D1271240

TOOL AND MANUFACTURING ENGINEERS HANDBOOK

VOLUME III
MATERIALS, FINISHING AND COATING

SOCIETY OF MANUFACTURING ENGINEERS

OFFICERS AND DIRECTORS, 1985-1986

TOOL AND MANUFACTURING ENGINEERS HANDBOOK

FOURTH EDITION

VOLUME III
MATERIALS, FINISHING AND COATING

A reference book for manufacturing engineers, managers, and technicians

Charles Wick, CMfgE
Editor-in-Chief

Raymond F. Veilleux
Staff Editor

Revised under the supervision of the SME Publications Committee in cooperation with the SME Technical Divisions

Society of Manufacturing Engineers
One SME Drive
Dearborn, Michigan

TMEH ™

ISBN No. 0-87263-176-1

Library of Congress Catalog No. 82-60312

Society of Manufacturing Engineers (SME)

First edition published 1949 by McGraw-Hill Book Co. in cooperation with SME under earlier Society name, American Society of Tool Engineers (ASTE), and under title *Tool Engineers Handbook*. Second edition published 1959 by McGraw-Hill Book Co. in cooperation with SME under earlier Society name, American Society of Tool and Manufacturing Engineers (ASTME), and under title *Tool Engineers Handbook*. Third edition published 1976 by McGraw-Hill Book Co. in cooperation with SME under current Society name and under title *Tool and Manufacturing Engineers Handbook*.

Printed in the United States of America.

PREFACE

The first edition, published as the *Tool Engineers Handbook* in 1949, established a useful and authoritative editorial format that was successfully expanded and improved upon in the publication of highly acclaimed subsequent editions, published in 1959 and 1976 respectively. Now, with continuing dramatic advances in manufacturing technology, increasing competitive pressure both in the United States and abroad, and a significant diversification of informational needs of the modern manufacturing engineer, comes the need for further expansion of the Handbook. As succinctly stated by Editor Frank W. Wilson in the preface to the second edition: "...no 'bible' of the industry can indefinitely survive the impact of new and changed technology."

Although greatly expanded and updated to reflect the latest in manufacturing technology, the nature of coverage in this edition is deeply rooted in the heritage of previous editions, constituting a unique compilation of practical data detailing the specification and use of modern manufacturing equipment and processes. Yet, the publication of this edition marks an important break with tradition in that this volume, dedicated to materials, finishing, and coating, is the third of five volumes to be published in the coming years, to comprise the fourth edition. Volume I, *Machining*, was published in March 1983, and Volume II, *Forming*, in April 1984. The final two volumes of this edition will be *Quality Control and Assembly* and *Management*.

The scope of this edition is multifaceted, offering a ready reference source of authoritative manufacturing information for daily use by engineers, managers, and technicians, yet providing significant coverage of the fundamentals of manufacturing processes, equipment, and tooling for study by the novice engineer or student. Uniquely, this blend of coverage has characterized the proven usefulness and reputation of SME Handbooks in previous editions and continues in this edition to provide the basis for acceptance across all segments of manufacturing.

The scope of this volume encompasses engineering materials, heat treatment, surface and edge preparation, and both inorganic and organic coatings. Materials discussed include the various steels, cast irons, nonferrous metals, plastics, advanced composites, and powdered metals. The heat treatment of steel and other metals, surface hardening, and heat treating furnaces are included.

A comprehensive section on surface and edge preparation includes individual chapters on mechanical and abrasive deburring and finishing; thermal, chemical, and electrochemical finishing; and cleaning. The section on inorganic coatings includes details on conversion coatings and anodizing, plating, electroless plating, thermal spraying and hard facing, porcelain enameling and hot dipping, vapor deposition, and special processes.

A separate section on organic coatings includes chapters on coating materials, application methods, curing methods, coating systems, and testing, troubleshooting, and safety. In-depth coverage of all subjects is presented in an easy-to-read format. A comprehensive index cross references all subjects, facilitating the quick access of information. The liberal use of drawings, graphs, and tables also speeds information gathering and problem solving.

The reference material contained in this volume is the product of incalculable hours of unselfish contribution by hundreds of individuals and organizations, as listed at the beginning of each chapter. No written words of appreciation can sufficiently express the special thanks due these many forward-thinking professionals. Their work is deeply appreciated by the Society; but more important, their contributions will undoubtedly serve to advance the understanding of manufacturing technology throughout industry and will certainly help spur major productivity gains in the years ahead. Industry as a whole will be the beneficiary of their dedication.

Further recognition is due the members of the SME Publications Committee for their expert guidance and support as well as the many members of the SME Technical Activities Board—particularly the members of the Composites Council and the Burr, Edge and Surface Conditioning Technology (BEST) Division—and the Association for Finishing Processes of SME (AFP/SME).

The Editors

SME staff who participated in the editorial development and production of this volume include:

EDITORIAL

Thomas J. Drozda
Director of Communications

Charles Wick
Manager, Reference Publications

Raymond F. Veilleux
Staff Editor

Darlene P. Copp
Technical Copy Editor

Shirley A. Barrick
Editorial Secretary

Judy A. Justice
Word Processor Operator

TYPESETTING

Shari L. Rogers
Typesetter Operator

GRAPHICS

Kathy J. Haas
Project Leader

Yvonne Haddix
Keyliner/Illustrator

Gail E. Salazar
Assistant Keyliner/Illustrator

SME

The Society of Manufacturing Engineers is a professional engineering society dedicated to advancing manufacturing technology through the continuing education of manufacturing managers, engineers, and technicians. The specific goal of the Society is "to advance scientific knowledge in the field of manufacturing engineering and to apply its resources to research, writing, publishing, and disseminating information."

The Society was founded in 1932 as the American Society of Tool Engineers (ASTE). From 1960 to 1969, it was known as the American Society of Tool and Manufacturing Engineers (ASTME), and in January 1970 it became the Society of Manufacturing Engineers. The changes in name reflect the evolution of the manufacturing engineering profession and the growth and increasing sophistication of a technical society that has gained an international reputation for being the most knowledgeable and progressive voice in the field.

As a member of the World Federation of Engineering Organizations, SME is the universally acknowledged technical society serving the manufacturing community. Among SME's activities are the following:

Associations of SME—The Society provides complete technical services and membership benefits through a number of associations. Each serves a special interest area. Members may join these associations in addition to SME. The associations are:

Association for Finishing Processes of SME (AFP/SME)

Computer and Automated Systems Association of SME (CASA/SME)

Machine Vision Association of SME (MVA/SME)

North American Manufacturing Research Institute of SME (NAMRI/SME)

Robotics International of SME (RI/SME)

Members and Chapters—The Society and its associations have some 80,000 members in 70 countries, most of whom are affiliated with SME's 380-plus senior chapters. The Society also has some 10,000 student members and more than 140 student chapters at colleges and universities in the United States and Canada.

Publications—The Society is involved in various publication activities encompassing handbooks, textbooks, videotapes, and magazines. Current periodicals include:

CIM Technology

Manufacturing Engineering

Manufacturing Insights (a video magazine)

Robotics Today

Technical Digest

Certification—This SME program formally recognizes manufacturing managers, engineers, and technologists based on experience and knowledge. The key certification requirement is successful completion of a two-part written examination covering (1) engineering fundamentals and (2) an area of manufacturing specialization.

Educational Programs—The Society annually sponsors over 200 conferences, expositions, and workshops throughout the world. It also operates the Center for Professional Development at its Dearborn, Michigan world headquarters.

CONTENTS

VOLUME III—MATERIALS, FINISHING AND COATING

SYMBOLS AND ABBREVIATIONS

The following is a list of symbols and abbreviations in general use throughout this volume. Supplementary and/or derived units, symbols, and abbreviations that are peculiar to specific subject matter are listed within chapters.

A

A	Ampere
ABS	Acrylonitrile butadiene styrene
a-c	Alternating current
A/dm^2	Ampere per square decimeter
AFM	Abrasive-flow machining
A/ft^2	Ampere per square foot
Ag	Silver
A·h	Ampere hour
AISI	American Iron and Steel Institute
AJM	Abrasive-jet machining
Al	Aluminum
AMS	Aerospace Material Specification (of SAE)
ANSI	American National Standards Institute
AOD	Argon-oxygen decarburization
As	Arsenic
ASM	American Society for Metals
ASME	American Society of Mechanical Engineers
ASTM	American Society for Testing and Materials
Au	Gold

B-C-D-E

B	Boron
Ba	Barium
bcc	Body-centered cubic
Be	Beryllium
Bhn	Brinell hardness number
Bi	Bismuth
Btu	British thermal unit
C	Celsius or Carbon
Ca	Calcium
CAD/CAM	Computer-aided design/computer-aided manufacturing
Cb	Columbium
CBN	Cubic boron nitride
Cd	Cadmium
Ce	Cerium
cfm	Cubic foot per minute
cm	Centimeter
CNC	Computer numerical control
Co	Cobalt
CPVC	Cellular polyvinyl chloride or Critical pigment volume concentration
Cr	Chromium
Cu	Copper
CVD	Chemical vapor deposition
dB	Decibel
dB(A)	Decibel ("A" scale)
d-c	Direct current
diam	Diameter
DNC	Direct numerical control
DPH	Diamond-pyramid hardness
EB	Electron beam
ECD	Electrochemical deburring
EDM	Electrical discharge machining
EPA	Environmental Protection Agency
Eq.	Equation
eV	Electron volt

F-G

F	Fahrenheit
fcc	Face-centered cubic
FCC	Federal Communications Commission
Fe	Iron
Fe_3C	Iron carbide or cementite
Fig.	Figure
fl oz	Fluid ounce
fpm	Foot per minute
fps	Foot per second
FRP	Fiberglass-reinforced plastics
ft	Foot
ft-lb	Foot-pound
ft-lb/in.	Foot pound per inch
ft/s	Foot per second
g	Gram or Gravity value
Ga	Gallium
gal	Gallon
g/cm^3	Gram per cubic centimeter
GP	General purpose
GPa	Gigapascal
gpm	Gallon per minute
GRN	Glass-reinforced nylon

H-I

H	Hydrogen
H_2	Hydrogen gas
HAZ	Heat-affected zone
HCl	Hydrochloric acid
hcp	Hexagonal close-packed
HDPE	High-density polyethylene
HDT	Heat deflection temperatures
HF	High frequency
Hg	Mercury
HIP	Hot isostatic pressing

HIPS	High-impact polystyrene
HMS	High-modulus strength
H_2O	Water
hp	Horsepower
hr	Hour
HSLA	High-strength low-alloy
HSS	High-speed steel
HV	Vickers hardness
Hz	Hertz (cycles per second)
IACS	International Annealed Copper Standard
ID	Inside diameter
In	Indium
in. or ″	Inch
in./in.	Inch per inch
in./min	Inch per minute
ipm	Inch per minute
IPN	Interpenetrating polymer networks
ipr	Inch per revolution
ips	Inch per second
ISO	International Organization for Standardization
IVD	Ion vapor deposition

J-K-L

J	Joule
K	Potassium
keV	Kilo electron volt
kg	Kilogram
kHz	Kilohertz
kJ	Kilojoule
km	Kilometer
kN	Kilonewton
kPa	Kilopascal
ksi	1000 pounds per square inch
kV	Kilovolt
kW	Kilowatt
L	Liter
lb	Pound
lb-in.	Pound-inch (torque)
lb/in.	Pound per inch
lb/in.^{3}	Pound mass per cubic inch
LDPE	Low-density polyethylene
Li	Lithium

M

m	Meter
max	Maximum
MEQ	Milliequivalent
MeV	Mega electron volt
Mg	Magnesium
mg	Milligram
mg/in.^{2}	Milligram per square inch
mHz	Millihertz
MIG	Metal inert gas
mil	Milli-inch (0.001 in.)
MIL	Military
min	Minimum or Minute
mL	Milliliter
mm	Millimeter
Mm	Megameter
m/min	Meter per minute
mm/m	Millimeter per meter
mm/mm	Millimeter per millimeter
mm/rev	Millimeter per revolution
mm/s	Millimeter per second
Mn	Manganese
MN-m^2	Meganewton meter squared
Mo	Molybdenum
MPa	Megapascal
MPIF	Metal Powder Industries Federation
m/s	Meter per second

N-O

N	Newton or Nitrogen
N_2	Nitrogen gas
Na	Sodium
NaOH	Sodium hydroxide
Nb	Niobium
NC	Numerical control
NFPA	National Fire Protection Association
Ni	Nickel
nm	Nanometer
N/m	Newton per meter
No.	Number
NV	Nonvolatile
O or O_2	Oxygen
OD	Outside diameter
OSHA	Occupational Safety and Health Administration
oz/yd^2	Ounce per square yard

P

P	Phosphorus
Pa	Pascal
PA	Polyamide
Pb	Lead
PC	Polycarbonate or Programmable controller
PE	Polyethylene
PES	Polyether sulfone
pH	Acidity measure
PH	Precipitation hardening
PID	Proportional, integral, derivative
PM	Powder metallurgy
PMMA	Polymethyl/methacrylate
PO	Phenoxy
PP	Polypropylene
ppm	Parts per million
PPO	Polyphenylene oxide
PS	Polystyrene
PSF	Polysulfone
psi	Pounds per square inch
pt	Pint
PVC	Polyvinyl chloride or Pigment volume concentration
PVD	Physical vapor deposition

R-S

$R_{A,B,C,\text{ or }H}$	Rockwell hardness, A, B, C, or H scale
rad	Radius
Re	Rhenium
RF	Radio frequency
rms	Root mean square
rpm	Revolution per minute
RT	Room temperature
RTM	Resin transfer molding or Resin injection molding
RTV	Room temperature vulcanizing
s or sec	Second
S	Sulfur
SAE	Society of Automotive Engineers
SAN	Styrene acrylonitrile
Sb	Antimony
SBQ	Salt-bath quench
SCR	Silicon-controlled rectifier
Se	Selenium
sfm	Surface feet per minute
Si	Silicon
SI	International System of Units
SLI	Starting, Lighting, Ignition
Sn	Tin
SO_2	Sulfur dioxide
SPE	Society of Plastics Engineers
SPI	Society of the Plastics Industry
Sr	Strontium

T

t	Metric ton
Ta	Tantalum
tan	Tangent
te	Transfer efficiency
Te	Tellurium
TED	Thermal energy deburring
TEM	Thermal energy method
Temp.	Temperature
TFE	Tetrafluoroethylene
Ti	Titanium
TIG	Tungsten inert gas
Tl	Thalium
TTT	Time-temperature transformation

U-V-W

UL	Underwriters Laboratories
UNS	Unified Numbering System
URTRI	Ultimately reinforced thermoset reaction injection
UV	Ultraviolet
V	Vanadium or Volt
VHN	Vickers hardness number
VOC	Volatile organic compound
W	Watt or Tungsten

Z

Zn	Zinc
Zr	Zirconium

α	Alpha
$^\circ$	Degree
Δ	Delta
μA	Microampere
μin.	Microinch
μm	Micrometer
μmhos	Micromhos
$\mu\Omega$	Microhm
μs	Microsecond
Ω	Ohm
%	Percent
ϕ	Phi
$\pm$	Plus or minus
θ	Theta

ENGINEERING MATERIALS

SECTION

1

ENGINEERING MATERIALS, CARBON AND ALLOY STEELS

ENGINEERING MATERIALS

CHAPTER CONTENTS:

Engineers today have at their disposal a multitude of materials that they can use to produce the products used in our rapidly developing society. These materials range from the ordinary materials that have been available for several hundred years to those that have recently been developed, and new materials are being introduced continuously.

This section of the Handbook, Volume III, contains nine individual chapters and is designed to provide the manufacturing engineer and other manufacturing professionals with the information they need to efficiently and effectively accomplish their respective roles. Information presented includes definitions of terms frequently encountered, brief discussions on how materials are produced, various material classifications, typical properties, current and potential applications, and descriptions of machinability, formability, and weldability.

A comprehensive discussion of machinability and machining of various materials can be found in Volume I, *Machining*, of this Handbook series; the forming of these materials is discussed in Volume II, *Forming*; and the assembly of these materials through various welding and joining processes can be found in Volume IV, *Quality Control and Assembly*. Heat treatment of ferrous and nonferrous alloys are discussed in subsequent chapters of this volume.

Throughout this section, the mechanical properties and some chemical compositions of the most commonly used materials are presented in tabular form for comparison. Tabular data are not intended for use in specifications or design. For specific values and compositions, consult the material manufacturer.

The materials discussed in this section include carbon and alloy steels; stainless and maraging steels; cast steels and irons; nickel alloys and superalloys; titanium and molybdenum alloys; aluminum, copper, and magnesium alloys; lead, tin, and zinc alloys; plastics and composites; powdered metals; and some refractory metals. Not included in this discussion of materials are precious metals, pure metals, and glasses and ceramics. Cutting tool materials are discussed in Volume I, *Machining*, and tool and die materials are discussed in Volume II, *Forming*, of this Handbook series.

GLOSSARY OF TERMS

Terms commonly used with respect to engineering materials are presented in this section and are adapted from the glossary in the third edition of the *Tool and Manufacturing Engineers Handbook*, *Modern Steels and Their Properties*,[1] and *Heat Treater's Guide*.[2] Additional terms common to the heat treatment of metals appear in Chapter 10.

acid steel Steel melted in a furnace with an acid bottom and lining and under a slag containing an excess of an acid substance such as silica.

alloy A metal containing additions of other metallic or nonmetallic elements to enhance specific properties such as strength and corrosion resistance.

basic steel Steel melted in a furnace with a basic bottom and lining and under a slag containing an excess of a basic substance such as magnesia or lime.

Bayer process A process for extracting alumina from bauxite ore before the electrolytic reduction to aluminum metal.

blast furnace A shaft-type furnace using solid fuel (coke), air, and oxygen to smelt ore in a continuous operation.

Brinell hardness number (Bhn) A measure of hardness determined by the Brinell hardness test, which forces a hard steel or carbide ball of specific diameter into a material under a specified load.

brittleness The quality of a material that leads to fracture without appreciable plastic deformation.

capped steel A steel similar to rimmed steel in which the application of a mechanical or chemical cap renders the rimming action incomplete by causing the top metal to solidify.

carbon steel Steel that derives its properties from the presence of carbon without substantial amounts of other alloying elements.

cast iron A generic term for a large group of cast ferrous alloys containing over 2% carbon and 1% silicon.

cast steel Steel in the form of castings, characterized by a grain structure produced by solidification.

***Contributors of sections of this chapter are: Calvin Cooley**, Metallurgical Engineer, American Iron and Steel Institute; **W. C. Leslie**, Professor, Dept. of Materials and Metallurgical Engineering, University of Michigan; **Tom Oakwood**, Manager, Wrought Alloy Steel Development, Climax Molybdenum, AMAX Inc.*

***Reviewers of sections of this chapter are: Calvin Cooley**, Metallurgical Engineer, American Iron and Steel Institute; **Dr. Lee Cuddy**, Associate Research Consultant, Heavy Products Div., U. S. Steel Tech Center;*

CHAPTER 1

ENGINEERING MATERIALS

cold working Deforming metal plastically under conditions that induce strain hardening; usually performed at room temperature.

compacted graphite cast iron Cast iron having a graphite shape intermediate between the flake form of gray cast iron and the spherical form of ductile cast iron.

compressive strength The maximum compressive stress that a material can withstand without significant plastic deformation or fracture.

consumable electrode remelting A process for refining metals in which an electric current passes between an electrode made of the metal or alloy to be refined and an ingot of the refined metal under the protection of a vacuum, inert atmosphere, or slag covers.

continuous casting A casting technique that produces steel in the form of slabs, blooms, or billets directly from the ladle.

creep The flow or plastic deformation of metals held for long periods of time at stresses lower than the normal yield strength.

creep limit The maximum stress that will result in creep at a rate lower than an assigned rate.

damping capacity The ability of a metal to absorb vibrations, changing the mechanical energy into heat.

dead soft A temper of nonferrous alloys and some ferrous alloys corresponding to the condition of minimum hardness and tensile strength produced by full annealing.

density ratio The ratio of the apparent density of a powder metallurgy compact to the absolute density of metal of the same composition. Usually expressed as a percentage.

ductile cast iron Cast iron that has been treated with magnesium or cerium while molten to spheroidize the graphite and thereby impart ductility to the cast metal.

ductility The ability of a metal to undergo substantial amounts of plastic deformation before fracture.

elastic limit The maximum stress that a metal can withstand without exhibiting a permanent deformation upon release of the stress.

elongation The amount of permanent extension within a specified gage length, measured after fracture in the tension test; usually expressed as a percentage of the original gage length, such as 25% in 2" (50 mm). Elongation may also refer to the amount of extension at any stage in any process that elongates a body continuously, as in rolling.

endurance limit The maximum stress that a metal can withstand without failure after a specified number of cycles of stress. If the term is employed without qualification, the cycles of stress are usually such as to produce complete reversal of flexural stress.

extra hard A temper of nonferrous and some ferrous alloys characterized by tensile strength and hardness about one third of the way from full-hard to extra-spring tempers.

extra spring A temper of nonferrous and some ferrous alloys corresponding approximately to a cold worked state above full hard beyond which further cold work will not measurably increase the strength and hardness.

fatigue The tendency of a metal to break under conditions of repeated cyclic stressing considerably below the tensile strength.

file hardness Hardness as determined by the use of a file of standardized hardness, on the assumption that a material cannot be cut with the file if the material is as hard as, or harder than, the file. Files covering a range of hardnesses may be employed.

flakes Short, discontinuous internal fissures in ferrous metals caused by localized internal stresses during cooling after hot working. Flaking may be associated with the presence of hydrogen in the steel.

free machining A term used to describe metals having alloying additions, such as lead, manganese, or sulfur, that reduce the tool force required in machining operations.

full hard A temper of nonferrous alloys and some ferrous alloys corresponding approximately to a cold worked state beyond which the material can no longer be formed by bending. In specifications, a full-hard temper is commonly defined in terms of minimum hardness or minimum tensile strength corresponding to a specific percentage of cold reduction following a full anneal.

grain An individual crystal in a metal or alloy.

grain size The average cross-sectional area or volume of grains in polycrystalline metals; usually expressed as average diameter or number of grains per unit of area or volume.

gray cast iron Cast iron that contains a large percentage of carbon in the form of flake graphite.

half hard A temper of nonferrous alloys and some ferrous alloys characterized by tensile strength about midway between that of dead-soft and full-hard tempers.

hardenability The relative ability of a ferrous alloy to form martensite when quenched from above the upper critical temperature.

hardness The resistance of a metal to indentation, defined in terms of the method of measurement.

heat treatment A sequence of controlled heating and cooling operations applied to a solid metal to impart desired properties.

inclusions Particles of nonmetallic compounds present in as-cast metals and carried over into wrought products. The shape and distribution of inclusions are changed by plastic deformation and contribute to directionality of mechanical properties.

ingot A casting intended for subsequent rolling, forging, or extrusion.

killed steel Steel treated with additions of silicon or aluminum to the melt to minimize the oxygen content so that no reaction occurs between carbon and oxygen during solidification.

machinability The relative ease with which materials can be shaped by cutting, drilling, or other chip-forming processes.

malleable cast iron A cast iron made by annealing white cast iron to eliminate some or all of the cementite.

mechanical properties The characteristics of a material that are displayed when a force is applied to the material. They

__Reviewers, cont.:__ __Gene Curry__, Manager, Metallurgical Laboratory, McLouth Steel Products Corp.; __Professor A. J. DeArdo__, Dept. of Metallurgical and Materials Engineering, University of Pittsburgh; __Dr. Robert M. Fisher__, Associate Director—Industry Participation, Center for Advanced Materials, Lawrence Berkeley Laboratories, University of California; __J. Klein__, Librarian, Jones and Laughlin Steel Corp.; __Dr. Peter J. Koros__, Senior Research Associate, Jones and Laughlin Steel Corp.; __W. C. Leslie__, Professor, Dept. of Materials and Metallurgical Engineering, University of Michigan; __Arnie Marder__, Senior Scientist, Research Dept., Bethlehem Steel Corp.; __Conrad Mitchell__, Manager, Bar, Rod, Wire & Semi-Finished Products, Product Metallurgy, U.S. Steel Corp.;

usually relate to the elastic or inelastic response of the material.

modulus of elasticity The ratio of stress to strain within the elastic range of a material; a measure of stiffness and the ability to resist deflection when loaded. Also referred to as Young's modulus.

modulus of rigidity The ratio of the shear stress to the shear strain in the elastic range. Also called the shear modulus.

notch impact strength A measure of the ability of a material to sustain high-velocity loading in the presence of a notch.

percent elongation See elongation.

percent reduction See reduction in area.

permanent set Plastic deformation.

physical properties Properties that pertain to the physics of a material, such as melting point, density, electrical and thermal conductivity, specific heat, and coefficient of thermal expansion.

plasticity The ability of a metal to undergo permanent deformation without rupture. (See also ductility.)

Poisson's ratio (μ) The ratio of the lateral expansion to the longitudinal contraction under a compressive load, or the ratio of the lateral contraction to the longitudinal expansion under a tensile load, provided the elastic limit is not exceeded.

powdered metals Metals and alloys in the form of fine particles, usually in the range of 39 μin. to 0.039″ (1-1000 μm), or 1-1000 microns.

proportional limit The greatest stress that a material can sustain without any deviation from proportionality of stress to strain.

quarter hard A temper of nonferrous alloys and some ferrous alloys characterized by tensile strength about midway between that of dead-soft and half-hard tempers.

recrystallization The formation of new, strain-free grains by annealing a cold worked metal. Also called primary recrystallization.

reduction in area The difference between the original cross-sectional area and that of the smallest area at the point of rupture of a tensile test specimen. Usually stated as a percentage of the original area. A measurement of the material's ability to deform plastically in a localized manner.

residual stress Stresses present in a free metal body, usually as a result of prior, nonuniform plastic deformation, severe temperature gradients during quenching, or chemical differences as in carburized surfaces.

rimmed steel Low-carbon steel that does not contain significant percentages of easily oxidized elements such as aluminum, silicon, or titanium.

semikilled steel Steel that is incompletely deoxidized and contains sufficient dissolved oxygen to react with carbon to form carbon monoxide and offset solidification shrinkage.

shear Deformation in which parallel planes within the metal are displaced by sliding but retain their parallel relation to each other.

shear strength The maximum stress that a material can withstand before fracture when the load is applied parallel to the plane of stress.

strain, engineering The change in length divided by the original length, L/L_o. Expressed as a dimensionless number.

strain rate The rate at which deformation occurs.

stress, engineering Internal force reactions set up in a body when it is subjected to a load. Calculated by dividing the load by the original cross-sectional area.

temper The degree of ductility and toughness produced in a hardened metal by reheating to a temperature below the transformation range and then cooling at a suitable rate.

tensile strength The maximum engineering stress in tension that a material can withstand before rupture. Calculated by dividing the maximum load by the original cross-sectional area of a specimen pulled to failure in a tensile test.

three-quarters hard A temper of nonferrous alloys and some ferrous alloys characterized by tensile strength and hardness about midway between those of half-hard and full-hard tempers.

torsion modulus See modulus of rigidity.

torsional strength The maximum stress that a material can withstand before fracture when subjected to a torque or twisting force. Stress in torsion involves shearing stress, which is not uniformly distributed in a conventional torsion test bar.

toughness The ability of a material to absorb energy without failure when a load is applied rapidly, such as in an impact. Represented by the area under a stress-strain curve.

ultimate strength See tensile strength.

vacuum refining Melting and/or casting in vacuum to remove gaseous contaminants from a metal.

yield point The stress at which a pronounced increase in strain is shown without an increase in load; observed in low and medium-carbon steels.

yield strength The stress at which a material exhibits a specific amount of permanent deformation. In tensile tests, it is usually measured as the stress at 0.2% offset on a stress-strain diagram.

yield/tensile ratio The ratio of the yield point stress or yield strength to the tensile strength. Annealed low-carbon steels may have a ratio of only 40%, while heat-treated alloy steels and cold worked steels may have a ratio of 90% or higher. Low-carbon steels usually have a ratio of 50 to 70%.

MATERIAL PROPERTIES

Materials are usually selected for a particular application based on the properties that they possess and/or display under certain circumstances. The three main material property categories are chemical, physical, and mechanical.

Chemical Properties

It is not possible to make a sharp distinction between the chemical and physical properties of a metallic material. Both are dependent upon the crystal structure, the strength of interactive bonds, defects in the crystal lattice, and the amount

Reviewers, cont.: ***Tom Oakwood**, Manager, Wrought Alloy Steel Development, Climax Molybdenum, AMAX Inc.;* ***Dr. Gordon Powell**, Professor, Dept. of Metallurgy Engineering, Ohio State University;* ***Dr. Arnie Preban**, Senior Research Metallurgist, Inland Steel;* ***Karl B. Rundman**, Professor, Dept. of Metallurgical Engineering, Michigan Technological University;* ***Everett E. Shields**, Product Metallurgist, Technical Services, Republic Steel Corp.;* ***Gilbert R. Speich**, Professor and Chairman, Metallurgical and Materials Engineering, Illinois Institute of Technology;* ***R. D. Stout**, Professor, Dept. of Metallurgy and Materials Engineering, Lehigh University;* ***Dr. Brian Taylor**, Research Scientist, APMES, General Motors Corp.*

and distribution of other phases within the body. Manufacturing engineers will be interested primarily in the interaction between metals and their environments. These include relative resistance to oxidation, galvanic corrosion caused by electrical contact between disimilar metals, general corrosion, and, especially, stress-corrosion cracking, which occurs in the simultaneous presence of a corroding agent and stress. The stress can be either residual or applied.

Physical Properties

The physical properties of a metal or alloy can be classified as thermal, electrical, magnetic, or optical. The thermal properties include melting temperature (or melting range), specific heat, thermal conductivity, and thermal expansion. The most important electrical properties are conductivity (and its inverse, resistivity) and superconductivity. Magnetic materials are either soft or hard. In the former, the direction of magnetization can be altered with little expenditure of energy. In the latter, the magnetization is permanent, which means that a large expenditure of energy is required to alter the direction. Soft magnetic materials are required for generators, motors, and transformers. Hard magnetic materials are used wherever permanent magnets are needed. The optical properties of a solid depend upon the type of interatomic bonding. Metallic bonding is responsible for the color, reflectivity, and emissivity of metallic surfaces.

Mechanical Properties

Mechanical properties are the characteristics of a material that are displayed when a force is applied. The mechanical properties of a material are usually the primary factors in selecting a material for a specific application.

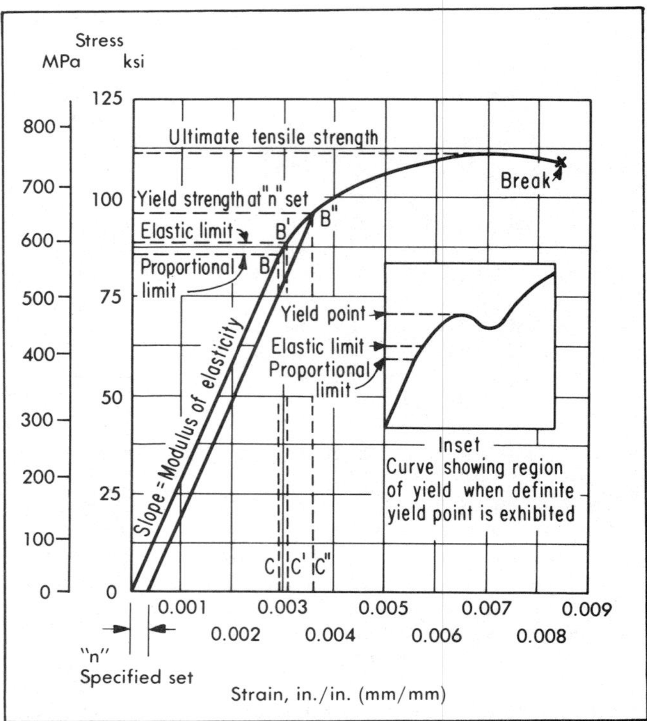

Fig. 1-1 A stress-strain diagram is a plot of the stress required to produce a given strain during a tensile test. The tensile strength, yield strength, and modulus of elasticity of a given material can be obtained from this diagram.

TABLE 1-1
Comparison of Density and Modulus of Elasticity of Commonly Used Engineering Materials

Material	Density, lb/in.^{3} (g/cm^3)	Modulus of Elasticity, psi x 10^6 (GPa)
Wrought steels	0.280-0.290 (7.76-8.03)	28.0-31.0 (193-215)
Cast steels		
Carbon and low alloy	0.282-0.285 (7.81-7.90)	29.3-31.4 (202-217)
High alloy*	0.272-0.294 (7.53-8.14)	24.0-29.0 (165-200)
Cast irons	0.251-0.280 (6.95-7.76)	9.6-28.0 (66-193)
Nickel alloys	0.282-0.334 (7.81-9.25)	24.0-30.0 (165-207)
Molybdenum alloys	0.369 (10.22)	46 (317)
Titanium alloys	0.158-0.175 (4.38-4.85)	15.0-17.9 (103-123)
Superalloys		
Iron based	0.286-0.291 (7.92-8.06)	
Nickel based	0.280-0.319 (7.76-8.84)	28.6-31.6 (197-218)
Cobalt based	0.290-0.333 (8.03-9.23)	29.5-32.6 (203-225)
Aluminum alloys	0.097-0.102 (2.69-2.83)	10.0-11.4 (69-79)
Magnesium alloys	0.064-0.067 (1.77-1.86)	6.5 (45)
Copper alloys	0.301-0.323 (8.34-8.95)	14.0-22.0 (97-152)
Lead alloys	0.351-0.410 (9.72-11.36)	2.0 (14)
Tin alloys	0.262-0.304 (7.25-8.42)	6.03 (41.6)
Zinc alloys	0.188-0.259 (5.21-7.18)	6.2-14.0 (43-97)
Powdered metals		
Ferrous	0.209-0.267 (5.80-7.40)	10.5-23 (72-159)
Nonferrous	0.047-0.296 (1.30-8.20)	
Plastics		
Thermosetting	0.0397-0.0632 (1.1-1.75)	1.0-1.5 (6.9-10.3)
Thermoplastic	0.0328-0.0769 (0.91-2.13)	0.025-4.0 (0.17-28)

Note: Values are based on materials at room temperature.
* The modulus of elasticity is sensitive to the material's crystalline structure and grain orientation. The range given is for equiaxed structures; columnar structures are lower.

Mechanical properties provide the engineer with information regarding strength, formability, rigidity, toughness, and durability. In the subsequent chapters, the tensile strength, yield strength, total elongation, and hardness are given in tabular form for commonly used materials. Figure 1-1 shows the tensile strength, yield strength, yield point, and modulus of elasticity of a material on a typical stress-strain diagram. The modulus of elasticity range and density are given in Table 1-1 to permit comparison of these values for various materials. Table 1-2 shows the approximate relations between various hardness scales for steel having uniform chemical composition and heat treatment.[3] The approximate tensile strengths for corresponding hardnesses are also given. These relations usually apply to only steels with uniform chemical composition and heat treatment.

The mechanical, physical, and chemical properties of a metal are sensitively dependent upon crystal structure. Engineering materials almost invariably have one of three structures: body-centered cubic (bcc), face-centered cubic (fcc), or hexagonal close-packed (hcp).

The mechanical properties of bcc metals are dependent upon temperature and the strain rate. As temperature is reduced and the strain rate increased, the strength of the metal increases rapidly and the ductility and toughness decrease. In contrast, strain rate and temperature changes have little effect on the mechanical properties of fcc metals. Often they are more easily formed than bcc metals. Hexagonal close-packed metals are generally more difficult to form than either bcc or fcc metals owing to the small number of slip systems that can operate in the hcp structures.

CARBON STEELS

Iron and carbon are the predominant elements in steels. Carbon content ranges from a few hundredths to about one percent. The amount of additional alloying elements determines whether the steel is considered to be a carbon or an alloy steel.

Steel is considered a carbon steel when no minimum content is specified or required for aluminum (except for oxidation or to control grain size), chromium, cobalt, columbium, molybdenum, nickel, titanium, tungsten, vanadium, zirconium, or any other element to obtain a desired alloying effect; when the specified minimum for copper does not exceed 0.40%; or when the addition of manganese, silicon, and copper is limited to a maximum of 1.65%, 0.60%, and 0.60% respectively.

On the basis of carbon content, carbon steels can be divided into three groups. The first group contains 0.001-0.30% carbon and is considered low-carbon steel. The second group contains 0.30-0.70% carbon and is considered medium-carbon steel. The third group contains 0.70-1.30% carbon and is considered high-carbon steel.

Certain grades may also specify the addition of boron to improve hardenability and aluminum for deoxidation and to control grain size. Carbon steels also contain small quantities of residual elements or impurities from the raw material such as copper, nickel, molybdenum, chromium, phosphorus, and sulfur, which are considered incidental.

Carbon steels may be classified according to chemical composition, deoxidation practice, quality, and end-product forms. Common end-product forms include bar, sheet/strip, plate, wire, tubing, and structural shapes. Carbon steel may also be classified as hot rolled or cold drawn (cold rolled when referring to sheets). Cold finished steels are produced from hot rolled steel by several cold finishing processes, resulting in improved surface finishes, dimensional accuracy, alignment, or machinability; elongation and yield and tensile strengths are increased. Cold rolled sheets are available in different tempers and can be precoated with zinc, aluminum, terne (lead-tin alloy), tin, and organic coatings.

STEELMAKING PRACTICE

Steelmaking may be described as the process of refining pig iron or ferrous scrap by removing undesirable elements from the melt and then adding desired elements in predetermined amounts. The additions are often the same elements that were originally removed, the difference being that the elements present in the final steel product are in the proper proportion to produce the desired properties.[4]

The various practices employed in steel production have a direct influence on the type and quality of the finished product. To ensure a quality finished product, it is necessary to exercise control over the raw materials used and to employ the proper melting, refining, and casting techniques.

Melting and Refining Techniques

The melting of pig iron in the production of steel is performed in basic oxygen, electric, and open-hearth furnaces. Before the molten metal is poured, it is sometimes subjected to a vacuum treatment that lowers the content of hydrogen and oxygen gases, minimizes slow cooling time, and improves alloy distribution and mechanical properties. The two types of vacuum treatments used are vacuum degassing and vacuum carbon deoxidation (VCD).

In recent years, ladle metallurgy has been adopted for refining steel. Ladle metallurgy permits the steelmaking processes to operate in a lower cost and higher productivity mode, while simultaneously ensuring production of high-quality steels.[5] The various treatments in ladle metallurgy are synthetic slag treatments, gas stirring or purging, direct immersion of reactants, lance injection of reactants, and wire feeding of reactants for the purpose of removing undesired elements such as oxygen, hydrogen, sulfur, and phosphorus. The treatments may be performed separately or in combination to achieve the desired results.

Casting Techniques

In wrought steel production, the molten steel is poured into tapered molds or into a strand-casting machine. Techniques for producing steel castings are discussed in Volume II, *Forming*, of this Handbook series.

Ingot casting. In ingot casting, the molten metal is poured into tapered, cast iron molds and allowed to solidify. The solidified metal is referred to as an ingot. Ingots can be square, rectangular, or round in cross section, with round corners and corrugated sides. The size of the ingot ranges from a few hundred pounds to several hundred tons.

TABLE 1-2
Comparison of Commonly Used Hardness Scales for Steel[3]

Brinell Indentation Diam, mm	Brinell Hardness No., * 10 mm Ball, 3000 kg Load			Diamond Pyramid Hardness No.	Rockwell Hardness No. **				Rockwell Superficial Hardness No., Superficial Brale Penetrator			Shore Scleroscope Hardness No.	Tensile Strength (approximate), ksi (MPa)
	Standard Ball	Hultgren Ball	Tungsten-Carbide Ball		A-Scale, 60 kg Load, Brale Penetrator	B-Scale, 100 kg Load, 1/16 in. Diam Ball	C-Scale, 150 kg Load, Brale Penetrator	D-Scale, 100 kg Load, Brale Penetrator	15-N Scale, 15 kg Load	30-N Scale, 30 kg Load	45-N Scale, 45 kg Load		
--	--	--	--	940	85.6	--	68.0	76.9	93.2	84.4	75.4	97	--
--	--	--	--	920	85.3	--	67.5	76.5	93.0	84.0	74.8	96	--
--	--	--	--	900	85.0	--	67.0	76.1	92.9	83.6	74.2	95	--
--	--	--	767	880	84.7	--	66.4	75.7	92.7	83.1	73.6	93	--
--	--	--	757	860	84.4	--	65.9	75.3	92.5	82.7	73.1	92	--
2.25	--	--	745	840	84.1	--	65.3	74.8	92.3	82.2	72.2	91	--
--	--	--	733	820	83.8	--	64.7	74.3	92.1	81.7	71.8	90	--
--	--	--	722	800	83.4	--	64.0	73.8	91.8	81.1	71.0	88	--
2.30	--	--	712	--	--	--	--	--	--	--	--	--	--
--	--	--	710	780	83.0	--	63.3	73.3	91.5	80.4	70.2	87	--
--	--	--	698	760	82.6	--	62.5	72.6	91.2	79.7	69.4	86	--
--	--	--	684	740	82.2	--	61.8	72.1	91.0	79.1	68.6	--	--
2.35	--	--	682	737	82.2	--	61.7	72.0	91.0	79.0	68.5	84	--
--	--	--	670	720	81.8	--	61.0	71.5	90.7	78.4	67.7	83	--
--	--	--	656	700	81.3	--	60.1	70.8	90.3	77.6	66.7	--	--
2.40	--	--	653	697	81.2	--	60.0	70.7	90.2	77.5	66.5	81	--
--	--	--	647	690	81.1	--	59.7	70.5	90.1	77.2	66.2	--	--
--	--	--	638	680	80.8	--	59.2	70.1	89.8	76.8	65.7	80	--
--	--	--	630	670	80.6	--	58.8	69.8	89.7	76.4	65.3	--	--
2.45	--	--	627	667	80.5	--	58.7	69.7	89.6	76.3	65.1	79	--
2.50	--	601	--	677	80.7	--	59.1	70.0	89.8	76.8	65.7	--	--
	--	--	601	640	79.8	--	57.3	68.7	89.0	75.1	63.5	77	--
2.55	--	578	--	640	79.8	--	57.3	68.7	89.0	75.1	63.5	--	--
	--	--	578	615	79.1	--	56.0	67.7	88.4	73.9	62.1	75	--
2.60	--	555	--	607	78.8	--	55.6	67.4	88.1	73.5	61.6	--	--
	--	--	555	591	78.4	--	54.7	66.7	87.8	72.7	60.6	73	298 (2053)
2.65	--	534	--	579	78.0	--	54.0	66.1	87.5	72.0	59.8	--	292 (2013)
	--	--	534	569	77.8	--	53.5	65.8	87.2	71.6	59.2	71	288 (1986)
2.70	--	514	--	553	77.1	--	52.5	65.0	86.7	70.7	58.0	--	278 (1917)
	--	--	514	547	76.9	--	52.1	64.7	86.5	70.3	57.6	70	274 (1889)
	495	--	--	539	76.7	--	51.6	64.3	86.3	69.9	56.9	--	269 (1855)
2.75	--	495	--	530	76.4	--	51.1	63.9	86.0	69.5	56.2	--	265 (1827)
	--	--	495	528	76.3	--	51.0	63.8	85.9	69.4	56.1	68	264 (1820)

TABLE 1-2—*Continued*

Brinell Indentation Diam, mm	Brinell Hardness No.,* 10 mm Ball, 3000 kg Load: Standard Ball	Hultgren Ball	Tungsten-Carbide Ball	Diamond Pyramid Hardness No.	Rockwell Hardness No.**: A-Scale, 60 kg Load, Brale Penetrator	B-Scale, 100 kg Load, 1/16 in. Diam Ball	C-Scale, 150 kg Load, Brale Penetrator	D-Scale, 100 kg Load, Brale Penetrator	Rockwell Superficial Hardness No., Superficial Brale Penetrator: 15-N Scale, 15 kg Load	30-N Scale, 30 kg Load	45-N Scale, 45 kg Load	Shore Scleroscope Hardness No.	Tensile Strength (approximate), ksi (MPa)
	477	--	--	516	75.9	--	50.3	63.2	85.6	68.7	55.2	--	258 (1779)
2.80	--	477	--	508	75.6	--	49.6	62.7	85.3	68.2	54.5	--	252 (1738)
	--	--	477	508	75.6	--	49.6	62.7	85.3	68.2	54.5	66	252 (1738)
	461	--	--	495	75.1	--	48.8	61.9	84.9	67.4	53.5	--	244 (1682)
2.85	--	461	--	491	74.9	--	48.5	61.7	84.7	67.2	53.2	--	242 (1667)
	--	--	461	491	74.9	--	48.5	61.7	84.7	67.2	53.2	65	242 (1667)
	444	--	--	474	74.3	--	47.2	61.0	84.1	66.0	51.7	--	231 (1593)
2.90	--	444	--	472	74.2	--	47.1	60.8	84.0	65.8	51.5	--	230 (1586)
	--	--	444	472	74.2	--	47.1	60.8	84.0	65.8	51.5	63	230 (1586)
2.95	429	429	429	455	73.4	--	45.7	59.7	83.4	64.6	49.9	61	219 (1510)
3.00	415	415	415	440	72.8	--	44.5	58.8	82.8	63.5	48.4	59	212 (1462)
3.05	401	401	401	425	72.0	--	43.1	57.8	82.0	62.3	46.9	58	202 (1393)
3.10	388	388	388	410	71.4	--	41.8	56.8	81.4	61.1	45.3	56	193 (1331)
3.15	375	375	375	396	70.6	--	40.4	55.7	80.6	59.9	43.6	54	184 (1269)
3.20	363	363	363	383	70.0	--	39.1	54.6	80.0	58.7	42.0	52	177 (1220)
3.25	352	352	352	372	69.3	(110.0)	37.9	53.8	79.3	57.6	40.5	51	171 (1179)
3.30	341	341	341	360	68.7	(109.0)	36.6	52.8	78.6	56.4	39.1	50	164 (1131)
3.35	331	331	331	350	68.1	(108.5)	35.5	51.9	78.0	55.4	37.8	48	159 (1096)
3.40	321	321	321	339	67.5	(108.0)	34.3	51.0	77.3	54.3	36.4	47	154 (1062)
3.45	311	311	311	328	66.9	(107.5)	33.1	50.0	76.7	53.3	34.4	46	149 (1027)
3.50	302	302	302	319	66.3	(107.0)	32.1	49.3	76.1	52.2	33.8	45	146 (1007)
3.55	293	293	293	309	65.7	(106.0)	30.9	48.3	75.5	51.2	32.4	43	141 (972)
3.60	285	285	285	301	65.3	(105.5)	29.9	47.6	75.0	50.3	31.2	--	138 (952)
3.65	277	277	277	292	64.6	(104.5)	28.8	46.7	74.4	49.3	29.9	41	134 (924)
3.70	269	269	269	284	64.1	(104.0)	27.6	45.9	73.7	48.3	28.5	40	130 (896)
3.75	262	262	262	276	63.6	(103.0)	26.6	45.0	73.1	47.3	27.3	39	127 (876)
3.80	255	255	255	269	63.0	(102.0)	25.4	44.2	72.5	46.2	26.0	38	123 (848)
3.85	248	248	248	261	62.5	(101.0)	24.2	43.2	71.7	45.1	24.5	37	120 (827)
3.90	241	241	241	253	61.8	100.0	22.8	42.0	70.9	43.9	22.8	36	116 (800)
3.95	235	235	235	247	61.4	99.0	21.7	41.4	70.3	42.9	21.5	35	114 (786)
4.00	229	229	229	241	60.8	98.2	20.5	40.5	69.7	41.9	20.1	34	111 (765)
4.05	223	223	223	234	--	97.3	(18.8)	--	--	--	--	--	--

TABLE 1-2—*Continued*

Brinell Indenta-tion Diam, mm	Brinell Hardness No., *, 10 mm Ball, 3000 kg Load			Diamond Pyramid Hardness No.	Rockwell Hardness No. **				Rockwell Superficial Hardness No., Super-ficial Brale Penetrator			Shore Sclero-scope Hardness No.	Tensile Strength (approxi-mate), ksi (MPa)
	Standard Ball	Hultgren Ball	Tungsten-Carbide Ball		A-Scale, 60 kg Load, Brale Penetrator	B-Scale, 100 kg Load, 1/16 in. Diam Ball	C-Scale, 150 kg Load, Brale Penetrator	D-Scale, 100 kg Load, Brale Penetrator	15-N Scale, 15 kg Load	30-N Scale, 30 kg Load	45-N Scale, 45 kg Load		
4.10	217	217	217	238	--	96.4	(17.5)	--	--	--	--	33	105 (724)
4.15	212	212	212	222	--	95.5	(16.0)	--	--	--	--	--	102 (703)
4.20	207	207	207	218	--	94.6	(15.2)	--	--	--	--	32	100 (690)
4.25	201	201	201	212	--	93.8	(13.8)	--	--	--	--	31	98 (676)
4.30	197	197	197	207	--	92.8	(12.7)	--	--	--	--	30	95 (655)
4.35	192	192	192	202	--	91.9	(11.5)	--	--	--	--	29	93 (641)
4.40	187	187	187	196	--	90.7	(10.0)	--	--	--	--	--	90 (620)
4.45	183	183	183	192	--	90.0	(9.0)	--	--	--	--	28	89 (614)
4.50	179	179	179	188	--	89.0	(8.0)	--	--	--	--	27	87 (600)
4.55	174	174	174	182	--	87.8	(6.4)	--	--	--	--	--	85 (586)
4.60	170	170	170	178	--	86.8	(5.4)	--	--	--	--	26	83 (572)
4.65	167	167	167	175	--	86.0	(4.4)	--	--	--	--	--	81 (558)
4.70	163	163	163	171	--	85.0	(3.3)	--	--	--	--	25	79 (545)
4.80	156	156	156	163	--	82.9	(0.9)	--	--	--	--	--	76 (524)
4.90	149	149	149	156	--	80.8	--	--	--	--	--	23	73 (503)
5.00	143	143	143	150	--	78.7	--	--	--	--	--	22	71 (490)
5.10	137	137	137	143	--	76.4	--	--	--	--	--	21	67 (462)
5.20	131	131	131	137	--	74.0	--	--	--	--	--	--	65 (448)
5.30	126	126	126	132	--	72.0	--	--	--	--	--	20	63 (434)
5.40	121	121	121	127	--	69.8	--	--	--	--	--	19	60 (414)
5.50	116	116	116	122	--	67.6	--	--	--	--	--	18	58 (400)
5.60	111	111	111	117	--	65.7	--	--	--	--	--	15	56 (386)

(Society of Automotive Engineers)

Note: Some of the values shown in this table correspond to values in the corresponding joint SAE-ASM-ASTM Committee on Hardness Conversions as printed in ASTM E 140.

* Brinell numbers are based on the diameter of impressed indentation. If the ball distorts (flattens) during test, Brinell numbers will vary in accordance with the degree of such distortions when related to hardnesses determined with Vickers Diamond Pyramid, Rockwell Brale, or other penetrator which does not sensibly distort. At high hardnesses, therefore, the relationship between Brinell, Vickers, or Rockwell scales is affected by the type of ball used. Steel balls (Standard or Hultgren) tend to flatten slightly more than carbide balls, resulting in larger indentation and lower Brinell number than shown by a carbon ball.

** Values in ( ) are beyond normal range and are given for information only.

After the ingots are stripped from the molds, they are held in a furnace or soaking pit to equalize the temperature throughout. When the suitable temperature has been attained, the ingot is rolled or forged into blooms, billets, or slabs through a series of mill operations.

Strand (continuous) casting. Strand casting is the direct casting of steel from the ladle into slabs, blooms, or billets. The process can be performed vertically or horizontally. In operation, the molten steel is poured from the ladle into an intermediate vessel, called a tundish, at the top of the strand-casting machine. The tundish acts as a reservoir for the molten steel and regulates the rate at which the molten steel flows into one or more oscillating, water-cooled molds. The water-cooled, open-ended molds incorporate the desired cross section of the slab, bloom, or billet.

Solidification of the steel begins in the mold and is completed by cooling the moving steel surface. The steel produced from one or more ladle through one mold is commonly referred to as a strand. Several strands may be cast simultaneously, depending on the size of the ladle and the cross section of the strand. A reduction in strand size may be carried out by hot working before cutting the strand into proper lengths. When two or more ladles are cast without interruption, the process is called continuous casting.

Strand or continuous casting techniques are being increasingly employed in the steel industry. Several advantages of the process include more uniform chemical composition and mechanical properties in the semifinished product, and increased productivity over ingot casting. Continuous casting methods are also being used in the production of copper and aluminum alloys, and gray and alloy-type cast irons.

TYPES OF STEEL

The principal reaction in steelmaking is the removal of excess carbon by the combination of carbon and oxygen to form a gas. If the extra oxygen remaining after this reaction is not removed prior to or during casting, the gaseous products continue to evolve during solidification. The type of steel produced is determined by the amount of deoxidation that takes place before casting. The four types of carbon steels produced are killed, semikilled, rimmed, and capped.

Killed Steels

Killed steels are strongly deoxidized by the addition of aluminum and/or silicon to the ladle before pouring. These elements combine with the oxygen; thus, only a negligible evolution of gases occurs during solidification. Killed steels are characterized by a high degree of uniformity in chemical composition and mechanical properties, which render them suitable for applications requiring forging, extrusion, severe cold forming, carburizing, and heat treatment. However, there may be variations in composition depending on the steelmaking practice used.

Semikilled Steels

Semikilled steels have characteristics intermediate between those of killed and rimmed steels. During solidification, the evolved gas is entrapped within the body of the ingot and counteracts the shrinkage.

Rimmed Steels

Rimmed steels are generally low-carbon steels that do not contain significant percentages of easily oxidized elements such as aluminum, silicon, or titanium. Since deoxidation is minimal, the carbon and residual oxygen react during solidification. The reaction stirs the liquid metal causing the metal that solidifies at the outer rim of the ingot to be lower in carbon, phosphorus, and sulfur than the average composition, whereas the inner portion, or core, is higher than average in those elements. The rimming action may continue until the reactions stop and the top of the ingot solidifies, or it may be stopped mechanically or chemically.

Rimmed steels have good surface and ductility characteristics. Because of their ductility, rimmed steels are suitable for moderate cold forming applications.

Capped Steels

Capped steels have characteristics similar to those of rimmed steel. The rimming action is controlled when the steel is cast so that gas produced during solidification causes the metal to rise in the mold. Capping occurs when the rising metal contacts a heavy metal cap placed on the bottle-top mold (mechanical capping). Adding ferrosilicon or aluminum to the ingot top after the ingot has rimmed for the desired period of time is another method of producing capped steel (chemical capping).

GRADES OF STEEL

Grade usually denotes the chemical composition of a particular steel. The grades may vary in chemical composition from almost pure iron to a material of complex constitution. A particular grade of carbon steel usually has specified limits for various elements, but the properties of products made from that grade can be diverse.

Lists of standard steels designed to serve the needs of fabricators and users of steel products are published by the American Iron and Steel Institute (AISI) and the Society of Automotive Engineers (SAE). The general acceptance and use of standard steels since their inception in 1941 have demonstrated that these steels have, in most cases, successfully replaced the many steels of specialized compositions previously used. The list is altered from time to time to accommodate steels of proven merit and to provide for changes that develop in industry. There are still specialized steels being produced, however, for particular applications.

Grade Designation

A four-numeral series, adopted by the AISI and the SAE, is used to designate standard carbon steels specified to chemical composition ranges. It is important to note that these designations do not indicate specifications. The prefix M is used to designate a series of merchant-quality steels and the suffix H designates standard hardenability steels.

The first two digits indicate the steel type and identifying elements as shown in Table 1-3. The last two digits indicate the approximate mean of the carbon range. For example, in the grade designation 1035, 35 represents a carbon range of 0.32 to 0.38%. It is necessary to deviate from this system and to interpolate numbers in the case of some carbon ranges and for variations in manganese, phosphorus, or sulfur with the same range. Special-purpose elements such as lead and boron are designated by inserting the letter L or B, respectively, between the second and third numerals.

In 1975, the Unified Numbering System (UNS) for Metals and Alloys was established by the American Society for Testing and Materials (ASTM) and the SAE. The UNS number consists of a single letter prefix followed by five digits. The letter G indicates standard carbon steels, and H indicates standard

TABLE 1-3
Grade Designations of Standard Carbon Steels

Series Designation*	Type and Approximate Percentages of Identifying Elements
10XX	Nonresulfurized, 1.00% manganese maximum
11XX	Resulfurized
12XX	Rephosphorized and resulfurized
15XX	Nonresulfurized, over 1.00% manganese

* XX indicates carbon content in hundredths of a percent.

hardenability steels. The first four digits usually correspond to standard AISI, ASTM, or SAE steel designations, and the last digit usually indicates that an additional element such as lead or boron is specified. The number four indicates that lead is added, the number one indicates boron, and the number six indicates that an electric furnace is used for melting.

Hardenability Grades

Hardenability is a term used to designate that property of steel that determines the depth and distribution of hardness induced by quenching from the austenitizing temperature.[6] Hardenability is mainly determined by alloying elements in the steel, whereas maximum attainable hardness is dependent upon carbon content and cooling rate. Hardenability is also discussed in Chapter 10, "Heat Treatment of Steel," of this volume.

Methods of specifying hardenability requirements. The recommended method and equipment used to determine the hardenability of steel is described in SAE Standard J406, "Methods of Determining Hardenability of Steels." The hardenability bands are tabulated in SAE Standard J1268, "Hardenability Bands for Carbon and Alloy H Steels." Rockwell hardness C-scale (R_C) is used to designate the minimum and maximum hardnesses of the test bar at specified distances.

Hardenability bands. In the AISI/SAE grade designation system, steels specified to hardenability band limits are identified by the suffix letter H. In the UNS, the prefix letter H indicates steels specified to hardenability band limits. The chemical composition limits of these steels have been modified somewhat from those in the same grade of steel without specified hardenability band limits. The modifications permit adjustments in chemical composition to reflect individual plant melting characteristics that may influence the level and widths of the hardenability bands. The hardenability bands are applicable to killed, fine-grain carbon steels.

CARBON STEEL QUALITY

The term quality is indicative of internal soundness, relative uniformity of composition, relative freedom from detrimental surface imperfections, and finish for any given steel. Steel quality also relates to general suitability for particular applications. For example, cold rolled sheet steel is available in classes for either exposed or unexposed applications. Exposed applications require a good painted surface, whereas the surface finish is not important in unexposed applications.

Carbon steels can be obtained in a number of qualities that reflect various degrees of the conditions mentioned above. These quality designations are summarized in Table 1-4. Some of the qualities may be modified by such requirements as austenite grain size, special discard, macroetch test, special hardenability, maximum incidental alloy elements, restricted chemical composition, and nonmetallic inclusions. In addition, several products have special qualities that are intended for specific end uses or fabricating practices.[7]

ALLOYING ELEMENTS

Alloying elements added to carbon steel influence both steelmaking practice and the mechanical properties of the finished steel. The effect of any given element depend on the quantities of other elements present.

Some of the commonly specified alloying elements include carbon, manganese, phosphorus, sulfur, silicon, and aluminum. Other elements are also specified on occasion to obtain desired properties.

Carbon

Carbon is the principal alloying element in steel. As the carbon content increases, the tensile strength, yield strength, and hardness increase, whereas ductility, weldability, and toughness decrease. In rimmed steels, surface quality is impaired as carbon content increases. In contrast, the surface quality of killed steels is poorer in the low-carbon grades.

Manganese

The addition of manganese contributes to the strength and hardness of steel, but to a lesser degree than carbon. The amount of increase in these properties depends on the carbon content; for example, high-carbon steels are affected more by manganese than are low-carbon steels. Increasing the manganese content decreases weldability, but to a much lesser extent than carbon. Manganese tends to increase the rate of carbon penetration during carburizing and increases the hardenability of the steel. Manganese is beneficial to surface quality in all types of steel except low-carbon, rimmed steels, and is particularly beneficial in high-sulfur steels.

Phosphorus

Phosphorus increases the strength and hardness of carbon steels, but reduces ductility and impact toughness, particularly in high-carbon steels that are quenched and tempered. Phosphorus content is generally held well below the specified maximum. In some free-machining steels, however, phosphorus content is greater because it improves machinability.

Sulfur

Sulfur lowers the transverse ductility and notch-impact toughness of carbon steel, but has only a slight effect on longitudinal tensile properties. The weldability of a steel also decreases with increasing amounts of sulfur. In addition, sulfur is very detrimental to the surface quality of steel, particularly in low-carbon and low-manganese steels. To minimize these negative effects, a maximum sulfur content is usually specified. For certain steels, however, sulfur content may be increased to improve machinability, as in the resulfurized grades of steel.

Silicon

Silicon is one of the two principal deoxidizers used in steelmaking; therefore, silicon content is directly related to the type of steel being produced. Silicon increases the strength and hardness of steel, but generally impairs machinability and cold forming.

TABLE 1-4
Summary of Quality Designations for Carbon Steel Bars and Sheets

Quality Designations	Characteristics	Applications
Hot Rolled Bars		
Merchant quality	Produced to chemical composition (ladle analysis only) within limits of 0.50% maximum carbon, 0.60% maximum manganese, 0.05% maximum sulfur, and 0.04% maximum phosphorus. May contain pronounced chemical segregation, internal porosity, surface seams, and other surface irregularities. Size ranges are limited; and the type of steel may be rimmed, capped, semikilled, or killed.	Used in structural and similar applications requiring mild cold bending, mild hot forming, punching, and welding when producing non-critical parts of bridges, buildings, ships, agricultural implements, road-building equipment, railway equipment, and general machinery. This quality is not suitable for applications requiring hot forging, heat treating, cold drawing, or other operations that require internal soundness or relative freedom from surface imperfections.
Special quality	Basic or standard quality for carbon steel bars. Produced using rimmed, capped, semikilled, or killed deoxidizing practices.	Special-quality bars are used when end use, method of fabrication, or subsequent processing treatment requires quality characteristics not available in merchant quality. Typical applications include hot forged, heat treated, cold drawn, and machined parts, and many structural uses.
Scrapless-nut quality	Steel must be of controlled soundness and free from detrimental surface imperfections. Annealing or spheroidize annealing may be necessary to obtain proper cold forming characteristics.	Production of scrapless nuts by piercing, upsetting, and forming round bars.
Axle shaft quality	Special rolling practices, special billet and bar conditioning, and selective inspection are employed to minimize injurious surface imperfections.	Production of power-driven axle shafts for automotive or truck applications.
Cold extrusion quality	Characteristics vary with application. Heat treatment may be required to obtain cold-forming characteristics.	Used in the production of solid or hollow shapes by means of severe, cold plastic deformation involving forward extrusion, backward extrusion, or both, with or without expansion.
Cold heading and cold forging quality	Steel must be free from surface imperfections. Annealing or spheroidize annealing may be required to achieve forming characteristics.	Used in applications requiring severe cold forming.
Cold Finished Bars		
Standard quality	Produced from special-quality, hot rolled carbon steel bars. May contain surface imperand require stock removal for finishing. fections Restrictive requirement bars are also available.	
Cold heading and cold forging quality	Steel produced by closely controlled steel-making practices. Grades over 0.30% carbon may require heat treatment to obtain proper hardness and microstructure.	For applications involving severe, cold plastic deformation by upsetting, heading, or forging.
Cold extrusion quality	To minimize age hardening, a fully killed, fine-grained steel is used. A sound internal structure is required. Heat treatment may be required to obtain proper forming characteristics.	For applications that require steel to flow in a die to produce a desired shape.

(continued)

TABLE 1-4—*Continued*

Quality Designations	Characteristics	Applications
Hot Rolled Sheet		
Commercial quality	Produced from rimmed, capped, or semikilled steel and does not contain a high degree of uniformity in chemical composition and mechanical properties. May require heat treatment before drawing operations and may contain coil breaks, stretcher strains, and fluting.	Suitable for applications when the presence of oxides and surface imperfections are not objectionable. Not recommended for exposed parts that require good surface finish.
Drawing quality	Production requires the special selection of raw materials, the use of specially produced or selected steels, and exacting control of the processing operations. Subject to coil breaks, stretcher strains, and fluting.	Used in drawing applications when surface finish and surface disturbances are not objectionable. Used when fabricating parts that are too difficult to form with commercial-quality steel.
Drawing quality, special killed	Supplied as a low-carbon, aluminum-killed steel. Production requires special selection of raw materials, the use of specially produced or selected steel, and exacting control of processing operations. Subject to coil breaks, stretcher strains, and fluting.	Used when fabricating parts that are too difficult to form from drawing-quality steel, when delays between draws detrimentally affect the drawing performance, or when inherent qualities of special killed steel are required.
Structural quality	Mechanical properties influenced by chemical composition, thickness, and variables in mill design, mill practice, or both. Improper heat treatment may adversely affect steel properties.	Used when fabricating parts that have specified mechanical properties. Not recommended for exposed parts that require good surface finish.
Cold Rolled Sheet		
Commercial quality	Produced from low-carbon grade of rimmed, capped, or semikilled steel. Does not have a high level of ductility or a high degree of uniformity of chemical composition and mechanical properties. Has matte finish.	Suitable for exposed parts requiring good surface finish.
Drawing quality	Produced from specially processed steel but not commonly specified to chemical composition. Improper heat treatment may adversely affect steel properties.	Suitable for fabricating parts requiring a more severe deformation than permissible with commercial-quality steel.
Drawing quality, special killed	Normally produced with a matte finish. Chemical composition is homogenous, and mechanical properties are stable over time.	Used when fabricating parts that require severe drawing and forming.
Structural quality	Formability of structural-quality steel decreases with increasing yield strength and/or hardness. Surface characteristics are the same as commercial quality.	Used when special mechanical properties are required.

Aluminum

Aluminum is the other principal deoxidizer in steelmaking; however, it also performs other functions. Aluminum combines with nitrogen in the solid steel to minimize austenite grain growth and to eliminate or minimize the effects of strain aging. This combination also helps control the plastic strain ratio of sheet products, discussed under mechanical properties next in this chapter.

Other Elements

The metallurgical characteristics of a steel are also influenced by the addition of various elements that are not commonly specified. These elements are added to improve fabricating characteristics or influence service behavior.

Some of the elements occasionally specified include copper, boron, lead, nitrogen, selenium, tellurium, bismuth, and calcium. Copper is normally added to improve resistance to atmospheric corrosion; however, copper is detrimental to surface quality and hot working behavior. Boron is added to improve the hardenability of steel; it may also improve machinability and formability in relation to other steels of the same hardenability without boron. Lead, nitrogen, selenium, tellurium, and bismuth can be added to improve the machinability of carbon steels. Nitrogen is also a low-cost strengthening agent. Tellurium and calcium may be added to control the shape of the sulphide inclusions, thus improving formability and fracture toughness.

MECHANICAL PROPERTIES

Mechanical properties are those properties of the material that are associated with the material's reaction when a force is applied. Mechanical properties are usually determined from tension, bend, and hardness tests. The properties most commonly specified are tensile and yield strengths, total elongation, reduction in area, and hardness.

Hot rolled and cold drawn bars are usually produced to meet mechanical property requirements as well as limited compositional requirements. The tensile characteristics of hot rolled bars are mainly influenced by chemical composition, thickness or cross-sectional area, and variables in hot rolling and cooling practices. The effect of cold working on cold drawn bars depends on chemical composition, cross-sectional area, amount of cold reduction, and thermal treatment. During cold working, the yield strength of a material increases more than the tensile strength. Table 1-5 lists mechanical properties of both hot rolled and cold drawn bars.[8]

Data from tension tests are used to determine the mechanical properties of sheet steel that influence drawing and stretching. The two main properties are the plastic strain ratio (r) and the work-hardening exponent (n).

The plastic strain ratio is indicative of the ability of a sheet to resist thinning during drawing and is defined as the ratio of width strain to thickness strain in the tensile test. Since the properties of the sheet are different in different directions, the average strain ratio ($\bar{r}$) is given. As the $\bar{r}$ value increases, the depth of permissible draw increases. Typical r and $\bar{r}$ values for low-carbon steels are given in Table 1-6.

The work-hardening exponent (n) is a measure of the ability of the sheet to resist localized straining and thus increase uniform deformation. A metal with a high n value tends to strain uniformly even under nonuniform stress conditions. Typical n values for low-carbon steels are 0.20-0.22. Figure 1-2 can be used to approximate the n value of low-carbon sheet from its yield strength.

APPLICATIONS

The selection of a carbon steel for a particular application is largely determined by its carbon content. As previously stated, carbon steels can be divided into three main groups: low, medium, and high-carbon steels.

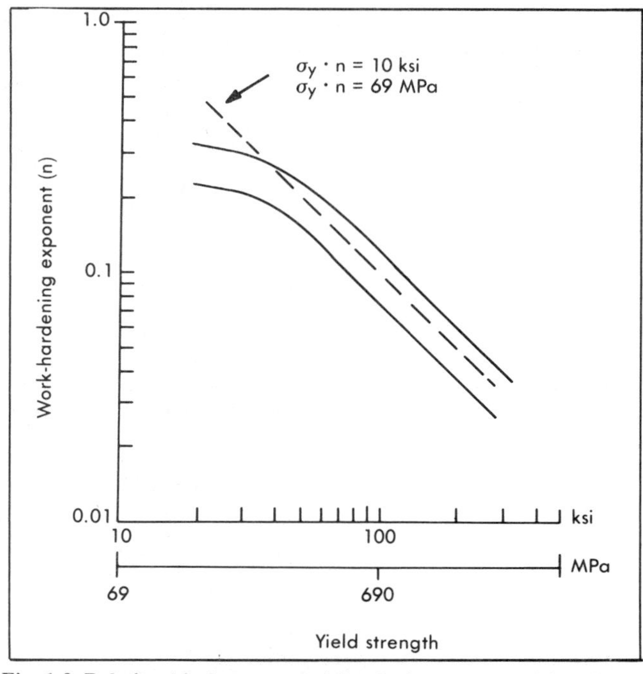

Fig. 1-2 Relationship between work-hardening exponent (n) and the yield strength. (*National Steel Corporation*)

Low-Carbon Steels

In general, low-carbon steels are used for industrial products such as nuts, bolts, sheet, strip, plates, shapes, tubes, and many machined components that are subject to low stresses. An important group of low-carbon steels are free-cutting or free-machining steels. In many instances, the products made from this class of steel are machined from hot or cold formed bars; products requiring a hard, wear-resistant surface can be subsequently surface (case) hardened.

Medium-Carbon Steels

The medium-carbon grades of steel are used when the strength and hardness requirements are greater than can be adequately met by low-carbon steels. The mechanical properties of this class of steel can be improved by quenching and tempering.

TABLE 1-5
Mechanical Properties and Machinability Ratings of Carbon Steel Bars[8]

Grade			Tensile	Estimated Minimum Values				
UNS No.	SAE and/or AISI No.	Condition*	Strength, ksi (MPa)	Yield Strength, ksi (MPa)	Elongation in 2″ (50 mm), %	Reduction in Area, %	Hardness, Bhn	Average Machinability Rating**
Nonresulfurized Carbon Steels, 1% Manganese Maximum								
G10060	1006	Hot rolled	43 (300)	24 (170)	30	55	86	50
		Cold drawn	48 (330)	41 (280)	20	45	95	

(continued)

TABLE 1-5—*Continued*

Grade			Tensile	Estimated Minimum Values				
UNS No.	SAE and/or AISI No.	Condition*	Tensile Strength, ksi (MPa)	Yield Strength, ksi (MPa)	Elongation in 2″ (50 mm), %	Reduction in Area, %	Hardness, Bhn	Average Machinability Rating**
G10080	1008	Hot rolled	44 (303)	24.5 (170)	30	55	86	55
		Cold drawn	49 (340)	41.5 (290)	20	45	95	
G10100	1010	Hot rolled	47 (320)	26 (180)	28	50	95	55
		Cold drawn	53 (370)	44 (300)	20	40	105	
G10150	1015	Hot rolled	50 (340)	27.5 (190)	28	50	101	60
		Cold drawn	56 (390)	47 (320)	18	40	111	
G10160	1016	Hot rolled	55 (380)	30 (210)	25	50	111	70
		Cold drawn	61 (420)	51 (350)	18	40	121	
G10180	1018	Hot rolled	58 (400)	32 (220)	25	50	116	70
		Cold drawn	64 (440)	54 (370)	15	40	126	
G10200	1020	Hot rolled	55 (380)	30 (210)	25	50	111	65
		Cold drawn	61 (420)	51 (350)	15	40	121	
G10300	1030	Hot rolled	68 (470)	37.5 (260)	20	42	137	70
		Cold drawn	76 (520)	64 (440)	12	35	149	
G10350	1035	Hot rolled	72 (500)	39.5 (270)	18	40	143	65
		Cold drawn	80 (550)	67 (460)	12	35	163	
G10370	1037	Hot rolled	74 (510)	40.5 (280)	18	40	143	65
		Cold drawn	82 (570)	69 (480)	12	35	167	
G10400	1040	Hot rolled	76 (520)	42 (290)	18	40	149	60
		Cold drawn	85 (590)	71 (490)	12	35	170	
G10450	1045	Hot rolled	82 (570)	45 (310)	16	40	163	55
		Cold drawn	91 (630)	77 (530)	12	35	179	65
		ACD	85 (590)	73 (500)	12	45	170	
G10460	1046	Hot rolled	85 (590)	47 (320)	15	40	170	55
		Cold drawn	94 (650)	79 (540)	12	35	187	65
		ACD	90 (620)	75 (520)	12	45	179	
G10490	1049	Hot rolled	87 (600)	48 (330)	15	35	179	45
		Cold drawn	97 (670)	81.5 (560)	10	30	197	55
		ACD	92 (630)	77 (530)	10	40	187	
G10500	1050	Hot rolled	90 (620)	49.5 (340)	15	35	179	45
		Cold drawn	100 (690)	84 (580)	10	30	197	55
		ACD	95 (660)	80 (550)	10	40	189	
G10600	1060	Hot rolled	98 (680)	54 (370)	12	30	201	60
		SACD	90 (620)	70 (480)	10	45	183	
G10700	1070	Hot rolled	102 (700)	56 (390)	12	30	212	55
		SACD	93 (640)	72 (500)	10	45	192	
G10800	1080	Hot rolled	112 (770)	61.5 (420)	10	25	229	45
		SACD	98 (680)	75 (520)	10	40	192	
G10950	1095	Hot rolled	120 (830)	66 (460)	10	25	248	45
		SACD	99 (680)	76 (520)	10	40	197	
Resulfurized Carbon Steels								
G11170	1117	Hot rolled	62 (430)	34 (230)	23	47	121	90
		Cold drawn	69 (480)	58 (400)	15	40	137	

TABLE 1-5—*Continued*

Grade UNS No.	SAE and/or AISI No.	Condition*	Tensile Strength, ksi (MPa)	Estimated Minimum Values: Yield Strength, ksi (MPa)	Elongation in 2″ (50 mm), %	Reduction in Area, %	Hardness, Bhn	Average Machinability Rating**
G11370	1137	Hot rolled	88 (610)	48 (330)	15	35	179	70
		Cold drawn	98 (680)	82 (570)	10	30	197	
G11410	1141	Hot rolled	94 (650)	51.5 (360)	15	35	187	70
		Cold drawn	105 (720)	88 (610)	10	30	212	
G11440	1144	Hot rolled	97 (670)	53 (370)	15	35	197	80
		Cold drawn	108 (740)	90 (620)	10	30	217	
Rephosphorized and Resulfurized Carbon Steels								
G12110	1211	Hot rolled	55 (380)	33 (230)	25	45	121	95
		Cold drawn	75 (520)	58 (400)	10	35	163	
G12120	1212	Hot rolled	56 (390)	33.5 (230)	25	45	121	100
		Cold drawn	78 (540)	60 (410)	10	35	167	
G12144	12L14	Hot rolled	57 (390)	34 (230)	22	45	121	160
		Cold drawn	78 (540)	60 (410)	10	35	163	
Nonrephosphorized Carbon Steels, over 1% Manganese								
G15410	1541	Hot rolled	92 (630)	51 (350)	15	40	187	45
		Cold drawn	102.5 (710)	87 (600)	10	30	207	60
		ACD	94 (650)	80 (550)	10	45	184	
G15520	1552	Hot rolled	108 (740)	59.5 (410)	12	30	217	50
		ACD	98 (680)	83 (570)	10	40	193	
G15480	1548	Hot rolled	96 (660)	53 (370)	14	33	197	45
		Cold drawn	106.5 (730)	89.5 (620)	10	28	217	50
		ACD	93.5 (640)	78.5 (540)	10	35	192	

(*Society of Automotive Engineers, Inc.*)

Note: The values of the mechanical properties listed are obtained from bars 3/4 to 1 1/4″ (20 to 30 mm) diam and should only be used for general information rather than for design guidelines. For specific values, contact the steel manufacturer.

* ACD = annealed cold drawn SACD = spheroidized, annealed cold drawn

** Based on cold drawn AISI 1212 = 100%

TABLE 1-6
Typical Values of Plastic Strain Ratio

Type of Steel	Plastic Strain Ratio,* r_0	r_{45}	r_{90}	$\bar{r}$
Rimmed, normalized	0.9	1.1	0.9	1.0
Rimmed, annealed	1.3	1.0	1.4	1.2
Aluminum killed (CR 40%, annealed)	1.4	1.2	1.6	1.4
Aluminum killed (CR 70%, annealed)	1.6	1.4	1.9	1.6

* The value r_0 is obtained in the rolling direction, r_{45} is in the diagonal direction, r_{90} is in the transverse direction, and $\bar{r}$ is the average of the three values based on the equation $\bar{r} = 1/4\,(r_0 + 2\,r_{45} + r_{90})$.

Medium-carbon grades are used in producing rails, railroad equipment, parts for lathes and presses, machined parts requiring moderate-to-high strength, heavy stamped or pressed products, crankshafts, connecting rods, axles, gears, and many other automotive parts. In addition, many items in the agricultural equipment and petroleum industries are made from medium-carbon steels.

High-Carbon Steels

High-carbon steels are used for manufacturing products that require high strength, high hardness, and, in certain instances, good wear resistance. Typical applications for high-carbon steels include cutting tools such as drills, reamers, taps and dies, and cutlery. High-carbon steels are also used for high-strength rope, cable, music wire, and springs. High-carbon steels are generally purchased in the annealed condition; the manufactured parts are then heat treated to obtain the desired properties.

MACHINING AND FABRICATING CHARACTERISTICS

The machining and fabricating characteristics of carbon steels depend on the properties of the particular grade of steel being machined or fabricated, as well as the specific equipment and tooling employed. The following sections provide general information to assist the manufacturing engineer in machining, forming, welding, and heat treating carbon steels. Detailed information on the various operations can be obtained from the references mentioned in each section.

Machinability

Machinability concerns the relative ease with which a steel can be cut in turning, drilling, milling, broaching, threading, reaming, or sawing. Machinability is influenced by machine and work material variables. Some common machine variables are cutting speed, dimensions of the cut, tool geometry and material, cutting fluid, condition of the machine, and type of tool engagement with the workpiece. Work material variables include hardness, tensile properties, chemical composition, microstructure, degree of cold work, strain hardenability, shape and dimension of workpiece, and rigidity of the workpiece.

Hot rolled carbon steels containing less than 0.25% carbon tend to be tough and gummy in machining. Increasing carbon and manganese content increases strength and hardness and results in improved surface finish and chip character. Increasing sulfur, phosphorus, or nitrogen content and adding lead also improves the machinability of carbon steels.

If carbon content is approximately 0.20-0.25%, machinability is improved over lower carbon grades in both hot rolled and cold drawn steels. Carbon content greater than 0.25% decreases machinability. Most carbon steels containing less than 0.35% carbon are machined in the as-rolled or as-rolled, cold drawn condition. Cold drawn grades containing greater amounts of carbon are usually annealed to improve machinability. In comparison to hot rolled bars of similar composition and microstructure, cold drawn bars have improved machinability because of the higher yield-tensile strength ratio. Machinability ratings of carbon steels are given in Table 1-5.

When machining both hot rolled and cold drawn carbon steels, it is necessary to allow for surface finishing. Resulfurized grades have a poorer surface finish and require more material to be removed for a proper surface finish than nonresulfurized grades. Table 1-7 gives the recommended minimum machining allowances, per side, for hot rolled and cold drawn carbon steel bars. To calculate the recommended allowance per side, multiply the diameter or thickness by the percentage allowed.

The total allowance calculated for nonresulfurized steels should always be greater than 0.010″ (0.25 mm) per side. If it is less, allow 0.010″ per side for machining. For resulfurized steels, the total allowance should always be greater than 0.015″ (0.38 mm) per side. If it is less, allow 0.015″ per side for machining. Carbon steels made to higher qualities may permit lesser amounts of surface removal. Additional information on machinability and machining processes for carbon steels can be found in Volume I, *Machining*, of this Handbook series.

Formability

Carbon steel bars and sheets are readily formed by a variety of processes. For bars and wires, these processes include forging, wire drawing, extruding, heading, and swaging. Sheet metal forming processes include bending, flanging, hemming, drawing, expanding, shrinking, stretch forming, roll forming, spinning, and several special forming processes. A discussion of these processes, including their advantages and applications, can be found in Volume II, *Forming*, of this Handbook series.

Low-carbon steels are the most easily formed because they contain less carbon and fewer alloying elements. Medium-carbon steels are usually not formed cold but can be successfully formed warm or hot. Both bar and sheet carbon steels are produced in special qualities that facilitate forming.

Weldability

Weldability is the capacity of a metal or combination of metals to be welded under fabrication conditions into a specific, suitably designed structure, and to perform satisfactorily in the intended service.[9] The weldability of carbon steel depends primarily on the carbon content or carbon equivalent, which in turn controls hardenability and the susceptibility of the welded structure to cracking or to hardening during thermal cycles induced by welding. Carbon equivalent is determined by the combined amount of carbon and other alloying elements present in steel.

Carbon steels with up to 0.30% carbon or with a carbon equivalent not over 0.40% are easily welded by arc, resistance, flash, oxyfuel gas, solid state, electron beam, or laser processes. The selection of the process is usually determined by the section thickness and the quality requirements of the weld. For carbon content over 0.15% and section thicknesses over 1.0″ (25 mm), it may be necessary to preheat the workpiece, control interpass temperature, and stress relieve the workpiece after welding. Resulfurized carbon steels have poor weldability due to their high sulfur content.

Carbon steels containing more than 0.30% carbon are weldable, but special techniques must be employed to prevent weld cracking. Preheating the workpiece from 300 to 600° F (150 to 316° C) and postheating between 1000 and 1200° F (550 and 650° C) helps to avoid any brittle microstructure. Low-hydrogen-type electrodes are recommended for welding these steels. Steels containing more than 1.00% carbon are not recommended for high-temperature welding processes. For additional information on welding processes and techniques, refer to Volume IV, *Quality Control and Assembly*, in this Handbook series.

TABLE 1-7
Recommended Machining Allowances for Hot Rolled and Cold Drawn Carbon Steels

Quality	Machining Operation	Nonresulfurized, %	Resulfurized, %
Hot rolled	Turned on centers	3.0	3.8
	Centerless turned or ground	2.6	3.4
	Other methods	1.6	2.4
Cold drawn		1.6	2.4

Heat Treatment

The versatility of steel can be attributed to its response to heat treatment. While the major percentage of steel is used in the as-rolled condition, heat treatment greatly broadens the spectrum of properties attainable.

Heat treatments fall into two general categories: (1) those that increase the strength, hardness, and toughness by virtue of rapid cooling from above the transformation range, and (2) those that decrease hardness and promote uniformity by slow cooling from above the transformation range, or by prolonged heating within or below the transformation range, followed by slow cooling. The first category can involve through hardening by quenching and tempering, or a variety of specialized treatments undertaken to enhance surface hardness to a controlled depth. The second category encompasses normalizing and various types of annealing to improve machinability, toughness, or cold forming characteristics.[10] Annealing after cold forming relieves stresses and restores ductility. Details of heat treating practices for steels are discussed in Chapters 10, 11, and 13 of this volume.

ALLOY STEELS

Simply stated, an alloy steel is a steel that has one or more alloying elements added to it to obtain properties not obtainable in carbon steels. Steel is considered to be an alloy steel when the maximum range for manganese, silicon, or copper exceeds 1.65, 0.60, and 0.60% respectively. A steel is also considered an alloy when a definite range or a minimum quantity is specified or required for aluminum, chromium (up to 3.99%), cobalt, columbium, molybdenum, nickel, titanium, tungsten, vanadium, zirconium, or any other alloying element.

The alloy steels discussed in this section are the low-alloy steels. These steels may be divided into the structural grades and those listed by the American Iron and Steel Institute (AISI) and the Society of Automotive Engineers (SAE). In the structural grades, the alloying elements are the principal means of strengthening the ferrite matrix. The structural grade alloy steels are generally used in the as-rolled condition, the quenched and tempered condition, and in the normalized or annealed condition. In the AISI/SAE grades, the alloying elements serve primarily to improve the mechanical properties over equivalent carbon steel and to enhance the response of the steel to heat treatment.

Alloy steels not included in this section are high-strength low alloy (HSLA) steels, stainless steels, and tool steels. The HSLA steels are discussed in the next section of this chapter, stainless steels are discussed in Chapter 2 of this volume, and tool steels are discussed in Volume I, *Machining*, and Volume II, *Forming*, of this Handbook series.

STEELMAKING PRACTICE

Alloy steel is made by basic open-hearth, basic oxygen, or basic electric furnace practices. In addition to the conventional melting practices, there are several methods of treating molten steel under vacuum, including vacuum-arc remelting, vacuum-induction remelting, and vacuum degassing. These methods improve the soundness, cleanliness, and mechanical properties (particularly transverse ductility, toughness, and fatigue life) of the steel by reducing its gaseous impurities and lowering inclusion content.

The casting techniques for alloy steels are the same as those discussed previously for carbon steels. Alloy steels are always produced as killed steels and are made using a fine-grain practice as described in American Society for Testing and Materials (ASTM) Standard E112.

Alloy steel, cold finished bars are produced from hot rolled steel by several cold finishing processes for the purpose of improving surface finish, dimensional accuracy, alignment, or machinability; also, in the case of cold drawn or cold rolled bars, to increase the yield strength and tensile strength. Cold finishing processes and surface improvement processes used singly or in combination include cold drawing, cold rolling, turning, grinding, polishing, and straightening. Cold finishing processes are frequently employed in conjunction with thermal treatments such as annealing, normalizing, quenching and tempering, and stress relieving when special properties are required in the finished bar.

Hot rolled steels, from which cold finished bars are produced, normally contain a decarburized, peripheral zone. The decarburized zone can be eliminated by sufficient mechanical surface removal. On bars that are only cold drawn, the decarburized, peripheral zone is not removed. Carbon can be restored to that zone by a carburization thermal treatment sometimes known as carbon restoration or carbon correction.

ALLOY STEEL GRADES

The grade of a particular alloy steel is commonly indicated by the percentage of the various elements that comprise its chemical composition. The composition may be specified by a maximum limit, a minimum limit, or by both minimum and maximum limits, which are referred to as the range. Lists of standard alloy steels designed to serve the needs of fabricators and users of steel products are published by the AISI and SAE. Specialized grades, steels not on the standard alloy steel lists, are also produced.

Grade Designation

As with low-carbon steels, a four-numeral series designates alloy steels specified to chemical composition ranges. For certain grades, a five-numeral series is used.

The last two digits of the four-numeral series indicate the approximate middle of the carbon range; for example, 20 represents a range of 0.18 to 0.23% carbon. In the five-numeral series, the last three digits represent the carbon range. The first two digits of both the four and five-numeral series indicate the primary alloying elements used in the grade, along with their approximate percentages. Table 1-8 defines the grade designation system as established by the AISI and the SAE.

The prefix letter E is used to designate steels normally made by the basic electric furnace practice. Steels without the prefix are normally manufactured by the basic open-hearth or basic oxygen processes.

ALLOY STEELS

TABLE 1-8
Alloy Steel Grade Designations

Series Designation		Type and Approximate Percentages of Elements
UNS	AISI/SAE*	
G13xxx	13xx	Manganese 1.75
G31xxx	31xx	Nickel 1.25 and chromium 0.65
G33xxx	33xx	Nickel 3.50 and chromium 1.55
G40xxx	40xx	Molybdenum 0.20 or 0.25; or molybdenum 0.25 and sulfur 0.042
G41xxx	41xx	Chromium 0.50, 0.95, or 1.05 and molybdenum 0.12, 0.20, or 0.30
G43xxx	43xx	Nickel 1.83, chromium 0.50 or 0.80, and molybdenum 0.25
G44xxx	44xx	Molybdenum 0.40 or 0.53
G46xxx	46xx	Nickel 0.85 or 1.83 and molybdenum 0.20 or 0.25
G47xxx	47xx	Nickel 1.05, chromium 0.45, and molybdenum 0.20
G48xxx	48xx	Nickel 3.50 and molybdenum 0.25
G50xxx	50xx	Chromium 0.28 or 0.40
G51xxx	51xx	Chromium 0.80, 0.88, 0.93, 0.95, or 1.00
G5xxxx	5xxx	Carbon 1.00 and chromium 1.03 or 1.45
G61xxx	61xx	Chromium 0.60, 0.80, or 0.95 and vanadium 0.10 and/or 0.15 min
G71xxx	71xx	Chromium 1.60, molybdenum 0.35, and aluminum 1.13
G81xx1	81Bxx	Nickel 0.30, chromium 0.45, and molybdenum 0.12
G86xxx	86xx	Nickel 0.55, chromium 0.50, and molybdenum 0.20
G87xxx	87xx	Nickel 0.55, chromium 0.50, and molybdenum 0.25
G88xxx	88xx	Nickel 0.55, chromium 0.50, and molybdenum 0.35
G92xxx	92xx	Silicon 2.00; or silicon 1.40 and chromium 0.70
G93xxx	93xx	Nickel 3.25, chromium 1.20, and molybdenum 0.12
G94xx1	94Bxx	Nickel 0.45, chromium 0.40, and molybdenum 0.12
G98xxx	98xx	Nickel 1.00, chromium 0.80, and molybdenum 0.25
Gxxxx1	B	Denotes boron steel, as in 86B45 (G86451) and others

(*American Iron and Steel Institute*)

Note: All values shown are mean values.
* xx indicates carbon content in hundredths of a percent.

In 1975, the Unified Numbering System (UNS) for Metals and Alloys was established by the ASTM and the SAE. The UNS designates the various alloy steel grades using a single letter prefix followed by five digits. The letter G indicates standard alloy or carbon steels, and the first four digits usually correspond to the AISI/SAE steel designations. The fifth digit indicates an additional element, such as boron (indicated by the number one), or a particular manufacturing practice.

Hardenability Grades

As a result of cooperative work done by the SAE and the AISI, hardenability bands have been developed for many of the constructional alloy steels. The hardenability limits were determined from data obtained by conducting standard 1" (25.4 mm) Jominy end-quench hardenability tests (ASTM Standard A 256) on many heats of each composition. Hardenability is also discussed in Chapter 10, "Heat Treatment of Steel," later in this volume.

As a means of identifying steels specified to hardenability requirements, the suffix letter H has been added to the conventional series number. The UNS designates these steels with the prefix letter H instead of G which is used to designate standard alloy or carbon steels.

ALLOY STEEL QUALITY

Alloy steels are made with more than ordinary care throughout their manufacture. They are more sensitive to thermal and mechanical operations, the control of which is complicated by the varying effects of different chemical combinations.

The quality characteristics of alloy steel include, among others, internal soundness, uniformity of chemical composition, and freedom from injurious surface imperfections. The degree to which these characteristics can be obtained is limited by existing raw materials, manufacturing methods, and the technological nature of the alloy steel. Quality characteristics are related to the suitability of the steel to make a particular part.

The qualities of alloy steels that are summarized in Table 1-9 concern characteristic properties that are adapted to the particular conditions encountered in the fabrication or use for which the steels are intended.

ALLOYING ELEMENTS

Alloying elements are added to ordinary steels for the purpose of modifying their behavior during heat treatment, which in turn results in improvement of the mechanical and physical properties. Specifically, the additions are made for one or more of the following reasons:

- Improve tensile strength without lowering material ductility.
- Improve toughness.
- Increase hardenability, which permits the hardening of larger sections than possible with plain carbon steels or allows successful quenching with less drastic cooling rates, reducing the hazard of distortion and quench cracking.
- Retain physical properties at elevated temperatures.
- Obtain better corrosion resistance.
- Improve wear resistance.
- Impart a fine grain size to the steel.
- Improve surface (case) hardening characteristics.

The effects of carbon, manganese, phosphorus, sulfur, silicon, aluminum, copper, and boron are discussed in the earlier section on carbon steels.

Nickel

Nickel is one of the common steel alloying elements. When present in appreciable amounts, it provides simplified and more economical heat treatment, increased hardenability, less distortion in quenching, improved corrosion resistance, and improved toughness, particularly at low temperatures.

TABLE 1-9
Summary of Quality Designations for Hot and Cold Rolled Alloy Steel Bars

Quality Designations	Characteristics	Applications
Regular quality	Basic or standard quality for alloy steel. These steels are killed and are produced as fine grain. May contain surface imperfections.	Used for regular constructional applications.
Axle shaft quality	Special rolling practices, special billet and bar conditioning, and selective inspection are employed to minimize surface imperfections.	Used for power-driven axle shafts for automobiles and trucks.
Ball and roller bearing quality	Subjected to restricted melting and special teeming, rolling, cooling, and conditioning practices. Thorough examinations for internal imperfections are performed.	Used for antifriction bearings.
Cold heading quality	Bars are supplied from steel produced by closely controlled steelmaking practices and are subject to testing and inspection to determine internal soundness, uniformity of chemical composition, and freedom from detrimental surface imperfections. Hardness and microstructure controlled by heat treatment.	Used in the production of fasteners, studs, anchor pins, bearing rollers, and cap screws by cold plastic deformation.
Special cold heading quality	Produced by closely controlled steelmaking practices to provide uniform chemical composition and internal soundness. Surface imperfections removed at intermediate stages by grinding or equivalent surface preparation. Hardness and microstructure controlled by heat treatment.	Used for applications requiring severe, cold plastic deformation such as for front suspension studs, socket screws, and valves.
Aircraft quality	Produced using exacting steelmaking, rolling, and testing practices. Phosphorus and sulfur limited to 0.025% maximum.	Used for highly stressed aircraft, missile, and rocket parts.

Chromium

Chromium is used in constructional alloy steels primarily to increase hardenability, provide improved resistance to abrasion and wear, and to promote carburization. It also contributes to corrosion and heat resistance. A maximum content of 3.99% chromium has been established for constructional alloy steels. Heat-resisting and stainless steels have much higher chromium percentages.

Molybdenum

Molybdenum is a nonoxidizing element that promotes hardenability of steel and is useful when hardenability control is important. Molybdenum provides hardenability with minimal detrimental effects on cold forming characteristics. It widens the temperature range of effective heat treating response since it has a strong tendency to form stable carbides. It also increases the tensile and creep strengths at high temperatures.

Vanadium

Vanadium increases the hot rolled mechanical properties of steel and may be used to enhance hardenability, provided that it is not combined into carbides. It is a deoxidizer and forms carbides and more stable carbo-nitrides. Vanadium inhibits grain growth and promotes a fine-grain structure that imparts strength and toughness to HSLA steels. It also provides secondary hardening during tempering through precipitation hardening.

Tungsten

Tungsten increases hardness, promotes a fine-grain structure, and is excellent for resisting heat. At elevated tempering temperatures, tungsten forms tungsten carbides, which are very hard and stable. The tungsten carbide helps prevent the steel from softening during tempering.

Columbium/Niobium

Columbium (or niobium) decreases the hardenability of steel by promoting a fine-grain structure when it is precipitated as a carbide or nitride. Columbium in solid solution increases hardenability. The fine grain size results in increased strength and impact resistance in the hot rolled condition of HSLA steels.

Titanium

Titanium is primarily used as a deoxidizer and helps to limit grain growth. When added to boron steels, it helps increase hardenability by preventing the loss of boron through reaction with nitrogen in the steel. It can also provide precipitation hardening by forming titanium nitrides.

Cobalt

In alloy steels containing more than 0.40% carbon, cobalt decreases hardenability. In low-carbon chromium steels, however, cobalt has improved the hardenability of the steel.

Zirconium

Zirconium inhibits grain growth and is a better deoxidizer than boron, silicon, titanium, vanadium, or manganese when precipitated as zirconium nitride. Its primary use is to improve hot rolled properties in HSLA steels. Zirconium in solution also improves hardenability slightly.

MECHANICAL PROPERTIES

Alloy steels are not directly produced to specific mechanical properties, but are usually heat treated to achieve desired properties. Cold finished alloy steel bars usually require thermal treatments in order to meet definite limitations for tensile or hardness values. Alloy steels in the annealed and cold finished condition can be produced to specified maximum hardness limits. For steels in the normalized and cold finished condition, minimum hardness or minimum tensile strength may be specified. If the steels are normalized and tempered before cold finishing, either maximum and minimum hardnesses or maximum and minimum tensile values can be produced to a range that varies with the tensile strength level and is equivalent to a Brinell indentation diameter range of four-tenths of a millimeter (e.g., 4.0 to 4.4) at any specified location. If the steels are quenched and tempered before cold finishing, either maximum and minimum hardnesses or maximum and minimum tensile strength values can be produced to a range that varies with the tensile strength level and is equivalent to a Brinell indentation diameter range of three-tenths of a millimeter (e.g., 3.6 to 3.9) at any specified location.

When both hardness and tensile values are specified at the same position, the limits should be consistent with each other. In many cases, when the Brinell limits are specified as surface values, the tensile test results, which are of necessity obtained below the surface, and the surface hardness results will not be consistent because they vary according to the size of bar and the hardenability of the steel involved. For that reason the purchaser should recognize inconsistencies between the two and specify limits accordingly. In either case, it is essential that the position at which Brinell hardness values are taken be specified by the purchaser.

Generally the yield, elongation, and reduction of area are specified as minimums for steel in the quenched and tempered or normalized and tempered conditions, and they should be consistent with the tensile strength or Brinell hardness. Table 1-10 lists the mechanical properties of commonly used grades of alloy steel bars.

TABLE 1-10
Representative Mechanical Properties of Various Alloy Steels*

Grade					Total		
UNS	AISI/SAE	Condition	Tensile Strength, ksi (MPa)	Yield Strength (0.2% Offset), ksi (MPa)	Elongation in 2″ (50 mm), %	Reduction of Area, %	Hardness, Bhn
G13400	1340	Annealed, 1475°F (800°C)	102 (703)	63.3 (436)	25.5	57.3	207
		Normalized, 1600°F (870°C)	121.3 (836)	81 (558.5)	22	62.9	248
		Oil quenched, 1525°F (830°C), and tempered, 1100°F (595°C)	118 (814)	98.3 (677)	21.7	60.1	241
G40270	4027	Annealed, 1585°F (865°C)	75 (517)	57.3 (326)	30	52.9	143
		Normalized, 1660°F (905°C)	93.3 (643)	61.3 (422)	25.8	60.2	179
		Water quenched, 1585°F (865°C), and tempered, 1000°F (540°C)	139.3 (960)	122.3 (843)	18.8	60.1	285
G41180	4118	Annealed, 1600°F (870°C)	75 (517)	53 (365)	33	63.7	137
		Normalized, 1670°F (910°C)	84.5 (583)	56 (386)	32	71	156
		Carburized, 1700°F (925°C), and tempered, 300°F (150°C)	119 (820.5)	64.5 (445)	21	37.5	241
G41300	4130	Annealed, 1585°F (865°C)	81.3 (560)	52.3 (360)	28.2	55.6	156
		Normalized, 1600°F (870°C)	97 (669)	63.3 (436)	25.5	59.5	197
		Water quenched, 1575°F (855°C), and tempered, 1000°F (540°C)	144.5 (996)	129.5 (893)	18.5	61.8	293
G41400	4140	Annealed, 1500°F (815°C)	95 (655)	60.5 (417)	25.7	56.9	197
		Normalized, 1600°F (870°C)	148 (1020)	95 (655)	17.7	46.8	302
		Oil quenched, 1550°F (845°C), and tempered, 1100°F (595°C)	140.3 (967)	135 (931)	19.5	62.3	285

TABLE 1-10—*Continued*

Grade		Condition	Tensile Strength, ksi (MPa)	Yield Strength (0.2% Offset), ksi (MPa)	Total Elongation in 2″ (50 mm), %	Reduction of Area, %	Hardness, Bhn
UNS	AISI/SAE						
G41500	4150	Annealed, 1525° F (830° C)	105.8 (729)	55 (379)	20.2	40.2	197
		Normalized, 1600° F (870° C)	167.5 (1155)	106.5 (734)	11.7	30.8	321
		Oil quenched, 1525° F (830° C), and tempered, 1100° F (595° C)	165.5 (1141)	150 (1034)	15.7	51.1	331
G43200	4320	Annealed, 1560° F (850° C)	84 (579)	61.6 (425)	29	58.4	163
		Normalized, 1640° F (895° C)	115 (793)	67.3 (464)	20.8	50.7	235
		Carburized, 1700° F (925° C), and tempered, 300° F (150° C)	152.5 (1052)	107.3 (739)	17	51	302
G43400	4340	Annealed, 1490° F (810° C)	108 (745)	68.5 (472)	22	49.9	217
		Normalized, 1600° F (870° C)	185.5 (1279)	125 (862)	12.2	36.3	363
		Oil quenched, 1475° F (800° C), and tempered, 1100° F (595° C)	164.8 (1136)	159 (1096)	16.5	54.1	331
G44190	4419	Annealed, 1675° F (915° C)	64.8 (446)	48 (331)	31.2	62.8	121
		Normalized, 1750° F (955° C)	75.3 (519)	51 (352)	32.5	69.4	143
		Carburized, 1700° F (925° C), and tempered, 300° F (150° C)	97.3 (670.5)	62.8 (433)	24.2	66.4	201
G46200	4620	Annealed, 1575° F (855° C)	74.3 (512)	54 (372)	31.3	60.3	149
		Normalized, 1650° F (900° C)	83.3 (574)	53.1 (366)	29	66.7	174
		Carburized, 1700° F (925° C), and tempered, 300° F (150° C)	98 (676)	67 (462)	25.8	70	197
G48200	4820	Annealed, 1500° F (815° C)	98.8 (681)	67.3 (464)	22.3	58.8	197
		Normalized, 1580° F (860° C)	109.5 (755)	70.3 (484)	24	59.2	229
		Carburized, 1700° F (925° C), and tempered, 300° F (150° C)	169.5 (1169)	126.5 (872)	15	51	352
G51400	5140	Annealed, 1525° F (830° C)	83 (572)	42.5 (293)	28.6	57.3	167
		Normalized, 1600° F (870° C)	115 (793)	68.5 (472)	22.7	59.2	229
		Oil quenched, 1550° F (845° C), and tempered, 1100° F (595° C)	127.3 (878)	105 (724)	20.5	61.7	262
G51500	5150	Annealed, 1520° F (827° C)	98 (676)	51.8 (357)	22	43.7	197
		Normalized, 1600° F (870° C)	126.3 (870)	76.8 (529)	20.7	58.7	255
		Oil quenched, 1550° F (845° C), and tempered, 1100° F (595° C)	137 (945)	115.3 (795)	20.2	59.5	277
G51600	5160	Annealed, 1495° F (815° C)	104.8 (722)	40 (276)	17.2	30.6	197
		Normalized, 1575° F (855° C)	138.8 (957)	77 (531)	17.5	44.8	269
		Oil quenched, 1525° F (830° C), and tempered, 1100° F (595° C)	145.3 (1001)	126 (869)	18	53.6	302
G61500	6150	Annealed, 1500° F (815° C)	96.8 (667)	59.8 (412)	23	48.4	197
		Normalized, 1600° F (870° C)	136.3 (939)	89.3 (615)	21.8	61	269
		Oil quenched, 1550° F (845° C), and tempered, 1100° F (595° C)	158.3 (1091)	150.5 (1038)	16	53.2	311
G86200	8620	Annealed, 1600° F (870° C)	77.8 (536)	55.9 (385)	31.3	62.1	149
		Normalized, 1675° F (915° C)	91.8 (633)	51.8 (357)	26.3	59.7	183
		Carburized, 1700° F (925° C), and tempered, 300° F (150° C)	126.8 (874)	83.8 (577)	20.8	52.7	255
G86300	8630	Annealed, 1550° F (845° C)	81.8 (564)	54 (372)	29	58.9	156
		Normalized, 1600° F (870° C)	94.3 (650)	62.3 (429)	23.5	53.5	187
		Water quenched, 1550° F (845° C), and tempered, 1000° F (540° C)	134.8 (929)	123 (848)	18.7	59.6	269

(continued)

TABLE 1-10—*Continued*

Grade UNS	Grade AISI/SAE	Condition	Tensile Strength, ksi (MPa)	Yield Strength (0.2% Offset), ksi (MPa)	Total Elongation in 2″ (50 mm), %	Reduction of Area, %	Hardness, Bhn
G86500	8650	Annealed, 1465° F (795° C)	103.8 (715)	56 (386)	22.5	46.4	212
		Normalized, 1600° F (870° C)	148.5 (1024)	99.8 (688)	14	40.4	302
		Oil quenched, 1475° F (800° C), and tempered, 1100° F (595° C)	153.5 (1058)	142.8 (984)	17.7	57.3	311
G87400	8740	Annealed, 1500° F (815° C)	100.8 (695)	60.3 (415)	22.2	46.4	201
		Normalized, 1600° F (870° C)	134.8 (929)	88 (607)	16	47.9	269
		Oil quenched, 1525° F (830° C), and tempered, 1100° F (595° C)	149.3 (1029)	134.5 (927)	18.2	59.9	302
G92550	9255	Annealed, 1550° F (845° C)	112.8 (777)	70.5 (486)	21.7	41.1	229
		Normalized, 1650° F (900° C)	135.3 (933)	84 (579)	19.7	43.4	269
		Oil quenched, 1625° F (885° C), and tempered, 1100° F (595° C)	150 (1034)	118 (814)	19.2	44.8	293

Note: the values of the mechanical properties listed are for comparison purposes only and are not intended for specifications or design purposes. For specific information, consult the steel manufacturer.
* Based on 1″ (25.4 mm) diam bars.

APPLICATIONS

As was stated previously, low-alloy steels can be divided into the structural and AISI/SAE groups. The structural group is produced according to ASTM specifications.

Structural Grades

The high-strength structural steels are used principally in the transportation and construction industries for applications that require moderately high strength and weight reduction. An alloy combination for a common low-alloy structural steel is usually balanced to produce a minimum tensile strength of about 70 ksi (483 MPa), with a corresponding minimum yield strength of about 55 ksi (379 MPa).

AISI/SAE Grades

The AISI alloy steels are used particularly in the automotive and aircraft industries for highly stressed members such as gears, studs, and axles and moving engine parts such as cams, crankshafts, and valves. Certain combinations of the various alloying elements, after appropriate heat treatment, can impart to any one steel certain specialized characteristics for use in specific applications. For example, carbon-molybdenum and several other molybdenum-bearing steels possess good creep characteristics and therefore find useful application for moderately high-temperature service, when oxidation is not too severe. Typical applications are found in piping for steam and oil refineries.

The nickel-chromium steels as a group exhibit excellent hardenability, high strength, good wear resistance, and toughness. The various nickel-chromium combinations, properly heat treated, demonstrate tensile properties embracing the entire range available with alloy steels.

The chromium-vanadium steels, after heat treatment, show remarkable toughness and good fatigue resistance. As such, they find wide application when the part is subjected to reversing cycles such as with leaf and coil springs.

The low-alloy machinery steels are generally characterized by high tensile strength, good ductility, and excellent toughness when appropriately heat treated. The alloy content in these steels imparts good hardenability to the steel and permits the steel to be oil quenched to obtain these characteristics and air quenched when the mass of the section is small enough. This combination of characteristics is also desirable from the standpoint of preventing serious distortion during heat treatment.

MACHINING AND FABRICATING CHARACTERISTICS

To secure the most satisfactory results, purchasers normally consult with the steel producers regarding the working, machining, heat treating, or other operations to be used in fabricating the steel ordered, the mechanical properties to be obtained, and the conditions of service for which the finished parts are intended. Particular attention should be given to informing the producer regarding the details of the first operation to which the steel will be subjected and subsequent operations when significant.

Alloy steels containing over 0.38% carbon are customarily given a thermal treatment prior to cold finishing. For best results when machining, cold heading, or performing other fabricating operations, thermal treatment of alloy steels having lower carbon content may be required.

Machining

Machinability concerns the relative ease with which a steel is cut by sharp tools in various operations, such as turning, drilling, milling, broaching, threading, reaming, or sawing. Machinability involves the concepts of tool life and surface finish and is influenced in an important way by cutting speed, tool geometry, cutting fluid, rigidity of the workpiece, and mechanical condition of the machine tool.

The characteristics of steel that influence machinability are composition, special additives, treatment, and structure. The chemical composition has a major influence since it affects the microstructure and mechanical properties.

Most low-carbon alloy steels are machined in the as-rolled or as-rolled and cold drawn condition. Higher carbon alloy steels and high-hardenability, low-carbon alloy steels may be conditioned for machining by annealing, either for softening or for producing a specified microstructure.

Average machinability ratings for cold finished alloy steel bars are given in Volume I, *Machining*, of this Handbook series. Cutting speeds for the various machining operations are also given in appropriate chapters of Volume I.

Forming

Alloy steels are not widely used in forming operations other than forging and heading. Special quality designations are assigned to those alloy steels that are used in different forming operations. The various forming operations are discussed in Volume II, *Forming*, of this Handbook series.

Welding

Alloy steels with a carbon content lower than 0.10% can be readily welded by most welding techniques. Since their carbon content is low, preheating or postheating is not required. Alloy steels containing between 0.10 and 0.30% carbon are slightly more difficult to weld than the steels with lower carbon content. Preheating and postheating of these steels are recommended to reduce internal stresses. Alloy steels containing more than 0.30% carbon are difficult to weld, and preheating and postheating techniques are required.

Heat treating the alloy steels with a higher carbon content helps to produce a uniform structure in the weld metal and the parent metal. Low-hydrogen electrodes are recommended when arc welding these steels to reduce brittleness in the weld. For additional information on welding techniques and processes, refer to Volume IV, *Quality Control and Assembly*, in this Handbook series.

Heat Treatment

Alloy steels are usually heat treated to achieve the required properties for a given application. Heat treating practices for steel are discussed in Chapter 10, "Heat Treatment of Steel," later in this volume.

HIGH-STRENGTH LOW-ALLOY STEELS

High-strength low-alloy (HSLA) steels are a group of steels that exhibit and develop strengths significantly higher than carbon steels owing to the addition of small amounts of alloying elements, coupled with special steel processing methods. The carbon content of these steels is usually less than 0.30% by weight. Small amounts of manganese, silicon, phosphorus, copper, aluminum, chromium, niobium, vanadium, titanium, molybdenum, nickel, zirconium, nitrogen, calcium, and rare earth elements are used singly or in combination to increase strength, toughness, formability, and corrosion resistance.

The total alloy content of a few grades of HSLA steel is high enough to qualify them as alloy steels. However, HSLA steels are considered distinct from traditional alloy steels, such as constructional alloy steels, since, with a few exceptions, they achieve their high strength without separate heat treatment after finishing.

PRODUCT FORMS AND APPLICATIONS

High-strength low-alloy steels are produced in a variety of product forms. Most HSLA steels are produced as hot rolled products including sheet and strip, plates, structural shapes, and bars. A few grades are also produced as cold rolled sheet and strip. Flat rolled product forms are also available with protective coatings, such as zinc, for corrosion resistance.

Sheet and strip products, either in the form of coils or cut lengths, are used in automobiles, trucks and trailers, agricultural equipment, mining equipment, railway equipment, tanks and containers, and in other miscellaneous industrial applications. Plate products are also used in these applications, and, in addition, they are used in bridges, ships, buildings, line pipe, and other structural applications. Structural shapes, which include I-beams, H-beams, channels, angles, tees, and zees, find a wide range of structural applications. Bars are used in applications that require cold and hot forming for the manufacture of structural parts for diverse industrial equipment and machinery.

PRODUCTION OF HSLA STEELS

High-strength low-alloy steels are produced using special rolling and finishing practices to achieve their high strength.

Melting and Casting

The steelmaking practices used for HSLA steels are in most respects the same as those for carbon steels. A description of these steelmaking practices is discussed in the section on carbon steels. The vacuum treatments and ladle metallurgy processes described for carbon steels are also applicable to HSLA steels. Both ingot and strand (continuous) casting methods are used. Killed, semikilled, and nonkilled grades are produced.

Rolling and Finishing

The rolling and finishing practices used for HSLA steels are also similar to those used for carbon steels. However, since the level of alloy addition is kept low, HSLA steels often rely on special processing methods in order to obtain optimum properties. Hot rolled products, particularly flat rolled products, use combinations of low reheating temperatures prior to rolling and low, finish hot rolling temperatures (compared to carbon steels), as well as accelerated cooling practices, to obtain maximum strength and toughness. These practices are often referred to as controlled rolling. A few hot rolled grades are normalized or quenched and tempered rather than controlled rolled to obtain good combinations of strength and toughness. Certain grades are given low-temperature aging treatments following hot rolling, normalizing, or quenching to achieve optimum properties.

Cold rolled HSLA sheet and strip steels are annealed after cold rolling to obtain specific combinations of strength and ductility. In most cases, annealing develops a recrystallized ferrite plus carbide microstructure that results in higher strength along with good ductility. Low-temperature annealing (recovery

HIGH-STRENGTH LOW-ALLOY STEELS

annealing) is used for some grades to obtain an unrecrystallized microstructure that exhibits somewhat higher strength but with limited ductility.

Dual-phase HSLA steels achieve high strength along with superior ductility through the use of high-temperature intercritical annealing. During the high-temperature anneal, a ferrite plus austenite microstructure is obtained. This microstructure transforms to ferrite plus martensite during rapid cooling following the anneal.

TYPES OF HSLA STEELS

High-strength low-alloy steels have been categorized or grouped largely on composition.[11,12] Several types can often achieve a given strength level, but with varying degrees of toughness, formability, weldability, and corrosion resistance. In addition, the strength, toughness, and formability of a given type of HSLA steel can vary depending on the rolling and finishing practices used during production.

HSLA steels are typified by a high strength-to-weight ratio, and, as a result, yield strength is an important consideration. Hot rolled grades exhibit yield strengths ranging from 42 to 90 ksi (290 to 620 MPa). Cold rolled sheet and strip grades develop yield strengths from 40 to 140 ksi (276 to 965 MPa).

Table 1-11 gives the chemical compositions for the various types of HSLA steels. Table 1-12 lists the various types of HSLA steels, along with a summary of their mechanical properties and manufacturing characteristics. Comparisons of HSLA properties with those of carbon steels are discussed subsequently.

Niobium (Columbium)/Vanadium Steels

This category of HSLA steels contains niobium and/or vanadium additions in amounts of approximately 0.1% by weight or less. These alloying elements combine with carbon and/or nitrogen to form fine precipitates in the microstructure during controlled rolling with the result that fine ferrite grain size is obtained. Subsequent precipitation hardening during cooling results in high strength. The steels are available in all product forms, are readily weldable, and have good formability. Toughness of hot rolled products is poor when this type of alloy steel is semikilled and conventionally rolled. When fully killed, toughness, along with formability and fatigue resistance, is significantly improved. These steels can also be controlled rolled to obtain excellent toughness along with high strength.

Manganese-Copper Steels

This group of HSLA steels contains higher amounts of manganese along with additions of copper to improve strength and corrosion resistance. Resistance to corrosion is about twice that of carbon steels. This type is produced largely as plates; however, a few grades are available as sheet,

TABLE 1-11
Chemical Compositions for HSLA Steels[11, 12]

Type	%C (max)	%Mn (max)	%Si (max)	Other
Niobium/ vanadium	0.09-0.26	0.8-1.65	0.1-0.9	Nb (0.005-0.01% min) and/or V (0.01-0.1% min). Nitrogen may be added with V.
Manganese-copper	0.25-0.28	1.6	0.3	Cu (0.2% min)
Manganese (heat treated)	0.14-0.24	1.35-1.6	0.3-0.6	Some grades use Ni, Cr, Mo, Nb, V, and Cu (opt.).
Managanese-vanadium	0.22	1.5	find	V (0.4% min). Some grades use nitrogen or Nb (0.01-0.05%).
Manganese-titanium	0.12-0.15	0.9-1.0	find	Ti (0.05-0.07% min)
Manganese-vanadium-copper	0.15-0.22	1.25	0.3-0.4	Cu (0.2% min), V (0.02% min)
Multiple alloy plus copper	0.12-0.25	0.9-1.6	0.3-1.0	Cu (0.2% min). Cr, Ni, Mo, V, Ti, and Nb may be added in various combinations.
Multiple alloy plus copper and phosphorus	0.12-0.22	1.0-1.25	0.2-0.9	Cu (0.2-0.55% min), P (0.12-0.15% max). Cr,

(continued)

TABLE 1-11—*Continued*

Type	%C (max)	%Mn (max)	%Si (max)	Other
				Ni, and Mo added in various combinations.
HSLA special formability	0.09-0.18	0.6-1.65	0.1-0.9	Nb (0.005-0.01% min) or V (0.01-0.02% min) or Ti (0.05% min) added. Nitrogen may be used with V, Ce, and Zr, and Ca used for sulfide shape control and desulfurization.
HSLA precipitation hardening	0.07-0.2	0.65-1.0	0.15-35	Cu (0.75-1.0% min), Ni (1.0-2.2% max). Cr, Mo, Nb, and V used in some grades.

Note: The ranges given for the maximum and minimum percentages for various elements represent the variations in these maximums and minimums for different manufacturers.

TABLE 1-12
Mechanical Properties and Manufacturing Characteristics for HSLA Steels[11, 12]

Type	Tensile Strength, ksi (MPa)	Yield Strength, ksi (MPa)	Formability	Weldability
Niobium/vanadium	60-90 (415-620)	42-80 (290-550)	Good	Weldable by filler metal and resistance methods.
Manganese-copper	70 (485)	50 (345)	Fair	Not recommended for welding.
Manganese (heat treated)	63-80 (435-550)	42-60 (290-415)	Fair	Weldable by filler metal methods.
Manganese-vanadium	80 (515)	60 (415)	Good	Weldable by filler metal and resistance methods.
Manganese-titanium	65-95 (448-655)* 65-140 (448-965)**	50-80 (345-350)* 50-140 (345-965)**	Good	Weldable by filler metal and resistance methods.
Manganese-vanadium-copper	60-70 (415-485)	45-50 (310-345)	Good	Weldable by filler metal and resistance methods.
Multiple alloy plus copper	60-90 (485-620)	45-80 (310-550)	Good	Weldable by filler metal and resistance methods.
Multiple alloy plus copper and phosphorus	70 (485)	50 (345)	Fairly good	Weldable by filler metal and resistance methods.
HSLA special formability	50-95 (345-965)	40-80 (310-550)	Excellent	Weldable by filler metal and resistance methods.
HSLA precipitation hardening	72-100 (500-690)	65-85 (450-585)	Very good	Weldable by filler metal and resistance methods.

Note: The values for tensile and yield strengths are minimums. The strength ranges shown are variations in minimum strength levels produced by various manufacturers.
* Hot rolled
** Cold rolled

HIGH-STRENGTH LOW-ALLOY STEELS

strip, bars, and shapes. Weldability is relatively poor, and, as a result, this group of HSLA steels is not recommended for use in welded structures.

Manganese Steels (Heat Treated)

These steels contain additional amounts of manganese for increased strength. Some grades also contain various combinations of nickel, chromium, molybdenum, niobium, and vanadium. In some cases copper is an optional addition. They are available as plates, and, in some cases, as hot rolled sheets. Plate grades are usually normalized to obtain optimum toughness, or, in some instances, quenched and tempered to maximize both strength and toughness. These steels are weldable by filler metal methods.

Manganese-Vanadium and Manganese-Titanium Steels

This group of HSLA steels contains additions of vanadium or titanium in order to improve strength by precipitation hardening and grain refinement. Manganese-vanadium grades may also contain additions of nitrogen or niobium. Manganese-vanadium steels are plate grades, and manganese-titanium steels are available as hot rolled and cold rolled sheet and strip. These steels have good formability and toughness, and are weldable by both filler metal and resistance methods.

Manganese-Vanadium-Copper Steels

These steels contain additions of manganese and vanadium for increased strength, as well as copper for increased corrosion resistance. They have good notch toughness and formability, and exhibit a corrosion resistance about twice that of carbon steel. These steels are weldable by both filler metal and resistance methods. They are available in all product forms.

Multiple Alloy Steels with Copper

These grades contain additions of silicon, copper, chromium, molybdenum, and nickel for improved strength and corrosion resistance. Niobium and vanadium are also added. These steels are weathering steels with corrosion resistance two to six times that of carbon steel. They are primarily plate steels, but a few grades are available as bars, sheets, or shapes. They have good formability, excellent notch toughness, and are readily weldable.

Multiple Alloy Steels with Copper and High Phosphorus

These HSLA steels are also weathering steels that contain additions of copper and phosphorus along with nickel, chromium, and molybdenum. Corrosion resistance is four to eight times that of carbon steel. They exhibit fairly good formability and are weldable; however, notch toughness is poor. These grades are available in most product forms.

HSLA Steels with Special Formability

These steels contain very low carbon contents (0.18% max), are usually fully killed, and often use additions of niobium or vanadium to increase strength. To achieve excellent formability and notch toughness, ladle desulfurization and/or sulfide inclusion shape control using calcium, zirconium, or rare earth elements are employed. These grades are produced primarily as hot and cold rolled sheet and strip; however, a few plate grades are available.

Precipitation-Hardening Steels

This category of HSLA steels contains significant amounts of copper, nickel, and molybdenum. Niobium or vanadium is also added in some instances. These steels develop their strength as a result of low-temperature aging treatments following hot rolling, normalizing, or quenching. They exhibit excellent corrosion resistance (four to six times that of carbon steel) and notch toughness. Product forms include plates, structural shapes, and bars.

Dual-Phase Steels

Dual-phase steels are a special category of high-strength steels that have been recently developed. Their properties result from a mixed ferrite plus martensite microstructure. They are characterized by relatively low yield strengths, 40-50 ksi (275-345 MPa), but work harden rapidly during straining, and develop high tensile strengths on the order of 85 to 100 ksi (585 to 690 MPa). As a result, very high strengths are obtained in fabricated parts. They exhibit significantly higher ductility and formability than more conventional HSLA steels. Dual-phase steels are produced as hot and cold rolled sheet and strip with very small amounts of alloy additions using intercritical annealing. A more highly alloyed version can be produced as directly hot rolled sheet and strip using additions of silicon, chromium, and molybdenum. Where needed, the low yield strengths of dual-phase steels can be increased by low-temperature aging or straining and aging treatments.

Rephosphorized and Renitrogenized Steels

The strength of carbon steels can be increased to the levels exhibited by many HSLA steels through additions of phosphorus or nitrogen. These types of steels are produced as hot and cold rolled sheet and strip products. Nitrogen-containing steels respond to straining and aging treatments resulting in high strength following fabrication. These steels can be obtained with yield strengths ranging from 35 to 140 ksi (240 to 965 MPa).

SPECIFICATIONS

The ASTM, the SAE, and the American Petroleum Institute (API) have developed specifications applicable to high-strength low-alloy steels. In addition, the AISI has developed a designation system applicable to high-strength sheet steels. It is important to note, however, that not all HSLA grades produced by all steel producers are covered by the various specifications. Many grades that are produced as proprietary grades by various manufacturers do not qualify for coverage by any specification.

ASTM Specifications

Table 1-13 lists 14 ASTM specifications that deal with HSLA steels. Each specification describes the composition limits, required properties, and applicable product forms for one or more grades of HSLA steels. Further information can be obtained through review of the individual specifications.

SAE Recommended Practices

The SAE has two recommended practices for HSLA steels, J410c and J1392. Practice J410c is currently under revision.

Practice J410c. SAE recommended practice J410c covers 14 grades of HSLA steel intended primarily for automotive applications. Table 1-14 gives the grade designations of these

TABLE 1-13
ASTM Specifications for HSLA Steels

Specification No.	HSLA Group(s)	Specification Title
A 242	Multiple alloy plus copper, multiple alloy plus copper and phosphorus	High-strength, low-alloy structural steel
A 440	Manganese-copper	High-strength structural steel
A 441	Manganese-vanadium-copper	High-strength, low-alloy, structural manganese-vanadium steel
A 572	Columbium/vanadium	High-strength, low-alloy columbium/vanadium steels of structural quality
A 588	Multiple alloy plus copper	High-strength, low-alloy structural steel with 50,000 psi (345 MPa) minimum yield point to 4" (100 mm) thick
A 606	Multiple alloy plus copper, multiple alloy plus copper and phosphorus	Steel sheet and strip: hot rolled and cold rolled, high-strength, low-alloy, with improved corrosion resistance
A 607	Columbium/vanadium	Steel sheet and strip: hot rolled and cold rolled, high-strength, low-alloy columbium and/or vanadium
A 633	Columbium/vanadium, manganese, manganese-vanadium, manganese-titanium	Normalized high-strength, low-alloy structural steel
A 699	---	Low-carbon manganese-molybdenum-columbium alloy steel plates, shapes, and bars
A 709	Columbium/vanadium	Structural steel for bridges
A 710	Precipitation hardening	Low-carbon, age-hardening nickel-copper-chromium-molybdenum-columbium and nickel-copper-columbium alloy steels
A 715	HSLA special formability	Steel sheet and strip: hot rolled, high-strength, low-alloy, with improved formability
A 736	Precipitation hardening	Pressure vessel plates: low-carbon, age-hardening nickel-copper-chromium-molybdenum-columbium alloy steel
A 737	Manganese	Pressure vessel plates: high-strength, low-alloy steel

steels, along with product forms and mechanical properties. The second and third digits of these grade designations give the specified minimum yield strength in ksi. (To obtain the yield strength in MPa, multiply strength in ksi by 6.895.) Only the maximum carbon, manganese, phosphorus, sulfur, and silicon contents are specified.

The steels designated by the suffix X contain strengthening elements such as niobium, vanadium, or nitrogen, added singly or in combination. These steels are usually made semikilled; however, if killed steel is desired, it may be specified by use of the K suffix, such as SAE 950XK. The killed grade should only be selected when improved low-temperature notch toughness is important.

Practice J1392. SAE recommended practice J1392 covers high-strength hot rolled, cold rolled, and coated sheet steels, and will eventually replace J410c for these particular product forms. The mechanical properties for the various grades covered by practice J1392 are given in Table 1-15 and Table 1-16.[14]

A six-character code is used to describe strength level, general chemical composition, general carbon level, and the deoxidation/sulfide inclusion control system. The first, second, and third characters give the minimum yield strength in ksi; for example, 035 is 35 ksi, 040 is 40 ksi, etc.

The fourth character describes the general chemical composition. The letter A means carbon and manganese only; B means carbon, manganese, and nitrogen; C means carbon, manganese, and phosphorus; S means carbon and manganese with nitrogen and/or phosphorus added at producer option; W refers to weathering compositions that include silicon, phosphorus, copper, nickel, and chromium in various combinations; X refers to HSLA compositions containing niobium, chromium, copper, molybdenum, nickel, silicon, titanium, vanadium, and zirconium added singly or in combination (along with nitrogen and/or phosphorus if desired), and that exhibit a 10 ksi (70 MPa) spread between specified minimal values of yield and tensile strengths; Y refers to the same compositions as X except with a 15 ksi (100 MPa) spread between specified minimal

TABLE 1-14
Mechanical Properties of HSLA Steels Specified by SAE J410c[13]

Grade	Form and Size	Tensile Strength, ksi (MPa)	Yield Strength,* ksi (MPa)	Elongation, %	
				2″ (50 mm)	8″ (200 mm)
942X	Plates, shapes, and bars to 4″ (100 mm) inclusive	60 (414)	42 (290)	24	20
945A, C	Sheet and strip	60 (414)	45 (310)	22	---
	Plates, shapes, and bars				
	to 1/2″ (12.7 mm)	65 (448)	45 (310)	22	18
	1/2 to 1 1/2″ (12.7 to 38 mm)	62 (427)	42 (290)	24	19
	1 1/2 to 3″ (38 to 76 mm)	62 (427)	40 (276)	24	19
945X	Sheet and strip	60 (414)	45 (310)	25	---
	Plates, shapes, and bars to 1 1/2″ (38 mm)	60 (414)	45 (310)	22	19
950A, B, C, D	Sheet and strip	70 (483)	50 (345)	22	---
	Plates, shapes, and bars				
	to 1/2″ (12.7 mm)	70 (483)	50 (345)	22	18
	1/2 to 1 1/2″ (12.7 to 38 mm)	67 (462)	45 (310)	24	19
	1 1/2 to 3″ (38 to 76 mm)	63 (434)	42 (290)	24	19
950X	Sheet and strip	65 (448)	50 (345)	22	---
	Plates, shapes, and bars to 1 1/2″ (38 mm)	65 (448)	50 (345)	---	18
955X	Sheet and strip	70 (483)	55 (379)	20	---
	Plates, shapes, and bars to 1 1/2″ (38 mm)	70 (483)	55 (379)	---	17
960X	Sheet and strip	75 (517)	60 (414)	18	---
	Plates, shapes, and bars to 1 1/2″ (38 mm)	75 (517)	60 (414)	---	16
965X	Sheet and strip	80 (552)	65 (448)	16	---
	Plates, shapes, and bars to 3/4″ (19 mm)	80 (552)	65 (448)	---	15
970X	Sheet and strip	85 (586)	70 (483)	14	---
	Plates, shapes, and bars to 3/4″ (19 mm)	85 (586)	70 (483)	---	14
980X	Sheet and strip	95 (655)	80 (552)	12	---
	Plates to 3/8″ (9.5 mm)	95 (655)	80 (552)	---	10

(*Society of Automotive Engineers*)

Note: The values of the mechanical properties listed are determined in accordance with ASTM A370 and should only be used for comparison purposes, not for design guidelines. For specific values, contact the steel manufacturer.
* Yield strength to be measured at 2% offset.

values of yield and tensile strengths; and Z refers to the same compositions as X except with a 20 ksi (140 MPa) spread between specified minimal values of yield and tensile strengths.

The fifth character describes the general carbon level. The letter H refers to the maximum carbon level, and L means 0.13% carbon maximum. The sixth character describes deoxidation/sulfide inclusion control practices. The letter K means killed and made to a fine-grain practice; F means sulfide inclusion controlled, killed, and made to a fine-grain practice; and O refers to other than K or F.

TABLE 1-15
Mechanical Properties of Hot Rolled HSLA Sheet and Strip Specified by SAE J1392[14]

Grade	Yield Strength, ksi (MPa) min	Tensile Strength, ksi (MPa) min	% Elongation* (2″ or 50 mm), min
035 A, B, C, S	35 (240)	**	21
035 X, Y, Z	35 (240)	**	28
040 A, B, C, S	40 (280)	**	20
040 X, Y, Z	40 (280)	**	27
045 A, B, C, S	45 (310)	**	18
045 W	45 (310)	65 (450)	25
045 X	45 (310)	55 (380)	25
045 Y	45 (310)	60 (410)	25
045 Z	45 (310)	65 (450)	25
050 A, B, C, S	50 (340)	**	16
050 W	50 (340)	70 (480)	22
050 X	50 (340)	60 (410)	22
050 Y	50 (340)	65 (450)	22
050 Z	50 (340)	70 (480)	22
060 X	60 (410)	70 (480)	20
060 Y	60 (410)	75 (520)	20
070 X	70 (480)	80 (550)	17
070 Y	70 (480)	85 (590)	17
080 X	80 (550)	90 (620)	14
080 Y	80 (550)	95 (650)	14

(*Society of Automotive Engineers*)

* Elongation values are dependent upon specimen geometry (cross-sectional area). Thicker and wider specimens normally result in higher percentages.

** Minimum tensile strength normally does not apply.

TABLE 1-16
Mechanical Properties of Cold Rolled and Coated HSLA Sheet and Strip Specified by SAE J1392[14]

Grade	Yield Strength, ksi (MPa) min	Tensile Strength, ksi (MPa) min	% Elongation* (2″ or 50 mm), min
035 A, B, C, S	35 (240)	**	22
035 X, Y, Z	35 (240)	**	27
040 A, B, C, S	40 (280)	**	20
040 X, Y, Z	40 (280)	**	25
045 A, B, C, S	45 (310)	**	18
045 W	45 (310)	65 (450)	22
045 X	45 (310)	55 (380)	22
045 Y	45 (310)	60 (410)	22
045 Z	45 (310)	65 (450)	22
050 A, B, C, S	50 (340)	**	16
050 X	50 (340)	60 (410)	20
050 Y	50 (340)	65 (450)	20
050 Z	50 (340)	70 (480)	20

(*Society of Automotive Engineers*)

* Elongation values are dependent upon specimen geometry (cross-sectional area). Thicker and wider specimens normally result in higher percentages.

** Minimum tensile strength normally does not apply.

API Specifications

The American Pipe Institute Specification 5LX covers high test-line pipe and includes both seamless and welded pipe. This specification includes yield strengths ranging from 42 to 70 ksi (290 to 483 MPa); for example, X42 to X70. Specification 5LS for spiral-weld line-pipe covers these same grades and two lower strength grades. Steel compositions other than those shown in the specifications may be supplied by agreement between purchaser and manufacturer. Niobium and vanadium are often used for higher strength grades, and the pipe skelp is often controlled rolled. Further information may be obtained from the specifications.

AISI Sheet Designation System

The AISI designation system of high-strength sheet steels contains three basic components: (1) the minimum yield strength, (2) the chemical composition, and (3) the deoxidation practice. A five-character code is used to describe these components.

The first three characters give the yield strength of a given grade. Yield strength is categorized in 5 ksi (35 MPa) increments from 35 to 60 ksi (241 to 414 MPa), in 10 ksi (70 MPa) increments from 60 to 80 ksi (414 to 550 MPa), and in 20 ksi (140 MPa) increments from 80 to 140 ksi (550 to 965 MPa). Thus, the designation "050" refers to a steel with a yield strength of 50 ksi (345 MPa).

The chemical composition of each grade is designated by a letter classification: S, X, W, or D. The letter S refers to structural-quality steels that contain carbon plus manganese; carbon plus manganese and phosphorus; carbon plus manganese and nitrogen; or carbon plus manganese, phosphorus, and nitrogen. Recovery-annealed steels, except those with the designation X, are included in this category. The letter X refers to low-alloy steel grades containing niobium, chromium, copper, molybdenum, nickel, silicon, titanium, vanadium, and zirconium either singly or in combination. Weathering steels containing silicon, phosphorus, copper, nickel, and chromium in various combinations are indicated by the letter W. Dual-phase steels containing martensite or other transformation products in a ferrite matrix are designated by the letter D. Dual-phase steels exhibit very high work-hardening rates, and, as a result, formed parts have significantly higher strengths than the original flat rolled sheets. Consequently, the yield strength of a dual-phase steel is designated as the strength after a 5% strain; for example, an 80D grade exhibits an 80 ksi (550 MPa) yield strength after 5% strain.

Deoxidation practice is also designated by a letter classification. The letter F means killed plus sulfide inclusion controlled, K means killed, and O means nonkilled. For example, the steel designation 040SF would mean a minimum yield strength of 40 ksi (275 MPa), structural quality, killed.

SELECTION FACTORS

Selection of an HSLA steel for use in a given application involves an evaluation of (1) properties in relation to the requirements for the application, and (2) manufacturing characteristics needed for the production of the part. Properties to be considered include strength, toughness, weldability, and corrosion resistance. In some instances, fatigue behavior also becomes important. Manufacturing characteristics include formability and weldability.

HIGH-STRENGTH LOW-ALLOY STEELS

Strength Characteristics

The strength of an HSLA steel depends on steel composition and production processing. However, a characteristic common to all HSLA steels is their high strength-to-weight ratio. As was mentioned earlier, yield strength is an important criterion for selection. Figure 1-3 shows partial stress-strain diagrams for a low-carbon mild steel (SAE 1010) and an HSLA steel (SAE 950X).[15] The higher yield strength and greater elastic range of the HSLA grade permit its use in thinner gages, resulting in weight savings. It should be noted, however, that the modulus of elasticity (Young's modulus) is the same for both grades. Thus, where stiffness, deflection, or buckling is a design consideration, it may not be possible to take full advantage of the increased strength of HSLA steels.

Toughness Characteristics

Toughness of a steel is the ability of the steel to absorb impact loads by plastically deforming prior to fracture.[16] The toughness of HSLA steels can vary considerably depending on steel composition and processing. In general, toughness decreases as strength increases. Toughness can be improved by using fully killed steels and through the use of desulfurization and sulfide inclusion shape control. The use of controlled rolling practices or heat treatment also results in good toughness. The relative notch toughness of HSLA steels specified in SAE recommended practice J410c is given in Table 1-17.[17]

Corrosion Characteristics

The corrosion resistance of HSLA steels depends primarily on alloy content as well as on the environment. With the exception of steels containing copper and weathering steels, HSLA steels exhibit a corrosion resistance approximately equal to that of carbon steel.

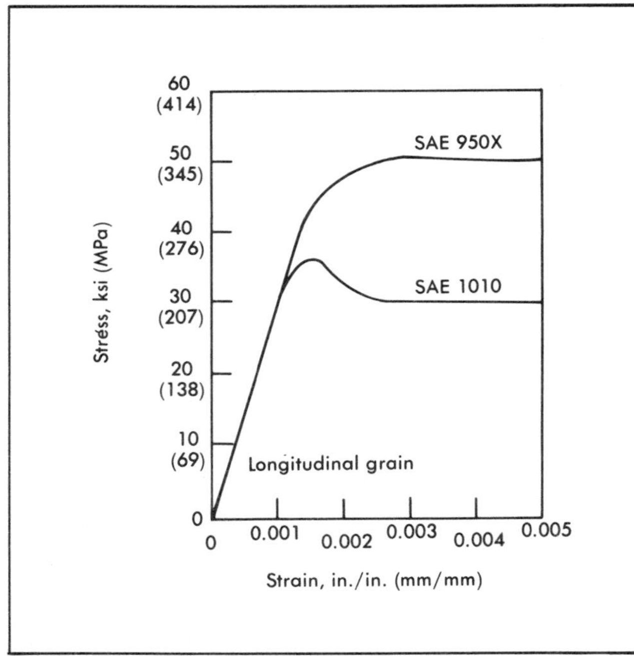

Fig. 1-3 Partial stress-strain diagrams for a low-carbon mild steel (SAE 1010) and an HSLA steel (SAE 950X). (*American Society for Metals*)

Fatigue Strength

Fatigue performance depends on a number of factors including loading cycle, material properties, design of the part, surface condition, and environment.[18] In general, fatigue strength increases as tensile strength increases.

Formability

Formability of a steel may be defined as the capability of the steel to be formed into a useful shape.[19] In general, formability increases as uniform and total elongation and reduction in area increase. As strength increases, formability generally decreases. Thus, while HSLA steels exhibit good formability for their strength, they are not as formable as ordinary low-carbon steels. Table 1-18 presents the relative formability for the various grades of HSLA steel as described in SAE recommended practice J410c.[20]

The formability of steel is usually evaluated by the forming limit diagram (FLD). The strain-hardening exponent (n) and the average plastic strain ratio ($\bar{r}$) are also measurements of a steel's formability. The FLD, strain-hardening exponent, and plastic strain ratio are discussed in greater detail in Volume II, *Forming*, of this Handbook series. It should be noted that the formability of HSLA steels is significantly improved by the use of steelmaking practices that incorporate killing, desulfurization, and sulfide inclusion shape control.

TABLE 1-17
Relative Notch Toughness of HSLA Steels Specified by SAE J410c[17]

Order of Increasing Toughness	
980X	Least tough
970X	
965X	
960X	
955X	
945C, 950C, 942X	
945X, 950X	
950D	
950B	
950A	
945A	Toughest

(*Society of Automotive Engineers*)

TABLE 1-18
Relative Formability of HSLA Steels Specified by SAE J410c[20]

Order of Increasing Formability	
980X	Least formable
970X	
965X	
960X	
955X	
950C	
950D	
950B, 950X, 942X	
945C, 945X	
950A	
945A	Most formable

(*Society of Automotive Engineers*)

Weldability

The weldability of HSLA steels is measured in many different ways. It takes into consideration the susceptibility of a steel to cracking during welding and the properties of the weld and heat-affected zone. In structural applications, most welding is carried out by metal arc welding.

When welding HSLA steels, five serious problems may occur. Three of these—hydrogen cracking (cold cracking), solidification cracking (hot cracking), and lamellar tearing—are fabrication problems. The other two—weld metal toughness and heat-affected zone toughness—are service problems.[21]

The susceptibility of a steel to hydrogen cracking, resulting from absorption of hydrogen into the weld, can be assessed through the use of carbon equivalent (CE). As total alloy content increases, the susceptibility to hydrogen cracking increases. Table 1-19 gives the carbon equivalent values for several HSLA steels and indicates a qualitative estimate of their weldability.[22] Solidification cracking results during solidification of the weld metal whenever the last portion to freeze has a significantly lower freezing temperature than the bulk solidified material. Shrinkage stresses then produce cracks. The presence of low-melting-point iron and alloy sulfides that wet interdendritic regions during solidification are principal causes of this type of cracking. High carbon and phosphorus levels are also detrimental. Thus, control of the amount and type of sulfides aids in prevention of this type of cracking. Lamellar tearing results from the action of shrinkage stresses in the through thickness (for example, plate thickness and direction). These stresses act on thin sulfide inclusions or other lamination-type defects. Control of the amount and shape of inclusions is the best means of preventing this type of cracking. The toughness of a weld and heat-affected zone vary with steel composition and welding procedure. Each application must be considered on an individual basis.

Resistance spot welding is used for many sheet applications, particularly in the automotive industry. Because of their higher hot strength, HSLA steels usually require somewhat higher weld pressures and weld times than low-carbon mild steels of the same thickness. This factor is usually offset, however, by the use of lighter gage HSLA grades. The higher resistivity of HSLA steels reduces to some degree the required welding current. Generally, higher carbon and alloy contents result in harder welds that are more susceptible to failure.[23] Thus, lower carbon, lower alloy HSLA steels are used when spot welding is an important fabrication consideration.

With a few exceptions, HSLA steels are generally weldable by filler metal and resistance methods. Preheat and postheat are usually not required. Table 1-20 ranks the relative weldability of the HSLA steels covered by SAE recommended practice J410c.[24] Additional information on welding methods and practices can be found in Volume IV, *Quality Control and Assembly*, of this Handbook series.

TABLE 1-19
Carbon Equivalent Values and Manufacturer's Weldability Rating for Six Types of HSLA Steel[22]

Steel Type	Specifications		Carbon Equivalent Value Range	Weldability Rating
	ASTM	SAE (J410c)		
Multiple alloy, low phosphorus	A242, Type 2 A606, Type 4	Grade 950B	0.28-0.34	Excellent
Multiple alloy, high phosphorus	A242, Type 1 A606, Type 4	Grade 950D	0.27-0.33	Excellent
Columbium	A572 A607, Type 1			
Min YS = 50 ksi (345 MPa)	Grade 50	Grade 950X	0.27-0.33	Excellent
Min YS = 60 ksi (415 MPa)	Grade 60	Grade 960X	0.27-0.33	Excellent
Min YS = 70 ksi (485 MPa)	Grade 70	Grade 970X	0.34-0.40	Very good
High formability	A715, Type 6	---	---	---
Min YS = 50 ksi (345 MPa)	Grade 50	---	0.18-0.24	Excellent
Min YS = 60 ksi (415 MPa)	Grade 60	---	0.23-0.29	Excellent
Manganese-vanadium	A441 A572, Type 2 A607, Type 2	Grade 950B	0.34-0.40	Very good
Manganese-copper	A440	Grade 950C	0.44-0.50	Fair

REFERENCES

TABLE 1-20
Relative Weldability of HSLA Steels Specified by SAE J410c[24]

Order of Increasing Weldability	
980X	Difficult to weld
970X	
965X	
960X	
955X, 950C, 942X	
945C	
950B, 950X	
945X	
950D	
950A	
945A	Easy to weld

(*Society of Automotive Engineers*)

References

1. *Modern Steels and Their Properties* (Bethlehem, PA: Bethlehem Steel Corp., 1980), pp. 191-198.
2. Paul M. Unterweiser, Howard E. Boyer, and James J. Kubbs, eds., *Heat Treater's Guide* (Metals Park, OH: American Society For Metals, 1982), pp. 447-457.
3. *1983 SAE Handbook*, vol. 1 (Warrendale, PA: Society of Automotive Engineers, 1983) p. 3.04.
4. *Modern Steels and Their Properties*, *op.cit.*, p. 6.
5. Robert D. Pehlke, "An Overview of Contemporary Steelmaking Processes," *Journal of Metals* (May 1982), p. 62.
6. *Modern Steels and Their Properties*, *op. cit.*, p. 43.
7. *1983 SAE Handbook*, *op.cit.*, p. 2.02.
8. *Ibid.*, pp. 2.15-2.17.
9. American Welding Society, *Welding Handbook*, 7th ed., vol. 1 (Miami, FL: American Welding Society, 1976), p. 137.
10. *Modern Steels and Their Properties*, *op. cit.*, p. 61.
11. E. E. Fletcher, *A Review of the Status, Selection, and Physical Metallurgy of High-Strength, Low-Alloy Steels,"* Battelle Report MCIC-79-39 (Columbus, OH: Battelle Columbus Laboratories, March 1979), pp. 5-110.
12. *Metals Progress 1978 Databook* (Metals Park, OH: American Society for Metals, 1979), pp. 47-59.
13. *1983 SAE Handbook*, *op. cit.*, p. 1.75.
14. *1983 SAE Handbook*, *op. cit.*, p. 1.77.
15. D. G. Younger, "How Ford Evaluates Conversions to HSLA," *Metal Progress* (May 1975), pp. 43-47.
16. Fletcher, *op. cit.*, p. 40.
17. *1983 SAE Handbook*, *op. cit.*, p. 1.76.
18. Fletcher, *op. cit.*, p. 93.
19. *Ibid.*, p. 75.
20. *1983 SAE Handbook*, *op. cit.*, p. 1.76.
21. Fletcher, *op. cit.*, p. 49.
22. *Dofascoloy High Strength Low Alloy Steels* (Hamilton, Ontario: Dominion Foundries and Steel Limited).
23. Fletcher, *op. cit.*, p. 59.
24. *1983 SAE Handbook*, *op. cit.*, p. 1.76.

Bibliography

American Society for Metals. *Metals Handbook*, vol. 4, 9th ed. Metals Park, OH: American Society for Metals, 1981.

Avitzur, B. *Handbook of Metal-Forming Processes*. New York: John Wiley and Sons, 1983.

Bardes, Bruce P., ed. *Metals Handbook*, vol. 1, 9th ed. Metals Park, OH: American Society for Metals, 1978.

Barrett, C. S., and Marsalski, J. B. *Structure of Metals*, 3rd ed. New York: McGraw-Hill, 1966.

Dieter, G. E. *Mechanical Metallurgy*, 2nd ed. New York: McGraw-Hill, 1976.

Flinn, Richard A., and Trojan, Paul K. *Engineering Materials and Their Applications*. Boston: Houghton Mifflin Company, 1975.

Hosford, W.F., and Caddell, R. M. *Metal Forming*. Englewood Cliffs, NJ: Prentice-Hall, 1983.

Leslie, W. C. *The Physical Metallurgy of Steels*. New York: McGraw-Hill, 1981.

Smith, W. F. *Structure and Properties of Engineering Alloys*. New York: McGraw-Hill, 1981.

Steel Founders' Society of America. *Steel Castings Handbook*, 5th ed. Des Plaines, IL: Steel Founders' Society of America, 1980.

Van Vlack, L. H. *Elements of Materials Science*, 4th ed. Reading, MA: Addison-Wesley, 1980.

STAINLESS AND MARAGING STEELS

STAINLESS STEELS

Stainless steels are iron-based alloys containing 10.50% or more chromium. These steels achieve their "stainless" characteristics as a result of the invisible and adherent, chromium-rich oxide film that forms on the material's surface. The oxide film is self-forming and self-healing in the presence of oxygen. Other elements added to improve corrosion resistance, fabricating and machining characteristics, or strength include nickel, molybdenum, copper, titanium, silicon, manganese, columbium, aluminum, nitrogen, and sulfur. Carbon is normally present in amounts from 0.03% to over 1.00% in certain martensitic grades, which contributes to improvements in the alloy's strength.

The selection of stainless steels is based on corrosion or heat resistance, mechanical properties, fabrication characteristics, availability, and the total product cost. Generally, corrosion resistance and mechanical properties are the predominant factors in selecting the appropriate grade of stainless steel for a given application.

MANUFACTURING PRACTICE

Stainless steels are usually melted in electric furnaces, followed by additional refining steps to adjust carbon content and remove impurities. Refining may be performed using oxygen-inert gas injection (argon oxygen decarburization), oxygen injection under vacuum (vacuum oxygen decarburization), consumable electrode (vacuum-arc or electroslag) remelting, or electron-beam melting or refining. Stainless steels have also been produced using the basic oxygen process. The choice of a particular melting or refining method depends on the level of purity and nonmetallic inclusion content necessary for the application. Special melting and refining methods, such as air or vacuum induction, may be employed if material demands justify the additional costs.

The molten metal is poured into solid molds to form ingots, followed later by blooming or slabbing, or is poured directly into a continuous casting machine to form slabs, blooms, or billets. Figure 2-1 illustrates the mill processes for making various stainless steel end products. Commonly produced finished forms include plates, sheets, strips, bars, structural shapes, wire, pipe, and tubing; semifinished forms include rods, blooms, billets, slabs, and tube rounds. Table 2-1 gives the dimensions of the various stainless steel product forms. Stainless and heat-resisting steel castings are discussed in the following chapter.

Material finishes are obtained by cold drawing; rolling between polished or textured rolls; grinding, polishing, and/or buffing with abrasive wheels, belts, or pads; blasting with abrasive grit or glass beads; or by chemical descaling and pickling techniques.

TYPES OF STAINLESS STEELS

Stainless steels possess resistance to attack by many corrosive media at room and elevated temperatures, and are produced in a variety of grades to cover a wide range of mechanical and physical properties for specific applications. Currently, over 57 standard grades of stainless steels are produced as well as proprietary stainless steels with special characteristics. The standard grades are those identified in the American Iron and Steel Institute (AISI) products manual entitled *Stainless and Heat Resisting Steels.*

The AISI classifies the different types of stainless steels according to a three-digit numbering system. The first digit indicates the type of stainless steel that is suggestive of the material's microstructure. The last two digits indicate the specific grade in the group. Letters following the last two digits indicate modifications of a specific grade. The Unified Numbering System (UNS), developed by the American Society for Testing and Materials (ASTM) and the Society of Automotive Engineers (SAE), uses six characters to designate a particular material type and grade. All stainless steels in this system are identified by the letter "S" and followed by five digits. In this Handbook, material designations are given using the three-digit and six-character numbering systems whenever they are applicable.

The five main types of stainless steels include austenitic, ferritic, martensitic, precipitation-

CHAPTER CONTENTS:

***Contributors of sections of this chapter are: D. C. Agarwal**, Market Development Engineer, New Products and Applications Engineering, Cabot Corp.; **Dr. George Aggen**, Manager, Stainless and Alloy Metallurgy Dept., Research Center, Allegheny Ludlum Steel Corp.; **Alan M. Bayer**, Technical Director, Teledyne Vasco; **Calvin J. Cooley**, Metallurgical Engineer, Manufacturing Research, American Iron and Steel Institute; **Ted Kosa**, Supervisor, Stainless Alloy Research and Development Dept., Carpenter Technology Corp.*

***Reviewers of sections of this chapter are: D. C. Agarwal**, Market Development Engineer, New Products and Applications Engineering, Cabot Corp.; **Dr. George Aggen**, Manager, Stainless and Alloy Metallurgy Dept., Research Center, Allegheny Ludlum Steel Corp.; **Alan M. Bayer**, Technical Director, Teledyne Vasco;*

CHAPTER 2

STAINLESS STEELS

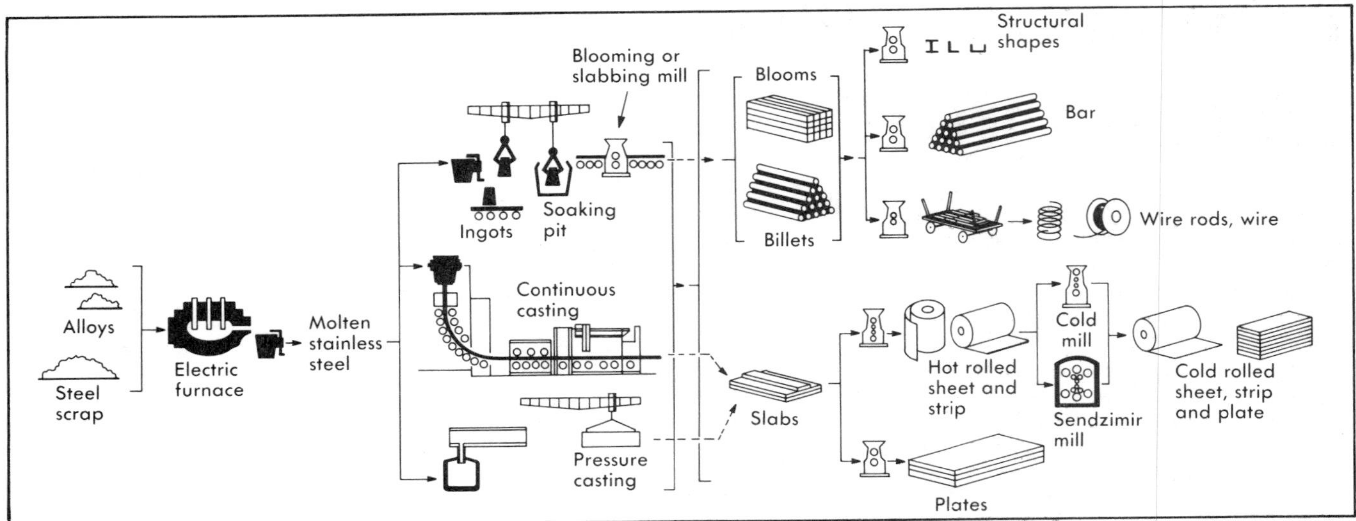

Fig. 2-1 Stainless steel is produced by pouring the molten metal into ingots which are then rolled into the final product form. Continuous casting techniques are used widely in the production of sheet and plate end products and also for casting blooms or billets. *(American Iron and Steel Institute)*

TABLE 2-1
Dimensions of Various Stainless Steel Product Forms

Product	Description	Dimensions, in. (mm) Thickness	Width	Diam or Size
Sheet	Coils and cut lengths:			
	Mill finishes	Under 3/16 (4.8)	24 (610) and over	---
	Polished finishes	Under 3/16 (4.8)	All widths	---
Strip	Cold finished, coils or cut lengths	Under 3/16 (4.8)	Under 24 (610)	---
Plates	Flat rolled or forged	3/16 (4.8) and over	Over 10 (254)	---
Bars	Hot finished rounds, squares, octagons and hexagons	---	---	1/4 (6.4) and over
	Hot finished flats	1/8 (3.2) and over	1/4-10 (6.4-254)	---
	Cold finished rounds, squares, octagons and hexagons	---	---	Over 1/2 (12.7)
	Cold finished flats	---	3/8 (9.5) and over	---
Wire	Cold finished only: Round, square, octagon, hexagon, flat wire	0.010-3/16 (0.25-4.8)	1/16-3/8 (1.6-9.5)	1/2 (12.7) and under
Pipe and tubing	Several different classifications, with differing specifications are available. For information on standard sizes consult the Committee of Stainless Steel Producers, American Iron and Steel Institute.			
Extrusions	Not considered "standard" shapes, but of potentially wide interest. Currently limited in size to approximately 6 1/2" (165 mm) diam circle, or structurals to 5" (127 mm) diam.			

(American Iron and Steel Institute)

Reviewers, cont.: ***Herb L. Black****, Manager, Technical Services, Universal-Cyclops Specialty Steel Div., Cyclops Corp.;* ***Calvin J. Cooley****, Metallurgical Engineer, Manufacturing Research, American Iron and Steel Institute;* ***Charles Divine****, Product Development, Al Tech Specialty Steel Corp.;* ***Tom Frappier****, Vice President of Sales and Marketing, Joslyn Stainless Steels;* ***Raymond M. Hemphill****, Supervisor, Research and Development Dept., Carpenter Technology Corp.;* ***Ted Kosa****, Supervisor, Stainless Alloy Research and*

hardening, and duplex. Austenitic, ferritic, martensitic, and duplex stainless steels are classified according to the three-digit and six-character number systems. Precipitation-hardening stainless steels are generally classified according to the six-character numbering system. Table 2-2 gives the three-digit designation for austenitic, ferritic, and martensitic stainless steels. Mechanical properties of the various stainless steel grades at room temperature are tabulated in Table 2-3 for comparison purposes. Mechanical properties for stainless steels at elevated temperatures can be found in ASTM Data Series Publications DS 3, DS 5, DS 5-S1, DS 5-S2, and DS 18.

Austenitic

Austenitic stainless steels are characterized by their austenitic (face-centered cubic) structure. Austenitic stainless steels are essentially nonmagnetic in the annealed condition and can only be hardened by cold working. They possess excellent cryogenic characteristics and the greatest corrosion resistance and best high-temperature strength of all stainless steels produced. However, some of the newer duplex stainless steels possess better corrosion and strength properties.

Austenitic stainless steels are composed of iron-chromium-nickel and iron-chromium-manganese-nickel alloys. Chromium content is from 16 to 26%, nickel 6 to 22%, and manganese 1 to 15%. Nickel improves the corrosion resistance of the alloy in some environments. The 2xx stainless steels contain nitrogen and 4-15.5% manganese to promote the austenitic structure and reduce the amount of nickel required. The 3xx stainless steels may contain up to 2% manganese. Carbon and, more importantly, nitrogen are also added to promote the austenitic structure. Molybdenum improves the high-temperature strength and increases the resistance to chloride pitting and crevice corrosion. Sulfur or selenium may be added to certain grades to improve machinability.

Cold working austenitic stainless steels significantly increases their tensile and yield strengths. In many—but not all—austenitic steels, cold working partially transforms the austenitic structure to a martensitic structure. The actual rate of work hardening is determined by the total alloy content. At room temperature, yield strengths range from 30 to 200 ksi (200 to 1379 MPa) depending on composition and the amount of cold working performed. Table 2-3 compares the mechanical properties of the austenitic alloys at room temperature.

TABLE 2-2
Types of Stainless Steels

Designation	Type and General Characteristics
2xx	Chromium-nickel-manganese steels: nonhardenable, austenitic, and nonmagnetic
3xx	Chromium-nickel steels: nonhardenable, austenitic, and nonmagnetic
4xx	Chromium steels: nonhardenable, ferritic, and magnetic
4xx	Chromium steels: hardenable, martensitic, and magnetic
5xx	Chromium steels: low chromium and heat resisting

(*American Iron and Steel Institute*)

Ferritic

Ferritic stainless steels are chromium alloys with body-centered cubic microstructures. Chromium content is usually from 10.5 to 27%, and some grades may contain up to 4% molybdenum to improve pitting and crevice corrosion resistance. Ferritic alloys are magnetic, have good ductility, and resist corrosion and oxidation; however, toughness may be limited. High-temperature strength is poor, and hardening can only be performed by cold working. Table 2-3 compares the mechanical properties of the standard ferritic grades at room temperature.

Martensitic

Martensitic stainless steels are chromium alloys that possess a martensitic microstructure in the hardened condition. These alloys are magnetic, resistant to corrosion in mild environments, and hardenable by heat treatment. They are also less ductile than the other types of stainless steels, but their ductility is improved in the fully solution annealed condition.

Chromium content ranges from 10.5 to 18% and carbon content may be as high as 1.20%. However, a proper ratio of carbon to chromium must be maintained to ensure a fully austenitic structure during hardening. Chromium increases corrosion resistance, but increasing carbon lowers it. High-carbon, high-chromium alloys provide increased hardness for cutting and wearing applications, as well as improved strength for highly stressed parts such as bolts and nuts. However, toughness will not be as high as that obtained with lower-carbon martensitic stainless steels. Adding sulfur and selenium to the alloy improves machinability, and adding nickel improves corrosion resistance in some media but may hinder full hardening.

Mechanical properties of martensitic stainless steels fall into two groups. The first group consists of grades containing up to 0.15% carbon and are generally referred to as low-carbon alloys. This group can attain a maximum hardness of R_C45. High-carbon alloys, with carbon content greater than 0.15%, can attain a maximum hardness of R_C60. The maximum hardness of both groups in the annealed condition is R_C26. The tensile strength of heat-treated martensitic stainless steels may exceed 200 ksi (1379 MPa). Table 2-3 compares the mechanical properties of martensitic stainless steels in various forms and conditions.

Precipitation-Hardening

Development of precipitation-hardening (PH) stainless steel alloys began in the 1940s for use in the aerospace industry. Currently, PH alloys are used in a wide variety of applications

---

__Reviewers, cont.__: Development Dept., Carpenter Technology Corp.; __Edward A. Loria__, Research Metallurgist, Universal-Cyclops Specialty Steel Div., Cyclops Corp.; __Conrad Mitchell__, Manager of Product Metallurgy—Bar, Rod, Wire and Semi-Finished Products, U.S. Steel Corp.; __Randall A. Oertel__, Product Metallurgist, Al Tech Specialty Steel Corp.; __Michael L. Schmidt__, Associate Metallurgist, Research and Development Dept., Carpenter Technology Corp.; __Joe B. Young__, Metallurgist, Eastern Stainless Steel Company, Eastmet Corp.

TABLE 2-3
Mechanical Properties of Commonly Used Stainless Steels at Room Temperature

Grade UNS	Grade AISI/SAE	Form and Condition	Tensile Strength, ksi (MPa)	Yield Strength (0.2% Offset), ksi (MPa)	Total Elongation in 2" (50 mm), %*	Hardness**
Austenitic:						
S20100	201	Plate, sheet and strip				
		Annealed	95 (655)	38 (262)	40	90 R_B
		1/4 hard	125 (862)	75 (517)	20	25 R_C
		1/2 hard	150 (1034)	110 (758)	9-10	32 R_C
		3/4 hard	175 (1207)	135 (931)	3-7	37 R_C
		Full hard	185 (1276)	140 (965)	3-5	41 R_C
S20200	202	Plate, sheet and strip				
		Annealed	90 (621)	38 (262)	40	90 R_B
		1/4 hard	125 (862)	75 (517)	12	25 R_C
		1/2 hard	150 (1034)	110 (758)	10	---
S30100	301	Sheet and strip				
		Annealed	75 (517)	30 (207)	40	92 R_B
		1/4 hard	125 (862)	75 (517)	25	25 R_C
		1/2 hard	150 (1034)	110 (758)	18	32 R_C
		3/4 hard	175 (1207)	135 (862)	12	37 R_C
		Full hard	185 (1276)	140 (965)	9	41 R_C
S30200	302	Sheet and strip				
		Annealed	75 (517)	30 (207)	40	92 R_B
		Cold rolled	125 (862)	75 (517)	12	---
		Bars				
		Annealed	75 (517)	30 (207)	50	---
		Cold finished, high tensile, 7/8" (22 mm) diam	125 (862)	100 (689)	12	---
S30215	302B	Sheet and strip				
		Annealed	75 (517)	30 (207)	40	95 R_B
		Bars				
		Annealed	75 (517)	30 (207)	40	---
S30300, S30323	303, 303 Se	Bars				
		Annealed	85 (586)	35 (241)	50	262 Bhn
		Annealed and cold drawn, 1" (25.4 mm) diam	100 (689)	60 (414)	40	228 Bhn
		Cold drawn, high tensile, 7/8" (22 mm) diam	125 (862)	100 (689)	12	---
S30400	304	Sheet and strip				
		Annealed	75 (517)	30 (207)	40	92 R_B
		Bars				
		Annealed	75 (517)	30 (207)	40	---
		Cold finished, high tensile, 7/8" (22 mm) diam	125 (862)	100 (689)	12	---
S30403	304L	Plate, sheet and strip				
		Annealed	70 (483)	25 (172)	40	88 R_B
		Bars				
		Annealed	70 (483)	25 (172)	40	---

(continued)

TABLE 2-3—*Continued*

Grade UNS	Grade AISI/SAE	Form and Condition	Tensile Strength, ksi (MPa)	Yield Strength (0.2% Offset), ksi (MPa)	Total Elongation in 2″ (50 mm), %*	Hardness**
S30451	304N	Plate, sheet and strip				
		Annealed	80 (552)	35 (241)	30	95 R_B
		Bars				
		Annealed	80 (552)	35 (241)	30	---
S30500	305	Plate, sheet and strip				
		Annealed	70 (483)	25 (172)	40	88 R_B
S30900, S30908	309, 309S	Plate, sheet and strip				
		Annealed	75 (517)	30 (207)	40	95 R_B
		Bars				
		Annealed	75 (517)	30 (207)	40	---
S31000, S31008	310, 310S	Plate, sheet and strip				
		Annealed	75 (517)	30 (207)	40	95 R_B
		Bars				
		Annealed	75 (517)	30 (207)	40	---
S31600	316	Plate, sheet and strip				
		Annealed	75 (517)	30 (207)	40	95 R_B
		Bars				
		Annealed	75 (517)	30 (207)	40	---
S31620	316F	Plate, sheet and bars				
		Annealed	85 (586)	35 (241)	40	20-28 R_C
S31603	316L	Sheet and strip				
		Annealed	70 (517)	25 (172)	40	95 R_B
		Bars				
		Annealed	75 (517)	30 (207)	40	149 Bhn
S31651	316N	Plate, sheet and strip				
		Annealed	80 (552)	35 (241)	30	95 R_B
		Bars				
		Annealed	80 (552)	35 (241)	30	---
S31700	317	Plate, sheet and strip				
		Annealed	75 (517)	30 (207)	35	95 R_B
		Bars				
		Annealed	75 (517)	30 (207)	40	---
S31703	317L	Plate, sheet and strip				
		Annealed	75 (517)	30 (207)	35	95 R_B
S32100	321	Plate, sheet and strip				
		Annealed	75 (517)	30 (207)	40	88 R_B
		Bars				
		Annealed	75 (517)	30 (207)	40	---
		Annealed and cold finished, 1″ (25.4 mm) diam	90 (621)	45 (310)	40	---
N08330	330	Plate, sheet and strip				
		Annealed	70 (483)	30 (207)	30	80 R_B

(continued)

TABLE 2-3—*Continued*

Grade UNS	Grade AISI/SAE	Form and Condition	Tensile Strength, ksi (MPa)	Yield Strength (0.2% Offset), ksi (MPa)	Total Elongation in 2″ (50 mm), %*	Hardness**
N08330	330	Bars Annealed, 1-3″ (25.4-76 mm) diam	70 (483)	30 (207)	30	---
S34700, S34800	347, 348	Plate, sheet and strip Annealed	75 (517)	30 (207)	40	88 R_B
		Bars Annealed	75 (517)	30 (207)	40	---
		Annealed and cold finished, 1″ (25.4 mm) diam	90 (621)	45 (310)	40	---
Ferritic:						
S40500	405	Plate, sheet and strip Annealed	60 (414)	25 (172)	20	88 R_B
		Bars Annealed	60 (414)	25 (172)	20	207 Bhn
S40900	409	Plate, sheet and strip Annealed	55 (379)	30 (207)	22	80 R_B
		Bars Annealed	65 (448)	35 (241)	25	---
S42900	429	Plate, sheet and strip Annealed	65 (448)	30 (207)	22	88 R_B
		Bars Annealed	70 (483)	40 (276)	20	---
S43000	430	Plate, sheet and strip Annealed	65 (448)	30 (207)	22	88 R_B
		Bars Annealed	70 (483)	40 (276)	20	---
		Annealed and cold finished	70 (483)	40 (276)	16	185 Bhn
S43020, S43023	430F, 430FSe	Bars Annealed	80 (552)	55 (379)	25	262 Bhn
S43400, S43600	434, 436	Plate, sheet and strip Annealed	77 (531)	53 (365)	23	83 R_B
S44600	446	Plate, sheet and strip Annealed	75 (517)	40 (276)	20	95 R_B
		Bars Annealed	70 (483)	40 (276)	20	---
		Annealed and cold finished	70 (483)	40 (276)	16	185 Bhn, 90 R_B
Martensitic:						
S40300, S41000	403, 410	Plate, sheet and strip Annealed	70 (483)	30 (207)	25	88 R_B
		Bars Annealed	70 (483)	40 (276)	20	---

(continued)

TABLE 2-3—*Continued*

Grade UNS	Grade AISI/SAE	Form and Condition	Tensile Strength, ksi (MPa)	Yield Strength (0.2% Offset), ksi (MPa)	Total Elongation in 2″ (50 mm), %*	Hardness**
S40300, S41000	403, 410	Tempered, 1200° F (649° C)	100 (689)	80 (552)	15	---
		Tempered, 600° F (315° C)	180 (1241)	140 (965)	15	375 Bhn, 39 R_C
S41400	414	Bars				
		Annealed	115 (793)	90 (621)	20	298 Bhn
		Tempered, 1200° F (649° C)	115 (793)	90 (621)	15	---
		Tempered, 600° F (315° C)	190 (1310)	145 (1000)	15	400 Bhn, 41 R_C
S41600, S41623	416, 416Se	Bars				
		Annealed	75 (517)	40 (276)	30	262 Bhn
		Tempered, 1200° F (649° C)	110 (758)	85 (586)	18	293-352 Bhn, 97 R_B
		Tempered, 600° F (315° C)	180 (1241)	140 (965)	13	375 Bhn, 39 R_C
S42000	420	Bars				
		Annealed	95 (655)	50 (345)	25	241 Bhn
		Tempered, 600° F (315° C)	230 (1586)	195 (1344)	8	500 Bhn, 50 R_C
S43100	431	Strip				
		Annealed	95 (655)	50 (345)	20	92 R_B
		Bars				
		Annealed	125 (862)	95 (655)	20	285 Bhn
		Tempered, 600° F (315° C)	195 (1344)	150 (1034)	15	400 Bhn, 41 R_C
S44002	440A	Bars				
		Annealed	105 (723)	60 (414)	20	269 Bhn
		Annealed and cold finished	115 (793)	90 (621)	12	285 Bhn
		Tempered, 600° F (315° C)	260 (1793)	240 (1655)	5	510 Bhn, 51 R_C
S44003	440B	Bars				
		Annealed	107 (738)	62 (427)	18	269 Bhn
		Annealed and cold finished	120 (827)	95 (655)	9	285 Bhn
S44004	440C	Bars				
		Annealed	110 (758)	65 (448)	14	269 Bhn
		Annealed and cold finished	125 (862)	100 (689)	7	285 Bhn
		Tempered, 600° F (315° C)	285 (1965)	275 (1896)	2	580 Bhn, 57 R_C
S50100	501	Plate				
		Annealed	60 (414)	30 (207)	18	---
		Bars				
		Annealed	70 (483)	30 (207)	28	160 Bhn
S50200	502	Plate				
		Annealed	60 (414)	30 (207)	18	---
		Bars				
		Annealed	65 (448)	25 (172)	30	150 Bhn

(continued)

STAINLESS STEELS

TABLE 2-3—*Continued*

Grade UNS	Grade AISI/SAE	Form and Condition	Tensile Strength, ksi (MPa)	Yield Strength (0.2% Offset), ksi (MPa)	Total Elongation in 2" (50 mm), %*	Hardness**
Precipitation Hardening:						
S13800		Plate, sheet and bar				
		Solution treated	160 (1103)	120 (827)	17	38 R_C
		Hardened, 950° F (510° C)	225 (1551)	210 (1448)	12	47 R_C
S15500		Plate, sheet, strip and bar				
		Solution treated	160 (1103)	145 (1000)	15	38 R_C
		Hardened, 900° F (480° C)	200 (1379)	185 (1276)	14	44 R_C
S17400		Plate, sheet, strip and bar				
		Solution treated	160 (1103)	145 (1000)	5	38 R_C
		Hardened, 900° F (480° C)	200 (1379)	185 (1276)	14	44 R_C
S17700		Plate, sheet, strip and bar				
		Solution treated	150 (1034)	65 (448)	25	100 R_B
Duplex:						
S32900	329	Plate				
		Annealed	90 (621)	70 (483)	15	28 R_C
		Bars				
		Annealed	105 (724)	80 (552)	25	230 Bhn
S32550		Plate and sheet				
		Annealed	130 (896)	100 (689)	25	25 R_C

(*American Iron and Steel Institute*)

Note: The values of the mechanical properties listed are for comparison purposes only and are not intended for specifications or design purposes. The values will vary depending on temperature for annealing, tempering, and solution treating. For specific information, consult the steel manufacturer.

* Elongation may be affected by Sheet or strip thickness.
**R_B = Rockwell hardness number B Scale
R_C = Rockwell hardness number C Scale
Bhn = Brinell hardness number

because of their high strength-to-weight ratio, ductility, and corrosion resistance at temperatures up to their precipitation temperature.

Precipitation-hardening stainless steels are chromium-nickel alloys with the addition of elements such as copper, aluminum, titanium, or molybdenum. Precipitation-hardening stainless steels can be classified as martensitic or austenitic based on the microstructure of the steel. The predominant microstructure is martensitic.

Hardening PH stainless steels to high strengths is accomplished by solution treatment and an aging process. In the precipitation-hardened condition, PH stainless steels are magnetic and attain tensile strengths up to 260 ksi (1790 MPa). Mechanical properties of the standard PH grades are given in Table 2-3.

Duplex

Another type of stainless steel that has been widely used in Europe and is now becoming popular in the U.S. and Canada is classified as duplex stainless steel and has a mixed structure of both austenite and ferrite. The exact amount of each phase is varied by the introduction of austenite and ferrite stablizers, but most compositions usually contain 50% austenite and 50% ferrite. The principal alloying elements are chromium and nickel with the addition of molybdenum, copper, and nitrogen in certain grades.

The corrosion resistance of duplex stainless steels is similar to austenitic stainless steels. However, they have improved resistance to stress-corrosion cracking and higher tensile and yield strengths in the annealed condition. They also offer good fabricability and toughness. Mechanical properties of common duplex grades are given in Table 2-3.

APPLICATIONS

Since their introduction over 50 years ago, stainless steels have been used in many different industries for a variety of applications. Some of the industries include chemical, petroleum, textile, nuclear power, pulp and paper, marine, fertilizer, and transportation. Stainless steels are also being used in hospitals, laboratories, cafeterias, dairies, breweries, food processing plants, and residential homes.

Austenitic alloys are the most widely used of the stainless steel alloys due to their resistance to corrosion, ease of fabrication, and wide range of mechanical properties in both elevated temperature and cryogenic environments. Ferritic alloys are frequently used in nonstructural applications that require good corrosion resistance and bright, highly polished finishes. Structural applications include automotive exhaust systems and farm equipment. Martensitic alloys are used in applications that require not only moderate corrosion resistance but high strength, high hardness, and good fatigue properties such as in cutlery. Precipitation-hardening alloys are used predominantly in aerospace and aircraft structural components. Duplex stainless steels are finding increasing use in industries where the 300 series stainless steels are susceptible to localized forms of corrosion such as pitting/crevice corrosion and chloride stress-corrosion cracking. A list of some typical applications for the standard stainless steel grades is found in Table 2-4.

TABLE 2-4
Typical Applications of Standard Stainless Steels

Grade	Application
	Austenitic Alloys
201, 202, 301, 302, and 304	Aircraft industry: screws, fuel lines, engine parts, firewalls
	Architectural: unlimited applications such as store facings, frames, rails, and trim walls
	Chemical processing: tanks, piping towers, heat exchangers, bubble caps
	Food processing: unlimited applications such as kettles, impellers, tanks, refrigerator cars, fermentation vats, pasteurizers, buckets, shovels, and barrels
	Household items: tableware, cooking utensils, mixing bowls, refrigerators, evaporators, washing-machine tubs
	Dairy industry: milk cans, pasteurizers, heat exchangers, freezers, coolers, tanks, separators
	Textile industry: dye and bleach equipment
	Transporation industry: trailer bodies, railway passenger cars, structures, passenger-car finish, automobile hubcaps, bumpers, trim
302B	Annealing covers, burner sections, furnace parts
303	Free-machining 18-8 screws, bolts and nuts, shafts, carburetor parts, and other machined parts with corrosion resistance of 18-8
305	Spun parts, drawn parts, cold-heading operations

TABLE 2-4—*Continued*

Grade	Application
308	Welding rods
309 and 310	Chemical and photographic processing equipment. Annealing boxes, covers, welding rods, furnace parts, resistors, and retorts for resistance to oxidation up to 2000° F (1090° C).
316 and 317	For chemical processing equipment requiring greater corrosion resistance than that of 18-8. Chemical processing, paper processing, and soap manufacturing equipment.
321 and 347	Similar applications to those for 301, 302, and 304 when fabrication by welding would cause harmful carbide precipitation. Also used when considerable corrosion resistance is required and regular or intermittent heating in the 800-1500° F (425-815° C) range occurs.
	Ferritic Alloys
405	Cast-in-place turbine blades, linings for the petroleum industry, heat-exchanger tubes, and welded assemblies operating below 1400° F (760° C) that cannot be annealed after welding.
409	Automotive mufflers, converters, and tail pipe parts. Farm equipment, structural supports, transformer cases.
430	Automotive trim and molding, interior decorative work, refrigerator trays, chemical and processing towers, condensers, bubble caps, tanks and screens. Furnace parts subject to temperatures up to 1550° F (815° C). Trim on irons, cabinets, mixers, toasters.
430F	Screws, nuts, bolts, fittings, and other machined parts requiring corrosion resistance similar to that of 430.
446	Primarily for heat-resisting applications such as annealing boxes, baffle plates, glass molds, heaters, oil-burner parts, stirring rods, valves and fittings, X-ray tube bases, and lead-in wires.
	Martensitic Alloys
403	Turbine buckets and valves
410 and 414	Nuts and bolts, kitchen tools, tableware, springs, rules and tapes, furnace parts and heat-resisting applications up to 1200° F (649° C), cutlery, pump parts, fastenings, screws, steam-turbine parts, gage parts, oil-burner parts, scissors, shipbells

(continued)

STAINLESS STEELS

TABLE 2-4—*Continued*

Grade	Application
416	Bolts and nuts, screws, intricate machined parts such as pump and valve parts, outboard motor parts
420	Cutlery, heat-treated springs, pump parts, knives, dental and surgical instruments, self-tapping screws
431	Pump parts, steering wheel spokes, windshield wiper arms, marine propeller shafting, aircraft stress members
440A, 440B, and 440C	Bushings and bearings, cutlery, valve seats and trim, heat-treated springs, gages, surgical and dental instruments
	Precipitation-Hardening Alloys
S13800	Valve parts, fittings, cold headed and machined fasteners, shafts, landing gear parts, pins, lockwashers, aircraft components, nuclear reactor components
S15500	Valve parts, fittings and fasteners, shafts, gears, engine parts, chemical processing equipment, paper mill equipment, aircraft components, nuclear reactor components
S17400	Oil field valve parts, chemical processing equipment, aircraft fittings, pumpshafts, nuclear reactor components, gears, paper mill equipment, missile fittings, jet engine parts
	Duplex Alloys
329 and S32550	Shell and tube heat exchangers, centrifuges, pump parts, wastewater treatment, cooling coils

MACHINING AND FABRICATING CHARACTERISTICS

Care must be exercised in the fabrication of stainless steels because they are sensitive to thermal and mechanical operations. The following sections give general information to assist the manufacturing engineer when machining, forming, welding, or heat treating stainless steels. To obtain the best results, it is advisable to consult with the steel producer regarding the specific stainless steel grade.

Machining

The machining characteristics of austenitic stainless steels are substantially different from those of carbon and alloy steels. Generally, most austenitic stainless steels are tough, gummy, and tend to seize and gall. However, recent modifications in melting and refining practices have improved the machinability of stainless steels.

Austenitic stainless steels are the most difficult to machine because they work harden at a very rapid rate. The 400 series, ferritic and martensitic alloys, are the easiest to machine but the stringy chip produced by these alloys can slow productivity. Machining stainless steels in a slightly hardened condition results in improved machinability and productivity.

Machinability improvements have also been made by modifying the chemical compositions of certain grades referred to as free-machining stainless steels. Sulfur, selenium, lead, copper, aluminum, or phosphorus are added to improve machinability by a variety of mechanisms. These mechanisms include reducing friction between the tool, workpiece, and chip; reducing the tendency for pressure welding of chips to the tool; and increasing the brittleness of the chip. However, the free-machining alloying elements can adversely affect corrosion resistance, transverse ductility, and other qualities, such as weldability.

Forming

Stainless steels can be successfully formed by hot and cold processes. Hot forming processes include forging, extruding, and heading. Cold forming processes include blanking, brake forming, bending, drawing, embossing, heading, punching, roll forming, and spinning. Detailed information on each process is given in Volume II, *Forming*, of this Handbook series. Table 2-5 presents the relative forming characteristics of various grades of stainless steel.

Hot forming. Most of the stainless steel grades have a moderately restricted hot-working temperature range. The temperature is largely determined by the alloy's composition. In most applications, the lowest feasible temperature is recommended. Since stainless steels possess greater yield strengths than carbon and low-alloy steels, the equipment employed must be capable of delivering greater force. The number of blows must also be increased when using hammers. Heat treatment is generally necessary after hot forming to obtain optimum corrosion resistance and mechanical properties. Austenitic alloys should be reheated to approximately 1900° F (1040° C) and water quenched for optimum corrosion resistance. To prevent cracking, martensitic alloys require slow cooling after forging followed by annealing. Ferritic alloys also require annealing for best corrosion protection.

Cold forming. Stainless steels exhibit maximum ductility in the fully annealed condition. Austenitic alloys have greater ductility than ferritic and martensitic alloys. The formability of the particular grade can be estimated from the room temperature ductility, which is often measured as percent elongation or percent reduction of area. As ductility increases, the formability of the material increases. In general, higher carbon alloys have reduced formability. All stainless steel grades can be slightly heated to improve their formability.

Since the shear strength of most stainless steels is 50-75% greater than mild steel, the press or shear must be capable of delivering the additional force required. Generally press or shear speed is reduced to 75% of normal operating speeds. Proper lubricants are also required to obtain the best cold forming results. Austenitic alloys work harden during cold forming and often require annealing between deep drawing operations.

Welding

All stainless steels can be welded employing the various welding methods available today. The two main methods employed are arc welding and resistance welding. However, the

TABLE 2-5
Relative Forming Characteristics of Stainless Steels

AISI Grade	Blanking	Brake Forming	Coining	Deep Drawing	Embossing	Forging, Cold	Forging, Hot	Heading, Cold	Heading, Hot	Punching	Roll Forming	Sawing	Spinning
Austenitic Alloys													
201	B	B	B-C	A-B	B-C	C	B	C-D	B	C	B	C	C-D
202	B	A	B	A	B	B	B	C	B	B	A	C	B-C
301	B	B	B-C	A-B	B-C	C	B	C-D	B	C	B	C	C-D
302	B	A	B	A	B	B	B	C	B	B	A	C	B-C
302B	B	B	C	B-C	B-C	B	B	D	B	B	---	C	C
303 and 303Se	B	D	C-D	D	C	D	B-C	D-C	C	B	D	B	D
304	B	A	B	A	B	B	B	C	B	B	A	C	B
304L	B	A	B	A	B	B	B	C	B	B	A	C	B
305	B	A	A-B	A	A-B	A-B	B	B-A	B	B	A	C	A
308	B	B	D	D	D	D	B	D	B	---	---	C	D
309	B	A	B	B	B	B-C	B-C	C	C	B	B	C	C
309S	B	A	B	B	B	B-C	B-C	C	C	B	B	C	C
310	B	A	B	B	B	B	B-C	C	C	B	A	C	B
310S	B	A	B	B	B	B	B-C	C	C	B	A	C	B
314	B	A	B	B-C	B-C	B-C	B-C	C-D	C	B	B	C	C
316	B	A	B	A	B	B	B	C	B	B	A	C	B
316L	B	A	B	A	B	B	B	C	B	B	A	C	B
317	B	A	B	A	B	B	B-C	C	B-C	B	B	C	B-C
321	B	A	B	A	B	B	B	C	C-B	B	B	C	B-C
347 and 348	B	A	B	A	B	B	B	C	C-B	B	B	C	B-C
384 and 385	B	B	A	B	A-B	A	B	A	B	B	A	C	A

(continued)

TABLE 2-5—*Continued*

AISI Grade	Blanking	Brake Forming	Coining	Deep Drawing	Embossing	Forging, Cold	Forging, Hot	Heading, Cold	Heading, Hot	Punching	Roll Forming	Sawing	Spinning
Duplex Alloys													
S32550	B	B	B	C	B-C	B	A-B	B	B	B	B	C	B
Ferritic Alloys													
430	A	A*	A	A-B	A	B	B	A	B	A-B	A	B	A
430F and 430FSe	B	B-C*	C-D	D	C	D	C	D	C	A-B	D	A-B	D
405	A	A*	A	A	A	B	B	A	B	A-B	A	B	A
429	A	A*	A	A-B	A	B	B	A	B	A-B	A	B	A
434	A	A*	A	B	A	B	B	A	B	A-B	A	B	A-B
436	A	A*	A	B	A	B	B	A	B	A	A	B	A-B
442	A	A*	B	B	B	B-C	B-C	B	B-C	A-B	A	B-C	B-C
446	A	A*	B	B-C	B	C	B-C	C	B-C	B	B	B-C	C
Martensitic Alloys													
410	A	A*	A	A	A	B	B	A	B	A-B	A	B	A
403	A	A*	A	A	A	B	B	A	B	A-B	A	B	A
414	A	A*	B	B	C	C	B	D	B	B	C	C	C
416 and 416Se	B	C*	D	D	C	D	C	D	C	A-B	D	B	D
420	B	C*	C-D	C-D	C	C-D	B	C	B	B-C	C-D	C	D
420F	B	C*	C-D	C-D	C	C-D	B	C	B	B-C	C-D	C	D
431	C-D	C*	C-D	C-D	C-D	C-D	B	C-D	B	C-D	C-D	C	D
440A	B-C	C*	D	C-D	C	C-D	B	C	B	---	C-D	C	D
440B	---	---	D	---	D	D	B	C-D	B	---	---	C	D
440C	---	---	D	---	D	D	B	C-D	B	---	---	C	D

(*American Iron and Steel Institute*)

Note: Ratings are for making comparisons of alloys within their own metallurgical group. They should not be used to compare 300 series with 400 series.
A = excellent B = good C = fair D = not generally recommended

*Severe sharp bends should be avoided.

actual techniques followed when using these methods are modified slightly to preserve corrosion resistance in the weld and heat-affected zone (HAZ), to maintain optimum mechanical properties in the joint, and to minimize heat distortion.

When welding joints made with a stainless steel and a low-alloy steel or with dissimilar stainless steels, a highly alloyed welding rod should be used to counteract dilution effects and preserve corrosion resistance. The slow cooling of heavy sections of austenitic stainless steels often results in chromium-carbide precipitation, which reduces corrosion resistance. This precipitation of carbides or sensitizing process can be corrected by solution annealing. Optimum mechanical properties in the joint can be achieved by using the proper weld rod. Following recommended practices by the American Welding Society or consulting with the rod manufacturer will ensure proper rod selection.

The heat generated by the welding process is dissipated at a slower rate in stainless steels than in other steels. The slow cooling rate may result in workpiece distortion. Distortion can be minimized by lowering weld current settings, employing skip-weld techniques, using backup chill bars or other cooling techniques, or by incorporating bevel joints in the design instead of square-end butt joints.

Preheating and postheating martensitic alloys slows the cooling rate. If the cooling rate is too fast, metallurgical changes may occur in the material that could result in cracking. Ferritic alloys are usually heat treated following welding operations to minimize sensitization and losses in ductility. Austenitic alloys do not generally require pre or postheating. In heavy sections that cannot be solution annealed, stabilized or low-carbon grades such as 321, 347, 304L, or 316L are recommended. Postheating is also performed to stress relieve components that may contain high residual stresses. In some chemical environments, residual stresses increase susceptibility to stress-corrosion cracking. Postheating treatments are normally employed for precipitation-hardening alloys to restore or improve their mechanical properties.

Stainless steels can also be joined together by employing brazing and soldering techniques. Brazing is usually preferred when joining stainless steel to another metal. Chloride fluxes must be avoided, or chlorides must be completely removed after brazing to avoid pitting or stress-corrosion problems. Phosphoric acid type fluxes are recommended when preparing stainless steel surfaces for soldering. Volume IV, *Quality Control and Assembly*, has additional information on stainless steel welding practices.

Heat Treating

Annealing and other forms of heat treatment are usually performed on all stainless steel grades to improve their corrosion resistance and mechanical properties. In certain high-temperature applications, stainless steels are used in the as-rolled or forged condition. The chemical composition influences how each particular grade responds to the treatment. Austenitic, ferritic, and duplex alloys are not capable of being hardened by heat treatment. These alloys are hardened by cold working. Martensitic alloys and precipitation-hardening alloys are hardened through heat treatment.

Heat-treating methods for stainless steels are similar to those used with other steels and special equipment is not required. Conventional electric, gas-fired, oil-fired, salt bath, or induction furnaces are used to heat the steel. The steel should not be directly exposed to the flame, and slightly oxidizing furnace atmospheres are preferred. For detailed information on the heat treatment of stainless steels, refer to Chapter 14 in this volume.

MARAGING STEELS

Maraging steels, developed by the International Nickel Company in the 1960s, comprise a special class of high-strength steels that use nickel as the main alloying element. The term "maraging" is derived from "martensite age hardening" and denotes age hardening of a low-carbon, iron-nickel martensitic matrix.

The annealed microstructure of maraging steel is essentially a carbon-free, iron-nickel lath martensite. In the annealed condition, the material is soft and can be readily machined or formed. The material is martensitic at room temperature but reverts to an austenitic, face-centered cubic structure when heated to 1500° F (815° C). During cooling, martensite starts to form at 310° F (155° C) and is 99% complete at 210° F (100° C). The material is heated to 900° F (480° C) for aging and then cooled to room temperature. During aging, the martensite is strengthened by short-range ordering and subsequent precipitation of nickel-molybdenum and nickel-titanium intermetallic compounds.

Maraging steels are produced by a double-vacuum melting process to maintain high purity and to reduce residual elements. The first process is usually vacuum induction melting which is then followed by a vacuum-arc remelting process. Maraging steels are produced in wrought steel compositions. Common wrought forms are bar, plate, and sheet.

ALLOYS

Currently only 18% nickel maraging steel alloys are being produced. The 20% and 25% nickel alloys were the two original maraging alloys developed, but they were discontinued because of their brittleness at high-temperature strength levels and the complexity of the annealing and aging treatments.

Table 2-6 shows that nickel rather than carbon is the principal alloying element in maraging steels. Cobalt, molybdenum, and titanium are used to develop the high strength. The use of more common elements like carbon, sulphur, phosphorous, silicon, and manganese is held to a minimum since they serve no significant purpose, and in some cases may promote embrittlement of the alloy. The balance of the analysis is iron. The numerical designations associated with each grade are generally indicative of the ultimate tensile strength of that grade in ksi. Table 2-7 lists the mechanical properties for the various grades of maraging steels.

APPLICATIONS

The 18% nickel maraging steels were developed to meet the exacting requirements of the aerospace industry. This class of steels filled the need for ultrahigh-strength steels that had high fracture toughness and high fatigue, tensile, and yield strengths.

TABLE 2-6
Nominal Chemical Composition of 18% Nickel Maraging Steels, %

Elements	Grade 200	250	300	350
Nickel	18.50	18.50	18.50	18.50
Cobalt	8.50	7.50	9.00	12.00
Molybdenum	3.25	4.80	4.80	4.80
Titanium	0.20	0.40	0.60	1.40
Aluminum	0.10	0.10	0.10	0.10
Silicon	0.10 max	0.10 max	0.10 max	0.10 max
Manganese	0.10 max	0.10 max	0.10 max	0.10 max
Carbon	0.03 max	0.03 max	0.03 max	0.03 max
Sulfur	0.01 max	0.01 max	0.01 max	0.01 max
Phosphorous	0.01 max	0.01 max	0.01 max	0.01 max
Zirconium	0.01	0.01	0.01	0.01
Boron	0.003	0.003	0.003	0.003
Iron	balance	balance	balance	balance

(*Teledyne Vasco*)

Initial maraging steel applications in the aerospace industry included solid-propellant rocket motor casings, load cells used to measure the thrust of large rockets, flexures used in the guidance mechanism of missiles, helicopter driveshaft components, mid-fan driveshafts in jet engines, and aircraft wing members. More recently, the materials have been included in the space program in the production of torsion bars for the Apollo 15 lunar roving vehicle as well as for a critical component on the space shuttle Columbia.

Shortly after their development, the great potential of the maraging steels in tooling applications was realized. Some of the initial applications in the tooling area included aluminum die casting dies, aluminum die casting die core pins, plastic molding dies, extrusion tooling, punches, blanking dies, and cold forming dies. These materials continue to be used in tooling areas because of:

1. Excellent machinability.
2. Good polishability.
3. A simple, precipitation-hardening, aging heat treatment.
4. Uniform, predictable shrinkage during heat treatment.
5. Through hardening without quenching.

TABLE 2-7
Nominal Mechanical Properties of 18% Nickel Maraging Steels

Grade	Form and Condition	Tensile Strength, ksi (MPa)	Yield Strength (0.2% Offset), ksi (MPa)	Total Elongation in 2″ (50 mm), %	Hardness, Rockwell C
200	1″ (25.4 mm) diam bars after aging	210 (1450)	206 (1420)	12	43/48
250	1″ (25.4 mm) diam bars after aging	260 (1790)	255 (1760)	11	48/52
300	1″ (25.4 mm) diam bars after aging	294 (2025)	290 (2000)	9	50/55
350	1″ (25.4 mm) diam bars after aging	350 (2415)	340 (2345)	7	55/60

(*Teledyne Vasco*)

Note: The values of the mechanical properties listed are for comparison purposes only and are not intended for specifications or design purposes. For specific information, consult the steel manufacturer.

6. Minimal distortion during heat treatment.
7. Total freedom from decarburization.
8. Good weldability without preheating.

FABRICATING CHARACTERISTICS

Maraging steels can be readily processed by machining, forming, welding, and heat-treating methods.

Machining

Machinability of maraging steels in the annealed condition is approximately comparable to steels such as prehardened AISI 4340. However, after aging, the choice of cutting tools and machining conditions becomes increasingly important. During machining operations, rigid equipment, firm tool supports, sharp tools, and an adequate coolant are essential.

Forming

Maraging steels can be readily formed using cold, warm, and hot-working methods. Cold working is performed in the annealed condition, and large reductions can be made without significantly increasing the tensile strength or hardness of the material. When necessary, annealing can be performed between forming operations. Warm forming can be performed below 600° F (315° C). Forming above 600° F may cause maraging to occur. Hot forming may be performed by forging, forming, or rolling operations between 1500 and 2100° F (815 and 1150° C).

Welding

The 18% nickel maraging steels have high weldability in both the solution annealed and the full-aged conditions. Most conventional welding procedures achieve sound, crack-free welds.

Best results for structural welding have been obtained by either using tungsten inert gas (TIG) or electron-beam welding. Although the metal inert gas (MIG) process may be employed, it does result in a loss of ductility. Preheating is not only unnecessary but should not be performed because of the hardening mechanism of these grades. Gas shielding with pure argon is recommended for either TIG or MIG welding. Thorough cleaning of the as-welded deposit and a maximum interpass temperature of 250° F (120° C) are essential.

To obtain the best possible weld microstructure, and thus the highest joint efficiency, the material should be solution annealed after welding and prior to aging. However, joint efficiencies of better than 90% have been obtained with the 200, 250, and 300 level material by merely aging after welding.

Heat Treatment

Maraging steels are supplied in the solution annealed condition by the steel supplier in the hardness range of R_C30-35. Aging is performed to develop the material's high strength, toughness, and hardness. The aging process may be performed in an air atmosphere or in a liquid salt-bath solution.

Bibliography

American Iron and Steel Institute. *Stainless and Heat Resisting Steels*. Washington, DC: American Iron and Steel Institute, December 1974.

American Iron and Steel Institute. *Design Guidelines for the Selection and Use of Stainless Steels*. Washington, DC: American Iron and Steel Institute, April 1977.

American Society for Metals. *Metals Handbook*, Vol. 3, 9th ed. Metals Park, OH: American Society for Metals, 1980.

Carpenter Technology Corporation. *Carpenter Stainless Steels—Working Data*. Reading, PA: Carpenter Technology Corporation, January 1983.

Peckner, Donald, and Bernstein, I.M., eds. *Handbook of Stainless Steels*. New York: McGraw-Hill Book Company, 1977.

Wick, Charles. "Cold Forming Stainless Steels." *Manufacturing Engineering* (February 1978), pp. 44-49.

CAST STEELS AND IRONS

CAST STEELS

In wrought steel production, the various alloying elements are melted together in a furnace and then poured into ingots to cool and subsequently be hot and/or cold worked until the desired form and size has been achieved. When steel castings are produced, the various alloying elements are melted together and then poured directly into a mold cavity having the proper design. The cast part is allowed to cool and then removed from the mold.

The compositions of steel castings are similar to those of wrought steels, with the exception of higher silicon and manganese content to ensure thorough deoxidation. Cast steels may contain alloying elements such as nickel, chromium, vanadium, and copper to give desirable combinations of hardness, tensile strength, and toughness not readily available in plain carbon grades. The total alloy content may be as high as 30% or greater. Chemical compositions of the various alloys are usually based on specifications of the American Society of Testing and Materials (ASTM). Cast steels are also available in American Iron and Steel Institute (AISI) designations, but the silicon and manganese percentages are higher than in wrought steels. The hardness of cast steels is measured by the Brinell test method because of the coarseness of their microstructure.

MANUFACTURING PRACTICE

The direct-arc electric process is predominantly used in the melting of carbon and low-alloy steel for castings. The open hearth, converter, and crucible processes are practically extinct. However, the AOD (argon-oxygen decarburization) process is gaining acceptance for melting of both low and high-alloy steels due to the process' ability to control sulfur and gas levels and minimize oxidation losses of chromium. The basic oxygen melting process has not been adapted to cast steels. High-frequency induction melting is used for high-alloy steels and especially for carbon steels containing less than 0.10% carbon. The high-frequency induction process is also used when melting small quantities of steel in a variety of chemical compositions. However, the high-frequency induction process is at an economical disadvantage with the direct-arc method for carbon and low-alloy steels. A discussion on the various casting and molding processes is found in Volume II, *Forming*, of this Handbook series.

CAST STEEL ALLOYS AND APPLICATIONS

Cast steels are available in carbon, low-alloy, corrosion-resistant, and heat-resistant alloys. The following material briefly describes the more commonly used alloys and includes the various areas and industries where these alloys are being used. Also included are tables of the mechanical properties that are most useful for manufacturing engineers.

Carbon and Low-Alloy Cast Steels

Cast carbon steels have carbon as the main alloying element, although other alloying elements are also present. They are usually classified by the amount of carbon contained in the steel; low-carbon steel castings contain up to 0.20% carbon, medium-carbon steel castings contain from 0.20 to 0.50% carbon, and high-carbon steel castings contain more than 0.50% carbon. Cast steels containing more than 1.00% manganese, 0.80% silicon, 0.50% nickel, 0.50% copper, 0.25% chromium, 0.10% molybdenum, 0.05% vanadium, and 0.05% tungsten are normally considered alloy steel castings. When the percentage of alloying elements, including carbon, is 8% or less, the cast steel is considered low-alloy cast steel. In low-alloy cast steel, carbon content is generally less than 0.45%.

Carbon steel castings. Medium-carbon steel is used predominantly in carbon steel castings for a variety of applications. Some of these applications include equipment for machinery and tools, rolling mills, mining, road building, and building construction, and in the railroad and other transportation industries. Low-carbon steel castings are

CHAPTER CONTENTS:

***Contributors of sections of this chapter are: Lyle R. Jenkins**, Technical Director, Ductile Iron Society; **Roy W. Lobenhofer**, Vice President—Technology, American Foundrymen's Society; **Paul J. Mikelonis**, Vice President—Technical Services, General Casting Corp.; **Wayne M. Riggle**, Technical Director, Castwell Div., Wells Manufacturing Co.; **John Svoboda**, Technical and Research Director, Steel Founders' Society of America; **Charles F. Walton**, Specialist in Iron Castings.*

***Reviewers of sections of this chapter are: Harold Coolidge**, Vice President—Technical Services, Wells Manufacturing Co.; **Dr. Robert C. Creese**, Associate Professor, West Virginia University; **William J. Hayes**, Manager of Metallurgy, C.W.C. Castings Div., Textron, Inc.; **Lyle R. Jenkins**, Technical Director, Ductile Iron Society; **Roy W. Lobenhofer**, Vice President—Technology, American Foundrymen's Society; **Paul J. Mikelonis**, Vice President—Technical Services, General Casting Corp.; **Raymond W. Monroe**, Research Manager, Steel Founders' Society of America; **Wayne M. Riggle**, Technical Director, Castwell Div., Wells Manufacturing Co.; **Donald Rupert**, Vice President of Sales, Erie Malleable Iron Co.; **John Svoboda**, Technical and Research Director, Steel Founders' Society of America; **Charles F. Walton**, Specialist in Iron Castings.*

CHAPTER 3

CAST STEELS

TABLE 3-1
Mechanical Properties of Most Commonly Used Carbon Grades of Cast Steels[1]

ASTM Standard	Grade	Condition*	Tensile Strength, ksi (MPa)	Yield Strength (0.2% Offset), ksi (MPa)	Total Elongation in 2″ (50 mm), %	Reduction of Area, %
A27	U-60-30	---	60 (415)	30 (205)	22	30
	60-30	A, N, NT, or QT	60 (415)	30 (205)	24	35
	65-35	A, N, NT, or QT	65 (450)	35 (240)	24	35
	70-36	A, N, NT, or QT	70 (485)	36 (250)	22	30
	70-40	A, N, NT, or QT	70 (485)	40 (275)	22	30
A216	WCA	A, N, or NT	60-85 (415-585)	30 (205)	24	35
	WCB	A, N, or NT	70-95 (485-655)	36 (250)	22	35
	WCC	A, N, or NT	70-95 (485-655)	40 (275)	22	35
A352	LCA	NT or QT at 1100° F (595° C)	60-85 (415-585)	30 (205)	24	35
	LCB	NT or QT at 1100° F (595° C)	65-90 (450-620)	35 (240)	24	35
	LCC	NT or QT at 1100° F (595° C)	70-95 (485-655)	40 (275)	22	35
A487	A, AN	NT at 1100° F (595° C)	60-85 (415-585)	30 (205)	24	35
	B, BN	NT at 1100° F (595° C)	70-95 (485-655)	36 (250)	22	35
	C, CN	NT at 1100° F (595° C)	70-95 (485-655)	40 (275)	22	35
	DN	NT at 1100° F (595° C)	80 (550)	40 (275)	17	25
	AQ	QT at 1100° F (595° C)	70-90 (485-620)	30 (205)	24	35
	BQ	QT at 1100° F (595° C)	80-105 (550-725)	36 (250)	22	35
	CQ	QT at 1100° F (595° C)	80-105 (550-725)	40 (275)	22	35

(*Steel Founders' Society of America*)

Note: The values of the mechanical properties listed should only be used as a general guide and not for design purposes. For specific information, contact the supplier of the steel castings.

* A = annealed
N = normalized
NT = normalized and tempered
QT = quenched and tempered

used in the automotive industry for components that are case carburized and also in manufacturing electrical components, which require good magnetic properties. Table 3-1 lists the mechanical properties of the most commonly used grades of carbon steel castings.

Low-alloy steel castings. Several types of low-alloy steel castings are currently being produced. Carbon-manganese steel castings are widely used for producing parts that are subject to dynamic loading and abrasion. Manganese-molybdenum steel castings are used when producing parts for agricultural machinery and equipment, construction equipment, road building machinery, and mining equipment. Manganese-nickel-chromium-molybdenum steel castings are used in structural and dynamic applications that require high yield-strength steels. Nickel steel castings are being used in the production of components operating in low-temperature environments. Nickel-chromium-molybdenum steel castings are used when producing large castings and parts that are subject to high static and dynamic stresses as in the aircraft industries. Chromium-molybdenum steel castings are used extensively for oil refinery equipment castings. Copper alloy steel castings are used in the logging and excavating industries. Table 3-2 lists the mechanical properties of the various grades of cast low-alloy steels that are commonly used.

Corrosion-Resistant Cast Steels

Corrosion-resistant, high-alloy steel castings are commonly referred to as cast stainless steels. Alloy composition is based on the alloy designation system adopted by the Alloy Casting Institute (ACI), and the various alloys produced are covered by ASTM A743, A744, A747, and A494 standards.

The principal grades are martensitic, ferritic, precipitation-hardening, austenitic-ferritic, and austenitic. These steels are generally used when manufacturing chemical processing and power generating equipment to resist corrosion in aqueous or liquid-vapor environments at temperatures below 600° F (315° C). The mechanical properties of the commonly used corrosion-resistant alloys are listed in Table 3-3.

Martensitic grades. Martensitic cast stainless steels are those alloys beginning with the designation *CA*. This type of steel has good resistance to atmospheric corrosion and organic media in mild service. Resistance to seawater corrosion increases with the addition of molybdenum. Martensitic cast stainless steels are used when manufacturing pumps, compressors, valves, hydraulic turbines, propellers, and machinery components.

Ferritic grades. Ferritic cast stainless steels are usually those alloys beginning with the designations *CB* and *CC*. However, there are also precipitation-hardening alloys designated with CB. The ferritic grade is nonhardenable.

Alloys beginning with CB have a greater resistance to corrosives than alloys beginning with CA, but have a low impact strength. They are generally used when manufacturing valve bodies and trim for chemical production and food processing industries. Alloys beginning with CC have good resistance to oxidizing corrosives, mixed nitric and sulfuric acids, and alkaline liquors. They are frequently used when manufacturing components that are in contact with acid mine waters and are used in nitrocellulose production.

Precipitation-hardening grades. Precipitation-hardening cast stainless steels are those alloys designated by *CB* and *CD* that are hardenable. Alloy-type CB-7Cu has better corrosion resistance than CA alloys and is generally used when manufacturing components that require high strength and good corrosion resistance, such as components in the aircraft and food processing industries.

Alloy CD-4MCu has approximately twice the strength of the austenitic-ferritic grade of cast stainless steels as well as equal or better corrosion resistance. This alloy is used for pumps, valves, and stressed components in the marine, chemical, textile, and paper industries.

Austenitic-ferritic grades. Cast stainless steel alloys in this grade are designated by *CE, CF,* and *CG*. The CF alloys comprise the majority of the corrosion-resistant cast steels currently being produced and are used in a variety of general purpose applications as well as in applications requiring resistance to nitric acid and other oxidizing agents.

Alloys designated by CE have good resistance to sulfurous acid and to stress-corrosion cracking in polythionic acid. The CG alloys have better corrosion resistance to sulfuric and sulfurous acid solutions than certain CF alloys but are not suitable for applications requiring resistance to nitric acid or other strongly oxidizing environments.

Austenitic grade. Cast stainless steel alloys in the austenitic grade are designated by *CH, CK,* and *CN*. Alloys CH and CK are high-chromium, high-carbon alloys and are used for components that are in contact with paper pulp solutions and nitric acid. The CN-7M alloy contains molybdenum and copper and is used for components in steel mills handling nitric-hydrofluoric pickling solutions and other components that operate in severe service environments.

Heat-Resistant Steel Castings

Cast steels discussed in this section are divided into two groups: (1) those that can be used for service up to 1150° F (620° C) and (2) those that can be used for service above 1150° F (620° C). The steels for use below 1150° F (620° C) are made up

TABLE 3-2
Mechanical Properties of Most Commonly Used Grades of Low-Alloy Cast Steels[2]

ASTM Standard	Grade	Condition*	Tensile Strength, ksi (MPa)	Yield Strength (0.2% Offset), ksi (MPa)	Total Elongation in 2″ (50 mm), %	Reduction of Area, %
A352	LC1	NT or QT at 1100° F (595° C)	65-90 (450-620)	35 (240)	24	35
	LC2	NT or QT at 1100° F (595° C)	70-95 (485-655)	40 (275)	24	35
	LC2-1	NT or QT at 1100° F (595° C)	105-130 (725-895)	80 (550)	18	30
	LC3	NT or QT at 1100° F (595° C)	70-95 (485-655)	40 (275)	24	35
	CA6NM	NT at 1050° F (565° C)	110-135 (760-930)	80 (550)	15	35
A487	1N	NT at 1100° F (595° C)	85-110 (585-760)	55 (380)	22	40
	2N	NT at 1100° F (595° C)	85-110 (585-760)	53 (365)	22	35
	4N	NT at 1100° F (595° C)	90-115 (620-795)	60 (415)	20	40
	6N	NT at 1100° F (595° C)	115 (795)	80 (550)	18	30
	8N	NT at 1250° F (680° C)	85-110 (585-760)	55 (380)	20	35
	9N	NT at 1100° F (595° C)	90 (620)	60 (415)	20	35
	10N	NT at 1100° F (595° C)	100 (690)	70 (485)	18	35
	11N	NT at 1100° F (595° C)	70-95 (485-655)	40 (275)	20	35
	12N	NT at 1100° F (595° C)	70-95 (485-655)	40 (275)	20	35
	13N	NT at 1100° F (595° C)	90-115 (620-795)	60 (415)	18	35
	16N	NT at 1100° F (595° C)	70-95 (485-655)	40 (275)	22	35
	1Q	AT at 1100° F (595° C)	90-115 (620-795)	65 (450)	22	45
	2Q	QT at 1100° F (595° C)	90-115 (620-795)	65 (450)	22	40
	4Q	QT at 1100° F (595° C)	105-130 (725-895)	85 (585)	17	35
	4QA	QT at 1100° F (595° C)	115 (795)	95 (655)	15	35
	6Q	QT at 1100° F (595° C)	120 (825)	95 (655)	12	25
	7Q	QT at 1100° F (595° C)	115 (795)	100 (690)	15	30
	8Q	QT at 1250° F (680° C)	105 (725)	85 (585)	17	30
	9Q	QT at 1100° F (595° C)	105 (725)	85 (585)	16	35
	10Q	QT at 1100° F (595° C)	125 (860)	100 (690)	15	35
	11Q	QT at 1100° F (595° C)	105-130 (725-895)	85 (585)	17	35
	12Q	QT at 1100° F (595° C)	105-130 (725-895)	85 (585)	17	35
	13Q	QT at 1100° F (595° C)	105-130 (725-895)	85 (585)	17	35
	14Q	QT at 1100° F (595° C)	120-145 (825-1000)	95 (655)	14	30

(*Steel Founders' Society of America*)

Note: The values of the mechanical properties listed should only be used as a general guide and not for design purposes. For specific information, contact the supplier of the steel castings.

*NT = normalized and tempered
QT = quenched and tempered

CAST STEELS

TABLE 3-3
Mechanical Properties of Most Commonly Used Corrosion-Resistant Cast Steels[3]

ASTM Standard	Grade	Condition*	Tensile Strength, ksi (MPa)	Yield Strength (0.2% Offset), ksi (MPa)	Total Elongation in 2″ (50 mm), %	Reduction of Area, %	Hardness, Bhn
A743	CF-8	ST at 1900°F (1040°C)	70 (485)	30 (205)	35	---	---
	CG-12	ST at 1900°F (1040°C)	70 (485)	28 (195)	35	---	---
	CF-20	ST at 1900°F (1040°C)	70 (485)	30 (205)	30	---	---
	CF-8M	ST at 1900°F (1040°C)	70 (485)	30 (205)	30	---	---
	CF-8C	ST at 1900°F (1040°C)	70 (485)	30 (205)	30	---	---
	CF-16F	ST at 1900°F (1040°C)	70 (485)	30 (205)	35	---	---
	CH-20	ST at 2000°F (1093°C)	70 (485)	30 (205)	30	---	---
	CK-20	ST at 2000°F (1093°C)	65 (450)	28 (195)	30	---	---
	CE-30	ST at 2000°F (1093°C)	80 (550)	40 (275)	10	---	---
	CA-15	NT or A	90 (620)	65 (450)	18	30	241 max
	CA-15M	NT or A	90 (620)	65 (450)	18	30	241 max
	CB-30	N or A	65 (450)	30 (205)	---	---	241 max
	CC-50	N or A	55 (380)	---	---	---	241 max
	CA-40	NT or A	100 (690)	70 (485)	15	25	269 max
	CF-3	As cast or ST	70 (485)	30 (205)	35	---	---
	CF-3M	As cast or ST	70 (485)	30 (205)	30	---	---
	CG6MMN	---	75 (515)	35 (240)	30	1	---
	CG-8M	ST at 1900°F (1040°C)	75 (520)	35 (240)	25	---	---
	CN-7M	ST at 2050°F (1120°C)	62 (425)	25 (170)	35	---	---
	CN-7MS	ST at 2050°F (1120°C)	70 (485)	30 (205)	35	---	---
	CA-6NM	NT at 1100°F (590°C)	110 (755)	80 (550)	15	35	285 max
	CD4MCu	ST at 1900°F (1040°C)	100 (690)	70 (485)	16	---	---
	CA-6N	NT at 1500°F (815°C)	140 (965)	135 (930)	15	50	---
A747	CB7Cu-1	Hardened at 900°F (480°C)	170 (1170)	145 (1000)	5	---	375
		Hardened at 925°F (495°C)	175 (1205)	150 (1035)	5	---	375
		Hardened at 1025°F (550°C)	150 (1035)	140 (965)	9	---	311
		Hardened at 1075°F (580°C)	145 (1000)	115 (795)	9	---	277
		Hardened at 1100°F (590°C)	135 (930)	110 (760)	9	---	269
		Hardened at 1125°F (605°C)	125 (860)	97 (670)	10	---	269
	CB7Cu-2	Hardened at 900°F (480°C)	170 (1170)	145 (1000)	5	---	375
		Hardened at 925°F (495°C)	175 (1205)	150 (1035)	5	---	375
		Hardened at 1025°F (550°C)	150 (1035)	140 (965)	9	---	311
		Hardened at 1075°F (580°C)	145 (1000)	115 (795)	9	---	277
		Hardened at 1100°F (590°C)	135 (930)	110 (760)	9	---	269
		Hardened at 1150°F (620°C)	125 (860)	97 (670)	10	---	269

(*Steel Founders' Society of America*)

Note: The values of the mechanical properties listed should only be used as a general guide and not for design purposes. For specific information, contact the supplier of the steel castings.
* ST = solution annealed
NT = normalized and tempered
A = annealed

of carbon and low-alloy cast steels, and the alloys for service above 1150°F (620°C) are made up of high-alloy cast steels.

Carbon and low-alloy steel castings. The two elements common to this group of cast steels and that contribute to creep resistance are molybdenum and chromium. These steels are covered by ASTM A216, A217, A356, and A389 standards. Table 3-4 lists the mechanical properties of the most commonly used cast steels in this category.

High-alloy steel castings. To provide effective resistance to oxidation (scaling) or to corrosive gases, these alloys contain chromium content in excess of 12%. Except for their higher carbon content, these cast steels are similar to corrosion-resistant cast steels. The three principal grades in this group are iron-chromium, iron-chromium-nickel, and iron-nickel-chromium. The various alloys are covered by the ASTM A297 standard. Table 3-5 presents the mechanical properties of the most commonly used alloys in this category.

Iron-chromium. The alloys belonging to this grade are designated by *HC* and *HD*. Alloy HD has greater strength because of its high nickel content. Iron-chromium alloys can be used for components in load-bearing applications up to 1200°F (649°C) and in lighter load-bearing applications up to 1900°F

TABLE 3-4
Mechanical Properties of Most Commonly Used Carbon and Low-Alloy, Heat-Resistant Cast Steels[4]

ASTM Standard	Grade	Condition*	Tensile Strength, ksi (MPa)	Yield Strength (0.2% Offset), ksi (MPa)	Total Elongation in 2" (50 mm), %	Reduction of Area, %
A216	WCA	A, N, or NT	60-85 (415-585)	30 (205)	24	35
	WCB	A, N, or NT	70-95 (485-655)	36 (250)	22	35
	WCC	A, N, or NT	70-95 (485-655)	40 (275)	22	35
A217	WC1	NT at 1100° F (595° C)	65-90 (450-620)	35 (240)	24	35
	WC4	NT at 1100° F (595° C)	70-95 (485-655)	40 (275)	20	35
	WC5	NT at 1100° F (595° C)	70-95 (485-655)	40 (275)	20	35
	WC6	NT at 1100° F (595° C)	70-95 (485-655)	40 (275)	20	35
	WC9	NT at 1250° F (675° C)	70-95 (485-655)	40 (275)	20	35
	WC11	NT at 1250° F (675° C)	80-105 (550-725)	50 (345)	18	45
	C5	NT at 1250° F (675° C)	90-115 (620-795)	60 (415)	18	35
	C12	NT at 1250° F (675° C)	90-115 (620-795)	60 (415)	18	35
	CA-15	NT at 1100° F (595° C)	90-115 (620-795)	65 (450)	18	30
A356	1	NT at 1100° F (595° C)	70 (485)	36 (250)	20	35
	2	NT at 1100° F (595° C)	65 (450)	35 (240)	22	35
	5	NT at 1100° F (595° C)	70 (485)	40 (275)	22	35
	6	NT at 1100° F (595° C)	70 (485)	45 (310)	22	35
	8	NT at 1100° F (595° C)	80 (550)	50 (345)	18	45
	9	NT at 1100° F (595° C)	85 (585)	60 (415)	15	45
	10	NT at 1100° F (595° C)	85 (585)	55 (380)	20	35

(*Steel Founders' Society of America*)

Note: The values of the mechanical properties listed should only be used as a general guide and not for design purposes. For specific information, contact the supplier of the steel castings.

* A = annealed
N = normalized
NT = normalized and tempered

(1038° C). Some typical components are rabble arms and blades for ore-roasting furnaces, salt pots, and grate bars.

Iron-chromium-nickel. The alloys in this grade are partially or completely austenitic and have higher strength and ductility than iron-chromium alloys. These alloys are designated by *HE, HF, HH, HI, HK*, and *HL*. Satisfactory results are obtained in either oxidizing or reducing atmospheres.

Typical applications for HE alloys are ore-roasting furnaces and steel mill furnaces. Alloy HF is used for tube supports and beams in oil refinery heaters, in cement kilns, and in ore-roasting and heat-treating furnaces. Alloy HH is used for manufacturing furnace parts that are not subjected to severe temperature cycling. Alloy HI is used in cast retorts for calcium and magnesium production. Alloy HK is used in the production of jet engines, gas turbines, hydrogen reformer tubes, and furnace parts. Alloy HL exhibits the best resistance to high-sulfur environments up to 1800° F (980° C) and is used in gas dissociation equipment.

Iron-nickel-chromium. These alloys—HN, HP, HT, HU, HW, and HX—are high-nickel steels and normally constitute about 40% of the total production of heat-resistant castings. Nickel is either the predominant alloying element or, in some cases, the base metal. The alloys can be used for most applications up to 2100° F (1150° C) and give excellent service life when subject to rapid heating and cooling. Resistance to thermal fatigue is excellent, but they are not recommended in atmospheres with high sulfur content.

Typical applications for alloy HN are brazing fixtures and highly stressed parts. Alloy HP is used for heat-treat fixtures, radiant tubes, and coils for ethylene pyrolysis heaters. Alloy HT is used for parts in heat-treating furnaces, glass rolls, enameling racks, and radiant heater tubes. Alloy HU is used for manufacturing burner tubes, lead and cyanide pots, retorts, and furnace parts. Alloy HW is used for hearths, mufflers, retorts, trays, boxes, burner parts, enameling fixtures, quenching fixtures, and containers for molten lead. Alloy HX finds the same applications as HW, particularly when improved resistance to hot gas corrosion is required.

PROCESSING STEEL CASTINGS

In order to produce a finished casting with specified dimensions and mechanical properties, several secondary processes are employed. These processes include machining, welding, and heat treating.

Machinability

In general, steel castings have the same machinability as comparable wrought steels. Machinability of carbon and low-alloy steel castings can be improved by altering the microstructure through heat treatment. However, this is only recommended when the production quantity and tool and time savings offset the cost of heat treatment. Plain carbon steel castings usually possess better machining properties than low-alloy steel castings.

The oxide scale or skin of the casting should be removed by abrasive blasting prior to any machining operation. The initial cut should be as deep as possible; for large castings the depth should be 1/4 to 3/8" (6 to 9.5 mm).

CAST STEELS

TABLE 3-5
Mechanical Properties of High-Alloy, Heat-Resistant Cast Steels at Room Temperature[5]

ASTM Standard	Grade	Condition*	Tensile Strength, ksi (MPa)	Yield Strength (0.2% Offset), ksi (MPa)	Total Elongation in 2" (50 mm), %	Hardness, Bhn
A297	HC	As cast	110 (760)	75 (515)	19	223
		Aged	115 (790)	80 (550)	18	---
	HD	As cast	85 (585)	48 (330)	16	90
	HE	As cast	95 (655)	45 (310)	20	200
		Aged	90 (620)	55 (380)	10	270
	HF	As cast	92 (635)	45 (310)	38	165
		Aged	100 (690)	50 (345)	25	190
	HH-type 1	As cast	85 (585)	50 (345)	25	185
		Aged	86 (595)	55 (380)	11	200
	HH-type 2	As cast	80 (550)	40 (275)	15	180
		Aged	92 (635)	45 (310)	8	200
	HI	As cast	80 (550)	45 (310)	12	180
		Aged	90 (620)	65 (450)	6	200
	HK	As cast	75 (515)	50 (345)	17	170
		Aged	85 (585)	50 (345)	10	190
	HL	As cast	82 (565)	52 (360)	19	192
	HN	As cast	68 (470)	38 (260)	13	160
	HP	As cast	71 (490)	40 (275)	11	170
	HT	As cast	70 (485)	40 (275)	10	180
		Aged	75 (515)	45 (310)	5	200
	HU	As cast	70 (485)	40 (275)	9	170
		Aged	73 (505)	43 (295)	5	190
	HW	As cast	68 (470)	36 (250)	4	185
		Aged	84 (580)	52 (360)	4	205
	HX	As cast	65 (450)	36 (250)	9	176
		Aged	73 (505)	44 (305)	9	185

(*American Society for Metals*)

Note: The values of the mechanical properties listed should only be used for comparison purposes and not for design guidelines. For specific information, contact the supplier of the steel castings.
*The temperature and duration of the aging treatment depends on the particular grade.

Weldability

Weldability of steel castings is mainly determined by the composition of the steel and the heat treatment performed. Generally, steel castings can be welded by the same processes used when welding wrought steels. These processes include shielded metal arc welding, gas tungsten arc welding, gas metal arc welding, flux-cored arc welding, submerged arc welding, and electroslag welding.

Preheating is usually recommended for carbon or low-alloy steels containing over 0.30% carbon to reduce the rate at which heat is extracted from the heat-affected zone (HAZ), relieve mechanical stress, prevent underbead cracking, and minimize hardening in the HAZ. Preheat temperatures are given in the ASTM specifications for the particular grade of steel. Corrosion-resistant cast steels do not generally require preheating; but in many cases, the weld is cooled between passes. Reheat treatment is usually required after welding to restore corrosion resistance at the welds.

Heat Treatment

Steel castings are heat treated to improve the as-cast structure and the mechanical properties. The customary heat treatments are annealing, normalizing, and normalizing and tempering. It is also common to quench and temper steel castings when the size, shape, and composition are not prone to serious distortion and cracking during quenching.

CAST IRONS

The term "cast iron" is a generic term that designates an entire family of cast ferrous metals. These metals possess a wide variety of properties that distinguish them from the family of steels. In composition, both steels and cast irons are primarily iron that is alloyed with carbon. However, steels always contain less than 2% combined carbon (and usually less than 1%), while cast irons contain more than 2% carbon. The carbon in cast iron is generally in the free state except for a maximum 0.65% combined carbon. Cast irons must also contain appreciable amounts of silicon, usually from 1 to 3%. These differences are not arbitrary, but have a metallurgical basis and effect the differing useful properties of these two families of ferrous alloys.

Because of the high carbon and silicon content, cast irons possess excellent casting characteristics and can be melted more easily than steels. Molten cast iron also flows better than molten steel and is less reactive with the molding material because of a lower pouring temperature. Shrinkage and contraction of cast iron during solidification are nominal and easily compensated for. High-strength parts can be cast close to machine dimensions with minimum material to machine off and discard. Machinability is very good since most of the carbon is in the free state. Since most cast irons are not as ductile as steels, they are not usually rolled or forged.

The tables supplied in the cast iron section have been compiled from tables in the *Iron Castings Handbook* and the *Tool and Manufacturing Engineers Handbook*, 3rd edition. The specification standards given refer only to the basic specification number and not the latest revision. When additional information is required, refer to the most current revision of the appropriate specification given by the specifying body.

MELTING AND CASTING TECHNIQUES

The equipment used for melting cast irons depends on the type of iron produced, quantity of iron required, and availability of the necessary energy. The two types of melting equipment most commonly used are the coreless induction furnace and the cupola.

Most iron is cast directly by the foundry into special shapes established by the customer's pattern. The common casting processes are green sand, resin-bonded sand, and shell mold. Processes that are less common include vacuum mold, permanent mold, ceramic mold, expendable pattern, and die casting. The selection of the casting process is usually determined by the part design and dimensions, the production quantities, and the accuracy required. Centrifugal and continuous casting processes are also employed in cast iron production. Centrifugal casting is used to produce hollow castings without cores such as pipe, and continuous casting is used in the production of bar stock having various cross sections. In the continuous casting of cast iron as opposed to steel, cast iron is always cast horizontally, the dies are made of graphite rather than copper, and cooling is only performed in the dies; any additional cooling is by radiation. For additional information on the various casting methods, refer to Chapter 16, "Casting," in Volume II, *Forming*, of this Handbook series.

TYPES OF CAST IRON

In most irons, an appreciable portion of the carbon content precipitates during solidification and appears as a separate constituent in the microstructure of the iron. The form and shape in which the excess carbon occurs determine the type of cast iron and establish the nature of its properties. The structure of the matrix metal around the carbon-rich constituent establishes the class of iron within each category.

The five basic types of cast iron are white iron, malleable iron, gray iron, ductile iron, and compacted graphite iron. In white iron, the majority of carbon occurs as the compound iron-carbide, which is a very hard constituent. Malleable iron is characterized by having most of the contained carbon present in irregularly shaped nodules of temper carbon that forms after annealing. Gray iron has the carbon occurring as graphite flakes. In ductile iron, the graphite occurs in spheres; and in compacted graphite iron, the graphite occurs primarily as stubby flakes with some spheres possible.

A sixth type of iron is composed of the high-alloy irons. High-alloy irons are white, gray, or ductile irons containing appreciable amounts of alloying elements, generally in excess of 3%. Their properties are not just modified, but may be essentially different from those of the base iron. Because of the high alloy content, special facilities are usually required for producing high-alloy iron castings.

It is not practical to designate cast irons by chemical analysis because the ranges of the chemical composition for the different cast irons overlap. The typical range of chemical analysis for different types of unalloyed cast irons is given in Table 3-6. Cast irons are usually specified by their mechanical properties and microstructure.

White Cast Iron

White cast irons differ from other irons in that they contain little or no graphite due to their chemical composition and rapid cooling rate during solidification. Virtually all the carbon present is chemically combined with the iron as iron carbide, a very hard and brittle substance. Since the iron carbide, also called cementite, dominates the microstructure, white iron is essentially hard and brittle and has a white crystalline fracture.

White irons are very high in compressive strength, excellent in wear resistance, and retain their hardness even up to a red heat. However, they are brittle and are not machinable under normal circumstances. The properties can be varied within a limited range by the amount of iron carbide in the structure and the nature of the matrix structure that surrounds it.

Castings can be made entirely of white iron in regular sand molds, or white iron can be formed on the surface of a gray iron by chilling. During chilling, the metal near the surface solidifies with sufficient rapidity to prevent graphite from forming. When a block of iron or graphite is used as part of the mold, it is called a chill, and the iron that is formed is called chilled iron. If white iron is chilled at an intermediate rate and contains both iron carbide and graphite, it is called mottled iron.

TABLE 3-6
Range of Compositions for Typical Unalloyed Cast Irons

Element	Gray Iron, %	White Iron, %	Malleable Iron (Cast White), %	Ductile Iron and Compacted Iron, %
Carbon	2.5-4.0	1.8-3.6	2.00-2.60	3.0-4.0
Silicon	1.0-3.0	0.5-1.9	1.10-1.60	1.8-2.8
Manganese	0.25-1.0	0.25-0.80	0.20-1.00	0.10-1.00
Sulfur	0.02-0.25	0.06-0.20	0.04-0.18	0.03 max
Phosphorus	0.05-1.0	0.06-0.18	0.18 max	0.10 max

CAST IRONS

Applications. White cast iron is used in applications requiring good wear and abrasive-resistant characteristics. Certain compositions of white iron are used as base metals in the production of malleable cast iron.

Specifications. Standard specifications for unalloyed and low-alloy white cast irons have not been established. Generally, unalloyed and low-alloy white irons can be specified by hardness ranging from 300 to 600 Bhn. The lower hardness values are characteristic of the lower carbon content irons that are a little tougher, but not as wear resistant as the harder irons. Maximum hardness and wear resistance can be obtained with the high-alloy white irons.

Malleable Iron

Malleable iron has the major portion of its carbon content occurring in the microstructure as irregularly shaped nodules of temper carbon. The structure is obtained by first producing a white cast iron casting having the correct composition. The casting is then heat treated at a temperature over 1750° F (950° C) for an extended period of time. During the heat treatment, the iron carbide dissociates and graphite precipitates in the iron matrix. This form of graphite has been called "temper carbon" because it is formed in the solid state during heat treatment. Heat treatment is done in a controlled atmosphere to prevent loss of carbon from the surface. Depending on the type of heat treatment performed, the matrix obtained after heat treatment gives ferritic or pearlitic malleable iron, each of which serves specific purposes and possesses a variety of typical characteristics.

Malleable iron is comparatively strong, ductile, and tough and has machining characteristics comparable to gray iron. It has higher toughness and fatigue strength than gray iron and can be surface hardened to produce good wear resistance. The rapid solidification required to form the preliminary white iron limits the thickness that is practical for a malleable iron casting to about 3" (75 mm).

Malleable iron castings are widely used in the automotive, agricultural equipment, and railroad industries. Some typical applications include universal joint yokes, rear axle housings, crankshafts, connecting rods, rocker arms, transmission gear components, hand tools, and plumbing fittings. Specifications for the composition and heat-treating practices when producing malleable iron can be found in the American Society for Testing and Materials (ASTM) Standards A47, A197, A220, A338, and A602.

Gray Iron

Because most of the iron castings produced are of gray iron, the generic term cast iron is commonly used when gray iron is intended. The name comes from the characteristic gray color of the metal on a fractured surface. The gray color is caused by the presence of graphite flakes in the iron that are formed when the carbon in the iron separates during solidification. Special foundry practices are necessary to obtain the correct form of graphite.

The amount of graphite present, as well as its size and distribution, is important to the properties of gray iron. The different sizes of graphite tend to enhance certain properties of gray iron. The large (Type C) graphite flakes improve thermal shock resistance by increasing the thermal conductivity and lower the modulus of elasticity to minimize thermal stresses. Coarse flakes, on the other hand, are not conducive to a good machine finish or to strength. A very small size of flake graphite (Type D) promotes a fine machine finish by minimizing surface pitting, but it is difficult to obtain a pearlitic matrix with this type of graphite. There are, of course, many casting applications in which the type of graphite is of lesser consequence as long as the mechanical property requirements are met.

The size and shape of the graphite have no direct influence on the soundness of the casting. An iron with coarse graphite, often referred to as open-grained iron, may have a pitted surface when machined because of the breakout of material in machining, but this condition is not an indication of unsoundness or porosity.

A classification system for graphite in iron castings has been established by the ASTM. Specification A-247 is entitled "Standard Method for Evaluating the Microstructure of Graphite in Iron Casting."

Applications. Gray cast irons are used in a variety of structural steel applications that do not require impact resistance and high tensile strength. Some of these applications include engine blocks, cylinder heads, housings, manifolds, hydraulic valve bodies, and other components having complex design. Gray cast irons are also used in manufacturing components requiring vibration damping characteristics and resistance to heat checking and heat shock. Typical examples of these components are clutch plates, brake drums, ingot and pig molds, machine tool members, and piano plates.

Specifications. Gray irons are usually specified by their hardness or tensile strength. Chemical analysis is specified only for special types of irons such as those used in elevated temperatures.

The most commonly used gray iron specification for general engineering application is the ASTM A-48. This specification designates the minimum tensile strength of the iron as determined in a test bar of the size that is comparable with the critical section of the casting. Table 3-7 lists the tensile strengths of the various gray iron classes, and Table 3-8 lists the standard test bar casting sizes.

The strength and hardness of gray irons are directly influenced by the total carbon content and the rate at which the metal cools after it is poured into the mold. The cooling rate is dependent upon the size of the casting, the thickness of the metal, and its complexity in shape. Specification ASTM A-48

TABLE 3-7
Classes of Gray Iron

Class*	Minimum Tensile Strength, ksi (MPa)
20 A, B, C, or S	20 (138)
25 A, B, C, or S	25 (172)
30 A, B, C, or S	30 (207)
35 A, B, C, or S	35 (241)
40 A, B, C, or S	40 (276)
45 A, B, C, or S	45 (310)
50 A, B, C, or S	50 (345)
55 A, B, C, or S	55 (379)
60 A, B, C, or S	60 (414)

* This class designation, which indicates the minimum tensile strength in thousands of psi, must be followed by the letters A, B, C, or S to designate the appropriate test bar size as given in Table 3-8.

TABLE 3-8
Test Bar Casting Sizes

Test Bar	Diameter, in. (mm)	Length, in. (mm) Minimum	Length, in. (mm) Maximum	Typical Related Casting Section,* in. (mm)
A	0.88 (22.4)	5.0 (127)	6.0 (152)	0.25-0.5 (6.4-12.7)
B	1.20 (30.5)	6.0 (152)	9.0 (229)	0.5-1.0 (12.7-25.4)
C	2.0 (51)	7.0 (178)	10.0 (254)	1.0-2 (25.4-51)
S	As agreed	---	---	Under 0.25 (6.4) and over 2 (51)

* The relative section effect (cooling rate) of the test bar and casting is a complex function and is not specified. A proper relation can be indicated by the relative hardness of test bar and casting.

lists three standard-sized test bar castings and also provides for other sizes to be agreed upon to represent the casting properly. Under this specification, the foundry is required to make tensile tests for each lot of castings produced, and the lot size is also stipulated.

The Society of Automotive Engineers (SAE) Standard J431c is the specification for different gray cast iron grades used in automotive as well as nonautomotive applications. This specification is based on the Brinell hardness of the iron in the casting at a designated location. Specifications for gray cast irons used in elevated temperature applications are also available.

Ductile Iron

Ductile iron was discovered in the 1940s and is one of the more recent additions to the family of cast irons. It is sometimes referred to as nodular iron and is called spherulitic graphite iron or SG iron in England. Intensive development work was conducted in the 1950s on ductile iron, resulting in a rapid increase in commercial application that continues today. The tonnage of ductile iron produced exceeds that of malleable iron and cast steel combined.

Ductile iron is similar in composition to gray iron but has a very restricted content on minor elements. A very small but definite amount of magnesium with cerium is added to the molten iron, which causes the graphite to nucleate in spheres or spherulites rather than individual flakes. Special control procedures must be employed when processing ductile iron to prevent excessive losses of magnesium. Magnesium prevents the free carbon from precipitating in nonspheroidal shapes.

The high carbon and silicon contents of ductile iron retain the advantages of the casting processes and the excellent machinability of gray iron; but the number, shape, and size of the graphite spheroids have an important influence on the functional properties of the casting. Ductile iron has a higher modulus of elasticity than gray iron, with a somewhat linear stress/strain relationship; a very good range of yield strength, fatigue strength, and impact resistance; and, as its name implies, excellent ductility. Castings are made in a wide range of sizes and in very thin to very thick sections.

Applications. Ductile iron castings combine strength, durability, and toughness with good machinability and low cost. In the automotive and allied industries, ductile iron is used in the production of crankshafts, connecting rods, gears, steering knuckles, idler arms, disc brake calipers, and rocker arms. In the chemical industry, ductile iron is used in both subzero and elevated temperature environments to produce valves, fittings, and pump bodies. Ductile iron castings are also used in the production of machinery for a variety of industries. The controllability and reliability of ductile iron has produced successful applications in safety-related products. Low-cost silicon alloys are used for high-temperature oxidation resistance and thermal stability such as in turbo-charger housings.

Specifications. The various grades of regular, unalloyed ductile iron are designated by their tensile properties. Different grades are produced by obtaining different matrix microstructures in the iron as cast or by subsequent heat treatment. The chemical analyses of the different grades are essentially the same, but may be varied to ensure the formation of the desired matrix microstructure. Alloying elements may be added to ductile iron to enhance as-cast properties and hardenability and to facilitate heat treatment. High-alloy ductile irons are also produced for special requirements.

The common grades of ductile iron based on tensile strength are covered by ASTM A536 and several other specifications. The SAE specification for ductile iron casting, J434c, covers similar grades but specifies them only by Brinell hardness and microstructure. There are a number of other specifications for ductile iron in special service castings such as at elevated temperatures or with a maximum impact resistance at low temperatures.

Compacted Graphite Iron

Compacted graphite (CG) iron is the newest member of the cast iron family. Although it was originally identified as a distinct type of cast iron at the same time ductile iron was discovered, it was not until improved production techniques became available in the late 1970s that the material attracted widespread commercial interest. Other names for CG iron are compacted flake graphite (CFG) iron, quasi-flake graphite iron, vermicular graphite iron, and compacted vermicular graphite iron (CVI). These names refer to the stubby, interconnected graphite flake structure.

Compacted graphite irons possess lower strength and ductility levels than ductile irons, but have machinability characteristics and thermal conductivity values that are close to gray irons. The casting properties of CG iron, such as shrinkage and mold yield, are superior to those of ductile iron if the amount of spheroidal graphite is restricted.

Techniques for producing CG iron are more difficult than they are for ductile iron. The major difference results from the alloying elements added to control the graphite formation. A minimum of 80% compacted-type graphite is required to optimize the soundness, strength, thermal conductivity, and machinability of the iron.

CAST IRONS

Applications. Compacted graphite irons are being used in the production of engine parts, hydraulic components, and ingot molds. Typical engine components include cylinder heads and blocks, flywheels, bearing caps, and exhaust manifolds.

Specifications. While compacted graphite irons are being used for a number of casting applications, the material does not, as yet, have any specifications by national organizations. Both the SAE and the ASTM have committees working on the formulation of specifications for this material at the time of this writing.

High-Alloy Irons

High-alloy irons are a special group of irons consisting of high-alloy gray irons, high-alloy ductile irons, and high-alloy white irons. Alloy content of these irons ranges from 3 to 40%, modifying the properties of the base iron. Malleable irons are not highly alloyed because this would interfere with the metallurgy of the malleable process. Currently, compacted graphite iron has not been highly alloyed.

High-alloy iron castings are produced by foundries that specialize in their production because the alloy content of these materials requires special melting facilities and techniques. High-alloy irons are classified by the type of unusual service conditions for which they are used.[6]

Applications. High-alloy irons are used in applications requiring excellent corrosion resistance, high heat resistance, and maximum wear resistance. They are also chosen when special magnetic and electrical properties or low thermal expansion are needed.

Specifications. Because the special property requirements of high-alloy irons are often too difficult to establish and verify for specification, these irons are usually specified by their chemical analysis. Mechanical property requirements may also be included when they are critical.

Eight different grades of high nickel content, austenitic gray iron are specified by ASTM A436. The gray irons with high silicon content used for extreme corrosion resistance are specified by ASTM A518. Eight different grades of high nickel content, austenitic ductile irons are specified by ASTM A439. Castings for low-temperature and nonmagnetic services are specified in ASTM A571. Seven different grades of high-alloy white irons are specified by ASTM A532.

MECHANICAL PROPERTIES

Hardness and tensile strength are the most commonly specified properties for iron castings. While hardness and tensile strength relate directly to many useful characteristics in metals, there are two aspects of mechanical properties in general that should be discussed.

1. Dynamic properties relate closer to part function than static tensile properties. Low-temperature capability cannot be measured by tensile tests nor can high strain rate applications be indicated by tensile tests.
2. The mechanical properties of metal, especially iron, are not specific to a particular batch or heat as is the chemical analysis of metal. Properties are also influenced by the section thickness in which the metal solidifies and the manner in which the metal cools.

As an example of the first qualification, hardness is a relatively good indication of machinability; however, gray iron and ductile iron with the same hardness can exhibit appreciable differences in tool life. That is, if the microstructure of either contains some free carbides, machinability is reduced much more than indicated by the small increase in hardness.

The second qualification results from the fact that the properties of iron are directly influenced by the rate of solidification and subsequent cooling. Appreciably different properties in various portions of a casting are apt to occur if the sections have sufficiently large differences in thickness or shape to cause a significant variation in cooling rate. With modern technology, however, castings can be more uniform throughout variously sized sections. Thus, both large and small castings from the same ladle of metal will have similar mechanical properties.

Cast iron test bars should have a cooling rate and composition that is relatively similar to the casting sections they represent. If the casting is of sufficient size, test bars should be cut from the critical areas of the casting and then tested. Test bars that are different in cooling rate than the castings they represent can be used to establish the relative quality of the metal being poured rather than to indicate the actual properties to be obtained in the casting. Often the tensile properties of the metal in a casting can be related to the properties of the test bar by the hardness of each. This is a valid but not a precise relationship when the microstructures are similar.

As a basic concept, hardness is resistance to abrasion or scratching. As it applies to metals, the measurement of hardness is based on the relative resistance to the penetration of an indenter. The most common testing methods for iron—Brinell, Rockwell, Knoop, and Vickers—use this principle. Hardness is the most frequently used test for metal because it is convenient, it is usually nondestructive, and the test results can be related to a number of other properties.

Hardness is not an absolute value, and test results can vary even under ideal conditions. Some impressions of Brinell diameters can vary 0.15 mm in reading when measured by different individuals. In specifying a required hardness, it is important to stipulate where on the casting the hardness is to be determined and a range to allow for variations when reading the impression. Because of the difference in solidification and cooling rate, different portions of the casting may vary in hardness. Edges and thin sections may be of a higher hardness but usually cannot be measured with conventional instruments.

Mechanical Properties of Gray Iron

The mechanical properties of gray iron castings are influenced by their microstructure, composition, cross section, and secondary processing such as heat treatment. As previously mentioned, there is a general relationship between Brinell hardness, tensile strength, and other properties of gray iron. However, it is not a precise relationship because the amount and shape of the flake graphite present influences some properties more than others. The general variation in properties and related characteristics for the various classes of gray iron is shown in Fig. 3-1. Improved properties or characteristics can be obtained by selecting the next class of iron as indicated by the direction of the arrows. The classes of gray iron vary primarily because of the quantity of carbon present in the iron. The tensile strength of the iron is only specified when it is critical. It is important to realize that different parts of the same casting or two differently sized castings from the same metal can have different properties because of differences in the cooling rate of the metal in the mold. This effect is illustrated in Fig. 3-2. Table 3-9 lists the typical properties of gray irons used in automotive applications.

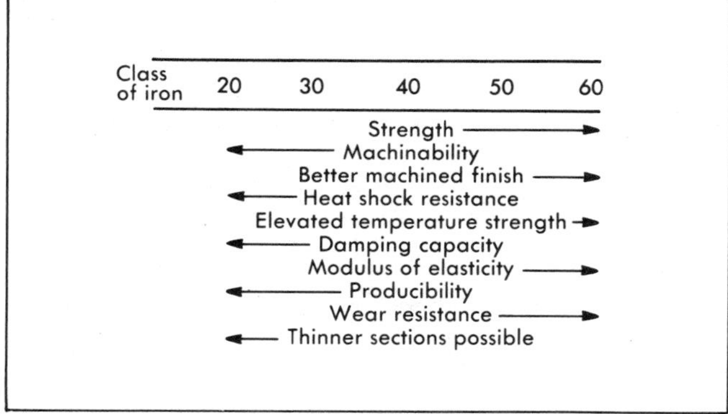

Fig. 3-1 General relation of gray iron properties and the class designated by ASTM Standard A-48.

The yield strength of gray iron is not usually specified because it is not a significant property of this material. The conventional offset method to designate yield strength can be used on the stress-strain curve with a 0.1% or 0.2% offset in strain. The latter will give values close to the tensile strength. The proof stress, or load, that produces 0.1% permanent strains will range from 65 to 80% of the tensile strength in gray iron. The compression strength of gray iron is exceptionally high, from 3 to 4 times its tensile strength. The larger ratio applies to the lower tensile strength irons, and the lower ratio is for the higher strength gray irons. Shear strength is from 1.1 to 1.5 times its tensile strength. Gray irons also retain their mechanical properties at temperatures up to 450° F (230° C).

Ductile Iron

The mechanical properties of ductile iron are primarily influenced by the microstructure of the metal around the graphite, which is referred to as the matrix. The matrix varies with the composition and the cooling rate of the casting.[7] Heat treatment can also be used to modify the matrix. However, a significant loss in fatigue strength occurs when ductile iron is oil quenched and tempered. A ferrite matrix provides good ductility, a ferrite and pearlite matrix is intermediate in strength, and a pearlite matrix provides high strength. The highest strength is with a martensitic matrix. Ductile iron can be austempered to produce more than double the tensile strength

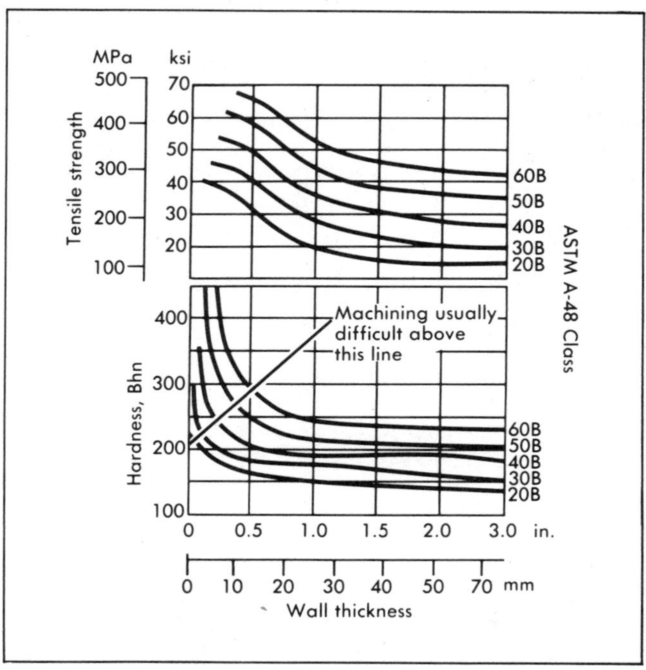

Fig. 3-2 The influence of casting section thickness on the tensile strength and hardness for classes of gray iron cast in 1.2" (30.5 mm) diam bars.

TABLE 3-9
Tensile Strength and Hardness of Common Gray Iron Castings Used in Automotive Applications

Specification Standard	Grade or Class	Microstructure	Minimum Tensile Strength, ksi (MPa)	Hardness, Bhn	Minimum Carbon, %
ASTM A-159	G 1800	Ferrite with some pearlite	18 (124)	143-187	
	G 2500	Pearlite with some ferrite	25 (172)	170-229	
	G 3000	Pearlite	30 (207)	187-241	
	G 3500	Pearlite	35 (241)	207-255	
	G 4000	Fine pearlite	40 (276)	217-269	
SAE J431	G 2500a	"A" graphite size 2-4 15% max ferrite	25 (172)	170-229	3.40
	G 3500b	"A" graphite size 3-5 5% max ferrite or carbide	35 (241)	207-255	3.40
	G 3500c	"A" graphite size 3-5 5% max ferrite or carbide	35 (241)	207-255	3.50

Note: The tabulated properties are for comparison purposes only and are not intended for design purposes. For specific values, contact the castings supplier or refer to the latest revision of the specification standard.

and yet increase fatigue strength, wear resistance, and toughness.

The tensile strength, yield strength, and percent elongation are the most commonly established properties for ductile iron. The minimums of these properties are generally established by the specification or implied by the hardness. Because the spherical graphite in ductile iron has a nominal effect on its properties, there is a graphical relationship between the hardness and tensile properties as shown in Fig. 3-3. The left side of the graph represents ductile irons possessing a completely ferritic matrix. The right side represents the very fine pearlitic matrix obtained from normalizing. The central portion represents a matrix of ferrite and pearlite or coarse pearlite. Ferrite content around the nodules (bulls-eyes) indicates slightly lower yield to ultimate strength ratios, but the bulls-eyes contribute to increased fatigue strength, ultimate strength, wear resistance, and elongation. The shear strength and torsional strength of ductile iron is 0.9 times the tensile strength. Table 3-10 lists the tensile properties and hardness values of the most commonly used grades of ductile iron.

Compacted Graphite Iron

The mechanical properties of compacted graphite cast iron are influenced by the percent of nodular graphite present in the microstructure. The properties presented here are for CG irons having a maximum of 20% nodular graphite present with the balance of graphite in compacted form.

Compacted graphite iron has tensile strength, yield strength, and elongation values between those of ductile iron and gray iron. The hardness of CG irons is also dependent on the microstructure of the iron. The hardness value of a particular grade is similar to the value of an equivalent ductile iron grade. Table 3-11 summarizes the tensile properties of the three most commonly used CG irons.

The compressive strength of a CG iron with a pearlitic microstructure is approximately three times the tensile strength. The compressive yield strength determined from the 0.2% offset method is slightly higher than the yield strength in tension.

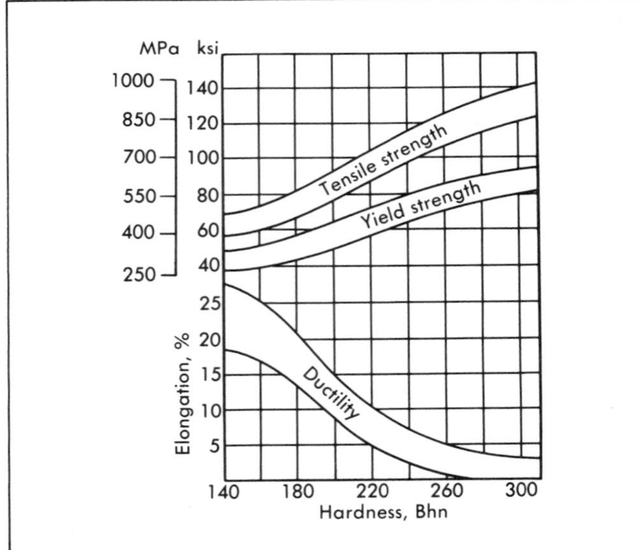

Fig. 3-3 Graphical relationship between hardness and tensile properties of ductile irons in the as-cast, annealed, or normalized condition with a microstructure of ferrite and/or pearlite.

Malleable Iron

A wide range of mechanical properties can be produced in malleable iron by controlling the matrix structure around the graphite with various heat treatments and alloy additions. The three common matrices are ferritic, pearlitic, and tempered martensitic.

The hardness of malleable iron is directly related to the other mechanical properties. Usually these irons are specified by either the hardness or tensile properties but should not be specified by both. Table 3-12 lists the tensile properties and hardness values of common malleable iron grades for comparison purposes only.

The ferritic grades have lower strengths but higher ductility than the pearlitic grades. Hardness values range from 110 to 156 Bhn for the ferritic grades, whereas the pearlitic grades range from 149 to 321 Bhn.

White and High-Alloy Irons

White cast irons and high-alloy cast irons are a group of irons that have controlled solidification times and/or composition to produce specific properties for selected applications. The white irons are used principally for their excellent wear and abrasion resistance. Higher hardness values in the white cast irons are indicative of better wear resistance caused by greater carbide content. A martensitic matrix provides a white iron with the maximum hardness of 500 to 700 Bhn; a pearlitic white iron has a lower hardness of 450 to 550 Bhn. The white irons retain their hardness at elevated temperatures with almost no decrease in hardness up to 800° F (425° C) and a stable hardness over 1000° F (540° C).

The high-alloy irons may be of the white iron, gray iron, or ductile iron types. The principal alloying elements may be silicon, nickel, chromium, molybdenum, or aluminum in amounts from 3 to over 30%. The high-alloy irons are used for special applications to resist corrosion, to retard oxidation at elevated temperatures, to be nonmagnetic, or to have an extremely low coefficient of thermal expansion. The matrix structures in the irons may be ferritic as in the 5% silicon irons, martensitic as in the 28% chromium irons, or austenitic as in the 22% nickel irons. Because of the specialized nature of these various alloys, it is impractical to review their individual properties here. Table 3-13 lists the mechanical properties of commonly used high-alloy irons.

PROCESSING CHARACTERISTICS

One or more secondary operations or processes are usually performed before the casting is placed in service. The most common are machining, welding, and heat treatment. Coining is performed on malleable and ductile iron castings to eliminate or reduce machining. Iron castings can also be coated successfully using organic and inorganic coatings. For additional information on the various coatings and techniques used, refer to the appropriate section in this volume.

Machining

The machinability of iron relates specifically to its microstructure whether the evaluation is tool life, surface finish, or power. The presence of graphite provides the free-machining characteristics of iron, and the shape and amount of graphite establish the potential surface finish obtainable with a cutting process and the necessary cutting force. The microstructure of the metal around the graphite determines the tool life and

TABLE 3-10
Tensile Properties of Most Commonly Used Ductile Irons

Specification Standard	Grade or Class	Microstructure and Condition	Tensile Strength, ksi (MPa)	Yield Strength (0.2% Offset), ksi (MPa)	Total Elongation in 2″ (50 mm), %	Typical Hardness, Bhn
ASTM A536	Common Grades of Ductile Iron					
	60-40-18	All ferrite, annealed	60 (414)	40 (276)	18	149-187
	65-45-12	Ferrite ---	65 (448)	45 (310)	12	170-207
	80-55-06	Ferrite and pearlite ---	80 (552)	55 (379)	6	187-255
	100-70-03	All pearlite, normalized	100 (690)	70 (483)	3	217-269
	120-90-02	Martensite, quenched and tempered	120 (827)	90 (621)	2	240-300+
SAE J434c	Automotive Ductile Iron Castings*					
	D-4018	Ferrite	60 (414)	40 (276)	18	170 max
	D-4512	Ferrite and pearlite	65 (448)	45 (310)	12	156-217
	D-5506	Ferrite and pearlite	80 (552)	55 (379)	6	187-255
	D-7003	Pearlite	100 (690)	70 (483)	3	241-302
	D, Q, & T	Martensite	A wide variety of properties will result from liquid quenching and tempering.			As specified

Note: Mechanical properties listed are for comparison purposes only and are not intended for specifications or design purposes. For specific values, contact the castings supplier or refer to the latest revision of the specification standard.
* Tensile strength, yield strength, and elongation are usually not specified. The graphite structure must be at 80% spheroidal.

TABLE 3-11
Mechanical Properties of Compacted Graphite Cast Iron*

Grade	Minimum Tensile Strength, ksi (MPa)	Minimum Yield Strength, ksi (MPa)	Elongation (at 0.2% Offset), %	Hardness, Bhn
265**	38 (265)	28 (190)	3	179 max
310	45 (310)	35 (240)	1.5	143-207
345	50 (345)	40 (275)	1	163-229
380	55 (380)	45 (310)	1	170-241
415†	60 (415)	49 (340)	1	207-269

Note: The tabulated values are for comparison purposes only and are not intended for design purposes. For specific values, contact the castings supplier.
* Matrix is compacted graphite with a maximum of 20% graphite in nodular form.
** The 265 grade is a ferritic grade. The castings producer usually determines type of heat treatment performed to obtain required mechanical properties and microstructure.
† The 415 grade is a pearlitic grade usually produced without heat treatment. Certain alloys are added to promote pearlite as a major part of the matrix.

TABLE 3-12
Mechanical Properties of Commonly Used Malleable Irons

Standard Specification	Class or Grade	Microstructure and Condition	Tensile Strength, ksi (MPa)	Yield Strength (0.2% Offset), ksi (MPa)	Elongation in 2" (50 mm), %	Hardness, Bhn
ASTM A-47	32510	Ferrite and temper carbon	50 (345)	32 (221)	10	156 max
	35018	Ferrite and temper carbon	53 (365)	35 (241)	18	156 max
ASTM A-220	40010	Pearlite	60 (414)	40 (276)	10	149-197
	45008	Pearlite	65 (448)	45 (310)	8	156-197
	45006	Pearlite	65 (448)	45 (310)	6	156-207
	50005	Pearlite	70 (483)	50 (345)	5	179-229
	60004	Pearlite	80 (552)	60 (414)	4	197-241
	70003	Pearlite	85 (586)	70 (483)	3	217-269
	80002	Pearlite	95 (655)	80 (552)	2	241-285
	90001	Pearlite	105 (724)	90 (621)	1	269-321
ASTM A-602 and SAE J-158a	M3210	Ferrite, annealed	50 (345)	32 (224)	10	156 max
	M4504	Ferrite and pearlite, normalized or quenched and tempered	65 (448)	45 (310)	4	163-217
	M5003	Ferrite and pearlite, air quenched and tempered	75 (517)	50 (345)	3	187-241
	M5503	Martensite, liquid quenched and tempered	75 (517)	55 (379)	3	187-241
	M7002	Martensite, liquid quenched and tempered	90 (621)	70 (483)	2	229-269
	M8501	Martensite, liquid quenched and tempered	105 (724)	85 (586)	1	269-302

Note: Mechanical properties listed are for comparison purposes only and are not intended for specifications or design purposes. For specific values, contact the castings supplier or refer to the latest revision of the specification standard.

establishes the most advantageous cutting speeds and feeds.

The relative ease with which iron castings can be machined is one of their important advantages. The annealed grades can be cut at very high speeds without generating burrs and without chip breakers. Coated carbide inserts and ceramic inserts can be used at rates as high as 3000 sfm (915 m/min) with adequate tool life. The harder grades of iron can also be machined economically and at reasonable rates of speed. White irons and high-alloyed irons are usually machined abrasively.[9] For additional information on machining iron castings, refer to Volume I, *Machining*, of this Handbook series.

Welding

Iron castings are successfully welded to assemble parts, to correct casting discontinuities, and to make repairs in-service. Welding is also used to apply wear or corrosion-resisting surfaces to iron castings and to rebuild worn or corroded surfaces.[10] Because welding of cast iron is difficult, the proper techniques must be carefully followed.

Gray, ductile, and malleable iron castings are weldable by standard welding processes including shielded metal-arc, oxy-fuel gas, braze, and gas metal-arc welding. White iron castings are considered unweldable. The selection of the welding process and the welding filler metal depends on the type of weld properties desired and the service life that is expected.[11]

The welding techniques employed when welding iron castings are designed to restrict penetration to the minimum depth required for proper fusion. This practice minimizes or prevents the base metal from forming a brittle zone at the weld due to the rapid freezing and cooling of the weld metal in the heat-affected zone (HAZ).[12] Preheating is desirable for welding iron castings using any of the welding processes in order to reduce the thermal gradient between the weld and the remainder of the casting. The selection of a preheat temperature is based on the welding process, type of filler metal, and the mass and complexity of the casting. After welding, the casting should be

TABLE 3-13
Mechanical Properties of Commonly Used High-Alloy Irons[8]

Specification Standard	Class or Grade	Minimum Tensile Strength,* ksi (MPa)	Minimum Yield Strength (0.2% Offset),* ksi (MPa)	Total Elongation in 2″ (50 mm), %	Hardness, Bhn
High-Alloy White Irons					
ASTM A-532	IA				550 Sand cast 600 Chill cast
	IB				550 Sand cast 600 Chill cast
	IC				550 Sand cast 600 Chill cast
	ID				550 Sand cast 600 Chill cast
	IIA				550 As cast 600 Hardened
	IIB				450 As cast 600 Hardened
	IIC				550 As cast 600 Hardened
	IID				450 As cast 600 Hardened
	IIE				450 As cast 600 Hardened
	IIIA				450 As cast 600 Hardened
High-Alloy Gray Irons					
ASTM A-436	1	25 (172)			131-183
	1b	30 (207)			149-212
	2	25 (172)			118-174
	2b	30 (207)			171-248
	3	25 (172)			118-159
	4	25 (172)			149-212
	5	20 (138)			99-124
	6	25 (172)			124-174

(continued)

TABLE 3-13—*Continued*

Specification Standard	Class or Grade	Minimum Tensile Strength,* ksi (MPa)	Minimum Yield Strength (0.2% Offset),* ksi (MPa)	Total Elongation in 2″ (50 mm), %	Hardness, Bhn
High-Alloy Ductile Irons					
ASTM A-439	D-2**	58 (400)	30 (207)	8	139-202†
	D-2B	58 (400)	30 (207)	7	148-211†
	D-2C	58 (400)	28 (193)	20	121-171†
	D-3**	55 (379)	30 (207)	6	139-202†
	D-3A	55 (379)	30 (207)	10	131-193†
	D-4	60 (414)	---	---	202-273†
	D-5	55 (379)	30 (207)	20	131-185†
	D-5B	55 (379)	30 (207)	6	139-193†

Note: The mechanical properties listed are for comparison purposes only and are not intended for specifications or design purposes. For specific values, contact the castings supplier or refer to the current specification standard.
* Test bar machined from 1″ keel block or a "Y" block in 1/2, or 1, or 3″ size by option of purchaser.
** Additions of 0.7-1.0% molybdenum will increase mechanical properties above 800° F (425° C).
† Lower hardness value based on minimum chemical composition and higher value based on maximum composition.

cooled slowly or stress relieved to prevent cracks from occurring and to improve the machinability of the HAZ. Additional information on welding cast irons can be found in Volume IV, *Quality Control and Assembly*, of this Handbook series.

Heat Treatment

Most iron castings respond to heat treatment, and several types are heat treated in the course of their manufacture. Iron castings are heat treated to relieve internal stresses, improve machinability, and increase toughness, ductility, strength, and wear resistance. The three categories of heat treatment for iron castings are stress relieving, annealing, and hardening. Localized surface hardening can also be performed by induction, flame, laser, or electron-beam hardening techniques. Additional information on heat-treating practices for cast irons can be found in Chapter 14 of this volume.

Stress relief. Castings are often of complex shape and can contain, under some circumstances, appreciable internal or locked-in stress. If allowed to cool in the sand mold in which it is made, a casting will usually be quite free of internal stresses because molding sand is a good insulator. However, this arrangement is not always possible because of the production method or the need for air cooling to obtain a desired hardness range. Stresses can also result from subsequent processing.

Stress relieving involves heating the casting to a high enough temperature to permit the stresses to relax. For the lower classes of gray iron, 900-1100° F (480-590° C) is generally satisfactory. The high-strength irons and alloyed irons usually require a temperature of 1200° F (650° C) for satisfactory relief of the stresses. After a suitable time and temperature, usually one hour after complete equalization of temperature, the casting must be slowly cooled so that the stresses are not reintroduced. Stress relieving can decrease the hardness and strength of higher hardness castings.

Annealing. Annealing of irons provides minimum hardness and maximum machinability. Some types of iron are fully annealed when they are produced. The annealing of other grades will, of course, reduce their mechanical properties as well as their hardness.

Hardening. The hardness of iron castings can be increased by heat treatment to provide excellent wear resistance or higher strength. The hardening of iron is similar to that for steel except that the critical temperature for iron is increased by its silicon content. Low-silicon irons, such as some malleable irons, should be quenched from temperatures high enough to redissolve the carbon and to provide good depth of quench, varying from 1600 to 1650° F (870 to 900° C). Austempering may require temperatures up to 1700° F (925° C). Ductile iron with 2.5% silicon may require a hardening temperature over 1650° F (900° C). Because of their high carbon content, irons have relatively high hardenability, which can be further increased by alloy additions. Malleable iron requires a lower temperature of quenching media than ductile iron since it possesses a lower hardenability.

Surface hardening by induction heating, laser heating, electron-beam heating, or flame heating plus quenching of working surfaces is commonly performed on iron castings to provide a full, hard working surface with a minimum of internal stress and distortion. Surface hardening creates surface stresses on the workpiece that help to improve fatigue strength. For this process, the casting should not be of the fully annealed type but should have a pearlitic or tempered martensite matrix. Although ferritic irons can be satisfactorily furnace hardened with special processing techniques, the short time at temperature in the flame or induction heating does not allow sufficient time for adequate carbon to diffuse across the ferrite matrix.

Air hardening or normalizing of small and moderately sized castings after furnace heating usually results in a pearlitic matrix with higher hardness and strength than castings in the as-cast or annealed condition. Some castings, depending on size and shape, may require a higher manganese content or other alloy additions to achieve these characteristics with air cooling. Complex castings should generally be stress relieved after normalizing. Oil quenching is practical for crankshafts, gear blanks, and similar castings. More complex castings may be subject to quench cracking. Quenching should be followed immediately by tempering.

References

1. *Steel Castings Handbook,* Supplement 2 (Des Plaines, IL: Steel Founders' Society of America, 1982), pp. 2 and 4.
2. Steel Castings Handbook, *loc. cit.*
3. *Ibid.*, pp. 10 and 14.
4. *Ibid.*, pp. 2, 4, and 12.
5. *ASM Metals Reference Book* (Metals Park, OH: American Society for Metals, 1981), p. 222.
6. Charles F. Walton, ed., *Iron Castings Handbook* (Des Plaines, IL: Iron Castings Society, Inc., 1981), p. 129.
7. *Ibid.*, p. 323.
8. *Ibid.*, pp. 100-102.
9. *Ibid.*, pp. 667-668.
10. *Ibid.*, p. 509.
11. Howard B. Cary, *Modern Welding Technology* (Englewood Cliffs, NJ: Prentice-Hall, Inc., 1979), p. 463.
12. Henry Horwitz, *Welding: Principles and Practice* (Boston: Houghton Mifflin Company, 1979), p. 570.

Bibliography

Bardes, Bruce P.; Baker, Hugh; and Benjamin, David, eds. *Metals Handbook*, Vol. 1, 9th ed. Metals Park, OH: American Society for Metals, 1978.

Wieser, Peter F., ed. *Steel Castings Handbook*, 5th ed. Des Plaines, IL: Steel Founders' Society of America, 1980.

HIGH-PERFORMANCE ALLOYS

High-performance alloys are a group of alloys used in applications requiring high strength and/or corrosion resistance over a wide range of temperatures. The high-strength, heat-resistant alloys are commonly referred to as superalloys in the aircraft and aerospace industries.

High-performance alloys achieve some of their high-temperature strength characteristics through solid-solution strengthening or a combination of solid-solution strengthening and precipitation hardening. However, high-temperature strength as well as ambient temperature strengths are also improved by second-phase particle strengthening. Carbides and gamma-prime-type precipitates function as second-phase strengtheners, with carbides playing a minor role in alloys of the gamma prime precipitate type and a major role in the solid-solution type.

The solid-solution strengthening in high-performance alloys comes from the presence of cobalt, chromium, iron, molybdenum, columbium, and tungsten elements in the face-centered cubic structure of the matrix. Of these elements, tungsten, molybdenum, cobalt, and chromium have the greatest strengthening effect.

The carbides in high-performance alloys are of various types depending on the particular alloy composition and the thermal conditions to which the material has been exposed. The carbides change from one form to another as the thermal conditions vary with time. Titanium carbides tend to be very stable; molybdenum, tungsten, and columbium carbides are moderately stable; and chromium carbides tend to be somewhat unstable. In some alloys used primarily for corrosion-resistant applications, carbon is kept at very low levels because carbides tend to lower corrosion resistance; in particular, intergranular corrosion resistance is lowered when the carbides precipitate at grain boundaries.

Gamma-prime-type precipitates are used in iron, iron-nickel, and nickel-based alloys that contain from 1 to 10% by weight of combinations of aluminum, titanium, and/or columbium. The alloys with the higher amounts of gamma prime formers can be strengthened most.

In high-temperature applications, the oxidation resistance of high-performance alloys is as important as high-temperature strength. Oxidation behavior is complex due to the number of elements involved. In simple terms, the oxidation resistance is primarily due to the adherent surface oxide film that forms as a result of the added elements. Chromium plays a large role in the high-temperature oxidation resistance. Chromium levels of over 15% by weight are normally required to achieve acceptable oxidation resistance. Aluminum and titanium can add to the oxidation resistance at any given chromium level. However, high-performance alloys with the highest strengths above 1200° F (650° C) must have the highest chromium contents in order to have adequate oxidation resistance at such high temperatures. Excessively high chromium can cause reversals in gamma-prime-type strengthening. Therefore, chromium levels above 20% are not used in these alloys.

For uniform corrosion resistance, molybdenum, chromium, and nickel contents are increased. Molybdenum improves resistance in nonoxidizing acids, chromium improves resistance in oxidizing environments, and nickel improves resistance in alkaline environments. Pitting corrosion can be improved by additional amounts of molybdenum and chromium. Increased amounts of nickel and molybdenum improve resistance to stress-corrosion cracking.[1]

CHAPTER CONTENTS:

MANUFACTURING PRACTICE

High-performance alloys are multiply melted in most cases.[2] The austenitic high-nickel alloys and nickel-based, solid-solution alloys are generally electric furnace melted followed by argon-oxygen decarburization (AOD processing). The primary melt and refine steps may sometimes be followed by electroslag remelting.

Precipitation-hardening, iron-based alloys are electric furnace or vacuum-induction melted followed by vacuum-arc or electroslag remelting. Double-vacuum melting may be used to improve fatigue strength, ductility, and impact strength. Precipitation-hardening, iron-nickel and nickel-based alloys are usually vacuum-induction melted followed by vacuum-arc or electroslag remelting. The number and type of sequential melting operations used is determined by the level of ductility, toughness, and fatigue life expected in service.

Vacuum-induction melting lowers the gas content and evaporates some trace elements that can adversely affect service life. Vacuum-arc and electroslag remelting processes help to refine the metal by reducing gases, nonmetallic and metallic impurities, and inclusions. With secondary

Reviewers of this chapter are: ***H. M. Butler****, Product Process Engineer, Special Metals Corp.;* ***Richard L. Kennedy****, Vice President, Teledyne Allvac;* ***Dwaine L. Klarstrom****, Manager, Product Development, Cabot Corp.;* ***L. W. Lherbier****, Director of Research and Development, Universal Cyclops;* ***Dr. Paul E. Manning****, Corrosion Market Development Engineer, Cabot Corp.;* ***Phil Ranson****, Technical Editor, Huntington Alloys, Inc.;* ***Donald J. Tillack****, Industry Manager, Huntington Alloys, Inc.*

remelting, larger ingots with a more uniform composition and a more dense homogeneous structure can be produced.

Powder metallurgy (PM) processes are also used when producing precipitation-hardening, nickel-based alloys. The two processes most commonly used are the rotating electrode method and the inert-gas atomization method. By using the PM processes when producing the more segregation-prone high-performance alloys, component fabrication costs can sometimes be reduced. Regardless, components from these alloys generally cannot be made with more conventional fabrication processes.

Two recently developed PM processes are rapid solidification processing (RSP) and mechanical alloying. In RSP, the molten metal is quenched at rates varying from 1800 to 1,800,000° F (1000 to 1 000 000° C) per second. The rapid cooling produces an extremely high homogeneity and a very fine grain structure, which strengthens the alloy.[3] In mechanical alloying, elemental powder blends are beaten into a uniform and homogeneous composition in a ball mill or attritor. Additional information on powder metallurgy can be found in Chapter 8, "Powdered Metals," in this volume and in Volume II, *Forming*, of this Handbook series.

High-performance alloys are available in both wrought and cast forms. Common wrought products include bars, billets, plates, sheet and strip, wire, tubes, pipes, and forgings. In general, cast forms perform better than wrought forms only for some applications above 1500° F (815° C). Precision casting is used to produce intricate shapes.

TYPES OF HIGH-PERFORMANCE ALLOYS

High-performance alloys encompass a wide range of complex materials.[4] It is difficult to define the term "high performance" precisely because it has broader or narrower limits depending upon who is using the word. In this chapter, a high-performance alloy is defined as one developed for elevated temperature, above 1000° F (535° C), and/or corrosion plus oxidation resistance service, usually based on Group VIII A elements. The high-strength, heat-resistant alloys encounter relatively severe multidirectional, often cyclical fatigue-type mechanical stressing and require high surface stability as well as internal structure quality.

High-performance alloys are predominantly iron-nickel-based, nickel-based, or cobalt-based, with a few iron-based alloys included in the group. They tend to be of complex composition with relatively large percentages of three or more alloying elements. Some high-performance alloys contain as many as 10 or more important elemental constituents, with precisely controlled quantities. Some of the more important elements added include chromium, molybdenum, tungsten, columbium, aluminum, cobalt, titanium, carbon, and nitrogen.

Metallurgically, high-performance alloys are somewhat like austenitic stainless steels. They have the face-centered cubic (fcc) crystal structure, and many of their physical characteristics resemble stainless steels. High-performance alloys sometimes display a reduction in ductility in the intermediate temperature range of 1200 to 1560° F (650 to 850° C). In addition, they are relatively poor conductors of heat. These alloys generally melt at lower temperatures than do steels, typically in the range of 2200 to 2550° F (1200 to 1400° C).

NICKEL-BASED ALLOYS

Nickel-based alloys generally have greater resistance to high temperatures than low-alloy steels and stainless steels. These alloys contain 30-75% nickel and up to 35% of both chromium and cobalt. Iron content ranges from relatively small amounts to as much as 35% in certain alloys. Aluminum, titanium, niobium, molybdenum, and tungsten are added to enhance either strength and/or corrosion plus oxidation resistance.

Nickel alloys that are strengthened by solid-solution additions are used in a variety of aerospace applications. Typical parts include jet engine ignitors, combustion can liners, diffuser assemblies, heat shields, exhaust systems, and turbine shroud rings. Precipitation-hardening nickel alloys are frequently used in forged components such as gas-turbine blades, discs, rings, shafts, and various compressor and diffuser components.

COBALT-BASED ALLOYS

Cobalt-based alloys contain over 38% cobalt, 20% chromium, usually some nickel, and substantial percentages of molybdenum and/or tungsten for strengthening. These alloys are divided into three main groups. The first group consists of those alloys used in high-temperature applications. They maintain their strength at temperatures from 1200 to 2100° F (650 to 1150° C). Some of the common applications of these alloys include nuclear reactor components, surgical implants, and combustors and transition ducts for gas turbines.

The second group consists of those alloys specifically designed to be work hardened. These alloys are generally used for producing fasteners. The third group consists of those alloys used for wear-resistant applications. Common applications are erosion shields in steam turbines and wear pads in gas turbines.

IRON-BASED ALLOYS

Iron-based alloys contain over 12% chromium, at least 9% nickel, and varying amounts of molybdenum, tungsten, cobalt, and titanium. Strengthening of these alloys is usually by precipitation hardening; however, some of these alloys are solid-solution strengthened. Typical applications for precipitation-hardened alloys are blades, discs, and fasteners in gas turbine engines. Iron-based alloys are generally used in the wrought form.

APPLICATIONS

High-performance alloys are relatively expensive and are therefore used only in applications where their special properties are essential to the satisfactory performance of the component. Their special properties include excellent strength and toughness at elevated temperatures and superior resistance to oxidation and corrosion. The maximum temperature for high-performance alloys usually does not exceed 1740° F (950° C). Each alloy has different levels of properties, and selection for a particular application is based on the match between its properties and the needs of the service to be performed.[5]

High-performance alloys are used for chemical and petrochemical processing equipment, industrial furnaces and related heat treating equipment, jet engines, rocket engines, and a variety of other applications involving strenuous service conditions imposed by oxidizing and corrosive atmospheres over a wide range of temperatures. Table 4-1 summarizes the characteristics and major applications of the most commonly used high-performance alloys.

TABLE 4-1
Applications of Commonly Used High-Performance Alloys

UNS No.	Alloy[a]	Description	Major Applications
N02200	200	Commercially pure wrought nickel, good mechanical properties, excellent resistance to many corrosives.	Food processing equipment, chemical shipping drums, caustic handling equipment and piping, electronic parts, aerospace and missile components, rocket motor cases, magnetostrictive devices.
N04400	400	High strength, good weldability, excellent corrosion resistance over wide range of temperatures and conditions.	Valves, pumps, shafts, marine fixtures, fasteners, electrical and electronic components, processing equipment, petroleum refining and production equipment, feedwater heaters and other heat exchangers.
N04405	R-405	Similar to alloy 400. Controlled sulfur added for improved machining characteristics.	Water meter parts, screw machine products, fasteners, valve parts.
N05500	K-500	Age-hardened version of alloy 400 for increased strength and hardness.	Pump shafts, impellers, doctor blades and scrapers, oil well drill collars and instruments, electronic components, springs, valve trim, fasteners.
N06600	600	High nickel, high chromium content for resistance to oxidizing and reducing environments. For severely corrosive environments at elevated temperatures.	Furnace muffles, electronic components, chemical and food processing equipment, heat treating equipment, nuclear steam generator tubing and other equipment.
N06601	601	Excellent high-temperature properties. Resistance to oxidizing, carburizing, and sulfur-containing atmospheres.	Heat exchangers, heat treating baskets and fixtures, radiant tubes, thermocouple tubes, furnace muffles and retorts, combustion cans, aircraft engine parts.
N06617	617	Optimum high-temperature mechanical stability, oxidation, and corrosion resistance. Excellent cyclic oxidation and carburization resistance at 2000° F (1095° C). Good stress-rupture properties above 1800° F (980° C).	Aerospace and engine components, afterburners, flame holders, spray bars, combustion can liners, turbine seals, heat treating equipment, nitric acid catalyst supports.
N06625	625	High strength and toughness from cryogenic temperatures to 1800° F (980° C), good oxidation resistance, exceptional fatigue strength, and good resistance to many corrosives.	Chemical and pollution control equipment, ash pit seals, nuclear reactors, marine equipment, ducting, thrust reverser assemblies, hot brine handling equipment, fuel nozzles, afterburners, spray bars.
N06690	690	A high-chromium modification of alloy 600. Good resistance to oxidizing chemicals and sulfur-containing gases. High mechanical properties.	Used for various applications involving nitric or nitric hydrofluoric acid solutions as in tail-gas reheaters in production of nitric acid and steam-heating coils in nitric hydrofluoric solutions for pickling stainless steels; reprocessing of nuclear fuels.
N09706	706	Similar to alloy 718 but with considerably improved machinability. High strength from cryogenic temperatures to 1300° F (705° C). Good weldability.	Gas turbine components.

(continued)

TABLE 4-1—*Continued*

UNS No.	Alloy[a]	Description	Major Applications
N07718	718	Excellent strength from -423° F to 1300° F (-253° C to 705° C). Age hardenable and may be welded in fully aged condition. Excellent oxidation resistance to 1800° F (980° C).	Jet engines, pump bodies and parts, rocket motors and thrust reversers, nuclear fuel element spacers, hot extrusion tooling.
N07750	X-750	Age-hardenable alloy with good corrosion and oxidation resistance. Excellent relaxation resistance.	Gas turbine parts, steam service and nuclear reactor springs, bolts, vacuum envelopes, extrusion dies, bellows, forming tools.
N08800	800	Strong and resistant to oxidation and carburization at elevated temperatures. Resists sulfur attack, internal oxidation, scaling, and corrosion in wide variety of atmospheres.	Heat exchangers, process piping, carburizing fixtures and retorts, heating element sheathing, nuclear steam generator tubing and other components.
N08810	800H	Similar to alloy 800 with better high-temperature strength. Higher design strength values for use above 1150° F (620° C). Improved creep and stress-to-rupture properties in 1100° F to 1800° F (595° C to 980° C) range.	Chemical and power plant superheater and reheater tubing; steam methane reformer pigtails, headers, and furnace tubing; process piping.
	Incoloy alloy 802[b]	High-carbon version of alloy 800 with improved mechanical properties at high temperatures. Excellent oxidation resistance.	Titanium hot forming dies, connecting pins for cast link furnace belts, tubing for ethylene furnaces, iron ore sintering equipment.
N08825	825	Excellent resistance to wide variety of corrosives. Resists pitting and intergranular type corrosion, reducing acids, and oxidizing chemicals.	Pickling tank heaters, hooks, etc.; spent nuclear fuel element recovery, chemical tank trailers, evaporators, other processing equipment; ash pit seals, hydrofluoric acid production, pollution control and radwaste systems.
N09925	925	An age-hardenable, nickel-iron-chromium alloy that possesses high strength at temperatures to 1000° F (540° C).	Used in oil production tubular products, tool joints, and equipment for surface and downhole hardware for intermediate sour-gas wells. Also used in fasteners, marine and pump shafting, and high-strength seamless tubing where resistance to general corrosion and pitting are required.
	Inconel alloy MA754[b]	An oxide-dispersion-strengthened, nickel-chromium alloy. High-temperature strength is developed by controlled thermomechanical processing.	Gas turbine components and other extreme service applications.
	Inconel alloy MA6000[b]	A nickel-based alloy produced by mechanical alloying and strengthened by oxide dispersion strengthening and precipitation hardening. Possesses high creep and rupture properties to 2000° F (1095° C). Possesses excellent resistance to oxidation and sulfidation.	Blade material for advanced gas turbines.
	Incoloy alloy MA956[b]	An oxide-dispersion-strengthened, iron-chromium-aluminum alloy produced by mechanical alloying. Possesses strength at high temperatures with excellent resistance to oxidation, carburization, and hot corrosion.	Gas turbine combustion chambers, components of advanced energy-conversion systems, and other applications involving rigorous conditions.
N10665	B2	A nickel-based wrought alloy with excellent resistance to hydrochloric and sulfuric acids at all concentrations and temperatures.	Chemical process applications in the as-welded condition.

TABLE 4-1—*Continued*

UNS No.	Alloy[a]	Description	Major Applications
N06455	C-4	A nickel-chromium-molybdenum alloy with outstanding high-temperature stability.	Chemical process applications that are exposed to hot contaminated mineral acids, solvents, chlorine and chlorine-contaminated media, dry chlorine, formic and acetic acids, acetic anhydride, and seawater and brine solution.
	Hastelloy alloy C-22[c]	A nickel-chromium-molybdenum alloy with exceptional resistance to oxidizing acids and chlorides. Good thermal stability.	Formic acid reactors, HF furnaces, SO_2 cooling towers, cellophane manufacturing, dye manufacturing, chlorine spargers, nuclear fuel reprocessing, plate and tubular heat exchangers, pickling system components, chlorination and sulfonation systems.
N10276	C-276	A nickel-molybdenum-chromium wrought alloy. Possesses outstanding resistance to localized corrosion and to both oxidizing and reducing media.	Flue-gas desulfurization systems and scrubbers. Chemical process applications involving corrosive chemical mixtures.
N06007	G	A nickel-based alloy possessing excellent resistance to hot sulfuric and phosphoric acids.	Flue-gas desulfurization systems, pollution control equipment, pulp and paper industry equipment.
N06985	G-3		Flue-gas desulfurization systems and chemical process industry applications.
	Hastelloy alloy B[c]	A nickel-molybdenum alloy combining high-temperature and corrosion-resistant properties.	Hydrochloric acid handling equipment.
N10003	N	A nickel-based alloy with good oxidation resistance, excellent resistance to aging, and good dimensional stability and high-temperature strength.	Material for molten-fluoride-salt containers.
	Hastelloy alloy S[c]	A nickel-based, high-temperature alloy possessing excellent thermal stability, low thermal expansion, and outstanding oxidation resistance to 2000° F (1095° C).	Applications involving severe cyclic heating conditions.
N06002	X	A nickel-based alloy that possesses exceptional strength and oxidation resistance to 2200° F (1205° C).	Furnace rolls, jet engine tailpipes, afterburner components, cabin heaters, other aircraft parts.
	Cabot No. 214[c]	A nickel-based, high-temperature alloy with excellent resistance to oxidation, carburization, and chlorination to 2200° F (1205° C).	Furnace parts and hardware, high-temperature test racks, heating elements, radiant tubes, gas turbine parts, brazing fixtures, roller hearths, retorts, support systems for ceramics.
	Multimet[c]	An economical, mixed-base, high-temperature alloy with excellent oxidation resistance, good ductility, and good formability. High-temperature properties are inherent and are not dependent upon age hardening.	Aircraft related parts such as tailpipes and tailcones, afterburner parts, exhaust manifolds, combustion chambers, and turbine blades, buckets, and nozzles.
	Haynes No. 556[c]	A lanthanum modified, iron-based alloy with excellent oxidation resistance to 2000° F (1095° C), good ductility, and good formability.	High-temperature environments.

(continued)

TABLE 4-1—*Continued*

UNS No.	Alloy[a]	Description	Major Applications
	Haynes No. 25[c]	A cobalt-based alloy that possesses good formability and excellent high-temperature properties.	Jet engine parts that include turbine blades, combustion chambers, afterburner parts, and turbine rings. Furnace muffles and liners in high-temperature kilns.
	Haynes No. 188[c]	A cobalt-based alloy with excellent high-temperature strength, oxidation resistance to 2100° F (1150° C), and good post-aging ductility.	Transition ducts, combustor cans, spray bars, flame holders, afterburner liners in jet engines.
	René 41	A vacuum-melted, nickel-based alloy that has excellent high strength at temperatures in the range of 1200 to 1800° F (650 to 980° C). Strength is developed by various solution and aging heat treatments.	Afterburner parts and nozzle diaphragm partitions in gas turbines.
	Cabot No. 263[c]	A nickel-based, vacuum-melted and electroflux-refined, gamma-prime-strengthened alloy. Possesses good strength up to 1600° F (870° C), good ductility, and good formability.	Combustors and combustor liners for use in gas turbines.
	Haynes Stellite No. 6B[c]	A cobalt-based alloy with good resistance to most types of wear and to a variety of corrosive media.	Screw conveyors, rock-crushing rollers, tile-making machines, and cement and steel mill equipment. Also food handling machinery and chemical equipment requiring both wear and corrosion resistance.
	Haynes Stellite No. 6K[c]	A cobalt-based alloy with good resistance to most types of wear.	Shaft sleeves, valve parts, pump plungers, doctor blades, feed screws, knives, wear plates, bearings.
	Allcorr[d]	A nickel-chromium-molybdenum-tungsten, solid solution hardened alloy with outstanding corrosion resistance to oxidizing and reducing environments. Readily hot and cold worked and welded.	Deep sour-gas and geothermal wells. Nuclear fuel processing, flue gas desulfurization plants, and scrubbers.
	Astroloy[d]	A nickel-chromium-cobalt-molybdenum, precipitation-hardened alloy with excellent high-temperature capabilities.	Turbine discs and blades, spacers, seals, and bolting.
	Waspaloy[e]	Precipitation-hardenable alloy with good strength and oxidation resistance for service to temperatures up to 1600° F (870° C).	Turbine and compressor blades and discs, shafts, spacers, fasteners, and miscellaneous jet engine hardware.

[a] Some high-performance alloys are made by more than one manufacturer.
[b] Inconel and Incoloy are registered trademarks of Huntington Alloys, Inc.
[c] Hastelloy, Haynes, Multimet, and Cabot are registered trademarks of Cabot Corp.
[d] Allcorr and Astroloy are registered trademarks of Teledyne Allvac.
[e] Waspaloy is a registered trademark of Pratt and Whitney Aircraft.

MECHANICAL PROPERTIES

A wide range of mechanical properties is obtainable with high-performance alloys. The properties vary with composition, product form, and heat treatment. Some representative room temperature mechanical properties are given in Table 4-2. Representative mechanical properties at elevated temperatures are presented in Table 4-3.

TABLE 4-2
Mechanical Properties of High-Performance Alloys at Room Temperature

UNS No.	Alloy[a]	Form and Condition	Tensile Strength, ksi (MPa)	Yield Strength (0.2% Offset), ksi (MPa)	Elongation in 2″ (50 mm), %	Hardness[b]
		Nickel-Based:				
N02200	200	Rod and Bar				
		Hot finished	60-85 (414-586)	15-45 (103-310)	55-35	45-80 R_B
		Cold drawn	65-110 (448-758)	40-100 (276-690)	35-10	75-98 R_B
		Plate, hot rolled	55-100 (379-690)	20-80 (138-552)	55-35	55-80 R_B
		Sheet, hard	90-115 (620-793)	70-105 (483-724)	15-2	90 R_B max
N02201	201	Rod and Bar				
		Hot finished, and hot finished, annealed	50-60 (345-414)	10-25 (69-172)	60-40	75-100 Bhn
		Cold drawn	60-100 (414-690)	35-90 (241-620)	35-10	125-200 Bhn
N02270	270	Rod and Bar				
		Hot finished	50 (345)	16 (110)	50	40 R_B
		Sheet	50 (345)	16 (110)	45-50	30-45 R_B
N03301	301	Rod and Bar				
		Hot finished	90-130 (620-896)	35-90 (241-620)	55-30	75 R_B
		Cold drawn	110-150 (758-1034)	60-130 (414-896)	35-15	90 R_B
		Strip, annealed	90-120 (620-827)	35-60 (241-414)	50-30	90 R_B max
	290	Strip, cold rolled	86-90 (595-623)	80-86 (558-590)	6	90 R_B
N04400	400	Rod and Bar, annealed	75-90 (517-620)	25-50 (172-345)	60-35	60-80 R_B
		Plate, hot rolled	75-95 (517-655)	40-75 (276-517)	45-30	70-96 R_B
		Sheet, annealed	70-85 (483-586)	25-45 (172-310)	50-35	73 R_B max
N04405	R-405	Rod and Bar				
		Annealed	70-85 (483-586)	25-40 (172-276)	50-35	60-76 R_B
		Cold drawn	85-115 (586-793)	50-105 (345-724)	35-15	85-23 R_C
N05500	K-500	Rod and Bar				
		Hot finished	90-155 (620-1069)	40-110 (276-758)	45-20	75R_B
		Cold drawn	100-140 (690-965)	70-125 (483-862)	35-13	88 R_B
		Sheet, cold rolled and annealed	90-105 (620-724)	40-65 (276-448)	45-25	85 R_B max
N06600	600	Rod and Bar				
		Cold drawn	105-150 (724-1034)	80-125 (552-862)	30-10	90 R_B
		Hot finished	85-120 (586-827)	35-90 (241-620)	50-30	75-95 R_B
		Sheet, cold rolled and annealed	80-100 (552-690)	30-45 (207-310)	55-35	88 R_B max
		Plate, hot rolled	85-110 (586-758)	35-65 (241-448)	50-30	80-95 R_B
N06601	601	Rod and Bar, hot finished	85-120 (585-825)	35-100 (240-690)	60-15	65-95 R_B
		Sheet, cold rolled	115-190 (790-1310)	100-175 (690-1205)	20-2	---
		Plate, annealed	80-100 (550-690)	30-45 (205-310)	65-45	60-75 R_B
N06617	617	Bar, solution annealed	111.5 (734)	51.5 (355)	56	181 Bhn
		Sheet, solution annealed	109.5 (755)	50.9 (351)	58	173 Bhn
		Plate, solution annealed	106.5 (734)	52.3 (361)	62	172 Bhn

(continued)

TABLE 4-2—*Continued*

UNS No.	Alloy[a]	Form and Condition	Tensile Strength, ksi (MPa)	Yield Strength (0.2% Offset), ksi (MPa)	Elongation in 2″ (50 mm), %	Hardness[b]
N06625	625	Rod, bar, and plate				
		As rolled	120-160 (827-1103)	60-110 (414-758)	60-30	175-240 Bhn
		Annealed	120-150 (827-1034)	60-95 (414-655)	60-30	145-220 Bhn
		Sheet and strip, annealed	120-150 (827-1034)	60-90 (414-620)	55-30	145-240 Bhn
N06690	690	Rod, hot rolled and annealed, 2″ (51 mm) diam	100 (690)	48.5 (334)	50	---
		Sheet, cold rolled and annealed, 0.15″ (3.8 mm) thick	105 (724)	50.5 (348)	41	---
N09706	706	Cold rolled sheet, 0.04″ (1.0 mm) thick				
		Solution treated	109.8 (757)	55.5 (383)	47	---
		Heat treated	186-193.5 (1282-1334)	148.5-161.3 (1024-1112)	22-24	---
		Cold rolled sheet, 0.062″ (1.6 mm) thick, heat treated	189 (1303)	153.5-159 (1058-1096)	18-12	---
		Hot finished rod, 0.562 (143) diam, heat treated	186-193 (1282-1331)	144-158 (993-1089)	19-21	---
		Forging stock, annealed at 1750° F (954° C) and aged	185 (1276)	145 (1000)	15	---
N07718	718	Sheet, strip, or plate: annealed and aged	180 (1241)	150 (1034)	12-15	36-38 R_C
		Bars, annealed and aged	185 (1276)	150 (1034)	10-12	331 Bhn
		Forging stock, annealed at 1750° F (954° C) and aged	195 (1345)	165 (1138)	12	---
N07750	X-750	Rod, bars, and forgings, heat treated	165-170 (1138-1172)	105-115 (724-793)	15-20	302-363 Bhn
N08825	825	Rod and bar, annealed, hot finished or cold drawn	85 (586)	35 (241)	30	---
		Plate and sheet, hot rolled and annealed	85 (586)	35 (241)	30	---
		Sheet and strip, cold rolled and annealed	85 (586)	35 (241)	30	---
N09925	925	Octagons, 8.125″ (206 mm), as rolled	150 (1034)	83.3 (574)	34.5	26 R_C
	Inconel MA754[c]	Thermomechanically processed bar, annealed at 2400° F (1315° C)				
		Longitudinal	140 (965)	85 (586)	21	---
		Long transverse	122 (841)	82 (565)	27	---

TABLE 4-2—*Continued*

UNS No.	Alloy[a]	Form and Condition	Tensile Strength, ksi (MPa)	Yield Strength (0.2% Offset), ksi (MPa)	Elongation in 2″ (50 mm), %	Hardness[b]
	Inconel MA6000[c]	Extruded plus hot rolled bar, 0.5 x 1.0 (12 mm x 25 mm)				
		Longitudinal	187.6 (1294)	186.2 (1284)	3.5	47.8 R_C
		Long transverse	184.8 (1274)	179.3 (1236)	5.5	48.7 R_C
	Inconel MA 956[c]	Thermomechanically processed sheet, heat treated				
		Longitudinal	93.5 (645)	80.2 (553)	10	---
		Transverse	95.3 (657)	79.9 (551)	9	---
N10665	B-2	Sheet, heat treated at 2100° F (1149° C) and hydrogen cooled	132.5 (914)	57.5 (396)	55	98 R_B
		Sheet and plate, heat treated at 1950° F (1066° C) and rapid quenched, 0.100-0.350″ (2.5-8.9 mm) thick	129.7 (894)	59.8 (412)	61	95 R_B
		Plate, heat treated at 1950° F (1066° C) and rapid quenched, 0.360-2″ (9.1-51 mm) thick	130.9 (902)	59 (407)	61	94 R_B
N06455	C-4	Sheet, 0.125″ (3.2 mm) thick, aged	114.6 (790)	54.6 (376)	56	90 R_B
		Plate, 3/8″ (9.5 mm) thick, aged	111.8 (771)	48.7 (336)	62	90 R_B
		Plate, 3/8″ (9.5 mm) thick, heat treated	114.7 (791)	51.6 (356)	59	90 R_B
	Hastelloy alloy C-22[d]	Bar, solution annealed, 1/2 - 2″ (12-50 mm) diam	111.6 (769)	52.5 (362)	70	90 R_B
		Sheet, solution annealed, 0.028-0.125″ (0.71-3.18 mm) thick	117.6 (810)	58 (400)	56	90 R_B
		Plate, solution annealed, 1/4-3.4″ (6-19 mm) thick	114.2 (787)	53.4 (368)	65	90 R_B
N10276	C-276	Sheet, heat treated at 2050° F (1121° C) and rapid quenched, 0.078″ (2.0 mm) thick	114.9 (792)	51.6 (356)	61	90 R_B
		Plate, heat treated at 2050° F (1121° C) and rapid quenched, 1″ (25 mm) thick	113.9 (785)	52.9 (365)	59	90 R_B
N06007	G	Sheet, solution annealed, up to 0.125″ (3.2 mm) thick	102 (703)	46.2 (319)	61	---

(continued)

TABLE 4-2—*Continued*

UNS No.	Alloy[a]	Form and Condition	Tensile Strength, ksi (MPa)	Yield Strength (0.2% Offset), ksi (MPa)	Elongation in 2″ (50 mm), %	Hardness[b]
		Plate, solution annealed, 3/8-5/8″ (9.5-16 mm) thick	99.6 (687)	45 (310)	62	---
N06985	G-3	Plate, heat treated at 2100° F (1149° C) and rapid cooled	100.4 (692)	45.1 (311)	58	---
	Hastelloy alloy B[d]	Sheet, 0.078″ (2 mm) thick, heat treated at 2000° F (1093° C) and rapid air cooled	134.1 (925)	67 (462)	51	---
		Sheet, 0.063″ (1.6 mm) thick and cold reduced 10%	147.3 (1016)	110 (758)	33	---
		Sheet, 0.063″ (1.6 mm) thick and cold reduced 20%	164 (1131)	138.5 (955)	22	---
		Bar, solution annealed	127 (876)	56 (386)	52	---
N10003	N	Sheet, 0.063″ (1.6 mm) thick, heat treated at 2150° F (1177° C)	115.1 (794)	45.5 (314)	50.7	---
		Sheet, 0.045″ (1.14 mm) thick, heat treated at 2150° F (1177° C)	114.4 (789)	44.7 (308)	50	---
		Cast	87 (600)	37.3-49.7 (257-343)	16.8-24[e]	---
	Hastelloy alloy S[d]	Sheet, 0.045-0.063″ (1.1-1.6 mm) thick	128.8 (888)	72 (496)	51	---
		Plate, 3/8-1″ (9.5-25 mm) thick	123.1 (849)	55.6 (383)	55	---
N06002	X	Sheet, 0.012-0.090″ (0.30-2.3 mm) thick, heat treated at 2150° F (1177° C) and rapid cooled	110.3 (760)	55.1 (380)	44	---
		Sheet, 0.091-0.312 (2.3-7.9 mm) thick, heat treated at 2150° F (1177° C) and rapid cooled	109.5 (755)	55.9 (385)	45	---
		Plate, 3/8-2″ (9.5-51 mm) thick, heat treated at 2150° F (1177° C) and rapid cooled	107.7 (743)	49.1 (339)	51	---
	Cabot No. 214[d]	Sheet, 0.040″ (1 mm) thick, bright annealed at 2050° F (1120° C)	133.6 (921)	77.1 (532)	39	---

TABLE 4-2—*Continued*

UNS No.	Alloy[a]	Form and Condition	Tensile Strength, ksi (MPa)	Yield Strength (0.2% Offset), ksi (MPa)	Elongation in 2″ (50 mm), %	Hardness[b]
		Sheet, 0.109″ (2.8 mm) thick, bright annealed at 2050° F (1120° C)	136.3 (940)	85.7 (591)	37	---
	René 41	Sheet, 0.050″ (1.27 mm) thick, heat treated	194.8 (1343)	155.2 (1070)	14	---
		Forging stock, annealed at 1950° F (1066° C) and aged	180 (1241)	140 (965)	12	---
	Cabot No. 263[d]	Sheet, 0.028-0.140″ (0.71-3.6 mm) thick, solution heat treated, aged and air cooled	144 (993)	87 (600)	37	---
	Waspaloy[f]	Forging stock, annealed at 1850° F (1010° C) and aged	190 (1310)	135 (931)	15	---
		Cobalt-Based:				
	Haynes No. 25[d]	Sheet, 0.109″ (2.78 mm) thick, solution heat treated	135 (931)	65 (448)	60	---
		Plate, solution heat treated	140.7 (970)	---	58	---
		Bar, 1″ (25 mm) diam, solution heat treated	150 (1034)	70 (483)	65[e]	---
	Haynes No. 188[d]	Sheet, 0.029-0.050″ (0.74-1.27 mm) thick, bright annealed	137.2 (946)	67.3 (464)	53	---
		Cold reduced sheet	158-246.4 (1089-1699)	129.6-220.1 (894-1518)	4-43	---
	Haynes Stellite No. 6B[d]	Sheet, 0.063″ (1.6 mm) thick, solution heat treated	146 (1007)	91.6 (632)	11	---
		Plate, 1/2″ (12.7 mm) thick, solution heat treated	148 (1020)	88 (607)	7	---
		Bar, 5/8″ (16 mm) diam, solution heat treated	154.1 (1063)	92.6 (638)	17[e]	---
	Haynes Stellite No. 6K[d]	Sheet, 0.063″ (1.6 mm) thick, solution heat treated	176.5 (1217)	102.7 (708)	4	---
		Plate, 1/2-7/8″ (12-22 mm) thick, solution heat treated	146.2 (1008)	108.2 (746)	1[e]	---
		Iron-Based:				
N08800	800	Annealed at 1800° F (980° C)	85.5 (590)	36.2 (250)	---	138 Bhn
		Hot rolled	96.4 (665)	64.6 (445)	---	198 Bhn

(continued)

TABLE 4-2—*Continued*

UNS No.	Alloy[a]	Form and Condition	Tensile Strength, ksi (MPa)	Yield Strength (0.2% Offset), ksi (MPa)	Elongation in 2″ (50 mm), %	Hardness[b]
N08810	800H	Annealed plate, 0.813″ (20.7 mm) thick	77.8 (536)	21.7 (150)	---	126 Bhn
N08801	801	Strip				
		Annealed	91.8 (63)	34.3 (236)	43.5	76 R_B
		Age hardened	114.5 (789)	56.3 (388)	40.5	93 R_B
		Cold rolled 10%	114.5 (789)	56.3 (388)	30.5	---
	Multimet[d]	Sheet, 0.052″ (1.32 mm) thick, heat treated and rapid air cooled	116 (800)	57 (393)	43	92 R_B
		Sheet, 0.063″ (1.6 mm) thick, heat treated and rapid air quenched	118.1 (814)	58 (400)	49	92 R_B
		Hot rolled bar, 1-2″ (25-50 mm) diam, heat treated and water quenched	111.3 (767)	54 (372)	57[e]	190 Bhn
		Sand cast, heat treated	97.9 (675)	54 (372)	23	89 R_B
		Investment cast, as cast	98 (676)	57.8 (399)	27[e]	21 R_C max
	Haynes No. 556[d]	Sheet, 0.019-0.180″ (0.48-4.57 mm) thick, heat treated and rapid cooled	116-121.6 (800-838)	551-58.9 (380-406)	51-58	---

Note: The values of the mechanical properties listed are for comparison purposes only and are not intended for specifications or design purposes. The data for these properties were supplied by Cabot Corp., Special Metals Corp., Teledyne Allvac, Universal Cyclops, and Huntington Alloys, Inc. For specific information, consult the material manufacturer.

[a] Some high-performance alloys are made by more than one manufacturer.

[b] R_B = Rockwell hardness number B scale.
R_C = Rockwell hardness number C scale.
Bhn = Brinell hardness number.

[c] Inconel and Incoloy are registered trademarks of Huntington Alloys, Inc.

[d] Hastelloy, Haynes, Multimet, Stellite, and Cabot are registered trademarks of Cabot Corp.

[e] Elongation in 1″ (25 mm).

[f] Waspaloy is a registered trademark of Pratt and Whitney Aircraft.

TABLE 4-3
Mechanical Properties of High-Performance Alloys at Elevated Temperatures

UNS No.	Alloy[a]	Form and Condition	Temperature, °F (°C)	Tensile Strength, ksi (MPa)	Yield Strength (0.2% Offset), ksi (MPa)	Elongation in 2″ (50 mm), %	Reduction of Area, %
		Nickel-Based:					
N07718	718	Hot rolled bar, 5/8″ (16 mm) diam, annealed & aged	1200 (650)	160-164.5 (1103-1134)	140-145 (965-1000)	15-28	25-59
		Cold rolled sheet, 0.065″ (1.65 mm) thick, annealed & aged	120 (650)	166 (1145)	149.5 (1031)	13.5	---

TABLE 4-3—*Continued*

UNS No.	Alloy[a]	Form and Condition	Temperature, °F (°C)	Tensile Strength, ksi (MPa)	Yield Strength (0.2% Offset), ksi (MPa)	Elongation in 2″ (50 mm), %	Reduction of Area, %
		Forging stock, annealed at 1750° F (950° C) & aged	1000 (540)	175 (1207)	145 (1000)	12	---
			1200 (650)	155 (1069)	135 (931)	12	---
			1400 (760)	120 (827)	110 (758)	17	---
N07750	X-750	Hot rolled bars, 3/4″ (19 mm) diam					
		Equalized and precipitation treated	1000 (538)	163.5 (1127)	115 (793)	20	25
			1200 (650)	143 (986)	110 (758)	7	7.8
		Solution treated and precipitation treated	1000 (538)	168.5 (1162)	128 (883)	13	18
			1200 (650)	143 (986)	122.5 (845)	6	8
		Cold rolled sheet, 0.062″ (1.6 mm) thick, annealed	1200 (650)	83 (572)	54.5 (376)	23	---
			1500 (816)	57 (393)	32 (221)	11	---
			1600 (871)	35 (241)	27.5 (190)	45	---
		Cold rolled sheet, 0.050″ (1.27 mm) thick, annealed & precipitation treated	1000 (538)	154 (1062)	112 (772)	26	---
			1200 (650)	123 (848)	105.5 (727)	6	---
			1500 (816)	80.3 (554)	76.4 (527)	11	---
N08825	825	Cold drawn bars, annealed	800 (427)	88.5 (610)	33 (228)	43.5	59.7
			1600 (871)	19.6 (135)	17 (117)	102	94.8
	Inconel MA754[b]	Thermomechanically processed bar, annealed at 2400° F (1315° C)					
		Longitudinal	1200 (650)	87 (600)	69 (476)	25	44
			1600 (871)	36 (248)	31 (214)	31	58
			2000 (1093)	21 (148)	19.5 (134)	12.5	24
		Long transverse	1200 (650)	69 (476)	58 (400)	24	27
			1600 (871)	32 (221)	49 (200)	11	16
			2000 (1093)	19 (131)	17.5 (121)	3.5	1.5

(continued)

TABLE 4-3—*Continued*

UNS No.	Alloy[a]	Form and Condition	Temperature, °F (°C)	Tensile Strength, ksi (MPa)	Yield Strength (0.2% Offset), ksi (MPa)	Elongation in 2″ (50 mm), %	Reduction of Area, %
	Inconel MA 6000[b]	Extruded and hot rolled bar, heat treated					
		Longitudinal	1400 (760)	141.6 (976)	113.3 (781)	5.5	12.5
			2000 (1093)	32.2 (222)	27.8 (192)	9	31
		Long transverse	1400 (760)	130.1 (897)	116.6 (804)	3.5	2.5
			2000 (1093)	25.7 (177)	24.7 (170)	2	1
	Incoloy MA 956[b]	Thermomechanically processed sheet, heat treated					
		Longitudinal	1472 (800)	20.2 (139)	17.7 (122)	12	---
			1832 (1000)	14.5 (100)	14.1 (97)	4.5	---
			2192 (1200)	11.5 (79)	11 (76)	2	---
		Transverse	1472 (800)	19.7 (136)	17.7 (122)	12	---
			1832 (1000)	14.6 (101)	13.6 (94)	6	---
			2192 (1200)	11.3 (78)	10.4 (72)	2	---
N10665	B-2	Sheet and plate, 0.100-0.350″ (2.5-8.9 mm) thick, heat treated & rapid quenched	400 (204)	123.2 (849)	50.8 (350)	59	---
			600 (316)	119.3 (823)	47.5 (328)	60	---
			800 (427)	116.9 (806)	44.9 (310)	60	---
		Plate, 0.360-2.0″ (0.1-51 mm) thick, heat treated & rapid quenched	400 (204)	126.2 (871)	52.3 (361)	60	---
			600 (316)	121.8 (840)	48.8 (336)	60	---
			800 (427)	119.3 (823)	46.3 (319)	61	---
N06455	C-4	Sheet, 0.065-0.156″ (1.7-4.0 mm) thick, heat treated & rapid quenched	400 (204)	98.3-102.4 (678-706)	39.9-58.5 (275-403)	49-55	---
			800 (427)	93.4-95.2 (644-656)	36.2-46.4 (250-320)	62-68	---
		Sheet, 0.125″ (3.2 mm) thick, aged	400 (204)	103.2 (712)	47.1 (325)	54	---
			800 (427)	97 (669)	40.6 (280)	60	---
		Plate, 3/8″ (9.5 mm) thick, aged	400 (204)	100.6 (694)	39.5 (272)	51	---
			800 (427)	97.2 (670)	37.1 (256)	57	---

TABLE 4-3—*Continued*

UNS No.	Alloy[a]	Form and Condition	Temperature, °F (°C)	Tensile Strength, ksi (MPa)	Yield Strength (0.2% Offset), ksi (MPa)	Elongation in 2″ (50 mm), %	Reduction of Area, %
	Hastelloy alloy C-22[c]	Solution annealed sheet, 0.028-0.125″ (0.71-3.2 mm) thick	400 (204)	103.7 (714)	43.9 (303)	57	---
			800 (427)	96.7 (666)	37.5 (258)	67	---
		Solution annealed bar, 1/2-2″ (12-50 mm) diam	400 (204)	96.9 (668)	38 (262)	74	---
			800 (427)	89.7 (618)	31.2 (215)	79	---
		Solution annealed plate, 1/4-3/4″ (6-19 mm) thick	400 (204)	99.3 (684)	41.3 (285)	68	---
			800 (427)	91.3 (630)	33.6 (232)	72	---
N10276	C-276	Sheet, 0.063-0.187″ (1.6-4.7 mm) thick, heat treated & rapid quenched	400 (204)	100.8 (695)	42.1 (290)	56	---
			800 (427)	95 (655)	34.8 (240)	65	---
		Plate, 3/16-1″ (4.8-25 mm) thick, heat treated & rapid quenched	400 (204)	98.9 (682)	38.2 (263)	61	---
			800 (427)	91.5 (631)	32.7 (225)	60	---
N06007	G	Sheet, up to 0.125″ (3.2 mm) thick, solution heat treated	400 (204)	90.9 (627)	37.4 (258)	74	47
			800 (427)	85 (586)	33.4 (230)	84	47
		Plate, 3/8-5/8″ (9.5-16 mm) thick, solution heat treated	400 (204)	87.9 (606)	33.9 (234)	63	52
			800 (427)	82 (565)	29.1 (201)	70	52
N06985	G-3	Plate, heat treated & rapid cooled	350 (177)	87.9 (606)	34.7 (239)	61	---
			800 (427)	79.4 (547)	27 (186)	67	---
	Hastelloy B[c]	Sheet, 0.078″ (2 mm) thick, heat treated & rapid air cooled	1200 (650)	106.6 (735)	50.4 (348)	50	---
			1600 (871)	71.6 (494)	41.1 (283)	22	---
			2000 (1093)	25.4 (175)	10.1 (70)	20	---
		Sand cast, solution heat treated	1000 (538)	75 (517)	---	15	---
			1200 (650)	63 (434)	---	15	---
			1500 (816)	56 (386)	---	19	---
		Investment cast, as cast	1000 (538)	77.4 (534)	---	15[d]	---
			1200 (650)	65.2 (450)	---	15[d]	---

(continued)

TABLE 4-3—*Continued*

UNS No.	Alloy[a]	Form and Condition	Temperature, °F (°C)	Tensile Strength, ksi (MPa)	Yield Strength (0.2% Offset), ksi (MPa)	Elongation in 2″ (50 mm), %	Reduction of Area, %
			1500 (816)	58.5 (403)	---	19[d]	---
N10003	N	Sheet, 0.063″ (1.6 mm) thick, heat treated & rapid air cooled	1100 (593)	86.9 (599)	32.9 (227)	45.3	---
			1300 (704)	69.6 (480)	31.6 (218)	30	---
			1500 (816)	55.9 (385)	29.5 (203)	24.3	---
		Sheet, 0.045″ (1.14 mm) thick, heat treated & rapid air cooled	1000 (538)	93 (641)	28.3 (195)	46	---
			1200 (650)	82.4 (568)	27.5 (190)	37	---
			1400 (760)	61.8 (426)	26.2 (181)	21	---
		Investment cast, heat treated & rapid air cooled	1200 (650)	53.6 (370)	24.2 (167)	23.5[d]	---
			1600 (871)	29.3 (202)	22.3 (154)	7[d]	---
			1800 (982)	22.1 (152)	20.6 (142)	4[d]	---
	Hastelloy S[c]	Sheet, 0.045-0.063″ (1.1-1.6 mm) thick	1200 (650)	98.8 (681)	51.9 (358)	51	---
			1400 (760)	82 (565)	48.3 (333)	56	---
			1600 (871)	52.7 (363)	27.8 (192)	43	---
		Plate, 3/8-1″ (9.5-25 mm) thick	1200 (650)	99.1 (683)	39.8 (274)	59	---
			1600 (871)	52.7 (363)	33.8 (233)	57	---
			2000 (1093)	17 (117)	8.8 (61)	69	---
N06002	X	Sheet, 0.109″ (2.8 mm) thick, heat treated & rapid cooled	1200 (650)	83 (572)	39.5 (272)	37	---
			1600 (871)	36.5 (251)	25.7 (177)	51	---
			2000 (1093)	13 (90)	8 (55)	40	---
			2200 (1204)	5.4 (37)	3.7 (26)	31	---
	Cabot alloy No. 214[c]	Sheet, 0.040″ (1 mm) thick, bright annealed	1112 (600)	100.3 (692)	68.3 (471)	20	---
			1652 (900)	55.1 (380)	39.2 (270)	11	---
			2012 (1100)	8 (55)	3.1 (21)	77	---
		Sheet, 0.109″ (2.8 mm) thick, bright annealed	1112 (600)	106.3 (733)	72.8 (502)	23	---

TABLE 4-3—*Continued*

UNS No.	Alloy[a]	Form and Condition	Temperature, °F (°C)	Tensile Strength, ksi (MPa)	Yield Strength (0.2% Offset), ksi (MPa)	Elongation in 2″ (50 mm), %	Reduction of Area, %
			1652 (900)	61.6 (425)	47.7 (329)	10	---
			2012 (1100)	8.9 (62)	2.7 (18)	116	---
	René 41	Sheet, three-stage heat treatment, 0.025″ (0.64 mm) thick	1200 (650)	158.8 (1095)	128.4 (885)	7	---
			1400 (760)	140.1 (966)	119 (821)	4	---
			1600 (871)	100.7 (694)	76.8 (530)	2	---
		0.050″ (1.27 mm) thick	1200 (650)	171 (1179)	134.9 (930)	8	---
			1600 (871)	106.4 (734)	81.6 (563)	5	---
			2000 (1093)	11.4 (79)	5.7 (39)	47	---
		Forging stock, annealed at 195° F (91° C) & aged	1000 (538)	170 (1172)	125 (862)	12	---
			1200 (650)	160 (1103)	120 (827)	12	---
			1400 (760)	140 (965)	110 (758)	8	---
	Cabot alloy No. 263[c]	Cold rolled sheet, 0.018-0.097″ (0.46-0.21 mm) thick, solution treated & aged	1436 (780)	84 (579)	70 (483)	20	---
		Hot rolled sheet, 0.128-0.212″ (3.2-5.4 mm) thick, solution treated & aged	1436 (780)	86 (593)	69 (476)	22	---
		Plate, 3/8″ (9.5 mm) thick, solution treated & aged	1436 (780)	86 (593)	66 (455)	22	---
	ALLCORR[e]	Solution annealed bar, 2″ (50 mm) diam	400 (204)	94.4 (651)	38.3 (264)	70.5[d]	71.7
			600 (316)	84.5 (583)	31.7 (219)	69.7[d]	67.2
			800 (427)	78.4 (541)	26.4 (182)	76.4[d]	64.1
	Waspaloy[f]	Bar, solution treated 1865° F (1018° C) & double aged	1000 (538)	178 (1227)	123 (848)	23[d]	31
		Forging stock, annealed at 1850° F (1010° C) & aged	1000 (538)	165 (1138)	115 (793)	15	---
			1200 (650)	150 (1034)	110 (758)	20	---
			1400 (760)	110 (758)	100 (690)	15	---

(continued)

TABLE 4-3—*Continued*

UNS No.	Alloy[a]	Form and Condition	Temperature, °F (°C)	Tensile Strength, ksi (MPa)	Yield Strength (0.2% Offset), ksi (MPa)	Elongation in 2″ (50 mm), %	Reduction of Area, %
	Astroloy	Bar, solution treated 2070°F (1132°C) plus four-step aging treatment	1400°F (760°C)	160 (1103)	132 (910)	21[d]	26
		Solution treated 2125°F (1163°C) plus four-step aging treatment	1800°F (982°C)	53 (365)	39 (269)	15[d]	16
		Cobalt-Based:					
	Haynes alloy No. 25[c]	Sheet, 0.109″ (2.77 mm) thick, solution heat treated	1200 (650)	103 (710)	35.4 (244)	35	---
			1800 (982)	33.5 (231)	18.2 (125)	40	---
			2200 (1204)	11.1 (77)	4.7 (32)	38	---
		Bar, 1″ (25 mm) diam, solution heat treated	1200 (650)	97 (669)	---	37[d]	---
			1500 (816)	65.7 (453)	---	24[d]	---
			1800 (982)	32.9 (227)	---	21[d]	---
	Haynes alloy No. 188[c]	Sheet, 0.029-0.050″ (0.74-1.27 mm) thick, bright annealed	1200 (650)	103.3 (712)	39.7 (274)	59	---
			1600 (871)	60 (414)	35.9 (248)	64	---
			2000 (1093)	18.7 (129)	9.3 (64)	32	---
		Cold reduced sheet 10% reduction	1000 (538)	125.5 (865)	86.1 (594)	60	---
		30% reduction	1000 (538)	173.5 (1196)	151.3 (1043)	7	---
		50% reduction	1000 (538)	230 (1586)	213.4 (1471)	2	---
	Haynes Stellite No. 6B[c]	Sheet, 0.063″ (1.6 mm) thick, solution heat treated & air cooled	1500 (816)	73.9 (509)	45.4 (313)	17	---
			1800 (982)	32.6 (225)	19.8 (137)	36	---
			2100 (1149)	13.3 (92)	7.7 (53)	22	---
		Plate, 1/2″ (13 mm) thick, solution heat treated & air cooled	1000 (538)	133 (917)	58.5 (403)	9	---
			1250 (677)	115 (793)	60.6 (418)	9	---
		Bar, 5/8″ (16 mm) diam, solution heat treated & air cooled	600 (316)	147.8 (1019)	74.5 (514)	30[d]	---
			1000 (538)	129.1 (890)	67.3 (464)	28[d]	---

TABLE 4-3—*Continued*

UNS No.	Alloy[a]	Form and Condition	Temperature, °F (°C)	Tensile Strength, ksi (MPa)	Yield Strength (0.2% Offset), ksi (MPa)	Elongation in 2″ (50 mm), %	Reduction of Area, %
			1600 (871)	58.3 (402)	37.9 (261)	34[d]	---
	Haynes Stellite No. 6K[c]	Sheet, 0.063″ (1.6 mm) thick, solution heat treated & air cooled	1200 (650)	146 (1007)	---	8	---
			1800 (982)	34.1 (235)	19.3 (133)	28	---
			2000 (1093)	17.4 (120)	8.6 (59)	53	---
		Plate, 1/2″ (12.7 mm) and 7/8″ (22.2 mm) thick, solution heat treated & air cooled	600 (316)	113.9 (785)	80.5 (555)	2[d]	---
			1000 (538)	105.7 (729)	75 (517)	3[d]	---
		Iron-Based:					
N08800	800	Cold drawn rod, 1″ (25 mm) diam	900 (480)	96.3 (664)	90 (621)	15	52.5
			1200 (650)	78.5 (541)	66.8 (461)	19.5	42.5
		Hot rolled material As rolled	1200 (650)	65.3 (450)	48.3 (333)	---	---
			1400 (760)	44.5 (307)	41.2 (284)	---	---
		Annealed	1200 (650)	58.7 (405)	25.5 (176)	---	---
			1400 (760)	34.5 (238)	21.6 (149)	---	---
N08810	800H	Plate, 0.813″ (20.7 mm) thick, annealed	1200 (650)	54.8 (378)	13.5 (93)	---	---
			1400 (760)	34.2 (236)	13.1 (90)	---	---
N08801	801	Cold rolled strip, reduced 10%, heat treated	1200 (650)	77.6 (535)	44.3 (305)	25.5	---
			1500 (815)	31.9 (220)	18.3 (126)	99	---
	Multimet[c]	Sheet, 0.063″ (1.6 mm) thick, heat treated at 2150° F (1177° C) & rapid air cooled	1200 (650)	73.5 (507)	37.6 (259)	28	---
			2000 (1093)	13 (90)	8.4 (58)	38	---
			2350 (1288)	2.6 (18)	---	7	---
		Sand cast, heat treated at 2140° F (1171° C)	1200 (650)	64.7 (446)	---	23	24
			1650 (900)	33.3 (230)	---	24	29
		Investment cast, as cast	1200 (650)	63.7 (439)	---	30[d]	28
			1650 (900)	30 (207)	---	36[d]	58

(continued)

TABLE 4-3—*Continued*

UNS No.	Alloy[a]	Form and Condition	Temperature, °F (°C)	Tensile Strength, ksi (MPa)	Yield Strength (0.2% Offset), ksi (MPa)	Elongation in 2″ (50 mm), %	Reduction of Area, %
	Haynes alloy No. 556[c]	Sheet, 0.019-0.180″ (0.48-4.57 mm) thick, heat treated & rapid cooled	1200 (650)	85.4-88.9 (589-613)	31-38.5 (214-265)	45-62	---
			1800 (982)	25.6-30.3 (177-209)	14.7-20.2 (101-139)	37-56	---

Note: The values of the mechanical properties listed are for comparison purposes only and are not intended for specifications or design purposes. The data for these properties were supplied by Cabot Corp., Special Metals Corp., Teledyne Allvac, Universal Cyclops, and Huntington Alloys, Inc. For specific information, consult the material manufacturer.

[a] Some high-performance alloys are made by more than one manufacturer.

[b] Inconel and Incoloy are registered trademarks of Huntington Alloys, Inc.

[c] Hastelloy, Haynes, Multimet, Stellite, and Cabot are registered trademarks of Cabot Corp.

[d] Elongation in 1″ (25 mm).

[e] ALLCORR is a registered trademark of Teledyne Allvac.

[f] Waspaloy is a registered trademark of Pratt and Whitney Aircraft.

MACHINING AND FABRICATING CHARACTERISTICS

High-performance alloys are difficult to machine and fabricate because they are designed to have high strengths and excellent surface stability even in aggressive environments.[6] The mechanical properties that make them particularly useful in such environments also make high-performance alloys difficult to machine, form, or weld. However, if careful attention is paid to the metallurgical factors involved, these special materials can be fabricated effectively.

MACHINING

In comparison to steel, high-performance alloys have low machinability ratings (Fig. 4-1). Some of the characteristics that contribute to difficulty in machining include high strength at elevated temperatures, high resistance to shear loads, presence of abrasive particles in the microstructure, low thermal conductivity, and rapid work hardening. Because of severe work hardening, very high surface stresses may remain in machined components. The stresses can cause distortion and/or premature service failure. Therefore, careful attention to the recommended machining practices is important. Machining high-performance alloys typically requires high torques, slow cutting speeds, and machine tools with a high degree of rigidity.

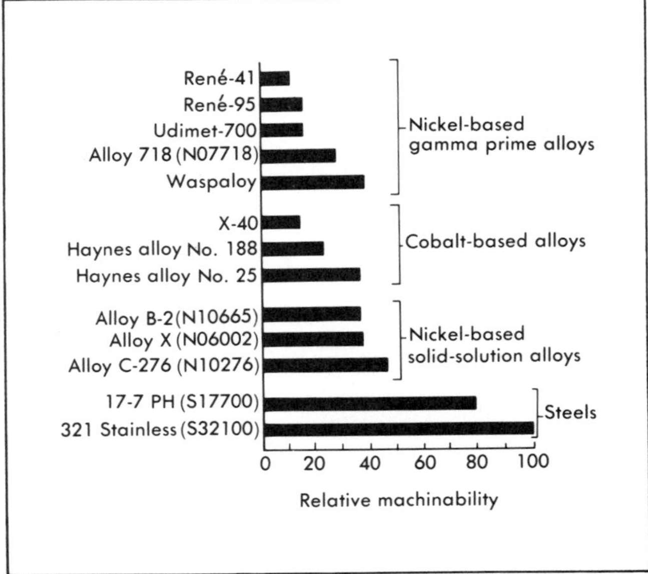

Fig. 4-1 Relative machinability ratings of high-performance alloys.

Machinability can be improved by a solution anneal followed by a rapid cool down to ambient temperature. Cutting tools made from high-speed steels containing generous amounts of cobalt are recommended for drilling, tapping, milling, and broaching. Carbide cutting tools tend to be more satisfactory for turning, reaming, and face milling. The carbide grades containing modest amounts of tantalum carbide (TaC) have better abrasion resistance than straight carbide grades. Ceramic (aluminum oxide) cutting tools are not normally used because they are too fragile and chips adhere to the aluminum oxide, thus shortening tool life.

Machining operations on high-performance alloys should be continuous because they work harden, which accelerates tool deterioration for interrupted machining. For most alloys, cutting depths of 0.040 to 0.080″ (1.0 to 2.0 mm) and cutting speeds of 32 to 130 sfm (10 to 40 m/min) are generally satisfactory. Premature tool failure results from too fast a cutting speed.

High-performance alloys can also be ground using bonded aluminum oxide or silicon carbide wheels. Wheel wear can be from 3 to 30 times greater than for steels, depending upon the specific alloy and other operating parameters.

Nontraditional machining methods such as electrochemical machining, electrical discharge machining, and electrochemical grinding are also used when machining high-performance alloys. For additional information on the various machining processes and their operating parameters, refer to Volume I, *Machining*, in this Handbook series.

MACHINING AND FABRICATING

FORMING

High-performance alloys are formed using both hot and cold forming methods. For additional information on the different forming methods, refer to Volume II, *Forming*, in this Handbook series.

Hot Working

High-performance alloys are quite resistant to hot deformation and thus require large forces to work them. In the initial breakdown of ingots to make wrought products, these alloys tend to have relatively narrow ranges of temperatures in which they may be worked. The range is usually around ±270° F (150° C), the recommended temperature. Too high a temperature causes incipient melting, and too low a temperature results in poor workability. Some precipitation-hardenable, nickel-based alloys are occasionally encased in other less sensitive alloys during primary breakdown to minimize surface cracking. In general, most high-performance alloys require more reheating operations during forming than do the more common alloys.

Once the cast structure is thoroughly broken up by deformation processing, the working temperature range is broader. However, it is important to keep the workpiece within relatively narrow temperature limits because the required working forces rapidly increase as the temperature decreases.

Hot working formability depends upon achieving both uniform temperature and a uniform metallurgical structure throughout the workpiece. Generally a longer hold time at temperature is needed for working high-performance alloys than is needed for steels. Since the thermal conductivity of high-performance alloys is low, it takes a longer time to heat them through to the center than for steels. Once the working temperature has been achieved throughout the workpiece, additional time is usually required for microstructural changes to come to equilibrium.

Induction heating methods are usually not recommended for low thermal conductivity high-performance alloys because the outside can be overheated while the center areas are still below the working temperature; however, induction heating is used in some isothermal forge equipment. Rapid heat input may result in cracking because of thermally induced stresses.

Cold Working

Cold working of high-performance alloys is common, and there are no major difficulties associated with this popular fabrication technique. These alloys work harden more rapidly than steels, and some care is necessary to ensure that they are not worked extensively before annealing. High forces are generally required to form these materials. A high-temperature anneal, followed by a rapid cool down to ambient temperature, is usually recommended for best cold working results. The anneal temperature, however, must be carefully selected; too high a temperature will dissolve carbides that can cause cracking in the heat-affected zone if the component is to be subsequently welded.

WELDING

Welding is an essential step in the fabrication of high-performance alloys because it permits the joining of subcomponents into high-integrity fabricated parts with very little increase in weight, and at reasonable cost. In joining high-performance alloys, it is important that the joint area have properties as nearly like those of the base material as practical. Although some degradation of properties from those of the base material is to be expected in a weld, the overall properties of a welded joint are generally as good as, or better than, the properties of joints made by nonwelding methods.

In general, solid-solution-strengthened alloys have welding characteristics similar to those of austenitic stainless steels and can be welded by a variety of methods; shielded metal arc is the most popular. Gas metal arc, gas tungsten arc, and resistance welding techniques can also be employed. Oxyacetylene welding is not normally used because of the susceptibility of these alloys to carbon and oxygen pickup.

Precipitation-hardening alloys are usually welded by the gas tungsten arc method, but several other methods are also used. Heat input should be held to a moderately low level to obtain the highest joint efficiency. For multiple-bead or multiple-layer welds, several small beads should be used instead of a few large, heavy beads. The oxide films should be removed to ensure good fusion and to prevent laminar oxide inclusions.

Rigid or complex structures must be assembled and welded with care to avoid excessively high stress levels. Units or subassemblies should be given sufficient annealing treatments to ensure a low level of residual stress when they are precipitation hardened. Any part that has been subjected to severe bending, drawing, or other forming operations should be annealed before it is welded.

High-performance alloys can also be successfully brazed and soldered together. For additional information on welding methods and practices, refer to Volume IV, *Quality Control and Assembly*, in this Handbook series.

HEAT TREATMENT

The heat treatment of high-performance alloys depends on their composition, the size and shape of the products, and the application. Some of the most common treatments performed include solution treating, solution treating and aging, stress relieving, and annealing.

Precipitation-strengthened alloys are solution treated prior to and after cold working to reduce the strength and increase the formability. Aging after a solution treatment strengthens these alloys. Optimum strengthening of the heavily alloyed, solid-solution-strengthened alloys is achieved by a full solution treatment and rapid cooling. Stress relieving is performed on alloys that are not age hardenable to remove residual stresses caused by hot or cold working. Annealing is generally performed on high-performance alloys to improve machinability. For additional information on heat treating procedures, refer to Chapter 14, "Heat Treatment of Other Metals," later in this volume.

References

1. Juri Kolts, James B.C. Wu, and Aziz I. Asphahani, "Highly Alloyed Austenitic Materials for Corrosion Resistance," *Metal Progress* (September 1983), p. 34.
2. F. R. Morral, ed., *Metals Handbook*, Vol. III, 9th ed. (Metals Park, OH: American Society for Metals, 1980), pp. 214-215.
3. Mary Helen Johnson, "Superalloys Take Flight," *High Tech* (July/August 1982), pp. 29-30.
4. Eugene W. Kelley, *The Fabrication of Superalloys*, SME Technical Paper MF78-642 (Dearborn, MI: Society of Manufacturing Engineers, 1978), p. 1.
5. *Ibid.*
6. Kelley, *op. cit.*, pp. 3-10.

TUNGSTEN, MOLYBDENUM AND TITANIUM

Tungsten and molybdenum are included among the refractory metals, all of which have exceptionally high melting points. A drawback is that tungsten and molybdenum oxidize readily above 1400° F (600° C). Other common characteristics of tungsten and molybdenum (and other refractory metals) are:

- Excellent strength at high temperatures.
- Low coefficient of expansion.
- High elastic modulus.
- Exceptional resistance to corrosion.
- High resistance to thermal shock.
- Good electrical and heat conducting properties.
- High density and specific gravity (tungsten, not molybdenum).
- High hardness properties at elevated temperatures.

In general terms, "refractory" means resistant to heat or capable of enduring high temperatures. The principal areas of use for the five most common refractory metals are shown in Table 5-1.[1] Alloys containing varying amounts of refractory metals are vital to virtually every major industry, including automotive, aerospace, mining, chemical and petroleum processing, electrical and electronics, metal processing, nuclear technology, and ordnance.

Titanium is not classified as a refractory metal; however, it does have a unique set of characteristics and properties, including a high melting point and a high strength-to-weight ratio at elevated temperatures. It also has good stiffness properties, having a much higher modulus than the other light metals, aluminum and magnesium. Titanium alloys generally exhibit good creep strengths, over a wide range of temperatures.

The principal uses of titanium and its alloys are in the aerospace industry, although its exceptional corrosion resistance has led to increasing applications in the chemical industry and in the medical profession, which uses titanium alloy prostheses for implanting in the human body. The major titanium alloy usage is in the form of wrought products; although titanium alloy casting usage, which accounts for about 10% of the total, is growing.

The basic production method for tungsten, molybdenum (and other refractory metals), and titanium is similar. The metals are extracted from ore concentrates, processed into chemicals, and then into sponge or powders. These primary forms are consolidated into finished products or mill shapes and ingots for further processing. Because of their high melting points and ease of oxidation, the metals in the refractory group are typically worked in powder form. Titanium products are usually made from consumable vacuum-melted ingots.

CHAPTER CONTENTS:

TUNGSTEN

Tungsten (W) has a silver-gray metallic luster. Its melting point, 6150° F (3400° C), is the highest of any metal; and, as illustrated in Fig. 5-1, tungsten is the heaviest engineering material. Other unusual physical properties include the highest modulus of elasticity of all metals (59 x 10^6 psi; 40.7 x 10^4 MPa) and extreme hardness. Tungsten is the hardest pure metal; annealed, the hardness can be as high as 425 Bhn. Tungsten also exhibits superior high-temperature strength characteristics, but is expensive and difficult to fabricate.

PRODUCTION

The commercially important tungsten-bearing minerals are wolframite, scheelite, hubernite, and ferberite. Deposits are found in various places, such as Burma, Portugal, and China, as well as in the United States. The pure metal is obtained by the reduction of tungstic acid with aluminum filings, and also by reduction of tungstic oxide by heating with charcoal.

PROPERTIES

Chemically, tungsten is relatively inert. It is not readily attacked by the common acids, alkalies, or aqua regia (mixture of nitric and hydrochloric acids). It reacts with a mixture of concentrated nitric acid and hydrofluoric acid. Molten oxidizing salts such as sodium nitrite attack tungsten rapidly. Gaseous chlorine, bromine, iodine, carbon dioxide,

***Contributors of sections of this chapter are: John T. Benedict**, Consultant; **Kurt H. Miska**, Supervisor, Technical Information, Climax Molybdenum Company of Michigan; **Dr. Harry W. Rosenberg**, Manager, Technical Services and Sales Development, TIMET.*

***Reviewers of sections of this chapter are: P. W. Blackburn**, President, The Rembar Co., Inc.; **C. E. Hulswitt**, Vice President, Astro Metallurgical Div., Harsco Corp.; **Stephen C. McCrossan**, Sales Dept., Marketing Div., Schwarzkopf Development Corp.; **Dr. Marvin Pesses**, President, The Pesses Co.; **Dr. Harry W. Rosenberg**, Manager, Technical Services and Sales Development, TIMET; **Stan R. Seagle**, Vice President, Research and Development, RMI Co.; **George A. Timmons**, Consultant, AMAX Specialty Metals Corp.*

CHAPTER 5

TUNGSTEN

TABLE 5-1
Principal Uses of Refractory Metals

	Tungsten	Molybdenum	Tantalum	Columbium	Rhenium
Electronics	•	•	•	•	•
Alloying	•	•	•	•	•
Nuclear Power	•	•		•	
Aerospace	•	•	•	•	
Chemicals/Catalysts	•	•	•		•
Metal Cutting & Forming	•	•	•		
Mechanical Parts	•	•			
Mining/Oil Drilling	•				

(*Refractory Metals Association*)

carbon monoxide, and sulfur react with tungsten only at high temperatures. Carbon, boron, silicon, and nitrogen also form compounds with tungsten at elevated temperatures; hydrogen does not.

APPLICATIONS

When added to iron or steel, tungsten improves high-temperature strength and hardness. More than 75% of the tungsten produced is used in ferrous and aluminum alloys and in tungsten carbide for cutting tools. (See Volume I, *Machining*, of this Handbook series.)

A well-known use for tungsten is in the filament material for electric light bulbs. Pure tungsten metal in the form of wire, rod, and sheet is important in the electric lamp and electrical industries. Industrial applications include electrodes for inert gas welding and electrical discharge machining.

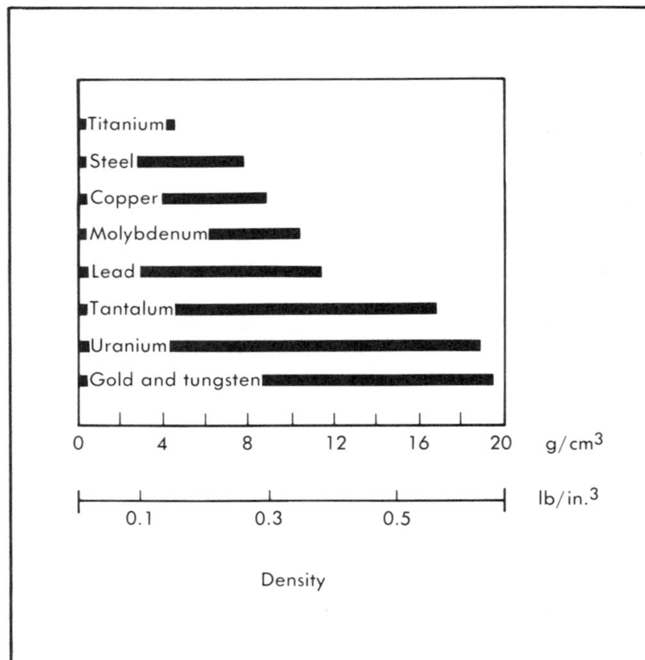

Fig. 5-1 Metal density comparison shows tungsten as the heaviest engineering material.

Small shapes of tungsten sintered to various densities and porosities are used as filters and in probes for ultrasonic nondestructive testing. Tungsten is also used for high inertia devices and for balancing masses. Because of tungsten's high density, a flywheel or balance weight made from tungsten requires only one-third as much space as a steel component.

Some inertial devices use a tungsten-nickel-copper alloy that is more machinable than pure tungsten. Several of the commercially available alloys of tungsten are listed in Table 5-2. The rhenium (Re) alloys are more ductile than unalloyed tungsten at room temperature. Thoriated tungsten is stronger than unalloyed tungsten at temperatures up to the recrystallization temperature.

WELDING AND BONDING

Manganese containing 16% (by weight) of nickel, 16% cobalt, and 1% boron is an effective tungsten-to-tungsten vacuum-brazing alloy. If tungsten parts are to be soft-soldered, they should be precoated with copper. To minimize loss of ductility in tungsten-to-tungsten spot welds, the spot welding is sometimes done with the parts under water. Tungsten surfaces to be joined should be mechanically cleaned, chemically etched, or preheated in hydrogen at 1200°F (650°C).

Arc welds or electron-beam welds in powder-metallurgy tungsten have considerable porosity. Welding should be done at a high speed to avoid excessive heat input.

Tungsten is embrittled more by melting during welding than by high-temperature exposure in a protective atmosphere or under a protective coating. Pressure-diffusion bonding can therefore join the metal without damaging it. Two pieces of tungsten can be united in the fraction of a second necessary for resistance welding or for impulse bonding, or joining by autoclave pressure bonding may require 24 hours.

FABRICATION

Deformation processing of tungsten usually begins with primary reduction and ingot breakdown by a compressive process, such as extrusion, before secondary fabrication, machining, and surface finishing are feasible. The reduction process is usually the only means of working an ingot of tungsten without cracking it. The crude metal is converted into a shape such as tube stock or sheet, or bar stock suitable for secondary working by rolling, swaging, or drawing into final shape.

TABLE 5-2
Typical Properties of Tungsten and Tungsten Alloys

	Nominal Composition, %	Form and Condition	Yield Strength (0.2% Offset), ksi (MPa)	Tensile Strength, ksi (MPa)	Elongation in 2 in., %	Hardness	Density, lb/in.^{3} (g/cm^3)	Melting Point, °F (°C)	Tensile Modulus of Elasticity, 1000 ksi (1000 MPa)
Tungsten-Nickel-Copper Alloy UNS R07030 ASTM B 459 Grade 4	W - 97.5 N + Cu - bal	Sintered powder shapes	80 (552)	100 (690)	3	---	0.67 (18.5)	---	52 (358)
Tungsten-Nickel-Copper Alloy UNS R07100 ASTM B 459 Grade 1	Ni - 7.5 Cu - 2.5 W - balance	Forged from sintered powder	85 (586)	120 (827)	17	285 Bhn	0.61 (16.9)	---	45 (310)
Tungsten-Rhenium Alloy	W - 75 Re - 25	Wire wrought	---	310 (2137)	---	---	0.71 (19.7)	5612 (3100)	57 (393)
		Recrystallized	---	190 (1310)	15	450 Vickers	---	---	---
Tungsten-Rhenium Alloy	W - 50 Re - 50	Sintered powder wrought	210 (1448)	240 (1655)	---	600 Vickers	0.50 (13.8)	4622 (2550)	52 (359)
		Recrystallized	123 (848)	150 (1034)	18	350 Vickers	---	---	---
Tungsten UNS R07004 ASTM B 410 Bars, rods, billets	W - 99.5 min	Sheet hard Cold rolled	360 (2482)	400 (2758)	---	---	0.70 (19.4)	6170 (3410)	50 (345)
		Annealed	---	270 (1862)	---	---	---	---	---

MOLYBDENUM

Molybdenum (Mo) melts at 4730° F (2610° C) and retains useful strength at temperatures of 2000° F (1093° C). The metal is characterized by high thermal conductivity, high modulus, low specific heat, and low coefficient of thermal expansion. Unalloyed molybdenum is produced commercially at the level of purity of 99.95% Mo. In the stress-relieved condition, it has a room temperature hardness of about 240-250 Brinell and a tensile strength of about 100 ksi (689 MPa). Molybdenum has a higher density than steel. Its specific gravity is 10.2 compared with 7.9 for steel. The chemical composition of molybdenum and its alloys is given in Table 5-3.[2]

Molybdenum is similar to tungsten in most of its properties. Its most serious limitation is the ready formation of a volatile oxide at a temperature of approximately 1400° F (760° C). In the worked form, molybdenum is inferior to tungsten in melting point, tensile strength, vapor pressure, and hardness. But, in the recrystallized condition, molybdenum's ultimate strength and elongation are higher.

PRODUCTION

The basic processes for the production of metallic tungsten and molybdenum are similar. The metals are recovered from their ores as high purity ammonium salts that are converted to oxides by calcination and reduced to fine metal powders (particle size 79-236 μ in., 2-6 μ m) by hydrogen at temperatures above 2000° F (1090° C). The powders are consolidated to ingots by two methods: (1) the powder metallurgy process in which the powders are compacted under high pressure and the compacted shapes are sintered at high temperatures and (2) vacuum-arc melting. Powder metallurgy consolidations may be extruded, forged, or rolled to accomplish the initial working of the metal. Vacuum-arc melted ingots are always extruded to break down the cast structure and to refine the grain size.

ALLOYS

Molybdenum can be alloyed with several other metals to enhance its special properties, but only two alloy groups are produced commercially: (a) TZM, which is molybdenum alloyed with 0.5% titanium and 0.1% zirconium, for high-temperature structural applications and (b) an alloy comprising 70% molybdenum and 30% tungsten, which is resistant to attack by molten zinc.

Unalloyed molybdenum and its two principal alloys cannot be hardened by heat treatment. The hardness and strength of mill products and forgings are developed by working the metal below its recrystallization temperature. To develop an optimum

TABLE 5-3
Chemical Composition of Molybdenum and Alloys

Element	Composition, % — Material Designation, UNS Number (ASTM Number)					
	RO 3600 (Mo 360)	RO 3610 (Mo 361)	RO 3620 (Mo 362)	RO 3630 (Mo 363)	RO 3640 (Mo 364)	RO 3650 (Mo 365)
Carbon	0.010-0.040	0.010 max	0.010-0.040	0.010-0.040	0.010-0.040	0.010 max
Oxygen, max	0.0030	0.0070	0.0030	0.0030	0.030	0.0030
Nitrogen, max	0.0010	0.0020	0.0010	0.0010	0.0020	0.0010
Iron, max	0.010	0.010	0.010	0.010	0.010	0.010
Nickel, max	0.005	0.005	0.005	0.005	0.005	0.005
Silicon, max	0.010	0.010	0.010	0.010	0.005	0.010
Titanium			0.40-0.55	0.40-0.55	0.40-0.55	
Zirconium				0.06-0.12	0.06-0.12	
Molybdenum	balance	balance	balance	balance	balance	balance
Production Method	Unalloyed Arc-Cast	Unalloyed PM*	Ti Alloy Arc-Cast	TZM Arc-Cast	TZM PM	Unalloyed Arc-Cast Low-Carbon

* Powder Metallurgy

(*Climax Molybdenum Company*)

combination of strength and toughness, the technically "cold worked" products are usually stress relieved by heating them to temperatures below the recrystallization temperature. Unalloyed molybdenum and its two commercial alloys are available in the various common wrought, mill product forms: forging billets, tubes, bars, rods, wire, plate, sheet, strip, and foil.

Although molybdenum and its alloys can endure at very high temperatures in vacuum, inert, or hydrogen atmospheres indefinitely, rapid oxidation of molybdenum at temperatures above 1400° F (760° C) precludes application at high temperatures for extended periods of time when exposed to air. The metal reacts with oxygen to form the volatile molybdenum trioxide (MoO_3), which evolves in the form of copious, nontoxic, white fumes. Special silicide coatings can protect unalloyed molybdenum and TZM from oxidation in air at temperatures as high as 2200° F (1200° C) for up to 50 hours.

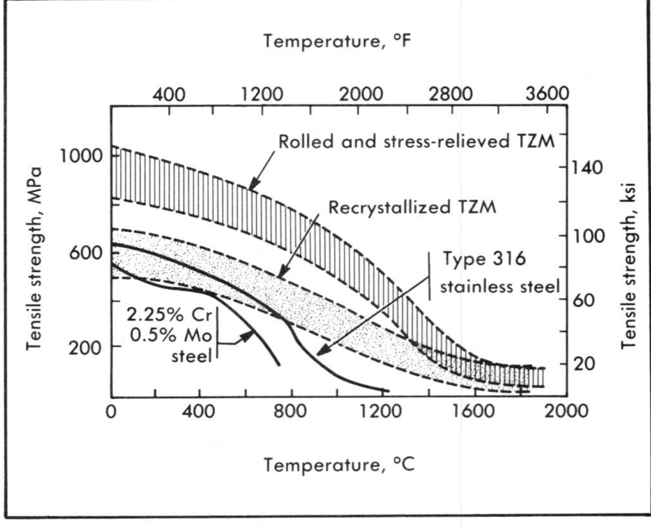

Fig. 5-2 Scatter bands of tensile strength versus temperature for rolled TZM molybdenum bar and curves for two steels that are commonly used at elevated temperatures.

APPLICATIONS

The high-temperature strength of molybdenum is responsible for most of the metal's industrial applications. As shown in Fig. 5-2, TZM molybdenum alloy retains good tensile strength properties at temperatures in excess of 2000° F (1093° C). Steels and nickel-based superalloys are inferior to molybdenum in this respect. As stated previously, molybdenum and its alloys do, however, oxidize rapidly in air at temperatures above 1400° F (760° C); hence, they are suitable for high-temperature usage only in vacuum or inert atmospheres.

Unalloyed Molybdenum

The principal uses for unalloyed molybdenum are in mandrels for the production of coiled tungsten filaments and the support wires for these filaments in incandescent light bulbs; semiconductor support discs; electrodes for heating molten glass; reflectors (heat shields) and heating elements for vacuum furnaces or special atmosphere furnaces; magnetron components; resistance welding electrodes; electrodes for electrical discharge machines; anodes, grids and supports in vacuum tubes; X-ray targets; and mirrors for laser beams.

Because it has a higher modulus of elasticity (47 x 10^6 psi, 33,000 MPa) than steel (30 x 10^6 psi, 21,000 MPa), molybdenum is often used for special boring bars and for the quills of internal grinding machines where high rigidity and absence of vibration are required. Unalloyed molybdenum has also been used successfully for nozzles and guidance systems for rockets.

Unalloyed molybdenum wire and powder can be flame sprayed (oxy-acetylene flame) or plasma sprayed onto irons, steels, and other substrates. (See Chapter 22.) Because the sprayed deposit is hard, and sufficiently porous to retain lubricating oil, it provides a wear-resistant surface for piston rings for internal combustion engines, for special bearing journals such as those used in surface grinders, and for dies used in deep drawing and forming. The flame-sprayed deposit also provides an excellent bond between steel substrates and other flame-sprayed deposits, eliminating the necessity of developing a special surface to be built up on the steel part for salvage purposes.

Molybdenum is an excellent material for trueing many grades of resinoid-bonded diamond grinding wheels under dry and wet grinding conditions. The metal is not satisfactory for dressing vitrified or metal-bonded diamond wheels.

Alloy Applications

Where high strength and resistance to thermal fatigue are required at high temperatures, the TZM alloy is used. The elevated temperature tensile strength of TZM is compared to the elevated temperature strengths of two popular steels recommended for high-temperature applications in Fig. 5-2. It can be observed that TZM in the fully recrystallized (annealed) condition is superior to the 2.25% Cr-1% Mo steel at temperatures over 400° F (200° C) and the austenitic Type 316 stainless steel at temperatures above 1500° F (815° C).

The improvements in hardness and tensile properties at 1600° F (871° C), and 100-hour rupture stress at 2000° F (1095° C), are presented in Table 5-4. The improved properties of the alloy are attributed in large measure to the higher recrystallization temperature imparted to molybdenum by the Ti-Zr combination. The alloy is regularly used for dies in the isothermal forging of superalloys and cores in the die casting of aluminum. TZM is used in several plants in Europe for cores and die inserts in the pressure casting of brass and bronze, and for die inserts used for extruding steel into special shapes. It is also used in special heating elements for vacuum or hydrogen atmosphere furnaces and for boats (open boxes) for high-temperature sintering. Hot gas valves for rockets and missiles are machined from arc-cast TZM bar stock.

The 70% molybdenum-30% tungsten alloy has excellent resistance to corrosive attack by molten zinc, but it exhibits no significant improvement over unalloyed molybdenum with respect to mechanical properties at elevated temperatures. For this reason its principal commercial applications are limited to the processing and transfer of molten zinc. Zinc processors and industries concerned with galvanizing use the 70-30 alloy for stirrers, pump impellers and shafts, and piping.

FABRICATION

Sheet products can be formed, deep drawn, or spun to produce the desired shapes. Forming, drawing, and spinning are best performed with the metal at temperatures in the range of 300 to 800° F (149 to 427° C). For spinning operations that require large amounts of plastic deformation, or for roll forming of thin-walled tubes, temperatures near 1600° F (871° C) may be advantageous.

If sheet is to be cut to shape by shearing, punching, or blanking, it is recommended that the dies be kept sharp. Thin sheet can be handled at room temperature, but for thick sheet the metal should be heated to about 800° F (427° C) to prevent delamination at the edges. If molybdenum sheet is to be cut to shape with a bandsaw, friction saw, or abrasive wheel, the operation can be accomplished readily at room temperature, but it will be found advantageous to back up the workpiece with a piece of wood or mild-steel plate to reduce chattering and delamination.

MACHINING

Molybdenum can be machined readily by conventional operations such as sawing, turning, boring, shaping, milling (face milling and end milling), drilling, topping, and threading.

General Considerations

No special machines are required to produce parts to accurate dimensions with excellent finishes. With respect to depth of cuts and feeds and speeds, molybdenum and TZM, at hardness levels of 220 to 300 HV50, may be characterized as machining like medium-carbon, low-alloy steel (such as AISI 4340), heat treated to the hardness of R_C 30-35 (300-345 HV50); the chips, however, are generally like the chips generated in the machining of gray cast iron. The hardness values of unalloyed molybdenum and its alloys fall within a fairly narrow range, 160-320 HV; powder metallurgy products and arc-melted products machine similarly, although the arc-cast products have less tendency to chip out at edges and usually develop smoother finishes.

TABLE 5-4
Typical Properties of Unalloyed Molybdenum and TZM Alloy
[1 in. (25 mm) diam bars, stress relieved]

	Molybdenum, Unalloyed	TZM Alloy
Room temperature:		
Diamond pyramid hardness	245	290
Yield strength, psi (MPa)	91,000 (627)	104,000 (717)
Tensile strength, psi (MPa)	102,000 (703)	122,000 (841)
Elongation, %	26	24
1600° F (871° C):		
Diamond pyramid hardness	115	179
Yield strength, psi (MPa)	37,000 (255)	52,000 (359)
Tensile strength, psi (MPa)	52,000 (359)	83,000 (572)
Elongation, %	24	15
Stress to rupture in 100 hr, psi (MPa)	37,000 (255)	75,000 (517)
2000° F (1093° C):		
Stress to rupture in 100 hr, psi (MPa)	11,000 (76)	52,000 (359)

CHAPTER 5

MOLYBDENUM

A perception of the machinability of molybdenum is conveyed by an industrial operation in which wrought bars 4″ (100 mm) diameter are turned on a conventional lathe using a carbide-tipped tool (BL-6-2A5) with a depth of cut of 0.100″ (2.54 mm), a feed rate of 0.005″ (0.13 mm) per revolution, and a cutting speed of 150 fpm (0.8 m/s) without any cutting fluid.

It is important when machining molybdenum that the machines be rigid and free from vibration. The workpieces should be gripped tightly in chucks or vises. For turning and face milling operations, positive rake angles are recommended: near 0° for carbide tools, and up to 20° for high-speed steel tools.

Tooling Factors

All molybdenum products may be characterized as "abrasive" due to the generation of fine particles (dust or grit) that separate from the chips during machining. These fine particles wear the cutting edges of high-speed steel tools; therefore, tool life is substantially shorter for a given machine operation on molybdenum than it is for steel. Removal of the chips by an air jet, soluble oils, or cutting fluids can mitigate the decreased tool life, but will not eliminate it. High-speed steel tools usually require more frequent regrinds when they are used to machine molybdenum.

All machining operations may be performed satisfactorily with high-speed steel tools. For heavy and intermittent cuts on lathes, and for most shaper operations, high-speed steel tool bits are usually recommended. However, for long continuous cuts on lathes and for face-milling operations, tools with carbide tips or disposable carbide inserts are preferred because they require fewer regrinds, thereby increasing productivity. The straight tungsten-carbide-cobalt types of tips or inserts are recommended for machining molybdenum.

Cutting Fluids

Molybdenum mill products can be machined "dry" in all operations, but soluble oils can be used effectively to improve tool life. The soluble oil emulsions are particularly effective in hacksaw and bandsaw operations, and for milling and boring. Highly chlorinated oils and sulfur-containing oils improve tool life for drilling, tapping, thread chasing, and for special fine finishes on turned parts. Much of the advantage attributed to the use of cutting fluids lies in the ability to remove fine particles of molybdenum before they can grind away the cutting edge of the tool.

DRILLING

Molybdenum is drilled with ordinary high-speed steel drills using the conventional 118° point angle. It is important in drilling molybdenum to maintain a continuous feed of at least 0.005″ (0.13 mm) per revolution and to provide for the removal of chips and the fine particles by a flow of air, soluble oil, or chlorinated oil. Drilling of small holes may be performed most effectively in the inverted position (the drill moving upward into the workpiece) to assure the removal of chips by gravity, thereby avoiding abrasion at the cutting edge of the drill.

THREADING

External threads over 1/2″ (13 mm) pitch diam should be machined on a lathe using a chlorinated cutting oil to preserve the integrity of the thread. Smaller threads can be produced effectively by thread rolling or by thread chasing on a lathe; threading with dies is not recommended. Internal threads up to 1-1/4″ (32 mm) diam are produced by tapping with high-speed steel taps; larger internal threads are produced on lathes with preground carbide inserts using a chlorinated cutting oil.

GRINDING

Excellent finishes and close-tolerance dimensions can be developed for molybdenum parts by grinding; surface finishes of 10 to 30 μ in. (0.3 to 0.8 μ m) can be developed. The selection of grinding wheels is important because some grades tend to load rapidly and require more frequent dressing. One grade that works well for surface grinding is 32A46L8VBE; a solution of 5% potassium nitrite (KNO_2) is used as the grinding fluid.

Cylindrical grinding on the outside surface ("0.0. grinding") can be accomplished successfully using wheels designated 32A60K5VBE with a soluble oil coolant.

TITANIUM

Titanium (Ti) is a light, strong, silvery gray metal. Pure titanium has a high melting point, 3047°F (1675°C). Its modulus of elasticity (related to stiffness) is 16.5×10^6 psi (11.4×10^4 MPa), midway between that of steel and aluminum.

Titanium has a lower coefficient of expansion and lower thermal conductivity than either steel or aluminum alloys and is nonmagnetic. Titanium exhibits outstanding corrosion resistance to chlorine and its derivatives in oxidizing or neutral aqueous solutions. It resists organic compounds, oxidizing acids, and nonoxidizing acids in low concentrations.

Titanium is abundant in nature; about 1% of the earth's crust is titanium. Almost all common rocks contain some percentage of titanium dioxide, TiO_2. Deposits of titanium minerals are found in many places throughout the world. Large deposits are located in the United States and Australia.

BASIC PRODUCTION

The most widely used method for producing metallic titanium from ore is the Kroll process, which uses titanium tetrachloride ($TiCl_4$) that is also produced for the titanium oxide pigment industry.

The Kroll process involves reducing titanium tetrachloride to titanium with molten magnesium (or sodium) under an inert atmosphere in a sealed retort. At this stage, the titanium "sponge" contains residual salts (NaCl or $MgCl_2$), and it is further refined by leaching with dilute hydrochloric acid or by vacuum distillation. The sponge, along with any necessary additives, is melted in a cold-wall, consumable vacuum-arc furnace to form ingots, which subsequently are remelted to form the final homogeneous titanium ingots that can be drawn, forged, or rolled.

ALLOYS

In metallurgical terms, the two crystalline forms of titanium are called alpha and beta. The alpha form, which exists at temperatures below 1620° F (882° C), is a close-packed hexagonal crystal structure. The beta phase, which is formed at temperatures above 1620° F, is body-centered cubic.

Pure titanium is soft, weak, and extremely ductile, hence is easily fabricated. With appropriate additions of other elements (mostly metals), the titanium base metal is converted into an engineering material that has unique characteristics, including high strength and stiffness, corrosion resistance, and ductility.

The addition of other metals to a titanium base will favor (stabilize) one or the other of the two crystalline forms, as listed in Table 5-5. The transition temperature (from alpha to beta phase) is raised or lowered, depending on whether a particular alloying element favors the alpha phase or the beta phase. Iron, for example, favors (stabilizes) the beta structure; hence, iron reduces the temperature at which alpha transforms to beta. Aluminum, on the other hand, favors the alpha phase, raising the temperature at which alpha changes to beta.

For convenience in differentiating between the various titanium and titanium alloy compositions, the available commercial grades are classified (depending upon the phases that are present in the metal's microstructure at room temperature) as commercially pure (CP) titanium, all-alpha (single phase) weldable alloys, alpha-beta (two phase) weldable alloys, alpha-beta nonweldable alloys, all-beta alloys, beta-lean alpha alloys, and corrosion-resistant alloys. Table 5-6 lists chemical compo-

TABLE 5-5
Common Titanium Alloy Additions

Alpha Stabilizers	Beta Stabilizers	Neutral
Aluminum	Vanadium	Zirconium
Oxygen	Tantalum	Tin
Nitrogen	Molybdenum	
Carbon	Chromium	
	Iron	
	Nickel	

TABLE 5-6
Nominal Chemical Composition of Titanium and Alloys, %

Material Commercial Designation	C	Mn	O	Cr	Si	H	Sn	Mo	Ta	Cb	Ti	Al	Fe	V	Zr	Cu	N	Ni
Commercially Pure																		
Ti-35A (ASTM Grade 1)	0.03	---	0.08	---	---	0.007	---	---	---	---	Bal	---	0.06	---	---	---	0.010	---
Ti-50A (ASTM Grade 2)	0.03	---	0.13	---	---	0.007	---	---	---	---	Bal	---	0.08	---	---	---	0.010	---
Ti-65A (ASTM Grade 3)	0.03	---	0.20	---	---	0.007	---	---	---	---	Bal	---	0.12	---	---	---	0.015	---
Ti-75A (ASTM Grade 4)	0.03	---	0.30	---	---	0.007	---	---	---	---	Bal	---	0.12	---	---	---	0.015	---
Corrosion Alloys																		
Ti-0.2Pd (ASTM Grade 7)	0.03	---	0.13	---	---	0.007	---	---	---	---	Bal	---	0.08	---	---	---	0.010	---
TiCode-12 (ASTM Grade 12)	0.03	---	0.12	---	---	0.007	---	0.3	---	---	Bal	---	0.08	---	---	---	0.010	0.8
Alpha Alloys																		
Ti-5Al-2.5Sn	0.03	---	0.17	---	---	0.007	2.5	---	---	---	Bal	5.0	0.30	---	---	---	0.015	---
Ti-5Al-2.5Sn ELI*	0.03	---	0.08	---	---	0.007	2.5	---	---	---	Bal	5.0	0.17	---	---	---	0.015	---
Alpha-Lean Beta Alloys																		
Ti-8Al-1Mo-1V	0.03	---	0.09	---	---	0.007	---	1.0	---	---	Bal	8.0	0.07	1.0	---	---	0.015	---
Ti-679	0.03	---	0.14	---	0.20	0.007	11.0	1.0	---	---	Bal	2.25	0.08	---	5.0	---	0.015	---
Ti-6Al-2Sn-4Zr-2Mo	0.03	---	0.11	---	---	0.007	2.0	2.0	---	---	Bal	6.0	0.08	---	4.0	---	0.015	---
Ti-5Al-6Sn-2Zr-1Mo-Si	0.03	---	0.10	---	0.20	0.007	6.0	1.0	---	---	Bal	5.0	0.12	---	2.0	---	0.010	---
Ti-6Al-2Cb-1Ta-1Mo	0.03	---	0.08	---	---	0.007	---	0.8	1.0	2.0	Bal	6.0	0.10	---	---	---	0.010	---
Alpha-Beta Alloys																		
Ti-8Mn	0.03	8.0	0.12	---	---	0.007	---	---	---	---	Bal	---	0.25	---	---	---	0.015	---
Ti-3Al-2.5V	0.03	---	0.12	---	---	0.007	---	---	---	---	Bal	3.0	0.12	2.5	---	---	0.015	---
Ti-6Al-4V	0.03	---	0.15	---	---	0.007	---	---	---	---	Bal	6.0	0.12	4.0	---	---	0.015	---
Ti-6Al-4V ELI*	0.03	---	0.10	---	---	0.007	---	---	---	---	Bal	6.0	0.10	4.0	---	---	0.015	---
Ti-6Al-6V-2Sn	0.03	---	0.17	---	---	0.007	2.0	---	---	---	Bal	5.5	0.6	5.5	---	0.6	0.015	---
Ti-7Al-4Mo	0.03	---	0.12	---	---	0.007	---	4.0	---	---	Bal	7.0	0.15	---	---	---	0.015	---
Ti-6Al-2Sn-4Zr-6Mo	0.03	---	0.12	---	---	0.007	2.0	6.0	---	---	Bal	6.0	0.08	---	4.0	---	0.010	---

(continued)

TABLE 5-6—*Continued*

Material Commercial Designation	C	Mn	O	Cr	Si	H	Sn	Mo	Ta	Cb	Ti	Al	Fe	V	Zr	Cu	N	Ni
Beta-Lean Alpha Alloys																		
Ti-10V-2Fe-3Al	0.03	---	0.10	---	---	0.007	---	---	---	---	Bal	3.0	1.8	10.0	---	---	0.010	---
Ti-17	0.03	---	0.10	4.0	---	0.007	2.0	3.8	---	---	Bal	5.2	0.12	---	1.8	---	0.010	---
Beta Alloys																		
Ti-15V-3Cr-3Al-35n**	0.03	---	0.10	3.0	---	0.007	3.0	---	---	---	Bal	3.0	0.12	15.0	---	---	0.010	---
Ti-13V-11Cr-3Al	0.03	---	0.11	11.0	---	0.007	---	---	---	---	Bal	3.0	0.15	13.5	---	---	0.02	---
Ti-3Al-8V-6Cr-4Mo-4Zr	0.03	---	0.10	6.0	---	0.007	---	4.0	---	---	Bal	3.5	0.2	8.0	4.0	---	0.01	---
Ti-11.5Mo-6Zr-4.5Sn	0.10	---	0.18	---	---	0.020	4.5	11.5	---	---	Bal	---	0.35	---	6.0	---	0.05	---

* ELI = Extra Low Interstitials
** Also designated Ti-15-3

sitions of pure titanium and various alloys. Table 5-7 provides data on mechanical properties. Tensile strength estimates from individual Rockwell hardness numbers may be in significant error, on the order of 20 ksi (138 MPa), 10% of the time.

Commercially Pure Titanium

This group consists of the unalloyed compositions containing over 99% titanium. The remaining percentage is made up of carbon, nitrogen, oxygen, hydrogen, and iron. Strength levels are influenced significantly by the amounts of oxygen and nitrogen that are present. All of the grades are available in billets, bars, wire, sheet, strip, and tubing, and some are offered in extruded forms.

Commercially pure titanium alloys, containing minor amounts of oxygen and nitrogen for added strength, are alpha alloys, and they find wide use in applications where maximum corrosion resistance and ease of fabrication are essential. Among the commercially pure titanium grades, two metals are listed in a "corrosion alloy" subgroup because that term is descriptive of their principal applications. Both of these alloys have improved corrosion resistance in reducing media, and they are weldable.

All-Alpha Weldable Alloys

The hexagonal structure compositions generally have the highest strength at elevated temperatures, 600-1100°F (316-590°C); the best weldability; and, depending on type, the best corrosion resistance. However, in general, these alloys have the lowest room temperature strength and are not heat treatable.

There is only one commercial alloy in this group. This 5% Al 2 1/2% Sn alloy is available as sheet, bar, and wire and may be fusion welded with close to 100% joint efficiency. This alloy is used extensively in aircraft applications.

Alpha-Beta Weldable Alloys

The majority of the titanium alloys are of the alpha-beta type. As a group, the alpha-beta alloys have higher strength and respond to heat treatment but are less formable than the commercially pure titanium. All of the alpha-beta alloys are available in bars and billets, and most of the alloys also are offered in sheet form. This class of alloy accounts for more than half of all titanium metal products. Refer to Chapter 14 for information on heat treating titanium.

The alpha-beta alloys vary widely in their composition and in their characteristics such as strength, heat treatability, and ductility. The titanium alloy containing 6% Al and 4% V was developed as a forging alloy and is now available in all wrought mill shapes. By far the most commonly used alpha-beta alloy, it is readily weldable (with some sacrifice of joint strength) and is the basic alloy for jet engines and airframes. Through heat treatment, the tensile strength of the 6% Al 4% V alloy can be varied from 120,000 to 180,000 psi (827 to 1241 MPa).

Other alpha-beta weldable alloys include Ti-8Al-1Mo-1V, Ti-6Al-2Sn-4Zr-2Mo, Ti-6Al-2Cb-1Ta-1Mo, Ti-3Al-2.5V, and the new beta-lean alpha alloys.

Alpha-Beta Nonweldable Alloys

The remainder of the alpha-beta and the alpha-lean beta alloys are nonweldable by fusion welding, although flash or resistance welding techniques may be used successfully in some cases.

Beta Alloys

The titanium beta alloy group formulations contain large amounts of alloying metals. As a result, the beta alloys typically are denser than other titanium alloys. The titanium beta class of alloys is fully heat treatable and cold workable in the solution-treated condition. Welding feasibility has been established for the newer beta alloys, and technical information is available from the suppliers.

APPLICATIONS

The unique properties of titanium make it the most cost-effective and satisfactory engineering material for many applications. Titanium product uses range from heat exchangers to geothermal power equipment, and from liquefied natural gas condensors to piping and instrumentation for oil exploration. Additional uses include steam turbine blades and nuclear waste containers.

Titanium oxide, TiO_2, a white compound, is used in the production of paint pigment, paper, plastics, glass, and ceramics. Titanium is also added to other metals, such as steel, aluminum, and copper, to attain certain desired properties. About 94% of the titanium raw material produced or imported into the United States is processed into titanium oxide for use as pigment and filler. Only about 6% is used in the production of titanium metal and alloys.

TITANIUM

Titanium and its alloys are widely used in aircraft airframes and engines for their strength and fatigue resistance. Since both light weight and medium temperature strength retention are important factors in alloy selection for aerospace, these qualities, plus corrosion resistance, have led to increasing usage of titanium in the aerospace field. Compared with other structural metals, the superior resistance of titanium to the chloride ion accounts for much of the use of titanium in corrosion-resistant applications. This quality, combined with titanium's electrochemical characteristics, has led to many applications in electrochemical devices.

On industrial machines, titanium finds application where high strength is needed and mass effects must be reduced. Examples include high-speed rolls, frames for flying shears, pump shafting, and quick-acting latches. Increasing applications are being made of a recently developed nickel-titanium alloy (Nitinol), which has the unusual property of regaining its previous shape when heated.

FABRICATION

The use of various metalworking processes for titanium fabrication into end products is handled much like the processing of other high-performance engineering materials, with due regard for titanium's distinctive properties. There are several important differences between titanium and steel or nickel-based alloys. Awareness of these differences enables titanium to be fabricated by techniques similar to those used for stainless steel and nickel-based alloys. Titanium's distinctive characteristics include:

- Lower modulus of elasticity.
- Higher melting point.
- Lower ductility.
- A tendency to gall.
- Sensitivity to welding contamination.

The fabrication of titanium demands close attention to cleanliness. It is not uncommon for metalworking shops that handle several metals to isolate an area solely for titanium. Welding, in particular, requires complete freedom from contaminants that could degrade the properties of titanium. Thus, the work area set aside for titanium should be free of air drafts, moisture, dust, grease, and other contaminants that might affect the weld metal.

Machining

Machining techniques for titanium are similar to those for other high-performance metals such as the austenitic stainless steels. Careful consideration should be given to the following problems and characteristics:

- The unusual chip-forming tendency and low thermal conductivity of titanium tends to cause a buildup of heat on the edge and face of cutting tools.
- The reactivity of titanium with cutting tools contributes to seizing, galling, abrasion, and pickup on cutting edges and faces.
- Titanium's low elastic modulus permits greater deflections of workpieces and, therefore, may require proper backup.

Guidelines. Machining conditions can be selected to minimize or circumvent the adverse effects of these characteristics, thereby allowing good tool life at acceptable production rates. The following are guidelines for machining titanium:

1. Use low cutting speeds. Tool tip temperature is strongly affected by cutting speed, and a low cutting speed will help to minimize tool edge temperature and maximize tool life. Lower speeds are required for titanium alloys than for pure titanium.
2. Maintain high feed rates. Tool temperature is affected less by feed rate than by speed, and the highest rate of feed consistent with good practice should be used. The depth of cut should be greater than the work-hardened layer resulting from the previous cut.
3. Use a generous quantity of cutting fluid. The coolant carries away heat in addition to washing away chips and reducing cutting forces.
4. Maintain sharp tools. Tool wear results in buildup of metal on cutting edges and causes poor surface finish, tearing, and deflection of the workpiece.
5. Do not stop feeding while tool and work are in moving contact. Permitting a tool to dwell in moving contact with titanium can cause work hardening and promote smearing, galling, and seizing. This could lead to total tool breakdown.
6. Use rigid setups. Rigidity of machine tool and workpiece ensures a controlled depth of cut.

Cutting tools. Cutting tools for titanium require abrasion resistance and adequate hot hardness, and general-purpose high-speed tool steels (such as Grades M1, M2, M7, and M10) often are suitable. Best results can generally be obtained with more highly alloyed grades (such as Carbide Grades C-2 and C-3).

Fluids. Large quantities of chemically active cutting fluid are needed to cool the titanium workpiece and the cutting tool during high-speed machining operations. Water-based fluids are more efficient than oils. A weak solution of rust inhibitor and/or water soluble oil (5-10%) is the most practical fluid for high-speed cutting operations. Chlorinated or sulfurized oils normally are avoided, but may be required for slow speed and complex operations to minimize friction and reduce the galling and seizing tendency of titanium. Carefully controlled post-machine cleaning operations must be followed if chlorinated cutting fluids are used on alloys that may be subject to hot halide stress-corrosion cracking, when they are subsequently heated above 500° F (260° C) under stress.

Forming

The conversion of titanium ingots into mill products such as forging billet, plate, sheet, and tubing is usually done on conventional metalworking equipment. Mills designed to roll and shape stainless or alloy steel are used with only slight modifications. Consequently, titanium and its structural alloys are available in most of the same forms and shapes as stainless steel.

Heating. Fabricating titanium mill products into finished parts is performed on conventional metalworking machinery. During titanium heating operations, it is necessary to avoid the contaminating and embrittling effects of hydrogen, nitrogen, and oxygen. It is critically important to maintain close control of furnace temperatures and atmospheres for heating titanium prior to forging and forming, or for heat treating. During welding operations, the hot and molten titanium metal must be shielded from the atmosphere to avoid brittle welds. Argon or

CHAPTER 5

TITANIUM

TABLE 5-7
Typical Mechanical Properties of Titanium and Alloys

Material: Commercial Designation	and Condition	Ultimate Tensile Strength: Temp.	ksi (MPa)	°F (°C)	ksi (MPa)	Yield Strength, 0.2%: Temp.	ksi (MPa)	°F (°C)	ksi (MPa)	Elongation, 2″ (51 mm): Temp.	%	°F (°C)	%	Hardness
Commercially Pure														
Ti-35A (ASTM Gr. 1)	Annealed	RT	48 (331)	600 (316)	22 (152)	RT	35 (241)	600 (316)	14 (97)	RT	30	600 (316)	32	120 Bhn
Ti-50A (ASTM Gr. 2)	Annealed	RT	63 (434)	600 (316)	28 (193)	RT	50 (345)	600 (316)	17 (117)	RT	28	600 (316)	35	200 Bhn
Ti-65A (ASTM Gr. 3)	Annealed	RT	75 (517)	600 (316)	34 (234)	RT	65 (448)	600 (316)	20 (138)	RT	25	600 (316)	34	225 Bhn
Ti-75A (ASTM Gr. 4)	Annealed	RT	96 (662)	600 (316)	45 (310)	RT	85 (586)	600 (316)	25 (172)	RT	20	600 (316)	25	265 Bhn
Corrosion Alloys														
Ti-0.2Pd (ASTM Gr. 7)	Annealed	RT	63 (434)	600 (316)	27 (186)	RT	50 (345)	600 (316)	16 (110)	RT	28	600 (316)	37	200 Bhn
Ti-Code-12 (ASTM Gr. 12)	Annealed	RT	80 (551)	600 (316)	50 (345)	RT	60 (413)	600 (316)	35 (241)	RT	24	600 (316)	30	R_C27
Alpha Alloys														
Ti-5Al-2.5Sn	Annealed	RT	125 (862)	600 (316)	82 (565)	RT	117 (807)	600 (316)	65 (448)	RT	16	600 (316)	18	R_C36
Ti-5Al-2.5Sn ELI	Annealed	RT	117 (807)	-320 (-196)	180 (1241)	RT	108 (745)	-320 (-196)	168 (1158)	RT	16	-320 (-196)	16	R_C35
Alpha-Lean Beta Alloys														
Ti-8Al-1Mo-1V	Duplex annealed	RT	145 (1000)	600 (316)	115 (793)	RT	138 (952)	600 (316)	90 (621)	RT	15	600 (316)	20	R_C35
				1000 (538)	90 (621)			1000 (538)	75 (517)			1000 (538)	25	
Ti-679		RT	160 (1103)	600 (316)	130 (896)	RT	144 (993)	600 (316)	110 (758)	RT	15	600 (316)	20	R_C36
				1000 (538)	110 (758)			1000 (538)	85 (586)			1000 (538)	24	
Ti-6Al-2Sn-4Zr-2Mo		RT	142 (979)	600 (316)	112 (772)	RT	130 (896)	600 (316)	85 (586)	RT	15	600 (316)	16	R_C32
				1000 (538)	194 (1338)			1000 (538)	71 (490)			1000 (538)	26	
Ti-5Al-6Sn-2Zr-1Mo-Si	Annealed	RT	152 (1048)	600 (316)	115 (793)	RT	140 (965)	600 (316)	82 (565)	RT	13	600 (316)	15	R_C36
				1000 (538)	100 (690)			1000 (538)	73 (503)			1000 (538)	19	
Ti-6Al-2Cb-1Ta-1Mo	As rolled	RT	132 (910)	600 (316)	85 (586)	RT	189 (1303)	600 (316)	67 (462)	RT	13	600 (316)	15	R_C30
				1000 (538)	70 (483)			1000 (538)	55 (379)			1000 (538)	15	
Alpha-Beta Alloys														
Ti-8Mn	Annealed	RT	137 (945)	600 (316)	104 (717)	RT	125 (862)	600 (316)	82 (565)	RT	15	600 (316)	18	R_C32
Ti-3Al-2.5V	Annealed	RT	100 (690)	600 (316)	70 (483)	RT	85 (586)	600 (316)	50 (345)	RT		600 (316)	25	R_C28

TABLE 5-7—*Continued*

Ti-6Al-4V	Annealed	RT	144 (993)	600 (316)	105 (724)	RT	134 (924)	600 (316)	95 (655)	RT	14	600 (316)	14	R_C36
				1000 (538)	77 (531)			1000 (538)	62 (427)			1000 (538)	35	
	Solution + Age	RT	170 (1172)	600 (316)	125 (862)	RT	160 (1103)	600 (316)	102 (703)	RT	10	600 (316)	10	R_C41
				1000 (538)	95 (655)			1000 (538)	70 (483)			1000 (538)	22	
Ti-6Al-4V ELI	Annealed	RT	130 (896)	600 (316)	98 (676)	RT	120 (827)	600 (316)	87 (600)	RT	15	600 (316)	15	R_C35
Ti-6Al-6V-2Sn	Annealed	RT	155 (1069)	600 (316)	135 (931)	RT	145 (1000)	600 (316)	117 (807)	RT	14	600 (316)	18	R_C38
	Solution + Age	RT	185 (1276)	600 (316)	142 (979)	RT	170 (1172)	600 (316)	130 (896)	RT	10	600 (316)	12	R_C42
Ti-7Al-4Mo	Solution + Age	RT	160 (1103)	600 (316)	127 (876)	RT	150 (1034)	600 (316)	108 (745)	RT	16	600 (316)	18	R_C38
				800 (427)	123 (848)			800 (427)	104 (717)			800 (427)	20	
Ti-6Al-2Sn-4Zr-6Mo	Solution + Age	RT	184 (1269)	600 (316)	148 (1020)	RT	170 (1172)	600 (316)	122 (841)	RT	10	600 (316)	18	R_C42
				1000 (538)	123 (848)			1000 (538)	95 (655)			1000 (538)	19	
Beta-Lean Alpha Alloys														
Ti-10V-2Fe-3Al	Duplex annealed 1400° F (760° C)	RT	186 (1282)			RT	176 (1214)			RT	7			R_C42
	910° F (488° C)			600 (316)	160 (1103)			600 (316)	142 (979)			600 (316)	13	
Ti-17	Duplex annealed 1550° F (843° C)	RT	176 (1214)			RT	163 (1124)			RT	8			R_C40
	1475° F (802° C)			600 (316)	151 (1041)			600 (316)	125 (862)			600 (316)	12	
Beta Alloys														
Ti-15-3	Duplex annealed 1450° (788° C)	RT	114 (786)			RT	110 (758)			RT	12			R_C32
	1000° F (538° C)		169 (1165)				156 (1076)							R_C39
	Duplex annealed 1450° F (788° C)	RT	182 (1255)			RT	167 (1151)			RT	10			
	900° F (482° C)			600 (316)	156 (1075)			600 (316)	135 (930)			600 (316)	9	R_C41
Ti-13V-11Cr-3Al	Solution + Age	RT	185 (1276)	800 (427)	160 (1103)	RT	175 (1207)	800 (427)	120 (827)	RT	8	600 (316)	12	R_C40
Ti-3Al-8V-6Cr-4Mo-4Zr	Solution + Age	RT	177 (1220)	400 (204)	165 (1138)	RT	167 (1151)	400 (204)	148 (1020)	RT	10	400 (204)	13	R_C42
				700 (371)	159 (1096)			700 (371)	139 (958)			700 (371)	12	
Ti-11.5Mo-6Zr-4.5Sn	Solution + Age	RT	172 (1186)			RT	163 (1164)			RT	12			R_C40

helium gas-shielding techniques are usually used. Titanium and its alloys cannot be welded in air. They cannot be welded to steel, nickel-based alloys, or aluminum. Titanium alloys, however, can be welded to the other reactive metals and to some of the refractory metals for which suitable technical information is available.

Formability. The lowest strength unalloyed grades of titanium have moderately good formability; however, the high-strength unalloyed grades and the alloys require bend radii up to five times the metal thickness. Forming operations must often be done in the temperature range of 400 to 1200° F (204 to 650° C). The beta alloys are an exception; they combine formability and high strength. The low-strength grades of pure titanium have annealed hardnesses in the range of soft steels. Titanium alloys, however, typically have a hardness greater than R_C30 and are difficult to machine. Turning operations are performed readily, but milling and drilling are more difficult.

Powder Metallurgy

The relatively high cost of titanium metal has generally restricted its use to products that require exceptional corrosion resistance or a high strength-to-weight ratio. Titanium's chemical reactivity necessitates specialized equipment and technology to produce conventional castings or wrought shapes. In addition, numerous processing steps usually are required.

The powder metallurgy (PM) process has been instrumental in expanding the applications for titanium. The PM process is attractive from a raw material cost viewpoint because the basic ore refining process produces a sponge product from the magnesium or sodium reduction of $TiCl_4$. Powder metallurgy processing temperatures are a fraction of the melting point of the metal, thus reducing the reactivity problem. A further PM advantage is that substantially fewer processing steps may be needed.

The PM process offers the generic attributes of powder metallurgy products, including fine grain size and homogeneous composition. A further significant advantage of the PM titanium process is its ability to produce near-net-shape components. This advantage may result in lower initial material cost than for ingot metallurgy material and also reduced machining cost.

References

1. Refractory Metals Association, Metal Powder Industries Federation, *Refractory Metals: Five Metals with Unique Characteristics* (Princeton, NJ).
2. American Society for Testing and Materials, *Annual Book of ASTM Standards*, Part 8 (Philadelphia: 1982).

Bibliography

American Society of Metals. *ASM Metals Reference Book*. Metals Park, OH: 1981.

American Society of Metals. *Metals Handbook*, Vol. 2, 9th ed. Metals Park, OH: 1979.

Bonini, J. J., and Sankaran, K. K. "Evaluation of Blended Elemental Powders for Titanium Missile-Structural Applications." National Powder Metallurgy Conference, Metal Powder Industries Federation. Montreal, Canada: May 1982.

Budinski, Kenneth. *Engineering Materials Properties and Selection*, 2nd ed. Reston, VA: Reston Publishing Co., 1983.

Facts About Machining Titanium. Niles, OH: RMI Company, 1978.

Froes, F. H.; Eylon, D.; Eichelman, G. E.; and Burte, H. M. "Developments in Titanium Powder Metallurgy." *Journal of Metals*. (February 1980), pp. 47-54.

__________; Eylon, D.; Wirth, G.; Grundhoff, K. J.; and Smarsly, W. "The Fatigue Properties of Hot Isostatically Pressed Ti-6Al-4V Powders." National Powder Metallurgy Conference, Metal Powder Industries Federation. Montreal, Canada: May 1982.

Kocinski, Edward J. "The Mechanical Properties of Titanium P/M Parts Produced from Superclean Powders." National Powder Metallurgy Conference, Metal Powder Industries Federation. Montreal, Canada: May 1982.

Poulsen, E. R., and Hall, J. A. "Extractive Metallurgy of Titanium: A Review of the State of the Art." *Journal of Metals*. (June 1983), pp. 60-65.

Properties of Some Metals and Alloys. New York: The International Nickel Company, 1981.

Society of Automotive Engineers. *SAE Handbook*, Part 1. Warrendale, PA: 1981.

Society of Automotive Engineers. *Unified Numbering System for Metals and Alloys*, 3rd ed. Warrendale, PA: 1983.

Wood, Nat. "Fabricating Breakthrough for Titanium." *Machine and Tool Blue Book*. (October 1978), pp. 98-107.

ALUMINUM, COPPER AND MAGNESIUM

CHAPTER CONTENTS:

ALUMINUM

Aluminum (Al) is the most abundant metallic element, estimated to form about 8% of the earth's crust. It is a light metal with a bright silvery luster, and the pure metal melts at 1220° F (660° C). With a specific gravity of 2.8, aluminum weighs only about 0.1 lb/in.^{3} (2.77 g/cm^3), as compared with 0.28 lb/in.^{3} (7.75 g/cm^3) for iron and 0.32 lb/in.^{3} (8.86 g/cm^3) for copper. Among the structural metals, only magnesium is lighter. The weight of a given volume of ferrous metals is almost three times that of an equal volume of aluminum.

Although pure aluminum is very ductile, it can be hardened by cold working and alloying. Some aluminum alloys can also be hardened by heat treatment. Pure aluminum is relatively soft and low in strength, but it can be alloyed with other elements to increase its strength and impart other useful properties. The most commonly used alloying elements are copper, magnesium, manganese, and zinc. Iron and silicon are commonly present as impurities.

Aluminum is an excellent conductor of both heat and electricity. Of all the metals, only sodium conducts heat and electricity better on a pound-for-pound basis; and, on a volume basis, only gold, silver, and copper are better conductors. At room temperature, the electrical conductivity of pure aluminum is over 64% of the International Annealed Copper Standard (IACS). Aluminum is also highly reflective of both light and heat.

A unique combination of properties makes aluminum a versatile engineering and construction material. It is light in weight, yet some of its alloys have strengths greater than that of structural steel. It has high resistance to corrosion under the majority of service conditions, and no colored salts are formed to stain adjacent surfaces or discolor products that it comes into contact with, such as fabrics in textile operations and solutions in chemical equipment. It has no toxic reaction. The metal can easily be worked into many forms and readily accepts a wide variety of surface finishes.

BASIC METAL PRODUCTION

Aluminum is a constituent of all common rocks except limestone and sandstone. Aluminum has a strong affinity for oxygen, and therefore it is not found in the purely metallic state. It occurs naturally in chemical combination with silicon or oxygen and in ore deposits of aluminum combined with sulfur and phosphorus.

The production of primary aluminum involves the assembly of large quantities of several raw materials close to a dependable source of continuous electric power. The most widely used technology for producing aluminum involves two steps: extraction and purification of alumina from ores, and electrolysis of the oxide after it has been dissolved in fused cryolite. In basic terms, aluminum ore is chemically processed to yield its content of aluminum oxide (alumina) that is, in turn, subjected to powerful electrolytic action to produce metallic aluminum.

The commercial ore and chief raw material of aluminum is bauxite, which varies in color from creamy white to reddish brown and contains large percentages of aluminum oxide. Specifically, bauxite contains about 50-60% alumina (aluminum oxide, Al_2O_3). It takes about two pounds of bauxite to make a pound of alumina, and two pounds of alumina to make a pound of aluminum. The principal sources of bauxite in the United States are in Arkansas and nearby Southern states. Large deposits are also found in Jamaica, the USSR, Greece, Guinea, France, Surinam, Guyana, Brazil, Hungary, Yugoslavia, and Indonesia.

There are a number of methods for refining bauxite, but the most common is the Bayer process. In this process, crushed bauxite is mixed in large pressure tanks with a solution of hot, caustic soda made from soda ash and lime. The caustic soda dissolves the alumina content of the bauxite, while the insoluble impurities—such as iron oxide, titania, and silica—are removed by settling and filtering and are left behind as "red mud."

The filtrate containing the alumina in solution is pumped into towering precipitation tanks, which are the distinctive architectural feature of an alumina plant, and is slowly cooled, with aluminum trihydrate settling in the form of fine crystals. After separation from the mother liquor, these crystals

Contributors of sections of this chapter are: ***James W. Barr****, Director, Standards, The Aluminum Association, Inc.;* ***John T. Benedict****, Consultant;* ***Arthur Cohen****, Manager, Standards and Safety Engineering, Copper Development Association, Inc.;* ***V. Samuel Hill****, Project Leader, Technical Service—Magnesium, Dow Chemical Co.;* ***Daniel Lea****, Director of Communications, International Magnesium Association.*

Reviewers of sections of this chapter are: ***M. D. Ballain****, Staff Process Engineer, Northwest Alloys, Inc.;* ***James W. Barr****, Director, Standards, The Aluminum Association, Inc.;* ***Arthur Cohen****, Manager, Standards and Safety Engineering, Copper Development Association, Inc.;* ***Donald D. Destito****, Manager, Technical Services and Product Development, Revere Copper Products, Inc.;* ***L. E. Gibson****, Technical Manager, Northwest Alloys, Inc.;*

CHAPTER 6

ALUMINUM ALLOY AND TEMPER DESIGNATION

are fed into large, revolving kilns at a temperature of approximately 2000° F (1100° C); and the chemically combined water is driven off, leaving aluminum oxide or alumina, the material actually used in the smelting process.

In the second half of the reduction process, the oxygen in the alumina is separated by the use of electricity, and pure aluminum remains. In this electrolytic reduction, called the Hall-Heroult process, the alumina is dissolved in a molten electrolyte bath of cryolite at a temperature of about 1800° F (982° C). Cryolite is a chemical salt consisting of sodium, fluorine, and aluminum that is mined commercially only in Greenland, but is also produced synthetically in the United States. The molten electrolyte is contained in a carbon-lined cast iron shell that serves as the cathode. Carbon blocks suspended in the electrolyte serve as the anode. The current passing through the electrolyte separates the dissolved aluminum oxide into metallic aluminum, which is deposited on the bottom of the cell, and oxygen, which is deposited on the carbon anodes and gradually consumes them. The cryolite remains substantially unaltered, while alumina is periodically stirred into the bath and dissolved to maintain the continuous operation of the process.

ALLOY AND TEMPER DESIGNATION[1]

The aluminum alloy and temper designation systems are completely set forth in the American National Standard ANSI H35.1 and in various publications issued by The Aluminum Association. Two numbering formats are employed for designating the individual alloys: one for wrought alloys and another for casting alloys.

In format, the two numbering systems differ only slightly. Designations for wrought alloys consist of four-digit numbers, while casting alloys are designated by a three-digit number followed by a decimal point and a fourth digit. In both systems, the basic temper designation (consisting of capital letters and numerals) follows the alloy designation and is separated from it by a hyphen.

These systems provide a standard means for designating aluminum and its alloys in all product forms—wrought, cast, and ingot—and the tempers in which they are produced.

Wrought Aluminum and Alloys

In the four-digit system used to identify wrought aluminum and aluminum alloys, the first digit indicates the alloy group or principal alloying element. As shown in Table 6-1, the 1xxx series is for minimum aluminum purities of 99.00% and greater, and the 2xxx through 8xxx series group aluminum alloys by major alloying elements. The last two digits identify a specific alloy within a given series or indicate the aluminum purity. The second digit indicates modifications of the original alloy or impurity limits. For codification purposes, an alloying element is any element that is intentionally added for any purpose other than grain refinement and for which minimum and maximum limits are specified.

Standard limits for alloying elements and impurities listed in Table 6-2 are expressed in the following sequence: silicon, iron, copper, manganese, magnesium, chromium, nickel, zinc, titanium, other elements (individual), other elements (total), and aluminum. Additional specified elements having limits are inserted in alphabetical order of their chemical symbols between zinc and titanium, or are specified in footnotes. Aluminum is specified as minimum for unalloyed aluminum and as a remainder for aluminum alloys.

Aluminum. In the 1xxx group for aluminum purities of 99.00% and greater, the last two of the four digits in the designation indicate the minimum aluminum percentage. These digits are the same as the two numerals to the right of the decimal point in the minimum aluminum percentage when it is expressed to the nearest 0.01 percent. The second digit in the

TABLE 6-1
Designations for Wrought Aluminum Alloy Groups

Group	Alloy Number
Aluminum, 99.00% min and greater	1xxx
Aluminum alloys grouped by major alloying elements:	
Copper	2xxx
Manganese	3xxx
Silicon	4xxx
Magnesium	5xxx
Magnesium and silicon	6xxx
Zinc	7xxx
Other element	8xxx
Unused series	9xxx

TABLE 6-2
Standard Limits for Wrought Aluminum Alloying Elements and Impurities

Composition is stated in weight percent maximum unless shown as a range or a minimum.

Standard limits for alloying elements and impurities are expressed to the following places:

Less than 0.001%	0.000x
0.001 to 0.01%	0.00x
0.01 to 0.10%	
Unalloyed aluminum made by a refining process	0.0xx
Alloys and unalloyed aluminum not made by a refining process	0.0x
0.10 through 0.55%	0.xx
(It is customary to express limits of 0.30% through 0.55% as 0.x0 or 0.x5.)	
Over 0.55%	0.x,x.x, etc.
(except that combined Si + Fe limits for 1xxx designations must be expressed as 0.xx or 1.xx)	

*Reviewers, cont.: **David L. Hawke**, Materials Specialist, Magnesium Div., Amax, Inc.; **V. Samuel Hill**, Project Leader, Technical Service—Magnesium, Dow Chemical Co.; **John P. Laughlin**, Manager, Metallurgical and Process Development, Oregon Metallurgical Corp.; **Don K. Lewis**, Staff Engineer, Aluminum Smelting and Refining Co., Inc.; **Clarence S. Lorenz**, Manager, Customer Service Engineering, Alloy Div., Brush Wellman, Inc.; **J. P. Lyle**, Senior Technical Consultant, Metallurgical Research & Development, Alcoa Laboratories, Aluminum Company of America; **J. Howard Mendenhall**, Technical Associate, Marketing, Olin Corp., Brass Group;*

designation indicates modifications in impurity limits. If the second digit in the designation is zero, it indicates that there is no special control on individual impurities; integers 1 through 9 indicate special control of one or more individual impurities.

Aluminum alloys. In the 2xxx through 8xxx alloy groups, the last two of the four digits in the designation serve only to identify the different aluminum alloys in the group. The second digit in the alloy designation indicates alloy modifications. If the second digit in the designation is zero, it indicates the original alloy. Integers 1 through 9, which are assigned consecutively, indicate alloy modifications.

Experimental alloys. Experimental wrought alloys are also designated in accordance with this system, but they are indicated by the prefix X. The prefix is dropped when the alloy is no longer experimental. During the early development period, before they are designated as experimental, new alloys are identified by serial numbers assigned by their originators. Use of the serial number is discontinued when the X prefixed number is assigned.

Cast Aluminum and Alloys

A system of four-digit numerical designations is used to identify aluminum casting alloys. The first digit indicates the alloy group, as shown in Table 6-3. The 1xx.x series is for aluminum purities of 99.00% and greater, and the 2xx.x through 9xx.x series group aluminum alloys by major alloying elements. The second two digits identify the aluminum alloy or indicate the aluminum purity. The last digit, which is separated from the others by a decimal point, indicates the product form, i.e., castings or ingot. A modification of the original alloy or impurity limits is indicated by a serial letter before the numerical designation. The serial letters are assigned in alphabetical sequence starting with A but omitting I, O, Q, and X—the X being reserved for experimental alloys. For codification purposes, an alloying element is any element that is intentionally added for any purpose other than grain refinement and for which minimum and maximum limits are specified.

Standard limits for alloying elements and impurities listed in Table 6-4 are expressed in the following sequence: silicon, iron, copper, manganese, magnesium, chromium, nickel, zinc, titanium, other elements (individual), other elements (total), and aluminum. Additional specified elements having limits are inserted in alphabetical order (of their chemical symbols) between zinc and titanium, or are specified in footnotes. Aluminum is specified as minimum for unalloyed aluminum and as a remainder for aluminum alloys.

Aluminum castings and ingot. In the 1xx.x group for minimum aluminum purities of 99.00% and greater, the second two of the four digits in the designation indicate the minimum aluminum percentage. These digits are the same as the two numerals to the right of the decimal point in the minimum aluminum percentage when it is expressed to the nearest 0.01%. The last digit, which is to the right of the decimal point, indicates the product form: 1xx.0 indicates castings and 1xx.1 indicates ingots.

Product form. In the 2xx.x through 9xx.x alloy groups, the second two of the four digits in the designation have no special significance but serve only to identify the different aluminum alloys. Alloy modifications are indicated by a serial letter before the numerical designation. Limits for alloying elements and impurities for xxx.1 ingot are the same as for the alloy in the form of castings, except as shown in Tables 6-5 and 6-6. The serial letters are assigned in alphabetical sequence starting with A but omitting I, O, Q, and X—the X being reserved for experimental alloys.

Experimental alloys. Experimental cast alloys are also designated in accordance with this system. They are identified by the prefix X and their numbering is administered by a procedure similar to that described previously for wrought experimental alloys.

TABLE 6-3
Designations for Aluminum Casting Alloy Groups

Group	Alloy Number
Aluminum, 99.00% min and greater	1xx.x
Aluminum alloys grouped by major alloying elements:	
Copper	2xx.x
Silicon, with added copper and/or magnesium	3xx.x
Silicon	4xx.x
Magnesium	5xx.x
Zinc	7xx.x
Tin	8xx.x
Other element	9xx.x
Unused series	6xx.x

TABLE 6-4
Standard Limits for Cast Aluminum Alloying Elements and Impurities

Composition is stated in weight percent maximum unless shown as a range or a minimum.

Standard limits for alloying elements and impurities are expressed to the following places:

Less than 0.001%	0.000x
0.001 to 0.1%	0.00x
0.01 to 0.10%	
Unalloyed aluminum made by a refining process	0.0xx
Alloys and unalloyed aluminum not made by a refining process	0.0x
0.10 through 0.55%	0.xx
(It is customary to express limits of 0.30% through 0.55% as 0.x0 or 0.x5.)	
Over 0.55%	0.x,x.x, etc.*

* Magnesium percent for some alloys is an exception to this rule.

Reviewers, cont.: ***B. A. Riggs****, Senior Research Metallurgist, Kaiser Aluminum & Chemical Corp.;* ***John D. Rutherford****, Manager, Market Development, Aluminum Smelting and Refining Co., Inc.;* ***Dr. Brian Taylor****, Research Scientist, Advanced Product and Manufacturing Engineering Staff, General Motors Corp.;* ***Dr. David S. Thompson****, Director, Dept. of Metallurgy and Metallurgical Services, Reynolds Aluminum, Reynolds Metals Co.;* ***Ralph D. Venen****, Materials Manager, Brass Products Div., Parker Hannifin Corp.;* ***Joseph S. Viland****, Vice President, Sales and Marketing, Magnesium Div., Amax, Inc.;* ***A. H. Wheeler****, Chief Engineer, North American Magnesium, Inc.*

ALUMINUM ALLOY AND TEMPER DESIGNATION

TABLE 6-5
Aluminum Alloy Ingot

Maximum Iron Percentage	
For sand and permanent-mold castings:	For ingot:
Up through 0.15	0.03 less than castings
Over 0.15 through 0.25	0.05 less than castings
Over 0.25 through 0.6	0.10 less than castings
Over 0.6 through 1.0	0.2 less than castings
Over 1.0	0.3 less than castings
For die castings:	
Up through 1.3	0.3 less than castings
Over 1.3	1.1 max

TABLE 6-6
Aluminum Alloy Ingot

Minimum Magnesium Percentage	
For all castings:	For ingot:
Less than 0.50	0.05 more than castings*
0.50 and greater	0.1 more than castings*
Maximum Zinc Percentage	
For die castings:	
Over 0.25 through 0.6	0.10 less than castings
Over 0.6	0.1 less than castings

* Applicable only when the specified magnesium range for castings is greater than 0.15%.

Temper Designation

The temper designation system is used for all forms of wrought and cast aluminum and aluminum alloys except ingot. Significant information about the characteristics and properties of an aluminum alloy is given by the alphanumeric coding that is based on the sequences of basic treatments used to produce the various tempers. The temper designation follows the alloy designation, the two being separated by a hyphen. Basic temper designations consist of letters. Subdivisions of the basic tempers, where required, are indicated by one or more digits following the letter. These designate specific sequences of basic treatments, but only operations recognized as significantly influencing the characteristics of the product are indicated. Should some other variation of the same sequence of basic operations be applied to the same alloy, resulting in different characteristics, then additional digits are added to the designation. The principal temper designations for aluminum alloys are as follows:

F: As fabricated. Applies to the products of shaping processes in which no special control over thermal conditions or strain hardening is employed. For wrought products, there are no mechanical property limits.

O: Annealed. Applies to wrought products that are annealed to obtain the lowest strength temper, and to cast products that are annealed to improve ductility and dimensional stability. The O may be followed by a digit other than zero.

H: Strain hardened (wrought products only). Applies to products that have their strength increased by strain hardening, with or without supplementary thermal treatments to produce some reduction in strength. The H is always followed by two or more digits.

W: Solution heat treated. An unstable temper applicable only to alloys that spontaneously age at room temperature after solution heat treatment. This designation is specific only when the period of natural aging is indicated—for example, W 1/2 hr.

T: Thermally treated to produce stable tempers other than F, O, or H. Applies to products that are thermally treated, with or without supplementary strain hardening, to produce stable tempers. The T is always followed by one or more digits.

Subdivisions of basic tempers (H, strain hardened). The first digit following the H indicates the specific combination of basic operations, as follows:

H1: Strain hardened only. Applies to products which are strain hardened to obtain the desired strength without supplementary thermal treatment. The number following this designation indicates the degree of strain hardening.

H2: Strain hardened and partially annealed. Applies to products that are strain hardened more than the desired final amount and then reduced in strength to the desired level by partial annealing. For alloys that age-soften at room temperature, the H2 tempers have the same minimum ultimate tensile strength as the corresponding H3 tempers. For other alloys, the H2 tempers have the same minimum ultimate tensile strength as the corresponding H1 tempers and slightly higher elongation. The number following this designation indicates the degree of strain hardening remaining after the product has been partially annealed.

H3: Strain hardened and stabilized. Applies to products that are strain hardened with mechanical properties that are stabilized either by a low-temperature thermal treatment or as a result of heat introduced during fabrication. Stabilization usually improves ductility. This designation is applicable only to those alloys that, unless stabilized, gradually age-soften at room temperature. The number following this designation indicates the degree of strain hardening after the stabilization treatment. The digit following the designations H1, H2, and H3 indicates the degree of strain hardening.

Numeral 8 has been assigned to indicate tempers having an ultimate tensile strength equivalent to that achieved by a cold reduction [temperature during reduction not to exceed 120° F (49° C)] of approximately 75% following a full anneal. Tempers between 0 (annealed) and 8 are designated by numerals 1 through 7. Material having an ultimate tensile strength about midway between that of the 0 temper and that of the 8 temper is designated by the numeral 4, about midway between 0 and 4 tempers by the numeral 2, and about midway between the 4 and 8 tempers by the numeral 6. Numeral 9 designates tempers with minimum ultimate tensile strength exceeding that of the 8 temper by 2.0 ksi (13.8 MPa) or more. For two-digit H tempers with a second digit that is odd, the standard limits for ultimate tensile strength are exactly midway between those of the adjacent two-digit H tempers with second digits that are even.

For alloys that cannot be cold reduced an amount sufficient to establish an ultimate tensile strength applicable to the temper (75% cold reduction after full anneal), the 6 temper tensile strength may be established by a cold reduction of approximately 55% following a full anneal, or the 4 temper tensile strength may be established by a cold reduction of approximately 35% after a full anneal.

The third digit, when used, indicates a variation of a two-digit temper. It is used when the degree of control of temper or the mechanical properties are different from, but close to, those for the two-digit H temper designation to which it is added, or when some other characteristic is significantly affected.

Subdivisions of T temper: thermally treated. Numerals 1 through 10 following the T indicate specific sequences of basic treatments, as follows:

- T1 Cooled from an elevated temperature shaping process and naturally aged to a substantially stable condition. Applies to products that are not cold worked after cooling from an elevated temperature shaping process, or in which the effect of cold working during flattening or straightening operations may not be recognized in mechanical property limits.
- T2 Cooled from an elevated temperature shaping process, cold worked, and naturally aged to a substantially stable condition. Applies to products that are cold worked to improve strength after cooling from an elevated temperature shaping process, or in which the effect of cold working in flattening or straightening is recognized in mechanical property limits.
- T3 Solution heat treated, cold worked, and naturally aged to a substantially stable condition. Applies to products that are cold worked to improve strength after solution heat treatment, or in which the effect of cold working in flattening or straightening is recognized in mechanical property limits.
- T4 Solution heat treated and naturally aged to a substantially stable condition. Applies to products that are not cold worked after solution heat treatment, or in which the effect of cold work in flattening or straightening may not be recognized in mechanical property limits.
- T5 Cooled from an elevated temperature shaping process and then artificially aged. Applies to products that are not cold worked after cooling from an elevated temperature shaping process, or in which the effect of cold work in flattening or straightening may not be recognized in mechanical property limits.
- T6 Solution heat treated and then artificially aged. Applies to products that are not cold worked after solution heat treatment, or in which the effect of cold working in flattening or straightening may not be recognized in mechanical property limits.
- T7 Solution heat treated and overaged/stabilized. Applies to wrought products that are artificially aged after solution heat treatment to carry them beyond a point of maximum strength in order to provide control of some significant characteristic. Applies to cast products that are artificially aged after solution heat treatment to provide dimensional and strength stability.
- T8 Solution heat treated, cold worked, and then artificially aged. Applies to products that are cold worked to improve strength, or in which the effect of cold working in flattening or straightening is recognized in mechanical property limits.
- T9 Solution heat treated, artificially aged, and then cold worked. Applies to products that are cold worked to improve strength.
- T10 Cooled from an elevated temperature shaping process, cold worked, and then artificially aged. Applies to products that are cold worked to improve strength, or in which the effect of cold work in flattening or straightening is recognized in mechanical property limits.

Additional digits, the first of which cannot be zero, may be added to designations T1 through T10 to indicate a variation in treatment that significantly alters the characteristics of the product. (See ANSI H35.1-1982 for additional three-digit H tempers and T tempers.)

A period of natural aging at room temperature may occur between or after the operations listed for tempers T3 through T10. Control of this period is exercised when it is metallurgically important.

PROPERTIES

For aluminum, as for other metals, effective product engineering and manufacturing require an understanding of the material's physical and mechanical properties, its corrosion resistance, and its fabrication, joining, and finishing characteristics.

Typical tensile strengths of aluminum range from as low as 5 ksi (34.5 MPa) for some commercially pure aluminum grades to 90 ksi (620 MPa) or more for high-strength alloys. In the automotive field, commonly used alloys have yield strengths from 18 to 60 ksi (124 to 414 MPa). Young's modulus, an important factor in a material's stiffness and bending strength, is 10×10^6 psi (68.9 GPa) for aluminum.

Aluminum, like steel and the various nonferrous metals, is not a single metal but a family of alloys, each of which has been formulated and developed to provide properties suitable for separate classes or fields of applications.

Effect of Alloying Elements

1000 Series. The aluminum is of 99% or higher purity. This type has many applications, especially in the electrical and chemical fields. These alloys are characterized by excellent corrosion resistance, high thermal and electrical conductivity, excellent workability, but low strength and mechanical properties. Moderate increases in strength may be obtained by strain hardening. Iron and silicon are the major impurities.

2000 Series. Copper is the principal alloying element in this group. These alloys require solution heat treatment to obtain optimum properties. In the heat-treated condition, mechanical properties are similar to, and sometimes exceed, those of mild steel. In some instances, artificial aging is employed to further increase the mechanical properties. This treatment materially increases yield strength, with attendant loss in elongation; its effect on ultimate tensile strength is not as great. The alloys in the 2000 series do not have corrosion resistance as good as most other aluminum alloys, and under certain conditions they may be subject to intergranular corrosion. Therefore, these alloys in the form of sheet are usually clad with a high-purity alloy or a magnesium-silicon alloy of the 6000 series that provides galvanic protection to the core material and thus greatly increases resistance to corrosion. Alloy 2024 is perhaps the best known and most widely used aircraft alloy.

3000 Series. Manganese is the major alloying element of alloys in this group, which are nonheat-treatable. Because only about 1.5% manganese can be effectively added to aluminum, it is used as a major element in only a few instances. One of these, however, is the popular 3003, which is widely used as a general-purpose alloy for moderate-strength applications requiring good workability. Alloy 3004 is widely used for aluminum can body stock, formed by drawing and ironing operations.

4000 Series. The major alloying element of this group is silicon, which can be added in sufficient quantities to cause substantial lowering of the melting point without producing brittleness in the resulting alloys. For this reason, aluminum-silicon alloys are used in welding wire and as brazing alloys where a melting point lower than that of the parent metal is required. Most alloys in this series are nonheat-treatable; but when used in welding heat-treatable alloys, they pick up some of the alloying constituents of the latter and respond to heat treatment to a limited extent. The alloys containing appreciable amounts of silicon become dark gray when anodic oxide finishes are applied and hence are in demand for architectural applications.

5000 Series. Magnesium is one of the most effective and widely used alloying elements for aluminum. When it is used as the major alloying element or with manganese, the result is a moderate-to-high-strength, nonheat-treatable alloy. Magnesium is considerably more effective than manganese as a hardener, about 0.8% magnesium being equal to 1.25% manganese, and it can be added in considerably higher quantities. Alloys in this series possess good welding characteristics and good resistance to corrosion in a marine atmosphere. However, because of possible stress corrosion susceptibility, the higher magnesium content alloys (more than 3% Mg) should not be used in the as-rolled (-H1X) tempers when service temperatures above approximately 150° F (66° C) are expected. For such applications, special -H3X tempers are necessary to avoid susceptibility to stress corrosion.

6000 Series. Alloys in this group contain silicon and magnesium in appropriate proportions to form magnesium silicide, thus making them heat treatable. The major alloy in this series is 6061, one of the most versatile of the heat-treatable alloys. Although less strong than most of the 2000 or 7000 alloys, the magnesium-silicon (or magnesium-silicide) alloys possess good formability and corrosion resistance, with medium strength. Alloys in this heat-treatable group may be formed in the T4 temper (solution heat treated but not artificially aged) and then reach full T6 properties by artificial aging.

7000 Series. Zinc is the major alloying element in this group. When coupled with a smaller percentage of magnesium, the result is a heat-treatable alloy of very high strength. Usually other elements such as copper and chromium are also added in small quantities. An outstanding member of this group is 7075, which is among the highest strength alloys available and is used in airframe structures and for highly stressed parts.

Cast Alloys. The cast alloys are essentially of two types: (1) alloys wherein the desired mechanical properties are secured by virtue of alloy additions alone and (2) alloys that are subsequently heat treated to improve the properties. Chapter 14, "Heat Treatment of Other Metals," in this volume gives the typical conditions for heat treatment of most of the cast aluminum alloys.

The alloys containing silicon are not only characterized by excellent casting qualities but also exhibit good weldability and freedom from hot-shortness. They are somewhat difficult to machine and, from the standpoint of tool wear, are similar to cast iron.

Many of the sand-casting alloys can be successfully used in permanent-mold casting, although precise alloy combinations have been developed for this and for die-casting purposes. Typical mechanical properties for sand and permanent-mold-casting aluminum alloys are given in Table 6-7.

TABLE 6-7
Typical Mechanical Properties of Sand and Permanent-Mold Casting Aluminum Alloys*

AA Number	Temper	Product**	Ultimate Tensile Strength, ksi (MPa)	Elongation in 2″ (51 mm), %
208.0	F	S	19.0 (131)	1.5
208.0	T55	S	21.0 (145)	
213.0	F	S	19.0 (131)	
213.0	F	P	23.0 (159)	
222.0	T2	S	23.0 (159)	
222.0	T61	S	30.0 (207)	
222.0	T551	P	30.0 (207)	
222.0	T65	P	40.0 (276)	
242.0	T21	S	23.0 (159)	
242.0	T571	S	29.0 (200)	
242.0	T571	P	34.0 (234)	
242.0	T61	P	40.0 (276)	
295.0	T4	S	29.0 (200)	6.0
295.0	T6	S	32.0 (221)	3.0
295.0	T62	S	36.0 (248)	
295.0	T7	S	29.0 (200)	3.0
296.0	T4	P	33.0 (228)	4.5
296.0	T6	P	35.0 (241)	2.0
296.0	T7	P	33.0 (228)	3.0
308.0	F	P	24.0 (165)	
319.0	F	S	23.0 (159)	
319.0	T5	S	25.0 (172)	

TABLE 6-7—*Continued*

AA Number	Temper	Product**	Ultimate Tensile Strength, ksi (MPa)	Elongation in 2″ (51 mm), %
319.0	T6	S	31.0 (214)	1.5
319.0	F	P	28.0 (193)	1.5
319.0	T6	P	34.0 (234)	2.0
328.0	F	S	25.0 (172)	1.0
328.0	T6	S	34.0 (234)	1.0
336.0	T551	P	31.0 (214)	
336.0	T65	P	40.0 (276)	
332.0	T5	P	31.0 (214)	
333.0	F	P	28.0 (193)	
333.0	T5	P	30.0 (207)	
333.0	T6	P	35.0 (241)	
333.0	T7	P	31.0 (214)	
355.0	T51	S	25.0 (172)	
355.0	T6	S	32.0 (221)	2.0
355.0	T7	S	35.0 (241)	
355.0	T71	S	30.0 (207)	
355.0	T51	P	27.0 (186)	
355.0	T6	P	37.0 (255)	1.5
355.0	T62	P	42.0 (290)	
355.0	T71	P	34.0 (234)	
C355.0	T61	P	40.0 (276)	3.0
356.0	T51	S	23.0 (159)	
356.0	T6	S	30.0 (207)	3.0
356.0	T7	S	31.0 (214)	
356.0	T51	P	25.0 (172)	
356.0	T6	P	33.0 (228)	3.0
356.0	T7	P	29.0 (200)	4.0
A356.0	T61	P	37.0 (255)	5.0
357.0	T6	P	45.0 (310)	3.0
A390.1	F	S	26.0 (179)	
	T5	S	26.0 (179)	
	T6	S	40.0 (276)	
	T7	S	36.0 (248)	
	F	P	29.0 (200)	
	T5	P	29.0 (200)	
	T6	P	45.0 (310)	
	T7	P	38.0 (262)	
B443.0	F	S	17.0 (117)	3.0
B443.0	F	P	21.0 (145)	5.0
514.0	F	S	22.0 (152)	6.0
513.0	F	P	22.0 (152)	2.5
512.0	F	S	17.0 (117)	
520.0	T4	S	42.0 (290)	12.0
535.0	F	S	35.0 (241)	9.0
535.0	T2	S	35.0 (241)	9.0
705.0	F or T5	S	30.0 (207)	5.0
705.0	T5	P	37.0 (255)	10.0
707.0	F or T5	S	33.0 (228)	2.0
707.0	T5	P	42.0 (290)	4.0
707.0	T7	P	45.0 (310)	3.0
710.0	T5	S	32.0 (221)	2.0
712.0	F or T5	S	34.0 (234)	4.0
713.0	F or T5	S	32.0 (221)	3.0
713.0	T5	P	32.0 (221)	4.0
771.0	T6	S	45.0 (310)	5.0
850.0	T5	S	16.0 (110)	5.0

(continued)

TABLE 6-7—*Continued*

AA Number	Temper	Product**	Ultimate Tensile Strength, ksi (MPa)	Elongation in 2″ (51 mm), %
850.0	T5	P	18.0 (124)	8.0
851.0	T5	S	17.0 (117)	3.0
851.0	T5	P	17.0 (117)	3.0
852.0	T5	S	24.0 (165)	
852.0	T5	P	27.0 (186)	3.0

(*The Aluminum Association, Inc.*)

* Values represent properties obtained from separately cast test bars. The customer should keep in mind that (1) some foundries may offer additional tempers for the above alloys; and (2) foundries are constantly improving casting technique, and, as a result, some may guarantee minimum properties in excess of the above.

** S = sand-cast.
P = permanent-mold cast.

Characteristics of Aluminum

Commonly used commercially pure aluminum has a tensile strength of about 13 ksi (90 MPa). Its usefulness as a structural material in this form is thus somewhat limited. By working the metal, as by cold rolling, its strength can be approximately doubled. Much larger increases in strength can be obtained by alloying aluminum with small percentages of one or more other metals such as manganese, silicon, copper, magnesium, or zinc. Like pure aluminum, the alloys are also made stronger by cold working. Some of the alloys are further strengthened and hardened by heat treatments so that aluminum alloys having tensile strengths approaching 100 ksi (690 MPa) are available. Typical mechanical properties of commonly used wrought aluminum alloys are given in Table 6-8.

TABLE 6-8
Typical Mechanical Properties of Wrought Aluminum Alloys*

Alloy and Temper	Tension: Strength, ksi (MPa)		Tension: Elongation, Percent in 2 in. (51 mm)		Hardness	Product Forms**
	Ultimate	Yield	1/16 in. (1.6 mm) Thick Specimen	1/2 in. (12.7 mm) Diameter Specimen	Brinell Number, 500 kg Load 10 mm Ball	
1060-0	10 (69)	4 (28)	43	---	19	SPT
1060-H12	12 (83)	11 (76)	16	---	23	SPT
1060-H14	14 (97)	13 (90)	12	---	26	SPT
1060-H16	16 (110)	15 (103)	8	---	30	SPT
1060-H18	19 (131)	18 (124)	6	---	35	SPT
1100-0	13 (90)	5 (34)	35	45	23	SPTEBR WFOMN
1100-H12	16 (110)	15 (103)	12	25	28	SPTEBR WFOMN
1100-H14	18 (124)	17 (117)	9	20	32	SPTEBR WFOMN
1100-H16	21 (145)	20 (138)	6	17	38	SPTEBR WFOMN
1100-H18	24 (165)	22 (152)	5	15	44	SPTEBR WFOMN
1350-0	12 (83)	4 (28)	---	---†	---	SPTER BWC
1350-H12	14 (97)	12 (83)	---	---	---	SPTER BWC
1350-H14	16 (110)	14 (97)	---	---	---	SPTER BWC
1350-H16	18 (124)	16 (110)	---	---	---	SPTER BWC
1350-H19	27 (186)	24 (165)	---	---‡	---	SPTER BWC
2011-T3	55 (379)	43 (296)	---	15	95	BWR
2011-T8	59 (407)	45 (310)	---	12	100	BWR
2014-0	27 (186)	14 (97)	---	18	45	SPTEBFR
2014-T4, T451	62 (427)	42 (290)	---	20	105	SPTEBFR

TABLE 6-8—*Continued*

Alloy and Temper	Tension: Strength, ksi (MPa) Ultimate	Yield	Elongation, Percent in 2 in. (51 mm): 1/16 in. (1.6 mm) Thick Specimen	1/2 in. (12.7 mm) Diameter Specimen	Hardness: Brinell Number, 500 kg Load 10 mm Ball	Product Forms**
2014-T6, T651	70 (483)	60 (414)	---	13	135	SPTEBFR
Alclad 2014-0	25 (172)	10 (69)	21	---	---	SP
Alclad 2014-T3	63 (434)	40 (276)	20	---	---	SP
Alclad 2014-T4, T451	61 (421)	37 (255)	22	---	---	SP
Alclad 2014-T6, T651	68 (469)	60 (414)	10	---	---	SP
2017-0	26 (179)	10 (69)	---	22	45	BWRN
2017-T4, T451	62 (427)	40 (276)	---	22	105	BWRN
2018-T61	61 (421)	46 (317)	---	12	120	F
2024-0	27 (186)	11 (76)	20	22	47	STEPRBW
2024-T3	70 (483)	50 (345)	18	---	120	STEPRBW
2024-T4, T351	68 (469)	47 (324)	20	19	120	STEPRBW
2024-T361	72 (496)	57 (393)	13	---	130	STEPRBW
Alclad 2024-0	26 (179)	11 (26)	20	---	---	S
Alclad 2024-T3	65 (448)	45 (310)	18	---	---	S
Alclad 2024-T4, T351	64 (441)	42 (290)	19	---	---	S
Alclad 2024-T361	67 (462)	53 (365)	11	---	---	S
Alclad 2024-T81, T851	65 (448)	60 (414)	6	---	---	S
Alclad 2024-T861	70 (483)	66 (455)	6	---	---	S
2025-T6	58 (400)	37 (255)	---	19	110	F
2036-T4	49 (338)	28 (193)	24	---	---	S
2117-T4	43 (296)	24 (165)	---	27	70	WRN
2218-T72	48 (331)	37 (255)	---	11	95	F
2219-0	25 (172)	11 (76)	18	---	---	SEFR BPT
2219-T42	52 (359)	27 (186)	20	---	---	SEFR BPT
2219-T31, T351	52 (359)	36 (248)	17	---	---	SEFR BPT
2219-T37	57 (393)	46 (317)	11	---	---	SEFR BPT
2219-T62	60 (414)	42 (290)	10	---	---	SEFR BPT
2219-T81, T851	66 (455)	51 (352)	10	---	---	SEFR BPT
2219-T87	69 (476)	57 (393)	10	---	---	SEFR BPT
3003-0	16 (110)	6 (41)	30	40	28	All forms
3003-H12	19 (131)	18 (124)	10	20	35	All forms
3003-H14	22 (152)	21 (145)	8	16	40	All forms
3003-H16	26 (179)	25 (172)	5	14	47	All forms
3003-H18	29 (200)	27 (186)	4	10	55	All forms
Alclad 3003-0	16 (110)	6 (41)	30	40	---	S
Alclad 3003-H12	19 (131)	18 (124)	10	20	---	S
Alclad 3003-H14	22 (152)	21 (145)	8	16	---	S

(continued)

TABLE 6-8—*Continued*

Alloy and Temper	Tension: Strength, ksi (MPa) Ultimate	Yield	Elongation, Percent in 2 in. (51 mm): 1/16 in. (1.6 mm) Thick Specimen	1/2 in. (12.7 mm) Diameter Specimen	Hardness: Brinell Number, 500 kg Load 10 mm Ball	Product Forms**
Alclad 3003-H16	26 (179)	25 (172)	5	14	---	S
Alclad 3003-H18	29 (200)	27 (186)	4	10	---	S
3004-0	26 (179)	10 (69)	20	25	45	SPT
3004-H32	31 (214)	25 (172)	10	17	52	SPT
3004-H34	35 (241)	29 (200)	9	12	63	SPT
3004-H36	38 (262)	33 (228)	5	9	70	SPT
3004-H38	41 (283)	36 (248)	5	6	77	SPT
Alclad 3004-0	26 (179)	10 (69)	20	25	---	S
Alclad 3004-H32	31 (214)	25 (172)	10	17	---	S
Alclad 3004-H34	35 (241)	29 (200)	9	12	---	S
Alclad 3004-H36	38 (262)	33 (228)	5	9	---	S
Alclad 3004-H38	41 (283)	36 (248)	5	6	---	S
3105-0	17 (117)	8 (55)	24	---	---	
3105-H12	22 (152)	19 (131)	7	---	---	
3105-H14	25 (172)	22 (152)	5	---	---	
3105-H16	28 (193)	25 (172)	4	---	---	
3105-H18	31 (214)	28 (193)	3	---	---	
3105-H25	26 (179)	23 (159)	8	---	---	
4032-T6	55 (379)	46 (317)	---	9	120	F
5005-0	18 (124)	6 (41)	25	---	28	SWNPR
5005-H12	20 (138)	19 (131)	10	---	---	SWNPR
5005-H14	23 (159)	22 (152)	6	---	---	SWNPR
5005-H16	26 (179)	25 (172)	5	---	---	SWNPR
5005-H18	29 (200)	28 (193)	4	---	---	SWNPR
5005-H32	20 (138)	17 (117)	11	---	36	SWNPR
5005-H34	23 (159)	20 (138)	8	---	41	SWNPR
5005-H36	26 (179)	24 (165)	6	---	46	SWNPR
5005-H38	29 (200)	27 (186)	5	---	51	SWNPR
5050-0	21 (145)	8 (55)	24	---	36	STPRWB
5050-H32	25 (172)	21 (145)	9	---	46	STPRWB
5050-H34	28 (198)	24 (165)	8	---	53	STPRWB
5050-H36	30 (207)	26 (179)	7	---	58	STPRWB
5050-H38	32 (221)	29 (200)	6	---	63	STPRWB
5052-0	28 (193)	13 (90)	25	30	47	STBWPO RN
5052-H32	33 (228)	28 (193)	12	18	60	STBWPO RN
5052-H34	38 (262)	31 (214)	10	14	68	STBWPO RN
5052-H36	40 (276)	35 (241)	8	10	73	STBWPO RN
5052-H38	42 (290)	37 (255)	7	8	77	STBWPO RN
5056-0	42 (290)	22 (152)	---	35	65	BWRON
5056-H18	63 (434)	59 (407)	---	10	105	BWRON
5056-H38	60 (414)	50 (345)	---	15	100	BWRON
5083-0	42 (290)	21 (145)	---	22	---	SEPFT
5083-H321, H116	46 (317)	33 (228)	---	16	---	SEPFT
5086-0	38 (262)	17 (117)	22	---	---	SETP
5086-H32, H116	42 (290)	30 (207)	12	---	---	SETP

TABLE 6-8—*Continued*

Alloy and Temper	Tension: Strength, ksi (MPa) Ultimate	Yield	Elongation, Percent in 2 in. (51 mm): 1/16 in. (1.6 mm) Thick Specimen	1/2 in. (12.7 mm) Diameter Specimen	Hardness: Brinell Number, 500 kg Load 10 mm Ball	Product Forms**
5086-H34	47 (324)	37 (255)	10	---	---	SETP
5086-H112	39 (269)	19 (131)	14	---	---	SETP
5154-0	35 (241)	17 (117)	27	---	58	STERBWP
5154-H32	39 (269)	30 (207)	15	---	67	STERBWP
5154-H34	42 (290)	33 (228)	13	---	73	STERBWP
5154-H36	45 (310)	36 (248)	12	---	78	STERBWP
5154-H38	48 (331)	39 (269)	10	---	80	STERBWP
5154-H112	35 (241)	17 (117)	25	---	63	STERBWP
5252-H25	34 (234)	25 (172)	11	---	68	S
5252-H38, H28	41 (283)	35 (241)	5	---	75	S
5254-0	35 (241)	17 (117)	27	---	58	SP
5254-H32	39 (269)	30 (207)	15	---	67	SP
5254-H34	42 (290)	33 (228)	13	---	73	SP
5254-H36	45 (310)	36 (248)	12	---	78	SP
5254-H38	48 (331)	39 (269)	10	---	80	SP
5254-H112	35 (241)	17 (117)	25	---	63	SP
5454-0	36 (248)	17 (117)	22	---	62	STPE
5454-H32	40 (276)	30 (207)	10	---	73	STPE
5454-H34	44 (303)	35 (241)	10	---	81	STPE
5454-H112	36 (248)	18 (124)	18	---	62	STPE
5456-0	45 (310)	23 (159)	---	24	---	SETP
5456-H112	45 (310)	24 (165)	---	22	---	SETP
5456-H321, H116	51 (352)	37 (255)	---	16	90	SETP
5457-0	19 (131)	7 (48)	22	---	32	S
5457-H25	26 (179)	23 (159)	12	---	48	S
5457-H38, H28	30 (207)	27 (186)	6	---	55	S
5652-0	28 (193)	13 (90)	25	30	47	SP
5652-H32	33 (228)	28 (193)	12	18	60	SP
5652-H34	38 (262)	31 (214)	10	14	68	SP
5652-H36	40 (276)	35 (241)	8	10	73	SP
5652-H38	42 (290)	37 (255)	7	8	77	SP
5657-H25	23 (159)	20 (138)	12	---	40	S
5657-H38, H28	28 (193)	24 (165)	7	---	50	S
6061-0	18 (124)	8 (55)	25	30	30	STPEBWF
6061-T4, T451	35 (241)	21 (145)	22	25	65	STPEBWF
6061-T6, T651	45 (310)	40 (276)	12	17	95	STPEBWF
Alclad 6061-0	17 (117)	7 (48)	25	---	---	S
Alclad 6061-T4, T451	33 (228)	19 (131)	22	---	---	S

(continued)

TABLE 6-8—*Continued*

Alloy and Temper	Tension: Strength, ksi (MPa) Ultimate	Yield	Elongation, Percent in 2 in. (51 mm): 1/16 in. (1.6 mm) Thick Specimen	1/2 in. (12.7 mm) Diameter Specimen	Hardness: Brinell Number, 500 kg Load 10 mm Ball	Product Forms**
Alclad 6061-T6, T651	42 (290)	37 (255)	12	---	---	S
6063-0	13 (90)	7 (48)	---	---	25	TEC
6063-T1	22 (152)	13 (90)	20	---	42	TEC
6063-T4	25 (172)	13 (90)	22	---	---	TEC
6063-T5	27 (186)	21 (145)	12	---	60	TEC
6063-T6	35 (241)	31 (214)	12	---	73	TEC
6063-T83	37 (255)	35 (241)	9	---	82	TEC
6063-T831	30 (207)	27 (186)	10	---	70	TEC
6063-T832	42 (290)	39 (269)	12	---	95	TEC
6066-0	22 (152)	12 (83)	---	18	43	TEF
6066-T4, T451	52 (359)	30 (207)	---	18	90	TEF
6066-T6, T651	57 (393)	52 (359)	---	12	120	TEF
6070-T6	55 (379)	51 (352)	10	---	---	RBTC
6101-H111	14 (97)	11 (76)	---	---	---	RBTC
6101-T6	32 (221)	28 (193)	15	---	71	RBTC
6262-T9	58 (400)	55 (379)	---	10	120	WBRTE
6463-T1	22 (152)	13 (90)	20	---	42	E
6463-T5	27 (186)	21 (145)	12	---	60	E
6463-T6	35 (241)	31 (214)	12	---	74	E
7001-0	37 (255)	22 (152)	---	14	60	TE
7001-T6, T651	98 (676)	91 (627)	---	9	160	TE
7075-0	33 (228)	15 (103)	17	16	60	SPRB TCFP
7075-T6, T651	83 (572)	73 (503)	11	11	150	SPRB TCFP
Alclad 7075-0	32 (221)	14 (97)	17	---	---	S
Alclad 7075-T6, T651	76 (524)	67 (462)	11	---	---	S
7178-0	33 (228)	15 (103)	15	16	---	SPER BWN
7178-T6, T651	88 (607)	78 (538)	10	11	---	SPER BWN
7178-T76, T7651	83 (572)	73 (503)	---	11	---	SPER BWN
Alclad 7178-0	32 (221)	14 (97)	16	---	---	S
Alclad 7178-T6, T651	81 (558)	71 (490)	10	---	---	S

(*The Aluminum Association, Inc.*)

* The indicated typical mechanical properties for all except 0 temper material are higher than the specified minimum properties. For 0 temper products, typical ultimate and yield values are slightly lower than specified (maximum) values.

** Below is a list of abbreviations appearing in the Product Forms column. These various product forms may or may not be available in all tempers listed for the respective alloys. For a detailed listing of product forms available, contact the Aluminum Association.

S = Sheet	O = Foil	F = Forgings and forging stock
P = Plate	N = Rivets	R = Rod
C = Pipe	M = Fin stock	B = Bar
T = Tube	E = Extruded wire, rod, bar and shapes	W = Wire rolled or cold finished

† 1350-0 wire will have an elongation of approximately 23% in 10 inches.

‡ 1350-H19 wire will have an elongation of approximately 1 1/2% in 10 inches.

Tempers. A wide variety of mechanical characteristics, or tempers, is available in aluminum alloys through various combinations of cold working and heat treatment. In specifying the temper for any given product, the fabricating process and the amount of cold working involved should be kept in mind. In general, the temper specified should be such that the amount of cold working the metal will receive during fabrication will develop the desired characteristics in the finished products.

Temperature effects. Aluminum and its alloys lose part of their strength at elevated temperatures, although some alloys retain good strength at temperatures from 400 to 500° F (204 to 260° C). Table 6-9 lists typical tensile strength data for aluminum alloys at a temperature of 300° F (149° C). At subzero temperatures, aluminum alloys retain their strength without loss of ductility, making aluminum a particularly useful metal for low-temperature applications.

Corrosion Resistance. When aluminum surfaces are exposed to the atmosphere, a thin, invisible oxide skin forms immediately and protects the metal from further oxidation. This self-protecting characteristic gives aluminum its high resistance to corrosion. Unless exposed to some substance or condition that destroys this protective oxide coating, the metal remains fully protected against corrosion. Aluminum is highly resistant to weathering, even in industrial atmospheres that often corrode other metals. It is also corrosion resistant to many acids. Alkalies are among the few substances that attack the oxide skin and therefore are corrosive to aluminum. Although the metal can safely be used in the presence of certain mild alkalies with the aid of inhibitors, in general, direct contact with alkaline substances should be avoided.

TABLE 6-9
Typical Tensile Properties of Aluminum Alloys at 300° F (149° C)*

Alloy and Temper	Tensile Strength, ksi (MPa)		Elongation in 2″ (51 mm), %
	Ultimate	Yield**	
1100-0	8 (55)	4.2 (29)	55
1100-H14	14 (97)	12 (83)	23
1100-H18	18 (124)	14 (97)	20
1145-0	11 (75)	5 (35)	---
1145-H19	24 (165)	21 (145)	---
2011-T3	28 (193)	19 (131)	25
2014-T6, T651	40 (276)	35 (241)	20
2017-T4, T451	40 (276)	30 (207)	15
2024-T3 (sheet)	55 (379)	45 (310)	11
2024-T4, T351 (plate)	45 (310)	36 (248)	17
2024-T6, T651	45 (310)	36 (248)	17
2024-T81, T851	55 (379)	49 (338)	11
2024-T861	54 (372)	48 (331)	11
2036-T4	49 (340)	28 (195)	---
X2037-T4	45 (310)	25 (170)	---
2117-T4	30 (207)	17 (117)	20
2219-T62	45 (310)	33 (228)	17
2219-T81, T851	49 (338)	40 (276)	17
2618-T61	50 (345)	44 (303)	14
3003-0	11 (76)	5 (34)	47
3003-H14	18 (124)	16 (110)	16
3003-H18	23 (159)	16 (110)	11
3004-0	22 (152)	10 (69)	35
3004-H34	28 (193)	25 (172)	22
3004-H38	31 (214)	27 (186)	15
4032-T6	37 (255)	33 (228)	9
5050-0	19 (131)	8 (55)	---
5050-H34	25 (172)	22 (152)	---
5050-H38	27 (186)	25 (172)	---
5052-0	23 (159)	13 (90)	50
5052-H34	30 (207)	27 (186)	27
5052-H38	34 (234)	28 (193)	24
5083-0	31 (214)	19 (131)	50
5086-0	29 (200)	16 (110)	50
5154-0	29 (200)	16 (110)	50
5254-0	29 (200)	16 (110)	50
5454-0	29 (200)	16 (110)	50
5454-H32	32 (221)	26 (179)	37
5454-H34	34 (234)	28 (193)	32
5456-0	31 (214)	20 (138)	50
5652-0	23 (159)	13 (90)	50
5652-H34	30 (207)	27 (186)	27
5652-H38	34 (234)	28 (193)	24
6053-T6, T651	25 (172)	24 (165)	13
6061-T6, T651	34 (234)	31 (214)	20
6063-T1	21 (145)	15 (103)	20
6063-T5	20 (138)	18 (124)	20
6063-T6	21 (145)	20 (138)	20
6101-T6	21 (145)	19 (131)	20
6151-T6	28 (193)	27 (186)	20
6262-T651	34 (234)	31 (214)	20
6262-T9	38 (262)	37 (255)	14
7075-T6, T651	31 (214)	27 (186)	30
7075-T73, T7351	31 (214)	27 (186)	30
7178-T6, T651	31 (214)	27 (186)	40
7178-T76, T7651	31 (214)	27 (186)	40

(*The Aluminum Association, Inc.*)

* This data is based on a limited amount of testing and represents the lowest strength during 10,000 hr of exposure at testing temperature under no load; stress applied at 5000 psi/min (34 MPa) to yield strength and then at strain rate of 0.05 in./in./min (1.3 mm/mm) to failure. Under some conditions of temperature and time, the application of heat will adversely affect certain other properties of some alloys.

** Offset equals 0.2%.

ALUMINUM APPLICATIONS

Some alloys are less resistant to corrosion than others, particularly certain high-strength alloys. Such alloys in some forms can be effectively protected from the majority of corrosive influences, however, by cladding the exposed surface or surfaces with a thin layer of either pure aluminum or one of the more highly corrosion-resistant alloys.

A word of caution is necessary in connection with the corrosion-resistant characteristics of aluminum. Direct contact with certain other metals should be avoided in the presence of an electrolyte; otherwise, galvanic corrosion of the aluminum may take place in the vicinity of the contact area. Where other metals must be fastened to aluminum, the use of a bituminous paint coating or insulating tape is recommended.

Conductivity. Aluminum is one of the two common metals having an electrical conductivity high enough for use as an electric conductor. The conductivity of electric conductor grade 1350 is about 62% of the International Annealed Copper Standard (IACS). Because aluminum has less than one-third the specific gravity of copper, however, a pound of aluminum will go about twice as far as a pound of copper when used for this purpose. Alloying lowers the conductivity somewhat, so that wherever possible the 1350 aluminum is used in electric conductor applications.

Other properties. Aluminum is also an excellent reflector of radiant energy in the entire range of wavelengths from ultraviolet through the visible spectrum to infrared and heat waves, as well as for electromagnetic waves of radio and radar. Aluminum has a light reflectivity of over 80%, which has led to its wide use in lighting fixtures. Aluminum roofing reflects a high percentage of solar radiation; consequently, buildings roofed with this material may be cooler in summer.

Not so well known are the nonmagnetic characteristics of aluminum. Nevertheless, these properties are of great importance for some uses. For example, its nonmagnetic properties make the metal useful for electrical shielding purposes such as in bus-bar housings or enclosures for other electrical equipment.

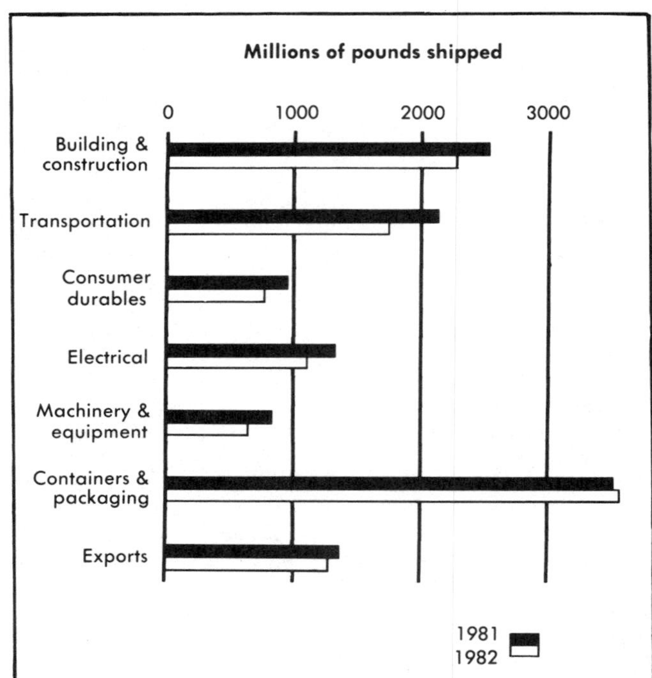

Fig. 6-1 Aluminum net shipments summarized by major markets. (*The Aluminum Association, Inc.*)

APPLICATIONS

As illustrated in Figure 6-1, containers and packaging represent the largest market for aluminum. Next in usage of aluminum are the fields of building construction and transportation. Aluminum's excellent resistance to atmospheric corrosion and the attractive appearance it maintains when anodically treated account for its many architectural uses. These include curtain walls for buildings, siding for houses, window frames, doors, and screen frames. It is also used for highway signs and bridge railings.

Aluminum is used extensively in the construction of motor vehicles, railway cars, aircraft, and aerospace equipment. In automobiles, aluminum is used in exterior and interior sheet metal and for trim, grilles, wheels, bumpers, radiators, and air conditioners. A major application is in automatic transmissions. Aluminum alloy castings are used for engine cylinder heads and blocks, crankcases, oil pans, pistons, and many other components of internal combustion engines, including jet engines. Aluminum alloy forgings are widely used in aircraft parts such as propellers and landing gear struts. Toughness, light weight, and heat reflective characteristics have led to aluminum's use for space satellites, moon rockets, and vehicles such as the lunar rover.

In electrical applications, aluminum wire and cable are major products. Underground electrical cables account for a large amount of aluminum usage. Aluminum wiring in residential, commercial, and industrial buildings is also gaining acceptance.

As stated previously, the packaging industry is a large and fast-growing field of application for aluminum. It is used in processing and storage equipment for foods and beverages, as well as in cans, foil, pouches, bags, and other containers. Consumer use includes kitchen utensils, hardware, tools, home appliances, and sporting equipment such as skis, baseball bats, and tennis rackets. Information on applications and comparative characteristics of wrought aluminum alloys is given in Table 6-10.

FABRICATION

The ease with which aluminum may be fabricated into any form is one of its most important assets. Often it can compete successfully with lower cost materials having less workability. The metal can be cast by any method known to foundries; it can be rolled to any desired thickness down to foil thinner than paper; aluminum sheet can be stamped, drawn, spun, or roll formed. The metal can also be hammered or forged. Aluminum wire, drawn from rolled rod, can be stranded into cable of any desired size and type. Aluminum's inherent characteristics make it well suited to the extrusion processes, and there is almost no limit to the different shapes in which the metal can be extruded.

Machining

The ease and speed with which aluminum can be machined significantly contributes to the low cost of finished aluminum parts. The metal can be turned, milled, bored, or machined in other ways at the maximum speeds of which most machines are capable. Another advantage of its flexible machining characteristics is that aluminum rod and bar can readily be used in the high-speed manufacture of parts by automatic screw machines.

TABLE 6-10
Comparative Characteristics and Applications for Wrought Aluminum Alloys

Alloy and Temper	Resistance to Corrosion				Weldability[f]			
	General[a]	Stress-Corrosion Cracking[b]	Workability (Cold)[e]	Machinability[e]	Brazeability[f]	Gas	Arc	Resistance Spot and Seam	Typical Applications
1060-0	A	A	A	E	A	A	A	B	Chemical equipment, railroad tank cars
H12	A	A	A	E	A	A	A	A	
H14	A	A	A	D	A	A	A	A	
H16	A	A	B	D	A	A	A	A	
H18	A	A	B	D	A	A	A	A	
1100-0	A	A	A	E	A	A	A	B	Sheet metal work, spun hollowware, fin stock
H12	A	A	A	E	A	A	A	A	
H14	A	A	A	D	A	A	A	A	
H16	A	A	B	D	A	A	A	A	
H18	A	A	C	D	A	A	A	A	
1350-0	A	A	A	E	A	A	A	B	Electrical conductors
H12, H111	A	A	A	E	A	A	A	A	
H14, H24	A	A	A	D	A	A	A	A	
H16, H26	A	A	B	D	A	A	A	A	
H18	A	A	B	D	A	A	A	A	
2011-T3	D[c]	D	C	A	D	D	D	D	Screw machine products
T4, T451	D[c]	D	B	A	D	D	D	D	
T8	D	B	D	A	D	D	D	D	
2014-0	---	---	---	D	D	D	D	B	Truck frames, aircraft structures
T3, T4, T451	D[c]	C	C	B	D	D	B	B	
T6, T651, T6510, T6511	D	C	D	B	D	D	B	B	
2017-T4, T451	D[c]	C	C	B	D	D	B	B	Screw machine products, fittings
2018-T61	---	---	---	B	---	---	---	---	Aircraft engine cylinders, heads and pistons
2024-0	---	---	---	D	D	D	D	D	Truck wheels, screw machine products, aircraft structures
T4, T3, T351, T3510, T3511	D[c]	C	C	B	D	C	B	B	
T361	D[c]	C	D	B	D	D	C	B	
T6	D	B	C	B	D	D	C	B	
T861, T81, T851, T8510, T8511	D	B	D	B	D	D	C	B	
T72	---	---	---	B	---	---	---	---	
2025-T6	D	C	---	B	D	D	B	B	Forgings, aircraft propellers
2036-T4	C	---	B	C	D	---	B	B	Auto body panel sheet
2117-T4	C	A	B	C	D	D	B	B	Rivets
2218-T61	D	C	---	---	---	---	---	C	Jet engine impellers and rings
T72	D	C	---	B	D	D	C	B	
2219-0	---	---	---	---	D	D	A	B	Structural uses at high temperatures [to 600° F (316° C)], high-strength weldments
T31, T351, T3510, T3511	D[c]	C	C	B	D	A	A	A	
T37	D[c]	C	D	B	D	A	A	A	
T81, T851, T8510, T8511	D	B	D	B	D	A	A	A	
T87	D	B	D	B	D	A	A	A	
2618-T61	D	C	---	B	D	D	C	B	Aircraft engines

(continued)

ALUMINUM APPLICATIONS

TABLE 6-10—*Continued*

Alloy and Temper	Resistance to Corrosion					Weldability[f]			
	General[a]	Stress-Corrosion Cracking[b]	Workability (Cold)[c]	Machinability[e]	Brazeability[f]	Gas	Arc	Resistance Spot and Seam	Typical Applications
3003-0	A	A	A	E	A	A	A	A	Cooking utensils,
H12	A	A	A	E	A	A	A	A	chemical equipment,
H14	A	A	B	D	A	A	A	A	pressure vessels,
H16	A	A	C	D	A	A	A	A	sheet metal work,
H18	A	A	C	D	A	A	A	A	builder's hardware,
H25	A	A	B	D	A	A	A	A	storage tanks
3004-0	A	A	A	D	B	B	A	B	Sheet metal work,
H32	A	A	B	D	B	B	A	A	storage tanks
H34	A	A	B	C	B	B	A	A	
H36	A	A	C	C	B	B	A	A	
H38	A	A	C	C	B	B	A	A	
3105-0	A	A	A	E	B	B	A	B	Residential siding,
H12	A	A	B	E	B	B	A	A	mobile homes,
H14	A	A	B	D	B	B	A	A	sheet metal work
H16	A	A	C	D	B	B	A	A	
H18	A	A	C	D	B	B	A	A	
H25	A	A	B	D	B	B	A	A	
4032-T6	C	B	---	B	D	D	B	C	Pistons
5005-0	A	A	A	E	B	A	A	B	Appliances,
H12	A	A	A	E	B	A	A	A	utensils,
H14	A	A	B	D	B	A	A	A	architectural,
H16	A	A	C	D	B	A	A	A	electrical conductor
H18	A	A	C	D	B	A	A	A	
H32	A	A	A	E	B	A	A	A	
H34	A	A	B	D	B	A	A	A	
H36	A	A	C	D	B	A	A	A	
H38	A	A	C	D	B	A	A	A	
5050-0	A	A	A	E	B	A	A	B	Builder's hardware,
H32	A	A	A	D	B	A	A	A	refrigerator trim,
H34	A	A	B	D	B	A	A	A	coiled tubes
H36	A	A	C	C	B	A	A	A	
H38	A	A	C	C	B	A	A	A	
5052-0	A	A	A	D	C	A	A	B	Sheet metal work,
H32	A	A	B	D	C	A	A	A	hydraulic tube,
H34	A	A	B	C	C	A	A	A	appliances
H36	A	A	C	C	C	A	A	A	
H38	A	A	C	C	C	A	A	A	
5056-0	A[d]	B[d]	A	D	D	C	A	B	Cable sheathing,
H111	A[d]	B[d]	A	D	D	C	A	A	rivets for magnesium,
H12, H32	A[d]	B[d]	B	D	D	C	A	A	screen wire,
H14, H34	A[d]	B[d]	B	C	D	C	A	A	zippers
H18, H38	A[d]	C[d]	C	C	D	C	A	A	
H192	B[d]	D[d]	D	B	D	C	A	A	
H392	B[d]	D[d]	D	B	D	C	A	A	

TABLE 6-10—*Continued*

Alloy and Temper	Resistance to Corrosion: General[a]	Resistance to Corrosion: Stress-Corrosion Cracking[b]	Workability (Cold)[e]	Machinability[e]	Brazeability[f]	Weldability[f]: Gas	Weldability[f]: Arc	Weldability[f]: Resistance Spot and Seam	Typical Applications
5083-0	A[d]	B[d]	B	D	D	C	A	B	Unfired, welded pressure vessels, marine, auto, aircraft equipment, cryogenics, TV towers, drilling rigs, transportation equipment, missile components
H321, H116	A[d]	B[d]	C	D	D	C	A	A	
H111	A[d]	B[d]	C	D	D	C	A	A	
5086-0	A[d]	A[d]	A	D	D	C	A	B	
H32, H116	A[d]	A[d]	B	D	D	C	A	A	
H34	A[d]	B[d]	B	C	D	C	A	A	
H36	A[d]	B[d]	C	C	D	C	A	A	
H38	A[d]	B[d]	C	C	D	C	A	A	
H111	A[d]	A[d]	B	D	D	C	A	A	
5154-0	A[d]	A[d]	A	D	D	C	A	B	Welded structures, storage tanks, pressure vessels, salt water service
H32	A[d]	A[d]	B	D	D	C	A	A	
H34	A[d]	A[d]	B	C	D	C	A	A	
H36	A[d]	A[d]	C	C	D	C	A	A	
H38	A[d]	A[d]	C	C	D	C	A	A	
5252-H24	A	A	B	D	C	A	A	A	Automotive and appliance trim
H25	A	A	B	C	C	A	A	A	
H28	A	A	C	C	C	A	A	A	
5254-0	A[d]	A[d]	A	D	D	C	A	B	Hydrogen peroxide and chemical storage vessels
H32	A[d]	A[d]	B	D	D	C	A	A	
H34	A[d]	A[d]	B	C	D	C	A	A	
H36	A[d]	A[d]	C	C	D	C	A	A	
H38	A[d]	A[d]	C	C	D	C	A	A	
5454-0	A	A	A	D	D	C	A	B	Welded structures, pressure vessels, marine service
H32	A	A	B	D	D	C	A	A	
H34	A	A	B	C	D	C	A	A	
H111	A	A	B	D	D	C	A	A	
5456-0	A[d]	B[d]	B	D	D	C	A	B	High-strength welded structures, pressure vessels, marine applications, storage tanks
H321, H116	A[d]	B[d]	C	D	D	C	A	A	
5457-0	A	A	A	E	B	A	A	B	
5652-0	A	A	A	D	C	A	A	B	Hydrogen peroxide and chemical storage vessels
H32	A	A	B	D	C	A	A	A	
H34	A	A	B	C	C	A	A	A	
H36	A	A	C	C	C	A	A	A	
H38	A	A	C	C	C	A	A	A	
5657-H241	A	A	A	D	B	A	A	A	Anodized auto and appliance trim
H25	A	A	B	D	B	A	A	A	
H26	A	A	B	D	B	A	A	A	
H28	A	A	C	D	B	A	A	A	
6053-0	---	---	---	E	A	A	A	B	Wire and rod for rivets
T6, T61	A	A	---	C	A	A	A	A	

(continued)

TABLE 6-10—*Continued*

Alloy and Temper	Resistance to Corrosion: General[a]	Resistance to Corrosion: Stress-Corrosion Cracking[b]	Workability (Cold)[e]	Machinability[e]	Brazeability[f]	Weldability[f]: Gas	Weldability[f]: Arc	Weldability[f]: Resistance Spot and Seam	Typical Applications
6061-0	B	A	A	D	A	A	A	B	Heavy-duty structures requiring good corrosion resistance, truck and marine, railroad cars, furniture, pipelines
T4, T451, T4510, T4511	B	B	B	C	A	A	A	A	
T6, T651, T652, T6510, T6511	B	A	C	C	A	A	A	A	
6063-T1	A	A	B	D	A	A	A	A	Pipe railing, furniture, architectural extrusions
T4	A	A	B	D	A	A	A	A	
T5, T52	A	A	B	C	A	A	A	A	
T6	A	A	C	C	A	A	A	A	
T83, T831, T832	A	A	C	C	A	A	A	A	
6066-0	C	A	B	D	D	D	B	B	Forgings and extrusions for welded structures
T4, T4510, T4511	C	B	C	C	D	D	B	B	
T6, T6510, T6511	C	B	C	B	D	D	B	B	
6070-T4, T4511	B	B	B	C	B	A	A	A	Heavy-duty welded structures, pipelines
T6	B	B	C	C	B	A	A	A	
6101-T6, T63	A	A	C	C	A	A	A	A	High-strength bus conductors
T61, T64	A	A	B	D	A	A	A	A	
6151-T6, T652	---	---	---	---	---	---	---	---	Moderate-strength intricate forgings for machine and auto parts
6201-T81	A	A	---	C	A	A	A	A	High-strength electric conductor wire
6262-T6, T651, T6510, T6511	B	A	C	B	A	A	A	A	Screw machine products
T9	B	A	D	B	A	A	A	A	
6463-T1	A	A	B	D	A	A	A	A	Extruded architectural and trim sections
T5	A	A	B	C	A	A	A	A	
T6	A	A	C	C	A	A	A	A	
7001-0	C[c]	C	D	B	D	D	D	B	High-strength structures
T6, T6510, T6511									
7075-0	---	---	---	D	D	D	C	B	Aircraft and other structures
T6, T651, T652, T6510, T6511	C[c]	C	D	B	D	D	C	B	
T73, T7351	C	B	D	B	D	D	C	B	
7178-0	---	---	---	---	D	D	C	B	Aircraft and other structures
T6, T651, T6510, T6511	C[c]	C	D	B	D	D	C	B	

(*The Aluminum Association, Inc.*)

[a] Ratings A through E are relative ratings in decreasing order of merit, based on exposures to sodium chloride solution by intermittent spraying or immersion. Alloys with A and B ratings can be used in industrial and seacoast atmospheres without protection. Alloys with C, D, and E ratings generally should be protected at least on faying surfaces.

[b] Stress-corrosion cracking ratings are based on service experience and on laboratory tests of specimens exposed to the 3.5% sodium chloride alternate immersion test.

A = No known instance of failure in service or in laboratory tests.

B = No known instance of failure in service; limited failures in laboratory tests of short transverse specimens.

C = Service failures with sustained tension stress acting in short transverse direction relative to grain structure; limited failures in laboratory tests of long transverse specimens.

TABLE 6-10—*Continued*

D = Limited service failures with sustained longitudinal or long transverse stress.

[c] In relatively thick sections the rating would be E.

[d] This rating may be different for material held at elevated temperature for long periods.

[e] Ratings A through D for workability (cold), and A through E for machinability, are relative ratings in decreasing order of merit.

[f] Ratings A through D for weldability and brazeability are relative ratings defined as follows:

A = Generally weldable by all commercial procedures and methods.

B = Weldable with special techniques or for specific applications that justify preliminary trials or testing to develop welding procedure and weld performance.

C = Limited weldability because of crack sensitivity or loss in resistance to corrosion and mechanical properties.

D = No commonly used welding methods have been developed.

Joining

Almost any method of joining is applicable to aluminum—riveting, welding, brazing, or soldering. A wide variety of mechanical aluminum fasteners simplifies the assembly of many products. Adhesive bonding of aluminum parts is widely employed, particularly in joining aircraft components.

Finishing

For the majority of its applications, aluminum needs no protective coating. Mechanical finishing methods—such as polishing, sandblasting, or wire brushing as discussed in Chapter 16, "Mechanical and Abrasive Deburring and Finishing," of this volume—meet the majority of needs. In many instances, the surface finish supplied is entirely adequate without further finishing. Where the plain aluminum surface does not suffice, or where additional protection is required, a wide variety of surface finishes may be applied. Chemical, electrochemical, and paint finishes are all used. Many colors are available in both chemical and electrochemical finishes. If paint, lacquer, or enamel is used, any color possible with these finishes may be applied. Vitreous enamels have been developed for aluminum, and the metal may also be electroplated. For additional information, refer to Chapter 19, "Conversion Coatings," and Chapter 20, "Plating," in this volume.

COPPER

Copper is a comparatively heavy metal, with a specific gravity of 8.96 at 68° F (20° C). The melting point is 1981° F (1083° C). The pure element is salmon pink in color and has a bright metallic luster when polished. Copper is nonmagnetic and has very high thermal conductivity and electrical conductivity. Among metals, only silver has a greater electrical conductivity. On a relative basis, with silver rated 100, copper is 95, aluminum 57, and iron 16. The usefulness of copper is derived from its combination of chemical, physical, electrical, and mechanical properties and its abundant supply.

The copper alloys also have the advantages of corrosion resistance, ease of forming, ease of joining, and are available in colors. On the other hand, copper and its alloys have relatively low strength-to-weight ratios and low strengths at elevated temperatures. Some alloys are susceptible to stress-corrosion cracking unless they are stress relieved. Copper and its alloys tend to work harden, and they can be hot or cold worked to increase strength. Ductility can be restored by annealing or in the heating that accompanies welding or brazing operations.

BASIC METAL PRODUCTION

The majority of the world's copper is obtained from the sulfide ores chalcocite, Cu_2S; covellite, CuS; chalcopyrite, $CuFeS_2$; bornite, Cu_5FeS_4; and enargite, $Cu_3(As\ Sb)S_4$. Native copper, once widespread in the United States, is now mined principally in Arizona, Utah, and Michigan's northern peninsula.

There are many variations in the basic production processes for copper. Starting with low-grade sulfide (with 0.3-1.0% copper) and oxide ores, copper is produced by a series of processing steps:

1. Concentration of crushed ores by flotation and other physical separation techniques.
2. Roasting of the ore in special furnaces to remove volatiles and to change the ore into a form more suited to further concentration. The ore does not melt in this process. It merely changes in chemical nature.
3. Smelting or melting the roasted ore in a reverberatory furnace to produce a matte, which is about 30% copper. The smelting operation is simply melting in a controlled atmosphere. This step further purifies the ore by, for example, converting iron in the ore to a sulfide that can be removed by subsequent oxidation.
4. Converting the molten copper matte to metallic copper is usually accomplished by blowing air or oxygen through the matte in converter furnaces, not unlike those used in steelmaking. The purification mechanism is the oxidation of copper sulfides to form metallic copper. The conversion produces blister copper with some dissolved oxygen and other impurities. Typically, the product is 98% to 99% copper.
5. Electrolytic refining is used to further reduce impurities. This process essentially involves making anodes from blister copper and plating in a form that is 99.9% pure copper. When the electrodeposited copper electrodes are melted and cast, the product is called electrolytic tough-pitch (ETP) copper.

ALLOY DESIGNATION SYSTEM[2]

The Unified Numbering System for Metals and Alloys (UNS) applied to wrought and cast copper and copper alloys evolved from the three-digit system developed by the U.S. copper and brass industry. The original three-digit designations were expanded to five digits, following a prefix letter C.

COPPER ALLOY DESIGNATION SYSTEM

UNS System Summary

The standard UNS designation system is summarized in Table 6-11. The UNS numbers are simply expansions of the former numerical system used by the Copper Development Association (CDA). For example, Copper Alloy No. 377 (forging brass) in the original three-digit system became C37700 in the UNS system. The overall UNS is jointly managed by the Society of Automotive Engineers (SAE) and the American Society for Testing and Materials (ASTM). For copper, the UNS designation system is administered by the CDA.

Alloy Designation Groups

The UNS designation system is an orderly method of identifying and defining coppers and copper alloys. It is not a metallurgical specification system, and the alloy designation numbers have no direct significance with regard to composition and properties. Numbers from C10000 through C79999 denote wrought alloys. Cast alloys are numbered from C80000 through C99999. Within these two categories, the compositions are grouped into the following families of coppers and copper alloys:

Coppers. Pure copper is metal that has a designated minimum copper content of 99.3% or higher and is essentially unalloyed copper.

High-copper alloys. For the wrought products, these are alloys with a designated copper content less than 99.3% but more than 96% that do not fall into any other copper alloy group. The cast high-copper alloys have a designated copper content in excess of 94%, to which silver may be added to create special properties.

Brasses. These alloys contain zinc as the principal alloying element with or without other designated alloying elements such as iron, aluminum, nickel, and silicon.

TABLE 6-11
Compositions and Standard Designations of Copper Alloy Families

UNS Number*	Description	Major Constituents
	Wrought	
C10100-C15500	Coppers	Cu
C16200-C19500	High-copper alloys	Cu plus Cd, Be, Cr or Fe
C20500-C28200	Brasses	Cu, Zn
C31400-C38600	Leaded brasses	Cu, Zn, Pb
C40500-C48500	Tin brasses	Cu, Zn, Sn
C50100-C52400	Phosphor bronzes	Cu, Sn
C53400-C54800	Leaded phosphor bronzes	Cu, Sn, Pb
C60600-C64200	Aluminum bronzes	Cu, Al
C64700-C66100	Silicon bronzes	Cu, Si
C66400-C69800	Special brasses	Cu, Zn plus Mn, Si, Al, etc.
C70100-C72500	Copper nickels	Cu, Ni
C73200-C79900	Nickel silvers	Cu, Ni, Zn
	Cast	
C80100-C81100	Coppers	Cu
C81300-C82800	High-copper alloys	Cu plus Cr, Be, Co, Ni, Si
C83300-C83800	Red brasses, leaded red brasses	Cu, Sn, Zn, Pb
C84200-C84800	Semi-red brasses, leaded semi-red brasses	Cu, Sn, Zn, Pb
C85200-C85800	Yellow brasses, leaded yellow brasses	Cu, Sn, Zn, Pb
C86100-C86800	Manganese bronzes, leaded manganese bronzes	Cu, Zn, Al, Mn, Pb
C87200-C87900	Silicon bronzes and brasses	Cu, Zn, Si
C90200-C91700	Tin bronzes	Cu, Sn
C92200-C92900	Leaded tin bronzes	Cu, Sn, Pb
C93200-C94500	High-leaded tin bronzes	Cu, Sn, Pb
C94700-C94900	Nickel-tin bronzes	Cu, Sn, Ni
C95200-C95800	Aluminum bronzes	Cu, Al, Fe, Ni
C96200-C96600	Copper nickels	Cu, Ni, Fe
C97300-C97800	Nickel silvers	Cu, Ni, Zn
C98200-C98800	Leaded coppers	Cu, Pb
C99300-C99700	Special alloys	Cu, Ni, Fe, Al, Zn

(*Copper Development Association, Inc.*)

* The Unified Numbering System (UNS) names an alloy using five digits that follow the letter C, for copper. It is based on the former three-digit CDA (Copper Development Association) system; e.g., UNS C10100 was formerly CDA 101.

Wrought. The wrought alloys comprise three main families of brasses: copper-zinc alloys, copper-zinc-lead alloys (leaded brasses), and copper-zinc-tin alloys (tin brasses).

Cast. The cast alloys contain four main families of brasses: copper-tin-zinc alloys (red, semi-red, and yellow brasses); "manganese bronze" alloys (high-strength yellow brasses); leaded "manganese bronze" alloys (leaded high-strength yellow brasses); and copper-zinc-silicon alloys (silicon brasses and bronzes).

Bronzes. Broadly speaking, bronzes are copper alloys in which the major alloying element is not zinc or nickel. Originally "bronze" described alloys with tin as the only or principal alloying element. Today the term is generally used not by itself but with a modifying adjective.

Wrought. For wrought alloys, there are four main families of bronzes: copper-tin-phosphorus alloys (phosphor bronzes), copper-tin-lead-phosphorus alloys (leaded phosphor bronzes), copper-aluminum alloys (aluminum bronzes), and copper-silicon alloys (silicon bronzes).

Cast. The cast alloys have four main families of bronzes: copper-tin alloys (tin bronzes), copper-tin-lead alloys (leaded and high-leaded tin bronzes), copper-tin-nickel alloys (nickel-tin bronzes), and copper-aluminum alloys (aluminum bronzes). The family of alloys known as "manganese bronzes," in which zinc is the major alloying element, is included in the brasses.

Copper nickels. These are alloys with nickel as the principal alloying element, with or without other designated alloying elements.

Copper-nickel-zinc alloys. Known commonly as "nickel silvers," these are alloys that contain zinc and nickel as the principal and secondary alloying elements, with or without other designated elements.

Leaded coppers. These include a series of cast alloys of copper with 20% or more lead, usually with a small amount of silver present, but without tin or zinc.

Special alloys. Alloys with chemical compositions that do not fall into any of the above categories are combined in the category "special alloys."

COPPER AND COPPER ALLOYS

More than 300 standard coppers and alloys are produced by the United States copper and brass industry. As a group, these alloys encompass a wide range of wrought and cast materials that are available in virtually all of the commercial mill and product forms. The fabricated forms include strip, plate, sheet, pipe, tube, rod, forgings, wire, bar, foil, extrusions, and castings. Table 6-12 lists the standard size ranges for wrought copper and copper alloy mill products. Cast products are available in varied shapes and sizes as needed for specific applications.

Mill Product Terminology

Copper and copper alloy mill products in all their forms are defined in ASTM General Requirements Specifications. ASTM B248 defines plate, sheet, strip, and rolled bar; ASTM B249 defines rod, bar, and shapes; ASTM B250 defines wire; ASTM B251 defines tube; and ASTM B224 contains a classification of copper refinery products. The terminology for copper and copper alloys is unique to the industry and its products.

In some instances, conflicting terminology is employed by various industries. For example, the product known as *bar* in the steel industry is, in specific cases, designated as *rod* by the brass mill industry. In the latter industry, a *rod* is a round, hexagonal, or octagonal solid section in a straight length. A *bar* is a square or rectangular solid section in a straight length. And a *shape* denotes a solid section that is oval, half oval, half round, triangular, pentagonal, or of any special cross section in a straight length. *Wire* is a solid section furnished in coils, on spools, or on reels.

Commonly Used Alloys

For applications requiring maximum electrical conductivity, the most widely used copper is C11000, "tough pitch," which contains approximately 0.03-0.06% oxygen and a minimum of 99.90% copper (including silver). In addition to high electrical conductivity, oxygen-free grades C10100 and C10200 provide immunity to embrittlement at high temperature. The addition of phosphorus produces grade C12200—the standard water-tube copper.

The high-copper alloys contain small amounts of alloying elements that improve strength with minimal loss in electrical conductivity. For example, cadmium in amounts up to 1% can increase strength by 50%, with a decline in conductivity to 85%. Small amounts of cadmium raise the softening temperature in alloy C14300, which is used widely for radiator fin stock. Tellurium, lead, or sulfur—present in small amounts in grades C14500, C14700, and C18700—increase machinability.

For machined products, the most used material is copper alloy C36000, which is called free-cutting brass. This alloy has a combination of good physical, mechanical, and fabrication properties, plus an acceptable cost level for a majority of screw machine product applications. A wide variety of alloys is available to provide other properties and characteristics for various applications. For example, alloy C46400, naval brass, provides higher strength and ductility, but has a lower machinability rating than free-cutting brass.

Pure Copper

Pure copper is alloyed with many other elements to produce minor changes in properties. Tellurium, sulfur, and lead are added to improve machinability. These elements are insoluble in small amounts, and machinability is improved by the action of dispersed tellurides, sulfides, and lead inclusions. Chromium is added for strengthening, as are other elements such as zirconium, cadmium, and tin. Some American copper ores contain silver as a natural impurity. Silver has excellent solubility with copper, and it remains with the copper during refining to produce silver-bearing pure copper, which has improved resistance to softening and grain growth at elevated temperatures. Total alloy additions in pure copper are usually less than 1%.

Commercially pure copper is available in several grades, all of which have essentially the same mechanical properties. The three most commonly used are all of the same purity but vary in some respects.

1. Electrolytic tough-pitch copper, containing nominally about 0.03-0.06% oxygen, has high electrical conductivity. It is susceptible to embrittlement when heated in reducing atmospheres.
2. Deoxidized copper, containing nominally about 0.02% phosphorus as a residual deoxidant, has substantially lower electrical conductivity than electrolytic tough-pitch copper. It has improved cold working characteristics

COPPER AND COPPER ALLOYS

TABLE 6-12
Standard Size Ranges for Wrought Copper and Copper Alloy Mill Products

	Thickness,		Width,	Outside Diameter,		Length,	Weight,
Product Form	Minimum in. (mm)	Maximum in. (mm)	Maximum in. (mm)	Minimum in. (mm)	Maximum in. (mm)	Maximum ft (m)	Maximum lb (kg)
Bar	0.188 (4.78)	5 (127)	12 (305)	---	---	60 (18.3)	2000 (907)
Extrusion	0.062 (1.57)	6 (152)	8 (203)	0.625 (15.88)	5 (127)	60 (18.3)	350 (159)
Foil	0.0005 (0.013)	0.005 (0.13)	17 (178)	---	---	---	4600 (2087)
Forging	0.094 (2.39)	6 (152)	12 (305)	0.375 (9.53)	12 (305)	12 (3.7)	12,000 (5443)
Pipe	0.062 (1.57)	2.5 (64)	---	0.405 (10.29)	10.75 (273.1)	95 (29.0)	20,000 (9072)
Plate	0.188 (4.78)	5 (127)	156 (3962)	---	---	40 (12.2)	20,000 (9072)
Plumbing Fittings	---	---	---	0.30* (7.62)	7.55* (191.8)	---	---
Rod							
Hexagonal	0.050 (1.27)	4.5 (114)	---	---	---	**	**
Octagonal	0.050 (1.27)	4.5 (114)	---	---	---	**	**
Round	---	---	---	0.050 (1.27)	4.5 (114)	**	**
Sheet	0.005 (0.13)	0.188 (4.78)	50 (1270)	---	---	10 (3.0)	5600 (2540)
Strip	0.005 (0.13)	0.188 (4.78)	24 (610)	---	---	---	8000 (3629)
Tube	0.0005 (0.013)	0.400 (10.16)	---	0.010 (0.25)	8.625 (219.08)	20 (straight) (6.1) 2000 (coils) (610)	900 (408)

(*Copper Development Association, Inc.*)

Note: Maximum and minimum sizes may be different from above for some specific alloys and tempers.
* Inside diameter.
** Consult mill.

and is not susceptible to embrittlement at elevated temperatures. It has better welding and brazing characteristics than the other grades.

3. Oxygen-free copper is thoroughly deoxidized but contains no residual deoxidants. It possesses exceptional plasticity and is not prone to embrittlement when heated in reducing atmospheres. It has the same high electrical conductivity as electrolytic tough-pitch copper.

Modified Copper

Copper can be modified by the addition of small amounts of other elements to achieve special characteristics. The two principal types of modified copper are:

1. Tellurium copper, a free-cutting copper containing 0.4-0.7% tellurium. It is somewhat less ductile and plastic than the commercially pure coppers. Selenium is occasionally used instead of tellurium. Leaded copper, another free-cutting copper, has been largely superseded by tellurium copper.
2. Tellurium-nickel copper, an age-hardenable alloy combining high strength, high conductivity, and excellent machinability.

COPPER AND COPPER ALLOYS

Copper-Based Alloys

Binary alloys of copper and zinc are known as brasses, while alloys of copper and tin are bronzes. From long established usage, some true brasses are called bronzes solely because their color is similar to that of the copper-tin alloys. Likewise, the term "bronze" is also used in modern metallurgy to refer to copper in which elements other than tin are the principal alloying material but that has the characteristic bronze color.

Nonleaded Copper-Zinc Alloys

The brasses are the most widely used and least expensive of the copper-based alloys. They possess relatively good corrosion resistance, moderately high strength, and in some compositions exceptionally good ductility and excellent forming characteristics when shaped by pressing, deep drawing, rolling, machining, etc. Cold working results in improved tensile properties. After cold working, they can be softened and recrystallized by appropriate annealing.

The color of the brasses varies with the copper content, starting with a color closely approximating that of copper, through a bronze at 90% copper, a reddish color at 85%, and a brassy yellow at 65%. Further reduction in the percentage of copper results in increased reddish color.

Alpha. Brasses with copper content above approximately 64% are called alpha brasses and are best known for their ability to withstand considerable cold working without annealing. These brasses may also be hot worked with reasonable ease.

Beta. The beta brasses, with copper content of 60% and less, are characterized by excellent hot working characteristics. They are not so adaptable to cold working as are the alpha brasses; and at the lower part of the range, about 55% copper, the strengths and elongations of the alpha and beta brasses vary according to the zinc and copper content as shown in Fig. 6-2.

Leaded Brass

Adding lead to the brasses in amounts from 0.5 to 4% results in free-cutting or free-machining alloys in which the elemental lead is present as uniformly dispersed particles. When the lead content is high, these brasses have relatively low ductility and plasticity. The mechanical properties of corrosion resistance and color are not materially affected by the presence of lead.

Casting Brass

While the wrought copper-zinc alloys are essentially simple binary alloys, the casting brasses contain substantial amounts of other elements, added principally to enhance the casting qualities of the alloy and to impart suitable strength properties. The usual elements, added either alone or in combination with one another, are tin (1-6%), lead (1-10%), iron (0.5-3%), and varying amounts of nickel, antimony, and aluminum.

Tin Bronze

Copper-tin alloys (phosphor bronzes) containing from about 1.25 to 10% tin possess good strength and cold forming characteristics and the typical bronze color.

The most important copper-tin alloys are those that have been deoxidized with phosphorus during the refining process and hence are known as phosphor bronze. The amount of residual phosphorus may range from a trace to about 0.35% or even higher in some special grades. The excess phosphorus, which exists in solid solution, materially increases the hardness and strength of the alloy, but it does so at the expense of ductility and electrical conductivity. It improves the toughness, elevated temperature properties, and resistance to corrosion. In amounts greater than 1.0%, phosphorus causes excessive brittleness and impairs surface appearance but affords a good bearing surface, as is evident by the use of high-phosphorus bronze compositions for gears and other machine parts subject to wear.

The phosphor bronzes include a variety of distinctive metals, characterized by strength, high resistance to corrosion and fatigue, and high yield strength.

Casting bronzes frequently contain as much as 10% lead, and some bearing compositions as much as 25%. The high lead content precludes the use of cold forming operations and limits the use of such alloys to temperatures below the melting point of lead.

The substitution of zinc for tin in the tin bronzes improves casting qualities, but zinc is seldom added in amounts greater than about 5% because of unattractive color and lower corrosion resistance to some media. A phosphor bronze containing approximately 4% each of tin, lead, and zinc has excellent free-cutting characteristics.

Nickel Silver

The nickel-silver alloys (once known as "German silver") are essentially brasses containing substantial amounts of nickel in addition to the copper and zinc. The designation of nickel silver—for instance, "65-18"—indicates the nominal percentages of copper and nickel. The addition of small amounts of lead produces nickel silvers with free-cutting characteristics. The color of the alloy varies from white with a yellowish tinge to a silvery white as the amount of nickel is increased.

Nickel-silver alloys are generally characterized by a pleasing color, high strength and ductility, good corrosion resistance, and ease in working by stamping, rolling, drawing, etc. Some compositions may be cold worked; others are best formed by hot working.

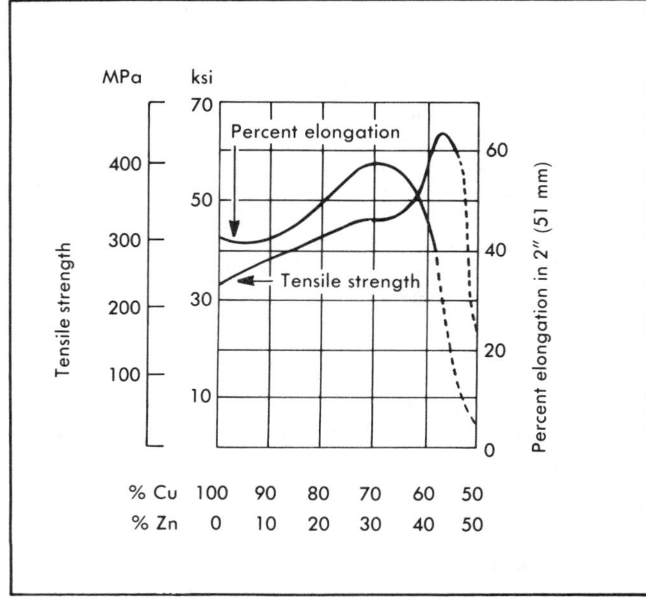

Fig. 6-2 Influence of composition on strength and elongation properties of annealed wrought brass.

CHAPTER 6

COPPER AND COPPER ALLOYS

Copper Nickel

Copper-based alloys containing nickel but no zinc are known as copper nickels. They have good strength and plasticity and are particularly noted for their high resistance to corrosion. They can be cold worked to a considerable degree and are readily hot worked.

Manganese Bronze

Manganese bronzes are essentially high-strength, modified copper-zinc casting alloys containing 55-60% copper, 38-42% zinc, 0-1.5% tin, 0-2% iron, 0-1.5% aluminum, and up to about 3.5% manganese. These alloys are available in various wrought forms and are characterized by good mechanical properties and corrosion resistance. Manganese bronzes have poor cold forming characteristics, but they can be readily hot worked. Their principal uses are for ship propellers, rudders, and other marine fittings.

Silicon Bronze

These alloys possess high strength (similar to mild steel) and good toughness, plus they exhibit excellent resistance to corrosion by brine and sulfite solutions, nonoxidizing inorganic acids, alkalies, and other media. They are readily hot worked, and the low-silicon alloys particularly have good cold working characteristics. The high-silicon alloys have excellent casting qualities and are superior in this respect to other high-strength nonferrous alloys such as aluminum bronze.

The composition of the silicon bronzes may vary over rather wide limits and may contain about 1-4% silicon, 0.25-1.5% manganese, 0.5-1.0% iron when present, and zinc as high as 22%. The addition of about 0.5% lead produces alloys with good machining characteristics.

Aluminum Bronzes

Aluminum bronzes have high strengths (comparable to medium-carbon steel), high hardnesses in the as-cast state, and excellent corrosion-resistance properties. They possess good antifrictional characteristics and resist scaling and oxidation at elevated temperatures. Aluminum bronzes can be hot worked readily, and some grades possess good cold forming characteristics and respond to a form of precipitation hardening.

The aluminum bronzes are essentially copper-aluminum alloys, containing up to about 13.5% aluminum, small amounts of manganese and nickel, and up to 4% iron for the purpose of hardening the alloys. The presence of iron in these alloys, in the form of an intermetallic compound ($FeAl_3$), contributes to wear resistance and hardness.

The good antifrictional characteristics of these alloys suit them for bearings, bushings, rollers, gears, etc. They resist scaling and oxidation at high temperatures and can be hot worked readily.

Beryllium Coppers

The addition of small amounts of beryllium to copper creates a family of high-copper alloys with strengths as high as alloy steel. The high strength is obtained by precipitation hardening, similar to the PH stainless steels. The commercial grades of beryllium-copper alloys contain from 0.4 to 2.0 weight percent beryllium in the wrought products, and up to 2.75 weight percent in the casting alloys. Small amounts of cobalt or nickel are also added to aid in the precipitation-hardening process and in grain refinement.

There are basically two families of beryllium-copper alloys: (1) high strength with moderate conductivity and (2) high conductivity with moderate strength. The principal characteristics of these alloys are their excellent response to precipitation-hardening treatments, excellent corrosion and fatigue resistance, moderate-to-good electrical and thermal conductivity, and resistance to stress relaxation. Elevated temperature performance is very good; however, the maximum long-time operating temperature should not exceed 400° F (204° C).

Beryllium copper is available in all forms: strip, rod, bar, plate, tube, wire, cast or forged billet, special shapes, and casting ingot. The high-strength beryllium coppers can be purchased in the solution annealed or cold worked tempers and can be easily formed or machined. A low-temperature heat treatment [600° F (316° C) for 2-3 hours] produces maximum strength and hardness. The high-conductivity alloys are also available in the annealed and cold worked tempers, and can be easily formed; however, due to the lower beryllium content (lower strength), these materials can also be formed in the heat-treated condition, thus saving heat treating and cleaning costs. A temperature of 900° F (482° C) for 2-3 hours is recommended for heat treating the high-conductivity alloys. Beryllium copper work hardens when formed or machined; however, it is not generally necessary for the fabricator of the material to solution anneal it to remove the work-hardening effects. For additional information, refer to Chapter 14, "Heat Treatment of Other Metals," in this volume.

Their high strength and ease of manufacture make beryllium-copper alloys particularly suited for bellows, bourdon tubing, diaphragms, fasteners, lock washers, springs, switch parts, relay parts, electrical and electronic components, retaining rings, roll pins, valves, pump parts, spline shafts, rolling mill parts, welding equipment, and instrument housings.

PROPERTIES

Copper is among the toughest of pure metals. It is moderately wear resistant and highly malleable and ductile. As shown in Fig. 6-3, the principal alloy groups have hardness and tensile strength values that are higher than those for pure copper.

Copper alloys do not have a sharply defined yield point. Yield strength is reported either as 0.5% extension under load or as 0.2% offset. On the most common basis (0.5% extension), yield strength of annealed material is approximately one-third the tensile strength. As the material is cold worked or hardened, yield strength approaches tensile strength.

Temper Specification

Wrought coppers and copper-based alloys are available and specified in various degrees of temper, which are established by cold working or annealing. Typical temper levels for flat products are soft, half hard, hard, spring, and extra-spring. The yield strength of a hard temper copper is approximately two thirds of its tensile strength. In the annealed or soft condition, the temper is based on grain size, which is a key factor in successful forming by deep drawing operations.

For brasses, phosphor bronzes, or other commonly cold worked alloy grades, the hardest available tempers are also the strongest. Ductility is sacrificed to gain strength. Beryllium-copper alloys such as C17000 to C17200 can be precipitation hardened to increase strength, as discussed in Chapter 14, "Heat Treatment of Other Metals," in this volume. Beryllium-copper temper designations AM through XHMS represent various

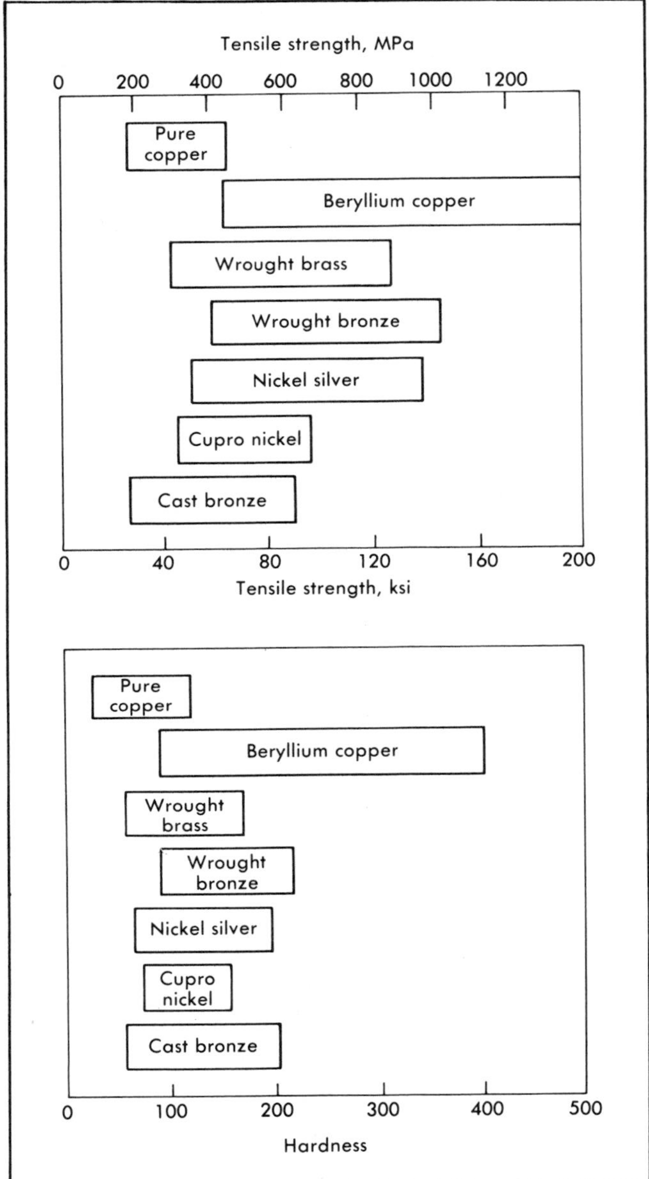

Fig. 6-3 Typical hardness and tensile strength ranges for various copper alloy groups. (*Engineering Materials, Kenneth Budinski, Reston Publishing Co.*)

strength levels available in mill-hardened strip. The process is proprietary; hence, the term AM does not necessarily signify annealing and heat treating but, rather, a set of properties. AM temper is the softest and XHMS is the hardest in this particular beryllium-copper alloy series.

Wrought Copper Properties

Table 6-13 lists the properties and characteristics of wrought copper and its alloys. Tensile strengths, yield strengths, and elongations vary somewhat with the shape of the section. For flat products, the section is taken at 0.040″ (1 mm) thick or, if that is not available, the nearest thickness for which data is available. For rod, the section is taken at 1.0″ (25 mm) diam or, if that is not available, the nearest diameter for which data is available. Yield strength is the stress corresponding to an extension of 0.5%. Data listed under "soft" is for 0.002″ (0.05 mm) average grain size or, if that is not available, the nearest grain size or anneal that is available.

The role of the various alloying elements used in copper systems is usually to increase strength and hardness. All of the ductile copper alloys can be cold worked to improve strength, but beryllium copper is the strongest. In the age-hardened condition, the tensile strength can be as high as 200 ksi (1379 MPa).

The ductility of wrought copper alloys depends on the amount of cold work. In the annealed condition, they all have high elongations and good formability. Wrought beryllium copper after heat treatment has poor ductility, but forming can usually be done in the annealed or cold worked condition. The ductility of some of the cast bronzes can be as low as 5% elongation.

Low temperatures have little effect on the ductility of copper alloys; that is, they do not get brittle. Most copper alloys soften at temperatures over 400° F (200° C), and the oxygen-bearing coppers embrittle in reducing atmospheres. Use at elevated temperatures requires careful consideration of elevated temperature property data. The ability to cold work the wrought alloys extensively produces a wide variety of strength ranges within each major alloy system.

Compared to high-carbon steels, copper alloys are not very hard, as shown in the lower graph in Fig. 6-3. Beryllium coppers, however, can achieve a hardness of R_C40 (Bhn 400), which approaches the hardness of some hardened steels.[3]

Cast Copper Alloy Properties

Copper alloy castings of irregular or complex, internal and external shapes can be produced by sand casting, die casting, and various other casting methods. Alloy and process selection are guided by the need for castings that have superior corrosion resistance, good electrical conductivity, good bearing quality, or other useful properties.

Table 6-14 lists typical mechanical properties of cast copper alloys selected from the more than 100 that are available. The greatest tonnage of copper castings is made by the sand-casting process. Alloy C83600 is widely used for general-purpose applications. Brass die castings are made when greater dimensional accuracy or a better surface finish are desired. High-copper alloys with varying amounts of tin, lead, and zinc account for a large percentage of all copper alloys used in cast form.

Corrosion Resistance

Copper and its alloys resist corrosion. They are inherently noble (chemically inert) because of their low driving force toward the oxidized state, as measured by the electromotive force or galvanic series. Their corrosion products form adherent surface films in various media.

Copper's performance in the atmosphere, in contact with both potable water and seawater, and in underground applications is well known. In all these media, copper performs well because its noble character keeps corrosion to a minimum even when the surface is disturbed.

Under outdoor atmospheric conditions, the protective oxide film eventually takes on the familiar blue-green color or patina; under indoor and manufacturing plant conditions, the film may consist of oxides, sulfides, or other salts. In potable water, protective films of oxides or of basic copper carbonates generally are formed.

CHAPTER 6

COPPER PROPERTIES

TABLE 6-13
Properties and Characteristics of Wrought Copper and Its Alloys

UNS No.	Name	Tensile Strength (average), ksi (MPa)				Yield Strength (average), ksi (MPa)				Elongation (average), % in 2 in. gage length				Rockwell Hardness			
		Flat Products		Rod		Flat Products		Rod		Flat Products		Rod		Flat Products		Rod	
		Hard	Soft	Hard	Soft	Hard	Soft	Hard	Soft	Hard	Soft	Hard	Soft	Hard B	Soft F	Hard B	Soft F
C10200	Oxygen-free copper	50 (345)	32 (221)	48 (331)	32 (221)	45 (310)	10 (69)	44 (303)	10 (69)	6	45	16	55	50	40	47	40
C11000	Electrolytic tough-pitch Cu	50 (345)	32 (221)	48 (331)	32 (221)	45 (310)	10 (69)	44 (303)	10 (69)	6	45	16	55	50	40	47	40
C12200	Phosphorus deoxidized Cu	50 (345)	32 (221)	---	---	45 (310)	10 (69)	---	---	6	45	---	---	40	40	---	---
C14500	Tellurium copper	---	---	48 (331)	32 (221)	---	---	44 (303)	10 (69)	---	---	20	50	---	---	48	40
C14700	Sulfur-bearing copper	---	---	48 (331)	32 (221)	---	---	44 (303)	10 (69)	---	---	12	52	---	---	48	40
C15100	Zirconium copper	58 (400)	38 (260)	---	---	55 (380)	10 (70)	---	---	5	36	---	---	58	49	---	---
C17200	Beryllium copper[g]	185 (1280)	70 (480)	85 (585)	75 (520)	170 (1170)	40 (280)	170 (1170)	30 (210)	3	45	3	50	40	77	40	94
C17500	Beryllium copper[g]	120 (830)	45 (310)	120 (830)	45 (310)	110 (760)	25 (175)	110 (760)	25 (175)	10	25	15	5	100	77	100	77
C18700	Leaded copper	---	---	48 (331)	32 (221)	---	---	42 (290)	10 (69)	---	---	15	45	---	---	48	40
C19100	Cu-Ni-P-Te	---	---	78 (538)	---	---	---	68 (469)	---	---	---	27	---	---	---	48	40
C19400	Cu-Fe-P	67 (460)	50 (345)	---	---	63 (425)	30 (210)	---	---	4	29	---	---	73	---	---	---
C19500	Cu-Fe-Sn-Co-P	86 (590)	55 (380)	---	---	83 (570)	30 (210)	---	---	2	28	---	---	84	---	---	---
C22000	Commercial bronze, 90%	61 (421)	37 (255)	45[a] (310)	40 (276)	54 (372)	10 (69)	---	---	5	45	25[a]	50	70	53	42[a]	55
C23000	Red brass, 85%	70 (483)	40 (276)	---	---	57 (393)	12 (83)	---	---	5	47	---	---	77	59	---	---
C26000	Cartridge brass, 70%	76 (524)	47 (324)	70[b] (483)	48 (331)	63 (434)	15 (103)	52[b] (359)	16 (110)	8	62	30[b]	65	82	64	80[b]	65
C28000	Muntz metal	70[b] (483)	54 (372)	72[c] (496)	54 (372)	50[b] (345)	21 (145)	50[c] (345)	21 (145)	10[b]	45	25[c]	50	75[b]	80	78[c]	80
C31400	Leaded commercial bronze	---	---	52[b] (359)	37 (255)	---	---	45[b] (310)	12 (83)	---	---	18[b]	45	---	---	58[b]	55
C34000	Medium-leaded brass	74 (510)	47 (324)	55[c] (379)	50 (345)	60 (414)	15 (103)	42[c] (290)	19 (131)	7	60	40[c]	60	80	64	60[c]	70
C35300	High-leaded brass	74 (510)	49 (338)	58[b] (400)	---	60 (414)	17 (117)	45[b] (310)	---	7	52	25[b]	---	80	68	75[b]	---
C36000	Free-cutting brass	56[c] (386)	---	58[b] (400)	49 (338)	45[c] (310)	---	45[b] (310)	18 (124)	20[c]	---	25[b]	53	62[c]	---	78[b]	68
C37700	Forging brass	---	---	---	52[d] (359)	---	---	---	20[d] (138)	---	---	---	45[d]	---	---	78[d]	---
C38500	Architectural bronze	---	---	---	60[d] (414)	---	---	---	20[d] (138)	---	---	---	30[d]	---	---	65[d]	---

TABLE 6-13—*Continued*

UNS No.	Name	Melting Point, °F (°C)	Capacity for Being Cold Worked[e]	Capacity for Being Hot Worked[e]	Hot-Forgeability Rating (Forging Brass = 100)	Hot-Working Temperature, °F (°C)	Annealing Temperature (30-60 min.), °F (°C)
C10200	Oxygen-free copper	1981 (1083)	E	E	65	1400-1600 (760-871)	700-1200 (371-649)
C11000	Electrolytic tough-pitch Cu	1981 (1083)	E	E	65	1400-1600 (760-871)	700-1200 (371-649)
C12200	Phosphorus deoxidized Cu	1981 (1083)	E	E	65	1400-1600 (760-871)	700-1200 (371-649)
C14500	Tellurium copper	1967 (1075)	G	G	65	1400-1600 (760-871)	800-1200 (427-649)
C14700	Sulfur-bearing copper	1969 (1076)	G	E	---	1400-1600 (760-871)	800-1200 (427-649)
C15100	Zirconium copper	1976 (1080)	E	E	---	1650-1740 (900-950)	
C17200	Beryllium copper[g]	1800 (980)	G	E	65	1200-1425 (650-775)	
C17500	Beryllium copper[g]	1960 (1020)	G	G	65	1350-1600 (700-870)	
C18700	Leaded copper	1976 (1080)	G	P	---	1400-1600 (760-871)	
C19100	Cu-Ni-P-Te	1980 (1082)	G	G	65	1400-1600 (760-871)	
C19400	Cu-Fe-P	1990 (1090)	E	G	---	1400-1600 (760-871)	
C19500	Cu-Fe-Sn-Co-P	1950 (1065)	E	G	---	1400-1600 (760-871)	
C22000	Commercial bronze, 90%	1910 (1043)	E	G	---	1400-1600 (760-871)	800-1450 (427-788)
C23000	Red brass, 85%	1880 (1027)	E	G	---	1450-1650 (788-899)	800-1350 (427-732)
C26000	Cartridge brass, 70%	1750 (954)	E	F	---	1350-1550 (732-843)	800-1400 (427-760)
C28000	Muntz metal	1660 (904)	F	E	90	1150-1450 (621-788)	800-1100 (427-593)
C31400	Leaded commercial bronze	1900 (1038)	G	P	---	---	800-1200 (427-649)
C34000	Medium-leaded brass	1700 (927)	G	P	---	---	800-1200 (427-649)
C35300	High-leaded brass	1670 (910)	F	P	---	---	800-1100 (427-593)
C36000	Free-cutting brass	1650 (899)	P	F	---	1300-1450 (704-788)	800-1100 (427-593)
C37700	Forging brass	1640 (893)	P	E	100	1200-1500 (649-816)	800-1100 (427-593)
C38500	Architectural bronze	1630 (888)	P	E	90	1150-1350 (621-732)	800-1100 (427-593)

(continued)

TABLE 6-13—*Continued*

UNS No.	Name	Tensile Strength (average), ksi (MPa)				Yield Strength (average), ksi (MPa)				Elongation (average), % in 2 in. gage length				Rockwell Hardness			
		Flat Products		Rod		Flat Products		Rod		Flat Products		Rod		Flat Products		Rod	
		Hard	Soft	Hard	Soft	Hard	Soft	Hard	Soft	Hard	Soft	Hard	Soft	Hard B	Soft F	Hard B	Soft F
C46400 C46700	Naval brass	70[c] (483)	62 (427)	75[b] (517)	63 (434)	58[c] (400)	30 (207)	53[b] (365)	30 (207)	17[c]	40	20[b]	40	75[c]	56B	82[b]	55B
C48500	Leaded naval brass	---	---	75[b] (517)	57 (393)	---	---	53[b] (365)	25 (172)	---	---	15[b]	40	82[b]	55B	---	---
C50500	Phosphor bronze, 1.25% E	65 (448)	40 (276)	---	---	50 (345)	14 (97)	---	---	8	48	---	---	75	68	---	---
C51000	Phosphor bronze, 5%A	81 (558)	47 (324)	70[b] (483)	---	75 (517)	19 (131)	58[b] (400)	---	10	64	25[b]	---	87	73	78[b]	---
C52100	Phosphor bronze, 8% C	93 (641)	55 (379)	80[b] (551)	---	72 (496)	---	65[b] (448)	---	10	70	33[b]	---	93	75	85[b]	---
C52400	Phosphor bronze, 10% D	100 (690)	66 (455)	---	---	---	28 (193)	---	---	13	68	---	---	97	55B	---	---
C54400	Leaded phosphor bronze	58[b] (400)	44 (303)	68 (469)	---	40[b] (276)	19 (131)	57 (393)	--	24[b]	50	20	---	68[b]	65	80	---
C61400	Aluminum bronze, D	85 (586)	80 (551)	82 (565)	---	58 (400)	40 (276)	40 (276)	---	35	40	35	0	86	83B	90	---
C62300	Aluminum bronze	---	---	95[b] (655)	---	---	---	50[b] (345)	---	---	---	25[b]	---	---	---	88	---
C63000	Aluminum bronze	---	---	118[b] (814)	---	---	---	75[b] (517)	---	---	---	15[b]	---	---	---	98	---
C63800	Cu-Al-Si-Co	120 (830)	82 (565)	---	---	101 (700)	54 (370)	---	---	7	36	---	---	99	86	---	---
C64200	Aluminum-silicon bronze	---	---	102 (703)	90 (621)	---	---	68 (469)	55 (379)	---	---	22	28	---	---	94	---
C65100	Low-silicon bronze, B	---	---	70 (483)	40 (276)	---	---	55 (379)	14 (97)	---	---	15	50	---	---	80	55
C65400	Cu-Si-Sn-Cr	114 (790)	72 (500)	---	---	102 (705)	32 (220)	---	---	5	48	---	---	97	94	---	---
C65500	High-silicon bronze, A	94 (648)	60 (414)	92 (634)	58 (400)	58 (400)	25 (172)	55 (379)	22 (152)	8	60	22	60	93	85	90	60B
C66400	Cu-Zn-Fe-Co	88 (610)	63 (435)	---	---	85 (590)	45 (310)	---	---	5	25	---	---	88	95	---	---
C67500	Manganese bronze, A	85 (586)	58 (400)	70[b] (483)	56 (386)	74 (510)	25 (172)	60[b] (414)	25 (172)	3	40	20[b]	42	87	85	78[b]	---
C68800	Cu-Zn-Al-Co	114 (790)	82 (565)	---	---	102 (700)	69 (465)	---	---	5	30	---	---	97	100	---	---
C70600	Copper nickel, 10%[f]	60 (414)	44 (303)	---	---	57 (393)	16 (110)	---	---	10	42	---	---	72	65	---	---
C71500	Copper nickel, 30%	55 (379)	---	75[b] (517)	---	20 (138)	---	70 (483)	---	45	---	15	---	35	---	80	---
C72500	Cu-Ni-Sn	83 (570)	---	55 (380)	---	81 (560)	22 (150)	---	---	3	35	---	---	42	80	---	---
C75200	Nickel silver, 65-18	85 (586)	58 (400)	70[b] (483)	56 (386)	74 (510)	25 (172)	60[b] (414)	25 (172)	3	40	20[b]	42	87	85	78[b]	---

TABLE 6-13—*Continued*

UNS No.	Name	Melting Point, °F (°C)	Capacity for Being Cold Worked[e]	Capacity for Being Hot Worked[e]	Hot-Forgeability Rating (Forging Brass = 100)	Hot-Working Temperature, °F (°C)	Annealing Temperature (30-60 min.), °F (°C)
C46400 C46700	Naval brass	1650 (899)	F	E	90	1200-1500 (649-816)	800-1100 (427-593)
C48500	Leaded naval brass	1650 (899)	P	G	70	1200-1400 (649-760)	800-1100 (427-593)
C50500	Phosphor bronze, 1.25% E	1970 (1077)	E	G	---	1450-1600 (788-871)	900-1200 (482-649)
C51000	Phosphor bronze, 5% A	1920 (1049)	E	P	---	---	900-1250 (482-677)
C52100	Phosphor bronze, 8% C	1880 (1027)	G	P	---	---	900-1250 (482-677)
C52400	Phosphor bronze, 10% D	1830 (999)	G	P	---	---	900-1250 (482-677)
C54400	Leaded phosphor bronze	1830 (999)	G	---	---	---	900-1250 (482-677)
C61400	Aluminum bronze, D	1905 (1041)	G	G	---	1450-1700 (788-927)	1100-1650 (593-899)
C62300	Aluminum bronze	1915 (1046)	G	G	75	1300-1600 (704-871)	
C63000	Aluminum bronze	1930 (1054)	P	G	75	1450-1700 (788-927)	
C63800	Cu-Al-Si-Co	1885 (1030)	G	G	---	---	
C64200	Aluminum-silicon bronze	1840 (1004)	P	E	80	1300-1600 (704-871)	
C65100	Low-silicon bronze, B	1940 (1060)	E	E	---	1300-1600 (704-871)	
C65400	Cu-Si-Sn-Cr	1865 (1020)	G	G	---	---	
C65500	High-silicon bronze, A	1880 (1027)	E	E	40	1300-1600 (704-871)	
C66400	Cu-Zn-Fe-Co	1930 (1054)	G	G	---	---	
C67500	Manganese bronze, A	1630 (888)	P	E	80	1150-1450 (621-788)	
C68800	Cu-Zn-Al-Co	1765 (965)	G	G	---	---	
C70600	Copper nickel, 10%[f]	2030 (1110)	E	P	---	---	
C71500	Copper nickel, 30%	2100 (1149)	G	G	---	1550-1750 (843-954)	
C72500	Cu-Ni-Sn	2065 (1130)	E	E	---	---	
C75200	Nickel silver, 65-18	2260 (1238)	G	G	---	1700-1900 (927-1038)	

(continued)

TABLE 6-13—*Continued*

UNS No.	Name	Tensile Strength (average), ksi (MPa)				Yield Strength (average), ksi (MPa)				Elongation (average), % in 2 in. gage length				Rockwell Hardness			
		Flat Products		Rod		Flat Products		Rod		Flat Products		Rod		Flat Products		Rod	
		Hard	Soft	Hard	Soft	Hard	Soft	Hard	Soft	Hard	Soft	Hard	Soft	Hard B	Soft F	Hard B	Soft F
C76200	Nickel silver, 59-12	97 (670)	63 (435)	---	---	89 (610)	29 (200)	---	---	4	46	---	---	92	89	---	---
C78200	Leaded nickel silver	85 (586)	53 (365)	---	---	73 (503)	23 (159)	---	---	4	40	---	---	87	78	---	---

[a] Eighth hard.
[b] Half hard.
[c] Quarter hard.
[d] As extruded.
[e] E, excellent; G, good; F, fair; P, poor.
[f] Tube.
[g] Properties are before (soft) and after (hard) precipitation heat treatment.

All of the alloys have high resistance to fresh water and steam. Most of the bronzes, copper nickels, and nickel silvers have high resistance to corrosion by salt water but are subject to limitations at high water velocities due to breakdown of the passivating film.

Copper alloys have good resistance to alkalies and organic acids but not to inorganic acids. Alloys with high zinc content may corrode by dezincification, selectively removing zinc from the alloys and leaving the remaining metal with a spongy structure. Copper alloys are attacked by even trace quantities of mercury and by moist ammonia or ammoniacal compounds. Several of the alloy systems are subject to stress-corrosion cracking, which can be avoided in part by the stress relieving of fabricated parts prior to exposure to the corrodent.

Electrical Conductivity

In the electronic and electrical industry, a large user of copper, high electrical conductivity is the single most important material property; although for industrial use, conductivity must be accompanied by other suitable characteristics and properties.

The conductivity of commercial copper is commonly rated on a percentage basis. This method affords a convenient standard for buying and selling, for comparison of quality for electrical purposes, and for comparison with other metals. The rating is based on percentage of a mass conductivity, adopted as a standard by the International Electrotechnical Commission in 1913 and subsequently by the American National Standards Institute, American Society for Testing and Materials, and other organizations.

Resistivity units are based on the International Annealed Copper Standard (IACS), which is $1/58\ \Omega \cdot mm^2/m$ at 68° F (20° C) for 100% conductivity. The value of $0.017241\ \Omega \cdot mm^2/m$ and the value of $0.15328\ \Omega \cdot g/m^2$ at 68° F (20° C) are respectively the international equivalent for volume and weight resistivity of annealed copper equal to 100% conductivity. The latter term means that a copper wire one meter in length and weighing one gram would have a resistance of $\Omega \cdot 0.15328$.

The resistivity of copper increases with temperature, the amount of change being given by the temperature coefficient 0.00393/° C at 20° C. Commercial electrolytic copper of the highest purity has a conductivity of about 102 IACS, and most of the copper sold averages between 100.5 and 101.8.

Wear Resistance

Copper alloys have been used in wear applications for centuries. Today these applications remain among the most important uses of copper alloys. Plain bearings of copper alloys are widely used in fractional horsepower motors, in marine applications, and in very large, heavily loaded journals such as paper mill rolls, railroad train bearings, and steel mill roll bearings. Wrought copper alloys are used for stamped gears, cams, and escapements in clocks, timing mechanisms, switch gears, and cameras.

Brasses, bronzes, and beryllium coppers are the most widely used copper alloys for wear components. Copper alloys are often subjected to cavitation and other forms of corrosive wear, but in machine applications the most important modes of wear are abrasion and metal-to-metal.

APPLICATIONS

Copper's corrosion resistance is an important property that adapts the metal to widespread commercial use. Copper resists oxidation while carrying water, and it does not pick up a mineral deposit from the water. This advantage accounts for the extensive use of copper for water pipes and valves and other fittings in plumbing systems. Copper alloys are widely used in marine applications, specified for propellers, bushings, hardware, and heat exchangers that must resist attack by salt water. In the chemical process industry, copper alloys withstand a wide range of acids, bases, and organic solutions. About one third of all the copper alloy material produced is used for tubing and pipe that carry corrosive fluids of one form or another.

Low Temperature

Copper and copper alloys were the first metals used in the construction of low-temperature equipment for liquefaction and storage of cryogenic fluids. Copper and many copper alloys retain excellent ductility at low temperatures. This characteristic plus good thermal conductivity provides a combination of properties needed for heat exchangers and other components in cryogenic plants and in low-temperature processing and storage facilities.

High Temperature

For applications involving elevated temperatures, the American Society of Mechanical Engineers (ASME) Boiler and

TABLE 6-13—*Continued*

UNS No.	Name	Melting Point, °F (°C)	Capacity for Being Cold Worked[e]	Capacity for Being Hot Worked[e]	Hot-Forgeability Rating (Forging Brass = 100)	Hot-Working Temperature, °F (°C)	Annealing Temperature (30-60 min.), °F (°C)
C76200	Nickel silver, 59-12	1902 (1039)	G	G	---	---	
C78200	Leaded nickel silver	1830 (999)	G	P	---	---	

Pressure Vessel Code is used to specify critical copper alloy components. The code recommendations make allowances for a falloff in creep or rupture strength, a relatively low strength on a strength-to-weight basis, and susceptibility to stress-corrosion cracking in certain environments. Silicon bronzes, aluminum brass, and copper-nickel alloys are widely used in high-temperature equipment applications.

End Use

The breakdown of 1983 market data on end uses for primary brass mill shipments includes the following major fields for use of copper and copper alloys: building products, 18%; transportation equipment, 10%; electrical and electronic products, 8%; and industrial machinery and equipment, 7%.

In the electrical equipment field, the four largest end uses in descending order are motors and generators, transformers, switchgears and switchboards, and current-carrying devices.

For fabricated metal products, the largest uses of copper are in valves and pipe fittings (leading with more than 9% of all copper use), and—in descending order—plumbing fixtures and trim, metal stampings, screw machine products, gears, bearings, and hardware.

One of the most severe mechanical applications of a copper alloy is the use of beryllium copper for cavities in plastic injection molds. The cavity must withstand high compressive loads and injection pressures often as high as 20 ksi (138 MPa). The hardness is not as important in this application as compressive and tensile strength.

In other types of machinery, refrigeration equipment accounts for about 4% of all copper used, followed by pumps, compressors, and general machine shop use.

In transportation equipment, motor vehicles claim almost 10% of all copper consumed. About 40 lb (18 kg) of copper is used per vehicle. A large airplane uses about 3000 lb (1361 kg) of copper, and a Pullman railroad car uses about 2000 lb (907 kg).[4]

Industry Users

Among the largest industry users of copper is the construction industry, where copper is used for building wire, roofing products, plumbing goods, builder's hardware, gutters, flashing, and fittings. In transportation, copper is used by many manufacturers: by automobile makers, in electric motors, radiators, heaters, defrosters, and oil lines; by railroad equipment makers, in locomotives, passenger trains, and signal devices; and by aircraft makers, in wiring systems.

The appliance industry is a large consumer of copper, particularly for washing machines, air conditioners, refrigerators, and radio and television sets. Copper is used extensively for telephone wire and cable, household cookware, and numerous other consumer and industrial products.

Copper is one of the principal metals used for electroplating. Not only can the outside coating be of copper, but in chromium plating on steel, the usual method is to plate the steel first with thin undercoatings of copper and nickel. Electroplating is sometimes done with brass or bronze. Copper is also used for cladding. In this process, copper sheet is hot rolled onto the surface of steel or other metal.

FABRICATION

Copper and copper alloys have always been noted for excellent formability and ease of fabrication into utensils and products. For additional information on the various sheet and bulk metalforming processes and equipment, refer to *Forming,* Volume II in this Handbook series.

Melting

Alloys are initially prepared by melting the various elements in induction furnaces. Usually the melt is protected from oxidation by covering it with charcoal. Copper-rich alloys are also deoxidized by adding elements such as phosphorus, lithium, or boron. The melt is usually poured at 100 to 200° F (38 to 93° C) above its melting point into oil and graphite-dressed, water-cooled copper molds of suitable shape.

Continuous casting of copper and its alloys is common practice and is accomplished by pouring the melt into the top of an open-ended mold as the solidified casting is withdrawn from the bottom. Various mold designs are used, but all use water to cool the mold itself in addition to use in cooling sprays and tanks below the mold. Casting can be performed in either a vertical or a horizontal plane.

Mechanical properties of castings can be affected by such problems as large grain size, directional grain growth, and segregation. Either hot working of the casting by rolling, forging, piercing, or extrusion, or cold working by rolling or drawing with necessary intermediate annealing can correct these conditions and improve properties.

Processing

Processing varies with alloy, form, and size. Flat products are cast in heavy cakes for initial hot rolling or in bars for cold rolling. Copper and lead-free brasses are readily hot rolled in a single heating from a 9″ (229 mm) thick cake to a 1/2″ (13 mm) slab. Further rolling is done cold with intermediate anneals. Some leaded brasses, tin bronzes, and other hot-short alloys (alloys that are brittle when heated above red heat) must be cold rolled from the casting to final size, using intermediate anneals that heat the particular alloy above its recrystallization tem-

TABLE 6-14
Typical Mechanical Properties of Cast (Heat-Treated) Copper and Alloys

UNS Number	Name	Tensile Strength, ksi (MPa)	Yield Strength,[a] ksi (MPa)	Elongation, % in 2 in.	Modulus of Elasticity, 1000 ksi (GPa)	Brinell Hardness, 500 kg Load, 10 mm ball	Compressive Strength, ksi (MPa)	Pattern-makers Shrinkage, in./ft (mm/m)	Electrical Conductivity, % IACS	Melting Range, °F (°C)	Relative Suitability for Welding[e]: Oxyacetylene	Gas-Shielded Arc	Coated-Metal Arc
Copper:													
C80100	High-conductivity copper	25 (172)	9 (62)	40	17 (117)	44	---	1/4 (20.8)	100	1948-1981 (1064-1082)	N	C	N
High-copper alloy:													
C81500	Chromium copper	---	---	---	16.5 (114)	---	---	1/4 (20.8)	82	1967-1985 (1075-1085)	N	C	N
Beryllium coppers:													
C82400	Beryllium copper[f]	150 (1030)	140 (970)	3	18.5 (128)	R_C36	---	3/16 (15.6)	23	1650-1825 (899-996)	N	C	C
C82500	Beryllium copper[f]	160 (1100)	135 (930)	2	18.5 (128)	R_C40	---	3/16 (15.6)	22	1575-1800 (857-982)	N	C	C
C82800	Beryllium copper[f]	190 (1310)	170 (1170)	2	18.5 (128)	R_C44	---	3/16 (15.6)	21	1625-1710 (885-932)	N	C	C
C82200	Beryllium copper[f]	100 (690)	75 (520)	10	20 (138)	175	---	3/16 (15.6)	45	1830-1980 (1000-1080)	N	C	C
Leaded red brasses:													
C83600	85-5-5-5	37 (255)	17 (117)	30	13.5 (93)	60	14 (97)	3/16 (15.6)	15	1570-1850 (854-1010)	N	N	C
C83800	Commercial red brass	35 (241)	16 (110)	25	13.3 (92)	60	11.5 (79)	3/16 (15.6)	15	1550-1840 (843-1004)	N	N	C
Leaded semi-red brasses:													
C84400	Valve composition	34 (234)	15 (103)	26	13 (90)	55	---	3/16 (15.6)	16.4	1549-1940 (843-1060)	N	N	C
C84800	Leaded semi-red brass	36 (248)	14 (96)	30	15 (103)	55	12.8 (88)	3/16 (15.6)	16.4	1530-1750 (832-954)	N	N	C
Yellow brasses and leaded yellow brasses:													
C85200	High-copper yellow brass	38 (262)	13 (90)	35	11 (76)	45	9 (62)	3/16 (15.6)	18	1700-1725 (927-941)	C	N	N
C85400	No. 1 yellow brass	34 (234)	12 (83)	35	12 (83)	50	9 (62)	3/16 (15.6)	19.5	1700-1725 (927-941)	C	D	D

TABLE 6-14—*Continued*

UNS Number	Name	Tensile Strength, ksi (MPa)	Yield Strength,[a] ksi (MPa)	Elongation, % in 2 in.	Modulus of Elasticity, 1000 ksi (GPa)	Brinell Hardness, 500 kg Load, 10 mm ball	Compressive Strength, ksi (MPa)	Patternmakers Shrinkage, in./ft (mm/m)	Electrical Conductivity, % IACS	Melting Range, °F (°C)	Relative Suitability for Welding[e]: Oxyacetylene	Gas-Shielded Arc	Coated-Metal Arc
C85700		50 (345)	18 (124)	40	14 (97)	75	---	3/16 (15.6)	22	1675-1725 (913-941)	N	N	N
C85800	Die-casting alloy	55 (379)	30[b] (207)	15	15 (103)	B55	---	---	20	1600-1650 (871-899)	N	N	N
Manganese and leaded-manganese bronzes:													
C86200		95 (655)	48[b] (331)	20	15 (103)	180[c]	50 (345)	1/4 (20.8)	7.5	1650-1725 (899-941)	B	C	B
C86300		119 (821)	83[b] (572)	18	14.2 (98)	225[c]	60 (414)	1/4 (20.8)	8	1625-1693 (885-923)	D	D	B
C86400		65 (448)	25[b] (172)	20	14 (97)	90	23 (159)	1/4 (20.8)	19	1583-1616 (862-880)	D	D	D
C86500		71 (490)	28[b] (193)	30	15 (103)	100	24 (165)	1/4 (20.8)	22	1583-1616 (862-880)	D	D	D
C86700		85 (586)	42 (290)	20	15 (103)	B80	---	1/4 (20.8)	16.7	1583-1616 (862-880)	N	N	N
Silicon bronzes and silicon brasses:													
C87200		55 (379)	25 (172)	30	15 (103)	85	18 (124)	1/4 (20.8)	6	1580-1780 (860-971)	B	C	C
C87400		55 (379)	24 (165)	30	15.4 (106)	70	---	3/16 (15.6)	6.7	1510-1680 (821-916)	C	C	N
C87500		67 (462)	30 (207)	21	15.4 (106)	115	26.5 (183)	3/16 (15.6)	6.7	1510-1680 (821-916)	C	C	N
C87800	Die-casting alloy	85 (586)	50[b] (345)	25	20 (138)	B85	---	3/16 (15.6)	6.7	1510-1680 (821-916)	N	N	N
Tin bronzes and leaded tin bronzes:													
C90300	88-8-4	45 (310)	21 (145)	30	14 (97)	70	13 (90)	3/16 (15.6)	12	1570-1832 (854-1000)	C	C	C
C90500	88-10-0-2 or G bronze	45 (310)	22 (152)	25	15 (103)	75	---	3/16 (15.6)	11	1570-1830 (854-999)	C	C	C

(continued)

TABLE 6-14—*Continued*

UNS Number	Name	Tensile Strength, ksi (MPa)	Yield Strength,[a] ksi (MPa)	Elongation, % in 2 in.	Modulus of Elasticity, 1000 ksi (GPa)	Brinell Hardness, 500 kg Load, 10 mm Ball	Compressive Strength, ksi (MPa)	Patternmakers Shrinkage, in./ft (mm/m)	Electrical Conductivity, % IACS	Melting Range, °F (°C)	Relative Suitability for Welding[e]: Oxyacetylene	Relative Suitability for Welding[e]: Gas-Shielded Arc	Relative Suitability for Welding[e]: Coated-Metal Arc
C92200	Composition M valve bronze	40 (276)	20 (138)	30	14 (97)	65	---	3/16 (15.6)	14.3	1518-1810 (826-988)	N	N	N
C92300		40 (276)	20 (138)	25	14 (97)	70	35 (241)	3/16 (15.6)	12	1570-1830 (854-999)	N	N	N
C92600	88-10-02 commercial bronze	44 (303)	20 (138)	30	15 (103)	70	12 (83)	3/16 (15.6)	9	1550-1800 (843-982)	N	N	N
C92700	Leaded bearing bronze	42 (290)	21 (145)	20	16 (110)	77	---	3/16 (15.6)	11	1550-1800 (843-982)	N	N	N
High-leaded tin bronzes:													
C93200	83-7-7-3	35 (241)	18 (124)	20	14.5 (100)	65	---	7/32 (18.2)	12	1570-1790 (854-977)	N	N	N
C93500	85-5-9	32 (221)	16 (110)	20	14.5 (100)	60	13 (90)	3/16 (15.6)	15	1570-1830 (854-999)	N	N	N
C93700	80-10-10	35 (241)	18 (124)	20	11 (76)	60	13 (90)	1/8 (10.4)	10	1403-1705 (762-929)	N	N	N
C93800	78-7-15	30 (207)	16 (110)	18	10.5 (72)	55	12 (83)	1/8 (10.4)	11.5	1570-1730 (854-943)	N	N	N
C94300	70-5-25	27 (186)	13 (90)	10	10.5 (72)	48	11 (76)	1/8 (10.4)	9	---	N	N	N
Nickel-tin bronzes:													
C94700	88-5-5-2	50 (345)	23 (159)	35	15 (103)	85	---	3/16 (15.6)	12	1660-1880 (904-1027)	C	B	B
C94800		50 (345)	23 (159)	35	15 (103)	80	---	3/16 (15.6)	12	1660-1880 (904-1027)	N	N	N
Aluminum bronzes:													
C95200		80 (552)	27 (186)	35	15 (103)	125	27 (186)	1/4 (20.8)	11	1907-1913 (1042-1045)	N	A	B
C95300		75 (517)	27 (186)	25	16 (110)	140	---	3/16 (15.6)	13	1904-1913 (1040-1045)	N	A	B

TABLE 6-14—*Continued*

UNS Number	Name	Tensile Strength, ksi (MPa)	Yield Strength,[a] ksi (MPa)	Elongation, % in 2 in.	Modulus of Elasticity, 1000 ksi (GPa)	Brinell Hardness, 500 kg Load, 10 mm Ball	Compressive Strength, ksi (MPa)	Pattern-makers Shrinkage, in./ft (mm/m)	Electrical Conductivity, % IACS	Melting Range, °F (°C)	Relative Suitability for Welding[e]		
											Oxya-cetylene	Gas-Shielded Arc	Coated-Metal Arc
C95400		85 (586)	35 (241)	18	15.5 (107)	170	---	3/16 (15.6)	13	1880-1900 (1027-1038)	N	A	B
C95500		100 (690)	44 (303)	12	16 (110)	195	---	3/16 (15.6)	8.5	1900-1930 (1038-1054)	N	B	B
C95600		75 (517)	34 (234)	18	15 (103)	140[c]	---	3/16 (15.6)	8.5	1800-1840 (982-1004)	N	B	C
C95800		95 (655)	38 (262)	25	16.5 (114)	159[c]	---	3/16 (15.6)	7.1	1910-1940 (1043-1060)	N	B	B
Copper nickels:													
C96200	90-10	45[d] (310)	25[d] (172)	20[d]	18 (124)	---	---	3/16 (15.6)	11	2010-2100 (1099-1149)	N	C	B
C96400	70-30	68 (469)	37 (255)	28	21 (145)	140[c]	---	7/32 (18.2)	5	2140-2260 (1171-1238)	N	B	B
C96600	Beryllium copper-nickel	---	70[b] (483)	7	22 (152)	---	---	7/32 (18.2)	4.3	2010-2160 (1099-1182)	B	C	C
Nickel silvers:													
C97300		35 (241)	17 (117)	20	16 (110)	55	---	3/16 (15.6)	5.7	1850-1904 (1010-1040)	N	N	N
C97600		40 (276)	24 (165)	20	19 (131)	80	30 (207)	1/8 (10.4)	5	2027-2089 (1108-1143)	N	N	N
C97800		55 (379)	30 (207)	16	19 (131)	130[c]	---	3/16 (15.6)	4.5	2084-2156 (1140-1180)	N	N	N
Special alloys:													
C99300	Incramet	95 (655)	55 (379)	2	18 (124)	200[c]	---	3/16 (15.6)	9	1955-1970 (1068-1077)	N	B	B

(*Copper Development Association, Inc.*)

[a] Taken at 0.5% elongation in 2 in. gage length under load.
[b] Taken at 0.2% offset.
[c] 3000 kg load.
[d] Minimum.
[e] Relative weldability ratings: A, excellent; B, good; C, fair; D, poor; N, not recommended.
[f] Properties after solution annealing and precipitation hardening.

COPPER FABRICATION

perature. These alloys are often cast continuously in thin sections that are cooled and coiled as they are cast.

Round copper rods and wire are started by hot rolling castings (wire bar) in grooved rolls, but most copper alloys are extruded. Both wire and rods are subsequently reduced in size by cold rolling, drawing, or both. In extrusion, a billet placed in a press is forced through a lubricated die by a ram.

Tin and tin-lead bronzes that are hot-short are continuously cast in rod form and are cold rolled or drawn to finished sizes using intermediate anneals as necessary.

Copper and high-copper, lead-free brass tubes are started by hot piercing wherein the solid billet is driven with a helical motion over a rotating mandrel to form a seamless tube. Tubes of most other alloys are formed by extrusion because the alloys do not stand piercing. In extrusion of tube, a mandrel is pushed through a billet forming the inside of the tube, and a ram pushes the billet over the mandrel and through a die to form a seamless tube.[5]

Casting

Pure copper has poor casting qualities caused mainly by the release of gases from the molten metal when it solidifies. Machining is also difficult. Castings of intricate shapes are seldom made of pure copper, although there are exceptions where high thermal or electrical conductivity is a primary consideration. Wirebars, cakes, billets, and similar objects of regular or simple shape intended for working are readily cast if the oxygen content of the molten metal is controlled. In such applications, some porosity is allowable because it will be closed up during working.

Most copper alloys have excellent casting qualities, and some of the bronzes and brasses are outstanding in their suitability for the casting processes. Die castings are made with some compositions of brass.

Forming

Pure copper is highly workable, as are many of the wrought copper alloys. The commonly used methods of working include rolling, extrusion, wire drawing, piercing, and stamping. Other processes that are used include forming by forging, pressing, deep drawing, flanging, spinning, and cupping, as well as various machining operations.

General data. Copper can be hot worked extensively at 1400 to 1600° F (760 to 871° C) and has excellent ductility in cold work. Sheet, rod, and tube are fabricated after cold rolling or drawing to increase strength and hardness as needed for a particular application. These materials can also be softened by annealing at 700 to 1400° F (371 to 760° C).

Copper wire about 0.04" (1.0 mm) diam is commonly made by drawing from a hot rolled rod without annealing, but smaller sizes may involve intermediate anneals. Copper shapes for switch parts are made by extrusion; brushes and commutator sections, by rolling and drawing. Copper for electrical purposes must be very pure; the presence of even traces of certain impurities (particularly phosphorus, arsenic, iron, titanium, and silicon) decreases the conductivity considerably. Such copper should have a minimum copper content of 99.90% (silver being counted as copper).

Copper containing small amounts of silver retains the effect of cold working to a higher temperature than pure copper. This characteristic is useful when comparatively high temperatures are involved, as in soldering operations or for stressed conductors that operate at moderately elevated temperatures.

Beryllium copper. Beryllium-copper strip is easily formed. The tooling requirements for stamping and forming it are the same as those for any other copper-based material of similar hardness. Tools should be kept sharp to minimize edge burr formation. The burrs should be removed prior to age hardening if solution annealed or cold rolled material is used. Die lubricants containing sulfur cause staining of copper alloys and hence their use should be avoided.

The beryllium-copper alloys provide significantly increased strength while retaining most of copper's desirable properties. In the electrical and electronic industries, a combination of strength, formability, and conductivity makes beryllium copper the appropriate material for application in high-performance sockets, terminals, connectors, switches, and relays.

In other areas of application, beryllium copper is used where there is need for high material strength along with corrosion resistance. In addition, beryllium copper's hardness leads to its use as a bearing material. Figure 6-4 provides a comparison of strength and formability between beryllium copper and phosphor bronze. In the AM temper, alloy C17200 has better formability than extra-spring temper phosphor bronze at the same strength level. For some forming applications, a further advantage of beryllium copper is that its formability is relatively uniform in both the longitudinal and transverse directions.

Machining[6]

Most copper alloys are readily machined by usual methods using standard tools designed for steel, but at higher speeds. Consideration of the wide range of characteristics presented by various types of copper alloys and the adaptation of the machining practice to the particular material concerned will provide greatly improved results.

Most machining involves operations in which either the workpiece or the tool revolves, and the machining operation is performed on the circumference and at right angles to the longitudinal axis. Exceptions are generally secondary operations, such as milling and cross drilling.

Characteristics. Copper and each of its alloys have individual characteristics such as ductility, hardness, tensile strength, and type of chip produced. Metal to be machined must be rigid enough to stand up against the turning tool without distortion. This demand requires a temper in the wrought copper alloys for automatic screw machining, yet the alloys may need to be soft enough to withstand subsequent cold working. Because of their varying characteristics, care must be taken with copper alloys to clearly define speed, rate of feed, and form of tool to be used in machining, whether it be a single-point tool, form tool, milling cutter, drill, tap, chaser, reamer, or saw. Additional information is available in *Machining,* Volume I, of this Handbook series.

Groupings. Copper and its alloys can be grouped into classifications for which broad ranges of speeds, feeds, and tool geometry can be established. These general groupings may be summarized as follows: Tellurium and sulfur are added to enhance the machinability of copper with minimum sacrifice in conductivity. The effect of adding controlled quantities of lead may be illustrated by considering C27000 (yellow brass, 65%), an alloy of 65% copper with 35% zinc. When cut, this alloy produces long, stringy chips. By replacing a like amount of copper with lead, a leaded brass results that speeds machining by producing an easily broken chip. This alloy—61.5% copper, 35.5% zinc, 3% lead, known as C36000 (free-cutting brass)—ranks as the finest machining alloy. It is given a rating of 100

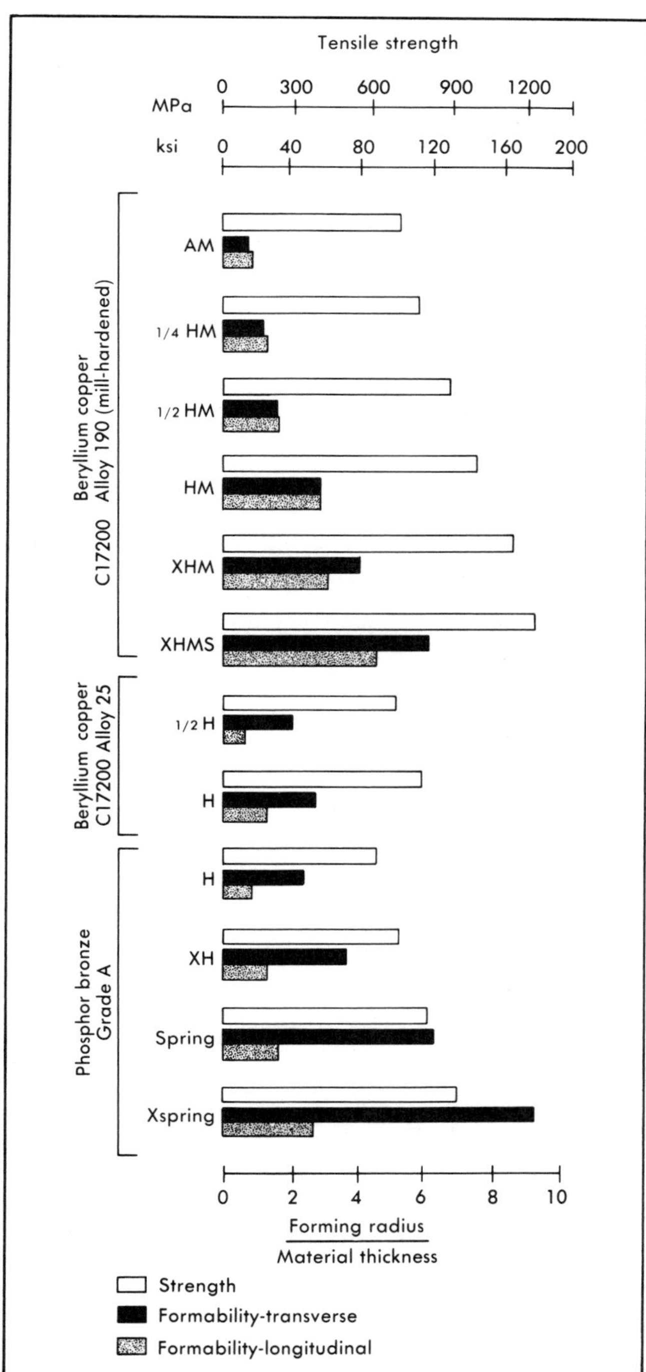

Fig. 6-4 Strength and formability comparison of beryllium copper to phosphor bronze. (*Brush Wellman, Inc.*)

and provides the standard to which other alloys are compared when discussing machinability.

Ratings. Machinability ratings in Table 6-15 are comparisons with free-cutting brass and are arbitrary values that should be considered only as approximate indications. While they provide a reasonable guide to suitable cutting speed and the amount of power required for cutting, many variables may materially affect the ratings.

In rating machinability of the metals listed, consideration was given to the following factors:

- Speed of machining.
- Tool wear.
- Finish.
- Accuracy.
- Power required.

A machinability rating of 20 was arbitrarily assigned to those metals that are difficult to machine and produce long, tough, stringy turnings.

Speed of machining. Machining speed is limited by many factors, principally the heat that is generated at the contact point of tool and workpiece. This heat must be carried away by the coolant/lubricant and to some extent by the chip itself. The efficiency of the cutting fluid affects the temperature of the work and the tool. Such factors as the shape of the tool and its sharpness and depth of cut affect the manner and speed with which the chip is carried away from the tool. These factors, in turn, control the temperature of the work and the tool.

With free-cutting alloys, and even with some of the moderately machinable alloys, the chip breaks up at a rapid rate and is only in momentary contact with the tool. This situation calls for a cutting fluid that principally serves as a coolant, with little regard for lubrication. Rake angles and clearances can be held to the minimum specifications. The small rake angle breaks the chip almost immediately, and the reduced clearance provides greater support for the tool's cutting edge.

The nonleaded alloys tend to weld on tools, and therefore a lubricant is required to build up a protective film between the work and the tool. This requirement diminishes the cooling characteristic of the cutting medium, necessitating a reduction in turning speed; and rake angles must be increased to start the chip flowing off the tool as rapidly as possible. The greater rake angle tends to prevent buildup on the face of the tool from which a rubbing or burnishing condition would result. Under this condition, a higher (better) finish attained when grinding the tool tends to reduce the friction produced by the chip with the tool.

High-strength alloys produce much heat due to the toughness of the material itself. Being intrinsically harder, these alloys produce considerably more heat by friction, and, at the same time, actually wear the cutting tool material much faster than do the free-cutting alloys.

Tool wear. The tool wear ratings in Table 6-15 are approximate, since they are based on user and producer reports of shop experience and represent an overall average of a wide variety of types of products and operations. The values shown represent the comparative number of tool dressings required for a given quantity of production. Carbide-tipped tools require fewer dressings than standard high-speed steel tools. A generally outstanding characteristic of copper alloys as compared with other metals is their ability to be machined in long runs with a minimum of tool dressing.

Heat Treatment

Pure copper and the copper alloys are usually used in the as-received condition; however, in some applications, heat treatments are used to modify the properties. Annealing may be done

COPPER FABRICATION

TABLE 6-15
Machinability Ratings and Tool Wear Ratings

Copper or Copper Alloy Number	Trade Name	Approximate Machinability Rating*	Approximate Tool Wear Rating** Number of Tool Dressings per Basic Production	
			Carbide-Tipped	Standard High-Speed Steel
	Machinability Rating Group 70 to 100			
C36000	Free-cutting brass	100	1	4
C35600	Extra-high leaded brass	90 to 100	1	5
C18700	Leaded copper	80 to 90	1 1/2 to 2	6
C34500, C35300	High-leaded brass	85	1 1/2	6
C14500	Phosphorus deoxidized tellurium-bearing copper	85	2	6
C14700	Sulfur-bearing copper	85	1 1/2 to 2	6
C19100	Tellurium nickel copper	85	2 to 3	7†
C31400	Leaded commercial bronze	80	1 1/2 to 2	5
C54400	Phosphor bronze, B-2	80	3	8†
C79800	Extruded leaded nickel copper	80	2 to 3	6
C33500	Low-leaded brass	70	2	5
C34000, C35000	Medium-leaded brass	70	2	5
C48500	Naval brass—high leaded	70	3	5
	Machinability Rating Group 30 to 60			
C63900	Aluminum silicon bronze	60	4	9†
C53400	Phosphor bronze, B-1	50	4	9†
C79200	Leaded nickel silver	50	3	9†
C26000	Cartridge brass, 70%	30	3	6
C46400	Naval brass	30	4	8†
C65100	Low-silicon bronze, B	30		
C67500	Manganese bronze, A	30	5	10
	Machinability Rating Group 20			
C10100, C10200	Oxygen-free copper	20	5	10†
C11000	ETP copper	20	3	6
C22000	Commercial bronze, 90%	20	3	6
C51000	Phosphor bronze, 5% A	20		

(*Copper Development Association, Inc.*)

* Approximate machinability rating based on No. 360 (free-cutting brass) = 100.

** The approximate tool wear rating can be defined in the following example:

If it is assumed that 100,000 units of any given shape and weight of No. 360 (free-cutting brass) can be made per tool dressing, a similar piece made from No. 464 (naval brass, uninhibited) using a carbide-tipped tool is expected to require four (4) dressings for the same quantity; or the expected quantity of No. 464 (naval brass, uninhibited) pieces per tool dressing = $\frac{100,000}{4}$. Similarly, using standard high-speed steel tooling, an equivalent piece is expected to require eight (8) dressings for the same quantity; or the expected quantity of No. 464 (naval brass, uninhibited) pieces per tool dressing = $\frac{100,000}{8}$.

† Carbide-tipped tools or super-high-speed steels are strongly recommended where practical when the tool wear rating is 7 or larger.

to soften wrought alloys when extensive drawing is required.[7]

Stress relieving is performed to provide dimensional stability for parts that are subjected to significant amounts of machining, or to reduce residual stress level on parts that may be exposed to atmospheres that cause stress-corrosion cracking.

Beryllium copper and certain other alloys can be solution treated and age hardened. Included in this group are aluminum bronzes, nickel-copper alloys, chromium copper, and zirconium copper. All of the cast age-hardenable alloys (except C82000 and C82200) attain a hardness of about R_C40 or more when age hardened to maximum strength. For additional information, see Chapter 14, "Heat Treatment of Other Metals," in this volume.

MAGNESIUM

Magnesium (Mg) is a silvery white metal with a specific gravity of 1.74. In the pure state, magnesium's melting point is 1202° F (650° C), which is approximately the same temperature required to melt aluminum.

Most pure metals, including magnesium, are soft and are not suitable for structural use. However, strength properties comparable to those of many aluminum alloys are obtained by alloying magnesium with other metals, and, in some cases, by heat treating or cold or hot working.

With a unit weight of 0.063 lb/in.^{3} (1.74 g/cm^3), magnesium is the lightest of the commonly used metals. For engineering applications, magnesium is usually alloyed with one or more elements, including aluminum, manganese, rare earth metals, lithium, thorium, zinc, and zirconium. The resultant alloys have very high strength-to-weight ratios.

While its light weight is magnesium's best known characteristic (aluminum weighs 1 1/2 times more than magnesium; iron and steel, 4 times more; copper, 5 times more), there are also other desirable properties. Magnesium's excellent machinability, for example, makes it economical in parts where weight saving may not be of primary importance, but where much costly machining is required. Parts made of magnesium can be machined at higher speeds, with fewer cuts, and with greater economies than are possible with most other metals.

BASIC METAL PRODUCTION

Magnesium occurs widespread in nature in the form of various compounds. The principal ores of magnesium are dolomite (calcium magnesium carbonate), magnesite, and carnallite. Magnesium also exists in nature as the chloride in seawater, in underground natural brines, and in salt deposits.

Metallic magnesium can be made by a number of processes. However, only two processes are in large-scale commercial use. The earliest method, and the one used to produce the most magnesium, is the electrolytic method. Another method, introduced during World War II, is a thermal reduction method called the ferrosilicon (Magnetherm or Pidgeon) process.

Thermal Reduction Processes

For thermal reduction, dolomite is generally used as the source of magnesium oxide. In the original process developed by Dr. L. M. Pidgeon, powdered dolime and ferrosilicon (an alloy of iron and silicon) are mixed, briquetted, and then heated for several hours in a steel retort at about 2000° F (1100° C) under a high vacuum. The magnesium volatilizes from the briquettes and is condensed to a solid "crown" at the cold end of the retort.

In the more recently developed Magnetherm process, ferrosilicon, bauxite or aluminum, and calcined dolomite are added to a molten bed. Energy requirements are provided by electrical resistance heating through the molten bed. This process allows the reduction of the magnesium oxide content of the calcined dolomite to proceed more rapidly and on a larger scale than the Pidgeon process does. The magnesium vapor emitted from the reduction furnace is condensed in a water-cooled crucible that can be detached from the reduction furnace. The crude magnesium requires processing in a foundry to produce a finished product.

Electrolytic Processes

The bulk of world magnesium production is by electrolytic processes wherein magnesium compounds are first precipitated from sea water by means of lime, ground oyster shells, or calcined dolomite. Seawater contains about 0.13% magnesium. The seawater is mixed with dolime and placed in large settling ponds where insoluble magnesium hydroxide settles to the bottom of the pond. The magnesium hydroxide slurry is pumped from the bottom of the ponds to filters where it is concentrated for further processing.

In one method, the precipitated magnesium hydroxide is converted to magnesium chloride with hydrochloric acid, dried, and fed to the electrolytic cells. The chlorine formed in the electrolysis is converted to hydrochloric acid and recycled. Another method is to dry and calcine the magnesium hydroxide to form magnesium oxide, which is reacted with carbon and chlorine from the cells to form the anhydrous magnesium chloride for feeding the cell.

In the electrolytic cell, the magnesium chloride is decomposed by an electric current into molten magnesium and chlorine gas. The magnesium metal is drawn off and cast into primary ingots. It can also be alloyed with metals, such as aluminum and zinc, and cast into magnesium alloy ingots. The chlorine is collected and recycled in gaseous form.

More recently developed electrolytic processes use concentrated magnesium chloride from natural brines as the raw material. Both magnesium and chlorine are produced as products.

ALLOY DESIGNATION SYSTEM

Primary magnesium, like most metals, lacks sufficient strength in its elemental state to be used as a structural metal. Therefore, it must be alloyed with various other metals, such as aluminum, manganese, thorium, rare earth metals, lithium, tin, zinc, and zirconium. Magnesium alloys are most commonly designated by a system established by the American Society for Testing and Materials (ASTM), which covers both chemical compositions and tempers.

The ASTM designations for alloys are based on chemical composition and consist of two letters representing the two alloying elements specified in the greatest amount, arranged either in decreasing percentages or—if of equal percentage—alphabetically. The letters are followed by the respective percentages rounded off to whole numbers, with a serial letter at

MAGNESIUM ALLOY DESIGNATION SYSTEM

the end. The serial letter indicates some variation in composition. Experimental alloys have the letter X between the alloy and serial numbers.

The following letters designate various alloying elements: A—aluminum, B—bismuth, C—copper, D—cadmium, E—rare earths, F—iron, G—magnesium, H—thorium, K—zirconium, L—lithium, M—manganese, N—nickel, P—lead, Q—silver, R—chromium, S—silicon, T—tin, Z—zinc.

Primary magnesium metal and alloys have also been assigned UNS (unified numbering system) designations according to the Unified Numbering System for Metals and Alloys, SAE HS1086a and ASTM DS-56A. The UNS designation for a metal or alloy consists of a letter followed by five numbers. The UNS system is intended to provide a nationally accepted means of correlating the many alloy designation numbers used by various organizations, and an improved system for indexing, record keeping, data storage and retrieval, and cross referencing. The numbers M10001 through M19999 have been reserved for magnesium and magnesium alloys.

ASTM Specification B 296 designates tempers for magnesium alloys. Temper designation is separated from alloy designation by a dash. Table 6-16 describes the ASTM tempers commonly used for magnesium cast and wrought products.

Magnesium alloys in their various forms are covered by government and national society specifications. The various specifications for ingot, cast, and wrought forms of magnesium alloys are given in Table 6-17.

TABLE 6-16
ASTM Temper Designations for Magnesium*

Description	Temper Designation
As fabricated	-F
Annealed recrystallized (wrought products only)	-O
Strain hardened	-H
Strain hardened only	-H1
Strain hardened and then partially annealed	-H2
Thermally treated to produce stable tempers other than -F, -O, or -H	-T
Annealed (cast products only)	-T2
Solution heat treated and naturally aged to a substantially stable condition	-T4
Artificially aged only	-T5
Solution heat treated and then artificially aged	-T6
Solution heat treated and then stabilized	-T7
Solution heat treated, cold worked, and then artificially aged	-T8

* ANSI/ASTM B296 "Temper Designations of Magnesium Alloys"

TABLE 6-17
Specifications Covering Selected Magnesium Products

Product	Alloy	ASTM	Federal	Military	AMS & SAE
Extruded rods, bars & shapes	AZ10A	---	QQ-M-31	---	---
	AZ31B	B107	QQ-M-31	---	---
	AZ31C	B107	---	---	---
	AZ61A	B107	QQ-M-31	---	4350
	AZ80A	B107	QQ-M-31	---	---
	HM-31A	---	---	MIL-M8916	4388 4389
	M1A	B107	QQ-M-31	---	---
	ZK21A	---	---	MIL-M-46039	4387
	ZK40A	B107	---	---	---
	ZK60A	B107	QQ-M-31	---	4352
	(P)ZK60B	---	---	MIL-M-26696	---
Extruded tubes	AZ31B	B107	WW-T-825	---	---
	AZ31C	B107	---	---	---
	AZ61A	B107	WW-T-825	---	4350
	AZ80A	B107	---	---	---
	M1A	B107	WW-T-825	---	---
	ZK21A	---	---	---	4387
	ZK40A	B107	---	---	---
	ZK60A	B107	WW-T-825	---	4352
Sheet and plate	AZ31B	B90	QQ-M-44	---	4375 4377 4376
	AZ31B				
	Tread plate	---	---	MIL-F-46048	---
	Tooling plate	---	---	---	4382
	AZ31C	B90	---	---	---
	HK31A	B90	---	MIL-M-26075	4384 4385

TABLE 6-17—*Continued*

Product	Alloy	ASTM	Federal	Military	AMS & SAE
	HM21A	B90	---	MIL-M-8917	4383 4390
	LA141A	B90	---	MIL-M-46130	4386
	LS141A	---	---	MIL-M-46130	---
	LZ145A	---	---	MIL-M-46130	---
	ZE10A	B90	---	MIL-M-46037	---
Forgings	AZ31B	B91	QQ-M-40	---	---
	AZ61A	B91	QQ-M-40	---	4358
	AZ80A	B91	QQ-M-40	---	4360
	HM21A	B91	QQ-M-40	---	4363
	LA141A	---	---	MIL-M-46130	---
	LS141A	---	---	MIL-M-46130	---
	LZ145A	---	---	MIL-M-46130	---
	M1A	---	QQ-M-40	---	---
	TA54A	---	QQ-M-40	---	---
	ZK60A	B91	QQ-M-40	---	4362
Welding rod	AZ61A	---	---	MIL-R-6944	---
	AZ92A	---	---	MIL-R-6944	4395
	AZ101A	---	---	MIL-R-6944	---
	EZ33A	---	---	MIL-R-6944	4396
	LA141A	---	---	MIL-R-6944	4397
Brazing filler metal	AZ91A	---	QQ-B-655	---	---
	AZ125A	---	QQ-B-655	---	---
Powder		---	---	MIL-M-382	---
		---	---	JAN-M-454	---
		---	---	MIL-P-14067	---
Primary ingot and stick	9980A	B92	---	---	---
	9980B	B92	---	---	---
	9990A	B92	---	---	---
	9995A	B92	---	---	---
	9998A	B92	---	---	---
Alloy ingot	AM100A	B93	---	---	---
	AZ63A	B93	---	---	---
	AZ81A	B93	---	---	---
	AZ91A	B93	---	---	---
	AZ91B	B93	---	---	---
	AZ91C	B93	---	---	---
	AZ92A	B93	---	---	---
	AM60A	B93	---	---	---
	AS41A	B93	---	---	---
Sand castings	AM100A	B80	---	---	---
	AZ63A	B80	QQ-M-56	---	4424 4420 4422
	AZ81A	B80	QQ-M-56	---	---
	AZ91C	B80	QQ-M-56	MIL-M-46062	4437
	AZ92A	B80	QQ-M-56	MIL-M-46062	4434
	EK41A	---	---	---	4434 4440 4441
	EZ33A	B80	QQ-M-56	---	4442
	HK31A	B80	QQ-M-56	MIL-M-46062	4445

(continued)

TABLE 6-17—*Continued*

Product	Alloy	ASTM	Federal	Military	AMS & SAE
	HZ32A	B80	QQ-M-56	---	4447
	K1A	B80	---	---	---
	LS141A	---	---	MIL-M-46143	---
	LZ145A	---	---	MIL-M-46143	---
	QE22A	B80	QQ-M-56	MIL-M-46062	4418
	QH21A	B80	---	---	---
	ZE41A	B80	---	---	4439
	ZE63A	B80	---	MIL-M-46062	4425
	ZH62A	B80	QQ-M-56	MIL-M-46062	4438
	ZK51A	B80	QQ-M-56	MIL-M-46062	4443
	ZK61A	B80	QQ-M-56	MIL-M-46062	4444
Permanent-mold castings	AM100A	B199	QQ-M-55	MIL-M-46062	4483
	AZ63A	---	QQ-M-55	---	---
	AZ81A	B199	QQ-M-55	---	---
	AZ91C	B199	QQ-M-55	MIL-M-46062	---
	AZ92A	B199	QQ-M-55	MIL-M-46062	4484
	EZ33A	B199	QQ-M-55	---	---
	HK31A	B199	QQ-M-55	MIL-M-46062	---
	HZ32A	---	QQ-M-55	---	---
	QE22A	B199	QQ-M-55	MIL-M-46062	---
Die castings	AZ91A	B94	QQ-M-38	---	4490
	AZ91B	B94	---	---	---
	AM60A	B94	---	---	---
	AS41A	B94	---	---	---
Investment castings	AM100A	B403	---	---	4455
	AZ81A	B403	---	---	---
	AZ91C	B403	---	---	4452
	AZ92A	B403	---	---	4453
	EZ33A	B403	---	---	---
	HK31A	B403	---	---	---
	K1A	B403	---	---	---
	QE22A	B403	---	---	---
	ZE41A	---	---	---	---
	ZE63A	---	---	---	---
	ZK61A	B403	---	---	---

(*Dow Chemical Company*)

ALLOYS AND PRODUCT FORMS

The common alloying elements for magnesium are aluminum, zinc, and manganese. The addition of 3 to 10% aluminum increases the hardness and strength about in proportion to the amount added. Magnesium castings with 5 to 10% aluminum respond well to heat treatment. Manganese added in small amounts improves corrosion resistance but has little effect on mechanical properties. Zinc is used up to 3% in the magnesium aluminum alloys, improving strength and saline corrosion resistance. Magnesium-zinc-zirconium alloys with up to 6% zinc provide high strength with good ductility. Where strength and creep resistance are required at moderately elevated temperatures, rare earth or thorium is added, along with zinc and zirconium. For high-temperature strength, a silver addition (2.5%) is used in combination with rare earth or thorium and zirconium.

General forms

Magnesium alloys are available in virtually all of the usual metal forms, including cast ingots, slabs, and billets; sand, permanent-mold, die, and investment castings; forgings, extruded bars, rods, tubes, structural shapes, and special hollow and solid shapes; and rolled sheet and plate.

Wrought Products

Wrought products of various magnesium alloys are available in plate, sheet, strip, rods, bars, irregular extruded shapes, and forgings. Wrought forms can be cold worked to a limited degree, but are usually heated before they are shaped, drawn, forged, or spun. Significant weight and structural savings are gained by using magnesium sheet, plate, and extrusions. High rigidity with reduced weight is possible, permitting diminished lateral and horizontal bracing and reduced fastening.

PROPERTIES

Magnesium offers significant advantages in strength-weight and stiffness-weight ratios. Table 6-18 compares bending strength and stiffness for beams of equal weight. Mechanical properties of magnesium alloys are given in Table 6-19.

Magnesium alloys do not have the sharp yield point characteristic of carbon steels. Instead, the metal yields gradually when stressed, and the term "yield strength" is used. Yield strength has been defined as the stress at which the stress-strain curve deviates 0.2% from the modulus line. In cast form, the tensile and compressive yield strengths of most magnesium alloys are substantially equal. In most wrought alloys, however, the compressive yield strength is less than that of the tensile yield. Magnesium alloys have a modulus of elasticity of approximately 6.5×10^6 psi (45 GPa), and Poisson's ratio is 0.35.

Magnesium provides high-speed machinability, good thermal and electrical conductivity, fatigue resistance and damping properties of a high order, high impact and dent resistance, and the ability to be cast and formed by a variety of standard metalworking processes. These factors allow the design of large yet light and rigid tools, and jigs and fixtures of ready portability. Moving machine elements made of magnesium have low inertia forces compared with those of steel or aluminum.

APPLICATIONS

The principal industrial and commercial uses of magnesium and its alloys are in nonstructural metallurgical, chemical, and electrochemical applications, as well as in structural components for aircraft, missiles, motor vehicles, portable tools, machinery parts, computers, and office equipment. Table 6-20 gives percentage estimates for areas of magnesium usage.

Nonstructural

In the metallurgical field, magnesium is an important constituent in aluminum alloys; in fact, the largest usage of magnesium is for alloying with aluminum. Additions of up to 10% magnesium result in aluminum alloys having favorable combinations of strength properties, formability, and corrosion resistance.

Magnesium also improves the properties and stability of zinc die-casting alloys and is used as a reducing agent in the manufacture of titanium and zirconium and as a desulfurization agent for iron and steel. Nodular or ductile cast iron is made by treatment of the molten alloy with magnesium, leaving a few hundredths of a percent magnesium in the nodular iron.

In electrochemical applications, magnesium anodes are used to prevent galvanic corrosion of steel in certain installations, such as underground pipelines, storage tanks, and domestic water heaters. Magnesium is also used in dry cell batteries.

Organic chemistry applications of magnesium include industrial synthesis processes such as the production of tetraethyl lead, an antiknock additive in gasoline.

In the printing industry, magnesium alloy plates are used for photoengraving. Magnesium etches rapidly to provide a sharp impression, and the byproducts do not require extensive effluent treatment for removal. When ignited, finely divided magnesium burns with intensive blue-white light. The powder is used in the pyrotechnical industry in making fireworks, military flares, and various incendiary devices.

TABLE 6-18
Relative Strength and Stiffness in Bending of Some Structural Metals*

Material	Thickness	Bending Strength	Stiffness
1025 Steel	100	100	100
Titanium 55A sheet	175	340	280
Aluminum 6061-T6 sheet & extrusion	290	820	840
Magnesium AZ31B-F extrusion	445	1180	1940
Magnesium AZ31B-H24 sheet	445	1590	1940
Magnesium ZK60A-T5 extrusion	430	2050	1750

(*International Magnesium Association*)
* Rectangular beams of constant width and equal weight.

Structural

In structural applications, magnesium is used primarily because of its low density. When stiffness or elastic buckling is the main design criterion, a structure will have the least weight when constructed of magnesium.

Structural uses of magnesium generally take advantage of the weight savings over other metals and alloys, especially in aircraft and aerospace applications, and military and electronic equipment. Special magnesium alloys containing zinc, zirconium, thorium, silver, and rare earth metals are used for components operating up to 572° F (300° C). Typical applications include gear boxes, jet engine housings, aircraft landing wheels, and cockpit canopy frames.

A large number of magnesium alloys are used. The most common contain up to 9% aluminum, up to 2% zinc, and small amounts of manganese. The alloys of lower aluminum content are used for the production of sheet and extrusions, while those of higher aluminum content are used mainly for castings.

While sand and permanent-mold castings are used, the major application is in the form of pressure die castings, such as for a variety of automobile and truck components where weight reduction is an important factor. Examples include air-cooled engine blocks and heads, clutch and transaxle housings, wheels, brackets, and grilles. Other examples of pressure die-casting applications include chain saw housings, rotary lawn-mower decks, computers and peripheral equipment, housings for office machines, loudspeakers, luggage, tennis racquets, and archery bows.

In addition to strength and lightness, properties desirable for structural applications include damping capacity, thermal and electrical characteristics, dent resistance, and the ability to be formed and joined. Rods, tubing, and other sections are produced by extrusion, and magnesium is also rolled into sheet and plate. Applications for these mill products range from fuel element containers to materials handling ramps, hand trucks, ladders, and garden tools.

TABLE 6-19
Typical Properties and Applications of Selected Magnesium Alloys

		Room Temperature			400° F (204° C) Short Time		
Product Form	Alloy*	Minimum Tensile Strength,** ksi (MPa)	Minimum Yield Strength,** ksi (MPa)	Minimum Elongation in 2 in. (51 mm),** %	Typical Tensile Strength, ksi (MPa)	Typical Yield Strength, ksi (MPa)	Typical Elongation in 2 in. (51 mm), %
			Casting Alloys				
Sand and permanent-mold castings	AM100A-F	20 (138)	10 (69)				
	AM100A-T61	34 (234)	17 (117)		17	6.5	25
	AZ63A-T6	34 (234)	16 (110)	3	18	12	17
	AZ81A-T4	34 (234)	10 (70)	7	20	11	30
	AZ91C-T6	34 (234)	16 (110)	3	17	12	40
	AZ92A-T6	34 (234)	18 (124)	1	17	12	36
	EZ33A-T5	20 (139)	14 (97)	2	21	11	20
	HK31A-T6	27 (186)	13 (90)	4	24	14	17
	HZ32A-T5	27 (186)	13 (90)	4	17	10	33
	K1A-F	24 (165)	6 (41)	14	8	5	71
	QE22A-T6	35 (241)	25 (172)	2	27	23	25
	ZE41A-T5	29 (200)	19 (131)	2	21	17	31
	ZH62A-T5	35 (241)	22 (152)	4	10 (69)†		
	ZK51A-T5	34 (234)	20 (138)	5	17	13	17
	ZK61A-T6	40 (276)	26 (179)	1.3			
Die castings	AM60A-F	32★ (221)	19★ (131)	8★			
	AS41A-F	31★ (214)	20★ (138)	6★			
	AZ91A and AZ91B-F	34★ (234)	23★ (159)	3★			
			Wrought Alloys				
Sheets and plates†	AZ31B-H24	34-39 (234-269)	18-29 (124-200)	6-8	15	8.5	55
	HK31A-H24	33-34 (228-234)	24-26 (165-179)	4	24	21	21
	HM21A-T8	30-33 (207-228)	18-21 (124-145)	6	14 (97)‡	12	12
	LA141A-T7	18-19 (124-131)	13-15 (90-103)	10			
	ZM21	40 (276)	26 (179)	8			
Extruded bars, rods, shapes, tube, and wire†	AZ21A-F						
	AZ31B-F	32-35 (221-241)	16-22 (110-152)	4-8			
	AZ61A-F	36-40 (248-276)	16-24 (110-165)	7-9	21	14	49
	AZ80A-T5	45-48 (310-331)	30-33 (207-228)	2-4	10 (69)‡		
	HM31A-T5	37 (255)	26 (179)	4	25	24	31
	ZK60A-T5	44-46 (303-317)	33-38 (228-262)	4	6 (41)‡		
	ZM21	37 (255)	27 (186)	17			
Forgings	AZ31B-F	34 (234)	19 (131)	6			
	AZ61A-F	38 (262)	22 (152)	6			
	AZ80A-T5	42 (290)	28 (193)	2			
	HM21A-T5	33 (228)	25 (172)	3			
	TA54A	36 (248)	22 (152)	7			
	ZK60A-T5	42 (290)	26 (179)	7	15	12	84

* Temper: F = as fabricated; H24, H26 = strain hardened then partially annealed; O = annealed; T4 = solution heat treated; T5 = artificially aged; T6, T61 = solution heat treated and artificially aged; T7 = solution heat treated and stabilized; T8 = solution heat treated, cold worked, and artificially aged.

** Properties of separately cast bars. All are minimums or ranges except those typical properties marked with a star (★). Young's modulus for LA141A = 6,100,000 psi (42 GPa); for others, 6,500,000 psi (45 GPa). It is important to note that the common practice of using hardness as an index to ultimate strength is not applicable or feasible for magnesium alloys. Hardness test results tend to exhibit a wide scatter.

† For magnesium alloy sheet, plate, extrusions, and wire, the actual minimum strengths vary with metal thickness. The ranges given are for thicknesses from 0.016″ (0.41 mm) through 3″ (76 mm).

‡ Tested at 600° F (316° C).

TABLE 6-19—*Continued*

Approximate Room Temperature Hardness,** Brinell	Applications
53	Pressure-tight sand and permanent-mold castings. Good combination of room temperature properties.
69	
73	
55	Tough, leakproof sand castings.
70	
81	
50	Pressure-tight sand and permanent-mold castings. For applications at 350-500° F (177-260° C).
55	Sand castings used at 400° F (204° C) and higher.
55	Sand castings used at 400° F (204° C) and higher.
---	Sand castings. Excellent damping capacity.
78	Sand castings. Good room and elevated temperature properties.
62	Sand castings. Good strength at room temperature. Improved castability over ZK alloys.
70	Sand castings. Good strength at room temperature. Improved castability over ZK alloys.
65	Sand castings. Good strength and ductility at room temperature.
---	Sand castings. Good strength and ductility at room temperature.
---	Automobile wheels.
---	Automobile engines and housings.
63	General-purpose alloy. Parts for cars, lawnmowers, luggage, business machines, tools.
73	
57	
---	Missile and aircraft uses up to 800° F (427° C).
54	Formed parts requiring a lightweight alloy.
---	Good damping capacity; stress relief not needed after welding.
---	Anode material for primary batteries.
47	General-purpose extrusions.
55	Extrusions requiring higher properties than AZ31B.
72	
---	Missile and aircraft uses as high as 800° F (427° C).
82	
---	Good damping capacity; stress relief not needed after welding.
50	For low-stressed forgings.
55	For forgings with higher strength than AZ31B.
72	
55	Aircraft and missile forgings up to 800° F (427° C).
52	General-purpose hammer forging alloy.
65	For forgings of maximum strength in aircraft and military applications.

(*Amax Magnesium Corp.*)

FABRICATION

Magnesium alloys are fabricated by common methods, including melting followed by casting, rolling, extrusion, and forging. Further fabrication includes forming, joining, and machining, after which standard assembly and finishing methods are used.

Magnesium alloys are highly formable at elevated temperature, and they can be readily cast into complex forms with a high degree of accuracy. Drilling, turning, and other machining operations can be performed at higher rates than with other metals; and strong joints between magnesium parts are possible through welding, adhesive bonding, and the use of mechanical joints such as rivets and threaded fasteners.

MAGNESIUM FABRICATION

TABLE 6-20
Estimated World Consumption of Magnesium

Nonstructural Applications	
Aluminum alloying	42%
Nodular iron and steel desulphurization	10%
Cathodic protection (anodes)	5%
Other	12%
Total	69%
Structural Applications	
Castings	26%
Mill products	5%
Total	31%

(*International Magnesium Association*)

Die Castings

Magnesium is ideally suited for die casting. The low latent heat allows rapid solidification in the die, with resultant high production rates. Molten magnesium does not react with or solder to die steels, resulting in longer die life and increased productivity. Draft and tape requirements are 20-25% less than those for aluminum die castings. Shrinkage rates are consistent and predictable, and parts are ejected from the die with minimum distortion and residual stress. Maximum mechanical properties of magnesium die-casting alloys are achieved in wall thicknesses ranging from 0.075″ to 0.150″ (1.91 to 3.81 mm), but walls thinner than 0.050″ (1.27 mm) are practicable in localized and unstressed areas.

Processes. Magnesium is suitable for production in both the cold chamber and hot chamber processes. Cold chamber machines, with greater locking force and injection pressure capability, allow the production of larger parts. Hot chamber machines are well suited for smaller parts, thinner wall castings, and automated systems. Both processes are viable and undergoing refinement in melting, handling, and metal metering. A recent development is the horizontal hot chamber machine, which offers increased injection pressures and metal velocities over the standard vertical machine, as well as ready convertibility from cold chamber to hot chamber mode.

A significant advance in magnesium die casting is the trend toward reduction or elimination of chloride-based cover fluxes in favor of dilute gas mixtures. The elimination of flux inclusions greatly improves the corrosion resistance of magnesium alloy die castings.

Alloys. The principal magnesium die-casting alloy has been AZ91B, with a lower copper version, AZ91A, available for better corrosion resistance. A recent major advance has been the introduction of high-purity AZ91, with strict limits on nickel, iron, and copper impurities. This high-purity alloy provides corrosion resistance in salt environments greatly superior to that previously associated with magnesium alloys. Other magnesium die-casting alloys in use are AM60A, with better elongation and toughness than AZ91B but lower tensile and yield strength, and AS41A for improved creep strength up to 350° F (177° C).

Sand and Permanent-Mold Castings

Sand and permanent-mold castings in magnesium alloys are produced in a large variety of sizes and shapes for many uses. Typical alloys with good casting qualities are ZE41, AZ63A, AZ91C, and AZ92A. The last provides the optimum combination of high yield strength and moderate elongation with good pressure tightness. Casting alloys containing zirconium, such as ZK51A and ZK61A, have been developed for their improved properties. Alloys containing thorium or rare earths, such as HK31A and EZ33A, respectively, are specified for elevated temperature service [for example, 345 to 600° F (175 to 315° C)], depending on alloy composition. Another relatively new casting alloy, QE22A, contains silver and didymium (cerium-free rare earth metals). This composition provides good room temperature properties, tensile yield, and creep properties up to about 600° F (315° C).

Sand and permanent-mold castings are usually heat treated (solution treated and age hardened, or age hardened alone) in order to achieve the desired properties. Heat treatment details are provided in Chapter 14, "Heat Treatment of Other Metals," in this volume.

Extrusions

The extrusion process starts with cast extrusion ingots that can be made by casting in thick-walled iron or steel molds, or preferably by means of the direct chill (DC) casting process, which gives a fine grain and less compound segregate.

Ingots that have been machined to remove the casting skin are preheated at 600 to 800° F (315 to 425° C), depending upon the alloy, placed in the container of the extrusion press, and then forced through a steel die by a hydraulically driven ram to provide the desired shape. Certain low-alloy magnesium alloys may be extruded at a rate as high as 100 fpm (30.5 m/min). Other alloys and more difficult configurations may have a limiting extrusion speed of only a few feet per minute. The reduction in cross-sectional area from the original ingot to the final extruded section should be relatively high (10:1 or greater) to obtain desired properties in the extrusion. Additional improvements in the properties of certain alloys can be attained after extrusion by a heat treatment such as artificial aging, which causes precipitation of compound. This heat treatment improves yield strength, with some sacrifice in ductility.

For ordinary temperature applications, an alloy of the AZ type, such as AZ31B, is used. The heat-treatable alloy ZK60A provides high strength and good toughness. If hot strength and creep resistance is needed, the 3% thorium alloy HM31A can be selected.

Forging

Magnesium forgings are usually made by the press forging process in closed dies, less often by hammer forging. The size of magnesium forgings seems to be limited only by the size of available equipment.

The principal forging alloys are AZ31B, AZ61A, AZ80A, and ZK60A. HM21A, a sheet and plate alloy, is also a good forging alloy. It is used where creep strength is required at temperatures as high as 800° F (427° C).

Rolling

The rolling of magnesium sheet and plate, like the extrusion process, starts with an ingot, currently made by the direct chill process. Rolling of magnesium alloy sheet requires several separate operations. First there is "breakdown" rolling in which slabs are heated in a two-stage preheating oven to a temperature of 800 to 900° F (427 to 482° C). The slab is then reduced in thickness by repeated passes through the rolls of a hot mill. In breakdown rolling, the original slab may be made as thin as about 1/4" (6.4 mm) without reheating. Following breakdown, the sheet, after reheating, can be semifinished on the same mill to as thin as 0.051" (1.3 mm). From the hot mill the sheet is transferred to a finishing mill and rolled, with intermittent annealing steps, to the desired thickness and temper. A number of additional operations—such as shearing at various stages in production, acid cleaning, wire brushing, chrome pickling, or oil finishing—are required to various extents in different mills.

The principal sheet and plate alloy is AZ31B for applications up to 200° F (93° C). For special applications requiring hot strength and creep resistance, HK31A and HM21A can be selected.

Forming

The forming of magnesium sheet or extrusions can be done by practically all the methods commonly used. Forming operations involve bending around generous radii. Relatively mild deformation can be performed on magnesium at room temperature. The formability at elevated temperatures of 400 to 600° F (204 to 316° C) is so greatly improved, however, that most magnesium forming is done in this temperature range.

This improved formability at elevated temperature is due to magnesium's hexagonal crystalline structure. When the metal is heated, additional slip planes become available, and plastic deformation characteristics improve. High-speed mechanical presses can be used to make draws in one step up to 59% reduction and at speeds up to 80 fpm (24.4 m/min). Magnesium alloys are also formed by bending, stretch forming, spinning, impact extrusion, drop hammer forging, and other common methods.

Joining

All the standard methods of joining are used with magnesium, including arc welding, electric resistance welding, brazing, riveting, bolting, and adhesive bonding.

Welding. For general welding, the inert-gas-shielded arc welding processes are used. Helium or argon can be used as the inert protective gas. In gas tungsten arc welding, a tungsten electrode is used to maintain the arc that melts the separate magnesium filler rod. Another process called gas metal arc welding uses a coil of magnesium wire that functions as both an electrode and filler rod. This method is sometimes described as consumable-electrode arc welding.

Inert-gas-shielded arc welding is preferred over gas flame welding processes because it requires no corrosive flux. Magnesium can be welded with oxyacetylene, oxyhydrogen, and oxyhydrocarbon gas, but a chloride-based flux is necessary to prevent oxidation. This flux is very corrosive and difficult to completely remove after welding. As a consequence, gas welding of magnesium alloys is usually limited to emergency repair to put broken parts back into service until replacement parts can be obtained.

Other methods. Spot welding is the most frequently used method of electric resistance welding. In addition, forge welding and stud welding can be used. Magnesium alloys can also be joined by furnace, flux-dip, and torch brazing. Soldering is not used as a method of joining because of the brittle joint that is formed; its use is limited to making electrical connections and to the filling of imperfections in the surface. Before soldering electrical connections to magnesium, the metal can be prepared by plating with electroless nickel, or with zinc followed by copper and tin.

Riveting. Riveting is a commonly used method of making mechanical joints in magnesium. Aluminum alloy 5056 rivets in the H32 temper are commonly used, although rivets of 6053-T61 or 6061-T6 can be substituted. These aluminum alloys minimize the possibility of galvanic corrosion. Rivets of steel, brass, copper, and certain aluminum alloys should not be used in magnesium because of the possibility of serious galvanic corrosion. Rivet holes in magnesium should be drilled rather than punched, and squeeze riveters are preferred over pneumatic riveting hammers. The latter are more likely to damage the magnesium sheet by overdriving. Optimum rivet-joint design calls for a minimum spacing in any direction of three times the rivet diameter. Similarly, a minimum edge distance of 2.5 times the rivet diameter is suggested. Riveted joints should preferably be protected with a sealing compound to prevent water entrapment; one or more coats of a chromate-pigmented primer are desirable.

Adhesive Bonding. Adhesive bonding has become a popular method of joining magnesium sheet in certain applications. It is often desirable when joining material too thin to be effectively riveted or welded. A large variety of adhesives are available. Most of them require heat, although in some cases the temperature can be as low as 200° F (93° C). Curing times will range from a few minutes to an hour or more. Certain adhesives require the application of pressure, but many of them require only contact. The shear strength of adhesive-bonded joints in magnesium will range from about 1000 to 4000 psi (6.9 to 27.6 MPa). Most bonded joints are useful at temperatures up to about 150° F (66° C). There is, however, an epoxy-phenolic adhesive that has shown good joint strength retention up to 500° F (260° C).

Machining

Magnesium is the easiest of all structural metals to machine. Its excellent machinability has often been the reason for its use in applications where a large number of machining operations were required. Appropriate procedures and controls should be applied to avoid a fire hazard in high-speed machining operations, especially in the presence of water liquid or vapor. Magnesium is normally machined dry; but, where very high cutting speeds are involved and there is a possibility of igniting fine turnings, it may be necessary to use a coolant. Mineral oils are used because water-based coolants may react chemically with the swarf. Table 6-21 compares the machinability of magnesium alloys with other metals and shows the relative power required to remove a unit quantity of metal, with magnesium rated as one.

Some advantages of magnesium's excellent machinability include reduced machining time (which can mean fewer machine tools and thus lower capital investment and less floor space to do a given job); four to five times greater tool life; an excellent surface finish with only one cut; well-broken chips that minimize handling cost; and less tool buildup. High-speed routers are standard tools for cutting and trimming magnesium. For additional information, see Chapter 1, "Principles of Metalcutting and Machinability," in Volume I of this Handbook series.

CHAPTER 6

FINISHING MAGNESIUM

FINISHING

The finishing requirements for magnesium are established by the end use of the part and may range from none up to highly protective paints, powder resin coatings, and metallic plating systems. Finishing operations may be divided into cleaning, pretreatment, and coating steps. For additional information, see Chapter 18 ("Cleaning"), Chapter 20 ("Plating"), and Chapter 26 ("Coating Materials") in this volume.

Cleaning

Cleaning may be mechanical or chemical. Sandblasting, grit blasting, and hydroblasting are used on sand and permanent-mold castings. Dry blasting with steel shot or with any media containing iron contaminates the magnesium surface with iron and drastically lowers corrosion resistance. Acid pickling with recommended solutions is mandatory after such operations. Grinding, sanding, and wire brushing are other mechanical means of cleaning. Vibratory finishing is a preferred method of mechanical preparation of magnesium surfaces for subsequent finishing.

A number of acid pickling solutions have been developed for cleaning and removing surface contamination and surface segregation from magnesium alloys. The particular formulation to be used varies with the alloy, wrought or cast form, and nature of surface alteration to be accomplished.

Organic contamination, oils, and other foreign matter can be removed by suitable solvent cleaning followed by alkaline cleaning. Unlike aluminum, magnesium is not attacked by hot, strong caustic solutions; and, with the exception of alloy ZK60A, can be cleaned with strong alkaline cleaners without etching.

Pretreatment

Chemical pretreatments are used to provide magnesium with shelf storage protection and, more importantly, to improve paint adhesion and performance. Pretreatments make the surface more receptive to paint by reducing its natural alkalinity and by providing a mechanical key for paint adhesion. Chemical pretreatments are of two main categories: chrome pickle (etching type) and dichromate (nonetching). A less effective class of pretreatment for magnesium is the phosphate type, which is adequate for some coatings in mildly corrosive environments.

Anodic pretreatments are used either as superior paint bases where the cost can be justified, or as thick coatings (1 mil or 0.025 mm) to provide abrasion resistance. Anodic coatings on magnesium are porous and should not be exposed to corrosive environments without supplementary sealing.

TABLE 6-21
Relative Power Required to Machine Metals

Metal	Relative Power Required (1 = lowest)
Magnesium alloys	1.0
Aluminum alloys	1.8
Brass	2.3
Cast iron	3.5
Mild steel	6.3
Nickel alloys	10.0

Coatings

Various organic coatings can be applied successfully to magnesium alloys. Epoxy, vinyl, polyurethane, acrylic, and other alkali-resistant vehicles are suitable. Baked finishes are superior to air-drying types. An electrophoretic, cationic epoxy primer has demonstrated good performance on magnesium. Epoxy, polyester, and polyurethane powder resin coatings have also performed well when applied over standard pretreatments.

Plating

The technology for electroplating of all standard metallic platings on magnesium is established. The process involves a zinc immersion plate followed by a copper strike as preparatory steps. Electroless nickel can be applied directly to suitably prepared magnesium from a specially formulated bath, but commercial baths require the standard zinc and copper strike procedure. Plating on magnesium is used for special purposes and is not cost effective as a general corrosion-protective measure.

References

1. The Aluminum Association, Inc., *Aluminum Standards and Data*, 7th ed. (Washington, DC: 1982).
2. Copper Development Association, Inc., *Standard Designations for Copper and Copper Alloys*, Application Data Sheet 101/3.
3. Kenneth Budinski, *Engineering Materials, Properties and Selection*, 2nd ed. (Reston, VA: Reston Publishing Co., Inc., 1983).
4. *Encyclopedia Americana*, international edition (Danbury, CT: Americana Corp., 1980).
5. *Encyclopedia of Science & Technology*, vol. 3 (New York: McGraw-Hill, 1977).
6. Copper Development Association, Inc., *Machining Rod Handbook: Copper, Brass, Bronze*, 702/9.
7. Budinski, *op. cit.*

Bibliography

Abernathy, John L. *Manufacturing Processes for Aluminum-Steel Hoods*. SAE Technical Paper No. 760056. 1979.

Aluminum Association, Inc., The. *T8—Design for Aluminum—A Guide for Automotive Engineers*, 1st ed. Washington, D.C.: 1980.

———. *T9—Data on Aluminum Alloy Properties and Characteristics for Automotive Applications*, 3rd ed. Washington, D.C.: 1979.

———. *Standards for Aluminum Sand and Permanent Mold Castings*, 8th ed. Washington, D.C.: 1980.

Aluminum Smelting & Refining Co., Inc. *390 Aluminum Alloys*. Maple Heights, OH.

———. *Data Book*. Maple Heights, OH.

Amax Magnesium Division of Amax, Inc. "Safety," "Joining," and "Machining." Technical bulletins. Salt Lake City, UT.

Armstrong, G. Leslie, and Luckett, Robert D. "Aluminum Die Casting Alloys." *Die Casting Engineer* (January/February 1984), pp. 10-16.

Brush Wellman, Inc. *Wrought Beryllium Copper*.

Cohen, Arthur. "Metric Material Standards." *Metal Progress* (June 1978).

Copper Development Association, Inc. *Copper Alloys for the Chemical Processing Industries.* Technical Report No. 114/9. New York: 1979.

———. *Copper and Copper Alloys for the Process Industries.* Application Data Sheet No. 143/8.

———. *Design Handbook: Sheet Copper Applications*. No. 401/10.

BIBLIOGRAPHY

___________. *How Pb-Sn Solders React with Copper.* Technical Report CDA No. 804/7.

___________. *Low-Temperature Properties of Copper and Copper Alloys.* Application Data Sheet No. 104/5.

___________. *Standards Handbook: Cast Products, Alloy Data/7.* No. 109/8. New York: 1978.

___________. *Standards Handbook: Wrought Mill Products, Tolerances/1.* No. 105/3. New York: 1983.

___________. *Standards Handbook: Wrought Mill Products, Alloy Data/2.* No. 101/3. New York: 1973.

___________. *Standards Handbook: Wrought Mill Products, Terminology/3.* No. 103/8-R. New York: 1978.

___________. *Standards Handbook: Wrought Mill Products, Engineering Data/4.* No. 103/3. New York: 1983.

___________. *Standards Handbook: Wrought and Cast Products, Sources/5 and Specifications Index/6.* No. 137/3. New York: 1983.

Dow Chemical Co. *Magnesium Mill Products and Alloys.* No. 141-481-83. Midland, MI: 1982.

___________. *Digest of Specifications for Magnesium Products.* No. 141-452-82. Midland, MI: 1982.

George, R. A.; Swenson, W. E.; and Adams, D. G. *Aluminum in Automobiles: Why and How It's Used.* SAE Technical Paper No. 760164. 1976.

Graham, Robert H. "Putting More Muscle in Aluminum Alloys." *Machine Design* (January 12, 1984), pp. 125-129.

International Magnesium Association. "Magnesium and Magnesium Alloys," 3rd ed. Excerpted from the *Kirk-Othmer Encyclopedia of Chemical Technology.* Dayton, OH: 1982.

___________. *Magnesium: Light, Strong, Versatile.* Dayton, OH: 1982.

___________. *Magnesium: More Die Castings per Pound.* Dayton, OH: 1982.

___________. *Proceedings: 39th World Magnesium Conference & Exposition.* Held in Dearborn, MI, 1982. Dayton, OH: 1982.

Hatschek, R. L., and Mayfield, J. "Machining Aluminum." *American Machinist* (March 1978), pp. SR-1 to SR-16.

Kellog, H. H. "The State of Nonferrous Extractive Metallurgy." *Journal of Metals* (October 1982), pp. 35-42.

Magers, Dwain. "Magnesium: Where's Its Future?" *Die Casting Management* (July/August 1983).

Miller, John C., and DeVor, Richard E. *An Experimental Approach for Machinability Testing.* SAE Technical Paper No. 800487. 1980.

Olin Corporation, Brass Group. *Olin Alloy 654 Engineering Guide,* 1st ed. 5M 1282. 1982.

Paton, N.E.; Hamilton, C.H.; Wert, J.; and Mahoney, M. "Characterization of Fine-Grained Superplastic Aluminum Alloys." *Journal of Metals* (August 1982), pp. 21-27.

Pollmear, I. J. *Light Alloys - Metallurgy of the Light Metals.* Metals Park, OH: American Society for Metals, 1982.

Reynolds Metals Company. *Aluminum Mill Products.* No. 700-1-2. Richmond, VA: 1982.

Smith, Karl A.; Riemer, Steven C.; and Iwasaki, Iwao. "Carbochlorination of Aluminum from Non-Bauxite Sources." *Journal of Metals* (September 1982), pp. 59-60.

Traulsen, H. R.; Taylor, J. C.; and George, D. B. "Copper Smelting—An Overview." *Journal of Metals* (August 1982), pp. 35-40.

Tschamper, Otto. "The New Alusuisse Process for Producing Coarse Aluminum Hydrate in the Bayer Process." *Journal of Metals* (April 1982), pp. 36-39.

"USAF Studies SST Aluminum Alloys." *Aviation Week* (December 6, 1982), pp. 113-114.

Vaccari, John A. "Forming Limits Set for Copper Metals." *American Machinist* (April 1983), pp. 99-104.

Venen, Ralph. "Free-Machining Brass Pays Off." *Tooling & Production* (September 1980).

LEAD, TIN AND ZINC

CHAPTER CONTENTS:

LEAD

Pure lead (Pb) is a bluish gray metal that takes on a silvery gray patina with atmospheric exposure. In industrial atmospheres, it may change to a dark gray to black color. The density of cast lead is 0.409 lb/in.^{3} (11.34 g/cm^3), and the melting point is 621° F (327° C).

Lead is readily and inexpensively fabricated into a great variety of useful forms. The diversity of available forms combined with lead's advantageous properties—high density, low melting point, corrosion resistance, chemical stability, malleability, lubricity, electrical properties, and ability to form useful alloys and chemical compounds—provide a material suitable for a wide range of applications.[1]

Although lead is widely known as a heavy metal, only about 10% of lead applications are based primarily on its high density. In many other applications, lead is selected because it melts at low temperature, is easy to cast and form, is useful for generating electric current in electrochemical reactions, is a good absorber of sound and vibration, or is among the easiest of metals to salvage from scrap. Lead also resists attack by many corrosive chemicals and acids, most types of soil, and marine and industrial environments. It is the most impervious of all common metals to X-rays and gamma radiation.

Nearly three fourths of all U.S. lead consumption is for chemical applications such as paint pigments, gasoline additives, and storage batteries.

BASIC METAL PRODUCTION

Lead is found usually associated with other metals, notably silver and zinc. Its association with silver is especially noteworthy; historically, lead was considered mainly a by-product of silver mining. Although many minerals containing lead are known, the most important is *galena* (lead sulphide), an easily recognized, brightly metallic mineral. Of the other lead minerals, the carbonate (*cerrusite*) and the sulphate (*anglesite*) are of commercial importance, but they are not as frequently found as galena.

Lead Production

Once the lead is mined, the ore is crushed, ground, and prepared for flotation, which is the method generally used to separate the lead sulfide from the rock matrix and other minerals and to concentrate it to about 60-70% of the original mass. Briefly, the flotation process involves taking finely crushed ore diluted with four times as much water and agitating the mass violently with air in a tank to which a fraction of one percent of pine oil, with or without small amounts of other suitable chemicals, has been added. As a result of the agitation, a froth containing most of the metallic constituents of the ore is formed on top of the tank, while the valueless gangue matter remains unaffected at the bottom of the tank. The froth flows from the tank and is easily dried after washing and filtering.

The concentrated material serves as feed for the lead smelter, its treatment consisting of sintering, smelting, drossing, and refining. The sintering step agglomerates the fine concentrates and removes most of the sulfur as sulfur dioxide, which at most smelters goes to an acid plant to be converted to sulfuric acid.

The sinter product, along with coke, fluxes, and internal by-products (dross), is smelted in a blast furnace to produce impure lead bullion, slag, and fume. The slag is crushed and can be treated in a slag fuming plant to recover zinc as an oxide fume. A drossing kettle is used to remove copper and nickel from the bullion by controlled cooling to the point where the copper becomes insoluble. The copper-containing dross is treated in a reverberatory furnace to make a copper matte, and the lead bullion is ready for refining. Other, less common processes treat the dross with sodium and make the matte directly in the drossing kettle. The lead from the blast furnace goes to the drossing kettle.

Contributors of sections of this chapter are: ***John T. Benedict***, *Consultant;* ***William B. Hampshire***, *Assistant Manager, Tin Research Institute, Inc.;* ***Anthony D. Ippolito***, *Manager, Metal Applications and Development, Lead Industries Association, Inc.;* ***Hugh Morrow III***, *Manager, Market Development, Zinc Institute, Inc.*

Reviewers of sections of this chapter are: ***Paul E. Davis***, *Manager, West Coast, Tin Research Institute, Inc.;* ***Carl R. DiMartini***, *General Superintendent, Metals Technology, Central Research Dept., ASARCO, Inc.;* ***Barry P. Dugan***, *Technical Service Engineer, St. Joe Resources Co.;* ***William B. Hampshire***, *Assistant Manager, Tin Research Institute, Inc.;* ***Richard A. Hicks***, *Manager, Technical Services, Zinc Institute, Inc.;* ***Anthony D. Ippolito***, *Manager, Metal Applications and Development, Lead Industries Association, Inc.;* ***Andrea Korcan***, *Market Research Assistant, St. Joe Resources Co.;* ***Donald K. Lewis***, *Metallurgist, Certified Alloys Div., Aluminum Smelting and Refining Co., Inc.;* ***Dr. Colin A. MacKay***, *Director, Research & Development, Semi-Alloys, Inc., Allied Corp.;* ***William Mihaichuk***, *Market Development Manager, Eastern Alloys, Inc.;* ***Hugh Morrow III***, *Manager, Market Development, Zinc Institute, Inc.;* ***R. David Prengaman***, *Vice President, Research & Development, RSR Corp.;* ***John D. Rutherford***, *Manager, Market Development, Certified Alloys Div., Aluminum Smelting and Refining Co., Inc.;* ***Russell B. Schaffer***, *Sales Representative, St. Joe Resources Co.;* ***Edward J. Schmidt***, *Market Development Engineer, St. Joe Lead Co.;* ***Mark Taylor***, *Metallurgist, Seitzinger Div., TaraCorp Industries.*

CHAPTER 7

LEAD ALLOY DESIGNATION

Lead Refining

In the most common lead refining process, lead is first melted and allowed to cool. Copper crystallizes and is removed by treating with sulfur or sulfide, and the dross is removed by skimming. The lead then passes to a reverberatory or softening furnace where the temperature is raised and the molten lead stirred. A blast of air oxidizes any antimony, tin, or arsenic, and the oxides are skimmed off because these impurities harden lead (hence the term "softening" furnace).

The Parkes process removes silver and any remaining copper through the addition of zinc. The zinc is removed by vacuum dezincing, and caustic is added to the mixture to remove the last traces of zinc and minor amounts of arsenic and antimony. Bismuth is removed from lead bullion, when required, by the Betterton-Kroll process through the addition of calcium and magnesium. As a result of these processes, the refined lead product contains more than 99.9% lead.

In an alternative electrolytic refining process, called the Betts process, anodes of lead bullion are electrolyzed in a solution of lead fluosilicate and free fluosilicic acid. The lead is electrodeposited on the cathode, while the impurities remain on the anode as a slime. The cathodes are melted, drossed, and cast into refined lead pigs.

The Imperial smelting process was developed in England to smelt sintered, mixed lead-zinc concentrates from complex lead-zinc ores. This blast furnace technique produces zinc metal and lead bullion containing gold and silver.

LEAD ALLOY DESIGNATION

The basic specification covering pig lead, as produced by most primary and secondary lead smelters in the United States, is the ANSI/ASTM B29-79 Standard Specification for Pig Lead. Used widely in the plumbing trade and for chemical, radiation protection, and sound attenuation applications, this specification is summarized in Table 7-1.

Pure lead is available with a purity of 99.999%. By agreement between purchaser and supplier, limits are established for specified elements, and the percentages of unspecified elements may be raised or lowered, depending on the particular application.

The Unified Numbering System (UNS) is a means of describing metal and alloy composition in a logical, numerical manner that is useful to industry and government operations. The system is jointly managed by the American Society for Testing and Materials (ASTM) and the Society of Automotive Engineers (SAE). The numbers and compositions shown in Table 7-2 use the 50,000 to 55,799 numbers and were developed by a subcommittee of lead and lead alloy suppliers and users. The format Lxxxxx is used in the UNS number series for "Low Melting Metals and Alloys."

LEAD PRODUCT FORMS

Lead forms a wide range of low-melting alloys. Lead and the various low-melting-point lead alloys are widely available in a large number of product forms, including castings, extrusions, shot, powder, rope, coatings, and sheet, strip and plate products. When lead is added to other alloys, such as steel, brass and bronze, it promotes machinability, corrosion resistance, or other special properties.

Castings

The low melting point of lead, 621° F (327° C), makes it one of the simplest metals to cast, at approximately 700° F (371° C). It is used for many applications, discussed later in this section. Small die castings can be produced with 0.05″ (1.3 mm) wall thickness and as-cast dimensions that are reproducible to 0.001″ (0.03 mm). Antimony, tin, calcium, or arsenic and other elements may be alloyed with lead to produce certain properties such as castability, strength, or hardness.

TABLE 7-1
Lead Chemical Requirements*

	Type			
	Corroding	Common	Chemical	Copper Bearing
	UNS Number			
	L50042	L50045	L51120	L51121
Element	Lead Composition (Weight %)			
Silver, max	0.0015	0.005	0.020	0.020
Silver, min	---	---	0.002	---
Copper, max	0.0015	0.0015	0.080	0.080
Copper, min	---	---	0.040	0.040
Silver and copper together, max	0.0025	---	---	---
Arsenic, antimony, tin together, max	0.002	0.002	0.002	0.002
Zinc, max	0.001	0.001	0.001	0.001
Iron, max	0.002	0.002	0.002	0.002
Bismuth, max	0.050	0.050	0.005	0.025
Lead, min (by difference)	99.94	99.94	99.90	99.90

* ANSI/ASTM B29-79
Standard Specification for Pig Lead

TABLE 7-2
UNS Number Ranges for Lead and Its Alloys

L50000-50099	Pure Leads	L53300-53399	Sb (9.0-10.99%)
L50100-50199	Pb-Ag		Pb-Sb
L50200-50299	Pb-Al	L53400-53499	Sb (11.0-12.99%)
L50300-50399	Pb-As		Pb-Sb
L50400-50499	Pb-Au	L53500-53599	Sb (13.0-15.99%)
L50500-50599	Pb-Ba		Pb-Sb
L50600-50699	Pb-Bi	L53600-53699	Sb (16.0-19.99%)
L50700-50799	Pb-Ca		Pb-Sb
L50800-50899	Pb-Ca	L53700-53799	Sb (>20%)
L50900-50999	Pb-Cd	L53800-53899	Pb-Se
L51000-51099	Pb-Co	L53900-53999	Pb-Si
L51100-51199	Pb-Cu		Pb-Sn
L51200-51299	Pb-Fe	L54000-54099	Sn (<1.0%)
L51300-51399	Pb-Ga		Pb-Sn
L51400-51499	Pb-Hg	L54100-54199	Sn (1.0-1.99%)
L51500-51599	Pb-In		Pb-Sn
L51600-51699	Pb-K	L54200-54299	Sn (2.0-3.99%)
L51700-51799	Pb-Li		Pb-Sn
L51800-51899	Pb-Mg	L54300-54399	Sn (4.0-7.99%)
L51900-51999	Pb-Mn		Pb-Sn
L52000-52099	Pb-Na	L54400-54499	Sn (8.0-11.99%)
L52100-52199	Pb-Ni		Pb-Sn
L52200-52299	Pb-O	L54500-54599	Sn (12.0-15.99%)
L52300-52399	Pb-P		Pb-Sn
L52400-52499	Pb-S	L54600-54699	Sn (16.0-19.99%)
	Pb-Sb		Pb-Sn
L52500-52599	Sb (<1.0%)	L54700-54799	Sn (20.0-27.99%)
	Pb-Sb		Pb-Sn
L52600-52699	Sb (1.0-1.99%)	L54800-54899	Sn (28.0-37.99%)
	Pb-Sb		Pb-Sn
L52700-52799	Sb (2.0-2.99%)	L54900-54999	Sn (38.0-47.99%)
	Pb-Sb		Pb-Sn
L52800-52899	Sb (3.0-3.99%)	L55000-55099	Sn (48.0-57.99%)
	Pb-Sb		Pb-Sn
L52900-52999	Sb (4.0-4.99%)	L55100-55199	(Sn>58%)
	Pb-Sb	L55200-55299	Pb-Sr
L53000-53099	Sb (5.0-5.99%)	L55300-55399	Pb-Te
	Pb-Sb	L55400-55499	Pb-Ti
L53100-53199	Sb (6.0-6.99%)	L55500-55599	Pb-Zn
	Pb-Sb	L55600-55699	Pb-Zr
L53200-53299	Sb (7.0-8.99%)	L55700-55799	*Miscellaneous alloys not included above.*
	Pb-Sb		

(Lead Industries Association, Inc.)

Sheet, Strip, Plate Products

The malleability of lead and most of its alloys allows it to be rolled to any desired thickness from 2″ (51 mm) to 0.005″ (0.127 mm). Lead sheet or strip is also produced by direct continuous casting to the desired thickness. Sheets are produced in standard widths up to 8 ft (2.4 m). Sheet is defined as a product up to 0.187″ (4.75 mm) thick and over 24″ (610 mm) wide. Strip thickness is 0.187″ (4.75 mm) or less, and width is 24″ or less. Plate is any product over 0.187″ thick and over 10″ (254 mm) wide.

Extrusions

Lead is extruded in the form of pipe, rod, wire, or practically any cross section, such as H-shaped window frames, hollow stars, and rectangular ducts. Commercially available extrusions are available from 12″ (305 mm) diam pipe down to 0.010″ (0.25 mm) solder wire. Lead is also extruded around steel bars, as well as around soft materials, such as rubber or plastic-covered power cables.

Shot

The surface tension of lead is such that when it is molten and poured through a sieve and allowed to free-fall, it forms spheres before solidifying. The size of the shot formed is controlled by the sieve size. The shot is cooled in water and collected at the bottom of the tower where it is graded for size. Typical shot diameter range is from 0.04″ (1.0 mm) to 0.23″ (5.8 mm). Larger size shot is cast in permanent molds.

CHAPTER 7

LEAD PROPERTIES

Powder

Lead powder, particles, and flakes are produced usually by atomization, in sizes from 157 μin. (4μm) and up. Lead particles impart useful properties when added to grease and pipe joint compounds. They are also used as constituents of powder metallurgy products.

Wool (Rope)

Lead wool, a loose rope of fibers weighing about 0.5 lb/ft (744 g/m), is produced by passing molten lead through a fine sieve and allowing it to solidify. The fibrous rope, when forced into a crevice under considerable force, cold welds into a homogeneous mass and forms a solid metal seal. This caulking process is useful where temperature or explosion hazards prohibit the use of flame heating. The gaps between lead shielding sheets in nuclear submarines are often caulked with lead wool.

Coatings

Coatings of lead and lead alloys over steel and other metals are available in thicknesses from a few thousandths of an inch to several inches. Terne (8% tin, 92% lead), other hot-dipped coatings, electroplate, and flame-sprayed lead and solder are some of the common forms.

Other Forms of Lead

Mixtures. Lead powder and paraffin mixtures combine the gamma shielding ability of lead with the neutron capturing quality of an organic material.

Lead and plastics mixtures show unusually effective sound attenuation and X-ray shielding. A mixture of lead and polyethylene has been used to mold precision parts with substantial mass, and lead shot in neoprene forms a combination acoustical, antifouling, antiturbulence coating for submarines.

Laminates. Originally developed for X-ray protection, a large family of lead laminates is now used for noise control. Typical examples include lead-plywood, lead-gypsum-board, lead-cinder-block, and leaded plastics and fiberglass combinations.

Bonded sheath. Lead in thicknesses from 1/64" (0.4 mm) to a foot (305 mm) or more may be bonded to other metals to impart special qualities. For example, lead may be bonded to or combined with steel, copper, or aluminum for corrosion resistance and strength, or for gamma shielding and heat transfer.

PROPERTIES

The properties of lead that make it useful in a wide variety of applications are its density, malleability, lubricity, flexibility, and coefficient of thermal expansion, all of which are high; and elastic modulus, elastic limit, strength, hardness, melting point, and electrical conductivity, all of which are low. Lead also has good resistance to corrosion under a wide variety of conditions, is easily alloyed with many other metals, and is easily cast. Table 7-3 lists some general physical and mechanical properties for unalloyed lead that is 99.9% pure. Table 7-4 presents data on mechanical properties for lead alloys.

Density

The high density of lead makes it very effective in shielding against X-rays and gamma radiation. In large installations, it is often used in conjunction with concrete structures to reduce the mass of concrete that otherwise would be required.

The combination of high density, high "limpness" (low stiffness), and high damping capacity makes lead an excellent material for deadening sound and for isolating equipment and structures from mechanical vibrations.

Because of its high density, lead is generally excluded from use in applications where light weight is important. However, even when light weight is desirable, the high density of lead can sometimes be used to advantage. For example, the use of lead in sailboat keels, aircraft counterweights, and as cabin linings for noise control is often advantageous because lead's high density allows more mass to be concentrated at the point of greatest effect. In addition, lead's low melting point and ease of casting make it possible to fit lead weights into irregular and out-of-the-way spaces.

Malleability, Softness and Lubricity

These three related properties account for the extensive use of lead in many applications. For example, high malleability is largely responsible for the value of lead as a caulking material, enabling it to fill caulked joints completely. The softness and self-lubricating properties of lead account in substantial part for its use in bearing alloys, gaskets, washers, and lead-headed nails. As a coating on wire or sheet metal, lead acts as a drawing lubricant; and in the form of powder or wire, it imparts lubricity to friction materials, such as brake linings. The commercial application in which the malleability of lead is used to greatest advantage is the manufacture of foil; lead foil is often rolled as thin as 0.0005" (0.013 mm). On the other hand, the softness of lead requires that care be taken in many applications. For example, excessive liquid velocity in lead pipes may result in severe erosion if precautions are not taken.

Strength

The fact that lead has low tensile strength and low creep strength must always be considered in lead components. However, even when good strength is an essential design criterion, the low strength of lead does not necessarily preclude its use. Lead products can be designed to be self-supporting, or inserts or supports of other materials can be provided. Alloying with certain other metals, notably antimony and calcium, is a common method of strengthening lead for many applications. In general, consideration should always be given to supporting lead structures by lead-covered steel straps. When lead is used as a lining in a structure made of a stronger material, the lining can be supported by bonding it to the structure.

Thermal Expansion

The relatively high coefficient of thermal expansion of lead is another important property. In lead roofing and flashing, thermal expansion is a key factor and is provided for by using small sheets and loose-locking each sheet to the next, thus minimizing both individual and cumulative expansion. In pipelines subject to wide variations in temperature, provision must be made for free expansion. The flexibility of lead can be used to advantage in such systems.

Corrosion Resistance

Lead is highly resistant to corrosion by the atmosphere, by water, and by a wide range of chemicals in common use. Where resistance to corrosion must be combined with long service life,

TABLE 7-3
Representative Physical and Mechanical Properties for Unalloyed Lead

Weight	
Cast lead, 68° F (20° C), calculated	0.4092 lb/in.^{3} (11.33 g/cm^3)
Rolled, 68° F (20° C) (density 11.37), calculated	0.4103 lb/in.^{3} (11.36 g/cm^3)
Liquid, 621° F (327.4° C), calculated	0.3854 lb/in.^{3} (10.67 g/cm^3)
Sheet lead, 1 ft^2 (0.09 m^2) by 1/64 in. (0.4 mm) thick	approximately 1 lb (0.45 kg)
Volume of 1 lb (0.45 kg) cast lead, 68° F (20° C), calculated	2.44 in.^{3} (40 cm^3)
Density	
Cast lead, 68° F (20° C)	0.41 lb/in.^{3} (11.35 g/cm^3)
Rolled, 68° F (20° C)	0.41 lb/in.^{3} (11.35 g/cm^3)
Just solid, 621° F (327.4° C)	0.40 lb/in.^{3} (11.07 g/cm^3)
Just liquid, 621° F (327.4° C)	0.39 lb/in.^{3} (10.80 g/cm^3)
Thermal Properties	
Melting point, common lead	621° F (327.4° C)
Melting point, chemical lead	618° F (325.6° C)
Casting temperature	790-830° F (421-443° C)
Mechanical Properties	
Hardness, Moh's scale	1.5
Brinell no., 1 cm ball, 30 s, 100 kg load	
Common lead	3.2 to 4.5
Chemical lead	4.5 to 6.0
Influence of temperature on Brinell hardness (chemical lead)	
Temperature °F (°C)	77 (25) 212 (100) 302 (150)
Hardness	5.3 3.6 2.6
Ultimate tensile strength	
Common lead	1400-1700 psi (9.6-11.7 MPa)
Chemical lead	2300-2800 psi (15.9-19.3 MPa)

Effect of Temperature on Tensile Properties, Lead annealed at 212° F (100° C)

Temperature, °F	°C	Tensile Strength, psi (MPa)	Elongation, %
68	20	1920 (13.1)	31
180	82	1140 (7.9)	24
302	150	710 (4.9)	33
383	195	570 (3.9)	20
509	265	280 (1.9)	20

Low-Temperature Properties

Temperature, °F (°C)	Tensile Strength, psi (MPa)	Elongation, %	Brinell Hardness
Cast Lead			
Room	3000 (20.7)	33	4.3
-300 (-184)	6200 (42.7)	40	9.0
Rolled Lead			
59 (15)	3600 (24.8)	52	---
-4 (-20)	7200 (49.6)	40	---
-40 (-40)	13,300 (91.7)	31	---
-103 (-75)	15,200 (104.8)	24	---

(Lead Industries Association, Inc.)

TABLE 7-4
Typical Mechanical Properties for Lead and Alloys at Room Temperature

UNS No.	Ultimate Tensile Strength, psi (MPa)	Elongation, %	Yield Strength, psi (MPa)	Hardness, Bhn
L50042	1740 (12)	30	7978 (55)	3.2-4.5
L50605	5511 (38)	220		9
L50640	6300 (43)	50		14
L50737	6500-7000 (45)	15		
L50765	6500 (45)	25		
L50790	7542 (52-55)	20-35		
L51110 L51120 L51121	2320 (16-19)	30-60	808 (6-8)	4-6
L51123 L51124	3000 (21)	40		5.8
L52605	2900 (20)	50		7
L52901	4020 (27.6)	48		8.1
L53105	6817 (47)	24		13
L53230	4650 (32.1)	31.3		9.5
L53346	10,000 (69)	5		19
L53565	10,000 (69)	5		20
L54321	4060 (218)	55	1450 (10)	8
L54520	4350 (30)	10		10
L54711	5800 (40)	16	3625 (25)	11.3
L54820	4930 (34)	18		12
L54915	5367 (37)	25		12
L55133	6235 (43)	7		
L55141	7500 (51.7)	32		14

(Lead Industries Association, Inc.)

the limitations imposed by the mechanical properties of lead must be carefully considered.

For chemical corrosion resistance, lead is usually chosen from the ASTM grades of chemical and acid-copper lead. Generally, either of these grades may be used up to approximately 450°F (232°C). Improved fatigue resistance may be obtained by the use of lead containing approximately 0.05% tellurium. Where high strength is desired and a lower maximum operating temperature is permissible, antimony-lead alloys (hard lead) are employed. Alloys containing up to 12% antimony are used for specific physical properties and sometimes for specific corrosion resistance. They are seldom used above 250°F (121°C), but below this temperature they have strength superior to unalloyed lead. The alloy with 6% antimony is most commonly used. The addition of small amounts of calcium (0.03% to 0.10%) and tin (up to 1.5%) to lead creates alloys that have significantly increased mechanical strength and that age harden at room temperature. In addition, the corrosion resistance of these alloys is higher than that of antimonial alloys in many applications.

APPLICATIONS FOR LEAD AND LEAD ALLOYS

In the metallic form, lead is used extensively in building construction as both sheet and pipe and in solder. Lead is also widely used in alloys such as type metal, antimonial lead, calcium lead for storage batteries, bearing alloys, foil, and collapsible tubes; as a coating for other materials; as lead weight, caulking lead, gaskets, seals, cable sheathing, bullets, and shot. A general summary of lead usage in the United States is given in Table 7-5.

Pure Lead

Lead in various forms and combinations is recognized as an effective material for controlling sound and vibration, and, as such, lead has a significant role in noise control. Its high internal damping characteristics make it one of the most efficient sound attenuators for industrial, commercial, and residential applications. Sheet lead, lead-loaded vinyls, lead composites, and lead laminates are frequently used to reduce machinery noise. Also, lead is important as a shielding material against X-rays and, in the nuclear industry, against gamma radiation.

In addition to its physical and mechanical properties, lead's chemical properties, mainly corrosion resistance, account for many of its uses. Lead is durable under varying weather conditions and when exposed to most types of soil, marine and

TABLE 7-5
Consumption of Lead in the United States in 1982

SIC Code*	Use	Tons	Metric Tons
	Metal Products:		
3482	Ammunition—shot and bullets	48,762	44 237
	Bearing metals:		
35	Machinery, except electrical	1340	1216
36	Electrical and electronic equipment	106	96
371	Motor vehicles and equipment	2227	2020
37	Other transportation equipment	3088	2801
	Total bearing metals	6760	6133
3351	Brass and bronze—billets and ingots	12,513	11 352
36	Cable covering—power and communication	16,734	15 181
15	Calking lead—building construction	4471	4056
	Casting metals:		
36	Electrical machinery and equipment	884	802
371	Motor vehicles and equipment	724	657
37	Other transportation and equipment	26,018	23 603
3443	Nuclear radiation shielding	**	**
	Total casting metals	27,626	25 062
	Pipes, traps, and other extruded products:		
15	Building construction	9099	8255
3443	Storage tanks, process vessels, etc.	467	424
	Total pipes, traps, and other extruded products	9567	8679
	Sheet lead:		
15	Building construction	11,011	9989
3443	Storage tanks, process vessels, etc.	138	125
3693	Medical radiation shielding	5561	5045
	Total sheet lead	16,710	15 159
	Solder:		
15	Building construction	7430	6740
341	Metal cans and shipping containers	8222	7459
367	Electronic components and accessories	6577	5967
36	Other electrical machinery and equipment	2978	2702
371	Motor vehicles and equipment	6208	5632
	Total solder	31,416	28 500

(continued)

TABLE 7-5—Continued

SIC Code*	Use	Tons	Metric Tons
	Storage battery grids, posts, etc.:		
36911	Storage batteries—SLI automotive	313,889	284 758
36912	Storage batteries—industrial and traction	30,670	27 824
	Total storage battery grids, posts, etc.	344,559	312 582
	Storage battery oxides:		
36911	Storage batteries—SLI automotive	410,146	372 082
36912	Storage batteries—industrial and traction	21,670	19 659
	Total storage battery oxides	431,816	391 741
371	Terne metal—motor vehicles and equipment	3624	3288
27	Type metal—printing and allied industries	3049	2766
34	Other metal products†	7820	7094
	Total metal products	965,427	875 830
	Pigments:		
285	Paints	14,743	13 375
32	Glass and ceramic products	38,058	34 526
28	Other pigments‡	14,291	12 965
	Total pigments	67,093	60 866
2911	Chemicals—petroleum refining	131,432	119 234
	Miscellaneous uses	21,471	19 478
	Total reported	1,185,422	1 075 408

(U.S. Bureau of Mines, Mineral Industries Service)

* Standard Industrial Classification
** Included in "Other transportation" to avoid disclosing company proprietary data.
† Includes lead consumed in foil, collapsible tubes, annealing, galvanizing, plating, and fishing weights.
‡ Includes lead content of leaded zinc oxide, color, and other pigments.

industrial atmospheres, and the action of many corrosive chemicals. Its resistance to sulfuric acid is used advantageously in the manufacture of the acid and in the most common method of storing electricity—the storage battery.

Lead Alloys

Lead alloyed with other metals is used commercially in large tonnages. Antimony, calcium, and tin are the alloying metals most commonly used. Antimony and calcium provide greater hardness and strength, as in storage battery plates, sheets, pipes, and castings. Tin also increases hardness slightly, but the lead-tin alloys are most commonly used for their low melting characteristics and ability to join other metals, as in solder. In solder, terne metal, and other lead-alloy coatings, tin imparts the ability to bond with metals like steel and copper. High-purity lead has no such bonding ability. Both tin and antimony improve the casting qualities of lead. Bismuth and tin are often alloyed with lead to obtain extremely low-melting fusible alloys.

Although lead does not alloy readily with copper, it is combined in rather high percentages in bronzes and free-machining brasses. Also, while lead is difficult to alloy with steel, fractional percentages of lead are used in the production of free-machining steel. Calcium, magnesium, and sodium in fractional percentages are sometimes alloyed with lead in special bearing alloys and cable sheathing. Cadmium additions prove helpful in coating copper wires with lead.

Soldering alloys. One of the most common of the lead alloys is solder. Numerous soldering alloys of lead are available for almost any metal joining purpose. The most widely used solders are the lead-tin alloys, most of which begin to melt at 361°F (183°C). Lead-tin solder is compatible for use with nearly all base metal cleaners and fluxes and heating methods. The melting characteristics of lead-tin solders are given in Table 7-6.

Other lead solders are lead-tin-antimony alloys. These alloys are frequently used where greater mechanical strength and higher operating temperatures are necessary. Lead-silver alloys are suitable for application temperatures up to 350°F (177°C). Minor additions of tin (less than 1.5%) are sometimes made to the silver solders to reduce corrosion susceptibility.

Fusible lead alloys. Many lead alloys melt at low temperatures and are used for electric fuses, automatic sprinkler systems, and boiler plugs. These generally contain a combination of lead, tin, bismuth, and cadmium or indium. To achieve better properties, minor additions of certain other elements are sometimes made. One group of these alloys is based on the ternary eutectic, containing 15.5% tin, 52.5% bismuth, and

TABLE 7-6
Melting Characteristics of Tin-Lead Solders

UNS No.	Composition (Weight %) Tin	Lead	Temperature °F (°C) Solidus	Liquidus	Pasty Range	Uses
L54210	2	98	601 (316)	611 (322)	10 (5.6)	Side seams for can manufacturing.
L54320	5	95	581 (305)	594 (312)	13 (7.2)	For automobile radiators.
L54520	10	90	514 (268)	576 (302)	62 (34.4)	For coating and joining metals.
L54560	15	85	440 (227)	551 (288)	110 (61.1)	
L54710	20	80	361 (183)	531 (277)	170 (94.4)	For coating and joining metals. For filling dents or seams in automobile bodies.
L54720	25	75	361 (183)	511 (266)	150 (83.3)	For machine and torch soldering.
L54820	30	70	361 (183)	491 (255)	130 (72.2)	
L54850	35	65	361 (183)	477 (247)	116 (64.4)	General purpose and wiping solder.
L54915	40	60	361 (183)	460 (238)	99 (55.0)	Wiping solder for joining lead pipes and cable sheaths. For automobile radiator cores and heating units.
L54950	45	55	361 (183)	441 (227)	80 (44.4)	For automobile radiator cores and roofing seams.
L55030	50	50	361 (183)	421 (216)	60 (33.3)	For general purpose. Most popular.
L13600	60	40	361 (183)	374 (190)	13 (7.2)	Primarily used in electronic soldering applications where low soldering temperatures are required.
L13630	63	37	361 (183)	361 (183)	0	Lowest melting (eutectic) solder for electronic applications.

(Lead Industries Association, Inc.)

32% lead. Wood's and Lipowitz's alloys are typical of the quaternary eutectic, containing lead, tin, bismuth, and cadmium; they melt at about 158°F (70°C). Tensile strengths for the fusible alloys range from 4000 psi (28 MPa) to more than 13,000 psi (90 MPa). Mechanical properties tend to change over a period of time, and care should be taken in selection of properties and alloys.

Bearing alloys. Lead-based (white metal) bearing alloys vary widely in composition. A common composition for a lead-based bearing used in railroad cars is 86% lead, 5% tin, and 9% antimony. Copper-lead alloys for bearings contain as much as 40% lead. Leaded bronzes contain smaller amounts of lead and are essentially mixtures of lead, copper, and tin, with a lead content of 4% to 25%. Numerous alloys of lead and alkaline earth metals, such as calcium and sodium, are used as bearing materials.

Type metals. Type metals are a class of metals used in the printing industry and generally are alloys of lead, antimony, and tin; small amounts of copper and arsenic are added to increase hardness for some applications. The lead base provides low cost, a low melting point, and ease of casting—properties desired in all type metals. Additions of antimony not only harden the alloy and make it more resistant to compressive impact and wear, but also lower the casting temperature and minimize contraction during freezing. Tin adds fluidity and greater ease of casting, reduces brittleness, and imparts a finer structure, a characteristic helpful in reproducing fine detail.

Hard lead (lead-antimony alloys). Hard lead contains from 1% to 12% antimony. The antimonial lead alloys have wide application in making castings where high strength is not required and high temperature is not encountered. Where optimum strength and fatigue resistance are required, 4-6% antimony grades are commonly used, at temperatures up to 250°F (121°C). Manufacturers of storage batteries use great quantities of antimonial lead containing approximately 2-6% antimony and frequently 0.25% tin, with small amounts of arsenic and copper. Recently, calcium has begun to be alloyed with lead for use in battery plates and lead roofing. The lead-calcium alloys have improved tensile strengths and corrosion resistance.

Antimonial lead sheet (6% antimony) is used widely in the United States for building construction. Hard lead is also used in cable sheathing and collapsible tubes. In addition, large quantities of antimonial lead are used in ammunition for both military and sporting purposes. The alloy used may contain up to 1% arsenic for shot and up to 2% antimony for bullet cores.

Protective coatings. Lead alloys are used to protect other metals, particularly sheet steel and copper. The coating is most commonly applied by hot dipping, but it may also be applied by spraying or electroplating. Terne plate—steel sheet hot-dip coated with a lead alloy containing from 7% to 20% tin—finds wide use in applications such as the manufacture of gasoline tanks for cars, fuel tanks for oil heaters, and car radiator supports. Thin steel strip in narrow widths is lead coated by hot dipping on a continuous basis, and the coating alloy may be lead-tin or 6% antimony-lead alloy. This material is also used in the manufacture of heater fins on some motor cars instead of the more expensive, traditional copper.

Anodes. Anodes made of lead or lead alloys are used in electrolytic refining and plating of metals, usually because of their high resistance to the sulfuric acid used in electrolytic solutions. Lead anodes also have high resistance to corrosion by seawater, making them economical to use in systems for cathodic protection of ships and offshore oil rigs. Anodes for these purposes are usually made of unalloyed lead but sometimes of lead alloyed with silver, tin, antimony, or calcium. These anodes are supplied in cast form and also as extruded bars or supported sheet.

Other uses of lead alloys. There are many uses for lead alloys, each use consuming only a small amount of lead; but when added together, they make a sizeable total. Typical of these uses are fire safety devices, low-cost jewelry, ornamental castings, seals, collapsible tubes, and seatings for manhole covers.

Compounds

Lead monoxide or Litharge (PbO) is the most important compound of lead. It is used in storage batteries, crystal glass, ceramics, and leaded-glass radiation-shielding windows, as well as in pigments and to vulcanize rubber.

Lead tetraoxide (Pb_3O_4) or "red lead" is prepared by the controlled oxidation of Litharge. It is commonly used for corrosion resistant pigments and storage batteries. Lead chromates—yellow, orange, and red—are used as corrosion-resistant, colored pigments. Lead carbonate, the familiar white lead, is sometimes used in plastics and nacreous coatings; lead azide is a standard explosive detonator; lead arsenate is an insecticide and herbicide; and lead silicates are paint pigments and enameling frits. Lead diamyldithiocarbonate (LDAC) is being investigated as an antioxidant for asphalt roads and roofing shingles. Some lead compounds are used to stabilize plastics resins.

Lead tetraethyl and tetramethyl have been added to gasoline since 1924 to control the flame propagation rate in internal combustion engines, resulting in greater efficiency and uniform combustion. Since 1974, however, new American cars have catalytic converters to control engine emissions and require unleaded gasoline, resulting in a substantial reduction in the use of leaded gasoline.

TIN

Tin (Sn) is a soft, ductile, silvery-white metal. It is highly malleable (similar to silver and gold) and can be hammered into very thin, flat sheets or drawn into thin wire. Tin does not tarnish in air, and it melts at a relatively low temperature, 449.4° F (231.9° C). The specific gravity is 5.77 in alpha form (gray tin), and 7.29 in the normal beta form (white tin). The Brinell hardness at room temperature is 3.9, which is somewhat harder than lead.

The earliest use of tin was in the copper-based alloy, bronze. The use of bronze is known to have been prevalent in Egypt and Mesopotamia before 3000 B.C. Tin was mined in Spain from the beginning of the Bronze Age, while the working of tin mines in France and England began about 500 B.C. Examination of ancient furnaces indicates that tin of 99.9% purity was produced.

The largest use of tin is for tin-coated steel containers used to preserve foods. The next largest uses are in solder alloys, babbitt (bearing metal), bronzes, fusible alloys, type metals, pewter, and dental amalgams. Tin chemical compounds, both inorganic and organic, find extensive use in the electroplating, ceramic, plastics, pesticidal, and antifungal industries.

BASIC METAL PRODUCTION

The important tin-producing countries are Malaysia, Indonesia, Bolivia, Thailand, Zaire, Nigeria, China, and Australia. Tin is relatively expensive because the commercial deposits are widely scattered and the ores are sparse in tin content (0.15% or less).

Only one tin-bearing mineral, cassiterite (SnO_2), is of commercial importance. There are no high-grade tin ores. The bulk of the world's tin ore is obtained from low-grade alluvial deposits averaging approximately ½ lb (¼ kg) of cassiterite per cubic yard (3000 lb), 0.76 m^3 (1361 kg).

The cassiterite is commonly recovered from alluvial deposits by dredging, by using water jets and gravel pumps, by hydraulic methods where a head of water permits it, and by open-pit mining. The fine grains of cassiterite have a density 2½ times that of the gravel, and concentration is readily accomplished by screening and gravity separation. The concentrates contain 70-77% tin.

Tin is comparatively simple to reduce to metal from the oxide by pyrometallurgy. However, this operation differs from the smelting of most common metals because it requires retreatment of the slag to obtain efficient metal recovery.

Smelting of tin concentrates is usually done in reverberatory furnaces, using coke or coal as the reducing agent. Electric-arc furnaces and rotary, reverberatory furnaces are also used in some smelters, and there are a few blast furnaces used to treat low-grade ores and to reduce tin slags. The reverberatories and rotary furnaces are fired with oil, natural gas, or coal, depending on availability and cost.

Secondary tin from metal scrap amounts to as much as one third of the total tin consumed in the United States. Most of the secondary tin comes from tin-bearing alloys, and secondary smelters rework the material into other alloys and chemicals. In addition, several thousand tons of high-purity tin is recovered from the detinning of tinplate scrap. The tin is dissolved in a hot caustic solution and converted to metal electrolytically.

TIN ALLOY DESIGNATION

Tin alloys are included in the Unified Numbering System (UNS) for Metals and Alloys under the "L" designation for "Low Melting Metals and Alloys." Many of these alloys are solders and often are lead based. Other tin-containing alloys appear in the aluminum alloy section of the UNS, throughout the copper section, and in the reactive and refractory metals section, as well as under miscellaneous alloys.

GRADES OF TIN

Most applications for tin begin with the metal in ingot form. The purity of the ingot usually conforms to the ASTM B339 Standard Classification "Grade A," with 99.80% minimum tin, as shown in Table 7-7. The requirements for Grade A tin have remained virtually unchanged for more than forty years and represent a compromise between the needs of the major tin-consuming industries and the cost to the producers of removing impurities. Since its major use is as a protective coating in the food industry (for example, for cans), there has been strong emphasis on high purity levels.

PROPERTIES

Because tin is mechanically weak, there are few uses for the pure metal. Tin easily forms alloys with many metals, however, and is used to increase resistance to corrosion and fatigue and to improve malleability. The metals most commonly alloyed with tin are antimony, copper, lead, zinc, and silver to produce the common tin alloys of bronze, pewter, solder, type metal, and babbitt metal. Some brasses also contain a small amount of tin.

TABLE 7-7
Tin Grades*

Element	Composition, %						
	AAA	AA	A	B	C	D	E
Tin, min	99.98	99.95	99.80	99.80	99.65	99.50	99.00
Antimony, max	0.008	0.02	0.04	---	---	---	---
Arsenic, max	0.0005	0.01	0.05	0.05	---	---	---
Bismuth, max	0.001	0.01	0.015	---	---	---	---
Cadmium, max	0.001	0.001	0.001	---	---	---	---
Copper, max	0.002	0.02	0.04	---	---	---	---
Iron, max	0.005	0.01	0.015	---	---	---	---
Lead, max	0.010	0.02	0.05	---	---	---	---
Nickel + cobalt, max	0.005	0.01	0.01	---	---	---	---
Sulfur, max	0.002	0.01	0.01	---	---	---	---
Zinc, max	0.001	0.005	0.005	---	---	---	---

* ASTM B-339, "Standard Classification of Pig Tin"

The typical physical and mechanical properties of tin are presented in Table 7-8. Tin has a highly ordered crystalline structure. When a bar of tin is bent, a distinctive sound called "tin cry" is produced by the reorientation of the crystals.

Nontoxic

The useful properties of tin that figure prominently in the commercial uses of the metal include nontoxicity, which allows tin coatings to be used on surfaces that come in contact with food or beverages, or in certain applications that require contact with the human body.

Corrosion Resistance

An equally important property for commercial application is the good corrosion resistance of tin. A suitable coating thickness of tin on steel, as in tinplate, gives sufficient protection for most applications.

TABLE 7-8
Typical Physical and Mechanical Properties for Tin

Property	Value
Melting point	449.4°F (231.9°C)
Boiling point	4118°F (2270°C)
*Shear strength (room temperature)	1817 psi (12.4 MPa)
*Tensile strength 59°F (15°C)	2135 psi (14.7 MPa)
*Elongation on 0.86" (22 mm), 59°F (15°C)	75%
*Hardness 68°F (20°C)	3.9 Bhn

* Beta form

Low Melting Point

The low melting point of tin makes it suitable for coating a variety of metals by simple hot dipping. Where greater control of thickness uniformity is required and where hot dipping is not feasible, tin can be readily electroplated onto a variety of surfaces. Overall, including the worldwide production of tinplate, tin is the most commonly electroplated metal.

Alloying and Reactive Ability

Another important property of tin is its ability to alloy with or react with many other metals. For this reason, tin coatings adhere well when applied to metallic surfaces, allowing forming, spinning, and deep drawing operations to follow coating without degradation of the coating protection. Furthermore, tin is the active ingredient in solders because of its low melting point and its ability to wet a wide range of surfaces. The high-tin solders (60-63% tin) are the standard alloys for electronics soldering in which reliable joints must be quickly made to minimize the exposure of the components to the heat source for soldering. A 50% tin-50% lead alloy is widely used and is favored for a variety of general engineering joining applications. This alloy is also used for plumbing, although 40% tin and 30% tin alloys are also widely used.

Other Properties

Other properties of tin are important for certain applications. For example, the ability of tin and many tin alloys to hold a film of lubricant leads to their use in a variety of commercial bearing alloys. Tin provides useful, improved properties when alloyed with copper (to form bronze) or when added to various titanium and titanium-aluminum alloys.

Limitations

A phenomenon with tin may require precautions in some circumstances. As is the case for several other metals, tin-plated surfaces can grow whiskers, which are single-crystal, thin filaments of metal that grow spontaneously. The exact mechanism is not fully understood, but it appears that the growth is stress driven. Tin whiskers occur most often on thin electroplated layers, especially on brass or when plating organics are codeposited in large amounts. Hot dipping or reflowing an electroplated coating greatly lowers the risk of tin whiskers, as does the codeposition of small amounts of certain metals, such as lead.

Another precaution is for tin use at low temperatures. The allotropic transformation (beta to alpha) is detrimental when it occurs. Several common impurities retard this transformation, even at the low levels found in commercially pure tin, so that the transformation in practical terms is little more than a laboratory curiosity.

APPLICATIONS

Tin is rarely used alone, but usually in metallurgical combination with another metal or as a coating. Coatings, alloys, and compounds are the largest uses for tin.

Pure tin and alloys of tin can be applied as coatings to all of the common metals by hot dipping or electrodeposition. Tin coatings give protection to metal surfaces that oxidize or corrode readily. They also aid in fabricating and joining metals, and provide a clean, adherent base for paints or lacquers.

Tinplate

About one third of United States tin consumption and about 40% worldwide goes into the production of tinplate. Over ten million tons of tinplate have been produced worldwide each year for the past two decades, mostly for food cans. Mild-steel strip is cold rolled, annealed, then coated with tin by electroplating. The resulting tinplate product is typically 0.006 to 0.010″ (0.15 to 0.25 mm) thick, with a thickness 0.000016 to 0.000032″ (0.0004 to 0.0008 mm) of tin on each surface. It can be delivered in large coils, slit into narrower coils, or cut into sheets.

Despite the growing use of other materials, tinplate remains the major material for food packaging. About 90% of the world tinplate production is used for container manufacture, with the traditional tinplate can being fabricated from three pieces—namely, a body and two ends. The can's body is formed into a cylinder starting with a flat sheet of tinplate. Then a hooked seam is created and soldered. Traditionally, this application has made use of the nontoxicity, corrosion resistance, and solderability of tin.

More recently, welding has supplanted some of the soldering of cans due to concerns over the possible lead contamination of foodstuffs. Tinplate has been found to be readily weldable using a continuous copper electrode system. The presence of tin improves the quality of the electrical contact of the electrode to the tinplate. As a result, the change from soldering to welding is relatively simple to accomplish.

A significant recent development has been the introduction of a method to produce two-piece cans. A circular blank of tinplate is drawn into a shallow cup making use of the metallic lubricity of the tin surface. The walls are then ironed, thinning and extending them upward to form the container. After filling, the can is typically sealed by seaming on an easy-open end. A substantial savings in material is gained with the two-piece drawn-and-ironed can, but long production runs of a single-size can are required to justify the capital equipment expense. Additionally, thinning and stretching of the tin layer during drawing tends to expose the steel surface; hence, internal lacquering is usually required.

Other uses for tinplate include containers for paint, petroleum products, household cleaners and polishes, cosmetics, and tobacco products. Battery cases, automotive filters and gaskets, baking trays, and a wide range of other cases, cabinets, and smaller parts are fabricated from tinplate.

Solder

About one third of United States tin consumption goes into the production of solder alloys, roughly as much tin as is consumed for tinplate. While solders for food container and radiator manufacture represent large tonnage markets, the tin content of the solder alloys used in those applications is low, so that the most important solders from a tin consumption viewpoint are used for electrical and electronics soldering. The low melting point of tin that minimizes thermal exposure of delicate components and the wetting ability of tin that speeds soldering are both particularly important properties in the latter applications. Molten solder readily penetrates and fills the capillary space of a well-designed joint.

Soldering process. Whether the solder is used for a plumbing joint or in a mass soldering operation, such as wave soldering an electronic printed circuit board, the same basic steps are followed. The surfaces to be joined are cleaned so that any dirt, grease, or oxide films are removed. In some cases cleaning may be done well in advance of soldering. A tin, solder, or other solderable coating can be applied to protect the surface until the time of soldering.

The next step is applying a soldering flux, a chemical of corrosivity appropriate to the materials to be joined. The flux prevents the cleaned surface from reoxidizing. Heat is applied by iron, torch, infrared heating, or other methods; then the solder alloy is introduced and melted by the heat, unless it is already molten as in wave or dip soldering. The solder displaces the flux and wets the surfaces to produce the joint itself.

Although the basic process of soldering is simple, the multiplicity of metals that may form part of the final assembly introduces complications that make soldering a process requiring careful design and control. More detailed information on soldering is provided in Volume IV, *Quality Control and Assembly*, of this Handbook series.

Solder alloys. Solder alloys typically contain tin and lead. Their variety includes pure tin, the electronics solders composed of about 60% tin and 40% lead, the plumbing and engineering solders of 50% tin and 50% lead or 30% tin and 70% lead, and the can and radiator solders of about 2% tin and the balance lead. Antimony is sometimes added to solder to lower the cost and increase the strength, but only where the reduced wetting of the solder can be tolerated. Silver is also added for strength and fatigue resistance.

Bearing Alloys

Tin-based bearing alloys are also called tin-based babbitt alloys, named for the 19th century inventor. They are primarily tin with 5-12% antimony and 1-5% copper. The alloying elements form a hard particulate framework, while the soft tin-based matrix offers excellent resistance to seizure, high corrosion resistance, embeddability, and conformability. This combination of properties has led to diverse uses for these alloys. The best examples of such applications are for large, slow-speed diesel engines and machinery, where temperatures are not very high and conformability is particularly important.

Alloyed with aluminum, tin is added in the 6-20% range to produce two types of bearing alloys for automotive crankshaft bearings, with the final product alloy roll-bonded onto a steel backing. A special alloy of 40% tin in aluminum has been introduced for marine diesel engine applications. A bronze containing 10% tin in copper (with 0.5% phosphorus) is useful for heavy loads and high temperatures at slow speeds. Lead is also added to improve the performance of the bearing alloy under poor lubrication conditions. Dry bronze bearings can be made (usually by continuous casting) by starting with bronze powder that is sintered into the bearing shape. Solid lubricity can be provided by polytetrafluorethylene (Teflon) powder intermixed or impregnated into the porous bronze structure after sintering. Sometimes the infiltrant is graphite; but, in either case, the infiltrant acts as a built-in lubricant for situations where a conventionally lubricated bearing is not possible. This enables production of a "sealed-for-life" bearing.

Bronzes

In addition to bearing applications, the uses of tin bronzes are quite varied, some of these predating recorded history. Tin additions improve the mechanical and corrosion properties of copper, while maintaining good workability. Chemical hardware, friction surfaces, mechanical hardware, and gear parts are some of the uses of bronzes containing up to 12% tin. Bell metal, which is used in bell founding, is typically 23% tin. Cymbals and glockenspiels are also made from bronzes with a high tin content.

Additions of zinc and/or lead produce the so-called gun-metals. Leaded bronzes are used to cast valves, gears, pump parts, and various fittings. Adding nickel to copper-tin produces a useful family of alloys with excellent corrosion resistance and good solderability.

Pewter

Modern pewter is nearly pure tin, typically 92% tin with 6% antimony and 2% copper. The alloy is easily spun or cast to make a variety of decorative and useful items, such as drinking vessels, candlesticks, trays, and jewelry. Unlike silver items, pewterware requires only occasional polishing in normal use, although regular rinsing in warm soapy water is recommended.

Other Uses

Smaller metallurgical alloying applications of tin include type metals for printing, and fusible alloys that melt at fairly low temperatures and are therefore valuable in safety devices such as fire sprinklers, safety valves, and fire door latches. Tin is added elementally in small amounts to produce and stabilize a pearlite in cast irons. Added to titanium or titanium-aluminum alloys, tin improves mechanical strength without reducing ductility. Minor additions of tin to zirconium give improved corrosion resistance in nuclear fuel packaging. Tin powder is used as a sintering aid in iron powder metallurgy where, in conjunction with copper, it gives excellent volume stability to reduce the subsequent machining and coining operations. In the electronics industry, tin roll cladding is used to produce solid state devices.

Tin Coatings

As exemplified by tinplate, tin coatings are useful because they are nontoxic, corrosion resistant, solderable, and easy to apply. Coatings can be electroplated from acid or alkaline electrolytes or applied by dipping in molten tin. They are used on electrical and electronic parts, food handling equipment, automotive parts, and a wide range of consumer goods. For additional information, refer to Chapter 20, "Plating," and Chapter 23, "Porcelain Enameling and Hot Dipping," of this volume.

Alloy Coatings

Tin-lead alloys are electroplated in various compositions, including as overlays on bearings surfaces, as corrosion-resistant coatings on steel wire or strip, and as solderable coatings. Tin-lead coatings for these same applications can also be produced by hot dipping. Terneplate is the name for lead-tin coatings containing 4-20% tin. These rely on the wetting ability of tin for good adhesion and on the high-lead alloy for improved corrosion resistance. Terneplate can also be produced electrolytically.

Tin-copper in the form of red bronze plating (7-20% tin) is used for its high wear resistance and excellent corrosion resistance. It has been used in hydraulic mining equipment where such properties are most useful, and for jewelry coatings.

Tin-nickel can be electroplated in an alloy of 65% tin and 35% nickel. It is hard and corrosion resistant with good frictional and lubrication-holding properties. Applications include watch parts, drawing instruments, automobile brake pistons, costume jewelry, and electrical contacts.

Tin-zinc (65-80% tin) also exhibits excellent corrosion resistance as needed for specific situations, and has been used in applications where cadmium can no longer be used. For example, tin-zinc's resistance to corrosion by mineral oil has led to its use as a coating for automotive brake-fluid reservoirs. Tin-cobalt (about 80% tin) has been considered a competitor for chromium plating and does have several advantages in the application process, although the plate itself is not quite so wear resistant. Still other binary and ternary tin alloys are plated, but only for specialized applications.

Chemicals

Inorganic tin chemicals have several important uses in the ceramics and glass industries. Stannic oxide is an opacifier in glass and enamels, as well as one of the oxides that constitute some common pigments. Stannic oxide is also used as a coating on glass in various thicknesses for different purposes. Very thin coatings strengthen glass and improve its abrasion resistance, important in extending the life of glass items that are subjected to rigorous use as in restaurant and catering businesses. Thicker coatings, greater than 39 μin. (1 μm), provide electrical conductivity and retain optical transparency. These properties are used for display signs and lighting, for aircraft deicing windows, and for controls in cathode ray tubes. Since these coatings reflect infrared radiation, they are also useful for heat insulating windows. Stannic oxide is used for glass-melting electrodes and as an electrical capacitor dielectric material.

Other inorganic tin chemical applications include the use of chlorides as reaction catalysts and stannous fluoride as a decay-preventative in toothpastes. Other chemicals find use as flame retardants and for certain biocidal applications.

Organotin compounds are compounds in which tin is linked directly to one or more carbon atoms. These compounds find their largest application as PVC stabilizers to maintain the clarity and color fastness of plastics and prevent embrittlement under long-term exposure to light and/or heat. Where PVC clarity is needed and color changes cannot be tolerated (as for packaging or for outdoor application), the organotin stabilizers give optimum performance. The nontoxicity of inorganic tin makes these stabilizers particularly suitable for PVC pipe for drinking water supplies. Related organotin compounds serve as catalysts for the production of polyurethane foams and for the curing of room temperature vulcanizing silicones.

Another group of organotin compounds has markedly different properties and are very useful as biocides. They are the active ingredients in formulations for wood preservatives, algicides, slimicides, disinfectants, marine antifouling paints, and agricultural biocides.

CHAPTER 7

ZINC PRODUCTION

ZINC

Zinc (Zn), in its unalloyed form, is a bluish-white metal with a specific gravity of 7.1, a melting temperature of 787°F (419°C), and a boiling temperature of 1661°F (905°C). It is readily cast and crystallizes in a hexagonal close-packed structure. The tensile strength of as-cast, unalloyed zinc is about 9000 psi (62 MPa), with an elongation of 1%. It can also be readily fabricated into a variety of wrought products such as rolled zinc sheet, which exhibits a tensile strength of 24,000 psi (165 MPa) with 35% elongation. Zinc is highly active in the electromotive series and forms a thin, complex oxide-carbonate-hydroxide film when exposed to air. This property accounts for its wide use in the corrosion protection of steel, in processes such as galvanizing and metallizing, and for zinc-rich paints.

Historically, the first known reference to zinc as a metal was made by Strabos in a passage describing Andriera in Mysia (now the northwestern section of modern day Turkey). The Romans used zinc in brasses as early as 200 B.C., and a zinc alloy idol was found at a prehistoric Dacian settlement in modern day Rumania. The commercial use of zinc in Europe dates roughly from the beginning of the sixteenth century, and the birth of the American zinc industry (in terms of commercial production volume) is generally placed at the time of the U.S. Civil War.

Spelter is an old name for slabs of cast zinc. The first spelter or metallic zinc was produced in the United States in 1838. An early grade of zinc, Sterling Spelter, was 99.5% pure. Today, zinc, its alloys, and its chemical compounds represent the fourth most industrially used metal after iron, aluminum, and copper.

BASIC METAL PRODUCTION

Zinc is obtained chiefly from sphalerite (ZmS) and is contained in ore deposits widely distributed throughout the world. Zinc has been extracted from a number of ores, but is now nearly all obtained from sulfide materials that also usually contain lead or copper and often silver. The most plentiful of these is zinc blende or sphalerite, although marmatite, containing iron, and zincite are also important.[2]

Zinc is removed from sulfide, silicate, or carbonate ores by a process involving concentration and roasting. One of two methods may then be used to transform the ore into metal. Either the zinc ore is reduced by carbon and simultaneous distillation of the zinc in batch or continuous retorts, or the oxide is leached out with sulfuric acid and the solution is electrolyzed after purification.

Zinc ores generally need to be concentrated before the metal can be extracted. Two processes are in use: a wet gravity method, which takes advantage of the difference in density between mineral particles in the ore and the worthless sandy gangue material; and the more widely used and more efficient flotation process, which relies mainly on the wetting characteristics of zinc sulfide particles.

Flotation Method

In the flotation process, the finely crushed ore is agitated with water containing chemicals and a suspension of air bubbles. The mineral particles—which tend to attach themselves to the air bubbles—are carried to the surface to form a froth that is skimmed off, while the accompanying worthless material sinks to the bottom. By adding suitable reagents, it is possible to make some constituents float and others sink, thus separating the lead and zinc sulfide ores from each other.

Roasting

The first stage in the treatment of zinc sulfide concentrates is a roasting process. Most of the sulfur is thereby converted to sulfur dioxide, which is used for the production of sulfuric acid. The crude zinc oxide product is then treated either by thermal smelting or electrolytic refining.

Leaching

As an alternative to roasting sulfide concentrates, which produces sulfur dioxide as a by-product, direct pressure leaching is being tried commercially in Canada.

Thermal Smelting

In the thermal method developed by the New Jersey Zinc Co., a briquetted mixture of roasted concentrates and anthracite coal is heated in a large vertical retort, which allows the reduction to proceed during the steady descent of the charge.

Electrothermic Process

An electrothermic refining process developed by the St. Joe Minerals Co. is a vertical shaft process in which a mixture of roasted concentrates and coke is heated internally by passing a heavy electric current through the charge. The zinc vapor that is formed may be mixed with air to form zinc oxide or condensed to produce metal or dust.

Electrolytic Process

Over 80% of the world's zinc is refined by an electrolytic process. For electrolytic zinc production, the sulfide concentrates are first roasted, most often in a flash or fluidized bed roaster, to yield a crude oxide in a form that will dissolve easily in dilute sulfuric acid to form zinc sulfate.

The purified electrolyte enters the electrolytic cells that have anodes consisting of lead alloyed with up to 1% of silver and cathodes made of aluminum sheets with wooden or plastic edge protectors. The zinc is deposited at the cathodes, which are hoisted out of the cells every 24-72 hours, stripped, and subsequently melted and cast into slabs. At the insoluble silver-lead anodes, oxygen is liberated into the air, and the sulfuric acid is regenerated and returned to the leaching plant.

The minimum purity of ordinary electrolytic zinc is 99.95%. It is possible to make metal of greater than 99.99% purity, suitable for the production of die-casting alloys.

ALLOY DESIGNATION SYSTEM

The Unified Numbering System (UNS) numbers and American Society for Testing and Materials (ASTM) cross-reference specifications for zinc and zinc alloys are given in Table 7-9.[3]

PRODUCT FORMS AND ALLOY TYPES

Cast zinc reaches the market primarily in the form of nominal 50 lb (22.7 kg) slabs, 1 to 1½″ (25.4 to 38.1 mm) thick, 8½ to 10″ (210 to 254 mm) wide, and 18 to 20″ (457 to 508 mm) long. This form of cast zinc is called zinc slab or spelter. Wrought zinc is produced in sheets, plates, ribbon, wire, extrusions, and forgings.

TABLE 7-9
UNS Numbers, Compositions, and Specifications for Zinc and Zinc Alloys

UNS No.	Description	Chemical Composition,* %								Specification
		Al	Cd	Fe	Pb	Cu	Mg	Sn	Zn	
Z1300	Zinc anodes, Type II	0.005 max	0.003 max	0.0014 max					Remainder	ASTM B418
Z13001	Zinc metal		0.003 max	0.003 max	0.003 max			0.001 max	99.990 min	ASTM B6, special high grade
Z15001	Zinc metal		0.02 max	0.02 max	0.03 max				99.90 min	ASTM B6, high grade
Z16001	Zinc metal		0.40 max	0.03 max	0.20 max				99.5 min	ASTM B6, intermediate
Z17001	Zinc metal		0.05 max	0.03 max	0.6 max				99.0 min	ASTM B6, brass special
Z19001	Zinc metal		0.20 max	0.05 max	1.4 max				98.0 min	ASTM B6, prime western
Z21210	Rolled zinc		0.005 max	0.010 max	0.05 max	0.001 max			Remainder	ASTM B69
Z21310	Rolled zinc		0.005 max	0.012 max	0.05-0.12	0.001 max			Remainder	ASTM B69
Z21540	Rolled zinc		0.20-0.35	0.020 max	0.30-0.65	0.005 max			Remainder	ASTM B69
Z25630	Zinc casting alloy	8				1			Remainder	ASTM B669 (ZA-8)
Z32120	Zinc anodes, Type I	0.10-0.4	0.03-0.10	0.005 max					Remainder	ASTM B418
Z33520	Zinc alloy AG40A	3.5-4.3	0.004 max	0.100 max	0.005 max	0.25 max	0.02-0.05	0.003 max	Remainder	AMS 4803, ASTM B86 SAE J468 (903), Federal QQ-Z-363
Z33521	Zinc alloy AG40A	3.9-4.3	0.003 max	0.075 max	0.004 max	0.10 max	0.025-0.05	0.002 max	Remainder	ASTM B240, SAE J468 (903)
Z35530	Zinc alloy AC41A	3.9-4.3	0.003 max	0.075 max	0.004 max	0.75-1.25	0.03-0.06	0.002 max	Remainder	ASTM B240, SAE J468 (925)
Z35531	Zinc alloy AC41A	3.5-4.3	0.004 max	0.100 max	0.005 max	0.75-1.25	0.03-0.08	0.003 max	Remainder	ASTM B86, SAE J468 (925), Federal QQ-Z-363
Z35630	Zinc casting alloy	11				1			Remainder	ASTM B669 (ZA-12)
Z35840	Zinc casting alloy	7				2			Remainder	ASTM B669 (ZA-27)
Z44330	Rolled zinc		0.005 max	0.012 max	0.05-0.12	0.65-1.25			Remainder	ASTM B69
Z45330	Rolled zinc		0.005 max	0.015 max	0.05-0.12	0.75-1.25	0.007 0.02		Remainder	ASTM B69

* Chemical compositions are for identification purposes and should not be used in lieu of the specifications.

ZINC PRODUCT FORMS AND ALLOY TYPES

Pure Zinc

Unalloyed zinc containing normal trace impurities from its ores or its smelting and refining processes is generally available in the United States in three grades that are defined by the ASTM Specification B6 for zinc (slab zinc) (see Table 7-10). The international specification for zinc (ISO 752) lists five grades for slab zinc.

Zinc Alloys

The zinc alloy family—which includes pressure die cast, gravity cast, and sheet alloys—provides alloys suitable for the production of cast and wrought components.[4]

Casting alloys. There are two basic groups of zinc casting alloys. The first consists of the traditional zinc-4% aluminum alloys, designated as alloy numbers 3, 5, and 7. These are also commonly known as Zamak 3, 5, and 7, and are cast by pressure die casting. The second group is made up of the new zinc-aluminum ZA alloys ZA-8, ZA-12, and ZA-27, which can be successfully cast by a variety of gravity casting processes, as well as by cold chamber die casting. ZA is a registered trademark of the Zinc Institute. The properties of these two groups of zinc alloys provide a broad flexibility in the design and processing of cast products.

Conventional alloys. The conventional zinc die-casting alloys are numbers 3, 5, and 7. Alloy 7 is essentially a high-purity form of alloy 3.

- Alloy 3 is by far the most widely used zinc die-casting alloy. It is usually the first choice for designers who are considering die casting. This alloy is characterized by outstanding castability and a good balance of properties, including excellent retention of impact strength and long-term dimensional stability.
- Alloy 5, usually specified as an alternative to alloy 3, is widely used in Europe. It offers improved mechanical properties and has demonstrated greater resistance to creep, but with a resultant decrease of impact strength after aging.
- Alloy 7 is a high-purity, low-magnesium form of alloy 3, having higher ductility and slightly better fluidity. It is the alloy of choice when optimum elongation and casting definition are needed in special designs.

ZA casting alloys. The ZA alloys are a series of zinc-based casting materials with much higher aluminum content than conventional die-casting alloys. The group is currently comprised of three alloys: ZA-8, ZA-12, and ZA-27. Each is alloyed with varying amounts of aluminum, copper, and magnesium, with the alloy number representing nominal aluminum content. As a reference, the traditional zinc die-casting alloys contain about 4% aluminum. In combination, aluminum, copper, and magnesium tend to increase alloy strength and hardness.

- ZA-8 was originally developed as a permanent-mold alloy with excellent finish and plating characteristics; now it is also produced by die casting. It is particularly suitable for decorative applications and can be die cast by the hot chamber process. As cast, the material has properties superior to alloy 3.
- ZA-12 is a general-purpose casting alloy for both gravity and cold chamber pressure die-casting processes. It is also recommended for use as an appropriate prototype alloy for components intended for production as traditional zinc die castings. Its combination of low cost, strength, castability, and resistance to wear have made it the most popular of the three ZA alloys. It can be finished and plated.
- ZA-27 may be sand, permanent-mold, or die cast (cold chamber). It provides the highest mechanical properties of the ZA alloys; it has some design restrictions, however, in heavy sections. It is the lightest zinc alloy, and sand-cast parts can be heat treated to increase ductility. ZA-27 provides the highest design stress capability at elevated temperatures of all the commercially available cast zinc-based alloys.

Slush-casting alloys. A family of zinc-aluminum alloys has been used for many years in slush and permanent-mold slush metal casting. These alloys center around the zinc-aluminum eutectic at 5% aluminum and generally range in composition from 4.75 to 5.5% aluminum. Tensile strengths are in the range of 25 to 28 ksi (175 to 195 MPa). Aging has little effect on these alloys.

Forming-die alloys (Kirksite alloys). The forming-die alloys are a family of zinc-aluminum-copper alloys used largely in plaster and sand-mold casting for the preparation of forming dies for other metals and plastics, injection molds, and similar items requiring high impact and compressive strength. Typical compositions contain 4% aluminum and 3% copper, with or without the addition of a small amount (0.1%) of magnesium. The tensile strengths are of the order of 30 to 40 ksi (210 to 280 MPa). The dimensional stability of these alloys may be a problem when die cast, but not if sand or permanent-mold cast.

Wrought alloys. Wrought zinc alloy products can be divided into three main categories of semifinished materials according to the deformation techniques used to shape them:

TABLE 7-10
Grades and Compositions of Slab Zinc (ASTM B6)

	Composition,* weight %			
Grade	Pb, max	Fe, max	Cd, max	Zn, min
Special high grade**	0.003	0.003	0.003	99.990
High grade	0.03	0.02	0.02	99.90
Prime western†	1.4	0.05	0.20	98.0

* When specified for use in the manufacture of rolled zinc or brass, aluminum shall not exceed 0.005%.
** Tin in special high-grade zinc shall not exceed 0.001%.
† Aluminum in prime western zinc shall not exceed 0.05%.

- Flat rolled products (rolled zinc).
- Wire drawn products.
- Forged or extruded products.

Rolled sheet and strip are the predominant forms of wrought zinc. They offer economic advantages for high production quantities, sometimes reaching levels of hundreds of millions of parts per year. Forgings add further flexibility to production with zinc and zinc alloys, and are another alternative when maximum strength and precise tolerances, as well as other well-known forging advantages, are required.

Rolled zinc. Sheet zinc alloys fall into three basic composition groups: the zinc-copper group, offering a range of mechanical properties; the creep-resistant zinc-titanium alloys; and the zinc-lead-cadmium alloys. Some of these compositions are covered in ASTM B69, "Rolled Zinc," while others are listed in the Federal Specification QQ-Z-100A, "Zinc Alloy Sheet and Strip." A large number of compositions are available in the United States for rolled zinc alloys, depending upon the specific application requirements. An even wider variety is available in Europe, where rolled zinc alloys have generally been more extensively used.

Sheet alloys generally require the same levels of purity as their die-cast and gravity-cast counterparts. An additional benefit is that their metallurgical structures cannot be adversely affected in fabrication because they are not remelted. Sheet zinc has good-to-excellent forming properties and little tendency to work harden; consequently, there is no need for intermediate annealing during multiple-stage forming operations.

Forging alloys. Four alloys, two of the zinc-titanium family and two of the zinc-aluminum family, have been developed especially for forging applications. Long die life and precise tolerances are two major advantages of zinc alloy forging because forging temperatures are only 580° F (304°C). The other well-known advantages of forging also apply, including the ability to orient grains in the direction requiring maximum strength, a low scrap rate, and complete lack of porosity. These alloys are exclusively press forged.

- The zinc-titanium alloys are recommended for general forging applications and possess moderate tensile properties with excellent ductility. Creep resistance and intergranular corrosion are excellent and are combined with an unusual resistance to grain growth at elevated temperatures. These alloy compositions are easily soldered or brazed.
- The zinc-aluminum forging alloys are lightweight alloys possessing excellent strength and impact properties at low temperatures, -60° F (-50° C). Strength is comparable to that of other high-strength, nonferrous alloys. The machinability and bearing properties of these alloys are excellent.

Superplastic Zinc-Aluminum Alloys

The term "superplastic," as applied to alloys in general, refers to a metallurgical condition in which the material's properties resemble those of the plastic state of thermoplastic polymers. In this condition, such alloys can be uniformly and plastically elongated, employing technology resembling that used in forming molten glass and plastics. These superplastic properties are found in many alloy systems, and usually occur within a narrow range of composition and depend upon the processing history.

The useful alloys are those that are superplastic at moderately elevated temperatures but that can be heat treated to exhibit good mechanical and physical properties at room temperature. At their forming temperatures, superplastic alloys exhibit no elastic behavior and do not work harden. The commercially available superplastic alloys have compositions with 78% zinc and 22% aluminum, often with minor additions.

PROPERTIES

Aside from zinc coatings, the casting alloys are the most widely used form of zinc. Zinc-based die-casting alloys were introduced in the late 1920s to meet the demand for strong, stable die castings. These alloys, familiarly known as the Zamak or Mazak alloys, show a unique combination of properties that permit rapid, economic casting of strong, durable, accurate parts, and hence have dominated the market since their inception. They have many advantages over other die-casting materials, such as aluminum and magnesium alloys. The zinc alloys are more easily cast, are stronger and more ductile, require less finishing, can be held to closer tolerances, and can be cast in thinner sections. Because of low casting temperatures, die life for zinc die castings exceeds that for other die-cast metals. The production rate (shots per hour) is also higher.

Die casting with zinc-based alloys is one of the most efficient and versatile production methods for the manufacture of accurate, complex metal components. In general, the mechanical properties of zinc alloy die castings used at normal temperatures are superior to those of sand-cast gray iron, brass, and aluminum, particularly fracture toughness and impact strength. Zinc alloy die castings are stronger, tougher, and more dimensionally stable than some injection molded plastics, and can be produced at a higher rate.

Overview of Characteristics

The properties and characteristics of zinc alloys include:

- A broad range of mechanical properties. As indicated in Table 7-11, the tensile, compressive, yield, and impact strengths of the zinc-based alloys compare favorably with those of competing materials and are often higher. Strength-to-weight ratios also compare favorably, especially for the relatively low-density zinc-aluminum alloys. Alloys ZA-8, ZA-12, and ZA-27 are being used as direct replacements for aluminum, iron, and copper-based alloys.
- Bearing and wear properties. The ZA alloys possess excellent bearing properties. Bearing studies and case histories comparing ZA-12 and ZA-27 bearings to SAE 660 bronze bearing alloys show the ZA alloys to possess lower coefficients of friction, higher load-bearing capacities, greater wear resistance, and superior lubrication retention characteristics.
- Ease of fabrication. Zinc die castings traditionally have been easy to cast, machine, and assemble in the field of low-cost, high-precision component manufacturing. When machining operations are required, their excellent machinability results in minimum processing costs. The ZA casting alloys exhibit similar capabilities when they are precision cast in steel or graphite molds.
- Improved die life and reduced maintenance. Replacement and maintenance costs for dies are typically 50-90% lower with zinc die castings than with those of other casting alloys. When products are fabricated from zinc

TABLE 7-11
Comparison of Typical Casting Alloy Properties

	Zinc Alloys											
	ZA-8*			ZA-12*			ZA-27*			No. 3 Zinc Die Cast Alloy (AG-40A)	No. 5 Zinc Die Cast Alloy (AG-41A)	No. 7 Zinc Die Cast Alloy
Mechanical Properties	Sand Cast	Perm Mold	Die Cast	Sand Cast	Perm Mold	Die Cast	Sand Cast	Sand Cast, heat treat**	Die Cast			
Ultimate Tensile Strength, ksi (MPa)	36-40 (248-276)	32-37 (221-255)	53-56 (365-386)	40-46 (275-317)	45-50 (310-345)	57-60 (393-414)	58-64 (400-440)	45-47 (310-325)	59-64 (407-440)	41 (283)	48 (383)	41 (283)
Yield Strength, 0.2% Offset, ksi (MPa)	29 (200)	30 (208)	42 (290)	30 (208)	30 (208)	46 (317)	53 (365)	37 (255)	53 (365)	---	---	---
Elongation, % in 2″ (51 mm)	1-2	1-2	6-10	1-2	1-2	4-7	3-6	8-11	1	10	7	13
Young's Modulus, psi x 10^6 (GPa)	---	12.4 (85.5)	---	12 (83)	---	---	10.9 (75)	11.5 (80)	---	---	---	---
Shear Strength, ksi (MPa)	---	35 (241)	---	36-38 (248-262)	---	---	41-43 (283-297)	32-33 (221-228)	---	31 (214)	38 (262)	---
Hardness, Brinell	82-88	85-90	95-105	92-96	88-90	95-110	110-120	90-100	105-120	82	91	80
Impact Strength, ft-lb (J)	13-18[a] (18-24)	13-18[a] (18-24)	24-35[a] (33-47)	17-22[a] (23-30)	---	15-27[a] (20-37)	25-40[a] (34-54)	35-55[a] (47-75)	2-5 (3-7)	43[b] (58)	48 (65)	43 (58)
Fatigue Strength, ksi (MPa), 5 x 10^8 cycles	---	7.5 (51.8)	---	15 (103.5)	---	---	25 (172.5)	15 (103.5)	---	6.9 (47.6)	8.2 (56.5)	---
Compressive Yield Strength, ksi (MPa)	28-30[h] (193-207)	29-32[h] (200-221)	---	32-34[h] (221-234)	33-35[h] (228-241)	---	47-48[h] (324-331)	37-38[h] (255-262)	---	60[f] (414)	87 (600)	---
Physical Properties												
Density, lb/in.^{3} (g/cm^3)		0.227 (6.3)			0.218 (6.03)			0.181 (5.00)		0.24 (6.6)	0.24 (6.6)	---
Melting range, °F (°C)		707-759 (375-404)			710-810 (377-432)			708-903 (376-484)		718-728 (381-387)	717-727 (381-386)	---

(continued)

TABLE 7-11—*Continued*

Mechanical Properties	Aluminum Alloys 380 Die Cast	Aluminum Alloys 319 Sand Cast	Aluminum Alloys 356-T6 Sand Cast	Brass/Bronze Alloys SAE 660 C93200 Sand Cast	Brass/Bronze Alloys SAE 40 C83600 Sand Cast	Brass/Bronze Alloys SAE 64 C93700 Sand Cast	Iron Alloys Class 30 Cast Iron	Iron Alloys 32510 Malleable Iron
Ultimate Tensile Strength, ksi (MPa)	47 (325)	27 (185)	33 (228)	35 (240)	37 (255)	35 (240)	31 (214)	50 (345)
Yield Strength, 0.2% Offset, ksi (MPa)	24 (165)	18 (125)	24 (165)	18[c] (125)	17[c] (117)	18[c] (125)	18 (125)	32 (224)
Elongation, % in 2" (51 mm)	2	3.5	3.5	20	30	20	---	10
Young's Modulus, psi x 10^6 (GPa)	10.3 (71)	10.7 (74)	10.5 (72.4)	14.5 (100)	13.5 (93)	11 (76)	13-16 (90-110)	25 (170)
Shear Strength, ksi (MPa)	27 (185)	22 (150)	26 (179)	---	---	---	43 (296)	45 (310)
Hardness, Brinell	75	70	70	65	60	60	170-269	110-156
Impact Strength, ft-lb (J)	3[a] (4)	4[a] (5)	8[a] (11)	6[d] (8)	11[e] (15)	11[e] (15)	---	40-65[a] (54-88)
Fatigue Strength, ksi (MPa), 5 x 10^8 cycles	20 (138)	10 (69)	8.5 (58.6)	16 (110)	11 (75.8)	13 (90)	14 (97)	28 (193)
Compressive Yield Strength, ksi (MPa)	---	19 (130)	25 (172)	46[g] (315)	37.5[g] (259)	47[g] (324)	109[f] (752)	---
Physical Properties								
Density, lb/in.^{3} (g/cm^3)	0.098 (2.74)	0.101 (2.80)	0.097 (2.69)	0.322 (8.93)	0.318 (8.80)	0.32 (8.9)	0.25 (6.9)	0.26 (7.2)
Melting Range, °F (°C)	1000-1100 (538-593)	960-1120 (516-604)	1035-1135 (557-613)	1570-1790 (854-977)	1570-1850 (854-1010)	1403-1705 (762-929)	>2150 (1177)	>2250 (1232)

* Complies with ASTM Specification B669-82.
** Heat for 3 hours at 610° F (320° C) and furnace cool.
[a] 10 mm (0.4") unnotched Charpy
[b] 1/4" (6.4 mm) unnotched Charpy
[c] At 1/2% elongation
[d] Izod
[e] Notched Charpy
[f] Compressive strength
[g] 0.1 set/in.
[h] 0.1% offset

sheet, tool wear may be minor, even when decorative features with fine details are required. Fabricators routinely form millions of sheet zinc parts with no die maintenance required.

- Low energy consumption and nonpolluting. Zinc alloys require less energy to convert from ore to ingot than many other metallic materials; they also require less energy in melting to produce castings. Equally important, the zinc alloys do not produce noxious airborne pollutants or toxic wastes when they are properly handled; therefore, applications are not likely to be adversely affected by government-imposed health and environmental regulations.
- Excellent surface finishing capability. Zinc alloys readily accept a wide variety of decorative and corrosion-resistant surface finishes. Low-cost zinc components are painted to match adjacent parts; chrome-plated to offer a durable luster; and barrel-plated or electroplated and brush finished to take on the rich appearance of brass, bronze, or stainless steel. Zinc die castings may also be dyed in a variety of colors to give them a distinct appearance. The good corrosion resistance of zinc coatings is enhanced by chromate or phosphate treatment, and further improved by anodizing.
- Joining capability. Zinc alloy components are commonly joined to similar or dissimilar metals with little or no problem of galvanic corrosion. Castings are internally threaded, externally threaded, or through-bolted, and they readily accept standard push-on, spin-on, or self-tapping fasteners. Alloys 3, 5, and 7 have sufficient ductility to permit riveting, staking, crimping, and swaging. Soldering and welding are possible on all zinc alloys that contain no aluminum. When metallurgically bonding zinc alloys that contain aluminum, however, it is important not to use solders or welding rods containing lead, tin, or cadmium. These elements are serious contaminants that can cause intergranular corrosion. Zinc sheet can be spot-welded on equipment designed for sheet steel.
- Corrosion resistance. Zinc-based alloys have good corrosion resistance under normal atmospheric conditions, in aqueous solutions, and when used with petroleum products. Their corrosion resistance can be further improved with a variety of surface treatments.
- Electrical and thermal conductivity. Because the electrical and thermal properties of pure zinc do not equal those for pure copper and aluminum, zinc alloys are often overlooked when selecting materials for applications in which these properties are important. However, a zinc alloy can sometimes be substituted for an aluminum or copper-based alloy in a particular application at equivalent electrical and/or thermal conductivity.
- Nonmagnetic and nonsparking. The nonmagnetic properties of zinc make it useful in electronics and for delicate moving parts that would otherwise be adversely affected by magnetic disturbances. Zinc alloys are generally nonsparking by nature, and this inherent characteristic is an advantage in the presence of inflammable or explosive substances.

Creep Behavior

When specifying and processing materials, it is customary to rely on tensile strength as a significant and useful property. Specifying this property is usually adequate for most metal alloys. However, when zinc alloys are involved and are expected to perform under constant tensile-load conditions or at elevated temperatures, the creep strength will be the determining factor for specifying the alloys.

Foundry alloys. Creep limitations exist with respect to conventional zinc die-casting alloys. However, the new ZA foundry alloys offer improved creep strengths for higher strength designs. Many applications do not impose constant stress at elevated temperatures, and so creep may not be their limiting factor.

Wrought alloys. Rolled unalloyed zinc creeps readily at room temperature; and although the addition of copper is beneficial, the improvement in creep resistance is small. However, the addition of a small amount (0.1-0.2%) of titanium to zinc or zinc-copper alloys provides a substantial improvement in the creep resistance, particularly in the case of the zinc-copper alloys. Because of its significant increase in creep resistance, compared with zinc or zinc-copper alloys, zinc-copper-titanium can be used in thin sheets for a wide range of architectural applications, where zinc and zinc-copper have had only limited use owing to the heavy gage and sheer bulk required to obtain the necessary resistance to creep.

The excellent creep resistance of the zinc-copper-titanium alloys can only be obtained (or maximized) if the alloys have been given the proper mechanical and thermal treatments. Improper mechanical or thermal processing can lead to low creep resistance because these alloys are very sensitive to rolling reduction, rolling direction, and rolling temperature, as well as composition. In addition, the as-rolled sheet is anisotropic in its creep behavior—a characteristic, however, that can be minimized by proper heat treatment after rolling.

Mechanical Properties

Typical mechanical properties of representative zinc casting alloys, wrought (rolled) alloys, superplastic alloys, drawn wire products, and forged and extruded zinc alloy products are presented in Tables 7-12, 7-13, 7-14, and 7-15.[5]

Superplastic alloys. The commercially available superplastic zinc alloys can be divided into three general groups, based on their compositions and mechanical properties. These groups have been arbitrarily designated A, B, and C, since no generally accepted nomenclature has been established to date. Alloy A is the basic binary zinc-aluminum alloy; alloy B contains copper, which has been added to improve the tensile and creep strengths; alloy C has magnesium added as well as copper to further improve these properties. In Table 7-13, the mechanical properties are tabulated under two conditions:

1. As rolled—the alloy in the superplastic condition.
2. Annealed and air cooled—the alloy in the finished condition after heat treatment to remove the superplasticity and reestablish good room temperature mechanical properties.

Wire drawn products. Zinc alloys in the form of wire are largely used for flame or arc-spray metallizing. Certain zinc alloy wire grades are also used either as filler metal for soldering or for their mechanical properties.

Forging and extruded alloys. Although the applications of extruded and forged zinc products have been relatively limited, the inherent advantages of the extrusion and forging processes are applicable to zinc alloys. Zinc extrusions and forgings can be readily machined; can be joined by soldering, welding, or

TABLE 7-12
Room Temperature Mechanical Properties of Selected Wrought (Rolled) Zinc Alloys

Alloy Family	Trade Name	Tensile Strength, ksi (MPa) Longitudinal	Tensile Strength, ksi (MPa) Transverse	Elongation, % Longitudinal	Elongation, % Transverse	Hardness Brinell	Hardness Vickers	Hardness Rockwell B
Zn	Z9-VM	17.4 (120)	21.8 (150)	60-80	40-60	---	30	---
	Zintane 5	17.4 (120)	21.8 (150)	60-80	40-60	---	30-35	---
Zn-Cu	Microzinc Alloy 190	25-30 (172-207)*	31-38 (214-262)*	25-45*	20-36*	---	---	55-65*
	Zilloy 15	37-28 (255-193)**	48-36 (331-248)**	20-25**	2-10**	80-61 **	---	---
	NJZ Alloy No. 15	24-31 (167-217)	31-38 (214-265)	26-39	12-29	---	---	71-82, 65-76**
Zn-Ti	Zilloy 20	32 (221)**	42 (290)**	38**	21**	---	---	---
	NJZ Alloy No. 25	24-30 (165-209)	32-44 (218-300)	28-54	16-47	---	---	69-80, 60-76**
	Microzinc Alloy 700	24-30 (165-207)*	30-38 (207-262)*	25-45*	20-36*	---	---	54-66*
	S.T.Z.	25 (170)	---	40	---	---	40	---
	RCA-Titane	22-25 (150-170)	26-29 (180-200)	60-80	30-50	46-50	53-58	---
Zn-Pb-Cd-Fe	Microzinc Alloy 100	18-24 (124-165)*	22-28 (152-193)*	30-65*	24-52*	---	---	41-58*
	NJZ Alloy No. 10	20-24 (139-163)	25-32 (174-220)	42-65	30-51	---	---	55-75, 47-58**
	NJZ Alloy No. 55	21-26 (143-176)	27-32 (183-222)	29-60	21-46	---	---	57-64, 51-74**
	Photo-engraving Zinc	23-28 (160-190)	33-41 (230-280)	30-45	25-45	---	45	---
Zn-Al-Mg	Microzinc Alloy 300	25-35 (172-241)	30-40 (207-276)	15-30	12-24	---	---	58-70
	Flash/Ultra	26-28 (180-190)	29-32 (200-220)	30-35	20-25	---	60	---

* Value dependent on thermomechanical treatment.
** Hot rolled condition.

(*Zinc Institute, Inc.*)

adhesives; and can be finished with paints, polymers, or with electroplated coatings. Because of the ductility of zinc extrusions and forgings, secondary operations—such as bending, swaging, flaring, stamping, and coining—can be readily performed, as required. Zinc extrusions and forgings can be produced to meet the requirements for a variety of applications.

The zinc-based forging alloys can be divided into two groups, the zinc-titanium family and the zinc-aluminum family. The zinc-titanium alloys are recommended for most forging applications. The zinc-aluminum alloys were developed for specialty applications, particularly where high impact strength at low temperatures is required.

Zinc-titanium. The zinc-titanium alloys possess good strength and ductility, are easily forged and machined, are dimensionally stable, have excellent creep characteristics, and offer unusual resistance to dimensional growth at elevated temperatures. Joining can be accomplished by any of the standard methods, and finishing characteristics are good.

Zinc-aluminum. The zinc-aluminum alloys are low density, two-phase materials having superior strength and impact properties at very low temperatures, -60° F (-50° C). They have somewhat better machine turning characteristics and have better bearing properties. Like the zinc-titanium alloys, joining properties and finishing characteristics are excellent.

APPLICATIONS

Unalloyed zinc is widely used for protective coatings and in engineering applications as anodes for the cathodic protection of structures immersed in an electrolyte, such as seawater or moist soil. Zinc alloys, on the other hand, traditionally have

TABLE 7-13
Room Temperature Mechanical Properties of Superplastic Zn-Al Alloys

Alloy	Heat Treatment	Tensile Strength, ksi (MPa)	0.2% Yield Strength, ksi (MPa)	Elongation, %	Hardness, Rockwell T-15
Zn-22Al (Alloy A)	As rolled	28 (193)	23 (159)	98	46
	Annealed at 660° F (350° C) and air cooled	51 (352)	45 (310)	13	79
Zn-22Al-0.5Cu (Alloy B)	As rolled	45 (310)	37 (255)	27	70
	Annealed at 660° F (350° C) and air cooled	58 (400)	51 (352)	11	84
Zn-22Al-0.5Cu-0.02Mg (Alloy C)	As rolled	55 (379)	43 (297)	25	79
	Annealed at 660° F (350° C) and air cooled	64 (441)	56 (386)	9	85

TABLE 7-14
Typical Compositions* and Room Temperature Mechanical Properties of Wire Drawn Zinc Alloys

Trade Name	% Pb	% Fe	% Cd	% Cu	% Al	% Ti	% Mn	% Cr	Tensile Strength, ksi (MPa)	Elongation, %	Vickers Hardness
Zinacor 100	< 0.003	< 0.0015	< 0.003	< 0.001	---	---	---	---	14-15 (95-105)	100-110	31.5
Zinacor 101	< 0.003	< 0.0015	0.03	< 0.001	---	---	---	---	20-23 (140-160)	80-100	---
Zinacor 850	< 0.003	< 0.0015	< 0.003	< 0.001	15	---	---	---	15-20 (105-140)	120-150	35
Zinacor 960	< 0.003	< 0.0015	< 0.003	< 0.001	4	---	---	---	11-12 (75-85)	150-200	---
Zinacor 980	< 0.003	< 0.0015	< 0.003	< 0.001	2	---	---	---	15-16 (100-110)	150-200	---
Zinacor 42	< 0.003	< 0.0015	< 0.003	0.4	---	0.1	---	---	28-29 (190-200)	60-70	70
Zinacor 432	< 0.003	< 0.0015	0.18	2	---	0.15	0.3	0.07	44-51 (300-350)	40-60	---

* Weight percent, balance zinc

TABLE 7-15
Typical Compositions* and Room Temperature Mechanical Properties of Forged or Extruded Zinc Alloys

Trade Name**	% Al	% Mg	% Cu	% Ti	% Mn	Tensile Strength, ksi (MPa)	0.2% Yield Strength, ksi (MPa)	Elongation, %	Brinell Hardness
KORLOY 2570	11.0	0.02	0.75	---	---	50-72 (344-496)	45-60 (310-414)	18-13	100-120
KORLOY 2573	14.5	0.02	0.70	---	---	50-75 (345-517)	45-65 (310-448)	20-14	100-125
KORLOY 3130	---	---	1.0	0.1	---	35 (241)	23 (159)	30	65
KORLOY 3330	---	---	1.0	0.1	1.0	48 (331)	35 (241)	20	90

* Weight percent, balance zinc
** Designations of Cominco Ltd.

been used extensively in decorative, nonload-bearing applications, but typically were not specified for engineered load-bearing components and structures. However, recent advances in fabricating technology, combined with active and fruitful programs in alloy development, have established the viability of zinc alloys in the field of engineering load-bearing materials.

The principal end uses for zinc are: (1) unalloyed zinc coatings or anodes for corrosion protection; (2) casting alloys; (3) as the alloying element in copper, aluminum, magnesium, or other-based alloys; (4) wrought alloys; and (5) zinc chemicals. In the corrosion protection category, hot-dip or continuous galvanizing accounts for the majority of zinc consumption. In casting alloys, it is the die-casting compositions that consume most of the zinc. In the zinc-containing alloys, the copper-based materials, such as the brasses, are the largest zinc consumers. Rolled zinc is the major form in which wrought zinc products are supplied, although drawn zinc wire for metallizing is gaining increasing usage. In the zinc chemical division, zinc oxide represents the compound of major importance.

Recent statistics indicate the following percentages for the consumption of zinc in the United States:

Galvanizing	44.5%
Die casting	25.6%
Brass alloys	13.5%
Rolled zinc	5.7%
Zinc oxide	9.1%
Miscellaneous	1.6%

For galvanizing applications, the less pure or higher lead content grades are generally used; however, the impurities (lead, aluminum, and iron) must be controlled for the various galvanizing processes. The grade of zinc used for die-casting alloys and anodes is usually special high grade, while both high grade and special high grade are employed for brasses and rolled alloys, depending on the specific application.

Zinc oxide is used in rubber, paints, ceramics, chemicals, agriculture, photocopying, floor coverings, and coated fabrics and textiles. Both the French process and the American process grades are used in most of these applications, depending upon the specific character of the end product. Zinc oxide's largest use is found in rubber products where it is an activator for the accelerators used to speed up the vulcanization process. It is a pigment in paints and ceramics, a soil nutrient in the agricultural field, a stabilizer in plastics, and provides the photosensitive character for coated papers used in some photocopying techniques.

Corrosion Protection

Almost half of the zinc consumed in the world goes into coatings for the corrosion protection of irons, mild steels, and low-alloy steels. Zinc corrodes at much lower rates than steel in atmospheric exposure and, in addition, will corrode sacrificially when the coated steel is exposed to moisture by scratches or cut ends. Zinc anodes are also used purely to provide galvanic sacrificial protection in underwater and underground applications.

Zinc coatings. The specific coating techniques by which zinc is applied to provide corrosion protection include:

- Hot-dip galvanizing after fabrication (see Chapter 23).
- Continuous line galvanizing (see Chapter 23).
- Electrogalvanizing (see Chapter 23).
- Electroplating (see Chapter 20).
- Metallizing (see Chapter 22).
- Zinc dust/zinc oxide painting (see Chapter 26).
- Mechanical plating (see Chapter 20).

The first three categories are the most important commercially, and, according to a recent Zinc Development Association estimate, account for over 90% of the total zinc used for coatings.

The major product forms of hot-dipped, zinc-coated materials include sheet, strip, wire, and tube. Galvanized sheet is used mainly for building and construction, automotive underbody panels, and household appliances. Electroplating and mechanical plating are normally used on fasteners and other relatively small objects where thin, uniform coatings are required. Zinc spraying is used on large structures, such as bridges, that are too large to conveniently hot-dip galvanize or are already in place. Zinc-rich or zinc oxide paints are effective across a wide range of products and are often used as the primers for painting storage tanks, ships, and other large structures.

Zinc anodes. If a structure requiring protection is not, or cannot be, coated with zinc, corrosion can still be controlled by using zinc as a sacrificial galvanic anode in a cathodic protection system. Cathodic protection is effective in such diverse media as saltwater, fresh water, and moist soil. The structures most frequently protected by this method are underground pipelines, steel piers, and ship hulls. The source of the protective cathodic current can be either impressed (rectified a-c or specially generated d-c) or galvanic.

Casting Alloys

Zinc-based casting alloys fall into three categories: (1) die-casting alloys, (2) foundry alloys, and (3) other casting alloys. Of these three categories, the die-casting alloys are by far the most important, accounting for more than 25% of all zinc consumption in recent years. The foundry alloys are a relatively new development, are just beginning to achieve significant production, and are commercially available from numerous sources. The slush-casting and forming-die alloys are produced in only small quantities, but have been utilized for some time for items such as lighting fixtures, lamp bases, and casket hardware.

Die-Casting Alloys. The major use of zinc as a structural material is in alloys for pressure die casting. Zinc alloy die castings are highly castable, easily finished, economical materials with good mechanical properties. They are used for a wide range of decorative and lightweight structural parts, and lend themselves readily to rapid mass production techniques such as those required in the automotive industry. These alloys may be cast into very complex shapes, in intricate detail, to tight dimensional tolerances, with excellent surface finish, and often in very thin section sizes.

Foundry Alloys. The zinc foundry alloys, commonly referred to as the ZA alloys, are experiencing growing commercial usage. The ZA alloys are suitable for casting by sand, permanent-mold, graphite permanent-mold, shell-mold, and high pressure die-casting methods. The graphite permanent-mold process was specifically developed for the good castability and low casting temperature of the ZA alloys (see ASTM B669).

The ZA alloys exhibit mechanical properties equal to or exceeding those of the conventional zinc die-casting, cast iron, aluminum, or copper alloys. In addition, they have excellent bearing properties, wear resistance, and machinability. Advantageous foundry properties include low melting temperatures, increased die life, and mold stability. They may be readily cast in thin sections in sand molds and require no fluxing or degassing.

Wrought Alloys

Wrought zinc and zinc alloy products may be obtained in a number of different forms such as rolled strip, sheet, or foil; forged or extruded bar, rod, and other shapes; and drawn wire or rod. They are readily fabricated and welded, and exhibit good corrosion resistance for many types of service. Since pure zinc will creep under load at room temperature, alloying additions are necessary for alloys used in any structural applications. Furthermore, zinc is anisotropic in its deformation behavior and will exhibit textures and preferred orientations. Roofing and siding are typical applications of these products.

Rolled zinc alloys. The majority of wrought zinc alloys are used in architectural and building materials, dry cell batteries, photoengraving plates, coinage, electrical contacts, superplastic alloy applications, and miscellaneous applications similar to those wrought brasses are used for. Rolled zinc products are popular because they can readily be formed by a number of secondary deformation techniques, exhibit good corrosion resistance, and are readily joined by conventional welding or soldering methods.

Superplastic alloys. Included in the zinc-aluminum alloy group are the superplastic rolled zinc alloys. They can be easily fabricated from sheet form into a desired shape and then heat treated to restore their high-strength properties. Annealing at 660° F (350° C) and air cooling also produces increases in strength. The superplastic zinc and 22% aluminum alloys are normally processed into sheet by conventional methods, and then fabricated into complex shapes by processes such as stretch forming, deep drawing, compression molding, or thermoforming.

Forged or extruded zinc alloys. The superplastic zinc and 22% aluminum alloys may also be processed by standard forging or extrusion methods. However, the two alloy groups most often used for forgings and extrusions are zinc-aluminum alloys with 11.0 to 14.5% aluminum and zinc-copper-titanium compositions containing 1% copper and 0.1% titanium.

Drawn wire and rod zinc alloys. As discussed under mechanical properties, the drawn wire and rod products are used mainly for thermal-spray, metallizing wire, filler metal wire for soldering, and for wire that is subsequently fabricated into nails, screws, slate hooks, or wire gauze. These alloys range from virtually pure zinc to multialloying element compositions. Other zinc-aluminum alloys with much higher aluminum contents (10-20%) have also been produced in wire and rod form for specialty applications.

Zinc-Containing Alloys

Zinc-containing alloys account for almost 15% of zinc consumption, with the largest group being the copper-zinc alloys known as the brasses or bronzes. Zinc is also used, in varying degrees, as an alloying element in aluminum-based, magnesium-based, and tin-based systems, and is often one of the constituents in brazing and soldering alloys.

Zinc Dust and Powder

Zinc dust refers to material condensed from zinc vapor, while zinc powder normally refers to atomized molten zinc. Zinc dust and powder are employed in the production of zinc-rich paints and for powder-fed thermal spraying or coating operations. Because of their reactivity, they are employed for many chemical engineering applications, particularly in the pulp, paper, textile, and linoleum industries. They are also used in several metal refining operations, in dry cell batteries, and in a large number of other applications.

Zinc Chemicals

The zinc chemicals of major importance are zinc oxide, zinc sulfide, zinc sulfate, zinc chloride, zinc chromate (zinc yellow), zinc stearate, and zinc phosphate. Zinc oxide is the most important and accounts for about 10% of total zinc consumption. It is used in rubber, paints, ceramics, general chemical applications, agriculture, and photocopying. Rubber applications represent about half of all zinc oxide usage. Zinc oxide is also present as a pigment in many paint and ceramics formulations.

FABRICATION

Zinc alloys can be fabricated into parts using a broad range of production processes that are capable of supplying components in quantities anywhere from millions a year down to a single part. Zinc alloys are used in more production casting processes than any other ferrous or nonferrous metal and, hence, fill a major role in industrial parts manufacture.

Die casting, permanent-mold casting, graphite permanent-mold casting, sand casting, and shell-mold casting are the methods most widely used in commercial zinc alloy components manufacture. Plaster-mold casting is used for prototyping zinc castings to allow testing of new component designs. An overview of process features and dimensional tolerances for these six casting methods is found in Table 7-16. Investment, low-pressure permanent-mold, centrifugal, continuous, and less traditional rubber-mold, bronze-mold, and aluminum-mold casting processes are also used to produce zinc alloy parts. In addition, wrought zinc alloys are fabricated into components from sheet zinc and by forging and extrusion.

Process Selection

A number of general process selection guidelines have evolved. Lowest casting cost per piece will be achieved by die casting, followed by permanent-mold casting, and then sand casting; however, tooling costs run in the opposite direction. A simple pattern is used to make an inexpensive sand mold, used just once. Reusable metal (or graphite) permanent molds are more costly, but easily allow automation and provide improved casting tolerances and detail compared to sand casting. Die-casting tooling is the most expensive initially, but provides intricate detail, excellent dimensional tolerances, and, by far, the fastest production rate. Thus, for high-volume requirements, when tooling cost can be amortized over many castings, die casting will result in the lowest overall piece price. Conversely, for low production quantities, sand casting results in the lowest total piece price, with permanent-mold casting fitting in between die casting and sand casting in cost per piece.

Generally, die casting is used when production requirements are 50,000 pieces per year or greater, up to millions of pieces per year. However, many die castings are produced in annual quantities of 10,000 or less because die casting can eliminate or minimize machining and reduce weight.

Permanent-mold casting is normally used for intermediate annual volume requirements, in the 5000 to 50,000 piece range. Surface finish and detail are better than with sand casting (but not as good as with die casting), and can justify process selection for production quantities of 1000 per year or less. Graphite permanent molds are less costly than metal molds, and their use is gaining acceptance for medium-volume precision component production in the 500 to 20,000 per year volume category.

Sand casting is used with zinc alloys for low-to-medium volume requirements ranging from a single piece up to a few thousand per year. However, there are exceptions, with production quantities in the tens of thousands not uncommon when used to exploit other advantages of sand casting.

Factors other than casting and tooling costs have a bearing on process selection. A component's structural requirements may dictate process selection; for example, when the traditional die-casting alloys do not fully satisfy mechanical property requirements. In such cases, the zinc-aluminum alloys ZA-12 and ZA-27 provide higher strength engineering alternatives when they are permanent-mold or sand cast.

Wrought Zinc Alloys

A variety of zinc alloys are available to produce wrought components. The wrought alloys have advantages, especially in die life, that should be considered in evaluating product design and processing factors.

Rolled zinc. Wrought (rolled) zinc may be formed readily by many techniques including bending, spinning, stamping, deep drawing, roll forming, coining, and impact extrusion. For joining operations, zinc alloys can be soldered and resistance welded with ease.

The tables of property data for zinc alloys normally do not contain data on the drawability of rolled zinc alloys. This omission is due to the fact that, unlike most other industrial metals and alloys, the drawability of zinc and its alloys is highly strain-rate dependent. Industrial drawing conditions for rolled zinc alloys are characterized by a high strain rate (high speed of drawing); hence, drawability test data are not applicable because the conventional tests are usually conducted at a low strain rate (low speed of drawing).

Superplastic alloys. In sheet form, these alloys can be vacuum or blow molded in low-cost dies, taking advantage of their 1000% elongation capability. Small amounts of copper and magnesium may be added to improve mechanical properties in the resulting products, which have found uses in casings and decorative wall tiles.

Processes. Thermoforming is the process most commonly used for superplastic zinc alloys. Other fabrication techniques being used or developed include compression molding, deep drawing, stretch forming, and powder metallurgy hot pressing of superplastic zinc alloy powders.

Finishing. Parts made from superplastic zinc alloys may be finished by any of the conventional techniques used for finishing other zinc alloys. However, because of their high aluminum content, their surface reactivity is somewhat different from that of the standard alloys, and modified procedures have been developed for anodizing and for surface preparation prior to plating.

Gravity Casting Alloys

Historically, production of cast zinc alloy components has been limited to the die-casting process. The largest market for zinc alloy die casting has been the automotive industry. The appliance, builders' hardware, electrical fittings, and toy industries are other important markets.

The new ZA alloys represent a significant departure from conventional zinc alloy technology. Unlike the zinc die-casting alloys, ZA alloys can be fabricated in virtually all major casting processes including sand, permanent mold, shell, investment, continuous, and die casting.

Zinc is not as widely used in the gravity casting/foundry alloy field as in the die-casting area in spite of its long history and wide usage. Prior to 1967, the only zinc gravity casting alloys of note were the slush-casting and forming-die alloys. The foundry market was dominated then, and still is, by the long-established copper, aluminum, and iron-based alloys. In that year, the first high-strength zinc gravity casting alloy, ILZRO 12 (International Lead Zinc Research Organization, Inc.), was introduced. ILZRO 12 was originally marketed as a prototype alloy to be used in sand or permanent-mold casting of parts that would later become die castings. However, it became apparent that ILZRO 12 had much wider application in the foundry industry.

Subsequent to the introduction of ILZRO 12, two other alloys were developed containing 8% aluminum (ZA-8) and 27% aluminum (ZA-27). Experience with these zinc alloys has shown that they can be cast in many types of molds, including sand, plaster, silicon rubber, graphite, and cast iron. Bronze, aluminum, and beryllium-copper molds have also been used successfully in many applications.

Alloy ZA-12 (ILZRO 12) is well suited to casting in graphite molds. The ZA-27 alloy is intended primarily for sand castings and has a tensile strength comparable to many of the higher strength ferrous and nonferrous casting alloys. The ZA-8 alloy can be sand cast but is generally recommended for permanent-mold casting. It has excellent finishing characteristics and presents an alternative to brass or bronze castings.

Advantages. In comparison with aluminum, the zinc foundry alloys melt at lower temperatures, have better fluidity, and do not require fluxing or degassing. They are also stronger and harder, are less brittle, and show improved galling resistance and superior platability. Compared to brass and bronze alloys, the zinc alloys have better castability at a lower temperature, lower density, improved strength and hardness, and show less wear in service.

Cast iron and the zinc foundry alloys can be cast with similar equipment and have similar properties. The zinc alloys, however, offer the advantage of speed of melting and have fewer pollution control problems. Since the zinc alloys are homogeneous, they do not have hard spots, and hence machinability is greatly improved.

Prototyping. The preparation of prototype components by die-casting alloys 3 and 5 is usually a costly and time-consuming process; therefore, these alloys are often gravity cast to produce prototypes of decorative (nonload-bearing) parts. However, with gravity casting, the mechanical properties of the castings are much poorer than those of corresponding die castings. Hence, gravity casting is not a suitable method for producing functional (load-bearing) parts from these alloys. The mechanical properties of gravity-cast ZA-12 and heat-treated ZA-27 approximate those of die-cast alloys 3 and 5; they are therefore suitable, in many cases, for making prototypes of functional parts designed to be subsequently produced by die casting.

Graphite-mold casting. Compared with iron or steel, the attributes of graphite as a mold material include fast, easy machining, superior dimensional stability, and greater thermal conductivity to promote rapid solidification and improved productivity. However, graphite deteriorates rapidly at the temperatures normally required for casting aluminum, bronze, and ferrous alloys. The low casting temperatures of the ZA alloys minimize attack on graphite, resulting in mold lives of 25,000 to 50,000 casting cycles. Consequently, graphite molds provide a low-cost, precision, gravity casting process that minimizes machining costs for intermediate production quantities of 500 to 20,000 pieces per year.

Graphite-mold casting has evolved into a sophisticated system using semiautomatic casting machines and die design concepts. ZA-12 is the alloy preferred for graphite-mold casting; however, both ZA-8 and ZA-27 may be cast in graphite molds for specialty applications.

Capabilities of graphite-mold casting fall between those of die casting and conventional permanent-mold casting. Graphite molds answer the need for situations requiring die-casting precision, but where the cost of die-casting tooling cannot be justified because of a low production volume. Features such as holes, countersinks, ribs, and contours that would normally require machining of a sand casting can be cast to size in graphite molds. Tolerances and surface finish are similar to those of investment castings.

Die-Casting Alloys

Zinc-based die-casting alloys were introduced in the late 1920s to meet the demand for strong, stable die castings. Alloys 3, 5, and 7 are die cast in a hot chamber die-casting machine; that is, one in which the injection plunger sleeve and gooseneck are continuously immersed in molten zinc in the machine pot or furnace. When die casting an alloy with a low aluminum content, the molten metal will slowly dissolve iron if left in contact with the steel plunger sleeve and cast iron gooseneck. Hence, the alloy is cast in a cold chamber machine in which the

TABLE 7-16
Casting Process Comparison

Casting Process	Process Description	Advantages	Limitations
Die casting	Molten metal is forced into closed steel dies at high velocities by the application of pressure. When the metal solidifies, the die is opened and the casting ejected.	Extremely smooth surfaces, excellent dimensional accuracy, and rapid production rates.	High initial die costs. Limited to nonferrous metals and size of parts limited. Porosity may be encountered as a result of entrapped air in the die.
Permanent-mold casting	Mold cavities are machined into metal die blocks and designed for repetitive use. Cores made of an expendable material. The mold halves are clamped and molten metal is gravity fed to the cavity (sometimes low pressure is applied). Molds open and castings are ejected.	Good surface finish and grain structure. Good dimensional accuracy. Repeated use of molds. Rapid production rates. Low scrap loss and low porosity.	High initial mold cost. Shape, size, and intricacy limited. High melting metals, such as steel, unsuitable.
Graphite permanent-mold casting	This process is very similar to conventional metal permanent mold. Mold cavities are machined into structural graphite block and metal is gravity fed without pressure assist.	Combined low cost tooling, precision details, and good surface finish (close to die-casting finish and accuracy).	Low to medium production runs. Zinc alloys are used due to their low pouring temperature.
Sand casting	Moist, bonded sand is packed around a wood or metal pattern. The pattern halves are removed, and the mold is assembled with or without cores. Resin-coated sand also used.	Almost any metal can be used. Almost no limit on size, shape, or weight of part. Low-cost, most direct route from pattern to casting.	Some machining always necessary. Large castings have rough surface finish. Close tolerance difficult to achieve. Long thin projections not practical. Some alloys develop defects.
Shell-mold casting	Resin-coated sand is applied to heated metal patterns, forming shell-like mold halves. Pattern halves are bonded together with or without cores.	Faster production rate than sand castings and good dimensional repeatability.	Requires expensive pattern equipment. Size of part limited.
Plaster-mold casting	Slurry of special gypsum plaster mixed with water and other ingredients is poured over pattern and allowed to set. Pattern is removed and mold halves baked.	High dimensional accuracy, smooth surface finishes, and almost unlimited intricacy.	Limited to nonferrous metals. Moldmaking time is relatively long. Some limit to maximum size. Limited to prototypes and low volume requirements.

furnace pot holding the molten metal is not an integral part of the machine, and metal must be ladled from a refractory holding pot into the injection cylinder either manually or mechanically, prior to each shot. Aluminum alloys are usually die cast from a similar machine.

Thin wall. During the 1970s, thin-wall zinc die casting techniques were developed. The term "thin wall," when applied to zinc alloy die castings, is used to identify the minimum wall thickness required for an intended application. For example, these techniques enable large castings up to about 6.5 lb (3 kg)

TABLE 7-16—*Continued*

Typical Tolerances	Metal Used	Size Range	Typical Yearly Quantity Range	Relative Tooling Costs
Zinc: ±0.0015 in./in. (mm/mm). Aluminum and magnesium: ±0.002 in./in. (mm/mm). Across parting line, add ±0.001″ (0.03 mm) to the 0.0015″ tolerance. Zinc: 0.025-0.040″ (0.64-1.02 mm) is the normal minimal thickness. Aluminum: 0.050-0.080″ (1.27-2.03 mm) is the normal minimal thickness.	All zinc alloys, aluminum, magnesium, and some brass alloys.	Not normally over 3 ft^2 (0.3 m^2). Under 15 lb (6.8 kg) is typical.	High volume: 20,000 to millions	High
Aluminum and zinc: ±0.015″ (0.38 mm) basic. Add ±0.002 in./in. (mm/mm). Across parting line, add ±0.010 to 0.030″ (0.25-0.76 mm). Normal minimal thickness is 0.100 to 3/16″ (2.54 to 4.8 mm).	All zinc alloys, aluminum, and some brass, bronze, magnesium, and iron.	Castings to 2 x 2 ft (0.6 x 0.6 m) are common. Ounces to 150 lb (68 kg).	Moderate volume: 5000-50,000	Medium
±0.006″ (0.15 mm) first inch (25 mm). ±0.002 in./in. (mm/mm) additional. Across parting line, add ±0.005″ (0.13 mm). Normal minimal thickness is 0.100″ (2.54 mm).	ZA zinc alloys.	Sizes up to 12 x 14 x 7″ (305 x 356 x 178 mm). Ounces to 10 lb (4.5 kg).	Low to moderate volume: 500-20,000	Low to medium
Nonferrous: ±1/32″ to 6″ (0.8-152 mm). Add ±0.003″ (0.08 mm) for each additional inch (25 mm). Across parting line, add ±0.020 to ±0.090″ (0.51-2.29 mm). Normal minimal thickness is 1/8″ (3.2 mm).	ZA zinc alloys and most ferrous and nonferrous casting alloys.	Unrestricted size and weight.	Low to unlimited volume: one to thousands (zinc)	Very low to medium
Nonferous: ±0.008 in./in. (mm/mm). Across parting line, add 0.005 to ±0.010″ (0.13-0.25 mm). Normal minimal thickness is 3/32″ (2.4 mm).	ZA zinc alloys and most ferrous and nonferrous casting alloys.	Maximum usable mold area of 550 in.^{2} (0.4 m^2). To 50 lb (22.6 kg).	Low to moderate to high volume: 500 to thousands	Low to medium
±0.005″ (0.13 mm) to 2″. ±0.002 in./in. (0.05 mm/mm) additional. Across parting line, add ±0.010″ (0.25 mm). Normal minimal thickness is 0.070″ (1.78 mm).	ZA zinc alloys, aluminum, brass, and bronze casting alloys.	Normally up to 500 in.^{2} (0.3 m^2) area. Under 20 lb (9.1 kg).	Low volume and prototyping: one to hundreds	Very low

(*International Lead Zinc Research Organization, Inc. and Zinc Institute, Inc.*)

to be made with about 0.03″ (0.8 mm) wall thickness, instead of the 0.08″ (2 mm) or more that was previously considered necessary. Thin-wall casting technology is particularly valuable when weight must be reduced while still retaining the advantages of zinc alloy die castings.

Processing considerations. Zamac zinc alloy die castings will undergo aging and slight shrinkage, about 0.0005 in./in. (mm/mm), if placed in service in the as-cast condition. Room temperature aging for five weeks will permit about 70% of the shrinkage to occur. If required, stabilization heat treatments may be applied to accelerate the aging process and thereby avoid dimensional changes in service. The stabilization anneals may be carried out at temperatures from 150 to 300°F (65 to 150°C) for corresponding times of 12 hours to 30 minutes. Lower temperature, longer time annealing is preferable since it results in less shrinkage after treatment, but the higher temperature anneal is more practical from a production standpoint.

The ZA alloys can shrink and then grow. The ZA-27 alloy is particularly susceptible to growth when held at elevated temperatures, especially when die cast. Alloys ZA-8 and ZA-12 are relatively stable and exhibit minimal growth.

Machining. Zinc alloy die castings are invariably cast within quite close dimensional limits; but some machining is commonly required to remove flash, and to drill, ream, or tap holes. Alloys 3, 5, and 7 may be machined with conventional tools and methods. Generally, only light cuts at high speeds are necessary. High-speed steels are usually satisfactory, but cemented tungsten carbide tools produce better surface finishes, closer dimensional control, and less tool grinding time.

Finishing. The zinc die-casting alloys are amenable to a wide range of finishing operations that are applied for decorative purposes, improved corrosion resistance, or better abrasion performance. Finishing may be accomplished through mechanical techniques, electrodeposition, chemical conversion, or the application of organic coatings or specialty coatings. Mechanical finishing may include buffing, polishing, brushing, or tumbling. Metals electrodeposited on zinc die castings include copper, nickel, chromium, brass, silver, and gold. The copper-nickel-chromium plating system is the most widely used; however, a variety of other electrodeposited coatings are available to produce different colors and textures on zinc alloy die castings.

The chemical conversion coatings include chromates, phosphates, and other inorganics that may be applied directly to the fresh zinc surface or to one of the other subsequent coatings to produce certain colors and surface effects, and to improve corrosion resistance. The organic coatings include paints, enamels, lacquers, varnishes, and plastics. Some of the specialty techniques employed for finishing zinc die castings include vacuum metallizing, hot foil stamping, and anodizing, which is also called "Iridizing."

Joining. Zinc alloy die castings may be joined into assemblies or components by many of the techniques employed for other engineering materials. Most often they are fastened by mechanical means such as rivets, studs, rollovers, bosses, or threaded fasteners. Adhesives have also been used successfully to join zinc die castings. Alloys 3, 5, and 7 are weldable and solderable, but these two joining techniques are not normally employed except in special circumstances. Soldering with the normal lead-tin solders may only be done if the die casting is first electroplated with copper or nickel. Gas welding of zinc alloy die castings may be accomplished under reducing conditions using a filler rod similar in composition to the die-casting alloy. Resistance and arc welding have also been successfully used to join zinc alloy die castings.

Cold chamber process. The conventional zinc die casting alloys have been used successfully for decades to produce all types of products via the hot chamber die-casting process. The new ZA alloys, in addition to production by virtually all major gravity casting processes, may also be successfully cast by cold chamber (pressure) die casting. These alloys are normally selected for a combination of higher strength and superior bearing, creep, and wear resistance characteristics.

Both ZA-12 and ZA-27 can be die cast using conventional cold chamber die-casting technology. These alloys are specified for high-performance applications when conventional zinc and aluminum alloys are inadequate. They may be substituted directly for aluminum using existing tooling. They can also be continuously and centrifugally cast for bearings and machine components.

High temperatures. The ZA-12 and ZA-27 alloys are cold chamber die-casting alloys, which means they must be processed with die-casting machines and methods normally used with aluminum because they are cast at temperatures that would attack the submerged ferrous metal injection system of hot chamber machines. Alloy ZA-8, however, can be die cast in hot chamber machines.

Future application. Compared to traditional zinc and aluminum die castings, higher metal costs preclude consideration of the ZA alloys on a purely economic basis. Instead, they are expected to extend the capability of the die-casting process into higher performance applications.

References

1. Lead Industries Association, Inc., "Properties of Lead and Lead Alloys" (New York: 1983).
2. Zinc Development Association, "Zinc Production, Properties, and Uses" (London: 1982).
3. Society of Automotive Engineers, Inc., *Metals and Alloys in the Unified Numbering System*, 3rd ed. (Warrendale, PA: 1983).
4. International Lead Zinc Research Organization, Inc. and Zinc Institute, Inc., *Designing in Zinc*, Ist ed. (New York: 1982).
5. International Lead Zinc Research Organization, Inc., *Engineering Properties of Zinc Alloys*, 2nd ed. (New York: 1981).

Bibliography

Botta, Mike. "Radiation Shielding Market Awaits Nuke Plant Construction." *American Metal Market* (August 29, 1983).

Certified Alloys Company. "Zinc-Aluminum Alloys." Maple Heights, OH.

Eastern Alloys, Inc. "ZA Casting Alloys." Technical Data 5M-10/82. Maybrook, NY: 1982.

International Lead Zinc Research Organization, Inc. "Lead Research Digest," no. 38. New York: 1980.

International Tin Research Institute. *Annual Report for 1980*. Greenford, Middlesex, England.

Kidd Creek Mines Ltd. "Zinc Casting Alloys." Manual No. 231. Toronto, Canada: 1983.

Lead Industries Association, Inc. "Annual Review, U.S. Lead Industry." New York: 1980.

__________. "Follow the Safety Line." New York.

__________. "Lead Acid Batteries." New York: 1981.

__________. "Modern Uses of Lead." New York.

Mihaichuk, William. "Near-Net-Shape Castings from Zinc-Aluminum Alloys." *Machine Design* (September 8, 1983), pp. 98-103.

Obrzut, John J., and Alman, Pat S. "Why Everyone's Shouting About Noise Control." *Iron Age* (July 24, 1978).

Reichenecker, W. J. "Shear Strength of Solder Alloys." *Materials Engineering* (June 1983), pp. 12-13.

Rutherford, John D., and Hoskin, Thomas A., eds. *Future Cast*, vol. 1, no. 2. Maple Heights, OH: Aluminum Smelting & Refining Co., 1983.

Thwaites, C. J. "Reliability in Modern Soldering Techniques for Electronic Assemblies." Review No. 166. Columbus, OH: Tin Research Institute, Inc., 1972.

PLASTICS AND COMPOSITES

PLASTICS

Chapter contents:

Plastics are nonmetallic materials that can be formed and shaped by many methods. Plastics can be made from such natural resins as shellac; however, most plastics used in industrial applications are produced from man-made synthetic resins. Additional information on the various thermoset and thermoplastic materials and their product manufacturing processes and equipment is provided in "Plastics Forming," Chapter 18, Volume II of this Handbook series.

Plastics have become one of the most common classes of engineering materials in the past decade. For the last five years, the production of plastics, on a volume basis, exceeded steel output. Engineering plastics, those grades devised to resist severe service conditions or structural loads are in widespread use and their applications are increasing rapidly.[1]

INDUSTRY STRUCTURE

Plastics are an outgrowth of the petrochemical industry; hence, the language and terms of plastics are usually expressed in organic chemistry terms. Metals and plastics behave quite differently. To understand plastics materials and their production parts processing, therefore, some of the language of plastics must be learned.

The plastics industry is organized differently than the metals industry. Raw plastics, called resins, are made from oil and gas by large high-technology firms and then sold to custom molders or captive shops for processing into plastics components. Resins are made by three types of firms: Several large oil companies manufacture resins. Corporations with substantial chemical knowledge apply their expertise to produce plastics resins. And, a few sophisticated plastics consuming firms have developed resins and produce engineering plastics.

A resin manufacturer usually produces only a few different types of plastics, with numerous variations in some instances. Because the industry is relatively new, patents still play a major role in industrial organization; and, in general, firms manufacture only those types of plastics for which they have a strong patent position.

The industry's organization tends to isolate the end-user from the resin manufacturer. The designers and manufacturing engineers responsible for developing and applying plastics parts usually deal with a custom molder. The molder is the part maker and is frequently considered an expert in plastics. A molder has considerable expertise in designing for ease of processing and selecting the type and grade of plastics that is easiest to process. However, molders generally have limited knowledge of mechanical or corrosion design factors. Molders also usually limit their work to plastics they are familiar with and to the amount of engineering time they can invest in quoting a part. To assure selection of the most suitable plastics material from among hundreds that are available, it is advisable to develop direct sources of information, including contacts with resin manufacturers. Table 8-1 lists representative properties for selected thermoplastics and thermosets commonly used in industrial applications.

PLASTICS DESIGNATION SYSTEMS

The youthfulness of the plastics industry is reflected in the conflicting and competing specification, designation, and nomenclature systems currently in use. Most of these systems evolved during the earliest days of plastics when issues related to the organic chemistry of producing resins and creating new resins were of great importance. These systems have minimal appeal to designers and engineers who tend to specify plastics by the manufacturer's trade name/grade—such as Du Pont's Delrin 100 ST, acetal plastics resin—or use one of the ASTM standards covering a specific plastics type. However, as the competition between resins increased and plastics product applications grew, a broader, standardized approach to selecting plastics and processes was needed.

Contributors of sections of this chapter are: ***John T. Benedict***, *Consultant;* ***Frank D. Diodato***, *Market Development Engineer, Engineering Resins Div., Celanese Corp.;* ***Jack Hill***, *President, J. Hill Associates;* ***Roy S. Klein***, *Manager, Materials Engineering, American Sterilizer Co.;* ***Blaise A. LeWark, Sr.***, *Owner, Polytech, Ltd.*

Reviewers of sections of this chapter are: ***John C. Foster***, *Press Relations, Media Services Group, General Electric Co.;* ***Dr. David J. Goldwasser***, *Engineer, D. S. Gilmore Laboratory, Upjohn Co.;* ***Jack Hill***, *President, J. Hill Associates;* ***John T. Hoggatt***, *Technology Manager, Boeing Aerospace Co.;* ***R. J. Juergens***, *Branch Chief—Technology, McDonnell Douglas Corp.;* ***Roy S. Klein***, *Manager, Materials Engineering, American Sterilizer Co.;* ***P. D. Kohl***, *Processing Consultant, Polymer Products Dept., Du Pont Co.;* ***Dr. Joseph K. Lees***, *Manager, Advanced Composites, Du Pont Co.;* ***Blaise A. LeWark, Sr.***, *Owner, Polytech, Ltd.;* ***R. A. Lofland***, *Manager, Research & Development Composites Lab, Hughes Helicopter, Inc.;* ***Roy L. Manns***, *President, Boston Plastics Group, MEIPEC;* ***Francis B. McAndrew***, *Manager, Laboratory Technical Services, Celanese Engineering Resins Div., Celanese Corp.;* ***Tilak M. Shah***, *Market Development Engineer, D. S. Gilmore Laboratory, Upjohn Co.;* ***Wallace Wannlund***, *Consulting Engineer;* ***Charles F. Woodward***, *Professor, Engineering Graphics and Industrial Design, Engineering Technology Dept., Western Michigan University.*

PLASTICS INDUSTRY STRUCTURE

In response to this need, the ASTM has created a designation system applicable to all plastics. The system includes strengths and properties, as well as chemical information. The significant nomenclature, classification, specification, and designation systems currently in use are summarized here to show the trend toward systems based on a thorough understanding of plastics' structures, properties, and applications.

Behavior During Heating

Thermoplastics denote a class of plastics that "melt" or, more accurately, soften and flow under loads when heated. Cured (fully reacted) thermosets, in contrast, char and degrade at elevated temperatures but do not melt. Thermoplastics can be injection molded, but thermosets, once chemically cured or set as in epoxy, generally cannot be injection molded. The thermoplastics can be remelted and reprocessed. The thermosets cannot be reprocessed, except for use as filler materials. Until the 1970s, thermosets were used for all high-temperature applications. However, high-temperature thermoplastics have now been developed that rival the thermosets for many uses.

Production Volume

In terms of annual production volume, four types of plastics constitute about 70% of plastics production. Thus, polyethylene (PE), polystyrene (PS), polypropylene (PP), and polyvinyl chloride (PVC) are the principal commodity plastics. Other plastics are termed engineering or specialty plastics. A recent trend, however, is blurring this distinction.

TABLE 8-1
Properties of Selected Industrial Plastics

			Mechanical Properties (Room Temperature)			
Type of Plastics	Molecular Packing	Specific Gravity	ASTM D-638 Tensile Strength, psi (MPa)	ASTM D-638 Elongation, percent	ASTM D-695 Compressive Strength, psi (MPa)	ASTM D-256 Impact Strength (Izod), ft · lb/in. (J/cm)
Polystyrene	Amorphous	1.10	7500 (51.7)	2	14,000 (96.5)	0.3 (0.2)
High-impact polystyrene	Amorphous	1.15	5000 (34.5)	10	7500 (51.7)	0.6-10.0 (0.3-5.3)
Acrylics	Amorphous	1.15	10,000 (69.0)	6	15,000 (103.4)	0.4 (0.2)
Polycarbonate	Amorphous	1.20	9000 (62.1)	100	10,000 (69.0)	15.0 (8.0)
ABS	Amorphous	1.05	6000 (41.4)	30	8000 (55.2)	6.0 (3.2)
Acetal (homopolymer)	Crystalline	1.40	10,000 (69.0)	40	18,000 (124.1)	1.8 (1.0)
Nylon 6/6 at 50% RH*	Crystalline	1.15	11,000 (75.8)	400	10,000 (69.0)	2.1 (1.1)
Polypropylene	Crystalline	0.91	4500 (31.0)	500	7000 (48.3)	1.0 (0.5)
Polyethylene (high density)	Crystalline	0.95	4000 (27.6)	600	3000 (20.7)	10.0 (5.3)
Polyethylene (medium density)	Crystalline with amorphous regions	0.93	2400 (16.5)	600	3000 (20.7)	8.0 (4.3)
Polyethylene (low density)	Semi-crystalline	0.91	1500 (10.3)	700	3000 (20.7)	No break
Epoxy	Cross-linked network	1.25	10,000 (69.0)	3	20,000 (137.9)	0.8 (0.4)
Phenolic	Cross-linked network	1.35	7000 (4-8.3)	2	10,000 (69.0)	0.4 (0.2)

*RH = relative humidity

PLASTICS DESIGNATION SYSTEMS

The commodity plastics, as well as most other plastics, are dependent on oil and gas supplies. Declining petroleum prices induced oil-producing nations in the Middle East and North America to develop their own commodity plastics production capabilities as outlets for oil and gas surpluses. Resin producers in industrially developed countries are responding by inventing modified grades of commodity plastics that compete with some grades of engineering plastics. Industry analysts predict that new plastics in the 1980s will actually be improved existing grades. This development will tend to further complicate the classification and application of the new, modified commodity plastics and blur the distinction between commodity and engineering plastics.

Generic Class

The generic or family name used for plastics is the chemical designation of the structural organic molecule upon which a type of plastics is based. At present this nomenclature has several limits. Many engineers and designers have difficulty with names like polymethyl/methacrylate (PMMA). As a consequence, short nicknames, such as acrylic for PMMA or nylon for polyamide, are commonly used, further complicating the use of this designation system. While these names are well known within the plastics industry, engineers from other fields often cannot readily recognize and interpret them.

Chemical Characteristics

Unlike generic names, a system based on chemical characteristics groups plastics according to similarity in chemical analysis. For example, most plastics are "hydrocarbons," meaning that hydrogen and carbon atoms in different arrangements make up the mers or primary molecular units for a particular polymer resin. Silicon can replace carbon and form polymer chain molecules, as in the silicones. Hydrogen atoms can be replaced by other atoms with a valence of one as in chlorinated (chlorine) plastics like PVC or fluorinated (fluorine) plastics like Du Pont's Teflon TFE (tetrafluoroethylene).

Commercial Trade Names

Plastics manufacturers have developed different commercial trade names and designation systems for similar types of plastics. For example, polyamide (PA) or nylon resins are called Zytel by Du Pont, Thermocomp by LNP, Maranyl by ICI-Americas, and Akulan by Schulman. Du Pont also markets a mineral-filled nylon as Minlon. Within each of these commercial designations are different numbering systems to indicate specific grades. As a result, there is no simple way to locate comparable grades from competing manufacturers. Two reference sources for tracking a plastic's trade name back to the manufacturer and generic type are the *International Plastics Selector* and the *Modern Plastics Encyclopedia.* The strongest advantage in using trade names is the considerable technical support offered by resin manufacturers. Specifying a manufacturer's resin gives the user access to both published data and consulting services from the manufacturer.

ASTM Classification System

The American Society of Testing and Materials (ASTM) is a technical organization bringing manufacturers, specifiers, and users together to standardize specifications and test methods. Several individual types of plastics have been covered by specific ASTM standards; D 789 for polyamide (nylon) and D 788 for acrylic are two examples. Standards are useful for two reasons: First, the specifier can be assured of minimum strengths and properties for design calculations. Second, competitive manufacturers' resins can be used. Unfortunately, the rapid growth of the plastics industry also limits the usefulness of single standards. As mentioned previously, there is considerable competition between different types of resins. New and modified grades of resins are being developed, more quickly than ASTM standards can be set.

To remedy this situation by establishing an industry-wide designation system, in 1982 the ASTM issued the Standard D 4000 "Guide for Identification of Plastic Materials" to "adequately identify plastic materials in order to give industry a system that can be used universally." In D 4000, ASTM is attempting to establish a single designation system for all types and grades of plastics. In essence, D 4000 combines generic designations with a unified system of identifying important modifications to the generic resin and significant engineering property minimums. Figure 8-1 describes a typical line callout for ASTM Standard D 4000 application. A regular designation for a glass-filled polyamide (PA) or nylon resin is the ASTM D 4000 specification for 33% glass-filled nylon (polyamide) resin grade, that is, ASTM D 4000 PA120G33A53380GA140 where:

ASTM D 4000 = Plastics material
PA120G33 = Basic—generic resin and modifications
A53380 = Cell—mechanical (physical) properties
GA140 = Suffix—special properties and tests

Further interpretation of this example is provided in Fig. 8-2. It should be noted that "cells" of property information are

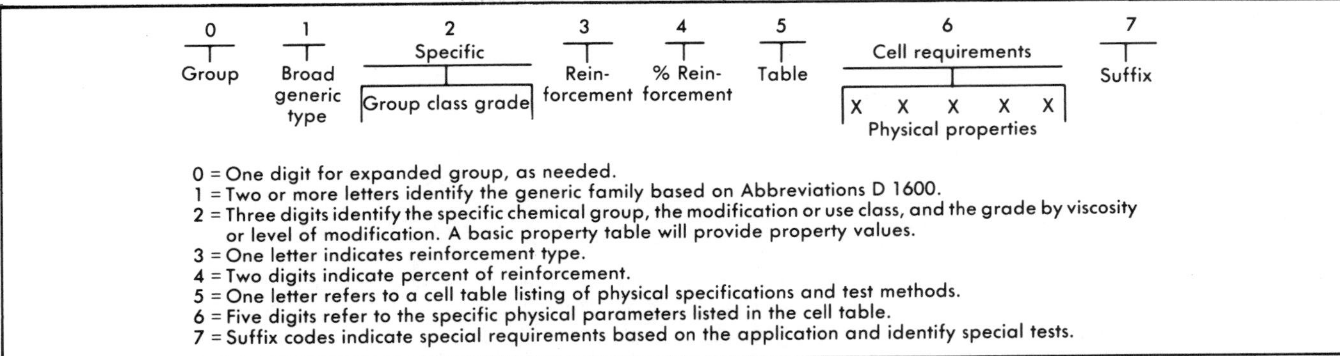

Fig. 8-1 ASTM D 4000 coding system. Line callout to designate a plastics material. (*ASTM*)

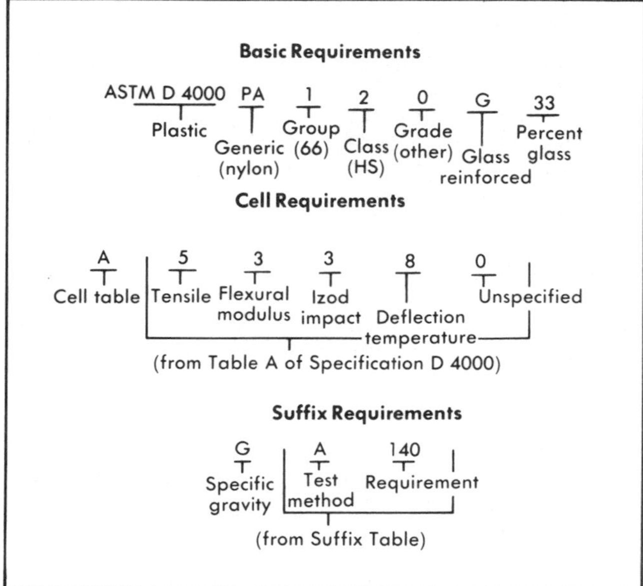

Fig. 8-2 Explanation of ASTM D 4000 designation PA120G33A533-80GA140 for 33% glass-filled nylon (polyamide) resin grade. (*ASTM*)

included in the designation. The specifier must be familiar with generic names, structural modifications, and engineering properties, as well as having access to reference data found in D 4000. Standard D 4000 recognizes the complexity of specifying plastics; however, at this time, the universal adoption of D 4000 is not assured. If this system is not accepted, one that is just as complex and thorough is likely to come into use.

BASIC TERMINOLOGY

The term "polymer" is commonly used interchangeably with the term "plastics." Neither term is entirely accurate in its delineation. Plastic means pliable, yet most engineering polymers are not plastic at room temperature. Polymer, on the other hand, can include every kind of material made by polymerization with repeating molecules. The ASTM definition (D 883) of a plastic is: "A material that contains as an essential ingredient an organic substance of large molecular weight, is solid in its finished state, and, at some stage in its manufacture or in its processing into finished articles, can be shaped by flow."

In broad terms, plastics are man-made polymers. Polymer is the generic name for all materials composed of long, chainlike molecules. Most living tissue and cells are polymeric. Plastics are created either by modifying natural polymers, such as cellulose fibers, or by causing small synthetic molecules to bond together into a chain. Compared to other classes of materials, the plastics molecular chain is enormous, giving it the term "macromolecule." Millions of macromolecular chains must be put together to make industrially useful quantities.

Mers or Repeating Units

The basic, repeating structural unit of a plastics chain is termed its "mer." In essence, the mer is a typical link of the molecular chain. The chemical name of a mer is also the generic name for that type of plastics. Before being reacted to form a chain, the small molecule that becomes the mer is termed a "monomer."

The atoms within a mer are covalently bonded by sharing valence electrons. Polyethylene (PE) is an example of a simple structure. The hydrogen atoms can be replaced either by single atoms with a valence of 1, as in polyvinyl chloride (PVC), or by small organic molecules, as in polypropylene (PP) or polymethyl/methacrylate (PMMA). The small molecules replacing hydrogens are termed "side groups" or "pendant groups." Many mers have carbon atoms as the backbone of the chain-link. However, oxygen can be inserted (as in polyacetal), which then changes a number of properties. New engineering plastics have very complex mers, frequently including small molecules.

Polymerization

Polymerization is the chemical reaction that bonds monomers into plastics chains. Addition polymerization, the most common method, is exemplified in the manufacture of polyethylene (PE). The monomers, molecules of ethylene gas synthesized from petrochemicals, are mixed under precisely controlled temperature and pressure conditions. Initially, the double bond between the carbon atoms of the monomer is broken down to a single, covalently shared electron pair. Using catalysts, the carbons are driven to fill their missing valence positions by covalently bonding with carbons in neighboring monomers, not by rebonding within the same monomer. As polymerization continues, monomers continue to add together, forming a chain. In this form, plastics is termed "resin," indicating it is not a finished product but ready to be used for molding into a plastics part.

During polymerization, chains generally grow to somewhat different lengths and, potentially, different shapes. Chain length is a significant factor in determining molecular weight and ease of processing. Using viscosity or thermal techniques, the average molecular weight and weight distribution are determined for each batch. Molecular weight is directly related to chain length. Since each mer is the same, the molecular weight is the weight of one mer times the number of mers per chain (chain length). Frequently, manufacturers designate specific grades of a plastics as best for molding, indicating that the chain length and distribution of chain lengths have been controlled for maximum flow during molding. Flowability decreases with increased molecular weight. Shapes can be varied by using catalysts to induce monomer additions at chain positions other than the chain ends. Shape modification is generally reserved for special applications. In chemistry, a polymer is described by drawing its physical structure of atoms, as illustrated in Fig. 8-3.

Plastics Chains

Although the mer is the principal chemical unit, the chains determine the unique properties of plastics by class and help differentiate between plastics. The behavior of the molecular chains is a key factor in determining the characteristics of a particular plastics material.

Motion. Plastics chains are in motion, even when the plastics material appears to be a solid. Temperature and applied load determine the rapidity and type of motion. As the temperature increases, chain motion increases until, at the molecular level, the plastics resembles a can of wriggling, intertwining worms. At the "glass transition stage" in amorphous polymers such as polystyrene and polycarbonate, and at the melting point for crystalline polymers such as Celanese Celcon acetal copolymer, chain motion becomes so marked that intertwining has little effect and chains can slide past one another. This condition is

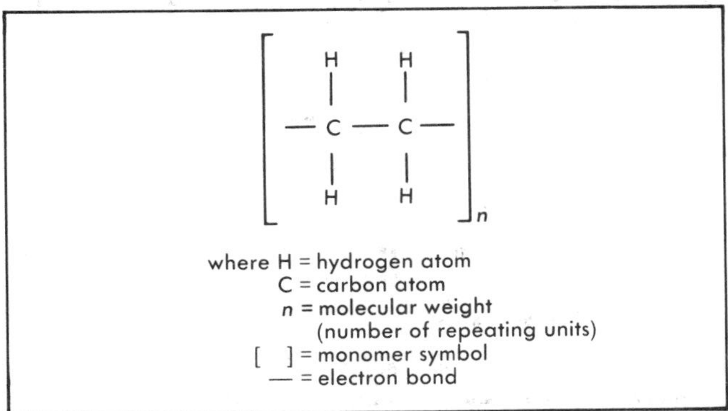

Fig. 8-3 Diagram representing arrangement of atoms in polyethylene is an example of the method used to describe polymers by a schematic drawing of the molecular structure.

evident by a drop in the plastics' viscosity and by the plastics becoming a tarlike liquid. Each type of plastics has a distinctive glass transition temperature range or melting point range.

Strength. The strength of unreinforced plastics is dependent primarily on the forces holding the chains together. These forces arise from the natural intertwining of chains and from structural manipulations made by resin manufacturers. One of the oldest strengthening mechanisms is "crosslinking," or using a small molecule to bond resin chains together as in epoxy. Crosslinked plastics are strong because the chains cannot move. Most crosslinked plastics are thermosets. Plastics that melt are thermoplastics.

Elastic flow. Visco-elastic behavior, a combination of flow and elastic response to applied load, is another characteristic property of plastics. Chain motion is related to the internal structure of a plastics. While visco-elastic behavior is not unique to plastics, plastics are one of the few engineering materials to behave visco-elastically in the room temperature range and under low loads. The design principles governing visco-elastic plastics are different than those for elastic metals. In the next section of this chapter, the fundamentals of visco-elastic properties will be presented as related to plastics materials applications and parts manufacturing processing considerations.

MECHANICAL PROPERTIES

Both metals and plastics are characterized by similar types of mechanical properties. Metals, however, tend to be consistent in the sense that their behavior is adequately characterized by stress-strain relationships. In contrast, while the individual plastics materials also display distinctive stress-strain characteristics, the mechanical properties of plastics are more dependent on the additional factors of temperature and time (under load). In the design application, and to some extent in processing, creep data are of significant importance in the field of plastics materials.

The engineering plastics materials have ultimate tensile and compressive strengths and stiffness properties that are significantly lower than those of metals. This difference is especially true when comparing plastics to tool steels and high-strength steels. However, the differential in ultimate mechanical strength is much less when plastics are compared with metals such as aluminum, magnesium, zinc, and copper. Strength ranges of typical plastics are compared in Fig. 8-4. Tables 8-2 and 8-3 list data for rigid PVC and polyurethane thermoplastics.

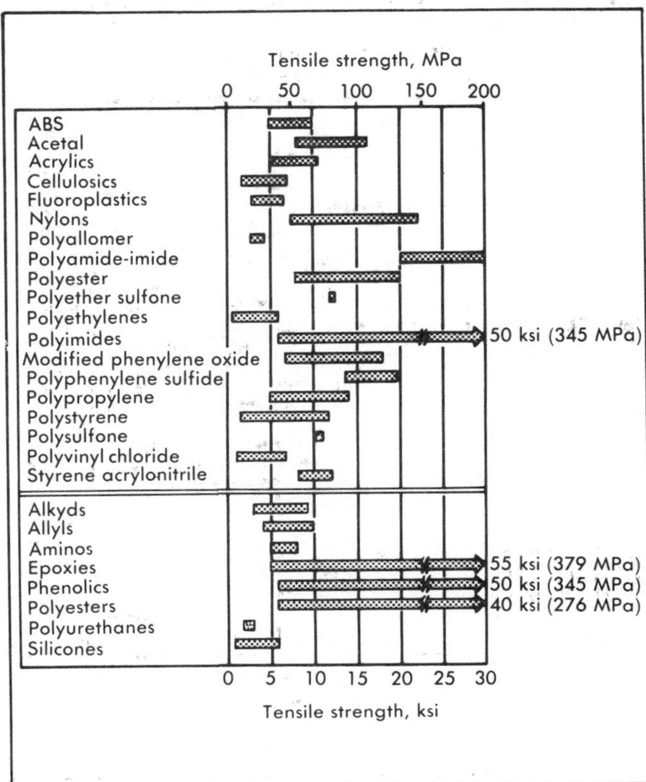

Fig. 8-4 Tensile strength of selected plastics at 68° F (20° C) per ASTM D 638. (*"Engineering Materials, Properties and Selection," Kenneth Budinski*)

TABLE 8-2
Typical Properties of Rigid Polyvinylchloride (PVC)

	ASTM Test Methods	Geon 87237[1]	Geon 87239
Specific gravity, g/cm^3	D-792	1.40	1.36
Hardness, Durometer D, pts.	D-2240	79	80
Hardness, Rockwell R, pts.	D-785	107	112
Tensile strength, ksi (MPa)	D-638	6.0 (41.3)	6.7 (46.2)
Elongation, %	D-638	135	140
Izod impact strength, notched at 72° F (22° C), ft-lb/in. (J/cm)	D-256	2.0 (1.1)	1.0 (0.5)
Coefficient of linear expansion, 10^{-5}8 in./in., °F (10^{-5} cm/cm, °C)	D-696	3.7 (6.6)	3.4 (6.1)

(*The B.F. Goodrich Co.*)

[1] Lead stabilized

CHAPTER 8

MECHANICAL PROPERTIES

Evolving Tests and Standards

Plastics are too new to have a well-defined, standard set of generic properties. As with metals, tensile, compressive, and impact tests are routine. Unlike metals, at room temperature visco-elastic effects dominate the behavior of plastics in service and are an important consideration in parts processing and manufacture. A second dissimilarity is the wide range of differences in property values between metals and plastics. Thus, special tests had to be devised for plastics. To account for visco-elasticity, creep testing data and thermal properties are routinely shown. For standard mechanical tests, special procedures have been developed to account for both the differences in value ranges and for visco-elasticity. The temperature, humidity, and rate of deformation must be carefully controlled to attain meaningful test results.

Reliable information on mechanical properties and other important characteristics of plastics can be obtained from various sources. The resin manufacturers publish technical information and data for the polymers they produce. Publications and reports are available from the Society of Plastics Engineers (SPE) and the Society of the Plastics Industry (SPI). Additional data sources are the tabulations published by technical periodicals and trade journals in the plastics field.

Tension and Compression Tests

Plastics data tabulations usually list ultimate tensile strength, yield strength, compressive strength, and elongation. However, since the tests for these properties tend to produce single point, very rapid loading rate, and short duration results, they are not good predictors of plastics behavior under prolonged, constant load. Further, plastics are visco-elastic and generally exhibit neither a well-defined modulus (stress-strain relationship) nor yield point. These general characteristics are illustrated by the stress-strain curves shown in Fig. 8-5. Because plastics test data is very dependent on strain rate (loading rate) and temperature, creep data (flow under load) must also be considered in design.

Tensile data provides useful information in several specific cases. First, at temperatures well below the glass transition (T_g) and for small deflections, the stress-strain relationship is elastic and linear. Second, if creep data indicates slight flow at the temperature of interest, then the tensile curve is a good approximation of behavior. Finally, applications like a snap-fit are a single, rapid load application and thus can be treated as a tensile (compressive) load modeled by the stress-strain curve.

TABLE 8-3
Typical Physical Properties of Polyurethane Engineering Thermoplastic*

	Test Method	Data
Density, g/cm^3	ASTM D-792	1.2
Hardness, Rockwell	ASTM D-7850	
R		105
M		48
Tensile strength, ksi (MPa)	ASTM D-638	
at yield		7.6 (52.4)
at break		8.5 (58.6)
Elongation, %	ASTM D-638	
at yield		6
at break		180
Water absorption, % (24 hours)	ASTM D-570	0.17
Notched Izod impact strength, ft-lb/in. (J/cm)	ASTM D-256	
1/8″ thickness (3.175 mm)		
at 73° F (23° C)		14 (7.5)
at -20° F (-29° C)		6 (3.2)
at -40° F (-40° C)		3 (1.6)
Taber abrasion, mg loss	ASTM D-1044	11

* Isoplast 201, Upjohn Co.

Temperature and Properties

Visco-elastic behavior becomes more pronounced as temperature, stress, and time increase. Therefore, elevated temperature processing and service require careful planning. Indeed, many plastics will flow under relatively low stress at room temperature. For prolonged service, plastics must be designed using creep data (in the same manner turbine blades are designed) because creep occurs at service temperatures that exceed half the melting point on the absolute temperature scale. The absolute temperature scale is, numerically, the same as the Rankine temperature scale, and °R = 459.7 ±°F.

Plastics at 70° F (21° C) exceed half their melt point in °R, just as nickel and cobalt alloys at 1500° F (816° C) exceed half their melting points, °R. Creep data requires considerable manipulation and may not be available because of the long time spans required for these tests. Therefore, several tests have been devised to provide guideposts for comparing plastics. The commonly used temperature-related tests include Vicat softening point (ASTM D 1525), heat deflection temperature (ASTM D 648), and Underwriters Laboratories (UL) relative temperature indexes for various thermoset and thermoplastic materials.

Creep Data

Creep is deformation or flow under constant applied load. Creep occurs in plastics, even at room temperature, as molecular

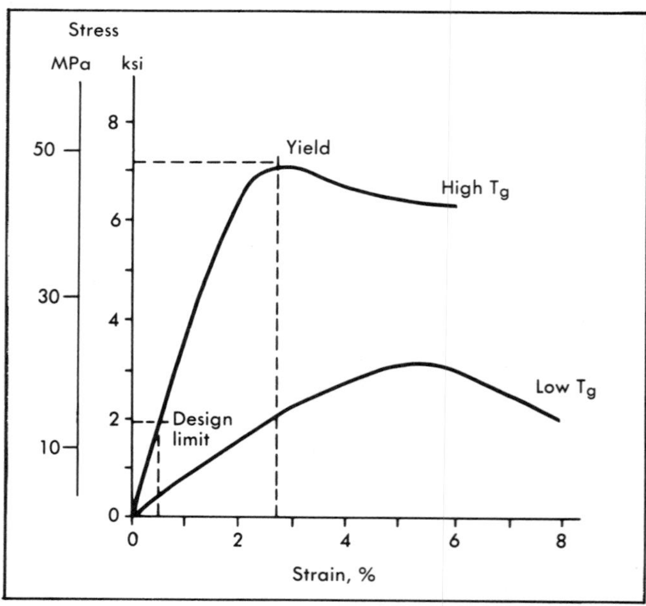

Fig. 8-5 Representative tensile stress/strain curve for plastics. (*Roy S. Klein*)

chains are untangled or move in the direction of the load. Creep in solids is usually evident at temperatures in excess of one-half the melting point, °R. In plastics, the glass transition can be used in place of melting. In terms of absolute temperature, creep will be evident in all plastics, since room temperature is about 525° R, and no industrial plastics resins (except the thermoset polyimide and the thermoplastic polymide) have a T_g higher than 500° F (260° C), or 960° R.

Creep occurs in three continuous stages. First, a rapid, large initial deformation is noted. In the second stage, the rate of deformation drops markedly to a lower, relatively constant value. Rupturing, the final stage, occurs either as a sudden failure for a brittle plastics or after a rapid, large deformation for a ductile plastics.

Creep can be regarded as viscous flow. The factors affecting viscous flow—time, temperature, and stress or pressure—also control creep behavior. Creep behavior is measured for plastics by creep strain (the deformation) or creep modulus, which is essentially the design modulus for creep conditions and is based on the slope of the rate of deformation versus time at a given load in second-stage creep.

ENGINEERING PLASTICS[2]

The thermoplastic engineering resins are usually characterized as those resins having the following combination of properties:

- Thermal, mechanical, chemical, corrosion resistance, and fabricability.
- Ability to sustain high mechanical loads, in harsh environments, for long periods of time.
- Predictable, reliable performance.

The term "engineering plastics" is neither rigorously defined nor restrictive in the sense that implies there is also a well-defined group of nonengineering plastics. Instead, some industry experts advocate a practical definition of engineering plastics that includes not only the property/performance criteria, but also market/pricing criteria.

The above two sets of criteria, taken together, place certain resins in the engineering category to the exclusion of others. The principal resins of the past decade that meet both sets of criteria are: nylon, acetal, thermoplastic polyester, modified phenylene oxide, and polycarbonate. In the 1980s, new materials, including some grades of acrylonitrile butadiene styrene (ABS), are being developed that can be categorized as engineering plastics.

When referring to engineering plastics, some knowledgeable people simply apply a broad definition based on property/performance criteria. This interpretation includes special grades and compounds of the commodity thermoplastics such as isocyanate-based resins and a variety of polymer alloys and copolymers, as well as polysulfone (PSO) and polyphenylene sulfide (PPS). The new types of rigid polyvinyl chloride (PVC) are also sometimes included. In addition, the growing scope of engineering plastics is commonly recognized to include the specialty plastics that offer high strength along with high-temperature performance. However, common usage of the term generally does not include the thermoset resins that were the forerunners of the engineering thermoplastics.

Production Volumes

In terms of annual production and sales volumes, nylon and polycarbonate are the most widely used engineering thermoplastics. Their combined total usage accounts for more than one half of the nearly 500,000 ton (453 550 metric ton) annual production of engineering plastics.

Balance of Properties

The engineering plastics have a good balance of high tensile properties, stiffness, compressive and shear strength, as well as impact resistance, and they are easily moldable. Their high physical strength properties are reproducible and predictable, and they retain their physical and electrical properties over a wide range of environmental conditions (hot, cold, chemicals). The engineering plastics can resist mechanical stress with good retention of properties for long periods of time. Flame retardance, which formerly had not been an essential requirement, has now become an important attribute for many applications. Typical mechanical properties of major families of engineering plastics are listed in Table 8-4.

The individual physical properties of a resin can be quantified and compared to other resins through established testing procedures, but the balance of properties essential to a true engineering resin requires a broader view. A balance of properties exists when the achievement of one property does not demand a tradeoff with another (for example, stiffness for low-temperature impact strength). Certain properties of commodity thermoplastics can be improved through the use of stabilizers, fibrous reinforcements, and particulate fillers to produce grades that directly compete with engineering plastics, but these improvements invariably cause a corresponding reduction in other properties.

The introduction of the term "commodity" leads to the market/pricing criteria that place the engineering plastics in a class of their own. The load-bearing engineering plastics form a distinct group, as compared to the high-volume/low-price commodity plastics and the low-volume/high-price specialty plastics. The chart in Fig. 8-6 shows where the various resins fall together and how they are grouped with respect to relative costs and production volume levels.

Advantages and Limitations

The advantageous characteristics of engineering plastics include high strength per unit weight, inherent corrosion resistance, little or no maintenance, and good retention of mechanical properties. Depending on the reinforcements or fillers added (such as glass or graphite), stiffness, lubricity, and other properties can be enhanced.

Because the end products typically are molded, design flexibility enables multiple functions to be combined into a single part, thus reducing the total parts count (as well as assembly time and labor), while eliminating finish grinding and machining. Also, because resins can be pigmented, painting is often not required.

The main inherent problem with engineering plastics is creep deformation under load. Glass reinforcement is commonly used to control and limit the creep characteristics of some plastics. In some applications, abrasion resistance and wear resistance are other limitations that must be taken into consideration.

Overview of Properties

Du Pont introduced nylon as the first engineering thermoplastic in the 1950s. It has since been joined by four other engineering resin families that have broadened the capabili-

TABLE 8-4
Representative Properties of Major Engineering Thermoplastics

Property	ASTM Method	Material							
		Nylon 6*	Nylon 6/6**	PC*	Modified Phenylene Oxide	Acetal†	ABS**	PSF	PPS
Specific gravity	D-792	1.12-1.14	1.13-1.15	1.2	1.06	1.41-1.42	1.03-1.06	1.24	1.34
Tensile strength, ksi (MPa)	D-638	10 (69.0)	11 (75.8)	9.5 (65.5)	9.6 (66.2)	8.8-10.5 (61-72)	6-7.5 (41-52)	10.2 (70.3)	10.8 (74.5)
Tensile modulus, ksi (GPa)	D-638	100 (0.7)	410 (2.8)	350 (2.4)	355 (2.5)	390-520 (2.7-3.6)	300-400 (2.1-2.8)	380 (2.6)	480 (3.3)
Elongation to break, percent	D-638	300	300	110	20-60	25-75	5-25	50-100	3
Izod impact strength, notched, ft-lb/in. (J/cm)	D-265	3.0 (1.6)	2.1 (1.1)	16 (8.5)	1.8-5.0 (1.0-2.7)	1.3-2.3 (0.4-1.2)	3-6 (1.6-3.2)	0.3-0.5 (0.2-0.3)	0.2 (0.1)

* Moisture conditioned
** Medium-impact grade
† Property range includes homopolymer (delrin) and copolymer (celcon)

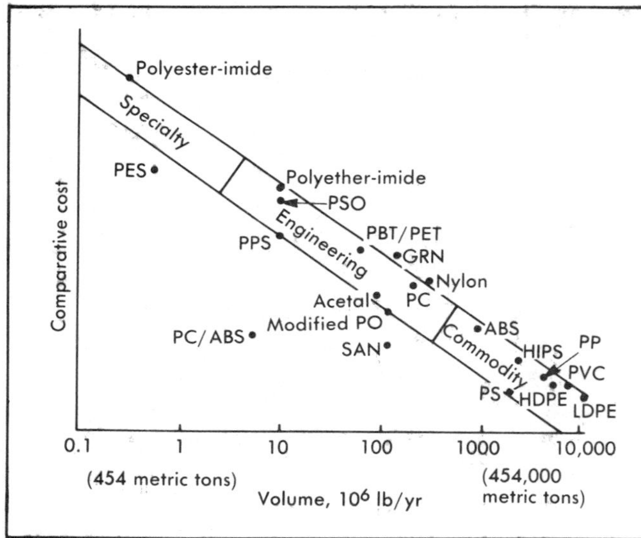

Fig. 8-6 Engineering thermoplastics charted by group, based on annual production volume and relative cost. This chart broadly depicts the market-oriented distinction between commodity, engineering, and specialty plastics resins. (*John Hill, Jr.*)

ties of thermoplastics in replacing metals, glass, and thermosetting plastics.

General-purpose (GP) nylon-6,6 has a heat deflection temperature (HDT) of 220°F (104°C) dry as molded; but at equilibrium, the HDT drops to 167°F (75°C). Its Underwriters Laboratories' (UL) thermal index is 167°F (75°C). The flexural modulus is 410 ksi (2.8 GPa) dry, but at equilibrium it is only 175 ksi (1.2 GPa) for nylon-6,6 and 140 ksi (965 MPa) for nylon-6. The notched Izod impact strength for nylon-6,6 is 2.1 ft-lb/in. (1.12 J/cm) of notch at equilibrium, but only 1.0 ft-lb/in. (0.53 J/cm) dry as molded. The Izod test, with its severe notch, is not necessarily the best method for comparing impact resistance, but it is frequently used in comparisons of engineering plastics.

Nylon's performance in any given application is dependent on moisture content; and, while the impact resistance improves as equilibrium is reached, the modulus drops off. Through modifications, all of these properties have been improved, but they are the key properties on which engineering plastics are based.

Acetals were introduced in 1960 in the form of Du Pont's Delrin homopolymer, followed, in 1962, by the Celanese copolymer Celcon. These resins brought a new set of properties to engineering plastics applications. The acetals offered somewhat higher heat deflection temperatures (HDT) and higher use temperatures; and, most important, their resistance to moisture pickup made physical properties independent of moisture content. The flexural modulus of acetals as molded is the same (homopolymer) or slightly lower (copolymer) than nylon-6,6 as molded, but it is relatively constant.

Nylon and acetal, both crystalline resins, surpassed amorphous ABS in such properties as tensile strength, long-term heat resistance, and chemical resistance; but neither of them could equal ABS in notched impact resistance. In the 1960s, two amorphous resins were introduced with impact resistance equal or superior to ABS.

With the appearance of polycarbonate (PC) in the early 1960s, a substantial gain was made in the UL thermal index rating. Because of its retention of properties at high temperatures, PC can be used continuously at 239°F (115°C), which is 45-72°F (8-40°C) higher than for previously available engineering resins. The HDT value of 270°F (132°C) for PC and its flexural modulus of 340 ksi (2.3 GPa) were similar to the properties of other materials. However, polycarbonates provided a substantial gain in impact resistance, with a notched Izod value of 14.0 ft-lb/in. (7.5 J/cm). In addition, PC was the first resin to offer the advantage of transparency. This property, combined with impact resistance and extrudability, brought to the engineering resins the potential for application as a glass replacement. As with the modified PPO, however, the tradeoff was in diminished chemical resistance.

General Electric's Noryl, a styrene-modified polyphenylene oxide resin, was introduced in 1966. Noryl's HDT is 265°F (129°C), with a UL thermal index of 194°F (90°C) and a flexural modulus of 360 ksi (2.5 GPa). These property values are similar to those for the acetals; but, with a notched Izod value of 5.0 ft-lb/in. (2.7 J/cm), the impact resistance was significantly improved. The chemical solvent resistance, however, is poor and is significantly less than that for the crystalline polymers such as nylon and acetal.

With the introduction of polybutylene terephthalate resins (PBT) in 1970, the potential for engineering resins in high-temperature applications opened up, excluding the unreinforced resins. Unreinforced PBT, as a crystalline resin, offers the typical chemical resistance of nylon and acetal, along with certain common properties such as lubricity. But unreinforced PBT's tensile and impact strengths are relatively low, and its HDT of 130°F (54°C) is very low.

High-temperature thermoplastic resins include the polysulfones (PSO) as a class of engineering thermoplastics with high thermal, oxidative, and hydrolytic stability; good resistance to aqueous mineral acids, alkali, and salt solutions; and fair resistance to oil and grease. Polyphenylene sulfide (PPS), Phillips Petroleum Company's Ryton, has good chemical resistance and electrical properties, along with excellent flame retardance, low coefficient of friction, and high transparency to microwave radiation. It can be injection and compression molded at high temperatures, 572-698°F (300-370°C). Among the high-temperature resins, polyetherimide, polyphenylene sulfide, and polysulfone compete with the basic engineering resins, but are classified at the high end of the five major engineering resin groups.

Reinforced Resins

As revealed by a comparison of property values, the various resin groups are affected by glass reinforcement in some markedly different ways. Glass reinforcement raises the flexural modulus for all of them. The heat deflection temperatures all go up as well, especially for nylon and PBT. Nylon's 480°F (249°C) HDT qualifies it for short-term high-heat exposure, and its long-term capability increases as well. In the case of PBT, glass reinforcement makes a true engineering resin out of a material that is not outstanding in its unfilled properties. The thermal index of 284°F (140°C) (the highest of all resins in this group) is important in relation to retention of electrical properties.

Glass reinforcement improves the various properties of acetal, compared to the unfilled resin; but, in comparison with other glass-reinforced materials, it is not outstanding. Glass reinforcement of modified phenylene oxide resin and polycarbonate results in decreased impact resistance; and, in the case of polycarbonate, it also causes a loss of transparency. Properties that are generally improved by the use of glass-reinforced resins include dimensional stability, moisture resistance, and flammability ratings.

Glass reinforcement reduces the ductility and unnotched Izod impact resistance of most thermoplastic resins. The use of reinforced resins is growing rapidly, and they are being increasingly specified for engineering applications that replace metal stampings, die castings, and, in some instances, cast iron.

Other Plastics Used for Engineering Applications

In some product applications, the five principal groups of engineering plastics compete with other plastics such as ABS, polystyrene, and various thermosetting plastics. Glass and/or mineral-filled polypropylene and flame-retardant ABS are widely used where their special characteristics are needed, and modified phenylene oxide (Noryl) must compete against flame-retardant ABS in some electronic cabinetry where a relatively low thermal index is acceptable. Ultrahigh molecular weight polyethylene (UHMWPE) offers excellent wear resistance, but limited processability.

Thermosets. Thermosets have long been available as insulators in electric/electronic applications, offering a wide range of capabilities in resistance to heat and other environmental conditions. Thermoset molding compounds may be formulated to satisfy one or more important uses. Typical distinctive properties of thermosets include dimensional stability, low-to-zero creep, low water absorption, maximum physical strength, good electrical properties, high heat deflection temperatures, high heat resistance, minimal values of coefficient of thermal expansion, low heat transfer, and specific gravities in the 1.35 to 2.00 range.

Processing advantages have allowed thermoplastics to replace thermosets in many markets, but competition remains in some of the more demanding uses. Thermoplastics in general offer faster molding, lighter weight, the possibility for thinner walls and more complex design, and greater impact resistance. Thermosets do not commonly exhibit as much creep at elevated temperatures as thermoplastics, including the reinforced grades.

The engineering resins do not compete directly with epoxies, but some of the high-temperature resins are being used for that purpose. The engineering resins rarely compete with polyurethanes or silicones. Among the major engineering thermoplastics, polybutylene terephthalate (PBT) and polyethylene terephthalate (PET) come closest to the thermosets in balance of properties. Polycarbonate comes closest in dimensional stability; but PBT is the first of the major engineering plastics to be considered for some of the most demanding thermoset applications, including industrial machinery and electrical/electronic usage.

Fluoropolymers. Fluoropolymers are often categorized along with the engineering plastics, but the two groups seldom directly compete. As a class, fluoropolymers do not offer the load-bearing capability of the engineering plastics, and load-bearing is generally one of the demands placed on plastics in product engineering specifications. In nonload-bearing uses, however, fluoropolymers have outstanding and unique properties, including resistance to very high and low temperatures, exceptional electrical properties, and a low coefficient of friction.

Specialty Plastics

Specialty plastics include a mixed group of materials sold at relatively high prices, compared to the engineering plastics, and in relatively low volumes. The members of this group generally have high-temperature capability, but this capability involves complex, costly synthesis and usually some processing difficulty. In this group, the polyimides (thermoset) and polyamides (thermoplastic) can be used continuously at temperatures in the 500°F (260°C) range.

There is no question that such materials can be used in engineering applications, but there are two principal materials in this group that compete directly with the engineering plastics: polyphenylene sulfide (PPS) and polysulfone (PSO).

The relative heat resistance of plastics materials is generally measured either by ASTM D 648, Deflection Temperature Under Load, or UL 746B, Polymeric Materials—Long-Term Property Evaluation (Relative Thermal Index). These two testing procedures are often referred to in connection with the main engineering plastics, but they become particularly relevant in attempting to define a high-temperature plastics. No comprehensive body of data exists to compare the mechanical, chemical, and electrical stress behavior of all materials under long-term elevated temperature conditions.

The UL thermal index is, however, the most useful indicator because heat deflection temperature merely indicates the temperature at which a bar (test specimen) begins to deflect. It provides no information about time to failure or mode of failure. For example, while glass-reinforced nylon and polyester have very high heat deflection temperatures, they are crystalline resins with sharp melting points, and this characteristic must be taken into account when considering prospective applications.

The UL thermal index is a valid indicator of continuous performance ability. An acceptable dividing line between the major engineering thermoplastics and the high-temperature thermoplastics can be drawn at the UL 302° F (150° C) thermal index. This temperature is the generic rating for phenolic molding compounds, and most thermosets can be used continuously above it. For thermosets, unlike thermoplastics, 302° F (150° C) is usually taken for granted when considering functional applications that involve moderately hot environments.

The significance of thermoplastics that can operate over 302° F (150° C) does not lie in their role as replacements for thermosets in the low-price range. With the exception of PPS, they are too expensive. They can function as replacements for metals, glass, epoxies, fluoropolymers, and specialty thermosets in areas where thermoplastic processing advantages make the cost worthwhile.

Polyphenylene sulfide (Phillips Chemical Co.'s Ryton) is known as PPS and is available for molding only in glass and/or mineral-reinforced compounds. It offers UL thermal index ratings of 338° F (170° C) at a relatively low cost. PPS also has outstanding chemical resistance, with no known solvents below 401° F (205° C). An additional asset is that PPS is inherently flame retardant. It is considered a thermosetting thermoplastic because optimal high-temperature properties can be obtained through annealing, in which some crosslinking takes place. PPS is also being used in high-temperature alloys.

Most PPS applications are in structural electric/electronic parts, where PPS competes against phenolic resins on the basis of cost savings through scrap reuse, and against other more expensive thermosets like diallyl phthalate (DAP) and diallyl isophthalate (DAIP) resins. Among the engineering thermoplastics, its closest competitor is PBT.

Polysulfone, Union Carbide's Udel, is a transparent, amorphous copolymer resin with a UL thermal index of 302° F (150° C). Polysulfone is selected instead of polycarbonate when the higher use temperature is required, and sometimes for better stress-crack resistance at lower temperatures. Its markets are often in glass and stainless steel replacement based on the advantages of transparency, heat resistance, hydrolysis resistance, suitability for food contact, and resistance to acids and alkalis.

Polysulfone applications, often for corrosion resistance, are in medical hardware, food processing and handling equipment, and automotive, electric/electronic, and industrial parts. There is some use of polysulfone as a substrate for circuit boards instead of the customary epoxy-glass composites for aircraft parts, in competition with carbon-epoxy composites at 230° F (110° C).

Plastics with still higher heat-resistance capabilities include some thermoplastics (other polysulfones), some thermosets (including certain polyimides and aramids), and some resins that can be melt-processed but require crosslinking for optimal property development. Flame-retardant plastics, such as Du Pont's Rynite thermoplastic polyester, are becoming increasingly important as replacements for thermoset plastics electrical components.

Plastics Alloys and Blends

Because of the high cost to develop and introduce a new plastics and provide the necessary marketing and technical support, new materials tailored for specific sets of properties are often made by chemical or physical modifications of existing resins. Desirable characteristics and properties can be obtained by blending or alloying resins, or by adding inorganic or organic fillers and reinforcements. Such modification is being done by primary resin suppliers who use these methods to produce special grades, by custom compounders, and—to an increasing extent—by end-users who tailor-make resins to fulfill their own specific needs.

Description. The distinction between alloys and blends is not clearly defined, but both terms are used for physical mixtures of two or more structurally different polymers. As compared to copolymers, in which the components are linked by strong chemical bonds, the components in alloys adhere primarily through Van der Waals forces, dipole interactions, and/or hydrogen bonding.

The process of alloying to improve certain desired characteristics of a polymer is not limited to adding only one other polymer. There are also terpolymers (three monomers in a chain) and plastics alloys with several polymer additives.

Example. Table 8-5 lists properties of General Electric's Xenoy, an alloy that consists primarily of polycarbonate (PC) and polybutylene terephthalate (PBT). This thermoplastic polymer alloy was developed specifically to fulfill requirements for automotive exterior parts. Xenoy's characteristics include:

- High impact strength over a wide temperature range.
- Dimensional stability at high and low temperatures and humidity levels.
- Flexural modulus of approximately 300 ksi (2 GPa).
- Resistance to gasoline (spillage) and automotive waxes and cleaners.
- Ultraviolet (UV) stability.
- Wide latitude in processing.

Polymer networks. A new technology that combines incompatible plastics to form interpenetrating polymer networks (IPNs) has been developed by Shell Chemical and other companies. This technique produces a new type of alloy consisting of intimate mixtures of two or more polymer networks held together by permanent entanglements. Unlike conventional alloys, the polymers need not be miscible, and the networks can be devised for optimum properties when they are needed, while using lower cost materials as the predominant ingredient when the property requirements are less severe.

TABLE 8-5
Typical Properties of a Thermoplastic Polymer Alloy*

Property	ASTM Test Method	Value
Mechanical		
Impact strength, notched Izod, ft-lb/in. (J/cm)	D-256	
73° F (23° C)		14.8 (7.9)
32° F (0° C)		12.2 (7.0)
-4° F (-20° C)		4.7 (2.5)
-40° F (-40° C)		3.0 (1.6)
Tensile strength	D-638	
at yield, ksi (MPa)		7.4 (51)
at break, ksi (MPa)		8.1 (56)
Elongation, at break, %		145
Other		
Coefficient of thermal expansion,	D-696	
in./in./° F		5.3×10^{-5}
(mm/mm/° C)		9.5×10^{-5}
Mold shrinkage, %	D-955	0.9
Specific gravity	D-792	1.22

*Xenoy 1100 resin, General Electric Co.

PROCESSING AND APPLICATIONS

The melt-processability of thermoplastic resins is a basic characteristic that distinguishes them from thermosets. This fact pertains not only to the advantages of injection molding as compared to compression or transfer molding, but also to the variety of processing alternatives that extend the utility of the thermoplastics. There are thermosets that can be injection molded; but only the thermoplastics offer the options of extrusion into sheet, film and profiles, or blow molding. For additional information on processing, refer to "Plastics Forming," Chapter 18, Volume II, *Forming*, of this Handbook series.

The degree to which each of the engineering plastics is amenable to alternative processing methods varies, and the relative potential of each of them depends also on their potential in alternative processes, not just on their utility in injection molding. The diagram shown in Fig. 8-7 generally relates the thermosets and thermoplastics and their respective parts production processes.

Applications Overview

In addition to its use in injection molding, nylon is extruded into monofilament and brush filament. Nylon-6 is used for sewing thread, fishing line, household/industrial brushes, and level-filament paint brushes. Nylon-6,6, stiffer than nylon-6, is used for sewing thread and household/industrial brushes. Nylon-6,12 dominates in personal-care brushes, although polyethylene terephthalate (PET) competes in these applications. In tapered-filament paint brushes, nylon-6,12's leading position has been taken over by PBT, a more expensive but more versatile filament.

Nylon is used as a wire coating, primarily as a protective abrasion-resistant coating over PVC-insulated wire. Nylon film can be cast or blown, or extrusion coated onto various substrates. Most nylon film is cast, and virtually all is sold to converters who add a sealant layer of low-density polyethylene (LDPE), ethylene-vinyl acetate copolymer (EVA), or ionomer. Its major market is vacuum packages of processed meats and cheese, usually combined with a PET-sealant cover web. Nylon film is also used for fresh-meat packaging, and a new market has opened in medical device packaging using techniques similar to those for formed-meat packaging. The most important properties in these applications are formability and heat resistance.

Nylon strapping began replacing steel strapping in the early 1960s, even at higher cost, because of the general advantages of nonmetallic strapping. In recent years, nylon has met increasing competition in this market from polypropylene and PET.

Nylon is also extruded into rods, tubes, and shapes for machining, an important option for low-volume runs. The blow molding of nylon has been restrained partly by cost and partly by the difficulties inherent in crystalline resins because of their sharp melting point. Nylon blow-molding resins have been developed with high melt strengths for parison forming and are used to some extent for monolayer and coextruded bottles and for gas tanks in small equipment. Nylon-6 is also cast to produce very large bearings.

Nylon-11 is used for powder coatings and for flexible tubing. Nylon-12 is used for the same purposes, but to a greater extent in Europe than in the U.S. These resins have exceptional moisture resistance, but they are considerably less stiff than nylon-6 or -6,6. They are used to some extent in rotational molding.

In contrast to nylon, acetal offers few options outside of the injection molding category. An acetal terpolymer is available for injection blow molding; but, apart from some carburetor floats and rod extrusions, it has found little usage. Although acetal is difficult to extrude, it is extruded into shapes, as is nylon, for subsequent machining. Almost all acetal consumption is in injection molding, a factor that limits its total consumption.

The PET thermoplastic polyesters used for film, sheet, and blow molding are not the same as those used for injection-molded engineering applications. PBT can be blow molded, but rarely is. While used almost entirely in injection molding, PBT does find some use in tapered brush filaments and in extruded strip for small electrical parts.

Noryl resin's use in extrusion is relatively minor compared to injection molding, but it is used to some extent for stock shapes and to an increasing extent for sheet and profiles. Noryl (General Electric's PPO) sheet competes with flame-retardant ABS as it does in injection molding, and it can compete with less expensive resins like ABS and PVC where its properties permit the extrusion of thinner walls.

The transparency of polycarbonate, combined with its extrudability and impact resistance, makes it a strong competitor for acrylic sheet in replacement of flat glass. Extruded sheet for glazing, lighting, and signs accounts for approximately 25% of polycarbonate's volume. Its use in extruded profiles is minor, but polycarbonate is widely used in blow molding for water bottles, milk bottles, baby nursing bottles, and miscellaneous packaging.

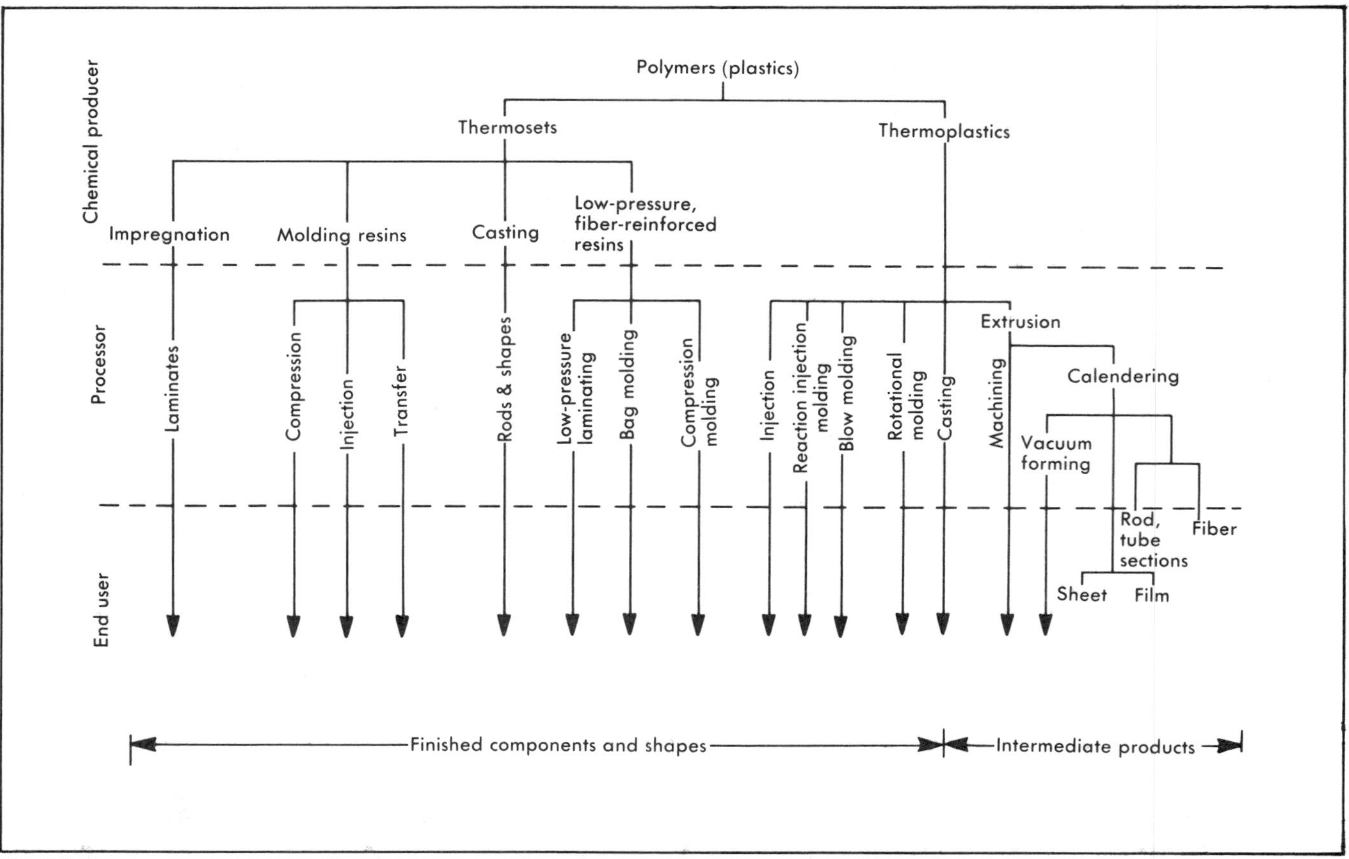

Fig. 8-7 Thermoset and thermoplastic polymers grouped by parts production processes. (*Roy Manns & Associates*)

Special Processing Considerations

When processing plastics resins to produce end products, it is important to avoid inadvertent creation of hazards. During parts production, the processing parameters and operations should not subject the resin to conditions that would alter its properties significantly. Information on the general behavior of plastics and the specific characteristics of the particular materials involved is essential for sound planning and performance of the manufacturing operations.

Flammability

Few areas of plastics technology have received so much attention and are as confused as the combustion properties of plastics. Like most other organic materials, plastics burn. But burning behavior is difficult to measure because the type of combustion, the stacking or packing of the plastics, the ratio of surface area to volume, the effects of the products of combustion, and the rate of combustion all interact during any specific fire. Tests vary and some experts disagree on the basic knowledge of flammability of plastics. For example, in the summer of 1983, the state of New York concluded that several existing flammability test protocols were scientific and could be used to begin setting fire standards, whereas the State of California concluded that these same tests were not sufficiently scientific to set meaningful standards.

Flammability Tests

Some of the major flammability tests and standards for plastics are:

- ASTM D 635 Burn Rate Test.
- ASTM D 2863 Oxygen Index.
- ASTM E 84 Tunnel Test.
- ASTM E 162 Radiant Panel Test.
- Underwriters Laboratories UL-94 Flame Class.
- Underwriters Laboratories UL-746c Enclosure of Live Electrical Parts.
- United States Department of Transportation MVSS-302 Flame Test.
- Canadian Standards Association CSA C-22.2, No. 6-1982.

Plastics burn because they contain carbon and hydrogen atoms, both capable of reacting with atmospheric oxygen. To increase fire resistance, chlorinated polymers have been developed, such as PVC and CPVC. At issue, however, is the potential effect of burning these plastics and producing HCL (hydrogen chloride gas), a hazardous substance. Other polymers with large "blocky" molecules in the backbone burn very slowly and tend to give off very little smoke. Unfortunately, these polymers are costly. Several flame retardants, usually halides, can be added to plastics. For any operation involving possible flammability, it is advisable to contact the appropriate agency, read the standards thoroughly, and talk to resin suppliers before selecting plastics or establishing processing conditions.

ADVANCED COMPOSITES

A composite material is created by the combination of two or more materials—a reinforcing element and a compatible resin binder (matrix)—to obtain specific characteristics and properties. The components do not dissolve completely into each other or otherwise chemically merge, although they do act synergistically. Normally, the separate components can be physically identified, as well as the interface between components.

A common example of a composite material is fiberglass. Glass fibers are very strong. If notched, however, they fracture readily; or if put in compression, they buckle easily. By encapsulating the glass fibers in a resin matrix, they are protected from damage; at the same time, the resin matrix transfers applied loads to the unified fibers so that their stiffness and strength can be fully utilized in both tension and compression.

The more advanced structural composites use fibers of glass, carbon/graphite, boron, Kevlar (aramid), and other organic materials. These fibers are very stiff and strong, yet lightweight. The strengthening effects of the fiber reinforcements in composites are derived from (a) the percentage of fibers (fiber-resin ratio), (b) the type of fibers, and (c) the fiber orientation with respect to the direction of the loads.

While advanced fiber, resin matrix composites are classified as reinforced thermosets, a special technology has developed involving these materials that sets them somewhat apart from other reinforced thermosets. Called "advanced composites," resin matrix composites can include hybrids, mixtures of fibers in various forms in the resin (usually epoxy) matrix.

Continuous-fiber reinforcements can be directionally oriented; short or chopped fibers can only be randomly oriented. Each fiber form has its advantages and limitations. In general, short fibers cost more than continuous fibers, yet fabrication costs are lower for the short fibers. The properties of composites from chopped fibers are weaker than those obtainable with longer or continuous fibers.

The four basic areas of composites technology are:

- Organic (resin) matrix composites.
- Metal matrix composites.
- Carbon-carbon composites.
- Ceramic matrix composites.

The majority of aerospace, military, and commercial product applications for advanced composites use the carbon or graphite fibers with organic matrices; hence, these are the principal composites discussed in this section of the chapter. Commercial composite materials, such as laminates and glass-reinforced plastics, are discussed previously in this chapter and also in Volume II, *Forming*, of this Handbook series.

Composite structural materials have evolved as a class of engineering materials that offer some unique properties and combinations of characteristics not exhibited by the more traditional materials systems such as "pure" metals, ceramics, and polymers. Figure 8-8 offers a schematic comparison of homogeneous metal and fiber-resin composite material. The industrial appeal of advanced structural composites is based on their inherent ability to replace conventional high-strength metals with lighter weight, higher strength material.

In the manufacture of aircraft, the predominant advanced composite material contains a reinforcing fiber that is almost pure carbon. The term *carbon* correctly describes the fiber since

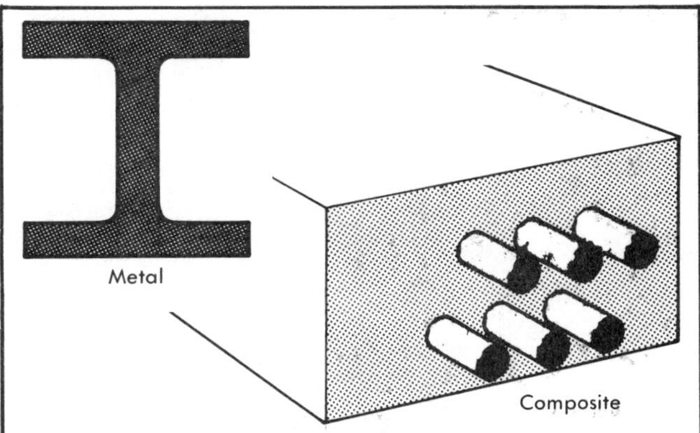

Fig. 8-8 Schematic comparison of homogeneous metal and a fiber-resin composite material. (*Polytech, Ltd.*)

it contains little or no graphitic structure; however, in common practice, the terms *carbon* and *graphite* are often used interchangeably to denote a particular group of fibers used in advanced composites. Generally, the modulus of elasticity of a carbon fiber increases as the degree of graphitization increases.

STATE OF THE ART

Generically, a composite can include metal, wood, foam, or other material layers, in addition to the fiber and resin components. The term *advanced composites* came into use in the late 1960s to designate certain composite materials with properties considerably superior to those of earlier composites. The term is, however, imprecise because it does not identify specific material combinations, nor does it indicate their arrangement or configuration in the composite, the strength level, or other qualities that distinguish an advanced composite from other composites. Currently, the industry defines advanced composites as composites that contain a fiber-to-resin ratio of greater than 50% fiber, with the fibers having a modulus of elasticity greater than 16 x 10^6 psi (110.3 GPa).

General Description of Advanced Composites

To engineers in the field, an advanced composite has come to denote a resin matrix material that is reinforced with high-strength, high-modulus fibers of carbon, aramid, or boron, and is usually fabricated in layers to form an engineered component. More specifically, the term is applied principally to epoxy-resin matrix materials reinforced with oriented, continuous fibers of carbon and fabricated in a multilayer form to make extremely rigid, strong structures. Another characteristic that distinguishes composites from reinforced plastics is the fiber-to-resin ratio. This ratio is generally greater than 50% fiber by weight; however, the ratio is sometimes indicated by volume since the weight and volume in composites are similar.

Composites Materials

Organic matrix composites are the most common, least expensive, and most widely used of all advanced composite materials. For purposes of discussion and reference, organic composites may be grouped according to their chronological development and commercialization, as presented in Table 8-6. It

CHAPTER 8

COMPOSITES NOMENCLATURE

TABLE 8-6
Groups of Organic Composites

Fibers	Resins
Group 1:* Established Composites Used on a Broad Scale	
E-glass	Polyester
S-glass	Vinyl ester
	Phenolic
Group 2:* Advanced Composites Primarily for Aerospace Applications	
Carbon/graphite (5 grades, 2 types)	Epoxy
Aramid (3 grades)	Phenolic
Boron (2 grades)	Bismaleimide
Glass (2 types)	

Group 3: Composites and Processes in the Developmental Stage

- Advanced bismaleimide resin matrix series for high-temperature service.
- Polyether etherketone (PEEK) thermoplastic matrix series for higher temperature service.
- Hybrid reinforcements and knitted/stacked ply fabrics and "3-D" (special shape) woven fabric reinforcements.
- Selective stitching of collated ply kits.
- URTRI (ultimately reinforced thermoset reaction injection) process for advanced composites using resin injection.

(*Polytech, Ltd.*)

* Fibers may be used separately or in combination with other fibers in the same group. Resins cannot be mixed together.

should be noted that while any combination of fibers can be used, only one resin is generally used in any given composites material.

The materials in Group No. 1 (see Table 8-6) have been widely used throughout the world for many years. The advanced composites (Group No. 2) are currently used in both commercial and military aerospace systems. The developmental composites (Group No. 3) include materials, processing methods, and controls that are still in the experimental stages and have not reached significant production, as well as new materials that have considerable potential for future applications. It is anticipated that emphasis on automation of organic matrix composites fabrication may lead to the development of new or modified materials, as well as processing methods and techniques for quality control of both the materials and processes.

NOMENCLATURE[3]

A-stage The condition of low molecular weight of a resin during which the resin is readily soluble and fusible.

anisotropic Exhibiting different properties when tested along axes in different directions.

aspect ratio The ratio of the length to either the width or the diameter of a fiber.

autoclave A closed pressure vessel for inducing a resin cure or other operation under pressure and heat.

axial winding In filament-wound reinforced plastics, a winding with the filaments parallel to the axis.

B-stage The condition of a resin-hardener mixture when it exhibits an initial degree of crosslinking while being still plastic and fusible and still somewhat soluble in selective solvents.

bag molding A technique of molding reinforced plastics composites using a flexible cover (bag) over a rigid mold. The composites material is placed in the mold and covered with the bag; pressure is applied by vacuum, autoclave, press, or by inflating the bag.

bidirectional laminate A reinforced plastics laminate in which the fibers are oriented in two directions in the plane of the laminate; typically, 90° opposed.

binder The resin or cementing constituent of a plastics compound that holds the other components together. Also, the agent applied to glass mat or preforms to bond the fibers before laminating or molding.

bismaleimide Thermoset resin used in producing advanced composites for elevated temperature applications, up to 700° F (370° C) continuously.

bulk molding compound (BMC) Thermosetting resins mixed with chopped reinforcements or fillers into a viscous compound for compression molding.

C-stage The condition of a resin when it is in the solid state, with high molecular weight, being insoluble and infusible (cured).

carbon A general term that includes any form of graphite or amorphous carbon.

carbon fiber Fibers produced by pyrolysis of an organic precursor fiber in an inert atmosphere at temperatures higher than 1800° F (982° C). Used as reinforcement for lightweight, high-strength, and high-stiffness structures.

carbon/graphite Fibers possessing high strength that are used with a resin matrix to form advanced composites structures.

cellular plastics Materials with cell structure throughout their mass and/or with integral skins (foams).

circumferential winding In filament-wound reinforced plastics, a winding with the filaments essentially perpendicular to the axis of rotation.

composite A homogeneous material created by the judicious assembly of two or more materials (selected reinforcing elements and compatible matrix) to obtain specific characteristics and properties.

compression molding A technique of thermoset molding in which the molding compound (generally preheated) is placed in the open mold cavity, the mold is closed, and heat and pressure is applied until the material has cured.

continuous filament An individual fiber strand of small diameter that is flexible and of great or indefinite length.

continuous filament yarn Yarn formed by twisting two or more continuous filaments into a single, continuous strand.

copolymer A compound resulting from the chemical reaction of two chemically different monomers.

cure To change the physical properties of a material by chemical reaction through the action of heat and catalysts, alone or in combination, with or without pressure.

delamination The separation of layers in a laminate through matrix cohesive failure or through interfacial failure (interlaminar shear).

E-glass A borosilicate glass, most commonly used for glass fibers in reinforced plastics.

endothermic reaction A reaction that is accompanied by the absorption of heat.

exotherm The temperature-versus-time curve of a chemical reaction; maximum temperature occurs at peak exotherm.

COMPOSITES NOMENCLATURE

exothermic reaction A reaction in which heat is given off; typical of thermoset resins.

FRP Fiberglass reinforced plastics. A general term for plastics that are reinforced with cloth, mat, strands, or any other form of fibrous glass.

fiberglass reinforcement A major material used to reinforce commercial grade plastics. Available as mat, roving, or fabric, it is incorporated into both thermosets and thermoplastics. The glass increases the mechanical strength, impact resistance, stiffness, and dimensional stability of the matrix.

fiber orientation Fiber alignment in a woven fabric laminate in which the majority of fibers may lie in the same direction, resulting in a higher strength in that direction.

fiber-resin ratio Ratio by weight or volume of fiber percent to resin percent, based on 100% of a composite material. Advanced composites are greater than 50% fiber.

filament Fiber of extreme length, used in yarns and other compositions. Generally used for large-diameter reinforcing materials such as boron or glass.

filament winding Process in which resin-impregnated strands are applied over a rotating mandrel to produce high-strength, reinforced cylindrical shapes.

graphite A crystalline form of carbon either found in natural deposits or formed by heating amorphous carbon.

hand lay-up Method of positioning successive layers of reinforcement mat or web (which may or may not be preimpregnated with resin) on a mold by hand. Resin is used to impregnate or coat the reinforcement, followed by curing the resin to permanently fix the formed shape.

homopolymer A polymer formed from a single monomer.

honeycomb A manufactured product of material formed into hexagonal-shaped cells; used as a core material in sandwich construction.

isotropic The ability to react the same regardless of direction of measurement. Isotropic materials will react consistently even if stress is applied in different directions. The stress-strength ratio is uniform throughout the material, and strength properties are equal in all directions.

laminated plastics A class of standard structural shapes, plates, sheets, angles, channels, rods, tubes, and zees that are produced by combining layers of resin-impregnated materials in a press under heat and pressure.

lay-up The reinforced material, sometimes resin-impregnated, that is positioned in the mold.

mandrel The core around which fiberglass that is impregnated with plastics resin is wound, as in filament winding. Also, the portion of an extrusion die that forms the hollow center in an extruded tube.

mer The repeating structural unit of a polymer (plastics resin).

MOHS hardness A measure of scratch resistance of a material; the higher the number, the greater the scratch resistance on a scale where diamond = 10.

monomer A single molecule that can join with another monomer or molecule to form a polymer or molecular chain.

polyacrylonitrile (PAN) A precursor for carbon fiber.

polyester resins A family of resins produced by the reaction of dibasic acids with dihydric alcohols. Polyethylene terephthalate (PET) is a thermoplastic that may be extruded or injection or blow molded. Unsaturated polyesters are thermoset and used in the reinforced plastics industry for applications such as boats and auto components.

polyimide resin A heat-resistant resin with composite forms that are stable to temperatures as high as 930° F (500° C). It also resists heat distortion to temperatures as high as 680° F (360° C).

polymerization A chemical reaction in which the molecules of monomers are linked together to form polymers.

polyurethane resin, thermoset A resin produced by reacting diisocyanate with organic compounds containing two or more active hydrogens to form polymers with free-isocyanate groups. These groups will react with each other under heat or by a catalyst to form thermosets.

postcure An operation on a thermoset molded part in which the part is subjected to elevated temperatures for a period of time to effect full cure or stress relaxation and to enhance properties.

precursor A textile product from which carbon fibers are made, such as polyacrylonitrile.

prepreg A ready-to-mold material in sheet form, generally a fabric preimpregnated with B-stage resin and stored for use. The prepeg, in frozen form, is supplied to the fabricator, who lays up the finished shape and completes the cure with heat and pressure.

pressure bag molding A process for molding reinforced plastics in which a tailored, flexible bag is placed over the contact lay-up on the mold, sealed, and clamped in place. Compressed air forces the bag against the part to apply pressure while the part cures.

pultrusion The extrusion of resin-impregnated roving in the manufacture of rods, tubes, and structural shapes of a constant cross section. After passing through the resin dip tank, the roving is drawn through a die and cured to form the desired cross section as it continuously runs through the machine.

reinforced plastics Molded, formed, filament-wound, or shaped plastics parts consisting of resins to which reinforcing fibers, mats, or fabrics have been added before the forming operation.

reinforcement A material used to reinforce, strengthen, or give dimensional stability to another material.

S-glass A magnesia-alumina-silicate glass that provides very high tensile strength reinforcement and is superior to E-glass.

sandwich construction A composite of three or more layers in which the outer layers form functional skins over a core.

sizing The surface treatment of fibers to enhance the fiber-resin interface.

thermoset A plastics that changes into a substantially infusible and insoluble material when cured by application of heat or chemical means.

tooling resins Resins that have applications as tooling aids, coreboxes, prototypes, hammer forms, stretch forms, or foundry patterns. Epoxy and silicone are common examples.

URTRI Ultimately Reinforced Thermoset Resin Injection (Polytech, Ltd.).

vacuum bag molding A process for molding reinforced plastics in which a sheet of flexible, transparent material is placed over the lay-up on the mold and sealed. A vacuum is created between the sheet and the lay-up. The entrapped air is next mechanically worked out of the lay-up and removed by the vacuum; finally, the part is cured.

wet lay-up The application of a resin-saturated reinforcement in the mold (hand lay-up). Also, the first stage in wet-wrap compression molding.

FIBER TYPES AND PRODUCTION

whisker A very short fiber reinforcement, usually of crystalline material.
Young's modulus Modulus of elasticity.

THE MATRIX

The matrix serves two important functions in a composite: (1) it holds the fibers in place; and (2) under an applied force, it deforms and distributes the stress to the high-modulus fibrous constituent. The matrix material for a structural fiber composite must have a greater elongation at break than the fibers for maximum efficiency. Also, the matrix must transmit the force to the fibers and change shape as required to accomplish this, placing the majority of the load on the fibers. Furthermore, during processing, the matrix should encapsulate the fibrous phase with minimum shrinkage, which places an internal strain on the fibers. Other properties of the composite, such as chemical, thermal, electrical, and corrosion resistance, are also influenced significantly by the type of matrix used.

The two main classes of polymer resin matrices are thermoset and thermoplastic. The principal thermosets are epoxy, phenolic, bismaleimide, and polyimide. Thermoplastic matrices are many and varied, including nylon (polyamide), polysulfone, polyphenylene sulfide, and polyether etherketone. The matrix material must be carefully matched for compatibility with the fiber material and for application requirements. The selection process should cover factors such as thermal stability, impact strength, environmental resistance, processability, and surface treatment of the reinforcing fibers (sizing).

Depending on the application, it is possible to view the role of the matrix in two different ways: either as the binder that contains the major structural elements (the fibers) and transfers load between them, or as the primary phase that is merely reinforced by the secondary fiber phase. The first concept is traditional since most composites have used a relatively soft matrix (such as a thermosetting plastics of the polyester, phenolic, or epoxide type). The strength of such composites is almost entirely that of the fibers; hence, for efficiency, it is desirable to optimize the fiber content. Thus, in most instances, small improvements in the structural properties of the matrix are of little value; its adhesion and processing characteristics, however, are of paramount importance.

Most structural composite components are produced with thermosetting resin matrix materials. In metal matrix composites, the most frequently used matrix material is aluminum, while alloys of titanium, magnesium, and copper are being developed. Boron and graphite fibers are generally used as the reinforcement for metal matrix materials.

FIBER TYPES AND PRODUCTION

The unique geometry of a fiber provides many of the advantages in an advanced composite. In their fiber form, materials such as carbon/graphite and boron (which are also known as polycrystalline ceramic fibers) show near-perfect crystalline structure. Parallel alignment of these crystals along the filament axis provides the superior strengths and stiffnesses that characterize advanced composites. Various production methods are used for the different fiber types.

Carbon/Graphite

Carbon or graphite fibers (these two terms are often used interchangeably to refer to a specific class of fibers), generally accepted as the most desirable materials for advanced structural composite applications, are produced from either polyacrylonitrile (PAN) or pitch. When superheated to temperatures of 4900° F (2700° C), various carbon grades can be manufactured into a synthetic graphite fiber. It is this graphite fiber, not the graphite mined naturally from the earth, that is woven into tape or fabric and bonded in a matrix of epoxy or other resin to form a composite. The carbon fibers are strong but brittle; hence, they may break rather than bend when forced into tight curves. Within a resin matrix, however, the fibers are positioned and protected. The loads are transmitted and distributed uniformly when grouped and held together by a matrix.

Polyacrylonitrile (PAN). The most widely used carbon fibers are formed by extruding polyacrylonitrile (PAN) through a spinnerette containing approximately 10,000 holes. The resulting fiber goes through ultrahigh heating, oxidation, carbonization, and graphitization processes to produce a filament with extremely high tensile strength and modulus (or stiffness). By varying process temperatures, engineers at Hercules, Inc. have produced both high-strength fibers [e.g., their AS4 graphite fiber at 520 ksi (3 585 MPa) and 32,000 ksi (221 x 10^3 MPa) modulus] and high-modulus fibers [e.g., their HMS at 50,000 ksi (345 x 10^3 MPa) with 320 ksi (2 206 MPa) tensile strength].

Pitch. The carbon or graphite fiber may also be manufactured by starting with petroleum pitch. With this method, the pitch must first be stabilized, and then a process is used to spin it into a filament, which is next made into the final carbon/graphite fiber. Pitch-based fibers are only about two thirds as strong as PAN fibers, but have a greater potential for developing high modulii.

Recent developments at Union Carbide's Carbon Products Division include a petroleum pitch-based carbon fiber with a 100,000 ksi (690 x 10^3 MPa) modulus. Under their trade name, Thornel, a typical epoxy matrix composite, grade VS-0054, exhibits a Young's modulus of 60,000 ksi (413 x 10^3 MPa) and a tensile strength of 165 ksi (1 138 MPa). By comparison, the typical Young's modulus and tensile strength of PAN-based carbon fibers in a similar epoxy matrix are 20,000 ksi (138 x 10^3 MPa) and 225 ksi (1 551 MPa), respectively. This new pitch-based fiber is expected to be cost effective in the fabrication of lightweight structures in which stiffness is critical, such as space hardware and driveshafts.

Aramid

As a result of its high tensile strength properties, rated at 525 ksi (3 620 MPa), Du Pont's Kevlar aramid has also gained acceptance in the field of advanced composites. Aramid is an organic fiber that is produced from a watery polymer solution extruded through a spinnerette. Du Pont's proprietary spinning process yields a super-rigid molecular chain with a modulus of 12,000 ksi (83 x 10^3 MPa) and 18,000 ksi (124 x 10^3 MPa) respectively for Kevlar grades 29 and 49. Aramid fibers exhibit one of the highest tensile strength to weight ratios of any commercially used fiber; however, the compressive strength is one of the lowest. These properties must be taken into consideration in the design stage when using the composite. Blending the aramid fibers with other fibers, referred to as hybridization, optimizes the properties obtained.

Boron

Boron begins as a high-priced substrate known as boron trichloride. Through a chemical vapor deposition process, boron is deposited on a heated, moving tungsten filament. The resulting boron filament exhibits an average minimum tensile strength of 400 ksi (2758 MPa), with a tensile modulus of 55 to 60 x 10^6 psi (379 to 414 GPa). Boron fibers exhibit the highest level of stiffness in resin matrices of all commercial fibers used today.

FIBER FORMS AND FABRICS

Composites can be classified in a number of different ways. The accepted classification types are fibrous (composed of fibers in a matrix), laminar (made from layers of materials), and particulate (made from particles in a matrix). Within the particulate type are flake and skeletal subcategories.

Continuous Fibers

Continuous fibers in yarns or tows (untwisted yarns) are used in filament windings and in unidirectional tape form. Tapes as wide as 48″ (1220 mm) are formed by collimating continuous fibers, applying a resin-compatible sizing, and impregnating the fibers with resin. The materials are partially cured (B-staged), then separated with a backing material to prevent them from sticking, and are referred to as prepregs. Prepreg tapes are usually processed by laminating them together in a desired configuration, then final curing them with heat and pressure.

Woven-fiber fabrics, or broadgoods, can be easily laid atop complex mold structures. Weaves of carbon/graphite fibers are available in unidirectional orientations that are sometimes bound with nonstructural tie yarns. Hybrid forms may include Kevlar as a locking element. Hybridizing can also be used to combine the impact resistance of Kevlar or cost savings of fiberglass with the superior strength and stiffness of carbon or boron. Multidirectional weaves of single-material fibers and/or multimaterial fibers (hybrids) are available in dry or prepregged forms.

Fabrics

Generally, the fibers are woven into fabrics that come in roll form. Fabrics can be woven for specific part requirements. There are various ways of placing the fibers and fabrics in the matrix; when the fibers are oriented to run in one direction, the resulting material is anisotropic in its strength properties.

Fiber Science

In the evolving technology of advanced composite materials for high-performance structural components, the term *fiber science* is applied to a materials-tailoring discipline that includes the type of fibers, their percentage of the whole composite, and the oriented placement of fibers in the matrix during parts fabrication and production processing.

Versatility. Fiber science in the field of advanced composites is not limited to off-the-shelf items and concepts. Fiber science may be treated as a separate subscience of the composite structure. Figure 8-9 illustrates the freedom and latitude available during part design and manufacturing operations. The materials are literally devised and produced through techniques that allow the tailoring of materials performance and properties to suit the intended application.

Fiber directionality. Fibers can run longitudinally, called the warp in a fabric; or they can run transversely, called the weft. Traditional bidirectional fabrics are called warp and weft. However, with the new weaving mill technology, fabrics can now run on the bias, as desired. The bias relates to the directionality and distribution of the fibers. Research is currently under way to produce three-dimensional hybrid fiber weaves; the ultimate hybrid fiber weave will be a "3-D" or cartesian weave.

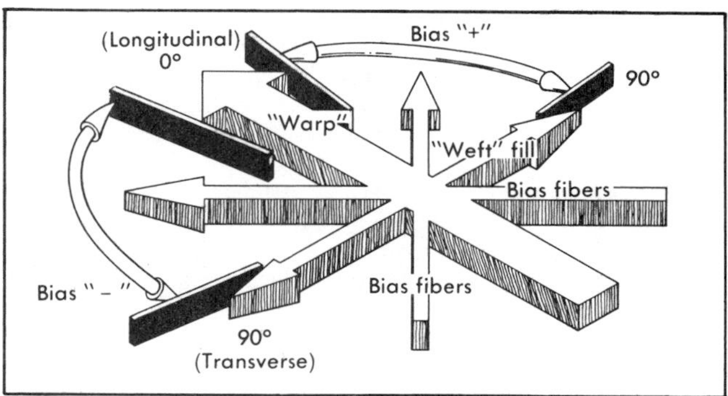

Fig. 8-9 Advanced composites fiber-orientation diagram. (*Polytech, Ltd.*)

Variable percentage. The first variable in fiber directionality is the percentage of fibers in each direction. The various directional fiber percentages can be specified and controlled to provide the desired strength levels and orientations. The bias fibers can be greater or less than the percentage of longitudinal fibers, which, in turn, may be greater or less than the percentage of weft or transverse fibers.

Fiber type. The other principal variables in the fiber reinforcement package are the type of fiber, and, especially, the growing capability for hybridization of fiber reinforcements for advanced composites.

Hybrid Composites

Hybrid composites, which combine two or more different fibers in a common matrix, greatly expand the range of properties that can be achieved with advanced composites. They also increase the potential for cost-effective applications since hybrids may cost less than materials reinforced only with graphite, aramid, or boron.

Characteristics. The term *hybrid* generally applies to advanced composites and refers to the use of various combinations of continuous graphite, boron, aramid, and glass filaments in thermoset matrices. Hybrids have unique features that can be used to meet diverse design requirements in a more cost-effective way than either advanced or conventional composites. Some of the advantages of hybrids over conventional composites are balanced strength and stiffness, optimum mechanical properties, thermal-distortion stability, reduced weight and/or cost, improved fatigue resistance, reduced notch sensitivity, improved fracture toughness, improved impact resistance, and, most of all, optimum cost as related to performance.

Fibers. Various types of graphite, boron, glass, and aramid fibers are used in hybrids, as are cloth and fabric woven from the fibers. Fibers are available with the following ranges of mechanical properties: tensile strength from 2500 to 5000 ksi (17.2 to 34.4 GPa) and tensile modulus from 10 to 60 x 10^6 psi (69 to 414 GPa).

Graphite fibers (five different types are commercially available) offer high stiffness and strength and low density; however, impact resistance is low when compared to glass or aramid fibers. Aramid polyamide fibers combine high tensile strength, low modulus of elasticity, and high impact resistance, but have low compressive strength; with aramid fibers, producing chopped fibers on standard equipment is difficult. Of the two types of glass fibers currently used, S-glass is more rigid and stronger than E-glass. The main limitation of the glass fibers is their high density compared with that of graphite and aramid fibers.

Resins. Both thermoset and thermoplastic resins are used for hybrid composites. The epoxies remain the chief thermoset, but the use of thermoset polyester is growing, especially for automotive applications. The epoxies are available with a wide range of properties, but intermediate-modulus epoxies are used in most hybrids. Epoxies have good-to-moderate elevated temperature properties, but their properties decline when exposed to temperatures near 350° F (177° C) because of moisture absorption. Bismaleimides are used as matrices for composites where extended operation at temperatures near 500° F (260° C) and higher are required.

Forms. The main forms of hybrid composites are interply, intraply, interply-intraply, selective placement, and interply knitting.

Interply. The interply hybrids, illustrated in Fig. 8-10, consist of plies from two or more different fiber types stacked in alternate layers to obtain the desired properties.

Intraply. Figure 8-11 depicts the principle of intraply hybrids, which consist of two or more different fiber strands intermixed in the same ply.

Interply-intraply. The interply-intraply hybrids are made up of plies of interply and intraply hybrids stacked in a specific sequence.

Selective placement. Figure 8-12 illustrates selective placement, which, from the viewpoint of cost-effective tailoring of material properties to design requirements, is regarded as a significant hybridization technique. As an example, the I-beam in the illustration shows the bulk of the fibers made of glass for economy, and places the costlier, higher strength graphite fibers at the extremities of the skins and flanges where the higher modulus (stiffness) qualities are most significant for the type of loading that is applied. Under deflection, the graphite carries the principal loads.

Interply knitting. A recently developed concept in composite fabric assembly called interply knitting or stitching is described in Fig. 8-13. This technique applies a vertical interply stitching with a polyester or aramid strand, thus connecting 2-5 plies together and strengthening the composite against interlaminar shear. Conceptually, the next step involves using a package of the interply-stitched fibers to make a part and selectively applying the stitching or interply knitting at designated places in the structure. Interply knitting enhances resistance to interlaminar shear by placing a vertical stitch through the multiple plies. Interlaminar shear occurs when the resin matrix fractures and the individual layers or plies of fabric peel apart.

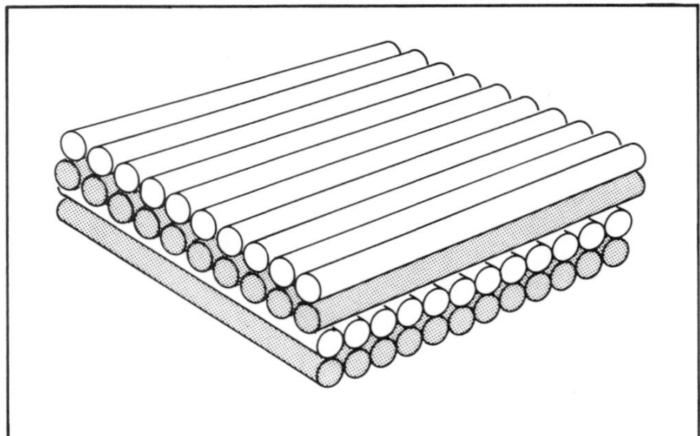

Fig. 8-10 Basic arrangement of interply hybrid composites. (*Owens Corning*)

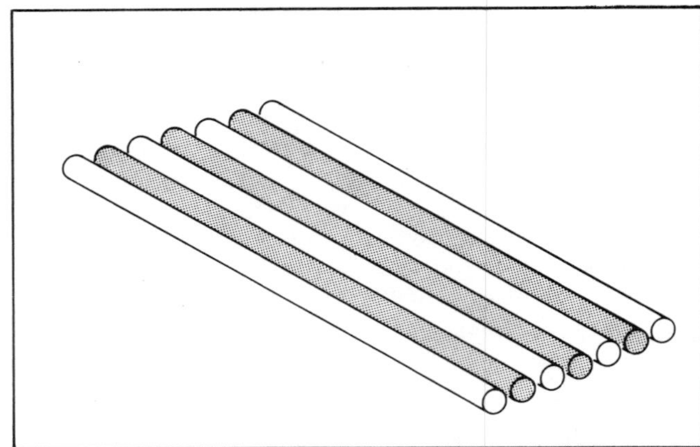

Fig. 8-11 Placement of fibers in intraply hybrid composites. (*Owens Corning*)

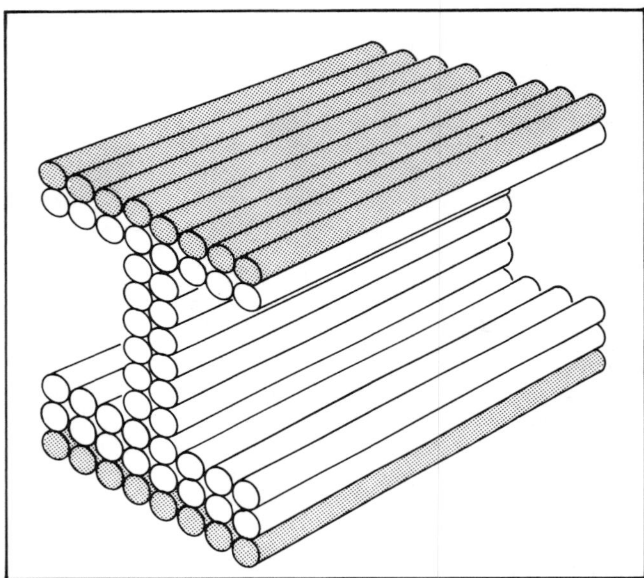

Fig. 8-12 Typical I-beam cross section illustrates selective placement of hybrid composite fibers to optimize strength and cost. (*Owens Corning*)

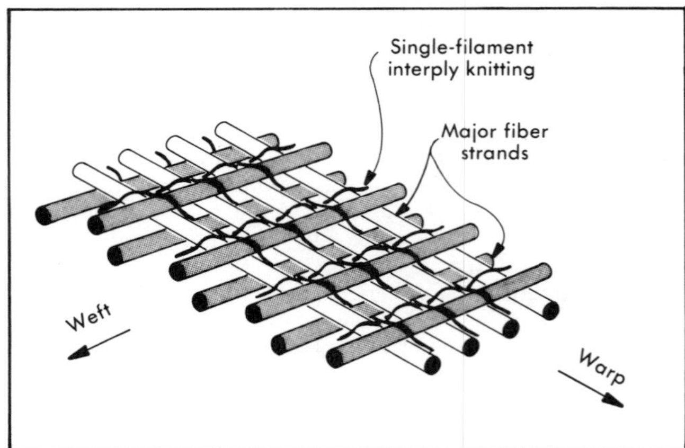

Fig. 8-13 Schematic illustration shows interply knitting to enhance fabric interlamellar shear strength of composite structures. (*Polytech, Ltd.*)

Combinations. The potential number of fiber-resin combinations for hybrids is great. With just two resins (epoxy or bismaleimide) and three types of fibers (graphite, aramid, or glass), many different hybrid composites can be produced, depending on the fiber content and orientation of each fiber in the matrix. With two or even three different fibers in a common matrix, it is possible to make the most effective use of each fiber. For example, a hybrid of graphite and aramid is a natural combination, to which each fiber contributes its best properties.

COMPOSITE CONSTRUCTION

Composites can be divided into laminates and sandwiches. Laminates are composite materials consisting of two or more layers bonded together. Sandwiches are multiple-layer structural materials that contain a low-density core between thin faces (skins) of composite materials. In some applications, particularly in the field of advanced structural composites, the constituents (the individual layers) may themselves be composites (usually of the fiber-matrix type).

Laminates

In the context of this Handbook section, laminates are the general form in which component parts and end products are fabricated from advanced high-strength structural composite materials. Theoretically, there are as many different types of laminates as there are possible combinations of two or more materials. If materials are divided into metals and nonmetals, and if nonmetals divide into organic and inorganic, there are six possible combinations in which laminates can be produced: metal-metal, metal-organic, metal-inorganic, organic-organic, organic-inorganic, and inorganic-inorganic. In laminates containing more than two layers, there are considerably more possibilities, and one or more of the layers may be a composite.

Sandwiches

As was previously stated, sandwiches consist of a relatively thick, low-density core (such as a honeycomb or foamed material) between thin faces of a material with higher strength and density. Although this distinction between laminates and sandwiches is more descriptive than technical, some general observations can be made. In sandwich composites, for example, a primary objective is improved structural performance, or, more specifically, high strength-to-weight ratio. The core serves to separate and stabilize the faces against buckling under edgewise compression, torsion, or bending, and provides a rigid and highly efficient structure. Other considerations, such as thermal insulation, heat resistance, corrosion resistance, and vibration damping, dictate the particular choice of materials used. While the choice of materials is an important consideration in sandwich composites, it is the configuration of the structure that controls the essential properties. In laminates, however, properties depend much more upon the combination of materials used than in sandwiches.[4]

Honeycomb. Honeycomb construction is a special type of sandwich core. The skin consists of two thin layers of strong composites between which is sandwiched a thick layer of weaker but very lightweight honeycomb core material.

As illustrated in Fig. 8-14, for beams of equal weight and the same material, the honeycomb is the most efficient load-carrying design configuration. The principle is similar to that of an I-beam, which is an efficient structural shape because as much of the material as possible is placed in flanges located farthest from the center of bending (neutral axis). Only enough material is used in the connecting web to cause the flanges to act in concert and to resist shear and buckling. In a structural honeycomb sandwich, the facings take the place of the I-beam flanges, and the core replaces the I-beam web.

The core is made of a different material than the facings, and it is dispersed in the cellular construction, instead of being concentrated in a narrow web. The facings act together to form an efficient internal stress couple or resisting moment that counteracts the externally imposed bending moment. The honeycomb core resists the shear stresses, and it has the additional important function of stabilizing the facings against wrinkling or buckling.

The honeycomb core must be rigid enough perpendicular to the faces to prevent crushing, and its shearing rigidity must be sufficiently high to prevent shearing deformations of a magnitude that would destroy the advantages gained through the increase in the bending rigidity of the sandwich wall. Through efficient design, each material can be stressed to its practical limit.[5]

Construction. Figure 8-15 illustrates the construction of a honeycomb beam. The faces do not buckle because they are supported by the multitude of small cells. To accommodate the requirements of a particular application, the core is made in various strengths by changing cell size and the material and thickness of the foil. The balance between strength, stiffness, and weight properties available in honeycombs is depicted in Fig. 8-16, which also compares the specific strength and stiffness characteristics of various structural materials.

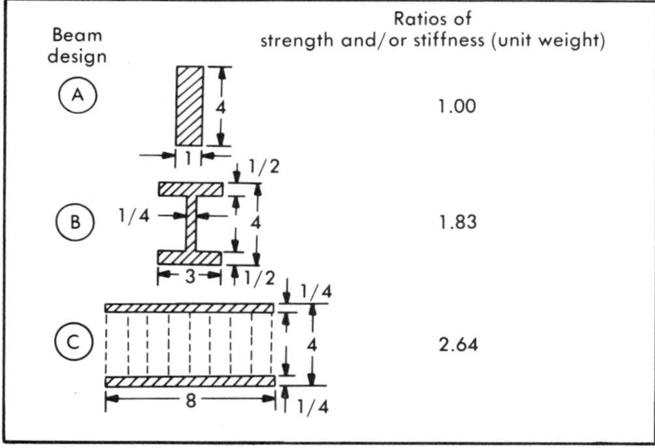

Fig. 8-14 Comparison based on ratios for moment of inertia indicates that honeycomb construction is 2 ½ times stronger and stiffer than a solid, rectangular cross section and 40% better than the typical I-beam configuration. (*American Cyanamid Co.*)

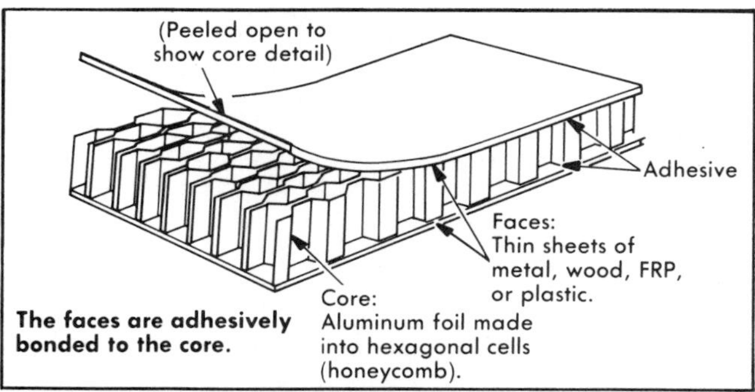

Fig. 8-15 Illustration of typical honeycomb construction identifies the faces and cellular core material. (*American Cyanamid Co.*)

CHAPTER 8

COMPOSITE PROPERTIES

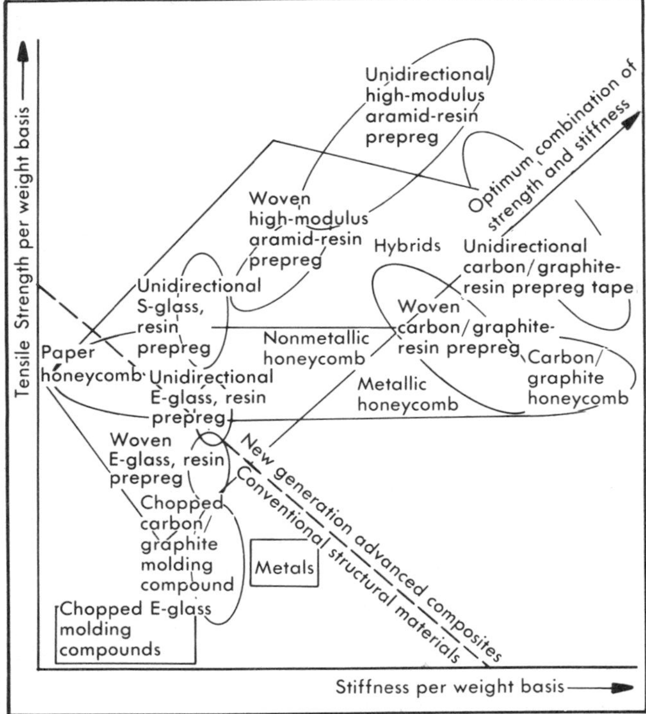

Fig. 8-16 This comparison of specific strength and stiffness characteristics of various structural materials depicts the general, relative position of composites in a graphic property matrix that includes conventional materials. (*Hexcel, Structural Products Div.*)

PROPERTIES

Composite structures, with their capability for tailoring of properties, open new dimensions of freedom not available with isotropic materials. Modulus of elasticity, tensile strength, and thickness can be varied within a single component; parts can be "softened" locally so that loads are transferred from highly stressed areas; and large one-piece structures can be produced without the need for a multitude of fasteners and parts.

Design and production considerations for composites are different from those for homogeneous, isotropic materials having well-defined elastic and plastic stress-strain behavior. For structural parts, all failure modes must be investigated, and particular attention must be paid to interlaminar tensile and shear stresses because the strength of composites depends principally on the matrix resin, not on the reinforcing fiber.

Components made from fiber-resin combinations, often called advanced composites, are usually of laminated construction. They can be essentially isotropic, quasi-isotropic, or anisotropic, depending on material form, lay-up configuration, and fabrication method. Most parts are designed to be anisotropic to exploit the directional properties of the fibers.[6]

Key Factors

Carbon, glass, and aramid fiber composites have outstanding stiffness and strength properties, along with low weight. These composite materials afford a new, wider latitude in the parameters of design, production, and performance.

Stiffness. A major advantage of using various forms of reinforcing fibers in structural parts is that these composite systems can be tailored to suit specific needs. Stiffness, for example, can be varied significantly in different areas of a composite part by selecting the type and form of fiber, by judicious orientation, and by controlling local concentrations of fibers.

Ductility. Like most rigid materials, carbon-reinforced plastics are relatively brittle because carbon fibers are brittle. This characteristic requires close attention to cutouts and areas involving fasteners, where stress concentrations can cause failure. Carbon-fiber composites have a low yield strength, and resistance to impact is low.

Conductivity. Incorporating carbon fibers into plastic resins makes the compounds conductive, both electrically and thermally. Electrical conductivity provides the benefits of static drain and radio-frequency suppression in such applications as electronic equipment enclosures, automotive ignition system devices, and small-appliance housings. This property is also an advantage in providing electrostatic paintability. Thermal conductivity provides the benefits of heat dissipation in such components as gears, bearings, brake pads, composites tooling, and other friction-related products and cryogenic processing equipment.

Thermal expansion. The coefficient of thermal expansion of a composite depends not only on the orientation of the fibers, but also on the fiber content and on the thermal behavior of the matrix material. Thus, care must be exercised to ensure that volumetric expansion of the composite in the usually unrestrained direction does not exceed a tolerable limit. Relative to metals, however, the thermal characteristics of composites are more stable.

Overview of Attributes

Composite materials can be made stronger than steel, lighter than aluminum, and stiffer than titanium. These advantages are possible through the careful selection and use of high-strength fibers such as carbon/graphite, aramid, or boron bound in a matrix of epoxy.

In fabricated aircraft structures, a graphite-epoxy composite offers about the same strength and stiffness as aluminum. The principal advantage is that the composite structure is up to 45% lighter. Composites exhibit a low thermal conductivity, good heat resistance, and improved fatigue life compared to metals. They are also generally noncorrosive and highly wear resistant. However, a corrosive chemical (galvanic) reaction can occur when graphite and aluminum are placed in direct contact in the presence of moisture; this reaction does not occur when titanium and graphite are in contact.

The two properties that are most important in a discussion of advanced, high-performance structural composite materials are tensile strength and Young's modulus of elasticity (which is related to stiffness). Aramid has a slightly higher tensile strength than carbon/graphite fibers, but a much lower Young's modulus.

Carbon filaments. Continuous carbon fibers provide high strength and weight reduction. Continuous fibers are available in a number of forms including yarns or tows containing 400-160,000 individual filaments; unidirectional, impregnated tapes up to 48″ (1200 mm) wide; multiple layers of tape having individual layers or plies at selected fiber orientation; and fabrics of many weights and weaves.

The individual carbon filaments, which are the basic elements of the reinforcement, are usually 200-600 μin. (5-15 μm) in diameter. Tensile strength can exceed 400 ksi (2.8 GPa), and modulus of elasticity can exceed 70 x 10^6 psi (483 GPa).

Commercial fibers are available in four modulus groups: 30-35 million psi (207-241 GPa), 50-55 million psi (345-379 GPa), 70-75 million psi (483-517 GPa), and 100 million psi (690 GPa).

The outstanding properties of carbon fiber-resin matrix composites are their high strength-to-weight ratio and stiffness-to-weight ratio. With proper selection and placement of fibers, the composites can be stronger and stiffer than steel parts of equivalent thickness and can weigh 40-70% less. The fatigue resistance of continuous-fiber composites is excellent, and chemical resistance is better than that of glass-reinforced composites, particularly in alkaline environments. The stiffness of a component can be varied significantly in different areas by the selection of various types and forms of fiber, by judicious orientation, and by controlling local concentrations of fibers. The following are some key basic properties of carbon/graphite fibers:

- High tensile modulus—up to 80 x 10^6 psi (550 GPa).
- High strength—up to 450 ksi (3.1 GPa).
- Medium density—.063 to 0.70 lb/in.^{3} (1.74 to 1.94 g/cm^3).
- Low coefficient of thermal expansion.
- High vibration-damping factor.
- High thermal shock resistance.
- Chemical inertness.
- Low coefficient of friction.
- High thermal conductivity.
- High electrical conductivity.

Aramid fibers. Aramid is a manufactured fiber in which the fiber-forming substance is a long-chain synthetic polyamide that has at least 85% of the amide linkages attached directly to two aromatic rings. Kevlar is the Du Pont trade name for a family of aramid fibers, which are offered in three types. Kevlar yarn is used in motor vehicle tires and other commercial applications. Bulletproof vests are an example of applications for Kevlar 29 fabric. This aramid fiber also has industrial uses, such as ropes, cables, ballistic fabrics, and asbestos replacement products. Kevlar 49 is the grade used in making advanced composites. This high-modulus aramid fiber is produced as a plastic resin reinforcement for parts ranging from airplane propellers and boat hulls to motor vehicle bodies and snow skis. The following are basic properties and characteristics of Kelvar 49 fibers:

- Tensile modulus—up to 19 x 10^6 psi (131 GPa).
- Density—.052 lb/in.^{3} (1.44 g/cm^3).
- Tensile strength—up to 525 ksi (3.62 GPa).
- Stiffness—intermediate between glass and carbon/graphite.
- High thermal stability.
- Impact strength exceeds carbon/graphite.
- Retains a high degree of room temperature properties at moderate temperatures.
- Good chemical resistance.
- Dielectric properties superior to glass composites and equivalent to quartz-fiber composites.

Tables of Properties

Table 8-7 compares data for the strength properties of various high-strength fibers. The relative strengths of various fibers in epoxy-resin matrices are shown in Table 8-8. Figure 8-17 is a graphic plot of specific tensile strength and specific tensile elastic modulus comparisons for various reinforcing fibers. The aramid fibers have the highest specific tensile strengths of any continuous fiber commercially available. They are 2 1/2 times as strong as E-glass, 5 times as strong as steel, and over 10 times as strong as aluminum. The specific tensile modulus of Kevlar 49 is 3 times that of E-glass and 75% that of high-strength graphite. Figure 8-18 compares the specific tensile strength and specific tensile modulus of various reinforcing fibers to conventional materials.

Beam strength. Table 8-9 presents a comparison of stiffness and load-carrying capability for beams of equivalent strength made of steel, titanium, aluminum, and the graphite-epoxy composite. The most significant data is on the bottom line, showing the approximate values of pounds per linear foot as follows: steel, 5; titanium, 3; aluminum, 2; and graphite-epoxy composite, 1. The composite graphite beam has the greatest strength-to-weight ratio.

TABLE 8-7
Strength Comparison for Commercially Available High-Strength Fibers

Fiber Type	Density, lb/in.^{3} (g/cm^3)	Tensile Strength, ksi (MPa)	Tensile Modulus, 10^6 psi (GPa)
Boron (4 mils) (0.0001 mm)*	0.10 (2.65)	495 (3410)	56 (386)
Boron (8 mils) (0.0002 mm)*	0.09 (2.38)	540 (3720)	56 (386)
Graphite (AS)	0.06 (1.75)	410 (2820)	30 (208)
Graphite (HTS)	0.06 (1.77)	410 (2820)	34-37 (234-253)
Graphite (HMS)	0.07 (1.91)	340 (2340)	50-55 (345-375)
Graphite (UHMS)(GY-70)	0.07 (1.96)	270 (1860)	70-75 (485-517)
Aramid (Kevlar 49)	0.05 (1.45)	400 (2760)	20 (138)
Glass (E)	0.09 (2.54)	500 (3440)	10.5 (72.5)

(*Polytech, Ltd.*)

* Tungsten core

COMPOSITE PROPERTIES

TABLE 8-8
Properties of Fibers in Epoxy-Resin Matrix

Fiber Type	S-Glass	Aramid	Hercules AS Carbon
Fiber content, %	45	53	46
Tensile strength, ksi (MPa)	126 (867)	158 (1089)	147 (1014)
Tensile modulus, 10^3 ksi (GPa)	3.9 (26.9)	9.4 (64.8)	13.0 (89.6)
Flexural strength, ksi (GPa)	142.0 (979)	61.0 (421)	177.0 (1220)

(*Polytech, Ltd.*)

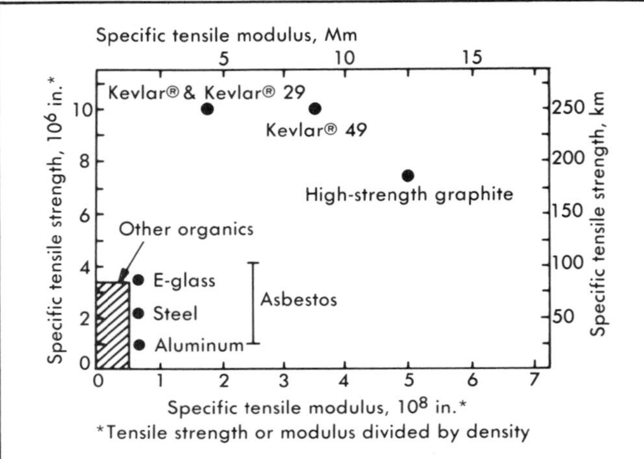

Fig. 8-17 Graphic representation of the specific tensile strength and specific tensile modulus for various reinforcing fibers, tested per resin-impregnated strand test, ASTM D 2343. (*Du Pont Co.*)

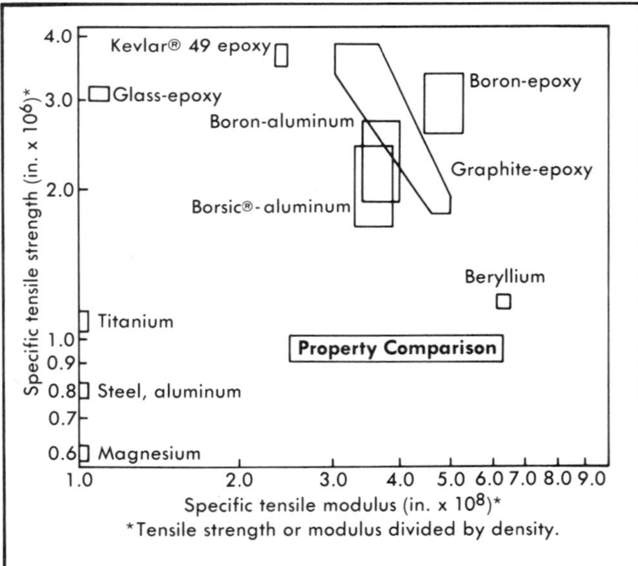

Fig. 8-18 Graphic plot illustrates that typical properties of common fiber composites exhibit significant strength and stiffness advantages over conventional materials. (*Composite Technology, Inc.*)

Carbon and graphite cloths. Graphite cloths and most carbon cloths are fair conductors of electricity. At room temperature, the volume resistivity of graphite cloth is about 40 times that of Nichrome wire. The electrical resistance of graphite cloth decreases with increases in temperature; the resistance at 2400° F (1316° C) is one half the resistance at room temperature. Typical properties of a representative carbon cloth are listed in Table 8-10. Table 8-11 lists typical properties of epoxy composites made from carbon cloth.

General description. Carbon and graphite cloths, consisting entirely of flexible filaments, are produced by the pyrolysis of rayon cloth at high temperatures, to make products with a high degree of purity. The cloth flexibility results from small carbon and graphite filaments, each having a diameter of 350 μin. (9 μm). Single filaments from carbon or graphite cloths have tensile strengths of 50 to 100 ksi (345 to 690 MPa) and a Young's modulus of about 4 x 10^6 psi (27.6 GPa). The average breaking strength of carbon and graphite cloths is 7-24 ksi (48.3-165.5 MPa), depending on grade. Typical laminate properties of carbon and graphite cloths are listed in Table 8-12.

Reactivity. Graphite cloth is stable in vacuum and inert atmospheres up to temperatures approaching the sublimation temperature, 660° F (3649° C). Constituents of neutral or reducing atmospheres may react with graphite cloth at high temperatures. For example, pure hydrogen begins to react appreciably with graphite at about 1000 ° F (538° C). Graphite begins to oxidize in steam at about 1350° F (732° C), and in carbon dioxide at !650° F (899° C). Graphite cloth reacts with air, and at 510° F (266° C) can lose as much as 1% of its weight in 10,000 hours; at 680° F (360° C), it can lose as much as 1% of its weight in 100 hours.

APPLICATIONS

By tailoring the materials and fabrication methods, and by modifying structural designs to accommodate their unique properties, advanced composites can be used for applications requiring high strength, high stiffness, or low thermal conductivity. The common feature of many aerospace uses is that the new materials weigh less than the metallic materials that they replace.

Applications Overview

Although advanced composites containing such materials as carbon/graphite or aramid fibers in an organic resin matrix are currently used mainly by the aerospace industries, these stiff, strong, lightweight materials are also used in various other commercial and industrial applications, ranging from aircraft structures to automobiles and trucks, from spacecraft to printed circuit boards, and from prosthetic devices to sports equipment. Products run the gamut from boat hulls and hockey shinguards to an advanced composites hinge for the retractable arm of the space shuttle.

The space shuttle cargo bay doors are the largest graphite-epoxy structures built to date. Filament-wound aramid forms the first, second, and third-stage motor housings of the Trident missile as well as the round pressure bottles aboard the space shuttle. Extensive use is made of composites in military aircraft, with applications also appearing in commercial jetliners and general aviation aircraft. For example, 3% of the total weight of the new Boeing 767, including floors, ceilings, bulkheads, and door panels, is comprised of composites. The DeHaviland Dash-7 aircraft uses honeycomb flooring that combines high-strength aramid skin with a lightweight core. The Dash-8 aircraft uses composites in the nose bay, elevator tabs, tail cone,

TABLE 8-9
Properties of Structural Beams

	Steel A36	Titanium 6 Al-4V	Aluminum 7075-T6	Graphite-Epoxy Composite
Moment of inertia, in.^{4} (cm^4)	12.51 (520.7)	12.51 (520.7)	12.51 (520.7)	12.51 (520.7)
Modulus of elasticity, 10^6 psi (GPa)	27 (186)	17 (117)	10 (69)	26 (179)**
Stiffness, 10^8 lb-in.^{2} (MN-m^2)	3.38 (0.97)	2.13 (0.61)	1.25 (0.36)	3.25 (0.93)
Ultimate tensile stress, ksi (MPa)	80 (552)*	160 (1103)	83 (572)	140 (965)***
Weight, lb/ft (kg/m)	5.2 (7.7)	2.9 (4.3)	1.9 (2.8)	1.0 (1.5)

(*Hercules, Inc.*)

* Heat treated.
** Approximate properties for beam using HMS fibers.
*** Approximate properties for beam using AS4 fibers.

TABLE 8-10
Typical Properties of Carbon Cloth*

Weave	8 harness satin
Count, ends/in. (ends/5 cm)	14 x 13 (28 x 26)
Weight, oz/yd^2 (g/m^2)	21 (712)
Fiber density, lb/in.^{3} (g/cm^3)	0.068 (1.88)
Thickness, in. (mm)	0.037 (0.94)
Break strength:	
Warp, lb/in. (N/2.5 cm)	200 (890)
Fill, lb/in. (N/2.5 cm)	175 (780)
Carbon assay, %	98+
Ash, %	0.1
pH	8
Moisture, %	<1
Filament count/yarn bundle	4000
Fiber strength (strand), ksi (MPa)	200 (1380)
Fiber modulus (strand), 10^6 psi (GPa)	20 (138)
Surface area, m^2/g	1
Thermal conductivity (room temperature):	
Across ply, Btu/hr · ft/°F (W/m°C)	1.45 (2.5)
Fiber (longitudinal), Btu/hr · ft/°F (W/m°C)	8.7 (15)
Specific heat (room temperature), Btu/lb/°F (J/kg · °C)	0.24 (1002)

* Thornel carbon cloth VC-0160, Union Carbide Corp.

TABLE 8-11
Typical Properties of Epoxy Composites*
Made With Carbon Cloth**

Density, lb/in.^{3} (g/cm^3)	0.058 (1.60)
Tensile strength, ksi (MPa)	40 (275)
Tensile modulus, 10^6 psi (GPa)	6.0 (41)
Flexural strength, ksi (MPa)	55 (380)
Flexural modulus, 10^6 psi (GPa)	5.5 (38)
Compressive strength, ksi (MPa)	50 (345)
Short-beam shear strength, 10^6 psi (GPa)	5.5 (38)
Cured ply thickness, in. (mm)	0.023 (0.59)
Coefficient of thermal expansion, $\Delta l/l \cdot °F$ ($\Delta l/l \cdot °C$)	5 x 10^{-6} (9 x 10^{-6})

* Nominal 55% fiber volume.
** Thornel carbon cloth VC-0160, Union Carbide Corp.

and the leading edges of horizontal and vertical stabilizers. Lear Fan's business airplane model 2100 uses graphite-fiber reinforced epoxy for almost the entire aircraft structure. In combat aircraft, the F15 Eagle has a 17% composite structure, the F/A-18 Hornet has 10%, and the AV-8B Harrier has 30%.

Carbon/graphite cloth reinforced plastics are used in a variety of applications requiring thermal stability, high-temperature strength, good ablation characteristics, and good insulating capability. Since graphite possesses greater heat resistance than carbon and is physically stable at elevated temperatures, graphite reinforcements are used in rocket nozzle throats and ablation chambers. Carbon-carbon composites are

TABLE 8-12
Typical Laminate*
Properties of Carbon and Graphite Cloths**

Properties†	Carbon Fabric	Graphite Fabric
Tensile strength, ksi (MPa)	16 (110)	15 (103)
Compressive strength, ksi (MPa)	24 (165)	16 (110)
Flexural strength, ksi (MPa)	25 (172)	24 (165)
Tensile modulus, 10^6 psi (GPa)	2.2 (152)	1.9 (131)
Compressive modulus, 10^6 psi (GPa)	1.6 (110)	2.2 (152)
Flexural modulus, 10^6 psi (GPa)	2.0 (138)	1.9 (131)
Specific gravity	1.4	1.4
Specific heat, Btu/lb/°F (J/kg · °C)	0.29 (1214)	0.24 (1005)
Barcol hardness	65	60

* MIL R-9299 resin
** Thornel, Union Carbide Corp.
† Federal Test Method 406

used when greater strength and lower conductivity is required, such as in re-entry vehicles, rocket nozzle entrance sections and exit cones, and critical insulation areas.

Figure 8-19 shows a new composites bicycle frame as an exemplary structure in which composites hybridization is used. This structure is a tubular sandwich with a foam core, hybrid fabric, and resin skins. The upper members are made out of glass and graphite fibers because they are always in compression; graphite and glass fibers possess good compressive properties. The frame combines graphite and aramid for the tensile loads on the lower members since aramid has higher tensile strength and is lighter in weight than graphite.

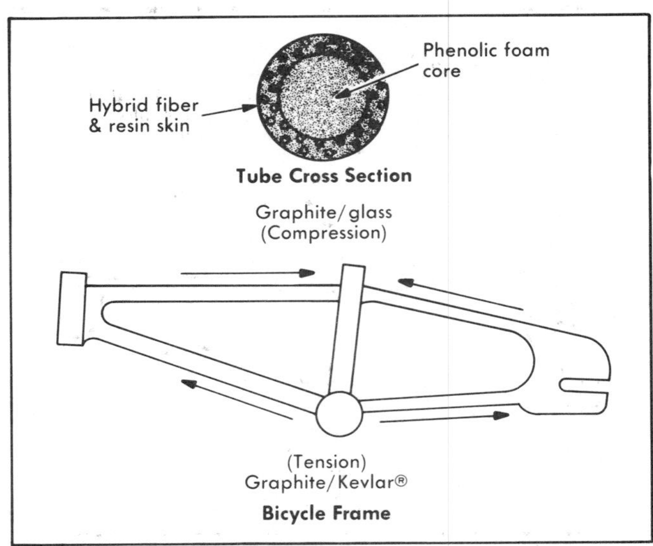

Fig. 8-19 Bicycle frame uses hybrid composite materials, including glass, graphite, aramid, and phenolic foam. (*Polytech, Ltd.*)

Tailoring

Composites are often considered as lightweight alternatives for more traditional structural materials. Innovation with composites can lead to higher production efficiency and lower life cycle costs. For example, creativity in the use of raw materials allows the product design-to-production team to build the composite at the same time the structure is fabricated and to use the anisotropy of the materials to tailor the properties of the composite to meet specific structural requirements.

Proper tailoring of material properties to provide greater local strength and stiffness around fastener holes, for example, increases the load-carrying efficiency of the structure and, at the same time, reduces the amount of material and time needed to manufacture a part. Selective arrangement of plies in a composite laminate may also be used to introduce prestrain during the fabrication process, which produces desired shapes and curvatures in panels and parts while eliminating secondary forming operations.

The desirable features of tailorability and fabrication efficiency are illustrated by the use of composites materials in space structures. Graphite-epoxy tubes are used as structural members in space structures because they are lightweight, stiff, strong, and relatively easy to fabricate. They are selected because many space structures must be dimensionally stable, even when exposed to extreme thermal gradients. Because graphite fibers have a negative coefficient of thermal expansion, they can be embedded in matrix materials having a positive coefficient of thermal expansion and selectively oriented in the laminate to produce a structure with a thermal expansion coefficient of zero to satisfy a stringent functional requirement of the space structure. The dimensional and orientational stability of the communications platform allows the space antennae to receive and send signals to stations on earth precisely and accurately.

Database Limitation

Despite more than 30 years of developmental and production applications activity, the database on composite materials does not approach the available accumulated repository of experience and data on the various metals. There is still much less hard data on composites regarding properties, durability, and failure mechanisms than on metals.

Scientific foundation. The state of the art in composites material technology does not yet have an extensively developed scientific foundation. For example, the method of determining stresses, which assumes that the fibers carry the load and the matrix mainly holds the fibers together, is limited in its capability to fully analyze the complex loadings encountered in aircraft structures. Also, the interface bond between the plastics resin matrix and the reinforcing fiber is known to be crucially important, yet the expertise is still evolving to fully understand these bonds and to determine when they are about to fail.

Impact resistance. Basic information on the reaction of composites to impact is also in a developmental stage. When a metal is hit sufficiently hard, it may dent or crack. However, when a composite sustains an impact, there may not be any apparent defect on the outside, but delamination may have been induced internally within the composite's layers.

Property prediction. Composites property prediction is another evolving area. The raw materials that are used in composites production are undergoing continuous review and research. Until recently, for example, thermosetting resins were thought to be mandatory because of their high-temperature properties. Recently, however, several thermoplastics have become available for use in high-performance advanced composites applications, and others are likely to emerge from current work in the area of engineering thermoplastics such as nylon, polyphenylene sulfide, polysulfone, and polyether etherketone (PEEK). As with other composites, performance data is limited and must be obtained mainly from the resin producers.

FABRICATION

Current composites fabrication techniques are similar to those used for producing fiberglass, including hand and automated tape lay-up resin injection, compression molding, vacuum bag and autoclave molding, matched die molding, pultrusion, and filament winding. Organic matrix composites are made primarily by molding in autoclaves, while metal (aluminum, titanium, etc.) matrix composites are formed mainly by diffusion bonding.

Lamination, filament winding, pultrusion, and resin transfer (injection) molding are four widely used methods of producing continuous-fiber composites with closely controlled properties. The shape, size, and type of part and the quantity to be manufactured determine construction techniques. The lamination method is used for comparatively flat pieces. Filament winding is a powerful and potentially high-speed process for making tubes and other cylindrical structures. So-called *pultrusion* and *pulmolding* can be used for parts with constant cross-sectional shapes. Injection molding can be used for small, nonload-bearing parts. A new epoxy injection process, called URTRI (Ultimately Reinforced Thermoset Reaction Injection), is used for making load-bearing structures, sandwiches, and torsion boxes.

Laminating

Advanced composites are typically used in the form of laminates and are processed by starting with a prepreg material (partially cured composite with the fibers aligned parallel to each other). A pattern of the product's shape is cut out, and the prepreg material is then stacked in layers into the desired laminate geometry.

A final product is obtained by curing the stacked plies under pressure and heat in an autoclave. Graphite-epoxy composites are cured at approximately 350° F (175° C) and with a pressure of 100 psi (690 kPa). The new high-temperature composites, such as bismaleimides, are cured at 600° F (316° C). The tooling is essentially a mold that follows a part through the lay-up and autoclaving processes. Tooling materials commonly used when manufacturing composite parts include aluminum, steel, electroplated nickel, a high-temperature epoxy-resin system casting, and fabricated graphite composite tools.

Conventional method. The two principal steps in the manufacture of laminated fiber-reinforced composite materials are (1) lay-up, which consists of arranging fibers in layers, and (2) curing, which is the polymerization of the resinous matrix material to form permanent bonds between fibers and between layers. Curing may occur unaided or under heat and/or pressure, which speeds the process. When curing is performed under pressure, the processes are referred to as vacuum bag and autoclave, hydroclave, or tool press molding.

In Figs. 8-20, 8-21, 8-22, and 8-23, schematic drawings illustrate preplying and direct-on-tool lay-up, hand lay-up, vacuum bag molding, and compression molding. Figure 8-24 illustrates a variation of compression molding that uses a

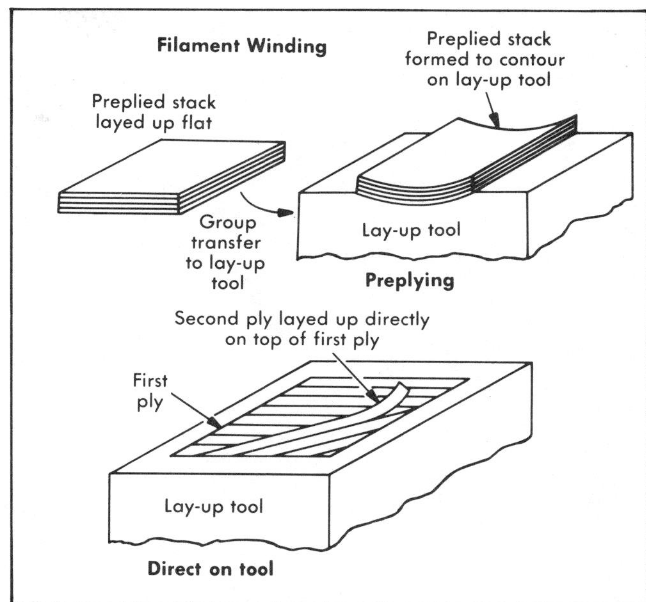

Fig. 8-20 Preplying and direct-on-tool lay-up techniques for composites product fabrication.

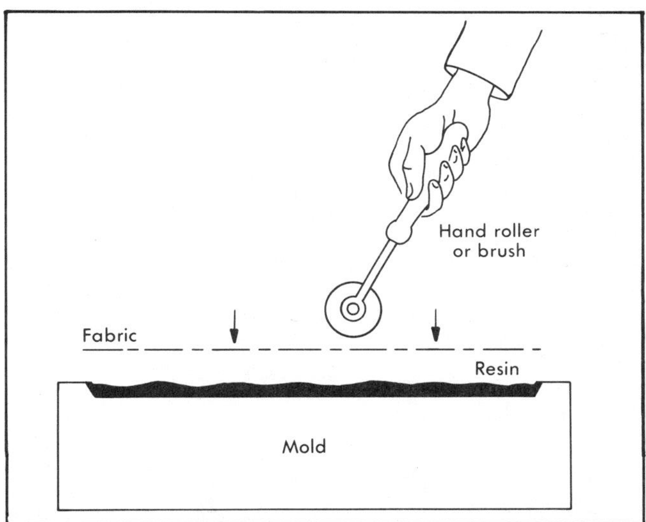

Fig. 8-21 Schematic illustration of basic hand lay-up technique, which is often followed by a room temperature curing stage. (*Polytech, Ltd.*)

premolded core between the composite layers. This wet wrap process, used in the manufacture of skis, is faster than conventional compression molding because resins are catalyzed and then mixed with the fabric and placed wet into the mold. An advantage of this process is higher production rates, at relatively low pressures and temperatures, due to faster cure times (as short as five minutes). For additional information on manufacturing processes for commercial and industrial laminates and reinforced plastics, see "Plastics Forming," Chapter 18, Volume II of this Handbook series.

Heated mold. The heated mold process begins in the conventional manner with prepreg fabrics of graphite or aramid that have been impregnated with resin (usually epoxy) and then

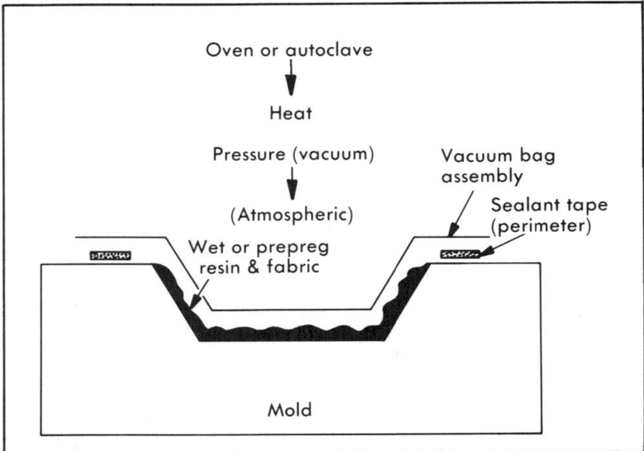

Fig. 8-22 The principles of conventional vacuum bag molding for advanced composites. (*Polytech, Ltd.*)

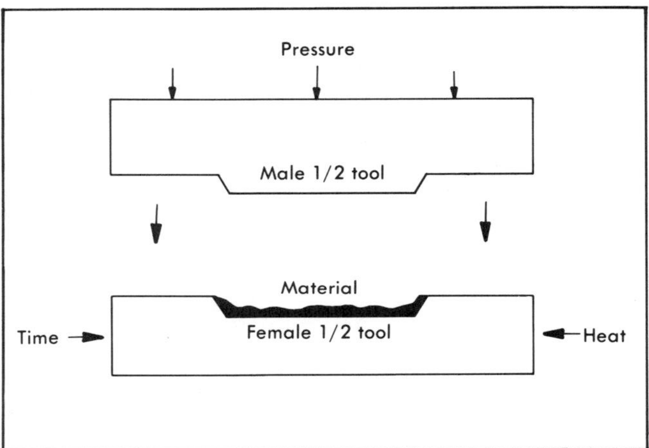

Fig. 8-23 Schematic illustration of basic compression molding process. (*Polytech, Ltd.*)

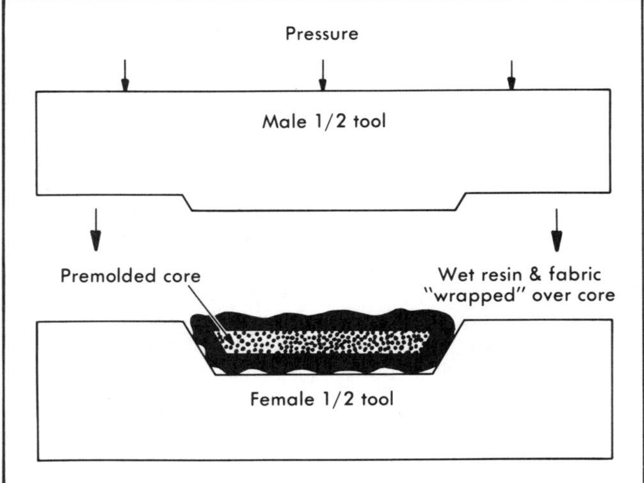

Fig. 8-24 Illustrating the wet-wrap compression molding process, which uses a premolded core. (*Polytech, Ltd.*)

cured to the B-stage (tacky) and frozen. Polymerization can be postponed for up to several months, allowing the material to be stored and shipped. At the fabrication site, the fabric is thawed, layed up, vacuum bagged, and conventionally cured. The overall process is slow and labor intensive.

A recent development eliminates the need for the use of an oven or autoclave curing stage in vacuum bag molding. The mold is built up of composite material, with an integral heating element. Pressure is created on the part through a vacuum bag and/or pressure bag. Since the mold has an integral heating element, faster cure rates are attained, and the need for autoclave/hydroclave curing is eliminated.

Filament Winding

In the filament winding process, illustrated in Fig. 8-25, fibers or tape are drawn through a resin bath and wound onto a rotating mandrel. Filament winding is a relatively slow process; but the fiber direction can be controlled, and the diameter can be varied along the length of the piece. In some versions of the process, the fiber bundle, which may be made up of several thousand carbon fibers, is first coated with the matrix material to make a prepreg tape. The tape is an endless strip with a width that may vary from an inch to a yard (several centimeters to a meter). With both the fiber and tape winding processes, the finished part is cured in an autoclave and later removed from the mandrel.

For strength-critical aerospace structures, carbon fibers are usually wound with epoxy-based resin systems. The polyesters, phenolics, and bismaleimides are limited to special applications. Filament winding is used to produce round or cylindrical objects such as pressure bottles, missile canisters, and industrial storage tanks. It also has been used to make automobile driveshafts.

Pultrusion

In composites technology, pultrusion is the equivalent of metals extrusion. Pultrusion (also called pultruding), depicted in Fig. 8-26, consists of transporting a continuous fiber bundle through a resin matrix bath and then pulling it through a heated die. The process can be used to make complex shapes; however, it has been limited to items with constant cross sections, such as tubing, channels, I-beams, Z-sections, and flat bars. Developmental activity is progressing on variable-section pultrusion, in which the geometry can be controlled by an articulating die. "Pulmolding" is a process variation that begins with pultruding; then the part is placed in a compression mold.

Resin Transfer Molding

Filling a niche between hand manufacturing lay-up or spray-up of parts and compression molding in matched metal molds is resin transfer molding (RTM), also called resin injection molding.

Conventional process. In the conventional RTM process, two-piece matched cavity molds are used with one or multiple injection points and breather holes. The key to RTM is low pressure in the mold, which allows the use of low-cost tooling. The reinforcing material, which is either chopped or continuous-strand mat, is cut to shape and draped in the mold cavity. The mold halves are clamped together, and a polyester resin is pumped through an injection port in the mold. Polyester, glassmat, and conventional RTM are not considered to be in the family of composites.

Compared with the spray-up method (discussed in Volume II, *Forming*, of this Handbook series), RTM permits faster cycle times and usually requires less labor. The RTM cycle times are longer than for compression molding, but the low cost of RTM tooling often compensates for the differential when the production run is less than 50,000 parts per year.

Advanced RTM. In the advanced RTM process, called URTRI (Ultimately Reinforced Thermoset Reaction Injection), reinforcements are placed in the mold, and cores can be handled as inserts. In this process, shown schematically in Fig. 8-27, a core is cast from syntactic foam (high-temperature epoxy with hollow glass microspheres) around a fitting. The core is then wrapped with multiple layers of bidirectional graphite fabric in 7 and 14 mil (0.18 and 0.36 mm) thicknesses and placed in the mold. Epoxy resin is next injected into the heated mold. Curing time is five minutes.

Potential benefits of URTRI include reduced part weight and cost, and improved quality for items such as aircraft wings, fins, elevons, passenger seat shells, landing-gear beams, and other high-performance structures. An example of a part made by the advanced RTM process is shown in Fig. 8-28.

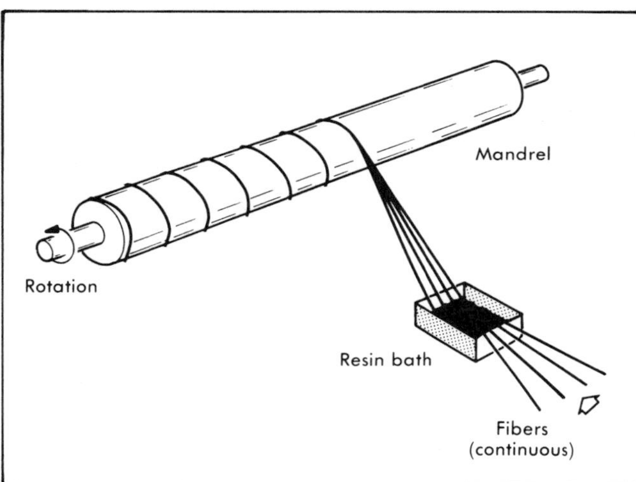

Fig. 8-25 Fundamental elements of the filament winding process for fabricating advanced composites structures. (*Polytech, Ltd.*)

Machining, Cutting, and Joining

The advanced composites materials generally are unsuited for the direct application of equipment, tools, and techniques used for machining, cutting, trimming, and joining metals; hence, special methods are applied to the final processing operations.

Machining. Acceptable-quality machining of composite materials must ensure that there is no splintering, cracking, fraying, or delamination of cured composite edges. Standard machining equipment can often be used with appropriate modifications. In general, the spindle speeds and feeds depend upon the type of laminate material, its thickness, and the cutting method. Cutting tools for machining operations include countersinks, cutoff wheels, router bits, high-speed steel drills, and reamers. While tungsten carbides and high-speed steel tools are generally used, a polycrystalline diamond insert tool performs satisfactorily and is cost effective.

Tools must be kept sharp to provide quality cuts and avoid delamination. Careful attention must be paid to factors such as tool selection and tool geometry because a composite material comprises two separate types of materials (fiber and matrix) with differing mechanical properties and different machining characteristics. Selective stitching around areas for holes and cutouts prevents delamination of plies while drilling, cutting, or router trimming.

Cutting. The conventional methods for cutting uncured composite materials, such as prepreg plys, involve manual cutting with a carbide disk cutter, scissors, or power shears. For cutting cured composites, principal techniques for production applications include reciprocating knife cutting, high-pressure waterjet cutting, ultrasonic knife cutting, and laser cutting.

The particular choice of cutters must be balanced between required performances and the problems created by advanced composites in each operation. Lasers perform the best for cured composites, but may burn or carbonize uncured materials; they are also limited by ply thicknesses. Knife cutters may create rough edges and may become clogged, while water cutters may lead to moisture problems and delamination.

Power cutters and saws. Rotary power cutters can effectively cut one layer of Kevlar fabric. Multiple plies of fabric or prepreg can be cut with abrasive grit, toothless bandsaw, or saber saw blades.

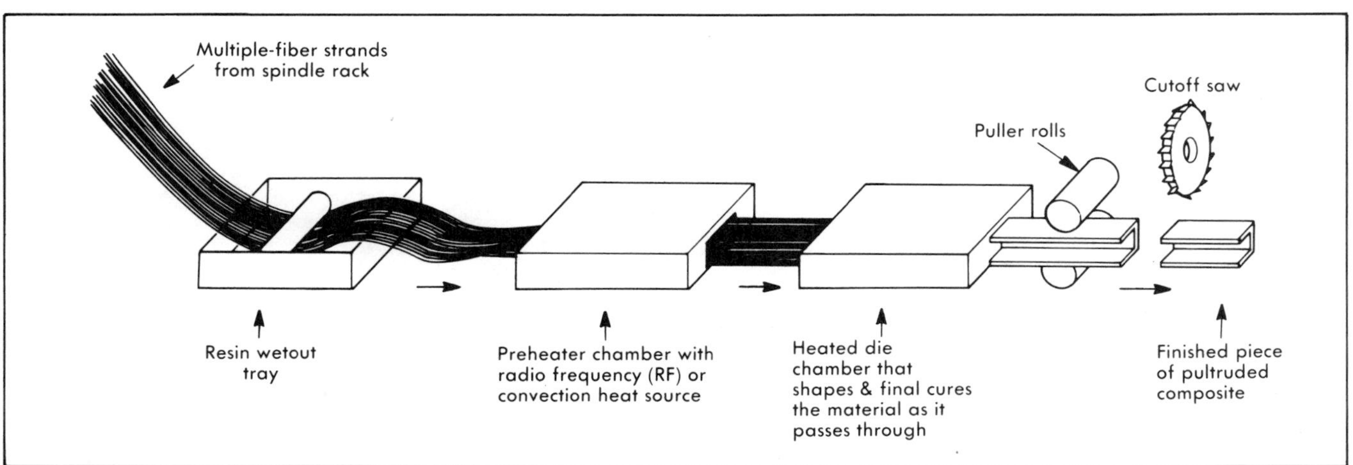

Fig. 8-26 In the pultrusion process, resin-impregnated fibers are pulled through a heated die to produce shapes of constant cross section. (*Polytech, Ltd.*)

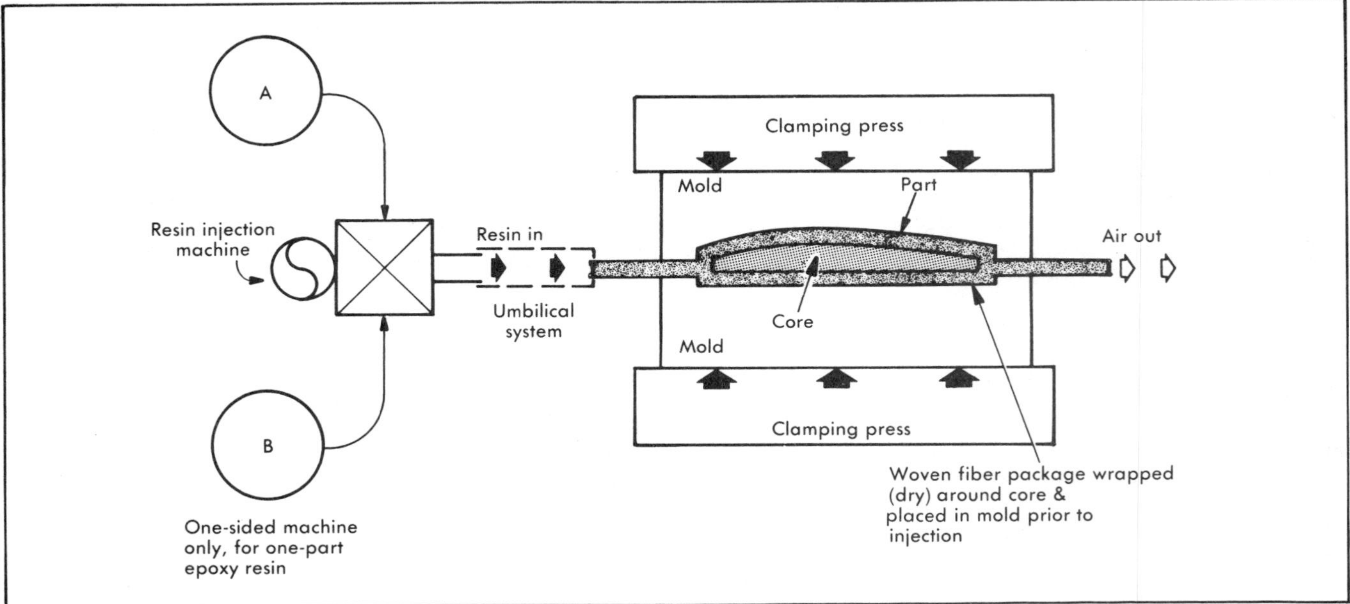

Fig. 8-27 Key features of the Ultimately Reinforced Thermoset Resin Injection (URTRI) process are identified in this schematic diagram. (*Polytech, Ltd.*)

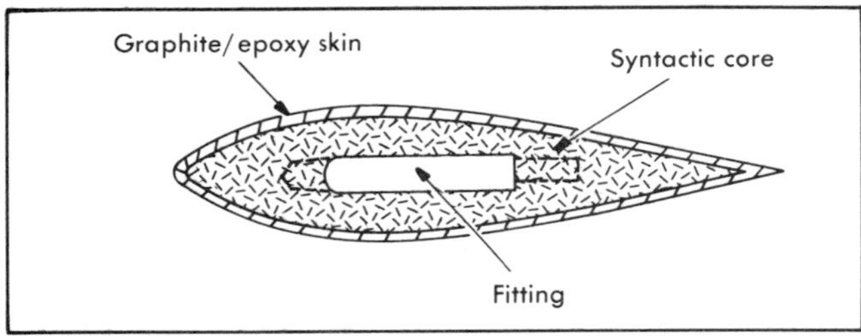

Fig. 8-28 A typical product of the URTRI process for advanced composites fabrication. (*Polytech, Ltd.*)

Laser beam. The laser provides another, newer method for cutting cured composites. Continuous-wave lasers (250 W CO_2) have produced cutting speeds up to 400 in./min (169 mm/s) in single-ply aramid prepregs and up to 300 in./min (127 mm/s) in single-ply boron-epoxy composites. Woven graphite broadgoods are readily cut at 300 in./min (127 mm/s) with nitrogen-assisted gas pressure.

Reciprocating knife. While the laser produces clean through-cuts for single-ply material thicknesses, a reciprocating knife cutter is often more suitable for making multiple-ply cuts. There are two basic types of reciprocating knife cutting machines, both of which use high-speed reciprocating knives that are driven through the material to be cut by a computer-controlled positioning system.

In one system, the cutting knife penetrates through the composite material into closely packed plastics bristles that constitute the cutting table surface. The knife cuts through the material onto the table using a chopping type of action.

The second system can cut patterns in a continuous line at high speeds. Curves, notches, and corners can be cut without lifting the knife from the material. The knife can be raised, as required, to start new cutting lines, to pass over sections without cutting, or to cut holes. The system uses a blade 0.25″ (6.4 mm) wide and cuts either by chopping or slicing. In the slicing mode, the blade remains buried in the material after the first stroke. Computer-controlled rotation of the knife about the C axis keeps the blade properly positioned.

Waterjet process. The waterjet process cuts material by forcing water through a small-diameter jet at high velocities. A typical waterjet cutter cuts from one to ten plies of carbon, aramid, or glass prepregs in cured form. Maximum cutting speed is typically 13″ (330 mm) per second. Cutting performance of waterjet equipment is affected by factors such as jet pressure, nozzle orifice diameter, traverse speed, and the types and thicknesses of materials that are being cut. In some operations, the cutting head is mounted on automated indexing and positioning equipment. Because of water contamination, the waterjet process has been used only on cured parts.

Joining. Advanced composites introduce engineering and production problems different than those common to metal and conventional molded plastics construction, particularly where components must be fastened together. Because most types of composite joints—whether they are adhesively bonded or mechanically fastened—involve some cutting or machining of the strength-providing fibers, joint configurations and joining operations require careful planning to minimize the possibility of failure.

Since structural composites are usually made with thermosetting resins, they cannot be joined by welding methods as can thermoplastics and metals. The choice for joining composites is between mechanical methods and adhesive bonding. Each technique produces a joint with significant differences in both function and production; each method has advantages and limitations. The best permanent joint incorporates both mechanical fasteners and adhesive bonding.

Bolted joints. Bolted mechanical joints are generally not as adversely affected by thermal cycling or humidity as are bonded joints. The joints permit disassembly without destroying the substrate, and they can be readily inspected for joint quality. Mechanical joints also require little or no surface preparation, and they do not call for clean-room production conditions. On the other hand, mechanical fasteners add weight and bulk to a joint; and since they require holes for installation, the members are weakened. In some instances, stress concentrations produced by mechanical fasteners can cause joint failure.

Bonded joints. The advantages of adhesive-bonded joints include light weight, distribution of the load over a larger area than with mechanical joints, and avoidance of drilled holes that weaken the structural members. Drawbacks to bonded joints are the difficulty of inspection for bond integrity, the possibility of degradation in service from heat and humidity cycling, and their relative permanence, which does not permit disassembly without destruction of the joined members. In addition, bonded joints require a rigorous cleaning and preparation of the adhering surfaces.

Quality Assurance

The advanced composite materials introduce special requirements for inspection and quality assurance. For metallic structures, inspection and quality control often focus on detection of a crack or discontinuities that could lead to a crack. Defects in composites range from those that are easily detected, such as delaminations, to those that are difficult to locate, such as improperly cured resin. Common flaws include voids, delaminations, disbonds, missing layers, contamination, inclusions, fiber breakage, improper lay-up, and ply slippage. While most metals fail through the fatigue mechanism of cracking, composites break under load (catastrophic failure), and the resin matrix shears (fractures), allowing the fabric plies to peel apart (interlaminar shear).

Since the composite structures may vary in makeup from point to point, the removal of samples for destructive analysis would be neither practical nor useful. Traditional X-ray inspection is sometimes rendered ineffective by fiber reinforcements and matrices that appear the same on film. Magnetic methods are marginally useful because of the nonconductive nature of most composite materials. Therefore, detecting and diagnosing the potential defects of composites requires a number of specialized nondestructive inspection techniques.

Radiology. Conventional radiography provides good resolution when the attenuation characteristics of the fiber and the matrix are quite different, such as in boron-epoxy composites. In graphite-epoxy or aramid-epoxy composites, however, where differences are small, defect detection is difficult due to lack of contrast. Computerized axial tomography and collimated conical radiography are two variations of radiography used increasingly in the composites industry with good results.

Ultrasonic. Ultrasonics are useful in inspecting the skin and its bond to the substrate and are effective in locating flaws in the thickness direction of the part. Liquid-coupled ultrasonics are currently the most widely used, but dry-coupled ultrasonics are the state of the art. The C scan and pulse echo methods of ultrasonic detection are also used widely. The pulse echo can lose resolution if the laminate is thin due to the echo from the back surface becoming superimposed on the defect echo. Also, if there are many defects, the echoes are difficult to resolve.

In acoustic emission, elastic stress waves are produced by flaws in the material under stress. With this method, however, the part must be under stress; also, current databases on emission characteristics of various flaws are not sufficient to provide a high level of confidence in distinguishing between various defects.

Other methods. Acousto-ultrasonics, thermography, optical holography, and eddy current testing techniques are also used in certain applications, with each method having its own advantages and limitations. Nuclear magnetic resonance is another new technique that offers promise in validating the use of proper materials, evaluating fabrication methods, and determining the chemical nature of composites.[7]

Automated Fabrication Methods

Composites technology is advancing rapidly, with a great deal of effort applied to improvement and automation in producing, cutting, applying, and handling the composite materials. The main compression molding, pultrusion, filament winding, and advanced RTM processes exhibit high-volume production potential.

CAD/CAM. In end product fabrication, research and development programs are making the composite structures increasingly amenable to computer-aided design and numerical control equipment. With CAD/CAM, design changes from metal to composite can be made with minimal risk, since the design can be evaluated as it develops. While the manufacture of advanced composite components is still highly labor intensive and expensive, plans are being implemented by at least one aircraft manufacturer to build an automated factory for producing composite structures of carbon-epoxy construction.

State of the art. Although isolated pieces of automated equipment are now becoming prevalent in composites fabrication, a workcell at the Northrop Corporation shows the extent to which automation can be applied to the composites industry. The cell automates each step of a complexly woven carbon-fiber prepreg lamination process.

In operation, an automated broadgoods spreader lays the prepreg across a cutting table. A computer-controlled reciprocating knife cutter then cuts six to twelve plies of the material. Next, a unique handling system known as the "flying carpet" transports the cutting surface between workstations along 230 ft (70 m) of electrified monorails. The precut prepreg is then carried to a robotic workstation, where a robot with a foam-faced vacuum head lifts the ply and places it in the proper orientation. The robot employs a vision system to search and locate the appropriate ply according to a programmed sequence. Lastly, the plies are laid into a contoured cure mold for delivery to a final cure station. The cell was developed as part of the Air Force Technology Modernization Program; and while not yet the norm in this rapidly evolving industry, it points to a trend of the future.[8]

References

1. Roy S. Klein, "Manufacturing Technology for Designing Plastic Parts," seminar materials (Dearborn, MI: Society of Manufacturing Engineers, 1983).
2. John Hill, Jr., *Engineering Thermoplastics*, Business Opportunity Report P-015R (Stamford, CT: Business Communications Co., Inc., 1983).
3. Blaise A. LeWark, Sr., "An Introduction to Advanced Composites," workshop lecture notes (Dearborn, MI: Society of Manufacturing Engineers, 1984).
4. Mel H. Schwartz, *Composite Materials Handbook* (NY: McGraw-Hill Book Co., 1984), p. 1.40.
5. George Lubin, ed., *Handbook of Fiberglass and Advanced Plastics Composites* (NY: Van Nostrand Reinhold Co., 1969), p. 494.

BIBLIOGRAPHY

6. Schwartz, *op. cit.*, p. 3.1.
7. Robert Waterbury, "Advanced Composites Aid Performance and Assembly," *Assembly Engineering* (July 1983), pp. 26-30.
8. Bruce Krauskopf and Patrick Burgam, "Advanced Composites: High-Tech Materials for High-Tech Industry," *Manufacturing Engineering* (August 1983), pp. 69-71.

Bibliography

ABS Product Design Manual. Borg-Warner Chemicals Publication P-407-8004-20M. Parkersburg, WV: Borg-Warner Corp.

Acetal Copolymer Design Manual. Chatham, NJ: Celanese Plastics Co., Celanese Corp.

"Advanced Composite Materials from Hexcel." Publication 113-41-0978-10M. Dublin, CA: Hexcel, Structural Products Div.

Bettner, Timothy J. "Computer Integrated Composites Manufacturing." *AUTOFACT 5 Conference Proceedings*. Held November 1983, Detroit, MI. Dearborn, MI: Society of Manufacturing Engineers, 1983.

Brown, R.L.E. *Design and Manufacture of Plastic Parts*. NY: John Wiley & Sons, 1980.

Cast Polyurethanes for Industry, Application Guide. Addison, IL: Polyurethane Products Corp.

Crosby, J.M., and Talley, K.L. "New Data for Flame-Retardant Reinforced Plastics." *Machine Design* (January 7, 1982), pp. 94-99.

Designing with Engineering Plastics. Polymer Products Dept. Publication E-45714. Wilmington, DE: Du Pont Co.

Designing with Lexan. Plastics Div. Publication CDC-536B. Pittsfield, MA: General Electric Co., 1982.

Designing with Plastics. Polymer Products Dept. Publication E-59937. Wilmington, DE: Du Pont Co.

DeYoung, H. Garrett. "Plastic Composites Fight for Status." *High Technology* (October 1983), pp. 63-67.

Dreger, Donald R. "New Thrust for Polymer Technology." *Machine Design* (June 24, 1982), pp. 51-55.

__________. "Design Guidelines for Joining Advanced Composites." *Machine Design* (May 8, 1980), pp. 89-93.

DuBois, J. Harry, and John, Frederick W. *Plastics*, 5th ed. NY: Van Nostrand Reinhold Co., 1976.

Elastomers in the Real World. Publication CE-54. Torrance, CA: The Upjohn Co.

Engineering Guide to Elastomers. Elastomers Div. Publication E-41875. Wilmington, DE: Du Pont Co.

Engineering Resins Product Data Bulletins CD-10, CD-14, 15M/04/78, C3B, 10M/583, and Product Information ND-08, 10M/583, 10M/582, and 3M/0283. Chatham, NJ: Celanese Plastics Co., Celanese Corp.

Engineering Thermoplastic Resins. Data Sheet MPR-2-215. St. Louis, MO: Monsanto Polymer Products Co.

Engineering Thermosets. Manchester, CT: Rogers Corp., Molding Materials Div.

Fried, Joel R. "Polymer Technology." *Plastics Engineering* (June 1982 to May 1983 series).

Guide to Material Properties, Design, Processing, and Secondary Operations, Ultem Polyetherimide Resin. Plastics Operations Publication ULT-201. Pittsfield, MA: General Electric Co.

Grayson, Martin, ed. *Encyclopedia of Composite Materials and Components*. NY: John Wiley & Sons, 1983.

Hercules, Inc., Plastics Technical Center. Product Data Bulletins MG-540A, MG-549B, and P-277B. Wilmington, DE.

Holt, Daniel J. "Graphite Composites: A Closer Look." *Aerospace Engineering* (November 1983), pp. 15-18.

__________. "Composite Part Production Techniques Reviewed." *Aerospace Engineering* (September 1983), pp. 9-13.

Johnson, Robert O., and Teutsch, Eric O. "Thermoplastic Aromatic Polyimide Composites." *Polymer Composites* (July 1983), pp. 162-166.

Krieger, Raymond B. *On the Beam: A Honeycomb Primer*. Wayne, NJ: American Cyanamid Co., Chemical Products Div., 1981.

Mock, John A. "Advancing Technology to Spur Growth for Plastic Composites During the '80s." *Plastics Engineering* (February 1983), pp. 13-19.

__________. "Automotive Plastics—1983 Status Report." *Plastics Engineering* (August 1983), pp. 19-25.

Noryl Engineering Plastics. Noryl Products Div. Publication CDX-80. Selkirk, NY: General Electric Co.

Plastics Handbook. Chicago, IL: Engineered Plastic Products Corp.

Plastics Properties Guide. Plastics Operations Publication PBO-140. Pittsfield, MA: General Electric Co.

Roder, Thomas M. *Injection Molding Productivity*. Technical Report E-49723. Wilmington, DE: Du Pont Co.

Simon, Robert M. "EMI Shielding Can Be Made of Conductive Plastics." *Industrial Research & Development* (June 1982).

Society for the Advancement of Material and Process Engineering (SAMPE). *Materials and Processes—Continuing Innovations*. Proceedings from the 28th National SAMPE Symposium. Azusa, CA: 1983.

Standard Tests on Plastics, 9th ed. Celanese Plastics & Specialties Co. Bulletin G1C. Chatham, NJ: Celanese Corp.

Stobart, Phil. "Designing Thermoplastics for Function and Performance." *Design News* (January 23, 1984), pp. 49-58.

Swanson, Frank O. "ASTM D 4000: A Classification System for Plastics." *ASTM Standardization News* (October 1982), pp. 31-34.

"Thermoplastics for the Engineer." *Chartered Mechanical Engineer* (May 1982), pp. 27-31.

Union Carbide Corp., Specialty Polymers and Composites Div. Technical Information Bulletins F-6355, F-7013, and F-7029. Danbury, CT.

Valox Injection Molding. Plastics Operations Publication VAL-15C. Pittsfield, MA: General Electric Co.

Valox Resin Design Guide. Plastics Operations Publication VAL-50A. Pittsfield, MA: General Electric Co.

Wehrenberg, Robert H. "Progress in Reinforced Plastics." *Materials Engineering* (April 1984), pp. 46-53.

Whitney, James; Daniel, Isaac; and Pipes, Byron. *Experimental Mechanics of Fiber Reinforced Composite Materials*. Brookfield Center, CT: The Society for Experimental Stress Analysis, 1982.

POWDERED METALS

Powder metallurgy (PM) is defined broadly as the technology of manufacturing articles from metal powders and of producing those powders, which range in diameter from 4 μin. to 0.04" (0.1 to 1000 μm). The powders are smaller than shot and larger than dust. The PM technique is one of the oldest kinds of metallurgy, going back 5000 years to the manufacture of Egyptian implements.

Modern powder metallurgy evolved in the last half of the 18th century and became commercially feasible in the 20th century with the development of tungsten wire for incandescent lamp filaments. The idea of producing shapes by powder metallurgy was described in a 1902 German patent; however, serious consideration of this technique has only been recent. A resurgence of research and development in metals, generated by competition from nonmetals and particularly from plastics, has enhanced the interest in potential new methods for fabrication of metals. For instance, rolling powder of special, high-priced compositions to produce strip or sheet is feasible and has become an accepted commercial process.

Powder metallurgy has two broad areas of application: the working of refractory metals (such as tungsten, molybdenum, columbium, and tantalum, which have high melting points and are difficult to process by conventional methods); and the fabrication of parts that are uneconomic, difficult, or impossible to produce by other techniques. The industries most interested in powder metallurgy are aerospace (particularly for the use of refractory metals and alloys); and producers of durable consumer goods and capital goods including automobiles and other transportation equipment, home appliances, business machines, industrial equipment, power tools, and hardware. For additional information, refer to Chapter 17, "Powder Metallurgy," in Volume II, *Forming*, of this Handbook series.

CHAPTER CONTENTS:

ATTRIBUTES OF POWDER METALLURGY

During the past 20 years, considerable progress has been made in advancing the science as well as the art of powder metallurgy. With a greater basic understanding of the materials and the processing steps in PM technology, metal powder parts no longer simply replace existing cast iron parts, sheet metal stampings, etc.

Product and production engineers and materials and machinery suppliers now collaborate to design and manufacture PM parts and assemblies and specify them for critical engineering applications. The available PM equipment, techniques, and powders, such as the high-strength and high-compressibility powders, make it feasible to design and produce PM parts that are stronger, or larger, or both.

Properties of PM forgings are comparable to wrought steel, yet production costs may be lower. PM forgings have improved detail and surface finish, hence minimal final machining is required. Accurate control of preform weight results in a minimum of flash and reduces material losses. A graphic comparison of tensile strengths for various materials, including PM, is shown in Fig. 9-1.

DISTINCTIVE OPERATIONS

Powder metallurgy is a highly developed method of manufacturing reliable ferrous and nonferrous parts. Made by mixing elemental or alloy powders and compacting the mixture in a die, the resultant shapes are then sintered or heated in a controlled atmosphere furnace to metallurgically bond the particles. Basically a "chipless" metalworking process, PM typically uses more than 97% of the original raw material in the finished part; therefore, the PM process conserves both energy and materials.

Powder metallurgy is cost effective in producing simple or complex parts at or very close to final dimensions at production rates that can range from a few hundred to several thousand parts per hour. As a result, only minor, if any, machining is required. PM parts can also be sized for close dimensional control that essentially eliminates secondary fabrication steps and/or coined for both higher density and strength.

UNIQUE CAPABILITIES

Ferrous and nonferrous PM parts can be oil-impregnated to function as self-lubricating bearings, resin-impregnated to seal interconnecting porosity, infiltrated with a lower melting point metal for greater strength and shock resistance, and heat treated and plated when required.

Most PM parts weigh less than 5 lb (2.3 kg), although parts weighing as much as 35 lb (15.9 kg) can be fabricated in conventional PM equipment. Many of the early PM parts, such as bushings and bearings, were very simple shapes, as contrasted with the complex contours and multiple "stepped" levels that are often produced economically today. In many cases, functions that normally would require intricate multiple parts and several assembly

***Reviewers of sections of this chapter are: Henry C. Adams**, Development Engineer, Dixon Sintaloy, Inc.; **John T. Benedict**, Consultant; **Peter K. Johnson**, Director, Marketing and Public Relations, Metal Powder Industries Federation; **Robert K. Owens**, Manager, Product Promotion and Sales Training, Hoeganaes Corp.; **Kempton H. Roll**, Executive Director, Metal Powder Industries Federation.*

ATTRIBUTES OF POWDER METALLURGY

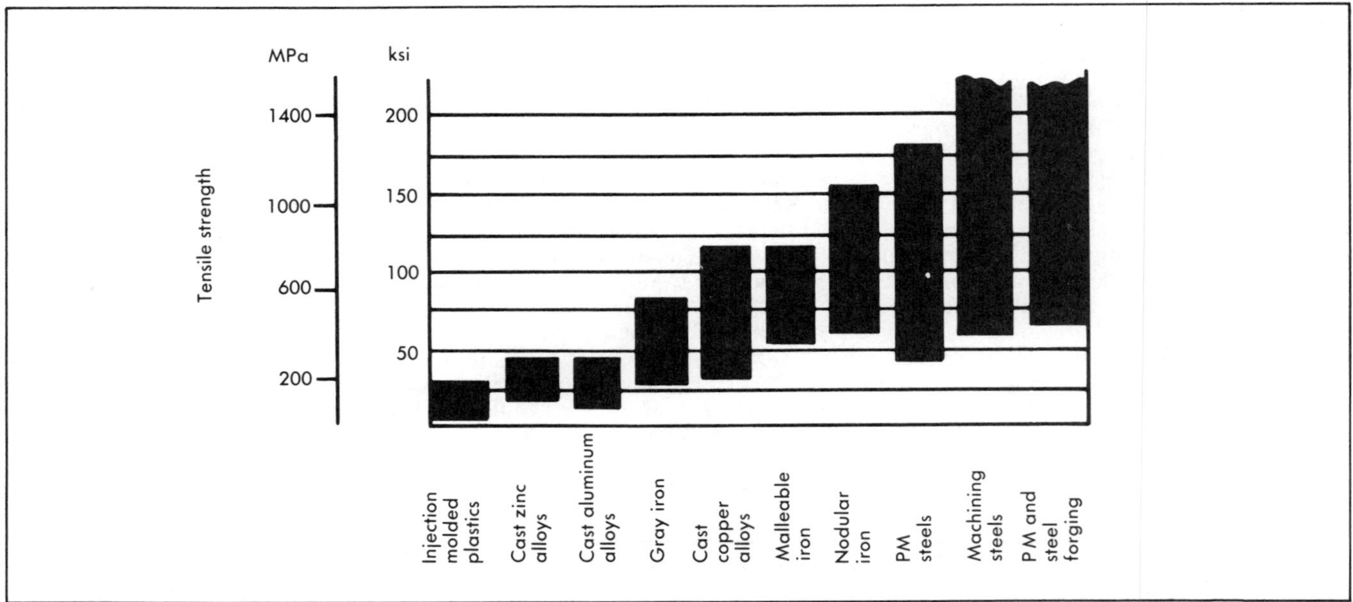

Fig. 9-1 General comparison of tensile strengths for various engineering materials. (*Metal Powder Industries Federation*)

steps are consolidated into a single PM part, thereby minimizing production procedures and reducing cost.

Since the PM process is not shape-sensitive and normally does not require draft, parts like cams, gears, sprockets, and levers are economically produced. In many cases, designs require that parts such as a cam and gear or a spur gear and a pinion gear be joined together by a secondary assembly method. These additional assembly steps can often be eliminated with multiple-level designs that combine the separate shapes into a one-piece PM part. In other instances, two PM parts may be assembled after pressing, then bonded into a one-piece part during sintering.

TYPICAL APPLICATIONS

The versatility of PM is valuable in numerous industries, including automotive, business machines, aerospace, electrical and electronic equipment, small and major appliances, agricultural and garden equipment, and hand and power tools. Thousands of different PM designs serve these industries in a wide range of engineering applications that fall into two main groups. In one group are parts that are difficult to make by any other production method. For example, parts of difficult-to-fabricate materials such as tungsten and molybdenum or tungsten carbide cannot be made efficiently by other processes. Porous bearings, filters, and many types of hard and soft magnetic parts are also exclusive products of the PM process.

The second and larger group applies PM as an effective alternative to machined parts, stampings, castings, and forgings. PM technology has expanded into precision hot forging from sintered blanks and preforms, thus providing PM parts with increased strength and PM forgings with closer tolerances.[1]

METAL POWDERS

Metal powders are precisely engineered materials, available in numerous types and grades to meet a wide range of performance requirements. Mechanical properties of PM parts compare favorably with metal parts made by other metal-working methods.

Most metal powders are produced by atomization, electrolysis, or chemical or oxide reduction. In powder form, the elemental materials include iron, tin, nickel, copper, aluminum, and refractory and reactive metals. These metals can be mixed and then sintered to produce alloy compositions. Sintering bonds the powder particles together metallurgically.

Molten, prealloyed compositions of low-alloy steels, bronze, brass, nickel-silver, and stainless steel, in which each powder particle is itself an alloy, can also be atomized. In addition, metal and nonmetal powders can be combined to provide composite materials with specialized properties.

The engineering properties of the PM part are significantly determined by the metal powder processing and fabrication techniques. Among the controlling elements are particle size and shape, size distribution or sieve analysis, apparent or bulk density of the powder, rate of powder flow into the die cavity, and powder compressibility in the die.

FERROUS MATERIALS

Iron powder, the most widely used PM material for structural parts, is sometimes used alone but most frequently used with small additions of other powders, singly or in combination, to improve mechanical properties of the pressed and sintered part. Among the powders added are carbon, copper, nickel, or molybdenum. Properties data, code designations, compositions, microstructure, and other information on PM structural parts is available in Metal Powder Industries Federation (MPIF) Standard 35.

Ferrous Alloys

Carbon steel. Plain carbon steels are made from mixtures of iron and graphite powders. Sintering the parts diffuses the graphite into the iron to produce a steel structure with carbon content up to about 0.90%. Beyond this point, carbides begin to form.

Low, medium and high-density parts produced from these powders range in tensile strength from 16 ksi to almost 60 ksi (110 to 414 MPa). Strengths in high-density parts can be increased to over 90 ksi (621 MPa) by heat treatment. Prealloyed low-carbon steel powder is also available. The MPIF designation for an iron-carbon steel is F-0008, the last two digits indicating the combined carbon in tenths of a percent (in this case 0.8%).

Copper steel. The addition of copper powder to iron powder increases strength and hardness. Copper steels contain from 1.5 to 5.0% copper and up to 0.9% carbon, depending on the application. Low, medium and high-density parts and bearings produced with these powders have tensile strengths from 35 ksi to more than 65 ksi (241 to 448 MPa); heat-treated, high-density parts can have strengths up to 105 ksi (724 MPa). The MPIF coding for copper steel is FC-OXOY, where *X* is the percent copper and *Y* the tenths-of-a-percent combined carbon (e.g. FC-0208: 2% copper, 0.8% carbon).

Nickel steel. Additions of 2 to 4% nickel powder to iron powders, with or without copper powder, result in tough, high-strength parts with excellent fatigue strength. Tensile strength for nonheat-treated parts ranges from 25 to 90 ksi (172 to 621 MPa); heat treatment increases strength to as high as 195 ksi (1344 MPa). These materials are designated FN-OXOY, *X* representing the percent nickel and *Y* the tenths-of-a-percent combined carbon (e.g. FN-0205: 2% nickel, 0.5% carbon).

Infiltrated iron and steel. Porous ferrous parts can be infiltrated with several lower melting materials, such as copper and various brasses, in order to approach 100% density and provide improved impact resistance. Copper content ranges from about 8 to 25%, depending upon the pressed density of the part, reducing residual porosity to 4 to 8%. Standard copper-infiltrated parts have tensile strengths ranging from about 65 to 90 ksi (448 to 621 MPa) in the infiltrated condition; heat treatment increases strengths up to 130 ksi (896 MPa). Besides improving tensile strength, infiltration can provide uniform part density, otherwise difficult to attain in the pressing operation. The absence of interconnected porosity permits plating without resin impregnation. The MPIF description of infiltrated steel is FX-ZZOY, the *Z*'s being the percent copper and *Y* the percent carbon (e.g. FX-1008: 10% copper and 0.8% combined carbon).

Phosphorous iron. Small additions of phosphorous to iron powder accelerate the sintering cycle, enhance electromagnetic properties, and provide a unique combination of good ductility and high yield and tensile strengths.

Low-alloy steels. Nickel-molybdenum-manganese (Ni/Mo/Mn) powders are used for production of parts that require heat treatment. The homogeneity of the atomized, prealloyed powder results in substantially greater hardenability than is obtainable with blended elemental alloying powders. Strengths range from 90 to 155 ksi (621 to 1069 MPa). The MPIF designations are FL-2205 and FL-4605.

Stainless steels. Powder metallurgy stainless steels are used in automotive, appliance, hardware, marine, and food industry parts and as porous filters. More costly than iron powders, their use is justified when strength and corrosion resistance are needed. MPIF designations are SS-316 and SS-410.

Ferromagnetic Materials

Direct current (d-c) and low-frequency, soft magnetic applications require materials offering good permeability, purity, and density. Sintered "soft" magnets, because of porosity, show lower flux densities than wrought magnets of the same size and shape. Using pure, high-compressibility iron powders or phosphorous irons and designing the magnetic parts somewhat larger can overcome the comparative flux deficiency. The cost of the small additional amount of material is usually more than offset by PM's elimination of scrap.

Various PM materials for soft magnetic parts are available:

- Low-density iron, characterized by a wide hysteresis loop with a low flux density (B max) and high coercive force (Hc), is the least expensive.
- High-density iron parts, with characteristics approaching wrought iron, have a higher B max and lower Hc, resulting in lower core losses.
- Phosphorous-iron offers markedly improved permeability and more precisely controlled coercive force. The phosphorous also minimizes the effect of contaminants such as oxygen and nitrogen.
- Silicon-iron PM parts, with 3% silicon content, are used where high electrical resistivity results in lower core losses and fast magnetic response. These materials have a flux density (B max) somewhat lower than iron, a low coercive force (Hc), and considerably higher permeability (μ) than iron.
- Nickel-iron provides fast magnetic response for a minimum applied field and has the fastest rise in permeability of any PM material. Powder metallurgy parts are generally made from 50/50 prealloyed powder and have very low coercivity (Hc).
- High-frequency cores, another type of PM magnetic part, are produced by insulating iron powder particles with a plastic coating. They are then pressed by conventional PM methods and the plastic binder is cured at low temperatures. These iron powder cores are used in high-frequency coils for numerous electronic applications. They are closely allied to ferrite cores and ceramic magnetic materials in electronic inductive components. The materials most commonly used in iron powder cores are the carbonyl irons, hydrogen-reduced, electrolytic, and flake iron, as well as magnetite powder.
- Permanent magnetic materials take advantage of powder metallurgy to produce sintered Alnico, rare earth, and bonded magnets.

NONFERROUS MATERIALS

Copper and aluminum-based powders are the most widely used nonferrous PM materials, although most nonferrous metals and alloys can be converted to powder and fabricated into engineering components. Copper-based materials include the brasses, bronzes, and nickel-silver.

Copper

Pure copper is selected when its properties are needed to meet high electrical and/or thermal conductivity requirements.

METAL POWDERS

Bronze

Bronze, made from mixtures of elemental copper and tin powders, is the most widely used product of nonferrous PM materials, notably as oil-impregnated, self-lubricating bronze bearings. PM bearings can provide between 18 and 30% interconnecting porosity as an oil reservoir. Tensile strengths of 8 to 18 ksi (55 to 124 MPa) are achieved. High-strength bronze structural parts with tensile strengths up to 50 ksi (345 MPa) are usually made from prealloyed bronze powders.

Brass

Brasses, in the form of prealloyed powders, are available in numerous compositions, with zinc content ranging from about 10% to as high as 30%. Leaded brasses contain from 70 to 90% copper and from 1 to 2% lead. Sintered brass parts have tensile strengths up to 40 ksi (276 MPa).

Machinability of the materials is comparable to cast and wrought brass stock of the same composition. Brass PM parts are well suited for applications requiring good corrosion resistance, good machinability, attractive appearance, and ductility. Brass is designated as CZ-00YY, the *YY* being the percent zinc. (CZ-0010 has 10% zinc.) The CZP designation applies to machining grade brass with 1 to 2% lead.

Nickel-Silver

Nickel-silver powders are copper-nickel-zinc alloys with 16 to 19% nickel. Nickel-silver PM parts have properties similar to brass, but with improved corrosion resistance. The MPIF material code for nickel-silver is CZN-1818, with 18% each of zinc and nickel in the copper alloy.

Titanium

Commercially pure titanium powders can be used for PM parts, with or without the addition of alloying elements. Providing a cost-effective means of using this comparatively expensive and difficult-to-machine material, the PM process expands the potential applications of titanium, capitalizing on its excellent corrosion resistance and strength-to-weight ratio.

Aluminum

Aluminum alloy powders for PM structural parts and bearings offer good corrosion resistance, light weight, and good electrical and thermal conductivity. Major alloying elements are copper and magnesium. Tensile strengths range from about 15 to 50 ksi (103 to 345 MPa) depending on composition, density, and heat treatment.

GENERAL ENGINEERING PROPERTIES

A wide range of powder metallurgy materials that meet the design requirements of particular applications is listed in Table 9-1. The following representative physical and mechanical properties data can be augmented when necessary with additional information and data from original sources, standards and specifications, and from PM powder and parts producers.

Density

Most properties of a PM part are closely related to the final density. This density is the weight per unit volume of the part expressed in grams per cubic centimeter (g/cm^3). Normally, density of mechanical and structural parts is reported on a dry unimpregnated basis, while density of bearings is reported on a fully oil-impregnated basis. Density may be calculated by any of several means. A commonly used method is given in Metal Powder Industries Federation (MPIF) Standard 42.

Density is also expressed as percent of theoretical density, which is defined as the ratio of a PM part's density to that of its wrought metal counterpart. In practice, PM parts less than 75% of theoretical density are considered to be low density; those above 90% are high density; and those between these two ranges are considered as medium density. In general, structural and mechanical parts have densities ranging from 80% to above 95%. Many self-lubricating type bearings have densities on the order of 75%, and filter parts usually have densities of 50%.

Porosity

Porosity is the percentage of void volume in a part. It is the converse of density. A part that is 85% of theoretical density will have 15% porosity. Porosity in PM parts can be present as a network of interconnected pores that extend to the surface like a sponge or as a number of closed holes within the part. Interconnected porosity is important to the performance of self-lubricating bearings and is included in the specification for these types of materials.

As a unique structural characteristic of PM parts, porosity is controllable and a function of the raw material and processing techniques. Parts can be produced either with uniform porosity or with variations in porosity (and also density) from one section to another in order to provide different properties. For example, parts such as gears can be made self-lubricating in one area and dense and strong in other areas. The method for calculating pore volume or oil content of self-lubricating PM components in terms of interconnected porosity is given in MPIF Standard 35.

Permeability

The ability to pass fluids or gas, as in filters, is another property that can be designed into PM products. Depending on the forming and sintering techniques, a PM part can provide permeability ranging from highly restricted to highly open flow. The part can be produced with permeabilities that will separate materials selectively, diffuse the flow of gases or liquids, regulate flow or pressure drop in supply lines, or act as flame arresters by cooling gases below combustion temperatures. Filters can be produced in almost any configuration, including sheets and tubes.

Mechanical Properties

As with wrought and cast metals, the chemical composition of PM parts strongly influences the mechanical properties. However, in PM parts there are the additional factors of density; particle size; pore size, shape, and distribution; and extent of sintering, upon which the mechanical properties also depend.

Because of these influences, mechanical property data are commonly given in graphs showing the relationship between the property and density (or percent of theoretical density). The graphs shown in Figs. 9-2 through 9-4 represent typical values and relationships depicting the trend of properties versus density from test specimens conforming to chemical and property requirements specified in MPIF Standard 35. (For more complete property data, refer to Standard 35.) Typical properties of representative iron and steel powders are shown in Table 9-2.

TABLE 9-1
MPIF Designations for Typical PM Materials

Designation	Chemistry	Applications Characteristics
F-0000	Pure Fe (Carbon below 0.03%)	Low strength at high density, a soft magnetic material
F-0005	Fe-0.5% C	Moderate strength, mild steel
F-0008	Fe-0.8% C	Moderate strength, mild steel
FC-0208	Fe-2% Cu-0.8% C	Higher strength structural parts
FN-0205	Fe-2% Ni-0.5% C	High strength when heat treated, good impact resistance
FN-0208	Fe-2% Ni-0.8% C	High strength when heat treated
FX-2008	Fe-20% Cu-0.8% C	Infiltrated steel, high strength, sintered or heat treated
SS-316	316 stainless steel	Good toughness and corrosion resistance
SS-410	410 stainless steel	Hardens during sintering, good abrasion resistance
CZ-0220	Brass: Cu-20% Zn, 2% Pb	Good toughness, elongation, easy machining, good corrosion resistance
CT-0010	Bronze: Cu-10% Sn*	Structural and bearing applications, excellent corrosion resistance
CZN-1818	Nickel silver: Cu-18% Ni-18% Zn	Decorative, tough and ductile, superior corrosion resistance.

Note: For more complete materials standards information, contact the Metal Powder Industries Federation. MPIF Standard 35 provides minimum strength values for design purposes. Designers can select materials based on the physical and mechanical properties of each material. Density need not serve as the basis for material selection.
*Diluted bronze available.

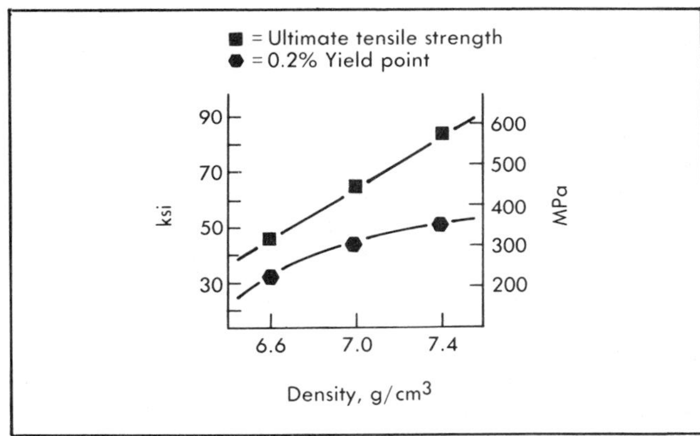

Fig. 9-2 Tensile and yield strength (at 0.2% elongation) vs. density for powdered metal FN-0208 (2% Ni, 0.8% C), as sintered. (*Metal Powder Industries Federation*)

Since variations of density in the part and test specimen may not be similar, test data obtained on standard specimens does not necessarily accurately represent the performance of the actual part. Thus, test specimen data should be considered only as an approximate predictor of part performance.

Strength. Figure 9-2 shows ultimate tensile and yield strengths of a 2% nickel and 0.8% carbon, pressed and sintered PM material as a function of density. Yield strength, generally 62-98% of ultimate strength, is closer to the tensile strength than with wrought metals. Also, the yield strength of many PM materials, particularly stainless steels, may be higher than the wrought forms.

Figure 9-3 shows impact strength of a typical PM nickel steel as a function of density, with impact strength rising significantly at higher densities and lower carbon contents.

Ductility. Ductility, the amount of plastic deformation prior to tensile fracture, is relatively low in PM materials, chiefly due to the presence of pores. Ductility as a function of density is illustrated in Fig. 9-4. Elongation is generally less than 10% for

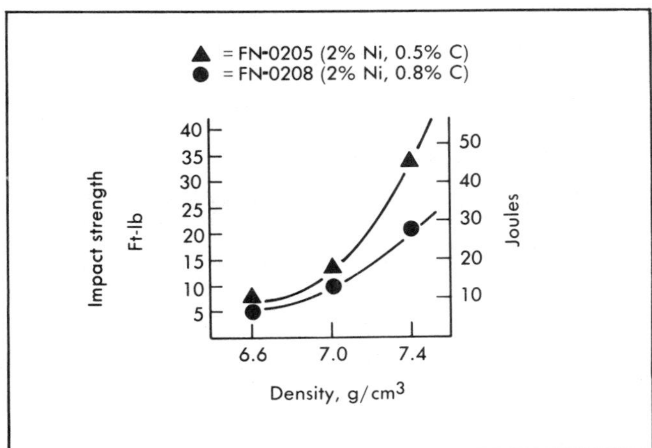

Fig. 9-3 Unnotched Charpy impact strength vs. density for powdered metal nickel steels, as sintered. (*Metal Powder Industries Federation*)

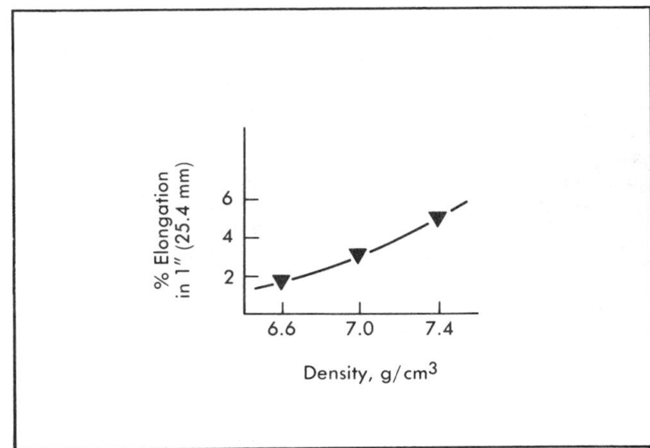

Fig. 9-4 Elongation vs. density for as-sintered powdered metal nickel steel, FN-0205 (2% Ni, 0.5% C). (*Metal Powder Industries Federation*)

TABLE 9-2
Typical Properties of Iron and Alloy Steel Powders

	Powder Properties				Compacted,* 30 tons/in.^{2} (413.7 MPa)		Sintered Properties					
Powder Type	C%	H_2 Loss %	Apparent Density, lb/in.^{3} (g/cm^3)	Flow seconds	Density, lb/in.^{3} (g/cm^3)	Green Strength, psi (MPa)	Additions C%		Additions Cu%		Density, lb/in.^{3} (g/cm^3)	Transverse Rupture Strength, ksi (MPa)
Sponge iron MH-100	0.01	0.25	0.09 (2.55)	30	0.23 (6.4)	2000 (13.8)	1.0		2		0.22 (6.1)	78 (538)
Atomized steel 1000	0.01	0.15	0.10 (2.90)	26	0.24 (6.6)	1400 (9.6)	0.9		2		0.24 (6.6)	127 (876)**
Atomized low-alloy steel							Alloy Content (0.5% C added) Ni%	Mo%	Mn%	Cu%		
2000	0.01	0.17	0.11 (2.98)	25	0.24 (6.6)	1100 (7.6)	0.45	0.6	0.30	---	0.24 (6.6)	86 (593)†
4600V	0.01	0.15	0.11 (2.97)	25	0.23 (6.5)	1000 (6.9)	1.77	0.5	0.18	---	0.24 (6.6)	95 (655)†
Atomized high-alloy steel							Cr%	Ni%	Mo%			
AISI 316L	0.03 max	---	0.10 (2.72)	31	0.23 (6.25)	420 (2.9)	17	13	2.5		0.25 (6.8)	140 (965)†
AISI 410L	0.03 max	---	0.10 (2.70)	29	0.21 (5.9)	350 (2.4)	13	---	---		0.23 (6.4)	140 (965)†

(*Hoeganaes Corp.*)

* Additions: 1% zinc stearate in sponge and atomized powders and 1% lithium stearate in high-alloy powders.
** Sintered in endothermic atmosphere at 2050° F (1120° C) for 30 minutes.
† Sintered in dissociated ammonia at 2050° F (1120° C) for 30 minutes.

ferrous materials. However, elongations for some PM brasses and stainless steels are up to 15 to 20%. Ductility of most PM materials can be increased considerably by hot or cold repressing followed by resintering. The resulting properties are improved by the use of materials with low hardness and high density.

Hardness. Because of differences in structure, gross indentation hardness values of wrought metals and PM parts cannot be directly compared. The hardness value of a PM part, when obtained using a standard tester and a scale, is referred to as "apparent hardness," a combination of powder particle hardness and porosity. (See MPIF Standard 43.) Figure 9-5 shows how an indenter can compress the surface between particles or displace powder particles in low-density parts. However, microhardness tests, such as Knoop, if carefully performed, will measure true particle hardness.

Because of the natural porosity of PM materials and the possible density variations, the area in which hardness measurements are made should be specified. Surface hardness values on case-hardened parts should be carefully interpreted because the atmosphere penetration during heat treating can cause case depth variations, and the hardness test indentations sometimes penetrate the case. (See MPIF Standard 37.)

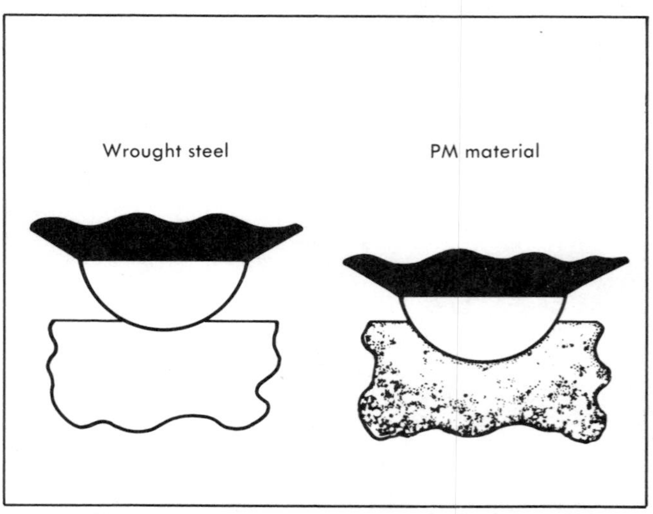

Fig. 9-5 This illustration shows that in PM materials a hardness tester penetrates deeper than in wrought steel because of the lower density that is characteristic of PM materials.

Corrosion Resistance

Porosity in PM parts significantly affects corrosion resistance due to possible entrapment of corrosive media. Higher density improves corrosion resistance, as it does most other properties. Stainless steel PM parts have relatively good corrosion resistance in the atmosphere and in weak acids. Nonferrous PM materials have corrosion-resistant properties similar to the wrought forms. For further information on corrosion resistance, see the sections on heat treating and finishing later in this chapter.

Surface Finish

Excellent surface finish is an inherent feature of PM parts. The overall smoothness and surface reflectivity depend on density, tool finish, and secondary operations. Conventional profilometer readings give an erroneous impression of surface smoothness (except at densities approaching 100%) because a different surface condition exists than is the case with machined or ground surfaces. The conventional root mean square (rms) readings take into account the peaks and valleys of machined surfaces, but PM parts have a series of very smooth surfaces that are interrupted by varying sized pores.

To get a true indication of the functional surface finish of PM parts, a chisel stylus (Fig. 9-6, *a*) is more effective than a standard cone stylus because it bridges the gaps caused by the pores and will measure any variation on the surface of the particle (Fig. 9-6, *b*). When this characteristic of PM surfaces is considered, the effective surface smoothness of PM ferrous parts compares favorably with ground or ground and polished surfaces. Surface smoothness can be further improved by secondary operations such as repressing, honing, burnishing, or grinding.

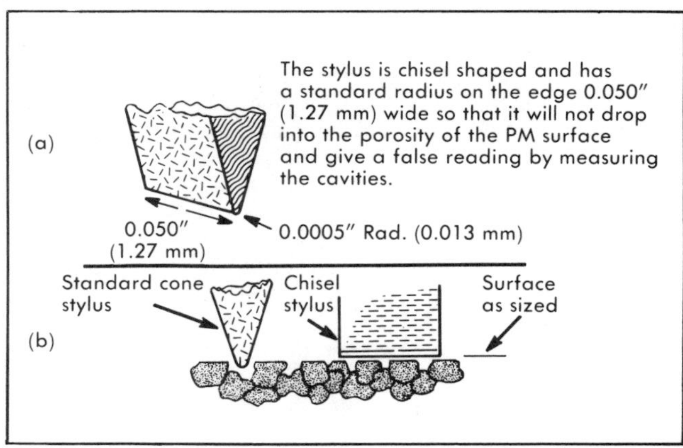

Fig. 9-6 View a shows a chisel stylus used for PM part surface measurement. View b shows the differing effects of chisel and cone styli on a PM part surface.

Sound Damping

The porous nature of PM parts provides good sound damping. Ringing, common with wrought steel gears and other parts, is reduced due to the controllable density in PM products. This characteristic is important in sound recording equipment and other business machines, as well as in air-conditioning blowers and similar products. Damping characteristics can be further improved by infiltration or impregnation with sound-damping materials. The controllable density of PM parts is also used to dissipate and muffle the noise of air-driven power tools.

POWDER METALLURGY PARTS PRODUCTION

The PM process, similar to other part fabrication methods, has its own set of design and manufacturing guidelines for producing soundly engineered, economical products. The particular aspects and requirements of PM production should be kept in mind. Design advantages can then be gained that are unique to the PM process, and the limitations of the process will have been taken into account.

Powder metallurgy product design and production require close cooperation between the part user, or buyer, and the producer, especially in the initial design stages. The PM part should be designed in the context of the whole assembly, with the manufacturer kept informed throughout the process. An improved, lower cost PM part can often be achieved through small changes in an assembly. Not infrequently, early designer-manufacturer interaction results in an expansion of the PM concept that nets overall production design simplification and cost reductions.

TOOLING CONSIDERATIONS

The PM part design and the complexity of tooling to produce it are closely related. Consultation between designers of the parts and the tools can often lead to minor design changes in the part and more simple, economical tooling.

PM Process Feasibility

The feasibility of making a part by powder metallurgy depends on being able to economically press the metal powder in the die to obtain the desired shape, dimensions, detail, and density. Two major factors in the compacting operation influence or control part design: the flow behavior of metal powders and the pressing action.

Metal powders do not flow hydraulically because of friction between the particles and the dies; therefore, the design should assure that adequate powder will be placed in the die cavity to allow adequate compaction and to attain the desired density. Because metal powders have limited lateral flow, there are also some limitations on the contours that can be produced.

The pressing action is applied only from the top and bottom, largely governing the shape, dimensional details, and length of part that can be made with the desired density. Another consideration of the pressing action is that the part shape must allow ejection from the mold.

Tolerances

The tolerances that can be held in the PM process compare favorably with those for many other conventional parts fabricating processes. In most cases the tolerances can be held closer than in parts made by sand casting, die casting, stamping, and forging. As with other processes, tolerances no smaller than necessary should be specified to reduce production cost. Tolerances on PM parts depend upon a number of factors, including the metal powder and its characteristics, the dimensions and size of the part, tool wear, and the runout tolerance.

Heat treatment. As with wrought materials, heat treatment of PM parts can cause dimensional changes. Size changes tend to be larger in low-density parts because the heat-treating atmosphere penetrates deeply and quickly.

Coining and repressing. The closest tolerances are obtained by coining or repressing parts. Sizing does not necessarily block the pores in medium-density and low-density parts that are to be self-lubricating. An increase in density reduces the oil-holding capacity. Table 9-3 lists typical tolerances for various PM materials when the parts are heat treated and/or sized.

HOW PM PARTS ARE MADE

Typical parts can be produced at rates of several hundred to thousands per hour with the PM process. Although normally associated with high-volume production, PM is also feasible for economic part production in lower volumes. For maximum efficiency, however, lower volume parts should be comparatively simple, permitting low tooling and maintenance costs. Also, extending tooling to produce more than one part, with varying thicknesses, or a common part with different sized holes, can enhance the low-volume economic rationale for PM. Another consideration is that the part design should take advantage of the ability of the PM process to minimize secondary processing operations compared with overall processing by other methods. When these guidelines are followed, PM parts can offer cost and performance advantages in runs as low as 1000 pieces.

Basic PM Steps

The three basic steps for producing conventional density parts by the powder metallurgy process are mixing, compacting, and sintering, shown schematically in Fig. 9-7. For additional information, see Chapter 17, "Powder Metallurgy," in Volume II, *Forming*, of this Handbook series.

Secondary Operations

Generally, PM parts are ready for use after sintering. However, to provide special properties, the parts can be repressed, impregnated, machined, tumbled, plated, or heat treated.

Heat Treating

Like wrought products, PM materials can be annealed, quenched, and surface hardened. For porous parts, a neutral, dry atmosphere should be maintained in the heat-treating furnace. High-density (7.2 g/cm^3 minimum) and infiltrated steel parts can be heat treated by most conventional methods. However, liquids—such as carburizing salts, brine or water, which can become entrapped within the part and cause corrosion—should be avoided with lower density parts. In general, oil should be the fast-quenching medium for low to medium-density parts.

Ferrous PM parts are generally surface hardened by carburizing or carbonitriding, with the depth of case depending on the carburizing medium, time, and temperature. A well-defined, hardened surface is obtained with high-density parts. With lower density parts, rapid carbon diffusion results in through-carburization or a case of indefinite thickness. The porosity of powdered metal enables full-depth hardening throughout the part when desirable.

Machining

Powder metallurgy parts are normally produced to finished dimensions. However, a machining operation is sometimes necessary for special shapes, threads, holes, crossholes, and for closer tolerances. Machining of PM parts and conventional castings is similar, except that materials to improve machinability can be added readily to the powder mix. Small additions of lead, sulfur, copper, or graphite are common in ferrous parts, and lead is frequently added to nonferrous parts. Oil or resin impregnation improves machinability of porous PM parts.

Machining feeds and speeds for high-density PM parts (above 92% of theoretical density) are similar to those for wrought metals. Lower density parts require feed and speed adjustments for optimum results.

Machining of self-lubricating, porous parts should avoid smearing the surface porosity. Use of sharp tools and light cuts in single point machining, such as turning or boring, helps maintain surface porosity. While cutting fluids are preferred for most machining operations, fluid pickup increases with greater part porosity. Ideally, all machining, except grinding, should be done before deburring because retained deburring abrasives can cause excessive tool wear.

Finishing

Practically all common finishing methods apply to PM parts. While procedures in many applications are similar to those for wrought and cast parts, the structural characteristics and porosity, particularly in low-density parts, require some modifications in finishing operations.

TABLE 9-3
Typical Tolerances for PM Parts Processing*

Material	As Sintered	As Sized	As Heat Treated
Brass	±0.0035″ (0.089 mm)	±0.0005″ (0.013 mm)	NA
Bronze	±0.0035″ (0.089 mm)	±0.0005″ (0.013 mm)	NA
Aluminum	±0.002″ (0.051 mm)	±0.0005″ (0.013 mm)	±0.0005″ (0.013 mm)
Iron	±0.001″ (0.025 mm)	±0.0005″ (0.013 mm)	NA
Copper alloy steel	±0.0015″ (0.038 mm)	±0.001″ (0.025 mm)	±0.0015″ (0.038 mm)
Nickel alloy steel	±0.0015″ (0.038 mm)	±0.001″ (0.025 mm)	±0.0015″ (0.038 mm)
Stainless steel	±0.001″ (0.025 mm)	±0.0005″ (0.013 mm)	NA

* Up to 0.50″ (12.7 mm) PM (other than length)
Length tolerance: ±0.004″ (0.102 mm) unless machined or ground
NA = Not applicable

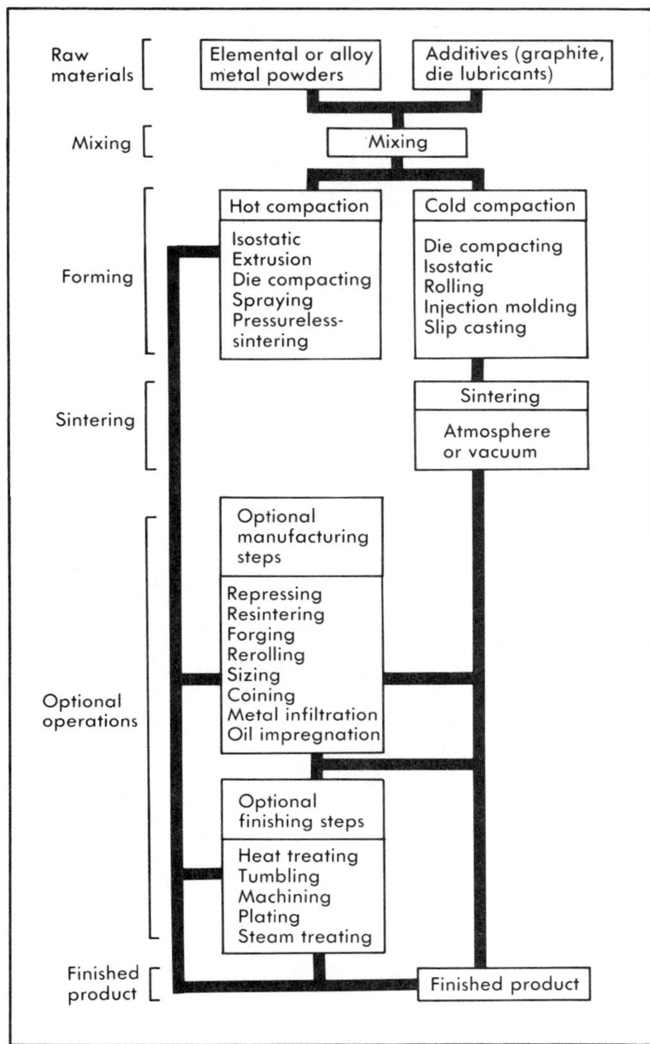

Fig. 9-7 Schematic diagram of the powder metallurgy process.

Deburring. Powder metallurgy parts are rolled in rotating barrels or agitated in vibrating tubs—usually with some form of abrasive media and water—for cleaning, deburring, shine rolling, or burnishing. Rust inhibitors should be added to the water. After the operation, parts may be spun dry; in addition, heat may be used to evaporate water from the pores.

Burnishing. Burnishing can improve part finish and dimensional accuracy, or it can work harden surfaces. Tumbling, roller burnishing, ball sizing, and stick burnishing—using a broach-like tool with buttons or ridges—are common burnishing techniques. Compared to wrought parts, closer tolerances can be held on most PM parts because porosity allows the metal to be displaced more easily. As with machining, closing of desired surface pores should be avoided.

Blueing (Blackening). Ferrous PM parts can be colored by several methods. To give parts indoor corrosion resistance, they can be blackened by heating in a furnace. Oil dipping gives a deeper color as well as slightly greater corrosion resistance. A dry-to-touch oil, which leaves a dry film on the parts, may be used for oil dipping.

Steam treating. As a commonly used process for PM, steam treating increases corrosion resistance of ferrous PM parts. Filling some of the interconnecting pores and much of the surface increases the PM part density and compressive strength. An oxide coating provides additional hardness and wear resistance. However, the process tends to cause a slight size change and makes the parts somewhat less ductile and more difficult to machine.

Welding/brazing. Ferrous PM parts can be welded or brazed. Additionally, some parts can be assembled during sintering, thus often avoiding the need for welding, brazing, or joining by mechanical fastening systems. If welding is needed, however, best results are achieved when the PM part has a density of 6.8 g/cm^3 or higher and carbon content is between 0.2% and 0.8%. Most of the conventional welding methods (TIG, MIG, electron beam, resistance, projection, and friction) are applicable.

Since porosity in PM parts could cause excessive absorption of costly brazing alloy, brazing requires a more specialized procedure. A brazing system designed for PM limits the penetration of the brazing alloy to the immediate area of the adjoining PM surfaces.

Plating. All types of plating in general use—including copper, nickel, chromium, cadmium, and zinc—can be applied on PM parts. High-density (7.2 g/cm^3) and infiltrated parts can be plated using the same methods as for wrought parts. Lower density parts should have the porosity sealed, usually by resin impregnation, to avoid entrapment of plating solutions in the pores. Parts that have been oil-impregnated or oil quenched must have all oil removed from the pores and surface prior to resin impregnation and/or plating. Electroless nickel plating can also be used, and peen (mechanical) plating is applicable to nonimpregnated parts.

HIGH-PERFORMANCE PM PARTS PRODUCTION

The powder metallurgy process has extended into the production of full-density material and parts. Different compacting methods are used, and the new processes may often involve greater emphasis on high-alloy materials. Some of the specialized processes involved in this new and growing technology are hot forging, isostatic pressing, injection molding, and extrusion and rolling.

HOT FORGED PM PARTS

Powder metallurgy components with wrought properties are now available. A popular material is the 4600 series of steels with varying carbon levels and conventional heat treatment when necessary. Various systems within the PM industry produce millions of parts annually, primarily for automotive original equipment.

CHAPTER 9

HIGH-PERFORMANCE PM PARTS

The basic process common to most higher density systems is to manufacture a sintered compact called a preform, heat, and then restrike/forge the preform to the required final density.

Two popular processes are shown in Fig. 9-8. Typical parts made by these processes include a steel chain drive sprocket and the race and cam for an automatic transmission in an automobile.

ISOSTATIC PRESSING

Isostatic pressing is generally used to produce PM parts to near-net sizes and shapes of varied complexity. Unlike conventional press compaction or molding, isostatic pressing is performed in a pressurized fluid such as oil, water, or gas. A flexible membrane or hermetic container surrounds the powder mass and provides a pressure differential between its contents and the pressurizing medium. For further information, see Chapter 17, "Powder Metallurgy," in Volume II, *Forming*, of this Handbook series.

Cold Compaction

Cold or room temperature compaction is carried out in liquid systems at pressures commonly reaching 60 ksi (414 MPa). Metal powder can be packed into complexly shaped rubber or other elastomeric molds before compacting. The resulting green shape, although smaller after compaction, closely duplicates the internal mold configuration. Free of die frictional forces, the powder compact reaches a higher and more uniform density at its external surfaces. Sintering can be performed by any of the conventional processes.

Hot Isostatic Pressing (HIP)

Hot isostatic processing is performed in a gaseous (inert argon or helium) atmosphere contained within the pressure vessel. In most installations, the gaseous atmosphere as well as the part to be pressed are heated by a furnace within the vessel. In some cases, however, heat is supplied externally to the part, which is kept hot by heat shields when it is in the pressure vessel.

Common pressure levels extend upward to 15 ksi (103 MPa) with combined temperatures to about 2300° F (1260° C). Processing volumes currently reach to 4 ft (1.2 m) diameter by 9 ft (2.7 m) long. Higher pressures and temperatures are in use but with considerably reduced work volumes.

Densification to full density is achievable with most materials. The resulting mechanical properties are equivalent to those of wrought parts in similar structural condition. In some materials, the properties of the HIP product are superior because of reduced anisotropy.

HIGH ALLOYS

Powder metallurgy processing of high-alloy materials, notably superalloys, is most common for compaction of near-net shapes and forging preforms for aircraft turbine engines.

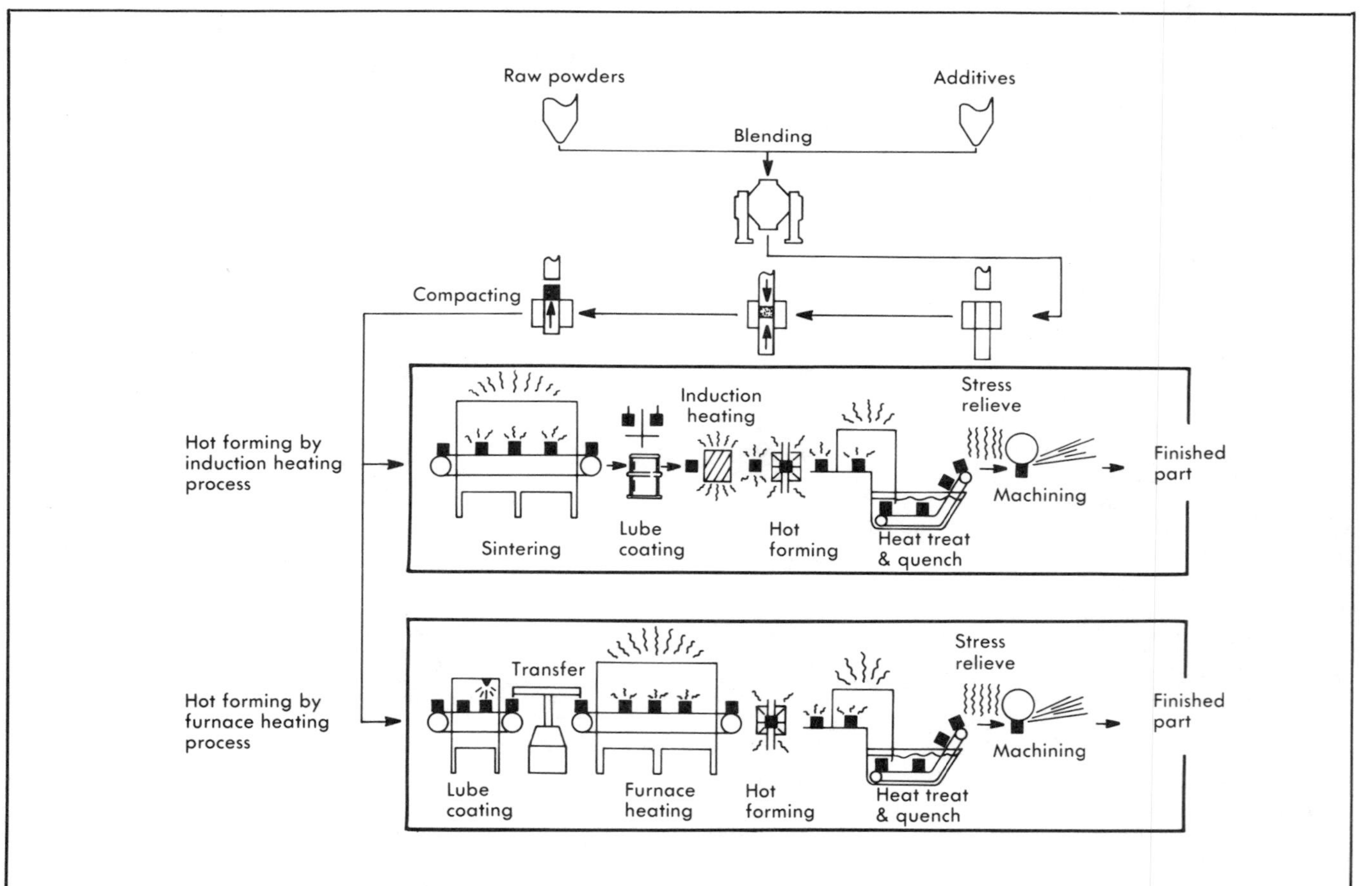

Fig. 9-8 Two versions of the PM hot forming (hot forging) process.

HIGH-PERFORMANCE PM PARTS

Economic benefits have been the prime driving force in the use of PM to process these high-cost alloys. Other benefits of PM processing of high-alloy materials include:

- Fine grain size and metallurgical homogeneity.
- Uniformity of mechanical properties.
- Improved hot workability and machinability.
- Ability to create shapes not practical by processing methods for ingot-based materials.
- Reduction of costly input material and difficult-to-machine excess stock.
- Reduction of processing steps.

Powder metallurgy superalloys are processed from highly controlled, metallurgically clean powders by techniques selected to maintain purity and achieve defect-free products at theoretical density. The alloy powders are produced by inert gas processes that minimize surface oxidation during processing.

The compaction processes in common use are:

- Extrusion.
- Hot isostatic pressing (HIP).
- Consolidation at atmospheric pressure (CAP).

The products produced by these processes can be used directly in machined form for selected applications. Extrusion produces axisymmetric blanks that may require extensive machining; HIP and CAP produce complexly shaped blanks that approximate the final machined shape more closely. In the CAP process, powder is placed in a shaped glass container that is then evacuated and heated at atmospheric pressure. Upon cooling, the glass spalls off, leaving a near-net shaped component ready for machining to final dimensions.

Hot forging has been employed as a secondary metalworking operation. It can change the shape of the billet or preform, and has achieved, in certain alloys, a refinement in the material structure. Forging under isothermal conditions has been most successful in gaining this improvement.

TOOL STEELS

Tool and high-speed cutting grades of steels provide a growing application area of PM technology. This use of PM was fostered by the development of both improved powder manufacturing and processing techniques. Finer grain structure, improved homogeneity, and distribution of secondary and carbide phases are definite advantages of the structure of the gas-atomized powder particles. The common stringering of secondary phases in worked cast material is also eliminated. The improvements in structure are manifested in the improved grindability and wear life of fabricated tools.

Hot isostatic pressing (HIP) has been used extensively in the commercial production of PM tool steels to near-net shapes of full density. Simple cylindrical billets up to 9 ft (2.7 m) in length and hollow cylinders of similar length measuring up to 24" (610 mm) in diameter have been produced commercially. Near-net shapes for the production of hobs and shaper cutters are produced by HIP techniques.

PM INJECTION MOLDING

With this PM process, fine powders, 39-787 + μin. (1-20 + μm), are coated with a thermoplastic to bind the particles into a uniform mass for injection molding. A part with complex geometry can be injection molded at relatively low pressures and temperatures. The parts are formed in a modified plastics injection molding machine at approximately 300° F (150° C) and 1000 psi (6.89 MPa). After forming, the organic material is removed and the green part is sintered. Densities equal to 96% of theoretical can be achieved. Ductility of as-sintered parts is high, with an elongation of 30% not uncommon.

EXTRUDING AND ROLLING

Metal powders can be hot extruded either with or without presintering. Extrusion of unsintered powders starts with a billet, prepared by placing the powder inside a steel or copper can. The assembly is evacuated and sealed, then heated and extruded as a unit.

Powder extrusion has been useful in fabricating parts from steels requiring extreme cleanliness. Another use is in making billets from alloys of magnesium, aluminum, beryllium, nickel, and titanium-aluminum-vanadium.

Another technique proven to be technically feasible is powder rolling. In the process, ferrous, nickel alloy, or aluminum powder is fed between rolls, squeezed into a porous green strip, and sintered.

HOW TO SPECIFY PM PARTS PROCESSING

It is important to discuss a proposed PM part application, design, and processing method with the potential manufacturer. To take advantage of the latest technology when requesting a parts production quotation, accurate and complete product information must be provided. It is helpful to refer to MPIF Standard 35 for PM materials, properties, and specifications. The necessary information typically includes:

1. The initial production quantity, annual usage, and estimate of future needs in order to take advantage of high-volume cost reductions.
2. Detailed drawings of the part, including assembly drawings and samples of existing parts or prototypes. Identify materials that have performed satisfactorily for the application.
3. Can part design be modified without affecting function? If so, where?
4. Will the PM part replace one currently in production, or is this a new application? Is the application commercial, military, aerospace, or medical?
5. Service conditions: heat, moisture, impact, corrosive, etc.
6. Necessary physical, mechanical, corrosion, or special properties (tensile, elongation, hardness, flatness, conductivity, impact, fatigue, etc.).
7. The finish required (plating, oxide coating, surface finish).
8. If any machining operations are required, will they be performed by the PM supplier?

SPECIFYING PM PARTS PROCESSING

9. For bearings, the load, shaft material and finish, speed, and diameter should be noted.
10. For gears, splines, sprockets, etc., specific data are required: (a) number of teeth, (b) theoretical pitch diameter, (c) pressure angle, (d) measurement over wires, (e) tooth thickness, (f) backlash, (g) helix angle, and (d) American Gear Manufacturers Association (AGMA) quality class.

SIZES AND SHAPES

Although there is no known theoretical limit, maximum practical size is governed chiefly by powder characteristics, part density, and available press capacity.

Part Size

The majority of conventionally pressed PM parts range in projected area from about 0.006 to 25 in.^{2} (3.9 mm^2 to 0.016 m^2) and between 1/32 and 6″ (0.8 and 152 mm) in height (direction of pressing). The practical height limit is closer to 3″ (76 mm). For a given metal powder, the lower the density required, the larger the part that can be produced on a particular press.

Parts that are relatively long in the direction of pressing are more difficult to produce with adequate density throughout the total length of the part, particularly in the center section. Another factor that limits part length is the apparent density of metal powders. In general, the compression ratio is at least 2 to 1, meaning that the die depth must be at least twice the height of the pressed part.

Part Shapes

The shapes most suitable for the PM process should have uniform dimensions in the direction of pressing.

Examples. Appropriate shapes include simple cylindrical, square and rectangular shapes, and those with the contour in a plane at right angles to the direction of pressing. Examples are parts with radial projections and contours, like cams and gears, and parts with no changes in thickness, which are relatively simple to press.

Since the tooling is subject to very high stresses, shapes that require fragile tools should be carefully considered. Under normal circumstances, the contour of parts must allow ejection of the green compact from the die with an upward motion of the bottom punch.

Complete spheres cannot be made by the PM process since powders do not flow hydraulically and must be compressed. Spherical PM parts are designed with straight or flat areas around the circumference. Parts that must fit into ball sockets are repressed after sintering to produce a more spherical shape. Hemispheres, such as those used in automotive ball joints, can be readily compacted. Spherical depressions up to a hemisphere contour are also possible.

Multilevel shapes. Simple variations or levels in part thickness can often be provided with face forms on upper and/or lower punches. Applications include such details as countersinks, steps not exceeding 15% of the overall part thickness, numbering or lettering, and other similar features.

More complex, multiple-motion tooling is required to maintain uniform density throughout parts with pronounced variations in steps or levels. Sophisticated pressing equipment with the ability to move tooling components independently is used to assure a proper vertical transfer of the metal powder before the compacting sequence begins.

PROCESS SELECTION

New PM applications can be evaluated with minimum dollar expenditure by testing parts machined from recommended PM sample material. There is no low-cost temporary tooling for die pressed parts. Materials for machining prototypes can be supplied by PM fabricators. Tooling does not have to be committed until after prototype samples are tested.

Material Variation

Customers frequently request that the PM material have properties that are identical to the material specified. Most properties usually can be met; but identical properties are not always possible, often may not be necessary, and can increase cost.

The powder metallurgy process can allow some cost variation if specific part requirements are not rigidly defined. Widely differing costs may result from differences in quality levels, changes in tolerance or design, or failure to specify minimum tensile properties. There may be more than one MPIF standard material that can satisfy functional requirements.

Quality Requirement

Due to high-volume production capability, repeatability, and attractive cost benefits, an acceptable quality level (AQL) should be agreed upon with the PM parts manufacturer. To achieve an acceptable quality level for critical dimensions or functional tests (torque, bending, etc.), it is recommended that identical sets of inspection fixtures, gages, etc. be provided by the customer (one to the PM supplier and one retained by the customer). This procedure will ensure that both supplier and customer evaluate the parts in the same manner. The acceptable quality level can significantly affect cost of a PM part.

More than a dozen PM processes are available for consideration. Some processes produce PM billets or preforms. Others consolidate the powders to full density in net shape. Selecting the appropriate production method requires consideration of metal properties, performance required, functional quality level, production quality, and cost.

References

1. Metal Powder Industries Federation (MPIF), *P/M Design Guidebook* (Princeton, NJ: 1983).

Bibliography

Balshin M. Yu, Kiparisov, S. S., and trans. I. V. Savin. *General Principles of Powder Metallurgy*. Moscow: MIR Publishers, 1980.

Bradbury, Samuel, ed. *Source Book on Powder Metallurgy*. Metals Park, OH: American Society for Metals, 1979.

Hirschhorn, Joel S. *Introduction to Powder Metallurgy*. New York: American Powder Metallurgy Institute, 1969.

Hoeganaes Corp. *Creating with Metal Powders*, 5th ed. Riverton, NJ: 1979.

James, W. Brian, and Powell, Robert A. "Powder Metallurgy Moves Ahead." *Machine Design* (Sept. 22, 1983), pp. 54-58.

Lenel, Fritz V. *Powder Metallurgy Principles and Applications*. Princeton, NJ: Metal Powder Industries Federation, 1980.

Lenel, F. V., and Ansell, G. S. "The State of the Science and Art of Powder Metallurgy." *Journal of Metals* (February 1982), pp. 17-27.

Materials Standards and Specifications, MPIF Standard 35. Princeton, NJ: Metal Powders Industries Federation, 1981.

HEAT TREATMENT

SECTION

2

HEAT TREATMENT OF STEEL

Heat treatment is an operation or combination of operations involving the controlled heating and cooling of solid metals and alloys to obtain a required microstructure with resultant desired properties. The properties desired vary widely, depending upon the applications for the metals. With current know-how and equipment, heat treatment is very versatile and can provide many different and predictable properties or combinations of properties.

Most heat treating operations can be classified into the two following types of processes:

1. Processes such as through hardening and surface (case) hardening that increase the strength, hardness, and toughness of metals. Surface hardening processes are discussed in Chapter 11, "Surface (Case) Hardening," and Chapter 13, "Selective Hardening," of this volume.
2. Processes such as annealing and normalizing that decrease the hardness of metals in order to improve their homogeneity, machinability, and formability, or to relieve stresses.

While heat treatment is applied to a wide variety of metals, this chapter is confined primarily to the heat treatment of steels. The heat treatment of cutting tool materials is discussed in Chapter 3, "Cutting Tool Materials," of Volume I, *Machining*, and die and mold materials is covered in Chapter 2, "Die and Mold Materials," of Volume II, *Forming*, of this Handbook series. The heat treatment of other common metals—including cast irons, cast steels, stainless steels, and nonferrous materials—is discussed in Chapter 14, "Heat Treatment of Other Metals," in this volume.

Proper heat treatment of any metal requires the combined input of design engineers, metallurgists, and manufacturing engineers. Essential information required for optimum heat treatment includes the following:

1. The composition and condition of the metal to be heat treated, and the intended applications.
2. The critical time-temperature transformation relationship for the specific metal to be heat treated.
3. The response of the metal to quenching and the method of cooling or quenching to be used.
4. The desired hardness and/or strength.

CHAPTER CONTENTS:

GLOSSARY OF HEAT TREATING TERMS

aging A time-temperature dependent change in the properties of certain metals occurring at room or slightly elevated temperatures following hot working or cold working, or following quenching after thermal treatment.

age hardening Hardening by aging. Also called precipitation hardening.

age softening A decrease in strength and hardness that occurs at room temperature in certain strain-hardened alloys, especially those of aluminum.

artificial aging Aging above room temperature.

interrupted aging Aging at two or more different temperatures, with cooling to room temperature after each heating.

natural aging Aging at room temperature.

quench aging Aging by rapid cooling after solution heat treatment.

step aging Aging at two or more different temperatures, without cooling to room temperature after each step.

air-hardening steel A steel containing sufficient carbon and other alloying elements to harden fully during cooling in air or other gaseous mediums from a temperature above its transformation range.

alpha iron The body-centered cubic form of pure iron. (See ferrite.)

annealing A heat treatment process to reduce hardness or brittleness, relieve stresses, improve machinability, facilitate cold working, or produce a desired microstructure or properties. The process consists of heating to a suitable temperature, which is dependent upon the type of annealing, followed by slow cooling. Common types of annealing include:

black annealing Box, close, or pot annealing of ferrous alloy sheet, strip, or wire.

blue annealing A process for softening hot rolled ferrous sheet by heating in an open furnace and then cooling in air. The formation of a bluish oxide on the surface is incidental.

*Contributors of sections of this chapter are: **Gary E. Armour**, Project Engineer/Metallurgist, Sunbeam Equipment Corp.; **D. Michael Donovan**, Director of Technology, Lindberg Corp.; **Dr. Robert W. Foreman**, Director of Research and Development, Park Chemical Co.; **William A. Leeper**, Chief Metallurgical Engineer, GH-Thornhill Craver; **Professor Karl B. Rundman**, College of Engineering, Dept. of Metallurgical Engineering, Michigan Technological University; **Daniel S. Zamborsky**, Manager of Metallurgy, Bendix Automation Group, Research Div., Bendix Corp.*

*Reviewers of sections of this chapter are: **Zoltan Ambrus**, Atmosphere Systems Group, Applied Research and Development, Industrial Gas Div., Air Products and Chemicals, Inc.; **Richard B. Bertolo**, Section Chief, Bar and Tubular Products, Metallurgical Div., Republic Steel Corp.; **C. R. Campbell**, General Manager, Gas Atmospheres Div., Modern Equipment Co.; **D. Michael Donovan**, Director of Technology, Lindberg Corp.; **Jon L. Dossett**, Vice President, Heat Treat Corporation of America; **Dennis J. Giancola**, Heat Treating Specialist, Industrial Market Development Dept., East Ohio Gas Co.; **G. G. Hoeft**, Manufacturing Engineering, Caterpillar Tractor Co.;*

CHAPTER 10

HEAT TREATING TERMS

box, close, or pot annealing A process of slow heating in a sealed container to minimize oxidation, followed by slow cooling. (See black annealing.)

bright annealing Annealing in a furnace containing a protective atmosphere to minimize surface oxidation and prevent discoloration of the bright surface.

cycle annealing Annealing with a controlled time-temperature cycle to provide a specific microstructure or properties.

flame annealing Annealing (softening) by heat from a high-temperature flame.

full annealing A term denoting annealing to produce minimum strength and hardness. For precise meaning, the condition of the starting material and the time-temperature cycle should be specified. The time-temperature cycle must produce complete transformation to austenite.

graphitizing Annealing of ferrous alloys to precipitate some or all of the combined carbon as graphite; generally applicable only to cast irons.

isothermal annealing Austenitizing a ferrous alloy and then cooling to and holding at a temperature at which the austenite transforms into a softer ferrite and carbide microstructure or a pearlite microstructure.

malleablizing Annealing white cast iron to transform some or all of the combined carbon to graphite.

partial annealing A process for stress relieving cold worked material or for reducing its strength.

process annealing Processes for softening metals to facilitate further cold working. The subcritical temperatures usually employed do not involve austenitization.

quench annealing Annealing of austenitic ferrous alloys by solution heat treatment (heating followed by rapid quenching).

recrystallization annealing Annealing cold worked metals to provide refined grain structures, primarily new, strain-free grains.

spheroidize annealing (spheroidizing) A process to produce a rounded or globular form of carbide in the structure. This process generally requires austenitization, followed by heating to a subcritical temperature for an extended period of time.

subcritical annealing Annealing of ferrous alloys at temperatures below those at which austenite begins to form. (See also partial and process annealing.)

austempering See quenching.

austenite A solid solution of carbon and other elements in face-centered cubic iron (gamma iron).

austenitizing Forming austenite by heating ferrous alloys into their transformation ranges (partial austenitizing) or above their transformation ranges (complete austenitizing).

bainite A transformation product (an aggregate of ferrite and cementite) developed from austenite at temperatures between those where pearlite and martensite form.

carbonitriding A surface hardening process consisting of heating ferrous metals in an atmosphere that permits absorption of both carbon and nitrogen into a shallow surface layer.

carburizing A process for introducing carbon into ferrous metals by heating above the transformation temperature while in contact with a carbonaceous medium (solid, liquid, gas, or plasma). Subsequent quenching produces a hardened surface. Homogeneous carburizing is the use of a carburizing process to convert a low-carbon ferrous alloy to one of uniform and higher carbon content throughout the section.

case hardening See surface hardening.

cementation A process for introducing elements into the surfaces of metals by high-temperature diffusion.

cementite A compound of iron and carbon known as iron carbide (Fe_3C), consisting of small, hard particles distributed throughout the ferrite.

chromizing A surface treatment at elevated temperature in which an alloy is formed by the diffusion of chromium into the base metal.

cold treatment Exposure to suitable subzero temperatures for the purpose of obtaining desired conditions or properties such as dimensional or microstructural stability.

critical cooling rate The minimum rate of cooling required to prevent the undesirable transformation of austenite, usually the transformation of austenite to pearlite.

critical temperature ranges See preferred term "transformation ranges."

cyaniding A surface hardening process consisting of heating ferrous metals to above their critical temperatures in molten salt containing cyanide, resulting in the simultaneous absorption of carbon and nitrogen. Subsequent quenching produces a hardened surface.

decarburization The loss of carbon from the surface of a ferrous metal as the result of heating in a medium, such as air, that reacts with the carbon.

drawing See tempering.

eutectic An alloy having the composition with the lowest melting point on an equilibrium diagram. Also, a minimum transformation temperature effecting a phase change from liquid to solid.

eutectoid A plain carbon steel containing 0.80% carbon. Also, a minimum temperature at which a phase change takes place in the solid state.

ferrite A solid solution of carbon in body-centered cubic iron (alpha iron), limited to a maximum carbon content of 0.02%.

flame hardening A surface hardening process consisting of heating ferrous metals with a flame to above the Ac_3 temperature, followed by quenching.

gamma iron The face-centered cubic form of pure iron. (See austenite.)

hardenability Comparative ability of metals to be hardened. For steels, hardenability refers specifically to the ability of martensite to be formed to a certain depth.

homogenizing A high-temperature heat treatment process to decrease or eliminate chemical segregation by diffusion.

hydrogen embrittlement A condition of low ductility in metals resulting from the absorption of hydrogen.

hypereutectoid Plain carbon steels containing more than 0.80% carbon.

Reviewers, cont.: ***L. E. Jones****, Division Manager, Lindberg Heat Treating Co.;* ***F. D. Lauricella****, Marketing Manager, Protective Atmospheres, Linde Div., Union Carbide Corp.;* ***Inna Lazarev****, Manager, Heat Treating Technical Service, E. F. Houghton & Co.;* ***William A. Leeper****, Chief Metallurgical Engineer, GH-Thornhill Craver;* ***Norman C. McClure****, Chief Metallurgist, Commercial Steel Treating Corp.;* ***Edward R. Mueller****, Director of Research, Tenaxol, Inc.;* ***Elliot S. Nachtman****, Manager, Industrial Technology Div., Tower Oil & Technology Co.;*

HEAT TREATING TERMS

hypoeutectoid Plain carbon steels containing less than 0.80% carbon.

induction hardening A surface hardening process consisting of heating ferrous metals by electromagnetic induction to above the Ac_3 temperature, followed by immediate quenching.

Jominy end quench A standardized technique, discussed subsequently in this chapter, that produces a hardness gradient as a measure of the hardenability of the steel.

ledeburite The eutectic of the iron-carbon system, the constituents being austenite and cementite.

malleablizing See annealing.

maraging A hardening treatment for maraging steels to precipitate one or more intermetallic compounds in a matrix of low-carbon martensite.

marquenching (martempering) See quenching.

martensite A transformation product of austenite having a body-centered, tetragonal crystal structure that appears needle-like in the microstructure.

McQuaid-Ehn test A test identifying grain size when metals are heated to a specific austenitic temperature.

neutral hardening The heat treatment of steel parts without affecting their surfaces with respect to carburization or decarburization.

nitriding A surface hardening process consisting of heating a ferrous alloy to a subcritical temperature in an atmosphere of ammonia or in contact with nitrogenous material. Hardening is attained by the formation of nitrides. Quenching is not required.

nitrocarburizing See carbonitriding.

normalizing A heat treatment process for ferrous metals consisting of heating to a temperature above the transformation range and then cooling in air. The process refines the microstructure and provides a carbide size and distribution favorable for subsequent heat treatment. Normalizing is also frequently used to treat large sections to provide a moderate increase in strength without an undue increase in stress and is often followed by tempering to reduce stresses further and to slightly modify the mechanical properties.

patenting A heat treatment process for wire and rod prior to drawing or between drafts. It consists of heating to a temperature above the transformation range and then cooling in air or in a bath of molten lead or salt.

pearlite A lamellar aggregate of ferrite and cementite resulting from the transformation of austenite at temperatures above the bainite range but below the temperatures at which transformation of ferrite to austenite begins during heating.

precipitation hardening See aging.

quenching Rapid cooling after heating by contact with liquids, gases, or solids.

austempering Quenching of ferrous alloys into a molten salt or lead bath from their austenitizing temperatures to temperatures below that of pearlite formation and above that of martensite formation, and maintaining at temperature until transformation to bainite is complete.

differential quenching A quenching process in which only certain portions of the heated part are quenched and hardened.

fog quenching Quenching in a vapor or mist.

hot quenching Quenching in a medium having a temperature substantially higher than room temperature. (See also isothermal quenching.)

interrupted quenching A quenching process in which parts are removed from one quench and placed in a second quench having a different cooling rate.

isothermal quenching The cooling of heated metal after austenitizing and holding at a reduced, constant temperature in the pearlite transformation range of 1000-1325° F (540-720° C). (See also hot quenching.)

marquenching (martempering) Quenching of austenitized ferrous metals in a medium having a temperature just above that at which the transformation of austenite to martensite starts, holding in the quench bath until the temperature is uniform throughout the metal, and then cooling slowly—generally in air—through the temperature at which transformation of austenite to martensite is substantially completed.

selective quenching See differential quenching.

slack quenching Incomplete hardening of steel as the result of quenching from the austenitizing temperature at a rate slower than the critical cooling rate for martensite formation.

spray quenching Quenching in a spray of liquid.

time quenching Interrupted quenching during which the times in the quenching mediums are controlled.

siliconizing Diffusing silicon into solid metal, usually steel, at an elevated temperature.

sintering A heat treatment process by which the adjacent surfaces of powder metal particles in a compacted part are bonded to develop strength and adhesion.

soaking Holding a load of parts in a furnace at a fixed temperature for sufficient time to allow equalization of the temperature throughout the load.

solution heat treatment A treatment in which an alloy is heated to a suitable temperature and held at this temperature for a sufficient length of time to allow a desired constituent to enter into solid solution, followed by rapid cooling to hold the constituent in solution.

spheroidite An aggregate of iron or alloy carbides of essentially spherical shape dispersed throughout a matrix of ferrite.

spheroidizing See annealing, spheroidize.

stress relieving A process to reduce internal residual stresses in metal parts by heating the parts to a suitable temperature, holding for a proper time at that temperature, and then cooling slowly to minimize the development of new residual stresses.

surface hardening Various processes for producing surface layers on ferrous alloys that are harder or more wear resistant than the softer, tougher cores.

***Reviewers, cont.:* *Dr. Harbhajan S. Nayar*, *Technical Center, BOC Group;* *Morton Perle*, *Manager of Gas Applications, Industrial Gasses Div., Liquid Air Corp.;* *Thomas R. Risbeck*, *Metallurgical Div., Research Center, Republic Steel Corp.;* *James A. Smith*, *Vice President, 3X Instruments and Tooling;* *J. F. Warchol*, *Technical Manager, Organic Additives, E. F. Houghton & Co.;* *Kenneth R. Watts*, *Heat Treat Staff Engineer, Manufacturing Engineering, Caterpillar Tractor Co.*

temper brittleness Loss of notch toughness that results when certain steels are held within, or cooled slowly through, a specific range of temperatures below the transformation range, usually 600-950° F (315-510° C).

tempering Reheating of previously hardened or normalized material for the purpose of decreasing the hardness, minimizing stresses, improving ductility, and increasing toughness.

transformation ranges The temperature range within which austenite forms when heating and the temperature range within which austenite disappears when cooling ferrous metals. The two ranges may overlap, but never coincide. Limiting temperatures of the ranges vary with the composition of the material and the rate at which the temperature changes.

transformation temperatures The temperatures at which changes in phase occur. The term is sometimes used to denote the limiting temperatures of transformation ranges. The symbols for transformation temperatures of primary interest with respect to iron and steels are as follows:

Ac_{cm} The temperature at which the solution of cementite in austenite is completed during the heating of hypereutectoid steels (steels containing more than 0.80% carbon).

Ac_1 The temperature at which transformation of ferrite to austenite begins during heating.

Ac_3 The temperature at which transformation of ferrite to austenite is completed during heating.

Ar_1 The temperature at which transformation of austenite to ferrite or to ferrite plus cementite is completed during cooling.

Ar_3 The temperature at which transformation of austenite to ferrite begins during cooling.

M_s The temperature at which transformation of austenite to martensite begins during cooling.

M_f The temperature at which transformation of austenite to martensite is substantially completed during cooling. Note: All these changes (except the formation of martensite) occur at lower temperatures during cooling than during heating and depend upon the rate of change of the temperature.

HEAT TREATMENT PRINCIPLES

The three major operations performed in the hardening of steel by heat treatment are: austenitizing, equalizing, and cooling.

AUSTENITIZING

As steel is heated and cooled, its structure changes in certain predictable steps that must be recognized for each alloy. Initially, steel that has not been previously heat treated is usually composed of a mixture of ferrite and carbides, often present in a lamellar microstructure known as pearlite.

When steel with this ferrite-pearlite structure is heated, it reaches a temperature at which the carbides in the lamellar pearlite begin to dissolve into the iron. As the temperature is raised, more of the carbides are dissolved until the steel reaches a point at which all the carbides are dissolved and the steel consists completely of a solid solution of carbon in iron called austenite. The temperature at which pearlite begins to transform into austenite is identified as the lower critical temperature, Ac_1, the temperature at which the steel becomes composed completely of austenite is called the upper critical temperature, Ac_3, and the temperature range between is the critical range or transformation range for the particular alloy.

The lower critical temperature is shown by the line A_1 in the iron-carbon equilibrium diagram (see Fig. 10-1). In actual practice, this temperature varies slightly depending upon whether the pearlite is beginning to be transformed to austenite or the austenite has completed transformation to pearlite—in other words, whether the steel is being heated or cooled. This difference is designated by the letters Ac_1 upon heating and Ar_1 upon cooling. These temperatures vary with the chemical composition of the steel.

The upper critical temperature varies in the same way. Shown as A_3 in the equilibrium diagram, it is known as Ac_3 upon heating and Ar_3 upon cooling. It is also designated as A_{cm} when the carbon content of the steel is above 0.80%. The temperature at which the steel is completely converted to austenite or at which pearlite just begins to form is partially dependent upon the alloying element content of the steel but primarily upon the carbon content.

As shown in Fig. 10-1, a plain carbon steel of 0.80% carbon will have the lowest full transformation temperature because, at this composition, the upper critical temperature A_3 is the same as the lower critical temperature A_1. This 0.80% carbon composition is called eutectoid steel; it is 100% pearlite below the transformation temperature and is characterized by the complete change from pearlite to austenite (and back to pearlite) at a single temperature. All other compositions of various percentages transform over a range of temperatures.

Hypoeutectoid Steels

Steels containing less than 0.80% carbon are called hypoeutectoid steels. Upon heating, the pearlite in such steels begins to transform to austenite at A_1; but because of the excess ferrite (alpha iron) in the composition, the transformation is not completed until the temperature reaches A_3. Holding at temperatures above the upper critical temperature increases the rate at which carbon and other alloying elements go into solution, but it also increases the austenitic grain size. Although large-grained steels harden to a greater depth, large grains decrease the steel's fatigue and rupture strengths. If the steel is held below A_3, some of the ferrite will not go into solution. Because the undissolved ferrite contains only a minimal amount of carbon, it does not harden when cooled, and the steel cannot reach its maximum hardness.

Hypereutectoid Steels

Reaching a full solution is not as great a problem with hypereutectoid steels (steels with a carbon content greater than 0.80%) because the undissolved materials are cementite (iron carbides and other carbides), which is harder than ferrite.

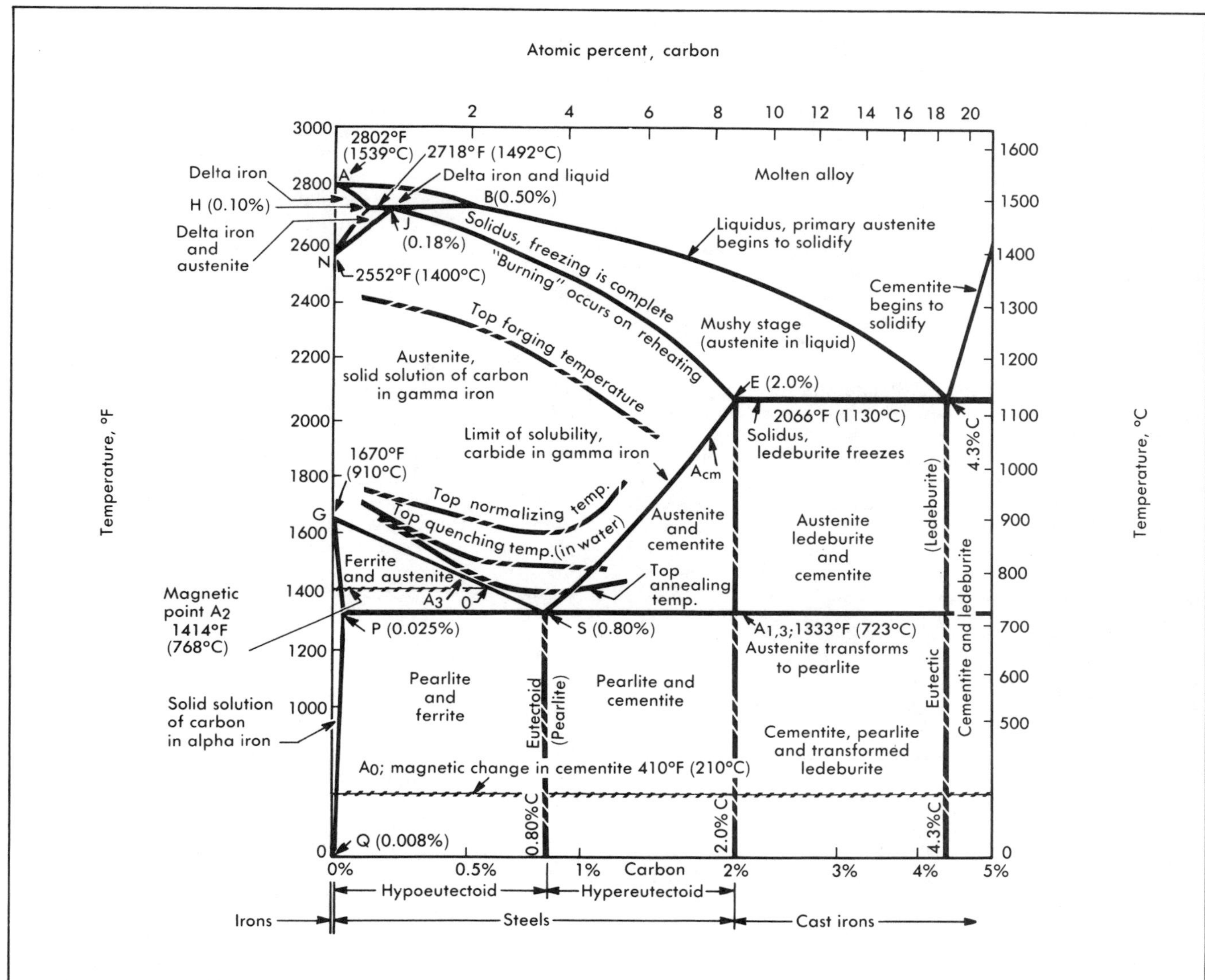

Fig. 10-1 The iron-carbon phase diagram. The critical temperature lines A_1, A_2, and A_{cm} represent transformation upon both heating and cooling. Because of the hysteresis of steel transformation and the effects of elements other than carbon, the transformations upon heating and cooling actually occur at different temperatures.

However, to achieve maximum depth of hardening, the steel must be held above the upper critical temperature (A_{cm}). The steepness of the A_{cm} line means that, for steels of 1.00% carbon or more, rapid growth of large grains will occur. Therefore, a compromise temperature between the upper and lower critical temperatures is generally chosen in order to leave some carbides to retard austenite grain growth and to serve as sites at which phase changes can initiate during cooling.

EQUALIZING

The rate of heating is usually not significant unless the steel is highly stressed initially, and then it should be heated slowly. Preheating in another furnace is often necessary during production when a furnace temperature cannot be altered because other parts are being processed at the same time or because productivity would decline. For equalizing or soaking (the operation that holds the steel at Ac_3, the upper critical temperature, until all sections are uniformly at heat), an adequate soak time is generally about ½-1 hour for each inch (25 mm) of the thickest cross section.

QUENCHING

Quenching, or the rate of cooling, is important in heat treating operations because it dictates the structure and properties of the steel. This present discussion is limited to the microstructural transformations during cooling; details of the various processes and media for quenching various steels are presented later in this chapter.

Pearlite Formations

Upon very slow cooling, ferrite begins to form at the austenite grain boundaries when the temperature reaches the A_3

line in the equilibrium diagram (Fig. 10-1) and continues to grow in amount until the temperature reaches the A_1 line. While the temperature is changing and ferrite is forming, the carbon content of the austenite increases from its nominal value to about 0.80%. The A_{cm} line also represents the saturation limit of carbon in the austenite. Therefore, in steels having a carbon content greater than 0.80%, cementite will precipitate on cooling, and the carbon content of the austenite will again approach 0.80% as the temperature approaches 1333° F (723° C).

This decomposition of austenite to proeutectoid ferrite or carbide may occur when equilibrium conditions are approached by slow cooling. The nonequilibrium condition associated with the hardening of steel by quenching can best be illustrated by considering the isothermal decomposition of austenite at temperatures below 1333° F (723° C): a number of thin wafers of a 0.80% carbon steel are heated to a temperature at which their microstructure becomes 100% austenite, and are then quenched in a molten-lead bath maintained at some temperature T_1, which is below 1333° F. Individual wafers are next removed from the lead bath after varying time intervals and are then quenched in water. This process is illustrated in Fig. 10-2. The time period P_1 is not long enough for the austenite to begin to decompose into ferrite and cementite; so on subsequent quenching to room temperature, the first wafer will show a structure of 100% martensite. Time period P_2 is long enough for some austenite to transform to pearlite, and time periods P_3 and P_4 show an increasing amount of pearlite, while P_5 is long enough for all the austenite to transform to pearlite. This process depends upon nucleation and growth, with the nuclei of pearlite forming at the austenite grain boundaries. As the temperature T_1 is lowered, the pearlite becomes increasingly finer in texture.

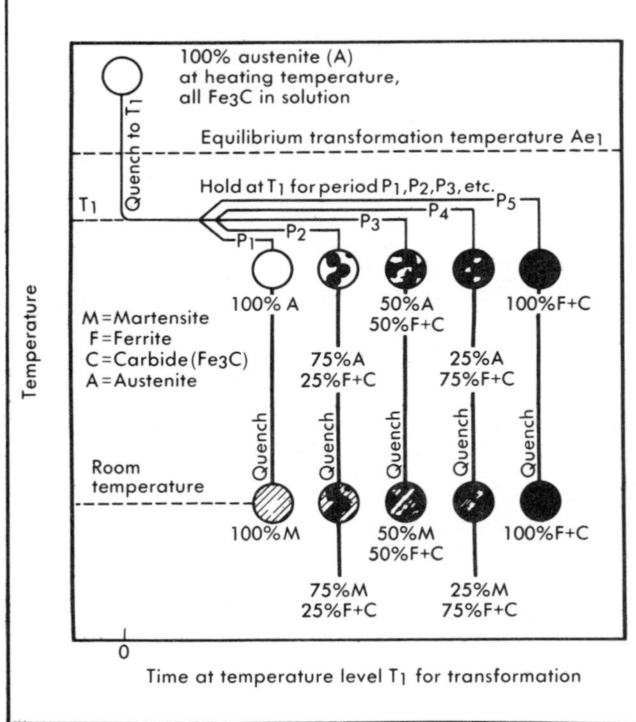

Fig. 10-2 Different phase percentages in steel for increased isothermal time periods. (*Pitman Publishing Corp.*)

Repeating this experiment at various temperature levels results in the accumulation of data for the beginning and ending of austenite decomposition at those temperatures. The data is then plotted as a conventional time-temperature transformation (TTT) diagram, sometimes called an S-curve. The TTT diagram shown in Fig. 10-3 is schematic for carbon steels and includes the proeutectoid line for ferrite and cementite, which applies to steels of less than or greater than 0.80% carbon, respectively. Such TTT diagrams are available for various steels and are used to roughly predict structures formed and mechanical properties obtained by isothermal heat treatment. It is important to note that TTT diagrams are drawn from isothermal transformation data. Since most parts are not cooled in accordance with TTT diagrams, better guides are the so-called continuous-cooling transformation diagrams.

Each steel has a temperature range in which transformation takes place most rapidly. That range of the TTT diagram is often referred to as the nose of the TTT curve. Above or below this rapid transformation range, the time required for transformations is considerably longer. In hardening steel, it is essential to quench at such a rate that transformation at a higher temperature is avoided. Quenching to a temperature above that of the nose of the curve results in a structure (after completion of transformation and subsequent cooling to room temperature) that is relatively soft compared to the martensitic product. However, as T_1 decreases to the nose of the curve, the pearlitic structure formed becomes finer and therefore much stronger.

Bainite Formation

The proeutectoid line joins the beginning of the pearlite transformation line at approximately 1000° F (540° C). This means that only pearlite would be present when fully transformed at this temperature, regardless of the carbon content of the steel. The transformation of austenite by isothermal decomposition below the knee of the diagram, 1000° F, results in the formation of bainite, a microstructure containing ferrite and carbide phases that looks distinctly different from pearlite. The structure of bainite is said to be acicular and like pearlite; the degree of fineness of the bainite structure increases as the temperature of formation is lowered from 1000° F. Pearlite is thought to be nucleated by cementite and bainite is thought to be nucleated by ferrite. Like pearlite, as the fineness of the bainite structure increases, so does the strength of the bainite.

Martensite Formation

As the temperature of the quenching bath is lowered, a temperature level is reached at which austenite will undergo a spontaneous transformation to martensite—a single-phase, body-centered, tetragonal structure. Martensite continues to form as the temperature is lowered, forming athermally; and it will not form isothermally. The temperature at which martensite first begins to form from austenite is termed the M_s point, and the temperature at which the austenite is fully transformed to martensite is called the martensite finish or the M_f point. The M_s temperature depends upon the composition of the steel and can be estimated in degrees Fahrenheit from the following calculation:

$$M_s = 1000 - (650 \times \%C) - (70 \times \%Mn) - (35 \times \%Ni) - (70 \times \%Cr) - (50 \times \%Mo) \quad (1)$$

The M_f temperature cannot be determined from the composition of the steel, but subtracting 385° F from the calculated M_s temperature will give an estimated working guide for that point.

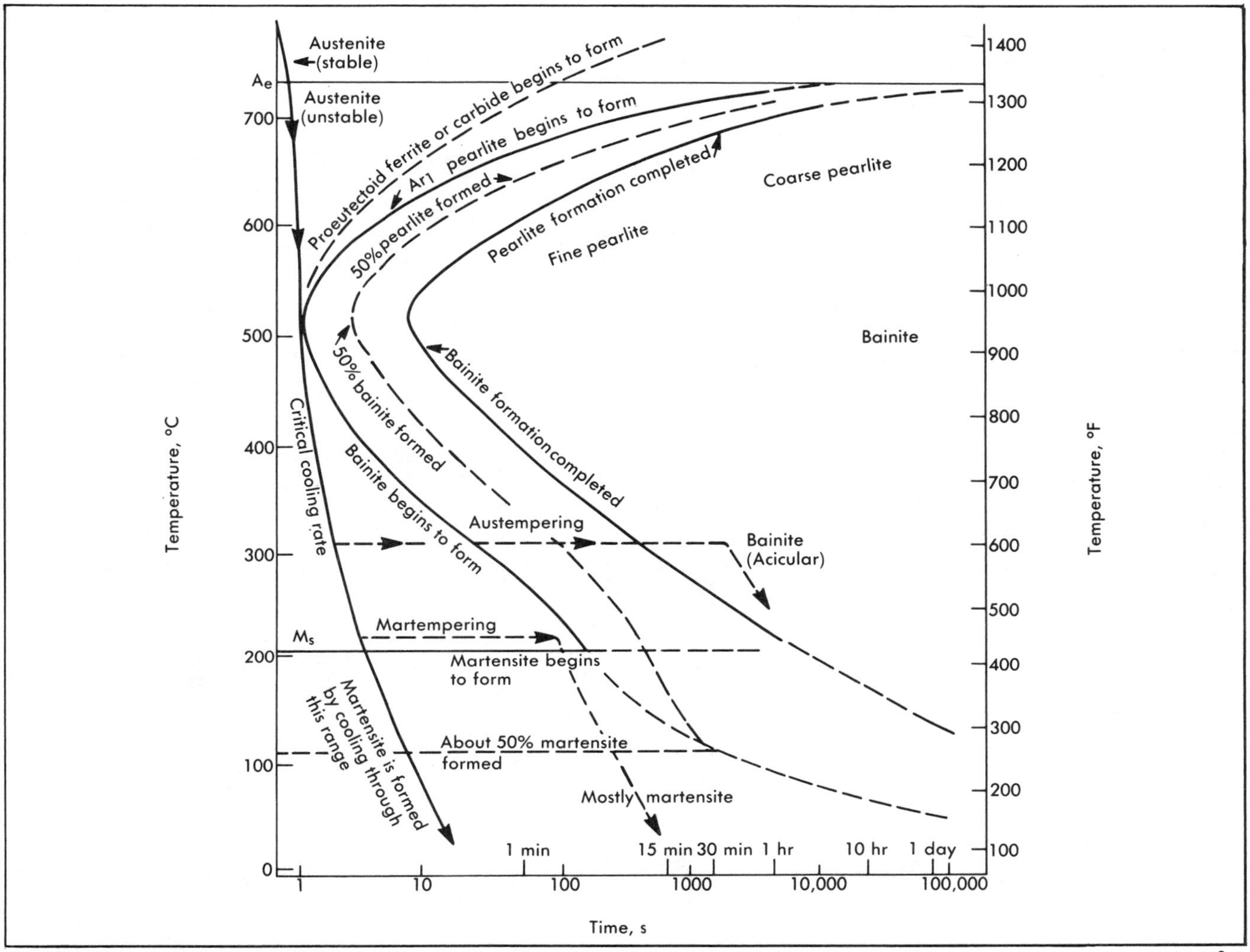

Fig. 10-3 Time-temperature transformation diagram for carbon steel showing critical cooling rate curve, M_s temperature, and temperatures for austempering and martempering. (*Pitman Publishing Corp.*)

The transformation of austenite to martensite depends not upon carbon diffusion, as is the case for pearlite and bainite formation. Instead, a temperature level is attained where the face-centered, cubic lattice structure of austenite transforms to a more stable type of lattice—the tetragonal, body-centered type of martensite, in which the carbon atoms are trapped as an atomic dispersion in its interstices. In this transformation, atomic movements are small and coordinated in a way which can be described as a shear process.

The primary purpose of quenching a steel is to produce martensite. Therefore, it is necessary to employ fast enough cooling rates to prevent the steel from dwelling in the higher temperature ranges where pearlite and/or bainite will form. Alloying elements shift the TTT diagrams to the right, indicating longer times for the beginning of transformation and permitting slower cooling rates for the formation of 100% martensite.

The precise hardness of martensite is apparently influenced primarily by carbon content. As shown in Fig. 10-4, a maximum hardness of about R_C65 may be attained under proper cooling conditions when the carbon content is in excess of about 0.55%.

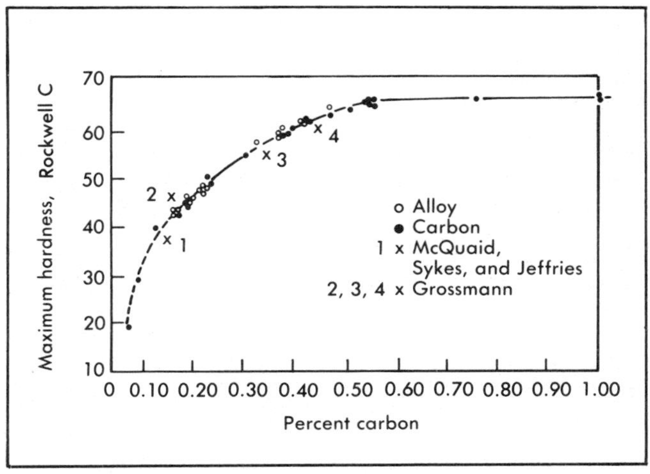

Fig. 10-4 The maximum as-quenched hardness of steel as influenced by carbon content.[1]

HARDENABILITY OF STEELS

Hardenability is the relative ability of ferrous metals to harden. It is the property that determines the depth and distribution of hardness induced by quenching or the size of a workpiece that can be hardened under given cooling conditions. It does not refer to the maximum hardness that can be attained in a given steel.

Hardenability is commonly measured as the distance below a quenched surface where the metal has a specific hardness or a specific percentage of martensite in the microstructure, as discussed later in this section. Metals having low hardenability—called shallow hardening—require faster cooling rates and can only be hardened to relatively shallow depths. Those with high hardenability can be hardened more deeply or completely through the material with slower cooling rates.

EFFECT OF CARBON AND ALLOY CONTENT

Unalloyed iron is soft and ductile and cannot be hardened by heat treatment. The maximum as-quenched surface hardness attainable in any steel is dependent primarily upon the carbon content of the steel and the cooling rate. The more carbon present in the steel, the higher hardness possible.

Hardenability, however, is largely governed by the nature and percentage of alloying elements, as well as the carbon content, grain size of the austenite, time and temperature during austenitizing, and prior structure.

Low-carbon, unalloyed steels—those generally with less than 0.20% carbon—are generally not heat treated. Exceptions exist when carbon is added to the surface, as discussed in Chapter 11, "Surface (Case) Hardening," of this volume. Medium-carbon, unalloyed steels (containing 0.30-0.60% carbon) and high-carbon steels (containing over 0.60% carbon) generally require rapid cooling rates to produce optimum mechanical properties. The addition of small amounts of boron (0.0005-0.003%) to carbon steels, particularly those having carbon contents ranging from 0.20 to 0.40%, increases their hardenability.

All alloying elements except cobalt and vanadium increase the hardenability of steels. Alloy steels transform more slowly from austenite than unalloyed steels, thus permitting an increased hardening effect with slower quenching rates.

Hardenability requirements depend upon many factors, including the specific material to be heat treated and the size, design, and application of the workpiece. For highly stressed parts, through hardening, followed by tempering, is generally preferred. For parts stressed primarily at or near their surfaces, or for applications requiring wear resistance or resistance to shock, shallow or surface hardening is often satisfactory.

Lean steels are generally carbon steels that may be water quenched. Alloy steels, generally those containing a total of 1 to 2% alloy additions, are often through hardened with an oil, water, or molten-salt-bath quenchant. Alloy additions of 5% or more produce metals that achieve their maximum properties by cooling in air; most tool and die steels are in this category. Most alloy steels have a practically linear relationship between hardness and tensile strength, with the tensile strength (in psi) generally being about 500 times the Brinell hardness number.

MEASURING HARDENABILITY

The most commonly used method of determining hardenability is the Jominy end-quench test. This test consists of heating a standard specimen, a 1" (25 mm) diam x 4" (102 mm) long round bar, above the upper critical temperature of the metal. The hot specimen is supported vertically in a fixture, and its lower end is quenched with a stream of cold water. After the specimen has cooled to room temperature, flat surfaces are ground on opposite sides and hardnesses are measured at 1/16" (1.6 mm) increments along the length of the specimen. Details of this laboratory test procedure are presented in SAE Standard J406c, "Method of Determining Hardenability of Steels," and ANSI/ASTM Standard A255, "End-Quench Test for Hardenability of Steel."

Data from these tests is plotted as hardness versus distance from the quenched end, the hardness decreasing as the distance from the quenched end increases. Published data, called hardenability bands (H-bands), is used to predict the hardening characteristics of various grades of steels, parts of different shapes, and parts having various cross-sectional shapes.

H STEELS AND HARDENABILITY BANDS

For the identification of steels specified to hardenability limits, the suffix letter H is added to the conventional series number. In the unified numbering system (UNS), the H appears as a prefix. To permit steel producers to meet the standards for hardenability limits, the chemical compositions of some steels have been modified somewhat from the same grades of steels that do not have a specified hardenability band.

Hardenability bands for carbon and alloy H steels are specified in SAE Standard J1268. The band graphs and tabular values show maximum and minimum hardenability limits, in Rockwell C-scale values, for various steels, based on standard end-quench tests. Steels are available with narrower (about one third) bands than standard hardenability bands. Such steels are sometimes used to help control distortion when heat treating certain parts, such as gears.

PROBLEMS IN HARDENING

Many problems can occur in hardening, most of which are the result of improper heat treatment specifications or poor heat treating practices. Some of the more common problems are due to decarburization, scaling, quench cracking, residual stresses, retained austenite, and dimensional changes.

Decarburization

Steel parts are often decarburized to some extent during heating for hot forming operations, such as rolling, extruding, and forging. The loss of surface carbon may prevent attaining full hardness in the finished parts. This situation may necessitate the removal of the softer surfaces of the parts by grinding or other means, preferably before heat treatment.

Decarburization during heating can be eliminated by using protective atmospheres in the furnaces. Decarburization can also be corrected by carbon restoration (carburizing), but it is better prevented than corrected.

Scaling

The formation of surface layers of oxidation products, called scaling, is another undesirable result of hardening. Such a condition often requires machining, grinding, cleaning, or other methods of descaling. Scaling can be minimized by using protective atmospheres in the heating furnaces.

Quench Cracking

The formation of cracks when quenching heated steels is a major problem. As the carbon content of a steel increases, the tendency to crack increases. Higher austenitizing temperatures also increase the tendency toward quench cracking. Steels with coarser grain size are more susceptible to cracks than fine-grained steels because the latter have more grain boundary area to block the movement of cracks, and grain boundaries help to absorb and redistribute residual stresses, as discussed next in this section. Nicks, scratches, too small a radius between section changes, and impurities on the surfaces of workpieces to be hardened are also possible sources of cracks. Careful machining, grinding, and cleaning prior to heat treatment is generally advisable.

Methods and media used for quenching heated parts, discussed later in this chapter, have a major influence on the tendency toward cracking. Direct quenching is generally more likely to cause cracks than isothermal quenching. Also, the use of more severe cooling media, such as water or brine, may sometimes result in more cracking than quenching in a slower cooling medium, such as oil.

Residual Stresses

Residual stresses (stresses that exist in a part that is free of external forces or thermal gradients) can cause distortion, warpage, cracking, or breakage, especially after quenching and before tempering. Major reasons for the formation of residual stresses are the nonuniform contraction of heated parts during cooling, expansion during transformation, or a combination of the two, as well as variations in heating and cooling rates.

Fast quenching produces higher stresses, and therefore the slowest cooling rate for producing the desired microstructure and hardness should be used. Subsequent tempering reduces the hardness and tensile strength, but minimizes stresses and improves ductility and toughness.

Distortion or warping is generally greater for more complexly shaped workpieces or when there is a large thickness change in the workpieces. Preheating, special cooling methods, or the use of workholding fixtures are sometimes necessary. Straightening of bars and other parts after thermal treatment can introduce stresses, which can cause distortion in subsequent machining or processing operations. Stress relieving prior to heat treatment is often required to minimize distortion, but some distortion and dimensional change are inevitable as the result of heat treatment.

Retained Austenite

Austenite that is not transformed during the heat treating cycle is soft and weak, and lowers the overall hardness of the steel if excessive in amount. It may also cause dimensional instability because of the possible transformation of the austenite at room temperature over a long period of time. In addition, retained austenite will often transform to martensite with the application of stresses.

The amount of retained austenite is affected by the alloy content of the steel, the austenitizing temperature, the cooling rate, and, in some cases, the amount and distribution of the stresses in the quenched parts. The effect of alloy content is so strong that some steels are entirely austenitic at room temperature. High austenitizing temperatures can result in increased amounts of retained austenite. Slow cooling rates also increase the tendency to retain austenite.

One method of eliminating retained austenite is cold treating—cooling the quenched parts to temperatures below freezing where transformation of austenite to martensite proceeds to completion. Another method, applicable to certain steels in which the austenite has not stabilized for too long a period prior to treatment, consists of heating the quenched parts to a temperature of about 300°F (150°C) or more to permit transformation to ferrite plus carbide.

HARDENING PROCEDURES

Optimum heat treatment requires the use of proper furnaces, accurate use of the time-temperature transformation relationships, and the right quenchant and cooling method. For many applications, controlled atmospheres in the furnaces are essential.

Heat treating procedures—with respect to temperatures, times, quenchants, and cooling methods—vary with the composition and condition of the metal, and the microstructure and properties desired. Typical procedures for heat treating some common carbon steels are presented in Table 10-1, and for alloy steels in Table 10-2.

HEATING METHODS

Many different types of furnaces are available, as are discussed in Chapter 12, "Heat Treating Furnaces," of this volume. Selection of a specific type depends primarily upon the workpieces to be heat treated, the production rate required, and the type of heat treating operation to be performed.

CONTROLLED ATMOSPHERES

The use of controlled atmospheres in heat treating processes is often for protection of the metal surfaces from oxidation. Suitable atmospheres for this purpose are those that do not discolor the surface of the metal and that allow the surface chemistry to retain its original composition. Surface oxides are also reduced, if present, to produce bright surfaces. The most notable use of controlled atmospheres for other purposes occurs in special heat treatments of ferrous alloys, such as in carburizing, nitriding, and carbonitriding as discussed in Chapter 11, "Surface (Case) Hardening," of this volume.

Atmospheres Used

Controlled atmospheres are usually a mixture of a number of gases. Most furnace atmospheres are prepared by the reaction of a fuel gas with air in generators, although specially

HARDENING PROCEDURES

TABLE 10-1
Typical Heat-Treating Procedures for Some Common Carbon Steels[2]

	Hardening				Annealing	
		Through				
Steel Grade, AISI/SAE	Surface (Case)*	Austenitizing Temperature, °F (°C)	Quench	Normalizing Temperature, °F (°C)**	Temperature, °F (°C)	Cooling
Nonresulfurized carbon steels:						
1008, 1010	Carburizing, nitriding, or carbonitriding	---	---	1700 (925)	1670 (910)	Slow, preferably in furnace.
1015,1018		---	---	1700 (925)	1625 (885)	
1020		---	---	1700 (925)	1600 (870)	
1030	---	1580 (860)	Water or brine. Thin sections may be quenched in oil.	1675 (915)	1600 (870)	Furnace cool to 1200° F (650° C) at a rate of not more than 50° F (28° C) per hour. From 1200° F to ambient temperature, the cooling rate is not critical.
1035, 1037	---	1575 (855)	Same as 1030 steel.	1675 (915)	1600 (870)	Same as 1030 steel.
1040, 1045, 1046	---	1550 (845)	Same as 1030 steel.	1650 (900)	1550 (845)	Same as 1030 steel.
1049, 1050	---	1525 (830)	Same as 1030 steel.	1650 (900)	1525 (830)	Same as 1030 steel.
1060, 1070	---	1500 (815)	Same as 1030 steel.	1625 (885)	1525 (830)	Same as 1030 steel.
1080	---	1500 (815)	Same as 1030 steel.	1600 (870)	1500 (815)	Same as 1030 steel.
1095	---	1475 (800)	Same as 1030 steel.	1575 (855)	1475 (800)	Same as 1030 steel.
Resulfurized carbon steels:						
1117	Carburizing, nitriding, or carbonitriding	---	---	1650 (900)	1575 (860)	Slow, preferably in furnace.
1137	---	1550 (845)	Same as 1030 steel.	1650 (900)	1625 (885)	Same as 1030 steel.
1141, 1144	---	1525 (830)	Same as 1030 steel.	1625 (885)	1550 (845)	Same as 1030 steel.
High-manganese carbon steels:						
15B21, 1522	Carburizing, nitriding, or carbonitriding	---	---	1700 (925)	1600 (870)	Slow, preferably in furnace.
15B35	---	1575 (855)	Varies with section thickness.	1675 (915)	1600 (870)	Same as 1030 steel.
1541, 15B41, 1548	---	1550 (845)	Varies with section thickness.	1650 (900)	1525 (830)	Same as 1030 steel.
1552	---	1525 (830)	Varies with section thickness.	1650 (900)	1525 (830)	Same as 1030 steel.

Note: Temperatures for tempering are selected to obtain desired hardness.
* Surface hardening processes are discussed in Chapter 11.
** Heating for normalizing is followed by air cooling.

prepared cylinder or bulk-storage gases are also commonly used. The more common gases are:

N_2 — Nitrogen
O_2 — Oxygen
H_2O — Water vapor
CO_2 — Carbon dioxide
CO — Carbon monoxide
H_2 — Hydrogen
CH_4 — Methane (natural gas)

Selection of a specific atmosphere depends primarily upon the application, as well as the availability and cost of the gases in the geographic location where the heat treating operations are performed.

Ferrous alloys. Molecular nitrogen (N_2) can be considered an inert gas with reference to its reaction with iron or common steels. Molecular nitrogen is also inert in reaction with other gases. Atomic nitrogen (N), however, will react with iron and certain alloys of iron to form a nitride. Atomic nitrogen does not occur in furnace atmospheres unless it is purposely introduced by adding ammonia (NH_3), which dissociates at elevated furnace temperatures to form molecular nitrogen (N_2) and hydrogen (H_2).

During the process of dissociation of ammonia (NH_3), atomic nitrogen is available for a short time before the two atoms of nitrogen form a molecule of nitrogen (N_2). With atomic or nascent nitrogen present at the iron surface, part of the atomic nitrogen will react with the iron to form iron nitride (FeN). However, the greatest percentage of atomic nitrogen reacts with itself to form molecular nitrogen, which is then essentially inert to iron and the other furnace gases.

Some iron alloys high in chromium act as a catalyst and dissociate part of the molecular nitrogen into atomic nitrogen, with the atomic nitrogen being subsequently absorbed into the metal. The absorbed or dissolved nitrogen affects the physical properties of the metal. However, this phenomenon is an exception rather than the rule, and molecular nitrogen can be considered an inert gas in the basic furnace atmosphere.

Free oxygen (O_2) oxidizes iron and steel, as shown in the following reaction:

$$2\,Fe + O_2 \rightarrow 2\,FeO_2 \qquad (2)$$

This reaction, which also produces Fe_2O_3 and Fe_3O_4, may be considered irreversible since it is uncontrollable; no product besides metal oxide is formed and no equilibrium is involved.

TABLE 10-2
Typical Heat-Treating Procedures for Some Common Alloy Steels[2]

Steel Grade, AISI/SAE	Hardening			Normalizing Temperature, °F (°C)	Annealing					
	Surface (Case)*	Through			Temperature, °F (°C)	Cooling				Hold Time, hr
		Austenitizing Temperature, °F (°C)	Quench			Method	From, °F (°C)	To, °F (°C)	Maximum Rate per Hour, °F (°C)	
1340	---	1525 (830)	Oil, water or brine for heavy sections.	1600 (870)	For a predominantly pearlitic structure:					
					1525 (830)	Furnace	1350 (730)	1130 (610)	20 (11)	
					or,	Rapid	1350 (730)	1150 (620)		4 1/2
					For a predominantly spheroidized structure:					
					1380 (750)	Furnace	1350 (730)	1130 (610)	10 (6)	
					or,	Rapid	1380 (750)	1180 (640)		8
4027	Carburizing	1575 (855)	Oil	1650 (900)	For a predominantly pearlitic structure:					
					1575 (855)	Rapid		1380 (750)		
							then,	1180 (640)	20 (11)	
					or, 1600 (870)	Rapid		1225 (660)		5
					For a predominantly spheroidized structure:					
					1425 (775)	Furnace	1370 (745)	1180 (640)	10 (6)	
					or,	Rapid		1225 (660)		8
4130	---	1600 (870)	Oil	1650 (900)	For a predominantly pearlitic structure:					
					1575 (855)	Rapid		1400 (760)		
							then,	1230 (665)	35 (19)	
					or,	Rapid		1250 (675)		4
					For a predominantly spheroidized structure:					
					1380 (750)	Furnace		1230 (665)	10 (6)	
					or,	Rapid		1250 (675)		8
4140, 4142	Nitriding	1575 (855)	Oil	1600 (870)	For a predominantly pearlitic structure:					
					1550 (845)	Rapid		1390 (755)		
							then,	1230 (665)	25 (14)	
					or,	Rapid		1250 (675)		5
					For a predominantly spheroidized structure:					
					1380 (750)	Furnace		1230 (665)	10 (6)	
					or,	Rapid		1250 (675)		9

(continued)

TABLE 10-2—*Continued*

	Hardening				Annealing					
		Through				Cooling				
Steel Grade, AISI/SAE	Surface (Case)*	Austenitizing Temperature, °F (°C)	Quench	Normalizing Temperature, °F (°C)	Temperature, °F (°C)	Method	From, °F (°C)	To, °F (°C)	Maximum Rate per Hour, °F (°C)	Hold Time, hr
4340	Nitriding	1550 (845)	Oil, but air for thin sections.	1600 (870)	For a predominantly pearlitic structure:					
					1525 (830)	Rapid		1300 (705)		
							then,	1050 (565)	15 (8)	
					or,	Rapid		1200 (650)		8
					For a predominantly spheroidized structure:					
					1380 (750)	Rapid		1300 (705)		
							then,	1050 (565)	5 (3)	
					or,	Rapid		1200 (650)		12
4620	Carburizing, carbonitriding			1700 (925)	1425 (775)	Rapid		1200 (650)		6
4720	Carburizing, carbonitriding			1700 (925)	1500 (815)	Rapid		1200 (650)		8
4820	Carburizing			1700 (925)	1200 (650)	Hold for 1 hr per inch of thickness, cooling rate not critical, or:				
					1370 (745)	Rapid		1125 (610)		8
5140		1550 (845)	Oil	1600 (870)	For a predominantly pearlitic structure:					
					1525 (830)	Rapid		1360 (740)		
							then,	1240 (670)	20 (11)	
					or,	Rapid		1250 (675)		6
					For a predominantly spheroidized structure:					
					1380 (750)	Rapid		1275 (690)		8
5150, 5155, 5160		1525 (830)	Oil	1600 (870)	For a predominantly pearlitic structure:					
					1525 (830)	Rapid		1300 (705)		
							then,	1200 (650)	20 (11)	
					or,	Rapid		1250 (675)		6
					For a predominantly spheroidized structure:					
					1380 (750)	Rapid		1300 (705)		
							then,	1200 (650)	10 (6)	
					or,	Rapid		1250 (675)		10
51B60		1550 (845)	Oil	1600 (870)	For a predominantly spheroidized structure:					
					1380 (750)	Rapid		1290 (700)		
							then,	1210 (655)	10 (6)	
					or,	Rapid		1200 (650)		12
E52100		1550 (845)	Oil	1625 (885)	For a predominantly spheroidized structure:					
					1460 (795)	Rapid		1380 (750)		
							then,	1250 (675)	10 (6)	
							or,	1275 (690)		16
6150		1600 (870)	Oil	1650 (900)	For a predominantly pearlitic structure:					
					1525 (830)	Rapid		1400 (760)		
							then,	1250 (675)	15 (8)	
					or,	Rapid		1250 (675)		6
					For a predominantly spheroidized structure:					
					1400 (760)	Furnace		1250 (675)	10 (6)	
					or,	Rapid		1200 (650)		10
8615, 8620	Carburizing, carbonitriding			1700 (925)	1625 (885)	Rapid		1225 (660)		4
					or, 1450 (790)	Rapid		1225 (660)		8

TABLE 10-2—*Continued*

Steel Grade, AISI/SAE	Hardening: Surface (Case)*	Hardening: Through: Austenitizing Temperature, °F (°C)	Hardening: Through: Quench	Normalizing Temperature, °F (°C)	Annealing: Temperature, °F (°C)	Annealing: Cooling: Method	Annealing: Cooling: From, °F (°C)	Annealing: Cooling: To, °F (°C)	Annealing: Cooling: Maximum Rate per Hour, °F (°C)	Annealing: Hold Time, hr
8630		1600 (870)	Oil	1650 (900)	For a predominantly pearlitic structure:					
					1550 (845)	Rapid		1350 (730)		
							then,	1180 (640)	20 (11)	
					or,	Rapid		1225 (665)		6
					For a predominantly spheroidized structure:					
					1400 (760)	Rapid		1350 (730)		
							then,	1200 (650)	10 (6)	
					or,	Rapid		1225 (665)		8
8640	Nitriding	1575 (855)	Oil	1600 (870)	For a predominantly pearlitic structure:					
					1525 (830)	Rapid		1340 (725)		
							then,	1180 (640)	20 (11)	
					or,	Rapid		1225 (665)		6
					For a predominantly spheroidized structure:					
					1380 (750)	Rapid		1340 (725)		
							then,	1180 (640)	10 (6)	
					or,	Rapid		1225 (665)		8
8720	Carburizing, carbonitriding			1700 (925)	1625 (885)	Rapid		1225 (660)		4
				or,	1450 (790)	Rapid		1225 (660)		8
9260		1600 (870)	Oil	1650 (900)	For a predominantly spheroidized structure:					
					1400 (760)	Furnace		1300 (705)	10 (6)	
					or,	Rapid		1225 (660)		10

Note: Temperatures for tempering are selected to obtain desired hardness.
* Surface hardening processes are discussed in Chapter 11.
** Heating for normalizing is followed by air cooling.

Oxygen compounds also react to form metal oxide, but the reactions are reversible and reach metal-metal oxide equilibrium as a function of temperature. In the reactions

$$Fe + H_2O \rightleftarrows FeO + H_2 \quad (3)$$

$$Fe + CO_2 \rightleftarrows FeO + CO \quad (4)$$

oxidation is carried out by water vapor and carbon dioxide. Water (H_2O) more often oxidizes to Fe_3O_4. Carbon dioxide (CO_2) and oxygen (O_2) usually oxidize to FeO or Fe_2O_3. These reactions differ from those involving free oxygen in that other products, the reducing gases hydrogen (H_2) and carbon monoxide (CO), are formed in addition to the iron oxides.

As the reaction continues, the relative concentrations of oxidizing gas and reducing gas reach a point of equilibrium. If an excess of reducing gas is supplied, the reaction is reversed to produce pure metal and water or carbon dioxide. Reactions of this type may thus be controlled to be oxidizing, reducing, or in equilibrium, depending upon the relative amounts of the constituent gases present and the metal being heat treated.

Figure 10-5 shows carbon monoxide-carbon dioxide equilibrium ratios, together with hydrogen-water vapor ratios, for a wide range of atmospheric proportions. The curves indicate that while a reducing atmosphere will tolerate increasing amounts of water vapor with increasing temperature, the

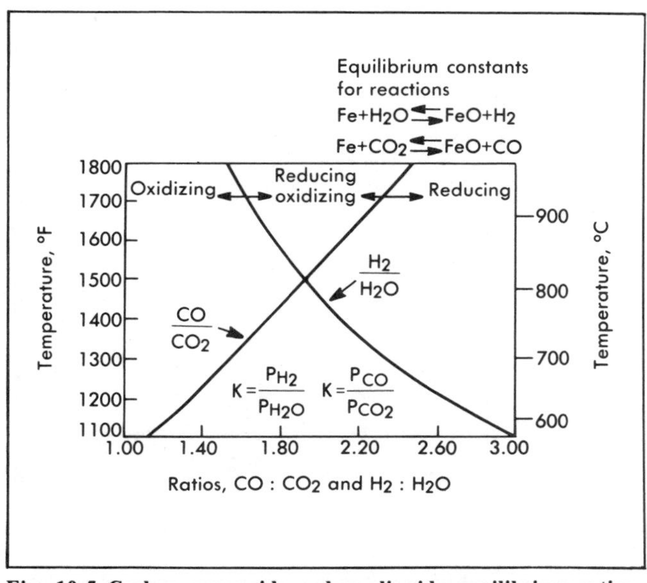

Fig. 10-5 Carbon monoxide-carbon dioxide equilibrium ratios, together with hydrogen-water vapor ratios, for a wide range of atmospheric proportions.

opposite is true when carbon dioxide is the oxidizing gas. The opposing actions of these oxidizers may also be controlled according to the following reversible, water-gas shift reaction:

$$CO + H_2O \rightleftarrows CO_2 + H_2 \quad (5)$$

In iron and steel, the carbon as well as the metal is subject to oxidizing/reducing reactions. Carburizing or decarburizing reactions occur if carbon is either absorbed or given up by the ferrous metal in the presence of a suitable gas atmosphere. Since most of the unique physical properties of iron and steel are derived from control of the carbon content, these reactions are of particular importance. From the following reactions

$$3\,Fe + 2CO \rightleftarrows Fe_3C + CO_2 \quad (6)$$

$$3\,Fe + CH_4 \rightleftarrows Fe_3C + 2H_2 \quad (7)$$

$$3\,Fe + CO + H_2 \rightleftarrows Fe_3C + H_2O \quad (8)$$

it is evident that carbon monoxide and methane (CH_4) are carburizing gases and that carbon dioxide, water vapor, and hydrogen are decarburizing gases, with water vapor being the most significant. The reaction of $CO + H_2$ is much faster than CO or CH_4 alone.

Carburizing/decarburizing reactions are reversible, and the equilibrium ratios may be expressed as $CO : CO_2$, $CH_4 : H_2$, and $CO : H_2 : H_2O$. Figures 10-6 and 10-7 graphically show iron carbide's (Fe_3C) relationship with water vapor (measured in terms of dew point) and carbon dioxide. Control or regulation of the atmosphere dew point or CO_2 level to match the carbon content of a steel workpiece is generally accomplished by the addition of an enriching gas, usually a straight hydrocarbon, such as methane or propane. Only hydrocarbons, such as methane, are used to control carbon potential. The addition of CH_3OH, which dissociates into $CO + H_2$, enhances the rate of carburization but has no effect on carbon potential.

With high-chromium stainless steels, dry hydrogen, nitrogen, or dissociated ammonia consisting of 75% hydrogen and 25%

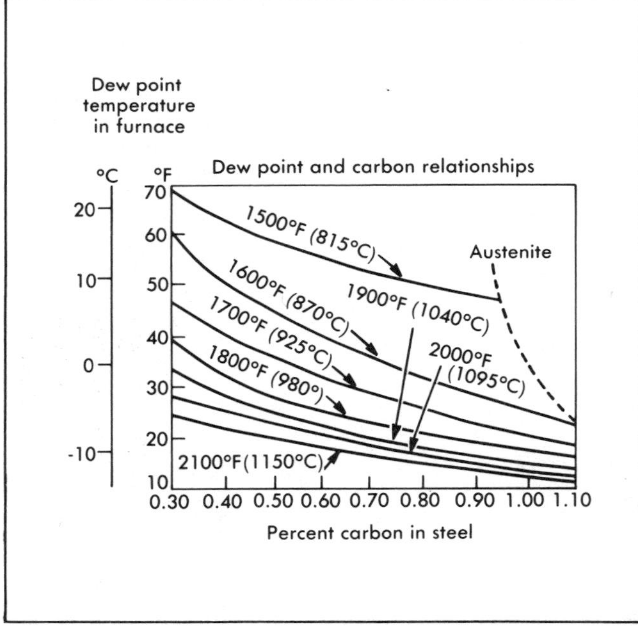

Fig. 10-6 Dew-point temperature equilibrium related to carbon content. (*After Koebel*)[3]

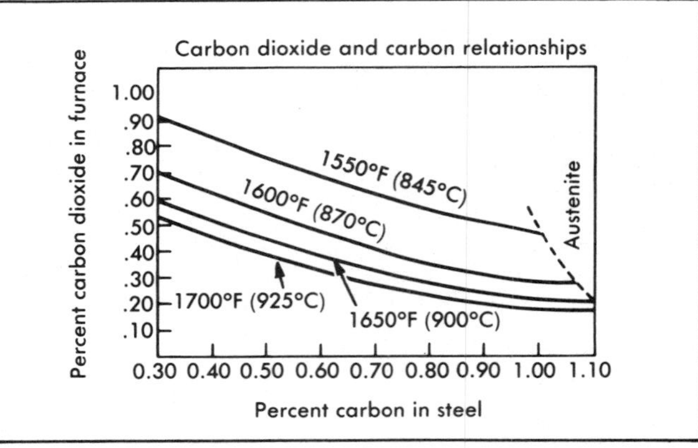

Fig. 10-7 Carbon dioxide percentage equilibrium related to carbon content.[4]

nitrogen can be used to obtain bright results. Traces of carbon monoxide, carbon dioxide, water vapor, or air will cause discoloration and oxidation. When absorption of nitrogen affects the physical properties, inert helium or argon can be used. Nitrogen absorption in stainless steels can be controlled with lower hydrogen levels and a protective oxide layer.

Nonferrous alloys. As with iron and steel, protection from free oxygen is necessary for heat treating most nonferrous materials, such as copper, copper alloys, nickel, nickel alloys, and refractory metals. All metals (with the possible exception of gold) react with oxygen to form a metal oxide by the following reaction:

$$M + 1/2\,O_2 \rightarrow MO \quad (9)$$

Even trace amounts of free oxygen will oxidize copper. However, carbon dioxide and water vapor, which are strongly oxidizing to steel, will not oxidize copper. In fact, steam can be used as an atmosphere for the bright annealing of copper. Likewise, products of combustion and nitrogen can be used as a bright annealing atmosphere, provided there are no traces of free oxygen or hydrogen sulfide.

With respect to atmosphere, alloys of copper with silver or nickel behave in a manner much like copper itself. However, alloys with tin, lead, aluminum, and beryllium will oxidize in atmospheres containing high carbon dioxide and water vapor. Best results are obtained on these alloys by keeping the hydrogen content high and the water vapor and carbon dioxide content low, as is done in bright annealing of steel.

The alloys of copper with zinc, such as yellow brass, present far greater difficulties in preserving their bright surface during heat treatment. Zinc or alloys containing a fair percentage of zinc oxidize when heated in air, carbon dioxide, or water vapor. Therefore, an atmosphere to prevent oxidation must exclude these constituents, and preferably have a high percentage of hydrogen to counteract any traces of the undesirable constituents. Another factor with respect to the bright annealing of yellow brass is that zinc has a tendency to vaporize and form a haze on the surface. The higher the temperature and the longer the annealing time, the greater is this reaction. As a result, even if pure dry hydrogen is used for the atmosphere, the brass color and brightness cannot be fully controlled when annealing yellow brass.

Figure 10-8 shows equilibrium ratios of water vapor to hydrogen throughout a range of temperatures for four common

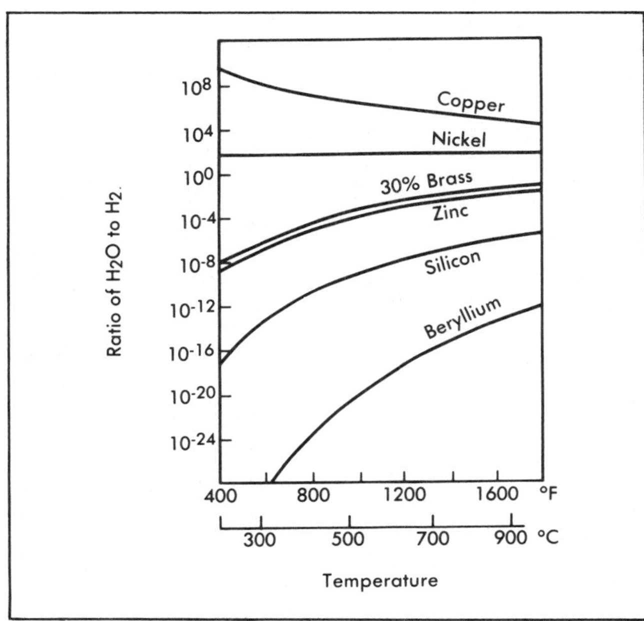

Fig. 10-8 Equilibrium ratios of water vapor to hydrogen at various temperatures for various metals.[5]

and two uncommon metals. Above each equilibrium curve, a mixture of water and hydrogen is oxidizing; and below each curve, the mixture is reducing. The curve for copper illustrates that this metal may be annealed in an atmosphere of pure steam without oxidation. The curves also indicate why nickel seems relatively easy to anneal without oxidation, whereas brass and zinc are difficult to keep bright or clean without dry hydrogen atmospheres in special furnaces. The equilibrium curve for iron is close to that for nickel; and in their reactions, chromium and stainless steel behave similarly to brass and zinc.

Titanium and some refractory metals will either react or absorb nearly every known gas or combination of gases, except argon and helium, resulting in severe embrittlement and loss of ductility. Dry air will not cause embrittlement, but will, of course, oxidize the metal.

Classification of Prepared Atmospheres

Mixtures of gases commonly used as protective atmospheres in the heat treatment of metals have been classified by the American Gas Association. The gases generated for furnace atmospheres are divided into six categories, as follows:

Class 100—Exothermic Base: An atmosphere composed of products of partial or complete combustion of an air-fuel mixture. The base may be modified by various degrees and methods of moisture removal, and the use of various air-fuel ratios.

Class 200—Prepared Nitrogen Base: An exothermic base (Class 100) with most of the carbon dioxide and water vapor removed.

Class 300—Endothermic Base: Formed by partial reaction of a gaseous fuel and air mixture in an externally heated catalyst-filled chamber.

Class 400—Charcoal Base: Formed by passing air through incandescent charcoal in an externally heated vertical retort. Popularity and acceptance of other atmosphere systems have made this class essentially obsolete.

Class 500—Exothermic-Endothermic Base: Formed by complete combustion of a mixture of fuel gas and air, removing most of the water vapor, and reforming most of the carbon dioxide to carbon monoxide by reaction with fuel gas in an externally heated catalyst reactor.

Class 600—Ammonia Base: Any atmosphere produced using ammonia as the feed stock, including raw ammonia, dissociated ammonia, or partially or completely combusted ammonia with most of the water vapor removed.

Class 100, 200, 300, and 500 atmospheres are prepared by the controlled reaction of a fuel gas with air. These atmospheres are normally used to produce an environment that is void of free oxygen. For example, fuel plus air produces CO_2, CO, H_2, H_2O, and N_2. Class 600 atmospheres are prepared by dissociating anhydrous ammonia to produce a gas consisting of H_2 and N_2. The six broad areas of atmosphere classification are subclassified by three-digit designations to indicate variations in the method by which they are prepared.

Nitrogen-Based Atmospheres

Gas mixtures equivalent to those produced from fuel gases and ammonia can be produced by blending bottled or bulk-storage gases and enriching nitrogen with a fuel gas (usually methane) or liquid alcohol (usually methanol). These are simple to apply as they basically require only regulation of the flow of component gases. Both the composition and flow rate of the atmospheres can be independently controlled. Safety purging with inert nitrogen can eliminate dangers associated with handling combustible gases at elevated temperatures. Industrially supplied nitrogen is most commonly produced through air separation (liquefaction and fractional distillation). The air is filtered, purified, compressed to drive it through the system, and cooled to remove the water and carbon dioxide. After being liquefied, it is distilled into its major constituents, the most abundant being nitrogen.

Hydrogen is also frequently blended with nitrogen, producing an atmosphere used as a replacement for either the exothermic base or dissociated ammonia atmospheres. Users of dissociated ammonia often do not require high hydrogen levels and can use a replacement nitrogen-based system with low hydrogen additions.

Atmosphere Generators

The two major types of generators employed to produce controlled atmospheres for heat treating furnaces are the exothermic and endothermic types.

Exothermic generators. Exothermic (Class 100) generators are among the simplest of all atmosphere producers. This generator's operation is based on the reaction or partial combustion of hydrocarbon gas with a limited amount of air to form an atmosphere containing carbon monoxide, carbon dioxide, hydrogen, water vapor, and nitrogen in controlled quantities to obtain the following reaction:

$$\begin{gathered} 2\,CH_4 + 2\,O_2 + N_2 \rightarrow \\ N_2 + CO + CO_2 + 3\,H_2 + H_2O + \text{heat} \end{gathered} \tag{10}$$

Because this reaction requires partial burning of the gas, with heat given off as the result, the reaction and the generator are termed exothermic, or simply exo. This atmosphere is still one of the most widely used for bright or clean annealing of copper-based alloys and low-carbon steels. For low-carbon steels, however, the dew point of the atmosphere should be kept low to prevent possible decarburization.

The exothermic generator consists basically of a mixer for the hydrocarbon gas and air, a combustion chamber, and a water cooler of either direct or indirect type to chill the gas after combustion has taken place. (See Fig. 10-9.) Exothermic generators are generally equipped to operate in two different air-fuel gas mixture ranges, rich and lean, using different air-gas mixing systems and a catalyst to promote reaction in the rich unit. A catalyst is not required for this type of generator when producing a rich exothermic gas containing 12% hydrogen or less. Figure 10-10 illustrates the variations in atmosphere constituents based on the air-fuel gas ratio.

Another type of exothermic generator is the nitrogen generator (Class 200). Again a mixture of fuel and air is burned, as in the exothermic generator. After the products cool, they pass through a carbon dioxide absorber, which may be a wet chemical such as monoethanolamine (MEA) or a dry, molecular sieve material. The dry, molecular sieve method precludes the danger of corrosive chemical action that may occur with a wet chemical absorber. The corrosiveness of MEA can be controlled by using sodium vanadate. Figure 10-11 illustrates the atmosphere constituents based on the air-gas ratio.

Endothermic generators. Endothermic (Class 300) generators produce atmospheres used to protect and control the carbon level of steels during heat treatment. The endothermic type of atmosphere contains substantially none of the oxidizing gases carbon dioxide and water vapor, but does contain amounts of the reducing gases carbon monoxide and hydrogen. In its operation, a source of external heat is required in the presence of a catalyst to bring about the following reaction:

$$2\,CH_4 + O_2 + N_2 + \text{heat} \rightarrow N_2 + 2\,CO + 4\,H_2 \qquad (11)$$

As shown in Figure 10-12, the endothermic generator consists essentially of an air-gas mixer and one or more externally heated retorts in which the air-gas mixture reacts in the presence of a nickel catalyst. The gaseous reaction is completed in the retort and the cracked gas is generally cooled by an external cooling system for transmission to the point of use.

Endothermic gas, sometimes called simply endo, is strongly reducing to iron and has varying carbon potential, as shown by the pertinent ratios:

$$CO/CO_2 = 20/0.1 = 200:1$$
$$H_2/H_2O = 40/0.35 = 110:1$$

These ratios are for the usual composition, with a dew point of 20°F (-7°C) and 0.1% CO_2. A high ratio of reducing to oxidizing components can be obtained with this system, with the nitrogen-based (Class 200) atmospheres previously discussed, or from industrially supplied nitrogen atmospheres. Figure 10-13 illustrates the atmosphere constituents based on the air-gas ratio.

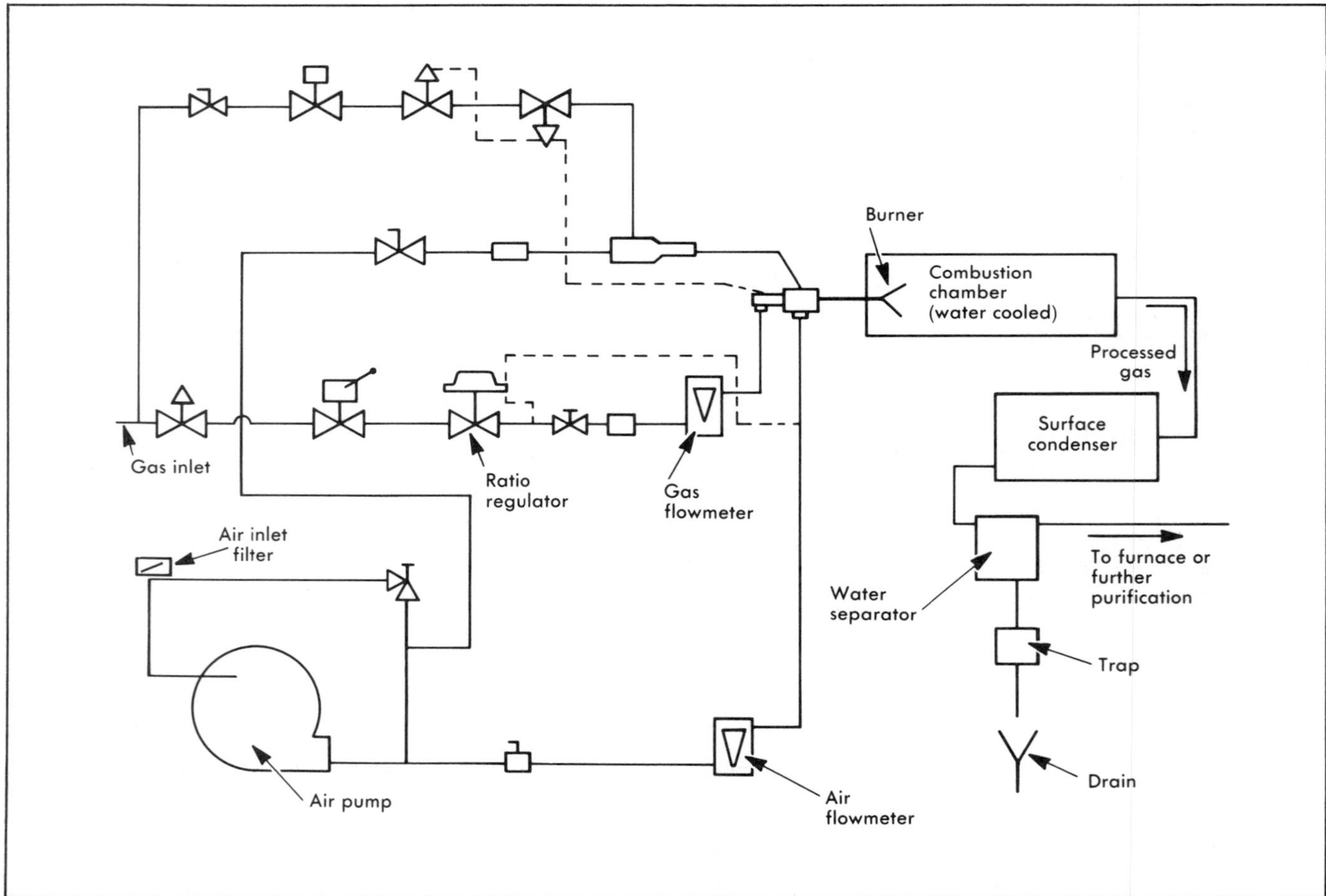

Fig. 10-9 Schematic flow diagram of a typical exothermic-type atmosphere generator.

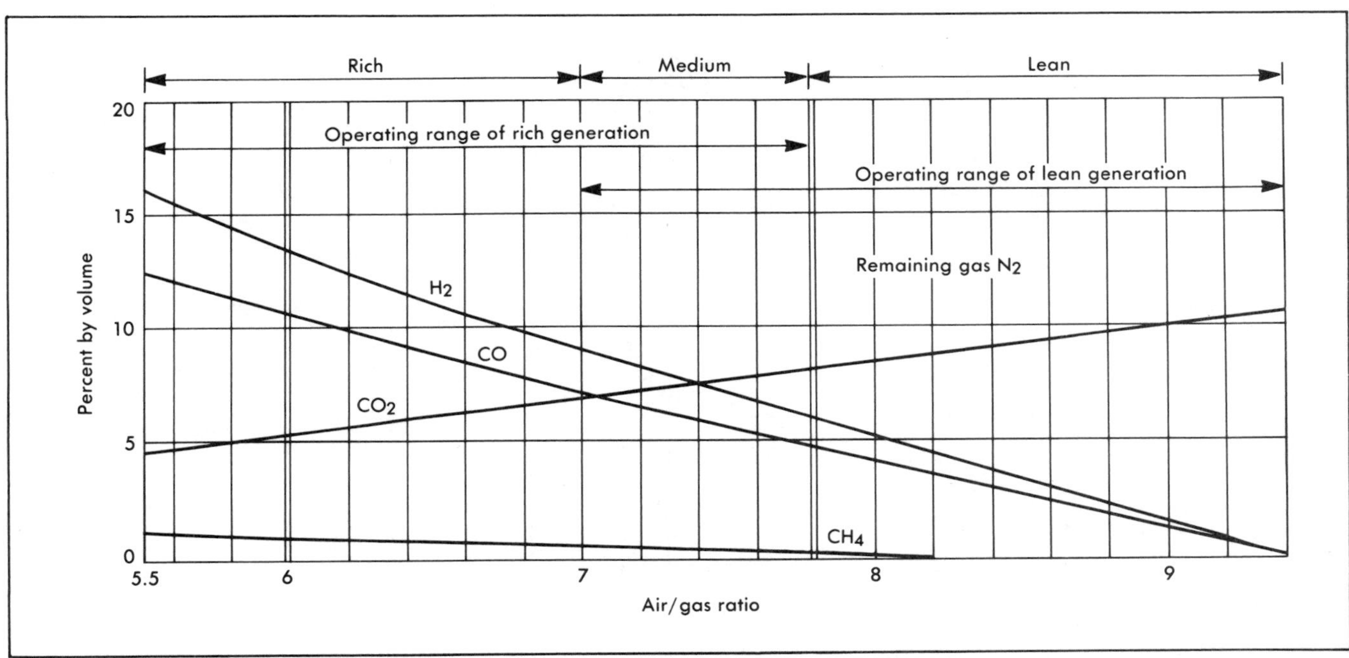

Fig. 10-10 Constituents of the atmosphere from an exothermic generator vary with the air to fuel gas ratio. Fuel gas is natural gas: 90% CH_4, 5% C_2H_6, and 5% N_2.

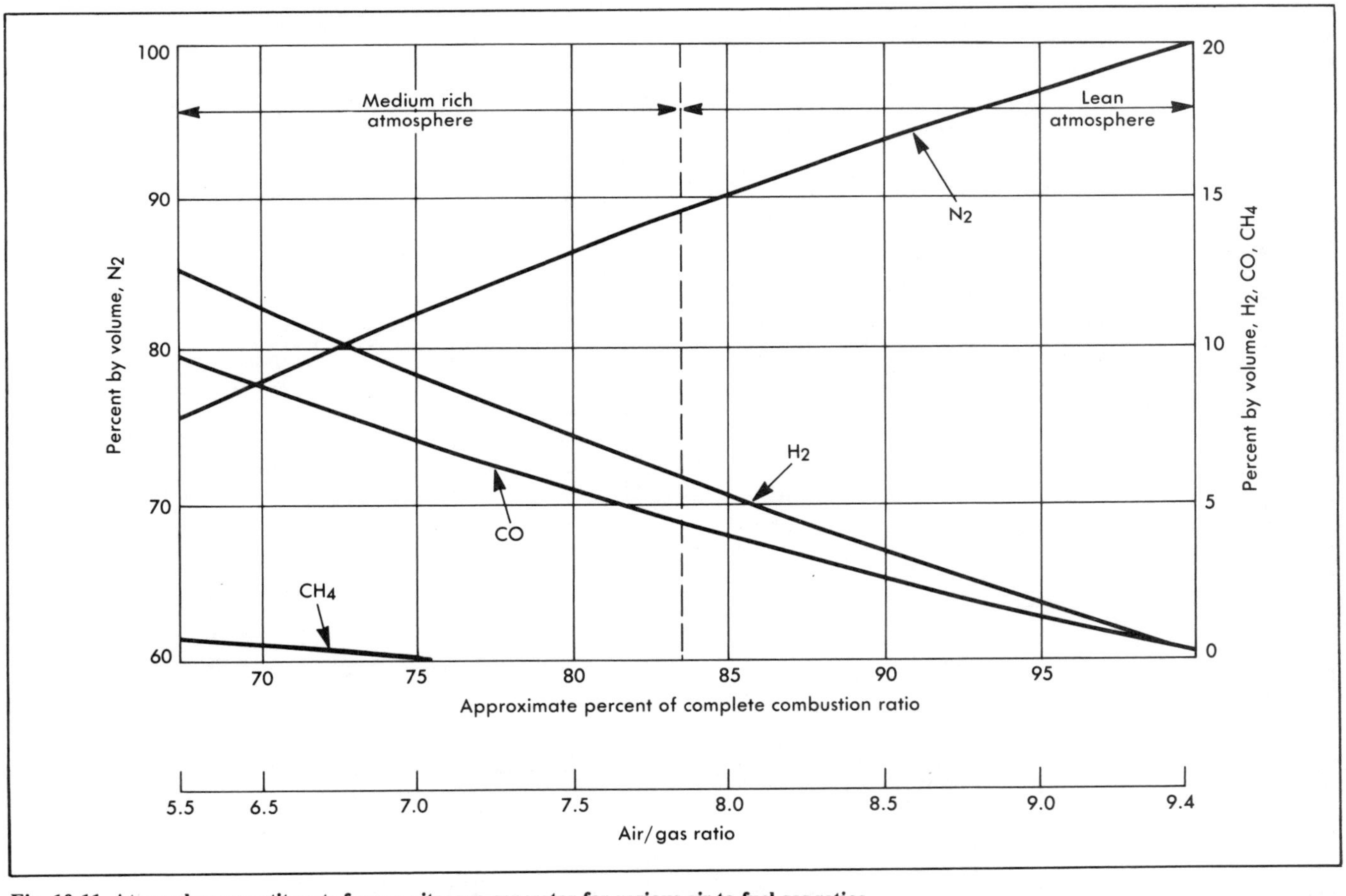

Fig. 10-11 Atmosphere constituents from a nitrogen generator for various air to fuel gas ratios.

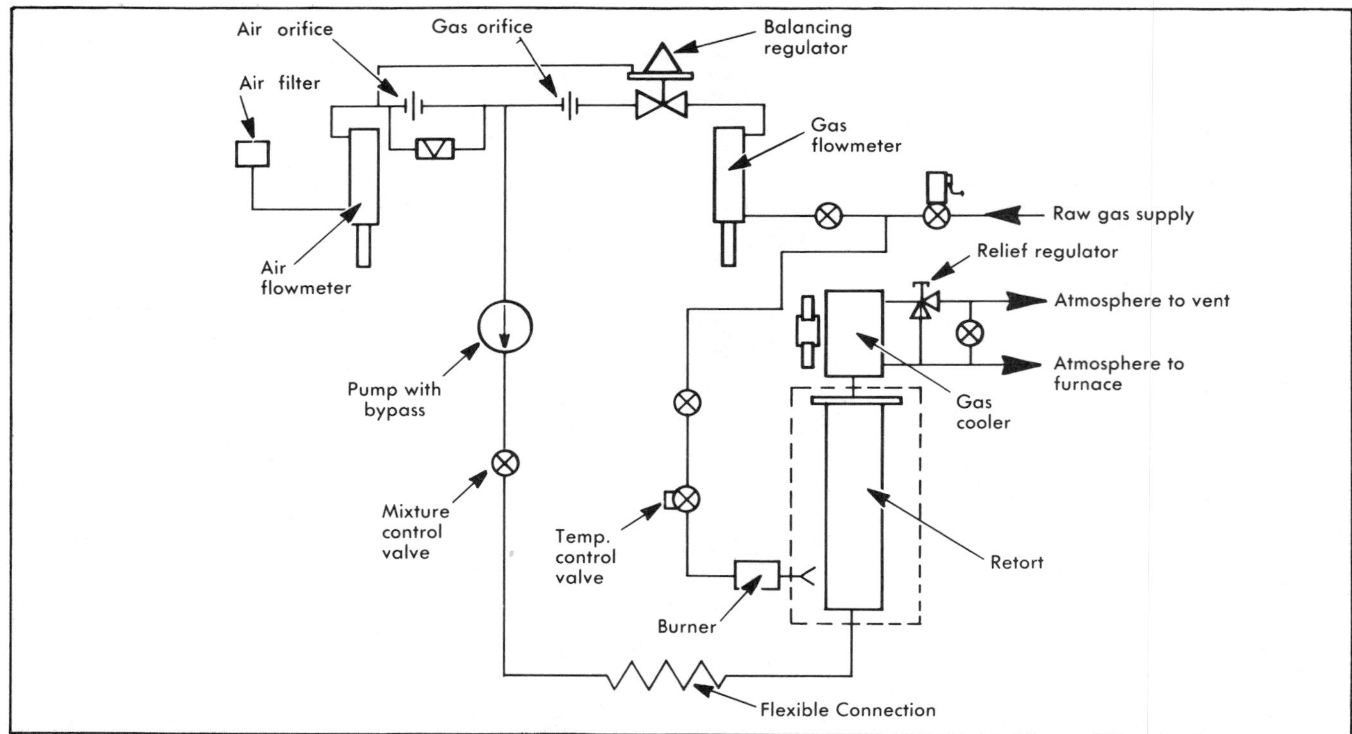

Fig. 10-12 Schematic flow diagram of a typical endothermic-type atmosphere generator.

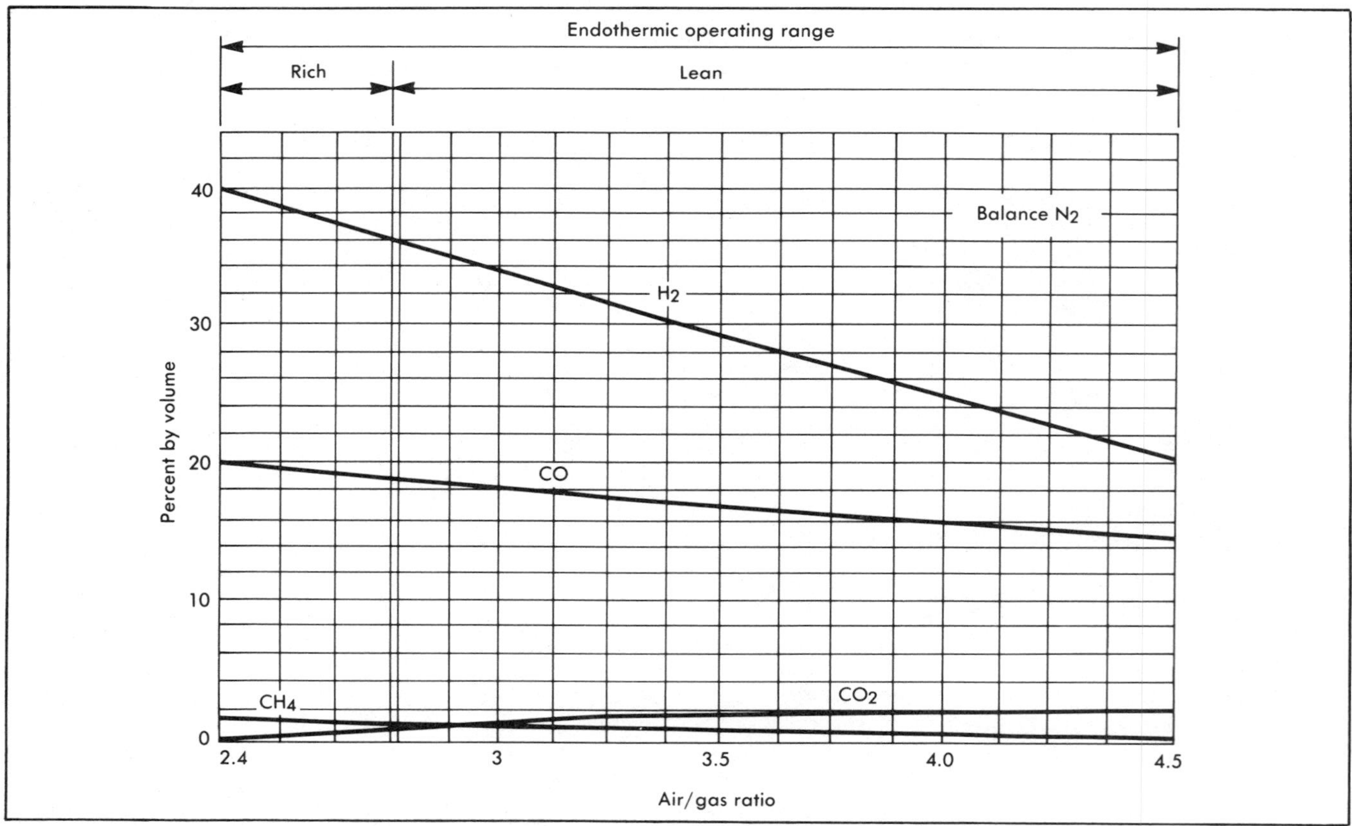

Fig. 10-13 Constituents of the atmosphere from an endothermic generator vary with the air to fuel gas ratio.

When using endothermic atmospheres, the atmospheres have to be adjusted for each different grade of steel treated and for the temperature of treatment. Generators are available with microprocessor-based control systems to automatically regulate the composition of the atmosphere supplied to the furnace.

Exothermic-endothermic. The exothermic-endothermic (Class 500) atmosphere system is a compromise between the strong reducing and carburizing effect of an endothermic atmosphere and the weaker effect of an exothermic atmosphere. After an exothermic gas (Class 100) has been cooled and dried, it is heated and reacted with additional fuel on a catalyst. The fuel combines with the oxygen present in the form of carbon dioxide to produce additional hydrogen and carbon monoxide. This atmosphere combines some of the qualities of the endothermic atmosphere with a lower fuel requirement.

Ammonia-Based Atmospheres

Ammonia-based (Class 600) atmospheres are those using ammonia as the feedstock (raw material), including raw ammonia, dissociated ammonia, combusted-dissociated ammonia, and hydrogen produced from ammonia. Ammonia dissociation is an endothermic reaction producing 75% hydrogen and 25% nitrogen as follows:

$$2NH_3 + \text{heat} \rightarrow N_2 + 3H_2 \quad (12)$$

Hydrogen-nitrogen generator. As shown in Fig. 10-14, a hydrogen-nitrogen generator consists essentially of an ammonia vaporizer, an externally heated retort filled with a nickel catalyst, and a gas cooler. The product gas is often purified in an absorption dryer to remove any traces of dissociated ammonia and water vapor.

Combustion of dissociated ammonia produces a nitrogen atmosphere with a controlled amount of hydrogen. The combustor is essentially the same equipment as that used for a Class 100 exothermic generator. Figure 10-15 illustrates the atmosphere constituents based on the air-gas ratio.

Diffusion cell system. High-purity hydrogen can also be produced from passing dissociated ammonia through a diffusion cell system. For applications requiring it, an ultrapure hydrogen stream can be produced by passing the 75% hydrogen mixture from the ammonia dissociator through a platinum-silver or palladium diffusion cell. In such a system, hydrogen diffuses through a membrane matrix, leaving the larger nitrogen molecules behind.

Atmosphere Controls

There are many types of instruments available to analyze the various constituents of a controlled atmosphere, individually or in groups (such as combustibles). These analyzers are generally mated to auxiliary equipment to serve a variety of functions, such as recorders, alarms, and controllers. The control of the furnace atmosphere can be either manual or automatic, depending upon the process, equipment size, and atmosphere used. Since most controlled atmospheres are prepared from a reaction of a fuel gas, their control may generally be accomplished with several types of analyzers, such as combustible analyzers, dew point analyzers, infrared analyzers, and oxygen probe systems. Additional information on furnace controls is presented in Chapter 12, "Heat Treating Furnaces," of this volume.

Combustibles analyzers. Most combustibles analyzers operate on the principle of catalytic combustion. A sample drawn from the gas generator or from the furnace is mixed with air (to supply oxygen) and then is passed over a detector. Any combustibles in the gas burn catalytically, raising the temperature of the detector and increasing the electrical resistance. The

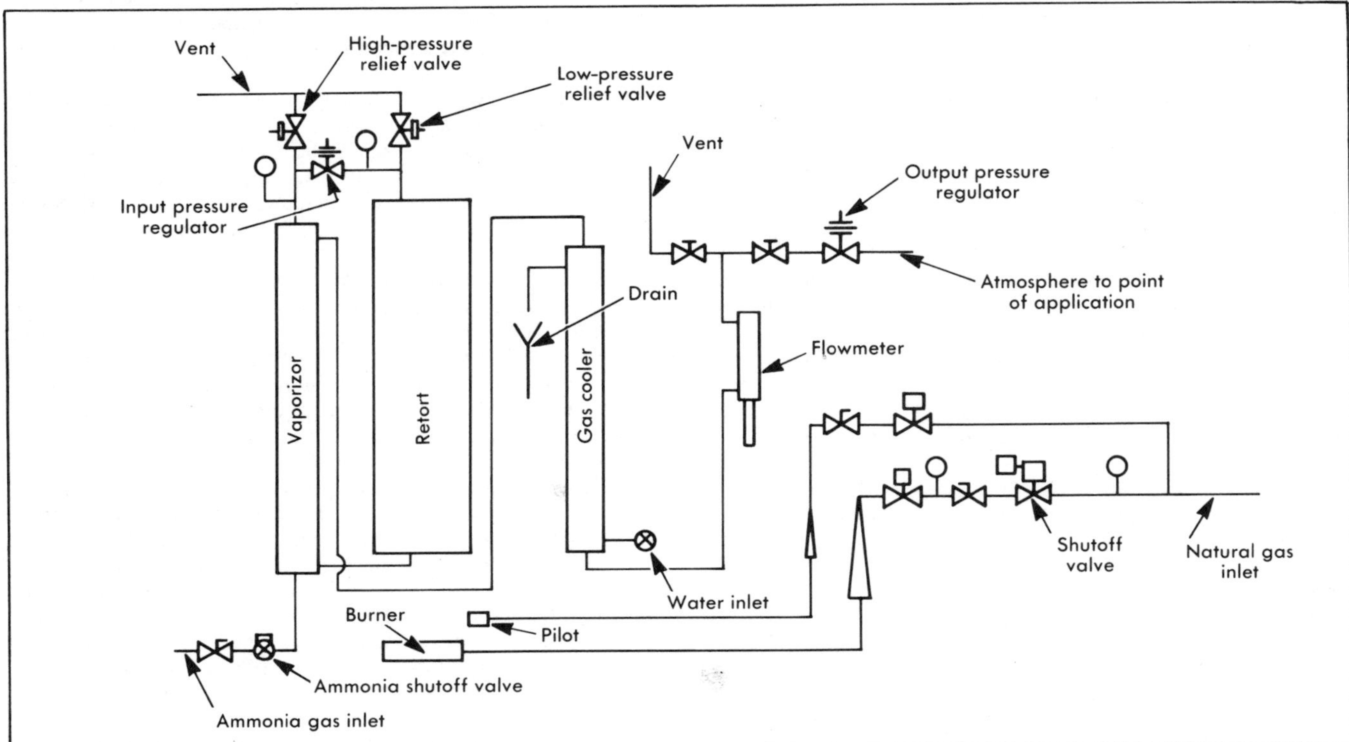

Fig. 10-14 Schematic flow diagram of a typical dissociated-ammonia atmosphere generator.

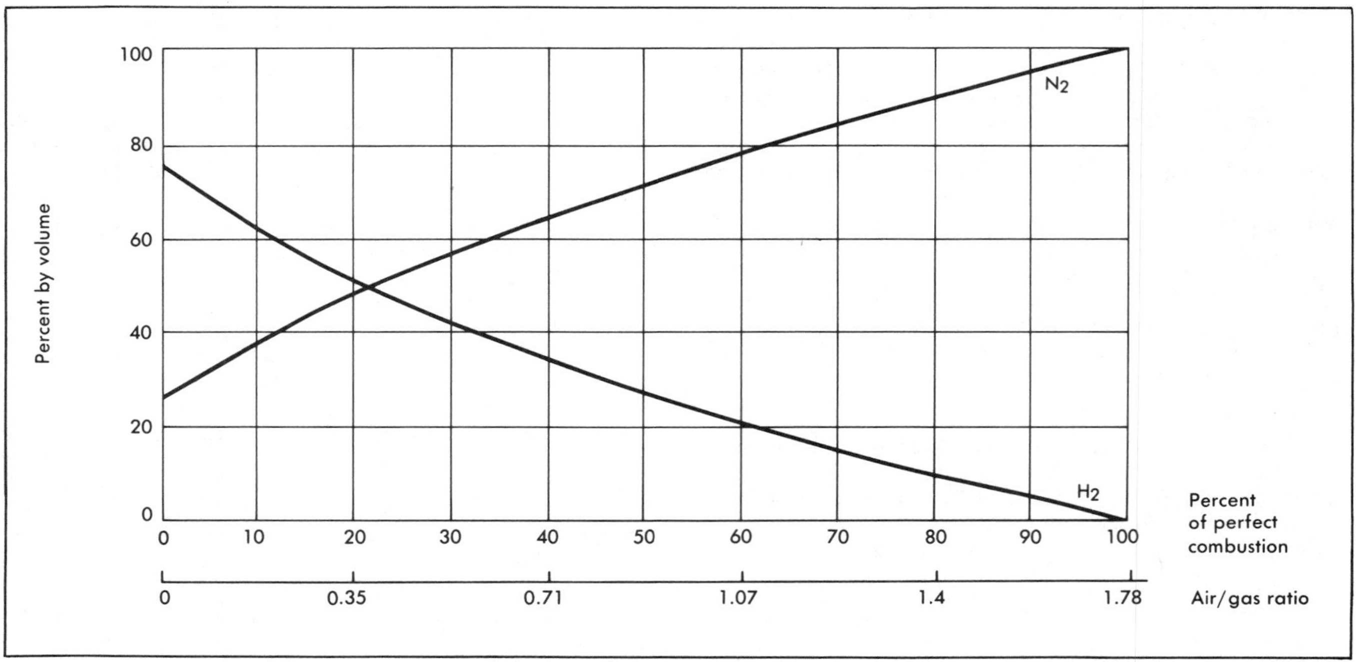

Fig. 10-15 Atmosphere constituents from a hydrogen-nitrogen generator for various air to fuel gas ratios. Fuel gas is dissociated ammonia: 75% H_2 and 25% N_2.

detectors in the catalytic chamber make up half of the balanced electric circuit. A corresponding imbalance in the bridge resistance develops and the resultant electrical output voltage, proportional to the concentration of total combustibles in the sample, is read on the panel meter.

Dew point analyzers. Dew pointers are among the most commonly used instruments for monitoring and controlling endothermic atmospheres. They are relatively inexpensive and simple to operate. Their major limitation, however, is in respect to accuracy. For example, an error of 5 to 6° F (2.8 to 3.3° C) in dew point measurement can represent a variation of 0.10 to 0.20% in carbon potential for a given furnace atmosphere.

Infrared controllers. Infrared controllers represent a popular method of measuring carbon monoxide, carbon dioxide, and methane content, rather than measuring the dew point. They provide the necessary sensitivity and stability required to correlate carbon dioxide concentration with the carbon potential of the furnace atmosphere, and a readable measurement of as low as 0.01% carbon dioxide may be obtained.

Infrared gas analyzers work on the principle that most substances have a unique absorption pattern of infrared energy. The infrared absorption of a compound is a characteristic of the type and arrangement of the atoms composing the molecule. The frequency or wave length at which absorption occurs is determined by the mass of the atoms and the strength of the bond joining them. Nearly all gases absorb infrared energy at characteristic wave lengths, except those few having no infrared absorption band, such as oxygen, hydrogen, and nitrogen.

Oxygen probes. The oxygen probe is the latest and most universal technique being used in the heat processing industry. The probes are employed for monitoring and controlling different atmospheres, such as endothermic gas, exothermic gas, steam with small traces of hydrogen, direct-firing (fuel-rich) atmospheres, and hydrogen-bearing gases with low to medium water content.

The oxygen probe is based in theory on a hot, ceramic electrochemical cell. The probe responds to oxygen, hydrogen, carbon monoxide, water vapor, and carbon dioxide. Thus, it can determine the oxidation potential of a gas. The output of the oxygen probe is a direct measurement of the oxidation potential of the atmosphere at the process temperature of the furnace. Therefore, when the probe temperature is close to the furnace temperature, the response of the probe is a direct indication of whether the atmosphere will oxidize or reduce steel, provided that the composition of the atmosphere with regard to the proportions of carbon gases and hydrogen is known. Under such conditions, the probe will give a reliable indication of the oxidation/reduction situation for all furnace temperatures. Hence, with simple mechanical methods, additive gases or liquids can be introduced into the furnace in order to control the oxidation/reduction potential of the atmosphere.

Specific gravity analyzers. A specific gravity analyzer is a simple mechanical instrument that is very sensitive to any changes in the specific gravity of the gas being measured. A common application is determining the complete purge of furnace retorts or bells to prevent explosions. The specific gravity analyzer is also well suited to measure and record ammonia dissociation in nitriding applications.

Thermal conductivity analyzers. Gas analyzers of the thermal conductivity type operate from the principle that each gas constituent has its own thermal conductivity characteristic. Typical applications include measuring hydrogen-nitrogen mixtures and impurities in hydrogen, helium, nitrogen, and argon. Operation of the analyzer is based upon the effect a gas sample has on the temperature of a heated wire filament. The filament is placed in a cylindrical well of a cell block and is one arm of a Wheatstone bridge circuit. The filament is heated and heat is lost to the cell walls at a rate depending upon the thermal conductivity of the gas. A change in concentration of the gas changes the rate of heat loss and upsets the bridge circuit, resulting in an analog signal proportional to the concentration of the component of interest.

Oxygen analyzers. Some applications require the use of analyzers to determine the oxygen content of a given processed gas. The most common practice for oxygen analysis in heat treating furnaces is to determine the lack of oxygen. Depending upon the level of oxygen desired or permitted in the process gas, specific ranges of concentration may have to be determined. Magnetic oxygen analyzers are commonly used. Oxygen has an affinity for magnetic fields; most other gases do not. By adding a sample gas to a magnetic field and a detector, the change in resistance can be measured in a cell. The output from this cell is fed to a meter or a recorder, and quite often is used as a permissive circuit in an atmosphere control system.

QUENCHING

The optimum combination of strength and ductility in a given steel is generally achieved by heat treatments that yield a homogeneous microstructure. One of the most effective means of accomplishing homogeneity is by first quenching to a martensitic structure and then tempering to obtain the desired hardness and ductility.

Effective heat treatment involves not only critical heating rates and temperatures, but also critical quenching rates and temperatures. These critical factors depend primarily upon the carbon and alloy content of the material to be heat treated, and the section size and shape of the workpiece. For optimum results in quenching, it is necessary that the quenching medium provide heat extraction fast enough to avoid pearlite transformation, but slow enough to ensure uniformity of heat extraction throughout the workpiece.

Rate of Heat Extraction

The rate of heat extraction varies widely depending upon the section size and geometry of the part, the amount of surface area available for heat transfer, the type of steel, the type of quenching medium, its temperature, and the degree of circulation or agitation of the medium. Quenchants are agitated in tanks by the use of propellers to move the medium with relation to the workpieces or by pumps that force the medium through appropriate orifices. In some cases the workpieces are moved through the medium, and for some applications a spray quench is used. Agitation of the quenching medium is sometimes accomplished through the use of ultrasonic energy. Some parts, such as gears, are often quenched in presses with the workpieces held in dies, where the medium is introduced.

Table 10-3 shows the comparative severity of some quenching media under various quenching conditions. Table 10-4 contains center and surface quenching rates for 1" (25 mm) diam steel rods cooled as much as 1000° F (540° C). It should be noted from Table 10-4 that cooling in air results in nearly simultaneous rates for both the surface and the center of the rods.

Table 10-5 lists the center cooling rates measured at temperatures of 1300 and 400° F (705 and 205° C), using a stainless steel probe to eliminate influences resulting from heats of transformation. These temperatures are critical in the quenching of many steels. To avoid transformation to pearlite, high cooling rates are required at a temperature of 1300° F. Slow cooling rates are desirable at 400° F to avoid rapid, nonuniform martensitic transformation and consequent distortion or cracking.

As shown in Table 10-5, water cools rapidly at 1300° F (705° C), but it also cools at a relatively high rate at 400° F (205° C). Oil cools somewhat slower than water at 1300° F and considerably slower at 400° F. Different types of water-soluble polymer quenchants, discussed subsequently in this section, come close to oil with respect to cooling rates or can provide rates between oil and water.

TABLE 10-3
Severity of Quench for Various Media

	Quenching Media			
	Air	Oil	Water	Brine
Quenching Conditions	Severity of Quench, H*			
No circulation of media or agitation of workpiece	0.02	0.25-0.30	0.9-1.0	2
Mild circulation or agitation	---	0.30-0.35	1.0-1.1	2-2.2
Moderate circulation	---	0.35-0.40	1.2-1.3	---
Good circulation	---	0.4-0.5	1.4-1.5	---
Strong circulation	---	0.5-0.8	1.6-2	---
Violent circulation	---	0.8-1.1	4	5

* Nominally, the severity of quench, H, is based on a value of 1.0 for still water.

TABLE 10-4
Surface and Center Quenching Rates for 1" (25 mm) Diam Steel Rods [ΔT = 1000° F (540° C)]

Temperature of Rods, °F (°C)	Quenchant	Quenchant Temperature, °F (°C)	Quenching Rate, °F/s (°C/s) Surface	Center
Austenitizing temperature	Air	30 to 110 (-1 to 43)	5 (2.8)	4 (2.2)
	Lead	800 to 1200 (427 to 649)	60 (33.3)	20 (11.1)
	Salt	950 to 1150 (510 to 621)	30 (16.7)	9 (5)
	Salt	400 to 600 (205 to 316)	50 (27.8)	10 (5.6)
1050 (566)	Air	30 to 110 (-1 to 43)	2 (1.1)	2 (1.1)

TABLE 10-5
Center Cooling Rates for a 1" (25 mm) Diam, Type 304 Stainless Steel Probe, Quenched from 1600° F (870° C), with Quenchant Circulation at the Rate of 100-150 fpm (30.5-45.7 m/min)

Quenchant	Quenchant Temperature, °F (°C)	Quenching Rate, °F/s (°C/s)	
		at 1300° F (705° C)	at 400° F (205° C)
Water	90 (32)	137 (76.1)	31 (17.2)
Accelerated oil	140 (60)	83 (46.1)	9.8 (5.4)
16% PVP	125 (52)	103 (57.2)	22.5 (12.5)
25% PVP	130 (54)	82.4 (45.8)	12.9 (7.2)
16% PAG	125 (52)	60 (33.3)	27.5 (15.3)
30% PAG	130 (54)	26.8 (14.9)	14.2 (7.9)

(*Park Chemical Co.*)

Mechanism of Heat Transfer

Any quenching medium cools only the surface of the metal. Heat inside the metal must flow to the surface by conduction and must be carried away by the quenchant. As a result, cooling takes longer for thicker parts and sections.

The rate of cooling in a workpiece during quenching is determined chiefly by the surface cooling rate and the thermal diffusivity of the workpiece material. This surface cooling rate varies with and is characterized by three distinct stages: vapor-blanket, vapor-transport, and liquid.

Vapor-blanket stage. The part being quenched is at a high temperature compared with the quenching medium. The liquid in immediate contact with the part vaporizes and surrounds the part with an insulating vapor film, and this film keeps additional coolant from contacting the part. The total heat loss is that which is necessary to vaporize the liquid plus the small amount of heat lost by radiation through the vapor film. Therefore, cooling in this stage is relatively slow.

Vapor-transport stage. This stage of quenching begins when the liquid breaks through the vapor film and makes contact with the part. This results in boiling of the liquid that contacts the part. The heat is rapidly transferred during this stage because fresh cooling liquid continually replaces the vapor film that tends to form at the surface of the part. For liquids having a high heat of vaporization, such as water, this is an especially rapid process.

Liquid stage. This stage begins when the rate of coolant flow is such that the liquid does not reach its boiling point while passing over the work, even though the surface temperature of the work is higher than the boiling point of the liquid. The boiling action ceases and the entire surface of the part is in intimate contact with the liquid. Cooling of the part then takes place by conduction and convection.

Quenching Media

As discussed previously, the choice of a quenching medium is determined by the metal to be treated, the size and shape of the parts being treated, and the desired physical properties. These factors also establish requirements for the cooling rates and method of quenching. The first requirement of a quenchant is that it provide the desired speed of cooling without distorting, cracking, or breaking the part. Other factors influencing the selection are cost (both initial and in-use), ease of control and maintenance, safety and cleanliness in use, ecological considerations (pollution and disposal), and stability in extended use.

A wide variety of quenchants are used for various heat treating processes. Those most readily satisfying the requirements discussed include air, water, oils, aqueous polymer solutions, brine, caustic solutions, and molten-salt and lead baths.

Cooling in air. Air is used preferably whenever the degree of distortion by oil quenching is objectionable or whenever the high alloy content of the workpiece material will permit full hardening by air. Both direct and indirect costs are low when the desired structure can be achieved by air cooling.

Water. Plain water is probably the most widely used of all quenchants except air and fully meets the requirements of low cost and availability. It is easily handled and safe, but its cooling characteristics change more with agitation and variations in temperature than do those of many oils and polymer solutions. The high cooling rate of water may result in excessive distortion or cracking of some workpieces.

With vigorous circulation and with maintenance of substantially constant bath temperatures, water may be considered a satisfactory quenchant for many applications. However, whenever the flow of water is slow and there is no provision for keeping it cool, water may prove wholly unsatisfactory. Also, its use is limited to parts of simple configuration made from water-hardening steels and certain nonferrous compositions.

Water quenching is effective for breaking scale from the surfaces of steel parts that are heated in furnaces without protective atmospheres. The water is generally maintained at a temperature of about 65° F (18° C) for most applications. As the temperature increases and/or agitation decreases, there is a tendency for an envelope of steam to form around the workpiece, thus reducing the cooling rate.

Quenching in oil. Oils comprise one of the most important groups of quenching liquids. In addition to straight mineral oils, many different types of oils are available in a wide variety of blends compounded to accelerate cooling rates by shortening the vapor-blanket stage. Heavier oils are available that have flash and fire points in excess of 500° F (260° C). These oils are used for quenching at elevated temperatures, such as 300-400° F (150-205° C) in modified martempering (marquenching), discussed subsequently in this section.

Most quenching oils have good stability and chemical inactivity with hot steel, plus little change in the cooling rate with minor variations in temperature. Cooling coils are sometimes used to control the temperature of the oil.

Polymer quenchants. Water-based polymer quenchants, sometimes called synthetics, are being used increasingly because they offer quenching characteristics intermediate between water and oil and, in some cases, very similar to oil. Because aqueous solutions are used, these quenchants are nonflammable, clean to use, offer fewer ecological problems, and are often more economical. On the other hand, they require more careful selection of quenching conditions, control, and maintenance. The principal types of polymer quenchants are polyglycols (PAG's), which are most widely used because they've been available longer, the newer polyvinylpyrrolidones (PVP's), and sodium polyacrylates (SPA's). These water-soluble polymers work by somewhat different mechanisms than other quenchants, but all act to some degree to slow down both the vapor transport and liquid cooling stages, compared with water. Polymer type, concentration, quenchant temperature, and agitation all strongly influence their cooling rates. Agitation is particularly important in comparison with quenching with oil.

Quenching in brine. Some brine quenchants include aqueous, sodium chloride solutions. Proprietary brines, which eliminate rusting and prevent corrosion, are also widely used in place of water when more uniform and faster cooling in the vapor blanket/vapor transport stages are desired. Control of agitation and temperature are important as with water. Disadvantages of brines include higher cost than plain water, possible corrosiveness, and the need for fume vent hoods in some applications.

Caustic solutions. Caustic solutions, such as sodium hydroxide, with strengths of about 5% are sometimes used to give a brine-like quench. These quenchants, however, are losing favor because of their corrosivity and handling problems, plus their tendency to pick up carbon dioxide. Carbon dioxide changes quench characteristics gradually and makes control difficult. High alkalinity of the solutions also make them harmful to the skin.

Molten salt. Baths of molten salt are used predominantly for quenching high-speed steels and for the interrupted quenching methods discussed subsequently in this section. Because of the high thermal conductivity of molten salts and the high temperatures at which they can be maintained, to 950-1050° F (510-565° C), these quenchants provide rapid initial cooling with low distortion and cracking. For interrupted quenching operations, molten-salt baths can also be used at temperatures ranging from 300 to 1000° F (150 to 540° C).

Lead baths. Lead baths are the only metal baths that have proved suitable as quenching media for heat treating. Lead has a working temperature range of 650 to 1700° F (345 to 925° C) and a high level of heat conductivity; thus, its quenching rates are high. However, the extreme toxicity of lead has greatly reduced its use in recent years.

Gas quenching. Quenching with pressurized gas provides cooling rates more rapid than with still air, but slower than with oil. Gas quenching is done in gas-tight chambers or zones of furnaces. Relatively cold gases are directed onto the workpiece surfaces, and after absorbing heat they are cooled and recirculated. Pressurized gases used include air, nitrogen, and mixtures of various gases. Cooling rates can be adjusted by varying the type, pressure, and velocity of the gas.

Fog quenching. In this process, streams of gas containing droplets of water are used to increase the cooling rate with minimal distortion.

Quenching Specific Steels

Production-grade steels and tool steels can be classified into three groups with respect to the most suitable quenchant for use with each group. The groups are: plain carbon, low-to-medium alloy, and high-alloy steels.

Plain carbon steels. These metals, such as SAE 1030 and 1045 steel or W1 tool steel, are often quenched in water at a temperature of about 70° F (20° C) or in brine (a 10% salt solution or an inhibited brine) at 80 to 100° F (25 to 40° C). For workpieces having variable section sizes that are susceptible to distortion or cracking, dilute solutions of polymers or fast-quenching oils are more suitable.

Low-to-medium alloy steels. These metals, such as SAE 4140 and 4340 or 01 tool steel, are usually quenched in oil or polymer quenchants that approach oil in quench severity. Molten-salt baths are also used for interrupted quenching processes.

High-alloy steels. These metals, such as A2 and M2 tool steels, are either cooled in air (as is the case with A2 tool steel) or quenched in a molten-salt bath at a temperature of 1000 to 1100° F (540 to 595° C), with subsequent air cooling. Interrupted quenching using oil or polymer quenchants is sometimes feasible but requires close control of quenching conditions.

Interrupted Quenching

Interrupted quenching is used to improve the ductility and toughness of steel, while retaining high hardness levels with minimum distortion and residual stresses. In interrupted quenching, the rapid cooling of the metal is stopped and held at points above the M_s temperature for specified lengths of time, followed by cooling in air. Three types of interrupted quenching are: austempering, marquenching (martempering), and isothermal quenching. For each of these methods, the temperature at which the quenching is interrupted, the length of time the steel is held at temperature, and the rate of air cooling are varied for different types of steels and workpiece sizes.

Austempering. The process known as austempering consists of a rapid quench through the pearlite formation range down to a temperature of about 450 to 750° F (230 to 400° C), depending upon the transformation characteristics of the particular steel involved; a hold at temperature for a length of time sufficient for isothermal transformation; and air cooling. Subsequent tempering is not required.

The generalized TTT diagram shown in Fig. 10-16 provides an example of austempering and isothermal quenching, compared with the form of interrupted quenching employed in marquenching, discussed next in this section. The product of austempering transformation is essentially bainite, which possesses hardness values of R_C 50 to 55, accompanied by a higher degree of toughness than can be obtained by quenching to martensite and tempering to the same degree of hardness.

The lower the temperature of transformation (as the transformation temperature approaches M_s), the harder the product of transformation. Therefore, if the rate of cooling in the upper region can be sufficiently rapid to avoid the transformation to pearlite, there is a fairly wide choice of temperatures at which transformation can be allowed to take place, depending upon the desired mechanical properties. It is usually best to use a quenching bath held at the temperature where the beginning of transformation is the most sluggish for the particular steel involved. Probably the most common temperature for austempering is 550-600° F (290-315° C). In any case, the

QUENCHING

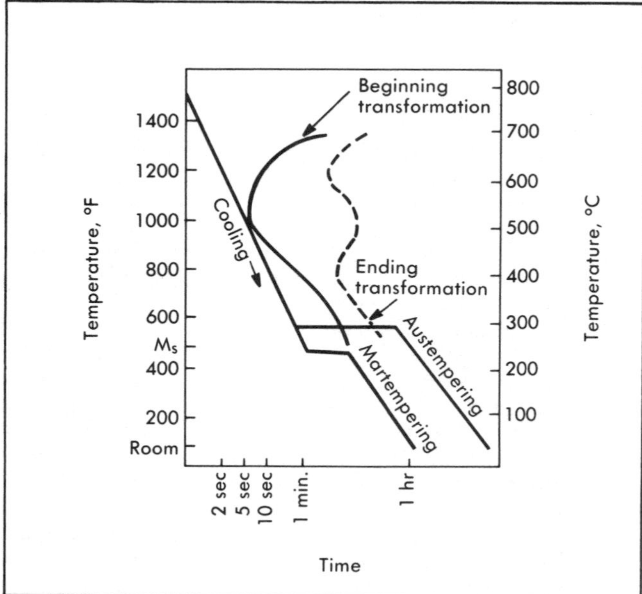

Fig. 10-16 Cooling procedure in austempering and marquenching.[6]

temperature should be above M_s. However, transformation to bainite is often very slow just above M_s. As a result, somewhat higher temperatures, 50-150° F (28-83° C) above M_s, are commonly used.

Molten-salt baths have been found most practical and economical for practically all isothermal quenching applications. Oils have been developed that suffice in some cases; but, in general, molten salt possesses better heat-transfer properties. Irrespective of the medium used, vigorous agitation is necessary.

Austempering is applicable to most medium-carbon steels and low-to-medium alloy steels, with limitations on workpiece thickness. The less time available during quenching for avoiding the transformation to pearlite, the thinner the section size that can be successfully austempered. Lower alloy steels are usually restricted to 3/8″ (9.5 mm) or thinner sections, while more hardenable steels can be austempered with sections up to 2″ (51 mm) thick. Recent work on austempering of cast irons shows promising results.

Marquenching (martempering). This process is similar to austempering in that the work is quenched rapidly from the austenitizing range into an agitated salt bath or similar medium held near the M_s temperature. It differs from austempering in that the work is allowed to remain at temperature only long enough for the temperature to be equalized throughout all sections of the workpiece. The process takes advantage of the time allowed in the temperature range just above M_s, where the beginning of transformation is quite sluggish. When the temperature has become equalized, but before the beginning of any transformation, the work is removed from the salt bath and allowed to cool in air to room temperature. During this time, martensite forms at a reasonably uniform rate throughout all sections.

In marquenching, it is absolutely necessary that the austenitizing time and temperature be closely controlled, since the beginning of transformation and the M_s temperature are both dependent on the composition of the austenite. It is also necessary to know the M_s temperature for the steel being treated. Theoretically, the quenching bath should be held at a temperature slightly above M_s; but for practical purposes, in order to obtain greater cooling power, the quenching bath is held at or slightly below M_s. Since the transformation to martensite takes place slowly at temperatures as low as 25-50° F (14-28° C) below M_s, it is usually not harmful if the transformation has started prior to removal of the work from the bath. Only a small fraction of austenite transforms to martensite under these conditions, but the transformation begins immediately on cooling.

Cooling from the marquenching bath to room temperature should usually be conducted in still air. The deeper hardening types of steel are especially sensitive and susceptible to cracking during the time martensite is forming. As a result, it is inadvisable to hasten cooling by further quenching or air blasting. The alloy carburizing steels, which also lend themselves to marquenching because of their softer core, are insensitive to cracking during martensite formation; consequently, the rate of cooling from the M_s temperature is not critical.

Marquenching does not remove the necessity for subsequent tempering. The structure of the metal is essentially the same as that formed by quenching directly in oil or polymer quenchants; hence the subsequent tempering treatment should be the same. If the process has been carried out as planned, its only difference from oil or polymer quenching lies in the fact that the hardened structure has been formed with far greater uniformity. Therefore, unbalanced residual stresses resulting from the formation of martensite (and the corresponding volumetric increase) are greatly decreased.

Some of the limitations imposed on austempering also apply to marquenching. Steels must contain some alloy to slow down the beginning of transformation, and carbon steels must be relatively thin in section to lend themselves to the process. Even for SAE 1060 steel, less than one minute is available before transformation starts at 550° F (290° C). Additions of water to molten salt at 350 to 600° F (175 to 315° C) produce a pronounced increase in the rate of cooling and are being used to allow marquenching of heavier sections.

Oils are used successfully for marquenching; but molten salt, because of its better heat-transferring properties, is usually preferred. Figure 10-17 shows a comparison in cooling rates induced by water, oil, polymer quenchants, and salt, as well as a comparison of temperature gradient in the sections, as the workpieces pass into and through the martensite formation range.

Regardless of the quench medium selected, a close temperature control must be maintained to ensure uniformity in the work. Obviously, a method of controlled heating must be provided to obtain the desired temperature. Then, after quenching starts, the major concern is to provide the highest rate of cooling possible to avoid transformation to other than martensite. Agitation is important and is usually done by motor-driven stirrers or pumps. Water additions are sometimes used in salt baths to increase heat extraction. Table 10-6 indicates the approximate amounts of water that can be maintained for many hours in molten-salt baths.

Under carefully controlled conditions, steels that are hardenable by oil quenching are hardenable by marquenching. The higher carbon steels, however, are best adapted to this process. Some of the medium-carbon, low-alloy steels may be similarly treated, provided that all their sections are thin. Carburized alloy steels are also martempered.

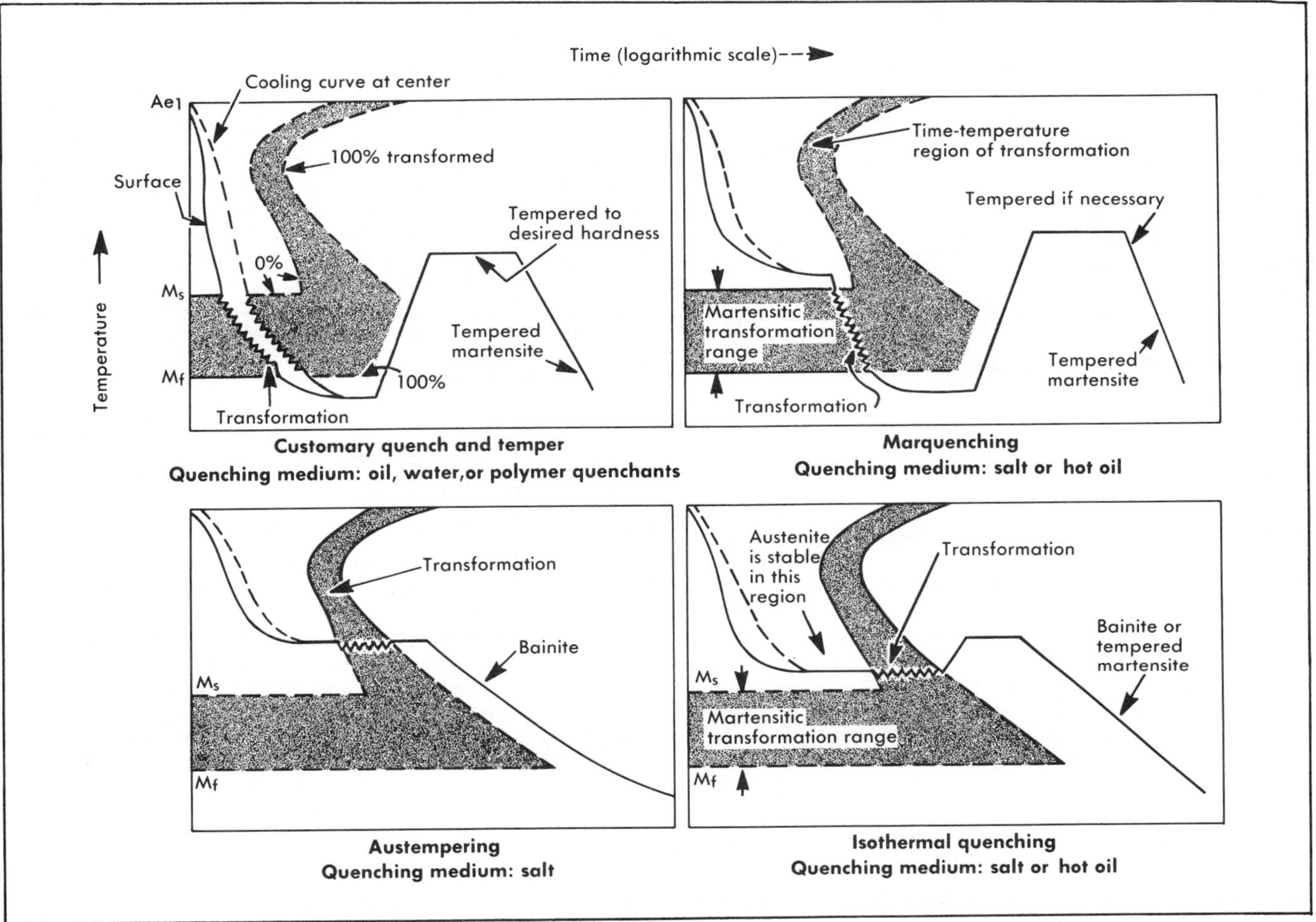

Fig. 10-17 Comparisons of quenching in various media.[7]

Isothermal quenching. Isothermal quenching is similar to austempering and, as such, is often not classified as a different type of quenching. The steel is rapidly quenched through the pearlite formation range to a temperature just above M_s where bainite is formed. However, isothermal quenching differs from austempering in that a two-bath quench is employed. After the first quench, and before transformation has time to begin, the work is quenched in another bath at a somewhat higher temperature. It is allowed to transform isothermally, after which it is allowed to cool in air. This method of quenching must be tailored for each specific application and must be closely controlled. As a result, it is not used extensively.

TABLE 10-6
Approximate Percentages of Water Usable in Molten-Salt Baths at Various Temperatures

Temperature of Molten Salt Bath, °F (°C)	Percentage of Water, by Weight, %
400 (205)	2.0
500 (260)	1.0
600 (315)	0.5
700 (370)	0.25

SOLUTION TREATING AND AGING

Solution treating and aging are processes employed to harden and strengthen many nonferrous alloys and some steels. The procedure generally requires the following two consecutive operations:

1. Solution heat treatment to produce a homogeneous solid solution that is retained by rapid quenching to room temperature.
2. Aging to produce fine precipitates in the solid solution.

SOLUTION TREATING AND AGING

SOLUTION HEAT TREATMENT

In solution heat treatment, an alloy is heated to a suitable temperature, held at this temperature for sufficient time to allow a desired constituent to enter into solid solution, and then rapidly cooled to hold the constituent in solution. Most of these solid solutions are comparatively soft and ductile. They are also structurally unstable and have a tendency to return to the form of an aggregate with aging. Solution annealing is discussed in a subsequent section of this chapter under the subject of annealing.

Metals with solid solutions that age harden at room temperature can be prevented from doing so by cooling below the freezing temperature after quenching and holding at a subzero temperature for as long as required. Such a delay in precipitation facilitates forming or other operations prior to aging. For example, rivets made from heat-treatable aluminum alloys are often refrigerated after solution treating, removed from low-temperature storage for assembly, and then allowed to age at room temperature.

AGING

Aging, also called age hardening or precipitation hardening, is a time-temperature dependent change in the properties of certain metals (higher strength and hardness) occurring at room or elevated temperatures. Aging at room temperature is called natural aging; aging above room temperature, which requires less time, is termed artificial aging. Quench aging is aging by rapid cooling after solution heat treatment.

Interrupted aging is aging at two or more different temperatures, with cooling to room temperature after each heating; step aging is also aging at two or more different temperatures, but without cooling to room temperature after each step. Double aging refers to a solution annealing followed by reheating, a second quenching, and a second reheating. Some alloys will precipitation harden after a double-aging cycle with no prior solution annealing.

Carbon Steels

When quenched, untempered, low-carbon steels are allowed to rest at room temperature for more than about 10 hours, their hardness will increase. A smaller increase in hardness is produced with high-carbon steels. Such increases in hardness resulting from precipitation in steels having medium-to-high carbon content are not as high as the increases attainable with transformation hardening. However, aging does provide an alternative when transformation hardening is not feasible or possible. Also, it is often possible to harden and strengthen the centers of large forgings by aging.

Stainless and Alloy Steels

The precipitation-hardening grades of stainless steels, maraging steels, and some heat-resistant, high-strength alloys are often hardened and strengthened by aging. Details of the solution treating and aging of these materials are presented in Chapter 14, "Heat Treatment of Other Metals," in this volume.

Nonferrous Metals

Many aluminum alloys containing elements such as copper, magnesium, zinc, and silicon are treated by natural or artificial aging. Magnesium alloys are solution heat treated to improve strength, toughness, and shock resistance. Artificial aging after solution treatment provides maximum hardness and strength, but toughness is reduced. Some titanium alloys are also solution treated and age hardened. Overaging is sometimes performed to increase toughness and ductility, with only a small loss in tensile strength. Data on the solution treating and aging of some of the more common nonferrous metals is presented in Chapter 14, "Heat Treatment of Other Metals," in this volume.

TEMPERING

In its hardened, as-quenched, fully martensitic form, steel is hard and brittle, thus restricting its usefulness. Hardened steel must be subsequently tempered to relieve quenching stresses and to provide a limited but necessary degree of toughness and ductility, protecting the part from cracking in storage, installation, and use. Impact resistance and improved elongation and area reduction qualities are afforded by tempering, but are brought about by a sacrifice in hardness, tensile strength, and wear resistance. Tempering consists of reheating previously hardened metal to a relatively low temperature (below the transformation range), generally followed by slow cooling. During this treatment, the martensite phase transforms to the more stable, two-phase mixture of ferrite and carbide. Certain steel parts, tempered in the temperature range of 400 to 700° F (205 to 370° C) may be water quenched to increase their impact strength.

Machinability is a function of hardness and toughness and is seldom changed appreciably by hardening and tempering at low temperatures. With tempering at low temperatures, grinding is often the only machining possible on some tempered steels. After tempering operations at 900 to 1200° F (480 to 650° C), turning or milling of medium-carbon steels is possible.

The higher the tempering temperature, or the longer the time at that temperature, the softer and more ductile the steel. Alloy steels containing carbide-forming elements show greater resistance to softening during tempering than do plain carbon steels. Therefore, higher temperatures and longer times should be used in tempering the higher alloy steels, as shown in Fig. 10-18. Tempering may be performed to give almost any desired combination of properties by correct selection of the time-temperature cycle, because the strength of steel is proportional to its hardness, as shown in Fig. 10-19. The composition of the steel, its condition before quenching, and the effectiveness of the quench all affect the tempering temperature required to produce a specified hardness.

AVOIDING BRITTLENESS

Tempering of martensitic steel to ferrite and carbide can be done at any temperature up to Ac_1, above which austenitization begins. The size of the carbide particles precipitated during tempering becomes coarser with higher tempering temperatures, and the steel becomes correspondingly softer. Parts made from carbon and low-alloy steels that will be subjected to impact stresses in service should not be tempered in the temperature range of 400 to 600° F (205 to 315° C). Impact toughness decreases in this temperature range—a phenomenon called blue brittleness because of the surface color developed.

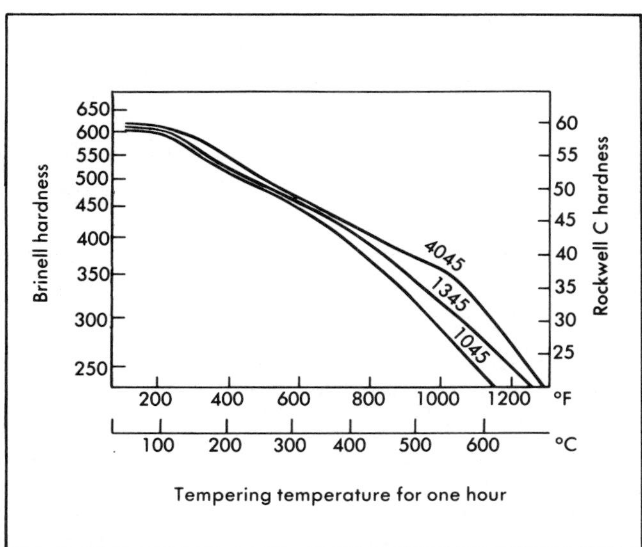

Fig. 10-18 Tempered hardness as affected by alloy content.

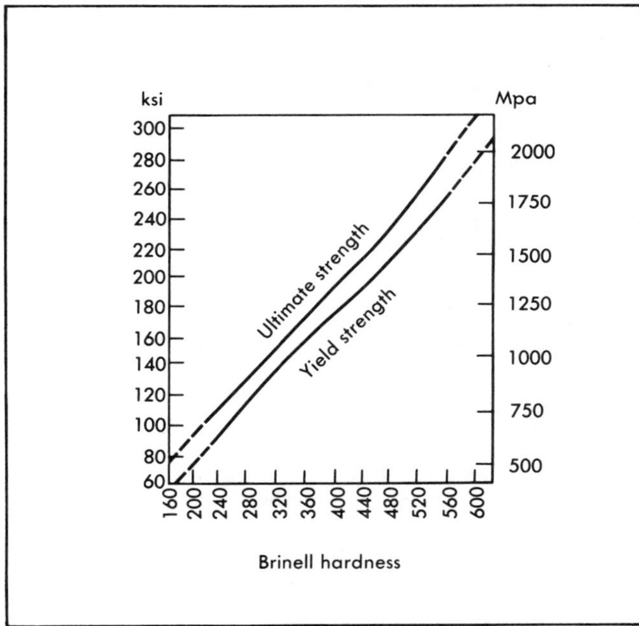

Fig. 10-19 Increase in strength of steel with increased hardness.

Another phenomenon, called temper brittleness, can occur when parts, primarily those made from alloy steels, are tempered in the temperature range of about 700 to 1000° F (370 to 540° C), or when they are cooled slowly through this temperature range from a higher tempering temperature. Temper brittleness can sometimes be avoided by quenching from tempering temperatures above about 1000° F (540° C). Alloy steels containing molybdenum, which require high tempering temperatures, have less tendency toward temper brittleness. To reduce the possibility of temper brittleness, impurity elements, such as phosphorous, tin, and antimony, should be kept at a minimum.

TEMPERING TOOL STEELS

Hardened tool steels are often in a highly and nonuniformly stressed condition. Since the danger of cracking is most pronounced just after quenching, the tools should be tempered just as soon as they can be handled comfortably with bare hands and then cooled slowly to room temperature. The rate of heating should be slow to obtain uniform distribution of temperature within the tool and prevent cracking. A cold quenched tool should never be placed in a molten bath for tempering for this could readily cause it to crack or warp. Tool steels should be cooled in still air after tempering to prevent residual stresses from forming. The transfer from quenchant to cleaner to tempering furnace should be within 20 minutes for crack-sensitive and high-carbon grades of steel, particularly when fully hardened. Sections up to 1" (25 mm) thick should be tempered 1-2 hours at temperature. The center temperature will lag behind furnace temperature readings, so one hour of temper per inch of steel section should be used to assure penetration of heat to the center.

MULTIPLE TEMPERING

Multiple tempering will very often correct poor hardening practices and may be metallurgically essential. Of necessity, the presence of variable amounts of retained austenite must be reduced by multiple tempering, a method more effective than a single tempering operation for highly alloyed steels. Double or triple tempering is advisable for air-hardening, hot-work, and high-speed steel tools. The first treatment tempers the martensite formed during quenching and also conditions some of the retained austenite for transformation to martensite during cooling to room temperature from the tempering temperature. The second treatment is then necessary to temper the martensite formed during the first tempering operation. The third tempering, when used, further decreases the amount of untempered martensite still remaining, thereby improving mechanical properties. Additional information on heat treatment is presented in Chapter 3, "Cutting Tool Materials," of Volume I, *Machining*, and Chapter 2, "Die and Mold Materials," of Volume II, *Forming*, of this Handbook series.

TEMPERING MEDIA

Media for tempering are usually simple baths or furnace atmospheres. The four most commonly used are:

1. *Hydrogen-nitrogen*. Hydrogen-nitrogen atmospheres for tempering are prepared by the dissociation of ammonia. Such atmospheres can leave bright surface finishes when the work is allowed to cool in a low dew-point hydrogen atmosphere.
2. *Air*. Air is a slightly oxidizing tempering medium and may produce scaling at temperatures above 1000° F (540° C).
3. *Molten nitrite/nitrate salts*. These media produce oxidizing effects very similar to those of air.
4. *Steam*. Dry steam is used to purge tempering furnaces heated to 700° F (370° C). The addition of steam is continued at the tempering temperature, and then the work is cooled in sulfonated, emulsified oil to enhance the resulting lustrous black finish.

Multiple tempers may be obtained in the following ways:

1. *Selective tempering*. Localized areas of the steel can be heated to tempering temperature by a torch, salt bath, or

induction heater to produce special localized impact or tensile properties without affecting the properties of the part as a whole.

2. *Cold treatment*. Tempered steel with over 10% austenite retained after quenching can be cooled to -120° F (-84° C) and then retempered to convert the remaining austenite to martensite, thereby reducing brittleness. Cold treatment is also useful for stabilizing closely toleranced parts. Additional information on cold treatment is presented in a subsequent section of this chapter.
3. *Nitriding*. Nitriding in dissociated ammonia or in molten cyanide salt provides wear-resistant nitride cases. Details about the nitriding process are presented in Chapter 11, "Surface (Case) Hardening," of this volume.

ANNEALING

Annealing is a heat treatment process that consists of heating metal to a suitable temperature, followed by slow cooling. The purposes of annealing include one or more of the following:

1. To reduce hardness or brittleness.
2. To relieve stresses.
3. To improve machinability or facilitate cold working.
4. To produce desired microstructure or properties.
5. To remove gases.
6. To alter the electrical or magnetic properties.

There are many different types of annealing, some of which are defined in the glossary of heat treating terms presented at the beginning of this chapter. Typical cycles for annealing some common carbon steels are presented in Table 10-1, and for alloy steels in Table 10-2. Continuous annealing of sheet steel is now being done by some steel producers. This method ensures more uniform physical properties and replaces the slower, less consistent annealing of coil stock in batch furnaces.

EFFECTS OF HOT AND COLD WORKING

If a pure metal or alloy is stressed beyond its yield point, it will deform plastically. Deformation at an absolute temperature greater than one-half the material's melting point will generally result in the spontaneous formation of new strain-free grains that can be repeatedly deformed to a high degree without cracking, spalling, or breaking. This elevated temperature deformation is called *hot working* and the formation of new strain-free grains is known as *recrystallization*.

Mechanical deformation at temperatures less than approximately one-half the melting point results in both macroscopic regions of compression and tension and microscopic disorientations of the atoms from their equilibrium or unstressed positions, which may persist for long times at the deformation temperature. This type of deformation is called *cold working*. The temperature at which cold working (also called *strain hardening*) begins is dependent on the metal composition, the rate of deformation, and the amount of total deformation.

The fields of stress within cold worked metal store energy that tends to make the material return to the unstressed condition by dissipating that energy. In other words, the cold worked metal is unstable because of the residual stresses. The amount of residual stress depends on the composition of the metal, the temperature at which the deformation takes place, the amount of cold work, and the deformation process.

Cold working affects the mechanical, electrical, and magnetic properties of the metal. Hardness, tensile and yield strengths, electrode potential, general tendency to corrode, and coercive forces are all increased. The coefficient of thermal expansion is also slightly increased. On the other hand, the elongation to fracture, reduction of area at fracture, drawing ability, and impact strength are all decreased. Slight decreases in density, electrical conductivity, and maximum magnetic permeability are also evident.

If the deformation is slight and thus the atomic mechanisms by which it occurred are simple, the residual stresses may relieve themselves at the deformation temperature over a moderate period of time. For example, single crystals of zinc simply deformed 10-20% at room temperature will return to the unstressed state in 24 hours. After larger and more complex deformations, on the other hand, the material must be heated to an elevated temperature for a period of time to recover its original properties and then cooled at a rate that will prevent the reintroduction of stresses. This heating and cooling cycle must be varied in both time and temperature for the different materials treated and the different material qualities desired. However, all heat treatments of this type are considered varying types of annealing.

STRESS RELIEVING

Stress relieving by heat treatment is a method of reducing the stresses that may be introduced with such operations as cold working, machining, welding, drawing, heading, and extrusion, without resorting to complete annealing or normalizing and without greatly affecting other properties. The reduction of such stress is important to avoid excessive distortion during hardening and to avoid cracking, which could result from the addition of the residual stress to the thermal stress produced in heating to the hardening temperature.

Stress relieving of steel basically involves the heating of the part to a subcritical temperature below Ac_1, holding at that temperature long enough to assure uniformity, and slow cooling to room temperature, usually in air, which will prevent the reintroduction of stresses. Microstructures are essentially unchanged. The temperatures most commonly used range from about 1100 to 1300° F (595 to 705° C). Stress relieving is commonly performed on structures where usual annealing or normalizing methods would be prohibitive because of part size or shape, warpage, or scale.

Stress relieving is not as effective as annealing in reducing residual stresses. However, annealing is a more costly and time-consuming operation; and because of the higher temperatures in annealing, decarburization, scaling, and grain growth are more of a problem. Because of the relatively low temperature involved in stress relieving, surface protection against scale or decarburization is usually not required. The temperature used is always below the transformation range of the steel. Therefore, the cooling rate is unimportant except that it must not be such as to cause thermal stress. Rapid cooling produces higher stresses than slow cooling.

Up to Ac_1 temperature, the higher the stress-relieving temperature, the more completely the stress is removed. The degree of stress relief afforded in this manner is usually sufficient to avoid the difficulties mentioned; but if for any reason complete stress relief is required, an annealing treatment must be employed.

Stress relieving will eliminate most of the stresses in parts prior to hardening and will reduce warpage and the risk of cracking. It is commonly used in the wire industry to relieve stresses introduced by drawing of the wire through draw dies, and it is also employed to partially relieve stresses introduced in the cold straightening of bar stock. This application is particularly effective for relatively long parts such as broaches, shafts, spindles, and leadscrews. In addition to relieving stresses, this type of thermal treatment can induce softness, spheroidize the cementite, reorient the grains, eliminate gases, and alter the mechanical and electrical properties.

After the steel has been stress relieved, the dimensions should be remeasured. Dimensional changes are an indication that stress has been relieved. It may therefore be necessary to correct some dimensions before proceeding with hardening. If no dimensional changes result from stress relieving, either the procedure was ineffective or no appreciable stresses were present.

Whenever a high degree of residual stress is involved, the stress-relieving operation can result in warpage and distortion. If straightening is required, it is often advisable to stress relieve again to reduce straightening stresses. Also, on intricate parts, it is often desirable to correct for dimensional changes that occur during stress relieving by machining or grinding before proceeding to the hardening operation.

FULL ANNEALING

Full annealing involves heating steel to above the upper critical temperature, Ac_3, for hypoeutectoid steels and just above the lower critical temperature, Ac_1, for hypereutecoid steels, followed by sufficient slow cooling through the transformation range to under 1000° F (540° C). This operation results in the softest pearlitic structure. Steel may be annealed to reduce hardness, to improve machinability, to improve cold working characteristics, to produce a desired microstructure, or to obtain desired physical properties. Ordinarily, full annealing is carried out in a batch-type furnace rather than a continuous furnace, and usually the work is cooled in the furnace. Slower rates of cooling may be obtained by progressively lowering the furnace controls to regulate the cooling at preselected rates. More recently, programmed heating and cooling is being accomplished accurately by the use of microprocessor-based temperature controls.

There are two important steps in full annealing: the initial formation of austenite and the subsequent transformation of the austenite to pearlite. Steel in the as-rolled or as-forged condition usually consists of ferrite and carbide. As described earlier in this chapter, these structures are converted to austenite upon heating to above the critical temperature, Ac_3. A fully austenitized microstructure produced at high austenitizing temperatures will most likely transform to a pearlitic constituent. Conversely, at lower annealing temperatures—which are closer to the lower critical temperature, Ar_1—a heterogeneous mixture of austenite plus ferrite *or* austenite plus cementite prevails, and the metal tends to form spheroidized structures upon transformation.

The temperature at which transformation occurs is also important. If the austenite transforms just below Ar_3, the lamellar pearlite or spheroidized carbide formed will be relatively coarse. Again, the conditions which determine whether the end product will be spheroidized or lamellar depend upon the composition of the steel and the annealing temperature. Generally, transformation just below the upper critical temperature results in the softest end product. At the lower transformation temperatures [about 100° F (55° C) below Ar_1], the transformation products are harder and less coarse. Also, the tendency toward a lamellar structure is greater. The time for transformation is shorter as compared with transformation just below the critical temperature. For a better understanding of the transformation of austenite, refer to the TTT curves for the steel under consideration.

In order to develop the softest condition during annealing, austenitization should usually be at less than 100° F (55° C) above Ac_3 to ensure a fine austenitic grain size, and transformation of austenite should usually be completed within 100° F of the Ar_3. Due to the relatively long time required to accomplish transformation in this temperature range, however, it is often expedient to allow most of the transformation to occur at temperatures within the range of 100° F below the Ar_3 temperature, and to finish the transformation at a lower temperature range where transformation proceeds at a faster rate. After complete transformation has taken place, the steel may be cooled to room temperature at a relatively rapid rate to decrease the total time of the annealing cycle.

It is difficult to generalize about the optimum structures for best machinability. Softer structures will often result in highest production rates, while harder structures tend to produce a better finish. As a guide, the common grades of alloy steels respond fairly well in machining at the following hardnesses with the following structures:

1. Low-carbon steel of 0.10 to 0.25% carbon should have a Brinell hardness of 183 to 207, with a microstructure of coarse, blocky ferrite plus uniformly distributed pearlite.
2. Medium-carbon steel of 0.25 to 0.50% carbon should have a Brinell hardness of 187 to 212 and a blocky, lamellar pearlite structure.
3. High-carbon steel, having a carbon content of 0.50 to 1.00%, should have a Brinell hardness of 197 to 217 and a completely spheroidized structure.

ISOTHERMAL (CYCLE) ANNEALING

Isothermal annealing may be defined as the austenitizing of a steel by heating it above its critical temperature Ac_3, followed by cooling it to and holding it for 4 to 8 hours at a preselected temperature, usually in the range of 1050 to 1300°F (565 to 705° C). In this temperature range, the austenite transforms to a relatively soft ferrite-carbide aggregate in accordance with the transformation diagram of the steel.

Isothermal annealing takes less time and is less costly than conventional full annealing with continuous cooling. In addition, the uniformity of hardness and structure is better with isothermal annealing. However, very close temperature control must be maintained. The equipment for isothermal annealing consists of a continuous furnace that includes one zone for austenitizing and another zone for holding and transformation. Salt baths can be ideal for isothermal annealing, depending upon cost. Properly rectified baths will protect steel from decarburization, surface attack, and scale formation. Salt baths permit close control of temperature for austenitization as well as for transformation, which is more important in cycle annealing than in full annealing.

ANNEALING

The generalized TTT curves in Fig. 10-20 illustrate a comparison of conventional full annealing and isothermal annealing. It should be noted for full annealing that transformation occurs over a range of temperatures, resulting in a variation in structure and hardness in the annealed product. Isothermal or cycle annealing occurs substantially at a single temperature or over a much narrower range, resulting in greater uniformity of structure and hardness. That range of temperatures may be found in the TTT curves published for many steels. The temperature at which the lowest hardness is obtainable should be noted because it is the proper temperature to which the metal should be cooled from the regular annealing heat; it is held for the length of time noted at the base of the chart or on the log-log table.

An alternate method of annealing steels that are normally difficult to full anneal—such as high-carbon, high-chrome tool and die steels, hot-work steels, and high-nickel alloy steels—follows:

1. Heat to 1550° F (845° C).
2. Soak thoroughly.
3. Furnace cool to 400° F (205° C).
4. Air cool.
5. Reheat to 1250-1300° F (675-705° C).
6. Soak four hours.
7. Furnace cool to a black heat.
8. Air cool.

Average hardnesses of 201 to 225 Bhn can be obtained in this way.

SPHEROIDIZING

Spheroidize annealing is a modified annealing process that results in a spheroidized, rather than lamellar, microstructure. This microstructure consists essentially of small globules or spheroidized cementite particles in a ferrite mixture, which is the softest condition for steels.

Spheroidized structures are generally preferred for cold heading, wire drawing, and cold forming operations where greater resistance to work hardening and cracking is desired. They are also generally preferred for subsequent hardening of tool and die steels. Low and medium-carbon steels that have thoroughly spheroidized structures, however, are slow to austenitize for hardening, and a blocky ferrite/pearlite structure is preferred. For best results, the spheroidized carbides should be small and uniformly distributed.

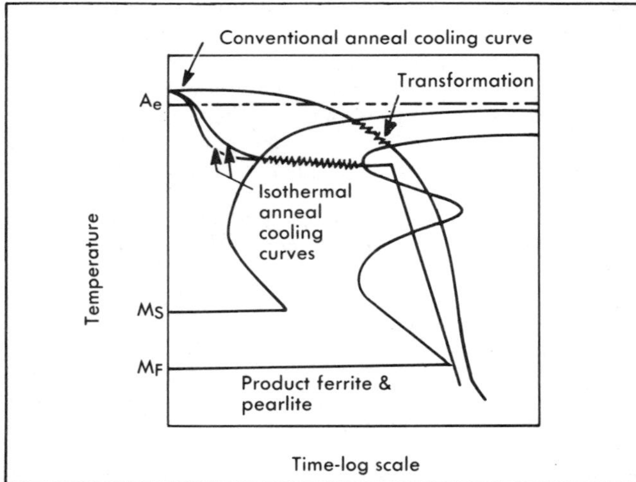

Fig. 10-20 Time-temperature transformation curves for conventional annealing and isothermal annealing.

Spheroidize annealing is often applied to tool steels, SAE 52100 steels, and high-carbon steels to improve machinability, and to other lower alloy, high-carbon steels to improve formability and drawability.

Four methods of spheroidizing are frequently used:

1. Prolonged holding at a temperature just below Ac_1.
2. Alternately heating and cooling between temperatures that are just above and just below Ac_1.
3. Heating to a temperature above Ac_1 or Ac_3, as applicable, and then cooling very slowly in the furnace or holding at a temperature just below Ar_1.
4. Cooling at a suitable rate from the minimum temperature at which all carbide is dissolved to prevent the reformation of a carbide network, and then reheating in accordance with methods 1 or 2. (Applicable to hypereutectoid steel containing a carbide network.)

The temperature range commonly used for spheroidizing is between 1325 and 1375° F (720 and 745° C). This temperature range is held very carefully for 4 to 8 hours, followed by furnace cooling.

As explained in the preceding section on full annealing, lower annealing and transformation temperatures close to the critical temperature promote the formation of a spheroidized structure. In all cases, low annealing temperatures are essential for spheroidizing.

LOW-TEMPERATURE ANNEALING

Heating of a severely cold worked ferrous alloy in the 200 to 500° F (95 to 260° C) range results in a 10 to 20% increase in the alloy's yield strength due to the segregation of carbon atoms to defects in the structure. In production, this effect is usually masked by variations in processing and composition from lot to lot. Figures 10-21 and 10-22 show the effects of low-temperature

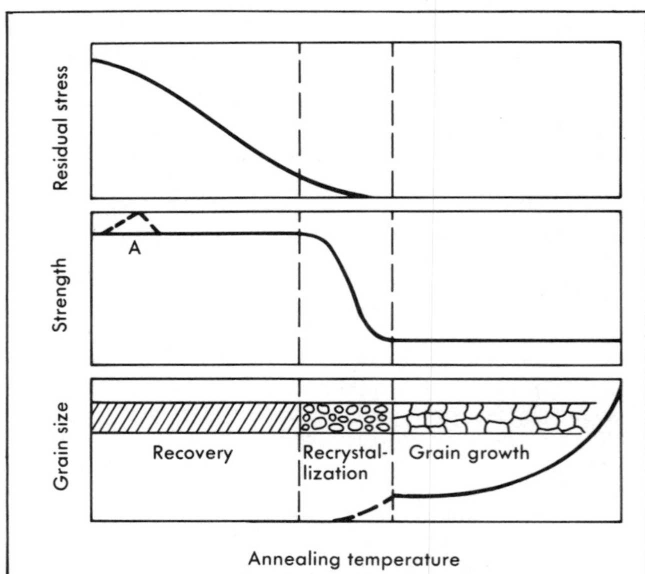

Fig. 10-21 Schematic diagram of recovery, recrystallization, and grain growth of cold worked steel. Curve at "A" represents a 10-20% increase in yield strength at 200 to 500° F (95 to 260° C). (*After Sachs*)[8]

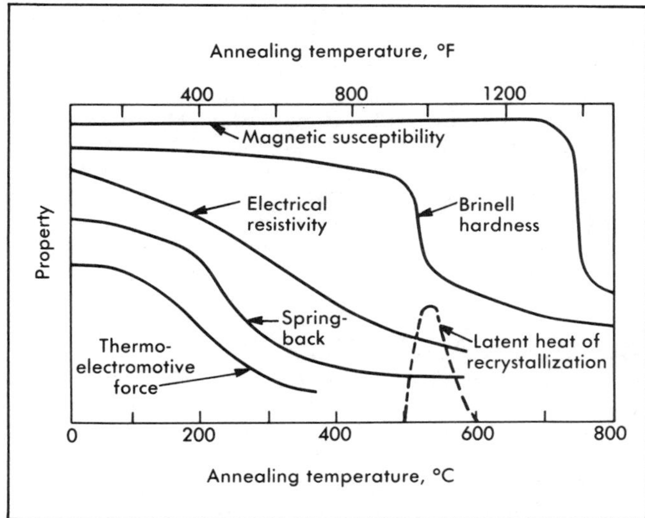

Fig. 10-22 Low-temperature recovery and recrystallization in cold worked iron. (*After Tammann*)[9]

annealing on metal yield strength. Heating into the critical temperature range results in lowering mechanical properties, altering electrical properties, removing stresses, reorienting the grains, and causing grain growth.

RECRYSTALLIZATION ANNEALING

At a temperature determined by the purity of the metal, its cold worked grain size, and the amount of cold work done on it, recrystallization will begin with nuclei of strain-free (defect-free) metal forming at regions of high stress. These nuclei grow at a rate that increases to a maximum and then decreases as the cold worked material is eliminated. This rearrangement is not a phase change but an atomic rearrangement in which the defects (dislocations) generated during cold work are eliminated. The rate of recrystallization increases with an increase in temperature.

The recrystallization temperature is defined as the temperature at which recrystallization is complete in one hour. This temperature is very sensitive to impurities and to the initial amount of cold work. For example, one source lists the recrystallization temperature of pure copper as 437° F (225° C), while another gives temperatures of 554 and 338° F (290 and 170° C) for purities of 99.967 and 99.988% respectively. Various batches of iron also show different recrystallization temperatures for the same method of purification due to differences in the amounts and types of foreign atoms present.

The more severely cold worked a metal is, the lower is its recrystallization temperature. In addition, a fine-grained metal has a lower recrystallization temperature than a coarse-grained metal subject to the same amount of cold work. An initially fine-grained metal also yields a finer recrystallized grain size. As the amount of cold work is decreased, the size of the recrystallized grains increase until a critical value of cold work is reached below which the metal will not recrystallize. Thus, the final recrystallized grain size is sensitive to the initial grain size, as well as the other variables already discussed.

Large, localized grain sizes developed because of small amounts of surface cold work are commercially objectionable. For example, in subsequent forming operations, large grains are known to produce a surface roughening known as "orange peel effect." This problem can be minimized by refining the recrystallized grain size, thus producing a starting stock which can be severely deformed and still yield a smooth surface, free of "orange peel."

Recrystallization annealing is suitable for both batch and continuous furnaces and both large and small production. If recrystallization is desired without grain growth, the recrystallization temperature for the desired cycle time must be determined experimentally, using test samples and metallographic examination techniques. A good starting point for such testing is 180° F (100° C) above the high end of the stress-relieving temperature range for the material. Generally, some grain growth can be tolerated in practical heat treatment. Temperatures for recrystallization annealing a specific alloy to a specific grain size are available in the heat treating literature.

Copper and aluminum alloys are not generally recrystallized to a set hardness or tensile strength value because of the steepness of their annealing curves in the recrystallization range. An alloy is usually fully recrystallized, then reworked to the desired hardness or tensile strength. The terminology "quarter hard," "half hard," and "full hard" is derived from the amount of cold work after recrystallization.

The grain boundaries in an alloy are a source of internal energy. There is a natural tendency toward reducing the internal energy of the system by reducing the number of grain boundaries. Thus, over longer times at recrystallization temperatures, especially at elevated temperatures, larger grains grow and consume smaller grains, thereby reducing the total area of grain boundaries. During grain growth, the square of the resulting grain diameter is proportional to the time at temperature. Grain growth may be impeded or eliminated by the presence of insoluble particles or even certain soluble alloying elements. While very large grains are generally objectionable, moderate grain growth in an alloy that has recrystallized into a fine-grained structure increases the alloy's formability.

RECOVERY ANNEALING

There is a region between room temperature and the recrystallization temperature wherein a large portion of residual stress is relieved, but no grains start to grow. In the low end of this range, the stresses decrease because of rearrangement of point defects (vacancies and solute atoms dissolved in the grain structure) into more stable configurations. At higher temperatures in the recovery range, the line defects (dislocations) introduced during cold working are free to move to more stable rearrangements in which small strain-free subgrains are formed. This rearrangement of crystal defects to form more perfect subgrains is called polygonization. While recovery decreases the residual stresses of cold work, it does not appreciably decrease the tensile strength. Likewise, hardness and magnetic susceptibility are not decreased. However, electrical resistivity and springback are sensitive functions of the distribution of point and line defects, and they decrease with decreasing stress level. Thus, for situations where it is desirable to retain most of the benefits of cold working, but the residual stress level is not wanted, recovery annealing is used.

PROCESS ANNEALING

In *nonferrous* metal production, process annealing refers to a quick annealing—heating without regard to rate, holding for a short period at temperature, and cooling without regard to rate—in the temperature range from the high end of the stress-relieving range through the top of the full-annealing range. In the lower end of the range, recovery, recrystallization, and grain growth take place without solution of impurities or secondary phases. As the temperatures come up into the full-annealing range, impurities and secondary phases begin to dissolve in the base metal or alloy. At these temperatures, the alloy elements rejected into the interdentritic volumes of a billet or casting during slow solidification can be taken back into solution.

In *ferrous* metal technology, process annealing refers both to quick annealing in the transformation temperature range and to subcritical annealing. The quick annealing involves heating without regard to rate into the full-annealing temperature range, holding for a short time (15 to 60 minutes) at temperature, and cooling at a moderate rate. Subcritical annealing involves heating at a moderate rate to just below the lower critical temperature, holding for a reasonable time [for example, two hours plus one hour for each inch (25 mm) of section thickness in plain carbon steel], and cooling at a moderate rate (such as still-air cooling for a plain carbon steel).

All the process annealing methods are used as intermediate softening treatments during a series of forming operations, such as in the multiple deformations of deep drawing, to allow forming without tearing or splitting. As the temperature increases, the hardness and tensile strength are decreased. A balance of physical properties must be determined experimentally to fit the part specifications. Alloy producers can generally supply guidelines to aid in pinpointing the time and temperature necessary to produce the desired result. Process annealing is suitable for either batch or continuous furnace equipment, and production is limited only by the furnace capacity.

HOMOGENIZATION ANNEALING

In castings, especially those of copper and copper alloys, coring and severe segregation of certain alloying elements and different melting temperature phases can take place. In hot rolled alloy shapes, moderate or rapid cooling may result in segregation of a constituent or phase at the grain boundaries. Subsequent cold or hot rolling of a segregated material can result in a laminated structure, as shown in Fig. 10-23. If the pressures and temperatures are high enough, the segregated material may soften and smear or even melt, allowing the various matrix layers to tear, wrinkle, or pull apart. Heating from 100 to 300° F (55 to 165° C) above the maximum annealing temperature for an extended period of time reduces segregation by diffusion of the segregated species into the matrix, a process called homogenization. The disadvantage of the process is the resulting large grain size. However, the homogenized structure can be cold worked to refine the grain.

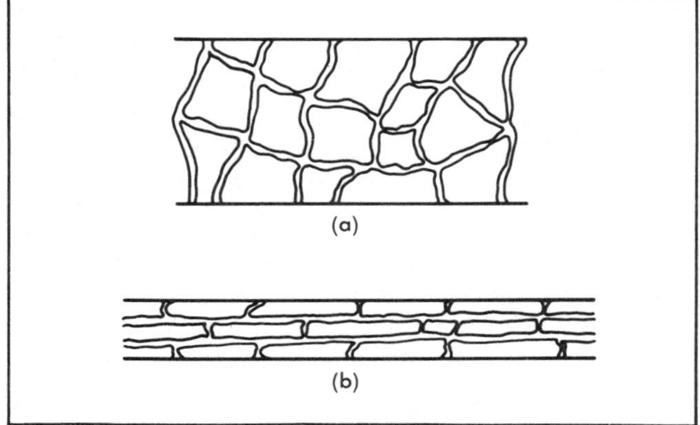

Fig. 10-23 Material segregated at grain boundaries: (a) forms seams that separate laminar, elongated grains after rolling; (b) high temperature can diffuse segregated layers into matrix.

Because the process requires long furnace times at elevated temperatures, large batch furnaces are needed to obtain even moderate homogenization annealing production rates. Since slow cooling is required for softness in the annealed condition, continuous furnaces would generally be prohibitively large or long.

SOLUTION ANNEALING

Solution annealing is the heating of a multiple-phase alloy into a temperature range in which only one homogeneous phase exists at equilibrium, holding at this temperature until the desired degree of homogeneity is achieved, and then rapidly cooling to freeze the atoms in their fixed positions. The atoms are not allowed time enough at the intermediate temperatures to reform the separate phases, which would be in equilibrium at room temperature. The trapped solute atoms in the frozen solution create a situation that provides a source of energy that tends to rearrange the atoms into separate phases when atom mobility is increased by reheating into an intermediate temperature range. Depending on the relative solid solubilities of the phases, this rearrangement can be a hardening and strengthening mechanism. Other alloys are solution annealed to obtain a uniform microstructure that will respond uniformly to subsequent cold work.

While austenitizing can be considered a solution annealing method, the term is usually applied to nonferrous alloys such as those of aluminum, magnesium, nickel, copper, and titanium, for which the upper solution annealing temperature is often within 5° F (3° C) of the eutectic melting temperature. If the eutectic melting temperature is exceeded in any region of the part being treated, grain boundary melting will occur, resulting in drastic embrittlement in some alloys when cooled. The solution annealing temperature range may also be very narrow, as little as a 20° F (11° C) spread. As shown in Fig. 10-24,

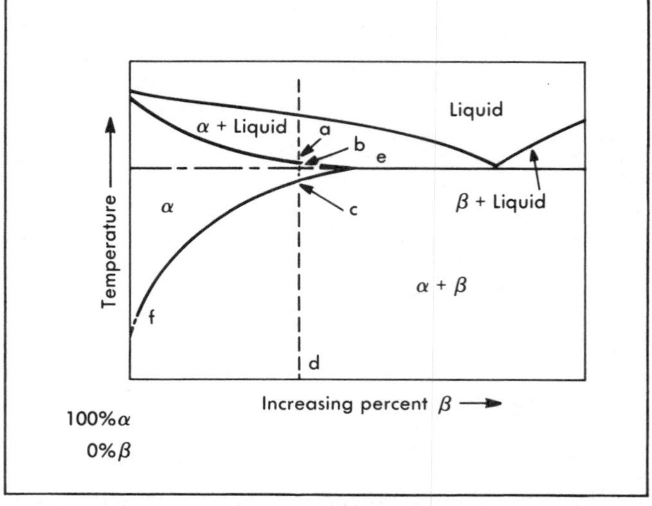

Fig. 10-24 Increasing solubility of material B in material A with increased temperature, showing sensitivity of solution-annealing range.

because of the flatness of the solubility curve *fce* near a typical alloy composition *d*, a small change in temperature can appreciably affect the degree of solubility and hence the strength level of the alloy when it is age hardened.

Solution annealing furnaces need ±5° F (3° C) controllability with a 10 to 15° F (6 to 8° C) temperature spread throughout the furnace work zone. The alloy soak times depend on the maximum section sizes being treated. Solution treating of some common metals is discussed in Chapter 14, "Heat Treatment of Other Metals," of this volume.

While small parts may be processed in continuous furnaces such as shaker hearths, rotary drums, or cast-link conveyors, medium to large parts are generally processed in batch equipment such as drop-bottom, quick-quench furnaces. Quick transfer from the heating chamber to quench immersion is necessary for most water-quenched materials. For example, 0.015" (0.38 mm) thick 2024 aluminum sheet has a maximum transfer time of six seconds. This short time requires positive, precise door opening and rapid fixture acceleration. As the load just touches the quenchant, the fixture is decelerated so that the work enters the quenchant at a linear rate, minimizing part distortion. Control of quenchant conditions, particularly temperature and agitation, is critical for consistent process results.

Double aging refers to a solution annealing followed by reheating, a second quenching, and a second reheating. Alternatively, a double-aging cycle may include cold working. For example, an alloy may be solution annealed, reheated, cooled, cold worked, reheated again, and recooled. Some alloys will precipitation harden after a double-aging cycle with no prior solution annealing. For example, an as-cast alloy may be heated into the precipitation temperature range, cooled, cold worked, reheated, and recooled. In addition, cold working can change some alloys from a soft condition to a precipitation-hardenable condition. Typically these alloys are solution annealed, cold worked, and reheated to accomplish the precipitation reaction. A wide range of values for the various alloy properties can be obtained by the above techniques. Solution treating and aging are discussed in a preceding section of this chapter.

ANNEALING FOR MAGNETIC PROPERTIES

Magnetic metals are divided into two classes: hard magnetic alloys from which permanent magnets are made and soft magnetic metals from which electromagnet cores, motor laminations, solenoids, and other electromagnetic devices are made.

Hard Magnetic Metals

The best known hard magnetic alloys are Alnico V (8% aluminum, 14% nickel, 24% cobalt, 3% copper, and the balance iron) and Alnico VI (8% aluminum, 15% nickel, 24% cobalt, 3% copper, 1.25% titanium, and the balance iron). Both alloys are annealed in the same manner. They are heated to 2200-2400° F (1205-1315° C) and held until homogenized. Thus, the initial heating is a solution treatment. From 2200 to 2400° F, the magnets are cooled at a controlled rate to 900° F (480° C). When the magnets are uniformly heated at 900° F, the temperature is raised to 1100-1200° F (595-650° C), and the magnets are stabilized. The magnets are then cooled without regard to rate through the Curie point, that temperature below which the magnetic domains or regions in the magnets are thermally stable. During this cooling period, which continues until the magnets reach room temperature, the magnets are subjected to an electromagnetic field of 3000 oersteds, oriented parallel to the field desired in the permanent magnets.

Because of the problems associated with establishing a strong, correctly oriented electromagnetic field for each magnet processed, magnetic annealing is limited to small production runs. However, either batch or continuous equipment can be used, generally with a pure dry hydrogen, dissociated ammonia, or vacuum atmosphere.

Soft Magnetic Metals

The soft magnetic metals include iron containing less than 0.005% carbon and iron-silicon, iron-aluminum, iron-aluminum-silicon, iron-nickel, and iron-cobalt alloys. Pure iron is annealed for several hours at 2000 to 2200° F (1095 to 1205° C) in hydrogen, dissociated ammonia, or refrigerated lean-exothermic atmospheres to give a structure with large grains and, consequently, increased magnetic permeability without carburizing or scaling. Iron plus silicon (up to 6%) is hot rolled close to size, pickled to remove the scale, and then cold rolled to final dimension for stackable lamination sheets. The sheets are coated with a refractory and then annealed several hours at 1350 to 2100° F (730 to 1150° C) to obtain a large grain size. The exact annealing temperature is determined by the percentage of silicon, the grain size desired, and the impurities present in the sheet. Either gas or electric batch furnaces, with hydrogen, dissociated ammonia, nitrogen, or refrigerated lean-exothermic atmospheres, are used.

The iron-aluminum and iron-aluminum-silicon alloys are not popular because they are difficult to fabricate, but powder metallurgy has made these materials economical. Iron plus 40-60% nickel and iron-cobalt alloys have very low hysteresis losses. That is, the alloys retain and dissipate as heat very little of the energy of an applied fluctuating or alternating electromagnetic field. They are annealed at 1850 to 2200° F (1010 to 1205° C) for several hours in pure dry hydrogen of -60° F (-50° C) dew point. A -60° F dew point represents 56 ppm by volume of water vapor in the gas.

NORMALIZING

Normalizing involves the heating of steel to above the critical temperature Ac_3, followed by still-air cooling to room temperature to obtain a uniform, fine-grained pearlitic structure. Normalizing differs from stress relieving, annealing, and quenching in that it is carried out from temperatures of about 1600 to 1700° F (870 to 925° C)—approximately 100-200° F (55-110° C) higher than the regular hardening temperature and as much as 200-250° F (110-140° C) over the regular annealing temperature (fully air-hardened steels are exceptions, however)—and in that the cooling rate is neither restricted nor accelerated.

Normalizing was originally applied to forgings of medium-carbon steels to refine their coarse grain structure. Normalizing results in finer grained, harder, more homogeneous structures than annealed structures because of the more rapid air cooling involved in the process. This finer microstructure (usually pearlitic) often coincides with higher mechanical properties for the normalized steels. In fact, normalized materials sometimes provide the specified properties without further heat treatment. Ferrous castings are frequently normalized to remove the strains of solidification and improve toughness or impact strength.

Higher alloy, medium-carbon steels may be given good strength and ductility by normalizing them and then tempering at 1200 to 1250° F (650 to 675° C). High alloyed, air-hardening steels, however, are never treated by other than a stress-relieving process at approximately 1200° F. A true normalizing would cause these steels to reaustenitize and then harden fully, thus defeating the primary purpose of the operation. Typical cycles for normalizing some common carbon steels are presented in Table 10-1, and for alloy steels in Table 10-2.

COLD TREATMENT

Cold (cryogenic) treatment consists of exposing metal parts or tools to subzero temperatures for the purposes of obtaining desired conditions or properties. Such treatment can provide improved strength, dimensional or microstructural stability, greater resistance to wear, stress relieving, and retarded aging.

For steels, proper cold treatment ensures a more uniform and completely transformed microstructure. Primarily, soft, retained austenite is transformed into hard, more stable martensite, which can be subsequently tempered.

RETAINED AUSTENITE

Depending upon the intended applications for heat-treated steel parts, it is often necessary to subject them to cold treatment in order to reduce or eliminate retained austenite. Some metals, especially nickel-containing alloy steels or steels having carbon contents above about 0.60%, have a tendency to retain austenite after hardening. In these instances, the M_f temperature is below room temperature, and the resulting retained austenite yields a reduced hardness. The higher carbon content in the surfaces of carburized parts can have as much as 30% or more of the soft, retained austenite, which greatly reduces the wear and fatigue resistance.

If a file is used and it indicates that the parts are file hard, but the Rockwell hardness reading is low, the problem is probably due to retained austenite, and cold treatment will usually increase the hardness. To ensure that there is no austenite left in the structure and that there will be no subsequent change in the size of the workpiece, a series of cold treatments are sometimes performed.

When subzero treatment is performed to transform retained austenite, care must be taken to avoid thermal shock, which can result in cracking. Slow, progressive cooling is preferred. It is often desirable to stress relieve (snap temper) before refrigeration and temper immediately after cold treatment.

STRESS RELIEVING AND AGING PREVENTION

Cold treatment is also used for stress relieving of castings and machined parts made from steels and aluminum alloys. This process is accomplished by alternately cooling and heating the parts a number of times to achieve dimensional stability.

Cold treatment also prevents or retards the aging of some metals. For example, rivets made from aluminum alloy are solution treated, quenched, cold treated, and held at the subzero temperature until ready for riveting. After the cold working during riveting, the rivets naturally age harden at room temperature to the required strength over a period of a few hours.

WEAR RESISTANCE

The improved wear resistance and longer life sometimes resulting from cold treatment are the major reasons for many of its applications. Many cutting tools and punches made from high-speed steels, cast alloys, and carbides are cold treated for longer life. The treatment stabilizes structures and dimensions and forces transformations closer to completion.

When copper welding electrodes are cold treated, electrical conductivity is improved and wear of the tools is reduced. Other cold-treated components include ball and roller bearings, gears, splines, springs, and powder metallurgy parts.

THE COLD TREATMENT PROCESS

Cold treatment immediately after heat treatment is generally the best procedure. However, care must be taken to prevent cracking. Workpieces especially susceptible to cracking may require tempering prior to cold treatment. Some firms specializing in cold treatment recommend controlled soaking for longer times at lower temperatures (see Table 10-7). Most parts are tempered after cold treatment.

EQUIPMENT USED

A variety of equipment is used for the cold treatment of metals. For some applications, a simple freezer, which produces a temperature of about 0° F (-18° C), is satisfactory. Dry ice (solid CO_2) placed on workpieces in an insulated container

TABLE 10-7
Comparison of Wear Resistance Ratios for Various Steel Samples Soaked at Different Cold Treatment Temperatures

	Cold Treatment Soak Temperature, ° F (° C)	
	-120 (-84)	-310 (-190)
Steel	Wear Resistance Ratio	
52100	2.0	6.5
D3	1.4	3.9
A2	2.0	6.5
M2	1.2	2.0
01	1.7	4.2

(3X Instruments and Tooling)

produces a temperature of about -75° F (-60° C). Some firms use a liquid bath, such as alcohol, plus dry ice to reduce the temperature to -110° F (-79° C). Mechanical refrigeration systems are available to achieve a temperature of -140° F (-96° C) or lower. Liquid nitrogen, with a temperature of -320° F (-196° C), is also used, but less frequently because of the cost and danger of thermal shock.

References

1. J. L. Burns, T. L. Moore, and R. S. Archer, "Quantitative Hardenability," *ASM Transactions*, Vol. 26 (1938).
2. Paul M. Unterweiser, Howard E. Boyer, and James J. Kubbs, *Heat Treater's Guide, Standard Practices and Procedures for Steel* (Metals Park, OH: American Society for Metals, 1982).
3. Lawrence H. Seabright, *The Selection and Hardening of Tool Steels*, 3rd ed. (New York: McGraw-Hill Book Co., 1968).
4. *Ibid.*
5. "Furnace Atmospheres for Modern Heat Treating," *Metal Progress*, Vol. 87, No. 3 (March 1965), pp. 91-105.
6. *Practical Metallurgy for Engineers*, 5th ed. (Philadelphia: E. F. Houghton & Co., 1952).
7. *Ibid.*
8. *Ibid.*
9. Robert F. Mehl, "Recrystallization," *Metals Handbook* (Metals Park, OH: American Society for Metals, 1948).

Bibliography

Beck, Robert I. *Nitrogen-Based Heat Treating Atmospheres*. SME Technical Paper EM78-454. 1978.

Haga, L. J. *Principles of Heat Treating*. Kentwood, MI: L. J. Haga Co., 1980.

____________. *Practical Heat Treating*. Kentwood, MI: L. J. Haga Co., 1981.

Heins, Robert W., and Mueller, Edward R. "Characterization of Polymer Quenchants by Cooling Curves." *Metal Progress* (September 1982).

McGannon, Harold E. *The Making, Shaping, and Treating of Steel*, 8th ed. Pittsburgh: United States Steel Corp., 1964.

Nayar, Harbhajan S., and Drew, John D. "Heat Treating Atmospheres: Feedstock Makeup, Properties." *Heat Treating* (July 1980).

Rainey, Paul E.; Williams, Roy L.; and Greenslate, Robert N. *The Hardenability of Plain Carbon and AISI Alloy Steels*. SME Technical Paper MR78-129. 1978.

Trucks, H. E. "How Cryogenics is Used for the Treatment of Metals." *Manufacturing Engineering* (December 1983), pp. 54-55.

Vaccari, John A. "Fundamentals of Heat Treating." *American Machinist* (September 1981), pp. 185-200.

Watters, James L. *Experience Replacing Generated Atmospheres with Nitrogen*. SME Technical Paper EM78-458. 1978.

Zamborsky, Daniel S. *Heat Treating: A Brief Introduction to Fundamentals*. SME Technical Paper CM81-317. 1981.

SURFACE (CASE) HARDENING

CHAPTER CONTENTS:

Surface hardening (also called case hardening) processes discussed in this chapter are thermochemical treatments in which the chemical composition of steel surfaces is altered. The processes involve adding carbon, nitrogen, or both to the steel surfaces in order to provide a hardened layer or "case" having a definite depth. Common processes used to create hardened cases include carburizing, nitriding, and carbonitriding.

Surface-hardening thermochemical processes are used extensively for low-to-medium carbon steel parts requiring high hardness or fatigue strength primarily at their surfaces, as in wear-resistant applications. The processes also provide sufficient core strength and toughness to withstand tensile or impact stresses and fatigue.

Processes employed to harden the surfaces of steel parts without changing their chemical composition, which are purely thermal treatments, are discussed in Chapter 13, "Selective Surface Hardening," of this volume. These processes include induction, flame, electron beam, and laser hardening.

CARBURIZING

Carburizing is a heat treating process for increasing the carbon content of exposed surfaces on steel parts, usually low-carbon grades with carbon contents below 0.30%. This procedure is accomplished by heating the steel above its upper critical temperature under controlled conditions and in contact with a suitable carbonaceous medium. Carburizing is normally done in the temperature range of 1600 to 1800°F (870 to 980°C). With proper conditions, an enriched carbon case is produced to the desired depth.

The basis for any carburizing operation is the equilibrium of the system. Given the proper conditions, the carbon content of the steel and the carbon potential of the surrounding gases—in the case of gas carburizing—will try to equalize. If the atmosphere has a higher potential than the steel, the steel will absorb the carbon. With atmospheres having extremely high carbon potential, iron carbides may form on the surface of the steel. If the carbon potential of the atmosphere is too low, the steel may yield carbon to the atmosphere and become decarburized.

Being a diffusion process, carburizing is affected by the amount of alloying elements in the steel. The case depth produced is temperature and time dependent, as shown in Fig. 11-1. For any given set of carburizing conditions, the case depth can be determined by using the following formula:

$$CD = k\sqrt{t} \qquad (1)$$

where:

CD = case depth
k = a constant based on the temperature and the chemical composition of the steel
t = time

It is important to distinguish between total case depth and effective case depth. The effective case depth is typically about two thirds to three fourths of the total case depth. It is critical that the required effective depth be specified so that the heat treater can process the parts for the correct time at the proper temperature. Most carburizing is done in a single or two-cycle process. For total case depths less than 0.030" (0.76 mm), a single carburizing cycle is used. For deeper cases, two cycles (carburize and diffuse) are generally used.

Typically, carburizing and quenching produce a case hardness of R_C 60-63, with a core hardness of R_C 10-40, depending upon carbon content. If low-carbon steels, such as SAE 1020, are carburized, it may be difficult to attain this hardness, owing to the low hardenability of the metals, and a more severe quenching medium is generally specified. Case and core hardnesses are dependent upon the composition of the steel, the section size of the workpiece, and the heat treating procedure.

Most steels that are carburized are killed steels (deoxidized by the addition of aluminum), which maintain fine grain sizes to temperatures of about 1900°F (1040°C). Steels made to coarse grain

*Contributors of sections of this chapter are: **Alan M. Bayer**, Technical Director, Teledyne Vasco; **Robert L. Chaney**, Metallurgist, Industrial Furnace Div., Wellman Thermal Systems Corp.; **D. Michael Donovan**, Director of Technology, Lindberg Corp.; **James R. Easterday**, Manager, Applications Engineering, Kolene Corp.; **L. E. Jones**, Division Manager, Lindberg Heat Treating Co.; **William A. Leeper**, Chief Metallurgical Engineer, GH-Thornhill Craver; **Q. D. Mehrkam**, Senior Vice President, Ajax Electric Co.*

*Reviewers of sections of this chapter are: **Alan M. Bayer**, Technical Director, Teledyne Vasco; **Richard B. Bertolo**, Section Chief, Bar and Tubular Products, Metallurgical Div., Republic Steel Corp., Research Center; **Robert L. Chaney**, Metallurgist, Industrial Furnace Div., Wellman Thermal Systems Corp.; **Jim G. Conybear**, Surface Combustion Div., Midland-Ross Corp.; **D. Michael Donovan**, Director of Technology, Lindberg Corp.; **James R. Easterday**, Manager, Applications Engineering, Kolene Corp.; **Robert T. Frazier**, Marketing Manager, Coating Technology, Chromizing Co. Div., Chromalloy; **L. E. Jones**, Division Manager, Lindberg Heat Treating Co.; **Bill Kovacs**, Heat-Vac Systems, Inc.; **William A. Leeper**, Chief Metallurgical Engineer, GH-Thornhill Craver; **Q. D. Mehrkam**, Senior Vice President, Ajax Electric Co.; **Dr. Daryle W. Morgan**, Professor, Dept. of Engineering Technology, Texas A & M University; **Professor Karl B. Rundman**, College of Engineering, Dept. of Metallurgical Engineering, Michigan Technological University.*

CHAPTER 11

CARBURIZING

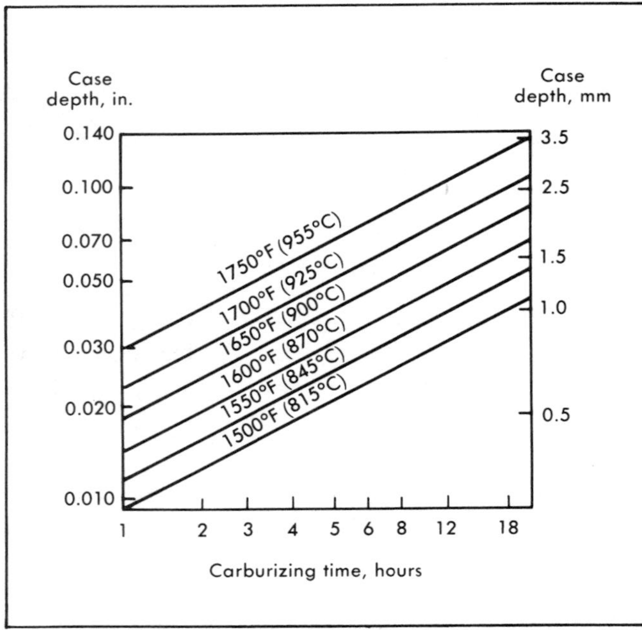

Fig. 11-1 Relation of case depth to carburizing time.

practices (semikilled, rimmed, or capped) should not be carburized. For coarse-grained steels, carbonitriding (discussed later in this section) is more appropriate because of the lower temperatures and shorter cycle times.

Parts carburized to a carbon content over 1% can have a brittle structure if slowly cooled from the carburizing temperature, due to the formation of proeutectoid carbides on the grain boundaries. For this reason, surface carbon is generally limited to about 0.9%. Typical carburizing temperatures and cooling media for various grades of carbon steels are presented in Table 11-1, and for alloy steels, in Table 11-2.

The three major methods of carburizing are liquid, pack, and gas. Gas carburizing is the predominant method, with liquid and pack carburizing being used less frequently.

LIQUID CARBURIZING

Liquid carburizing uses a molten salt primarily consisting of controlled percentages of sodium cyanide (8-12%), barium chloride, and small percentages of other salts. This method of surface hardening is used to produce case depths of 0.020-0.062″ (0.51-1.57 mm), although it can be used for deeper cases, to 0.250″ (6.35 mm). For general use, the most economical operating temperature for liquid carburizing is 1650-1700° F (900-925° C). Temperatures to 1750° F (955° C) provide more rapid penetration, but at some added expense for increased material loss (due to oxidation of the salt) and a slight increase in equipment deterioration.

Furnaces Used

Many liquid-carburizing setups use ordinary steel or alloy pots when operated at temperatures to 1700° F (925° C). These pots are conventionally fired with oil or gas. For operation to about 1750° F (955° C), steel, alloy, or ceramic pots with internal electrodes are common. More detailed information on salts and salt-bath furnaces is presented in Chapter 12, "Heat Treating Furnaces," in this volume.

Applications

Liquid carburizing is generally done in batches, but continuous installations may be justified where production requirements are high. The liquid carburizing process has the flexibility to handle a wide range of parts of varied design and to produce different case depths simultaneously by removing parts at specified times. It is relatively rapid because of the short time required for the work to heat to the carburizing temperature. It also has the advantage of producing clean work, after a rinse in a molten-cyanide bath to remove carbonates.

Precautions Necessary

Cyanide salts are highly poisonous when taken internally or when in contact with open wounds. Fumes from cyanide baths are highly objectionable, and the baths should be hooded and well vented for worker comfort and safety. If water inadvertently comes in contact with the hot molten salt, the water vaporizes with explosive force. All work should be dried carefully before it is placed in the hot salt pot. Additional information with respect to safety and environmental considerations is presented in Chapter 12, "Heat Treating Furnaces," of this volume.

CYANIDING

Cyaniding, sometimes referred to as liquid carbonitriding, is similar to the liquid carburizing process in that it employs a molten-sodium-cyanide bath to provide the hardening elements. However, cyaniding differs from carburizing in both its operating principles and its effects. In cyaniding, 20% or more of the salt bath is composed of sodium cyanide, and the balance, sodium carbonate and sodium chloride in specific percentages. This mixture, when used at temperatures ranging from 1450 to 1650° F (790 to 900° C), decomposes to free carbon and nitrogen, which are then absorbed into the steel to form a hardened carbide-nitride case. Thus, the resulting case is similar to that produced by carbonitriding, discussed later in this chapter. In fact, the term gas cyaniding is a misnomer for carbonitriding.

Applications

Cases produced by cyaniding are usually held to 0.001-0.015″ (0.03-0.38 mm) in depth. Small stampings and screw machine parts made from carbon or free-machining steels are commonly cyanided. The addition of nitrogen and carbon to the surface provides sufficient hardenability that the surface will harden, even in an oil quench, to full file-hard condition. The superficial nitrided case, referred to as hot hardness, is harder than a carbon case and resists tempering, thus providing an advantage for parts operating without lubrication.

A cyanided part may not require a finishing operation. This type of case has high wear resistance. Also, the molten bath lends itself to partial heating of parts where overall hardening is not required or where excessive distortion is a factor.

Parts heated in cyanide may be either oil or water quenched, depending upon the composition, size, and shape of the parts. Quenched parts, when washed in a hot, alkaline water solution, will have clean and file-hard surfaces.

Furnaces Used

Equipment used for cyaniding, as well as liquid carburizing, depends upon the economics of available fuels and power. Gas or oil-fired furnaces may be used economically where the temperature is not in excess of 1600° F (870° C). Pots may be

CARBURIZING

TABLE 11-1
Typical Heat Treatments for Case-Hardening Grades of Carbon Steels[1]

SAE Steels*	Carburizing Temperature, °F (°C)	Cooling Medium	Reheat Temperature, °F (°C)	Cooling Medium	Carbonitriding Temperature,** °F (°C)	Cooling Medium	Tempering Temperature,† °F (°C)
1010, 1015	---	---	---	---	1450-1650 (790-900)	Oil	250-400 (120-205)
1016	1650-1700 (900-925)	Water or caustic	---	---	1450-1650 (790-900)	Oil	250-400 (120-205)
1018, 1019, 1020, 1022, 1026, 1030	1650-1700 (900-925)	Water or caustic	1450 (790)	Water or caustic††	1450-1650 (790-900)	Oil	250-400 (120-205)
1109	1650-1700 (900-925)	Water or oil	1400-1450 (760-790)	Water or caustic††	---	---	250-400 (120-205)
1117	1650-1700 (900-925)	Water or oil	1450-1600 (790-870)	Water or caustic††	1450-1650 (790-900)	Oil	250-400 (120-205)
1118	1650-1700 (900-925)	Oil	1450-1600 (790-870)	Oil	---	---	250-400 (120-205)
1513, 1518, 1522, 1524 (1024), 1525, 1526, 1527 (1027)	1650-1700 (900-925)	Oil	1450 (790)	Oil	---	---	250-400 (120-205)

* Generally, it is not necessary to normalize these carbon steel grades for fulfilling either dimensional or machinability requirements; although, where dimensions are of vital importance, normalizing temperatures of at least 50°F (28°C) above the carburizing temperatures are sometimes required.

** The higher manganese steels, such as 1118 and the 1500 series, are not usually carbonitrided. If carbonitriding is performed, care must be taken to limit the nitrogen content because high nitrogen will increase the tendency of the steels to retain austenite. Retained austenite can be eliminated or minimized by putting the parts in a freezer after tempering.

† Tempering is not necessary for many applications. Tempering is generally employed for partial stress relief and improves resistance to cracking from grinding operations. Higher temperatures than those shown may be employed where the hardness specification for the finished part permits.

†† 3% sodium hydroxide.

made of cast or fabricated steel, or of cast or fabricated high-temperature alloys. Good housekeeping and upkeep are necessary to assure satisfactory pot life. Refer to Chapter 12, "Heat Treating Furnaces," of this volume for additional information.

Precautions Necessary

The hazards of using cyanide salts in the cyaniding process are the same as those for the liquid carburizing process. Consequently, the same precautions should be taken. Environmental laws require close control of all cyanides and cyanide wastes. A permit is required to use and dispose of cyanide, including rinse waters containing cyanides. Systems, including the chemical methods, are available for treating wastes. One system consists of a reactor, a d-c (direct current) power supply, a storage tank, and a pump. Waste is pumped through the reactor where an applied d-c potential transforms the cyanide and cyanate into nitrogen, carbon dioxide, and trace amounts of ammonia, which are safely vented to the atmosphere. Chapter 12, "Heat Treating Furnaces," in this volume contains additional information on safety and environmental considerations.

NONCYANIDE LIQUID CARBURIZING

Liquid carburizing in a molten-salt bath containing a special grade of carbon has been growing in commercial use. In this bath, carbon particles are dispersed in the molten salt by mechanical agitation, which is achieved by means of one or more simple propeller agitators that occupy a small fraction of the bath.

The chemical reaction involved is not fully understood but is thought to involve adsorption of carbon monoxide on carbon particles. The carbon monoxide is generated by reaction between the carbon and carbonates and the air. Then the adsorbed carbon monoxide is presumed to react with steel surfaces, similar to gas or pack carburizing.

Operating temperatures for this type of bath are generally higher than those in cyanide-type baths. A range of about 1650 to 1750°F (900 to 955°C) is most commonly used. Temperatures below about 1600°F (870°C) are not recommended and may even lead to decarburization of the steel. The case depths and carbon gradients produced are in the same range as for high-temperature cyanide-type baths, but there is no nitrogen in the case.

CARBURIZING

TABLE 11-2
Typical Heat Treatments for Carburizing Grades of Alloy Steels[2]

	Pretreatments							
SAE Steels[a]	Normalize[b]	Normalize and Temper[c]	Cycle Anneal[d]	Carburizing Temperature,[e] °F (°C)	Cooling Medium	Reheat Temperature, °F (°C)	Quench Medium	Tempering Temperature,[f] °F (°C)
4012, 4023, 4024, 4027, 4028, 4032, 4118	Yes	---	---	1650-1700 (900-925)	Quench in oil[g]	---	---	250-350 (120-175)
4320	Yes	---	Yes	1650-1700 (900-925)	Quench in oil[g]	---	---	250-350 (120-175)
				1650-1700 (900-925)	Cool slowly	1525-1550 (830-845)[i]	Oil	250-350 (120-175)
4419, 4422, 4427	Yes	---	Yes	1650-1700 (900-925)	Quench in oil[g]	---	---	250-350 (120-175)
4615, 4617, 4620, 4621, 4626, 4718	Yes	---	Yes	1650-1700 (900-925)	Quench in oil[g]	---	---	250-350 (120-175)
				1650-1700 (900-925)	Cool slowly	1500-1550 (815-845)[i]	Oil	250-350 (120-175)
				1650-1700 (900-925)	Quench in oil	1500-1550 (815-845)[h]	Oil	250-350 (120-175)
4720	Yes	---	Yes	1650-1700 (900-925)	Quench in oil	1500-1550 (815-845)[h]	Oil	250-350 (120-175)
4815, 4817, 4820	---	Yes	Yes	1650-1700 (900-925)	Quench in oil[g]	---	---	250-325 (120-165)
				1650-1700 (900-925)	Cool slowly	1475-1525 (800-830)[i]	Oil	250-325 (120-165)
				1650-1700 (900-925)	Quench in oil	1475-1525 (800-830)[h]	Oil	250-325 (120-165)
5015, 5115, 5120	Yes	---	---	1650-1700 (900-925)	Quench in oil[g]	---	---	250-350 (120-175)
6118	Yes	---	---	1650 (900)	Quench in oil[g]	---	---	325 (165)
8115, 8615, 8617, 8620, 8622, 8625, 8627, 8720, 8822	Yes	---	---	1650-1700 (900-925)	Quench in oil[g]	---	---	250-350 (120-175)
				1650-1700 (900-925)	Cool slowly	1550-1600 (845-870)[i]	Oil	250-350 (120-175)
				1650-1700 (900-925)	Quench in oil	1550-1600 (845-870)[h]	Oil	250-350 (120-175)

TABLE 11-2—*Continued*

SAE Steels[a]	Pretreatments: Normalize[b]	Pretreatments: Normalize and Temper[c]	Pretreatments: Cycle Anneal[d]	Carburizing Temperature,[e] °F (°C)	Cooling Medium	Reheat Temperature, °F (°C)	Quench Medium	Tempering Temperature,[f] °F (°C)
9310	---	Yes	---	1600-1700 (900-925)	Quench in oil	1450-1525 (790-830)[h]	Oil	250-325 (120-165)
				1600-1700 (900-925)	Cool slowly	1450-1525 (790-830)[i]	Oil	250-325 (120-165)
94B15, 94B17	Yes	---	---	1650-1700 (900-925)	Quench in oil[g]	---	---	250-350 (120-175)

[a] These steels are fine grained. Heat treatments are not necessarily correct for coarse-grained steels.
[b] Normalizing temperatures should be at least as high as the carburizing temperatures followed by air cooling.
[c] After normalizing, reheat to temperature of 1100-1200° F (595-650° C) and hold at temperature about 1 hr per in. (25 mm) of maximum section or 4 hr minimum time.
[d] Where cycle annealing is desired, heat to at least as high as the carburizing temperature, hold for uniformity, cool rapidly to 1000-1250° F (540-675° C), hold 1-3 hr, and then air or furnace cool to obtain a structure suitable for machining and finish.
[e] It is general practice to reduce the carburizing temperature to about 1550° F (845° C) before quenching to minimize distortion and retained austenite. For the 4800 series steels, the carburizing temperature is reduced to about 1500° F (815° C) before quenching. (Carburizing temperature should be a minimum of 50° F (28° C) above the critical temperature of the steel.)
[f] Tempering is optional and is generally employed for partial stress relief and improved resistance to cracking from grinding operations. Temperatures higher than those shown are used in some instances where application requires.
[g] This treatment is most commonly used and generally results in a minimum of distortion.
[h] This treatment is used when the maximum grain refinement is required and/or when parts are subsequently ground on critical dimensions. A combination of good case and core properties is secured with somewhat greater distortion than is obtained by a single quench from the carburizing treatment.
[i] In this treatment, the parts are slowly cooled, preferably under a protective atmosphere. They are then reheated and oil quenched. A tempering operation follows if required. This treatment is used when machining must be done between carburizing and hardening or when facilities for quenching from the carburizing cycle are not available. Distortion is at least equal to that obtained by a single quench from the carburizing cycle, as described in note [e].

Because of the lack of nitrogen, slowly cooled noncyanide carburized parts are more easily machined than cyanide carburized parts. Also, retained austenite is less likely to be present in quenched cases that contain no nitrogen.

The noncyanide carburizing process is limited to applications where the parts to be carburized are individually spaced or suspended. Success of the process depends on free circulation of the mechanically agitated salt around all surfaces to be carburized. Shallow basket loading may be used if the geometry of the parts allows free flow of the salt through the load. Control of the carbon potential is also difficult with noncyanide carburizing.

PACK CARBURIZING

Pack carburizing is the carburizing process that requires the least specialized equipment and techniques. It requires a source of carbon that will be diffused into the surface of the heated low-carbon steel to form a carburized case.

Commercially available packing compounds include a catalyst or activator that permeates the compound. This activator deteriorates with use, so additions of at least 10% to as much as 50% of new compound should be made after each cycle of heating.

Equipment Used

The metal containers used for holding the compound and the work need only be thick enough to withstand handling after prolonged heating. They may be made of mild steel, although scaling may deteriorate the containers and make them short-lived. Other good box materials are 80-20 chrome nickel, 35-15 nickel chrome, and 18-8 stainless steel, either cast or welded. A loose-fitting cover, luted or sealed with fire clay, is normally provided for each container to keep the furnace atmosphere from directly contacting the compound. Pack carburizing can be done in a wide variety of furnaces, the major requirement being uniform temperature.

Processing

The compound used for pack carburizing should be dust-free; otherwise, soft spots in the work may result. The work to be carburized should also be reasonably free from foreign material. The bottom of the container is first covered to a depth of at least 1″ (25 mm) by compound. The parts to be carburized are then placed in the box so that they are at least 1/2″ (13 mm) apart and 1″ from the side of the box. Several layers of parts may be placed in one box, to within 2 to 3″ (51 to 76 mm) of the top. The container is next filled with compound and shaken or rapped sharply to settle its contents. The cover is then put on and sealed with fire clay, and the container is ready for the furnace.

Applications

Pack carburizing is normally not used for light case work, less than 0.050″ (1.27 mm) deep, but is more often used for deep cases, such as 0.100″ (2.54 mm) or more. With pack carburizing, little control over the surface carbon content is possible as saturated austenite is produced at the carburizing temperature. For example, at 1700° F (925° C) the surface carbon would be approximately 1.25%. This may be an advantage on parts requiring high wear resistance with consequent low impact resistance.

Limitations

Pack carburizing is a labor intensive, low production, deep case process used for such items as rolling mill rolls, earth drilling bits, and large-diameter wear pins. Some of the disadvantages of pack carburizing are:

1. The compound is dirty and difficult to handle when hot.
2. It is difficult to remove parts from the pack for quenching.
3. Case depths may be uneven because of the insulating qualities of the compound. This factor causes the edges and thin sections of the work to heat first and carburize deeper than the slower heating heavy sections.
4. The compound is becoming more expensive and difficult to obtain because the process to produce it is more expensive as environmental standards are raised.

GAS CARBURIZING

Most carburizing is performed with gaseous atmospheres in sealed furnaces where the products of combustion and air do not come in contact with the work. The gaseous atmosphere used can vary from natural gas to endothermic to nitrogen (plus additives), and each, when controlled, can be used successfully. Gas chemistries and selection of atmospheres are discussed in Chapter 10, "Heat Treatment of Steel," of this volume. It is sufficient to know that a source of carbon must be available (CO, CH_4,C_2H_6, etc.), and the furnace must be of the sealed type, capable of maintaining the correct atmosphere. The atmosphere pressure in the furnace must be positive to exclude air from infiltrating. Various furnaces available are described in Chapter 12, "Heat Treating Furnaces," of this volume.

With gas carburizing, the atmosphere can be controlled by monitoring the moisture (dew point), carbon dioxide, methane, oxygen, or a combination of these, and adjusting the entering gases to provide the carbon potential desired, as discussed earlier in Chapter 10. Instrumentation is available to program the time, temperature, and carbon potential to produce the desired case properties and to duplicate the process from batch to batch. Information on furnace controls is presented in Chapter 12.

Work may be hardened by quenching from the carburizing atmosphere. Generally, the temperature is reduced to 1450-1550° F (790-845° C) prior to quenching to reduce warpage. Cooling to a lower temperature before quenching is also necessary to ensure a proper microstructure. By lowering the temperature of the steel but keeping it above the critical temperature (Ar_3), less heat has to be removed in quenching, and a relatively uniform microstructure can be achieved.

In some instances, parts are slowly cooled from the carburizing temperature and then reheated and quenched to refine the grain and reduce the retained austenite. Other reasons for slow cooling include the need to keep the part soft so that the case can be machined away in specific areas. After reheating and quenching, those areas that have had the case machined away will be relatively soft and can be further machined.

The steels used for carburizing range from low-carbon steels—such as SAE 1018, 1020 and 1022—to low-carbon alloy steels, such as SAE 8620, 4615, and 9310 (see Tables 11-1 and 11-2). It is not desirable to carburize through-hardening steels, such as SAE 4140 or 4340, as the core may be too brittle to withstand any impact loading.

Vacuum Carburizing

Vacuum carburizing, sometimes called partial-pressure carburizing, is a form of gas carburizing. The requirements of temperature, time, and carbon potential are essentially the same as in gas carburizing to produce a given case depth. Normally, nonoxygen-bearing gases are used in vacuum carburizing to provide the carbon, at a pressure of 300 to 400 torr (760 torr equals 1 atmosphere). The advantages of vacuum carburizing include versatility, closely controllable operations, and shorter cycle times if carburizing temperatures are raised. Vacuum furnaces, discussed in Chapter 12, are designed for high-temperature operation. Therefore, unless other reasons prohibit, temperatures to 1900° F (1040° C) may be used. Parts carburized in vacuum furnaces are bright and scale-free. However, the initial cost of vacuum furnaces is high, and maintenance costs are substantial.

In vacuum carburizing, parts are loaded into the furnace, the heating chamber is evacuated, and the parts are radiantly heated to a temperature of 1700 to 1900° F (925 to 1040° C). A controlled amount of hydrocarbon gas (much less than used in conventional gas carburizing) is then introduced into the chamber. After a predetermined time, the gas flow is stopped, excess gas is evacuated, and the diffusion cycle begins. Surface carbon produced depends upon the amount of time allowed for diffusion of the carbon inward from the surface and may vary from 0.7 to 1.0%.

Fluidized-Bed Carburizing

Carburizing is also done in fluidized-bed furnaces, discussed in Chapter 12, "Heat Treating Furnaces." A nitrogen atmosphere, with the addition of natural gas, propane, or vaporized methanol, is used. Control of carburizing potential is achieved from the time that the atmosphere is switched from nitrogen to the carburizing atmosphere. Vaporizers are available to convert liquid methanol to gas for deep-case carburization. This gas is blended with nitrogen, and additions of natural gas permit controlling the carbon potential.

Ion (Plasma) Carburizing

In ion carburizing, carbon is imparted to steel surfaces by the impingement of carbon ions escaping from an ionized gas (plasma). The process is performed in vacuum furnaces equipped with a high-voltage power supply and conventional radiant heating elements. Energy for the absorption reaction is achieved by a high-voltage glow discharge between the cathodic workpiece and an anodic electrode in the furnace. Carbon ions accelerate from the ionized hydrocarbon gas toward the workpiece and impinge on the surfaces.

The advantages of ion carburizing include energy conservation and reduced requirements for gas, compared to atmosphere and vacuum carburizing. Processing time is shorter as well. This method produces more uniform case depths, especially on irregular surfaces. It can also provide deeper cases in blind holes and cavities. Equipment and maintenance costs, however, are high compared to conventional methods of carburizing.

High-Temperature Carburizing

If higher temperature carburizing can be used, the time required to produce a given case is reduced substantially. The risks involved include greater furnace maintenance, increased usage of alloy trays and fixtures, less load weight per tray, possibly more warpage, and, sometimes, increased grain size. Careful consideration must be given to these factors before the advantages of higher temperature carburizing can be realized.

Since the dew point is inversely proportional to carbon potential and the carbon potential is inversely proportional to temperature, there is a distinct relationship between temperature and dew point variables. An important factor, sometimes not considered, is that as temperature increases, the measured dew point changes.

A typical chart of case depths for various furnace temperatures and for a constant atmosphere is shown in Fig. 11-2. This chart also shows the carbon potential/dew point data that apply to the temperature ranges considered practical for carburizing. These curves are plotted from test data selected by sample processing and production rounds. A temperature of 1900° F (1040° C) has been reported as the highest practical for use in many applications. A normal 0.85-1.00% carbon specification would require dew point control within an estimated 2° F (-17° C) range at this temperature. A comparative 4° F (-16° C) deviation would be allowed at 1700° F (925° C). This fact points out the critical need for precise automatic control at higher temperatures. Similar considerations exist for carbon dioxide and oxygen control. Numerous control systems have been devised for automatically compensating for changing temperature and gas compositions.

Carbon Correction

Carbon correction or restoration is the process of increasing or decreasing the surface carbon content of steel parts, generally to restore carbon content to a decarburized case equal to that of the core of the part. For example, if an SAE 4140 steel were being heat treated and simultaneously carbon corrected, a hardening temperature of 1600° F (870° C) would be used. A furnace dew point of 52° F (11° C) would be in equilibrium with the carbon content of the core.

If steel bolts were soaked at 1600° F (870° C), the surface carbon would be restored to 0.50%. There would be no danger of exceeding this amount regardless of time and temperature. Usually a soak time of about twice that required for normal hardening is used, but this time also depends upon the depth and degree of decarburization.

Homogeneous Carburizing

Homogeneous carburizing is a process for raising the carbon content of a low-carbon steel to a uniform medium or high-

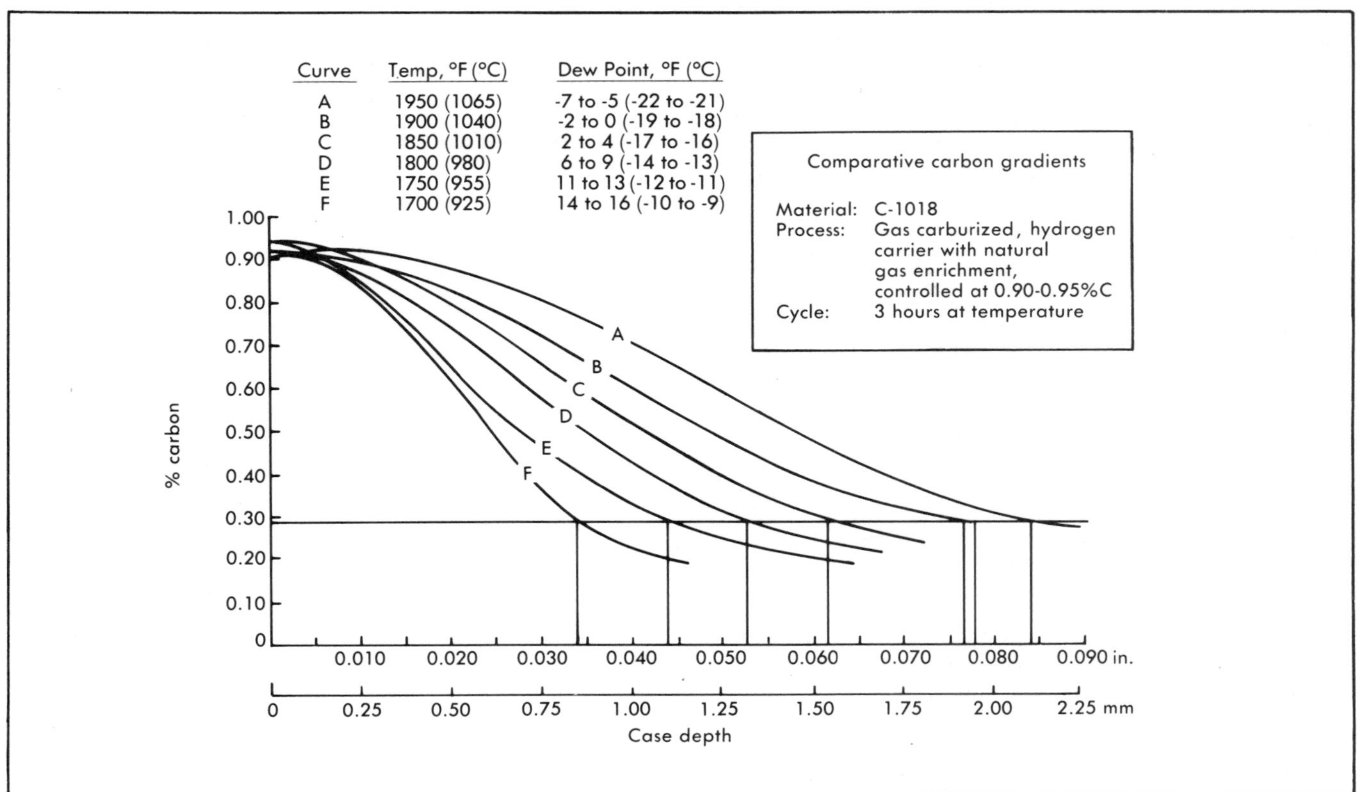

Fig. 11-2 Carbon potential and carburized case depth at various temperatures in constant atmosphere.

carbon composition throughout its section. As long as the carbon potential of the atmosphere is at the desired level, thin sections and thick sections can be carburized at the same time without overcarburizing. The principal advantage of this process is that intricate parts can be formed from low-carbon steel, with attendant savings in die life, steel cost, and scrap that would be impossible to achieve in the cold forming of medium and high-carbon steel. It also has an advantage in allowing a single low-carbon steel to be used to meet several specifications in the finished products. As an example, springs can be made from low-carbon steel by homogeneous carburizing.

Selective Carburizing

If it is necessary to prevent carburization on certain areas of a part, stopoff compounds or copper plating may be used. Preventing carburization requires attention to cleanliness and handling to achieve the desired stopoff. Copper plate thicknesses of 0.001″ (0.03 mm) minimum are required. Stopoff compounds are proprietary, and instructions should be closely followed as it is difficult to achieve 100% stopoff. Carburization prevention may be necessary on areas to be further machined after heat treating or to prevent a thin area from being carburized all the way through its section and thereby becoming brittle.

NITRIDING

Nitriding is a surface hardening process in which nitrogen is diffused into the surfaces of ferrous alloys at subcritical temperatures to produce a shallow case of nitrides without requiring quenching. The purposes of nitriding include the following:

1. To increase the surface hardness.
2. To improve wear resistance.
3. To increase fatigue life.
4. To possibly increase corrosion resistance slightly, except in the case of stainless steels.

The low temperatures employed and the lack of quenching in nitriding minimize distortion, but the process may require considerable time, especially for producing deep cases with gas nitriding. Also, allowance must be made for a small amount of dimensional growth, generally less than 0.001-0.002″ (0.03-0.05 mm), resulting from the formation of the nitrides. Distortion and growth are much less when using ion (plasma) nitriding, discussed later in this section. The amount of growth with all methods of nitriding is dependent upon the depth of the nitride layer.

STEELS NITRIDED

Steels that are to be nitrided should contain at least one of the following alloying elements: aluminum, chromium, vanadium, tungsten, or molybdenum. These elements form the nitride precipitates that are necessary for diffused cases. Commercial alloy steels contain varying amounts of these elements, with aluminum, chromium, and molybdenum being the most desirable. Steels that are commonly nitrided include AISI 4140 and 4340.

Aluminum is the strongest nitride former, and the Nitralloy series of steels, containing 0.85-1.50% Al, have been developed specifically for nitriding because of this property. Chromium steels can approach the high hardnesses of nitrided aluminum-bearing steels if the level of chromium content is high enough, about 5% or more.

With certain nitriding processes, discussed later in this section, plain carbon steels and cast irons can be successfully nitrided. Typical steels that are nitrided and the hardnesses attained are presented in Table 11-3.

Maraging steels are routinely nitrided, with best results obtained when the parts to be nitrided are in the solution-annealed condition. In gas nitriding with a 25 to 30% dissociated-ammonia atmosphere, the steels are simultaneously aged and nitrided at a temperature of 825 to 850° F (440 to 455° C) when held at temperature for 40 to 48 hours. This produces an effective case depth of 0.005-0.007″ (0.13-0.18 mm). The surface hardness is about R_C 65, and the core hardness is generally 2-4 points higher on the Rockwell C scale than normally achieved during conventional aging.

Case depths to 0.030″ (0.76 mm) are obtainable with low-alloy steels, such as AISI 4140. Nitriding of this steel to a case depth of about 0.008″ (0.20 mm) and hardness of R_C 50 is common, using a temperature of 1000° F (540° C) for 24 hours. Dimensional growth is less than 0.0006″ (0.015 mm). For high-alloy steels, such as stainless steels, case depths are generally limited to 0.010″ (0.25 mm) or less because of slow case development. Deep cases may require 70 hours or more at temperature. Major methods of nitriding are: liquid (molten-salt bath), gas, and plasma (ion nitriding). Some pack nitriding is also done.

PRIOR TREATMENT

Unlike some other surface hardening processes, most parts must be hardened and tempered before being successfully

TABLE 11-3
Steels Nitrided and Typical Surface Hardnesses Attained

Steels Nitrided	Typical Maximum Surface Hardness	Usual Surface Hardness, R_C
Nitralloy 135M	R_C 70	63-65
AISI Series:* 4100, 4300, 5100, 6100, 8600, 8700, 9300, 9800	R_C 60	52-60
AISI Tool Steels: H11, H12, H13, M2, M42	R_C 70	64-68
Stainless steels**	1200 Knoop	64-70

* These should be medium-carbon steels.

** Generally, stainless steels cannot be nitrided in a conventional system. Stainless steels also lose some of their corrosion resistance when nitrided.

nitrided. The final tempering temperature before nitriding should be at least 50° F (28° C) above the nitriding temperature. This treatment produces the proper metallurgical structure for nitriding and minimizes dimensional distortion during the nitriding process.

Any decarburization on functional surfaces of the workpieces should be removed before nitriding. Decarburized surfaces tend to become brittle and will spall after nitriding. Workpieces should also be carefully cleaned before nitriding.

LIQUID NITRIDING

Liquid nitriding is done in molten-salt baths—discussed in Chapter 12, "Heat Treating Furnaces"—in the temperature range of 950 to 1075° F (510 to 580° C). During this thermal/chemical treatment, workpieces are exposed to a nitriding environment generated by salts containing cyanides and cyanates or salts containing no cyanides. Nitrides of iron and alloying elements are formed to depths within 0.001" (0.03 mm). The specific temperature used and the immersion time (generally 5 minutes to 2 hours) vary with the material being nitrided and the application of the workpieces.

Advantages of liquid nitriding include precise uniformity of temperature throughout the bath, minimal distortion of the workpieces due to the buoyancy effect, and the capability of nitriding plain carbon steels. The principal disadvantage is in the use of salts containing cyanides and cyanates, requiring precautions discussed previously in this chapter under the heading of liquid carburizing. The availability of nontoxic salts containing no cyanides, however, has minimized this problem.

The treatment of cutting tools and forming dies is a major application for nitriding because of the significant improvement in performance attained. Nitrides that are formed improve properties, including high surface hardness, and provide excellent lubricity, with resulting increases in tool and die life ranging from 200 to 1000%.

Applications include the following:

- For die-casting dies made from H series steels, improved resistance to heat checking, soldering, galling, and seizing.
- For plastic molds and dies made from SAE 5120 and 6120 steels, reduced manufacturing costs and risks.
- For blanking and forming dies made from carbon steel, improved fatigue properties and wear resistance, and minimum distortion during heat treatment.
- For cutting tools (drills, reamers, taps, end mills, form tools, hobs, and shaper cutters) made from high-speed steels, improved performance resulting from increased resistance to wear and galling, and minimized welding of chips to the cutting edges of the tools.

Nitriding of maraging steels in a liquid salt bath for 90 minutes at 1000° F (540° C) has successfully produced a case hardness of R_C 64-66, with a total effective case depth of about 0.003" (0.08 mm). When using salt-bath nitriding, aging of maraging steels may be performed prior to or simultaneously with the nitriding operation.

GAS NITRIDING

Most nitriding is done with gaseous atmospheres in retort furnaces where the products of combustion and air do not contact the workpieces. Both single and double-stage processes are employed. Nitriding is also done in fluidized-bed and vacuum furnaces and by the ion (plasma) process.

In gas nitriding, the term "dissociation rate" is commonly used to describe the percentage of gas volume in the furnace that is not molecular ammonia. However, dissociated ammonia cannot nitride steel. It is the availability of usable nitrogen in undissociated ammonia (NH_3) that controls the nitriding process.

Single-Stage Nitriding

In the single-stage nitriding process, ammonia with a dissociation rate of 15 to 30% is added to the chamber of the furnace and the load is heated to the temperature range of 950 to 1000° F (510 to 540° C). A byproduct of this process is a brittle white layer that forms on the surface of steel. This layer is made up of iron nitrides, and its thickness ranges from 0.0001-0.002" (0.003-0.05 mm), depending upon the temperature, the time at processing temperature, and the nitriding potential employed. This layer is sometimes removed by chemical or mechanical means to prevent it from spalling when the part is used. In most cases, the thickness of the white nitride layer is simply controlled to a reasonable level, about 0.0005" (0.013 mm) by use of the two-stage process.

Double-Stage Nitriding

In double-stage nitriding, also known as the floe process, a second cycle is added to the single-stage method. During the second cycle, the nitriding potential is decreased by reducing the ammonia concentration to 15-35% by volume. This reduction is accomplished with a carrier gas, either dissociated ammonia (25% N_2, 75% H_2), produced by cracking the ammonia outside the furnace, or nitrogen gas. The temperature used during the second cycle can be the same as the single-stage process, or it may be increased to as high as 1050° F (565° C). Advantages of double-stage nitriding include reduced ammonia consumption, minimizing of the white layer, and more rapid case development.

Fluidized-Bed Nitriding

Nitriding is also done in fluidized-bed furnaces, discussed in Chapter 12, "Heat Treating Furnaces." Ammonia and natural gas or nitrogen is passed through the bed, and the fluid bed acts as the heating medium. Most of the dissociation takes place on the parts being processed.

Vacuum Nitriding

Controlled case depths can be produced on critical parts with energy and time savings by nitriding in vacuum furnaces (also discussed in Chapter 12) using a partial-pressure cycle. After the heating chamber is evacuated, the workpieces are heated radiantly, and the furnace is back-filled with a controlled amount of ammonia and natural gas or propane. After a predetermined time, the gas flow is stopped and diffusion begins. This process has had limited application.

Ion (Plasma) Nitriding

Ion nitriding is similar to ion carburizing, discussed previously in this chapter, with two important differences:

1. The gas mixture used is a mixture of nitrogen and hydrogen to which a small amount of methane is sometimes added.
2. Radiant heating elements are not generally required in the vacuum vessels because the energy from the glow discharge and current passing through the workpieces is sufficient for heating. However, radiant elements are sometimes provided for faster heating and better temperature uniformity.

The vacuum chamber is the anode, and the workload (electrically isolated from the chamber) is the cathode. After a controlled amount of gas mixture has been introduced into the chamber, a d-c potential difference is established between the chamber and workpiece to ionize the gas. Positively charged nitrogen ions bombard and are diffused into the surfaces of the metal and combine with the alloying elements to form nitrides.

An important advantage of ion nitriding is the versatility of the process. It provides a wide temperature range; and by varying the gas mixture, an extensive range of metals—including stainless steels—can be nitrided, and different cases produced. Reliable control is provided with respect to case depth and composition. Processing times are generally shorter, and the method is more energy and gas efficient than conventional processes. Since no ammonia gas or cyanide salts are needed, the process is environmentally safe. A possible limitation is that workpieces require racking or fixturing to expose all surfaces to be nitrided.

Masking for selective nitriding can be done mechanically (with sheet covers, plates, or screws), by plating, or with painting. Care is necessary, however, with plating or painting, as the ion bombardment has a tendency to remove these coatings.

CARBONITRIDING

Carbonitriding is similar to cyaniding, discussed previously in this chapter under the subject of carburizing, except that the simultaneous absorption of carbon and nitrogen into the steel surfaces is accomplished by heating in a gaseous atmosphere. The process is sometimes referred to as dry cyaniding or gas cyaniding. The composition of the case produced depends upon the atmosphere, temperature, time, and steel composition. While carbonitriding is done above the phase-transformation temperatures, the process called nitrocarburizing is performed at temperatures below phase transformation. However, both processes are performed at temperatures higher than required for nitriding.

Temperatures of 1450 to 1650° F (790 to 900° C) are commonly used for carbonitriding steel parts to be quenched, generally in oil (see Table 11-1). Lower temperature processing, such as nitrocarburizing, is sometimes used where a liquid quench is not required. Endothermic atmospheres, however, are explosive at temperatures below 1400° F (760° C). Most heat treating furnaces have safety devices that prevent the introduction of endothermic gases into the furnace unless the chamber is at a temperature of at least 1400° F.

Atmosphere cooling is valuable whenever there is danger of distortion and core properties are of no concern. The case resulting from atmosphere cooling, however, is brittle and not generally satisfactory when parts are subjected to impact or bending in service. Case depths may range from 0.001-0.030″ (0.03-0.76 mm) or more, but generally vary from 0.003-0.015″ (0.08-0.38 mm).

ADVANTAGES

An important advantage of carbonitriding over carburizing is the effect of nitrogen in improving the hardenability of the case. This process makes possible the use of low-carbon steel to achieve surface hardness equivalent to high-alloy carburized steel without the need for drastic quenching, resulting in less distortion and reduced danger of cracking the work.

Carbonitriding is sometimes combined with carburizing. A thin carbonitrided case is added to the surface of a previously carburized part to provide high surface hardness. Such surfaces can be highly polished to give better frictional properties and longer wear life than the original carburized surfaces.

ATMOSPHERES USED

Atmospheres required for carbonitriding are similar to those used for carburizing except for the addition of ammonia. Hydrocarbon-gas additions to the atmosphere are generally maintained between 2 and 10%. Ammonia additions vary from 0.5 to 25%, the lower percentages being recommended when liquid quenching is used. The higher percentages of ammonia are necessary to produce a nitrogen case for full hardening without benefit of quenching. More detailed information on atmospheres is presented in Chapter 10, "Heat Treatment of Steel," in this volume.

FURNACES EMPLOYED

Furnaces often used for carbonitriding are of gas-tight batch or continuous designs, discussed in Chapter 12, "Heat Treating Furnaces." The type of conveying mechanism determines the application of continuous furnaces because of the time factor required in obtaining the proper case depths. Batch furnaces are more flexible, but their use is determined by the cost of the furnace versus case depth and the quantity of parts to be treated.

Fluidized-Bed Furnaces

Carbonitriding is done in fluidized-bed furnaces, described in Chapter 12, by using mixtures of ammonia, natural gas, and nitrogen. Since the furnaces are usually left uncovered, larger volumes of gases are required and flammable gases burn at the surface. Workpieces, loaded on racks or in baskets, are immersed in the fluid bed for the required time and then generally quenched.

Vacuum Furnaces

Carbonitriding of critical parts is being done in vacuum furnaces, also discussed in Chapter 12, using a partial-pressure cycle. Back-filling of the vacuum container is generally done with natural gas mixed with nitrogen or propane and nitrogen. Some ammonia is sometimes added, but this requires special equipment design because ammonia attacks copper, especially at high temperatures.

Ion Carbonitriding

Some work is also being done with fast, high-temperature ion carbonitriding, using vacuum vessels similar to those employed for ion carburizing and nitriding. For ion carbonitriding, the ionized gas (glow-discharge plasma) consists of a mixture of hydrocarbon (usually methane) and nitrogen gases, thus permitting both carbon and nitrogen to be added to steel surfaces simultaneously.

CHROMIZING

Chromizing is a high-temperature process in which chromium is transported to and diffused into the surface of metals to provide a chromium-enriched alloy layer with high corrosion, heat, and wear-resistance properties. Current commercial techniques that provide for the simultaneous transportation and diffusion of chromium are:

1. Powder-pack cementation process. In this process, parts to be chromized are placed in a retort in direct contact with a powder mixture of a chromium source, chromium halide salts, and inert refractory oxides. The retort may be sealed, or it may be designed so that a reducing or inert atmosphere can be maintained in it. It may also be placed under a vacuum.
2. Fused salt-bath process. In this process, parts are placed in a salt bath containing chromium halide salts and elemental chromium, and the bath is protected by an inert atmosphere such as argon. The process may be carried out with or without the use of an external current that, if used, deposits chromium on the base part, which serves as the cathode. Faster deposition is possible with the fused salt-bath technique, but its major disadvantages include the need for increased facilities for production and the need for special handling fixtures.
3. Granular-pack process. The granular-pack process uses only porous granules of chromium containing halides within the pores; no inert refractories are used. Workpieces may be either in direct contact with the granules or out of contact if a smoother finish is required. Parts and granules are packed in a retort; and upon heating, the chromium halides vaporize and pass to the parts, where chromium is deposited and diffused into the surfaces. The gaseous reaction products are then transported back to the chromium-source granules to generate more chromium halides. Upon completion of the diffusion, the retort is cooled, and the halides are reabsorbed into the granules, thereby providing primed granules for the next chromizing heat.
4. Gas convection process. In this method, also called chemical vapor deposition (CVD), the parts are placed apart from the chromium sources. Gaseous halogens or halogen acids are then made to react with the chromium to form a gas that is conveyed by forced circulation to the parts. The gaseous reaction products are then exhausted or returned to the source, where a reverse reaction generates additional chromium halides. More detailed information about the CVD process is presented in Chapter 24, "Vapor Deposition Processes," of this volume.

Differences among these processes are chromizing potential, heating and cooling rates, surface finish, transporting power, and economy. For example, the simultaneous chromizing and heat treatment of air-hardening tool steels and austenitic stainless steels require a subsequent rapid cooling cycle to develop their respective martensitic and austenitic structures, free of chromium carbides in the grain boundaries. This freedom from carbides can be achieved by the granular-pack process, which employs solid, porous chromium granules with good heat conductivity, but not with the powder-pack cementation process, which uses refractory oxides having poor heat conductivity. The gas convection process provides smooth chromized surfaces not attainable by any of the other processes. The chromium potential, which is dependent upon the activity of the chromium, dictates the chromium gradient and the surface chromium content in the alloy layer of metals containing little or no carbon.

In the chromizing of iron and low-carbon steels, such as AISI 1008, case depth is proportional to the square root of time. The case depth versus time curve (see Fig. 11-3) becomes parabolic after about the first two hours of chromizing. Case depth is an exponential function of temperature, as shown in Fig. 11-4.

Carbon is the most important single element in determining chromium case depth and composition on steel. Carbon diffuses rapidly from the base-metal core to form chromium carbides that act as a barrier to further chromium diffusion. Generally, steels containing more than 0.20% carbon are limited to a 0.001″ (0.03 mm) case depth.

Growth due to chromizing varies with the process and the base metal. Growth on a cross section can be as little as 0.0002″ (0.005 mm) on tool steels, one-half the chromized case depth on low-carbon steels, and as great as the equivalent of the case depth on nickel and its alloys. The growth can be predicted once the base metal, case depth, and process have been established.

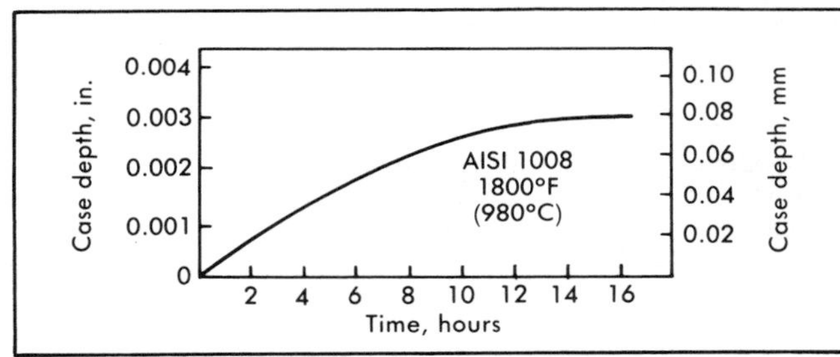

Fig. 11-3 Chromized case depth versus chromizing time.

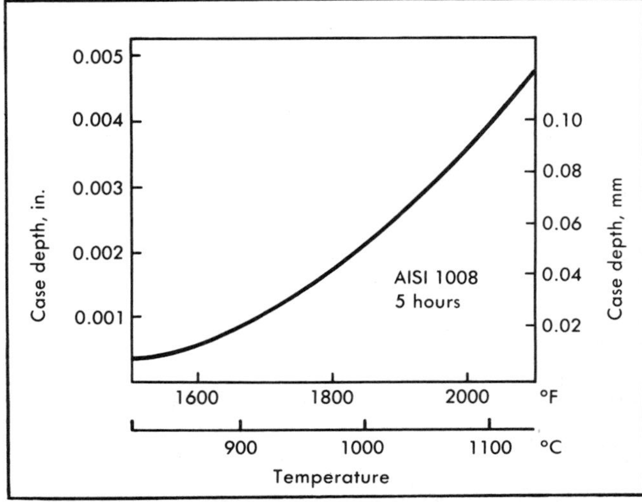

Fig. 11-4 Chromized case depth versus chromizing temperature.

Chromizing offers a variety of advantages. It purifies the steel by removing carbon and nitrogen, causes coarsening of the ferrite grain, removes the stresses and strains of fabrication, provides air-gap stability in electrical components, and increases wear and corrosion resistance. Because chromizing entails high temperatures and relatively long processing times, however, possible distortion and dimensional change of workpieces that have thin cross sections or those containing residual stresses must be considered. Excessive decarburization because of chromizing can occur in thin cross sections, since carbon diffuses outward from the base steel to the surface to combine with the chromium. Bulky parts with large surface area to weight ratios are sometimes costly to chromize because of the volume they occupy in the furnace.

BORONIZING

Boronizing is another diffusion process for the surface hardening of steels. Compatible materials for this process include mild steels, tool steels, cast irons, nickel-based alloys, and cobalt-based alloys. When a boride layer is applied to the appropriate substrate, the layer will provide additional wear and abrasion resistance, often comparable to that of sintered carbide.

Boronizing can be carried out in a variety of media: gases, molten-salt mixtures or powders, and by plasma processes. The temperature at which the boron is diffused depends upon the material and case depth required. Usually the temperature is between 1550 and 1850°F (845 and 1010°C).

The case depth that can be achieved with boronizing depends primarily upon the material being processed. For example, case depths of 0.005″ (0.13 mm) can be produced on low-alloy and carbon steels. Usually, case depths greater than 0.002″ (0.05 mm) are not economical for high-alloyed materials, such as stainless steels and tool steels. The most common case depths are 0.001-0.002″ (0.03-0.05 mm) for low-alloy and carbon steels, and 0.0005″ (0.013 mm) for high-alloyed materials.

The Vickers (diamond pyramid) hardness of boride cases ranges from 1300 to 2300. This range exceeds hard chrome electroplate or hardened tool steels, but approximates the hardness of tungsten carbide. In general, the hardness will increase with an increasing amount of alloy in the substrate.

Size changes result from the processing parameters of boronizing. Typically, a growth can be expected in the range of 15 to 30% of the boride case thickness per surface. This growth may vary with the composition of the substrate but is reproducible for a given combination of material and thermal cycle.

Boronizing provides good wear-resistant properties and reduces the coefficient of friction. The final properties are comparable to carbide. The thermal stability of the boride compounds is good at subcritical temperature, and the dependent factor for performance is the hot strength of the substrate material. Boride case hardness is unaffected by exposure to subcritical temperatures in service.

It is important to provide a core material with sufficient strength to support loads applied to the case. This is not of particular concern in sliding wear situations, but it is crucial when high unit loads are applied. The core material can yield under such high loads and cause the case to spall. Boronized cases are brittle and should not be used for impact loading applications.

Another advantage of boronizing is additional corrosion resistance. While each corrosive situation is unique, low-alloy and carbon steels usually increase in corrosion resistance, while the stainless steels are unchanged or decreased in corrosion resistance, depending upon the environment.

References

1. *1983 SAE Handbook*, Vol. I, SAE Standard J 412h (Warrendale, PA: Society of Automotive Engineers).
2. *Ibid.*

Bibliography

Johnson, Paul, and Nayar, Harbhajan S. "Gas Carburizing Systems: Choosing the Best One for You." *Heat Treating* (February 1981).

Verhoff, Steve H., and Grube, William H. "Benefits of Ion Processing in Heat Treating Practice." *Heat Treating* (September 1981), pp. 52-58.

HEAT TREATING FURNACES

Many different types of furnaces are used for heat treating. Some of the more common types, their control, and their safety and environmental requirements are discussed in this chapter. The subject of controlled atmospheres used in furnaces is presented in Chapter 10, "Heat Treatment of Steel." The equipment used for induction, resistance, flame, electron beam, and laser heating is described in Chapter 13, "Selective Surface Hardening."

CHAPTER CONTENTS:

METHODS OF HEATING FURNACES

Furnaces are either fuel fired, with gas or oil, or electrically heated. The advantages and disadvantages of each type are summarized in Table 12-1. The metal to be heated and the type of treatment to be performed are important considerations in selecting the method of heat application. Surfaces of metals absorb heat transmitted by radiation, convection, or conduction, or a combination of these; and the heat is transferred through the metals by conduction.

DIRECT-FIRED FURNACES

In direct-fired furnaces, the work being processed is directly exposed to the products of combustion, normally referred to as flue products. The analysis of the flue products can be varied to minimize scaling (oxidation) of the work. This control is accomplished by adjusting the fuel-air ratio of the combustion system. Adjustment can be manual or, for a more precise analysis, can be controlled automatically by a variety of fuel-air ratio control systems available.

When direct-fired burner equipment is used in a heat treat furnace, the parts being processed are generally in some primary or intermediate stage of manufacture. As a result, the oxide formed is not usually detrimental to the part and is generally removed subsequently in the manufacturing process. Such is the case in the annealing of cold formed parts between successive drawing operations and in the annealing of rough castings prior to machining operations.

Gas-Fired Furnaces

Gaseous fuel can be natural gas, straight propane, a propane-air mixture, or a manufactured (producer) gas. With the proper selection of burners, controls, orifices, and pipe sizes, a single heating system can be designed to operate on propane gas having a heating value of 2500 Btu (2638 kJ), on natural gas with a value of 1000 Btu (1055 kJ), and on producer gas with 160 Btu (169 kJ). Manual adjustments are required for converting from one gas to another.

Oil-Fired Furnaces

Almost any grade of oil that can be satisfactorily atomized can be burned in direct-fired furnaces. However, in many cases, consideration must be given to the sulfur, vanadium, sodium, and potassium content of the fuel oil. These elements can dramatically affect the life of internal furnace components, particularly the nickel-chromium, heat-resisting alloys. Lower viscosity oils, such as diesel fuel and No. 2 fuel oil, can be easily atomized with low-pressure air. These oils are probably the most commonly used fuel oils for heat treating processes. Even with easily atomized oils, caution should be taken in using them in furnaces operating at temperatures below 1400° F (760° C) with interrupted pilots. At low oil-flow rates and with burners operating with high excess air, nuisance shutdowns can occur from the flame supervision devices. In certain instances, as dictated by the National Fire Prevention Association, constant pilots may be used to eliminate these shutdowns. It is important to note, however, that the insuring body must approve the use of constant pilots for each particular application.

Heavier grades of oil that are burned have to be atomized by sources other than low-pressure air. Typical alternate methods of atomization are high-pressure air and steam. Burners are available that are dual-fuel fired for use with gas and oil. In most instances, oil is used as the standby fuel for peak periods when the availability of natural gas is reduced.

ELECTRICALLY HEATED FURNACES

Electrically heated furnaces are used for practically all temperature ranges, from low-temperature tempering to high forging temperatures. A basic consideration in the use of electric furnaces is

Contributors of sections of this chapter are: ***Steve Balme***, *Chief Engineer, C. I. Hayes Inc.;* ***Anthony G. Fennell***, *Vice President, Fennell Corp.;* ***Edward F. Grady***, *Manager of Marketing, C. I. Hayes Inc.;* ***W. James Laird, Jr.***, *Vice President, Upton Industries, Inc.;* ***Q. D. Mehrkam***, *Sr. Vice President, Ajax Electric Co.;* ***John E. O'Neil***, *Fabricated Materials Market Manager, Leeds & Northrup Co., A Unit of General Signal Corp.;* ***John W. Smith***, *Technical Director, Holcroft/Loftus Div., Thermo Electron Corp.;* ***Ronald W. Sustich***, *Applications Engineer, Fennell Corp.*

Reviewers of sections of this chapter are: ***E. Black***, *Manager of Engineering, A. F. Holden Co.;* ***Frank Boyle***, *Industrial Heat Treating Equipment Group, Selas Corporation of America;* ***Richard K. Brown***, *Director of Research and Development, Consarc;* ***Ed R. Byrnes, Jr.***, *Vacuum Section, Marketing, Ipsen Industries;*

CHAPTER 12

METHODS OF HEATING FURNACES

TABLE 12-1
Advantages and Disadvantages of Fuel-Fired and Electrically Heated Furnaces

Fuel-Fired Furnaces		Electrically Heated Furnaces	
Advantages	Disadvantages	Advantages	Disadvantages
1. Easy adjustment. Input can be easily varied, normally with a simple orifice change. 2. Recuperator-type heat-saving devices can be added. 3. Easily controlled cooling in furnace, with properly designed combustion system. 4. Generally faster heatup. 5. Lower operating and maintenance costs.	1. Requires extensive ventilation system. 2. Potential hazards of fire or explosion. 3. Requires more labor for startup and shutdown. 4. Easier to get out of adjustment, resulting in excessive fuel use.	1. Systems are clean and free of pollution. 2. Cooler plant environment, without need for exhaust hoods and stacks. 3. Quieter operation—no blower or combustion noise. 4. More uniform heat pattern. 5. No exhaust or makeup air systems required. 6. Purge and flame safety systems not required. 7. Availability of electric power.	1. System is inflexible with respect to changing heating capacity. 2. Higher initial, operating, and maintenance costs. 3. Longer cooldown times.

selecting the type of heating element. Elements available include the open type, which are exposed to the furnace's environment, and the indirect type, which are protected from the furnace's internal environment by some means, such as a radiant tube, muffle, or retort.

Factors affecting the selection of the type of heating element include the furnace atmosphere, the need for protection, and the available space.

Effect of Furnace Atmosphere

Almost all furnace atmospheres, other than air, will in some way affect the overall performance and life of each type of heating element material. Manufacturers of heating element materials provide charts that allow an engineer to predict their performance within any standard atmosphere.

Each heating element material will operate with varying degrees of success when exposed to different furnace atmospheres. An exception exists when using a carburizing or an enriched reducing atmosphere. The conventional nickel-chromium strip heating element does not perform well in a carburizing atmosphere as the element itself is rapidly carburized, thus affecting element performance. Generally, with a carburizing atmosphere present, the heating elements are placed inside radiant tubes. However, there are some alternative heating elements specifically designed to operate efficiently when exposed to a carburizing atmosphere.

Protecting the Heating Elements

It is necessary to protect the heating elements against mechanical damage from the parts being heated and from accumulations of metallic scale or broken refractories. In a furnace where bottom heat is mandatory and scale can readily be formed on the parts, or where parts can fall off the tray or conveyor, the use of electric elements protected in radiant tubes below the hearth is recommended, even though open elements may be used throughout the upper portion of the furnace.

Available Space

The space available to locate the heating elements is an important consideration. There are occasions where the physical space available will dictate the design of the element. For example, high watt-density-release elements are sometimes

Reviewers, cont.: ***David L. Checkley****, General Manager, Atmosphere Furnace Co.;* ***Kenelm W. Doak****, Marketing, Vacuum Industries Div., GCA Corp.;* ***Dennis J. Giancola****, East Ohio Gas Co.;* ***G. G. Hoeft****, Manufacturing Engineering, Caterpillar Tractor Co.;* ***Carl Hotchkiss****, Manager of Field Service, Sunbeam Equipment Corp.;* ***L. E. Jones****, Lindberg Heat Treating Co.;* ***William R. Jones****, President, Vacuum Furnace Systems Corp.;* ***Richard J. Kawka****, National Sales Manager, Lindberg, A Unit of General Signal;* ***C. B. Kentnor, III****, W. S. Rockwell Co.;* ***Q. D. Mehrkam****, Senior Vice President, Ajax Electric Co.;* ***Kenneth M. Minoletti****, Engineering Dept., Upton Industries, Inc.;* ***Wayne F. Parker****, Product Assurance Manager, General Heat Treat Equipment, Surface Combustion Div., Midland-Ross Corp.;* ***W. A. Paulson****, Principal Engineer/Scientist, Industrial Instruments Div., Barber-Colman Co.;* ***Burt B. Roens****, Manager, Marketing Communications, Fischer & Porter Co.;* ***Fred Smith****, General Manager, Marketing Operations, Wellman Furnaces Inc.;* ***John W. Smith****, Technical Director, Holcroft/Loftus Div., Thermo Electron Corp.;* ***Ronald W. Sustich****, Applications Engineer, Fennell Corp.;* ***Theodore K. Thomas****, Business Unit Advertising Manager, Process Control Div., Honeywell Inc.;* ***Rajat Verma****, Applications Engineer, ABAR Corp.;* ***William H. Wiley****, Engineering Supervisor, Industrial Instruments Div., Barber-Colman Co.;* ***M. R. Winter****, Manager, Proposal Engineering, Sauder Furnace Div., Sauder Energy Systems.*

METHODS OF HEATING FURNACES

used to get the required amount of heat release in a limited area. Care must be taken in their design so that hot spots created by the high release elements do not affect the temperature uniformity of the workload.

Heating Element Materials

While metallic, nickel-chromium strips and rods are the most common materials for heating elements in electric furnaces, nonmetallic materials are also used. Silicon carbide (Globar-type) elements or molybdenum disilicide rod elements are used with success, directly exposed to various atmospheres. Silicon carbide elements, however, are normally not recommended for use in a carburizing atmosphere; they have been used inside radiant tubes for protection against carburizing atmospheres.

Metallic Resistance-Type Heating Elements

Metallic resistance-type heating elements are used in several types of electric furnaces, including low and high-temperature units.

Low-temperature furnaces with open elements. The temperature range of these heat treat furnaces varies from about 300 to 1250° F (150 to 675° C), and they are normally of the recirculated-wind type. The simplest type of heating element used in these furnaces is a commercially available duct heater. Such heaters are usually full-voltage (440V or 220V) and are useful when employed within their limitations. Maximum temperature is normally limited to 750° F (400° C).

With regard to physical size, the heater should cover the entire cross section of the recirculating duct; however, the heater sizes commercially available are limited. The watt density for commercial duct heaters is normally 22 W/in.^{2} (0.034 W/mm^2).

As an alternative to commercial units, custom-built duct heaters are used. The steel or alloy frame of such units can be designed to completely fill the air-duct cross section. The units are easily removed in a single piece through a sidewall or roof bulkhead. The nickel-chromium ribbon material is strung in a serpentine pattern on ceramic insulators, mounted in tiers. A common element material contains 35% nickel, 18% chromium, and 44% iron, and is normally selected in light gages and narrow widths. A typical cross section for the ribbon material is 1/2 x 0.030" (12.7 x 0.76 mm). Depending upon the temperature and flow velocity, custom-built duct heaters are designed with a watt density range of 15 to 30 W/in.^{2} (0.023 to 0.046 W/mm^2).

High-temperature furnaces with open elements. The temperature range of these furnaces varies from about 1250 to 1750° F (675 to 955° C), and they are normally of the radiant heating type.

When large wall areas are available inside a furnace, a common method of mounting the nickel-chromium strip element is to hang it in a serpentine pattern from ceramic or electrically isolated alloy anchors, normally on the vertical walls only. With this design, especially for higher temperatures, the structural strength of the element material and the configuration must be considered. The element must be capable of supporting itself at temperature without excessive droop or warping. Warping, which causes the elements to touch at various points, can shorten the effective length of the element, decrease the resistance, and cause premature failure due to excessive currents and watt densities from the increase in generated power.

On larger furnaces with accessible wall areas, maintenance of such elements is relatively easy. On smaller furnaces, accessibility for replacement of wall elements becomes a problem. Other types of modular or drawer-type radiant heating elements are available, which make element removal and maintenance easier.

The element strip material used in these high-temperature furnaces is generally one of the following types: 80% nickel-20% chromium, 68% nickel-20% chromium, or 35% nickel-18% chromium-44% iron. The elements are generally thicker and wider than for low-temperature furnaces for increased structural stability. A typical cross-section range would be from 0.050 to 0.090" (1.27 to 2.29 mm) thick and 3/4 to 1 1/2" (19 to 38 mm) wide. Depending upon temperature and the location of the elements, the watt density would be in the range of 8 to 15 W/in.^{2} (0.01 to 0.02 W/mm^2).

An alternative to the nickel-chromium strip element is a cast nickel-chromium heating element. This element has good structural strength, stability, and resistance to atmosphere attack. The quality control necessary in the manufacture of this element, however, has made it slightly less flexible and popular than strip. The casting must be of uniform density and cross section to ensure good resistance without danger of hot spots. Castings made with the lost-wax process generally meet the quality requirements.

Low voltages at the cast elements and the resultant high currents tend to make control hardware and wiring expensive. A common cast element material contains 35% nickel and 15% chromium. The watt density range is from 8 to 10 W/in.^{2} (0.012 to 0.015 W/mm^2).

Nonmetallic Resistance-Type Heating Elements

In general, nonmetallic heating elements are used for high-temperature furnaces, those above 1850° F (1010° C). The two major types of nonmetallic elements are silicon carbide and molybdenum disilicide.

Silicon carbide. Silicon carbide (Globar) elements are generally used in the higher temperature ranges, 1850° F (1010° C) and above. Because they tend to be fragile, care should be taken in design to allow for proper support of the element and freedom for it to expand and contract as the element or furnace is heated and cooled.

Silicon carbide elements go through a resistance increase as they age. To maintain constant power over the life of the element, it is necessary to have a voltage adjustment available (usually with a multitap transformer). A rule of thumb is to set the useful life of the element at the point where its resistance has increased four times. To maintain constant power over this resistance range would mean that a total voltage increase of twice the initial voltage would be available from the step transformer.

Silicon carbide elements are available in various diameters and lengths with rated resistances. Watt densities vary with temperature, atmosphere, and other factors. Conservative watt densities result in better element life. In a sintering furnace operating at 2100° F (1150° C) with an endothermic atmosphere, a watt density of 30 to 50 W/in.^{2} (0.05 to 0.08 W/mm^2) is considered reasonable.

Molybdenum disilicide. These elements are commonly formed in U-shaped rod configurations and should be mounted vertically. The maximum temperature, in air, is over 3000° F (1650° C), which covers most furnace requirements.

METHODS OF HEATING FURNACES

Element location is an important consideration as these elements are designed to operate at high watt densities. Molybdenum disilicide elements go through a high resistance change from cold to hot temperatures (resistance increases with temperature). The control hardware and wiring must be properly designed to handle high initial currents. A typical selection for a 1750° F (955° C) furnace, with an endothermic atmosphere, would be a rod element having a diameter of 0.35″ (9 mm), rated at 158 W/in.^{2} (0.24 W/mm^2), and with an element temperature of 2610° F (1430° C).

RADIANT-TUBE FURNACES

With fuel-fired, radiant-tube heated furnaces, the work chamber is protected from the products of combustion. Normally, the work chamber contains a manufactured controlled atmosphere as dictated by the process. There are, however, cases where the chamber remains filled with air, and the only purpose of the radiant tubes is to protect the work from the high dew point and/or corrosive flue products. Electrically heated radiant tubes are normally used to protect the heating element material from attack by the furnace atmosphere.

Gas-Fired Radiant Tubes

Gas-fired radiant tubes (see Fig. 12-1) are the most common type of fuel-fired indirect heating, due mainly to the wide availability of natural gas. The proper selection of burner components, controls, orifices, and piping allows the same radiant tube to be fired with a wide variety of gases. Fuels used include natural gas, a propane-air mixture, straight propane, and certain manufactured gases. A radiant-tube burner arranged to use natural gas, propane, or producer gas as the fuel is shown in Fig. 12-2.

Radiant-tube burners are of two basic types: sealed head and open type. A particular advantage of the sealed-head burner is that it is adaptable to heat recuperation. This adaptation is accomplished by using the products of combustion and an air-to-flue-gas heat exchanger to preheat the combustion air prior to its entering the burner. Preheating of the air can result in considerable fuel savings.

Radiant tubes are designed for structural strength at temperature and are normally constructed from centrifugal-cast tubing of various diameters, with wall thicknesses from 3/16 to 5/16″ (5 to 8 mm). In many cases, tubes fabricated from wrought alloy, with wall thicknesses of 1/8 and 3/16″ (3 and 5 mm), are adequate. With the high temperatures attained in radiant tubes in most heat treating furnaces, high-grade alloys are commonly used. These include *cast* HT, HK, and NA22H alloys, and *wrought* types 330, 601, and Incoloy 800.

Oil-Fired Radiant Tubes

Straight, oil-fired radiant tubes are not too common and are used mainly when an adequate supply of gaseous fuel is not available or would be too expensive. Dual-fuel (oil and gas) radiant tubes are more common, with oil, usually No. 2, being used as the standby fuel.

When oil-fired radiant tubes are used and corrosive attack from flue gas is not a problem, the construction and types of materials used are similar to those used for gas-fired radiant tubes. Heat recuperation is possible with some sealed-head oil burners. Burner manufacturers should be consulted about possible damage to or problems with the atomizing system from the preheated combustion air, normally supplied at a temperature of 700 to 1000° F (370 to 540° C).

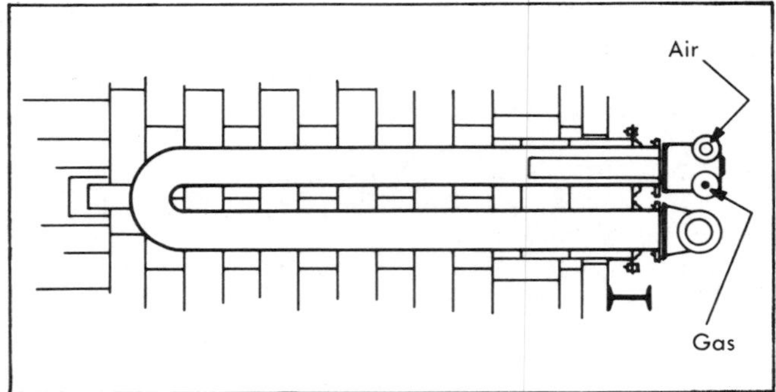

Fig. 12-1 Gas-fired radiant-tube furnace heater.

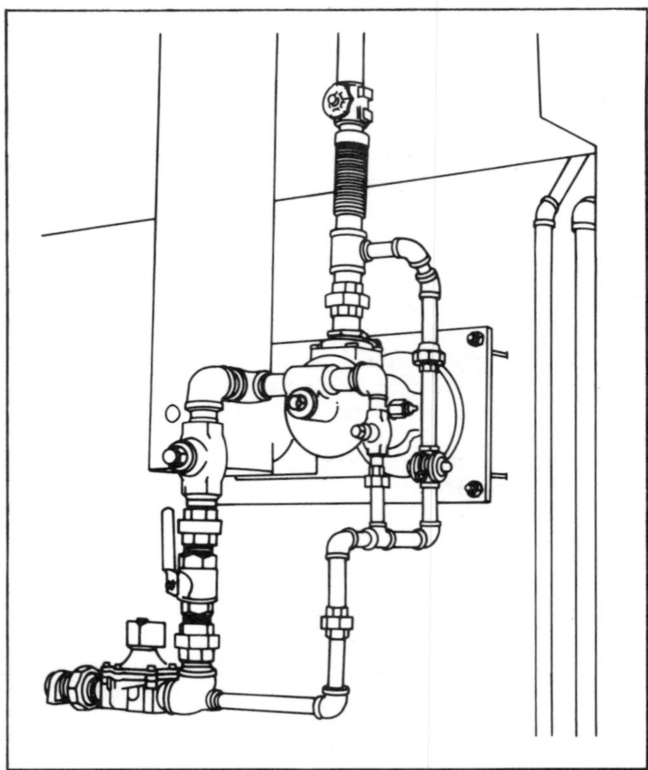

Fig. 12-2 Radiant-tube burner arranged to use natural gas, propane, or producer gas as the fuel.

Electrically Heated Radiant Tubes

With an electrically heated design, the radiant tube is used to protect the resistance heating element from the effects of the furnace atmosphere. A common design is to use rods, made from a nickel-chromium alloy, inside the radiant tube. The rods are formed into hairpin shapes that are supported and contained by ceramic spacer discs.

With the heating element contained in a tube, it is important to conservatively select the watt density of the element. The watt density is a direct function of furnace temperature and varies from 12 to 30 W/in.^{2} (0.02 to 0.05 W/mm^2). Some element designs use strip material, rather than rod, that is formed into similar hairpin patterns.

If a user wishes to convert a gas-fired U-tube furnace to electrically heated radiant tubes, it is desirable to replace the U-tube with two straight tubes for improved uniformity and reduced watt density. With this type of conversion, it is possible to change from natural gas to electricity without loss in production capacity.

Other types of electric radiant tubes are available, in which the radiant tube itself becomes the resistive heating element. The quality and condition of the radiant tube, however, will determine if it will function properly as the resistance element. Silicon carbide elements have also been used inside radiant tubes to protect them from carburizing-type atmospheres.

ENERGY CONSERVATION

In the combustion of fuel in a furnace, using room temperature combustion air, a large portion of the energy in the fuel is used to heat the nitrogen in the air to furnace temperature. This energy, absorbed by the nitrogen, reduces the amount of energy available to heat the work in the furnace. Considerable energy savings can be realized by using heat-saving recuperators.

Stack-Type Recuperators

The typical metallic, stack-type recuperator is a single or multiple-pass, air-to-flue-gas heat exchanger that recovers waste heat from hot flue products normally exhausted into the atmosphere. The exhausted flue products can be from a radiant tube or from a flue opening on a direct-fired furnace. A single-pass, metallic, stack-type recuperator is shown in Fig. 12-3, mounted on a natural gas, radiant-tube burner.

Normally, a stack-type recuperator is a direct replacement for an exhaust stack, and few, if any, exhaust system modifications are generally required when adding recuperation to existing equipment. The energy from the hot exhaust gases is transferred to the incoming room temperature air used for combustion, elevating the temperature of the combustion air. As a result, less gas is needed to raise the burner flame and/or furnace to operating temperature.

A stack-type, metallic recuperator normally consists of two or more concentric cylinders, assembled and welded gas-tight, but that remain free to expand with respect to one another. Hot exhaust gases flow upward through the inner cylinder, and room temperature combustion air flows between the outer cylinder(s). Heat from the exhaust gases is transferred through the thin inner cylinder wall to the combustion air by a combination of convection, radiation, and turbulent heat transfer. In a typical heat treat furnace, the net effect is to create preheated combustion air temperatures as high as 900° F (480° C), with exhaust gases generally entering the recuperator at 1800° F (980° C).

Although higher flow velocities can increase the heat transfer coefficient inside the recuperator, they also create excessive, undesirable pressure drops in the recuperator. With low pressure-drop recuperators, considering both the combustion air and exhaust gas sides, furnaces are able to operate with relatively low-pressure combustion air blowers [1-1.5 psi (7-10 kPa)] and at normal internal furnace pressures. If combustion air is preheated to a temperature over 900° F (480° C), special high-temperature burners are required. The cost of such burners initially offsets the fuel savings on new furnaces and makes retrofits to existing furnaces more expensive. Internal gas pilots or direct electric-spark ignition of main burners are normally recommended with recuperation to optimize the fuel savings.

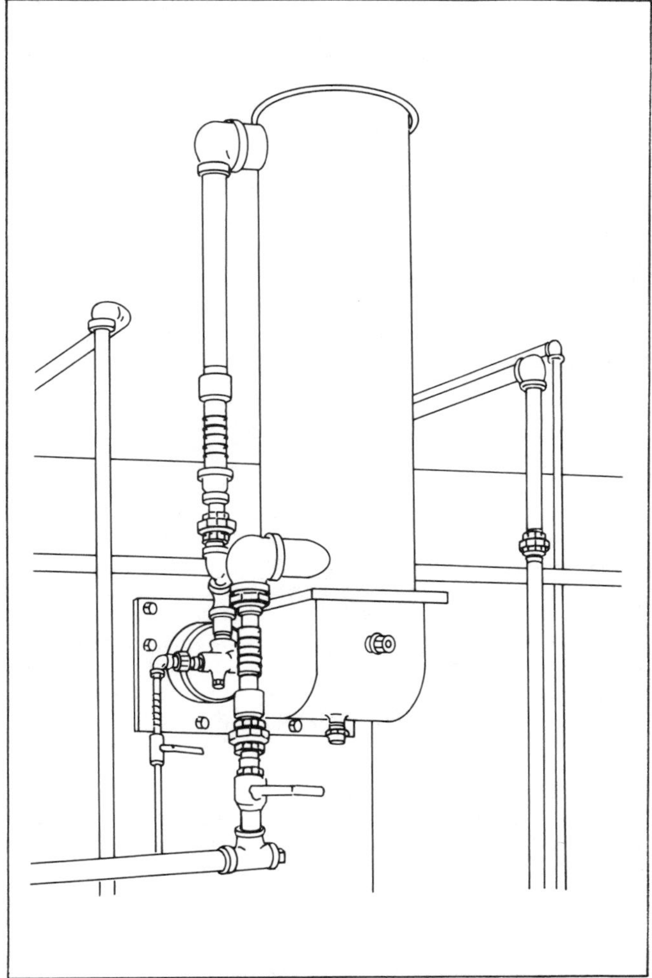

Fig. 12-3 Single-pass, metallic, stack-type recuperator mounted on a natural gas radiant-tube burner.

Savings obtained with radiant-tube applications are normally greater than those with direct-fired applications because the preheated combustion air increases flame temperature and actually improves the combustion efficiency and heat transfer characteristics inside the radiant tube. In instances where a recuperated, sealed-head, radiant-tube burner is compared to an open-type, radiant-tube burner with eductor, the fuel savings can be 40% or more. Normal savings on furnaces operating between 1400 and 1900° F (760 and 1040° C) are 13% for stack-type recuperators and 25% for sealed-head burners. In most instances, the use of a core buster, placed inside the exhaust gas stream travelling through the recuperator, can improve these values.

For maximum efficiency on furnaces that have recuperators and operate at different temperatures, a mass flow system is often added to the cold side of the combustion air supply of each control zone. This addition serves to keep the combustion system on proper ratio for varying combustion air preheat temperatures. The system also improves efficiency during startup of a furnace until maximum combustion air preheat temperatures have been attained.

Other Type Recuperators

Other types of recuperators, in addition to the stack type just described, are readily adaptable to the combustion systems of heat treating furnaces. Some use a ceramic core or matrix as the heat transfer surface to absorb heat from the exhaust gas stream and transfer it to the room temperature combustion air. Ceramic core recuperators can normally accept flue-gas exhaust temperatures in excess of 2000° F (1095° C) and deliver higher preheats than metallic types. However, they tend to have high pressure drops on the flue-gas side, which normally requires an external means of powering the exhaust gases through the recuperator.

Radiant-tube recuperators are also available with the heat-exchange device inserted into the exhaust leg of the radiant tube. This arrangement may offer an advantage where space limitation is a problem. A self-recuperated, direct-fired burner has the flue gas exhausted through the individual burner housing assembly, where the heat exchange to the combustion air takes place. This type of self-contained unit may offer an advantage on larger input burners or, again, where space is limited. A single-stack-type recuperator on a direct-fired furnace can be manifolded to serve more than one or all the burners in a single zone.

Each fuel-fired installation must be analyzed separately when considering the addition of recuperation. With the large variety of heat-saving devices presently available, however, it is difficult not to justify recuperation using hot exhaust flue gases, unless they are extremely corrosive, difficult to capture, or have low heat content.

TYPES OF FURNACES

Heat treating furnaces are heat-holding enclosures that can be classified as either batch or continuous types. There are many variations in design and construction within each type. Selection of a specific furnace depends primarily upon the size, weight, and shape of the parts to be heat treated; the type of heat treatment to be performed; production rates required; and the initial and operating costs. The productivity of furnaces can be rated in pounds-per-hour production per dollar of capital expenditure (including installation), per dollar of energy cost, and per dollar of labor costs (including maintenance).

Direct-fired, batch-type furnaces are the simplest kind. Continuous furnaces, with automatic control for high-production requirements, are much more sophisticated. Many furnaces are equipped for controlling the atmospheres in their work chambers. While standard furnaces are available in a wide variety of sizes, many are of special design to suit a specific application.

FURNACE CONSTRUCTION

Furnaces should be well insulated and tightly built to conserve energy. Cast steels and high-alloy castings are used for furnace construction whenever possible because of their low cost, but these materials are generally not satisfactory when exposed to high temperatures. Ductile cast irons, however, are sometimes used for hearth bars and plates, depending upon operating temperature, but provision must be made for expansion and growth. Special alloy materials are used extensively for many furnace components to minimize maintenance and prolong life.

Insulating Materials

Insulating firebrick has been used extensively in furnace construction because it enables the furnace chamber to contain the heat and minimize heat loss. Various grades of material are available with different hot-face temperature values and thermal properties. More recently, however, insulating materials in the form of blankets or modular blocks have replaced many firebrick applications. Ceramic fiber is now a commonly used insulating material. Advantages include low thermal conductivity, lightness, and resistance to thermal shock. The low heat retention of ceramic fiber also permits faster response to desired temperature changes. Care must be taken, however, to avoid subjecting ceramic fiber to mechanical damage.

Heat Distribution and Control

Various methods of heating furnaces are discussed previously in this chapter. An important design requirement for any furnace is to provide uniform heat distribution and temperature throughout the work chamber. The method and position of heat application, circulation of gases and heat in the furnace, and efficient flow of heat around the workpieces are important considerations to attain this goal.

Since the heating and cooling rates required for different metals vary, it is generally advisable to provide flexible means for controlling these rates. Details of various furnace controls are presented later in this chapter.

Furnace capacities must be sufficient to maintain the time-temperature relationships required for various heat treatments and to meet production needs. Convenient means for providing material handling—into and out of the furnace, and to quench tanks or other furnaces—are also essential. Some furnaces are constructed with integral quench tanks, and complete automation can be provided.

Direct-Fired Furnaces

In direct(open)-fired furnaces, the work being processed is directly exposed to the products of combustion. For many applications, the oxides formed on the workpiece surfaces are not detrimental because they are generally removed in subsequent manufacturing operations. When protection of workpiece surfaces is necessary, semimuffle or muffle (retort) construction (see Fig. 12-4), radiant-tube designs, or controlled-atmosphere furnaces are used.

Semimuffle furnaces. The construction of these furnaces basically uses a direct-fired technique, but the work is protected from the direct effects of the burner firing by a solid hearth and sidewalls, with combustion taking place under the hearth. The tile hearth plate joins the front and back walls but does not touch the sidewalls. The fuel, usually gas or oil, is burned in the space below the tile hearth plate, causing the combustion products to rise through the space between the tile and the sidewalls and to vent through the holes in the roof. The atmosphere surrounding the parts heat treated in such furnaces consists of combustion products, but the parts do not lie in the direct path of the flame.

The amount of roof vent allowed provides a rough control over the flame travel. For example, if there are two or three

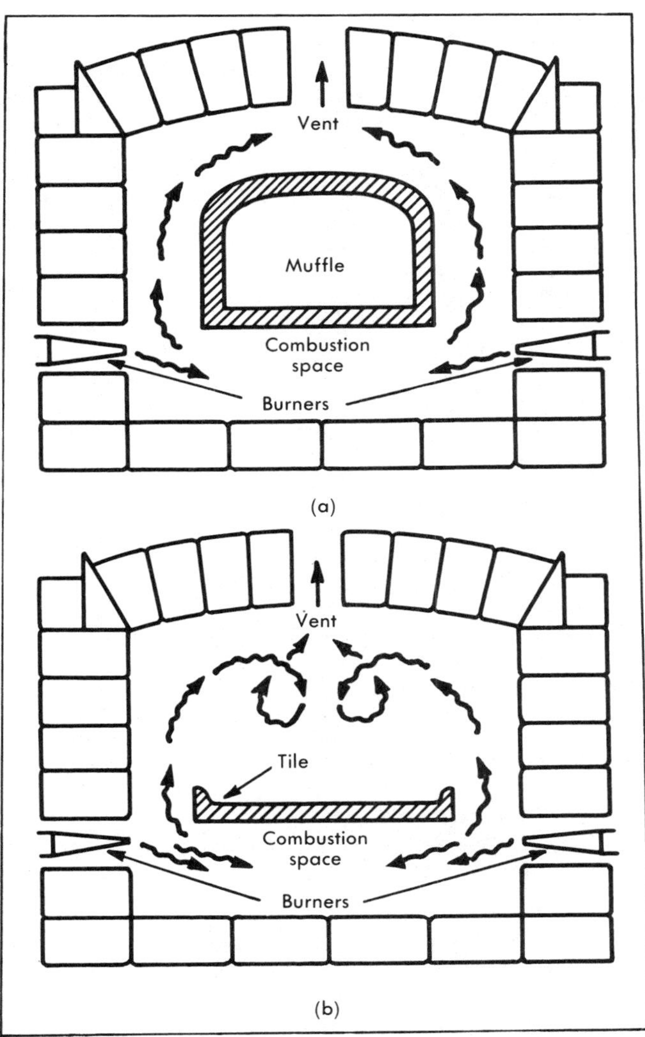

Fig. 12-4 Direct-fired heat treating furnaces of (a) muffle and (b) semimuffle construction.

vents, certain vents can be partially covered, forcing the flame toward other parts of the furnace. If all vents are covered, there is a pressure buildup in the furnace that forces the furnace gases to escape around the door. This is not a desirable situation because it can be dangerous, but it does prevent room atmosphere from being sucked in the door to cool the furnace and perhaps damage the work. A pressure sensor can be provided in the furnace to open or close a flue damper, thus regulating furnace pressure.

Full-muffle furnaces. Wherever gas or oil-fired furnace equipment is to be used for processes requiring protective atmospheres or accurate control of surface chemistry, the products of combustion must be isolated from the work chamber. This isolation is accomplished by a full-muffle design. The transfer of heat energy in these designs is one of radiation through the muffle wall. In a full-muffle design, the work is passed through or treated within a cast or fabricated muffle that takes various cross-sectional configurations, such as a square, rectangle, D-shape, or cylinder. The burners are mounted on the outside, thus excluding the products of combustion from the muffle and the workpiece. The use of muffles, however, is limited with respect to size, temperature, and cost.

Radiant-Tube Furnaces

Radiant-tube furnaces, discussed previously in this chapter, are gas or electrically heated or, less commonly, oil fired. They are more economical than muffle furnaces and are generally used when controlled atmospheres are required. To achieve temperature uniformity and the required heating rates, radiant tubes can be placed at various locations in the heating chambers and mounted horizontally or vertically. Recuperators can be fitted to the tubes to preheat the air for combustion.

Controlled-Atmosphere Furnaces

While muffle, radiant-tube, and electric furnaces do not permit combustion products to contact the workpieces being heat treated, they do allow air to do so. Air contact can result in oxidation of the workpiece surfaces and decarburization of steels. Oxidation and decarburization is avoided by the use of controlled atmospheres, discussed in Chapter 10, "Heat Treatment of Steel."

Controlled-atmosphere furnaces require special construction features to prevent loss of the atmosphere and to eliminate the entry of air. Furnace housings are generally welded gas-tight, and batch-type furnaces are provided with sand or other type seals. Continuous furnaces may have flame curtains to burn oxygen from any infiltrating air, or the furnaces are operated under atmospheric pressure to prevent air from infiltrating. Sometimes gases in the furnaces contain hydrocarbons that burn any infiltrating oxygen.

When doors are required on the furnaces, they must fit tightly and be provided with means for clamping. Continuous furnaces are sometimes provided with vestibules and inner doors to minimize furnace contamination during loading and unloading of workpieces.

BATCH FURNACES

Batch furnaces consist essentially of an insulated chamber with a reinforced steel shell, a heating system, and one or more access doors for loading and unloading the heated chamber. They are available in many different types, the more common being box, car-bottom, elevator, bell, pit, and salt-bath furnaces. While most batch furnaces are loaded and unloaded manually, they can be automated for transferring parts or loads to and from the furnace chambers and integral quench tanks.

Batch furnaces are used extensively to heat treat limited quantities of small and large parts, workpieces difficult to handle automatically, and precision-machined components. Other common applications include the heat treating of parts requiring long cycle times and processes requiring a variety of heating and cooling cycles.

Box Furnaces

Box furnaces are batch furnaces that are shaped like boxes (see Fig. 12-5). They have solid hearths and are loaded and unloaded through vertically operating doors at one or both ends of the furnaces. Box furnaces are available in a wide variety of sizes, with electrically heated or fuel-fired systems. For convection heating, the furnaces have recirculating fans mounted either internally or externally.

Common applications for box furnaces include laboratory operations, the heat treating of single parts or small lots, and for limited production requirements. Some box furnaces are used for annealing and also for hardening when a fast quench is not required. Workpieces are loaded onto the furnace hearth or into baskets that are charged into the furnace with a forklift truck. Slot furnaces are a special type of box furnace used to heat the ends of stock to be forged.

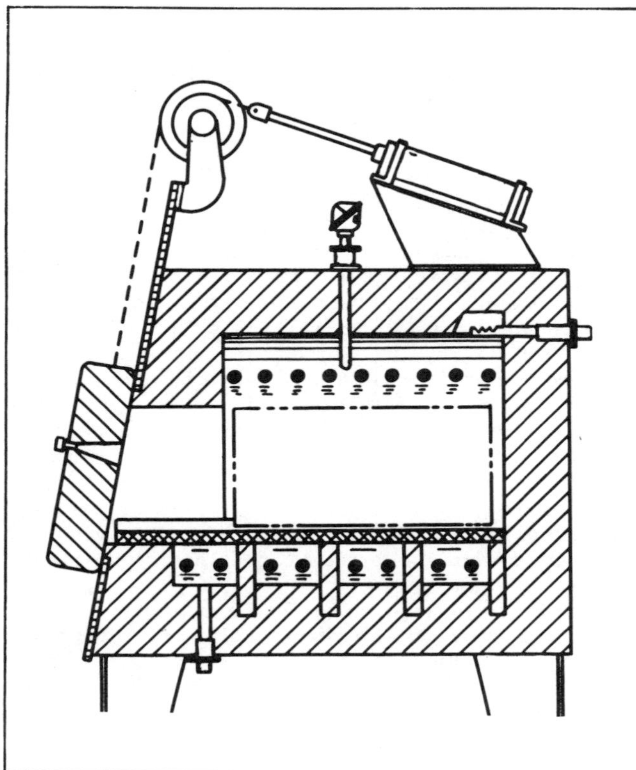

Fig. 12-5 Typical batch-type box furnace for heat treating.

Car-Bottom Furnaces

Car-bottom furnaces are essentially large box furnaces designed for loading and unloading outside the furnace. The furnace hearth is built on a car (see Fig. 12-6) that is moved into and out of the furnace through a door or doors. The car, mounted on wheels or rollers, moves on tracks by means of a power drive. When in position inside the furnace, the car is sealed to the furnace structure. For better heat circulation and more uniform heating, the workpieces are generally supported on the car with castings of heat-resistant materials or on refractory piers.

Car-bottom furnaces are available with fuel firing or electric heating and in many different sizes. They are most commonly used to heat treat large parts, such as castings, forgings, weldments, bar stock, and heavy plates. Some car-bottom furnaces have two heating chambers side by side, with a common dividing wall, thus permitting two heat treating operations to be performed. Auxiliary cooling systems are sometimes provided for faster cooling of the heated parts.

Elevator Furnaces

Elevator furnaces are similar to car-bottom furnaces except that the car and hearth are rolled under a furnace shell and raised into the furnace, or the shell is lowered over the car and hearth. These furnaces generally have built-in lifting or lowering mechanisms, thus eliminating the need for crane facilities. Gas firing or electric heating are commonly used, with oil firing being employed less frequently. The temperature range for these furnaces is generally about 600 to 2200° F (315 to 1205° C).

Bell-Type Furnaces

Bell-type furnaces, sometimes called cover or gantry-type furnaces, have removable retorts or covers called "bells." The retort is removed by a crane for loading the hearth, and then the retort is replaced. For heat treating operations requiring protection of the workpiece surfaces from oxidation or decarburization, inner retorts (or bells) are used (see Fig. 12-7). The inner retort is placed over the loaded hearth, sealed at the bottom, and provided with a constant supply of protective atmosphere; then the outer heating shell is lowered over the assembly.

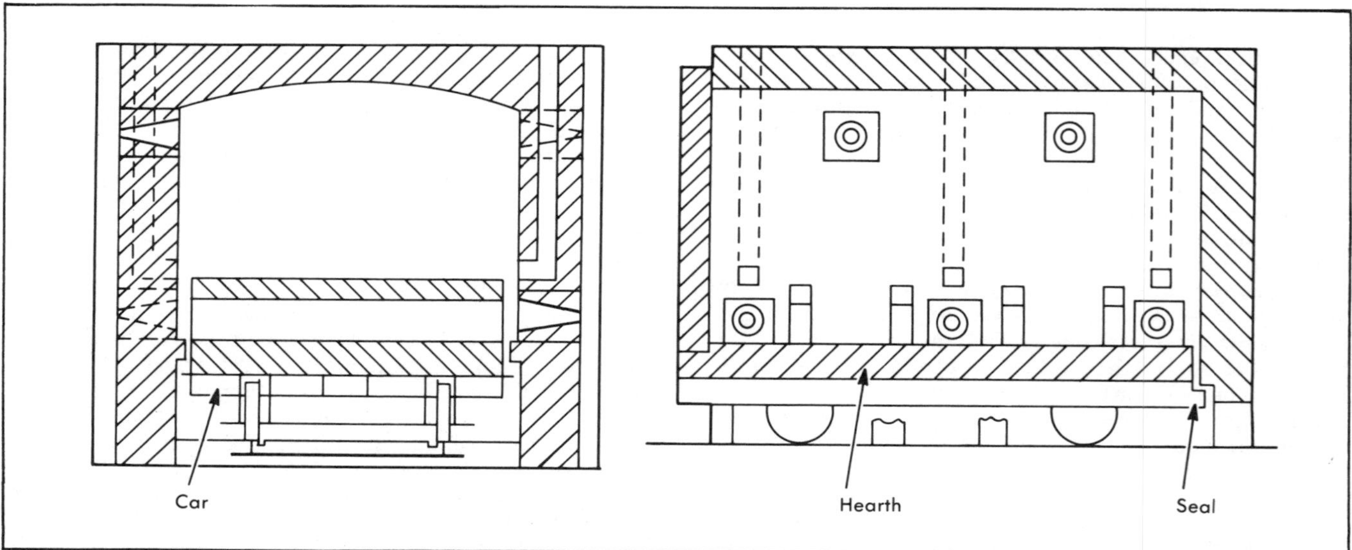

Fig. 12-6 Car-bottom furnace designed for loading and unloading outside the furnace. (*National Fire Protection Assn.*)

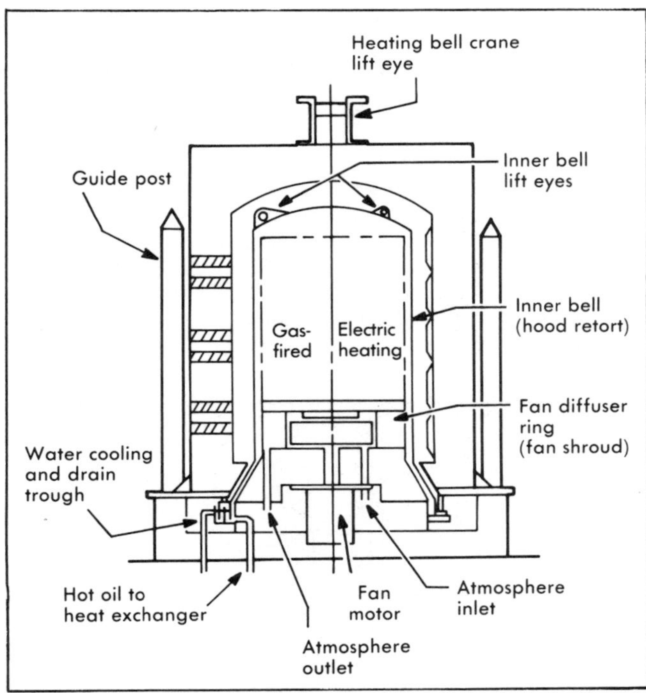

Fig. 12-7 Bell-type furnace with inner retort for atmosphere control. (*National Fire Protection Assn.*)

Pit (Pot) Furnaces

Pit furnaces, sometimes called pot furnaces, are cylindrical or rectangular in shape and have openings, usually with covers that swing aside, for loading and unloading workpieces. Large pit furnaces are generally installed with at least part of their heating chambers below floor level. Smaller furnaces are usually mounted on the floor. Workpieces are suspended from fixtures, held in baskets, or placed on bases in the furnaces. These furnaces can also be fitted with a removable muffle or retort to contain the workpieces.

Other Batch Furnaces

Other common types of batch furnaces include salt-bath, vacuum, and fluidized-bed furnaces, discussed in subsequent sections of this chapter.

CONTINUOUS FURNACES

Continuous furnaces contain the same basic components as batch furnaces, but are used for high-production requirements when consistent cycles are required. Advantages of continuous furnaces include uniform quality of the heat-treated parts, reduced labor requirements, and the capability of integrating the furnaces into production systems. Workpieces are moved through the furnaces while they are being heated, and the furnaces operate in uninterrupted cycles. Rate of travel through such furnaces and the zoning required must be coordinated to obtain the desired heat treating cycle.

Two typical kinds of continuous furnaces are circular types with rotating hearths and those having a single, long, straight chamber or series of chambers (see Fig. 12-8). Straight-chamber furnaces can be classified according to the way parts are moved: rotary retort, roller hearth, pusher type, conveyor type, shaker hearth, walking beam, strand type, and overhead monorail.

Rotary-Hearth Continuous Furnaces

In rotary-hearth (rotating-table) furnaces (see Fig. 12-9), the walls and roof are stationary. The floor of the heating chamber is formed by the rotating hearth, the periphery of which contacts the stationary wall through a sand or liquid seal. Loading and unloading are often done at one station generally equipped with a door, but double-door designs are also used.

Continuous rotation or indexing of the hearth is accomplished with electric, pneumatic, or hydraulic systems. Heating sources may be electric, gas, or oil. While most rotary-hearth furnaces operate with an air atmosphere and products of combustion, protective atmospheres are also used for some applications.

Rotary-hearth furnaces are often preferred for heat treating parts that have to be handled individually. They are also used for small parts that are loaded into baskets or trays.

Straight-Chamber Furnaces

Rotary-retort furnaces. Rotary-retort furnaces consist essentially of large alloy drums that rotate around their horizontal axes. The furnaces can be equipped with automatic feeders, multiple heating zones, quench tanks, and automatic unloaders to provide completely automated processing lines. These furnaces are used extensively for the continuous hardening of relatively small parts, such as fasteners, washers, and bearing balls, rollers, and needles.

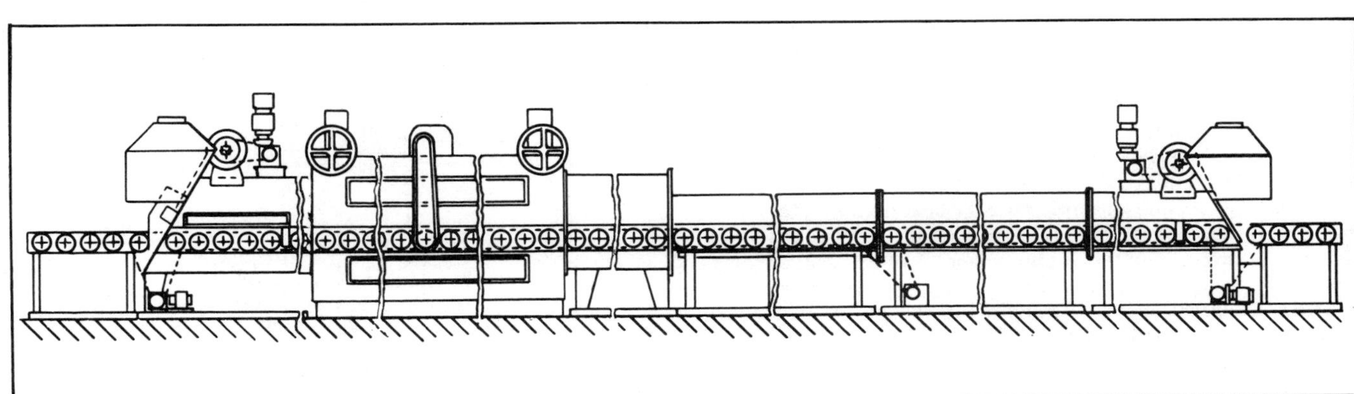

Fig. 12-8 Continuous type of heat treating furnace.

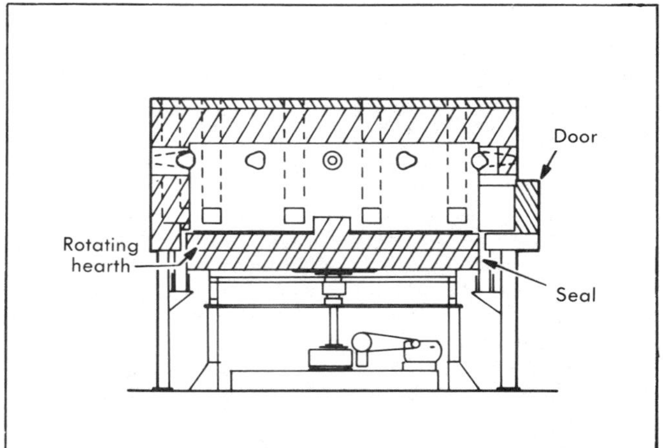

Fig. 12-9 Rotary-hearth furnace. (*National Fire Protection Assn.*)

Roller-hearth furnaces. These furnaces (see Fig. 12-8) are available as single units, as a line of furnaces for zone heating and cooling, and with an intermediate quench tank if desired. Workpiece movement through the furnace is accomplished with powered, shaft-mounted rollers that contact the workpieces or trays holding the workpieces. Sheets of alloy steel are sometimes placed between the rollers and the workpieces to protect the surfaces of the workpieces. Fast-drive sections may be incorporated to facilitate charging or discharging, or for negotiating quench units.

Pusher-type furnaces. In furnaces of this type, the workpieces are pushed against each other, or the parts are loaded in trays, fixtures, or other carriers that are pushed. Pusher furnaces are equipped with ceramic skids or roller rails. They are also available with multiple chambers connected by gas-tight passageways with barrier doors for zone heating or carburizing.

Conveyor-type furnaces. Furnaces of this type are similar to roller-hearth furnaces except that mesh or cast-link belts are used to move the parts. Such furnaces are preferred for small parts that cannot be moved satisfactorily with rollers. Conveyors used include belts of suitable material, chains with projecting lugs, pans or trays connected to roller chains, and mesh or woven chains.

Shaker(shuffle)-hearth furnaces. Furnaces of this type are similar in design to conveyor furnaces except that workpieces are carried through the furnace on a long pan. The pan is vibrated or shaken to advance the parts, which generally fall off the far end of the pan into a quench tank. Such furnaces are commonly used for heat treating small, lightweight parts required in large quantities. For some applications, workpieces are fed by gravity from a hopper onto the vibrating pan. In other cases, automatic parts feeders are used.

Walking-beam furnaces. Furnaces of this type have rows of support beams placed in longitudinal slots in the hearth and staggered throughout the length of the furnace. Toggles or cams below the hearth intermittently raise the beams, move them forward, and then lower them, thus moving the workpieces onto the beams ahead. This type of movement is often used for long and heavy parts, such as bars, slabs, tubes, and structural shapes.

Strand-type furnaces. Continuous strand-type furnaces for heat treating uncoiled strip reduce the handling and cycle times required with batch-type furnaces for sheet in coil form. Furnaces of this type also permit combining other operations, such as cleaning and/or coating. These furnaces usually have high production rates.

Overhead-monorail furnaces. In these furnaces, workpieces to be heat treated are suspended from rods attached to carriers on a monorail.

Furnaces with Integral Quench Tanks

Many heat treat furnaces are provided with integral, enclosed quench tanks. These permit keeping the workpieces under atmosphere control from the time they enter the furnace until they have been quenched and removed from the furnace. Three basic types of integral quench tanks used on furnaces are illustrated in Fig. 12-10.

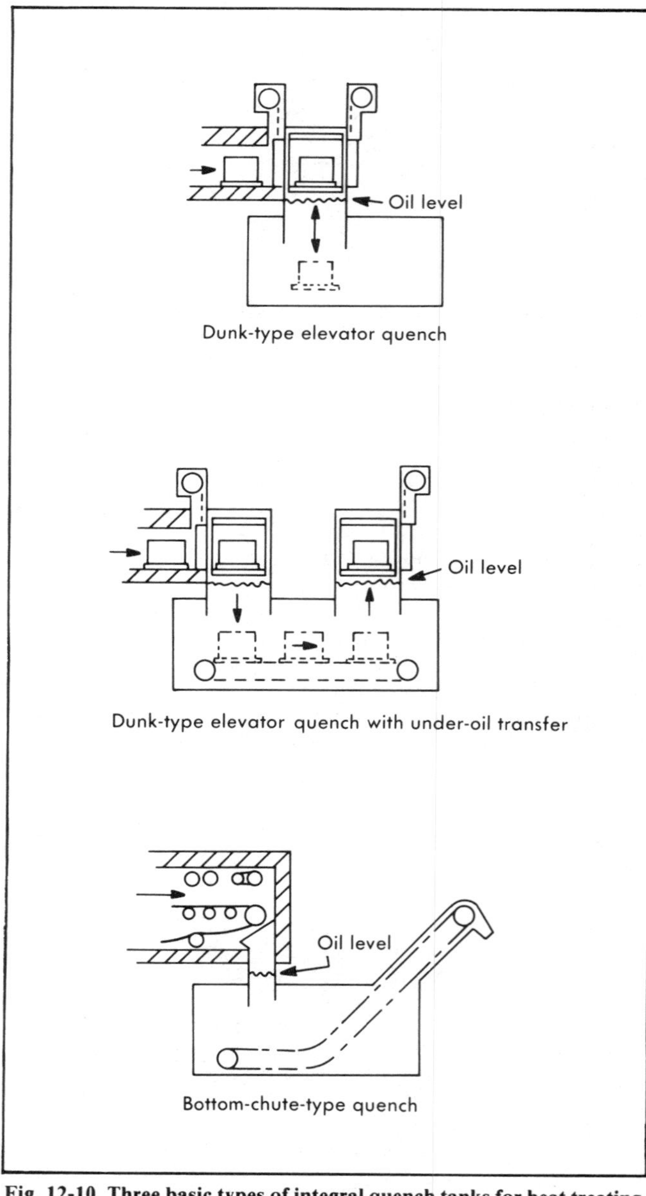

Fig. 12-10 Three basic types of integral quench tanks for heat treating furnaces. (*National Fire Protection Assn.*)

SALT-BATH FURNACES

Salt-bath furnaces are basically ceramic or metal containers with molten salt in which workpieces are immersed for either heating or cooling within a temperature range of 300 to 2400° F (150 to 1315° C). The molten media generally consists of one or more salts, such as nitrates, chlorides, carbonates, cyanides, or hydroxides. The furnaces are heated by gas, oil, or electricity.

ADVANTAGES

A major advantage of salt-bath furnaces is rapid heating of the workpieces. The conduction method of heating with molten salts is up to six times faster than convection or radiation methods of heating. The heating rate is limited only by the thermal conductivity of the metal being heated and the ability of the furnace to supply energy at a rate fast enough to maintain the bath temperature. Rapid heating permits increased production rates and/or the use of a smaller furnace.

Protection of the workpiece surfaces from the atmosphere during heating is another important advantage of salt-bath furnaces. Decarburization of steel parts and oxidation and scaling are generally eliminated. While these advantages are also obtained with controlled atmosphere and vacuum furnaces, capital expenditure requirements for salt-bath furnaces of similar production capacities are lower.

Another advantage of salt-bath furnaces is the uniformity of heating. The temperature throughout any size of internally heated bath averages ±5° F (3° C) of the preset temperature, thus heating all surfaces and sections simultaneously. For high-temperature applications, temperature conformity of ±1 1/2° F (0.8° C) is not unusual. The uniformity of workpiece heating and precise control of temperature minimize distortion, prevent cracking, and ensure consistently high-quality results. The density of the molten bath helps support the workpieces, thus minimizing sagging, bending, or distortion. Salt-bath furnaces can also be used for selective hardening by using fixtures to immerse only the portions of the workpieces to be hardened. Parts of different sizes and shapes may be treated simultaneously if desired.

LIMITATIONS

Salt-bath furnaces are best suited for production work requiring daily or continuous operation. Applications requiring intermittent use are best served with externally heated pot furnaces.

Buoyant workpieces are difficult to handle in salt baths, and it is desirable to avoid such applications by redesign of the workpieces if possible. Blind cavities, which can trap air or salt, present problems in handling and salt removal. These problems can sometimes be solved by redesigning the parts or repositioning them for satisfactory drainage.

PRECAUTIONS

Workpieces must be thoroughly dry before they are immersed in molten-salt baths to avoid splashing of the salt. Hollow fixtures and tooling that has been capped must not be used. Bright surfaces cannot be maintained on workpieces when both heating and cooling operations are performed in salt baths.

APPLICATIONS

Salt-bath furnaces are being used for a wide variety of heat treating operations. These include austenitizing, solution treating, age hardening, tempering, annealing, normalizing, liquid carburizing, cyaniding, liquid nitriding, carbonitriding, quenching, and dip brazing. The furnaces are used extensively for heat treating ferrous and nonferrous alloys; ceramics, such as optical glass; and for curing some polymers.

One major application of salt-bath furnaces is the hardening of high-speed steel tools that require maximum surface protection and uniformity. Reasons for the predominant use of salt-bath furnaces for this application include simplicity and versatility, uniform and rapid heating, precise control, and freedom from scaling.

Typically, hardening of high-speed steels in salt-bath furnaces consists of four steps: (1) preheating, (2) austenitizing, (3) quenching or air cooling, and (4) drawing or tempering. Preheating is done to lessen thermal stress and safeguard against cracking and distortion. Intricately shaped workpieces may require more than one preheating operation. Austenitizing, the most critical step, is often performed at temperatures close to the melting point of the high-speed steel and therefore requires accurate temperature control. Drawing or tempering of the quench-hardened steels modifies the microstructure to provide the desired strength, hardness, and toughness.

FURNACE TYPES

The two major kinds of salt-bath furnaces are ceramic lined and metal pot types. Ceramic-lined furnaces, used primarily for neutral chloride applications, are generally heated electrically. They are available with submerged electrodes or immersed electrodes that conduct electrical current to heat the bath internally for maximum temperature uniformity. Pot furnaces are externally heated and have less temperature uniformity, typically ±10° F (6° C).

Submerged Electrode Furnaces

Submerged electrode salt-bath furnaces are typically rectangular boxes (see Fig. 12-11), but can be designed to almost any shape or size. Electrodes are located on one or both sides and are spaced from top to bottom to provide uniform heating. Electrode life is much longer than with immersed electrodes.

Typically, submerged electrode, salt-bath furnaces are constructed with a 6 to 9" (150 to 230 mm) thick ceramic pot, surrounded by castable refractory material and/or firebrick. This construction provides a temperature gradient that causes the molten salt to solidify within the ceramic pot, thus reducing furnace distortion and lengthening life.

Immersed Electrode Furnaces

Ceramic-lined salt baths heated by immersed (over-the-top) electrodes (see Fig. 12-12) were developed for ease of electrode replacement and continuity of operation. Another advantage of immersed electrodes is easy startup. Since the electrodes penetrate the surfaces of the baths, it is only necessary to melt puddles of salt between the electrodes with a gas torch or resistance heating element to start the furnaces. However, due to corrosive attack at the air-to-salt interface, service life is only a fraction of the service life for submerged electrodes.

CHAPTER 12

SALT-BATH FURNACES

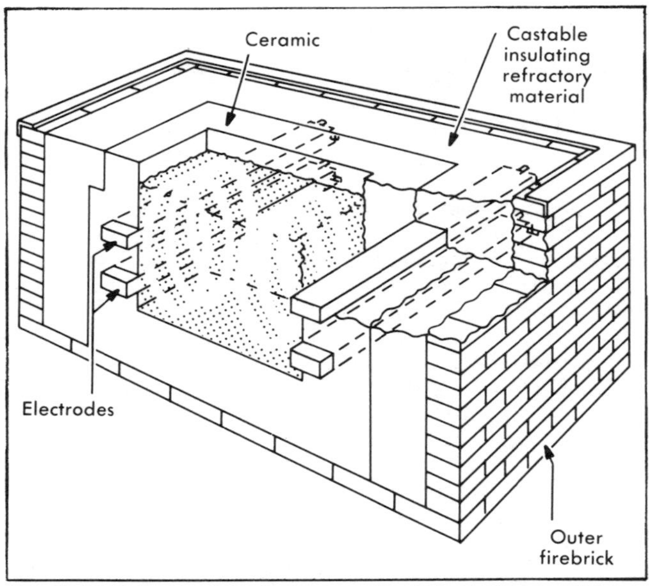

Fig. 12-11 Submerged electrode type of salt-bath furnace. (*Upton Industries*)

Externally Heated Pot Furnaces

Externally heated pot furnaces—heated by gas, oil, or electrical resistance elements—have been used for many years. Gas or oil-fired furnaces have two or more self-cooling burners that fire into a combustion chamber surrounding the round or rectangular pot. Electrical furnaces are heated by a series of resistance elements surrounding the pot. Round pots for these furnaces usually range from about 9 3/4 to 35 1/2″ (250 to 900 mm) diam x 8 to 29 1/2″ (200 to 750 mm) deep, but larger pots have been built. Improved temperature uniformity is obtained by internal heating.

Due to rapid deterioration, externally heated pot furnaces are limited to a temperature of about 1700° F (925° C). Although these furnaces are used at higher temperatures, pot life is greatly shortened. If the steel or alloy pot fails, with salt leaking into the heating chamber, a complete furnace rebuild will be required. Pot life seldom exceeds one year, depending upon operating temperature. Pot materials are selected according to application and may be steel, aluminized steel, alloy, or ceramic-coated alloy.

Externally heated pot furnaces are now available for operation at temperatures to 1850° F (1010° C), and furnaces with maximum operating temperatures to 2350° F (1290° C) are being developed. Pot furnaces provide all the advantages of ceramic-lined salt bath furnaces plus lower initial capital cost

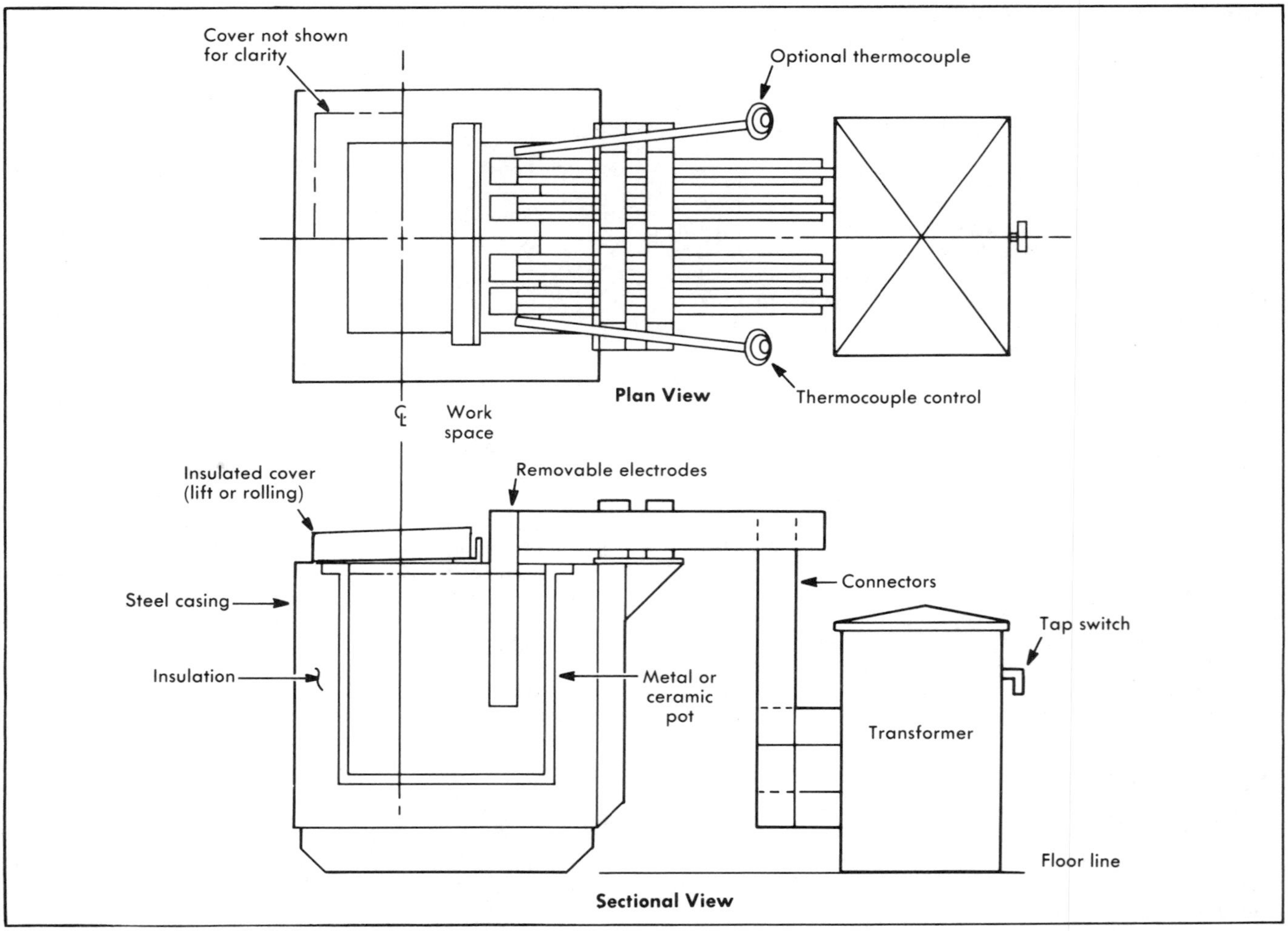

Fig. 12-12 Typical over-the-top, immersed electrode salt-bath furnace. (*Ajax Electric Co.*)

and the ability to start and stop quickly. Startup from room temperature to 1550° F (845° C) generally takes only about 3 1/2 hours, compared to about 12 hours for submerged or immersed electrode ceramic furnaces, depending upon workpiece size and section thickness. However, operating and maintenance costs should be carefully considered; externally heated pot furnaces have a higher operating cost in dollars per pound of metal treated.

Energy efficiency of pot furnaces is less than ceramic-type furnaces, but considerably more than fluidized-bed furnaces, discussed later in this chapter. Useful energy delivered to the work is about half that attainable with submerged electrode furnaces. The cost of operating an externally heated pot furnace is higher than internally heated furnaces due to higher energy use and more frequent pot replacement. If used continuously, this type of furnace loses its advantage due to higher cost per pound of metal treated.

FURNACE AUTOMATION

Any salt-bath heat treating process can be arranged in line with a quench tank or furnace, wash and rinse tanks, and other equipment to form an automated system. Closed-loop chain systems are used for some high-speed-steel hardening applications because of the short cycle times. Figure 12-13 illustrates a system using a programmable hoist operated from a monorail. A transfer-arm mechanism quickly moves workpieces from the austenitizing furnace to the quench furnace.

Programmable hoists incorporate two dual-speed, heat-shielded electric motors that use a positive drive to transport the load. Control of the hoist is achieved through a master console having a microprocessor-based programmable controller. Any work-transfer cycle can be easily programmed from the console, but manual-mode stations can be provided at desired locations along the line.

Safety devices available with programmable hoist systems include overdrives and interlocks to prevent jamming, occupied station detectors, and emergency stop systems. Other devices include a full-bath detector to prevent overloading any furnace or tank and provisions for missed workstations.

SALTS USED

Many different salts are used for various heat treating applications, and details about them are described in Military Specification MIL-10699B. The salt to be used depends upon the specific application.

In selecting a salt for a given application, the following factors should be considered:

1. The salt must have the proper working range to suit the operating temperature requirements.
2. The salt must have the proper melting point to avoid prolonged heatup times for large workloads.
3. The salt must be compatible with the furnace materials.
4. The salt must be compatible with other salts, oils, and chemicals used in the same heat treating line.
5. The salt must be easy to remove from the workpieces by washing, have a low affinity for moisture, and be versatile for different applications.

By balancing these considerations, a salt best suited for a particular application can be chosen.

For preheating and hardening high-speed steels, a ternary-eutectic chloride mixture is often used. These salts, which melt at about 1006° F (540° C) and are usable at temperatures of 1100 to 1650° F (595 to 900° C), are formulated to preheat or harden without decarburization. The heating rate for a salt at 1650° F is rapid, thoroughly heating a 1" (25 mm) diam tool in three minutes or less. A binary chloride mixture is commonly used for austenitizing hardenable steels within the temperature range of

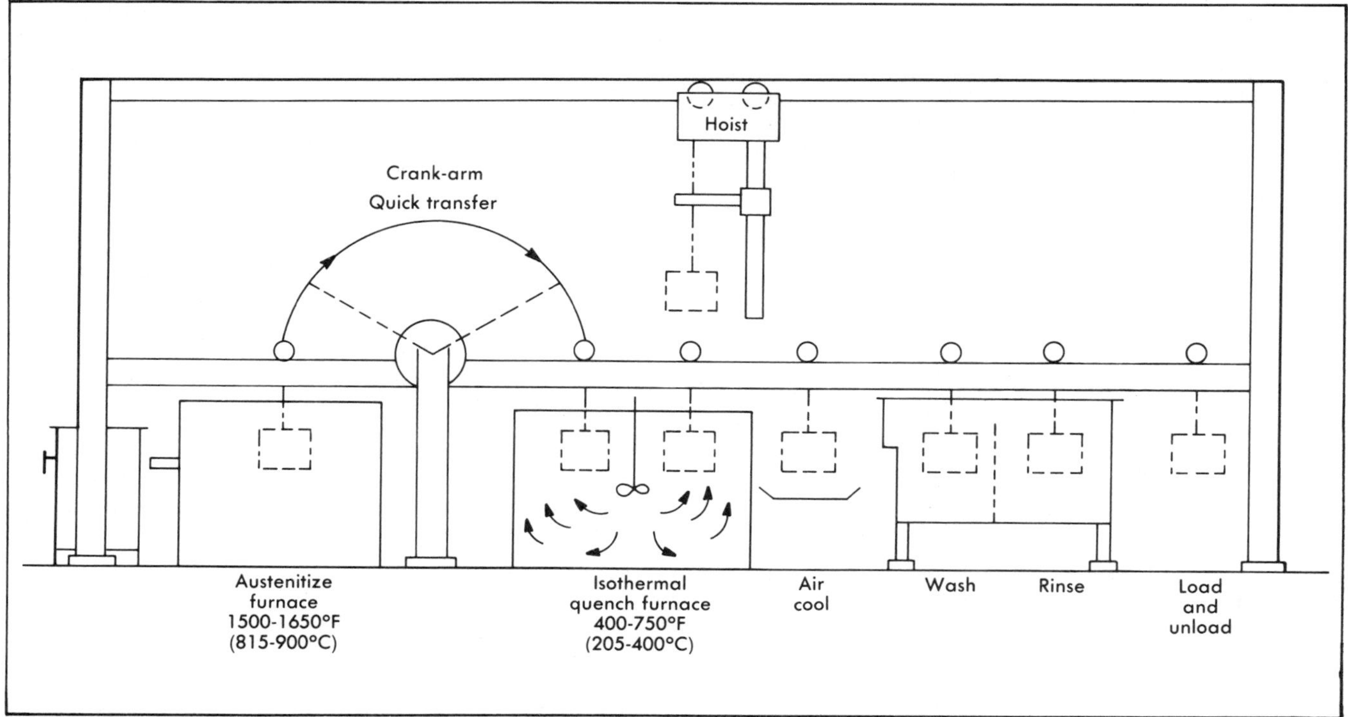

Fig. 12-13 Single-rail, programmable hoist provides flexibility for automating salt-bath heat treating. (*Ajax Electric Co.*)

1450 to 1650° F (790 to 900° C) and is compatible with other chlorides. Neutrality is maintained by the use of solid or gaseous rectification practices. Replenishment after mechanical dragout is usually sufficient to control bath chemistry. Various rectification practices are available to maintain neutrality and freedom from decarburization. Periodic desludging of the furnace bottom is necessary to remove accumulated contaminants and workpieces.

Austenitizing baths for high-speed steels are generally anhydrous barium chloride, which begins to melt at about 1760° F (960° C) and has a working range of 1800 to 2400° F (980 to 1315° C). In maintaining the neutrality of the barium chloride, two procedures are used, each at different bath temperatures.

The first procedure involves barium chlorides at temperatures above 2000° F (1095° C). The immersion of carbon rods in the furnace bath allows a chemical reaction that removes metal oxides, consequently depositing the oxides on the carbon rods. The rods are immersed for about one hour for every 5-8 hours of heat treating operation.

The second rectification procedure is for use with methyl chloride. Upon completion of the first rectification procedure, the furnace-bath temperature should be reduced below 1900° F (1040° C). Methyl chloride is then introduced below the surface of the bath. When broken down, chloride ions combine with harmful metallic oxides to form neutral metallic chlorides.

Various salts and other media are used for quenching high-speed steels from the austenitizing temperature. Water is sometimes used to quench cold-heading tools, but a triple-eutectic chloride salt is recommended for quenching other high-speed steels.

Tempering is generally done in nitrate-nitrite mixtures that melt at a temperature of about 290° F (145° C) and are usable at temperatures of 325 to 1100° F (165 to 595° C). Neutral chloride salts can be used to temper small parts and light workloads, but they are not recommended for production unless the tempering temperatures are above 1100° F.

SAFETY CONSIDERATIONS

The use of heated baths of molten salt presents potential hazards requiring safety precautions. These precautions are discussed in a subsequent section of this chapter.

VACUUM FURNACES

Vacuum furnaces basically consist of a container that is evacuated to create a vacuum in which workpieces are thermally treated by electric radiant heat. Substantial advances have been made during recent years in the design, operation, control, and versatility of vacuum furnaces. As a result, they are being used extensively as an alternative to atmosphere-controlled and salt-bath furnaces.

ADVANTAGES

A major advantage of vacuum furnaces for many heat treating applications is in eliminating the need for costly protective atmospheres, controls, and venting systems. The use of a vacuum prevents unwanted gases from contaminating or oxidizing the metal being treated. By eliminating oxygen, workpieces are kept bright and scale-free, and chemistry of the metal, including carbon content, remains unchanged. Dissolved gases, such as hydrogen, water vapor, and air, are evaporated from the workpieces; and other surface contaminants, such as films, solvents, and lubricants, are removed. In many cases, superior surface properties are attained with vacuum heat treatment.

Precise control of vacuum and temperature ensures consistently high quality of heat-treated parts. Versatility is another important benefit of vacuum furnaces. The furnaces are used for many different heat treating processes, discussed later in this section. Energy efficiency of vacuum furnaces is high. Also, energy is only required when the furnace is in use; the furnace can be completely shut down when not needed.

Increased worker safety and an improved working environment are other advantages of vacuum furnaces. By eliminating flammable or explosive gases and toxic substances, less elaborate safety precautions are required. Reduced heat and the elimination of smoke, dust, scale, and other pollutants permit placing the furnaces anywhere in the plant.

LIMITATIONS

The initial cost of vacuum furnaces is a possible limitation to their use. However, the operating savings possible for many applications permit rapid write-off of the investment. In some cases, the cost of operating a vacuum furnace is much less than for a controlled-atmosphere furnace.

Evaporation of metals having high vapor pressures can be a problem when using vacuum furnaces. Metals such as zinc, cadmium, aluminum, and silver are a few with vapor pressures that, at normal heat treating temperatures, are in excess of the vacuums produced in heat treating equipment. Since these metals can be vaporized from the solid state, alloys of these metals cannot be processed in a vacuum unless the level can be controlled to prevent the vaporization. Removal of zinc from brass during annealing is a typical problem. With sufficient time, temperature, and low pressure, zinc is depleted until only spongy copper remains after such processing. Chromium, too, vaporizes out of alloys at processing temperatures. This vaporization problem can be prevented by using higher pressures, obtained by back-filling the vacuum furnace with inert gases.

Many alloying elements of commercial alloys have relatively high vapor pressures; but the alloy contents are low, and they are tied to compounds that are slow to dissociate or migrate to the surface. For example, manganese in steel is usually present in amounts to 1% and is not readily evaporated. Consequently, vacuum heat treatments may be specified for such alloys when a low percentage of manganese is present and the treatment time is short. Nitrided parts are an example of material that should not be treated in vacuum.

Condensed metal vapors will deposit in the cold zones of vacuum furnaces. In cold-wall furnaces (discussed later in this section), this condensation will cause shorts in the electrical connections and will also cause darkening of radiation shields. When metallic radiation shields are used, it is important to keep the shield bright and reflective in order to prevent an increase in heat loss. One technique to retard evaporation is to bleed an inert gas into the vacuum chamber to produce pressures in excess of the vapor pressure concerned.

Most organic and many inorganic compounds tend to break down to their element constituents in a vacuum furnace. Some of these constituents are readily removed as gases by the pumping system, while others will deposit in hot and cold areas of the furnace. These deposits can also cause shorts in the electrical system and decrease the reflective properties of the metallic shields. To avoid these problems, workloads should be cleaned to prevent large amounts of contaminants from being placed in the vacuum furnace.

The cleaning action of a vacuum can be a problem for some infrequent applications. With ultraclean surfaces in contact, diffusion bonding between the parts and the fixtures can occur. This condition sometimes necessitates the application of a parting compound to the metallic points in contact. Relatively inert, paintlike ceramics are commercially available for this purpose.

Properly designed furnaces, using special oils and having low vapor pressures, are available for hot oil quenching. Vacuum furnaces with pressurized-gas quenching are used for high-speed steel parts having large sections that require maximum hardness. Parts treated in this way include long broaches, drills, dies, and similar components.

Highly polished work that is being vacuum processed to maintain the polish may also come out of vacuum processing with a rough surface—a typical pitted, alligator, or orange-peel surface—because too much energy is applied during polishing. The polishing operation must not cold work the part excessively, or it will cause the surface to anneal and recrystallize during heat treating and result in surface roughening. Work that is expected to remain bright after vacuum processing must be polished much more gently than it would be if it were not to be heat treated.

APPLICATIONS

Vacuum furnaces were originally used primarily for the high-temperature processing of highly reactive and refractory metals, such as columbium, tantalum, tungsten, molybdenum, and titanium alloys. With advances in design and control, these furnaces are no longer limited to special heat treating operations on small, expensive parts, but are now used extensively for hardening and annealing many alloys previously heat treated in atmosphere or salt-bath furnaces.

Operations Performed

Thermal treatments currently being done with vacuum furnaces include stress relieving, homogenization, annealing, hardening, tempering, solution treating, and aging. By incorporating gas or oil-quenching features in vacuum furnaces, most metals and alloys can be heat treated in a vacuum. Quenching or cooling in inert gases preserves the surface metallurgy of the heated metal because there is no reaction. The best guarantee of purity in inert gases is to use a gas obtained from a liquid storage source, although welding-grade or high-purity cylinder gases may also be pure enough for high-quality results.

Tempering

Although the low temperatures, under 1000°F (540°C), required for some tempering operations cause response-time problems with respect to temperature, minor adaptations permit these operations to be performed in a single vacuum furnace. Heat transfer in vacuum is essentially accomplished by radiation that is greatly reduced at temperatures below 1100°F (595°C). However, introducing an inert gas to achieve a partial pressure in the vacuum chamber results in convective and radiant heating, and better response at lower temperatures. Vacuum tempering furnaces are available with forced convection heating and cooling for improved uniformity.

Surface Hardening

Techniques have been developed for producing controlled case depths when vacuum carburizing, nitriding, and carbonitriding, with some savings in time and energy requirements compared to atmosphere furnaces. Some vacuum furnaces have been developed for plasma ion carburizing and ion nitriding processes used to obtain surface (case) hardening.

Some surface hardening is done in vacuum furnaces by using a partial-pressure type of heating cycle. The vacuum container is generally back-filled with natural gas, natural gas mixed with nitrogen, or propane and nitrogen during part of the total heat processing cycle.

Other Operations

Heat treating in a vacuum is also used to remove dissolved gases from metals. The removal of embrittling hydrogen from titanium and titanium alloys is one example of this treatment. Other common applications for vacuum furnaces include brazing, diffusion bonding, and sintering.

Workpieces and Materials Treated

Machined parts, weldments, and assemblies ranging from watch parts to assemblies over 5 ft (1.5 m) long are heat treated in vacuum. Stainless steel heat treatment examples include the processing of investment castings of several grades. These castings are usually precipitation-hardening stainless or nickel alloys and are heat treated in vacuum furnaces to maintain their high surface finish. In addition to a fast heating rate, the furnace must also have the capability of rapidly and uniformly cooling the workload in order to produce the desired physical properties and to prevent distortion. Net loads to 3000 lb (1360 kg) have been successfully processed in batch-type vacuum furnaces.

Air-hardening tool steels used for die-casting dies have been hardened and double tempered in a single vacuum furnace. Since a considerable variation in section thickness is common in such dies, these furnaces must also be able to cool the load rapidly and uniformly in order to produce the desired physical properties and prevent distortion.

Reduced Costs

The higher cost sometimes wrongly associated with heat treating in a vacuum because of the higher initial investment cost, as compared with controlled-atmosphere or molten-salt methods, can usually be more than offset by the minimization of subsequent finishing operations and work spoilage. An example of this economy is the heat treatment of processing dies for pressing watch glasses. Mechanical polishing generally requires two days after an atmosphere heat treatment to achieve the required properties in these dies. The same dies heat treated in vacuum, however, require only four minutes of polishing to produce the required surface finish.

Annealing Applications

In recent years, the vacuum annealing of high-permeability magnetic materials has gained in popularity. Alloys such as iron-silicon, iron-nickel, and iron-cobalt are now being annealed in vacuum furnaces by a number of manufacturers. Results are substantially the same or superior to those that have been achieved by annealing in hydrogen. In the annealing of laminated stock, parts are stacked closely together and are supported on fixtures that have been previously sprayed with an aluminum-oxide paint to prevent the parts from diffusion bonding to them. The iron-silicon material is annealed at about 1825° F (995° C) in a vacuum furnace. After 40 minutes at heat, the parts are vacuum cooled for a few minutes, after which they are cooled to room temperature by forced convection with an inert gas. High-purity nitrogen is normally used for the cooling gas. If the iron-silicon laminations have a tendency to stick together, they can be dusted with powdered magnesium oxide before heat treatment. Another major application is the in-process annealing of 300 and 400 Series stainless steel tubing for the medical, aerospace, and petrochemical industries.

TYPES OF FURNACES

Vacuum furnaces are similar in design to other types of furnaces, the major difference being the addition of a vacuum chamber or work container. Vacuum levels are usually maintained as high as the particular pumping system used is capable of producing. These furnaces can be used to heat treat almost any ferrous metal that can be heat treated in conventional furnaces. For alloys containing elements with low vapor pressures, the cycles can be modified to include partial-pressure operation at higher temperatures. It is essential that the proper pressure, temperature, and cycle time be carefully selected for the specific metal workpieces to be processed.

Furnace Classification

Vacuum furnaces are classified into two types: cold wall and hot wall. The two types differ only in respect to the method of heating and shell design.

Cold-wall furnaces. These furnaces have the heating source inside the vacuum chamber. The hot zone is inside a double-wall, jacketed chamber, with cooling water circulating in the jacket. Hot zones can be of various configurations, but round zones provide better radiation heating than rectangular zones. Maximum temperatures in these furnaces can be to approximately 4000° F (2205° C), depending upon the intended application. Furnaces are available with external blowers and heat exchangers (see Fig. 12-14) for uniform, fast cooling and increased efficiency. This design eliminates potential problems associated with movable baffles and water leakage when the heat exchanger is located inside the furnace chamber.

Hot-wall furnaces. These furnaces have a single container or wall that is heated externally, without water cooling. As a result, they have a slow response to temperature changes and are generally preferable only as continuous production units. With single-wall construction and no water cooling, hot-wall furnaces are limited to a maximum temperature of about 1800° F (980° C).

Heating Methods

Various types of heating elements and insulation systems are used for the electrical heating of vacuum furnaces, depending upon the specific application and furnace manufacturer. One type of construction uses metallic heating elements and radiation shields. Radiation shields consist of layers of reflective metal sheets arranged to reflect heat back into the heating zone of the vacuum chamber. The number of shields and the material used depend upon the operating temperature required. All-metallic construction permits low vacuum levels and rapid heating and cooling, and can withstand high-velocity gas flows for gas quenching. Repairs, however, are sometimes difficult and costly, and contamination of the radiation shields must be kept at a minimum to prevent a decrease in reflectivity.

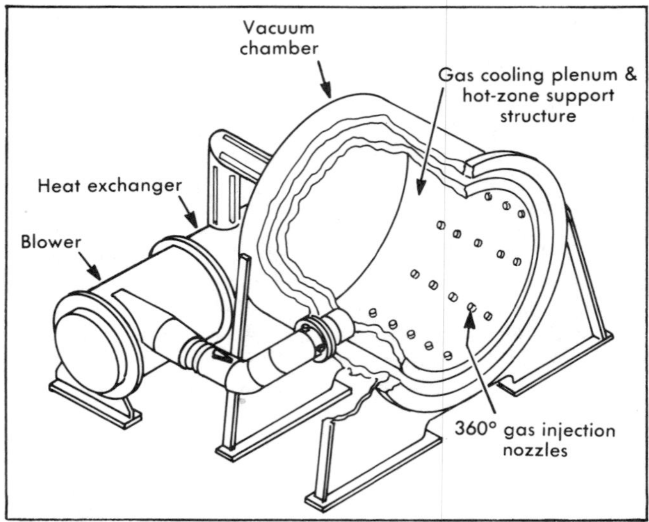

Fig. 12-14 Vacuum furnace with gas cooling and external blower and heat exchanger. (*ABAR Corp.*)

A sandwich-type insulation system is used for some vacuum furnaces. Metallic shields are used on both the inner and outer surfaces of the heating chamber, with a lightweight blanket of ceramic insulation between the two shields. Graphite rod or bar-type heating elements are generally used with this design. One disadvantage of this construction is that the fibrous insulating material tends to slow down the heating, cooling, and evacuation rates.

Some furnace designs use all-graphite construction. Graphite-fiber material is used for insulation, and graphite cloth, rods, or flat bars for the heating elements. Graphite, bar-type heating elements with bolted design minimize problems of maintenance and temperature uniformity. When using a hydrogen atmosphere with graphite construction, care must be taken to ensure that the furnace is leak-free to eliminate chemical reaction at high temperatures.

Furnace Configurations

The four basic configurations of vacuum furnaces are single, dual, or three-chamber design, or a continuous type with multiple in-line chambers.

Single-chamber design. Vacuum furnaces of single-chamber design are used extensively for batch-type heat treating operations. The vacuum chamber can be in either a vertical or a horizontal position, and the workpieces are not moved during the heating cycle. Vertical furnaces are available with bottom-load elevator bases. The furnaces can be equipped with internal fans to recirculate the atmosphere through a heat exchanger to cool the workload and furnace. A single-chamber, batch-type vacuum furnace designed for gas pressure quenching is illustrated in Fig. 12-15.

Dual-chamber design. In vacuum furnaces of dual-chamber design, heating is usually done in one chamber, and cooling by gas or oil quenching in the other chamber. These furnaces are

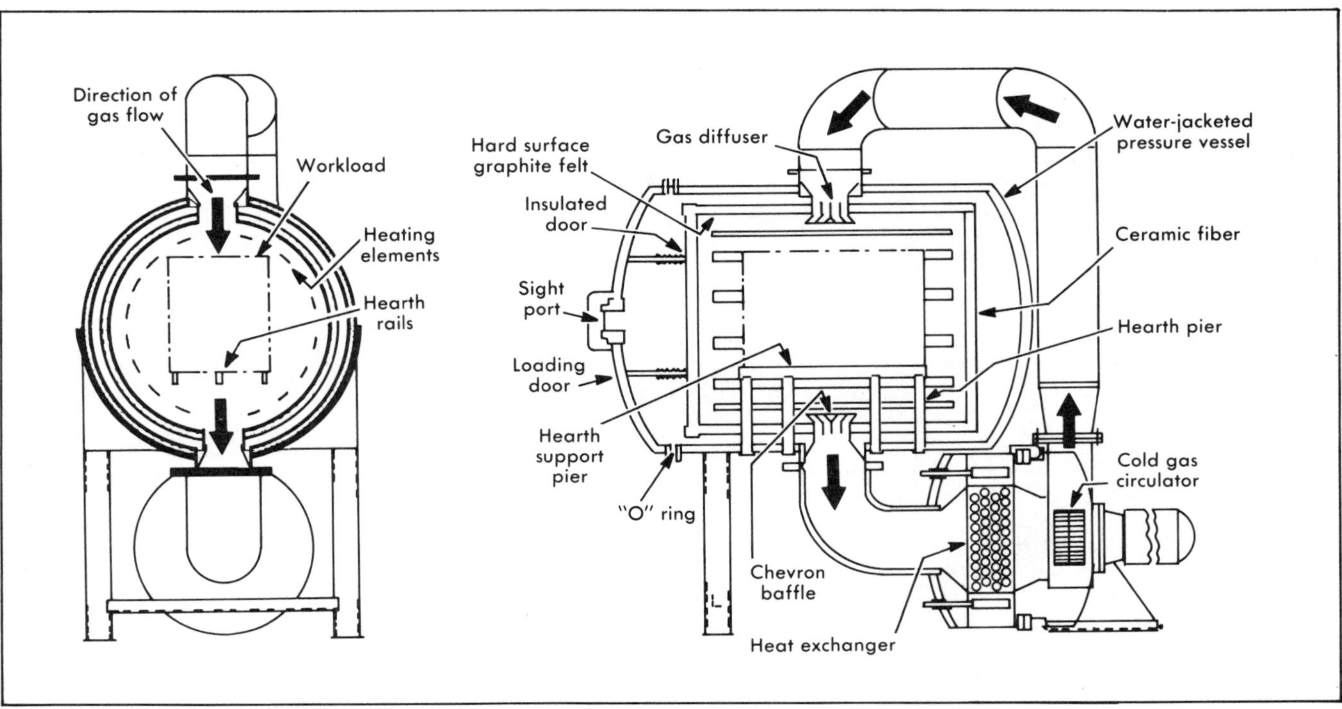

Fig. 12-15 Single-chamber, batch-type vacuum furnace designed for gas pressure quenching. (*C. I. Hayes, Inc.*)

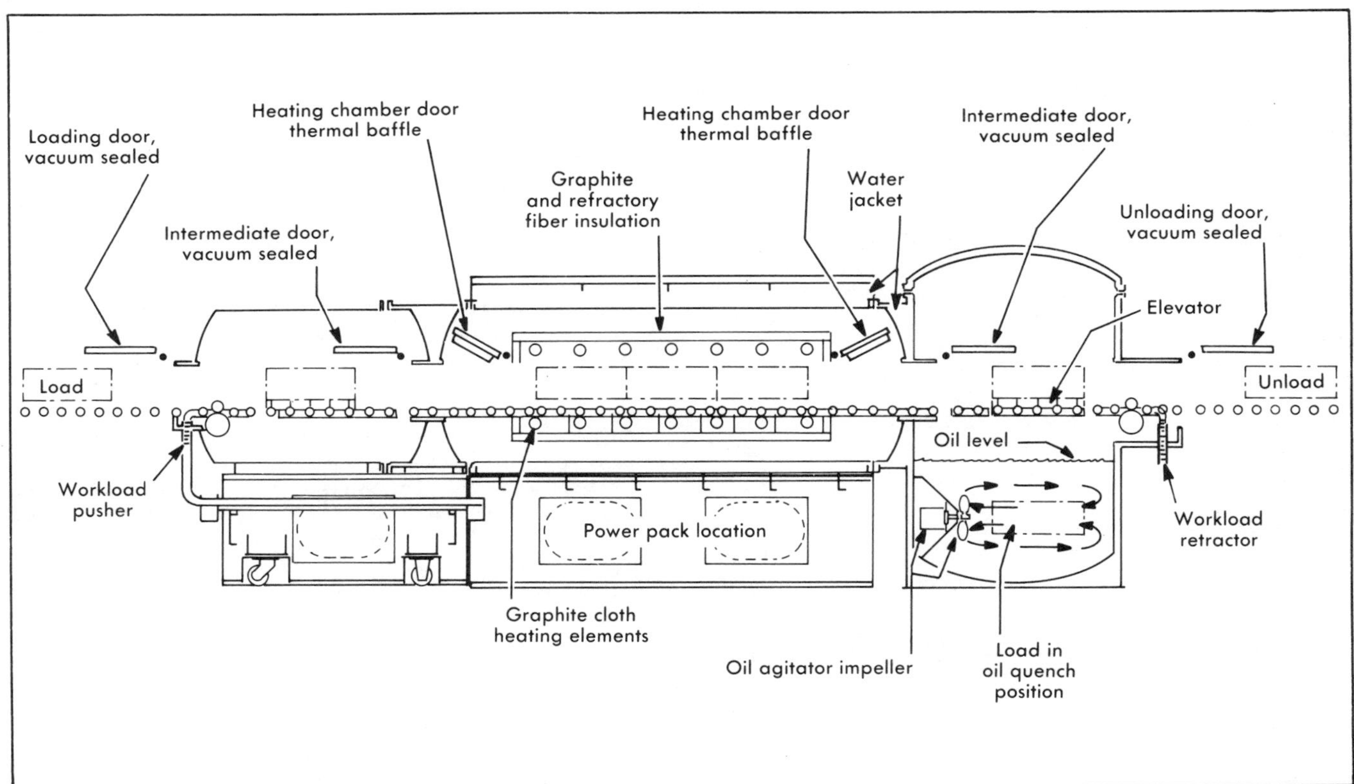

Fig. 12-16 Continuous vacuum furnace designed for oil quenching. (*C. I. Hayes, Inc.*)

available with the chambers arranged either horizontally or vertically. Oil quench tanks can be exposed to the vacuum environment and isolated from the heating chamber by an internal insulating door. The tanks can be of cold-wall design to cool the oil, or the oil can be removed from the tank and recirculated through a heat exchanger.

Three-chamber design. In vacuum furnaces of this type, the first chamber is generally a vestibule with a vacuum lock and is usually equipped for gas pressure quenching. The second chamber is generally used for heating. The third chamber is provided with a vacuum lock and generally incorporates a means of oil and/or gas quenching.

Continuous type. Vacuum furnaces of the continuous type are being used for the large-volume production of similar workpieces. These furnaces consist of a vacuum lock at their front end, a segmented heating chamber in the middle section, and quenching facilities (oil, gas, slow cooling, or a combination of these) at the end of the furnace. A continuous vacuum furnace designed for oil quenching is shown in Fig. 12-16.

Ion Processing Equipment

The glow-discharge ion processes for carburizing, nitriding, and carbonitriding are discussed in Chapter 11, "Surface (Case) Hardening." Vacuum equipment used for these processes is generally referred to as vessels rather than furnaces. For ion (plasma) carburizing, the vacuum vessels are equipped with a high-voltage power supply and conventional radiant heating elements. Energy for the absorption of carbon into the workpiece surfaces is achieved by a high-voltage glow discharge between the cathodic workpieces and an anodic electrode in the vessel.

For ion (plasma) nitriding, radiant heating elements are not required in the vacuum vessels because the energy from the glow discharge is sufficient to heat the workpieces. In this process, the grounded wall of the vacuum chamber is the anode, and the workpieces (electrically isolated from the chamber) are the cathode. A d-c potential difference between the chamber and workpieces ionizes the gas.

Controlling Temperature and Vacuum

Controls used for vacuum furnaces are similar to those used for other furnaces and discussed in another section of this chapter. Microprocessor-based programmable controllers, many with digital readouts, are being used extensively. Gas-pressure, sensor-type gages and relay-operated valves control the vacuum system to provide the required level of vacuum. Complete computer control is available to provide fully automatic operation, data accumulation, and functional diagnostic information.

Vacuum Systems

Vacuum furnaces normally use an automatic valve, high-vacuum pumping system with a direct rough-pumping line (see Fig. 12-17). An oil-vapor, fractionating diffusion pump, a blower with mechanical backing pump, and a holding pump may also be used. In order to obtain pressures of 1×10^{-5} torr (1.33×10^{-3}Pa) and lower, pumping systems capable of removing very large volumes of gas in a relatively short period of time are necessary. A specialized type of pump known as a sorption pump is used in conjunction with an ion pump for relatively small vacuum chambers. Turbomolecular pumps are also being used for some laboratory applications.

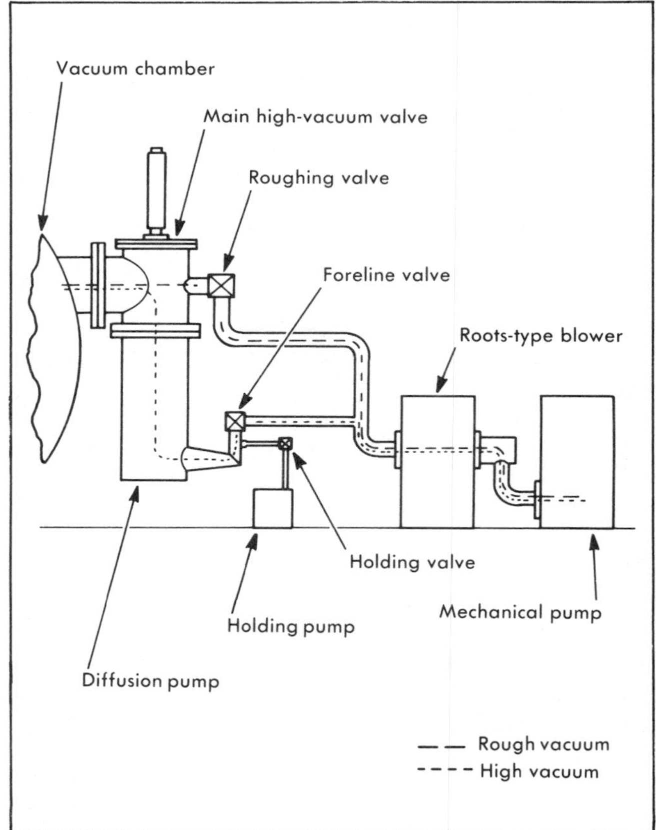

Fig. 12-17 Vacuum pumping system with diffusion pump and mechanical blower pump. (*ABAR Corp.*)

Several types of pumps are capable of operating against atmospheric pressure, but the mechanical vacuum pump is most commonly used. One type of mechanical pump, known as the booster pump, is used to increase pumping speeds at intermediate vacuum levels. It is used in series with a rotary-vane or rotary-piston pump. High vacuum levels are provided by the oil-vapor diffusion pump, which operates by vaporizing a low-vapor-pressure oil by means of a boiler within the pump base. The oil vapor is driven upward through a central stack to a series of angular jets. The jets direct the stream outward and downward at supersonic speeds to collide with and entrain gas molecules, compressing the gas to a level at which it can be removed by the mechanical vacuum pump.

The efficiency of a pumping system at the high vacuum levels attained by a diffusion pump can be increased by interposing a cold-trap type of pump between the furnace chamber and the inlet to the diffusion pump. Such a trap consists of a series of baffles, usually arranged in chevron configuration, that are cooled either by chilled water or by liquid nitrogen. Condensable vapors and gases are collected on the chilled baffles, thereby further reducing the pressure by this gas pumping means.

SAFETY CONSIDERATIONS

Precautions necessary to ensure safety in the operation of vacuum and other type furnaces are discussed in a subsequent section of this chapter.

FLUIDIZED-BED FURNACES

Fluidized-bed heat processing is performed in a retort containing mobile, inert particles of uniform size that are suspended in a flowing stream of gas. When properly fluidized, the bed attains liquid-like properties, and products are heated by direct immersion. A variety of materials may be used for the bed media, such as aluminum oxide, silicon carbide, or zirconia sand. The primary requirements of the medium are that it be uniform in size and remain inert at operating temperature.

OPERATING PRINCIPLES

The general nature of fluidized-bed heat transfer is illustrated in Fig. 12-18 with a graph of the heat transfer coefficient versus fluidizing-gas velocity. Heat transfer characteristics of a fluidized bed fall into three distinct areas. In Section 1, the bed is in a static state, with the gas velocity insufficient for fluidization. The heat transfer coefficient rises only slightly with relation to gas velocity and is at a low value.

The minimum velocity for the start of fluidization is indicated by V_1 (Fig. 12-18). Across Section 2, the heat transfer coefficient rises rapidly until the optimal velocity V is reached. Beyond this velocity, the bed becomes separated and attains gas-like properties. Thus, in Section 2, the heat transfer coefficient decreases with increasing velocity.

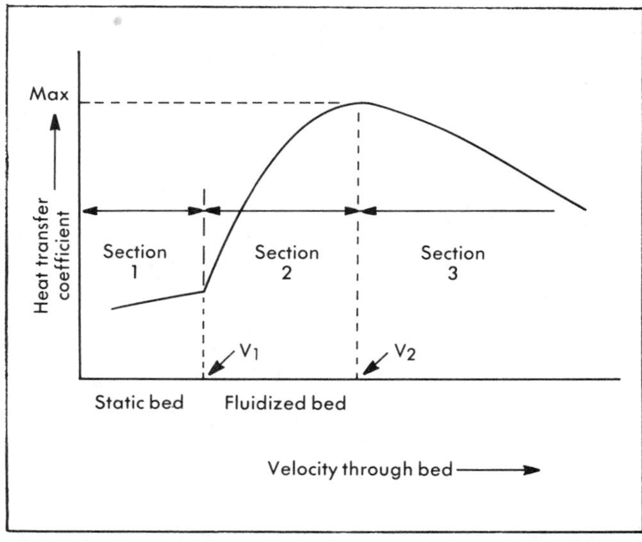

Fig. 12-18 Heat transfer coefficient versus gas velocity in fluidized-bed furnace.

Pressure Drop

Pressure drop across the bed increases linearly with gas flow in Section 1 (Fig. 12-18), attaining a maximum at velocity V_1. The pressure then quickly drops slightly to a value that remains constant across Section 2. For media commonly in use, this pressure value is approximately two inches of water column per inch of bed depth (0.5 kPa per cm of bed depth). In Section 3, the pressure drop decreases with increasing gas velocity.

Media Used

In general, the majority of heat treating is done using calcined, aluminum-oxide media of 99% purity, with a size range of 46 to 100 mesh. This type of medium remains inert to a temperature of 3300° F (1815° C). Operating velocities are in a range of 0.3 to 2 fps (0.09 to 0.61 m/s), approximately double V_1 for the particles being used.

Typical operating heat transfer rates are 100 Btu/hr/ft^2/°F (560 J/s/m^2/°C) for 80 mesh media, and 85 Btu/hr/ft^2/°F (476 J/s/m^2/°C) for 46 mesh media. These rates may vary somewhat depending upon the geometry and structural design features of the furnace, the thermal conductivity of the fluidizing gas, and the thermal/physical properties of the fluidized-bed medium.

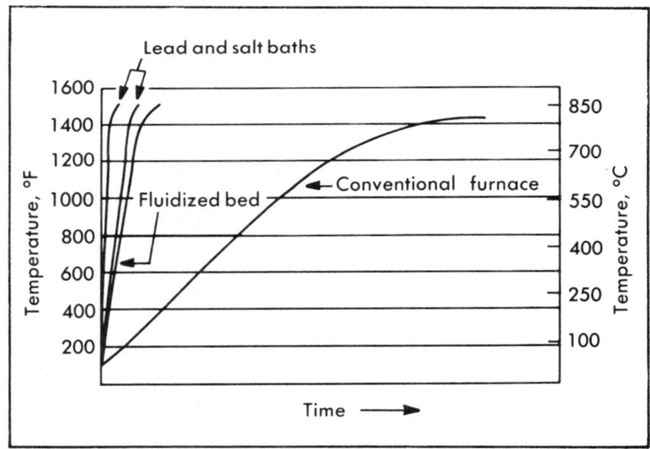

Fig. 12-19 Heating rates for steel in various type furnaces. (*Fennell Corp.*)

Comparative Heating Rates

A comparison of the heating rates for fluidized-bed furnaces versus standard furnace types is shown in Fig. 12-19. Although the fluidized bed exhibits a lower heat transfer rate than lead or salt-bath furnaces, it has the flexibility, under certain circumstances, to be operated at a higher temperature without appreciable detrimental effects. This advantage is due to the inertness of the fluid-bed media. Process times may thus be equalized, although at some cost in fuel efficiency.

FURNACE DESIGN

Modern fluidized-bed furnaces are used in three configurations. Two types are fuel fired, using products of combustion as the fluidizing gas, and are known as internally and externally fired types. The third type is indirectly fuel fired or electrically heated, independent of the source of fluidization, and may use any fluidizing gas desired.

Internally Fired Beds

An internally fired fluidized-bed furnace is illustrated in Fig. 12-20. Within the system, air and gaseous fuel are premixed and passed through a porous, ceramic distributor tile that supports the bed at the bottom. The air/gas mixture is burned at the top of the distributor plate, and the products of combustion then fluidize the bed.

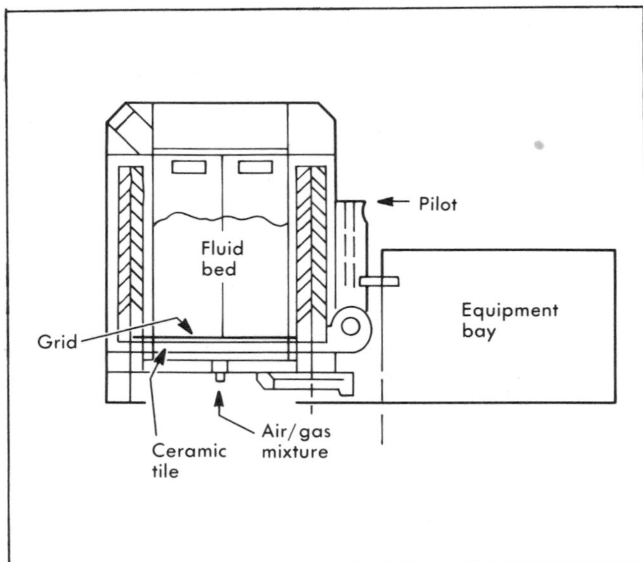

Fig. 12-20 Internally fired fluidized-bed furnace.

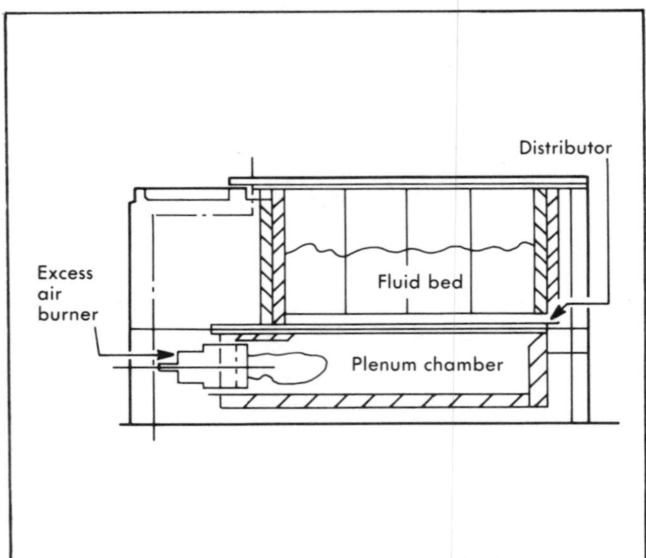

Fig. 12-21 Externally fired fluidized-bed furnace.

Combustion takes place within about 1″ (25 mm) of the top of the tile. The ceramic tile is designed to effect a uniform flow of gas through the bed for its full operating temperature range. Temperature of the bed is controlled by modulating the gas at a constant air flow to the distributor; this produces a modulation of thermal input at a relatively constant fluidizing-gas velocity. Normal operating temperatures for internally fired fluidized-bed furnaces are in the range of 1250 to 2200° F (675 to 1205° C), with either propane or natural gas used for fuel.

Externally Fired Beds

At temperatures below 1250° F (675° C), combustion begins to become unstable in internally fired fluidized-bed furnaces. Therefore, for such lower temperatures, externally fired fluidized-bed furnaces are used. In such furnaces, an excess-air burner fires into a plenum, either on the side of the bed or at the bottom, as shown in Fig. 12-21. The products of this combustion then fluidize the bed. The bed is supported on a metallic distributor plate, which may be built in a variety of configurations. The temperature of the bed is controlled simply by varying the gas supply to the burner with a constant air input.

As is the case with internally fired fluidized-bed furnaces, combustion in externally fired furnaces effects a varying thermal input with a relatively constant fluidizing-gas velocity. The operating temperature range of externally fired fluidized-bed furnaces is from 250 to 1200° F (120 to 650° C). Natural gas, propane, or light oil may be used as fuel.

Indirectly Heated Beds

For applications requiring special atmospheres, such as in neutral hardening, carburizing, nitriding, or carbonitriding, internally and externally fired fluidized-bed furnaces are not appropriate. Instead, indirectly heated, controlled-atmosphere fluidized-bed furnaces are used. The general configuration of this type furnace is shown in Fig. 12-22.

Fluidizing gas is introduced to the bottom of the work chamber and is regulated independent of the means of heating.

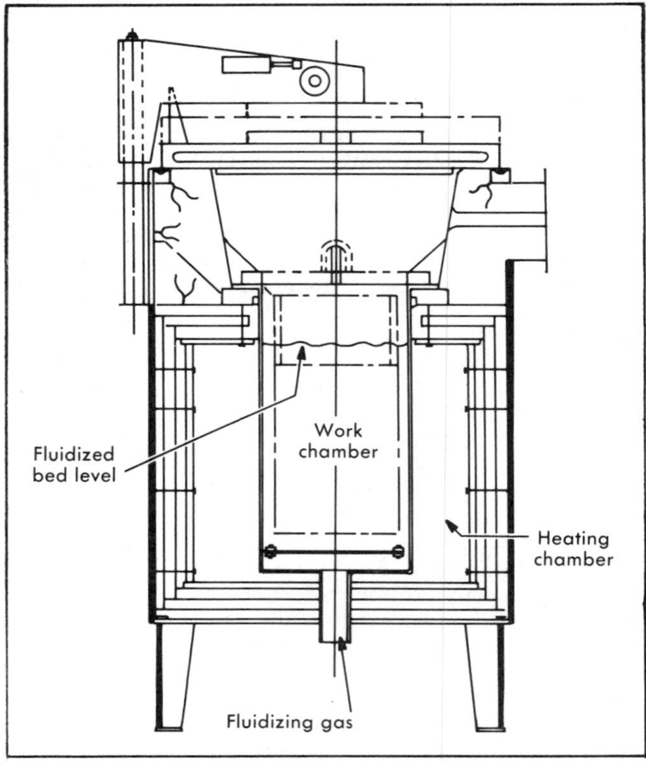

Fig. 12-22 Indirectly heated fluidized-bed furnace.

A metallic distributor plate, of the same general configuration as that for externally fired furnaces, is used. The heating chamber is completely separate from the work chamber and may contain an electric or fuel-fired heating system. Fuels used include natural gas, propane, and light oil.

In the furnace configuration shown in Fig. 12-22, all process heat is transferred to the fluidized bed through the side walls of the retort. In another configuration, the heating source is placed above the fluidized bed, and process heat is transferred to the bed through its surface. In yet another configuration, electric heating elements are immersed directly in the fluidized bed.

Operating temperatures for indirectly heated fluidized-bed furnaces can range from subambient to 2200° F (1205° C). Since any gas will properly fluidize the bed and since the fluidizing gas is also the furnace atmosphere, the same furnace can be used for a variety of processes.

TEMPERATURE AND ATMOSPHERE CONTROL

Temperature control in fluid-bed furnaces is generally accomplished by use of a thermocouple immersed in the fluid bed. Since the temperature fluctuations within the bed (or within each zone of a multizone continuous furnace) are minor, a thermocouple placed anywhere in the control zone is satisfactory for accurate temperature control. Additional thermocouples are installed in the annulus (the space surrounding the retort) and/or the plenum chamber, depending upon the furnace design, to provide over-temperature safety control.

With operating temperatures to 2200° F (1205° C), the temperature may be held to within ±5° F (3° C) any place within the control zone. Furnaces are normally controlled with proportioning-type controllers. Contactor or silicon-controlled rectifier (SCR) electrical systems are used for power control on electrically heated units. Fuel/air-proportioning burner systems are used on gas and oil-fired furnaces. Microprocessor-based control systems are available that can be programmed to automate furnace operations. Microprocessor control of cycle parameters during batch-type processing can maximize furnace output and reduce operator error.

Atmospheres for various heat treating and surface conditioning operations performed in externally heated furnaces are controlled by the use of flow-control valves and indicators. Additional information on furnace controls, both for temperature and atmospheres, is presented later in this chapter.

In addition to atmosphere analysis, test pieces of material such as shim stock can be heat treated along with the workpieces. Since fluid-bed furnaces can be operated in the closed or open-top position, test samples can be removed for metallographic evaluation at any time during the processing cycle, without disturbing the atmosphere composition within the bed.

FLUIDIZED-BED QUENCHING

With the high heat transfer rates inherent in fluidized-bed furnaces, these furnaces can be applied to quenching operations. Both gas and air quenching of workpieces can be accelerated within a fluidized bed. By incorporating both heating and cooling of the fluidized bed, quenching temperatures from subambient to over 1000° F (540° C) can be achieved.

MATERIAL HANDLING EQUIPMENT

Batch-type fluid-bed furnaces, with an opening at their tops, lend themselves to easy work loading, using fixtures of various designs. Because fluidization is dependent upon the gas movement upward through the medium, it is important that the cross-section area of the furnace not be completely obstructed. It is considered good practice if the total cross-section area of the load (including fixture) does not exceed 50% of the cross-section area of the furnace.

Geometry of the workpiece is important in fixture design. Cylindrical parts, shafts, drill bits, and similarly shaped parts should be loaded vertically, in order to minimize distortion and media carry-over to the quench process. Parts with complex shapes should be positioned to minimize the number of horizontal surfaces, thus minimizing carry-over.

Fixtures and baskets should be constructed of a material capable of withstanding the furnace temperature and atmosphere composition called for in the processing cycle. When baskets are used, they should have a coarse mesh with openings no smaller than 1/8″ (3.2 mm), to minimize media carry-over. In general, the largest mesh size available (consistent with the geometry of the workpiece) should be used for baskets.

Batch furnaces can be grouped together in tandem or with appropriate quenching facilities to provide a complete fluidized-bed heat treating system. To optimize material handling operations, an overhead conveyor can be used. Industrial robots are also used, either with stand-alone control or for operation in conjunction with a central computer system. The fast process times possible with fluidized-bed furnaces make them particularly suitable for continuous heat treating systems.

Rotary-drum fluid-bed furnaces have been developed to heat treat small parts, such as washers and fasteners. The parts are conveyed through the fluidized-bed furnace by a rotary scroll within the drum. The outer skin of the drum is perforated to permit the fluidized media to pass through the skin, but the diameter of the perforation is small enough to contain the parts. Various groupings of rotary-drum furnaces in conjunction with conveyorized quenching equipment can be assembled to provide continuous, multistage processing of parts.

In addition to rotary-drum processing, continuous, fluid-bed heat treating systems including an overhead conveyor can be constructed. Heat treating of continuous material, such as wire and strip, can also be accomplished in fluidized-bed furnaces.

PROCESS CAPABILITIES

Fluidized-bed furnaces are being used for many different heat treating operations. Common applications include neutral hardening, carburizing, carbonitriding, nitriding, nitrocarburizing, and steam tempering. When using generated atmospheres, however, care has to be exercised to ensure that the pressure and flow rate of the atmosphere generated is consistent with the fluidization requirements of the bed.

FURNACE CONTROLS

Heat treatment of metals always requires the control of temperatures and, for many applications, also necessitates the control of time, atmospheres, flow rates, and pressures. In addition to the regulation of such process variables, control of the process logic is sometimes needed for functions such as material handling and safety interlocks. Time is a critical element in the application of both process and logic control systems. Efficient control can provide significant cost savings in heat treatment, as well as more predictable, consistently accurate results.

CONTROL REQUIREMENTS

The control of a process variable requires a sensor, a controller, and a final control element or actuator connected to the process in a control loop. This basic control loop is used for both on-off and proportioning or modulating types of control. The basic control loop for a heat treating process concerned with temperature control is illustrated schematically in Fig. 12-23. A temperature sensor detects the process temperature and generates a signal proportional to this temperature, which is transmitted to the controller. The controller set-point (the desired temperature) is compared to this signal (the actual temperature).

The difference or error signal between the actual and desired temperature determines the controller's action. In heating processes using on-off control, the energy source is turned off when the actual temperature is above the set-point and turned on when it is below the set-point. Consequently, the process temperature varies above and below the desired set-point. A common misconception is that this temperature cycling arises because such controllers exhibit a dead-band between the power-off and power-on switching points, and therefore cycling can be reduced or eliminated by reducing the dead-band. In fact, cycling is caused by process dynamics and only in exceptional cases will changing the dead-band have any effect.

Continuous control systems adjust the process energy input as required to maintain process temperature. Proportional controllers, which produce an output proportional to the current error, are easy to adjust, but allow some offsets to exist. Adding integral and derivative terms to the error-signal computation results in the popular proportional-integral-derivative (PID) controller. Although more expensive and more difficult to adjust, PID controllers are generally preferred when very close control is required.

The basic control loop is frequently provided with auxiliary devices, such as recording instruments and set-point programmers. The measurement instruments may be recorders or indicators, showing the process temperature for guidance to the process operator and to maintain process records. Computer-based data acquisition and microcomputer control systems are advanced methods, which use printers or typewriters to log the process temperature.

With some applications of multiple devices, the temperature sensor signal will be measured by a temperature transmitter, which amplifies the signal. This amplification helps to avoid interaction between the measurement and controlling instruments, and minimizes pickup of electrical noise when these signals are transmitted long distances. The set-point programmer automatically varies the controller set-point to provide a temperature versus time cycle, which is called a recipe or program. The rate of heating or cooling and the time at specific temperatures can be automatically maintained as required by the process. Many modern controllers combine the set-point programmer and controller in a single unit, in both single and multiple-loop forms. Temperature controllers also often use directly connected thermocouples and provide a process signal for recording instruments.

Sensors and transmitters used to determine process variables must be accurate, reliable, and properly installed and maintained. As part of the basic control loop, these sensors and transmitters may represent only 1% of the instrumentation cost, but their performance is vital to the operation of the control loop.

TEMPERATURE CONTROL

Heat treating applications in the metals industry require temperatures ranging from -150 to 2500° F (-100 to 1370° C), but the majority of applications are from 200 to 2200° F (95 to 1205° C). Temperature sensors are classified as contact or noncontact types and as either electrical or nonelectrical.

Electrical sensors are most commonly used in heat treating applications because of the high temperatures involved and the dominant use of electronic instrumentation. Contact sensors, such as thermocouples and resistance temperature devices (RTD's), are used for over 95% of the measurements, with thermocouples employed for about 90% of these measurements. Noncontact sensors, which operate as a function of radiant energy, are more expensive than contact sensors and serve applications not suited for contact measurements.

Thermocouples

Thermocouples are contact sensors consisting of two wires of dissimilar metals joined at one end, the measuring junction. They are rugged, inexpensive, and accurate. They also cover a wide temperature range and are fast in response. Thermocouples generate an electrical signal (millivolts) in proportion to the difference in temperature between their measuring or hot junctions and their cold or reference junctions.

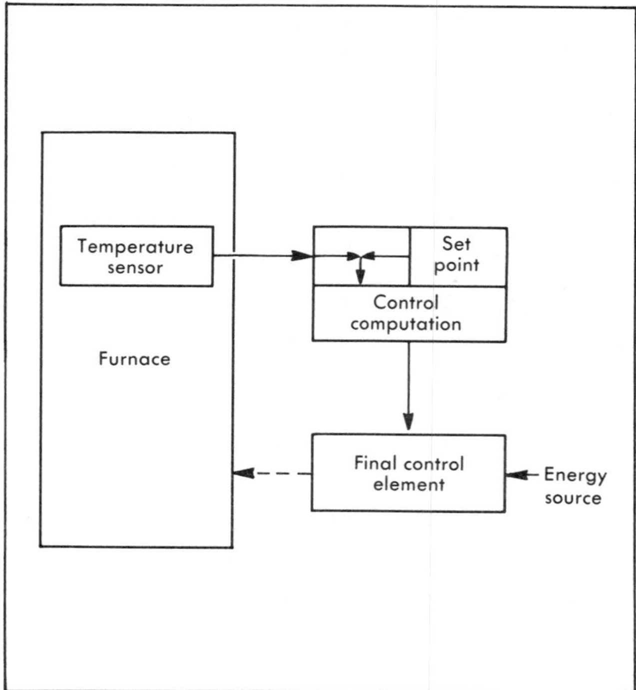

Fig. 12-23 Basic control loop for temperature control of a heat treating furnace.

Different types of thermocouples, classified by their metallurgical composition, have different temperature-millivolt calibrations and are selected based on application requirements. The thermocouple's measuring junction is exposed to the temperature to be measured, and the other end is connected with extension lead wire (which has the same calibration) to the copper wire of the instrumentation. The junction with the copper wire is the reference junction.

The circuitry and assembly of a typical thermocouple are shown in Fig. 12-24. The assembly is mounted on the furnace with its tip exposed to the furnace temperature. The metal or ceramic protection tube protects the thermocouple element, and insulators separate the thermocouple wires. The terminal junction in the head provides for connection to the extension lead wire. The lead wire, installed in conduit, is connected to the instrumentation. The conduit provides mechanical protection and some shielding for the extension lead wire to prevent picking up electrical signals from high-voltage electrical wiring and machinery.

Thermocouple assemblies have mounting fittings that must be gas-tight for atmosphere, vacuum, or high-pressure furnaces. The fittings, however, can have adjustable flanges for other applications, permitting changes in the depth of insertion. Thermocouple calibration errors are frequently due to external ambient contamination, producing a lower output signal and an erroneously low temperature measurement. If improper thermocouples are used, parts being heat treated may be overheated, resulting in wasted energy and damage.

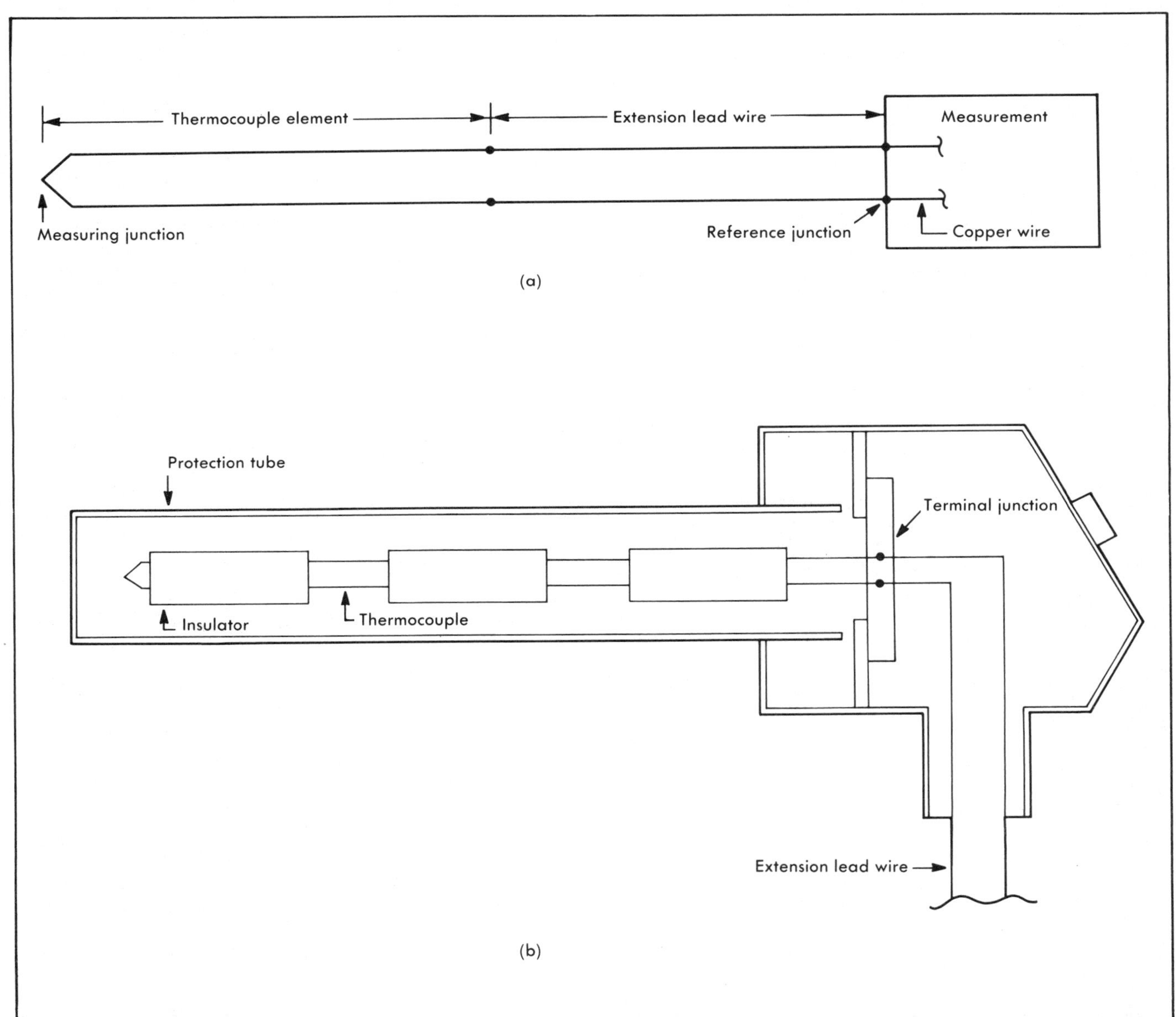

Fig. 12-24 A typical thermocouple: (a) circuitry and (b) assembly.

FURNACE CONTROLS

Calibration may be checked with a second, standard thermocouple in an adjacent location or inserted in the same protective tube if space permits. Contamination may be due to process gases penetrating the protection tube. Cutting oils on the pipe threads of the protection tube or oils and dirt from handling components during assembly also cause contamination. Replacement protection tubes should be degreased before use. Insulators should not be reused as they absorb contaminants during use. Gloves should be worn when assembling thermocouples. Thermocouple wires should not be repeatedly flexed; and the extension lead wire should not be strained with excessive pulling force when installed in electrical conduit, as this cold works the wire and can cause calibration errors.

Installation of thermocouples. The tip or measuring end of the thermocouple must be exposed to the temperature to be measured or controlled. The most common error is not inserting the thermocouple far enough into the furnace, resulting in conduction errors caused by heat flowing away from the tip to the colder furnace wall and producing a low temperature measurement. Thermocouples should not be located near burners or heater elements because radiant heat may cause measurement errors.

The speed of response of the thermocouple to heating or cooling should be similar to or faster than the change in the temperature of furnace or workload to achieve good control. The controller actually controls the temperature of the thermocouple and, indirectly, the furnace and workload temperature. Rugged, heavy construction may be used for longer thermocouple life on large car-type furnaces for processing large workpieces; but a light construction is required on vacuum furnaces because their rapid heating rate requires a fast response. For fast processes, a minimum of physical protection should be used, in spite of shortened life, to achieve satisfactory control. Furnace temperature uniformity should be measured with test thermocouples to determine the best location for the control thermocouples.

Thermocouple types. Thermocouples of various compositions are available to satisfy installation requirements of temperature ranges, oxidation, accuracy, and cost. Tables 12-2, 12-3, and 12-4 provide performance criteria for selecting thermocouples. The U.S. National Bureau of Standards publishes nomographs documenting the performance and calibration data for these thermocouples.

The major thermocouples used for heat treating are types J and K, and the platinum types S, R, and B. Type J (iron-constantant) is used extensively because of its price-performance characteristics. However, its useful maximum temperature of about 1500° F (815° C) is not adequate for many heat treating applications. Consequently, Type K (nickel, chromium-nickel, aluminum) is used at higher temperatures because of its superior oxidation resistance. Types S, R, and B (the platinum-type thermocouples) are used for high to very high temperature applications because of their stability and resistance to oxidation.

Thermocouple assemblies. Assemblies are used to protect and provide mechanical support for the thermocouple elements. Protection from mechanical damage, contaminating atmospheres, and liquids used in quenching applications is required. Mechanical support is needed to locate and maintain the position of the thermocouple tip where the temperature measurement is desired. Mounting features include flanges, threaded bushings, and threaded thermowells used to mount the thermocouple assemblies in furnace and oven walls or quench tanks.

Protection tubes and thermowells are available in various materials for different temperature ranges and environments, with ceramic tubes being used extensively for high-temperature applications. Compacted ceramic-type thermocouples are used as shown in Fig. 12-25 or mounted in protection tubes and wells. The thermocouple wires are separated by a compacted ceramic powder inside a metal sheath. This thermocouple design can be small in diameter, fast in response, and can be bent to suit installation requirements. The grounded construction shown in Fig. 12-25 has the thermocouple element in contact with a metal plug in the tip to provide fast temperature response. Ungrounded construction is used to provide electrical isolation.

Resistance Temperature Detectors

Resistance temperature detectors (RTD's) are an alternative to thermocouples for applications up to 1000° F (540° C), which limits their use for heat treating applications. Their resistance changes with temperature, thus providing accurate, stable detectors, which may be of platinum, copper, or nickel construction. They are mounted in assemblies similar to those used with thermocouples. Within their temperature limits, they are much more precise than thermocouples.

Noncontact Sensors

Noncontact sensors measure temperatures based on radiant energy emitted from hot targets that may be stationary or moving. Consequently, furnace or product temperatures can be measured. Typical heat treating applications are in the 1200 to 2400° F (650 to 1315° C) range. Accurate, reproducible measurements can be made with a minimum of sensor maintenance in continuous, high-temperature and reducing-atmosphere applications.

Typical detectors used are thermopile types (a series of thermocouples) or photosensitive types like silicon. The hot-target radiant energy is focused on the detectors, which develop a millivolt output proportional to the target temperature. Sighting conditions involving dust and smoke or energy-absorbing gases, and the emissivity of target materials are application considerations. These sensors are factory calibrated by sighting on a black-body target (emissivity of 1.0) at a known temperature. In heat treating applications when the target emissivity is less than 1.0, an optical pyrometer or test thermocouple measurement may be used to calibrate the measurement instrument used with the sensor.

Optical-path interference from energy-absorbing gases such as steam or carbon dioxide from combustion processes requires the use of spectrally selective detectors not affected by the absorption bands of these gases. An alternative, also useful with interference from dust and smoke, is the use of an open-ended sighting tube purged with dry air and mounted on the sensor. The purge air is used to clear the optical path between the sensor and the hot target.

Control Actuators

The temperature controller regulates an actuator that controls the electricity or combustion system used to provide heat. Actuators for systems heated by electrical resistance are contactors that are turned on and off, or power regulators, such as silicon controlled rectifiers (SCR's), that proportion the energy. Induction heat treating is normally done by timing the power-on cycle, but product temperature measured with a noncontact sensor may be used to turn the power off. Actuators for combustion systems are electrical motors or pneumatic devices used to position valves regulating fuel or air flow to combustion burners.

TABLE 12-2
Comparison of Thermocouple Types

ANSI Type	Usable Temperature Range °C	°F	Advantages	Restrictions
J (iron-constantan)	-185 to 870	-300 to 1600	Comparatively inexpensive; suitable for continuous service to 1600° F (870° C) in neutral or reducing atmospheres.	Maximum upper limit in oxidizing atmosphere is 1400° F (760° C) due to the oxidation of the iron; protection tubes should be used above 900° F (480° C); protection tubes should always be used in a contaminating medium.
K (nickel, chromium-nickel, aluminum)	-20 to 1370	0 to 2500	Suitable for oxidizing atmospheres; in higher temperature ranges, provides a more mechanically and thermally rugged unit than platinum or rhodium-platinum, and longer life than iron-constantan.	Especially vulnerable to reducing atmospheres, requiring substantial protection when used.
T (copper-constantan)	-185 to 370	-300 to 700	Resists atmosphere corrosion; applicable in reducing or oxidizing atmospheres below 600° F (315° C); its stability makes it useful at subzero temperatures; has high conformity to published calibration data.	Copper oxidizes above 600° F (315° C).
E (nickel, chromium-constantan)	-185 to 870	-300 to 1600	Has high thermoelectric power; both elements are highly corrosion resistant, permitting use in oxidizing atmospheres; does not corrode at subzero temperatures.	Stability is unsatisfactory in reducing atmospheres.
S (platinum, 10% rhodium-platinum) R (platinum, 13% rhodium-platinum)	-20 to 1480	0 to 2700	Usable in oxidizing atmospheres; provides a higher usable range than type K; frequently more practical than noncontact pyrometers; has high conformity to published calibration data.	Easily contaminated in other than oxidizing atmospheres.
B (platinum, 30% rhodium-platinum, 6% rhodium)	870 to 1650	1600 to 3000	Better stability than types S or R; increased mechanical strength; usable to higher temperatures than types S or R; reference-junction compensation is not required if junction temperature does not exceed 150° F (65° C).	Available in standard grade only; high temperature limit requires the use of alumina insulators and protection tubes; easily contaminated in other than oxidizing atmospheres.

(*Leeds & Northrup Co.*)

TABLE 12-3
Temperature Limits for Thermocouples in Oxidizing Atmospheres*

Thermocouple			AWG (American Wire Gage) 8		14		16		20		24		30	
ANSI			Temperature Limits											
Type	Material	Condition	°F	°C	°F	°C	°F	°C	°F	°C	°F	°C	°F	°C
J	Iron-constantan	Bare	1200	650	900	480	900	480	800	425	650	345	600	315
		Protected	1400	760	1100	595	1100	595	900	480	700	370	700	370
K	Nickel, chromium-nickel, aluminum	Bare	2000	1095	1700	925	1700	925	1600	870	1400	760	1300	705
		Protected	2300	1260	2000	1095	2000	1095	1800	980	1600	870	1500	815
T	Copper-constantan	Bare	600	315	600	315	500	260	400	205	400	205	400	205
		Protected	700	370	700	370	600	315	500	260	400	205	400	205
E	Nickel, chromium-constantan	Bare	1400	760	1100	595	1100	595	900	480	700	370	700	370
		Protected	1600	870	1200	650	1200	650	1000	540	800	425	800	425
S and R	Platinum, rhodium-platinum	Protected	---	---	---	---	---	---	2800	1540	2700	1480	2400	1315
B	Platinum, 30% rhodium-platinum, 6% rhodium	Protected	---	---	---	---	---	---	---	---	3100	1705	---	---

(Leeds & Northrup Co.)

* Upper temperature limits are a function of wire diameter; because heat tends to have deleterious effects on thermocouples, the more material in their cross sections, the longer they can be expected to last.

TABLE 12-4
Limits of Error for Thermocouples

ANSI Type	Thermocouple Material	Temperature Range °F	Temperature Range °C	Limits of Error* Standard	Limits of Error* Special
J	Iron-constantan	-310 to -100	-190 to -75	---	±2%
		-100 to +600	-75 to +315	±4° F (±2° C)	±2° F (±1° C)
		600 to 800	315 to 425	±4° F (±2° C)	±0.33%
		800 to 1400	425 to 760	±0.75%	±0.33%
K	Nickel, chromium-nickel, aluminum	32 to 530	0 to 275	±4° F (±2° C)	±2° F (±1° C)
		30 to 2300	275 to 1260	±0.75%	±0.38%
T	Copper-constantan	-300 to -75	-185 to -60	---	±1%
		-150 to -75	-100 to -60	±2%	±1%
		-75 to +200	-60 to +95	±1.5° F (±1° C)	±0.75° F (±0.5° C)
		200 to 700	95 to 370	±0.75%	±0.38%
E	Nickel, chromium-constantan	32 to 600	0 to 315	±3° F (±2° C)	±2° F (±1° C)
		600 to 1600	315 to 870	±0.5%	±0.38%
S	Platinum, 10% rhodium-platinum	0 to 1000	-15 to +540	±2.5° F (±1.5° C)	±1.5° F (±1° C)
		1000 to 2700	540 to 1480	±0.25%	±0.15%
R	Platinum, 13% rhodium-platinum	0 to 1000	-15 to +540	±2.5° F (±1.5° C)	---
		1000 to 2700	540 to 1480	±0.25%	---
B	Platinum, 30% rhodium-platinum, 6% rhodium	1600 to 3100	870 to 1705	±0.5%	---

(Leeds & Northrup Co.)

* When expressed as a percentage, the limit of error is a percentage of the temperature reading, not of the range.

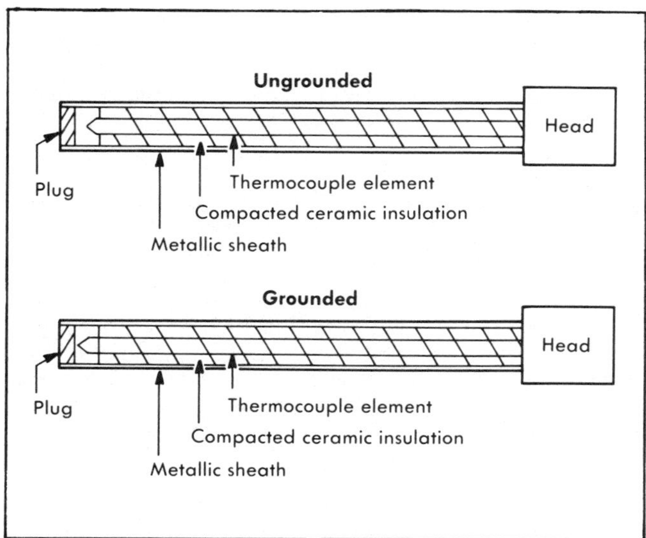

Fig. 12-25 Ungrounded and grounded ceramic-type thermocouples.

ATMOSPHERE CONTROL

Major atmosphere-controlled heat treating applications include carburizing, hardening, carbonitriding, and nitriding. Carburizing and hardening using carbonaceous atmospheres are the largest volume applications. In hardening, the metal has the desired carbon concentration, and the objective is to maintain it while the part is brought to temperature prior to quenching. In carburizing, the objective is to increase the carbon content of the metal at its surface. Consequently, carbon-rich atmospheres (atmospheres with a high carbon potential) are used for carburizing, and atmospheres with a lower carbon potential are used for hardening. Carbonitriding involves the additions of nitrogen to a carbonaceous atmosphere, while nitriding involves only nitrogen. This discussion is limited to carburizing and hardening applications.

Atmospheres produced by generators and furnace atmospheres are controlled. A variety of components exist in a carboneacous atmosphere, and their concentration, with allowance for temperature, may be related to carbon potential. Oxygen measured with a zirconia probe and carbon dioxide measured with an infrared gas analyzer are the most common measurements for continuous control. Water measured with a dew point analyzer is the most common basis for manual control.

Oxygen Sensors

An oxygen sensor, similar in appearance to a thermocouple assembly, can be inserted through the furnace wall into the furnace atmosphere. The components of the sensor include platinum electrodes, a zirconia tube, and an air supply that provides a reference gas of constant oxygen content. The electrical output, corrected for temperature, may be related to carbon potential in percent carbon (see Fig. 12-26). This sensor is fast in response and easy to install. Its major disadvantage is the need for replacement, normally at intervals of a year or longer, due to component failures or damage because of mechanical or thermal shock. Controllers are available that accept oxygen sensor and thermocouple outputs, and read directly in carburizing potential (%C).

Carbon Dioxide Analyzers

A carbon dioxide analyzer extracts a sample of the atmosphere from the furnace or generator to determine its content. Frequently it is multiplexed to several furnaces and/or generators, thus reducing the cost of measurements. These measurements, corrected for temperature, may also be related to carbon potential in percent carbon. Infrared analyzers are accurate and easy to calibrate. Their major disadvantages are high initial cost, maintenance of the multiplexed sampling system, and more complex electronics.

Atmosphere Controllers

An atmosphere controller regulates the enrichment of carrier gases, with on/off or proportional controllers used to regulate the flow of enrichment gas. The controller regulates flow by controlling a solenoid valve or proportioning valve. The carrier gas provides sufficient volume to fill the furnace and the required flow rate of fresh gas over the parts being carburized or hardened. The enrichment is used to maintain the desired carbon potential of the gas mixture. The mixture of carrier and enrichment gases is burned off as it exits the furnace.

Carbon potential of the gas from the generator is maintained by regulating the air-to-fuel ratio of the combustion process in the generator. Frequently generators are operated at fixed air-to-fuel ratios for specific applications. The gases generated are monitored for carbon potential to detect any changes in the combustion process, catalyst operation, or fuel quality.

VACUUM MEASUREMENT AND CONTROL

In vacuum furnaces, the vacuum replaces the inert gas atmospheres used in other furnaces to protect the work from oxidation at high temperatures. Mechanical and diffusion pumps are operated continuously to evacuate (pump down) the furnace from atmospheric pressure to the desired vacuum. A common method of maintaining the desired vacuum level is to regulate the flow of an inert, back-fill gas, such as nitrogen, while continuing operation of the vacuum pumping. A solenoid-operated gas valve, regulated by an on-off controller, is normally used to control the flow.

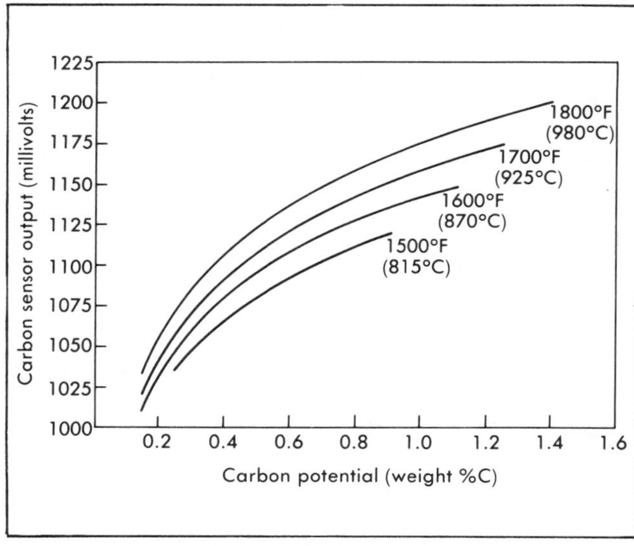

Fig. 12-26 Sensor output versus carbon potential. (*Corning Glass Works*)

FURNACE CONTROLS

Vacuum measurements are referenced to standard atmospheric pressure at sea level. Vacuum instruments are usually calibrated in microns and torrs. Multiple pumps and multiple instruments are commonly used. A diaphragm vacuum gage, calibrated from 800 to 10 torr, is often used with a thermocouple vacuum gage, calibrated in microns down to 1 micron (0.001 torr). The millivolt output from the micron gage is used by an on-off controller to regulate the flow of back-fill gas to maintain the desired vacuum.

PROCESS CONTROLLERS

A controller compares the process variable value to the desired value (the controller's set-point) and regulates the controlled variable based on this comparison. In heat treating applications, the process variable can be temperature, percent carbon, or vacuum. The controlled variable can be electricity or fuel for temperature control, an enriching gas or fluid for carbon control, or an inert back-fill gas for vacuum control. Controllers are classified as either on-off or continuous types.

On-Off Controllers

On-off controllers are inexpensive, simple to operate, and easy to maintain. The on-off control action may cause the process variable to cycle above and below the controller's set-point, but cycling usually occurs below the set-point and the bottom of the dead band of the controller. On temperature controlled processes, this cycling wastes energy, increases furnace and oven maintenance, and may damage the product.

For example, during startup, the heat remains on until the temperature reaches the controller's set-point and then turns off. Consequently, the thermal inertia of the furnace and product causes the temperature to overshoot the set-point, wasting energy and possibly damaging the furnace and the work.

Continuous Controllers

Continuous controllers are more expensive and complex than on-off controllers. However, they provide superior control, resulting in reduced operating costs and improved product quality. The controller proportions the controlled variable to maintain the process variable at set-point. For example, the heat input is continuously matched to the heat demand of the process by adjusting a burner to provide the amount of heat needed to maintain the desired temperature.

The controller computes the required output based on monitoring any deviation between its set-point and the process variable. The most common computation is a PID algorithm, so called because it incorporates proportional, integral, and derivative terms. The corresponding tuning adjustments are proportional band or gain, reset, and rate. Tuning adjustments are required to match the controller to the dynamics of the process. They may be adjusted manually or automatically.

Control Actuators

The controller output selected depends upon the control actuator used. Electrical energy, proportioned with a silicon controlled rectifier (SCR), normally requires a current output (4-20 milliamperes) from a current-adjusting controller. Gas flow to a burner or an atmosphere enrichment can be regulated with a proportioning valve operated by an electric motor. The electric motor may be positioned using raise and lower signals from a position-adjusting controller.

Electrical contactors used for control of electrical energy and solenoid valves used for control of gas or liquid flow require a time-proportioning controller that produces timed electrical pulses. The proportion of on-time to off-time is continuously regulated. For example, a pulse cycle with a solenoid valve of 75% on-time and 25% off-time provides a net flow to the process equivalent to a three-quarters-open proportioning valve.

Controller Circuitry

Controllers may use analog or digital circuitry; however, digital electronics is used on most new products. The preference for digital electronics is based on superior man/machine interface features, flexibility, performance features, and the ability to conveniently interface with computers and their peripherals. The readability and resolution of numeric digital displays, the alpha displays for word messages, and push-button controls are preferred by operators and maintenance personnel.

The accuracy and stability provided by digital computations also provide superior measurement and control functions. Microprocessors and digital memories can be used to provide a wide variety of features in each controller. Consequently, one basic controller can be used to serve a wide variety of applications, permitting standardization throughout a plant. If process requirements change, some controllers may be reprogrammed rather than replaced to meet the new conditions.

PROGRAMMABLE CONTROLLERS

Programmable controllers, also called programmable logic controllers, are used to replace relay control logic, electromechanical timers, and counters. In addition, they can perform sequencer functions, computations, and information handling. Digital electronics are used for all operational and display functions.

The programmable functions enable the user to readily modify the control to meet changing requirements. The use of digital circuitry provides high-speed, reliable systems that may implement complex control strategies, including some mathematical computations beyond the capabilities of relay systems. Programming may be done in ladder logic, corresponding to relay ladder diagrams, or by using Boolean expressions. The ladder diagram approach has greater acceptance by electrical maintenance personnel because of familiarity and troubleshooting convenience. Boolean entry is preferred by engineering personnel as a more efficient programming method. Alphanumeric displays and video displays are used in a variety of programs and operating displays.

COORDINATED CONTROLS

Control products that incorporate both process-variable (loop) and programmable logic functions provide the advantage of interaction between these functions, as they are both based on digital electronics. A block diagram of this type of system, with operator's console and video terminal, is shown in Fig. 12-27.

Individual process-variable (loop) controllers, providing the advantage of single loop integrity and using digital communications, can interface with programmable logic controllers or digital computers. The computer can monitor the performance of the controllers, using the data collected for high-level computation, report generation, and alarms. In addition, the computer can download programs to the controllers.

Programmable logic controllers can provide, in addition to logic functions, process-variable control functions. However, the process-variable functions provided are limited, compared to the power and flexibility of the logic functions.

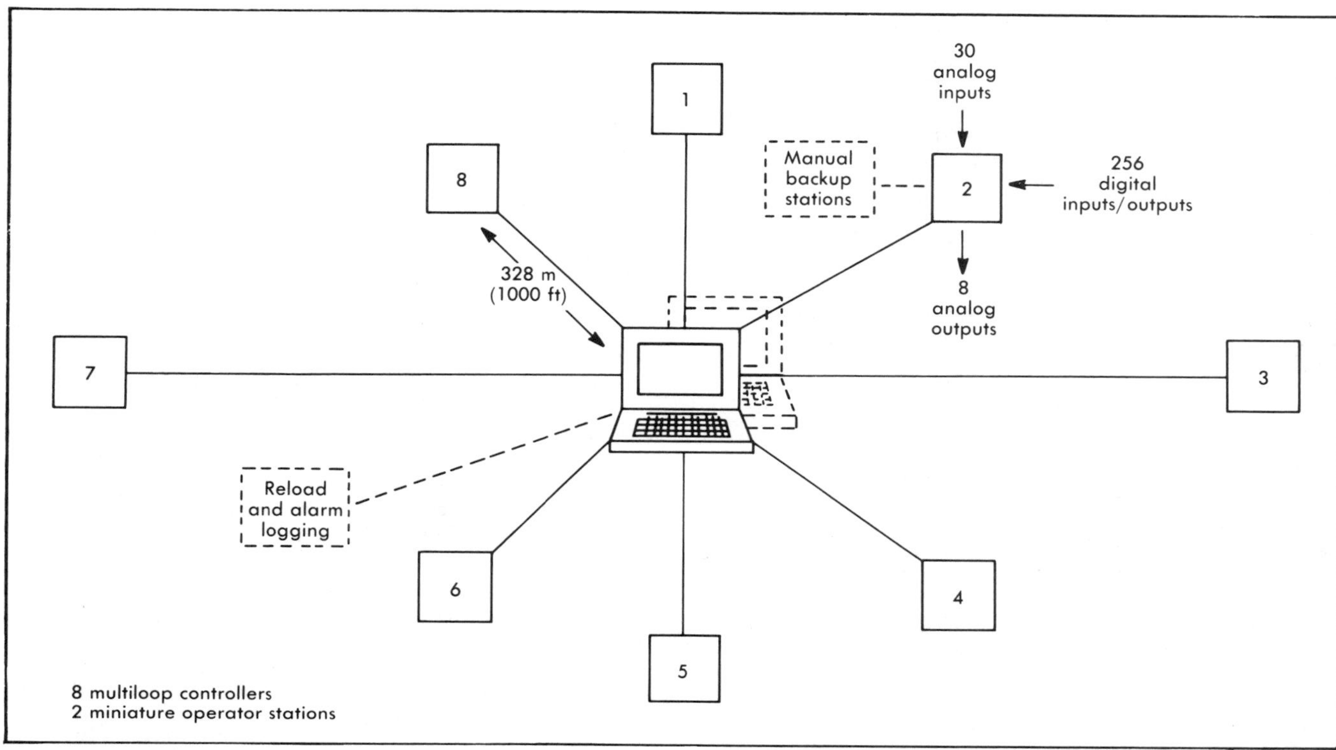

Fig. 12-27 Block diagram of coordinated control system with operator's console and video terminals. (*Leeds & Northrup Co.*)

Multiple-loop and logic control products, and programmable logic control products use digital communications to interface between individual installations of these products and a central supervisory system. These installations are referred to as distributed control systems. The central supervisory facility can be interfaced to a computer.

DATA CENTERS

Data center functions vary with the complexity and requirements of different heat treating installations. Traditional requirements are continuous or multipoint strip-chart and round-chart recorders. Digital technology has replaced the indication recording and alarm features of strip-chart recorders. Some trend records are useful for operators during process startup and changeover. Hard copy records or reports are frequently required to confirm that the heat treating cycle meets specifications. Data acquisition and computer processed data are more commonly displayed in log formats, which facilitate summary records of process variables, alarms, and computed information.

SAFETY CONSIDERATIONS

Since heat treating presents many potential hazards, a number of precautions are necessary to ensure safety of personnel, equipment, and buildings. Proper design of equipment and control systems, as well as periodic inspection and preventive maintenance are essential. The National Fire Protection Association (NFPA) Standard 86C contains many requirements with respect to furnace design, operation, and safety. Adequate training programs for personnel in proper procedures for operating furnaces are critically important. Furnaces that are out of service for extended periods should be properly maintained and serviced.

SAFETY PROGRAMS

Safety in heat treating requires management commitment and qualified personnel to execute an effective program. The program must also be accepted and supported by well-trained employees. All safety programs should be designed to minimize the exposure of personnel to hazardous conditions. A detailed discussion of safety programs is presented in Chapter 20, "Safety in Forming," in *Forming*, Volume II of this Handbook series.

Safeguards provided to protect workers against hazards include barriers, shields, guards, and other safety devices, as well as warning signs. Thorough training in material handling safety is essential. Various methods of safeguarding, as well as fire prevention and control, are described in Chapter 18, "Safety and Noise Control," in *Machining*, Volume I of this Handbook series. Adequate ventilation is essential to prevent heat stress, and first aid and medical assistance should be immediately available.

CHAPTER 12

SAFETY CONSIDERATIONS

PERSONAL PROTECTIVE EQUIPMENT

Safety glasses and/or shields should be provided for eye protection of personnel when they are using molten-salt baths, especially where the surface of the molten salt is exposed; and the wearing of hard hats is recommended. Suitable gloves and special protective clothing are recommended for protecting the skin from burns, cuts, abrasions, and dermatitis.

When workers are subjected to toxic, irritating, or asphyxiating pollutants, air-supply breathing apparatus or filter devices to cover the face and mouth are necessary. A good ventilating system can minimize problems from these breathing hazards. Helmets for head protection are recommended where overhead handling is involved. Safety footwear may also be needed, depending upon the hazards present.

FUEL, ATMOSPHERE AND CARBURIZING GASES

Most fuel, atmosphere, and carburizing gases used in heat treating furnaces are explosive when mixed with air. Many of the gases are highly toxic and flammable, and some are asphyxiating and poisonous. These gaseous hazards necessitate proper design of furnaces, equipment, and controls, plus cautious handling. Furnace design as specified in NFPA 86C requires a positive source of ignition where the flammable gas mixes with air. The design of various types of furnaces and control systems available is discussed previously in this chapter, and details with respect to atmospheres are presented in Chapter 10, "Heat Treatment of Steel."

The greatest potential danger from combustible gases is present during the startup and shutdown of furnaces, or during electrical power or gas failures. If power or gas flow fails, the furnace is generally shut down and purged with an inert gas. Purging procedures are not required if burnout is possible. Burnout can normally be used with furnaces operating at temperatures above 1400° F (760° C). Thorough venting of gas-fired furnaces is necessary before lighting the burners.

Mixtures of endothermic gas and air in practically all proportions are explosive at temperatures below about 1400° F (760° C); therefore, such mixtures should generally not be supplied to furnaces heated below this temperature. However, endothermic gas can be used in furnaces operating below 1400° F if air is removed prior to the introduction of endothermic gas.

QUENCHING OILS AND TANKS

Another potential hazard that can cause fires is the improper use of quenching oils. Overheating of the oils by hot workpieces can raise the temperatures of the oils above their flash points. The temperature of the oil should be checked periodically, and a controller that sounds an alarm when the temperature exceeds a preset limit is recommended. Good precautions include the use of large quench tanks; providing cooling coils, heat exchangers, or agitation for the tanks; and/or using oils having higher flash points. Agitation supplies cooler oil to the area of a fire, lowers the flash point of burning oil, and extinguishes the fire.

Partial submersion of hot workpieces into the quench tank can cause ignition of the oil surface. The surface oil will continue to burn until the workpieces are completely submerged or removed from the tank, or until the oil is drained from the tank. In furnaces equipped with integral quench tanks and special atmospheres over the quenching oil, however, internal fires should not occur because the atmosphere is free of oxygen. Pneumatic or hydraulically powered elevators permit lowering hot loads into the tanks, even in the event of a power failure, thus minimizing the possibility of a fire with a hot load sitting over a tank of oil.

Contamination of the oils with water must be guarded against because it can cause violent expulsion of the oils and consequent fires. Shell or tube-type heat exchangers in the area of quench tanks should use oil or air, rather than water. Systems that detect water in quench oils are recommended.

Overflow of oil from a quench tank, resulting from an increase in volume during work immersion and heating of the oil, can also cause fires, and thus adequate drainage is essential. It is recommended that all oil quench tanks be equipped with a properly sized, quick-dump bottom drain to remove, by gravity, oil from the tank to a remote, safe location in the event of an uncontrollable fire in the quench tank. The drain pipe to the remote location should contain a trap to prevent the fire from following the oil along the pipe. The drain valve in the bottom of the tank can be triggered from a remote location or by a thermal link that would be melted by a fire. Fire protection is discussed in Chapter 18, "Safety and Noise Control," in Volume I, *Machining*, of this Handbook series.

The design of enclosed quench systems in furnaces with combustible atmospheres requires particular care. Control of the liquid level—which serves to seal the quench system, maintain a readily purgeable area, and prevent the infiltration of air into the quench enclosure—is extremely important. Since some liquid is continuously lost from the tank as drag-out with the workpieces, it is necessary to add liquid periodically, either manually or automatically.

MOLTEN-SALT BATHS

The use of heated baths of molten salt as the heat transfer medium for the thermal treatment of immersed parts presents many potential hazards. Many of the salts used for heat treatment contain toxic and/or corrosive chemicals. Most are poisonous if taken internally and can produce dermatitis or serious burns if they come in contact with the skin. In addition to care with the baths themselves, precautions are also necessary in handling and storing the salts.

Workpieces and material handling equipment must be clean and dry. Immersion of wet or contaminated parts or devices can create steam or gases and cause violent expulsion of the molten salt. The introduction of combustible materials can cause fires. Air trapped in hollow parts will expand during heating and cause splashing of the salt.

The addition of incorrect salt can cause expulsion from the bath. For example, nitrate-nitride mixtures are incompatible with cyanide and cyanide-bearing salts. Overheating of nitrate-nitrite salt baths above a temperature of about 1100° F (595° C) can cause an explosion or fire and may create toxic fumes, requiring adequate ventilation by means of hoods, ducts, and fans. Waste salts must be disposed of in conformance with local, state, and national regulations. However, some salts, such as nitrates or nitrates/nitrites, can be recovered from wash waters by a specially designed and patented process that approaches a zero discharge efficiency.

VACUUM FURNACES

Operating and maintenance personnel must be thoroughly familiar with the vacuum equipment used with these furnaces and fully aware of the manufacturer's precautions. Preventive maintenance is required to avoid malfunctioning equipment, such as valves, pumps, controls, and interlocking devices. Serious malfunctioning can lead to implosion and explosion accidents.

If it is necessary to enter the interior of the vacuum chamber for loading/unloading or maintenance, ventilation is required to remove residual gases that may not be life supporting. Ventilation is also necessary when using solvents or cleaners to remove contaminants, such as oil, from the interior surfaces, resulting from the use of an integrated oil-quench system. Additional safety considerations are presented in ANSI/NFPA Standard 86D, "Industrial Furnaces Using Vacuum as an Atmosphere."

FLUID-BED FURNACES

The variety of gases employed as the fluidizing media in fluid-bed furnaces necessitates the same precautions for the use of gases discussed previously in this section. Many of these gases are highly toxic and flammable, as well as asphyxiating and poisonous. In addition, fluidization of the bed material, usually a sand or alumina, invariably causes fine dust particles to become airborne. Suitable hooding of the area and well-designed exhaust handling systems are required to capture these gases, vapors, and airborne particulates; and they must be disposed of in conformance with local, state, and national regulations.

The area surrounding any atmosphere furnace should be well ventilated. Ventilation is particularly necessary when using heavier-than-air gases, such as propane, that could accumulate in a lower level area and result in a serious explosion. At all times, operators of fluid-bed furnaces should wear protective equipment, such as a face shield, heat-resistant clothing, and gloves. It is characteristic of fluidized-bed furnaces for bursts of hot bed material to erupt up and out of the furnace retort. Operators must take suitable precautions, and combustible materials should not be stored in the immediate area.

FURNACE MECHANISMS

Furnaces are equipped with a wide variety of powered mechanisms—such as pushers, transfer units, elevators, fan drives, door operators, and other devices—that require safety precautions. All such mechanisms should be properly guarded before they are operated. For some equipment, a limit switch or plug-in connection can be electrically interlocked in series with the power supply of the mechanisms, making it necessary for the guard to be in place before the unit will operate. Proper warning signs are also essential.

Emergency pushbuttons should be provided to stop all mechanisms instantly. However, pneumatic or hydraulically powered mechanisms, without special protection, will continue their movements as long as pressure is available. Consequently, means should be provided to instantaneously remove such pressure from the system.

The use of shear pins should be limited to drives that will stop instantly and not cause a safety hazard. Shear pins should never be used on elevators, dumpers, or noncounterweighted door drives where a shear pin failure could cause the mechanism to fall and possibly cause damage or injury.

Fire-resistant hydraulic fluids are recommended for all furnace applications. Hydraulic power units should be supplied with pressure switches to stop the system in case of a line breakage.

Bibliography

American Society for Metals. *Metals Handbook*, 9th ed., vol. 4, "Heat Treating." Metals Park, OH: 1981.

Japka, Joseph E. "High Temperature Controlled Carburization of Steel in Fluidized Bed Furnaces." *Industrial Heating* (August 1983).

Lehr, Thomas A. *Industrial Energy Management—A Cost-Cutting Approach*. Dearborn, MI: Society of Manufacturing Engineers, 1983.

McGannon, Harold E. *The Making, Shaping and Treating of Steel*, 8th ed. Pittsburgh: United States Steel Corp., 1964.

Nayar, Harbhajan S., and McKinley, William. "Training and Maintenance: Keys to Atmosphere Safety." *Heat Treating* (December 1980), pp. 44-46.

Petrarca, Dale M. *Vacuum Heat Treating and Equipment*. SME Technical Paper CM80-395. Dearborn, MI: Society of Manufacturing Engineers, 1980.

"Using the Fluidized Bed for Nitriding-Type Processes." *Metal Progress* (February 1983).

Wolfson Heat Treatment Centre. *Guidelines for Safety in Heat Treatment: Part 1, Use of Molten Salt Baths* and *Part 2, Health and Personal Protection*. Birmingham, England: The University of Aston in Birmingham, 1981.

SELECTIVE SURFACE HARDENING

CHAPTER CONTENTS:

Processes discussed in this chapter involve surface hardening by thermal treatments—heating and quenching with no chemical change occurring on the surfaces of the work. They thus differ from the thermochemical surface-hardening processes described in Chapter 11, "Surface (Case) Hardening," in which the chemical composition of steel parts is altered prior to quenching and tempering.

Methods used for surface hardening without any chemical change of the metals are induction, high-frequency resistance, flame, electron beam, and laser hardening. For most applications, these methods are used to harden selected surface layers of parts, but they are also used to harden all surfaces, as well as for through heating of workpieces. Like the processes described in Chapter 11, these methods are used to provide high surface hardness, strength, and wear resistance. The electron beam and laser hardening processes eliminate the need for quenching the heated workpieces.

INDUCTION HARDENING

Induction heating is based on the concept of electromagnetic inducement of electrical energy into an electrically conductive part. The electrical considerations involve the phenomena of hysteresis and eddy currents, with the major factor being eddy currents. However, the only requirement of a material in order to respond to induction heating is electrical conductivity. An induction heating circuit is fundamentally a transformer wherein the inductor, also called the induction heating coil, carrying the alternating current is the primary of a transformer, and the part to be heated is made the secondary by merely placing it within the confines of or in close proximity to the inductor loop or coil.

There is no contact or connection between coil and workpiece. The current, usually flowing circumferentially through the inductor, sets up magnetic lines of flux in a circular pattern, which link or thread through the material being heated and induce a circumferential flow of electrical current to flow in the workpiece. It is important to note that the current flow pattern in the workpiece is essentially a mirror image of inductor current flow (see Fig. 13-1).

Inductor coil and workpiece configurations are not limited to an encircling arrangement, with the workpiece surrounded by a single-turn or multiturn (solenoid) coil. Many other configurations are possible: internal, pancake, and other shaped coils (see Fig. 13-2). Additional designs, such as hairpin, channel, nonencircling, single shot, and others, are described later in this section.

Encircling-type coils are usually most efficient electrically. Other types, however, can be very effective in heating applications and are used to suit specific application requirements or part handling considerations. Current patterns must be closed loop.

ADVANTAGES OF INDUCTION HEATING

Induction systems provide a fast, efficient, and economical method of heating any electrically conductive material to a precise temperature. The equipment uses readily available electric power to heat the entire surface of the workpiece or selected areas. Heat depth can be limited to just the surfaces or can include the entire cross section.

Induction heating can be equally efficient for job shops or high-production operations. The same basic equipment can be used to heat a wide range of sizes and shapes of parts, as well as different materials. Since radiated and secondary heating is eliminated, process heating can be installed directly in a production line or general work area. The equipment is compatible with existing in-plant material handling systems and can be automated to meet specific production requirements.

Important advantages of induction heating include increased production, reduced costs, and improved products. Cycle times are a matter of seconds, and machines can be completely automated. Precise controls reduce or eliminate scrap, distortion is minimized, and the need for more costly alloy steels is sometimes eliminated. Floorspace requirements are reduced, and working conditions improved.

Hardened areas can be accurately controlled with respect to depth, width, location, and hardness.

Contributors of sections of this chapter are: ***Mary Anton****, Applications Metallurgist, Leybold-Heraeus Vacuum Systems, Inc.;* ***Richard E. Haimbaugh****, Induction Heat Treating Corp.;* ***Ronald C. Hanson****, Manager, E. B. Systems, Sciaky Bros., Inc., Allegheny International;* ***C. N. Hubbard, Jr.****, Vice President, Domestic Sales, Thermatool Corp.;* ***Gary LaFlamme****, Research & Development Engineer, Leybold-Heraeus Vacuum Systems, Inc.;* ***George D. Pfaffmann****, Director of Marketing and Technical Services, TOCCO Div., Park-Ohio Industries, Inc.;* ***Ole A. Sandven****, Chief Metallurgist, Avco Everett Research Laboratory, Inc.;* ***George P. Welch****, Product Manager, Heat Treating, Ajax Magnethermic Corp.;* ***Stanley Zinn****, Ferrotherm Inc.*

CHAPTER 13

INDUCTION HEATING

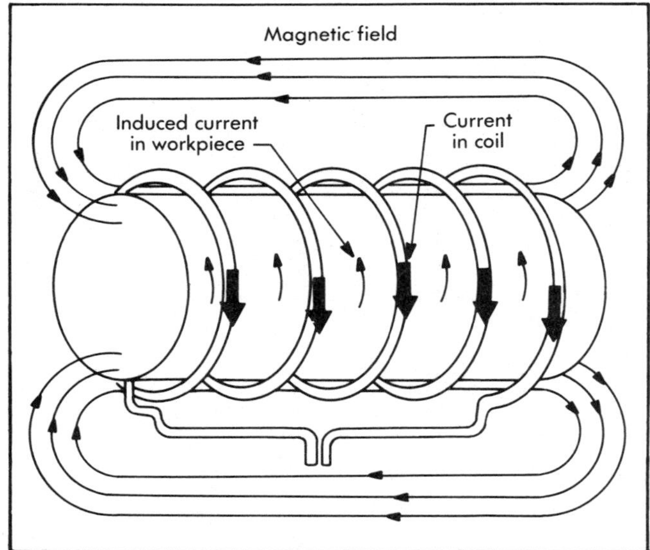

Fig. 13-1 Current flow in workpiece is opposite to current flow in coil.

The original ductility of the cores is retained, and the rapid heating reduces scale formation. Compressive residual stresses generated in selective, localized hardening enhance fatigue life.

OTHER APPLICATIONS

The use of induction heating for selective or through hardening, stress relieving, tempering, normalizing, and annealing is the primary purpose of this discussion. However, induction heating is employed for many other purposes, including joining, brazing/soldering, bonding, curing, and drying. Another important application is warm/hot densification and near-net-shape forming of powdered metal parts. Induction heating is also used for melting and warm and hot forging.

PRINCIPLES OF INDUCTION HEATING

Since the workpiece or material that carries the induced current is acting as a conductor, it offers resistance to the flow of electrical current. Thus, induction heating can be compared to ordinary resistance heating and established as the heat that is liberated as a result of I^2R losses, where I equals the current in amperes and R is the resistance in ohms.

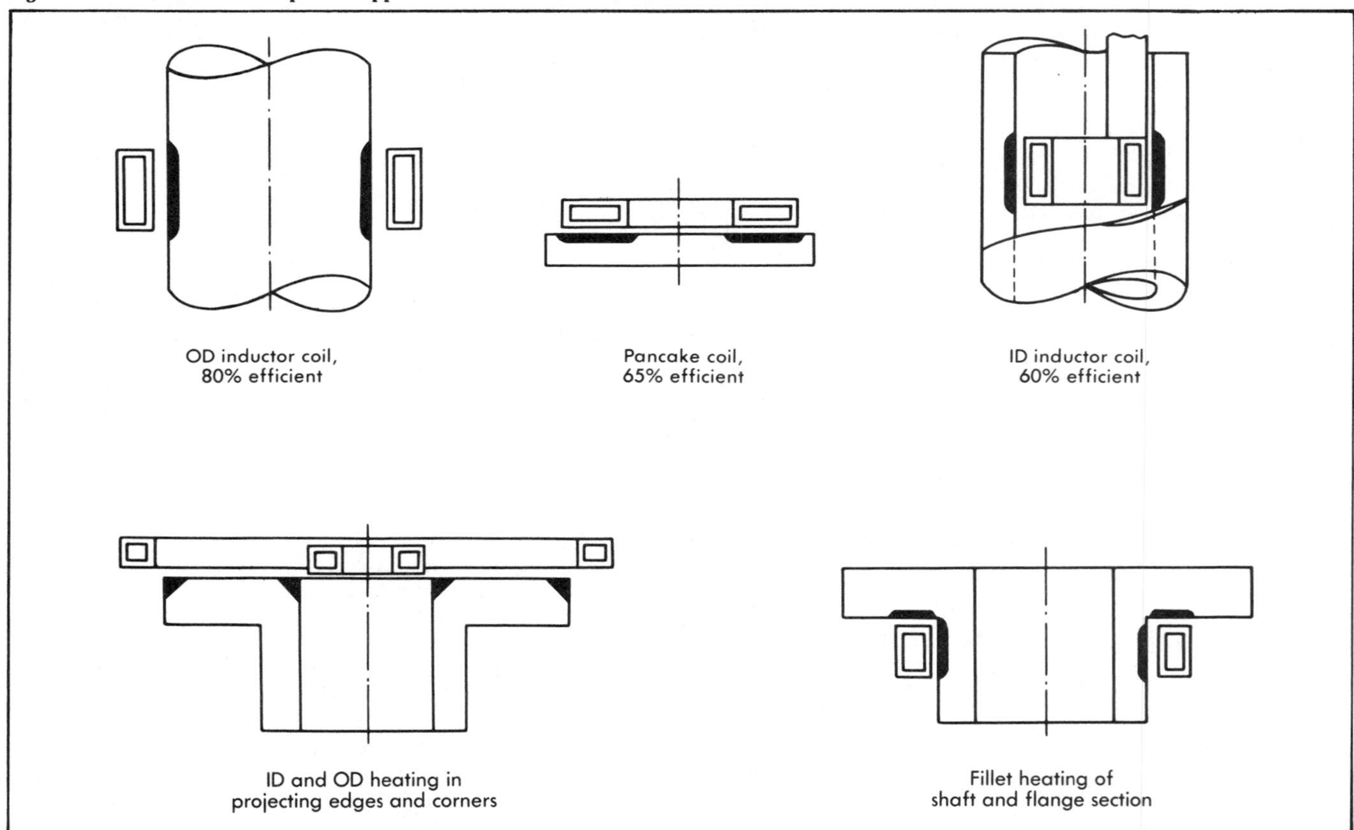

Fig. 13-2 Typical heat patterns and efficiencies (below Curie temperature) for ferrous metals with various inductor coil configurations. (*TOCCO Div., Park-Ohio Industries*)

Reviewers of sections of this chapter are: ***Richard Barber****, Director of Applications, Photon Sources, Inc.;* ***Kenneth M. Barrette****, Chief Engineer, Quantum Laser Corp.;* ***Donald W. Bennett****, Vice President/General Manager, Laser Div., Coherent, Inc.;* ***David E. Carnahan****, Vice President, Pennsylvania Flame Hardening Co., Inc.;* ***Mike Faber****, Senior Project Engineer, Heat Treating, Ajax Magnethermic Corp.;* ***Terry R. Gonser****, Production Superintendent, Allison Gas Turbine, General Motors Corp.;* ***Richard D. Green****, Marketing Manager, MAPP Products;* ***Jack LeClaire****, Michigan Flame Hardening Co.;* ***Gregory G. Pawlowski****, Supervisor, Marketing Communications, Photon Sources, Inc.;* ***George D. Pfaffmann****, Director of Marketing and Technical Services, TOCCO Div., Park-Ohio Industries, Inc.;*

INDUCTION HEATING

The magnetic lines of force, or flux, generated by the inductor, which induce the flow of current in the workpiece, are more concentrated at the midpoint of the width of the inductor and near its inside face. The unusual characteristic of high-frequency heating, upon which all surface hardening applications depend, is its tendency to concentrate on the surface of the conductor through which it flows, in both inductor and workpiece. This phenomenon, called skin effect, is a function of frequency. Other factors being equal, the higher the frequency, the shallower will be the depth of penetration of the heated zone. Details of this relationship are discussed later in this section.

In induction heating, the original core properties of the workpiece are maintained, with surface hardness being secured upon subsequent, proper quenching. Continued application of power causes an increase in depth of heating. As each layer of steel increases in temperature, additional depths of heat penetration result from the flow of energy by thermal conduction and the change in electrical resistivity.

The use of high power at a given frequency for a short period of time permits raising the temperature in shallow depths to hardening temperatures, the heat flow by conduction being limited. Conversely, a small amount of power induced into the surface of a bar for a long period of time produces greater depths of heating since heat flow by conduction is permitted to continue. Thus, the selection of the proper frequency and the control of power and heating time make possible the fulfillment of any desired specification for surface hardening, as well as through hardening and other operations.

There are certain fundamental relationships between frequency and diameter or thickness of stock to be heated, upon which may depend the selection of the frequency to be used for a particular application. With the ease of adjustment for induction heating equipment, many applications can be successfully performed with not only various frequencies, but various types of equipment. Careful consideration of all the variables is necessary in selecting a particular system.

FREQUENCY SELECTION

A convenient equation can be used for selecting frequency from a scientific basis. The depth at which most of the induced current flows and, therefore, the metal in which the I^2R losses occur and heating takes place is mathematically related to the frequency used, the relative magnetic permeability, and the electrical resistivity of the material being heated in accordance with the following formula:

$$D = K \frac{R}{PF} \qquad (1)$$

where:

D = depth of current flow
R = resistivity
P = relative magnetic permeability
F = frequency
K = factor which depends on the units of the above variables

The resistivity of a material increases gradually with temperature. Relative magnetic permeability of magnetic materials decreases with temperature but at an increasing rate from a certain value at room temperature to a constant value of one at a temperature of about 1350° F (730° C), the Curie temperature, at which point the steel becomes nonmagnetic. All nonmagnetic materials have a value of one at all temperatures. Values for relative magnetic permeability depend upon the strength of the magnetic field in which the material exists, and the magnetic field depends upon the power density used.

For surface hardening applications, the power density must be high in order to have short heat cycles and to be able to control the heat flow by conduction. For through-heating applications, the power density is lower; but in order to maintain production requirements, it is usually no lower than 1/2-1 kW/in.^{2} (0.08-0.16 kW/cm^2) of workpiece surface area. At these levels, the material may be considered to have a relative magnetic permeability of 100. If the magnetic field is substantially reduced, however, this number can be increased several times, and it is important in connection with the heating of strip.

For steel at room temperature (having a relative magnetic permeability of 100), the approximate depth can be determined from the following formula:

$$D_1 = \frac{2.3}{F} \qquad (2)$$

where:

D_1 = depth, in.
F = frequency (Hz)

For steel at temperature above the Curie point, 1350° F (730° C), the approximate equation is:

$$D_1 = \frac{23}{F} \qquad (3)$$

The value D_1 is referred to as the reference depth and is defined as the depth in which there is uniform current density (uniform distribution of eddy currents). Below this depth, the current density falls off sharply, and pure induction heating can be ignored. About 80% of the current flows in the reference depth area.

The ratio of workpiece diameter to reference depth is often called workpiece resistance or the electrical size of the workpiece. To heat with good efficiency, this ratio is usually 3:1 when heating for forging or through heating. For surface hardening applications, this value is usually higher than 4:1.

EQUIPMENT USED

An induction heating system consists essentially of a power supply, a workstation, a heating coil or inductor, controls, workpiece handling units, and ancillary equipment. After the required frequency has been determined, equipment selection should be dictated by an evaluation of the cost of the equipment, operating and maintenance costs, and, where applicable, flexibility of the equipment for job changeover.

Reviewers, cont.: ***Ronald H. Podolak****, Detroit Flame Hardening Co.;* ***Ole A. Sandven****, Chief Metallurgist, Avco Everett Research Laboratory, Inc.;* ***Fred D. Seaman****, Manager, Laser Center, IIT Research Institute;* ***Donald R. Shank****, Director of Product Marketing, Photon Sources, Inc.;* ***Richard Sommer****, Vice President of Research & Development, Ajax Magnethermic Corp.;* ***Richard E. Trillwood****, President, E. B. Engineering, Inc.;* ***George P. Welch****, Product Manager, Heat Treating, Ajax Magnethermic Corp.;* ***Dennis Werth****, Manager of Applications, Photon Sources, Inc.;* ***Stanley Zinn****, Ferrotherm Inc.*

INDUCTION HEATING

Power Supply

A power supply unit converts the available line frequency, such as 60 Hz, to the required frequency for the specific application. The types of power units in prominent use today are:

1. Vacuum-tube oscillators (radio-frequency generators): 1-4 mHz and 200-500 kHz.
2. Motor generator sets: 1, 3, and 10 kHz.
3. Frequency multipliers: 180 and 540 Hz.
4. Frequency inverters (solid state): 0.2, 0.5, 1, 3, 6, 10, 25, and 50 kHz. Many modern solid-state power supplies are capable of operating over a range of frequencies.

Although there are many motor generator units in service today, the economics of present day manufacturing, operating efficiency, flexibility, and dependability strongly favor the solid-state inverter.

Workstation

The work or heat station contains electrical hardware, such as output transformers, capacitors, high-frequency contactors (with medium-frequency range), protective devices, and cooling-water manifolds.

The function of the workstation is to provide the proper electrical impedance match between the output of the power supply and the inductor (induction heating coil) for optimum power transfer into the workpieces to be heated. The coil is normally mounted on the front of the workstation. Several workstations can be operated in sequence from a single power supply, and some power supplies can operate with workstations in parallel.

Workstation components can be combined into the power supply, as is done on small solid-state, medium-frequency units, and particularly on vacuum-tube oscillator units (radio-frequency generators) where the only component that may be needed for matching is the output transformer.

Heating Coil (Inductor)

The heating coil (inductor) is the component that produces specific heating results in the workpieces. It is specially designed to properly couple the induction heating energy into the workpiece to produce the required results. Design of the heating coil is discussed later in this section.

Workpiece Handling

Workpiece handling equipment for induction heating systems takes many forms. For a job shop application, the equipment can consist of a simple, manually operated holding device. For a production application, sophisticated, precision automation equipment can be used for workpiece handling.

Control Systems

The high speed of heating and selectivity of heat placement possible with induction heating requires precise control. Many different control systems are used, depending upon the application, including infrared temperature controls, NC/CNC units, programmable controllers, computers, programmable automation systems, and robotics.

Ancillary Equipment

It is common practice to include other components with induction heating installations. These can include water cooling systems and a quench recirculator, particularly for hardening operations.

A properly designed and operating water cooling system is critical to successful induction heating. Because of the high electrical currents carried in many parts of the system, satisfactory water pressure and flow at reasonable temperatures is a must. In addition, the water-cooled electrical parts may operate at substantial potential (voltage); therefore, the electrical conductivity of the water must be low, requiring a minimum of additives. Other desirable water characteristics include limits on the amount of calcium carbonate ($CaCO_3$), total solids (dissolved and suspended), and the pH valve. Water systems generally include facilities for recirculation (pumping), heat removal (heat exchangers, chillers, towers, etc.), and a piping network, normally recommended to be a nonferrous material to prevent contamination.

INDUCTION SURFACE HARDENING

It is common practice to stress relieve parts after induction heating, accomplished by heating to 300° F (150° C) or lower to stabilize the internal stresses caused in the selective hardening process. This treatment prevents problems in subsequent manufacturing operations such as grinding. With few exceptions (certain grades of stainless steels), all ferrous materials that are hardened by induction heating are magnetic at the time the heating starts and become nonmagnetic before the hardening temperature is reached, usually at a temperature of about 1350° F (730° C).

As previously explained, the depth in which 80% of the heating initially takes place is expressed by D_1, determined from equation (2) in this chapter. Additional depth of heating results from heat flow by conduction from the heated surface to the colder core and the increase in depth of current flow as the resistivity of the material increases with temperature. The additional depth due to conduction can be determined approximately from the following formula:

$$D_2 = \sqrt{0.0015T} \quad (4)$$

where:

D_2 = additional depth of heating
T = heating time, s

Therefore, the total depth of heating is $D_1 + D_2$.

In discussing any phase of depth control, it is important that consideration be given to an analysis of the material and its structure. These factors control the rate of solution of the microconstituents necessary to permit thorough austenitizing prior to quenching for hardening. The inherently short heating time of induction heating permits the use of higher austenitizing temperatures than with conventional heating practices. Consequently, it is generally possible to obtain satisfactory hardness with lower carbon steels using such higher temperatures.

Surface condition of the steel to be hardened is also important, and all decarburized material must be removed. It is frequently necessary to remove some stock from cold-drawn steel bars to eliminate detrimental imperfections such as seams or pitting.

With respect to carbon steels, plotting the austenitizing temperature against carbon content produces practically a straight-line function. When the carbon content of the steel exceeds about 0.50%, additional carbon content has no effect on the hardness obtained; however, there is a pronounced effect on the ease of obtaining full hardness.

Carbon in excess of 0.50% may not be placed completely in solution; but from the standpoint of wear resistance, this may not be necessary. A bar of normalized SAE 1090 steel shows very rapid response to austenitizing and requires only moderate temperature requirements. Further, a low-carbon steel that has been previously heat treated to a fine-grained pearlitic structure shows similar tendencies for ease of austenitization.

Assuming the presence of structures or analyses that respond readily to heat, Table 13-1 shows the minimum depths of hardness that should be considered for production work in the hardening of steel. For comparative purposes, the approximate theoretical depths of penetration are also shown. Depths of hardness data in Table 13-1 represent actual results obtained and are values noted with both single-shot and progressive methods of treatment, generally with power inputs considerably greater than the accepted minimums. Diameter of the stock or wall thickness of the workpiece must be sufficient to provide a reasonable core. It is not considered practical to surface harden 1/8" (3.2 mm) diam stock regardless of the frequency, nor any small tube when the wall thickness is much less than twice the depth of hardness anticipated. There are other factors, such as the relationship of wall thickness to diameter, that assume importance in such instances. Small-diameter stock, however, can be through heated in selected areas by induction.

As shown in Table 13-1, the depth of hardness is always considerably in excess of the theoretical depth of electrical energy penetration. The *skin effect* (the tendency for heat to be developed on the surface of a part) assumes less significance as the heating time increases. The fullest advantage of this phenomenon is obtained only if the surface area can be brought to hardening temperature and quenched within a few seconds. To accomplish this, a sufficient amount of energy must be introduced into the inductor to induce adequate current generation in the surface of the part. Therefore, power densities used for surface hardening applications are generally of a high order of magnitude.

TABLE 13-1
Depths of Hardness for Production Applications

Frequency, Hz (cycles/s)	Approximate Theoretical Depth of Hardness of Electrical Energy, in. (mm)	Minimum Practical Depth of Hardness,* in. (mm)	Working Depth of Hardness,** in. (mm)
1000	0.060 (1.52)	0.136 (3.45)	0.200-0.350 (5.08-8.89)
3000	0.050 (1.27)	0.090 (2.29)	0.150-0.200 (3.81-5.08)
10,000	0.020 (0.51)	0.060 (1.52)	0.080-0.120 (2.03-3.05)
50,000	0.010 (0.25)	0.035 (0.89)	0.050-0.080 (1.27-2.03)
500,000	0.003 (0.08)	0.020 (0.51)	0.040-0.080 (1.02-2.03)
1,000,000	0.002 (0.05)	0.010 (0.25)	---

* High power densities, a minimum of 15kW/in.^{2} (2.33 kW/cm^2), and for steels with optimum prior structure.

** Medium power densities, about 10 kW/in.^{2} (1.55 kW/cm^2), and for steels with a prior structure of medium-fine grains.

For static or single-shot hardening, the power density depends upon the frequency used and depth required. These values usually range from 5 to 20 kW/in.^{2} (0.78 to 3.10 kW/cm^2). However, for shallow depths on small diameters, it is necessary to use power densities considerably higher. The influence of the prior structure (fineness of microstituents) of steel parts on power requirements for surface hardening is illustrated in Fig. 13-3.

High power cannot be used arbitrarily to obtain high production. A coarse pearlitic or spheroidized structure will require more heat time or higher temperature than a tempered martensitic structure. Coarsening of grain structure is not likely to be a factor when using relatively short heat time in induction hardening.

QUENCHING

A successful hardening process must properly apply a suitable cooling medium to the heated area to ensure that the particular material cools faster than the critical cooling rate. Low-carbon steels must be quenched with a minimum of delay and in a medium like water to obtain maximum hardness. With high-carbon or alloy steel grades, a longer delay and a slower quench medium, like oil or any of the polymer types, can be tolerated. For some applications, the use of polymer quenchants provides an intermediate control of quenchability between that of water and oil. These materials can be adjusted for quench capability by controlling their concentration in a water solution or by controlling their temperature.

It is possible to use no liquid medium at all if the heated volume of metal is very small compared to the total part size or volume. This operation is termed *mass quenching.*

With induction heating, the medium and method used for quenching are generally not critical as long as they provide sufficient speed to meet critical cooling requirements. The part

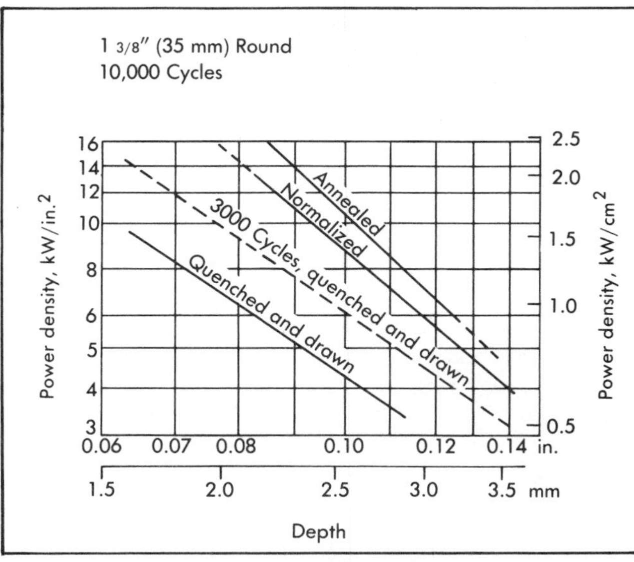

Fig. 13-3 Influence of prior structure of steel on power requirements for surface hardening.

geometry can help determine the quench medium. The more complex the part surface geometry, the more susceptible the process is to nonuniform cooling rates, which can contribute to cracking. For instance, a smooth shaft (no tool marks) of SAE 1090 steel can be induction heated and quenched in cold water with no cracking. Conversely, a shaft with several snap-ring grooves will require an oil or polymer quench, which may necessitate a high-carbon or alloy steel to meet hardness requirements.

The period of time between the shutoff of power and the application of the quench is important from a control standpoint. When the part processed has a reasonably large cross section and the limitation of the depth of hardness is not important, the delay period between heating and quenching is not critical, other than from a standpoint of reproducing itself each time a part is processed.

A delay period permits additional time for diffusion of the metallurgical structure of the area to be hardened, without the need for using higher temperatures, and is therefore advantageous. However, the short heating cycles generally associated with induction heating permit the use of higher temperatures for obtaining the same results, thereby limiting heat flow by conduction without introducing deleterious coarsening of the microstructure.

When a part has an area to be hardened that is exceedingly long, heating is accomplished by passing the part progressively through the magnetic field of an inductor. The heated surface area is quenched progressively as it emerges from the effect of the high-frequency energy. This technique makes it possible, with limited power available, to maintain high power densities for control of depth since the power can be introduced into a narrow band around the surface of the part.

While the depth of penetration is a function of frequency, it is primarily controlled by power density and the rate of travel through the inductor, which is comparable to time of heating. Quenching, while not involving a delay period in terms of time, is controllable by varying the angle at which the quench holes are drilled; this controls the distance between the point at which the part leaves the effect of the high-frequency current and the quench. The longer the time interval between leaving the inductor and reaching the quench, the greater the heat flow by conduction. The volume of quenchant used is also critical with respect to heat removal.

SCAN HARDENING VERSUS SINGLE-SHOT HARDENING

Both the scan and single-shot systems of hardening have certain advantages. The following comparison is based on hardening a 1½" (38 mm) diam shaft with a 10" (254 mm) heated length, requiring an approximate hardened depth of 0.150" (3.81 mm) containing 50% martensite. The material is SAE 1040 steel, and a water quench is used. Gross production required is 300 pieces per hour.

For scan hardening, the power required is about 122 kW, scanning at the rate of 0.5 in./s (12.7 mm/s). A typical cycle for manually loaded scanners is presented in Table 13-2. Based on this cycle, two dual-spindle scanners, hardening two parts at a time and operating alternately, would be required to effectively produce 300 shafts per hour. The two scanners would need 244 kW of power each or this same amount of power to switch between the two.

Considering the single-shot heating approach for the same part, about 300 kW would be required, with a heating time of five seconds. With this type of heating, a walking-beam hardening machine with three rotating spindles would be used. The part is heated in the first spindle and transferred to the second and third spindles for quenching. Even though the heat time is relatively short, the quench time of the second station is almost three times longer to adequately quench without any effect of tempering during the transfer delay between quenches. A typical machine cycle is shown in Table 13-2. The production rate with one walking-beam machine, using 300 kW of 10 kHz power, would be 300 pieces per hour.

An alternative to either scan or single-shot hardening would be continuous hardening. This process requires that the shafts be uniform in diameter to permit feeding with a set of feed rolls, and it excludes parts with flanges or shoulders. With continuous hardening, the same parts could be processed at the rate of one every 20 seconds, or 180 pieces per hour, using 122 kW of 10 kHz power and eliminating load, unload, and transfer times.

INDUCTOR COIL DESIGN

Inductor coils have a variety of shapes and are of different types. Their geometry is based upon the geometry of the workpiece, heat pattern configuration, and processing technique. Computer-based mathematical models are used exten-

TABLE 13-2
Comparative Cycle Times for Scan and Single-Shot Hardening

Manually Loaded Scanner		Single-Shot	
Function	Time, s	Function	Time, s
Load	5	Walking-beam index	2*
Index to heat	6	Raise part and center	1*
Scan harden	20	Delay to heat	3
Quench out	8	Heat, station no. 1	5
Index to load	4	Quench, station no. 2	8*
Unload	5	Quench, station no. 3	8
		Open centers and lower	1*
Total cycle time	48	*Index cycle time	12
Production, parts per hr	75	Production, parts per hr	300

sively to design long, multiturn, solenoid-type coils for applications such as heating for forging, through heating, and similar uses. The design of most other coils, however, is evolved and refined largely by an empirical process of trial and error. Ceramic or refractory-coated coils are commonly used when possible to reduce radiation losses, improve inductor life, and reduce contact damage and scale-related problems.

The simplest form of inductor coil is the single-turn type shown in Fig. 13-4. Normally the workpiece is rotated since the flux density is weaker at the gap where the leads connect it to the transformer.

Single-turn inductors can be made wider for longer heated areas and can be equipped with an integral quench to provide a combination heating and quenching tool (see Fig. 13-5). As shown on the lower half of the inductor section in Fig. 13-5, the coil can be contoured to flatten the flux pattern and redistribute the part current and flow to provide a uniform hardened depth.

Long areas along the length of a shaft can be scan hardened with a progressive, integral-quench inductor as shown in Fig. 13-6. Workpieces are rotated essentially for two reasons:

1. To cancel out the weaker flux at the fishtail gap, thus promoting more uniform heating.
2. To provide a more uniform quenching action on workpiece surfaces.

Figure 13-7 illustrates a progressive integral quench for hardening bores. With internal coils, heat generation in the workpiece is slower because the flux paths are concentrated in the center of the inductor. Current flows on the inside diameter (ID) of the inductor, thus reducing the effectiveness of heating on the outside diameter (OD). With such inductors, it is normally necessary to use flux magnetic cores—such as iron, C-type laminations, or a powdered-iron material—to improve flux linkage to the workpiece. The cores force coil current to flow on the OD of the inductor face for improved performance and efficiency. Flux magnetic cores are especially beneficial for small internal coils, one or two-turn coils, channel-type coils, some fillet-heating coils, and special coils, such as those used in scanning gear teeth.

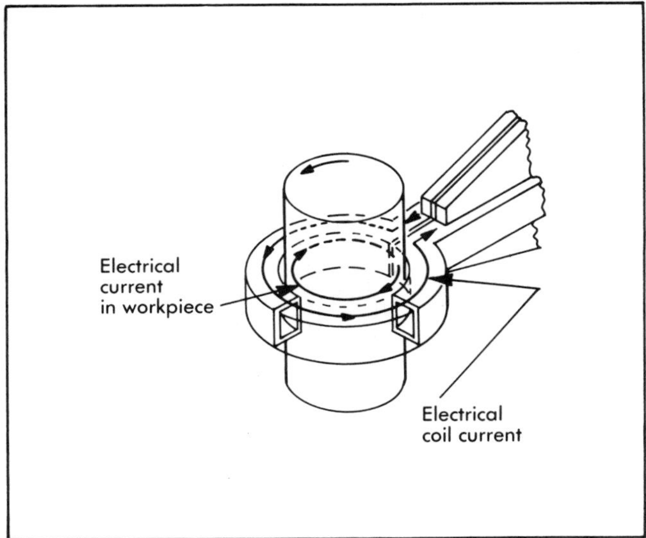

Fig. 13-4 Simple, single-turn inductor coil.

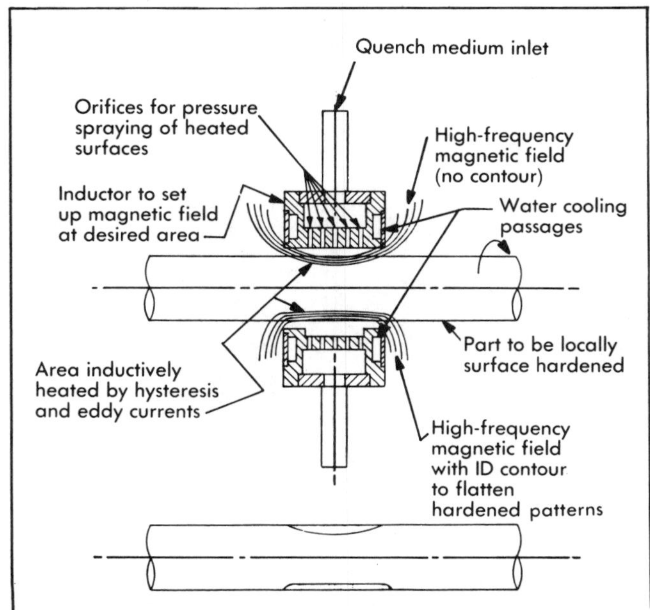

Fig. 13-5 Wide, single-turn inductor with integral quench.

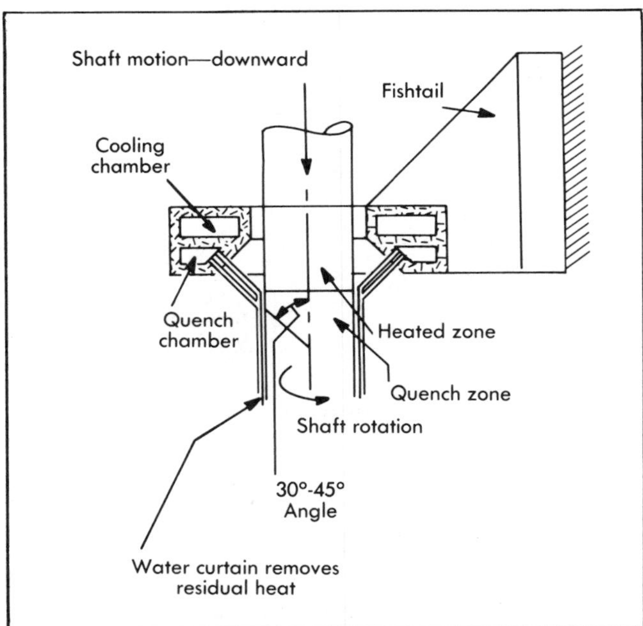

Fig. 13-6 Progressive-type inductor with integral quench.

For surface heating, two common variations of inductors are hairpin and pancake coils (see Fig. 13-8). For hairpin-type inductors, performance is improved with the use of flux magnetic cores. For both hairpin and pancake inductors, the coupling or distance between the bottom of the coil and top of the workpiece is critical. The current generated in the workpiece is inversely proportional to the square of this coupling distance. Also, because of the inductive effect, these type of coils are more responsive to the higher frequencies, such as 450 kHz.

INDUCTION HEATING

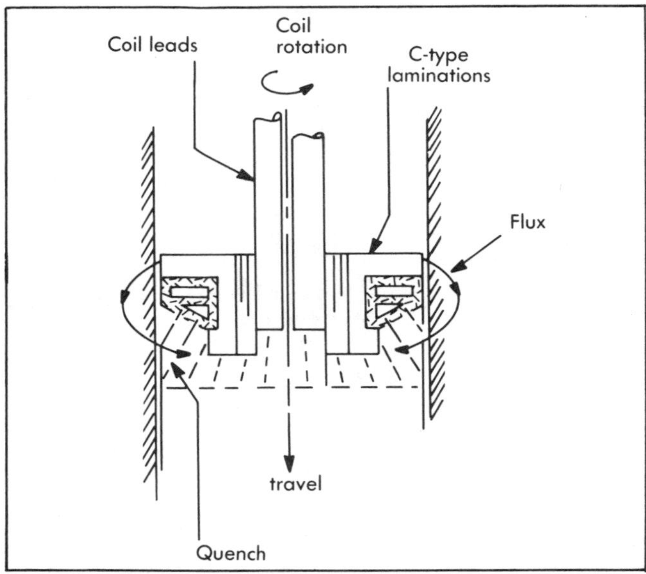

Fig. 13-7 Progressive inductor with integral quench for hardening bores.

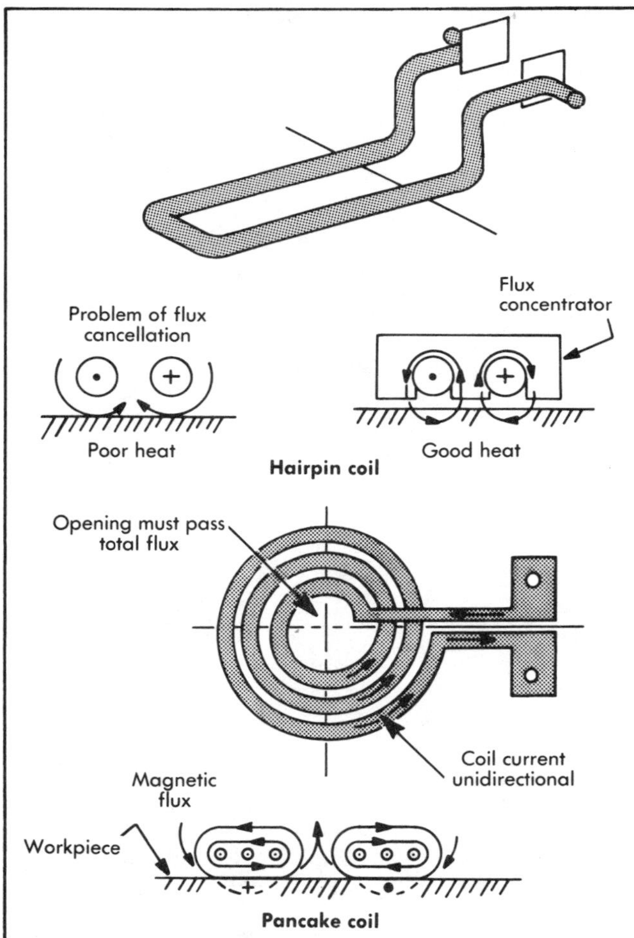

Fig. 13-8 Two variations of inductors for surface heating are the hairpin and pancake coils.

Folding hairpin or pancake coils over onto themselves essentially produces channel coils. By changing the circular configuration of a pancake-type inductor to a rectangular shape and then folding, the resulting form is a multiturn channel coil normally used for heating a continuous flow of workpieces, such as the ends of bars and the threaded ends of parts for annealing.

The basic difference in current flow through a workpiece is that the flow is circumferential with encircling heating and longitudinal with nonencircling heating. With the single-shot technique, the current flows axially along the part, and the part must be rotated (see Fig. 13-9) to distribute the heat uniformly around the circumference. Flux magnetic cores are generally used to improve performance and efficiency. A particularly important advantage of the current flow in a workpiece with the

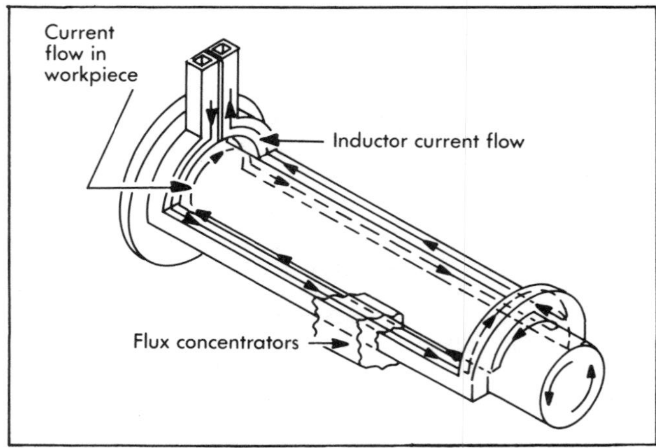

Fig. 13-9 Nonencircling, single-shot inductor for heating a long shaft.

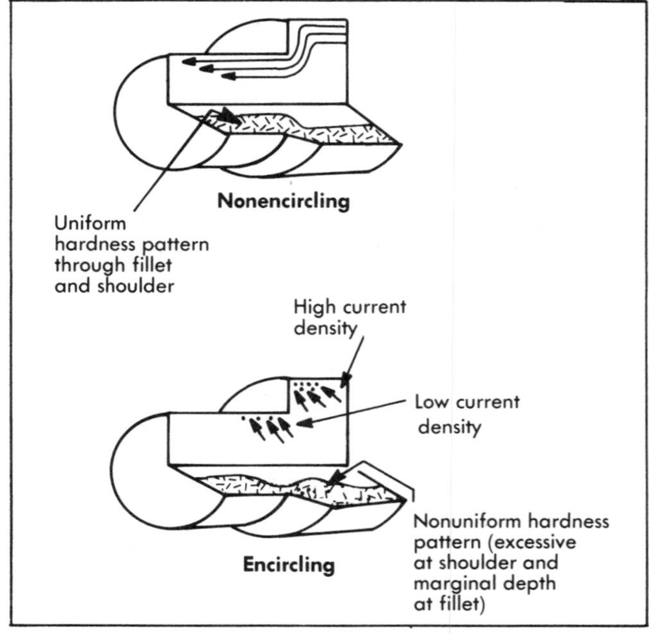

Fig. 13-10 Comparison of effect on hardening pattern resulting from current flow with nonencircling and encircling heating.

single-shot technique is the ability to obtain adequate hardened depth at shoulders, grooves, and similar surface geometries. A comparison of the effect on hardening patterns resulting from current flow with nonencircling and encircling heating is presented in Fig. 13-10.

WORKPIECE HANDLING

A wide variety of methods are employed for processing, handling, and transferring workpieces for induction heating (see Fig. 13-11). Manufacturing engineers are limited only by their experience and innovative ability.

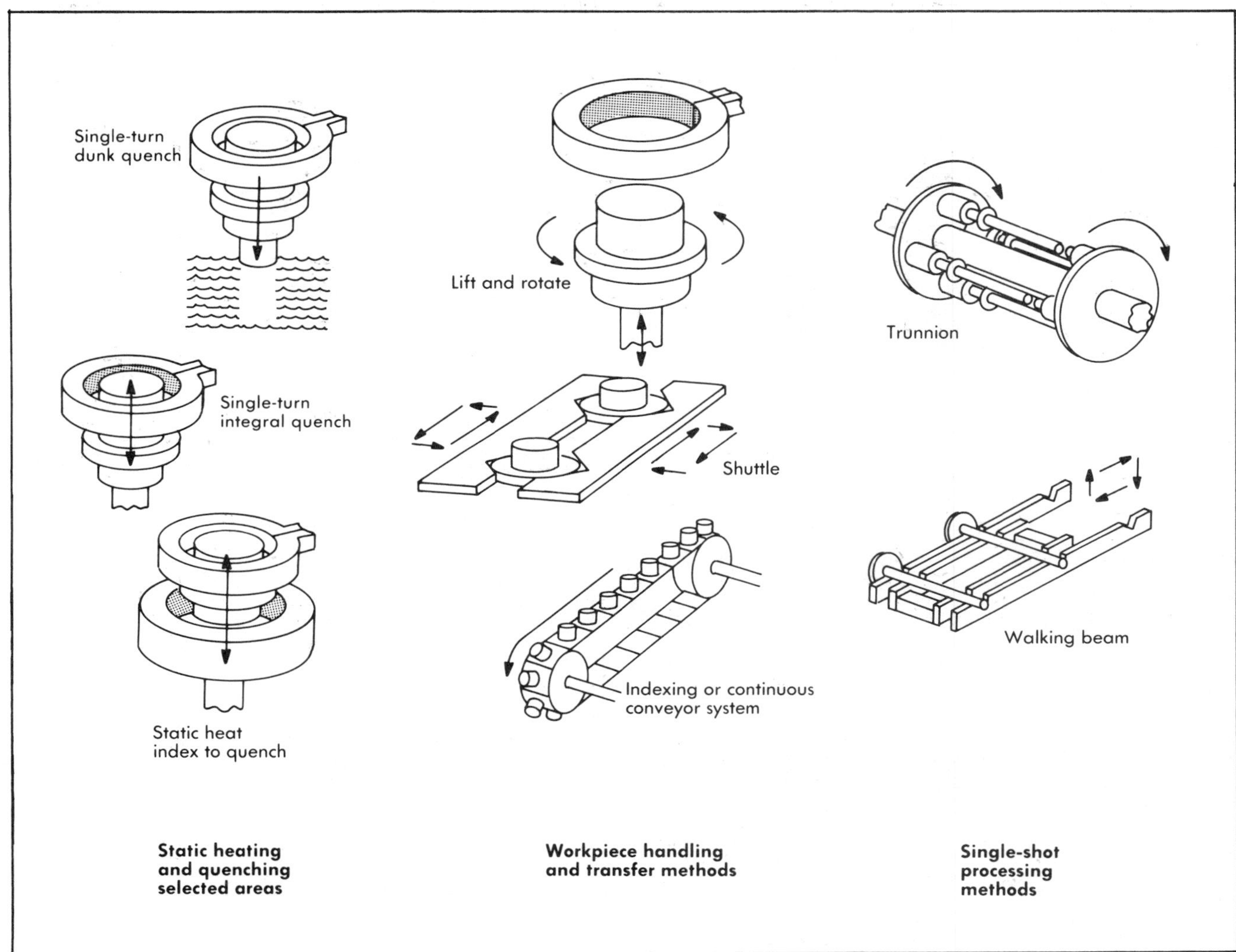

Fig. 13-11 Different methods for processing, handling, and transferring workpieces.

HIGH-FREQUENCY RESISTANCE HARDENING

High-frequency (400 kHz) resistance heating is used to selectively harden specified areas on the surfaces of workpieces, rather than entire surfaces. In this way, only those areas most subject to wear are hardened, thus considerably reducing energy requirements and distortion of the workpieces. The fast heating cycles make this process especially suitable for high-production applications.

PROCESS DETAILS

An advantage of resistance heating with high-frequency current is that a closed loop of current is not required, as is the case with high-frequency induction heating, discussed previously in this chapter. A line or stripe of almost any reasonable shape can be heated between two points. The basic principle is illustrated in Fig. 13-12 where a block of hardenable steel is

HIGH-FREQUENCY RESISTANCE HARDENING

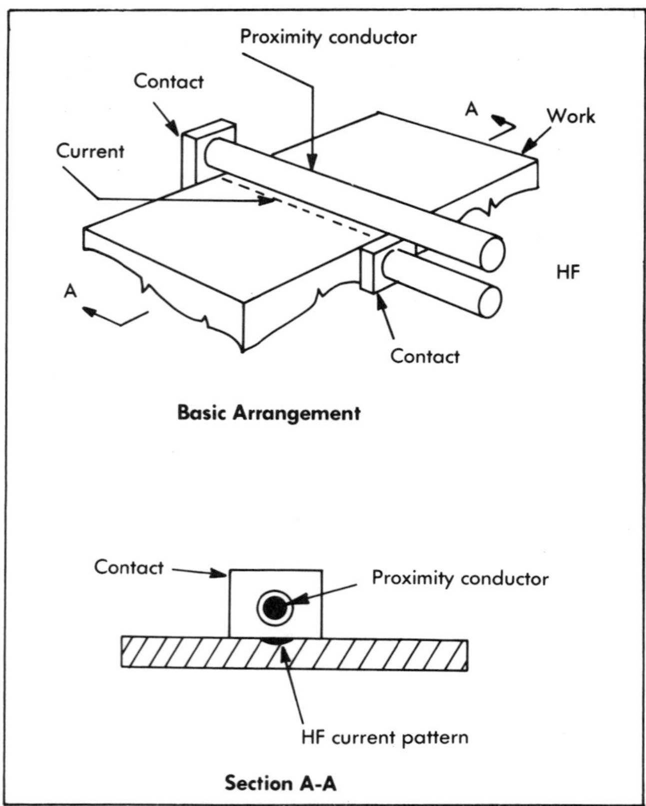

Fig. 13-12 Basic principle of high-frequency resistance hardening. (*Thermatool Corp.*)

arranged with a pair of contacts on its outer edges. A water-cooled proximity conductor is placed close to the surface to be heated and connected to the contacts. The source of 400 kHz power for heating is typically 50-300 kW, depending upon the area to be hardened.

When 400 kHz current is applied, a shallow line of heat is rapidly generated on the surface of the workpiece, immediately underneath the proximity conductor. In hardening steel, the metal is heated to about 1600° F (870° C), and the power is then turned off. The high power densities generated by this process heat very rapidly (typically in ½ second or less), thus resulting in hardening through self-quenching by the cold steel surrounding the hot stripe.

The depth of the hardness varies from 0.025 to 0.035″ (0.64 to 0.89 mm) in steel, depending upon the time of heating and the power level. Also, two, three, or even four stripes can be hardened simultaneously from a single power source.

The contacts used for high-frequency resistance heating are normally very small, about ⅜″ (9.5 mm) square, and are water cooled. Contact pressure is light, and there is no pitting or marking of the work surfaces under the contacts, although the surfaces must be smooth to begin with.

TYPICAL APPLICATIONS

For many applications, hardened lines are produced on the bores of cylindrical parts (see Fig. 13-13). Terminated hardened lines can be produced on both internal and external surfaces by placing the contacts at desired locations on the surfaces. The hardened lines produced extend from one contact to the other (see Fig. 13-14). Skewed (angled) and curved lines are hardened in a similar fashion.

Semicircular, finite-length lines are often produced on the end surfaces of cylindrical parts, leaving sections unhardened where the contacts rested (see Fig. 13-15). Two or more such hardened lines can be made sequentially or, under certain conditions, simultaneously. If the entire end surfaces require hardening, the sequential lines would be overlapped. The overlapped areas would have a slightly reduced hardness. The second operation tempers portions of the first hardened line in the overlap area, reducing the hardness by one to five points on the Rockwell C scale, depending upon the width of the line.

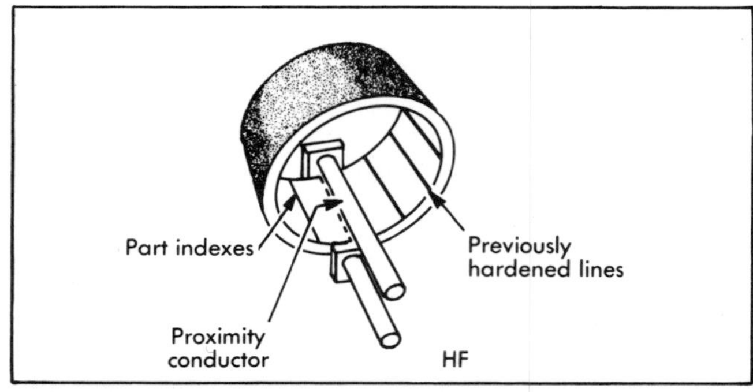

Fig. 13-13 Setup for high-frequency induction hardening of bores. (*Thermatool Corp.*)

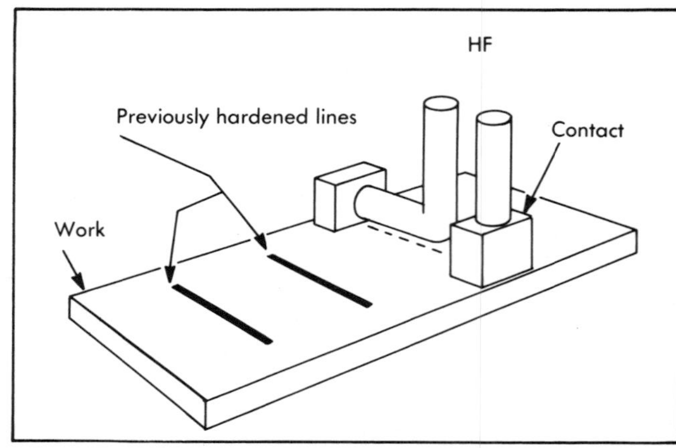

Fig. 13-14 Hardened surface lines extend from one contact to the other. (*Thermatool Corp.*)

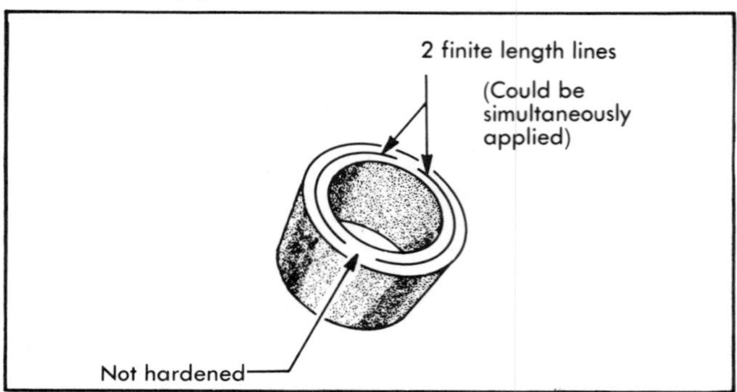

Fig. 13-15 Semicircular lines can be produced on end surfaces of cylindrical parts. (*Thermatool Corp.*)

HIGH-FREQUENCY RESISTANCE HARDENING

Interrupted (pulsed) hardened lines can be produced in a single operation by several special techniques. For some special applications, line hardening is done continuously by moving the workpieces through a contact system. In this way, continuous hardened lines of any desired length are produced. A typical application is hardening continuous lines on steel coil stock.

Hardened lines are sometimes produced on parts previously hardened by other methods. As an example, a spring made from AISI 1075 steel is first hardened and tempered conventionally to provide a hardness of R_C47. Two hardened lines, 0.17" (4.3 mm) wide x 0.032" (0.81 mm) deep and having a hardness of R_C60, are then added by high-frequency resistance heating to provide a wear-resistant area at the contact points. Because of the fast heating and self-quenching inherent in the high-frequency process, there is no significant loss of hardness in the heat-affected transition zone.

Where lines of reduced hardness are desirable, back-tempering can be achieved with the same equipment and without moving the contacts or proximity conductors. Back-tempering is accomplished by simply applying a second pulse of high-frequency current at a much lower power level than the original hardening pulse. The second, lower-level heating reduces the hardness of the line.

Some recent applications of high-frequency resistance hardening include ring grooves in diesel engine pistons, blades for hedge clippers, and the teeth on wrench jaws. The hardening of constant-velocity universal joints for front-wheel-drive transmissions is another major application.

MATERIALS HARDENED

Many different types of steel can be hardened by high-frequency resistance heating. Stripes having a hardness of R_C43 are produced in this way on the wearing faces of levers made from AISI 1117 low-carbon steel. The hardness of the parent metal in these parts is R_C26 as the result of cold working.

Stripes, 0.19" (4.8 mm) wide x 0.033" (8.4 mm) deep, are hardened to R_C56 on parts made from AISI 1045 steel. The high-frequency resistance hardening process produces a predominately martensitic microstructure with some partially dissolved, but mainly unaffected, grain-boundary ferrite.

A wide variety of tool steels are being hardened by this process. In one application, stripes 2" (51 mm) long x 0.57" (14.5 mm) wide x 0.038" (0.97 mm) deep are being hardened to R_C62 on parts made from AISI 01 tool steel.

Cast irons and similar metals are also hardened with the high-frequency process. In hardening gray cast iron, the pearlitic matrix is replaced by a fine martensitic structure having a hardness of R_C62.

Many parts made by the powder metallurgy (PM) process are also hardenable by high-frequency resistance heating. Parts made from iron powder with 2% copper and 0.8% carbon develop a fine martensitic matrix when hardened in this way. Parts made from iron powder with 2% nickel and 0.8% carbon also develop a martensitic matrix, but have patches of retained austenite associated with the nickel-rich areas.

OPERATING PARAMETERS

Power requirements for high-frequency resistance hardening depend upon the length, width, and number of lines to be hardened, and generally range from 50 to 300 kW. Typically, 50 kW of 400 kHz current are required to produce a 4" (102 mm) long x 1/4" (6.4 mm) wide hardened line in steel in 1/3 second. Depth of hardening is shallow, generally being in the 0.025 to 0.035" (0.64 to 0.89 mm) range.

Lines to 42" (1067 mm) long can be hardened and typically require 50-100 kW/in.^{2} (7.8-15.5 kW/cm^2), depending upon line size and heating time. Typical line widths vary from 1/8 to 5/8" (3.2 to 15.9 mm).

FLAME HARDENING

Flame hardening of steels and cast irons is a practical method of developing general or local surface hardness to increase resistance to wear under abrasive conditions and to increase resistance to surface breakdown under concentrated unit loading. Basically, the hardening of structural carbon and alloy steels, tool steels, and cast irons is accomplished by rapid heating of surface areas followed by suitable quenching. Flame-heating equipment may be a single torch with a specially designed head or an elaborate apparatus that automatically indexes, heats, and quenches parts.

Flame hardening has become a standard method of surface hardening as a result of the development of equipment and techniques that have refined the process. Large parts such as gears and machine-tool ways, with sizes or shapes that would make furnace heat treatment impractical, are easily flame hardened. With improvements in gas-mixing equipment, infrared temperature measurement and control, and burner design, flame hardening has been accepted as a reliable heat treating process that is adaptable to many varied applications.

APPLICATIONS

This method of surface hardening is suitable for small, odd-lot quantities, as well as for medium-to-high production requirements. Its initial cost is low, it has flexibility for complex shapes, and it can be adapted to automatic control. Flame hardening competes favorably with the other surface hardening methods for most applications, and is generally preferable for the surface treatment of massive parts that would be beyond the capacity of induction hardening equipment. The original equipment cost and setup costs are also lower for flame hardening than for the other methods, and the price differential increases with part size.

Material intended for flame hardening may or may not have had previous heat treatment. As a rule, parts requiring high-strength applications are first heat treated conventionally to obtain the desired core properties. The surface areas are then heated as rapidly as possible to the desired temperature and are quenched, either by immersion or by a spray quench trailing the flame head or burner. The hardness range when flame hardening carbon steels to a depth of 1/8" (3.2 mm) is illustrated in Fig. 13-16.

FUEL GASES

Fuels used for heating are listed in Table 13-3. Either the torch or the part is kept moving slowly to prevent localized overheating and surface burning. The time necessary for

TABLE 13-3
Fuel Gases Used for Flame Hardening

Gas	Flame Temp in 0_2, °F (°C)	Oxygen-Fuel Ratios*		
		Carburizing	Neutral	Oxidizing
Acetylene	5589 (3087)	1.2 to 1	1.3 to 1	1.4 to 1
Natural gas	4600 (2538)	1.6 to 1	1.7 to 1	1.8 to 1
Propane	4579 (2526)	3.5 to 1	4.0 to 1	4.5 to 1
Propylene	5193 (2867)	3.5 to 1	4.0 to 1	4.5 to 1
MAPP gas	5301 (2927)	3.0 to 1	3.5 to 1	4.0 to 1

(*MAPP Products*)

* Usual ratios for single-port tips.

heating varies with the size of workpiece and the method of flame hardening. Case depths to 0.500" (12.70 mm) are possible, depending upon the workpiece material, the fuel, and the method used; but depths of 0.090 to 0.125" (2.29 to 3.18 mm) are more commonly specified. Such cases are deeper than those obtained with other surface hardening methods, which is an important advantage of flame hardening when deep cases are required. A typical hardness curve for flame-hardened, medium-carbon steels is presented in Fig. 13-17.

METHODS

The three methods of flame hardening (see Fig. 13-18) are:

1. Spot hardening. Spot or stationary flame hardening is accomplished by positioning a flame or group of flames over a given area of the part and applying heat until the part's surface is above the critical temperature. The part is then quenched.
2. Spin hardening. In spin hardening, a torch or group of torches is arranged around the periphery of the workpiece. The part is then revolved at approximately 100 sfm (30.5 m/min) while flames are applied. After the surface is heated above the critical point, it is quenched.
3. Progressive hardening. Progressive flame hardening may be applied to either flat surfaces or cylindrical parts. In this method, the torch is passed over the work surface, and the quench is applied immediately after the last flames have passed. The heat input and the motion must be coordinated so that the work surface is above the critical temperature when the quench is applied. The quench most frequently used is water; but soluble oil, air, and polymer quenches are also used, especially for alloy steels having a high susceptibility to cracking.

WORKPIECE DESIGN

Although uniform sections lend themselves most readily to flame hardening, intricate shapes and patterns are best hardened by this method. For such applications, precautions must be taken to prevent hardening of areas in which a hard surface is undesirable.

Whenever flame hardening is being considered for use, parts should be of the best design to take advantage of the method. Potential problem areas, such as holes close to edges and thin sections, should be eliminated from the part design if possible. Whenever possible, all sections should be uniform in thickness.

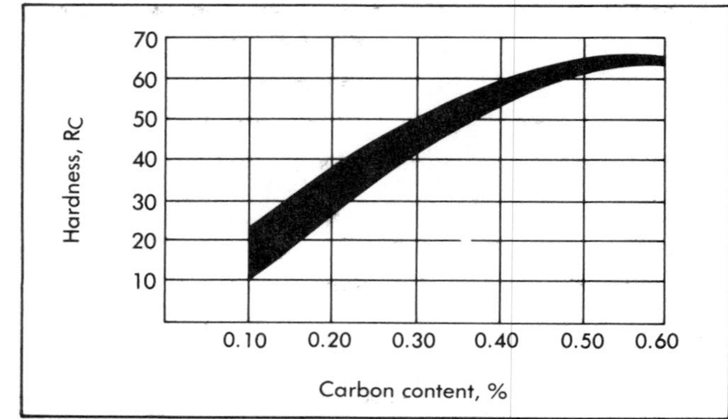

Fig. 13-16 Hardness range when flame hardening carbon steels to a depth of 1/8" (3.2 mm). (*Detroit Flame Hardening Co.*)

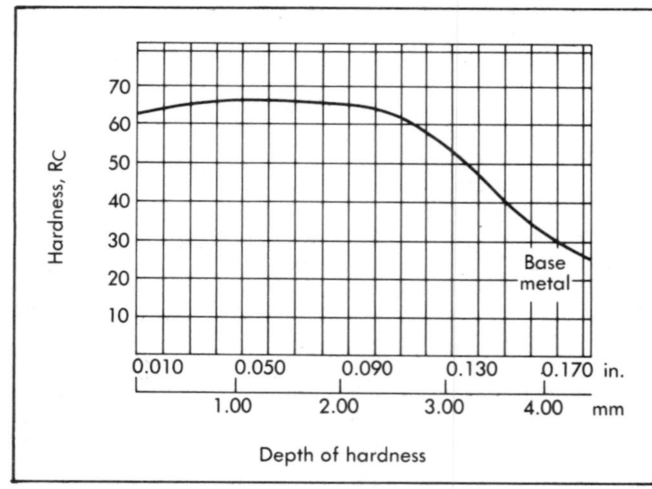

Fig. 13-17 Typical hardness curve for flame-hardened, medium-carbon steels. (*Detroit Flame Hardening Co.*)

Variations in mass cause variations in surface temperature and depth of hardness, resulting in unequal cooling, a nonuniform structure, and possible distortion. Large surfaces containing projections greater than 1/16" (1.6 mm) should be avoided

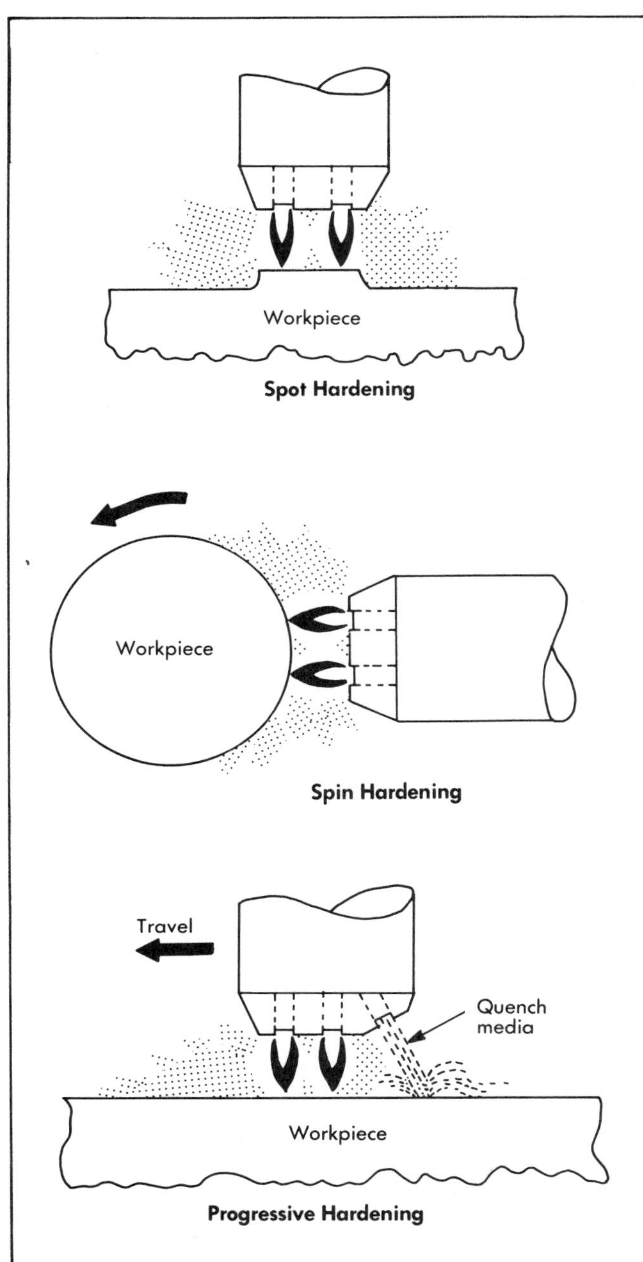

Fig. 13-18 Three basic methods of flame hardening. (*MAPP Products*)

when the progressive hardening method is used; the projections heat faster than the lower surface and will overheat. When hardness in corners or grooves is required, generous radii must be allowed. Consultation between the design and heat treating engineers during the design stage is recommended.

EQUIPMENT USED

The basic flame head used for flame hardening is the oxygen-fueled type. Oxy-fuel-gas flame heads consist of a mixer tube into which two gases are metered and a flame tip or head made of brass or copper. The mixer tube must have adequate capacity to handle the size and number of orifices in the flame head to prevent flashback. Large flame heads must be water cooled.

Flame heads must be used with pressure regulators, valves, flowmeters, and protection devices. Oxy-fuel-gas burners must be cleaned periodically to remove carbon deposits in the orifices, and they must be replaced when erosion or corrosion results in oversized burner openings.

TEMPERATURE CONTROL

Temperature measuring devices, such as thermopiles, are often used instead of conventional devices to control the heat cycle. When thermopiles are used, it must be possible to expose the heated surface to them without interference from the flame itself in order to attain accurate temperature measurements. The lens of the thermopile must also be kept clean to avoid erratic results. Infrared temperature measurement is also being used successfully for flame hardening, especially when the workpieces are rotated during processing.

WORKHOLDING FIXTURES

The reproducibility of the flame heating process is as dependent on the accuracy of the work locating fixture as it is on the flame control. As the flame temperature varies with proximity to the tip, it is essential that the surface of the part be positioned accurately in the flame each time. In a progressive operation, the scanning rate, as well as the position, must be constant and reproducible by the fixture.

SAFETY PRECAUTIONS

All fuel gases are explosive when mixed with air or oxygen. Personnel involved in their use should be properly trained in starting and operating the equipment to avoid accumulations of gas by improper ignition, leaks, or line-purging operations. Information on safe operating procedures is available from the National Fire Protection Association, American Gas Association, American Insurance Association, and the Compressed Gas Association.

ELECTRON-BEAM HARDENING

Recent technological advancements have accelerated the use of electron-beam (EB) heat treating (transformation hardening), a relatively new process for localized surface hardening of components made from carbon and alloy steels.

THE PROCESS

The EB heat treating process uses a concentrated beam of high-velocity electrons as an energy source to selectively heat desired surface areas of ferrous parts. Electrons are accelerated and formed into a directed beam by an electron-beam gun. After exiting the gun, the beam passes through a focus coil, which precisely controls beam density levels (spot size) at the workpiece surface, and then passes through a deflection coil.

The deflection coil allows the beam spot to be moved about on the workpiece surface at speeds to 400,000 ips (10 160 m/s), forming geometric patterns appropriate for the configuration of the zone to be hardened. A typical arrangement of system components is shown in Fig. 13-19.

Heating patterns are applicable to both the static and traveling methods of EB heat treating (see Fig. 13-20). Static methods employ no relative motion between the workpiece and the heating pattern. Traveling patterns generally use workpiece motion to accomplish surface hardening of large areas. Both methods permit an infinite variety of geometric shapes.

The surface of the component being bombarded by the electron beam heats rapidly to a high temperature, confined to the target area of the beam; the rest of the component remains relatively cold. This rapid and precise buildup of heat, which austenitizes the surface and metal immediately under the surface, normally occurs in 0.5 to 3.0 seconds.

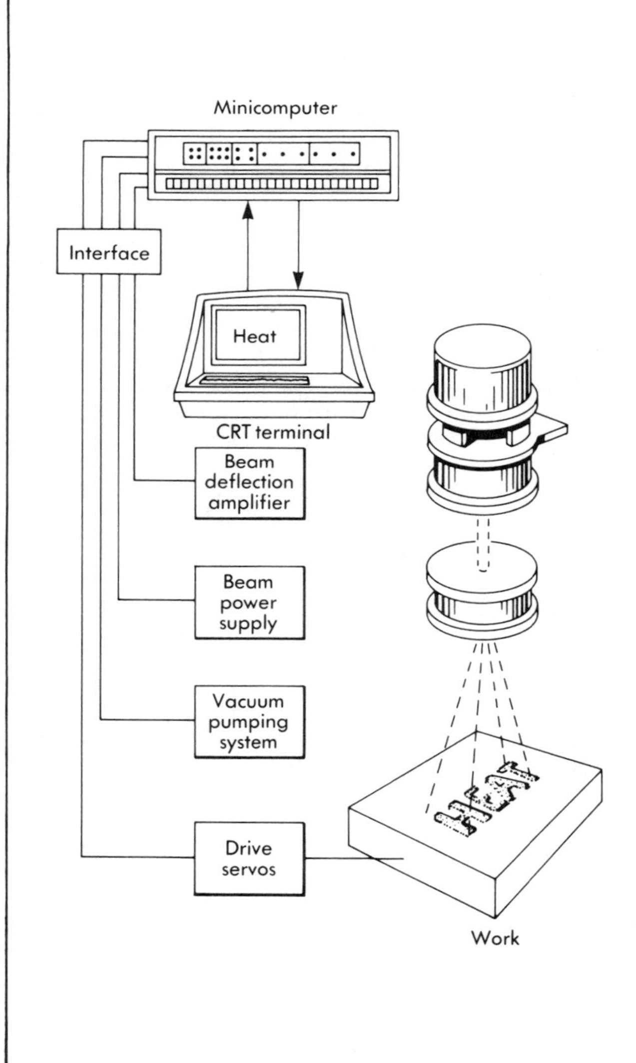

Fig. 13-19 Typical arrangement of system components for the electron-beam (EB) heat treating process. (*Sciaky Bros., Allegheny International*)

When the beam is turned off, heat flows from the high temperature zone to the region that is still cold. The result of this flow is called self-quenching and is more rapid and just as effective as immersion into a liquid.

Self-quenching is controlled by the mass of the workpiece relative to the volume of metal austenitized, hardenability of the metal, initial temperature of the workpiece, and the rate of heating. Self-quenching produces a martensitic structure with beneficial surface compressive stresses. Surface hardness values are typically one or two Rockwell C points higher than achieved with conventional heat treating processes.

The EB heat treating process therefore consists essentially of the following two primary functions:

1. Distribute the energy delivered by the electron beam over the precise geometrical area to be hardened, thus achieving a uniform surface temperature.
2. Control the beam power as a function of temperature magnitude and time to produce the maximum possible surface hardness to the depth desired.

A typical relationship between energy input by an electron beam and the resulting surface temperature, with subsequent case depth approximations, is illustrated in Fig. 13-21.

Computerized control of the process provides correct parameter adjustment of energy input as a function of time and position. Compensation for such nonuniformities as heat transfer characteristics of the workpiece material and the configuration of the workpiece can be accommodated.

VACUUM ENVIRONMENT

To produce an electron beam, a high vacuum of 10^{-5} torr (1.3 x 10^{-3} Pa) is needed in the region where the electrons are emitted and accelerated. This vacuum environment protects the emitter from oxidizing and avoids scattering of the electrons while they are still traveling at a relatively low velocity.

Surfaces to be heat treated, however, can be located in any of three different environments, as follows: high vacuum, partial vacuum, and nonvacuum. These basic types of EB systems and their approximate ranges of operating pressures are shown schematically in Fig. 13-22.

Although a high vacuum provides the cleanest environment, it requires the longest pumping time and therefore results in the lowest production rates. A partial-vacuum level shortens the evacuation and production times while still protecting the workpieces from oxidation during cooling. A nonvacuum EB system can be used for surface hardening when accurate control of energy distribution is not needed. The beam is not electromagnetically deflected with the nonvacuum process, and workpiece motion is often used.

PROCESS ADVANTAGES

Surface hardening with the EB process offers many advantages, including eliminating the need for quenchants. Processing in a vacuum eliminates oxidation and scaling; and there is only minimum, if any, distortion of the workpieces. Also, the process does not disturb the surface finishes on the workpieces, and extensive testing has shown that no cracks are produced.

Because of minimal distortion and undisturbed surface finish, cost savings are often realized by the elimination of subsequent operations, such as straightening or finishing. Additional savings also result from the overall energy efficiency of the EB heat treating process. With the EB process, up to 65% of the electrical input power is converted to usable heating energy, compared to about 15% or less conversion efficiency with laser heating and approximately 20% or more with other heat treating methods.

Unlike laser heat treatment, discussed next in this chapter, no energy-absorbing coatings are required on the workpiece surfaces. Since energy input and self-quenching are rapid and repeatable, hardness areas and depths of hardening are consistent.

LIMITATIONS

Workpiece materials to be heat treated by the EB process must contain sufficient carbon, alloying elements, and hardenability to produce the required hardness. Areas on the workpieces requiring heat treatment must be capable of being exposed to a beam of electrons, with a minimum beam impingement angle of 25°. Workpieces that have been magnetized in previous processing must be demagnetized prior to hardening to prevent deflection of the beam.

An important requirement for successful EB heat treatment is that the workpiece mass must be sufficient to permit self-quenching of the heat-treated areas. A mass of up to eight times that of the volume to be hardened is required around and beneath the heated surfaces.

MACHINES USED

Workpieces can be surface hardened by the EB process on a variety of standard machines, including single and multiple-station types. Selection of a specific machine depends primarily upon production requirements. All machines can be adapted to the use of robots for automatic loading, unloading, and workpiece handling. In many instances, the machines can be linked to centralized computer control systems.

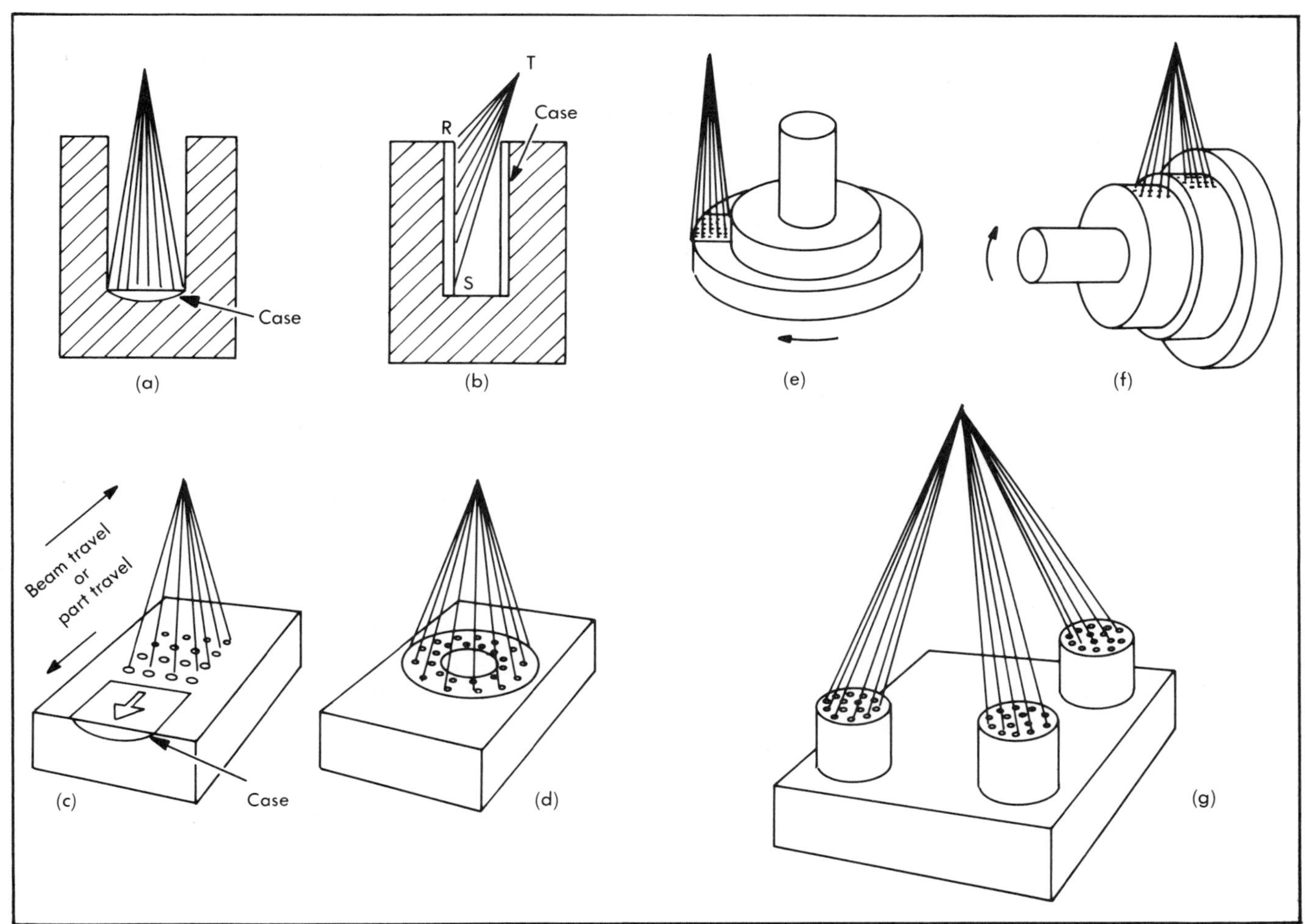

Fig. 13-20 Examples of static and traveling methods of EB heat treating: (a) display static pattern within cavity in workpiece; (b) maintaining angle at 25° minimum; (c) display static pattern with workpiece or pattern moving to heat treat large areas; (d) display static pattern of annular configuration; (e) display static pattern with workpiece rotating; (f) display of more than one pattern with workpiece rotating; (g) display of multiple patterns on one or more workpieces for simultaneous hardening; patterns may be similar or dissimilar in geometric shape.[1]

ELECTRON-BEAM HARDENING

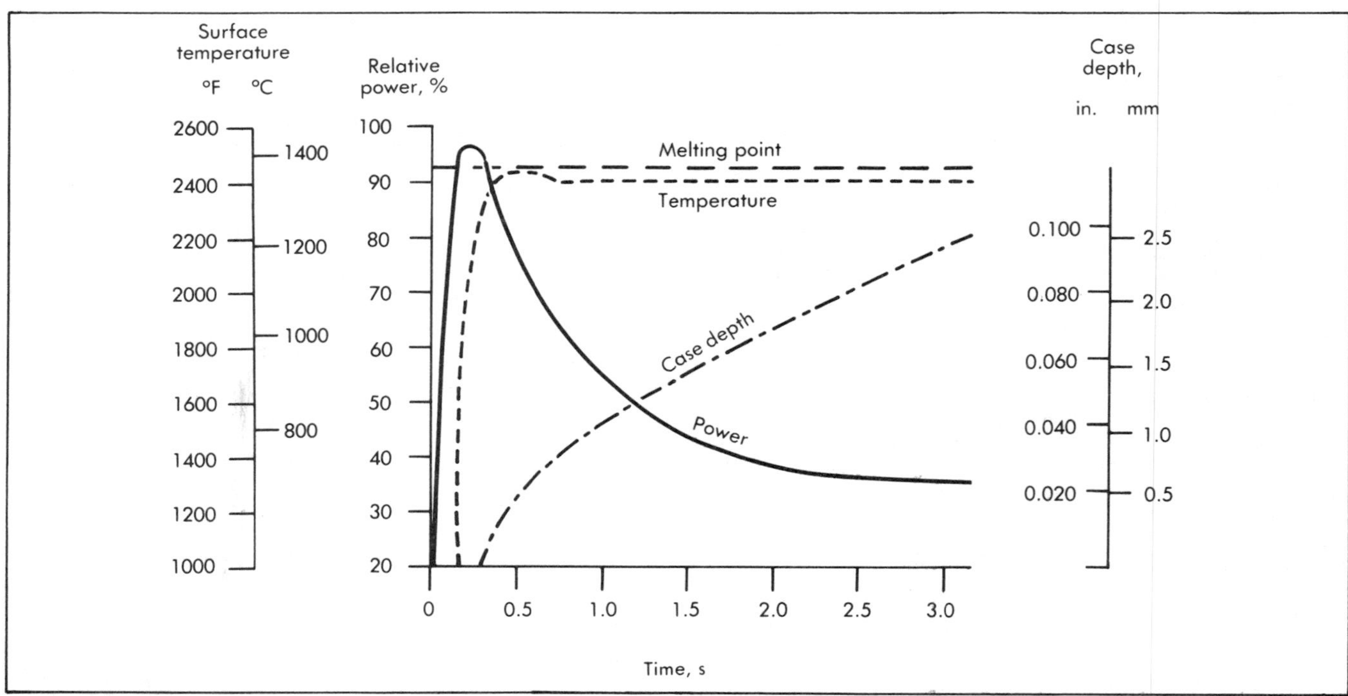

Fig. 13-21 Typical relationship between energy input by an electron beam and resulting surface temperature, with subsequent depth approximations.[2]

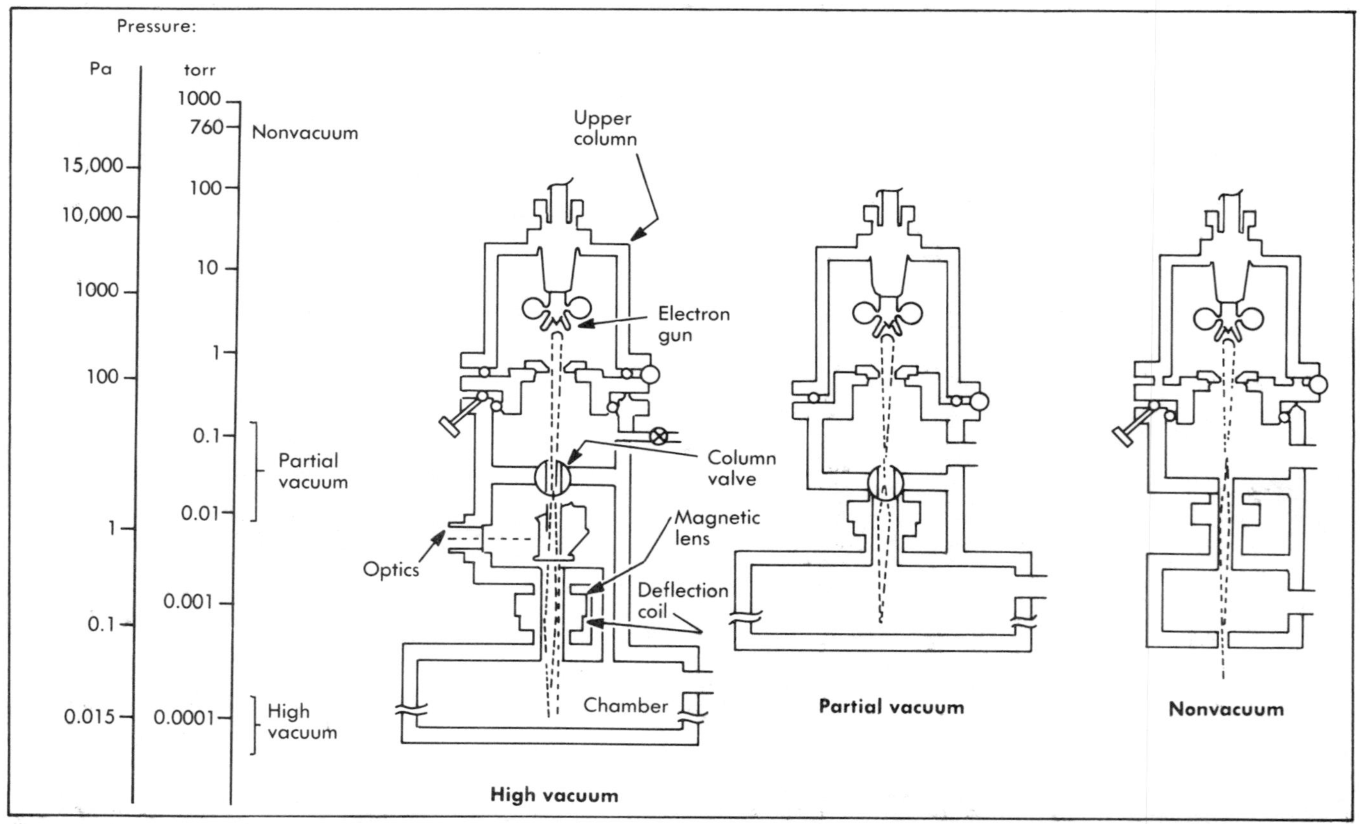

Fig. 13-22 Schematic of the three basic types of EB systems, with their ranges of operating pressures at the workpiece. (*Leybold-Heraeus Vacuum Systems*)

Single-Station Machines

A single-station machine for EB heat treating (see Fig. 13-23) offers moderate production rates, using a small vacuum chamber and high pumping speeds. The machine and chamber size provide flexibility for manually or automatically loading and unloading single parts or batch loads.

A typical production cycle for a single-station, partial-vacuum machine requires 5 seconds for loading and unloading, 20 seconds for the machine to cycle, and 1.5 seconds for heat treatment. This total cycle of 26.5 seconds permits a production rate of 136 parts per hour. A nonvacuum, single-station machine requires less time for the machine to cycle because of the elimination of chamber pumping.

Four-Station Machines

Standard four-station machines for EB heat treatment, such as the one illustrated in Fig. 13-24, are usually dedicated to the production of a single type of part. The four-station, rotary-index table, the small chamber size, and the fast pumping speed make the machine ideal for moderate-to-high production rates. Optional attachments are available for automatic loading/unloading and/or inspection.

Typical cycling on a four-station, partial-vacuum machine requires 10.5 seconds for the machine cycle and 1.5 seconds for heat treatment. No time is required for loading and unloading because it is done during the machine cycle. The total cycle of 12 seconds results in a gross production rate of 300 parts per hour. As with single-station machines, a nonvacuum mode requires less machine cycle time.

Six-Station Machines

Standard six-station machines, similar to the four-station machines just described, are available with seal plates and staged pumping systems. Staged pumping permits prepumping of the tooling cavity, thus eliminating pump downtime. Machine cycle time (for indexing only) is reduced to 1.5 seconds, and heat treatment time remains at 1.5 seconds. Loading/unloading adds no time to the cycle since these functions are performed during the dwell, when heat treatment is taking place. The total cycle of 3 seconds permits a production rate of 1200 parts per hour.

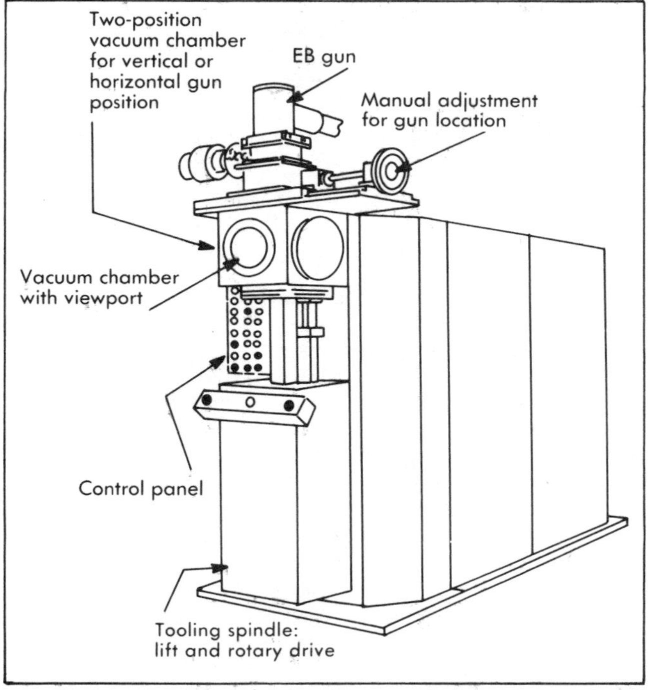

Two-position vacuum chamber for vertical or horizontal gun position
EB gun
Manual adjustment for gun location
Vacuum chamber with viewport
Control panel
Tooling spindle: lift and rotary drive

Fig. 13-23 Single-station EB heat treating machine. (*Sciaky Bros., Allegheny International*)

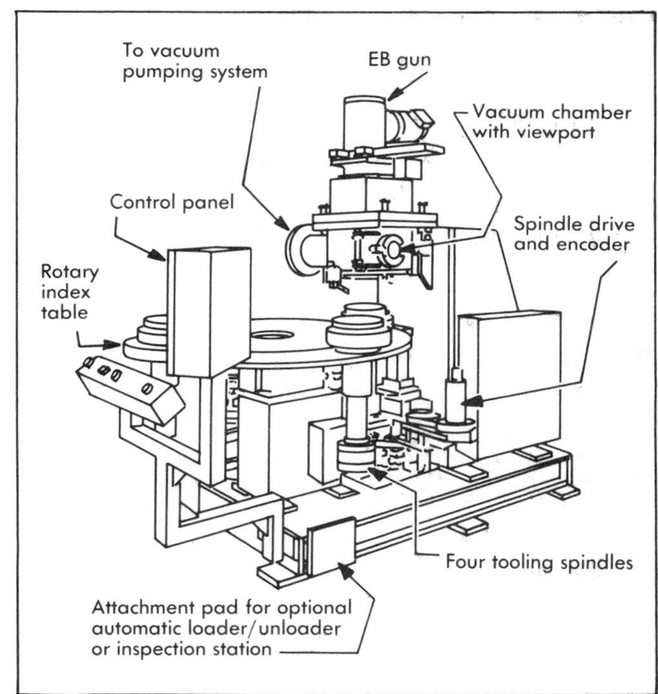

Fig. 13-24 Four-station machine for EB heat treating. (*Sciaky Bros., Allegheny International*)

LASER HARDENING

Surface-transformation hardening with lasers (*laser* is an acronym for "light amplification by stimulated emission of radiation") is a relatively new process made possible by the development of high-power industrial lasers. As in other surface-transformation hardening processes, a relatively thin surface layer is generated in which the material has undergone

LASER HARDENING

transformation to martensite. The process is limited to materials, such as hardenable cast irons and steels, that are capable of undergoing such a transformation. A carbon content of at least 0.30% is generally necessary to attain any significant hardness in steels.

In surface hardening, only the surfaces of the workpieces need to be heated to the austenitizing temperature, and it is desirable that the surface heating be as rapid as possible. This requirement makes the laser an ideal heat source because it can easily produce the energy fluxes (power densities) needed. In fact, lasers are capable of heating surfaces so rapidly that the required, subsequent quenching occurs by fast heat conduction to the still-cold interiors of the workpieces. This so-called self-quenching is a major advantage of laser surface hardening.

THE HARDENING PROCESS

For surface-transformation hardening with lasers, the required power density is much lower than for laser welding and cutting, but the exposure time is longer. The power densities for laser hardening generally range from 15.5 to 1550 W/in.^{2} (2.4 to 240 W/cm^2). Laser output power is spread uniformly over a relatively large area (spot), typically 0.4" (1 cm) square, depending upon the area to be hardened. Laser beam machining is discussed in Volume I, *Machining,* and laser welding in Volume IV, *Quality Control and Assembly,* of this Handbook series.

Hardening is performed by moving the spot over the workpiece surface at a controlled speed. With the proper power density and speed, the desired hardened strip is produced. Output power of the laser can be shaped or raster scanned into other forms to cover broad-area spots, but the principle is the same. When more than one path is required to cover an area, consideration must be given to overlap of the paths. Insufficient overlap produces a thinning of the case. Overlap may also cause a tempered zone in the previous path.

Absorption Coatings

Absorption of laser radiation in cold metals is very low; and because the power densities used in surface hardening are relatively moderate, it is necessary to use energy-absorbing coatings on the workpiece surfaces. Many different substances, including manganese, phosphate, and paints containing graphite, silicon, and carbon black, are used for this purpose. With such coatings, the absorbed power is 60% or more of the incident laser power. Other considerations influencing the choice of a coating include effect on the heated metal and the ease of application and postprocess residue removal.

Depths of Hardening

The ease with which laser beams can be controlled permits the surface hardening of specific areas of workpieces to controlled depths. Case depths obtainable in practical applications depend upon the metal to be hardened. Typically, depths of 0.040 to 0.060" (1.02 to 1.52 mm) can be produced in medium-carbon steels and a depth of 0.080" (2.03 mm) or more in low-alloy steels. The ability to accurately control case depths to less than 0.010" (0.25 mm), when applicable, is a distinct process advantage.

Laser-hardened parts do not generally require an external quench. Heat input is so rapid and localized that the part stays cool enough to quench itself. However, in order to take advantage of self-quenching, the part should be 5-8 times thicker than the case at the point where the laser beam strikes.

Subsequent Tempering

Laser hardening produces a surface layer of untempered martensite; but post-treatment heat treating operations, such as tempering, are seldom required because laser hardening is very localized. For certain applications, however, this layer is sometimes tempered to increase the toughness and reduce the brittleness of the martensite. In some cases, this can be done by using a laser with a reduced power output; however, since tempering requires time at temperature, it is generally best performed using conventional heating techniques. The laser, by its nature, is not well suited for steady-state heating processes, such as tempering and annealing.

Advantages and Limitations

Major advantages of laser hardening include:

- Easy and accurate control of power input to the workpiece.
- Rapid heating and self-quenching, with reduced distortion.
- Ability to case harden specific areas to controlled depths.
- High power flux reduces total energy input.
- Ability to harden normally inaccessible areas on workpiece surfaces.
- Ability to harden specific areas on large and irregularly shaped workpieces because no protective atmosphere is required and the distance from the last optical element can be long.

Possible limitations of laser hardening include:

- High cost of capital equipment.
- Maximum depth of case obtainable is limited.
- Need for absorption coatings, entailing costs for application and removal.

METALLURGICAL CONSIDERATIONS

The rapid heating rate and short time the metal is kept at an elevated temperature result in several metallurgical effects that must be considered. These concerns include high transformation and metal temperatures and limitations on case depths.

Transformation Temperatures

The rapid heating of laser hardening causes an increase in the transformation temperature, resulting in a shallower hardened case than would be expected. This effect becomes more pronounced as the processing speed increases and exposure time decreases. Also, if the metal to be treated has a structure in which the carbon is unevenly distributed, the short time spent above the transformation temperature may not be sufficient to allow the carbon to redistribute uniformly by diffusion. As a result, the austenite has a variable carbon content, and the structure obtained after self-quenching is not uniform.

Nonuniformity of the structure is generally not too important for medium and high-carbon steels and low-alloy steels, but can cause problems when low-carbon steels, high-alloy steels, or nodular cast irons are hardened. If the structure contains a large amount of free (proeutectoid) ferrite, a high volume of alloy carbides, and the grain size is large, the metal may not be hardenable by the laser process. In such cases, the metal can be through hardened by conventional means, tempered, and then surface hardened with a laser. This approach is possible because the carbon is more evenly distributed after tempering.

Metal Temperatures

Because of the unsteady-state heat flow in laser hardening, it is necessary to heat the surfaces of workpieces to temperatures considerably above those necessary for austenitizing to obtain a sufficient temperature rise at the desired case depth. Care must be taken to prevent the surface temperature from reaching the melting point of the metal, particularly if a deep hardened case is desired. In some cases, high surface temperatures may result in local grain growth and possibly an increased amount of retained austenite in the vicinity of the hardened surface.

Small amounts of surface melting are unavoidable in the hardening of cast iron, which has graphite flakes at or just below the surface. Surface melting is not generally a problem when the hardened parts are only subjected to sliding wear, but should be considered if the surfaces are stressed in service. Light grinding or honing operations are often performed after hardening to improve lubrication, and such processing may also remove the effects of melting.

Case Depths

The depth of hardened case that can be obtained in laser hardening depends upon the hardenability of the metal to be processed. Materials having high hardenability can be processed at relatively low speeds because a high rate of self-quenching is not required. In such cases, sufficient time is available for relatively deep heat penetration, and deep cases can be obtained. With steels such as SAE 4140, for example, case depths in excess of 0.10″ (2.5 mm) can be produced.

For steels with lower hardenability, such as carbon steels, the processing speed must be higher to obtain rates that permit self-quenching. Heat penetration is therefore shallower, and the depth of case obtainable is less than for alloy steels.

EQUIPMENT USED

Most metalworking lasers used for heat treating are of the carbon dioxide (CO_2) gas type, but solid-state, yttrium-aluminum-garnet (YAG) types are also employed for shallow cases. The output power of these lasers ranges from 50 W to 15 kW. Lasers with outputs up to 400 W are usually of the YAG type. For power outputs in excess of 400 W, CO_2 lasers are used. Most surface hardening is done with CO_2 lasers in the 500 to 5000 W range.

Power Requirements

Because the cost of metalworking lasers is approximately proportional to their output power (although the cost per watt generally decreases), laser size should be determined on the basis of production rate requirements, the minimum practical spot size, and necessary power densities for the applications contemplated. However, in addition to surface hardening, metalworking lasers can be used for other processes, such as cutting, drilling, welding, and hardfacing. If a laser is to be used for a variety of applications, it may be beneficial to have as much power available as possible.

Output beams from lasers can be projected a considerable distance without much loss of power. This projection allows the laser power to be used for several workstations on a time-sharing basis or simultaneously by splitting the output beam into two or more separate components. Each specific application can then be assigned to a separate workstation with its own optical beam-shaping system, thus making effective use of the laser. Some manufacturers separate the laser head (optical portion) from the power supply. As a result, an appropriate laser head can be assigned to each separate workstation, and all the laser heads supplied with power from a central source.

Optical Systems

The output beam from a laser is shaped into usable form by means of lenses or mirrors. Because the wavelength of the laser radiation is long, any transmission optical elements must be made from special materials that do not readily absorb the laser beam. For high power densities, reflective optics, such as water-cooled copper or molybdenum mirrors, must be used. The laser beam is usually collimated near the output window of the laser and directed through ducts to the workstation. The collimated beam is then shaped into the desired form for a specific application in the workstation.

Optical integrators. For laser hardening, an optical integrator is frequently used. This device (see Fig. 13-25) will shape the collimated beam into a square or rectangular beam with uniform power density in the focal plane of the device. The integrator consists of an array of square or rectangular molybdenum mirrors mounted on a spherical surface. The broad-area spot is formed at the center of the sphere. Such a spot is well suited to the hardening of flat surfaces but can also be used for more complicated surfaces. Figure 13-26 shows an arrangement for surface hardening a flat plate using an optical integrator.

A square or rectangular laser spot can also be created by scanning a focused laser beam over the desired area. Scanning can be done conveniently through the use of a vibrating mirror. If the frequency of scanning is sufficiently high, about 100 Hz, the time-averaged power density is nearly uniform over the scanned area.

An optical integrator offers the advantage of easily changing the size and shape of the working spot by varying the vibration amplitude of the reflecting mirror. However, the effective power density tends to be somewhat higher at the edges of the spot than in the middle because of the reversal of scan directions.

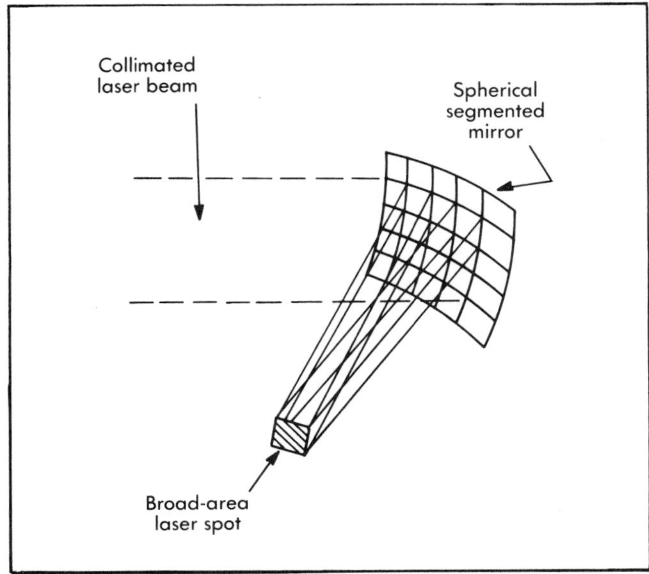

Fig. 13-25 Optical integrator frequently used for laser hardening.

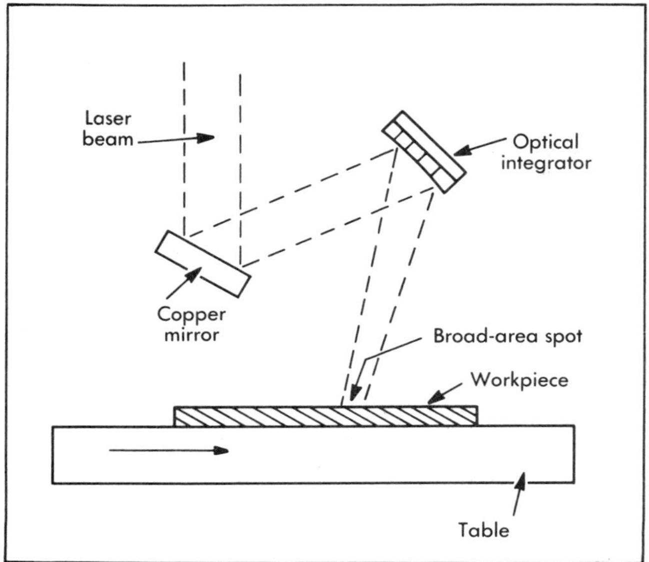

Fig. 13-26 Arrangement for surface hardening a flat plate with an optical integrator.

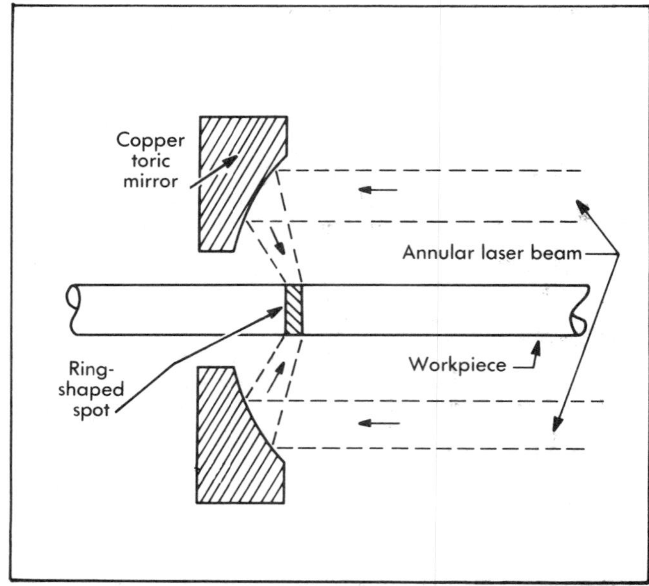

Fig. 13-27 Use of toric mirror for hardening OD of cylindrical part.

Toric mirrors. The hardening of cylindrical workpieces can be achieved by means of so-called toric mirrors. For some applications, a collimated output beam that is annular in shape is employed, and a dish-like mirror is used to reflect the laser beam onto the workpiece surface so that a ring-shaped laser spot is formed around the periphery of the workpiece surface. Figure 13-27 shows the system used for hardening the outer surface of a cylinder, and the use of a toric mirror for hardening the inner wall of a hollow cylinder is illustrated in Fig. 13-28. In both cases, the workpiece is rotated rapidly in order to ensure uniform power density, and the entire surface can be hardened by axial translation of the workpiece.

Toric mirrors are useful for obtaining continuous hardened cases on cylindrical workpieces, but they require relatively high laser output because the laser radiation is distributed over a large area. About 5 kW of output power is required for a 1″ (25 mm) diam workpiece.

Beam splitting. Another laser hardening technique used for nonplanar surfaces is beam splitting. The laser beam is split into two or more separate components and directed to the workpiece surface by means of suitable reflective optical components. A system used for hardening the teeth of large spur gears, shown in Fig. 13-29, splits the laser beam into two equal parts by means of a copper prism. The two beams are then shaped into square beams and directed to the workpiece surface by optical integrators. In this way, both flanks of a gear tooth can be hardened simultaneously. If only a single beam was used, only one flank could be hardened at a time, and thermal distortion and back-tempering of previously hardened material on the other side of the tooth would complicate the process. Some laser heads actually generate two separate beams independently, thereby eliminating the need for a copper prism.

Optic maintenance. All optical components, whether transmissive or reflective types, cause some loss of power. This loss can be kept small, however, if the optical components are properly maintained. The loss involved in a single reflection

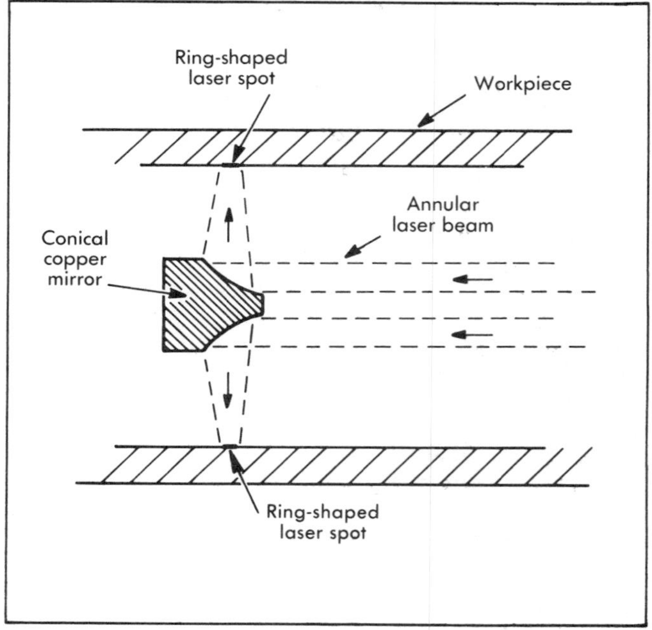

Fig. 13-28 Use of toric mirror for hardening ID of hollow cylinder.

from a copper mirror, for example, can be limited to 1-2% if the mirror surface is kept clean and free from mechanical damage.

The output beams from metalworking lasers are invisible due to their long wavelength. In order to aid in the alignment of the workpiece to the power beam, a visible, low-power alignment laser beam can be used. A helium-neon laser with visible red output, coaxial with the power beam, is frequently used for this purpose.

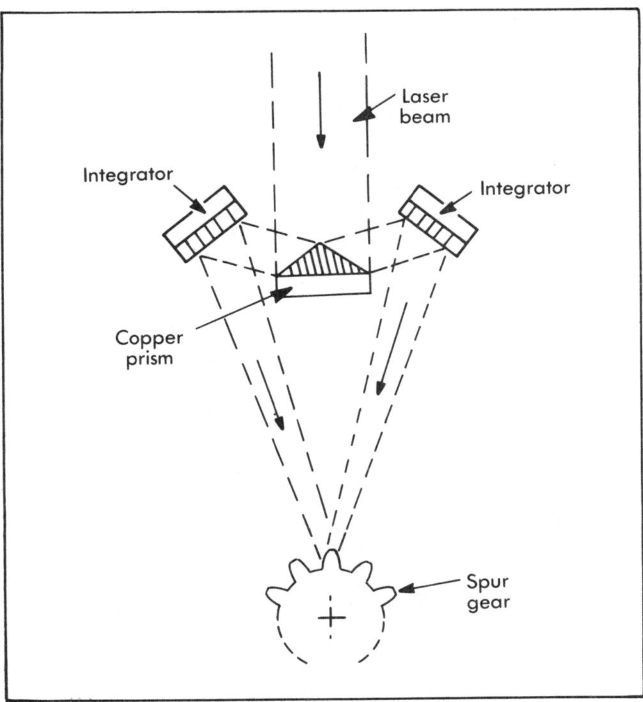

Fig. 13-29 Beam-splitting system for hardening spur gears.

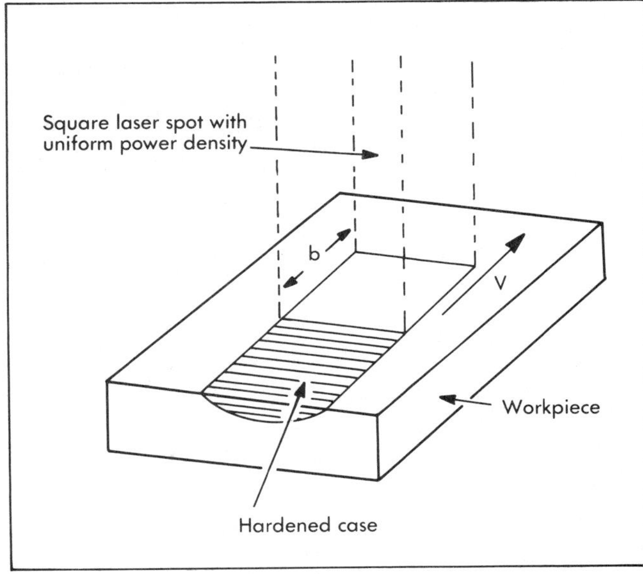

Fig. 13-30 Square laser spot having a length b in the direction of movement. The speed at which the spot moves over the workpiece surface is indicated by v.

OPERATING PARAMETERS

Operating parameters that determine the depth of case in laser hardening are the power density used and the processing speed. The relationship between these two variables is illustrated in Fig. 13-30, showing a square laser spot moving over a workpiece surface. The time that a given point on the workpiece surface is exposed to the laser spot is determined by the following formula:

$$t_d = \frac{b}{v} \tag{5}$$

where:

t_d = dwell (exposure) time, s
b = length of laser spot in the direction of movement, cm or in.
v = speed of spot movement, cm/s or in./s

The longer the dwell time, the deeper the heat penetration. Temperatures generated also depend upon the laser power, with the maximum temperature occuring at the trailing edge of the spot.

Transition temperature can be calculated from the following formula:

$$T_T = T_R + \frac{2Q}{B}\sqrt{Dt_D}\,(ierfc)\left(\frac{C}{2\sqrt{Dt_D}}\right) \tag{6}$$

where:

T_T = transition temperature, °C
T_R = room temperature, °C
Q = power density, W/cm^2
B = thermal conductivity, W/cm/°C
D = thermal diffusivity, cm^2/s
t_D = dwell time, s
$ierfc$ = integrated complimentary error function
C = depth of case, cm

Integrated complimentary error functions (*ierfc*) are tabulated in various reference works, such as *Conduction of Heat in Solids.*[3]

It is also necessary to have data for the thermophysical values B and D for the metal being processed to use formula (2), given earlier in this chapter. For steels, the thermal conductivity, B, is in the range of 0.25 to 0.40 and the value for thermal diffusivity, D, is generally 0.04 to 0.1. These are average values for conditions from room temperature to maximum operating temperature. Table 13-4 presents this data for some common steels. Transition temperatures given in this table are for the temperatures at which transformation to austenite is complete.

The analysis just discussed is only valid for relatively fast processing. For slower processing speeds, a more complicated analysis is necessary. The results of such an analysis for a medium-carbon steel are shown in Fig. 13-31, and for a low-alloy steel in Fig. 13-32. The good hardenability of low-alloy steels allows slow processing without loss of self-quenching if the workpieces are not too small or their sections too thin. Data given in these two illustrations has been corrected for incomplete absorption of the incident laser energy. An absorption coefficient of 0.85% was used, making the real absorbed power density 85% less than the applied power density indicated.

Surface temperatures should be kept well below the melting points of the metals being processed to avoid possible localized melting due to variations in compositions of the metals (for example, in cast irons) or to uneven power distribution. Too close an approach to the melting point can also cause grain growth and a possible increase in retained austenite after quenching.

TABLE 13-4
Thermophysical Values for Various Steels

SAE Steel	Thermal Conductivity, B, W/cm/°C* (W/in./°F)	Thermal Diffusivity, D, cm^2/s* ($in.^2/s$)	Transition Temperature, T_T, °C (°F)
1045	0.36 (0.078)	0.068 (0.011)	790 (1455)
4140	0.31 (0.068)	0.049 (0.008)	780 (1435)
4340	0.28 (0.061)	0.047 (0.007)	750 (1380)
8620	0.28 (0.061)	0.050 (0.008)	840 (1545)

* To convert values to SI units W/m/K and mm^2/s, multiply by 100.

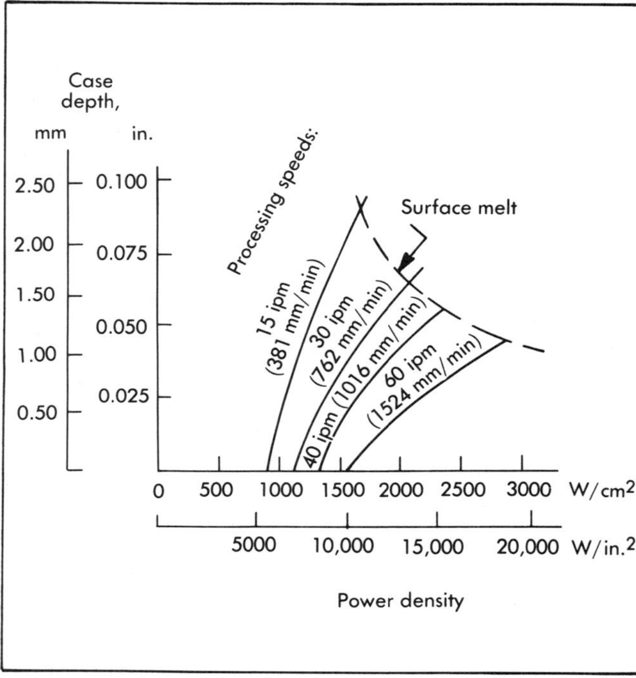

Fig. 13-31 Depth of case vs. power density in laser hardening SAE 1045 steel with a 1/2" (12.7 mm) square laser spot.

SAFETY CONSIDERATIONS

Safety problems associated with metalworking lasers are not severe, but certain precautions must be taken to protect the laser operator. Basically, precaution consists of preventing exposure to the laser beam, accomplished by employing workstations having positive interlocks that prevent the laser beam from being admitted to the enclosed work area unless the access doors are closed.

Thin sheets of Lucite or similar material between the workpieces and the operators are sufficient to absorb reflected radiation if the laser is of the CO_2 type. With YAG lasers, having output radiation of relatively short wavelength, goggles with appropriate lenses to suit the wavelength should be used by the operator for additional protection.

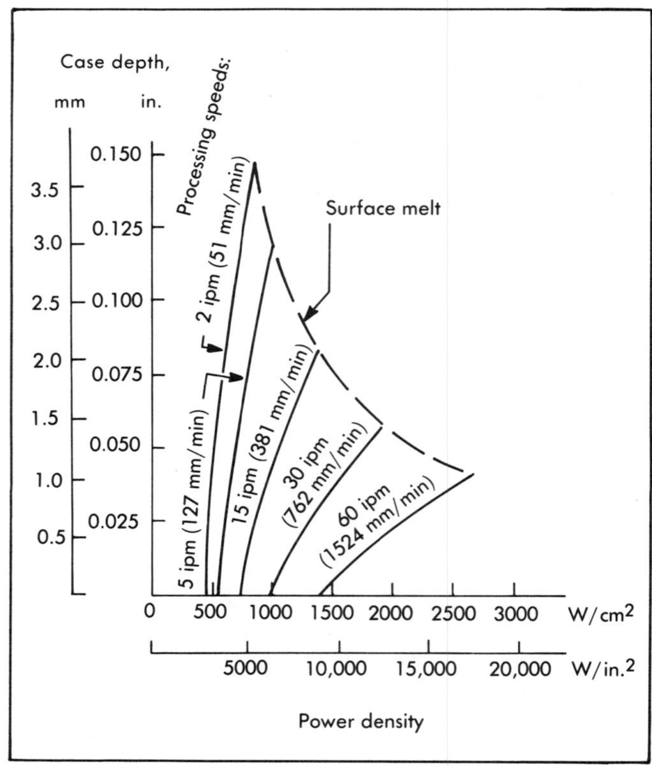

Fig. 13-32 Depth of case vs. power density in laser hardening SAE 4340 steel with a 1/2" (12.7 mm) square laser spot.

Details with respect to safety precautions for laser operation are presented in ANSI Standard Z136.1, "Safe Use of Lasers."

References

1. *Metals Handbook*, 9th ed., vol. 4, "Heat Treating" (Metals Park, OH: American Society for Metals, 1981), p. 519.
2. Terry R. Gonser, *Selective Surface Hardening with Electron Beams*, SME Technical Paper AD81-318 (Dearborn, MI: Society of Manufacturing Engineers, 1981).
3. H. S. Carslaw and J. C. Jaeger, *Conduction of Heat in Solids*, 2nd ed. (New York: Oxford University Press, 1959), p. 485.

Bibliography

Engel, Simon L. *Basics of Laser Heat Treating*. SME Technical Paper MR76-857. Dearborn, MI: Society of Manufacturing Engineers, 1976.

Mazumder, J. "Laser Heat Treatment: The State of the Art." *Journal of Metals* (May 1983), pp. 18-26.

Rudd, Wallace C., and Udall, Humfrey N. "Selective Surface Hardening by High Frequency Resistance." ASM 6th Heat Treating Conference/Workshop. Held in Cincinnati, September 23, 1981.

Stauffer, Robert N. "Hardening to Precise Patterns with Numerical Control." *Manufacturing Engineering* (May 1977), pp. 44-45.

Wick, Charles. "Laser Hardening." *Manufacturing Engineering* (June 1976), pp. 35-37.

HEAT TREATMENT OF OTHER METALS

Many other metals beside steels are heat treated. The heat treatment of carbon and alloy steels is discussed in previous chapters of this section. This chapter provides information on the heat treatment of some of the other, more common metals: cast irons; cast steels; stainless and maraging steels; titanium alloys; heat-resisting, high-strength alloys; and aluminum, magnesium, and copper alloys.

CHAPTER CONTENTS:

CAST IRONS

The majority of commercially produced cast irons, discussed in Chapter 3, "Cast Steels and Irons," do not receive any heat treatment. Instead, they are used as produced in the mold, called the as-cast condition. Most gray irons are generally used as cast, except for intricate castings that are often stress relieved. Some of the high-alloy gray irons do receive some hardening heat treatments when specialized service conditions require increased strength and wear resistance. Other castings are annealed to improve machinability and tool life.

The free carbon in cast irons redissolves into the matrix when heated to a temperature above 1450° F (790° C). The amount of carbon redissolved depends upon the temperature level and time at heat, which also determines the hardness, strength, and wear resistance of the metal.

At one time, ductile irons were usually heat treated to obtain required mechanical properties. Now, however, many of the ductile grades are used as cast. High-strength grades of ductile irons are available with tensile strengths of 100 ksi (690 MPa) or more as cast. Oil quenched and tempered ductile irons have higher yield strengths, but lower tensile strength, elongation, and long-cycle fatigue strength. The wear resistance of as-cast ductile irons having hardnesses to 260 Bhn is better than heat-treated irons of the same hardness. Austempering treatments for ductile irons give tensile strengths from 160 to 260 ksi (1103 to 1793 MPa), depending upon the final hardness.

Compacted graphite cast iron, a recently developed engineering material, is used as cast or receives heat treatment similar to ductile iron. Much research and development work, however, is still necessary to realize the full benefits of heat treating compacted graphite cast iron.

The ferritic, pearlitic, and martensitic grades of malleable irons are heat treated in the manufacturing process. Subsequent heat treatments are usually for surface hardening.

In general, heat treatment of cast irons is performed for one or more of the following reasons:

1. To relieve residual stresses.
2. To reduce hardness, which improves machinability but reduces strength.
3. To produce the necessary microstructure for desired mechanical or physical properties.
4. To harden the entire casting and change the mechanical properties.
5. To increase surface hardness.

STRESS RELIEVING

Stress relieving of cast irons is a relatively low-temperature thermal processing to remove residual stresses. Such stresses may form in castings as the result of a differential cooling rate within the casting due to changes in casting cross sections, or when the casting is restrained within the mold, limiting contraction of the metal as it cools. Machining, shot cleaning, grinding, or other processing may also produce residual stresses. Shot cleaning induces surface compressive stresses that aid fatigue strength in ductile or malleable irons, but induces harmful stresses in gray cast irons. Some gray iron cylinder blocks are not shot cleaned after stress relieving due to induced stresses.

Gray Iron Castings

Gray cast irons may be heated to a temperature between 900 and 1200° F (480 and 650° C) for stress relieving. With temperatures to 1100° F (595° C), the casting hardness will not be affected. At temperatures of 1000 to 1050° F (540 to 565° C), the

Contributors of sections of this chapter are: ***Alan M. Bayer***, *Technical Director, Teledyne Vasco;* ***Charles A. Divine***, *Director of Product Development, AL Tech Specialty Steel Corp.;* ***V. Sam Hill***, *Technical Service and Development, Lake Jackson Research Center, Dow Chemical Co.;* ***J. Howard Mendenhall***, *Technical Associate, Marketing, Olin Brass;* ***Paul J. Mikelonis***, *Vice President, Quality and Technology, General Casting Corp.;* ***Randall A. Oertel***, *Product Metallurgist, AL Tech Specialty Steel Corp.;* ***Harry W. Rosenberg***, *Manager, Technical Services, TIMET;* ***Michael J. Rothman***, *Group Leader, High Temperature Alloys, Cabot Wrought Products Div., Cabot Corp.;* ***Donald J. Tillack***, *Industry Manager, Huntington Alloys, Inc.*

Reviewers of sections of this chapter are: ***L. C. Andrews***, *Stainless Metallurgy, Allegheny Ludlum Steel Corp.;* ***James W. Barr***, *The Aluminum Assn.;* ***Edward J. Berger***, *Associate Metallurgist, Tool and Alloy Research and Development, Carpenter Steel Div., Carpenter Technology Corp.;* ***Emmett Bossing***, *Chief Metallurgist, Teledyne Cast Products;* ***Alan H. Braun***, *Vice President, Engineering, Wellman Dynamics Corp.;*

maximum relief from stresses occurs (75-85% removed), with no decomposition of pearlite. At 1100° F (595° C), a greater percentage of stresses will be relieved, but strength and hardness will be lowered.

If carbide or pearlite-stabilizing alloying elements such as chromium, or other alloying elements such as molybdenum or manganese, are present in the casting material, a higher stress-relieving temperature, possibly 1100 to 1200° F (595 to 650° C), will be required. Table 14-1 compares an unalloyed gray iron and a chromium-nickel-molybdenum alloyed gray iron for percentage of stress relief at various temperatures for two and four hours at temperature.

The rate of heating for stress relieving gray cast irons can be up to 300° F (165° C) per hour. The rate of cooling should be slow, generally under 100° F (55° C) per hour to 600° F (315° C) before removing the castings from the furnace into still air.

TABLE 14-1
Percentage of Stress Relief Obtained by Heating Unalloyed and Alloyed Cast Irons to Various Temperatures

Stress Relieving Temperature, °F (°C)	Time at Temperature, hr	Unalloyed Gray Iron	Alloyed Gray Iron
		Percentage of Stress Relief	
900 (480)	2	56	20
	4	60	22
1000 (540)	2	70	54
	4	75	58
1100 (595)	2	82	74
	4	86	77
1200 (650)	2	96	89
	4	98	93

(*General Casting Corp.*)

Ductile Iron Castings

The practice for stress relieving ductile iron castings is the same as for gray iron castings. Normally, unalloyed ductile irons are held at 1000 to 1050° F (540 to 565° C). For alloy ductile irons, the stress-relief temperature is raised to 1050 to 1150° F (565 to 620° C). Austenitic grades of ductile iron generally require a temperature of 1200 to 1250° F (650 to 675° C), and even higher temperatures, 1550 to 1650° F (845 to 900° C), for increased dimensional stability.

ANNEALING

Annealing softens cast irons by the decomposition of pearlite and/or carbides. Annealing also reduces strength. A number of different annealing treatments are used for gray iron, each of which is accomplished at different temperature ranges.

Low-Temperature Annealing

With a low-temperature anneal, often called a ferritizing anneal, the castings are held at temperatures between 1200 and 1400° F (650 and 760° C). This type of anneal is used for unalloyed or low-alloyed gray iron when no primary iron carbides are present and the intent is to convert pearlite to ferrite and graphite. The castings should be heated to temperature at a rate not exceeding 200° F (110° C) per hour to minimize any large thermal gradients, particularly where thick and thin sections are present in the castings. Furnace cooling at a rate not exceeding 100° F (55° C) per hour to 600° F (315° C) is recommended before air cooling.

Full Annealing

Full annealing, often referred to as medium-temperature annealing, is used to ferritize the matrix for gray irons that may be moderately alloyed with molybdenum, vanadium, or chromium, and that contain no primary carbides. This same treatment is sometimes used for unalloyed gray irons when some small amounts of carbide may be present and complete ferritization is desired. Temperatures for full annealing vary from 1550 to 1650° F (845 to 900° C). The same heating and cooling rates are recommended for low-temperature annealing. Unalloyed gray iron must be cooled slowly in the furnace through the critical range to prevent the formation of pearlite.

High-Temperature Annealing

High-temperature annealing of gray cast irons is used to remove primary carbides from the matrix by decomposing them at a temperature range of 1650 to 1750° F (900 to 955° C). By cooling slowly in the furnace through the critical temperature range, a soft ferritic structure results. When a higher hardness is desired, normalizing (discussed subsequently in this section) is required.

As in the other annealing processes, heating at a rate not exceeding 200° F (110° C) per hour and furnace cooling at a rate not exceeding 100° F (55° C) per hour to 600° F (315° C) before withdrawing to still air is recommended. Figure 14-1 shows the tensile strengths before and after annealing gray iron castings of various diameters.

Annealing Ductile Irons

Ductile irons are annealed in the same manner as gray irons with some slight modifications in the annealing temperatures. For unalloyed ductile irons requiring ferritizing, an annealing temperature of 1250 to 1350° F (675 to 730° C) is used. Medium-temperature annealing is used for low-alloyed and low-silicon ductile irons to accomplish complete ferritizing, employing a

Reviewers, cont.: ***Arthur Cohen***, *Manager, Standards and Safety Engineering, Copper Development Assn.;* ***S. L. Couling***, *Research Scientist, Process Metallurgy Section, Battelle Columbus Laboratories;* ***Donald D. Destito***, *Manager, Technical Services and Product Development, Revere Copper Products, Inc.;* ***Engineering Dept.***, *Cabot Wrought Products Div., Cabot Corp.;* ***Louis P. Flesch***, *Foundry Metallurgist, Electric Steel Castings Co.;* ***Mark D. Gorman***, *Engineer, Stainless Steel Research, Armco Research and Technology;* ***John B. Guernsey***, *Vice President, Technical Services, Jessop Steel Co.;* ***David L. Hawke***, *Materials Specialist, Magnesium Div., AMAX Inc.;* ***Roy R. Heffner***, *Plant Metallurgist, Pennsylvania Steel Foundry & Machine Co.;* ***Raymond M. Hemphill***, *Supervisor, Tool and Alloy Research and Development, Carpenter Steel Div., Carpenter Technology Corp.;* ***V. Sam Hill***, *Technical Service and Development, Lake Jackson Research Center, Dow Chemical Co.;* ***James G. Hoag***, *Waukesha Div., Abex Corp.;* ***Lyle R. Jenkins***, *Technical Director, Ductile Iron Society;* ***John F. Kane***, *Senior Information Scientist, Selective Information Services, Alcoa Laboratories, The Aluminum Company of America;*

CAST IRONS

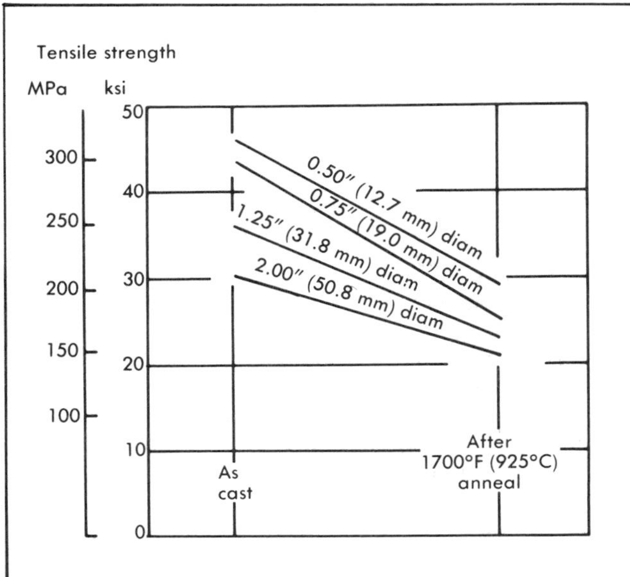

Fig. 14-1 Tensile strengths of gray iron castings before and after annealing. (*General Casting Corp.*)

temperature of 1600 to 1650° F (870 to 900° C), followed by slow cooling through the temperature range of 1250 to 1350° F (675 to 730° C).

For ductile iron that may have primary carbides, a temperature of 1650 to 1750° F (900 to 955° C) is used. By furnace cooling from the upper temperature range to 1300° F (705° C) and holding for two hours, a more complete ferritizing can be accomplished. After holding at this temperature, cooling to 600° F (315° C), with a cooling rate of 100° F (55° C) per hour, is recommended before withdrawing to still air. An alternative is slow cooling through the critical range of 1300 to 1400° F (705 to 760° C) at a rate of 50° F (28° C) or less per hour.

Annealing Malleable Irons

Annealing of malleable irons is used in producing both the ferritic and pearlitic grades. Ferritic malleable iron is produced using a three-stage annealing. The first stage is heating the castings to a temperature of 1650 to 1780° F (900 to 970° C), which causes nucleation of the temper carbon. The castings are then held at this temperature until all carbides are eliminated. After this stage, the castings are cooled rapidly to a temperature of about 1365° F (740° C). The third stage consists of cooling the castings slowly at a rate of 30° F (15° C) per hour, resulting in a totally ferritic matrix.

Pearlitic malleables undergo varying combinations of thermal processing to obtain a pearlitic matrix, but all grades start with a first-stage anneal at a temperature of 1650 to 1780° F (900 to 970° C). Air cooling or liquid quenching is used after the first-stage annealing, followed by tempering to bring the castings to the desired hardness range. Tempering temperatures vary from 1150 to 1320° F (620 to 715° C) for times of two to eight hours.

NORMALIZING

Gray irons are normalized to increase tensile strength and hardness. The castings are heated to about 100° F (55° C) over the critical temperature range, usually 1550 to 1650° F (845 to 900° C), and held long enough to ensure a uniform temperature throughout the casting section. The normalizing or cooling action can be in still air for thin-section castings, providing a uniform cooling of the entire casting. Heavy-section castings, however, will require faster cooling through the use of a blower that can distribute cool air quickly and evenly over the entire casting.

Castings with drastic changes in section size may not be suitable for normalizing treatment because uneven cooling can result in nonuniform hardness, as well as induce high stresses. There may be a need to stress relieve these castings after normalizing. Gray irons that are unalloyed will not develop increased strength and hardness when normalized in still air. However, gray irons with moderate amounts of alloys, such as chromium, molybdenum, and/or nickel, will have their Brinell hardness increased 10-30 points and their tensile strength increased 5-6 ksi (34-41 MPa) when cooled in still air from 1700° F (925° C).

Ductile irons are usually normalized in combination with tempering to provide the desired microstructure and mechanical properties. Normalizing is performed from temperatures of 1600 to 1725° F (870 to 940° C), and tempering is done at temperatures varying between 900 and 1200° F (480 and 650° C). Normally, the higher the tempering temperature, the lower the hardness and the lower the tensile strength. Tempering also acts as a stress-relief treatment.

HARDENING

Hardening of cast irons is accomplished by quenching the castings in a liquid medium that can be oil, water, brine, or a water-soluble polymer, after they have been held at a high enough temperature to austenitize the structure. The higher the temperature of austenitizing, the higher the hardness after quenching, but the greater the tendency to crack. After quenching, tempering is required to reduce residual stresses, reduce the amount of retained austenite, and lessen the possibility of cracking. Tempering temperatures range from

Reviewers, cont.: ***Robert L. Kane***, *President, Titanium Industries;* ***Ted Kosa***, *Supervisor, Stainless Alloy Research and Development, Carpenter Steel Div., Carpenter Technology Corp.;* ***Clarence Lorenz***, *Manager, Customer Service Engineering, Brush Wellman Inc.;* ***J. Howard Mendenhall***, *Technical Associate, Marketing, Olin Brass;* ***Glen Morgan***, *Titanium Alloy Operations, Howmet Turbine Components Corp.;* ***Daniel Rapoport***, *Supervisor, Materials and Process Control, Technical Center, Howmet Turbine Components Corp.;* ***William E. Royer***, *Senior Product Metallurgist, Specialty Metals Div., Crucible Inc.;* ***Stan R. Seagle***, *Vice President, Research and Technical Development, RMI Titanium;* ***William P. Shulhof***, *Technical Director, Central Foundry Div., General Motors Corp.;* ***Richard V. Stumph***, *Director of Engineering and Reliability, Central Foundry Div., General Motors Corp.;* ***Joseph S. Viland***, *Vice President, Magnesium Div., AMAX Inc.;* ***James D. Voss***, *Vice President, Engineering and Technical Services, Hamilton Foundry Div., Hamilton Allied Corp.*

450° F (230° C) for maximum hardness to 1050° F (565° C) for maximum machinability.

While normalizing or air quenching produces a pearlitic microstructure in cast irons, liquid quenching should produce a martensitic microstructure. Alloy cast irons, both gray and ductile, require higher tempering temperatures than unalloyed irons for obtaining equivalent hardnesses. It is recommended that time-temperature transformation (TTT) diagrams be used as a means to predict the microstructure and hardness that can be obtained for cast irons having a given chemical composition. Such diagrams define the time and temperature before transformation will occur. Cast irons of differing chemical composition each have their own individual TTT diagram.

Effect of Temperature

Varying the austenitizing temperatures affects the hardness of both unalloyed and alloyed gray irons. Low hardnesses are generally the result of failure to heat above the critical temperature, usually 1450 to 1500° F (790 to 815° C). Subsequent tempering increases the strength and toughness, but reduces the hardness. The higher the tempering temperature, the greater the decrease in hardness. The presence of alloys in the material, however, reduces the decrease in hardness. There are specialized hardening treatments for attaining specific microstructures and properties, the most common of which are austempering and martempering.

Austempering

In austempering, the matrix attained is retained austenite and bainite. The processing cycle consists of heating to a temperature between 1500 and 1650° F (815 and 900° C), and holding for ½ hour per inch (25 mm) of casting section. The casting is then quenched in molten salt or oil baths and maintained at a temperature of 450 to 800° F (230 to 425° C). The castings are generally held at the quenching temperature for 2 to 3 hours before cooling to room temperature. Higher hardnesses are attained at the lower end of the quench temperature range.

Alloyed cast irons require longer holding times in the quench bath because alloying elements increase the transformation time. As the section size of the casting increases, it becomes more difficult to get a fast enough quench to prevent austenite transformation to pearlite and to produce a uniform structure.

Martempering

Martempering develops a matrix of martensite. The thermal cycling is similar to that for austempering except that quenching is performed at a temperature of 400 to 500° F (205 to 260° C) in molten salt or oil quenchant. A tempering treatment is given to the castings after quenching. For tempering, the castings are then cooled to 400° F (205° C) and held at this temperature for two hours. When molten-salt baths are used, the castings must be washed to remove any adhering salt.

Surface (Case) Hardening

Surface hardening is accomplished either by flame or induction methods, as well as by other techniques, discussed in Chapter 13, "Selective Surface Hardening." With gray irons, the presence of alloying elements enhances the ease of hardening and provides a greater depth of the formed martensite. With all cast irons, a high percentage of pearlite is desirable to allow quick response to the hardening treatments and maximum hardness without excessive case depths. For gray irons, hardnesses to 550 Brinell (about R_C 54) are attainable. For pearlitic ductile irons and pearlitic malleable irons, hardnesses to about 690 Brinell (R_C 62) have been obtained.

HEAT TREATING PRACTICES

To maintain good heat treating operations, it is important to know the limitations of the furnaces used with relation to the loads they are called upon to process, and casting loads must be restricted to the limitations of the furnaces. Heat treat cycles must be determined by the type and amount of castings to be processed, how they are located in the furnace, and the exact time and temperature of the cycle required.

Temperature Control

In thermal treatments, the required temperature refers to the temperature of the castings and not the furnace temperature. Thus, it is necessary to know the temperature of the castings during the heat treating cycle and not rely on the furnace thermocouple for monitoring temperatures. Thermocouples are frequently placed within the castings or throughout the casting load to record the temperature of the castings. If this is done and related to the furnace thermocouple, fairly good control of heat treating cycles can be established. Such procedures ensure that castings attain the correct temperatures and cycles during treatment.

In addition, frequent calibration and checking of the furnace temperature and the recording devices should be performed to ensure that the records obtained are correct. Performing calibrations at least once a month is recommended. Graphic records of the heat treatment cycles should be obtained and retained as evidence of performance.

Furnace Conditions

Furnace conditions are also extremely important for good heat treating operations. Tightness of doors and bottom seals should be checked frequently, daily in most cases, to avoid heat loss or cold spots in the furnace. In larger furnaces, it may be found after temperature checks have been made that fans have to be used to circulate the heat to avoid hot and cold spots. The use of excess-air burners can minimize energy costs. Because malleable irons require long heating times at high temperatures, controlled-atmosphere furnaces are often used to prevent surface decarburization, which reduces machinability.

Hardness Checks

Another control measure for heat treating operations consists of taking hardness readings of the castings and, in some cases, making microstructure checks. In using hardness testing, the checks should be made on different castings from several areas of the load. In the case of large castings, several tests should be taken from different areas and thicknesses of the castings. Hardness tests should be made on castings after they have cooled to room temperature. If the indentation is made on a hot casting, an erroneous reading will be obtained.

Quality Control

Most commercial heat treating firms belong to the Metal Treaters Institute (MTI) and adhere to the MTI 200 quality document, a quality control guide to be followed in heat treating operations. It is recommended that foundries using heat treating vendors obtain a copy of this control agreement and make a quality assurance surveillance of heat treaters' capabilities and performance. When using outside heat treaters, it is also important to supply them with as much information about the castings as possible, especially if there is a wide variation in chemical composition.

CAST STEELS

Most steel castings,[1] discussed in Chapter 3, "Cast Steels and Irons," are heat treated for one or more of the following reasons:

1. To improve the as-cast structure and attain the desired mechanical properties.
2. To relieve stresses in the castings.
3. To improve machinability, facilitate processing, and avoid difficulties during finishing of the castings.
4. To produce optimum corrosion resistance of high-alloy steels and austenitic stainless steels.

HEAT TREATING PROCEDURES

This discussion is a general overview of the heat treatment of steel castings. Specific data with respect to heating rate, temperature, time at temperature, and cooling rate for a specific material should be obtained from the material supplier or the casting producer. The properties of some cast steels in various heat-treated conditions are presented in Chapter 3, "Cast Steels and Irons." The heat treatment of steel castings is essentially the same as that for wrought steel products. Heat treating processes and equipment are similar to those used for similar steel components produced by welding, rolling, or forging.

For increased productivity and maximum economy, steel castings are generally placed in a hot furnace, allowed to soak until their centers reach the temperature in the furnace, maintained at this temperature for 15 minutes or more (as required by the specifications), and then furnace cooled, liquid quenched, or air cooled. This type of short heating procedure will not usually cause distortion or cracking except in infrequent cases with some casting designs or highly alloyed steels, such as some tool steels.

Controlled-atmosphere or molten-salt-bath furnaces are used when it is desirable to prevent oxidation scaling or other surface reactions during heat treating. Such furnaces are also used when it is necessary to produce a compositional change on the surfaces of the castings, such as in carburization. Some steels that are investment cast are heat treated in vacuum furnaces.

ANNEALING

The annealing of castings made from carbon and low-alloy steels provides a soft, stress relieved, readily machinable structure. While the strength of annealed castings is low, ductility is high. Many annealed castings are given a final heat treatment for increased strength after machining is completed.

NORMALIZING

Normalizing of steel castings produces a uniform structure having low residual stresses, minimum distortion, and good machinability. Tensile strengths to 95 ksi (655 MPa) can be obtained by normalizing. Higher strengths are possible with high-carbon and low-alloy steels. Some castings are tempered after normalizing.

It is essential that the castings be placed in the furnace so that air can circulate freely around each casting. If the air flow is restricted during cooling, the heat treating operation will be more like annealing. Accelerated cooling by the use of fans or forced-air flow may produce results more like quenching.

HARDENING BY QUENCHING

Quenching of carbon and low-alloy steel castings is always followed by tempering to obtain the optimum combination of strength and toughness. Martempering, isothermal, and austempering methods of quenching are used only infrequently for steel castings.

Water is generally used as the quench medium whenever possible, but high-carbon and deep-hardening steels require oil quenching. Some complexly shaped castings may also require the slower cooling of oil quenching to minimize the possibility of cracking. For safety reasons, polymeric solutions that simulate oil may be used.

Some casting grades of stainless steels, such as CA-15 containing 13% chromium and CA-6NM containing 13% chromium and 4% nickel, are considered air hardening and do not require accelerated cooling by quenching. However, higher hardnesses may be obtained by accelerated cooling of these alloys. Certain low-alloy steels are also air hardened when cast in thin sections.

TEMPERING

The tempering of steel castings following quenching or normalizing will reduce residual stresses and alter the strength and other mechanical properties of the castings. Tempering is accomplished at a temperature below the transformation range and immediately following the hardening cycle. Tempering below 1100° F (593° C) may cause temper embrittlement of certain steels.

STRESS RELIEVING

Since stress relief also occurs during tempering, there is often no need for special stress relieving after tempering. For steel castings that are not tempered, stress relief is obtained by heating the castings to temperatures above 500° F (260° C). There is the possibility, however, of blue brittleness occurring in some steels when they are heated to a temperature between 500 and 1100° F (260 and 593° C). When it is necessary to heat castings in this temperature range, the danger of blue brittleness can be lessened by quenching the castings from the stress-relieving temperature into water.

Stress relief is sometimes required when operations such as straightening, welding, grinding, or induction hardening performed after heat treatment leave residual stresses in the castings. The maximum temperature for such stress relieving is generally limited to 50° F (28° C) below the tempering temperature used in heat treatment.

SOLUTION HEAT TREATMENT

Austenitic stainless steels, age (precipitation)-hardening steels, and wear-resistant, austenitic manganese steels are often solution heat treated to dissolve second phases and to produce a homogeneous structure. For austenitic stainless steels and austenitic manganese steels, this heat treatment produces austenite and dissolves carbides. A quench or other accelerated cooling is required after heating to retain carbon in solution and prevent the precipitation of carbides.

Temperatures for the solution heat treatment of austenitic stainless steels range from 1900 to 2100° F (1038 to 1149° C), depending primarily upon alloy content. Austenitic manganese steels are commonly heated to temperatures between 1850 and 1950° F (1010 and 1066° C) for best toughness properties, although higher temperatures are necessary for high-carbon and alloy steel grades.

For age-hardening steels, solution heat treatment is necessary to develop the required structure prior to aging. Solution heat treatment creates a soft condition that makes the castings easy to machine compared to their hardened condition after aging. The cooling rate from the solution treatment temperature must be rapid enough to develop the desired structure for subsequent precipitation hardening.

AGE (PRECIPITATION) HARDENING

Manganese-copper cast steels are a family of low-alloy cast steels that are age hardened. The usual heat treating practice for these steels is to normalize at 1700° F (927° C) and then age harden for about two hours in the temperature range of 900 to 950° F (482 to 510° C). This procedure produces an increase of 20 ksi (138 MPa) in tensile strength with a minimum decrease in ductility.

Age-hardenable stainless steels for casting include CB-7Cu-1 and CB-7Cu-2. The CB-7Cu alloys can be hardened by heating to 1925° F (1052° C), followed by cooling to room temperature and aging at a temperature of 900 to 1150° F (482 to 621° C), depending upon desired mechanical properties.

Proper heat treatment of steel castings is dictated by the specific material from which the castings are made and/or process specifications. Use of concensus specifications simplifies the heat treating process. However, specialized heat treating processes are used for many steel castings.

STAINLESS AND MARAGING STEELS

Manufacturing practices, types, applications, and machining and fabricating characteristics of stainless and maraging steels are discussed in Chapter 2, "Stainless and Maraging Steels," of this volume. The five main types of stainless steels are austenitic, ferritic, martensitic, duplex (austenitic-ferritic), and precipitation hardening (PH).

AUSTENITIC STAINLESS STEELS

The austenitic stainless steels are not hardenable by heat treatment but can be moderately hardened by cold working. These alloys, however, are often annealed, heat treated to eliminate sensitization, or stress relieved.

Annealing

Annealing of austenitic stainless steels is performed either to restore maximum corrosion resistance or to soften or relax stresses in strained or cold worked materials. Most austenitic stainless steels can be annealed successfully in the 1900 to 2050° F (1040 to 1120° C) temperature range. However, special modifications in the composition of these alloys can require the use of higher temperatures to 2200° F (1205° C) or lower temperatures to 1700° F (925° C). Time at temperature depends upon workpiece thickness. Soaking at temperature for only a few minutes is generally adequate provided the entire section is at the specified temperature and all carbides are dissolved from the grain boundaries. A time at temperature of 30 minutes per inch (25 mm) of thickness is sometimes used as a guideline, but shorter times are often successfully used with adequate metallurgical checking.

Cooling from the annealing temperature should be as rapid as possible to prevent sensitization and precipitation of carbides if maximum corrosion resistance is to be obtained. Water quenching is most often used. When practical considerations of workpiece distortion prohibit the use of water quenching, slower oil or air cooling may be used if the metal composition permits. Such slower cooling is almost always metallurgically acceptable for the low-carbon or stabilized grades and is often acceptable for the higher carbon standard grades for thin workpieces.

Sensitization

Sensitization is the precipitation of chromium carbide particles from the austenitic matrix. It can be caused by welding, improper heat treatment, or service at elevated temperatures. Cold working increases the tendency to sensitize. Precipitated carbides form on the grain boundaries and leave the area adjacent to the boundaries depleted of chromium, thus lowering corrosion resistance in this area. The rate of precipitation, which is time and temperature dependent, occurs over a temperature range of 850 to 1600° F (455 to 870° C), but is most rapid in the temperature range of 1100 to 1300° F (595 to 705° C). For this reason, water quenching is used after annealing whenever possible to minimize the length of time in this temperature range.

Three commonly used methods of eliminating sensitization from austenitic stainless steels or preventing its harmful effects are:

1. The use of stabilized alloys.
2. The use of low-carbon grades.
3. Solution annealing.

Stabilized alloys. These alloys have controlled amounts of titanium or columbium (niobium) added, which combine with carbon to form titanium or columbium carbides. These carbides are more stable and preferentially form chromium carbides. Types 321 and 347 are the most well-known stabilized grades.

To achieve maximum resistance to sensitization when using the stabilized grades, it may be necessary to use a stabilizing anneal after the normal solution anneal is employed.

By heating Type 347 alloy to approximately 1550 to 1650° F (845 to 900° C) for several hours, all the carbon is converted to columbium carbide.

Similarly, with Type 321 grade, carbon can be converted to titanium carbide by heating at 1550 to 1650° F (845 to 900° C) for the same period of time. Subsequent exposure to temperatures in the sensitizing range will not lower corrosion resistance because chromium carbide cannot form.

Low-carbon grades. Another approach to at least minimizing sensitization is to lower the carbon content of the alloy. The lower the carbon content, the lower the amount of chromium carbide that will be formed. Types 304L and 316L are two common low-carbon grades having a maximum carbon content of 0.030%. These grades can be used successfully when the material will be held in the sensitization temperature range for only a short time, such as in arc welding. Types 321 or 347 should be used when material is heated for extended periods of time in the sensitizing range.

Solution annealing. Sensitization can also be eliminated through the use of a solution annealing treatment. By heating well above the sensitizing range, usually at a temperature of 1900 to 2050° F (1040 to 1120° C), and cooling rapidly, all chromium carbides can be taken into solid solution and retained there. The metal is susceptible to resensitization if again heated into the sensitization temperature range.

Stress Relieving

Many of the austenitic stainless steels are heavily cold worked to provide the required strength and hardness ranges needed for the final part. This cold working can impart large and uneven stresses to the material that can result in distortion when machining the material. To minimize this effect, a stress-relieving heat treatment is sometimes used. This treatment consists of heating the material in the temperature range of 600 to 900° F (315 to 480° C) for 1 to 3 hours (sometimes a shorter time), followed by air cooling. Stress relieving in this manner causes some distortion, which often requires some subsequent straightening. The use of temperatures above 850° F (455° C) can cause some sensitization to occur. However, the rate of sensitization in the temperature range of 850 to 900° F (455 to 480° C) is quite slow. Stress relieving for short periods of time in this range, therefore, will normally not reduce corrosion resistance.

Highly Alloyed Austenitic Grades

Highly alloyed, austenitic stainless steels have relatively large amounts of chromium, nickel, and molybdenum to provide excellent corrosion resistance. They are fully austenitic in the solution-annealed condition, but may form sigma or delta ferrite phases under certain conditions of heat treatment or service. These phases can be detrimental to corrosion resistance and mechanical properties. Solution annealing temperatures are held to a narrow range to avoid sigma-phase formation at lower temperatures or delta ferrite at higher temperatures.

FERRITIC ALLOYS

The ferritic family of stainless steels cannot be hardened by heat treatment, except for a partial hardening that can take place in certain modifications of 430 and 434 grades. Therefore, annealing is the only heat treating process commonly used.

Annealing

The purpose of annealing the ferritic grades is usually to relieve stresses developed during cold working or welding and to provide maximum corrosion resistance. Table 14-2 shows recommended annealing treatments for some common ferritic grades. Soaking times at temperature need only be long enough to ensure that the work is heated uniformly throughout.

TABLE 14-2
Recommended Annealing Treatments for Ferritic Stainless Steels*

Grade		Temperature Range,	
UNS	AISI/SAE	°F (°C)	Quench**
S40500	405	1200-1500 (650-815)	Air or water
S40900	409	1400-1650 (760-900)	Air or water
S43000	430	1300-1550 (705-845)	Furnace cool for temp above
S43020	430F	1300-1550 (705-845)	1450° F (790° C); air cool for
S43400	434	1300-1550 (705-845)	lower temp
S44600	446	1400-1650 (760-900)	Water or air

* There are many high-chromium, low interstitial ferritic alloys produced for special applications. The producers of the alloys should be consulted for specific annealing recommendations.

** Cooling rates should be as rapid as possible.

Grain Growth

Ferritic alloys, when heated above 1650° F (900° C), are subject to grain coarsening, which greatly reduces notch toughness as measured by such tests as the Charpy impact test. With most ferritic alloys, this coarse-grained structure cannot be refined by heat treatment because there is no ferrite-to-austenite transformation, as with the martensitic alloys (discussed next in this section).

Because large grain size has such a detrimental effect on notch toughness, it should be avoided. The only methods that reduce the grain size in most ferritic alloys are hot working or cold working, followed by annealing, and such working is not always possible. In Type 446, nitrogen is used as a grain refiner. However, this alloy is still subject to grain growth if heated above 1650° F (900° C).

Embrittlement

An embrittlement phenomenon may occur when the ferritic grades (and martensitic grades to a lesser degree) are held for prolonged periods of time in the temperature range of 750 to 1000° F (400 to 540° C). The maximum rate of embrittlement occurs at a temperature of about 885° F (475° C). Slow cooling through the temperature range of 1000 to 750° F can also cause embrittlement.

Embrittlement is most noticeable in the high-chromium (16% or more) ferritic alloys. When annealing the ferritic grades, air cooling is usually sufficient to avoid embrittlement, except when large sections are involved. Embrittlement can be eliminated by annealing the material using the treatments listed in Table 14-2.

CHAPTER 14

STAINLESS AND MARAGING STEELS

MARTENSITIC STAINLESS STEELS

As a result of their high carbon content, the hardenability of the martensitic stainless steels is high. Drastic quenching is not required.

Hardening Treatments

Hardening treatments for the martensitic stainless steels are listed in Table 14-3. Preheating is frequently used prior to austenitizing these alloys and is almost mandatory for the higher carbon grades (Types 420 and 440 series). Cracking or warpage of parts can occur if preheating is not used. Preheating is accomplished by heating slowly and holding for about one hour at 1400 to 1450° F (760 to 790° C) prior to austenitizing. Extremely large parts must be held at temperature until all portions of the part have reached the preheat temperature. After hardening, martensitic stainless steels must be either stress relieved or tempered, discussed subsequently in this section.

In order to obtain optimum toughness in martensitic stainless steels, the hardening temperature should be selected based upon the subsequent heat treatment. A hardening temperature on the high side of the allowable range should be used if the material will be subsequently stress relieved. Conversely, a hardening temperature on the low side of the range should be used if the metal will be tempered.

TABLE 14-3
Hardening Treatments for Martensitic Stainless Steels

Grade		Austenitizing Temperature,*	Typical Hardness,
UNS	AISI/SAE	° F (° C)	Bhn
S40300	403	1700-1850 (925-1010)	410
S41000	410	1700-1850 (925-1010)	410
S41400	414	1800-1900 (980-1040)	420
S41600	416	1700-1850 (925-1010)	410
S42000	420	1800-1950 (980-1065)	500
S43100	431	1800-1950 (980-1065)	430
S44002	440A	1850-1950 (1010-1065)	560
S44003	440B	1850-1950 (1010-1065)	580
S44004	440C	1850-1950 (1010-1065)	610

(*AL Tech Specialty Steel Corp.*)

* Time at temperature: 15-60 minutes
Air or oil quench.

Quenching. In most cases, air cooling or oil quenching is adequate for martensitic stainless steels. Oil quenching is often chosen because it usually produces a slightly higher hardness than air cooling. Water quenching is never recommended because of the possibility of quench cracks being formed. Oil quenching is preferred for the low-carbon grades (Types 403, 410, and 416), while air cooling may be used for the high-carbon and high-nickel grades (Types 414, 420, 431, and 440 series). For extremely large sections, oil quenching may be required but may cause cracking or distortion, especially with complex shapes. If air cooling cannot achieve the required as-quenched hardness in the center, marquenching may be used successfully.

Bright hardening. Bright hardening techniques may be used successfully on the martensitic stainless steels provided the cooling rate is at least as fast as air cooling. A cooling rate that is too slow can reduce corrosion resistance and ductility, particularly in the high-carbon grades. Since most parts that are bright hardened are machined to, or near, final dimensions, consideration must be given to the small size change that takes place during hardening as a result of the transformation of ferrite to austenite upon heating and austenite to martensite upon cooling. This size change is also affected by the initial condition of the metal and subsequent stress-relieving or tempering operation but will generally be less than 0.001 in./in. (mm/mm). Bright hardening must be done in gas-tight furnaces or retorts with special gases of high purity and dryness.

High-carbon and high-nickel alloys. A complication in hardening the high-carbon grades (Types 420 and 440 series) and the high-nickel alloys (Type 431) is a tendency for austenite retention after quenching. This tendency will be aggravated by using hardening temperatures that are higher than necessary. If retained austenite is present in the as-quenched condition, it may be eliminated by subzero cooling to -100° F (-75° C) and/or by double tempering. These cooling and/or tempering methods are similar to those used for high-speed steels.

When quenching complex parts made from high-carbon grades, such as Type 440C, it may be necessary to prevent them from reaching a temperature below 150° F (65° C) prior to stress relieving or tempering. Quenching to room temperature increases the probability of distortion or cracking of some parts, such as those with sharp angles.

Stress Relieving

Stress relieving, as applied to as-quenched steels, relaxes internal stresses without permitting important microstructural changes to occur that can affect mechanical properties. In general, this process is carried out at temperatures below those used for tempering. Common stress-relief treatments for martensitic stainless steels employ temperatures of about 300 to 700° F (150 to 370° C) for times of 1 to 4 hours.

Tempering

Tempering involves higher temperatures than those used for stress relieving, usually 400-1400° F (205-760° C) for 1 to 4 hours. Tempering treatments, unlike stress relieving, result in marked changes in microstructure and mechanical properties. The effect of tempering temperature on the properties of martensitic stainless steels is similar to that produced in tempering low-alloy and carbon steels. Higher tempering temperatures reduce hardness, tensile and yield strengths, and corrosion resistance, but increase elongation, reduction of area, and impact strength (see Fig. 14-2). A distinction between stress relieving and tempering is commonly not made, with both types of treatments often referred to as tempering.

Retained Austenite

Because the high-carbon, martensitic stainless steels, particularly Type 440C, have rather low M_f temperatures (the temperatures at which transformation of austenite to martensite finishes during cooling), double tempering is recommended to eliminate retained austenite. Subzero cooling to -100° F (-75° C) following the first tempering treatment will ensure maximum transformation. The second treatment will then temper the newly formed martensite.

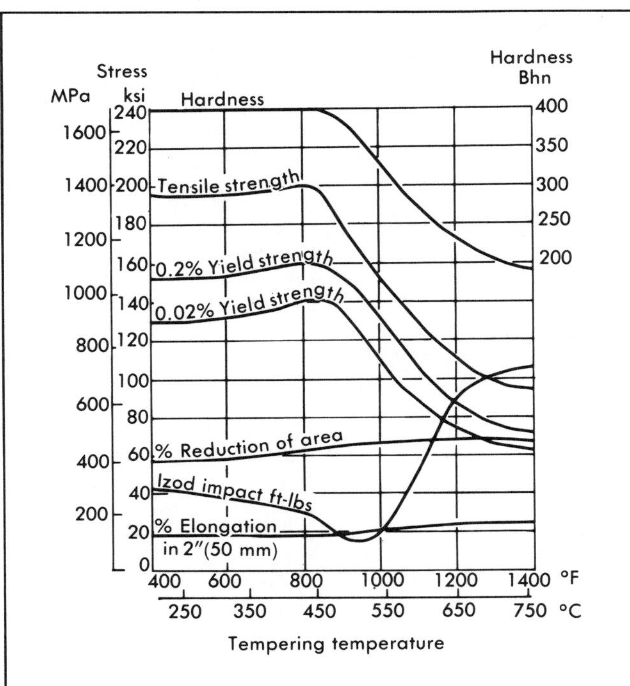

Fig. 14-2 Variations in mechanical properties for different tempering temperatures in heat treating types 403, 410, and 416 martensitic stainless steels. Hardening was done at a temperature of 1750° F (955° C) for one hour, followed by an oil quench. Tempering was done for four hours, followed by air cooling. (*A L Tech Specialty Steel Corp.*)

Embrittlement

Of considerable importance in the stress relieving and tempering of martensitic stainless steels is the phenomenon of embrittlement. The brittleness is accompanied by some loss of corrosion resistance and is caused by exposure in the temperature range of 750 to 1000° F (400 to 540° C). Hence, prolonged exposure in this temperature range should be avoided when stress-relieving or tempering operations are performed.

Annealing

The martensitic stainless steels are annealed or softened by two principal methods: full annealing and subcritical or process annealing.

Full annealing. In full annealing, the metals are heated into the austenitizing range, followed by slow cooling to below the critical temperature. This treatment results in a structure of carbide particles in a ferrite matrix, a microstructure that possesses the best ductility for many fabrication processes.

Subcritical annealing. In the second softening method, called subcritical or process annealing, the material is heated to just below the critical temperature, held for a specified length of time, and then cooled in air. An oil or water quench can be used but is generally not necessary. The resultant microstructure is tempered martensite or ferrite with carbide particles, depending upon the temperature and time at temperature. This heat treatment yields sufficient ductility for many fabrication processes without the complications of entering the austenitic region. Table 14-4 presents typical annealing cycles for the martensitic types, as well as the hardness values to be expected from each treatment.

TABLE 14-4
Annealing Treatments for Martensitic Stainless Steels*

Grade		Type of		Time at	Hardness,
UNS	AISI/SAE	Anneal	Temperature, ° F (° C)	Temperature, hr	Bhn
S40300 S41000 S41600	403 410 416	Process Full	1200-1400 (650-760) 1550-1650 (845-900)	2-4 1-3	165-195 135-160
S41400	414**	Process	1200-1300 (650-705)	4-8	235-265
S42000	420	Process Full	1350-1450 (730-790) 1550-1650 (845-900)	2-4 1-3	200-230 170-200
S43100	431**	Process	1150-1250 (620-675)	4-8	255-285
S44002	440A	Process Full	1300-1400 (705-760) 1550-1650 (845-900)	2-4 1-3	230-250 200-230
S44003	440B	Process Full	Same as for Type 440A		235-255 215-235
S44004	440C	Process Full	Same as for Type 440A		255-275 230-255

* Cooling from annealing temperature:
Process annealing—air cool.
Full annealing—slowly (furnace) cool at rate of 25-50° F (14-28° C) per hour to a temperature of 1100° F (595° C); then air cool.

** These alloys do not respond to slow cooling from above the critical temperature.

PRECIPITATION-HARDENING ALLOYS

The group of precipitation-hardening stainless steels discussed in Chapter 2, "Stainless and Maraging Steels," contain many alloys with marked differences in composition, mechanical properties, and microstructure. In most cases, the final heat treatment performed is an aging treatment that causes a second phase to precipitate from the martensitic or austenitic matrix, and the precipitate serves to strengthen the material.

Generally, the martensitic, precipitation-hardening alloys are not recommended for use in the unaged condition. In this condition they have lower ductility than material in the aged condition. Because of the complexity of these alloys, it is recommended that the user consult published standards or the producer of the material for detailed heat treatment recommendations.

Alloy 17-4PH

Alloy 17-4PH (UNS S17400) has an essentially martensitic structure in the solution-annealed condition (called Condition A). After fabrication, the material is precipitation hardened by aging in the temperature range of 900 to 1150° F (480 to 620° C), as shown in Table 14-5A. For maximum softness and optimum machinability, this alloy may be supplied in the H1150M condition (see Table 14-5B); however, the material must then be solution annealed prior to the final aging treatment.

Alloy 15-5PH

Alloy 15-5PH (UNS S15500) is a compositional variation of grade 17-4PH that is normally melted in a vacuum arc or consumable electrode furnace for increased cleanliness and greater homogeneity. It also has a martensitic structure in the solution-annealed condition (Condition A). However, 15-5PH has its composition varied to provide a structure essentially free of delta ferrite. When compared with 17-4PH, this feature provides improved transverse ductility and hot workability.

After fabrication, Condition A material is hardened by a single aging treatment in the temperature range of 900 to 1150° F (480 to 620° C), as shown in Table 14-5A. This alloy can be supplied in the H1150M condition (see Table 14-5B) for maximum softness and optimum machinability; however, the material must then be solution annealed prior to the final aging treatment.

Alloy 17-7PH

Alloy 17-7PH (UNS 17700) has a structure consisting of an austenitic matrix with some delta ferrite in the solution-annealed condition (Condition A). In this condition, the metal has good formability and can be subsequently heat treated to high strength levels. Solution annealing consists of heating to a temperature of 1900° F (1040° C) for 30 minutes and then water quenching.

Age hardening of alloy 17-7PH is accomplished by either of two treatments: TH1050 or RH950. In the TH1050 treatment, austenite conditioning (also called trigger annealing) is done by heating to a temperature of 1400° F (760° C) for 90 minutes and then air cooling within one hour to below 60° F (15° C). This procedure results in transformation of the austenite to martensite. The workpiece is held at the cooled temperature for not less than 30 minutes and then age hardened at 1050° F (565° C) for 90 minutes, followed by air cooling. The minimum tensile strength will then be 170 ksi (1172 MPa).

In the RH950 treatment, austenite conditioning is accomplished by heating to a temperature of 1750° F (955° C) for not less than 10 minutes, but not more than one hour, followed by air cooling. The workpiece is then cooled as soon as possible, within one hour, to a temperature of -100° F (-73° C), held at this temperature for at least eight hours, and then warmed in air to room temperature. The alloy will then be martensitic. Precipitation hardening (aging) is done by heating to a temperature of 950° F (510° C) for one hour and then air cooling, which produces a minimum tensile strength of 185 ksi (1276 MPa).

Alloy 17-7PH is also supplied in the solution-annealed and cold worked condition (Condition C). Martensitic transformation takes place during cold working and, as a result, only precipitation hardening (aging) is required after fabrication. Aging is accomplished by heating to a temperature of 900° F (480° C) for one hour, followed by air cooling, which produces a tensile strength of 230 ksi (1586 MPa) or more, depending upon workpiece section.

Alloy PH 15-7Mo

Alloy PH 15-7Mo (UNS S15700) is similar to grade 17-7PH except that its composition is varied to yield higher elevated temperature properties in the aged condition. It is supplied in the same conditions as grade 17-7PH, and identical heat treating procedures are used. With the TH1050 treatment, a minimum tensile strength of 180 ksi (1241 MPa) is obtained. The RH950 treatment provides a minimum tensile strength of 200 ksi (1379 MPa).

Alloy PH 13-8Mo

Alloy PH 13-8Mo (UNS S13800) is a double-vacuum-melted alloy that provides excellent toughness at high strength levels. It is normally supplied in the solution-annealed condition and has essentially a martensitic structure. After fabrication, the material is hardened by a single aging treatment in the temperature range of 950 to 1150° F (510 to 620° C) (see Table 14-5A). Condition H1150M, identical to the heat treatment for 17-4PH and 15-5PH alloys, can also be used for maximum softness and machinability.

Custom 450

Custom 450 alloy (UNS S45000) is normally supplied in the solution-annealed condition (Condition A) and has an essentially martensitic structure. After fabrication, the material can be used in the solution-annealed condition or hardened by a single aging treatment in the temperature range of 900 to 1150° F (480 to 1150° C) (see Table 14-5A).

Alloy AM-355

The AM-355 (UNS S35500) grade of precipitation-hardening stainless steel is generally supplied in the equalized and overtempered condition and is primarily designed for bar production. In this condition, the structure is an overtempered martensite, which provides maximum softness and is best for machining. After machining, the material is heat treated to attain high strength levels.

Specific heat treating recommendations should be obtained from the supplier of the material. However, parts in the equalized and overtempered condition can be hardened by austenite conditioning, subzero cooling, and tempering. The austenite condition is achieved by heating to 1750° F (955° C), holding at this temperature for 10-60 minutes, and then quenching in water. Subzero cooling is accomplished by lowering the temperature of the workpiece to -100° F (-73° C), holding at this temperature for at least three hours, and warming in air to room temperature. Tempering consists of heating to 850° F (455° C) or 1000° F (540° C), holding at this

TABLE 14-5A
Heat Treating Procedures for Some Precipitation-Hardening Stainless Steel Grades

Grade		Solution Annealing			Precipitation Hardening			
UNS	Other Designation	Temperature, °F (°C)	Time,* hr	Cooling	Temperature, °F (°C)	Time, hr	Cooling	Min Tensile Strength, ksi (MPa)
S17400 S15500	17-4PH 15-5PH	1900 (1040)	1/2	Air or oil quench**	900 (480)	1	Air	190 (1310)
					925 (495)	4	Air	170 (1172)
					1025 (550)	4	Air	155 (1069)
					1075 (580)	4	Air	145 (1000)
					1100 (595)	4	Air	140 (965)
					1150 (620)	4	Air	135 (931)
S13800	PH13-8Mo	1700 (925)	1/2	Air or oil quench†	950 (510)	4	Air	220 (1517)
					1000 (540)	4	Air	205 (1413)
					1025 (550)	4	Air	185 (1276)
					1050 (565)	4	Air	175 (1207)
					1100 (595)	4	Air	150 (1034)
					1150 (620)	4	Air	135 (931)
S45000	Custom 450	1900 (1040)	1/2	Water†† or oil quench	900 (480)	4	Air	180 (1241)
					950 (510)	4	Air	170 (1172)
					1000 (540)	4	Air	160 (1103)
					1025 (550)	4	Air	150 (1034)
					1050 (565)	4	Air	145 (998)
					1100 (595)	4	Air	130 (896)
					1150 (620)	4	Air	125 (862)

* Depends upon section size.

** Thick sections should not be oil quenched because cracks may form. Air cooling is satisfactory for most applications. Material must be cooled to below 90° F (30° C) before the precipitation-hardening treatment.

† Thick sections should not be oil quenched because cracks may form. Air cooling is satisfactory for most applications. Material must be cooled to below 60° F (15° C) before the precipitation-hardening treatment.

†† Forced air cooling may be required for thicker sections. Material must be cooled to below 90° F (30° C) before the precipitation-hardening treatment.

TABLE 14-5B
H1150M Heat Treatment for Some Precipitation-Hardening Stainless Steel Grades

Grade				Cooling		Reheating			Min Tensile
UNS	Other Designation	Temperature, °F (°C)	Time, hr	Media	To Below °F (°C)	Temperature, °F (°C)	Time, hr	Cooling	Strength, ksi (MPa)
S17400 S15500	17-4PH 15-5PH	1400 (760)	2	Air	90 (30)	1150 (620)	4	Air	115 (793)
S13800	PH13-8Mo	1400 (760)	2	Air	60 (15)	1150 (620)	4	Air	125 (862)

temperature for at least three hours, and then cooling in air. Typical, minimum tensile strengths attained are 190 ksi (1310 MPa) when tempering is done at a temperature of 850° F and 170 ksi (1172 MPa) when tempering at 1000° F.

Another method used to heat treat parts made from AM-355 stainless steel consists of solution treating, subzero cooling, austenite conditioning, subzero cooling, and tempering. For solution treating, parts are heated to 1900° F (1040° C), held at this temperature for 1 to 3 hours, and then water quenched. The first subzero treatment consists of cooling the parts to -100° F (-73° C) for at least three hours, and then warming to room temperature in air. For austenite conditioning, parts are heated to 1750° F (955° C) for 10-60 minutes and quenched in water. The second subzero treatment involves again cooling the parts to -100° F, holding for at least three hours, and warming in air to room temperature. Tempering is done by heating to 850 or 1000° F (455 or 540° C), holding for at least three hours, and air cooling. Typical, minimum tensile strengths produced are 190 ksi (1310 MPa) for the lower tempering temperature and 170 ksi (1172 MPa) for the higher temperature.

MARAGING STEELS

Maraging steels are strengthened by the precipitation of intermetallic compounds produced by age hardening a matrix of low-carbon martensite. Because of the low-carbon, iron-nickel, martensitic matrix, these alloys possess excellent toughness and ductility, both in the solution-annealed and aged conditions. Aging develops high strength and hardness. At the present time, only maraging steels containing 18% nickel are being produced.

Annealing

All maraging steels are supplied in the solution-annealed condition. In this condition, the soft metal, usually having a hardness of R_C30 to 35, can be machined or formed easily.

Maraging steels are generally double annealed from temperatures of 1700 and 1500° F (925 and 815° C), with heating times of about one hour per inch (25 mm) of section thickness. Workpieces are cooled in air to room temperature between the two annealing cycles and after the final annealing treatment. The cooling rate is not critical, but the workpieces must be cooled to room temperature before age hardening.

Because of the low carbon content of the metals, a protective furnace atmosphere is generally not required for solution annealing of maraging steels to prevent decarburization. Atmosphere controls are sometimes used, however, to minimize possible surface damage.

Age Hardening

Age hardening of maraging steels is accomplished by heating to a temperature of about 900° F (480° C) for 3 to 5 hours, thus developing the desired high strength and hardness. Table 14-6 shows the recommended aging temperatures and heating times, as well as resulting hardnesses, for the four grades of maraging steels.

Air is commonly used as the furnace atmosphere for age hardening of maraging steels. These steels are also often surface (case) hardened by nitriding, using either gas nitriding or the molten-salt-bath method, discussed in Chapter 11, "Surface (Case) Hardening." Best results are obtained when the parts to be nitrided are in the solution-annealed condition.

During aging, maraging steels shrink uniformly and predictably in all dimensions. Table 14-7 shows the typical shrinkage, which varies with the aging temperature and heating time employed.

TABLE 14-6
Age-Hardening Cycles for Maraging Steels

Grade	Aging Temperature, °F (°C)	Aging Time, hr	Resulting Hardness, R_C
200	900-925	6	44-48
250	900-925	6	48-52
300	900-925	6	50-55
350	900-925	6	55-60
	or 950	3	56-60

(*Teledyne Vasco*)

TABLE 14-7
Shrinkage of Maraging Steels Resulting from Age Hardening

	Heating Time at Aging Temperature of 900° F (480° C)	
	6 hr	9 hr
Grade	Shrinkage, in./in. (mm/mm)	
200	0.0006	0.0007
250	0.0009	0.0011
300	0.001	0.0012
350	0.001	0.0013

(*Teledyne Vasco*)

DUPLEX STAINLESS STEELS

These alloys have two phases present in their microstructure, austenite and ferrite. Depending upon their composition, the alloys can be predominantly ferrite, predominantly austenite, or have approximately equal amounts of each phase present. These grades are not hardenable by heat treatment, but can be moderately hardened by cold working. The only heat treatment commonly used is annealing. Annealing treatments are very similar to those used for austenitic stainless steels. Temperatures of 1850 to 1950° F (1010 to 1065° C) are often used, followed by water quenching. However, it is recommended that the user contact the producer for exact annealing recommendations.

TITANIUM AND ITS ALLOYS

Titanium and its alloys, discussed in Chapter 5, "Tungsten, Molybdenum, and Titanium," are readily heat treated. The response to heat treatment, however, depends upon the composition of the metal. In general, the more highly alloyed the alloy, the more heat treatable it will be. Not all types of heat treatment are applicable to all titanium alloys. Alpha stabilizers, such as aluminum, in the alloys raise the alpha-to-beta transformation temperature; beta stabilizers, such as vanadium, lower the transformation temperature.

The grain size of titanium and its alloys cannot be refined by heat treatment alone. Either hot or cold working of the metal is necessary for grain refinement. The degree of cold working performed on single-phase alloys also affects the recrystallization temperature and time of heating required. Mixed-phase alloys are recrystallized during hot working and are generally not cold rollable to a degree sufficient to cause grain refinement during recrystallization. The final properties of the heat-treated materials thus reflect the variables introduced during prior cold or hot working.

TYPES OF HEAT TREATMENT

Many types of heat treatment are performed on titanium and titanium alloys to suit specific requirements. The more common treatments include stress relieving, annealing, stabilizing, betatizing, solution treating, and age hardening. Solution treating and age hardening are usually combined in a two-stage thermal treatment to provide high strength. Another heat treatment, recrystallizing, is done at intermediate-to-high temperatures to produce a uniform, dislocation-free microstructure. Recrystallizing may or may not induce grain refinement or a new texture, depending upon prior working of the metal.

Stress Relieving

Stress relieving of titanium and titanium alloys is a low-temperature heat treatment to relax or remove any residual stresses existing in the material as a result of machining, forming, welding, or quenching. The process is most commonly used when full annealing is undesirable.

Various temperatures and heating times are used for stress relieving to suit specific requirements. Higher temperatures are generally employed with shorter times and lower temperatures with longer times. The rate of cooling from the stress-relieving temperature is not critical, but the uniformity of cooling is important. Oil or water quenching is not recommended because of possible uneven cooling, which can produce residual stresses.

Annealing

Annealing is a heat treatment that provides good ductility to the metal, but the strength of the metal is only low to moderate. The treatment is generally performed at an intermediate temperature of 1200 to 1500° F (650 to 815° C). Common treatments include mill (full), duplex, and beta annealing.

Stabilizing

Stabilizing is a low-temperature heat treatment commonly performed on creep-resistant alloys. The purpose is to stabilize the structure and phases present so that little or no further change occurs in service. Stabilizing is generally done in the temperature range of 1100 to 1200° F (595 to 650° C).

Betatizing

Betatizing is a high-temperature heat treatment often applied to alpha-lean (super-alpha), beta-lean alpha (near beta), and alpha-beta grades of titanium alloys. The purpose is to impart maximum toughness or creep resistance to the metal.

Solution Treating

Solution treating is a high-temperature process followed by cooling at a rate sufficient to retain age-hardening capability. Too slow a rate of cooling can prevent effective subsequent strengthening. This process may or may not retain the phases existing at the solution treating temperature.

Age Hardening

Age hardening is a low-temperature heat treatment following solution treating and is generally performed in the temperature range from 800 to 1200° F (425 to 650° C). This treatment significantly increases the strength of titanium alloys. Aging at or near the annealing temperature of the metal results in overaging. Overaging is sometimes done to provide a smaller increase in strength while maintaining toughness and increasing the ductility of the material.

HEAT TREATMENT SCHEDULES

Typical temperatures, times at temperature, and cooling methods for various heat treatments of titanium and titanium alloys, as well as some of the properties attained, are presented in Table 14-8. Many other heat treatment schedules are used to develop specific property requirements. Consultation with the supplier or producer of the material is recommended for special needs.

In general, increasing the temperature for solution treating will increase the strength produced by age hardening. However, increasing the aging temperature will decrease the strength attained. Low aging temperatures may require additional time at temperature to produce full strength. Measuring the hardness of the material is generally not a very reliable method for verifying the heat treatment of titanium alloys.

TITANIUM AND ITS ALLOYS

TABLE 14-8
Heat Treatment Schedules and Properties for Titanium and Titanium Alloys

Metal Grade Designation	Stress Relieving[a] Temperature, °F (°C)	Stress Relieving[a] Time, hr	Annealing[a] Temperature, °F (°C)	Annealing[a] Time, hr	Annealed Properties, minimum: Tensile Strength, ksi (MPa)	Annealed Properties, minimum: Yield Strength, ksi (MPa)	Annealed Properties, minimum: Elongation, %
Unalloyed, commercially pure:							
Ti-35A (ASTM Grade 1)	1000 (540)	1	1300 (705)	1/4	35 (241)	25 (172)	24
Ti-50A (ASTM Grade 2)	1000 (540)	1	1300 (705)	1/4	50 (345)	40 (276)	20
Ti-65A (ASTM Grade 3)	1000 (540)	1	1300 (705)	1/4	65 (448)	55 (379)	18
Ti-75A (ASTM Grade 4)	1000 (540)	1	1300 (705)	1/4	80 (552)	70 (483)	15
Corrosion alloys:							
Ti-0.2Pd (ASTM Grade 7)	1000 (540)	1	1350 (730)	1/4	50 (345)	40 (276)	20
Ti-Code 12 (ASTM Grade 12)	1050 (565)	1	1350 (730)	1/4	70 (483)	50 (345)	18
Alpha alloys:							
Ti-5Al-2.5 Sn	1100 (595)	2	1400 (760)	1/4	120 (827)	115 (793)	10
Ti-5Al-2.5 Sn ELI	1100 (595)	2	1400 (760)	1/4	105 (724)	95 (655)	10
Alpha-lean beta alloys:							
Ti-8Al-1Mo-1V	1200 (650)	2	1250 (675)	1	130[d] (896)	120[d] (827)	10[d]
Ti-6Al-2Sn-4Zr-2Mo	1200 (650)	2	1250 (675)	1	130[d] (896)	120[d] (827)	10[d]
Ti-6Al-2Sn-4Zr-6Mo	1200 (650)	2	1250 (675)	1	150[e] (1034)	140[e] (965)	10[e]
Alpha-beta alloys:							
Ti-3Al-2.5V	1100 (595)	1	1350 (730)	1/4	90 (621)	75 (517)	15
Ti-6Al-4V	1100 (595)	1	1350 (730)	1/4	130 (896)	120 (827)	10
Ti-6Al-4V ELI	1100 (595)	1	1350 (730)	1/4	125 (862)	115 (793)	10
Ti-6Al-6V-2Sn	1100 (595)	1	1350 (730)	1/4	155 (1069)	145 (1000)	8
Beta-lean alpha alloys:							
Ti-10V-2Fe-3Al	1200 (650)	1	1250 (675)	1	[c]	[c]	[c]
Ti-17	1020 (550)	4	1300 (705)	1	[c]	[c]	[c]

TABLE 14-8—*Continued*

Stabilizing[a]		Betatizing[a]		Solution Treating[b]		Age Hardening[a,b]		Aged Properties, minimum		
Temperature, °F (°C)	Time, hr	Temperature, °F (°C)	Time, hr	Temperature, °F (°C)	Time, hr	Temperature, °F (°C)	Time, hr	Tensile Strength, ksi (MPa)	Yield Strength, ksi (MPa)	Elongation, %
c		*c*		*c*		*c*		*c*	*c*	*c*
c		*c*		*c*		*c*		*c*	*c*	*c*
c		*c*		*c*		*c*		*c*	*c*	*c*
c		*c*		*c*		*c*		*c*	*c*	*c*
c		*c*		*c*		*c*		*c*	*c*	*c*
c		*c*		*c*		*c*		*c*	*c*	*c*
c		*c*		*c*		*c*		*c*	*c*	*c*
c		*c*		*c*		*c*		*c*	*c*	*c*
1100 (595)	8	1950 (1065)	1/4	1800 (980)	1[a]	*c*		*c*	*c*	*c*
1100 (595)	8	1850 (1010)	1/4	1750 (955)	1	*c*		*c*	*c*	*c*
1100 (595)	8	1800 (980)	1/4	1625 (885)	1	*c*		*c*	*c*	*c*
c		*c*		*c*		*c*		*c*	*c*	*c*
c		1875 (1025)	1/4	1700 (925)	1[f]	1000 (540)	4	165[g] (1138)	155[g] (1069)	10
c		1875 (1025)	1/4	*c*		*c*		*c*	*c*	*c*
c		*c*		1625 (885)	1[f]	1050 (565)	8	175[g] (1207)	160[g] (1103)	6
c		1550 (845)	1	1400 (760)	1[h]	925 (495)	16	180 (1241)	165 (1138)	4
c		1700 (925)	1/4	1475 (800)	4[f]	1175 (635)	8	170[i] (1172)	160[i] (1103)	12[i]

(continued)

TABLE 14-8—*Continued*

Metal Grade Designation	Stress Relieving[a] Temperature, °F (°C)	Stress Relieving[a] Time, hr	Annealing[a] Temperature, °F (°C)	Annealing[a] Time, hr	Annealed Properties, minimum: Tensile Strength, ksi (MPa)	Annealed Properties, minimum: Yield Strength, ksi (MPa)	Annealed Properties, minimum: Elongation, %
Beta alloys:							
Ti-13V-11Cr-3Al	1100 (595)	1/2	1350 (730)	1	130 (896)	125 (862)	8
Ti-15V-3Cr-3Al-3Sn	1200 (650)	1/4	1450 (790)	1/2	105 (724)	100 (690)	15
Ti-3Al-8V-6Mo-4Sn-4Zr	1200 (650)	1/4	1400 (760)	1/4	125 (862)	120 (827)	6
Ti-Beta III	1200 (650)	1/4	1400 (760)	1/4	122[i] (841)	107[i] (738)	20[i]

HEAT TREATING PRACTICES

Care is necessary in heat treating titanium and titanium alloys, and several precautions are essential. It is recommended that workpieces be cleaned and dried before heat treatment. Precise temperature control in the furnaces used is necessary, especially for solution treating and aging.

Titanium, being a reactive metal, readily oxidizes in air at intermediate and high temperatures. Oxygen produces oxide scale and an alpha-phase case on the workpieces. Alpha cases are oxygen-enriched, brittle layers that develop on the surface or beneath the scale. Scale can be removed by mechanical or chemical means, but alpha cases are more difficult to remove.

Alpha cases can be effectively removed by machining or pickling in a solution, such as one containing 35% nitric acid (HNO_3), 5% hydrofluoric acid (HF), and the balance water. A problem with such processing is knowing when removal of the alpha case is complete, and the only sure method of determining this is to examine the surface metallographically with a microscope. The recommended practice to eliminate or minimize scale and alpha cases is to use an argon or helium atmosphere in the furnace, or to heat treat in a vacuum furnace.

Another precaution is to prevent halide contamination by cleaning the workpieces prior to heat treatment and preventing contact with halides thereafter. Halide contamination causes no problems unless residual or applied stresses are present during heat treatment at low-to-intermediate temperatures. However, the presence of halides, such as chlorides in the form of salts or organic compounds, can induce hot, stress-corrosion cracking in most titanium alloys. Such cracking is most troublesome in hot forming operations, but in rare instances can occur during stress relieving, stabilizing, or age-hardening treatments. If the cracks are not deep, they can be removed mechanically or chemically.

HEAT-RESISTING, HIGH-STRENGTH ALLOYS

Heat-resisting, high-strength alloys are used extensively in the aircraft and aerospace industries, chemical and petrochemical processing equipment, industrial furnaces and related heat treating equipment, and a variety of other applications involving strenuous service conditions imposed by oxidizing and corrosive atmospheres over a wide range of temperatures. These materials consist essentially of iron, nickel, and cobalt-based alloys, and are available in many special, proprietary compositions.

Heat treatment of these alloys varies with their compositions, the size and shape of the products, and the applications. Typical solution treating, aging, stress relieving, and annealing cycles for a few of the more common heat-resisting alloys are presented in Table 14-9. Specific recommendations for heat treating a particular alloy and part should be established by consultation with the producer of the material and the designer of the component.

Some heat-resisting alloys are strengthened by means of aging treatments designed to promote precipitation hardening, using second phases; these are usually referred to as precipitation-strengthened alloys. Other alloys are normally used in the annealed or solution heat-treated condition and develop strength from in-service precipitation of carbides or from solid-solution strengthening. These alloys are usually called solution-strengthened or carbide-strengthened alloys. Still a third group of alloys are oxide dispersion strengthened (ODS) alloys, which derive their strength from a fine dispersion of inert oxide particles. These alloys are also typically used in the annealed condition.

TYPES OF ALLOYS

Examples of wrought, iron-based, solution-strengthened alloys include iron-nickel-chromium alloys such as alloys 800H and 330, and Type 310 stainless steel. Also included are materials of the more complex iron-nickel-chromium-cobalt family, such as alloy N-155. Precipitation-strengthened, iron-based, heat-resisting alloys usually contain some amount of titanium (to

TABLE 14-8—*Continued*

Stabilizing[a]		Betatizing[a]		Solution Treating[b]		Age Hardening[a,b]		Aged Properties, minimum		
Temperature, °F (°C)	Time, hr	Temperature, °F (°C)	Time, hr	Temperature, °F (°C)	Time, hr	Temperature, °F (°C)	Time, hr	Tensile Strength, ksi (MPa)	Yield Strength, ksi (MPa)	Elongation, %
[c]		[c]		1350 (730)	1[a]	900 (480)	48	180[i] (1241)	160[i] (1103)	5[i]
[c]		[c]		1450 (790)	1/2[a]	950 (510)	8	180 (1241)	165 (1138)	6
[c]		[c]		1500 (815)	1/2[a]	950 (510)	8	180 (1241)	170 (1172)	6
[c]		[c]		1400 (760)	1/2[a]	900 (480)	8	175[i] (1207)	165[i] (1138)	6[i]

(*TIMET*)

[a] Air cool.
[b] Schedules shown are typical commercial practice. Most heat-treatable titanium alloys can be custom heat treated to produce a variety of properties.
[c] Not applicable.
[d] Stabilized condition.
[e] Duplex annealed: 1600° F (870° C) for 1/2 hr and air cooled; then 1300° F (705° C) for 1/4 hr and air cooled.
[f] Water quench.
[g] Limited with respect to section size.
[h] Oil quench or water quench.
[i] Typical values.

3%) and sometimes aluminum. Examples are alloys A-286 and 901, both of which also contain molybdenum. Inco Co.'s Incoloy alloy MA 956, which contains a dispersion of yttrium oxide particles, is a good example of an iron-based ODS alloy.

As in the case of iron-based alloys, nickel-based, heat-resisting alloys fall into several categories. Solution-strengthened alloys, designed principally for low stress/high-temperature applications, usually contain chromium, molybdenum and/or tungsten, sometimes cobalt or iron, and possibly a variety of other minor elements. The more basic materials in this category, such as alloy 600, Inco's Inconel alloy 601, and alloy 214 made by the Cabot Corp., contain no refractory metals. More sophisticated alloys, such as RA333 made by Rolled Alloys, Hastelloy alloys S and X made by Cabot Corp., and Inconel alloy 617 made by Inco Co., contain as much as 15% molybdenum or tungsten.

A great many nickel-based, heat-resisting alloys are precipitation strengthened. These alloys, which usually contain appreciable amounts of aluminum, titanium, and columbium, are designed for high-stress/high-temperature applications. They usually contain chromium, often contain molybdenum and/or tungsten, sometimes contain cobalt or iron, and also contain a variety of minor elements. Basic alloy examples include alloy X-750 and Inco's Nimonic alloy 80A, with relatively low aluminum and titanium contents (to about 3.5% total). Intermediate alloys containing 3.5-7% aluminum, plus titanium and columbium, include Udimet 500, made by Special Metals Corp.; alloy No. 718; and René 41 alloy, made by Teledyne-Allvac Corp. Very high-strength, precipitation-strengthened alloys containing over 7% aluminum and titanium include Nimonic alloy 115 and Udimet 700. Many other alloys are used in cast form, often containing even higher, strengthening element content.

Cobalt-based, heat-resisting alloys for the most part are either solution strengthened or carbide strengthened and, as such, are normally not subject to age hardening by heat treatment. Most cobalt-based alloys contain chromium, some amount of nickel, and major amounts of tungsten, tantalum, and—less often—molybdenum. Examples include Haynes alloy No. 188 made by Cabot Corp., alloy L-605, and alloy S-816 as wrought materials; and alloy X-40 and Martin Metal's Mar-M 509 as cast materials.

Presently available refractory metals and alloys are not heat treated to achieve high strength. These materials are used in the stress-relieved or annealed (recrystallized) condition.

SOLUTION TREATING AND AGING PRECIPITATION-STRENGTHENED ALLOYS

Increasing the strength of precipitation-strengthened, heat-resisting alloys usually requires solution treatment and aging, but alternative heat treatments are sometimes used to improve specific properties. In some instances, the solution treating temperature employed depends upon the properties desired.

Quenching is done in oil or water or by air cooling after solution treatment, depending upon the alloy. The cooling rate for some alloys is critical. Cold working of heat-resisting alloys

TABLE 14-9
Typical Heat Treatments for Some Heat-Resisting Alloys

Alloy	Solution Treating Temperature, °F (°C)	Cooling/Quenching	Annealing Temperature,* °F (°C)
Solution-Strengthened Alloys**			
Iron-based:			
800H	2100 (1150)	Air cool	1800 (980)
330	2000 (1095)	Air cool	1875 (1025)
N-155	2150 (1175)	Rapid cool	2150 (1175)
Haynes 556	2150 (1175)	Rapid cool	2150 (1175)
Nickel-based:			
600	2050 (1120)	Air cool	1850 (1010)
Hastelloy X (HX)	2150 (1175)	Rapid cool	2150 (1175)
Inconel 617	2150 (1175)	Rapid cool	1900 (1040)
Hastelloy S	1950 (1065)	Rapid cool	1950 (1065)
RA 333	2000 (1095)	Rapid cool	1925 (1050)
Inconel 601	2100 (1150)	Rapid cool	1800 (980)
625	2050 (1120)	Rapid cool	1925 (1050)
Cobalt-based:			
Haynes 188	2150 (1175)	Rapid cool	2150 (1175)
S-816	2150 (1175)	Rapid cool	2200 (1205)
L-605	2200 (1205)	Rapid cool	2200 (1205)

Alloy	Solution Treating Temperature, °F (°C)	Cooling/ Quenching	Aging Temperature, °F (°C)	Cooling/ Quenching	Annealing Temperature,* °F (°C)
Precipitation-Strengthened Alloys**					
Iron-based:					
A-286	1800 (980)	Oil quench	1325 (720)	Air cool	1800 (980)
901	2000 (1095)	Rapid cool	1450 (790) +1325 (720)	Air cool Air cool	2000 (1095)
Nickel-based:					
X-750	1800 (980)	Rapid cool	1325 (720) +1150 (620)	Furnace cool Air cool	1800 (980)
René 41	1950 (1065)	Rapid cool	1400 (760)	Air cool	1975 (1080)
718	1750 (955)	Air cool	1325 (720) +1150 (620)	Furnace cool Air cool	1750 (955)
718 (cast)	2000 (1095)	Air cool	1150 (620) or 1600 (870)	Air cool Air cool	
Mar-M 200 (cast)			1600 (870)	Air cool	
C-263	2100 (1150)	Rapid cool	1472 (800)	Air cool	1965 (1075)

(continued)

TABLE 14-9—*Continued*

Alloy	Solution Treating Temperature, °F (°C)		Cooling/Quenching		Annealing Temperature,* °F (°C)
Nimonic 90	1975 (1080)	Air cool	1290 (700)	Air cool	1900 (1040)
IN 738 (cast)	2050 (1120)	Air cool	1550 (840)	Air cool	
Nimonic 115	2175 (1090)	Air cool	2010 (1100)	Air cool	
Udimet 700 (cast)	2100 (1150)	Air cool	1400 (760)	Air cool	
Udimet 700	2150 (1175) +1975 (1080)	Air cool Air cool	1550 (845) +1400 (760)	Air cool Air cool	2075 (1135)
Incoloy 903/909	1800 (980)	Air cool	1325 (720) +1150 (620)	Furnace cool Air cool	1800 (980)

* Minimum hardness is achieved by cooling rapidly from the annealing temperature to prevent precipitation of hardening phases. Water quenching is often preferred and is usually necessary for heavy sections. However, air cooling is generally preferred for heavy sections of some heat-resisting alloys, such as Udimet 700 and Waspaloy, because water quenching causes cracking. For complex shapes subject to excessive distortion, oil quenching is often adequate and more practical. Rapid air cooling is usually adequate for parts formed from strip or sheet. Rapid cooling from the annealing or solution treating temperature does not suppress the aging reaction of some alloys, such as Astroloy; these alloys become harder and stronger.

** Stress relieving at intermediate temperatures is not recommended for these alloys as exposure to intermediate temperatures can cause undesirable aging reactions or phase precipitation. When a stress relief is required, a full anneal treatment is normally employed.

is usually performed in their solution-treated condition because of the lower strength and increased ductility of the materials before aging.

Factors that influence the selection of the number of aging steps and the aging temperatures include:

1. The type and number of precipitating phases available in the alloy being treated.
2. The anticipated service temperature of the part treated.
3. The size of the precipitates.
4. The combination of strength and ductility desired.

SOLUTION TREATING SOLUTION-STRENGTHENED ALLOYS

Optimum strengthening of the more heavily alloyed, solution-strengthened, heat-resistant alloys can often be achieved through the use of a full solution treatment and rapid cooling. Alternate treatments are often used, however, to provide a balance with other properties, such as ductility and corrosion resistance. The cooling rate for most alloys is critical, although rates approaching a water quench are not typically required. Solution heat treatment of more basic, solution-strengthened alloys can cause lower strength due to excessive grain growth.

STRESS RELIEVING

Stress relieving of heat-resisting alloys and refractory metals frequently entails a compromise between the desirability of maximum relief from residual stresses and possible effects deleterious to high-temperature properties and corrosion resistance. True stress relieving of wrought materials is usually confined to alloys that are not age hardenable.

Some heat-resisting alloy castings are placed in service in the as-cast condition. However, castings are generally stress relieved when:

1. They are of complex shape that might cause cracking during initial heating in service.
2. Their dimensional tolerances are stringent.
3. They have been welded.

Iron-based, heat-resisting alloys that depend entirely upon hot and cold working to develop high strength frequently require stress relieving. Both stress relieving and annealing are generally applied to wrought nickel-based alloys of the solution-strengthened type, but castings are seldom stress relieved or annealed. Most of the wrought, cobalt-chromium-nickel-based alloys can be safely stress relieved when fabrication is completed, provided that excessive distortion does not occur. Stress-relief heat treatments are applied to cobalt-based alloys after welding.

Great care must be taken in the application of stress-relief treatments to avoid excessively high temperatures. Such exposures can cause undesirable aging reactions or carbide precipitation reactions to occur, which can reduce alloy strength and ductility.

ANNEALING

Annealing, as applied to heat-resisting alloys, implies full annealing (complete recrystallization) and the attainment of maximum softness. This treatment is usually applied to wrought alloys of the nonhardening type. For a majority of the hardenable alloys, annealing cycles are the same as those used for solution treating.

Severe cold forming of wrought, heat-resisting alloys may require several intermediate annealing operations. Full annealing must be followed by fast cooling. Annealing of weldments should immediately follow welding of age-hardenable alloys when highly restrained joints are involved. If the configuration of the weldment does not permit annealing, aging can be used for stress relieving the joints.

Some iron-based, heat-resisting alloys, such as A-286 and 901, cannot be stress relieved. Instead, ductility is restored and stresses reduced in cold formed parts and weldments made of these alloys by heating rapidly to the annealing temperature. Castings of iron-based, heat-resisting alloys are not annealed except for special reasons.

Some age hardenable, nickel-based alloys, such as René 41 alloy, are more crack sensitive than the iron-based alloys and must be annealed during fabrication to relieve forming and welding stresses. For some wrought, nickel-based alloys, annealing can have a marked effect on subsequent response to solution treating and aging.

Full annealing, rather than stress relieving, is recommended for most of the wrought, solution-strengthened alloys whenever high residual stresses are developed during fabrication. Annealing of cast, solution-strengthened alloys is sometimes avoided because of an undesirable precipitation of coarse carbides that can occur during slow cooling after annealing.

HEAT TREATING PRACTICE

Proper practices are essential for the successful heat treatment of heat-resisting alloys. Possible intergranular corrosion and surface contamination must be guarded against. The use of protective atmospheres and the proper selection of furnaces and fixtures must be given careful consideration.

Intergranular Oxidation

At temperatures used for solution treating, age-hardenable, heat-resisting alloys can be susceptible to intergranular oxidation, a defect that adversely affects mechanical properties. Prolonged exposure of age-hardenable alloys to air environments at temperatures above 2000° F (1095° C) is generally not desirable. Use of protective atmospheres, particularly for thin sections, is often recommended. Solution-strengthened alloys are generally less sensitive to this problem and require less attention in this regard, although there are exceptions.

Surface Contamination

All exposed surfaces of heat-resisting alloy parts should be free of dirt, fingerprints, oil, grease, forming compounds, lubricants, and scale before heat treatment. Carburization of these materials at the heat treating temperatures can reduce corrosion resistance. Carbon pickup can also occur if the solution treating atmosphere has a carburizing potential.

Protective Atmospheres

Protective atmospheres, as discussed in Chapter 10, "Heat Treatment of Steel," are used in annealing or solution treating if heavy oxidation or intergranular oxidation cannot be tolerated. If oxidation can be tolerated, because of subsequent stock removal, heat-resisting alloys can be solution treated in air or in the normal mixture of air and combustion products found in gas-fired furnaces. Refractory metals, however, must always be heat treated in a vacuum, an inert-gas atmosphere, or hydrogen. In some cases, ceramic coatings are used to prevent surface attack.

Furnaces

Considerations in selecting furnaces for heat treating heat-resisting alloys are essentially the same as for heat treating steels. A discussion of various types of furnaces available is presented in Chapter 12, "Heat Treating Furnaces." Batch heating for annealing or solution treating is usually done in box furnaces. Aging is also commonly done in box furnaces, with or without protective atmospheres. Vacuum furnaces are used for heating niobium, tantalum, and other heat-resisting alloys.

Fixturing

Fixtures for holding finished parts or assemblies during heat treatment may be of either the support or the restraint type. For alloys that must be cooled rapidly from the solution treating temperature, the best practice is to employ minimum fixturing during solution treating and quenching and to control dimensional relations by the use of restraining fixtures during aging.

Support fixtures are used when restraint is not required or when the part itself provides sufficient self-restraint. Long, narrow parts are most easily fixtured by hanging vertically. Asymmetrical parts can be supported by laying on a tray of sand or on a ceramic casting formed to the shape of the part.

Restraint fixtures may require machined grooves, lugs, or clamps. It is possible to perform some straightening of parts in aging fixtures by forcing slightly distorted parts into the fixtures and clamping them. The use of threaded fasteners for clamping is not recommended because they are difficult to remove after heat treatment. A slotted bar held in place by wedges is preferred.

ALUMINUM ALLOYS

In high-purity form, aluminum is soft and ductile. Most commercial uses, however, require greater strength than pure aluminum affords. Strengthening of aluminum is achieved first by the addition of other elements to produce various alloys,[2] which singly or in combination impart strength to the metal. Further strengthening is possible by means that classify the alloys roughly into two categories, nonheat-treatable and heat-treatable.

NONHEAT-TREATABLE ALLOYS

The initial strength of alloys in this group depends upon the hardening effect of elements such as manganese, silicon, iron and magnesium, singly or in various combinations. Effects of the various alloying elements are discussed in Chapter 6, "Aluminum, Copper and Magnesium." The nonheat-treatable alloys are usually designated in the 1000, 3000, 4000, or 5000 series. Since these alloys are work-hardenable, further strengthening is made possible by various degrees of cold working, denoted by the H series of tempers. The alloy and temper designation systems for aluminum are also discussed in Chapter 6. Alloys containing appreciable amounts of magnesium, when supplied in strain-hardened tempers, are usually given a final elevated temperature treatment called *stabilizing* to ensure stability of properties.

HEAT-TREATABLE ALLOYS

The initial strength of heat-treatable aluminum alloys is enhanced by the addition of alloying elements such as copper, magnesium, zinc, and silicon. Since these elements singly or in various combinations show increasing solid solubility in aluminum with increasing temperature, it is possible to subject them to thermal treatments that will impart pronounced strengthening.

The first step, called solution heat treatment, is an elevated temperature process designed to put the soluble element or elements in solid solution. Solution heat treatment is followed by rapid quenching, usually in water, which momentarily freezes the structure and for a short time renders the alloy very workable. It is at this stage that some fabricators retain this more workable structure by storing the alloys at below freezing temperatures until they are ready to form them.

At room or elevated temperatures the alloys are not stable after quenching, however, and precipitation of the constituents from the supersaturated solution begins. After a period of several days at room temperature, termed aging or room temperature precipitation, the alloy is considerably stronger. Many alloys approach a stable condition at room temperature; but some alloys, particularly those containing magnesium and silicon or magnesium and zinc, continue to age harden for long periods of time at room temperature.

By heating for a controlled time at slightly elevated temperatures, even further strengthening is possible, and properties are stabilized. This process is called artificial aging or precipitation hardening. By the proper combination of solution heat treatment, quenching, cold working, and artificial aging, the highest strengths are obtained. Aging for extended periods of time or at high temperatures can result in a reduction of mechanical properties. Restoration of the properties requires repeating the cycle of solution treatment, quenching, and aging.

ANNEALING CHARACTERISTICS

All wrought aluminum alloys are available in annealed form. In addition, it may be desirable to anneal an alloy from any other initial temper, after working, or between successive stages of working such as in deep drawing. Typical annealing treatments for aluminum alloy mill products are presented in Table 14-10. These treatments are typical for various sizes and methods of manufacture, and may not exactly describe the optimum treatment for a specific item. Castings, especially those with varying cross sections, are often annealed to homogenize their structures for uniform mechanical properties.

TYPICAL HEAT TREATMENTS

Typical heat treatments for aluminum alloy mill products are presented in Table 14-11. Recommended heat treatments for aluminum sand and permanent-mold castings are given in Table 14-12. The times and temperatures given for solution heat treatment are critical. The time required for the solution heat treatment of castings depends upon the section thicknesses of the castings, the alloys used, and the solidification rates of the castings. Less time is required for castings with thin, rapidly cooled sections.

Quenching should be accomplished by complete immersion of the workpieces with a minimum delay after they are removed from the furnace. Precipitation-hardenable aluminum alloys progressively lose their ability to develop maximum strength and corrosion resistance with decreasing quench rates. Limiting factors, however, with respect to quench rates are possible cracking, distortion, and the introduction of residual stresses.

Under certain conditions, complex castings that might crack or distort in a water quench can be quenched in oil, a synthetic medium, or by an air blast. When such quenching is done, however, the purchaser of the castings and the foundry making the castings must agree to the procedure used and the acceptable level of mechanical properties. Aging treatments can be varied slightly to attain the optimum treatment for a specific casting or to produce slightly different levels of mechanical properties.

HEAT TREATING FURNACES

Solution heat treatment of aluminum alloys is done in either molten-salt baths or air furnaces, depending primarily upon the size and shape of the parts to be treated, as well as the alloys from which they are made. Air furnaces are more common, generally because of their increased flexibility with respect to changing the operating temperatures and handling various size parts. Oil or gas-fired furnaces should be designed so that their combustion products do not contact the parts being treated because contact can cause oxidation.

Salt-bath furnaces generally permit faster heating of parts than is possible with air furnaces. They are also more suitable for handling small quantities of parts requiring different heating times. Other advantages of salt-bath furnaces include minimum distortion of the heated parts and uniform temperature control.

Regardless of the type of furnace used, precise temperature controls are essential. Each zone of the furnace should have its own control system. A detailed discussion of various types of heat treating furnaces and their controls is presented in Chapter 12, "Heat Treating Furnaces."

TABLE 14-10
Typical Annealing Treatments for Aluminum Alloy Mill Products*

Aluminum Alloys	Metal Temperature, °F (°C)	Approx. Time at Temp., hr	Temper Designation
1060, 1100, 1350	650 (345)	**	0
2014, 2017, 2024	775 (415)†	2-3	0

(continued)

TABLE 14-10—*Continued*

Aluminum Alloys	Metal Temperature, °F (°C)	Approx. Time at Temperature, hr	Temper Designation
2036	725 (385)†	2-3	0
2117, 2219	775 (415)†	2-3	0
3003	775 (415)	**	0
3004, 3105, 5005, 5050, 5052, 5056, 5083, 5086, 5154, 5254, 5454, 5456, 5457, 5652	650 (345)	**	0
6005, 6053, 6061, 6063, 6066	775 (415)†	2-3	0
7001, 7075, 7178	775 (415)††	2-3	0
Brazing sheet:			
No.'s 11, 12, 21, 22, 23, and 24	650 (345)	**	0

* Treatments listed are typical for various sizes and methods of manufacture, and may not exactly describe the optimum treatment for a specific item.

** Time in the furnace need not be longer than necessary to bring all parts of the load to annealing temperature. Rate of cooling is unimportant.

† These treatments are intended to remove effects of solution heat treatment and include cooling at a rate of about 50° F (30° C) per hour from the annealing temperature to 500° F (260° C). The rate of subsequent cooling is unimportant. Treatment at 650° F (345° C), followed by uncontrolled cooling, may be used to remove the effects of cold work, or to partially remove the effects of heat treatment.

†† This treatment is intended to remove the effects of solution heat treatment and includes cooling at an uncontrolled rate to 400° F (205° C) or less, followed by reheating to 450° F (230° C) for four hours. Treatment at 650° F (345° C), followed by uncontrolled cooling, may be used to remove the effects of cold work, or to partially remove the effects of heat treatment.

TABLE 14-11
Typical Heat Treatments for Aluminum Alloy Mill Products[a] [2]

Aluminum Alloy	Product	Solution Heat Treatment[b]		Precipitation Heat Treatment		
		Metal Temperature,[c] °F (°C)	Temper Designation	Metal Temperature,[c] °F (°C)	Approx. Time at Temperature, hr[d]	Temper Designation
2011	Rolled or cold-finished rod and bar	975 (525)	T3[e]	320 (160)	14	T8[e]
			T4	---	---	---
			T451[f]	---	---	---
2014[g]	Flat sheet	935 (500)	T3[e]	320 (160)	18	T6
			T42	320 (160)	18	T62
	Coiled sheet	935 (500)	T4	320 (160)	18	T6
			T42	320 (160)	18	T62
	Plate	935 (500)	T42	320 (160)	18	T62
			T451[f]	320 (160)	18	T651[f]

TABLE 14-11—*Continued*

Aluminum Alloy	Product	Solution Heat Treatment[b] Metal Temperature,[c] °F (°C)	Solution Heat Treatment[b] Temper Designation	Precipitation Heat Treatment Metal Temperature,[c] °F (°C)	Precipitation Heat Treatment Approx. Time at Temperature, hr[d]	Precipitation Heat Treatment Temper Designation
2014[g]	Rolled or cold-finished wire, rod, and bar	935 (500)	T4	320 (160)[h]	18	T6
			T42	320 (160)[h]	18	T62
			T451[f]	320 (160)[h]	18	T651[f]
	Extruded rod, bar, shapes and tube	935 (500)	T4	320 (160)[h]	18	T6
			T42	320 (160)[h]	18	T62
			T4510[f]	320 (160)[h]	18	T6510[f]
			T4511[f]	320 (160)[h]	18	T6511[f]
	Drawn tube	935 (500)	T4	320 (160)[h]	18	T6
			T42	320 (160)[h]	18	T62
	Die forgings	935 (500)[i]	T4	340 (170)	10	T6
	Hand forgings and rolled rings	935 (500)[i]	T4	340 (170)	10	T6
			T452[j]	340 (170)	10	T652[j]
2017	Rolled or cold-finished wire, rod, and bar	935 (500)	T4	---	---	---
			T42	---	---	---
			T451[f]	---	---	---
2018	Die forgings	950 (510)[k]	T4	340 (170)	10	T61
2024[g]	Flat sheet	920 (495)	T3[e]	375 (190)	12	T81[e]
			T361[e]	375 (190)	8	T861[e]
			T42	375 (190)	9	T62
				375 (190)	16	T72
	Coiled sheet	920 (495)	T4	---	---	---
			T42	375 (190)	9	T62
				375 (190)	16	T72
	Plate	920 (495)	T351[f]	375 (190)	12	T851[f]
			T361[e]	375 (190)	8	T861[e]
			T42	375 (190)	9	T62
	Rolled or cold-finished wire, rod, and bar	920 (495)	T4	375 (190)	12	T6
			T351[f]	375 (190)	12	T851[f]
			T36[e]	375 (190)	8	T86[e]
			T42	375 (190)	16	T62
	Extruded rod, bar, shapes and tube	920 (495)	T3	375 (190)	12	T81
			T3510[f]	375 (190)	12	T8510[f]
			T3511[f]	375 (190)	12	T8511[f]
			T42	375 (190)	16	T62
	Drawn tube	920 (495)	T3[e]	---	---	---
			T42	---	---	---
2025	Die forgings	960 (495)	T4	340 (170)	10	T6
2036	Sheet	930 (500)	T4	---	---	---
2117	Rolled or cold-finished wire and rod	935 (500)	T4	---	---	---
			T42	---	---	---
2218	Die forgings	950 (510)[k]	T4	340 (170)	10	T61
		950 (510)[l]	T41	460 (240)	6	T72
2219[g]	Flat sheet	995 (535)	T31[e]	350 (175)	18	T81[e]
			T37[e]	325 (165)	24	T87[e]
			T42	375 (190)	36	T62

(continued)

TABLE 14-11—*Continued*

Aluminum Alloy	Product	Solution Heat Treatment[b] Metal Temperature,[c] °F (°C)	Solution Heat Treatment[b] Temper Designation	Precipitation Heat Treatment Metal Temperature,[c] °F (°C)	Precipitation Heat Treatment Approx. Time at Temperature, hr[d]	Precipitation Heat Treatment Temper Designation
2219[g]	Plate	995 (535)	T31[e]	350 (175)	18	T81[e]
			T37[e]	350 (175)	18	T87[e]
			T351[f]	350 (175)	18	T851[f]
			T42	375 (190)	36	T62
	Rolled or cold-finished wire, rod, and bar	995 (535)	T351[f]	375 (190)	18	T851[f]
	Extruded rod, bar, shapes and tube	995 (535)	T31[e]	375 (190)	18	T81[e]
			T3510[f]	375 (190)	18	T8510[f]
			T3511[f]	375 (190)	18	T8511[f]
			T42	375 (190)	36	T62
	Die forgings and rolled rings	995 (535)	T4	375 (190)	26	T6
	Hand forgings	995 (535)	T4	375 (190)	26	T6
			T352[j]	350 (175)	18	T852[j]
2618	Forgings and rolled rings	985 (530)[k]	T4	390 (200)	20	T61
4032	Die forgings	950 (510)[l]	T4	340 (170)	10	T6
6005	Extruded rod, bar, shapes and tube	[m]	T1	350 (175)	8	T5
6053	Die forgings	970 (520)	T4	340 (170)	10	T6
6061[g]	Sheet	985 (530)	T4	320 (160)	18	T6
			T42	320 (160)	18	T62
	Plate	985 (530)	T4[n]	320 (160)	18	T6[n]
			T42	320 (160)	18	T62
			T451[f]	320 (160)	18	T651[f]
	Rolled or cold-finished wire, rod, and bar	985 (530)	T4	320 (160)[o]	18	T6
				320 (160)[o]	18	T89[e]
				320 (160)[o]	18	T93[p]
				320 (160)[o]	18	T913[p]
				320 (160)[o]	18	T94[p]
			T42	320 (160)[o]	18	T62
			T451[f]	320 (160)[o]	18	T651[f]
	Extruded rod, bar, shapes and tube	985 (530)[m]	T4	350 (175)	8	T6
			T4510[f]	350 (175)	8	T6510[f]
			T4511[f]	350 (175)	8	T6511[f]
		985 (530)	T42	350 (175)	8	T62
	Drawn tube	985 (530)	T4	320 (160)[o]	18	T6
			T42	320 (160)[o]	18	T62
	Die and hand forgings	985 (530)	T4	350 (175)	8	T6
	Rolled rings	985 (530)	T4	350 (175)	8	T6
			T452[j]	350 (175)	8	T652[j]
6063	Extruded rod, bar, shapes and tube	[m]	T1	360 (180)[q]	3	T5
		970 (520)[m]	T4	350 (175)[r]	8	T6
		970 (520)	T42	350 (175)[r]	8	T62

TABLE 14-11—*Continued*

Aluminum Alloy	Product	Solution Heat Treatment[b]: Metal Temperature,[c] °F (°C)	Solution Heat Treatment[b]: Temper Designation	Precipitation Heat Treatment: Metal Temperature,[c] °F (°C)	Precipitation Heat Treatment: Approx. Time at Temperature, hr[d]	Precipitation Heat Treatment: Temper Designation
6063	Drawn tube	970 (520)	T4	350 (175)	8	T6
				350 (175)	8	T83[e,m]
				350 (175)	8	T831[e,m]
				350 (175)	8	T832[e,m]
			T42	350 (175)	8	T62
6066	Extruded rod, bar, shapes and tube	990 (530)	T4	350 (175)	8	T6
			T42	350 (175)	8	T62
			T4510[f]	350 (175)	8	T6510[f]
			T4511[f]	350 (175)	8	T6511[f]
	Drawn tube	990 (530)	T4	350 (175)	8	T6
			T42	350 (175)	8	T62
	Die forgings	990 (530)	T4	350 (175)	8	T6
6070	Extruded rod, bar, shapes and tube	1015 (545)[m]	T4	320 (160)	18	T6
			T42	320 (160)	18	T62
6105	Extruded rod, bar, shapes and tube	[m]	T1	350 (175)	8	T5
6151	Die forgings	960 (515)	T4	340 (170)	10	T6
	Rolled rings	960 (515)	T4	340 (170)	10	T6
			T452[j]	340 (170)	10	T652[j]
6262	Rolled or cold-finished wire, rod, and bar	1000 (540)	T4	340 (170)	8	T6
				340 (170)	12	T9[p]
			T451[f]	340 (170)	8	T651[f]
			T42	340 (170)	8	T62
	Extruded rod, bar, shapes and tube	1000 (540)[m]	T4	350 (175)	12	T6
			T4510[f]	350 (175)	12	T6510[f]
			T4511[f]	350 (175)	12	T6511[f]
		1000 (540)	T42	350 (175)	12	T62
	Drawn tube	1000 (540)	T4	340 (170)	8	T6
				340 (170)	8	T9[p]
			T42	340 (170)	8	T62
6463	Extruded rod, bar, shapes and tube	[m]	T1	400 (205)[q]	1	T5
		970 (520)[m]	T4	350 (175)[r]	8	T6
		970 (520)	T42	350 (175)[r]	8	T62
6951	Sheet	985 (530)	T4	320 (160)	18	T6
			T42	320 (160)	18	T62
7001	Extruded rod, bar, shapes and tube	870 (465)	W	250 (120)	24	T6
				250 (120)	24	T62
			W510[f]	250 (120)	24	T6510[f]
			W511[f]	250 (120)	24	T6511[f]
7005	Extruded rod, bar, and shapes	---	---	---	---	T53[s]
7075[g]	Sheet	900 (480)[u]	W	250 (120)[t]	24	T6
						T62
				[v]	[v]	T76[w]
				[x,y]	[x,y]	T73[w]

(continued)

TABLE 14-11—*Continued*

Aluminum Alloy	Product	Solution Heat Treatment[b]		Precipitation Heat Treatment		
		Metal Temperature,[c] °F (°C)	Temper Designation	Metal Temperature,[c] °F (°C)	Approx. Time at Temperature, hr[d]	Temper Designation
7075[g]	Plate	900 (480)[u]	W	250 (120)[t]	24	T62
				[x,y]	[x,y]	T7351[f,w]
			W51[f]	250 (120)[t]	24	T651[f]
				[v]	[v]	T7651[f,w]
	Rolled or cold-finished wire, rod, and bar	915 (490)[u]	W	250 (120)	24	T6
						T62
				[x,y]	[x,y]	T73[w]
			W51[f]	250 (120)	24	T651[f]
				[x,y]	[x,y]	T7351[f,w]
	Extruded rod, bar, shapes and tube	870 (465)	W	250 (120)[z]	24	T6
						T62
				[x,y]	[x,y]	T73[w]
				[v]	[v]	T76[w]
			W510[f]	250 (120)[z]	24	T6510[f]
				[x,y]	[x,y]	T73510[f,w]
				[v]	[v]	T76510[f,w]
			W511[f]	250 (120)[z]	24	T6511[f]
				[x,y]	[x,y]	T73511[f,w]
				[v]	[v]	T76511[f,w]
	Drawn tube	870 (465)	W	250 (120)	24	T6
						T62
				[x,y]	[x,y]	T73[w]
	Die forgings	880 (470)[i]	W	250 (120)	24	T6
				[x]	[x]	T73[w]
			W52[j]	[x]	[x]	T7352[f,w]
	Hand forgings	880 (470)[i]	W	250 (120)	24	T6
				[x]	[x]	T73[w]
			W52[j]	250 (120)	24	T652[j]
				[x]	[x]	T7352[j,w]
	Rolled rings	880 (470)	W	250 (120)	24	T6
7178[g]	Sheet	875 (470)	W	250 (120)	24	T6
						T62
				[aa]	[aa]	T76[w]
	Plate	875 (470)	W	250 (120)	24	T62
			W51[f]	250 (120)	24	T651[f]
				[aa]	[aa]	T7651[f,w]
	Extruded rod, bar, shapes and tube	870 (465)	W	250 (120)	24	T6
				250 (120)	24	T62

TABLE 14-11—*Continued*

Aluminum Alloy	Product	Solution Heat Treatment[b] Metal Temperature,[c] °F (°C)	Solution Heat Treatment[b] Temper Designation	Precipitation Heat Treatment Metal Temperature,[c] °F (°C)	Precipitation Heat Treatment Approx. Time at Temperature, hr[d]	Precipitation Heat Treatment Temper Designation
7178[g]				[bb]	[bb]	T76[w]
			W510[f]	250 (120)	24	T6510[f]
			W511[f]	250 (120)	24	T6511[f]
			W510[f]	[bb]	[bb]	T76510[f,w]
			W511[f]	[bb]	[bb]	T76511[f,w]

[a] The times and temperatures shown are typical for various forms, sizes, and methods of manufacture, and may not exactly describe the optimum treatment for a specific item.

[b] Material should be quenched from the solution heat treating temperature as rapidly as possible and with minimum delay after removal from the furnace. Unless otherwise indicated, when material is quenched by total immersion in water, the water should be at room temperature and suitably cooled to remain below 100° F (40° C) during the quenching cycle. The use of high-velocity, high-volume jets of cold water is also effective for some materials.

[c] The nominal metal temperatures should be attained as rapidly as possible and maintained ±10° F (6° C) of nominal during the time at temperature.

[d] The time at temperature will depend on time required for load to reach temperature. The times shown are based on rapid heating, with soaking time measured from the time the load reaches within ±10° F (6° C) of the applicable temperature.

[e] Cold work subsequent to solution heat treatment and, where applicable, prior to any precipitation heat treatment is required to atain the specified mechanical properties for these tempers.

[f] Stress relieved by stretching. Required to produce a specified amount of permanent set subsequent to solution heat treatment and, where applicable, prior to any precipitation heat treatment.

[g] These heat treatments also apply to alclad sheet and plate in these alloys.

[h] An alternate treatment comprised of 8 hours at 350° F (175° C) may also be used.

[i] Quench after solution heat treatment in water at 140 to 180° F (60 to 80° C).

[j] Stress relieved by 1 to 5% cold reduction subsequent to solution heat treatment and prior to precipitation heat treatment.

[k] Quench after solution heat treatment in water at 212° F (100° C).

[l] Quench after solution heat treatment in air blast at room temperature.

[m] By suitable control of extrusion temperature, product may be quenched directly from extrusion press to provide specified properties for this temper. Some products may be adequately quenched in air blast at room temperature.

[n] Applies to tread plate only.

[o] An alternate treatment comprised of 8 hours at 340° F (170° C) may also be used.

[p] Cold working subsequent to precipitation heat treatment is necessary to secure the specified properties for this temper.

[q] An alternate treatment comprised of 1-2 hours at 400° F (205° C) may also be used.

[r] An alternate treatment comprised of 6 hours at 360° F (180° C) may also be used.

[s] No solution heat treatment, 72 hours at room temperature following press quench, followed by two-stage precipitation heat treatment of 8 hours at 225° F (105° C), plus 16 hours at 300° F (150° C).

[t] An alternate two-stage treatment comprised of 4 hours at 205° F (95° C) followed by 8 hours at 315° F (155° C) may also be used.

[u] With optimum homogenization, heat treating temperatures as high as 928° F (500° C) are sometimes acceptable.

[v] The aging practice will vary with the product, size, nature of equiment, loading procedures, and furnace control capabilities. The optimum practice for a specific item can be ascertained only by actual trial treatment of the item under specific conditions. Typical procedures involve a two-stage treatment comprised of 3 to 30 hours at 250° F (120° C), followed by 15 to 18 hours at 325° F (165° C) for extrusions. An alternate two-stage treatment of 8 hours at 210° F (100° C) followed by 24 to 28 hours at 325° F (165° C) may be used.

[w] The aging of aluminum alloys 7075 and 7178 from any temper to T73 (applicable to alloy 7075 only) or T76 temper series requires closer than normal controls on aging practice variables such as time, temperature, heating-up rates, etc., for any given item. In addition to the above, when re-aging material in the T6 temper series to the T73 or T76 temper series, the specific condition of the T6 temper material (such as its property level and other effects of processing variables) is extremely important and will affect the capability of the re-aged material to conform to the requirements specified for the applicable T73 or T76 temper series.

(continued)

TABLE 14-11—*Continued*

[x] Two-stage treatment comprised of 6 to 8 hours at 225° F (105° C) followed by a second stage of:
(1) 24 to 30 hours at 325° F (165° C) for sheet and plate.
(2) 8 to 10 hours or at 350° F (175° C) for rolled or cold-finished rod and bar.
(3) 6 to 8 hours at 350° F (175° C) for extrusions and tube.
(4) 8 to 10 hours at 350° F (175° C) for forgings in T73 temper and 6 to 8 hours at 350° F (175° C) for forgings in T7352 temper.

[y] An alternate two-stage treatment for sheet, plate, tube, and extrusions comprised of 6 to 8 hours at 225° F (105° C) followed by a second stage of 14 to 18 hours at 335° F (170° C) may be used providing a heating-up rate of 25° F (14° C) per hour is used. For rolled cold-finished rod and bar, the alternate treatment is 10 hours at 350° F (175° C).

[z] An alternate three-stage treatment comprised of 5 hours at 210° F (100° C), followed by 4 hours at 250° F (120° C), followed by 4 hours at 300° F (150° C) may also be used.

[aa] A two-stage treatment comprised of 3 to 5 hours at 250° F (120° C), followed by 15 to 18 hours at 325° F (165° C).

[bb] A two-stage treatment comprised of 3 to 5 hours at 250° F (120° C), followed by 18 to 21 hours at 320° F (160° C).

TABLE 14-12
Recommended Heat Treatments for Aluminum Sand and Permanent-Mold Castings[3]

			Solution Heat Treatment[b]		Aging Treatment	
Aluminum Alloy	Temper	Product[a]	Metal Temperature, °F (°C), ±10° F (6° C)	Time, hr	Metal Temperature, °F (°C), ±10° F (6° C)	Time, hr
201.0	T6	S	950-960 (510-515), then 980-990 (525-530)	2 14-20	310 (155)	20
201.0	T7	S	950-960 (510-515), then 980-990 (525-530)	2 14-20	370 (190)	5
204.0	T4	S or P	970 (520)	10	---	---
208.0	T55	S	---	---	310 (155)	16
222.0	0[c]	S	---	---	600 (315)	3
222.0	T61	S	950 (510)	12	310 (155)	11
222.0	T551	P	---	---	340 (170)	16-22
222.0	T65	---	950 (510)	4-12	340 (170)	7-9
242.0	0[d]	S	---	---	650 (345)	3
242.0	T571	S	---	---	400 (205)	8
242.0	T77	S	960 (515)	5[e]	625-675 (330-355)	2 min
242.0	T571	P	---	---	330-340 (165-170)	22-26
242.0	T61	S or P	960 (515)	4-12[e]	400-450 (205-230)	3-5
295.0	T4	S	960 (515)	12	---	---
295.0	T6	S	960 (515)	12	310 (155)	3-6
295.0	T62	S	960 (515)	12	310 (155)	12-24
295.0	T7	S	960 (515)	12	500 (260)	4-6
296.0	T4	P	950 (510)	8	---	---
296.0	T6	P	950 (510)	8	310 (155)	1-8
296.0	T7	P	950 (510)	8	500 (260)	4-6
319.0	T5	S	---	---	400 (205)	8
319.0	T6	S	940 (505)	12	310 (155)	2-5
319.0	T6	P	940 (505)	4-12	310 (155)	2-5
328.0	T6	S	960 (515)	12	310 (155)	2-5
332.0	T5	P	---	---	400 (205)	7-9
333.0	T5	P	---	---	400 (205)	7-9
333.0	T6	P	940 (505)	6-12	310 (155)	2-5
333.0	T7	P	940 (505)	6-12	500 (260)	4-6
336.0	T551	P	---	---	400 (205)	7-9
336.0	T65	P	960 (515)	8	400 (205)	7-9
354.0	---	[f]	980-995 (525-535)	10-12	[g]	[g]
355.0	T51	S	---	---	440 (225)	7-9
355.0	T6	S	980 (525)	12	310 (155)	3-5
355.0	T7	S	980 (525)	12	440 (225)	3-5
355.0	T71	S	980 (525)	12	475 (245)	4-6
355.0	T51	P	---	---	440 (225)	7-9
355.0	T6	P	980 (525)	4-12	310 (155)	2-5

TABLE 14-12—*Continued*

Aluminum Alloy	Temper	Product[a]	Solution Heat Treatment[b] Metal Temperature, °F (°C), ±10°F (6°C)	Solution Heat Treatment[b] Time, hr	Aging Treatment Metal Temperature, °F (°C), ±10°F (6°C)	Aging Treatment Time, hr
355.0	T62	P	980 (525)	4-12	340 (170)	14-18
355.0	T7	P	980 (525)	4-12	440 (225)	3-9
355.0	T71	P	980 (525)	4-12	475 (245)	3-6
C355.0	T6	S	980 (525)	12	310 (155)	3-5
C355.0	T61	P	980 (525)	6-12	Room, then 310 (155)	8 min 10-12
356.0	T51	S	---	---	440 (225)	7-9
356.0	T6	S	1000 (540)	12	310 (155)	3-5
356.0	T7	S	1000 (540)	12	400 (205)	3-5
356.0	T71	S	1000 (540)	10-12	475 (245)	3
356.0	T51	P	---	---	440 (225)	7-9
356.0	T6	P	1000 (540)	4-12	310 (155)	2-5
356.0	T7	P	1000 (540)	4-12	440 (225)	7-9
356.0	T71	P	1000 (540)	4-12	475 (245)	3-6
A356.0	T6	S	1000 (540)	12	310 (155)	3-5
A356.0	T61	P	1000 (540)	6-12	Room, then 310 (155)	8 min 6-12
357.0	T6	P	1000 (540)	8	350 (175)	6
A357.0	---	[f]	1000 (540)	8-12	[g]	[g]
359.0	---	[f]	1000 (540)	10-14	[g]	[g]
A444.0	T4	P	1000 (540)	8-12	---	---
520.0	T4	S	810 (430)	18[h]	---	---
535.0	T5[c]	S	750 (400)	5	---	---
705.0	T5	S	---	---	Room or 210 (100)	504 (21 days) 8
705.0	T5	P	---	---	Room or 210 (100)	504 (21 days) 10
707.0	T5	S	---	---	310 (155)	3-5
707.0	T7	S	990 (530)	8-16	350 (175)	4-10
707.0	T5	P	---	---	Room or 210 (100)	504 (21 days) 8
707.0	T7	P	990 (530)	4-8	350 (175)	4-10
707.0	T5	P	---	---	Room or 210 (100)	504 (21 days) 8
707.0	T7	P	990 (530)	4-8	350 (175)	4-10
710.0	T5	S	---	---	Room	504 (21 days)
711.0	T1	P	---	---	Room	504 (21 days)
712.0	T5	S	---	---	Room or 315 (155)	504 (21 days) 6-8
713.0	T5	S or P	---	---	Room or 250 (120)	504 (21 days) 16
771.0	T53[c]	S	775 (415)	5[i]	360 (180)	4[i]
771.0	T5	S	---	---	355 (180)	3-5[i]
771.0	T51	S	---	---	405	6
771.0	T52	S	---	---	[c]	---
771.0	T6	S	1090 (590)	6[i]	265 (130)	3

(continued)

TABLE 14-12—*Continued*

Aluminum Alloy	Temper	Product[a]	Solution Heat Treatment[b] Metal Temperature, °F (°C), ±10°F (6°C)	Time, hr	Aging Treatment Metal Temperature, °F (°C), ±10°F (6°C)	Time, hr
771.0	T71	S	1090 (590)	6[d]	285 (140)	15
850.0	T5	S or P	---	---	430 (220)	7-9
851.0	T5	S or P	---	---	430 (220)	7-9
851.0	T6	P	900 (480)	6	430 (220)	4
852.0	T5	S or P	---	---	430 (220)	7-9

[a] S = sand cast, P = permanent-mold cast.
[b] Unless otherwise noted, quench in water at a temperature of 150-212° F (65-100° C).
[c] Stress relieve for dimensional stability by: (1) holding at 775° F (415° C), ±25° F (14° C) for 5 hours; then (2) furnace cool to 650° F (345° C) for 2 or more hours; then (3) furnace cool to 450° F (230° C) for not more than 1/2 hour; then (4) furnace cool to 250° F (120° C) for about 2 hours; and then (5) cool to room temperature in still air outside the furnace.
[d] No quench required. Cool in still air outside furnace.
[e] Use air-blast quench.
[f] Casting process varies (sand, permanent mold, or composite) to obtain desired mechanical properties.
[g] Solution heat treat as indicated, then artificially age by heating uniformly for the time and at the temperature necessary to obtain the desired mechanical properties.
[h] Quench in water at a temperature of 150 to 212° F (65 to 100° C) for 10 to 20 seconds only.
[i] Cool to room temperature in still air outside furnace.

MAGNESIUM ALLOYS

Like many other metals, magnesium alloys are heat treated to improve their mechanical properties or to facilitate fabrication. The type of heat treatment used depends upon the specific alloy, the form of the product (wrought or cast), and the service requirements. Die castings are not usually heat treated; but sand, investment, and permanent-mold castings are generally heat treated to improve their properties. Details of various magnesium alloys, including temper designations and properties, are presented in Chapter 6, "Aluminum, Copper and Magnesium."

ANNEALING

Annealing of magnesium alloys is generally restricted to wrought products. This treatment is performed to achieve maximum ductility, but also results in lowered mechanical properties. The temperatures employed for annealing range from 550 to 850° F (290 to 455° C), depending upon the alloy.

Recommended temperatures for annealing some of the more common magnesium alloys are presented in Table 14-13. The time the workpieces are held at these temperatures is generally one hour or more, with no maximum time limit.

SOLUTION TREATING AND AGING

Solution heat treatment of the thermally responsive magnesium alloys provides maximum toughness and shock resistance. Artificial age hardening (precipitation heat treatment) after solution heat treatment usually results in maximum strength and hardness, with some reduction in toughness and ductility.

Recommended temperature/time cycles for solution treating and aging some of the more common magnesium alloys and product forms are presented in Table 14-14. The temperatures and times given are typical for various workpiece sizes and different production methods. As a result, they may not be optimum for a specific workpiece.

For cast magnesium alloys containing thorium, such as HK31A and HZ32A alloys, it is advisable to bring the furnace load to temperature as fast as possible to avoid grain coarsening. For other alloys, the furnace load can be raised to temperature more slowly, thus guarding against melting of eutectic compounds and minimizing condensation on workpiece surfaces, with resultant staining.

STRESS RELIEVING

A stress-relieving heat treatment is sometimes required for wrought magnesium alloy parts. Such a treatment will reduce

TABLE 14-13
Annealing Temperatures for Various Wrought Magnesium Alloys*

Magnesium Alloy	Starting Temper	Annealing Temperature, °F (°C)
AZ Commercial	F	650 (345)
AZ31B	F, H24, H26	650 (345)
AZ31C	F	650 (345)
AZ61A	F	650 (345)
AZ80A	F, T5, T6	725 (385)
HK31A	F, H24	750 (400)
HM21A	F, T5, T8, T81	850 (455)
HM31A	F, T5	850 (455)
ZK60A	F, T5, T6	550 (290)

* Time at temperature is one hour minimum, no maximum.

TABLE 14-14
Solution Heat Treatment and Age Hardening of Some Magnesium Alloys[a]

Magnesium Alloy	Temper	Solution Heat Treatment: Temperature, °F (°C)	Solution Heat Treatment: Soak Time, hr	Artificial Age Hardening: Temperature, °F (°C)	Artificial Age Hardening: Soak Time, hr	Ultimate Tensile Strength, ksi (MPa): Typical	Ultimate Tensile Strength, ksi (MPa): Minimum
Sand, investment, and permanent-mold casting alloys:							
AM100A	T61	775 (415)	18[b]	400 (205)	24	40 (276)	34 (234)
AZ63A	T6	725 (385)	12	425 (220)	5	40 (276)	34 (234)
AZ81A	T4	775 (415)	18[b]	---	---	40 (276)	34 (234)
AZ91C	T6	775 (415)	18[b]	335 (170)	16	40 (276)	34 (234)
AZ92A	T6	765 (405)	18[c]	425 (220)	5	40 (276)	34 (234)
EZ33A	T5	---	---	650 (345)	2	---	---
				then 420 (215)	5	23 (159)	20 (138)
HK31A	T6	1050 (565)	2[d]	400 (205)	16	32 (221)	27 (186)
HZ32A	T5	---	---	600 (315)	16	30 (207)	27 (186)
QE22A	T6	975 (525)	8	400 (205)	8	40 (276)	28 (193)
ZE41A	T5	---	---	625 (330)	2	---	---
				then 350 (175)	16	30 (207)	28 (193)
ZH62A	T5	---	---	625 (330)	2	---	---
				then 350 (175)	16	40 (276)	35 (241)
ZK51A	T5	---	---	350 (175)	12	40 (276)	34 (234)
ZK61A	T6	930 (500)	2	265 (130)	48	40 (276)	30 (207)
Extruded solid shapes:[e]							
AZ80A	T5	---	---	350 (175)	16	55 (379)	48 (331)
HM31A	T5	---	---	425 (220)	16	44 (303)	37 (255)
ZK60A	T5	---	---	300 (150)	24	52 (359)	45 (310)
Extruded tube:[e]							
ZK60A	T5	---	---	300 (150)	24	50 (345)	46 (317)
Sheet and plate:							
AZ31B	H24	---	---	275 (135)[f]	2	37-42[e] (255-290)	29-39[e] (200-269)
AZ31B	H26	---	---	250 (120)[f]	2	38-40[e] (262-276)	35-39[e] (241-269)
AZ31B	0	---	---	650 (345)	2	36-37[e] (248-255)	32[g] (221)
HK31A	H24	---	---	620 (325)[h]	1	37-39[e] (255-269)	33-34[e] (228-234)
HM21A	T8	---	---	715 (380)	1	33-37[e] (228-255)	30-33[e] (207-228)

[a] Heated magnesium alloy parts are generally cooled in air, with a fan or fans, after removal from the furnace. However, some alloys, such as QE22A, are sometimes quenched in water or other suitable media from the solution heat treating temperature.
[b] As an alternative, if necessary to prevent grain germination, soak for 6 hours at 775° F (415° C), 2 hours at 665° F (350° C), and then 10 hours at 775° F.
[c] As an alternative, if necessary to prevent grain germination, soak for 6 hours at 765° F (405° C), 2 hours at 665° F (350° C), and then 10 hours at 765° F.
[d] Furnaces should be heated to temperature before loading castings.
[e] Tensile strengths vary with metal thickness.
[f] Partial annealing after strain hardening; not age hardening.
[g] Maximum tensile strength of 40 ksi (276 MPa) for complete annealing.
[h] Temperature applies only to sheets. For plates, soak at 400° F (205° C).

Note: Magnesium-aluminum-zinc casting alloys, such as AZ63A, AZ81A, AZ91C, and AZ92A alloys, should be slowly heated (for about 2 hours) from 500 to 700° F (260 to 370° C).

the residual stresses present due to welding or cold working operations, such as forming or bending. Stress relieving is also required for some castings to reduce residual stresses caused by welding, contraction resulting from mold restraint, preferential cooling after heat treatment, or machining.

The temperatures and time cycles recommended for stress relieving in this section are intended to reduce stresses and provide practical dimensional stability, while not affecting mechanical properties.

Castings

For all magnesium-aluminum alloys, of all tempers, a stress-relieving temperature of 500° F (260° C) should be used for one hour. For zinc-zirconium alloys, stress relieving should be done at 625° F (330° C) for two hours, followed by 265° F (130° C) for 48 hours.

Wrought Products

Recommended temperatures and times for stress relieving some common wrought magnesium alloys are presented in Table 14-15.

REHEAT TREATMENT

Reheat treatment of magnesium alloys is not usually necessary. However, if the parts have been overaged or the compound ratings of the castings are too high, reheat treating can usually be accomplished without any detrimental effects. An exception to this may be with the AZ and thorium-containing (HK and HM) alloys, which are subject to grain coarsening. Products made from these alloys should be carefully examined after reheat treatments for possible coarse grain structures.

HEAT TREATING PRACTICES

Optimum heat treatment requires the use of proper furnaces and procedures. Since protective atmospheres are generally used for the solution heat treatment of magnesium alloys, gas-tight furnaces are necessary. The furnaces must have means for circulating the atmosphere and be equipped with suitable controls and safety devices. There should be no direct radiation from the heating elements or impingement of flames on the workpieces. Details about the various furnaces available are presented in Chapter 12, "Heat Treating Furnaces."

Atmospheres

Since magnesium castings are subject to excessive surface oxidation at temperatures of 750° F (400° C) or more, a protective atmosphere containing sulfur dioxide (SO_2) gas is generally maintained in the furnace for solution heat treatment. A minimum SO_2 content of 0.5% is necessary in the furnace atmosphere when solution heat treating AZ81A, AZ92A, AM100A, and AZ91C magnesium alloys.

Alloys containing rare earth metals and thorium require a minimum SO_2 content of 1%. Concentrations above these percentages are not harmful, so it is desirable to maintain an excess. Below a temperature of 750° F (400° C), a protective atmosphere is not necessary, but advisable. While SO_2 acts as a fire preventive, it does not extinguish a fire once started.

In areas where air pollution is a problem or furnace atmospheres are not collected in air control equipment, such as scrubbers, other inert gaseous atmospheres may be used. For example, 3% CO_2 may be substituted for the 0.5% SO_2 requirement, and 5% CO_2 for the 1% SO_2. Another possibility is the use of SF_6 (sulfur hexafluoride) at ¼ to ½%. Since SF_6 is

TABLE 14-15
Stress-Relieving Temperatures and Times for Various Wrought Magnesium Alloys

Wrought Magnesium Alloy	Temper/Product Form	Stress-Relieving Temperature, ° F (° C)	Soak Time, min
AZ Commercial	All tempers and forms	500 (260)	15
AZ31B	Extrusions, forgings, and annealed sheets	500 (260)	15
AZ31B	H temper sheets	300 (150)	60
AZ61A	Extrusions, forgings, and annealed sheets	500 (260)	15
AZ61A	H temper sheets	400 (205)	60
AZ80A	F temper	500 (260)	15
AZ80A	T5 and T6 tempers	400 (205)	60
HK31A	F and O tempers	650 (345)	60
HK31A	H24 temper	550 (290)	30
HM21A	F, T5, and T8 tempers	700 (370)	30
HM21A	T81 temper	750 (400)	30
HM31A	F and T5 tempers	800 (425)	60
ZK60A	F temper	500 (260)	15
ZK60A	T5 and T6 tempers	300 (150)	60

odorless, nontoxic, and an excellent retardant for magnesium oxidation, it can be used to create a useable furnace atmosphere.

Loading Furnaces

Magnesium castings to be heat treated must be clean, dry, and free from grinding dust and machining chips. Each furnace load must consist of parts having the same alloy composition because some magnesium alloys have lower eutectic points than others. The specific alloy and the size and shape of the casting determine the manner in which it is loaded into the furnace.

At temperatures used for solution heat treating, magnesium castings of long and slender or unusual shapes must be given adequate support. In some cases, special fixtures may be required to prevent distortion of the castings. Haphazard piling of castings and overloading of the furnace can impede circulation of the atmosphere and contribute to warpage. Castings made from alloys requiring higher temperatures for solution heat treating are more likely to warp than those treated at lower temperatures.

The recommended temperature for loading castings into a furnace is about 500° F (260° C). The maximum temperature for furnace loading must never exceed 670° F (355° C) when treating magnesium-aluminum-zinc alloys. For these alloys, the load should be uniformly and slowly heated so that about two hours elapse between temperatures of 500 and 700° F (260 and 370° C). Faster heating may cause eutectic melting, resulting in low mechanical properties. When solution heat treating HK31A alloy, the furnace should be heated to temperature before loading and then reheated (with the load) rapidly to temperature.

Temperature Control

The temperature range in which magnesium alloys can be heat treated is narrow. As a result, temperature variations in the furnace must be held to a minimum; +10° F (6° C) is generally recommended. Uniformity of temperature throughout the furnace is commonly achieved by using high rates of atmosphere recirculation.

Safety Considerations

In heat treating magnesium alloys, a potential fire hazard exists. When the furnace temperature exceeds the maximum solution heat treating temperature of the alloy, because of control failure or other reasons, the metal can ignite and burn. A safety device to shut off all power is essential for each furnace.

One method of extinguishing a magnesium fire is to introduce boron trifluoride into the closed furnace. Water should never be used to extinguish a magnesium fire. It is recommended that the furnace be allowed to cool to below 300° F (150° C) before opening the door and exposing the workpieces to air.

COPPERS AND COPPER ALLOYS

The wrought coppers and many copper alloys, discussed in Chapter 6, "Aluminum, Copper and Magnesium," are single-phase metals. As a result, heat treatment is limited to the annealing of cold worked metal for recrystallization or stress relieving. Castings are also sometimes annealed to soften them and increase their ductility and/or toughness. Annealing may be carried out in roller-hearth, batch-type, and strip or strand-annealing furnaces. Furnaces are discussed in Chapter 12, "Heating Treating Furnaces." Since some lubricants on the metals can cause staining, such lubricants should be removed before annealing. Furnace atmospheres should be controlled to suit the composition of the metal being annealed in order to minimize oxidation.

Castings and billets of copper alloys are sometimes given a homogenizing heat treatment to reduce chemical segregation and to attain the required hardness, ductility, or toughness. Homogenizing consists of heating to temperatures above the upper annealing range and soaking at temperature for up to 10 hours. Temperatures and times vary with the alloy, the size of the cast grains, and the desired results.

COPPERS

Wrought coppers, except for the special-purpose types, include oxygen-free coppers; oxygen-free, silver-bearing coppers; tough-pitch coppers; tough-pitch, silver-bearing coppers; and phosphorus-deoxidized coppers. Mill products may be cold reduced from 40 to 90% prior to annealing.

Oxygen-Free Coppers

Oxygen-free coppers without silver are generally annealed for one hour at temperature in the temperature range of 600 to 1200° F (315 to 650° C), depending upon the amount of cold work that has preceded the anneal. Lower annealing temperatures provide a finer grain size, which is generally beneficial if additional cold forming is to be done. Oxygen-free, silver-bearing coppers are generally annealed for one hour in the temperature range of 700 to 1400° F (370 to 760° C), depending upon both the silver content and the amount of cold work that has preceded the anneal.

Silver is added to oxygen-free coppers for the purpose of raising the annealing temperature, inhibiting grain growth, and increasing the endurance limit. The annealing temperature increases in proportion to the amount of silver present. Oxygen-free coppers can be bright annealed in reducing atmospheres containing free hydrogen.

Tough-Pitch Coppers

Tough-pitch coppers contain oxygen in the form of cuprous oxide, with up to 0.07% oxygen present. Wrought, tough-pitch copper that does not contain silver can be annealed for one hour in a temperature range of 600 to 1400° F (315 to 760° C). Silver-bearing, wrought tough-pitch coppers can be annealed in a temperature range of 700 to 1400° F (370 to 760° C), depending upon the silver content and amount of cold work that has preceded the anneal.

The cuprous oxide in tough-pitch coppers restricts grain growth. Within the temperature ranges given, it is unlikely that the grain size would exceed an average of 0.050 mm (0.002″), and the strength and ductility would be adequate for further cold working.

Tough-pitch coppers should not be annealed in reducing atmospheres that contain free hydrogen because they are subject to hydrogen embrittlement. Hydrogen diffuses through the hot copper and combines with the oxygen of the cuprous oxide to form steam fissures, which embrittle the metal.

COPPERS AND COPPER ALLOYS

Phosphorus-Deoxidized Coppers

Phosphorus-deoxidized coppers contain enough residual phosphorus to ensure complete deoxidation. These metals are not subject to hydrogen embrittlement, and their formability is enhanced by the absence of cuprous oxide. Their annealing temperatures are in the range of 700 to 1200° F (370 to 650° C) for one hour at temperature. Since the grain size of phosphorus-deoxidized coppers can be varied in a controlled manner, these metals may be annealed to meet a grain size range. The lower annealing temperatures produce a finer, recrystallized grain size, and the higher annealing temperatures produce a coarser grain size. The residual phosphorus in these coppers lowers the electrical conductivity.

BRASSES

Copper-zinc alloys, in general, are called brasses, although each has a commonly used trade name. When these alloys are annealed after cold working, they are annealed to meet a specific grain size or tensile strength requirement. Grain size is the average diameter in millimeters of the grains observed. For sheet and strip stock, grain size is measured on a plane parallel to the surface of the metal.

Gilding Metal

Alloy C21000 (95-5 copper-zinc), gilding metal, has nominal grain sizes given in ASTM Specification B36 of 0.015 mm (0.025 mm maximum), 0.025 mm (0.015-0.035 mm), 0.035 mm (0.025-0.050 mm), and 0.050 mm (0.035-0.090 mm). The approximate annealing temperatures that correspond to these nominal grain sizes for metal having been reduced 40-60% in thickness by cold rolling are: 930° F (500° C), 1040° F (560° C), 1165° F (630° C), and 1240° F (670° C)—for one hour at temperature.

Commercial Bronze

Alloy C22000 (90-10 copper-zinc), commercial bronze, has nominal grain sizes given in ASTM B36 of 0.015 mm (0.025 mm maximum), 0.025 mm (0.015-0.035 mm), 0.035 mm (0.025-0.050 mm), and 0.050 mm (0.035-0.090 mm). The approximate annealing temperatures that correspond to these nominal grain sizes for metal having been reduced 40-60% in thickness by cold rolling are: 840° F (450° C), 975° F (525° C), 1065° F (575° C), and 1150° F (620° C)—for one hour at temperature.

Red Brass

Alloy C23000 (85-15 copper-zinc), red brass, has nominal grain sizes given in ASTM B36 of 0.015 mm (0.025 mm maximum), 0.025 mm (0.015-0.035 mm), 0.035 mm (0.025-0.050 mm), 0.050 mm (0.035-0.070 mm), and 0.070 mm (0.050-0.100 mm). The approximate annealing temperatures that correspond to these nominal grain sizes for metal having been reduced 40-60% in thickness by cold rolling are: 885° F (475° C), 975° F (525° C), 1065° F (575° C), 1155° F (625° C), and 1220° F (660° C)—for one hour at temperature.

Low Brass

Alloy C24000 (80-20 copper-zinc), low brass, has nominal grain sizes given in ASTM B36 of 0.015 mm (0.025 mm maximum), 0.025 mm (0.015-0.035 mm), 0.035 mm (0.025-0.050 mm), 0.050 mm (0.035-0.070 mm), and 0.070 mm (0.050-0.120 mm). The approximate annealing temperatures that correspond to these nominal grain sizes for metal having been reduced 40-60% in thickness by cold rolling are: 860° F (460° C), 985° F (530° C), 1065° F (575° C), 1130° F (610° C), and 1200° F (650° C)—for one hour at temperature.

Cartridge Brass

Alloy C26000 (70-30 copper-zinc), cartridge brass, is one of the least expensive of the copper alloys and has excellent formability. It is by far the most popular copper alloy because of its great adaptability for many and diverse applications. It has nominal grain sizes given in ASTM B36 of 0.015 mm (0.025 mm maximum), 0.025 mm (0.015-0.035 mm), 0.035 mm (0.025-0.050 mm), 0.050 mm (0.035-0.070 mm), 0.070 mm (0.050-0.120 mm), and 0.120 mm (0.070 mm minimum). The approximate annealing temperatures that correspond to these nominal grain sizes for metal having been reduced 40-60% by cold rolling are: 795° F (425° C), 895° F (480° C), 970° F (520° C), 1040° F (560° C), 1095° F (590° C), and 1200° F (650° C)—for one hour at temperature.

High-Zinc Brasses

The high-zinc brasses are subject to stress-corrosion cracking if a sufficiently high residual or applied stress is present along with small amounts of ammonia, mercury, and their compounds and cyanides. In fact, any copper alloy containing more than 15-18% zinc is subject to dezincification and stress-corrosion cracking. Stress-corrosion cracks are typically intergranular but can be transgranular. Unless conditions conducive to stress-corrosion cracking are very severe, alloy C23000 with 15% zinc can be used to avoid stress corrosion when the stresses are applied stresses. For parts formed from alloys of higher zinc content, where the residual stresses are high enough to cause stress-corrosion cracking, a stress-relieving heat treatment should be used.

Stress-relieving heat treatments may also be used to relieve residual stresses in formed parts in order to improve their functionability in spring applications. Typical stress-relief heat-treating temperatures with one hour at temperature are given in Table 14-16.

Leaded Brasses

Lead is added to brasses in amounts from 0.25 to 3.0% to provide good machinability. The leaded brasses that are used in greatest amounts contain 60-65% copper, 0.25-3% lead, and the remainder zinc. They have annealing characteristics similar to the high-zinc brasses. In ASTM B121, their nominal grain sizes are given as 0.025 mm (0.015-0.035 mm), 0.035 mm (0.025-0.050 mm), 0.050 mm (0.035-0.070 mm), and 0.070 mm (0.050-0.100 mm). The approximate annealing temperatures that correspond to these nominal grain sizes for metal having been reduced 40-60% in thickness by cold rolling are: 840° F (450° C), 1005° F (540° C), and 1020° F (550° C). The grain size range for 0.070 mm nominal cannot be met unless the copper content exceeds 63% because the required temperature causes the alpha-to-beta transformation to begin to take place, refining the grain size. The leaded brasses are also subject to stress-corrosion cracking.

Tin-Bearing Brasses

Tin is added to the copper-zinc brasses to increase their strength and enhance their corrosion resistance. The low-zinc, tin-bearing brasses are used mostly in electrical connectors and terminals for wiring devices and communications equipment. These metals are seldom used in annealed tempers or annealed during forming operations. Alloys C40500, C40800, C41100,

TABLE 14-16
Stress-Relieving Treatments for Copper Alloys, One Hour at Temperature

Alloy	Trade Name	Stress-Relieving Temperature, °F (°C)
C21000	Gilding metal	390-450 (200-230)
C22000	Commercial bronze	390-450 (200-230)
C23000	Red brass	390-450 (200-230)
C24000	Low brass	390-450 (200-230)
C26000	Cartridge brass	390-450 (200-230)
C27000	Yellow brass	390-450 (200-230)
C28000	Muntz metal	390-450 (200-230)
C35000	Medium-leaded brass	390-450 (200-230)
C35300	High-leaded brass	390-450 (200-230)
C42200	Lubronze	390 (200)
C42500	Lubaloy X	390 (200)
C44300	Admiralty, arsenical	550 (290)
C44400	Admiralty, antimonial	550 (290)
C44500	Admiralty, phosphorized	550 (290)
C51000	Phosphor bronze, 5%	350-425 (175-220)
C52100	Phosphor bronze, 8%	300-400 (150-205)
C61300	Aluminum bronze	650 (345)
C61400	Aluminum bronze	650 (345)
C63800	Coronze	600 (315)
C65400	Ultronze	445 (230)
C65500	High-silicon bronze	650 (345)
C66400	Cobron	390 (200)
C68800	Alcoloy	540 (280)
C70600	Copper nickel, 10%	500-600 (260-315)
C71500	Copper nickel, 30%	500-600 (260-315)
C72500		540 (280)
C75200	Nickel silver, 18%	550-650 (290-345)
C76200	Nickel silver, 12%	525-625 (275-330)
C77000	Nickel silver, 18%	480-600 (250-315)

C41300, C41500, C42200, C42500, C43000, and C43400 are covered in ASTM B591. Nominal grain sizes of 0.015 mm (0.025 mm maximum), 0.025 mm (0.015-0.035 mm), and 0.035 mm (0.025-0.050 mm) are listed for annealed tempers. The annealing temperatures for one hour at temperature to meet these grain sizes are given in Table 14-17.

TABLE 14-17
Annealing Temperatures for Tin-Bearing Brasses, One Hour at Temperature

Alloy No.	Nominal Grain Size, mm: 0.015	0.025	0.035
	Annealing Temperature, °F (°C)		
C40500	950 (510)	1020 (550)	1130 (610)
C40800	1040 (560)	1130 (610)	1165 (630)
C41100	985 (530)	1110 (600)	1155 (625)
C41300	1005 (540)	1130 (610)	1165 (630)
C41500	1040 (560)	1130 (610)	1165 (630)
C42200	975 (525)	1095 (590)	1130 (610)
C42500	975 (525)	1065 (575)	1130 (610)
C43000	970 (520)	1040 (560)	1110 (600)
C43400	970 (520)	1040 (560)	1110 (600)

Phosphor Bronzes

The phosphor bronzes are copper-tin alloys that contain a phosphorous residual from the deoxidation of the molten metal. In general, the phosphor bronzes are used in cold worked tempers for functional springs, such as contact springs in electrical switches and connectors for electrical and electronic devices. Bellows are drawn parts that are frequently made from phosphor bronze and require annealing during fabrication. The commercially important phosphor bronzes are alloys C51000, C51100, and C52100. In ASTM B103, tensile strength ranges are given for soft temper. However, in commercial processing, alloys are annealed to meet grain size ranges. The nominal grain sizes frequently specified and the nominal corresponding one-hour annealing temperatures for the alloys are given in Table 14-18.

TABLE 14-18
Annealing Temperatures for Phosphor Bronzes, One Hour at Temperature

Alloy No.	Nominal Grain Size, mm: 0.015	0.030	0.050
	Annealing Temperature, °F (°C)		
C51000	970 (520)	1095 (590)	1220 (660)
C51100	950 (510)	1075 (580)	1165 (630)
C52100	950 (510)	1110 (600)	1200 (650)

Aluminum Bronzes and Brasses

The aluminum bronzes C60600, C61000, C61300, and C61400 are covered by ASTM Specification B169, and are annealed to meet minimum tensile strength, yield strength, and elongation requirements. The *ASM Metals Handbook*, 9th edition, volume 4, gives the annealing temperature ranges for these alloys (see Table 14-19).

Aluminum brass alloy C68700 is most often used in the form of tubes and plates for fresh-water power plant condensers. It must be stress-relief annealed after cold working to avoid stress-corrosion cracking. The annealing temperature range is 800-1100° F (425-595° C).

Alloy C68800 is a cobalt-bearing, high-strength aluminum brass produced in strip form and used in cold rolled tempers for wiring devices, blade receptacles in wall outlets, and in many other applications where spring properties are required. The alloy's applications seldom require annealing. After a cold

TABLE 14-19
Annealing Temperatures for Aluminum Bronzes

Alloy No.	Annealing Temperature Range, °F (°C)
C60600	1000-1200 (540-650)
C61000	1110-1250 (600-675)
C61300	1400-1600 (760-870)
C61400	1400-1600 (760-870)

reduction of 40 to 60% in thickness by rolling, alloy C68800 can be annealed in the temperature range of 750 to 1110°F (400 to 600°C).

In Table 14-16, a temperature of 540°F (280°C) is recommended for stress-relief annealing. When the cold worked alloy is annealed at this temperature, a reaction takes place that increases the strength about 10 ksi (69 MPa). Lower stress-relieving temperatures increase the alloy's susceptibility to stress corrosion. Other high-zinc, cold worked brasses and nickel silvers are also strengthened somewhat when they are stress-relief annealed.

Alloy C63800 is another alloy that is unique, containing 2.8% aluminum, 1.8% silicon, and 0.4% cobalt. Cobalt is present as a dispersion of intermetallic particles that strengthens the alloy. As a result, a minimum of cold rolling is required to produce high strength. A combination of strength and good formability results, and therefore the alloy is used extensively in spring applications. The alloy is also used occasionally in drawn parts. When annealing is required, a temperature range of 840 to 1200°F (450 to 650°C) for one hour at temperature is recommended.

Alloy C65400, containing 3.0% silicon, 1.5% tin, and 0.07% chromium, is a high-strength spring alloy that is heat treated for another reason. Springs are often required to function in environments where the temperature is higher than normal atmospheric; for example, under the hood of an automobile. Under these conditions, stress relaxation can occur, and alloy C65400 was designed to be an inexpensive, high-strength alloy that could resist stress relaxation at temperatures up to 220°F (105°C). After forming into parts, a stabilizing anneal at 445°F (230°C) for one hour enhances resistance to stress relaxation.

COPPER-NICKEL ALLOYS

The two most popular copper-nickel alloys are C70600 and C71500. They are used extensively in applications requiring resistance to corrosion in both fresh and salt water. In tubular form, they are used in power plant condensers where the coolant may be salt, brackish, or fresh water. They contain iron, and it is important that the iron be in solution in the copper-nickel alloy matrix. Annealing temperatures are high, in the range of 1100 to 1500°F (595 to 815°C), in order to put the iron into solution. Rapid cooling by quenching is desirable to keep the iron in solution.

NICKEL-SILVER ALLOYS

The nickel silvers are copper-nickel-zinc alloys, also called nickel brasses. Their nickel contents vary from 8 to 18%, and zinc contents from 10 to 27%. Their name is derived from their silvery color. They are used in the fabrication of silver-plated holloware and flatware, and also as relay springs, connectors, and terminals in electronic and communication devices. The grades used to make holloware are frequently supplied in annealed tempers. The high-zinc, high-strength grades are used in cold rolled tempers and seldom annealed during parts fabrication. Annealing, when done, is used to meet grain size requirements. ASTM B122 lists nominal grain sizes of 0.015 mm (0.025 mm maximum), 0.035 mm (0.025-0.050 mm), and 0.070 mm (0.050-0.100 mm). The nickel-silver alloys are listed in Table 14-20, with the nominal annealing temperatures for one hour at temperature to correspond with the nominal grain sizes.

Nickel-silver alloys are prone to fire cracking. When parts containing residual stresses are to be annealed, they should be brought to temperature slowly. Annealing temperatures are high; and if highly stressed parts are put into a hot furnace, fire cracking is likely to occur.

TABLE 14-20
Annealing Temperatures for Nickel-Silver Alloys, One Hour at Temperature

Alloy No.	Nominal Grain Size, mm		
	0.015	0.035	0.050
	Annealing Temperature, °F (°C)		
C73500	1110 (600)	1380 (750)	---
C74000	1095 (590)	1200 (650)	1240 (670)
C74300	970 (520)	1040 (560)	1110 (600)
C75200	1095 (590)	1290 (700)	---
C76200	1040 (560)	1130 (610)	1255 (680)
C77000	1110 (600)	1200 (650)	1290 (700)

MULTIPHASE ALLOYS

Besides the beryllium coppers, discussed next in this section, there are several other alloys that consist of more than one phase. The metallurgical structure of these alloys depends upon their heat treatment, and they contain alloying elements that are only partially soluble in the solid solution of the matrix at room temperature.

Two of the multiphase alloys are the zirconium coppers C15000 and C15100. C15000 is a premium alloy made by oxygen-free melting practices. Alloy C15100 is made by standard melting practices controlled to avoid excessive oxidation of the zirconium addition. Up to 0.15% zirconium can be dissolved in copper at a temperature of 1770°F (965°C). Upon slow cooling, copper-zirconium intermetallic particles precipitate out, strengthening the alloy and providing electrical conductivity of 95% IACS (International Annealed Copper Standard) in the annealed temper.

Alloy C15100 is made in strip form. The annealing practice during processing provides precipitation heat treatment needed to give maximum strength and conductivity. Solution heat treatments can be accomplished at 1380 to 1650°F (750 to 900°C) for 30 minutes, followed by quenching in water. Precipitation heat treatments are most effective when applied to cold worked metal; at temperatures of 750 to 840°F (400 to 450°C) for two hours, they are effective in developing maximum strength and conductivity combinations. Leadframes for semiconductor devices are a major application for alloy C15100.

Two other alloys that contain intermetallic particles in their structures, providing good combinations of strength, conductivity, and softening resistance, are alloys C19400 and C19500. Alloy C19400 has an electrical conductivity of 65% IACS in the annealed temper, as supplied by the mills. Although both the iron and phosphorus in the alloy form intermetallic particles with the copper, the alloy is not heat treatable. If subjected to a high temperature that causes solution of the alloying elements and loss of conductivity, the conductivity can be restored by annealing at a temperature of 840 to 930°F (450 to 500°C) for at least one hour. For recrystallization annealing of cold worked alloy C19400, a temperature of 880 to 970°F (470 to 520°C) for one hour at temperature is recommended.

Alloy C19500 is somewhat similar to alloy C19400 in metallurgical structure. Its alloying elements are iron, tin, cobalt, and phosphorus, which form intermetallic compounds in copper. Alloy C19500 has higher strength than C19400, but the conductivity is lower, at about 50% IACS. The majority of alloy C19500 applications are in electrical connectors and terminals, and leadframes for semiconductor devices where the combination of strength and conductivity provided by rolled tempers is desirable. When annealing cold worked metal for recrystallization, a temperature in the range of 1020 to 1290° F (550 to 700° C) for one hour at temperature is recommended.

Alloy C72500 contains 9% nickel and 2% tin. These elements can form intermetallic compounds in a matrix, which is a solid solution of the three elements. No strong effects on properties are occasioned by the presence of these precipitate particles. Both the elongation and strength of rolled temper alloy C72500 can be increased by annealing at 750° F (400° C) for one hour. Recrystallization to produce a grain size range of 0.010 to 0.035 mm can be achieved by annealing cold rolled metal in a temperature range of 1110 to 1290° F (600 to 700° C). Alloy C72500 is used extensively in telecommunications equipment.

In general, when questions of heat treatment or annealing of copper or copper alloys are faced by manufacturing engineers, the specific problem should be discussed with one of the metallurgists for the company that produces the metal. They have access to the data needed to tailor the process for a specific application.

HEAT TREATMENT OF BERYLLIUM COPPER ALLOYS

A major advantage of beryllium copper alloys is their ability to be strengthened by a low-temperature thermal treatment—called aging, age hardening, or precipitation hardening—after workpieces have been cast or formed. The properties of beryllium copper alloys are enhanced by cold working prior to age hardening. Another thermal treatment for these materials is solution heat treatment, frequently called annealing, which results in softening and in dissolving alloy constituents that cause precipitation hardening during subsequent aging. The thermal treatments vary for wrought and casting alloys.

Wrought Alloys

Fabricators of parts made from beryllium copper alloys are ordinarily concerned only with age hardening because solution heat treatment is generally done by the material supplier at the mill. Solution heat treating requires a water quench immediately after exposure to high temperature. If the process is performed correctly, beryllium coppers will be as soft as they can be. Parts that have been formed or machined should not be solution heat treated because they will most likely distort when water quenched.

By varying the age-hardening times and temperatures, different combinations of properties are obtainable. For example, elongation can be increased by either underaging or overaging (changing the time-temperature relationships), but with some sacrifice in hardness, as well as tensile and yield strengths. Overaging is generally preferred to underaging for obtaining special property combinations.

Temperature ranges and heating times for solution heat treatment and age hardening of several common alloys are presented in Table 14-21. The materials will shrink up to 0.003 in./in. (mm/mm) during age hardening.

Circulating-air furnaces are recommended for age hardening of beryllium copper alloys. Controlled atmospheres are generally not necessary because the discoloration, oxide film, or scale caused by heating in air does not significantly affect properties. If required, bright, scale-free surfaces can be restored after age hardening without a controlled atmosphere by chemically or abrasively cleaning the parts. Workholding fixtures are sometimes used for age hardening when shapes must be closely controlled.

Salt-bath furnaces are also used for age hardening because of their rapid, uniform, and precise heating with minimum distortion, especially for short cycle times. When salt baths are used, they must be kept clean and not used for other materials. Workpieces must be carefully cleaned after removal from the

TABLE 14-21
Temperature-Time Cycles for Solution Heat Treatment and Age Hardening of Beryllium Copper Alloys

UNS Alloy Designation No.	Solution Heat Treatment Temperature,* °F (°C)	Age Hardening**				
		Temperature Range, °F (°C)	To Obtain Highest Strength and Hardness			
			Cold Worked Material		Solution Heat Treated Material	
			Temperature, °F (°C)	Time, hr	Temperature, °F (°C)	Time, hr
Wrought alloys:						
C17000	1425-1475 (775-800)		600 (315)	2-3	600 (315)	3
C17200, C17300	1425-1475 (775-800)		600 (315)	2-3	600 (315)	3
C17500	1650-1750 (900-955)		900 (480)	2-3	900 (480)	3
C17510	1650-1750 (900-955)		900 (480)	2-3	900 (480)	3

(continued)

COPPERS AND COPPER ALLOYS

TABLE 14-21—*Continued*

UNS Alloy Designation No.	Solution Heat Treatment Temperature,* °F (°C)	Age Hardening** Temperature Range, °F (°C)	To Obtain Highest Strength and Hardness: Cold Worked Material Temperature, °F (°C)	Cold Worked Material Time, hr	Solution Heat Treated Material Temperature, °F (°C)	Solution Heat Treated Material Time, hr
Casting alloys:						
C82000, C82200	1650-1750 (900-955)	900 (480)	Time at temperature is 3 hours for both as-cast and solution heat treated materials.			
C82400, C82500, C82510, C82600, C82700, C82800	1425-1475 (775-800)	625 (330)				

* Time at temperature for solution heat treatment should be the minimum necessary for effective softening in order to minimize grain growth. In general, the time is one hour per inch or fraction of an inch of section thickness for heavy products, and a maximum of 20 minutes for thin strip and wire products. All alloys should be water quenched immediately from the solution treating temperature.

** Air cooling should be used to cool parts from the age-hardening temperature.

baths. Salt baths should never be used for the solution heat treatment of beryllium copper alloys.

Residual stresses can develop when beryllium copper alloys are fabricated after age hardening. Moderate stress relief can generally be obtained by heating the parts to temperatures between 300 and 400° F (150 and 205° C) for up to two hours. This treatment does not cause any loss of hardness, and ductility, yield strength, and dimensional stability may be improved.

Casting Alloys

Castings of beryllium copper alloys can be age hardened in the as-cast condition, or they can be solution heat treated first and then age hardened. Solution heat treatment, followed by age hardening, is recommended when maximum strength and hardness are required. However, omitting the solution heat treatment and aging the castings directly is satisfactory when maximum strength and hardness are not essential. Most castings are machined after aging, so some distortion resulting from thermal treatment can generally be tolerated.

Temperature ranges and heating times for the solution heat treatment and age hardening of several common casting alloys are presented in Table 14-21. When solution heat treating, the maximum temperatures shown in the table for each alloy should not be exceeded, and salt baths should not be used for solution heat treatment. While rapid quenching in water from the solution-treating temperature is recommended, castings subject to cracking can be cooled in hot water or oil, or by forced air. Slower cooling, however, can cause premature precipitation and less response to subsequent age hardening.

Age-hardening times and temperatures can be varied to enhance certain desirable properties. Circulating-air furnaces are preferable for age-hardening castings. As is the case with wrought materials, casting alloys will shrink up to 0.003 in./in. (mm/mm) during age hardening.

References

1. Peter F. Weiser, *Steel Castings Handbook* (Des Plaines, IL: Steel Founders' Society of America, 1980), pp. 21-15 to 21-23.
2. The Aluminum Association, *Aluminum Standards and Data*, 7th ed. (Washington, DC: 1982).
3. The Aluminum Association, *Standards for Aluminum Sand and Permanent Mold Castings*, 8th ed. (Washington, DC: 1980).

Bibliography

ANSI/ASTM Standard B661, "Recommended Practice for Heat Treatment of Magnesium Alloys."

Brush Wellman Inc. *Wrought Beryllium Copper*. Cleveland.

Cabot Wrought Products Div., Cabot Corp. *Beryllium Copper Heat Treatment*. Bulletin 304-TD10. Reading, PA.

Huntington Alloys, Inc. Alloy brochures. Huntington, WV.

Magnesium and Magnesium Alloys. Dayton, OH: The International Magnesium Association.

Metals Handbook, vol. 4, 9th ed. "Heat Treating." Metals Park, OH: American Society for Metals, 1981.

Rolled Alloys, Inc. Alloy brochures. Detroit.

Rosenberg, H. W. *Heat Treating of Titanium Alloys*. SME Technical Paper EM77-162. Dearborn, MI: Society of Manufacturing Engineers, 1977.

Walton, Charles F., and Opar, Timothy J. *Iron Castings Handbook*. Des Plaines, IL: Iron Castings Society, Inc., 1981.

Wickle, Keith G. "Beryllium Copper: An Overview of Heat Treat Techniques." *Heat Treating* (July 1981), pp. 28-34.

Wood, R. A., and Favor, R. J. *Titanium Alloys Handbook*. MCIC-HB-02. Columbus, OH: Battelle Institute, Metals and Ceramics Information Center.

SURFACE AND EDGE PREPARATION: DEBURRING, FINISHING AND CLEANING

SECTION

3

CHAPTER 15

SURFACE AND EDGE IMPROVEMENT

There are many different reasons why surface, edge, and corner preparation, conditioning, cleaning, or finishing are required. Most of these reasons can be classified with respect to functional or appearance requirements.

Improving the appearance of a product for purely aesthetic reasons can be important because it often increases the salability of the product, with appearance reflecting a concern for quality by the producer. The increased product performance and safety provided by proper edge and surface finishing is also important. The removal of burrs and sharp edges improves safety for both the worker and product user by eliminating the possibility of cuts and making parts easier to handle. For critical components, the surface condition and edge geometry can be a major influence on component performance and durability.

CHAPTER CONTENTS:

FUNCTIONAL REQUIREMENTS

Many functional requirements necessitate surface and edge preparation. These requirements include improved surface finishes and better surfaces for subsequent coating, easier assembly, improved operating performance, and increased strength.

IMPROVED SURFACE FINISHES

Improved surface finishes are often required for increased wear resistance, lubricant retention, and corrosion resistance. The efficiency of gears and other power transmission components is improved by smooth surface finishes. Proper finishing can eliminate surface defects, such as nicks, scratches, or tool marks, that can act as stress risers. Smooth, blended internal passages, as well as some external surfaces, reduce turbulence and increase the flow rates of gases or fluids.

HIGH-QUALITY COATINGS

Requirements for surface and edge conditioning are often dictated by subsequent processing, such as plating, painting, anodizing, or other coating, or by the intended applications for the components. For the subsequent application of organic and inorganic coatings, specifications with respect to surface finish and cleanliness vary with the specific coating method.

Surface finish and cleanliness affect the adhesion, appearance, quality, and performance of the coating. Any boundary layer, chemical or physical, between the coating and the substrate can interfere with proper bonding. Rough surfaces and the presence of flash or burrs can ruin the coating. High-quality coatings require proper preparation of surfaces and edges, as well as mechanical and/or chemical cleaning. The type of coating dictates the types of finishing and cleaning processes used prior to coating.

FACILITATING ASSEMBLY

Proper surface and edge finishing facilitates assembly, thus reducing costs, and is required for automatic assembly. By preventing interferences and damage to mating parts, the possibility of mechanisms jamming or assemblies failing is minimized.

IMPROVED PERFORMANCE

Proper conditioning often improves operating performance and lengthens product life. Burrs and flash not removed can break off during service and may cut wires or seals, or cause electrical shorts. The reduction of stress concentrations also helps lengthen product life.

INCREASED FATIGUE STRENGTH

Producing specified chamfers or radii on edges and corners can increase the resistance to fatigue failure of highly stressed components. With some finishing processes, surface integrity can be improved by reducing or removing tensile stresses, or by imparting compressive stresses to the surfaces, thus improving fatigue strength.

Contributor for this chapter is: ***L. K. Gillespie****, Senior Project Engineer, Kansas City Div., Bendix Corp.**

**Operated for the U.S. Department of Energy by The Bendix Corporation, Kansas City Division, under Contract No. DE-AC04-76DP00613.*

Reviewers for this chapter are: ***Ezra A. Blount****, Secretary, Metal Finishing Suppliers' Assn., Inc.;* ***L. K. Gillespie****, Senior Project Engineer, Kansas City Div., Bendix Corp.;* ***J. Bernard Hignett****, Vice President, Harper Co.;* ***John B. Kittredge****, Consultant;* ***Larry Rhoades****, President, Extrude Hone Corp.*

MINIMIZING COSTS

The cost of burr and flash removal, edge breaking, radius forming, descaling, cleaning, and surface finishing has been estimated to be more than $2 billion per year to U. S. industry alone. Actual costs are undoubtedly higher because of hidden costs due to reworking, reinspecting, replacing, and the intangible cost of loss of customers' goodwill.

REASONS FOR EXCESSIVE COSTS

High finishing costs are often the result of one or more of the following:

- Improper design of the component.
- Incomplete or improper specification of requirements.
- Incorrect processing during prior manufacturing operations.
- Not using efficient processes, machines, and tools.
- Not keeping tools sharp and machines properly maintained.
- Excessive finishing.
- Too many manual operations.
- Use of untrained or unqualified personnel.
- Lack of research and development.
- Improper or excessive inspection.

METHODS OF REDUCING COSTS

Finishing costs can often be reduced significantly by:

- Designing components to minimize or accommodate burrs. If possible, specify a workpiece material or material condition that will minimize burring tendencies.
- Developing and using suitable standards that reflect actual needs.
- Proper processing to minimize, prevent, or control the maximum size of burrs. Production machines, tooling, and workholding fixtures should be as rigid as possible; tools should be of proper geometry and kept sharp or changed frequently; and operating parameters should be selected for optimum conditions.
- Using the most efficient deburring and finishing techniques and equipment.
- Proper training of finishing personnel, inspectors, engineers, and management.

The cost of burr removal is directly related to burr size and location. Details regarding design and processing considerations for reducing finishing costs are presented later in this section. First, however, it is essential to understand the nomenclature, formation, and properties of burrs.

BURR NOMENCLATURE

One reason that burr removal has remained costly is that industry has not accepted uniform definitions relating to burrs and deburring. Such definitions are essential for production, inspection, and other manufacturing departments to easily and quickly determine whether a projection is a burr. A number of proposed definitions are as follows:

burr (verb) (1) To remove burrs from a part (preferred terminology is "to deburr"); (2) to form a short flange around holes (preferred terminology is "to flange").

burr (noun) A burr is an undesirable projection of material that results from a cutting, forming, blanking or shearing operation. (This definition has been adopted by the Society of Manufacturing Engineers' Burr, Edge and Surface Technology (BEST) Division.)

burr hardness Hardness of the burr in the vicinity of the base of the burr.

burr height The distance a burr projects above the surface of the workpiece.

burring (1) The process of removing burrs (preferred word is "deburring"); (2) the intentional production of a small flange around holes in a pressworking operation (preferred word is "flanging"); (3) the act of burr formation.

burr length Length of edge along which burr exists.

burr radius Radius on backside of burr at the junction of the burr and the workpiece surface.

burr thickness Thickness of a burr where it joins the parent metal when measured along the plane of the projected surface.

compressive burr The burr produced in blanking or piercing operations in which the slug separates from the stock in a compressive stress field. Burrs produced in this manner initially begin as tensile burrs and are typically identified as tensile-plus-compressive burrs.

corner Intersection of three or more surfaces.

corner break The radius or chamfer produced on the corners of a part.

cutoff burr A projection of material left when the workpiece falls from the stock before the separating cut has been completed.

deburring The process of removing burrs by any of several techniques. Unless otherwise indicated, the Class III definition is implied.

Class I: Complete removal of all burrs when viewed without magnification.

Class II: Removal of all burrs except those of a specified or given size.

Class III: Removal of all loose fragments and burrs so that no sharpness or loose particles are detected by an inspection method specified.

deflashing The removal of flash by any method from cast, molded, and forged parts. Unless otherwise indicated, the Class III definition is implied.

Class I: Complete removal of all flash when viewed without magnification.

Class II: Removal of all flash except that of a specified size.

Class III: Removal of all loose fragments and flash so that no sharpness or loose particles are detected and the parts fall within the required tolerance.

dross Molten metal that resolidifies on the workpiece. Dross occurs at the edges of material cut by such thermal processes as plasma arc, oxyacetylene, and Heliarc cutting.

edge Intersection of two surfaces.

edge break The amount of material removed from the theoretical intersection of two surfaces.

entrance burr Burr formed on the surface at which the cutting tool or its teeth enters the workpiece.

exit burr Burr formed on the surface at which the cutting tool or its teeth leaves the workpiece.

feather burr A very fine or thin burr.

feather edge Same as feather burr except that feather edge can also refer to the ends of a lead-in or lead-out thread, which is a very thin machined ridge. Sometimes called a wire edge or whisker-type burr.

fin Portion of flash that flows into the small gap between movable parts of a mold.

flash The excess material squeezed out of the cavity as a compression mold closes or as pressure is applied to a transfer or injection mold. Flash includes both the fin and fragments that are left in the mold.

flash extension Portion of flash remaining on a workpiece after trimming. Flash extension is measured from the intersection of the draft and flash at the body of the forging or molded part to the trimmed edge of the stock.

hanging burr Loose or flexible portions of a burr that are not firmly attached to the workpiece (i.e., hanging from the workpiece). Sometimes called a flag.

hinge area The juncture of a hanging burr with the more rigidly adhered portion of a burr.

Poisson burr A burr formed predominantly by the phenomenon that is responsible for Poisson's ratio. Sometimes called a flow-type burr.

rollover burr Burr formed by a cutter when it exits over a surface and allows the chip to be rolled away from the cutter, rather than sheared.

tear burr Burr formed by the sides of a cutter as the cutter tears a chip from the workpiece. Also a ragged form of the Poisson burr caused by a built-up edge on the cutting tool.

tensile burr The burr produced in blanking or piercing operations in which the slug separates from the stock as a result of tensile stresses.

tensile-plus-compressive burr The burr produced in blanking or piercing operations in which the slug separates from the stock in a stress field that is initially tensile, but changes to compressive at actual separation. This burr is usually associated with small die clearances and short die life.

trimming The mechanical shearing of flash from molded and forged parts.

trunk The thick base of a burr.

wire edge A very fine or thin burr that is typically sharp, looks like a wire, and is parallel to an edge. Sometimes called a feather edge or whisker-type burr.

Burr Values

Many individuals unconsciously include a value judgment of what an acceptable burr is as part of their definition, as in "a burr is material 0.0030″ (0.076 mm) or higher projecting from an edge." Such definitions are incompatible with the need for broad standards. A hump of material 0.0029″ (0.074 mm) high forms by the same mechanism as a 0.0030″ high hump. It makes little sense to define one as a burr and not the other. It is more logical to define both humps as burrs; and, in a separate standard or section of a standard, indicate that burrs 0.003″ (0.08 mm) or higher are not allowable. Defining burrs as loose material or sharp fragments ignores how the material is formed and requires an additional definition of what the material is below the fragments.

Burr Height

The terms *burr height* and *burr length* are sometimes used interchangeably since there is seldom any question about the extent of the edge that has a burr on it. For consistency, however, burr height should be used to indicate the size of a burr. Burr thickness (at the root of the burr) is the most important factor with respect to difficulty in removal.

BURR FORMATION

Burrs, flash, and related protrusions are formed by the six physical principles listed in Table 15-1. Burrs formed by one of the first three principles listed in this table involve plastic deformation of the workpiece material. Solidification of material on the working edges, the fourth principle of formation, forms a burr-like projection. The fifth type of burr occurs when the workpiece is allowed to fall from the part before the cut is completed. Flash forms whenever the pressure on molten material is sufficient to force the material between the two halves of a die or mold.

TABLE 15-1
Physical Principles in Forming Burrs, Flash, and Related Protrusions

Physical Principle of Formation	Name of Protrusion
Lateral flow of material	Poisson burr
Bending of material (such as chip rollover)	Rollover burr
Tearing of chip from workpiece*	Tear burr
Redeposition of material	Recast bead
Incomplete cutoff	Cutoff projection
Flow of material into cracks	Flash

* A tear burr also forms in stamping operations when the punch tears the part from the stock.

Poisson Burr

A Poisson burr occurs whenever the cutting edge extends past an edge of the workpiece (see Fig. 15-1). It is the result of lateral deformation that occurs whenever any solid is compressed. The extent of this deformation is a function of the workpiece material, the size and shape of the contacting components, and the applied load. The extent of the lateral

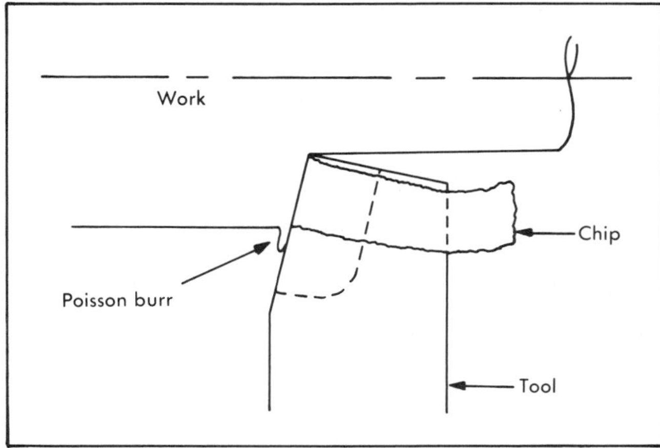

Fig. 15-1 Poisson burr formed when cutting edge of tool extends past edge of workpiece.

deformation (burr size) is usually relatively small. The size of the burr is proportional to the cutting edge radius and the applied pressure.

Entrance Burr

As shown in Fig. 15-2, when a cutting edge first indents a workpiece, it is possible for another type of burr to form. This entrance burr is material that has flowed opposite to the direction of the tool. It is similar to the ridge that forms around the indentation made by a Brinell hardness tester.

Whether or not a burr forms at this point depends upon the workpiece properties and, probably, the shape of the cutting edge. Strain hardening also plays an important role. A burr generally forms when the material has a low strain-hardening exponent (below 0.3). Materials having high strain-hardening exponents tend to cause a bulge, but not a sharp burr.

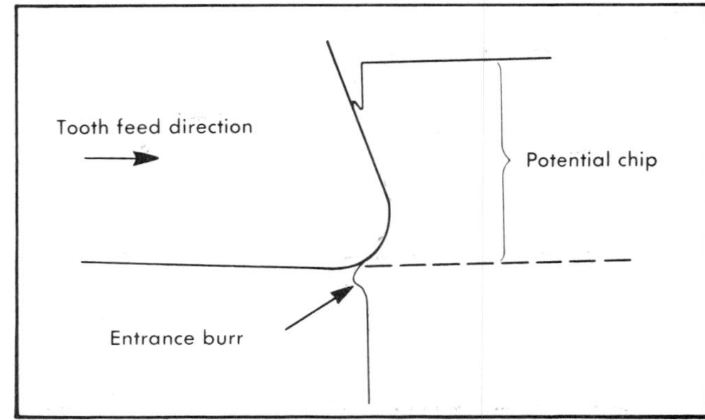

Fig. 15-2 Cutting edge produces indentation burr as it enters workpiece.

Rollover Burrs

When a cutting edge exits from a workpiece, a rollover burr occurs when it is easier to bend the chip than to cut it. As seen in Fig. 15-3, the thickness of a rollover burr in a continuous depth-of-cut process can be expressed by the following equation:

$$h = t\,(\csc A - 1) \tag{1}$$

where:

h = thickness of rollover burr, in. or mm
t = depth of cut, in. or mm
A = shear angle, expressed approximately from the following equation:

$$A = \tan^{-1} \frac{(t/t_2)\cos B}{1 - (t/t_2)\sin B} \tag{2}$$

where:

t_2 = chip thickness, in. or mm
B = rake angle of cutting tool

Shallow depths of cut and large shear angles minimize burr thickness. Large rake angles and small cutting forces minimize the chip thickness, thus reducing burr thickness.

The length of a rollover burr is a function of the cutting conditions and the plasticity of the workpiece material. With the exception of some milling operations, the length of the burr cannot be more than the total depth of cut. If, in bending, the strain exceeds the strain required to fracture, then the majority of the burr will break off, leaving only a short burr.

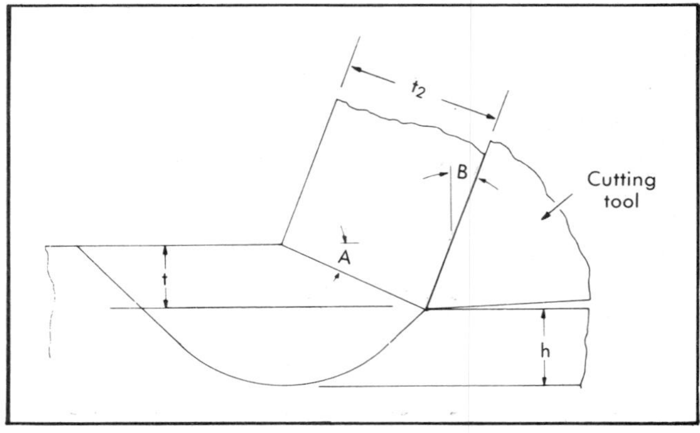

Fig. 15-3 Rollover burr thickness, h, as a function of idealized conditions. (*After Barash and Schoech*)[1]

Tear Burrs

Tear burrs form when chips are torn rather than being sheared from the workpiece. Although such burrs can form in most cutting processes, they are easiest to produce in side-milling operations. The milling cutter tooth forces the chip up and forward. As it does so, the sides of the chip are torn from the workpiece. The torn portion remaining on the workpiece is the tear burr.

Common Characteristic

The Poisson burr, the rollover burr, and the tear burr all have one common characteristic: a radius occurs on the back side of these burrs (see Fig. 15-4). The total shape of the burr cross section can be expressed by height and radius. When all of the burr must be removed, the burr thickness should be specified.

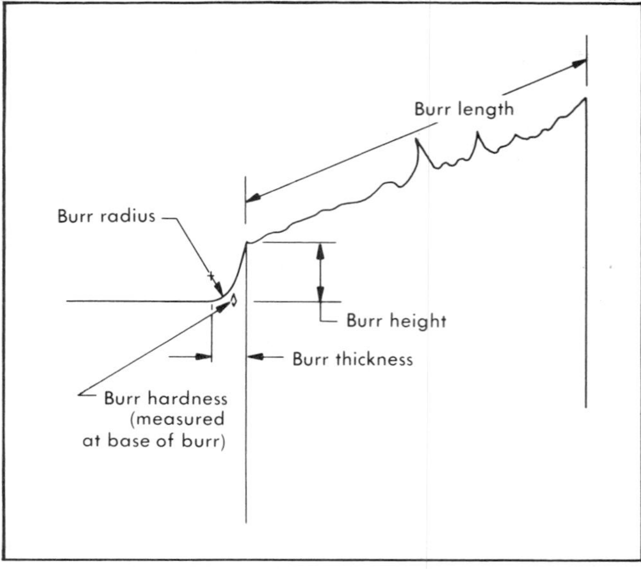

Fig. 15-4 Definitions of burr characteristics. (*Bendix Corp.*)

Recast Material

Recast material is formed when molten metal resolidifies on an edge of a workpiece. The most common occurrence of this type of burr is found on electrical discharge machined (EDM) features. The amount of recast material is a function of polarity, workpiece material, and frequency and amperage of the current. Details of the EDM process are presented in Volume I, *Machining*, of this Handbook series.

Cutoff Projections

Cutoff projections (see Fig. 15-5) occur when workpieces are severed from barstock. In such cases, the workpiece is allowed to fall from the stock before the parting tool completely severs the material. Projections can occur in any parting operation, although they are most common in turning operations.

Flash

Provided the clamping pressure between dies or molds is adequate, flash will only occur when die or mold halves are not perfectly matched. In such cases, flash thickness is equal to the thickness of the voids between die or mold halves, and flash height is equal to or shorter than the length of these voids. The hardness of the flash is approximately the same as the workpiece. Flash is generally more difficult to remove than burrs because it is often thicker next to the parent metal; it is usually removed with trimming dies in secondary press operations, discussed in Volume II, *Forming*, of this Handbook series.

Burrs Produced by Specific Operations

Turning operations. A Poisson burr typically occurs in a turning operation (see Fig. 15-1). When the tool passes over a groove, however, a rollover burr forms in the groove. In producing a groove, a tear burr forms. If sufficient rubbing occurs between the grooving tool and the walls of the groove, material will also flow from the sides of the groove, as the result of the Poisson effect. A small rollover burr also occurs when a facing cut is taken.

Drilling. A Poisson burr also occurs at the entrance of drilled holes. The burr formed as the drill exits from the bottom of the workpiece is initially formed as chip rollover; but if the corners and margins on the drill are sharp, the burr properties will actually be those of a Poisson burr. In this case, the burr is entirely formed by the leading edge of the margins on the drill. A rollover burr is first partially formed, and then the sharp corners and margins of the drill shear the material at the base of the rollover burr. Dull drills produce exit burrs by chip rollover only.

Side milling. In a side-milling slotting operation, burrs are produced on eight edges (see Fig. 15-6). A rollover burr occurs at the edges when the cutter exits from the workpiece. The height of the burr, b_L in Fig. 15-7, is equal to the depth of the cut, *d*, or less. An entrance burr occurs at edge 8; a hybrid Poisson tear burr occurs at edges 6 and 7; and a tear burr occurs at edges 1, 2, 3, and 4.

If the milling cutter has a helix angle, the properties of the burrs at each edge will be different from those on every other edge. There are eight different sets of thickness, height, radius, and hardness for the burrs with such a cut. For precision deburring, these variations are important because they mean that each edge will respond differently to a single deburring process. Properties of the burrs also vary along the length of all edges.

End milling. In end-milling operations, Poisson burrs are formed on edges 1, 2, 4, and 10 (see Fig. 15-6 and Fig. 15-8). An entrance burr occurs on edge 6, and rollover burrs are produced on edges 3, 7, and 9. On half of edges 5 and 8, an entrance burr is produced, while a rollover burr occurs on the other halves. The height of the rollover burr equals the radial depth of cut until the cut exceeds 0.6 times the tool diameter. For cuts deeper than this, the height tends to remain constant at 0.6 times the tool diameter, although it may vary with workpiece properties and tool forces.

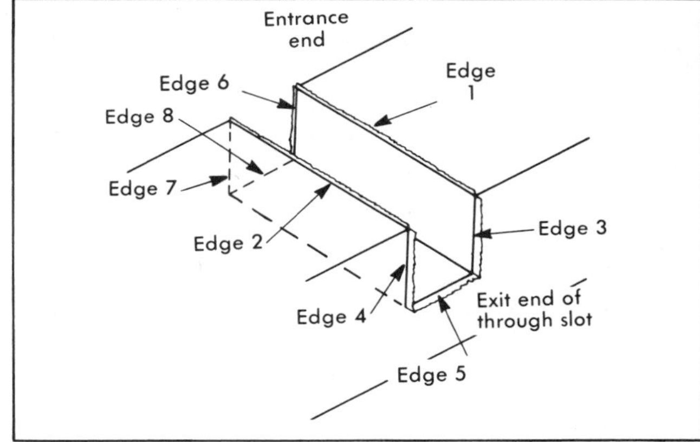

Fig. 15-6 Burr locations in slots produced by side or end mills. (*Bendix Corp.*)

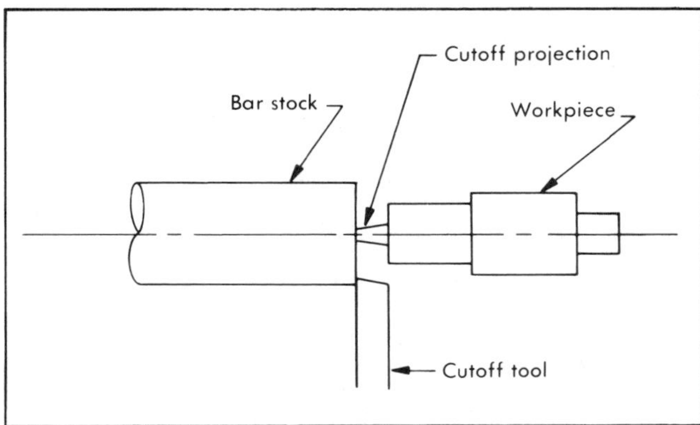

Fig. 15-5 Projection left on workpiece after cutoff operation.

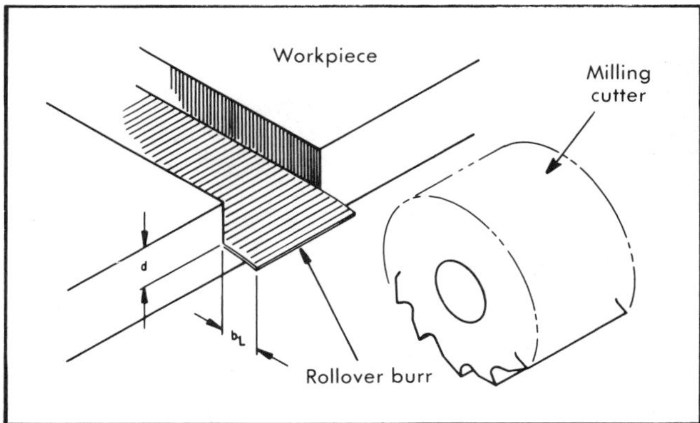

Fig. 15-7 Location of rollover burr in milling a slot. (*Bendix Corp.*)

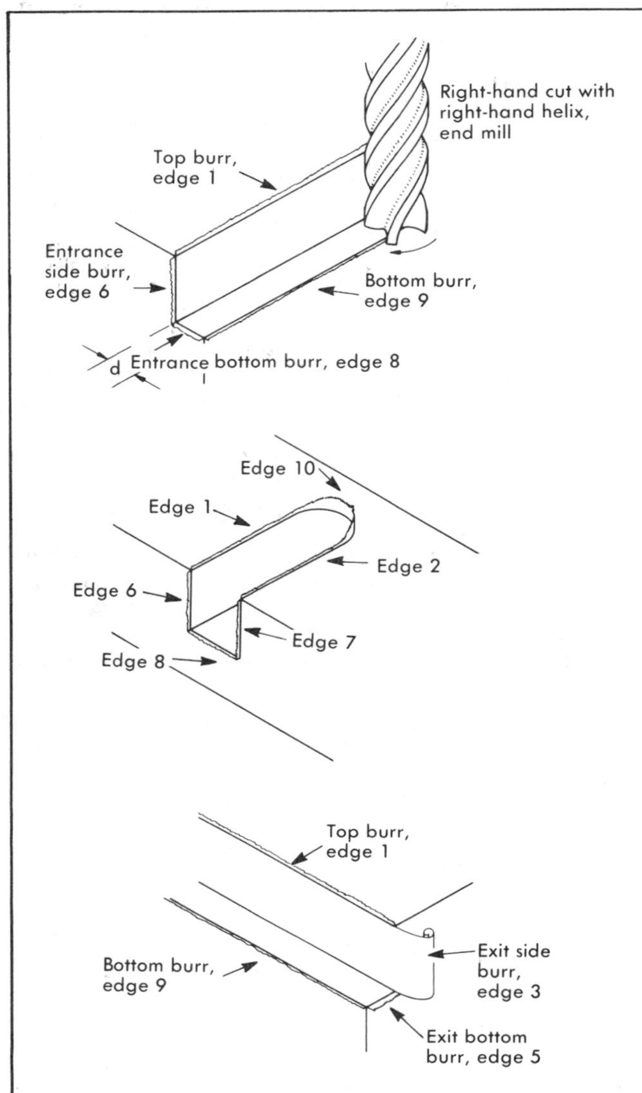

Fig. 15-8 Identification of burr locations in end-milling edges of workpieces. (*Bendix Corp.*)

BURR PROPERTIES

The difficulty in removing a burr varies with each situation, but the following list, in approximately the correct order of significance, is typical with respect to difficulty in removal:

- Burr thickness.
- Burr location.
- Burr toughness.
- Burr height (length).
- Workpiece tolerances.
- Burr radius.

In the past, measuring burrs meant measuring burr height, but burr thickness is the single most important size consideration for burr removal. In most situations, the combination of thickness and height, or thickness and location, rather than just a single measurement, is important. Burr properties depend on which metalcutting mechanism produced the burr.

It is burr thickness at the root of the burr, not burr height, that makes a burr difficult to remove. From a theoretical standpoint, burr height is directly related to burr thickness; but from a practical standpoint, the relationship is not readily evident.

Burrs assume many shapes. They can be triangular or rectangular in cross section; and they can be uniform or ragged (irregular), or feather-like in height. Each of these shapes indicates a different metalcutting phenomenon.

Burr properties vary considerably from part to part, and even within the same part. As an example, an abusive face-milling operation on 303Se stainless steel produced an average burr height of 0.0095″ (0.241 mm), as shown in Table 15-2. Measuring 100 edges, however, revealed heights varying from 0 to 0.020″ (0.51 mm).

DESIGN CONSIDERATIONS

Economical and proper finishing requires close coordination of design, manufacturing, and process engineers, supervisors, operating personnel, and inspectors. The critical starting point, however, is in product design. Designers must determine the best workpiece material and level of quality required, and design the components to eliminate or minimize finishing requirements and costs.

Level of Quality Needed

Determining the level of quality needed requires a knowledge of component and assembly functions, and how critical each edge and surface is to the application of the component and assembly. In most situations, all surfaces do not require the same degree of surface finish, and all edges do not require the same chamfer, radius, or burr-free condition. Many burrs are unnecessarily removed because of design specifications that are not essential. Edge conditions and surface finishes can sometimes be changed slightly to make subsequent processing easier and more economical.

With respect to burrs, some of the considerations a design engineer must make include:

1. Is a burr-free condition required? If so, why and where? How should it be measured? What would happen if the part were not burr-free? Should stock allowance be specified on some dimensions to compensate for stock to be removed in deburring operations?
2. Is a burr allowable? Will it be a safety hazard, cause unallowable stress concentrations, or accelerate wear beyond allowable limits? Will it cause misalignment, interference fits, jamming of mechanisms, or electrical short circuits? How can the part be designed to minimize or accommodate the burr?
3. Is an edge break or radius required? If so, why and where? How should it be measured? What would happen if the edge did not have a break or radius?

Manufacturing and process engineers, supervisors, production operators, and inspectors must all be informed of the actual quality required on every surface and edge. In some cases, quality can be controlled by the choice of deburring or finishing technique. For many parts, however, the design engineer must designate the quality for various edges and surfaces. One method of accomplishing this designation is by using codes on the part drawing, as illustrated in Fig. 15-9.

TABLE 15-2
Comparative Burr Sizes Produced in Milling Operations

Material	Milling Condition*	Burr Thickness, in. (mm) Average	Burr Thickness, in. (mm) Standard Deviation	Burr Height, in. (mm) Average	Burr Height, in. (mm) Standard Deviation	Hardness, Knoop** Parent Metal	Hardness, Knoop** Burr
Aluminum	Abusive face	0.0078 (0.198)	0.0011 (0.028)	0.0138 (0.351)	0.0045 (0.114)	115	129
	Gentle profile	0.0023 (0.058)	0.0003 (0.008)	0.0015 (0.038)	---	---	---
Low-carbon steel	Abusive face	0.0046 (0.117)	0.0010 (0.025)	0.0255 (0.648)	0.0191 (0.485)	230	275
	Gentle profile	0.0014 (0.036)	0.0002 (0.005)	0.0024 (0.061)	0.0009 (0.023)	---	---
Stainless steel, 303 Se	Abusive face	0.0049 (0.124)	0.0020 (0.051)	0.0095 (0.241)	0.0053 (0.135)	305	428
	Gentle profile	0.0020 (0.051)	---	0.0013 (0.033)	---	---	---
Beryllium copper	Abusive face	0.0045 (0.114)	0.0008 (0.020)	0.0189 (0.480)	0.0017 (0.043)	249	268
	Gentle profile	0.0007 (0.018)	0.0005 (0.013)	0.0016 (0.041)	0.0011 (0.028)	---	---

* Abusive face milling: dull cutter and 0.025" (0.64 mm) depth of cut. Gentle profile milling: burr produced by the bottom of a 1/4" (6.4 mm) diam end mill rotating at 2730 rpm and fed at the rate of 9 3/16" ipm (233 mm/min).
** Knoop hardness number with 100 g load (249 Knoop hardness number = R_C 24; 115 = R_B 73).

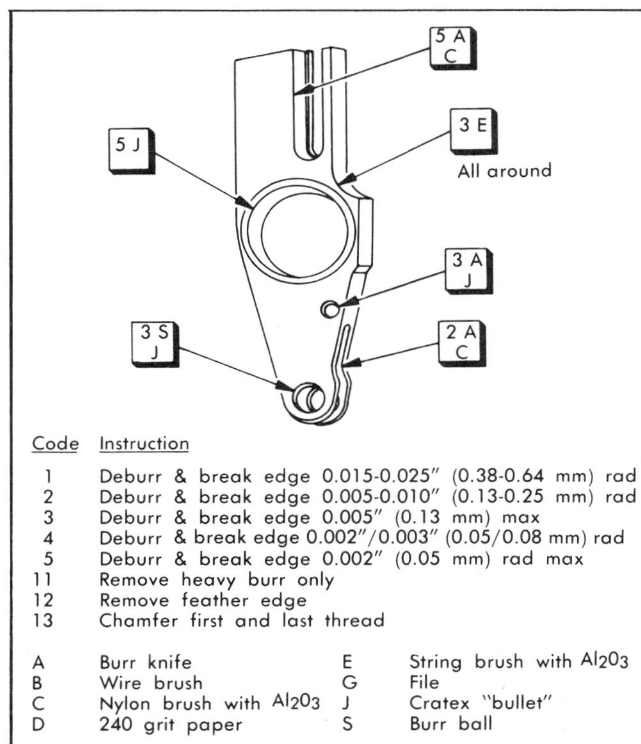

5 A C
3 E
All around
5 J
3 A J
3 S J
2 A C
Code Instruction
1 Deburr & break edge 0.015-0.025" (0.38-0.64 mm) rad
2 Deburr & break edge 0.005-0.010" (0.13-0.25 mm) rad
3 Deburr & break edge 0.005" (0.13 mm) max
4 Deburr & break edge 0.002"/0.003" (0.05/0.08 mm) rad
5 Deburr & break edge 0.002" (0.05 mm) rad max
11 Remove heavy burr only
12 Remove feather edge
13 Chamfer first and last thread
A Burr knife
B Wire brush
C Nylon brush with Al_2O_3
D 240 grit paper
E String brush with Al_2O_3
G File
J Cratex "bullet"
S Burr ball

Fig. 15-9 Instructions for edge requirements and deburring. (*Bendix Corp.*)

Design Standards

The existence or lack of edge standards can contribute significantly to deburring costs. A cardinal rule in any value engineering approach to reducing costs is to specify only what is actually required. Overspecification increases cost and production time; underspecification creates confusion and problems, and also increases cost. If standards do not exist, the product designer, in conjunction with the manufacturing engineer and the machining and inspection departments, should consider establishing plant-wide standards, as well as standard nomenclature for use on individual part prints.

Definition methods. Several approaches can be used to define allowable burrs or edge conditions:

- Define them on the part print.
- Define them in a process engineering (manufacturing) specification.
- Define them on the production traveler (routing sheets).
- Define them by an interpretative memorandum (including sketches, photos, and measuring techniques).
- Define them on the inspection traveler (routing sheets).
- Define them with photos of acceptable and unacceptable conditions.
- Define them by the use of comparative masters (each master is given a tool or gage number, a visual aid, or a visual standard number).
- Define them by go/no go gages (if the gages fit, the burrs are acceptable).
- Define them by taking specific exception to general workmanship specifications.
- Define them by special specifications.

- Define them by such phrases as "firmly adhered burrs or raised metal is allowable in this area provided a microtool with a 90° hook will not dislodge them."

Inadequate notes. While notes such as "small burr satisfactory" or "burr raised in slotting operation is acceptable" may be adequate for parts made within a specific plant, they should be avoided if parts are to be made by outside vendors. Sooner or later the product designer will be asked to define "small burr."

Preferred notes. The preferred practice for specifying edge quality, with allowable burr sizes described, is illustrated in Fig. 15-10. Chamfering produces a generally smaller burr than the burrs produced by other processes and may be the only deburring required. Either drawing notes or an in-plant standard should be used to indicate whether chamfering represents adequate deburring. When a smooth blend is required, the edge should be specified with a radius.

Edge breaks (chamfers) should be so specified that either a chamfered or a radiused condition is allowable. This specification permits the manufacturing engineer to determine whether a machining or a deburring process will provide the most economical edge condition. Typical corner breaks are 0.015" (0.38 mm) x 45°, or 0.010-0.015" (254-381 μm) x 45°. Radii should not normally be specified larger than 254 μm (0.010") nor smaller than 76.2 μm (0.003"), unless required by function or stress concentrations.

Burr direction. The direction a burr faces is sometimes more critical than its actual size. In these cases, the orientation of the burr should be noted on the part drawing. With symmetrical threaded parts, it is helpful to the manufacturer if the designer indicates which end of the part the thread is started from, eliminating the need to deburr both ends.

Specific notes. A burr always forms at the intersection of two holes. When a burr cannot be tolerated in one hole but can in the other, it should be noted on the part drawing (see Fig. 15-11). Defining where burrs can exist on formed parts may eliminate the need to deburr the sheet stock. With proper thought and communication between product designer, tool designer, and manufacturing engineer, forming dies can be designed so burrs on the blank will be in an out-of-the-way location in the finished part.

Removing sharp edges. On many parts, the only significant edge requirement is that all sharp edges be removed. In such cases, beating over burrs and dulling edges is adequate. Designers can handle this requirement in at least two ways:

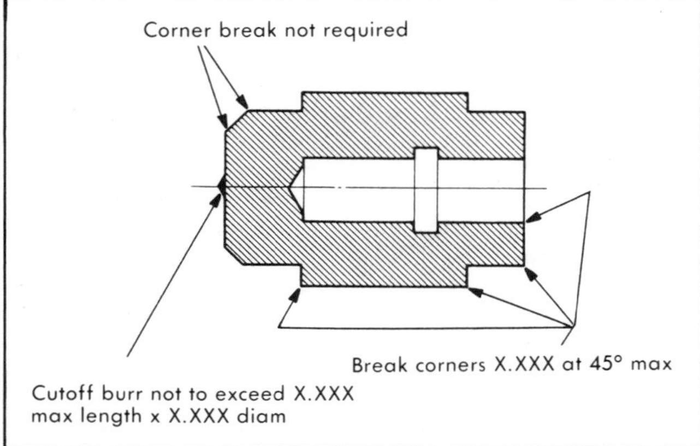
Corner break not required
Break corners X.XXX at 45° max
Cutoff burr not to exceed X.XXX
max length x X.XXX diam

Fig. 15-10 Typical burr notes for external edges.[2]

1. By specifying the process that gives an acceptable edge.
2. By defining the actual edge quality needed.

Vibrating parts in steel media will dull the edges of most parts economically. Similarly, on many materials, thermal energy deburring (TEM), discussed in Chapter 17, can be used to ensure that no loose burrs or particles will be present to jam assemblies. While specifying on the drawing that parts should be vibrated in steel media to dull edges is often done for parts made within a plant, such notes are not complete enough for work contracted to others. Outside vendors must know what size steel ball to use, how long to run the machine, and the amplitude and frequency of the machine to be used. These requirements can be specified by developing explicit processing standards and referring to them on the part drawings. Such an approach is easy and relatively problem-free when the majority of parts have similar requirements.

When a wide variety of parts is designed and manufactured every year, using drawing notes may restrict the manufacturing engineer's ability to make parts by the least expensive process. For example, a standard may specify that abrasive blasting be used. On some parts, a centrifugal barrel finisher might be more economical. Although drawing notes can be changed, the paperwork and delays involved add unnecessary costs. Different manufacturers have different approaches to providing the same edge quality, and each uses the lowest cost method available at the time.

Proposed system. At present, there are no American national standards for specifying allowable edge conditions. The examples discussed previously illustrate some logical approaches now in use. On complex parts having both precision and commercial features, however, a drawing can quickly become cluttered with notes describing allowable edge quality. In-house standards may not be adequate for such parts because of the variety of edge requirements. A system has been proposed that can be used in these situations.

The proposed system is based on the observation that allowable or desirable edge conditions can exist in any of the four quadrants defined by two perpendicular lines. Although a radius typically occurs after deburring, often a slight protrusion

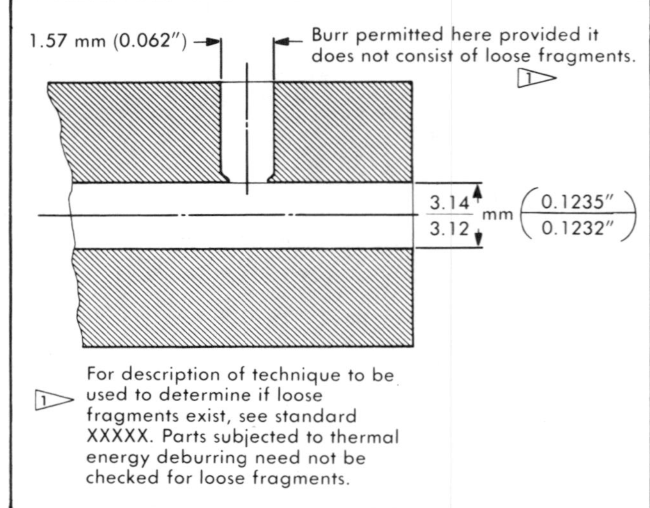
1.57 mm (0.062")
Burr permitted here provided it
does not consist of loose fragments.
1
3.14
3.12
mm
0.1235"
0.1232"
1
For description of technique to be
used to determine if loose
fragments exist, see standard
XXXXX. Parts subjected to thermal
energy deburring need not be
checked for loose fragments.

Fig. 15-11 Allowable burr and its location should be noted on drawing. (*Bendix Corp.*)

exists (see Fig. 15-12). The proposed system uses the two intersecting surfaces to define the four quadrants.

To simplify numerical definitions of edge condition, a series of deburring classes have been established (see Fig. 15-13). An edge radius of 0.16-0.315 mm (0.006-0.012") represents a Class 6. Since no material is allowed past the theoretical intersection of the two surfaces, a zero is indicated in those quadrants. In some instances, because of the view shown, the quadrant in which a radius occurs will vary (see Fig. 15-14).

A Class 1 edge allows 0.01 mm (0.0004") high projections or radii. A Class 9 allows 2.50 mm (0.100") conditions at edges. When edges intersect at angles other than 90°, quadrants are defined by the planes on the part and not by orthogonal planes.

This system requires the use of at least one of the following three notes (either on drawings or in in-plant standards):

1. Burrs need not be removed except as noted.
2. Tightly adhered burrs need not be removed. A sufficient check for adherence is defined in Standard XXXXX.
3. Sharp edges not permitted. Sharpness will be checked with Underwriters Laboratories sharpness monitor.

Note 1 applies to parts in which burr measurements are not required. This, of course, eliminates the need for the proposed system except on specified edges. Note 2 applies when loose burrs or chips will cause malfunction of the assemblies. Note 3 is normally placed on most prints or in in-plant standards. When a radius rather than a chamfer is desired, the letter r can be added after the class number, or the radius can be specified. When a chamfer is allowable, in-plant standards must indicate whether or not a small burr on the chamfer is allowable.

Available standards. Sheared sheet stock can be purchased by an AISI edge condition number. Edge Number 3 indicates that the shearing burr remains on the sheet. Edge Number 5 indicates a burr-free sharp edge, and Edge Number 1 indicates chamfered or rounded edges. The National Screw Machine Products Association has a standard indicating that its supporting members do not deburr screw machine parts unless specifically agreed upon between buyer and seller. Burr, edge, and surface standards are common in the aircraft and aerospace industries, but they are not used industry wide, being usually confined to individual companies. Failure to know the standards of a particular industry or company, or to specify requirements explicitly, can result in rejected parts and added costs.

Workpiece Material

Two factors with respect to workpiece material are directly linked to burr size:

1. Ductility of the workpiece material.
2. Strain-hardening exponent of the material.

Large burrs cannot form in brittle materials. Cast irons, for example, often have edges with no visible burr. These materials have values of elongation of 0.5 to 3.0% in a 2" (50 mm) gage length. Since the material has little capacity for plastic deformation, large burrs cannot form. If, however, the cutting tool heats the cast iron enough to change its structure and the material is no longer brittle at the edges or machined surfaces, a noticeable burr can form.

Burr size is also a function of the strain-hardening exponent (coefficient). Nonstrain-hardening materials will form burrs, but they will be considerably smaller than formed on materials having large strain-hardening tendencies. As the strain-hardening exponent increases, burr thickness generally increases, but the relationship is not usually directly proportional.

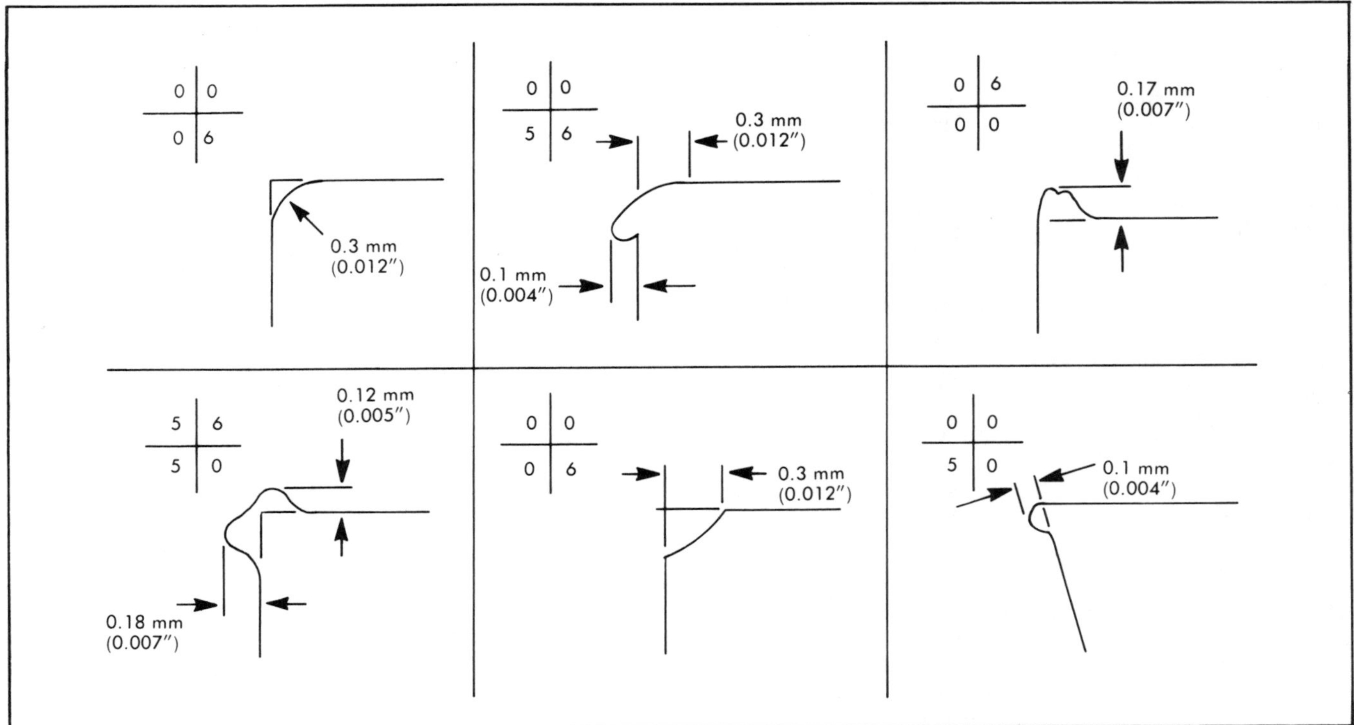

Fig. 15-12 Edge conditions and related notations.[3]

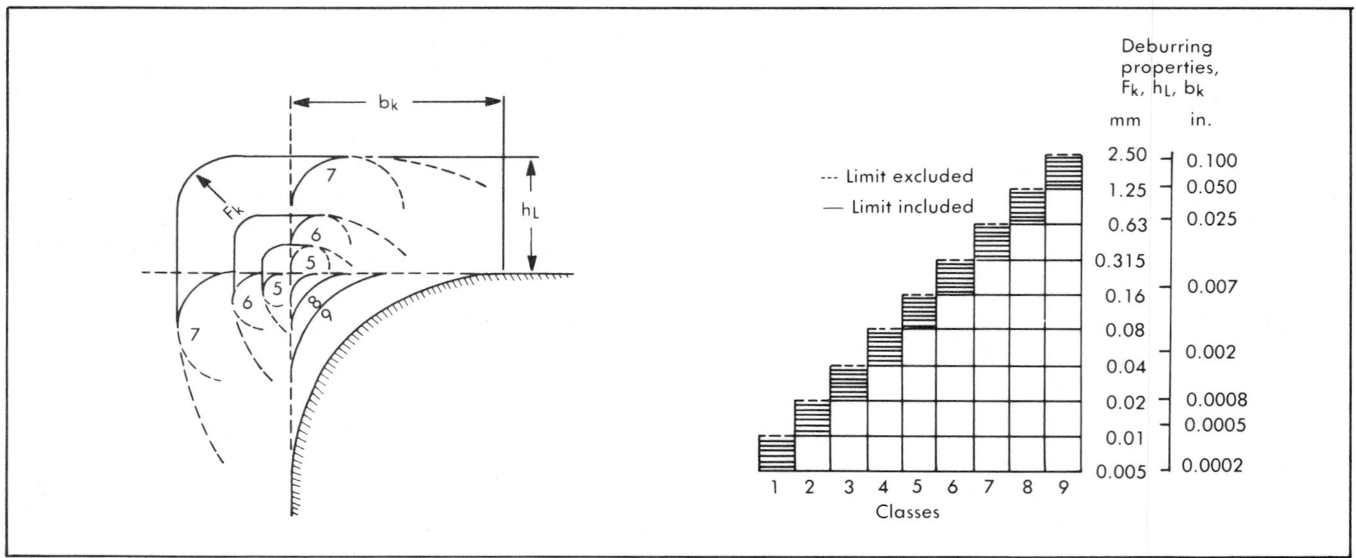

Fig. 15-13 Classes of allowable edge quality proposed by Schafer.[4]

Designing to Decrease Deburring

The need for deburring can be eliminated or minimized by the following two design approaches:

1. Design components with configurations that eliminate or minimize the need for deburring.
2. Specify the actual edge quality required.

Some components and assemblies can operate adequately without deburring. Sheet metal edges are often more aesthetic and trouble-free if a rolled edge is produced. In such cases, deburring is not required on the hidden edge; thus, the designer can use the geometry of the part to reduce deburring. The majority of assemblies may not lend themselves to such obvious design changes. However, if deburring can be eliminated from even one part in an assembly, there is a consequent cost savings. A common example in which burr removal is not required is shown in Fig. 15-15. Pins that are pressed into holes often do not have to be burr-free if provisions are made in the holes to accommodate the burrs.

Designers can often eliminate or minimize deburring by specifying large corner angles. When a cutter passes over an edge having an angle much larger than 90°, little or no burr forms on the edge. Conversely, when the edge angle is small, a large burr forms because there is no support for the metal being cut. Figure 15-16 illustrates the influence the workpiece edge angle has on burr size when a shell mill is used; as depicted, no noticeable burr is produced when edge angles exceed 150°. While the data presented is for low-carbon steel in an unhardened condition, this basic phenomenon can be observed in all materials.

From a practical standpoint, however, 90° edges are preferred on most machined components because they are easier to design, inspect, and machine. On cast parts, however, an angle larger than 90° is required to remove parts from the mold. Draft is required in forgings to increase tool life as well as for part removal. The draft angle generally used for castings is in the order of 1 to 5°. If the draft were 60°, the included edge angle would be 150°, and no burr would form when these features

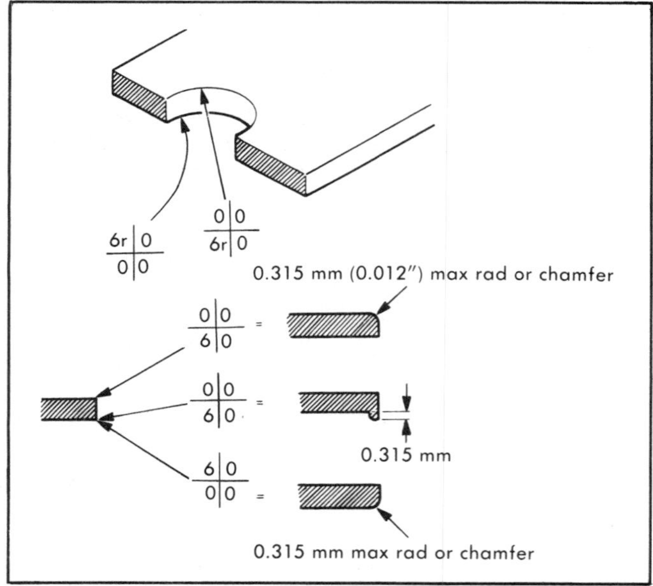

Fig. 15-14 Schafer's burr notations as a function of the view.[5]

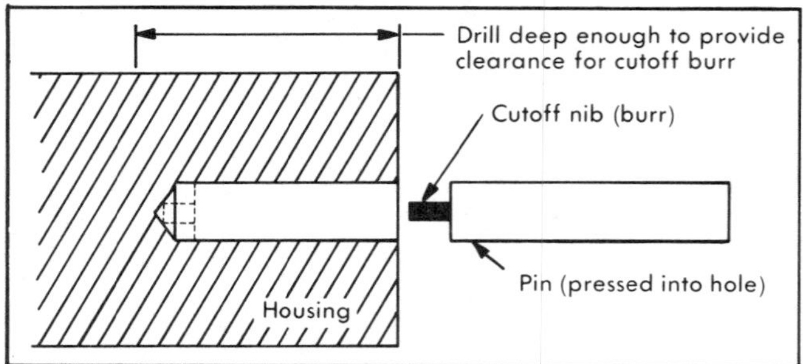

Fig. 15-15 An example of workpiece design that eliminates the need for deburring.

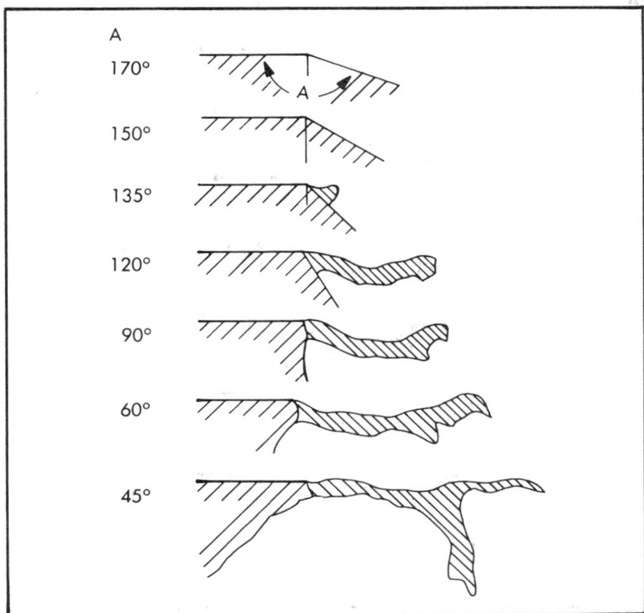

Fig. 15-16 Effect of edge angle on size of burr at end of cut.[6]

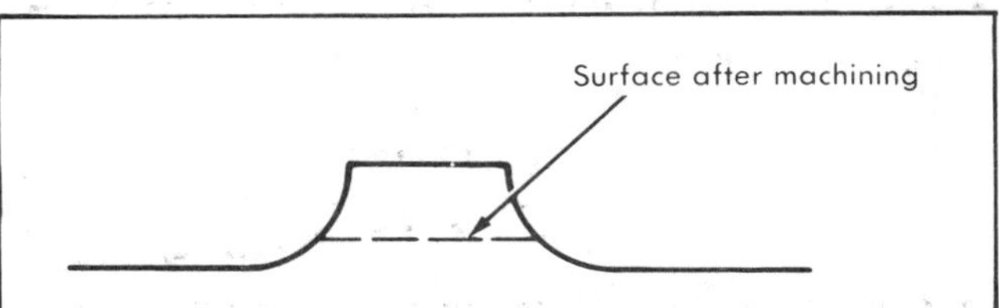

Fig. 15-17 Specifying a large fillet radius on castings can result in a large edge angle.

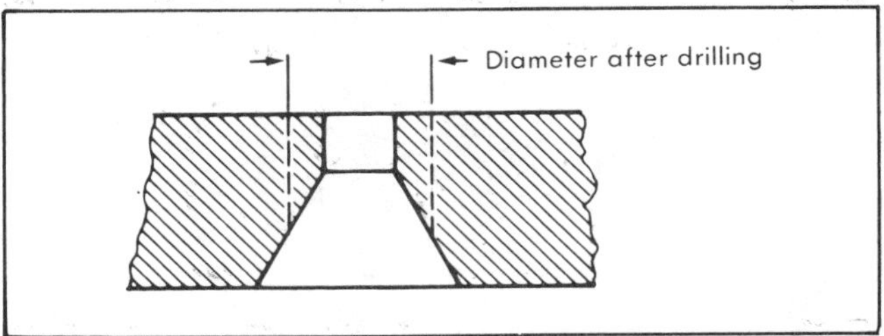

Fig. 15-18 Providing relief on cored holes helps reduce size of burrs when holes are finish machined.

were machined. Specifying a properly selected radius on castings can result in a large edge angle (see Fig. 15-17).

It is possible to eliminate rollover burrs from drilling, turning, grinding, and other processes, as well as from milling operations, by specifying a large exit angle. In areas to be drilled, cored holes should be provided with relief (see Fig. 15-18) to help reduce burr size when the holes are finish machined.

Figure 15-19 illustrates another significant aspect with respect to angles. The amount of radius that can be produced economically by vibratory finishing after removing the burrs is a function of the angle between intersecting surfaces. Large radii can be produced relatively quickly when the included angle is large. Finishing times are 20 times longer when the included angle is 30° than when it is 120° (see Fig. 15-20). Precision edge-radius tolerances are harder to maintain, however, when large angles are present. When a component has features involving several different edge angles, edge radii will vary significantly. Designers must recognize this variation when assigning tolerances to edge radii if they want to eliminate the extra costs required to produce equal radii.

Designing to Decrease Flash

It is easier to predict the location and size of flash from die casting and molding processes than for burrs produced in machining because flash corresponds to die configuration rather than the path of a cutting tool. As a result, designing to minimize flash is a technique that has been widely used for a number of years.

An aspect of molded parts that simplifies deflashing is that trimming equipment generally conforms to the part contour. Trimming attacks the burrs and not the entire workpiece surface, as is the case in many deburring processes. If flash cannot be prevented at a nonworking (aesthetic) surface, it should be on a rolled surface rather than a parting-plane edge. This procedure allows trimming to be used, and any mold offset is not visible after trimming.

When surfaces must be ground to remove caps of flash around holes, any projecting features should be designed with recesses to provide an unobstructed grinding path. While piercing dies can be used to remove these caps, their use can result in tearing of the edges.

Designing for Mass Finishing

While many designers do not know which deburring process will be used to remove burrs or flash, it is significant that at least 50% of all parts produced in the world are subjected to mass finishing processes, as discussed in Chapter 16. Because these processes all use the same basic metal removal mechanism, there are a few common considerations with respect to workpiece design. Essential factors in all mass finishing processes include the following:

1. The media used must have access to and be able to slide over all critical surfaces and edges that require deburring.
2. Large media deburrs faster than small media, except on very small parts. (Media used for mass finishing is discussed in Chapter 16 of this volume.)
3. The use of special preformed shapes, such as triangles, cones, and cylinders, have advantages over gravel-like media.

In the top view of Fig. 15-21, the step height H_1 is smaller than the radius R_S on the deburring media, which in this case is a triangle. As a result, the media will often bend the burr over rather than remove it. Complete removal can be obtained only with long cycle times or by using blends of media with smaller sizes present. If the designer can specify a larger step height, as shown in the bottom view, this problem is eliminated since the media can work over the edge and against both surfaces. For deburring slots with triangular media having a standard 60° configuration, the slot width *b* (see Fig. 15-22) must be greater than $3R_S$, and the depth must be greater than R_S.

When inner surfaces near fillet radii must be smoothed by mass finishing processes, it is essential that the fillet radii specified be equal to or exceed the major radius on the media (see Fig. 15-23). This is a particularly significant consideration

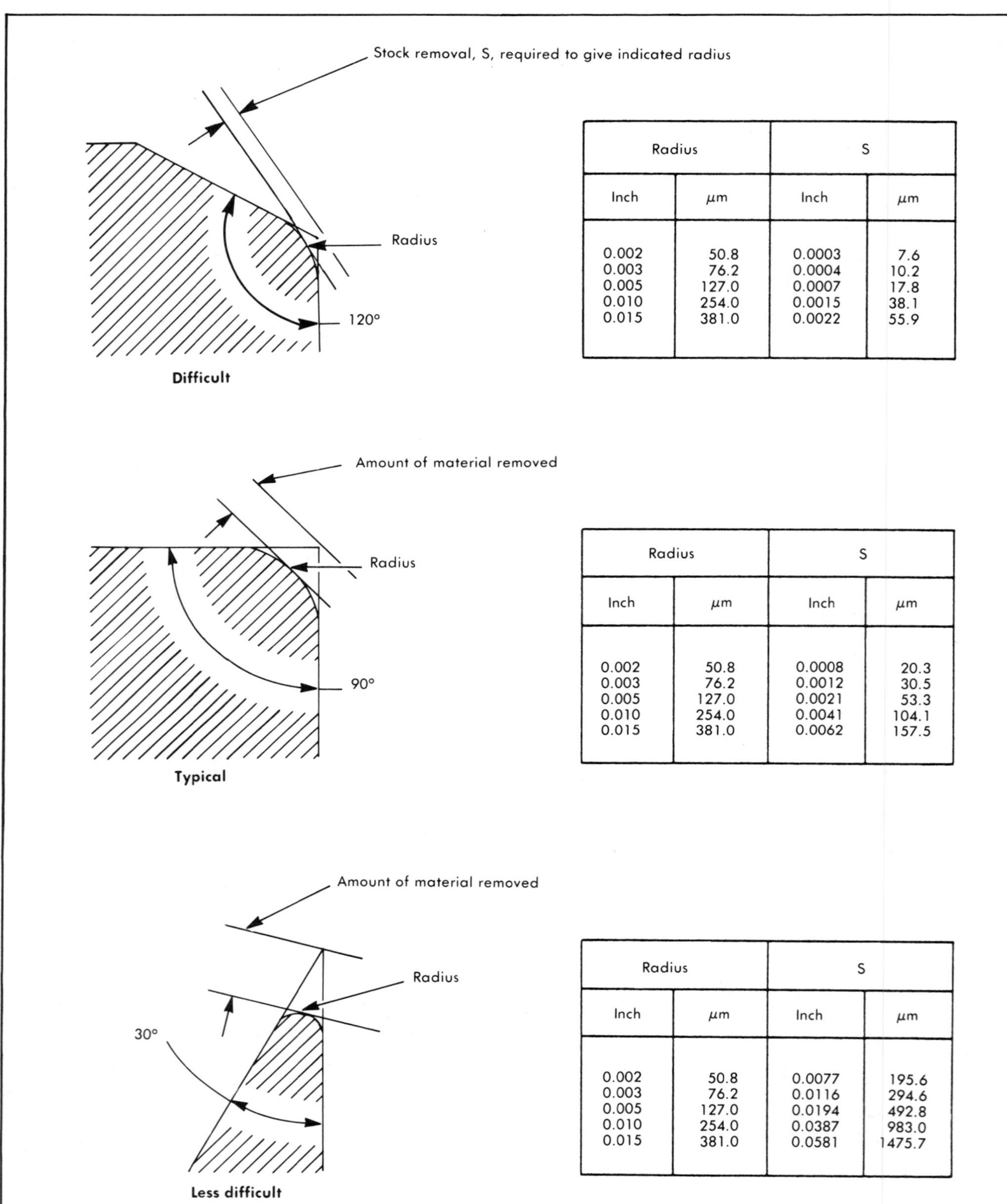

Radius		S	
Inch	μm	Inch	μm
0.002	50.8	0.0003	7.6
0.003	76.2	0.0004	10.2
0.005	127.0	0.0007	17.8
0.010	254.0	0.0015	38.1
0.015	381.0	0.0022	55.9

Radius		S	
Inch	μm	Inch	μm
0.002	50.8	0.0008	20.3
0.003	76.2	0.0012	30.5
0.005	127.0	0.0021	53.3
0.010	254.0	0.0041	104.1
0.015	381.0	0.0062	157.5

Radius		S	
Inch	μm	Inch	μm
0.002	50.8	0.0077	195.6
0.003	76.2	0.0116	294.6
0.005	127.0	0.0194	492.8
0.010	254.0	0.0387	983.0
0.015	381.0	0.0581	1475.7

Fig. 15-19 Effect of angle between intersecting surfaces on edge radiusing. (*Bendix Corp.*)

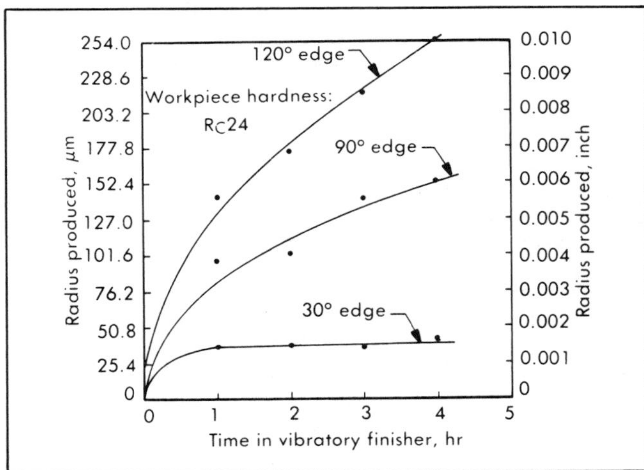

Fig. 15-20 Effect of edge angle and time of vibratory finishing on edge radiusing of phosphor-bronze workpieces. (*Bendix Corp.*)

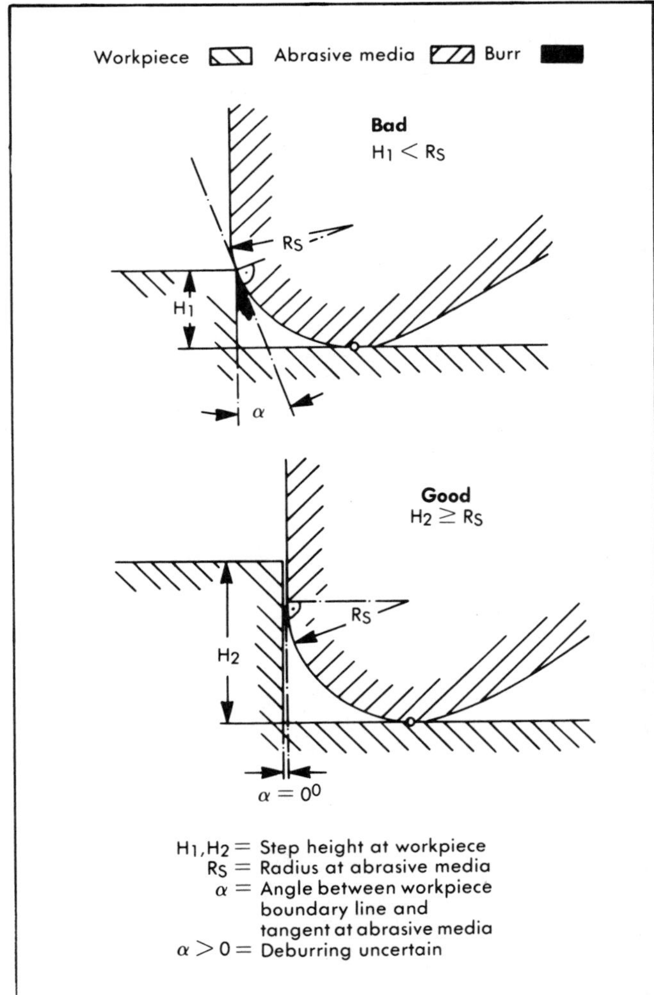

Fig. 15-21 Complete deburring requires that angle α = 0°.[7]

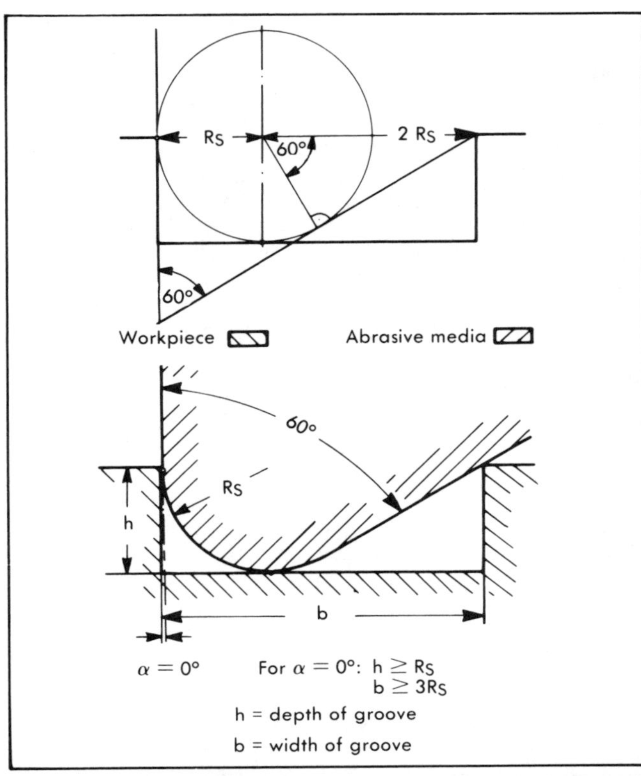

Fig. 15-22 For deburring slots with triangular media, slot width, b, must be greater than $3R_S$, and the depth must be greater than R_S.[8]

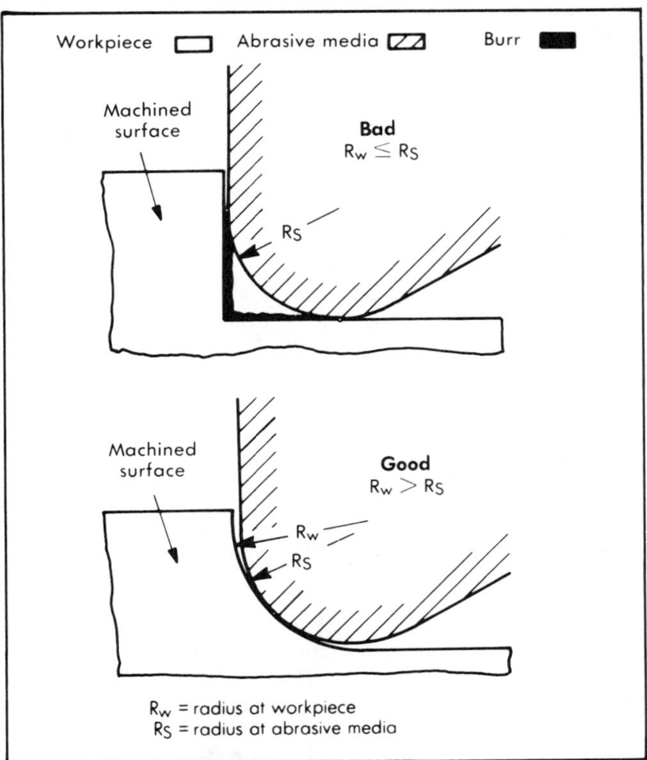

Fig. 15-23 Deburring action at fillets.[9]

when the feature is an extruded item, which is controlled by the die shape. Small steps in a workpiece will not receive significant deburring action, even with small media or balls, because size and geometry prohibit contact. Other media shapes, however, usually overcome these problems.

When the same edge radius and deburring quality is required on all edges of a part, smaller media sizes are required. Radiusing action is largest where larger media can most readily contact the part. While such differences are not critical on most parts, they are on many others. Complete deburring can occur only when the burr is totally accessible by the media.

Each of these factors should be considered while parts are still in the design phase. While smaller deburring media or various shapes can be used, they create other problems. Whenever possible, workpiece drawings should be reviewed by manufacturing engineers responsible for machining and deburring before the drawings are finalized.

Relief on Features Machined Through Threads

Burrs formed by machining through threads are extremely difficult to remove. This problem can be eliminated by turning a relief diameter that is smaller than the minor diameter of the thread (see Fig. 15-24). The designer must indicate that these conditions are allowable or desirable through drawing notes or in-house standards.

V-Grooves on Turned Parts

If a small burr is allowable on the outside diameter (OD) of a slotted part, an optional V-groove can be placed at the bottom of the slot. This groove permits the existence of a burr at the bottom of the slot without affecting OD size. Although a small burr also forms at the sides of the slots, it may not be large enough to require removal. If it does, it is much easier to remove than the burr normally left at the bottom of the cut.

When a V-shaped groove is placed on the inside diameter (ID) of a cylindrical part so that the cutoff tool will pass through the groove (see Fig. 15-25), the cutoff burr will not be as objectionable as on most parts. In such cases, the improvement is the result of both the edge angle created by the vee and the fact that when properly placed, the burr will form downward into the vee rather than upward on the ID. If total burr removal is required, a vibratory operation should prove adequate. In this case, the designer may have to specify the allowable chamfer that the V-groove can introduce to the completed part.

Designing to Reduce Die Wear

Part designs such as the one shown in the left-hand view of Fig. 15-26 lead to early die wear because of the sharp edges required. This design creates heavy burrs, particularly at the needle-like projections and cutouts. Designing in accordance with the guidelines shown in the right-hand view extend die life and minimize burrs significantly. When sharp corners are necessary, they can be provided by ordering more expensive progressive dies.

PROCESSING CONSIDERATIONS

Burrs and the cost of removing them can sometimes be eliminated and often minimized by optimum processing during the manufacture of the parts. Important factors that must be considered include process selection, operating parameters, burr placement, and the design and maintenance of tools, fixtures, dies, and molds. When many different workpieces are processed, the workpieces should be grouped into families of similar parts, which often permits deburring families of parts together on a single machine.

Burr Prevention

Methods of preventing burrs include optimum design of the workpieces (discussed previously in this section), the use of

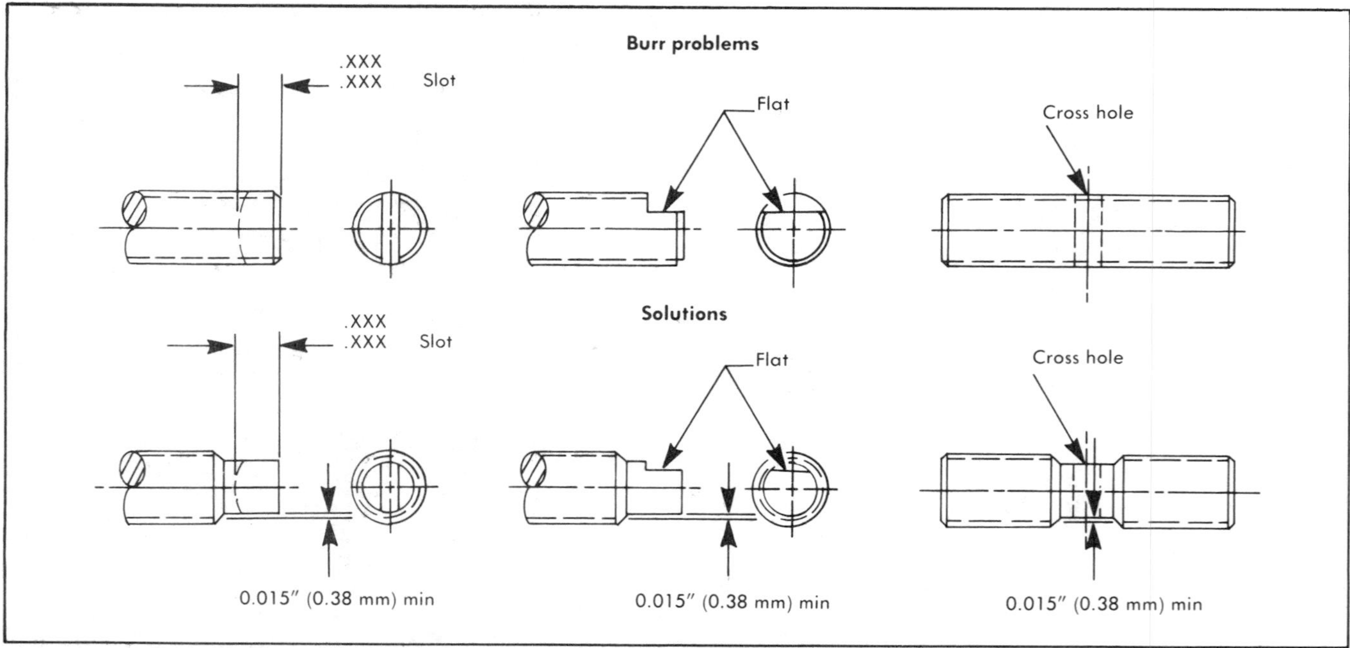

Fig. 15-24 Methods of eliminating deburring problems when machining through threads.[10]

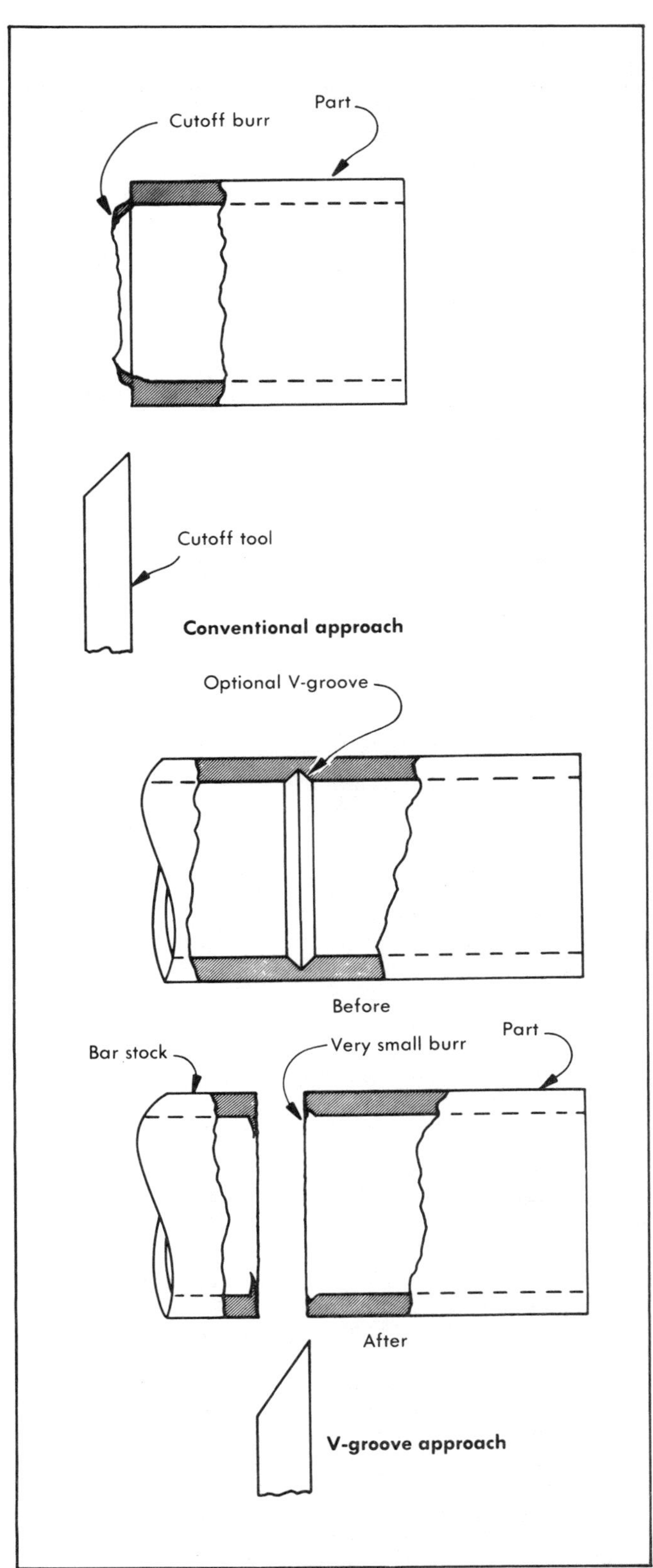

Fig. 15-25 Placing a V-groove on ID minimizes size of cutoff burr. (*Bendix Corp.*)

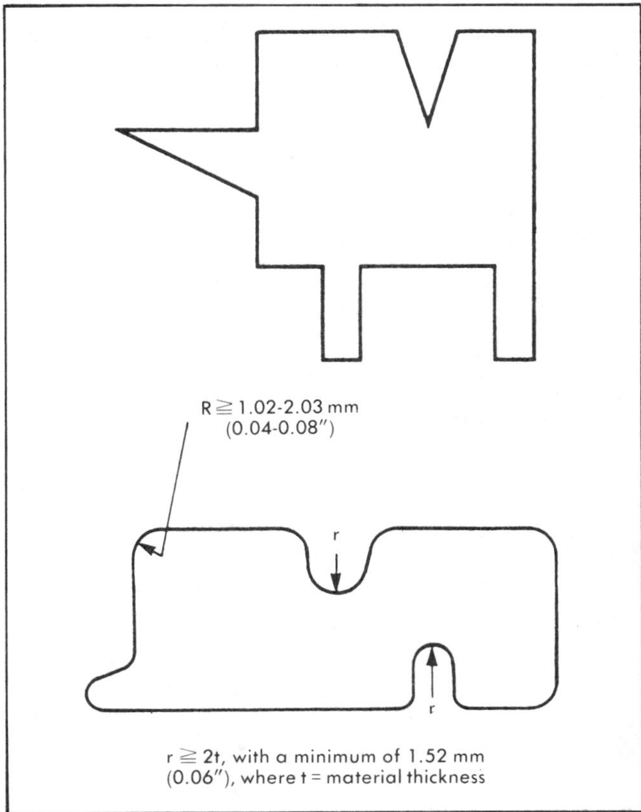

Fig. 15-26 Large radii are preferred for blanked parts to minimize burr size and lengthen die life.[11]

back-up material during machining, employing nontraditional processes, or using different production techniques.

Back-up materials. Theoretically, supporting a workpiece with back-up material should prevent burrs. From a practical standpoint, however, back-up material only helps to minimize burrs and is generally only economical when deburring costs are high. It is also important that the back-up material have the same properties as the workpiece to prevent burr formation.

Most operations produce burrs at more than one location. In drilling, for example, a burr occurs at both hole entrance and hole exit. In a milling operation, burrs can be produced on up to ten edges. At best, burrs on only one side of the workpiece are minimized by using back-up material.

Nontraditional processes. Burrs can be prevented by employing some of the nontraditional processes discussed in Volume I, *Machining*, of this Handbook series. As described in Table 15-3, most of the nontraditional processes do not produce burrs. The electrical discharge machining (EDM), electron beam machining (EBM), and laser beam machining (LBM) processes produce burr-like projections of recast material.

Processes such as chemical machining (CHM), electrochemical grinding (ECG), electrochemical machining (ECM), electrochemical honing (ECH), electropolishing (ELP), and electrostream machining (ESM) not only eliminate deburring costs, but they also provide excellent surface finishes and minimize welding, brazing, and plating problems caused by media impregnation or improper cleaning. In addition, the

TABLE 15-3
Nontraditional Machining Capabilities

Process	Typically Makes Burr?*	Typical Edge Radius Produced, in. (mm)	Typical Machining Tolerance, in. (mm)
Abrasive jet machining (AJM)	No	0.003 (0.08)	---
Chemical machining (CHM)	No	Unknown	±0.002 (0.05)
Electron beam machining (EBM)	Yes	---	±0.001 (0.03)
Electrochemical discharge machining (ECDM)	Unknown	Unknown	Unknown
Electrochemical grinding (ECG)	No	0.003 (0.08)	±0.002 (0.05)
Electrochemical machining (ECM)	No	0.001 (0.03)	±0.002 (0.05)
Electrochemical honing (ECH)	No	0.0005 (0.013)	±0.0002 (0.005)
Electrical discharge machining (EDM)	Yes	---	±0.0006 (0.015)
Electropolishing (ELP)	No	0.001 (0.03)	±0.0005 (0.013)
Electrostream machining (ESM)	No	0.002 (0.05)	±0.001 (0.03)
Hot chlorine gas (HCG)	No	0.002 (0.05)	±0.003 (0.08)
Ion beam machining (IBM)	No	0.00005 (0.0013)	±0.0001 (0.003)
Laser beam machining (LBM)	Yes	---	±0.001 (0.03)
Plasma arc machining (PAM)	Yes	---	±0.003 (0.08)
Ultrasonic machining (USM)	No	0.001 (0.03)	±0.001 (0.03)
Water jet machining (WJM)	No	Unknown	±0.003 (0.08)

* Where burr or burr-like projection is visible under 30x magnification.

elimination of unnecessary operations could reduce costs and shorten production time.

Disadvantages of using the nontraditional processes generally include high equipment costs, limitations to certain geometries and workpiece materials, slower processing, and workpiece tolerance and surface integrity problems. In addition, these processes are not always applicable.

High-speed machining. The use of cutting speeds in excess of 10,000 sfm (3048 m/min) has been observed to produce no burrs in some metals. In many cases, a claim of "no burr" actually means a much smaller burr or "no burr of consequence to most users."

Each metal has a critical machining velocity above which less energy is required to induce failure or separation. Lower energy requirements are the result of smaller amounts of plastic deformation, which is the cause of burrs. In the case of 17-7 PH stainless steel at 600 fps (183 m/s), its ductility is one tenth that at 100 fps (30 m/s).

Preventing plastic deformation. A key to burr prevention is to prevent plastic deformation from occurring macroscopically at part edges. Any environment that prevents plastic deformation will minimize or prevent burrs. One technique used to prevent burrs is to extrude complex sections rather than machine them. Extrusion saves greatly on machining costs, and the only burrs formed are at cutoff edges. Other processes being used to prevent or minimize burrs include flashless forging, hot forming on closed-die machines, and wedge (roll) rolling, all discussed in Volume II, *Forming* of this Handbook series.

Burr Minimization

Burrs can be minimized by:

- Proper design of the workpiece or a change in its configuration.
- Use of a more appropriate workpiece material.
- Changing the types and/or sequence of production operations. Deburring can sometimes be done on NC/CNC machines, along with other operations.
- Changing machine stiffness and/or minimizing cutting forces.
- Changing operating parameters (cutting speed, feed rate, depth of cut, etc.).
- Changing the cutting tool material, configuration, and/or sharpness.
- Using back-up materials.
- Using better workholding methods.
- Placing burrs in positions that facilitate removal.

Conventional machining methods always produce burrs. Altering the operating parameters (such as speeds, feeds, and depths of cut) or tool geometries will not prevent burrs. However, the use of sharp tools, correct tool geometry, and proper cutting conditions can minimize burr size and control their repeatability. Consistency of burrs simplifies burr removal. The cost of keeping tools constantly sharp can be prohibitive, and an economic balance must be established between the cost of resharpening or replacing the tools and the cost of deburring.

In some cases, burrs will be minimized at one location, but may increase at another location on the same part. Burrs formed in drilling and milling operations exhibit the most variability with respect to burr thickness. Grinding and honing burrs are normally more consistent, partly because they involve smaller forces at workpiece edges.

Burr Placement

Making a burr easy to remove involves two requirements:

1. Locating the burr in the best position for easy removal. Such location should be established before the machining process and tooling are finalized.
2. Controlling burr size by appropriate selection of machining sequences.

Figure 15-27 provides an excellent example of burr placement. With the cutoff burr on the large-diameter flat end of the part, burr removal by vibratory finishing is difficult. With the burr on the rounded end, the burr is more exposed to the media action of vibratory deburring and can be readily removed, provided tolerances are adequate.

On intersecting features, placing the burr on the most accessible edge can sometimes reduce burr costs up to 50%. The correct placement of burrs can change a hand deburring situation to a much more economical and repeatable vibratory deburring operation. In some operations, burr placement can be controlled by a simple change in the program of an NC/CNC machine. In some cases, climb milling rather than conventional milling will provide better burr location.

The decision with regard to the machining sequence to be used can affect deburring time. For example, a cylindrical part having a triangular cross section for part of its length would typically be turned, and then the three flats would be milled. This sequence, however, creates two undesirable conditions. First, milling typically makes a heavy burr; and, second, deburring would require a separate operation. The company producing this part chose to mill the flats before turning the stem diameter. This procedure produced a smaller burr and allowed the operator to brush the burrs off with a wire brush as a part of the lathe operation.

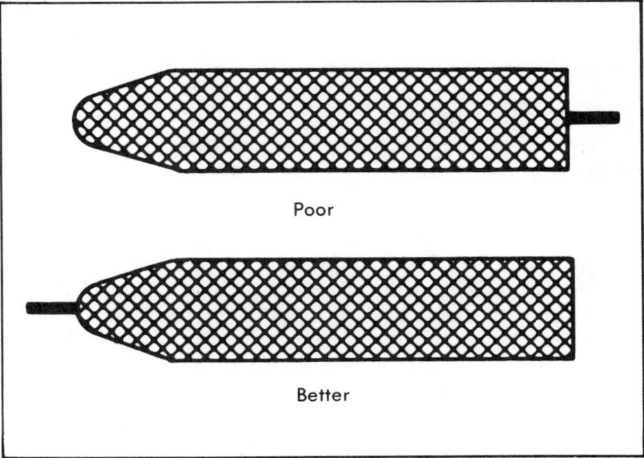

Fig. 15-27 Placing the cutoff burr on the rounded end facilitates removal by vibratory finishing.[12]

When a manufacturing engineer considers the machining sequence to follow, the burr should be considered as significant a factor as the desired surface finish and tolerance. The following checklist helps in processing and troubleshooting with respect to burrs:

1. Does the burr have to be removed?
2. Can the part be redesigned for less machining? for less deburring?
3. Will the burr be cut off in a subsequent machining operation? If the machining sequence were changed, would the burr be cut off?
4. Is the burr accessible?
5. Should the sequence of operations or the direction of the cut be changed?
6. Can a cutter be chosen that gives a smaller burr?
7. Will the feed rate selected form the smallest burr?
8. Will subsequent heat treating make the burr brittle?
9. Would a change in cutting fluid or method of application make the burr brittle?
10. Does the burr have to be removed now?

When a rollover-type of exit burr forms near a hub, it is difficult to remove by mechanized means. If the rollover burr is located at the opposite end of the part (see Fig. 15-28), no projection will interfere with the deburring operation.

If the hobbing exit burr shown in Fig. 15-29 is allowed to form on the hub side, precision deburring will be very time consuming. By fixturing the part so that the hob exits over the flat face of the gear, hobbing burrs can be removed easily by a quick hand-abrasive operation. Any small burrs remaining can then be removed and edges radiused by brushing, or by one of the mass finishing processes.

Even when backup material is used in multiple-part hobbing, the cutter should exit from the flat surface. Since direction of feed or rotation of the cutter cannot be changed when cutting ratchet teeth and nongear shapes, burr location must be chosen before hobs are ordered. When milling L-shaped configurations (see Fig. 15-30), exit burrs should be located on the back and top, rather than under ledges.

Turning Operations

The following recommendations will help to minimize or eliminate the formation of burrs in turning operations on lathes and other machine tools:

- When possible, use a form tool to cut diameters. No burr can occur when the tool produces the diameter and the adjoining face at the same time. Any intermediate burr formed is wiped off.
- Break all edges with a chamfering tool. Since tool position can generally be controlled accurately to 5.08 μm (0.0002″) on many materials, the chamfering tool can remove the burrs and still ensure small, final edge breaks.
- Generate corner radii rather than machining perpendicular, sharp edges by cam design or NC program.
- Reduce cutting forces and the weight of the workpieces to reduce the diameters of the cutoff projections. The use of sharp, narrow cutoff tools is also helpful.
- Remove built-up edges, increase rake angles, use sharp cutting edges, minimize flank wear, and use shallow cuts to minimize burrs since these factors reduce cutting forces when cutting all materials.

Drilling Operations

Burrs form on both the entrance and exit sides of drilled holes. The burrs on the entrance side are typically small, while those on the exit side are typically long and ragged. There are numerous combinations of helix angle, point angle, and feed rate that produce the same burr thickness. Significant reductions in burr thickness are possible by the correct selection of these parameters. Maximum burr height on any hole is proportional to the drill diameter.

A burr thickness of 0.0043″ (0.109 mm) can be expected when drilling AISI 1018 steel, having a hardness of 150 Bhn, with a ¼″ (6.35 mm) diam HSS twist drill having a 27° helix angle and 112° point angle, using a feed rate of 0.006 ipr (0.15 mm/rev). By increasing the helix angle to 36°, the burr

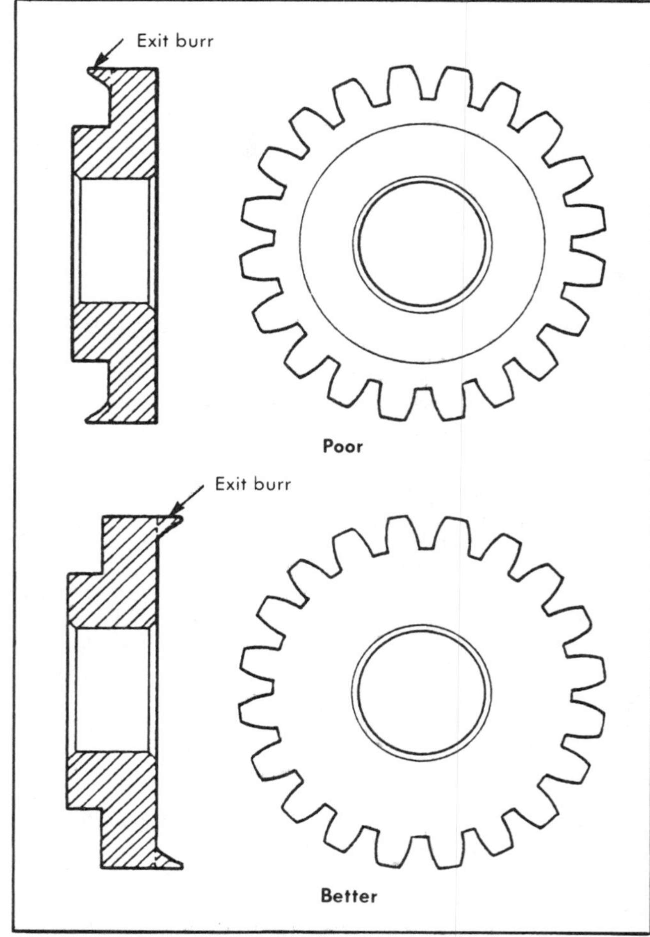

Fig. 15-29 Location of exit burr resulting from hobbing affects deburring time.

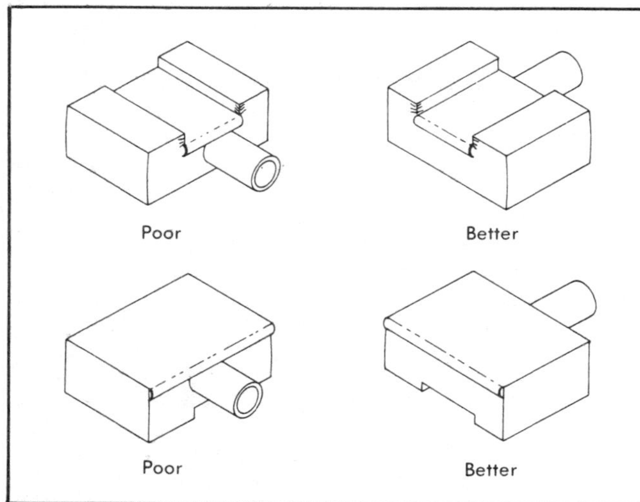

Fig. 15-28 Locations of rollover-type exit burrs. (*Bendix Corp.*)

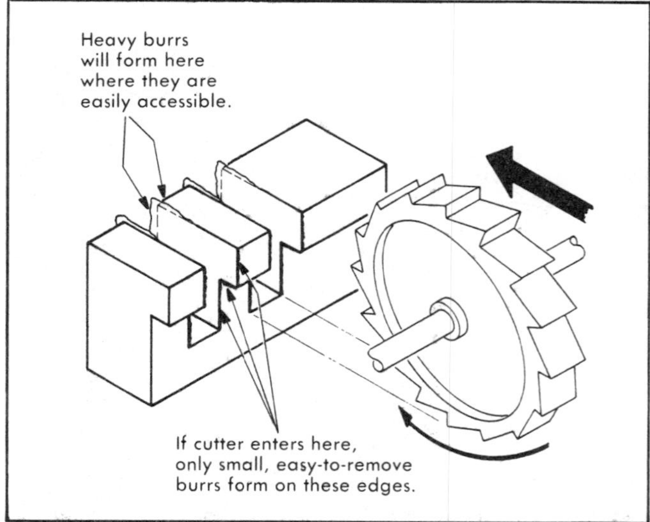

Fig. 15-30 Heavy exit burrs should be located on back and top when milling L-shaped workpieces. (*Bendix Corp.*)

thickness is reduced to 0.0026" (0.066 mm). If the feed is lowered to 0.0035 ipr (0.089 mm/rev) and the point angle reduced to 98°, the burr thickness is reduced to 0.0015" (0.038 mm). A burr this thin can normally be removed easily by barrel or vibratory deburring.

Burr thickness can be expressed by the following equation:

$$T = C_1 H^{-1.72} P^{0.84} L^{-0.36} f^{0.86} \quad (3)$$

where:

T = burr thickness
C_1 = 4.271 when T is in inches and f is given in ipr
= 6.797 when T is in mm and f is given in mm/rev
H = helix angle of twist drill
P = point angle of twist drill
L = lip clearance angle of twist drill
f = feed rate

This equation shows that increasing the helix angle and lip clearance angle decreases the burr thickness. Thickness is only slightly affected by the lip clearance angle. Decreasing the point angle and feed rate also decreases the burr thickness.

Burr thickness can also be expressed in terms of the spindle speed, hardness of the workpiece material, and stiffness of the drilling system by the following equation:

$$T = C_2 N^{-0.783} B^{-0.998} K^{-0.007} \quad (4)$$

where:

T = burr thickness
C_2 = 252 when T is in inches and K is given in lb/in.
= 7128 when T is in mm and K is given in N/m
N = spindle speed of the drilling machine, rpm
B = hardness of the workpiece material, Bhn
K = stiffness of drilling system

This equation shows that increasing the spindle speed or material hardness reduces the thickness of an exit burr. Because of the low exponent over the range of values examined, the system stiffness has an insignificant effect. The effect of spindle speed and workpiece hardness when K is 27,000 lb/in. is shown in Fig. 15-31.

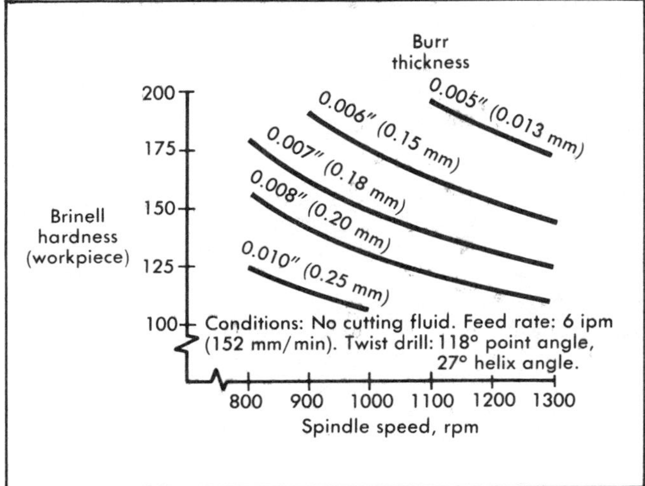

Fig. 15-31 Effect of spindle speed and hardness on burr thickness when drilling AISI 1018 steel with a ¼" (6.35 mm) diam HSS twist drill. (*Bendix Corp.*)

A practical aspect of the results shown in Fig. 15-31 is that for a given workpiece hardness, the burr thickness can be reduced by increasing the spindle speed. It can be seen that the thickness can be reduced from 0.010 to 0.007" (0.25 to 0.18 mm) by increasing the spindle speed from 800 to 1300 rpm. Changes in material hardness of 25 points on the Brinell hardness scale cause variations of 0.001 to 0.002" (0.03 to 0.05 mm) for the higher and lower Brinell numbers respectively. Use of high speeds for drilling aluminum, however, can cause mushroom-type burrs at the entrance sides of holes, and large burrs or breakout at the exit sides.

The effect of drill lip radius on exit burr thickness when drilling 6061-T6 aluminum alloy is illustrated in Fig. 15-32. Figure 15-33 shows the effect of drill point angle on exit burr thickness when drilling 1050-0 aluminum alloy. Increasing the point angle to 100° increases burr thickness. Above that angle, thickness decreases to a minimum at about 180°. The generally recommended point angle for drilling aluminum is 118 to 140°, but this angle results in comparatively large burrs.

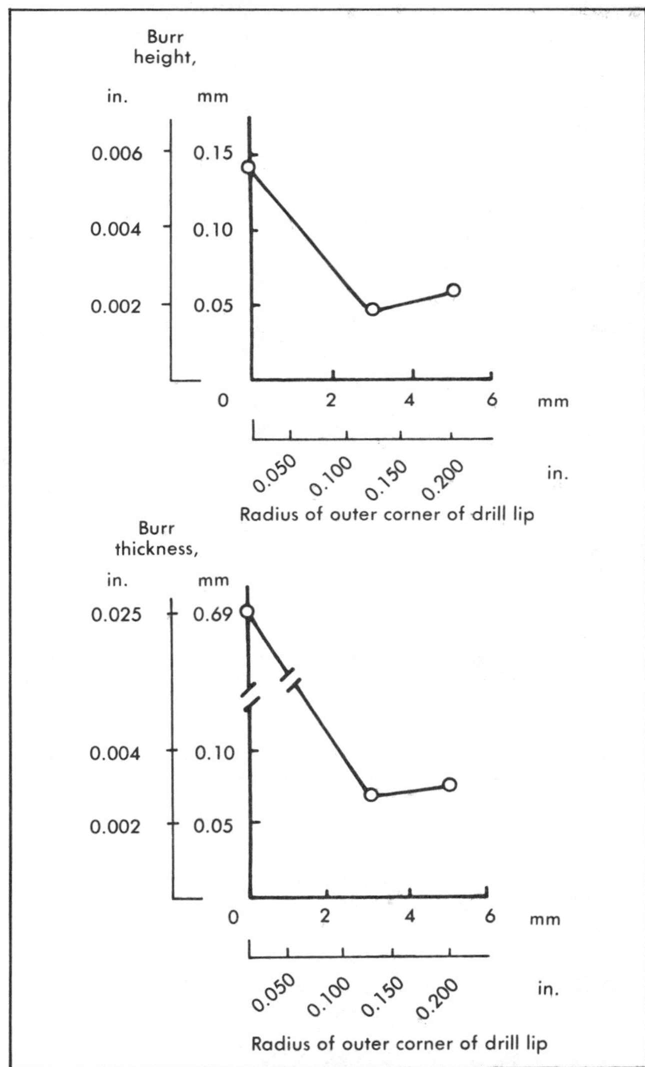

Fig. 15-32 Effect of lip radius of drill on burr height and thickness when drilling 6061-T6 aluminum alloy.[13]

There are two approaches to diminishing the thickness of burrs with respect to the point angle of the twist drills used. The first is to decrease the angle at the outer corner (increasing the point angle also does this), and the other is to make the point angle as large as 180°. For a small equivalent point angle, the zone of plastic flow diminishes at the hole exit owing to the small chip thickness. For a 180° point angle, the radial component of cutting resistance becomes small. Both methods reduce burrs. The latter is employed widely in drilling thin sheet metals using a fishtail point drill.

Drills with chamfered or dubbed corners, such as used for drilling cast iron, are used for decreasing the angle at the outer corner, as is the radial-lip drill geometry. In some cases, and for drilling certain materials, the radial-lip drill can reduce the chip thickness at the drill corners to a very small value, which produces the same effect as slowing the feed rate. Table 15-4 indicates the relative significance of variables on burr size for drilling operations. Some researchers have found that as silicon content is increased in aluminum, drill exit burrs become more brittle and are thus more easily removed. Additional information on twist drills, their geometries, and the drilling process is presented in Volume I, *Machining*, of this Handbook series.

One study of oil-hole drills indicated that the radial-lip design with a 3 mm (0.118") radius at the drill corners produced a smaller burr in aluminum than conventional drills. Cutting fluids with 20% lard content produced much shorter burrs than those with lower lard content. Forced application of the cutting fluid provided the smallest burr.

Large differences occur in the size of burrs when the feed rate employed for drilling is manually controlled. While drilling occurs under supposedly constant conditions, the operator cannot control the breakthrough thrust. Positive-feed drilling minimizes feed rate fluctuations. If there is flexibility with respect to the workpieces or fixture, end play in the spindle of the drilling machine, sloppy gears, or leaky hydraulics, the resultant lurch on breakthrough—in effect, a rapid feed rate increase—will cause burrs, even with optimum drill points. Rigid machinery, sturdy fixtures, well-supported workpieces, and regulated feeds are necessary for best results.

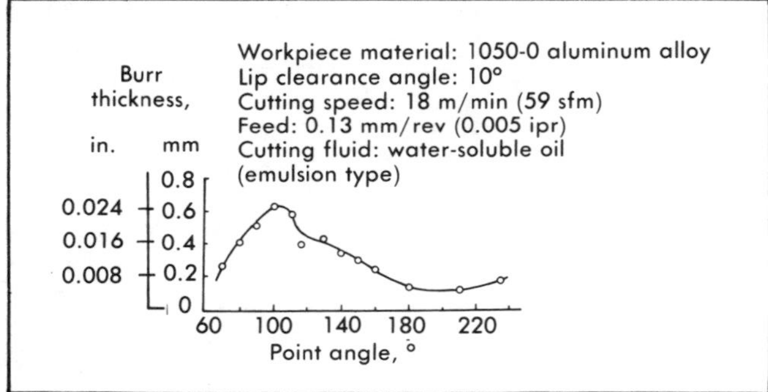

Fig. 15-33 Effect of point angle of drill on burr thickness.[14]

Exit burr size in drilling can be reduced by minimizing the size of the clearance hole in the fixture below the hole to be drilled. For commercial parts, a clearance hole 0.002" (0.05 mm) larger than the drill diameter has been recommended (see Fig. 15-34). Such a clearance should result in burrs only 0.001" (0.03 mm) thick. For precision holes, the clearance should be smaller, although tool life may be greatly shortened by this approach.

The use of a hard (R_C42), sacrificial back-up material in drilling also minimizes exit burr size. The effect of back-up material hardness on burr height and thickness when drilling Type 303 Se stainless steel is illustrated in Fig. 15-35. Hole saws will typically produce a smaller burr than twist drills in ductile material, partly owing to the fact that they greatly reduce the cutting forces that create the burrs.

Milling Operations

Minimizing end-milling burrs. The number of different burrs produced on various edges of workpieces in end-milling operations is discussed previously in this chapter (see Fig. 15-6 and Fig. 15-8). The height, the thickness, the radius on the back side of each burr, the hardness, and the appearance of each of these groups of burrs are different from the other groups—group indicating that the burr on each edge is the linear accumulation of many scallop-like cuts, each spaced one revolution apart (see Fig. 15-36).

The large number of edges and burr properties prohibit a full discussion of machining effects on burr size, but the following observations are significant:

- Fast feed rates, 0.002" (0.05 mm) vs. 0.0005" (0.013 mm)/rev/tooth, reduce burr height up to 50% but slightly increase burr thickness.
- Dull cutters significantly increase burr height and thickness.
- Helix angle changes increase some burr sizes while decreasing others.

TABLE 15-4
Effect of Variables on Burr Size in Drilling 303 Se Stainless Steel

Variable	Entrance Burr		Exit Burr		Ranked Significance
	Thickness	Height	Thickness	Height	
Helix angle	---	---	---	---	1
Feed rate	0	0	+	+	2
Diameter	0	---	+	+	3
Surface velocity	+	---	+	x	4
Corner angles*	x	x	---	+	5

--- : Increasing variable reduces burr property.
0 : No effect.
+ : Increasing variable increases burr size.
x : Conditions not studied.
* For conventional drills, the corner angle equals 180° minus half the point angle.

Minimizing side-milling burrs. Side-milling operations can also produce burrs at ten different edges. For conventional side-milling cutters, burr thicknesses can be minimized by reducing the feed rate and using a 0° helix angle. Full-radius cutters (cutters that produce a U-shaped slot cross section) should not be used unless essential on materials that elongate noticeably. These cutters produce thicker, as well as much longer burrs than do square-edge cutters. Table 15-5 provides some relative values of burr size along edges of one cut.

Minimizing face-milling burrs. While no data has been published on this process, it is similar to end milling. It is significant that burr height is a function of the angle at which the cutter teeth exit over the workpiece edge. Tests indicate that burr height can be controlled by varying the position of the cutter teeth with respect to the edge of the workpiece.

Geometry and tool sharpness play a major role in determining burr size. The burr formation force is that component of the resultant cutting force that is vertical to the exit edge, and depends on the angle between edge and cutting force direction. The burr size in the horizontal plane can be reduced by minimizing the angle, but the burr in the vertical plane may be enlarged by this approach. Any factor that reduces the burr formation force will reduce burr thickness. Such factors include slowing the feed rate, using sharp cutters, employing a cutting fluid, and using maximum rake angles.

Workholding Fixtures

Workholding jigs and fixtures should be designed with the realization that burrs may prevent easy removal of the workpieces. Providing clearances and corner reliefs (see Fig. 15-37) will generally eliminate some of the in-process deburring frequently required. In some cases, knockouts (such as the spring ejector shown in Fig. 15-38) are required.

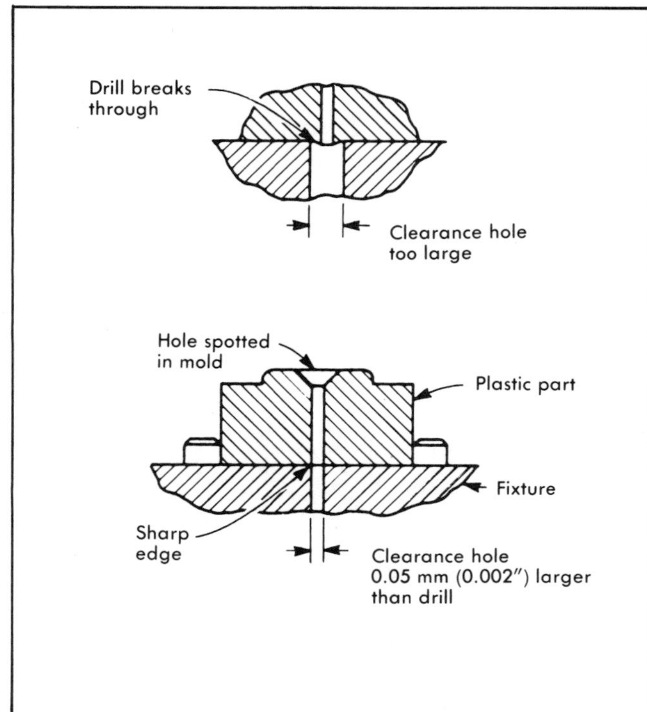

Fig. 15-34 Drill clearance in fixture for burr minimization. (*Bendix Corp.*)

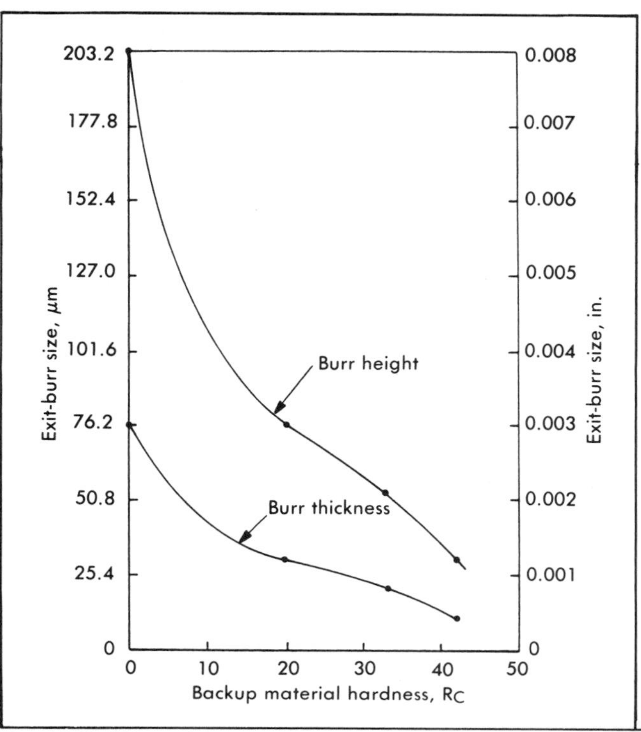

Fig. 15-35 Effect of backup material hardness on burr size when drilling Type 303 Se stainless steel. (*Bendix Corp.*)

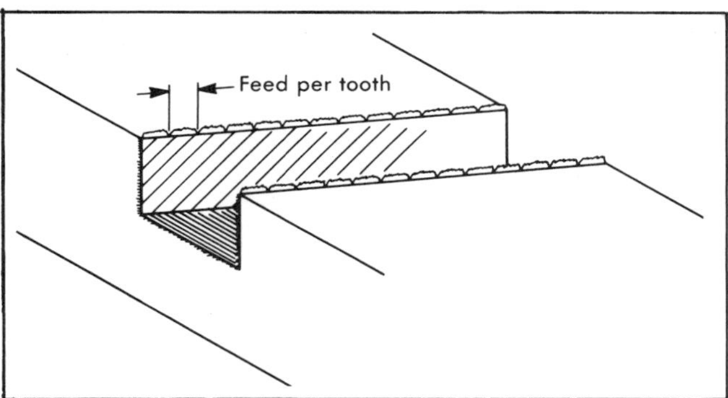

Fig. 15-36 Periodic nature of milling burrs.

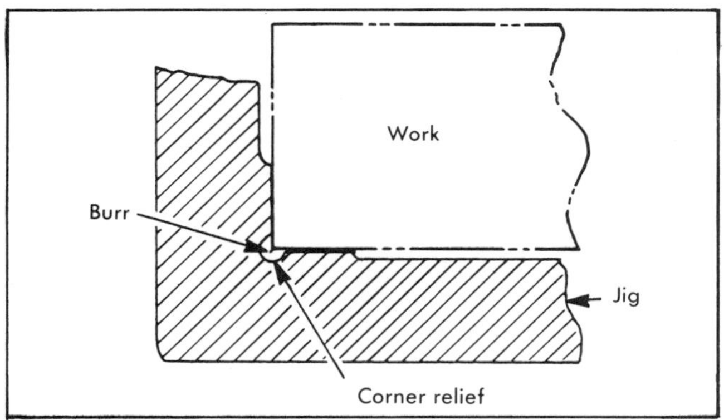

Fig. 15-37 Burr clearance provided by corner relief in jig.[15]

TABLE 15-5
Typical Burr Sizes Produced in Milling 303 Se Stainless Steel (R_C 32)

Milling Operation and Burr Size	Burr Location (See Fig. 15-6 and Fig. 15-8)								
	1	2	3	4	5	6	7	8	9
Side Milling:*									
Burr height, in. (mm)	0.0028 (0.071)	0.0020 (0.051)	0.0140 (0.356)	0.0073 (0.185)	0.0600 (1.524)	0.0001 (0.003)	0.0004 (0.010)	0.0002 (0.005)	---
Burr thickness, in. (mm)	0.0014 (0.036)	0.0016 (0.041)	0.0020 (0.051)	0.0016 (0.041)	0.0010 (0.025)	0.0014 (0.036)	0.0015 (0.038)	0.0011 (0.028)	---
End Milling:**									
Burr height, in. (mm)	0.0010 (0.025)	0.0020 (0.051)	0.0850 (2.159)	0.0001 (0.003)	0.0024 (0.061)	0.0002 (0.005)	0.0600 (1.524)	0.0007 (0.018)	0.0029 (0.074)
Burr thickness, in. (mm)	0.0007 (0.018)	---	0.0008 (0.020)	---	0.0009 (0.023)	0.0003 (0.008)	---	0.0003 (0.008)	---

* Using a 4″ (102 mm) diam, 1/2″ (12.7 mm) wide HSS cutter, with 18 teeth and 0-10° helix angle; operating at 98 rpm and fed at rates of 0.00057″ (0.0145 mm) to 0.002″ (0.05 mm)/rev/tooth.

** Using 0.125-0.375″ (3.18-9.52 mm) diam, four-flute end mills, operating at 1240 rpm and fed at rates of 0.0005″ (0.013 mm) to 0.0033″ (0.084 mm)/rev/tooth.

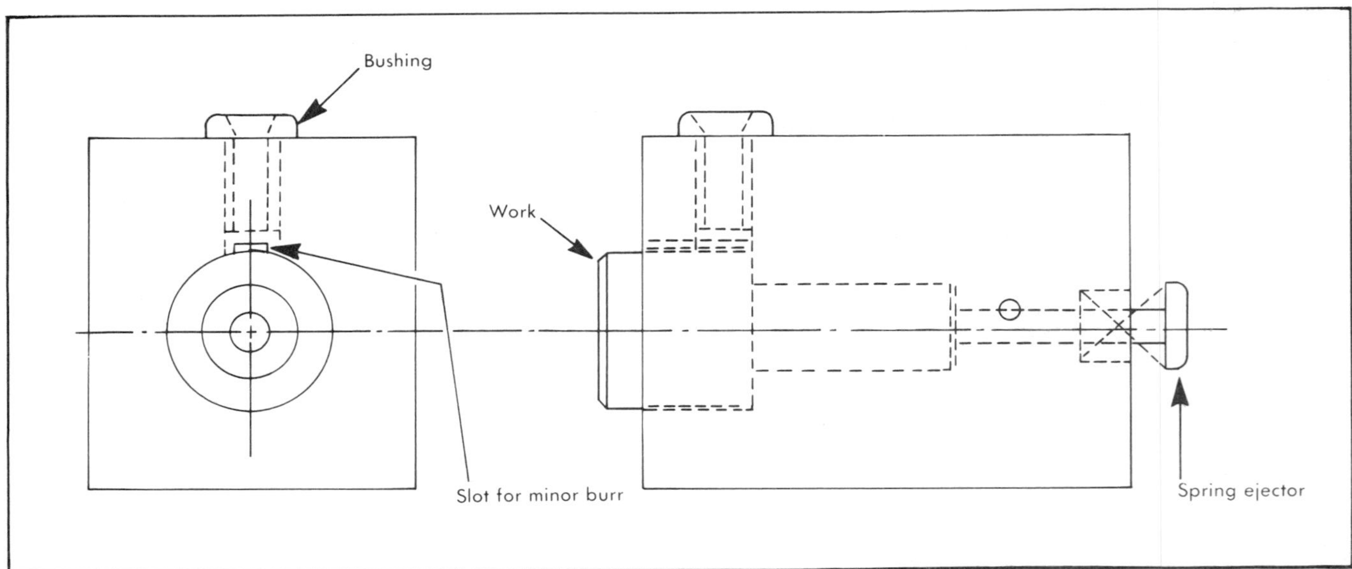

Fig. 15-38 Knockout (spring ejector) for removing workpiece having drill burr. (*Bendix Corp.*)

Pressworking

Clearance. In conventional blanking and shearing, clearance between the punch and die, workpiece clamping conditions, cutting edge sharpness, and workpiece properties are the factors that most affect burr size. Extremely large or small clearances will cause large burrs. Clearances of 5-15% are usually adopted in order to minimize burr height. Burr height is proportional to tool dullness.

Die radius. Die radius affects burr height more than clearance, punch speed, die taper, die throat length, punch slide taper, or punch bottom taper. Burr height at the beginning of a run is greatly influenced by the construction of the die button and is minimized when a straight land is used. When a tapered relief is used instead of a straight land, the burr height at the beginning of a run is inversely proportional to the clearance. The burr height increases in direct proportion to the number of punch strokes made. In conventional dies, burr height will generally increase at a rate of 0.002" (0.05 mm) per 100,000 press strokes, or faster, depending upon die construction and workpiece material.

Some data indicate that a large die radius reduces burr size in conventional blanking. Burr thickness and height also increase as punch-to-die clearance increases, but the relationship is not linear. For steels, a clearance of 2.5% of the stock thickness generally produces the smallest burr, but such tight fits greatly increase tool wear. Initial burr size is reduced as the finish on the punch face is improved.

Opposed-punch processes. When burr-free edges are essential, the opposed-punch processes known as reciprocating blanking, push-back blanking, opposed-die blanking, control cutting, or roll-slit methods should be considered. The basic principle behind these processes requires that die roll be generated on both the top and bottom of the workpiece surfaces. The separation of the slug from the blank occurs in the interior of the cross section rather than at the edges. More detailed information on blanking and shearing is presented in Volume II, *Forming,* of this Handbook series.

Die casting and plastics molding. Die and mold designs that result in rolls are depicted in Fig. 15-39. The labels of "poor design" and "good design" are only appropriate for aesthetic conditions. In many functional applications, the design shown in view *a* is the most applicable. Design considerations such as rolls and unobstructed surfaces are major factors in much of the aluminum and zinc die casting and plastics molding industries since deflashing costs represent 20-35% of the total manufacturing cost. The design of gating systems, ejector pins, and parting lines must consider deflashing approaches. Overflow wells, for example, should be designed to allow the use of strong trimming punches (see Fig. 15-40). Gates should also be designed to facilitate the removal of flash. In addition, gates should always have a shape that ensures fracture of the gate occurs at the edge of the part (see Fig. 15-41), and they should be as thin as possible to produce a clean fracture (see Fig. 15-42). Reportedly, with a shallow gate, the fracture is almost straight and follows the vertical face of the component. A wider gate breaks on an insweeping curve, which finishes at a point a few thousandths of an inch inside the correct line. Feather edges at the ends of threads should be avoided because they make mold fit more critical and promote flashing.

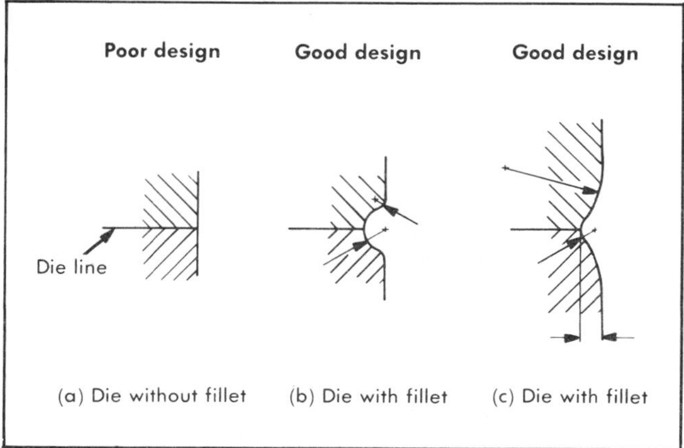

Fig. 15-39 Die and mold designs for aesthetic conditions and easy flash removal.[16]

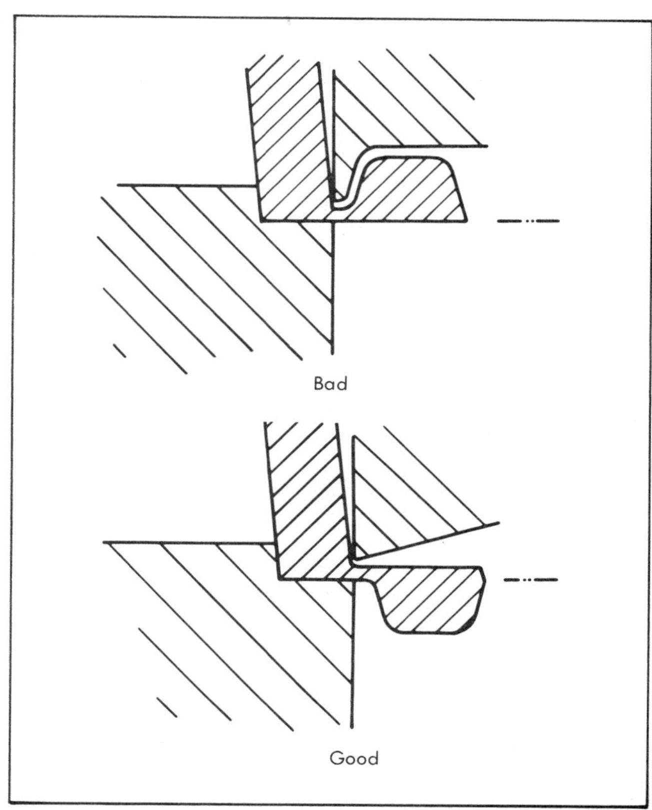

Fig. 15-40 Design of overflow wells for easy trimming and long tool life.

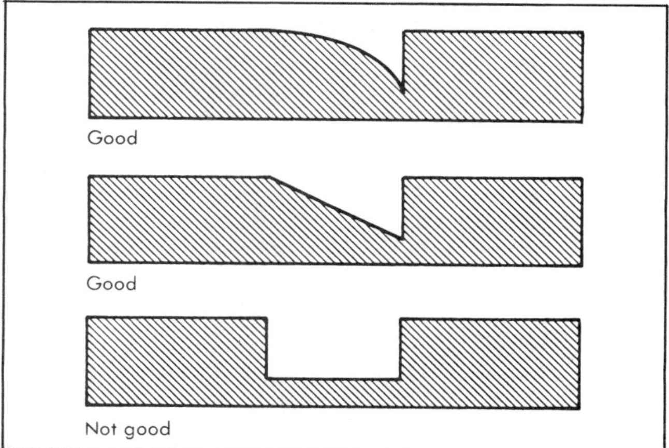

Fig. 15-41 Gate designs.[17]

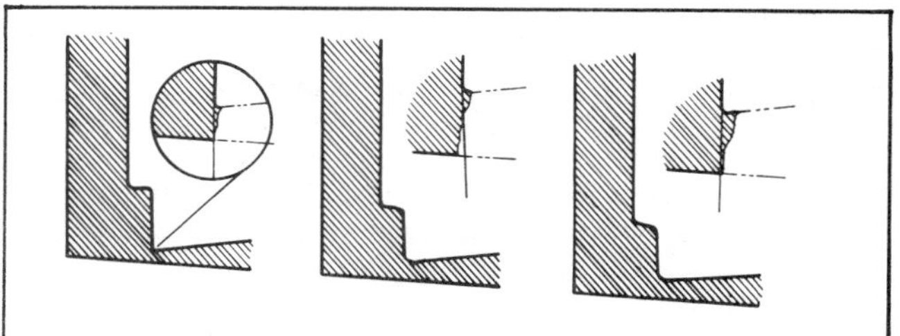

Fig. 15-42 Effect of gate thickness on edge quality.[18]

SURFACE AND EDGE FINISHING METHODS

There are many different processes employed for surface, edge, and corner preparation, conditioning, cleaning, and finishing. In recent years, there has been dynamic growth in the development and improvement of these processes, as well as the equipment, tooling, media, and compounds used.

Lack of knowledge about many of these processes and the improvements that have been made has limited their application. There is currently no formal undergraduate program available in the United States to train mechanical finishing engineers, nor is there any requirement for students of mechanical or manufacturing engineering to take any courses in these processes. Much has been accomplished in educating industry about these advances, however, by conferences, technical papers, and books prepared by the Burr, Edge and Surface Conditioning Technology (BEST) Division of SME.

PROCESSES USED

Of the many different processes used for finishing, some simply clean contaminants from surfaces. Others remove or form the surface material to produce the desired results. The finishing processes can be broadly classified into mechanical, abrasive, thermal, chemical, and electrochemical methods, but some combine several methods.

Mechanical and abrasive finishing methods, including hand deburring; brushing, buffing, and polishing; mass finishing; blasting; abrasive flow finishing; and grinding and honing are discussed in Chapter 16 of this volume. Thermal, chemical, and electrochemical finishing are the subjects of Chapter 17. Cleaning processes are reviewed in Chapter 18.

Advantages and Limitations

Advantages, typical applications, and possible limitations of some of the more common finishing processes are presented in Table 15-6. While this data specifically applies to miniature precision parts made from aluminum, beryllium copper, and stainless steel, the comments are generally pertinent for other parts and materials. Other details with respect to specific processes—such as cycle times, workpiece materials, burr sizes removed, typical radii and surface finishes produced, and resultant changes in workpiece dimensions—are discussed in Chapters 16 and 17 of this volume.

TABLE 15-6
Advantages and Limitations of Various Finishing Processes

Process	Advantages and Typical Applications	Possible Limitations
Abrasive flow	Removes hard-to-reach burrs. Polishes surfaces.	Blind features not deburred.
Abrasive blasting	Good for hard metals.	Produces matte finish. Dust control required. Burr must be accessible.
Barrel tumbling	Low cost. Suitable for all materials.	Edges must be exposed. Slow and not effective for interior surfaces and edges.
Brushing and buffing	All accessible burrs and edges. Polishes surfaces.	
Centrifugal barrel	Fast process. Suitable for all materials. Residual compressive stresses.	
Electro-chemical	Removes hard-to-reach burrs.	Possible stray etching. Limited to conductive metals.
Electro-polishing	Good for thin burrs. Polishes surfaces.	Possible pitting and streaking.
Hand deburring	For hard-to-reach areas and small volume requirements.	Usually expensive and inconsistent. Burrs must be accessible.
Spindle finishing	Fast process. For uniform shapes.	Fixturing needed.
Thermal energy	For thin burrs. Deburrs blind features.	Covers part with oxide film. Burr area must be free of oil and water.
Vibratory	Versatile, economical process. Many deburring and finishing applications.	Not usually suitable for removing internal burrs in intersecting holes.

Side Effects

Many different side effects that result from various finishing processes can affect subsequent processing, including machining, heat treating, assembly, plating, or painting. For example, inadequate finishing or incomplete burr removal can cause plating buildup, interference fits, excessive wear, or localized electrical or magnetic-field disturbances.

Changes in dimensions and surface finishes. Most finishing processes remove material from all exposed surfaces and edges. As a result, part dimensions change, and radii are formed on edges. Provisions for surplus stock or minimization of burr formation in prior machining and care in finishing are essential for critical components. Most finishing processes also change surface finishes, often improving them.

Residual stresses. Most mass finishing processes, such as barrel tumbling and vibratory finishing, reduce surface tensile stresses in workpieces. High-energy mass finishing, such as centrifugal barrel finishing, can induce compressive surface stresses. Compressive stresses are desirable for parts subject to fatigue loading because they tend to increase service life. However, compressive stresses can be a disadvantage for some applications, and can cause distortion, especially of thin, nonsymmetrical parts.

Color or luster. Some finishing processes change the color of surfaces, depending primarily upon the surface finishes, chemical changes, or contamination. For example, brushing operations and the compound used for tumbling control color; media types and compound control the luster produced.

Contaminated surfaces. A film is sometimes produced on the surfaces of parts subjected to incorrect mass finishing. This film is the result of using improper conditioners, agents, and abrasive particles. If allowed to dry, the film forms a barrier that inhibits proper plating and passivation. Films can be prevented by proper compound selection and solution use.

PROCESS SELECTION

Different finishing processes vary with respect to their capabilities, possible applications, side effects, and costs. Capability factors include production rates, repeatability, and finishes produced. Total costs include those for equipment, consumables, tooling and fixtures, labor, and overhead. For a specific requirement, there are generally only two or three processes suitable. In some cases, especially for the removal of large burrs, more than one finishing process is required.

Selection Factors

For the optimum selection of a finishing process, knowledge of the following factors is essential:

1. Burr locations.
2. Burr properties and size: workpiece material, hardness of burr in relation to that of the workpiece, thickness of burr at its base, radius of burr at its base, and burr height.
3. The amount of stock loss (dimensional changes in workpiece) that can be tolerated as the result of the finishing operation. In many cases, burr removal must be accomplished while maintaining specified workpiece dimensions.
4. Required and allowable results from finishing: how large a radius or chamfer, the surface finish produced, and whether side effects of the finishing process are tolerable.

Other factors that should be considered for each specific application include workpiece size and geometry, the sizes and locations of any holes in the workpieces, production requirements, and the capabilities and economics of the finishing processes. Conducting various deburring tests on sample workpieces may be the best method of determining the most economical and efficient process to be used.

Deep, intersecting features within the workpieces often make burrs inaccessible and limit the choice as to the finishing process that can be used. Low volume requirements and short lead times generally make single-purpose, special finishing machines impractical. Small workpieces having tiny holes, undercuts, slots, and similar features make workhandling and finishing difficult.

Computer-Aided Selection

Several manufacturing and consulting firms are developing computer-aided selection systems for determining the optimum finishing process for a specific application. With one system, numerically coded input data entered into the computer include workpiece materials, machining methods before deburring, the shapes of the burrs and their classifications, the weight and volume of the workpieces, and deburring requirements. Output data include the types of deburring methods that could be used, in order of their cost, and the deburring conditions resulting from the most economical method.

References

1. M. M. Barash and W. J. Schoech, *Advances in Machine Tool Design and Research* (Elmsford, NY: Pergamon Press, 1970), pp. 603-613.
2. National Screw Machine Products Association, *Designer's Guide* (Brecksville, OH: 1983).
3. F. Schafer, *Deburring Processes in Perspective*, SME Technical Paper MR 75-482 (Dearborn, MI: Society of Manufacturing Engineers, 1975).
4. *Ibid.*
5. *Ibid.*
6. F. Schafer, *Product Design Influences on Deburring*, SME Technical Paper MR75-483 (Dearborn, MI: Society of Manufacturing Engineers, 1975).
7. *Ibid.*
8. *Ibid.*
9. *Ibid.*
10. National Screw Machine Products Association, *loc. cit.*
11. Federico Strasser, "How Control of Burrs Aids Sheet Metal Stamping," *Iron Age* (January 21, 1960), pp. 90-92.
12. "Design of Screw Machine Products," *American Machinist* (December 7, 1964), p. 127.
13. Yoshio Hasegawa, Shigeo Zaima, and Akiyasu Yuki, *Burr Drilling Aluminum and Prevention of It*, SME Technical Paper MR75-480 (Dearborn, MI: Society of Manufacturing Engineers, 1975).
14. *Ibid.*
15. "Chip Control Saves Tool Lives," *Tooling and Production* (December 1973), pp. 44-45.
16. F. Schafer, *loc. cit.*
17. Robert Kramer, "Good Die Design Makes Barrel Trimming Easy," *Precision Metal* (September 1971).
18. H. K. Barton and L. C. Barton, "The Relationship of Runner Layout to Ease of Trimming," *Machinery*, vol. 90 (April 26, 1957), pp. 945-955

Bibliography

Blotter, P. Thomas; Huang, S. F.; Spear, Carl D.; and Canfield, Ronald V. *Deburring Costs Related to Cutting Speed*. SME Technical Paper MR80-661. Dearborn, MI: Society of Manufacturing Engineers, 1980.

Fickers, John J. *The Machinist's Look at Burrs*. SME Technical Paper MR82-266. Dearborn, MI: Society of Manufacturing Engineers, 1982.

Gillespie, LaRoux K. *Deburring Technology for Improved Manufacturing*. Dearborn, MI: Society of Manufacturing Engineers, 1981.

———. *Advances in Deburring*. Dearborn, MI: Society of Manufacturing Engineers, 1978.

———. *Deburring Capabilities and Limitations*. Dearborn, MI: Society of Manufacturing Engineers, 1976.

Hignett, J. Bernard. *Mechanical Finishing—The Future of This New Technology*. SME Technical Paper MR81-383. Dearborn, MI: Society of Manufacturing Engineers, 1981.

Miller, Emery P. *Substrate Preparation*. AFP Monograph FCR83-09. Dearborn, MI: Society of Manufacturing Engineers, 1983.

Murphy, James A. *Surface Preparation and Finishes for Metals*. New York: McGraw-Hill Book Co., 1971.

Rhoades, Lawrence J. *Cost Guide for Automatic Finishing Processes*. Dearborn, MI: Society of Manufacturing Engineers, 1981.

MECHANICAL AND ABRASIVE DEBURRING AND FINISHING

CHAPTER CONTENTS:

HAND (MANUAL) DEBURRING

Hand deburring, also called hand or bench work, benching, or manual deburring, is any operation in which a handheld deburring tool is used or in which a handheld part is placed against a fixtured tool. The process employs various cutters and motorized tools, such as those listed in Table 16-1. While this list is not complete, it is representative. An estimated 10,000 hand deburring tools are commercially available, and one tool manufacturing company lists over 1000 rotary burs in its catalog. Because of the tools used, hand deburring overlaps other processes such as brush deburring and finishing with coated abrasives, discussed later in this chapter.

WHY HAND DEBURRING IS USED

Hand deburring is still used extensively, even though it is slow, labor intensive, and costly, and often provides less consistent results than desired. Advantages of hand deburring include the versatility of the process and minimal capital investment. A number of reasons for the continued use of hand deburring are presented in Table 16-2. For specific applications, each of these reasons can represent the least expensive approach to deburring.

Hard-to-reach burrs are not in themselves a valid reason for using hand deburring because some of the processes discussed later in this and the next chapter can reach remote areas. A combination of burr accessibility and size, edge and surface requirements, available equipment, and cost usually dictates the selection of a specific process. The quantity of workpieces to be produced in itself may not justify machine setup for deburring.

EVALUATING HAND DEBURRING

Before using hand deburring, the following questions should be answered:

1. Is deburring necessary?
2. What are the actual requirements regarding workpiece specifications and production quantities?
3. Why should hand deburring be used?
4. What would be accomplished if hand deburring is specified?
5. How much would hand deburring cost?
6. What are the technical capabilities and economics of mechanized processes?
7. Is it possible to use an economically viable combination of hand deburring and mechanized finishing?

Accomplishments

The question as to what is actually accomplished with hand deburring is particularly significant. The answer requires a knowledge of the tools employed, the sequence in which the tools are used, the edges actually finished, worker dexterity and capability, and the criteria for determining the adequacy of finishing. A study of these factors often reveals many undesirable practices.

Limitations

Precision edge breaks are difficult to maintain consistently with hand deburring, especially when the workpieces have a number of edges that require breaking. It is almost impossible to consistently produce breaks of 0.002 to 0.003" (0.05 to 0.08 mm) on all portions of every edge of intricate workpieces. Larger breaks, typically 0.005-0.010" (0.13-0.25 mm), generally have to be specified to ensure that most parts will be acceptable. In the aerospace industry, edge radii to 0.062" (1.57 mm) are sometimes specified. Hand deburring, however, is very time consuming when edge breaks or radii greater than 0.015" (0.38 mm) are required.

Contributors of sections of this chapter are: ***Walter G. Bainbridge***, *Vice President, Roto-Finish Co., Inc.;* ***Davis L. Baughman***, *Senior Application Engineer, Pangborn Co., A Sohio Co.;* ***E. W. Bendziunas***, *Lea Manufacturing Co.;* ***Michael C. Burr***, *Manager of Engineering, Cogsdill Tool Products, Inc.;* ***James J. Daly***, *Vice President, Metal Improvement Co.;* ***Warren B. Depperman***, *Vice President, Engineering, Cogsdill Tool Products, Inc.;* ***Thomas R. Dombrowski***, *Manager, Market Development, S. S. White Industrial Products, Pennwalt Corp.;* ***Mike Ferrara***, *Sales Manager, Ultramatic Equipment Co.;* ***George A. Gazan***, *President, French Enterprises, Inc.;* ***L. K. Gillespie***, *Senior Project Engineer, Kansas City Div., Bendix Corp.*;* ***Richard Gillott***, *Sales Engineer, Spadone Machine Co., Inc.;* ***Tom Hankins***, *Product Engineering Manager, Sunnen Products Co.;* ***R. Z. Herr***, *Industry Manager, Airblast & Shot Peening, Pangborn Co., A Sohio Co.;* ***Frank J. Hettes***, *Vice President of Engineering, Weiler Brush Co., Inc.;* ***J. Bernard Hignett***, *Vice President, Harper Co.;* ***John H. Indge***, *Sales Manager, Peter Wolters of America, Inc.;* ***William J. Miller***, *President, Jackson Buff Co., A Div. of AMCA International Corp.;* ***Roy B. Pleiman***, *Vice President, United Technologies Elliott;* ***Lawrence J. Rhoades***, *President, Extrude Hone Corp.;* ***Richard R. Robinson***, *Technical Services Coordinator, Ex-Cell-O Corp., Micromatic Operations;* ***Robert C. Sasena***, *General Sales Manager, Anderson Operations, Dresser Industries, Inc.;* ***Alfred F. Scheider***, *Vice President, Research and Development, Osborn Manufacturing Corp., A Div. of AMCA International Corp.;* ***Dr. William J. Westerman***, *President, Cogsdill Tool Products, Inc.*

** Operated for the U.S. Department of Energy by The Bendix Corporation, Kansas City Division, under Contract No. DE-AC04-76DP00613.*

CHAPTER 16

HAND DEBURRING

TABLE 16-1
Typical Tools Used for Hand Deburring

Abrasive-filled: Cork products (bullets, ball-nose cylinders) Cotton products (cylinders, balls, bars) Nylon/synthetic materials Rubber products (bullets, cylinders, flat bars, discs, cups, dental bullets)
Abrasive paper products (discs, rolls, sheets, cord)
Brushes (wheel, end and cup, tube, crosshole deburring, side action)
Burs, bur balls, rotary files
Countersinks
Drills and reamers
Felt bobs (discs, bullets, cylinders)
Files (large, miniature, round, half round, triangular, curved or bent)
Hand stones (bars, triangles, cones, points)
Knives (triangular, oval, special shapes, scalpel blades)
Lapping compounds
Mandrels for tools
Miscellaneous tools (swivel blade tools, back side cutters, special design tools, vacuum probes)
Motorized tools (bench motors, air motors, dental tools, belt sanders, reciprocating files, jitterbug sanders)
Mounted points (balls, discs, cylinders, cones, special shapes)
Picks
Pin vises (dog nose, collet)
Scrapers

HAND DEBURRING COSTS

An analysis of hand deburring costs is relatively straightforward. A typical form used to analyze the costs is presented in Table 16-3. Cost, however, should not be the only criterion in determining if special efforts are required for improving manual efforts.

TABLE 16-2
Reasons for Using Hand Deburring

Increased flexibility of the process.
Burrs are hard to reach by any other process.
Only a small number of parts are needed and/or a short lead time is required, making hand deburring the least expensive method.
Parts are delicate and/or require a precision finish.
Parts require close tolerances.
Required edge radii are large, 0.060″ (1.52 mm) or more; or, allowable edge radii are small, 0.002″ (0.05 mm) or less.
Highly variable burr size prohibits use of other processes.
Deburring or finishing machine will not accept part size required, or machine is out of commission.
Hand deburring can be done while other machining operations are being performed.
To remove or reduce the size of burrs or extruded metal left by mechanized finishing process.
To prevent media impregnation, oxide formation, or other undesirable side effects from other finishing processes.
Inability to change the process because it has been accepted in conformance with military specification.

Percentage of Total Costs

One indicator that hand deburring costs may be out of line with normal costs is evident when the deburring man-hours or costs are more than 5% of total machining time or costs. Another cost problem is indicated when most of the production parts are being delayed at the burr benches. A third warning develops when a manufacturing firm is spending considerably less than 5-7% of their capital equipment dollars on finishing equipment, which is the industry average.

Outside Deburring

An inexpensive and effective approach to determine if hand deburring costs are excessive is to have a deburring job shop finish a representative sample of workpieces. If the quality of the finished parts is acceptable and the cost is less than the in-house cost, hand deburring methods being used in-house require careful scrutiny.

Reviewers of sections of this chapter are: ***Walter C. Bainbridge**, Vice President, Roto-Finish Co., Inc.;* ***Davis L. Baughman**, Senior Application Engineer, Pangborn Co., a Sohio Co.;* ***E. W. Bendziunas**, Lea Manufacturing Co.;* ***K. R. Bennett**, Managing Director, Lapmaster International, Ltd.;* ***Ezra A. Blount**, Secretary, Metal Finishing Suppliers' Assn., Inc.;* ***Bramwell W. Bone**, Sales Manager, Hegenscheidt Corp.;* ***E. L. Bradley**, Sales Engineer, R. Howard Strasbaugh, Inc.;* ***C. P. Butterfield**, Vice President, Sales, Jackson Buff Corp.;* ***Glen A. Carlson, Jr.**, President, Acme Manufacturing Co.;* ***Ronald I. Cosler**, Vice President and General Manager, Barnes Drill Co.;* ***James J. Daly**, Vice President, Metal Improvement Co.;* ***Thomas R. Dombrowski**, Manager, Market Development, S. S. White Industrial Products, Pennwalt Corp.;* ***George A. Gazan**, President, French Enterprises, Inc.;* ***L. K. Gillespie**, Senior Project Engineer, Kansas City Div., Bendix Corp.;* ***Richard Gillott**, Sales Engineer, Spadone Machine Co., Inc.;* ***Lars Gustafsson**, Product Manager, ASEA Robotics;* ***Tom Hankins**, Product Engineering Manager, Sunnen Products Co.;* ***Matthias H. Hermanns**, President, Thielenhaus Microfinish Corp.;* ***R. Z. Herr**, Industry Manager, Airblast & Peening Systems, Pangborn Co.;* ***Frank J. Hettes**, Vice President, Engineering, Weiler Brush Co., Inc.;* ***J. Bernard Hignett**, Vice President, Harper Co.;* ***Douglas O. Hoag**, Senior Product Engineer, Metal Products Section, Norton Co.;* ***John H. Indge**, Sales Manager, Peter Wolters of America, Inc.;* ***Michael F. Jarvis**, Director of Engineering, Empire Abrasive Equipment Corp.;* ***Richard L. Jette**, Western Regional Sales Manager, Automated Finishing;* ***R. W. Johannesen**, Consultant;* ***Ted M. Kahn**, President, Titan Tool Supply Co.;* ***Robert W. Kenagy**, Director, Marketing Communications, SWECO, Inc.;*

TABLE 16-3
Elements of Deburring Costs

Source of Cost	Estimated Hours per Year for Deburring	Salary, dollars	Overhead, dollars/hr	Yearly Cost, dollars
Hand deburr personnel	______	______	______	______
Deburring machine operators	______	______	______	______
Deburr supervisor	______	______	______	______
Engineering or management	______	______	______	______
Inspection	______	______	______	______
Deburr equipment depreciation				______
Maintenance of equipment				______
Sharpening of deburr tools	______	______	______	______
Deburr supplies				______
Energy costs				______
Deburr on machine cycle	______	______	______	______
Scrap due to inadequate deburr				______
Warranty work caused by burrs				______
Floor space costs				______
Total				______

Rejected Parts

While costs are a major concern for all manufacturing firms, the percent of rejects for burr-related reasons is also a significant consideration. Continual rejections are a sign of an out-of-control situation, implying that parts with burrs are passing inspection or that deburring operations are not satisfactory.

Effect of Burr Size

Both burr size (see Fig. 16-1) and workpiece size have a considerable effect on deburring time and costs. Washer-like parts with 0.005″ (0.13 mm) thick burrs require four times longer for hand deburring than similar parts with 0.0015″ (0.038 mm) thick burrs. Figure 16-2 illustrates typical edge

Reviewers, cont.: ***John B. Kittredge****, Consultant;* ***F. Peter Lynah, Jr.****, Director of Engineering, P. R. Hoffman Machine Products;* ***R. Bruce MacLeod****, Manager of Engineering, Taft-Peirce Manufacturing Co.;* ***R. Steven Marcus****, President, Markee Corp.;* ***Howard J. McAleer****, General Manager, Formax Manufacturing Co.;* ***Danny McCoy****, President, B & D Sencoy, Inc.;* ***William M. Meushaw****, Plant Manager, PPG Industries, Inc.;* ***Daniel E. Miller****, Metalworking Div., Sales Manager, Spitfire Tool and Machine Co.;* ***Douglas E. Miller****, Systems Marketing and Planning Manager, 3M Co.;* ***Russell Osborne****, Manufacturing Manager, Gorham Div., Textron, Inc.;* ***William M. Parker****, Technical Service and Methods Development Manager, 3M Co.;* ***Thomas H. Patton****, Gyromatic Mfg. Co.;* ***Winfield B. Perry****, President, Dynetics Corp.;* ***Henry B. Rand****, Senior Product Engineer, Materials Div., Norton Co.;* ***Lawrence J. Rhoades****, President, Extrude Hone Corp.;* ***Roger Rice****, President, Michigan Hone and Drill, Inc.;* ***Richard R. Robinson****, Technical Services Coordinator, Ex-Cell-O Corp., Micromatic Operations;* ***L. Brown Sanders****, Marketing Manager, Materials Cleaning Systems Div., Wheelabrator-Frye, Inc.;* ***Alfred F. Scheider****, Vice President, Research and Development, Osborn Manufacturing Corp., A Div. of AMCA International Corp.;* ***Jay N. Stricker****, National Sales Manager, Finishing Equipment Group, SWECO, Inc.;* ***John Taylor****, Wickman Corp.;* ***Edward H. Tulinski****, Regional Manager, Harper Co.;* ***John M. Tuma****, Chas. F. L'Hommedieu & Sons Co.;* ***George F. Weaver****, Vice President, Divine Brothers Co.;* ***Dr. William J. Westerman****, President, Cogsdill Tool Products, Inc.;* ***Bruce A. Yount****, Marketing Manager, Anderson Products, Div. of Wilton Corp.*

CHAPTER 16

IMPROVING HAND DEBURRING

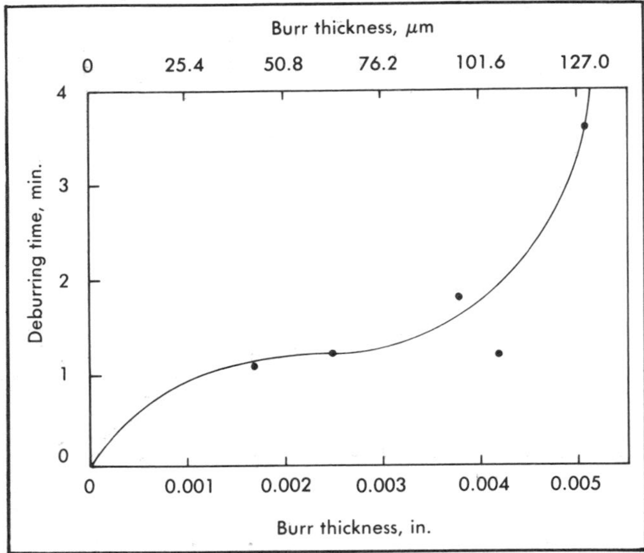

Fig. 16-1 Effect of burr thickness on hand deburring time. (*Bendix Corp.*)

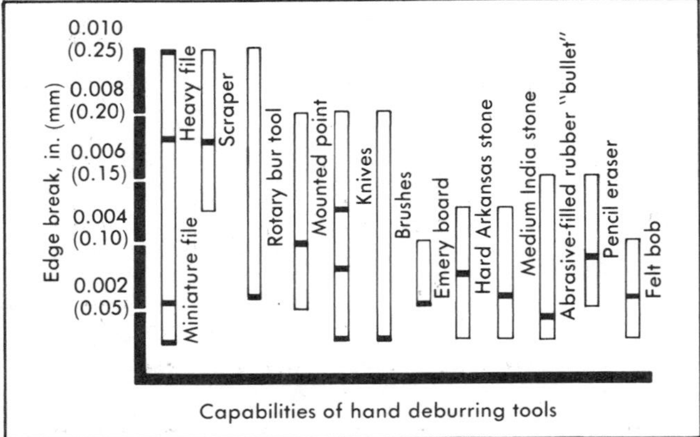

Fig. 16-2 Basic edge break capabilities using specific tools to remove 0.003″ (0.08 mm) thick by 0.003″ high burrs from 303Se stainless steel. Horizontal bars indicate measured results by one operator. (*Bendix Corp.*)

breaks produced by a variety of deburring tools, discussed subsequently in this section.

IMPROVING HAND DEBURRING

While the capabilities and limitations of most mechanized deburring processes have been defined reasonably well, the capabilities of hand deburring methods have not, which is a major problem in trying to improve hand deburring. For example, data should be available with respect to coated-abrasive products as to which grain size, structure, and bond can be used without damaging a specific surface finish or reducing a diameter more than an allowable amount in a certain time.

Quality Requirements

Clearly defining the quality requirements for various edges and surfaces is an important part of reducing costs for hand deburring. One system for accomplishing this objective is to use specific instructions and codes on the part drawings (see Fig. 15-9, Chapter 15). With this method, every worker knows the exact requirements, the tools employed, and the sequence in which the tools should be used. Another quality method consists of painting a sample part white and then color coding the edges requiring finishing (see Fig. 16-3). When more than one tool is required to finish an edge, the color closest to the edge indicates the first tool; the next color, the second tool; etc.

Training Programs

Formal training programs are effective in providing the skills and expertise necessary for precision hand deburring. Hand deburring by an unskilled operator often replaces a burr with a chamfer and two new burrs.

Mechanized Deburring

Since mechanized deburring is generally 80-90% efficient, it should be used whenever possible. For heavy burrs, it frequently costs less to remove them by machining than to use hand deburring.

FINDING BURRS

Many burrs, especially minute ones, are difficult to see. Cleanliness of the workpieces is essential in locating burrs. On small features, it is practically impossible to tell if a particle at an edge is dirt or a burr unless the part is clean. The use of special lighting may be necessary, augmented by microscopes and borescopes.

Looking in the Right Direction

Many burrs are almost impossible to see when looking in the wrong direction. For example, burrs on the top edges of holes may not be visible if viewed only from the bottoms of the holes; however, by looking through the holes from the top to the bottom surfaces of the workpieces, many of these burrs can be located. To verify that all burrs have been removed, first look at the top surface and then through the hole; next, turn the part over and look at that surface, and then through the hole.

On most parts, it is important to tip the part to an angle with relation to the line of eyesight in order to see burrs correctly. Observing edges at an angle of 30 to 60° is generally preferred. A flexible or miniature-lens borescope connected to a video system is helpful. It is also important to have proper lighting at the correct angle. Figure 16-4 illustrates effective angles.

Optimum Lighting

Various lighting techniques are used for microscopic deburring and inspecting. The more common methods include lights or illuminators on or off the microscopes, ring lights (fluorescent tubes around the lens), fiber-optic ring lights, coaxial illuminators (through the microscope), back lighting, polarizing accessories, color filters, and miniature illumination systems with flexible probe cords that enter openings as small as 0.110″ (2.79 mm) diam.

Probing for Burrs

Fingernails and toothpicks are only two of a wide variety of items used to determine tactilely if burrs exist. Dental picks are excellent tools for this purpose because of their small diameter and bent ends; electronic probes are also used. Use of a miniature borescope with video hookup is a good method of searching for hidden burrs. In addition, size (gage) wires are being used, particularly for probing through the teeth of small

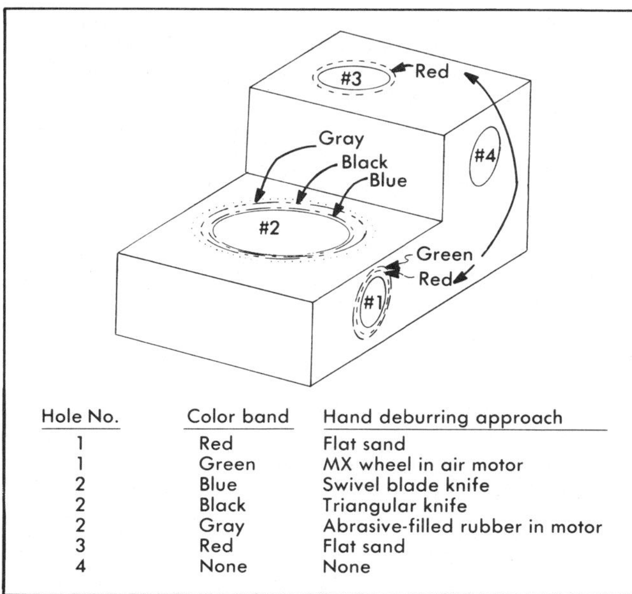

Hole No.	Color band	Hand deburring approach
1	Red	Flat sand
1	Green	MX wheel in air motor
2	Blue	Swivel blade knife
2	Black	Triangular knife
2	Gray	Abrasive-filled rubber in motor
3	Red	Flat sand
4	None	None

Fig. 16-3 Color coding of sample part indicates finishing requirements for various edges. (*Bendix Corp.*)

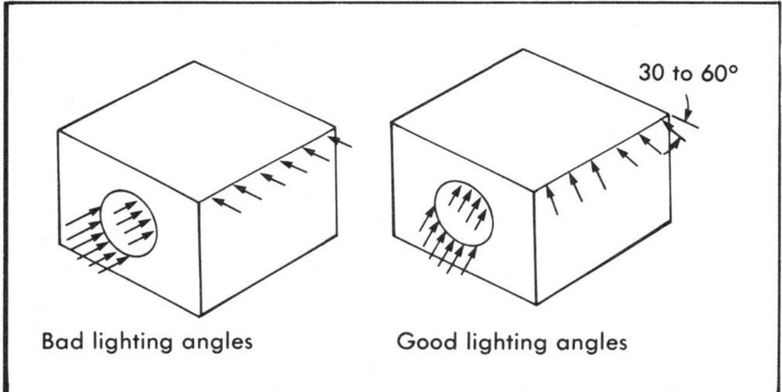

Fig. 16-4 Effect of lighting angles on burr detection. (*Bendix Corp.*)

gears. For some applications, pipe cleaners are dragged over edges; loose pieces of cotton indicate rough edges or burrs.

Tactile probing to detect burrs involves three major problems:

1. It may scratch precision parts.
2. It is time consuming when the entire part must be probed.
3. It is not foolproof. For example, it is not satisfactory for probing in a direction parallel to the burr.

Using Gages

Gages are used for some parts, particularly for those having holes, to determine if burrs are present or if the burrs are of excessive size. If a go-type plug gage fails to enter a hole, it indicates that the hole is undersized or that a burr has been thrown over into the hole. Similarly, if the gage enters the hole but does not leave the opposite end, the hole is tapered or undersized, or a burr has been thrown into the exit end of the hole.

Plug gages do not absolutely indicate that a burr is present, plus burrs can wear or scratch the gages. Another problem in using gages is that, in some cases, the burrs may lock the gage in the workpiece; forcing the gage either through the hole or back out can scratch both the gage and the workpiece.

DEBURRING TOOLS

There are basically 19 types of tools with 59 subcategories of these tools in everyday use for hand deburring. These tools (many are listed in Table 16-1) include knives, files, rotary burs, brushes, abrasive products, and various other items.

Knives

A wide variety of knives are used for hand deburring. While knives are inherently dangerous, they do not generally present a safety problem in normal use. Most knives are of special design for a specific application. They are typically made from American Iron and Steel Institute (AISI) Type M2 high-speed steel and generally have a hardness of R_C61.

Typical applications. Triangular knives are typically used to cut or scrape burrs from straight edges. Oval knives are normally used for hole edges or hard-to-reach areas. Specially shaped knives are designed to remove burrs from specific workpieces, but are often useful for other applications as well. Scalpel blades are used extensively to trim flash from parts made of plastics, but they are rarely successful for removing metal burrs.

Methods of holding. Knives can be used as fist tools or as finger tools. Fist tools, scrapers, large knives, and many other large tools are used with power supplied by the operator's hand and arm. While the fist-holding position is used extensively for deburring large parts, it is not a successful approach for small or precision parts because it does not permit a sense of feel for the burrs. With a knife held in a fist, it is difficult to stop the momentum or control the amount of edge break produced on a workpiece. This method can also cause cuts to the operator and gouges in the workpieces. As a result, knives should not be used in the fist position except when necessary.

For small and precision parts, the finger is used as a tactile indicator of the actual conditions. There are two approaches to the finger position; both approaches are widely used for hand deburring and are safer than the fist position. In one approach, the knife is held with two fingers and the thumb, like holding a pen or pencil, and is called the overhand or pencil position. In this position, the knife can be rotated to accommodate changes in intersecting planes on the workpiece. The middle finger can be used as a surface feeling probe. By using this method under a microscope, small burrs can be removed easily.

The second approach to holding knives with the fingers is the underhand position. In this position, the small and ring fingers can be used as a surface support, and the index finger as a surface feeling probe. The knife can also be rotated for easy burr removal. This method is particularly useful for deburring holes. It permits rotating the knife around contours and provides uniform force against edges. Interchangeable handles and bladeholders, including adjusting and telescoping types for deep recesses and holes, are available for different knife blades.

Triangular knives. As implied by its name, the triangular knife has three cutting edges. These knives can be ground either concave or convex (see Fig. 16-5). Concave or hollow ground knives have very sharp edges, but they dull quickly and are more likely to chip. Convex ground tools have a more rigid body behind each edge and, as a result, generally last longer. In addition, they can be reground much faster with conventional equipment than can concave tools.

Oval knives. The oval knife (see Fig. 16-6) has a cylindrical body and a flat surface ground at an angle, which provides a working edge having an oval shape. These knives are designed primarily for removing burrs from holes, but they are often used to chisel out burrs and to reach difficult areas.

HAND DEBURRING TOOLS

Angle hooked knives. The angle hooked knife (see Fig. 16-7) has two cutting edges at 90° to the shank. This design allows the user to remove the edge burrs along a slotted groove, sometimes removing burrs from both edges of the slot at the same time. Special angle hook knives are designed to clean out and deburr threads or to scrape along a single edge.

Flat pointed knives. These tools, also called chisels, are generally cylindrically shaped with a flat ground on each side, resulting in a flat, sharp edge similar to a screwdriver. The tool is used with a gentle pushing motion to scrape burrs. A typical application is to clean chips from the bottoms of blind tapped holes.

Cone-end tools. These tools have a double cone (back to back) with a flat surface ground to the centerline (see Fig. 16-8). This design enables the tool to remove burrs from hard-to-reach areas, such as blind holes. The tools are often used to clean and deburr threads, and to deburr slots in the walls of tubes. Some operators prefer this type of tool to an angle hook tool for deburring threads or tapped holes, particularly larger sizes.

Knife-workpiece relationship. Normally, workers should not hold workpieces down on a table or microscope base. It is important for both personal safety and workpiece quality to hold workpieces above a solid surface. If necessary, wrists or arms may rest on a tabletop.

When using a table to support workpieces, any sudden lunge with the hand holding the knife will gouge the workpiece or cut the other hand. Abrupt changes in burr sizes and locations can also cause lunges. By holding the workpiece in the air, the workholding hand can quickly and easily respond to sudden motions of the hand holding the knife.

Files

More than 2700 different files in a variety of sizes are available commercially. Common types include Swiss escapement, parallel machine, needle, and riffler. The files, in a variety of miniature sizes, are rectangular, triangular, round, curved, or bent. Some are made from high-speed steels, and others have carbide or diamond particles bonded onto steel shanks.

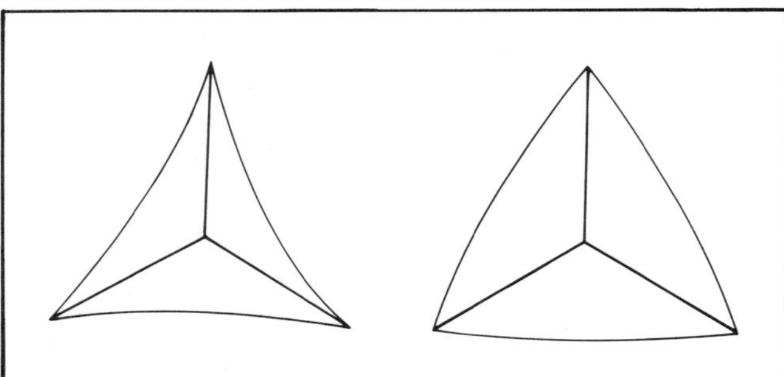

Fig. 16-5 Concave grind (left) and convex grind (right) on triangular knife, with concavity and convexity exaggerated for clarity. (*Bendix Corp.*)

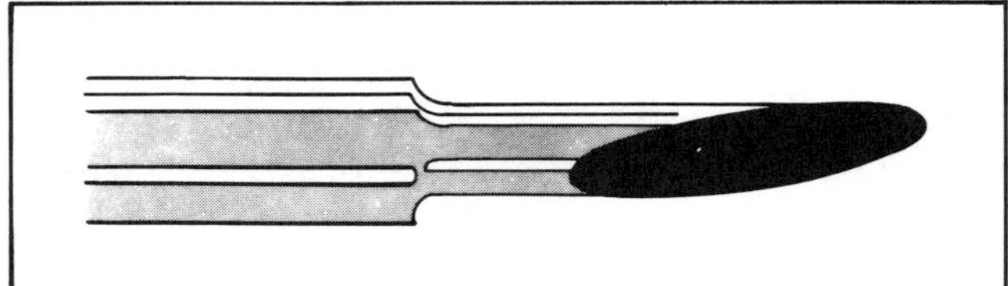

Fig. 16-6 Oval knife. (*Bendix Corp.*)

Most files can be used on any workpiece, although it is easy to exceed 0.003″ (0.08 mm) edge breaks with many of these tools. Needle files have almost sharp points, generally with a radius of about 0.030″ (0.76 mm). The narrowest rectangular file is about 0.015″ (0.38 mm) thick. Some triangular files have knife-like edges thinner than 0.015″. All files can be further thinned if required.

Rotary Burs

Rotary burs, bur balls, and related tools, sometimes called rotary files, are available in a variety of diameters, shapes, tooth coarseness, and materials. Shapes include round, oval, cone, inverted cone, cup, barrel, cylinder, wheel, tree, flame, knife edge, and round edge. Although these tools are sometimes called burrs, it is preferable to refer to them as *burs* or *rotary burs* to eliminate confusion between the tools and the material being removed.

For hand deburring of most miniature parts, a bur tool is placed in pin vises and rotated by hand. The tools, however, are also inserted in bench-mounted electric motors, air motors, and dental motors for faster action. They are also used on various machine tools.

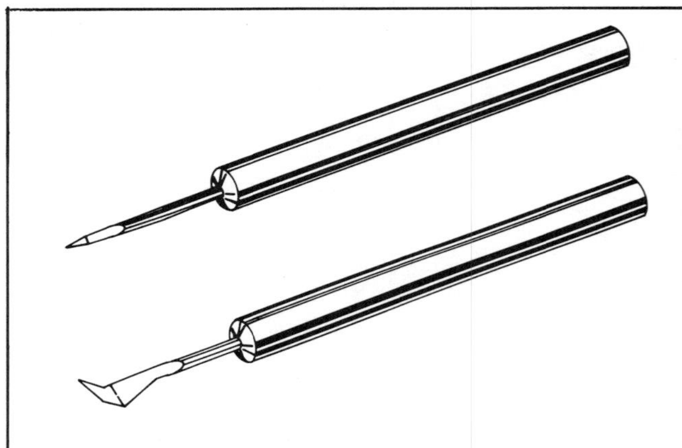

Fig. 16-7 Top and side views of angle hooked knife often used to deburr slots. (*Bendix Corp.*)

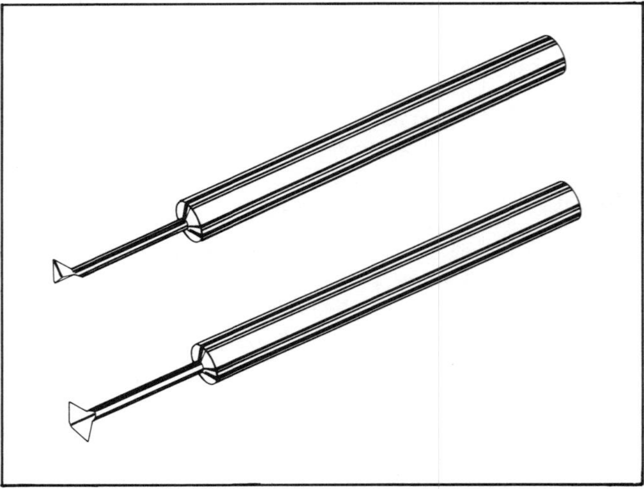

Fig. 16-8 Top and side views of cone-end tool. (*Bendix Corp.*)

The smallest commercially available tool has a 0.004″ (0.10 mm) diam ball, but the shanks are normally much larger, which may present a problem for internal deburring. Rotary burs tend to chatter or cut too deep unless very fine teeth are used. Diamond-plated burs are being used with good results. Several shapes of these tools are used to deburr the back sides of holes that may be inaccessible by other means.

Brushes

The seven basic types of brushes used for hand deburring are radial, cup, end, tube, crosshole, side action, and toothbrush. The sizes of commercially available brushes are shown in Table 16-4.

Radial brushes. These brushes, which resemble wheels with spokes extending from them, are the most commonly used for deburring. They are generally used in a bench motor but are also employed by hand. They should never be used with high-speed air motors because fibers will be thrown from the brushes and small-diameter shanks may bend.

Cup brushes. Inverted cup brushes are useful for deburring the edges of relatively large gears, parts having short shafts extending from them, and difficult-to-reach areas where a soft action is required. They are also handy for deburring parts that can be held in the hand, and for larger parts or features 1/2″ (13 mm) or more in diameter. Figure 16-9 illustrates several applications of these brushes.

End brushes. On these tools, the fibers all extend axially outward from the ends of the brushes. These brushes are relatively stiff and, because of this, are often used in hard-to-reach places. A typical application is shown in Fig. 16-10. End brushes are also used to remove recast metal from EDM operations, as well as to clean the workpieces.

Tube brushes. These tools are often used to deburr the insides of tubes, through holes, or relatively deep blind holes. They are available in a variety of stiffnesses and materials, as well as different diameters. In general, they are designed to be softer and more flexible than side-action brushes, discussed later in this section.

TABLE 16-4
Size Ranges of Commercially Available Standard Brushes for Hand Deburring

Brush Type	Filler Material	Size Range Available, in. (mm)
Radial	Stainless steel	3/4 to 15 (19 to 381)
	Brass	3/4 to 6 (19 to 152)
	Nylon	3/4 to 2 (19 to 51)
	Abrasive-filled nylon	1 3/8 to 14 (35 to 356)
	Nonwoven abrasive-filled nylon	2 to 6 (51 to 152)
	Tampico	2 to 16 (51 to 406)
	Cord filled	1 to 16 (25 to 406)
	Abrasive tipped	1/2 to 2 (13 to 51)
	Elastomer filled	1 1/4 to 12 (32 to 305)
	Flapwheel	1 to 16 (25 to 406)
	Felt bob	1/4 to 3 (6 to 76)
	Abrasive-filled muslin	3/16 to 1 (5 to 25)
Cup	Stainless steel	3/4 to 6 1/4 (19 to 159)
	Brass	3/4 to 3 1/4 (19 to 83)
	Elastomer filled	4 to 6 (102 to 152)
End	Stainless steel	1/4 to 1 (6 to 25)
	Elastomer filled	1/2 to 1 (13 to 25)
Crosshole or tube	Stainless steel	1/32 to 3 (0.8 to 76)
	Brass	1/4 to 2 (6 to 51)
	Nylon	1/4 to 2 (6 to 51)
	Elastomer filled	1/2 to 1 (13 to 25)
Side action or sibot	Stainless steel	1/4 to 1 1/4 (6 to 32)
	Brass	1/4 to 7/8 (6 to 22)
	Nylon	1/4 to 7/8 (6 to 22)
	Abrasive-filled nonwoven nylon	1/2 to 3 (13 to 76)
Flare	Steel	1/2 to 4 (13 to 102)

(*Bendix Corp.*)

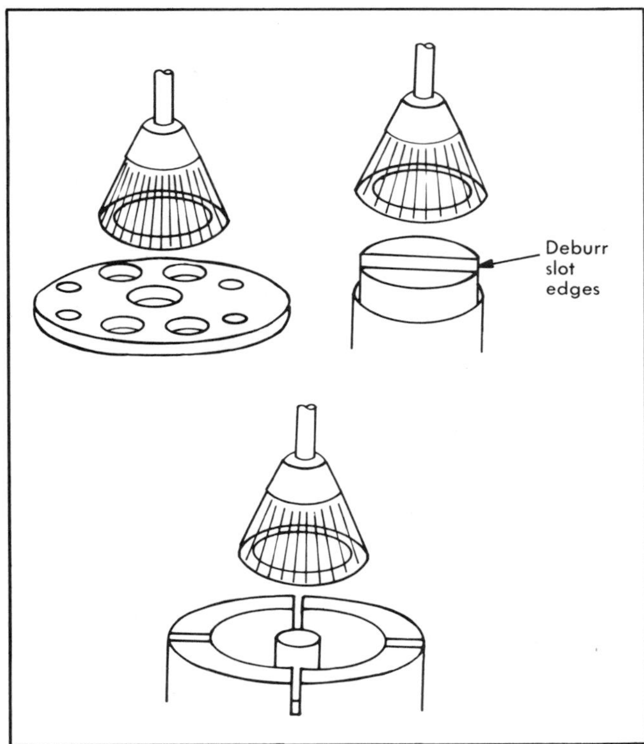

Fig. 16-9 Typical applications for cup brushes. (*Bendix Corp.*)

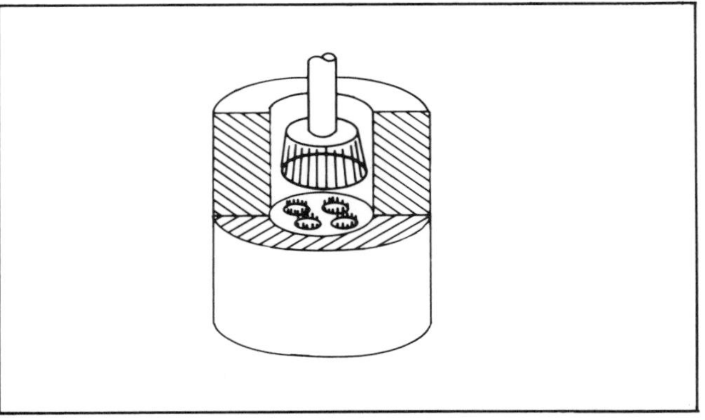

Fig. 16-10 An application of end brushing. (*Bendix Corp.*)

HAND DEBURRING TOOLS

Crosshole brushes. These tools have their stainless steel or nylon fibers extending outward in a helical path around the shanks of the brushes. They are available in sizes as small as 0.024″ (0.61 mm) diam, and can be made smaller. While these tools are visually the same as tube brushes, they are classified as a separate entity because of their minute sizes. Because of their small sizes, these brushes must be fairly soft acting. As a result, they are not capable of removing large burrs. Their most common use is in cleaning and removing fine burrs from small-diameter plain and threaded holes.

If the wrong size of brush is used in a threaded hole, the brush fibers can break loose and become lodged in the hole. In many cases, trying to remove these small fibers is a difficult task. Table 16-5 presents recommended brush sizes for various hole and thread sizes.

Side-action and related brushes. Side-action (butterfly) brushes are similar to sibot (square trim) brushes. An advantage of side-action brushes is that they can be obtained in sizes as small as 1/4″ (6.4 mm) diam. These brushes are very aggressive and are typically used to deburr plain or threaded holes. When precision dimensions are involved, the brushes work satisfactorily on stainless steels, but not on aluminum alloys. While these brushes can be used by hand, they are normally employed with some type of motorized or powered tool.

Brush filaments. A variety of bristle materials are used in the manufacture of brushes. However, only the following eight are used to any extent for deburring: carbon steel, stainless steel, brass, tampico fibers, nylon, abrasive-filled nylon, nonwoven synthetic material, and elastomer-filled wire bristles.

Carbon steel brushes are still used because of their low cost and long life, but they are not applicable for many parts. The steel rusts easily, and the brushes deposit steel particles onto the workpieces, which causes rusting. Stainless steel brushes require careful use because they can form large burrs and may weld to stainless steel workpieces, creating a rough surface.

Brushes having miniature brass bristles are excellent tools for finishing or radiusing edges. They are rigid enough to produce small edge breaks, yet soft enough to prevent forming new burrs. Brass, however, easily deposits on workpieces and can cause discoloration. Discoloration is not allowable on most finished parts, but it can be removed by proper cleaning procedures.

Brushes made with tampico fibers are too soft to produce much radiusing action, but they do provide good finishes. If these brushes are used with a deburring compound containing abrasive particles, they can produce fine edge finishes or aggressive deburring, if desired; such compounds, however, are hard to remove from the workpieces.

Nylon bristle brushes are too soft to remove large burrs, but they are useful for cleaning operations. Brushes having either aluminum oxide or silicon carbide abrasive embedded in the nylon fibers, however, are among the most widely used brushes for the deburring and edge finishing of precision miniature parts. As the nylon wears, new cutting edges are continually exposed. With gentle use, these brushes will not provide more than a 0.005″ (0.13 mm) edge break.

Nonwoven synthetic brushes, filled with abrasive, are used extensively by the aircraft industry. They can provide either gentle or aggressive action and leave an excellent surface finish. For more aggressive action without undue scratching of the surfaces, some firms use plastics-filled metal brushes. Plastics keep the metal fibers from spreading apart and reduce wear, thus making the brush a stiffer tool. These brushes are not applicable to precision or miniature parts.

Additional information about brushes and their operation, factors affecting aggressiveness, side effects of brushing, and safety considerations is presented in a subsequent section of this chapter under the heading of "Power Brushing."

TABLE 16-5
Crosshole Brush Sizes for Deburring Various Sized Holes

Hole Size	Brush Diameter, in. (mm)
Hole Diameter, mm (in.):	
0.5 (0.0197)	0.024 (0.61)
0.8 (0.0315)	0.032 (0.81)
1 (0.0394)	0.032 (0.81)
1.2 (0.0472)	0.047 (1.19)
Nominal Size and Threads per Inch:	
0-80	0.047 (1.19)
2-56	0.079 (2.01)
4-40	0.097 (2.46)
5-40	0.109 (2.77)
6-32	0.125 (3.18)
6-40	0.125 (3.18)
8-32	0.142 (3.61)
10-24	0.156 (3.96)
10-32	0.156 (3.96)

(*Bendix Corp.*)

Abrasive Tools

A variety of abrasive tools are used for deburring. The more common include abrasive-filled rubber, cotton, and cork products, abrasive paper, and hand stones.

Abrasive-filled rubber products. For deburring small holes, bullet-shaped rubber tools are used extensively. They are available in four different levels of aggressiveness, the degree of aggressiveness being determined by the size and number of silicon carbide particles molded into the rubber. To facilitate identification, the degree of aggressiveness is indicated by colors, the color codes varying with different manufacturers of the products.

If these products are used on machine tools, the particles that are worn off may score or scratch the machine ways. If the particles are blown into chuck jaws or collets, they can indent subsequent parts processed.

Short-shank dental tools are less aggressive than industrial tools of this type, but they permit use on smaller features of workpieces. The short lengths of dental tools allow them to be used crosswise, with dental motors, in holes as small as 0.750″ (19 mm) diam.

Abrasive-filled cotton products. When more aggressive cutting is required than provided by abrasive-filled rubber, abrasive-filled cotton or muslin tools are often used. By varying abrasive size and the type of bonding agent, it is possible to produce smooth radii and fast burr removal with these tools. While these tools are relatively unknown in the general metalworking industry, they are used extensively by aircraft firms.

Abrasive-filled cork products. Cork and abrasive particles, bonded together with special glues or resins, are also used for fast cutting with smooth radiusing.

Abrasive paper products. A variety of sandpaper-like products are used for deburring and finishing. Miniature discs, with adhesive backs for holding them to mandrels, are used for small workpieces. The smallest commercially available disc is 0.5″ (12.7 mm) diam, but smaller sizes can be fabricated. Abrasive belts for deburring and finishing are discussed later in this chapter.

Hand stones. A wide variety of handheld abrasive stones are used to remove burrs and improve finishes. They are available in sizes as small as 0.010″ (0.2 mm) thick or less.

Mounted points. Miniature grinding wheels, known as mounted points, are used for edge finishing by many firms. They are available in a wide variety of sizes, shapes, and materials. Mounted points, however, are too aggressive for many applications and often tend to chatter in operation.

The mounted points used in the dental industry, including diamond-coated types, quickly break edges in hard-to-reach areas, without chatter, and are employed extensively for miniature parts. Mounted points made of cotton or muslin fibers, rather than bonded abrasives or ceramics, are softer acting tools, used primarily by the aircraft industry.

Other Tools

There are many other tools used for deburring and finishing. These include drills and reamers, lapping compounds, felt bobs, pin vises, motorized tools, and special-design tools.

Drills and reamers. These tools are frequently used to deburr small intersecting holes. Reamers are generally used first to remove the burrs and, if possible, leave a sharp edge. When burrs are too thick for reamers to remove by hand, drills are used.

Lapping compounds. While lapping compounds are not typically used for removing burrs, they can remove small burrs and produce smooth, lustrous surfaces. They are most easily used on flat surfaces and in holes.

Felt bobs. Occasionally, specially formed felt shapes, called bobs, are used with a diamond lapping compound to remove burrs. Usually, however, the abrasive-filled rubber products will deburr faster without damaging parts.

Pin vises. A variety of pin vises are used to hold workpieces and tools for deburring and finishing. Some of these have brass collets or jaws to prevent indenting the workpieces while they are being grasped. Some vises locate on the OD of workpieces, while others grasp the ID. When necessary, the vises can be machined to hold small-diameter or specially shaped parts.

Motorized tools. More than 500 motorized tools are available for deburring and finishing, with one company offering 15 different dental-tool motors. All can be obtained with belt drive, integral electric, or air-powered motors, as well as flexible-shaft drives, and operate at speeds from 2000 to 400,000 rpm. For most applications, motors that operate at 20,000 rpm or less are designed to remove burrs by applying a measurable force to the workpieces; faster motors require only a light touch.

Special tools. Whenever conventional tools are not satisfactory for removing burrs, special-design tools should be considered. Some tools, such as several swivel-bladed types, are designed to remove burrs from the bottom edges of holes that are not normally accessible.

BRUSHING, POLISHING AND BUFFING

Brushing, polishing, and buffing are related but different processes. Brushing is used for both deburring and/or finishing, including hand deburring, discussed in the preceding section of this chapter. Polishing is an abrading operation using abrasives on resilient wheels, belts, or buffs, with the amount of stock removed dependent primarily upon the grit size and abrasive used, as well as on the pressure exerted. It is more aggressive on surfaces than brushing and leaves a defined pattern on workpiece surfaces. Chemical polishing and electropolishing, discussed in Chapter 17, do not leave lines on the surfaces. Buffing is a final surface finishing process that produces smooth, lustrous surface finishes with less-defined line effects and slower stock removal than polishing. All three processes can be used for descaling, deflashing, deburring, radius generating, surface improving, and stress-relieving operations.

POWER BRUSHING

Power-driven, rotary industrial brushes are widely used for deburring, cleaning, and finishing because they result in time and cost savings for many applications. Being flexible tools and available in an extensive range of types, sizes, and materials, industrial brushes are employed for many different purposes. Applications include edge blending, deburring, controlled surface roughening or refinement, cleaning, and finishing.

Stock removal with brushes varies from a minimum to substantial amounts. Fine finishes are produced on miniature components, and scale is removed from large parts. Depending upon brush filament size and speed, wire-wheel brushes can affect workpiece dimensions; being flexible, they are not suitable tools for achieving final dimensional size. Equipment used to rotate wheel brushes includes bench-mounted electric or air motors, flexible shaft tools, bench and pedestal grinders, portable tools, polishing and buffing lathes, and other machine tools. When improperly used, wire wheels can displace material rather than removing it.

Potential limitations to the use of power brushing include the possibility of contaminating the workpieces, changing the color or surface finish of the workpieces, the generation or turning over of burrs, and hardening of the workpiece surfaces.

Fill Materials

Depending upon the applications, various filament materials are used in the manufacture of brushes. Although some are man-made and others are derived from natural sources, they are known collectively as *brush fill material.* These materials include tempered and untempered high-tensile-strength steel wires; tempered and untempered stainless steel wires; a number of nonferrous wires such as brass, nickel silver, and beryllium copper; brass and/or zinc-coated steel, titanium, and zirconium; fiberglass-coated, abrasive-loaded plastics and heat and corrosion-resistant alloys; synthetics such as nylon and polypropylene; natural animal hairs such as horsehair; and a large number of vegetable fibers, including tampico, bahia, bassine, and other combination mixtures.

Natural and synthetic filaments. Natural brush filaments are selected on the basis of filament stiffness. Animal hair is flexible, with bassine being one of the stiffer natural fibers. Man-made synthetic filaments, available in sizes from 0.005 to 0.040″ (0.13 to 1.02 mm) diam, are selected for performance, their ability to resist abrasion, and their stability in withstanding the effects of industrial chemicals, such as alkaline or acidic cleaners.

Tempered carbon steels. Steel wire is used more than other fill materials in power brushes, with the tempered rather than the untempered variety being predominant. This steel wire is high-quality material similar to that used in springs. It is generally made of 0.65 to 0.75% carbon steel rod, with tensile strengths ranging from 320 to 380 ksi (2206 to 2620 MPa). Because of the millions of cycles of stress to which the filaments are subjected, it is highly important that the wires be able to withstand repeated flexing without premature breaking. The wires must be free of pits, die marks, rust, excessive scale, scrapes, splits, laps, cracks, seams, and excessive decarburization. A grade of tempered steel wire with 0.55 to 0.70% carbon and lower tensile strength than previously mentioned is also used when resistance to abrasion and service requirements are not as demanding.

Untempered carbon steel. Untempered steel wires are hard drawn to their highest tensile strength and are sometimes coated for corrosion protection. Three grades are normally used:

1. Low-carbon brush wire (0.14-0.20% carbon) with a tensile strength of approximately 140 ksi (965 MPa).
2. Scratch brush wire (0.45-0.60% carbon) with a tensile strength of 230 to 290 ksi (1586 to 2000 MPa), depending on the diameter of the wire.
3. High-strength wire (0.55-0.75% carbon) with a tensile strength of 300 to 380 ksi (2068 to 2620 MPa), depending on the diameter of the wire.

The low-carbon brush wire is commonly produced in sizes of 0.001 to 0.008″ (0.03 to 0.20 mm) diam, while scratch brush wire and high-strength wire are usually produced in diameters ranging from 0.005 to 0.035″ (0.13 to 0.89 mm) diam. Untempered wires do not have as great a cutting capability, nor are they as abrasion-resistant as tempered wires; however, they are adequate for many applications.

Stainless steel. Applications that involve elevated temperatures or corrosive environments, as well as those requiring nonsparking capabilities, prompt the use of stainless steel wire. Two types are used. The first and probably most common is standard untempered, bright-finished Type 302 stainless steel, which is hard drawn to tensile strengths of 320 to 360 ksi (2206 to 2482 MPa). This type of wire is frequently used in brushing or finishing stainless steels, specialty alloys, and aluminum or other nonferrous materials in an effort to preclude oxidation caused by iron deposits when carbon steel fill material is used.

Tempered stainless steels, with tensile strengths ranging from 240 to 255 ksi (1655 to 1758 MPa), are also used in some power brushes. If avoidance of sparking is the primary criterion, either beryllium-copper or silicon-bronze wire fill material is recommended. Tempered stainless steel has distinct advantages for some applications. It is superior to untempered stainless in its cutting capabilities, fatigue resistance, and abrasion resistance.

There are some applications for which it is desirable to use brushes with nonmagnetic stainless steel wire. For these nonmagnetic uses, Type 420 tempered stainless steel is not satisfactory, and Type 302 untempered should be used. It is not advisable, however, to check Type 302 wire with a magnet. Wire from all heats of Type 302, when drawn to spring quality as used in brushes, can be attracted by a magnet to varying degrees. This amount of magnetic attraction for a given size of wire in the spring tensile range is a function of the degree of cold working that the wire has undergone.

Cold work increases the tensile strength and, simultaneously, the magnetic permeability of the wire. Only in its fully annealed condition can Type 302 stainless steel be considered essentially nonmagnetic, because in this state it exhibits permeabilities of less than 1.02. A simple test to determine if the wire is Type 302 stainless is to heat a piece of the brush wire in question to a red color, which will anneal it; if it is not attracted to a magnet, it can be assumed to be Type 302 stainless.

Brush Selection

It is impossible to recommend a single type of brush for any given application because of the many variables that exist in brushing conditions. However, in choosing a brush for a particular application, some basic design considerations that influence the choice are:

1. Type and modulus of elasticity of the fill material, discussed previously in this section.
2. Diameter or cross-sectional area of the individual filaments.
3. Trim length (the length of the filaments from the retaining members).
4. Density of the fill material.
5. Speed of brush rotation, discussed subsequently.
6. Filament configuration (crimped, straight, or multistrand crimped).

Each of these factors contributes to the physical characteristic known as stiffness. The difference in stiffness of filament materials that differ in modulus of elasticity, diameter, or length is obvious. It is also true that the mass of filament ends at the periphery of a brush has a feel of stiffness depending on the density of the fill material. Centrifugal force is a less obvious factor that permits brush filaments to resist lateral displacement and to feel stiffer.

Filament size. The diameter of the filaments is referred to as the filament size. Wire filament size is generally designated by gage number: No. 20 = 0.035″ (0.89 mm) diam; No. 25 = 0.020″ (0.51 mm) diam; No. 30 = 0.014″ (0.36 mm) diam; No. 33 = 0.0118″ (0.300 mm) diam; No. 34 = 0.0104″ (0.264 mm) diam; and No. 38 = 0.008″ (0.20 mm) diam.

Large-diameter filaments result in a low-density brush that provides aggressiveness, good impact action, and coarse finishes, but a short brush life. Small-diameter filaments result in a high-density brush that produces fine finishes, with good flexibility and contour accommodation, and a long service life.

Trim length. Trim length is the length of wire exposed from the faceplate to the outside diameter (OD) of the brush. A short trim brush has a firm face and low flexibility, but produces aggressive action and fine finishes. A long trim brush has greater flexibility and good impact action and contour accommodation, but produces coarser finishes.

Fill density. The fill density is the number of filament ends per unit of area at the working face of a brush. Higher density brushes are more aggressive, produce finer finishes, and have longer life. Low-density brushes have more flexibility, good impact action, and good contour accommodations, but produce coarser finishes.

Filament configuration. Another basic factor of brush design affecting stiffness is the configuration of the filaments used. Every power-driven brush constructed of metallic filaments uses one of the following four basic forms: crimped, knot or twisted tuft, straight, or multistrand crimped.

Crimped filaments. Crimps are a repeating, approximately sine-wave shape formed into filaments (see Fig. 16-11). The amplitude and frequency of the crimp is controlled by the manufacturing specification. A brush with filaments having a shallow (low-amplitude) crimp has a high density and produces fast, strong action, but results in coarse finishes. A brush with a deep (high-amplitude) crimp has a low density, but produces fine finishes and has a long service life.

A crimped wire brush is illustrated in Fig. 16-12. Many types of crimped wire are used in brushes; however, the best crimping results are obtained when the wire is displaced over multiple planes. Although there are a number of crimping techniques that will produce multiple-plane displacement, the end result should be similar to the crimp configuration that is produced by running wire through two sets of gears perpendicular to each other. Each set of gears should have a different pitch. The number of crimps per inch is governed by the finer of the two sets of gears. The wavelength and frequency of the crimps should be uniform over the entire length of wire. The amplitude of the wire should be held to a tolerance of ±0.004″ (0.10 mm), with a ±0.002″ (0.05 mm) tolerance sometimes being required in special applications.

Knot or twisted tuft. In making knot or twisted tuft filaments, a bundle of straight wires of equal length is formed into a hairpin shape around a retaining member. The bundle is then twisted into a cable with a helix angle, generally about 45°. The standard twist has the tuft twisted for two thirds of its length; however, in applications requiring maximum impact, such as extremely severe cleaning or heavy incrustation removal, the tuft is twisted for its entire length, sometimes referred to as cable twist. Multiple rows of twisted tufts, instead of a single row, are sometimes used to give additional stiffness for the most severe operations. A knot or twisted tuft power brush is also shown in Fig. 16-12.

Straight filaments. In the fabrication of straight wire brushes, rows of holes are drilled in a radial pattern on the periphery of a circular wood, composite, or plastics hub. Small bundles of straight wire, bent into hairpin shape, are then pulled or driven into the holes and locked firmly into place with staples or retaining wire. The protruding ends of the filament wire, being relatively unsupported by other wire filaments, have a high degree of flexibility with a minimum amount of filament entanglement. This design permits them to penetrate narrow slots and shapes better than other filament configurations.

Multistrand crimped. Multistrand crimped filaments are made by stranding together approximately 5 to 25 straight fine wires into a steel cord, usually ranging from 0.020 to 0.050″ (0.51 to 1.27 mm) in cable diameter. This cord or cable is then crimped in multiple planes, just as is conventional crimped wire. The wire used in such a cable has a tensile strength of 360 to 400 ksi (2482 to 2758 MPa) and is used in applications that require the stiffness of a coarse wire with the fatigue strength and multiple-point advantages of a fine wire.

Brush Types

The primary types of power brushes are radial or wheel, both standard and centerless; cup; end; tube; strip; wide-face or cylinder; and miniature (see Fig. 16-13). In addition, elastomer-

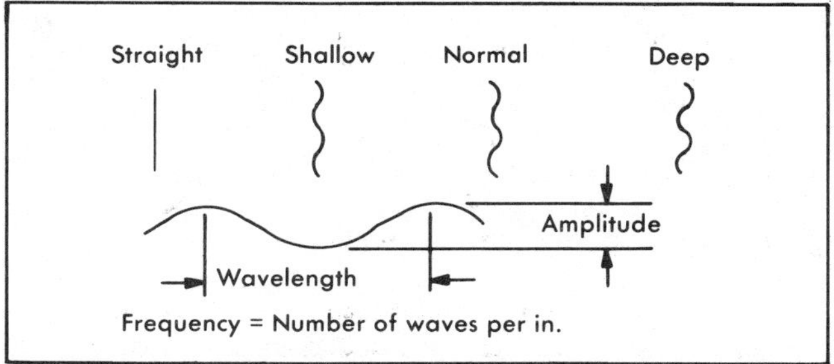

Fig. 16-11 Terminology of crimps formed in filaments. (*Anderson Products, Div. of Wilton Corp.*)

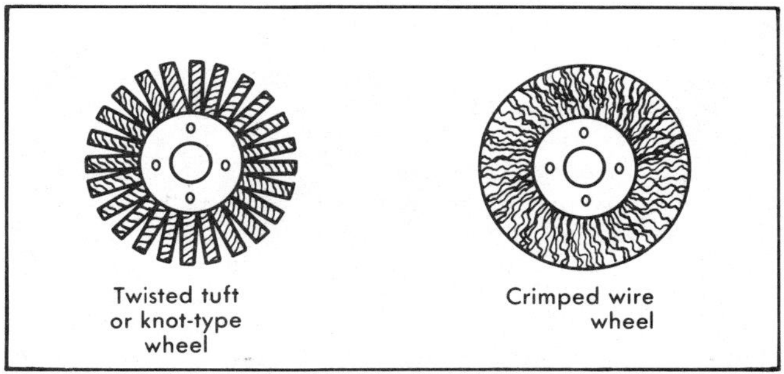

Fig. 16-12 Filament configurations of power brushes.

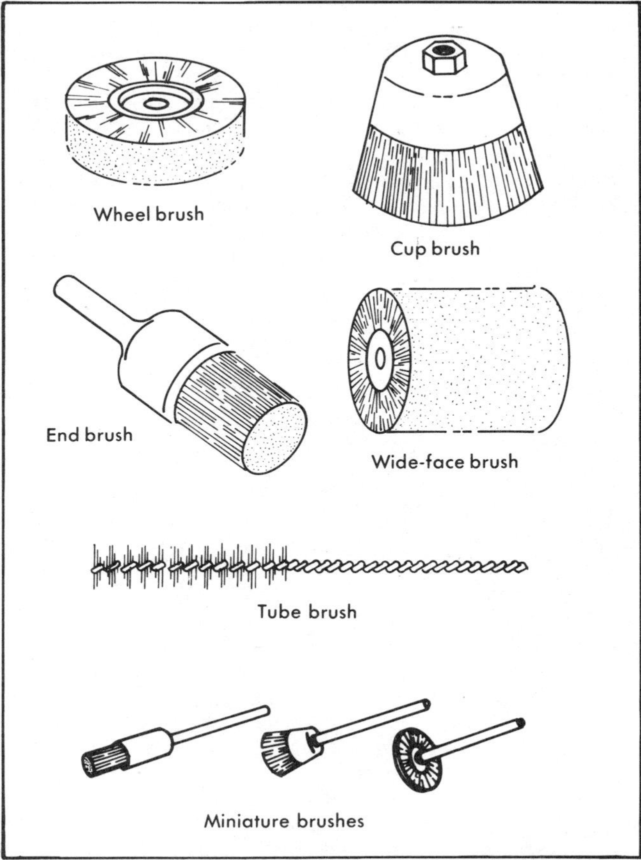

Fig. 16-13 Primary types of power brushes. (*Weiler Brush Co.*)

bonded brushes in most of these types are in common use. Each of these basic types varies in design and composition, and is otherwise modified to provide the versatility required in thousands of complex and sophisticated industrial applications. The dimensional terminology employed for power brushes is illustrated in Fig. 16-14.

Radial (wheel) brushes—metallic. Metal-filled radial or wheel brushes, used in almost every type of industry, are the most popular of all brush configurations. Brush diameters range from 5/8 to 18″ (916 to 457 mm), and face widths usually vary from 1/8″ (3.2 mm) on small-diameter brushes to 2″ (51 mm) on large-diameter brushes. Radial brushes are furnished in short, medium, or long trim lengths, with either a knotted or crimped wire construction.

Short trim. Radial brushes with short trim are densely filled and provide maximum cutting action. They can be furnished either in crimped or in knotted types. They are used extensively for the removal of burrs from gears in the automotive, agricultural implement, and aircraft industries. These short-trim brushes impart the proper radii and adequately blend the edges of gears, while at the same time removing all the minute fragments of metal that may remain following the use of some other techniques of burr removal.

One distinct advantage of the short-trim radial brush is its capability of removing the burr from an edge without changing the dimensional characteristics of the two intersecting surfaces. Short-trim radial brushes are used extensively for heavy-duty deburring and edge blending of stampings and metal parts, for removal of scale and surface encrustations, and for flash removal in the rubber industry. They are designed for efficient, fast, low-cost performance on automatic, semiautomatic, and manual machines.

Medium trim. Radial brushes with medium trim are furnished in both knot and crimped wire styles, and they are used more extensively than all other brushes combined. They lend themselves to use on portable equipment and bench grinders, and they may be ganged in multiples on a common arbor.

Because of their high-impact action, knotted medium-trim brushes are used rather universally for cleaning all kinds of welds and weld spatter. Their virtues also permit them to be used in applications such as roughening for adhesion, scale removal, cleaning, deburring, and flash removal.

Crimped wire, medium-trim brushes are generally conceded to be the best general-purpose wheels. The fact that they are not very stiff permits them to follow and conform to the contours of irregular surfaces. Relatively inaccessible places can be reached because of the flexibility of their filaments. Feather-type machining and grinding burrs can be removed from edges and light flash on rubber and plastics can be taken off with these wheels. They are also used extensively for cleaning, finishing, roughening for adhesion, and rust and other contaminant removal.

Long trim. Long-trim radial brushes are also furnished in either knot or crimped style but are more specialized in their uses. When the applications dictate a need for a narrow face or maximum flexibility, the long-trim brush is customarily used. A wide range of results can be secured by varying the diameter of the brush, as well as by modifying wire size, peripheral speed, and the wire configuration.

The knotted wheels efficiently remove rubber flashing. Because they develop a very narrow face at high speed, they are also used to clean the crevices between sections of roadways and airplane runways before the cracks are filled with mastic. At slower speeds they are much more flexible and are consequently adaptable to applications such as surface finishing, paint removal, and the cleaning of wire mesh and other types of conveyors.

Crimped wire, long-trim brushes are especially useful in the attainment of satin or brushed finishes on metals. Other major applications include very light deburring, surface finishing, and the brushing of uneven surfaces for scale or paint removal.

Radial brushes—nonmetallic. A substantial portion of applications requiring fine finishes of 4 to 30 μin. (0.10 to 0.76 μm) cannot be accomplished with wire-filled radial brushes. Similarly, applications for a reduction of stress concentrations can, in many instances, be best accomplished by means of brushes constructed of nonmetallic fill material. Three primary fill materials for this type of work are tampico, natural and synthetic strings or cords, and abrasive nylon monofilaments.

Tampico. Tampico is a cellular vegetable fiber used to make a flexible, resilient wheel brush. Used singly or in multiples, tampico wheels are used in many dusting, scrubbing, cleaning, and surface finishing applications. It has been found that tampico brushes can be treated and cured with a lacquer for improved stiffness and abrasive grit retention. The treatment minimizes the shedding, knifing, and weaving that sometimes occur at high speeds. Treated brushes can therefore perform an infinitely wider range of operations than can be accomplished with the natural, untreated fiber. They are used extensively in deburring and edge blending, as well as in smoothing or blending surface imperfections such as pits, scratches, parting lines, mold marks, tool and die marks, and stretched metal strains. The stiffness and tackiness attained by treatment make it easy to apply the grease stick and abrasive compounds used with these brushes for polishing and buffing, discussed later in this section.

The main reason that treated tampico brushes are used is to increase the life of the part being brushed. Over half of all failures of brushed parts are fatigue failures caused by the widening of microscopic cracks that result from stress concentrations. The radii and surface finishes produced by treated tampico brushes on parts subjected to repeated cyclic stresses can almost totally eliminate the scratches, burrs, and sharp corners that cause cracks and progressive fractures.

String and cord. String wheels, also called string brushes, are made of soft cotton yarn fastened to a hub and used with greaseless polishing compounds to produce fine finishes. Because abrasive compounds are commonly used with string wheels, their use is probably more closely related to polishing operations than to brushing.

Cord wheels or brushes are similar to string wheels, but they are made of special natural or synthetic cords that create a denser brush to provide less-smooth finishes. Cord wheels are also used with cutting compounds in polishing and buffing.

Abrasive monofilament. Since its introduction as a power brush fill material, abrasive monofilament, made into all types of brushes, has replaced a number of other conventional abrasive-style products, as well as light-to-medium gage, metallic and nonmetallic filled brushes. Brushes made with nylon monofilaments, in which silicon carbide, aluminum oxide, or pumice grit is permanently encapsulated, constitute the most rapidly growing new technology.

Between 30 and 40% of the abrasive particles, by weight, are uniformly distributed throughout the filament. Grit sizes range between 600 (fine) and 80 (coarse). The coarse grits are encapsulated in relatively large-diameter filaments, while the

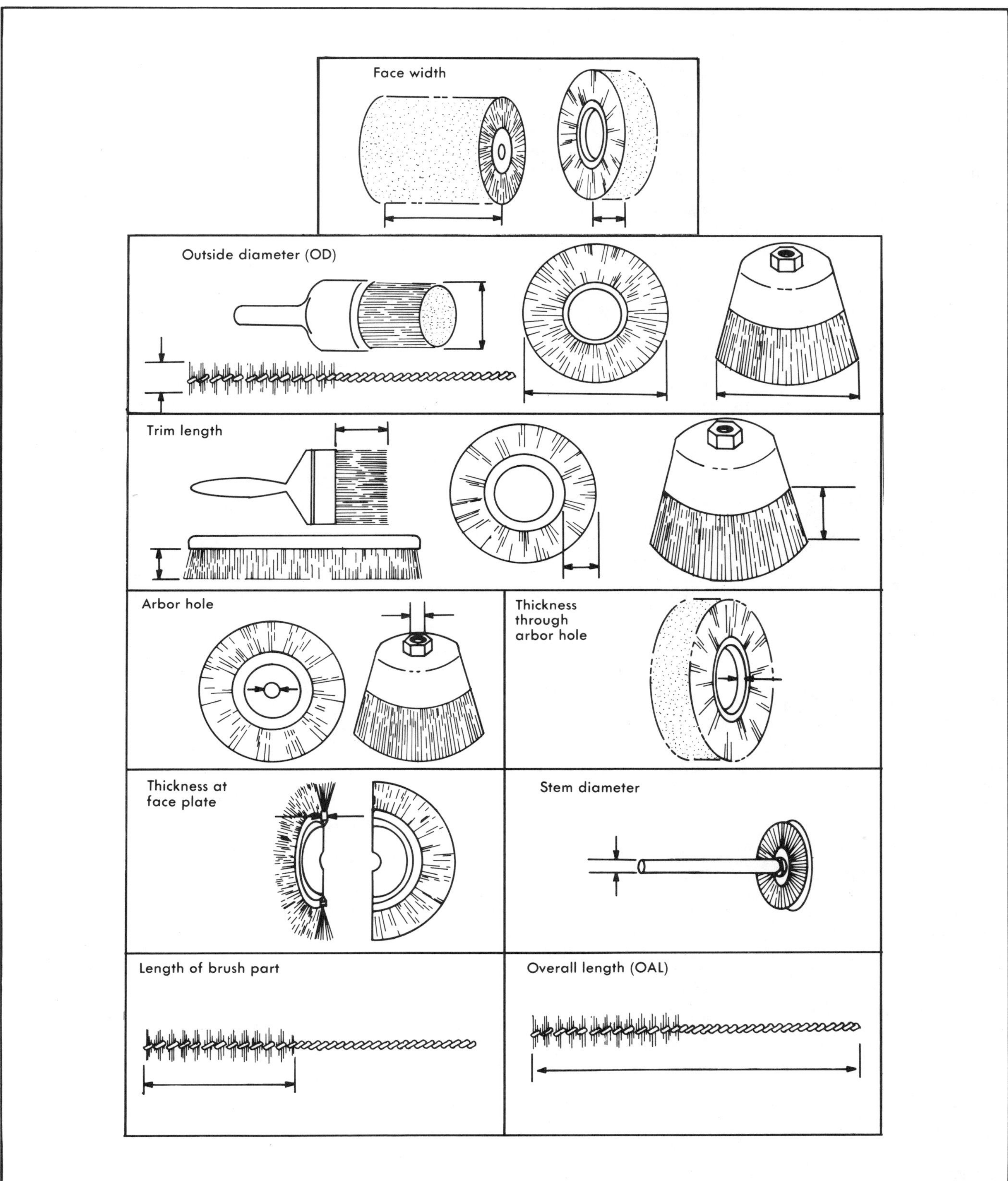

Fig. 16-14 Dimensional terminology for power brushes. (*Weiler Brush Co.*)

fine size are dispersed in small-diameter filaments. Usually the monofilament diameter is between 0.014 and 0.050″ (0.36 and 1.27 mm), depending upon the size of grit used.

Abrasive filament brushes can be used either wet or dry. Experience indicates some advantages of wet use when water, mineral oil, or water-soluble oil is used as the cooling medium. Grease sticks have also been used efficiently. Generally, these brushes can be used in all the applications for which treated tampico brushes are used. Their advantage lies in the fact that the need for compounds and slurries is eliminated; such compounds can be used with abrasive filament brushes, however, to accelerate the cutting action.

Abrasive filament brushes can also be used in applications for which the tackiness of treated tampico brushes would be objectionable, such as in the decorative finishing of wood, leather, plastics, cork, rubber, ceramics, and glass. These brushes have also exhibited long life in cleaning and finishing steel strip and in controlling oxide on steel rolls in the primary metals industry.

Long-string abrasive brushes are being used to remove burrs and to smooth corner surfaces on a wide variety of aircraft parts. The brush rolls contain rows of abrasive-impregnated nylon fibers (see Fig. 16-15). The flexibility of the long fibers and their wraparound effect permits deburring both sides of some parts, but stock removal is limited to about 0.001″ (0.03 mm). The brushes are rotated at slower speeds than wire brushes.

Centerless radial brushes. Centerless brushes fit standard centerless grinders. These large radial brushes have a wire fill, 0.005 to 0.014″ (0.13 to 0.36 mm) diam, or a fill of abrasive monofilament, treated tampico, or cord. When the fill is wire or abrasive monofilament, these tools remove grinding burrs and produce a minimum edge of 0.005″. Finishing work can be done either dry or (preferably) with normal grinding coolant.

Fine wire fill or abrasive monofilament brushes, working with conventional grinding coolant, remove feather grinding burrs and improve surface finishes on such parts as control valves for automatic transmissions. In such applications, they can improve the finish from 7 to 4 μin. (0.18 to 0.10 μm). If the fill is treated tampico or cord, applying a grease-based polishing compound to the wheel face will provide finishes in the range of 2 to 7 μin. (0.05 to 0.18 μm).

It is important to remember the following facts about centerless brushes:

- They will not remove appreciable amounts of metal from cylindrical surfaces. Parts must therefore be ground to size before brushing.
- The greatest improvements in finish are obtained when the unbrushed parts have a finish in the 25 to 30 μin. (0.64 to 0.76 μm) range. A part with a 25 μin. finish can be brushed rapidly to obtain a 10 to 15 μin. (0.25 to 0.38 μm) finish; a part with a 10 μin. finish will have a 4 to 6 μin. (0.10 to 0.15 μm) finish after brushing.
- Centerless brushing follows centerless grinding principles, with the following exception: accuracy in pressure and adjustment is not critical with brushes. Through-feed centerless operations may be arranged in tandem to feed parts directly from the last grinding operation into the brushing operation.

Cup brushes. Provided in both the crimped and twisted tuft construction, cup brushes are designed for use on high-speed air and electric tools, drill presses, pipeline machinery, and special-purpose equipment. The twisted tuft or knot-type brush affords maximum impact and is generally used for severe cleaning and the removal of weld scale or heavy incrustations. These brushes are most universally used in the structural steel and shipbuilding industries to clean welds, scale, and rust. Different diameters of brushes, ranging from 3 to 8″ (76 to 203 mm), and different wire fill diameters, from 0.0118 to 0.035″ (0.300 to 0.89 mm), are usually adequate for most requirements. Light brushing is usually accomplished with crimped wire cup brushes, which are especially valuable for the removal of lighter encrustations of rust, oxides, or paints, and when large surface areas have to be cleaned.

End brushes. A mounting shank is an integral part of most end brushes. The standard end brushes used are the solid-end type, in either crimped or knot configuration, and the circular flared brush. Hollow-center, pencil, and pilot-bonded types are also available. Diameters generally range from 5/32 to 1 1/8″ (4 to 28.6 mm).

Typical end brush applications include weld cleaning, mold cleaning, tool and die polishing, spotfacing, casting cleaning, and automotive carbon cleaning, mostly in corners, crevices, and other hard-to-reach places. These tools are often used in portable air and electric-powered tools.

Tube brushes. Designed for use in slow-speed portable tools or drill presses, most tube brushes are made by twisting the filament wires between two or more larger retaining wires. Tube brushes are excellent brushing tools for cleaning and finishing in restricted areas. They are also used for cleaning and deburring keyways and grooves in parts, and for internal hole deburring and thread cleaning. Diameters generally range from 3/16 to 1 1/4″ (4.8 to 32 mm), but other sizes are available.

Strip brushes. One of the most versatile brushes available is the strip brush, which can be customized by either the manufacturer or the user, depending upon the purpose of the

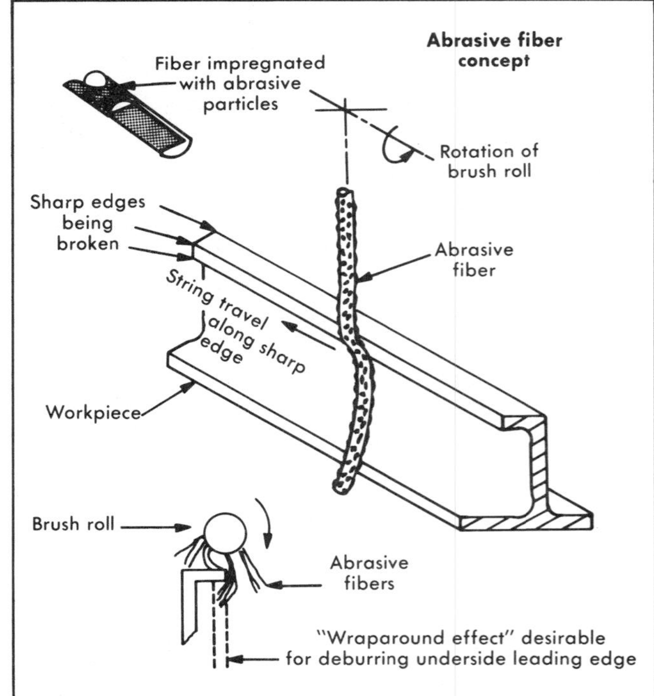

Fig. 16-15 Long-string abrasive brush used to deburr aircraft parts. (*Avco Aerostructures Div.*)

brush. For example, virtually any strip length, height, channel material, or channel size desired can be obtained. The choice of fill materials includes the entire spectrum of synthetic, vegetable fiber, hair, and abrasive monofilaments, or metallic filaments.

Applications for strip brushes are as extensive as the variety of brushes available. Strip brushes are used extensively as splash curtains or curtain walls on machine tools and ovens; as lag brushes, static electricity eliminators, and backups for coated abrasive wheels; for dusting rubber goods such as tires; as cotton pickers; and in many vegetable and fruit cleaning operations. One of the widest uses of strip brushes is for assembly on rotary hubs or extruded mountings to form rotary brushes for the cleaning of conveyor belts and screens, the scrubbing of continuous steel strip, and the spreading of various materials. The strips are mounted either straight, parallel to the axis of the brush, or helically around the axis to ensure continuous contact with the work. Because of the space between the strips, the brush does not load up, and its flexibility permits it to conform to the irregular surfaces on which it is used.

Wide-face or cylinder brushes. Probably 90% or more of all wide-face brushes are custom made to suit the particular requirements of the user. They are made in diameters ranging from 2 to 30" (51 to 762 mm) and in lengths up to 25 ft (7.6 m). Every conceivable type of fill material is used, including all the popular synthetics, hairs, natural fibers, abrasive monofilament, and metallic wires. Wide-face brushes, as well as other brushes that operate at speeds over 500 rpm, generally require dynamic balancing. Such brushes should also be designed for ventilation, especially when used for continuous and/or heavy-duty applications. Radial fin-type mountings with ends open to allow the free passage of air generally provide sufficient ventilation.

Although the customer can provide specifications of overall dimensional characteristics, a specialist in brush technology should provide the detailed specifications relating to the kind of fill density, trim length, and type of construction required. The brush manufacturer should also determine whether the brush should be made with wheel sections mounted on a common arbor or with a continuous strip helically wound into convolutions of sufficient number to produce the proper face width. The manufacturer should also recommend whether the brush should be used wet or dry and what pressure and speed should be used.

It is not uncommon for worn brushes, mounted on permanent arbors, to be returned to the manufacturer for replacement. The brush manufacturer removes the worn brush from the arbor; replaces, trims, and dynamically balances the brush assembly, including the arbor; and then returns it to the customer. For some less-sophisticated applications, a helically wound brush element or a group of sectional brushes may be ordered by customers to mount on their own arbors.

Another tool is the unitized section or throwaway can. For these tools, the brush manufacturer fabricates wide-face brushes on a thin-walled steel tube that contains inserts to reduce the tubing to the customer's arbor size. These units are forwarded to the customer, who merely slides them onto the arbor and locks them into position. Since the worn brush elements may be thrown away, the cost and time required to ship arbors back and forth between manufacturer and customer are eliminated. Minimal spare parts inventories are required when throwaway cans are used.

The throwaway-can brush is used extensively in the steel industry for scrubbing steel on galvanizing lines, electrolytic tinplate lines, cold reduction cleaning lines, paint coating lines, and roller levelers. In most of these applications, the brushes are the flow-through variety to facilitate a continuous flow of water or mild alkaline solution through the brush.

The circulation of the water, which is emitted through the face of the flow-through brush, serves a number of purposes in addition to dissipating the heat generated in brushing. It also dissolves the dirts and oils on the strip being cleaned; and, most important, it flushes the brush face free of dirt.

The flow-through principle is not confined to throwaway-can brushes, but is also frequently used for the refillable, permanent arbor types. In addition, brushes using the flow-through principle are found in aluminum, copper, and brass mills, and, in many instances, in washing lines for sheet glass products. Precision-ground metallic and abrasive monofilament brushes weighing hundreds of pounds each are used to control the aluminum oxide buildup on the steel work rolls of continuous and reversing hot mills in the aluminum industry.

Special stainless steel, wire-filled brushes, covering stainless steel conveyor rolls, are used directly inside of hearth furnaces to convey aluminum and alloy plate through the furnace for attainment of critical metallurgical properties and smooth, nonscuffed surfaces. To produce a matte finish on stainless steel, a pumice slurry is used in combination with tampico or abrasive monofilament wide-face brushes. Slurry is also used in dulling the finish on boards and panels such as formica.

There are more applications for wide-face brushes used dry, however, than for those used wet or with a slurry. The textile industry alone uses hundreds of different types of brushes, and tanneries use brushes for seasoning, dusting, buffing, polishing, and oiling. The paper industry requires large brushes for dampening, for buffing clay-coated stock to attain high-luster finishes, for coating and dusting, and for cleaning wire mesh.

Wire-filled elastomer brushes. Elastomer-bonded wire-filled brushes represent a significant development in the brush industry. These brushes have proved themselves to be extremely fast cutting, with excellent operator safety. Because the bonding material supports the wire to the very tip, the brush provides a maximum amount of cutting action. By periodically reversing the brush, optimum cutting efficiency is maintained throughout its life. Compared to conventional nonbonded wire-filled brushes, there is a minimal loss of wire through fatigue with wire-filled elastomer brushes.

Bonded wire-filled brushes are made in radial, cup-style, flared-end, and straight-end designs, and can be used in place of conventional wire-filled products. These brushes are effective for many operations because they maintain a uniform face throughout their life and perform at maximum efficiency. Examples of the types of jobs these brushes perform are as follows:

1. Removing oxide and weld scale from various material surfaces.
2. Removing burrs from keyslots and other hard-to-reach areas. The wire-fill material for these applications is generally 0.0118" (0.300 mm) in diameter, and the brushes are rotated at 5500 to 6500 sfm (1676 to 1981 m/min).
3. Removing stamping burrs and producing radii on ductile metals, without producing secondary rolled conditions. The wire-fill material is usually 0.008-0.0118" (0.20-0.300 mm) diam, and the brushes rotate at 5000 to 7000 sfm (1524 to 2134 m/min).
4. Removing burrs on powdered metal parts, aluminum die castings, and fine-pitch gears. The wire-fill material for these applications is usually 0.006-0.008" (0.15-0.20 mm)

diam, and the brushes are rotated at 5000 to 6500 sfm (1524 to 1981 m/min).

5. In the electrical industry, removing insulation from copper leads. The wire-fill material is 0.0025-0.008" (0.06-0.20 mm) diam, and the brush surface speed ranges from 4000 to 6000 sfm (1219 to 1829 m/min).

Two factors should be taken into consideration when elastomer-bonded brushes are used. First, periodic dressing of the brush face will extend brush life. Second, because the cutting action is so fast, brush adjustment is important; the part must be held precisely if the brush is to be used efficiently.

Miniature brushes. These brushes are small versions of end, wheel, and cup-type brushes. They are used for cleaning, deburring, and finishing of delicate or miniature components. Brush sizes range from 5/32 to 1 1/2" (4 to 38 mm) diam. Fill materials include steel, stainless steel, brass, bronze, and nickel-silver wire, as well as various types of animal hair and nylon filaments. The hair-filled types are sometimes used with abrasive media for deburring and polishing small and detailed parts. These brushes are generally mounted on a 3/32 or 1/8" (2.4 or 3.2 mm) diam shank.

Machine Requirements

It is essential that the correct equipment and the proper brush be used for the various applications because the correct brush on inadequate equipment would preclude efficient brushing. In selecting the equipment for brushing, the largest diameter brush permitted by the machine and workpiece size will usually provide the most efficient brushing and the lowest service cost. When large areas are to be brushed, the widest possible brush face undoubtedly will provide the maximum efficiency.

Various machines and equipment are used for power brushing. Equipment used includes bench motors, pedestal grinders, single and dual-spindle polishing lathes, and many different special machines. Brushing parts on the machines used to produce the parts is often more economical than performing a secondary brushing operation. Numerically controlled equipment and industrial robots provide excellent flexibility for on-machine deburring and finishing.

Several important factors should be considered in the design of brushing machines. These factors include the means for measuring and relieving pressure, and adjusting, reciprocating, and conditioning the brush. In addition, shaft sizes for brushing machines are important. The ability to reverse shaft direction may also be desirable for some applications.

Pressure measurement. The quality of the work produced and brush life are both significantly affected by brushing pressure. As a result, it is important to have some means for measuring the pressure on all semiautomatic and automatic brushing machines. One simple method of accomplishing this measurement is to connect an ammeter to the driving motor circuit for each brushing head as a permanent part of the machine. The meters can be easily calibrated with respect to pressure requirements for various applications.

Pressure relief. Brushing machines should also have some means of relieving pressure under conditions of pressure buildup resulting from misalignment, oversized parts, or improper location of the workpieces. Such a feature is especially important when using brushes having low flexibility (short trim and/or dense fill), when damage to the brushes and the machine could occur. For small workpieces, it is usually best to allow the parts to retract when possible. For large parts, the machine heads can be of floating design, thus allowing the brushes to rise and fall.

Brush adjustments. An important feature for machines, especially those on which wide-face or cylinder brushes are employed, is a means for adjusting the brushes to obtain predetermined pressures and to compensate for wear of the brushes. Good results are being obtained by using a single adjusting screw with a fine control for both bearing slides. This design ensures that wide-face brushes remain parallel to the pass line. Adjustments can be hand controlled or automated through power feed controls.

Brush reciprocation. A means for axial reciprocation, especially of wide-faced brushes, aids in maintaining uniform action across the full faces of the brushes. Axial reciprocation avoids the possibility of developing grooves and prevents streaking when finishing shapes such as sheets, continuous strip, and rods. Reciprocation, however, will not eliminate a streaking pattern caused by grooves or spacings in the brush face when such grooves or spaces are continuous and in a plane perpendicular to the rotational axis of the brush.

Reciprocating brushes at the rate of 1/2 to 2 reversals per minute is effective. When brushing multiple parts across the brush face, the reciprocation length must exceed the width of one part plus one half the space between parts.

Backup rolls. These rolls are often used for power brushing. They should be designed to provide proper support for the workpiece, but should not interfere with the brush. A brush cannot be used successfully as a backup roll for another brush. For thin strip or sheet material, 1/4" (6.35 mm) thick or less, two backup rolls (one on each side of the brush) are often used. For thicker materials or parts, a single backup roll is generally positioned directly opposite the brush.

Shaft sizes. While shaft sizes for brushing machines have become fairly standardized as the result of many years of use, it is still essential to calculate shaft diameters for special machines and operating conditions. In general, for semiautomatic machines having a shaft length requirement to provide 8 to 12" (203 to 305 mm) overhang, a 1 1/4" (32 mm) diam shaft has been found to be satisfactory. For similar service with a maximum overhang of 20" (508 mm), a 2" (51 mm) diam shaft is generally suitable. For brushing operations, a maximum deflection of 1/32" (0.8 mm) in 6 ft (1.8 m) is the standard recommended allowance.

Brush replacement. To minimize downtime, machines should be designed for easy and fast replacement of brushes. In some cases, spare heads are provided so that brushes can be replaced while the machine is in operation.

Operating Parameters

Power requirements. Inefficient brushing is sometimes encountered because of inadequate horsepower for driving the brush. The three primary factors that determine the amount of horsepower required for a given operation are as follows:

1. The brushing pressure required to accomplish the work.
2. The resistance developed between the work and the brush.
3. The speed of the brush.

The brush configuration itself and the type and size of filaments used are also influencing factors.

Table 16-6 provides a working guide to determine brushing horsepower requirements for a 1" (25 mm) wide brush face. This guide is based on experience and general application conditions. For conditions that are more complex, additional power consumption data is required to ascertain the correct horsepower.

Brush speed. Effective and low-cost brushing is also influenced by the speed at which the brush rotates. For each particular application, there is an optimum operating speed, and to go higher or lower will affect the efficiency of the brushing operation. As a generalization, Table 16-7 lists surface speeds for various types of brushing applications. Table 16-8 presents speed conversions from revolutions per minute to surface feet per minute for various brush diameters. For semiautomated and automated applications, brushes should be operated at the highest practical speed with the lightest possible pressure.

TABLE 16-6
Power Requirements for 1" (25 mm) Wide Brushes

Brush Diameter, in. (mm)	Recommended Motor Size, hp (kW)	Motor Speed, rpm
4 (102)	1/4 (0.2)	3450
6 (152)	1/2 (0.4)	3450
8 (203)	3/4 (0.6)	3450
10 (254)	1 (0.75)	1750
12 (305)	1 (0.75)	1750
15 (381)	1 1/2 (1.12)	1750

TABLE 16-7
Recommended Surface Speeds for Brushing Applications

Application	Surface Speed Range, sfm (m/min)
Burr removal	5500-7500 (1675-2285)
Scale removal	7500-10,000 (2285-3050)
Weld cleaning	7200-9400 (2195-2865)
Edge blending	4700-7500 (1435-2285)
Cleaning, dry (wire brushes)	4000-5500 (1220-1675)
Cleaning, wet	1900-4000 (580-1220)
Surface polishing	6400-8000 (1950-2440)
Surface buffing	8000-10,000 (2440-3050)

Brushing techniques. In any brushing application, it is generally the tips of the brush filaments that do the work. If excessive pressure is exerted, the filaments are displaced so that a wiping action rather than a cutting action occurs. Even worse, excessive pressure overstresses the filaments, due to excess displacement, and causes premature fatigue failure.

It is important to have the full face of the brush in contact with the workpiece to avoid grooving the brush. Operations in which the brush face cannot be in full contact with the work may require some provision for periodically dressing the brush face. In some cases, brush reciprocation can be used to avoid grooving. Some correct and incorrect brushing techniques are illustrated in Fig. 16-16.

In many operations, it has been found that more efficient brushing is accomplished if the brush is periodically reversed in its direction of rotation. The filament tips become dull with use, and reversing the direction of rotation of the brush tends to resharpen the tips or at least presents new cutting edges to the work. The self-sharpening action also tends to reduce operator fatigue by reducing the pressure required to do the work.

Troubleshooting in Brushing

Some of the possible problems that may be encountered in power brushing and their suggested corrections are presented in Table 16-9. Since power-driven brushing tools are expendable items, they must be replaced periodically. In analyzing the cause of rapid brush wear or failure, it has been found that two primary causes exist: abrasive wear and fatigue.

Abrasive wear. To increase the wear life of a brush, a number of steps can be taken. Since the abrasion resistance of a brush is decreased in proportion to the stiffness of the filaments, longer life requires a reduction of the stiffness of the filaments in the brush. This reduction can be made by altering any of the stiffness factors discussed previously, including modulus of elasticity, trim length, and rotational speed.

Brush life may also be increased simply by reducing the load on the brush—that is, by changing the relationship that exists between the brush and the work being brushed. A secondary

TABLE 16-8
Relationship of Surface Speed to Rotational Speed for Various Brush Diameters

Rotational Speed, rpm	Brush Diameter, in. (mm)					
	4 (102)	6 (152)	8 (203)	10 (254)	12 (305)	15 (381)
	Surface Speed, sfm (m/min)					
1000	1050 (320)	1575 (480)	2100 (640)	2625 (800)	3150 (960)	3925 (1125)
1500	1575 (480)	2350 (715)	3150 (960)	3925 (1195)	4725 (1440)	5900 (1800)
1750	1800 (550)	2750 (840)	3650 (1115)	4550 (1385)	5500 (1675)	6800 (2075)
2000	2100 (640)	3150 (960)	4200 (1280)	5250 (1600)	6275 (1915)	7850 (2395)
2500	2625 (800)	3925 (1195)	5250 (1600)	6550 (1995)	7850 (2395)	9825 (2995)
3000	3125 (950)	4725 (1440)	6275 (1915)	7850 (2395)	9425 (2875)	11,775 (3590)
3450	3600 (1095)	5400 (1645)	7200 (2195)	9000 (2745)	11,000 (3355)	13,500 (4115)
3750	3900 (1190)	5900 (1800)	7800 (2375)	9800 (2985)	11,800 (3595)	
4000	4175 (1275)	6275 (1915)	8375 (2555)	10,475 (3195)		
4500	4700 (1435)	7075 (2155)	9425 (2875)			
5000	5225 (1595)	7875 (2400)				
6000	6275 (1915)	9425 (2875)				
8000	8375 (2555)					

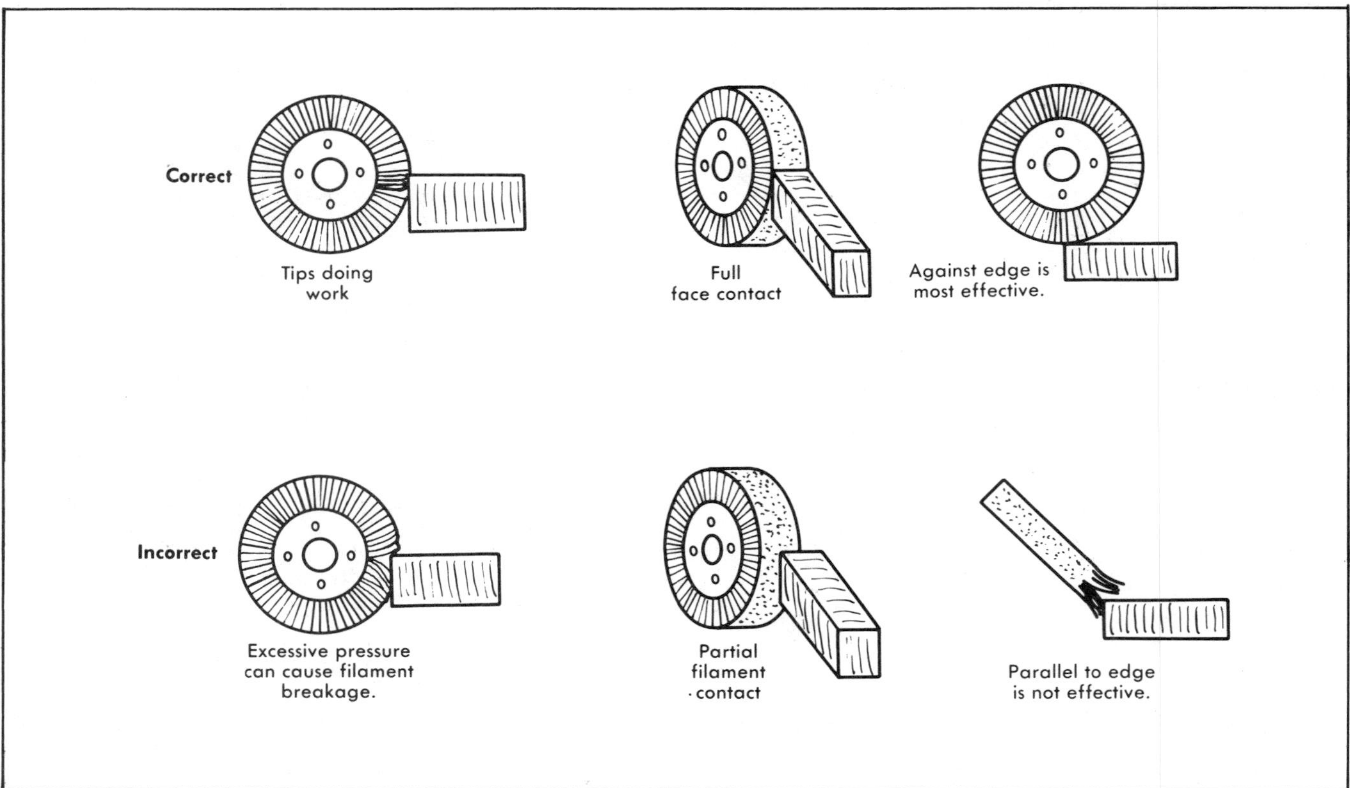

Fig. 16-16 Some correct and incorrect brushing techniques.

TABLE 16-9
Brush and Operating Adjustments to Improve Results

Observed Results	Corrections Suggested
Brush works too slowly.	1. Increase surface speed by increasing OD or rotational speed of brush. 2. Decrease trim length and increase fill density of brush used. 3. Increase filament diameter. 4. Use a heavier duty brush.
Brush works too fast.	1. Reduce surface speed by decreasing OD or rotational speed of brush. 2. Increase trim length and decrease fill density of brush used. 3. Reduce filament diameter.
Brush rolls or peens burr to adjacent surfaces instead of removing it.	1. Decrease trim length and increase fill density of brush used. 2. If wire brush tests indicate metal is too ductile, change to a nonmetallic brush, such as tampico, treated with abrasive burring compound.
Finer or smoother finish required.	1. Decrease trim length and increase fill density of brush used. 2. Reduce wire filament diameter. 3. Try treated tampico or cord brushes with suitable compounds at recommended speeds. 4. Use an auxiliary buffing compound with brush. 5. Increase brush speed.
Finish too smooth or lustrous.	1. Increase trim length and decrease fill density of brush used. 2. Reduce surface speed by decreasing OD or rotational speed of brush. 3. Increase filament diameter.
Brushing action not sufficiently uniform.	1. Increase trim length and decrease fill density of brush used. 2. Devise handheld or mechanical workholding fixture, or use semiautomatic or automatic machine, to avoid irregular offhand manipulation.
Unable to reach some areas of irregular surfaces.	1. Use brush with longer trim and lower fill density for increased flexibility.
Short brush life.	1. Increase trim length. 2. Reduce filament diameter.

(*Weiler Brush Co.*)

advantage of load reduction is a decrease in the amount of filament deflection. Wear increases more rapidly than the proportional increase in load when the filaments are wiping instead of cutting.

Fatigue. After abrasive wear, filament fatigue is the next most important factor leading to premature brush failure. The very nature of brushing produces forces that compress, flex, and sometimes overheat the brush filaments. These forces are conducive to fatigue when they reach an intensity that exceeds a certain tolerance called the endurance limit. When the endurance limit is exceeded, the brush failure is almost immediate; and the greater the mechanical stress, the sooner the damage to the brush filaments occurs. Although fatigue failure is possible for any size of filament, susceptibility varies directly with the diameter of the filament. Finer wires are capable of resisting the overstressing to a much greater degree than larger diameter filaments.

Brushing Safety

When properly used, power-driven brushes are safe, although they inherently have all the dangerous attributes of any rotating body. It is therefore essential that they be mounted, handled, and used carefully, observing all the necessary precautions. The operator and others in the work area must wear safety goggles or full-face shields over safety glasses, with side shields and other suitable protective equipment. A brush should only be used when it is guarded by an enclosure that protects the operator, as well as any other workers who are in the proximity of the brushing operation.

The machine on which a brush is being used should not be operated at an excessive speed, and the brush itself should never exceed the maximum safe speed that has been established for it. Before operation, the machine should be jogged to ensure that it is ready for use and that the brush is fastened securely and is concentric with the axis of rotation. After the machine has been started, the brush should be run at operating speed in a protected enclosure for at least one minute before applying the work; during this time, no one should stand in front of or in line with the brush.

Brushes are self-sharpening tools inasmuch as they continually renew their working face by a process of breaking off dulled tips of filament wires and exposing new, sharp cutting edges. The fragments of wire from this renewal process, in addition to the particles being removed from the work being brushed, are propelled at great velocities. It is essential, therefore, that the operator and others in the area be protected from them by guards and other protection equipment.

In operation, the work should never be pressed into the face of the brush far enough to slow the drive motor. Visual evidence of excessive pressure can be noted when the sides of the filaments are in contact with the work; only the tips should be in contact. Further, brushes should be stored in an environmental atmosphere that precludes rusting or oxidizing, and in a manner that protects the brushes from physical distortion, which might result in brush imbalance.

Excellent safety standard requirements and explanatory information are contained in the American National Standards Institute's (ANSI) Standards B165.1, "Safety Requirements for the Design, Care, and Use of Power Driven Brushing Tools," and B165.2, "Safety Requirements for the Design, Care, and Use of Power Driven Brushing Tools Constructed with Wood, Plastic, or Composition Hubs or Cores." These standards should be referred to and fully understood before setting up any new applications, and they should be periodically reviewed for present and existing brushing applications. With separate sections for the user, machine tool builder, and brush manufacturer, the standards contain a wealth of engineering and technical information.

POLISHING

Polishing in the metalworking industry, except for chemical polishing and electropolishing (discussed in Chapter 17), generally refers to an abrading process that can remove small or large amounts of stock and usually produces a defined line pattern on workpiece surfaces. Polishing uses abrasive grains that are carried to the workpieces by resilient wheels or belts. The availability of improved, bonded-abrasive belts and flap wheels has made polishing, in some instances, comparable to abrasive machining. Polishing is a more aggressive operation (removes stock more rapidly) than brushing and buffing.

Wheels or Belts?

Wheels and coated-abrasive belts can generally perform many polishing operations equally well. Some of their special features and advantages are summarized in Table 16-10. Precoated-abrasive belts are available in various grain sizes that are ready for polishing without the need for a setup room. The softness or stiffness of belts depends upon the type of cloth backing and the type of adhesive used to bond the abrasive to the belt surface. Belts are often best for polishing flat surfaces, and wheels best for concave and contoured surfaces.

Except for the abrasive grades used, belt polishing techniques are similar to coated-abrasive grinding. The use of coated-abrasive belts for general-purpose stock removal is discussed in Volume I, *Machining*, of this Handbook series. Details on coated-abrasive belts for polishing and finishing are presented later in this chapter.

Contact wheels for belt polishing. The performance and life of an abrasive belt depend upon the contact wheel used. Generally, the contact wheel, which is usually serrated or smooth rubber, should be as hard as the shape of the workpiece will permit. Durometer hardnesses of the wheels range from 15 (soft) to 90 (hard). The more uneven the shape of the workpiece, the softer the contact wheel should be. The larger the surface of the workpiece, the coarser the serrations should be, resulting in a faster rate of stock removal for a given abrasive grain size.

Pneumatic-type contact wheels are available, in which air pressure is introduced into a pump sleeve within the abrasive belt to increase the hardness. Another type of contact wheel consists of a series of buffing wheels made of rubber-coated fabric to provide the necessary resiliency. For some applications, conventional sewed cotton and bias-type buffs, sometimes sewed together to provide the desired face width, are preferred.

Diameters of the contact wheels are also important. Larger diameter wheels increase the contact area, which results in a better surface quality. Conversely, smaller contact wheels decrease the cutting area, thereby increasing the rate of stock removal. The diameter of the contact wheel also determines the cutting speeds used, all other conditions remaining constant, as discussed later in this section.

Comparative costs. Comparative costs with regard to expendable materials used are considerably less for setup wheels than belts. However, the labor cost of handling and heading the setup wheel, unless done carefully and efficiently, can make wheel polishing uneconomical. A belt can offer more abrasive than a polishing setup wheel without necessarily providing greater abrasive volume because the abrasive coating

TABLE 16-10
Polishing with Wheels and Belts

Criterion	Wheels	Belts
Adaptability		
Wide flat surfaces	Limited to a few inches per pass.	Available in widths of several feet.
Contours	Yes, by forming faces. No scuffing adjacent surfaces.	Mild contours only without special equipment.
Use in hand tools	Yes.	Requires idler wheel—sometimes awkward.
Automatic operations	Limited by factors of wheel wear and more frequent changes.	Yes.
Inner or closed surfaces	Yes.	Not generally.
Work on flat or side of wheel	Yes.	No, but can be used with platen.
Grit sizes available	Any.	Any.
Densities available	Any.	Softness limited by tension and body of belt. No limit on hardness.
Special equipment required	Setup room with drying facilities, wheel dresser, balancer, etc.	Back stand; idler wheel arrangement for deep work.
Convenience	Wheel changes plus reheading.	High; belt changes only on run of same work.
Uniformity of finish throughout life of abrasive	Good.	Tend to cut finer as the grains dull.
Quality of finish obtainable	Probably little difference when each is used within its proper limits.	

on a belt is normally one grain thick, whereas the abrasive head of a polishing wheel may be many grains thick. The only valid comparison is the total cost per unit of acceptable work produced, and this cost can vary in favor of wheels or belts according to circumstances.

Polishing Wheels

Polishing wheels, also called setup wheels or headed wheels, are usually made of circular plies of cloth or felt. The type of cloth, the number of plies, the number of stitches per linear unit of length, the design of the stitching, and the type of adhesive bond and compression all control the density of the wheel. Leather wheels were once used extensively, but they are now less common, generally being confined to the fine polishing required in cutlery and gun work. Brushes and coated-abrasive flap wheels are also used for polishing. The different materials and methods of construction allow great flexibility and wide variations in hardness to meet various requirements for specific applications.

In general, more rigid polishing wheels are used when there is a requirement for fast stock removal or when the workpieces have flat surfaces. Softer wheels are employed when workpiece surfaces have contours or irregularities and rapid stock removal is not necessary.

Cloth wheels. Cloth wheels are most often made of fabric buff sections glued or sewed together to provide the required wheel width. They range from extra-hard cemented canvas to soft sewed muslin. The cloth sections may be either full discs or pieced sections. Although wheels made up of sections must be balanced, they are usually more economical. Cloth wheels are used for practically all classes of polishing, from heavy snagging with coarse abrasives to oiling-out operations.

Felt wheels. Felt wheels are available in a wide variety of densities and several grades of felt. Because of their even texture throughout, these wheels are easily formed and are suitable for contoured parts. They cover the range of fine polishing with abrasives, ranging from fine to medium, on any material from stainless steel through the nonferrous metals; but, for economy, they are generally used only with finer abrasive sizes. Compressed-wool felt wheels have better balance than cotton or canvas wheels and have uniform density over their entire surface.

Flap wheels. Coated-abrasive flap wheels, which polish with a slapping action, have come into predominant use, especially for off-hand polishing. The wheels, made from a number of coated-abrasive flaps attached to a hub, are available in various designs, such as the one illustrated in Fig. 16-17. A major advantage of flap wheels is their conformance to the shapes of the workpieces (see Fig. 16-18).

Some flap wheels of small diameter have epoxy hubs, but most have metal hubs. They are made in many different diameters, widths, and grain sizes. Various designs have folded abrasive cloths or single-cut sheets glued into segments. Some wheels are precontoured to the shape of the workpiece, and others are slashed (prescored) to soften the brushing action and allow conformance with irregular surfaces. They are also available as combination wheels, with coated-abrasive sheets interleafed with three-dimensional abrasive sheets. This design offers a cushioning effect, absorbs shock, and reduces operator fatigue because less pressure is required. For very fine finishes, wheels are made from nonwoven synthetic material impregnated

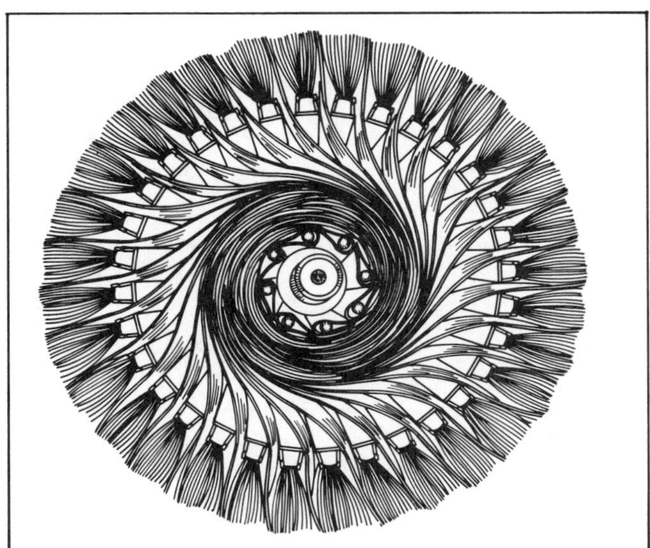

Fig. 16-17 Coated-abrasive flap wheel for polishing. (*Grinding and Polishing Machinery Corp.*)

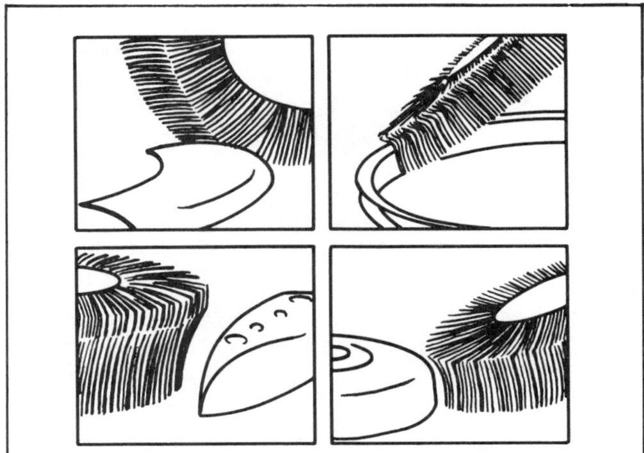

Fig. 16-18 Coated-abrasive flap wheel conforms to the shape of the workpieces.

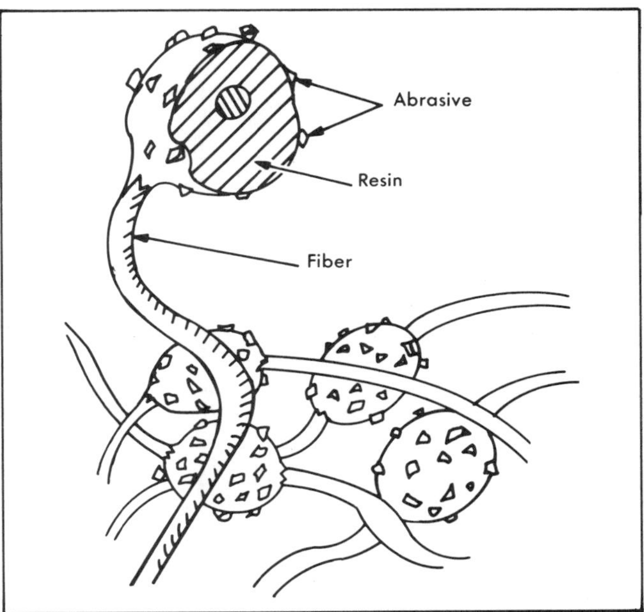

Fig. 16-19 Magnified diagram of three-dimensional abrasive media. (*3M Co.*)

with abrasive grains, used alone or combined with coated-abrasive material.

Polishing Abrasives

Natural abrasives such as emery and corundum are used occasionally in specialized operations that require the finest quality finishes. Turkish emery was once used as the standard natural abrasive, and its fast smoothing action still makes it valuable for fine polishing. Artificial abrasives, however, chiefly aluminum oxide and silicon carbide, will handle practically any polishing problem. Aluminum oxide is easier to bond to the wheel than silicon carbide. Aluminum oxide, which is tough and hard, is used on most carbon, alloy, and high-speed steels; wrought and malleable iron; and nonferrous metals except aluminum. Silicon carbide, which is sharp and hard but brittle, is recommended for finishing low-tensile-strength materials such as brass, copper, cast iron, and aluminum; it is also the preferred abrasive for polishing marble, granite, glass, and other ceramic products. Any abrasive used on stainless steel should be iron-free to prevent contamination and possible rust.

Three-dimensional abrasives. Three-dimensional or low-density abrasive media is made from fiber, resin, and abrasive grains such as aluminum oxide, silicon carbide, flint, or garnet, as illustrated in Fig. 16-19. The three components are oriented to permit controlled cutting of contoured surfaces. Product forms include cleaning and flap brushes, unitized wheels (discs of abrasive web compressed together and adhesively bonded), convolute wheels (abrasive web convolutely wrapped around a fiber core and adhesively bonded between layers), and pads.

Progressive polishing. The general practice is to polish the work with a succession of wheels set up with different grain sizes progressing from a coarse to a fine grit, followed by buffing. Grain progression and the number of grain sizes used depend on the work and are carried only to as fine a point as is necessary to obtain the desired finish. Progression is usually by about 30 grain numbers; for example, 60, 90, 120, 150, 180, 220. How far to carry the polishing process (that is, how fine the final grain size should be) depends on the finish required and whether or not buffing is to follow. If the article is to be plated, a final polishing with 220 grit abrasive will usually suffice on most ferrous metals prior to copper plating, which is followed by buffing. On softer metals, except aluminum, it is often possible to buff satisfactorily from a somewhat coarser finish. On aluminum, it is generally advisable to carry the polishing further, possibly to 320 grit finish, because anodizing, which usually follows buffing, will reveal rather than hide any surface imperfections. In many cases, the savings in buffing time makes an additional polishing step good policy.

The Abrasive Grain Association's (AGA) standard artificial grain sizes are listed in Table 16-11. Tables 16-12 and 16-13, also compiled by the AGA, serve as general guides to the best progression of abrasives.

TABLE 16-11
Abrasive Grain Association's Standard Artificial-Abrasive Grain Sizes

Abrasive	Screened Sizes	Unclassified Flours	Classified Flours
Aluminum oxide	4, 6, 8, 10, 12, 14, 16, 20, 24, 30, 36, 46, 54, 60, 70, 80, 90, 100, 120, 150, 180, 220, 240	F, 2F, 3F, 4F, XF	280, 320, 400, 500, 600
Silicon carbide	8, 10, 12, 14, 16, 20, 24, 30, 36, 46, 60, 70, 80, 90, 100, 120, 150, 180, 220, 240	F, 2F, 3F, 4F, XF	280, 320, 400, 500, 600

TABLE 16-12
Natural Polishing Grain Selection for Various Applications (Naxos Emery and Turkish Emery)

	Polishing Operation				
Parts	First	Second	Third	Fourth	Fifth
Aluminum, sand-cast (inside bottom)	36-46				
Aluminum, sand-cast (outside)	60-80	120-180	Buff		
Aluminum, die-cast	150*	Buff			
Aluminum, sheet	120*	180*	Buff		
Auto bumpers	60-90	120	150-180*	220*	
Auto headlights	180-220*	Buff			
Axes	46-60	70-90	120	150-180*	
Bandsaw steel	60-80	120-150			
Brass, sand-cast	60-80*	150-180*			
Brass, sheet	180-220*	Buff			
Electric irons	80	120*	150*	180-240*	
Glass beveling	70-90	120-150	220	Pumice	Rouge
Granite polishing	60-90	120-150	F	3F	Buff tin oxide
Gray iron, pickled	80	120-150			
Gray iron, not pickled	70	120-150			
Hammerheads	46-60	100-120*			
Knives, table and steel blades	80-90	120-150*	220F	Buff (special machines)	
Knives, table backs	46-60				
Knives, machete, edges	46-60				
face	80	120*			
Lenses					
Prescription	60-80	180-220	Optical flour	Rouge	
Telescope	60-80	180-220	Optical flour	Optical flour	Rouge flour
Locomotive side rods	36	60-70	120		
Monel metal, deep drawn	120	150	180*	Buff	
Monel metal, cast	80	120	150	150*	Buff
Monel metal, full-finish sheet	180	180*	220*	Buff	
Plows	24-36	80	120-150	180-220*	
Plowshares	36-46				
Plow disks	30-46	70-90			
Shears, tinsmith	46	60	120-150	180	
Shovels, blades	30-46	120			
Shovels, straps	36-70		120 (belt)		
Stainless steel					
Mirror finish	60-80	100-120*	150*	220-3F*	Buff
Commercial finish	80	100*	120*	150*	
Wrenches	30-60	80-90	120*		

* Denotes grease or oil wheel.

TABLE 16-13
Artificial Polishing Grain Selection for Various Applications (Aluminum Oxide, Unless Otherwise Specified)

Parts	Polishing Operation					
	First	Second	Third	Fourth	Fifth	Sixth
Aluminum, sand-cast (inside bottom)	36-46					
Aluminum, sand-cast (outside)	60-80	120-180	Buff			
Aluminum, die-cast	150*	Buff				
Aluminum, sheet	120*	180*	Buff			
Auto bumpers	60-90	120	150-180*	---	220*	
Auto headlights	180-220*	Buff				
Automobile fenders and sheet stock for enameling	90	120-150				
Automotive hardware	36-54	90	120	220*		
Axes	46-60	70-90	120	150-180*		
Bandsaw steel	60-80	120-150				
Brass, sand-cast	60-80*	150-180*	220-3F*			
Brass, sheet	180-220*	Buff				
Cutlery	80	120	Pumice	Buff with compound		
Electric irons	60-80	120*	150*	180-240*		
Forks, hay	60-70	100-120				
Forks, spade	24					
Glass beveling	70-90 SiC	120-150	220	Pumice	Rouge	
Granite polishing	60-80 SiC	120-150 SiC	F SiC	3F SiC	Buff tin oxide	
Gray iron, pickled	80	120-150				
Gray iron, not pickled	70	120-150				
Hammerheads	46-60	100-120*				
Hoes, first quality	36-46	70	100-120			
Hoes, second quality	36-46					
Knives, table and steel blades	80-90	120-150*	220-F (special machines)	Buff		
Knives, table backs	46-60					
Knives, machete, edges	46-60					
face	80	120*				
Lenses	60-80	180-220	Optical flour	Rouge		
Lenses, prescription	60-80 SiC	180-220	Optical flour	Rouge		
Lenses, telescope	60-80	180-220	320	Optical flour	Optical flour	Rouge
Locomotive side rods	36	60-70	120			
Monel metal, deep drawn	120	150	180*	Buff		
Monel metal, cast	80	120	150	150*	Buff	
Monel metal, full-finish sheet	180	180*	220*	Buff		
Plows	24-36	80	120-150	180-220*		
Plowshares	36-46					
Plow disks	30-46	70-90				
Shears, tinsmith	46	60	120-150	180		
Shovels, blades	36-46	120				
Shovels, straps	36-60					
			120 (belt)			
Stainless steel						
Mirror finish	60-80	100-120*	150*	220-3F*	Buff	
Commercial finish	80	100*	120*	150*		
Tools, small hand**						
Wrenches	30-46	80	120*			

* Denotes grease or oil wheel. ** Small tool polishing operations are too numerous to enumerate in detail. Apply to any abrasive grain manufacturer for recommendations.

Bonding Agents

In bonding setup polishing wheels, the choice of bonding agents is between hot animal-hide glue, one of the cold cements based on sodium silicate, or a plastics resin. Any of these agents can provide excellent bonding results, although hot glue is definitely preferred for use with finer grains from number 180 up. Proprietary manufactured adhesives and cement compounds have largely replaced hide glues because wheels set up with cement are practically unaffected by the heat of polishing, whereas glue bonds will tend to soften, cushioning the grain and reducing its cutting power. Cold silicate-based cements are being used increasingly. Most silicate cements are proprietary mixtures that can be obtained from the manufacturers in grades to suit any particular job. For the occasional or limited user, cold cement is preferable because of its greater convenience. Plastic resin-based cements now compete with glue for fine, flexible polishing applications.

Glue-bonded wheels—grain size and all other factors being equal—will produce a finer finish than cemented wheels, and subsequent buffing operations will be easier. The correct procedure for the setting up and handling of polishing wheels is of great importance. The use of hot glue requires utmost cleanliness in the glue room; bacteria, particularly in hot weather, can ruin a day's batch of wheels overnight.

Heading of Setup Wheels

New wheels require a thinned sizing coating of the adhesive to be used for bonding. After drying, if the wheel is to be headed with fine grit, it should be run on a shaft, and the face should be sanded lightly with medium-grade sandpaper. A coating of full-strength bonding material is then brushed on, and the wheel is immediately rolled in a trough containing abrasive grain—two or three times around the circumference is enough. Heavy pressure and pounding of the wheel into the grain are bad practices; they tend to open the face of soft wheels, driving the grain in between the sections. Machines give the most satisfactory results for heading. Glue wheels should be allowed to dry until the first coat loses its sheen before the second coat is applied; usually the first coat is dry within one to three hours. Cold cement wheels may be given a second coat within one-half hour or less.

Old wheels should have the worn head smoothed down on a wheel-dressing machine with an abrasive brick or diamond wheel-dressing tool. If the wheel is to be reheaded with the same grit, the previous head should not be completely removed unless the wheel face shows wear down through the sizing coat to the fabric. If the fabric does show through, the entire previous head must be removed, and any grease must be removed with lump pumice and, if necessary, a solvent. The wheel is then resized and treated as a new wheel.

Drying should not be hurried. Hot glue heads should be given 24 hours at about 80° F (27° C) to reach full strength. Cold cement heads will dry to satisfactory strength in 24 hours at room temperature; drying can be speeded up by the careful use of infrared lamps, or other artificial heat sources, combined with good air circulation. The relative humidity should be 40-45% to properly dry a wheel. A green wheel (not fully dried) will glaze. An overdried wheel will have a brittle coating, causing shedding of the grains.

After the wheel is thoroughly dried, the head should be "cracked" in order to restore the original density of the wheel. Cracking, particularly important in the case of soft wheels, is accomplished by striking the wheel with an iron rod or pipe at an angle of about 45° to the axis of the arbor hole while the wheel is being supported. After the entire circumference has been cracked, the operation should be repeated from the other side of the wheel. Cracking causes the head to separate into islands of abrasive, each resiliently supported by the fabric beneath and of a size depending on the density of the wheel.

The wheel should then be balanced. Any tendency of the wheel to turn while it is supported by a rod through its arbor suggests an out-of-balance condition. Static balance may be restored by screwing flattened lead weights to the side of the wheel as required. A better, much safer practice, however, is to procure wheels with built-in balancing tubes spaced at intervals; these permit easy balancing by the insertion of short lengths of lead wire into the tubes, which may then be crimped into the tubes without danger of flying out.

Wheel Speed

For most polishing operations with glue-bonded wheels, speeds will be in the range of 6000 to 8000 sfm (1830 to 2440 m/min), with the higher speeds favored for the stainless and tougher steels. The relation of speed to wheel density is important. It is useless to specify a supersoft wheel for contour finishing and then run it so fast that it cannot conform to the work. High speeds promote burning of the work, glazing, and short head life. Specifying the slowest wheel speed that will adequately do the job will ensure improved results and longer wheel life. Wheels set up with cold cement can be run at speeds up to 9000 sfm (2745 m/min). The heat tolerance of the metal being polished is also a factor in determining the proper polishing speed.

Lubrication

Proper lubrication makes a wheel run cooler, decreases its tendency to grab and hold particles of metal removed from the work, reduces the cut, and therefore makes for a smoother, finer finish. Lubrication is used principally with the finer grits when polishing is to be followed by buffing. Grease sticks are made of beef tallow and stearic acid, with various waxes and hardeners added to regulate the melting point. It is important to choose a grease stick of the right melting point for each job. The grease should melt, flow onto the running wheel, and give good coverage without thinning to the point at which it is thrown off. A 1:1 mixture of SAE 20 oil and kerosene is often applied on wheels of 150 grit and finer to keep them cutting freely. The wheels are cleaned of chips and grease with lump pumice held lightly against them as they run. Finishing wheels of 220 grit and finer are frequently treated with emery paste or emery cake. Sprayed liquid lubricants are used with automatic polishing equipment.

Flexible Polishing with Paste Wheels

For many fine polishing applications, it is better to have a complete mixture of the abrasive grit and the adhesive rather than a wheel face on which the abrasive is exposed. When the abrasive is thoroughly mixed with melted glue before it is applied to the wheel, the result is called a *paste wheel*. Silicate-based adhesives have been used with abrasives for paste wheels. Greaseless compounds applied to the wheels as they rotate are also employed.

In the latter type of paste wheel, the greaseless compound stick is applied to the wheel after power has been shut off and the wheel begins to decelerate, thus keeping the compound from being slung off the wheel. This technique allows the preparation of polishing wheels in less than 10 minutes, without taking the wheels from the machines. An adhesive sizing coat will increase the adhesion, the cutting properties, and the life of a wheel used with a greaseless compound. Adhesive sizing is available in stick form and is applied the same as greaseless compound.

Flexible polishing wheels have the following advantages over conventional cracked wheels: (1) wheels can be prepared rapidly; (2) wheel inventory can be reduced; (3) softer cloth can be used to allow polishing of irregular contours; and (4) part cleaning procedures can often be simplified. Flexible wheels operate best at speeds from 5000 to 6000 sfm (1524 to 1830 m/min).

Machines and Equipment Used

A wide variety of machines and equipment is used for polishing. Similar machines and equipment are used for brushing (discussed previously in this section) and buffing (discussed next), but with different media and tools. The machines and equipment used include portable air and electric tools, bench-mounted air and electric motors, bench and pedestal grinders, flexible-shaft tools, polishing and buffing lathes, abrasive-belt machines, and many different semiautomatic and automatic machines. All such tools and machines have a rotating shaft (spindle) to which the polishing or buffing wheel is attached.

Selection. The selection of a specific machine or piece of equipment for a particular application depends upon several factors, including the size, weight, and shape of the workpieces; production requirements; the surface and appearance specifications for the workpieces; the degree of versatility needed; floor space available; and costs.

In off-hand (manual) polishing or buffing, either the workpiece or the tool is held, forced into contact, guided, and moved by hand. This process is used extensively, especially for limited production requirements or when a wide variety of parts must be processed. To provide maximum operator safety, it is necessary that machines be properly guarded and that operating instructions from the machine manufacturer be followed carefully. Misuse of the equipment and/or operator carelessness can result in serious injuries.

Semiautomatic and automatic machines are generally limited to high-production requirements. When the design of the workpieces permits the use of such machines, advantages generally include improved and more consistent finishes, higher quality products with less rework required, and cost savings resulting from reduced labor needs.

Off-hand operations. Powered hand tools are used extensively for workpieces and assemblies that are too bulky or heavy to be mounted on a machine. They are also employed for intricate blending and smoothening operations that are not practical on machines.

Polishing and buffing lathes, sometimes referred to as hand lathes or polishing jacks, are commonly used for off-hand operations. These machines (see Fig. 16-20) have more of a forward overhang arrangement of the wheel spindle than floor-stand grinders. Also, their spindles extend farther horizontally to provide ample clearance for manipulation of the workpieces. A wide selection of machines is available with single or double-end spindles, having arrangements for speed changes by pulleys or by variable-speed drives. A back stand can be added for abrasive belt operation, as illustrated in Fig. 16-20.

Semiautomatic and automatic machines. These machines are available in many different designs (see Fig. 16-21). Some

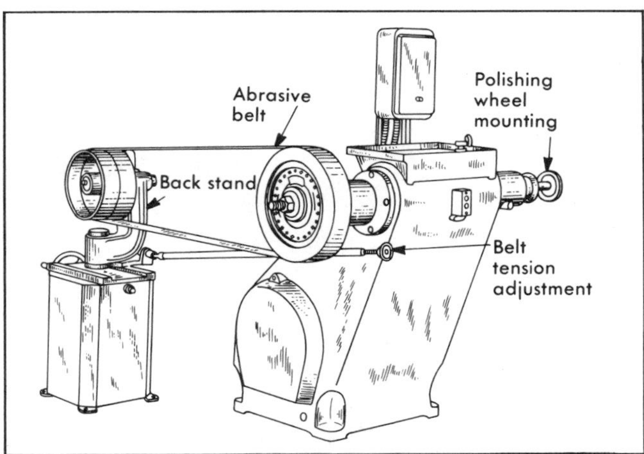

Fig. 16-20 Floor-type machine for off-hand polishing and buffing, with abrasive belt mounted on one end of spindle.

are single-station machines, generally for small production requirements. For higher production, machines have multiple operating stations and are built for specific applications. Work transfer units for multistation machines are generally of two types:

1. Rotary table, either indexing or continuous.
2. In-line conveyors, either straight-line (usually over-and-under conveyor design) or rectilinear.

Camming and positioning mechanisms are provided to control the location, orientation, and, in some cases, the motions of fixtured workpieces in relation to the polishing or buffing heads. The heads can incorporate traversing devices, retractors and/or lifters, wheel-wear compensators, spindle oscillators, variable-speed drives, and, in the case of buffing operations, automatic applicators for liquid or solid-bar compounds. Enclosures, collectors, and duct work are required for ventilation control, as specified in ANSI Standard Z43.1.

Abrasive-belt machines. Many abrasive-belt machines, discussed in Volume I, *Machining*, of this Handbook series, can be arranged for manual, semiautomatic, or fully automatic polishing operations. These machines include back stand, rotary, free-belt roll, swing-frame, platen, conveyorized, and centerless grinders.

Machines for long parts. Reciprocating table machines (see Fig. 16-21) are being used to simultaneously polish and buff long parts, as well as a number of small parts. Workpieces are mounted on the table and reciprocated under one or more heads. Pinch-roll machines use pinch or drive rolls to convey long workpieces under one or more heads. Special machines have been built with traveling heads or with multiple heads on a traveling carriage to polish or buff long parts.

BUFFING

Buffing is a surface finishing process that can produce smooth, lustrous finishes, with minimal pronounced line effects. The process is accomplished with rotating wheels, called buffs, to which abrasives and compounds are applied. The buffing wheels by themselves do little work; they are designed to be carriers for the abrasives and compounds. However, the density and construction of the buff, type of abrasive and compound, pressure exerted, and rotary speed of the buff all affect the aggressiveness (ability to do work) of the process.

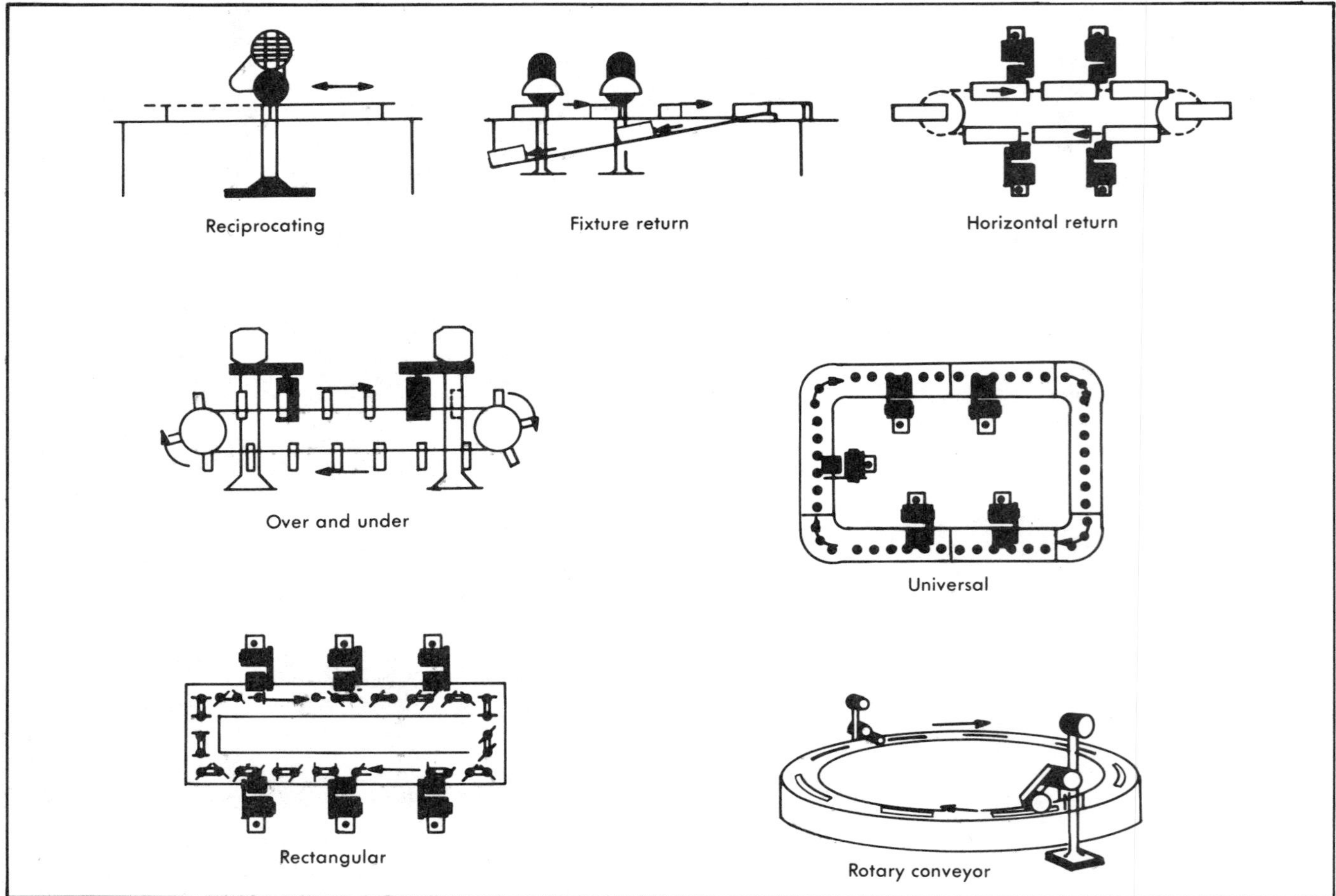

Fig. 16-21 Various types of semiautomatic and automatic machines for brushing, polishing, and buffing. (*Acme Manufacturing Co.*)

Buffing may be considered a two-stage operation: (1) cutting down and (2) coloring. Cutting down is done with the workpiece moving against the motion of the buff using medium-to-hard pressure. It is performed to refine a surface by removing scratch lines from polishing, stretch marks from forming, die marks, or other surface imperfections. It can make a relatively smooth surface smoother. Buffing can also deburr, blend edges, generate radii, and stress relieve. Coloring, with the workpiece traveling with the motion of the buff using medium-to-light pressure, refines the cut-down surface and brings out maximum luster or a mirror finish.

Product Design

Substantial reductions in finishing costs can frequently be realized through slight changes in product design. These savings are true particularly for parts produced in volumes sufficient to justify automatic polishing and buffing. In general, the following features may result in slower and more costly finishing: sharp corners, both inside or outside; deep recesses or grooves; compound curves; projections; and attachments, such as handles.

Buff Materials

Fabrics used in buffing wheels are designated by thread count and number of linear yards, with various widths available. The most often used fabrics are cotton muslin and cotton-polyester blends, designated 60/60 and 86/82, the numbers indicating the threads per inch in the warp and fill respectively. Weights are approximately 5 3/4 oz/yd^2 (3.12 kg/m^2) for 86/82 fabric, and 4 1/2 oz/yd^2 (2.44 kg/m^2) for 60/60 cloth. Fabrics are frequently hardened and stiffened by mill treatments to promote faster cutting and better compound retention. Most buff manufacturers now dip treat buffs after they are made, resulting in still more aggressive action.

Flannels, both single napped (canton) and double napped (domet), are used for coloring precious and nonferrous metals. Sisal, a natural fiber, makes a fast-cutting buff widely used on steel and stainless steel. Kraft paper, either used alone or alternated with cloth plies, has limited use, principally on regular surfaces to increase cutting power.

Although 100% synthetic fibers are not generally well adapted to buffing operations, nonwoven nylon, polyesters, and other synthetics may be used. Wheels made of synthetic materials hold abrasives evenly; and because of their water resistance, they may be used with wax or grease lubricants. They are usually operated at slow speeds of about 2500 sfm (760 m/min). Synthetic fibers are often used for buffing parts with deeply irregular contours or sharp edges because they tend to wear longer than 100% cotton buffs. However, when used at high speeds or under excessive pressure, synthetic fibers will streak the surface finish due to heat buildup.

Buff Construction

Buff construction controls the action of the buff by making it harder or softer, or more or less aggressive. The construction itself may be of more importance in determining final results on a particular job than the thread count.

Full-disc buffs. These are made of discs of sheeting cut full size and assembled into a number of plies (commonly 20) to make a buff section about 1/4" (6 mm) thick. Sections are assembled on the spindle to provide the required face width. Full-disc buffs may be sewed around the arbor hole only, spirally sewed over the entire surface, sewed in concentric circles, square sewed, radially sewed in a variety of patterns, or folded into wedge-shaped sectors and assembled radially before sewing (see Fig. 16-22).

Packed buffs. These are full-disc buffs, usually sewed only around the arbor hole, with small-diameter discs of cloth or cardboard inserted between the larger diameter plies; the result is a soft, open face that does not close when rotating. This construction is useful for reducing frictional drag and cutting on soft, easily deformed metals such as aluminum, and for obtaining extreme softness in wheels used with greaseless compounds.

Pieced buffs. These are made of remnants and strips of cloth assembled and matched to a more or less uniform thickness. Such buffs require one of the types of full-disc sewing in order to stay together in use. Their chief virtue is their low cost. They are used for the relatively rougher buffing operations and for assembly into polishing wheels.

Bias buffs. These are used when a very fast cutting and coloring buff is required. Strips of fabric (86/82 most commonly) are cut on the bias to minimize raveling and are assembled around a hub or core. The chief advantages of bias buffs are increased face width for a given amount of cloth, cool running, and resistance to burning. They are adaptable to operations from the heaviest cutting to coloring by varying the number of plies and amount of pucker in their faces. The puckered face tends to break up lines that may be left in the work. These buffs combine flexibility and cutting power, and are most generally used for automatic applications. They may be ventilated or nonventilated.

Ventilated buffs. This is a type of bias buff constructed around a perforated metal hub that draws in air near the arbor and blows it centrifugally outward through the layers of the buff. This air ventilation not only cools the buff and workpiece surfaces, but it also causes the buff to form pockets that hold buffing compound and increase buffing efficiency. Ventilated bias buffs are now the accepted standard for industrial use, both for off-hand and automatic machine applications.

Finger or spoke buffs. These are made of folds of cloth assembled radially around a hub. The folds may be left unsewed and soft, or may be sewed lengthwise to give greater aggressiveness. They combine great cutting power with the ability to flex and cover contours. Spoke buffs are made of buffing materials ranging from 60/60 (soft) to sisal for maximum cutting action.

Goblet buffs. These are designed to buff inside surfaces of hollowware or areas of other articles that cannot be reached by the usual methods. They are so constructed that the individual threads of the muslin fabric run toward the periphery and also toward the bottom or end; thus, buffing can be done on two surfaces at once. These buffs are adaptable to heading with glue and grain, with greaseless compound, or with regular grease-based compositions. Practical limitations in size are 2-6" (51-152 mm) diam and depths to 4" (102 mm). Goblet buffs are operated on tapered threaded spindles or screw points.

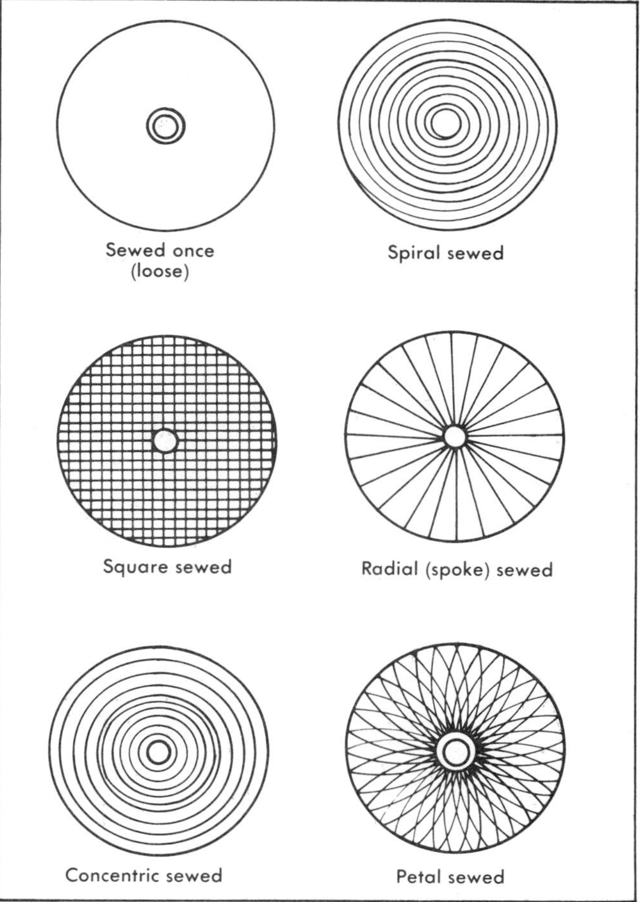

Fig. 16-22 Various types of sewing used for full-disc buffs. (*Divine Brothers Co.*)

Abrasive buffs. High-density, abrasive-impregnated bias buffs are treated with abrasive and binder. These buffs cut fast for increased metal removal and require less buffing compound. They also require less pressure. Buff wear is reduced because of the abrasive action at the contact area between the buff wheel and the workpiece. Performance and finish produced is uniform throughout the buff life.

Buffing Abrasives

Tripoli. This abrasive is often considered to be an amorphous silica that occurs in nature. Too soft and fragile for use on the ferrous metals, it is highly suitable for buffing aluminum, copper, brass, and die castings. Soft or amorphous silica also occurs in the natural state, but, unlike tripoli, is iron-free and white in color. It may be used alone for coloring nonferrous metals, or it may be mixed with tripoli to give added coloring ability to so-called cut-and-color compounds.

Recent evaluations of tripoli have shown the product to contain approximately 75% crystalline silica. Based on these findings, it should not be classified as an amorphous silica. It has also been determined that the so-called "white tripoli" contains about the same percentage of crystalline silica. Unlike amorphous silica, crystalline silica may cause delayed lung injury for operators over a period of time. Users of products containing these abrasives should be aware of this possibility.

Aluminum oxide and other powders. Aluminum oxide powders, fused and unfused, are the abrasives in most compounds for buffing hard metals. Chromium oxide is used for highest color buffing of stainless steel, chromium, and nickel plate. Rouge is the preferred compound for color buffing on copper, brass, gold, and silver.

Pumice. This is a natural product used with tampico brushes for scratch finishes, in compounds for buffing plastics, and with water for wet-ashing operations on plastics and glass that cannot tolerate the heat generated by normal buffing.

Buffing Compounds

Buffing compounds are supplied in either solid bar or liquid form. Both forms use the same abrasive agents, only the vehicles differ.

Solid bars. Bar compounds get their shape from the binder, consisting usually of a blend of tallow, stearic acid, waxes, or petroleum greases. These binders vary widely in their effect on the buffing operation and in their ease of cleaning from the work prior to plating. Greaseless compounds are also used, mainly for satin finishing. These compounds combine fast-cutting abrasives with glue binders, and they are generally used at speeds of 5000 to 6000 sfm (1525 to 1830 m/min).

Liquid compounds. Spray-type buffing compounds use the same abrasives as the bar compounds, but they are used in oils or water emulsions that remain fluid at ordinary temperatures. These mixtures are sprayed onto the buffing wheel under air pressure. Liquid compounds are more easily cleaned from workpieces than solid compounds and are highly suitable for machine or automatic buffing. Automatic spray guns are often a part of buffing equipment.

With recently developed high-pressure, airless spray systems having become commonplace, gravity application of liquid compounds has become obsolete. Advantages of airless spray include virtual elimination of overspray and dusting, reduced waste of compound and compressed air, and better compound penetration into the buffs. Spray guns that confine the high pressures to the guns are available, thus eliminating the need for high-pressure feed lines and associated equipment.

One automatic spray gun will serve a buffing wheel up to 5" (127 mm) wide. Spray guns are mounted near the wheels, but their distance from and angle to the wheels must vary with the rotating blanket of air surrounding each wheel. The air blanket varies with different buffs and with their speed, and it must be considered when the proper spray angles and pressures are being determined. The compressed-air supply is controlled by automatic valves.

Principal advantages of liquid spray buffing compounds include the following:

1. The right amount of compound is always supplied to the wheel.
2. Excessive wheel wear from contact with a solid abrasive-compound bar is eliminated.
3. The cleaning of buffing dirt from parts buffed with liquid compounds is usually simpler and quicker.
4. Production rates are increased because solid-bar compound changing is eliminated.
5. All the compound purchased can be used; there are no bar ends to be discarded or reclaimed.
6. The possible hazards of buffs catching fire from friction is minimized if the liquid compound is noncombustible.

Buff Selection

Of the factors that must be considered in buff selection, the shape (contour) of the workpiece may be as important as the material itself. Since it is impossible to combine maximum cutting power and flexibility in one buff, compromises are often required. Table 16-14 indicates the types of buffs best suited for various workpiece materials, especially for cutting-down operations. Table 16-15 presents speed conversions from rpm to sfm for various brush diameters. In some cases, because of a heavy cutting buff's inability to cover the surface efficiently, buffs of more flexible construction and of less cutting power will be required.

Wide-Face (Mush) Buffing

The use of wide-faced buffing heads, called mush buffing, is a recent development that has proven to be very effective in finishing a variety of parts, ranging from small die castings to large components such as steel and aluminum bumpers. Such wide heads cover larger areas than possible in contact buffing, providing more contact time between the buff and workpiece(s) and excellent coverage, while reducing the number of heads necessary. Buff life is generally long, and compound use low.

The buffs are generally spaced apart on shafts. Shafts to 12 ft (3.7 m) long are being used, with buffs mounted along the shafts, normally 1/2 to 3" (12.7 to 76 mm) apart. Workpieces travel across the faces of the buffs, approximately parallel to the axis of their rotation; and the buffs are generally rotated at slower speeds than for contact buffing. The buffs act as flexible tools, following the contours of the workpieces being buffed.

Buffing of Plastics

Thermosetting plastics. These materials are readily buffed with low-grease compounds on muslin buffs, operating from 2000 to 5000 sfm (610 to 1525 m/min) for coloring, and between 3000 and 6000 sfm (915 and 1830 m/min) for cutting down. Headed polishing wheels or belts in suitable grit sizes may be used for the cutting of molding gates or flash lines. The buff should be full disc, either loose or sewed, depending on the shape of the part. Tripoli is suitable for many jobs, but unfused aluminum oxide powders are recommended for fine luster.

Thermoplastics. These materials are more difficult to buff than thermosetting plastics because of their tendency to melt and flow under the heat developed by buffing. The cutting of molding gates or flash lines must be done carefully, either by a wet belt to prevent overheating, or by a soft buff and fine white silica or tripoli in grease binders. Further refinement and blending to the surrounding color may require careful experimentation to discover the right combination of buff and compound. Here the experience of the compound manufacturer may be invaluable and should be sought. The generalization that speeds should be low and the pressure light is about the only valid one for buffing thermoplastics. Buffing speeds should ordinarily be held between 1000 and 4000 sfm (305 and 1220 m/min).

Plastic laminates. As far as the surface is concerned, these may be handled in accordance with the suggestions previously given, depending upon whether the binder is thermoplastic or thermosetting. Cut edges, however, present peculiar problems. In many cases it is not possible to go further than the finish obtainable with a soft buff and one of the finer grades of greaseless compound, especially if the laminate contains long, hard fibers such as fiberglass.

Buffing Machines and Equipment

The machines and equipment previously described in this chapter for polishing are also used for buffing. Off-hand buffing, especially of small workpieces, is frequently done on polishing and buffing lathes having double-end spindles. One wheel on each spindle is used for cutting-down operations and the second wheel is used for coloring. Air-powered handheld tools are also used extensively for off-hand buffing operations, especially for complex contours.

TABLE 16-14
Suggested Buff Materials and Speeds for Buffing Various Metals

	Cutdown			Coloring		
Material	Compound	Buff	Speed, sfm (m/min)	Compound	Buff	Speed, sfm (m/min)
Aluminum	Tripoli bar	Bias sewed, loose ventilated, or sisal	6000-8000 (1830-2440)	Rouge, silica, or unfused aluminum oxide bar or liquid	Loose or low-density ventilated, bias, or sisal or liquid	7000-9500 (2135-2895)
Copper and alloys	Tripoli bar or liquid	Loose sewed or ventilated	5000-9000 (1525-2745)	Rouge, silica, or unfused aluminum oxide bar or liquid	Loose or low-density ventilated	5000-9000 (1525-2745)
Chromium and plate	Combined fused and unfused aluminum oxide bar (for burned areas)	Loose or ventilated	6500-8000 (1980-2440)	Chrome green oxide or unfused aluminum oxide bar or liquid	Loose or low density	Metal: 5000-6500 (1525-1980) Plate: 7000-9000 (2135-2745)
Nickel and alloys	Tripoli bar or liquid	Bias, loose sewed, or ventilated	5000-8000 (1525-2440)	Chrome green oxide or unfused aluminum oxide bar or liquid	Loose or ventilated	5000-9000 (1525-2745)
Plate				Lime bar, chrome green oxide, or unfused aluminum oxide bar or liquid	Loose or low-density ventilated	6500-7500 (1980-2285)
Steel and stainless steel	Aluminum oxide bar or liquid	Ventilated, bias sewed sisal or tampico	8000-10,000 (2440-3050)	Chrome green oxide and/or unfused aluminum oxide bar or liquid	Loose or low-density ventilated	3000-6000 (915-1830)
Zinc	Tripoli bar or liquid	Bias, loose ventilated, sewed, or sisal	7500-9000 (2285-2745)	Silica or unfused aluminum oxide bar or liquid	Loose or low density	6000-8000 (1830-2440)
Magnesium*	Tripoli or aluminum oxide bar or liquid		4000-8000 (1220-2440)	Lime bar	Loose or ventilated	

* Precautions should be taken to control the collection of flammable magnesium dust.

TABLE 16-15
Relationship of Surface Speed to Rotational Speed for Various Buff Diameters

Rotational Speed, rpm	Buffing Wheel Diameter, in. (mm)										
	4 (102)	6 (152)	8 (203)	10 (254)	12 (305)	14 (356)	16 (406)	18 (457)	20 (508)	22 (559)	24 (610)
	Surface Speed, sfm (m/min)										
800	837 (255)	1256 (383)	1675 (511)	2094 (638)	2513 (766)	2932 (894)	3351 (1021)	3770 (1149)	4189 (1277)	4608 (1405)	5026 (1532)
900	942 (287)	1413 (431)	1885 (575)	2356 (718)	2827 (862)	3298 (1005)	3770 (1149)	4241 (1293)	4712 (1436)	5184 (1580)	5655 (1724)
1000	1047 (319)	1570 (479)	2094 (638)	2618 (798)	3141 (957)	3665 (1117)	4189 (1277)	4712 (1436)	5236 (1596)	5760 (1756)	6283 (1915)
1200	1256 (383)	1884 (574)	2513 (766)	3142 (958)	3769 (1149)	4398 (1341)	5027 (1532)	5655 (1724)	6283 (1915)	6912 (2107)	7540 (2298)
1400	1466 (447)	2199 (670)	2932 (894)	3666 (1117)	4398 (1341)	5131 (1564)	5865 (1788)	6597 (2011)	7330 (2234)	8064 (2458)	8796 (2681)
1600	1675 (511)	2513 (766)	3351 (1021)	4189 (1277)	5026 (1532)	5864 (1787)	6703 (2043)	7540 (2298)	8378 (2554)	9216 (2809)	10,053 (3064)
1800	1885 (575)	2827 (862)	3770 (1149)	4713 (1437)	5654 (1723)	6597 (2011)	7540 (2298)	8482 (2585)	9425 (2873)	10,368 (3160)	
2000	2094 (638)	3141 (957)	4189 (1277)	5236 (1596)	6283 (1915)	7330 (2234)	8378 (2554)	9425 (2873)	10,472 (3192)		
2200	2304 (702)	3455 (1053)	4608 (1405)	5760 (1756)	6911 (2106)	8063 (2458)	9215 (2809)	10,367 (3160)			
2400	2513 (766)	3770 (1149)	5027 (1532)	6284 (1915)	7540 (2298)	8796 (2681)	10,053 (3064)				
2600	2722 (830)	4084 (1245)	5445 (1660)	6807 (2075)	8168 (2490)	9529 (2904)					
2800	2932 (894)	4398 (1341)	5864 (1787)	7331 (2234)	8796 (2681)	10,262 (3128)					

ROLLER AND BALL FINISHING/BURNISHING

Finishing or burnishing with rollers or balls is a surface finishing technique based on the cold forming of metal. The high forces generated in the contact zone between the tools and the workpieces results in plastic deformation of the work surfaces. Although the degree of metal flow in these finishing processes is low compared to more severe cold forming operations, such as extruding, heading, and swaging, they also result in the characteristic cold forming advantages of increased strength, improved size accuracy, and low manufacturing costs.

Roller finishing and roller burnishing are similar processes, and the terms are often used interchangeably. For the purposes of discussion in this section, however, roller finishing, including deep rolling, is considered a heavy-duty operation. Roller burnishing, discussed later in this section, employs multiroll tools for lighter duty applications. Ball burnishing, also called ballizing, produces dimensionally precise, smooth holes by forcing balls through the holes.

ROLLER FINISHING

Roller finishing, together with size rolling and deep rolling, is a process employed for cylindrical components. It has been used for many years by railroad car manufacturers for finishing sleeve-bearing seats on railcar axles.

Advantages and Limitations

Besides providing the desired surface finish and improvements in geometrical accuracy, roller finishing also has the following advantages:

- Improved surface finish and roundness will result in noise level reductions for shafts, such as motor armatures, that run in sintered-metal sleeve bearings with only microscopic, noncushioning oil films.
- Direct cost savings may be realized by the elimination of some and the replacement of other finishing methods,

such as grinding, polishing, lapping, and honing.

- Direct cost savings may also result from the possible elimination of surface heat treating operations such as induction hardening.
- Other cost savings may be achieved by reductions in tool costs, possible changes to lower grade materials, the elimination of grinding wheels, and the relaxation of quality control.

The major limitation of roller finishing is the initial hardness of the workpiece; however, roller finishing can still be performed on materials with hardnesses of R_C 40 to 45, which are slightly above the general limits for machinability. Some processing bypasses the hardness specification. For example, particular parts may be induction hardened after roller finishing and then passed through a chemical wash to remove the resultant oxides before chrome plating. Many applications, such as for shock absorber rods, permit the roller finishing of materials at hardnesses of approximately R_C 40, which will still provide hard enough surfaces to avoid scratching during handling or assembly.

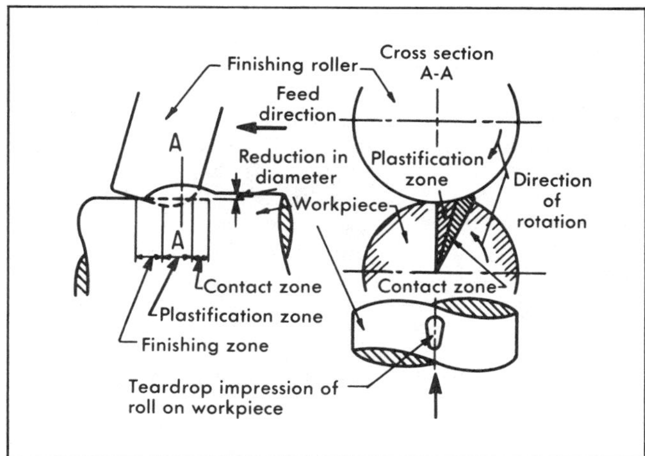

Fig. 16-23 Contact area zones in through-feed roller finishing.

Process Principles

Roller finishing theory defines three zones in the contact area between the roller and the workpiece. On the leading edge of the roller is the *contact zone* or *lead zone*, which is followed by the *plastification zone*. On the trailing edge of the roller is the *finishing zone*. These zones are illustrated in Fig. 16-23.

Roller finishing may be accomplished by either through-feed or plunge operations. For through-feed operations, the work roller axis is slightly inclined relative to the workpiece axis so that a teardrop-shaped contact area is generated. The length of the teardrop is determined mainly by hardness, the roughness or waviness of the surface, and the desired feed rate. Teardrop area lengths most commonly vary from 1/4 to 1 1/8" (6.4 to 28.6 mm).

For plunge rolling operations, as illustrated in Fig. 16-24, the work roller axis must be parallel to the workpiece axis in order to generate a line of contact. Line contact eliminates the finishing zone, and this, in turn, could require finer finish preparation than that for through-feed rolling. The setting of the line contact over the roller or part length should also compensate for machine deflection that results under the required rolling forces.

Plastic flow. As the work roller is forced against surface peaks, steadily increasing compressive stress is generated in the plastification zone. At a certain stress level, plastic flow of the material in the peaks will occur in the direction of least resistance—toward the surface valleys. This flow will then lift the level of the valleys up to the level of the work roller. This effect is illustrated in Fig. 16-25.

The magnitude of the plastic flow depends upon the rolling forces used. In some cases, it is desirable to select a rolling force that does not permit the surface valleys to close up entirely; such a surface may be desirable for its lubrication-holding capability combined with a high bearing-contact area. The volume of the workpiece is not reduced, but its diameter is reduced as a function of its surface roughness before and after rolling.

Rolling force. The rolling force required to create the plastification zone on the workpiece depends upon (1) the tensile strength of the workpiece material, (2) the length of the contact area form, and (3) the size of the workpiece and work roller diameters. Under known conditions of tool diameter and contact area length, the required rolling forces can be depicted

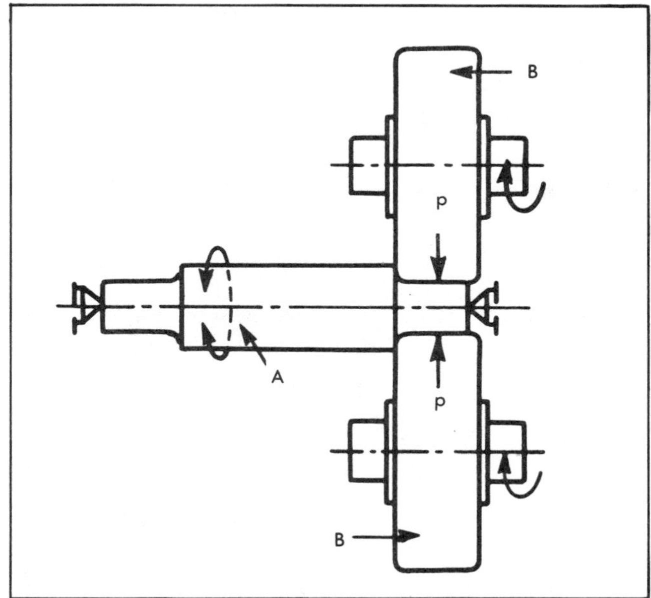

Fig. 16-24 Typical plunge-type roller finishing operation: center-held workpiece (A); plunge work rollers (B); direction of roller pressure (p).

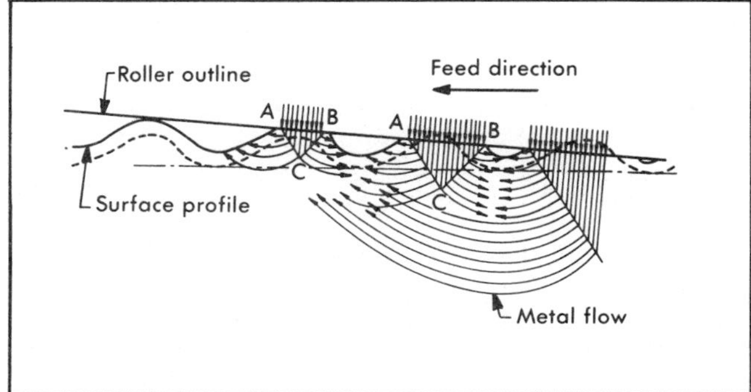

Fig. 16-25 Roller finishing plastification-zone cross section, showing the directions of compressive stress applied by the work roller and the plastic flow of metal from surface peaks to valleys.

ROLLER FINISHING

graphically in relation to workpiece diameter and tensile strength (see Fig. 16-26) or the carbon content of steels.

Effects of Roller Finishing

Surface hardness. The characteristic of steel to work harden under plastic deformation explains the fact that roller-finished components show an increase in surface hardness. Because of stress distribution, the greatest hardness increases actually occur below the surface. Conventional hardness tests cannot measure these increases because the testing tools break through the hardness scale under normal preloads. However, microhardness techniques, such as the Knoop test with 50 or 100 gram preloads, indicate a hardness increase of 2 to 8 points on the Rockwell C scale, with the lesser increases on materials with higher hardnesses before rolling.

Surface finishing. Roller finishing results primarily in a desired surface finish. When properly employed, the process will produce surface finishes between 1 and 5 μin. (0.025 and 0.127 μm) from a turned surface of 125 to 150 μin. (3.2 to 3.8 μm); thus, the best finishes for improved wear resistance can easily be obtained. The degree of finish depends upon the application. For example, the bearing diameter for a seal should not be roller finished below a surface value of approximately 15 μin. (0.38 μm); on the other hand, the stem of an engine valve could be roller finished to a surface value of 8 μin. (0.20 μm), provided the accompanying valve guide is manufactured with a surface roughness capable of retaining lubricating oil.

Diameter reduction. Roller finishing reduces workpiece diameter in proportion to the surface finishes before and after rolling. The diameter reduction will average approximately four times the difference in rms values before and after rolling, measured in μin. If the surface roughness before rolling and the rolling force are kept within a permissible limit of variation, the diameter reduction will remain constant.

The most accurate way to determine the diameter reduction is to make rolling tests after the machine is set up or the finishing has been done. However, workpieces that require close dimensional accuracy must be machined to a size that takes into consideration the change in diameter; and even though a finish-grinding operation to obtain the required dimensional accuracy may be necessary, roller finishing still offers considerable advantages. Another means of obtaining high dimensional accuracy is to premachine the workpiece to approximately 150 to 200 μin. (3.8 to 5.1 μm) and then roller finish it with a multiple-roller attachment. In this case, however, dimensional deviations resulting from previous operations will lead to a varying surface finish and bearing-contact area.

Product quality. The surface quality of the part is determined not only by the surface finish prior to rolling and the hardness of the material, but also by the form accuracy. Excessive roller finishing forces could result in undesirable geometry changes such as elongation, end tapering, and metal flow into grooves or small crossholes.

The plastic deformation that takes place during roller finishing creates compressive stress layers in the surface of the workpiece. These stresses are symmetrical to the workpiece axis and should not result in macrogeometric changes. Experience, however, indicates that long components can bend during roller finishing. Bending, at first interpreted as a deficiency in the process, has been shown to occur only if the workpiece has an unequal stress pattern caused by such operations as machining, cold forming, or heat treating before the roller finishing operation. The relief of these unbalanced stresses by finishing results in the bending.

Similarly, a certain type of surface flaking in a line pattern parallel to the part axis has been falsely attributed to the rolling process. Actually, such flaking is the result of minute surface cracks created during previous operations. Thus, roller finishing reveals manufacturing faults that, in the past, were probably never properly recognized.

Machines and Tooling Used

The relationship between the mass of the workpiece and the mass of the tooling will generally determine whether the rotational motion is delivered through the workpiece or the work rollers. If the moment of inertia of the workpiece is high in comparison with that of the tooling, the part is held between centers, and rotational motion is transferred from the work to the tools. This arrangement is generally used when large parts are being finished with a two-roll attachment. The opposed work roller arrangement eliminates bending or deflection, and it can be applied to either plunge or through-feed operations. Feed motion must be generated by the machine, however, because the workpiece is held between centers.

If the moment of inertia of the workpiece is small, the centerless rolling principle is applicable. Rotational motion is supplied by a drive roller to the workpiece and the work rollers. Neither plunge nor through-feed principles are applied in this case, and feed must be accomplished by inclination of the drive roller axis toward the workpiece axis.

Work rollers must be designed for the smallest possible size to minimize the rolling force needed. Multiple-roller attachments offer optimum conditions. For centerless roller finishing heads, small floating work rollers are backed up by properly supported rollers. The bearing life of the backup rollers and the three-point contact geometry between the work rollers and the part require roller heads to be designed only for a certain diameter range.

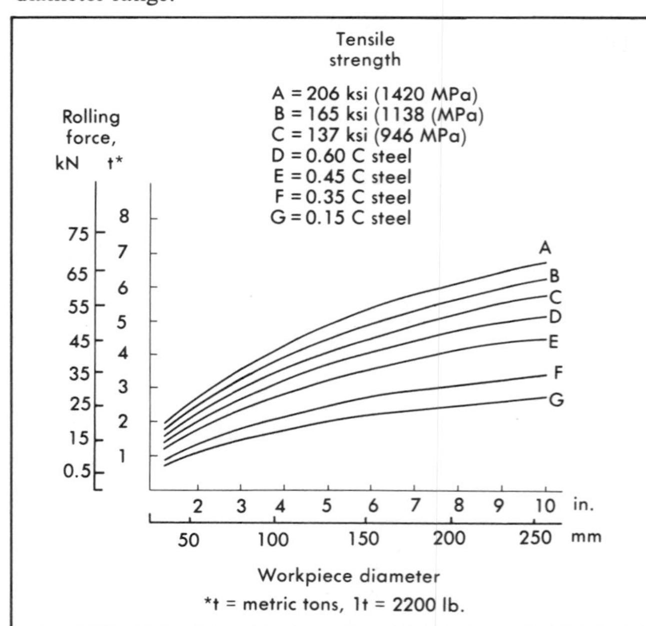

Fig. 16-26 Required rolling forces for roller finishing in relation to the workpiece diameter and tensile strength or the carbon content of steels.

Multiple-roller attachments are most commonly mounted on engine or turret lathes, but they are also used as standard tooling on special roller finishing machines. Diameter adjustments are relatively limited, and the setting remains constant during the operation. The work rollers are arranged symmetrically and are retained in such a way that they produce a teardrop contact area with the workpiece. These fixed-diameter attachments require good workpiece size control before roller finishing, but they are especially well suited to internal roller finishing. Their use for external rolling is restricted to short bearing extensions. Special attachments for tapers, chamfers, spheres, and fillets are also available.

General-purpose semiautomatic roller finishing machines and single-purpose automatic machines are available with capacities up to a maximum workpiece diameter of 4" (102 mm). These machines work on the centerless principle, and their tooling can be adjusted to any desired contact area form or line. Feed rates are infinitely variable within machine capability. Required rolling forces are generated by hydraulic pressure. The tooling and the machines are designed to provide optimum flexibility for manual or automatic operation, for adaptation of loading devices, and for transfer lines. The diameter range of the roller heads can be used without adjustments, making the finishing effect independent of any diameter changes. Machine setup and tool changes can be performed in minimum time to conform to high-production requirements.

Two-roll and single-roll attachments are compatible in design. Single-roll attachments, however, should be used only on diameters where there will be no deflection of workpiece or machine when rolling forces are applied. Table 16-16 shows workpiece diameter capacities, general applications, and rolling force requirements of roller finishing equipment.

Operating Parameters

The number of over-rollings to be given a particular part is an important factor in surface failure, especially for materials with low ductility. The number of work rollers used, the length of the contact area form, and the feed rate per workpiece revolution determine the number of over-rollings. Generally, a minimum of three over-rollings per part is desirable, but satisfactory surfaces have been achieved with only one over-rolling. The maximum number of over-rollings is not critical for steel, but it is for cast iron and screw machine materials, which require over-rolling control because of their sensitivity to subsurface shear failures.

Rolling speed and feed. Extensive tests performed to determine the optimum speed for roller finishing various materials indicate that the speed has no effect on surface finish. Neither are tool wear nor diameter reduction affected by changes in rolling speed. Other considerations, such as the required cycle time, workpiece acceleration, dynamic behavior of the workpiece, and tooling or bearing heat, all limit rolling speed to a practical value. Best results have been achieved with rolling speeds of approximately 200 sfm (61 m/min) for diameters up to 1 1/2" (38 mm). Very small diameters of 1/16 to 3/16" (1.6 to 4.8 mm) should be rolled below 100 sfm (30.5 m/min) since part revolution could otherwise be excessively high.

Rolling tools should always contact the work before rotation begins and should accelerate to rolling speed. Excessive slippage will cause wiping of the workpiece material, which might be visible or cause out of roundness. The feed rates that can be maintained are comparable with those of centerless grinding, and a maximum feed of 0.080 ipr (2.03 mm/rev) is used.

Lubrication. Lubrication during roller finishing does not affect surface quality. Lubricating oil may even be undesirable because of slippage, which causes a reduction in feed rate. Other factors, however, make a light spindle oil desirable. For example, a continuous stream of filtered and recirculated oil flushes dirt and metal particles away from the rolling tools, especially for stainless steel parts, which have a tendency to flake microscopic particles. A light spindle oil is also often used to lubricate the bearings holding the work rollers.

DEEP ROLLING

Deep rolling is a process related to roller finishing that results in increased bending or torsional fatigue strengths, as well as improved surface finishes. Although similar to roller finishing, deep rolling differs from that process in its objectives, results, applications, and tooling designs.

The primary objective of deep rolling is to impart a deeply penetrating compressive stress layer into highly stressed portions of workpieces. Any failure of a highly stressed component can

TABLE 16-16
Capacities, Applications, and Rolling Forces of Roller Finishing Equipment

Workpiece Diameter, in. (mm)	Rolling Equipment	Typical Applications	Available Rolling Force, lb (kN)
1/16 to 4 (1.6 to 102)	Roller finishing machines and internal rolling attachments	Valve stems, shock absorber rods, bolts, armature shafts, dental drills, pump plungers, seal diameters, rocker arm shafts, driveshafts, and piston rods	200 to 10,000 (0.89 to 44.5)
4 to 25 (102 to 635)	Two-roller attachments and internal rolling attachments [16" (406 mm) diam only]	Piston rods, tierods, turbine shafts, rolls for the paper and plastics industries	200 to 18,000 (0.89 to 80)
Above 25 (635)	Single-roll attachments	Rolls and roll bearings	1500 to 4000 (6.7 to 17.8)

be analyzed as a failure in tension. The compressive stresses provided by deep rolling oppose the tension stresses that occur during part operation. Deep rolling shows increased stress levels to a depth of more than 0.125" (3.18 mm) from the part surface. The average values of these stresses, measured by X-ray diffraction, are as high as 60,000 psi (414 MPa) near the surface of parts deep rolled with an average specific rolling force of 500,000 psi (3448 MPa). Sectional dimension changes of as much as 5 to 20% may occur under these pressures.

Higher specific rolling forces and the pressure control required demand a much more sophisticated machine tool design for deep rolling than for roller finishing. Deep-rolled workpieces are rotated positively in the machines and, in turn, rotate the tooling. Feed rates are generated by a feed screw or hydraulic feed cylinder, and rolling pressure is varied according to the tool position relative to the workpiece.

Advantages and Limitations

Deep rolling compares favorably with most processes commonly employed for increasing fatigue strength, such as nitriding, shot peening, and induction hardening. Further distinct advantages are the cleanliness of the process, its reliability, and its low cost.

Manufacturing cost savings are obtainable directly by eliminating grinding, polishing, or heat treating operations. Further savings are possible through savings on grinding wheels, changes to less costly or lower quality materials, and higher production line efficiency or increased tool life. Another advantage of deep rolling is the possibility of combining roller finishing and deep rolling.

Like roller finishing, deep rolling is most applicable to materials that have a percentage elongation of over 6%. Common candidates are plain carbon steel, alloy steel, stainless steel, and nodular cast iron. Materials such as low-grade steels or cast iron, bronze, aluminum, and brass may be finished by roller finishing, but they will not withstand the high specific rolling forces applied during deep rolling.

Another limitation concerns the hardness of the material to be deep rolled. Hardness values of R_C 40 are the top limit and require extremely high rolling forces, resulting in excessive bearing and tool wear. The high rolling pressures also endanger the form accuracy of the component. A practical value for surface hardness limits is R_C 35.

Principles

Deep rolling may be performed by either the contour rolling or the plunge rolling methods. Contour rolling, which is a contour tracing method, does not require close control of preliminary machining with regard to size, form, or surface finish. Fillet tolerances and surface finishes obtainable by a single-point tracer turning operation are adequate. The rolling tool is approximately 20% smaller than the fillet to be rolled, and it follows the existing part contour by using the part itself as a template. A normally permissible tolerance for a fillet up to 0.500" (12.70 mm) diam is ±0.020" (0.51 mm). The tool path and compressive stresses provided by contour deep rolling are depicted in Fig. 16-27.

The plunge deep rolling method requires a relatively close mating of part and rolling tool geometries. The plunge method therefore requires closer manufacturing tolerances in workpiece design in order to create a fatigue-strength increase. Required tolerances for plunge rolling a fillet radius up to 0.375" (9.52 mm) are ±0.005" (0.13 mm) for size, ±0.003" (0.08 mm) for form trueness, and a maximum of 90 μin. (2.3 μm) for surface finish. Figure 16-28 illustrates the tool paths for plunge deep rolling and the resulting stress pattern.

Plastic flow. Deep rolling results when the work rollers forced against the workpiece exceed the yield point of the material. Plastic flow occurring in the rolling direction causes a metal wave to build up near the leading edge of the tool-contact area and to fill the surface valleys in the part. This metal wave can reach a height of 0.008" (0.20 mm); resulting stresses in the outer layers of the material are very high and bear the danger of subsurface shear failure.

Contour rolling keeps this metal flow within acceptable limits by control of rolling forces exerted in the two directions of tool movement (lateral as well as rotational); any metal built up during deep rolling will be smoothened out at the appropriate location as rolling pressure is decreased. The plunge rolling method does not allow for this smoothening since tool movement occurs only in rotation; metal flow can be limited, however, by reducing any mismatches between tool and fillet geometries. Tool design features such as shoulder rollers provide only limited metal flow control.

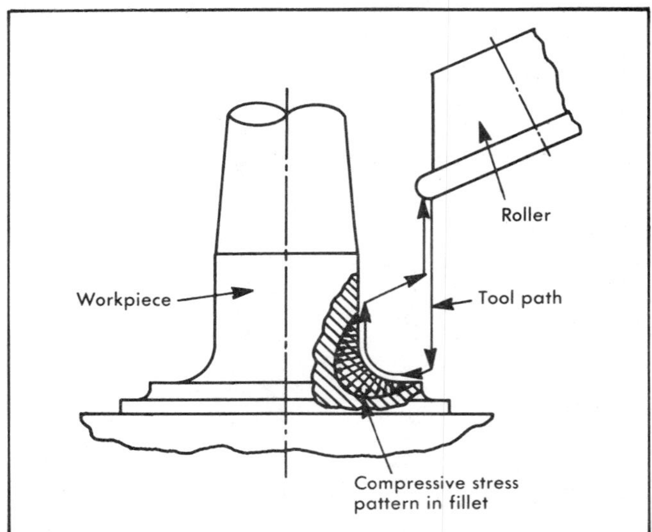

Fig. 16-27 Tool path and compressive stress pattern in contour deep rolling.

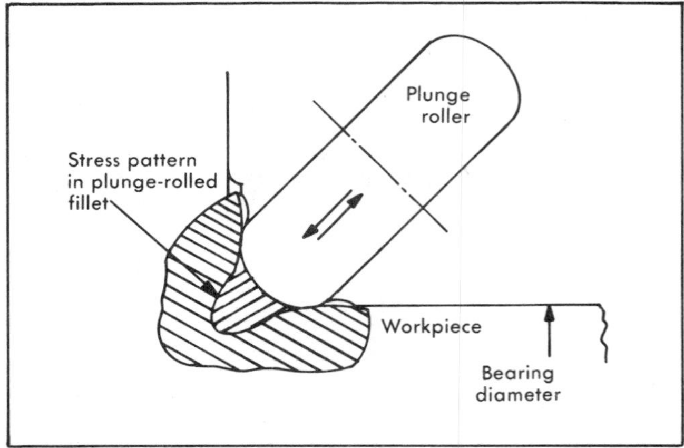

Fig. 16-28 Tool path and compressive stress pattern in plunge deep rolling.

Contact area. The contact area between the workpiece and the deep rolling tooling consists of a lead zone and a rolling zone. The lead zone carries the metal wave ahead of the rolling tool and ends at the point at which actual contact between the workpiece and the roller is made. The rolling zone is a contact area in which cold working occurs, and it is this zone for which specific rolling force calculations can be made. Following the rolling zone is another, substantially lower wave caused by backflow. The magnitude of the backflow wave is less than that of the front wave because the metal has been cold worked and prestressed in the contact zone. The backflow wave causes a spiral feed pattern on the part surface, especially on long components such as torsion bars. This spiral pattern is not detrimental to part fatigue strength; but if it is not desirable, a tool design that will avoid backflow, such as a combination of deep rolling and roller finishing tooling, will be more suitable.

Rolling force. The rolling forces required for deep rolling depend upon the desired compressive stresses in the fillet, the tensile strength of the material, the length of the rolling zone, and the contacting workpiece and work roller diameters. Deep rolling uses specific area pressures from 250,000 to 750,000 psi (1724 to 5171 MPa). A clear relationship between specific area pressure, compressive stress layer depth, and the resultant fatigue-strength increase has not been established because of the many influencing factors that vary with each application. Empirical data, however, provide a good starting point for the rolling force calculation. As a general rule, the specific area pressure used during the deep rolling should be three to four times the ultimate tensile strength of the material.

Effects of Deep Rolling

Surface hardness. The highest hardness increase in deep-rolled parts, according to stress distribution theory, should be measured below the part surface. As in roller finishing, conventional hardness testing methods such as the Rockwell or Brinell tests are not suitable for measuring deep-rolled hardness increases because the test media break through the hardness scale under normally used preloads. Microhardness tests such as the Knoop test, working with preloads of 50 or 100 grams, perform satisfactorily. The hardness increase can range from 10 to 35% of the base material hardness, with the lower values experienced on components high in hardness before rolling.

The increase in surface hardness provides a means of checking process reliability. This testing method is nondestructive and will leave no detrimental imprints or stress-rising notches in the surface of the component. Special microhardness testing equipment might be required, however, depending upon the part configuration.

The compressive stresses imparted to the rolled area are normally measured by X-ray diffraction. This process is destructive and not suitable for production. X-ray diffraction tests on deep-rolled C1046 steel steering knuckles (240-280 Bhn) with bearing fillets rolled at approximately 520,000 psi (3585 MPa) specific rolling pressure, for example, show a compressive stress layer of 60,000 psi (414 MPa). These same tests indicate a fatigue-strength increase of over 40%.

The surface hardness increase and compressive stress layer are always safe indicators of fatigue-strength increase in the deep-rolled component. Another method of predicting a strength increase without performing fatigue tests is microetching. A properly cold worked grain structure will show a disturbance and directional layer adjustment, which indicates that a fatigue-strength increase can be expected.

Surface finish. The surface finish of a deep-rolled part has a high-gloss appearance. Finish values of 5 μin. (0.13 μm) or better are reached from turned surfaces of 125 to 175 μin. (3.18 to 4.44 μm). The surface finish prior to rolling does not affect the deep rolling results or fatigue-strength increases.

Diameter reduction. Diameter reduction in deep rolling is due to closing of the surface and plastic flow. The reduction is subject to changes in hardness and surface finish. Therefore, an accurate determination of the diameter reduction or form change experienced as a result of deep rolling cannot be made without actual rolling tests.

Product quality. The compressive stresses resulting from deep rolling are symmetrical to the workpiece axis and do not result in any macrogeometric form changes of the component. Bending, as in roller forming, however, can result from the alteration of stress patterns caused in the work by previous operations.

Machines and Tooling Used

Even when working with high specific rolling forces, tool life with deep rolling is good. For example, a small crankshaft fillet roller measuring approximately 0.600″ (15.24 mm) diam and operating under 50,000 psi (345 MPa) contact pressure will produce approximately 20,000 crankshafts. Similarly, the ball bearings supporting steering knuckle rollers will last through approximately 15,000 rolling cycles.

Even though the deep rolling process is relatively new, highly efficient rolling equipment has been developed. The size or the configuration of a part to be deep rolled usually dictates the equipment or rolling procedures required. In the case of very large workpieces, it is possible to design special rolling attachments that can be mounted on existing equipment.

Table 16-17 shows the diameter ranges, equipment types, typical applications, and available rolling forces of deep rolling machine tools. The horizontal and vertical deep rolling machines are completely compatible in design and differ only in size capacity and floor space requirements. The crankshaft deep rolling machine is considered a special machine because of its special multiple-roller arrangement, which is entirely dependent upon the design of the crankshaft. Rolling attachments are often used on special applications for which a special rolling machine design would be too costly.

Operating Parameters

The number of over-rollings to be given a part during deep rolling is an important factor in the part's surface failure, particularly for low-ductility materials. The number of work rollers used, the length of the contact area, and the feed rate determine the number of over-rollings; normally a minimum of 10 are required per surface element. As a general rule, the number of over-rollings is greater for deep rolling than for roller finishing because of the greater rolling intensity required.

The maximum number of over-rollings for steel is not critical; and nodular cast iron, often used for crankshafts, is highly suitable for multiple rollings under high contact stresses. Materials with high sulfur content are sensitive to subsurface shear failure and are therefore not well suited to deep rolling.

Rolling speed and feed. Extensive tests performed to determine the best rolling speed indicate that not deep rolling results, but rather the rolling tools used or their bearing arrangement are limiting factors. Because high specific rolling forces cause heat or wear on the tooling or bearings, the rolling speed must be held within practical values. Best results for the

TABLE 16-17
Capacities, Applications, and Rolling Forces of Deep Rolling Equipment

Workpiece Diameter Range, in. (mm)	Workpiece Maximum Length, in. (mm)	Rolling Equipment	Typical Applications	Available Rolling Force, lb (kN)
1/4 to 2 (6.4 to 51)	12 (305)	Vertical machine	Steering knuckles, slip yokes, connecting-rod bolts, spider cross valves, high-tensile bolts	4000 (17.8)
1/4 to 10 (6.4 to 254)	48 (1219)	Universal, high-production machine	Driveshafts, piston rods, ballscrews, turbine shafts, cylinder-head studs, torsion bars	16,000 (71.2)
2 to 12 (51 to 305)	102 (2590)	Horizontal machine	Railroad axles, landing-gear rods, gun barrels, roll necks	22,000 (98)
1 to 6 (25 to 152)	60 (1524)	Crankshaft machine	Automotive crankshafts	8000 (35.6)

contour rolling process have been achieved with rolling speeds of approximately 100 to 200 sfm (30.5 to 61 m/min). Plunge rolling of crankshafts is an exception and is normally performed at a rolling speed of 60 rpm. The design of the tooling and the relatively large rotation mass are the reasons for this limitation with crankshafts. Feed rates for deep rolling range between 3 and 15 ipm (76 and 381 mm/min). Higher feed rates, to 0.080 ipr (2.03 mm/rev), are possible on straight workpieces.

Lubrication. Lubrication during the rolling process does not affect the degree of fatigue-strength increase. Lubrication, however, is generally desirable for cooling. The crankshaft deep rolling process requires a continuous stream of filtered oil flowing through the tooling onto the workpiece to flush dirt and metal particles out of the rolling area. The lubricating oil may be filtered and recirculated.

The contour deep rolling process, in which the work rollers are properly held on ball bearings, does not require a continuous stream of lubricating oil; a few drops of coolant oil are satisfactory. This oil is not recirculated but is considered waste after it is used.

ROLLER BURNISHING

Roller burnishing is a plastic deformation process employing a variety of rotary tools containing caged rollers. The process is used to improve the surface finish, tolerance control, surface hardness, and fatigue life of metal components. It is applied to the inside diameters of holes, outside diameters of shafts, flat surfaces of revolution, tapered and spherical surfaces, and fillets (radii at shoulders). Special tools burnish multiple surfaces simultaneously.

Process Advantages

There are four predominant reasons for roller burnishing parts: (1) improved surface finish, (2) improved tolerance control, (3) increased surface hardness, and (4) improved fatigue life.

Surface finish. Surface finishes typically obtained from various metalworking processes on steel and aluminum are compared in Fig. 16-29. Roller burnishing would normally follow a turning or reaming operation, where the preburnished finish would typically be in the 60 to 120 μin. (1.5 to 3 μm) range.

Since the attainable surface finish with roller burnishing is dependent upon the preburnished (machined) surface, as well as the workpiece material, it is important that the machined finish be free of tears and reasonably uniform. In general, a surface having a given microfinish will burnish better if it is generated with a cutting tool having a small nose radius and fed at a moderate rate, rather than by a tool having a large nose radius fed at a rapid rate. While the heights of the peaks may be the same in both cases, the peaks would be closer together in the former case than in the latter, and will burnish smooth with less subsurface material flow and less roll force.

Parts having preburnished finishes of 80 to 120 μin. (2 to 3 μm) can be roller burnished to 2-15 μin. (0.05-0.38 μm) in one pass and at a rate of roughly 2 to 4 in.^{2} (13 to 26 cm^2) of surface per second. Table 16-18 lists typical results for several materials.

Tolerance control. Consider a steel part in which it is required to machine a 1.000" (25.4 mm) diam hole with a tolerance of ±0.0005" (0.013 mm). This requirement would normally present a difficult problem, but one that is commonly encountered in wristpin holes in pistons or connecting rods, as well as in hydraulic valves and other components. If the hole is drilled and then reamed or bored slightly undersized, say to 0.9995" (25.387 mm), ±0.001" (0.25 mm), with a surface finish of 250 μin. (6.35 μm), it can be roller burnished to the required size and tolerance.

The roller burnishing process will deform the surface material, rolling the microscopic peaks into the valleys. The process also works the surface of the hole on the small side of the tolerance range more than on the large side. The smallest drilled, reamed, or bored hole could be expected to measure 0.9985" (25.362 mm), with a 250 μin. (6.35 μm) finish prior to burnishing. After burnishing, the hole size would become 0.9995" (25.387 mm) with perhaps a 5 μin. (0.13 μm) finish. A nominal hole would be reamed or bored to 0.9995" and could be expected to become 1.0000" (25.4 mm) after burnishing. The largest hole, at 1.0005" (25.413 mm) diam, would be essentially untouched by the burnishing tool. In any case, hole size variation would be cut in half.

This simplified discussion assumes that the part containing the hole is essentially infinite in any dimension and that the material is completely incompressible and unyielding, except for the material on the surface. In fact, this is not the case, and material movement can be accomplished somewhat beyond that which would be indicated from the consideration of just surface finish effects.

There are many cases where it is possible to open up a 1" (25.4 mm) diam hole by as much as 0.001 to 0.002" (0.03 to 0.05 mm) and hold ±0.0002" (0.005 mm) tolerances. In one application, stainless steel tubing having the required internal diameter of 2.115" (53.72 mm) was not available, and it was impractical to bore available tubing. A 2 1/8" (53.98 mm) diam burnishing tool was used to expand standard tubing almost 1/8" (3.2 mm), from an ID of 2.000" (50.80 mm) to 2.115" (53.72 mm), while holding tolerances of ±0.0015" (0.038 mm).

Typical experience indicates that consistently held tolerances of ±0.00025" (0.0064 mm) can be expected with the feeding-contact roller burnishing process discussed later in this section, given proper part preparation. If closer tolerances are required, a slightly different tool, the bearingizer, using the full-contact burnishing principle (also discussed subsequently), can be used to achieve tolerances of ±0.0001" (0.003 mm).

Increased surface hardness. Roller burnishing can be applied to any ductile or malleable material having a hardness to R_C40, although harder parts have been burnished on occasion. Since the metal is compressed past its yield point, the grain structure is changed, and the part becomes strain hardened. Through this granular dislocation and deformation, the grain size is decreased and the boundary area is increased after the cold working during roller burnishing.

Roller burnishing can increase hardness from 5 to 10%, with a surface penetration of 0.010 to 0.030" (0.25 to 0.76 mm). Bearingizing can increase hardness to 30%, with less surface penetration. Table 16-19 provides data from a typical hardness experiment.

Improved fatigue life. The fact that fatigue life can be improved through the elimination of surface imperfections, such as machining marks, is well known. In addition to improved surface finish, however, roller burnishing plastically deforms the material near the surface, leaving a compressive residual stress that extends into the material. This compressive stress has the effect of minimizing the maximum tensile stress that the material experiences through the stress reversal cycles during its lifetime.

In general, no matter in which direction the original machine marks are (parallel or perpendicular to the applied stress), about a 300% improvement in fatigue life, as a rule of thumb, can be expected in roller burnishing aluminum or steel. Some relatively new applications have taken a unique approach. For example, holes in aluminum plates are drilled and reamed undersized. The holes are then burnished well beyond what is required for a good finish, to the point that the metal surface fatigues and surface flaking occurs. Finally, the holes are reamed again, thus removing the surface imperfections. Fatigue life is reported to be improved up to 500%.

Other Applications

In addition to the many applications of roller burnishing due to the four major benefits already discussed, the process is also employed for other reasons, such as the reduction of noise and/or vibration. Roller burnishing the shafts of windshield wiper motors mounted inside a car's passenger compartment eliminated a noise problem. Burnishing the platter shaft and bushings for a high-fidelity turntable reduces vibration that can be transmitted to the tone arm and pickup unit.

Roller burnishing two diameters simultaneously on a previously ground stepped shaft provides a smoother finish that has doubled the life of sliding O-rings. Centerless grinding of motor shafts for small appliances sometimes produces a three-

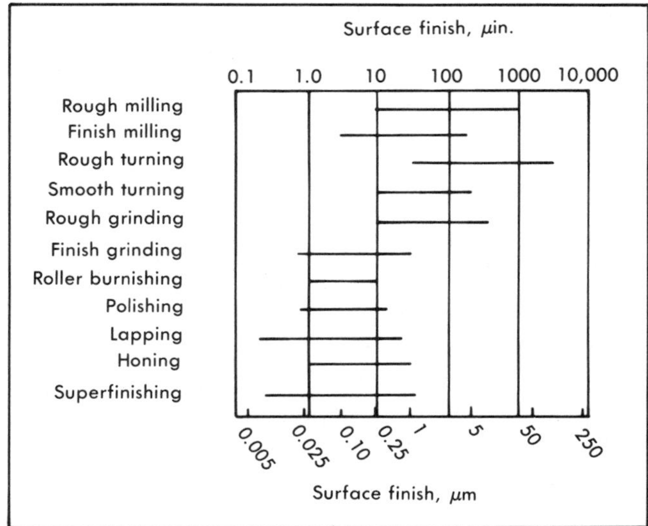

Fig. 16-29 Typical surface finishes produced by various metalworking processes.

TABLE 16-18
Surface Finishes Produced with Roller Burnishing

Workpiece Material	Surface Finish Range, μ in. (μm)
Cast irons	10-15 (0.25-0.38)
Steels, including stainless	4-8 (0.08-0.20)
Bronze and aluminum alloys	2-8 (0.05-0.20)

TABLE 16-19
Hardness Readings Below Surfaces

Surface Condition	Depth Below Surface, in. (mm)	Hardness,* R_B
Machined	0.002 (0.05)	75.5
	0.004 (0.10)	80
	0.010 (0.25)	80
	0.020 (0.51)	79
Burnished	0.002 (0.05)	92
	0.004 (0.10)	99.5
	0.010 (0.25)	97
	0.020 (0.51)	96
	0.040 (1.02)	88
	0.060 (1.52)	87
	0.100 (2.54)	78

* With 500 gram load and 20x objective magnification.

lobed diameter; this condition is eliminated by roller burnishing. Electric motor shafts are also roller burnished to reduce friction up to 35%, thus improving efficiency. Shafts to be chromium plated are roller burnished to improve the finish; this step reduces chromium requirements and lessens the need for polishing.

Additional applications for roller burnishing include the surfaces of disc-brake rotors to eliminate undesirable tool marks, brazed faucet assemblies to eliminate distortion, and holes for gear shafts in planetary carriers (consisting of two stampings welded together) to ensure axial alignment.

Process Principles

The roller burnishing process is a technique in which hard, smooth rollers are brought into pressure contact with a metal surface to be finished. The rollers cause the workpiece material to undergo localized cold flowing. In the process, the surface finish of the normally rough part is improved, the surface of the material becomes work hardened, and the material is left with a residual stress distribution that is compressive on the surface.

Figure 16-30 presents a schematic of the roller burnishing process. As illustrated, the roller comes in contact with the rough prefinished surface of the part; and, as the roller advances, the surface is progressively compressed through the elastic compression zone. At the beginning of the plastic deformation zone, the yield point of the material is exceeded, and cold flow takes place. After the material has been subjected to the maximum compressive strain under the bottom of the roller, it begins to elastically relieve through the elastic recovery zone, finally exiting from beneath the roller with a smooth surface and a residual compressive stress of significant peak value.

In actual practice, the material is subjected many times to the action described by a tool that has a multitude of rolls. While Fig. 16-30 illustrates the basic aspects of roller burnishing, there are essentially two different approaches to accomplish the finishing action. The difference is only apparent when viewing the roller along its length (see Fig. 16-31). In view *a*, a machined surface is being burnished by a roller along its whole length, L. In this case, the surface of the roller that contacts the workpiece is parallel to the workpiece surface. A force exerted on the roller drives the roller into the workpiece surface.

In view *b* of Fig. 16-31, the roller is being fed along the machined surface. The roller surface that contacts the part is not parallel to the part surface, but is designed to have a slight back relief. Thus, the nose radius of the roller does most of the work, while the roller is inducing a teardrop "footprint" on the part material at any given point. In addition, the three zones (see Fig. 16-30) can be shown to exist along the length of the footprint. Roller burnishing tools are designed using one or the other of these two principles, depending upon the specific application.

The improved surface finish produced by roller burnishing is accomplished by forcing the peaks downward into the base surface. The material flows outward, thereby raising the valleys in the machined surfaces. There is no scrubbing action, and the peaks are not bent or folded over into the valleys. The radii of burnishing rolls are so much larger than the roughness profiles of machined workpieces that the burnishing action is almost equivalent to pressing the rough profile with a flat plate.

Tools Employed

Rotary burnishing tools are used on a wide variety of machine tools. For small-quantity production, manually fed drill presses or hand-rotated toolholders are often used. For volume work, special machines are common. Burnishing is also frequently combined with machining and/or assembly operations on transfer machines.

Tools for burnishing the inside diameters of holes and the outside diameters of shafts are made in two different types. One type is adjusted to a fixed diameter and remains fixed during operation. These are called interference tools and generally operate on the feeding-contact principle. With the second type, the diameter can be changed to exert a predetermined burnishing pressure. These tools operate on the full-contact principle and

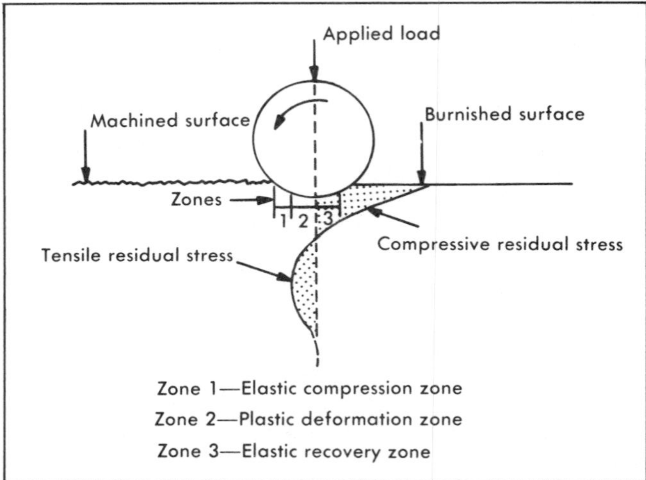

Fig. 16-30 Schematic representation of the roller burnishing process.

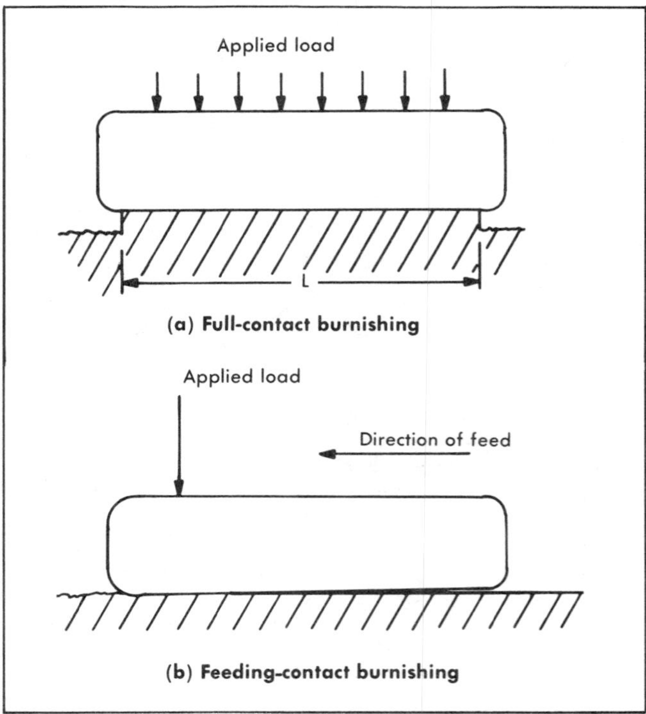

Fig. 16-31 Roller burnishing methods: (a) full contact and (b) feeding contact.

are called *expanders* for burnishing inside diameters and *contractors* for outside diameters.

In addition, there is a class of tools that operates on the feeding-contact principle and exerts a predetermined burnishing pressure by changing the working diameter during the cycle; these are called compensating tools. There is also a class of tools having a fixed working diameter that uses the full-contact principle; these are called bearingizers.

Fixed-working-diameter tools are the only type to provide tolerance control; however, they will not provide a constant surface finish unless the premachined workpieces are precisely the same diameter and have exactly the same prefinish from piece to piece. Tools with working diameters that are capable of changing provide a nearly constant surface finish, even though the parts may vary considerably in diameter from piece to piece. Table 16-20 summarizes the various types of tools.

Fixed-working-diameter tools. This type of tool is available for burnishing the inside diameters of holes and the outside diameters of shafts.

Inside-diameter tools. The most common roller tool designed to burnish a hole is illustrated in Fig. 16-32. The tool shown is for a 1″ (25.4 mm) diam hole, but tools are available from 1/8 to 10″ (3.2 to 254 mm) and larger. Precision tapered rollers are captured in a cage-like member. The rolls rotate around and bear upon an inversely tapered mandrel or race. The rolls are caused to turn by the rotation of the mandrel and apply a rolling pressure against the work surface. The cage serves to retain the rolls and to keep them properly spaced.

The combined diameter of the rollers and mandrel is set to be slightly greater than the size of the hole, hence the term *interference tools*. Tool size can be controlled by axial adjustments of the tapered mandrel within the roller and cage assembly. This adjustment is accomplished by pulling back on the castellated adjustment collar, which, since it is keyed to the shank, normally prevents the bearing collar from rotating. The bearing collar, which is threaded to the shank, is turned, thereby altering its position relative to the shank and mandrel. The bearing collar is again locked in place when the spring-loaded adjustment collar is released. Movement of one castellation changes the tool diameter by 0.0001″ (0.003 mm).

TABLE 16-20
Classes of Roller Burnishing Tools

Type of Tool	Full Contact	Feeding Contact
Fixed working diameter	Bearingnizers	Interference tools
Variable working diameter	Expanders	Compensators
	Contractors	
Other tools	Flat surface tools	Spherical tools
	Taper tools	
	Contour tools	

Total adjustment is typically 0.041″ (1.04 mm) for tools over 1/2″ (12.7 mm) diam. In certain cases, the rolls are held at a slight helix angle relative to the axis of the tool; this angle provides a self-feeding action. When the rolls are not set in the cage at an angle, the tool must be machine fed. When the tool is withdrawn from the hole, the cage and sleeve assembly, which has been forced up against the thrust bearing in the bearing collar, moves to the right. This movement allows the diameter of the tool to be reduced and the tool to be retracted effortlessly from the hole.

The tool shown in Fig. 16-32 is a through-hole tool and will not burnish to the bottom of a blind hole. Standard tools are available in which the cage is open ended, the rolls are provided with an undercut near their back end, and the rolls are retained in the cage with a tine to accommodate bottoming applications.

Outside-diameter tools. A typical tool for interference burnishing an outside diameter is shown in Fig. 16-33. The principle is the same as that of the internal interference tool, but the rolls bear against a race rather than a mandrel. The rolls are captured by a cage that is moved axially to adjust for diameter. Again, the adjustment is in 0.0001″ (0.003 mm) increments over a total of typically 0.021″ (0.53 mm) diam.

With this tool, the cage thrust bearing is moved indirectly by the adjustment collar through three dowel pins that extend through the shank member. In this way, it is possible to minimize any tool runout since the shank is one piece from the Morse taper to the recess into which the precision-ground race is fit.

Tools of this type are available to burnish diameters from 0.062 to 5.00″ (1.5 to 127 mm). When it is necessary to burnish a different diameter than that for which the tool was initially furnished, it is only necessary to replace the cage, race, and, sometimes, the rolls.

When the work length is insufficient in the standard tool, three alternatives are available. First, it is possible to order a special tool with increased work length. Second, tools having through-holes and capable of burnishing long lengths are available; however, the workpiece must usually be rotated while the tool is held still. The third alternative is to use self-powered

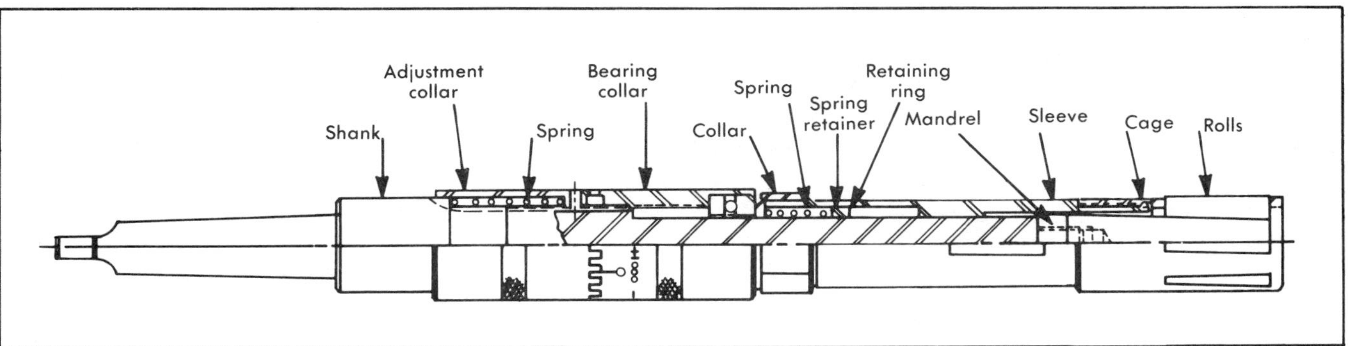

Fig. 16-32 Typical roller burnishing tool for internal applications. (*Cogsdill Tool Products*)

outside-diameter burnishing machines that can burnish shafts of any length.

Bearingizers. The bearingizer is an interference-type tool that works on the full-contact principle and also introduces a peening action to the rolls. It is similar to the roller burnishing tool, but the rollers are cylindrical, rather than tapered. The rollers operate against a multilobed, cylindrical (nontapered) mandrel or arbor within the tool, rather than against the smooth, round mandrel of the roller burnishing tool. Figure 16-34 is a schematic of the process.

The bearingizer tool is adjustable only about 0.004″ (0.10 mm), and this adjustment is accomplished by changing the complete set of rolls. Sets of rolls are available in 0.0001″ (0.003 mm) increments.

In operation, the bearingizing tool is rotated, and the rolls alternately expand and contract against the workpiece as they ride against the multilobed arbor, resulting in as many as 200,000 peening blows per minute. There is little kinetic energy involved, however, and the burnishing action is the result of the same principle already discussed. The cam action primarily allows the tool to be fed into the hole.

From a practical standpoint, the rolls are much smaller in diameter than those used in roller burnishing tools. Small-diameter rolls result in a much smaller "footprint" (total width of zones 1, 2, and 3 of Fig. 16-30) and a higher specific pressure for a given amount of force. Furthermore, for a given diameter tool, it is possible to have a greater number of rolls. For example, a 1″ (25.4 mm) diam bearingizing tool has 12 rolls, whereas the roller burnishing tool has 5-7 rolls.

The combination of small-diameter rolls and a large number of rolls makes the bearingizer preferable when burnishing a thin-walled part or a part in which the elasticity of the hole is nonuniform due to a variable cross section of material surrounding the hole. Such a situation is best illustrated by the example of the wristpin hole in a connecting rod.

Generally, it is considered normal to be able to maintain tolerances of ±0.00025″ (0.006 mm) with roller burnishing, given proper part preparation. Bearingizing will normally maintain tolerances, under the same circumstances, of ±0.0001″ (0.003 mm). Roller burnishing can typically increase the surface hardness of work-hardening materials 5-10%, whereas bearingizing will increase it 10-30%. Surface penetration of the hardening effect, however, is between 0.010 and 0.030″ (0.25 and 0.76 mm) with roller burnishing and only half this depth with bearingizing.

Variable-working-diameter tools. These tools are designed to operate on the full-contact burnishing principle. They are called expanders when used to burnish inside diameters and contractors when burnishing outside diameters. These special tools are generally recommended when one or more of the following conditions are present:

- The diameter variation from workpiece to workpiece is as great as ±0.003″ (0.08 mm).
- An interrupted surface is present, such as that created by a small crosshole or partial slot (for a Woodruff key, as an example). Existence of a full length keyway, however, would often preclude roller burnishing.
- It is desired to complete the burnishing job very quickly, say in one to two seconds, for a surface of perhaps 2″ (51 mm) diam by 2″ long.

Generally, expanders and contractors are limited to situations where the length to be burnished is less than 2 ½ times its diameter. There are many examples of variable-working-diameter tools, however, that contradict this rule of thumb.

The tool shown in Fig. 16-35 is in its fully expanded condition. It enters the workpiece by means of the spring, with the cage and the stop positioned towards the front of the tool. This design causes the rolls to ride on the small end of the tapered mandrel. The machine feeds the tool into the part until

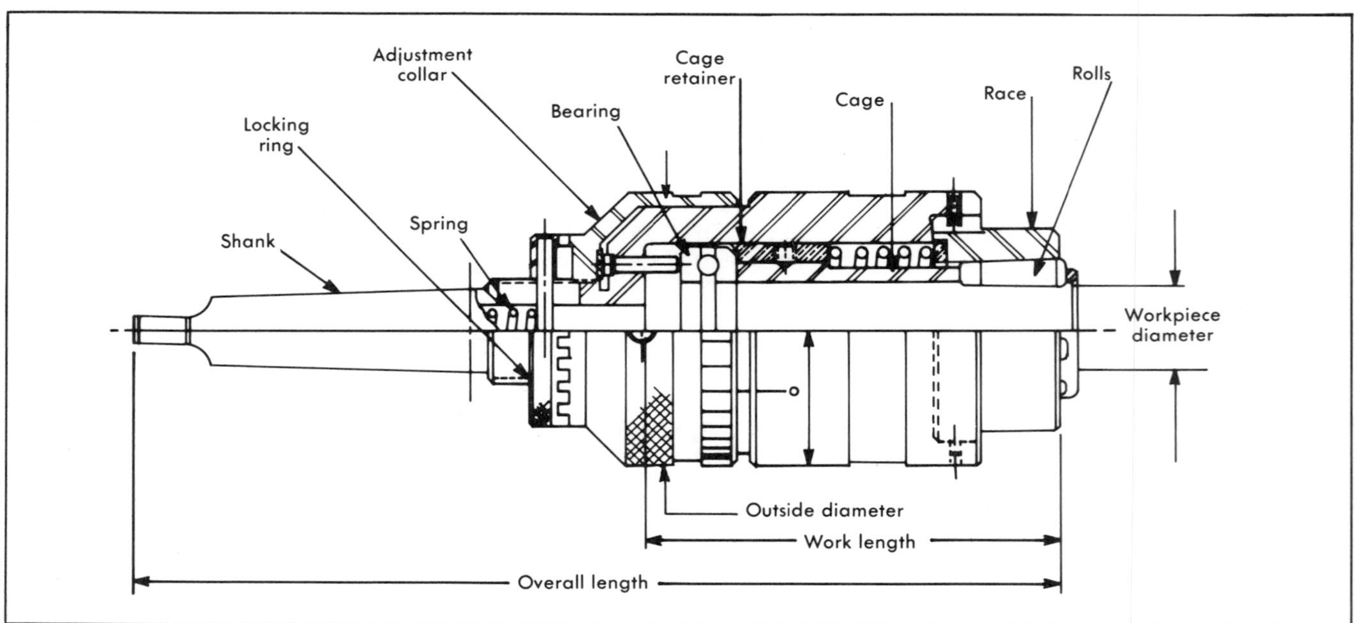

Fig. 16-33 Typical roller burnishing tool for external applications. (*Cogsdill Tool Products*)

the stop contacts a surface of the workpiece or a fixture plate. At that point, the stop ceases rotating and prevents the cage from moving axially. The tool with the mandrel continues to feed forward. This action causes the rolls to expand as they are forced up the taper of the mandrel. Expansion continues until the adjusting collar contacts the cage thrust bearing and the mandrel has advanced as far as it will go. Thus, the tool will expand to a specific diameter determined by the position of the adjustment collar.

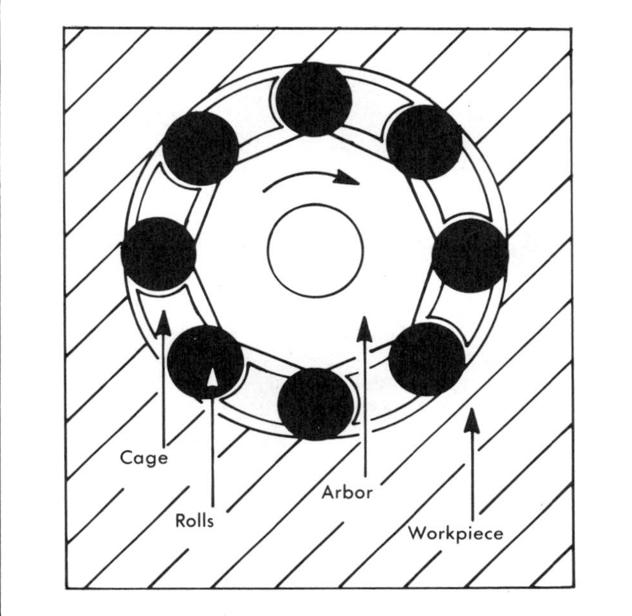

Fig. 16-34 Schematic representation of the bearingizing process.

In a variation of this type of tool, the adjustment collar is either not used or is set so far to the rear of the tool that the tool is free to expand to its largest diameter. The amount of burnishing is then determined by the amount of force exerted on the mandrel. This force can be controlled either by a machine that has a pressure-controlled head or by fitting an overtravel unit to the shank. The overtravel unit consists of a sleeve that fits over the shank and contains suitable springs (typically a stack of Belleville washers) and a drive mechanism.

In applications of overtravel units, the machine is adjusted so the head or quill feeds to a prepositioned stop. This design compresses the spring in the overtravel unit when the tool is in a workpiece, thereby exerting a specific force on the shank. Through the mechanical advantage of the tapered mandrel and rolls, a corresponding specific force is exerted on the burnishing rolls. The fact that a slightly oversized workpiece will experience slightly less force and a slightly undersized workpiece will experience slightly greater force, due to the spring constant of the overtravel unit, is normally not significant.

Compensating tools. These tools are designed to operate on the feeding-contact burnishing principle. They are available for outside and inside-diameter burnishing. Compensating tools are used when one or more of the following conditions exist:

- The diameter variation from piece to piece is ±0.003″ (0.08 mm) or more.
- The diameter variation is ±0.003″ (0.08 mm) or more, and the length to be burnished is greater than 2 ½ times the diameter.
- A part to be burnished has a diameter that is not constant along its length.

Compensating tools are designed so that a constant force, or nearly constant force, is exerted on the burnishing rolls by the mandrel or race over a fairly wide range of diameters. The force that is exerted on the mandrel (relative to the cage) is generated by air pressure acting on a pneumatic cylinder built into the tool

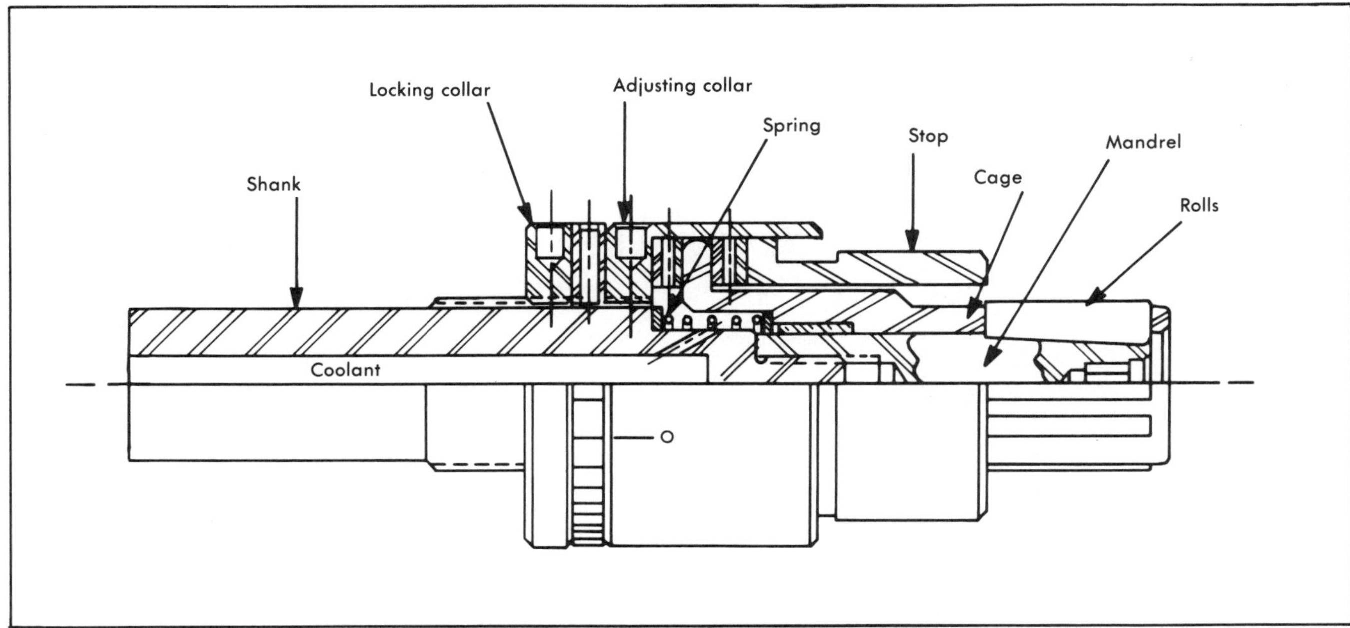

Fig. 16-35 Expander type of variable-working-diameter tool for roller burnishing. (*Cogsdill Tool Products*)

as an integral part, or by means of a spring. The pneumatic cylinder gives a constant burnishing force, totally independent of workpiece diameter, while the spring-operated tool gives a nearly constant force. The variations in force of the spring-operated tool are a function of the diametral change, the roll-mandrel taper, and the spring constant.

The diameter range over which a compensating tool can operate is a function of the diameter of the burnishing rolls and is typically 30% (±15%) of the roll diameter. Thus, as the part diameter becomes larger, a tool with larger rolls can be used, and a greater operating range can be achieved. Table 16-21 lists typical operating ranges for some selected nominal diameters.

Other tools. With the roller burnishing tools just described, very little torque is exerted on the workpieces. Feeding forces are nonexistent or very low for self-feeding tools, even with the contractors and expanders. Low forces result from the mechanical advantage of the wedgelike action of the tapered rolls and mandrel. Flat surface tools, tapered tools, and spherical tools, however, all require considerable force to be exerted against the workpiece. Therefore, the workpiece must be well supported, often in a floating fixture; and the machine head must be capable of generating the forces required.

Flat surface tools. A typical flat surface tool, the simplest type of roller burnishing tool, is illustrated in Fig. 16-36. Plain cylindrical rollers are held in a flat cage and run against a flat surface mandrel or cam. The tools are designed to burnish an annular surface, but cannot burnish to the center of a circle without using a sophisticated workholding device. Such devices must simultaneously rotate the workpiece as the tool is independently rotated on a different centerline. This procedure is sometimes referred to as planetary burnishing.

Tapered tools. Roller tools are available for burnishing both external and internal tapered surfaces. (A typical tool for burnishing external tapers is shown in Fig. 16-37.) Their construction is similar to flat surface tools. The tool illustrated in Fig. 16-37 uses cylindrical rolls, but tools are also built with tapered rollers. To use tapered rollers, the tools are designed so that the rollers roll around the cam without relative motion or slippage.

A common misconception is that a tapered tool will correct a taper that has been machined at the wrong angle. Such correction is not possible because these tools only move metal within the limits of the preexisting finish.

Contour tools. In contour tools, the inverse of the surface of revolution to be burnished is formed into the rollers, and the mandrel or cam is ground to properly support the rollers. These tools operate on the full-contact burnishing principle.

Spherical burnishing tools. It is possible to burnish convex spherical surfaces with a tool having cylindrical rolls, resembling the tapered tool shown in Fig. 16-37. Such tools are commonly used for automotive ball joints, valve balls, artificial hip joints, and similar components.

TABLE 16-21
Working Range of Compensating Tools for Roller Burnishing

Nominal Diameter, in. (mm)	Operating Range, in. (mm)
0.250 (6.35)	0.023 (0.58)
0.500 (12.70)	0.048 (1.22)
1.000 (25.40)	0.080 (2.03)
2.000 (50.80)	0.122 (3.10)
3.000 (76.20)	0.158 (4.01)

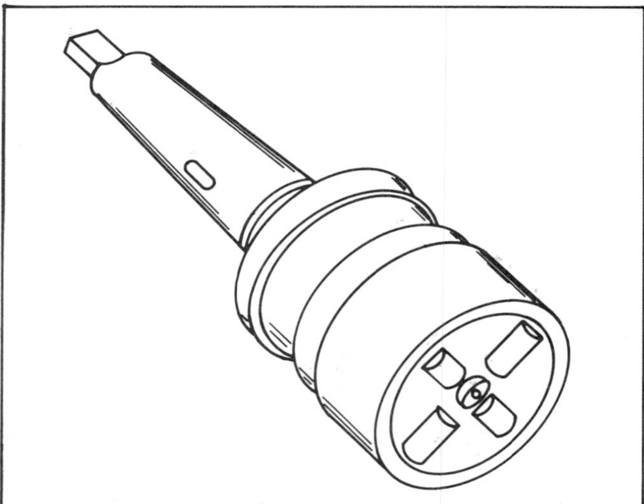

Fig. 16-36 Roller burnishing tool for flat surfaces.

The tapered tool is rotated about an angle to the centerline of the part. As the tool rotates, it burnishes a circular line around the workpiece. The diameter and location of this line is a function of the included angle between the cylindrical rollers and the angle between the centerline of the part and that of the tool. The part is slowly rotated as the tool is more rapidly rotated. When the part has made one revolution, it is completely burnished. A typical automotive ball joint can be burnished in three seconds.

Operating Parameters

Part preparation. Successful burnishing is dependent upon proper part preparation. In about 90% of the cases when an operator is experiencing difficulty with achieving the desired burnished finish, the problem can be traced back to an improper prefinish or excessive size variation (when using fixed-working-diameter tools).

Any suitably ductile material with hardness less than R_C 40-45 can be successfully burnished if the prefinish is relatively constant and free of tears. With steels, aluminum, copper, and most other materials, a prefinish to 150 μin. (3.8 μm) is satisfactory for burnishing.

For a reamed hole, the best success is obtained when the reamer has a relatively sharp front chamfer of about 20° and a back taper of approximately 0.0005 in./in. (mm/mm). The reamer should also have radial relief. Rose-type reamers without radial relief are not recommended.

For turning or boring operations, a tool with a small nose radius and operated at a moderate feed rate is preferred to one with a large radius and operated at a high feed rate, even though both may give the same finish. As a general rule of thumb, a surface suitable for roller burnishing can be produced by using the tool nose radii and corresponding feed rates shown in Table 16-22. These combinations will usually produce a surface finish of 80 to 120 μin. (2 to 3 μm).

The burnishing process changes the size of a component by reducing the height of the roughness profile. The size change during burnishing must be predictable so that the desired

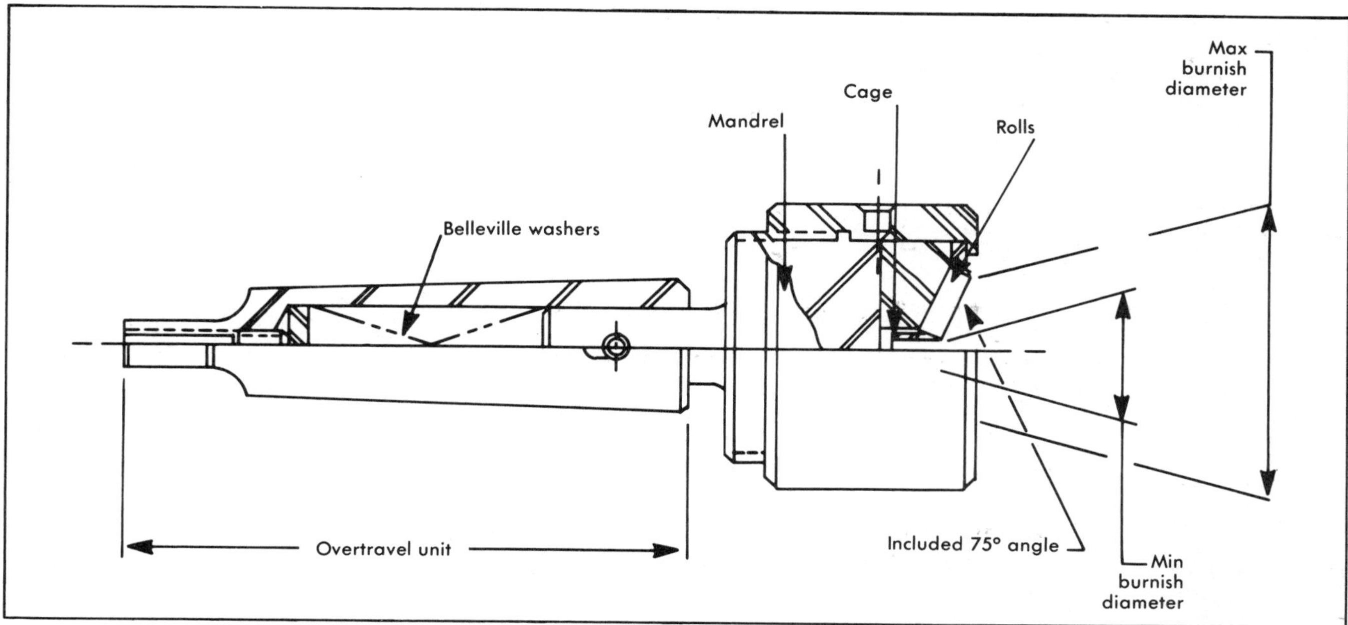

Fig. 16-37 Roller burnishing tool for external tapers.

TABLE 16-22
Recommended Tool Nose Radii and Feed Rates for Turning and Boring Operations Prior to Roller Burnishing

Tool Nose Radii, in. (mm)	Feed Rates, ipr (mm/rev)
0.010 (0.25)	0.003 to 0.005 (0.08 to 0.13)
0.015 (0.38)	0.007 to 0.010 (0.18 to 0.25)
0.030 (0.76)	0.012 to 0.015 (0.30 to 0.38)
0.045 (1.14)	0.017 to 0.020 (0.43 to 0.51)
0.060 (1.52)	0.022 to 0.025* (0.56 to 0.64)

* Not usually recommended.

postburnished size can be obtained. For a bore, the preburnished diameter must be smaller than the final diameter; and for a shaft, the reverse is true.

Actual diameter changes resulting from roller burnishing can be approximated from the following equation:

$$D_c = K\,(R_a \text{ before burnishing} - R_a \text{ after burnishing}) \qquad (1)$$

where:

D_c = actual diameter change
K = a constant having a value between 4 and 8
R_a = 1/2 the height of the peaks above the median line in the peak-valley surface profile

Speeds and feeds. Roller burnishing tools are very tolerant with regard to the effect of speed upon resultant surface finish and tolerance control. Tools are sometimes hand cranked through the parts to be burnished with satisfactory results. For the purposes of production, however, the speed ranges shown in Fig. 16-38 are generally recommended. The shape of the curve shows that the inertial effects of the rotating tool become the dominant consideration as tool size increases.

Bearingizing tools should be fed rapidly for longest tool life. A feed rate of 150 to 250 ipm (3810 to 6350 mm/min) is recommended, regardless of tool diameter. Figure 16-39 shows the feed rates that are obtained from one manufacturer's self-feeding tools. These are tools in which the rolls are held in the cage at a slight helix angle. The helix angle of the rolls changes at 2.5" (63.5 mm) diam. If self-feeding tools are used in a machine with power feed, the machine feed rate must exceed the feed rates shown in Fig. 16-39. The higher feed rates will prevent the tool from collapsing.

If tools in which the rolls are not held at a helix angle are used, the machine must have a power feed. Any feed rate from 0.010 ipr (0.25 mm/rev) up to the maximum rate shown for helix tools in Fig. 16-39 can be used.

Lubrication. Any standard grade of lightweight, low-viscosity lubricating oil, or any mineral, sulfur, or soluble oil that is compatible with the alloy or metal to be burnished, can be used. For aluminum or magnesium alloys, a highly refined, paraffin-based oil of low viscosity is recommended. For cast iron, a mineral-seal solution is best, and flooding the part is necessary.

It is essential that the part be clean and free of chips and that the lubricant be continuously filtered if a closed-loop system is used. If these precautions are not taken, chips, flakes, and other contaminants will be rolled into the part surface by the action of the burnishing tools.

Troubleshooting

To achieve the maximum benefits from roller burnishing, proper burnishing procedures must be followed. Care should be taken to follow the manufacturer's recommendations, such as using the proper speeds and feeds.

Unlike common machining operations, such as turning, boring, or grinding, roller burnishing operations are very dependent on the premachining operation for their success. For this reason, when problems arise the operator should first check his premachining operation to ensure proper part preparation.

ROLLER BURNISHING

Size and surface finish are related, and any change in one without a complimentary change in the other will affect the results of roller burnishing. As an example, a prefinish to 100 μin. (2.5 μm) and a stock allowance of 0.001″ (0.03 mm) will give the same results as a prefinish of 50 μin. (1.3 μm) and a 0.0005″ (0.13 mm) stock allowance, as long as the burnishing tool is adjusted appropriately.

Scratches or lines in the surface that appear after burnishing are referred to as tear marks. Tear marks occur when the cutting tool used prior to burnishing becomes worn and actually tears the surface, as opposed to cutting the surface.

A general and concise troubleshooting guide for roller burnishing is presented in Table 16-23. The problems and possible solutions are divided into four sections, based on the finish and geometry resulting from roller burnishing and the type of tool used (feeding contact or full contact).

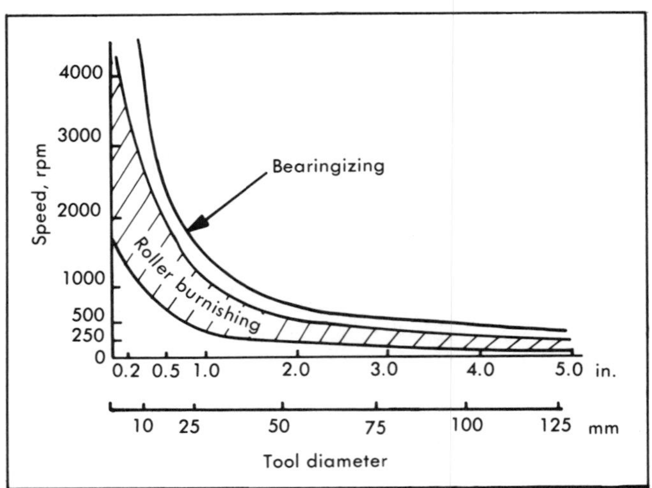

Fig. 16-38 Recommended speed ranges for production applications of roller burnishing tools of different diameters.

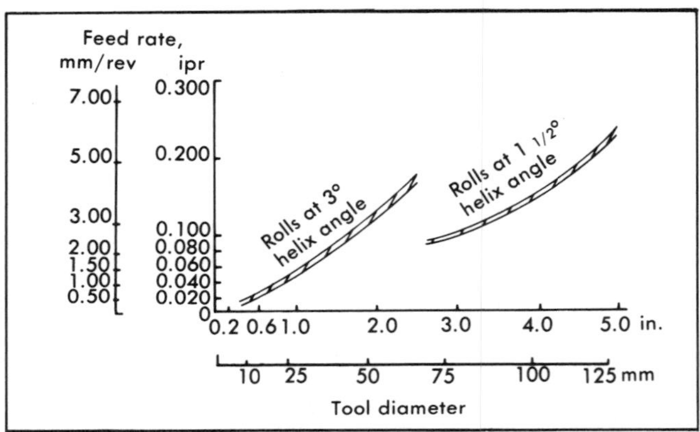

Fig. 16-39 Feed rates for roller burnishing with self-feeding tools of various diameters.

BALL BURNISHING

Dimensionally precise holes with smooth, high-density surfaces are produced by a process called ball burnishing or ballizing. In this process, an oversized, tungsten carbide ball is forced through an undersized hole. High pressure is used to force the ball through the hole at high speed. A lubricant is applied to prevent galling, seizing, or distortion of either the ball or the workpiece.

Unlike abrasive finishing methods, ballizing does not remove any metal or leave any residue. The process is similar to roller burnishing, discussed previously in this section, in that the workpiece surface is compressed by the tooling. However, roller burnishing involves cold working of the metal, while ballizing entails hot working.

Ball burnishing also differs from other types of burnishing in that a narrow, peripheral line on the ball creates a concentrated radial force that pushes the workpiece material into the surface of the hole. This narrow contact line between the ball and the wall of the hole work hardens the surface. There is no flaking of the material, and no ringlets are formed through overlapping the peaks and valleys of the machined surface.

TABLE 16-23
Troubleshooting Guide for Roller Burnishing

Problems	Possible Causes	Possible Solutions
1. Finish problems with feeding-contact tools:		
Workpieces have scratches after roller burnishing.	Prefinish R_a value is too large for stock displacement.	Reduce prefinish value or increase stock allowance.
	Peak-to-peak distance of surface profile too large.	Reduce nose radius and feed rate of cutting tool used prior to burnishing.
	Lubricant contains chips, dirt, or other foreign matter.	Clean and filter lubricant.
	Rollers excessively worn.	Inspect rollers; if discolored or marred, replace.
	Rollers stuck in cage.	Inspect and clean cage; replace if necessary.

(continued)

TABLE 16-23—*Continued*

Problems	Possible Causes	Possible Solutions
1. Finish problems with feeding-contact tools-cont.:		
Workpieces are flaked or appear scaly after roller burnishing.	Prefinish R_a value is too small for stock displacement.	Increase prefinish value or reduce stock allowance.
	Cage may be broken, preventing rollers from rotating freely.	Inspect cage; replace if necessary.
2. Geometry problems with feeding-contact tools:		
Workpieces are tapered or bellmouthed after roller burnishing.	Misalignment between the workpiece and the tool.	Correct alignment.
	Varying cross sections or improper support of thin walls.	Correct workholding fixture to fully support workpiece.
	Tool runout.	Inspect tool; repair if necessary.
3. Finish problems with full-contact tools:		
Workpieces have scratches after roller burnishing.	Prefinish R_a value is too large for stock displacement.	Reduce prefinish value and increase tool pressure.
	Peak-to-peak distance of surface profile too large.	Reduce nose radius and feed rate of cutting tool used prior to burnishing.
	Lubricant contains chips, dirt, or other foreign matter.	Clean and filter lubricant.
	Rollers excessively worn.	Inspect rollers; if scratched or worn, replace.
	Rollers stuck in cage.	Inspect and clean cage; replace if necessary.
Workpieces are flaked or appear burnt after roller burnishing.	Prefinish R_a value is too small for stock displacement.	Increase prefinish value or reduce stock allowance.
	Dwell time too long, causing overburnishing and excessive heat.	Reduce dwell time.
	Rollers stuck in cage.	Inspect and clean cage; replace if necessary.
Residual tool marks are present after roller burnishing.	Premachining operation is producing incorrect geometry.	Correct machining operation.
	Cutting tool tearing metal.	Sharpen or replace cutting tool.
	Burnishing pressure too low.	Increase pressure.
	Insufficient dwell.	Increase dwell time.
	Fixture deflecting.	Correct fixture.
4 Geometry problems with full-contact tools:		
Workpiece geometry incorrect after roller burnishing.	Improper fixture.	Correct fixture to ensure rigidity; replace if necessary.
	Burnishing pressure may cause workpiece distortion, especially with thin-wall components.	Reduce burnishing pressure and/or dwell time. Correct fixture to ensure rigidity; replace if necessary.

CHAPTER 16

BALL BURNISHING

Process Advantages

Hot working of the metal during ball burnishing refines the molecular structure of the hole surfaces. The refined structure increases bearing surface density and strength, and provides improved resistance to friction and/or heat stress. Also, to a limited degree, the process stabilizes the molecules so that workpiece distortion during subsequent heat treatment is minimized. If more precise tolerances are required after heat treating, the workpieces can be ball burnished first, then heat treated, and finally ball burnished again to correct any minor distortion that may occur during heat treatment.

Surfaces and Materials Burnished

Ballizing is used to produce smooth surfaces in blind, through, cross, and elliptical holes. It is also used to finish surfaces in elbows, S-bends, interrupted segments, and undercuts. A typical setup for ball burnishing a blind hole is illustrated in Fig. 16-40. Blind holes are also ballized by the step method, as depicted in Fig. 16-41.

Ball burnishing has been used successfully on a wide variety of metals, both soft and hard. These metals include the full range of low and high-carbon steels, tool steels, stainless steels and other nickel-bearing alloys, and superalloys. The process is also employed on sintered metal parts and many nonferrous materials, including aluminum, magnesium, and copper alloys. Cadmium and copper-plated parts and soft tungsten carbide are also finished by ball burnishing.

Limitations

There are certain limitations to the ball burnishing process. If, for example, specifications require parts to be case hardened to R_C 62, with a case depth of 0.010 to 0.020" (0.25 to 0.51 mm), the parts can be ballized after heat treating, but only for size. Through-hardened parts with the same hardness must be ballized for both size and finish prior to heat treating.

In some cases, especially with improper burnishing speeds, pressures, and lubricants, ballizing can produce small chips and burrs. Another possible limitation is that ball burnishing will generally improve out-of-round or tapered conditions by only about 50%. Workpieces must also be clean and dry before ballizing.

Typical Applications

Ball burnishing is an efficient method of forming, sizing, finishing, and deburring bores, all in one operation. Other applications include coining or staking parts together, expanding, and changing wall thickness. Figure 16-42 illustrates a swaging operation that eliminated the need for brazing in producing pressure vessels.

Deburring operations in which size or finish are generally of no concern are a common application for the ballizing process. For example, when a hole is drilled from both ends of a workpiece to maintain a sharp entry edge at both ends of the hole, there is often a mismatch and burr produced near the center of the hole (see Fig. 16-43, view *a*). Such burrs can interfere with the flow of liquids or gases in operation. Ballizing irons out such burrs (see view *b*), producing a small radius at the mismatch area but without significantly changing the sharp edges at the ends of the hole.

When crossholes are more than 25% the diameter of main bores being ballized, slight burrs are formed at the top edges of the crossholes, but about 90% of the existing burr is removed. A 100% burr removal can be attained by inserting fill plugs of exact configuration in the crossholes.

The 80 μin. (2.03 μm) finish of reamed holes can typically be improved to a 2 to 3 μin. (0.05 to 0.08 μm) finish with just one pass of a ball. On hard materials, such as heat-treated tool steels, ball burnishing can improve the finish from a previously ground 15 to 20 μin. (0.38 to 0.51 μm) to within the 5 to 6 μin. (0.13 to 0.15 μm) range. Prior to ball burnishing, holes can generally have tolerances more than double the design requirements and surface finishes more than 10 times those allowable.

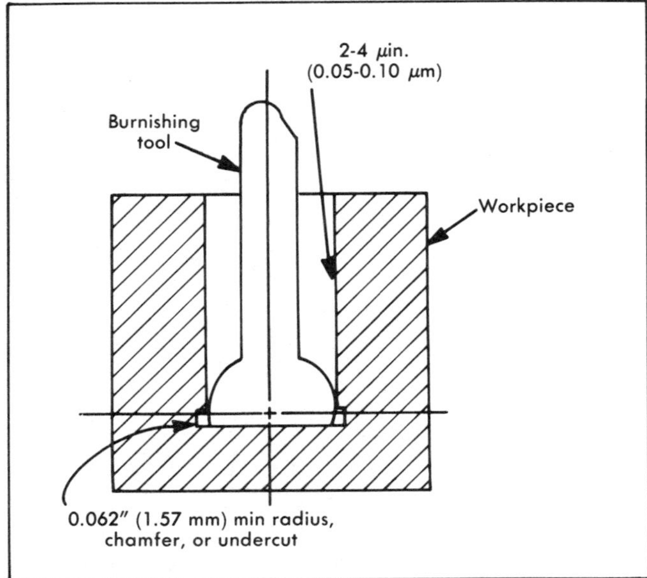

Fig. 16-40 Ball burnishing of a blind hole. (*French Enterprises*)

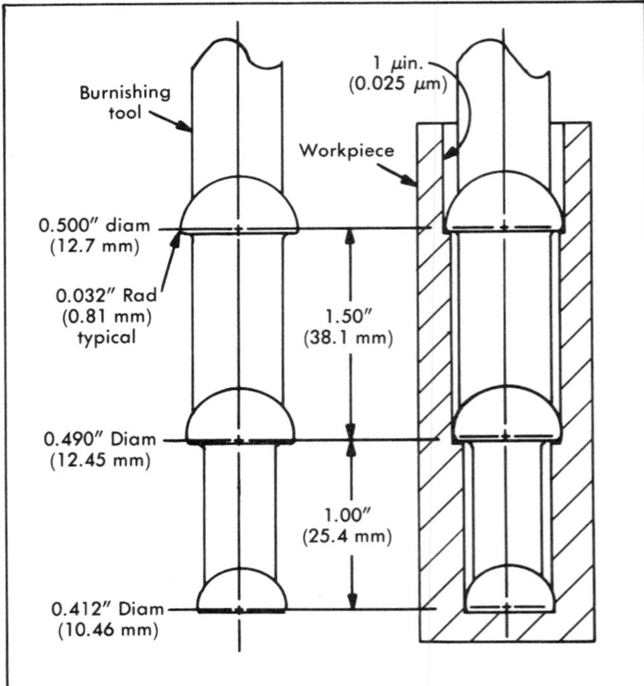

Fig. 16-41 Step ballizing of a blind hole. (*French Enterprises*)

In many cases, ballizing eliminates the need for finishing operations such as precision boring, honing, lapping, or other burnishing methods. With proper machines and precision balls, many jobs can be done faster, more precisely, and more economically.

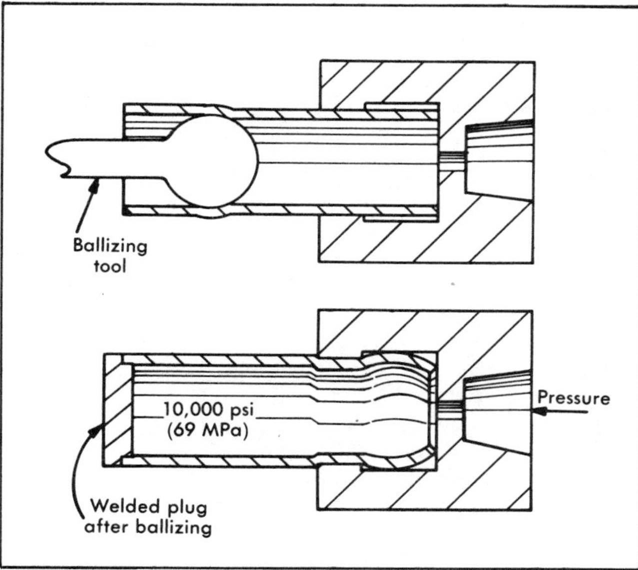

Fig. 16-42 Swaging by ballizing eliminates the need for brazing. (*French Enterprises*)

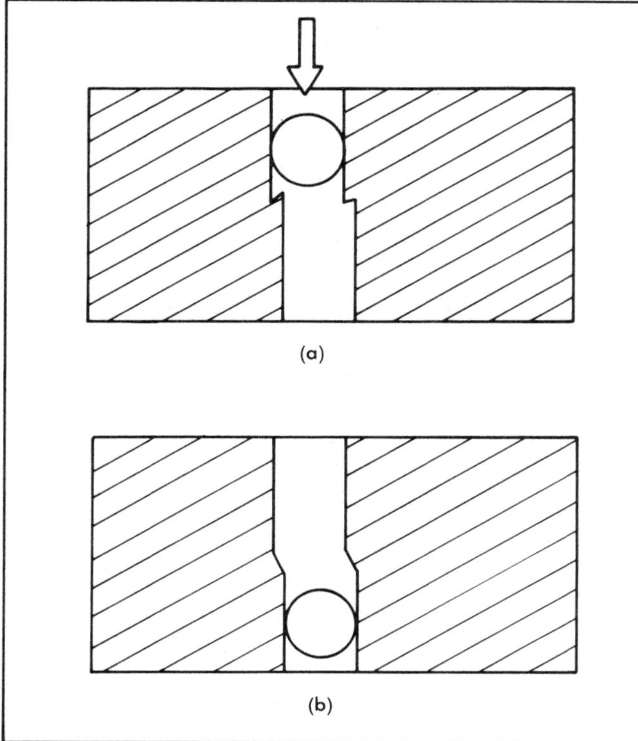

Fig. 16-43 Deburring by ballizing: (a) burrs near midpoint of hole produced by drilling from both ends and (b) burrs ironed out by pressing ball through hole.

Successful applications include parts for electric motors, grinders, precision machine tools, and automotive and aircraft assemblies. Other high-precision components finished by ballizing include parts for missiles, electrical and electronic instruments, medical and dental devices, air and hydraulic valves, fuel injection nozzles, carburetor jets, robot components, and computer linkages.

Machines and Equipment

Machines. While some ballizing is done by pressing balls through holes with arbor presses or other mechanical means, faster production and more precise results are being achieved with high-speed presses. Even higher production rates are attainable by the automatic recirculating-ball method, called the Ball-O-Matic process, patented by French Enterprises, Inc.

Ballizing production rates vary widely, depending upon the workpiece configuration and material, hole length, ball speed, and the machine used. With hand loading/unloading, the average production rate is 1000-1700 parts per hour. With automatic loading, the rate can reach 10,000 per hour.

Automatic machines incorporate a ball return track that allows a single sizing ball and several pusher balls to return to the top of the machine (see Fig. 16-44). During recirculation, the balls are cleaned, cooled, and lubricated. A special configuration in the ball track prevents irregular ricochet patterns or bounce-back as the balls return to their starting positions.

Controlled air pressure behind the balls, as well as controlled vacuum ahead of them, allows the balls to seat instantly when they reach the top of the track. The ball tracks can be quickly interchanged to handle different size balls. A modification for dual operation is illustrated in Fig. 16-45. In this arrangement, a smaller free-floating ball recirculates, and a larger ball is brazed to the end of a vertically reciprocating shank.

One manufacturer offers ten standard ballizing machines. Sizes range from a 1/4 ton (2.2 kN) capacity unit, with a ball size range of 0.015 to 0.062" (0.38 to 1.57 mm) diam, to a 10 ton (89 kN) machine, with a ball range of 1.000 to 5.000" (25.40 to 127.00 mm) diam. Smaller models operate on air, while larger machines are powered hydraulically. The force capacities of various machines generally used to ball burnish holes of different sizes are presented in Table 16-24. All standard machines are designed to accommodate hand, semiautomatic, or automatic tooling.

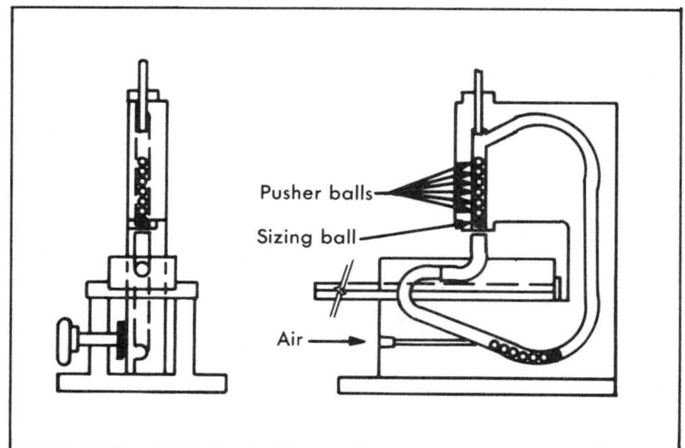

Fig. 16-44 Automatic burnishing machine incorporates ball-return track. (*French Enterprises*)

BALL BURNISHING

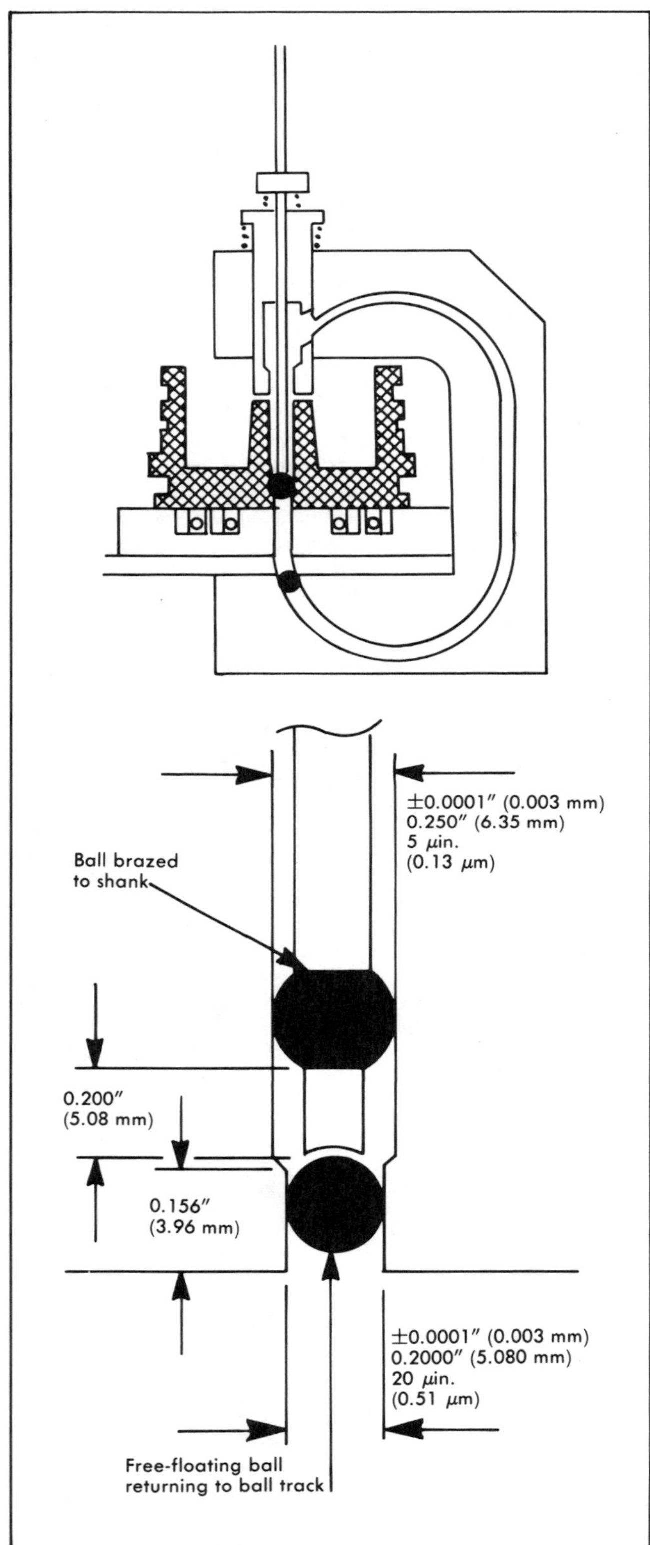

Fig. 16-45 Dual burnishing operation employs a small recirculating ball and a large ball brazed to the end of a reciprocating shank. (*French Enterprises*)

TABLE 16-24
Force Capacities of Machines Used to Ball Burnish Holes of Various Sizes

ID* of Hole Before Ball Burnishing, in. (mm)	Force Capacity of Machine, tons (kN)
0.015-0.062 (0.38-1.57)	1/4 (2.2)
0.020-0.140 (0.51-3.56)	1/2 (4.4)
0.062-0.375 (1.57-9.52)	3/4 (6.7)
0.062-0.625 (1.57-15.88)	1 1/4 (11.1)
0.062-1.140 (1.57-28.96)	6 (53.4)
1.000-5.000 (25.40-127.00)	10 (89)

(*French Enterprises, Inc.*)

* ID = inside diameter

Custom machines have been developed for special applications. One horizontal machine is capable of finishing 20 ft (6.1 m) long tubes having inside diameters ranging from 2.000 to 5.000″ (50.80 to 127.00 mm). A high-speed machine for sizing and finishing shock absorber tubing has a production rate of four tubes every 4.5 seconds.

Standard machines can burnish two, three, or four holes having different diameters simultaneously, providing the holes have a common centerline. Multiple holes in a single workpiece can also be ballized at the same time, with special tooling eliminating the need for repositioning the workpieces for each hole. Machines are controlled by an air logic device that prevents the machines from operating until the balls are in place at the tops of the tracks, seated, and ready for the next cycle.

Burnishing balls. The precision tungsten carbide balls used for burnishing have a cobalt content of 5.5 to 6.5% and are available in diameters from 0.002 to 5″ (0.05 to 127 mm). Diameter uniformity of the balls is maintained within 0.000050″ (0.00127 mm), and roundness within 0.000025″ (0.0006 mm). Burnishing balls have a compressive strength of 643,000 psi (4433 MPa), in conformance with standards of the Anti-Friction Bearing Manufacturers Association (AFBMA). Normal service life of the balls ranges from 500,000 to 2,000,000 holes, depending upon the hardness and/or abrasiveness of the material being worked. A mist of special lubricant, having a high flash point of 1500° F (815° C), is used to resist the high pressures and prevent material pickup or galling, even under the elevated temperature generated during ball burnishing.

Fixturing. Automatic fixturing, including workpiece hoppers, is available for single-part shuttle operation, permitting production rates of 2500 to 10,000 pieces per hour. Automatic operation includes the ejection of finished parts into holding receptacles. The machine automatically stops if either a workpiece or the ball tooling is not in correct position.

Workholding. Because the ball follows the centerline of the hole in most burnishing operations, workpieces need not be clamped or chucked. The workpieces are normally nested or allowed to float so that the ball follows the centerline of the hole automatically. Neither the workpiece nor the tool rotate. Each free-floating nest is generally designed to accommodate two workpieces or two different holes in a single workpiece.

In determining the nest diameter for ballizing a bushing or similar workpiece, the ID of the nest should be from 0.002 to 0.005″ (0.05 to 0.13 mm) larger than the OD of the workpiece, providing the wall thickness of the workpiece is not abnormally thin. For workpieces having very thin walls, the nest ID is generally not made more than 0.001" (0.03 mm) larger than the OD of the workpiece. The use of a fairly tight nest helps prevent distortion. Proper nesting can usually prevent expansion of the OD's of thin-walled tubular parts, if desired, and can also prevent camber or other distortion during ballizing, particularly on long parts.

Operating Parameters

There is no formula to determine the amount of material that a ball will push into the wall of a hole. With softer materials, such as screw machine stock, a normal amount of interference between hole ID and ball OD to produce a good finish is 0.003" (0.08 mm). Average interference ranges from 0.0002 to 0.0005" (0.005 to 0.0013 mm) for balls smaller than 1/32" (0.8 mm) diam, to 0.002 to 0.003" (0.05 to 0.08 mm) for balls up to 1" (25.4 mm) diam. The actual amount of interference depends upon springback of the workpiece material and size and finish requirements. When tolerances are not particularly tight and a very dense surface finish is desired, an interference up to 0.010" (0.25 mm) can be readily achieved.

Ball speeds generally range from 120 to 310 ips (3048 to 7874 mm/s), depending upon workpiece material and length of hole. The hardness of a ballized surface cannot be determined precisely. Estimates range from 2 to 15% harder than the surrounding material. Wear tests indicate an increase in wear life of from 20 to 30 times.

SPECIAL-PURPOSE MACHINES AND ROBOTIC DEBURRING

Specialized machines are being used extensively to remove burrs and finish parts when production requirements are high or families of similar parts are processed. Various machines available are manually, semiautomatically, or automatically operated, depending upon production needs.

Tools employed on the different machines include trimming dies, rotary cutters or pinch rolls, grinding wheels, brushes, and honing stones. Typical parts processed on such machines include stampings, castings, forgings, and extrusions; sheet and plate stock; many different types of workpieces; tubes and bars; gears; and cutting tool inserts.

TYPES OF SPECIALIZED MACHINES

Specialized machines for deburring and finishing include trimming presses, edgers, end finishers, gear tooth deburring/chamfering machines, and cutting-tool-insert finishing machines. Also, when feasible, industrial robots and NC machining centers are used. Robotic deburring is discussed later in this section.

Use of NC/CNC Machines

The NC/CNC machines that produce parts can often effectively deburr or finish the parts, or many features of the parts. These machines ensure the accuracy of features produced in the machining operations and can result in considerable savings by eliminating the need for subsequent operations. Because of more liberal tolerances on cast or forged surfaces, these machines may not be satisfactory. When wall locations vary from part to part, edges may be produced with too much or too little chamfering or rounding.

Deburring and/or finishing tools can often be stored in the magazines of NC/CNC machining centers that have automatic toolchanging capabilities. Macros are available for the NC/CNC units to remove burrs by automatic chamfering.

Several CNC machines are specifically designed for removing sprues, risers, gates, and runners from castings, a process called *fettling*. These are rigid, single-tool machines that often employ touch sensors to define surface orientations and then rotate the part to provide a straight-line path for the cutter. Such CNC machines are used on simple contours for nonprecision edge requirements.

When considering NC/CNC machines for precision deburring, it is important to consider the following factors:

- Rotary burs are designed to be used at high speeds and low forces. Many of the spindles on NC/CNC machines will not approach the recommended cutting speeds for these tools.
- When deburring and machining are performed on the same setup, the operator has better control or visibility with respect to appropriate toolchanging times. If dull tools cause too large a burr, the burrs will not be removed by chamfering.
- Abrasive stones or filled rubber tools generate a large number of fine abrasive particles that can become lodged in collets, holding fixtures, and machine ways.
- The rubber in some abrasive-filled rubber products reacts with coolants to produce sticky machine-way surfaces.
- Thin flanges may distort during machining to the extent that chamfering tools may not even touch the edges. This possibility is a probability when flanges are less than 0.030" (0.76 mm) thick and chamfers of 0.003" (0.08 mm) maximum are required.

Trimming Presses

Mechanical, hydraulic, or pneumatic presses are used extensively to remove excess material from stampings or drawn parts, or to remove flash from forgings, castings, or moldings. This method requires the use of cutting dies, such as pinch-trim, Brehm (shimmy), or notching dies. Shaving dies or broaching tools are employed for more precise requirements. Presses and dies are discussed in detail in Volume II, *Forming*, of this Handbook series.

For some applications, trimming is performed by roller cutters rather than a press-mounted die. This approach generally requires less expensive tooling than with trim dies. Roller dies can accommodate stock up to 1/2" (12.7 mm) thick, at production rates up to 500 parts per hour.

Edging Machines for Sheet Metal

Sheet metal edgers use either small grinding wheels or pinch rolls to remove burrs. Those employing grinding wheels are generally adjustable to vary the chamfer produced, and some will deburr both the top and bottom edges on one side of the sheet (see Fig. 16-46). Such machines are available for sheet stock ranging from 0.023 to 0.250″ (0.58 to 6.35 mm) thick and in widths of 1/2″ (12.7 mm) or more. Linear feed rates of 60 fpm (18.3 m/min) are possible.

Edging machines using pinch rolls to remove burrs or reroll the material into the edges can deburr both sides of sheet stock, as well as top and bottom, in a single pass (see Fig. 16-47). These machines can process sheets ranging from 0.023 to 0.120″ (0.58 to 3.05 mm) thick and in widths to 48″ (1219 mm). Feed rates up to 100 fpm (30.5 m/min) are possible.

On some machines, the stock passes through a set of wide pinch rolls that feed the stock. The next set of rolls, mounted on a vertical axis, set up the edges. Subsequent sets of power-driven rolls perform the edging and finishing operations. Either chamfered or radiused edges can be produced.

The rolls, similar to small V-belt pulleys, are hardened and ground. Pressure exerted by the rolls progressively forces the burrs and sharp edges over, under, or into the stock. While this method works well with steel, stainless steel, and most aluminum sheets, it will not deburr dead-soft copper or aluminum. The latter materials tend to buckle under roll pressure.

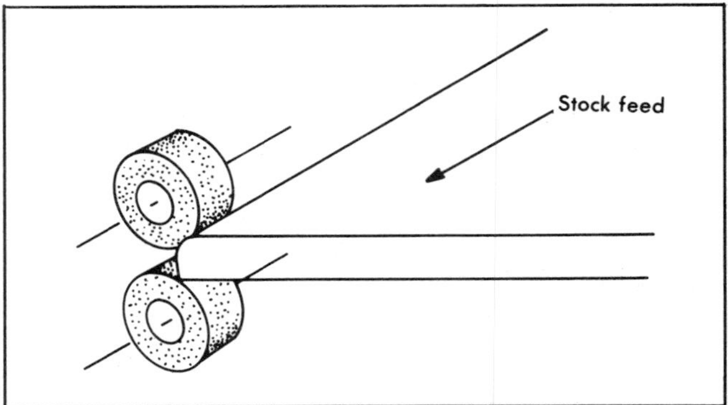

Fig. 16-46 Small grinding wheels deburr and chamfer top and bottom edges of sheet stock at feed rates to 60 fpm (18.3 m/min). (*Bendix Corp.*)

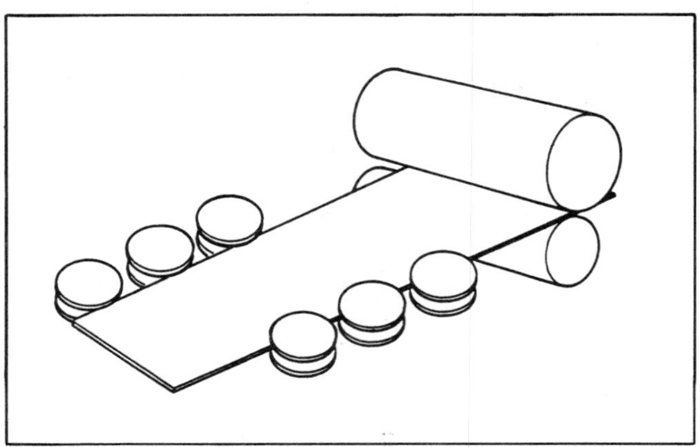

Fig. 16-47 Pinch rolls are often used to remove burrs from sheet stock ranging from 0.023 to 0.120″ (0.58 to 3.05 mm) thick. (*Bendix Corp.*)

Workpiece Edgers

Two basic approaches are used for general-purpose workpiece edging. With one approach, rectangular workpieces are placed in an angle-iron trough and hand fed over a small grinding wheel, providing a small, adjustable chamfer. The second approach is to use a piloted chamfering tool extending upward through the flat table of a machine (see Fig. 16-48, view *a*). Workpieces are moved into contact with the pilot on the cutter and then fed across the cutter, using the pilot as a guide or stop. Chamfer depth is controlled by adjusting the cutter upward or downward. Another approach is to install a permanent stop (view *b* of Fig. 16-48) or a magnetic device and use a standard chamfering tool. Straight and contoured edges are deburred with such machines.

End-Finishing Machines

Finishing machines for the ends of tubes and bars are available in various degrees of sophistication and automation. A manually operated machine capable of finishing up to 1500 ends per hour is illustrated in Fig. 16-49. Simultaneous inside-outside deburring, chamfering, and facing are possible on this machine. The operator places the workpiece through the self-centering chuck jaws, against an adjustable stop, and pulls the starting lever. This action closes the chuck jaws, pivots the stop clear, and feeds the work to the rotating cutters.

On some machines, the rotating cutter is mounted on the same spindle as an abrasive cutoff saw. In operation, the workpiece is sawed to length and then fed into the cutters. Double-end machines are available to deburr, chamfer, and face both ends of tubes, rods, or pipes at rates to 2000 parts (4000 ends) per hour. Some machines are integrated into transfer lines, using magazine or cradle loading, unloading, and storage devices.

Single-Purpose Machines

Single-purpose machines are available in a wide variety of types for deburring, brushing, grinding, polishing, and buffing. They are designed for manual, semiautomatic, or automatic operation. Many of the machines are designed to handle a specific workpiece, but some can accommodate different parts by changing the workholding fixtures and/or the finishing heads. A number of these machines are discussed in the preceding sections of this chapter under the subjects of brushing, polishing, and buffing, and some types are illustrated schematically in Fig. 16-21.

One type of special-purpose machine uses grinding wheels to deflash four different faces of cast engine blocks. A number of machines have been built to deflash, grind flat surfaces, and polish and buff hand tools to specific requirements. Such machines frequently use rotary or straight-line automatic transfer of the workpieces, with up to 30 stations per machine. Multistation machines have production rates up to 3000 parts per hour. Some machines rotate workpieces both clockwise and counterclockwise to reach recessed corner areas.

An automated machine for deburring flat sheet metal parts is illustrated in Fig. 16-50. It consists of two modules, each containing two media wheels that are webs of abrasive-impregnated fibers, and a hot air dryer. The four counter-rotating wheels are rotated at a surface speed of 4000 to 5000 sfm (1220 to 1525 m/min) and oscillate along their axis of rotation. Water is applied for cooling and rinsing, and then filtered before recirculating. Workpieces are fed through the machine by a chain-driven conveyor with rubber-covered rolls. For one-pass operation, conveyor speed is normally 20-30 fpm (6.1-9.1 m/min).

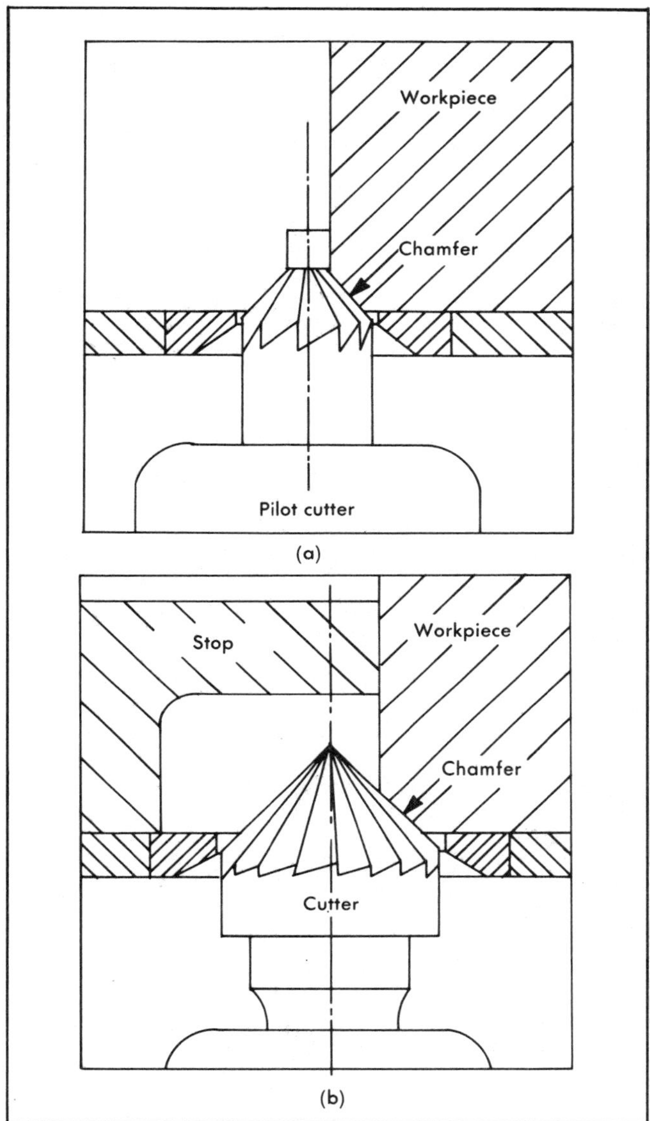

Fig. 16-48 Two concepts for locating edges on a general-purpose workpiece edger: (a) piloted chamfering tool and (b) permanent stop. (*Reishauer Corp.*)

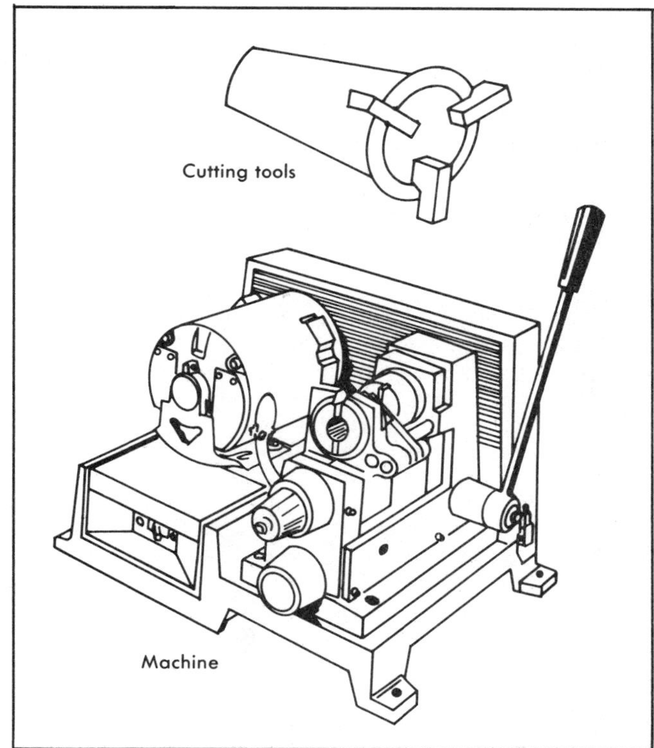

Fig. 16-49 Tube and rod end-finishing machine. (*Teledyne Pines*)

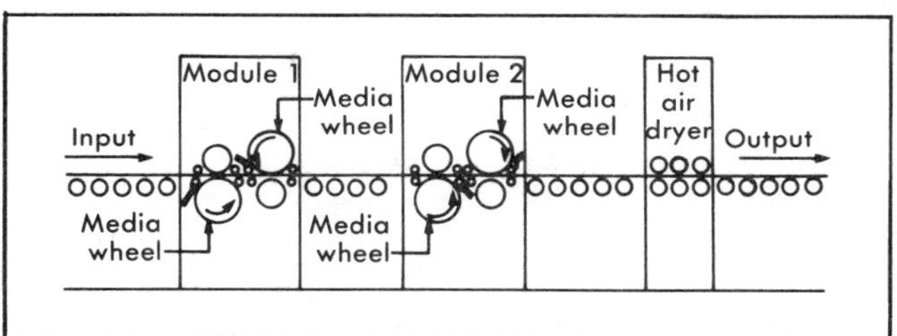

Fig. 16-50 Automated machine for deburring flat sheet metal parts. (*Lockheed-Georgia Co.*)

Special contouring machines are being used to deburr and finish compressor and turbine blades, and other parts having complex contours. In one design, the deburring head has three-dimensional motion, fully controlled by three plate cams mounted on the same shaft. The head can be equipped with a high-speed spindle for handling rotary burs or milling cutters. A fourth motion can be incorporated to maintain the cutting edge of a slowly rotating, single-point tool normal to the workpiece surface. Rotational speed of the camshaft can be varied to change the cycle time, usually 5 to 18 seconds per workpiece, depending upon the complexity of the operation.

Gear Deburring Machines

In gear hobbing or shaping, burrs are formed on the end faces of the teeth. Such burrs can seriously impair smooth and reliable gear operation. A wide variety of methods are used to deburr and chamfer gear teeth. These methods include using a rotating tool or small grinding wheel to traverse all tooth edges, brushes, a skiving tool and subsequent brushing, an abrasive-belt grinder (sometimes followed by brushing), and blasting with steel shot.

For heavy burrs, it is possible to combine a milling operation with tooth chamfering. Many gears, particularly those used by the watch industry, are being deburred on gear hobbing machines, thus eliminating a secondary operation. Small pinions are deburred by automatically holding a lathe tool against the hob exit side of the workpiece, the tool shearing the burrs. Some machines retract the burr cutting tools several rotations prior to the end of the hobbing cycle. This action allows the hob to remove the remaining burrs between the gear teeth produced by the tool. Rotating milling cutters are used on larger gears.

Several designs of special machines are available for deburring, chamfering, or rounding gear teeth. One type uses a rotary deburring tool (a driving gear and two cutters), with the workpiece driven by the tool. The workpiece and tool are in a parallel-axis relationship, and the tool automatically swings into engagement with the clamped workpiece. Stock removal along the edges of the workpiece teeth is accomplished by overlapping cuts of the edges on the rotary tool; simultaneously, stationary tools engage lateral surfaces of the workpiece to remove any projecting burrs. At the completion of the cycle, the rotary tool swings back and the workpiece is released for unloading. Clusters of up to four gears of different configurations can be chamfered and deburred simultaneously on these machines.

A cam-operated machine for deburring, chamfering, or rounding teeth can handle spur and helical gears, as well as splines, both internal and external. The horizontal cutter spindle of this machine is advanced in timed relationship with the rotation of the gear teeth for continuous cutting. The cutter spindle is mounted in an eccentric housing that can be offset to either side of the work spindle for operations on helical gears. One machine chamfers both ends of external gear teeth simultaneously at a rate of 400 teeth per minute.

Gear Burnishing Machines

Machines are available to remove nicks and burrs from the tooth surfaces of gears. One machine is of the centerless type and uses three hardened or abrasive-impregnated burnishing gears to finish the workpiece with adjustable burnishing pressure. Gears are automatically fed into the workstation, where they are engaged by the burnishing gears. The workpiece gear is constrained by pads supported by a series of leaf springs that are driven by an eccentric shaft. This motion causes the workpiece gear to oscillate as the burnishing gears revolve. Production rates are as high as 900 gears per hour.

Edge Honing of Cutting Tool Inserts

Machines are available for deburring and honing the edges of carbide or ceramic inserts for cutting tools. In addition, the machines are used for deburring flat and cylindrical components, and can also produce sharp corners with no burrs.

On one such machine, the inserts are finished between two rotating, resilient, abrasive discs. The discs consist of silicon carbide abrasive in a rubberlike bond and have a pressure-sensitive adhesive on their backs for application to the wheels. Each insert is held and guided in a separate opening of a carrier plate, which is part of an epicyclic sprocket holder. This design is similar to the two-wheel lapping machines discussed later in this chapter. The machine has three independent drives for the upper and lower wheels and for the carrier plate.

The required pressure is applied to the top wheel through an adjustable pneumatic device. Pressures and processing times vary with the sizes and shapes of the inserts, as well as with the materials from which they are made and the required finish. Depending upon requirements, insert edges can be rounded on one or both sides. For one side, the top disc is replaced with a pressure pad having a wear-resistant surface. The form of the honed edge is a multiradius compound curve having an average radius of about 0.0017" (0.043 mm).

ROBOTIC DEBURRING, FETTLING AND FINISHING

For many companies, the use of industrial robots for deburring, fettling, and finishing has considerably reduced costs and improved quality. Robots can operate three shifts a day, virtually untended; they reproduce the same motions accurately; they can process parts faster than humans; they can use heavier, higher powered tools for faster finishing; and they can work in dusty, noisy environments that are unsuitable and/or hazardous for humans. Robotic deburring, fettling, and finishing are not actually processes in themselves, but are simply automation added to mechanical cutting, grinding, brushing, blasting, and other operations.

Robotic deburring, fettling, and finishing are typically used for parts that have production rates falling between job shop quantities and automotive industry requirements. Robotic methods are not normally economical for one-time runs of very low quantities because the engineering and programming time exceed savings. Unlike many off-hand processes, the use of robotic systems requires considerable planning to ensure optimum results.

Typical Applications

Industrial robots are being used for many different deburring and finishing operations. Most applications are for large parts and are dedicated to a specific part or close geometric family of parts. Some precision and intricate parts are being deburred by robots, but they are the exception because many robots are not technically capable of performing precision deburring.

The foundry industry is a major robot user, with the robots handling the castings or tools, depending upon the size, weight, and configuration of the casting; the material from which the casting is made; the amount of stock to be removed; the type of operation; and production requirements. Operations performed include deburring and chamfering internal edges, and removing sprues, flash, risers, gates, parting lines, and runners—a process called *fettling*.

Because fettling requires cutting through thick sections of cast metal and smoothening wide sections of flash, high-powered tools are required and large forces occur, necessitating more powerful robots. While a human may use a 4 hp (3 kW) power tool in fettling, robots can handle 45 hp (34 kW) tools, which may weigh 50 lb (22.7 kg) and result in cutting forces of 100 lb (445 N) or more. In contrast, deburring will typically use 1/3 hp (248 watt) tools and involve forces of 1 to 10 lb (4 to 45 N).

Other robotic operations performed in many industries include brushing, polishing, buffing, and grinding. Automotive parts being deburred and finished with robots include transmission and steering knuckle housings, connecting rods, and plastic moldings. Oil holes in crankshafts are being deburred with robot-held tools. Laser-drilled holes, which have resolidified molten-metal burrs on their edges, are deburred by wet abrasive blasting; the blasting nozzle is attached to the robot arm.

Robot Requirements

Inherent limitations of robot design are often subtle. As an example, some of the radial or polar-motion robots cannot move in a straight line. Attempts to approximate straight lines require memories far in excess of conventional capabilities. At least one radial-motion robot has a wrist that allows it to move in a straight line; unfortunately, it cannot make circular motions without significantly extended memory. There may also be a limitation of too-limited vertical travel for some deburring applications, but the robots may still be appropriate for shallow parts.

For successful deburring, fettling, and finishing, robots should have the following characteristics: high accuracy and repeatability, continuous-path capability, the capacity for easy and quick tool or spindle changes, rigidity, and low inertia. Other desirable features include easy off-line programming, circular interpolation, and the ability to translate movements to similar features on the same workpiece.

Additional features often required for robots are a large-reach envelope and a small wrist/arm configuration. Quick servo response, no backlash, and calibration are desirable for most applications. Load capacity of the robot is generally specified to be three times the expected working force. Five-axis capability is generally the minimum requirement; six axes are necessary when nonrotational, nonsymmetrical tools are used.

Depending upon the accuracy requirements for a specific application, the specifications and costs of a robotic system can vary greatly. The closer the tolerances and accuracies that must be maintained on the workpieces, the higher the cost of the system. For applications involving workpieces having wide tolerance allowances, lower cost robots with less accuracy but good repeatability can often be used. Such applications are especially used with compliance tooling or fixturing.

Robot Accuracies

Positioning accuracy. Few manufacturers of industrial robots state the accuracy of their products. Many robots cannot go to a predefined point or location, such as x = 1 mm, y = 2 mm, and z = 0.00 mm. Because of this limitation, the traditional NC machine concept of programming to a point and then measuring how close the machine came to that point is not directly transferable to robots. Even when robot controls allow this predefinition of destination, precision robots may miss their targets. This limitation is not acceptable for deburring when a tolerance of ±0.005″ (0.13 mm) is required on parts. While compliance overcomes some of the accuracy limitations, it is desirable to have a robot with high accuracy for deburring. Less accuracy is acceptable for fettling operations.

Some robot manufacturers specify "accuracy to a point in space." In this case, the programmer can move to a point in space by using visual judgment or sensory devices that identify the point. That point, however, is not associated within the machine as a dimension; it is identified as *resolver counts*. Some emerging robot technology that uses visual or electrical signals now achieves accuracies of 0.002″ (0.05 mm) to a "point" in space. Accuracies to edge contour (continuous-path rather than isolated points) typically are ±0.025″ (±0.64 mm). Most continuous-path machines use hydraulics rather than servos, losing the inherent accuracies of servo resolver systems. Electric robots now available provide continuous-path capability and high accuracies, and eliminate the noise and other undesirable aspects of hydraulic units.

Repeatability. The repeatability of an industrial robot in deburring applications depends upon a combination of the following factors:

- The accuracy and positioning repeatability of the robot mechanism.
- Consistency in positioning the edges of the workpiece to a reference surface (clamping forces can significantly deflect some parts).
- Nature of the edge before deburring (burr uniformity, thickness, and location consistency).
- Repeatability of the deburring tool in the robot holder.
- Repeatability and accuracy of the cutting tool geometry.

Control Methods

Most precision robots are point-to-point machines with the capability of storing a maximum of 3000 points. They have no circular interpolation capability, and this effectively prohibits their use for deburring. If a machined casting has 100″ (2540 mm) of edge to deburr and points are identified every 0.003″ (0.08 mm) to ensure that a 0.010″ (0.25 mm) maximum break requirement is not exceeded, then more than 30,000 positions have to be stored—ten times the capacity of most robots. For complex parts, the only workable solution is to select a continuous-path robot.

Point-to-point robots. These are programmed by moving each axis of the robot to a position that yields the desired location of the robot end effector, and then recording the individual position of each of the robots axes into memory. In replaying these stored points, each axis runs at its maximum or some limited rate until it reaches its final position. As a result, some of the axes reach their destination prior to others. Because there is no coordination of motion between axes, the path and velocity of the end effector between points is not easily predicted.

Continuous-path robots. These are generally programmed by the operator who physically grasps the robot arm, or a similar teaching arm, and leads it through the desired path in the exact manner and velocity desired for the robot's motions. During this programming, the position of each axis is recorded into the robot's memory on a continuous-time basis, thus generating a continuous-time history of each axis position. The sampling frequency is typically in the range of 60 to 80 Hz.

Controlled-path robots. These are programmed by manipulating the robot arm to the desired tool center-point location and tool orientation. Instead of moving each axis individually, computer control allows coordination of all robot axes when teaching the program. The position of the robot's arm is recorded in the controller's memory at each desired program-point location. During program replay, the robot's computer control generates a mathematically definable path between the program points, including the control of velocity along the path and acceleration/deceleration for the start/stop points.

With the controlled-path type of robot motion control, several modes of generating the robot's path are possible, depending on the type of motion interpolation employed. These interpolation types include joint, linear, and curvilinear. Joint interpolation results in controlled-path motions in which the computer calculates the individual robot axes motions so that each arrives at the program-point location at the same time. Linear interpolation results in straight-line motion of the robot's tool center point. Curvilinear interpolation results in tool center-point motions that follow a desired curved path. Robots are available with one, two, or all three interpolation systems in a single unit. Selection depends upon specific requirements.

Programming Robots

On-line programming. Many robots in operation today use a teach pendant to program path moves (spray painting robots are one exception). The operator stands near the robot and uses a joy stick or directional buttons to cause the robot to traverse the path in a learning mode. It is a mode of try, correct, try, correct, and correct again. It is a convenient approach for simple parts because the results are immediately visible, as are the needed corrections. Because of robot inaccuracies and servo lag, it is also essential in many situations to at least make final adjustments to the program through on-line programming.

This approach has some limitations, including the following:

- It shuts down production until the program is developed.
- Because the deburring motor and cutter masks tool-to-workpiece contact, it may be difficult to make precision adjustments.
- Because the operator must be near the robot arm, it is not as safe as desired if the arm should suddenly fail.
- Considerable time may be required to obtain precision edge results on complex parts.

For continuous production of the same part for several weeks, however, the downtime incurred for teaching may not be significant.

Off-line programming. When cutter paths generated for NC machining exist, off-line programming has several advantages. The robot program can, at least theoretically, be generated quickly, and it can be done without having to face the problems of seeing the hidden cutter-workpiece interface. The deburring path can be generated and debugged without stopping production. The operator does not have to be in proximity to the robot arm for as long a time, which increases safety. In addition, computer-generated programs can produce smoother and more precise transitions around complex geometries.

Very few robots now in use have an off-line programming capability because postprocessors are just beginning to be developed, and the existing technology does not account for tool or robot arm deflections due to weights and cutting forces. As a result, off-line programming provides only a good starting point that must be corrected during a trial run. It has been predicted, however, that most robots for deburring, fettling, and finishing will soon have off-line programming capability.

Approaches to Robotic Deburring

At present, there are three philosophical approaches to accommodate robot inaccuracies and workpiece variations and to obtain more precise results. They are identified as compliant, fine-tuned robot, and force feedback approaches.

Compliant approach. This approach uses compliance to accommodate robot inaccuracies and workpiece variations. It can be used with any robot; but for deburring, finishing, and grinding operations, robots with continuous-path capability are generally used. Five or six-axis robots may be required.

Although every robot has some inherent compliance, the compliant approach adds extra compliance in one or more directions by using special-design toolholders. Single-axis compliance is normally necessary for smooth edges; although, occasionally, controlled two-direction compliance is needed. For some buffing applications, three-directional compliance is advisable.

If designed properly, programmed correctly, and if proper cutting tools are used, the compliant approach produces consistent edge breaks of 0.010" (0.25 mm) or less, even though part edge locations vary considerably between parts. Since a consistent volume of material is removed per unit of time and the burr volume is normally much less than the volume of the parent material, consistent edge breaks can be maintained even if the relative burr sizes vary slightly.

A major variance between NC machining centers and robots is the difference in their dynamic rigidity. The tool needs to move around a corner quickly, but it must move so that the surface finish on the workpiece meets requirements without chatter marks. Compliance is often needed in the direction perpendicular to the part surface, but it must be absolutely rigid in the feed direction to prevent chatter and skipping. As illustrated in Fig. 16-51, a robot is blind when the force measurements alone must determine whether any change in force is the result of a change in burr thickness, contour variation, or encountering an unexpected surface.

Rigidity is the key to the success of one robot versus another. If the tool can twist in any direction, it will climb, push away, dig into the surface, or chatter, and will not traverse the corner as accurately as desired. An ideal robot would have servos quick enough to compensate for the system, but ideal robots do not exist.

Fine-tuned robot approach. By fine tuning servos and modifying resolvers, it is possible to improve robot accuracy by a factor of two or three. This fine tuning slows the system slightly, and the servo may be pushed to an overload condition when the tool tries to go to an exact position. Theoretically, this approach could eliminate the need for compliance. In practice, however, because of workpiece variations, the need to minimize chatter, and the need for maximum reliability, the fine-tuned approach is not normally recommended.

Every time mechanical maintenance is performed on a compliantless, fine-tuned robot, every point in every program must again be retaught. Similarly, any accidents will require reprogramming of every program. Motions are defined from workpiece drawings (off-line programming). The fine-tuned approach requires a six-axis, continuous-path robot and depends upon the small amount of inherent compliance.

Force feedback. The concept of force feedback is well established, but the practicality of using this approach for robotic deburring has not been resolved for most parts because the dynamic effects of the robot system are so difficult to define and control. Force feedback, however, is seen as one eventual solution to the problems of robot inaccuracies, workpiece tolerance variations, and the use of on-line databases and off-line programming.

Force feedback requires the use of a five or six-axis robot having continuous-path capability. The six-axis force sensing unit and a computer interact to tell the robot what to do next. If necessary, the tool passes over an edge several times to achieve the desired force profile.

The following four constraints prevent the widespread use of force feedback:

- The robotic system cannot respond quickly enough for many needs.
- The dynamic response of the robot system is difficult to define for all arm positions.
- General algorithms that provide the robot logic are difficult to generalize for three-dimensional curvilinear geometries having variable burr size and fluctuating part geometries.
- Robot compliance affects sensor data.

Compliance of the robot is a significant factor in defining control logic. When the robot deflects, as a result of compliance, then the system never knows what the real cutting or interference forces are. Compliance may vary as a function of wrist position in space.

In maneuvering the robot arm on any complex workpiece, the reaction moments are not organized along the robot axis in any particular fashion. Determining what is going on at the cutting tool when a force sensor is placed on the wrist is a difficult problem. On complex workpieces, the sensing system may need another independent feedback system to close

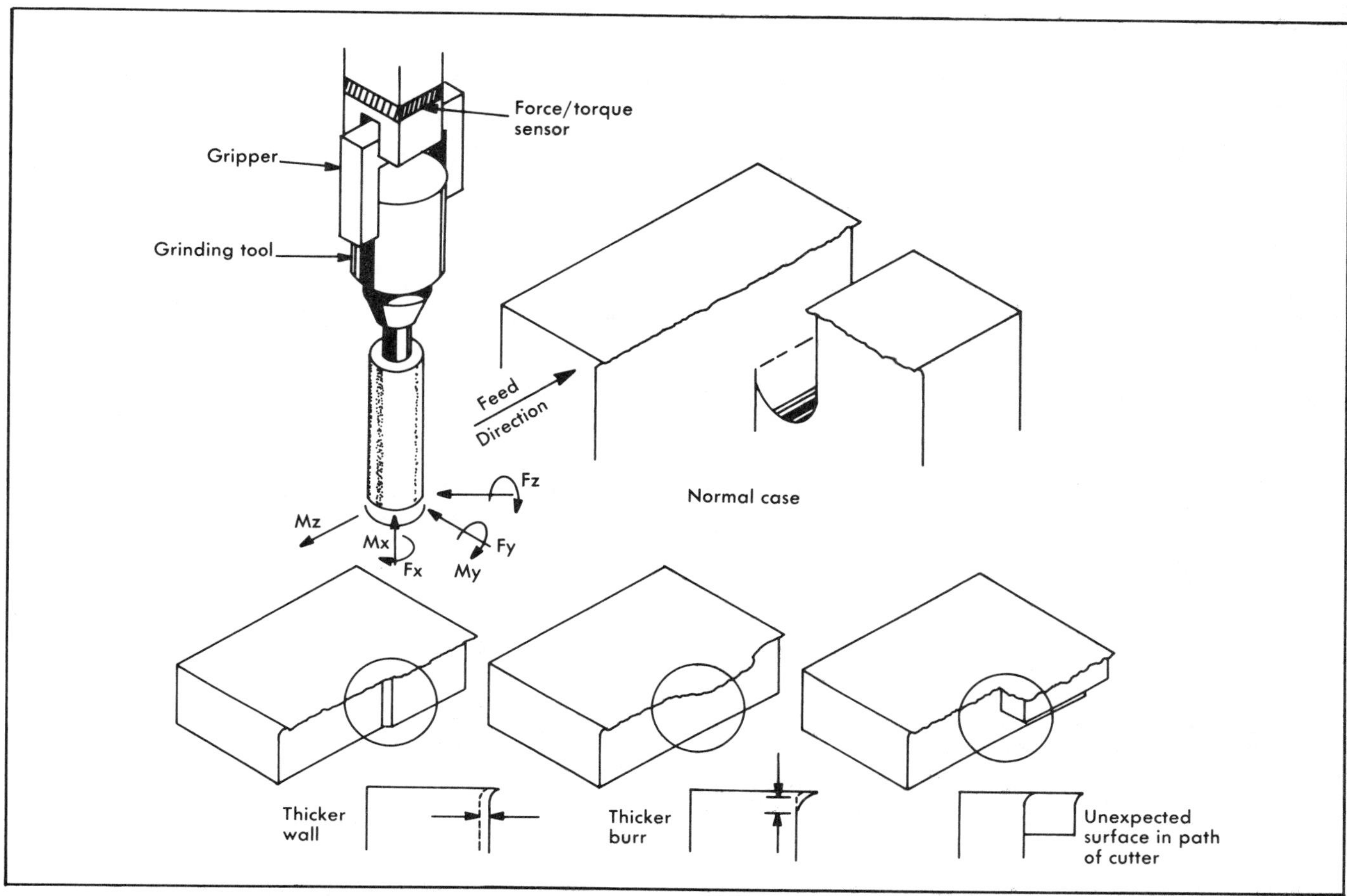

Fig. 16-51 Compliance is required for robotic deburring. (*Bendix Corp.*)

the loop around the force sensor compliance limitations. At fast feeds, the controller may have to accept data at megabaud rates to adequately provide the control for precision and complex workpieces.

Force sensing is being used for some simple, straight-line fettling operations. These systems work well as long as the burrs have the same dimensions at every location on the workpieces. Fettling operations get more complicated if the burrs are not in a straight line and if the dimensions of the burr vary. In such cases, the reaction force of the robot cannot be used to shift the program in the Z axis since it is not known whether the change in reaction force is caused by changing burr dimensions or by the contour of the part. The required calculations and computations are complicated.

With robots now available, encoders are located behind the robotic arm transmission train. As a result of losses through the train, the system cannot account for tool weight or process-induced deflections. A deburring force of 1 lb (4.45 N), for example, fed through transducers is overridden by system noise when the system must also be able to accommodate large cutting forces.

For a given workpiece material and set of operating conditions, the cutting force F_c is defined by the following equation:

$$F_c = \text{constant} \times \text{burr width} \times \text{burr height} \times \text{feed rate} \quad (2)$$

Thus, if the robot detects an increase in cutting force, it is logical to assume that either the width or the height of the burr increased. Because the burr volume increased, the cutting force increases, which causes greater robot deflections. If the robot keeps feeding the cutter at the same rate, the cutter will be deflected more from the part surfaces, thus leaving an area of incompletely removed material. By slowing the feed rate, the deflecting force can be maintained constant. As a result, the finished edge or surface should not have small islands of incompletely removed material.

If workpiece geometries vary significantly, then simple control logic is no longer sufficient. Increased forces may be the result of contacting a wall sooner than expected. Dull or chipped tools are similarly not accounted for by simple control logic.

For simple fettling operations and a spring-force servo, the normal stiffness of an electric robot can be adjusted to reduce the contact force values as required by the application. By using a proportional speed controller, the forces on the arm will be proportional to the position error. By using outputs from the robot, the robot stiffness can be adjusted as required in the program. The spring-force servo can be used to compensate for small variations in wheel size and small tolerance variations in the castings.

Force sensing systems can also be used with a search function to produce a transformation of the program. In this

ROBOTIC DEBURRING, FETTLING AND FINISHING

approach, the robot intermittently compensates for wear of a wheel, as well as detecting the edges of fins and the surface of castings. To compensate for wheel wear, the robot moves the wheel to a measuring hole and moves sideways in a search routine until a force results between it and the hole walls (see Fig. 16-52). An offset is calculated that accounts for the wheel wear, and then the offset is fed into the robot to define the new normal centerline position. A simpler approach, however, is to automatically offset a given dimension after a certain number of workpieces. It is desirable to have a robot that has tool center-point registers that change the offset automatically.

Researchers are investigating dynamic controls to improve robot accuracy. These controls have the potential for significantly improving deburring capabilities. Software programs, originally developed to design robots, can calculate the deflection and the compliance of the robot system. Such software is not being used yet for control, but the capability exists. If the forces are known, then the corrections can be made. If tool changes occur during the cycle, then the system must account for the difference in spindle or tool weights in order to ensure accuracy.

An NC tape made for one robot will not work as accurately in another robot of the same model because of robot manufacturing tolerance and control system tuning differences. Robots for high levels of accuracy, without sensory system programs, must have their programs generated on the robot that will use the tape. It is possible to get matched robots that can use the same tape, however, by specifying such needs on purchase orders and by building them at the same time.

Robot Capabilities

It is important to understand robot capabilities before selecting one for deburring. With the radial robot arm shown in Fig. 16-53, whenever the end of the arm is moved in an X or Y direction, the tool remains in the same angular orientation with respect to the arm, but changes significantly with respect to the workpiece. As a result, every movement (other than a radial motion into the horizontal plane) requires a reorientation of the tool to provide perpendicularity to the workpiece edges. Because off-line programming is not yet available for many robots, thousands of manual reorientations are required with such robots for deburring most parts.

Similarly, some robot wrists roll automatically with up-and-down movements of one of the principal axes. Some robots can take data for a feature in one position and then automatically provide the data for any other position, but most robot software does not allow this. An example is entering the data for deburring a 0.625″ (15.88 mm) diam hole in one horizontal plane and then translating that data to deburr the same size hole on a higher surface. Those robots that cannot automatically perform the translation and rotation require data reentry for each feature. Each of these limitations is a software restriction. Software may be developed to eliminate these restraints, but users should know that they exist.

Most deburring robots currently rely on an operator manually teaching the robot how to move to perform deburring. Such programming is the most convenient approach for many workpieces. For complex parts that already have an NC database, however, the ability to use off-line programming is desirable.

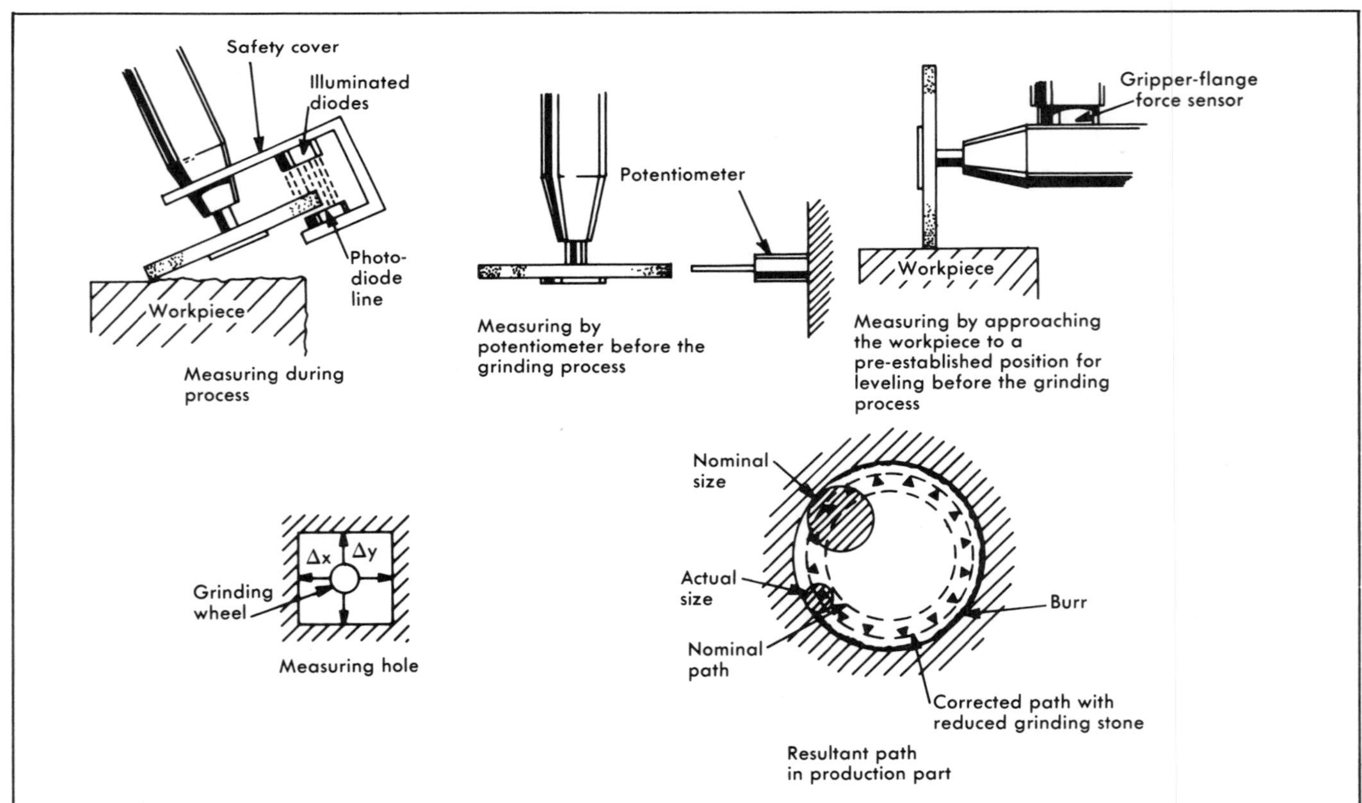

Fig. 16-52 Measurement compensation for wear by moving grinding wheel inside a known hole size until wheel contacts hole walls. (*Bendix Corp.*)

Off-line programming offers the advantage of minimizing machine downtime for teaching. In theory, with a preexisting NC database, off-line programming should be faster and more accurate than manual teaching. Because current robots are not precise, however, off-line programming with existing robots typically requires some manual final adjustment.

Few manufacturers of robots promote off-line capabilities and some specifically discourage its use. In practice, it is likely that off-line programming may require some postprocessor philosophies that are not written yet. If the robot cutter runs into a wall and exceeds the load capability of the robot, a permanent distortion of the robot arm occurs. Such accidents frequently require total reentry of data locations, a common problem on new installations. The problems can be overcome by manual programming.

In practice, deburring compliance requires offsetting the tool from the edge to be deburred (see Fig. 16-54) by as much as 0.24″ (6.10 mm). The tool is normally presented at an angle of 45° to the workpiece edges for contouring operations, as shown in Fig. 16-55. This method has at least two benefits. A cylindrical cutter can be used rather than a specially shaped cutter. In addition, programming is easier because all rotation is performed around a single axis, identified in Fig. 16-55 as "A." This approach also generally provides a smoother finish when the tool must pass through a fillet radius transition to another wall.

Offsets are also used in fettling, grinding, and finishing operations when wheels or abrasive belts are used. The offsets serve multiple purposes, such as wheel-wear compensation, permitting larger variations in workpieces and locations, and reducing programming time. The degree of offset depends upon the specific application and the compliance device used.

For tools that have rotational symmetry and are rigidly flange mounted to the robot, a five-axis capability usually allows any desired tool orientation within the working area of the robot. However, as soon as the working force on the tool is unidirectional (for example, when nonsymmetrical tools such as chisel hammers or grinding belts are used), six-axis capability is often required.

The usable working space of robots is far less than the volume that can be swept by the robot (see Fig. 16-56). The area swept by a robot's arm does not account for the fact that the robot cannot provide the desired tool orientation in all positions. In addition, many robots have singularity points within the working volume. When one attempts to drive them through these points, they lock up or perform undesirable movements. In addition, the presence of the workpiece itself forces the robot to crook its arm, wrist, and tool into difficult positions to reach all surfaces. The net result is that a robot that can sweep a volume of 10 or 20 ft^3 (0.28 or 0.56 m^3) may actually have a usable volume of only 2 to 9 ft^3 (0.06 to 0.25 m^3).

State-of-the-art capabilities of robotic deburring are summarized in Table 16-25. Edge chamfers of 0.005″ (0.13 mm) can be produced in deburring operations, and contours of 0.010″ (0.25 mm) are possible in some fettling applications. The heights of bosses or related features should exceed 0.080″ (2.03 mm).

Influence of Workpiece Geometry

Both the number of edges and their lineal lengths on flat parts affect deburring quality, cost, and programming difficulty. Shallow areas, such as counterbores and ledges having depths or heights less than 0.080″ (2.03 mm), make deburring with robots difficult. The points of chamfering tools tend to

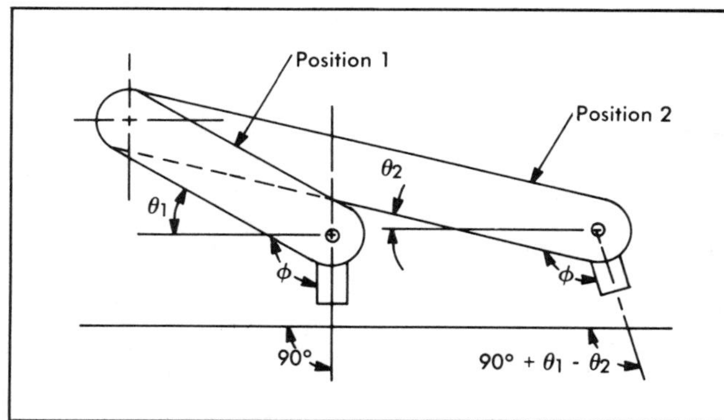

Fig. 16-53 When robot arm is moved from position 1 to position 2, the tool is no longer perpendicular to the workpiece surface. (*Bendix Corp.*)

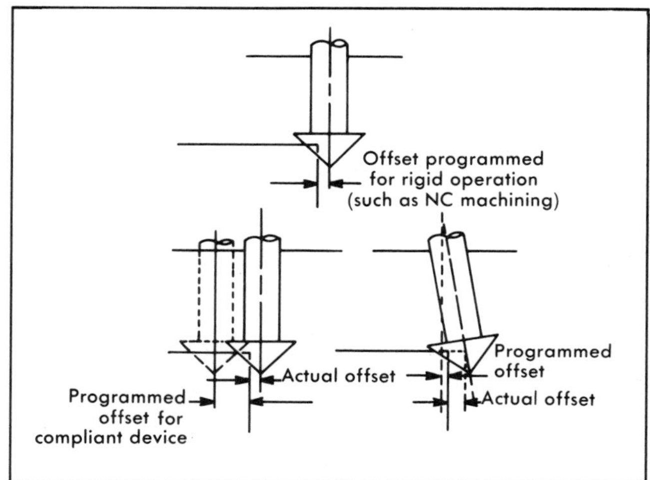

Fig. 16-54 Offsets for deburring. (*ASEA Robotics, Inc.*)

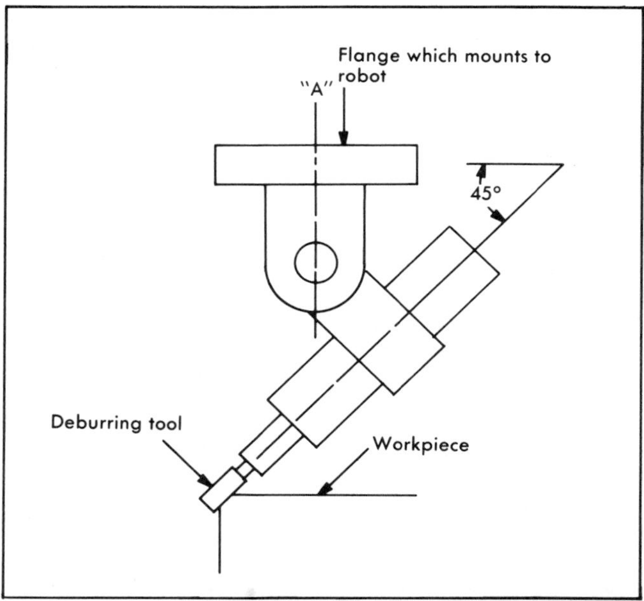

Fig. 16-55 Deburring tool presented at 45° to workpiece edge. (*ASEA Robotics, Inc.*)

scrape, resulting in tool marks that prevent the production of smooth chamfers. Also, the teeth of conical cutters wear rapidly near their points because of their small diameters.

The liberal tolerances for most castings make robotic deburring difficult. Variations in cast wall locations are especially troublesome. Edges that flare out into adjacent areas cannot be totally finished with robots without gouging the walls. Small, internal edge radii are difficult to deburr and require small tools and precision programming. Thin walls may distort under deburring forces. Holes close to walls may be impossible to finish because of tool motor, cutter, or spindle interferences.

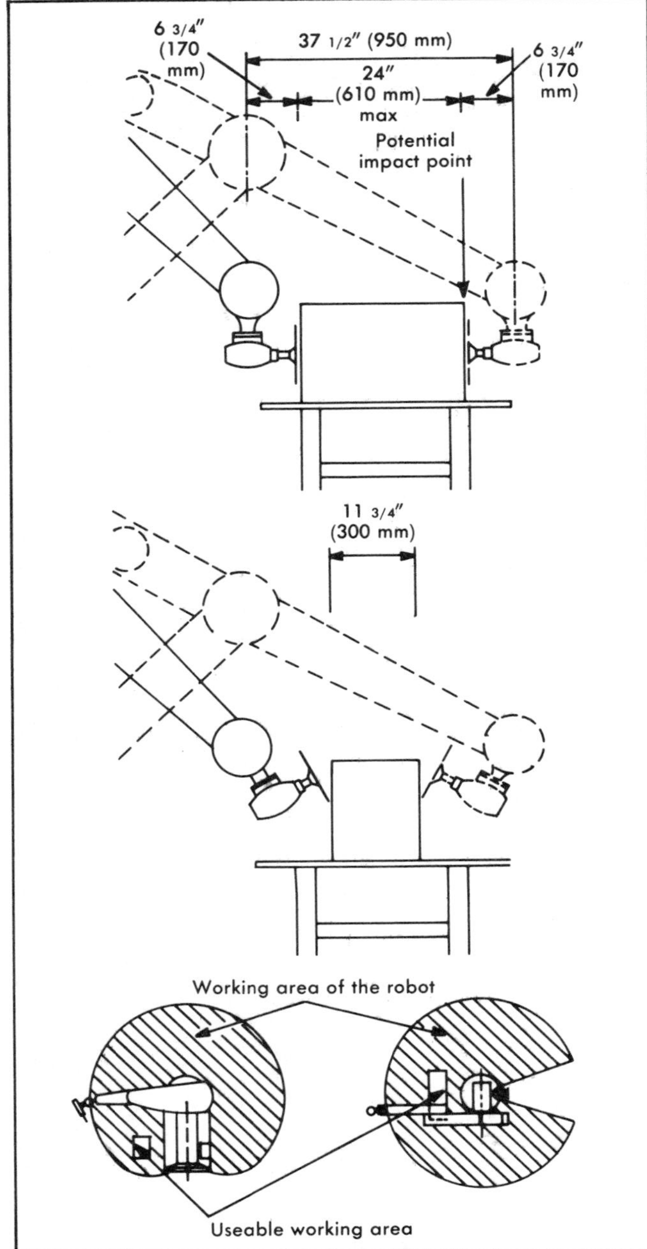

Fig. 16-56 Useable work area of a robot is less than total working area. (*Bendix Corp.*)

Burr size. Burr sizes and locations determine the parameters of the deburring process and influence the edge breaks obtainable. Burrs vary considerably along each edge, among edges, and among different workpieces, as discussed in Chapter 15 of this volume. Typical burr sizes produced in milling operations are presented in Table 15-2 of that chapter.

Deburring requirements. Incomplete or improper specification of requirements can increase deburring costs, as discussed in Chapter 15, and may become a limiting factor in robot applications. The development and consistent use of suitable standards are recommended. Allowable burrs or edge conditions can be defined on part prints, specifications, routing sheets, or memorandums with sketches or photos, or by the use of comparative masters or gages. The more stringent the requirements are for deburring, the less likely that robotic deburring can be easily and successfully implemented.

Economic Justification

Any economic evaluation of robotic deburring should compare it with other methods, including hand deburring, for both simple and complex parts. Savings generally result because a robot is faster than a human and an operator is only required part time. However, robot programming must be considered as a cost, and cutting tool or brushing costs may be higher as a result of more continuous use. Energy and space costs can also be significant.

Tool Versus Workpiece Handling

A robot can carry deburring tools to the workpiece or can carry the workpiece to the deburring tools. Manipulating the deburring or fettling tools is most frequently used when the workpiece is heavy or only two or three deburring spindles must be used. The workpiece is manipulated when several spindles are required, when it is desired to minimize workhandling devices, when it minimizes fixturing costs, or when the workpieces are light, typically less than 25 lb (11 kg).

Tool handling. When the robot manipulates the deburring tool, the workpieces must generally be brought to the robot. This procedure typically requires a rotary table and fixtures to orient the workpieces. For dedicated applications, fixture costs are relatively low, but two to six sets of each fixture may be required to keep the robot operating at a constant rate. Figure 16-57 illustrates a robot deburring a gearbox case presented to the robot on a conveyor.

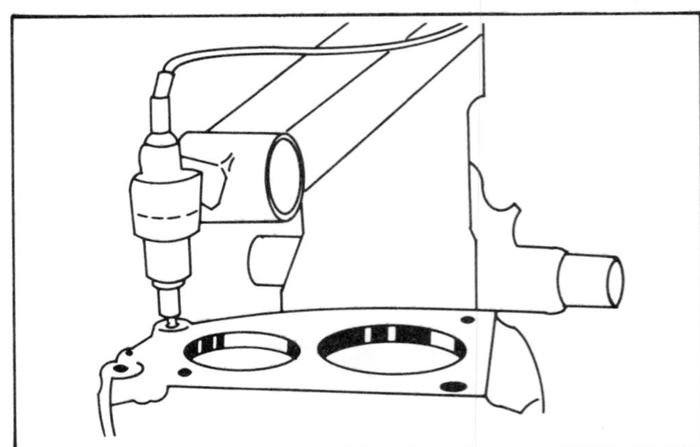

Fig. 16-57 Robotic arm with live spindle for deburring gearbox case. (*ASEA Robotics, Inc.*)

ROBOTIC DEBURRING, FETTLING AND FINISHING

In job shop applications, many workpieces will require their own fixtures. If a family of geometrically similar parts are involved, however, one fixture may accommodate all parts. Tooling costs in job shop applications can be expensive, and fixture storage space requirements can be high.

A robot can accommodate two live spindles on its wrist. When more spindles are required, the robot must move to a tooling rack, remove the spindle(s) on its wrist, and grasp another—generally requiring about 10 seconds per tool change.

This approach does have the advantage that programmers work in a "right side up" orientation. When the part is presented to the tool on fixed stands, programmers must normally contend with "upside down" programming.

The tool-holding robot is obviously independent of part weight. It also allows use in situations in which a future part might be heavier than the parts being run today. For deburring and fettling, robots are available that can accommodate almost all tool weights.

TABLE 16-25
Comparison of Automated Deburring Capabilities

Factor	Precision Robot	Machining Center	Fettling NC Machine	Fettling Robot
Memory capacity, bytes	32K	16K	480 steps	500 points
Repeatability, mm	±0.15	±0.0038	±0.2	±0.13
Accuracy, mm	3.68	0.0076	---	0.40
Number of axes	6	4	4	5
Maximum speed, mm/s	1000	128	100	2286
Tool capacity, number of tools	Unlimited	20 or 40	1	Unlimited
Maximum tool weight, lb	5	Not specified	Not specified	135
Floor space required, ft^2	120	120	50-100	120
Weight of equipment, lb	200	16,000	Not specified	1,650
Spindle speed, rpm	Unlimited	3,000	3,000	Unlimited
Work space, m^3	3.0	3.3	.25	3.0
Tool change time, s	10-20	5.5	---	17
Sensor controlled	Yes	No*	Yes	Yes
Continuous path	No	Yes	Yes	Yes
Normal data input mode	Teach/ computer disk	Magnetic tape	Stored program	Teach/ computer disk
Tool compensation	No	Yes	Yes	Yes
Suitable for DNC	---	Yes	No	Yes
Suitable for totally unmanned	Yes	No**	No	Yes
Dimensional drift	Yes	No	No	No
Off-line programming	Yes†	Yes	No	No
Manual overrides	Yes	Yes	No	No
Applicable to other tasks	Yes	Yes	No	Yes
Maintain 0.76 mm maximum break	No	Yes	No	No
Circular interpolation	No‡	Yes	Yes	Yes

* May be with hardware/software modifications.
** Part loading/unloading is required.
† Off-line programming may increase costs.
‡ Circular interpolation is feasible with off-line programming, but standard packages are not available. Some robots have a variation of circular interpolation.

ROBOTIC DEBURRING, FETTLING AND FINISHING

Tool-holding robots do present some tool power limitations since they must handle tools that have power umbilical cords (unless power inputs can be made through removable wrist connections). This limitation is normally only a concern when several spindles must be changed during the operation.

Deburring forces increase as the depth of cut and width of cut increases. In practice, for a vibration-free chamfer of 0.020 to 0.080" (0.5 to 2.0 mm), forces of 3 to 6 lb (13 to 26 N) occur.

Workpiece handling. When workpieces are light, they can be lifted from a magazine or conveyor and moved by the robot between two or more fixed tools. They can also be manipulated in such a way that the contours are passed over the tools. At the end of the cycle, the parts are placed on a conveyor or are palletized, and the robot grasps another part. Cycle times as low as 40 seconds can be achieved.

For universal applications, the robot can be surrounded by multiple pedestal-mounted grinders, files, rotary burrs, or brushes. The robot traverses the workpiece edges with whichever tools are appropriate for that part. This approach provides flexibility for handling many part geometries with no loss in toolchanging time.

With appropriate gripper design, the workpiece can be manipulated to expose most workpiece surfaces. If necessary, an intermediate regrip fixture can be used to allow access to all part edges. In contrast, when the robot holds the tools, the fixturing shields at least a portion of one part surface. With a family of similar part geometries, one gripper may accommodate many parts.

Tool Considerations

Toolchanging capability is an important consideration for deburring robots. Three approaches commonly used for toolchanging are:

- Design-integral multiple motors on the end effector (at 90° to each other) to eliminate the need for changing tools (see Fig. 16-58).
- Change preset spindles.
- Change grippers and tools.

In many cases, the most appropriate method for toolchanging is to exchange the motor, which contains the tool, for another motorized cutting unit. Systems that know when cutters are dull, clogged, broken, or worn excessively are desirable, but not generally available. As a result, users of cutters must expect to change the tools at predetermined intervals. The power cords or air hoses of motorized tools must be considered in tool design to ensure that they do not undesirably drape across the workpiece or other tooling.

It is important in installing a robot that the robot not be limited by the speed at which the operator can and will load workpieces and tools, or start and stop the machine. For maximum cost savings, it is appropriate that once the robot is turned on, the parts always be available to be finished. As a result, the material handling portion must be automated, as well as the deburring portion. If it is not, the deburring time is only going to be as fast as in any other operation in which a person must push program buttons.

Cutting tools. Most of the tools used for other deburring methods or mechanical cutting can be used for robotic deburring. Many of the tools used for hand deburring, discussed previously in this chapter (see Table 16-1), such as files, rotary burs, countersinks, brushes, abrasive-filled rubber and cotton products, abrasive paper products, and mounted points, are used for robotic deburring. Table 16-26 lists acceptable tools and tool sequences for robotic deburring.

Although most deburring is done with tools having tapered cutting edges, effective deburring demands a variety of tool configurations, tool axes, and motions. These variations include deburring with the flat end of some tools, with the cylindrical portion in either of two axes, or with brushes (see Fig. 16-59). Rotary burs have a life of 30 to 300 hours, depending upon application. It is significant to note, however, that most carbide cutters generate two smaller (secondary) burrs. By appropriate geometry selection, it is possible to minimize the size of these burrs.

Special motions can be used to deburr some features. The top surface of a bur can be used to deburr the bottom side of a hard-to-reach hole (see Fig. 16-60, view *a*). In this case, the tool must be used not as a countersink tool, but as a profiling tool for the hole edges. Double-sided chamfering tools are available to chamfer both the top and bottom of rectangular stock or related features (see Fig. 16-60, view *b*). The smallest standard cutters of this design are 0.76" (19.3 mm) diam.

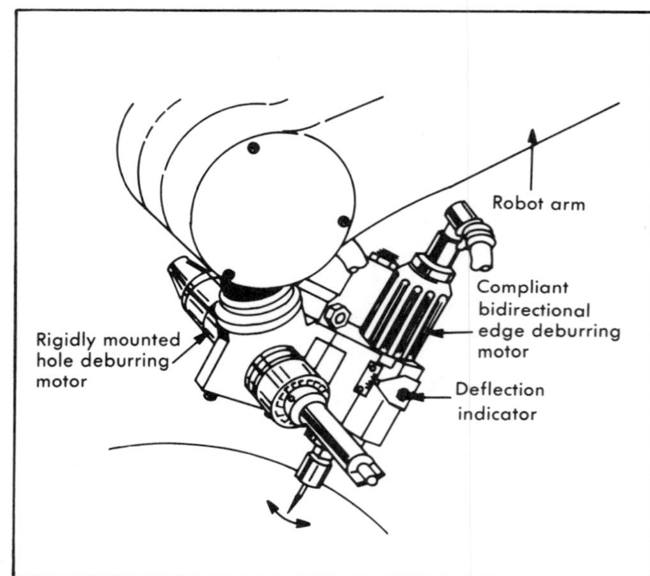

Fig. 16-58 Multispindle tools. (*ASEA Robotics, Inc.*)

TABLE 16-26
Combinations Appropriate for Robotic Deburring

Principal Approach	Subsequent Finishing Approaches: Abrasive Rubber	Brushing
Bur balls	X	
Chamfer tools	X	
Grinding wheels		X
Reverse radius cutters		X
Abrasive rubber		X
Abrasive-filled cotton		X
Brushing		
Grinding belts		
Reciprocating files		

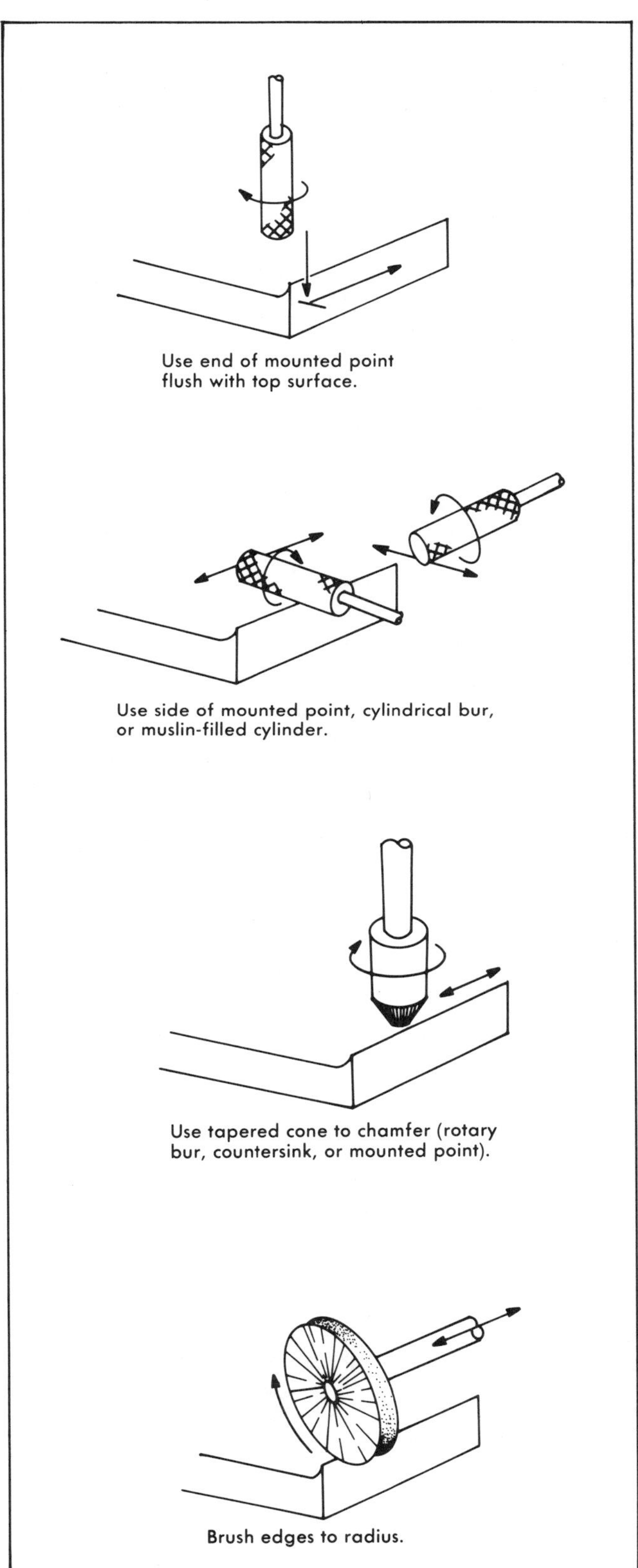

Use end of mounted point
flush with top surface.
Use side of mounted point, cylindrical bur,
or muslin-filled cylinder.
Use tapered cone to chamfer (rotary
bur, countersink, or mounted point).
Brush edges to radius.

Fig. 16-59 Tool orientations for robotic deburring. (*Bendix Corp.*)

Optimum use of cutters requires the following:

- Correct contact point (errors often lead to vibrations).
- Correct contact angle (correct angle results in the avoidance of secondary burrs).
- Correct path direction (incorrect direction of the tool motion relative to the rotation of the tool often results in vibrations).
- Correct path velocity (depends on type of burrs, desired quality, and other factors).
- Correct resilient mounting (compliance).

Files. Rotating files made of high-speed steel or carbide are the most common robotic deburring tools. A carbide cutter rigidly held by a robot can produce excellent surface finishes that are unattainable with hand deburring. Because chip-cutting tools blunt rather than undergo significant geometrical wear, wear compensation is not possible. Because a robotic system will not know whether increased forces are the result of tool wear or changed geometry, one must periodically change cutters.

Reciprocating files are suitable for applications in which only minor burrs are to be removed. The speed and stroke of the file can be adjusted, and files with different cuts can be used to vary the amount of material that the file removes. This control simplifies the programming necessary. It is possible to compensate for variations in tool and workpiece by mounting the motor in a compliant holder. Oscillating files are particularly suitable for hard or brittle materials. Soft materials such as aluminum tend to accumulate between the file teeth. Reciprocating files are also a good choice for reaching into sharp corners.

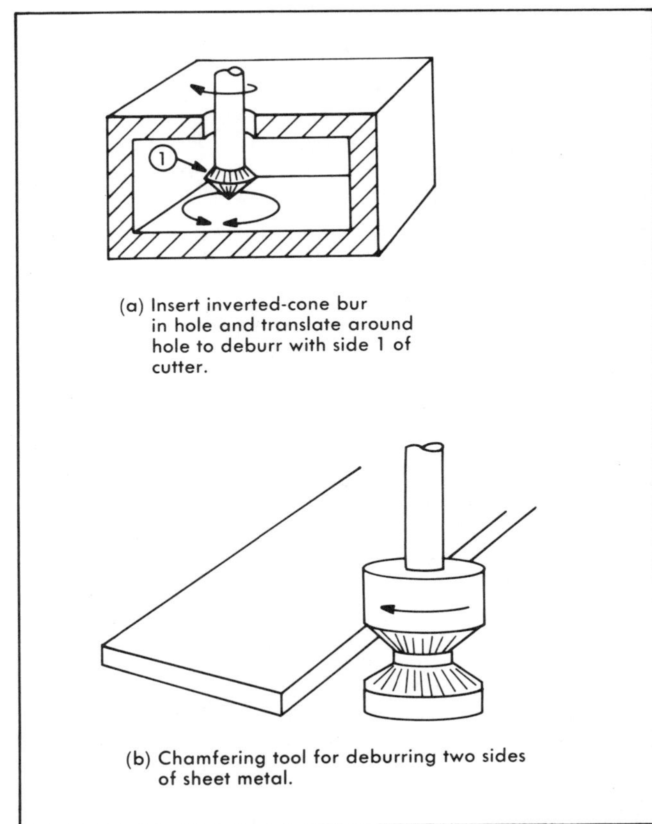

1
(a) Insert inverted-cone bur
in hole and translate around
hole to deburr with side 1 of
cutter.
(b) Chamfering tool for deburring two sides
of sheet metal.

Fig. 16-60 Special tools and deburring movements. (*Bendix Corp.*)

Abrasive-belt grinding. Edges easily accessible from the outside are suitable for deburring, fettling, or finishing with an abrasive-belt grinder, which removes material rapidly and leaves even surfaces. However, chamfer size can be difficult to control with this type of device. The grinding belt should be spring or air-loaded, or the driving equipment should be suspended in a compliant mount to provide flexibility. A longer belt can be used if the grinder is pedestal mounted. Long belts on pedestal-mounted grinders decrease the frequency of belt changes and allow the use of more powerful drive units for higher stock removal rates.

Brushing. For many commercial applications, a brush is used rather than a cutting tool. When an aggressive brush can be used, all normal burrs can be removed successfully by brushing. Brushes that will ensure a smooth blend not exceeding 0.010" (0.25 mm), however, will frequently beat small burrs flush against the surface. In precision applications, the fact that the burrs are not removed usually causes the parts to be rejected. Nevertheless, a judicious choice of brushing applications and approaches may eliminate the need for cutters. Details with respect to types of brushes, fill materials, brush selection, and operating parameters are presented earlier in this chapter under the subject of "Power Brushing."

Robotic brush deburring is suitable for all external surfaces when light abrasion on areas adjacent to the brushed edges is acceptable. Internal edges can be deburred if they are located so that the brush can reach the burrs at an angle between 25 and 50°. Programming for brush applications is simple because of the brush flexibility and because the deburring effect is not greatly dependent on the applied force. It is possible to compensate for brush wear by automatically offsetting the path, as is the case for compensating for grinding wheel wear.

Spindle Holders

One-way spring-loaded holders. In most cases, the spindle holder is designed to give compliance in just one direction, the direction perpendicular to the feed direction. Mounted on the robot's fifth axis (wrist), the holder is continuously turned by the robot as the part contour changes. This action maintains the direction of deflection perpendicular to the feed direction. In this instance, the spring force is adjustable, and compliance can be used with this tool. The major advantages of this type of device are that vibration is minimized and the tool follows complex contours with a minimum amount of programming.

Two-way spring-loaded holders. The holder shown in Fig. 16-61 can deflect in two directions (in the same plane) from a neutral position. To ensure that the tool center point is accurately located, two pneumatic cylinders on either side of the pivot point are activated. By activating the cylinder opposite the desired direction of compliance, the tool will always be firmly located, but not on a hard end stop.

The working range can be extended when using this type of holder. The holder can be offset from the robot wrist and does not have to be turned 180°, as does the one-way holder, to change the direction of compliance. This action minimizes robot wrist motions and total cycle time.

To further increase the accessibility of the tool, the direction of compliance can be turned continuously and integrated with the robot's control system. The tool is positioned with the robot's fifth axis, and the direction of compliance is turned to a right angle relative to the deburred surface by means of the robot's sixth axis.

Grippers can be designed to be multifunctional. They can hold two separate motors, provide air blowoff, or clean while deburring. As a result, they can significantly increase output when situations allow multiple, simultaneous operations. If necessary, they can stamp parts, verify the existence of hole patterns, and perform related tasks, or work with equipment that can perform these operations.

Spindles

For fettling operations, the spindle on the robot must have a stall force greater than the normal fettling forces. A 3/4 hp (0.56 kW) motor, for example, may have a stall force of 10 lb (44.5 N), but grinding or abrasive-belt finishing may require 50 lb (222 N) of force to finish in a reasonable cycle time. As a result, users should consider using 3 to 5 hp (2.2 to 3.7 kW) motors for rapid metal removal.

The main criterion for deburring and fettling spindles is a good power-to-weight ratio, such as high-frequency electric tools have. Pneumatic motors, however, can be overloaded without damage. High performance can be achieved with hydraulic drives, but only within certain speed ranges.

Operating Parameters

Most robot manufacturers advocate fast traverse rates that far exceed normal human movements. For applications in which high rates are possible, robotic deburring can greatly increase production, as well as minimize costs. For most nondeburring robotic applications, the robot works at about the same pace as manual operators.

For all deburring applications, the robot must move continuously at a constant feed rate. In most applications, climb (down) cutting provides a smoother and more even surface finish than conventional (up) cutting. At optimum feed rates, when moving around sharp corners, it may be necessary to start turning the holder before it reaches the corner. The change of compliant direction is then completed as the tool moves in the new direction.

When using rotating files for deburring, the cutting speed should be about 4200 ft/s (1280 m/s), and a feed rate of 300 ipm (7620 mm/min) is recommended. For deburring holes, the cutting speeds and feed rates used are about the same as for drilling or thread cutting. Low feed rates of 1 to 2 ipm (25 to 50 mm/min) are often used.

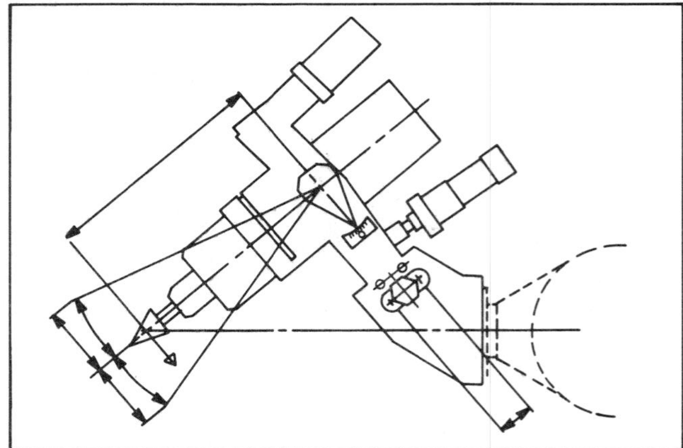

Fig. 16-61 Spindle holder for robotic deburring having dual-directional compliance. (*ASEA Robotics*)

Procedures for programming with a rigid holder versus compliant holders are different. The rigid holder requires accurate positioning of the robot and a great number of instructions. It is important that an open surface (minimum contact area) be used as the starting point. If more than 30° of the tool's radius is in contact with the surface, vibrations and chattering may occur. The rigid holder limits the maximum programmable feed rate due to servo lag created as the robot's speed increases.

Compliant toolholders decrease the number of instructions needed since they do not have to adjust for small changes in workpiece geometry. They also allow a relatively high feed rate when moving around complex contours, and the positioning instructions do not have to be exactly located along surfaces.

Safety Considerations

There are seven major safety hazards in robotic deburring, as follows:

1. Robot runaway.
2. Inadvertent human contact with the robot.
3. Robot's sudden release of workpiece or tooling.
4. High feed rates.
5. High spindle speeds.
6. Noise level.
7. Flying debris from broken wheels, burrs, flash, broken wire from brushes, and sparks from grinding.

The first four items are common concerns in all robot applications. If a robot loses its gripping power in a runaway condition, the workpiece in its gripper or the tooling on the wrist could become a flying missile. This situation creates a hazard of some concern because of the weight of the workpiece or tooling and the high travel speeds.

Feed rates of over 100 ipm (2540 mm/min) are considered high in the majority of metalcutting operations, but not in robotic deburring. The action of robots is so fast that it is important to ensure that individuals are not in the immediate vicinity of the robot when it operates at these rates because they cannot react fast enough to get out of the way. High spindle speeds are used in both mechanized and manual operations, but these speeds still require some care to ensure that chips do not fly wildly and that appropriate tools are inserted and correctly chucked.

In fettling operations, since higher power tools can be used than in normal operations, the noise levels can exceed allowable OSHA limits. As a result, these operations may require walled installations.

If grinding wheels are pinched or twisted as they bind, the wheels will fly apart. A walled area is desirable in all installations involving fettling to prevent injury from flying debris.

Off-line programming, as previously mentioned, provides some safety advantages in that the programmer is not near the robot during initial programming. If the programming is not well done, however, the robot can be damaged in the first test of the program.

MASS FINISHING

Mass finishing refers to several processes for cleaning, deburring, deflashing, edge and corner radiusing, and surface finishing. In general, mass finishing means handling a quantity of workpieces in bulk. A more precise definition is: the edge and surface conditioning of workpieces in a mass of media, normally, but not necessarily, involving a mass of workpieces.

All mass finishing processes are based on loading parts into a container, usually holding abrasive or nonabrasive media, water, and a compound. While mass finishing commonly refers to workpieces loosely loaded into a container, equipment is available in which the workpieces are either fixtured or located in individual compartments within the container.

Action of the container causes the media to rub against the workpieces or the workpieces to rub against one another, thus producing the desired results. The major mass finishing processes are rotary barrel (tumbling), vibratory, centrifugal barrel, centrifugal disc, and spindle finishing. These and other mass finishing techniques are discussed in this section.

For most mass finishing applications, workpieces are loaded into a container with some form of media. Most processes are also performed wet, using a solution of water and a compound within the container or flowing through the container. The different media and compounds used are discussed subsequently in this section.

PROCESS ADVANTAGES

Mass finishing is a simple, versatile, and low-cost means of conditioning the edges and surfaces of various components. Normally, individual handling or fixturing of workpieces is not required, thus eliminating costs associated with manual and most other mechanized finishing processes. With proper control, consistent results can be attained from workpiece to workpiece and batch to batch.

All metals and many nonmetals, as well as most sizes and shapes of workpieces, can be processed by mass finishing. Surface finishes smoother than 1 μin. (0.025 μm) are produced in some cases. In addition, the process can be highly automated for increased productivity.

The texture (surface irregularities) of machined or ground parts has a directional pattern characteristic of the specific cutting operation used. When surfaces are mass finished, this directional pattern is changed to a random one that has been found to be particularly advantageous for parts requiring lubrication because lubricants are more readily trapped by the random, nondirectional surface pattern. Also, product break-in periods can be substantially reduced when functional parts are mass finished prior to operation.

LIMITATIONS

A limitation to the use of mass finishing is that its action is generally effective on *all* surfaces, edges, and corners of workpieces that contact the media. It is not normally possible to give preferential treatment to specific areas; however, masking specific areas has been successfully used for some applications. The action is greater on the edges of workpieces than on equally exposed surfaces. The action in holes and recesses is signifi-

MASS FINISHING

cantly less than on exposed areas; and in small, deep recesses, it is unusual to be able to do any finishing unless the workpiece is fixtured.

A basic limitation to its use is that much of industry does not consider mass finishing to be a predictable and controllable technology. However, mass finishing methods can satisfy most requirements when adequate specifications and standards with respect to quality before and after finishing are used (as discussed in Chapter 15), and when the proper process and optimum operating parameters are selected.

There are applications that are not suitable for mass finishing. For example, the size of the burrs may be too large to be completely removed by mass finishing. In internal areas, burrs may be inaccessible to the media. Surfaces requiring smoothening may be too rough for finishing to the required degree. Or, long cycles may result in critical dimensions becoming out of tolerance.

The media may become lodged in workpieces with different sized openings, but this problem can generally be avoided by selecting a different size or shape of media or by sealing the openings. Mass finishing for surface finish can produce corner and edge radii. If such radii cannot be tolerated, mass finishing should not be attempted. Generally, the smoothening of a plane surface from 20 to 5 μin. (0.51 to 0.13 μm) will be accompanied by the generation of radii on the order of 0.015" (0.38 mm) on exposed edges, depending upon the workpiece material, size of the media, and other parameters.

APPLICATIONS OF MASS FINISHING

Mass finishing processes are being used for many different purposes. Major uses include deburring, deflashing, edge and corner radiusing, improving surface finish, and cleaning (removing rust, scale, and other surface contaminants). Other applications involve changing surface profiles, generating suitable surface textures for coating or painting, modifying surface stresses (normally to develop compressive stresses), inhibiting corrosion, removing discoloration, drying workpieces, and applying coatings, such as lubricants and wax.

PROCESS SELECTION

Many variables affect the selection of the best mass finishing process for a specific application. Major factors that should be considered prior to equipment selection include production requirements, quality requirements, process variables, and costs.

Production Requirements

Total production requirements, batch sizes, and the variety of workpieces to be finished are critical criteria with respect to process and equipment selection. Size, weight, and configuration of the workpieces, as well as the material from which they are made, must be analyzed. Any anticipated changes in future production requirements and automation needs must also be considered.

Quality Requirements

Specific requirements with respect to surface finish, edge and corner conditions, cleanliness, and uniformity must be known and considered before process and equipment can be judiciously selected. The consistency of workpiece quality entering the finishing department and post-treatment requirements and variables are important factors in selecting the process and equipment. Burr locations, properties, and sizes are critical in this selection. Cycle times for the process selected depend primarily upon the finishing requirements and the condition of the parts prior to finishing.

Process Variables

The various mass finishing processes are different with respect to capital investment requirements, capabilities, processing times, repeatability, the quantity of consumable materials used, energy and water needs, and operating and maintenance costs. Adaptability of the process and equipment to automation, the effect on subsequent operations, floor space requirements, and the effect on EPA and OSHA regulations are other important considerations. Table 16-27 presents some general comparisons of major mass finishing processes. Possible side effects of the mass finishing process, such as changes in dimensions and surface finishes, residual stresses, changes in color or luster, decreased cleanliness, and reduced contamination of the surfaces (all discussed in Chapter 15), must also be analyzed.

Process Costs

In comparing various processes for possible use, the total cost per finished workpiece must be a major consideration. Many elements enter into any cost comparison, including capital equipment, depreciation, maintenance, labor, and power costs. The cost of perishable items, such as media, compound, and water, must also be analyzed. Overhead costs include management, supervision, insurance, heat and light, floor space, and waste disposal. The possible need for subsequent cleaning or other processing can also affect costs. Scrap, rework, and inspection represent additional expenses.

Other Considerations

Since most machining and forming processes do not generally produce consistent burrs and surface and edge finishes, the versatility of the mass finishing process selected is important in handling variations from prior operations. It has been traditional in the metalworking industry to separate the mechanical finishing department from all other manufacturing processes and to have it handle all products and problems. However, when considering new or expanded mass finishing processes, it is recommended that the relative merits of departmentalized versus dedicated equipment be examined; finishing equipment already available should also be considered.

Computer-Aided Selection

As discussed in Chapter 15, computer-aided selection systems are being developed for determining the optimum finishing process for specific applications. Input data for such systems include workpiece materials, details of prior processing, burr and surface specifications, and finishing requirements. Some computer simulation systems recommend the correct type and size of machine, and the proper media and compound to use, as well as providing an analysis of capital and operating costs.

BARREL FINISHING/TUMBLING

Conventional rotary barrel tumbling was the original mass finishing technique. Ancient Chinese and Egyptians used tumbling barrels with natural stones as media to achieve smooth finishes on weapons and jewelry. The process was known as barreling, rattling, or tubbing.

TABLE 16-27
Comparisons of Major Mass Finishing Processes

Mass Finishing Process	Advantages	Limitations
Rotary barrel	Low initial, operating, and maintenance costs. Batch automation capability.	Slow process. High operator skill needed. Large floor space required. No in-process inspection. Wet working area. Little-to-no work in part recesses.
Vibratory tub	Faster than barrel. Can handle small to very large and very long parts. Open for in-process inspection. Full automation capability. Batch or continuous operation.	Noisy. Slower than high-energy processes (centrifugal barrel and disk). External material handling generally needed.
Vibratory bowl	Faster than barrel. Handles very small to large parts. Open for in-process inspection. Integral separation. Less noisy than vibratory tub. Generally lower initial cost and better media-parts mixing than vibratory tub. Economical for general-purpose work, heavy deburring, and continuous processing. Good automation capabilities. Batch or continuous operation.	Cannot handle very long parts.
Centrifugal barrel	Fast processing. Can handle precision and fragile parts. Produces smooth finishes. Batch automation capability. Versatile—can change from fine finishing to heavy stock removal. Improves fatigue strength. Low floor space needs.	No in-process inspection. Complex to automate.
Centrifugal disk	Fast processing. Open for in-process inspection. Batch automation capability. Versatile.	Limited part sizes. External material handling.
Spindle	Fast processing. No impingement of workpieces. Produces smooth finishes. Automation capability with robot reload.	Parts must be fixtured. Limited part sizes and geometries (typically used for cylindrical parts). Less versatile than other processes.

(*The Harper Co.*)

Barrel finishing is now a vastly improved process compared to the old-time tumbling operations, but it is generally slow. As a result, the more sophisticated mass finishing processes discussed next in this section are making the barrel obsolete for most applications. However, barrel finishing is a versatile means of edge and surface conditioning, equipment costs are generally low, operation is simple, and there are still applications where this process offers economies.

Operating Principle

Barrel finishing is a low-pressure abrading process generally performed by the controlled sliding and rolling action of workpieces, media, compound, and water. Not all these ingredients are used in all applications. In a rotary or tumbling barrel (see Fig. 16-62), the upper layer of the workload has a sliding movement. As the barrel rotates, the load moves upward in the barrel to a turnover point. The force of gravity overcomes the tendency of the mass to stick together, and then the upper layer slides toward the bottom of the barrel.

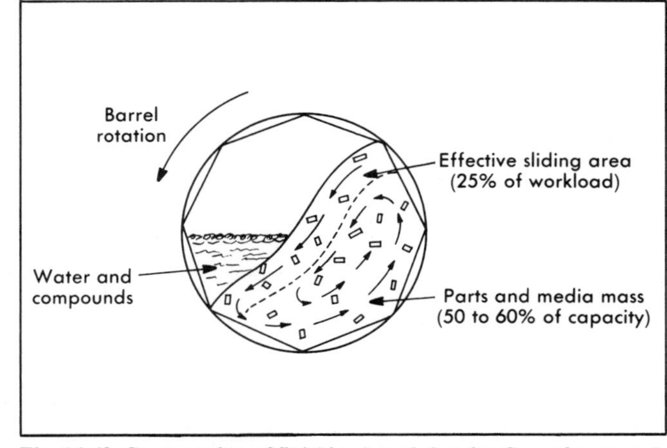

Fig. 16-62 Cross section of finishing barrel showing flow of parts and media.

BARREL FINISHING/TUMBLING

The barrel is normally loaded to about 60% of capacity with a mixture of workpieces, media, water, and compounds. Higher load levels (to 80%) are desirable for some workpieces, such as large or heavy components. Increasing the load level decreases the length of slide, thus reducing the probability of the workpieces contacting each other and reducing the force with which they contact, resulting in much longer cycles.

Although abrading action may occur as the workload rises in the barrel, about 90% of the rubbing action occurs during the slide. With horizontal barrels just over one-half filled, the most effective action occurs along the top surface of the mass. The faster the barrel rotation, the steeper the angle of slide and the faster the action. However, faster barrel rotation increases the tumbling action, increasing the likelihood of workpieces being damaged and reducing the quality of edge and surface conditions. For some applications, small barrel diameters and low rotational speeds may be required.

With barrel finishing, large radii are produced on the workpieces in relation to the amount of material removed from flat surfaces. This process also generally produces a higher luster and, often, a smoother finish on workpieces than other mass finishing methods. The peening action of barrel finishing can have a stress-relieving effect and produce denser workpiece surfaces. The process also permits usage of the media for extended times—about twice the time permissible for vibratory finishing, discussed next in this section—resulting in more efficient usage of the media.

Barrel finishing is generally a noisy operation, and this fact should be considered in selecting the locations of the machines. Sound enclosures may be required to meet OSHA regulations. Another possible limitation is that barrel rotation must be stopped before in-process inspection can be performed when using closed barrels. Barrel finishing is also intrinsically slow because only a small proportion of the workload is having any action applied to it at any given time. Nevertheless, the process is capable of developing surfaces of outstanding quality.

Equipment Used

Barrel finishing machines are simple in design, relatively low in cost, and available in a wide variety of sizes and types (see Fig. 16-63). Open-ended and horizontal types are the most popular. Horizontal units generally provide faster action, but open-ended units offer the advantages of inspection and making additions while in operation, plus easier unloading. Barrels are generally of multisided construction, with octagonal or hexagonal-shaped interiors being common. Most modern machines have flexible, wear-resistant, and replaceable linings in the barrels. Such linings, frequently polyurethane, extend barrel life, offer some protection to workpieces, and reduce noise levels. Machines are also available with individual compartments that allow workpieces to be processed singly.

Most machines are designed to permit changing the rotational speed of the barrel. Loading and unloading of horizontal barrels is done through wall openings that are closed during operation by watertight doors. During long processing cycles or when acidic compounds are used, heat can build up pressure within sealed horizontal barrels. This pressure must be released by a pressure-release device before opening the doors for unloading the barrels.

Material handling for barrel finishing machines is generally manual. Hoist pans and overhead hoists are normally required for servicing larger machines. Storage bins are also needed for media not in use whenever a variety of different materials, sizes, and shapes are required to produce desired results on a wide range of workpieces. While barrel finishing is almost always a batch processing operation, it is possible to automate the equipment by incorporating a material handling system. Feed troughs, chutes, timers, and automatic dumpers are common in such systems.

Process Variables

Process variables that must be properly selected for optimum barrel finishing include the shape and size of the barrel, the media used and its proportion to the workload, the compound used, the water level, rotational speed of the barrel, and processing time. Barrels are available for handling a wide variety of workpiece sizes, from small stampings or machined parts to large castings or forgings. Large or fragile workpieces are often fixtured in the barrels to eliminate part-to-part contact. Thin and long workpieces are generally not suitable for barrel finishing because of possible distortion or breakage. Media and compounds used are discussed subsequently in this section.

Rotational speeds generally range from as low as 4 rpm for large barrels to 60 rpm for small barrels, with surface speeds usually ranging from 20 to 200 sfm (6 to 61 m/min). For maximum productivity, barrel speed should be as high as possible without damaging the workpieces; however, the workpieces being finished and the required results may necessitate slower speeds. Burnishing is generally done at lower speeds, while deburring or cut-down operations are done at higher speeds.

Processing times vary from minutes to many hours, depending primarily upon the workload, the media used, the desired severity of action, and the required results. Separation of workpieces and media when unloading the barrels is generally done by dumping on vibrating screens. In some cases, the parts and media are discharged into a pan and transferred to a machine designed for separation. Trench drains are generally provided under barrel finishing machines to collect contaminants (liquids and solids), and the drains terminate in sumps that retain the solids.

VIBRATORY FINISHING

Vibratory finishing is now the most popular type of mass finishing and, next to hand deburring, the most common surface conditioning method used by industry. This versatile process is used for cleaning, deburring, deflashing, descaling, edge and corner radiusing, surface finishing, and stress relieving. Workpieces in a wide variety of sizes and shapes are handled, and all metals and many nonmetallic materials are processed. Large quantities of parts can be run in batch or continuous process setups without handling or fixturing, thus minimizing costs.

Vibratory equipment is made in two basic configurations: rectangular tub and round bowl types. The first tub-type vibratory finishing machine was introduced commercially in 1957, and the bowl type about five years later. Both types use an open-top work chamber containing an aggregate of media, compound, water, and the workpieces. While the chamber is vibrated, work is performed by the scrubbing or peening action of the media and compound on the workpieces.

Tub-Type Vibratory Equipment

With tub-type machines, workpieces are loaded into the open top of a container holding the media, compound, and water. The tub (container) is rectangular, generally has a U-

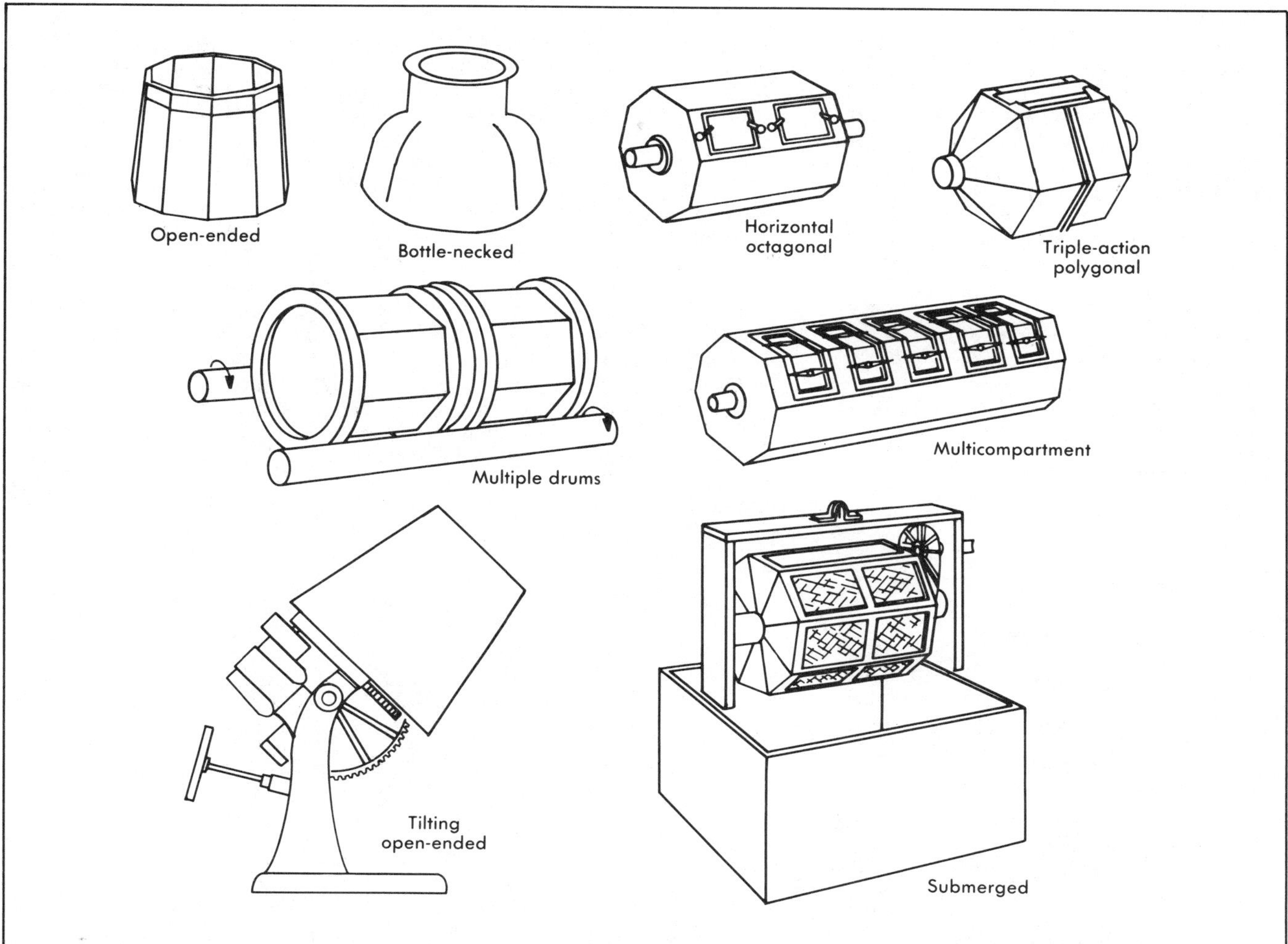

Fig. 16-63 Various types of barrel finishing machines. (*The Harper Co.*)

shaped cross section with flat parallel ends, and is usually mounted on coil or rubber springs (see Fig. 16-64). On some machines, the containers are suspended on air bags, and certain small units employ composite or metal strips for suspension. Processing containers other than U-shaped, such as keyhole shapes, and enclosed tubular units are sometimes used to improve unrestricted mass movement. Tub liners, generally polyurethane, are used to prevent wear, reduce noise, and resist chemical attack; they also transmit energy to the media. Removable separators can be installed to divide the rectangular tub into a number of compartments, thus permitting workpieces to be processed individually or the use of different media to finish various parts at the same time.

Vibration of the container is accomplished through several means, including the following:

- A vibratory motor having counterweights on its shaft and attached to the bottom of the container. This method is used on bowl-type machines.
- A single shaft or twin shafts, with eccentric weights, driven by a standard motor.
- A system of electromagnetic vibration generators.

Action of the media against the workpieces takes place throughout the whole load. The media rubs against the workpieces while the whole load is turning over within the container. As a result, cycle times are substantially shorter than for barrel finishing. Other advantages of tub-type vibratory finishing include the capability for in-process inspection, unloading and reloading without stopping the machine, and the possibility of automating for either batch or in-line processing.

Vibratory equipment can provide more finishing action than barrels in the recesses of workpieces. Vibratory machines can also process larger parts than barrels. Tub vibrators have been built to process components, such as aircraft wing spars, 30 ft (9 m) or more in length. Modular-type machines mounted in series can handle parts over 100 ft (30.5 m) long.

With long tub-type vibrators, fully automated, continuous processing of small parts is possible. Workpieces are loaded at one end of the tub and unloaded at the opposite end. At the unloading end, the media is separated from the workpieces and is generally returned to the tub by a conveyor (see Fig. 16-65), often through reclassification systems. Postprocessing operations, such as rinsing and drying, can be incorporated into such continuous systems.

VIBRATORY FINISHING

Round Bowls

Round bowl, or toroidal, vibratory finishing machines have a doughnut-shaped chamber (see Fig. 16-66) that permits a continuous circular flow of media and workpieces. The bowls may have either flat or spiral bottoms. As with tub-type machines, the chambers are provided with liners.

Vibration of the bowl is accomplished either by a vibratory motor or by an eccentric weight system mounted vertically in the center tube of the bowl. With either method, the amount of weight placed on the top and bottom of the eccentric system and the angular displacement between the two weights control the following:

- The finishing action (the amount of media vibration against the workpieces).
- The speed at which the mass rolls over within the bowl.
- The speed at which the mass rotates around the bowl.

Various degrees of automation are easily achieved with bowl vibrators. Figure 16-67 illustrates a machine equipped with a screen for parts and media separation and a discharge chute. A gate can be raised or lowered within the bowl to act as a dam, stopping the continuous flow of workpieces and media around the bowl when desired. When the gate is either raised or lowered, a screen separates the workpieces and media, the workpieces being discharged down the chute and the media being retained in the bowl.

Magnetic separators are sometimes used when finishing ferrous parts with nonferrous media. Machines can be equipped with magnetic parts collectors and demagnetization equipment, permitting the media to remain within the bowl.

Bowl vibrator machines have been built with indexing compartments that isolate one workpiece in each compartment, thus eliminating workpiece contact. Compartment dividers are indexed by means of a carousel mechanism, and no work-holding fixtures are required.

Machines can be placed in tandem. For example, one machine can be used for heavy deburring, discharging parts into a second machine for surface refinement or burnishing and rust inhibiting, and then discharging into a third machine for drying. Parts can be automatically transferred from the dryer onto a conveyor for subsequent operations or into containers.

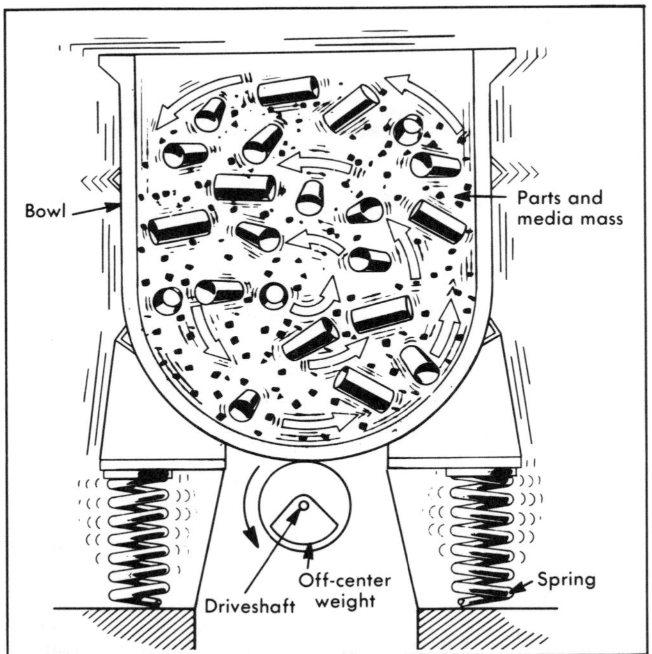

Fig. 16-64 Sectional view of tub-type vibratory finishing machine.

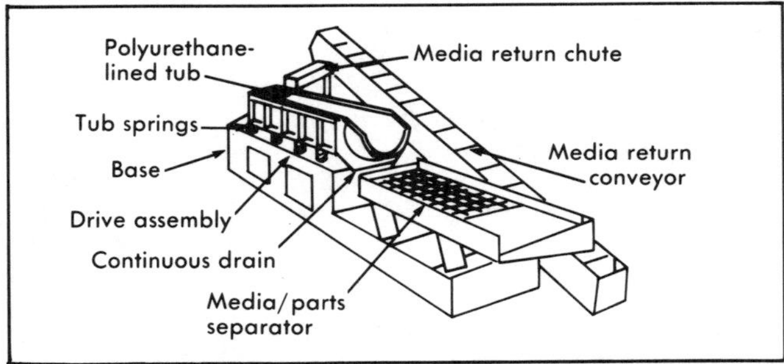

Fig. 16-65 Tub-type vibrator with automatic workpiece loading conveyor and another conveyor system to return media to the tub.

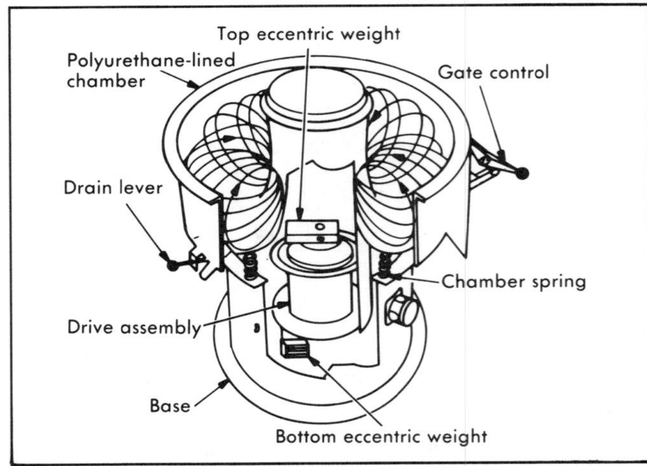

Fig. 16-66 Basic configuration of a round bowl vibratory finishing machine. (*Sweco, Inc.*)

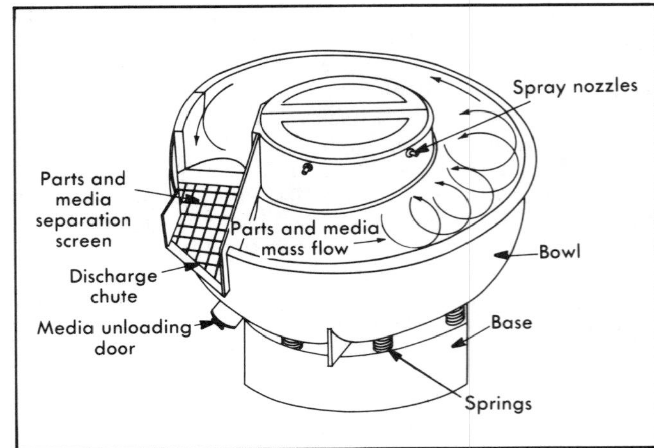

Fig. 16-67 Automated continuous-feed vibratory finishing machine with screen for separating parts and media, and conveyor and chute for returning the media. (*Gyromatic Mfg. Co.*)

Advantages of Vibratory Finishing

The basic merit of vibratory machines, both tub and bowl types, is that they offer a more convenient and faster means of finishing than do barrels. They are also more versatile, more easily automated, and provide cleaner operation. Processing cycles generally range from 20 minutes to several hours, but many applications are being done in 3 to 20 minutes with continuous processing.

Bowl vibratory equipment may be 5-10% slower than tub vibrators. However, the convenience and economy of integral separation of workpieces and media, along with the gentle-to-aggressive action, ease of adjustment, and versatility of the equipment make bowl vibrator machines desirable for finishing a wide range of products. Floor space requirements, capital investment, and noise are also minimized with internal media/parts separation. For all these reasons, bowl vibratory equipment is being used extensively for finishing.

Process Variables

Process variables for vibratory equipment are similar to those for barrels. For both types of equipment, the choice of the proper media, which must be sufficiently aggressive to remove burrs and to round edges while being sufficiently gentle to achieve desired surface finishes, is the variable that generally requires the most trial-and-error testing. The size and shape of the media are also important to ensure action on all critical areas of the parts, including any recessed areas of the workpieces, to prevent the media from lodging in such recesses or holes, and to facilitate simple separation of workpieces from the media. If the media can pass freely through holes in the workpieces, it will produce radii on the hole edges.

Water can have a considerable effect on the results obtained with vibratory equipment. Too much water will substantially dampen the action by flooding. For this reason, vibratory machines are designed for continuous flow-through of the water and compound solution. Compounds used serve the same functions in vibrators as in barrels: they enhance abrasive action, improve the color produced, inhibit corrosion, and, primarily, maintain cleanliness of the total load in the machine.

Other process variables for vibratory equipment are the amplitude of vibration, a variable provided on virtually all vibratory machines, and the frequency, which is variable on only some vibratory machines. Operational amplitudes generally range up to 0.375" (9.52 mm), and frequencies vary between 800 and 3600 vibrations per minute. Speed of rotation equals the frequency of the transmitted vibration. Higher speeds and/or amplitudes of vibration produce faster metal removal rates but rougher surface finishes for given media and compound; media wear rate is also increased. Vibratory tub and bowl machines are now available with high frequencies and/or amplitudes for higher energy mass finishing than previously possible.

HIGH-ENERGY MASS FINISHING

High-energy mass finishing encompasses processes in which the energy created within the mass in a container is greater than that obtained with standard vibratory methods. One advantage of the high-energy method is shorter processing cycles. Some high-energy systems permit easy adjustment of the energy level. Maximum energy can be used for fast deburring, edge radiusing, and metal removal, and reduced energy for gentler action to refine edges and surfaces. With some systems there is no workpiece impingement. High-energy mass finishing methods include centrifugal barrel, centrifugal disc, and spindle finishing, as well as chemically and electrochemically accelerated standard processes.

Centrifugal Barrel Finishing

Centrifugal barrel finishing, like other mass finishing processes, uses media, compound, and water to deburr and surface finish components. The difference between this and other processes is that centrifugal action results in a very fast, highly controllable operation. The process, sometimes called orbital barrel finishing, maintains a smooth rubbing action with no workpiece impingement, making it possible to produce fine finishes on precision and fragile parts. Another important advantage is the capability of imparting high compressive stresses in the surfaces of workpieces.

Operating principles. On centrifugal barrel machines, a number of drums are mounted on the periphery of a turret (see Fig. 16-68). The turret rotates in one direction, while the drums are rotated at a slower speed in the opposite direction, as illustrated in Fig. 16-69. Machines are available with the turret

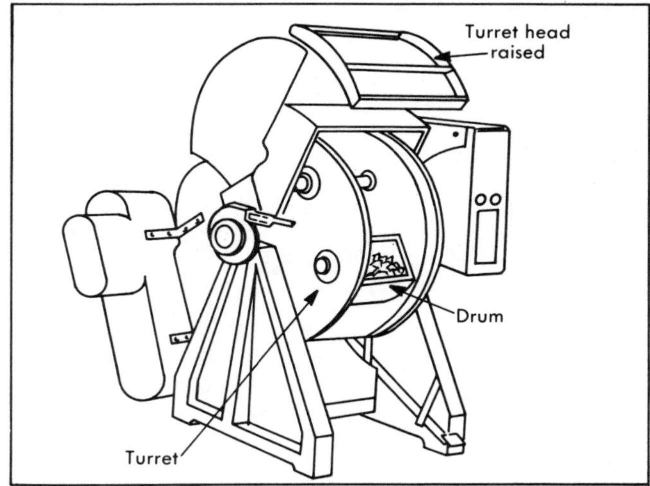

Fig. 16-68 Centrifugal barrel finishing machine with turret head raised for loading/unloading of drums. (*The Harper Co.*)

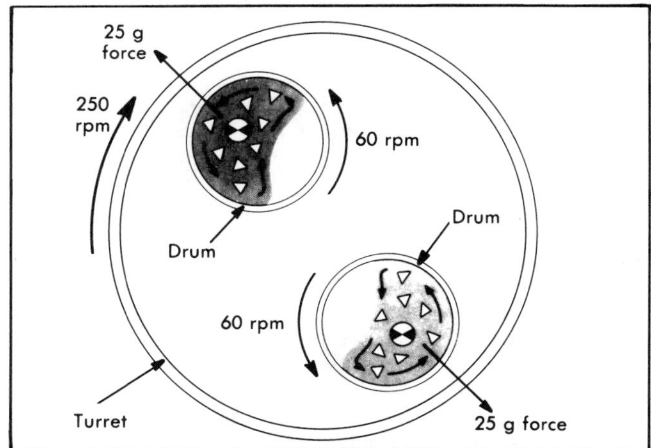

Fig. 16-69 Action of turret and drums in centrifugal barrel finishing. (*The Harper Co.*)

and drums rotating in a horizontal plane or in a vertical plane. Rotation in a horizontal plane is frequently used when the drums (work containers) are removable from the machine for reloading.

Drums are loaded and unloaded in a manner similar to that for normal barrel finishing. About 60-80% of each drum capacity is filled with a mixture of workpieces, media, compound, and water.

Turret rotation creates a high centrifugal force, as much as 50 times earth's gravity. This force compacts the loads within the drums into tightly packed masses. Rotation of the drums causes the media to slide against the workpieces, removing burrs and refining surfaces.

Process advantages. The abrading action under high centrifugal forces results in short processing cycles—20 to 50 times faster than conventional vibratory finishing. Another important advantage is that counter-rotation of the drums produces a smooth sliding action of media against workpieces, with no possibility of one workpiece impacting against another. As a result, consistent and reproducible results can be obtained; close tolerances can be maintained, even with fragile and precision components; and very smooth surface finishes can be produced.

Small-sized media can be used with centrifugal barrel finishing, thus ensuring uniformity of radii and edges, and permitting greater action in holes and recesses. Small parts can be handled as well. All equipment is quiet in operation and requires no special noise-abatement measures.

Economic considerations frequently dictate the choice among conventional tumbling barrels, vibratory machines, centrifugal barrel equipment, or other finishing machines. In general, if satisfactory results can be achieved in vibratory machines with process cycles of less than one hour, they are usually more economical than centrifugal barrel finishing. Small centrifugal barrel machines are less convenient to operate than vibratory units because their containers are normally closed and there is no means for integral workpiece/media separation. However, faster processing often offsets the higher cost and any inconvenience with centrifugal barrel finishing.

Applications. Originally, centrifugal barrel equipment was developed for finishing small, light components and workpieces having critical tolerances that could not be handled in other types of mass finishing machines. With the development of large machines, up to 100 ft^3 (2.8 m^3), and automation, the process is now being used for a broader range of applications.

Figure 16-70 illustrates one type of automatic centrifugal barrel machine. Automatic centrifugal barrel machines still involve batch operation, but they provide automatic loading, unloading, and separation of workpieces from the media. The equipment may be used for heavy stock removal and edge radiusing operations for both high and low-volume applications, as well as for precision finishing. Standard deburring and finishing operations seldom require more than 15 minutes.

Another important application for centrifugal barrel equipment involves imparting compressive stresses into the workpiece surfaces. This procedure improves resistance to fatigue failure and is being used for bearings, aircraft engine parts, springs, bearings, and pump components. Edge radiusing and surface finishing can be performed simultaneously, thus further enhancing the fatigue strength. This method generally costs less and is more effective than a combination of another finishing operation followed by shot peening.

Process variables. Process variables for centrifugal barrel finishing are similar to those for other mass finishing methods. Drums are loaded with workpieces and media, generally with a somewhat higher proportion of workpieces to media than in conventional barrel tumbling. Relative action on edges and surfaces can be controlled by the choice of media size and the amount of centrifugal force. It is not possible, however, to impart a radius on one edge while maintaining sharpness on another, equally accessible edge.

It is standard practice with centrifugal barrel finishing to operate at high speeds for maximum abrading action, and then automatically switch to lower speeds, possibly with a change from cutting compound to finishing compound, to achieve finely refined surface and edge conditions. As a result, two operations are combined in a single cycle.

Centrifugal Disc Finishing

The centrifugal disc process is the newest of the high-energy mass finishing methods. Machines consist of an open-top bowl having stationary sidewalls and a rotary disc for a base (see Fig. 16-71). Media, compound, and workpieces are contained in the bowl.

Operating principles. As the disc rotates with peripheral speeds up to 2000 sfm (610 m/min), the mass within the container is accelerated outwards. When the media and workpieces contact the stationary sidewalls of the container, the sidewalls act as a brake, and the load starts to decelerate as it is forced upwards by the mass behind it. The media and workpieces rise to the top of the load, then flow towards the center of the container, and from there downwards toward the disc.

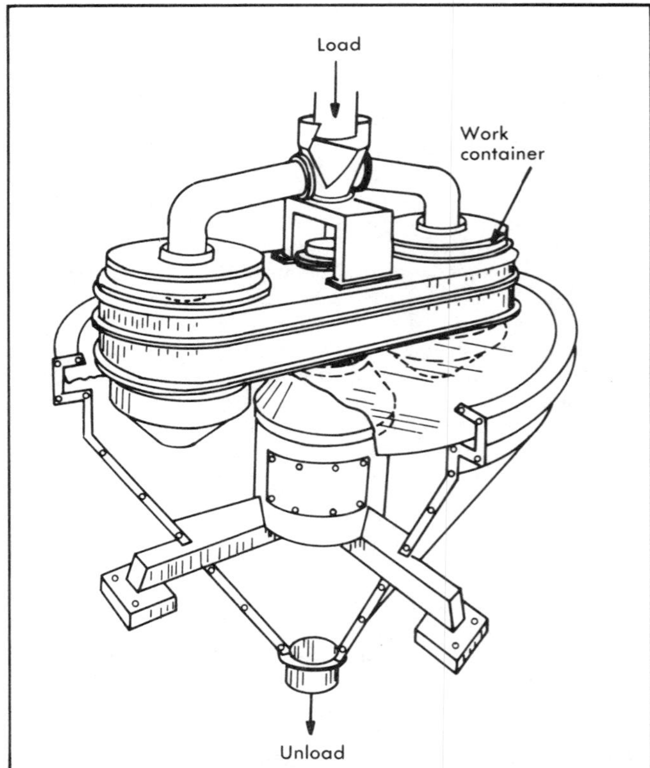

Fig. 16-70 Automatic centrifugal barrel finishing machine. (*The Harper Co.*)

Process advantages. The action in centrifugal disc machines is substantially faster than in vibratory equipment. Centrifugal force developed is as much as 10 times the earth's gravity, which presses the media against the workpieces. Processing cycles typically vary between one tenth and one twentieth those possible with vibratory finishing.

Short process cycles result in decreased costs, reduced floor space requirements, and less work in process and inventory. Centrifugal disc finishing is a fairly versatile process and easily adapted to batch automation. As with vibratory finishing equipment, workpieces can be readily inspected during the process cycle. With variable-speed equipment, it is frequently possible to combine heavy deburring and edge radiusing with a final, more gentle surface refinement operation.

Centrifugal disc equipment provides a smooth action. While centrifugal disc machines cost about five times that of vibratory equipment, processing cycles are typically one-tenth less, thus making them economically justifiable for many applications. The cost of media and compounds per part processed is comparable to their cost in vibratory finishing.

Limitations. The centrifugal disc process complements the barrel, vibratory, and centrifugal barrel processes. It is capable of producing very fine finishes, but cannot yet handle as wide a variety of parts as barrel, vibratory, and centrifugal barrel processes. Machines currently available cannot handle parts much larger than about 6 in.^{3} (98 cm^3), although parts to 12" (305 mm) long are being processed. Very small parts or media cannot be handled easily because there must be some gap, generally less than 0.010" (0.25 mm), between the disc and the cylinder walls. Also, centrifugal disc machines are not adaptable to continuous automation.

Equipment available. Centrifugal disc equipment is available with capacities (workpieces and media) ranging from about 1 to 25 ft^3 (0.03 to 0.71 m^3). Machines have been developed recently for fine finishing and burnishing operations. Abrasion-resistant materials for the cylinders and discs are essential to provide a reasonable service life.

Unloading and reloading is accomplished either by doors in the sidewalls of the cylinders or by turning the entire units over. Subsequent media separation, classification, washing, and return operations are external, similar to those used for tub-type vibratory equipment.

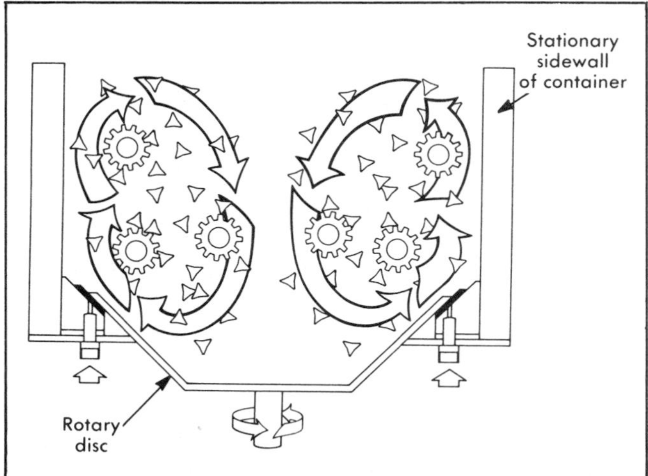

Fig. 16-71 Principle of centrifugal disc finishing. (*The Harper Co.*)

Spindle Finishing

Spindle finishing is another high-energy, mass finishing process that features fast and precise deburring or finishing. Good control, smooth finishes, and a high degree of reliability and uniformity are other characteristics of this process.

Processing cycles with spindle finishing rarely exceed five minutes and are frequently less than 30 seconds. A limitation is that the workpieces must be individually mounted on the spindles of the machines with workholding collets or fixtures. Mechanical arms or industrial robots are sometimes used to automate the loading and unloading.

Operating principles. Most spindle finishing machines consist essentially of a circular rotating tub that holds the media and one or more rotating spindles on which the workpieces are held and immersed in the media (see Fig. 16-72). With the workpieces rotating in one direction and the media moving rapidly in the opposite direction, the media flows swiftly over the edges and surfaces of the workpieces. Flow-through of the compound solution, using perforations in the tub, is typical practice.

Spindles are generally rotated at speeds from 10 to 3000 rpm. For some applications, however, the spindles are stationary; and in some cases, they are oscillated vertically. Multispindle heads are used on some machines, and there can be more than one head per tub. The tubs are held stationary or vibrated. Air is sometimes injected into the media to keep it loose.

Spindles are generally adjustable to vary the results obtained. For maximum action, the spindles are positioned to place the workpieces as close as possible to both the periphery and the bottom of the rotating tub. This arrangement stems from the fact that the velocity of the media is proportional to the radius from the tub's center, and hence there is an advantageous pressure effect created by compressing the media between the workpieces and the bottom and sidewalls of the tub. Air or hydraulic cylinders are generally used to pivot or lower the spindles into the tubs.

High-pressure spindle finishing. Higher velocity and confinement of the media generates greater finishing forces. These requirements can be met by fitting a standard tub with a baffle and center dome (see Fig. 16-73). The baffle tends to force the media downward toward the workpiece as the result of centrifugal force and permits the tub to be operated at a higher speed for faster processing. The center dome creates a pocket

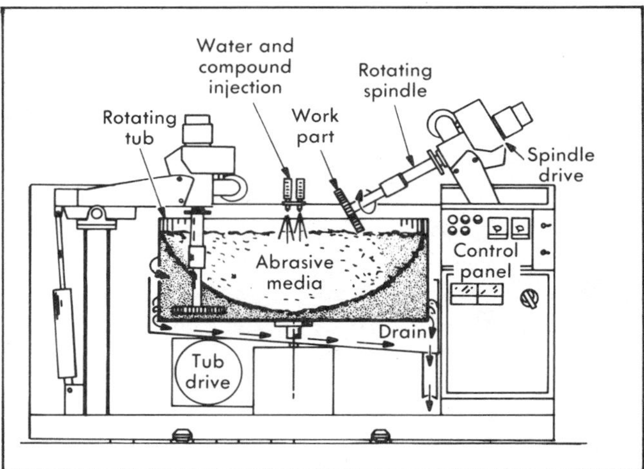

Fig. 16-72 Two-spindle mass finishing machine (*Almco Industrial Finishing Systems*)

effect that does not allow the media to flow into the center area of the tub. For small workpieces, the dome allows the media to recover faster and creates a dense, solid mass. For large workpieces, the confinement generates high finishing forces, resulting in short cycling times.

Continuous spindle finishing. To eliminate the unproductive time spent removing spindles from the tubs and unloading/reloading workpieces, continuous machines, such as the one illustrated in Fig. 16-74, have been developed. With this continuous machine, a head holding six rotating spindles indexes the workpieces into and out of the media in a continuously rotating tub. Unloading and reloading of the workpieces can be done manually or with a robot (as shown in Fig. 16-74) to increase productivity.

Typical applications. Spindle machines are well suited for deburring and finishing uniformly shaped cylindrical components, such as gears, sprockets, and bearing cages. Such components are easily fixtured, and the action of the media is uniform over all significant surfaces. Because the parts are fixtured, there is no possibility of impingement of one against another. Workpieces can often be stacked on a spindle so that several parts can be deburred or finished simultaneously.

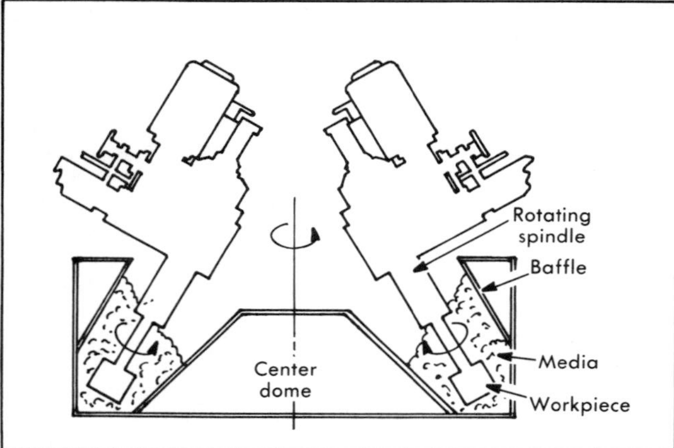

Fig. 16-73 Tub is equipped with a baffle and center dome for high-pressure spindle finishing. (*Almco Industrial Finishing Systems*)

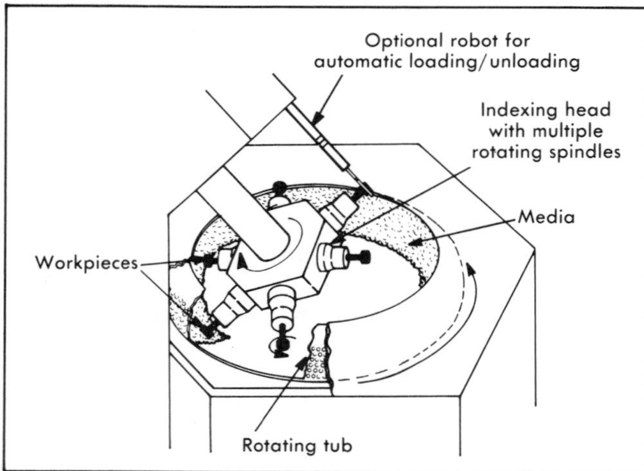

Fig. 16-74 Continuous spindle finishing with indexing multispindle head and robot for loading/unloading.

OTHER MASS FINISHING PROCESSES

There are many other mass finishing processes that have been developed. Some, such as resonant energy, chemically accelerated, electrochemically accelerated, and cryogenic finishing, are being employed and show promise for many applications. Other mass finishing methods having only limited, if any, commercial application are discussed briefly later in this section.

Use of Resonant Energy

Cleaning, deburring, and radiusing of intricate internal passages and recesses is being done with resonant power. Essentially, the process consists of submerging vibrating workpieces, fixtured at the ends of a resonating beam, in abrasive media for a predetermined time (see Fig. 16-75). The combination of resonant motion of the workpieces and orbital movement of the abrasive grains cleans and finishes the surfaces.

One or more workpieces are clamped in fixtures secured to both ends of a steel beam. High energy levels are produced by resonating the elastic beam near its natural frequency, using a mechanical oscillator with a counterweighted shaft, connected to a variable-speed drive motor. Processing is generally done at 75 to 100 Hz (4500 to 6000 vibrations per minute), depending upon the weight of the workpieces. Typical cycle times for cast iron workpieces range from 2 to 3 minutes.

Media containers are raised and lowered hydraulically, with a two-speed control to provide rapid advance and retraction of the containers, and a slower speed as the workpieces enter and leave the media. The abrasive media commonly used is randomly shaped, fused aluminum oxide; abrasive size depends upon the workpieces being processed. Water and a mild alkaline cleaning compound are sprayed on the abrasive in the containers during processing. When the containers are lowered to remove workpieces from the media after a preset time, beam vibration is continued for a short period to remove residual media from the parts and return it to the containers. Using similar principles, another machine oscillates the media container or the parts, without using a machine beam.

Use of Chemical Accelerators

Chemical accelerators are being used to reduce the cycle times of vibratory finishing processes. Investigations into the

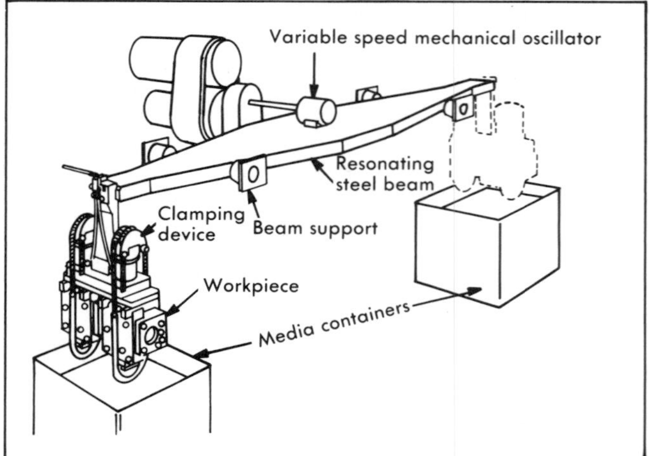

Fig. 16-75 Machine for internal finishing and cleaning of precision castings, which are vibrated at high frequencies by means of a resonant beam. (*Wheelabrator-Frye, Inc.*)

use of accelerators started in 1971 with a zinc alloy finishing program. Some research was done previously with the use of accelerators for barrel finishing.

When chemical accelerators are added to abrasive media in a vibratory machine, the chemicals react with the metal workpieces to remove metal. The vibrating media wipes away any passivating films that form and provides additional, mechanical cutting and burnishing action. Conventional vibratory finishing media and machines can be used, although linings in the tubs or bowls may be affected by the chemical solutions.

While accelerators are used for finishing steel and brass parts, they are also employed for zinc die castings. The chemical accelerators are most useful for vibratory deburring if the die castings have been trimmed closely to leave only small burrs; the large, jagged burrs left on many trimmed castings require the use of abrasive belts before vibratory finishing.

Aliphatic acids have been found useful as accelerators for barrel and vibratory finishing, with reductions in cycle times up to 90%. Maleic acid appears to be the most effective aliphatic acid and the least costly. A sodium bisulfate addition to the maleic acid solution provides increased metal removal rates for deburring. However, a solution [5 g/L (0.66 oz/gal)] of maleic acid with no sodium bisulfate generally produces smoother surface finishes.

Chemically accelerated processes are more difficult to control than standard mass finishing methods, and there are greater problems with effluent disposal. When using accelerators, care is required in handling the chemicals. Also, any residual chemicals left on the workpieces must be removed or neutralized. Information on chemical deburring and polishing processes is presented in Chapter 17, "Thermal, Chemical and Electrochemical Finishing," of this volume.

Cryogenic Deflashing

Cryogenic deflashing of molded rubber or plastics products and aluminum or zinc die castings is being done in barrels and vibratory machines by freezing the workpieces with liquid nitrogen or carbon dioxide. Flash on the workpiece becomes embrittled when its temperature is reduced, and mechanical means are used to remove the flash. When using liquid nitrogen, proper ventilation is essential, and care is required in handling the cold parts and media. The cost of liquid nitrogen may be a limitation to the use of this process.

Barrels and vibratory machines used for cryogenic deflashing require special double-walled, insulated work chambers to conserve the liquid nitrogen. The chambers are fitted with multiple nozzles for spraying the nitrogen. It is general practice to cool the media in the work chamber prior to adding the workpieces, which must then also be cooled prior to starting the machine. Temperature control during processing is essential. Temperatures as low as -320° F (-196° C) are used for some parts.

Electrochemical Mass Finishing

In this process, workpieces are placed in an electrically insulated drum filled with electrolyte and media, which may be graphite spheres. British patents, filed in 1968, indicate that phosphoric acid is used as the electrolyte and that either silicon carbide or copper-coated aluminum oxide can also be used as the deburring media. Electrodes, connected to an external power source, are also placed in the drum. When the drum is vibrated, the media acts as extensions of the electrodes to remove the burrs electrolytically. The action is similar to nonvibratory electrochemical deburring, discussed in Chapter 17, "Thermal, Chemical and Electrochemical Finishing," of this volume. The media also abrades the passivating surface oxides formed on the workpieces, thus exposing fresh surfaces to the electrolytic action.

The major advantage of this process is fast cycling; but the equipment required is comparatively expensive, and the process is limited to workpiece materials that conduct electricity.

Limited Application Methods

There are several other mass finishing methods that have only limited or no commercial application. Consequently, they are only discussed briefly in this section.

Vibratory rotary barrel finishing. Conventional rotary barrels that vibrate as they rotate have been developed for specific applications when problems are encountered with workpieces migrating out of the media in vibratory machines.

Reciprocal finishing. In this process, workpieces are moved back and forth in a stationary tub containing media. It is similar to spindle finishing in that the workpieces must be attached to a holding device, but the workpieces are oscillated rather than being rotated. The process permits handling parts too large or odd shaped for spindle finishing equipment. Finishing action can be concentrated on certain areas while shielding other areas. Processing cycles are normally much longer than with spindle finishing machines.

Reciprocal finishing can be fully automated. One design incorporates a traverse unit that reciprocates on rails fixed to an upper frame, fixture clamping arms pivoted to the traverse unit, a loading apparatus, and a transfer mechanism for loading and unloading. Several modifications of these machines are in use, including one type that employs spindle rotation.

Magnetic-abrasive mass finishing. Workpieces are placed in a container of magnetic, loose abrasive media, and an oscillating magnetic field is produced to vibrate the media. While the workpieces can be stationary, more effective finishing is obtained by tumbling or otherwise moving the parts.

When magnetic media is used, the workpieces must be nonmagnetic to obtain relative motion. However, it is conceivable that a magnetic field could oscillate miniature magnetic parts through nonmagnetic media. Magnetic media can be coated with ceramics or other metals to minimize contamination of the workpieces. Most iron-based components will not react well to high-frequency changes in a magnetic field.

In an experimental method of magnetic-abrasive finishing, cylindrical parts are vibrated, rotated, and fed axially in a magnetic field between two poles. Abrasive grains are combined magnetically between the poles and the workpiece to perform the cutting action. This process is also being used for noncylindrical parts.

MEDIA FOR MASS FINISHING

Media refers to the abrasive or nonabrasive, consumable elements used in mass finishing processes. The main function of media is to abrade or burnish the edges and surfaces of components to the desired finish. Media also helps keep the workpieces from impinging on each other and serves as a carrier for any compounds used.

Selection of Media

Selection of the proper media for a specific application is critical for optimum deburring and surface finishing results.

MEDIA FOR MASS FINISHING

Unfortunately, the choice of media is difficult because of the diversified requirements of numerous applications and workpieces and the wide variety of types, sizes, and shapes of media available. In many cases, selection must be made on a trial-and-error basis.

Major selection factors to be considered are the composition (type), size, shape, and weight of the media to be used. Other important considerations include the capability, versatility, economy, and availability of the media. Workpiece material, size, and geometry; burr sizes and locations; finishing requirements; finishing equipment used; and subsequent operations and functions of the workpieces are also critical factors. Some general recommendations with respect to media selection are presented in Table 16-28.

Media composition. The composition of a media determines whether it is a cutting or finishing type of media. Cutting types of media contain abrasives, while finishing types have no abrasives or only very fine abrasives. The workpiece material and its hardness, the size of the burrs, and the surface finish requirements all govern the type of media to be used.

Media with aggressive cutting capabilities are generally best for the removal of large burrs. Requirements for smooth surface finishes necessitate the use of slower cutting media with fine or no abrasives. Harder media can generally be used in the high-energy methods of mass finishing.

The composition of a media also determines its weight, which affects the rate at which work is performed. For fixed operating parameters, heavier media exerts more pressure on the workpieces than lighter media. Increased pressure can provide faster cutting and more polishing action, but may mar soft surfaces or distort fragile parts.

Media size. The size of the media used is important for several reasons. One function of size is helping to keep the workpieces separated during finishing. Small-sized media helps keep small parts separated. On the other hand, large media cuts faster and produces a rougher finish than small media. As a result, a mixture of media sizes is normally used for processing parts.

Another important consideration is that the size of the media must allow for reaching all surfaces to be finished without lodging in holes and recesses of the workpieces. The ease with which the media can be separated from the workpieces must also be analyzed. With separation by screening, the media must be either larger or smaller than the workpieces. However, with magnetic separation of ferrous parts, media size in relation to workpiece size is not a factor. Smaller media can generally be used in the high-energy methods of mass finishing without increasing cycle times significantly yet improving the surface finishes produced.

Media shape. Different shapes of media (see Fig. 16-76) have advantages for specific applications, depending primarily upon the configurations of the workpieces. Major considerations in selecting a shape are that it provides access to all workpiece surfaces requiring deburring or finishing, does not lodge in holes or recesses, and permits easy separation from the parts at the end of the cycle. A mixture of media shapes is sometimes used.

Triangles. Media of triangular shape provides uniform action and is effective in reaching corners and slotted areas. Angle-cut triangles have sharper corners and edges, and provide deeper penetration into remote areas. Notched triangle (arrowhead) media, in addition, provides sharper edges for reaching into slots, holes, or openings. The smallest standard triangle available has a corner radius of about 0.040" (1.02 mm), but it is frequently desirable to use larger triangles that have larger radii since they minimize cycle time.

Cylinders. Cylindrically shaped media works effectively for some applications. With its ends cut at angles of 22 to 60°, cylindrical media reaches recesses more easily.

Diamonds. Diamond-shaped media has sharp points that can reach into corners, slots, and recesses without lodging problems.

Cones. Conically shaped media is very versatile. It permits partially entering holes of various diameters, without lodging problems.

Spheres. Spherically shaped media has good flow action and surface contact. It is useful for uniform blending and smoothening of surfaces.

Media capability. A most important consideration in media selection is the ability of the media to remove burrs and produce the required corner, edge, and surface conditions with consistent uniformity. Cycle times required to achieve these results are also critical. Minimum break-in requirements, wear, and reclassification needs, as well as good cushioning action, are desirable features.

Versatility. Other features desirable in media include the

TABLE 16-28
Media Selection Guide for Mass Finishing

Effect Desired	Degree of Effect	Media Recommended*
Deburring	Light	Steel or ceramic
	Medium to heavy	Ceramic or plastic
Radiusing	Light to heavy	Ceramic or plastic
Surface roughness	Reduce to lower value.	Plastic or ceramic
	Preplate quality on softer alloys	Plastic, ceramic, plastic → plastic, plastic → steel, ceramic → steel, steel, or wood
Surface reflectivity	Brighten or highlight.	Steel or ceramic
	Best quality, hard alloys	Ceramic or ceramic → steel
	Best quality, soft alloys	Plastic → steel
	Best quality, plastics	Wood
Clean surfaces	All metals	Steel or ceramic
	Irregular surfaces	Randomly shaped
Least operating cost	Varies greatly with grade involved.	Steel, ceramic, plastic, or wood
Fastest processing speed	Knowledge of overall capabilities of media must be understood.	Steel, ceramic, plastic, or wood

(*John B. Kittredge, Consultant*)

* All pertinent factors must be evaluated before making a selection. A media may be inappropriate for overall part requirements. Two-step processes are denoted by "→" between media steps. Media not listed does not infer unsuitability.

MEDIA FOR MASS FINISHING

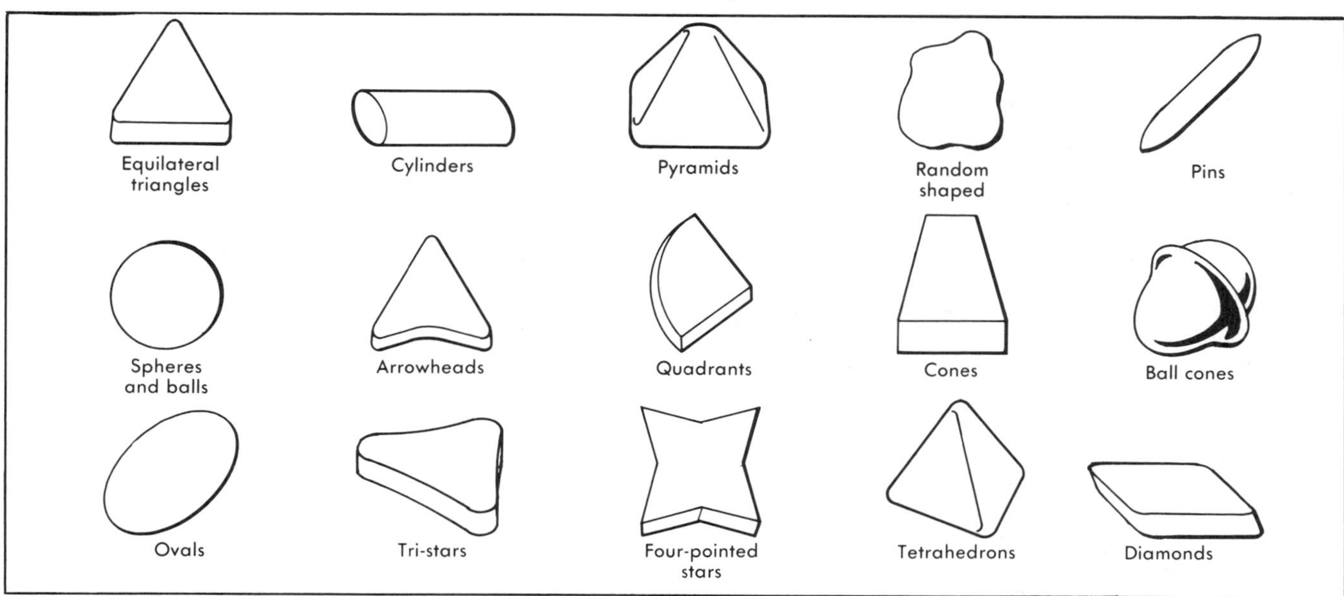

Fig. 16-76 Some of the many different media shapes available for mass finishing. (*Bendix Corp.*)

ability to finish a variety of workpieces in a given machine and the ability to achieve different finishes.

Economy. Cost per pound of different media is of little significance in overall economic considerations. Wear characteristics of the media and cycle times to attain the required results must be considered in determining the all-important cost per finished part.

Availability. The consistent availability of uniform, high-quality media from a reliable supplier is another important consideration.

Types of Media

The most important types of mass finishing media, from the standpoint of most common usage, include natural abrasives, agricultural products, synthetic random media, preformed ceramic and resin-bonded media, and metallic media.

Natural abrasives. Quarried, crushed, and graded stones were the original mass finishing media. They have, however, been largely replaced by synthetic materials primarily because natural stones are softer, wear more rapidly, are less consistent, and occur only in random shapes. Synthetic materials are harder, provide longer life, and have more consistent cutting action, wear rate, and dimensions.

Despite their limitations, natural abrasives still have some applications. Corundum is a natural crystalline oxide of aluminum (Al_2O_3) that can be economical for some uses. Novaculite is a fine-grained silica stone that maintains sharp particle shapes and can be of value for removing burrs from small holes and grooves. Soft limestone and hard but more friable granite can still be economically justified for some barrel finishing applications. Natural abrasives are sometimes used in the manufacture of bonded preforms.

Agricultural and wood products. Ground corncobs and walnut or other nut shells, as well as hardwood sawdust, are used in tumbling barrels and heated barrels for drying parts. These materials produce a good luster on some workpieces. When mixed with fine abrasives, they are suitable for fine polishing operations, particularly in the jewelry industry.

Wood shapes, such as cubes, pegs, or wedge-ended pegs, are made from hardwoods. They are being used to deburr and finish wood, plastic, ceramic, and metallic components. Sometimes they are coated with oils or waxes to transfer coatings to the workpieces or fine abrasives for increased cutting action. Other wood shapes used for finishing include dual wedge-ended pegs, and diamonds. Hardwood media can duplicate the effect of color buffing on plastics, wood, or soft metals.

Synthetic random media. Fused aluminum oxide (Al_2O_3), crushed and graded, has greater abrasion properties than natural materials. Other advantages include good wear resistance and consistency of size and quality. However, the physical characteristics of fused aluminum oxide vary with different producers. Sintered aluminum oxide is the longest lasting random-shaped abrasive material available for mass finishing. It is used extensively for fine deburring and finishing.

Fused silicon carbide (SiC) media is available, but sizes above about 1/4" (6 mm) diam fracture too readily to be well suited for mass finishing. This synthetic mineral may have to be used for components that have to be subsequently brazed or welded and when aluminum oxide cannot be used. However, preformed media is normally preferred to silicon carbide for such applications.

Preformed media. The development of media of controlled shape, size, and abrasion and finishing characteristics has been largely responsible for making mass finishing a precision process. Preformed media is available in ceramic, resin-bonded, and metallic types, all of which are used extensively. The abrasives used in preformed ceramic and resin-bonded media are often aluminum oxide, silica, or silicon carbide.

Preformed ceramic media. Ceramic-bonded media are produced by mixing porcelain, clay, or other vitreous materials with varying percentages of or with no abrasives, forming into desired shapes (usually by extrusion and cutting), and then firing to vitrify. Fusion bonding is used to make some media. Such media is available in a wide range of shapes and sizes. Also, selection of the proper grade of abrasive, the correct proportion of abrasive to binder, and the most suitable binder makes this media adaptable to a great variety of applications and workpieces. Ceramic media with about 50% aluminum oxide abrasive of 60 grit is the fastest material for deburring, edge generation, and stock removal.

Preformed resin-bonded media. This media is formed by bonding abrasives (generally 40-70% by weight) into polyester or urea-formaldehyde resins, usually by casting. This so-called "plastic media" is softer, has a lower density than preformed ceramic media, provides a cushioning effect, and sometimes has a shorter life. It is well suited for finishing softer metals and fragile parts, achieving preplating requirements, and producing smooth finishes free of impingement. Plastic media can also be produced in a wide variety of desirable shapes.

High-density polyester media. A more recent development, high-density polyester media is made the same way as low-density media except for a difference in composition: a polyester resin and heavier silicate filler are used. The additional weight of the media and sharper crystal facets improve the cutting performance compared with low-density media. It is also claimed that smoother surface finishes can be produced than with ceramic media.

Metallic media. Preformed, hardened steel media, free of abrasive, is available in a variety of shapes and sizes. It is used primarily for burnishing to achieve maximum luster for decorative purposes, for light deburring applications, and for heavy-duty cleaning. Advantages include uniformity of shape and size, the elimination of possible fracturing in use, and no wear. The high bulk density of steel media results in rapid peening and deburring, and supports the workpieces. Soft steel shapes are used as carriers for fine abrasive particles.

Media shapes preformed from zinc are occasionally used for burnishing operations as an alternative to steel. Tacks, pins, wire brads or clippings, and nails, mixed with fine abrasive, are sometimes used to remove burrs from small holes and recesses.

Other media. There are many other media used for special mass finishing applications. For example, pieces of leather or felt are sometimes used for light burnishing or wiping operations. Nylon media is used for cryogenic deflashing, and rubber-bonded abrasive media for light deburring and smoothening of soft metals.

Ratio of Media to Workpieces

Determining the proper ratio of media to workload to be used is essential for economical and successful mass finishing. High ratios reduce the chance for impingement. Any change in the optimum ratio can have a significant effect on the quality of the finishing achieved. Table 16-29 presents some typical media-to-part ratios for some commercial applications of barrel and vibratory finishing.

COMPOUND SOLUTIONS FOR MASS FINISHING

Compounds are combinations of chemicals that dissolve in water and form solutions to maintain consistency or to modify the action of media against the workpieces in mass finishing. The use of too little, too much, or an improper compound can adversely affect the cycle time and/or ability of the media to perform as intended.

There are some finishing applications where no compound solutions are used. For example, loads of workpieces are sometimes finished dry, perhaps with sawdust, ground corncobs, wood pegs, or other media, for burnishing purposes.

TABLE 16-29
Typical Media-to-Parts Ratios for Commercial Applications of Barrel and Vibratory Finishing

Media-to-Parts Ratio, by volume	Commercial Applications
0:1	No media for cutting or parts separation. Part-on-part contact sometimes used to beat off burrs.
1:1	Equal volumes of media and workpieces. Produces very rough surfaces.
2:1	More gentle action due to greater separation. Severe workpiece damage still possible.
3:1	About minimum ratio for nonferrous parts. Fair-to-good results with ferrous metals. Considerable workpiece contact.
4:1	Average ratio for nonferrous parts; good for ferrous metals. Produces fair-to-good surface finishes.
5:1	Minimal part-to-part contact. Good for nonferrous metals.
6:1	Very good for nonferrous metals. Common for preplate work on zinc alloys with plastic media.
8:1	For high-quality preplate finishes.
10:1 to 15:1 or more	Used for irregularly shaped parts or parts subject to tangling or bending.
One part per machine or compartment, or fixtured	Avoids part-to-part contact.

(*John B. Kittredge, Consultant*)

Media for dry finishing is often treated with an oil, grease, or wax-based formulation with finely powdered abrasives to produce smooth surface finishes; cycle times are much longer than for wet operations. There are other applications where the desired results are achieved with a suitable compound solution and no media; the workpieces themselves act as the media, tumbling against each other for the required action.

Functions of Solutions

There are many different functions served by various compound solutions. Major functions include cleaning of the workpieces and media, and control of the process. In some applications, the solution modifies the luster or color of the workpieces. The use of corrosion-inhibiting compounds is especially important when any metal is used.

Cleaning functions. All mass finishing operations require clean machine compartments, media, and workpieces for effective and efficient results. Compound solutions are often required to clean the workpieces by removing tarnish or scale, emulsifying oils, and suspended solids. The solutions are also expected to maintain the cleanliness of the workpieces and media during the finishing cycle. Buildup of dirt on the media causes glazing and reduces the cutting capability. Strongly alkaline formulations are often used for cleaning iron and steel parts, while milder alkaline to mildly acidic compositions are generally best for nonferrous metals and alloys. Some workpieces require cleaning and/or descaling prior to mass finishing.

Process control. Compound solutions are often required to control the hardness of the water used, the pH of the solutions, and foaming. In barrel finishing, foaming is often desirable; in vibratory finishing, it is not. Other control functions include wetting the surfaces of the workpieces, providing cooling and lubricity, cushioning the workpieces, and preventing corrosion. The cooling function is very important when using high-energy mass finishing methods.

The compound solution is sometimes required to modify the action of the media being used. For example, in some applications it may be necessary to change from maximum abrasion to maximum finish improvement conditions. Another important function of compound solutions is to produce effluents that meet EPA, OSHA, state, and local requirements. The development of phosphate-free, biodegradable types of compounds have made it easier to conform to these regulations.

Types of Compounds

Compound types may be classified basically according to function, as follows: (1) abrasive, (2) descaling or bleaching, (3) cleaning, (4) deburring and edge radiusing, and (5) burnishing and coloring. Many different compounds are available for specific functions. Selection depends upon numerous factors, including workpiece material, media type, the finishing process being used, water hardness, and cycle times. Compounds are available in liquid and powder forms. When a finishing application requires free abrasives to influence the cutting action of the media, the compounds containing free abrasives must necessarily be powders. Requirements for free abrasives, however, are limited; and liquid compounds are preferred for most mass finishing applications because they are easier to handle and to consistently use with precise results.

Most compounds contain several types of ingredients, such as chemicals to perform a certain function and other ingredients to facilitate finishing. Water conditioners minimize the deposit of hard salts on the media and workpieces by reducing the water hardness. If the available water has an exceptionally high mineral content (over 200 ppm), it may be necessary to use demineralized or preconditioned water. Detergents help keep the workload clean. Corrosion inhibitors minimize rusting. In general, the compound used must be tailored to the workpiece material and condition, and to the results required.

Applying Compounds

For batch-type operations, when workpieces and requirements vary from load to load, the manual method of adding compounds is the simplest. A correctly weighed amount of compound is added directly to the machine compartment, together with the media, workpieces, and required amount of water. Granular or powdered compounds should be added before the workpieces.

Semiautomated systems generally entail premixing of the compound with water in a separate tank. This solution is then pumped into the work container of the finishing machine. Proper control of compound concentration can be a problem with this method.

For most automated mass finishing applications, compound solutions are added by flow-through systems. With these systems, fresh compound solution is metered into the finishing machine, allowed to perform its functions, and is then permitted to drain from the machine. Flowmeters ensure precise control of the proportion of compound to water and the amount of solution that enters the machine, thus providing consistently uniform results.

The amount of compound added varies with the specific compound being used and the results desired. In general, from 1/2 to 4 oz (14.2 to 113.4 g) of compound are added to each gallon (3.8 L) of water. The compound solution is generally introduced at rates ranging from 1/2 to 2 gal (1.9 to 7.6 L) per cubic foot (0.03 m^3) of workload per hour.

Water Levels

The water levels used in mass finishing equipment are important with respect to cycle times and the results obtained. Barrels usually require much more water than vibratory machines. Low water levels in barrels, about 2" (51 mm) below the top of the mass, provide a harsher action, suitable for deburring operations; high levels, about 2" above the mass, result in a gentler action and more cushioning, suitable for burnishing.

TROUBLESHOOTING

There are so many variables (equipment, operating parameters, workpiece materials, media, compound solution, desired results, etc.) in mass finishing that all possible problems that may be encountered cannot be discussed in this section. Four major problems are unacceptable results, excessively long cycle times, high media/compound costs, and media lodging in the workpieces.

Unacceptable Results

When unsatisfactory finishes, incorrect corner radii, incomplete burr removal, or other problems are encountered in mass finishing, they are probably the result of using improper equipment, media, compound solution, or workpieces. Improper operating parameters for the machines, incorrect concentrations of compounds in the solutions, and trying to finish workpieces differing in quality, all contribute to poor results.

Long Cycle Times

Excessively long cycle times can result for many reasons. The time can sometimes be shortened by changing the compound solution used to reduce glazing of the media and improve the cutting action. The media itself may be the problem; it may be too large and impeding separation from the workpieces. Changing to a different shaped and/or faster cutting media can often reduce cycle times. Machines used should be kept in good condition and operated at proper speeds, or they may lengthen cycle times.

In some cases, long cycle times are the result of overfinishing the parts. Specifications for actual requirements should be checked, and then a shorter cycle tried to see if the requirements can be met. Sometimes, the preprocessing operations performed on the parts can be changed to reduce the amount of finishing needed. Attempts should be made to increase the uniformity of quality and cleanliness of the workpieces to be finished in order to reduce cycle times.

High Media and/or Compound Costs

The most obvious solution to this problem is to see if acceptable results can be obtained with lower cost media or compounds. Increasing the media size so that it is not carried out in the workpieces can often cut costs. Another possible solution is to try using media having greater wear resistance, especially for short-cycle operations. Sometimes reducing the machine speed cuts media and compound costs, often with no significant effect on cycle time.

Lodging of Media in Workpieces

Wedging of media in the holes, slots, or recesses of workpieces can result in the need for costly removal. Such lodging is sometimes caused by the breaking or fracturing of the media, improper choice of media size, or the media becoming contaminated with foreign material. Possible solutions include using a larger or differently shaped media, frequent screening and reclassifying of the media to maintain a uniform size, and removing contaminants from and keeping the media clean by using the proper compound.

SAFETY CONSIDERATIONS

Like all other metalworking processes, mass finishing does pose some potential health hazards. These hazards include the chance of injury from the machinery, potential hearing loss from exposure to high noise levels for extended periods of time, or possible skin diseases.

Machine Safety

Mass finishing machines should be equipped with lockout devices to prevent them from being accidentally turned on during loading, unloading, or maintenance procedures. All pinch points, belts, doors, and covers must be properly guarded. Adequate material handling equipment should also be provided. Good housekeeping practices to ensure dry, clean, and uncluttered floors are also important.

Noise Levels

Barrel finishing machines produce little noise, but vibratory equipment can be noisy, depending primarily upon the media being used. Noise levels can be reduced by using smaller media and/or noise-abatement covers for the equipment. Vibrating mechanical screen separators can also be coated with rubber or plastics. Current OSHA noise regulations are usually met by one of these means. When noise levels are excessively high, the operators may be required to wear ear plugs or muffs.

Skin Protection

Protection from contact with abrasive, chemical, or workpiece dusts and fluids can be provided by adequate shields and ventilation on the machines. Safety glasses and shoes for the operators are recommended. Other protective clothing, such as face shields, rubber gloves or protective creams, and aprons, may be required.

ABRASIVE-FLOW MACHINING

Abrasive-flow machining (AFM) is a process in which a semisolid abrasive media is forced, or extruded, through a workpiece passage. The three major elements required for the AFM process are the machine, the workpiece fixture (tooling), and the media.

Machines used (see Fig. 16-77) hydraulically clamp the workholding fixtures between two vertically opposed media cylinders. These cylinders extrude the media back and forth through the workpiece(s). Two strokes, one from the lower and one from the upper cylinder, comprise one process cycle, as illustrated in Fig. 16-78.

The AFM process can be thought of as the use of a self-forming abrasive tool that precisely removes workpiece material from those areas in which media flow is purposely restricted. In general, the media used determines the kind of abrasion that occurs, the fixture determines where the abrasion occurs, and the machine determines how much abrasion occurs.

PROCESS ADVANTAGES

By proper control of parameters and abrasive flow, the AFM process can perform a wide range of precision machining and finishing operations. These operations include deburring, edge radiusing, honing, polishing, and the removal of recast layers from workpiece surfaces. Important features of the process include selectivity, the capability of finishing inaccessible areas, and versatility.

Selectivity and Finishing Inaccessible Areas

Abrasion in the AFM process occurs only in areas where media flow is restricted; other areas are unaffected. Inaccessible areas and any restricted places through which the media can be forced to flow can be finished, including complex internal passages.

Versatility of the Process

The AFM process is being used on a wide range of workpiece and passage sizes. It can also process many selected areas on a workpiece simultaneously. Several to hundreds of holes, slots, or edges can be deburred, radiused, and/or polished in one operation.

A number of workpieces can also be processed simultaneously. Depending upon workpiece and machine sizes, several or even dozens of parts can be processed in one fixture load, resulting in production rates to hundreds of parts per hour.

Both small and large production quantities can be handled economically with the AFM process. Changeover from one job to another, including replacing the tooling and media, can normally be done in minutes.

LIMITATIONS OF THE PROCESS

Abrasive-flow machining is a versatile and controllable finishing process because the workpiece is held stationary and the abrasive media is directed to and often through the passages to be finished by the tooling. These characteristics, however, also impose some limitations on the process. One limitation is that most workpieces require a fixture, which involves some expense. Also, the workpieces must be loaded and oriented in the fixture, and then removed from the fixture after processing.

After processing, the abrasive media normally remains both within the part's interior and surrounding its exterior, making fully automated processing and handling difficult. Some recent, fully automated systems incorporate a machine with sufficient

ABRASIVE-FLOW MACHINING

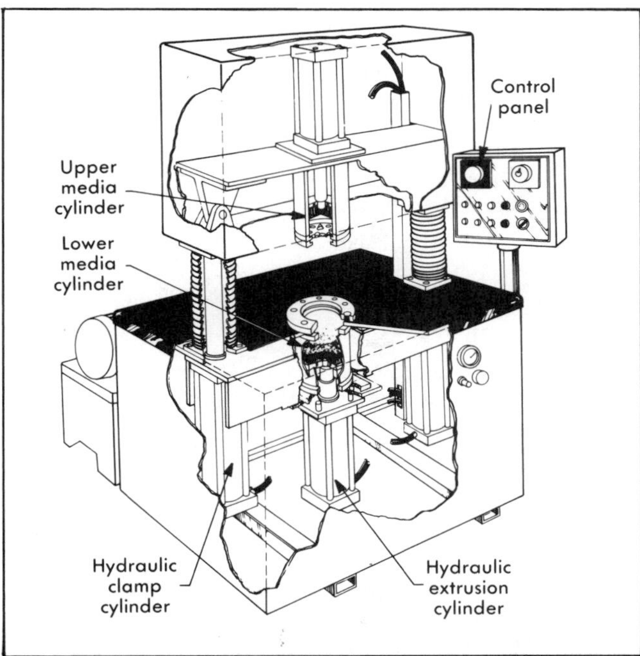

Fig. 16-77 Cutaway view of high-production machine for abrasive-flow machining (AFM). (*Extrude Hone Corp.*)

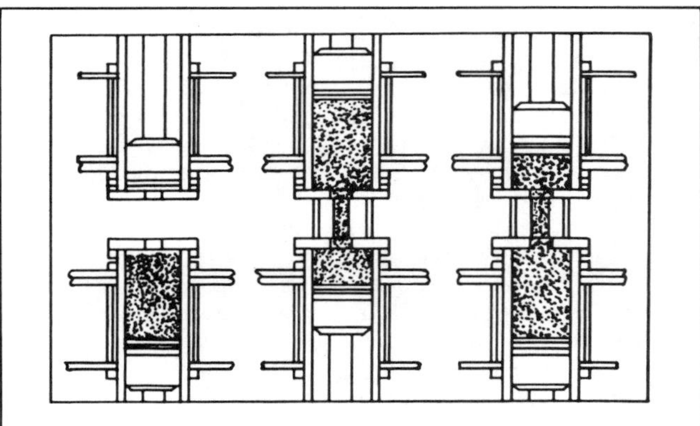

Fig. 16-78 One AFM process cycle consists of two cylinder strokes that extrude the media through the workpiece twice. (*Extrude Hone Corp.*)

capacity to completely process a fixture load of workpieces in one stroke. Flowing only from the interior of the workpieces, the abrasive media exits freely and falls into a receiving cylinder or collection hopper. This approach makes both tooling and automatic unloading/reloading easier because the outside of the workpieces are relatively free of media.

Removal and retrieval of media and final cleaning of the finished workpieces is another task required with abrasive-flow machining. Media can be removed from the workpieces by blowing it out with shop air. Fine particles and lubricant in the media are normally removed in an ultrasonic cleaning operation using commercial solvents. Such cleaning is essential because fine, loose particles left after blowoff can cause plug gages or mating components to seize in the holes. The careful choice of abrasive size and tooling will prevent abrasive particles from lodging in small holes.

The removal and retrieval of media and final cleaning of parts has recently been made more convenient for automated systems by using vacuum and water-based solvents. The solvents dissolve the media carrier and can be used as a spray or pulsed flush, without special venting.

TYPICAL APPLICATIONS

The AFM process is being used on workpiece sizes ranging from as small as 0.060″ (1.52 mm) diam to turbine discs nearly 4 ft (1.2 m) diam. Passages in the workpieces range from orifices as small as 0.004″ (0.10 mm) diam to splines 2″ (51 mm) or more across. The process can generate edge radii from less than 0.001″ (0.03 mm) to more than 0.060″ (1.52 mm).

For a given set of deburring conditions, the edge radius produced after removing the burr is a function of burr size. Thick, tall burrs reduce the radiusing action. With standard tooling, removing 0.005″ (0.13 mm) thick x 0.002″ (0.05 mm) high burrs in stainless steel results in a radius of about 0.005″. The same conditions in aluminum result in a radius of about 0.0075″ (0.19 mm). It is not possible with standard tooling to remove large burrs and produce small edge radii of 0.005″ or less.

Polishing with AFM can improve surface finishes of 30 to 300 μin. (0.76 to 7.6 μm) to one tenth or less of the original finishes. Material removal by the abrasion results in a dimensional change of about 25 to 35% more than the total roughness of the surface. For example, in reducing a surface finish of 100 μin. (2.54 μm) produced by EDM to a finish of 10 μin. (0.25 μm) requires the removal of about 0.001″ (0.03 mm) of stock. In most cases, stock removal can be held uniform throughout a passage within ±20%.

The finishing of dies (extruding, compacting, cold heading, upsetting, and others) is a major application of the AFM process. Advantages include reduced costs and longer die life. Costs are reduced by eliminating or minimizing the need for time-consuming hand polishing of the dies. Longer die life results from the directional finish produced by AFM and the improved uniformity and quality of the finished surfaces.

Other components commonly finished by the AFM process include gears, bearing races and cages, and splined parts. The manufacturers of jet engines and turbines use the process extensively for finishing components such as fuel swirlers, combustion liners, turbine discs and blades, and compressor wheels and vanes. Finishing by AFM is also used for critical components made by the aerospace, medical, and other industries in which demanding performance is essential.

PROCESS PRINCIPLES

Key to the AFM process is the special abrasive media used and its ability to precisely abrade only selected areas along its flow path. The media performs its selective abrasion by altering its physical characteristics, namely its viscosity (resistance to flow). In its normal semisolid state, the media is relatively pliable and yielding. However, the media temporarily increases its viscosity when it encounters a drastic restriction in its flow. This viscosity change is a special phenomenon of the polymer media carrier. When the polymer carrier experiences a sudden force, it reacts by becoming more viscous.

A drastic restriction in the media flow path creates a sudden force when the media mass attempts to flow through a smaller passage than previously encountered (see Fig. 16-79). Therefore, the portion of the media within this restriction becomes more viscous as it enters and flows through the passage. While in this more viscous state, the rigid media portion can be thought of as a slug. When the slug exits the restriction, its viscosity returns to normal.

Such momentary viscosity increases explain how the media used in the AFM process can selectively abrade local areas along its flow path. A fixture can be used to direct media so that the desired areas will form the narrowest passage in the flow path. If the area where abrasion is desired is not a sufficient restriction to flow, tooling can provide an insert or mandrel within this area to act as a restrictor, thus reducing the cross-sectional flow area so that it becomes a restriction.

The AFM process can simultaneously abrade two, and occasionally more than two, successive restrictions in the same flow path, providing that these areas are equal in cross-sectional area; otherwise, only the smaller of these passage areas will be abraded (see Fig. 16-80). Any number of parallel passages can also be abraded. If, however, two parallel passages have differing cross-sectional areas, then the larger of these restrictions will receive more flow and consequently more abrasion since it offers the least restriction to flow (see Fig. 16-81).

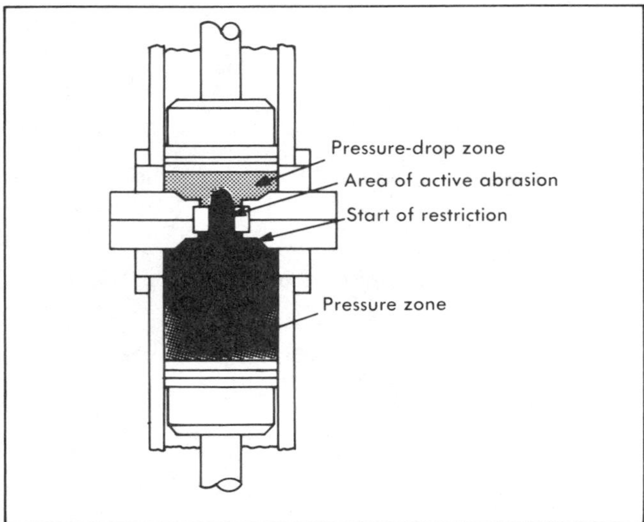

Fig. 16-79 Pressure changes of media as it flows through a straight passage in a workpiece. The restricted orifice formed by the workpiece causes the media to increase in viscosity, producing abrasion in this area. (*Extrude Hone Corp.*)

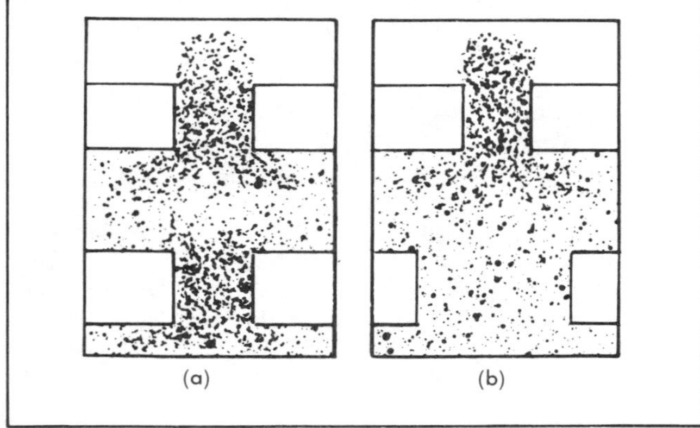

Fig. 16-80 Media viscosity while flowing through successive restrictions. View a: Both restrictions will be equally abraded if they are equal in cross-sectional area. View b: If the successive restrictions are of different cross-sectional areas, only the one with the smaller area will be abraded. (*Extrude Hone Corp.*)

Controlling the Abrasion

In AFM, the two main process factors that determine the abrasion produced are as follows:

1. The depth of cut made by each of the slug's abrasive grains. This cut depth depends on the size, hardness, and sharpness of the grains, and the machine extrusion pressure.
2. The number of cuts made by these grains within the process cycle. This number is a function of the length of media that actually passes through the restriction as a slug.

The maximum force initially applied to the abrasive slug is equal to the machine extrusion pressure multiplied by the cross-sectional area of the restricting passage. Some of this force is lost within the media mass due to its internal resistance to flow. The remainder is transferred through the suddenly stiffened media slug as it approaches and passes through the restricting passage, and, from there, to the abrasive grains contacting the workpiece edges and surfaces. The force against these grains is determined by the machine extrusion pressure, the media viscosity, and the abrasive grain size and type.

The number of cuts an abrasive grain will produce during a given time period depends upon how fast the media slug is extruded through the workpiece. Slug flow speed is also responsible, in part, for the type of flow pattern that will be produced. Media flow patterns and their effect on abrasion are discussed in the next section. The slug flow speed is controlled by extrusion pressure and media viscosity, in addition to passage area and length.

With all other things being equal, the final parameter determining the amount of abrasion will be the flow volume. A greater volume of media will bring a greater number of abrasive grains in contact with the workpiece, thereby producing more abrasion.

Process parameters and the two main factors they control cannot only determine the amount of abrasion produced by the media, but also the type of abrasive effect produced. Whereas

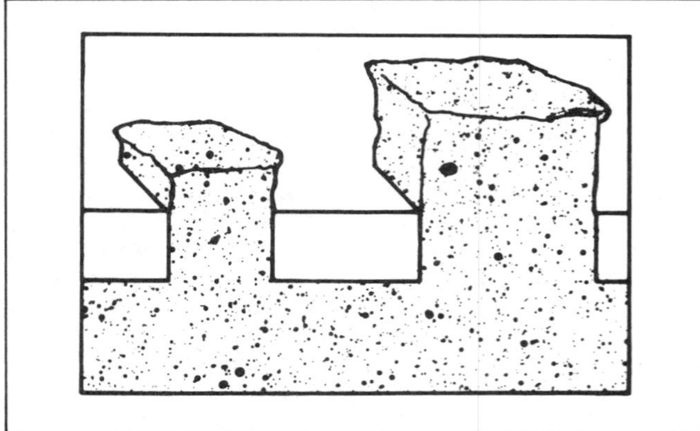

Fig. 16-81 When processing parallel restrictions, the larger passage allows more media flow and is therefore abraded the most. (*Extrude Hone Corp.*)

flow volume primarily determines the overall amount of abrasion, slug flow speed can be altered to affect the media flow pattern and the resulting type of machining action produced.

Wall Versus Edge Abrasion

The media flow pattern as it enters and extrudes through the restricting passageway determines whether the media's greatest abrasion will be concentrated at the passageway's edges or along its walls. If a polishing or honing effect is desired (in which a uniform amount of stock must be removed from the walls of the restricting passage), then a flow pattern is produced in which the media slug's internal flow is nearly uniform throughout the passageway. Slow speeds help the media slug flow evenly throughout the passageway (see Fig. 16-82, view *a*).

If the purpose is radiusing and deburring the edges of the passageway, higher flow speeds cause the edges at the passageway entrance to be abraded more than the passage walls. Higher slug flow speeds accentuate this edge radiusing effect (see Fig. 16-82, view *b*). Since one flow cycle is normally in both directions, both ends of the passage will be abraded.

The following parameters determine the media flow pattern: media viscosity, passage size (width and length), slug flow speed (extrusion pressure), and media rheology (the viscosity change of the polymer carrier under various stresses, due to its degree of internal bonding).

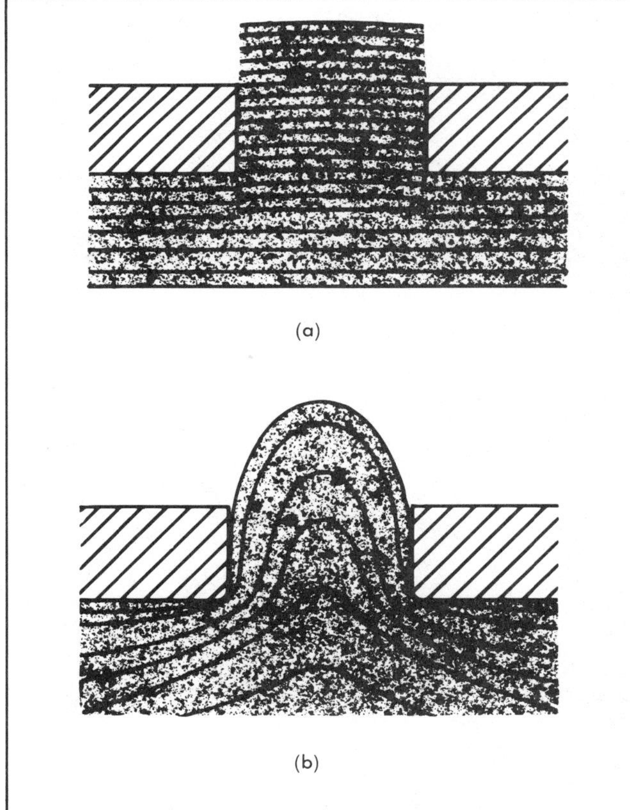

Fig. 16-82 Cross-sectional views of slug flow patterns. View a: Uniform pattern resulting from slower flow speeds and media of higher viscosity produces a honing or polishing effect. View b: Uneven pattern resulting from higher flow speeds and media of lower viscosity produces a deburring or edge radiusing effect. (*Extrude Hone Corp.*)

Surface Improvement

Surface improvement by AFM is a result of direct abrasion of the peaks and, to a lesser extent, the valleys of the microsurface. There is no smearing of the higher material into the lower valleys, which can happen with hand polishing.

The peaks of the microsurface are quickly abraded at first. However, as the peaks flatten into plateaus, the valleys likewise receive abrasion until a smooth and even profile is produced and no trace of the initial surface remains. This ability to remove undesirable, tensionally stressed surface material has made AFM ideal for improving the surface integrity and, consequently, the service life of critical components that are subjected to thermal or mechanical fatigue stress. This capability is particularly important to the growing numbers of workpieces being machined and/or treated by thermal methods such as electrical discharge, laser, and electron beam machining.

MACHINES USED

Machines used for the AFM process primarily control extrusion pressure, a crucial parameter for determining the amount of abrasion that will eventually be produced. Within the range of machines available, pressures from 100 to 3200 psi (690 to 22 060 kPa) can be produced, and at flow rates of up to 100 gpm (378 L/min) and higher. The ideal media flow rate depends upon the size and number of passages to be processed and the kind of flow characteristics desired. Flow rate is affected by media viscosity and extrusion pressure, and is generally not a crucial variable.

The dominant factors in controlling the amount of work done by a specific media composition are flow volume and extrusion pressure. These variables are controlled at the machine by presetting the displacement of each media cylinder stroke, the number of two-stroke cycles for the job, and the hydraulic extrusion pressure. Low-cost, low-production machines have media cylinder diameters ranging from 3 to 6" (76 to 152 mm). Machines for larger components or higher production rates have cylinders from 6 to 24" (152 to 610 mm) diam.

Microprocessor controllers can monitor and modify additional process parameters at the machine, such as media temperature, media viscosity, media wear, and flow speed. Accessories such as part cleaning stations, unload/reload stations, media refeed devices, and media heat exchangers are frequently included in installations for production applications.

WORKHOLDING FIXTURES

The workholding fixtures, or tooling, for AFM are primarily designed to hold the workpiece(s) in position on the machine and to direct the media to the desired areas. If necessary, the fixture can also restrict those passages that must be abraded but that do not restrict the media flow sufficiently.

Tooling is made of aluminum, steel, nylon, urethane, or combinations of these materials. Aluminum and nylon are easily machined, lightweight materials. Steel is used for its strength and durability. Urethane is sometimes preferred because of its high resistance to abrasion and its ability to be cast in complex shapes. For these reasons, passageway restrictors, which concentrate abrasion at those areas that are not sufficiently narrow to be abraded without them, are most often made of urethane. In many applications, the fixture will be machined of steel, with portions made of nylon or urethane inserted to fit into the more complex workpiece areas.

ABRASIVE-FLOW MACHINING

Fixture design can be very straightforward (see Fig. 16-83) or very complex, depending upon part configuration, media flow path(s), and processing rates. Many applications only require simple universal fixtures. For instance, a part with a straight-through passage (such as a tube, an internal spline, or a valve plate) only needs tooling to hold the part in place between the two media cylinders, allowing the workpiece passage itself to provide the greatest restriction in the flow path. In similar fashion, the great majority of extrusion die polishing applications require no tooling. If AFM is used to process external edges or surfaces, the tooling contains the part in a flow passage, restricting the flow between the exterior of the part and the interior of the fixture (see Fig. 16-84).

Internal passages, surfaces, and edges can be processed as long as these areas can be reached by the media flow. However, workpieces with complicated internal configurations may require processing in two or more operations, with some areas being abraded in one operation and other areas, requiring different flow paths, being abraded in the next. Very large parts may also be processed in sections. Counterbores, recessed areas, and even blind cavities can be processed by using a mandrel to fit inside the component and restrict the media flow at the desired areas. Tooling may also be used to protect critical workpiece areas that are adjacent to or in the flow path of a desired abrasion area.

High-production fixtures are designed to facilitate part handling operations, including unloading, loading, and cleaning. A common design for high-production processing uses two fixtures mounted on opposite ends of an index table. While one fixture is being processed, the other fixture is unloaded and reloaded with parts. After the first fixture-load of parts is completely processed, the table is indexed 180°, swinging the second fixture into position between the media cylinders. These and other types of high-production fixture arrangements are sometimes designed in conjunction with automated part handling systems to further increase production efficiency.

One of the more recent developments in AFM fixturing has enabled the process to accurately machine an external workpiece surface to a specific close-tolerance shape while also polishing or finishing it. Although large amounts of stock cannot be efficiently removed by AFM, precise forms and shapes can be produced on workpieces by placing specially shaped restrictors over the workpiece areas to be abraded.

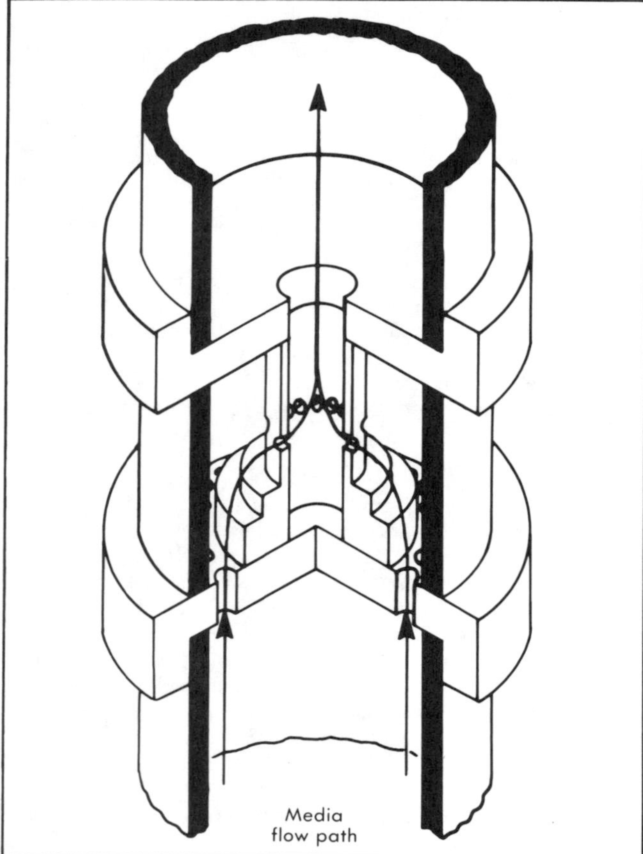

Fig. 16-83 Simple sleeve fixture directs the media through the holes and over the external edges of a workpiece. (*Extrude Hone Corp.*)

MEDIA FOR ABRASIVE-FLOW MACHINING

The media used for AFM is made of a polymer carrier and the abrasive grains it holds. Carriers are available in a number of different viscosities (see Table 16-30) and rheologies, according to the type of media flow desired. The abrasive grains vary in type, size, and concentration. Silicon carbide is the most commonly used abrasive, although other available types include boron carbide, aluminum oxide, and diamond. Most any material, from soft aluminum to rough nickel alloys and ceramics, can be abraded. Particle sizes range from 8 grit [0.180" (4.57 mm)] to 1000 grit [0.0002" (0.005 mm)] diam.

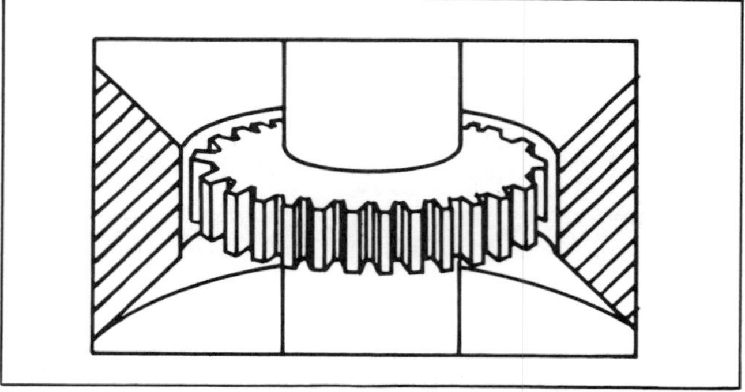

Fig. 16-84 Fixture for processing the external surfaces of a gear. The media flows between the gear teeth and the bore of the fixture. (*Extrude Hone Corp.*)

TABLE 16-30
Passage Sizes Processed by Five Basic Types of Media for Abrasive-Flow Machining

Media Viscosity	Passage Size,* in. (mm) Minimum	Maximum
Low	1/64 (0.4)	1/8 (3.2)
Low/medium	1/32 (0.8)	1/4 (6.4)
Medium	1/16 (1.6)	1/2 (12.7)
High/medium	1/8 (3.2)	1 (25.4)
High	1/4 (6.4)	2 (50.8)

* Indicated passage sizes are widths or diameters, and data is based on passage lengths being two times the width or diameter.

The effective lifetime of the media depends on initial batch quantity, abrasive size and type, flow speed while processing, part configuration, and the aggressiveness of the abrasive work performed. These factors combine to wear down the sharp corners of the abrasive. In addition, since the media mass absorbs all the material removed from the workpiece, performance will also begin to change as the media becomes loaded.

Lubricants in the media mixture are also carried off during processing. However, since the cutting tool in the process is a blend of millions of individual cutting elements with the specially compounded carrier (and some material removed from the workpieces), both the consistency and the average life of the abrasive particles in the tool can be kept constant by periodically adding fresh media.

BLAST FINISHING

Blast finishing refers to several versatile processes for the controlled cleaning and finishing of materials by the impact of media against the surfaces of workpieces. The original process, called sandblasting and often done outdoors, consists of introducing ordinary sand into a stream of high-velocity, pressurized air and manually directing it at the surfaces to be treated. Many different media and various processes are now being used for blasting indoors, using enclosures to contain the media.

The two major methods of modern blast finishing are dry and wet. Dry blasting is done with pressurized air or by airless processes. In air blasting, the media is propelled by a stream of pressurized air. In airless or mechanical blasting, the media is thrown against the work surfaces by centrifugal force, using a rotating bladed wheel. In wet blasting, the media is generally suspended in water that is forced by compressed air through nozzles to clean or finish the workpieces.

PROCESS ADVANTAGES AND LIMITATIONS

Blast finishing is generally an economical process, often requiring less labor than many other cleaning and finishing methods. The equipment can be used for many different operations and a wide variety of workpieces. The media is generally reused, thus reducing costs. Most blasting processes are suitable for low or high-volume production requirements.

APPLICATIONS OF BLAST FINISHING

Blast finishing is used extensively for cleaning operations. These include the removal of sand from castings, scale from forgings and heat treated parts, mill scale and rust, paint and other coatings, various soils, and other surface contaminants. Soils or dirt to be removed must be dry and workpiece surfaces free of grease or oil when using dry blasting.

Finishing applications of media blasting generally include either roughening or smoothening workpiece surfaces. Surface roughening (conditioning) is often required prior to the application of adhesives, paints, or other coatings. Smoothening operations include the removal of burrs, flash, directional machining or grinding lines, and other surface irregularities; the rounding of edges; and improvement in surface finishes. The use of media such as plastic pellets permits the removal of burrs and flash without affecting the surface finish.

A related process, called shot peening, discussed later in this section, cold works metals and improves stress characteristics in the parts blasted. This process is also used to form sheet and plate stock, as discussed in Volume II, *Forming*, of this Handbook series.

DRY BLASTING

There are two commonly used methods of dry blasting. In the air-blast system, the media is propelled by a stream of pressurized air. In the airless system, the media is fed into the center of a rotating vaned wheel, which hurls the media by centrifugal force. Although the air-blast system is the most versatile, it has the limitations of requiring a source of high-pressure air and being more energy intensive. However, air-blast systems accelerate the media to higher velocities, permitting the effective use of smaller media particles to produce smooth finishes.

Surface Preparation

Whenever parts are coated with grease or oils, they are not easily cleaned or finished by dry blasting. Not only do these substances resist removal, but they contaminate the media that is used. For these reasons, degreasing and drying should be performed prior to blasting. Metal surfaces prepared for painting or porcelain enameling by abrasive cleaning must have all traces of surface soil removed before blast cleaning. Any type of soil that will still adhere to the blasted surface or the shot or grit particles may produce surface defects in the subsequent phosphating, painting, or porcelain enameling operations.

Normally, materials that will readily flake and shatter under the blast impact will be reduced to fines and removed by separation and dust collection systems. Examples are rust, weld spatter compound, and dry dust such as sand or dirt.

Air-Blast Systems

Air-powered methods of dry blasting are divided into three basic types: direct pressure, gravity, and induction suction.

Direct pressure. In the direct pressure system of air blasting, the media, from a pressurized tank, is fed directly into the blast cleaning equipment hose and discharged through a nozzle. All types of media are used, the selection depending upon the material to be cleaned and the degree of cleaning necessary. This method is the workhorse of the air-blast systems and is generally used for exceptionally heavy cleaning, such as the removal of scale from castings and forgings. It is also ideal for cleaning interior cavities, blind holes, narrow recesses, and localized areas. This process generally provides a more uniform finish in a shorter time than other methods of air blasting. The media is ejected at higher velocities and with greater concentration than in the suction method. However, nozzles and lines are subjected to greater wear, and more media is consumed. Also, the force of the blast stream may warp thin material, except when two streams are used to hit both sides simultaneously.

DRY BLASTING

Gravity. In the gravity-fed type of air-blast system, a media feed hopper is located above the nozzle, and media flows by gravity to the nozzle. At the nozzle, air entering from a separate line mixes with the media to propel it. A bucket elevator, pneumatic system, or other mechanism returns the spent media to the feed hopper. A wider spray pattern results from this method, and the media has less force than that from the direct pressure type, resulting in slower cleaning. However, the slower speed results in less media breakdown, and a larger percentage of the media can be reused.

Gravity blast cleaning equipment is generally used for less severe cleaning applications, such as removing light scale or rust and other contaminants from castings, structural shapes, and various other parts. Unobstructed flow is sometimes difficult to maintain with traversing nozzles, and it is difficult to blast upward at surfaces.

Suction. In the induction suction type of blast cleaning equipment, the media is drawn from a collecting hopper into the nozzle by a partial vacuum. This vacuum is created in the suction line leading to the nozzle by high-velocity air flow. A larger proportion of air to media makes this method ideal for light applications on easily cleaned surfaces and for removal of burrs from machined parts. When low media velocities are sufficient, this method is the least expensive of the air-blast types. Induction suction systems will also propel media upward, and they can blast continuously without stopping for media refills.

Equipment used. Selection of the equipment to be used for blasting requires careful consideration of many factors. These include the number, size, variety, and condition of the parts to be blasted; the required production rate and finish specifications; the amount of floor space and ceiling height available; and environmental and safety regulations. Standard machines cost less than special machines and hence should be used whenever suitable. However, many applications necessitate the use of specially designed machines.

The most common type of air-blast equipment requires manipulation of the nozzle by an operator, who is located outside a cabinet or inside a room or other enclosure. Small workpieces are frequently processed in self-contained hand-blast cabinets. Large workpieces that must remain stationary are blasted in rooms or enclosures with the operator wearing special apparel and manipulating a flexible air-blast nozzle.

Cabinets. Standard hand-blast cabinets are relatively low in cost, very versatile, and available in various sizes. With a typical cabinet, the operator places his hands, protected by gloves, into the cabinet to manipulate a nozzle and/or the workpiece while viewing the operation through a window. Workpieces are loaded and unloaded through an access door. Cabinets are generally equipped with a media supply and recovery system, and a means for ventilation and dust collection.

Limitations to the use of cabinets are that they depend upon operator control and are not suitable for large parts. However, several means are available to increase production and improve control. A simple turntable, sometimes mounted on a wheeled dolly with tracks, can be provided for handling heavy or complex parts. Additional nozzles, supported by adjustable brackets, and vertical or horizontal oscillation of the nozzles are sometimes used. Timers can be employed to control the duration of the blast cycle, and programmable controllers are being used for increased versatility in some applications.

When production requirements for a specific workpiece are sufficiently high, conveyorized handling is often used. In-line conveyors are generally preferable for small parts that can be automatically loaded and unloaded. Rotary conveyors, usually of the indexing type, are more frequently used for larger or more complex parts, but they are slower than in-line conveyors. Industrial robots or programmable controllers are being used when a variety of parts are to be blasted with a single machine or when consistent results are required on complex parts.

Rooms/enclosures. Rooms or enclosures for blasting large parts provide space for positioning workpieces and retain the media for safety, reclaiming, and recycling purposes. Manual or power-driven turntables are used for blasting heavy parts. Because of the abrasiveness of most media, the rooms should have a minimum number of moving parts and be provided with wear-resistant liners. Loose-hanging rubber sheets are effective as protective liners. Polyurethane-lined cabinets are also becoming common.

Nozzles. Nozzles for air-blast equipment used in dry blasting are made of several different materials in a variety of shapes and sizes. Hard iron, boron carbide, silicon carbide, tungsten carbide, and ceramic are the principal wear-resistant materials used as nozzle liners. The nozzle material must be compatible with the wear-producing characteristics of the media being used. The wear of any type of nozzle is influenced by the hardness and sharpness of the media and its velocity. In general, a nozzle will last from two to four times longer when used with steel shot or grit than when used with sand.

Various nozzle shapes are available. Short, straight-bore nozzles are used for blasting close to the workpiece. Venturi nozzles are used for the most efficient cleaning at distances of 10 to 24" (254 to 610 mm) from the work. Venturi nozzles produce 20-50% faster cleaning than straight nozzles. Side-shooting nozzles are available for inside-diameter cleaning of such items as pipes, tubes, and valve-body bores.

A nozzle's shape does not influence its wear life to any great degree. However, the inlet must be well centered in the hose and must taper smoothly from the hose inner diameter to the nozzle's final size. The length of the straight section at the bore should be at least five times the bore diameter. Nozzles are considered to be worn out when the outlet has worn to 1 1/2 times its original size. Replacement at this point will save money by maintaining the efficiency of compressed air usage and the high terminal velocity of the media.

Compressed air requirements. Air consumption for various jet diameters used in air-blast systems is shown in Fig. 16-85. Compressed air is expensive to produce; and if excess plant air is not available, the cost of an air compressor must be considered when air-blasting equipment is purchased. Also, excess moisture in plant air sometimes leads to problems with rust on shot mixtures or workpieces. Water and oil traps, and, sometimes, air driers or moisture separators are often required. If plant air pressures vary with demand, care must be taken that the pressure drop will not be below that required for production.

Operating Parameters

Selection of the proper blasting pressure, impact angle, and standoff distance of the nozzle from the workpiece are essential for efficient and economical blasting.

Pressure requirements. When using the direct pressure method of air blasting, pressures of 40 to 60 psi (276 to 414 kPa) are generally the most economical. For the induction suction process of air blasting, pressures of 60 to 80 psi (414 to 552 kPa) are common. Excessive pressures accelerate breakdown of the media, with only minimal decreases in blasting times. Often, the pressure used is determined by that available in the plant.

The pressure used also depends upon the application and the media used. For blasting precision parts with fine, soft media, pressures as low as 10 psi (69 kPa) may be used. On the other hand, for heavy-duty operations, such as removing scale from forgings, pressures of 80 psi (552 kPa) or more are often used.

Impact angle. Maximum cutting action and the production of rough surfaces are obtained with cutting media when the air-blasting nozzle is held at an angle of 45° with respect to the workpiece surface. The surfaces thus produced are often desirable when subsequent coating operations are to be performed. However, when smooth surface finishes are required, an impact angle of about 30° is employed. Nozzle angles are sometimes adjustable, especially on automated, multinozzle machines.

Standoff distance. The distance that the nozzle is held from the workpiece depends upon the specific application and the media used, but is generally maintained constant during the operation. This distance generally varies from 3 to 24" (76 to 610 mm), with the shorter distances usually being employed with softer media. With a given media at a fixed velocity, longer standoff distances increase the diameter of the blast pattern; but with light abrasives, the impact force may be reduced.

Abrasive-Jet Machining

Abrasive-jet machining (AJM) is a specialized form of blasting for removing material by a high-speed stream of media particles carried from a nozzle by a gas, usually air. It is used to cut, deburr, and clean hard, brittle materials such as germanium, silicon, mica, glass, ceramics, titanium, and tantalum without heating or cracking. Cutting rates are generally low. However, since the tool does not contact the workpiece, the process is inherently free from chatter and vibration problems. Also, the cutting action is cool because the carrier gas serves as a coolant. This cool action makes the method ideal for cutting materials that are sensitive to heat damage, as well as thin sections of hard materials that might chip easily.

Material is removed by the impingement of fine media particles, usually about 0.001" (0.03 mm) diam, entrained in a high-velocity, 500 to 1000 fps (152 to 305 m/s), gas stream. By varying the nozzle diameter and configuration, as well as the standoff distance (see Fig. 16-86), a variety of results can be obtained. Nozzle-tip diameters generally range from 0.003 to 0.032" (0.08 to 0.81 mm). Rectangular nozzles are also used. The nozzles must be resistant to abrasive wear and are usually made from tungsten carbide or synthetic sapphire.

Variables influencing material removal rate, geometry of cut, surface roughness, and nozzle wear rate include the media (composition, strength, size, shape, and flow rate), gas used (composition, pressure, and velocity), and the nozzle (geometry, composition, and distance to and inclination with the work surface).

Aluminum oxide is the most widely used media for this process, but silicon carbide, glass beads, and other materials are used for special applications. Media powder is generally fed from a vibrating mixing chamber into an orifice chamber where it is entrained in the gas stream, and then passed through a hose to the nozzle. Good results have been obtained with media particles varying from 0.0006 to 0.0016" (0.015 to 0.041 mm) diam, with flow rates from 2 to 20 g/min (0.07 to 0.71 oz/min). Finer powders are used for cleaning and polishing. Increasing the flow rate tends to increase the material removal rate, but increasing the proportion of media in the jet lowers the stream velocity and tends to lower the material removal rate. Reuse of the media powder is not recommended because it will decrease the cutting action and contamination may clog the orifices.

Air, nitrogen, and carbon dioxide at pressures of 30 to 120 psi (207 to 827 kPa) have been used as the carrier gas. Commercial gases are generally sufficiently pure, but air should be filtered. Higher pressures provide more rapid material removal rates but increase nozzle wear. A typical material removal rate is about 0.001 in.^{3}/min (0.016 cm^3/min). Normal tolerance for production applications is ±0.005" (0.13 mm). Masking is sometimes used to confine the cutting action. Surface finishes produced generally range from 6 to 50 μin. (0.15 to 1.27 μm). A dust collection system is essential to protect operators, equipment, and workpieces.

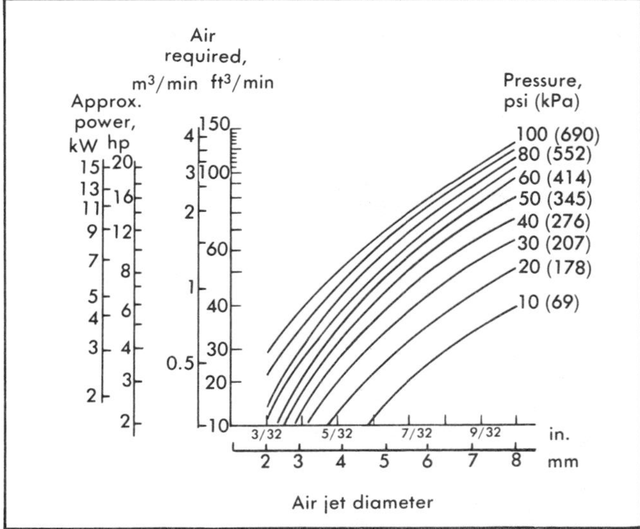

Fig. 16-85 Air required for air blasting with various jet diameters at different pressures; based on single-stage compressor at sea level, 5 ft^3 (0.14 m^3) equals 1 hp (0.75 kW), approximately. Data may be applied to either dry or wet blasting.

Airless Blast Systems

Centrifugal wheel (airless) blasting is a mechanical method by which media is propelled by centrifugal force from a rotating vaned wheel (slinger). The media forms an elongated, cone-shaped pattern, and the operator remains outside the machine during blasting. The economy of this type of equipment is apparent from the fact that a single wheel, with a 100 hp (75 kW)

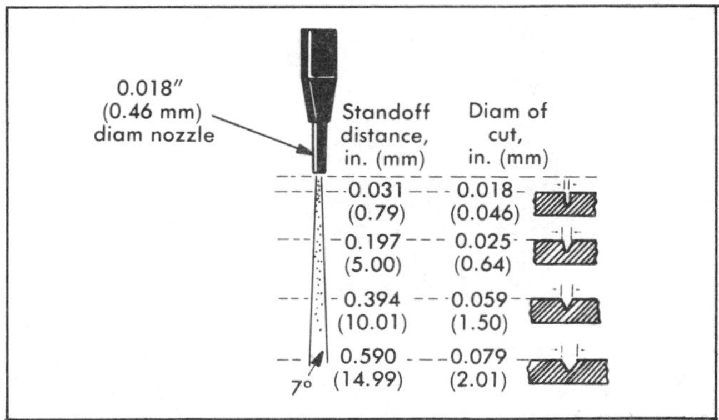

Fig. 16-86 Cutting actions in abrasive-jet machining (AJM) with various standoff distances. (*S. S. White Industrial Products*)

drive motor, can discharge up to about 2600 lb (1180 kg) of steel shot or grit per minute at velocities reaching 14,000 fpm (4267 m/min). In comparison, a single air nozzle is limited to about 100 lb (45 kg) of media per minute. Figure 16-87 illustrates this type of equipment. Media from an overhead storage hopper is fed through a spout to an impeller at the center of the wheel. From the impeller, the media enters the wheel vanes and is accelerated before being flung onto the work.

Airless blasting is used most commonly in batch and continuous equipment, especially when a large volume of media is required over a fairly large work area. The kind and size of unit will depend on the size, shape, and fragility of the metal parts to be cleaned. The mechanical wheels on single or multiple units are positioned on the equipment to expose the desired areas of the work to the proper volume of abrasive at the proper angle and velocity. The equipment must turn or roll the workpieces continuously to expose their surfaces completely.

Mechanical blast cleaning will effectively remove molding sands from iron or steel castings and iron oxide scale from forgings and heat-treated parts. Descaling of structural steel and plate is more readily performed with airless blasting than with an air-blast system. Removal of the carbonaceous film accumulated in oil quenches on heat-treated parts is another ideal application. Some of the more popular airless systems include barrels, tables, hanging machines, axial-flow units, rooms, cabinets, and special-purpose machines.

Blast-cleaning barrels. Batch-type barrels, sometimes called mills, are generally mechanically charged with a skip-hoist loader. Barrels continuously fed at one end and discharged at the other end are used for high-production, automated systems. The work is slowly rotated or rocked under the media stream, usually by a traveling, slat-type conveyor in a closed-container system. Timing or a visual inspection determines duration of the blast operation, which remains generally constant for a given type of scale removal.

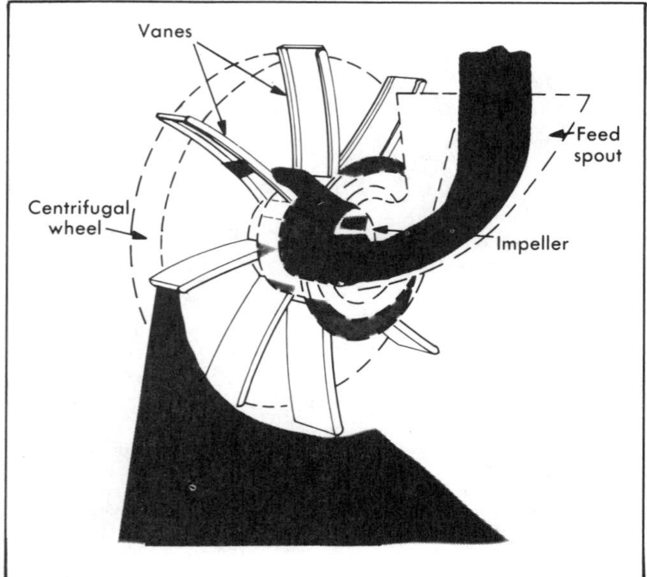

Fig. 16-87 Cutaway view of centrifugal dry-blasting wheel. The impeller picks up media from the feed spout. The media exits through openings in the impeller case and is deposited on the vanes of the wheel to be accelerated. The vanes throw the media in a targeted pattern. (*The Pangborn Co.*)

Table machines. Machines with rotating tables and one or more wheels can be used to clean parts that cannot be tumbled in barrels. The parts are positioned on the table in such a way as to expose the surfaces to be cleaned; repositioning may be necessary. These machines can employ a single large table or tables mounted on each of two doors so that loading and unloading can take place simultaneously with blasting. Other units employ auxiliary tables on the main table for smaller work. Partitions between the auxiliary tables are used to prevent media from escaping from the cabinet toward the operator.

Hanging fixtures. Hanging parts on fixtures or "tree hooks" is desirable from the standpoint of exposing all surfaces without repositioning. However, the difficulty of hanging and the number of different hanging fixtures that may be required must be taken into account.

Hanging fixtures can be mounted on various material handling devices for transporting into and out of the cabinet. Hand-pushed trolleys, self-powered carriers, and continuous or indexing trolley conveyors are all used. The conveyors can move in and out one side of the cabinet or pass through the cabinet. A method of rotating the work while in the cabinet is usually employed, along with multiple blast positions and multiple wheels.

Axial-flow units. A high-production, continuous blasting machine, specifically engineered to clean engine blocks, manifolds, heads, and similar parts, is illustrated in Fig. 16-88. It is fully automatic, including loading and unloading, and utilizes a barrel-type work support and axial work handling. The unit is capable of cleaning up to 1000 castings an hour effectively.

The single barrel-type work support is installed inside the blast cabinet. It is fabricated from special heat-treated steel bars for longer wear life. The barrel is built to accommodate given configurations and constructed to expose the entire part to thorough and rapid cleaning. As the parts pass through the machine, they are maintained in a fixed position. This workpiece orientation at the unload end also facilitates automatic work handling in subsequent operations.

Rooms and special-purpose machines. For cleaning large workpieces, such as forgings, fabrications, and railroad cars, rooms are used. These rooms have sufficient wheels to cover the part to be cleaned and material handling devices to get the work

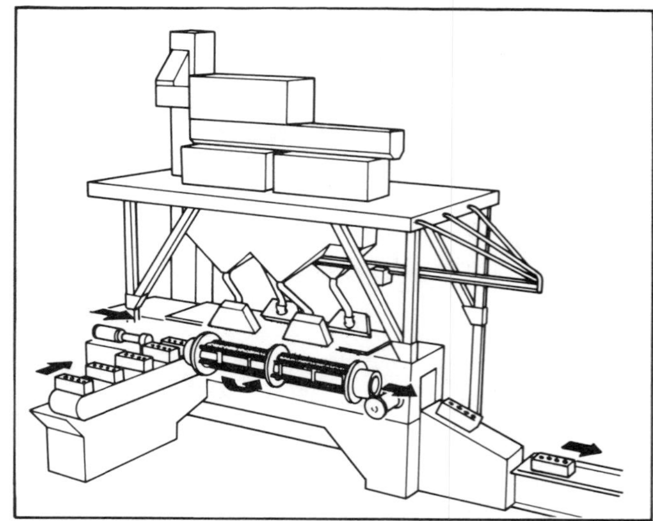

Fig. 16-88 Automated airless blast system with axial-flow conveyor.

in and out. Large self-propelled work cars with rotating tops and overhead monorails and cranes are commonly used. Rooms have been furnished up to 50 ft (15.2 m) square and 50 ft high for workpieces weighing up to hundreds of tons.

Special-purpose devices, such as boom wheels for cleaning the inside of items such as pipe, nuclear reactor fabrications, and railroad hopper-bottom cars, can be furnished as part of rooms or as stand-alone special-purpose machines.

Steel descaling machines. With the advent of highway building programs and sophisticated coating systems for steel, a family of machines for descaling steel has developed. This equipment can be furnished for bars, structural shapes, plates, pipe, or combinations of these. Roller conveyors, skewed roller conveyors, monorails, and cars are common methods of transporting the work through the machines. Care must be taken in the placement of the blast wheels so as to cover the entire shape of the workpiece without necessitating repositioning.

Descaling is also used at steel mills to replace or reduce the use of acid pickling. Machines capable of descaling steel strip at speeds matched to the rolling equipment are common. Wire in coiled or uncoiled form is also descaled by blasting.

Media for Dry Blasting

The many types of media available, both abrasive and nonabrasive, as well as the possible variations in the pressures used, make blasting a very versatile process. Different media are used for various applications to obtain the desired results and production rates, as well as to ensure an economical media life.

Media selection factors. The choice of media, as to both type and size, depends upon the type of material to be removed, the geometry and condition of the workpieces, the production rates required, media cost and wear rate, equipment wear (maintenance cost), and the intended application. Selection can affect both the efficiency and cost of the operation. Media costs can range from 30 to 75% of total cleaning or finishing costs, depending upon the results required. For heavy forging scale, large shot is used in order to break the scale from the work efficiently. Large shot is also commonly used for large castings from which scale and burned-in sand must be removed. Blasting with shot results in a peened appearance, varying from readily visible indentations when large shot is used to a very fine peened finish with the small shot. The smaller sizes are used in applications in which the scale is thinner and requires less impact. Provided the impact is sufficient for the particular type of scale or other material to be removed, or for the finish required, smaller shot cleans faster because of the greater coverage resulting from more pellets per unit weight of shot. Larger shot, however, provides greater impact.

Grit is used extensively whenever an etched finish is desired—for example, to obtain a better bond for painting or porcelain enameling. The texture of the etch varies with the different grit sizes. A coarse, deep etch is produced by large grit, while the smaller sizes produce a very fine etch. Nonferrous shot is used for the softer metals on which grit or steel shot would be damaging. Nonmetallic abrasives are used when it is desirable to remove a minimum of material from the parts or to impart a bright finish. Table 16-31 lists abrasives and equipment that have been used for various materials and purposes, but the table should be used only as a general reference.

Air-blast applications generally use a wider variety of media and smaller sized media than airless blasting. Fine powders of materials such as silicon carbide, aluminum oxide, glass beads, and bicarbonate of soda are used for some so-called "microabrasive" air-blast operations. Such fine media, which must be kept dry, can be accelerated to high speeds with air blasting, producing surface effects not attainable with airless blasting.

Cast steel shot. Shot has a usable life longer than any other type of abrasive media and reduces the wear and maintenance of equipment. Various hardness ranges are available for special peening and cleaning applications. Table 16-32 lists some of the sizes for cast shot as established by the Society of Automotive Engineers (SAE). Solid, spherical shot flattens and dimples the surfaces blasted, plus makes them shiny. The heavy weight of shot gives it high-impact action.

Metallic grit. Cast steel grit is durable and is particularly effective in applications where an etched finish is desired. Grit is made by crushing round shot and is angular in shape. Specifications for grit sizes have also been established by the SAE (see Table 16-33). Grit generally dulls the surfaces blasted and removes metal. For some applications, shot and grit are mixed to produce different finishes. Other metallic grit used for specific requirements includes copper, bronze, or aluminum nuggets for blasting some nonferrous materials; cut ferrous and nonferrous wire; and zinc.

Cutting abrasives. Cutting abrasives are used extensively for blasting or when fast cutting is a major consideration. The life of the equipment is shortened when using these abrasives because of rapid wear. Also, the cost of the abrasives is high in terms of first cost or reusability. Silica sand is still being used in some outdoor blasting applications. Sand is low in cost, but is not very efficient because it breaks down into dust after a single use and generally cannot be recycled. It also causes considerable wear of equipment and creates problems with respect to OSHA and EPA regulations. In many areas, sand is not permitted because of possible health hazards.

Aluminum oxide. Crushed and graded, fused aluminum oxide is available in a wide variety of sizes and shapes. This media is fast cutting and has good wear resistance, but is comparatively expensive. It is used extensively for cutting operations and produces a desirable surface pattern for subsequent coating operations, but is rarely used with airless blasting equipment. The physical characteristics of aluminum oxide vary with different producers.

Silicon carbide. This media is sharper than aluminum oxide and therefore faster cutting, but larger sizes tend to fracture readily. It is used to clean very hard surfaces, such as tungsten carbide, welded or plated parts, and components that are to be subsequently brazed or welded, when aluminum oxide is not suitable.

Nonmetallic media. A wide variety of these materials are available, including sawdust, crushed nut shells, fruit pits, rice hulls, corncobs, nylon, and plastic and glass beads. These media are used to remove light fins from aluminum castings or light scale and carbon from intricate parts, to deflash thermoset plastics parts, and whenever contamination from metallic media is not acceptable.

Agricultural products. Media such as crushed nut shells, corncobs, and fruit pits are used to remove foreign material, such as carbon, oil, grease, and dirt, without etching, scratching, or marring the surface. Possible limitations include the media being dirty or dusty, lack of uniformity from batch to batch, and high usage rates because of minimum recovery.

Nylon media. This media, made from cut nylon wire, is used for applications similar to those for media made from agricultural products. An advantage for nylon media is that it eliminates dust problems.

DRY BLASTING

Glass beads. These nonabrasive media are clear, colorless, transparent spheres generally made from a soda-lime-based, high-silicon glass. They are available in diameters from a maximum of 0.187" (4.75 mm) to 0.0015" (0.038 mm) or less. This type of media is used when ferrous contamination of the workpieces must be held to a minimum, when workpiece dimensions and tolerances must not be changed, when the small sizes available can reach areas inaccessible to other media, and when the special finishes attainable (bright matte to dull satin) are desirable.

Plastics media. Plastics with sharp, angular-surface particles, such as polycarbonate resins, are a relatively new media for

TABLE 16-31
Typical Abrasives and Equipment Used for Dry Blasting

Workpiece Material	Purpose	Abrasive Material	Abrasive Size	Equipment*
Cast iron	Cleaning	Steel	S230	C
	Coating preparation	Iron and steel	G25-G40	A,C
Gray iron	Cleaning	Malleable iron	S460	C
	Descaling	Malleable iron	S550	C
Steel:				
Cold rolled	Cleaning	Steel	G40	A,C
Tool	Descaling	Steel	G80	C
Hot rolled	Paint preparation	Steel	G25-G40	A,C
Rod, bar	Cleaning	Steel	G40	C
	Coating preparation	Steel	G40-G50	A,C
Structural	Paint preparation	Steel	G40	C
Welds	Descaling	Steel	S230-S280	C
Aluminum	Satin finishing	Steel	G80	A,C
	Paint preparation	Iron	G80	C
Bronzes	Satin finishing	Sand	50	A
Plastics:				
Clear	Frosting	Sand	50	A
Molded	Deflashing	Walnut shells, polycarbonate	---	C
Phenolic	Deflashing	Walnut shells, polycarbonate	---	A,C

* A = air blast; C = centrifugal

TABLE 16-32
Condensed SAE Specifications for Cast Shot

SAE	None on Screen		Maximum on Screen			Minimum on Screen			Maximum Through Screen		
	Screen			Screen			Screen			Screen	
No.	No.	Opening, in. (mm)	%	No.	Opening, in. (mm)	%	No.	Opening, in. (mm)	%	No.	Opening, in. (mm)
780	7	0.1110 (2.819)	---	---	---	85	10	0.0787 (1.999)	3	12	0.0661 (1.679)
660	8	0.0937 (2.380)	---	---	---	85	12	0.0661 (1.679)	3	14	0.0555 (1.410)
550	10	0.0787 (1.999)	---	---	---	85	14	0.0555 (1.410)	3	16	0.0469 (1.191)
460	10	0.0787 (1.999)	5	12	0.0661 (1.679)	85	16	0.0469 (1.191)	4	18	0.0394 (1.001)
390	12	0.0661 (1.679)	5	14	0.0555 (1.410)	85	18	0.0394 (1.001)	4	20	0.0331 (0.841)
330	14	0.0555 (1.410)	5	16	0.0469 (1.191)	85	20	0.0331 (0.841)	4	25	0.0280 (0.711)
280	16	0.469 (1.191)	5	18	0.0394 (1.001)	85	25	0.0280 (0.711)	4	30	0.0232 (0.589)
230	18	0.0394 (1.001)	10	20	0.0331 (0.841)	85	30	0.0232 (0.589)	3	35	0.0197 (0.500)
170	20	0.0331 (0.841)	10	25	0.0280 (0.711)	85	40	0.0165 (0.419)	3	45	0.0138 (0.351)
110	30	0.0232 (0.589)	10	35	0.0197 (0.500)	80	50	0.0117 (0.297)	10	80	0.0070 (0.178)
70	40	0.0165 (0.419)	10	45	0.0138 (0.351)	80	80	0.0070 (0.178)	10	120	0.0049 (0.124)

TABLE 16-33
Condensed SAE Specifications for Cast Grit

Grit No.	Through Screen		Minimum Through Screen			Maximum Through Screen		
	Screen			Screen			Screen	
	No.	Opening, in. (mm)	%	No.	Opening, in. (mm)	%	No.	Opening, in. (mm)
G-10	7	0.1110 (2.819)	80	10	0.0787 (1.999)	10	12	0.0661 (1.679)
G-12	8	0.0937 (2.830)	80	12	0.0661 (1.679)	10	14	0.0555 (1.410)
G-14	10	0.0787 (1.999)	80	14	0.0555 (1.410)	10	16	0.0469 (1.191)
G-16	12	0.0661 (1.679)	75	16	0.0469 (1.191)	15	18	0.0394 (1.001)
G-18	14	0.0555 (1.410)	75	18	0.0394 (1.001)	15	25	0.0280 (0.711)
G-25	16	0.0469 (1.191)	70	25	0.0280 (0.711)	20	40	0.0165 (0.419)
G-40	18	0.0394 (1.001)	70	40	0.0165 (0.419)	20	50	0.0117 (0.297)
G-50	25	0.0280 (0.711)	65	50	0.0117 (0.297)	25	80	0.0070 (0.178)
G-80	40	0.0165 (0.419)	65	80	0.0070 (0.178)	25	120	0.0049 (0.124)
G-120	50	0.0117 (0.297)	60	120	0.0049 (0.124)	30	200	0.0029 (0.074)
G-200	80	0.0070 (0.178)	55	200	0.0029 (0.074)	35	325	0.0017 (0.043)

blasting. They are being used to deflash plastics parts and to clean molds, dies, electronic circuit boards, some static-sensitive devices, and other parts.

Media mixtures. Some sort of systematic replacement program must be set up for media. If replacement is not regular, the result will be the loss of the unit's blasting capability. The production of uniform surfaces depends upon keeping a uniform working mix of media in the machine at all times. Performing a screen analysis of the working mix can cover a broader spread of sieve sizes than is used for as-purchased media.

A practical method of maintaining a fairly constant working mix is to make additions of new media periodically, usually at least once every eight hours, to keep the media level in the supply tank, bin, or hopper at a uniform level. If the required surface finish of the work is critical, media additions may have to be made as frequently as every hour. The overall amount of media required to fill the machine to operating capacity can also affect the frequency of additions. A machine having large excess capacity for the storage of media can run longer without additions because the working mix changes more slowly.

Media Recycling Systems

Reclaiming and recycling of the media are important steps in controlling blasting costs. It is generally desirable to remove fines and contaminants from the media before it is reused; and for some applications, it is necessary to reclassify the media so that only certain sizes are reused. Devices used for separation can be classified into the following three basic categories: air wash systems, centrifugal systems, and screening devices. Magnetic separators are often used to remove magnetic particles from nonmagnetic contaminants.

Air wash separators. These systems are efficient and the most commonly used for airless blast equipment in which large volumes of media are necessary. They generally consist of creating a curtain of media by pouring it down an inclined plane. Air is drawn through the curtain to deflect lighter pieces, and skimmer plates trap these particles. The size range of the working mix can be changed by adjusting the skimmer plates and the blast gates controlling the volume of air drawn through the curtain.

Centrifugal separators. These units, often called cyclone separators, are more common for air-blast machines in which smaller volumes and lighter media are used. The separators combine the air wash principle of removing fines and contaminants with the centrifugal principle of throwing heavier pieces from an air stream.

Media return systems. These units vary in design depending upon the requirements of the blasting system with which they are used. For most airless machines, the media is collected from the bottoms of the machines by one or more screw conveyors that carry it to the sides of the machines; here, vertical bucket conveyors carry the media upward to the top of the separator. Stationary, rotating, or vibrating screens are usually provided along the media transfer path to remove larger contaminants or, possibly, workpieces before the media passes through the air wash separator and returns to a storage hopper.

Air-blast machines generally use vacuum tubes to return the media to the separator, usually through stationary screens. The discharge end of the return vacuum line is generally an integral part of the centrifugal separator, which is also operated in a vacuum.

Dust collector. A dust collector of some type is generally used to remove dust from the separator air before it is discharged into the atmosphere or recirculated in the system.

WET BLASTING

Wet blasting, sometimes called vapor or hydraulic blasting, is a process using fine media particles in a slurry form. The process does not require a dust collector or ventilation equipment and, with the use of rust-inhibiting compounds, prevents immediate oxidation. Finer media, down to a particle size of 0.0001" (0.003 mm), can be used. The process is generally considered a precision finishing operation. It can be controlled to produce or avoid metal removal and permits maintaining dimensional tolerances to within 0.0001".

CHAPTER 16

WET BLASTING

Process Principles

In wet blasting, a media is mixed with a liquid, usually water, and then pumped to a nozzle as a slurry. At the nozzle, the slurry is introduced into an air stream and propelled against the workpiece. Additives of rust inhibitors, wetting agents, and anticlogging or antisettling agents, if required, may be included in the slurry. Petroleum distillates can be used in place of water; however, special design features are necessary in such installations.

Media particles are kept in suspension by pumping, air agitation, or mechanical means. The pumping principle has several advantages, including reduced consumption of compressed air, recirculation of the slurry to keep the particles in uniform suspension, and feeding of the slurry at a positive pressure into the blast gun. In machines using air or mechanical agitation, the slurry is brought to the blast gun by suction created by jet action. Unlike airless dry blasting, wet blasting is ordinarily limited to use with compressed air because of the great differences in specific gravity between the fluid and the media.

Advantages and Limitations

The advantages of wet blasting as a cleaning and finishing method are useful in many industries. In the preparation of metals for plating, it can be used instead of pickling. Hydrogen embrittlement is avoided, and a metallurgically clean surface is produced, with various matte finishes obtainable by the selection of finer or coarser media. If polishing is required, the time is reduced for polishing prior to plating, and also for polishing on many types of dies such as those for molding, drawing, stamping, forging, forming, die casting, and rubber and glass molding.

Wet blasting also provides a good surface for paint adhesion. The surfaces generated by wet blasting generally have a lower rms value than those generated by dry blasting, although laboratory tests indicate that stainless steel and titanium are exceptions to this statement. However, the surface generated on *any* metal by wet blasting will appear to have a smoother finish than that resulting from dry blasting.

Another advantage of the wet blast process is its suitability for use in toolrooms and tool grinding departments. A slurry of very fine particles is used to remove feather burrs from multitooth cutting tools such as milling cutters, hobs, broaches, and taps, thus eliminating hours of hand honing time.

Wet blasting also deburrs intersecting hole junctions in precision parts that are inaccessible by any other method. The use of glass beads for close-tolerance peening of relatively thin-walled parts has become a successful application of the wet blasting process. Other advantages of wet blasting include the ability to blast wet parts, the capability of removing oily or wet contaminants, and the beneficial cushioning or buffering effect of the liquid used, resulting in less wear of the equipment and longer media life. However, media reclamation is difficult.

Wet blasting does have certain limitations and is particularly not recommended for heavy work that can withstand sandblasting or shot peening, such as the preparation of castings and forgings for machining. Wet blasting is also not ordinarily used for the removal of heavy burrs adjacent to smooth finished areas or machined configurations on which close fits or tolerances must be held. Wet blasting is not considered to be a high-production process because it requires more time than dry blasting. It is also not as clean a process as dry blasting, and precautions must be taken against oxidation of the workpieces. In addition, wet media may pack into cavities, holes, and crevices, causing removal problems.

Precleaning

To prevent the contamination of the recirculated slurry, it may be necessary to preclean the parts; however, every case must be decided on its own merits. When contaminants such as grease, protective coatings, and excessive and heavy oils are present, degreasing is recommended. Rust that is thick enough to be scraped off should be removed. In some cases, soaking in a caustic cleaner is helpful in cleaning such items as forging or forming dies and glass molds from which the burnt and hardened residues of lubricating materials are to be removed.

If machined parts have a light layer of oxidation or heat treatment scale, the contaminated slurry can be either discarded or, in more complex machines, separated and filtered. The principal use of the blasting process in this case is to restore the parts to their intended use.

Equipment Used

Many types and sizes of wet blasting equipment are available. Machines do not require special rooms and can be placed in production lines. Whether a machine is a simple hand-operated cabinet, just large enough to handle the product, or a complicated multigun, automated machine, a properly designed unit based on the experience of the manufacturer is required. The lowest priced unit may be the most expensive in the long run. Because of the specific gravity of even the finest particles, these must be kept in suspension in a hopper, usually accomplished by air agitation, mechanical movement, or pump recirculation.

Air agitation. Air agitation has the advantages of reduced initial cost and elimination of pump maintenance. However, the nozzles that are immersed in the slurry will have to be maintained. Also, compressed air stirs the particles against one another with a high velocity, causing wear of the particles and erosion of the hopper, necessitating the use of heavy-gage steel. Additionally, the blast gun has the double duty of aspirating the slurry into the gun plus blasting the slurry against the object to be processed—all using compressed air, which is often in short supply and costly to produce.

Mechanical movement. Mechanical movement to keep the media particles in suspension presents several difficulties. The moving element must be powerful enough to move the total volume of slurry with sufficient force. Bearing sealing problems arise, paddles and propellers are eroded, and the gun has to do double duty, as is the case with air agitation.

Pump recirculation. Recirculation by pump has the disadvantages of higher initial cost and increased maintenance. However, there are worthwhile advantages, including reduced consumption of compressed air, recirculation of the slurry to keep the particles in uniform suspension, and the feeding of the slurry at a positive pressure into the blast gun. The basic requirements are a properly designed pump and a lifetime range of 1000 to 4000 working hours for the replaceable parts. The lifetime range is dependent on mesh sizes of the abrasives, and the pump is usually a relatively low-maintenance item. To avoid costly maintenance, a close-tolerance direct pressure pump should be used for only the very finest meshes.

A low-velocity pump of suitable volume not only will furnish the blast gun with a positive delivery of slurry at low pressure, but will also keep all particles in even suspension, thus preventing erosion of the hopper and excessive use of compressed air. The blasting results are uniform, and the active life of the media is increased. The form of the hopper is also of importance to the proper suspension of the media in order to maintain a uniform slurry delivery in the blast gun.

Hopper machines. One common type of wet blasting equipment available is the hopper-type machine used in conjunction with a pressurized tank. The slurry is preferably pumped into the tank and, by the proper application of valving, is then pressurized. In cases when it is necessary to deliver the slurry at high pressure and velocity, more fine particles are expelled, resulting in intense blasting of confined areas such as internal intersecting holes and similar configurations. It is important that the slurry be correctly suspended in the feeding hopper and the pressurized slurry tank.

Tumbling/blasting devices. Another type of machine is a combined tumbling and blasting device, consisting of a tumbling barrel mounted on a wheeled cart that moves on a railed stand. This device may be moved into a cabinet. The barrel is a self-contained unit with a driving mechanism that turns the barrel and oscillates the blast gun, or guns, inside the barrel. Commercial sizes range from 12 to 26″ (305 to 660 mm) diam and are perforated to allow the slurry to escape and recirculate. Up to 50 lb (23 kg) of parts can be simultaneously tumbled and wet blasted.

Cabinets. A third typical piece of equipment is a hand-operated cabinet large enough to clean forging dies up to 20 ft (6.1 m) long x 10 ft (3 m) wide x several feet (1 m) deep, and weighing up to 100 tons (90 t). The machine may incorporate a cabinet for precleaning and soaking with a heated solvent for removing carbonized lubricants. The wet blasting process minimizes the wear on costly forging dies, increasing the useful life and facilitating the flow of the forging material. These factors justify the outlay for an installation of such magnitude. However, the cleaning must be done manually owing to the many variables in form, contours, and depths of cavities in these dies.

Semiautomatic machines. Semiautomatic wet blasting machines for flat printed circuit boards or similar flat parts of various dimensions, compositions, and production requirements are also available. The lateral movements are screw operated, with built-in reciprocal motion and adjustable speeds. Blast guns are vertically oscillated, with variable speeds and strokes.

For higher and continuous production, conveyorized machines are used. Unitized wet blasting machines combining washing, rinsing, and drying with fluxing and soldering operations are not uncommon.

Turntable machines. Indexing-type turntable machines are still another variation of wet blasting equipment. This type of machine is suitable for a variety of production items. A number of blasting stations may perform internal or external blasting for deburring or final finishing. A separating system and a closed-circuit filtering system permit the parts to be rinsed and blown off to a high degree of cleanliness before removal from the holding fixture, thus eliminating the constant flow of rinse water and the wasteful carryout of media particles.

Nozzles for Wet Blasting

The form, shape, length, and diameter of nozzles for wet blasting vary with the size and application of the parts to be blasted and the type of media used. Some commonly used types are cylindrical, special, and internal nozzles.

Cylindrical nozzles. The most common nozzle form is a cylindrical nozzle about 1″ (25 mm) long, with about a 1/2″ (13 mm) inside diameter and roughly a 3/4″ (19 mm) outlet diameter, made of inexpensive mild steel. Such nozzles are used on hand-operated cabinets for most general cleaning purposes. The lifetime of these nozzles is about 40 working hours, with the use of silicate or quartz abrasives in the 100 to 300 mesh range and blasting pressures from 80 to 90 psi (550 to 620 kPa).

The changing of nozzles in hand-operated cabinets normally presents no problem and takes only 2 to 3 minutes. When fast-cutting abrasives such as 80 to 150 mesh aluminum oxide are used, the lifetime of the nozzle may be as little as 20 working hours. However, short nozzle life does not justify the use of more expensive, longer lasting nozzles in a hand-operated machine, unless, of course, maintenance personnel must be called in to frequently change short-lived nozzles. Ceramic nozzles, either plain or jacketed, are also used; although their lifetime is considerably longer, their price is higher.

Special nozzles. Special nozzles such as fan-shaped types are usually made from cast iron of good quality, such as Meehanite chilled to a high grade of hardness, or rubber.

In mechanized machines for which surface preparation beyond cleaning is of importance, as in the preparation of aluminum for anodizing or surface conditioning for phosphatizing, it is important to use higher grade nozzles made of carbides or similar alloys. Cylindrical nozzles of this kind are much higher priced, but worthwhile savings in maintenance may be achieved. Special forms such as fan shapes, however, are expensive, and careful consideration should be given to justify their cost.

The lifetime of cylindrical carbide nozzles can be several thousand hours of use with fine abrasives. After over 1000 hours of use with 140 mesh quartz, no measurable wear can be detected. For automatically cycled machines and complex configurations, the carbide nozzle is more economical. For intermittent hand use, a good grade of steel or case-hardened nozzle is satisfactory. Inside diameters of 1/8, 3/16, and 1/4″ (3.2, 4.8, and 6.4 mm) are not uncommon. For extremely small inside diameters, industrial diamonds are used.

Internal nozzles. Many types of internal blast nozzles are being used. The shapes and principles of nozzles having the greatest effectiveness are proprietary. Diameters and lengths are tailored to suit the customer's requirements. Nozzles up to 30 ft (9.1 m) in length have been developed.

Media and Additives

Any natural abrasive in screened or ground form, as well as any manufactured abrasive, may be used in wet blasting. In some cases, ground fruit pits and even sodium bicarbonate compounds are used, but most agricultural media cannot be used for wet blasting. There are several forms of silicon dioxide with different physical characteristics that are widely used, such as silica sand, ground silica sand, ground quartz, and novaculite. Table 16-34 presents characteristics and applications for some of the more popular compounds.

Novaculite. Novaculite compounds, ranging in size from 100 mesh down to a theoretical 5000 mesh, are very pure, are screened to particle-size combinations, and have excellent cutting qualities. These compounds are successfully used for finishing, corrosion removal, and surface preparation for phenolic coatings, painting, and plating. They are also being used for defuzzing cutting tools after grinding, blending grinding marks on tools, and magnetic-particle or fluorescent-penetrant inspection. When used with suitable antirust compounds added to the slurry, novaculite provides an unusual corrosion resistance.

Silica and quartz. Silica, in mesh sizes of 50, 80, and 140 (U.S. sieve), is used for general cleaning purposes. A 400 mesh

TABLE 16-34
Characteristics and Applications of Wet Blasting Abrasives

Grade of Abrasive	Applications
40 to 80 mesh silica	Deburring steel and cast iron; removal of oxides from all forms of steel. Fast cutting and will remove metal. No close tolerances can be observed.
50 mesh silica	Deburring steel and cast iron; removal of oxides from all forms of steel. Fast cutting and will remove metal. No close tolerances can be observed.
80 mesh silica	Deburring steel and cast iron; roughing of surfaces; plastics bonding or rough plating; peening action. Fast cutting and will remove metal. No tolerances can be observed.
80 mesh ground silica	Heavy burrs; light or medium scale; bad rust conditions. Use on nickel alloy steels also. Exceptionally fast cutting and will remove metal. No tolerances can be observed.
100 mesh novaculite	Cleaning of carbon from pistons and valve heads; deburring of brass, bronze, and copper. Can also be used on crankshafts. Fast cutting and will remove metal. No close tolerances can be observed.
100 mesh ground quartz	Blending in preliminary grind lines on steel, brass, and die castings. Removal of medium-hard carbon deposits. For small-radius requirements. Fast cutting and will remove metal.
140 mesh silica	Removal of small burrs on steel, copper, aluminum, and die castings. Can be used on rough cleaning of dies and tools. No close tolerances can be observed. Will remove metal.
325 mesh novaculite	Second stage for cleaning aluminum pistons, impellers, and crankshafts and valves. First stage in cleaning master rods and all glass. Will follow close tolerances to 0.0025. Cuts slowly.
400 mesh aluminum oxide	Excellent abrasive agent when working under oil contamination conditions. Fast cutting agent on stainless steel, zinc, and aluminum die castings. Tolerances can be held.
1250 mesh novaculite	Second stage for crankshafts, rods, pistons, impellers, valves, gears, and bearings. For polishing all metals; also dies, tools, and die castings. Tolerances can be held.
5000 mesh novaculite	Should be used on any parts where an extra-fine surface is needed.
Glass beads from 40 mesh to 325 and even finer	Use for removal of light heat-treated scale and/or discoloration; also light oxide removal for jet engines and electronic components. No metal removal.

silica is used as a carrier medium for other selected abrasives. Quartz, ground and graded to 60, 80, 100, and 140 mesh sizes, is a more aggressive material and has a greater resistance to breakdown than silica. It is used for blending existing grinding and machining lines on steel parts, for brass parts, die castings, and for the removal of semihard carbon deposits. It is fast cutting and will remove metal.

Aluminum oxide. Aluminum oxide is a manufactured abrasive that is well known for its fast cutting action and good resistance to breakdown. It is accurately graded, chemically pure, and does not contain any iron oxide or silicon dioxide, which is of considerable importance in certain industries. It is available in mesh ranges from 24 to 800. Ground ore is a natural product that is not as pure as manufactured aluminum oxide. It contains some iron oxide, compares favorably with aluminum oxide, and is widely used.

Glass beads. Glass beads are available under various names and descriptions, ranging from 40 to 3000 mesh and even finer sizes. Some are carefully graded, and others are a combination of mesh sizes. Beads are widely used to achieve a peening-quality finish. The particles are globular and create hemispherical depressions rather than sharp, angular cuts. No metal is removed, and the resulting finish gives a bright, almost polished appearance.

Related Processes

Several other wet blasting processes are being used for various applications. These include high-volume wet blasting, water-jet deburring, and liquid-hone deburring.

High-volume wet blasting. This process differs from other wet blasting processes in that media velocity is generated by a pump rather than by compressed air; small amounts of compressed air are used only to atomize the liquid as it exits the process nozzle. The process employs high-volume, variable-speed centrifugal pumps that develop pressures to 80 psi (552 kPa). The pumps, which have no packing or seals, are commercially available to recirculate up to 1200 gpm (4542 L/min) for multiple-nozzle applications. Advantages claimed include improved control of the process and the capability of providing a wide range of finishes with one type of media.

For removing small burrs and for gentle finishing operations, the pump is operated at high speeds to provide partial atomization of the spray. For removing large burrs and more aggressive finishing, low pump speeds and complete atomization are employed. The scrubbing action resulting from the high volumes of slurry accomplishes simultaneous degreasing of the workpieces without the need for chemicals. Rinsing, drying, and rust-inhibiting operations can be incorporated in the same machine.

Water-jet deburring. High-pressure water jets, without abrasives, are being used for the rapid and efficient cutting of many nonmetals, including paper, cloth, wood, and fiberglass. A discussion of this process, called hydrodynamic machining (HDM), is presented in Volume I, *Machining*, of this Handbook series.

A lower pressure version of the HDM process called water-jet deburring or machining (WJM) is being used to deburr and deflash both metallic and nonmetallic parts. Fan-shaped blades or conical streams of water under pressures to 30,000 psi (207 MPa) impinge workpiece surfaces at high velocities, removing burrs or flash and flushing them away. The process, using no abrasives, leaves sharp corners with no radiusing.

Since cycle times are longer for burrs and flash of larger size, the process is best suited for light operations. Applications include the finishing of air compressor cylinders and automatic transmission cases and components.

The water-jet deburring process is easily automated, using various designs to move the nozzles or workpieces. Production rates to 460 pieces per hour have been reported. Automation of wet blasting and water-jet deburring processes is also being accomplished with programmable industrial robots that can control the pump, pressure, and manipulation of the nozzle. One application of wet blasting involves the deburring of laser-drilled holes in vanes and blades for turbine engines. Figure 16-89 illustrates a system using a six-axis robot for removing material by water blasting, with the workpiece rotated on a turntable, which provides a seventh axis.

Liquid-hone deburring. This is a process in which fine abrasive particles suspended in water are forced over workpiece edges having burrs (see Fig. 16-90). The steady flow abrades the burrs and generates smaller radii. It differs from wet blasting in that it does not involve the impact of a high-velocity, concentrated stream. Instead, it is a gentler process, with the media being forced over the edges instead of at them. Tooling is generally used to contain and direct the flow of media.

Liquid-hone deburring has limited application, being used primarily to remove small burrs from drilled holes 0.062" (1.57 mm) or less in diameter, which requires a cycle time of 1 to 3 minutes. Larger holes would require longer cycles, primarily because most of the media would be in the centers of the holes rather than at the edges. Typically, silicon carbide or aluminum oxide abrasive, 120 grit or finer, is used in a stream of water at a pressure of 50 psi (345 kPa).

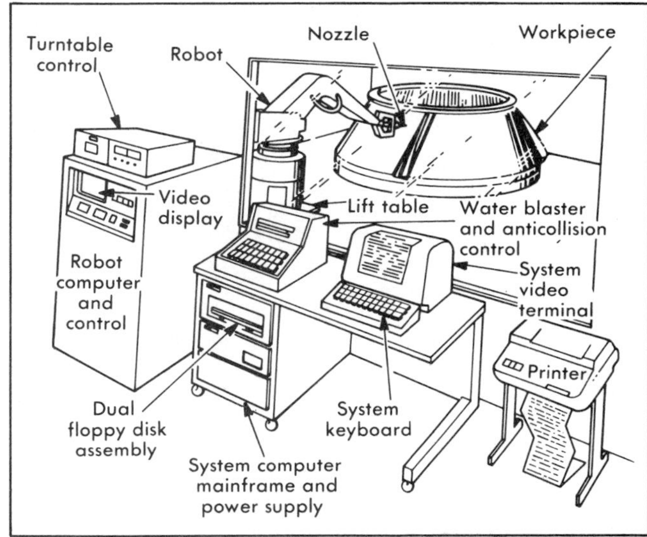

Fig. 16-89 Robot used for water blast cleaning of workpiece mounted on turntable. (*Marshall Space Flight Center*)

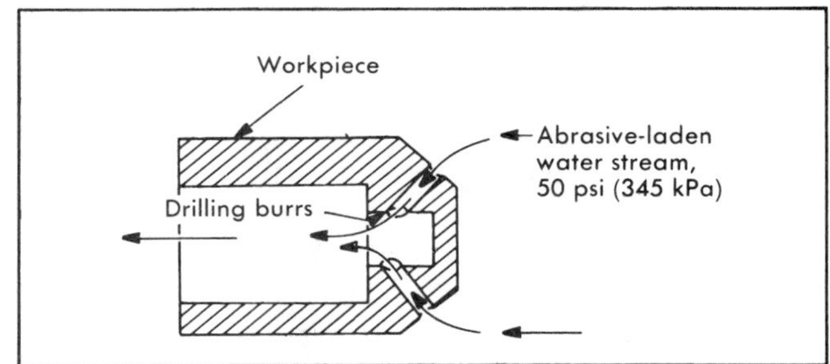

Fig. 16-90 Liquid-hone deburring.

SHOT PEENING

Shot peening is the cold working of a metal surface with a stream of spherical shot particles applied to the surface at high velocity under carefully controlled conditions.

EFFECTS OF SHOT PEENING

Shot peening is described as a process that provides the major advantage of inducing residual compressive stresses or work hardening. While stress relief may occur provided stresses are present, and while work hardening may also occur if the material is susceptible and the peening application is heavy enough and of long enough duration, shot peening is not generally used for either of these purposes. Its primary benefits are the generation of a uniform compressive stress pattern and the effective elimination of microscopic defects in the thin surface shell of the part.

Shot peening is most effective in reducing fatigue failures in parts subject to cyclic loading. Failures originate in surface areas under repeated tensile loading, and cracks will propagate from a surface defect or other stress riser. Shot peening prevents these failures by creating compressive stress layers in the surfaces of parts. As a part is loaded, its critical surface area will not develop tensile stresses until the shot-peen-induced com-

SHOT PEENING

pressive stresses are first overcome, thus permitting an increase in the allowable stress level and hence in the service life of the part. The effect of shot peening in improving the surface integrity of the part is also important. No matter how carefully a part is manufactured, it will exhibit some surface imperfections. These flaws may be localized areas of tensile stresses or phase transformations from machining or grinding, as well as pits, scratches, and other surface defects. As peening cold works the part surface, it blends these surface imperfections and effectively eliminates them as stress concentration points.

As each individual particle of shot strikes the metal surface, it produces a slightly rounded depression. Plastic flow and radial stretching of the surface metal occur at the instant of contact, and the edges of the depression may rise slightly above the original surface. In a completely worked part, the residual compressive stress layer usually extends to about 0.005 to 0.040" (0.13 to 1.02 mm) below the surface. Below this depth, a tensile stress layer will develop to achieve equilibrium. This tensile stress will be much lower than the surface compressive stress because it is distributed in a comparatively thicker core.

In thin materials peened on one side, a convex curve will develop with the peened surface on the outside. In this condition, the permanent compressive stress is low because the thin, undisturbed layer countering the compressive stress is too thin to completely resist the elongation of the peened side and curving results. This ability of shot peening to curve relatively thin materials is the basis of the standard test-strip method used for control of the peening process.

The magnitude of residual stress that can be induced by shot peening is a minimum of about half the yield strength in metals. This amount can be extended to nearly full yield strength by subjecting the surface to tensile stress while peening. Strain peening is commonly used on leaf springs and in the peen forming of aircraft wing contours, discussed in Volume II, *Forming*, of this Handbook series. Heat treatment above the stress relief temperature of the material or machining operations that would remove the induced compressive stress layer are not feasible after shot peening. Light honing or lapping is permitted on steels but is usually limited to 10% of the induced compressive layer. Charts are available from shot peening firms and machine manufacturers that correlate intensity, as measured by the Almen test, to the depth of the compressive layer for various materials.

SHOT PEENING APPLICATIONS

Peening has long been used to improve the fatigue characteristics of leaf and coil springs. Increases in fatigue life up to 800% are often obtained. The process is also used extensively on gears, driveshafts, crankshafts, torsion bars, axles, ball studs, high-strength fasteners, railroad wheels, and oil well drilling equipment. Often it is applied only at a critical area, such as a fillet radius. Overspray of shot is often permitted. If not, simple masking techniques can be used.

Peening will reduce the notch sensitivity of hard steels. Applied to highly stressed parts prior to chrome and electroless nickel plating, peening acts to prevent any cracks or imperfections in the chrome deposit from spreading into the parent part, and is required by federal specification QQC-320 for plated parts subject to dynamic loading.

The automotive industry uses peening extensively since the moving parts of the drivetrain are subject to cyclical loading. As the industry shifts to lighter weight cars, shot peening can give the industry lighter weight parts with excellent fatigue characteristics. Some special uses of shot peening include building high-performance racing engines from specially shot peened parts and peening aluminum die-cast transmission housings or gearboxes to prevent "leakers" or loss of lubricant through porous wall areas. Bearing surfaces of engine crankshafts are peened to retain lubricants, with the shot indentations acting as minute reservoirs. Fillet radii are also peened on most crankshafts.

The aerospace industry has long used shot peening for the improvement of the fatigue life of landing gear and other structural members. It is used extensively on jet engine parts, such as compressor and turbine blades. Wing and stabilizer skins are formed by peening and then saturation peened for fatigue resistance. Some of this work is done with glass beads, ceramic shot, cast steel shot, and shot produced from conditioned cut wire made of carbon or stainless steel.

Stress-corrosion cracking is a problem that has become critical in recent years as a result of the use of higher strength materials. Failures resulting from such cracks are caused by the complex interaction of corrosive (environmental) attack and sustained tensile stress (residual or applied) at the surface of a metal. Sustained static loadings are present in any thin-walled tank under a constant internal pressure. For example, the hydraulic reservoirs in aircraft landing gear are stressed by internal hydraulic pressure while at rest on the ground. Shot peening is highly effective in helping to prevent stress-corrosion cracking. Since most materials are susceptible to stress-corrosion cracking under specific conditions, parts should be designed for a stress level below the stress-corrosion threshold if possible. Shot peening is also effective in reducing fretting corrosion. It is widely used on bolt and fastener holes in aerospace structures when diameters permit.

Peening is sometimes used to straighten parts that may have become distorted in heat treatment or other operations. Rings that are out of round can sometimes be corrected by peening. Highly machined bulkheads and other large structural shapes that twist or deform as a result of machining operations can sometimes be straightened by judicious peening. The technique used in such cases is to select and peen a critical area that if expanded or elongated will straighten the part. Shot peening is also used to rework shafting to make diameters slightly larger and holes slightly smaller in diameter.

Parts requiring shot peening should be clean, dry, and free of any scale. Any material on top of the base material may absorb the energy of the controlled peening process. Degreasing, descaling, or blast cleaning operations are often performed prior to shot peening to ensure proper and beneficial results.

PROCESS PARAMETERS

The three variables used in the peening process are intensity, coverage, and shot size. Factors that affect intensity are shot size and hardness, velocity, and application angle. Good design data is available with respect to selecting shot sizes and intensities for shot peening applications. For example, the military specification MIL-S-13165B, amendment 2, presents some design data, and many of the commercial shot peening specifications also give recommendations. In all cases, however, notes on design drawings with respect to shot sizes and intensities take precedence because design engineers generally base their data on proven test results. Coverage in peening is a matter of industry practice.

Process Control

Unfortunately, methods discovered to date do not provide the user of the process a nondestructive way of determining if the desired compressive stress layer has been induced in a part. For this reason, control of the shot peen process is critical. Such control is essentially a system of documenting all of the variables on procedure sheets, which are approved by manufacturing, materials, and design engineers, including all of the qualifications of the machine and the setup.

Intensity

The term commonly used to describe the overall effect of a shot peening or mechanical prestressing application is *peening intensity*. The manufacturing engineer is most concerned with using the proper peening media and controlling the intensity produced by that media. The Almen test-strip method of measuring peening intensity displays the results of these factors as measurable curvature of the strip. Care must be taken to place standard Almen strips and blocks at every change in the part geometry to ensure uniform peening intensity. Qualification of the Almen test strips is usually made once per shift, as well as before the lot is run (when the setup is made) and at the end of the run (when the last part has been peened).

Test strips are furnished in three thicknesses (see Fig. 16-91) and are identified as N, A, or C strips according to their thickness. Test strips of type A are used for intensities that produce arc heights of 0.006 to 0.0224" (0.15 to 0.569 mm). For lesser intensities of peening, the N strip is recommended, and for greater intensities, the C strip. The test strips must be assembled to a standard Almen holding block (see Fig. 16-92). The arc height or bow induced in a strip by peening is read in thousandths of an inch by an Almen No. 2 dial-indicator gage, as shown in Fig. 16-93. The Almen No. 1 gage system is no longer used in shot peening. Figure 16-94 graphs the correlation of A, N, and C strips. The test strips should not be reused after they have been removed from the holder but should be marked and maintained as a record.

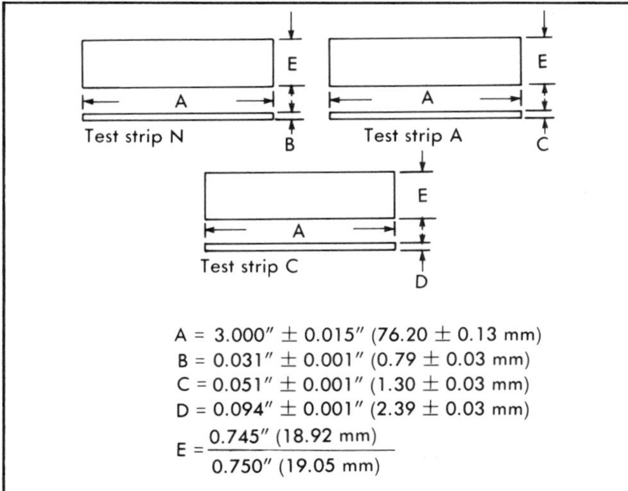

Fig. 16-91 Shot peening test-strip specifications: Analysis of stock: SAE 1070, cold rolled, spring-steel edge No. 1 [on 3" (76 mm) edges]. Finish: Blue temper (or bright), uniformly hardened. Heat set between flat plates under pressure for a minimum of 2 hr at 800° F (425° C), ±25° F (14° C). Hardness of R_C44-50. Flatness: ±0.001" (0.03 mm) arc height (strip A); ±0.0015" (strip C); ±0.001" (strip N) as measured on a standard Almen gage No. 2.

For good control, checks of peening intensity with a test strip should be made in addition to checks on shot size and application angle. Control of the required test-strip arc heights and coverage must be maintained, but even these factors can be misleading because only the correct size and shape of shot can produce an even layer of compressive stress to the required depth. Smaller shot at higher velocity may produce the required intensity and coverage but not the depth on soft parts, such as those made from aluminum.

Coverage

Surface coverage is a measure of how completely an area has been covered by the indentations of the individual shot particles. The proper coverage is a matter of industry practice that varies from 80 to 100% for spring applications and to 200% in the aircraft industry. Springs are peened to as low as 80% coverage since tests have shown additional coverage does not add enough to the life of the spring to justify the additional cost. Aircraft industry parts are peened to 200% coverage for safety reasons. The proper coverage should be determined by fatigue tests.

There are four accepted methods of reading coverage. The first method is visual, using optical magnification. This method, although not quantitative, is widely employed. It is difficult to estimate coverage visually above 98%. As a result, this 98% has been chosen to represent full coverage.

The second method involves the use of a polished Almen test strip. The strip is exposed to the shot and then magnified to 50 diameters in the field of a metallurgical camera. Using a piece of transparent paper as a ground glass, the indented areas are traced with a sharp pencil. The area of all the indentations can be measured with a planimeter, and the ratio of the indented areas to the total are determined. This method is time consuming and assumes the area selected for magnification is typical. Several areas on each test strip should be calculated, and the average used as the probable coverage.

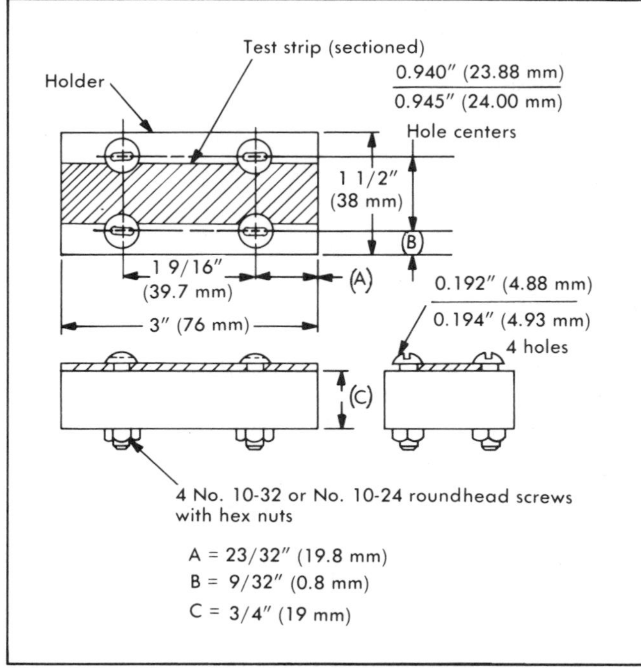

Fig. 16-92 Standard Almen block for holding test strips.

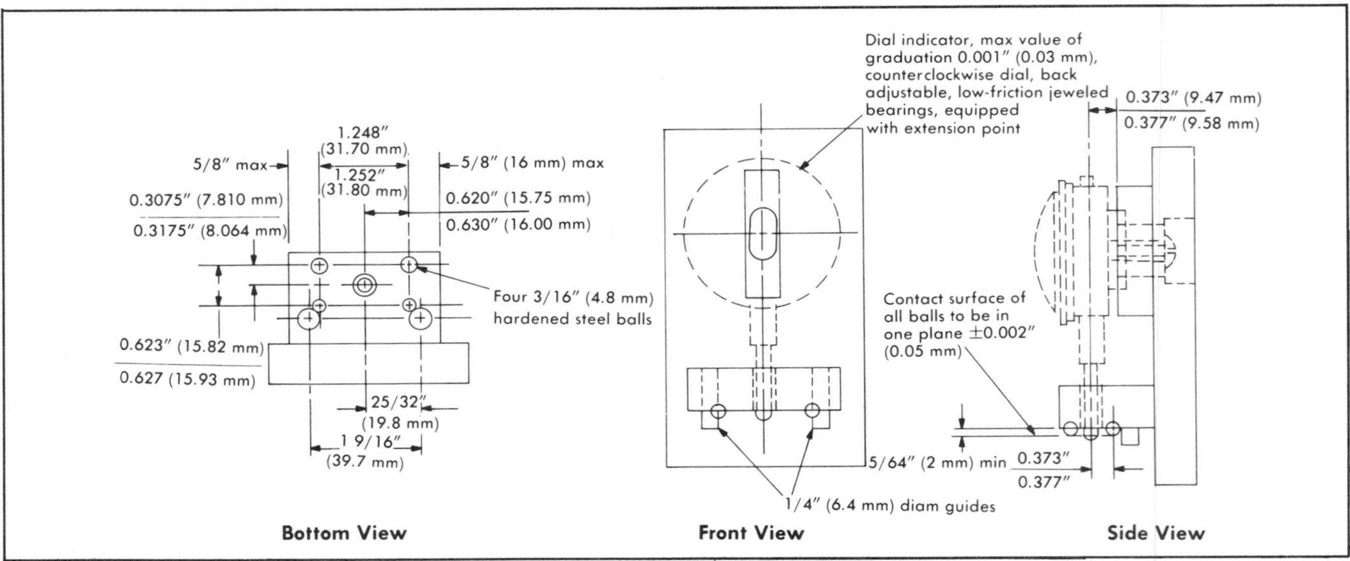

Fig. 16-93 Almen gage No. 2 for measuring test-strip arc height. Periodic checks should be made of the balls to determine if there are flat spots, which could give erroneous intensity readings. The balls should be rotated or replaced when flat spots occur.

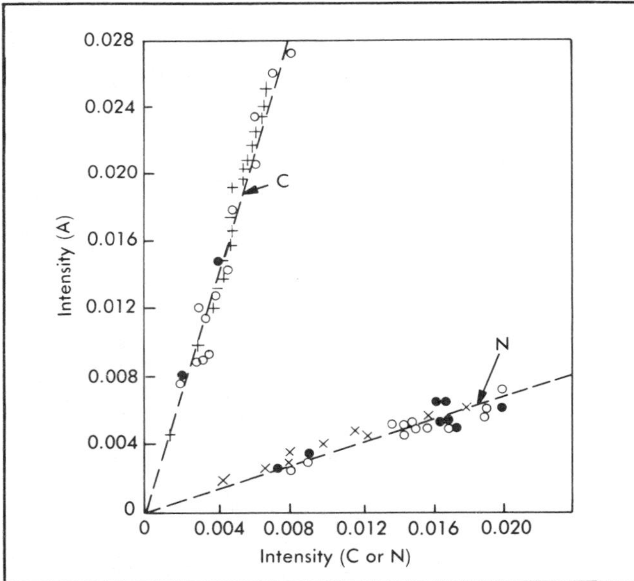

Fig. 16-94 Correlation of A, N, and C test strips as measured on Almen gage No. 2.

Figure 16-95 can be used with the above methods to guide the user as to the required exposure time to achieve a given coverage. As an example, suppose that after a given exposure time the part had 50% visual coverage, and 100% coverage is desired. Using the graph in Fig. 16-95, this example represents an exposure time factor of about 2 1/2, indicating an exposure time of approximately seven times the original exposure time will be required to reach 100% coverage, at which point the factor of exposure time is 18. The curve in Fig. 16-95 suggests that some areas are struck many times before all areas are struck at least once. Coverage requirements beyond 100% are expressed as a multiple of the exposure time required to produce 100% coverage.

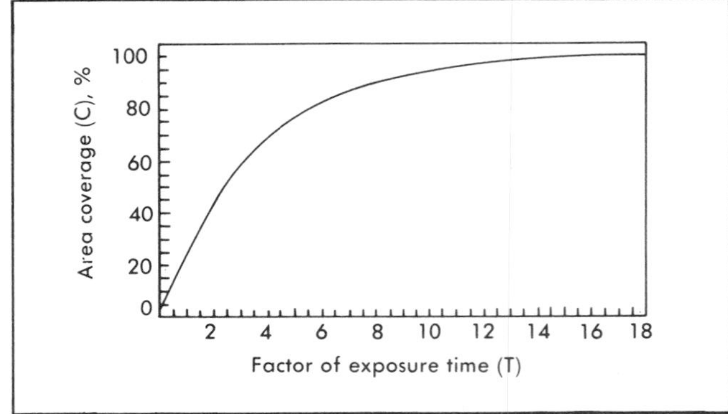

Fig. 16-95 Relationship of peening coverage to exposure time. Coverages above 100% are direct multiples of the time required for 100% coverage.

The third method of determining coverage involves the use of the Almen test strip. By this method, it is possible to make a fairly accurate and quick coverage determination if the desired coverage is 100% minimum. First, expose a test strip to the estimated exposure for 100% coverage and read the arc height. Then, expose another strip for two times the original exposure and read the arc height. If the second test strip's arc height is more than 10 to 15% of the first strip, it indicates that the first was not to saturation (100% coverage). Usually, several of these tests, adjusting exposure time between tests, can locate the position on the intensity-determination curve for the coverage desired.

When using the Almen strip method of checking coverage, it should be understood that the time required to saturate the test strip is not necessarily equal to the time required to visually saturate the part. If the hardness of the part is appreciably different from the R_C44 to 50 hardness of the test strip, the time required to saturate will vary, and a soft material will reach a point of saturation faster. The hardness of shot can also influence required time.

A fourth method involves the use of Dyescan tracer liquids, as used in the Peenscan process. A controlled specimen of the same material and hardness as the actual workpiece is coated with the tracer liquid by either dipping, spraying, or painting, and then the fluid is allowed to dry. The coating is checked under a black light to verify uniform coverage of the coated area to be shot peened. The controlled specimen is then shot peened under proper conditions for the required intensity and coverage, as prescribed in the peening procedure. A reexamination under the black light determines if the tracer liquid has been uniformly removed; full coverage is indicated by complete removal of the tracer liquid. Areas that do not produce full coverage will show a white color under the black light; areas of full coverage will give off a dark color. Coverage of actual production pieces can be established by employing the same procedure used for the controlled specimen; the fluorescent liquid is used for each part on a statistical sampling basis.

Peening Media

Shot peening is generally performed using cast steel shot, conditioned cut wire shot, or glass beads. Most controlled shot peening is done with cast steel shot. For automotive and general applications, shot conforming to SAE recommended practice J444 (see Table 16-35) is generally used. For aerospace and defense work, shot conforming to the military specification MIL-S-13165B is generally used. This specification requires a closer control of size, hardness, allowable nonrounds, and unacceptable shapes.

Conditioned cut wire shot, also allowable under the military specification, is a wrought material that breaks down more slowly than cast steel shot and may produce better surface finishes for some applications; it is also more expensive than cast steel shot. The MIL-S-13165B specifications for sizes of cast steel shot and cut steel wire are shown in Tables 16-36 and 16-37 respectively.

Selection of the shot size depends upon the thickness of the work, the appearance desired, the size of fillets, and the intensity desired. At a given velocity, smaller shot will provide a lower intensity, and larger shot, a higher intensity. Also, smaller shot, having more particles per pound, provides coverage more rapidly and is therefore more economical. Peen forming of heavy sections may employ steel balls up to 1/4" (6.4 mm) in diameter or larger. The use of the larger balls provides a smoother finish.

The use of glass beads as a media has good application for peening thin sections requiring low intensities. Steel shot will leave a ferric contamination on aluminum, which can be removed by acid dipping and, in some cases, by overpeening with glass beads; however, this overpeening is not as effective as acid dipping. Because glass beads are inert, they will not contaminate the aluminum. Glass is also effective in peening stainless steel and titanium, provided effective equipment is available to remove broken particles. Consideration should also be given to ceramic shot. Tests indicate ceramic shot can produce acceptable intensities without as much breakdown as glass beads.

The ideal peening application would use perfectly round media of uniform size, with each particle having the same hardness and density. Each particle would be applied at the same velocity and impact angle. Although these conditions are impractical, the use of high-quality steel shot in equipment designed to control shot size, velocity, and impact angle cannot be overemphasized.

Air wash separators, supplied with most machines, are capable of removing dust, fine particles, and small, broken pieces of shot. Vibrating screens are sometimes used to provide improved sizing of the media. Both air wash separators and vibratory screens can return large portions of broken shot or unacceptably shaped media to the machine for reuse. For more critical applications, such as aerospace component peening, spiral separators or inclined, vibrating-plate separators are used to control nonrounds. These devices use the rolling characteristic of round media to separate these particles from nonround particles. Automatic shot replenishers should be employed to maintain the required shot quantity.

Shot Velocity and Impact Angle

Shot velocities of 265 fps (80.8 m/s) or higher are possible with air-blast and centrifugal wheel systems. Air pressure or wheel speed is varied to change shot velocity. When a series of test strips are peened with a fixed machine setting for different exposure times, a saturation (full coverage) curve, as shown in Fig. 16-96, can be obtained. The time necessary to produce saturation on the test strip can be defined as the time required to achieve the specified arc height at which doubling the exposure time will not increase the arc height by more than 10 to 15%. An arc height is not properly termed intensity unless saturation has been achieved. If the specified intensity cannot be obtained in a reasonable time, it is indicated that the shot velocity is too low or that the shot size is too small. Conversely, excessive arc height indicates too high a shot velocity.

Shot particles from air nozzles have varying velocities. This phenomenon has been proven photographically by tests using two timed stroboscopic light flashes, each 3 μs in duration, which effectively stop the shot at two positions on the film. A third electronic flash of longer duration, timed between the other two flashes, causes a streak on the film that positively identifies the particle path. This varying shot velocity is probably caused by hose or nozzle discontinuities, air turbulence, or a combination of these. This phenomenon may be responsible for the lazy knee on the intensity-determination curve that is sometimes observed, but that cannot be attributed to a poor shot mix. This phenomenon has been observed in both suction and direct pressure systems. Shot from centrifugal wheels does not, by the same test, exhibit varying velocities.

The shot energy absorbed by the work will vary approximately as the sine of the angle between the plane of the work and the line of motion of the shot. Usually, parts with complex shapes require shot application angles varying from 90 to 45°, and the specifications should allow variation in test-strip intensity through this range, if possible. For example, if the maximum test-strip intensity is 0.012" (0.30 mm), the minimum would be 0.008" (0.20 mm), allowing a spread of 33 1/3%. An additional reason for the spreading of Almen intensity is the fact that the Almen strip has a tolerance of ±0.0015" (0.038 mm). The Almen gage has a tolerance of ±0.001" (0.003 mm) and a hardness tolerance range of R_C44 to 50.

SHOT PROPULSION

Peening media is propelled by two methods: compressed air or centrifugal force. For a discussion of the mechanics of these methods, refer to the preceding dry and wet blasting sections in this chapter.

TABLE 16-35
Cast Shot Specifications for Peening and Cleaning*

National Bureau of Standards Screen No.	Standard Screen Opening,** mm	Screen Size, in.	SAE Shot Number S1320	S110	S930	S780	S660
			Maximum and Minimum Cumulative Percentages Allowed on Corresponding Screens				
4	4.75	0.187	All pass	---	---	---	---
5	4.00	0.157	---	All pass	---	---	---
6	3.35	0.132	90% min	---	All pass	---	---
7	2.80	0.111	97% min	90% min	---	All pass	---
8	2.36	0.0937	---	97% min	90% min	---	All pass
10	2.00	0.0787	---	---	97% min	85% min	---
12	1.70	0.0661	---	---	---	97% min	85% min
14	1.40	0.0555	---	---	---	---	97% min
16	1.18	0.0469	---	---	---	---	---
18	1.00	0.0394	---	---	---	---	---
20	0.850	0.0331	---	---	---	---	---
25	0.710	0.0278	---	---	---	---	---
30	0.600	0.0234	---	---	---	---	---
35	0.500	0.0197	---	---	---	---	---
40	0.425	0.0165	---	---	---	---	---
45	0.355	0.0139	---	---	---	---	---
50	0.300	0.0117	---	---	---	---	---
80	0.180	0.007	---	---	---	---	---
120	0.125	0.0049	---	---	---	---	---
200	0.075	0.0029	---	---	---	---	---

* SAE Recommended Practice J444a.
** Corresponds to ISO Recommendations.

Shot Peening Using Compressed Air

The two most efficient compressed air systems in terms of air required per unit of shot moved are the direct pressure method and the gravity nozzle system. Of the two systems, direct pressure produces the highest shot velocity and is the only system that can move shot through long lances and side-shooting nozzles to peen deep holes or cavities.

Control of air pressure and flow is extremely critical for shot peening operations; variations can severely affect the resulting shot velocity and flow. Air pressure sensors and shot flow meters can be used to monitor these two variables and signal faults. Such monitoring can protect against nozzle plugging, nozzle wear, and air pressure variations. These two devices plus a microprocessor can also be used to set shot velocity. Velocity and the angle of impact should be confirmed using an Almen strip.

The air-blast tank used for shot peening can be a single chamber, an intermittent type, or a two or three-chamber continuous type. The intermittent type is used when the peening time cycle will not exceed the capacity of the tank. Also, adequate time must be available between cycles to depressurize, refill, and repressurize the tank.

Most direct pressure shot peening is done with continuous tanks that have more than one chamber. The bottom chamber is never depressurized while the machine is running and is refilled before it runs out of media. The top chamber is depressurized to receive the next load of media from a storage bin and then pressurized. The valve between the two chambers is next opened, and the media falls to the bottom chamber.

For critical shot peening, a three-chamber tank should be used. With such a system, the third chamber at the top measures, by volume rather than by time, the correct charge of shot for each cycle. The pressurizing and depressurizing valves never close on the shot since the charge is sized to drop completely through the valve without completely filling the chamber below. This system also employs a separate low-pressure air supply to activate the valves, avoiding the common

TABLE 16-35—*Continued*

SAE Shot Number								
S550	S460	S390	S330	S280	S230	S170	S110	S70
Maximum and Minimum Cumulative Percentages Allowed on Corresponding Screens								
---	---	---	---	---	---	---	---	---
---	---	---	---	---	---	---	---	---
---	---	---	---	---	---	---	---	---
---	---	---	---	---	---	---	---	---
---	---	---	---	---	---	---	---	---
All pass	All pass	---	---	---	---	---	---	---
---	5% max	All pass	---	---	---	---	---	---
85% min	---	5% max	All pass	---	---	---	---	---
97% min	85% min	---	5% max	All pass	---	---	---	---
---	96% min	85% min	---	5% max	All pass	---	---	---
---	---	96% min	85% min	---	10% max	All pass	---	---
---	---	---	96% min	85% min	---	10% max	---	---
---	---	---	---	96% min	85% min	---	All pass	---
---	---	---	---	---	97% min	---	10% max	---
---	---	---	---	---	---	85% min	---	All pass
---	---	---	---	---	---	97% min	---	10% max
---	---	---	---	---	---	---	80% min	---
---	---	---	---	---	---	---	90% min	80% min
---	---	---	---	---	---	---	---	90% min
---	---	---	---	---	---	---	---	---

TABLE 16-36
Cast Steel Shot Numbers and Screening Tolerances*

Peening Shot No.	All Pass U.S. Screen No.	All Pass U.S. Screen Size, in. (mm)	Maximum of 2% On: U.S. Screen No.	Maximum of 2% On: Size, in. (mm)	Maximum of 50% On: U.S. Screen No.	Maximum of 50% On: Size, in. (mm)	Cumulative Minimum, 90% On: U.S. Screen No.	Cumulative Minimum, 90% On: Size, in. (mm)	Maximum of 8% On: U.S. Screen No.	Maximum of 8% On: Size, in. (mm)	Maximum Deformed Shot Acceptable: No. of Shot	Maximum Deformed Shot Acceptable: Area, in.^{2} (cm^2)
930	5	0.157 (3.99)	6	0.1320 (3.353)	7	0.1110 (2.819)	8	0.0937 (2.380)	10	0.0787 (1.999)	5	1 (6.45)
780	6	0.1320 (3.353)	7	0.1110 (2.819)	8	0.0937 (2.380)	10	0.0787 (1.999)	12	0.0661 (1.679)	5	1 (6.45)
660	7	0.1110 (2.819)	8	0.0937 (2.380)	10	0.0787 (1.999)	12	0.0661 (1.679)	14	0.0555 (1.410)	12	1(6.45)
550	8	0.0937 (2.380)	10	0.0787 (1.999)	12	0.0661 (1.679)	14	0.0555 (1.410)	16	0.0469 (1.191)	12	1 (6.45)
460	10	0.0787 (1.999)	12	0.0661 (1.679)	14	0.0555 (1.410)	16	0.0469 (1.191)	18	0.0394 (1.001)	15	1 (6.45)
390	12	0.0661 (1.679)	14	0.0555 (1.410)	16	0.0469 (1.191)	18	0.0394 (1.001)	20	0.0331 (0.841)	20	1 (6.45)
330	14	0.0555 (1.410)	16	0.0469 (1.191)	18	0.0394 (1.001)	20	0.0331 (0.841)	25	0.0280 (0.711)	20	1/2 (3.22)

(continued)

TABLE 16-36—*Continued*

Peening Shot No.	All Pass U.S. Screen No.	All Pass Size, in. (mm)	Maximum of 2% On: U.S. Screen No.	Size, in. (mm)	Maximum of 50% On: U.S. Screen No.	Size, in. (mm)	Cumulative Minimum, 90% On: U.S. Screen No.	Size, in. (mm)	Maximum of 8% On: U.S. Screen No.	Size, in. (mm)	Maximum Deformed Shot Acceptable No. of Shot	Area, in.^{2} (cm^2)
280	16	0.0469 (1.191)	18	0.0394 (1.001)	20	0.0331 (0.841)	25	0.0280 (0.711)	30	0.0232 (0.589)	20	1/2 (3.22)
230	18	0.0394 (1.001)	20	0.0331 (0.841)	25	0.0280 (0.711)	30	0.0232 (0.589)	35	0.0197 (0.500)	20	1/2 (3.22)
190	20	0.0331 (0.841)	25	0.0280 (0.711)	30	0.0232 (0.589)	35	0.0197 (0.500)	40	0.0165 (0.419)	20	1/2 (3.22)
170	25	0.0280 (0.711)	30	0.0232 (0.589)	35	0.0197 (0.500)	40	0.0165 (0.419)	45	0.0138 (0.351)	20	1/2 (3.22)
130	30	0.0232 (0.589)	35	0.0197 (0.500)	40	0.0165 (0.419)	45	0.0138 (0.351)	50	0.0117 (0.297)	30	1/4 (1.61)
110	35	0.0197 (0.500)	40	0.0165 (0.419)	45	0.0138 (0.351)	50	0.0117 (0.297)	80	0.0070 (0.178)	40	1/4 (1.61)
70	40	0.0165 (0.419)	45	0.0138 (0.351)	50	0.0117 (0.297)	80	0.0070 (0.178)	120	0.0049 (0.124)	40	1/4 (1.61)

* Military Specification MIL-S-13165B.

TABLE 16-37
Size Classification of Cut Steel-Wire Shot*

Shot No.	Wire Diameter, in. (mm)	Length of Ten Pieces,** in. (mm)	Weight of 50 Pieces,† g (oz)
CW-62	0.0625 (1.59) ±0.002 (0.05)	0.620 (15.75) ±0.040 (1.02)	1.09 to 1.33 (0.038 to 0.047)
CW-54	0.054 (1.37) ±0.002 (0.05)	0.540 (13.72) ±0.040 (1.02)	0.72 to 0.88 (0.025 to 0.031)
CW-47	0.047 (1.19) ±0.002 (0.05)	0.470 (11.94) ±0.040 (1.02)	0.48 to 0.58 (0.017 to 0.020)
CW-41	0.041 (1.04) ±0.002 (0.05)	0.410 (10.41) ±0.040 (1.02)	0.31 to 0.39 (0.011 to 0.014)
CW-35	0.035 (0.89) ±0.001 (0.03)	0.350 (8.89) ±0.030 (0.76)	0.20 to 0.24 (0.007 to 0.008)
CW-32	0.032 (0.81) ±0.001 (0.03)	0.320 (8.13) ±0.030 (0.76)	0.14 to 0.18 (0.005 to 0.006)
CW-28	0.028 (0.71) ±0.001 (0.03)	0.280 (7.11) ±0.030 (0.76)	0.10 to 0.12 (0.003 to 0.004)
CW-23	0.023 (0.58) ±0.001 (0.03)	0.230 (5.84) ±0.020 (0.51)	0.05 to 0.07 (0.002 to 0.0025)
CW-20	0.020 (0.51) ±0.001 (0.03)	0.200 (5.08) ±0.020 (0.51)	0.04 to 0.05 (0.0014 to 0.0018)

* Military Specification MIL-S-13165B.

** Shot particles to be checked for length shall be mounted and ground and polished to expose a central longitudinal section. The combined length of ten random particles shall be within the tolerances shown.

† At the option of the contractor, the particles may be weighed instead of being mounted and measured. When weighed, the total weight of 50 randomly selected particles shall be within the limits specified.

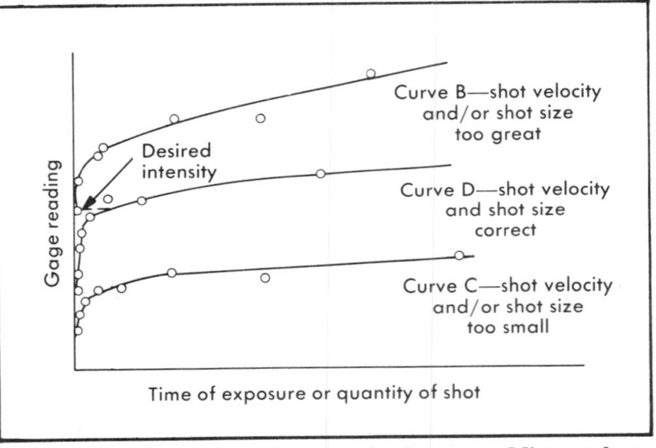

Fig. 16-96 Shot peening intensity-determination curves. Mixture of too many smaller shot sizes or varying shot velocities will extend the time to reach desired intensity.

air pressure drop at the nozzle when the valves operate. This design assures more even peening velocities and smooth operation at pressures from 125 down to 3 psi (862 to 20 kPa). This tank also utilizes an adjustable shot flow valve that automatically shuts off media to the mixing chamber when the air is shut off.

Because of their small size, air nozzles are readily aimed to reach difficult areas. Also, various nozzles can be adjusted to different angles and air pressures in one setup for a difficult part. An alternative to adjusting multiple nozzles is to automatically manipulate fewer nozzles, shortening the setup time, but lengthening the peening cycle time. In peening large areas, "barber poling" or striping must be carefully avoided by coordinating the part rotation and nozzle oscillation speeds to insure overlapping of the blast patterns.

Some peening is performed with wet systems using glass beads. These machines use suction-type nozzles; but the water-bead slurry is usually mechanically pumped to the nozzle, and a constant water-to-bead ratio is almost impossible to maintain. Wet peening also has other problems, such as the additional requirements for rinsing, drying, rust inhibiting, and disposing of drain-clogging spent shot.

Centrifugal Wheel Systems

Centrifugal wheels are efficient shot-throwing devices for peening. They have been used for many years for peening automotive springs and other parts for which high production was required, and for which the wheels could be fixed in location. More recently, manipulated wheels have been used to give the machines more flexibility.

Centrifugal wheels are most effective for peening large, flat surfaces. Their limitation is that they are not able to peen small crevices and internal surfaces. Another limitation is that the wheels generate hot spots on the workpieces, and these are the most affected areas of the peening. Large parts requiring peening must be brought under the hot spot of the centrifugal wheel in a uniform manner to insure maximum benefit from the process. This procedure is easily accomplished on large areas with minimum changes in geometry or by manipulating the wheel for changes in geometry. Single or multiple wheels can be manipulated to obtain much of the flexibility of air-blast nozzles.

Centrifugal wheels conserve energy. A 20 hp (14.9 kW) centrifugal wheel can project 36,000 lb (16 330 kg) of shot per hour, compared with the air nozzle system that would require a 420 hp (313 kW) air compressor for the same effect. Shot from centrifugal wheels does not vary in velocity. Variable-frequency or silicon-controlled rectifier wheel drives allow manual or programmable control of wheel speed and easily achieve variable Almen intensities. With centrifugal wheels, the velocity imparted to steel particles does not effectively diminish at a distance of 6 ft (1.8 m) or more. All shot sizes to 1/4" (6.4 mm) diam can be projected to high velocities by centrifugal wheels.

OTHER EQUIPMENT CONSIDERATIONS

In addition to a method for propelling shot, a peening machine must provide a means for collecting the spent shot and reclassifying it for use. As described previously, this recycling is done with a combination of screens, conditioning spirals, or vibrating plates, and an air wash system to maintain correct shot size and shape. A shot replenisher may be required on larger machines.

Workhandling systems may range from simple turntables to complex variable-speed work cars or monorail conveyors. Indexing table systems with one or more peening stations are adaptable to small and medium-sized parts and extensive production.

Microprocessor-based systems are available for:

- Wheel or air-blast manipulation. Typical is the centrifugal wheel system shown in Fig. 16-97, which provides five positioning movements for shot peening.

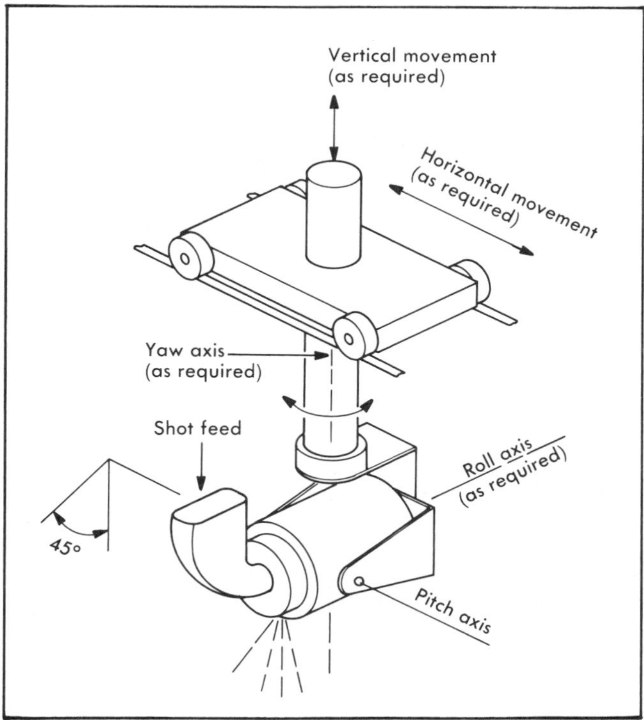

Fig. 16-97 Centrifugal wheel system provides five positioning movements for shot peening. (*The Pangborn Co.*)

- Shot flow monitoring and control.
- Shot velocity.
- Part manipulation.
- Documentation of process parameters and production information.
- Data transfer to host computers.
- Interfacing with peripheral devices in a manufacturing cell, such as with robots and conveying systems.

The systems may use programmable controllers (PC) or computer numerical control (CNC). Either type of control provides repeatable programs or routines, tailored to the requirements of the individual parts to be peened, stored in the memory or on tape. These systems shorten setup time and lessen the possibility of operator error. When microprocessor-based systems are not used, operator training in the technology of shot peening is paramount to obtaining good results. Operators should be certified prior to using the process.

GRINDING WITH BONDED ABRASIVES

Methods using bonded abrasives to finish parts include grinding with bonded wheels, coated-abrasive grinding, honing, and superfinishing.

GRINDING WITH BONDED WHEELS

Grinding with bonded wheels is a common process for both substantial stock removal and the fine finishing of flat, round, or curved surfaces to close tolerances and desired requirements. This process removes material by means of abrasive grains bonded into solid bodies that are referred to as wheels.

Advantages of grinding with bonded wheels include superior size control and the capability of varying the finish and lay of the surface produced. Since only a minimum force is exerted on the workpieces in finish grinding, there is little if any distortion.

The process is easily automated, and rough and finish grinding can often be combined in a single operation.

A comprehensive discussion of grinding is presented in Volume I, *Machining*, of this Handbook series. Subjects include the principles of grinding, grinding wheels and discs, grinding machines and processes (surface, cylindrical, and special), workholding and control methods, and troubleshooting.

COATED-ABRASIVE GRINDING

Coated abrasives are multipoint cutting tools available in sheet, disc, roll, belt, and other forms. They are used extensively for both heavy and light stock removal. Finishing operations performed include deburring, deflashing, blending, surface finishing, and polishing. Polishing operations with coated-abrasive belts and flap wheels are discussed in a preceding section of this chapter.

The versatility of coated-abrasive products results from their wide range in flexibility, sizes available, and adaptability to manual, semiautomatic, or automatic operation. The manufacture of coated-abrasive products and details with respect to different backings, abrasive grains, and bonding adhesives used are presented in Volume I, *Machining*, of this Handbook series. Applications of abrasive-belt machining for general-purpose and heavy stock removal are also discussed in Volume I.

Finishing Applications

Coated-abrasive products, especially belts, are being used for a wide variety of finishing applications. New abrasives, backings, and bonds, as well as improved machines, are continuously expanding the use of coated-abrasive products. Materials finished with coated abrasives include practically all metals (both ferrous and nonferrous), glass, stone, clay, concrete, wood, leather, ceramics, and plastics. Abrasives of finer grit size are used for finishing; but for maximum efficiency and economy, the coarsest grit that will produce the required finish should be used.

Coated-Abrasive Belts

Both narrow and wide coated-abrasive belts are being used extensively for finishing operations. Narrow belts are considered to be those 6″ (152 mm) or less in width, with lengths from 42 to 168″ (1069 to 4267 mm) long. Belts to 126″ (3200 mm) wide are being used to finish sheet and plate stock.

Deburring and deflashing operations are normally performed with intermediate (40 to 120 grit) grades of abrasive. For finishing operations, finer (150 to 400 grit) grades are used. For some mirror-finishing operations, abrasive grades as fine as P1200 are necessary prior to buffing to a mirror finish.

Film-backed coated abrasives. Fine finishes of 2 μin. (0.05 μm) or less are being obtained with fine abrasives that are electrostatically coated and resin bonded to a thin, flexible backing of polyester film. These products are available in belt, roll, sheet, and disc forms, with sheet being least commonly used. Aluminum oxide and silicon carbide abrasives are used, with aluminum oxide being more common. The tough yet flexible polyester film backing is impervious to water, allowing the use of water-based lubricants. The hardness of the film minimizes sinking of the abrasive into the backing, thereby preserving the cutting edges of the abrasive grains. Thickness of the film, 0.003 or 0.005″ (0.08 or 0.13 mm), is more constant than cloth or paper backing and has greater wear resistance for longer product life.

Rolls of film-backed coated abrasives are available in lengths to 450 ft (137 m). The rolls are used in setups such as the one shown in Fig. 16-98 to continuously or intermittently present fresh abrasive to the workpieces. Some machines are designed to oscillate the abrasive, thus producing a crosshatch pattern on surfaces.

Abrasive sizes. Abrasive grains of 50 to 15 micron size, 0.0020 to 0.0006″ diam, are commonly used for finishing operations with products made from film-backed coated abrasives. Grades are identified by both the U.S. CAMI (Coated Abrasive Manufacturers Institute) and the European P system (see Fig. 16-99).

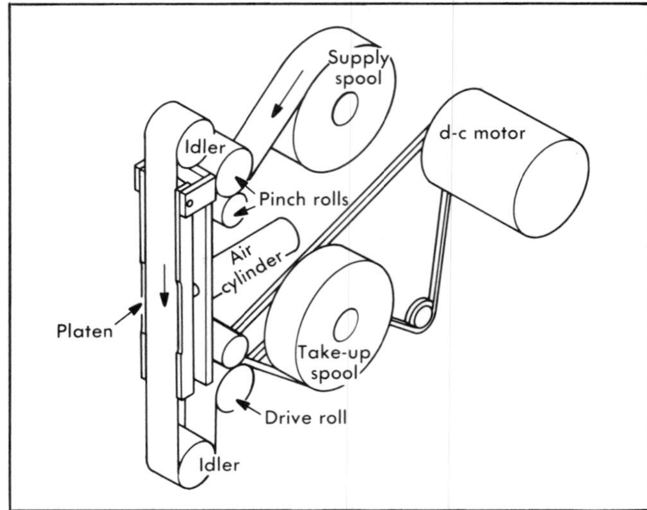

Fig. 16-98 Typical mechanical arrangement for using microabrasive tapes to produce fine finishes. (*3M Co.*)

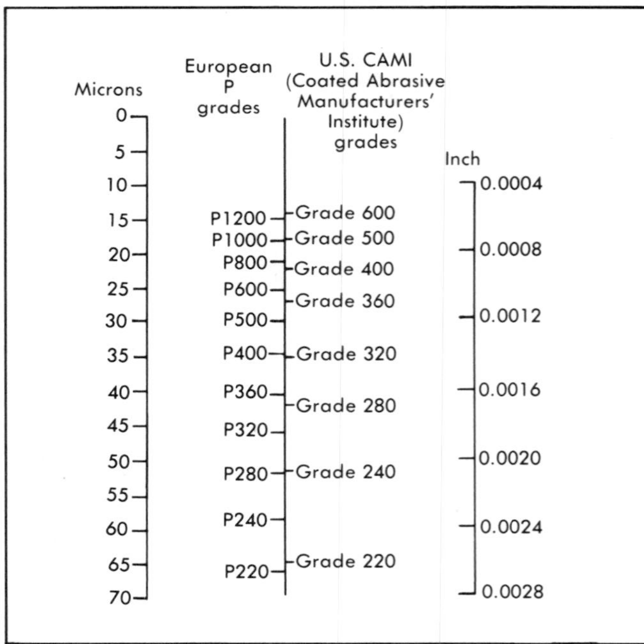

Fig. 16-99 Comparative sizes and grades of fine abrasive grains. (*Norton Co.*)

COATED-ABRASIVE GRINDING

Machines for Coated-Abrasive Finishing

Finishing operations with coated-abrasive products are performed on many different types of machines. These include back stand, swing frame, free-belt roll, vertical, centerless, and sheet and strip grinders, as well as portable tools, discussed in Volume I, *Machining*, of this Handbook series.

Many centerless grinders using bonded-abrasive wheels have been replaced by abrasive-belt machines for finishing bars, tubes, pipe, and other cylindrical parts. Conveyorized or pinch-roll-driven wide-belt machines are being used extensively to finish sheet and plate stock, with some having top and bottom belts to finish both sides in a single pass.

Three of the many types of multiple-head, conveyorized abrasive-belt machines for roughing and finishing in one pass are shown in Fig. 16-100. View *a* illustrates a continuous through-feed machine that is often equipped with a variable-speed drive unit. View *b* shows a two-head machine, one head equipped with an abrasive belt and the other head with a wheel or brush. The machine seen in view *c* has air or hydraulically controlled billy and pinch rolls for finishing coil or strip stock, as well as long workpieces.

Modern coated-abrasive machines are available with increased power, easy and fast belt-changing devices, and improved controls. Other features include motorized belt-head positioning, rapid head retraction, automatic belt tracking, workpiece gaging, and lubrication systems. Automatic feeders and loading/unloading devices, including industrial robots, are also in use.

Operating Parameters for Finishing

For maximum efficiency and economy in finishing with coated abrasives, careful consideration must be given to many variables. These include the workpiece material, hardness, and condition; the coated-abrasive product used (grit size and type, backing, bond, and coating method); speed of the coated-abrasive product; workpiece feed rate; backup support; belt tension; and lubricant used. An often overlooked condition for improved finishing with coated-abrasive belts is the direction of rotation of the belts in relation to workpiece feed. Rotation of the belt *with* the direction of workpiece feed (climb cutting) usually improves the finish produced. Rotation *against* the direction of workpiece feed (conventional cutting) will provide additional cutting action.

Belt speeds. Fast belt speeds generally produce smooth surface finishes, but slow speeds are often required for difficult-to-grind materials. Typical speeds for finishing various materials are presented in Table 16-38. Higher speeds are sometimes used, especially for high-pressure, power-assisted grinding, but they seldom exceed 10,000 sfm (3050 m/min).

Backup support of belt. While some light polishing operations are performed by pressing the workpieces against slack-of-belt locations, most operations require support in the area of workpiece contact. Such support is provided by contact wheels and rolls or platens.

Contact wheels and rolls. A variety of contact wheels and rolls, with respect to material, size, and shape, are available. For finishing operations, cloth wheels are common. Rubber wheels, both serrated and plain, are also used. Softer and larger diameter wheels generally produce smoother finishes.

Platens. Support with platens is provided for many types of flat surfacing operations. For precision work, the platens are often carbide faced and provided with a coolant system.

Belt tension. Tension maintained on the belt is generally lower, 4 to 10 lb/in. (0.07 to 0.18 kg/mm), for finishing operations than for stock removal requirements, which use tensions to 40 lb/in. (0.7 kg/mm) or more. For high-pressure abrasive-belt grinding operations, belt tensions approach 100 lb/in. (1.75 kg/mm).

Lubricants/coolants. Coated-abrasive belts are sometimes sprayed intermittently with a lubricant/coolant. Longer belt

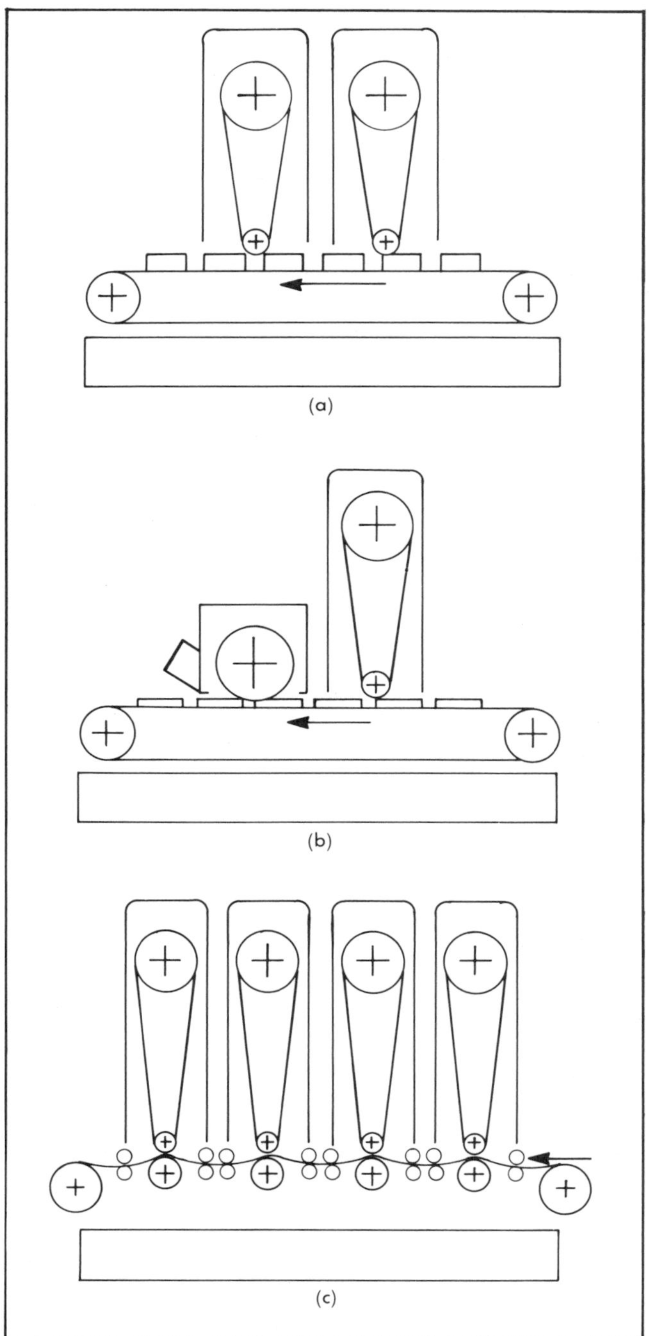

Fig. 16-100 Multiple-head, conveyorized abrasive-belt machines for roughing and finishing in one pass.

HONING

TABLE 16-38
Typical Speeds for Coated-Abrasive Belts in Finishing Various Metals*

Workpiece Material	Belt Speed, sfm (m/min)
Carbon steels	3500-6500 (1065-1980)
Stainless steels	3000-5500 (915-1675)
Nickel-chromium alloys	2500-3500 (760-1065)
Inconel	3000-5000 (915-1525)
Tool steels	4000-7000 (1220-2135)
Titanium	2000-3500 (610-1065)
Gray cast iron	4000-6000 (1220-1830)
Aluminum alloys	4500-7000 (1370-2135)
Brass/bronze alloys	5000-7000 (1525-2135)
Copper alloys	4500-5500 (1370-1675)

(3M Company)

* Contact belt manufacturer for recommended speed for specific operation.

life and better finishes can often be attained by wet machining, using a flood of filtered and recirculated lubricant/ coolant. A variety of lubricants and coolants are used, depending primarily upon the workpiece material and the operation being performed. Lubricants/coolants being used for various applications include water, water solutions, soluble oils, straight mineral oils, sulfurized and chlorinated oils, and grease or soap sticks. For some polishing operations, the lubricants/coolants are impregnated with abrasives.

HONING

Honing is a controlled, low-velocity abrasive process using bonded-abrasive stones, sometimes called sticks, to remove stock and improve surface finishes. The process is used for both heavy and light stock removal. Honing for fast and heavy stock removal, to 0.250″ (6.35 mm) or more on diameter, is discussed in Volume I, *Machining*, of this Handbook series. This section is confined to the use of honing for light stock removal and surface finishing operations.

In addition to removing stock, honing serves the important purpose of generating specified functional characteristics for surfaces and involves the correction of errors resulting from previous operations. Functional characteristics generated by honing include geometric accuracy (diametric roundness and straightness, and axial straightness), dimensional accuracy, and surface character (roughness, lay pattern, and integrity). Ten common bore errors caused by machining, heat treating, or workholding are illustrated in Fig. 16-101. Honing can correct all of these conditions with the least possible amount of material removal.

The most common application of honing is on internal cylindrical surfaces. For simplicity, such surfaces will be used throughout most of this section to explain the process. However, honing is also used to generate functional characteristics on external cylindrical surfaces, flat surfaces, truncated spherical surfaces, and toroidal surfaces (both internal and external). A

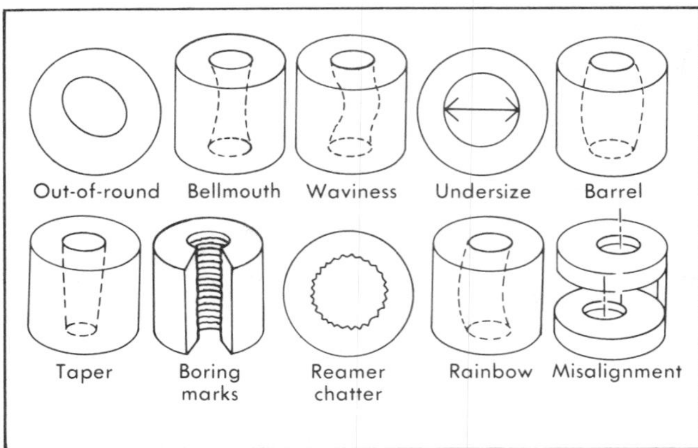

Fig. 16-101 Ten common bore errors that can be corrected by honing. (*Sunnen Products Co.*)

characteristic common to all these shapes is that they can be generated by a simple combination of motions.

Operating Principles

Control of the honing process is made possible by the manner in which the abrasive grains are used to shear off chips. The grains are bonded in the form of stones (sticks) by a vitreous (glass or clay), resinous, carbonaceous, or metallic material. The stones are presented to the work so their full cutting surface is in contact with the workpiece. In this way, thousands of small abrasive grains cut chips simultaneously, ensuring a substantial stock removal. Type of abrasive grain and bonding material used must be chosen to match the material being honed and the results desired. The most important consideration in choosing the stone specification, however, is that, under the specific operating conditions, the stone must be self-sharpening.

In honing cylindrical surfaces, speed of the abrasive is the result of two motions—rotation and reciprocation. These two motions combine to result in a crosshatch lay pattern with an included angle between 20 and 60° (see Fig. 16-102). The crosshatch angle (see Fig. 16-103) can be calculated from the following equation:

$$\tan \frac{A}{2} = 0.6366 \; \frac{L \times S}{D \times R} \qquad (3)$$

where:

A = average crosshatch angle, degrees
L = workpiece length, in. or mm
S = strokes per min of honing tool
D = workpiece diameter, in. or mm
R = revolutions per minute of honing tool

To keep honing stones cutting at all times and prevent glazing of their cutting surfaces, a steady, continuous, and consistent breakdown of the stone must be ensured. The aim is to keep sharp grits exposed to the work surface, thus ensuring minimum pressure and heat generation. The rotary speed, stone pressure and area, and bond hardness are factors in obtaining correct breakdown of the stones. In general, for all workpiece materials and hardnesses, honing speeds range from 150 to 250 sfm (45 to 75 m/min) when using aluminum oxide or silicon carbide abrasives, and from 200 to 300 sfm (60 to 90 m/min) when using cubic boron nitride (CBN) or diamond abrasives.

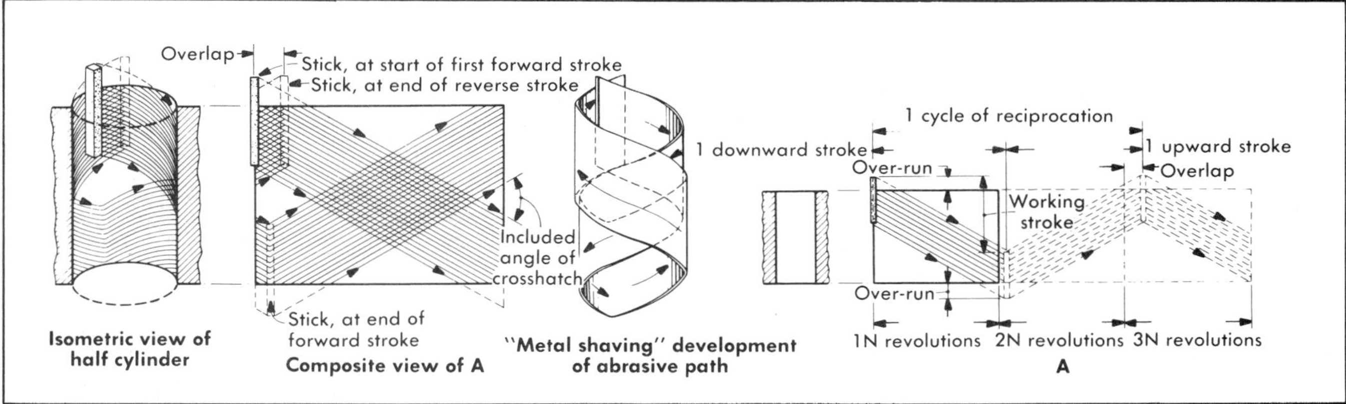

Fig. 16-102 Combined rotation and reciprocation result in a crosshatched surface finish, generated on a true cylindrical surface.

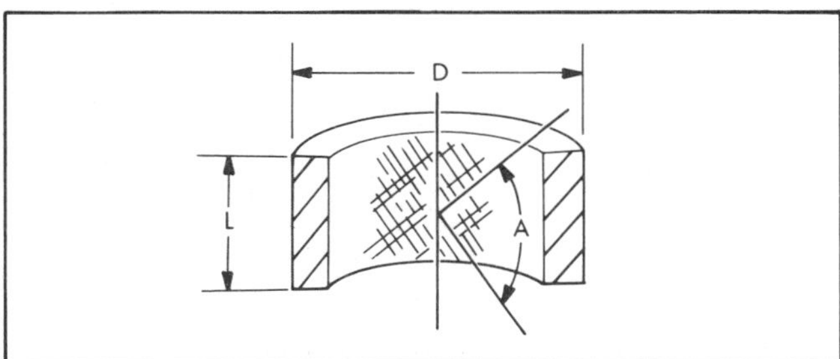

Fig. 16-103 Crosshatch angle (A) produced in honing. (*Sunnen Products Co.*)

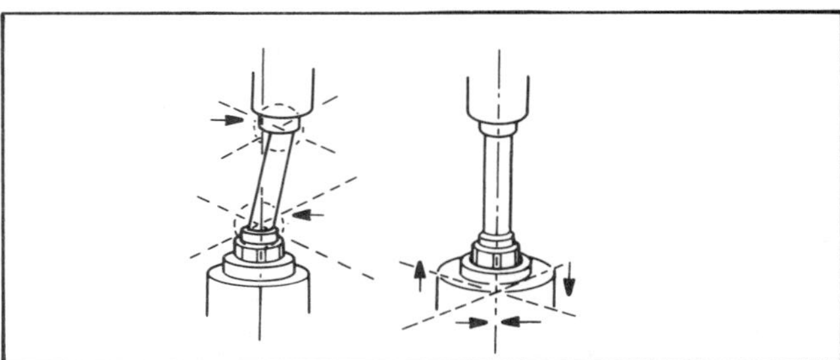

Fig. 16-104 Either the tool or the workholding fixture floats to permit the bore and the tool to align.

Reciprocation speed is less critical. However, an increase in the rate of reciprocation will sometimes improve the self-dressing characteristic of some abrasives and thus increase the stock removal rate; therefore, increased stock removal rates and fine finishes do not generally go together. If several thousandths of an inch of stock must be removed at a high production rate and less than a 20 μin. (0.51 μm) finish is required, two honing operations should be used.

A coarse abrasive grain (60 to 180 grain size) is usually used first to generate geometry and size. Then a second, finer abrasive (240 to 600 grain size) is used to refine the surface finish. The finish achieved with specific stones, however, depends upon the material being honed.

Developing Round, Straight Bores

A phenomenon of honing that makes it possible to develop a round, straight bore is the relationship of the cutting faces of the stones to the surface being honed, which is completely independent of the machine. The fact that either the tool or the workpiece floats (see Fig. 16-104) enables the tool to exert equal pressure on all sides of the bore, regardless of vibration in the machine or its environment.

Except in special cases, such as with the square axis requirement described next, honing will not change the axial location of a hole. The centerline of the tool follows the neutral axis of the bore as established by previous operations; the tool or workpiece must float so that the bore and the tool may align themselves. If practical, the workpiece should be mounted in a fixture that permits it to float. If the part is too large or unbalanced to float, however, the tool should be designed with universal ball joints (see Fig. 16-105). The fixture must be designed to locate the workpiece approximately in line with the machine spindle and take the torque and axial thrust of the tool without distorting the workpiece.

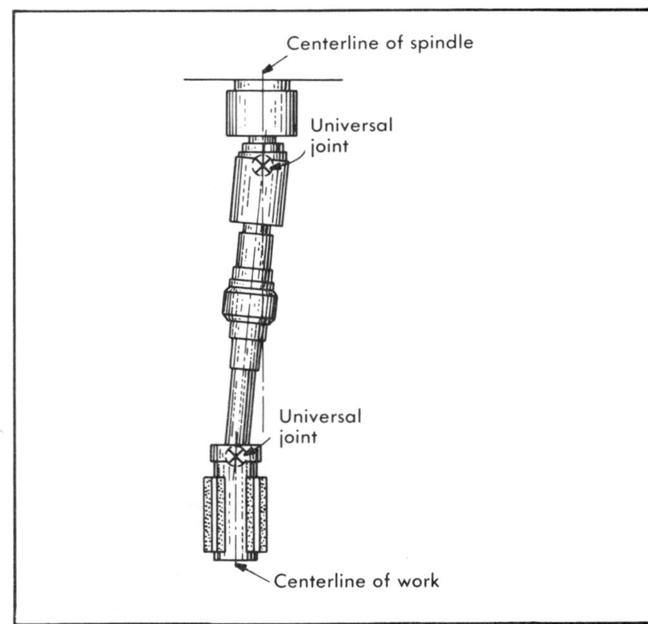

Fig. 16-105 Universal ball joint honing tool.

For bores that must be held square with an end surface, square axis honing has been developed. In this technique, both the tool and fixturing are rigid. The workpiece is slipped over the tool and automatically aligned with it before clamping.

As a tool is stroked through a bore, the pressure (and resulting penetration of the grit) will be greatest at the tight spots. The stones are long enough to bridge the low spots (see Fig. 16-106) and cut only on the crests of the waves until all snakiness or waviness is removed and the bore is made straight. After high spots have been leveled off, every section of the surface receives equal abrading action. The stones are stroked out of the ends of the bore approximately one fourth to one third of their length. This overstroke of the tool must be adjusted for each type of workpiece to provide maximum accuracy.

Because the full surface of the workpiece is abraded by thousands of grains on each reciprocation of the tool, the operation may be stopped at any time, and no step will be found in the bore. An operator can hold size by expanding the tool a known amount, which should be the sum of the stock removal and the stone wear.

Surface Finish and Integrity

Surface finishes generated by honing are a crosshatched lay pattern made up of a multitude of small, diamond-shaped plateaus, each surrounded by the cut pattern of the honing stone. The plateaus carry the load, and the valleys between act as oil reservoirs.

Because the cutting points of the grains are so small and there are so many grains cutting simultaneously, heat and stress generated in the workpiece are never concentrated. As a result, a minimum of damage is done to the surface, and the integrity of honed surfaces is especially good. With respect to surface character, the same surface roughness and overall quality are developed in every bore, a collateral benefit of honing, because the cutting faces of the stones always have the same degree of sharpness due to the self-dressing characteristic of the abrasive.

Materials and Parts Honed

Any metallic material can be honed including cast iron, all types of steel and carbides, brass, bronze, aluminum, chromium, and silver, as well as many nonmetallic materials such as glass, ceramic, and plastics. Hardness of the material does not limit the honing process, it only affects the rate at which the stock may be removed. Stock removals on diameters for holes of various diameters and lengths are presented in Table 16-39.

With fixtured honing equipment, the smallest part believed to have been honed to date is 0.060" (1.52 mm) diam. The maximum bore size honed to date is 60" (1524 mm) diam and 52 ft (15.8 m) long. The only honing limitation for part size is the machine on which the tool is mounted. On manual or power-stroked honing equipment (see Figs. 16-107 and 16-108), the diameter range is about 0.060 to 6.000" (1.52 to 152.4 mm).

Any amount of material can be removed by honing. However, if roughing operations used to locate the bore prior to honing can remove stock more economically, then a minimum of material should be left for honing. For parts such as long steel tubing, 4 to 30 ft (1.2 to 9.1 m) in length, honing is often the only practical and economical method of machining the bore.

Honing Machines

Honing machines may have spindles mounted vertically, horizontally, or at an angle. They may also have one or more spindles mounted on the same column. Figure 16-109 shows a typical vertical spindle machine of the type used for honing

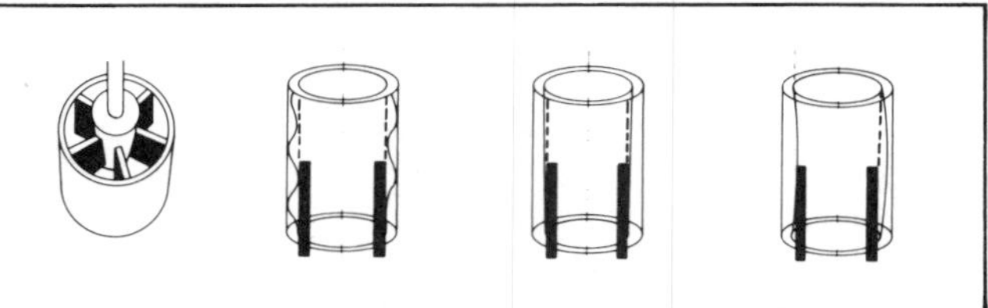

Fig. 16-106 Long, rigidly supported stones generate a true cylinder by abrading off all inaccuracies and then cutting equally on all sections of the bore.

short cylindrical surfaces. Figure 16-110 illustrates a heavy-duty, single-spindle vertical machine for honing long, large-diameter bores.

Adaptive control is available for use on high-production honing machines. Its purpose is to maintain the stone feed pressure at a rate that will ensure a controlled surface roughness, regardless of variations in the work material or in the environment. The control automatically senses how the stones are cutting, compares this performance with a previously established norm, and adjusts the pressure on the stones to keep them cutting at that norm.

Stone Holders

Honing stones are mounted in holders that adapt them to the tool. In small-diameter tools, these holders may be made of plastics, molded integrally with the stones, or they may be steel with the abrasive sections bonded to them. An angle that matches the cone or wedge angle is mounted into the holder. For larger diameter tools, the stones are cemented or clamped into steel holders, and the expanding angle plate may be a separate member of the tool.

Abrasives Used

Abrasive grains used in honing stones are silicon carbide, aluminum oxide, cubic boron nitride, or diamonds. Silicon carbide is generally used to hone cast iron and nonferrous materials. Aluminum oxide is widely used for honing steel. Cubic boron nitride (CBN) is used to hone all steels (both soft and hard), nickel and cobalt-based superalloys, stainless steel, and a variety of other metals, including beryllium copper and zirconium. Diamonds are used on chromium plate, carbides, ceramics, glass, cast irons, brass, bronze, and surfaces nitrided to depths greater than 0.001" (0.03 mm). Diamond hones are also being used for blind holes (where little overstroke is available), intermittent and interrupted bores, transfer-line honing, and the automatic honing of automotive parts in mass production.

Size of the grain used depends on the surface finish requirements (see Table 16-40). Grains coarser than 60 or finer than 800 are seldom used. An exception is the honing of ball tracks for antifriction bearings and other similar parts. These parts require a refinement of both geometry and surface finish, and levigated alumina has proven satisfactory for this type of application. Grain sizes and hardness grade numbers for manual or power-stroked honing are given in Tables 16-41 and 16-42.

Honing Tools

Honing tools (see Fig. 16-111) are as varied in configuration as the surfaces to be honed. The basic fixtured tool (see Fig. 16-112) used to hone internal cylindrical surfaces is made up of four functional elements: the body, cone or wedge, cone or

TABLE 16-39
Stock Removal in Honing Holes of Various Sizes

Diameter and Length of Hole	Brass and Soft Bronze	Cast Iron*	Aluminum	Soft Steel*	M42 Tool Steel,* R_C63	Carbide,* R_A44
	Stock Removal on Diameter, ipm (mm/min)					
1/16 x 5/16 (1.6 x 7.9)	0.029 (0.74)	0.022 (0.59)	0.015 (0.38)	0.006 (0.15)	0.005 (0.13)	0.005 (0.13)
1/8 x 7/16 (3.2 x 11.1)	0.051 (1.30)	0.030 (0.76)	0.032 (0.81)	0.016 (0.41)	0.014 (0.36)	0.013 (0.33)
1/4 x 7/8 (6.4 x 22.2)	0.081 (2.06)	0.060 (1.52)	0.056 (1.42)	0.038 (0.97)	0.018 (0.46)	0.022 (0.56)
3/8 x 1 1/4 (9.5 x 31.8)	0.087 (2.21)	0.074 (1.88)	0.064 (1.63)	0.044 (1.12)	0.016 (0.41)	0.017 (0.43)
1/2 x 1 5/8 (12.7 x 41.3)	0.072 (1.83)	0.048 (1.22)	0.048 (1.22)	0.041 (1.04)	0.015 (0.38)	0.014 (0.36)
5/8 x 1 13/16 (15.9 x 46)	0.053 (1.35)	0.037 (0.94)	0.030 (0.76)	0.038 (0.97)	0.014 (0.36)	0.012 (0.30)
1 1/4 x 2 3/8 (31.8 x 60.3)	0.041 (1.04)	0.017 (0.43)	0.019 (0.48)	0.015 (0.38)	0.009 (0.23)	0.007 (0.18)

(*Sunnen Products Co.*)

* Stock removal rates for these materials are normally obtainable only with metal-bonded CBN or diamond honing stones.

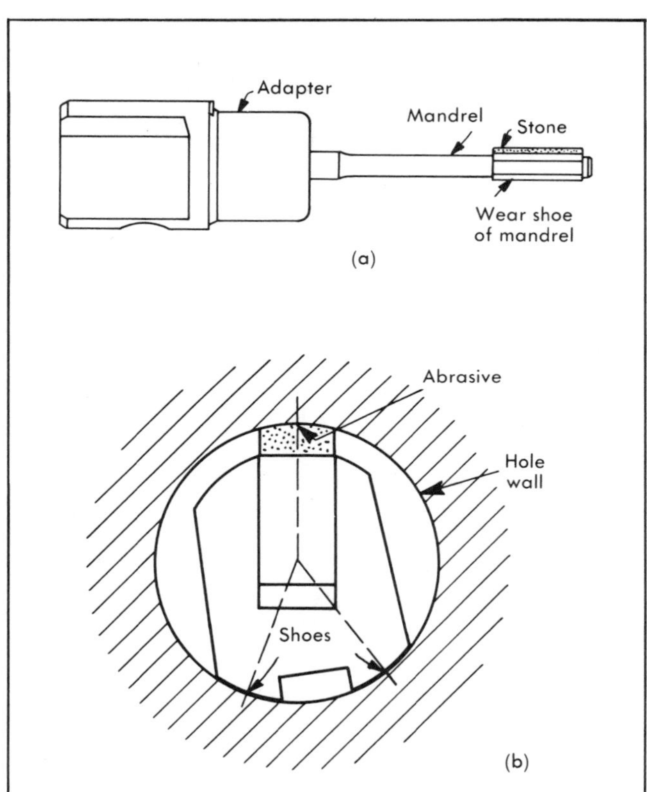

Fig. 16-107 (a) Unit for manual or power-stroked honing; (b) abrasive contact of the tool with the workpiece. (*Sunnen Products Co.*)

wedge rod, and driveshaft. The body supports the honing stones, to keep them in proper relation to each other and the work surface. The cone is a cone-shaped wedge that forces the stones out radially as it is moved axially in the body. The cone rod attaches the cone to the tool-expansion mechanism. The driveshaft attaches the tool body to the machine spindle. Multiple tools are available for honing a row of coaxial bores in one operation.

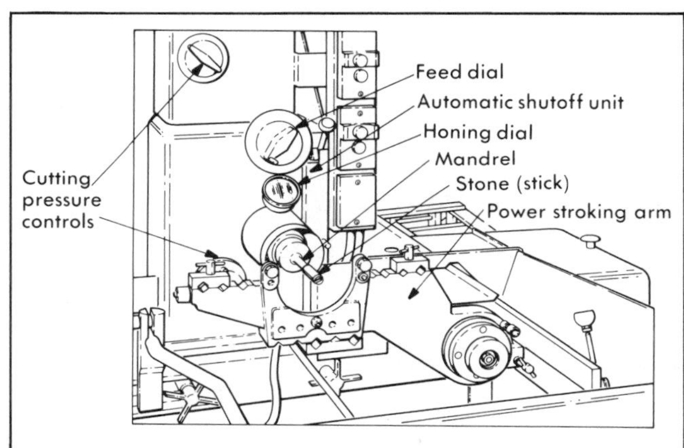

Fig. 16-108 Power-stroked honing machine. (*Sunnen Products Co.*)

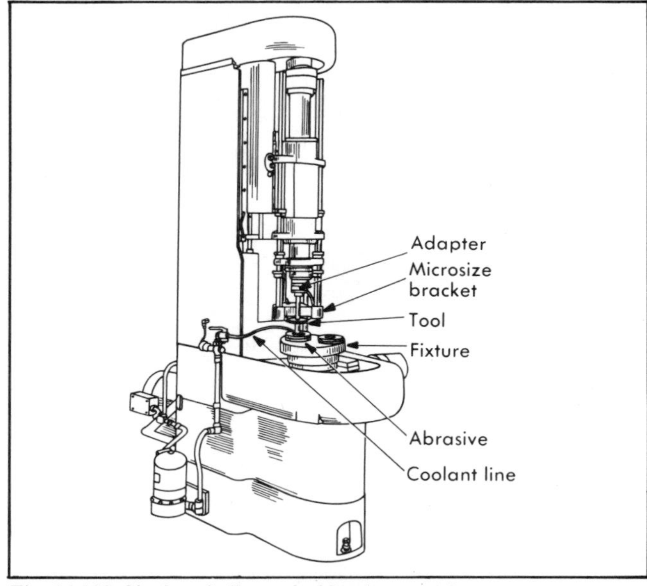

Fig. 16-109 Single-spindle vertical honing machine.

HONING

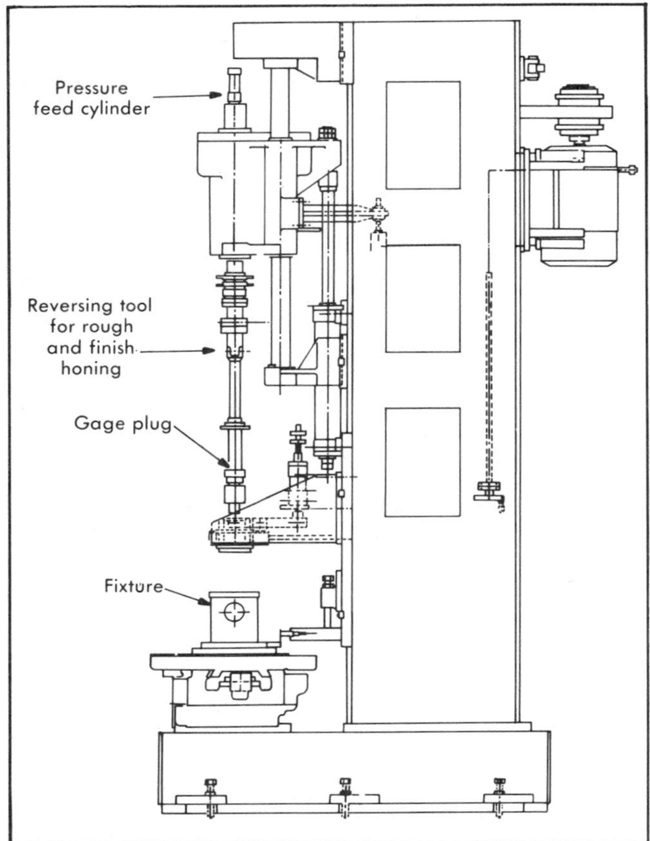

Fig. 16-110 Heavy-duty, single-spindle, vertical hydraulic machine for honing long, large-diameter bores.

Tool-expansion mechanisms are usually a part of the machine spindle. They may be actuated manually, mechanically, hydraulically, or pneumatically. Their purpose is to move the cone and force the stones outward radially with a pressure that will keep them cutting at the required rate.

Flex-Hone tools, made by Brush Research Manufacturing Co., Inc., are used on conventional internal honing machines, as well as with portable tools, to finish bores, but they will not affect geometry. They are resilient, flexible tools, resembling brushes, but having globules of abrasive at the outer ends of their filaments. They are available in diameters from 0.276 to 36″ (7 to 914 mm), with silicon carbide, aluminum oxide, boron carbide, or tungsten carbide, and in grit sizes from 40 to 800.

Cutting Fluids

Cutting fluids used in honing have a definite effect on the way the abrasive will act. They perform three important functions. One is flushing the abrasive stone surfaces. If chips or sludge collect on the faces of the stones, they will load and not cut, or act harder than they would under normal conditions. The second function is to act as a coolant to carry away heat caused by shear and friction. The third function is lubrication. Usually, a sulfurized, mineral-based or lard oil mixed with kerosene or other fluid of high flash point is used for honing. Water-soluble cutting fluids are successful with metal-bonded diamond hones on cast iron and ceramics, but are less successful with CBN hones.

Automatic In-Process Gaging

In modern high-production equipment, accurate diametric size is duplicated from piece to piece with automatic in-process gaging equipment. The diameter is automatically gaged continually throughout the honing cycle; and when the desired diameter is obtained, a "to-size" signal is initiated. Expansion of

TABLE 16-40
Approximate Surface Finishes Obtainable on Power-Stroked or Manual Honing Equipment

Workpiece Material	Type of Abrasive*	Grit Number (Size)								
		2 (80)	3 (100)	4 (150)	5 (220)	6 (280)	7 (320)	8 (400)	9 (500)	0 (600)
		Approximate Surface Finish, R_a (arithmetic average)**								
Hard steel	A or S	25	---	20	18	12	10	5	3	1
	CBN	---	55	45	30	28	---	20	---	---
Soft steel	A or S	85	---	35	25	20	16	7	4	2
	CBN	---	125	---	---	---	---	25	---	---
Cast iron	S	100	---	32	20	12	10	6	5	3
Aluminum, brass, and bronze	S	170	---	80	55	33	27	15	12	2
Carbide	D	---	---	30	20	---	---	7	---	3
Ceramic	D	---	---	50	40	---	---	20	---	15
Glass	D	---	---	95	70	---	---	30	---	15

(*Sunnen Products Co.*)

* A = aluminum oxide; S = silicon carbide; CBN = cubic boron nitride; and D = diamond.
** To convert to metric R_a in μm, divide by 40.

the tool is then stopped. A short dwell period follows, during which the pressure on the stones is gradually reduced, and then the tool is collapsed and withdrawn from the bore. Several types of sizing devices are used with honing.

Plug sizing. One method of automatic size control is by means of an adjustable, plug-type sensing tip (see Fig. 16-113). The sensing tip is set to the finished bore size and positioned to try to enter the bore on each stroke during the honing cycle. As long as the bore is undersized, the sensing tip is too large to enter the bore and will be depressed on each stroke, and the machine will hone for additional strokes. As soon as enough material is honed from the bore, the sensing tip enters. When this occurs, the honing cycle stops automatically with the bore to size, and the machine is ready for the next cycle.

Two-point sizing. Another plug method of automatic size control uses two sizing points (see Fig. 16-114). In this design, the gage plug is part of the honing tool, riding freely around the tool driveshaft and trying to enter the workpiece bore on each stroke. As the bore increases in size, the gage plug enters the bore and makes contact with the first size point. At this time, expansion of the abrasive stones is halted, thus decreasing the stock removal rate and resulting in sparkout. When the gage plug contacts the second size point, the stones collapse and the tool is withdrawn.

Air-to-electronic gage sizing. This method is similar in principle to plug sizing (see Fig. 16-115). An air gage enters the workpiece bore on each stroke of the honing tool. As the bore diameter increases, pressure in the air circuit decreases. When this back pressure drops, the air signal is turned into an electronic signal that is fed to a programmable controller. When the bore diameter reaches the programmed size, the honing cycle is automatically terminated.

TABLE 16-41
Grain Sizes for Manual Power-Stroked Honing

Mandrel Diameter, in. (mm)	Grit Sizes			
	Aluminum Oxide	Silicon Carbide	Cubic Boron Nitride	Diamond
0.060-0.100 (1.52-2.54)	280	280, 500, 600	280, 400	280, 600
0.100-0.185 (2.54-4.70)	280, 320	280, 400, 500, 600	280, 400	150, 220, 280, 400, 600, 1200
0.185-0.308 (4.70-7.82)	150, 220, 280, 320	150, 220, 280, 400, 500, 600	100, 150, 220, 280, 400	150, 220, 280, 400, 600, 1200
0.308-0.990 (7.82-25.15)	150, 220, 280, 320	150, 220, 280, 400, 500, 600	70, 100, 150, 220, 400	70, 150, 220, 400, 600
0.990-3.625 (25.15-92.08)	150, 220, 280, 320	80, 150, 220, 280, 400, 500, 600	70, 100, 150, 220, 400	70, 150, 220, 400, 600

(*Sunnen Products Co.*)

TABLE 16-42
Hardness Grade Numbers for Stones Used in Manual or Power-Stroked Honing

Stone Hardness No.*	Description	Recommended Uses
1	Very soft	For very hard, smooth holes, such as chrome plate. Not normally recommended.
3	Soft	For hardened metals and smooth, long holes.
5	Standard	For normal honing.
7	Hard	For rough holes and short bushings.
9	Very hard	For deburring operations. Not recommended for normal honing.

* Sunnen Products Company designation.

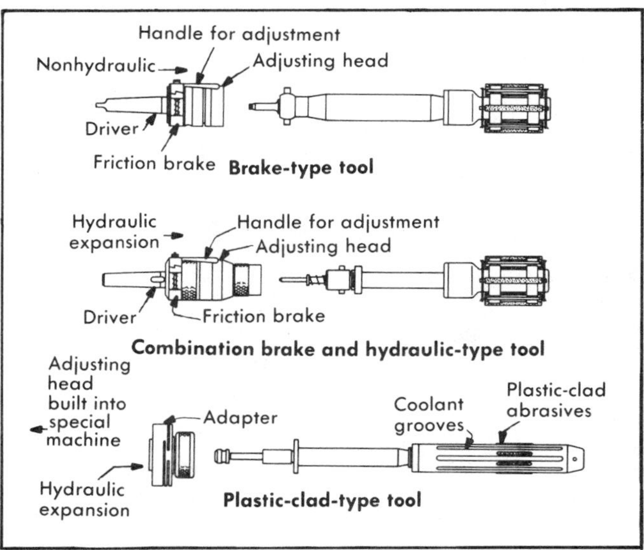

Fig. 16-111 Three of the many types of fixtured honing tools.

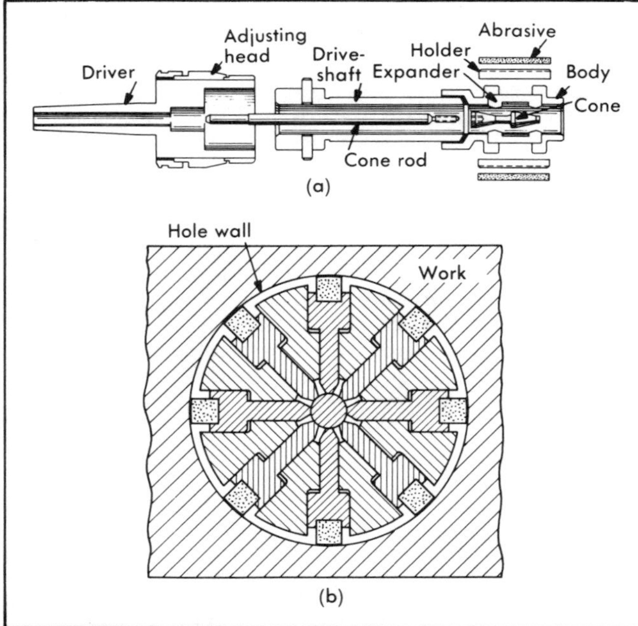

Fig. 16-112 (a) Fixtured honing tool; (b) abrasive contact of the tool with the workpiece.

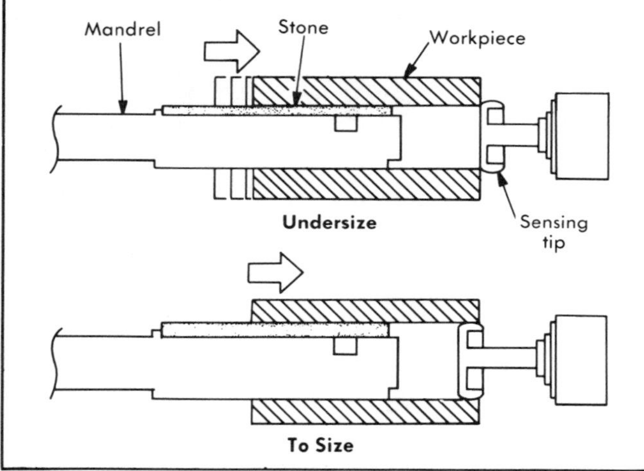

Fig. 16-113 Plug-type sensing tip controls stroking circuit and honing cycle. (*Sunnen Products Co.*)

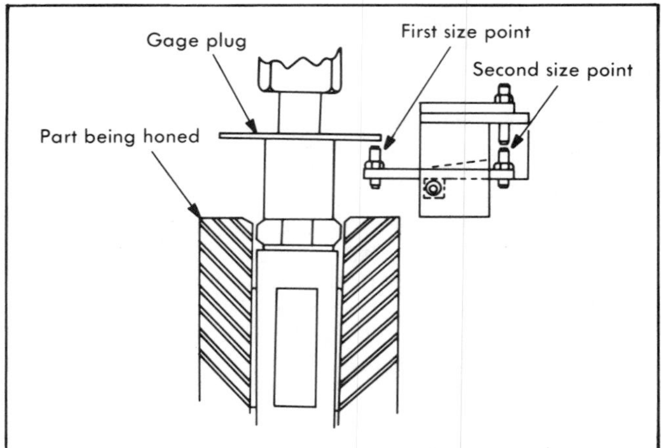

Fig. 16-114 Two-point sizing method of controlling honing cycle. (*Ex-Cell-O Corp., Micromatic Operations*)

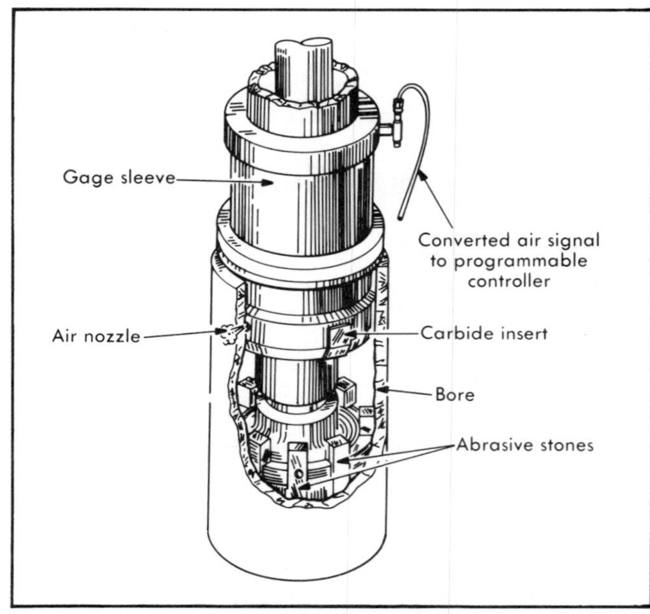

Fig. 16-115 Air-to-electronic gage sizing for automatic size control in honing. (*Micromatic Operations, Ex-Cell-O Corp.*)

Related Processes

There are several processes that are similar to yet different than conventional honing.

Plateau honing. This honing process is a method of producing a special plateau finish, having the surface peaks removed, but retaining the deep valleys. Such a finish has been found desirable in engine performance because the valleys act as oil reservoirs for improved lubrication, especially during engine break-in.

A plateau finish is produced by first rough honing to final size. Then the surface is finished with a finer grit stone for about 45 seconds, depending upon the amount of plateauing desired. The plateauing operation, with a 600 grit stone, removes so little stock that the bore diameter is not measurably increased.

Single-stroke process. This process is a fast and accurate method of sizing certain bores to final size. The tool used is an expandable, diamond-plated or CBN sleeve mounted on a tapered arbor. The sleeve is expanded only during setup, and no adjustments are necessary during honing. The rotating tool is pushed through the bore only one time, with the workpiece being removed after the return stroke.

This process is limited with respect to the types and volumes of material that can be removed. The size and volume of the chips produced must be no more than the spaces between the diamond grits on the sleeve; otherwise, the tool will seize in the bore of the workpiece. The process is best suited for interrupted or short bores that produce a low volume of chips, and is most successful for honing cast iron.

Hone-Forming. This procedure is a combination honing and electroplating process developed by Micromatic Operations, Ex-Cell-O Corporation. Details of this process are presented in Volume I, *Machining*, of this Handbook series.

Electrochemical honing. In this process, metal is removed by introducing an electrolyte into a gap between a cathodic honing tool body and an anodic workpiece. The process, discussed in Volume I, was also developed by Micromatic Operations, Ex-Cell-O Corporation, but equipment for the process is no longer offered.

External honing. This process is used to polish and improve the surface finish on the outside diameters of cylindrical workpieces. Instead of using reciprocating tools, as in conventional inside-diameter honing, cup or flat abrasive wheels or stock hones arranged in opposed vees are used. While surfaces can be finished faster with cylindrical or centerless grinding, honing is often preferred because of the desirable surface character (smoothness and lay pattern) obtained and the elimination of damage to surfaces through the use of lower speeds and gentler action.

External honing can be done with special tools on conventional internal honing machines or with machines designed specifically for the process. When cup wheels are used, the wheels and workpieces are rotated in opposite directions, and their axes are offset. In another method, workpieces roll between revolving flat wheels, which rotate in a direction opposite to that of the workpieces. The workpieces are nested in slotted workholders, with the individual parts free to revolve in their slots. Machines used are similar to those employed for flat honing, discussed next in this section.

Flat honing. This process produces burr-free, smooth, flat surfaces with multidirectional patterns. Workpieces are allowed to float while being guided over or between honing wheel faces. On two-wheel machines (see Fig. 16-116), both sides of the workpieces can be finished simultaneously, while maintaining precise tolerances with respect to flatness, parallelism, and thickness. Machines used differ from disc grinders (face grinding machines) in that the wheels move at a slower speed and the abrading action is more gentle, resulting in little if any surface damage.

Flat honing machines are available in a variety of sizes and types. On two-wheel machines, the top wheel, lower wheel, and workholder each have separate drives and controls. Machines are available with automatic controls to gradually increase pressure on the top wheel during the honing cycle. Automatic size control is also available. The workpiece carrier is part of an epicyclic sprocket holder.

Edge honing, rounding, and deburring. Machines are available to hone the edges of carbide and ceramic cutting tool inserts without chipping them. Inserts can be rounded on one side or, with a two-wheel machine, on both sides. The rounded edges produced are multiple-radius, compound curves that produce good results and long life in machining operations.

The machines used are similar to those for external and flat honing. On two-wheel machines, three independent drives are provided to operate the lower wheel, upper wheel, and workholder. Each insert is held and guided in a separate aperture of the workpiece carrier. Wheels used are generally fine-grit silicon carbide in a rubberlike, resilient bond. Pressure is applied through an adjustable pneumatic device.

Gear-tooth honing. This process is performed to correct the geometry and improve the surface finish of hardened spur or helical gears. Honing can correct errors in profile, lead, and eccentricity, as well as oversize conditions resulting from heat treating, but cannot correct tooth spacing errors. The process also removes nicks and burrs, and generally reduces the noise generated in the operation of the gears.

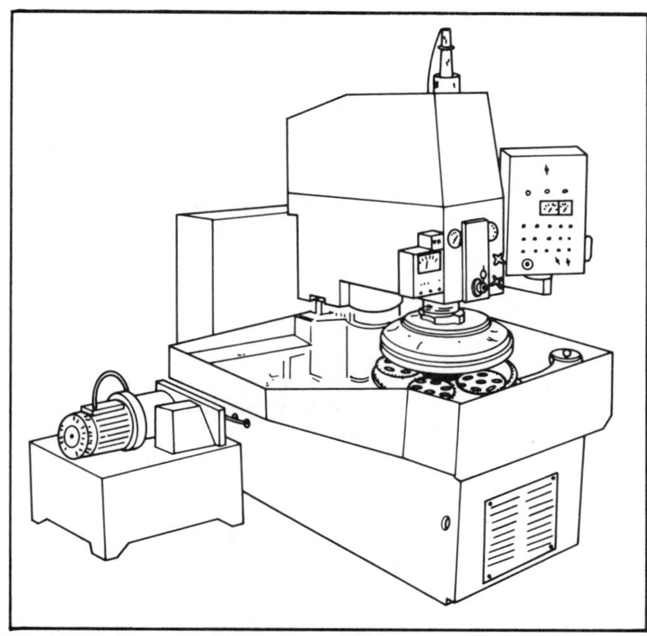

Fig. 16-116 Two-wheel flat honing machine. (*Peter Wolters of America*)

Gear-tooth honing consists of rotating the hardened gear in mesh with a gear-shaped honing tool. The tool can be abrasive-impregnated plastic or diamond, or carbide-coated steel. Gear-tooth honing can be performed on standard gear cutting machines or special equipment. Additional information on this process is presented in Volume I, *Machining*, of this Handbook series.

SUPERFINISHING

Superfinishing, also referred to as microfinishing, microstoning, superfinish honing, short-stroke honing, and microhoning, is a low-temperature abrading process, closely related to honing, for producing smooth, long-wearing surfaces. The process uses bonded-abrasive, stick-type stones for cylindrical work or cup wheels for flat and spherical work, with finer grit sizes (320 to 6000) than honing generally being used. When using cup wheels, the process is practically the same as flat honing, discussed in the preceding section of this chapter.

Superfinishing produces controlled surface conditions involving size, finish, geometry, and metallurgical structure. The resulting optically smooth surfaces are metallurgically free of any fragmented, amorphous, or smeared metal from previous operations. The process also corrects inequalities in workpiece geometries, such as grinding flats, and restores surface integrity by eliminating surface stresses and burns. Surface finishes of 0.5 μin. (0.013 μm) or less are being produced, with roundness held to 0.000040″ (0.00102 mm) and sphericity to 0.000010″ (0.00025 mm).

Superfinishing is efficient in the surface refinement of cylindrical, flat, spherical, and cone-shaped parts. Although it is not primarily a dimension-changing process, stock averaging 0.0002 to 0.001″ (0.005 to 0.03 mm) on diameter is removed.

SUPERFINISHING

With multistage operation and synthetic diamond or cubic boron nitride (CBN) abrasives, stock removal to 0.004″ (0.10 mm) is being achieved.

Almost any reasonable depth of scratch pattern can be produced, from practically none at all (mirror finish) to 30 μin. (0.76 μm) or more. The smoother finishes do not have sufficient scratches to exhibit any directional effect; they react the same to every direction of movement. When an intentional scratch pattern is produced, it will be a crosshatch at an angle or almost perpendicular to any normal direction of motion. Such patterns are desirable from the standpoint of reduction in friction or galling when surfaces rub together, as in bearing applications, or for better retention of lubricant.

Both general-purpose and high-production superfinishing machines are available for application of the process to almost any reasonably symmetrical surface. Multistep machines are also available for successive roughing and finishing. The high degree of surface improvement by superfinishing is often achieved at a lower overall cost than with other methods because only a short time is required to obtain the finish, there is a smaller percentage of rejection for reworking or complete loss, and the process usually offers a more reliable operation.

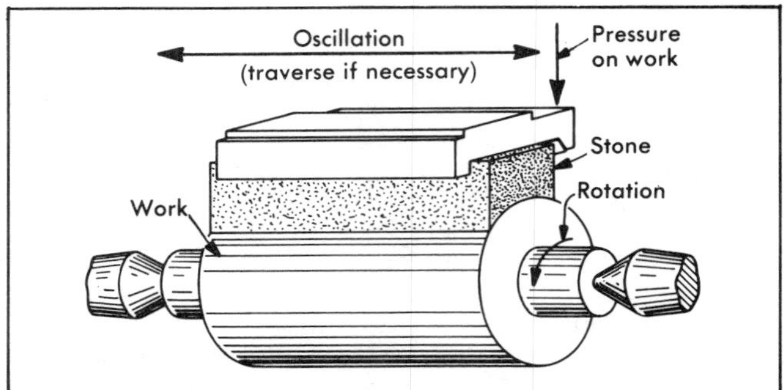

Fig. 16-117 Motions in superfinishing.

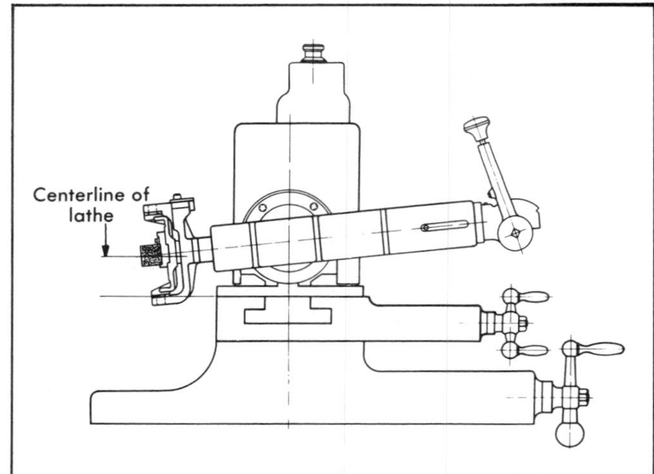

Fig. 16-118 Attachment for superfinishing on a lathe or cylindrical grinder.

Cylindrical Superfinishing

In grinding, only a line contact is in effect; while in superfinishing, a large area of abrasive contact is employed (see Fig. 16-117). For example, on small cylindrical work, the usual superfinishing method uses a stone that is 25 to 60% of the part diameter in width, and often the same length as that of the surface to be refined. With such a large area of contact, grinding imperfections, such as chatter and feed lines, are completely removed, together with any smearing or burning. In cylindrical superfinishing, the essential mechanism (see Fig. 16-118) is an arm containing a pressure-loaded quill on the end of which the abrasive stone is located. This assembly is mounted on a base having an oscillating mechanism by which the arm and the stone are oscillated a distance, usually about 3/16″ (5 mm), at a rate of up to 3000 complete strokes per minute. The arm and base can be mounted on machines such as lathes or cylindrical grinders for rotating the work, or they can be incorporated in a machine specially designed for cylindrical superfinishing. If the surface to be refined is not too long, 1-3″ (25-76 mm), depending upon the diameter, traverse will not be necessary since the stone will be the same length. If the surface is longer, traverse should then be incorporated, and the stone should be about 1 1/2 times as long as its width.

The stone is formed originally by the supplier or by one of two methods. With the quicker, but less accurate method, the part is placed between centers, and a strip of emery cloth is placed over the work surface (with the emery side toward the stone). The stone is brought into contact with the strip, which is then drawn back and forth until the approximate radius of the work surface has been formed on the stone. In use, the stone will soon wear to an exact shape. With the other method, a boring bar containing an adjustable diamond or a diamond-coated workpiece is placed between centers, and the stone is formed to the exact radius in the same location as it is to do the work.

After its first forming, a superfinishing stone needs no further dressing since the many tiny defects on the workpiece surface will themselves dress the stone to the required work-surface contour. The stone wears too slowly for any of the work-surface defects to have any appreciable effect on its accuracy of shape. Instead, the stone acts as a master in reproducing its true geometry on the work surface.

After the roughness of the unfinished workpiece surface is cut away, the stone face is no longer dressed. Instead, the stone dulls and glazes in production until a condition is reached where the viscosity of the lubricant will almost completely separate stone and work, and then the cutting ceases. Application of the stone to another unfinished surface will again form it, and the described cycle will be repeated.

After the stone is dressed, the part should be rotated at a surface speed of from 50 to 150 sfm (15 to 45 m/min). The stone should then be brought into contact with the work under a pressure of from 10 to 170 psi (70 to 1170 kPa) of stone area, with the stone oscillating and the work area flooded with lubricant. When traverse of the stone is necessary, faster feed rates, up to 1 ipr (25 mm/rev), should be used for the first part of the superfinishing cycle, and slower feed rates for the last part of the cycle.

The oscillating movement, which is parallel to the axis of the part, is necessary for several reasons. Through the continual change in direction of cut, the size of the chip is reduced, which prevents appreciable loading of the stone. If these chips were allowed to form, the resultant stone loading would cause deep scratches. By oscillating the stone, the direction of cut is changed continuously, and chips of consequential size seldom accumulate. Also, by reason of the oscillation, metal is removed by a shearing cut. The ridges left from previous operations (grinding or turning) are crossed, not followed, providing efficient stock removal and producing a greater smoothness without burnishing the surface. Due to the low speeds and pressures, superfinishing produces less heat and no distortion of the metallic structure.

Centerless Short-Stroke Honing

With centerless through-feed machines (see Fig. 16-119), one or more fine-grit stones contact about 60 to 70° of the total workpiece circumference. Stone pressure varies from 15 to 175 psi (103 to 1207 kPa), workpiece rotation from 500 to 6000 rpm, feed rates to 280 ipm (7 m/min), oscillation frequency from 500 to 3000 double strokes per minute, and stroke length from 0.039 to 0.315" (0.99 to 8.00 mm). These machines can be computer controlled and equipped with automatic preprocess and postprocess gaging.

Finishing Bearing Races

One common method of finishing bearing raceways uses the centerless shoe principle. Two carbide shoes are applied on the ID of an inner race (see Fig. 16-120, view *a*) or on the OD of an outer race (view *b*). The shoes have a highly polished surface to avoid cold working the races. The races may be driven by friction clamp or magnetic drivers. The centerline of the drive spindle is offset slightly from the centerline of the bearing race, preventing any eccentricity or runout of the driven spindle from being transmitted to the bearing race. This design also forces the race downward for more positive seating against the shoes.

In another method of superfinishing the ball and roller paths of bearing races, the races are held in hydrocentric chucks, where they rotate on a film of pressurized oil (see Fig. 16-121). For ball bearing races, the stone is pivoted; for roller bearing races, the stone is reciprocated. These operations are generally performed in two steps, first with a roughing stone and then with a finer polishing stone. During the last few revolutions of the races in the finishing step, stone motion is slowed. This slowing of the stone motion eliminates crosshatched surfaces, leaving only annular lines invisible to the naked eye and resulting in quiet bearing operation.

Flat and Spherical Superfinishing

Both flat and spherical superfinishing, similar to flat honing (discussed previously in this chapter), are performed with a machine (see Fig. 16-122) containing an upper and a lower vertical, rotating spindle. The upper spindle has a spring or hydraulically loaded quill on which the cup-shaped stone is located, while the lower spindle carries the work. Positions of work and stone can be reversed. As the work revolves, the end

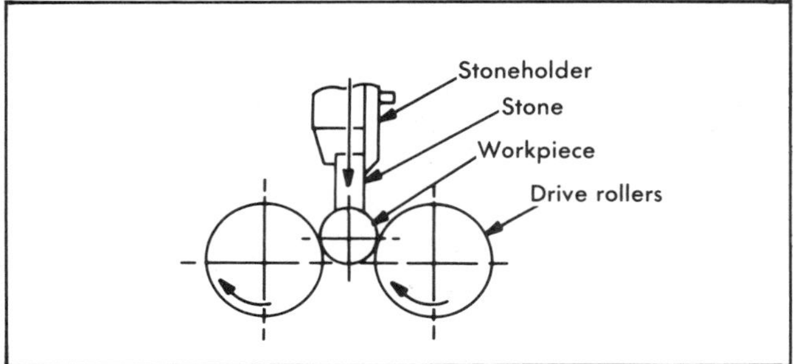

Fig. 16-119 Centerless short-stroke honing.

face of the cup is brought into contact with the work. Some degree of offset is given the cup with relation to the work position so that the path of any grit is not repeated.

When the two spindles are exactly parallel, the combined effect of both stone and work revolution results in the generation of flatness. If the upper spindle is adjusted to some angularity with the lower, a spherical shape is generated. As in cylindrical work, the stone automatically wears to the most desirable shape.

Surface Measurement of Superfinish

The usual, accepted methods of surface-quality comparison do not show the existence or degree of waviness. The superfinishing process, however, provides a superior method of investigation. A very short application of a worn-in stone will abrade only the high spots and reveal the pattern of waviness perfectly and instantly. The almost invariable defects of grinding are thus readily exposed.

Because superfinish reveals surface irregularities, superfinished surfaces can be measured with more consistent accuracy than ground surfaces. The dimension measured on ground parts is that of the highest ridges within the anvils of the measuring device. Unless extreme care is exercised to measure in exactly the same location, no two measurements are likely to be exactly the same since another pair of defects will then determine the indicated dimension. Further, since the major interest lies in what that dimension will be after a breaking-in period, it is well to recognize that such high points will be the first areas to be worn away. In that case, the exact running clearance will be impossible to predict.

Superfinish Speeds and Production Rates

The usual superfinishing speed is 50 to 60 sfm (15 to 18 m/min) under a flexible pressure of but 10 to 40 psi (70 to 275 kPa) of stone contact area and under a flood of lubricant. Under these conditions, there is no appreciable production of heat to alter the metallurgy or of violence to disturb or fragment the crystalline structure of the metal. Time consumed in a superfinishing operation depends somewhat upon the quality of the preceding operation. If the surface is rough and contains more than the average waviness, more metal must be removed with a greater length of time required (see Fig. 16-123). Nevertheless, the time required for a given dimension of a part is surprisingly consistent.

When a smoother surface is required for some special reason, a fairly definite relationship exists between the time necessary to obtain that quality and one of better quality. For example, when a stone has created a smoothness of 3 μin. (0.076 μm) in one minute, its cutting action has slowed up to the point that three minutes more are required for it to produce a 2 μin. (0.051 μm) finish. This approximate ratio will prevail on any size of surface. However, with multistep machines that produce finer finishes at each successive step, cycle times are reduced substantially.

Flat and spherical surfaces are superfinished in somewhat shorter times than cylindrical surfaces. A tappet head, for example, is corrected in geometry and finished to a smoothness of 5 μin. (0.127 μm) in no more than 30 seconds. Cast iron pump bearing faces are often completed in 20 seconds. A fully automatic, tappet finishing and chamfering machine has a cycle time of only 4.2 seconds, including loading and unloading.

Although no hard-and-fast rule can be laid down to indicate the amount of stock removal in superfinishing, Table 16-43 will

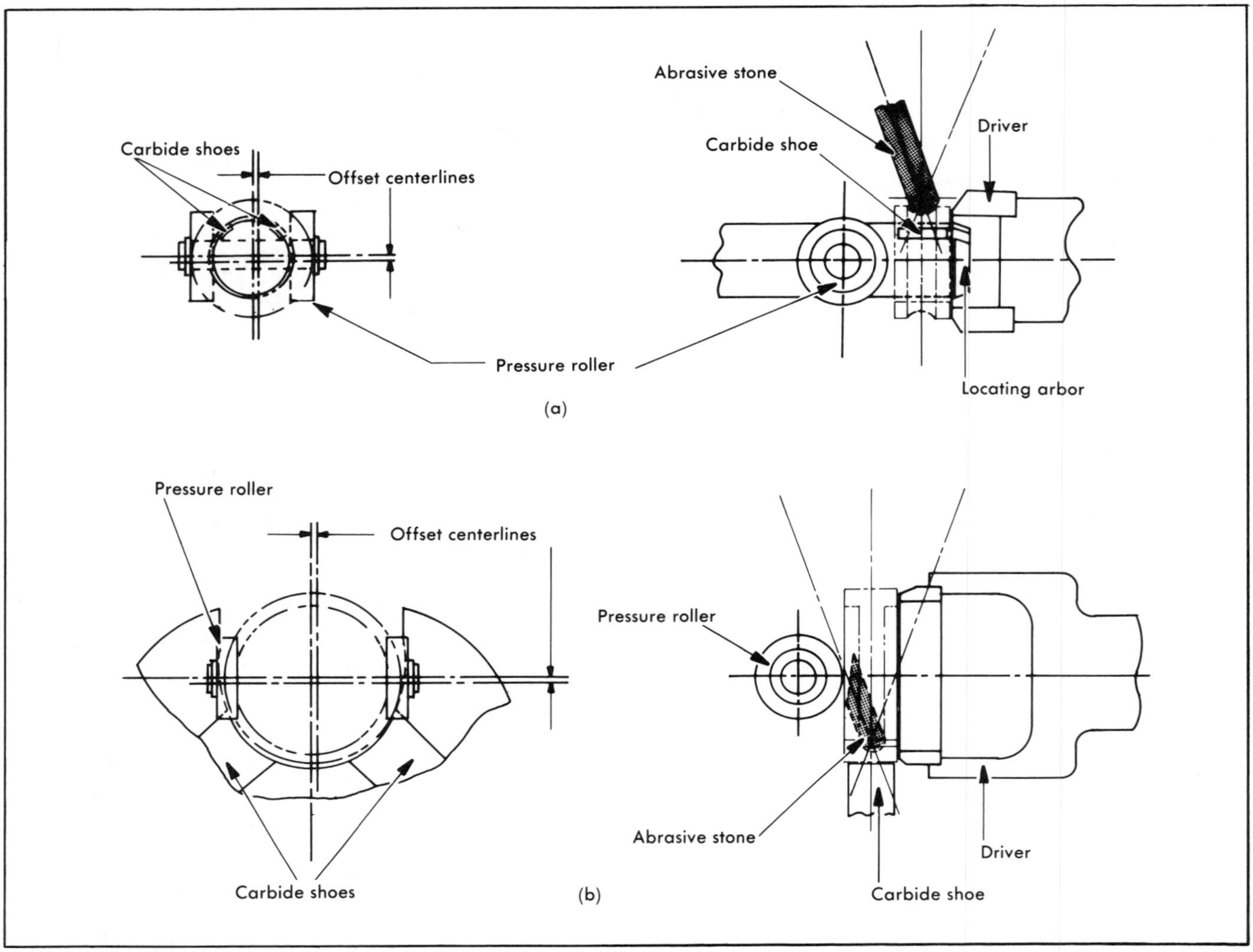

Carbide shoes
Offset centerlines
Pressure roller
Abrasive stone
Carbide shoe
Driver
Locating arbor
(a)
Pressure roller
Offset centerlines
Pressure roller
Abrasive stone
Carbide shoes
(b)
Driver
Carbide shoe

Fig. 16-120 Centerless shoe method of finishing bearing raceways. (*Ex-Cell-O Corp., Micromatic Operations*)

Rotating driver
Pressure rolls
Ball-bearing outer race
Oscillating stone
Pressurized oil
Hydrocentric chuck
(a)
Rotating driver
Pressure rolls
Reciprocating stone
Roller-bearing outer race
Hydrocentric chuck
(b)

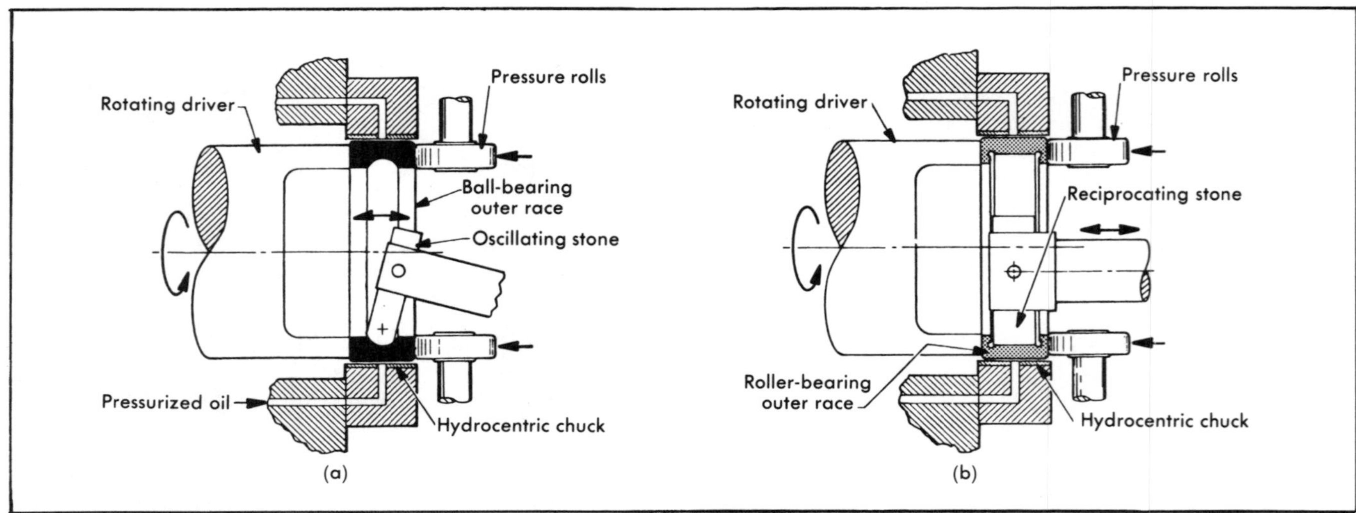

Fig. 16-121 Superfinishing of (a) ball bearing races and (b) roller bearing races. (*Thielenhaus Microfinishing Corp.*)

be found fairly reliable under average conditions for cylindrical workpieces. Actual stock removal will vary with stone pressure, the cutting capacity of the stone, oscillation amplitude, frequency, work speed, and cycle time. To indicate approximate production times for superfinishing operations, some representative figures for automotive parts are listed in Table 16-44.

Selection of Abrasives and Cutting Fluids

Abrasives. Aluminum oxide grit is generally used for superfinishing tough materials, such as alloy steels. Silicon carbide is usually best for hard, brittle materials and on cast iron and nonferrous metals. In addition, silicon carbide is often used on soft steels of R_C 30 or lower hardness and on diameters less than 1/2" (12.7 mm); it is also appropriate for very fine surfaces, 1.5 μin. (0.038 μm) or less.

Vitrified bond is generally used for superfinishing. Resin and shellac bonds will produce very smooth surfaces; however, their cutting action is so slow that their use is commonly restricted to small diameters, 1/4" (6.4 mm) or less, to aluminum alloys, or to jobs when a second operation is needed on such difficult metals as soft stainless steel or brasses. Where a scratch pattern of more than 3 or 4 μin. (0.076 or 0.102 μm) is desired, a coarser grit stone should be used, together with a slower surface speed of the work. For example, a 320 mesh stone and a speed of 35 sfm (10 m/min) should result in a profilometer reading of 7 to 12 μin. (0.178 to 0.305 μm) if the stone is withdrawn just as a complete cleanup of the grinding pattern is reached. Correction in geometry will be just as effective as when smoother surfaces are produced.

Cutting fluids. Special fluids having a higher flash point are replacing the mixtures of kerosene and oil formerly used for superfinishing. These special fluids sometimes have a chlorine additive.

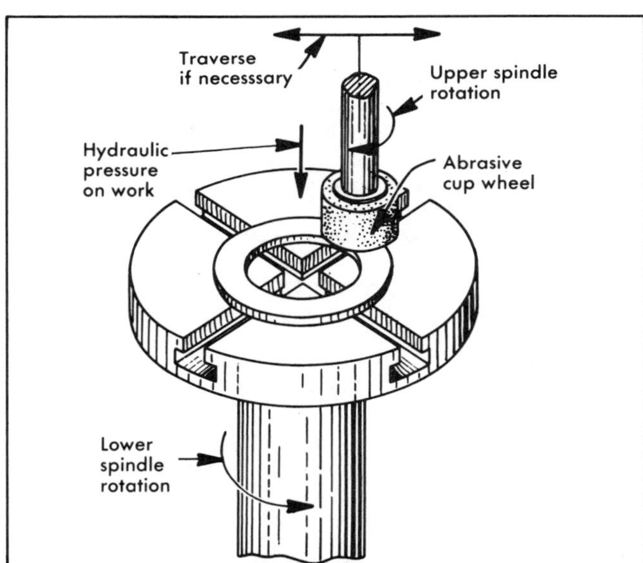

Fig. 16-122 Basic features of a flat superfinishing machine.

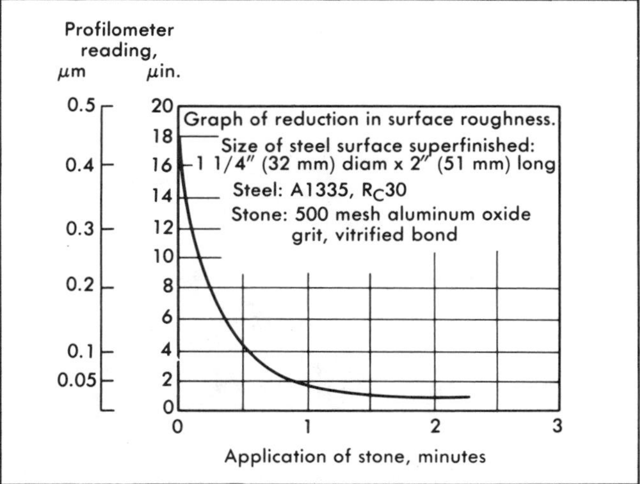

Fig. 16-123 Relation of surface roughness to superfinish time.

TABLE 16-43
Amount of Stock Removal by Superfinishing*

Finish of Ground Surface, μin. (μm)	Stock Removed in Superfinishing per Side or Radius, in. (mm)
10 (0.254)	0.00012 (0.0030)
15 (0.381)	0.00019 (0.0048)
20 (0.508)	0.00025 (0.0064)
25 (0.635)	0.00030 (0.0076)
30 (0.762)	0.00035 (0.0089)
35 (0.889)	0.00040 (0.0102)

* Stock removals given are for cylindrical workpieces only, with stones of only one grit size. For spherical and flat surfaces, stock removal is generally much higher and is normally controlled by in-process gaging.

TABLE 16-44
Representative Superfinishing Production Rates for Automotive Parts

Automotive Part	Ground Finish, μin. (μm)	Number of Spindles on Superfinishing Machine	Superfinishing Motion	Superfinish, μin. (μm)	Production Rate, parts per hour
Tappet head	30-40 (0.762-1.016)	1	Spherical or flat	5-8 (0.127-0.203)	900
Crankshaft	30-40 (0.762-1.016)	10	Cylindrical	5-8 (0.127-0.203)	80
Stem pinion bearing	15-25 (0.381-0.635)	2	Cylindrical	2-4 (0.051-0.102)	120

(continued)

TABLE 16-44—*Continued*

Automotive Part	Ground Finish, μin. (μm)	Number of Spindles on Superfinishing Machine	Superfinishing Motion	Superfinish, μin. (μm)	Production Rate, parts per hour
Distributor shaft	30-40 (0.762-1.016)	12	Cylindrical	3-5 (0.076-0.127)	720
Pressure plate	100-200* (2.540-5.080)	2	Flat	7-12 (0.178-0.305)	100
Brake drum	200-250* (5.080-6.350)	1	Internal cylindrical	15-25 (0.381-0.635)	150
Tappet body	10-20 (0.254-0.508)	12	Cylindrical	2-4 (0.051-0.102)	800
Camshaft main bearing	15-25 (0.381-0.635)	10	Cylindrical	2-4 (0.051-0.102)	80
Gear thrust face	10-20 (0.254-0.508)	9	Flat	2-4 (0.051-0.102)	500
Tapered bearing race	40-50 (1.016-1.270)	2	Cylindrical	2-4 (0.051-0.102)	450

* Starting finishes for pressure plates and brake drums are for turned surfaces rather than ground. In some cases, pressure plates are ground to a finish of 20 to 25 μin. (0.508 to 0.635 μm).

FINISHING WITH LOOSE ROLLING ABRASIVES

There are several processes, in addition to mass finishing and blasting (discussed previously in this chapter), that use unbonded, loose abrasives for finishing operations. These include lapping and free-abrasive machining (FAM). Polishing and buffing, other finishing processes using loose abrasives or compounds, are discussed in preceding sections of this chapter.

Lapping is a low-velocity abrading process that removes controlled, very small amounts of material. It is accomplished with loose abrasive grains (usually retained in a viscous or liquid media, called the vehicle) between a tooling plate or wheel, called the lap, and the work surface to be finished.

The terminology is confusing, however, because lapping is also sometimes used to designate finishing processes that operate with fine-grained, bonded abrasives. Such processes are similar to some honing and superfinishing methods discussed previously in this chapter. Some machines are designed for operation with either laps or bonded-abrasive wheels. This section will primarily discuss lapping with loose abrasives.

Lapping and free-abrasive machining use the same basic techniques, but there are several different designs of machines employed for the processes. Some manufacturing personnel feel that the distinction between the two processes is as follows: with lapping, material is removed by abrasive grains that have been embedded in the lap; with free-abrasive machining, material is removed by rolling abrasive grains. However, this distinction is not always the case. For example, abrasive grains do not become embedded in hardened steel laps.

Basic lapping machines are usually of open-face design with pressure between the lap and work surfaces applied by hand or weights; these machines can be equipped with liquid cooling of the lap. Other lapping and free-abrasive machines are equipped for applying pneumatic pressure and with pumps to constantly replenish the laps with abrasive.

LAPPING

Lapping is generally a final finishing operation that results in four major refinements in the workpiece: (1) extreme accuracy of dimension, (2) correction of minor imperfections of shape, (3) refinement of surface finish, and (4) close fit between mating surfaces. The life of moving parts that are subject to wear can be greatly increased by eliminating the hills and valleys on workpiece surfaces and creating a maximum percentage of bearing area. Besides developing a workpiece that meets the surface finish requirements and is correct for geometrical and dimensional accuracy, there is no distortion, as lapping procedures do not require the use of magnetic chucks or other holding or clamping devices.

In normal lapping operations, less heat is generated than in most other finishing operations, thus minimizing the possibility of rehardened and decarburized areas on hardened or heat-treated parts. When both sides of a flat piece are lapped in the same operation, extreme accuracy in flatness and parallelism can be accomplished; and by removing the same amount of stock from both sides of the part simultaneously, any inherent stresses in the piece are equally relieved. In both manual lapping operations and semiautomatic machine lapping operations, the end results depend on many factors, chief of which are: type of lap material, type of lapping medium, speed of lapping motion, and material to be lapped. Specific recommendations are given in Table 16-45.

Lapping with Loose Abrasive and Prepared Compounds

Lap materials. The most efficient material for machine lapping is cast iron. While cast iron laps are in common use for manual lapping, there are other materials used, such as hardened alloy steel, soft steel, bronze, brass, lead, leather, and cloth. The last two materials are actually used for polishing and do not correct the surface. For most operations, the lap should be softer than the workpiece material so that the abrasive material becomes temporarily embedded in the lap until it is dulled or fractured from the pressure of the lapping action. With hardened steel laps, however, there is no impregnation of the abrasive grains.

TABLE 16-45
Lap, Abrasive, and Vehicle Selection for Various Operations

Type of Lapping Operation	Lap Material	Type of Abrasive[a]	Grit Size	Vehicle[a]	Viscosity[a]	Solvent[a]
General and Toolroom Applications[b]						
General use	Cast iron	A	90	SO	H	W
	Cast iron	A	120	SO	H	W
	Cast iron	A	180	SO	H	W
Stock removal	Cast iron	C	220	O	M	K
Semifinish	Cast iron	A	400	SO	H	W
Fine finish	Cast iron	A	600	G	H	K
Polish	Cast iron	UA	800	SO	H	W
	Cast iron	A	1000	SO	H	W
Dies and molds:						
Roughing	Whitewood, lead	C	280	G	H	W
Semifinish	Whitewood, lead	UA	240	G	H	W
Finish	Leather	UA	400	G	H	W
Polish	Hard felt	UA	600	G	H	W
Gages, ring and plug:						
First lap	Cast iron	2A	240	G	H	W
Second lap	Cast iron	2A	400	G	H	W
Polish	Leather, felt	UA	800	G	M	W
Gages, thread:						
10P and coarser	Cast iron or soft	2A	600	G	M	W
Finer than 10P	steel	2A	800	G	M	W
Carbide tools and parts:[c]						
Matte finish	Cast iron	C	240	G	H	K
High finish	Cast iron	C	500	O	M	W
Draw dies	Cast iron	C	240	G	M	K
	Cast iron	C	320	G	M	K
	Cast iron	C	500	O	M	W
Gears[d]						
Cast iron spur and helical:						
Small	Cast iron	A	240	G	H	K
Medium	Cast iron	A	180	G	H	K
Large	Cast iron	A	90	G	H	K
Bevel and hypoid:[e]						
Bevel	Cast iron	C	280	O	L	K
Hypoid	Cast iron	C	320	O	L	K
Steel spur and helical:						
Coarse pitch	Cast iron	C	150	G	H	K
Fine pitch	Cast iron	C	280	G	H	K
Wormgears,[f] general use	Cast iron	2A	500	SO	M	W

[a] Abrasives: A, fused aluminum oxide; UA, unfused aluminum oxide; 2A, white fused aluminum oxide; C, silicon carbide. Vehicle: SO, soluble oil; O, oil; G, grease. Viscosity: H, heavy; M, medium; L, low. Solvent: K, kerosene; W, water.

[b] On all these operations, stock removal should be kept to a minimum. Normally, 0.0002″ (0.005 mm) is sufficient stock for final finish, provided the previous operation has left a proper finish. The polishing of the various parts does not remove stock but does bring out a high luster. On dies and molds and other irregular surfaces, the lapping usually follows the use of abrasive sticks and rubs. On many flat surfaces as well as on parts such as molds and die parts, constant pressure should be applied for stock removal and a diminishing pressure to bring up the finish. Both external and internal cylindrical surfaces require an adjustable lap that can be contracted for external surfaces and one that can be expanded for internal lapping.

[c] Carbide tools and wear parts are sometimes lapped with a diamond compound or diamond dust mixed with neat's-foot oil or olive oil. Both cast iron and whitewood laps are used. The cutting face of the lap is charged with the lapping compound and for higher finishes should be operated at 800 to 1200 sfm (245 to 365 m/min). The grit size of the diamond compound ranges from 400 to 800 grit, dependent on finish requirements.

[d] In all gear-lapping operations, a speed just below the point where the compound would be thrown from the gears is the most desirable.

[e] Gears in this classification are lapped together, such as pinion and ring or other mating gears.

[f] Most wormgears are lapped in assembled position with the driven worm.

LAPPING

Lapping media. Of the manufactured abrasives that are used for the majority of lapping operations, silicon carbide is generally used when rapid stock removal is a requirement. Fused aluminum oxide is softer than silicon carbide and is used to improve the finish, especially in lapping comparatively soft metals such as soft steel and nonferrous metals. Unfused aluminum oxide is very soft and breaks down rapidly under lapping pressures. It is inefficient for stock removal, but especially suitable for producing extremely fine finishes. Grit sizes used range from 90 to 600, and, sometimes, 800 mesh for light, fine finishing.

Diamond, being the hardest material, is generally used to lap precious stones and tungsten carbide. Natural abrasives, other than diamonds, that have been used for lapping are garnet, natural corundum, and Turkish emery. All of these are softer than aluminum oxide and will break down more rapidly. As a result, they are slower cutting, but tend to produce an improved finish.

It is possible to use loose grain with oils, greases, or soaps, but it is best to use a compound commercially prepared by manufacturers who best know how to compound. Compounding must take into consideration the susceptibility to temperature changes, provision for speed of cut without sacrificing finish, noncorrosive and nontoxic properties, easy cleanability without the use of highly inflammable cleaners, and capacity for holding abrasive particles in suspension. All compounds are mixed to various viscosities, commonly classified as light, medium, or heavy.

Lapping speeds and pressures. The most efficient lapping speeds range between 300 and 800 sfm (90 and 245 m/min), whether it be the rotation of the lap or movement of the work over a stationary lap. Speeds lower or higher than this range will lower the rate of cut, but higher speeds will also improve the surface finish.

The amount of pressure applied depends considerably upon the material being lapped, as well as the desired rate of cut and the finish specified. When lapping with loose abrasives or prepared compounds, the pressure will range from 1 to 3 psi (7 to 21 kPa) for soft materials, including nonferrous metals and alloys, and up to 10 psi (70 kPa) for hard materials.

Excessive pressures will cause rapid breakdown and possibly score the workpiece. A constant pressure provides the best results except when leather and cloth are used as lap materials, in which case a diminishing pressure helps to bring out the desired luster.

Lapping Finishes

In hand lapping operations, where the compound is usually brushed on the lap, the finish produced for a corresponding grit size will vary according to the operator's skill. On power-driven flat laps, cylindrical laps, and internal laps, the speed of lapping motion is more or less fixed. The pressure with power-driven flat lapping depends on the operator, while cylindrical and internal lapping depend on the adjusted tension of the lap on the workpiece.

The final finish achieved with various abrasive grit sizes is consistent for semiautomatic and fully automatic lapping machines, in which the abrasive compound is continuously fed to the work. In machine lapping with continuously fed compound, a predetermined stock removal cycle can be set up by using the proper compound under a known pressure. In such operations, the resulting surface finish will remain uniform. Table 16-46 lists approximate surface finishes obtained with continuously fed abrasives of different types and grit sizes. This table indicates that, for corresponding grit sizes, aluminum oxide produces a much finer surface finish than silicon carbide. The rate of stock removal with silicon carbide is much faster; and on parts where flatness and dimensional accuracy are the only requirements, the use of a silicon carbide compound will prove more economical. The type of material being lapped will also affect the surface finish. Softer, nonferrous metals require a finer grit size to produce surface finishes comparable with those produced on steel.

TABLE 16-46
Lapping Finishes Obtained with Continuously Fed Abrasives

Type of Abrasive*	Grit Size	Finish, μin. (μm)
S	220	30-40 (0.762-1.016)
S	320	25-30 (0.635-0.762)
S	400	18-25 (0.457-0.635)
S	500	15-18 (0.381-0.457)
S	600	10-15 (0.254-0.381)
S	800	5-10 (0.127-0.254)
A	400	3-6 (0.076-0.152)
A	800	2-3 (0.051-0.076)
A	900	1-2 (0.025-0.051)

* S, silicon carbide; A, fused aluminum oxide.

Lapping with Handheld Tools

For manual finishing of flat work, a lapping block or plate is used, on which the work is rubbed by hand. The plate is charged with the abrasive material or compound. By using figure-eight or similar motions and taking care to cover the entire surface of the plate, the plate will remain flat for a considerable amount of work. Abrasive should be used sparingly. The use of too much abrasive will increase the wear of the lap without doing any additional work.

Manual lapping demands specialized skill of the operator, upon whom the selection of the compound should largely depend. Skill is required especially when lapping flat surfaces since the proper lapping speed and pressure depend on feel and judgment. In all manual lapping, an ever-changing contact must be made between the lap and the work. This change in contact will result in uniform lap wear, uniform stock removal, and good size control. A typical lapping block is made of soft, close-grained cast iron, heavily ribbed and fairly thick so it will not distort. The surface may be serrated or plain faced. When serrated, it is easier to clean if the grooves are V-shaped.

Ring lapping is the original toolmaker's or gagemaker's method of finishing plug gages and other cylindrical objects. It relies upon a closely adjusted lap to correct errors of roundness and taper. As with machine lapping, the product and the lap mutually improve each other as the work progresses.

Ring laps are usually made of soft, close-grained cast iron. One type of ring lap (see Fig. 16-124) has several partial cuts, in addition to the complete slit. Closing screws and expanding or opening screws are provided to enable precision adjustment to

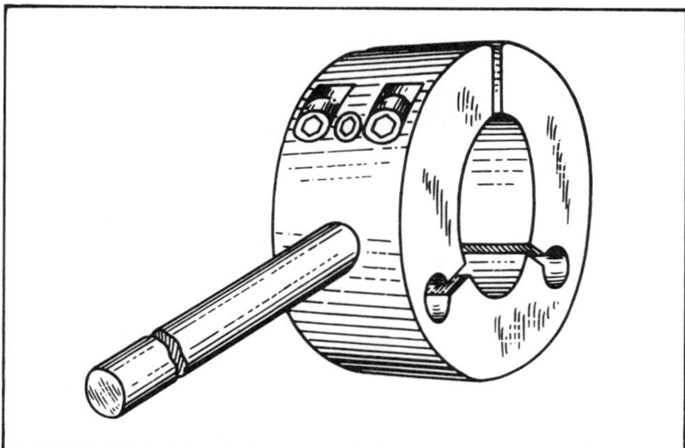

Fig. 16-124 Adjustable ring lap.

be easily maintained. The bore must be smooth, straight, and close to the size of the work.

The length of the lap should be slightly shorter than the length of the work. On long spindles, the lap length should not exceed three or four diameters. Greater length of lap would be difficult to handle as it is imperative that it be adjusted tightly to the work and reciprocated rather rapidly. The lap must be adjusted so that the work warms up somewhat; and, when reciprocated, the lap should overrun the work about one-third its length. The work should be cooled before measuring. The abrasive and vehicle should be fed through the slot to maintain a straight round hole in the lap.

Ring lapping is recommended for stepped plug gages or gages made in small quantities and for precision machine spindles when great accuracy for roundness is a factor. Skill must be acquired before quality work can be assured. Recommended stock removal is less than 0.0005″ (0.013 mm). Hardened steel is easy to lap and receives the highest finish by this method. Soft steels need specially selected abrasives, or the work will score.

Internal or Hole Lapping

Long holes with small bores and short holes having large bores are difficult to lap without encountering bellmouthing and errors in straightness. A loosely fitted lap will cause both these defects. There are two methods of lapping holes. The first calls for a series of solid laps of diameters ranging from the size of the unlapped hole to that of the finished product. The series must be in small increments, depending upon the size of the product, and care must be taken to use each size only until the next larger lap in the series will enter the hole. The other method utilizes an adjustable lap. When size permits, this latter method is quite satisfactory.

The length of the lap must be selected to suit the product. It is usually slightly longer than the product so that the hole is straightened as it is lapped. However, if bellmouthing exists before lapping, it may be necessary to select a short lap to rough out the hole, changing later to a longer lap.

Inasmuch as the accuracy resulting from hole lapping depends largely upon a close fit of the lap in the hole, any feature that will enable the operator to maintain that fit and take up the "slack" as the lap wears and the hole enlarges will contribute to good lapping. Therefore, great care must be taken in preparing and maintaining the laps.

Larger, solid laps are usually made of soft cast iron. For smaller sizes, this material is too brittle, and steel or hard brass may be used. The lap should be turned or ground straight and round. A coarse, helical groove, cut both right and left-handed, prevents the laps from pulling through and provides clearance for feeding the abrasive and oil.

Adjustable laps of various designs have been made and found suitable for all classes of hole lapping. Commercial laps with replaceable copper sleeves (see Fig. 16-125, view *a*) are available in sizes from 1/8 to 2 1/2″ (3.2 to 63.5 mm). These laps have an adjustable mandrel on which the sleeve is expanded to fit the hole.

Other types of adjustable laps are made (see view *b* of Fig. 16-125) with screw expansion devices that enlarge the body of a lap that has been partially split longitudinally. In view *c*, a split-sleeve type of lap sliding upon a tapered mandrel is illustrated. The lap in view *d*, recommended for large work, is designed to be used vertically.

In hole lapping, as with external manual lapping, the lap should be tight enough to create heat during the operation. Hole lapping is simplified when provision is made to collapse the lap before and after the lapping operation. Internal-lapping machines have this feature, with a micrometer adjustment for limiting the size of the finished hole. Semiautomatic machines are available with hydromechanical, oscillating-stroke control and an automatic stop position for tool expansion.

Cylindrical Lapping

Vertical lapping machines typically carry one rotating lap and one stationary lap. The lower lap rotates at a speed of not over 300 sfm (90 m/min). The upper lap does not rotate and is free to float and rest upon the work, while the latter rides upon the face of the lower lapping plate. The laps are of heavy ring-type construction, and the lapping pressure is provided by gravity or pneumatically.

Lapping machines of the type shown in Fig. 16-126 can handle both cylindrical and flat lapping operations with the same arrangement of the lapping plates. In flat lapping, a quantity of similar pieces are drawn between the parallel surfaces of the laps, with resulting parallelism imparted to the products.

The principle of lapping multiple parts at one setting results in remarkable size control because the rate of stock removal decreases as more and more of the pieces are finished. It is therefore possible to hold the final size to very close tolerance limits. The resultant accuracy can be as close as 0.00001″ (0.0003 mm) on gage work and 0.000025″ (0.00064 mm) on regular production.

The workholders used in machine lapping are guides rather than holders. They retain the work loosely and guide it through a complicated path between the lap faces. In cylindrical lapping, the work propels the workholder; while in flat lapping, the workholder propels the product. Figure 16-127 illustrates the spider or leg-type workholder suitable for hollow cylinders.

Figure 16-128 depicts the slotted-plate type of retainer, suitable for both solid cylinders and hollow parts. Work must be removed individually from this type of workholder, but the spider type can be removed from the machine with the work intact. It will be noticed that the work must fit freely on the leg or in the slot of the slotted-plate workholder. It is common practice to allow at least 1/16″ (1.6 mm) clearance.

Extremely high operational uniformity and dimensional accuracy can be obtained by a method called *transposing*. It consists of rearranging the work assembled in the workholder in

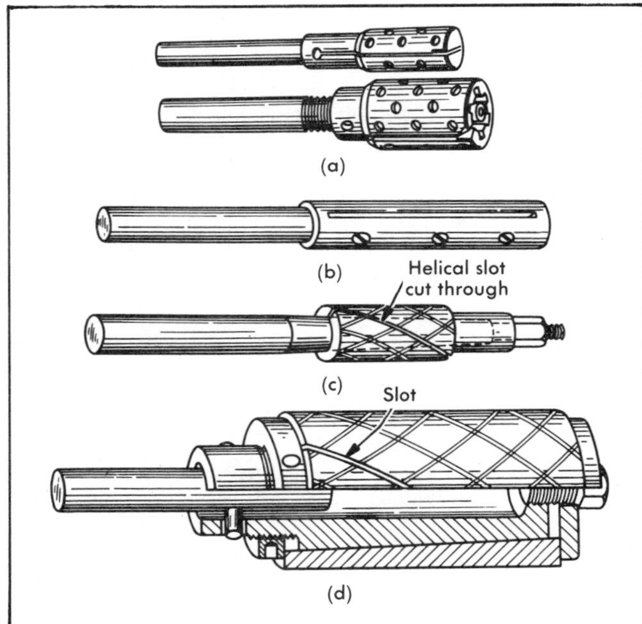

Fig. 16-125 Adjustable laps: (a) with replaceable copper sleeves; (b) with screw expansion device; (c) split-sleeve type; and (d) split-sleeve type for large work.

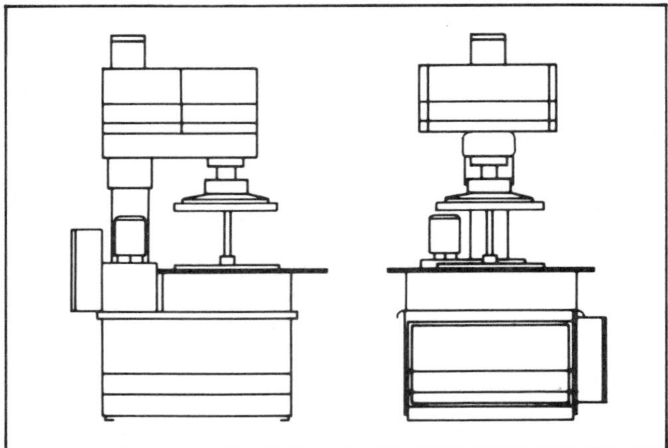

Fig. 16-126 Double-wheel lapping machine. (*Spitfire Tool and Machine Co.*)

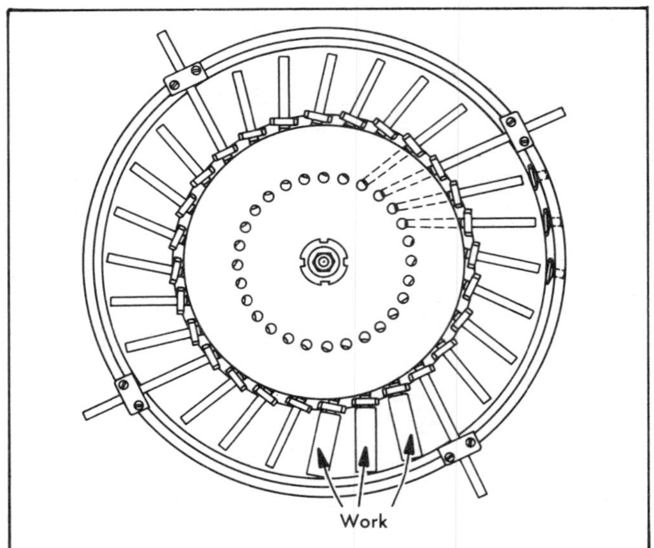

Fig. 16-127 Spider-type workholder for lapping hollow cylindrical workpieces on a vertical machine.

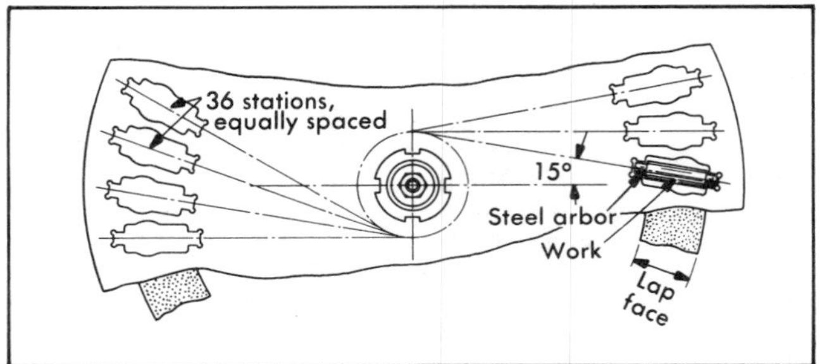

Fig. 16-128 Slotted-plate workholder for lapping cylindrical workpieces on vertical machine.

such a way as to distribute the error and relapping repeatedly until all measurable variation has been eliminated. The transposition may be deliberate, according to a predetermined plan, or the work can be unloaded and reloaded at random.

Gages and similar close-tolerance work should be lapped by transposing. If the laps are accurately flat and the conditions regarding lap dimensions, workholder, and size of the work are proportioned to maintain the accuracy of the laps, the refinement of machine lapping will practically equal the accuracy to which the product may be measured. This procedure, however, is not a high-production method and results in high costs.

Flat Lapping

Flat machine lapping is done on the same general type of vertical spindle machine as used for cylindrical lapping, except for certain modifications. The motion given to flat work is comparable to that given an individual piece when one face is being lapped by hand. However, because a large quantity of similar parts are being handled at one time and both sides of each piece are surfaced simultaneously, the resultant parallelism and uniformity of dimension are finer than can be obtained by hand lapping, and flatness is equal to the best that can be done by hand. Lapping time for the entire load is little more than that required for lapping one side of one piece. Therefore, the efficiency is proportionate to the quantity of parts that can be handled.

Workholders for flat lapping are of two types. The plate type of workholder is a simple plate made from sheet metal, Micarta, or plywood, and has rough apertures approximately the size of the work. The workholder is propelled by three drivepins that impart rotary and gyratory motions, causing the work to cover the entire surface of the lap. The planetary-type or two-piece workholder makes it possible to lap a large quantity of small pieces on a relatively wide-faced lap and yet make each piece cover the entire surface of the lap in a constantly changing path.

For manual flat lapping, vertical spindle machines are available with single horizontal lapping plates, with either cast iron laps or bonded-abrasive lapping wheels. Machines arranged for bonded-abrasive laps require a trueing device, a coolant pump, and a tank. Work is held by hand and moved at random across the face of the rotating lap. The weight of the work may be sufficient to provide adequate lapping pressure; if not, additional pressure must be applied by the operator. The pressure will vary with the degree of finish required and with the rigidity and size of the work. The pressure must be applied to suit the shape of the workpiece and must be directed to the section with the greatest area. Light pressures will provide the most accurate results, but higher pressures remove stock more rapidly.

Manual flat lapping machines of this type are useful for lapping flat surfaces on large and small articles. They are especially suited for lapping irregularly shaped pieces, when manipulation is needed to apply the pressure in a specific manner to ensure even lapping. They are not suited to lapping tall pieces, or parts where the area to be lapped is a small surface in a relatively large piece, or when dimensional accuracy is a critical factor. Parallelism cannot be controlled with manual flat lapping.

Mechanically held flat lapping dispenses with the manual effort of applying pressure and motion to the individual workpieces, and also permits loading the machine to capacity. Work is carried in a holder of the planetary type, consisting of a series of frictionally or positively driven circular adapters that revolve and rotate about the center of the lap. Openings, roughly the shape of the work, are cut in each adapter to accommodate as many pieces as possible. If the workholders move slowly, the individual workpieces can be removed and replaced without stopping the lap. There is no provision for applying pressure collectively. If the weight of the product itself is insufficient to apply the proper working pressure, additional weight must be added to each workpiece or to each group of pieces nested in the same adapter or retainer. This arrangement is more flexible than other forms of mechanical lapping inasmuch as work of different sizes can be placed in the machine at one time. One adapter can be arranged for one size of piece and another for a different size.

Certain work that has already been lapped for size and parallelism on conventional machines may still be unsatisfactory with respect to finish. Also, certain workpieces are of such shape that they cannot be lapped by the conventional methods and must be handled individually. Typical examples of hand lapping include lapping hardened steel to a practically scratch-free finish and individually finishing the seal faces of the components of rotary seals (such as siphon seals with bellows attached and carbon and bronze thrust washers), which must be lapped extremely flat and with the extremely high surface finish necessary for optical flat readings.

Flat lapping with nonrotating lapping plates is excellent for certain types of flat lapping, such as gage blocks and similar work, especially when small parallel surfaces on comparatively tall pieces are to be lapped, and also for lapping thin delicate pieces such as crystals. Machines for this type of operation carry a stationary lower lap and a removable, nonrotating upper lap of the same size. The weight of the upper lap is made to suit the class of work to be handled. Workpieces are carried between the nonrotating laps by the workholding retainers.

For quartz crystals and similar thin objects, a very light upper lap is made to be removed by hand, rather than with a lifting device. Heavy laps are provided for tool work such as gage blocks of hardened steel, when mechanical lifting devices are arranged to suit.

A positive work-drive mechanism is provided to rotate and gyrate the workholders, operated either by the eccentric rotary drive or by the planetary principle. One method uses a stationary outer ring, with an inner driving member to rotate the workholders. The other method (see Fig. 16-129) employs an inner toothed driving member and an outer toothed driving ring that are geared to each other to impart planetary motion to the work. In each case, motion given to the workholders causes each workpiece to cover the entire working surface of the lapping plates. The laps, being stationary, impart no tilting movement to the work.

Flat and Cylindrical Lapping with Bonded-Abrasive Wheels

Lapping on cast iron laps with loose abrasives will leave a film of abrasive and metal on the product that must be removed before measurements can be taken or the product is put to use. This film is sometimes difficult to remove, especially if there are channels or crevices where it may lodge. Furthermore, if the material from which the part is made is anything less than glass-hard, the use of loose abrasive will not produce the bright, smooth finish usually associated with a lapped surface. It is possible to remove the film or brighten the surface by an additional operation, but bright and clean surfaces can be obtained with proper lapping procedures.

In machines similar to those used for flat honing (see Fig. 16-116), two bonded-abrasive lapping wheels rotate on precision spindles having adjustable alignment. Hydraulic cylinders raise and lower the upper lap and actuate the diamond trueing device, which trues the laps while they are both rotating. Lapping pressure can be regulated to suit the product. Two-wheel lapping machines are available with electronically controlled hydraulic pressure systems, permitting infinitely variable working pressures during the preload portion of the cycle.

Workholders are identical in style with those used in other vertical lapping machines. However, there is one peculiarity in using these workholders. Turning the workholder over, thus reversing the tangent angle of the work, changes the rate at which the assembly will revolve. This change in relative speed is useful in the lapping of certain cylindrical parts and is one means of slightly changing the surface finish.

For flat lapping, the plate-type workholder and the two-piece or planetary type are used exactly as in machines using cast iron laps, except that the workholders traverse the work

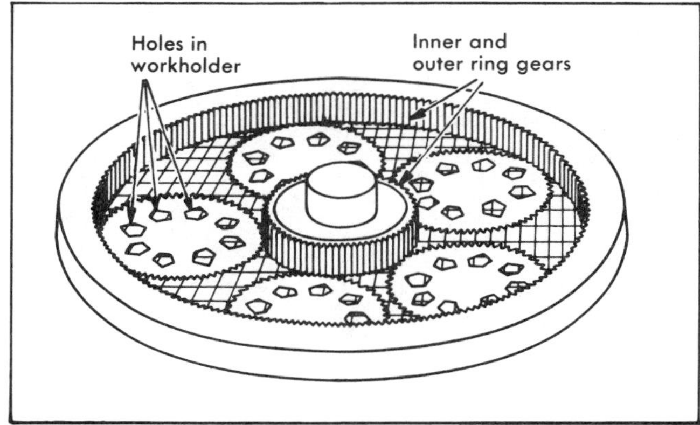

Fig. 16-129 Planetary workholder arrangement for lapping quartz crystals.

more slowly because both laps rotate. The surface finish obtained will vary with the product and the abrasive wheels used. Dimensional accuracies ranging between 0.00005 and 0.0001″ (0.0013 and 0.003 mm) for parallelism and between 0.0001 and 0.0005″ (0.003 and 0.013 mm) for thickness or diameter have been reported. Finer finishes can be obtained by a second operation, but accuracy is not necessarily improved.

Vertical-type lapping machines using bonded-abrasive wheels of the disc type are capable of producing finishes as fine as 2 μin. (0.051 μm), but their prime purpose is to produce flat surfaces such as on pressure plates and sealing parts. Stock removal should be limited to a few thousandths of an inch. All metals are being successfully finished on a production basis with this type of operation.

In the past, abrasive recommendations have been silicon carbide vitrified-bond wheels ranging from 150 to 500 grit in J to L grades. There have been recent developments, however, that indicate white aluminum oxide in the same grade range and grit size does a comparable job. Coolant recommended for all flat-metal lapping is mineral seal oil or a thin mineral oil similar to kerosene that is not as harsh on the operator's skin. For cylindrical work, a water coolant with soap flakes is commonly used, but mineral seal oil is also satisfactory. An exception is in the lapping of articles made of carbon, for which clear water is used for rough lapping and dry wheels for final finishing, when a polish is desired.

Centerless Lapping with Bonded-Abrasive Wheels

Machines of the type shown in Fig. 16-130 are a refinement of centerless grinding to provide higher surface finish on straight cylindrical objects. An extra-long grinding wheel and regulating wheel are provided to allow the work to remain in abrading contact longer than is necessary for regular centerless grinding. The wheel spindles are swiveled so that, when trued, they form a mild hourglass shape. This shape results in a line of contact at an angle with the axis of the work, with a wraparound effect upon the latter.

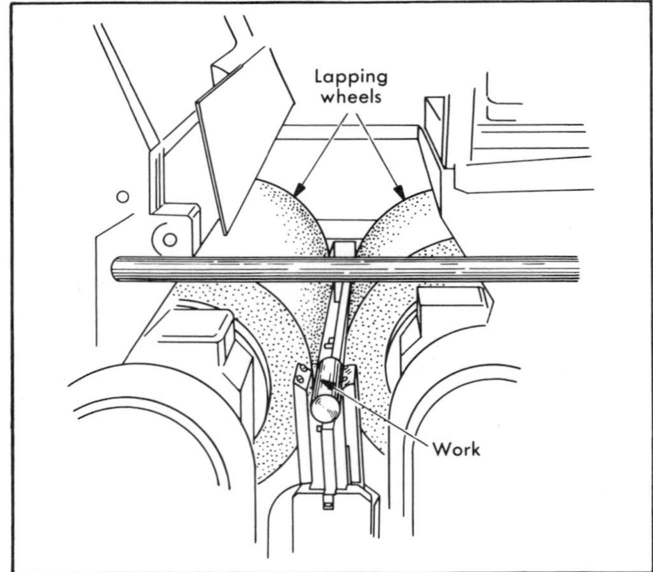

Fig. 16-130 Setup for centerless lapping.

Preceding lapping operations, the work must be ground as round as desired in the finished product. The finest finish provided by this method, comparable with that obtained from machines with flat lapping plates, requires three or more operations with progressively finer wheels. The three stages are identified as (1) grind lap, (2) first lap, and (3) final lap. The grind lap operation uses a vitrified-bonded aluminum oxide wheel of 180 or 220 grit size and O or P grade, and removes up to 0.0005″ (0.013 mm) of stock. The first lap removes approximately 0.0002″ (0.005 mm) of stock with a silicon carbide resinoid-bonded wheel of 320 grit. The final lap removes 0.0001″ (0.003 mm) of stock or less, producing a finish of 2 μin. (0.051 μm) or better, using a 500 grit wheel. The regulating wheel used for lapping operations is the same size as the lapping wheel, but is usually somewhat harder. A 320 grit, silicon carbide resinoid-bonded wheel is the most efficient.

The usual dimensional tolerances of parts from a centerless operation are within 0.000025″ (0.00064 mm) for concentricity and parallelism. Wheel speeds vary according to the finish and production requirements. The lapping or cutting wheel speeds are much lower than centerless grinding wheel speeds and will vary from 500 to 2000 sfm (150 to 610 m/min). The regulating wheel speeds vary from 200 to 500 sfm (60 to 150 m/min). With the higher regulating wheel speeds, the angle of draw can be reduced to a point where the sequence of lapping operations will maintain a uniform throughfeed from the previous grinding operations.

The type of machine illustrated in Fig. 16-131 is designed for lapping individual plug gages, measuring wires, and other straight or tapered cylindrical objects. The lapping roller is twice the diameter of the regulating roller; and both revolve slowly, at the same rpm, in the same direction. Work is laid between the two rollers and is caused to rotate at the speed of the regulating roller. Because of its increased surface speed over that of the work, the lap roller creates a rapid lapping action. The lapping pressure is applied manually through a notched fiber stick, held against the work and moved evenly over the entire surface. If it is desired to correct taper or other error, the stick is held longer on the portion requiring more stock removal. Blanks for gages to be lapped by this method must be ground to within 0.0005″ (0.013 mm) of finished size and well within the tolerance for roundness desired in the final product.

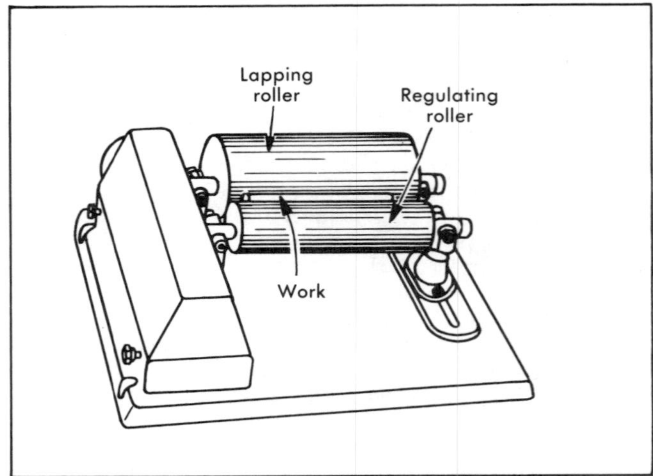

Fig. 16-131 Roller-type machine for centerless cylindrical lapping.

Spherical Lapping

The lapping of spherical surfaces other than balls, shown diagrammatically in Fig. 16-132, view *a*, uses a lap that is the counterpart of the surface to be finished. When used for lapping hardened steel, the lap may be of cast iron of suitable proportions, cut with a concave or convex surface to suit the work. The lap is carried in a machine similar to a drill press. A crank of suitable throw is held in the chuck. The crankpin is furnished with a ball end that enters freely into a blind hole in the back of the lap, on the centerline of the curved surface. With the workpiece in line with the drill spindle, the latter is rotated, thus gyrating the lap. The lap itself must be heavy enough to provide the lapping pressure.

Another method of lapping convex spherical surfaces, shown in Fig. 16-132, view *b*, consists of having a pair of rotatable spindles at an angle to each other. The lap is rigidly attached to one spindle, and the work to the other. Axial alignment must be accurate in order to generate a true radius. One spindle must be free to slide, to take up wear and apply the necessary pressure. The lap in this instance can be of ring formation, with only the rim concave. This method is used for generating spherical surfaces in the manufacture of lenses and makes use of diamond wheels as the abrading medium.

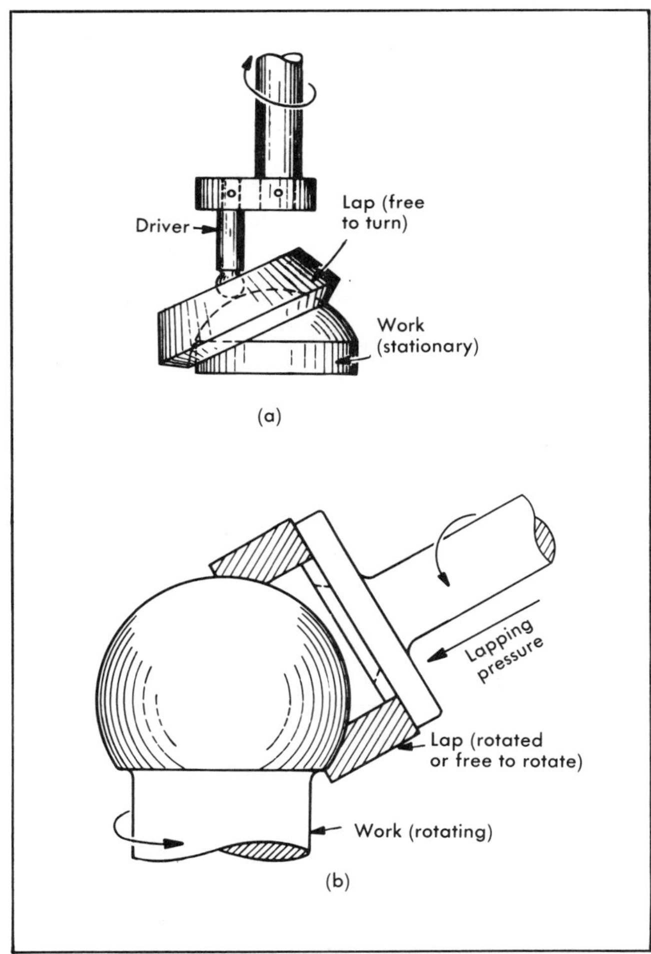

Fig. 16-132 Spherical lapping motions: (a) work stationary and (b) work rotating.

Gear Lapping

Small inaccuracies in tooth profile, distortion from heat treatment, chatter and cutter marks, tooth spacing, and surface defects will cause undue friction and noise when gears are in operation. Many of these faults can be corrected by lapping. (Table 16-45 offers lap, abrasive, and vehicle selection for various lapping operations.) Additional information on gear lapping is presented in Volume I, *Machining*, of this Handbook series.

Accuracy of Cast Iron Laps

Accuracy of external lapping operations depends upon the accuracy of the working surface of the laps. For precision work, the laps must be dressed frequently. Surface wear on cast iron laps, used in pairs for flat and cylindrical lapping, will usually be detected in the product, but the error will not be so obvious in the laps themselves. Upon inspection, the wear will usually be found to be one of the following types:

1. Surface undulated or wavy.
2. Both lap faces hollow.
3. Upper lap concave, lower lap convex.
4. Both laps concave.

Errors are usually very slight, but they can generally be detected with a long knife-edge straightedge, of a length equal to the diameter of the lap to be tested. The laps must be perfectly clean and dry before any examination can be made. Errors of types 1, 2, and 4 (as listed above) can be corrected by lapping the laps together. A type 3 error will not correct itself, but tends to increase if the laps are dressed in pairs. This type of error should be corrected by facing in a lathe or by grinding, using a vertical spindle, rotary-table type of grinding machine. In grinding, both laps must be made flat within 0.0003″ (0.008 mm) and slightly hollow, rather than crowned. After grinding or turning, they should then be lapped together.

For extreme precision, three laps of equal size should be lapped together in regular sequence in the order of 1 with 2, 2 with 3, and 1 with 3. In this order, the three laps alternately become "upper laps," thus providing all the necessary changes to ensure accuracy. Only laps of the same size should be dressed together by any of these methods.

For the finest work, the lap surfaces must be optically flat. For such a test, it is necessary to obtain a reflective finish before the check can be made. The laps should be prepared as described, with only the finest abrasive available. After a thorough cleaning with a good solvent, the laps should again be lapped together. This time, naptha or a similar thin fluid should be used, and the cleaning operation should be repeated until a reflective surface is built up.

Cleaning of Lapped Parts

All lapped parts should be thoroughly cleaned before being put in use. Thorough cleaning is especially true for soft materials upon which some of the particles of abrasive grain may become embedded in the finished surface. Water-soluble compounds may be cleaned by hand with a brush and hot or cold water. Soap and detergents may be used in a water solution for machine cleaning. For most compounds, when the vehicle is grease or oil, organic solvents are suitable. Solvents that might create a fire hazard are not necessary and not recommended.

FREE-ABRASIVE MACHINING

Free-abrasive machining, also called lapping, is a process for removing hard material by the action of abrasive grains, suspended in oil or water, that are interposed between a rotating wheel and the workpieces. As the wheel rotates, carrying the loose abrasive across the face of the work, it removes stock from the work—often as effectively and efficiently as a conventional grinding wheel. The process is frequently considered competitive with conventional grinding, especially for stock removal from flat surfaces. It is also used extensively for applications that are impractical for conventional grinding and machining. Typical application examples include the machining of solid materials, extremely thin work, and parts that might distort if clamped or magnetized. Plastics, glass, ceramics, and nonmagnetic material are also surface ground by free-abrasive machining.

One type of free-abrasive machine is arranged for workpieces to be held in four cylindrical retaining rings (see Fig. 16-133). Some machines have three rings, also called workstations. The wheel (actually an annularly shaped table, rotating in a horizontal plane under the rings) carries abrasive grains across the lower face of each workpiece. Pressure is applied to the parts by weights or by pneumatic cylinders acting upon pressure plates that fit inside the retaining rings. On some machines, the pneumatic cylinders also act as lifting devices for the retaining rings. In addition to retaining the workpieces, the rings help correct and maintain the flatness of the wheel.

In normal drive, the retaining rings are usually driven counterclockwise by friction with the wheel. This normal drive tends to develop a concave surface on the wheel, which is corrected by a reverse drive. In the reverse or dressing drive, a spur gear is placed over a shaft in the center of the wheel and forces the retaining rings (which have gear teeth on their peripheries) to rotate in a direction opposite to that of normal drive. This reverse drive quickly brings the wheel surface back to normal flatness. On some machines, correction of the wheel is accomplished by shifting the positions of the retaining rings relative to the annular face of the wheel.

Wheels may be of single piece or segmented construction, made from hardened steel, cast iron, brass, and other materials, depending upon workpiece material. Heat generated by friction is dissipated by water circulating through the wheel or through evaporation of the liquid used as the slurry vehicle. Wheels may

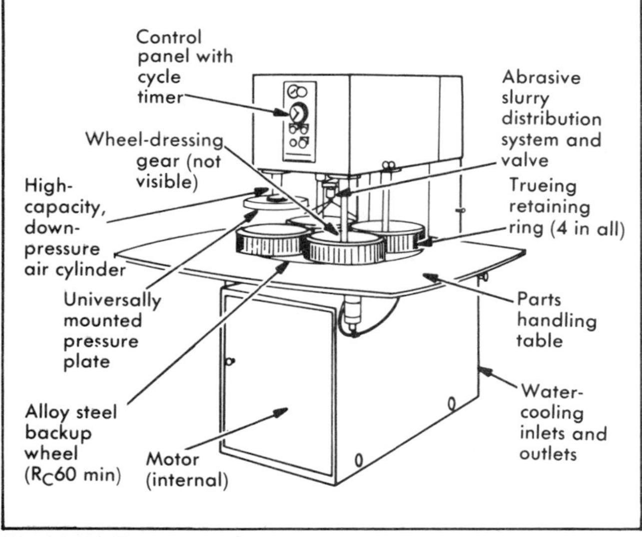

Fig. 16-133 Free-abrasive machine arranged for holding workpieces in four cylindrical retaining rings. (*Speedfam Corp.*)

or may not become impregnated with abrasive grains. The wheel usually sweeps the grains in a rolling slurry film across the faces of the parts being machined, constantly bringing a fresh supply of abrasive into action.

Retaining rings are generally made from the same material as the wheel, but are sometimes made from a different material. Production can be speeded up by using an extra set of retaining rings, which can be loaded with parts while other parts are being machined in the other set of rings.

The pressure plates may exert force on the workpieces by weights or by a variable-pressure pneumatic system. On some machines, the retaining rings are prevented from being ejected from the wheel by the pressure plates. Ejection is prevented by connecting each plate to the piston of an air cylinder with a universal joint, thus permitting finishing parts of varying heights. On other machines, the retaining rings are restrained by two yoke-supported rollers near wheel-top level. In one design, one roller at each station is powered to assist the rotation of the retaining ring.

Surface finishes as smooth as 2 μin. (0.051 μm) can be produced on parts, and flatness is easily attained with a properly maintained wheel. A slight concavity can be produced by making the wheel slightly convex. Parallelism is attained by transposing the workpieces between operations. Felt or rubber pads are sometimes placed between the pressure plates and top surfaces of the workpieces for the initial operation. With several transpositions, parallelism and part size can be held to a tolerance of 0.0001" (0.003 mm).

Free-abrasive machining is ideally suited for small workpieces, but some relatively large parts are also finished by the process. Stock removal rates (see Table 16-47) are much lower than conventional surface grinding when machining one or a few parts at a time, but production increases rapidly when large numbers of small parts are machined simultaneously. In general, stock removal rates in free-abrasive machining are proportional to the speed and pressure, and surface finish is proportional to the grain size and inversely proportional to the hardness of the work. The process has been found to be competitive with other methods of finishing when stock removal is in the 0.002 to 0.010" (0.05 to 0.25 mm) range on hard metals, and up to 0.040" (1.02 mm) on softer metals.

A semiautomated machine for automotive transmission housings differs from the standard machine previously described in several ways. Housings are palletized for conveyance to, through, and out of the free-abrasive machine. The weight of these pallets provides sufficient pressure, thus eliminating the need for pressure plates. Workpieces are loaded at the first of four stations and then indexed incrementally around the machine table. Loading and unloading are done with a toolchanger-type arm.

Another type of free-abrasive machine (see Fig. 16-134) resembles lapping machines in that abrasive materials are fed to a circular, rotating Meehanite plate. While some abrasive becomes embedded in the Meehanite, most of the loose abrasive particles are carried across the face of the rotating plate. Workpieces are held in three or four retainer rings, depending upon the size of the machine, and are subjected to a predetermined pressure by pneumatic arms applying force to pressure plates that fit inside the rings. The powered pressure arms are capable of exerting 900 lb (4000 N) of force and swing 360° to facilitate loading and unloading.

The rotating-plate wheels on these machines are usually Meehanite (especially for heavy-duty stock removal), but are

TABLE 16-47
Typical Stock Removal Rates and Surface Finishes in Free-Abrasive Machining Various Materials

	Abrasive*							
	#10		#30		#50		#60	
Workpiece Material	Stock Removal, in./hr (mm/hr)	Surface Finish, μin. (μm)	Stock Removal, in./hr (mm/hr)	Surface Finish, μin. (μm)	Stock Removal, in./hr (mm/hr)	Surface Finish, μin. (μm)	Stock Removal, in./hr (mm/hr)	Surface Finish, μin. (μm)
Aluminum	0.020 (0.51)	12 (0.305)	0.040 (1.02)	16 (0.406)	0.100 (2.54)	30 (0.762)	0.200 (5.08)	45 (1.143)
Bakelite	0.025 (0.64)	28 (0.711)	0.060 (1.52)	40 (1.016)	0.250 (6.35)	62 (1.575)	0.360 (9.14)	75 (1.905)
Brass	0.008 (0.20)	5 (0.127)	0.015 (0.38)	10 (0.254)	0.040 (1.02)	15 (0.381)	0.060 (1.52)	25 (0.635)
Carbon	0.240 (6.10)	18 (0.457)	0.500 (12.70)	28 (0.711)	1.00 (25.4)	45 (1.143)	1.50 (38.1)	55 (1.397)
Cast iron	0.005 (0.13)	6 (0.152)	0.008 (0.20)	12 (0.305)	0.012 (0.30)	18 (0.457)	0.018 (0.46)	30 (0.762)
Copper	0.030 (0.76)	18 (0.457)	0.048 (1.22)	26 (0.660)	0.100 (2.54)	52 (1.321)	0.120 (3.05)	60 (1.524)
Ferrite	0.060 (1.52)	17 (0.432)	0.150 (3.81)	25 (0.635)	0.500 (12.70)	35 (0.889)	0.600 (15.24)	45 (1.143)
Glass	0.100 (2.54)	15 (0.381)	0.150 (3.81)	25 (0.635)	0.235 (5.97)	35 (0.889)	0.300 (7.62)	50 (1.270)
Hard chromium	0.001 (0.03)	8 (0.203)	0.002 (0.05)	10 (0.254)	0.005 (0.13)	15 (0.381)	0.006 (0.15)	18 (0.457)
High-alumina (92%) ceramic	0.002 (0.05)	14 (0.356)	0.004 (0.10)	16 (0.406)	0.025 (0.64)	25 (0.635)	0.060 (1.52)	35 (0.889)
Inconel X	0.001 (0.03)	7 (0.178)	0.002 (0.05)	9 (0.229)	0.005 (0.13)	14 (0.356)	0.007 (0.18)	16 (0.406)
Molybdenum	0.001 (0.03)	6 (0.152)	0.002 (0.05)	10 (0.254)	0.004 (0.10)	18 (0.457)	0.006 (0.15)	22 (0.559)
Nickel	0.002 (0.05)	6 (0.152)	0.004 (0.10)	10 (0.254)	0.008 (0.20)	22 (0.559)	0.010 (0.25)	25 (0.635)
Nylon	0.030 (0.76)	14 (0.356)	0.060 (1.52)	18 (0.457)	0.095 (2.41)	26 (0.660)	0.120 (3.05)	30 (0.762)
Rubber, 60 durometer minimum			Not recommended				0.040 (1.02)	35 (0.889)
Silicon			Using 125A abrasive:				0.030 (0.76)	6 (0.152)
Sintered iron, soft	0.010 (0.25)	14 (0.356)	0.017 (0.43)	18 (0.457)	0.032 (0.81)	28 (0.711)	0.040 (1.02)	35 (0.889)
Stainless steel, 303	0.002 (0.05)	4 (0.102)	0.004 (0.10)	8 (0.203)	0.006 (0.15)	12 (0.305)	0.008 (0.20)	20 (0.508)
Stainless steel, 440	0.003 (0.08)	4 (0.102)	0.005 (0.13)	6 (0.152)	0.008 (0.20)	10 (0.254)	0.012 (0.30)	16 (0.406)

(continued)

TABLE 16-47—*Continued*

Workpiece Material	Abrasive* #10 Stock Removal, in./hr (mm/hr)	#10 Surface Finish, μin. (μm)	#30 Stock Removal, in./hr (mm/hr)	#30 Surface Finish, μin. (μm)	#50 Stock Removal, in./hr (mm/hr)	#50 Surface Finish, μin. (μm)	#60 Stock Removal, in./hr (mm/hr)	#60 Surface Finish, μin. (μm)
Steel, soft	0.004 (0.10)	5 (0.127)	0.006 (0.15)	8 (0.203)	0.012 (0.30)	14 (0.356)	0.020 (0.51)	25 (0.635)
Steel, R_C40-45	0.003 (0.08)	4 (0.102)	0.005 (0.13)	7 (0.178)	0.008 (0.20)	12 (0.305)	0.012 (0.30)	20 (0.508)
Steel, R_C60	0.002 (0.05)	3 (0.076)	0.004 (0.10)	6 (0.152)	0.006 (0.15)	10 (0.254)	0.008 (0.20)	15 (0.381)
Stellite	0.005 (0.13)	8 (0.203)	0.010 (0.25)	10 (0.254)	0.030 (0.76)	12 (0.305)	0.050 (1.27)	16 (0.406)
Teflon	0.025 (0.64)	12 (0.305)	0.052 (1.32)	16 (0.406)	0.090 (2.29)	25 (0.635)	0.110 (2.79)	28 (0.711)
Titanium	0.0005 (0.013)	6 (0.152)	0.001 (0.03)	8 (0.203)	0.002 (0.05)	12 (0.305)	0.003 (0.08)	14 (0.356)
Tungsten carbide, C2	0.001 (0.03)	10 (0.254)	0.002 (0.05)	10 (0.254)	0.003 (0.08)	12 (0.305)	0.006 (0.15)	15 (0.381)

(*Speedfam Corp.*)

* Abrasive numbers represent the average size of the grains in microns.

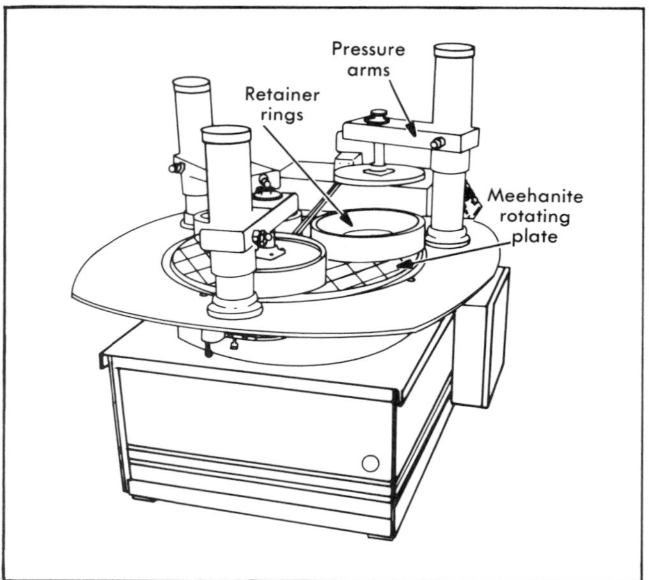

Fig. 16-134 Free-abrasive machine with three retainer rings and swinging pressure arms. (*Spitfire Abrasive Machining Systems*)

sometimes made of soft plates for light grinding of tungsten carbide and similar operations. The porosity of Meehanite facilitates pickup and retention of the abrasive particles. Cooling water is circulated immediately beneath the table.

Abrasives used are generally silicon carbide and aluminum oxide. Silicon carbide in the 90 to 600 grit range is used for heavy stock removal, and 800 mesh aluminum oxide is generally employed for light finishing operations. Abrasives are fed to the wheel by capillary action with a multiple-feed cup at the center of the wheel.

Free-abrasive machines, also called lapping machines, are available for double-side processing; finishing is accomplished on both sides of a quantity of workpieces simultaneously. Previously machined or as-cast surfaces can be precision finished flat and parallel on these machines, thus eliminating the need for grinding. A water slurry of free abrasives is automatically fed down through the upper lapping wheel to provide a film of rolling abrasive grains between the workpieces and faces of the lapping wheels.

At the start of a cycle, light pressure is applied until the high points on the workpiece surfaces are abraded. The pressure is gradually increased and then maintained constant until the workpieces are finished. All materials can be processed by free-abrasive machining. One example is the finishing of sintered carbide cutting tool inserts, accomplished at the rate of 1500 inserts per hour.

Bibliography

Belanger, James A. *Innovations in Flap Wheels.* SME Technical Paper MR81-404. Dearborn, MI: Society of Manufacturing Engineers, 1981.

BIBLIOGRAPHY

Estabrook, David T. *Vaqua High Volume Liquid Abrasive Process*. SME Technical Paper MR83-190. Dearborn, MI: Society of Manufacturing Engineers, 1983.

Fischer, Hans. *Honing*. SME Technical Paper MR82-939. Dearborn, MI: Society of Manufacturing Engineers, 1982.

Gazan, George A. *A "Hole" New Ball Game*. SME Technical Paper MR81-384. Dearborn, MI: Society of Manufacturing Engineers, 1981.

Gillespie, LaRoux K. *Advances in Deburring*. Dearborn, MI: Society of Manufacturing Engineers, 1978.

———. *A Guide to Deburring, Deflashing and Trimming Equipment, Supplies and Services*. SME Bibliography Series Report. Dearborn, MI: Society of Manufacturing Engineers, 1976.

———. *A Training Guide for Precision Hand Deburring*: Part 1, Bendix Kansas City Report BDX-613-2400, May 1980. Part 2, Bendix Kansas City Report BDX-613-2534, Nov. 1980. Part 3, Bendix Kansas City Report BDX-613-2572, Nov. 1980. Part 4, Bendix Kansas City Report BDX-613-2582, Nov. 1980. (Available from National Technical Information Service, NTIS.)

———. *Deburring Capabilities and Limitations*. Dearborn, MI: Society of Manufacturing Engineers, 1976.

———. *Deburring Technology for Improved Manufacturing*. Dearborn, MI: Society of Manufacturing Engineers, 1981.

———. *Tools Used For Hand Deburring*. Bendix Kansas City Report BDX-613-2588, March 1981 (available from NTIS).

Hignett, J. Bernard, and Coffield, John. *Automated High Energy Mass Finishing*. SME Technical Paper MR83-693. Dearborn, MI: Society of Manufacturing Engineers, 1983.

Indge, John H. *The Technology of Precision Flat Honing*. SME Technical Paper MR84-904. Dearborn, MI: Society of Manufacturing Engineers, 1984.

Kittredge, John B. *The Mathematics of Mass Finishing*. SME Technical Paper MR81-399. Dearborn, MI: Society of Manufacturing Engineers, 1981.

———. *The Vibratory Finishing Compound Solution System*. SME Technical Paper MR82-239. Dearborn, MI: Society of Manufacturing Engineers, 1982.

Laschy, Bernard. *Techniques of Barrel and Vibratory Finishing*, 2nd ed. St. Paul, MN: BJL Co., 1981.

McAleer, Howard J. *Polish Up Your Buffing Data*. SME Technical Paper MR79-755. Dearborn, MI: Society of Manufacturing Engineers, 1979.

Murphy, James A. *Surface Preparation and Finishes for Metals*. New York: McGraw-Hill Book Co., 1971.

Olsen, Kent R. *New Developments In High Pressure Spindle Finishing*. SME Technical Paper MR81-391. Dearborn, MI: Society of Manufacturing Engineers, 1981.

Rhoades, Lawrence J. *Cost Guide to Automatic Finishing Processes*. Dearborn, MI: Society of Manufacturing Engineers, 1981.

———. *What's New in Abrasive Flow Finishing*. SME Technical Paper MR83-200. Dearborn, MI: Society of Manufacturing Engineers, 1983.

Safranek, W. H.; Secrest, A. C.; and Turn, J. C. *Chemical Accelerators for Vibratory Finishing and Deburring of Zinc Die Castings*. SME Technical Paper MR76-832. Dearborn, MI: Society of Manufacturing Engineers, 1976.

Southem, Malcolm. *High Density Polyester Media for Mass Finishing*. SME Technical Paper MR83-680. Dearborn, MI: Society of Manufacturing Engineers, 1983.

Stauffer, Robert N. "What You Should Know About Vibratory Finishing." *Manufacturing Engineering* (July 1979), pp. 48-54.

Takazawa, Koya; Shinmura, Takeo; and Hatano, Eiju. *Development of Magnetic Abrasive Finishing and Its Equipment*. SME Technical Paper MR83-678. Dearborn, MI: Society of Manufacturing Engineers, 1983.

Westerman, William J. *An Overview of Roller Burnishing as a Surface Conditioning Technique*. SME Technical Paper MR81-401. Dearborn, MI: Society of Manufacturing Engineers, 1981.

Woelfel, Michael M. *Glass Bead Blasting for Surface Conditioning and Improvement*. SME Technical Paper MR83-686. Dearborn, MI: Society of Manufacturing Engineers, 1982.

THERMAL, CHEMICAL AND ELECTROCHEMICAL FINISHING

THERMAL ENERGY METHOD

Burrs and flash, both internal and external, are rapidly burned away by the thermal energy method (TEM), also called thermal energy deburring (TED). Parts to be deburred or deflashed are placed in the chamber (pressure vessel) of a machine in which a gas mixture is ignited to create intense heat.

ADVANTAGES AND LIMITATIONS

While the equipment required is more expensive than for more traditional deburring systems, the thermal energy method is the fastest known means of deburring most parts. Burning of the burrs takes less than 20 milliseconds. The usual cycle time for a high-production machine, including loading and unloading, is about 25 seconds. Results achieved are of consistent quality, reducing the need for inspection and subsequent part rejection or rework; and treated parts are free of contaminants.

The process is ideal for metal parts through which fluids or gases must flow. It has the unique ability of deburring blind and intersecting holes, while simultaneously deburring the outside surfaces of parts. Since the process does not use any abrasive media, there is no change in the dimensions of surfaces adjoining the burrs; also, these surfaces are not generally affected by heat. The bodies of most parts processed rarely exceed a temperature of 300° F (150° C). With a water-cooled chamber and the fast cycling, only thin parts, such as some stampings, get hot enough to disturb the parent metal. A recast layer that forms at the workpiece-burr interface is usually of negligible thickness—0.00008 to 0.001" (0.0020 to 0.03 mm); however, this layer may cause cracks in previously hardened steels.

Burrs or flash of uniform thickness, but limited size, are removed completely, down to their roots. With parting lines, however, which generally have a thicker root than a burr, a rounded rise is left on the surface of the part. When designing parts to be thermally deburred, the thinnest sections that have useful functions should be at least 15 times thicker than the burrs. This design limitation is beneficial in that the process will not remove threads, but only slightly round their extreme edges. For most stamped parts, however, which typically have triangular-shaped burrs, the process removes only the tips of the burrs.

Parts to be deburred by the thermal energy method must be free of oil. If oil remains on the parts, it will be burned into the surfaces, leaving a carbon smut that is difficult to remove. Also, blind and tapped holes must be free of compacted chips. If the gas cannot surround each chip, they will not be removed by the process. Another possible limitation is that it is not always possible to achieve specific radii on individual edges. Some edge radiusing is possible on steel and cast iron parts, but is seldom achieved on aluminum or stainless steel workpieces. In many cases, it is possible to remove burrs, but keep edges sharp.

Safety has been no problem with hundreds of machines in use for many years. Nevertheless, local and federal safety regulations require that these machines meet stringent safety codes.

APPLICATIONS

The thermal energy method is being used to deburr and deflash many different parts. Impact of the process on costs, production rates, and suitability of the parts varies with the specific applications. All metals can be treated by TED, but certain metals are more difficult to process. The oxidation resistance of stainless steels, for example, makes them less adaptable to the method. Metals with low thermal conductivity, such as steels and zinc alloys, however, give excellent results.

Iron and Steel Castings

Deburring and deflashing of iron and steel castings are major applications of the thermal energy method. The process is particularly suitable for and offers the most substantial savings with castings having internal intersections of recesses and bores that are difficult to reach. Hydraulic and

CHAPTER CONTENTS:

*Contributors of sections of this chapter are: **John F. Jumer**, President, Alchemize Corp. and Electro Glo Co.; **Jerome F. Miller**, President, Chemtool, Inc.*

*Reviewers of sections of this chapter are: **Donald Ball**, President, Anocut, Inc.; **Raymond A. Barry**, Supervisor, Technical Service, Globerite Dept., Ashland Chemical Co.; **Raymond G. Dargis**, Marketing Manager, Electroplating Processes, McGean-Rohco; **L. K. Gillespie**, Senior Project Engineer, Kansas City Div., Bendix Corp.; **J. Bernard Hignett**, Vice President, Harper Co.; **John F. Jumer**, President, Alchemize Corp. and Electro Glo Co.; **Horst Kissel**, Manager, ECD Engineering, SurfTran Div., Robert Bosch Industries; **Jerome F. Miller**, President, Chemtool, Inc.; **Robert Narins**, President, Deburring Laboratories, Inc.; **Roland R. Ricci**, President, Chemform, Inc.; **Ronald Sonego**, President, SurfTran Div., Robert Bosch Industries.*

pneumatic valve bodies are typical castings that are deburred and deflashed by TED.

Zinc Die Castings

A wide variety of zinc die castings are being deburred, deflashed, and cleaned by the thermal energy method. Carburetor components, automotive-lock cylinders, and many other automotive parts are being treated in this way. A major benefit is the consistently high quality obtained. Substantial savings also result because no tooling is necessary and no change in setup is required to handle parts of different shapes or sizes.

Machined Steel Components

Stainless steel valve spools and steel gears are just two of many machined components treated by the thermal energy method. Ring grooves and slots on the valve spools are deburred at the rate of 720 spools per hour, with six spools being handled per cycle. For the gears, burrs are removed from the teeth and from tapped and through holes at the rate of 120 gears per hour.

Plastics

Deflashing and cleaning of plastics parts is a recent development for the thermal method. Machines used for such parts operate at lower pressures and temperatures, and gases are monitored more precisely. The chamber is generally evacuated of air before gases are introduced to maintain precise mixtures and to minimize contamination. Temperatures are held to about 2000° F (1095° C); and because of the short duration of the heat, workpiece temperatures rise only approximately 30° F (15° C), with a maximum increase of 100° F (55° C).

Aluminum

Heat sinks used to dissipate heat from electronic modules are one of many aluminum components being deburred by the thermal energy method. Cycle time has been reduced from 15 minutes to 1 minute per unit, and the quality of the heat sinks has been improved.

PROCESS PRINCIPLES

Heat shock in the thermal energy method is applied by burning a mixture of oxygen and combustion gas (generally natural gas) (see Fig. 17-1), with a ratio of oxygen to combustion gas of 2 1/2 to 1. To release the necessary energy, the mixture of gases is compressed before entering the sealed and pressurized work chamber. The gas mixture distributes itself evenly throughout the chamber, and in all holes and recesses of the workpieces. The more the gases are compressed, the greater the energy and the stronger the deburring action.

The combustible gas mixture is ignited with a 30,000 volt spark, creating a temperature generally in the range of 2500 to 3500° C (4530 to 6330° F). Within two to three milliseconds, the fuel is burned and the fire extinguished; however, the burrs or flash burst into flame and burn in the excess oxygen. Burrs and flash continue to vaporize until the heat begins to dissipate into the body of the workpiece, when the flames extinguish themselves.

In the process of vaporizing, the burrs or flash become oxides of the metal being processed—for example, iron oxide from steel parts and aluminum oxide from aluminum parts. The oxide settles on the parts as a loose, powdery residue. Although the workpiece surfaces are discolored, they are not oxidized. The residue can be left undisturbed on aluminum or zinc components since it does not hinder further processing, such as enameling or plating. When necessary, the residue can be washed off with a suitable cleaner. Chemical post-treatment of ferrous components often eliminates the need for cleaning.

MACHINES USED

Machines for thermal energy deburring are available with C frames (see Fig. 17-2) or three posts to hold the water-cooled chamber halves together. A mechanical toggle mechanism is provided to clamp and lock the chamber. Machines are now available to handle parts up to 10″ (254 mm) in diameter and 30″ (762 mm) long.

A dial-type indexing table permits unloading and reloading a lower closure (bottom half of the chamber) while parts on

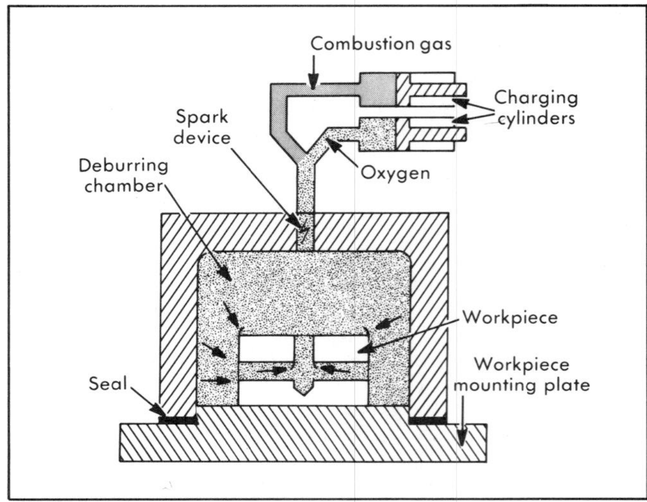

Fig. 17-1 Schematic of thermal energy method for deburring. Burrs on workpiece are indicated by heavy arrows. (*SurfTran*)

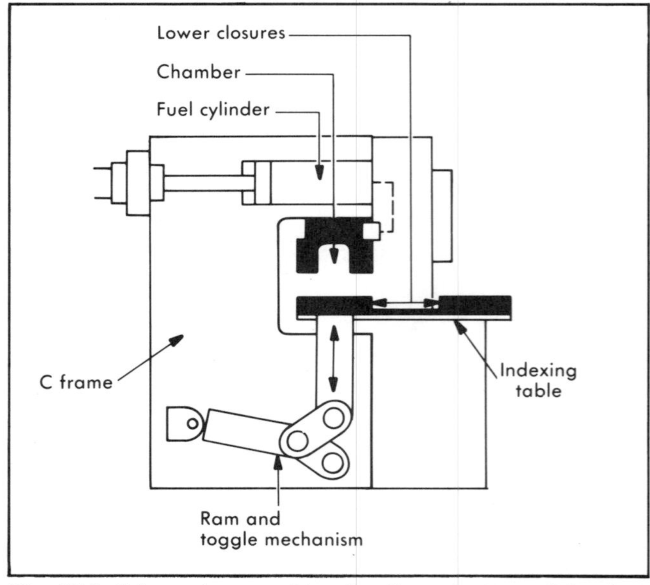

Fig. 17-2 C-frame-type machine for thermal energy deburring. (*SurfTran*)

another closure are being processed. The number of closure stations on the table depends upon the size of the workpieces and the production requirements.

Workholding Means

Several methods are used to hold parts in the chamber. Large and heavy parts are generally placed directly on the closure plates. Small parts that can be handled in bulk are loaded into pots or baskets that are placed on the closure plates.

Certain parts require insertion in perforated plates, mounting on pins, or restraining in a fixture. Such is the case for parts with external threads or sealing surfaces that must remain nick and scratch-free, and for thin die castings that have little ribbing support to prevent distortion. With certain other parts, the deburring action must be confined to specific areas, with other areas protected.

Gas Supply

Natural gas is supplied to a compressor that fills the fuel cylinder. Oxygen is generally supplied to its cylinder from pressurized bottles. The amount of gas charged into the cylinder is controlled by a system consisting of a series of spacers that can be added or removed to change the strokes of the cylinders. Gages are provided to indicate the fill and chamber pressures. Pressures in the chambers are normally adjustable between 50 and 300 psi (345 and 2070 kPa).

OTHER THERMAL METHODS

In addition to the thermal energy method discussed in the preceding section of this chapter, there are several other, less commonly used methods employing heat to deburr and finish parts. These processes include hot wire deflashing, and resistance heat, flame, electrical discharge, plasma, chlorine gas, and laser deburring. Most of these processes are used in very limited applications but are of interest for special requirements.

HOT WIRE DEFLASHING

External edges and the surfaces of large cutouts of thermoplastic parts can be deflashed with an electrically heated thin wire that is moved parallel to the edges and surfaces. The heat in the thin wire melts and removes the flash. Cycle time is proportional to the lineal distance to be trimmed. With some plastics, it should be possible to deflash at a rate of 24 ipm (610 mm/min) or more.

For this process, the heated wire passes over tensioning and drive pulleys (see Fig. 17-3). On most parts, a radius will not be produced on the edges. Typically, a small projection of flash is left. It is possible, however, to produce a chamfer if required.

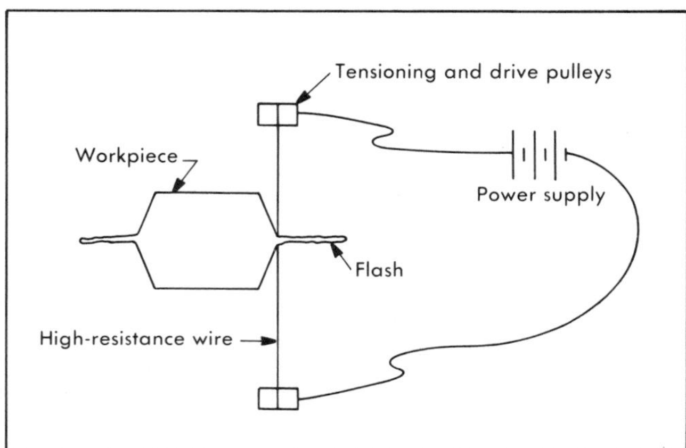

Fig. 17-3 Schematic of hot wire process for deflashing thermoplastic parts.

RESISTANCE HEATING

Heat generated by current passing through a burr can be sufficient to burn away the burr. Workpieces are attached to a power supply (see Fig. 17-4) that provides low-voltage, high-amperage current. The workpiece is attached to the negative side of the electrical source, and a positively charged electrode is placed in contact with the burr. The process can be used for any shape of part, provided electrical current connections can be made. In one application of this process, burrs are removed from gear teeth by using a meshing gear electrode.

Possible limitations of the resistance heating process include the need for the workpieces to be electrically conductive and for the electrode to be movable as the burrs shorten. In addition, the larger the burrs, the higher the current required.

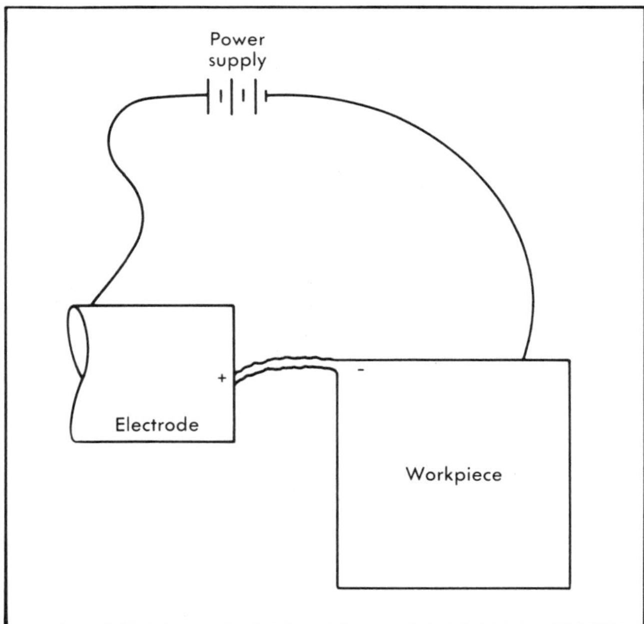

Fig. 17-4 Schematic of resistance heating method of deburring.

TORCH FLAME DEBURRING

The flames from oxyacetylene or similar torches can be used to melt burrs and the edges of parts. This process has been used to trim flash, sprues, and gates from large castings. Torch traverse rates of 20 to 40 ipm (508 to 1016 mm/min) are possible. Most edges exposed to this process will have a radius of 0.005" (0.13 mm) or more, but smaller radii are possible. Possible limitations of torch flame deburring include a heat-affected zone and possibly resolidified droplets around the edges, plus the need for workholding fixtures and some means for traversing the torches. In some cases, electric arcs are being used rather than flames. Robots are currently being applied to automate this process in foundries.

ELECTRICAL DISCHARGE DEBURRING

Burrs can be removed by the electrical discharge machining (EDM) process, discussed in detail in Volume I, *Machining*, of this Handbook series. An electrode and the workpiece are submerged in an electrolyte, and the electrode is advanced to within 0.001 to 0.003″ (0.03 to 0.08 mm) of the burr-laden edge. With power on, sparks melt the burrs, and the electrode continues to advance until it is below the burred edge. For circular parts with burrs projecting from their diameters, rod or tube electrodes can be used.

This process removes heavy burrs fairly quickly, but it leaves a recast surface at the edges having a roughness that is normally 30 to 100 μin. (0.76 to 2.54 μm). The recast surfaces must be removed if the parts are to be used for highly stressed, fatigue-prone applications. Most edges treated by this process will have a small radius, about 0.003″ (0.03 mm), but larger radii can be produced with special tooling.

PLASMA AND LASER DEBURRING

Burrs can be removed by placing the edges of parts in a plasma flame. Thermal energy from the flame is concentrated at the corners and edges of the parts, and the high temperature quickly melts the burrs. This process is limited in that both small and thin precision parts may be distorted. Loose fiber ends resulting from machining fiberglass-reinforced products are being removed by laser beams and glow-discharge plasma deburring.

CHLORINE GAS DEBURRING

On an experimental basis, burrs have been removed from heated steel workpieces by exposing them to a chlorine atmosphere in a closed chamber. The workpieces are heated to 315° C (600° F) or more before being exposed to the atmosphere. The chlorine combines with iron in the steel to form iron chloride, which vaporizes below 315° C. Because of the thin cross section and the high surface-area-to-mass ratio of the burrs, they become hotter than the parent metal and are therefore removed faster than material from the workpiece surfaces. The burr removal cycle time ranges from 20 to 60 seconds.

Possible limitations to the use of this process are that it is effective only on iron-based alloys and the process changes the dimensions of the workpiece. A one-minute cycle removes 0.004 to 0.010″ (0.010 to 0.25 mm) of stock from the workpiece surface, depending upon the workpiece temperature. Also possibly limiting the use of this process are the special necessary safety precautions due to the toxicity of chlorine. No known chlorine gas deburring units are commercially available, apparently because of safety considerations.

ELECTROPOLISHING

Electropolishing or, more correctly, electrochemical polishing has been used since the early 1930s, originally for metallography. The process is similar to electroplating with respect to processing, except that, in electropolishing, metal is selectively removed rather than deposited. Under proper conditions, metal is removed uniformly, thus causing smoothening and/or brightening. With proper techniques, the process can also be made burr selective—the burr being removed faster than adjacent stock. However, electropolishing cannot be confined to burr removal without stock removal unless the adjacent metal is electrically insulated or masked.

When the current is applied in electropolishing, a polarized film forms on the surface of the metal. The nature of this film is responsible for both the brightening and leveling actions. Film strength and viscosity of the electrolyte are responsible for the nature of this anodic film. The metal ions are converted to metallic salts and must diffuse through this film. The film is thinner over the projections and thicker over the depressions of the metal, allowing the projections to be more exposed to electrolytic action and to have lower electrical resistance than the depressions. The thinner film results in the projections being more rapidly dissolved by electrolysis than the depressions, which are more protected by the thicker film.

Atoms of the metal being removed in electropolishing pass through the polarized anodic film as metal salts and enter the electrolyte, either to be dissolved therein, deposited on the cathode, or to precipitate as sludge. Electropolishing solutions are therefore considered full sludging, semisludging, or nonsludging solutions.

APPLICATIONS

A number of specialized electrochemical processes have evolved over the years, all based on the mechanisms of electropolishing. Some of these processes are:

- Electrochemical brightening. In this process, surfaces are brightened, but not smoothened or leveled as in conventional electropolishing. A smoother surface is required initially for electrochemical brightening.
- Electrochemical deburring, discussed later in this chapter.
- Electrochemical machining (ECM), discussed in Volume I, *Machining*, of this Handbook series.

In addition to smoothening and brightening operations, electropolishing is used for many other applications. These include the reduction of friction, fatigue, and out-gassing. Electropolishing can eliminate the gases, vapors, and volatiles absorbed on surfaces, which would be released during pump-down for high-vacuum service. The process provides superior results in antistick and release properties, wear and corrosion resistance, and passivity. Electropolishing is also used as a means of inspecting metal surfaces and welds, and as a means of controlling size and/or weight. Electropolishing is applicable to a great variety of metals and alloys, the most popular being: aluminum and aluminum alloys, copper and copper alloys, iron, steel, and stainless steels.

PRETREATMENT REQUIREMENTS

Some type of pretreatment is usually required prior to electropolishing. If the surfaces of the workpieces are clean, however, this step is unnecessary. Grease, oil, soil, or combinations of these are removed either by solvent cleaning, degreasing, or aqueous alkaline cleaning (soak cleaning), all discussed in Chapter 18.

A variation of soak cleaning is electrolytic cleaning, also discussed in Chapter 18. In electrolytic cleaning, low-voltage direct current is applied to cause gassing. The gas released effects a scrubbing action for better contaminant release. Agitation of the solution, mechanical or by air, is beneficial. Heat tint, oxide, or scale, if present, should be removed chemically. Such removal is accomplished either by acid or alkaline descaling, sometimes with the aid of either ultrasonics or pickling for heavier or more tenacious deposits.

EQUIPMENT USED

The electropolishing tank may either be made of austenitic stainless steel or mild steel lined with polyvinyl chloride, polyethylene, polypropylene, fiberglass, or special types of rubber. For some applications, either lead-lined or plain steel tanks are used. However, steel tanks are electrically conductive, which may create problems and be undesirable. Selection of materials for tank construction depends mainly on the electropolishing solution, composition, and operating temperature. Electropolishing tanks are usually rectangular, greater in length than width.

The electropolishing tank is fitted with tank rods or bars, and the pieces to be electropolished are hung on the work (anode) rods, which carries positive current. On a single work-rod tank (see Fig. 17-5), the rod is centered with respect to the width of the tank and is parallel to the longer tank sides. The cathode rods are parallel to and equidistant from the anode or work rod, and are charged with negative current.

In some instances, tanks are equipped with two work rods and then require three cathode rods (see Fig. 17-6). The rods are arranged in the same manner as in a single work-rod tank.

Provisions should be made for heating and/or cooling the solution. Heating is achieved by either electric immersion heaters, or by hot water or steam passing through coils. The coils are made of lead, stainless steel, titanium, or, more recently, Teflon. The same coils can be used for cooling the solution with water, which can then be used in rinsing as well. Automatic temperature control is generally provided.

Low-voltage d-c is furnished by a rectifier, similar to that used in electroplating. The rectifier should be provided with a means of controlling d-c output, either by voltage control or constant-current control for specialized applications.

Rinse tanks are made of the same material as the electropolishing tanks, except when operating temperatures may differ. In such cases, lower cost materials and construction are used when possible. Rinse tanks are fitted with a means for draining and filling, and may be arranged for counterflowing to reduce effluents and water consumption.

The pieces to be electropolished are usually racked. The rack not only holds the pieces in the most advantageous positions for electropolishing, but also supplies the current to the pieces, either through one or several contact points. Any number of pieces can be put on a rack, providing the electropolishing tank is of sufficient size and volume. Usually several racks are put in the tank at the same time.

Some pieces may not require racking and can be hung on J-hooks, which may be less productive since the full depth of the solution may not be utilized. Small pieces that do not tangle or nest can be processed in bulk with special equipment at rates up to 8000 parts per hour. Bulk processing avoids much of the labor normally associated with hand racking. Equipment is also available for reel-to-reel or continuous electropolishing of wire and strip stock.

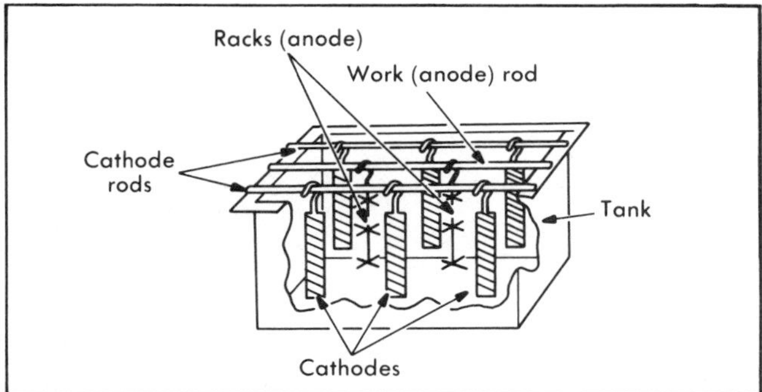

Fig. 17-5 Typical electropolishing tank with single work rod. (*Electro Glo Co.*)

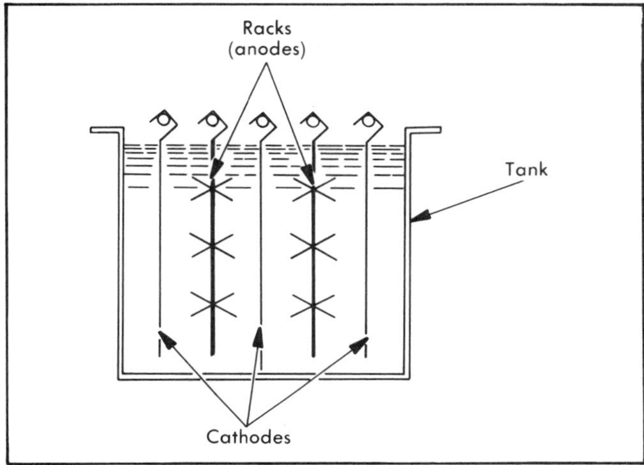

Fig. 17-6 Electropolishing tank with two work rods. (*Electro Glo Co.*)

ELECTROPOLISHING SOLUTIONS

There is no one electropolishing solution that can successfully electropolish all metals and alloys well; therefore, many specific types of electropolishing solutions exist. One supplier makes electrolytes (solutions) for over 55 different metals and alloys.

Electropolishing solutions are most often acid-based for the more commonly electropolished alloys; however, other electrolytes may be alkaline, cyanide, or metal salt systems. The characteristics of these various electrolytes are not only different in composition, but also in operating conditions—mainly operating temperatures and current densities. Temperatures may range from ambient to near their boiling points, and current densities range from 2.70 to 27.00 A/dm^2 (25 to 250 A/ft^2), depending upon the specific solution.

For many years, the most widely used electropolishing solutions for various metals and alloys were of a concentrated acid nature (see Table 17-1). These acids included acetic, citric, chromic, hydrofluoric, phosphoric, sulfuric, tartaric, and others. Some of these compositions are now considered toxic. Phosphoric acid is the major constituent in many of the acid-type solutions.

CHAPTER 17

ELECTROPOLISHING

TABLE 17-1
Typical Electropolishing Solutions

Material	Solution Components	Component Percentage (by weight)	Temperature, °F (°C)	Current Density, A/ft² (A/dm²)
Steel	Phosphoric acid* Trialkali metal phosphate Alkali metal sulfate	55-85 1-15 0.5 min	160-180 (70-80)	400-800 (43-86)
Stainless steel	Phosphoric acid** Glycerin	35 55 min	200 min (95)	25 min (2.7)
	Sulfuric acid Phosphoric acid Water	41 45 14	170-230 (75-110)	200-350 (22-38)
	Phosphoric acid Chromic acid Water	56 12 32	80-175 (25-80)	100-1000 (11-108)
Aluminum	Sulfuric acid† Hydrofluoric acid	1-60 0.2-1.5	140 (60)	100 (11)
	Sulfuric acid‡ Phosphoric acid Chromic acid Trivalent metals	4-45 40-80 0.2-9 6	160-200 (70-95)	25-950 (2.7-103)
	Sodium carbonate (anhydrous) Trisodium phosphate	(pH 10.5)	176-180 (80-82)	20-30 (2.2-3.2)
Copper and alloys	Sodium tripolyphosphate§ Boric acid	(pH 7-7.5)	125-135 (50-55)	100 min (11)
	Sulfuric acid Phosphoric acid Chromic acid Water	14 59 0.5 36.5	60-170 (15-75)	100-1000 (11-108)
Nickel and alloys	Sulfuric acid Water	70 30	Room	50-200 (5.4-22)

* Patented by Hammond, Edgeworth, and Bowman.
** Patented by H. Uhlig.
† Patented by Mason and Tosterud.
‡ Patent 2,550,544. 1951.
§ Patented by H. Wiesner.

More recently, organic acids and addition agents are used with acid-type electrolytes to reduce many of the disadvantages inherent in the straight acid-type electropolishing solutions. Such disadvantages include poor throwing power, high operating temperatures, narrow operating range, excessive current densities, and short useful life.

The majority of electropolishing solutions in use today are of a proprietary nature, supplied much in the same manner as in the plating industry. An important advantage of proprietary solutions is that considerable research and development go into the product. An added advantage is the technical and laboratory assistance available to the user when necessary.

Electropolishing solutions based on cyanides or solutions containing chromium compounds, halogens, or other noxious materials require exhausting with hoods or lateral exhaust systems. In concentrated phosphoric-acid-type solutions, a slight spray is liberated that rises only several inches; it is not harmful, other than its being corrosive. The gases produced for phosphoric-acid-type solutions are hydrogen and oxygen, and air circulation should be provided. Effluents resulting from rinsing after electropolishing and metallic salts in the tank sludge must be disposed of in accordance with EPA requirements.

OPERATING CONSIDERATIONS

The quality of surface finish is usually evaluated either visually or measured by a stylus-type instrument. With instrument measurement, surface finish is usually expressed as microinches (μin.). The average height of surface roughness is expressed in millionths of an inch [0.0256 microns (μm)]. These measurements are satisfactory for mechanically produced surfaces, but do not account for surface geometry, which is three dimensional. As a result, these measurements may not provide adequate data when appearance, reflectance, and geometry are significant on electropolished surfaces.

Surface conditions, regardless of metal or alloy, determine to a large extent the degree of successful electropolishing. The greatest contributing factors toward unsatisfactory results are:

large grain size, nonuniform structure, nonmetallic inclusions, directional roll marks, salt or scale contaminations, over-pickling, and insufficient or excessive cold reduction.

Most electropolishing is done with low-voltage d-c, with the voltage usually ranging from 5 to 18 V. Ordinarily, current densities vary with the type of solution and the purpose of the electropolishing. Each type of electropolishing solution has an ideal current density range.

The quantity of electropolishing done is determined by the current density (the number of amperes of current applied to a given area) and the time the current is applied. When the distance between the piece to be electropolished (the anode) and the electrode (the cathode) is small, the voltage required to produce the desired amperage is low. As the distance is increased, an increase in voltage is needed to overcome the additional resistance to maintain the same current density.

Since current takes the path of least electrical resistance, areas that are more distant or not receiving the same amount of current will not respond the same as in the high current density areas. The term expressing the quantity or quality of coverage of the lower current density areas is commonly referred to as *throwing power*.

Due to the nature of the electrolyte, corners, edges, and protrusions (including burrs) on workpieces are natural high current density areas, where electrolytic action is greatest. Conversely, areas more distant, such as recesses, holes, and inside corners, receive less current. Being in low current density areas, conforming cathodes (see Fig. 17-7) are required to attain a uniform current density over the entire surface. Cathode design to reduce high current density areas is illustrated in Fig. 17-8.

Factors affecting throwing power are: the size and shape of the workpiece, the size and shape of the processing tank, the number and location of the electrodes, and the size and shape of the electrodes. Also affecting the throwing power are the chemistry or composition of the electropolishing solution, the processing time, the temperature, and the current density employed.

POST-TREATMENT

Nearly all electropolishing solutions are aqueous systems, with water used for rinsing. Several rinses are usually required after electropolishing, unless the part is too large to be rinsed by immersion and must be hosed off. The number and type of rinses are dictated by the metal or alloy to be rinsed and the end use intended. Acid or alkaline neutralizing rinses may be necessary between water rinses. Overflow-type rinse tanks are being replaced by more efficient, counterflow rinsing tanks, and spray rinsing has gained wide use. In the case of static or overflow rinsing, agitation is beneficial, either by air or mechanical agitation of the solution or the workpieces.

Drying, staining, and spotting usually occur from the final rinse water, either contaminated from the carryover of the previous rinses or from the hardness of the water itself. If the final rinse water is hot, evaporation further concentrates the contaminants, which, when dried, results in staining or water spotting. This problem can be alleviated by more frequent changes of the final rinse or with the use of a suitable wetting agent. Wetting agents eliminate water beading by reducing surface tension, leaving a thin water film and less deposit. The ideal final rinse, although more costly, would use deionized or distilled water. Recently, the use of Freon for water displacement and rapid drying is becoming popular in specialized cases. Electropolished parts may be further processed with operations such as plating, anodizing, and conversion and organic coating.

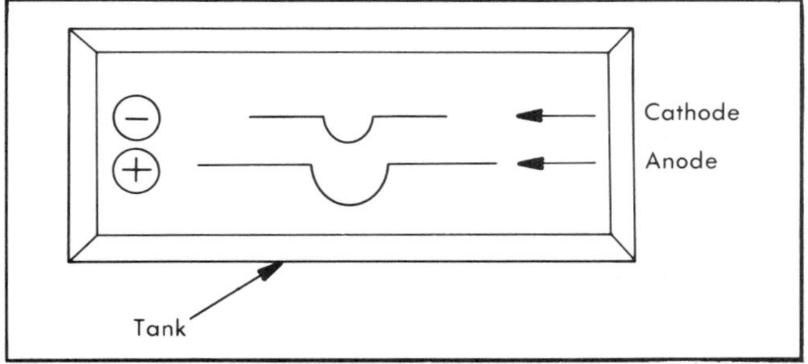

Fig. 17-7 Conforming cathode arrangement for improved throwing power. (*Electro Glo Co.*)

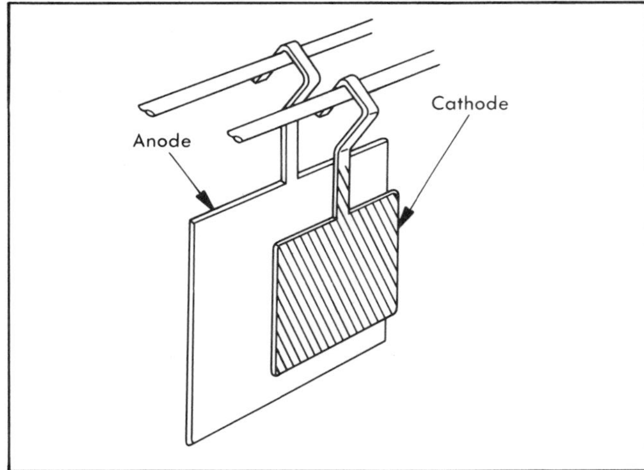

Fig. 17-8 Example of reduction of high current density areas by cathode design. (*Electro Glo Co.*)

CHEMICAL POLISHING AND BRIGHT DIPPING

Over the years, the progression of various chemical surface treatments for metals evolved into what has become known as chemical polishing. During the industrial revolution, the pickling, or acid dipping, of ferrous metals became important as a means of removing oxide and scale. At the same time, an etch or open structure resulted in the surfaces of the metals, which provided better bonding or adhesion for galvanizing and enameling.

Copper and copper alloys, such as brass and bronze, required a pickle for similar reasons. The pickle most generally used was a mixture of sulfuric and nitric acids with various additives. It was found (accidently) that, under the proper conditions, a brightening of the surface was attainable; hence, "bright dipping" became the accepted term.

CHAPTER 17

CHEMICAL POLISHING AND BRIGHT DIPPING

PROCESS DIFFERENCES

Chemical polishing and bright dipping are incorrectly used interchangeably. Bright dipping evolved from pickling in much the same way that chemical polishing followed bright dipping. While bright dipping is still in wide use, mainly due to its lower cost, it is less sophisticated and generally does not have the surface leveling, or microdeburring effect, that is achieved by chemical polishing. Chemical polishing is selective stock removal, somewhat like electropolishing (discussed in the preceding section of this chapter), whereas bright dipping is not.

As the name implies, bright dips are generally used after electroplating, such as on zinc or cadmium for improved corrosion resistance, and also on solid, nonplated metal surfaces for visual improvement or brightening. The process is also used for improved electroplate adhesion.

Chemical polishing will also brighten the surface of metals, but is more selective in nature than bright dipping. It is thus able, to some extent, to smooth or level the surface and also cause a deburring effect. Chemical polishing is somewhat electrochemical in nature because of the oxidizing agent in the solution; an electrochemical potential thus exists.

In almost all bright-dipping and chemical polishing solutions, an oxidizing agent must be present, usually furnished by the use of nitric acid, chromic acid, dichromates or peroxides, and persulfates. Unfortunately, regardless of the oxidizing agent used, there are disadvantages. The use of nitric acid evolves fumes that are toxic and must be exhausted. Chromic acid and dichromates are reduced from the hexavalent state to trivalent and therefore become exhausted, and the solution must be dumped. Chromium is objectionable in effluents. Peroxides and persulfates are useful only as long as they are able to liberate oxygen and decompose in the presence of iron and/or copper ions. Generally, neither chromic acid nor the peroxide types can be rejuvenated with any success. Recently, two different chemical polishing solutions have been experimentally developed for aluminum that do not require an oxidizing agent.

Bright-dipping and chemical polishing solutions fall into two categories, either dilute or concentrated acid systems. The concentrated acid systems can tolerate more metal concentrations and can usually be rejuvenated, thus extending the useful life, but are more expensive initially. Most dilute bright dips operate at ambient temperature. Concentrated acid systems may operate from ambient to elevated temperatures, sometimes about 200° F (95° C), depending upon the metal or alloy being processed.

Nearly all bright dips and chemical polishing solutions liberate vapors and/or fumes, and hence some degree of exhaust may be required. Also, since most systems are acidic in nature and contain metal salts, the disposal of solution and/or rinses must comply with EPA regulations.

Of the various metals and alloys that are bright dipped or chemically polished, aluminum is by far the most popular, followed by copper and its alloys. The brightening of zinc and cadmium on plated articles accounts for the greatest tonnage of bright-dip chemical consumption. Nickel and its alloys, carbon steels, and stainless steels are also applicable, but to a much lesser extent.

PRETREATMENT REQUIREMENTS

Some type of pretreatment is usually required prior to chemical polishing and bright dipping. If the surfaces of the workpieces are clean, this step is unnecessary. Grease, oil, soil, or combinations of these are removed either by solvent cleaning, degreasing, or aqueous alkaline cleaning (soak cleaning), all discussed in Chapter 18.

A variation of soak cleaning is electrolytic cleaning, also discussed in Chapter 18. In electrolytic cleaning, low-voltage direct current is applied to cause gassing. The gas released effects a scrubbing action for better contaminant release. Agitation of the solution, mechanical or by air, is beneficial. Heat tint, oxide, or scale, if present, should be removed chemically. Such removal is performed either by acid or alkaline descaling, sometimes with the aid of either ultrasonics or pickling for heavier or more tenacious deposits.

EQUIPMENT USED

Because most bright-dipping and chemical polishing solutions are acidic, they require corrosion-resistant equipment. Solution and rinse tanks are usually constructed of austenitic stainless steel, fiberglass, rubber, and plastics such as polyethylene, polypropylene, and polyvinyl chloride. Because of the variety of solution compositions and temperatures used, the various materials of construction must be evaluated to determine which are best suited for the particular application.

Parts may be processed in several different ways, depending upon their size and shape. Parts with flat surfaces or large workpieces are usually racked individually. Small parts that do not tangle or nest and that are not fragile can be processed in bulk in either a dipping basket or a tumbling barrel. When parts cannot be bulk processed and when individual hand racking is expensive, a compromise is the use of compartmentalized or partitioned baskets or containers.

Racks and fixtures are usually made of metal, such as austenitic stainless steel, aluminum, or titanium. Dipping baskets, compartmentalized containers, tumbling barrels, and exhaust equipment are made either of metal or one of the plastics used for tanks.

When either bright dipping or chemical polishing parts in production, a means of cooling or heating may be required, depending upon the particular solution involved. To regulate temperature in the bright-dip or chemical polishing tank, parts can be preheated or cooled in a water tank preceding the process. Most often, cooling is accomplished with water passing through tank coils made of stainless steel, titanium, lead, or Teflon. Heating can be accomplished by means of coils of the same materials, with hot water or steam, or electrically, with the same construction materials previously mentioned or with fused quartz. The materials selected must be evaluated for suitability in the particular solution. The water from heating or cooling can be discharged into the rinse tanks.

SOLUTIONS FOR VARIOUS METALS

Typical solutions for the chemical polishing and bright dipping of various metals are presented in Table 17-2. Operation and control of these solutions, which have been used for many years, is entirely dependent on the proficiency of the operator and is still more of an art than a science. Recent improvements in proprietary formulations for both bright dipping and chemical polishing have reduced many of the shortcomings inherent in these earlier solutions and provide a more scientific means of control and analysis.

TABLE 17-2
Typical Chemical Polishing and Bright Dipping Solutions

Workpiece Material	Solution Components	Component Amounts	Temperature, °F (°C)
Aluminum	Phosphoric acid Nitric acid Acetic acid Water	80% (by vol.) 5% (by vol.) 5% (by vol.) 10% (by vol.)	220 (105)
	Phosphoric acid Hydrogen peroxide Water	75% (by wt.) 3.5% (by wt.) 21.5% (by wt.)	195 (90)
Cadmium	Chromic acid Sulfuric acid	13.4 oz/gal (100 g/L) 1/8 to 1/4 oz/gal (0.9 to 1.9 g/L)	Room
	Hydrogen peroxide (30%) Sulfuric acid	7% (by vol.) 0.3% (by vol.)	Room
Copper and alloys	Sulfuric acid Nitric acid Water Hydrochloric acid	45% (by vol.) 22% (by vol.) 33% (by vol.) 1/2 oz/gal (3.7 g/L)	Room
Iron and steel	Oxalic acid Hydrogen peroxide Sulfuric acid	3.3 oz/gal (24.7 g/L) 1.7 oz/gal (12.7 g/L) 0.01 oz/gal (0.07 g/L)	Room
Stainless steel	Nitric acid Muriatic acid Phosphoric acid Acetic acid (First depassivated by immersion in hot 5% sulfuric acid)	36% (by vol.) 9% (by vol.) 9% (by vol.) 46% (by vol.)	160 (70)
Zinc and alloys	Chromic acid Sodium sulfate (5 to 30 s followed by cold water rinse)	30 to 40 oz/gal (225 to 300 g/L) 2 to 4 oz/gal (15 to 30 g/L)	Room

SOLUTION CONTROL AND ANALYSIS

The need for analytical control of many of the various bright-dipping solutions is not always necessary because some types cannot be rejuvenated and are discarded. However, with some user experience, only the addition of the oxidizing agent is necessary to prolong the useful life, since this constituent is depleted first.

Other considerations for solution control are the amount of metal salt or salts in solution and the water content. These constituents are usually determined and controlled by specific gravity, rather than by analysis. As the metal salt concentration increases, a corresponding increase in water content is usually required, but within stringent limits. Water content excesses, whether too little or too much, will aggravate instead of aid the situation. Exacting specific gravity determinations are necessary, and measurements must always be taken at the same temperature.

OPERATING CONSIDERATIONS

The degree of surface brightness and, in some instances, smoothening are controllable, the time and/or temperature being the variable factors. Stock removal increases with increases in time and temperature up to the point at which increases in either do not improve the brightness or surface smoothness. Excessive immersion time can cause a deterioration of brightness or smoothening and also contributes to shortened solution life due to excessive metal concentration.

Unlike electropolishing, throwing power is not a problem, providing all surfaces being treated are free of gas pockets. Should gas pockets be likely, either a different orientation of the workpiece or agitation of the solution will eliminate them in most cases.

The time required for removal of the piece or pieces from the bright-dipping or chemical polishing tank and movement to the first water rinse tank is called the transfer time. This time should be kept to a minimum because chemicals on the metal surface that are exposed to the air too long can either dull the surface brightness or cause chemical attack or etching.

Agitation of the solution is desirable, regardless of the type of processing, and mechanical agitation is preferred when possible; alternate methods are solution agitation with mechanical mixers or pumps and air agitation. Agitation provides fresh solution at the metal surface for improved chemical polishing; it can eliminate gas streaks on the processed parts and prevents temperature stratification.

Suitable ventilation should be provided to remove mist and/or acid fumes. Exhausting should be sufficient to draw off all mist and fumes during transfer to the first water rinse. Stainless steel, fiberglass, rigid polyvinyl chloride, or other plastic corrosion-resistant coatings are recommended for ducts and blowers, depending upon the nature of the solution and the operating temperature.

POST-TREATMENT

Several rinses are usually required, unless the part is too large to be rinsed by immersion and must be hosed off. The number and type of rinses are dictated by the metal or alloy to be rinsed and the end use intended. Acid or alkaline neutralizing rinses may be necessary between water rinses. Overflow-type rinse tanks are being replaced by more efficient, counterflow rinsing tanks, and spray rinsing has gained wide use. In the case of static or overflow rinsing, agitation is beneficial, either by air or mechanical agitation of the solution or the workpieces.

Drying, staining, and spotting usually occur from the final rinse water, either contaminated from the carryover of the previous rinses or from the hardness of the water itself. If the final rinse water is hot, evaporation further concentrates the contaminants, which, when dry, result in staining or water spotting. This problem can be alleviated by more frequent changes of the final rinse water or by the use of a suitable wetting agent. Wetting agents eliminate water beading by reducing surface tension, leaving a thin water film and less of a deposit. The ideal final rinse, although more costly, uses deionized or distilled water. Recently, the use of Freon for water displacement and rapid drying is becoming popular in specialized cases.

ELECTROCHEMICAL DEBURRING

In electrochemical deburring (ECD), burrs are dissolved from metallic workpieces electrochemically and are flushed away by pressurized electrolyte. In addition to deburring, edge and corner radii are generated. The process is a specialized, static adaptation of electrochemical machining (ECM), discussed in Volume I, *Machining*, of this Handbook series. The basic difference between ECM and ECD is that with ECM the electrode moves into the workpiece with a constant feed; with ECD, the electrode remains stationary.

With the ECD process, electrolyte flows through a gap between a tool (cathode) and the workpiece (anode), thus completing the electrical circuit needed for the d-c power to dissolve the burrs (see Fig. 17-9). The tool and workpiece do not contact each other, and the workpiece is not exposed to any mechanical or thermal stresses. As a result, there are no changes in the physical or chemical properties of the metal.

Special workholding fixtures and tools are used to ensure that the conductive surfaces of the tools conform to the areas or edges of the workpieces to be deburred. Proper insulation of the tools and protective shielding or masking of the workpieces limit the electrochemical action to the desired surfaces. Multiple tooling stations are commonly used for high-production requirements.

Optimum use of the ECD process for consistently uniform results requires control of the tool design, current density, electrolyte, and cycle time. Variables with respect to the electrolyte that must be controlled include its type, concentration, pH, temperature, clarification, pressure, and velocity.

ADVANTAGES OF THE PROCESS

Fast production and consistent results are major advantages of the ECD process. Electrochemical deburring is ideal for selective (controlled) deburring and the removal of burrs that are inaccessible with most other processes. For parts having surfaces that cannot be altered or scratched, ECD is often the only possible method of deburring. The process is used extensively to deburr cross, threaded, and intersecting holes, and parts made from metals that would work harden.

It has been estimated that electrochemical deburring is from 5 to 40 times faster than hand deburring, with cycle times generally ranging from 5 to 60 seconds, depending upon burr size. Smooth edge breaks of from 0.001 to 0.020" (0.03 to 0.51 mm) are consistently produced. Large or small workpieces of various shapes and materials can be handled on the same machine. Holes as small as 0.034" (0.86 mm) diam can be deburred. The absence of mechanical forces permits the deburring of thin sections and fragile parts without distortion or damage.

The process does not effect workpiece dimensions, except at the edges being deburred. It can be used as a manual or fully automatic operation, and ECD stations can be incorporated in transfer machines or other automated production systems.

PROCESS LIMITATIONS

One obvious limitation to the use of ECD is that parts made from nonconductive materials cannot be deburred in this way. Conductive parts to be deburred must be clean; and, in most cases, loose chips clinging to the workpieces from preceding operations should be removed to prevent short circuits during the ECD process. In addition, relatively close tolerances are required for the workpieces because of the small clearance required between the burrs and the tool, and orientation of the burrs must be consistent from part to part to prevent short circuits and ensure uniform burr removal.

The capability for close-tolerance edge radiusing is limited to small radii, typically not larger than 0.020" (0.51 mm). Another possible limitation is that the process may produce slight etching effects on areas immediately adjacent to edges being deburred. Etching is evidenced by some discoloration (slight darkening) but can be controlled by providing insulating masks in the tooling or eliminated by bright dipping the parts. Smut is the word commonly used to describe the darkened surfaces around deburred edges produced by the ECD process on steel components.

Another limitation of the ECD process is lack of versatility. Workholding fixtures and tooling are generally required for each workpiece to be deburred. However, universal and modular-type tooling, discussed later in this section, has been developed to permit the deburring of parts with a minimum number of changes with respect to tooling.

Parts processed by ECD require immediate rinsing with warm water, and any residue remaining can be easily removed by brushing or blasting with a soft abrasive. Rigid maintenance and cleaning schedules are required for the ECD process, including daily rinsing of equipment and fixturing with fresh water, and maintaining the pH of the electrolyte. Disposal of sludge produced by the ECD process must meet local, state, and Federal codes; this subject is discussed later in this section.

ELECTROCHEMICAL DEBURRING

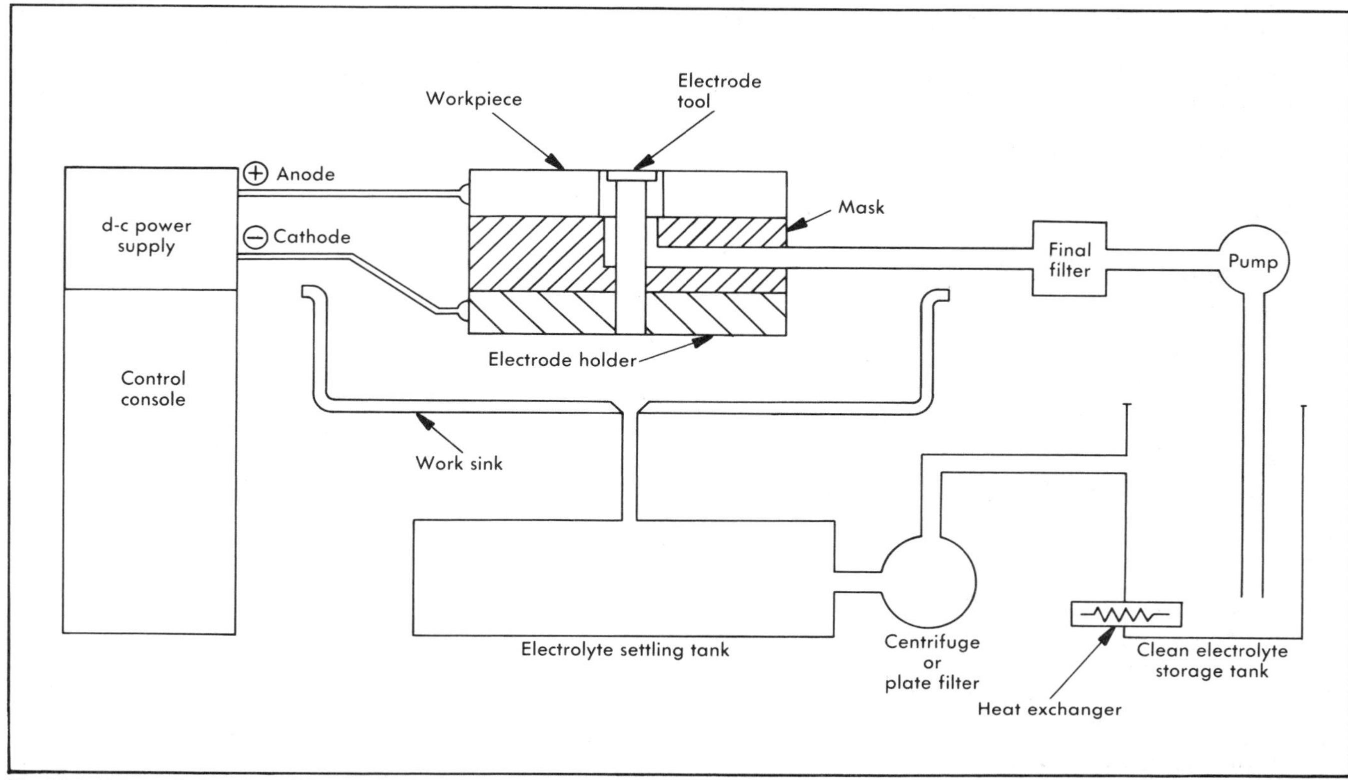

Fig. 17-9 Schematic drawing of the electrochemical deburring process. (*Chemtool, Inc.*)

TYPICAL APPLICATIONS

The ECD process is primarily used for removing specific burrs. It is also used extensively to remove recast (remelt) surfaces resulting from electrical discharge machining (EDM) and laser and electron beam machining (LBM and EBM). The process for polishing or finishing surfaces is referred to as static electrochemical machining, with no motion between the cathode and workpiece, and is usually done on a deburring bench.

Electrochemical deburring can be used for parts of any size or shape that are made from practically any conductive metal. The process is used extensively on parts made from most steels, stainless steels, aluminum, aluminum alloys, and exotic materials. Aluminum alloys with high silicon contents, however, result in a textured rather than a smooth surface. Nonuniform and possibly unsatisfactory results are obtained with some titanium alloys unless special additives are mixed in the electrolyte. Deburring of sintered tungsten requires the use of special chemicals, and zinc die castings often present processing problems.

There are many applications for the ECD process in automotive manufacturing, with many machines having multiple tooling. Deburring of pistons, connecting rods, and crankshafts is commonplace. Wristpin holes, snap-ring grooves, and oil-return slots are all deburred on pistons. Bearing lock slots, bolt holes, thrust faces, snap-ring grooves, and weight control bosses on connecting rods are deburred simultaneously.

Deburring the edges of pump gears is another major application. Many hydraulic and pneumatic components, including valve bodies, with interconnecting and crosshole passages are deburred with the ECD process to improve fluid and gas flow. The ECD process is also being used to deburr the cooling holes in turbine blades, as well as for many other aircraft engine components.

PROCESS PRINCIPLES

The ECD process is the reverse of electroplating, metal being removed rather than added. The removal of metal is in accordance with Faraday's law of electrolysis, the amount removed being proportional to the product of the current used and the time of exposure. Direct current at low voltage passes between a tool (the cathode) and the workpiece (anode) through a conductive, neutral salt electrolyte.

At the edges of the workpiece, electrons are removed by the current flowing through the electrolyte, which causes the metallic bonds of the molecular structure of the edges to be broken. Atoms removed go into the electrolyte solution as metal ions. Due to the flow of the electrolyte through the gap (distance between the tool and workpiece), the ions are prevented from becoming attached to the tool. Dissolved metal, in the form of hydroxides, is carried away by the controlled flow of electrolyte, which is then filtered by various methods discussed later in this section.

PRETREATMENT OF WORKPIECES

Workpieces should be cleaned prior to electrochemical deburring, and, in many cases, metal chips must be removed. Cleaning is necessary because many standard machine oils used in previous operations will contaminate the electrolyte. Some oils affect the ECD process by insulating the workpiece, resulting in uneven deburring. However, some of the newer

ELECTROCHEMICAL DEBURRING

water-soluble oils clean the workpieces, and hence washing before the ECD process is not required. Workpieces having internal chips can create short circuits and may damage plastic locators used in the ECD process.

EQUIPMENT REQUIREMENTS

The basic equipment required for the ECD process consists of a workstation, a d-c power supply, an electrolyte system, controls, and tooling.

Workstation

The workstation, often called the sink, of an ECD machine consists of an insulated work and tool-mounting plate, plus a trough to collect the electrolyte. Except for the mounting plate, the station is generally made of molded fiberglass or other chemically inert material, sometimes combined with a steel weldment. Stations are made in various sizes to suit the parts to be deburred, and multiple stations are often provided on a single machine to perform simultaneous or alternate operations.

Power Supply

Power supply units for electrochemical deburring are generally solid-state rectifiers, with air or water cooling. They are available in a wide range of capacities from 2 to 40 V d-c and 100 to 10,000 A. A rule of thumb for requirements is that 10 to 20 A are needed for every inch (25 mm) of edge from which burrs are to be removed. Electrical connecting cables should preferably be finely stranded, extra-flexible welding cable.

Electrolyte System

Electrolytes used. The electrolytes used for the ECD process are neutral salt solutions. Salts used include sodium nitrate, sodium chloride, sodium nitrite, sodium chlorate, sodium bromide, and potassium chloride, with sodium nitrate being the most common. A mixture of salts or proprietary solutions are sometimes used.

Sodium nitrate. The commercial grade of this salt is generally mixed with water in concentrations of 1 ½ to 2 lb/gal (0.18 to 0.24 kg/L). This mixture produces an excellent electrolyte that is easy to handle, minimizes stray machining, and retards the tendency of processed parts to rust. Sodium nitrate will produce satisfactory results on most ECD applications.

Sodium chloride. Sodium chloride is common table salt, which is generally mixed with water in concentrations of 1/2 to 3/4 lb/gal (0.06 to 0.09 kg/L). Electrolytes of this type are the least expensive, but they produce the greatest amount of corrosive damage and stray etching. Inhibitors are commonly used with sodium chloride to reduce stray etching.

Pumps. The pumps used to circulate electrolytes for the ECD process are generally of the centrifugal type. They are made from stainless steel, plastics, or bronze, and usually have output pressure capacities from 20 to 50 psi (138 to 345 kPa). Volume capacities vary with the requirements of the application.

Settling tanks. These tanks generally have divider baffles to allow for some clarification of the electrolytes by settling. They can be constructed from fiberglass or stainless steel, and capacities usually range from 100 to 1000 gal (378 to 3785 L).

Filters. Adequate filtration of the electrolyte is essential for successful ECD. Filtration removes foreign material that tends to plug the tooling and also reduces staining (the formation of smut) on the workpieces. Deburring cycle times increase when the electrolyte is allowed to become dirty.

Bag and cartridge filters. A polypropylene bag or cartridge-type filter is generally provided in series with the pump to remove foreign particles and metal hydroxides from the electrolyte in small ECD systems. For large systems, a centrifuge or plate-press filter may be added.

Centrifuges. These separators generally have stainless steel bowls with outer housings of reinforced fiberglass. They usually have a flow capacity in the range of 4 to 10 gpm (15 to 38 L/min) and are suitable for medium-size systems in which the disposal of wet sludge is not a problem. Centrifuges can be equipped for either manual or automatic sludge removal.

Plate-press filters. These units have a series of filter plates compressed together in a steel frame and remove metal hydroxides down to submicron size. This method of filtration produces solid cakes that are easier to dispose of than wet sludge. It is particularly desirable for workpieces where staining is a problem.

Canister filters. A fully automatic filtration system uses a metal canister having about 100 polypropylene, woven-mesh tubes suspended from a flexible shaker plate. Diatomaceous earth (a filter aid) is used to coat the tubes, which permits removing particles as small as 1 to 2 microns in size. A pneumatic cylinder automatically flexes the shaker plate to remove the filter aid and then allows recoating. After several recoatings, the system is shut down, and the material is dumped into a receiving tank where a vacuum or pump pulls the electrolyte back into the system, leaving a solid cake of filter aid and metal hydroxides.

Electrolyte temperature. It is recommended that the temperature of the electrolyte be controlled for steady production conditions, with this control considered essential for high-production applications. The proper electrolyte temperature to be used for the ECD process is a matter of disagreement. Some experts recommend a temperature of about 55° F (13° C) for steel parts, and 75° F (24° C) for workpieces made from stainless steel and aluminum. Others suggest 95 to 100° F (35 to 38° C) for all materials.

Heat exchangers. Controlling the electrolyte temperature through the use of plant water is an efficient control method. One recommended heat exchanger allows plant water to surround electrolyte-filled tubes made from stainless steel. The unit uses a maximum of 6 gpm (23 L/min) of electrolyte, and only about 3 gpm (11 L/min) are generally needed for ECD systems requiring approximately 200 A of current. If an enclosed work chamber is used, an exhaust system is necessary to remove the gas generated by the process.

Refrigeration units. One efficient design used to control electrolyte temperature is a multicoil unit, with Freon surrounding the electrolyte-filled stainless steel coils. The unit is air cooled and available in various capacities to suit specific requirements.

Fittings and hoses. Electrolyte supply fittings are frequently made from plastics because of the moderate temperatures and pressures used in the ECD process. Stainless steel fittings are also used. For some applications, brass fittings are more economical than plastics or stainless steel and are sufficiently corrosion resistant to require little or no maintenance.

Tubing for the electrolyte may be made from polyethylene, polyvinyl, nylon, or similar materials, depending upon availability and cost. Hoses incorporating metal reinforcements should be avoided because of possible problems with stray currents. Connections should be made with similar materials; for example, plastics to plastics or metal to metal.

Storage tanks. When different workpiece materials have to be deburred at frequent intervals, separate storage tanks should be provided for the clean electrolytes required for each different material. Each tank should have its own pump, valves, and filtering system. Separate work sinks are also desirable for deburring workpieces made from different materials.

Machine Controls

Machines for ECD generally have relay-type control logic. Some newer machines feature programmable controllers that permit reprogramming for different applications. Such machines can be used with industrial robots and connected to computer systems for maintenance, management information, or other functions.

TOOL DESIGN

Important considerations in the design of tools (cathodes) for ECD include the gaps between the tools and the workpieces, the flow of the electrolytes, and the materials from which the tools are made.

Gap Between Tool and Work

Based on considerable experience and extensive testing, recent improvements have been made in the ECD process with respect to the positioning of the tool in relation to the burr on the workpiece. For many years, this gap was maintained between 0.010 and 0.015″ (0.25 and 0.38 mm). Such small gaps, however, created problems, including shorts when burrs touched the tools and frequent overcutting on critical parts. Variations in the tools because of manufacturing tolerances caused some parts of the gap to be only 0.002 to 0.005″ (0.05 to 0.13 mm). Variations in the gap change the velocity of the electrolyte and cause irregular flow, and, in some cases, create air bubbles. Air acts as an insulator and thereby stops the deburring action.

Pulsating d-c power supplies and gaps of 0.025 to 0.040″ (0.64 to 1.02 mm) are being used for some applications. Advantages of the larger gaps include fewer shorting problems, longer tool life, and better finishes, but cycle times are generally longer. An even flow and uniform velocity of the electrolyte flushes the metal hydroxides away faster, which reduces smutting of the workpiece surfaces. Cathode wear can also be minimized by using spark-detection protection systems.

Electrolyte Flow Path

Practically all tooling for the ECD process is designed so that the tools have a corner-to-corner relationship to the burred edges of the workpieces (see Fig. 17-10), resulting in removal of the burrs and rounding of the edges. The gap, established by the X and Y dimensions shown, is varied to suit the application and to provide the desired corner break. The electrolyte flow path safely removes any particles of metal cut away during the process.

Modifications of the working gap, by increasing or decreasing the X and Y dimensions, will affect the type of corner break produced. However, both dimensions must be sufficient to allow for the maximum variations in workpiece geometry permitted by production tolerances. Even under the most adverse conditions of dimensional tolerance stacking, a gap must still be provided to prevent electrical contact and to permit sufficient electrolyte flow. The corner-to-corner relationship can be designed in different ways to accommodate workpieces of various configurations.

Through holes. In Fig. 17-11, view *a* shows a simple approach for removing the exit burr from a drilled hole. This method can only be used if the tool can enter the hole from the upper surface of the workpiece. To provide electrolyte flow in the correct direction to wash burrs out of the gap, the electrolyte supply has to be located on the side of the workpiece opposite the tool, requiring a more complex fixture than if both the tool and electrolyte supply could be located on the same side. View *b* of Fig. 17-11 illustrates how a workpiece can be located to have the tool and electrolyte flow come from the same side.

Figure 17-12 illustrates a through-drilled hole having a radius in place of an entrance burr, a tool with two conductive corners, and a plate made from plastic or other insulating material on which the workpiece rests. This design breaks the following two rules that are generally observed in the ECD process:

1. Never have two working gaps in series in an electrolyte stream because chips removed in the first gap can cause shorting in the second gap.
2. The direction of electrolyte flow should force chips out of the gap instead of into it.

These rules can be broken in this instance because the entrance burr in most drilled holes is so minimal that there is no danger of pieces breaking loose. Also, the function of this operation is more for corner radiusing than burr removal.

For the most uniform results in bores and the smoothest possible radiusing of edges, a circular electrolyte flow is desirable. Such flow prevents striations on the workpiece surfaces, and the centrifugal forces cause the electrolyte to follow the outer walls of cavities, thus producing radii (see Fig. 17-13) rather than conical corner breaks. This technique is helpful where large, smooth radii are required for stress-reduction purposes.

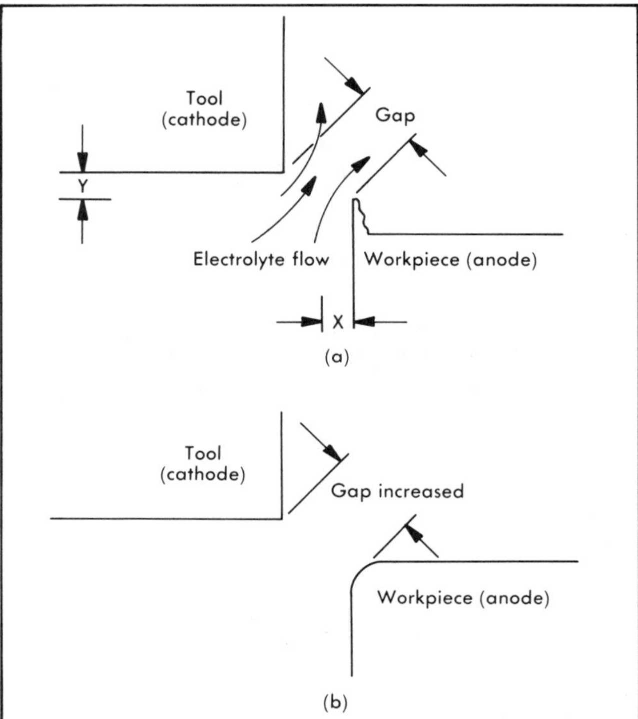

Fig. 17-10 Corner-to-corner relationship for the ECD process, (a) before and (b) after deburring. (*Chemform, Inc.*)

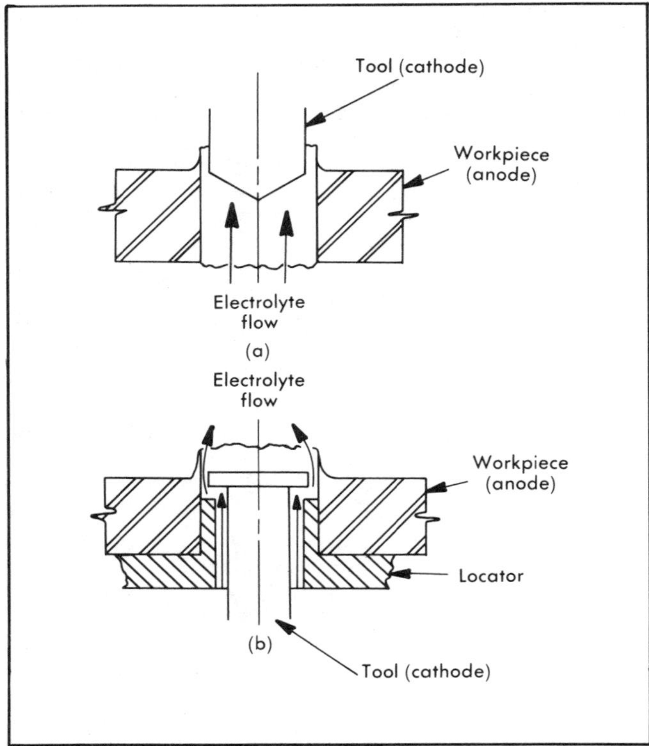

Fig. 17-11 Two methods of removing the exit burrs from through-drilled holes. (*Chemtool, Inc.*)

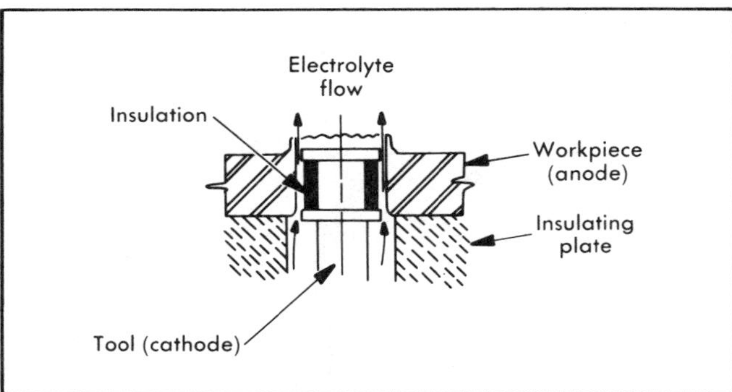

Fig. 17-12 Tool with two conductive corners for electrochemical deburring and edge radiusing. (*Chemtool, Inc.*)

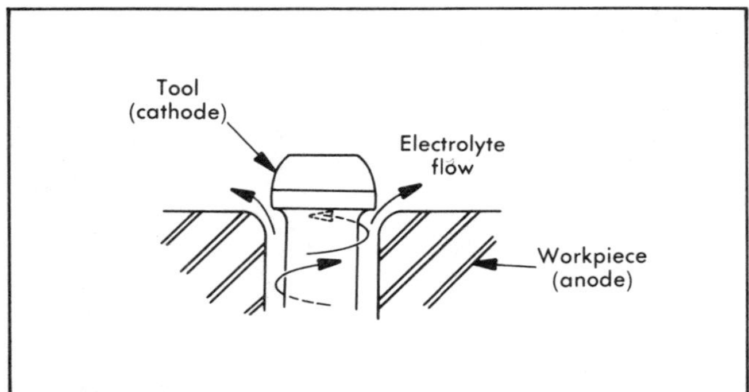

Fig. 17-13 Circular flow of electrolyte can produce large, smooth radii. (*Chemform, Inc.*)

Circular electrolyte flow can be accomplished by introducing the fluid tangentially into a distribution groove. A manifold for the electrolyte may be provided by milling slots into the face of a plate that is connected to an electrolyte supply hole or channel (see Fig. 17-14).

Blind holes. One method of deburring the edges of blind holes is illustrated in Fig. 17-15. The electrolyte exits from a central hole in the tool. This exit hole must be large enough to allow the passage of chips produced in the ECD process.

Intersecting holes. Two different approaches for deburring the intersection of a small hole (or holes) with a large center hole in a workpiece are illustrated in Fig. 17-16. View *a* shows a workpiece on a plastic locator, with a tool completely insulated except for the edge adjacent to the burr to be removed. This design produces smooth, uniform radii, the size of the radii depending upon cycle times.

With the method seen in view *b* of Fig. 17-16, the locator protects the complete bore of the workpiece. There is a slot or hole machined in the locator that is 0.020 to 0.040″ (0.51 to 1.02 mm) larger than the hole in the workpiece and at the same location. The tool is positioned inside the locator with its end opposite the hole in the workpiece; the tool is 0.020 to 0.050″ (0.51 to 1.27 mm) smaller in diameter than the hole in the workpiece. This design produces a slight counterbore, together with a small radius. It is useful when many holes in a part have to be deburred because it provides good control of electrolyte flow for all the holes, even when they are on different levels.

Intersecting holes, with one hole at an angle to the other, are generally deburred with portable-type ECD probes (see Fig. 17-17), especially in large workpieces. View *a* shows a probe pushed into the straight hole. A magnetic-type permanent stop is mounted on the outside of the workpiece to properly locate the tool portion of the probe. A plastic diffuser on the probe forces the electrolyte out the angular hole. View *b* shows a probe located in an angular hole. This method is generally used when the angular hole is drilled after the straight hole.

Five internal crossholes and two half-moon slots in the center bore of a cast iron turbine housing are deburred in one operation with the setup shown in Fig. 17-18. Three air-actuated tools are used, with a programmable controller providing proper sequencing of the tools.

Gear teeth. Radiusing of gear teeth and odd-shaped profiles is a common application for the ECD process. External gears are placed in plastic locators, positioning the gears radially by means of internal gear-shaped tools. The tools are made by using a gear as an electrode. The exact opposite shape of the gear to be radiused is produced by the electrical discharge machining method.

For producing small, closely controlled radii, 0.001 to 0.002″ (0.03 to 0.05 mm), electrolyte flow is generally upward through a series of small holes into a reservoir area above the tool. The electrolyte then turns 90° and flows inward against the edges of the gear teeth. For producing larger radii, the flow of electrolyte is usually in the opposite direction and then upward across the teeth. To prevent stray cutting of gear teeth, the electrolyte should be kept cold or adequate masking should be provided.

Tool Materials

Tools for the ECD process must be made from electrically conductive materials. A wide variety of such materials are used, depending primarily upon the specific applications. The use of copper tungsten for the working edges of the tools is recommended for many applications because this material is easily silver brazed to copper-based and ferrous metals. Whenever possible, it is generally preferable to provide mechanical mounting. For removing small burrs, free-machining brass is often an ideal tooling material.

Tool Insulation

A nonconductive shield or coating is applied to ECD tools adjacent to the areas of the workpieces where deburring is required. One commonly used insulating material for both tools and fixtures is a dense, epoxy-bonded, glass-cloth laminate called greenglass, conforming to the National Electrical Manufacturers Association's (NEMA) Specification G-10. This material is available in rounds, tubes, and sheets. Water absorption is low and dimensional stability is reasonably good, although occasional warpage may occur with sections unevenly exposed for long periods of time. Greenglass can be machined with conventional equipment, but the life of cutting tools is generally short.

Acetal resins, available under the trade names of Delrin and Celcon, are also used extensively as insulating materials. These materials machine well with longer cutting tool life than when machining greenglass.

When using thermosetting epoxy to make tools or fixture parts, an oven is required for preheating and curing. Powder-spray (flocking) guns and a spray booth with a filtered exhaust system are often used for the application of insulating materials.

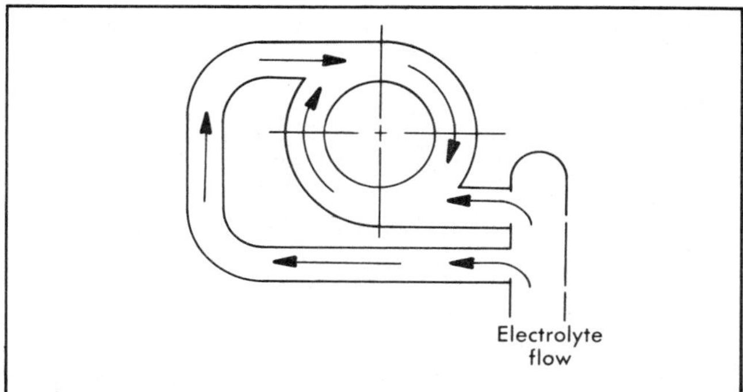

Fig. 17-14 Plan view of a plate used to obtain circular flow of electrolyte. (*Chemform, Inc.*)

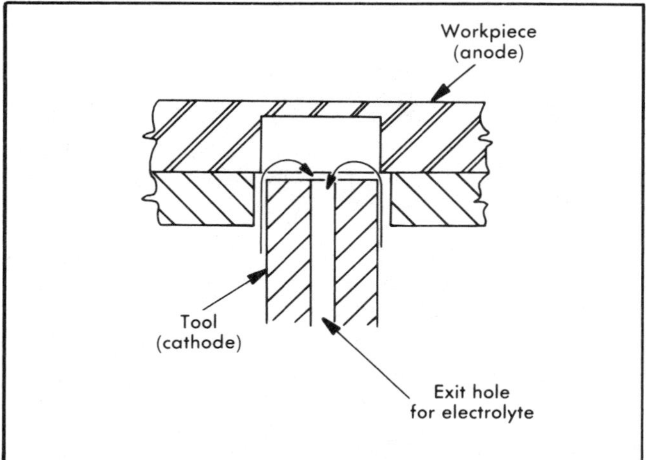

Fig. 17-15 In deburring blind holes, the electrolyte exits from a center hole in the tool. (*Chemtool, Inc.*)

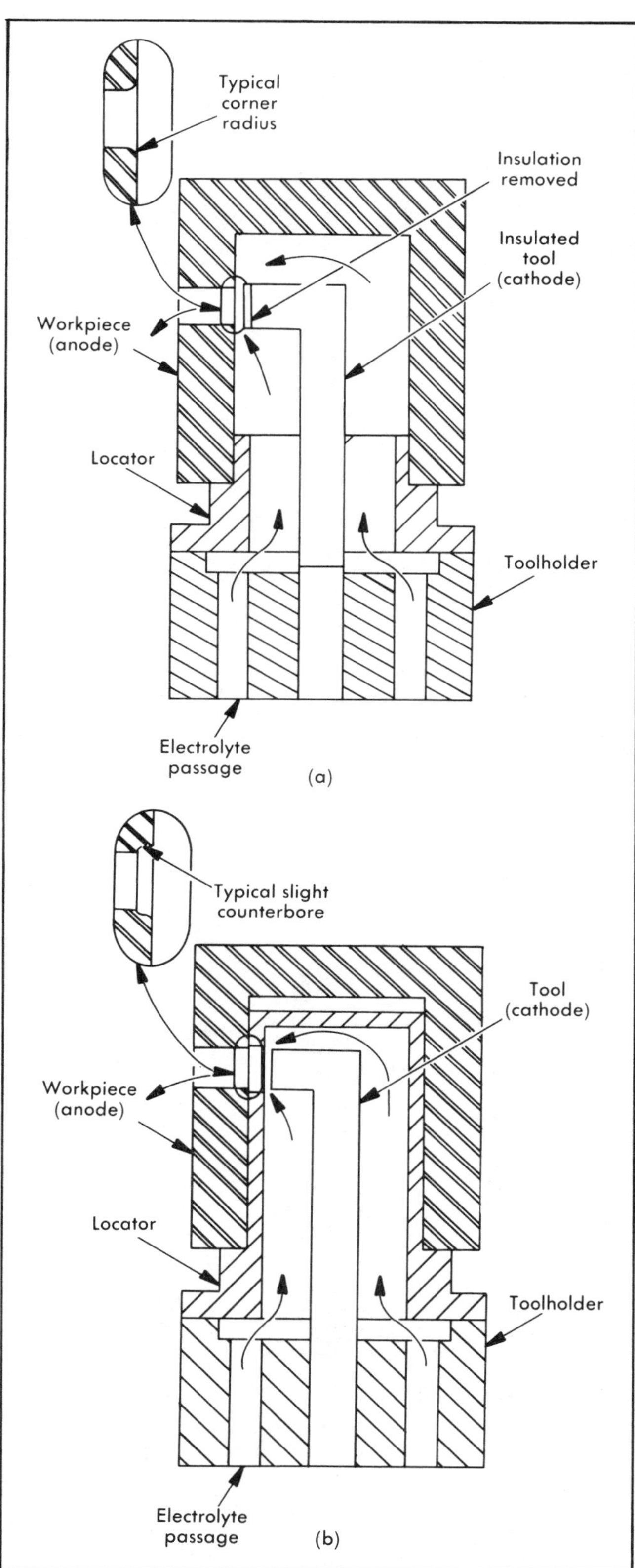

Fig. 17-16 Two methods of electrochemically deburring the intersection of a small hole (or holes) with a large center hole. (*Chemtool, Inc.*)

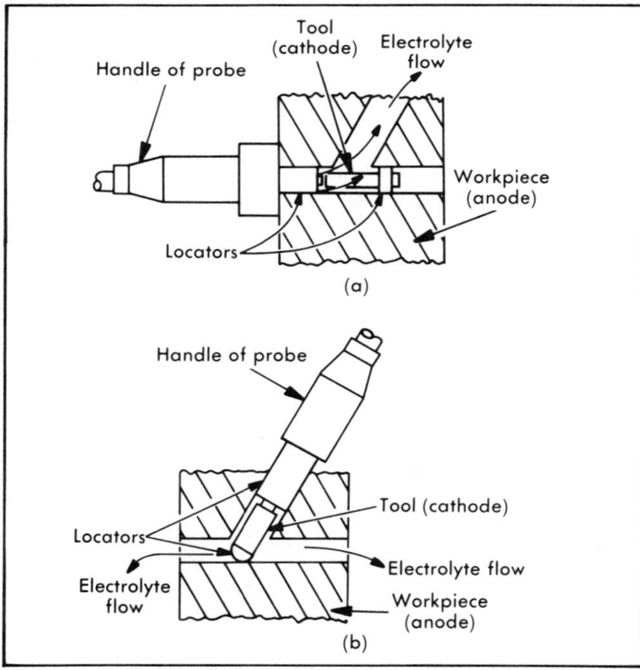

Fig. 17-17 Two methods of electrochemically deburring intersecting holes when one hole is at an angle to the other.

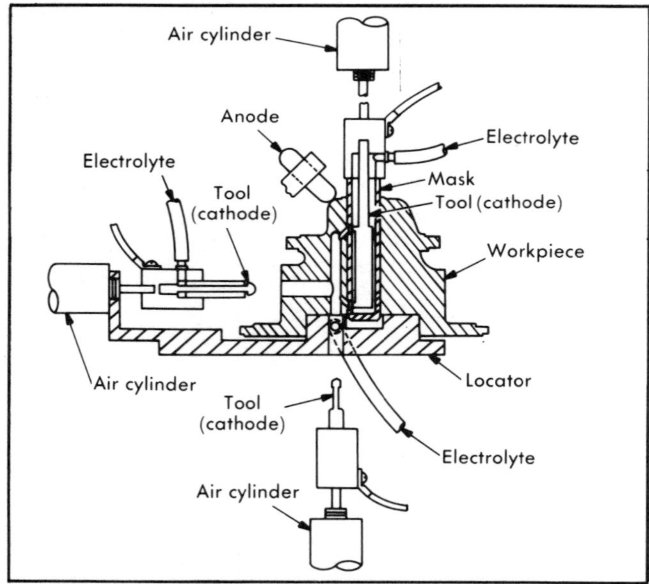

Fig. 17-18 Five internal crossholes and two half-moon slots are deburred in one operation with this setup. (*Chemtool, Inc.*)

FIXTURE DESIGN

In designing fixtures for the ECD process, an attempt should be made to have the anode contact the workpiece in a dry area. Contact pressures must increase with higher amperage requirements. Round anodes should have a minimum diameter of about 7/16" (11 mm) to serve as a heat sink for reducing hot spots. Copper is frequently used for the anodes; but where contact is in an area where electrolyte is flowing, platinum-tipped anodes may be required.

Fits and Tolerances

For most applications, the fits and tolerances used for conventional jig and fixture design will provide satisfactory performance in ECD tooling. However, where very porous materials are applied and are subject to liquid saturation, allowances must be made for anticipated expansion.

Fixture Materials

Brass is the most widely used metal for fixture components, primarily because it is easy to machine and a good conductor of electricity. Stainless steels of the 300 series are also used extensively, with Type 316 having a slight advantage over others in the series with respect to resistance to corrosion by the electrolyte.

Copper has a conductivity about 10 to 20 times better than stainless steel, depending upon the alloy, and is often used for contact surfaces that provide current to the workpieces. The strength of stainless steel can also be exceeded by using beryllium copper, which provides nearly the same conductivity as pure copper.

The closeness of fixture components, damp surfaces, and the possibility of trapped pools of electrolyte make insulation essential. Nonconductive shields or coatings of the type previously described for tool insulation are also used for fixtures. Clear acrylic (Plexiglas or Lucite) blocks and sheets are used extensively, especially when transparency is advantageous, but they do not have the wear resistance or mechanical strength of greenglass. Machining of acrylic is fast and easy, and bonding of acrylic to fixture components requires only clean, closely fit surfaces.

Other solid plastics materials, such as Lexan and Teflon, are used successfully for components of ECD fixtures. In fact, any dielectric material that has low water absorption and satisfactory physical strength may be used. In many cases, selection is based on ready availability rather than the physical characteristics of the material.

Fasteners and Dowels

Assembly of sections of the fixture that are permanently bonded is normally achieved by use of stainless steel bolts and cap screws. Coarse threads are preferred for tapped holes in the various dielectric materials. The use of socket-head cap screws allows counterboring and flush surfaces. Such screws also permit the use of temporary insulation to protect the fasteners, yet allow ready removal for disassembly.

Covering material for the fasteners may be ordinary silicon seal or, for a somewhat more solid but still removable cover, vinyl-based spackling paste. Where dowels are necessary, stainless steel will suffice; titanium is better; and in plastics materials, glass rods, ground to diameter, are best. Solid glass rods should be used, not laminates.

Electrical Connecting Cables

If any parts of the fixture move and conduct current, they should be bonded to the rest of the fixture electrically by the use of swedged and sealed conductors. This design minimizes the gradual disintegration that is inevitable when a voltage difference exists between two metal parts in the presence of an electrolyte.

Workpiece Clamping

Clamping of the workpieces, for positioning relative to the electrodes and/or to provide electrical conduction, should be fast, simple, and require minimal operator effort. For many applications, these criteria can be met by adapting simple air cylinder clamps.

The use of air cylinders is practical and offers operating economy. Air cylinders can be trouble-free if a few simple precautions are observed. Always select and install double-acting cylinders. This approach insures that any leakage across any of the cylinder's seals or joints will be from the inside to the outside, with no possibility of drawing in electrolyte, as can happen with a single-action, spring-return unit. Next, if it is impractical to mount the cylinder so that it does not get sprayed or soaked with electrolyte, a stainless steel or brass cylinder should be used.

Air cylinders may also be necessary to insert or move the tool into position, which by virtue of the workpiece configuration cannot be made an immovable feature of the fixture. With small, round tools requiring no orientation, it is possible to use the cylinder as both an actuating and clamping means. A double-ended rod is recommended so that the rear of the rod can have an adjustable stop mounted on it. For more stringent requirements, various stainless steel slides can supplement the cylinder to properly position the tools.

Cleaning the Tooling

The corrosive effects of the electrolyte on fixture parts after the fixture has been allowed to sit unused for a period of time can usually be overcome by a thorough rinsing of the entire fixture with fresh water. Ideally, this practice of rinsing the fixture is even more effective if applied immediately after shutting down the machine. The use of a flush system built into the ECD machine is desirable.

The tools used for ECD must also be kept clean when in use or when stored. All brass, stainless steel, or copper tungsten parts must have good electrical connections between each other. If connections are not clean, there will be a slight voltage drop. In some areas of the tool, this drop will result in dissolving parts of the tooling. A preventive maintenance program that allows for cleaning the tooling on a regular schedule is important; a bright dip is available that helps protect brass parts from tarnishing.

STANDARD MODULAR TOOLING

Considerable savings are being realized by the use of standardized modular tooling for the ECD process. Because many of the components can be used for deburring various workpieces, such tooling eliminates the cost of designing and manufacturing complete tools and fixtures for each specific application.

Modular tooling designed to deburr holes that break into the center holes of workpieces is shown in Fig. 17-19. Components called the mask and the electrode have to be made for each different workpiece, but the other components are suitable for many parts. The mask fits over the electrode, and a protrusion on the electrode is centered with respect to a hole in the mask. A workpiece is located on the mask, with the hole in the mask aligned with the crosshole in the workpiece. Electrolyte flows around the electrode and deburrs the edges of the crosshole. All other surfaces of the workpiece are protected by the mask.

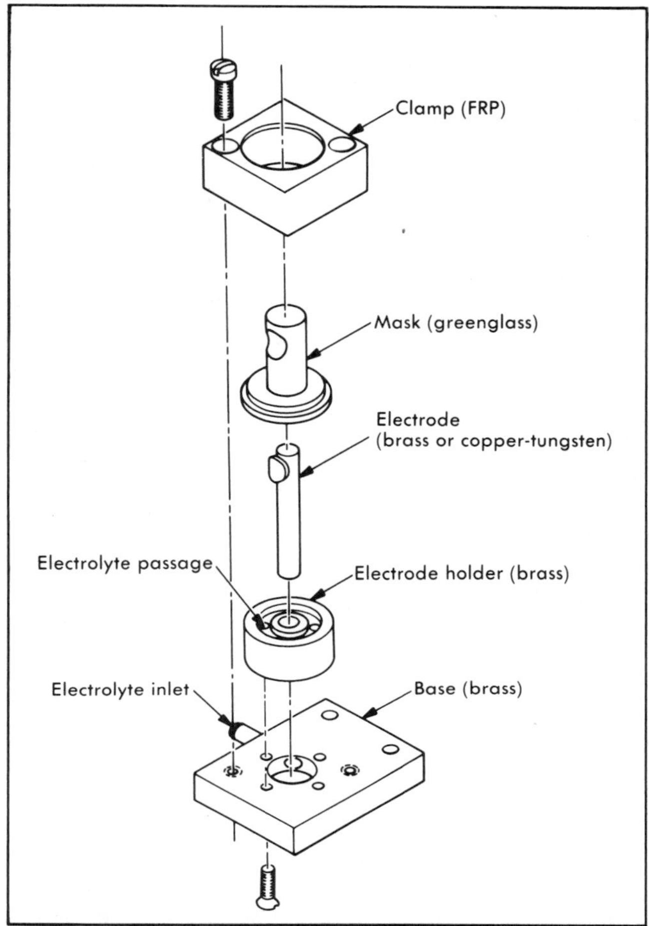

Fig. 17-19 Standard modular tooling designed to deburr holes that break into the center holes of workpieces. (*Chemtool, Inc.*)

OPERATING PARAMETERS

There are many operating parameters that must be carefully controlled for efficient and economical deburring with the ECD process. Important parameters discussed in this section include gap size, current density, operating voltage, cycle time, and the type, temperature, pressure, and velocity of the electrolyte.

Gap Size

The closer the tool and the workpiece are to each other, the faster burrs can be removed and the more consistent the deburring action. The size of the gap for a specific ECD application, however, must be determined by computing the effect of workpiece and tooling tolerances on the location of the edges to be deburred. Too small a gap will cause shorts and limit electrolyte flow, which is a critical factor in the process.

Normally, depending upon the tolerances, a minimum gap of 0.020" (0.51 mm) is used. It is generally desirable not to exceed 0.045" (1.14 mm), if practical. When production requirements are low and longer time cycles can be tolerated, larger gaps are sometimes used. For the shortest possible cycles and higher production rates, proper tool design is critically important.

Voltage and Current Densities

As the voltage used for ECD is increased, the time required for deburring is decreased. For example, doubling the voltage from 10 to 20 V will reduce the time required by one half.

ELECTROCHEMICAL DEBURRING

However, increasing the voltage can produce undesirable results, such as increasing the amount of stray machining and producing rough surface finishes. Increasing the voltage may remove metal too fast so that the electrolyte cannot carry the hydroxides away fast enough. This condition creates smut that may act as an insulator to slow or even stop the process.

The optimum operating voltage can only be determined by actual test cuts on representative parts. For a starting point, 15 V is generally recommended. If intolerable amounts of stray machining are produced, the voltage may be reduced to help overcome this condition; however, it may be desirable to investigate other solutions to this problem. If no adverse conditions are experienced, the voltage may be gradually increased, which will reduce the amount of time required to perform the operations.

Time Cycles

Time cycles for the ECD process vary from a few seconds to one minute or more. Most close-tolerance holes can be deburred in 5 to 10 seconds. Workpieces having more liberal tolerances or heavy burrs generally require cycle times of about 20 seconds. Parts having very loose tolerances, requiring gaps of about 0.030" (0.76 mm), may take one minute or more to produce the desired results. Aluminum and brass parts can be deburred about twice as fast as parts made from steel or stainless steel. Time cycles can only be determined accurately by making test cuts on parts representing the worst operating conditions.

Electrolyte Conditions

Selection of the proper type of electrolyte, maintaining the electrolyte at the correct temperature, and the importance of clarifying the electrolyte by filtration have been discussed previously in this section. Other important operating parameters for the electrolyte include the pressure and velocity used.

Electrolyte flow requirements for deburring involve maintaining sufficient velocity so that sludge will not accumulate and plug the flow path; this requires only moderate pressures, typically 20 to 50 psi (138 to 345 kPa). Low pressures minimize spray and the possibility of fixture leakage, and permit operation, in most cases, without enclosures, which can interfere with accessibility and decrease production.

POST-TREATMENT

Operations required on parts that have been deburred by the ECD process include the removal of salts and sludge. Another task is the disposal of electrolyte and sludge.

Removing Salt

All parts should be rinsed with warm water [120 to 140° F (50 to 60° C)] after the ECD process to remove salt. Fresh water should be added to the rinse tank at a sufficient rate to prevent contamination of the tank contents with excess salt concentrate. Standard agitating-type washing machines with heaters work well for removing salt from most parts. Many degreasing systems and special cleaning solutions do not normally remove salt from the parts as well as warm water. Parts subject to rust should be rinsed in a water-soluble rust inhibitor after rinsing in warm water.

Smut Removal

Smut produced by the ECD process does not always have to be removed. For example, aluminum parts can be anodized, painted, and finished by other processes over the smut without affecting quality. Brass parts can be plated or chromated with no problems. All types of plating and painting can be done on steel parts without removing the smut, and quality is not affected. In fact, the smut on steel parts is a carbon that serves as a lubricant, which is beneficial for some parts. Stainless steel parts can be passivated, electropolished, and painted without removing smut. Smut must be removed, however, from instrument components that are to be operated in ultraclean environments.

Removal of smut, when required, varies with the workpiece material. For aluminum parts, either acid or alkaline cleaning solutions are used without affecting the workpiece tolerances. Different types of acidic solutions are used to remove smut from steel parts. Either acid or alkaline solutions can be used to desmut brass and copper parts. Different types of acids are being used to remove smut from parts made of stainless steel, but electropolishing generally works best. In addition, some tumbling and vibratory finishing processes can be used to remove smut from certain steel and stainless steel parts.

Disposal of Electrolytes and Sludge

The disposal of solutions containing certain metals and pollutants considered hazardous is regulated by local, state, and federal codes. Pollutants are removed from electrolytes by the various methods of filtration discussed previously in this section.

Electrolyte can sometimes be reused after filtration or dumped into sanitary waste facilities, again depending upon applicable codes. It may be necessary to test for chemicals in the solution before disposal, and the pH value must be 7. Sodium nitrate, the most common electrolyte, is used extensively for treating industrial and domestic waste.

When deburring parts made of stainless steel or other metals having high chromium contents, it may be necessary to add sodium metabisulfite to the electrolyte. If there is any hexavalent chromium present, this chemical will reduce it to a trivalent form, which can generally be dumped in waste systems. Sludge remaining after filtration of electrolytes can be either stored or discarded in landfills, if regulations permit.

ULTRASONIC DEBURRING

Ultrasonic deburring is a little used process that could have been included in Chapter 14 of this volume because it employs mechanical and abrasive action. However, this method also uses a weak acid etching solution, which makes it also chemical in nature. Ultrasonic agitation of the abrasive-solution slurry also generates heat. Use of the process is limited to small lots of specialized parts having minimal burrs, and it is not recommended for high-volume requirements.

Ultrasonic deburring uses ultrasonic cavitation and a specially compounded acid slurry. The slurry consists of weak etching solutions and small quantities of abrasive particles. Ultrasound provides high-energy shock waves that cause the abrasives to remain suspended and to constantly bombard the workpiece; the shock waves also increase the etching action. Edges of the workpieces are attacked more than the surfaces because of strain-hardening microstructural changes in the areas of the burrs. Finishes produced are bright and oxide-free.

Little information is available on the ultrasonic deburring process, primarily because the contents of the slurries are considered secret. Apparently, fine-grained abrasives, such as silicon carbide, 0.0012 to 0.0079" (0.030 to 0.200 mm) diam, are used. Other slurry constituents that have been identified include calcinated soda, stearic acid, sulfuric acid, and glycerin, but the amounts are not known. Vibration frequencies to 30 kHz have been used. Because of the use of acids, tank linings are required.

Ultrasonic deburring is used primarily for thin burrs, those having a thickness of about 0.0005" (0.013 mm). It can only generate small edge radii, normally not larger than 0.005" (0.13 mm). The process can be used on a number of different materials, but is generally limited to small parts. Cavitation is directly proportional to the size of the transducer. Large parts require large tanks and powerful, costly transducers. Cycle times are directly proportional to burr sizes and have been reported to range from 10 to 25 minutes. Workpieces should be rinsed with water after ultrasonic deburring.

When workpieces are racked, allowing the slurry flow more penetration, the results are often more desirable. Racking increases production costs, but close-tolerance workpieces requiring this type of finish generally have to be handled delicately.

Applications include the removal of flash from complexly shaped parts having through or blind holes. The process is especially suitable for reaching inaccessible areas of workpieces. Other applications include deburring the teeth of precision gears, the ends of needles, and small precision stampings.

Bibliography

Gillespie, LaRoux K. *Deburring Technology for Improved Manufacturing*. Dearborn, MI: Society of Manufacturing Engineers, 1981.

__________. *Advances in Deburring*. Dearborn, MI: Society of Manufacturing Engineers, 1978.

__________. *Deburring Capabilities and Limitations*. Dearborn, MI: Society of Manufacturing Engineers, 1976.

Kidd, Thomas L. *Improving Productivity and Quality with TEM*. SME Technical Paper MR83-692. Dearborn, MI: Society of Manufacturing Engineers, 1983.

Koroskenyi, James B. *Thermal Energy Deburring for Improved Productivity*. SME Technical Paper MR81-296. Dearborn, MI: Society of Manufacturing Engineers, 1981.

Miller, Jerome F. *Electrochemical Deburring of High Technology Machine Parts*. SME Technical Paper MR83-197. Dearborn, MI: Society of Manufacturing Engineers, 1983.

Rhoades, Lawrence J. *Cost Guide to Automatic Finishing Processes*. Dearborn, MI: Society of Manufacturing Engineers, 1981.

Tuneski, Glenn E., and Hignett, J. Bernard. *Some Recent Developments in Electrochemical Deburring*. SME Technical Paper MR81-964. Dearborn, MI: Society of Manufacturing Engineers, 1981.

CLEANING

Cleaning is the process of removing objectionable matter from the surfaces of manufactured products. Drawing and stamping lubricants, cutting fluids, heat treatment scale and oxides, and fingerprints are typical of soils that must be removed.

Important reasons for adequate cleaning of components and products are as follows:

- As an intermediate step to facilitate manufacturing operations, like inspection and assembly.
- To prepare surfaces for subsequent operations such as coating.
- As a final finish, to improve performance, appearance, and salability.

The cleaning and surface preparation of manufactured components and products are accomplished by many different processes. Some of the mechanical and abrasive processes discussed in Chapter 16 and the thermal, chemical, and electrochemical finishing methods described in Chapter 17 are used for cleaning and surface preparation. Other common processes, discussed in this chapter, include a variety of chemical cleaning methods such as the application of solvents, aqueous cleaners, and acids by soaking or spraying; steam and flame cleaning methods; vapor degreasing; and ultrasonic cleaning.

CHAPTER CONTENTS:

SELECTING A CLEANING PROCESS

No one cleaner or cleaning method is best for all applications because of the many variables, including the workpiece, the contaminants, and the requirements. Successful surface preparation for subsequent coating operations often requires the use of more than one type of cleaner and operation. For example, a solvent or alkaline cleaner may be used to remove oil, and then an acid cleaner used to provide an oxide-free surface.

Important considerations in selecting a cleaning material and/or method include the following:

- Workpiece material, size, and shape.
- Contaminants to be removed and their adherence to the surfaces.
- Degree of cleanliness required.
- Subsequent operations to be performed, including the type of coating to be applied.
- Quality and hardness of the water to be used for cleaning.
- Production requirements.
- Floor space available.
- Efficiency and economy of the cleaning method and material.
- Environmental and safety considerations, including the disposal of spent solutions.

WORKPIECE MATERIAL, SIZE AND SHAPE

The type of material to be cleaned determines the chemical nature of a cleaner. Consideration of possible metal attack must also be considered. Ferrous metals, stainless steels, and magnesium alloys are not normally attacked by highly alkaline cleaning solutions. However, nonferrous metals, such as alloys of aluminum and zinc, may be attacked or tarnished by such solutions, and thus properly inhibited cleaners must be used.

Size and shape of the workpieces to be cleaned affect the selection of the process to be used. Large, cold parts act as heat sinks and decrease the temperature of the cleaning solutions, requiring increased solution temperature and/or concentration or longer cleaning cycles. Very large parts may be more effectively cleaned by spraying. The presence of recesses or blind holes in the workpieces also affects selection.

TYPES OF SOILS

The contaminants to be removed by cleaning can be a single substance or a complex mixture of materials. Proper identification of soils to be removed is essential in selecting an optimum cleaning system. Generally, cutting fluids, forming lubricants, and fingerprints are removed by solvents and detergent-type cleaners. Such cleaners are also effective for removing quenching and rustproofing oils. Drawing lubricants and buffing and polishing compounds are usually best removed with alkaline emulsifying cleaners that are specially formulated to saponify fatty acids and keep them in solution. Cutting fluids and lubricants should be compatible with both the workpiece material and all subsequent operations, including cleaning.

Additional materials to be removed are solid particles, such as metal chips, abrasive grains, pigments, carbon smut, and shop dirt. These materials are usually held to surfaces by oils or greases, or by forces such as static electricity,

---

Contributors of sections of this chapter are: ***Daniel E. Duffy****, Marketing, Automated Chemical Systems;* ***Joseph V. Otrhalek****, Director, Research & Development, Industrial Chemical Specialties Div., Detrex Chemical Industries, Inc.;* ***Graham Pendleton****, Vice President, Kleer-Flo Co.;* ***H. M. Sadwith****, President, Industrial Washing Machine Corp.;* ***Avery B. Smith****, President, Automated Chemical Systems.*

Reviewers of sections of this chapter are: ***Thomas C. Atkiss****, Regional Sales Manager, Van Straaten Chemical Co.;* ***Elizabeth A. Barisich****, Application Specialist, Process Equipment Dept., Industrial Equipment Div., Westinghouse Electric Corp.;* ***Charles M. Bessey****, Research Associate, Kolene Corp.;* ***Mike D'Angelo****, Marketing, MacDermid, Inc.;* ***Tona Del Prete****, Marketing Services Manager, Baron-Blakeslee, Inc.;*

magnetism, or other attractions. Alkaline cleaners are used to emulsify the oily or greasy binders and disperse the solid particles. In some cases, the physical forces must be neutralized by other means of cleaning.

Corrosion products, such as oxides, scale, tarnish, and rust, must also be removed. These products generally respond to precleaning with solvents or alkaline cleaners, followed by acidic pickling to remove the oxide films.

The cost and relative ease or difficulty of cleaning depends upon the adherence of the dirt. How tightly the dirt is bonded depends upon the workpiece material, the dirt, and the time and temperature at which they have been exposed. Some oils left on materials for a long time may be converted by oxidation to a varnish-like polymer that is difficult to remove.

DEGREE OF CLEANLINESS REQUIRED

All workpieces do not demand the same degree of cleanliness. Determining the degree of cleanliness required may sometimes be difficult, but it is essential for economical manufacturing. Parts cleaned between manufacturing processes to permit inspection or for storage can be protected from rusting or fingerprinting by leaving a light oily film after cleaning or by applying a dry-film rust preventive. A solvent emulsion is frequently used to leave a rust-preventive film after solvent evaporation. Parts that are to be subsequently coated require a high degree of cleanliness.

One method commonly used to determine cleanliness is the water-break test. After rinsing in flowing water, a continuous, unbroken film of water on the surface indicates good cleaning, whereas the presence of water beads indicates an unclean surface. Since results obtained in this way are not generally uniform, many other tests for cleanliness have been developed. These include fluorescent, residual pattern, wiping, Nielson, and radioisotope tracer techniques. Standard test methods for cleanliness are discussed in ASTM Standard B 322.

With a relatively new method developed by FMC Corporation, solutions of different known surface tensions are dropped on the cleaned surfaces. If the droplets bead, the solution's surface tension is greater than the free energy of the surfaces; if they spread, the surface tension is less. With this method, successively using different solutions of known surface tension, measurements can be used to quantify cleanliness.

SURFACE PREPARATION FOR COATING

The purpose of cleaning parts that are to be subsequently coated is to remove objectionable dirt or contaminants that could interfere with the application, adhesion, and corrosion resistance of the coating. However, excessive cleaning can increase production costs without improving the quality of the coating. Table 18-1 presents some metal cleaning processes typically used for various purposes.

Inorganic Coatings

The application of inorganic coatings, discussed in Section 4 of this volume, generally requires the removal of all traces of oils, greases, solid particles, oxides, and other corrosion products. However, the cleaning cycle necessitated varies with the specific inorganic coating to be applied.

Parts cleaned before phosphating or conversion coating require a clean surface free of oily deposits. Iron phosphating may be done with a single-step operation, followed by rinsing, since the baths can be designed to possess some detergent properties to clean as they phosphate. However, the removal of oil and oxide from the surface is essential prior to zinc phosphating. Chromate conversion coating of aluminum alloys requires a very clean surface with minimum oxides to obtain good quality performance. A high degree of cleanliness is required to ensure adhesion for surfaces to be brazed, galvanized, tinned, soldered, spot welded, plated, or finished with porcelain enamel. Parts to be electroplated usually need the additional cleaning steps of electrocleaning and acid activating to prepare their surfaces for plating. Typical cleaning sequences used prior to the application of various inorganic coatings are presented in Table 18-2.

Organic Coatings

The application of organic coatings such as paint, discussed in Section 5 of this volume, requires the removal of all visible surface dirt, scale, and rust, as well as oil, grease, and old coatings. Proper surface preparation is essential for good adhesion of the coating. The best performance in terms of paint adhesion and corrosion resistance is achieved by phosphatizing the metal prior to the application of the organic coating.

WATER USED

Impurities in water, such as calcium and magnesium salts, which cause hardness, impair the performance of alkaline cleaning solutions by forming insoluble compounds. Most proprietary cleaning compounds overcome this action by sequestering the hard water salts through the use of inorganic polyphosphate sequesterants or organic chelating agents and surfactants that are unaffected by water hardness.

Costs can often be reduced by taking steps to reduce the amount of water used for cleaning. Significant savings can result from the use of countercurrent rinses or by using rinse water for additions to cleaning tanks. Since the quality of the rinse water may have a significant bearing on the coating subsequently applied, deionized water may be required, especially for critical applications such as the sealing of phosphate coatings and the rinsing of electronic components.

PRODUCTION REQUIREMENTS

Production requirements for parts to be cleaned will determine the optimum cleaning methods and equipment to be used. More efficient and costlier cleaners and equipment can often be economically justified for high-volume and fast production needs. For example, fully automated equipment for cleaning, rinsing, drying, and unloading, with automatic control of the cleaning solution, can provide high production rates with consistently high quality.

Reviewers, cont.: ***M. Arthur Detrisac****, Technical Director, Surface Finishing Products, McGean-Rohco, Inc.;* ***T. R. Emerick****, National Sales Manager, Final Phase;* ***F. John Fuchs****, Blackstone Ultrasonics;* ***Jeffrey R. Hilgert****, Worldwide Product Manager, Branson Cleaning Equipment Co.;* ***Wallace L. Hilgren****, President, Kleer-Flo Co.;* ***Jack Horner****, Director, Technical Services, Allied-Kelite Div., Witco Chemical Corp.;* ***Q. D. Mehrkam****, President, Ajax Electric Co.;* ***W. Mooney****, Market Manager, E. F. Houghton & Co.;* ***Maurice O'Donoghue****, Applications Manager, Crest Ultrasonics;* ***Joseph V. Otrhalek****, Director, Research & Development, Industrial Chemical Specialties Div., Detrex Chemical Industries, Inc.;* ***Howard G. Pekar****, Technical Director, Globerite Div., Ashland Chemical Co.;* ***Joe Pokorny****, Chemist, Baron-Blakeslee, Inc.;*

SELECTING A CLEANING PROCESS

TABLE 18-1
Metal Cleaning Processes Typically Used for Various Purposes (Listed in Order of Decreasing Preference)[1]

Types of Production	In-Process Cleaning	Preparation for Painting	Preparation for Phosphating	Preparation for Plating
Removal of pigmented drawing compounds:[a]				
Occasional or intermittent	Hot-emulsion hand slush, spray emulsion in single stage, vapor slush degrease.[b]	Boiling alkaline, blow off, hand wipe. Vapor slush degrease, hand wipe. Acid clean.[c]	Hot-emulsion hand slush, spray emulsion in single stage, hot rinse, hand wipe.	Hot alkaline soak, hot rinse (hand wipe, if possible), electrolytic alkaline, cold water rinse.
Continuous high production	Conveyorized spray-emulsion washer.	Alkaline soak, hot rinse, alkaline spray, hot rinse.	Alkaline or acid[d] soak, rinse, alkaline or acid[d] spray, rinse.	Hot emulsion or alkaline soak, rinse, electrolytic alkaline, rinse.
Removal of unpigmented oil and grease:				
Occasional or intermittent	Solvent wipe. Emulsion dip or spray. Vapor degrease. Cold solvent dip. Alkaline dip, rinse, dry or dip in rust preventive.	Solvent wipe. Vapor degrease or phosphoric acid clean.[d]	Solvent wipe. Emulsion dip or spray, rinse. Vapor degrease. Alkaline soak or spray.	Solvent wipe. Emulsion soak, barrel rinse, electrolytic alkaline rinse, hydrochloric acid dip, rinse.
Continuous high production	Automatic vapor degrease. Emulsion, tumble, spray, rinse, dry.	Automatic vapor degrease.	Emulsion power spray, rinse. Vapor degrease. Acid clean.[c]	Alkaline soak, electrolytic alkaline rinse, hydrochloric acid dip, rinse, alkaline soak, rinse.
Removal of chips and cutting fluid:				
Occasional or intermittent	Solvent wipe. Alkaline dip and emulsion surfactant. Stoddard solvent or trichlorethylene. Steam.	Solvent wipe. Alkaline dip and emulsion surfactant. Solvent or vapor.	Solvent wipe. Alkaline dip and emulsion surfactant.[f] Solvent or vapor.	Solvent wipe. Alkaline dip, rinse, electrolytic alkaline,[g] rinse, acid dip, rinse.[h]
Continuous high production	Alkaline (dip or spray) and emulsion surfactant.	Alkaline (dip or spray) and emulsion surfactant.	Alkaline (dip or spray) and emulsion surfactant.	Alkaline soak, rinse, electrolytic alkaline,[g] rinse, acid dip and rinse.[h]
Removal of polishing and buffing compounds:				
Occasional or intermittent	Seldom required.	Solvent wipe. Surfactant alkaline (agitated soak), rinse. Emulsion soak, rinse.	Solvent wipe. Surfactant alkaline (agitated soak), rinse. Emulsion soak, rinse.	Solvent wipe. Surfactant alkaline (agitated soak), rinse, electroclean.[i]

(continued)

***Reviewers, cont.: Greg Poland**, Manager, Marketing Services, Graymills Corp.; **Greg Robertson**, Auto Sonics, Inc.; **B. Mitchel Robin**, Vice President, Technical, E. F. Houghton & Co.; **H. M. Sadwith**, President, Industrial Washing Machine Corp.; **Norman H. Schellenger, Jr.**, Group Leader, AMCHEM Products, Inc.; **John O. Sparks**, Product Manager, Industrial Finishing/Metalworking, DuBois Chemical Co., Div. of Chemed Corp.; **Richard O. Toles**, Vice President, Technology, Crest Ultrasonics Corp.; **Ken Watson**, Plant Operations Manager, Baron-Blakeslee, Inc.; **William J. Wittke**, Manager, Technical Service, Oakite Products, Inc.; **B. S. Yaffee**, Manager, Technical Service, Industrial Products Group, Diversey Wyandotte Corp.; **Nabil Zaki**, Technical Director, Frederick Gumm Chemical Co., Inc.*

TABLE 18-1—*Continued*

Types of Production	In-Process Cleaning	Preparation for Painting	Preparation for Phosphating	Preparation for Plating
Continuous high production	Seldom required.	Surfactant alkaline spray, spray rinse. Agitated soak or spray, rinse.[j]	Surfactant alkaline spray, spray rinse. Emulsion spray, rinse.	Surfactant alkaline soak and spray, alkaline soak, spray and rinse, electrolytic alkaline,[i] rinse, mild acid pickle, rinse.

[a] For complete removal of pigment, parts should be cleaned immediately after the forming operation, and all rinses should be spray when practical.
[b] Used only when pigment residue can be tolerated in subsequent operations.
[c] Phosphoric acid cleaner-coaters are often sprayed on the parts to clean the surface and leave a thin phosphate coating.
[d] Phosphoric acid for cleaning and iron phosphating. Proprietary products for high and low-temperature application are available.
[e] Some plating processes may require additional cleaning dips.
[f] Neutral emulsion or solvent should be used before manganese phosphating.
[g] Reverse-current cleaning may be necessary to remove chips from parts having deep recesses.
[h] For cyanide plating, acid dip and water rinse are followed by alkaline and water rinses.
[i] Other preferences: stable or diphase emulsion spray or soak, rinse, alkaline spray or soak, rinse, electroclean; or solvent presoak, alkaline soak or spray, electroclean.
[j] Third preference: emulsion spray rinse.

TABLE 18-2
Typical Cleaning Sequences Used Prior to the Application of Various Inorganic Coatings

Iron Phosphating	Zinc Phosphating	Plating
1. Simultaneously clean and phosphate.	1. Alkaline clean and descale.	1. Alkaline clean, soak and electroclean.
2. Rinse.	2. Rinse.	2. Rinse.
3. Seal (optional).	3. Acid descale (optional).	3. Acid activate.
	4. Rinse.	4. Rinse.
	5. Zinc phosphate.	5. Plate.
	6. Rinse.	
	7. Seal (optional).	

(*Frederick Gumm Chemical Co.*)

SAFETY AND ENVIRONMENTAL CONSIDERATIONS

The toxicity and flammability limitations set by OSHA and other regulatory agencies limit the use of solvents in cleaners, and toxic or flammable solvents should be avoided in cleaners wherever possible. Certain environmentally objectionable toxic chemicals, such as chromates, cyanides, and phenolic compounds, have practically been eliminated from new cleaners.

The use of new oil-displacing cleaners, rather than emulsifying types, allows effective separation of oil, thereby reducing oil loads in waste water. Air pollution standards for aromatic solvents are particularly stringent in some areas. Whenever possible, industry is replacing solvent cleaners with solvent emulsion cleaners or, still better, with straight alkaline cleaners to ease pollution problems.

Attention must be given to the disposal of spent cleaning solutions and other industrial wastes. Items that have to be essentially removed from the effluent of finishing operations include oil, heavy metals, and, in some areas, phosphates. Also, the pH of the solution has to be adjusted to meet regulatory requirements. Detergents normally contain biodegradable surfactants and reduced phosphates or no phosphates to meet waste disposal requirements. Detailed information with respect to proper handling and disposal of cleaning materials is available from manufacturers and suppliers of compounds.

CHEMICAL CLEANING

Chemical cleaning is the most widely used method of providing a suitable surface for subsequent finishing by phosphating, electroplating, organic coating, or other coating processes. Chemical cleaning usually depends upon the use of solvents or a chemical action between the cleaning material and the contaminant. Table 18-3 compares the most common of the basic chemical cleaning methods. The types of chemical cleaners generally used for removing contaminants from metal surfaces fall into the broad categories of solvent cleaners, solvent emulsion cleaners, alkaline cleaners, and acid cleaners.

REACTIONS IN CHEMICAL CLEANING

The following definitions help in understanding the different physical and chemical processes that take place during cleaning:

buffer A compound or mixture that causes a solution to resist a change in its pH value. High-quality alkaline cleaners are well buffered to maintain their alkalinity longer and thereby extend their effectiveness.

detergent A surface-active agent that possesses the ability to clean soiled surfaces.

TABLE 18-3
Comparison of Chemical Cleaning Methods

Factors	Emulsifiable Solvent Cleaning	Alkali Cleaning, immersion or spray	Alkali Electrocleaning	Acid Pickling	Molten-Salt Descaling
Equipment	Same as for alkali cleaning without current.	Still tanks or conveyorized spray washers of various sizes and capacities.	Open steel tanks with d-c current source, bus bars, and control equipment.	Appropriate corrosion-resistant tankage.	Low-carbon steel tanks with high-temperature heat source.
Cleaning medium	Mineral solvents mixed with suitable emulsifiers; used in water solution.	Appropriate proprietary cleaners in water solution properly inhibited for sensitive metals.	Same as for alkali cleaning without current.	Various acids.	Mixture of fused salts.
Operating temperature, °F (°C)	Depends upon flashpoint of solvent. Usually room temperature to 140 (60).	Room temperature 212 (100).	120-200 (50-95)	Room temperature to 180 (80).	400-950 (205-510)
Health hazards	Safe when body contact is avoided. Venting may be required.	Safe when body contact is avoided. Venting may be required.	Safe when body contact is avoided. Venting may be required.	Safe with protective clothing, face shields, goggles, etc. Venting may be required.	Safe with protective clothing, gloves, face shields, goggles, etc.
Fire hazard	Temperature must not exceed flashpoint of solvent.	None.	None.	None.	Slight.
Type soil removed	Both organic and inorganic matter.	Both organic and inorganic matter.	Normally used for final cleaning prior to plating to remove slight contamination and to activate surface.	Oxides, scale, rust, etc.	Scale, oxides, etc.
Effect on base metal	None.	None when properly inhibited, but caustic alkali will attack aluminum and zinc.	None when properly inhibited.	Very slight surface attack when properly controlled.	Sometimes slight etch.
Time required, minutes	1 to 15	1 to 30	1/2 to 3	1 to 30	Varies from seconds to minutes, depending upon the application.

dispersion To break big particles into small particles and suspend them in water so that they can be removed by rinsing. Alkaline silicates and phosphates are beneficial for dispersion.

emulsification To break large particles of oil and grease into small globules surrounded by a surfactant film so they can be rinsed away. Wetting agents are used in cleaning solutions to achieve this phenomenon.

etching (pitting) The localized attack of metal surfaces. Controlled etching of metals improves the adhesion of organic coatings. By contrast, uncontrolled etching of metals by an acid can cause damage by weakening the crystal structure.

flocculation The aggregation into larger particles to the point where precipitation occurs.

flotation The rising of soil particles to the surface of cleaning baths for removal by skimming.

SOLVENT CLEANERS

hydrogen embrittle The embrittlement of a metal by the absorption of hydrogen during a pickling, cleaning, or plating process.

inhibitor A substance used to reduce the rate of a chemical or electrochemical reaction, commonly corrosion or pickling.

neutralization The balancing of acidity and alkalinity by interaction. In the context of cleaning, the removal of acid soils by alkalies and alkaline soils by acids.

oxidation reduction To change the valence state of oxide scale and rust to soluble forms for removal from metal surfaces. Rust is chemically changed in this way to a more soluble form, easily dissolved by acids.

pH The symbol whose value is an index of the relative acidity or alkalinity of a solution. The pH scale extends from 0 to 14, with the midpoint 7 being neutral. Solutions having pH values less than 7 are acid; those having values greater than 7 are alkaline.

pickling The removal of fines and oxides from a metal surface by means of an acid solution.

precipitation The consolidation of soil particles (the opposite of dispersion) in a cleaning bath, permitting their removal as sludge and prolonging bath life. The hardness of water can be reduced by precipitation with soda ash or trisodium phosphate.

saponification The conversion of insoluble fats and fatty acids to water-soluble soaps by alkalies. The cleaning of buffing compounds from metallic surfaces can be achieved by saponification.

sequestration and chelation To inactivate calcium, magnesium, and iron salts in water so that they will not interfere with cleaning. Chelated alkaline compounds can remove oxides and rust from steel surfaces, eliminating the need for acids that may react with the metal to cause hydrogen embrittlement.

solvency The property of removal of soils by dissolving in a cleaning solution. For example, oils and fats are soluble in some solvents.

surfactant A surface-active agent (chemical compound) that has the property of altering the spreading conditions at boundaries between liquids and solids or liquids and liquids. Wetting of soiled surfaces is more effective with cleaners containing surfactants.

water break The appearance of a discontinuous film of water on a surface signifying nonuniform wetting and usually associated with an organic surface contamination.

wetting The penetration of soil by the cleaning solution. Soap and/or wetting agents increase the wetting action of water or solvents on a surface or soil by reducing surface tension. This wetting action helps in dislodging and removing soil.

SOLVENT CLEANERS

Solvents are derived from coal or petroleum and vary in napthenic and aromatic content. Glycol derivatives are water soluble and may be used to add solvency to water-based cleaners. While aromatic solvents are primarily used directly as solvent cleaners, some are also used in alkaline and acid cleaners.

Certain soils containing oils, fats, and waxes can be removed with organic solvents by the mechanism of solvency. The solvents used for this purpose are of three types:

1. Petroleum solvents such as kerosene, naphtha, or stoddard solvent, applied by wiping or immersion.
2. Nonflammable solvents, generally chlorinated, such as trichloroethane or perchlorethylene, used in vapor degreasers, discussed later in this chapter.
3. Specially formulated solvents, such as emulsion cleaners, emulsifiable solvents, or diphase cleaners, which allow the use of water-solvent mixtures.

Petroleum Solvents

Petroleum solvents are often used when removal of heavy soil is necessary as a precleaner before other cleaning operations, when water-based solvents cannot be used, or when rapid, light surface cleaning is required. They offer the advantages of high-speed soil penetration, a low cleaning temperature, being fast drying, and leaving behind a light solvent residue that can give some protection against rust.

Besides their relatively high cost, petroleum solvents can be disadvantageous in that the residue they leave on parts is often not tolerable for further operation, their fumes can be flammable or toxic, they are unsuitable for removal of certain soils such as soaps or pigments, and they often require expensive equipment.

Emulsifiable Solvents

The emulsifiable solvents have an emulsifier incorporated into kerosene or other aromatic solvents. After the soil has softened and/or dissolved, the workpiece is sprayed with water or is flushed with mineral spirits to remove soil and solvent. An emulsion forms when the soil and solvent combination is flushed with water. The solvents are effective in the removal of heavy oil films, waxes, greases, and dried-on, pigmented drawing compounds and rust preventives. Objectionable factors in the use of emulsifiable solvents can be their cost and the length of time required for cleaning.

Emulsion cleaners. These cleaners contain sufficient emulsifying agents to permit use at low concentrations. The emulsifiable solvent concentrate is commonly diluted with water from 1 to 5% for spray application or 8 to 12% for immersion application. Either method may be used from room temperature to about 140° F (60° C). Spray application is often preferred because it assists soil removal. Emulsion cleaning is commonly used for in-process removal of the bulk of soil to permit further processing and to provide temporary rust inhibition prior to the next processing step. Since emulsion cleaners have a neutral or mildly alkaline pH, they are sometimes referred to as neutral or rust-inhibitive cleaners.

Diphase cleaners. These cleaners have separate layers of a solvent phase and an aqueous phase. The aqueous phase removes dry soil, and the solvent phase removes oily soil by solvency. In some instances, the aqueous phase can serve as a seal for the solvent phase, retarding its evaporation. One major application for diphase cleaners is for precleaning buffed, zinc-based die castings.

ALKALINE CLEANERS

Alkaline cleaners are most widely used for the removal of metalworking compound soils in soak cleaning and spray cleaning operations. They are also used for barrel cleaning, electrocleaning, and ultrasonic cleaning (discussed later in this chapter). They are generally used to remove soils for in-process cleaning and to prepare metals for operations such as painting or plating.

Alkaline cleaners have the advantage of being economical. They can remove insoluble solid soils, water-soluble soils, and oily soils, and they can be rinsed with water. Alkaline cleaners can be formulated to operate at room or elevated temperature, depending upon the soil conditions. They can also be formulated for immersion or spray applications and for minimal or no attack on metals. Accurate control of the effective alkali concentration in the solution is essential for effective cleaning. This control is usually accomplished by simple titration of the alkali content.

The alkali cleaners available differ in composition depending upon the metal to be processed and the method by which they are applied. So-called "heavy duty," uninhibited, highly alkaline cleaners are not suitable for brass, zinc, and aluminum. Cleaners used in spray washers or electrocleaning tanks require special detergents, emulsifying agents, and low-foaming surfactants. These requirements are necessary to prevent excessive foaming caused by aeration in spray washers or the absorption of gases in electrocleaning. For cleaning steel, the cleaner may have a pH value of 9 to 14, depending primarily upon the type of soil.

Alkaline cleaners are formulated for optimum performance and economy. They usually contain alkaline builders, sequestering and chelating agents, and surfactants. They may also contain solvents and corrosion inhibitors for some applications. They are formulated to clean by a combination of mechanisms, including saponification, emulsification, dispersion, chelation, wetting, and solvency.

Alkaline Builders

Alkaline builders are usually the major components in alkaline cleaners. Chemical alkaline builders used include sodium hydroxide, potassium hydroxide, sodium carbonate, sodium silicates, potassium silicates, borax, and sodium or potassium phosphates. Such builders provide the alkalinity necessary for saponification and dispersion in cleaning, and also provide conductivity for electrocleaning.

Because of their better rinsability, phosphates are preferred over silicates when acid treatments follow cleaning. Silicates are favored for nonferrous metals, such as zinc and brass, to inhibit the corrosive action of alkalies. Borates are used in moderate alkalinity cleaners for sensitive nonferrous metals, such as aluminum. Organic amines and nitrites are used for cleaners that are not rinsed to provide in-process rust protection.

Sequestering and Chelating Agents

Sequestering and chelating agents, such as complex phosphates, sodium gluconate, and citrates, are usually added to alkaline cleaners in sufficient amounts to inactivate the calcium and magnesium water hardness so as not to interfere with cleaning and rinsing or cause excessive hard-water scale formation in processing equipment. The complex phosphates have the added advantage of acting as builders to provide detergency. The chelating agents are used when phosphates are restricted in the plant effluent and for special functions, such as the removal of scale, rust, or heavy metal soap soils. The type of chelating agent used is carefully selected so as not to tie up the heavy metals to the extent that they cannot be removed in the waste treatment plants by conventional methods.

Chelating agents consist of two basic types: soft and hard chelates. Soft chelates are identified as complexing agents and include gluconates, citrates, and alkanol amines. Modern waste treatment methods can tolerate small amounts of these soft agents. Chelating agents such as EDTA and NTA are referred to as hard chelates and cannot be removed in waste treatment plants by conventional methods.

Surfactants

Surfactants are used in cleaning formulations to penetrate, displace, and/or emulsify soil. They are generally classified as anionic, nonionic, cationic, and amphoteric. Among the anionic surfactants are soaps and alkyl benzene sulfonates or napthalene sulfonates. The nonionic surfactants are the most widely used, especially in spray cleaning formulations. The amphoteric surfactants are cationic at acid pH and anionic at alkaline pH; they are used for special functions, such as corrosion inhibition. Practically all the surfactants used in cleaners today are biodegradable, with little or no sacrifice of detergency.

Specific Cleaners

Specific cleaners (those developed for specific processes and applications) are usually formulated for ferrous and nonferrous metals to avoid undue etching or tarnishing. They are made for soak cleaning, electrocleaning, and spray cleaning. Spray cleaners are formulated with low-foaming or defoaming surfactants to prevent excessive foam in spray washers.

Soak cleaners are commonly used at a concentration of 4 to 12 oz/gal (0.03 to 0.09 kg/L), with a temperature of 75 to 190° F (25 to 88° C), and the parts are immersed for 5 to 15 minutes. Spray cleaners are used at a concentration of 1 to 4 oz/gal (0.008 to 0.03 kg/L), with a temperature of 75 to 160° F (25 to 70° C); the parts are cleaned from 1 to 3 minutes, with a spray pressure of 15 to 30 psi (103 to 207 kPa).

Low-Temperature Cleaners

The availability of heating can have a profound effect on cleaning. In general, the speed of the chemical reactions in cleaning about doubles with each 20° F (11° C) rise in temperature. This temperature effect is more pronounced in removing buffing compounds, heavy greases, and lubricant residues. In such instances, a temperature above the softening point of the soil should be maintained for effective removal. Lower temperatures can be more easily compensated for in spray cleaning applications owing to the impingement energy of the spray compensating for some loss of chemical cleaning activity. Low-temperature spray cleaners and soak cleaners are readily available.

Operating temperatures. A low-temperature cleaner operates between 70 and 120° F (20 and 50° C) and does require limited heat on occasion. A cold-temperature cleaner operates below 70° F and requires no external heat energy. Cold-temperature cleaners are effective for the removal of certain easy soils. A standard high-temperature cleaner normally operates above 140° F (60° C), well into the 170 to 190° F (75 to 90° C) range.

Requirements. Whether the metal finishing operation uses a cold, low, or high-temperature cleaner, the results must be the same. The processed part must be free of soil or contamination, be rinsed free of cleaner residues, and show good compatibility with the overall finishing line.

Advantages. The most apparent advantage of cold and low-temperature cleaners is a reduction or elimination of heating costs. Fringe benefits, also of considerable value, are a reduced water consumption due to less water evaporation loss (up the stack), less chemical wear on equipment, better working conditions (less heat and humidity), and a more efficient usage of counterflow rinsing.

Limitations. Low-temperature and cold cleaners are susceptible to certain production problems normally not encountered with high-temperature operations. The effectiveness of such cleaners is a function of the soil or lubricant to be removed.

They are not as universal in scope of application. For example, a change in the type of soil or an increase in soil load can suddenly reduce the cleaning efficiency of these compounds.

Electrolytic Cleaning

Another form of alkaline cleaning, electrolytic cleaning or electrocleaning, involves the use of electric current to release gas bubbles on the work surface. These bubbles produce a scrubbing action that, when combined with the selected electrocleaning detergent, results in a high degree of cleaning. The parts to be cleaned are immersed in a hot alkaline solution, 140 to 180° F (60 to 80° C), and current is applied. Basically there are three methods of electrocleaning: cathodic or direct current cleaning, anodic or reverse current cleaning, and periodic reverse current cleaning.

Cathodic or direct current electrocleaning. In this method, the work is negatively charged. Electrons flow from the work to the anode, and hydrogen gas evolves on the surface of the work as a result of water electrolysis. This method of cleaning is generally applied to brass, nickel, lead, and nickel silver. Other metals may also be cleaned with direct current, although infrequently. Cathodic cleaning solutions must be kept free of soil, particularly metallic fines that may plate onto the surface of the work.

Cathodic cleaning can produce hydrogen embrittlement in ferrous metals. It is therefore important to use cleaning materials that offer controlled foaming. A slight layer of foam should form on top of the electrocleaning solution, thick enough to prevent the escape of excess vapors and mist, yet brittle enough to prevent excessive entrapment of hydrogen with its consequent danger of explosion.

Typical variables in a direct current electrocleaning operation include a cleaner used at about 4 to 12 oz/gal (0.03 to 0.09 kg/L) of water, depending upon the type of metal being cleaned, and a direct current of 30 to 100 A/ft^2 (322 to 1075 A/m^2) at 4 to 9 V. Immersion time will vary depending upon the degree of soil present; it will normally take from 30 seconds to 2 minutes. After cleaning, parts should be removed from the tank and immediately rinsed.

Anodic or reverse current electrocleaning. In this method, the work is positively charged. Current flows from the cathode to the work, and oxygen gas evolves on the surface of the work. This method of cleaning is generally applied to steel, stainless steel, iron, copper, and sometimes zinc, but it is seldom used on brass because of its tendency to tarnish that metal.

With anodic cleaning, there is no danger of hydrogen embrittlement, and the deplating action produces a higher degree of cleanliness. In cathodic or direct current cleaning, particles or smut are positively charged and attracted to the work; these particles are repelled in reverse current cleaning. However, less agitation is provided since oxygen is given off at the surface of the work in only half the amount that hydrogen is given off in direct current cleaning. A reverse current electrocleaning operation proceeds in the same manner as a direct current operation, with the exception of the work being positively charged.

Periodic reverse current electrocleaning. Although this method of electrocleaning is not frequently employed, it is probably the most effective of the three methods. In this method, the procedure for direct current cleaning is followed, except that the current is applied directly for 10 seconds and then reversed for 20 seconds. At the end of the cleaning period, the current is shut off, preferably in the reverse part of the cycle. This method is somewhat more expensive in initial costs and is seldom required for most operations.

Alkaline Descaling

Alkaline descaling is a cleaning method in which soil, paints, oil, scale, and oxides can be removed from aluminum, zinc, steel, and tin in one operation, eliminating the need for acid descaling, which may introduce hydrogen embrittlement into the metals. This method is normally not recommended for stainless steel, magnesium, copper, brass, and certain other metals. Alkaline descaling operations may be accomplished in tanks or barrels, or with spray equipment or electric current; but they are most effective when used with periodic reverse current electrocleaning. Alkaline descalers (derusters) are often used to remove light rust from steel parts prior to plating. The solutions sometimes contain surfactants to aid in removing light soils or soak-cleaner films. Periodic reverse current is required to remove rust.

ACIDIC CLEANERS

Acidic cleaners are designed to remove oxides, rust, flux residues, corrosion products, tarnish films, perspiration stains, and heat scales resulting from forging, heat treating, extruding, welding, soldering, and brazing. Phosphoric acid is the most frequently used acid for cleaning and painting preparation; sulfuric and hydrochloric acids for rust and scale removal; and chromic acid for cleaning zinc, aluminum, and magnesium alloys. In preparing metals for plating, chromic acid is avoided in the pretreatment steps, unless the work is to be chromium plated directly (without intermediate plates of other metals). If plating is not done directly, precautions are required to remove any traces of chromic acid residues from the surface before subsequent activation and plating operations.

Nitric acid alone and/or with hydrofluoric acid is used to deoxidize and remove smuts from aluminum alloys. These two acids are also used to remove scale from stainless steels. Other acids such as gluconic, sulfamic, citric, oxalic, and acetic have specific applications. These acids may be compounded with surface-active agents to increase wetting and detergency, with solvents to permit one-step cleaning and oxide removal, and with inhibitors to prevent attack on base metals. While organic inhibitors are beneficial, they can cause problems in subsequent operations, especially in plating lines, if improperly controlled or inadequately rinsed.

Pickling

Pickling is the treatment of metal surfaces to remove oxides, heat scale, stains, weld discolorations, and smut. It can be achieved with a dilute acid product, generally sulfuric acid at 10 to 25%, or commercial hydrochloric acid mixed with an equal amount of water. Other acids used in industrial pickling operations are phosphoric acid, hydrofluoric acid, and nitric acid; these are inexpensive and do a relatively good job. There are instances, however, in which another acidic material may be desirable. In addition to their inherent ability to remove oxides, smut, and other contaminants, acids can be blended with wetting agents and solvents to provide soil-removing properties. Many pickling compounds include inhibitors that act after the soil has been removed by reducing or halting attack on the base metal. Proprietary acid salts are available to replace hydrochloric and sulfuric acids. The acid salts are easier and safer to handle than concentrated mineral acids.

Acids work by reacting with scale, rust, and oxides to form water-soluble salts that are readily rinsed away. An important benefit derived from using acidic materials prior to organic coating is improved paint adhesion. The treated metal surfaces are highly receptive to paint and other finishing operations, being both chemically compatible and etched for better bonding of organic coatings. Pickling operations should be followed with a water rinse to prevent attack on the metal prior to plating.

Bullard-Dunn Process

The Bullard-Dunn process is a proprietary, catalyzed-salt cleaning process used for steel and occasionally for copper and brass. It involves pickling the metal in a cathodic operation and minimizes hydrogen embrittlement. The metal is immersed in a water solution of 10% sulfuric acid (by volume) at 150° F (65° C), and a current density of 60 to 75 A/ft^2 (645 to 806 A/m^2) at 6 V is applied. As the scale is removed, tin plating is used on areas where bare metal is exposed to protect the base metal against attack and pitting. When all the scale is removed, the tin deposit may be unplated in an anodic or reverse current alkaline cleaning solution.

Salt-Bath Descaling

The salt-bath descaling processes are fused molten-salt operations accomplished at temperatures from 400 to 1000° F (205 to 540° C). They are safe on such metals as titanium, chromium, cobalt, manganese, molybdenum, nickel, and stainless steel. Salt-bath descaling is primarily designed for very difficult cleaning jobs. Three types of molten-salt baths are commonly used: reducing, oxidizing, and electrolytic.

Reducing type. This molten-salt operation is generally recommended for chromium-containing steel alloys of the ferritic and martensitic type. This operation is not recommended when glass, graphite, carbon, or other organic compounds are superimposed on the scale. Temperatures used range from 700 to 750° F (370 to 400° C); and the bath is composed of 1.5 to 2% sodium hydride, with the balance being fused liquid, anhydrous sodium hydroxide. The sodium hydride reduces metal oxides to pure metal. Following the molten-salt operation, a bright dip in a nitric-hydrofluoric acid solution is normally used.

Oxidizing method. This type of salt-bath descaling is more flexible than the reducing method, and temperatures may range from 400 to 1000° F (205 to 540° C). The base material is sodium hydroxide in combination with various activators, which react with the metal oxides to convert them to a more acid-soluble form. This is a particularly useful method for continuous-strip bath processing and batch operations, and is the most effective method for descaling all types of annealed stainless steels. It is frequently followed by a water quench and then a bright dip in a dilute nitric-hydrofluoric acid pickle solution.

Electrolytic descaling. In this method, sodium hydroxide containing catalytic agents is used at temperatures ranging from 800 to 900° F (425 to 480° C), with a 3 to 6 V direct current electrical source, normally of the periodic reversing type. Scale, soil, carbon, graphite, silica, and oils are removed quickly from cast iron and other ferrous materials. This process is particularly suited to cleaning materials for subsequent coating or joining operations.

RINSING

Factors that determine whether a rinse is necessary after cleaning include: the type of cleaning materials used, the next processing step, and the destination of the workpieces. If the next step is painting and the cleaning has been done with an alkaline solution, then rinsing and a neutralizing dip are necessary. If the parts are to be stored and have been cleaned in a solvent material, then rinsing is normally not required.

Medium Used

Most rinsing operations consist of water only, although proprietary rinse aids are available that may be used in conjunction with the water. Most water contains varying quantities of chlorides, sulfates, and carbonates of such metals as calcium, magnesium, and iron. These materials may cause spotting or contamination of plating solutions used in subsequent finishing. Water-softening equipment is suitable for preparing hard water for use as rinse water; but deionized water, although it is more expensive, may be desirable for rinses preceding bright plating, chromium plating, and anodizing solutions. Wetting agents or antispotting additives can be used to prevent spotting or rusting in recesses and blind holes during subsequent drying. Reclamation of used rinse water is often necessary for maximum economy of water use, as well as for compliance with waste disposal regulations.

Water used for rinsing may be hot, warm, or cold. Warm water is generally preferred because cold water is less efficient and hot water tends to promote the formation of oxide films. The water must also be kept relatively clean. The contamination limit (maximum allowable concentration of chemicals in the rinse solution) is generally based on experience and varies with the amount of cleaning solution on the parts during rinsing.

Rinsing Methods

Rinsing is accomplished by immersion or spray, with spraying generally being the most effective method. Spray cleaning is often incorporated into multizone machines that perform cleaning, rinsing, and drying. Immersion is accomplished by simply dipping the parts into a tank of water or through the counterflow or multiple rinsing methods.

With counterflow rinsing, a series of tanks, usually two or three, are used. Water flows into the first tank, overflows into the second tank, and—in three-stage operations—into the third tank. Workpieces to be rinsed flow in the opposite direction to the flow of rinse water: first into the second or third tank and then into the second or first tank. This method results in more efficient use of water and better rinsing. The tanks should be made as small as possible, and automatic control with respect to contamination limit is desirable.

A continuously running rinse or one that is constantly overflowing is desirable to prevent serious contamination of the rinse water and to aid in carrying off excess soil in the rinse. However, excessive overflow after alkaline or acid cleaning can result in flash rusting of the workpieces. Relative motion between the workpieces and water is also desirable for most rinsing applications.

DRYING

As previously mentioned, heating of the rinse water facilitates drying. For parts that do not readily dry upon removal from the final rinse stage, a dryoff station is often required. The two most common drying methods are a forced blowoff with room temperature air or a heated, recirculating air dryer. Heated air is generally preferable because it permits faster drying and reduces air and floor space requirements. The temperature of heated air for drying is usually limited to a maximum of 170° F (75° C).

CHAPTER 18

CHEMICAL CLEANING PROCESSES AND EQUIPMENT

Flat-surfaced parts that readily drain are often passed through a high-velocity, room temperature air curtain to blow off excess liquid prior to leaving the washing machine. Parts that are bulk loaded into baskets or that retain substantial cleaning solution generally require a high-temperature dryoff stage.

Typical heated dryers recirculate high-temperature air through a drying chamber to evaporate the solution. An exhaust fan is sometimes provided to remove excess moisture. The air is heated with steam, gas, oil, or electricity, depending upon the economic considerations dictated by the plant location and the costs of energy sources. A heated dryer should also include well-insulated outer walls and either well-insulated doors or curtains to contain the heat and reduce energy consumption.

Absorption drying is used to remove solution from some small parts. This method consists of tumbling the parts in dry sawdust or ground corncobs. A possible limitation is that the process leaves a residue on the workpieces. Flammable solvents, such as acetone and alcohol, can be used to dissolve the solution; but because of potential fire hazards and disposal problems, this method is seldom used.

Water-displacement systems using chlorinated solvents are also used for drying. The use of some fluorocarbon solvents for displacement drying eliminates spots on workpiece surfaces, and these solvents are not flammable.

PROCESSES AND EQUIPMENT

The primary methods of cleaning workpieces are by hand, by immersion (dipping or soaking), by spraying, or by a combination of immersion and spraying. Automated process lines incorporating cleaning, rinsing, and drying are often used for high-production requirements. Selection of the type of chemical cleaner to be used depends upon the method of application.

Process Selection

The choice of a particular cleaning process depends upon several factors, including the following:

- The number, size, and geometry of the workpieces to be cleaned.
- The workpiece material.
- Production requirements.
- Floor space limitations.
- Availability of heat, power, water, and other facilities.
- Type of labor available and prevailing labor rates.
- Processes preceding and following cleaning.
- Degree of cleanliness required and types of soil to be removed.
- Safety considerations.
- Cost of the cleaning equipment needed.
- Local regulations concerning waste disposal.

Hand Cleaning

Chemical cleaning by hand wiping or scrubbing is less commonly used than other methods because of high labor costs, but it has advantages for some applications. It requires only simple, low-cost equipment, such as common mops, sponges, brushes, and cloths; but it is labor intensive and restricted by OSHA regulations. When strongly acidic or caustic cleaners are used, resistant pails and protective clothing for workers are required.

Immersion Cleaning

Chemical cleaning by workpiece immersion (dipping or soaking) is a simple and efficient process that is used extensively. It is performed in tanks, barrels, or drums, with heated or unheated chemical solutions. Tanks are the predominant means for chemical cleaning, with barrels and drums being more commonly used in plating operations. Installations vary from a single, unheated tank to multistage systems (see Fig. 18-1) having heated and agitated solutions.

Workpieces can be handled individually or in bulk. Individual parts or baskets of workpieces are often hand loaded with an overhead hoist, or they can be mounted on racks suspended from a conveyor. Industrial robots are also being used for loading and unloading. If work baskets are used, some means of preventing workpieces from contacting one another must be provided, or incomplete cleaning of the surfaces will result. Barrels or drums that are immersed and rotated in tanks are used to clean large quantities of small parts. A common type of immersion washer automatically lowers a platform holding the workpieces into the solution and agitates the load for thorough flushing.

Advantages. When immersion tanks, barrels, and drums are constructed of proper materials, they can be used for emulsion, alkaline, electrolytic, and acidic cleaning, as well as for pickling. The immersion process is well suited to cleaning castings, weldments, machined parts, and some sheet metal components. Workpieces in a wide variety of shapes, sizes, and weights can be processed.

When heating of the chemical solution is required, operating temperatures are normally economically maintained. The process is adaptable to programmed, automatic operation. Properly designed immersion cleaning systems do a good job of cleaning both the interior and exterior surfaces of many parts, such as castings, weldments, and machined components.

Limitations. Drag-out losses of chemical solution from immersion equipment can be high, contaminating the water used for subsequent rinsing and increasing the cost of treating the effluent for disposal. Immersion cleaning often requires

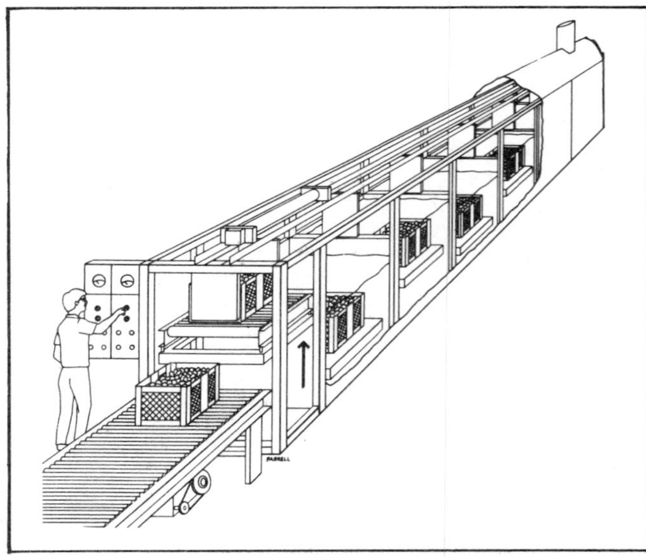

Fig. 18-1 Multistage processing line for chemical cleaning. (*Automated Chemical Systems Corp.*)

higher solution concentrations, higher operating temperatures, and longer cycle times than spray cleaning.

Air pockets can form during immersion cleaning, resulting in uncleaned areas on the workpieces. Skimming devices are often required to prevent the dragging of cleaned parts through a soil layer on top of the chemical solution.

Tank construction. Ideally, tanks for holding chemical solutions should be larger than the workpieces to be cleaned and have sufficient depth for settling of the sludge. A rule of thumb is that the tank dimensions should provide 2 to 3 gal (7.6 to 11.4 L) of solution for every square foot (0.09 m^2) of surface area of the workpiece load to be cleaned. This capacity assists in maintaining the proper operating temperature with less required energy by reducing the decrease in temperature when cold parts are immersed. An overflow weir is recommended to permit skimming of soil from the surface of the solution. A typical heated immersion tank is illustrated in Fig. 18-2.

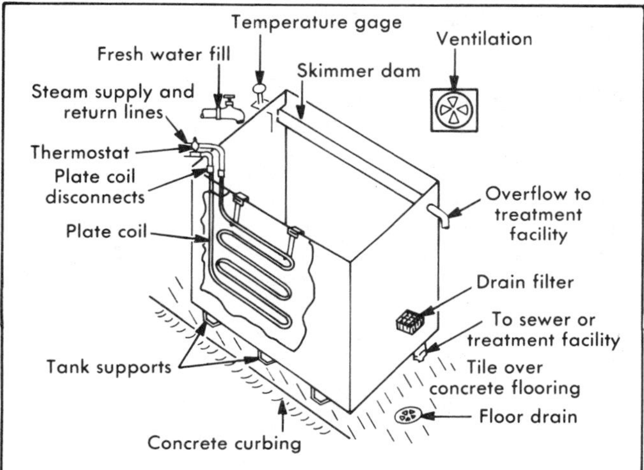

Fig. 18-2 Typical installation of a heated immersion tank for chemical cleaning. (*Detrex Chemical Industries, Inc.*)

Low-carbon steels, generally hot rolled, are used extensively for tanks that hold aqueous alkaline and emulsion cleaning solutions. Channel or angle-iron reinforcements are provided when required. All corners and joints are generally welded, both internally and externally. Heating coils and pumps are sometimes made of stainless steel, generally of the 300 series.

For acid cleaning and pickling, the tanks must be made from acid-resistant materials, again with double-welded corners and joints. Stainless steel is attacked by hydrofluoric and hydrochloric acids. Plastic, brick, concrete, or steel tanks with acid-resistant linings are sometimes used. Workholding baskets, racks, and hooks, as well as grates and conveyor systems, must also be resistant to chemicals.

Agitation. The workpieces and/or the chemical solution are agitated to improve the cleaning efficiency, to reduce cycle times and temperature requirements, to eliminate or reduce the formation of air pockets, and to facilitate the cleaning of multiple parts suspended in the tank. However, many dip tanks, especially those used for acid cleaning and pickling operations, operate with no induced agitation; they depend upon chemical action and/or convection movement of the solution to clean the parts.

The primary methods of agitation used for immersion cleaning are: workpiece movement, air pressure stirring, mechanical stirring, ultrasonics, and pumping. In electrocleaning, the electrical current used and the gas evolution agitates the solution at the work surface.

Workpiece movement. When parts are cleaned in baskets, agitation is accomplished by raising, lowering, and turning the baskets. With barrel cleaning, agitation is provided by rotation of the barrel. In automated systems, continuous movement of the workpieces often provides sufficient agitation.

A common method of agitating the workpieces is to use a vertically moving workholding platform that oscillates the parts in the solution. These platform washers can be equipped with an air cylinder to raise the platform to the top of the tank for easier loading and unloading. For multistage applications, conveyors can be provided to and from the platform. A tank cover can also be provided to automatically open when the platform is raised.

Air pressure stirring. Underwater air jets are commonly used for agitation, which is generally accomplished by passing blown or compressed air through a pipe, having small holes for the air to escape, located in the tank below the workpieces. The holes should be drilled at an angle of 30° below the horizontal centerline of the pipe to prevent clogging by dirt. Air directed upward in the tank also minimizes the formation of surface soil on the chemical solution. Compressed air should be filtered to prevent hydraulic oil from being introduced into the solution.

If the air agitation is too violent, there is the possibility of the operator being splashed during manual loading or unloading. Also, air agitation has a cooling effect on heated solutions, thus increasing energy requirements.

Air agitation is of questionable value in some applications. For example, when using heavy-duty, caustic-based cleaning solutions, air bubbles cause the buildup of carbonates, and caustic mist is emitted into the operator's breathing zone. Carbonates add to the total alkalinity of the cleaning solution, giving false results in titration tests for concentration. Under these conditions, it is advisable to use free-alkali control procedures.

Mechanical stirring. Electric-motor-driven propellers are sometimes provided in the tanks to agitate the chemical solution, especially for cleaning large, bulky parts. Care is necessary in properly locating and guarding the propeller blades so they will not contact the workpieces, workholders, or the operator.

Pump agitation. Small-capacity pumps are often preferred for agitating chemical solutions in the cleaning tanks. Pump intakes are generally located about halfway down from the top of the solution to the bottom of the tank to avoid disturbing and/or pumping the sludge that collects on the tank bottom. The closed-loop return is located about one-fourth the distance below the top of the solution. The use of vertical pumps minimizes packing-gland leaks and air entrainment, which can produce foam. The primary purpose of pumping is gentle movement of the solution, not "blasting" of soil from the work surfaces.

Ultrasonic agitation. This method of agitation is sometimes used, especially for cleaning crevices in small parts and blind holes; however, it requires higher capital investment than the other methods of agitation discussed. Ultrasonic cleaning is covered later in this chapter.

Solution heating. Considerable acid cleaning and pickling is done with unheated solutions, but heated solutions are often required in emulsion and alkaline cleaning. Solution heating is accomplished with steam, gas, oil, or electric energy sources, depending upon the availability and cost of energy sources at the specific plant location. Heating of cleaning solutions generally necessitates the insulation and use of covers for immersion tanks, and ventilation is often required to remove excessive moisture.

Steam heat. Steam as a source of heat energy is readily available and economical in many plants, often from plant heating facilities. Plate coils are normally used for steam heating. Coils should be mounted on the side walls or suspended from the top of the tank to permit sludge to settle on the tank bottom.

The coils must be checked periodically for surface buildup and leaks. If the coil surfaces are not kept clean, heat transfer to the solution is poor, requiring excessively long heatup times and causing problems in maintaining recommended operating temperatures. Also, scale deposits may be corrosive, creating pitting of the heating coil surfaces. Plate-and-frame, shell-and-tube, and similar heat exchangers are highly efficient, but cost more than coils.

Gas/oil heat. Gas or oil are used in conjunction with fire tubes in the tanks to heat the solutions. Open-flame heating is not recommended. The tube surfaces must be kept free of buildup for fast heating and good temperature control, and to prevent burnout of the tubes. Tube burnout can cause several problems, including carbonate buildup when using heavy-duty alkaline cleaners. Because of fluctuations in the availability of gas and oil, many installations are being designed so that either gas or oil can be used as the energy source, but gas is generally more efficient and desirable.

Electric heat. This source of energy for heating cleaning solutions is generally much more costly and is consequently often avoided when other energy sources are available. Nevertheless, electric heating is sometimes chosen because of ease of installation. All that is necessary is to connect the electrical power supply to the contactors of immersion heating elements and to a thermostat. Also, there are plant locations where electric heat is the only energy source available. Electricity is sometimes used as an auxiliary heat source and when tanks have been modified to increase their volume.

Rotary drum washers. In rotary drum washers, cleaning is basically an immersion operation with the rotating, perforated drums or barrels partially or fully immersed in a tank of solution. Cleaning is also done by placing the workpieces and solution in a conventional tumbling barrel, discussed in Chapter 16, "Mechanical and Abrasive Deburring and Finishing." Rotary drum washers are used extensively for cleaning large quantities of small parts, especially those that are not too heavily soiled. In some designs, scoops or paddles are used to pick up the solution and dump it onto the workpieces.

Electrolytic (electrocleaning) tanks. Tanks for electrolytic alkaline cleaning (see Fig. 18-3) are generally made from hot rolled, low-carbon steel. For electrolytic acid descaling, acid-resistant materials must be used for the tanks, and venting is generally required. The tanks are equipped with heating units, anodes, and busbars connected to a d-c power supply, generally a rectifier. For some applications, the tank itself is used as the anode. Making the workpieces the anodes prevents redeposition of metallic ions on the parts. Anodic alkaline cleaning also prevents hydrogen embrittlement.

Ample tank dimensions are essential so that the heating units and electrodes do not contact the workpieces during entry, cleaning, or exit from the solution. Good insulation between the busbars and tank is critical to prevent stray currents that can cause many problems. An overflow weir or other means to prevent a soil layer from collecting on top of the cleaning solution should also be provided.

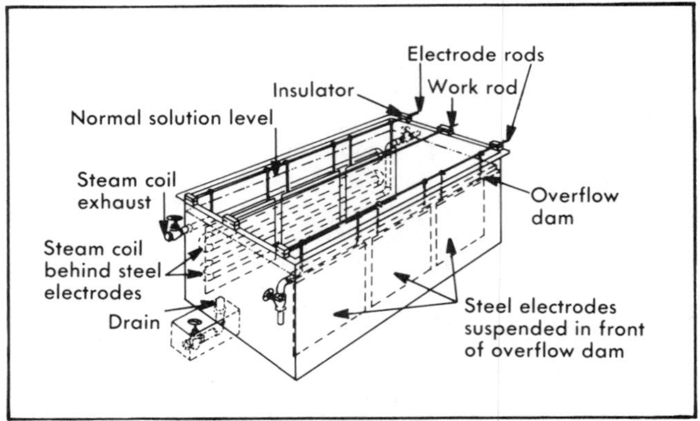

Fig. 18-3 Electrolytic cleaning tank equipped with steam heating coils and steel electrodes suspended from rods.

Spray Cleaning

Spray cleaning consists of impinging the spray of a cleaning solution upon the workpieces to remove unwanted soil. Machines used for spray cleaning, called spray washers, have a pump to pressurize the solution, a reservoir tank, connecting piping, spray nozzles, and, generally, some means for moving the workpieces or nozzles and for heating the solution. Multistage washers often have zones for cleaning, rinsing, and drying the parts. Drainage corridors should be provided between stations to facilitate drainage and reduce drag-out.

The recirculating solution should pass through screens having holes smaller than the spray nozzles. Both the screens and the nozzles should be easily accessible and removable for cleaning. Heating of the solution, when required, is done with steam, gas, oil, or electric energy, as discussed previously for immersion cleaning. Keeping the solution temperature at a minimum reduces heat losses due to aeration and evaporative losses from spraying. Construction materials for spray washers are similar to those for immersion tanks, with welding recommended for all corners and joints.

Advantages. An important advantage of spray cleaning is that high-production rates are possible, especially with similarly shaped workpieces. Clean solution is directed onto the work surfaces, and impingement of the pressurized solution assists in the cleaning operation.

Cleaner consumption is less than with immersion methods because the solution can be recirculated many times and solution concentration can often be lower. With proper draining time, drag-out losses of solution are minimal; and with proper workholding, uniform cleaning results are obtained. Spray cleaning is also ideal for light sheet metal parts that tend to float in immersion tanks.

Limitations. Equipment costs for spray washers are higher than for immersion tanks, and hence high-production requirements are generally necessary for economic justification. Also, more floor space and maintenance are generally required, and energy costs are usually higher. In addition, safety considerations limit the use of combustible petroleum solvents in spray washers.

Spray washers require regular maintenance. The nozzles must be free of all obstructions and properly aligned at all times. Pumps should be kept in good repair and provided with double screens between the cleaning solution and the pump intake to reduce plugging of the nozzles and to prolong pump life. Since foaming can be a problem with spray cleaning, the cleaners used should be specially formulated for low foaming.

Spray cabinets. Cabinets, or booths, provide for the simplest method of spray cleaning. They have doors for loading and unloading workpieces. In the cabinet, provisions are made for moving the nozzles, or a worktable that rotates or reciprocates is installed. Moving tables are generally preferred because the speed can be controlled and all surfaces of the workpieces can be exposed to the spray.

Figure 18-4 shows a plan view of a segmented cabinet-type machine in which workpieces placed on a mesh turntable are successively washed, rinsed, and dried before being returned to the load/unload position. Pressures at the spray nozzles vary from 2 to 200 psi (14 to 1380 kPa), depending upon the specific application and requirements.

Conveyorized spraying. For faster spray cleaning and higher production rates, parts are carried through washer tunnels by flat-belt, roller, or overhead conveyors. With mesh-belt conveyors (see Fig. 18-5), operators or automation equipment are required on both ends of the washers for loading and unloading. With overhead conveyors, workpieces can be loaded on racks or hooks at any location remote from the washers and carried from the washers to subsequent operations.

Positioning of the workpieces as they pass through the washers is of critical importance for efficient cleaning. Also, the workpieces must be positioned for proper draining to minimize carrying solution from the washers or from stage to stage and into the dryer.

The zones in multistage, conveyorized washers are separated to permit adequate draining and to prevent overspray from contaminating subsequent zones. Overall lengths of the washers depend upon the cycle times required in each zone, and the time in each zone is determined by zone length and conveyor speed.

Rotary drum washers. If the workpieces can be tumbled, rotating drum spray washers (see Fig. 18-6) are frequently used, especially for irregularly shaped stampings and screw machine parts. Workpieces are generally placed in a hopper at the loading end of the washer, and they tumble under the spray nozzles as they progress through the perforated drum. Solution drains into sumps below the drums for recirculation.

Combination immersion-spraying machines. These machines are available in two designs. In one, often used for large workpieces, both immersion and spray cleaning are performed at the same time. The second concept uses separate stages for each function, allowing the complete removal of heavy, tenacious soils, such as drawing compounds. The immersion cycle softens and partially removes the soil, and the spray cycle completes the soil removal.

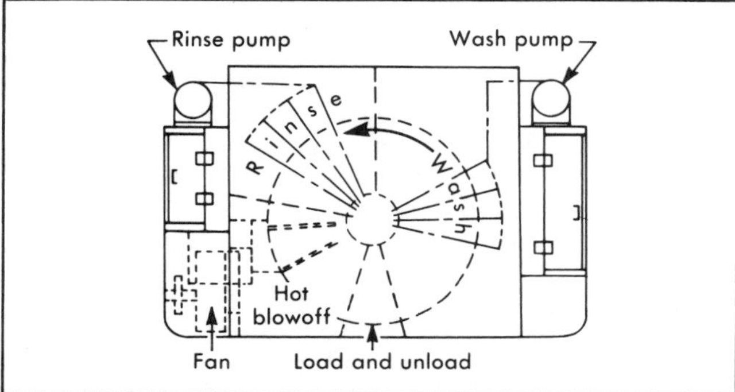

Fig. 18-4 Plan view of a segmented, cabinet-type machine for washing, rinsing, and drying parts. (*Industrial Washing Machine Corp.*)

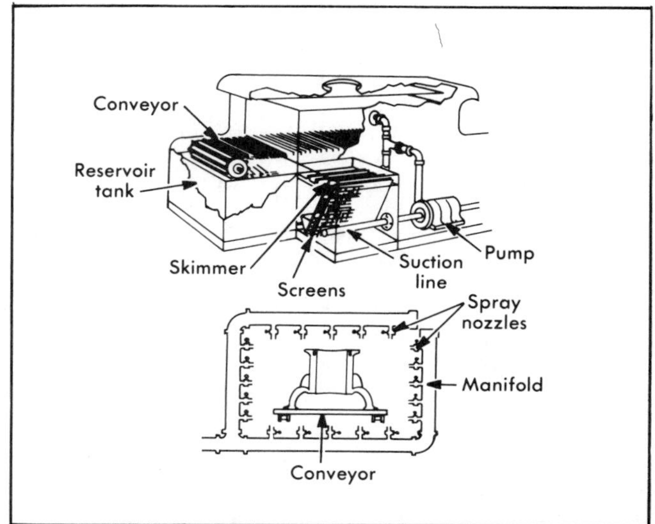

Fig. 18-5 (top) Typical conveyorized spray washer; (bottom) cross-sectional view showing arrangement of spray nozzles.

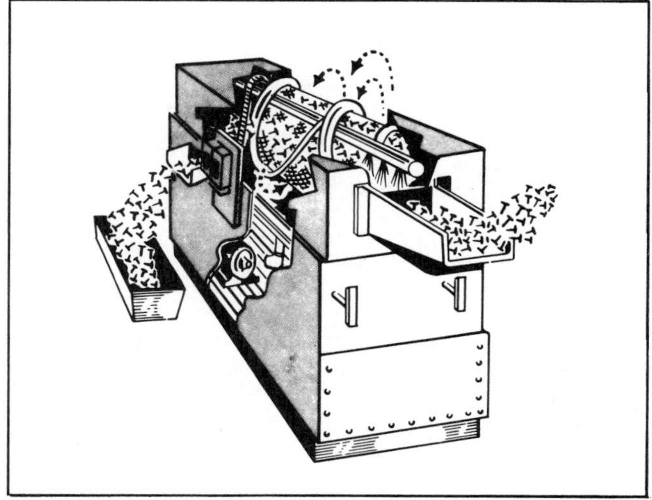

Fig. 18-6 Rotating drum spray washer for cleaning small parts. (*Manumatic International, Inc.*)

STEAM AND FLAME CLEANING

Steam and flame cleaning are less commonly used than the conventional chemical cleaning methods previously described. However, these processes are desirable for certain applications; for example, they are used to clean large and heavy parts that are difficult to move or that cannot be conveniently accommodated in immersion tanks or spray washers. These methods are also used to preclean some parts that are especially difficult to clean. Both steam and flame cleaning are faster than manual cleaning, and the portability of the equipment used is advantageous for some applications.

STEAM CLEANING

Steam cleaning, a variation of spray cleaning, is the application of a hot blast of steam and cleaning solution on parts to remove soil. The live, pressurized steam moves through an eductor, thereby creating a vacuum that picks up the cleaning solution. This type of cleaning offers several advantages over manual and immersion methods by combining three effective cleaning agents: heat, pressure, and chemical action. Heat speeds up and generally increases the cleaning action of any operation. Pressure provides force at the point of impact to loosen soils, making them easier to remove. Chemical action is provided by a material that chemically attacks soils, by either saponification or emulsification, also easing and speeding their removal.

Besides its obvious advantages over manual cleaning, steam cleaning is a highly portable operation when compared to tank immersion. In addition, many items that are simply too large or cumbersome to be cleaned in tanks may be cleaned easily with steam equipment. Another advantage of steam lies in its ability to get at otherwise inaccessible, trapped dirt—as in recessed areas of automobile engines, machinery, processing equipment, and molds—to clean areas that cannot be reached in any other way.

Steam cleaning does not offer nearly as strong an impact force as high-pressure spray cleaning equipment, nor is it as portable. However, steam cleaning equipment is relatively inexpensive compared to high-pressure spray units, and many jobs either do not require or cannot tolerate high-impact forces.

Equipment Used

Basically, the equipment required for an efficient steam cleaning operation is as follows:

1. A steam-generating boiler or generator capable of producing a minimum of 120 lb (54 kg) of steam per hour at 30 psi (207 kPa) and a maximum of 270 lb (123 kg) of steam per hour at 90 psi (621 kPa). Plant steam, if available, is often used. The recommended boiler horsepower range is from 3 1/2 to 8 hp (2.6 to 6 kW).
2. A solution holding tank in which the cleaning chemical is mixed with water, either hot or cold. The tank may be constructed of iron or steel; in some instances, the cleaner drum itself is used.
3. A steam eductor gun. Many varieties are available, but the best ones are insulated against heat buildup and offer rotating barrels that can clean the sides and bottom of equipment without the problem of handling a heavy hose.
4. Two hoses, one a steam hose and the other a heat-resistant hose.

Operation

With the generating device operating properly, the cleaning solution is added to the holding tank and mixed thoroughly. The steam hose is connected from the steam outlet to the gun, and the other hose from the holding tank to the solution inlet of the gun. For optimum results, the cleaning solution should first be heated in the holding tank. Steam cleaning can be accomplished simply by entering the barrel of the gun into the solution tank and slowly opening first the steam valve and then the solution valve. A partial vacuum is formed at the eductor valve by the steam, and the solution is drawn through the hose and delivered in a hot, pressurized spray to the point at which it is directed.

The point or nozzle of the gun is normally held from 2 to 4" (51 to 102 mm) away from the parts being cleaned. A systematic approach to the job is to keep the spray directed toward one area until it is thoroughly clean and then to move the gun slowly in one direction or pattern. A common procedure when cleaning large parts is to apply the spray from the bottom to the top of each part. After all surfaces have been sprayed once, they are given a second application. Next, the surfaces are rinsed thoroughly—first with steam, then with clear water, and finally with steam again to ensure quick drying. It should be remembered that steam cleaning involves water and thus can cause rusting on unprotected parts.

When work with a self-contained, steam-generating unit has been completed, the flame should be shut off first, allowing the solution to continue through the gun for another minute or two and ensuring that the coils cool down while they still contain the solution. Were the pump to be shut off first, the solution would evaporate, and solid particles and scale would be baked onto the barrel. One ounce of solution per gallon (7.5 g/L) used through a solution-lifting steam gun, with steam applied from a separate source, will provide a nozzle concentration of somewhat less than 1 oz/gal. The same solution passed through the coils of a self-generating steam unit will provide a nozzle concentration of slightly more than 1 oz/gal. The area used for steam cleaning generally requires venting.

Subsequent Processing

With most chemical cleaners, steam cleaning should be followed by rinsing or neutralizing and drying. Subsequent surface treatment may be required because paint adhesion is sometimes poor on steam-cleaned surfaces. When such treatment is not feasible, paints such as the asphalt types, which adhere well, should be used.

FLAME CLEANING

In flame cleaning, flames from an oxyacetylene torch are used to remove loose mill scale and rust. The flames are played over the metal surface, and the sudden temperature increase causes the scale to expand and flake away. Flame cleaning may also be used to burn off oils that will not leave heavy smut and to dry water from the metal surface. Any residue, loosened scale, or rust left after flame cleaning is usually removed by wire brushing. Flame cleaning must be closely controlled to prevent overheating of the metal surface with consequent surface hardening or changes in metallurgical properties. The flames used should be adjusted to have a neutral characteristic. Traverse of the flame on work surfaces should be fast enough to prevent fusing of the scale and foreign matter to the surfaces; however, a slow traverse is often necessary to clean heavily scaled or rusted surfaces. Flame cleaning is not generally applicable to thin sections that are subject to warping. The process is an alternative to blast cleaning, but results obtained are not usually as good.

One application of flame cleaning is the preparation of unpainted structural steel surfaces prior to painting or coating. Flame cleaning leaves a warm, dry surface to which a prime coat of paint can be applied before the surface cools. Painting while warm prevents condensation of moisture on the cleaned surfaces.

Most oils and greases are generally removed from work surfaces by solvent cleaning prior to flame cleaning. The solvent cleaning should be done a sufficient enough time before flame cleaning to ensure the vaporization and removal of all solvents before the application of flames. Welding flux and spatter should also be removed before flame cleaning. If fire or explosion hazards are present, proper precautions must be taken before flame cleaning, and ventilation provided during cleaning. Safety goggles should be worn by the operators during cleaning.

VAPOR DEGREASING

Vapor degreasing is a method of dissolving solvent-soluble and other soils, such as oils and greases, from the surfaces, crevices, and capillaries of metal, porcelain, and plastics products. Parts at ambient temperatures are brought into contact with hot solvent vapors, causing the vapors to condense on the parts. This condensation results in a liquid flow over the surface and throughout the part. The high rate of condensation results in an erosion effect that helps remove some nonsoluble soils like carbon black and lint. The solvents cannot dissolve inorganics such as oxides, salts, or smut.

The discussion of vapor degreasing in this section is a general overview of the process. For details required in specific applications, equipment and solvent manufacturers and suppliers should be consulted.

PROCESS ADVANTAGES

With the proper degreasing equipment and solvent, vapor degreasing is a proven, economical, and reliable method for cleaning many manufactured components. It produces a dry surface free of organic contamination and is often the preferred method for most finishing operations, without further treatment.

The vapor degreasing process can be used to clean all common industrial metals, including all types of cast iron, steels, copper, brass, bronze, zinc, aluminum, magnesium, nickel, lead, tin, titanium, and their alloys. As the process is primarily a physical rather than a chemical method, there is no detrimental chemical attack or etching of highly polished surfaces or delicate metal parts. Vapor degreasing also has the advantage of being able to clean assemblies and workloads containing parts of different metals. With the proper choice of solvents, many other materials, such as glass and plastics, can be cleaned as individual components or in assemblies.

Because vapor degreasers can be designed to suit the sizes of the parts to be cleaned, from miniature electronic components to diesel engine blocks and 100 ft (30.5 m) lengths of tubing, they usually result in minimum floor space requirements. A high degree of cleanliness is obtainable when the correct cleaning cycle is employed. Under normal conditions, the cleaned parts are ready for subsequent polishing, passivating, phosphatizing, or painting. Degreasing is frequently employed prior to final inspection and packaging. Radioactive and water-break tests indicate that a degree of cleanliness between 0.1 and 1.0 monomolecular layers of soil is attainable by solvent vapor degreasing.

Solvent vapor degreasing is often the most economical method of cleaning because of the low initial cost, minimum floor space requirements, and ease of installing the equipment. Degreasers used according to the equipment manufacturer's recommendations can provide low unit cleaning costs. The major cost of vapor degreasing is the solvent. The major costs of aqueous cleaning include utilities, equipment, and waste water treatment.

While utilities are drawn from the main supply lines, the cost of solvent is easily identified. As a result, misconceptions can result in erroneous conclusions that degreasing is more costly than aqueous cleaning. Therefore, the following factors should be considered for a fair overall comparison of vapor degreasing versus aqueous cleaning:

- Initial equipment and maintenance costs.
- Chemical costs.
- Cleaning ability.
- Floor space requirements.
- Electric power, steam, water, and heated air costs.
- Labor costs.
- Disposal costs versus waste treatment costs.
- Regulatory compliance costs.

Some studies have indicated that aqueous cleaning costs can be 12-25% higher than costs for vapor degreasing operations.

LIMITATIONS OF THE PROCESS

Vapor degreasing may not be effective in removing contaminants that are not soluble in solvent or that do not contain a sufficient portion of solvent-soluble material. These contaminants are materials such as metallic salts, oxides, heat treatment and welding scale, graphite or carbonaceous deposits, and certain inorganic soldering, welding, and brazing fluxes.

PRINCIPLES

There are many types of vapor degreasers, but all operate on the same principle of confining the vapors by the use of condensing coils. The vapors condense on the coils, and the liquid solvent is collected in a condensation trough. From the trough, the liquid solvent passes through a water separator to a reservoir or the clean dip tank of the vapor degreaser. Finally, the solvent flows back into a boiling sump through the overflow pipe of the reservoir or over the weir wall of the clean dip tank.

Cleaning is accomplished by suspending the parts to be cleaned in the vapor zone of the degreaser. Parts should be lowered vertically at a rate of less than 11 fpm (3.4 m/min), with several periods of pausing to minimize the piston-like effect of the descending mass of parts and thereby prevent solvent pushout from the degreaser. Solvent pushout can be minimized by limiting workpiece sizes to less than 50% of the surface area of the degreaser. As the solvent vapors condense on the cool part, the liquid flows across the surface imparting a mechanical or erosive effect that helps to remove insolubles, as well as soluble materials.

The removal of heavy contamination from surfaces requires more mechanical action, which can be accomplished by immersing the parts in the boiling sump or by spraying them in the vapor zone below the condensation trough. When necessary, other techniques can be used to obtain additional mechanical action.

VAPOR DEGREASING

Parts should be given a final rinse in the vapor zone. They should then remain there until liquid solvent no longer drips from the load. Dripping ceases when the temperature of the mass reaches the temperature of the vapors.

METHODS OF CLEANING

Basically, there are three variations of vapor degreasing: (1) vapor only, (2) vapor-spray-vapor, and (3) liquid-immersion vapor.

Vapor Only

The simplest vapor degreasing machine uses the straight vapor cycle. The work to be cleaned is lowered into the vapor zone (see Fig. 18-7). When the solvent ceases to drip from the parts, the workload is drawn into the freeboard area. Freeboard is the distance from the established vapor zone to the degreaser lip, where substantially all of the escaping vapors are confined before the cleaned parts are moved to the unloading area. Optimum equipment design and operation can minimize solvent loss; but even when vapor loss is controlled and no work is passing through the degreaser, some vapor will be lost when the degreaser is open.

Vapor-Spray-Vapor

When the workload contains a large surface area, recessed cavities, or blind holes, or when the solid portion of the soil is heavy and not readily removed by the vapors, a more thorough cleaning can be obtained by augmenting the vapor cycle with a forceful spray of clean liquid solvent, referred to as the vapor-spray-vapor cycle. Parts can be sprayed by a hand spray lance in an open-top degreaser (see Fig. 18-8, view *a*), or by fixed banks of multispray nozzles in conveyorized equipment such as a monorail (view *b*). Spraying pressures are normally in the 2 to 5 psi (14 to 35 kPa) range, but may be as high as 20 psi (138 kPa). The least amount of pressure produces the smallest vapor loss. Sealed liquid-spray units (see Fig. 18-9) allow for spraying pressures in the range of 50 to 75 psi (345 to 517 kPa) and produce a degree of cleanliness better than that obtained by hand wiping.

Liquid-Immersion Vapor

To clean either work of large dimensions—such as tubing, piping, or large heat treatment trays—or small items nested in work baskets, the parts can be immersed in warm solvent, followed by a vapor rinse and dry (see Fig. 18-10). This commonly used cycle produces satisfactory results, especially when the major portion of the soil to be removed is solvent soluble. Ultrasonic devices can also be incorporated in the degreaser, particularly in the clean dip tank, to aid in soil removal.

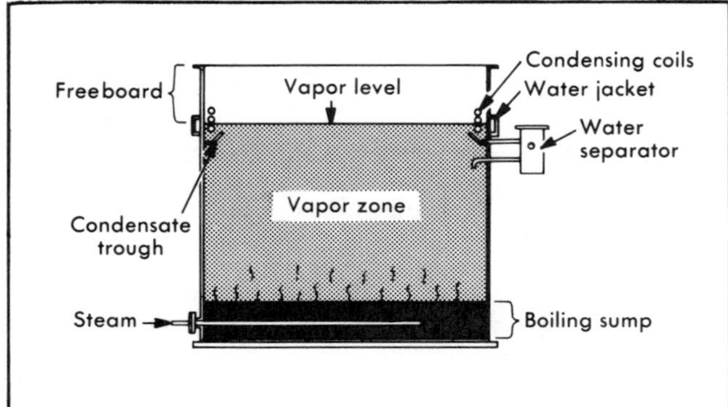

Fig. 18-7 Vapor-only degreaser.

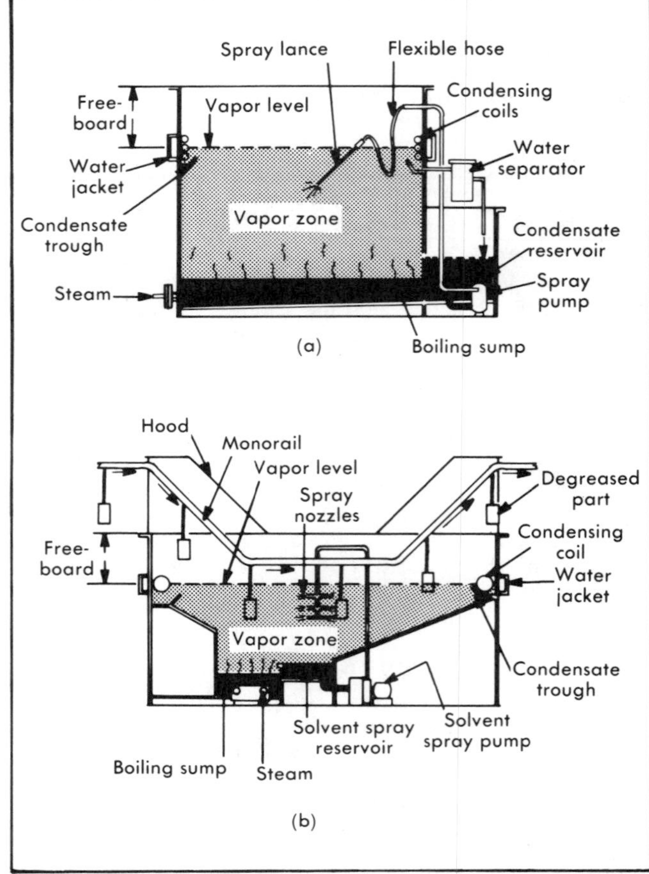

Fig. 18-8 Vapor-spray-vapor degreasers: (a) vertical loading/unloading and (b) equipped with monorail work carrier.

VAPOR DEGREASING SOLVENTS

Physical properties of typical solvents used for vapor degreasing are presented in Table 18-4. Applications of various solvents and factors affecting their selection are identified in Table 18-5.

Trichloroethylene

Trichloroethylene had been the most commonly used solvent for vapor degreasing. Its boiling point of 188° F (87° C) makes vapor hot enough to dissolve most soils. However, this liquid is now regulated as a volatile organic compound (VOC) under the EPA's Clean Air Act and is therefore a nonexempt solvent.

Perchloroethylene

Perchloroethylene is used when a hot vapor, 250° F (121° C), is required, as is the case in removing waxes and tar having high melting temperatures. It is the most costly solvent to heat, requiring steam at 50 to 60 psi (345 to 444 kPa). Any moisture trapped in the workpieces is vaporized, resulting in completely dry parts. However, use of this solvent is also limited by air pollution controls.

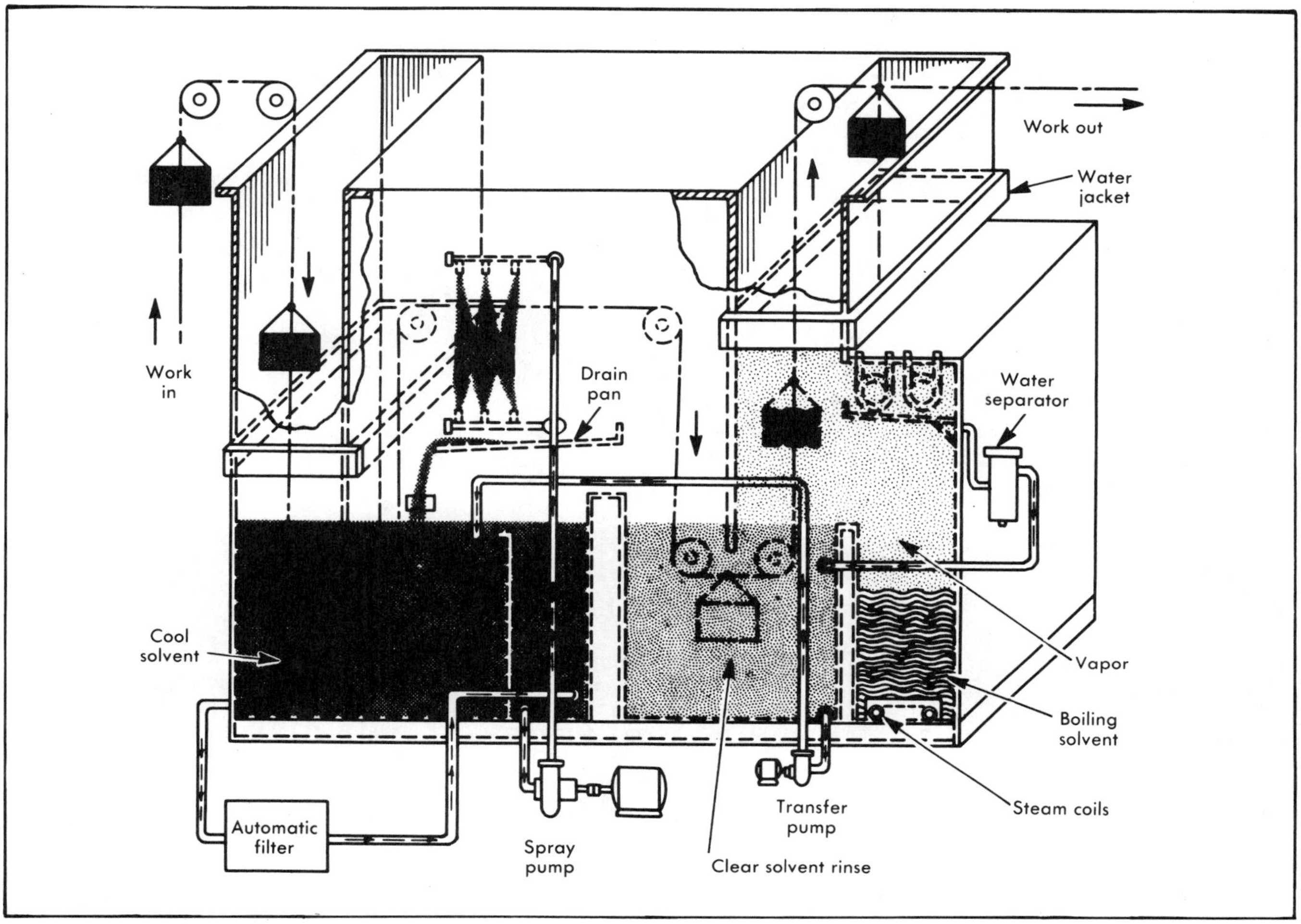

Fig. 18-9 Vapor degreaser with sealed liquid-spray zone.

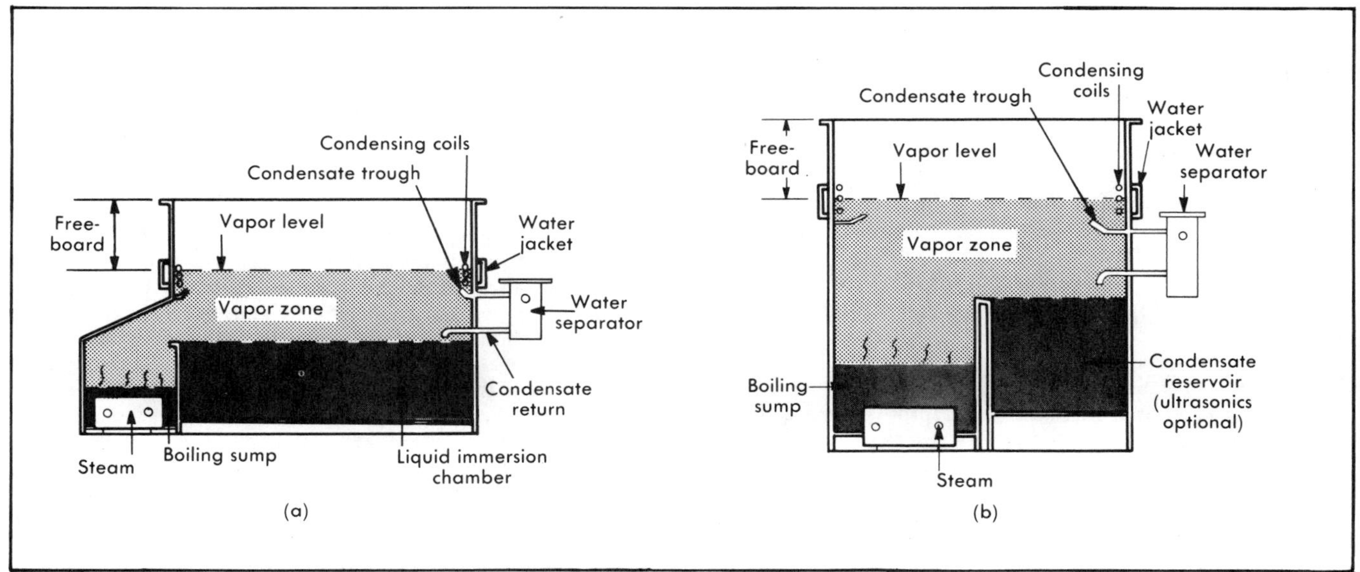

Fig. 18-10 (a) Warm-solvent immersion degreaser; (b) boiling-solvent vapor degreaser.

TABLE 18-4
Physical Properties of Solvents for Vapor Degreasing

	Methylene Chloride	1,1,1-Trichloroethane	Trichloroethylene	Perchloroethylene
Chemical structure	CH_2Cl_2	CH_3CCl_3	$CHClCCl_2$	CCl_2CCl_2
Molecular weight	84.94	133.42	131.40	165.85
Boiling point, °F (°C)	104 (40)	165 (74)	188 (87)	250 (121)
Specific gravity at 25°C (77°F)	1.316	1.322	1.456	1.613
Pounds per gallon (kg/L) at 25°C (77°F)	10.98 (1.32)	10.92 (1.31)	12.11 (1.45)	13.47 (1.62)
Water content, max ppm	100	100	63	75
Color, platinum-cobalt concentration, %, max	15	10	10	15
Nonvolatile residue, max ppm	10	10	10	25
Acid acceptance (as NaOH), min % by weight	0.23	0.20	0.17	0.10
Free halogens, ppm	None	None	None	None
Acidity (as HCl), max ppm	10	10	None	None
Flammability limits, volume:				
lower	14.8% (25°C,77°F)	7.5% (25°C,77°F)	8.0% (25°C,77°F)	None
upper	25% (50°C,122°F)	15.0% (25°C,77°F)	10.5% (25°C,77°F)	
Vapor density	2.93	4.55	4.54	5.83
OSHA standards, threshold limit values	500	350	100	100
Odor threshold, ppm	100-300	75-100	75-100	20-50

1,1,1-Trichloroethane

This solvent has become a replacement for trichloroethylene and perchloroethylene because it is exempt from current (1982) federal Clean Air Act regulations. It can be used to clean such metals as aluminum, magnesium, and zinc, as well as ferrous metals. This solvent is also appropriate when ultrasonic devices are used for vapor degreasing because it has low surface tension, vapor pressure, and specific gravity. For most applications, a frequency of 20 to 25 kHz is used.

Methylene Chloride

The comparatively cool vapors, 104°F (40°C), of methylene chloride solvent make it ideal for cleaning temperature-sensitive parts and lightly soiled components.

SOLVENT CONSERVATION

Vapor degreasers that are well designed, properly maintained, and sensibly operated will function with minimum solvent loss. Additional equipment and methods are available to further conserve solvents, as described in the following sections.

Chillers

In degreasers, chillers are refrigerated coils in the freeboard area above the vapor zone that can help reduce solvent loss. These devices are powered by a one to five-ton compressor, depending upon the size of the degreaser. Since a refrigerated chiller acts as a dehumidifier in the freeboard zone, it should be equipped with a separate trough to collect the water-solvent mixture and direct it to a water separator; otherwise, an accumulation of moisture condensation in the degreaser can lead to degradation of the solvent, resulting in corrosion and the generation of hydrogen chloride.

Carbon Adsorption

Carbon adsorption is a means for removing solvent vapors from a gas. This method has been in use for some time and, if properly adapted to the degreaser and maintained, can be effective in reducing solvent loss. Normally, beds of carbon pellets are located in two cylindrical tanks. Air containing solvent vapors from degreasing operations is directed through one bed until the carbon approaches saturation. When one bed of carbon is loaded, the degreaser exhaust is directed to the second bed, and the first bed is desorbed. Desorption is accomplished by steam or flue-gas stripping the carbon bed. The depletion of stabilizers in the solvent by the carbon adsorption process is well recognized and should be considered and monitored. Carbon adsorption recovery of 1,1,1-trichloroethane is not recommended.

TABLE 18-5
Typical Applications of Solvents for Vapor Degreasing

Application	Solvent	Factors Affecting Selection
Removal of soils from parts	Trichloroethylene or 1,1,1-trichloroethane	Most commonly used.
Removal of road grime from transmissions, etc.	Trichloroethylene	Higher operating temperature required.
Removal of high-melting waxes, oils, greases, and buffing compounds	Perchloroethylene	High operating temperatures are required to remove waxes having melting points greater than 175° F (80° C).
Removal of water films from metals	Perchloroethylene	Rapid and complete drying in one operation.
Clean thin sheets of metals, such as aluminum	Perchloroethylene	Requires high condensation to clean surfaces.
Cleaning coils and components for electric motors	1,1,1-trichloroethane or trichlorotrifluoroethane	Solvent must not damage wire coating or sealing agents and must have good dielectric properties.
Cleaning temperature-sensitive materials	Methylene chloride, 1,1,1-trichloroethane or trichlorotrifluoroethane	Used where part must not be exposed to high temperatures during cleaning.
Cleaning components for rockets or missiles	1,1,1-trichloroethane or trichloroethylene	Cleaned parts must be free of soils or residues that might react with oxidizers.
Lightly oiled parts	1,1,1-trichloroethane	Used when the oil can be removed because of the high solvency of the solvent.
Cleaning with ultrasonics	1,1,1-trichloroethane or fluorinated hydrocarbons	For cleaning efficiency beyond that obtained from standard vapor degreasing. Solvent must be kept clean by continuous

(continued)

TABLE 18-5—*Continued*

Application	Solvent	Factors Affecting Selection
Cleaning with ultrasonics, cont.		distillation and filtration during use. Selection should be based on preliminary trials.

Resolvers

The use of resolvers is a relatively new technology that employs refrigerated cooling coils stored in a bed of selected porcelain chips. The low temperature of the chips is retained by the freezing of residual moisture on their surfaces. The ice formed serves as a thermal bridge between the chips so that chips not in contact with the coil tubes are also cooled. This technique has little or no effect on removing valuable inhibitors from chlorinated solvents and is ideal for recovery of 1,1,1-trichloroethane.

Tank Covers

By Occupational Safety and Health Administration (OSHA) regulation, an open-top degreaser should be equipped with either a manual or powered roll or rigid cover. Such covers are of greatest value in minimizing solvent loss during idling periods or times of degreaser shutdown. A cover on an open-top degreaser can reduce solvent loss from 5 to 12%.

Freeboard Extensions

A freeboard extension is an effective means of lowering solvent losses. A study in which the freeboard height-to-width relationship increased from 0.50 to 0.75 showed a 46% reduction in the emissions rate. However, extending the freeboard height of a vapor degreaser will result in some additional capital costs.

In-House Reclamation

Reclaiming and extending the life of the solvent can best be accomplished by making use of the following procedure for boildown of a degreaser still. When the temperature of 1,1,1-trichloroethane in the boiling sump of the still reaches 174° F (79° C), the sludge should be concentrated in the still. The temperatures for other solvents are: methylene chloride, 107° F (42° C); trichloroethylene, 194° F (90° C); and perchloroethylene, 257° F (125° C). The transfer pump should be turned off, and the gate valve in the dirty solvent line closed. The solvent should continue to be distilled until the liquid level drops to the recommended minimum level of 2″ (51 mm) above the heating elements.

The heat in the boiling sump of the still should be turned off, and the solvent sludge allowed to cool to 90 to 100° F (32 to 38° C) before draining. The solvent sludge is then drained into 55 gal (210 L) drums. Parts, metallic fines, and chips or other insolubles may be removed from the sludge. The still should be cleaned thoroughly, giving special attention to the area under the heating elements. Next, the drain valve is closed, and the gate valve in the solvent line from the degreaser boiling sump to the boiling sump of the still is opened. The transfer pump is then turned on.

Care should be taken to prevent the heating elements from being exposed when activated. When the liquid level reaches 2″ (51 mm) above the heating elements, the heat is turned on. The necessary virgin solvent is added to the clean dip portion of the degreaser to properly maintain the liquid levels in all chambers of the degreaser and still.

Sludge accumulated in 55-gallon drums can be reclaimed in-house or by commercial reclaimers. When the oil concentration reaches 60 to 70%, the sludge material should be discarded in compliance with local, state, and federal regulations governing the disposal of hazardous waste products.

Stop-and-Go Technique

The following procedures, collectively called the stop-and-go technique, have been developed to reduce solvent vapor concentration in the operator's working area and to reduce the solvent loss of a degreasing operation from 7 to 12%. The workload should be lowered into the vapor zone at a slow speed to prevent an excessive wave formation of the vapors that would push an unnecessary amount of the vapors out of the degreaser. As the workload enters the vapor zone, the vapors will collapse. Whenever the vapors have dropped 2 to 4″ (51 to 102 mm), lowering of the load should be stopped until the vapors stabilize or start to recover, then the load can be lowered further until the vapors have dropped another 2 to 4″.

This stop-and-go method of entry prevents solvent vapors from being pushed out of the degreaser by the plunger effect of the workload. Once the workload is covered by the vapors, it should not be lowered further. Stop-and-go entry allows maximum vapor recovery with shorter cleaning cycles.

Removal of the workload should also be in increments of 2 to 4″, with pauses to allow the vapors to be entrapped in the freeboard area. This procedure for withdrawal decreases vapor drag-out. Once the workload has cleared the vapor zone, it should remain in the freeboard area until all parts are dry and no solvent drips from the workload.

WATER USE AND SLUDGE DISPOSAL

The vapor degreasing systems described in this section provide a partial solution to water pollution and waste treatment problems. There is no effluent discharge from the process; water is used only for condensation purposes. In most instances, the water can be reused for other purposes. Where the availability of water is limited, cooling towers and water chillers may be used to minimize consumption.

The final sludges produced by vapor degreasing must be disposed of according to applicable regulations. However, sludge volume is very small, and commercial disposal companies operate in all sections of the country. No storage ponds or treatment facilities are required at the plant site.

SAFETY PRECAUTIONS

The use of vapor degreasing solvents requires precautions with operating procedures and sometimes with appropriate clothing or equipment. The minimum recommendation is for operators to wear safety glasses and gloves if they contact the liquid. If solvent gets into the eyes, water should be used to flush them. If solvent soaks clothing, the clothing should be removed, and the skin washed. Prolonged and/or repeated skin contact may cause dermatitis. If solvent is accidentally swallowed, vomiting should not be induced; obtain immediate medical assistance.

The greatest potential hazard in the use of vapor degreasing is the inhalation of solvent vapors, but this hazard is minimized by providing adequate ventilation. When it is necessary to clean a tank, it should be thoroughly drained, exhausted, and ventilated before entering it; a positive-pressure breathing apparatus should be used. Instructions of the degreaser manufacturer should be followed carefully.

ULTRASONIC CLEANING

Ultrasonic cleaning uses high-frequency vibrations in a cleaning system. In the process, a cleaning solution is subjected to the rapid oscillation of longitudinal waves, identical to audible sound waves but of higher frequency.

Before ultrasonics can be used effectively, it is important to realize the following:

- Parts to be cleaned must be immersed in a liquid.
- Generally, with a few exceptions, the entire volume of liquid must be supplied with ultrasonic energy.
- The use of ultrasonics does not eliminate the need for cleaning chemicals.

Workpieces to be cleaned are placed in the solution, and the rapid oscillation of the solution resulting from the high-frequency sound waves creates minute vapor voids in the solution that implode against the workpiece and effectively clean its surfaces. Mechanically held contamination is released from the surfaces, soluble materials are rapidly dissolved, and oil and similar contaminants are easily emulsified.

CAVITATION PRINCIPLES

The effect created by the action of ultrasound waves in a liquid cleaning agent is a phenomenon known as cavitation. It consists of the formation and collapse of countless tiny cavities, or vapor voids, in the liquid. These voids occur throughout the liquid, even in recesses, cavities, and holes in the workpieces that the liquid penetrates.

Cavitation is produced by the alternating patterns of compression and rarefaction—high and low pressure points—generated during sound wave half-cycles. During rarefaction, the liquid is stretched beyond its tensile strength, and the pressure is lowered to allow instantaneous vaporization. The cavities formed grow from microscopic nuclei. During the subsequent compression phase, they implode violently. This phenomenon occurs at a rate proportional to the applied ultrasonic frequency, which can range from the ultrasonic threshold up to a practical limit of about 90 kHz.

Although cavitation vacuities are extremely small and release only minute amounts of energy individually, the cumulative effect of millions of implosions per second is intense. Theoretical localized temperatures of 20,000° F (11 093° C) and pressures of 10,000 psi (69 MPa) are created, providing a strong scouring action that can dislodge tenacious soils. Intimate mixing of the cleaning media with the contaminants speeds the rate at which soil is dissolved or dispersed. Cavitation prevents the formation of a neutral film on workpiece surfaces that could impede cleaning and raises the temperature of the liquid, thereby increasing the rate of chemical activity.

Effects of Cleaning Media

The point at which cavitation starts is called the cavitation threshold. It is reached when the applied energy is sufficient to drop the pressure within the liquid below its vapor pressure during rarefaction. The initial quantity of energy required to attain this point varies with different cleaning solutions. Frequency also plays a significant role in the onset of cavitation.

Physical properties of a cleaning agent affect not only its ability to achieve a cavitation threshold, but the intensity of the cavitation process as well. Thus, different cleaning media will cavitate at different levels according to their density, viscosity, surface tension, vapor pressure, and temperature.

Vapor pressure. A cleaner having a vapor pressure of medium value is most suitable for ultrasonic cleaning purposes. Low vapor pressure produces cavitation bubbles that implode with relatively greater force, but also results in the formation of fewer bubbles, as well as a higher cavitation threshold. High vapor pressure will lower the threshold and create more bubbles, but these will collapse with less intensity owing to a smaller differential between internal and external pressure. At either pressure extreme, cavitation ceases.

Surface tension. A cleaner with moderate surface tension gives best results. With high tension, bubbles have less elasticity and collapse with more intensity; too high a level impedes their formation. Low surface tension permits bubbles to grow larger, but also leads to less implosive force, lowering cavitation intensity.

Viscosity. The viscosity of a cleaner should be low to promote cavitation. The higher the viscosity, the more energy needed for transmission of ultrasonic waves.

Density. A cleaner having high density is desirable to create intense cavitation. Although high density liquids require somewhat more energy to expand the cavitation bubbles, they also produce greater implosive force.

Temperature. The temperature of a liquid affects both cavitation quality and cleaning action; both can usually be improved by increasing the operating temperature. There is, however, an ideal temperature at which cavitation intensity is greatest. This optimum temperature varies in liquids according to the physical properties previously discussed. Aqueous solutions usually cavitate best within the range of 140 to 190° F (60 to 88° C). Beyond this ideal temperature, cavitation steadily diminishes. Recent research indicates that there may be some advantages to cleaning near the boiling point of the liquid, but the physics are not as yet well understood.

Degassing Necessary

Before ultrasonics can be applied effectively to either an aqueous or solvent cleaning medium, dissolved gases (usually air) must be removed; otherwise, the vacuum bubbles formed will fill with gas, cushioning the force of cavitation. Degassing is accomplished by ultrasonically agitating a liquid while raising its temperature to within 10 to 20° F (6 to 11° C) of its boiling point. The time required for initial degassing of a liquid varies with its nature. To degas a water-based solution, only 15 minutes may be required; solvent degassing, however, may take up to an hour because organic solvents have a greater affinity for air than does water. Degassing must be repeated at the start of each subsequent day's cleaning operation, but the time required will be only a fraction of the original period. The degassing time can be reduced, if necessary, by decreasing the liquid depth or using greater power intensity with pulsed waves.

SYSTEM COMPONENTS

Industrial ultrasonic cleaning systems are composed of three basic components: (1) a generator, (2) a transducer, and (3) a tank containing the cleaning solution. The generator transforms standard line current of 50 or 60 Hz into a desired higher frequency. This high-frequency current, in turn, is converted into sound waves (mechanical energy) of a corresponding frequency via a transducer, which radiates the waves into a cleaning solution in the tank.

Generator (Power Supply)

Generators differ in the wave forms they emit and in their power output ratings. The best choice for a given system depends primarily on the type, number, and frequency of the transducer elements involved.

Three wave forms are used: half wave, full wave, and continuous wave. Of these, half-wave and full-wave forms are most common. For a full-wave unit, peak power is twice the average power output; for a half-wave generator, four times the average. In a continuous-wave system, peak and average power are the same.

Progressively increasing the power for ultrasonic cleaning tends first to accelerate cavitation intensity, then to quell the process if carried too far. The input of too much power creates a superagitated state in which the liquid near the radiating surface becomes elastic. This phenomenon, called surface cavitation, actually blocks sound wave transmission by preventing coupling.

Manufacturers of ultrasonic cleaning systems use solid-state full-wave generators for generating power at frequencies of 20 to 40 kHz. They are manufactured with various output ratings and are specifically designed to power either magnetostrictive or electrostrictive (piezoelectric) transducers. The high-frequency output is typically obtained by the use of a gate-driven silicon-controlled rectifier (SCR) or an array of transistors. The SCR generators use automatic frequency-tracking circuitry to optimize the frequency for varying acoustic loads. The transistor-controlled generator uses an electronic feedback loop for automatic fine tuning. Other standard components found in ultrasonic generators are fans, fuses, and radio-frequency interference filters as required under FCC regulations.

Transducers

Transducers most commonly used for ultrasonic cleaning are either magnetostrictive or electrostrictive (piezoelectric) elements. The magnetostrictive type contains ferromagnetic nickel laminations with an electrical winding. During operation, a varying magnetic field causes the laminations to alternately expand and contract.

Piezoelectric transducers have ceramic crystals that similarly expand and contract (vibrate) in a varying electrical field. Quartz crystals were tried originally, and then barium titanate, which could be used only at relatively low temperatures. Lead-zirconate-titanate ceramic is now used; it can withstand temperatures in excess of 200° F (93° C).

Magnetostrictive transducers are most often polarized by passing a d-c current through the electrical winding. The operational high-frequency current is simultaneously superimposed.

Like tuning forks, transducers are fashioned to generate sound waves of specific frequencies. Different frequencies produce distinct variations in the character and intensity of cavitation. Within the effective cleaning range of ultrasonics, the higher the frequency, the greater the volume of cavitation voids created and the smaller their size. The lower the frequency, the more intense the implosion. For industrial cleaning, frequencies are usually 20 to 40 kHz. Differences in properties and cleaning applications for 25 and 40 kHz frequencies are presented in Table 18-6.

Cleaning Units

There are two major types of ultrasonic cleaning units: integrated and modular. Integrated units have all components, including the tank for the cleaning solution, in a single enclosure. Modular systems consist of a separate generator linked to either tanks equipped with transducers or immersible transducers.

Integrated ultrasonic cleaners. These units are portable and self-contained, requiring only a power connection for use. Most are compact models designed to accommodate small workpieces. They are suitable for industrial laboratory cleaning needs and such specialized applications as cleaning optical goods, watch and clock movements, pieces of test equipment, electronic componentry, and plastics parts and assemblies.

One manufacturer offers a series of integrated ultrasonic cleaners ranging from 1 pt (0.47 L) to 8 gal (30 L) in capacity. Most are available with heaters to maintain the optimum cleaning temperature. Timers, covers, parts baskets, and other accessories are offered with these units for operating convenience. All use relatively high frequencies of 50 or 60 kHz, with cavitation that is more penetrating and less aggressive than that generated by lower frequencies. A larger integrated cleaner, having a capacity of 10 gal (38 L), is designed for industrial applications.

Modular systems. Tanks with transducers for modular systems are generally stainless steel cleaning vessels with transducer elements bonded to their exterior, usually on the bottom (see Fig. 18-11). Ultrasonic energy is directed upward through the tank. The transducer elements are driven through a connecting cable by a remotely located ultrasonic generator. Besides their modularity, these tanks are differentiated from integrated ultrasonic cleaners by their greater size and variety of available types. They also have heavier walls, which make them more adaptable to industrial needs. Tanks with transducers typically have lip flanges to allow out-of-the-way recessing into table or counter tops.

The number of transducer elements per tank is determined by the tank volume and its designated power rating. Standard intensity models are suitable for about 90% of all cleaning applications. Low-intensity tanks are sometimes needed for especially delicate workpieces. High-intensity units may be needed for liquids having high surface tensions (higher than tap water). Tanks generally vary in capacity from 2.6 to 34 gal (10 to 129 L) or may be made to order in larger sizes.

It is important that the selected tank and generator have matched ratings for power and frequency, otherwise serious damage will result, usually to the generator. Tank heating is a common option for tanks with transducers and is specified by a majority of users to keep the cleaning solution at its best temperature for cavitation and cleaning effectiveness. Typical accessories include stainless steel parts baskets and tank covers. Some tanks with transducers do not have water jackets or other means of maintaining a vapor blanket as they are intended for use with aqueous media only. Tanks intended for use with solvents have water jackets, peripheral cooling coils, and a separate vapor-generating sump.

Immersible transducers are hermetically sealed stainless steel containers holding a number of transducer elements. Fundamentally, they are unit sources of ultrasonic energy that can be placed inside cleaning tanks or presoak tanks to create an ultrasonic cleaning system. Like tanks with transducers, they are driven by separate ultrasonic generators; however, a single generator may power an entire bank of connected immersible transducers. Immersible transducers can be used in either aqueous or solvent systems.

Versatility is a major benefit of immersibles. During installation, they can be oriented to direct and concentrate ultrasonic energy where it is most needed, according to workpiece size and the location of its soiled surfaces. They can be used with almost any tank, regardless of size. Immersible transducers are available in various configurations for mounting on tank bottoms or sidewalls. Some can be installed without drilling holes in the tanks.

TABLE 18-6
Characteristics and Cleaning Applications of 25 and 40 kHz Ultrasonic Frequencies

Frequency, kHz	Characteristics	Recommended Applications
25	Relatively small numbers of large cavitation bubbles that implode with great force, providing powerful scrubbing action. Higher noise level than 40 kHz.	Large or massive workpieces with broad, accessible surface areas. Parts with stubborn soil residues.
40	Relatively large numbers of small cavitation bubbles that implode with less intensity. Relatively low operating noise.	Components with intricate surfaces, narrow crevices, and blind holes. Delicate materials and parts such as glass, printed circuit boards, and electronic components.

(*Branson Cleaning Equipment Co.*)

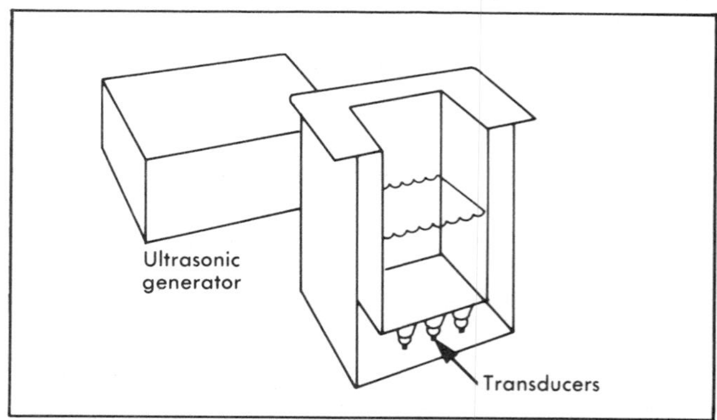

Fig. 18-11 Tank equipped with transducers and power supply generator for ultrasonic cleaning. (*Branson Cleaning Equipment Co.*)

Cylindrical ultrasonic modules are available for in-line cleaning of wire, rod, and strip, and for various liquid processing applications. These transducers are constructed of stainless steel and are hermetically sealed, allowing immersion in either aqueous solutions or degreasing solvents. Liquid-tight armored cables connect these transducers to ultrasonic generators.

Workhandling Methods

Loading fixtures for ultrasonic cleaning and vapor degreasing can range from simple hooks and wire hangers to racks and baskets for batch handling of small components. Whatever the type, their common requirements are: sufficient strength to support the workpiece during cleaning, adequate length to lower the workpiece into the cleaning zone without bringing workers into contact with the cleaning media or its hot vapor, and compatibility with the cleaning media. When ultrasonics are involved, other requirements are important to minimize loading fixtures interfering with the process of cavitation.

Parts baskets. These should have bottoms of open-mesh wire screen or solid, thin-gage sheet metal, preferably stainless steel in both cases. Small-mesh wire screen severely impedes cavitation, causing energy losses as high as 60%. A large, open mesh, at least 3/8″ (9.5 mm), is much less disruptive to cavitation, involving a loss of only about 5% of applied energy, but is unsuitable for workpieces smaller than the mesh size.

To hold very small items, solid-bottom baskets are recommended; the solid metal surface acts as a secondary radiating surface, with only about a 10% energy loss. Basket sides should be perforated or made of wire screen for drainage. It is also recommended that parts baskets have 2 to 3″ (51 to 76 mm) long legs at all four corners to hold workpieces off the tank floor.

Parts loading. In any ultrasonic cleaning system, aqueous or solvent, the *antinodal region* of cavitation should be recognized. It is the peak power area of ultrasonic vibrations, occurring 1 1/2 to 2 1/2″ (38 to 64 mm) away from the radiating face. Since the tank bottom is usually the radiating surface, placing parts directly on it can interfere with power transmission, thus lessening the cleaning intensity.

To receive the greatest cavitational force, soiled parts should be positioned 1 to 3″ (25 to 76 mm) above the tank floor. Even in the case of small ultrasonic aqueous cleaners, it is more effective to suspend items in the cleaning solution, rather than allowing them to rest on the bottom of the tank.

For best results in loading and removing work, the following additional points are recommended:

- Parts must be totally immersed in the cleaning media.
- Contaminated surfaces should be positioned to receive maximum ultrasonic energy. Stacking is not advisable, as it may obscure soiled surfaces and blind them to the cleaning process.
- Workers should not contact the cleaning media.

High-volume workhandling. Applications involving high-volume processing of components in repetitive, multistage cleaning cycles can be greatly aided by some means of automated cycle regulation and parts transfer. An assortment of equipment is available, including specialized, complex cleaning systems and separate, accessory transfer systems. Among the former are monorail, mesh-belt, and crossrod conveyors; programmed vertical lifts; and in-line lift and indexing systems.

Monorail conveyor systems. These are used to move large parts and assemblies through vapor degreasing cycles on a continuous basis.

Mesh-belt conveyor systems. These provide belt movement of parts through immersion, spray, and vapor cleaning cycles. They are especially useful for defluxing printed circuit boards and are also used in aqueous systems.

Crossrod conveyor systems. These are designed for baskets and racks of small parts, which are returned to their point of entry after various immersion and vapor cleaning steps.

Programmed vertical-lift systems. These carry racked parts through timed cleaning sequences within a vapor degreaser.

In-line lift and indexing systems. These consist of modular cleaning tanks aligned for a specific cleaning flow. Parts are automatically transferred through timed stages, which can be varied for different needs.

Cleaning Media

Aqueous ultrasonic systems use a variety of water-soluble detergents and cleaning agents that can be diluted with water to the concentration desired. The most common aqueous cleaning media are alkaline detergents, discussed previously in this chapter. Acidic solutions are seldom used in ultrasonic cleaning systems, but there are certain exceptions. One proprietary, mildly acidic cleaner, formulated especially for ultrasonic cleaning, attacks soil combinations such as oxides and light oil residues. Table 18-7 shows some detergents used for ultrasonic cleaning of various materials, parts, and soils.

REGULATIONS GOVERNING ULTRASONICS

Although ultrasound is, by definition, inaudible to the human ear, ultrasonic equipment does have an accompanying range of audible sounds, primarily caused by subharmonics. These sounds are created by frequencies that are integral multiples of the main frequency and that fall within audible range. Regulations established under the Occupational Safety

TABLE 18-7
Ultrasonic Cleaning Solutions Used for Various Materials, Parts, and Soils

Material of Construction	Types of Parts	Soils	Choice of Detergent
Iron, steel, stainless steel	Castings, stampings, machined parts, drawn wire, diesel fuel injectors	Chips, lubricants, light oxides	High caustic with chelating agents
	Oil-quenched, used automotive parts; fine-mesh and sintered filters	Carbonized oil and grease, carbon smut, heavy grime deposits	High caustic, silicated

(continued)

TABLE 18-7—*Continued*

Material of Construction	Types of Parts	Soils	Choice of Detergent
Iron, steel, stainless steel, cont.	Bearing rings, pump parts, knife blades, drill taps, valves, ceramics	Chips; grinding, lapping, and boning compounds; oils; waxes and abrasives	Moderately alkaline
	Roller bearings, electronic components that are affected by water or pose drying problems, knife blades, sintered filters	Buffing and polishing compounds; miscellaneous machining, shop, and other soils	Chlorinated-solvent degreaser (inhibited trichloroethylene, for example)
Aluminum and zinc	Castings, open-mesh air filters, used automotive carburetor parts, valves, switch components, drawn wire	Chips, lubricants, and general grime	Moderately alkaline, specially inhibited to prevent etching of metal, or neutral synthetic (usually in liquid form)
Copper and brass (also silver, gold, tin, lead, and solder)	Printed circuit boards, waveguides, switch components, instrument connector pins, jewelry—before and after plating, ring bearings, etc.	Chips, shop dirt, lubricants, light oxides, fingerprints, flux residues, buffing and lapping compounds	Moderately alkaline, silicated, or neutral synthetic (possibly with ammonium hydroxide for copper oxide removal)
Magnesium	Castings, machined parts	Chips, lubricants, shop dirt	High caustic with chelating agents
Glass and ceramics	Television tubes, electronic tubes, laboratory apparatus, coated and uncoated photographic and optical lenses	Chips, fingerprints, lint, shop dirt	Moderately alkaline or neutral synthetic
Plastics	Lenses, tubing, plates, switch components	Chips, fingerprints, lint, lubricants, shop dirt	Moderately alkaline or neutral synthetic
Various metals	Heat-treated tools, used automotive parts, copper-clad printed circuit boards, used fine-mesh filters	Oxide coatings	Moderately to strongly inhibited proprietary acid mixtures specific for the oxide and base metal of the part to be cleaned (except magnesium)
Various metals, plastics (nylon, Teflon, epoxy, etc.), and organic coatings when water solutions cannot be tolerated	Precision gears, bearings, switches, painted housings, printed circuit boards, miniature servomotors, computer components, clean room operations as in aerospace work	Lint, other particulate matter, and light oils	Trichlorotrifluoroethane (fluorocarbon solvent), sonic-vapor degreaser

and Health Act (OSHA) limiting worker exposure to noise are therefore applicable. The maximum steady-state exposure level currently established by OSHA is 90 dBA per eight-hour day.

Decibel levels produced by ultrasonic equipment will vary with the cleaning chemical used, equipment design, and transducer characteristics—configuration, type, and frequency. All ultrasonic equipment must also comply with radio-frequency interference standards listed under Part 18 of the FCC rules and regulations. Accordingly, ultrasonic generators and cleaners should be equipped with filters for radio-frequency interference.

References

1. William G. Wood, coordinator, *Metals Handbook*, vol. 5, *Surface Cleaning, Finishing, and Coating*, 9th ed. (Metals Park, OH: American Society for Metals, 1982), p. 5.

Bibliography

Otrhalek, Joseph V., and Sokalski, Stanley M. *Surface Preparation Via Chemical Applications*. SME Technical Paper FC83-634. Dearborn, MI: Society of Manufacturing Engineers, 1983.

INORGANIC COATINGS

SECTION

4

CONVERSION COATINGS AND ANODIZING

The coating processes discussed in this section include conversion coatings, plating, electroless plating, thermal spraying, hard facing, porcelain enameling, hot dipping, vapor deposition, and other special processes. Additional information can be obtained on the various coating processes from the references and bibliography listed at the end of each chapter.

CHAPTER CONTENTS:

CONVERSION COATINGS

Conversion coating, sometimes referred to as chemical reaction priming, is the formation of a coating on a ferrous or nonferrous metal surface as a result of controlled chemical or electrochemical attack. The converted surface is not superimposed on the underlying metal, such as a paint coating, but is rather a strongly adherent chemical entity formed at the interface by interaction between the chemical coating solution and the ions formed from the metallic surface immersed in the solution. The two most common methods of applying conversion coatings are spraying and immersion.

CONVERSION COATING TYPES

Conversion coatings may be classified as either natural or man-made depending on the environment that was used to cause the interaction. Man-made conversion coatings can be further broken into chemical and electrochemical processes.

Natural Conversion Coatings

Coatings, usually oxides, form on many metals in their natural environment. In the presence of moisture, anodic and cathodic areas develop all over a metal surface providing a corrosion mechanism. The process of converting outer metal atoms into soluble or nonadherent chemical compounds continues until all, or almost all, of the metal is consumed or combined with oxygen, as in the rusting of iron. However, if the corrosion products form an invisible, adherent film on the metal, as occurs naturally in the formation of an oxide on aluminum or copper, further attack is stopped or greatly retarded. The chemical reactions and thermodynamic relationships inherent in destructive corrosion are basically the same as those that result in beneficial surface layers.

Chemical Conversion Coatings

Chemical conversion coatings are produced by contacting the metallic surface with a chemical solution. The most common are phosphate, chromate, and oxide coatings. Table 19-1 lists the different chemical conversion coatings and their characteristics.

Phosphate conversion coatings. Phosphate conversion coatings are produced from chemicals containing phosphoric acid and its salts. The coatings formed from heavy metal phosphates are crystalline in nature and consist of layers of water-insoluble phosphorus compounds of iron, zinc, manganese, calcium, or a combination of these. Phosphate coatings formed from alkali metals of Group I elements in the periodic table and phosphoric acid are known as iron phosphates and are amorphous (noncrystalline) in nature. Crystalline phosphate coatings are usually different shades of gray; however, pretreatments or post-treatments may be performed to produce various colors including black, red, or green. Iron phosphates are light blue to red iridescent.

Chromate conversion coatings. Chromate conversion coatings are produced from compounds of chromium in combination with other water-soluble inorganic materials. During the conversion process, the base metal (aluminum, cadmium, or zinc) is converted on its outside surface to a complex chemical entity constituting salts of hexavalent chromium and trivalent chromium. Chromate conversion coatings are amorphorus and gelatinous when applied to the parts, but harden and become hydrophobic after they are dried. These coatings can be produced in shades of bright clear, yellow, bronze, green olive drab, or black. Some chromate conversion coatings can absorb certain organic dyes (similar to anodized aluminum coatings) to produce a spectrum of hues.

Oxide conversion coatings. Two types of oxide coatings are currently being used. The first is a black or bluish oxide coating produced on iron or steel when they are treated with hot caustic soda solutions containing accelerators. The second oxide coating is produced on cadmium, copper, iron, steel, and zinc alloys from acidic compositions

***Contributors of sections of this chapter are:* Mirza Baig**, *Technical Director, Aldoa Co.;* **Norman F. Callahan**, *President, Cidona, Inc.*

***Reviewers of sections of this chapter are:* Thomas C. Atkiss**, *Product Marketing Manager, Van Straaten Chemical Co.;* **Mirza Baig**, *Technical Director, Aldoa Co.;* **Calvin H. Biggar**, *Technical Director, Electrofilm, Inc.;* **P. W. Bolmer**, *Head, Finishing Section, Kaiser Aluminum & Chemical Corp.;* **Norman F. Callahan**, *President, Cidona, Inc.;* **Mike D'Angelo**, *Product Manager, Metal Finishing, MacDermid, Inc.;*

CHAPTER 19

TYPES OF CONVERSION COATINGS

TABLE 19-1
Characteristics of Chemical Conversion Coating Treatments

Process Type	Treated Metal	Principal Constituents of Coating	Coating Weight, mg/ft² (g/m²) Base for Paint	Corrosion	Appearance
Chromate	Steel	Iron and chromium oxides	15-50 (0.16-0.5)		Amorphous
	Zinc	Zinc and chromium oxides			Bluish gold, amorphous
Chromic oxide	Aluminum	Oxides of aluminum, chromic compounds	10-50 (0.1-0.5)		Colorless to gold, amorphous
Chromic oxide/ chromic phosphate	Aluminum	Chromic phosphate, aluminum oxides	4-500 (0.04-5.4)		Colorless to green, amorphous
Complex oxide	Zinc	Oxides of zinc and of heavy metals iron and cobalt			Pale yellow gold, amorphous
Iron phosphate	Steel	Iron phosphate, iron oxide	15-90 (0.16-0.9)		Bluish gold to reddish, amorphous
Manganese phosphate	Steel	Iron and manganese phosphates		800-2500 (8.6-27)	
Molybdate/ phosphate	Aluminum	Aluminum molybdates and phosphates			Golden blue
	Steel	Iron molybdate and phosphate, iron oxides	15-40 (0.16-0.4)		Bluish gold, amorphous
	Zinc	Zinc molybdates and phosphates			Bluish gold, amorphous
Zinc phosphate	Aluminum	Zinc phosphates, aluminum phosphates	150-500 (1.6-5.4)		Gray, crystalline
	Steel	Zinc and iron phosphates	100-600 (1-6.5)	800-2500 (8.6-27)	Gray, crystalline
	Zinc	Zinc phosphate, often with a trace of nickel	100-350 (1-3.8)		Gray, crystalline
Zinc-calcium phosphate	Steel	Calcium-zinc phosphate, zinc phosphate, iron phosphate	300-1000 (3.2-10.8)		Crystalline
	Zinc	Zinc phosphate	300-1000 (3.2-10.8)		
Zirconium/ tannin	Aluminum	Zirconium oxides			Colorless
Titanium/ tannin	Aluminum	Titanium oxides			Colorless to pale yellow

Reviewers, cont.: ***Jack Elliott***, *Vice President, Duralectra, Inc.;* ***David Gerstenkorn***, *General Manager, Nimet Industries, Inc.;* ***P. J. Ging***, *Laboratory Supervisor, Aluminum Company of America;* ***Arthur D. Godding***, *Manager, Tech Services and Equipment Engineering, Heatbath Corp.;* ***Dennis Hall***, *Director of Technical Services, Specialty Chemicals and Services, Inc.;* ***Drew Johnston***, *Vice President, Walgren Co.;* ***Carl LaBarbera***, *Vice President, DV Industries, Inc.;* ***Dr. Moisey Lerner***, *Director, Research and Development, Sanford Process Corp.;* ***Paul V. Mara***, *Vice President—Technical, Aluminum Association, Inc.;* ***James I. Maurer***, *Vice President, Technology & Commercial Development, Parker Chemical Co.;* ***Roger Miller***, *Midwest Regional Manager, Parker Chemical Co.;* ***Jeff Pernick***, *Plant Manager, International Hardcoat, Inc.;* ***John H. Powers***, *Technical Specialist, Product Engineering Div., Alcoa Laboratories, Alcoa Technical Center;*

at moderate temperatures. Unlike other conversion coatings, oxide conversion coatings contribute little corrosion protection to the part; they are generally produced for abrasion resistance, aesthetic, or identification purposes only. The colors range from gray to blue to black.

Electrochemical Conversion Coatings

Electrochemical conversion coatings are produced on the surfaces of aluminum, magnesium, and titanium alloys by the anodizing process. This type of conversion coating can also be applied to zinc and zinc alloys using proprietary chemical formulations. Electrochemical conversion coatings provide good corrosion and abrasion resistance. The color of the coatings ranges from clear to yellow to green to black. In some cases, it is necessary to seal the coating to obtain maximum benefit. Anodizing is covered in a subsequent section of this chapter.

APPLICATIONS

Material losses caused by corrosion are great, and conversion coatings serve a valuable role in limiting these losses. Metal fabrication problems in mass production suggest other uses for chemical treatments of metal surfaces. The application of a conversion coating substantially increases the quality and performance of a finished product by forming a stable and nonmetallic coating that chemically combines with the base metal; salt spray resistance and paint adhesion are improved.

The conversion coating choice is influenced by the ware and its finish, its application or purpose, its corrosion potential, the type of metal, the equipment available, and the cost of the treatment. If the coating is to be used as a base for paint, the paint itself must also be considered; and in the case of phosphate coatings, the choice of treatment will also depend upon the cleaning process used prior to treatment.

Some of the purposes of conversion coating can be illustrated by the following objectives for which coating provides economic value:

1. Corrosion protection.
2. Prepainting treatment.
3. Cold forming; lubrication carrier.
4. Wear reduction/resistance.
5. Electrical-resistance coating.
6. Decorative final finish.
7. Identification.

In analyzing the economics of a treatment, not only must the cost of application per unit area be considered, but also what this cost may bring about in possible savings. Savings may be related to reductions in material costs, changes in complexity of manufacturing procedures, and elimination of costly service failures.

Corrosion Protection

Conversion coatings provide resistance to corrosive environments in their own right; but when coupled with a rust-preventive oil or wax, they may greatly enhance the resistance to corrosion. The coating may act as a blotter to absorb the oil and increase the protection of articles that are not to be painted.

Prepaint Treatments

Paint adhesion and life are usually improved by conversion coatings because they (1) provide mechanical bonds such as capillary pores and cavities for the paint, (2) provide increased surface area on which the molecular forces contributing to adhesion can act, (3) minimize or inhibit the spread of corrosion if the organic finish is damaged, (4) result in a nonalkaline surface that is not harmful to paint or to sensitive metals, (5) ensure a clean surface before painting, and (6) in the case of zinc and other sensitive metal surfaces, prevent reaction between the paint and the metal surface.

The useful life of a paint finish depends upon the durability of the paint and its adhesion to the metal surface. Conversion of ferrous surfaces to a phosphate coating increases the adhesion and resultant durability of organic coatings. Conversion of nonferrous metals to a phosphate or chromate coating inhibits reaction of the paint vehicle with the metal and increases adhesion and durability of organic finishes. Various methods of preparing metal for painting are available, and the environment in which the paint is used should determine the preparation method used. The paint system must offer the maximum durability for the particular environment to which the metal product is to be subjected.

In any finishing system, the preparation of the metal prior to painting is as important as the paint used. Removal of oils and soils before application of organic coatings is essential to proper adhesion and corrosion protection. The paint must be formulated to protect the surface against blistering, cracking, checking, chipping, and loss of adhesion.

The selection of conversion coating depends upon many factors, but the end use of the product and the type of metal to be treated will frequently delimit the reasonable options. An excessively thin paint film coupled to an extra-heavy phosphate deposition will cause dulling of the film and early failure. An iron phosphate solution designed to clean and coat in one operation is not recommended when the highest performance is required of the finished product.

Almost all automotive and appliance manufacturers use conversion coating as a base for paint. By incorporating a conversion coating into their paint setup, they have been able to produce finished articles that meet their specifications at the lowest cost. Proper selection of the conversion coating and paint has permitted production of appliance and automobile finishes with greater life than was previously possible.

The treatment of sheet and strip metal in which zinc, steel, or aluminum coils are uncoiled, treated with a conversion coating, painted, and recoiled has developed into a substantial industry. The ability of the conversion coating and the paint formulation to accept severe deformation without rupture during recoiling and subsequent forming operations has permitted the extensive use of treated, painted metal in the building industry and in many other industries for which metal stampings are used.

The use of conversion coatings before electrodeposition is being heavily researched with excellent results. In electrodeposition, the treated parts are immersed in a water-reducible paint that is deposited by a controlled electric current. Because

Reviewers, cont: ***Roberta Priemon***, *Managing Standards Editor, ASTM;* ***Clint Rasmusson***, *Technical Service Director, ABA, Parker Chemical Co.;* ***Al Ryalls***, *President, Accurate Metal Finishing, Inc.;* ***D. J. Schardein***, *Supervisor, Surface Technology Section, Corporate Research Div., Reynolds Metals Co.;* ***Dr. Krish G. Sheth***, *Chief Chemist, Electrofilm, Inc.;* ***John Sparks***, *Product Manager, Metal Finishing Dept., DuBois Chemicals Co.;* ***Robert Stansbury***, *Industry Manager, Metals Industry Div./Metalprep Dept., Pennwalt Corp.;* ***Lester Steinbrecher***, *Director, Research and Development, Amchem Products, Inc.;* ***Edward Taylor***, *Manager, Technology Development Laboratory, Kolene Corp.;* ***Paul F. Wilson***, *Product Manager, Allied-Kelite Div., Witco Chemical Corp.;* ***William J. Wittke***, *Manager of Technical Services, Oakite Products, Inc.;* ***Joe Ziegeweid***, *Industrial Finishing Consultant.*

CONVERSION COATING APPLICATIONS

of the rapid development of paint formulations, the paint supplier and the manufacturer of the conversion coating should be consulted for advice about the best combination to use. Electrodeposition is discussed in greater detail in Chapter 27, "Application Methods," of this volume.

Cold Forming Lubrication

Some deep draws would be virtually impossible without the use of zinc phosphate coatings. These coatings are capable of reacting with or absorbing oil and soap-type lubricants and, in some cases, an insoluble lubricant film. The inherent heat-resistant characteristics of the conversion coatings permit their use in such applications that might otherwise destroy other types of drawing compounds or lubricants.

A large-scale industrial use of phosphatizing in the cold working of metals is in the production of automobile bumpers. The metal for the bumpers is polished in the flat, phosphated for resistance to stretch and cold forming marks, lubricated, and then cold formed. After forming, the phosphate coating is removed, and the bumper is cleaned and plated. Another area in which conversion coatings are used extensively to aid in metalforming is the drawing of mild and stainless steel tubing.

Wear Reduction/Resistance

Certain phosphate coatings, through their ability to produce parting layers and promote continuous oil films that are not subject to rupture, are used to reduce wear on bearing surfaces and permit uniform break-in of new parts. Most of the coatings used for this purpose are manganese phosphate coatings. However, zinc phosphate coatings are receiving increasing interest.

In metalforming, the coating acts in conjunction with the lubricant to form a separating layer. However, when used to aid wear resistance, the coating is rapidly removed by the interaction of the two surfaces rubbing together. Small, uniform pits produced in the base metal during the formation of the coating act as reservoirs to maintain a uniform oil film between the two surfaces and to prevent seizing of the surfaces.

Phosphate coatings that reduce wear have been used extensively in automotive engines and in refrigerator compressors. Typical parts treated include pinion gears, camshafts, pistons and rings, rocker arms, worm gears, tappets, and oil-distribution rods. However, coatings have not been successful in reducing wear in roller or ball bearings.

Applications in which close tolerances are involved must be specially considered. Conventional treatment provides a non-metallic buildup of approximately 0.00015-0.0003″ (4-8 μm) per surface depending on the specific cleaning method used. When tolerance specifications do not allow this great a buildup, the coating deposit can be controlled by either chemical or mechanical means. If tolerances allow only 0.0001-0.00015″ (2.5-4 μm), the acidity of the bath must be adjusted and activating agents added to the rinse preceding the treatment solution to produce a fine, dense coating. When the tolerance is less than 0.0001″ (2.5 μm), some mechanical operation such as burnishing or hand lapping must be used.

Electrical-Resistance Coating

These coatings are formed from baths of chromium, iron, alumina, and silica, with or without mica, and are applied by the roll-on technique. The coatings produced must: (1) prevent the adhesion of steel surfaces during gas-fired annealing for two hours at 1550° F (845° C); (2) have an electrical-resistance value before and after annealing no greater than 5/10 as measured by the Franklin tester (a lower measurement preferred); (3) not add more than 1% to the thickness of stacked laminates (although more is permissible in some instances); (4) not hold dust during handling; (5) not cause wear on stamping dies; and (6) upgrade low-carbon steel to a performance equal to low-silicon steel.

METALS TREATED

The four metals of major industrial importance in the automotive, electrical appliance, office equipment, housing, and related fields are steel, zinc, galvanized steel, and aluminum. Most metal products are painted to add decorative value, decrease corrosion, and increase service life. In addition, many are cold formed and/or subject to wear in use. However, almost any metal may be conversion coated for a useful purpose. The fabrication of quality metal products generally presupposes the use of conversion processes to serve one or more purposes. Some of the processes available for different metals are listed in Table 19-1. Chromate as well as phosphate coatings have been used to protect metals, and it is interesting to note that mixed chromate-phosphate baths have also been employed.

Conversion coatings are applied to cleaned articles by immersion or by spraying on solutions at the required temperature for the required length of time. Coating weights depend upon the manner in which the articles are cleaned, the composition of the processing immersion bath or spray, and the type and surface condition of the metal.

Aluminum

Most of the aluminum used, particularly in the missile and aircraft industries, is in the form of alloys that vary greatly in their physical and chemical properties. Aluminum alloys treated may be sheets, castings, forgings, or extruded or rolled structural forms. Common methods of converting aluminum surfaces include (1) electrolytic (anodizing) and chemical processes for developing aluminum oxide coatings, (2) acidic proprietary solutions to produce chromium chromate and chromium phosphate coatings, and (3) deposition of crystalline zinc phosphate coatings.

Crystalline zinc phosphate deposits form on aluminum from specially formulated baths. These baths are more difficult to control than those used for steel because the accumulation of aluminum ions in the bath quickly reduces the amount of coating formed. Aluminum ions are precipitated as sodium aluminum fluorides by maintaining an excess amount of sodium fluoride in the bath through frequent chemical analysis.

Zinc phosphate deposits are seldom used except when it is necessary to coat fabricated items made of both aluminum and steel or when they are required by military specifications. A notable exception is the use of zinc phosphate as a substrate for a lubricant in severe forming (drawing) of aluminum; sometimes the availability of the zinc phosphate bath for this purpose dictates the use of the same bath for prepainting treatment.

Two types of chromate paint-based coatings for aluminum are in general use. One consists of a mixture of chromic phosphate, aluminum phosphates, and oxides; it is green in color, nonfading, and suitable for both indoor and outdoor use. The other type is a chromium chromate film that is amorphous and gold to light brown or dark brown in color. These coatings are produced from sludgeless solutions in short process times by controlled spraying, immersion, and strip-line methods. Operations may be carried out at room temperature without the use of electricity, and the coating weight from thin films to heavy deposits can be controlled. In general, the nonheat-treatable, low-alloying constituent metals are easiest to treat and provide the maximum resistance to corrosion.[1]

Cadmium

Cadmium is painted rather infrequently and does not present troublesome oxide problems, so pretreatment is less vital than for aluminum. Bright dips are commonly used on cadmium to improve appearance and extend the shelf life of the coatings. Cadmium, usually in the form of cadmium-plated steel, is frequently chromated but may also be phosphatized in accordance with military specifications that require a phosphate coating on cadmium plate.

Copper

Copper alloys are buffed, bright dipped, or both. The passive, oxide-covered surface generated in bright dipping provides a good base for paint. To improve paint adhesion on copper and copper alloys, chromate coatings similar to the iridescent yellow coatings formed on zinc, cadmium, or aluminum may be applied.

Lead

Lead forms protective surface deposits by reaction with carbon dioxide in the air. Lead is seldom painted, but terneplate (lead-covered steel) sometimes requires painting. Although common practice is often to clean and paint without using a conversion coating, lead has been successfully treated with an oxalate bath, and an oxalate coating has been applied to terneplate ammunition boxes for military specifications.

Magnesium

Like zinc and cadmium, magnesium requires care in prepainting metal preparation. Magnesium is a very reactive metal that forms an alkaline hydrated oxide rapidly when exposed to moist conditions or atmospheres containing traces of salts. The growth of this oxide under paint destroys adhesion.

Magnesium parts may be immersed, sprayed, or brushed with solutions containing chromate, dichromate, acid fluoride, magnesium sulfate, nitric acid, calcium, magnesium, sodium, or ammonium ions. Some chemical-conversion treatments for magnesium are commonly used as paint bases; others are more frequently used without an organic finish. A dilute chromic acid solution is a popular commercial paint base, especially for magnesium luggage. Treatments applicable to magnesium alloys must be modified if aluminum is contained in the assembled part.

Chromate treatments are used in one way or another on almost all magnesium items such as wrought parts and sand or die castings, even if only a chromate pickle is used prior to shipment or storage. Reactions most probably involve chromic solution CrO_3 acid, which reacts with the magnesium metal and forms mixtures of reduced chromium oxides in complex arrangements with the magnesium surface.

Silver

Chromate conversion coatings may be applied to silver to reduce tarnishing if the silver is not painted.

Stainless Steel

On chromium and chrome-nickel steels, and for seamless alloy tubing, a coating of iron oxalate is produced as part of a lubricating system for cold forming. This coating is more easily applied than lime or the metallic coatings such as copper and lead, and it does not present the removal problems after drawing that the other preparations do.

Steel

Differences in steel composition can influence the quality of the product and affect the performance of organic coatings exposed to the atmosphere. Three types of coatings have been used to provide protection for steel or iron: iron, zinc, and manganese phosphates. Steel is frequently given a phosphate coating as a base for paint. Applications requiring modest final-film performance and salt spray resistance are normally coated with iron phosphate. Applications requiring good final-film performance are normally coated with zinc phosphate. Zinc phosphate coatings are more expensive and more difficult to apply than iron phosphate coatings.

Zinc phosphate coatings also act as lubricants and as carriers for oil and soap film lubricants. The crystalline, nonmetallic, and absorptive characteristics of these coatings (1) prevent welding of metals under load, (2) increase lubrication efficiency, (3) permit faster drawing speeds, (4) permit deeper draws, and (5) allow heavier cold extrusions.

A manganese dihydrogen phosphate solution, containing a small amount of iron phosphate and an oxidizing agent, protects properly cleaned surfaces even under pressure and after abrasion. The etching action of the manganese iron phosphate pits the base metal, and the pits serve as a carrier for lubrication even after the surface layer is worn away. Manganese phosphate coatings increase the wear resistance of moving parts, prevent metal from sticking where it is not wanted, and prevent seizing. However, zinc phosphate coatings are also used when the wear encountered is less severe.

Tin

An alkaline phosphatizing process has been used to treat tin to prevent sulfide blackening on the interiors of tin cans and to retard rust penetration of the outer surfaces during storage in damp conditions. Normally, only an alkaline cleaner is used on tin prior to lacquering or painting.

Titanium

Titanium requires careful cleaning prior to deep drawing, wire drawing, or tube drawing. It is usually pickled with hydrofluoric acid and nitrates. This pickling operation becomes absolutely necessary whenever titanium is superfically oxidized; for instance, at the time of heat treatment. An extremely fine coat on titanium is produced from a bath of trisodium phosphate, potassium fluoride, and hydrofluoric acid after 2 to 3 minutes immersion.

Zinc

A conversion coating is usually necessary when zinc is to be painted. As with cadmium and magnesium, zinc requires care in prepaint preparation. Zinc that is to be painted is usually in the form of zinc-based die castings, zinc-plated steel, and hot-dip galvanized steel.

Chromates are almost always applied to electroplated zinc and are frequently applied to zinc die castings to further protect the zinc from corrosion. It is common practice to phosphatize electrogalvanized (zinc-plated) steel at the mills. Galvanizing processes use a silicate chromate bath to prevent "bloom" or white rust during storage in humid conditions, although the silicate treatment presents particular problems in subsequent prepaint processing. At present the use of silicates is rare, but chromate baths are widely used.

Phosphatizing of zinc is desirable before painting and is the common pretreatment for galvanized steel. For the best appearance, spangle-free, hot-dip galvanized steel must be used. A zinc phosphate with nickel additions is most commonly chosen to give best all-around results; however, an amorphous, complex oxide coating produced from an alkaline solution has proved superior to commercially available surface treatment preparations and particularly excellent for the treatment of coils of galvanized metal that are to be painted and later formed.

One problem in phosphatizing galvanized steel is the "spangle" caused by a small amount of zinc-aluminum alloy in the hot-dip metal coating. Galvanizing flux also creates problems, as do the treatments used by the mills to prevent corrosion in storage. Fluoride in the bath may aid in reducing these problems.

Mixed Metals

For production of mixed steel, galvanized steel, and aluminum surfaces, zinc phosphate-based solutions are preferred, although iron phosphate solutions have been used. For aluminum and zinc surfaces in mixed production, chromium trioxide-based solutions containing fluoride and a mixture of tungstate and/or molybdates are employed. Steel, zinc, and aluminum mixtures are treated with solutions of zinc phosphates or solutions of acidic, alkali metal phosphate salts and wetting agents. Because plain steel and galvanized steel are often painted in the same process, phosphate coatings must be applied on both metals, often in the same tanks. Special baths are available for this purpose.

PHOSPHATE CONVERSION COATINGS

Phosphate conversion coatings bring about transformations of metal substrates into new surfaces having nonmetallic and nonconducting properties. The transformations occur in phosphating solutions containing divalent metal phosphates and, in some instances, in solutions containing monovalent metal phosphates. Generally, the solutions are prepared from liquid concentrates containing one or more divalent metals (zinc, magnesium, calcium, etc., phosphates), free phosphoric acid, and an accelerator.

Three types of phosphate conversion coatings are currently being used: zinc, iron, and manganese. Zinc phosphating is often used as a pretreatment for painted parts. It is also used to impart corrosion resistance and to aid in cold forming operations. Zinc coating weights are usually 200-500 mg/ft^2 (2.2-5.4 g/m^2). Iron phosphate coatings are primarily used to form a passive substrate under paint and have coating weights of 50 to 100 mg/ft^2 (0.5 to 1.0 g/m^2). Manganese phosphate coatings are used primarily on machined parts such as gears and internal combustion engine components as an antiscuff film for break-in wear. Coating weights are usually 1000-3000 mg/ft^2 (10.8-32.2 g/m^2).

Since the seventies, a trend to reduce heating costs, improve working conditions, prolong equipment life, reduce sludge, and reduce processing steps has resulted in low-temperature iron and zinc phosphate coatings and, to a limited degree, solvent phosphating solutions.

Chemical Reaction

When a steel part is immersed in a dilute phosphoric acid solution, the metal surface is attacked and hydrogen in the phosphoric acid is replaced by the metal ion, producing iron phosphate and hydrogen gas. Depending upon whether one, two, or three hydrogen atoms of the phosphoric acid molecule are replaced by metal ions, primary, secondary, or tertiary phosphates are formed as follows:

$$\underset{\text{iron}}{Fe} + \underset{\text{phosphoric acid}}{2H_3PO_4} \rightleftarrows \underset{\text{soluble primary ferrous phosphate}}{Fe(H_2PO_4)_2)} + H_2 \qquad (1)$$

$$Fe + Fe(H_2PO_4)_2 \rightleftarrows \underset{\text{insoluble secondary ferrous phosphate}}{2FeHPO_4} + H_2 \qquad (2)$$

Because of the reactions of Eqs. (1) and (2), a reduction in acid concentration or an increase in pH is particularly great and rapid at the interface between the metal and the phosphating solution; therefore, the precipitation of insoluble phosphates occurs at the metal surface. Part of the iron dissolved by the attack of phosphoric acid converts to secondary ferrous phosphate and enters the coating; part remains in solution as primary ferrous phosphate. Under certain conditions, air oxidation at the bath surface will be sufficient to oxidize the ferrous iron to tertiary ferric phosphate and precipitates as a sludge.

$$2FeHPO_4 + 1/2\,O_2 \rightarrow \underset{\text{ferric phosphate}}{2FePO_4} + H_2O \qquad (3)$$

On standing, the divalent iron of the secondary ferrous phosphate coating can convert to trivalent ferric phosphate in accordance with Eq. (3). This change destroys the protective value of the coating in a short time. Insoluble phosphates of manganese and zinc are not susceptible to oxidation.

When the phosphoric acid solution contains no primary ferrous phosphate, only a thin protective coating of a few milligrams per square foot of insoluble ferrous phosphate is produced. It is readily understood from Eq. (2) that if phosphoric acid solution contains primary ferrous phosphate, a heavier, insoluble, protective ferrous phosphate coating will be formed. If the solution contains primary manganese or zinc phosphates, it readily forms mixed crystals with secondary ferrous phosphate.

The primary phosphates of iron, manganese, calcium, zinc, etc., are water soluble. Their secondary and tertiary phosphates are not. However, the primary salts will dissociate into insoluble secondary and tertiary phosphates liberating phosphoric acid as follows:

$$\underset{\text{soluble, primary}}{M(H_2PO_4)_2} \rightleftarrows \underset{\text{insoluble, secondary}}{MHPO_4} + H_3PO_4 \qquad (4)$$

$$\underset{\text{soluble, primary}}{3M(H_2PO_4)_2} \rightleftarrows \underset{\text{insoluble, tertiary}}{M_3(PO_4)_2} + 4H_3PO_4 \qquad (5)$$

Here M stands for bivalent metal such as iron, calcium, manganese, or zinc.

Some of the insoluble metal phosphates are precipitated from the bath solution in the form of sludge in accordance with Eqs. (4) and (5). It is also evident from these equations that an excess phosphoric acid, above that which is bound by primary phosphate, must be maintained to keep primary phosphates in solution. The amount of this excess phosphoric acid required depends upon the type of primary metal phosphate present in the phosphating solution. Some excess phosphoric acid is also needed to dissolve a thin layer of the metal surface. For example, phosphating solutions containing primary zinc phosphate require more free (excess) phosphoric acid than those containing primary manganese phosphate, and phosphating solutions containing primary ferrous phosphate need less free acid than solutions containing primary manganese phosphate.

It is believed that the acid attack takes place at a large number of anodic sites occurring at grain boundaries, imperfections, projections, and discontinuities of the metal surface. Iron is oxidized to ferrous ions with the loss of electrons. These electrons move to the adjacent cathodic areas where hydrogen ions are reduced to hydrogen gas, consequently raising the pH at cathodic sites. Nuclei of the insoluble secondary and tertiary phosphate-crystalcline precipitates start to form at the cathodic sites on the metal surface. As the crystals grow to cover these areas with nonconductive coating, the pattern of the activation energy of the surface changes to create new anodic and cathodic areas adjacent to the crystalline deposit. The crystals continue to form until the entire metal surface is covered, barring micropores that maintain the galvanic cell.

Most modern phosphating solutions are prepared and replenished with concentrates in liquid form. On dilution, these concentrates produce a solution containing a divalent metal phosphate—usually zinc or manganese phosphates or a combination of the two, free phosphoric acid, and an accelerator. Iron phosphate solutions do not contain an accelerator. For example, in the case of manganese phosphating, the overall reaction that occurs at the metal/solution interface may be indicated, in simplified form, by Eq. (6).

$$H_3PO_4 + 3Mn(H_2PO_4)_2 + 5Fe \rightarrow Mn_3(PO_4)_2\downarrow + 5FeHPO_4\downarrow + 5H_2\uparrow \quad (6)$$

The manganese phosphate coating is composed of a mixture of tertiary phosphates and mixed crystals such as mixed iron-manganese phosphate.

Much can be done by varying the type of accelerator used and the proportions of the various ingredients in the bath to change the crystal structure in the coating as well as the coating weight. The addition of an oxidizing agent as an accelerator causes a finer coating to form more quickly, but the coating weight levels off at a moderate value in the neighborhood of 150 to 250 mg/ft^2 (1.6 to 2.7 g/m^2). With a fixed bath composition, some variation can be made in the coating by changing the processing time and the temperature.

Accelerators

The addition of heavy metal ions, such as cupric ions, to a conversion bath greatly reduces the coating formation time and the size and nonuniformity of the coating crystals. Copper, which is cathodic to the dissolving metals, deposits on the base metal to form many local cells, thus increasing the potential difference between the local anode and cathode areas. Hence, the rate of dissolution of the metal is greatly increased, with a proportionate reduction in hydrogen ion concentration in the solution layer next to the metal, resulting in faster coating precipitation as shown in Eqs. (1), (4), and (5). The myriad of local cells become centers for crystal growth, allowing rapid formation of small crystals on the metal surface. It should be noted, however, that an excess of copper in solution will plate out on the metal and inhibit coating formation.

Nickel salts behave differently than cupric ions, and their benefit may result from a catalytic action connected with the release of molecular hydrogen or from the control of certain reactions with other accelerators present in the bath. Nickel additions are frequently used when hardened materials are treated and when the activity of the metal surface being treated is low. Nickel is also used to improve a zinc phosphate coating on galvanized steel surfaces.

Besides copper and nickel, other heavy metal compounds act as promoters, catalysts, accelerators, or coating supplements. These accelerators include molybdenum, tungsten, vanadium, zirconium, and cerium compounds. However, adding most of these compounds to phosphate coating baths has been generally discarded because of the difficulty of maintaining the proper concentration in the processing solution and because of the reduced corrosion resistance of the coatings obtained. Some solution manufacturers have used molybdenum compounds as an integral part of proprietary iron phosphate compounds.

The addition of chemical substances that do not in themselves enter the coating also reduces the processing time. These compounds function as depolarizers and metal oxidizers at the cathode areas by removing the hydrogen formed as the metal dissolves, as indicated in Eq. (1). Thus, the rate of the anodic dissolution of the metal is increased, and coatings are subsequently formed more rapidly. These oxidizing agents are used primarily in zinc-based solutions. Of the various oxidants usually employed, only nitrate ions are compatible with ferrous ions. Chlorate, nitrite, bromate, and peroxide ions all oxidize ferrous ions to the insoluble ferric phosphate, thus producing substantially iron-free solutions.

Nitrate is used alone in solutions for the preparation of heavy phosphate coatings, and in combination with nitrite or chlorate in solutions for the preparation of lightweight base coatings. When a ferrous surface is treated with a phosphatizing bath containing nitrate as the sole oxidizing agent, not all the dissolved iron is oxidized to the ferric state, and the remainder accumulates in the bath as ferrous iron.

The nitrate ion also serves another useful purpose: ferric phosphate, the primary constituent of the sludge formed in phosphate coating baths, is more soluble in nitrate solutions. This is an important factor because ferric ion in acid solution reacts readily with hydrogen and is thereby reduced to ferrous ion; in this way, the iron is allowed to enter the coating instead of being lost as sludge. Because ferric ion solubility increases with nitrate ion concentration, it is advantageous to maintain the maximum practicable nitrate concentration in the phosphatizing solution.

Nitrate-containing zinc phosphate immersion baths are frequently controlled for concentration, temperature, and loading rates so that no ferrous iron accumulates in the bath. If the temperature of the solution drops lower than 170° F (77° C), a gradual buildup of ferrous iron will occur. The coatings produced under the proper operating conditions are lower in weight than those produced when the bath contains ferrous iron, and are normally used as a base for paint or for metalforming. Also, under these conditions, some of the nitrate breaks down to nitrite. In other cases, nitrite is added to the bath as sodium nitrite to ensure that no ferrous iron accumulates. The acceleration of coating formation is a function of the concentration of nitrous acid in the solution or at the interface and is not merely a simple result of hydrogen removal.

Nitrite-accelerated phosphate treatments are also applied extensively by spraying. Because nitrite is volatile, continuous additions of sodium nitrite are required. At temperatures greater than 170° F (77° C), nitrite is formed in immersion baths from the nitrate present.

With chlorate-accelerated baths, the type of coating obtained and the extreme rapidity of formation include a mechanism comparable with that of nitrite-accelerated baths. The chlorate-accelerated coating solution may be prepared by adding sodium chlorate to the concentrate. Thus, an advantage of the chlorate-accelerated bath is that the oxidizing agent is incorporated in the concentrate. However, chlorate-accelerated baths usually produce heavy scale on the heat tubes and require higher operating temperatures and more frequent desludging.

Coating Characteristics

As noted earlier, the reaction rate and crystal nucleation depend on the concentration of anodic and cathodic sites on the metal surface and its electrochemical activity. Hence, the crystal size and the structure of coating vary tremendously with the pretreatment given; the type (organic or inorganic crystal modifiers in the bath), pH, and temperature of the treating solution; and the time the surface remains in contact with the coating solution. Variations in coatings obtained from one type of phosphating solution may result from variations in pretreatment of the metal, the condition of cleanliness of the metal surface, and the degree of metal saturation of the phosphate coating solution.

The method used to remove the grease and oil from the metal surface creates the difference between a coating consisting of large crystals with uncoated areas between the crystals and a dense growth of small crystals located very closely together. When phosphate coatings are produced in the same solution at the same time on surfaces that have been given various types of preparation, their crystal structures will be different. A pickling or a caustic treatment will result in large crystals with uncoated areas between them, while a mild alkali treatment will result in much smaller crystals with fewer bare areas. Oxalic acid cleaning causes much smaller crystals, while hand-wiped surfaces produce the smallest crystals of all these cleaning methods. The use of titanium activators in the rinse prior to coating also produces small crystal coatings.

Thick phosphate coatings consist of relatively coarse crystals, whereas thinner coatings consist of finer crystallites. The coatings made up of relatively coarse crystals result from acid phosphatizing baths based on manganese, zinc, and ferrous ions. These coatings gradually increase in thickness and are usually produced in the range from 0.0001 to 0.002″ (2.5 to 50 μm), although thicknesses greater than 0.002″ have been observed in the case of certain coarse crystalline coatings. Since precise measurement of thickness is difficult, coating quantity is generally expressed in milligrams per square foot of coated area (with metric, it is expressed as grams per square meter). Coating weights usually vary from 50 to 4000 mg/ft^2 (0.5 to 43 g/m^2).

The coatings obtained from solutions exhibiting a pH of 4 to 6 are moderately thin, in the range of 25 to 100 mg/ft^2 (0.25 to 1.0 g/m^2), and are thought to consist of a mixture of ferrous and ferric phosphate with iron oxide. If phosphoric acid alone is used to treat the metal surface (as in acid wash), a very thin coating having a weight in the range of 2 to 8 mg/ft^2 (0.02 to 0.09 g/m^2) and consisting of a mixture of ferrous and ferric phosphate with ferric and ferrous oxides is formed.

Operating Conditions

As was mentioned previously, the crystal size and the structure of the coating vary according to the pretreatment, type, pH, and temperature of the phosphating solution, and the length of time the surface remains in contact with the coating solution.

Solution control. All the constituents of a phosphate coating solution are depleted during operation, so a replenishing solution is added to maintain the proper operating strength. Replenishing materials are compounded so that all constituents in the processing solution remain within a definite range. In addition to meeting the chemical requirements of keeping the bath in balance, additive materials must be formulated so that they may be handled easily.

Most phosphating solutions used in industry are of a proprietary nature that necessitates consultation with the supplier or vendor when seeking the appropriate analytical method. However, a zinc phosphate solution generally contains zinc with free and combined phosphoric acids. As the solution is used, the total acid concentration goes down causing a change in the reaction. For the sake of uniform production rate and quality coating, it is necessary to keep the solution in balance. A knowledge of the total acid, free acid, pH, temperature, and iron concentration is necessary every 4-8 hours of operation. For nitrite/nitrate-accelerated bath, a frequent accelerator titration is recommended.

The concentration of iron and acid is generally expressed by an arbitrary set of values called points. It is recommended that iron concentration be maintained below 11 to 12 points. As the concentration of iron in the phosphatizing solution increases, a greater number of acid points are required to maintain the working bath. The following formula may be used to determine the required number of total acid points:

$$3 \times \text{ferrous points} + 30 = \text{approximate number of acid points } (\pm 2) \qquad (7)$$

A 1% by volume of a heavy zinc phosphate solution is generally equivalent to 10 points total acid. Adding a specified amount of the solution concentrate lowers the pH and increases the number of total acid points.

Calculating ferrous points. The number of iron points in a phosphatizing solution can be determined using the following procedure:

1. Pipet a 10 mL sample of the phosphate solution into a 250 mL Erlenmeyer flask.
2. Add 10-15 drops of 50% sulfuric acid.
3. Titrate the sample with 0.2 Normal (N) potassium permanganate until a permanent pink color is obtained.

The number of milliliters of 0.2N potassium permanganate added represents the number of points.

Calculating total acid points. The number of total acid points in a phosphatizing solution can be determined using the following procedure:

1. Pipet a 10 mL sample of the phosphate solution into a 250 mL Erlenmeyer flask.
2. Add 3-4 drops of phenolphthalein indicator.
3. Titrate the sample with 0.1N sodium hydroxide until a permanent pink color is obtained.

The number of milliliters of 0.1N sodium hydroxide added represents the number of points.

Temperature. Although operating temperatures of different phosphating solutions may range from 90 to 210°F (32 to 99°C), individual solutions are compounded to operate at maximum efficiency within specific temperature limits.[2] The more recent low-temperature phosphating solutions can be applied by spray or immersion methods at temperatures from 60 to 120°F (16 to 49°C).

If the temperature is too low, the phosphate coating will be very thin or nonexistent. Too high a temperature causes an excessive coating buildup that is nonadhering and has a

TABLE 19-2
Troubleshooting Iron Phosphate Coatings[3]

Problem	Possible Causes	Proposed Solutions
Smut or inorganic soot on part surface	Excessively high pH.	Reduce pH.
	Spray nozzles blocked or out of adjustment.	Clean and adjust for proper spray pattern.
	Concentration too low.	Add materials to attain proper concentration.
	Poor quality steel or improper storage.	Check condition of cleaning system. Preclean with a solvent.
	Improperly racked parts.	Change racking.
Flash rusting	Poor phosphate development.	Check solution concentration and adjust as required.
	Low pH.	Adjust pH.
	Recessed areas catching moisture (cupping).	Arrange parts on rack to ensure complete drainage.
Water spotting	Cleaner or soil residue clinging to parts.	Check cleaner/phosphate and rinses for high solids or oil contamination.
	Improperly cleaned part.	Check cleaning system. Use cleaner additive.
	Parts are improperly racked.	Arrange parts on rack to ensure complete drainage.
Powdering	Excessively high pH.	Bring pH to desired range.
	Excessive sludge in bath.	Remove sludge, renew bath, or improve rinsing.
	Phosphating material concentration too high.	Dilute to the proper concentration.
	Excessive drying temperature.	Keep drying temperature below 300° F (150° C).
	Unsuitable system. Extremely dirty steel.	Preclean in solvent or remove smut by available techniques. Mild pickle after cleaning is recommended.
Mottling	pH too low.	Adjust pH.
	More easily cleaned areas develop heavier coating.	Reduce mottling by precleaning or with two phosphate baths in which the first bath has a high pH and acts as a cleaner and the second bath does the phosphatizing.
	Moist air oxidation may develop films of varying thicknesses.	
	Light splashback of chemicals on dried surface.	Adjust entrance nozzles on phosphate zone to prevent precoating.
Insufficient coating	Concentration too low or pH too high.	Add materials to attain proper concentration or lower pH.
	Contact time too short.	Raise temperature.
	Work too dirty and less time for coating.	Preclean or better cleaner.
	Temperature too low.	Bring up to suitable temperature.

powdery surface. High temperatures also result in excessive sludge and scale.

Time. The time required to phosphate a part is largely determined by the application method and the coating desired. The spraying method produces a given coating weight in a shorter period of time than does immersion. In general, sprayed phosphate coatings can be applied in less than 2 minutes. For some applications, the time may be as short as 3-5 seconds. Phosphate coatings applied by immersion usually require 3-5 minutes.

Troubleshooting

Many problems associated with phosphate coatings are due to improper cleaning of the workpieces or improper maintenance of the phosphating bath and rinse tanks. Poor cleaning can be detected as water beading on the surface of the part or loose particulate matter clinging to the workpiece. These water spots inhibit adhesion of organic coatings on the surface. An incorrect pH or solution concentration may cause several other surface-related problems (smut or inorganic soot, flash rusting, powdering, or mottling) and an insufficient phosphate coating.

Smut, or inorganic soot, appears as a black, gritty dust on the workpiece surface. Flash rusting creates reddish-brown patches on the surface, and powdering occurs as a white dust over the surface of the workpiece. Mottling is different color shades on the workpiece surface, ranging from light blue or gray to a golden iridescent color. An insufficient phosphate coating is normally difficult to see with the naked eye; but, on some occasions, an absence of color on the workpiece surface is an indication of an insufficient coating. Table 19-2 lists some of the common problems associated with iron phosphate coatings, along with possible causes and solutions.[3]

COATING EVALUATION

Since iron and zinc phosphate coatings are designed to bond paint to metal surfaces, testing the effectiveness of these coatings is done on the total system after painting is completed.[4] The two primary tests performed are tests for adhesion of the paint to the metal surface and corrosion resistance. Table 19-3 lists the standard specifications given by the American Society for Testing and Materials (ASTM) and the U.S. government.

Safety

Safety and care of workers and the environment are just as important as the safety and cleanliness of the equipment and the plant. The workers operating the phosphating equipment and handling the parts must wear safety equipment to protect their faces, eyes, noses, hands, and clothing. The combined effect of heat, free phosphoric acid, and the reaction of iron with nitric acid produces noxious fumes containing oxides of nitrogen. It is therefore necessary to ventilate the phosphating tank with an exhaust system that is equipped with fume scrubbers.

Effluent Treatment

The bath effluent or spillage should not be discharged directly to the sewer system, rivers, lakes, or any underground facility without being treated according to local and federal regulations. Phosphating solutions may contain phosphates, nitrates, chlorates, fluorides, zinc, manganese, magnesium, iron, nickel, or calcium. All of these materials, except nitrates and chlorates, can be removed from the solution by adjusting the pH to 7.0 with lime and alum and then followed by a final pH adjustment to 9.2 with caustic soda.

A typical treatment is as follows:

1. Collect waste phosphatizing bath in settling tank.
2. Add alum to bath and mix well [usually 3 lb (1.5 kg) per 100 gal (378L) of bath].

TABLE 19-3
Specifications for Evaluating Iron and Zinc Phosphate Coatings[4]

Specification Number	Title
D-522	Method of Test for Elongation of Attached Organic Coatings with Conical Mandrel Apparatus
D-714	Method of Evaluating Degree of Blistering on Paints
D-870	Method of Water Immersion Test of Organic Coating on Steel
B-117	Method of Salt Spray (Fog) Testing
B-287	Method of Acetic Acid-Salt Spray (Fog) Testing
D-1014	Method of Conducting Exterior Exposure Tests of Paints on Steel
D-1735	Method of Water Fog Testing of Organic coatings
MIL-C-46487	Cleaning, Preparation, and Organic Coating of Steel Cartridge Case
DOD-P-16232	Phosphate Coating Heavy, Manganese or Zinc Based (for Ferrous Metals)
MIL-T-12879A (MR)	Treatments, Chemical Prepaint and Corrosion Inhibitive for Zinc Surfaces
QQ-P-416	Plating, Cadmium (Electro-deposited)
QQ-Z-325	Zinc Plating (Electrodeposited)
TT-C-490	Cleaning Methods and Pretreatment of Ferrous Surfaces for Organic Coatings
No. 57-0-2	Finishes, Protective for Iron and Steel Parts
MIL-C-12968	Coatings, Phosphate Protective (for Iron and Steel)
AMS-2480	Phosphate Treatment, Paint Base
AMS-2481	Phosphate Treatment, Anti-Chafing

(*Metal Finishing*)

Note: Only the basic specification or standard number has been given. It is therefore necessary to refer to the most current revision given by the specifying body when following these procedures.

3. Add lime slurry to bath until pH reaches 7.0. Mix well.
4. Add liquid caustic soda to bath while mixing and adjust pH to 9.2.
5. Add flocculent as recommended and let stand.
6. Discharge clear liquid.
7. Remove and then dispose of sludge in accordance with local and federal regulations.

During this treatment, the soluble chloride is converted to nontoxic sodium chloride. If the nitrate concentration in the liquid is below the allowable limit, the liquid can be drained or used as rinse water. If the concentration is too high, the liquid can be diluted with fresh or rinse water, or the liquid can be heated to convert the nitrate to ammonia at high pH before discharging.

CHROMATE CONVERSION COATINGS

Chromate conversion coatings, often referred to as chromate coatings, are produced by chemically converting certain metal surfaces with aqueous solutions of chromic acid, chromates, dichromates, and certain other organic and inorganic chemicals. The treating bath compositions are generally proprietary, but all contain two basic ingredients—hexavalent chromium ions and enough acid to produce a desired pH. A few recent formulations, however, are based on trivalent chromium ions that are used to produce clear coatings on electroplated zinc and cadmium. Chromate-phosphate mixtures are also used to form combination conversion coatings on aluminum.

One of the oldest and still popular chromating processes for cadmium and zinc is based on a solution containing sodium dichromate slightly acidified with sulfuric acid. One chromating process for aluminum uses a bath of chromic, phosphoric, and hydrofluoric acids in well-defined proportions. Several chromating processes for magnesium have been developed by Dow Chemical Company.

Metals commonly treated include aluminum, cadmium, copper, magnesium, silver, zinc, and their alloys. The coatings are generally applied by immersion, although spraying, brushing, swabbing, or electrolytic methods are also used.

Chemical Reaction

The films in most common use are formed by the chemical reaction of hexavalent chromium with a metal surface in the presence of other components, or activators, in an acid solution. The hexavalent chromium is partially reduced to trivalent chromium during the reaction, with a concurrent rise in pH, forming a complex mixture consisting largely of hydrated chromium chromate and hydrous oxides of both chromium and the basis metal. The composition of the film is rather indefinite since it contains varying quantities of the reactants, reaction products, and water of hydration, as well as the associated ions of the particular systems.[5]

The mechanism of chromate coating formation from solutions containing hydrofluoric acid and its salts probably involves fluochromic acid ($HCrO_3F$), which hydrolyzes and forms mixtures of reduced chromium oxides in complex arrangements with the metal surface. Most chromate coatings absorb hexavalent chromium salts in varying amounts to give different shades of yellow. In the case of black chromate coatings on zinc or cadmium, for instance, a silver salt solution is employed. At the solution-metal interface, silver oxide is produced, giving the film a dark gray or black color.

Reactions. The reaction for a chromic-acid-based chromating solution for aluminum can be illustrated by the following equation:

$$6H_2Cr_2O_7 + 30HF + 12Al + 18HNO_3 \rightarrow$$
$$3Cr_2O_3 + Al_2O_3 + 10AlF_3 + 6Cr(NO_3)_3 + 30H_2O \qquad (8)$$

For the chromic-acid/phosphoric-acid-based solutions, a possible equation is:

$$2H_2Cr_2O_7 + 10H_3PO_4 + 12HF + 4Al \rightarrow$$
$$CrPO_4 + 4AlF_3 + 3Cr(H_2PO_4)_3 + 14H_2O \qquad (9)$$

The efficiency of the coating reaction depends on holding the following reaction to a minimum:

$$6HF + 2Al \rightarrow 2AlF_3 + 3H_2\uparrow \qquad (10)$$

Activators. Regulated concentrations of activators are added to the many proprietary formulations to promote the formation of the chromate film on the metal surface. Some of the common activators added include acetate, formate, sulfate, chloride, ferricyanide, fluoride, nitrate, phosphate, and sulfamate ions. Selection, separation, and control of the activator in a given formulation are performed by the chemical supplier.

Coating Characteristics

Chromate coatings are generally amorphous, nonporous, and gel-like when initially formed. As the coating dries, it slowly hardens or ages and becomes hydrophobic, less soluble, and more abrasion resistant. The coatings formed from solutions containing both hexavalent and trivalent chromium provide maximum protection by creating a mechanical barrier against corrosion and by allowing hexavalent chromium to leach out and resist corrosion.

The film formation begins at the interface between the chromate coating and the material surface and then grows outward. The amount of coating deposited is generally expressed in grams per unit of surface area, which is referred to as coating weight. Typical coating weights range from 15 to 150 mg/ft^2 (0.16 to 1.6 g/m^2). Generally, the coating weight or thickness increases in proportion to the immersion time and solution temperature.

Chromate coatings are available in a wide range of colors such as olive drab, bronze, iridescent yellow, blue bright, and clear. The actual color attained depends on the type of chromating solution used, the pH of the solution, and the thickness of the coating. For example, thinner coatings are usually lighter in color.

Applications

Chromate conversion coatings were widely used during World War II for the protection of cadmium and zinc-plated steel parts in tropical service. Since then they have been used in a variety of applications for decorative and functional purposes. Decorative chromate coatings are usually very thin and colorless, and are used as sealants over phosphate, oxide, or metallic coatings. The functional coatings are thicker, usually complete in themselves, and provide good corrosion resistance for the base metal that is exposed to an oxidizing environment. The degree of protection is proportional to the coating or film thickness. Chromate coatings can also be used as a nonporous bond for all paints that have good molecular adhesion.

CHAPTER 19

CHROMATE CONVERSION COATINGS

Chromate coatings are used over zinc and cadmium to simulate the appearance of bright nickel and chromium. They are also used to prevent the formation of white rust on zinc or cadmium-plated parts. Table 19-4 summarizes the common uses of chromate coatings for different materials.[6]

Operating Conditions[7]

In addition to the chemical makeup of the chromating solution, the treatment time, solution temperature, and solution agitation also govern film formation. Once these parameters are established for a given operation, they should be held constant.

Treatment time. Immersion time, or contact time of the metal surface with the solution, can vary from as little as one second to as much as one hour, depending on the solution being used and metal being treated. If prolonged treatment times are required to obtain desired results, a fault in the system is indicated and should be corrected.

Solution temperature. Chromating temperatures vary from ambient to boiling, depending on the solution being used and the metal being treated. For a given system, an increase in the solution temperature will accelerate both the film forming rate and the rate of attack on the metal surface, resulting in a change in the character of the chromate film. Temperatures should be adequately maintained, therefore, to ensure consistent results.

Solution agitation. Agitation of the working solution, or movement of the work in the solution, generally speeds the reaction and provides more uniform film formation. Air agitation and spray installations have been used for this purpose. There are, however, a few exceptions when excessive agitation will produce unsatisfactory films.

Coating Evaluation

Chromate conversion coatings are covered by many internal company standards and/or U.S. government and ASTM specifications. Tests are performed on chromated workpieces to determine coating presence, thickness, corrosion resistance, continuity, and adhesion. Table 19-5 lists the more commonly used specifications for chromate coatings on different materials.[8]

Solution Control

Maintenance and control of chromating solutions are generally done by controlling the acid level (pH) and the concentration of hexavalent and trivalent chromium. Each solution has a proper pH level at which it is to operate; a lower level causes the reaction products to remain in the solution rather than deposit as a coating on the metal surface. The pH is usually determined with a pH meter. Indicators and papers are not recommended because of discoloration by the chromate solution.[9] Typical pH levels range from less than zero to about 2.8.[10] The chromium concentration in the solution can be determined by titration. Because of the proprietary nature of chromating baths, the user must adhere to the recommendations of the supplier in order to obtain consistently good results.

Safety

The worker should wear appropriate safety equipment to protect eyes, nose, face, hands, and clothing. Since some chromating solutions emit toxic fumes and mists, a ventilation system is required. Solution spillage or effluents must not be discharged to the municipal sewer system, rivers, lakes, or underground. All effluents from chromating baths must be stored safely, properly treated, pH adjusted, filtered, and then disposed of in accordance with local and federal regulations.

OXIDE CONVERSION COATINGS

Oxide-type conversion coatings are chemically produced on iron, steel, stainless steel, aluminum, copper and copper alloy, zinc, and cadmium surfaces. These coatings may be used to provide color corrosion protection or abrasion resistance to the base metal.

The processes used to produce these coatings are referred to as blackening, nitriding, carbonitriding, sulfidizing, and QPQ™.

TABLE 19-4
Common Uses of Chromate Conversion Coatings[6]

	General Usage				
Metal	Corrosion Resistance	Paint Base	Chemical Polish	Metal Coloring	Remarks
Aluminum	X	X		X	• Economical replacement for anodizing if abrasion resistance is not required. • Used to "touch up" damaged areas on anodized surfaces.
Cadmium	X	X	X	X	
Copper	X	X	X	X	• Thin coatings prevent "spotting out" of brass and copper electrodeposits. • No fumes generated during chemical polishing.
Magnesium	X	X			
Silver	X				
Zinc	X	X	X	X	

(*Metal Finishing*)

TABLE 19-5
Reference Specifications for Evaluation of Chromate Coatings on Different Materials[8]

Material	Specification Number	Title
Aluminum	AMS 2473	Chemical Treatment for Aluminum Base Alloys—General Purpose Coating
	AMS 2474	Chemical Treatment for Aluminum Base Alloys—Low Electrical Resistance Coating
	ASTM D1730	Preparation of Aluminum and Aluminum Alloy Surfaces for Painting
	MIL-C-5541	Chemical Films and Chemical Film Material for Aluminum and Aluminum Alloys
	MIL-C-81706	Chemical Conversion Materials for Coating Aluminum and Aluminum Alloys
	MIL-W-6858	Welding, Resistance: Aluminum, Magnesium, etc.: Spot and Seam
Cadmium	AMS 2400	Cadmium Plating
	AMS 2416	Nickel-Cadmium Plating, Diffused
	AMS 2426	Cadmium Plating, Vacuum Deposition
	ASTM B201	Testing Chromate Coatings on Zinc and Cadmium Surfaces
	MIL-C-8837	Cadmium Coating (Vacuum Deposited)
	QQ-P-416	Plating, Cadmium (Electrodeposited)
Magnesium	AMS 2475	Protective Treatments, Magnesium Base Alloys
	MIL-M-3171	Magnesium Alloy, Process for Pretreatment and Prevention of Corrosion on
	MIL-W-6858	Welding, Resistance: Aluminum, Magnesium, etc.: Spot and Seam
Silver	QQ-S-365	Silver Plating, Electrodeposited, General Requirements for
Zinc	AMS 2402	Zinc Plating
	ASTM B201	Testing Chromate Coatings on Zinc and Cadmium Surfaces
	ASTM D2092	Preparation of Zinc-Coated Steel Surfaces for Painting
	MIL-A-81801	Anodic Coatings for Zinc and Zinc Alloys
	MIL-C-17711	Coatings, Chromate, for Zinc Alloy Castings and a Hot-Dip Galvanized Surface
	MIL-T-12879	Treatments, Chemical, Prepaint and Corrosion Inhibitive, for Zinc Surfaces
	MIL-Z-17871	Zinc, Hot-Dip Galvanizing
	QQ-Z-325	Zinc Coating, Electrodeposited, Requirements for

(*Metal Finishing*)

Note: Only the basic specification or standard number has been given. It is therefore necessary to refer to the most current revision given by the specifying body when following these procedures.

The abbreviation QPQ™ stands for quench-polish-quench, which is a process licensed by Kolene Corporation. Nitriding and carbonitriding are discussed in detail in Chapter 11, "Surface (Case) Hardening," of this volume.

QPQ™ Process

The QPQ™ process is based on a salt-bath nitriding process that also incorporates an oxidizing salt-bath quench and mechanical polishing. A typical cycle consists of heating the parts to 750° F (400° C) and then immersing them in an aerated nitriding salt bath at 1000 to 1075° F (540 to 580° C) for 10 to 180 minutes, depending on material and properties desired. The parts are then immediately transferred to the oxidizing salt-bath quench maintained at 650 to 750° F (345 to 400° C) for 5 to 10 minutes, followed by cooling and a water rinse. Light mechanical polishing, lapping, or vibratory finishing is performed to achieve the desired surface finish. After polishing, the parts are reimmersed into the oxidizing salt bath to improve corrosion resistance and develop the black surface finish. Figure 19-1 is a time-temperature profile of the complete cycle.

Advantages and limitations. The QPQ™ process provides better corrosion protection than conventional salt-bath nitriding and, for some applications, chromium plating. The surface finish of the parts is also improved. However, certain thin parts cannot be treated in this manner because of the high heat treatment temperatures.

Applications. The QPQ™ process can be used for applications requiring corrosion and wear resistance as essential properties. Many parts in the automotive industry can be treated with the QPQ™ process. Some of the applications include shock absorber piston rods, carburetor parts, engine

OXIDE CONVERSION COATINGS

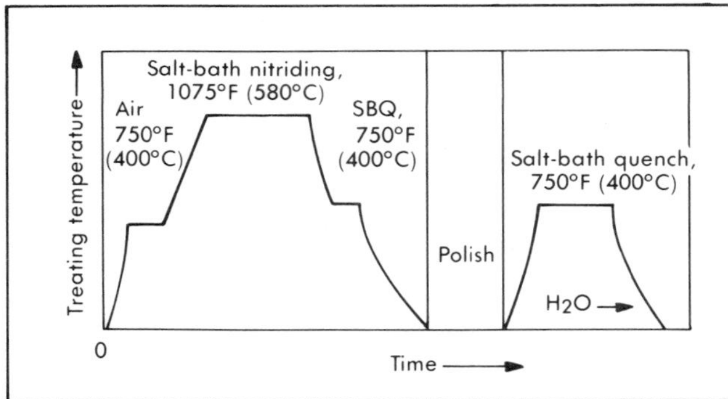

Fig. 19-1 Time-temperature cycle of the complete QPQ™ cycle. (*Kolene Corp.*)

valves, splined drive train couplings, and hydraulic components. Other applications include firearms, golf club heads, video recorder components, clamps and bolts, and camera parts.

Blackening Baths

Some blackening baths, especially those used for blackening iron and steel, are highly concentrated solutions of caustic soda. However, alkaline-chromate oxidizing processes and fused-salt oxidizing processes are also used for corrosion and abrasion-resistance applications on steel. The blueing or browning of steel also comes under the same category as blackening of steel. Table 19-6 lists the specifications for black oxide coatings.

Ferrous metals. The metal surface is attacked by caustic soda to produce metal hydroxide. Because of high temperatures and the presence of oxidizing agents such as nitrates, nitrites, or chlorates, the metal hydroxide is converted to oxides of varying oxygen contents. Some baths also contain activators such as cyanides, tannates, and tartrates that remove dissolved iron by complex formation.

The oxide coatings formed range from 0.00002 to 0.0002″ (0.5 to 5 μm) thick and have color hues of brown to gray to black. Oxide coatings formed on steel in alkaline solutions are not very resistant to corrosion. Oxide coatings are less porous than phosphate coatings but can serve as a suitable base for oil, wax, or paint. A black oxide coating possessing good corrosion resistance can be formed on stainless steel by immersion in fused sodium dichromate in the temperature range of 730 to 750° F (385 to 400° C).

Nonferrous metals. Some blackening processes produce a black coating on zinc and cadmium surfaces and are based on acidic compositions. Because of the nature of the base metals, these coatings are not abrasion resistant and provide very little corrosion protection. They are generally used for aesthetic and identification purposes. However, if they are subsequently coated with oils, waxes, or clear lacquer, their resistance to corrosion can be improved. Color of the coating ranges from dark brown to black.

The operating temperature and pH of these solutions are moderate and present little danger to workers. However, it is advisable to wear protective equipment for hands, face, and eyes in case of an accident. The effluent from these processes should be treated with lime or caustic solutions to precipitate and remove heavy metals before discharging.

PROCESSING

The complete process for chemical conversion treatment normally consists of the following steps: (1) cleaning, (2) water and/or conditioning rinsing, (3) treating with the coating solution, (4) rinsing, (5) post-treating or final rinsing, and (6) drying. In some cases the cleaning and conversion coating steps are combined. Depending on other factors, including the size and shape of the parts and the rates at which they are to be treated, additional cleaning and rinsing steps may be necessary. Typical treatment times for metal preparation are given in Table 19-7.

TABLE 19-6
Specifications Applying to Black Oxide Coating

Specification Number	Title
MIL-C-13924	Coating, Oxide, Black for Ferrous Metals
No. 57-0-2	Finishes, Protective for Iron and Steel Parts
MIL-HDBK-205	Phosphatizing and Black Oxide Coating of Ferrous Metals
AMS 2485	Black Oxide Treatment
MIL-F-495	Finish, Chemical, Black for Copper Alloys

Note: Only the basic specification or standard number has been given. It is therefore necessary to refer to the most current revision given by the specifying body when following these procedures.

Cleaning

It is important that the surface to be coated is clean and free of contaminants because contaminants can prevent the formation of a conversion coating. Physical characteristics of the conversion coating such as thickness, crystalline structure, and penetration of the base metal can be controlled by the method of cleaning. The choice of a cleaner will depend on the type and form of metal to be treated, the nature of soil present, the ease of soil removal, the completeness of the cleaning job required, and the type of coating desired. Strong alkali or acid cleaning produces a coarse crystalline coating and deeply etches the base metal, and it is used only in isolated applications.

Mechanical and chemical cleaning methods that may be used for conversion coating pretreatment are discussed in Chapter 16 and Chapter 18 respectively of this volume. Any of the mechanical cleaning methods are suitable for conversion coating pretreatment. However, caution should be exercised when selecting a chemical cleaner to prevent interference with the chemical reactions of the coating solution and the metal surface. The selected cleaner must also be compatible with the type of waste treatment used. In most cases, it is possible to select a chemical cleaning method that will properly condition the metal surface for conversion coating.

On some occasions, a smut may develop on the surface of parts made from aluminum, magnesium, copper, or brass during the cleaning operation. This smut may consist of oxides of copper and/or silicon that are present as alloying constituents. A pretreatment in a solution containing nitric acid, sulfuric acid, chromic acid, hydrofluoric acid, or a combination of these removes the smut and prepares the part for chromating.

When aluminum is treated, an alkali cleaner that will etch or dissolve the surface of the aluminum to remove nonuniform

TABLE 19-7
Typical Treatment Times for Conversion Coatings

Conversion Coating	Stage	Immersion, sec	Spray, sec
Iron phosphate (three stage)	Clean and phosphate	3-5 min.	60-120
	Warm water rinse	20-30	20-30
	Acidified rinse	20-30	20-30
	Drying		
Iron Phosphate (six stage)	Clean		60-120
	Warm water rinse		30
	Warm water rinse		30
	Iron phosphate		60-90
	Cold water rinse		30
	Acidified rinse		30
	Deionized water rinse (opt.)		15
	Drying		
Zinc phosphate	Clean	3-5 min.	60-90
	Warm water rinse	20-30	30
	Activating rinse	20-30	30
	Zinc phosphate	90-180	60-90
	Warm water rinse	20-30	20-30
	Acidified rinse	20-30	20-30
	Deionized water rinse (opt.)	15-30	15
	Drying		
Manganese phosphate	Clean	3-5 min.	
	Hot water rinse	20-30	
	Activating rinse		
	Manganese phosphate	10-30 min.	
	Cold water rinse	20-30	
	Acidified rinse	20-30	
	Drying		
Chromate (electroplated cadmium or zinc)	Plate		
	Cold water rinse		
	Acidified rinse (opt.)		
	Cold water rinse (opt.)		
	Chromate	15-60	
	Cold water rinse		
	Drying (opt.)	30-60	
	Cold water rinse		
	Drying		
	Lacquer application (opt.)		

oxides may be used. Following heavy etchant cleaners, it is often necessary to use an acid solution to remove the dark deposits that are not soluble in an alkali cleaner. An acidic chromate deoxidizes most aluminum alloys; nitric acid and hydrofluoric acid are used for high-silicon alloys. Alkali cleaners must be followed by a thorough rinse. Special conditioning materials based on colloidal titanium phosphate are used after alkali cleaners so that fine, hard, uniform phosphate coatings may be obtained. In some cases it is possible to blend the conditioner into an alkali cleaner.

Titanium activators are most necessary for producing consistent zinc phosphate coatings on steel or galvanized steel. A typical aqueous solution (pH 8.0-8.5) for pretreatment consists of 0.1 to 0.2% disodium phosphate and 0.001 to 0.005% titanium salt. The titanium phosphate compound is manufactured to form a colloidal dispersion in the pretreatment bath. Usually 10-15 seconds of contact with the bath is sufficient, although additional benefits can be obtained when treatment is extended to 45 seconds. The traces of colloidal titanium left on the metal surface activate the surface so that the time to form a phosphate coating is reduced and a fine, crystalline structure is obtained.

When producing chromate coatings on galvanized steel, the surfaces should be cleaned in a caustic solution or in a phosphoric acid solvent cleaner before the chromate dip. Zinc die castings are thoroughly cleaned in an appropriate cleaner, rinsed in cold water, and then dipped in a 1 to 2% sulfuric acid solution before chromating.

Water Rinsing

Whether used after the cleaning or after the conversion coating, the purpose of water rinsing is to remove the unreacted chemicals from the surface of the metal. Water rinsing minimizes the transfer of chemicals from one stage of the operation to another, prevents contamination of the baths that follow, and eliminates the possibility of undesirable chemical reactions. A thorough water rinse after alkali cleaning removes alkaline residue that would cause problems under paint; and after a hydrochloric acid pickle, the water rinse removes residual acid and iron salts that would interfere with subsequent operations.

The presence of soluble salts in most water can also cause problems. These salts can cause visible deposits that may detract from the appearance of the finished part. Chlorides and sulfates may cause premature corrosion of the finished part; and when paint is applied over parts with soluble salts dried on, blistering of the paint and loss of adhesion may result. To prevent such failures, it is important that these salts be absent from the water used in the final rinse or coating operation. In many locations it is desirable, if not necessary, to use deionized water in the final wet stage, and frequently it is desirable to use it in other stages to prevent any possibility of leaving soluble salts on part surfaces. A passivating final rinse prior to painting enhances the adhesion and corrosion resistance of painted metal.

Coating

Most conversion coatings are produced by immersing, spraying, brushing, or swabbing the work at temperatures from 70 to 350°F (20 to 175°C) and for processing times of 10 seconds to 30 minutes. Many of the solutions used contain wetting agents or detergents and are capable of removing soil while simultaneously coating the work. Processing solutions for treating steel that produce oxidation of ferrous to ferric ions are generally used as sprays for paint-bonding coatings, but may be used in immersion-type equipment.

Post-Treatment

When properly used, post-treatments enhance the quality of the final finish in terms of higher resistance to corrosion and humidity failures, and improved paint adhesion. Excess and unreacted coating chemicals on the workpieces after treatment are removed by cold water rinsing lasting 20-60 seconds. Then the articles are given a post-treatment with a dilute solution of chromic acid and dichromate solutions or a mixture of chromic and phosphoric acids. Solutions of hexavalent and trivalent chromium are also used. The use of a chromate treatment is not universal, and sometimes it is replaced by a second water rinse, preferably deionized water. The acid rinse may be applied hot, with temperatures ranging from 150°F (65°C) down to room temperature, depending upon the drying facilities available. The concentration of the chromate solution varies with the type of phosphate and chromate treatment and is in the range of 0.01 to 0.1% chromium trioxide. Thus, the concentration of the chemical is very low, and in most cases solution is left on the coated part. It is important to have just enough, and no more, of the final rinse chemical because the chemical accumulates at the lower edges of the coated part, at holes, and at seams. The pH of the final rinse should be adjusted to 3.5 to 5.0.

It has been theorized that the hexavalent chromium acts chiefly in passivating the metal exposed in the pores of the conversion coating. The chromate rinses have the ability to reduce the harmful effects of certain water salts, thus minimizing blistering under highly humid conditions.

A final rinse treatment based on hexavalent and trivalent chromium has been developed that may be rinsed off with deionized water without the loss of corrosion resistance to the final painted product. Before this development, corrosion resistance would fall drastically during the final rinse treatments with deionized water because the effect of chemicals in the rinse depended primarily upon mechanical deposition of the hexavalent chromium rather than reaction of the trivalent chromium with the conversion coating.

EQUIPMENT

The equipment required for conversion coating depends upon the method of application. Design of the equipment is dependent on the end use of the product, quality requirements, space availability, and other similar factors.

The most commonly used methods of coating are by immersion and spray. Workpiece handling is usually done by monorail, manual, automatic, and continuous strip methods. Immersion may also be performed in a barrel-type unit. On items too large for immersion, application by brushing steam or pressure spraying is used. The method of application chosen depends upon (1) the time for completing the process, (2) the size and shape of the work, and (3) the material handling method used. Some coating processes may require 30 minutes, while others may be completed in a matter of seconds.

Immersion Equipment

In the commercial processes that were first developed, as well as in modern operations in which a wide variety of objects are coated, immersion is used. Some shapes, such as tubes and long cylinders, can be treated by immersion only because of the inaccessibility of the various surfaces to a spray. Small tubular, cup-shaped, or irregularly shaped pieces can be immersed in the processing solutions in a tumbling barrel as required. In recent years, there has been a rapid increase in the treatment of automobile bodies by immersion.

For immersion, the equipment chosen depends upon the work handling method. These methods may vary from hand operations to more sophisticated indexing and computerized hoist systems. Solutions are held in gondola-shaped, open-topped or enclosed tanks arranged in the proper process sequence. The conveyor systems must include some way of lowering the work, which is carried on racks or in tumbling barrels, into the processing baths. The dimensions of the tanks depend upon the size of the rack or container for the work and the number of racks or units in the tank at one time.

On a continuously running conveyor, work is dipped successively into each tank and then drained between the stages. In another type of immersion system, used for treating wire or continuous metal strip, the work is held down in the processing baths by submerged rolls and carried over elevated rolls between stages. In these continuous immersion systems, the length of travel in the tank is determined by the speed of the conveyor or work and the time required in the different operational steps.

A more modern method of application for strip lines is the so-called reaction cell, which is essentially a modified type of immersion (see Fig. 19-2). The strip enters and exits a shallow container through slots. The processing solution from a holding tank is pumped into the compartmentalized trough and overflowed back into the working tank to maintain a head of ½ to 1½" (12-38 mm) above the strip. By proper valving, all or only part of the upper trough can be used for the application of a coating; and by proper valve positioning, the treatment time can be varied to suit line speed. Proper valving prevents overtreating or undertreating on lines that are varied widely in

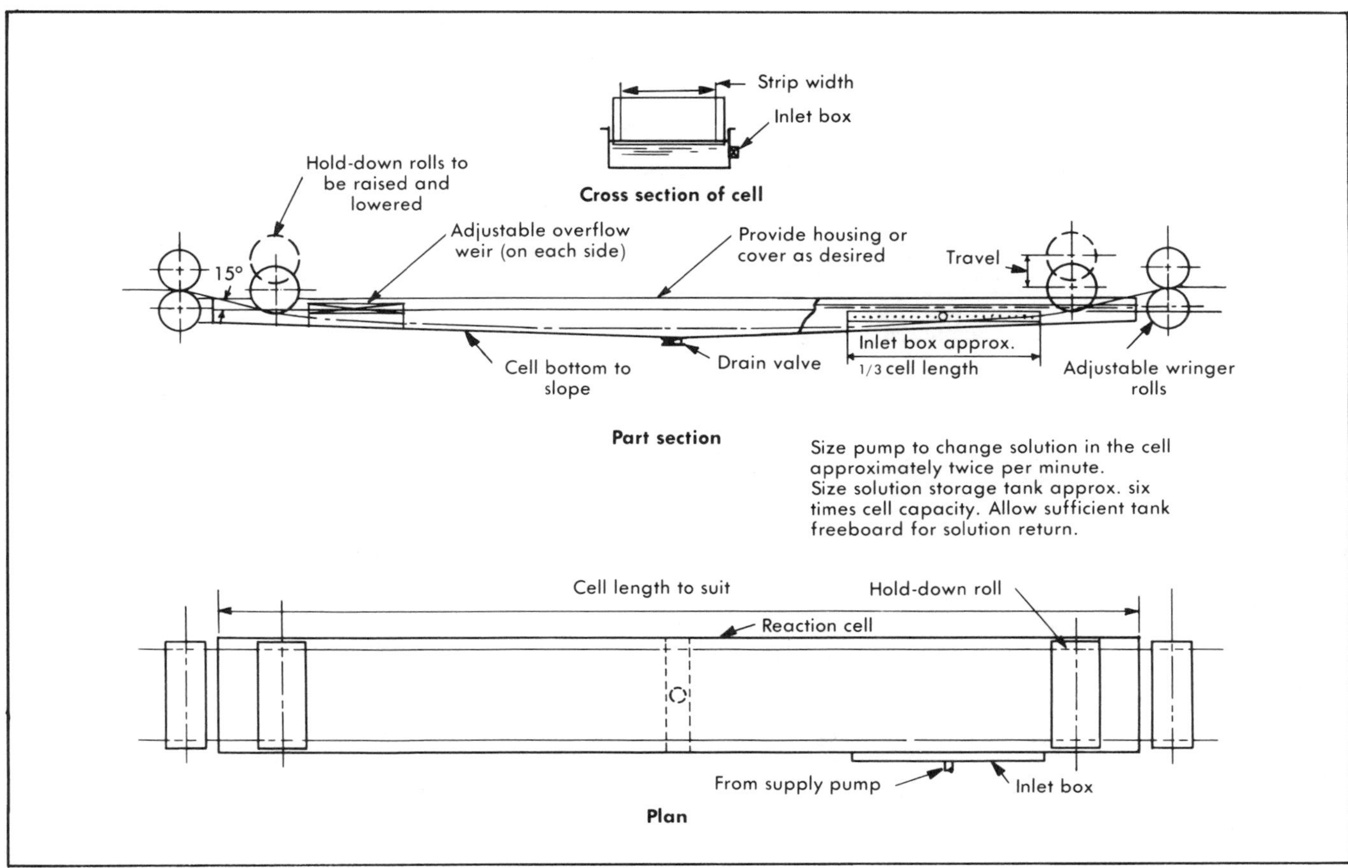

Fig. 19-2 Conversion coating reaction cell for strip lines. (*Parker Chemical Co.*)

speed and will result in greater product uniformity. Chromate coatings have been applied in this way to aluminum, galvanized steel, and cold rolled steel in as little as one second. This cell will also permit treatment of very thin gages that cannot be treated by conventional spray methods.

Immersion tanks are heated by steam coils, high-pressure hot water coils, gas-fired immersion or electric heaters. Gas-fired immersion heaters may be detrimental to phosphate baths because of localized heating. The types of materials used for various parts of the immersion system depend upon their corrosion resistance to the solutions. Recommendations of the chemical supplier should be obtained before specifying materials.

Sometimes a combination of immersion and spray systems is used, particularly for heavy phosphate coating applications that require immersion in the treatment stage but spraying in the cleaning and rinsing stages.

Spray Equipment

The spray technique has proved valuable in large-scale manufacturing operations that use conversion coatings. In most spray systems, the work is carried continuously through successive spray zones, sometimes referred to as spray washers, on overhead monorails, floor conveyors in which the work is carried on dollies, belt conveyors, and disk conveyors that are used particularly for flat sheets. For wire and metal strip, the work may be pulled through by a bridle and then recoiled. The lengths of various spray zones depend upon the processing time and the speed at which the work is traveling. Depending upon the process used, the time in each stage may vary from a few seconds to one or two minutes.

Conveyor. Figure 19-3 shows a typical layout of a spray washer for the cleaning, coating, and post-treatment of work to be carried on an overhead monorail conveyor. The monorail conveyor is a flexible chain that connects several hangers supported by rollers running on a beam. Splash guards protect the conveyor by reducing the amount of contact with the processing solutions, which may coat, corrode it, or wash away its lubricants.

Housing. The housing in which the spray operation takes place must be large enough to enclose the spray pipes and nozzles and provide clearance for the work. The work is hung on conveyor hangers and arranged so that it can be sprayed by stationary pipes and adequately drain between spray zones.

Note in Fig. 19-3 that the housing is a continuous piece, so ventilation is required only at both ends to prevent the escape of steam and fumes into the workroom. Exhaust from the housing should be avoided as much as possible. Access doors are provided in the various drain zones for servicing, and they may be opened to observe the spray in operation.

Spray piping and nozzles. Piping is generally arranged to support and feed the nozzles that spray the work. Spray nozzles are preferred over simple holes drilled in the pipe because they allow better control of spray pattern and capacity. Nozzles should be mounted so the spray is directed to all surfaces of the

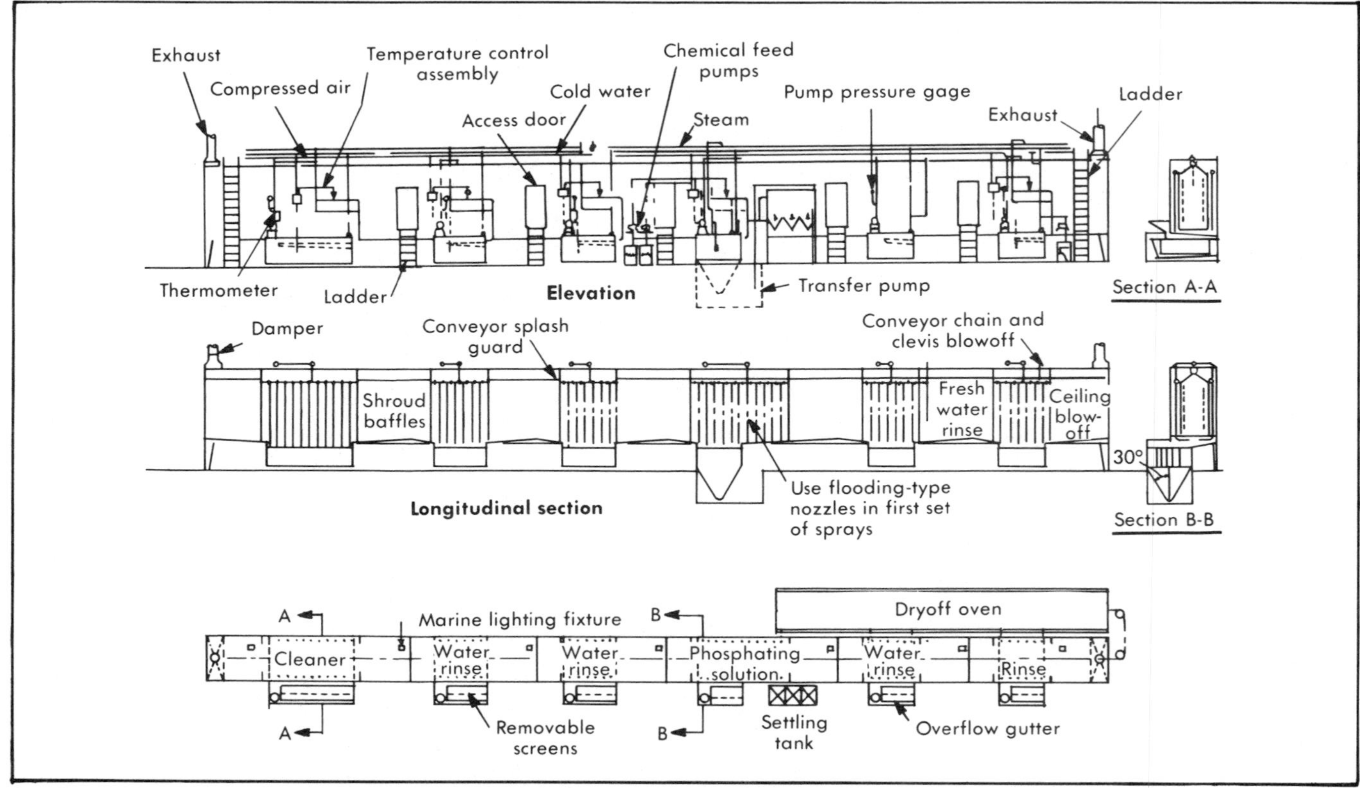

Fig. 19-3 Typical spray washer for cleaning, conversion coating, and post-treating of conveyed parts. (*Parker Chemical Co.*)

work. Nozzle pressure may vary from 5 to 20 psi (35 to 140 kPa) in most zones; however, in some designs higher pressures are used in the cleaning stage.

The most common nozzles used are flat-form, hollow-cone, wide angle, and low-impact types. The type of nozzle used depends upon the purpose of the particular spray zone. In the cleaning and rinsing stages (when mechanical scrubbing action is important), a solid stream or narrow angle, flat-fan spray is used to provide greater impingement. The impact of the spray helps to remove soil from the work. For the cleaning stage, flat-fan nozzles with a capacity of 5 gpm (19 L/min) at 20 psi (140 kPa) or greater are recommended because they have less tendency to clog. Nozzles with capacities less than 2 gpm (7.5 L/min) at 20 psi (140 kPa) should never be used unless the pump capacity is too small to support larger ones.

In the coating or treatment stage, hollow-cone nozzles are frequently used because they offer a large spray pattern with a minimum of nozzle restriction. The large opening reduces the tendency toward plugging by scale that may circulate with the solution. The nozzle with the largest orifice will stay open longer, and only stainless steel or plastic nozzles with a recommended capacity of 3½ gpm (13 L/min) at 20 psi (140 kPa) or greater should be used. However, if the work is difficult to spray, a flat-form nozzle may be preferred as in the cleaning stage.

Spray coverage is the most important function of phosphate washers, particularly in the phosphatizing stage. Coverage must be checked regularly because if it is inadequate the coating will suffer. Although there are several ways to clean nozzles (including the use of compressed air), it is best to replace nozzles on a regular schedule. In the phosphatizing stage, for example, at least 25% of the nozzles should be replaced each week. In the post-treating stage, wide angle, low-impact nozzles are used since excessive pressures or impact may tend to remove the coating that has been formed.

Pumps. Pumps circulate the solutions from the reservoir tank to the spray pipes and nozzles. Centrifugal pumps are generally used because of their dependability and flexibility in operation; they may be easily throttled if lower spray pressures are desired. Because shaft leakage is a frequent problem with centrifugal pumps, vertical, submerged types are often used.

Tank. The solutions are heated in tanks and maintained at the desired spray temperature. The method of heating is similar to that described for immersion tanks; however, for a given size of tank, the quantity of heat required is considerably greater because of some heat loss in spraying. Heat exchangers may be located outside the tank to heat the solution that is recirculating or flowing to the spray nozzles.

Dryer. A drying oven is the last piece of equipment in the post-treatment stage. The length or duration of drying time varies, but usually 3-5 minutes in circulating air from room temperature to 400° F (200° C) is adequate.

Sheet and Strip Equipment

Figure 19-4 shows a typical unit for the cleaning and conversion coating of continuous strip metal. Several components in Fig. 19-4 are similar to those described above; but because the processes used are much faster and greater strip

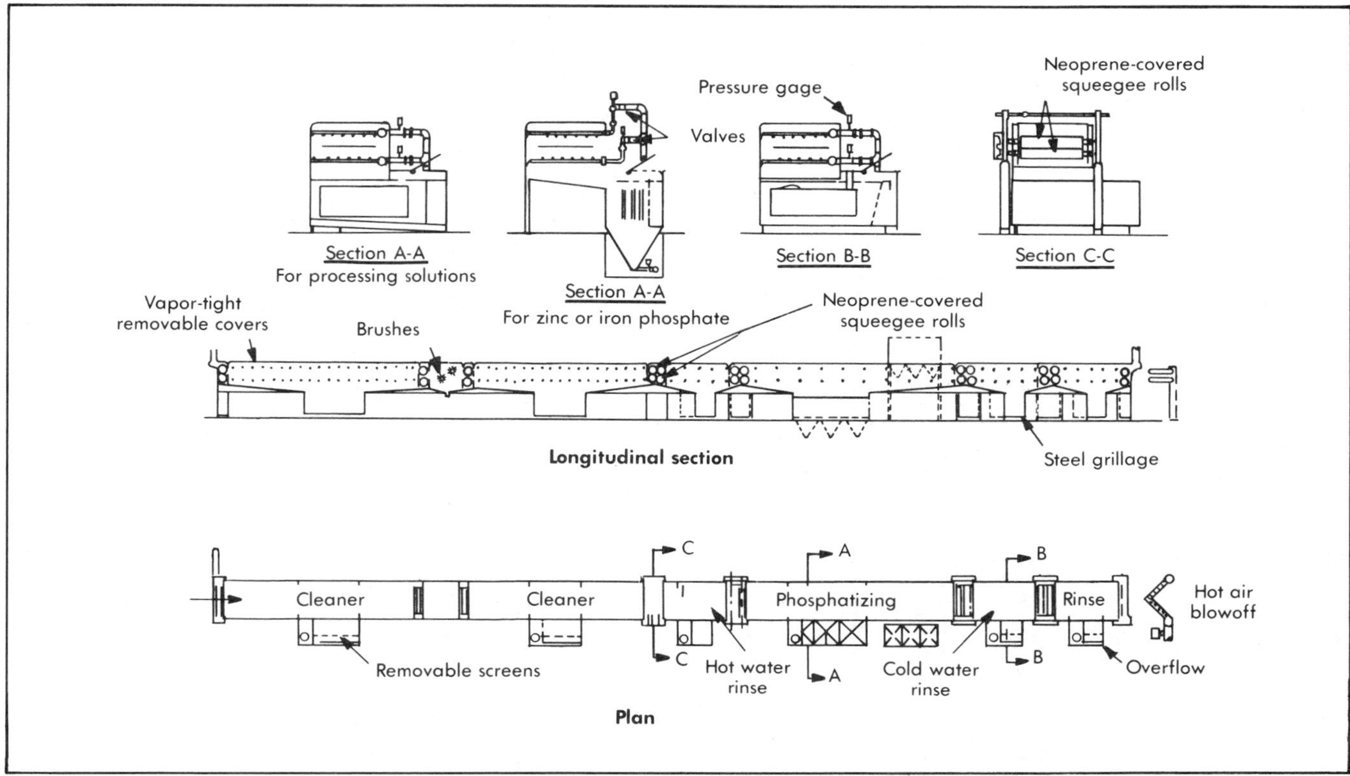

Fig. 19-4 Typical spray equipment for phosphate treatment of continuous steel strip at 200 fpm (60 m/min). (*Parker Chemical Co.*)

speeds are permitted, the spray zones may be longer. Because of the uniformity of the work going through, the nozzles can be placed for more efficient solution use.

Another feature of this unit is the presence of squeegee rolls between several spray stages to reduce carryout and overspray from one solution to the next; the squeegee rolls also reduce the drain zones between stages to a minimum. Power brushes are sometimes installed for more effective cleaning of the work as it passes through.

With this unit, dryoff equipment can be simpler since most of the final solution is removed by the squeegee rolls. Drying continuous strip is generally accomplished by blowing hot air at high velocity over the surfaces of the work.

Units designed for handling flat sheet are similar to those for continuous strip. The work, however, is carried on a disk conveyor consisting of rows of disks, mounted on driven shafts, that upon rotation carry the sheets along on top of them. The squeegee rolls must be driven at peripheral speeds comparable with those of the disks.

Automatic Control

Completely automatic electronic units for controlling the solutions used in conversion coating systems before painting, lamination, lubrication, and similar processes have been developed to add chemicals or water automatically as required. These automatic controlling and recording devices are essentially based on the electrical conductivity, pH, and redox potentials of solutions. Housed in a compact cabin, each unit is equipped with a controller/recorder, switches, and a series of visual indicators. From this single control point, the system (1) electronically monitors the chemical concentrations of all solutions, (2) energizes pumps for the controlled addition of chemicals in the amounts required, (3) compensates for temperature fluctuations, and (4) records production line conditions for future reference. Figure 19-5 is a schematic drawing of such a unit.

General Maintenance

In the operation of some conversion coating processes, insoluble residues are formed as by-products of the chemical reactions. These materials settle to the bottom of the tanks and should be removed regularly before they cause dusty coatings or interfere with the operation of spraying systems. Acidic phosphatizing baths, with or without accelerators, form a sludge of insoluble phosphates that must be removed if the phosphatizing solution is used for extended periods. The sludge can be removed by gravity through sludge settling tanks, by centrifuge, or by means of a conical, cyclone-type of liquid solid separator.

When a phosphate solution has been in service for some time, scale may form on the heating unit. This scale acts as an insulator and must be removed at regular intervals to ensure adequate heat transfer and the proper processing temperature. Scale buildup can be determined by checking the stack temperature; temperatures exceeding 700°F (370°C) indicate that an excessive amount of scale has formed on the immersion tubes and descaling is necessary. [Normal stack temperatures are usually 300-400°F (150-200°C).] Descaling can be performed with the heat exchanger either in the tank or out of the tank. The scale may then be removed by a suitable chemical or mechanical method.

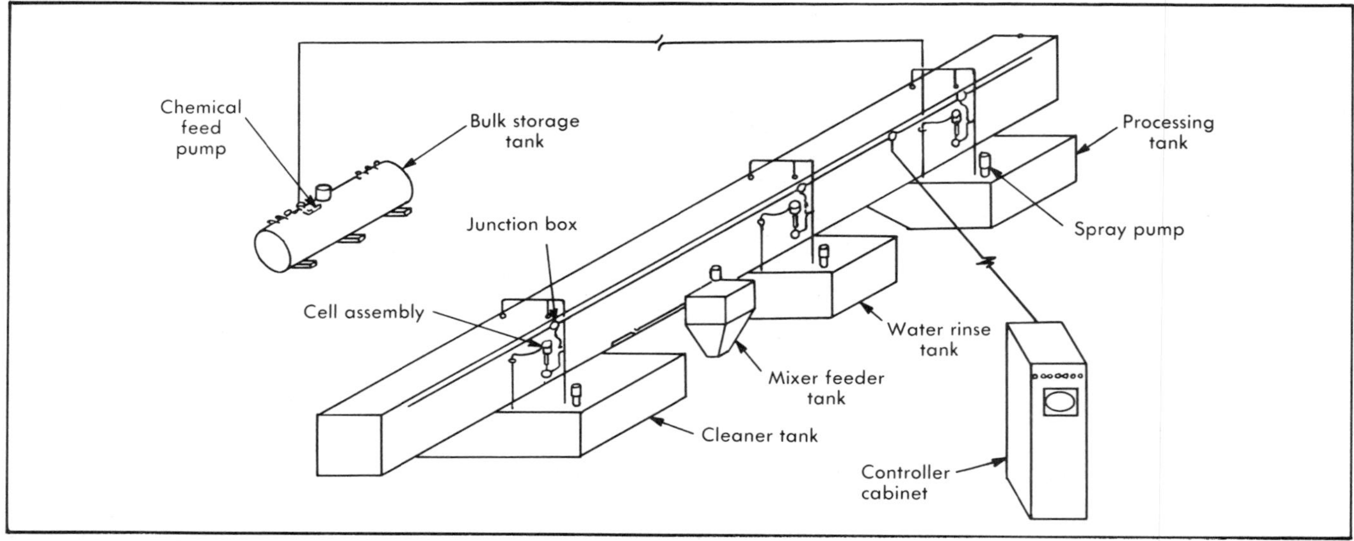

Fig. 19-5 Automatically controlled conversion coating system. (*Parker Chemical Co.*)

ANODIZING

Anodizing is the common commercial term for the electrolytic treatment of metals that forms stable films or coatings on the metal's surface. Aluminum and magnesium are anodized to the greatest extent on a commercial basis, but other metals such as zinc, beryllium, titanium, zirconium, and thorium also respond to anodic treatment to form films of varying thicknesses. The coatings on these metals are used primarily for decorative purposes, but anodic coatings on aluminum alloys are used for functional and decorative purposes or some combination of the two.

Anodizing differs from electroplating in two ways. In electroplating, the work is made the cathode, and the metallic coatings are deposited on the work. In anodizing, the work is made the anode, and its surface is converted to a form of its oxide that is integral with the metal substrate.[11] The metallic portion of the electrolytic container is made the cathode.

THEORY

Extensive work has been done on the mechanism of the oxide layer formation during the anodizing process. The theories developed are not in universal agreement, with the manner of oxide formation being complex and not fully understood. It is generally agreed that the oxide coating consists of two layers, with the initial layer forming against the surface of the part being anodized. This layer is termed the barrier layer, and it has distinct properties when compared with the succeeding layer. The barrier layer is relatively thin, very dense, nonporous, and may contain two forms of aluminum oxide, one crystalline and the other amorphous. The barrier layer thickness, cell diameter, and pore structure are dependent on the formation voltage and the nature of the electrolyte. The thickness has been determined to be 13-14 Angstroms (Å) (1.3-1.4 nm) per volt of applied formation voltage.

The outer, heavier layer of the anodic coating is assumed to be more porous with the pores stacked somewhat like parallel tubes extending through the layer from the surface nearly down to the barrier layer. It is important to recognize that, unlike plating where the metal ion travels from the bath to the part and builds up the plated layer by adding more plate, the anodic layer is always being formed from the underlying aluminum surface and forcing the existing layer outward. The oxide last formed is always between the metal surface and the previously formed oxide. The oxygen ion from the electrolyte, under the influence of the impressed voltage, travels through the pore channels to reach the aluminum-ion-rich area at or near the base of the barrier layer.

The major conflict between theories concerns the reaction of the barrier layer. Whether this layer has a rectification effect and allows the passage of the aluminum ion outward only, whether the barrier layer breaks down at weak points as the voltage is raised, or whether the layer is constantly breaking up and reforming, the sequence is not generally agreed upon. As the mechanism of the oxygen ion traveling through the pore structure toward the barrier layer continues, the scrubbing action on the pores near the outer part of the layer tends to develop a funneling of the pore channels, which then become larger near the surface and taper smaller inward. This tapering effect is also accelerated by the concurrent dissolution action of the electrolyte dissolving the surface of the oxide layer and reducing the thickness of the pore walls near the surface. As the thickness of the anodic film increases, the impressed voltage needed to sustain the action increases. The final thickness is determined by the terminal voltage, temperature, solution, time, and other operating variables.

Characteristics of Anodic Coatings

The first oxide layer formed is located at the interface of the anodic coating and the aluminum surface. Subsequent oxide formation is more porous than the coating formed underneath because it is in contact with the solvent action of the electrolyte longer. The type of electrolyte, its temperature, and the duration of the treatment determine the amount of solvent action. The porosity of the anodic coating located between the barrier layer and the outer surface of the coating is one of its outstanding features.

Anodic coatings have a definite cellular structure, and the size of the cell is determined by the type of the electrolyte and the formation voltage. There are millions of cells per square inch, with a pore in the center of each cell. The model shown in Fig. 19-6 represents the present concept of the cell and pore structure.[12] In this figure, "D" is the linear size of a cell and "a" is the pore diameter. Typical pore diameters range from 60 to 330 Å (6 to 33 nm), depending on the type of the electrolyte used.

The weight of anodic coating per unit area formed is dependent upon the thickness and porosity of the coating. The overall thickness of an anodic coating is generally determined by the total ampere minutes of current used during the anodic oxidation cycle. Figure 19-7 shows the relationship between solution temperature, anodic oxidation time, and coating weight on aluminum alloy 1100 in a 15% (by weight) sulfuric acid solution at different temperatures using a current density of 12 A/ft^2 (129 A/m^2).

By selecting the proper combination of electrolyte and operating conditions, anodic coatings with definite predetermined characteristics can be formed. The term *coating ratio* is an expression for overall anode efficiency with respect to anodic coating formation. This ratio is obtained by dividing the weight of the unsealed anodic coating by the weight of metal removed during the formation of the coating. For example, the theoretically perfect coating ratio for pure aluminum is 1.889, based on all the aluminum reacting electrochemically and being converted to aluminum oxide. A coating ratio lower than 1.889 indicates that the aluminum oxide has been dissolved by the electrolyte. Figure 19-8 indicates the coating ratio with varying electrolyte temperatures and concentration using a current density of 12 A/ft^2 (129 A/m^2) for 30 minutes on 99.9% pure aluminum, with a sulfuric acid solution as electrolyte. (Note the increase in coating ratio when the electrolyte temperature or electrolyte concentration is lowered.)

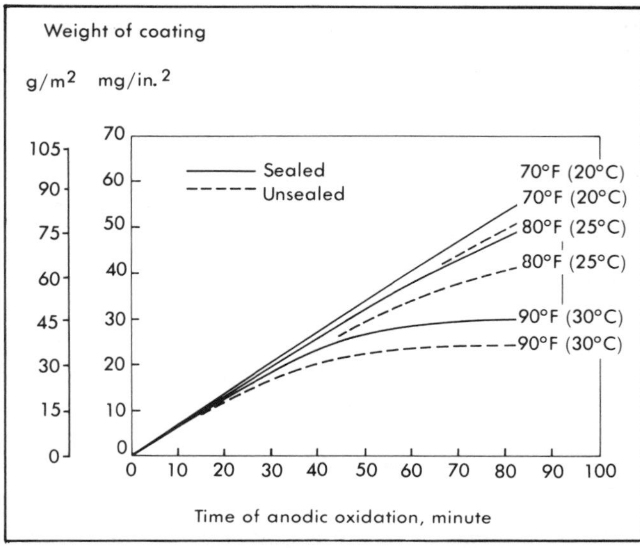

Fig. 19-7 Relationship between time of anodic oxidation and unsealed anodic coatings at various electrolyte temperatures. (*Aluminum Company of America*)

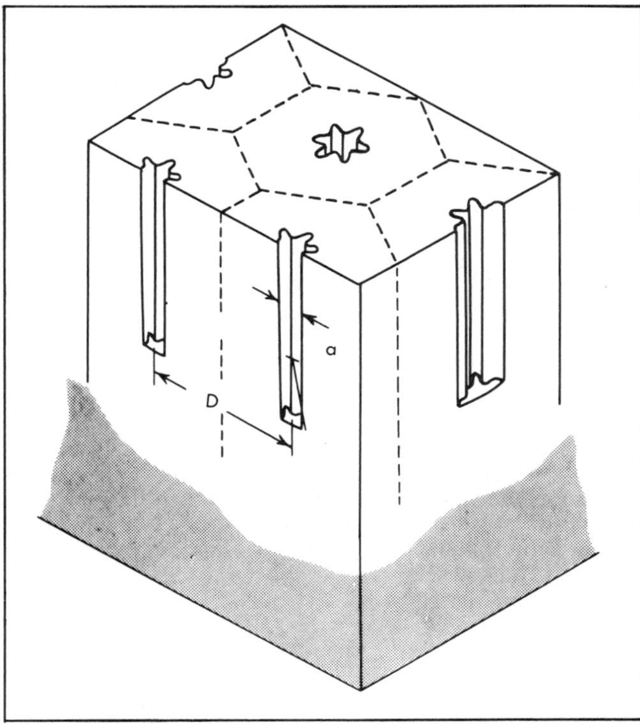

Fig. 19-6 Cell and pore structure of an anodic coating produced in a phosphoric acid electrolyte. Magnification is 65,000 x 10^3.[12] (*Aluminum Company of America*)

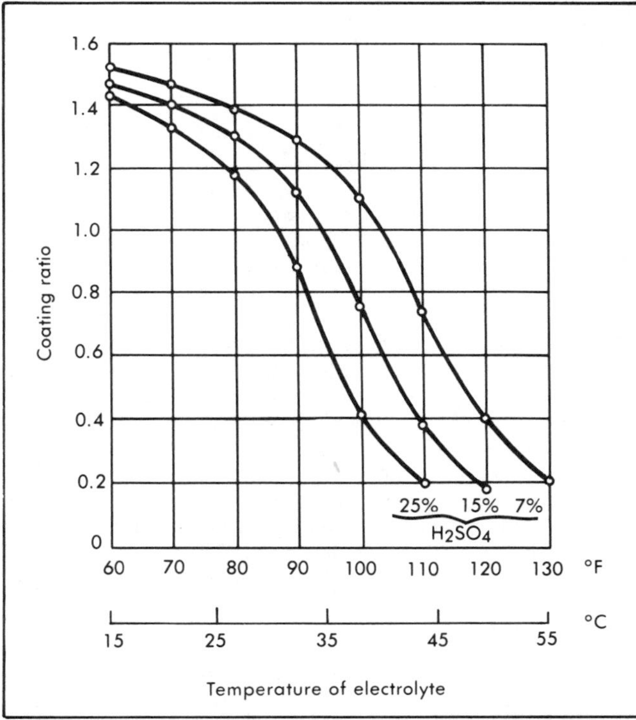

Fig. 19-8 Relationship between coating ratio and temperature of sulfuric acid electrolyte at various concentrations. (*Aluminum Company of America*)

COATINGS PRODUCED

Anodic coatings may be thin and dense or thick and porous. The degree of porosity determines the absorption characteristics of the coating. Denser coatings tend to have better abrasion resistance. Sealing the anodic coating can reduce the abrasion resistance up to 30%.

Alloy Effects

One of the most important factors that influences the oxide coatings produced by anodizing is the composition of the aluminum alloy. Therefore, any discussion of anodizing must include the response of the various aluminum alloys to the process.

Undissolved constituents can cause voids and disruptions in the coating. Some constituents of aluminum alloys will themselves oxidize under the conditions of anodizing, and the resultant oxidation products will color the coating. Manganese produces a brown opaque appearance from the presence of manganese dioxide. Chromium gives a yellow tint because the coating contains oxidation products of chromium.

Reaction of all the various possible alloying or impurity constituents is complex and, in many cases, not entirely understood; however, the alloy composition should be recognized as an important factor in determining the characteristics of the oxide coating. The purer the aluminum, the more continuous and transparent is the anodic coating formed. The so-called "super-purity" (99.99+ percent) aluminum produces the most transparent oxide coating. Matching appearances in assemblies comprised of various alloys can be a problem, and care must be used in selecting alloys for assemblies if a uniform overall appearance is desired. Castings and wrought products in the same assembly are undesirable for an appearance match; although with careful selection of alloys, tempers, and finishing procedures, the mismatch can be minimized. Since anodic coatings reproduce the surface on which they are formed, surfaces that have been textured either mechanically or by etching will, in many cases, assist in solving the matching problem for uniform overall appearance.

COATINGS PRODUCED

Many variations of anodizing have been developed to produce coatings for a wide range of applications. Anodic coatings produced in sulfuric acid electrolyte are generally transparent, and a wide range of properties may be obtained by varying operating techniques. Coating thicknesses range from 0.0001 to 0.003″ (2.5 to 75 μm), depending on the application. The coatings produced in a chromic acid electrolyte have an opaque, slightly iridescent appearance and are much thinner than coatings produced in sulfuric acid solutions, usually around 0.0001″ (25 μm) thick. Since the chromic acid coatings are thin, they also have a low resistance to abrasion, a high degree of flexibility, and are excellent as a base for paint, enamel, or lacquer. Anodic oxidation in oxalic acid electrolytes produces coatings that are essentially transparent but can vary in color from a light yellow to bronze. The thicknesses of these coatings are usually 0.0004-0.0015″ (10-40 μm).

Dyeing and Sealing

Anodic films are capable of being colored using a wide variety of organic dyestuff. Dense films do not dye as readily or as well as the more porous films; thus, a film formed in chromic acid is more difficult to dye than one formed in sulfuric acid. The clarity of the anodized layer is important since the brilliance and depth of color depend largely upon the ability of the clear film to reflect the underlying bright aluminum surface. Chromic anodize films, which by nature are pearly or procelain appearing, will have a muted or somewhat flat appearance after dyeing. Dyeing hard coatings black is satisfactory and results in a richer or deeper color than the natural black that can be obtained in the process.

Sealing treatments are generally used for porous anodic coatings. Barrier film coatings require no sealing treatment because they are essentially nonporous. Sealing processes make the coatings nonabsorptive and are generally performed by immersion in boiling deionized water, sodium bichromate, or nickel acetate solutions, or steam. Deionized water makeup is preferred for all sealing solutions because certain anions such as phosphates, silicates, or chlorides that retard the sealing or hydration reaction are removed. Although the exact mechanism of sealing is not completely understood, it is generally recognized that the anhydrous aluminum in the outer layers of the anodic film reacts with the water and hydrolizes the oxide to the monohydrate form. Since the monohydrate occupies a greater volume than the alumina from which it was formed, the reaction tends to close down and plug the pore structure. The dye intake and the sealing effect do not penetrate to the full depth of the anodic coating.

In addition to the use of dyes, it is also possible to treat the part immediately after anodizing and washing by immersion in other solutions for specific effects. Immersing in a Teflon dispersion results in some of the Teflon particles lodging in the pore structure. Immersing in oils may also improve the lubricity, and treatment in waxes and other materials improves the dielectric strength.

Another common treatment, especially for military and marine applications, is sealing in a 5% solution of sodium dichromate operating at 178 to 212° F (80 to 100° C) for ½ hour. The treatment deposits chromates in the pores and improves corrosion resistance. Duplex sealing involves a double seal: for example, a sealing in plain water at 160° F (70° C) followed by a sealing in the nickel acetate solution, or sealing in a dichromate solution followed by the nickel acetate seal.

It is general practice to seal all chromic and sulfuric anodized parts, whether they are dyed or not, to improve the general corrosion resistance and to prevent staining or absorption during service. Hard anodized parts are usually not sealed since this operation can reduce the abrasion resistance up to 30%. Typical operating conditions for some of the commercial sealing processes are listed in Table 19-8.

Integral and Electrolytic Color Anodizing

In the late 1950s, a new use of anodizing was introduced based in part on the observation that integral colors could be produced when hard coatings were applied to certain alloys used largely in architectural applications. As the coating thickness increased from 0.001 to 0.004″ (25 to 100 μm), the corresponding color changed from a silver gray to a bronze, deep bronze, and finally at 0.004″ (100 μm) to a blackish color. Since the color is inherent in the coating and is a function of the alloying elements and thickness of the coating, it has good permanency, fade and corrosion resistance, and durability. Later developments using different electrolytes enabled room temperature operations and reduced refrigeration costs. When bronze was used in building facades and curtain walls, integral-color anodized aluminum became a direct replacement.

In production, it was discovered that small amounts of impurities, especially iron, caused shading in parts run in the

TABLE 19-8
Typical Sealing Procedures and Operating Conditions for Anodic Coatings

Type	Concentration	Temperature, °F (°C)	pH	Time, min.	Remarks
Water		208-212 (98-100)	5.5-6.5	10-15	Use pure water (deionized) for best results.
Nickel acetate	5 g/L	208-212 (98-100)	5.5-5.8	5-10	Used for sealing anodic coatings colored with organic dyes.*
Nickel-cobalt acetate	Nickel acetate 5 g/L, cobalt acetate 1 g/L	208-212 (98-100)	5.0-6.0	5-10	Used for sealing anodic coatings colored with organic dyes.*
Potassium dichromate	50 g/L	208-212 (98-100)	5.0-6.0	15	Used for sealing all types of anodic coatings for maximum resistance to corrosion.
Phosphate chromic acid	20 g/L of KH_2PO_4, 2 g/L of CrO_3	170-180 (77-82)	5.0-6.0	Two times anodic oxidation time	Used for sealing anodized reflectors to minimize bloom.
Chromic acid	0.01 g/L	208-212 (98-100)	5.6-6.4	10	Used for sealing anodic coating formed in chromic acid electrolyte.
Potassium dichromate	0.03%	170-180 (77-82)	5.0-6.0	1 1/2 times anodic oxidation time	Used for sealing reflectors in coastal or marine service.

* Wetting agents and boric acid sometimes added to decrease sealing smudge and buffer solution.

same processing load but originating from different mill melt lots. This observation led to the use of specially controlled analysis melts of aluminum alloys that were designated for use in integral-color anodizing applications. Some of the trade names licensed for these processes are Duranodic-300 (Aluminum Company of America), Permalux (Reynolds Metal Co.), Veroxal (Vereinigte Aluminum Werke, Germany), and Permanodic (Kawneer Co.).

Recently, superior results have been obtained by a two-step process that originated in Europe and developed in Japan. The first step is the application of a conventional but heavy, clear anodized coating, usually sulfuric, in the order of 0.0007 to 0.001″ (18 to 25 μm) thick. The load is then transferred to a second tank containing any one of a number of aqueous solutions of metallic salts such as nickel, cobalt, tin, cadmium, silver, selenium, or tellurium. When single-phase alternating current is applied to the bath, color is obtained by the electrolytic deposition of chemically colored metal oxide deep into the anodic film. The colors attained in this process are shades of gray, bronze, gold, black, and red. The color imparted is independent of the coating thickness and can be accurately controlled by process instrumentation. Color can also be checked during processing by partially lifting the load from the tank and comparing the color obtained with a wetted color-match panel. If the color is too light, the load is lowered back into the tank, and added cycle time is applied. The two-step electrolytic process is capable of producing even color matches, which are necessary when building panels are assembled side by side. Some of the common trade names used in these two-step processes are Anolok (Aluminum Company of Canada), Colinol-2000 (Swiss Aluminum Limited), Reynolds Tru-Color (Reynolds Metal Co.), Eurocolor-800 (Pechiney CIE, France), and P3-Amecolor (Henkel-Amchem Products).

HARD COATING

Hard coating, often referred to as hard anodizing, is a relatively new process that was introduced in the United States in early 1952. The process is similar to conventional sulfuric anodizing but with reduced temperatures, higher unit amperages, and higher final voltages. These finishes are known commercially as Alumilite (Aluminum Company of America) hard coatings, Martin hard coatings, Sanford hard coatings, or Hardas coatings. Hard coatings have high resistance to abrasion, erosion, and corrosion, and can have thicknesses ranging from 0.001 to 0.012 (25 to 300 μm); the thicknesses of conventional chromic or sulfuric anodized finishes are in the range of 0.0001 to 0.0010″ (2.5 to 25 μm), depending on the application. In practice, however, most hard coatings are 0.002-0.004″ (25-200 μm) thick. This type of finish is becoming increasingly popular for parts requiring light weight in combination with high resistance to wear, erosion, and corrosion. Hard coating has also been used to salvage parts that have oversized dimensions.

The weights of anodic coatings on various wrought alloys formed by the conventional Alumilite procedure and the Alumilite hard coating procedure are compared in Fig. 19-9. Figure 19-10 is a comparison of the wear resistance of anodized hard coatings to that of other hard materials.[13]

Design Considerations

With the introduction of heavier anodic coatings, it becomes important to recognize the growth aspect involved in the formation of anodic coatings. As the aluminum oxide is being formed by dissolving aluminum from the surface of the part and combining with the oxygen from the bath, it both penetrates beyond the original surface and builds up above the original surface. While the theoretical growth is approximately equal to

HARD COATING

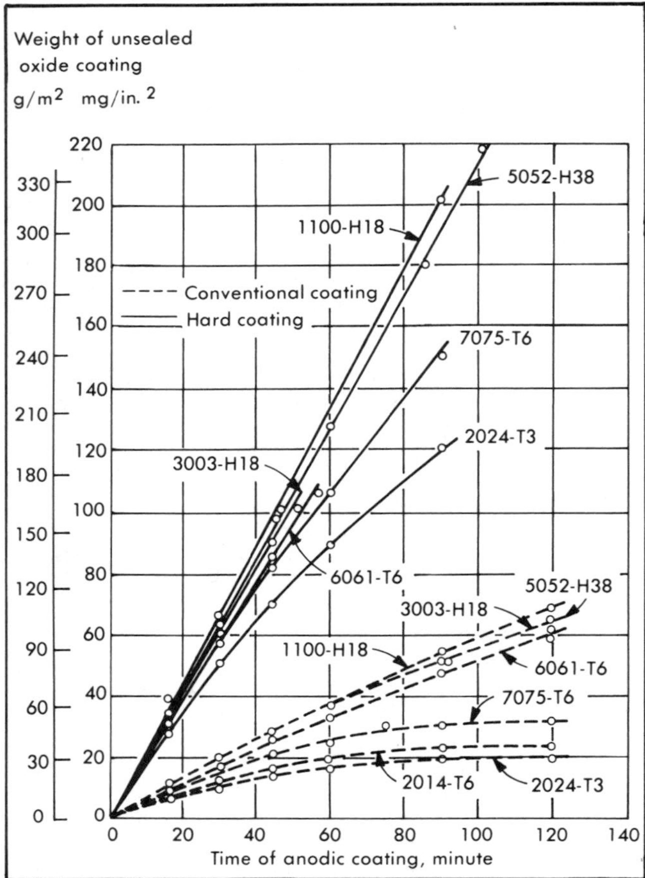

Fig. 19-9 Relationship between weight of anodic coatings produced under conventional conditions and hard anodic coatings produced at higher current densities and lower electrolyte temperatures. (*Aluminum Company of America*)

one half of the total coating thickness, the growth portion of the coating is reduced by the solvent action of the electrolyte; the solvent action, in turn, is dependent upon the temperature, strength of the electrolyte, and exposure time in the tank. In chromic anodizing, the film is so thin that the change in dimension is insignificant. In sulfuric anodizing, the coating growth is approximately 30% of the total coating thickness and is a consideration on closely machined parts. In hard coating, because of the refrigerated electrolyte, the coating growth is approximately 50% of the coating thickness; in many applications this becomes a design factor, especially on threads, dowel holes, close tolerance bores, and bearing diameters. A hard coating with a coating thickness of 0.002″ (50 μm), including buildup and penetration of the coating, would penetrate 0.001″ (25 μm) and build up 0.001″ (25 μm).

Another factor to consider in hard coating applications is that the buildup of the coating is perpendicular to the surface; at sharp edges, a corner effect results in a gap in the coating, producing areas that are prone to chipping, especially as the coating thickness increases. If the corner is well radiused, the coating will approximate the curvature. A recommended radius for a 0.001″ (25 μm) thick coating is 1/32″ (0.8 mm); for a 0.002″ (50 μm) coating, 1/16″ (1.6 mm); and for a 0.003″ (75 μm) coating, 3/32″ (2.4 mm).

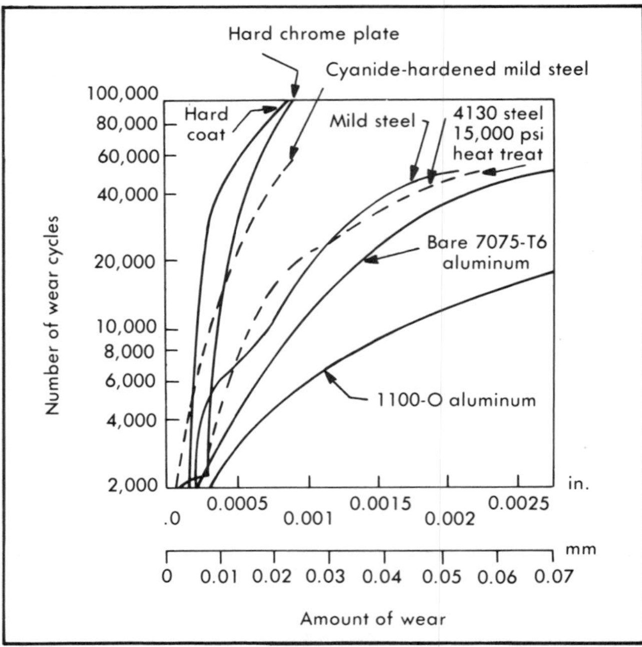

Fig. 19-10 Comparison of abrasion and wear-resistance tests of hard materials on a Taber Abrasiometer using CS-17 wheels and having a 1000 g load.[13]

Finishing Recommendations

As the hard coating builds up, the surface roughness increases considerably. However, it is possible to hone, lap, polish, or grind the coating to obtain a surface finish under five root mean square average (rms).

After applying a 0.002″ (50 μm) thick hard coat on a 6061 test piece lapped to a 3 rms, the rms increased to 60. However, lapping off 0.0003″ (7.5 μm) of the hard coat restored the surface to the original 3 rms finish. If a final rms of 16, for example, on a cylinder bore were to be required, it is not necessary or practical to go finer than a 16 or even a 20 rms in the precoat finishing operation. In the hard coat operation, the final roughness will increase to 60 rms or better and will necessitate a final sizing operation to obtain the 16 rms required. In production, where an rms finish under 50 is required, it is preferable to incorporate a final shop operation after hard coating to obtain the finish required.

Masking Considerations

In hard coating, since the temperature of the process is lower, it is easier to mask and selectively hard coat. For example, with an hydraulic cylinder, where the end use required a hardened bore but where other areas on the part had close tolerances and needed only corrosion protection or a paint-based film, by selectively masking, the bore could be stopped off, and the entire part conventionally anodized; next, the masking would be reversed to expose the bore for hard coating, while masking the balance of the part to protect the first anodized coating.

A variation of this procedure would use the chromic-anodized coating as a mask. In this case, the bore would be left undersized, and the entire part would be chromic anodized and well sealed. The bore would then be machined to remove the anodized coating and to obtain the precoat size necessary to accommodate the hard coat. The coating would then be applied by immersing the entire part in the electrolyte. Only the bare bore would accept the hard coat since the chromic film in the other areas would act like a maskant much in the same way as a coat of lacquer.

The advantage of the latter method is the elimination of the hand masking work and the neatness of the final result. The disadvantage is the double handling involved in shipping out for chromic anodizing, back for machining, and out again for the hard coat operation. In most cases, the double handling proves to be more desirable from a cost and end-quality standpoint. With average or small parts, the cost of masking operations usually exceeds the cost of coating the entire part; therefore, for many parts, especially where dimensions are not critical and where anodic film is either required or optional on the remaining surfaces, it is less costly to hard coat the entire part.

Post-Treatments

Hard anodic coatings are also used in conjunction with certain supplemental films to lower the coefficient of friction or to produce good release properties. After hard coating, the parts are immersed in a Teflon dispersion, removed, and then allowed to dry. The parts may be used either as dried, or they may be put through a sealing solution. The Teflon-lubricated surface is useful for break-in purposes. At present, the most popular processes are Anolube (International Hardcoat, Inc.), Tufram (General Magnaplate Corp.), Nituff (Nimet Industries), and Sandford Hardlube (Sanford Process Corp.). Another post-treatment similar to the processes previously mentioned is TEF-LOK (Forestik Plating and Manufacturing Co.), which mechanically locks some of the Teflon into the coating. These thick coatings with maximum density can be obtained on most aluminum alloys; however, the selection of alloy and temper must be given careful consideration.

APPLICATIONS

Anodizing is used for many purposes, usually with the application taking advantage of one or more of the specific properties of the anodic coating. The corrosion resistance of a painted part is largely derived from the paint system; however, paint adheres poorly to bare aluminum, and the tightly adherent anodized film acts as an intermediary for accepting the paint and binding it to the part. The anodic layer also provides a portion of the overall corrosion-resisting properties of the system. In aircraft applications, parts are usually chromic anodized, with the exterior parts given a primer and several top coats.

One limited use of chromic anodizing is based on the extraordinary penetrating ability of the chromic acid. The part to be inspected is anodized and lightly surface washed, then it is allowed to set for a period of time. If a crack is present, the chromic acid may bleed out of the crack and appear as a yellowish stain against the gray chromic-anodized background.

Commercially, sulfuric-anodized and dyed parts are used in automotive trim, cosmetic cases, aerosol caps, keys, picture frames, nameplates, and sporting goods, among other applications. The integral-color anodizing and variations of this process account for a large use of aluminum in outdoor architectural and building applications. Hard coating is increasing in usage mainly because of its sliding wear and corrosion-resistant properties. Although hard coating has a DPH of 400 to 600 hardness, the relatively thin layer does not add to the bearing strength of the underlying aluminum. Also, like most hard surfacing, hard coating reduces fatigue strength up to 50%, depending upon the alloy, thickness of coating, and the stress levels involved. In aircraft applications, where fatigue stresses are a design consideration, the hard coat is kept out of radii and other notch-sensitive areas.

Typical applications for hard coating include hydraulic cylinder bores, gun parts, and ordnance and missile components. Hard coatings are also used in laser targets, paper mill rolls, fire hose connectors, heat sinks, coarse screw threads, surveying instruments, aluminum gears, cams, torpedo hulls, textile bobbins, pots and pans, sailboat masks, davits, and pulleys.

Anodic coatings also have insulative properties. A breakdown voltage of 500V for a 0.001″ to 0.002″ (25 to 50 μm) thick coating when measured ground to surface is normal expectancy. The shape of the part is a factor in measuring breakdown voltage. On a flat surface, the results may be reasonably consistent; but around holes, threads, and edges, the breakdown voltage will be lower. The increase in breakdown voltage is not directly linear with anodic coating thickness. The coating thickness has a direct correlation to corrosion resistance, with tests indicating the coating thickness is the most important single factor in corrosion resistance of anodic films.

COATING PROCESS

The actual process or cycle followed in anodizing depends on the type of electrolyte being used. In general, the parts are cleaned, etched or pickled, anodized, colored, and sealed. Between each processing step, the parts are thoroughly rinsed in clean water. Figure 19-11 is a typical flow sheet for a chromic, sulfuric, or hard coat anodizing operation. Table 19-9 shows some typical operating parameters for the most common anodic coatings.[14]

Part Preparation and Precleaning

The anodized part surface must be thoroughly cleaned to ensure proper serviceability and appearance of the finish. Surface preparation can be performed mechanically or chemically.

In all anodizing operations the part preparation is substantially the same. The texture, pattern, surface roughness, matte, brightness, graining, or other desired final surface appearance is produced mechanically before the anodizing operation. A bright surface finish is accomplished by buffing or polishing. Matte or flat finishing is done by wire brushing, sanding, or blasting treatments. The type of finishes produced by chemical pretreatments is limited. On certain alloys, usually pure or low-alloyed aluminum, simulated polishing can be obtained by electropolishing or bright dipping. In these procedures, the mixed acids or chemicals used attack the peaks on the surface faster than the valleys between the peaks, producing a leveling effect that optically resembles a buffed surface. A flat or matte finish can be obtained by caustic or acid etching of the part; however, this procedure often brings out the crystalline structure of the surface and is not as satisfactory as mechanical methods for obtaining a consistent matte finish. In general, if a particular finish effect is required on a part, the finish is performed before and not as part of the anodizing process.

Precleaning is always performed as part of the overall processing cycle. Heavy preservatives are taken off in vapor degreasers or slush cleaning tanks. Light soils are removed from parts after they are racked by processing the loaded rack

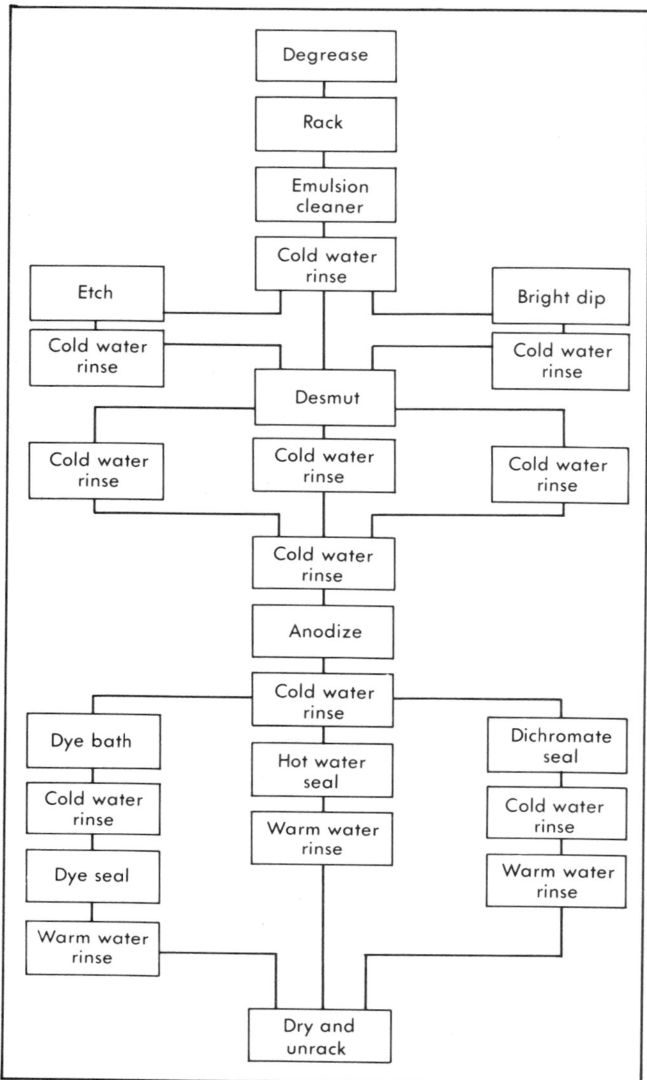

Fig. 19-11 Flow diagrams for anodic coatings produced in sulfuric and chromic acid electrolyte. (*Cidona, Inc.*)

through a hot, nonetching detergent-type cleaner. The rack is then deoxidized, washed several times, and placed in the anodizing tank. If the part dimensions are not critical, the racked parts may be dipped into a hot caustic etch to remove the existing surface and expose an active clean surface. The etch forms a smut that is removed in subsequent desmutting or deoxidizing treatments.

For additional information on mechanical cleaning and finishing methods, refer to Chapter 16 in this volume. Chemical cleaning is discussed in Chapter 18 and chemical finishing in Chapter 17.

Masking

It is possible to apply two or more kinds of anodized coatings on the same part, or to anodize certain areas and mask off the areas that are not to be anodized. Masking is done with pressure-sensitive electroplaters' tape, stopoff vinyls and lacquers, hot wax, adhesive-backed foil, and rubber corks, among other materials. Masking against chromic anodizing is difficult and hence not commonly done because of the heat involved in the process and the tendency of the chromic acid to undercut and penetrate. When paint is used as a maskant, the paint is removed after anodizing by immersion in a tank of methyl ethyl keystone, acetone, methylene chloride, or similar solvent. Vapor degreasers are sometimes used to remove maskants, especially wax that may be only partially removed initially in hot water. Masking operations are costly operations because of the materials and hand work involved.

Methods

Anodic coatings are applied to aluminum and its alloys by a variety of methods, including batch, bulk, and continuous strip methods. The batch method is similar to the conventional batch process employed in electroplating, except that the parts are anodic instead of cathodic. Bulk methods for applying anodic coatings use special perforated, nonmetallic cylindrical containers. Pressure is applied to the mass of parts in the container through a threaded center contact post. This pressure maintains the initial contact between the surfaces of the parts throughout the anodizing cycle.

The continuous strip process is used to apply anodic coatings to aluminum sheet that is subsequently roll formed into weather strip and other articles that require minimum forming. In Europe, this process is sometimes used to apply extremely thin [0.00005" (1.3 μm)] anodic coatings to aluminum sheet as a surface preparation for organic coatings. The coated sheet is then formed into special food containers.

Electrolyte

In Europe and Japan, oxalic acid and other special electrolytes are commonly used; but in the United States, virtually all of the basic commercial anodizing, excluding architectural, is performed with chromic or sulfuric acid electrolytes. When the electrolyte is sulfuric acid, the process is referred to as sulfuric anodizing or, in the trade, as "clear anodizing" because the coating is transparent and not readily distinguishable from the appearance of unanodized parts. When chromic acid is the electrolyte, the process is referred to as chromic anodizing, and the resultant coating is pearly to gray colored and tends to have a porcelain enamel-like appearance. Combination electrolytes of sulfuric and oxalic acids are sometimes used. Such combinations produce a denser coating with greater resistance to abrasion.

Because chromates are corrosion inhibitors and sulfates are conducive to corrosion, chromic anodizing is used largely in aircraft applications. Sulfuric anodizing is popular commercially because the more porous coating readily accepts dyes, the coating is heavier, and the operation itself presents less of a disposal problem than chromic anodizing.

Chromic anodizing. Chromic anodizing usually takes place in a plain carbon steel tank using a 3-10% by weight solution of chromic acid in deionized water, with the operating pH held between 0.5 and 1.0. The tank is air agitated and is usually shielded on the sides with glass plates to keep the ratio of anode area to cathode area about one to one. The tank contains heating and cooling pipes to hold the temperature within ±4° F (±2° C) of the selected operating temperature, which is usually 91-99° F (32-37° C).

A typical cycle would consist in bringing the voltage up to 40V in 5 to 8 minutes and then holding the voltage at 40V for 30 to 45 minutes. The rectifier usually has a 42V capacity, with the

TABLE 19-9
Typical Anodizing Operating Parameters for Various Electrolytes[14]

Electrolyte	Temperature, °F (°C)	Operating Conditions: Voltage, V	Operating Conditions: Current Density, A/ft^2 (A/m^2)	Time, min.
Sulfuric acid (10-20% by weight)	60-80 (15-25)	10-25[a]	10-15 (107.6-161.5)[a]	20-40[b]
Chromic acid	95 ± 3.6 (35 ± 2)	[c]	[c]	[c]
Hard coatings	25-50 (-4 to 10)	[d]	20-25 (215.3-269)	60-200[e]

[a] Depending on electrolyte concentration and temperature.
[b] Longer anodizing times produce thicker films.
[c] During the first 10 minutes the potential is increased from 0 to 40 volts. When the potential reaches 40 volts, anodizing is continued for 20 minutes. At this potential, the current density will range between 3-5 A/ft^2 (32.3-53.8 A/m^2). After 20 minutes, the potential is increased from 40-50 volts and then held at 50 volts for 5 minutes.
[d] To maintain this current density the potential rises from 20-25V to 40-60V.
[e] Thickness is approximately a linear function of time.

amperage output sized to the load requirements. For every square foot (0.09 m^2) of surface area, amperage input is approximately one ampere. The coating thickness of the chromic acid film obtained varies from 0.00005 to 0.0003″ (1.3 to 7.6 μm). The film, while somewhat fragile because of its thinness, is extremely dense and has excellent corrosion-resistant properties. Normally all chromic films are sealed, unless otherwise specified. Chromic anodizing is not recommended for alloys containing over 5% copper or over 7% silicon, or having a total alloying content of more than 7.5%.

Sulfuric acid anodizing. A typical sulfuric acid anodizing tank would be fabricated from plain carbon steel, completely lead-lined, and would contain the standard heating, cooling, and air agitation systems. The electrolyte is a 10-20% by weight sulfuric acid solution in deonized water. The operating temperature is usually 68-88° F (20-31° C), depending upon the alloy and whether the film is to be dyed. If the part is to be dyed, a temperature on the high side would be used to increase the film's porosity and its ability to accept the dye. The current density is 9-12 amperes per square foot (97-130 amperes per square meter) of surface area being anodized in the tank, and the cycle is from 15 minutes to 1 hour, with a resulting film thickness of 0.0002 to 0.001″ (5 to 25 μm). A typical cycle might be 10 A/ft^2(108 A/m^2) in a 15% by weight solution, run at 78° F (26° C), and brought up to and held at 18V for a total cycle time of 35 minutes to obtain a coating thickness of 0.0006″ (15 μm). The rectifier is sized by the load requirements in amperage and usually has a maximum voltage capacity of 24V.

Hard coating. The hard coating process involves reduced electrolyte temperatures, higher unit amperages, and final voltages up to 100V. In conventional sulfuric anodizing, operating at room temperature, the hot acid dissolves the film as it is being formed so that a counteraction of electrolytic formation and chemical dissolution is occurring simultaneously. By lowering the solution temperature, the chemical activity is retarded so that the reaction becomes more one-sided, and heavier oxide films are created electrochemically. The oxide is the same as that formed in other anodizing processes, but it is distinctive because it is much thicker than any hitherto obtainable. Typical electrolyte temperatures are from 25 to 50° F (-4 to 10° C), and current density is 25-50 amperes per square foot (270-538 amperes per square meter) of surface area.

Electric Current

Although direct current is the most popular type of electric current for forming anodic coatings, alternating current may also be used to anodize aluminum and its alloys with all the electrolytes mentioned; however, with alternating current the aluminum surface is anodic only half of the cycle so that an oxide coating is formed at only half the rate of coatings formed with direct current. Superimposing alternating current with direct current is used for high-speed anodizing and has been most successful for the production of hard, thick anodic coatings by the high-current-density Hardas process and also by the Sanford Plus process.

EQUIPMENT

The main components used in anodizing include tanks, racks, cathodes, agitators, a power supply, cooling coils or heat exchangers, and chillers, along with a ventilation system.[15]

Tanks and Cathodes

The tank proper has a dual function: it must be chemically resistant to the electrolyte being used, and it also serves as the cathode or ground of the electrical system. The tank can be made entirely of a suitable type of stainless steel, wood, acid-resistant brick, concrete, fiberglass, or plain carbon steel. Where required, the tank is lined with lead, koroseal, rubber, polyethylene, PVC, or plastisol. If the tank interior is non-conductive, either lead or aluminum plates are hung around the inside perimeter and connected in common to the ground or negative output of the power source.

Busbars for supporting the work racks are usually flat or round copper and rest on insulators on the top of the tank. The positioning of the busbars depends upon the size of the tank and the general type of work being processed. In small batch operations, the busbars are located for easy accessibility from the long sides of the tank. In large operations, the busbars are loaded in a separate racking area, and the entire busbar system with the racks attached is lifted as a unit by an overhead crane and placed on prefixed saddles on the tank. The busbars are connected in common to the positive side of the power source. The size of the tank is determined by consideration of the size and shape of the work to be processed. If, for example, drum-shaped parts were the only end product, the tank might be basically square. If jobbing of miscellaneous small parts were to be done on a batch basis, then the tank might be constructed long and narrow with a single, central busbar along the length to facilitate working from both sides of the tank, and to minimize the reach required to place the rack on and off of the busbar.

Racks

Anodizing of small parts with certain shapes, such as rivets, bolts, and similar parts that will not nest can be done in bulk using a perforated basket that is generally plastic. The baskets contain a central aluminum rod that runs through the top plate and the solution to make contact with the busbar. The work is randomly dumped into the basket and compressed lightly by holding the top down with a screw arrangement. The parts in contact with the central bar distribute the electrical current to the other parts by part-to-part contact. While many parts can be done per load with a minimum amount of handling, the coating thickness may not be uniform throughout the basket load. Cupped parts such as caps, covers, and lipstick cases may nest or be inverted in the basket, resulting in voids, bare areas, or linear contact marks. In basket work where appearance is a factor, 100% visual inspection is required. A 65% yield on certain types of parts might be considered satisfactory, and the rejects are then stripped and reprocessed.

Aside from bulk basket work, all anodized parts must be individually racked, usually by hand. The rack serves to conduct the current from the busbar to the work and also to hold the part firmly under the agitated electrolyte. Racks are constructed of commercially pure titanium or aluminum. By virtue of their good mechanical characteristics, 6063 and 6061-T6 are very good aluminum alloys for use as general racking material. Tantalum has also been used for rack contacts, but it is very expensive.

While titanium has the advantage of remaining conductive and is relatively inert chemically, it is a poor conductor. Rating copper at 100, aluminum has 60% of the conductivity of copper and titanium has 5%. Titanium racks are usually limited to smaller parts that require less unit amperage. Titanium should not be used with the higher alloy content parts, such as 7075 and 2024, because the galvanic couple in a conductive cleaner, rinse water, or electrolyte causes rapid pitting of the parts on the rack; titanium should also not be used for racks that are used in integral or electrolytic color anodizing. Titanium racks must be stripped periodically in hydrofluoric acid to remove the thin, discolored oxide film and to ensure uniform contact. Good practice is to bake the racks occasionally to drive off the hydrogen and lessen the tendency towards embrittlement.

When using aluminum racks, the rack itself anodizes, as well as the part, so it is necessary after each cycle to remove the insulative anodic layer from the contact areas. This removal is accomplished by caustic etching or by using special etches that attack the oxide but not the base aluminum. With repeated usage, the rack becomes thinner, loses its clamping strength, and must be discarded. To reduce the area of rack exposed during anodizing and to maintain spring strength, aluminum racks are often painted, taped, or entirely plastisol coated, after which the contact tips are cut back for metal contact.

Rack design, construction, and application require experience and are the forte of the anodizer. The area that contacts the part must be sufficient to carry the unit amperage required, and the part must be positioned to avoid gas pockets and to ensure positive solution sweep over the part. If the part contains tapped holes and is larger in size, the rack may be threaded into one or more of the tapped holes and locked with a jam nut. A commonly used rack is tree-shaped with a central bar going down the length of the rack and bent at the top to hook onto the busbar (see Fig. 19-12). Along the square or hexagonal cross-sectional length, sets of cross holes are drilled to accommodate clusters of three rods that extend out from either side of the rack and are staked where they pass through the central bar. By spreading the ends of the three rods, numerous types of parts can be accommodated. Tubular parts can be racked by placing

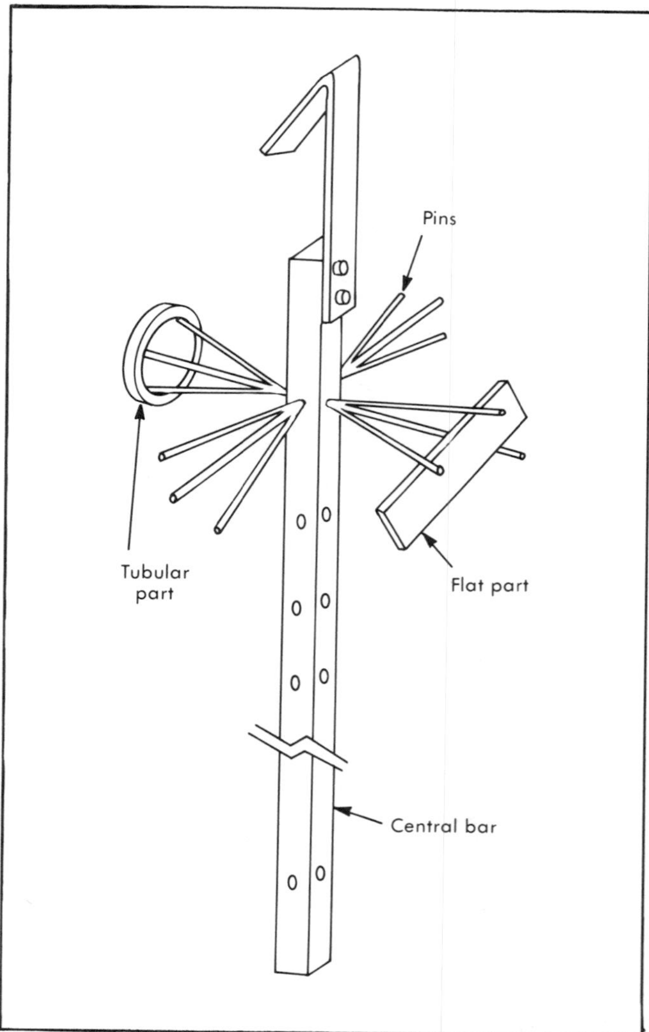

Fig. 19-12 Pin racks can be used for holding parts that are tubular and flat in shape. (*Cidona, Inc.*)

the three rods inside of the tube, while a flat part is held by placing two rods on one edge and the third rod on the opposing edge or by hooking the rod ends through holes in the part.

Racks are stamped and formed out of sheet, with fingers, tapered points, or clips for fastening the part (see Fig. 19-13). These stamped forms are then attached to a single aluminum bar that serves as a backbone for distributing the current evenly along the rack and provides an attachment method for connection to the busbar. A number of stamped racks can be attached in rows on an aluminum frame, with extended arms at the ends for busbar mounting. Individually stamped clips can be bolted with titanium fasteners to a basic spline. Standard types of racks are obtainable from manufacturers, or custom designs are fabricated by rack makers. In handling production parts that are repetitive in shape, the usual practice is to construct a specially designed rack to accommodate the particular part.

Depending upon the process, rack marks are made on the part in varying sizes. These contact points have no anodic film and often stand out as bare spots, in contrast to the color of the surrounding anodized coating. The contact marks are more noticeable if the part is dyed. If certain areas of a part require corrosion protection, electrical insulation, or other features that the anodic film provides, the designer must specify on the drawing or processing instructions where not to contact the part. Preferred practice is not to instruct the anodizer where to contact the part, but rather to specify the areas where not to contact since this allows the anodizer freedom to use his expertise in racking within the noncritical areas.

Agitators

To maintain uniform temperature and a supply of fresh electrolyte against the work, the solution is agitated by mechanical stirrers or, more commonly, by air agitation arising from a piping system mounted in the bottom of the tank. High-pressure plant air may be used for this agitation, but lower pressure blower air is preferred because its air supply is oil-free and the blower is independent from the system. Mechanical agitation of the busbar, which is common in plating, is seldom used in anodizing because movement of the rack in the agitated solution can loosen the work. A properly laid out air supply system takes the place of moveable busbar agitation.

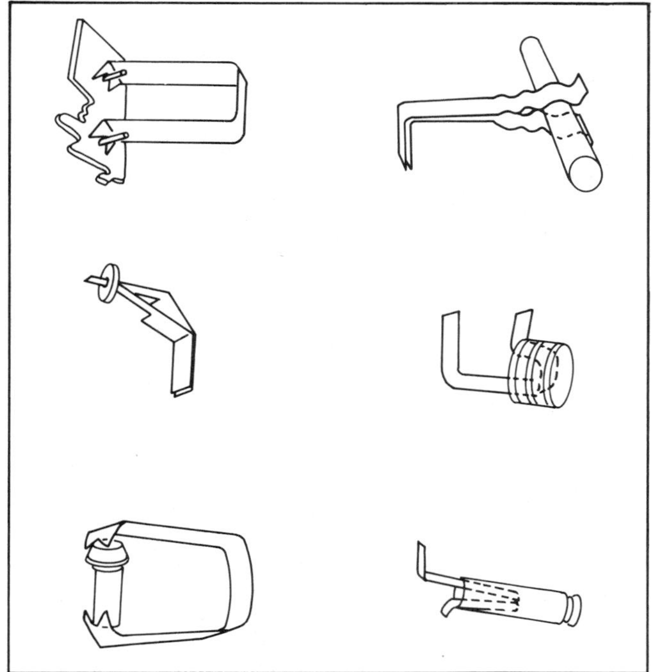

Fig. 19-13 Common stamped and formed clips used for holding different part designs during anodizing operations. (*Cidona, Inc.*)

Power Supply

Motor generator sets have largely been replaced by rectifiers that are quieter, have electronic circuitry, are rugged, require less maintenance, and are easier to program for control. The rectifier control panel can be removed and placed near the operating tank, while the rectifier proper is protected in an adjacent room. Alternating current has been used in some types of anodizing, but direct current is predominately standard. In some applications, a small percentage of alternating current is superimposed over the direct current, or the direct current itself can be pulsed.

The size of the power supply is a function of the anticipated area to be processed in a given load and the type of anodizing process being performed. Likewise, the volume of the tank is associated with area and amperage. A rule of thumb might be 1 gal (3.8 L) of electrolyte for each ampere of rectification, which in turn depends upon the type of anodizing being performed and the anticipated area per load.

For normal sulfuric acid anodizing, a rectifier or motor generator capable of 25V and 5 to 15 A/ft^2 (54.8 to 161.2 A/m^2) is usually sufficient. However, barrier films for electrolytic capacitors, chromic acid anodizing, phosphoric acid anodizing, and oxalic acid anodizing may require voltages as high as 150V. The variety of anodizing processes to be performed determines the voltage and power requirements of the power supply used. Integral-color anodizing requires a rectifier capable of 100V, and chromic acid anodizing a 40V rectifier. Electrolytic coloring typically requires a 25V a-c convertor.

Cooling Coils

In the anodizing operation, it is necessary to hold the electrolyte temperature within a certain range, depending upon the particular process requirements. To maintain temperature, it is necessary to employ heating and cooling systems. To achieve a working range of 75 to 85°F (24 to 30°C), the electrolyte may initially need to be heated to the working temperature. After the working temperature has been attained, refrigeration may be required to dissipate the heat acquired by the electrical energy wattage input from the power source. Heating is usually done with steam pipes, plate coils, or hot water pipes. For certain applications, electrical immersion heaters may be used. Cooling is accomplished by coils through which either plant water is piped and discharged, or a mixture of water and glycol is circulated in a closed circuit with an external chiller. For lower temperature requirements, the electrolyte itself may be pumped from the tank through the heat exchanger unit of an external refrigeration unit and then back into the tank. The pumping action of this method assists in the required agitation within the tank.

Ventilation

Ventilation hoods, ductwork, and exhaust fans are required to remove the fumes generated during anodizing.

TESTING AND SPECIFICATIONS

TESTING AND SPECIFICATIONS

There are several methods for testing the quality of anodic coatings. These include the American Society for Testing and Materials (ASTM) specifications B 117, B 136, B 137, B 244, B 457, B 487, B 538, B 680, and D 658.

The B 136 method is a stain test used to determine the effectiveness of the sealant in preventing stains. Method B 680 evaluates seal quality by measuring resistance to acid attack. Method B 137 is a chemical procedure used to determine the weight of the oxide coating.

One of the most important nondestructive tests for measuring coating thickness is the procedure described in B 244. This procedure measures coating thickness with an eddy current instrument that can produce results that closely check microscopic measurements; however, a flat, reasonably smooth surface is necessary for best results.

Method B 117 is a salt spray test that is sometimes used to determine resistance to corrosion; however, it is not always a reliable test for predicting actual performance. Some modifications of the salt spray tests are the B 287 acetic acid salt spray test, the B 368 copper-accelerated acetic acid salt spray test, and the B 380 test that employs a slurry consisting of kaolin, ammonium chloride, ferric chloride, and copper nitrate. After the slurry is dry, the test sample is exposed to high humidity for 24 hours.

The abrasimeter test, ASTM D 658, employing an abrasive air-blast mixture, is one method of determining resistance to abrasion. In this method, the amount of abrasive material required to wear through the coating under controlled conditions is accurately weighed. These results can be correlated with coating thickness so that the method can be used as a measurement of thickness. It is, of course, a destructive method and requires a flat surface of definite size. ASTM D 4060 measures the abrasion resistance of organic coatings using the Taber Abraser, which uses a rotating abrasive-impregnated wheel to contact the revolving anodized surface. The Taber Abraser can also be used on anodic coatings if specific test conditions are carefully selected and controlled.

Other tests to evaluate the quality of anodic coatings include the measurement of shade change for colored anodic coatings, appearance, and reflectance. Atmospheric exposure tests are, of course, the most reliable for evaluating colorfastness. ASTM G 23 defines accelerated fading tests that give reproducible results, but these results do not always correlate with atmospheric exposure. The degree of fading or color change can be determined by colorimeters using ASTM method D 244.

Appearance characteristics such as image clarity can be quantitatively measured by the Dori meter (Aluminum Company of America). The Dori-Gon Glossmeter (Hunter Associates Laboratory, Inc.), used in accordance with ASTM E 167, can be employed for quantitative evaluation of bloom on the diffusing characteristics of the anodized surface. Total reflectance is measured by a Taylor-Baumgartner integrating-sphere instrument.

Humidity tests check anodic coatings to be used in household refrigerators. One variation of the humidity test employs a mixture of sulfur dioxide and carbon dioxide in combination with controlled humidity. Another variation uses a sodium sulfite solution for immersion of the test specimen. In both tests, the development of a white bloom on the anodic coating indicates a poor seal.

As defined by ASTM B 457 and B 538, electrochemical tests are based upon contacting the anodic coating with a special solution, which serves as an electrolyte, and employ a special contact; they have been designed to determine the quality of an anodic coating from the standpoint of thickness and degree of seal.

Metallographic sectioning with microscopic measurement of coating thickness is considered to be the most accurate test available. This test is destructive and, in most cases, impossible to use on actual production parts as a routine quality test. The thickness of anodic coatings on curved surfaces can be measured nondestructively using a microscope fitted with an interference objective attachment. Another nondestructive test, described in ASTM B 588, measures coating thicknesses in the range of 0.000008 to 0.0004" (0.2 to 10 μm) using a double-beam interference microscope.

Test methods and anodic coating specifications are also covered in MIL-A-8625, Aeronautical Materials Specifications AMS-2468, AMS-2469, AMS-2470, AMS-2471 and AMS-2472, and in standards developed by the International Organization for Standardization (ISO). The standards by the ISO are particularly important for international trade. Table 19-10 lists the commonly used specification numbers and testing methods for anodic coatings.

TROUBLESHOOTING

Compared with other processes, the susceptibility to rejects in anodizing practice is relatively minor. When uniformity of coating thickness is required, depending upon the thickness tolerance agreed upon, variations in a given lot occur because of the uneven spring force in individually racked parts and also as a result of the positioning of the racks within a tank load. The rack in the center of the tank, being farther away from the side wall cathodes, may not draw as much current as a rack in the corner of the tank. Likewise, when the rack is in the center of the tank, parts on the bottom of the rack may draw more current than the upper parts on the rack, resulting in nonuniform coating thicknesses. Therefore, if a given tank load were to be checked 100% for thickness, a scatter would be found to exist, but would be acceptable within the normal range of operating tolerances. If a part is not racked tightly and movement on the rack occurs due to tank agitation during the anodizing process, electrical arcing can occur at one or more of the contact points.

In hard coating, a common problem is termed *burning*. Burning occurs when there is an excess amount of amperage in one area of the part, usually at a point, projection, edge, or in a pocket area where circulation of the electrolyte is inadequate. In a part having heavy and thin sections, burning can occur in the thin sections. Normally, if the burn is light or commences near the end of the cycle, the area appears chalky or powdery. If the part was initially polished, the coating will have a higher polish in the burn areas. If the burning is severe, there may be a loss in dimension or, in extreme cases, a cannibalization of the part. On a sheet or thin section, partial or localized burning has a sooty or moire pattern appearance that is identical on both sides of the sheet or section.

In any anodizing process, once the process is stopped and the part is washed, it is difficult and also poor practice to attempt to apply more coating. Interruption of the coating process may result in a laminated coating or in blotches, with the coating picking up in some areas but not in others. If a part has insufficient coating, the normal procedure is to strip and reanodize the part. While this procedure can readily be done with chromic anodized parts, and may be done with sulfuric anodized parts if the tolerance is not critical, it is

TABLE 19-10
Specifications Used in Anodizing Procurement

Specifying Body	Specification Number and Title
Aerospace Material Specifications	AMS-2468 Hard Coating Treatment of Aluminum Alloys
	AMS-2469 Hard Coating Treatment—Process and Performance Requirement of Aluminum Alloys
	AMS-2470 Anodic Treatment Aluminum Alloys, Chromic Acid Process
	AMS-2471 Anodic Treatment, Sulphuric Acid Process—Undyed Coating
	AMS-2472 Anodic Treatment, Aluminum Base Alloys Dyed Treatment (Sulphuric Acid Process)
Military Specifications	MIL-A-8625 Anodic Coating for Aluminum and Aluminum Alloys
	MIL-I-8474 Anodizing Process for Inspection of Aluminum Parts (Chromic Acid Process for Crack Detection)
Federal Specifications	Federal Test Method Standard No. 141—Paint, Varnish, Lacquer and Related Materials: Method of Inspection, Sampling and Testing
	Federal Test Method Standard No. 151—Metals, Test Method
	Federal Standard 595—Color Standards
American Society for Testing and Materials	B110 Standard Method of Test for Dielectric Strength of Anodically Coated Aluminum
	B117 Standard Method of Salt Spray (Fog) Testing
	B136 Standard Method of Test for Stain Resistance of Anodic Coatings on Aluminum
	B137 Standard Method of Test for Weight of Coating on Anodically Coated Aluminum
	B244 Method for Measuring Thickness of Anodic Coatings on Aluminum with Eddy Current Instruments
	B457 Method for Measuring the Impedance of Anodic Coatings
	C448 Abrasion Resistance Testing
	D2244 Method for Instrumental Evaluation of Color Differences of Opaque Materials

generally impossible to do with hard-coated parts because of the coating thickness.

Hard-coated parts may exhibit crazing (which is more obvious on the more transparent coating) caused by differential coefficients of expansion between the anodic coating and the base metal, but its occurrence is not considered detrimental to the end use of the coating. In color anodizing, rejections may be associated with uneven dye match in a lot. At a constant temperature, the depth of dye color varies with the thickness of the coating, condition of the dye bath, as well as time in the bath; hence, over a long run of parts, the color shading can vary. Streaking or washing of the color is associated with precleaning patterns or insufficient rinsing between processing steps. A powdery or velvet surface on dyed parts, which is removable by wiping, is usually caused by improper sealing practices. In chromic acid anodizing, yellow stains may result from the electrolyte draining from a cavity. Tapped holes or similar receptacles in the part can be corrected by forced water spraying and more intensive washing after anodizing. Similar problems occur in sulfuric or hard coating and appear as a light-colored streak. Table 19-11 lists some of the common symptoms of anodizing problems, along with the possible causes and suggested solutions.[16]

SAFETY AND ENVIRONMENTAL CONSIDERATIONS

Anodizing processes are relatively safe because of the limited group and low concentrations of chemicals involved and the low closed-circuit electrical requirements. Slipping on a wet platform is probably a more realistic danger than injury from the chemicals involved. Chromic acid and bright-dip operations are the most hazardous anodizing processes and require proper ventilation. Handling any concentrated acid used for tank makeups requires proper safety clothing, goggles, gloves, and safety-toe boots. Only experienced personnel should handle the acid concentrates. Strong deoxidizers are labeled by the manufacturer, and chromic acid flakes, nitric acid, and other deoxidizers used in anodizing operations must be segregated in storage and handled properly. Because of the EPA federal laws, all states are in the process of implementing controls to ensure conformance. These controls, which are becoming more constrictive and detailed, will ultimately result in isolation and sealing off the anodizing plant to make it, in effect, a leakproof container. Sealed drains, leakproof floors with curbs, and general containment are the aims of these requirements.

The anodizing facility, being a waste generator, is assigned a number as are the independent waste-hauling firms. All transactions involving waste removal must be done on a manifest that specifies, among other data, the chemicals involved, the quantity, and the waste hauler with identification numbers. Copies of the manifest are forwarded to the state environmental office.

In general, chromates and phosphates used in anodizing present the major disposal concern. Various methods are used to treat anodizing wastes. Evaporation systems operating under pressure to reduce evaporation temperatures, sedimentation,

TABLE 19-11
Common Symptoms and Possible Causes of Anodized Parts[16]

Symptom	Possible Causes	Suggested Solutions
Small pits or craters at contact points	Improper or loose electrical contact causing arcing.	Clean contacts well before racking.
No film in recesses	Trapped air or gases in recesses.	Position the recesses on the side to allow the electrolyte to circulate.
Variation of coating thickness	Poor electrical contact; nonuniform bath temperature; variation in alloy or heat treatment.	Clean contacts before racking; prevent rack movement during anodizing; increase bath agitation.
Patchy or streaky appearance	Improper or insufficient cleaning variation in alloy; variations in polishing or buffing; variation in heat treatment.	Rinse well between processing tanks and allow good draining.
Pitting or etching	Corrosion of surface before anodizing; reanodizing without stripping previous film; excess chlorides in bath; insufficient agitation; bath temperature too high.	Abrade or machine to new surface if tolerances permit; increase agitation; increase cooling.
Excess current density	Bath temperature too high.	Increase cooling; monitor current controls more closely.
White powdery "bloom"	Temperature too high; current density too high; local overheating; excessive sealing time; high calcium-salt content in sealer.	Reduce temperature, current, and sealing time; dump and remake seal bath and reduce seal time; increase agitation.
Failure to absorb dye	No film; film is sealed; dye too cold; dye too dilute; dye time too short; film too hard (bath temperature too low and voltage too high).	Renew bare aluminum surface if tolerance permits; increase dye temperature and strength and immersion time.
Variation in color between pieces	Alloys vary from the norm; variation in heat treating or mechanical processing of the same alloy; variations in part contact; variations in dyeing time; change of dyeing conditions (concentration, pH, temperature); failure to maintain dye balance.	Check for good contact; maintain dye bath at optimum conditions; use same dye time for all pieces.
Dark specks on surface	Particles of undissolved or precipitated dye in the dye bath; inclusions in the surface of the alloy.	Filter dye bath.
Film dull or colored slightly	Nature of the alloy; impure alloy; excess aluminum in solution.	Keep dissolved aluminum below 15 g/L.
	For Chromic Acid Anodizing Only	
Anodic film is a dirty yellow	Too much trivalent chromium.	Try dummying with aluminum panel twenty times the size of the cathode.
High current density	Copper-containing alloy with more than 5% copper; bath temperature too high.	Decrease voltage setting; decrease bath temperature.

(*Products Finishing*)

reverse osmosis, filtration, and ion and cation exchangers are among the methods commonly used. The anodizing baths contain metal and metal salts from the dissolution of the alloying elements in the work during anodizing. These metals are precipitated and separated from the acidic electrolyte, and the latter is neutralized before disposal. The sludge may be filter pressed before hauling away to be dried and briquetted prior to disposal in a state-approved site.

References

1. Fred W. Eppensteiner and Melvin R. Jenkins, "Chromate Conversion Coatings," *Metal Finishing—Guidebook Directory Issue* (January 1984), p. 490.
2. *Metals Handbook*, vol. 5, 9th ed. (Metals Park, OH: American Society for Metals, 1982), p. 441.
3. David L. Toyne, *Making It Stick With Pretreatment*, SME Technical Paper FC81-444 (Dearborn, MI: Society of Manufacturing Engineers, 1981).

REFERENCES

4. Albert M. Pradel and W. J. Wittke, "Phosphate Coatings," *Metal Finishing—Guidebook Directory Issue* (January 1984), p. 510.
5. Eppensteiner and Jenkins, *op. cit.*, pp. 488 and 490.
6. *Ibid.*, p. 488.
7. *Ibid.*, p. 491.
8. *Ibid.*, p. 495.
9. *Ibid.*, p. 493.
10. Frederick A. Lowenheim, *Electroplating* (New York: McGraw-Hill Book Co., 1978), p. 445.
11. *Ibid.*, p. 452.
12. F. Keller, M. S. Hunter, and D. L. Robinson, "Structural Features of Oxide Coatings on Aluminum," *Journal of The Electrochemical Society*, vol. 101 (1954), p. 481.
13. Charles F. Burrows, "New Finish Gives Aluminum Good Wear Resistance," *Iron Age* (August 24, 1950), pp. 73-75.
14. David Thomas, "Anodizing Aluminum," *Metal Finishing Guidebook and Directory Issue* (1983), pp. 510-516.
15. *Ibid.*, p. 512.
16. Lawrence J. Durney, "Trouble in Your Tank," *Products Finishing* (Cincinnati: 1983), p. 69.

Bibliography

Brace, A. W., and Sheasby, P. G. *Technology of Anodizing Aluminium*, 2nd ed. Gloucestershire, England: Technicopy Ltd., 1979.

Graham, A. K., ed. *Electroplating Engineering Handbook*, 3rd ed. New York: Van Nostrand Reinhold Publishing Corp., 1971.

Lorin, Guy. *Phosphating of Metals*. Hampton Hills, Middlesex, UK: Finishing Publications Ltd., 1974.

Magnesium Finishing. Midland, MI: Dow Chemical Co.

Practical Phosphate Coatings. Cleveland, OH: R. O. Hull and Co., Inc.

VanHorn, Kent R., ed. *Aluminum*, Vol. III. Metals Park, OH: American Society for Metals, 1967.

Wernick, S., and Pinner, R. *Surface Treatment of Aluminum*, 4th ed. Teddington, England: Robert Draper Ltd., 1972.

Wiederbolt, Wilhelm. *The Chemical Surface Treatment of Metals*. Teddington, UK: Robert Draper Ltd., 1965.

PLATING

ELECTROPLATING

Electroplating is an electrolytic process whereby a metal is cathodically deposited onto another metal or a surface that has been made conductive. Thus, a part made of metal, plastics, or other materials may be coated with a thin metal deposit to impart certain desirable properties while avoiding the prohibitive cost of fabricating the part entirely from the metal used as a coating. Plating provides desirable characteristics such as protection of the basis metal (substrate) from corrosion, improvement of the appearance of the substrate, and improved solderability, wear resistance, electrical conductivity, contact resistance, and lubricity. Depending on their use, electroplated coatings may be classified as either decorative or engineering, with some falling into both classifications.

THEORY

In electroplating, the workpiece is made cathodic in a solution containing the ions of the metal being deposited. Direct current is passed between the anode and the workpiece (cathode). The anode is usually constructed of the same material as the metal being plated, although some plating processes use insoluble anodes. As the current flows, the metal ions gain electrons at the cathodic workpiece and transform into a metal coating based on the following equation:

$$M^{+y} + ye^{-} \rightarrow M^{\circ} \text{ (metallic state)} \quad (1)$$

For nickel electroplating, the following equation describes the reaction at the cathode:

$$Ni^{+2} + 2e^{-} \rightarrow Ni^{\circ} \text{ (metallic state)} \quad (2)$$

Proprietary additives (brighteners) are usually incorporated in the plating solution to alter the deposit in a desirable fashion. These additives brighten or level the deposit as well as improve the uniformity of the deposit's thickness over the entire workpiece. The additives may also be used to alter physical properties such as hardness, ductility, internal stress, and corrosion resistance.

Faraday's Law

In 1883, Michael Faraday explained the nature of electrolysis. Faraday's law is basic to the understanding of all electrolytic processes and may be summarized by the two following statements:

1. The amount of chemical change (metal deposited) that is produced by an electric current passing through solution in a cell is directly related to the quantity of electricity that flows.
2. The amount of substance liberated (i.e., metal deposited) by a given quantity of electricity is directly related to the gram equivalent weight of the substance. More specifically, 96,500 ampere seconds (or coulombs) reduce one gram equivalent of metal ions to metal atoms.

Faraday's law is used by the plater to determine the plating time and current required to deposit a specified weight (and considering the specific gravity and surface area, also a specified thickness) of a metal. Table 20-1 gives the electrochemical equivalents for some typically electroplated metals. The amount of time required to deposit a certain thickness at a specific current density can be calculated from this data.

Cathode Efficiency

Faraday's law states that the total amount of chemical change at an electrode is directly related to the quantity of electricity (that is, the current flowing). In plating, however, the main concern is the quantity of metal deposited. Any other reactions, such as the liberation of hydrogen at the cathode, decrease the cathode efficiency. The average cathode efficiency percentage is determined by dividing the actual weight of metal deposited by the theoretical weight and then multiplying by 100. Changes in solution composition and other external factors can improve cathode efficiencies to some degree; however, the solution used is the major factor for determining efficiency. Table 20-2 presents approximate cathode efficiencies for typical electrolytes.

Current Distribution

Current density is the applied current divided by the total surface area of the workpiece. Current

CHAPTER CONTENTS:

Contributors of sections of this chapter are: ***Alan Brooks****, Chief Metallurgist, Waldes Kohinoor, Inc.;* ***Mike D'Angelo****, Product Manager, Metal Finishing, MacDermid, Inc.;* ***Max DiMarco****, Product Manager, Plating on Plastics, MacDermid, Inc.;* ***Dr. Otto Kardos****, Consultant, M & T Chemicals, Inc.;* ***Joe A. Miglionico****, Marketing Manager, Sifco Selective Plating Div., Sifco Industries, Inc.;* ***Joe C. Norris****, Metallurgist, Sifco Selective Plating Div., Sifco Industries, Inc.;* ***Richard F. Rapids****, Vice President, Rapid Electroplating Process, Inc.;* ***Hal Thrasher****, Marketing Manager, MacDermid, Inc.*

Reviewers of sections of this chapter are: ***Sidney Beach****, Technical Director, McGean Rohco, Inc.;* ***Alan Brooks****, Chief Metallurgist, Waldes Kohinoor, Inc.;* ***Robert R. Brookshire****, President, Brushtronics Engineering;* ***Mike D'Angelo****, Product Manager, Metal Finishing, MacDermid, Inc.;* ***Dr. George D. DiBari****, Product Manager, Technical Services, International Nickel, Inc.;* ***Max DiMarco****, Product Manager, Plating on Plastics, MacDermid, Inc.;*

CHAPTER 20

ELECTROPLATING THEORY

TABLE 20-1
Electrochemical Equivalents and Deposit Metal Weight and Thickness Calculated from Faraday's Law for Commonly Electroplated Metals*

Metal	Valence	Atomic Weight	Specific Gravity	Weight, oz/A·h (g/A·h)	Weight in 1 mil (1 μm) thick coating, oz/ft² (g/m²)	A·h to deposit 1 mil/ft² (μm/m²)
Cadmium (Cd)	2	112.40	8.65	0.074 (2.097)	0.71 (8.659)	9.73 (4.125)
Chromium (Cr)	6	52.01	7.1	0.011 (0.323)	0.59 (7.146)	51.8 (22.110)
	3		7.1	0.023 (0.646)	0.59 (7.146)	25.9 (11.060)
Cobalt (Co)	2	58.93	8.9	0.039 (1.099)	0.74 (8.719)	19.0 (7.926)
Copper (Cu)	1	63.54	8.96	0.084 (2.370)	0.74 (8.935)	8.84 (3.770)
	2	63.54	8.96	0.042 (1.186)	0.74 (8.935)	17.8 (7.540)
Gold (Au)	1	197.0	19.3	0.236 (7.348)**	1.47 (19.32)**	6.2 (2.631)
	3	197.0	19.3	0.079 (2.449)**	1.47 (19.32)**	18.6 (7.887)
Indium (In)	3	114.82	7.31	0.045 (1.428)**	0.56 (7.278)**	12.0 (5.092)
Iron (Fe)	2	55.85	7.86	0.037 (1.042)	0.65 (7.868)	17.9 (7.54)
Lead (Pb)	2	207.19	11.34	0.136 (3.865)	0.94 (11.350)	6.9 (2.936)
Nickel (Ni)	2	58.71	8.90	0.039 (1.095)	0.74 (8.880)	19.0 (8.044)
Palladium (Pd)	2	106.4	12.0	0.064 (1.985)**	0.86 (24.34)**	13.5 (6.045)
Platinum (Pt)	4	195.09	21.41	0.058 (1.819)**	1.60**	27.8 (11.77)
Rhodium (Rh)	3	102.91	12.4	0.041 (1.280)**	0.95 (8.29) **	22.9 (9.73)
Silver (Ag)	1	107.87	10.5	0.129 (4.024)**	0.79 (10.5) **	6.2 (2.605)
Tin (Sn)	2	118.69	7.30	0.078 (2.214)	0.61 (7.30)	7.8 (3.30)
	4	118.69	7.30	0.039 (1.106)	0.61 (7.30)	15.6 (6.604)
Zinc (Zn)	2	65.38	7.14	0.043 (1.219)	0.59 (7.15)	14.3 (5.863)

* Assumes 100% cathode efficiency for metal deposition.
** Weight is in 1 troy oz/ft².

density is usually measured in A/ft² or A/dm², depending on the system of units.

In electroplating, current is concentrated at edges and points as well as in areas closer to the opposite electrode (anode), and these latter areas consequently receive a greater deposit thickness (see Fig. 20-1). Normally, a part is plated to meet a minimum thickness specification. Since the excess metal thickness in high current density areas is usually not desired, the plater should select a bath with good throwing power. Throwing power is defined as the electrolyte's ability to minimize the difference in deposit thickness between high and low current density areas.

Plating solutions in which the cathode efficiency decreases with an increase in current density generally display better throwing power because less metal will be deposited in the less efficient high current density areas. Cyanide zinc is an example of such a bath. Chromium plating, on the other hand, becomes more efficient with an increase in current density and consequently has poor throwing power.

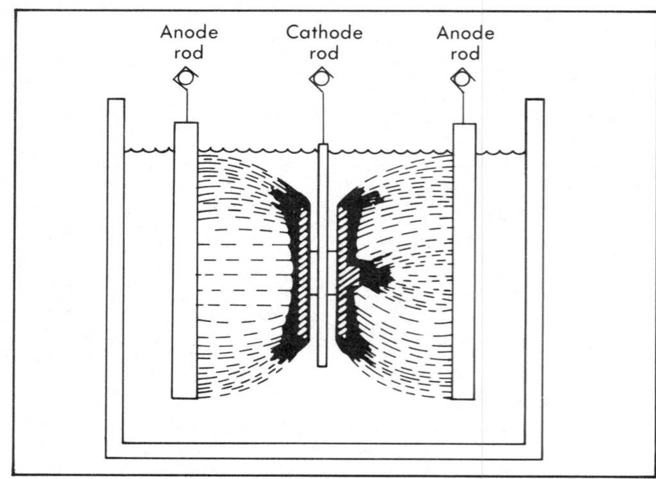

Fig. 20-1 The closeness of the dotted lines indicates the distribution of current and plating thickness.

Reviewers, cont.: ***Henry R. Friedberg**, Consultant, Henry R. Friedberg & Associates;* ***Carter A. Graff**, General Manager, Sifco Selective Plating Div., Sifco Industries, Inc.;* ***Joseph L. Greene**, Senior Development Chemist, Allied-Kelite Div., Witco Chemical Corp.;* ***Timothy M. Hawk**, Application Engineer, Engineered Products Div., Renold, Inc.;* ***Edwin W. Hoover**, Consultant, Udylite Plating Systems, OMI International Corp.;* ***Beldon Hutchinson**, Vice President, Liquid Development Co., Inc.;* ***Douglas Hutchinson**, Engineer, Liquid Development Co., Inc.;* ***Maynard L. Isabell**, Product Manager, Engineered Products Div., Renold, Inc.;* ***Dr. Otto Kardos**, Consultant, M & T Chemicals, Inc.;* ***Louis Klemm**, Assistant Technical Service Director, McGean Rohco, Inc.;* ***William Knapp**, Service Engineering, M & T Chemicals, Inc.;* ***Maurice LaFreniere**, Director of Sales and Marketing, Selectrons, Ltd.;* ***Gerald A. Laitinen**, Product Manager, Allied-Kelite Products, Witco Chemical Corp.;*

TABLE 20-2
Average Cathode Efficiencies of Common Plating Solutions

Metal	Type of Bath	Cathode Efficiency,* %
Cadmium	Cyanide	88-95
Chromium	Chromic acid-sulfate	12-16
Copper	Acid sulfate	97-100
Copper	Cyanide	30-95
Copper	Rochelle-cyanide	40-70
Cobalt	Acid sulfate	95-98
Gold	Cyanide	70-90
Indium	Cyanide	30-50
Indium	Fluoborate	30-50
Indium	Acid sulfate	70-80
Iron	Acid chloride	90-98
Iron	Acid sulfate	95-98
Lead	Fluoborate	100
Lead	Fluosilicate	100
Nickel	Acid sulfate	94-98
Silver	Cyanide	100
Tin	Acid sulfate	90-95
Tin	Stannate	60-90
Rhodium	Acid phosphate	10-18
Rhodium	Acid sulfate	10-18
Zinc	Chloride	95-98
Zinc	Cyanide	75-90
Zinc	Alkaline	50-75

* The cathode efficiencies are approximate. Efficiencies vary depending upon the electrolyte compositions and actual plating conditions.

GLOSSARY OF TERMS

The definitions for these selected terms are taken from the American Society for Testing and Materials' Standard B 374.[1]

activation Elimination of a passive condition on a surface to ensure adhesion.

activity (ion) The ion concentration corrected for deviations from ideal behavior. Concentration multiplied by activity coefficient.

addition agent A material added in small quantities to a plating solution for the purpose of modifying the character of the deposit.

adhesion The attractive force that exists between an electrodeposit and its substrate that can be measured as the force required to separate an electrodeposit and its substrate.

amorphous Noncrystalline or devoid of regular structure.

ampere The current that will deposit silver at the rate of 0.0011180 g/sec. Current flowing at the rate of one coulomb per second.

anode The positive electrode in electrolysis, at which negative ions are discharged, positive ions are formed, or other oxidizing reactions occur.

anodic coating A protective, decorative, or functional coating, formed by conversion of the surface of a metal in an electrolytic oxidation process.

anodizing An electrolytic oxidation process in which the surface of a metal, when anodic, is converted to a coating having desirable protective, decorative, or functional properties.

anolyte The portion of electrolyte in the vicinity of the anode; in a divided cell, the portion of electrolyte on the anode side of the diaphragm.

antipitting agent An addition agent for the specific purpose of preventing gas pits in a deposit. *See* wetting agent.

back emf (electromotive force) The potential set up in an electrolytic cell that opposes the flow of current, caused by such factors as concentration polarization and electrode films.

barrel plating (or cleaning) Plating or cleaning in which the work is processed in bulk in a rotating container.

base metal A metal that readily oxidizes or dissolves to form ions. The opposite of noble metal.

basis metal (or material) Material upon which coatings are deposited.

bipolar electrode An electrode that is not directly connected to the power supply but is so placed in the solution between the anode and the cathode that the part nearest the anode becomes cathodic and the part nearest the cathode becomes anodic.

bright dip (nonelectrolytic) A solution used to produce a bright surface on a metal.

bright plating A process that produces an electrodeposit having a high degree of specular reflectance in the as-plated condition.

bright plating range The range of current densities within which a given plating solution produces a bright plate.

bright throwing power The measure of the ability of a plating solution or a specified set of plating conditions to uniformly deposit bright electroplate upon an irregularly shaped cathode; in particular, recessed, low current density areas of the plated part.

brightener An addition agent that leads to the formation of a bright plate or that improves the brightness of the deposit.

brush plating A method of plating in which the plating solution is applied with a pad or brush, within which is an anode that is moved over the cathode to be plated.

buffer A compound or mixture that, when contained in solution, causes the solution to resist change in pH. Each buffer has a characteristic limited range of pH over which it is effective.

building up Electroplating for the purpose of increasing the dimensions of an article.

Reviewers, cont.: ***Joe L. Lester****, President, Metaplast Electrochemicals Corp.;* ***Joe A. Miglionico****, Marketing Manager, Sifco Selective Plating Div., Sifco Industries, Inc.;* ***Joe C. Norris****, Metallurgist, Sifco Selective Plating Div., Sifco Industries, Inc.;* ***Art O'Cone****, Marketing Manager, 3M Company;* ***Leslie W. Piddington****, Director of Special Services, Sifco Selective Plating, Sifco Industries, Inc.;* ***Richard Rapids****, Vice President, Rapid Electroplating Process, Inc.;* ***Carl M. Rodia****, Vice President, Audio Matrix, Inc.;* ***Gary Shawhan****, Product Manager, Electroless Nickel, Enthone, Inc.;* ***Gary W. Smith****, Research and Development Manager, Sifco Selective Plating Div., Sifco Industries, Inc.;* ***Gary Sovran****, Service Engineering, M & T Chemicals, Inc.;* ***Hal Thrasher****, Marketing Manager, MacDermid, Inc.;* ***Robert A. Tremmel****, Research Group Leader, Udylite Plating Systems, OMI International Corp.;* ***John White****, President, Liquid Development Co., Inc.*

CHAPTER 20

GLOSSARY OF TERMS

burnt deposit A rough, noncoherent, or otherwise unsatisfactory deposit produced by the application of an excessive current density and usually containing oxides or other inclusions.

bus (busbar) A rigid conducting section for carrying current to the anode or cathode bars.

catalyst An element or ion that promotes or facilitates a reaction without itself being affected or changed.

cathode The negative electrode in electrolysis at which positive ions are discharged, negative ions are formed, or other reducing actions occur.

cathode film The layer of solution in contact with the cathode that differs in composition from that of the bulk of the solution.

cation A positively charged ion.

colloidal particle An electrically charged particle, generally smaller in size than 0.2 μm, dispersed in a second, continuous phase.

coloring The production of desired colors on metal surfaces by appropriate chemical or electrochemical action.

complex ion An ion composed of two or more ions or radicals, both of which are capable of independent existence, that imparts the property of solubility necessary for electroplating.

complexing agent A compound that will combine with metallic ions to form soluble ions. *See* complex ion.

conditioning (plastics) The conversion of a surface to a suitable state for successful treatment in succeeding steps.

contact plating Deposition of a metal with the use of an internal source of current by immersion of the work in solution in contact with another metal.

conversion coating A coating produced by chemical or electrochemical treatment of a metallic surface that produces a superficial layer containing a compound of the metal; for example, chromate coatings on zinc and cadmium or oxide coatings on steel.

coulomb The quantity of electricity that passes any section of an electric circuit in one second when the current in the circuit is one ampere.

covering power The ability of a plating solution, under a specified set of plating conditions, to deposit metal on the surfaces of recesses or deep holes. (To be distinguished from throwing power.)

critical current density A current density above which a new and sometimes undesirable reaction occurs.

current density (cd) Current per unit area; usually expressed in amperes per square foot (A/ft^2) or amperes per square decimeter (A/dm^2).

diaphragm A porous or permeable membrane separating anode and cathode compartments of an electrolytic cell from each other or from an intermediate compartment.

diffusion coating An alloy coating produced by applying heat to one or more metal coatings deposited on a basis metal.

dispersing agent A material that increases the stability of a suspension of particles in a liquid medium.

drag-in The water or solution that adheres to the objects introduced into a bath.

drag-out The solution that adheres to the objects removed from a bath.

dummy (or dummy cathode) A cathode in a plating solution that is not to be used after plating. Often used for removal or decomposition of impurities.

electrochemical equivalent The weight of an element, compound, radical, or ion involved in a specified electrochemical reaction during the passage of a unit quantity of electricity, such as a faraday, ampere hour, or coulomb.

electrode A conductor through which current enters or leaves an electrolytic cell at which there is a change from conduction by electrons to conduction by charged particles of matter, or vice versa.

electrodeposition The process of depositing a substance upon an electrode by electrolysis. *See* electroforming, electroplating, electrorefining, and electrowinning.

electrode potential The difference in potential between an electrode and the immediately adjacent electrolyte.

electroforming The production or reproduction of articles by electrodeposition upon a mandrel or mold, which is subsequently separated from the deposit.

electrogalvanizing Electrodeposition of zinc coatings.

electroless plating Deposition of a metallic coating by a controlled chemical reduction that is catalyzed by the metal or alloy being deposited.

electrolyte A conducting medium in which the flow of current is accompanied by movement of matter. Most often an aqueous solution of acids, bases, or salts, but includes many other media, such as fused salts, ionized gases, and some solids.

electrolysis Production of chemical changes by the passage of current through an electrolyte.

electrolytic cell A unit apparatus in which electrochemical reactions are produced by applying electrical energy, or which supplies electrical energy as a result of chemical reactions and which includes two or more electrodes and one or more electrolytes contained in a suitable vessel.

electrophoresis The movement of colloidal particles produced by the application of an electric potential.

electroplating The electrodeposition of an adherent metallic coating upon an electrode for the purpose of securing a surface with properties or dimensions different from those of the basis metal.

electrorefining The process of anodically dissolving a metal from an impure anode and depositing it cathodically in a purer form.

electrowinning The production of metals by electrolysis with insoluble anodes, in solutions derived from ores or other materials.

faraday The number of coulombs (96,500) required for an electrochemical reaction involving one chemical equivalent.

flash (or flash plate) A thin electrodeposit that is less than 0.1 mil (2.5 μm). *See* strike.

free cyanide (1) *Calculated*—the concentration of cyanide or alkali cyanide present in solution in excess of that calculated as necessary to form a specified complex ion with a metal or metals present in solution. (2) *Analytical*—the free cyanide content of a solution, as determined by a specified analytical method.

foam blanket An additive that forms a layer on the surface of electroplating baths that have poor anode/cathode efficiency and prevents any mist or spray from escaping.

galvanic cell An electrolytic cell capable of producing electrical energy by electrochemical action.

galvanic series A list of metals and alloys arranged according to their relative potentials in a given environment.

galvanizing Application of a coating of zinc.

GLOSSARY OF TERMS

gram-equivalent weight The weight in grams of a substance involved in an oxidation-reduction reaction that is equivalent to one mole of electrons.

hard chromium Chromium plate for engineering rather than decorative applications. Not necessarily harder than the latter, but generally thicker or heavier.

hydrogen embrittlement Embrittlement of a metal or alloy caused by absorption of hydrogen during a pickling, cleaning, or plating process.

hydrophilic Tending to absorb water or to concentrate in the aqueous phase.

hydrophobic Tending to repel water or lacking affinity for water.

immersion plate A metallic deposit produced by a displacement reaction in which one metal displaces another from solution.

indicator (pH) A substance that changes color when the pH of the medium is changed. In the case of most useful indicators, the pH range within which the color changes is narrow.

inert anode An anode that is insoluble in the electrolyte under the conditions prevailing in the electrolysis.

inhibitor A substance used to reduce the rate of a chemical or electrochemical reaction.

ion An electrified portion of matter of atomic or molecular dimensions.

ion exchange A reversible process by which ions are interchanged between a solid and a liquid with no substantial structural changes of the solid.

leveling action The ability of an electroplated deposit to produce a surface smoother than that of the basis metal.

mandrel A form used as a cathode in electroforming; a mold or matrix.

mechanical plating The application of an adherent metallic coating by mechanical means involving the compacting of finely divided particles of such metal to form coherent coatings.

metallizing (1) The application of an electrically conductive metallic layer to the surface of nonconductors. (2) The application of metallic coatings by nonelectrolytic procedures such as spraying of molten metal and deposition from the vapor phase.

microthrowing power The ability of a plating solution or a specified set of plating conditions to deposit metal in tiny pores or scratches.

mil One thousandth of an inch (0.001″).

noble metal A metal that does not readily tend to furnish ions, and therefore does not readily dissolve nor easily enter into such reactions as oxidations. The opposite of base metal.

nucleation (plastics) The preplating step in which a catalytic material, often a palladium or gold compound, is absorbed on the surface.

nodule A rounded projection formed on a cathode during electrodeposition.

pH The negative logarithm of the hydrogen-ion activity or hydrogen-ion concentration in gram equivalents per liter that is used in expressing both acidity and alkalinity on a scale of values running from 0 to 14; 0 to 7 represents acidity, 7 neutrality, and 7 to 14 alkalinity.

periodic reverse plating A method of plating in which the current is reversed periodically. The cycles are usually no longer than a few minutes and may be much less.

plating range The current density range over which a satisfactory electroplate can be deposited.

primary current distribution The distribution of the current over the surface of an electrode in the absence of polarization.

pulse plating A method of plating that uses a power source capable of producing square-wave current pulses.

rack (plating) A frame for suspending and carrying current to articles during plating and related operations.

reducing agent A compound that causes reduction, thereby itself becoming oxidized.

ripple (d-c) Regular modulations in the d-c output wave of a rectifier unit or a motor generator set, originating from the harmonics of the a-c input system in the case of a rectifier or from the harmonics of the induced voltage of a motor generator set.

sacrificial protection The form of corrosion protection wherein one metal corrodes in preference to another, thereby protecting the latter from corrosion.

sealing of anodic coating A process that by absorption, chemical reaction, or other mechanism increases the resistance of an anodic coating to staining and corrosion, improves the durability of colors produced in the coating, or imparts other desirable properties.

sensitization (plastics) The surface absorption of a reducing agent, often a stannous compound.

specific gravity The density of a substance relative to some standard substance taken as unity (usually water = 1).

spotting out The delayed appearance of spots and blemishes on plated or finished surfaces that is most prevalent on porous basis metals or substrates.

strike (1) A thin film of metal to be followed by other coatings. (2) To plate for a short time, usually at a high initial current density.

substrate Surface material or electroplate upon which a subsequent electrodeposit or finish is made. *See* basis metal or material.

throwing power The improvement of the coating (usually metal) distribution over the primary current distribution on an electrode (usually cathode) in a given solution, under specified conditions. The term may also be used for anodic processes for which the definition is analogous.

total cyanide The total content of cyanide expressed as the radical CN^- or as alkali cyanide, whether present as simple or complex ions. The sum of both the combined and free-cyanide content of a solution.

trees Branched or irregular projections formed on a cathode during electrodeposition, especially at edges and other high current density areas. Shield, robbers, or other parts are used to prevent this formation from developing.

vapor deposition (1) Chemical—a process for producing a deposit by chemical reaction, induced by heat or gaseous reduction of a vapor condensing on the substrate. (2) Physical—a process for depositing a coating by evaporating and subsequently condensing an element or compound, usually in a high vacuum.

water break The appearance of a discontinuous film of water on a surface signifying nonuniform wetting and usually associated with a surface contamination.

wetting agent A substance that reduces the surface tension of a liquid, thereby causing it to spread more readily on a solid surface.

whiskers Metallic filamentary growths, often microscopic; sometimes formed during electrodeposition and sometimes spontaneously during storage or service after finishing. Common on electrical contacts plated with zinc or cadmium.

CHAPTER 20

METALS ELECTROPLATED

METALS ELECTROPLATED

Many different metals can be successfully electroplated onto other materials. Each metal possesses unique properties and characteristics that can improve the usefulness of the basis metal. Some of the more common electroplating metals are nickel, copper, chromium, zinc, tin, cadmium, and lead. Alloys and precious metals are also deposited.

Nickel Plating

Solutions. Nickel plating is documented as early as 1842, but modern nickel plating really began in 1916 with the introduction of the Watts bath. The Watts formulation, consisting of nickel sulfate, nickel chloride, and boric acid, is by far the most widely used electrolyte. Modern-day organic additives produce brilliant, level, low-stress deposits. These additives revolutionized conventional nickel plating in that they have virtually eliminated the laborious process of mechanical buffing. The more corrosion-resistant, sulfur-free, semibright nickel deposits have also become important as undercoatings for bright nickel.

A nickel sulfamate bath is used for engineering (usually nondecorative) applications to produce low-stress deposits. This bath is also useful in electroforming and for plating parts that are susceptible to fatigue failure. An all-chloride Woods nickel formulation is useful as a strike bath for treating passivated nickel and difficult-to-plate substrates such as stainless steel. There is also limited use of the nickel fluoborate bath for electroforming and other engineering applications. Table 20-3 presents the compositions of the typical nickel electrolytes.

Applications. Nickel plating is widely used for decorative purposes. Bright nickel plate—especially in combination with a lower layer of sulfur-free, semibright nickel and a much thinner upper layer of chromium—is very widely used over steel, brass, zinc die castings, aluminum, and chemically metallized plastics to provide a bright and corrosion-resistant finish with a nontarnishing and wear-resistant surface. Typical applications are decorative trim for automotive and consumer products. For best corrosion resistance, the chromium deposit should be microdiscontinuous (microcracked or microporous).

Nickel deposits are also used for nondecorative purposes to improve or modify surface properties such as corrosion resistance, hardness, wear, and magnetic characteristics. For example, in the automotive industry, nickel coatings, generally greater than 5 mil (1 mil = 0.001″) (125 μm) thick, are deposited on pistons, cylinder walls, ball studs, transmission thrust washers, and differential pinion cross-shafts to improve wear resistance. Nickel coatings are also deposited on pots and kettles in the food processing industry, on drying cylinders and rolls in the paper and pulp industry, on tape condensers and calender rolls in the textile industry, and on elbows, joints, and other piping components in the chemical and nuclear industries to improve corrosion resistance. Table 20-4 lists the properties of nickel coatings deposited from typical nickel electrolytes.

Depending on the substrate and the end use of the plated product, nickel may be applied directly to the substrate or over another metal coating such as copper and single or duplex nickel, with or without subsequent chromium. The thickness of the deposit depends on the substrate and the end use. For additional information, refer to American Society for Testing and Materials (ASTM) Specification B 456.

Copper Plating

Solutions. Copper is deposited from two main types of plating baths, the alkaline cyanide copper and acid sulfate copper baths. Among the alkaline baths, the pyrophosphate copper bath is used to some extent for electronic applications; among acid baths, the fluoborate bath is also used for electronic applications.

Despite waste disposal problems, the cyanide copper bath continues to be widely used owing to its superior throwing power and ability to satisfactorily plate steel and zinc die castings or zincate or stannate-treated aluminum substrates before nickel plating. For these applications, copper is generally first deposited from a cyanide copper strike, followed by a high-speed cyanide copper or an acid copper bath.

The acid copper bath is primarily used for decorative applications, especially for plating on plastics. Proprietary additives help achieve good microthrowing power (leveling)

TABLE 20-3
Basic Compositions (excluding additives) of Commonly Used Nickel Plating Baths, oz/gal (g/L)

Constituents	Watts	All Chloride	Sulfamate	Fluoborate
Nickel chloride	4-8 (30-60)	30-40 (225-300)	---	---
Nickel sulfate	30-50 (225-375)	---	---	---
Nickel sulfamate	---	---	60 (450)	---
Nickel fluoborate	---	---	---	40 (300)
Boric acid	4-6 (30-45)	4-6 (30-45)	5 (37.5)	4 (30)
Antipitting agent	yes	yes	yes	yes

TABLE 20-4
Representative Mechanical and Physical Properties of Nickel Deposits

Property	Watts	All Chloride	Sulfamate	Fluoborate
Tensile strength, ksi (MPa)	60 (413)	100 (690)	60 (413)	58 (400)
Elongation in 2″ (50 mm), %	28	14	30	30
Internal stress, ksi (MPa)	20 (138)	50 (345)	2 (14)	20 (138)
Hardness, Vickers	140-160	240	190	150

and brightness. Other uses of the copper sulfate and sometimes the fluoborate bath include pattern plating of printed circuit boards, electroforming, and plating of rotogravure printing plates. Common bath compositions for cyanide and acid copper baths are presented in Table 20-5.

Applications. Copper plating is commonly used:

- As a decorative and corrosion-resistant deposit under nickel and chromium.
- As a base deposit for nickel and/or chromium-plated plastics and zinc die castings.
- As a heat treat and nitriding stopoff.
- For electroforming applications (records and waveguides).
- As a decorative final finish (usually lacquered).
- For through-hole plating of printed circuit boards.

TABLE 20-5
Compositions of Commonly Used Copper Plating Baths

		Concentration, oz/gal (g/L)				
				High Efficiency Baths		Pryo-phosphate Bath
Type	Constituents	Strike Bath	Rochelle Bath	Sodium	Potassium	
Alkaline Cyanide	Copper cyanide	2.5-3.5 (18.5-26)	3.0-4.0 (22.5-30)	9-11 (67.4-82.4)	9-11 (67.4-82.4)	---
	Sodium cyanide	0.7-1.4 (5.2-10.5)	0.5-1.0 (3.75-7.5)	0.5-1.5 (3.75-11.2)	---	---
	Sodium carbonate	2.0-8.0 (15-60)	2-8 (15-60)	0-12 (0-90)	---	---
	Potassium carbonate	---	---	---	0-16 (0-120)	---
	Sodium hydroxide	---	---	3-5 (22.5-37.5)	---	---
	Potassium hydroxide	---	---	---	4-7 (30-52.4)	---
	Potassium cyanide	---	---	---	1.0-2.0 (7.5-15)	---
	Rochelle salt	---	4-8 (30-60)	---	---	---
	Copper (Cu)	---	---	---	---	3-5 (22.5-37.5)
	Pryophosphate (P_2O_7)	---	---	---	---	20-33 (150-247)
	Nitrate (NO_3)	---	---	---	---	2/3-1 1/3 (5-10)
	Ammonia (NH_3)	---	---	---	---	1/8-3/8 (1-2.8)
Acid Copper		Sulfate	Fluoborate			
	Copper sulfate	27-33 (200-250)	---			
	Sulfuric acid	6-10 (45-75)	---			
	Copper fluoborate	---	33 (250)			
	Fluoboric acid	---	5.3 (40)			
	Proprietary additives	1-2% by volume				

(*Metal Finishing*)

CHAPTER 20

METALS ELECTROPLATED

Chromium Plating

Solutions. Most chromium plating is done from hexavalent (CrO_3) baths, but trivalent systems are gaining in popularity. There has also been a fair amount of microdiscontinuous, either microcracked or microporous, chromium deposits used to improve corrosion resistance. In the galvanic couple, the underlying nickel acts as the anode, and the chromium behaves as the cathode. By exposing a greater surface area of nickel through the use of microdiscontinuous chromium, the corrosion current is spread over a large area; consequently, corrosion proceeds uniformly. As a result, corrosion protection in outdoor environments has been improved fivefold.

Hexavalent chromium plating baths consist of chromic acid and small amounts of a catalyst ($SO_4^{=}$). Recently, mixed catalyst baths containing fluoride compounds in addition to chromic acid and sulfate have been employed. Proprietary self-regulating baths control the concentration of the catalyst automatically. Some typical bath formulations are summarized in Table 20-6.

Applications. Chromium plating is divided into decorative and hard coatings. Decorative coatings are applied over a base deposit of nickel or copper plus nickel to provide color and tarnish resistance as well as a hard, protective finish. Coating thicknesses are usually less than 0.03 mil (0.75 μm). Decorative chromium coatings are most often found on automobiles, furniture, and kitchen appliances.

Hard chromium coatings are generally deposited directly on the base material without a nickel undercoat in thicknesses ranging from 0.1 to 20 mils (2.5 to 500 μm). Hard coatings provide resistance to wear, heat, abrasion, and/or corrosion. Typical applications for hard coatings include hydraulic pistons and cylinders, piston rings, wearing parts in business machines, aircraft engine parts, yarn and thread guides for textiles, plastics molds, and various parts of nuclear reactors where galling is a particular concern.

Zinc Plating

Solutions. Commercially, zinc is deposited from three different baths: the conventional cyanide bath, the acid chloride bath, and the alkaline noncyanide (or zincate) bath. Cyanide baths offer ease of control and normally trouble-free plating. However, the cost of cyanide destruction and the toxicity of the bath have prompted platers to install low-cyanide solutions or cyanide-free baths.

TABLE 20-6
Compositions of Commonly Used Decorative Chromium Plating Baths

Type of Bath	Constituents	Concentration, oz/gal (g/L)
Hexavalent	Chromic acid (CrO_3)	27-40 (200-300)
	Sulfuric acid	0.20-0.33 (1.5-2.5)
	Ratio of chromic acid to sulfate	100:1 to 125:1
Trivalent	Chromium (total)	2.4-3.3 (18-25)
	Conductivity salts	26.6-40 (200-300)
	Surfactant	0.13-0.26 (1.0-2.0)*
	Complexor	6.4-10.2 (50-80)*

(*MacDermid, Inc.*)

* fl oz/gal (mL/L)

Chloride zinc baths have been available since the late 1960s. The original baths were chelated or based on complexing agents such as ammonium chloride. Today, state-of-the-art chloride baths use potassium or ammonium chloride. The advantages of the chloride systems include brilliant deposits, high cathode efficiency, low electric power consumption, and nontoxic, easily treated electrolyte. The disadvantages are poor throwing power, a higher initial equipment investment, and higher brightener costs compared to the alkaline processes.

Alkaline noncyanidic electrolytes consist of zinc and sodium hydroxide. In the absence of cyanide, proprietary sequestering agents are sometimes used to yield grain refinement. Alkaline noncyanide electrolytes are simple and low cost. The solutions and rinse waters are easily treated, and metal hydroxide sludges are reduced owing to low zinc content in the bath. These baths, however, offer the lowest cathode efficiency, and the deposits may be yellowish in color. Blistering is common at higher thicknesses and may be related to the greater hydrogen occlusion.

There are approximately 88,000 tons (80 000 000 kg) of zinc currently consumed per year in the United States for electroplating. Approximately 40% is plated from cyanide-bearing baths, and another 40% is plated from chloride zincs, which are fast gaining popularity. The remainder is plated from alkaline noncyanide baths and, to a limited extent, from acid sulfate baths, which are used for high-speed strip and conduit plating. Table 20-7 identifies typical compositions of the three commercially employed electrolytes. Table 20-8 compares the baths with respect to some important characteristics.

Applications. Zinc is plated over iron and steel on a wide variety of parts where sacrificial corrosion resistance is required. The conventional zinc coating is dull gray in color with a matte finish, but whiter and more lustrous deposits can be produced by the use of proprietary addition agents. Zinc plate is almost always passivated with a chromate coating for added corrosion protection. Coating thickness ranges from 0.2 to 2.0 mils (5 to 50 μm) after chromate coating, depending on the particular application.

The formation of white corrosion products in marine environments makes zinc less desirable than cadmium; but because it is less toxic and less expensive and its electrolytes are more easily waste treated, zinc has replaced cadmium in many applications. Zinc coatings are also superior to cadmium coatings in industrial environments. Common applications for zinc plating include fasteners, wire goods, tools, and sheet metal parts.

Tin Plating

Tin deposit finishes may be bright as plated or matte. Matte deposits may be reflowed, a practice which is almost universally applied to continuously plated steel strip for can manufacturing and is referred to as tinplate or electrotinplate. In tinplating, the reflowing is accomplished by induction or resistance heating of the continuously moving strip. Small articles are often reflowed in hot oil or fat; however, many tin-plated articles are used without reflowing. Tin plating is covered by ASTM Specification B 545.

Solutions. Tin may be plated from either alkaline electrolytes (stannate) or acid electrolytes (sulfate, fluorborate), as described in Table 20-9. A special Halogen tin electrolyte containing stannous chloride, sodium fluoride, potassium bifluoride, sodium chloride, plus other additives is also used for tinplating rapidly moving strip at high current densities. Typical operating conditions are a temperature of 150° F (65° C), pH of 2.7, and

TABLE 20-7
Basic Compositions of Common Zinc Plating Baths

Constituents	Concentration, oz/gal (g/L) Mid-Cyanide	Chloride (Potassium)	Alkaline, noncyanide
Zinc cyanide	4.0 (30)	---	---
Zinc chloride	---	8.0 (60)	---
Zinc oxide	---	---	1.3 (9.5)
Sodium cyanide	2.7 (20)	---	---
Potassium chloride	---	28.0 (210)	---
Sodium hydroxide	10.0 (75)	---	14.0 (105)
Sodium carbonate	2.0 (15)	---	2.0 (15)
Boric acid	---	4.3 (32)	---
Sodium polysulfide	0.3 (2)	---	---
Addition agents	As required	As required	As required

(*MacDermid, Inc.*)

TABLE 20-8
Zinc Electrolyte Comparison

Consideration	Mid-Cyanide	Chloride	Alkaline Noncyanide
Appearance	Bright	Brilliant/level	Bright
Cathode efficiency	75-85%	95-100%	50-75%
Throwing power	50%	25%	45%
Covering power	Good	Superior	Fair
Equipment required	Minimum expense	Acid-resistant materials	Minimum expense
Waste disposal	Destroy cyanide, precipitate metal, and then neutralize solution.	Metal precipitation and solution neutralization only.	Metal precipitation and solution neutralization only; low sludge generation.
Chromate receptivity	Excellent	Good	Fair

(*MacDermid, Inc.*)

TABLE 20-9
Compositions of Common Tin Plating Baths

Type of Bath	Constituents	Concentration, oz/gal (g/L) A*	B*	C*
Alkaline stannate tin	Potassium stannate	14 (105)	56 (420)	---
	Sodium stannate	---	---	14 (105)
	Potassium hydroxide	2 (15)	3 (22)	---
	Sodium hydroxide	---	---	1.3 (10)
		Sulfate: Matte	Sulfate: Bright	Fluoborate
Acid tin	Stannous sulfate	8-13 (60-100)	2-6 (15-45)	---
	Tin (as metal)	4-6.5 (30-50)	1-3 (7.5-22.5)	3.25-10.7 (24.4-81)
	Free sulfuric acid	5.3-9.3 (40-70)	18.7-28 (140-210)	---
	Stannous fluoborate	---	---	8-26.7 (60-200)
	Free fluoboric acid	---	---	8-26.7 (60-200)
	Free boric acid	---	---	2.7-5.4 (20-40)
	Proprietary additives	As required	As required	As required

* A = Low-potassium stannate bath
B = High-metal potassium bath
C = Sodium stannate bath

METALS ELECTROPLATED

current density of 4.8 A/ft^2 (45 A/dm^2). The factors influencing the choice of electrolytes are coating appearance, operation, equipment, anodes, and power requirements.

Coating appearance. Alkaline baths provide a matte coating that is sensitive to fingerprints. Acid baths can be used with proprietary brighteners to produce a bright coating that is less sensitive to fingerprinting.

Operation. Alkaline baths are easier to operate than acid baths. The preplate cycles are also simpler and require fewer steps than acid baths.

Equipment. Tanks for alkaline baths can be made from unlined steel, while acid baths require lined tanks. Alkaline baths need to be heated to 170 to 190°F (77 to 88°C), while bright acid baths need to be cooled to 60 to 80°F (16 to 27°C).

Anodes. Alkaline baths can be operated with inert anodes for plating internal threads or steel couplings, as well as soluble tin anodes or tin anodes alloyed with 1% aluminum (high-speed anodes). Acid baths cannot be operated with inert anodes.

Power requirements. The metal deposition rate at the same current density is at least twice as high in acid baths as in alkaline baths because the electrochemical equivalent of Sn^{2+} (acid baths) is twice as high as Sn^{4+} (alkaline baths), and because acid baths have a higher current efficiency. Operating temperatures and bath voltages are lower in the acid baths.

Applications. Tin, as a metal, displays some very desirable properties. Tin is compatible with foods, is nontoxic, has a relatively low melting point [450°F (232°C)], and is readily soldered. Consequently, tin electroplate finds use in a multitude of products, including tinplate (tin-plated steel strip) for use in consumer food containers, buttons, gas-tank hardware, semiconductors, copper wire, electronic devices, and printed circuit boards.

Advantages and limitations. Tin provides sacrificial protection for copper, nickel, and other nonferrous metals; it does not provide similar protection to steel. A drawback with tin is the possibility of whisker growth, which is the formation of fine metal slivers that can short-circuit electronic devices. Tin is also a relatively expensive metal.

Cadmium Plating

Cadmium is a soft, silver-white metal with unique engineering properties that is obtained as a by-product of zinc smelting. Some of these properties include good conductivity, solderability, lubricity, ductility, and adhesive retention; excellent corrosion resistance and porous surface coverage; and minimal hydrogen embrittlement. Chromating after plating provides additional corrosion resistance. However, cadmium is extremely toxic and costly and is consequently being replaced by zinc wherever possible. About one third of the cadmium produced today is used in electroplating.

Solutions. Cadmium is primarily plated from a cyanide electrolyte. Acid cadmium baths are also used to a limited extent, and are fluoborate, sulfate, or chloride in nature. The acid-type baths are more desirable if hydrogen embrittlement is a problem, and their waste treatment is simplified. However, the cyanide baths are easier to control than the acid baths. Typical cadmium plating compositions are summarized in Table 20-10.

Applications. Cadmium is used as a protective coating over steel, cast iron, and malleable iron. It also provides galvanic protection on iron and steel. The government is by far the largest specifier of cadmium, for military applications. Deposit thicknesses range from 0.2 to 1 mil (5 to 25 μm), depending on the degree of exposure to corrosives and wear. Typical applications include springs, lock washers, fasteners, electronic and electrical parts, washing machine parts, and military hardware. Cadmium plating should never be used on parts that will come in contact with food or beverages.

Lead

Lead is electroplated to a limited extent to provide corrosion resistance or a good bearing surface, and is more apt to be plated as an alloy than as a pure metal. Lead baths may be fluoborate, fluosilicate, perchlorate, or sulfamate. The fluoborate is the most often used for storage battery parts and the linings of tanks and chemical apparatus.

Alloy Electroplating

Although hundreds of alloys have been successfully electroplated in the laboratory, few have been commercially used. The more important processes are described subsequently.

Brass. The main brass alloy consists of approximately 70% copper and 30% zinc, and has been plated for many years for decorative purposes as a substitute for solid brass. Most

TABLE 20-10
Typical Cadmium Electroplating Baths

	Concentration, oz/gal (g/mL)			
Constituent	Cyanide	Neutral, noncyanide	Fluoborate	Acid Sulfate
Ammonium chloride	---	---	---	---
Ammonium fluoborate	---	---	8.0 (60)	---
Ammonium sulfate	---	10.0-15.0 (75-112)	---	---
Boric acid	---	---	3.6 (27)	---
Cadmium	2.6 (19.5)	0.5-1.5 (3.7-11.2)	12.6 (94)	---
Cadmium fluoborate	---	---	32.2 (241)	---
Cadmium oxide	3.0 (22.5)	---	---	1.0-1.5 (7.5-11.2)
Sodium carbonate	4.0-8.0 (30-60)	---	---	---
Sodium cyanide	13.3 (100)	---	---	---
Sodium hydroxide	1.87 (14)	---	---	---
Sulfuric acid	---	---	---	4.5-5.0%
Brighteners	As required	As required	As required	As required

deposits are rather thin, from a mere flash up to 1 mil (25 μm) thick, and are generally protected by a clear lacquer coating. Brass is often plated over bright nickel to improve the luster of the nickel coating.

Brass has been plated over steel to promote rubber adhesion and to provide lubricating characteristics. White brass, containing 30% copper and 70% zinc, is also used as a substitute for bright nickel on toys, tubular furniture, and automotive interior trim and hardware; this practice is decreasing, however, because of the superior performance of decorative nickel/chromium coatings. The main brass plating bath is cyanide.

Bronze. The main electroplating alloy in this group contains from 8 to 15% tin, with the remainder copper. Bronze may be deposited from a potassium stannate and copper cyanide bath that contains potassium hydroxide and potassium cyanide. Relatively little bronze plating is done, but bronze plating has been used as an undercoating in the preparation of aluminum alloys for plating with nickel-chromium coatings.

Tin alloy deposits. Tin-lead alloy plating is used to protect steel from corrosion and for etch resistance. However, the major purpose for its use in the plating of wire, electronic devices, and printed circuit boards is to facilitate solderability. The most widely used tin alloy electrolyte is the fluoborate bath. The deposit normally has a matte appearance as plated, but the deposit can be reflowed, as in tin plating, to improve the appearance. Recently, proprietary organic additives have been developed to produce brilliant deposits. Proprietary, fluoborate-free electrolytes have also been introduced and appear to offer advantages such as ease of waste disposal and reduction in the rate of tin oxidation.

Among the additives used to improve the density of tin deposits by eliminating porosity and treeing are bone glue, gelatin, resorcinol, and peptone. These additives also improve throwing power. Peptone has frequently been used owing to its wide availability as a stabilized solution. Proprietary addition agents, however, are more effective grain refiners and minimize the need for frequent carbon treatments to remove reduction products.

Tin-cobalt alloys are used extensively in Europe, and to some degree in the U.S., as an alternative for chromium deposits. Tin-zinc alloys can be plated from the zincate-cyanide-stannate system, as can tin-cadmium alloys from the stannate-cadmium-cyanide system. Tin-nickel alloys, on the other hand, are plated from a completely different electrolyte containing stannous chloride, nickel chloride, and ammonium bifluoride, at a pH of 2.0 to 2.5. The tin-nickel deposit is hard, corrosion resistant, and has an attractive pink color; however, it has had little commercial use.

Nickel alloys. Nickel-iron alloys, containing 10-40% iron, have been deposited for over 10 years as a substitute for decorative bright nickel, including those demanding high corrosion protection. Deposit appearance is identical to bright nickel, and reasonable improvements in ductility, adhesion, and chromium receptivity are achieved.

Nickel-cobalt alloys have improved high-temperature properties, are harder than pure nickel deposits, and are also available commercially; however, they are not extensively used.

Plating with Precious Metals

The electrodeposition of precious metals for decorative and engineering purposes is an important part of the metal finishing industry. The high cost of each gallon of solution requires exceptionally good housekeeping. Metal recovery, security, personnel capability, and accounting are critical when operating a precious metal plating facility.

Silver. Commercial silver electroplating has been practiced since the middle of the nineteenth century. The plating bath contains silver in the form of potassium-silver cyanide, $KAg(CN)_2$, and free-potassium cyanide. Sodium cyanide may be used, but the potassium cyanide formulation is preferred. Usually a small amount of potassium carbonate and/or potassium hydroxide is also added. Silver baths are generally operated at room temperature, although high-speed plating has been done at temperatures as high as 120°F (50°C).

When hard, bright silver deposits are desired, proprietary additives containing metals or organic brighteners are generally used. Some additive combinations increase the tarnish resistance of the silver deposit. As with all bright solutions, the metal content of the bath must be closely controlled. The free cyanide is also monitored regularly.

The items to be electroplated are normally cleaned, conditioned, and cathodically activated with a potassium cyanide solution, and then given a strike in a bath with low potassium-silver cyanide and high free cyanide at 4 to 6V for 10 to 25 seconds before they are transferred to the silver plating tank.

The largest use of silver deposits, despite the popularity of stainless steel, is in the flatware/holloware trade. The second largest use is in the electronics industry where large amounts are plated onto conductors, waveguides, and similar items to take advantage of the unsurpassed electrical conductivity of silver. In most of these applications, silver is plated over copper and copper alloys.[2] Silver has also been plated over steel and used as bearings in reciprocating aircraft engines.

Gold. There are four types of gold plating solutions: alkaline, neutral, acid, and noncyanide baths. The alkaline cyanide baths were used for over a century. Because of the complexing action of cyanide, however, it was difficult to obtain consistent codeposition of other metals with the gold, except at high current densities and temperatures; as a result, deposits were limited to flash deposits.

At the midpoint of the twentieth century, bright baths were developed that used silver and selenium as alloying agents. The jewelry trade, however, desired light-colored alloys. Some success has been experienced with neutral baths, free of cyanide at the start, that build up potassium cyanide by virtue of adding potassium gold cyanide, $KAu(CN)_2$, to replenish the gold. Potassium gold cyanide is stable at a pH as low as 3.5; as a result, cobalt, nickel, indium, tin, and many other base metals can be plated with the gold in wide alloy variations. Gold deposits can therefore be tailored to the need with respect to hardness, ductility, and color.

Until recently, gold plating was primarily used for decorative purposes in jewelry, flatware, holloware, and similar items. In the last 20 years, however, the use of gold plating has widely expanded in the electronics industry because of its good electrical contact properties and corrosion and oxidation resistance. Typical applications include printed circuit boards, contacts, connectors, transistor bases, and integrated circuit components. Gold plating is also used in the chemical industry for reactors and heat exchangers. The high cost of gold in recent years has made conservation critical and has led to a search for substitutes.

Rhodium. Rhodium is plated from sulfate (rhodium sulfate plus sulfuric acid), phosphate (rhodium phosphate plus phosphoric acid), and phosphate-sulfate (rhodium phosphate plus sulfuric acid) baths. Low-stressed deposits are obtained from sulfate baths containing proprietary stress-reducing agents. The bright bath (phosphate or phosphate-sulfate) deposits are highly stressed and cracked, but have an attractive color that sets off diamonds well. The rhodium sulfate baths yield relatively heavy deposits that are hard and wear resistant. The low-stressed deposits are used when heavy plate [up to 10 mils (254 μm)] is required for good wear resistance with no cracks. Low-stressed deposits are not bright.

Platinum. A number of platinum plating solutions are available. Titanium anodes plated with platinum are used as auxiliary anodes for gold and other precious metal plating baths when plating plastics, die castings, and steel parts with complex shapes. However, an anode of mechanically clad tantalum and titanium is generally preferred. Platinum can be plated to very heavy thicknesses, but the cost is so high that there is little demand for these deposits. Heavy platinum deposits are not bright, although thin deposits can be.

Palladium. Palladium plating is more common in Europe than in the United States. Palladium baths are of the pH 9.0 to 10.0 variety and based upon palladium P salt (a palladium diamino dinitrite). Palladium coatings are not used for decorative finishes because they are dark in color and tarnish. They have been used as an undercoating for rhodium. Two common applications are watch cases and moving watch parts. The use of palladium and palladium-nickel alloy deposits appears to be growing in the electronics industry.

Ruthenium. Processes for electrodeposition of ruthenium are available. This relatively low-cost precious metal is finding use in switching devices and other electronics-related applications.

METAL SUBSTRATES

The three most common basis metals are steel, brass, and zinc (die castings). The physical properties and composition of these metals have a major influence on the platability of the part or selection of the specific finish. Adherent coatings can be deposited upon hard-to-plate surfaces by using specially developed preplating cycles or intermediate coatings differing from the final coating specified.

An example for variable platability is in the plating of cast iron. Cadmium can be plated on cast iron both in the barrel plating operation or in rack plating, but it is difficult to plate cast iron or malleable iron with zinc or nickel. If zinc is desired as the final coating, a strike plate of cadmium is usually recommended before plating, except for parts that could come in contact with food. If, however, a part is not too complicated in design or shape, it can be covered with zinc from an acid zinc plating bath, followed by an alkaline-cyanide zinc plating bath for buildup to the final thickness. This sequence is selected because the acid zinc bath will cover the cast iron more readily, and the alkaline zinc will give better metal distribution over a significant area of the part. The neutral chloride zinc bath will plate on cast iron effectively without the need for a prestrike bath. Case-hardened or carbonitride materials are similar to cast iron in platability.

With the exception of leaded brasses, the copper alloys of zinc and/or tin are not considered difficult to plate. Substrates containing lead require a special pretreatment step in order to obtain adequate and proper adhesion of the electrodeposited coating.

Because of its physical properties, such as the strength-weight ratio, beryllium is becoming a more important construction material. Many of the applications require that it be covered with another metal more resistant to service conditions. Pretreatment cycles are usually recommended to ensure adhesion of the electrodeposited coating.

If the basis metal is readily soluble in the plating solution of the selected finish, it is necessary to interpose another metal, not as soluble, before applying the final finish. As an example, aluminum is not plated in an acid nickel bath without first copper plating it; here a low alkaline copper strike bath is used prior to the high-efficiency copper bath. Some aluminum alloys have been anodized and then coated with a nickel strike.

The condition of the surface of zinc die castings influences the quality and durability of the finished electroplated product. Considerable effort has been expended in the development of proper die construction and proper casting conditions. The porosity of the surface and the presence of fissures are particularly troublesome. If the porosity is too severe, the coating may blister after processing. Preplating inspection of zinc die castings under good lighting conditions is strongly recommended.

Polishing and buffing zinc die castings to improve the porosity usually results in penetration and removal of the harder, more dense surface skin, aggravating the problem rather than relieving it. Research has shown that vibratory finishing can deburr as well as salvage defective zinc alloys before plating. After the coating has been applied, however, there is no economical way of salvaging the part should it become a reject. An acid copper deposit is now specified by automotive companies after a cyanide copper strike or plate to improve the zinc casting porosity.

If chromium is to be plated directly over copper or copper alloys, the operating conditions must be altered. Compared to nickel, the initial voltage required to plate over copper or copper alloys is usually higher to obtain rapid coverage, followed by a lowering of the voltage to one that will give the required thickness in the time allowed without burning at high current density areas.

High-tensile-strength steel alloys and steel spring stock are subject to hydrogen embrittlement. In theory, and as verified by practice, susceptible steels develop a highly brittle condition when subjected to a hydrogen atmosphere, which may occur during acid pickling as well as plating. The sensitive steels are most commonly used for engineering parts or industrial applications rather than for decorative purposes. For this reason, zinc or cadmium coatings are more widely used for corrosion protection of the basis metal. Of these two, the cathode efficiency of cyanide zinc is lower, and therefore more hydrogen is evolved during the plating period. The embrittlement is relieved by a baking operation at 350 to 450°F (175 to 230°C) for three hours, during which the hydrogen that penetrated into the steel migrates to the steel surface and is removed by permeation through the zinc coating. Since cadmium has a very low permeability factor, the baking step is not as effective as on zinc. It is believed that elimination of all hydrogen evolution at any point in the plating cycle will yield the best results. If the part can be plated in a bath having 100% cathode efficiency and no hydrogen deposited, then acid pickling should be eliminated in favor of the alkaline-type descaling process.

DESIGN CONSIDERATIONS

The laws of current distribution over the surface of the part being plated present a serious problem to both the designer and the production plater. The amount of metal deposited during electroplating is a function of the current density and time. The thickness of the deposit over any area is governed by the current density and current efficiency. The term *average current density* is commonly used to describe the plating condition used for a shaped part. Sharp corners, edges, and protruding sections or areas receive a greater portion of the current than the flat areas, and recesses receive a lesser amount of the current. Areas of high current density receive an excessive metal buildup at the expense of thinner deposits in the areas of the recesses, or low current density areas.

Correct placement of the part on the plating rack, so that edges of workpieces are shielded by adjacent workpieces or the frame of the rack, will greatly aid in improving the metal distribution. The use of auxiliary anodes placed in proper position with respect to the recessed areas is also very helpful. Any decrease in thickness of metal on areas of high current density while plating the areas of lower current density to meet specifications is a definite cost savings for the finished part.

Some of the more important design considerations affecting plating are illustrated in Fig. 20-2. Although the illustration is primarily concerned with die castings, the same principles apply to other types of parts. The plating of assembled parts can also present the problem of proper distribution of deposited metal. Often the individual parts can be more economically plated prior to assembly, especially if the individual pieces could be barrel plated while the assembled part would have to be rack plated.

Crimped or spotwelded assemblies usually have crevices under the crimp, along the weld line, or between spots that trap the processing solutions. The solution is almost impossible to rinse away because of the very small opening and may cause a number of problems throughout the processing sequence. After final rinse, the solution bleeds out and stains the finished piece. The unrinsed plating solution is also a source of contamination and may cause problems in post-treatment baths.

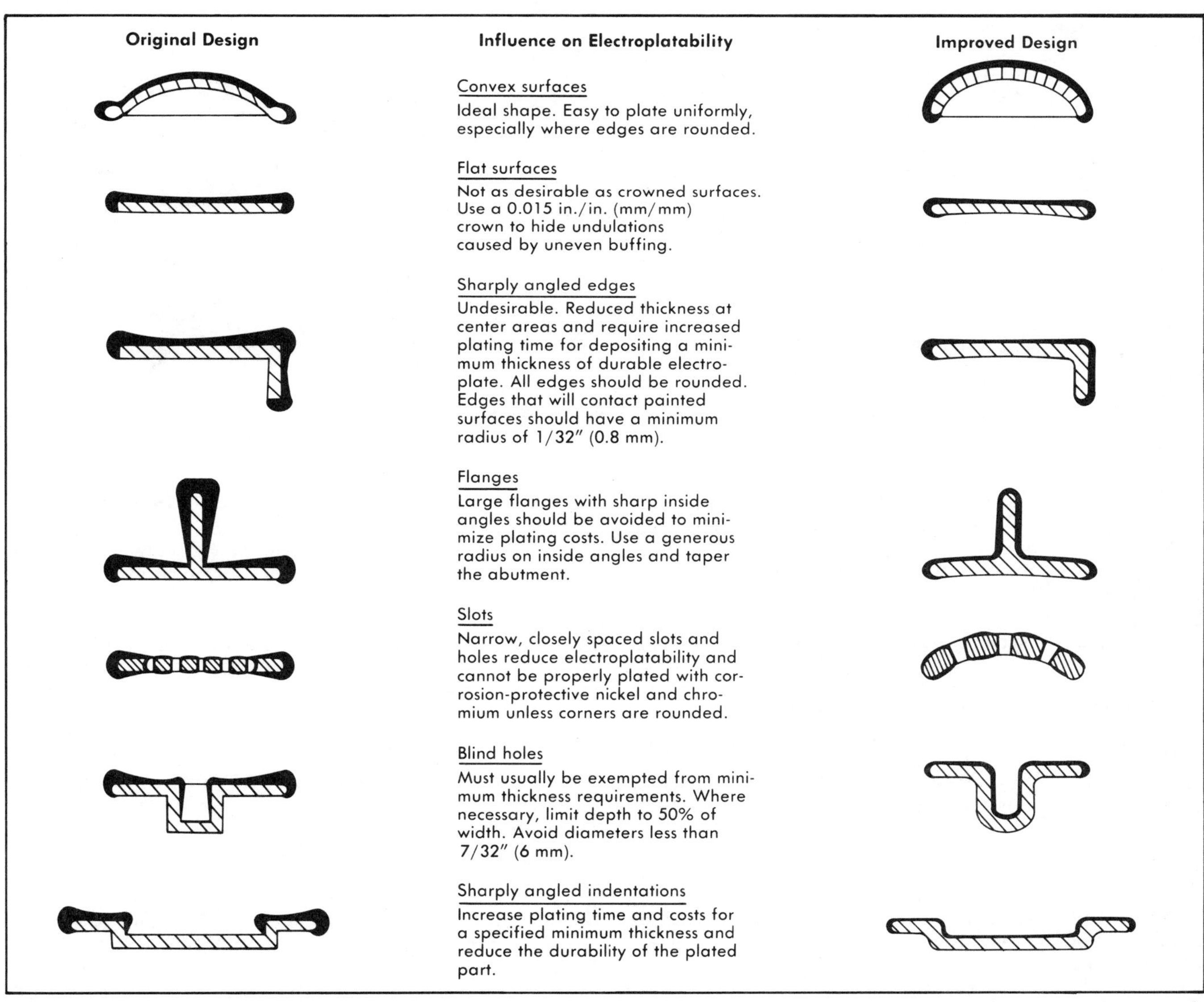

Original Design
Influence on Electroplatability
Improved Design
Convex surfaces
Ideal shape. Easy to plate uniformly, especially where edges are rounded.
Flat surfaces
Not as desirable as crowned surfaces. Use a 0.015 in./in. (mm/mm) crown to hide undulations caused by uneven buffing.
Sharply angled edges
Undesirable. Reduced thickness at center areas and require increased plating time for depositing a minimum thickness of durable electroplate. All edges should be rounded. Edges that will contact painted surfaces should have a minimum radius of 1/32" (0.8 mm).
Flanges
Large flanges with sharp inside angles should be avoided to minimize plating costs. Use a generous radius on inside angles and taper the abutment.
Slots
Narrow, closely spaced slots and holes reduce electroplatability and cannot be properly plated with corrosion-protective nickel and chromium unless corners are rounded.
Blind holes
Must usually be exempted from minimum thickness requirements. Where necessary, limit depth to 50% of width. Avoid diameters less than 7/32" (6 mm).
Sharply angled indentations
Increase plating time and costs for a specified minimum thickness and reduce the durability of the plated part.

Fig. 20-2 Improved part design permits better electroplating. *(continued)*

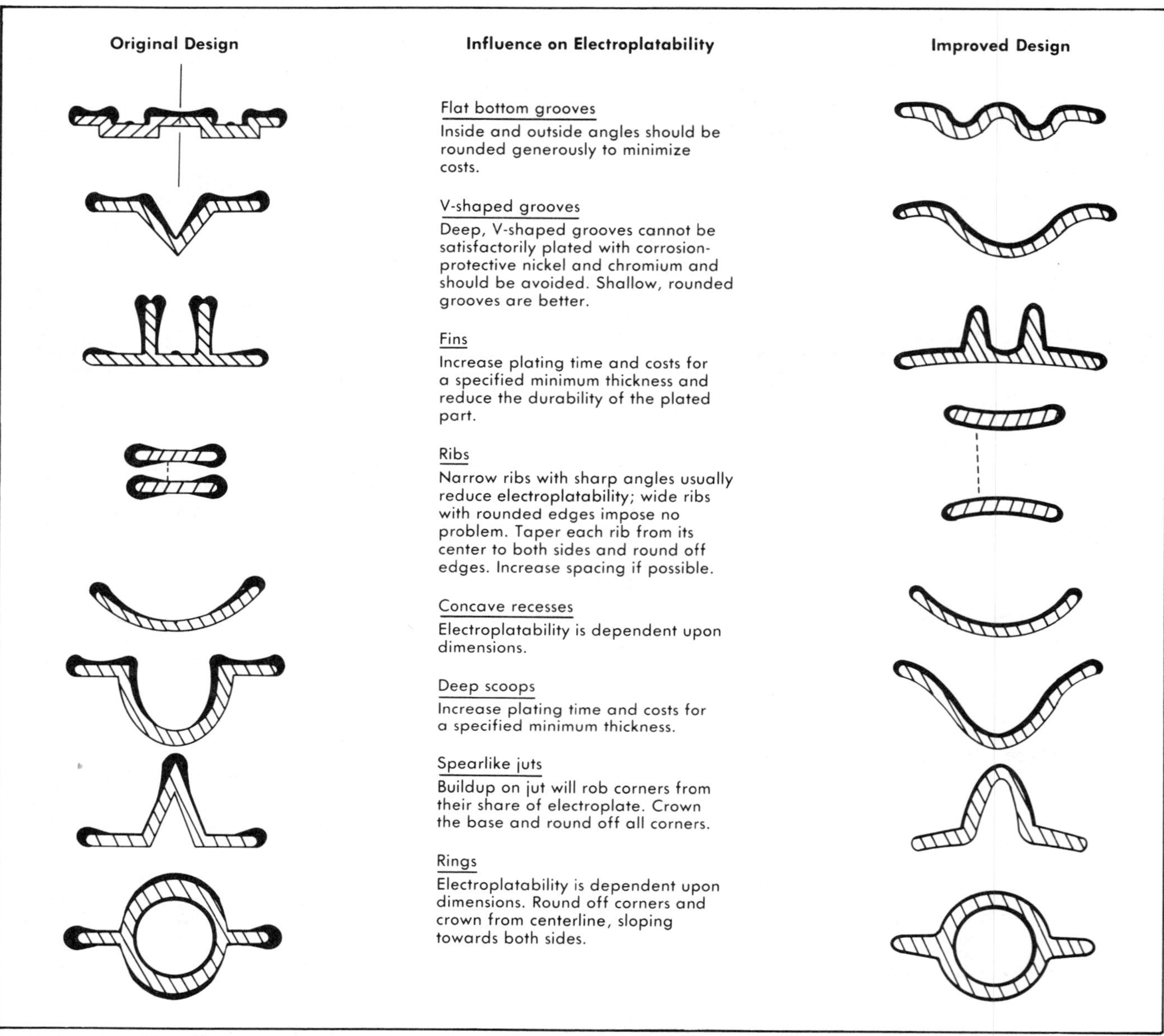

Original Design
Influence on Electroplatability
Improved Design
Flat bottom grooves
Inside and outside angles should be rounded generously to minimize costs.
V-shaped grooves
Deep, V-shaped grooves cannot be satisfactorily plated with corrosion-protective nickel and chromium and should be avoided. Shallow, rounded grooves are better.
Fins
Increase plating time and costs for a specified minimum thickness and reduce the durability of the plated part.
Ribs
Narrow ribs with sharp angles usually reduce electroplatability; wide ribs with rounded edges impose no problem. Taper each rib from its center to both sides and round off edges. Increase spacing if possible.
Concave recesses
Electroplatability is dependent upon dimensions.
Deep scoops
Increase plating time and costs for a specified minimum thickness.
Spearlike juts
Buildup on jut will rob corners from their share of electroplate. Crown the base and round off all corners.
Rings
Electroplatability is dependent upon dimensions. Round off corners and crown from centerline, sloping towards both sides.

Fig. 20-2 *continued* Improved part design permits better electroplating. (*Zinc Institute, Inc. and Metal Finishing Suppliers' Association, Inc.*)

PLATING METHODS AND EQUIPMENT

Electroplating may be divided into barrel, rack, and strip (continuous reel-to-reel) plating. Barrel plating is used for plating smaller parts in some electrolytes. Rack plating is used for larger parts and for chromium plating.

Barrel Plating

Barrel plating is usually performed in either horizontal or oblique barrels constructed of polypropylene or other suitable plastics. The walls of the barrels are perforated, and the barrel is rotated during plating. Electrical contact is obtained via a flexible conductor known as a dangler. Some barrels are not perforated, but contain the plating solution and an anode.

Mixed loads are not recommended in barrel plating, unless the plater is not too concerned about plating thickness and distribution, or unless the parts require just a flash plate. Longer or larger parts in the mixture will usually receive the greater deposit, and buildup on high current density areas will be increased. In addition, sorting the parts after plating is time consuming.

Large pieces weighing more than 1 lb (0.5 kg) and containing sharp edges should not be barrel plated because they will be damaged in the tumbling action of the barrel and the barrel itself may be damaged. Flat and lightweight parts should not be barrel plated because they tend to stick together and do not tumble properly. Wire forms are more easily rack plated than barrel plated.

Rack Plating

Rack plating is usually employed in the processing of parts that are too heavy, too large, or too complex in shape to be barrel plated. The parts can vary from a small knob that is to be nickel-chromium plated, to a large roller, weighing a ton or more, for hard chromium plating. Rack plating is used with manual, semiautomatic, and fully automatic machines.

One of the most important considerations in designing a plating rack is that the rack must have adequate current-carrying capacity. Racks are usually constructed of copper because of its high current-carrying capacity, ease of fabrication, and relatively high strength. After fabrication, the racks are covered with an inert insulating material to protect them and prevent metal plating on them, and to keep the plating solutions from becoming contaminated. Parts are hung or clipped to the insulated rack. The contact to the part should be made on a non-critical area of the part, such as the back, inside, or through a hole.

Another important factor in designing a plating rack is the position of the work on the rack. It must be positioned to obtain the most uniform current distribution possible and to prevent entrapment of air or gas in holes or pockets, which will restrict deposition of metal in these areas, resulting in nonuniform thicknesses. The position of the plating rack in the processing tank is also important for obtaining as uniform a plating deposit as possible. If the work is located too close to the anodes, excessive buildup of plating thicknesses, known as burning, occurs in the high current density areas of the part. The parts are usually racked so that they tend to shield the edges of adjoining parts, and are frequently racked back to back to increase the capacity of the rack and at the same time reduce the deposition of the metal in noncritical areas. If a particular part has a deep recess, an auxiliary anode can be used to ensure the deposition of the required thickness in the recessed area.

Strip Plating

Strip plating is a plating process whereby the workpiece is a continuous strip being pulled through each process station (tank) by a take-up roll. Wire and lead frames are commonly strip plated with tin, tin lead, nickel, and precious metals. Steel sheet may be continuously zinc, tin, chromium, copper, brass, nickel-iron, or nickel-zinc plated. The strip may be plated at specific points as it goes through the cycle; such selective plating is very common with precious metals.

Process Tanks

Tanks may be fabricated of hot rolled low-carbon steel and protected with a rubber or other suitable liner rated to withstand the bath's corrosive action and to keep the steel tank from contaminating the solution. Polypropylene and other high-strength plastics may be used for smaller tanks. Alkaline-cleaner tanks and some plating solutions do not require liners. Rinse tanks may be equipped with spray nozzles and the water counter-flowed back into the previous rinse tank to reduce water usage.

Temperature Control

Temperature control is important in most plating solutions. Some require cooling, heating, or both. For cooling, cold water piped through cooling coils of suitable construction may be sufficient, or heat exchangers may be required. Heat may be provided by steam through heating coils or by electric immersion heaters. Instruments are commonly used to automatically control plating bath temperatures.

Agitation

Many plating processes, such as bright nickel and bright acid copper, may specify air agitation. Agitation increases the permissible current density and often enhances the effectiveness of the brighteners and extends the bright current density range. Low-pressure high-volume blowers are the preferred source of air as compressed air may contain oils and other foreign contaminants. In certain processes, cathode-rod agitation is used. In barrel plating, the rotation of the barrel provides sufficient solution agitation.

Filtration and Ventilation

Filtration is usually incorporated in electroplating tanks to remove organic substances and suspended solids that can cause rough deposits. A preferred solution turnover rate is two to four times per hour, but the rate depends on the plating solution. Bright nickel and copper require continuous filtration, but cyanide zinc and cadmium require less frequent filtration.

Because irritating chemical vapors are often formed from plating solutions, tanks should either be provided with adequate ventilation or should have inhibitors or surface tension-reducing agents added to form a foam blanket that reduces the amount of toxic mist escaping into the atmosphere. However, foam blankets are not totally effective and can result in processing problems.

Power Supplies

Alternating current is converted to direct current by means of a rectifier or motor generator set, with rectifiers being preferred. Regulated and unregulated power supplies are available that provide filtered direct current with good reliability. A regulated power supply may be of the silicon-controlled or saturable reactor type. These power supplies maintain constant current or voltage under varied loads and are generally used in large operations. The more common unregulated types include the tap switch and manual powerstat. Both air and liquid-cooled rectifiers are available. The major advantage of the more expensive liquid-cooled type is that the unit is completely sealed so that rectifier components are not exposed to the atmosphere, which may be very corrosive in a plating shop. As a general rule, a rectifier should be operated at a minimum of 50% of its maximum rated output current to provide low-ripple direct current.

Anodes

The anode carries a positive charge and completes the electrolytic circuit. In most cases, the anode is fabricated of the metal being plated and serves to replenish the solution with metal ions. However, insoluble (inert) anodes are used in some applications, such as chromium plating and with precious metals. Anodes may be fabricated metal slabs or balls, or chips contained in an inert basket. The metallurgical structure and composition are important considerations for proper anode performance. Titanium is often used for the baskets because of its inert oxide film.

Most modern nickel plating installations use nickel chips in titanium baskets, shot, or cathode-deposited shapes. Zinc is usually plated from ball anodes, and copper from ball or bar anodes.

Automated Control

Electroplating is subject to a wide variety of variables that frequently change. Automatic controllers add stability and consistency to the operation and should be used wherever possible. Automatic ampere hour feeders monitor the plating time and may be used to approximate thickness as well as automatically feed addition agents. Solution pH may be controlled automatically with a relatively inexpensive piece of equipment; such control is especially useful in electrolytes where the pH tends to rise due to generation of the hydroxyl ion, such as in nickel and chloride zinc solutions. Rinse waters may be controlled by automatic conductivity meters connected to solenoid valves; these ensure good rinsing with reduced water consumption.

SUBSTRATE PREPARATION

In preparing the substrate for plating, it is important to properly select the correct pretreatment method. Pretreatment influences the adhesion, appearance, composition, and corrosion resistance of the deposit.

Some of the factors that should be considered when selecting the pretreatment cycle are type of substrate, nature of the contamination, how the part is used, and part geometry.[3] Each basis metal may require a different pretreatment. Aluminum, for example, cannot be properly processed in solutions formulated for steel. Even variations in alloy may cause the finisher to change pretreatments. Table 20-11 lists typical processing cycles for different substrates. Table 20-12 identifies the various pretreatment practices published by the American Society for Testing and Materials (ASTM).

TABLE 20-11
Typical Processing Cycles for Substrate Preparation for Electroplating

Low-carbon steel—oily, not rusted

1. Soak clean, 180° F (82° C), 2-3 minutes.
2. Cold water rinse, 30 seconds.
3. Anodic electroclean, 180° F (82° C).
4. Rinse.
5. Acid dip,[a] 30 seconds.
6. Cold water rinse, 30 seconds.
7. Plate.

Low-carbon steel—rusted

1. Soak clean, 180° F (82° C), 2-3 minutes.
2. Cold water rinse.
3. Periodic reverse electroclean, room temperature, 30-60 A/ft^2 (3.2-6.4 A/dm^2).
4. Plate.

Stainless steel and high-nickel alloys

1. Soak clean, 1-5 minutes.
2. Cold water rinse, 30 seconds.
3. Periodic reverse electroclean (10 seconds anodic, 7 seconds cathodic), room temperature, 30-60 A/ft^2 (3.2-6.4 A/dm^2), 3-5 minutes.
4. Cold water rinse, 30 seconds.
5. Hydrochloric acid dip, 30% by volume, room temperature, one minute.
6. Wood's nickel strike, 20-60 A/ft^2 (2.2-6.4 A/dm^2), 1-3 minutes.
7. Cold water rinse, 30 seconds.
8. Plate.

TABLE 20-11—*Continued*

Cast iron

1. Soak clean, 1-3 minutes.
2. Cold water rinse.
3. Periodic reverse electroclean (7 seconds anodic, 5 seconds cathodic), 5 minutes.
4. Warm water rinse.
5. Cold water rinse.
6. Hydrochloric acid, 20% by volume, room temperature, 10 seconds.
7. Cold water rinse.
8. Plate.

Cast iron—alternate method

1. Cathodic electroclean, 1-3 minutes.
2. Warm water rinse.
3. Cold water rinse.
4. Hydrochloric acid dip, 10% by volume, 5-10 seconds.
5. Cold water rinse.
6. Plate.

Wrought aluminum alloys—most alloys

1. Soak clean, nonetch cleaner, 1-3 minutes.
2. Cold water rinse.
3. Alkaline etch clean.
4. Cold water rinse.
5. Desmut in a solution containing 50% by volume nitric acid, 25% by volume sulfuric acid, 25% by volume water, and 1 lb/gal (120 g/L) fluoride salt at room temperature for 20-45 seconds.
6. Cold water rinse.
7. Zincate,[b] 5-30 seconds.
8. Rinse.
9. Plate, preferably first in a cyanide copper strike.

Wrought aluminum alloys—5000 and 6000 series

1. Soak clean, nonetch cleaner, 1-3 minutes.
2. Cold water rinse.
3. Acid etch, 160° F (70° C), 1-3 minutes.
4. Cold water rinse.
5. Nitric acid, 50% by volume, room temperature, 30 seconds.
6. Cold water rinse.
7. Zincate,[b] 5-30 seconds.
8. Cold water rinse.
9. Plate, preferably first in a cyanide copper strike.

Sand or die-cast aluminum alloys[c]

1. Nonetch soak clean, 1-2 minutes.
2. Cold water rinse.
3. Alkaline etch clean.
4. Cold water rinse.
5. Desmut in a solution containing 50% by volume nitric acid, 25% by volume sulfuric acid, 25% by volume water, and 1 lb/gal (120 g/L) fluoride salt at room temperature for 20-45 seconds.
6. Cold water rinse.
7. Zincate,[b] 5-30 seconds.
8. Rinse.
9. Plate, preferably first in a cyanide copper strike.

TABLE 20-11—*Continued*

Copper and copper alloys

1. Soak clean, 3-5 minutes.
2. Cold water rinse.
3. Electroclean (cathodic or anodic depending on formulation), 30-90 seconds.
4. Cold water rinse.
5. Acid dip, 10% sulfuric acid or dry acid salt, 30 seconds.
6. Cold water rinse.
7. Plate.

Zinc-based die castings

1. Soak clean, 3-5 minutes.
2. Cold water rinse.
3. Spray alkaline clean, 30-60 seconds.
4. Cold water rinse.
5. Anodic electroclean, 10-25 A/ft^2 (1.1-2.7 A/dm^2), 25-50 seconds.
6. Cold water rinse.
7. Acid dip,[d] 30 seconds.
8. Rinse.
9. Plate, preferably first in a cyanide copper strike.

(*Products Finishing*)

[a] The acid dip may be 10% sulfuric acid at 122-180° F (50-82° C), 40% hydrochloric acid at room temperature, or 16 oz/gal (120 g/L) dry acid salt at 77-140° F (25-60° C).
[b] The two most widely used processes for the pretreatment of aluminum alloys before electroplating are the zincate process and the stannate process.
[c] On difficult-to-plate alloys, use double zincate with an intermediate nitric acid dip to improve deposit adhesion.
[d] Acid dip may be 1% hydrofluoric acid, 1% fluoboric acid, or a dry acid salt.

TABLE 20-12
American Society for Testing and Materials' Recommended Practices for Preparation of Substrates to be Electroplated

Metal Substrate	Standard Number
Low-carbon steel	B 183
High-carbon steel	B 242
Zinc alloy die castings	B 252
Aluminum alloys	B 253
Stainless steel	B 254
Copper and copper-based alloys	B 281
Lead and lead alloys	B 319
Iron castings	B 320
Nickel	B 343
Magnesium and magnesium alloys	B 480
Titanium and titanium alloys	B 481
Tungsten and tungsten alloys	B 482
Nickel alloys	B 558

Several stages are generally required to provide adequate cleaning of the substrate and activation: precleaning, intermediate alkaline cleaning, and electrocleaning.[4] Precleaning is designed to remove a large excess of soil, especially deposits of buffing compound or grease. It is also useful in reducing the viscosity of waxes and heavy oils to enable later cleaning stages to be more effective, or to surround fingerprints and dry dust with an oily matrix to facilitate removal by alkaline cleaners.

Intermediate alkaline cleaning removes solvent residues and residual soil that has been softened or conditioned by precleaning. Spray or soak alkaline cleaning may also be used as a precleaning stage, followed by additional alkaline cleaning, if the soil and metal lend themselves to this treatment. Electrocleaning is soak cleaning with agitation provided by the upward movement of bubbles of hydrogen or oxygen formed by the electrolytic decomposition of water in the solution. Additional information on cleaning methods can be found in Chapter 18, "Cleaning," of this volume.

Some parts cannot be etched because surface finish must be maintained, just as parts used in structural applications should not be subjected to pretreatments that may cause hydrogen embrittlement. The design of the work may require special handling and surface treatment. For example, a large part may require external manual finishing or parts with deep recesses or blind holes may require special handling and drainage techniques to avoid excessive drag-out and cross-contamination.

OPERATING PARAMETERS

The four main concerns in electroplating are temperature, pH, and chemistry of the plating bath as well as current density. Most plating solutions have an optimum temperature range for producing best results, and close control of temperature is important for proper current control. As the temperature of the solution increases, conductivity increases, and therefore the current increases for a fixed applied voltage; the converse is also true. Overplating or underplating occurs if the temperature is not maintained properly.

The pH control of plating solutions is necessary to maintain the acidity or alkalinity that has been determined to produce the best results. Appearance, stress, leveling, electrode efficiency, and coating hardness are influenced by the pH of the solution.[5] Current density is a very important variable in all electroplating operations. The character of the deposit, its distribution, the current efficiency, and perhaps whether a deposit forms at all may depend on the current density employed.[6] Table 20-13 lists temperature, pH, and current density for typical electroplating solutions.

PLATING PLASTICS

Plating can be successfully performed on many plastics, including ABS, polypropylene, polysulfone, modified polyphenylene oxide, polycarbonate, polyester, and nylon, to provide a decorative finish or a hard surface for wear and corrosion resistance. Plating can improve physical properties of the plastics part, such as tensile and flexural strength and the heat deflection temperature. Because of their light weight and ease of design, plastics have been used in many applications to replace zinc die castings, brass, and steel. The total cost to plate plastics is competitive with metals.

Product design of parts to be plated is particularly critical in determining the success of the plating operation. Basic plastics design practices should be followed to achieve a good molding in the unplated product, and it is advisable to have the design reviewed by the plater. Because the proper choice of resin for products to be plated is of basic importance, the resin supplier should be consulted while the product is in the design stage.

The mechanism by which the electroplated metal adheres to the plastics substrate is a subject that is widely debated. One

TABLE 20-13
Operating Parameters for Typical Electroplating Solutions

Electroplating Bath	Temperature, °F (°C)	pH	Current Density, A/ft² (A/dm²)
Watts, nickel	110-150 (44-65)	2-4.5	10-100 (1.1-10.8)
All chloride, nickel	130 (55)	2	50 (5.4)
Sulfamate, nickel	105-140 (40-60	3-5	10-250 (1.1-26.9)
Fluoborate, nickel	105-175 (40-80)	2-3.5	40-100 (4.3-10.8)
Nickel iron	110-150 (44-65)	2.5-3.6	10-150 (1.1-16.5)
Cyanide copper, strike bath	70-90 or 100-120 (21-32 or 38-50)	11.0-12.2	5-10 or 10-20 (0.5-1.1 or 1.1-2.2)
Cyanide copper, Rochelle bath	140-150 (60-66)	12.2-12.8	30-40 (3.2-4.3)
Cyanide copper, high-efficiency potassium	140-175 (60-75)	13 and above	20-60 (2.2-6.5)
Copper pyrophosphate	122-140 (50-60)	7-8	9-74 (1.0-8.0)
Acid copper, sulfate	68-120 (20-50)	<1	20-100 (2.2-10.8)
Acid copper, fluoborate	68-120 (20-70)	0.8-1.5	70-100 (7.5-10.8)
Decorative hexavalent chromium	115-125 (45-52)		75-175 (8.0-18.8)
Trivalent chromium	72 (23)	3-3.5	75-125 (8.0-13.5)
Zinc, mid-cyanide	Room		20-80 (2.2-8.6)
Zinc, chloride	65-85 (18-30)	5-5.8	10-150 (1.10-18.2)
Zinc, alkaline (noncyanide)	Room		20-80 (2.2-8.6)
Alkaline stannate tin, A*	150-190 (66-88)		30-100 (3.2-10.8)
Alkaline stannate tin, B*	170-190 (77-88)		400 (43.0)
Alkaline stannate tin, C*	140-180 (60-82)		5-30 (0.54-3.2)
Acid tin, sulfate	60-120 (16-49)		10-100 (1.1-10.8)
Acid tin, fluoborate	60-100 (16-38)		10-200 (1.1-2.2)

(continued)

TABLE 20-13—*Continued*

Electroplating Bath	Temperature, °F (°C)	pH	Current Density, A/ft² (A/dm²)
Cadmium, cyanide	60-100 (15-38)		5-90 (0.54-9.7)
Cadmium, neutral (noncyanide)	60-100 (15-38)		2-15 (0.2-1.6)
Cadmium, fluoborate	70-100 (21-38)		30-60 (3.2-6.5)
Cadmium, acid sulfate	60-90 (15-32)		10-60 (1.1-6.5)

(*MacDermid, Inc.*)

* A = Low-potassium stannate bath
B = High-potassium stannate bath
C = Sodium stannate bath

group proposes that the adhesion is mainly mechanical. Anchoring points for the metallic coating result from undercuts or shallow pits from preplate etching procedures. Another group proposes that a chemical bond occurs as a result of the conditioning treatment. However, it is not in the scope of this Handbook to elaborate on these theories.

Various specifications and tests have been standardized by the American Society for Testing and Materials (ASTM) and the International Organization for Standardization (ISO). These standards help facilitate world trade, improve productivity, make mass production techniques possible, and lead to consumer satisfaction. Table 20-14 lists the standards published by the ASTM for electroplated plastics.

Preplate Cycle

Since plastics are nonconductive, they must first be processed through a preplate cycle, during which a metallic coating is deposited by an electroless plating process to make the plastics part conductive. The preplate cycle consists of etching, neutralizing, catalyzing, accelerating, and electroless plating. A typical cycle for plating ABS plastics is shown in Fig. 20-3. The actual cycle is dependent on the type of plastics being processed and the end application.

TABLE 20-14
American Society for Testing and Materials' Standards for Electroplated Plastics

Standard Number	Title
ASTM B 532	Specification for the Appearance of Electroplated Plastics Surfaces
ASTM B 533	Test Method for Peel Strength of Metal Electroplated Plastics
ASTM B 553	Test Method for thermal Cycling of Electroplated Plastics
ASTM B 554	Measurement of Thickness of Metallic Coatings on Nonmetallic Surfaces
ASTM B 604	Specification for Decorative Electroplated Coatings of Copper/Nickel/Chromium on Plastics
ASTM B 727	Preparation of Plastics Materials for Electroplating

Depending on their condition, parts may require alkaline cleaning and/or conditioning before etching. If these two preliminary steps are performed, multiple rinses are recommended between each step.

Etching. The etch bath consists of a highly concentrated acid solution of equal concentrations of chromic acid and sulfuric acid. This solution chemically oxidizes selective areas on the plastics part. The holes produced by the oxidizing action are absorbing sites that hold small metallic particles that serve as activators for electroless plating. The hole size influences adhesion and other physical properties. The type of etchant used is determined by the type of plastics being etched. After etching has been completed, the part is thoroughly rinsed.

To obtain the best etching possible, it is necessary to carefully monitor the temperature, time, and water levels of the bath. In general, the etch bath should be maintained at a temperature of 145 to 155°F (60 to 70°C), the parts should remain in the bath for 5 to 8 minutes depending on the part geometry, and the water level should be consistently maintained.

Neutralizing (sensitizing). The neutralizing bath, containing mild acids or alkaline solutions along with complexing or reducing agents, chemically neutralizes the acid used in the etching bath. A common solution for ABS plastics parts contains a mixture of an acid and a reducing agent, such as sodium bisulfite. The bath is usually maintained at 100°F (40°C), and the parts are immersed for 1 to 2 minutes.

Catalyzing (activating). In the catalyzing step, a catalytic film is deposited on the oxidized part surface to prepare it for subsequent electroless metal deposition. Palladium is generally used as the catalytic film and may be deposited in either a one or two-step procedure; the one-step procedure is more popular.

One-step. In the one-step catalyzing procedure, the plastics parts are immersed in a solution of colloidal stannous chloride and palladium chloride containing excess hydrochloric acid. The solution is maintained at a temperature of 68 to 85°F (20 to 30°C), and the parts are immersed for 1 to 3 minutes. Rinsing in

PLATING PLASTICS

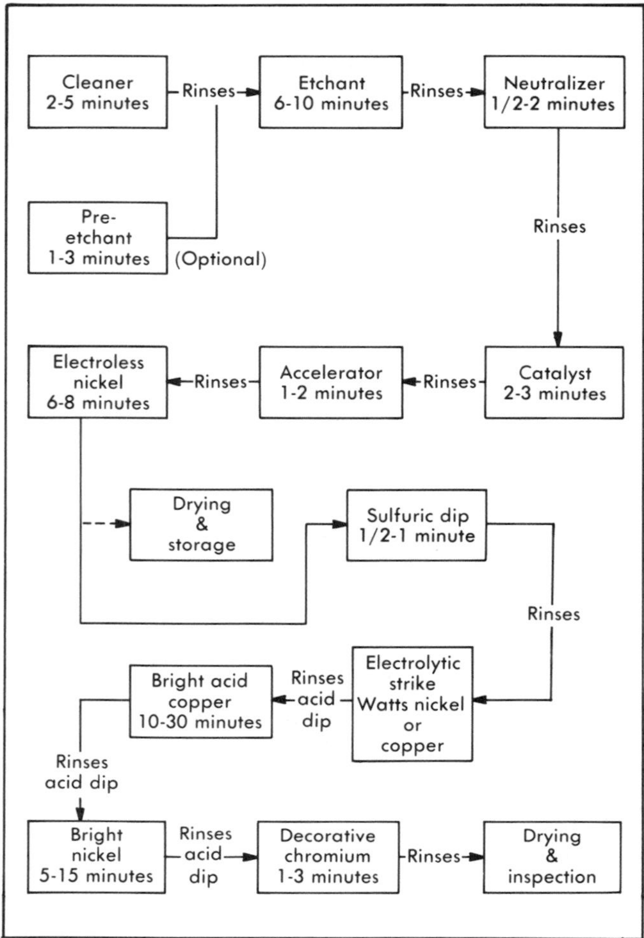

Fig. 20-3 Typical flow chart for plating ABS plastics. (*McGean-Rohco, Inc.*)

water leads to the surface formation of metallic palladium nuclei surrounded by stannic hydroxide, which is removed by accelerating.

Two-step. In the two-step catalyzing procedure, the parts are initially immersed in a solution of stannous chloride at a temperature of 68 to 77° F (20 to 25° C) for 1 to 3 minutes. After a second thorough rinsing, the parts are immersed in a solution containing palladium chloride. The palladium ions react with the stannous ions to form palladium metal and stannic chloride. After a thorough rinsing, the parts can be coated with nickel or copper by autocatalytic deposition.

Accelerating. The accelerator bath removes the stannic and or stannous hydroxide or excess stannous chloride that remains on the surface of the part after the one-step catalyzing procedure. It also chemically accelerates the catalytic film to ensure rapid and complete coverage with electroless deposits. Generally, the bath contains hydrochloric acid or a solution of an acid salt.

The accelerator bath should be maintained at a temperature of 100 to 120° F (38 to 50° C). The optimum time for processing is one minute because skip plating, due to a loss of absorbed palladium, could result from leaving the part in the bath for too long.

Electroless plating. The electroless plating bath is the final step in the preplate cycle. A thin metallic film, 0.01-0.02 mils (0.25-0.50 μm), is deposited on the plastics parts. The deposit can be either nickel or copper, depending on the intended application. Electroless-plated copper may improve the performance of electroplated plastics in wet corrosive environments.

The electroless nickel bath is usually maintained at 20 to 50° F (68 to 122° C). The parts should be left in the bath for 5 to 10 minutes, depending on the quality and thickness of the coating. Thicker coatings may be applied by low current density electroplating from nickel or copper strike baths.

Plating Cycle

After the plastics parts have been electroless plated in the preplate cycle, they can then be electroplated. A typical electroplating cycle consists of a strike bath, acid copper bath, semibright nickel bath, bright nickel bath, and a chromium bath.

Strike bath. The copper or nickel strike bath is used to electroplate a more conductive coating over electroless copper or nickel. In this bath, the plastics parts are given a negative charge, and the anode baskets filled with copper or nickel are given a positive charge. Copper strike baths are generally of the copper sulfate or pyrophosphate type, and nickel strike baths are generally of the Watts or sulfamate type.

The strike bath can be operated over a wide range of temperatures. The length of time that the parts are immersed in the bath is dependent on the thickness requirements. Plating time is generally 5-10 minutes.

Acid copper bath. In this bath, the plastics part gets its first layer of decorative metallic film. The copper plated on the part should be bright and have a high luster. The brightness achieved in this bath determines to a great extent the overall appearance of the plated plastics part. This copper layer acts as a cushion between the nickel and chromium layers, and absorbs the stress that occurs when the plastic expands under those layers.

The electrolyte used is copper sulfate with sulfuric acid added to improve electrical properties. Organic brighteners are also added on a regular basis. The acid copper bath is usually kept at 70° F (21° C). Plating time is normally 25-30 minutes.

Semibright nickel bath. The semibright nickel bath is similar to a nickel strike bath in operation and is generally required in automotive applications. Coating parts with a semibright plating followed by bright nickel plating is referred to as double-layer (duplex) nickel. The double-layer nickel coating is usually specified for outdoor decorative work that is exposed to corrosive environments. The semibright nickel bath is operated at a temperature of 130 to 135° F (55 to 60° C) for periods of 30 to 60 minutes, depending on the application.

Bright nickel bath. The bright nickel bath gives plastics parts the required decorative finish; and in certain applications, this is the only nickel coating that is applied. In most applications, however, the parts are then given a final coating of chromium to add tarnish and wear resistance to the nickel coating.

The bright nickel baths may be operated at a temperature between 125 and 150° F (50 and 65° C) for periods of 10 to 15 minutes for duplex deposits. Plating time is increased when the bright nickel is employed as a single nickel coating.

Chromium plating bath. The chromium plating bath is the final step in the process. Chromic acid with a sulfuric acid catalyst is used as the basic electrolyte. Various additives are also used to improve plating bath characteristics. The anodes used in this bath are either lead or lead-tin alloys.

Plating time is usually 2 1/2 to 3 1/2 minutes. The bath temperatures are generally maintained at 110 to 120°F (42 to 50°C). Plating voltage should be started at the minimum and then slowly increased until gas is liberated at the part locations; at this point, the voltage setting can be slowly increased to the maximum.

SOLUTION CONTROL AND TROUBLESHOOTING

Proper solution and process control along with preventive maintenance is the best insurance against plating problems. Alkaline soak and electrocleaners should be tested for alkalinity (cleaner concentration) before the start of each work shift. Temperature, amperage, and voltage readings should also be noted regularly. A regular discard schedule for cleaners and acid pickles should be established. This schedule is established by actual production experience, bearing in mind that the cost of a few loads of plating rejects may far outweigh the cost of fresh solutions.

The plating bath should be analyzed for its constituents regularly. Variables such as pH, temperature, or brightener level fluctuate and may require more careful manual control or, in some instances, automatic controllers. The Hull cell is a useful piece of test equipment that is employed to evaluate the plating bath performance over the range of current density used in production. This device indicates current density range, the general appearance of the deposit, the presence of impurities if such are detrimental, and the effect of additions to the bath. Many possible corrective measures can be tried in the Hull cell and then incorporated in the bath. Additions to plating baths should be small and frequent as opposed to large and infrequent additions. Permanent records of all additions should be kept. Hull cell tests should be run on a daily basis, and samples of the plating solution should be sent to the brightener supplier periodically to verify analytical results. Most major metal finishing suppliers offer this service free of charge.

Troubleshooting electroplating processes requires common sense and the use of the process of elimination. It is important to start with the basic aspects of the process and then proceed to the more complex aspects. The basic aspects of plating that should be checked include temperature, voltage, amperage, polarity, process cycle time, pH, and chemistry of the solution. The next step would be to determine if all the parts display the problem or only a select few. If all the parts are not affected, current density should be checked, and/or substrate preparation procedures should be monitored. If cleaning problems appear evident, an additional cleaning step can be manually performed before placing the parts in the plating solution.

Metal finishing suppliers can also help as they generally maintain a staff of technically trained service engineers to provide troubleshooting service to customers. Table 20-15 lists the common problems of various plating processes, along with the possible causes and suggested solutions.[7]

TABLE 20-15
Troubleshooting Guide for Electroplating

Problem	Possible Causes	Suggested Solutions
	Nickel Plating	
Poor adhesion	Inadequate cleaning.	Check age, concentration, and temperature of cleaner. Use standards or established methods of precleaning specific metals or plastics.
	Parts not acidified before entering nickel bath.	Pass parts through clean, dilute solution of hydrochloric or sulfuric acid salts.
	pH out of range (too high or too low).	Check and adjust pH of nickel bath.
	Nickel too stressed (may be due to metallic contamination or excess brightener).	Check for metallic contamination. Use high-pH treatment followed by low current density dummying to remove metallic impurities. Check for organic contamination. Add activated carbon to bath before high-pH treatment. Adjust brightener level as required.
	Chromium contamination.	Reduce hexavalent chromium with careful addition of sodium bisulfate and low current density dummying.
	Nitrate contamination.	Use high current density dummying to destroy nitrates.
Poor adhesion or laminated deposit	Current interrupted.	Check rectifier meters for output reliability. Check busbars for cleanliness and tight connections. Use established method for plating nickel on nickel when current is interrupted.

(continued)

TABLE 20-15—*Continued*

Problem	Possible Causes	Suggested Solutions
Poor adhesion or laminated deposit	Nickel over nickel without proper activation.	Check nickel activation.
	Replating without complete stripping of chromium.	Check chromium stripping procedure and check for complete stripping before plating.
	Deposit highly stressed.	Make stress determinations on the nickel deposit. May require purification of the bath to remove inorganic and organic impurities.
	Excess iron in solution.	Use high-pH treatment followed by low current density dummying.
Poor adhesion involving peeling from copper	Failure to remove brightener film from copper plate.	Cathodically clean alkaline copper deposit and rinse carefully. Treat in 2-5% sulfuric acid before nickel plating.
	Copper tarnished in rinsing process.	Use room temperature rinse water before nickel plate. Shorten rinse time after copper deposit.
Deposit too highly stressed	Contamination with zinc, lead, or cadmium.	Use high-pH treatment followed by low current density dummying.
	Iron content too high.	Use high-pH treatment followed by low current density dummying.
	Brighteners out of balance or too concentrated.	Perform Hull cell test to check brightener levels or send to brightener supplier. To reduce brightener level, add activated carbon to bath and perform high-pH treatment. Adjust brightener level as required.
	pH out of range.	Check pH and adjust accordingly.
	Chloride content too high.	Check for chloride content. Remove required amount of solution and replace with water.
Pitting of deposit	Solution contaminated with grease or oil.	Perform carbon and/or high-pH treatments followed by dummying. Adjust brightener level as required.
	Solid particles in solution.	Check anode bags. Filter bath through carbon bags.
	pH too low.	Check and adjust pH to proper value.
	Metal too low.	Analyze for metal content and adjust accordingly.
	Organic contamination.	Check for organic contamination. Add activated carbon to bath and perform high-pH treatment.
	Inadequate agitation.	Increase agitation.
	Temperature too low.	Check temperature and adjust.
	Boric acid low.	Analyze for boric acid and adjust concentration.

(continued)

TABLE 20-15—*Continued*

Problem	Possible Causes	Suggested Solutions
Deposit dark	Metallic contamination, especially copper. (Zinc and cadmium first produce a bright, stressed deposit, then, at higher levels, black streaky deposits.)	Analyze bath for contaminant metals. Use low current density electrolysis to reduce metallic impurities.
Deposit streaky	See "deposit dark."	
	Chromium contamination	Reduce chromium with careful addition of sodium bisulfate and low current density dummying.
	Nitrate contamination.	Use high current density dummying to destroy nitrate.
	Brightener imbalance.	Perform Hull test to determine brightener level. Add activated carbon to bath and perform high-pH treatment to reduce brightener level.
	pH out of range.	Check and adjust pH to proper value.
	Inadequate agitation.	Increase agitation.
	Low wetting agent concentration.	Determine surface tension and adjust wetting agent as required.
	Wrong wetting agent. (Wetting agents for mechanical agitation are usually different than those for air agitation.)	Analyze bath for proper wetting agent. Obtain recommendations from supplier.
Deposit rough	Solid material in solution.	Examine anode bags carefully and correct. Increase filtration rate.
	Smut not removed in cleaning.	Wipe workpieces with clean tissue to check for smut before plating. Use proper procedures for smut removal.
	Excessive current density.	Check current density and correct.
	Boric acid content too high (usually also associated with low temperature).	Analyze bath for boric acid content. Check and correct bath temperature.
Deposit hazy	Organic contamination.	Check for organic contamination. Add activated carbon to bath and then perform high-pH treatment. Adjust brightener level as required.
	Immersion deposit on basis metal (usually from processing multiple alloys through the same acid dip).	Analyze acid dip for contaminant metals and replace if necessary. Do not use same acid dip for multiple alloys.
	Iron content too high.	Analyze bath for iron content. Use high-pH treatment followed by low current density dummying.
	pH out of range.	Check and adjust pH to proper level.
	Inadequate cleaning.	Check precleaning procedures. Correct and improve where necessary.
Failure to plate	Faulty electrical system.	Check rectifier output meters and current delivery to anode/cathode bars. Clean corrosion from anode/cathode bars.

(continued)

TABLE 20-15—*Continued*

Problem	Possible Causes	Suggested Solutions
Failure to plate	Contamination with chromium or nitric acid.	Analyze bath for chromium or nitric acid contamination. Refer to suggestions under "poor adhesion."
	Cyanide Copper Plating*	
Insufficient adhesion	Inadequate cleaning and/or oxide removal.	Check cleaning bath and correct accordingly. Inspect parts for complete removal of oxides and correct accordingly.
	Free cyanide too high or too low.	Analyze bath for free and/or total cyanide. Adjust accordingly.
Copper "hard"	Current density too low.	Check actual versus recommended current density values and correct.
	Metal content too low.	Analyze bath for copper metal content. Correct by proper addition of copper cyanide if necessary.
Rough, granular deposit	Current density too high.	Check actual versus recommended current density. Correct if variance occurs.
Shelf roughness	Suspended solids in solution.	Check condition of anode bags. Replace if damaged or unsound. Increase filtration rate.
Solution not clear	Free cyanide too low.	Analyze bath for free cyanides. Adjust accordingly.
	Suspended material in solution.	Check condition of anode bags. Increase filtration rate or reduce filter pore size.
Solution blue (where tartrates are also used)	Free cyanide too low.	Analyze bath for free cyanides and adjust. Add cyanide (sodium or potassium) until blue color disappears.
Solution crystallizes when cold	Carbonates too high.	Filter or decant solution.
Poor efficiency	Free cyanide too high. Metal too low. Chromium contamination. (Chromium may also result in scattered blistering or failure to obtain coverage, resulting in patchy deposits especially in the low current density areas.)	Analyze bath for free cyanides, copper metal, and other contaminating metals. Make corrections for deficient components. If chromium is present, carefully add sodium bisulfite in Hull cell until patchy condition in low current density area clears. Make necessary corrections to plating bath.
Anodes too bright	Excess free cyanide.	Analyze bath for free cyanides. Add copper cyanide to complex excess.
Anodes polarized	Lack of free cyanide.	Add small amounts of potassium or sodium cyanide to Hull cell until desired results are obtained. Correct bath. Analyze bath for free cyanide and adjust accordingly.
Anodes black (Film does not always dissolve when current is off.)	Metallic contamination, especially lead.	Add controlled amounts of dilute sodium sulfide in Hull cell to precipitate lead. Correct bath on basis of Hull cell determination. Filter bath.

(continued)

TABLE 20-15—*Continued*

Problem	Possible Causes	Suggested Solutions
Staining or spotting out	Insufficient rinsing.	Improve rinsing procedure. Increase rinsing time or temperature.
	Porosity of basis metal.	Examine basis metal and reject if necessary.
	Porosity of plate.	
No plate	Chromium contamination.	Analyze bath for chromium contamination. Add sodium bisulfite in Hull cell until corrected. Make corrections to plating bath.
	Free cyanide too high.	Analyze bath for free cyanides. Add copper cyanide to complex excess.
	Inadequate cleaning and/or oxide removal.	See "insufficient adhesion."
	Bright Acid Copper Plating**	
Burning at high current density	Low copper content. Low chloride levels.	Analyze bath for copper and chloride content. Correct as required.
Dullness in low current areas	Excess chloride.	Analyze bath for chloride content. If chloride content is too high, precipitate excess chloride with silver sulfate and then filter.
	Brightener imbalance.	Carbon treat to remove brighteners, then add proper brightener balance.
Loss of brightness	Lack of brightener.	Preform Hull cell test to determine brightener level. Correct brightener level in Hull cell and then correct entire bath proportionately.
	Organic contamination.	Carbon treat to remove contaminants and then correct brightener level.
Ridging or gas streaking	Brightener imbalance.	Carbon treat to remove brighteners and then adjust brighteners to proper level.
	Insufficient agitation.	Increase solution agitation.
Brown deposits at low current	Excess brightener.	Carbon treat to remove excess brighteners and adjust brighteners to proper level.
Rough deposit	Suspended matter in solution.	Inspect anode bags and replace if necessary. Increase filtration rate or decrease filter pore size.
Coarse, granular deposit	Low acid content.	Analyze bath for acid content. Add sulfuric acid as required.
	Temperature too low.	Check bath temperature and adjust as required.
	Current density too high.	Check current density.
Deposit soft	Low acid content.	Analyze bath for acid content. Add sulfuric acid as required.
	Bath too warm.	Check bath temperature and adjust.
	Current density too low.	Check current density and adjust to recommended level.

(continued)

TABLE 20-15—*Continued*

Problem	Possible Causes	Suggested Solutions
Deposit hard and/or brittle	Too much free acid.	Analyze bath for free-acid concentration. Add copper carbonate to decrease concentration.
	Solution too cold.	Check bath temperature and adjust accordingly.
	Organic contamination. Excess brightener.	Treat solution with activated carbon and then filter. Adjust brightener level.
	High iron contamination.	Analyze bath for iron content. Raise pH and then blow clean air through solution to precipitate iron. Filter bath.
Poor throwing power	Low acid.	Analyze bath for acid content. Add sulfuric acid to increase acid.
Low conductivity	Temperature too low.	Check bath temperature and adjust as required.
	Low acid.	See "poor throwing power."
	Low copper.	Analyze bath for copper content and increase as required.
Anodes bright and crystalline	Free acid too high.	Analyze bath for free-acid content. Add copper carbonate as required.
Anodes polarized	Low acid content.	Analyze bath for acid content. Add sulfuric acid to increase content.
	Insufficient anode area.	Compare actual anode area versus recommended area and correct.
Anodes polarized with green deposit	High chloride contamination. (Note: plate will also be seriously affected.)	Analyze bath for chloride content. Add silver sulfate to reduce chloride content as required and then filter solution to remove silver chloride.
	Decorative Chromium	
Poor chromium coverage	Rectified failure.	Check rectifier meters for stable operation. AC ripple should not exceed 5%. Rectifier should operate at 75% rated amperage.
	Poor electrical connections.	Check electrical connections for tightness and contact.
	Inactive anodes.	Check coating on anodes. A yellowish lead chromate coating indicates that the anode is inactive; a black-brown lead peroxide coating is on active anodes. Make sure proper current is used. When plating inside diameters, use largest anode possible.
	Incorrect bath temperature.	Check bath temperature. Check bath agitation.
	Incorrect bath concentration.	Mix bath thoroughly and take bath sample. Analyze chromium and sulfate content. Adjust accordingly.

(continued)

TABLE 20-15—*Continued*

Problem	Possible Causes	Suggested Solutions
Poor chromium coverage	Improper racking.	Check rack-to-fixture connections. Incorporate shields or robbers in racking to direct current to specific areas.
	Poor cleaning or rinsing.	Check temperature of rinsing water and maintain between 70 and 75° F (21 and 24° C).
	Copper or nickel roughness.	Check copper or nickel plate for roughness.
	Passive nickel.	Dip nickel-plated part in 50% by volume solution of hydrochloric acid. If part brightness increases, correct nickel plating bath.
	Chloride contamination.	Dummy with high anode current density and agitation or treat with silver oxide.
	Improper current density.	Adjust current density to solution temperature and concentration.
	Incorrect chromic acid to sulfate catalyst ratio.	Add sulfuric acid to lower ratio and barium carbonate to increase ratio.
	Impurities.	Limit metallic impurities to less than 0.5 oz/gal (3.75 g/L). Increase chromic acid concentration.
Burned chromium deposits	Passive nickel deposit.	Correct nickel brightener concentration. Increase density in nickel bath.
	Incorrect bath temperature.	Correct bath temperature.
	Improper bath concentration.	Analyze chromium and sulfate content. Correct if necessary.
	Inactive anodes.	Check anode color. Inactive anodes are coated with a yellowish lead chromate.
	Incorrect anode length.	Anode should be 2-4" (50-100 mm) shorter than cathode. Mask anodes that are too long with suitable insulator or cut to proper length.
	Current interruption.	Check racking of parts and electrical contacts.
	Improper racking.	Check rack to fixture connections. Incorporate shields or robbers in racking to direct current to specific areas.
	Improper solution level.	Place work 3" (75 mm) below surface of plating solution.
	High current density.	Adjust current density to solution temperature and concentration.
	Sulfate concentration or other catalyst too high.	Add barium carbonate to bath to lower sulfate concentration. Ratio should be 2:1.
	High or low concentration ratio.	Check ratio and adjust accordingly.
Dull chromium deposits	Dull or passive nickel.	Correct nickel brightener concentration. Increase curent density in nickel bath.

(continued)

TABLE 20-15—*Continued*

Problem	Possible Causes	Suggested Solutions
Dull chromium deposits	Excessive ripple in current.	Check rectifier operation. Ripple should not exceed 5%.
	Current interruption.	Check racking of parts and electrical contacts.
	Incorrect bath temperature.	Correct bath temperature.
	Improper bath concentration.	Take Baume reading of bath and correct.
	Low current density.	Adjsut current density to bath temperature and concentration. Check rinsing procedures.
	Impurities.	Limit metallic impurities to less than 0.5 oz/gal (3.75 g/L). Increase chromic acid concentration.
	Excess fluoride catalyst or fluoride contamination.	Add boric acid to reduce catalyst or use high anode current density dummy.
	High or low concentration ratio.	Check ratio and adjust accordingly.
White blotching	Flim of immersion nickel on workpiece.	Rinse workpiece thoroughly.
	Passive nickel.	Check nickel plating bath for proper concentration.
	Excessive ripple in current.	Check rectifier operation. Ripple should not exceed 5%.
	Bipolarity caused by rapid work movement in automatic machines.	Reduce speed of work movement through machine.
	Bipolarity caused by dead entry into chromium.	Use live lead at reduced current.
	Bipolarity caused by dead exit from nickel tanks.	Use live exit-lead from nickel.
	Impurities, especially chloride.	Check rinsing procedures.
Deposit roughness	Poor basis metal finish.	Reject parts with rough finish.
	Copper or nickel roughness.	Check copper or nickel plating for roughness.
Poor chromium adhesion	Poor nickel adhesion.	Refer to solutions for bipolarity conditions under "white blotching."
No chromium plate	Rectifier failure.	Check operation of rectifier. Rectifier should operate at no less than 75% rated amperage.
	Poor electrical connections.	Check connections for good contact.
	Missing anodes.	Check for replacement of anodes after they have been removed.
	Passive nickel.	Check nickel bath.
	Low current density.	Adjust current density to bath temperature and concentration.
	Incorrect catalyst concentration.	Analyze chromium plating solution and adjust accordingly.

(continued)

TABLE 20-15—*Continued*

Problem	Possible Causes	Suggested Solutions
	Hard Chromium	
Poor chromium brightness	Poor basis metal finish.	
	Excessive reverse etching.	
	Improper bath temperature.	Correct bath temperature. Check bath agitation.
	Improper bath concentration.	Take Baume reading of bath and correct.
	Improper current density.	Adjust current density to bath temperature and concentration.
	High concentration ratio.	Check ratio and adjust accordingly.
Chromium roughness	High trivalent chromium.	Check color of chromium solution; a black color indicates trivalent chromium contamination. Electrolyze chromium plating solution.
	Magnetic particles.	Attach a magnet to a piece of wood and then drag through solution to remove particles.
	Excessive reverse etching.	
	High current density.	Adjust current density to bath temperature and concentration.
	Improper concentration ratio.	Check ratio and adjust accordingly.
Uneven chromium deposit.	Improper bath temperature.	Correct bath temperature.
	Poor fixturing.	Check rack design and contacts. Use shields to direct current to specific areas.
	Incorrect anode length.	Anode should be shorter than cathode. Mask anodes that are too long with suitable insulator.
	Improper anode-cathode relationship.	For outside-diameter plating, anode-cathode distance should be approximately 4″ (100 mm). For inside-diameter plating, the distance should be from 1/2 to 1″ (12-25 mm).
Poor chromium adhesion	Improper cleaning or rinsing.	Follow recommended cleaning procedures.
	Improper reverse etching.	Do not reverse etch in plating bath. Current used in reverse etching should be proportionated to work area. Work should be at bath temperature before applying current.
	Current interruption.	Check racking of parts and electrical contacts.
Burned chromium deposits	Etching in alkaline cleaner.	Check cleaner concentration.
	Improper fixturing.	Racks should be made from copper for best results. Check rack design and contact.
	Improper anode-cathode relationship.	See "uneven chromium deposit."
	Improper solution level.	Work should be 3″ (75 mm) below surface of plating solution.

(continued)

TABLE 20-15—*Continued*

Problem	Possible Causes	Suggested Solutions
Burned chromium deposits	Improper anode length.	Anode should be shorter than cathode.
	Inadequate stopoff anodes.	
	Improper bath temperature.	Correct bath temperature.
	Low bath concentration.	Mix bath thoroughly and take sample to obtain Baume reading. Adjust accordingly.
	High current density.	Adjust current density to bath temperature and concentration.
	Impurities.	
	Excess catalyst or chloride contamination.	Adjust sulfate and/or fluoride ratio to proper level.
Lack of chromium hardness	Incorrect bath temperature.	Check bath temperature and agitation.
	Improper bath concentration.	Take Baume reading of bath and correct.
	High ratio.	Check ratio and adjust.
	Insufficient deposit thickness.	Check deposit thickness.
Pitting	Poor basis metal and rework.	Check metal prior to plating.
	Improper cleaning.	Check cleaning procedures.
	Incorrect catalyst concentration.	Analyze chromium plating solution and adjust accordingly.
	Incorrect fume suppressant.	Eliminate use of fume suppressant that lowers surface tension.
No chromium plate	Rectifier failure.	Check operation of rectifier.
	Poor electrical connection.	Check connections for good contact.
	Missing anodes.	Check if anodes are properly placed.
	Inactive anodes.	Check anode color. Inactive anodes are coated with a yellowish lead chromate.
	Incorrect catalyst concentration.	Analyze chromium plating solution and adjust accordingly.
	<u>Cyanide Zinc Plating</u>	
Poor adhesion	Inadequate cleaning and/or oxide removal.	Examine parts for water breaks and oxides after rinsing but before plating; oxides appear as stains or blotches. Correct acid pickle time or concentration and/or cleaner time or condition if necessary.
	Hydrogen absorption (blistering is usually delayed).	Bake parts for 1 hour at 250° F (120° C) immediately after plating to remove absorbed hydrogen or strike with cadmium before zinc plate.
	Contamination with chromium or nitrates.	Analyze bath for chromium and nitrates. Add sodium hydrosulfite to reduce hexavalent chromium to trivalent chromium and agitate bath vigorously. Add high-purity zinc dust to reduce nitrates. Filter bath.

(continued)

TABLE 20-15—*Continued*

Problem	Possible Causes	Suggested Solutions
Poor adhesion	Solution out of balance.	Analyze bath for proper balance and correct.
Burned deposit	Excessive current.	Reduce current to level recommended by supplier.
	Low cyanide content.	Check cyanide content and correct as necessary. The cyanide-to-zinc ratio is based on bath temperature.
	Low caustic content.	Analyze caustic content. Add sodium hydroxide to adjust to proper level.
Deposit dark	Metallic contamination.	Use low current density dummying treatment overnight. Add sodium sulfide to precipitate lead or calcium. Treat bath with zinc dust and then filter to remove copper.
	Brighteners out of balance.	Perform Hull cell test to determine brightener level. Correct brightener level in Hull cell and then correct whole bath proportionately.
	Solution out of balance.	Analyze bath and restore to proper balance.
Deposits hazy or stained	Solution out of balance.	Analyze bath and restore to proper balance.
	Organic contamination.	Treat bath with potassium permanganate or circulate solution through a carbon-packed filter.
	Metallic contamination.	See "dark deposits."
	Insufficient rinsing.	Check rinsing procedures and correct.
	Lack of brightener.	Perform Hull cell test to determine brightener level and correct.
Shelf roughness	Solids suspended in solution.	Check anode bags and replace if used. Filter solution through appropriate filter.
Low conductivity	Lack of caustic. Lack of cyanide. Low metal.	Analyze caustic, cyanide, and metal content. Correct if necessary.
	Low temperature.	Increase bath temperature.
Poor throwing power	Cyanide too high. Caustic too high. Metal too high.	Analyze cyanide, caustic, and metal content. Correct if necessary.
Anodes polarized	Low cyanide. Low caustic.	Analyze cyanide and caustic level. Adjust as required.
	Insufficient anode area.	Increase anode area.
Metal content increasing	Anode area (zinc) too high. Increase proportion of steel to zinc.	Reduce zinc anode area. Replace with steel anodes.
Solution crystallizes when cold	Carbonates too high. (Note: conductivity may also be low, and anodes tend to polarize.)	Cool bath to allow sodium carbonate to precipitate. Filter or decant bath. Dilute to full bath volume and readjust bath chemistry.

(continued)

TABLE 20-15—*Continued*

Problem	Possible Causes	Suggested Solutions
Steel embrittled	Hydrogen absorption.	Bake parts for 1 hour at 250° F (120° C) immediately after plating and rinsing. Deposit cadmium or copper strike before zinc plating.
Staining in storage	Failure to use proper post-treatment (for example, chromates). Poor rinsing.	Rinse parts thoroughly after plating. Dip parts in 1/4% nitric acid rinse and then give appropriate chromate treatment.
	Spotting out.	Check parts for unacceptable porosity.
	Acid Zinc Plating	
Dull deposits	Lack of brightener.	Perform Hull cell test to determine brightener level. Correct brightener level in Hull cell and then correct entire bath.
Brittle deposits	Solution imbalance.	Analyze bath and correct.
	Excess brightener.	Dummy bath overnight to reduce brightener level, or carbon treat and filter bath. Readjust brightener level.
Pitting	Solution imbalance.	Analyze bath and correct.
	Lack of wetting agent.	Determine surface tension of the bath and add appropriate wetter agent.
Deposit generally dull with burning in high current area	Low metal concentration	Add metal ion concentrate based on lab analysis recommendations. Add metal ion concentrate based on experimental Hull cell test.
Deposit brownish	Excess chloride.	Check chloride content. Dummy bath with inert anodes to suitable level or dilute bath.
	Temperature too low.	Check and adjust operating temperature.
	Brightener imbalance.	Check brightener balance in Hull cell and adjust accordingly.
Poor throwing power	The pH is too low.	Check pH and adjust.
	Zinc too high.	Dummy bath with inert anodes or dilute bath and then readjust. Remove zinc anodes when not in use.
Deposit darkens in chromating or nitric dipping	Metallic contamination.	Check for metallic contamination. Use appropriate technique to remove contaminant.
	High iron contamination.	Treat with hydrogen peroxide and then filter insoluble materials.
Shelf roughness	Suspended matter in solution.	Check anode bags and replace if necessary. Increase filtration rate or filter size.
	pH too high.	Check pH. Decrease pH with small amounts of hydrochloric acid.
Barrel work shows print of barrel holes	Current density too high.	Check current density and adjust.
	Barrel speed too low (parts sticking to side of barrel).	Increase barrel speed.

(continued)

TABLE 20-15—*Continued*

Problem	Possible Causes	Suggested Solutions
Barrel work shows print of barrel holes	High iron contamination.	Check iron concentration. Add 1.2 fl oz/100 gal (9.3 mL/100L) of 30% hydrogen peroxide and the filter insoluble materials.
Spongy, dark deposit	Acid concentration too low.	Check acid concentration and adjust as necessary or adjust pH.
Low efficiency	Temperature too low.	Check temperature and correct if necessary.
	Metal content too low.	Analyze for zinc metal and adjust with appropriate zinc salt. Increase zinc anode area versus current density relationship.
	Solution out of balance (deficiency of conducting salts).	Add conducting salts on basis of lab analysis.
	Tin Plating—Stannate Solutions	
Poor conductivity with anodes gray	Low caustic content and/or low metal content.	Analyze bath and adjust by adding caustic and/or sodium or potassium stannate.
Rough, dark, or spongy deposits	Stannous tin in solution (anodes must be kept properly polarized).	Use hydrogen peroxide regularly as a preventive. Reduce tin anode area to the extent that a yellow-green film is maintained on the anodes during use. Remove anodes from bath when not in use.
Anodes black	Anode current density too high. Anodes improperly polarized.	Increase anode area to the point where the yellow-green film is maintained on the anodes. Remove black film by acid treatment, then polarize to yellow-green film.
Solution crystallizes when cold (sodium stannate bath)	Carbonates too high.	Cool and decant bath or filter.
Staining after plating	Improper rinsing.	Improve rinsing.
Anodes lose polarization or will not polarize properly	If anode area is right, caustic level is too high.	Analyze bath for free alkali. If it is beyond recommended level, add solution of 10% acetic acid to neutralize.
	Tin Plating—Acid Sulfate Solution	
Coarse-grained deposit	Addition agent concentration too low. (Voltage required for normal current densities will also be low; use this as a signal of approaching problems.)	
Shelf roughness	Suspended solids in solution.	Check anode bags. Increase filtration rate.
Burned deposits	Current density too high.	Check actual versus recommended current density values and correct.
Slow deposition rate	Low temperature.	Increase bath temperature.
	Low metal content or low acid concentration.	Analyze for metal content or acid concentration and correct.
Polarized anodes	Anode current density too high.	Reduce current density.
	Cadmium Plating	
Poor adhesion	Inadequate cleaning.	Improve cleaning cycle.

(continued)

TABLE 20-15—*Continued*

Problem	Possible Causes	Suggested Solutions
Poor adhesion	Low cyanide. (Anodes will also be polarized.)	Increase cyanide content.
	Failure to neutralize after acid dipping.	Correct preplate cycle.
	Hydrogen absorption (blisters will develop some time after plating).	Use mechanical cleaning techniques. Avoid using strong acid pickling solutions. Bake parts at 350-400° F (175-205° C) after pickling and before plating. Use higher current densities during plating. Bake parts at 350-400° F (175-205° C) after plating.
Dark, granular deposit	Current density too high.	Decrease current density.
	Cyanide content too low.	Add cyanide to solution.
	Lack of brightener or imbalance.	Analyze solution for brightener and adjust accordingly.
Dark deposit	Metallic contamination (copper, lead, tin, nickel, or silver).	Analyze solution for contaminants. Use zinc dust or sulfide treatment to remove metallic contaminants.
Shelf roughness	Suspended material in solution.	Check anode bags and replace if necessary. Increase filtration of solution.
Poor efficiency	Low metal.	Analyze bath for metal content and increase if necessary.
	High cyanide.	Decrease cyanide content.
	Chromium contamination.	Reduce hexavalent chromium to trivalent chromium with sodium hydrosulfate and then filter out trivalent chromium.
Poor throwing power	Excess cyanide.	Decrease cyanide content.
	Low metal.	Analyze bath for metal content and increase if necessary.
Anodes too bright	Excess cyanide.	Decrease cyanide content.
Anodes polarized	Low cyanide.	Add cyanide.
	Lack of sufficient anode area.	Increase anode area.
Metal content climbs	Excess anode area.	Decrease anode area.
Poor conductivity	Lack of caustic.	Increase caustic content to 2-3 oz/gal (15-22.5 g/L).
	Low temperature.	Increase bath temperature.
Solution crystallizes when cold	Excess carbonates.	Reduce solution temperature to 25-40° F (-4 to +4.4° C) and allow sodium carbonate to settle. Decant or filter solution.
Staining after plating	Poor rinsing.	Alternate cold and hot rinses. Rinse in boiling deionized water after tap water rinses.
	Failure to use neutralizing dip or chromate conversion coating.	Dip in a 2-10% solution of sodium hypochlorite.
Failure to plate	Excess cyanide.	Decrease cyanide content.

(continued)

TABLE 20-15—*Continued*

Problem	Possible Causes	Suggested Solutions
Failure to plate	Chromium contamination.	Reduce hexavalent chromium to trivalent chromium with sodium hydrosulfite and then filter out trivalent chromium.
	Inadequate cleaning.	Check cleaning cycle.
	Parts may be magnetized.	Check parts for magnetization. Demagnetize if necessary.
	Lead Plating	
Lack of adhesion	Improper part preparation.	Correct preplate cycle.
Rough deposit	Current too high.	Reduce current density to manufacturer's recommendation.
	Agitation too low.	Increase solution agitation.
	Temperature too low.	Increase solution temperature.
Shelf roughness	Suspended material in solution.	Check anode bags and replace if necessary. Improve solution filtration.
Deposit crystalline or feathery, treed	Lack of colloid in solution.	
Deposit bright but thin	Current too low.	Increase current density.
	Metal concentration too low.	Check metal concentration and increase if necessary.
Deposit dark and nonadherent	Organic contamination.	

(*Products Finishing*)

* Some proprietary bright cyanide-copper solutions also require the use of sodium or potassium hydroxide. Failure to maintain the hydroxide content at the proper level can strongly affect the performance of the brighteners. Bright copper solutions are also affected by the presence of organic contamination. This contamination usually produces a haze in the mid-to-low current density ranges. Consult your supplier for the exact effect to be expected.

** These problems are typical of one widely used bright acid copper. Other proprietary solutions may respond somewhat differently. Consult your supplier for exact information on the bath you are using.

SAFETY

The precautions taken when handling other chemicals should also be taken when handling plating chemicals. Protective clothing—such as goggles, aprons, rubber boots, and rubber gloves—is essential in handling strong acids and caustics. Storage and mixing containers should not be used interchangeably to prevent bath contamination and possibly the formation of hydrocyanic gas. It is just as important not to interchange containers used for transferring dry chemicals. For example, mixing dry chromic acid with reducible organic compounds, such as alcohol or acetone, usually results in a fire. Adequate ventilation should be provided to transfer any fumes or heat generated by the process to the outside.

EFFLUENT TREATMENT

The constituents of the plating baths contaminate the water used to rinse plated parts and their fixtures. These rinse waters must be treated prior to discharge into ground water or municipal sewer systems. State, local, and federal regulations govern the quantity of pollutants that are acceptable for discharge, and it is the responsibility of the metal finishing firm to ensure these guidelines are met. The effluent limitations and standards for existing sources (PSES), using the best practicable control technology (BPT) and the best available technology (BAT), are listed in Table 20-16.[8] Table 20-17 presents the new source performance standards (NSPS) and pretreatment standards for new sources (PSNS).[9]

In addition to rinse waters, other products requiring treatment prior to discharge are alkaline cleaners, acid pickles, chromate solutions, and metal strippers. The regulatory agencies establish limitations for various pollutants including metals, cyanides, hexavalent chromium, fluorides, and ammonia. Metals are usually removed from effluent streams by adjusting the pH, which causes the metals to precipitate as their hydroxide. Cyanide must be destroyed by one of a number of methods, the most common being chlorination. Hexavalent chromium must be reduced to the trivalent form prior to precipitation of the metal hydroxide, most often accomplished by reduction with sulfites. Other pollutants are handled by various treatment procedures. Recovery of metals in rinse waters by evaporation and ion exchange has become more popular. Closed-loop plating lines recycle and reuse the rinse water instead of discharging it. Other equipment has been developed to remove metal from the effluent stream. Metal finishing suppliers have and will continue to develop more ecologically feasible processes that will command a premium in the marketplace.

As regulatory legislation becomes more stringent, the cost to the finisher will increase. Investment in waste treatment equipment will not generate any profits for the plater; however, it will ensure that the installation can meet the regulations to protect our water and environment in the most cost-efficient manner.

TABLE 20-16
BPT/BAT/PSES Effluent Limitations and Pretreatment Standards

		Concentration, mg/L	
Standard*	Parameter	Maximum per Day	Maximum Monthly Average**
BPT/BAT/PSES	Cadmium	0.69	0.26
BPT/BAT/PSES	Chromium	2.77	1.71
BPT/BAT/PSES	Copper	3.38	2.07
BPT/BAT/PSES	Lead	0.69	0.43
BPT/BAT/PSES	Nickel	3.98	2.38
BPT/BAT/PSES	Silver	0.43	0.24
BPT/BAT/PSES	Zinc	2.61	1.48
BPT/BAT/PSES	Cyanide (T)†	1.20	0.65
BPT/BAT/PSES	Cyanide (A)†	0.86	0.32
BPT/BAT/PSES	Total toxic organics	2.13	---
BPT	Oil and grease	52	26
BPT	Total suspended solids	60	31
BPT	pH††	6.0-9.0	6.0-9.0

* BPT = Best practical control technology; BAT = Best available technology; PSES = Pretreatment standards for existing sources.

** Monthly average based on 10 samples per month.

† Industrial facilities with cyanide treatment may, upon agreement with the pollution control authority, apply the amenable (A) cyanide limit in place of the total (T) cyanide limit. Cyanide monitoring must be conducted after cyanide treatment and before dilution with other wastewater streams.

†† pH range, standard units.

TABLE 20-17
NSPS and PSNS Effluent Limitations and Pretreatment Standards

		Concentration, mg/L	
Standard*	Parameter	Maximum per Day	Maximum Monthly Average**
NSPS/PSNS	Cadmium	0.11	0.07
NSPS/PSNS	Chromium	2.77	1.71
NSPS/PSNS	Copper	3.38	2.07
NSPS/PSNS	Lead	0.69	0.43
NSPS/PSNS	Nickel	3.98	2.38
NSPS/PSNS	Silver	0.43	0.24
NSPS/PSNS	Zinc	2.61	1.48
NSPS/PSNS	Cyanide (A)†	1.20	0.65
NSPS/PSNS	Cyanide (T)†	0.86	0.32
NSPS/PSNS	Total toxic organics	2.13	---
NSPS	Oil and grease	52	26
NSPS	Total suspended solids	60	31
NSPS	pH††	6.0-9.0	6.0-9.0

* NSPS = New source performance standards; PSNS = Pretreatment standards for new sources.

** Monthly average based on 10 samples per month.

† Industrial facilities with cyanide treatment may, upon agreement with the pollution control authority, apply the amenable (A) cyanide limit in place of the total (T) cyanide limit. Cyanide monitoring must be conducted after cyanide treatment and before dilution with other wastewater streams.

†† pH range, standard units.

ELECTROFORMING

Electroforming[10] is a special type of electroplating in which a part is fabricated by the deposition of the desired metal on a form called a mandrel or matrix. The electrodeposited metal is built up to the desired thickness on the mandrel, and then the two are separated (see Fig. 20-4). Much of the equipment and many of the techniques that are employed in electroforming are the same as those used in the production of electrodeposited coatings. Electroforms differ from electrodeposited coatings, however, in that they are used as separate structures and are therefore usually substantially thicker than plated coatings. Electrodeposited coatings for decorative use are normally less than 2 mils (50 μm) thick. Requirements for some electroforms dictate that the thickness exceeds 1/4″ (6 mm).

New techniques of electroforming make it possible to duplicate complex forms at lower costs by using high-quality, precision mass-production methods. In many applications, electroforming eliminates machining and joining, producing one-piece forms of continuous, uniform, nonporous, stress-free metal. The most intricate surface detail and the most irregular internal designs can be reproduced with tolerances as small as 0.0001″ (2.5 μm). The quality of the surfaces is limited only by the quality of the mandrel.

Electroforming offers a wide range of control over strength, density, porosity, and purity of the deposited metal to meet individual product needs. While electroforming in the past was considered to be a high-cost production method, its current costs compare favorably with mechanical reproduction in many applications and are often lower.

APPLICATIONS

Extremely fine molds and dies can be electroformed from a variety of metals. The virtually perfect surface reproducibility offered by electroforming processes make them ideal for such dimensionally critical applications as lens molds, phonograph record stampers, embossing plates, fine printing plates, and, most recently, high-resolution video disk and optical disk molds. The video disk, read by a pinpoint laser beam, is the greatest testimony to the accuracy of electroforming as a process that provides virtually absolute replicability and accuracy. Information bits encoded on the average optically read disk consist of impressions having an average mean diameter of 0.000008″ (0.2 μm), a tolerance well within the range of electroforming process accuracy. Nickel stampers for replicating holograms have also been produced by electroforming. Like optical disks, these holograms have microsurfaces with tolerances in the submicronic range.

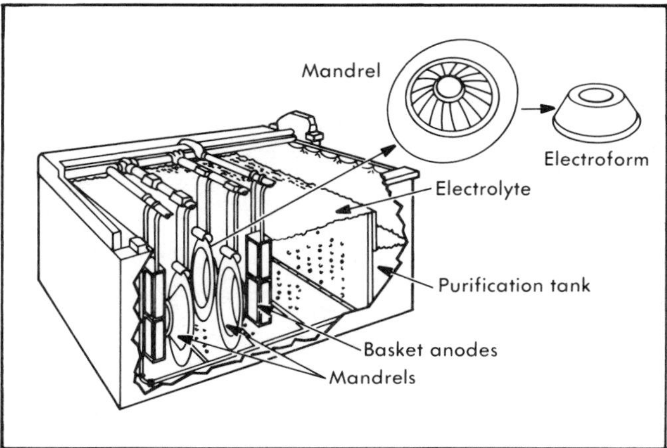

Fig. 20-4 Typical electroforming operation. (*The International Nickel Co., Inc.*)

ELECTROFORMED METALS

A variety of metals are available having properties suitable for thin metal-mold electroforming. Since many of the molding processes that use thin metal molds involve heat transfer through the mold to a thermoplastic or heat-setting material, the thermal conductivity of the metal selected is important. The metal's durability and abrasion and scratch resistance are also important when considering the quantity and quality of the moldings being manufactured. The metals most commonly used in electroforming are copper, nickel, and iron.

Copper

Copper is used when a highly conductive mold is required and when the oxidation commonly associated with copper surfaces can be tolerated by the process. Because of its softness, copper is generally not recommended for use in compression molding operations since deformation of the mold can occur at the high pressures, 30-40 ksi (207-276 MPa), normally associated with compression molding. Although the surface of a copper electroform is not chemically passive, it will readily form a passive separating film onto which subsequent generations of electroforms can be plated. The passive layer effectively separates the copper from the electroform grown on it, and the two are easily separated when plating is complete. Copper electroformed molds are not recommended for applications using thermoplastic molding compounds containing abrasive fillers since copper's softness leads to rapid wear and scratching when exposed to these compounds.

Nickel

Much research and development has been done in nickel plating, particularly in the sulfamate process, over the past 25 years. The low-stress characteristics of the metal as deposited from this bath along with inherent ductility, thermal conductivity, relative hardness, and corrosion resistance make it an attractive choice for thin metal molds. Lens manufacturers, phonograph record and optical disk manufacturers, and fine printing plate makers have all found nickel to be ideally suited for the high-pressure, high-temperature conditions.

Nickel Cobalt

Considerable effort has been put into the development of nickel-cobalt electroforms in recent years. In particular, the phonograph record industry has been interested in the physical properties of this hard, durable, scratch-resistant alloy as a substitute for pure nickel phonograph record stampers. The plating bath used for nickel-cobalt electroforming is conventional, and the anode system requires using a cobalt metal and nickel metal system. Results are impressive, but the material costs are many times that of conventional nickel plating process systems.

ELECTROFORMING BATHS

Iron

The primary use for iron has been as a substitute for nickel during periods of extreme shortages or restricted use during wartime. Iron baths are classically operated at extremes of high temperature and low pH, and hence are objectionable. Iron electroforms are brittle and must be annealed before use. They also are prone to immediate rust formation and must be chromium plated right away to preserve the iron surface for molding applications.

ELECTROFORMING BATHS

Plating solutions for electroforming do not differ in principle from ordinary plating baths for decorative and protective purposes. But because the physical properties of electroforms are usually of much more concern than those of ordinary electroplates, the composition and operating conditions of electroforming solutions tend to be more critical than those for the thinner deposits used in standard electroplating. Many agents, added for brightness, leveling, and other purposes, are not allowable in electroforming solutions because they introduce trace contaminants into the deposits that have deleterious effects on the properties of the electroforms. Also, because thicker deposits are involved, speed of plating is of greater importance; plating times, which are normally long, should be minimized.[11] Tables 20-18 to 20-20 list the compositions of the plating baths used for the commonly electroformed metals, along with the operating parameters and mechanical properties of the metals.

TABLE 20-18
Composition, Operating Parameters, and Mechanical Properties of Common Copper Electroforming Baths

Composition	Concentration, oz/gal (g/L)	
	Sulfate	Fluoborate
Copper sulfate	30-32 (224-240)	---
Copper fluoborate	---	30-60 (224-449)
Fluoboric acid	---	to pH 0.3-1.4
Sulfuric acid	6-10 (45-75)	---
Temperature, °F (°C)	80-110 (27-43)	80-120 (27-49)
Voltage, V	<6	4-12
Current density, A/ft^2 (A/dm^2)	30-150 (3.2-16)	75-300 (8-32)
Baume at 80° F (27° C)	---	29-31°
Hardness, diamond pyramid (VHN$_{100}$)	51-170	40-75
Tensile strength, ksi (MPa)	85-42 (241-290)	17-32 (117-220)
Elongation in 2″ (50 mm), %	5-25	3-14
Internal stress, psi (MPa)	0-5000 (0-34.5)	---

(*Metal Finishing*)

TABLE 20-19
Composition, Operating Parameters, and Mechanical Properties of Common Nickel Electroforming Baths

Composition	Concentration, oz/gal (g/L)		
	Sulfamate	Fluoborate	Watts
Nickel sulfamate	40-60 (300-450)	---	32-44 (240-330)
Nickel fluoborate	---	40-60 (300-450)	---
Nickel chloride	0-2 (0-15)	---	5-7 (37-52)
Boric acid	4-6 (30-45)	3-5 (22-37)	4-6 (30-45)
Temperature, °F (°C)	100-140 (38-60)	120-130 (49-54)	115-140 (46-60)
pH (electrometric)	3.4-4.5	3.0-4.0	1.5-4.0
Current density, A/ft^2 (A/dm^2)	25-200 (2.7-21.5)	25-100 (2.7-10.8)	25-100 (2.7-10.8)
Hardness, diamond pyramid (VHN$_{100}$)	140-240	130-250	140-160

(continued)

TABLE 20-19—*Continued*

Composition	Concentration, oz/gal (g/L)		
	Sulfamate	Fluoborate	Watts
Tensile strength, ksi (MPa)	60-110 (414-758)	56-80 (386-552)	55-65 (379-448)
Elongation in 2″ (50 mm), %	5-30	10-32	20-30
Internal stress, ksi (MPa)	1-6 (7-41)	15-24 (103-165)	18-22 (124-152)

(*Metal Finishing*)

TABLE 20-20
Composition, Operating Parameters, and Mechanical Properties of Common Iron Electroforming Baths

Composition	Concentration, oz/gal (g/L)	
	Chloride	Fluoborate
Ferrous chloride	30-60 (225-450)	---
Calcium chloride	15-20 (112-150)	---
Iron fluoborate	---	20-30 (150-225)
Sodium chloride	---	1-1.5 (7.5-11)
Temperature, °F (°C)	190-200 (88-93)	135-145 (57-63)
pH	0.5-1.5 (electrometric)	2.7-3.0 (colorimetric)
Current density, A/ft^2 (A/dm^2)	20-80 (2-8.6)	20-100 (2-10.8)
Baume at 80°F (27°C)	---	19-21°
Hardness, Bhn	125-220	---
Tensile strength, ksi (MPa)	50-65 (345-448)	---
Elongation in 2″ (50 mm), %	10-30	---
Internal stress, ksi (MPa)	15-40 (103-276)	---

(*Metal Finishing*)

MANDRELS

By the proper selection of mandrels, it is possible to electroform complicated shapes in one piece with extreme accuracy. The design of mandrels and the choice of materials from which to make them involve several considerations. They should be impervious to various types of cleaning and plating solutions, and should be dimensionally stable. They should also have the proper surface finish required for the applications and be capable of being separated or dissolved from the electrodeposited metal.

Types of Mandrels

Mandrels used in electroforming can be completely reusable or completely destroyed in the process, depending on the particular applications.

Permanent mandrels. Permanent mandrels, also known as matrices, are used to produce many electroformed parts and are usually machined, cast, or electroformed. A permanent mandrel must be treated to form a parting film on its surface to ensure separation from the electroform. Common methods used to form this film include passivation in dichromate solutions,

MANDRELS

anodic treatment, or by applying an absorbed film from a colloidal material. A part intended for production on permanent mandrels must be designed with sufficient draft (1 to 3°) to permit removal of the mandrel from the part without damage. If the design is such that these conditions cannot be met, expendable mandrels will probably be required. Common substrates for permanent mandrels are stainless steels, electroplated metals, copper, brass, nickel, and rigid plastics.

Stainless steels (Types 302 and 304). Stainless steels are commonly used for the production of permanent mandrels because of their inherently passive surface and good corrosion resistance. Stainless steels can also be machined to close dimensional tolerances, and the surface finish is not readily destroyed. A few examples of electroforms produced on stainless steel mandrels are high-precision nickel tubes, waveguides, surface roughness standards, and screens.

Electroplated metals. Steel, copper, or brass mandrels plated with a flash of either nickel or chromium to facilitate separation are also used for electroforming seamless, tube-type parts. Other typical electroforms produced on electroplated metal mandrels are consumer items such as cream pitchers, sugar bowls, pen caps, and decanters. Unless suitable dimensional allowances can be made for the thickness of the flash coating, these mandrels should not be used for the production of parts requiring high precision, such as those mentioned under stainless steel.

Copper and brass. Mandrels of these materials are recommended for reproducing engraved or etched surface textures. Engravers prefer to work with copper and brass, especially when chasing or photoengraving operations are required. An electroform can also serve as the original from which the production tooling mandrels are prepared.

Nickel. Nickel mandrels are frequently employed for large-volume production of consumer items or molds and dies. Individual electroformed mandrels are less expensive than those produced by the usual toolmaker's methods, which often require a great deal of manual finishing. Flat mandrels for repeated use are produced readily by electroforming. Typical products manufactured from such mandrels are phonograph record stampers, builders' hardware such as switch covers and door escutcheons, surface roughness standards, and molds.

Plastics. Mandrels made of epoxy resins, vinyl plastisols, and rigid vinyl sheets are generally used for producing surface textures and intricate shapes. Epoxy resins are used specifically for mandrels that can be cast in master mandrel molds prepared from flexible organic materials. Vinyl plastisol mandrels are made by either casting or spraying, and are used for applications similar to flat or cylindrical embossing dies. Rigid vinyl sheets are used in the preparation of mandrels for electrotype printing plates. Under heat and pressure, the vinyl sheet is molded into the original printing material to produce the required mandrel. The application of a silver spray will make plastics mandrels conductive.

Expendable mandrels. Expendable mandrels are completely destroyed during separation from the electroform and are usually made of nonmetallic, nonconductive materials that generally require metallizing after preparation in order to accept the electrodeposit. Some common materials used in producing expendable mandrels include aluminum alloys (2024, 6061, and 7075), plastics, fusible metals, waxes, glass, wood, and plaster.

Mandrel Design

Mandrels for electroforming are designed to produce the desired pattern and texture on either the inside or the outside surface of the part. When the outside surface of a part is the significant area, the internal surface of the mandrel upon which it is formed must be carefully designed. When the inside surface of a part is important, the external surface of the mandrel requires special care. In the former case, the mandrel is called a negative mandrel; in the latter, a positive mandrel.

Negative mandrels. The following suggestions apply only to mandrels in which the internal surfaces are significant:

- Sharp internal angles must be avoided to prevent weak corners.
- Fillets should be used if acute angles are necessary.
- Grooves should be made as shallow as possible; the width should exceed the depth.
- Corners should have a 1/32″ (0.8 mm) radius.

Positive mandrels. The suggestions for negative mandrels also apply to positive mandrels, but to a lesser degree because external surfaces are easier to prepare than internal ones.

EQUIPMENT

Certain features are commonly included in electroforming equipment systems that differ from conventional plating equipment. Electroforming tanks are usually designed to accommodate the particular shape and size of electroforms being produced. Agitation systems are generally included to enhance the high-speed deposition. Power supplies must have output ripple characteristics consistent with the mechanical properties desired on the formed deposit. Special rectifier controls that automatically raise the plating current from an initially low starting current to high-speed levels are desirable in most production applications where close monitoring of plating conditions is essential. Microprocessor controls have been integrated into electroforming systems used in the manufacture of disks and holograms. These controls allow a wide range of process variations to be programmed into a system, adding flexibility, stability, and reproducibility when manufacturing specialized electroforms.

SELECTIVE PLATING

Selective plating, also referred to as portable, brush, contact, or spot electroplating, is a method of depositing metal from a concentrated electrolyte solution without using immersion tanks. The first brush system was patented in the United States in 1925, with subsequent development and refinement in both the U.S. and France. Evolving from plating shop "jerry rigs" used for touch-up, the early systems used actual brushes with integrated anodes. True brush plating systems are still available, most commonly used to provide thin platings over small areas. Selective plating systems have been available in the U.S. for over 40 years and can deposit a variety of metals with industrially useful thicknesses for corrosion protection, surface buildup, electrical contacts, and plating repairs.

The principle of selective plating is the same as that of tank electroplating—metal ions in a liquid electrolyte are chemically reduced by an electric current and deposited as a metal at the cathode (workpiece). The main difference is that a large electrolytic bath is not used. Rather, the selective plating electrolyte is carried in an absorbent anode covering and applied directly to the area of the workpiece to be plated. As with tank plating, the workpiece must be electrically conductive, either intrinsically (e.g., a metal workpiece) or artificially (e.g., covered with a conductive paint).

PROCESS

Since electroplating is normally used to provide a relatively thin coating, the surface of the base material must be properly prepared both for cleanliness and finish prior to plating. Heavy oxides, grease, wax, or other debris prevent (mask) deposition or degrade adhesion of the plating; consequently, the workpiece must be cleaned thoroughly to expose its bare metal surface. In addition, the properties of some metals require a thin intermediate deposit or an underplating in order to ensure good adhesion and performance of the final plating. Finally, the conformal nature of plating means that any scratches or chips on the workpiece will be reflected in the plating. Unless a thick plating with subsequent grinding and polishing is intended, preplating surface finishing is imperative.

In operation, the cathode (electrically negative) lead from the direct current power supply is connected to the workpiece being plated, and the anode (electrically positive) lead is connected to an insulated plating tool handle. An anode, covered by absorbent material, is attached to the insulating handle to complete the basic equipment (see Fig. 20-5). The actual plating is a hand operation accomplished by alternately dipping the covered anode into a plating solution and then rubbing it over the area to be plated. When the plating solution comes into contact with the workpiece, the electrical circuit is completed, a low-voltage current flows, and metal ions begin depositing on the workpiece as described by Faraday's law.

As with all hand operations, operator technique plays an important part in job quality; with practice and thoughtful evaluation of the results, quality deposits can be routinely applied. In addition to anode pressure, which controls current density and electrolyte volume available for plating, anode movement determines the uniformity of the deposit. The best motion on rectangular surfaces is an up-and-back circular (orbital) movement. Circular components should be rotated in a lathe or low-speed turning head whenever possible to maintain concentricity; and the anode may be clamped, held steady in contact with the rotating surface, or moved slightly back and forth.[12]

Solutions

As mentioned earlier, surface preparation is important, and various preparatory and stripping solutions are available for

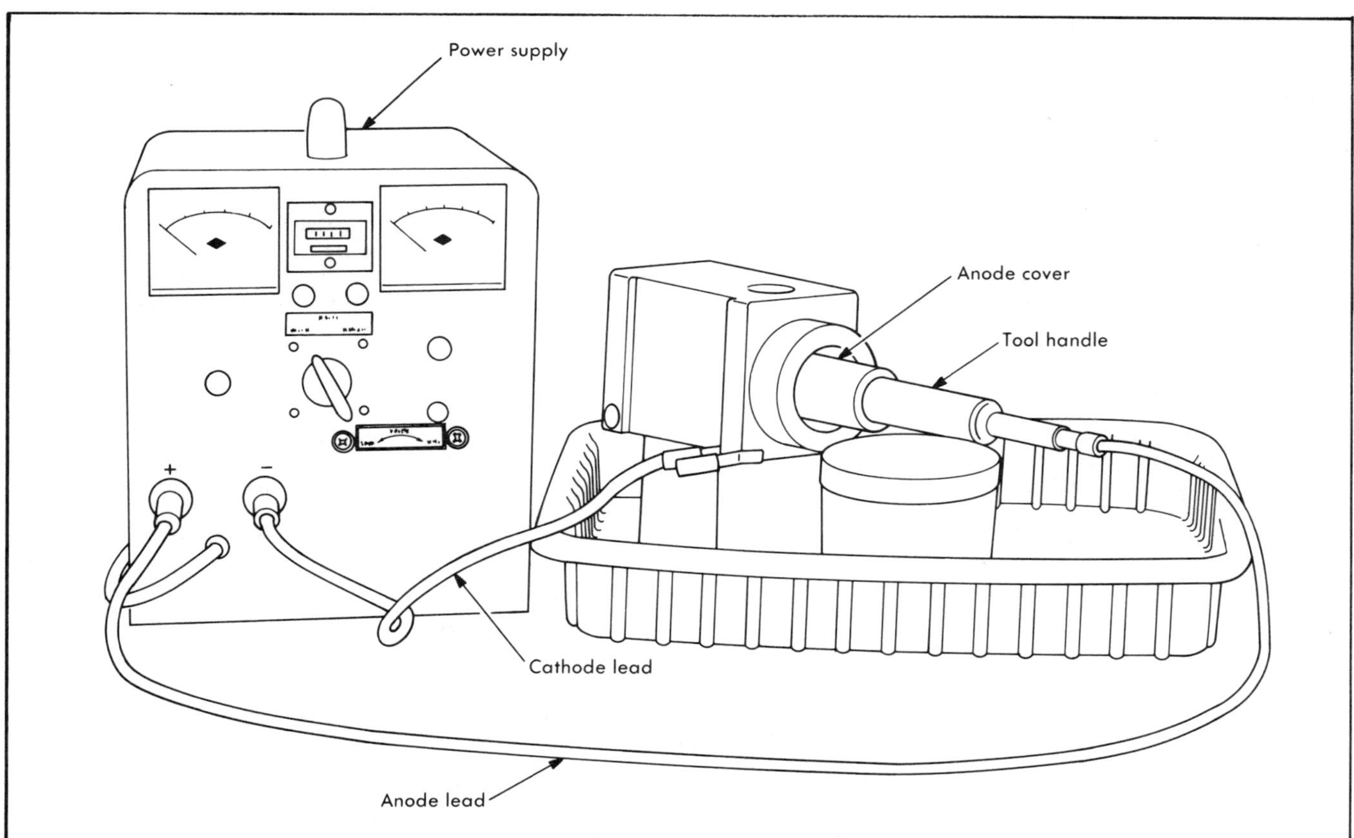

Fig. 20-5 Selective plating the inside diameter of a mismachined workpiece. (*SIFCO Selective Plating*)

SELECTIVE PLATING

cleaning the work surface. Preparatory solutions are used to remove minor surface contaminants such as fingerprints, light oils, or oxides. Heavy amounts of oil, grease, rust, or surface defects should be removed with an appropriate solvent or abrasive before using the preparatory solutions. Various stripping solutions are also available to chemically (dissolve) or electrochemically (reverse plate) remove defective electroplated coatings.

In contrast to the controlled nature of tank plating, the dynamics of selective plating (as a hand operation) require the electrolyte (plating solutions) to be capable of depositing metal over a wide range of current densities and temperatures. The compositions of these solutions are usually proprietary, involve higher metal concentrations than normal tank plating solutions, and limit the use of toxic materials like cyanides.

The common metals that can be deposited on workpieces include antimony, bismuth, cadmium, chromium, cobalt, copper, iron, lead, nickel, tin, and zinc. Alloy plating solutions are commercially available for brass, bronze, nickel cobalt, tin indium, tin lead nickel, cobalt tungsten, babbitt, and cadmium tin. Precious and semiprecious metal electrolytes such as gold, gallium, indium, palladium, platinum, rhodium, and silver are also available, as are special solutions for anodizing, applying black optical coatings, and electropolishing.

Specifications

A variety of specifications for selective plating have been issued by several manufacturing companies and the government, and, if cited, the latest revision from the specifying organization should be consulted. Although they may have been written around particular selective plating processes, the essence of the specifications is to describe the quality, final finish, and thickness of metal platings considered acceptable for various purposes. Table 20-21 lists some of the active specifications.

ADVANTAGES AND LIMITATIONS

The equipment for selective plating is portable and hence permits the plating or repair to be performed at the work location. Parts may also be plated without having to be extensively masked or disassembled; however, since liquids are used, the ability to completely clean and dry the workpiece will dictate the amount of masking and disassembly required. The selective plating process is simple and does not require nearly the same degree of control over electric current and solution chemistry as does tank plating. Yet, selective plating can provide greater deposition rates than tank plating. The basic equipment configuration provides fast and simple plating for applications that do not require close tolerances for thickness or flatness; for applications having close tolerances or requirements to minimize mechanical finishing (grinding, lapping, etc.), more complex equipment with current control, electrolyte pumps, and mechanically controlled anode movement is available. With frequent anode dipping or adequate electrolyte flow, atomic hydrogen deposition can be minimized to reduce the danger of hydrogen embrittlement to high-strength workpieces. Finally, workpieces that are large and of complex shape can be successfully selectively plated.

Selective plating is generally a labor-intensive process, and the plating solutions can be more costly than equal quantities of tank plating solutions because of their complex formulations and higher metal content. However, since smaller quantities of solutions are used in selective plating, the capital investment is not prohibitive. Industrial use is usually limited to low production rates because the parts are plated one at a time. In general, high-production volumes can be more efficiently tank plated, unless the workpieces are large and the areas to be plated relatively small, making tank plating impractical.

TABLE 20-21
Specifications for Selective Plating

Specification No.	Specifying Body
BAC5849	Boeing Aircraft Co.
DPS9.89	Douglas Aircraft Co.
N.M70-031	General Electric Co.
CP6-625-S6	General Electric Co.
SS8494	Sikorsky Aircraft Div., United Technologies Corp.
FPS1046	General Dynamics Corp.
MIL-Std-865A	U.S. Air Force
NAVSHIPS0900-038-6010	U.S. Navy

APPLICATIONS

Selective plating has diverse uses and should be considered any time that a surface requires modification to improve appearance, dimension, or chemical and mechanical properties. Mismachined or worn parts can often be salvaged with a buildup plating of nickel, while contact resistance on copper busbar joint areas can be reduced with a plating of silver. Steel parts can be protected from corrosion with a zinc or cadmium plating, while printed circuit board contact fingers can be plated (originally or as a repair) with gold. Bearing surfaces can be repaired with tin or babbitt metal, while steel parts can be brass plated to improve appearance. A plating thickness of 2 mils (50 μm) can routinely be applied; and with proper technique and solution flow, 10-70 mils (250-1780 μm) can be applied.[13] The application of selective plating is limited only by the imagination, availability of a particular plating solution, and labor considerations.

In general, conflicting requirements for surface and bulk properties can be resolved with the proper selection of a plating solution. Limitations on the area to be plated, disassembly, or transportation problems that would make conventional tank plating impractical can usually be overcome by the selective nature and portability of selective plating.

EQUIPMENT

In addition to the solutions already discussed, the equipment used in selective plating includes a variable, direct current power supply, a plating tool with anodes, and auxiliary materials such as anode covers and masking material. Equipment variations include semiautomated to fully automated configurations involving pumps to feed electrolyte to the anode, and actuators to apply the anode and replace workpieces in an assembly line operation.

Power Supplies

Since selective plating requires direct current, variable power supplies that provide rectified (filtered or unfiltered, full-wave or half-wave d-c) current are available in a variety of output ratings, up to 500 A at 25 V d-c, with inputs of 110 or 220 V a-c. Although workstations can be built in a shop area, a versatile selective plating system uses a power unit that can be taken to the work location. Small units are suitable for plating small areas, while large capacity power supplies are required to plate large areas at maximum plating rates.

Variable transformers in the power supply permit voltage and current adjustments from zero to maximum output. Voltmeters and ammeters with analog or digital readout are placed in the power unit to monitor operation conditions. Ampere hour meters are used to monitor the total current consumed; and with controlled electrolyte flow and assumptions about electrolyte metal ion depletions and cathode efficiency, the total amount of metal deposited on the workpiece can be estimated. However, uniformity of the plating thickness is controlled by the anode movement and operator technique. Forward-reverse switches permit quick changes to be made in the polarity of the voltage, which can be useful in preparing the workpiece for plating. Circuit breakers in high-current supplies can minimize damage to the workpiece or harm to the operator if the anode or tool shorts against the part.

With the availability of microprocessors, some manufacturers have included programmable calculators to assist the operator in estimating plating times and currents. The added cost of computational capability, however, must be weighed against the usefulness of precision calculations and complexity of graphs.

Plating Tool and Anode

The plating tool consists of a handle or stylus with electrical input connectors and a removable anode. Handles are made from various materials, insulate the operator from the plating circuit, and can incorporate fins to dissipate the heat generated during heavy plating operations. Large handles, used for high-current applications, can also be cooled by pumping the plating solution through them (see Fig. 20-6).

The anodes used in selective plating can be soluble or insoluble. Soluble anodes are made of the metal to be plated; they contribute metal ions to the plating solution to prolong the life of the solution, and thus are slowly consumed. In contrast, insoluble anodes, usually made from graphite but occasionally from relatively inert metals, do not play an active part in the plating process other than supplying current; the plated metal comes solely from the plating solution. Regardless of the type, the anode must always be clean and appropriate for the solution in order to avoid contamination.

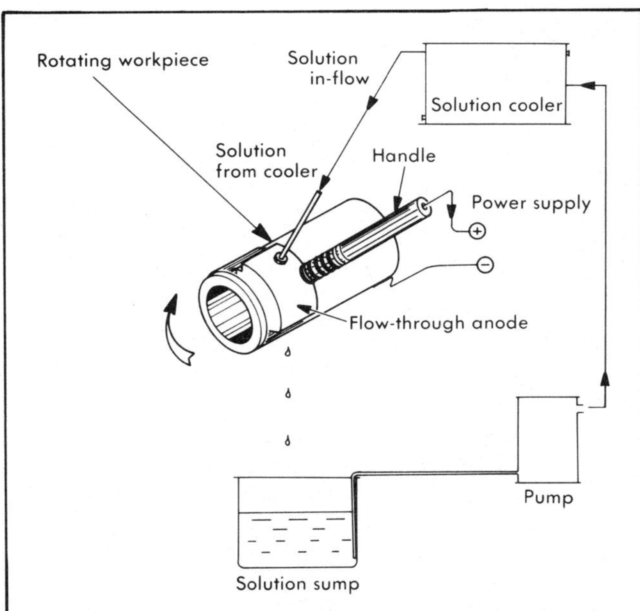

Fig. 20-6 Typical equipment used in flow-through selective plating. (*Selectrons, Ltd.*)

Anodes are commercially available in a variety of sizes and shapes and should be chosen to accommodate the different workpieces (see Fig. 20-7). A stylus/anode that approximates the surface shape and covers approximately one third to two thirds of the total area to be plated is efficient; but the anode size and, hence, current draw should be matched to the power supply outlet. Flat anodes are used as general-purpose tools for flat or curved surfaces. Concave anodes are used for plating shafts as they rotate in a turning device, while round anodes are normally used for plating holes or inside diameters of housings or bearings.

Auxiliary Materials

The anode cover or sleeve carries the plating solution to the workpiece and then distributes the solution over the plating area. The anode cover also separates the anode from the workpiece, thereby creating the electrolytic cell and reducing the chance of a short circuit between the anode and workpiece. During the plating cycle, the cover mechanically scrubs the surface of the part to disperse the depleted electrolyte layer and allow fresh solution to reach the workpiece. The cover may be

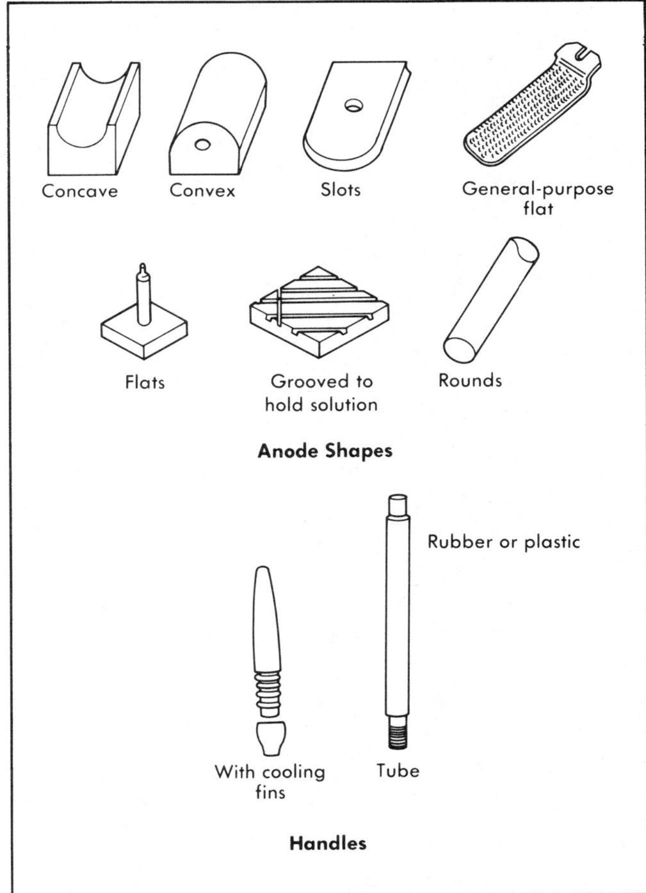

Fig. 20-7 Typical selective plating handle styles and anode shapes. (*Metal Finishing magazine and Rapid Electroplating Process, Inc.*)

SELECTIVE PLATING

either one or two piece. Two-piece covers have an inner cover to hold and distribute the solution and an outer cover to provide wear resistance.

Cost, wear resistance, absorbency, and purity are all factors governing the choice of anode coverings. Cotton batting and surgical sleeving are commonly used for the inner cover, and tube gauze is used as the outer cover. Covers are also made from presewn cotton or polyester materials. For high-volume applications, white Scotchbrite™ can be used because of its wear resistance and abrasiveness. Care must be exercised during the plating operation to prevent the abrasive coverings from damaging the thin coating of soft materials.

Masking is applied to the workpiece to prevent metal from depositing on areas where it is not desired, to sharply define the plating area, and to reduce the potential for solution contamination. Masking materials include conductive and nonconductives tapes and paints. When plating with a solution that can form a displacement reaction with the workpiece metal (e.g. silver on copper or copper on iron), masking may be required to prevent excess plating solution from depositing a poorly adherent layer on the base metal adjacent to the plating area.

OPERATING PARAMETERS

As discussed earlier, using some assumptions, it is possible to estimate the total amount of metal deposited on the workpiece. If the operator technique is good (i.e. uniform motion over the work and even pressure across the anode-workpiece interface), it is possible to estimate the plating thickness from the total electrical charge (ampere hours) used. It may also be advantageous to estimate the time and average current required to plate a desired thickness. All these computations are straightforward and listed in Table 20-22.

In addition to the dimensions of the area to be plated, it is necessary to know the following three parameters, which can be obtained from the solution manufacturer:

1. The plating solution factor is used to compute the ampere hours required for a particular thickness and area of plating. The factor is a calibration factor that is determined by measuring the number of ampere hours of plating required to deposit 0.0001″ (2.5 μm) of metal on an area of 1 in.^{2} (645 mm^2). Consequently, it is necessary to express the desired plating thickness in terms of multiples of the factor reference thickness (for example, if 0.0005″ is desired, t = 5 for use in Eq. 3, which is found in Table 20-22). The factor can be constant over a broad range of plating conditions, or it may vary with the plating conditions. Generally, solution manufacturers provide the plating factors, along with their optimal operating conditions.
2. The average current density that can be tolerated by the solution (again, a factor supplied by the solution manufacturer) must be used to estimate the current (Eq. 4, from Table 20-22), and also plating time (Eq. 5, from Table 20-22), needed for a particular plating application.
3. The relative anode-workpiece movement speed (recommended by the solution manufacturer in order to provide more uniform deposits) can be converted to an equivalent

TABLE 20-22
Operating Parameter Estimating Formulas

Electrical charge:

$$EAH = f \times a \times t \quad \text{(Eq. 3)}$$

Current:

$$EPA = CA \times acd \quad \text{(Eq. 4)}$$

Plating time:

$$EPT = \frac{EAH \times 60}{EPA} \quad \text{(Eq. 5)}$$

Rotational speed:

$$RPM = \frac{s \times C_0}{d} \quad \text{(Eq. 6)}$$

where:

	Values or Units	
	English	(Metric)
EAH = Estimated ampere hours	A · h	(A · h)
f = Plating solution factor*	A · h/in.^{2}/10^{-4} in.	(A · h/cm^2/micron)
a = Area to be plated	in.^{2}	(cm^2)
t = Thickness to be plated	multiples of 0.0001″	(multiples of 1 μm)
EPA = Estimated plating amperage	A	(A)
ca = Contact area (anode-workpiece)	in.^{2}	(cm^2)
acd = Average current density for solution*	A/in.^{2}	(A/cm^2)
EPT = Estimated plating time	minutes	(minutes)
RPM = Rotational speed of workpiece	rpm	(rpm)
s = Recommended anode-workpiece speed*	ft/min	(cm/min)
d = Diameter of workpiece	in.	(cm)
C_0 = Conversion factor	3.81	(0.318)

(*Rapid Electroplating Process, Inc.*)

* Plating solution operational parameters supplied by manufacturer.

rotational speed by means of Eq. 6 found in Table 20-22. In order to maintain concentricity, circular cross-section workpieces (e.g., shafts) should be plated while uniformly rotating (e.g., on a lathe); the equivalent rotational speed will preserve the recommended linear speed at the rotating surface.

TROUBLESHOOTING AND TESTING

The metallurgical properties of the plating, surface smoothness, hardness, brittleness, and internal stress (either compressive or tensile) are controlled by the rate of deposition and nonmetallic inclusions dragged from the solutions. The rate of deposition, in turn, is controlled by the current density used, as well as by the concentration of metal in the plating solution. In general, hardness, stress, and surface irregularities are increased with high current densities. Critical applications may require a trial and error process to develop the right combination of anode pressure and control voltage to routinely deposit a plating with the desired properties.

Beyond the metallurgical properties, adhesion of the deposit to the base metal is the most important consideration in evaluating deposit quality. Ideally, the bond between the plating and base metal is as strong as the inherent strength of the weakest material. However, poor cleaning and preparation of the base metal and/or improper plating conditions contribute to poor bonding. A simple tape test is performed by many users to evaluate the bond. In this test, a piece of tape with a strong adhesive is placed on a clean, dry portion of the plated deposit and removed with a quick pull. Other, more destructive tests such as chiselling, grinding, or bending may be performed to make a qualitative judgement of the deposit.

The appearance of the deposit also reveals a great deal about its quality. Shiny or milky deposits are usually dense, while matte (dull) deposits that cannot be easily buffed are porous or powdery and may compress under load. Too great a current density (attempting to plate faster than the solution will allow) can result in a "burned" (dark) or rough, unevenly crystallized deposit. The detailed appearance of platings varies from the pure metal because of plating conditions and microsmoothness of the resulting surface. However, most platings (with the possible exception of extremely hard platings) can be polished to reveal the expected metallic luster. Most manufacturers describe the appearance of the deposit in their instruction manuals.

Since selective plating deposits are usually applied manually, the operator can judge the quality of the deposit while plating. By knowing the desired appearance of the deposit, the operator may take corrective action, such as changing voltage, anode-to-workpiece speed, or solution supply rate to maintain proper deposit appearance and quality.

With proper operator training, selective plating may be accomplished without difficulty. However, situations do arise when the operator may encounter a problem with a plated deposit. Problems such as poor adhesion or quality, or low thickness can usually be traced to one or more of four basic causes:

1. Improper basis metal preparation—Foreign matter or oxide/corrosion remaining on the workpiece is interfering with deposition. The intermediate coating or underplating was improperly applied.
2. Wrong current density— Too low a current results in no or slow metal deposition. Too high a current results in fast solution depletion and/or burned or poor deposits with significant changes in the solution chemistry.
3. Depleted electrolyte— Little metal remains in the solution, and/or the solution chemistry has been modified to the point that poor plating results.
4. Poor anode technique— Too great a pressure on the anode squeezes the electrolyte out of the anode area; too slow a motion can allow the electrolyte to quickly deplete; nonuniform motion causes nonuniform deposits.

Table 20-23 describes some of the common problems associated with selective plating along with possible causes and suggested solutions.

TABLE 20-23
Troubleshooting Brush Plating Deposits

Problem	Cause	Corrective Action
Poor adhesion 1) Deposit coming off base metal	Base material incorrectly identified.	Correctly identify base material and follow manufacturer's recommendations for preparing base metal.
	Surface has foreign coating (chrome plate, metal spray, etc.).	Remove foreign coatings and replate or prepare foreign coating for deposit instead of base metal.
	Did not thoroughly, properly or quickly carry out preparatory procedure.	Consult instruction manual for proper instructions.
	Contamination of preparatory solutions, preparatory or preplate tools or covers.	Locate and correct contamination problem.
	Selected wrong plating solution.	Consult instruction manual or manufacturer for proper recommendation.
	No preplate used.	Consult instruction manual for preplate required on base metal.

(continued)

TABLE 20-23—*Continued*

Problem	Cause	Corrective Action
2) Final deposit coming off preplate	Did not prewet surface per instructions.	Prewet with final plating solution before applying current per manufacturer's instructions.
	Selected wrong plating solution.	Consult instruction manual or manufacturer for recommendations.
	Surface was prewet.	Corrosive plating solutions such as copper, silver, etc., should not be prewet.
3) Final deposit coming off itself	Surface dried during plating.	Do not interrupt plating cycle.
	Burned deposit.	Consult instruction manual for proper plating recommendations.
	Anode/solution contamination.	Locate and correct contamination problem.
	Wrong anode cover used.	Consult instruction manual for proper cover material to be used.
Poor deposit quality (roughness, stress cracking, stress crack lifting)	Selected wrong plating solution.	Consult instruction manual or manufacturer for recommendations.
	Anode/solution contamination.	Locate and correct contamination problem.
	Wrong anode cover used.	Consult instruction manual for proper cover material to be used.
	Incorrect procedures followed for solution.	Consult instruction manual for correct plating conditions (voltage, amperage, anode/cathode speed, etc.).
Low thickness deposit achieved	Did not properly calculate area or ampere hours required.	Recheck calculations.
	Insufficient solution supply.	Dip more frequently for solution or increase pump flow to anode.
	Plating occurs in tool cover because of overheating or graphite contamination.	Increase solution supply to dissipate heat. Remove graphite contamination.
	Plating solution factor applied incorrectly.	Consult instruction manual for correct plating conditions for solution.
	Solution overused (exceeded maximum recommended use for solution).	Change to fresh plating solution.
	Overetched base material during preparation.	Consult instruction manual for correct preparatory procedures.
	Excessive plating on masked edges.	Concentrate plating on area to be plated.
Nonuniform thickness of deposit	Tool too small for area.	Use large tool to cover area if power supply is adequate.
	Incorrect tooling selected (does not match contour of area).	Evaluate tooling requirements. If necessary, make special tool for area.
	Insufficient solution supply.	Dip more frequently for solution or increase pump flow to anode.
	Tool cover thickness varied.	Rewrap tool for more even contact and solution distribution.
Took too long to finish job	Solution overused.	Change to fresh plating solution.
	Plating tool too small.	Use larger tool to increase contact area if power supply is adequate.
	Power supply too small.	Increase size of power supply to match job requirements.

(continued)

TABLE 20-23—*Continued*

Problem	Cause	Corrective Action
Took too long to finish job	"Variable factor" plating solution applied incorrectly.	Consult instruction manual for correct plating conditions necessary (plating voltages, amperage, etc.).

(*SIFCO Selective Plating*)

SAFETY

Selective plating is a relatively safe process. However, it is necessary to exercise common sense when operating the equipment and handling the solutions. Specifics of the hazards associated with the materials can be obtained from the manufacturers via their material safety data sheets (OSHA Form 20 or equivalent). With proper knowledge, even cyanides can be safely handled in a shop environment.

Despite the low voltages employed, safe electrical practices must be followed. The power supply and outlet must be properly grounded to protect the operator from shock. The operator must be aware of and respect the fact that large currents can cause burns.

All electroplating solutions are potential hazards in varying degrees. The solutions containing cyanides, alkalines, and toxic metals pose the greatest risks to the operator and should always be segregated from other chemicals that could create an adverse reaction (in particular, cyanides must always be separated from acids). The risk of injury can be minimized by adhering to the instructions on the label of the solution bottle and maintaining operator awareness of and respect for the hazards. The solutions should only be used in well-ventilated areas and never mixed with other chemicals. Safety goggles and liquid-repellent gloves should be used when handling the solutions. On some occasions, face shields, aprons, and, for extreme situations, rubber boots may be appropriate. Personal hygiene is important to reduce the chance of spreading or ingesting the solutions; as with all chemical operations, the hands should be thoroughly washed after plating operations and especially before smoking or eating. If the solution contacts the skin, the affected area should be thoroughly rinsed with water. Eye contact and ingestion are always matters of serious concern that warrant a physician's attention.

WASTE DISPOSAL

By its nature, selective plating produces little waste, even though many of the plating solutions contain heavy metals, toxic chemicals, or otherwise hazardous materials. In addition, preparatory solutions are often highly alkaline or acidic (near the extremes of the pH scale). Depending on local pollution standards, even the rinse water and wiping/drying materials may be considered hazardous. Since the solutions are normally used in small amounts and much of the water carrier evaporates during the plating process, confinement and collection of the waste material is easily accomplished with trays placed beneath the plating operation.

In general, no selective plating solution should be casually discarded or discharged into the sewer system. The ideal disposal method is the use of a central treatment facility to remove or neutralize hazardous waste. However, alternatives exist for operations not large enough to warrant in-house treatment. Arrangements can be made with similar but larger volume waste producers (e.g., tank plating companies) to handle disposal; or, waste may be accumulated until use of a commercial treatment company would be worthwhile. With adequate dilution, determined by solution concentration from the manufacturer's OSHA Form 20 and local sewer regulations, waste products can be flushed through an industrial drain. Whatever mode of disposal is chosen, it is imperative to follow the guidelines of federal, state, and local regulations governing pollutants and hazardous waste products.

MECHANICAL PLATING

Mechanical plating is a process used to deposit malleable metals onto the surface of metallic parts. Mechanical plating has also been referred to as peen plating and impact plating. The term *mechanical galvanizing* is used in reference to deposits of zinc thicker than 2 mils (50 μm).

Mechanical plating is a room temperature process in which metal coatings are applied to parts without electricity, which is used in electroplating, and without high heat, as used in hot-dip galvanizing. The parts to be coated are tumbled in a lined barrel with water, metal powder, special promoters or accelerators, and glass impact beads. Most metal powders are approximately 0.0002″ (5 μm) in size. The promoter or accelerator provides the proper chemical environment for the plating process to take place. The glass beads or media are usually spherical in shape with a diameter of 0.006 to 0.25″ (0.15 to 6.4 mm). The media may also be made from ceramic materials and may be angular in shape. The mechanical energy generated from the barrel's rotation is transmitted through the media and causes the metal particles to be peened flat into a metallurgical bond.

PROCESS DESCRIPTION

The process sequence for mechanical plating and galvanizing is a straightforward sequence of soil and/or scale removal, surface preparation, addition of promoters or accelerators, and addition of metal powders. Descaling and/or soil removal can be accomplished in either the plating barrel or in an off-line cleaning system. After cleaning, the parts, glass beads, water, and surface conditioners are added to the rotating barrel. The surface conditioners remove residual traces of metal scale and oxides, producing a lightly coppered workpiece. The slurry of glass beads exerts a scrubbing action that facilitates oxide removal during surface preparation and provides a cushion between heavy parts, which minimizes damage to edges and sharp corners. Flat parts are also kept from clinging together owing to surface tension created by the slurry mixture. Finally, the slurry carries the plating material into the holes and recesses of the parts providing the mechanical energy necessary for plating. After the accelerator and metal powder are added, the metal particles are cold welded onto the parts by the many impingements of the small glass beads. Figure 20-8 shows the procedures for plating three different categories of parts.

The time cycle for the process, exclusive of workhandling, loading, unloading, rinsing, and dichromating, is approximately 40-75 minutes. Table 20-24 provides an approximate cycle breakdown when applying a 0.3 mil (7.6 μm) zinc coating to a batch of parts.

Advantages and Limitations

Mechanical plating eliminates the hydrogen embrittlement of steel parts, commonly associated with electroplating, since no electric current is required in the plating process. Thicker coatings can be deposited at little increased cost; and corners, recesses, threads, and the inside of tubular parts can be successfully plated. Using just one barrel, coatings of different metals (sandwich coatings) can be applied merely by adding powders of different metals in the proper sequence. Composite coatings can be deposited by using mixtures of two or more metal powders.

The chemical environments encountered in both surface preparation and plating are relatively mild. The materials are sold as concentrates; however, in use, the concentration of alkaline or acid surface preparation solutions is approximately 5%. Plating environments are even milder. The chemicals used are depleted and discarded after the cycle. Hazardous materials, such as fulminates, are not used, nor are highly toxic materials, such as cyanides, employed. Consequently, personnel safety problems are minimized, and waste disposal is simple when required. In addition, the relatively mild nature of the chemical system makes it possible to plate most powdered metal parts without initially filling or sealing the parts with organic materials.

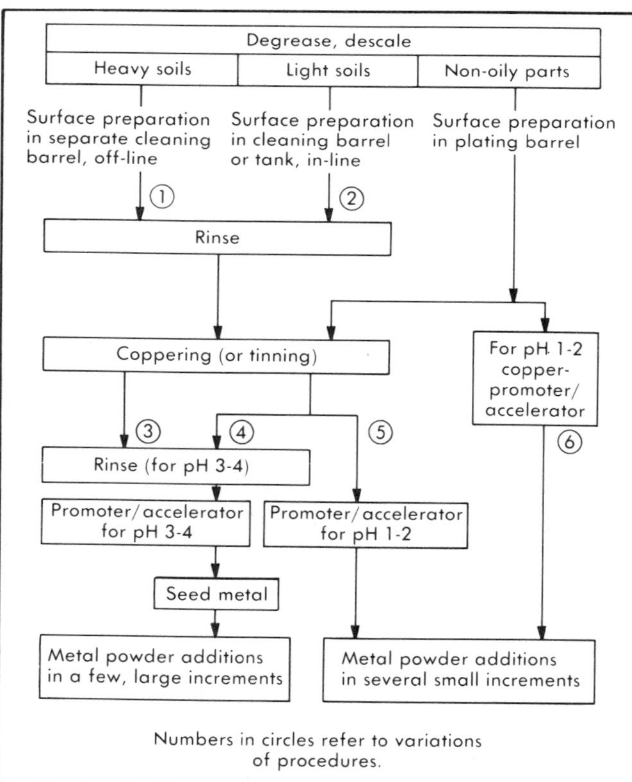

Fig. 20-8 Mechanical plating procedure or cycle for parts in a variety of conditions. (*Waldes Kohinoor, Inc.*)

Applications

The parts to be mechanically plated are most commonly made from ferrous metals (including highly alloyed and heat-treated steels), brass, bronze, and copper. Powdered metals can also be successfully plated.

The size and shape of the part determines how suitable it is for plating or galvanizing. Generally, parts range in size from 3/8 to 8" (9.5 to 200 mm) and weigh less than 1 lb (0.5 kg). Threaded rods in excess of 40" (1000 mm) long have been plated, but the cost is usually prohibitive for parts of that size.

Mechanical plating and galvanizing are performed on parts that are normally handled in batches. Parts that are normally rack plated electrolytically cannot be mechanically plated.

Typical parts that are mechanically plated include screws, bolts, nuts, washers, J-nuts, U-clips, self-tapping screws, nails, and chain links. Parts that are frequently mechanically galvanized include those previously mentioned, as well as parts used by the construction and highway maintenance industries. Through sandwich or composite procedures, the coatings can be tailored to meet special requirements such as enhanced corrosion protection for the base metal, improved surface appearance, resistance to tarnish, and improved lubricity of the surface.

Metals Deposited

The plating metals used, in order of current volumes, are zinc, cadmium, tin, lead, and their alloys. Copper, silver, and gold have also been deposited by this process. Corrosion resistance imparted by a mechanical zinc or cadmium plate is essentially the same, thickness for thickness, as that applied by other methods.

The surfaces of mechanically plated parts exhibit a bright nonspecular appearance. The mirror-bright surface possible with electroplate is not achieved, but the appearance is much brighter and smoother than is produced by hot-dip galvanizing. In addition, postplating treatments such as dichromating can be applied.

The thickness of the metals deposited by mechanical plating is usually thinner than by mechanical galvanizing. Coating thickness for mechanically plated parts varies from 0.2 to 1 mil (5 to 25 μm), and for mechanically galvanized parts the coating thickness ranges from 1 to 3 mil (25 to 75 μm).

Specifications

The United States government and various industries either accept mechanical plating under existing electroplating specifi-

TABLE 20-24
Typical Plating Cycle Time for Producing a 0.0003" (7.6 μm) Thick Coating of Zinc*

Cycle Step	Time Required, minutes
Cleaner	5-10
Copper flash	5
Accelerator	3
Zinc additions	20-30
Total	33-48

(*Waldes Kohinoor, Inc.*)

*Following procedure number 5 as shown in Fig. 20-8.

cations (for the same metal) or have written new specifications to take advantage of the special properties of the mechanical plating process. For example, aircraft manufacturers may specify mechanical plating as the only acceptable finish on parts that tend to be embrittled by electroplating. Government specifications are given in MIL-C-81562 for cadmium, tin cadmium, and zinc. The American Society for Testing and Materials' (ASTM) specification is B 695 for zinc coatings in thicknesses from 0.2 mil (5 μm) to equivalents for hot-dipped galvanized coatings of 2 oz/ft^2 (610 g/m^2) or 4 mil (102 μm) thick. The ASTM specification for cadmium is B 696, and B 635 for cadmium tin. In addition, there are hundreds of specifications in the original equipment manufacture (OEM) market of the automotive and construction industries for both mechanical plating and mechanical galvanizing.

OPERATING PARAMETERS

The three main parameters to consider in mechanical plating or galvanizing are the plating thickness, operating cost, and the energy necessary to plate the parts.

Plating Thickness

The thickness of coatings deposited by mechanical plating is independent of plating time but controlled by the amount of metal powder used per load. A thick coating can be deposited in only a little more time than a thin coating. Table 20-25 lists the amount of metal powder required to obtain a coating thickness of 0.1 mil on a surface of 100 ft^2 or a coating thickness of 1 μm on a surface of 1 m^2.

Uniformity of thickness from part to part generally falls within ±10%. Variation in plating thickness on a given part exhibits a reverse pattern from that experienced with electroplating. On mechanically plated parts, the plating is somewhat thinner on sharp edges and projections. Holes as small as 1/32″ (0.8 mm) diam, grooves, and recesses (areas of low current density in electroplating) generally receive adequate thickness by mechanical plating if the proper mix of media sizes is used.

Any problems in attaining the desired thickness, adhesion, normal smoothness, or coverage usually stem from inadequate cleaning of the parts initially and/or insufficient impacting energy. Flaking may result from an excess amount of impact energy.

TABLE 20-25
Amount of Metal Powder Required to Produce a Specified Coating Thickness on a Given Surface Area*

Type of Metal Powder	Weight of Metal Powder, lb/0.0001″/100 ft^2 (g/μm/m^2)
Zinc	0.40 (7.7)
Cadmium	0.50 (9.6)
Cadmium/zinc (50/50)	0.45 (8.7)
Cadmium/zinc (25/75)	0.43 (8.3)
Tin	0.42 (8.1)
Cadmium/tin (50/50)	0.46 (8.8)
Cadmium/tin (25/75)	0.43 (8.3)
Copper	0.51 (9.8)
Lead	0.65 (12.5)

(*MacDermid, Inc.*)

* For U.S. customary units, the metal powder weight is based on a coating thickness of 0.0001″ and a surface area of 100 ft^2. For metric units (SI), the metal powder weight is based on a coating thickness of 1 μm and a surface area of 1 m^2.

Operating Costs

The first value to determine when calculating operating costs is the surface area of the parts because chemical and metal dosage is based on the surface area. The dosage amounts are given as factors of the surface area in a given load. The load, in turn, depends on the shape, rolling characteristics, and bulk density of the part, the barrel chosen, and the lot size. The operating costs can be calculated by the equation:

$$C = A + BX \qquad (7)$$

where:

C = cost of a given barrel load
A = fixed costs (time, labor, and overhead) plus the cost of chemicals used for the load
B = desired thickness as an integer or multiple of 0.0001″ (2.5 μm)
X = cost of the metal powder required to cover the surface area in the load to the desired thickness

Impact Energy

The impact energy is the energy delivered by the glass beads or media as they flow over the surfaces of the parts. This energy is a function of both physical and chemical factors. The physical factors include:

- A balance between the weight and rolling characteristics of the parts and the weight and size distribution of the media. The media should be screened periodically to maintain desired size distribution.
- The rotation speed and configuration of the barrel in balance with the previous factors and the liquid quantity.
- The cleanliness of the media. The beads must be cleaned periodically since the metal powder tends to plate onto the beads.

The chemical factors include:

- The rate of metal deposition. This rate is controlled by the amount of promoter/accelerator and the dosage rate of the metal powder. Too fast a rate will cause the powder particles to stick to each other rather than distribute evenly on the part surface. A fast deposition rate may also cause an uneven surface on the part.
- The pH of the liquid, which must not rise above the desired range. A high pH will cause too slow a reaction and too thin a deposit, leaving metal powder suspended in the liquid.

Electric power requirements generally run about 5% of what would be required for the same load done electrolytically since the power is only used to rotate the barrel.

MECHANICAL PLATING EQUIPMENT

The equipment used in mechanical plating and galvanizing consists of several individual components. The main component is a rotating barrel having a metal parts capacity of 1 to 20 ft^3 (0.003 to 0.6 m^3); the actual barrel volume is approximately 4.5 times greater than the metal parts capacity. The barrel may be of the open-ended inclined or closed horizontal types, and is made from stainless or mild steel with a rubber or polypropylene liner. The depth to diameter ratio is approximately 1.4:1. Barrel actuation is achieved either hydraulically or electrically, with hydraulic actuation being the most popular. The speed of the barrel is usually adjustable from 70 to 200 sfm (20 to 60 m/min) to permit plating of both heavy and light parts.

Other components include a media supply hopper, media separator, and a dryer, all of which may be used as separate units or integrated and mechanized together. Figure 20-9 shows an integrated mechanical plating system incorporating a plating barrel, surge hopper, feeder and media separator, chromator, and dryer. Weighing, loading, media handling, and chemical feeding systems are also included to complete the plating system. The surge hopper/separator feeder receives the load of plated parts from the barrel and then separates the parts from the glass media with a vibrating screen separator or magnetic belt, or a combination of these components. The media is pumped back to an overhead reservoir to be used in the next coating run. After separation, the parts are either dried or chromated and dried.

POLLUTION TREATMENT

The chemicals used in mechanical plating are usually nontoxic and are essentially completely consumed in the plating process. In some instances, the effluent from a given process can be reused. For example, in the nonrinse procedures (refer to Fig. 20-8), the effluent can be used for up to ten successive barrel loadings before it must be treated.

When it is necessary to treat the effluent, the liquid is removed from the system and an alkali (such as caustic soda or soda ash) is added. The alkali raises the pH of the liquid, which is then decanted and re-acidified to the proper pH level. If desired, the liquid may be discharged into the sewer, providing the pH has been brought to a level acceptable by local, state, and federal regulations. The precipitated solids can be filtered and removed for disposal as solid waste. Chromating solutions are handled in the same manner as in electroplating.

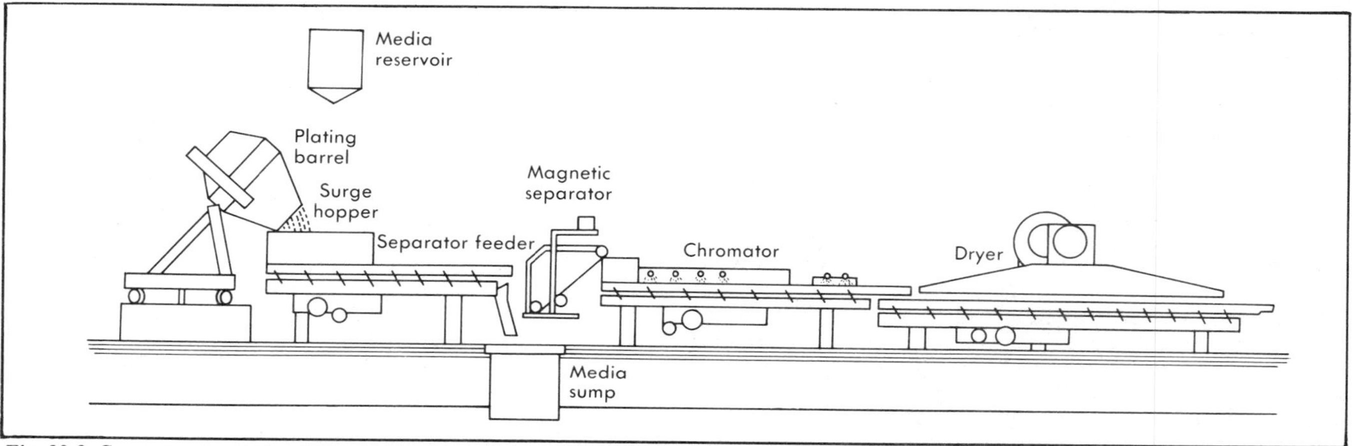

Fig. 20-9 Components used in a typical mechanical plating system. (*Renold, Inc.*)

References

1. American Society for Testing and Materials, *Annual Book of ASTM Standards*, vol. 02.05 (Philadelphia, PA: 1984).
2. Frederick A. Lowenheim, *Electroplating* (New York: McGraw-Hill, Inc., 1978), p. 257.
3. Carmine P. Nargi, "Preparing Metals for Plating," *Products Finishing Directory* (1984), pp. 87-89.
4. American Society for Testing and Materials, *Cleaning Metals Prior to Electroplating*, ASTM Standard B 322 (Philadelphia, PA: 1979).
5. Lowenheim, *op. cit.*, p.516.
6. Lowenheim, *op. cit.*, p. 14.
7. Lawrence J. Durney, *Trouble in Your Tank?* (Cincinnati, OH: *Products Finishing*, 1983), pp. 71-80.
8. Environmental Protection Agency, *Federal Register*, vol. 48, Part III (Washington, DC: July 15, 1983), pp. 32486-32487.
9. Environmental Protection Agency, *loc. cit.*
10. Carl M. Rodia and J. L. Lester, "Electroforming," *Metal Finishing—Guidebook Directory Issue* (1984), pp. 379-387.
11. Lowenheim, *op. cit.*, p. 436.
12. Marv Rubinstein, "Selective Plating," *Metal Finishing—Guidebook Directory Issue* (January 1983), p. 433.
13. Joe C. Norris, "New Developments in Brush Plating," *Products Finishing* (May 1983), p. 65.

Bibliography

American Society for Testing and Materials. *Preparation of Plastics Materials for Electroplating*, B 727-83. Philadelphia, PA: 1983.

American Society of Electroplated Plastics. *Standards and Guidelines for Electroplated Plastics*, 3rd ed. Washington, DC: 1984.

Blum, W., and Hogaboom, G. B. *Principles of Electroplating and Electroforming*, 3rd. ed. New York: McGraw Hill Book Co., 1949.

Brooks, Allan. "Mechanical Plating." *Metal Finishing* (August 1983).

DiBari, G. A. "Decorative Electrodeposited Nickel-Chromium Coatings." *Metal Finishing* (June 1977).

———. "Electroforming." *Electroplating Engineering Handbook*, 4th ed. New York: Van Nostrand Reinhold Company, 1984.

Graham, Kenneth A. *Electroplating Engineering Handbook*, 3rd ed. New York: Van Nostrand Reinhold Publishing, 1971.

Groshart, Earl. "Brush Plating Techniques." *Metal Finishing—Guidebook Directory Issue* (January 1984), pp. 358-369.

Kruger, Lennard G., ed. *Metals Handbook*, vol. 5, 9th ed. Metals Park, Ohio: American Society for Metals, 1982.

Lowenheim, Frederick A., ed. *Modern Electroplating*, 3rd ed. New York: John Wiley & Sons, 1974.

MacDermid, Inc. Numerous internal documents provided by, with permission to use. Waterbury, CT.

Maitland, Douglas W., and Deitsch, Marshall J. "Selective Plating." *Metals Handbook*, vol. 5, 9th ed. Metals Park, OH: American Society for Metals, 1982.

ELECTROLESS PLATING

The electroless plating process, also called autocatalytic deposition, deposits a uniform coating onto catalytic surfaces, regardless of the shape of the part. Once a primary layer of metal has formed on the substrate, that layer, as well as each subsequent layer, becomes the catalyst that causes the reaction to continue. Electroless plating, in contrast to conventional plating, does not use external electric current to produce a deposit. Deposition occurs in an aqueous solution containing metal ions, a reducing agent, and a catalyst (part). Chemical reactions on the surface of the catalytic part being plated cause deposition of the metal or alloy. Table 21-1 lists the different components in the plating solution, along with their respective functions.

Nickel is the most common metal deposited autocatalytically. The chemical reactions that occur when using sodium hypophosphite as the reducing agent are as follows:

$$(H_2PO_2)^- + H_2O \rightarrow (H_2PO_3)^- + H_2 \qquad (1)$$

$$Ni^{++} + H_2PO_2^{\circ} + H_2O \xrightarrow{\text{catalyst}} Ni^{\circ} + (H_2PO_3)^- + 2H^+ \qquad (2)$$

$$(H_2PO_2)^- + H^+ \rightarrow P + OH^- + H_2O \qquad (3)$$

Both nickel and phosphorus are reduced simultaneously, and the coatings produced are metastable solutions of phosphorus in nickel. When the coatings are properly heat treated, nickel-phosphide particles, which increase hardness and improve wear resistance, form within the coating. The coating thickness is not limited, except by practical considerations.

Electroless plating provides several unique characteristics superior to electrodeposition processes:

- The process produces a uniformly thick coating on both simple and complexly shaped workpieces.
- The process may be applicable to a variety of substrates ranging from metals and semiconductors to nonconductors.
- The use of an auxiliary power supply and the need for electrical contact is eliminated, except for precleaning and surface activation purposes.
- Some electroless deposits have unique and controlled chemical, mechanical, and magnetic properties.

CHAPTER CONTENTS:

METALS DEPOSITED

Electroless plating processes are used by industry to alter the surface of a part, providing a uniform, conductive metallic coating. Nickel, copper, cobalt, and gold are the most commonly deposited metals. The commercially produced electroless alloys are listed in Table 21-2.

Composite coatings have also been successfully deposited. These coatings consist of particles of such materials as synthetic diamonds, silicon carbide, aluminum oxide, and polytetrafluorethylene (PTFE), codeposited with nickel or cobalt. Composite coatings enhance the wear and friction characteristics of the metal deposit.

PROPERTIES

The properties of electroless coatings are varied and are determined by the specific metal and process employed. In the case of electroless nickel, the corrosion and wear resistance, conductivity, solderability, and magnetic properties of the deposit find useful application.

Producing specific properties in the deposit from a particular plating process requires careful control. For example, the hardness of the base metal determines which heat treatments are required to prevent hydrogen embrittlement caused by the pretreatment processes. It may be necessary to control the surface finish, elongation, structure of the deposit, reflectivity, conductivity, melting point, composition, coefficient of thermal expansion, density, internal stress, and elasticity to achieve specified properties. The producer must therefore have a working knowledge of how the process parameters affect the deposit. Some of the properties of electroless metal deposits are summarized in Table 21-3.

APPLICATIONS

The applications of electroless metal deposits are many and diverse. Each metal has specific applications for which it can be used. Typical metal deposits include copper, gold, nickel, and nickel alloys.

Contributors of sections of this chapter are: Dr. George D. DiBari, Product Manager, Technical Services, International Nickel, Inc.; Phil Stapleton, Technical Director, Stapleton Company & Associates.

Reviewers of sections of this chapter are: Teri L. Arney, Senior Research Chemist, Elnic, Inc.; Edward G. Buckley, Technical Service Manager, Technic, Inc.; Barry R. Chuba, Technical Marketing Specialist, Enthone, Inc.; Dr. George D. DiBari, Product Manager, Technical Services, International Nickel, Inc.; Ronald Duncan, Director of Research and Engineering, Elnic, Inc.; Dr. Nathan Feldstein, Surface Technology, Inc.; Gerald A. Laitinen, Product Manager, Allied Kelite Products, Witco Chemical Corp.; Jovona L. McDowell, Manager, Laboratory Services, Elnic, Inc.; Gary Shawhan, Electroless Nickel Product Manager, Enthone, Inc.; Phil Stapleton, Technical Director, Stapleton Company & Associates.

CHAPTER 21

METALS DEPOSITED

Copper

Electroless copper deposits are used in engineering applications to provide conductivity, such as in the through-hole plating of printed circuit boards and in the preparation of plastics for decorative plating. The deposits are uniform, conductive, and can be applied over many substrates that have been properly activated. Electroless copper deposits are also being used, in combination with electroless nickel, to solve electromagnetic shielding problems.

TABLE 21-1
Typical Components in Electroless Plating Solutions

Component	Function
Metal salt (e.g., nickel sulfate)	Provide the metal ions to be plated.
Reducing agent (e.g., sodium hypophosphite)	Provide the reducing power at the catalytic interface.
Complexing agent (chelators) (e.g., lactic acid)	Complexing of the metal ions and preventing bulk decomposition.
Buffering agent (e.g., acetic acid)	To resist the pH changes caused by the hydrogen released during deposition.
Accelerators (Exultants)	To help increase the speed of the reaction.
Inhibitors (Stabilizers)	To control reduction reaction in the plating solution.

Gold

Electroless gold deposits are used for improving solder shelf life by preventing oxidation of the surfaces. Gold deposits are normally thin, under 0.1 mil (1 mil = 0.001″) (2.5 μm), and are applied over electroless nickel, which prevents diffusion of the gold into copper. Electroless gold is used in electronic applica-

TABLE 21-2
Commercially Produced Electroless Alloys

Electroless Process	Au	Cu	Ni	Co	P	B	Tl	Trace
Copper		99.9						0.1
Gold	99.98							0.02
Nickel phosphorus low			98		2			0.3
Nickel phosphorus medium			94		6			0.3
Nickel phosphorus high			89		11			0.1
Nickel cobalt phosphorus			75	20	5			0.3
Nickel cobalt phosphorus			16	78	6			0.3
Nickel boron			99.5			0.5		0.1
Nickel thallium boron			92			3.5	4.5	0.3

(*Stapleton Company*)

TABLE 21-3
Summary of Deposit Properties

Property	Cu	Au	NiP9	NiCoP	NiB	NiTlB
Hardness,* Vickers (100 g load)	---	---	500-550	570	650-700	750
Modulus of elasticity, 10^6 psi (GPa)	---	---	17-28.4 (118-196)	---	17 (118)	17 (118)
Tensile strength, ksi (MPa)	---	---	35.5-128 (245-883)	42.7-99.5 (294-686)	---	15.6 (108)
Internal stress, ksi (MPa)	---	---	-4.3 to +42.7 (-29 to +294)	---	+42.7 (+294)	+15.6 (+108)
Density, g/cm^3	8.9	19.3	7.95	8.1	8.3	8.25
Thermal expansion, μm/m · °C	16	14	13	13	10	12.1
Melting point, °F (°C)	1980 (1080)	1940 (1060)	1635 (890)	1740 (950)	2640 (1450)	1980 (1080)
Magnetic, oersteads	0	0	1	up to 400	>5	>5
Resistivity, μ Ω · cm	1.6	2.1	20-110	20-110	<10	30-70

(*Stapleton Company*)

* Hardness values given are before heat treatment.

tions to provide coverage on isolated circuits where conventional electroplating is not possible. Electroless gold deposits are also being used on printed circuit boards to improve wear.

Nickel Phosphorus

Electroless nickel-phosphorus deposits are widely used to reduce corrosion and wear. Corrosion protection is provided by isolating the base material from the environment and by the natural oxide layers that form on the coating surface. Wear resistance is provided by the natural lubricity and hardness of the coating. Electroless nickel can be deposited onto many materials in thicknesses ranging from 0.05 to 100 mils (0.0013 to 2.5 mm). A typical engineering coating would be approximately 1 mil (25 μm) thick. The amount of phosphorus is from 1 to 13% and can generally be controlled to ±0.5%. The properties of these coatings depend on the phosphorus content, which should be specified when developing applications (see Table 21-4).

When electroless nickel deposits are heated to above their transition temperature, 390-660° F (200-350° C), they begin to precipitation harden and crystallize. Coatings have been produced that have hardnesses greater than 950 Vickers using a 100 g load. During this process, the coating changes from an amorphous structure to a mixture of nickel phosphide and crystalline nickel. When mid-phosphorus deposits are heat treated at 1380° F (750° C), they increase in strength; at the same time, high-phosphorus deposits decrease in strength due to greater shrinkage and the microcracking that may occur. Heat treating high-phosphorus deposits at lower temperatures, 230-400° F (110-200° C), and for a longer period of time helps to increase hardness while minimizing microcracking.

Strength, elasticity, elongation, and ductility are improved in deposits with high phosphorus, providing superior performance in many engineering applications including electroless-formed bellows, hydraulic cylinders, and pressure vessels. Electroless nickel-phosphorus deposits are also being used to provide electromagnetic shielding for plastic cabinets and enclosures.

Nickel Cobalt Phosphorus

Electroless nickel-cobalt coatings are used in magnetic applications where uniform, thin deposits are required. These coatings are in the thickness range of 0.003 to 0.02 mils (0.07 to 0.5 μm) and contain between 20 and 80% cobalt, depending on the application.

Nickel Boron

Electroless nickel-boron deposits are used in place of electroless gold, providing good wire bonding and solderability. These coatings are hard, uniform, and conductive and contain about 0.5-5.0% boron, the balance being nickel. When they contain between 3 and 4% boron, these deposits possess good wear and erosion characteristics for engineering applications.

Nickel Thallium Boron

Electroless nickel-thallium-boron deposits are used in high-wear applications, providing excellent lubricity and high hardness at high temperatures and high loads. Deposits containing 5% thallium and 5% boron are applied over a wide range of base materials in thicknesses of 0.5 to 2.0 mils (13 to 50 μm).

TABLE 21-4
Types of Electroless Nickel-Phosphorus Deposits

Deposit Type	Phosphorus Content, %		Comments
Low phosphorus	1.5	2.5	These processes provide low-temperature operations for plating on plastics. The coatings are high in stress and produced from alkaline solutions.
Medium phosphorus	3.5	8.5	These processes provide conventional electroless nickel coatings. The deposits are used for preventing wear and corrosion in engineering applications. Coatings may be bright depending on the chemistry. Solutions are operated in a pH range of 4 to 5.
High phosphorus	9.3	13	These processes provide increased protection in aggressive environments over medium-phosphorus coatings. The amount of trace metals plated into the deposit are reduced, producing a more pure deposit. These processes operate in a pH range of 4 to 5. Coatings have higher elongation and strength than medium-phosphorus coatings in the as-plated condition.

(*Stapleton Company*)

ELECTROLESS PLATING PROCESS

Electroless plating processes depend on chemical reduction reactions to deposit uniform metallic coatings on parts, eliminating the need for external current. The coating is uniform on all wetted surfaces. Thickness is determined by the length of time the article or part is kept immersed in the solution. The entire process involves several different steps as discussed in this section.

PRECLEANING

Precleaning removes heavy soils and scales on the surface and exposes the bare base material. The precleaning can be accomplished by wet or dry blasting, vapor degreasing, mechanical cleaning, and descaling, among other methods. In almost all cases, the base material will not be clean enough after precleaning to achieve adequate adhesion. The surface will still

ELECTROLESS PLATING PROCESS

have oxides and oils present. For additional information, refer to the various cleaning methods discussed in Chapter 16, "Mechanical and Abrasive Deburring and Finishing," and Chapter 18, "Cleaning," of this volume.

CLEANING

Cleaning removes the oils and organic material on the surface of the parts and is first accomplished by soaking in a solubilizing solution at an elevated temperature. Some base materials require electrocleaning as a second step to remove carbon or grinding material. The last step is to completely remove the oxides in an acidic solution.

The cleaning of metals prior to electroplating is discussed in American Society for Testing and Materials (ASTM) Standard B 322. Most of the information in that standard is applicable to the preparation of metals for electroless deposition. Additional information can be obtained in those standards that deal with specific metals and alloys.[1]

INITIATING DEPOSITION

Autocatalytic nickel will usually deposit on clean, wetted surfaces. Aluminum, beryllium, platinum metals, iron, cobalt, nickel, titanium, and their alloys can be plated directly. Certain base metals cannot be plated directly; these include zinc, lead, cadmium, tin, bismuth, arsenic, antimony, and alloys containing large proportions of these metals. The metals that cannot be plated directly should be electroplated with a thin copper or nickel strike prior to being immersed in the electroless plating solution.

Preliminary electrolysis or contact with a catalytic metal like iron or nickel is required for deposition on copper, silver, gold, carbon, vanadium, molybdenum, tungsten, chromium, selenium, and uranium. In some cases, immersion deposition processes may be used to initiate electroless deposition. Special heat treatments may be required on aluminum, beryllium, titanium, and other metals to obtain maximum adhesion.[2]

PLATING

When a properly prepared material is placed into the electroless plating solution, the surface potential reaches a value where metal deposition is possible. The deposition potential is dependent on the pH, temperature, ionic concentration, and chemical composition of the solution.

As the potential changes, a charge between the solution and the base material is created. The charge causes the cations in the solution to organize themselves and produce what is called the matrix. This matrix is like a blanket that covers the surface and controls the deposition process. The reducing agents in the solution provide electrons at the interface. Through several sequential reactions within the matrix, the metal ions in solution are reduced, producing the coating.

Each electroless plating process uses unique chemistry to achieve this reduction reaction. Powerful reducing agents like sodium borohydride and titanium trichloride are used to plate out thallium and gold, while sodium hypophosphite and formaldehyde will reduce nickel and copper, respectively. The concentrations of these reducing agents and the metal salts change as the plating process proceeds and by-products are produced. To sustain the reduction reactions, additions of fresh chemicals must be made. In addition, the special additives used to provide brightness, stability, wetting, controlled cloud points, and controlled transition temperatures must be monitored and kept at optimum levels.

Proprietary electroless processes are available, and suppliers of these processes provide instructions for maintaining and controlling individual processes. Electroless processes can be maintained by performing simple chemical analyses from which chemicals required to replenish the solution can be determined.

CONTROLLING THE PLATING PROCESS

Impurities in the solution can affect the ductility, corrosion resistance, and appearance of electroless deposits while contaminants in the solution can cause pitting, adhesion, and roughness problems. Hence, the solution must be kept relatively free of impurities to control the properties of the deposit and avoid producing defectively coated parts that must be scrapped or reprocessed at great expense to the metal finisher.

A deliberate, sustained campaign to keep impurities out of solution is better than having to resort to time-consuming and expensive treatments to remove contaminants. For this effort to be successful, everyone involved in processing the parts should become familiar with the sources of impurities, which are one or more of the following:

- Contamination from alkaline cleaners, acid pickling solutions, and other solutions by direct transfer (drag-in), or by airborne spray.
- Airborne particulate matter; for example, from grinding and polishing operations.
- Improper removal of grease, oil, and buffing and polishing compounds from the work.
- Defective filters, pumps, and other equipment.
- Defective exhaust ducts.
- Tank leachates and defective rack coatings.
- Grease and oil from pumps, hoists, and overhead equipment.
- Dissolution of metals or parts that accidentally fall into the solution.
- Hard water (calcium sulfate).

Assuming that all potential sources of contamination can be identified, it should be possible to compile a list of actions to maintain the purity of the solution and, consequently, avoid generating rejects. Some recommended actions are as follows:

- Remove metals or parts that have fallen into the solution as quickly as possible.
- Use adequate rinses and electrocleaning in the preparation process.
- Monitor the conductivity of rinse tanks to check cleanliness of the water and monitor excessive drag-out levels.
- Exhaust grinding and polishing equipment properly, or isolate these processes from the plating area.
- Use de-misting agents to prevent airborne spray.

- Inspect and maintain filters, pumps, and other auxiliary equipment on a regular schedule.
- Inspect and repair racks and rack coatings.
- Maintain solution at near optimum levels, maintain temperature, and check pH on a regular basis.
- Follow the instructions of the supplier of the process for bath makeup, replenishment, and operation.

Despite all precautions, problems may occur. Table 21-5 lists common problems that occur in electroless nickel plating, gives the possible causes, and suggests solutions to these problems.[3]

TABLE 21-5
Troubleshooting Electroless Nickel Deposits[3]

Problem	Possible Causes	Suggested Solutions
Skip plating, poor coverage, edge pullback and frosted edges	Improper cleaning.	Temperature should be checked as well as purity and concentration of cleaner and other pretreatment solutions.
	Improper activating.	Activators should be checked. Some metals and alloys, such as leaded steels, brasses, copper, aluminum and magnesium, require special preparation.
	Improper rinsing.	Rinse temperature and rinsing time should be checked. Too long a time in a rinse may cause an oxide film to form; too short a time may not remove residual films. Rinses must be kept clean.
	Metallic contamination.	Bath should be dummied or discarded and replaced.
	Organic contamination.	Bath should be cooled, carbon treated, and filtered. Rinses should be checked for residual drag-in.
	Too much air agitation.	Air agitation should be reduced.
Roughness in deposit	Contamination from solid particles; i.e., dust, loose nickel, or metal chips.	Solution should be filtered through #5 micro filter at high flow rates.
	Turbidity of solution.	May be result of high pH, which aids in formation of metallic hydroxides. pH should be checked and adjusted. Complexers may be added.
	Heavy or too rapid EN solution replenishment while work is being plated.	Replenisher should be added slowly and mixed thoroughly. Additions should be made in area as far away from work as possible.
	Contaminated makeup water.	Quality of water should be checked. Water may require carbon treatment and filtration.
	Localized overheating.	Air agitation should be checked or temperature adjusted. Derated heaters should be used.
	Particles of metal or soil on work.	Cleaning and rinsing should be improved.
	Contaminated filter cartridges, liners, or bags.	Liners should be leached prior to makeup. Filters should be changed.
	Only one side of work affected.	Agitation around work should be increased. Work rod or solution agitation should be provided.

(continued)

TABLE 21-5—*Continued*

Problem	Possible Causes	Suggested Solutions
	High sodium-phosphate concentration.	All or part of the solution should be discarded.
Streaks in deposit	Gas streaks from position of work.	Either solution or work rod agitation should be provided. Work should be repositioned periodically. Agitation should be increased.
	Silicate drag-in.	Nonsilicated cleaners should be used. Rinsing should be improved.
	Poor rinsing.	Process cycle should be checked; rinsing improved.
	Improper cleaning.	Cleaning and/or rinsing should be improved.
	High concentration of metals.	Bath should be dummied to remove metals. Bath may have to be discarded and replaced with new solution.
	Organic contamination.	Bath should be carbon treated and filtered, or discarded and replaced with new solution.
	Poor agitation.	Air agitation should be increased, or a different air pattern developed.
	Low surface area.	Surface area should be increased to recommended range.
	Low reducer content.	Reducer should be checked and adjusted.
	Too much agitation.	Agitation should be reduced.
	Too much complexer.	Complexer should be reduced.
Pitting	Heavy metal contamination.	Bath should be dummied or discarded and replaced.
	Ethylene glycol contamination from jacketed tank.	Bath should be discarded and replaced. Ethylene glycol must be kept from bath.
	Basis metal pitted.	Basis metal should be checked after each step in the cycle.
	Organic contamination.	Bath should be cooled, carbon treated, and filtered.
	Improper cleaning.	Cleaning and rinsing should be improved.
	Excessive bath activity.	pH should be lowered.
	Copious evolution of hydrogen.	Surface area being plated should be reduced. Reducing agent concentration should be checked and adjusted. If problem is being caused by plating of tank walls and equipment, it may be necessary to clean tank and equipment.
	Low complexer concentration.	Complexer should be added in small increments.
	Improper reducer concentration.	Bath should be analyzed and adjusted.
	Bath very old (more than 15 regenerations).	Bath should be discarded and new bath prepared.

(continued)

TABLE 21-5—*Continued*

Problem	Possible Causes	Suggested Solutions
Dull or matte deposit	Organic contamination.	Bath should be cooled, carbon treated, and filtered.
	Improper cleaning.	Cleaning should be improved. Electrocleaning may be required.
	High or low pH.	pH should be adjusted with acid or alkali.
	Bath imbalance.	Nickel and reducer should be checked and bath adjusted.
	Poor quality substrate.	Substrate should be improved.
	Low temperature.	Temperature should be adjusted.
	Excessive brightener drag-out.	Solution chemistry should be balanced.
	Metallic contamination.	Bath should be dummied onto a large surface.
	Low reducer concentration.	Reducer should be added.
	Low metal content.	Metal should be added.
Poor adhesion on aluminum	Metal contamination.	Bath should be dummied or discarded.
	Improper surface preparation.	Cleaning and rinsing should be improved.
	Improper zincate or other pretreat condition.	Concentration and aluminum substrate should be checked.
	Reoxidation.	Transfer times should be reduced.
	Drag-in of acid inhibitors.	Uninhibited acid should be used; rinsing improved.
	Organic contamination.	Bath should be carbon treated and filtered.
	Improper heat treatment.	Heating time and temperature should be corrected.
	Quality of alloy.	Alloy quality should be checked.
Frosted deposits	Low workload.	Workload should be increased.
	Metallic contamination.	Bath should be dummied or discarded and replaced.
Step plating	Metallic contamination.	Bath should be dummied or discarded and replaced.
	Certain alloys, such as lead, steel, brass, or copper.	A copper or nickel strike should be used prior to plating, followed by galvanic activation for copper.
	Bath imbalance.	Bath should be analyzed and adjusted.
Laminar deposits	Poor temperature, pH, and/or bath control.	Uniform temperature, pH, and bath control must be maintained.

(continued)

TABLE 21-5—*Continued*

Problem	Possible Causes	Suggested Solutions
Poor wear resistance	Low heat treatment temperature and/or short time.	Temperature and time cycle should be adjusted.
	Low phosphorus content.	Phosphorus content should be increased by lowering pH and temperature. Bath should be analyzed and chemical balance restored.
Poor corrosion and/or chemical resistance	Low phosphorus.	Phosphorus content should be increased by lowering pH and increasing reducer concentration.
	Metallic contamination.	Bath should be dummied or discarded and replaced.
	Poor bath control.	Uniform temperature should be maintained; pH and replenishment controlled.
Dark to black deposits	High stabilizer content.	Bath should be dummied or diluted.
	Metallic contamination.	Bath should be dummied or diluted.
	Bath imbalance.	Bath should be analyzed and adjusted.
	Poor activation.	Reactivate.
	Organic contamination.	Bath should be carbon treated and filtered.
	Improper surface preparation.	Cleaning, pickling, and rinse cycles should be improved.
	Inadequate rinse.	Rinsing should be improved.
	High workload.	Workload should be reduced.
	Low workload.	Workload should be increased.
	Low reducing agent.	Reducing agent should be increased.
	Low metal concentration.	Metal content of the bath should be increased.
Blistering	Inadequate surface preparation.	Cleaning, pickling, and rinsing should be improved.
	Improper zincating of aluminum.	Concentration of zincate and processing time should be checked.
	Organic contamination.	Bath should be carbon treated and filtered.
	Metallic contamination.	Bath should be dummied or discarded and replaced.
	Improper pH.	pH should be checked and adjusted.
	Improper heat treatment.	Heat treating temperature and time should be checked and corrected.

(*Metal Finishing Supplier's Association*)

PROPERTIES OF ELECTROLESS DEPOSITS

The appearance, adhesion, porosity, hardness, and thickness of the deposit can be controlled by following the prescribed procedures for solution makeup and by following the recommended operating parameters. To obtain reproducible properties, impurities must be kept to a minimum, as previously discussed.

Appearance of the deposit is affected by the metallurgical condition of the surface being coated; hence, the surface should be as uniform and as smooth as possible. Adhesion to the base material is affected by the surface finish of the base material and how it is cleaned and activated. Impurities in the solution can also affect adhesion.

The porosity and protective value of the coating are influenced by the uniformity, smoothness, porosity, and other surface defects of the substrate and by how the substrate is prepared for coating. In the case of electroless nickel, the phosphorus content affects corrosion resistance, and the nature of the stabilizer and complexing agents may also play a role in corrosion resistance.

Hardness of the deposit is controlled by the pH, temperature, composition, and agitation of the solution, as well as by post heat treatment of the deposit. Hardness is measured by microhardness testing with Knoop indenters on the cross section. Thickness is a function of immersion time in the solution. Coupons are plated along with the base material to determine the thickness of the deposit and the rate of deposition. The base material should never be removed from the plating solution until the desired thickness has been achieved. Additional information on hardness and thickness measurements can be found in ASTM Standard B 578.

Solderability is controlled by the surface condition of the deposit. Most metals oxidize with time, producing a thin film that interferes with the soldering process. By removing this film prior to soldering, the coatings can be soldered. Gold coatings oxidize very slowly and can be readily soldered, whereas coatings of nickel phosphorus require cleaning in a mild rosin flux to achieve solderability. Heat treating of electroless nickel deposits may make soldering difficult, and a highly active, acid flux may be required.

EQUIPMENT

The choice of electroless plating equipment influences the life of the plating bath and the quality of the deposits. Plating tanks are generally rectangular in shape and may be made from a variety of materials. Alternating the plating operation between two tanks minimizes downtime and allows more time for stripping the tank not being used.

Most electroless baths operate at approximately 200° F (90° C) and require heaters and temperature controllers for proper control; the maximum temperature variation is ±5.0° F (3.0° C). Equipment should be checked frequently to ensure accuracy. Steam heat exchangers are the most efficient for tanks having a capacity of greater than 200 gal (760 L), while electric immersion heaters work well in smaller tanks. Electric heaters should have a safety interlock that would prevent a fire in case the heater elements are exposed.

Pumps are required for solution transfer and filtration. Vertical centrifugal pumps are most commonly used and are usually made from CPVC plastic or stainless steel. Cartridge filters and filter bags are used to filter out particles larger than 0.2 mils (5 μm) in size. Filter bags are less expensive than cartridges and minimize the restriction at the discharge side of the pump. Mechanical or low-pressure air agitators are used to provide a fresh supply of solution to the part and to remove the hydrogen produced during deposition.

Rectifiers are used for some pretreatments, such as electrocleaning, and are also used when the activation process cannot employ immersion techniques. Rectifiers must be sized to achieve the specified current density of the electrocleaner. Typically their size range is between 50 and 300 A/ft^2 (5.4 and 32 A/dm^2).

TEST METHODS

The properties of electroless coatings can be determined by the test methods described below. Table 21-6 lists the test methods that apply to electroless coatings.[4]

APPEARANCE AND ADHESION

The appearance of electroless coatings should be uniform and free of pits, blisters, cracks, patterns, and pimples. These defects are visible on the surface and can be evaluated with normal corrected vision or under some level of low magnification. Chemical tests can also be performed to check the coating appearance.

Four tests commonly used to evaluate adhesion are the tape test, punch test, bend test, and quench test. These and other qualitative methods of measuring adhesion are discussed in ASTM Standard B 571. The coating's adhesion to the base material should be significant enough to maintain the coating's integrity in the most adverse service conditions.

TABLE 21-6
Standards and Specifications for Electroless Nickel Plating

Specifying Body	Specification No.	Title of Subject
ASTM	B 571	Adhesion of Metallic Coatings
	B 567	Thickness of Coating by Beta Backscatter Method
	B 568	Thickness of Coating by X-ray Spectrometry
	B 499	Thickness of Coating by Magnetic Method for Nonmetallic Coatings on Magnetic Base Metals
	B 287	Acetic Acid Salt Spray (Fog) Testing
	G 43	Acidified Synthetic Sea Water (Fog) Testing
	B 368	Copper-Accelerated Acetic Acid Salt Spray (Fog) Testing (CASS)
	B 117	Salt Spray (Fog) Testing
	B 656	Standard Guide for Autocatalytic Nickel Deposition on Metals for Engineering Use
	B 733	Standard Specification for Autocatalytic Nickel-Phosphorus Coatings on Metal
Aerospace Materials Specifications (AMS)	2404	Electroless Nickel Plating
	2405	Electroless Nickel Plating—Low Phosphorus
Military	MIL-C-26074B	Coatings, Electroless Nickel, Requirements for
International Standards Organization (ISO)	ISO 4527	Specification for Autocatalytic Nickel/Phosphorus Coatings (Including Test Methods)

The tape test method is used on gold and copper deposits when high loads are not present. A strong adhesive tape is applied to a coated surface and then removed. If the plating is removed by the tape, the adhesion is inadequate.

The punch test method is used on high-strength coatings to determine if the coating will separate from the base material upon deformation. In this method, a standard spring-loaded punch with a 2-3 mm radius-rounded tip makes three small indents at 120° intervals. The area between the indents is examined for flaking; loss of plating indicates inadequate adhesion.

The bend test method is used on coupons processed with the parts. A strip is plated and then bent over a mandrel. The diameter of the mandrel should be four times the thickness of the strip. After bending the coupon through 180°, the coating should remain attached on the radius. Small cracks at the radius are permissible.

The quench test method can be used on the part, which is heated to 400°F (205°C) in an oven and next quenched in ambient temperature water. The coating is then examined for poor adhesion at four times magnification.

POROSITY

The ability of the coatings to provide a pore-free barrier is important in applications where corrosion resistance is required. Porosity is the condition that consists of small holes in the deposit, exposing the base material. This condition is different from the pitting that occurs in service as a result of corrosion. Porosity is controlled by agitation and proper racking.

Measuring porosity is accomplished by immersing the part into a reagent solution that will react with the base material and create some visible salt. Many different reagents can be used. The selection of a suitable reagent is dependent upon the base material and deposit. Copper sulfate will deposit on iron and can be used for thin [0.1 mil (2.5 μm)] electroless nickel deposits. Ferroxyl will also work well with electroless nickel and copper deposits on iron. Alizarin sulfonate can be used for aluminum-based materials and potassium ferricyanide reagents with copper-based materials.

HARDNESS

To measure the hardness of the coating, a deposit is plated onto a coupon and metallurgically mounted, polished, and indented with a Vickers or Knoop diamond. Measurements are made on the indent scar, and the hardness is calculated. Hardness measurements made on the cross section of the deposit are usually more accurate than those made on the surface. The ASTM Standard B 578 describes the procedures for determining the hardness of electroless plating deposits.

THICKNESS

Coating thickness is an important property and may be measured destructively or nondestructively for most base materials and coatings. The density of the coating will affect the results obtained with certain test methods.

Beta back-scatter techniques can be used on coatings that have greater than a five atomic number difference between the base material and coating. Beta particles are reflected from the interface, causing a count on a geiger-muller tube. As the coating increases in thickness, the beta particles slow down and more become reflected, increasing the count rate.

X-ray fluorescence techniques can be used on all coatings, but only on certain thicknesses. With these techniques, the surface is bombarded with X rays of a significant electron voltage that will create the fluorescence of X rays specific to the elements in the deposit. If the density is known, the counts of X rays will be proportional to the thickness at specific electron voltages.

Magnetic techniques can be used on certain coatings that are nonmagnetic over magnetic-based materials. The magnetic flux of the base material is used to calculate the distance between the probe and the base material.

COMPOSITION

The precise composition of a deposit is important in specific applications. It can be determined by an inductively coupled plasma (ICP) test, atomic absorption (AA) test, or by conventional wet analysis. A coupon or sample of the deposit should be dissolved in nitric acid and then aspirated into the ICP or AA. All the major elements should be analyzed to determine the percentage of each element in the deposit. Autocatalytic deposits can also be analyzed nondestructively by X-ray fluorescence.

CORROSION RESISTANCE AND PROTECTION

The corrosion resistance of the deposit may be important in certain aggressive applications. If some measurement of corrosion performance is required, the parts or coupons can be evaluated by tests such as copper acetic-acid salt spray (CASS), Corrodkote, acetic-acid salt spray, and SO^2 salt spray. Neutral salt spray tests are commonly used on steel panels at regular intervals to evaluate the protective value properties of the coating. See Table 21-6 for standard methods of evaluating corrosion performance.

WEAR RESISTANCE

Many of the electroless nickel deposits are hard and have natural lubricity. To evaluate the wear resistance of a coating, both abrasive and adhesive tests can be used. Abrasive tests cut the coating with hard particles. In adhesive tests, the surfaces are point welded together and then pulled apart, causing cohesive failure of the deposit. In both test methods, the amount of material removed at a given velocity, load, and area per unit of time is reported.

SAFETY AND ENVIRONMENTAL FACTORS

Many of the plating processes use corrosive chemicals. Some processes also use hazardous and toxic materials, and care should always be exercised when handling these materials. An OSHA Form 20 should be reviewed and kept on file for every chemical used. In addition, all equipment should be designed to code and maintained. Eye wash equipment should be installed; and articles such as boots, aprons, gloves, and eye protection should be used to protect against splashing of chemicals on personnel. Ventilation of the process area is necessary to remove vapors from strong acids and hot process tanks.

All plating processes produce some metal waste. When operating efficiently, metals removed from the base material, drag-out of plating solution, and spent solutions will be less than 10% of the total metal in salts used. Some of these metals require removal from the solution, while others are permitted to be discharged directly into the sewer.

Local public-owned treatment works (POTW) are mandated by federal law to maintain strict environmental standards on their discharge to ground water systems. If the local POTW has developed a recovery plan, it may offer removal credits that can be used to increase the single generator limits. Present limits require several methods of treatment by the plating facility in order for the POTW to comply with federal law.

Rinse waters require neutralization to an alkaline pH and the removal of trace metals near to the 1 ppm level. Spent plating solutions must also be treated to the same levels before they can be discharged to the sewer. The chemistry of each process has unique features that determine the treatment method required. When choosing a method, the disposal of the waste sludge that is produced in the treatment process, the distance to the local landfill, the cost of transportation, and the nature of the sludge being produced must all be considered.

Several new techniques are now available that minimize the production of waste sludge and provide an economical means of recycling the metals. Ion exchange can be used to extract small amounts of metals from waste waters. Gold is recovered in this manner using several rinse stations. Nickel and copper may also be removed from waste water by ion exchange, providing clean water for recycling. By using this technique, the total water usage can be reduced to less than 10,000 gal (37 800 L) per day for a moderately sized facility. At this level of water usage, the facility is exempt from many of the strict reporting laws mandated by the federal government.

Plating and destruction techniques can be applied to the spent solutions that produce a solid metal sludge, which is suitable for smeltering. Methods that produce hydroxide sludges or sediment sludges should be avoided because of leaching in landfills.

References

1. American Society for Testing and Materials, *1984 Annual Book of ASTM Standards*, vol. 02.05 (Philadelphia, PA: 1984).
2. *Ibid.*
3. Metal Finishing Suppliers' Association, *Electroless Nickel Plating* (Birmingham, MI: 1983), pp. 25-28.
4. American Society for Testing and Materials, *op. cit.*

Bibliography

Gawrilov, G. G. *Chemical (Electroless) Nickel-Plating*. Surrey, England: Portcullus Press Ltd., 1979.

International Nickel, Inc. *Engineering Properties of Electroless Nickel Plating*. Saddle Brook, NJ: 1971.

Kruger, Lennard G., ed. *Metals Handbook*, 9th ed., vol. 5. Metals Park, OH: American Society for Metals, 1982.

Stapleton Company. *Electroless Nickel*. Long Beach, CA: 1983.

THERMAL SPRAYING AND HARD FACING

THERMAL SPRAYING

CHAPTER CONTENTS:

Although the concept of depositing a molten metal or ceramic coating on a substrate is not new, much of the technology is new, including the precision and range of both processes and applications. The ever-increasing emphasis on cost reduction, high productivity, and more efficient manufacturing has made thermally sprayed coatings attractive for providing superior resistance to wear, corrosion, or erosion, as well as for reclaiming worn or mismachined components.

Thermal spraying is the process of depositing molten or semimolten materials such as metals, alloys, or ceramic coatings on substrate materials so that they solidify and bond to the substrate (see Fig. 22-1). The process is also called metallizing and flame or metal spraying. The coating improves surface characteristics, but does not usually change the properties of the structural component. Thus basic structural design considerations must be maintained.

The spray materials can be in the form of wire, rod, cord, or powder. As the materials pass through the spray unit, they are heated to a molten or semimolten state and then atomized and/or projected onto the substrate. Heating can be accomplished by an electric arc, gas flame, or detonation of a combustible gas mixture. In some devices, the hot particles are conveyed from the spray equipment to the substrate by an air jet, which also accomplishes atomization and particle acceleration. As the sprayed particles impinge on the substrate, they cool and build up, particle by particle, into a cast-like structure. As they strike the surface, the particles flatten and form thin platelets that conform to the irregularities of the previously prepared surface, as well as to each other (see Fig. 22-2).

MATERIALS DEPOSITED

Most metals, oxides, cermets (ceramic plus metal), and metallic compounds, some carbides and organic plastics, and certain glasses can be deposited by one or more of the thermal spraying processes. Table 22-1 lists the different materials that can be deposited by thermal spraying. The substrates onto which the coatings can be applied include metals, oxides, ceramics, glass, most plastics, wood, and even some disposables such as paper and relatively thin films. Not all sprayable materials can be applied to all substrates, however, and some materials require special application techniques.

The deposited structure of thermally sprayed coatings differs from that of the same material in the wrought or cast form because of the incremental nature of the coating buildup as well as the reaction with the hot process gases or the surrounding atmosphere. In the case of metals, because of intermetallic and other valency compounds, the deposited coating tends to be harder, more brittle, and more porous than the original material.

The as-applied structure of all thermally sprayed coatings is generally lamellar in nature but may vary depending on the particular thermal spraying process used, parameters and techniques employed, and the material applied. Some processes do not heat the material particles sufficiently to provide a complete fusion between the metal deposit and the substrate. It is then necessary to perform a subsequent fusion process with an oxyacetylene torch to melt and diffuse the metal particles into an intricate coating. If the coating is not fused to the substrate and subsequent coating particles, the coating will spill and crack.

The density of coating varies with working temperature and particle impingement velocity. Figure 22-3 depicts the range of particle impact velocities developed by each of the commonly used spray techniques. For example, the oxy-fuel powder spray gun with an impact velocity of 80 to 400 fps (24 to 122 m/s) produces the lowest density and bond strength of the thermal spray processes.

Contributors of sections of this chapter are: ***Doug H. Harris****, President, APS-Materials, Inc.;* ***Ken Moyer****, Product Development Engineer, Hoegenaes Corp.;* ***Merle L. Thorpe****, President, TAFA, Inc.*

Reviewers of sections of this chapter are: ***Dick Baker****, Technical Services Engineer, Technology Dept., Cabot Corp.;* ***T. E. Barnard****, Vice President, Marketing, Hitemco;* ***K. Barrette****, Director of Engineering, Quantum Laser Corp.;* ***Bob Creech****, Vice President, Manufacturing Dept., Rexarc International, Inc.;* ***Dr. Paul Crook****, Wear and Welding Group Leader, Technology Dept., Cabot Corp.;* ***Dennis Daugherty****, Sales Manager, Rexarc International, Inc.;* ***Robert D. Dowell****, Vice President, Marketing, Plasma Technology, Inc.;* ***Samuel C. DuBois****, Vice President, Research and Development Lab., Wall Colmonoy Corp.;* ***Doug H. Harris****, President, APS-Materials, Inc.;* ***C. R. McKinsey****, Manager, Marketing Services, Coatings Service, Union Carbide Corp.;* ***Ken Moyer****, Product Development Engineer, Hoegenaes Corp.;* ***Sy Priceman****, President, Hitemco;* ***Gilbert A. Saltzman****, Technical Director, Metallurgical Industries, Inc.;* ***Fred Smith****, Manager, Alloy Rods, Hardfacing Dept., Rexarc International, Inc.;* ***Merle L. Thorpe****, President, TAFA, Inc.;* ***Bob Zounek****, Product Planning Manager, METCO, Inc.*

CHAPTER 22

MATERIALS DEPOSITED

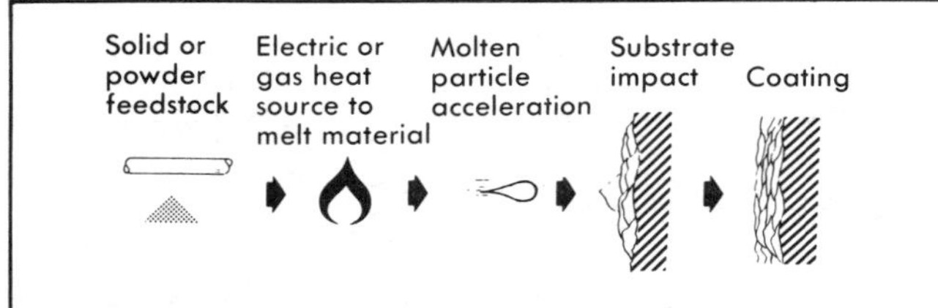

Fig. 22-1 The thermal spray process. (*TAFA, Inc.*)

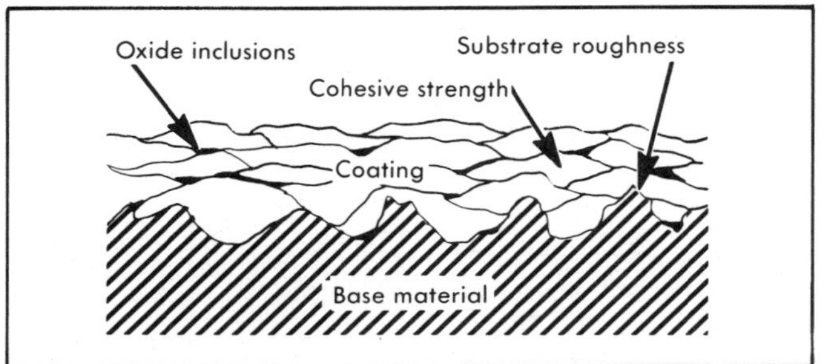

Fig. 22-2 Cross section of typical thermal spray coating. (*TAFA, Inc.*)

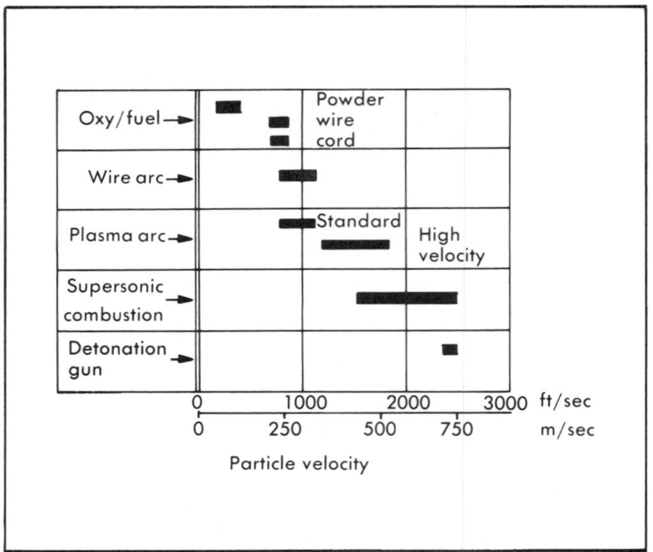

Fig. 22-3 Particle impact velocities of common thermal spraying processes. (*TAFA, Inc.*)

TABLE 22-1
Coating Materials Applied by Thermal Spraying

Elements		
Ag	Fe	Sb
Al	Ge	Si
Au	Ir	Sn
B	Mg	Ta
Be	Mn	Ti
Bi	Mo	U
Cd	Nb	V
Co	Ni	W
Cr	Pt	Zn
Cu	Re	Zr

Beryllides*
$MoBe_{12}$
$NbBe_{12}$
Nb_2Be_{17}
$TaBe_{12}$
Ta_2Be_{17}
$TiBe_{12}$
Ti_2Be_{17}
UBe_{13}
VBe_2
$ZrBe_2$
$ZrBe_{13}$

Silicides
$CrSi_2$
$MoSi_2$
$TaSi_2$
Ta_5Si_3
U_3Si_2
WSi_2
$ZrSi_2$

Carbides
B_4C
Cr_3C_2
HfC
MoC
Mo_2C
NbC
TaC
Ta_2C
TiC
UC
UC_2
VC
WC
W_2C
ZrC

Borides
CrB_2
HfB_2
Mob_2
NbB_2
TaB_2
TiB_2
VB_2
WB
ZrB_2

Sulfides
BeS
CeS
MoS_2
ThS
US
US_2

Plastics
Epoxies
Nylon
Penton
Polyethylene
Polyimide
Polyvinyl chloride
Teflon

Other Inorganics
NiAl
TiH_4
Glass

(continued)

TABLE 22-1—*Continued*

Alloys	Oxides	Blends (typical)
Ag	Al_2O_3	Al_2O_3-Cr_2O_3-Cr
Al-Si	$Al_2O_3 \cdot TiO_2$	Al_2O_3-Mo
Babbitt	$Al_2O_3 \cdot Cr_2O_3$	Al_2O_3-Ni
Bronze	$3Al_2O_3 \cdot 2SiO_2$	Al_2O_3-Nichrome
Co-Cr-W	$BaO \cdot TiO_2$	Al_2O_3-Nb
Co-Cr-W-B	$BaO \cdot ZrO_2$	Al_2O_3-Hastelloy C
Hastelloy B	BeO	Al_2O_3-stainless
Hastelloy C	$CaO \cdot ZrO_2$	Al_2O_3-TiO_2
Hastelloy D	CeO_2	Au-SeTe
Haynes	CoO	BeO-Hastelloy C
Haynes Stellite #21	Cr_2O_3	BeO-stainless
Inconel	$FeO \cdot TiO_2$	BeO-$TiBe_{12}$
Invar	HfO_2	Cr_3C_2-Co
Ir-Re	$MgO \cdot Al_2O_3$	Cr_2O_3-Al
Monel	$MgO \cdot ZrO_2$	Cr_2O_3-Cr
Nichrome	$MoO \cdot ZrO_2$	Cu-Fe
Ni-Cr-Si-B	NiO	HfO_2Cr
Ni-Si-B	$SrO \cdot TiO_2$	HfO_2-Hastelloy C
Ni-Fe	$SrO \cdot ZrO_2$	HfO_2-Ir
Rene 41	ThO_2	HfO_2-Mo
Rene 80	TiO_2	HfO_2-stainless
4340 steel	UO_2	HfO_2-ZrO_2
Ni-Cr steel	WO_3	Rare earth oxide mixtures
Low-C steel	Y_2O_3	
Medium-C steel	ZrO_2	ThO_2-Hastelloy C
High-C steel	$ZrO_2 \cdot SiO_2$	ThO_2-Hastelloy C-Nb
304 stainless		ThO_2-Ir
316 stainless		ThO_2-Mo
410 stainless		ThO_2-Nb
Ti-Al-V		ThO_2-stainless
Tungsten-rhenium		ThO_2-W
U		ThO_2-ZrO_2
W-Re		WC-Co
Zircaloy		WC-Co-Cr
Germanium-silicon		ZrO_2-Co
		ZrO_2-Cr
		ZrO_2-Hastelloy C
		ZrO_2-Mo
		ZrO_2-Nichrome
		ZrO_2-stainless
		ZrO_2-Zr

* These compounds are highly toxic and must be deposited with extreme caution. Controls have been instituted by the Occupational Safety and Health Act and the Environmental Protection Agency.

In general, the higher the particle impact velocity, the better the coating properties.

With most thermally sprayed deposits, adhesion to the substrate is mainly mechanical and surface preparation by roughening is necessary for good bonding. Important exceptions are found in some metal coatings such as molybdenum and certain composites and alloys containing aluminum and titanium, which form a metallurgical bond with certain metal and alloy substrates. Metallurgical bonds are often higher than the interparticle strength of the sprayed deposit; consequently, when coatings are tested to destruction, failure occurs within the coatings.

Bond coats are especially useful when hard metals, which cannot be roughened effectively, need to be coated. In some cases, bond coatings can be applied directly to relatively smooth, technically clean surfaces, thus eliminating the need for surface roughening, which is customarily a costly step. After initial preparation, a thin coating, normally about 0.002-0.005" (0.05-0.12 mm) thick, of a suitable bond material is thermally sprayed prior to the application of the primary deposit.

Arc-sprayed bond coatings usually have significantly higher bond strengths than flame-sprayed coatings because of the higher temperature at which the atomized particles hit the substrate; at higher temperatures, some microwelding and diffusion also occurs. For example, aluminum sprayed on mild steel with a gas flame gun yields 1400 psi (9.6 MPa) tensile bond strength, while arc spray yields 4900 psi (34 MPa).

PROCESSES

Several processes are used to thermally deposit the coating material on the substrate, and they can be divided into two basic categories: (1) combustion flame spray and (2) electric (thermal) arc spraying. Each process has specific advantages and limitations that require the process to be carefully matched to the application.

Table 22-2 lists the different thermal spray processes, along with the particle velocity and bond strength that is associated with each process. The processes listed have a generic similarity in that the deposited materials are melted and separated in a high-temperature moving gas stream and impinged on a substrate. They vary widely, however, in their ability to produce well-bonded, low-porosity deposits.

Combustion Flame Spraying

The combustion flame spraying process is the oldest of the thermal spray techniques and includes specific equipment to apply metallic and ceramic coatings from wire, powder, or rod materials. Acetylene is generally employed as the gas, but propane, Mapp, hydrogen, or natural gas have also been used. The first four techniques mentioned in Table 22-1 use oxyacetylene fuel, which generally produces a 5000° F (2760° C) combustion flame. The coating material is injected into the flame, where it is melted and plasticized by a compressed air blast. This compressed air blast accelerates the particles to the maximum velocity, at which point they are impinged on the substrate. The compressed air also serves to cool the substrate during the coating process, thus maintaining the part temperature below 400° F (200° C). On some equipment and for specific applications where oxidation may be a problem, protective gases such as nitrogen may be used in place of compressed air. In all the oxyacetylene processes, coating thicknesses greater than 1/8" (3 mm) can be obtained, although most coatings are much thinner; thicknesses under 0.003-0.005" (0.07-0.12 mm) are not recommended because their uniformity is difficult to maintain.

The tensile bond strength of coatings applied by combustion flame processes generally do not exceed 1000 psi (7 MPa) on steel substrates, unless the material is exothermic or an intermediate bond coat is employed. Most often, a nickel-aluminum exothermic material or molybdenum is used to ensure adequate bond strength.

Powder flame spray. Powder flame spraying is a type of flame spraying that involves the application of metals and other materials in the powder form. The powder materials may be held in a hopper on top of the spray gun and then gravity fed into the gun where they are picked up by the oxyacetylene (or hydrogen) gas mixtures and carried to the gun nozzle (see Fig.

TABLE 22-2
Thermal Spray Coating Process Characteristics

Process	Particle Velocity, fps (m/s)	Tensile Bond Strength, psi (MPa)	Porosity, %
Combustion Flame Spray			
Powder flame spray	80 to 120 (24 to 36)	600 to 4000* (4 to 27.6)	6 to 15
Powder flame spray/fused	80 to 120 (24 to 36)	>>20,000 (138)	1
Wire flame spray	800 (244)	2000 to 4000* (13.8 to 27.6)	6 to 15
Ceramic rod flame spray	800 (244)	2000 to 4000* (13.8 to 27.6)	6 to 15
Detonation spray	>2400 (730)	>12,000** (82.7)	1/4 to 1
Supersonic combustion spray	2500 (760)	>12,000** (82.7)	1/4 to 1
Electric-Arc Heating			
Two-wire-arc gun	800 to 1100 (244 to 335)	6000 to 10,000 (41 to 69)	2 to 8
Air plasma spray	400 to 2400 (122 to 730)	4000 to >12,000** (27.6 to 82.7)	1/4 to 5
Vacuum plasma spray	>3000 (900)	>12,000** (82.7)	1/4 to 1

(APS - Materials, Inc.)

* Values achieved with bond coat.
** Cannot be measured with the standard ASTM C633 to 69 Method.

22-4). Another gun design has the hopper located in a separate location, and an air aspiration carburetor system is used to feed the powder to the gun nozzle.

Powder spraying is widely used for the hard facing of steel and cast iron components. A number of powders have been developed for specific applications such as salvage and machine element buildup, abradable seals, and ceramic hard facings. With these materials, an operator can achieve relatively high deposition rates with acceptable bond strengths. Spray rates are usually 1-20 lb (0.5-9 kg) per hour. Major shortcomings result from the high proportion of heat that is transferred into the substrate during the spraying, which may promote oxidation or eventual stresses in the coating.

Powder flame spray/fusing. Powder flame spray/fusing operations incorporate the addition of a high-temperature fusion treatment in the overall process. Nickel or iron-based metals, containing small amounts (approximately 1.0-5.0%) of boron and silicon, are deposited on metallic substrates and simultaneously or subsequently fused. These materials develop hardnesses ranging from R_C20 to R_C68 and often contain moderate percentages of crushed tungsten carbide as an additive. The fusing operation results in a totally metallurgical-bond coating that is usually impervious to any corrosive liquid.

Wire flame spray. In wire spraying or metallizing, the flame is only used to melt the wire material. Spraying is accomplished by surrounding the flame with a coaxial stream of air in order to disintegrate the molten material and propel it onto the workpiece (see Fig. 22-5). Spray rates for materials like stainless steel are in the range of 1 to 20 lb (0.5 to 9 kg) per hour.

Wire spraying is a fast and low-cost method of applying a wide variety of metal coatings and some ceramics. (Ceramic coatings are formed into wire using plastic binders.) Carbon steels, stainless steels, brass, bronze, zinc, aluminum, and other metals can be applied over a grit-blasted, nickel-aluminum bond-coated substrate. Such coatings are often used for shaft buildup and general restoration of worn or mismatched parts. The macrohardness of such deposits may range from the basic material value to a few points higher; while the microhardness and roughness of the deposit may be substantially improved by the dispersion of oxides that are formed during the spray deposition process. Very often, machined parts that are restored by wire flame spraying will outlast the original parts by several times, and thus some original equipment manufacturers specify finished surfaces employing this process.

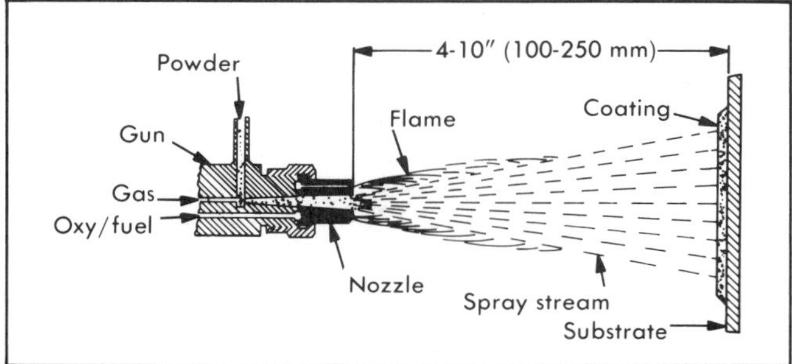

Fig. 22-4 Cross section of powder-fed combustion flame spraying torch. (*Metco, Inc.*)

Combustion rod spraying. Combustion rod spraying generally uses a 1/4" (6 mm) diam rod and will cover approximately 60-65 in.^{2} (385-420 cm^2) of substrate in about four minutes. Larger diameter rods, 3/8" (9 mm), are available for higher deposition rates. This process is often used to apply ceramic coatings such as alumina, alumina-titania, and chromium oxide. Since these coating materials are porous, the microstructure can be filled with tetrafluoroethylene (PTFE) and other organic lubricants.

Detonation gun spraying. Detonation gun spraying employs the controlled detonation of an oxyacetylene gas mixture to produce high temperatures and extremely high particle velocities. In this process, a mixture of oxygen and acetylene is fed into a combustion chamber in the rear of a long-barreled gun (see Fig. 22-6). The coating material is added to the gases, and the gaseous powder mixture is then ignited by a spark plug. The resulting detonation produces a high-velocity shock front that travels down the length of the barrel at 10 times the speed of sound. The powder particles are heated to a plastic state by the detonation and are accelerated to a velocity of approximately 2600 fps (790 m/s). The high kinetic energy of each powder particle is converted to additional heat upon impact with the work surface, thereby producing a very strong bond.

Unlike other combustion and arc spray torches that operate continuously, the detonation gun fires four to eight times a second, thereby forming a lamellar coating on the workpiece. The successive detonations can build up coatings to maximum thicknesses approaching 0.020-0.030" (0.5-0.8 mm), but 0.005-0.010" (0.12-0.25 mm) is more common. As with other thermal spraying techniques, auxiliary cooling is employed to maintain workpiece temperatures between 150 and 300° F (65 and 150° C).

A variety of coatings may be applied with the detonation gun process, including tungsten carbide, aluminum oxide, and chromium carbide. However, because of the high velocity of this coating technique, it is limited to the plating of metallic substrates. Nonmetallic surfaces such as plastics and graphite or ceramics may be eroded by the high velocity of the particles.

The bond strength of coatings produced with the detonation gun process is comparatively high and cannot be measured by the standard tensile bond strength method (ASTM C 633)

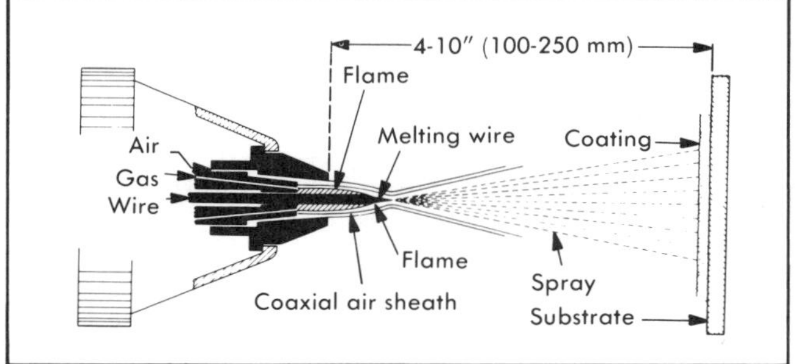

Fig. 22-5 Cross section of combustion rod spray torch. (*TAFA, Inc.*)

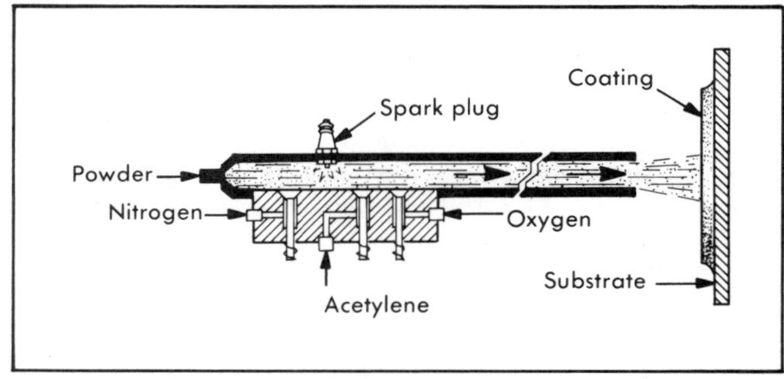

Fig. 22-6 Schematic of detonation gun thermal spraying.[1]

because epoxy adhesives fail at about 12,000 psi (83 MPa). The hardness values for these coatings can range up to 1200 Vickers with a 300 g load (approximately R_C 72). The surface porosity of some detonation coatings is as low as 0.5%, and they can be finish lapped to as low as 1 μin. (25 μm). Detonation gun coatings have gained widespread acceptance in precision machine part manufacture. However, they can only be applied at specialized third-party contract facilities.

Supersonic combustion. The supersonic combustion system is the most recent combustion flame spraying process. It reportedly yields hypersonic flame velocities and particle velocities estimated at about 2500 fps (760 m/s). The concept is essentially that of a rocket engine in which powders are injected into the burning gases and accelerated in a manner similar to the principles stated for the detonation gun. Unlike the detonation gun, however, the supersonic system is a continuous process that is easily used manually or in automatic fixtures. A high noise level, approximately 150 dB, is produced during operation, and large quantities of gas are used.

The supersonic combustion system has been used for depositing carbide coatings where microhardness in excess of 1200 Vickers with a 300 g load have been reported. Coating thicknesses are usually up to 0.015" (380 μm); thicker coatings tend to peel away from the base substrate due to extensive cold working. At this time, supersonic combustion does not appear to be a practical process for depositing ceramic materials since the combustion temperature is not hot enough to melt the ceramic material. It does appear to lend itself to other metallic coatings; and because of its compact size, may be a versatile system for on-site applications of various hard facings and corrosion-control coatings.

Electric-Arc Spraying

Electric-arc spraying processes include a wire-arc and a nontransferred plasma arc process. In the nontransferred plasma arc process, the inert gas is heated to the plasma state by an arc within the plasma arc gun. (In the transferred plasma arc process, the arc occurs between a nonconsumable cathode and the workpiece.) The nontransferred arc process can be performed in the natural environment or in a controlled environment depending on the application.

Wire-arc gun. The two-wire-arc gun is used for electric-arc spraying of metals or other conductive materials and achieves high deposition rates. Two wires or rods of the material to be deposited are fed through electrical contacts to form positive and negative electrodes (see Fig. 22-7). These wires are driven through the unit using compressed-air turbines or electric motors. When the wires form a narrow gap, an arc is created, resulting in temperatures of 4000 to 10,000° F (2200 to 5530° C), depending on the material sprayed. The wires are melted, and then the liquid drops are atomized and carried to the substrate by a blast of compressed air or inert gas. Spray-stream geometry and spray particle size can be changed with different atomizing heads and wire intersection angles.

By using two wires of different compositions, alloy-type coatings can be deposited. The high temperatures in the arc zone and resulting high droplet temperatures produce coatings that have excellent adhesive strength and high cohesive strength. High melting efficiency and air-stream cooling limit the substrate temperature to 125 to 150° F (50 to 65° C) and permit arc spraying even on heat-sensitive substrates such as plastics. Deposition rates up to 60 lb (27 kg) per hour for stainless steel are reported, and even higher values are attainable for the deposition of aluminum and zinc coatings as used in atmospheric and marine environment corrosion protection. Because of the high deposition rates and the improved bonding, the two-wire arc system can be used to produce very thick overlays, up to 1/2" (12 mm), and to cover very large surface areas, such as bridges and ship decks. The wire-arc gun can also be used to apply precision, high-density coatings on small electrical components.

Nontransferred plasma arc. Nontransferred plasma arc spraying is an extremely versatile thermal spraying process. Figure 22-8 shows the cross section of a plasma arc gun. In operation, a nontransferred, high-intensity arc is struck between the gun body (anode) and a tungsten cathode. An inert gas, such as argon, is passed through this arc, where it is heated to the plasma state [up to 30,000° F (16 650° C)] and sometimes accelerated to supersonic speed. Other gases such as helium, nitrogen, and various additions to argon are used in the processing of materials. The coating material in powder form is injected into the plasma stream, where it is heated to a molten state, accelerated, and carried to the workpiece. Auxiliary cooling can maintain the workpiece temperature at less than 150° F (65° C), reducing the chance for distortion or change in the metallurgical properties of the substrate material. Air or liquid carbon dioxide is most often used for cooling, although water-cooled substrate mounts are sometimes beneficial. A wide range of substrates can be coated using the nontransferred plasma arc, including graphite, carbon, glass, ceramic, some plastics, cement, wood, and even certain papers.

Rather stringent process control is required for the nontransferred arc process as compared to the oxyacetylene and wire-arc processes. The plasma torch is much more sensitive to powder flow, powder particle size, and process variables such as standoff distance, feed rate, arc-gas flow, and arc currents. The traverse velocities between the workpiece and the gun must be closely controlled to prevent variation from pass to pass, which would significantly affect coating quality. Electrodes are apt to be short-lived, and even minor wear after a few hours of spraying may require adjustments in the spraying parameters to maintain coating quality. Spray rates range from 1 to 50 lb (0.5 to 23 kg) per hour, depending on power setting and melting difficulty of the powder.

Subsonic plasma. Subsonic plasma provides velocities ranging from 400 to 1100 fps (120 to 335 m/s), is the most commonly used plasma process, and has the lowest operating costs. Various material suppliers offer one-coat material systems that incorporate metallurgical properties to enhance the bonding of the coating. The area to be coated must be grit blasted to ensure high bond strength. Many coating materials do not require an interface material to perform well; however, it is often beneficial

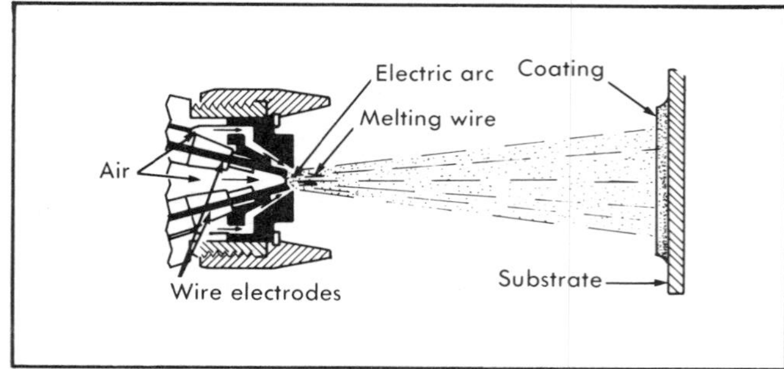

Fig. 22-7 Cross section of two-wire-arc gun.[2]

and preferable to use a bond coat to ensure high integrity. Material hardnesses up to 1300 Vickers with a 300 g load (approximately R_C 75) can be achieved with this process.

High-energy/high-velocity equipment. High-energy/ high-velocity equipment employs higher kinetic energies through the use of appropriately shaped arc chambers and higher gas flow rates and velocities. High-energy levels are used up to 120 kW. Such equipment can generate gas velocities of approximately 2200 fps (670 m/s) when operating under atmospheric conditions, and molten particles are accelerated to velocities greater than 1400 fps (425 m/s). Coatings are thereby produced that have higher integrity through greater bond strength, density, and uniformity. In general, no bond coat is required for this process. Because of the higher kinetic energy, the powder parameters must be more stringently controlled. At these higher velocities, the deposition rates are usually lower, and the collection efficiency may be reduced. Coatings are generally thin to minimize residual stresses. For these reasons, high-energy/high-velocity processes are generally not used for machine element salvage or other applications that require thicker overlays.

Inert chamber plasma or wire arc. Inert chamber spraying places the plasma or wire-arc gun and workpiece in a chamber containing an inert atmosphere. The gun can be manipulated either by hand or mechanically. This technique is generally used for refractory metals and other material systems where oxygen atmospheres are detrimental. The process characteristics remain the same as previously described for nontransferred plasma arc spraying.

Vacuum plasma arc. Vacuum plasma spraying further enhances the characteristics of the deposit. Both the workpiece and gun are mounted within the confines of an evacuated chamber. Low chamber pressures, in the range of 15 to 60 torr (2 to 8 kPa), enhance the velocity of the molten particles and remove much of the heat loss associated with operation of the plasma stream in an air environment. The workpiece-to-gun distance is much longer because of the nature of the plasma, and the overall effective diameter of deposition expands to several inches. In this environment, the motion between the gun and the workpiece must be completely mechanically controlled. Molten particle velocities in excess of 3000 fps (900 m/s) are reported, and metallurgical bonding is often noted.

The vacuum plasma arc process is generally used for coatings of a critical nature, when maximum density and bond strengths are essential. Since the process is confined in an evacuated chamber, transferred arcs between the anode and the workpiece are sometimes employed to bring about additional heating of the substrate, which enhances the bond and can improve deposit metallurgy. Some reports, however, indicate potential damage to substrate properties through the use of transferred arcs. Low-pressure plasma generally puts more heat into the substrate and is more difficult to control if lower

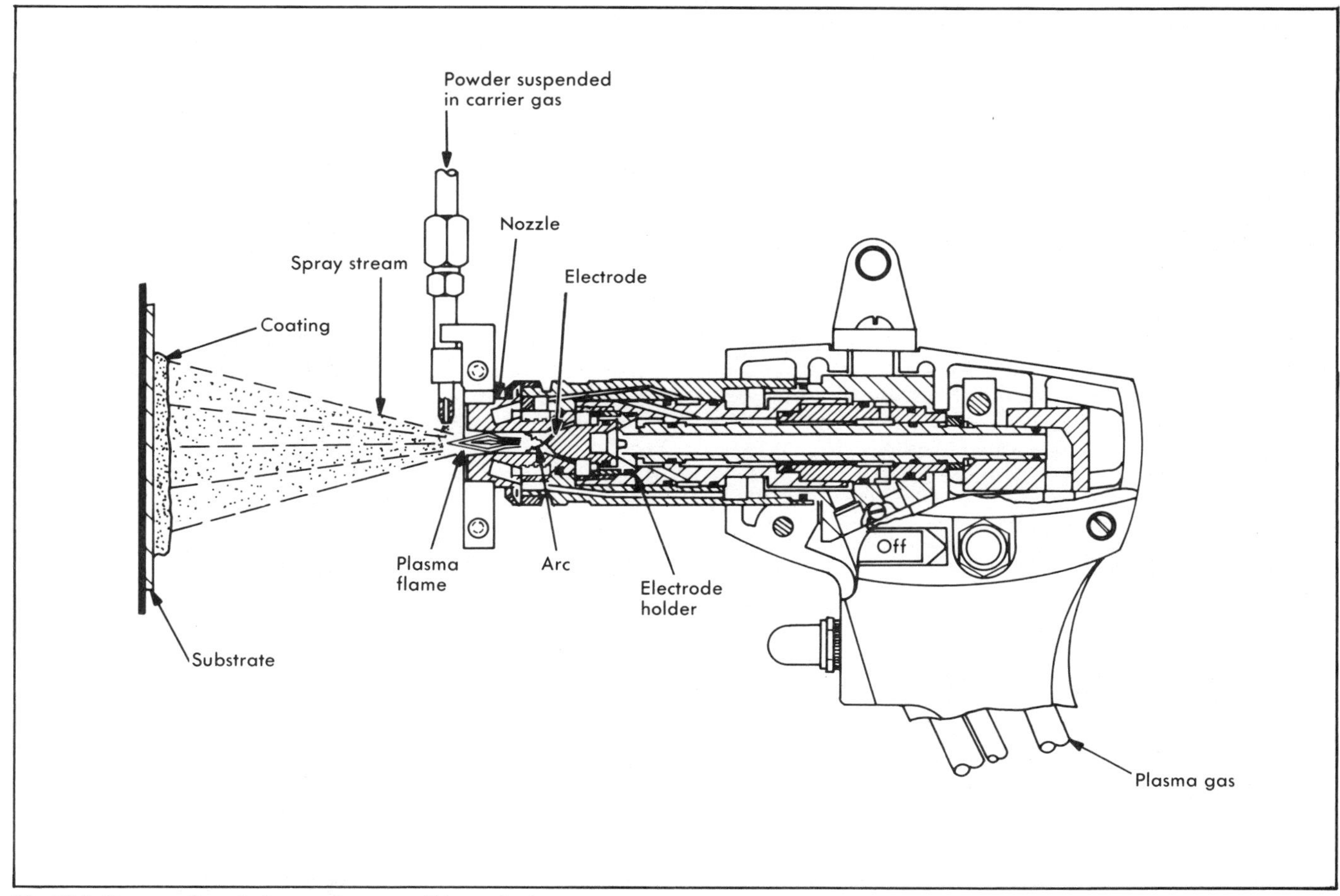

Fig. 22-8 Cross section of plasma spray gun. (*Metco, Inc.*)

substrate temperatures are required. The process lends itself to batch methods, rather than continuous production, and is therefore more expensive to use.

Process Selection and Economics

Process selection is usually made on the basis of technical merit, cost, and the specific advantage of a process. Some of the advantages include strength and adhesion of the coatings, coating thicknesses, and the materials that can be sprayed.

The powder flame spraying process is characterized by low equipment costs, low operating noise, and low fume generation. The process is widely used despite reduced coating properties and moderate operating costs. The wire flame spraying process is the oldest of the thermal spraying processes, with medium equipment costs; the main limitations are reduced coating properties and the limited materials available in wire or rod form.

Among the electric-arc spraying processes, wire-arc spraying offers medium equipment costs, low operating costs, and high bond and coating strengths. The direct conversion of electrical energy into molten material is considerably more efficient than the other methods, approximately 1/9th the energy consumption of flame spraying and 1/15th that of plasma arc spraying. Wire-arc spraying also offers high deposition rates. However, it is uncommon for a specific process to be selected on the basis of energy costs or production rates alone.

Plasma arc spraying is the most versatile of the thermal spraying processes, permitting the application of almost all materials with good coating properties. Currently, high equipment costs, complicated support requirements, and high production costs are the main limitations of this process.

The detonation gun process produces good coating properties. However, it is available only as a vendor service, is relatively expensive, and is limited to a few materials such as some oxides and carbides.

For original equipment manufacturer applications, the major economic and/or technical merit of thermal spraying is its ability to produce composite structures with longer life or improved physical properties. For example, in corrosion protection, the initial cost of spraying might be higher than painting, but the greatly increased life expectancy yields an overall savings. In electromagnetic shielding, superior signal attenuation is achieved with spraying at a cost comparable with competing processes. In rebuilding, the cost of resurfacing is usually less than half the cost of a new component, with a significant reduction in machine downtime.

The suitability of the thermal spraying process for any particular application may be determined according to the following fundamental principles:

1. No strength is imparted to the base material by the sprayed deposit. It is essential that the component to be sprayed should be able to withstand, in its prepared form, any mechanical loading to which it will be subject in service. In a few applications, some strength can be added by thermal spraying; however, such applications are unusual, and they should be carefully tested.
2. Parts or any section of the total area subject to shear loading in service should not be thermally sprayed. Examples of these parts include gear teeth, splines, and threads.
3. Point loading with line contact on a metal sprayed deposit will eventually spread the deposit, causing detachment. If the deposit is on a moving component with point loading, the deposit failure will occur rapidly. Examples are needle and roller bearing seatings where the bearing elements are in direct contact with the sprayed deposit.
4. If the base metal of the component to be treated has been nitrided, thermal spraying cannot be performed. Other forms of hardening also require special treatment.

APPLICATIONS

The original use of thermal spraying was for rebuilding worn parts and for mismachined components. In recent years, the refinement of techniques and modern equipment have permitted the manufacture and rebuilding of highly stressed components such as gas turbine aircraft engines. Now these refined techniques and equipment have been used to extend the process into production engineering to impart desirable surface characteristics, such as wear and corrosion resistance, and electrical properties to new components. Table 22-3 lists some of the typical applications for the thermal spraying processes, as well as some of their coating characteristics.

Mechanical Properties

The success of thermal-sprayed coatings for component reclamation has led to the extensive use of thermal spraying in original equipment manufacture. Many paper mill rolls, steel mill roll journals, wire-drawing cones, and marine propeller and winch shafts are thermally sprayed to provide improved performance. In the oil industry, oil rig drilling mandrels are coated with a 13% chromium steel deposit as original equipment. Various stainless steels and oxides are sprayed as reinforcing agents and wear points for nonstick Teflon-coated surfaces.

For the transportation industry, piston rings have been sprayed with a variety of materials. A unique capability of the wire-arc process is to apply composite coatings of a strong, wear-resistant steel and weaker, high-conductivity copper to provide lightweight disc brakes for racing motorcycles. These coatings are produced by using one steel wire and one copper wire as the feedstock to the wire-arc system or by using mixed powders in plasma units.

In the aircraft engine industry, abradable coatings are used to produce the rotor-to-shroud clearance by spraying a sacrificial material on the engine shroud. During operation, the upper blades machine away the coating to produce a minimum clearance, thus reducing bypass gas flows and increasing engine performance.

Thermal barrier coatings are also used on gas turbine and diesel engine combustion chambers to increase the temperature within the chamber, which, in turn, increases its operating efficiency. A typical coating is the duplex type, consisting of a metallic bond coat overlaid with a ceramic.

Wear can be reduced significantly in fretting, abrasion, and erosion applications. In some cases, self-fluxing alloys of iron with nickel and cobalt are used on engine valve seats to achieve resistance to chemical attack as well as erosion. In cases where materials cannot be fused, hard metals, carbides, or ceramics are used. In other cases, sprayed metal coatings are sealed to provide good chemical resistance.

For the agricultural industry, grain-crushing rolls and sugar cane crushers have been sprayed with 13% chromium steel when new and may be reclaimed many times. After severe damage, sugar crushers have been sprayed with up to 3" (75 mm) of steel.

Most coatings are ground before use; but, in some cases, the rough as-sprayed surface texture is required. Typical examples

TABLE 22-3
Typical Thermal Spray Process Applications

Part Name	Process	Coating Type	Coating Thickness, in. (μm)	Surface Finish (rms), μin. (μm)	Operating Conditions
Feed rolls	Plasma	Tungsten carbide	0.002-0.004 (0.05-0.10)	75 (2)	Low load, dry rubbing
Compressor rods	Detonation gun and plasma	Tungsten carbide	0.005-0.008 (0.12-0.20)	3-8 (0.08-0.2)	Sliding wear
Turbine and compressor vanes	Plasma	Chromium carbide	0.003-0.005 (0.08-0.12)	150 (4)	High temperature and fretting wear
Seals	Detonation gun	Chromium carbide	0.008-0.010 (0.20-0.25)	1-3 (0.02-0.08)	High temperature and sliding wear
Pump sleeves	Powder flame/ fused	Nickel chromium boron	0.020 (0.50)	10 (0.25)	High abrasion
	Plasma	Chromium oxide	0.015-0.018 (0.38-0.46)	16 (0.4)	Corrosion/abrasion
Carbon-graphite seals	Plasma	Aluminum oxide	0.008-0.010 (0.20-0.25)	10 (0.25)	High load, fretting wear
Shaft sleeves	Plasma	Chromium oxide	0.015-0.018 (0.38-0.46)	16 (0.4)	Ambient temperature, various chemical solutions
Process vessel	Plasma	Hastelloy "C"	0.035 (0.90)	---	HCL environment
Gate valves	Detonation gun	Tungsten carbide	0.008 (0.20)	1-3 (0.08-0.2)	Sliding wear
	Supersonic combustion	Tungsten carbide	0.008 (0.20)	1-3 (0.08-0.2)	Sliding wear
Thermal barrier coatings	Plasma	Zirconia	0.008-0.013 (0.20-0.33)	---	High temperature, hot gases
Abradable seals	Powder flame	Nickel graphite	0.100 (2.5)	---	Temperatures up to 800° F (425° C)

(APS-Materials, Inc.)

of the latter include the surfaces of dynamometer barrels, the surfaces of rolls used in backing carpets with latex, paper feed grippers, and antiskid surfaces on shipdecks, metal walkways, and treads.

Electrical/Electronic

The electrical engineering and electronics industries are also large users of thermal spraying. Conductors and resistors are sprayed to provide electrical circuits for surface heating elements on aircraft control surfaces and industrial and domestic hot-plates. The ability of sprayed coatings to adhere to nonmetallic substrates at low temperatures has led to the application of conductive terminals on carbon and ceramic electrical resistors, silicon carbide resistors, varistors, newer electronic oxides, and foil capacitors.

One rapidly expanding application of thermal spraying is that of providing electrically conductive surfaces on plastic housings, plugs, and other large and small enclosures for such items as computers, electronic and medical equipment, and shielded rooms. The coatings shield the devices from or prevent emission of electromagnetic or radio frequency interference. Zinc is typically used in these applications, but copper and aluminum have also been used. Usually a coating of zinc, 0.001-0.003" (25-75 μm) thick, is applied to the inside of a plastic housing, such as for a word processor. The resultant continuous metal coating produces signal attenuations in the range of 60 to 90 dB to meet design and recent federal specifications.

Thermal spraying is also used to produce thick-film electrical circuits. These types of circuits can carry higher currents than printed circuits, yet they are more flexible than stamped circuits. Conductor rolls for tinning sheet steel are sprayed with thick copper coatings to provide adequate conductivity at a reduced cost. Copper conductors have also been sprayed on the axles of electric railway locomotives to reduce contact resistance.

Corrosion Protection

Corrosion is one of the most expensive natural phenomena in that it reduces the effectiveness of engineering components and structures. Sprayed coatings up to 0.020″ (0.5 mm) thick of zinc or aluminum will protect structural steels for over 20 years in most natural environments.

Sprayed metals are more effective than painted metals because they are galvanic; and, unlike galvanizing, sprayed deposits are not restricted to coating bath size and can be applied in-plant during manufacture without the risk of material distortion or embrittlement. When galvanized components are used in fabrication, the sheared edges and welded areas can be recoated by thermal spraying. When the coatings are expected to bear loads in marine environments, nickel-chromium or copper-nickel alloys can be used along with a suitable deep-penetrating sealant.

Thermally sprayed deposits may also be used in certain circumstances to resist chemical corrosion; however, material and sealant selection is critical. Aluminum coatings provide steels with excellent protection from high-temperature oxidation and corrosion, and have been extensively applied to exhaust ducts, chimneys, furnace parts, parts prior to rolling or forging, and automotive valve seats.

PROCESS VARIABLES

Consistent success in the use of thermally sprayed coatings depends on the surface preparation, substrate temperature, and application techniques. The substrate temperature is determined by the type of coating applied, the substrate material, and the process used. For specific application techniques, follow the recommendations of the equipment manufacturer.

Surface Preparation

Thermal spray processes require extensive surface preparation for maximum adherence of the coating. Preparation of the substrate prior to spraying is virtually the same for all processes. Grit blasting, using aluminum oxide or chilled iron shot with mesh sizes ranging from 20 to 80, is most commonly employed. For proper roughness and maximum efficiency, blasting pressures in excess of 80 psi (550 kPa) are required. Prior to grit blasting, surfaces are degreased using chlorinated solvents or alcohol. Chemical cleaning is used on parts contaminated or impregnated with material that otherwise cannot be removed. It is advisable to deposit the coatings as soon as possible after the substrate is prepared to prevent surface oxidation, which impairs bonding.

Wire flame spraying often employs the use of threading on shafts in place of, or in addition to, grit blasting. The correct procedure is to mash the top of the threads so as to effect a keying action. As was previously discussed, some newly developed thermally sprayed bond coatings can be deposited on smooth, clean surfaces without any surface preparation.

Masking

The surfaces to be coated can be controlled by the use of various masking techniques. High-temperature tapes, special masking paints, rubber, plastics, or machined-metal can be used for coatings applied by flame spraying and electric-arc processes. Metal masking is generally the only acceptable technique for masking coatings applied with the detonation gun because of the erosive effect of the coating particles.

Surface Finish

The coated surface finish of materials applied by flame spray techniques varies from 100 to 400 μin. (2.5 to 10 μm) rms. The pure metal coatings may be machined to much lower surface finishes, but the harder carbide and oxide materials must be diamond ground. For example, the use of a 220 grit diamond wheel will produce a surface finish in the range of 10 to 20 μin. (0.25 to 0.50 μm) rms on wire arc, plasma arc, and detonation gun coatings, and 20 to 40 μin. (0.50 to 1.0 μm) rms on flame-sprayed coatings. Diamond lapping can reduce the finish to less than 1 μin. (0.025 μm) rms on the denser detonation gun coatings. Nondimensional finishing with silicon carbide paper has also been successful in obtaining surfaces in the range of 30 to 40 μin. (0.75 to 1.0 μm) rms.

Sealing

The porosity present in thermally sprayed coatings renders all coatings but zinc and aluminum, which are galvanic, relatively ineffective as corrosion barriers. Organic sealing materials have been used, including fluorocarbons and epoxy, to seal the pores in thermally sprayed coatings. Regardless of the sealant used, it is important that it penetrate deeply. Laboratory tests have shown some sealants to increase the substrate corrosion resistance 10 times; however, these organic sealants are limited to maximum temperatures of 400 to 500° F (200 to 260° C). Generally, sealants do not significantly affect the wear life of the coating.

TOOLING AND FIXTURING

Routine machine element buildup in salvage is generally accomplished by manual operation of the spray torch to provide transverse motion. Shafts and other cylindrical shapes can be rotated using lathes and rotary tables. For production, however, complete automation is recommended to maintain proper standoff, torch speed, and substrate temperatures. Equipment is currently on the market to perform all manner of X, Y, and Z-axis motions of the thermal spray stream relative to the substrate. Standard welding-type workholders are used to hold parts; however, the rotational and travel speeds required are usually higher for thermal spraying than for welding.

Robots are available that can perform complex manipulations and thereby improve the quality of production coatings. Users of such equipment must take precautions to shield expensive robots from the surrounding dust and hot gas environments. The type, size, and degree of automation of any system selected must be tailored to the requirements of the parts to be coated, environmental considerations, efficiency, and production rate.

SAFETY

Because most of the thermal spray processes generate intense light, eyes must be protected by using the appropriate welding shades. For plasma spray processes particularly, the intense ultraviolet rays dictate shades that are at least No. 9 or higher.

Many of the thermal spray processes generate high noise levels; the most significant offenders are the detonation gun, supersonic combustion, and plasma processes. Current safety requirements dictate that hearing protection be provided for all operators and other personnel in the spray area. Users are cautioned to check with current OSHA regulations regarding the length of time an operator can be exposed to the intensity generated by a specific process.

Combustion flame gases do not generally present a serious exhaust problem, but ozone in significant amounts can be created by plasma arc processes. In addition, many of the material systems that are thermally sprayed generate toxic fumes. For environmental control, spraying operations are usually isolated and often performed in enclosed areas.

Dust, fumes, and overspray must be removed from the work area as rapidly as possible. For specific safety recommendations, refer to American Welding Society Publication C2-73. Waste disposal of most overspray is quite conventional being treated like metallic dust. However, the overspray from coatings containing expensive alloying elements, such as nickel or cobalt hard-facing alloys, should not be discarded like inexpensive iron dust, but should be reprocessed. When spraying special materials, it is necessary to follow the disposal recommendations for those materials and be aware that fine powders can be explosive. Dry collection should only be used on the basis of experienced field recommendations. Proper ventilation is, of course, essential.

HARD FACING

Surfacing is defined as the deposition of filler material on a metal surface to obtain desired properties or dimensions. It is usually employed to extend the life of a part that may not otherwise have all the properties necessary for an engineering application, or to replace metal that has worn or corroded away. The overlay may contribute corrosion resistance, wear resistance, toughness, or antifriction properties exactly where they are needed most.[3] When the filler material is primarily used to improve wear resistance, the process is generally referred to as hard facing. The filler material adheres to the base metal by fusion or metallurgical bonding. Some of the advantages of hard facing include increased productivity through less downtime for repair and replacement, increased efficiency by permitting use of higher applied loads, reduced maintenance costs through reclamation of worn parts, and optimum compromise between wear and toughness through the use of less expensive, tougher base metals.[4]

Wear-resistant coatings are also applied by thermal spraying; however, the bond between the substrate and filler is sometimes mechanical rather than metallurgical. The flexible overlay, discussed in Chapter 29, is another recently developed process for applying wear-resistant surfaces to the workpiece.

APPLICATIONS

Hard-facing overlays are used in a wide variety of applications requiring wear resistance on particular surfaces. The wear may be due to hard particles or projections forced against and moved relative to a surface (abrasive wear), such as occurs in rock crushers and pulverizers, or may be metal-to-metal wear as takes place in control valves. Hard facing is also used to control combinations of wear and corrosion as encountered by mud seals, plows, knives in the food processing industry, valves, and pumps handling corrosive liquids or slurries. In most instances, parts are made of either plain carbon steel or stainless steel, materials that provide little wear resistance on their own.[5]

HARD-FACING MATERIALS

Many alloys, ceramics, and combinations of these are available for hard facing.[6] Conventional hard-facing materials are normally classified as steels or low-alloy ferrous materials, chromium white irons or high-alloy ferrous materials, carbides, nickel-based alloys, or cobalt-based alloys. A few copper-based alloys are sometimes used for hard-facing applications. Another group of alloys that is becoming increasingly popular are the self-fluxing alloys. Table 22-4 classifies the various alloys used in hard facing, along with their characteristics and typical applications.[7]

Selection of these alloys is based primarily on wear and cost considerations. In addition, it is necessary to take into consideration manufacturing and environmental considerations such as base metal composition, application method, and impact, corrosion, oxidation, and thermal requirements.

Hard-facing alloys are available in the form of rods (bare and coated), cored wire, strip, or powders. Fluxes may be in the

TABLE 22-4
Classification of Hard-Facing Alloys

Classification by Basic Types	Important Features	Successful Applications
Tungsten carbide deposits	Maximum abrasion resistance.	
Granules or inserts	Best design approach.	Oil well rock drill bits and tool joints
Coarse-granule-tube rods	Worn surfaces become rough.	A wide range of severely abrasive conditions
Fine-granule-tube rods	Best performance when gas welded.	Nonskid horseshoes
High-chromium irons	Excellent erosion resistance.	
Multiple-alloy type	Hot hardness from 800-1200° F (425-650° C) with tungsten and molybdenum.	Abrasion by hot coke
Martensitic type	Can be annealed and rehardened.	Erosion by catalysts in refineries
Austenitic type	Oxidation resistant.	Agricultural equipment in sandy soil
Martensitic alloy irons	Excellent abrasion resistance.	General abrasive conditions with light impact
Chromium-tungsten type	High compressive strength.	Machine parts subject to repetitive metal-to-metal wear and impact

(continued)

TABLE 22-4—*Continued*

Classification by Basic Types	Important Features	Successful Applications
Chromium-molybdenum type	Good for light impact.	A wide variety of abrasive conditions
Nickel-chromium type, austenitic alloy irons	More crack-resistant than martensitic irons.	General erosion conditions with light impact
Chromium-cobalt-tungsten alloys	Hot strength and creep resistance.	
High-carbon type, 2.5%	Brittle and abrasion-resistant.	Hot wear and abrasion above 1200°F (650°C)
Medium-carbon type, 1.4%	Tough and oxidation-resistant.	Exhaust valves of gasoline engines, valve trim of steam turbines, chainsaw guides
Nickel-based alloys		
Nickel-chromium-boron type	Good hot hardness and erosion resistance.	Oil well slush pumps
Nickel-chromium-molybdenum-tungsten type	Corrosion resistance but no abrasion resistance.	Many corrosive environments
Nickel-chromium-molybdenum type	Resistant to exhaust-gas erosion.	Exhaust valves of trucks, buses, and aircraft
Nickel-chromium type	Oxidation-resistant.	---
Copper-based alloys	Antiseizing, resistant to frictional wear.	Bearing surfaces
Martensitic steels	---	General abrasive conditions with medium impact
High-carbon type, 0.65-1.7%	Fair abrasion resistance.	---
Medium-carbon type, 0.30-0.65%	Good resistance to medium impact.	Hot working dies
Low-carbon type, below 0.30%	Tough, economical.	Tractor rollers
Semiaustenitic steels	Tough, crack-resistant.	General low-cost hard facing
Pearlitic steels	Crack-resistant and low in cost.	Base for surfacing or a buildup to restore dimensions
Low-alloy steel	Suitable for buildup of worn areas.	---
Simple carbon steel	A good base for hard facing.	---
Austenitic steels	Tough, excellent for heavy impact.	General metal-to-metal under heavy impact
13% manganese—1% molybdenum	Fair abrasion and erosion resistance.	---
13% manganese—3% nickel type	Lower yield strength.	Railway trackwork
13% manganese-chromium type	High yield strength for austenitic types.	Many conditions involving heavy impact with or without abrasion; also used for joining austenitic manganese steel parts.
High-carbon, nickel-chromium stainless type	Oxidation and hot-wear resistance.	Frictional wear at red heat, furnace parts
Low-carbon, nickel-chromium stainless type	Oxidation and corrosion resistance.	Corrosion-resistant surfacing of large tanks

central core portion of wire, on the surface of covered electrodes, introduced as a granular blanket, or mixed with powdered filler metal.[8]

Cobalt-Based Alloys

There are two types of cobalt-based alloys: those that contain carbides and those that contain the hard, intermetallic Laves phases. Resistance to abrasive wear is imparted primarily by the carbides or Laves phases, and resistance to corrosion and/or elevated temperature hardness is imparted by the matrix. The matrix also controls metal-to-metal wear and cavitation erosion properties. Alloys containing the Laves phases are less abrasive to mating materials than carbide-containing alloys in metal-to-metal wear applications.

Nickel-Based Alloys

Most nickel-based hard-facing alloys are divided into boride-containing or carbide-containing alloys, and alloys containing the Laves phases. The boride-containing alloys were first commercially produced as spray-and-fuse powders. These alloys are currently available from most manufacturers of hard-facing products under various trade names and in a variety of forms such as bare cast rod, cored wires, and powders, to suit a variety of welding processes.

The use of carbide-containing alloys has been extremely limited. However, these alloys are gaining popularity as low-cost alternatives to cobalt-based alloys. Only one alloy containing the Laves phases is available; it possesses good metal-to-metal wear properties and moderate abrasive wear resistance, but poor impact strength.

Iron-Based Alloys

Iron-based hard-facing alloys are more widely used than cobalt and/or nickel-based hard-facing alloys and constitute the largest volume of hard-facing alloys. Iron-based alloys offer low cost and a broad range of desirable properties. They do not wear as well as nickel-based alloys of equal hardness, but thicker deposits can be applied to compensate for the reduction in wear resistance. Iron-based alloys can be divided into pearlitic steels, austenitic steels, martensitic steels, and high-alloy irons.

Pearlitic steels are useful as buildup overlays, primarily to rebuild machinery parts to their original size. Austenitic steels are further divided into low-chromium and high-chromium alloys; low-chromium alloys contain up to 4% chromium, 12-15% manganese, and some nickel or molybdenum, while high-chromium alloys contain 12-15% chromium and approximately 15% manganese. Low-chromium austenitic alloys are generally used to rebuild machinery parts subject to high impact. High-chromium austenitic alloys are typically used to rebuild manganese steel and carbon steel parts subjected to high metal-to-metal pounding. Martensite steels are formed upon air cooling and resemble tool steels with hardnesses in the range of R_C40 to 45. They are primarily used in applications that have unlubricated metal-to-metal rolling or sliding parts. High-alloy irons are similar to cast irons and contain large amounts of carbides. Compared with the cobalt and nickel-based alloys, iron-based alloys lack resistance to oxidation and aqueous corrosion. However, stainless-steel-based alloys resemble the cobalt alloys in that they resist all forms of wear and are moderately resistant to corrosion. The stainless steel alloys are suited to a variety of welding and spraying processes.

Carbides

The amount of carbides used for hard-facing applications is small compared to iron-based alloys, but carbides are important for severe abrasion and cutting applications. Historically, tungsten-based carbides were used exclusively; but, recently, carbides of other elements such as titanium, molybdenum, tantalum, vanadium, and chromium have also proved useful in hard-facing applications. The carbide hard-facing alloys are often inserted in a steel or alloy tube, which is used as the weld-consumable material, and are also available in composite powders for plasma transferred arc, spray-and-fuse, and puddle-torch applications.

Copper-Based Alloys

The copper-based hard-facing alloys, which are similar to bronzes, are used as overlays of low-carbon steel instead of having the part made entirely of the copper-based alloy. They are used for applications requiring resistance to corrosion, cavitation erosion, and metal-to-metal wear, as in bearing materials. These hard-facing alloys have poor resistance to corrosion by sulfur compounds, poor abrasive wear, and poor elevated temperature creep; they are also more difficult to weld.

Self-Fluxing Alloys

Self-fluxing alloys are generally nickel-based powdered alloys, although carbide-based alloys are sometimes used, containing chromium, carbon, and boron to provide hard-phase carbides, nitrides, and corrosion resistance. Silicon is also added to provide a wide range of fusion temperatures so that the alloys behave like glass and not like metallic material, which has a true melting point. The compositions of these powders provide coatings ranging in hardness from R_C15 to R_C65; the higher the alloy content, the harder the coating. Fusion temperatures vary between 1900 and 2050°F (1040 and 1120°C); the lower the alloy content, the higher the fusion temperature. Self-fluxing alloys are usually applied by oxyacetylene welding methods and are typically used in the glass and oil well industries.

APPLICATION METHODS

Hard-facing alloys are usually applied by a variety of welding methods such as oxyacetylene, shielded metal arc, open arc, submerged arc, gas tungsten arc, plasma arc, and laser.

The selection of the method is based on the size of the workpiece, metallurgical properties of the substrate, the form and composition of the metal being deposited, weld characteristics desired, welder skill, and the cost of the welding operation. Table 22-5 lists the various hard-facing processes and characteristics in order of capital cost, with the first process costing the most. When determining the total hard-facing cost for a given application, it is necessary to take into consideration the cost of any machining or finishing operations required.

Manual, semiautomatic, and automatic techniques may all be used. In most semiautomatic welding, the deposit is positioned by means of a handheld gun, and the wire is fed by rollers on a variable-speed motor. In some machines, the motor is controlled by the voltage drop across the arc. In automatic welding, both the arc and the work are positioned mechanically. For detailed information on these methods, refer to Volume IV, *Quality Control and Assembly*, in this Handbook series.

Oxyacetylene Welding

Gas welding is used widely for hard facing because of its flexibility and portability, precision deposit placements, minimum dilution from the base metal, and minimum thermal stress due to slower heating and cooling rates. Operator skill is important for high-grade results, however; and it is not the lowest cost procedure. The overlay metal is usually in the form of a conventional welding rod, steel wires filled with carbide granules, or powders. Deposit thicknesses are generally between 1/32 and 1/8" (0.8 and 3.2 mm).

Hard facing by gas welding is usually done with a reducing flame on a steel base; the flame is adjusted to a feather-to-cone ratio of 3:1. The flame is applied to the base until it shows a wet, glistening surface. The rod tip is then melted by the flame until it

HARD-FACING APPLICATION METHODS

TABLE 22-5
Hard-Facing Processes and Characteristics[9]

Process	Deposit Thickness, in. (mm)	Dilution of Single Layer Deposit, %	Heat Input to Component	Distortion of Component	Deposition Rate, lb/hr (kg/hr)
Laser	0.010 to 0.100 (single pass)	1	Very low	Low	1 to 10 (0.5 to 4.5)
Plasma arc	0.02 to 0.25 (0.50 to 6.4)	5	High but local	Medium*	1 to 15 (0.5 to 6.8)
Gas tungsten arc	1/16 to 3/16 (1.6 to 5)	5 to 10	Medium	Medium*	10 to 20 (4.5 to 9.0)
Open arc	3/32 (2.5)	Up to 30	Medium	High*	5 to 25 (2.3 to 11.3)
Submerged arc	1/8 and greater (3.2)	20 to 30	Low	High*	10 to 25 (4.5 to 11.3)
Powder welding	1/32 to 5/64 (0.8 to 2.0)	Up to 5	Medium	Medium	1 to 15 (0.5 to 6.8)
Oxyacetylene (rod)	1/16 to 3/16 (1.6 to 5)	Up to 5	High	High*	1 to 6 (0.5 to 2.7)
Shielded metal arc	1/8 and greater (3.2)	10 to 25	Low	High*	1 to 6 (0.5 to 2.7)

(*Chartered Mechanical Engineer*)

* When used on large masses, distortion may be limited, but cracking may occur with harder grades.

flows over the surface. When powdered material is used, the powder feed is opened to spray a thin layer on the workpiece.

Shielded Metal Arc and Open Arc

Shielded metal arc (SMAW) and open arc welding are the most economical methods for hard facing. High deposition rates can be attained, but thermal stresses are also high, resulting in distortion or residual stresses. The first layer of arc weld may be diluted as much as 50% by melted base metal. Dilution is the interalloying of the base metal with the hard-facing alloy. A dilution of 50% means that the deposit contains 50% of the base metal and 50% of the hard-facing alloy. Generally, hardness and wear resistance decrease as dilution increases.

Coated electrodes are normally used for applying hard facing with manual processes. The coating contains arc-stabilizing ingredients, fluxing ingredients, protecting ingredients, and alloying elements. Semiautomatic and automatic processes may use coiled tubular electrodes with powdered alloying ingredients encased in a soft steel sheath.

Submerged Arc Welding

Submerged arc welding (SAW) with a single electrode is the most widely used automated technique for hard facing. Because of its high deposition rate, submerged arc hard facing can result in deposits made at lower costs than by other methods; however, distortion may occur due to the high heat input. Because of dilution, two or three layers are generally required to achieve good wear-resistant properties. Hard facing by submerged arc welding is limited to flat or cylindrical workpieces. For some applications, as many as six electrodes can be used; however, single or double-electrode methods are more common.

Gas Tungsten Arc

Gas tungsten arc welding (GTAW), also referred to as tungsten inert gas (TIG), is a nonconsumable-electrode arc method that shields both the molten hard-facing alloy and the nonconsumable electrode with an inert gas. The TIG method is commonly used for applying hard facing on titanium-stabilized stainless steels and aluminum-bearing, nickel-based alloys. It is also more adaptable to hard facing small, intricate parts, and generally produces higher quality deposits than several other hard-facing processes. Gas tungsten arc welding generally produces hard-facing deposits with minimal base metal dilution.

Plasma Arc

In the plasma arc process, an arc is struck between a nonconsumable cathode and a conductive workpiece. The hard-facing material, usually in the form of powder or wire, is fed into the plasma stream, where it is melted and then puddled on the surface of the workpiece. Deposits are from 0.020 to 0.250" (0.50 to 6.4 mm) thick in a single pass. Low dilutions are achievable at relatively high deposition rates; however, most deposits are generally applied in at least two-layer thicknesses. Up to 0.75" (19 mm) can be deposited with multiple passes. This hard-facing process offers the greatest flexibility in the selection of the coating material and also offers excellent robotics control for hard facing complex shapes.

Lasers

Laser hard facing is a method in which a high-intensity beam of light is focused on the surface of the part, then rapidly scanned over the area while injecting the powder into the interaction zone. This method allows for the hard facing of irregular shapes and deposits powder only where it is needed. Deposits of 0.010 to 0.10" (0.25 to 2.5 mm) are attainable in a single pass. Due to the speed of the process and the localized heat, distortion and dilution are minimized. Rapid solidification of the hard facing permits the formation of previously unattainable microstructures.

References

1. Merle L. Thorpe, "Thermal Spraying Becomes A Design Tool," *Machine Design* (November 24, 1983), p. 72.
2. *Ibid*.
3. Stanley T. Walter, ed., *Welding Handbook*, 6th ed., sec. 3A (Miami, FL: American Welding Society, 1970), p. 44.3.
4. Joseph R. Davis, ed., *Metals Handbook*, 9th ed., vol. 6 (Metals Park, OH: American Society for Metals, 1983), p. 772.
5. *Ibid*., p. 771.
6. *Ibid*., pp. 773-777.
7. Walter, *op. cit.*, p. 44.34.
8. Walter, *op. cit.*, p. 44.6.
9. T. H. Shailes, "Coating Parts With Hardfacing Alloys—the Cost Effective Solution to Wear," *Chartered Mechanical Engineer* (September 1983), pp. 56-58.

PORCELAIN ENAMELING AND HOT DIPPING

PORCELAIN ENAMELING

CHAPTER CONTENTS:

Porcelain enamels can be defined as highly durable, alkaliborsilicate glass coatings that are bonded by fusion to various metal substrates at temperatures above 800° F (425° C).[1] They are distinguished from other ceramic coatings by their predominantly vitreous nature and by the types of applications for which they are used, and from paint by their inorganic composition and the temperature at which the coating matrix is fused to the substrate metal.

Porcelain enamels were used initially in making jewelry and ceremonial objects. Today these coatings are widely used for industrial products, household appliances, plumbing fixtures, signs, and architectural applications. They are also used in jet engine components as a protective metal coating to extend service life and for many types of industrial and chemical process equipment requiring resistance to extreme heat and corrosion. Normally, porcelain enamels are selected for products or components requiring chemical resistance, corrosion protection, weather resistance, specific mechanical or electrical properties, appearance or color needs, cleanability, or thermal shock capability. Table 23-1 lists the major applications of porcelain enamels today.

GLOSSARY OF TERMS

The definitions of the following terms are taken from Standard C 286, published by the American Society for Testing and Materials (ASTM).[2]

bisque A coating of wet process porcelain enamel that has been dried, but not fired.

blister A defect caused by gas evolution consisting of a bubble that forms during fusion and remains when the porcelain enamel solidifies.

boiling A defect caused by gas evolution that is visible in the fired porcelain enamel and results in the formation of blisters, pinholes, black specks, dimples, or a spongy surface.

bubble structure Size and spatial distribution of voids within the fired porcelain enamel.

chipping Fracturing and breaking away of fragments of a porcelain-enameled surface.

color oxide A material used to impart color to a porcelain enamel.

continuity of coating The degree to which a porcelain enamel or ceramic coating is free of defects, such as bare spots, boiling, blisters, or copperheads, that could reduce its protective properties.

copperhead A defect occurring in sheet metal ground coat that appears as a small freckle or pimple-like spot, reddish brown in color.

cover coat A porcelain enamel finish applied to and then fused over a ground coat or applied directly to the metal substrate and then fused.

deflocculating Thinning the consistency of a slip by adding a suitable electrolyte.

dipping The process of coating a metal shape by immersion in slip, removal, and draining. In dry process enameling, the method of coating by immersing the heated metal shape for a short time in powdered frit.

direct fire A method of maturing porcelain enamel so that the products of combustion come in contact with the ware.

dredge, dredging In dry process enameling, (1) the application of dry, powdered frit to hot ware by sifting; or (2) the sieve used to apply powdered porcelain enamel frit to the ware.

dry process enameling A porcelain enameling process in which the metal article is heated to a temperature above the maturing temperature of the coating, usually 1600-1750° F (870-955° C). The coating materials are applied to the hot metal as a dry powder and fired.

dry weight The weight per unit area of the bisque.

film strength The relative resistance of the bisque to mechanical damage.

fineness of enamel A measurement of the degree to which a frit has been milled in wet or dry form, usually expressed in grams residue retained on a certain type of mesh screen from a 50 cm^3 or a 100 g sample.

Contributors of sections of this chapter are: ***John Alexander***, *Product Specialist, Coated Product Services, Eastern Steel Div., Armco, Inc.;* ***Ed Meyer***, *Vice President, Sales, Barrett Centrifugals, Inc.;* ***Gary T. Satterfield***, *Technical Director, American Hot Dip Galvanizers Assn., Inc.;* ***George Updike***, *Product Manager, Porcelain Enamel Coatings, Ferro Corp.*

Reviewers of sections of this chapter are: ***J. W. Barr***, *Director of Standards, The Aluminum Assn., Inc.;* ***John H. Collins***, *General Manager, Daytona Finishing Corp.;* ***John E. Cox***, *Laboratory Manager, O. Hommel Co.;* ***J. Donald Gardner***, *Metallurgical Engineer, The Aluminum Assn., Inc.;* ***Wayne L. Gasper***, *Chief Process Engineer, The Maytag Co.;* ***Cullen L. Hackler***, *Manager, Research & Development, Inorganic Chemicals Div., Mobay Chemical Corp.;* ***William B. Hampshire***, *Assistant Manager, Tin Research Institute;*

CHAPTER 23

GLOSSARY OF TERMS

TABLE 23-1
Porcelain Enamel Applications

Industrial Products	Architectural
Chemical reactors	Awnings
Commercial heat exchangers	Baseboards
Food processing vessels	Chalkboards
Induction heating coils	Exterior building panels
Ion gun parts	Floor, roof, and wall tiles
Jet engine components	Tunnel panels
Microcircuitry boards	Wall murals
Mufflers	**Plumbing Fixtures**
Transformer cases	Bathtubs and lavatories
General Products	Kitchen sinks
Camping equipment	Laundry tubs
Cooking and serving utensils	Tub enclosures
Grills	**Signs**
Hospital ware	Advertising signs
Kitchen cabinets	Highway and traffic signs
Lighting reflectors	House numbers
Meat scales	Street signs
Silos	
Tabletops	
Telephone booths	
Venetian blinds	
Household Appliances	
Air conditioners	
Broilers	
Dishwashers	
Freezers	
Home laundry equipment	
Ranges, gas and electric	
Refrigerators	
Space heaters	
Water heaters	

firing The controlled heat treatment of ceramic ware in a kiln or furnace to develop the desired final properties.

firing time The period during which the ware remains in the firing zone of the furnace to mature the coating.

fishscaling A defect appearing as small half-moon-shaped fractures, somewhat resembling the scales of a fish.

flocculating The thickening of the consistency of slip by adding a suitable electrolyte.

flow coating The process of coating a metal shape by causing the slip to flow over its surface and then allowing the excess slip to drain.

flux A substance that promotes fusion in a given ceramic mixture.

frit, porcelain enamel The small friable particles produced by quenching a molten glassy material.

glass A term sometimes used for porcelain enamel or frit.

gloss The shine or luster of a porcelain enamel. See ASTM Test Method C 346 for "45° Specular Gloss of Ceramic Materials."

ground coat (1) A porcelain enamel applied directly to the base metal to function as an intermediate layer between the metal and the cover coat. (2) On sheet steel, a porcelain enamel coating containing adherence-promoting agents that may be used either as an intermediate layer between the metal and the cover coat or as a single coat over the base metal.

one-coat ware, one-coat work (1) Articles finished in a single coat of porcelain enamel. (2) Sometimes a contraction of one-cover-coat ware in which the finish consists of a single cover coat applied over the ground coat.

opacifier A material that imparts or increases the diffuse reflectance of porcelain enamel.

opacity The property of reflecting light diffusely and non-selectively, properly defined in ASTM Test Method C 347, for "Reflectivity and Coefficient of Scatter of White Porcelain Enamels," under the term *contrast ratio*.

orange peel A surface condition characterized by an irregular waviness of the porcelain enamel, resembling an orange skin in texture; sometimes considered a defect.

pickup The amount of slip retained per unit area on dipped workpieces.

pinhole, pinholing A porcelain enamel surface defect caused by gas evolution and characterized by a small hole resembling a pin prick that may extend to the base metal.

pit A defect similar to a dimple but slightly smaller.

porcelain enamel A substantially vitreous or glassy inorganic coating bonded to metal by fusion at a temperature above 800° F (425° C).

primary boiling The evolution of gas during the initial firing of porcelain enamel; sometimes considered a defect.

reboiling Gas evolution occurring and recurring during repeated firing of the ground coat; sometimes considered a defect.

reclaim Overspray that is removed from the spray booth and reconditioned for further use.

sagging (1) A defect characterized by a wavy line or lines appearing on those surfaces of porcelain enamel that have been fired in a vertical position. (2) A defect characterized by irreversible downward bending in an article insufficiently supported during the firing cycle.

screen test A standard test for the fineness of porcelain enamel slip or powder.

scumming A defect characterized by areas of poor gloss on the surface of porcelain enamel.

set A flow property of porcelain enamel slip affecting the rate of draining, residual thickness, and uniformity of coating.

setting-up agent or setup agent An electrolyte used to increase the measured pickup of slip.

slip, slurry A suspension of finely divided ceramic material in liquid.

slump test A test to determine consistency of slip whereby measurement is made of the spreading of a specified volume of slip over a flat plate.

smelt A specific batch or lot of frit.

Reviewers, cont.: ***David E. Horan****, Director of Metallurgical Services, Thomas Steel Strip Corp.;* ***Robert K. Laird****, Ceramic Engineer, American Standard, Inc.;* ***Daniel J. Maykuth****, Manager, Tin Research Institute;* ***Dennis E. McCloskey****, Factory Manager, White Consolidated Industries;* ***William M. McClure****, Plant Manager, Cleveland Div., Magic Chef, Inc.;* ***John C. Oliver****, Executive Vice President, Porcelain Enamel Institute, Inc.;* ***Marvin B. Pierson****, General Manager, Coatings Research, Armco, Inc.;* ***William Planick****, Technical Writer;* ***James F. Quigley****, Manager of Porcelain Enamels, Ferro Corp.;* ***James S. Roden****, Senior Process Engineer, Rheem Manufacturing Co.;* ***Gary T. Satterfield****, Technical Director, American Hot Dip Galvanizers Assn.;* ***Donald R. Sauder****, Division Finishing Manager, Appliance Div., Tappan Co.;*

smelter A furnace in which the raw materials of the frit batch are melted.

spall, spalling, or spontaneous spalling A defect characterized by chipping that occurs without apparent external causes.

stippled finish A pebbly textured porcelain enamel, often multicolored.

tearing A defect in the surface of porcelain enamel characterized by short breaks or cracks that have been healed.

wet milling The grinding of porcelain enamel materials with sufficient liquid to form a slurry.

wet process enameling A method of porcelain enameling in which slip is applied to a metal article at ambient temperature, dried, and fired.

TYPES OF PORCELAIN ENAMELS

Porcelain enamels for sheet steel and cast iron are classified as either ground coat or cover coat enamels. A ground coat enamel contains oxides that promote adherence of the enamel to the metal substrate and may be used as a single functional coat or as a base for additional cover coats. Cover coat enamels are applied over ground coats to improve the appearance and chemical and physical properties of the coating. Cover coats may also be applied directly to properly prepared decarburized steel substrates. The color of ground coats is limited to various shades of blue, black, brown, and gray. Cover coats, however, may be clear, semiopaque, opaque, or pigmented to take on a great variety of colors; colors may also be smelted into the basic coating material. Opaque cover coats are usually white.

For aluminum, neither ground coats nor adherence-promoting oxides are required. Single-coat systems are used for most applications. When two coats are desired, the first coat can be of any color. Porcelain enamels for aluminum are usually transparent and can be pigmented and opacified inorganically to produce the desired appearance.

Porcelain enamels are not designated by composition because all are varieties of silicate glass. Selection is made on the basis of end use and processing requirements. Some common designations for porcelain enamels with particular characteristics are acid resistant, alkali resistant, heat resistant, glossy, low gloss, and matte. Specifications usually include the main requirements for quality and the processing parameters.

Preparation of Porcelain Enamels

The basic material of the porcelain enamel coating is called frit, a special glass of small friable particles produced by quenching a molten glassy mixture. Because porcelain enamels are usually designed for specific applications, the compositions of the frits from which they are made vary widely. Frits consist of five to twenty or more components that are thoroughly mixed together and melted into a glassy system. The molten glass is then quenched to a friable (easily broken up) condition by being either poured into water or rolled into a thin sheet between water-cooled rolls. If quenched in water, the frit is dried before use. If quenched in sheet form by water-cooled rolls, the sheet is ordinarily shattered into small flakes by mechanical means before shipment or use.

Porcelain enamel is usually applied as a suspension of finely milled frit in water. However, it may also be applied on sheet steel as a dry powder via electrostatic spraying or on cast iron by dredging (sifting). The wet process frit is reduced to a fine powder in a ball mill. After loading the frit charge and mill additions such as clay, bentonite, electrolytes, and coloring oxides, water is added. Frits for dry electrostatic application are ground without water by the frit supplier and furnished to the porcelain enameler in a ready-to-use form. Small amounts of proprietary additives are included during grinding to aid in the electric charge retention during application and handling. Particle size is fitted to the appropriate level for each particular application.

The enamel slip is unloaded either by gravity flow or by applying air pressure. Centrifugal or vibratory screening and magnetic separation of the slip as it is transferred from the mill to the storage tank is a recommended practice. Separate mills are used for grinding frits for ground coats, white cover coats, and colored cover coats.

Frit ground to proper particle size for some single-coat dark colors can be purchased from the manufacturer. This frit is combined with the usual mill additions and water. The mixture is blunged (blended) by the porcelain enameler into a suitable slip in a high-speed blunger operating at 900 to 2000 rpm. Frit applied by the dry process is milled in a ball mill without water.

Clays and electrolytes are used with frit in producing enamels applied by the wet process to control rheological properties of the slip. Clays or electrolytes are not used when preparing frits for aluminum. Refractory materials and pigments may also be added to the frit to impart desired properties to the fired porcelain enamel.

The soluble electrolytes that are highly alkaline, such as potassium carbonate and sodium aluminate, are responsible for deflocculation of the clays. Deflocculation is necessary to permit proper suspending action by the clays. Minor adjustments in the flow properties of the slip are made by varying the content of the clay and the electrolytes.

Mill additions to wet process enamel frits for aluminum include boric acid, potassium silicate, and sodium silicate. These materials are used to control the wet suspension of the frits and to contribute to the characteristics of the fired porcelain enamel. Titanium dioxide and ceramic pigments are also added to produce opacity and the desired color, respectively.

Properties of Porcelain Enamels

Porcelain enamels are prepared to ensure satisfactory properties for specific environments.

Chemical resistance. Porcelain enamel is extensively used because of its resistance to household chemicals and foods. Mild alkaline or acid environments are generally involved in household applications. Table 23-2 presents examples of corrosive environments in which porcelain enamels are widely used for long periods of service. Special enamels or glass compositions are available to resist most acids—except for hydrofluoric or concentrated phosphorics—to temperatures of 450° F

*Reviewers, cont.: **William C. Schieferstein**, Ceramic Engineer, Clyde Div., Whirlpool Corp.; **Arthur V. Sharon**, Laboratory Manager, Chi-Vit Corp.; **Dr. Robert R. Shuck**, Senior Research Manager, LTV Steel Co.; **Jerome Smith**, Vice President, Lead Industries Assn., Inc.; **Lester N. Smith**, Vice President, Cherokee Porcelain Enamel Corp.; **Larry L. Steele**, Research Metallurgist, Research & Technology, Armco, Inc.; **James D. Sullivan**, Manager, Ceramic Lab, Protective Coatings Div., A. O. Smith Corp.; **Dr. Charles Tennant**, Manager, Technical Services, Zinc Institute, Inc.; **Donald A. Toland**, Metallurgical Consultant; **George Updike**, Product Manager, Porcelain Enamel Coatings, Ferro Corp.; **Floyd J. Williams**, Manager, Enamel Engineering, Porcelain Metals Corp.*

TABLE 23-2
Applications in Which Porcelain Enamels are Used for Resistance to Corrosive Environments

Application	Corrosive Environment		
	Temperature, °F (°C)	pH	Corrosive Media
Bathtubs	to 120 (49)	5-9	Water, cleansers
Chemical ware	to 212 (100)	12	Alkaline solutions
	to 212 (100)	1-2	All acids except hydrofluoric
	350-450 (175-230)	1-2	Concentrated sulfuric acid, nitric acid, and hydrochloric acid
Home laundry equipment	to 160 (71)	11	Water, detergents, and bleach
Dishwashers	to 180 (82)	8-12	Water, strong detergents
Range exteriors	70-150 (21-66)	2-10	Food acids, cleaners
Range oven liners:			
Conventional	70-600 (20-315)	2-10	Food acids, cleaners
Pyrolytic	70-1000 (20-540)	2-10	Food acids, cleaners
Range burner grates	70-1100 (20-595)	2-10	Food acids, cleaners
Refrigerators	0-70 (-18 to 20)	2-10	Food acids, cleaners
Kitchen sinks, lavatories	to 160 (71)	2-10	Food acids, water, and cleansers
Water heaters	to 160 (71)	5-8	Water

(*American Society for Metals*)

(230° C); they also resist alkali concentration to a pH of 12 at 200° F (93° C).

Weather resistance. The important factors that determine the weather resistance of porcelain enamels are chemical durability, color stability, cleanability, and continuity of coating. Gloss and enamel texture do not necessarily affect weather resistance.

Appearance for indoor exposure. Where corrosive attack is unlikely to limit the life of a given part, and attractive appearance is the principal requirement, enamel selection and processing are directed toward providing reproducible color matching along with optimum gloss and smoothness. Different porcelain enamels are frequently used on different parts of the same product to ensure the best balance of properties and cost, particularly if high volume is involved. For example, range tops and side panels or refrigerator food compartment liners and crisper pans have somewhat different enamel compositions because of differing property and appearance requirements, even though processed in the same plant. Appearance standards, in particular, are established according to the component, its use, and its location as the end product. Small surface defects may be tolerated in areas not heavily used or readily seen in the finished part, provided they do not affect basic serviceability.

Parts exhibiting defects may be recoated one or more times as required, but recoating increases the enamel thickness and hence decreases resistance to chipping. Therefore, every effort is made to achieve realistic standards and high quality during the first enameling cycle.

Service temperatures. The temperature to which porcelain enamels can be exposed is limited by the softening of the glassy matrix. The softening releases gases remaining from reactions between the enamel and ferrous metal, producing random defects known as *reboil.* Service temperature limits for porcelain enamels are listed in Table 23-3.

Thermal shock intensifies the effect of elevated temperature, as does operation under severe temperature gradients. Enamels are formulated so that expansion characteristics place the

TABLE 23-3
Service Temperature Limits for Porcelain Enamels

Service Temperature, °F (°C)	Limiting Conditions
800 (425)	Usual limit for enamels maturing at about 1500° F (815° C).
1000 (540)	Maximum for enamels maturing at about 1500° F (815° C), without reboil.
1400 (760)	Operating limit for special high-temperature enamels.
2000 (1095)	Refractory enamels useful for short periods for protection of stainless steels and special alloys.

(*Porcelain Enamel Institute*)

enamel in compression under service conditions. If combinations of mechanical stress and elevated temperature place the enamel in tension, a pattern of fine cracks perpendicular to the tensile stress are formed. The forming of these cracks is commonly referred to as *crazing*.

Mechanical properties. The hardness of porcelain enamels ranges from 3.4 to 6.0 on the Moh's scale. Porcelain enamels show a high degree of abrasion resistance. Under abrasive test conditions where plate glass retains 50% specular gloss, porcelain enamel compositions retain from 35 to 85% specular gloss. Subsurface abrasion resistance varies with processing variables that affect the bubble structure of the enamel; that is, gas bubbles frozen in during cooling of the enamel. A decrease in abrasion resistance occurs with an increase in the number or size of gas bubbles. Abrasion resistance can be increased by adding crystalline particles to the enamel composition or by a devitrification heat treatment.

Electrical properties. Porcelain enamels are electrical insulators. The electrical resistance per unit area is a function of thickness, enamel composition, and temperature. In addition to resistance (usually expressed as resistivity), other electrical properties of interest include the dielectric constant, dissipation factor, and dielectric strength. As with many glassy materials, these properties vary with temperature. In general, as the temperature increases, the resistivity and dielectric strength decrease, while the dielectric constant and dissipation factor increase. The dielectric constant and the dissipation factor also vary with frequency.

When porcelain enamel is used for its electrical properties, the selection of the enamel composition and the enameling process requires careful attention. For electronic applications, such as porcelain-enameled substrates for hybrid circuits, special electronic grade enamels are used. These electronic grade compositions have considerably higher resistivities and dielectric strengths and are less sensitive to temperature changes than conventional porcelain enamels. Such specialty porcelain coatings are currently being used increasingly in sophisticated electronic circuitry.

METAL SUBSTRATES

Porcelain enamels are primarily applied to products made of sheet iron or steel, cast iron, aluminum, or aluminum-coated steel to impart selected chemical, physical, and aesthetic properties.

Steels

Typical compositions of the various grades of low-carbon sheet iron or steel that are commercially available for porcelain enameling are listed in Table 23-4. Sheet steels used in porcelain enameling can be divided into three groups: (1) those having ultralow carbon contents, a maximum of 0.008% carbon, including those in which the carbon is stabilized by the addition of titanium or niobium (these steels do not exhibit primary boiling and are suitable for direct cover coat enameling applications); (2) those having low-carbon contents of about 0.02% (these steels are suitable for ground or two-coat enameling); and (3) conventional cold rolled sheets with higher carbon contents of about 0.06% (such sheets have a tendency to primary boiling and sagging and are used in less critical ground and two-coat enameling applications). Hot rolled steels are generally not used in porcelain enameling because of their tendency for fishscaling; where thickness requirements dictate their use, they should be coated on one side only to minimize the fishscaling hazard.

Cast Iron

Cast iron for enameling usually has a composition within the limits given in Table 23-5. Total carbon and silicon should vary in opposite directions within the ranges shown. If both are low, the iron tends to be brittle and blisters during porcelain enameling. If both are high, the iron is soft and warps easily when reheated for porcelain enameling.

Manganese and sulfur should range in the same direction so that all of the sulfur is converted to manganese sulfide. Within the normal range, phosphorus has a negligible effect on the strength of the iron at porcelain enameling firing temperatures.

Aluminum

The common porcelain enameling alloys for the various forms of aluminum are:

Sheet:	1100, 3003, and 6061
Extrusion:	6061
Casting alloys:	43, 344, and 356

TABLE 23-4
Composition of Low-Carbon Sheet Iron and Steel for Porcelain Enameling

Enameled Metal	Element, weight %						
	Carbon	Manganese	Phosphorus	Sulfur	Aluminum	Titanium	Niobium
Enameling iron	0.03	0.05*	0.01	0.02	---**	---	---
Decarburized enameling steel	0.005	0.20-0.30	0.01	0.02	---**	---	---
Titanium-stabilized enameling steel	0.05	0.30	0.01	0.02	0.05	0.30	---
Interstitial-free enameling steel	0.005	0.20	0.01	0.02	---	0.04	0.09
Cold rolled steel	0.06	0.35	0.01	0.02	---**	---	---

(*American Society for Metals*)

* Some enameling iron may have manganese contents of 0.20 weight %.
** Some steels may be supplied as aluminum-killed products.

TABLE 23-5
Typical Composition Ranges of Cast Irons Used in Porcelain Enameling

Constituent	Amount, %
Total carbon	3.20-3.60
Silicon	2.30-3.00
Manganese	0.30-0.60
Sulfur	0.05-0.12
Phosphorus	0.40-0.80

(*American Society for Metals*)

Of wrought alloys, only 6061 is heat treatable. Because of its higher strength, 6061 alloy has better handling characteristics before and during porcelain enameling and is stronger after porcelain enameling. The nonheat-treatable alloys are easier to form before porcelain enameling and are used for small parts in which the amount of distortion and low strength encountered after firing are acceptable. However, nonheat-treatable alloys are unsuitable for more than one coat of porcelain because of crazing after a second firing.

DESIGN CONSIDERATIONS

Design, product, and manufacturing engineers should not hesitate to consult with suppliers of metal, enameling materials, and processing equipment as lack of experience does not preclude a successful adaptation of a porcelain enamel finish for new or existing products. The outstanding qualities of porcelain enameling are worth considering whenever a high-quality finish is desired.

The glasslike nature of porcelain enamel and the high firing temperatures used in its application impose limitations on the design of articles to be enameled to ensure that finished work is within dimensional tolerances and durable enough for intended service. In general, the size of products that can be porcelain enameled is limited only by the ability of facilities to accommodate them.

Steel Parts

Distortion during firing is minimized by uniformity of stress and temperature during the cycle. Nonuniform thicknesses heat at different rates and cause distortion and variations in enamel maturity. Specific aspects of design that should be considered are:

- Bend and corner radii should be at least 3/16" (4.7 mm). In drawn parts, the use of symmetrical embossed ridges and panels increases resistance to distortion caused by uneven residual forming stresses.
- Flanges increase strength and flatness but can cause irregular stresses in firing. Flanges on one side only may require welded braces. Flanges meeting at a corner must be welded there and should not vary in depth by a factor of more than three. Flanges may sag during firing if not properly designed or supported by braces.
- Welded lugs and ears for attachment and assembly result in double metal thickness. They should be of the same or lighter gage as the main part and as small as possible. Double metal thickness can result in color variations, hairlining, or poor adherence.
- Spot and seam welds also result in double metal thickness. Spot welds are difficult to enamel because of movement during firing and entrapment of solutions used in preparing the metal for enameling. Seam welds should be flattened to prevent burrs, rough projections, and protruding edges, all of which enamel poorly.
- Fusion welds must be free of crevices and oxide seams. Weld spatter must be removed to avoid coating defects.

Cast Iron Parts

Uniformity of section, simplicity of design, and the minimizing of lugs and braces are desirable characteristics for cast iron parts to be enameled. Radii of curved sections should be as generous as design limitations permit. The minimum radius may be 1/4" (6.3 mm) for decorative beading on a flat or slightly curved surface, or as large as 1 1/2" (38 mm) for one of the components of a compound curve on a large casting.

Aluminum Parts

Because enamel is ordinarily applied to aluminum at only about half the thickness as applied to steel, freedom from surface scratches, burrs, and irregularities is doubly important for aluminum. Aluminum is usually formed before enameling, but the thin coating permits some bending, shearing, punching, and sawing of the enameled part.

Surfaces to be enameled should have generous inside radii of not less than 3/16" (4.8 mm) and outside radii of not less than 1/16" (1.6 mm), with 1/8" (3.2 mm) preferred for dark and light colors. Attachments should be welded to the coated back side of enameled heavy-gage aluminum sheet or extrusions. The visible metal surfaces must not be overheated as overheating causes the aluminum to blister, while altering the color and gloss of the enamel. Welding can be done before enameling, provided the weld area is cleaned properly before coating.

METAL PREPARATION

The adhesion and appearance of porcelain enamel depend on closely controlled cleaning and roughening of the metal surface. Complete removal of oil, sand, drawing compounds, weld oxides, and other surface contaminants is required. Additional information on metal preparation can be obtained in Chapter 16, "Mechanical and Abrasive Deburring and Finishing," and in Chapter 18, "Cleaning," of this volume.

Steel

Steel may be prepared for porcelain enameling by chemical or mechanical procedures.

Chemical preparation. Chemical preparation involves cleaning, pickling or etching, nickel flashing, and neutralizing; however, for some applications, it may only include cleaning and neutralizing. Alkaline cleaners are generally used, and pickling is usually done in a 6 to 8% sulfuric acid solution at temperatures of 140 to 165° F (60 to 74° C). When a cover coat is to be applied directly to carburized steel, a minimum of 2 g/ft^2 (22 g/m^2) of metal surface should be removed using sulfuric acid; or a sequence of sulfuric acid, ferric sulfate, and sulfuric acid; or a more aggressive acid such as mixtures of nitric and sulfuric acid. A flash of nickel by a nickel sulfate solution promotes adherence of the enamel coating. The surface is neutralized in a weak alkali solution to prevent rusting before the coating is applied. Thorough rinsing between the various steps in the process is necessary.

Metal preparation for no nickel/no pickle enameling requires at least the same amount of cleaning as conventional metal preparation does for conventional enameling. There is, however, no acid etching or nickel deposition required with the no nickel/no pickle system. An advantage of this system lies in the reduction of wastewater treatment problems.

Equipment may be chosen for either tank immersion or spraying. Hand-operated traveling cranes or mechanized systems similar to those in electroplating operations may be used to transfer the workpieces. Spray units use monorail, mesh belt, or live roller conveyors to transfer the parts between workstations.

Mechanical preparation. Mechanical preparation consists of abrasive blasting using steel shot or steel grit. Before blasting, oil and drawing compounds are removed by alkaline cleaning or by heating at 800 to 850° F (425 to 455° C) to burn off the organic contaminants.

Flat areas of parts made from sheet steel thinner than 0.06" (1.5 mm) are usually not prepared mechanically due to distortion. Abrasive blasting is used particularly for preparing hot rolled steel and parts that are to be enameled on one side only. Stainless steels and other metals that may pick up surface contamination from steel abrasives are blasted with sand or other nonmetallic grit.

Cast Iron

Cast iron is prepared by blasting to remove adhering mold sand and the thin surface layer of chilled iron. Quartz sand, propelled by compressed air, is commonly used; however, steel shot, steel grit, and chilled cast iron grit, propelled centrifugally from rotating wheels, are generally used for cleaning sanitary ware.

After blasting, the castings are inspected for cracks, sand holes, slag holes, blow holes, fins, and washes. Cracks and large holes are filled by welding, followed by grinding. Fins and washes are removed by grinding. The repaired casting is then blasted a second time before enameling. Small holes are usually filled with a ceramic paste after final blasting.

Aluminum

Soil contaminants can be removed from nonheat-treatable aluminum alloys with alkaline cleaners or vapor degreasing, followed by a thorough rinse. If the contaminants are light organic substances, a prefire at 1000° F (540° C) can be used.

Heat-treatable aluminum alloys, such as Type 6061, require a more extensive pretreatment. This process involves cleaning, deoxidizing, and chromatizing. Cleaning is performed in a nonsilicated, mildly etching cleaner with free-rinsing characteristics. The deoxidizer is a mixture of hot chromic and sulfuric acids. The chromate dip is sodium chromate. Thorough rinsing is required between steps and after the sodium chromate dip.

THE PORCELAIN ENAMELING PROCESS

Processing methods for porcelain enameling parallel those used for organic coatings, which are discussed in Chapter 27, "Application Methods," of this volume. Dipping, flow coating, spraying, electrostatic spraying (wet or dry), and electrodeposition methods are used to apply the porcelain enamel to the base metal. However, because of the need for firing along with the use of water suspensions and thicker films in porcelain enameling, the equipment requirements and process control are quite different. The best method of application for a particular part is determined by quantity and quality requirements, the type of material being applied, units produced per hour, capital investment, labor cost, and, ultimately, part cost.

Regardless of the method of application, good porcelain enameling techniques must be used to ensure uniform coverage in the areas requiring the porcelain enamel protection. Excessive thickness, beads, or pooling of the porcelain enamel reduce product quality and make the product more prone to chipping. At the same time, the protection and decorative capabilities of porcelain enamel are reduced in areas where the coating is too thin.

Porcelain enamel is applied to steel by dipping, flow coating, spraying of wet slip, electrostatic powder spraying, and electrodeposition. Cast iron is porcelain enameled by a dry process (nonelectrostatic) or by the same wet process as used for enameling sheet steel. The wet process permits easier part handling since the part is cold, and more uniform coats of enamel can be applied than with the dry process; however, it is difficult to enamel large articles, such as sanitary ware, by the wet process. Porcelain enamel slips for aluminum are usually applied by spraying.

Dipping

Dipping is widely used as a method to apply porcelain enamel, particularly when both sides of the parts require coverage. Dipping can be used for both ground coat application and cover coat application. It is performed by immersing the part in the prepared porcelain enamel slip and then withdrawing it, allowing the excess material to drain from the part (see Fig. 23-1). Sometimes it is necessary to rotate, tilt, spin, or shake complex shapes to ensure uniform coverage. In areas where excessive porcelain enamel slip is retained on the part after draining, the excess enamel can be removed with a siphon or a wiping device before the part is dried.

Dipped porcelain enamel films are normally applied at a thickness of 0.002 to 0.004" (50 to 100 μm) to provide adequate coverage. Automatic dipping equipment is available when part configuration allows and production volume warrants the investment.

Flow Coating

In flow coating, the porcelain enamel slip flows onto the surface of the part. The process is applicable to high-volume continuous operations for parts requiring the same porcelain enamel. In automatic flow coating, the parts are placed on hangers at the correct angle for draining and carried by conveyor through the flow coating chamber (see Fig. 23-2). The porcelain enamel slip is pumped at a high volume, 150 gpm

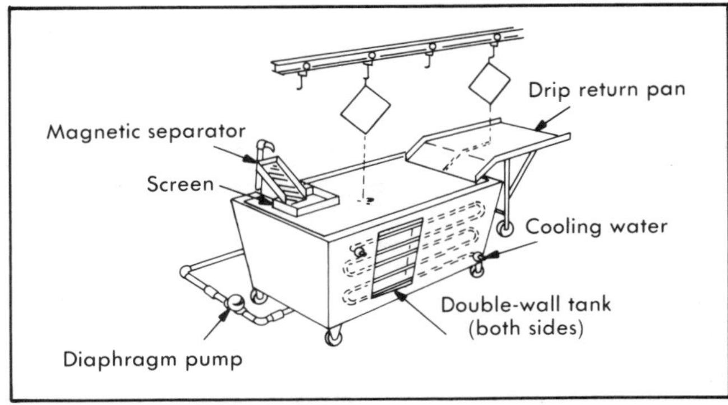

Fig. 23-1 Hand dipping equipment used for applying ground coat enamel coating to automatic washing machine lids.[3]

(570 L/min), and a low pressure, 10-15 psi (70-105 kPa), through a series of nozzles that are directed at various areas of the part to ensure complete coverage. Flow coating may also be used for single-side applications. Automatic flow coating is favored over hand dipping because it offers higher rates of production, improved coating quality, and reduced cost for the applied film.

Spraying

Spraying of the porcelain enamel slip is done primarily for one-side coverage. It is also used for reinforcing the bisque and for making repairs on enameled surfaces. Spraying is ideal for parts that are too large for hand or mechanical manipulation, particularly where service and appearance requirements do not permit beading due to draining or buildup of the porcelain enamel.

Wet electrostatic spraying of porcelain enamel is used to reduce losses in material by charging the porcelain enamel slip during atomization to a potential of 100,000 to 120,000 V. The electrostatically charged droplets are attracted to the grounded parts being sprayed. A well-operated electrostatic unit can deposit up to 85% of the sprayed material on the part as compared to 30 to 50% in conventional spraying operations. Figure 23-3 depicts a typical production installation for the spray application of porcelain enamels.

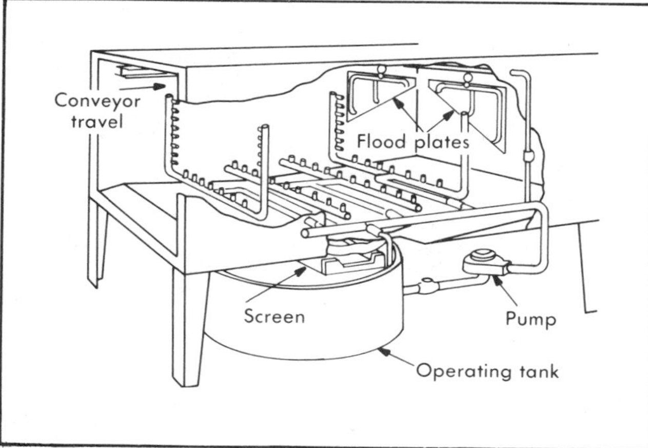

Fig. 23-2 Automatic equipment for applying enamel slip by flow coating.[4]

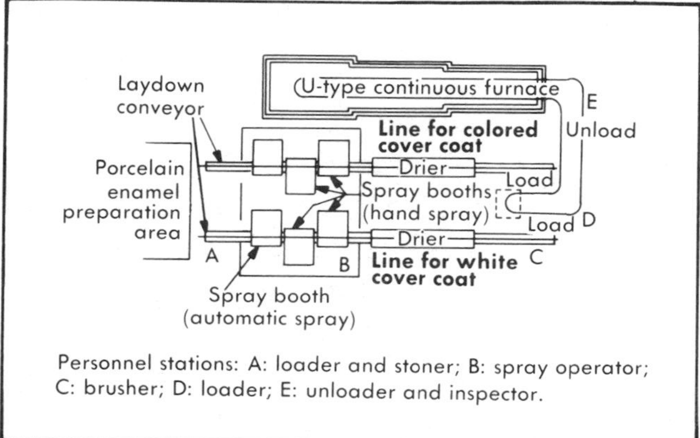

Fig. 23-3 Typical production installation for spraying porcelain enamel coatings.[5]

Powder Coatings

Electrostatic powder spraying is used when a large volume of parts is being produced. Most parts can be properly and evenly coated by this process. Furthermore, it is a very efficient method of applying porcelain enamel; up to 99% of the material is used, with little or no direct labor required for the operation. Smooth-running conveyors and careful handling are required to prevent loss of powder prior to firing the parts.

Powder is delivered to the spray guns from a feeder unit where it is diffused by clean and dry compressed air into a fluid-like state. The fluidized powder is then siphoned by the movement of high-velocity air flowing through a venturi and is propelled through powder feed tubes to the spray gun. The powder feeder provides a steady, controlled flow of powder to the guns. Independent control of powder and air volume ensures the proper ratios to provide the desired thickness coverage on the product.

The powder leaves the spray gun in the form of a diffused cloud being propelled toward the workpiece. A high-voltage, low-amperage power unit supplies current to the charging electrode, causing powder to seek out and attach itself to the grounded workpiece. By using a variable voltage power unit, the operator can regulate the voltage to the type of powder being sprayed, improving the wrap-around coating characteristics.

The recovery equipment booth serves to collect and return the powder that is not held on the workpiece; the workpiece moves through a closed-loop system with the use of filters so that none of the airborne powder escapes into the environment. The humidity and temperature of the powder application room must be carefully monitored.

Electrodeposition

Electrodeposition uses a series of tanks in which the parts are submerged, with the enamel being deposited electrophoretically. This process is basically limited to direct-on enameling, but can be considered for two-coat/one-fire applications. The main advantages of electrodeposition are a very uniform appearance and exceptionally thin enamel layers.

Porcelain Enameling of Aluminum

Porcelain enamel slips for aluminum are usually applied by spraying, using either manual or automatic equipment with agitated pressure tanks. Slips for aluminum are not self-leveling and, therefore, must be deposited smoothly in an even thickness and without runs or ripples. Many aluminum parts are coated satisfactorily by the one-coat/one-fire method. Although the heat-treatable alloys can be recoated one or more times, opacity and color of the coating changes with the thickness of the porcelain and with repeated firing.

Dry Process for Cast Iron

In the dry process, a very thin coat of ground coat enamel slip (slush coat) is applied to the cold casting, generally by spraying but sometimes by dipping or other methods. After the ground coat is dry, the casting is put in a furnace and heated to a bright red heat. After being withdrawn from the furnace and while the casting is still hot, the cover coat, in the form of dry powder, is sprinkled over the surfaces to be covered by means of a vibrating sieve; the powder melts as it falls on the hot surface. The application of powdered enamel continues until the temperature of the casting drops to where the powder will not melt; the casting is then returned to the furnace and heated until the enamel is properly fused. For some types of products, such as lavatories, one application of powder is sufficient, but other products may require several applications; bathtubs and combination sinks require two or more.

Drying

Coated parts may be air dried or placed in dryers using radiant heating or heated circulating air. Air drying is an inefficient procedure for production operations. Drying by radiant heating is at least 20% faster than by convection.

Parts coated with porcelain enamel slip are dried before firing to:

- Permit the application of additional porcelain enamel slip when required, without disturbing the previously applied coating.
- Permit brushing of the coated parts for decorative or functional purposes, if required.
- Allow parts to be handled more easily for transfer to the holding fixture used during firing.
- Reduce the amount of water vapor introduced into the enameling furnace; a high moisture level in the furnace atmosphere should be avoided.

ENAMELING FURNACES

Firing is accomplished in continuous, intermittent, or batch furnaces heated by oil, natural gas, propane gas, or electricity. With oil heating, a muffle furnace is used to prevent the products of combustion from contaminating the enamel coating. Gas-fired furnaces are either muffle, radiant-tube, or luminous-wall types. Continuous furnaces are of either straight-through or U-type design (refer to Fig. 23-3); both designs use air curtains to prevent heat loss through the end openings.

A laydown wire-mesh-belt conveyor is used for products such as small rings, dials, microcircuitry, and other flat pieces. Most continuous furnaces, however, are equipped with overhead monorail conveyors that are located above the furnace roof. Alloy hook or drop rods extend down from the conveyor trolleys through a narrow slot in the furnace roof to transport the parts. This slot is sealed by the use of alloy steel plates carried by the conveyor.

Intermittent furnaces are equipped at both ends with split side-opening doors, are air cylinder operated, and have an overhead monorail conveyor similar to that of the continuous furnaces. With this type of furnace, however, the conveyor moves in increments so that when one load is discharged, a new load enters the furnace, where it remains until the firing cycle is completed. Doors and conveyor are electronically interlocked so that the firing cycle can be time controlled.

Forced convection is the preferred method of heating furnaces for firing porcelain enamel on aluminum. The heat is provided by electric package heaters, quartz-tube electric heaters, or metal-sheath heaters, all specially designed for operation at high air temperature. Quartz-tube and metal-sheath heaters are adapted to the furnace so that radiant heat is available in the firing zone along with forced circulation. Package heaters are remote from the firing zone; this placement is the most effective means of eliminating direct radiation and hot spots. Heat imparted to the work from the package heater is derived completely from air circulation to maintain a uniform temperature throughout the furnace of $\pm 1\%$ of the nominal operating temperature.

Furnace construction for porcelain enameling of aluminum generally requires the use of stainless steel inner-liner sheets, low-density wall insulation, and a plain carbon steel exterior shell. This type of fabrication eliminates long heatup and cooldown periods.

PROCESS VARIABLES

The thickness of the applied porcelain enamel layer, firing time, and firing temperature markedly affect the properties of the coating.

Coating Thickness

Increasing the thickness of the coating increases the resistance to burnoff and produces truer colors; however, thin coatings have the greatest flexibility. The optimum thickness of porcelain enamel depends on the substrate metal and the service requirements of the part.

On sheet steel, a ground coat is generally applied 0.002 to 0.004" (50 to 100 μm) thick to promote adherence and metal protection. A finish coat, of white or pastel color, is applied 0.003 to 0.006" (75 to 150 μm) thick to provide chemical, physical, and aesthetic properties. Thus, a product with a two-coat finish has a thickness ranging from 0.005 to 0.010" (125 to 225 μm).

Brightly colored porcelain enamels are produced by applying less-opaque coats with more saturated colors over a white intermediate coating. The total thickness of brightly colored porcelain enamels usually ranges from 0.010 to 0.015" (255 to 380 μm). For some decorative textured finishes, the thickest parts are 0.025" (635 μm) thick; a thickness of more than 0.025" (635 μm) is usually undesirable. Coating thickness is about 0.005 to 0.006" (125 to 150 μm) for cover coat porcelain enamels applied directly to decarburized or specially stabilized steels.

Coatings for cast iron products are much thicker than those for sheet steel or aluminum. Dry process coatings on cast iron products, such as sanitary ware, range from 0.040 to 0.070" (1020 to 1780 μm) in thickness. Thinner than dry process coatings, wet process coatings range from 0.010 to 0.025" (255 to 635 μm) thick.

On aluminum, porcelain enamel is applied to produce a fired enamel thickness ranging from 0.0025 to 0.005" (65 to 125 μm). A tolerance of $\pm 0.0005''$ (± 13 μm) is required for a white enamel coating 0.0045" (115 μm) thick to maintain uniform opacity.

An enamel thickness of 0.0025 to 0.007" (65 to 180 μm) is desirable for aluminum architectural panels. When white or light-colored enamel is used, however, the enamel thickness is above 0.003" (75 μm) to produce acceptable opacity. Two coats, with a total thickness of about 0.005" (125 μm), result in more uniform opacity than one coat of the same total thickness. Additional details on coating thickness are available from the Porcelain Enamel Institute Bulletin P-302.

Firing Time and Temperature

The firing of porcelain enamels involves the flow and consolidation of a viscous coating and the escape of gases through the coating during its formation. Within limits, time and temperature are varied in a compensating manner. For example, similar properties and appearance develop when firing liners for household refrigerators at 1480° F (805° C) for 2 1/2 minutes or at 1450° F (790° C) for 4 minutes. In all instances, there is a minimum practical temperature for the attainment of complete fusion, acceptable adherence, and desired appearance. Most ground coat enamels for high-production steel parts exhibit acceptable properties when fired to within 100° F (55° C) of the specified temperature and at an optimum firing time; however, control within 20° F (11° C) is usually maintained to produce uniform appearance and allow interchangeability of parts.

Firing of enamel on aluminum is accomplished between 980 and 1020° F (525 and 550° C). To control the color and gloss of the enamel within acceptable limits, the temperature throughout the work should be held to ±2 1/2° F (±1 1/2° C). Time cycle may range from 5 to 15 minutes, depending on design of the part.

In-Process Repairs

Processing conditions limit the probability of producing 100% acceptable parts in one cycle; recoating parts is common, depending on quality specifications.

Repair of ground coat. Steel parts rejected at the ground coat stage are repaired and refired, if necessary, before the cover coat is applied. It is good practice to confine repair techniques to the defective area and not to grind or stone adjacent surfaces indiscriminately.

Dirt particles, scale, and similar contaminants are removed with a sharp-pointed instrument. The surrounding area should then be stoned lightly with an alundum rubbing stone. The dust generated in this operation should be blown off. Ground coat enamel is spotted-in at the repair area, and the entire piece is lightly dust coated. The piece is then dried and fired.

Repair of cover coat. Defects that are missed during inspection of the ground coat usually become visible in the cover coat. Specks, blisters, copperheads, and dents are typical of such transmitted defects and can be removed by grinding the ground coat. The ground coat is then repaired as previously described before another cover coat is applied. Cover coat defects such as lumps, handling defects, chips, and thin coating areas are repaired by stoning, respraying lightly, and then refiring.

HOT DIPPING

Hot dipping is a process by which a metal is coated onto the surface of a metal product by immersion in a bath of molten metal. During the coating process, an intermetallic compound layer is formed that provides the adherence to join the coating metal to the base metal. There are a number of coating metals used to hot dip steel, the most important being zinc, aluminum, tin, and lead. The corrosion rates of these metals are substantially less than that of steel. Thus, metallic coatings on steel preserve the useful properties of the base metal while producing a composite material having greater utility than either base or coating alone.

The two major corrosion protection mechanisms provided by hot-dip coatings are barrier and sacrificial protection. In barrier protection, the coating isolates the substrate from the corrosive environment. A major limitation of barrier protection is that the protection ceases the instant the barrier is broken; barrier damage can be caused by a scratch, abrasion, impact, shaping, forming, or other mechanical removal. Coatings providing only barrier protection should generally be used in areas where mechanical damage is not likely to occur; these coatings are also painted to provide additional protection. Tin and lead are two common metallic coatings that provide barrier protection to the substrate.

In sacrificial protection, the substrate is protected by a coating that corrodes preferentially to the substrate metal. As corrosion proceeds, the coating is removed electrochemically, thus preventing attack of the base metal. This electrochemical process continues until so much of the coating is removed that it can no longer protect the nearby substrate from corrosion.

Some metallic coatings on steel provide both barrier and sacrificial protection. The most popular metals used for dual protection are zinc and aluminum. A primary function of the zinc coating is to isolate the substrate from a corrosive environment. Since zinc is corrodible, the length of time that the coating can provide protection depends on coating thickness and the environment. Zinc also provides sacrificial protection when the steel substrate is exposed due to scratches, machining, or mechanical damage. Aluminum provides barrier protection by forming a tight oxide on its surface that reduces its rate of corrosion. If the oxide is removed, it usually reforms quickly in an atmospheric environment. However, the rapid formation of the aluminum oxide limits the amount of sacrificial protection in atmospheric environments. When the aluminum-coated part is exposed to a salt solution, the oxide layer does not form and the aluminum coating protects the substrate sacrificially.

ZINC HOT DIPPING (GALVANIZING)

Zinc hot dipping or galvanizing is a process by which a coating of zinc and zinc-iron alloy is developed on the surface of steel products by immersing the properly prepared base metal in a bath of molten zinc. It is the most widely used of the zinc coating processes and has been employed commercially for almost two hundred years. Another process used for depositing zinc onto steel substrates is referred to as mechanical plating, which is discussed in Chapter 20, "Plating." Research is also being conducted on developing a zinc-aluminum coating containing approximately 95% zinc and 5% aluminum.

Zinc hot dipping is applied primarily to finished parts and semifabricated parts in batch galvanizing; whereas sheet, strip, wire, and tube are commonly galvanized on the continuous, automated lines of the steel producers. The amount of zinc used in continuous galvanizing in comparison to batch galvanizing is approximately 4:1. Standard specifications for both batch and continuous galvanizing processes have been established by the American Society for Testing and Materials and are identified in Table 23-6.

Steels containing less than 0.25% carbon, 0.05% phosphorus, and 1.35% manganese individually or in combination can be galvanized as long as the silicon content is less than 0.05%. Steels with carbon and silicon contents greater than these amounts have been successfully galvanized, but their composition greatly influences the rate of the galvanized coating buildup and the final coating structure.

Coating Characteristics

The batch galvanized coating consists of a series of layers. Starting from the steel surface, each successive layer contains a higher proportion of zinc until the outer layer is pure zinc. There is no real line of demarcation between the iron and the zinc, but a gradual transition occurs through the series of zinc-iron alloys that provides a powerful bond between the basis metal and the coating. The temperature of the bath has little effect on the nature of the coating as long as the temperature is maintained between 810 and 880° F (430 and 470° C).

The number and extent of the alloy layers and the coating thickness depend on the composition and physical condition of the steel being treated, as well as on the dwell time in the zinc bath and the bath temperature. The surface coating layer, also known as the free-zinc layer, is largely pure zinc and gives the galvanized part its characteristically bright, shiny appearance. With some steel compositions, reaction between molten zinc and the steel is extremely rapid and can produce coatings made entirely of a zinc-iron compound. These coatings may be gray or mottled and will almost always present a dull, matte surface, instead of the more characteristic bright, spangled galvanized surface. Coatings on reactive steels are usually thicker than coatings on other steels. Heavier coatings also tend to be deposited on rough-surfaced and coarse-grained steel. The corners of the coated parts have a heavier coating than adjacent flat surfaces.

The total thickness of the coating may be controlled by varying the time that the parts are immersed in the molten zinc and the speed at which they are removed. The parts may be centrifuged or mechanically wiped after being removed from the bath to remove excess zinc and provide smoother surfaces. Generally, galvanizing does not significantly change the mechanical properties of structural steels. The service life of galvanized steel is directly dependent on the coating thickness and the environment in which the galvanized part is placed. Figure 23-4 suggests the estimated service life of zinc coatings in a variety of environments.

Batch Galvanizing

Batch galvanizing is hot-dip galvanizing applied to prefabricated steel items. The advantage to galvanizing after fabrication is that the zinc completely seals edges, overlaps, rivets, and welds; establishes liquid tightness; and prevents corrosion from starting. Iron and steel in all shapes and sizes can be coated with zinc by batch galvanizing. The process is simple, extremely versatile, and has been used to provide protection to articles ranging from small items, such as bolts, nuts, and miscellaneous hardware, to large items like structural beams for bridges or buildings. The virtually unrestricted size range of parts that can be galvanized and the ability to bolt or weld prefabricated sections after galvanizing enables almost any structure to be built from galvanized steel. Shape is not a restriction to batch galvanizing. Tubes, open vessels, drums, tanks, and complicated shapes such as large heat exchangers are readily galvanized on the inside and outside in one operation.

TABLE 23-6
ASTM Standards for Hot-Dipped Zinc Coatings

Type of Process	Standard Number	Title
Batch	A 123	Zinc (Hot-Galvanized) Coatings on Products Fabricated from Rolled, Pressed, and Forged Steel Shapes, Plates, Bars, and Strip
	A 153	Zinc Coating (Hot Dip) on Iron and Steel Hardware
	A 386	Zinc Coating (Hot Dip) on Assembled Steel Products
Continuous	A 444	Zinc-Coated (Galvanized) Iron or Steel Sheets for Culverts and Underdrains
	A 446	Zinc-Coated (Galvanized) Steel Sheets of Structural Quality, Coils and Cut Lengths
	A 525	General Requirements for Delivery of Zinc-Coated (Galvanized) Iron or Steel Sheets, Coils, and Cut Lengths Coated by the Hot Dip Method

Coating thickness. The thickness of the coating is controlled by the composition of the steel substrate and the immersion time. Part withdrawal rate and any postgalvanizing treatments, such as centrifuging, also influence the coating thickness.

The zinc coating on batch galvanized parts is generally specified in ounces or grams per unit of surface area, measured in either square feet or square meters on a single surface or on one side of the part. With proper coating techniques, the coating weight can usually be controlled between 2 and 4 oz/ft^2 (610 and 1220 g/m^2), equivalent to a coating thickness of approximately 0.0017 to 0.0034" (43 to 86 μm) per side.

Immersion time. The time of immersion is determined by the weight, mass, and configuration of the part being galvanized. Heavy parts require a longer time than light parts. Parts containing internal venting spaces, such as hollow work, also require a longer immersion time. Although timing is to some extent dependent on ease of handling and must be established for each design of part being coated, the duration of immersion is usually in the range of 1 to 5 minutes.[6] The minimum time required allows the immersed part to reach the zinc bath temperature.

Process description. Batch galvanizing consists of surface preparation, fluxing, and galvanizing. Small or threaded parts are generally centrifuged after galvanizing to remove excess zinc and produce a more uniform coating. The parts are ordinarily placed in a wire-mesh or perforated basket before immersion into the molten zinc. Upon withdrawal, the basket is immediately placed in a low-speed centrifuge.

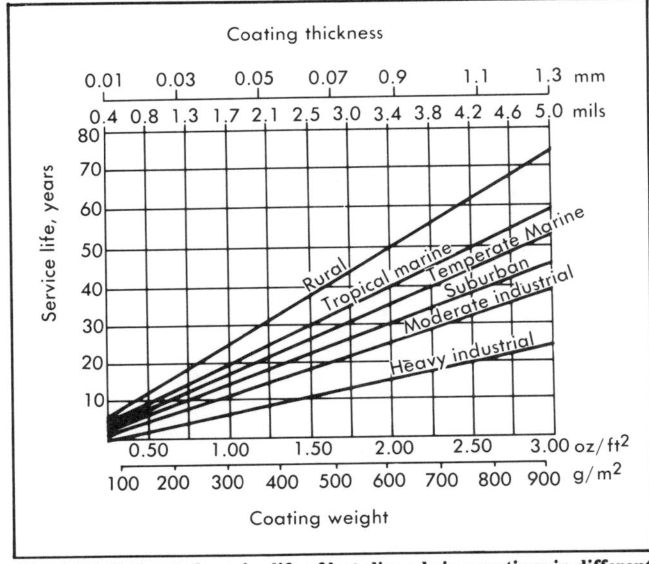

Fig. 23-4 Estimated service life of hot-dipped zinc coatings in different environments based on coating weight. (*American Hot Dip Galvanizers Association*)

ZINC HOT DIPPING (GALVANIZING)

Surface preparation. A properly cleaned surface is extremely important when galvanizing to enable the molten zinc to wet the steel, thus forming a metallurgical bond. The presence of any foreign substance on the steel prevents the metallurgical bonding and the formation of the proper protective coating.

The actual sequence of surface preparation is caustic (alkaline) cleaning, water rinsing, acid pickling, and another water rinse. The caustic cleaner frees the steel of organic contaminants, such as dirt, paint markings, grease, and oils, not readily removed by acid pickling. Scale and rust are generally removed by pickling in hot sulfuric acid or room temperature hydrochloric acid. Immersion time during pickling is as short as possible to minimize hydrogen absorption and undesirable attack on the work surface. Grit, shot, or sand blasting may also be used to clean heavily rusted articles or ferrous castings. Additional information on metal preparation can be found in Chapter 16, "Mechanical and Abrasive Deburring," and Chapter 18, "Cleaning," of this volume.

Fluxing. The final cleaning of the steel before actual galvanizing is performed by a flux that dissolves any oxide film left on the steel after pickling and facilitates the wetting of the steel by the molten zinc. In the dry galvanizing process, the parts are dipped in a flux solution of zinc ammonium chloride and then transferred to a low-temperature drying oven. In wet galvanizing, the parts are dipped through a flux blanket that floats on the molten zinc. The flux blanket is made of zinc ammonium chloride along with foaming agents and fume-suppressing compounds. The flux blanket may cover the entire surface of the zinc bath or be confined to one side.

Galvanizing. After the parts are thoroughly cleaned and fluxed, they are immersed in the molten-zinc bath, which is normally maintained at a temperature of 835 to 855° F (445 to 460° C). The zinc used in the bath must meet the American Society for Testing and Materials' (ASTM) Standard B 6. The amount of zinc in the galvanizing kettle is about 15 to 40 times the weight of parts being coated. In addition to molten zinc, the bath may contain aluminum and lead in small amounts. Aluminum improves coating flow and drainage and increases the brightness of the coating. Lead promotes proper spangle formation and also improves drainage.

After galvanizing has been completed and the parts are withdrawn, light-gage parts may be cooled in air, but heavier sections are often quenched in water to solidify the coating. An adequate time for complete drying must be allowed before stacking or storage to prevent wet-storage stain.

Equipment. Equipment for batch galvanizing consists of tanks for cleaning and fluxing, a kettle for the molten zinc, associated heating systems, and material handling equipment. Alkaline cleaning and rinsing tanks typically are made from hot rolled mild-steel plate. Pickling and fluxing tanks are generally made of steel or concrete and lined with acid-proof brick; a lining of rubber or an acid-resistant plastics is placed between the tank sides and brick lining.

The galvanizing kettle is made from mild steel of controlled composition for resistance to molten zinc, usually between 3/4 and 2" (19 and 50 mm) thick. The kettles are heated either by oil or gas combustion, or by electric-resistance heaters. Firebrick linings protect the kettle from direct contact with burner flames.

The dimensions of the galvanizing kettles are determined by the anticipated sizes of items to be galvanized and the amount of heat transfer required to bring immersed work up to galvanizing temperature, to maintain a constant bath temperature, and to melt any added zinc. These factors rest on economics and market requirements, varying considerably from galvanizer to galvanizer.

Handling equipment consists of overhead cranes and hoists, with their accompanying chain slings and lifting fixtures, and centrifuge baskets for small items. Special jigs and racks are often constructed to simultaneously galvanize a large number of similar articles.

Continuous Galvanizing

Hot-dipped galvanized coatings are also applied to a variety of mill products by highly mechanized mass production methods at speeds of over 300 fpm (90 m/min). Several designs of galvanizing lines have been developed for commercial use by the steel suppliers. Most steel suppliers can produce galvanized coils (sheets) in widths of 10 to 72" (250 to 1830 mm) and thicknesses of 0.017 to 0.164" (0.43 to 4.17 mm). Typical applications for mill coated sheets are roofing and siding panels, guardrails, appliance cabinets, automotive body parts, and ductwork. Galvanized pipe and tubing is used for fencing, sign poles, playground equipment, and plumbing.

The coatings applied by continuous galvanizing contain more aluminum than batch galvanized coatings to minimize the formation of the brittle zinc-iron alloy and permit deep drawing and bending without damage to the coating. Coating weights vary from 0.5 to 2.75 oz/ft^2 (152.5 to 840 g/m^2). The zinc coating may be on one side of the sheet only, of equal weight on both sides of the sheet, or differentially applied (one side has a thicker coating than the other side).

Forming. Continuous zinc-coated steels can generally be formed with the same tools and procedures as the base metal without peeling or flaking. As with any sheet steel, slow double-acting presses provide the best results when making deep draws. Die clearances are normally increased slightly to avoid scraping or galling the coating. Forming tools should be kept clean, polished, and lubricated to avoid zinc pickup. Additional information on forming processes and methods can be found in Volume II, *Forming*, of this Handbook series.

Welding. Galvanized steels can be joined by resistance, arc, and oxyacetylene welding methods. Careful control must be exercised to avoid minute cracks during continuous resistance seam welding, especially for watertight or gastight containers. Resistance welding should be used whenever possible because less of the protective coating is burned off. Higher welding currents, greater pressure, and longer times are used in spot welding galvanized steel than in spot welding cold rolled steel of the same gage. Truncated-cone electrode tips are also used in place of radius face tips, and tip cooling is recommended. A good ventilation system is required to remove the zinc oxide fumes created during arc or gas welding. Since welding burns off the zinc coating, the welded area may require recoating to prevent corrosion. Additional information on welding processes can be found in Volume IV, *Quality Control and Assembly*, of this Handbook series.

Painting. Galvanized coatings require a special surface treatment before the application of most paints. When paint is applied to a fresh zinc coating without treatment, there is a reaction between the paint vehicle and the zinc causing a loss of adhesion and the separation of the paint film from the zinc coating. Untreated galvanized steel can be painted with a minimum of preparation if paints designed specifically for galvanized steels are used. Zinc-dust, latex-type, and cement-in-oil paints provide the best results, but other proprietary paints can also be used. Additional information on paints can be found in Chapter 26, "Coating Materials," of this volume.

ALUMINUM HOT DIPPING

Although zinc is well established as a coating material, aluminum is gaining in popularity. In normal or mild industrial environments, aluminum hot-dipped coatings provide good barrier protection against corrosion. The process of applying a hot-dipped aluminum coating to various substrates is also referred to as aluminizing. The coating process creates a strong metallurgical bond between the coating and steel base, and provides a uniform coating thickness and matte surface finish.

Two types of aluminum coatings are commercially available and can be applied using both batch and continuous processes. The specifications and standards for these coatings are given in ASTM Specification A 463. The first type of coating (Type 1) contains 5-10% silicon with the remainder being aluminum. These coatings are generally used in applications requiring good corrosion and heat resistance, and can generally be used to temperatures as high as 1250° F (675° C). The other coating (Type 2) is pure aluminum and is used in applications requiring good corrosion resistance. In most corrosive environments, aluminum coatings perform better than zinc coatings; research has shown that Type 2 aluminum coatings will outlast zinc coatings 5:1 in bold atmospheric exposure.

An aluminum-zinc alloy coating has also been developed by Bethlehem Steel to combine the sacrificial protection of zinc with the barrier protection of aluminum. This alloy coating contains 55% aluminum, 43.4% zinc, and 1.6% silicon; silicon controls the growth of the intermetallic layer. Aluminum-zinc coatings are applied by continuous lines and are used for products such as metal roofing and siding, awnings, ductwork, water collection tanks, and highway culverts.

Batch Aluminizing

The batch aluminizing process is similar to the batch galvanizing process (see Fig. 23-5). The process consists of cleaning, heating, fluxing, and coating. Organic soils can be removed from the parts to be coated by alkaline cleaning and water rinsing. Steel parts are then descaled by abrasive blasting or acid pickling, followed by rinsing and drying. Gray or malleable iron parts are cleaned in molten salt to remove carbon smut.[7] Additional information on metal cleaning can be found in Chapter 18, "Cleaning," of this volume.

Fluxing dissolves any oxide film left on the part surface after cleaning and is carried out in either molten-salt baths or aqueous solutions prior to hot dipping. The salt bath is usually maintained at approximately 1100° F (595° C), and the parts are immersed in the bath from 30 seconds to several minutes. Immediately after fluxing, the parts are immersed in the molten-aluminum bath at 1300° F (705° C). A layer of molten salt floats on top of the molten aluminum to prevent oxidation.

Temperature control of the molten aluminum is important in order to control the coating thickness. Temperatures that are too high increase the thickness of the intermetallic layer and decrease the thickness of the pure aluminum coating. Immersion time must also be closely controlled. Coating thickness is also affected by substrate composition and aluminum coating bath composition.

Parts that can be aluminum hot dipped include fasteners, poppet valves, and blades and nozzle vanes for gas turbines. Parts with small threads are not considered practical, nor are parts that may nest during the coating process. Continuous, uniform coatings are difficult to produce on items with complicated configurations involving blind holes and reentrant angles in which air can be entrapped.[8]

Continuous Aluminizing

The aluminum dip coating of continuous steel strip, sheet, or tubing is produced on continuous in-line equipment similar to that used for galvanizing. The process consists of surface preparation, heat treatment, and immersion coating with aluminum. Surface preparation may take place in an oxidizing furnace or in a nonoxidizing preheater; if an oxidizing furnace is used, the surface oxides are reduced in a suitable atmosphere. The immersion time, the temperature of the steel before and after immersion, and the temperature of the molten aluminum must be controlled to prevent the formation of an excess of iron-aluminum interfacial alloy.

Aluminum coating on steels is generally specified in weight per surface area. For steels with the pure aluminum coating, the coating weight is approximately 0.60 oz/ft^2 (183 g/m^2), or 1.3 mils (33 μm) thick per side. Typical applications for products made from these steels include roofing, siding, and roof decks for commercial and industrial buildings; sheathing for air conditioners and other equipment located outdoors; metal insulation lagging; building panels and rolling doors; and other applications that require resistance to atmospheric corrosion.

Aluminum-silicon coatings are available in weights of either 0.25 oz/ft^2 (76 g/m^2) or 0.40 oz/ft^2 (122 g/m^2). Typical applications for products made from steels with this coating include reflectors and housings for industrial heater panels, interior panels and heat exchangers for residential furnaces, microwave ovens, automotive and truck exhaust systems, and heat shields for catalytic converters.

Forming. Moderate forming, drawing, and spinning operations can be performed with aluminum-coated steel without causing the coating to flake or peel. The aluminum-coated steel is generally available in several different quality designations, with the selection of the steel based on forming requirements and end-use requirements. The same die metal, draw rings, and generally the same lubricants may be used in drawing aluminum-

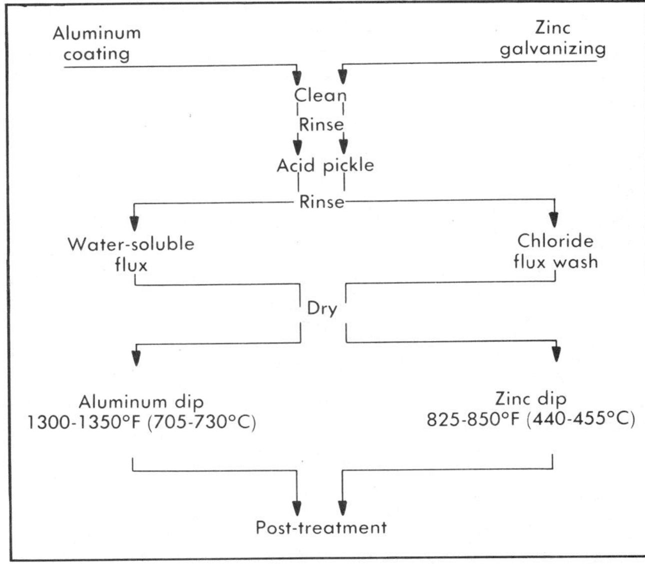

Fig. 23-5 Comparison of aluminum hot-dip coating with galvanizing.

coated steels as in drawing uncoated steels of the same quality designation, such as commercial quality, drawing quality, special killed drawing quality, or high-strength steel.

An aluminum-coated steel can generally be cold bent 180° over a radius equal to its own thickness; larger radii are recommended when parts will be operated in high-temperature environments, and especially when they are cooled and then heated cyclically. Since the aluminum oxide is abrasive, cutting edges and tooling surfaces should be inspected more frequently than with uncoated steel. Additional information on forming processes and practices can be found in Volume II, *Forming*, of this Handbook series.

Painting. In most applications, the appearance of the aluminum surface is quite satisfactory and requires no paint for corrosion resistance; however, aluminum-coated steel is relatively easy to paint. Pretreatment is nonessential, but it will provide better adherence and longer paint life for products requiring such added benefits. Chromate conversion coatings give the best results for pretreatment. The best type of paint coating consists of an epoxy primer and a baked-on alkyd enamel coat. Other satisfactory coatings include solvent-type plastics paints, zinc-dust zinc oxide paints, and cement-based paints. Additional information on conversion coatings can be found in Chapter 19, "Conversion Coatings and Anodizing," and on paint coatings in Chapter 26, "Coating Materials," of this volume.

Joining. Aluminum-coated steel can be joined by any of the welding processes suitable for uncoated steel. Good joint efficiency can be expected, provided the welding procedures are adapted to the special properties of the coating material. Resistant spot, projection, and seam welding processes are especially suited for aluminum-coated steels. The mechanical strength of the welds can be as good or better than for uncoated steel. Fusion processes such as gas tungsten arc, gas metal arc, shielded metal arc, and flux-cored arc welding require proper procedures and techniques because of the following reasons:

- Aluminum oxide formation may cause weld porosity and slag entrapment.
- Aluminum entering the weld metal as an alloy may decrease ductility.
- Loss of aluminum coating along the weld joint by some processes may require postweld operations to restore corrosion resistance.

The use of stainless steel electrodes or filler rod in any of the arc welding processes is an excellent way to ensure corrosion resistance without subsequent treatment of the weld area. Oxyacetylene processes are not normally recommended for aluminum-coated steel because of the damage to the coating during the joining process.

Aluminum-coated steels can also be joined with mechanical fasteners such as bolts, screws, rivets, or clips. The fastener selected should have heat and corrosion resistance at least equal to that of the aluminum coating to avoid failure in service or unsightly discoloration and bleeding onto the steel surface. Welding and joining processes and practices are discussed further in Volume IV, *Quality Control and Assembly*, of this Handbook series.

HOT-DIP TINNING (TINPLATING)

In hot-dip tinning or tinplating, a thin coating of molten tin is applied to the surface of iron or steel substrates. The tin provides a nontoxic, protective, or decorative coating for food handling, packaging, or dairy equipment; facilitates the soldering of a variety of components used in electronic and electrical equipment; and assists in bonding another metal to the basis metal.[9] The production and consumption of tinplate for food cans in the U.S. has dropped dramatically over the years due to the more popular electrolytic processes. A modern hot-tinning line may produce approximately 1100 tons (1000 t) of hot-dipped tinplate per month, whereas an electrolytic line may produce that much in a day. Changes in effluent standards for electroplating wastes, however, may revive hot dipping to some extent.[10]

Coating Characteristics and Specifications

Three grades of tin are used in hot-dip tinning: standard tin containing 99% tin, refined tin with over 99.8% tin, and high-purity tin with over 99.9% tin. The amount of impurities that can be present in these grades is covered by ASTM Specification B 339; lead is the most common impurity and may cause the coating to be spangled. Military Specification MIL-T-10727-B covers pretreatment processing of base metals and the coating characteristics of hot-dip tinned coatings. The majority of the hot-dip tinning applications use grade A tin (99.8% tin). Standard tin is used in applications that only require soldering and do not come in contact with food. Coating weights range from 0.05 to 0.15 oz/ft^2 (15 to 45 g/m^2).

The term *hot tin dip* has also been used generically when a hot solder dip coating is applied to a workpiece; the coating generally consists of 60% tin and 40% lead. While this type of coating offers corrosion protection similar to pure tin, it is used primarily on electronic parts that must be soldered after assembly and when a corrosive-type flux cannot be tolerated.

Substrates[11]

Low-carbon steels, containing less than 0.2% carbon, are intrinsically well suited for hot-dip tinning. Medium-to-high carbon steels, 0.3 to 1.0% carbon, are also used, but may require greater care during pickling. The higher alloy steels, especially those with high-chromium content (such as 18-8 stainless), are difficult to hot-dip tin satisfactorily, and special procedures may have to be used. To ensure good tinning characteristics, steel mill products should be purchased to a specification that includes suitability for hot-dip tin coating.

Cast irons having chemical compositions within the ranges given in Table 23-7 may also be successfully hot-dip tinned. Annealed irons are less suitable than the as-cast-type irons;

TABLE 23-7
Chemical Composition Ranges for Cast Irons Suitable for Hot-Dip Tinning

Element	Composition, %
Total carbon	3.2-3.5
Silicon	1.7-2.7
Manganese	0.5-0.8
Sulfur	0.05-0.12
Phosphorus	up to 1.3

(*American Society for Metals*)

substrate preparation and methods are different than those for steel. Refer to Chapter 18, "Cleaning," for detailed information.

Process Description[12]

In a typical hot-tinning process, individual steel sheets are pickled with hydrochloric acid or sulfuric acid, possibly with cathodic polarization, and then rinsed. On entering the bath of molten tin, the steel passes through a zinc chloride/ammonium chloride, aqueous flux cover that floats on the bath. It then passes through the molten tin, which is held at about 450° F (230° C) in an indirectly heated pot, and leaves the bath through the flux cover. Thicker coatings may be achieved by inserting the steel in a second pot of tin. The tin thickness is regulated by passage through wipers, where about 90% of the excess molten tin is wiped off the part and drops back into the bath. The tin coating thickness is related to a number of factors including contour of the tinning wipers, the wiper pressure, speed, bath temperature, and surface condition of the steel sheet before tinning. The drainage of the coating during solidification may result in a "list" or "drip" edge on the sheet, which must normally be removed by trimming.

LEAD ALLOY HOT DIPPING (TERNE COATING)

The term *terneplate* or *terne coating* is used to describe a hot-dipped coating of an alloy of lead and tin on a steel substrate. Since lead alone does not adhere to steel, a small amount of tin in the molten bath forms a thin layer of $FeSn_2$ compound, thus providing an adherent base for the terne coating. Tin also assists in the formation of a relatively continuous coating and adds to the solderability of the coating. Tin content is usually in the range of 2 to 15%. Terne coatings conform to ASTM Specification A 308.

Terne coatings are the most economical of the metallic coatings applied to steel; and, unlike the other coatings, the coating adherence does not deteriorate as the coating weight increases. However, since the terne coating is not sacrificial, it will show "pinhole" corrosion more rapidly than zinc or aluminum-coated steel in accelerated tests and in harsh atmospheric exposure; but as the steel rusts, iron oxide fills the pinholes and resists further corrosion until the terne in the area is consumed.

Terne coatings are usually applied over carbon steel substrates having commercial quality, drawing quality, or drawing quality special-killed steel designations. Terne-coated stainless steel is also available.

Batch Lead Alloy Hot Dipping

In hot dipping formed and fabricated parts with a lead alloy, the steel is first cleaned of oils and greases using the appropriate solvents or detergents, followed by pickling in a sulfuric acid or hydrochloric acid solution. Before hot dipping, the parts are immersed in a zinc chloride or zinc-ammonium chloride flux. The lead alloy bath is usually maintained at 620 to 675° F (325 to 390° C). The bath temperature and immersion time are determined by part weight. After the parts are withdrawn from the bath, they may be placed in a centrifuge or shaken to remove any excess coating. The part may then be quenched in water to solidify the coating.

Hot-dipped lead alloy coatings on fabricated parts vary from 0.2 to 0.6 mil (5 to 15 μm) thick, depending on the application; thicker coatings are required for parts exposed to salt or aggressive chemicals that may remove the coating. Coating thickness is controlled by the withdrawal rate, base metal surface roughness, and the amount of centrifuging or shaking. Typical parts that are batch coated include bolts, nuts, washers, plates, brackets, and other small fixtures used in industry.[13]

Continuous Lead Alloy Hot Dipping

The production of terneplate is essentially the same process as that of hot-dipped tinplate.[14] Two general classifications of terneplate are produced: short ternes and long ternes. Short ternes are used only for roofing applications and are always painted to provide additional corrosion protection. Short ternes are heavier than long ternes and are produced from sheet material rather than coils. Long terne is produced on a continuous coating line at speeds ranging from 30 to 150 fpm (9 to 45 m/min), with coating weights averaging from 0.25 to 0.55 oz/ft^2 (76 to 168 g/m^2). Coils are processed in widths from 18 to 54" (460 to 1370 mm) and in thicknesses from 0.017 to 0.080" (0.43 to 2.03 mm). Sheared sheets are available in lengths of 30 to 156" (760 to 3960 mm).

Terne-coated sheet products possess good corrosion-resistant properties as well as good formability, paintability, weldability, and solderability, permitting them to be used in the transportation, electrical appliance, and building industries. Typical applications include gasoline tanks, radiator parts, gaskets, valve covers, and air filter containers; chassis for radios, television sets, and tape recorders; fire door and screen construction; and cabinet, file drawer, and track manufacturing.

Forming. The lead-tin coating is very ductile and adherent, enabling it to be formed and drawn to the limit of the steel substrate. The same procedures and equipment can be used in drawing long terne steel as in drawing cold rolled steels of the same quality. Drawing lubricants recommended for cold rolled steels are also used with long ternes; only those lubricants that are corrosive need to be removed immediately after drawing. Long terne steels can be roll or brake formed with the same equipment, rolls, and guides used in forming mild steel. Dies and forming rolls should be kept highly polished to prevent coating pickup. Forming processes and practices are discussed further in Volume II, *Forming*, of this Handbook series.

Painting. The terne coating has good paintability, requiring no special surface treatments or primers. However, the surface should receive normal cleaning with a mild alkaline cleaner or vapor degreaser. Painted parts should be handled carefully because the alloy coating is soft.

Joining. Parts made of terne steel are readily welded by high-production spot and seam welding processes. Electrode force and weld current should be slightly higher for coated steels than for uncoated steels, and weld time should be longer so that the coating is forced from the weld interface and a good steel-to-steel weld nugget is formed. Electrode-wheel cooling is recommended in high-production seam welding. The lead-tin coating is also readily solderable with noncorrosive fluxes. Since the alloy coating is a solder itself, only small amounts of additional solder are required to form a good bond between equally receptive metals. Lead vapor, which is extremely toxic, is produced in quantities above safe limits when terne steels are arc welded. High-volume ventilation systems or approved respirators are required to meet the air-to-lead concentration standards established by the Occupational Safety and Health Act (OSHA). Additional information on welding processes and practices can be found in Volume IV, *Quality Control and Assembly*, of this Handbook series.

CHAPTER 23

REFERENCES

References

1. Lennard G. Kruger, ed., *Metals Handbook*, 9th ed., vol. 5 (Metals Park, OH: American Society for Metals, 1982), pp. 509-531.
2. American Society For Testing and Materials, *Definition of Terms Relating to Porcelain Enameling*, ASTM Bulletin C-286, (Philadelphia, PA).
3. Kruger, *op. cit.*, p. 517.
4. Kruger, *op. cit.*, p. 517.
5. Kruger, *op. cit.*, p. 520.
6. Kruger, *op. cit.*, p. 327.
7. Kruger, *op. cit.*, p. 337.
8. Kruger, *op. cit.*, p. 338.
9. Daniel Maykuth, "Hot Dip Tin Coating of Steel and Cast Iron," *Metals Handbook*, 9th ed., vol. 5 (Metals Park, OH: American Society for Metals, 1982), p. 351.
10. Martin Grayson, ed., *Encyclopedia of Composite Materials and Components* (NY: John Wiley & Sons, Inc., 1983), p. 804.
11. Maykuth, *loc. cit.*
12. International Tin Research Institute, *Guide to Tinplate*, ITRI Publication No. 622 (Columbus, OH: 1983), p. 15.
13. Kruger, *op. cit.*, p. 358.
14. American Welding Society, Inc., *Soldering Manual*, 2nd ed., (Miami, FL: 1978).

VAPOR DEPOSITION PROCESSES

CHEMICAL VAPOR DEPOSITION

Chemical vapor deposition (CVD) is a distinctly different coating process than vacuum evaporation, ion plating, or sputtering. A heat-activated process, CVD relies on the reaction of gaseous chemical compounds with suitably heated and prepared substrates.

The primary reactive vapor can be either a metal halide (chloride, bromide, iodide, or fluoride) or a metal carbonyl, $M(CO)_x$, although some hydrides and organometallic compounds are used. In general, deposition of a pure metal from the halide can occur either by hydrogen reduction at a given temperature or by direct pyrolytic decomposition at a higher temperature. The reduction method is illustrated in Fig. 24-1.

Whenever carbides, nitrides, or borides are the desired CVD products, the metal halide vapor is accompanied by an additional reactive species such as methane, nitrogen, or boron trichloride respectively. The addition of a reactive species lowers the free energy of the reaction products so that the compound is formed in preference to the pure metal. However, manipulation of reactor pressure, temperature, and/or reactant composition can also influence the deposition products.

ADVANTAGES AND LIMITATIONS

The versatility of the CVD process has been demonstrated by the wide variety of geometrical configurations into which it forms refractory metals, alloys, and refractory compounds. Users of CVD maintain that the process compares favorably with other methods of plating and forming in its ability to produce high-density materials, high-purity materials, high-strength materials, and complex shapes. The extraordinary throwing power of the CVD technique is such that very intricate coatings may be formed on substrates of complex geometry.

Materials in excess of 99.9% of theoretical density are commonly produced by this process. Thin structures made by CVD are used for applications in which vacuum tightness is required; in some cases, such structures could not be made by any other process. For example, as little as 0.002" (50 μm) of a vapor-deposited material is commonly used to seal a vacuum of 10^{-8} torr (13.3 x 10^{-6} Pa), even at temperatures of 3630° F (2000° C).

Engineering materials can be deposited at considerably higher purity levels than is possible with other manufacturing means because of CVD's ability to purify the precursors. The strength of CVD materials is dependent on crystal structure and size, purity, density, and internal stress, much the same as that of materials formed by other processes. Generally the CVD materials are more ductile because of their purity and can have comparable or higher mechanical strength than wrought material, providing that they are dense and have a fine equiaxed grain structure.

Not all deposited materials exhibit high strength; certain materials, such as pure tantalum and pure columbium, which are weak in their fully annealed condition, have low yield strengths and high ductility in the vapor-deposited condition. However, when pure rhenium, pure columbium, tantalum alloys, and most rhenium alloys are formed by CVD, they have higher strengths.

Metals that are readily electroplated or electroformed commercially are generally not well suited to chemical vapor deposition. In the case of the former, CVD processes are precluded owing to the unsuitable or hazardous properties of potential reactants. For example, electroplated metals such as lead, tin, and copper, and alloys such as brass are available as cyanides and fluoroborates and not as CVD-compatible halide salts.[1] With electroforming, in general, the electroformed part is the product desired and not the coating-substrate combination. However, molds for forming plastics that are vapor formed from nickel carbonyl are available for commercial use when their longer life justifies the higher costs.

CHAPTER CONTENTS:

Contributors of sections of this chapter are: ***C. H. Alexander****, Consultant, Technical Enterprises, Inc.;* ***Charles A. Baer****, President, Charles A. Baer Associates;* ***Clark Bergman****, Staff Scientist, Multi-Arc Vacuum Systems, Inc.;* ***Russell J. Hill****, Vice President, Technology, Temescal, Div. of the BOC Group, Inc.;* ***Frederick C. Hornbeck****, Vice President, Advanced Technology, Southwall Technologies, Inc.;* ***Dr. B. L. Kindberg****, Senior Technical Service Representative, American Hoechst Corp.;* ***Thomas C. Leister****, Senior Project Engineer, SPS Technologies;* ***Lee H. Miller****, President, Elitine Corp.;* ***Dennis R. Nichols****, Manager, Coating Applications Laboratory, Thin Film Technology Div., Varian Associates;* ***Dr. Stefan Reineck****, Product Manager of Sputtering Equipment, Leybold-Heraeus Technologies;* ***George E. Stephens****, President, General Engineering Services USA, Ltd.;* ***Kenneth E. Steube****, Unit Chief, Ivadizing Technology, McDonnell Aircraft Co.;* ***Richard P. Vento****, Manager of CVD Coatings, Bernex Div., Sylvester & Co.*

Reviewers of sections of this chapter are: ***C. H. Alexander****, Consultant, Technical Enterprises, Inc.;* ***Charles A. Baer****, President, Charles A. Baer Associates;* ***Clark Bergman****, Staff Scientist, Multi-Arc Vacuum Systems, Inc.;*

CHAPTER 24

CHEMICAL VAPOR DEPOSITION

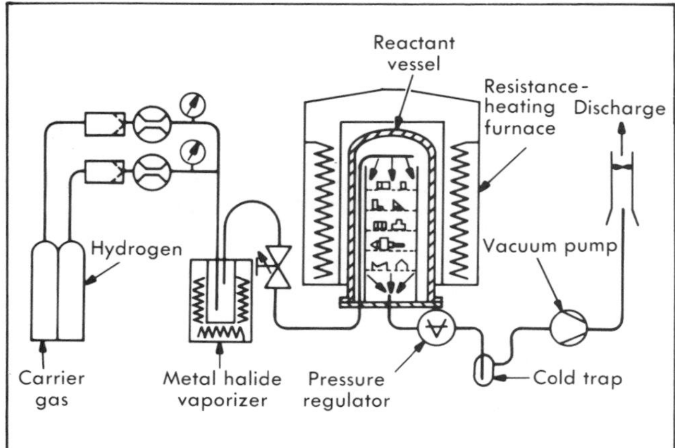

Fig. 24-1 Typical components used in the reduction method of chemical vapor deposition. (*Sylvester-Bernex Corp.*)

A closed system is usually required when depositing the materials because of the corrosive, toxic, or moisture-sensitive characteristics of most reactants. Within the control volume, low pressures reduce the likelihood of homogeneous (gas-phase) nucleation and induce conditions favorable for heterogeneous (gas-to-solid) nucleation. Moreover, closed systems tend to normalize gas compositions while preventing the formation of undesirable species. Material utilization can be low, and reactant cost may be high, which may increase the overall cost of the process. In addition, the high temperature of the reaction can cause distortion of the parts being coated.

COATINGS DEPOSITED

A variety of pure metals or carbides, nitrides, borides, silicides, and oxides of metals can be deposited by CVD processes.[2] Coating thicknesses are usually in the range of 0.0002 to 0.050″ (5 to 1270 μm). The codeposition of titanium carbide (TiC) and titanium nitride (TiN) to form titanium carbonitride (TiC,N) by the Bernex Moderate Temperature Chemical Vapor Deposition (MTCVD) process is rapidly expanding in use.[3] Chromium deposition on steel substrates, resulting in carbides of chromium and iron, $(CrFe)_7C_3$, provides exceptional resistance to corrosion and cold welding.[4]

The selection of the base material and the coating depends on their mutual interaction and on the operating conditions in the field. Moreover, these compounds have shown very good oxidation resistance in hot forging applications. Tables 24-1 and 24-2 list the refractory compounds and metals that are commonly deposited by CVD, along with the various operating conditions. Operating conditions for additional refractory compounds and special materials are given in Table 24-3.

APPLICATIONS

The major applications of CVD take advantage of the unique characteristics of the process, such as good throwing power, the ability to deposit refractory materials at temperatures far below the normal ceramic processing temperatures, and the capability of producing materials of exceptionally high purity. Typical uses for the CVD process include the fabrication or coating of tubing, tungsten boride crucibles, decorative trim, and dinnerware.

A substantial field of CVD application exists for the hard coating of tools fabricated from high-speed steels, air-hardenable tool and stainless steels, and the cemented carbides. In particular, both titanium carbide (TiC) and titanium nitride (TiN) are regularly used to enhance the life of cemented carbide cutting tools and high-speed steel punches and dies. In the fastener industry, the vertical integration of high-temperature, low-pressure CVD of TiC and TiN into the tool manufacturing sequence has shown the greatest benefit.[15] Similarly, the incorporation of TiC and TiN in the carbide cutting tool industry has proven to be of significant commercial value. Table 24-4 compares the life of uncoated tools to that of coated tools. Tools coated by CVD have a significantly longer life span.

More exotic depositions include the silicon and silicide coating of tungsten, and the platinum and rhenium coating of electronic hardware. Emitting surfaces and structures for cathodes as well as structural hardware for grids and anodes can be created by this process. Chemical vapor deposition is also the basis for several steps in the manufacture of integrated circuits for microelectronic applications and for solar cell fabrication.[16]

Chemical vapor deposition is used for depositing a refractory metal on jet aircraft turbine blades and for joining tungsten to other refractories at low temperature. Many applications are found in the nuclear power field, such as the coating of nuclear fuel particles for fission product retention or for matrix compatibility. Nuclear fuel can be contained by direct deposition of refractory metals and/or pyrolytic carbon on the surface of the fuel itself.[17] Ceramic-to-metal seals have been made by the deposition of metals, ceramics, or both; and there are many applications for composite materials that use deposited high-density oxides and metals. Tungsten and/or carbon fibers have been encapsulated with both boron carbide and silicon carbide by CVD processes.[18] The filaments have been subsequently used as reinforced composites in structural components.

The protection of chemical equipment is an obvious application of CVD. Steel with less than a 0.002″ (50 μm) deposited coating of tantalum will frequently outperform stainless steel, or Monel or Inconel (the International Nickel Co., Inc.) at a competitive cost. Cladding for abrasion resistance also warrants serious consideration. Many hard refractory-metal compounds cannot readily be formed or finished by mechanical means. On the other hand, hard coatings such as silicon carbide, tungsten carbide, tantalum carbide, or titanium diboride are readily

__Reviewers, cont.: John Blocher, Jr.__, Consultant; __Melvin F. Browning__, Research Leader, Coatings and Electronic Materials Section, Battelle-Columbus Div.; __William A. Bryant__, Staff Engineer, P. M. McKenna Laboratory, Kennametal, Inc.; __Philip J. Clough__, President, Gorham International, Inc.; __Robert B. Goodman__, Vice President, Gomar Manufacturing Co., Inc.; __Don Griffin__, General Manager, Manufacturing, Southwall Technologies, Inc.; __Jack Griswold__, Research & Development Engineer, Varian Associates; __Russell J. Hill__, Vice President, Technology, Temescal, Div. of the BOC Group, Inc.; __Frederick C. Hornbeck__, Vice President, Advanced Technology, Southwall Technologies, Inc.; __Dr. Richard W. Kidd__, Principal Research Scientist, Coatings and Electronic Materials Section, Battelle-Columbus Div.; __Dr. B. L. Kindberg__, Senior Technical Service Representative, American Hoechst Corp.; __John Laurilliard__, Chief Chemical Engineer, Aerospace and Industrial Products Div., SPS Technologies; __Lee H. Miller__, President, Elitine Corp.; __Walter Miller__, Applications Specialist, Stokes Div., Pennwalt Corp.;

applied by chemical vapor deposition to resist abrasion and corrosion on devices such as pump impellers, valves, and nozzles. Superficial coatings of such materials may be particularly adherent and of high density, and they have been measured at hardnesses above 2500 Vickers hardness using a 200 g weight.

TABLE 24-1
Conditions for Chemical Vapor Deposition of Refractory Compounds[5]

Deposit	Metal Reactants	Nonmetal Reactants	Other Reactants	Temperature, °F (°C)	Pressure, torr (kPa)	Deposition Rate,* mil/min (μm/min)
Al_2O_3	$AlCl_3$	CO_2, H_2O and O_2 combined	H_2	1650-1830 (900-1000)	76-760 (10-100)	0.01-0.05 (0.25-1.3)
SiO_2	$SiCl_4$	CO_2, H_2O and O_2 combined	H_2	1470-2010 (800-1100)	76-760 (10-100)	0.01-0.05 (0.25-1.3)
TiC**	$TiCl_4$	Volatile hydrocarbon	H_2	2550-3270 (1400-1800)	76-760 (10-100)	0.005-0.1 (0.13-2.5)
SiC**	CH_3SiCl_3	---	---	3630-4350 (2000-2400)	76-380 (10-50)	0.005-0.1 (0.13-2.5)
SiC**	CH_3SiCl_3	---	H_2	2550-3270 (1400-1800)	7.6-760 (10-100)	0.05-0.7 (1.3-18)
W_2C**	$W(CO)_6$	---	---	570-750 (300-400)	0.01-5 (0.001-.67)	0.01-0.05 (0.25-1.3)
WC**	WCl_6	Volatile hydrocarbon	H_2	1470-1830 (800-1000)	1-10 (0.13-1.3)	0.1-0.5 (2.5-13)
ZrN**	$ZrCl_4$	N_2	H_2	2730-4530 (1500-2500)	76-380 (10-50)	0.005-0.02 (0.13-0.50)
Si_3N_4	$SiCl_4$	N_2, NH_3	H_2	1830-2730 (1000-1500)	76-760 (10-100)	0.005-0.05 (0.13-0.50)
(AlB_x)**	$AlCl_3$	BCl_3	H_2	1830-2190 (1000-1200)	76-760 (10-100)	0.2-0.3 (5-7.5)
TiB_2	$Ti(BH_4)_3$	---	---	570 (300)	0.1-10 (0.013-1.3)	0.2-0.3 (5-7.5)
(MoB_x)**	$MoCl_5$	BCl_3	H_2	2190-2730 (1200-1500)	1-10 (0.13-1.3)	0.2-0.3 (5-7.5)
$MoSi_2$	$MoCl_5$	$SiCl_4$	H_2	1650-2730 (900-1500)	1-10 (0.13-1.3)	0.01-0.1 (0.25-2.5)
$(ZrSi_x)$**	$ZrCl_4$	$SiCl_4$	H_2	2190-2910 (1200-1600)	1-10 (0.13-1.3)	0.005-0.05 (0.13-1.3)

* 1 mil = 0.001″
** Exact stoichiometry of deposit varies with operating conditions.

***Reviewers, cont.: Donald E. Muehlberger**, Program Manager, Ivadizer Technology, McDonnell Douglas Corp.; **John R. Newton**, New Products and Applications Development Manager, Films Division, ICI Americas, Inc.; **Dennis R. Nichols**, Manager, Coating Applications Laboratory, Thin Film Technology Div., Varian Associates; **Dr. Stefan Reineck**, Product Manager, Sputtering Equipment, Leybold-Heraeus Technologies; **Joseph J. Rizzo**, Marketing Manager, Bee Chemical Co.; **Stan R. Roth**, Development Engineer, Eastman Kodak Co.; **Paul Rothman**, Senior Chemist, Technical Coatings Co.; **Jan Spanjer**, Manufacturing Manager, Jonergin, Inc.; **Robert L. Sproat**, Vice President of Manufacturing and Engineering, SPS Technologies; **George E. Stephens**, President, General Engineering Services USA, Ltd.; **Richard P. Vento**, Manager of CVD Coatings, Bernex Div., Sylvester & Co.; **Robert Wade**, Project Leader, Coatings Development, General Electric; **Richard P. Welty**, Research & Development Manager, Vac-Tech Systems, Inc.*

TABLE 24-2
Conditions for Chemical Vapor Deposition of Refractory Metals[6]

Deposit	Metal Reactants	Other Reactants	Temperature, °F (°C)	Pressure, torr (kPa)	Deposition Rate,* mil/min (μm/min)
W	WF_6	H_2	480-2200 (250-1200)	1-760 (0.13-100)	0.005-2.0 (0.13-50)
	WCl_6	H_2	1560-2550 (850-1400)	1-20 (0.13-2.7)	0.01-1.5 (0.25-38)
	WCl_6	---	2550-3630 (1400-2000)	1-20 (0.13-2.7)	0.1-2.0 (2.5-50)
	$W(CO)_6$	H_2	350-1110 (180-600)	0.1-1 (0.01-0.13)	0.005-0.05 (0.13-1.3)
Mo	MoF_6	H_2	1290-2200 (700-1200)	20-350 (2.7-47)	0.05-1.2 (1.3-30)
	$MoCl_5$	H_2	1200-2200 (650-1200)	1-20 (0.13-2.7)	0.05-0.8 (1.3-20)
	$MoCl_5$	---	2280-2910 (1250-1600)	10-20 (1.3-2.7)	0.1-0.7 (2.5-18)
	$Mo(CO)_6$	H_2	300-1110 (150-600)	0.1-1 (0.01-0.13)	0.005-0.05 (0.13-1.3)
Re	ReF_6	H_2	750-2550 (400-1400)	1-100 (0.13-13)	0.05-0.6 (1.3-15)
	$ReCl_5$	---	1470-2200 (800-1200)	1-200 (0.13-27)	0.05-0.6 (1.3-15)
Nb	$NbCl_5$	H_2	1470-2200 (800-1200)	1-760 (0.13-100)	0.003-1.0 (0.08-25)
	$NbCl_5$	---	3415 (1880)	1-20 (0.13-2.7)	0.1 (2.5)
	$NbBr_5$	H_2	1470-2200 (800-1200)	1-760 (0.13-100)	0.003-1.0 (0.08-25)
Ta	$TaCl_5$	H_2	1470-2200 (800-1200)	1-760 (0.13-100)	0.003-1.0 (0.08-25)
	$TaCL_5$	---	3630 (2000)	1-20 (0.13-2.7)	0.1 (2.5)
Zr	ZrI_4	---	1470-2910 (1200-1600)	1-20 (0.13-2.7)	0.05-0.1 (1.3-2.5)
Hf	HfI_4	---	2550-3630 (1400-2000)	1-20 (0.13-2.7)	0.05-0.1 (1.3-2.5)
Ni	$Ni(CO)_4$	---	300-840 (150-450)	100-760 (27-100)	0.1-1.5 (2.5-38)
Fe	$Fe(CO)_5$	---	300-840 (150-450)	100-760 (27-100)	0.1-2.0 (2.5-50)
V	VI_2	---	1830-2200 (1000-1200)	1-20 (0.13-2.7)	0.05-0.1 (1.3-2.5)
Cr	CrI_3	---	1830-2200 (1000-1200)	1-20 (0.13-2.7)	0.05-0.1 (1.3-2.5)
Ti	TiI_4	---	1830-2550 (1000-1400)	1-20 (0.13-2.7)	0.05-0.1 (1.3-2.5)

* Deposits made on mandrels of comparable geometry; 1 mil = 0.001″.

TABLE 24-3
Operating Conditions for Depositing Refractory Compounds and Some Special Materials

Deposit	Metal Reactants	Nonmetal Reactants	Other Reactants	Temperature, °F (°C)	Pressure, torr (kPa)	Deposition Rate,* mil/min (μm/min)	Chapter 24 Reference Number
TiN	$TiCl_4$	N_2	H_2	1600-1710 (870-930)	230 (30)	0.0008-0.009 (0.020-0.023)	**
TiC	$TiCl_4$	CH_4	H_2	1800-1960 (980-1070)	76 (10)	0.0007 (0.018)	**
Ti (C,N)	$TiCl_4$	C-N compounds	H_2	1290-1650 (700-900)	150-600 (20-80)	0.0001-0.0002 (0.003-0.005)	7
NbC	$NbCl_5$	CH_4, H_2	Ar, HCl	1700-2280 (930-1250)	---	3.78 gm/hr	8
NbC	$NbCl_5$	H_2	Ar, HCl	3270-3810 (1800-2100)	---	2.34 gm/hr	8
NbC	$NbBr_5$	CH_4, H_2	Ar, HBr	1700-2280 (925-1250)	---	2.14 gm/hr	8
ZrC	$ZrCl_4$	CH_4, H_2	Ar	1710-2550 (930-1400)	---	1.3-7.2 gm/hr	8
B	---	Diborane	Ar, H_2	750-1650 (400-900)	760 (100)	---	9
PBN	BCl_3	NH_3	---	3270 (1800)	450 (60)	0.1-0.3 gm/hr	10
TiB_2	$TiCl_4$	BCl_3	H_2	1290-1830 (700-1000)	15-25 (2-3.3)	---	11
B_4C	BCl_3	C_3H_6	H_2	1650 (900)	15-25 (2-3.3)	---	12
PyC	---	C_2H_2	---	2010-2190 (1100-1200)	---	0.0001 (0.003)	13
FeB_2	$FeBr_2$	BBr	---	2460 (1350)	0.04 (0.005)	0.4 g/hr	14

(*Compiled by Sylvester-Bernex*)

* 1 mil = 0.001″
** Data furnished by Sylvester-Bernex & Co. of Beachwood, OH.

TABLE 24-4
Comparison of CVD Coated Tools With Uncoated Tools

Part	AISI Grade	Coating	Application	Pieces Machined or Operations Before Resharpening: Uncoated	Coated
Trim die	M2	TiN	Cold heading stainless steel hex-head bolts	10,000	40,000
Punch	CPMT15*	TiN	Cold heading AISI 8630 steel sockets	20,000	40,000
Class C hob	M3-2	TiN	Hobbing AISI 4630 steel gears	1500	4500

(continued)

TABLE 24-4—*Continued*

Part	AISI Grade	Coating	Application	Pieces Machined or Operations Before Resharpening Uncoated	Coated
Thread roll die	D2	TiC	Rolling threads on low-carbon steel bolts	500,000	2,000,000
1/2 in. (13 mm) pipe tap	M2	TiN	Tapping holes in gray iron	3000	9000
5/16 in. (8 mm) drill	---	TiN	Drilling low-carbon steel	1000	4000
Form tool	T15	TiC	Screw machine application	4950	23,000
Cutoff tool	M2	TiN	Cutting low-carbon steel	150	1000
Wiping ring	D2	TiN	Forming transmission parts	5000	40,000
Piercing punch	---	TiC/TiN	Piercing nuts	25,000	175,000
Saw	---	TiC/TiN	Slotting screws	2 hours	10-12 hours
First-blow upsetting punches and dies	---	TiC/TiN		10,000	100,000
6 mm taps	---	TiC/TiN	Metalcutting	18,000	60,000

(*Scientific Coatings, Inc.*)

* Trade name of Crucible Specialty Metals Div., Colt Industries, Inc., for its powder metallurgy T15.

PROCESS FACTORS

Some important factors in CVD processes are the thermodynamic combinations of gas pressure, temperature, and velocity, and reactant composition; composition material; substrate cleanliness and temperature; the gas storage, flow, and recovery systems; scrubber systems for the by-products; and the composition and construction of the reaction vessel.

Substrates

When the process is used for plating or coating, the substrates must be free of greases and oils for good adhesion. This surface preparation is accomplished by ultrasonic and/or vapor degreasing before loading. Substrate surfaces are also frequently vapor honed to improve adhesion. During heatup, oxide layers may be removed by insitu scrubbing with reducing gases or mild acids. The temperature of the substrate promotes the reaction; ideally, it should be uniform and vary as little as possible because variations tend to affect the uniformity of structure, thickness, composition, and properties of the film.

Substrate materials must accept heat without deforming, since deposition temperatures may vary from 400 to 3600° F (200 to 2000° C). However, compensation for deformation during deposition is practical in many cases. Conducting heat into the substrate is simple if the workpiece is small and regular in shape, but difficult if it is large and complex. Small workpieces may have to be racked or rotated. The irregular geometry of some workpieces will call for individual designs for injecting and exhausting the gas for film uniformity; but some substrates can be tumbled, vibrated, or fluidized for uniform deposition, especially if the transport properties of the reactant species are low.

Because one of the driving potentials for deposition is the temperature of the substrate surface, and because cracks, crevices, and recesses in the surface frequently maintain a higher temperature due to reduced losses by radiation and conduction of the plating gases, deposits will be at least as easily formed on such higher temperature areas. However, there are limitations on the ability of the reactant gases to penetrate crevices, which tend to offset the effect of temperature on the deposition rate.

Gas Systems

A gas feed system whereby the active material can be injected into the deposition chamber and the exhaust products continuously removed is a necessary part of the CVD process. Several variations of gas feeding are used. The active compound can be prepared separately and stored in a separate container. A gaseous feed may then be introduced by expanding, distilling, or subliming this reactant from the container into the deposition chamber. Highly volatile materials may be distilled or sublimed by simply dropping the pressure by opening a valve to a lower pressure chamber; but less volatile materials must be heated and passed through a transfer system maintained at a high temperature to prevent condensation. Another method of transferring less volatile reactants is to use a carrier gas that may be passed over or percolated through the reactant storage vessel. This method provides a means of metering the flow of the reactant (at saturated pressure) without the problem of metering hot corrosive gases. Moreover, liquids can be metered-in directly, then flash vaporized in controlled portions to supply the reactant species.

Because most of the suitable reactants are metal halides that reduce to halide acids, inert materials must be used for reactant containment and transfer. In many cases, Pyrex and Vycor (Corning Glass Works) or quartz is used, but these materials are unsuitable for systems utilizing metal fluorides, particularly if moisture is present. When the reactants are fluorides, pure copper feed systems are used at modest feed temperatures. Stainless steels, such as 316 and 304, and graphite are acceptable materials whenever chlorides are present. The most suitable metallic materials for many applications, however, are nickel-based alloys such as Monel and Inconel.

Reducing gases such as hydrogen are easily fed into the system; but if the hydrogen must be heated, as is sometimes the case to prevent condensation of the metal chemical after mixing, precautions must be observed. These precautions include adequate vacuum tightness in the feed lines, proper ventilation, installation of hazardous gas detectors, and spark-proof fans to remove combustibles.

Reaction Temperature

The purpose of the CVD process is to apply a thin film on the substrate only. To prevent uneconomical coating buildup on tanks or reaction vessels, their temperatures should usually be maintained well below the catalyzing temperature. Also, because of the corrosive and catalyzing effects some tank construction materials have on the basic reaction, the tank material must be carefully chosen.

To decompose or reduce the metal compound, a transfer of heat energy is involved, and the substrate is usually held at a substantially higher temperature than any other part of the system. For this reason, the reaction chamber may present more of a high-temperature problem than any other part of the system. Most reactions are also conducted in an anhydrous and anaerobic environment, and frequently at subatmospheric pressures.

Typical deposition temperatures range from 1500 to 2200° F (800 to 1200° C). The selection of temperature and pressure is dictated by the properties desired in the deposit, the degree of thickness and compositional uniformity, and the deposition rate, which may vary from 0.000003 in./min (0.07 μm/min) at low temperatures to significantly more than 0.001 in./min (25 μm/min) at higher temperatures. The ability to provide uniformly thick coatings with refined grain is also infuenced by the deposition temperature. In both cases, low-temperature processing is frequently desirable, although a tradeoff with rate of deposition must often be made. Fewer CVD reactions are available for use at temperatures below 1500° F (800° C) than above. However, the temperature required for a given reaction can, in selected cases, be lowered by exposing the substrate to an electrical plasma in the gas phase during deposition. This procedure is referred to as plasma-assisted CVD or simply plasma CVD.

Reaction chambers for low deposition temperatures may be made of polymeric materials such as Teflon (Du Pont) or metal, such as stainless steel. Deposition is also frequently conducted in Pyrex or stainless steel chambers at moderate temperatures. High temperatures require the use of Vycor, quartz, high-temperature metals such as Inconel, refractories such as alumina, or graphite. The choice of the chamber material may be influenced by the means of providing heat to the substrate, as well as by the temperature or the chemical corrosion problem involved.

Simple part configurations such as tubing may be heated either by passing a current directly through the tubing or by placing a resistance heater inside the part. Irregular shapes may be heated by placing the entire deposition chamber in a furnace and permitting the deposition of the vapor over the chamber as well as over the part to be coated; but, more frequently, induction heating is used to selectively heat irregular parts. Induction coils may be placed within the chamber with gastight seals; in this case, metallic chamber materials may be used. It is frequently desirable to have the induction coils on the outside of the chamber; nonconducting chamber materials must then be used.

The type of induction heating to be used, particularly the frequency, is dictated by the geometry and certain physical properties of the substrate. Thick sections of highly conductive material are typically heated at frequencies of 10 kHz or less, whereas semiconductors, or extremely thin sections of conductive materials, are usually heated in the megahertz frequency range.

Reaction Pressure

Most CVD films can be improved in uniformity of thickness, grain size, composition, and properties by deposition at subatmospheric pressure. Special pumping system designs are necessary since the compound gases are not only hot, but frequently corrosive to ordinary vacuum pumps. One solution to these problems is to cool the gases and trap the corrosive compounds before they reach the pumps. Cooling equipment most often employs cold traps ranging from cool tap water to liquid nitrogen temperatures. For many applications, simple untrapped pumping systems are adequate. If pressures not lower than approximately 50 torr (6.7 kPa) are used, water-jet exhausters or water-ring pumps can be used for evacuation. More elaborate systems use Monel vacuum equipment. Satisfactory performance has been obtained from polyvinyl chloride exhausters, which are inexpensive and readily available, when the gases are cooled before they enter the equipment. The toxic by-products from the process should be thoroughly neutralized or eliminated.

ION VAPOR DEPOSITION

In the early 1960s, Sandia Corporation reported on an ion plating process for aluminum.[19] Based on that work, McDonnell Aircraft Company developed the ion vapor deposition (IVD) process as a method of coating aircraft components with aluminum. The IVD aluminum coating was originally developed as a corrosion protection coating to replace cadmium. An IVD aluminum coating is soft, ductile, and adherent, with properties virtually identical to those of pure aluminum.

The IVD process takes place in an evacuated chamber in which an inert gas (usually argon) is added to raise the pressure and to become ionized when a high negative potential is applied to the parts to be coated (see Fig. 24-2). The positively charged argon ions bombard the negatively charged parts, providing a final cleaning. The metal that will coat the parts is melted and vaporized. Some metal vapor is ionized, accelerating it to the parts where the coating is formed.

ION VAPOR DEPOSITION

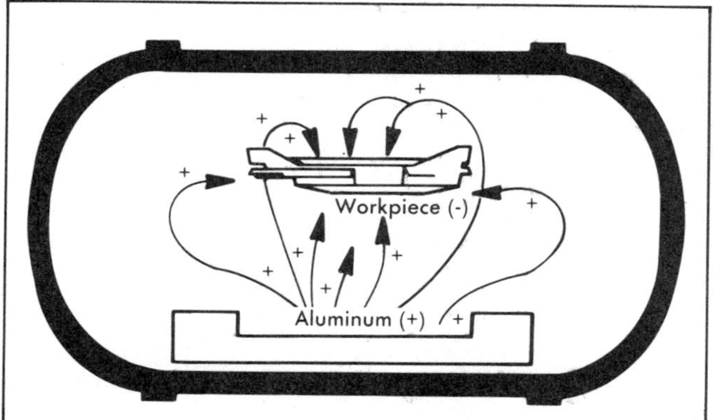

Fig. 24-2 Schematic diagram of the ion vapor deposition process. (*McDonnell Aircraft Co.*)

ADVANTAGES AND LIMITATIONS

The IVD process offers a number of advantages over other coating methods. It produces a dense, adherent, and uniform coating that is not limited to line-of-sight deposition. Thus, coating uniformity is maintained even on complex shapes. It is a nontoxic and nonpolluting process, eliminating the need for pollution control and waste disposal equipment and procedures.[20]

Since the IVD process does not introduce hydrogen, hydrogen embrittlement is not a consideration, and therefore the process can be used on steels at any strength level. Ion vapor deposition also eliminates solid-metal embrittlement of titanium and does not reduce the fatigue properties of aluminum alloys—both problems related to other coating processes.[21] With a chromate conversion coating, the IVD aluminum coating provides excellent electrical conductivity.

Use of the IVD process is limited by the initial expense of the equipment and its capacity, which is similar to other vacuum coatings or metallizing processes. However, recent developments have increased product throughput significantly.

APPLICATIONS

The major use of the IVD process is to coat alloy steel, titanium, or aluminum alloys with pure aluminum. The aluminum coating is generally applied per military standard MIL-C-83488,[22] which provides for three classes of thicknesses, from 0.0003 to 0.0010" (8 to 25 μm) minimum, and two types of finished coatings, either chromated or nonchromated.

Some specific applications for IVD aluminum are as follows:

- Aircraft applications: alloy steel details of all strength levels; fatigue-critical aluminum alloy structures; alloy steel, corrosion-resistant steel, and titanium fasteners.
- Aluminum alloy fuel and pneumatic line fittings.
- Electromagnetic interference (EMI) applications.
- Powder metallurgy details.
- High-temperature applications to 925° F (496° C).
- Depleted uranium substrates.
- Marine applications.

PROCESS DESCRIPTION

A typical coating cycle consists of surface preparation, loading, glow-discharge cleaning and coating, unloading, and supplemental processing. Table 24-5 lists details of process time, work throughput, and labor requirements for various coating cycles.[23]

Surface Preparation

Parts to be coated by the IVD process must be thoroughly cleaned, and their surfaces must be properly conditioned. The first step is vapor degreasing to remove oil and grease. Dry blasting with aluminum oxide grit or chemical cleaning is used to remove metal oxides or other scale. A grit size of 220 mesh is recommended to produce the surface texture required for optimum coating adhesion.

Loading

Large parts are individually racked. Standard racks can be used for many applications, and special configurations can be fabricated if needed. Smaller parts, such as fasteners, are fed from a hopper into a vacuum-locked chamber and then into two counter-rotating, perforated barrels. The racks and barrels are coated along with the work; they are made of a material from which the aluminum can be stripped periodically.

Glow-Discharge Cleaning and Coating

After loading, the vacuum chamber is sealed and evacuated to 10^{-4} torr (0.013 Pa) pressure; this takes approximately 25 to 40 minutes, depending on moisture conditions inside the chamber. The chamber is then backfilled with argon to a pressure of 10^{-2} torr (1.3 Pa) within two minutes. At this pressure, a high negative potential is applied between the parts to be coated and the evaporation source, ionizing the argon and creating a distinctive purple glow discharge around the parts to be coated. The positive gas ions are attracted to the negatively charged parts, bombarding their surfaces and giving them a final cleaning that is essential for coating adhesion. The glow discharge lasts 10 to 15 minutes.

After the glow discharge, the coating metal in wire form is fed continuously into resistance-heated crucibles that melt and evaporate the wire. As the metal vapor passes through the glow discharge, a portion is ionized and accelerated to the parts, where it condenses to a dense, adherent coating. The length of this stage of the process is ten or more minutes, depending on the thickness of coating required. Approximately ten minutes of vaporization is required to deposit 0.0003" (8 μm) aluminum on alloy steel fasteners in the barrel coater. Depositing 0.0010" (25 μm) of aluminum on racked parts requires 15 to 25 minutes for the coating cycle. Longer coating cycles must be interrupted to allow cooling of temperature-sensitive parts, such as aluminum alloy structures.

Unloading

When coating racked parts, the cycle is completed by backfilling with air to atmospheric pressure and opening the chamber for unloading. In the barrel process, the parts are poured into vacuum-locked cannisters from which they are unloaded; uncoated parts are then loaded into the barrels, and the cycle is repeated without breaking the vacuum.

The entire cycle for coating racked parts with 0.0010" (25 μm) aluminum takes about 60 to 75 minutes, while one cycle in the barrel coater for a 0.0003 to 0.0005" (8 to 13 μm) thick coating requires 40 to 60 minutes. Development of the derivative barrel system is aimed at increasing the capacity of the barrel coating process. A barrel accessory is also available for the rack coater, permitting batch barrel coating of small parts such as fasteners.

TABLE 24-5
Typical Equipment Throughput and Labor Requirements

Type IVD Coater	Coater Capacity	Cycle (hour)	Throughput	Labor Requirement, man-hour*
Standard pack	50 ft^2 (4.65 m^2) work area	1.25	40 ft^2/hr** (3.75 m^2/hr)	2.0
Standard barrel	60 lb (27.2 kg) steel fasteners	0.67	90 lb/hr (41 kg/hr)	2.0
Derivative barrel	300 lb (136 kg) steel fasteners	1.00	300 lb/hr (136 kg/hr)	2.5
Derivative barrel	1200 details aluminum alloy, 1 1/4″ (3.175 cm) diam x 1 1/2″ (3.81 cm) long	0.77	1550 parts/hr	2.5

(*McDonnell Aircraft Co. and the Society of Automotive Engineers*)

* Includes part precleaning, processing, and post-treatment.

** The maximum flat plate area that can be coated per cycle is 50 ft^2 (4.65 m^2). The total surface area of parts that can be coated depends on part configuration and the number of parts conveniently racked in a 50 ft^2 (4.65 m^2) area.

Supplemental Processing

Parts coated with aluminum are allowed to cool to room temperature and are then glass-bead peened for a smooth finish. In addition, the peening provides a check on the coating adhesion on each part. A supplemental chromate treatment per MIL-C-5541[24] may be applied for additional corrosion protection and for adhesion of subsequent painting applications.

EQUIPMENT

The basic equipment needed for the IVD process consists of a vacuum chamber, a vacuum pumping system, an evaporation source, a parts-holding rack, and a high-voltage power supply capable of 2 kV and 2.5 A. In addition, conventional supporting equipment may include a vapor degreaser, grit blasting equipment, inert gas storage and supply, glass-bead peening equipment, a chromate treatment line, and a system for stripping aluminum coating.

The first production system was used by the Naval Air Rework Facility in San Diego, California in 1974.[25] Since then, a number of systems have been fabricated for production use. These systems are available as racked parts coaters, barrel coaters, or combinations that can be converted for either racked parts or barrel coating. Schematic diagrams of the two basic systems are shown in Fig. 24-3 and Fig. 24-4.

Rack Coater

The rack coater has a 5 ft (1.5 m) wide by 10 ft (3 m) long parts-holding rack from which parts are suspended above the evaporators by hangers during the coating process. With the rack and parts in place inside the chamber, the evaporation sources traverse the length of the chamber and return while aluminum is being evaporated. Since wire is continually fed into the evaporative sources, the coating of aluminum is uniformly applied to the suspended parts. The thickness of the applied coating is controlled by the speed of the sources and the number of passes up and down the chamber. For parts of small dimensions, the coating wraps around and coats the top surfaces away from the evaporative sources. Larger parts may require rotation of 180° and recoating to obtain optimum uniformity on both sides. For many parts, rotation can be accomplished without removal from the vacuum chamber or breaking the vacuum. Coating thickness can range from 0.0003 to 0.002″ (8 to 50 μm), with a uniformity of $\pm$15%.

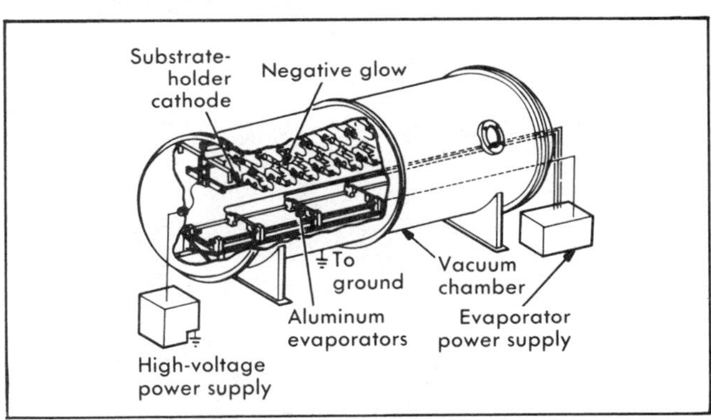

Fig. 24-3 Schematic diagram of large-parts rack coater. (*McDonnell Aircraft Co.*)

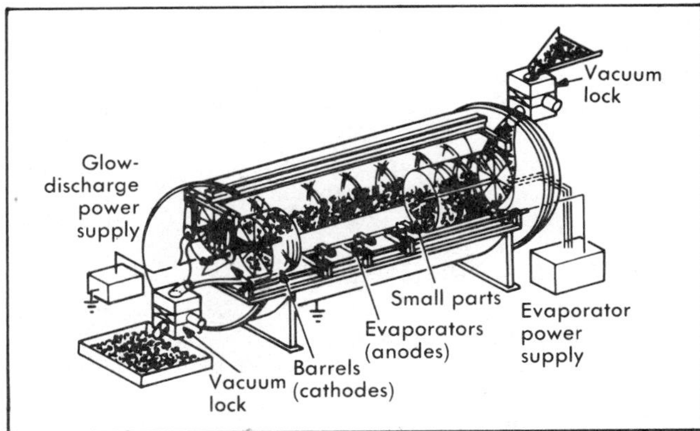

Fig. 24-4 Schematic diagram of small-parts barrel coater. (*McDonnell Aircraft Co.*)

Barrel Coater

The standard barrel coater features two 14" (350 mm) diam by 4 ft (1.2 m) long mesh barrel fixtures to coat small parts by tumbling. The barrels are commonly made from stainless steel. Rotating, rack-type parts holders are also available for larger parts that cannot be tumbled. These fixtures, using the same rotary drive mechanism as that used for the barrels, increase the versatility of this type of coater.

Both barrel and rack fixtured parts are coated in the barrel coater by being rotated above stationary evaporation sources that are positioned longitudinally inside the chamber. The coater operation using barrels is semicontinuous, with parts being loaded and discharged through airlocks that prevent loss of vacuum and thereby decrease process time.

PHYSICAL VAPOR DEPOSITION OF HARD COATINGS

Physical vapor deposition (PVD) covers a broad class of vacuum coating processes in which material is physically removed from a source by evaporation or sputtering, transported through a vacuum or partial vacuum by the energy of the vapor particles, and condensed as a film on the surfaces of appropriately placed parts or substrates. Chemical compounds are deposited by either using a similar source material, or by introducing a gas (nitrogen, oxygen, or simple hydrocarbons) containing the desired reactants, which react with metal(s) from the PVD source.

A myriad of PVD processes have evolved over the last twenty years. They are known by various phrases or acronyms that are sometimes more confusing than descriptive. Most are named for the physical vapor source; for example, diode or triode sputtering, planar or cylindrical magnetron sputtering, direct current (DC) or radio frequency (RF) sputtering, electron beam evaporation, activated reactive evaporation, and arc evaporation. Despite any name confusion, all PVD processes can be separated into three distinct phases:

1. *Emission* from a vapor source.
2. Vapor *transport* in vacuum.
3. *Condensation* on substrates to be coated.

Historically, PVD processes have been relatively expensive as the result of slow deposition rates, expensive vacuum equipment, and throughput limitations. Painting, electroplating, thermal spraying, and chemical vapor deposition (CVD) are typically less expensive coating processes. Additional information on painting, electroplating, thermal spraying, and CVD can be found in the designated chapters of this volume. In recent years, the use of PVD methods has expanded at an extremely rapid rate owing to reduced costs and, more importantly, because of increased demand for high-performance materials and coatings that cannot be produced by other methods.

This section describes the common reactive physical vapor deposition processes being used to deposit hard coatings. The characteristics and applications of the coatings deposited by these processes are also discussed. Other sputtering methods are discussed subsequently. Additional information on vacuum physics and PVD processes can be located by using the references and bibliography listed at the end of this chapter.

COATINGS DEPOSITED

A broad spectrum of metals, alloys, and electrically conductive, semiconductive, and insulating compounds are deposited commercially using PVD methods. Physical vapor deposition methods are used almost exclusively to deposit films that are 1-200 μin. (0.03-5 μm) thick. Thicker films are sometimes deposited, but the cost-benefit ratio usually acts as a barrier, dictating the use of films thinner than 200 μin. (5 μm).

The use of reactive PVD hard coatings, especially titanium nitride (TiN), to improve the performance of metalcutting tools, as well as a wide variety of other tools and wear components, has grown at an exceptionally high rate since 1980. Titanium nitride is a refractory material that has a hardness of greater than R_C80, which is approximately three times harder than typical high-speed tool steels. The increased hardness, in addition to antiwelding properties, reduces the rate of wear for most cutting surfaces.

A TiN coating provides resistance to chemical deterioration because it is a stable (almost inert) material. It also prevents chip welding in cutting tools owing to the antigalling properties of the coating. Titanium nitride has a lower coefficient of friction than hard chromium coatings, thus improving chip flow and reducing friction between the tool and workpiece. Most of the heat generated by the cutting action flows into the chips rather than into the cutting tool. The physical properties of TiN and other transition-metal carbides and nitrides that are used as hard coatings can be found by consulting Volume 7 of *Refractory Metals.*[26]

APPLICATIONS

Physical vapor deposition coatings are an integral part of many products ranging from conductive, resistive, or insulating coatings for microelectronic devices to heat-reflective and decorative coatings on architectural glass; from photoelectric and antireflective coatings in solar cells to corrosion-resistant coatings on aircraft fasteners; and from decorative coatings on pens and watchcases to wear-resistant coatings on metalcutting tools.

Titanium nitride hard coatings are widely used in the automotive, tractor, gear, bearing, fabrication, aircraft, and oil industries. These coatings are most commonly used on metalworking tools and plastic injection molds.

Metalworking Tools

Chemical vapor deposition (CVD) processes for depositing wear-resistant refractory coatings were introduced on metalworking tools in 1969. Due to the high application temperature, initial work was done exclusively on carbide tools. In recent years, CVD has been applied to tool steels; coated parts must be rehardened with care to minimize dimensional changes. For many critical parts, the high-temperature processing and rehardening alter dimensions more than specifications allow.

Cutting tools coated with TiN by PVD were introduced in Japan and Europe in the late 1970s and have subsequently achieved a high degree of acceptance worldwide. Some of the tools that are currently being coated include drills, end mills, broaches, hobs/shaper tools, punches, reamers, taps, blades/knives, and forming tools. Life increases of two to tenfold and sometimes higher are common in many applications, even after resharpening. For hobs and shaper cutters, recent tests have demonstrated economic benefits for tools that are recoated after resharpening; this is especially true for large end users with in-house coating capabilities. Physical vapor deposition is also competitive with CVD in coating carbide tools for applications involving interrupted cutting; the CVD process sometimes reduces the transverse rupture strength of carbides. Cutting tools and tool materials are discussed further in Volume I, *Machining*, of this Handbook series.

Plastic Injection Molds

The use of PVD TiN on plastic injection mold parts is in its infancy compared to its use on metalcutting and metalworking tools. Early results indicate, however, that the impact of PVD coatings in this new field of application will have as much or greater impact as that experienced with cutting tools.

Many injection molds are made from A, O, and P series steels that begin to soften at temperatures of 400 to 500°F (200 to 260°C). Since most injection mold parts are made to very tight tolerances, typically $\pm 0.0001''$ (2.5 μm), rehardening after coating at higher temperatures is undesirable because of distortions that would likely occur.

The optimum coating thickness for typical injection mold parts is about 80 μin. (2 μm). Unlike electroplated parts, PVD-coated molds do not require post-treatment to restore proper mechanical tolerances. Early test results indicate that TiN-coated injection molds exhibit useful life increases approximately three times that of uncoated molds due to reduced wear, abrasion, and corrosion, while showing dramatic improvements in release characteristics and a reduction in mold maintenance and cleanup requirements.

Other Applications

In addition to metalworking tools and plastic injection molds, TiN coatings have been used successfully in soldering, electrical, and erosion-resistant applications. For example, TiN-coated soldering bars have been found to resist corrosion pitting 500 times longer than uncoated bars when using chloride-activated fluxes. Electrical connectors coated with TiN exhibit lower contact resistance after many insertion cycles than do gold-plated connectors. Finally, titanium nitride coatings on titanium alloys have been observed to significantly reduce erosion rates.

REACTIVE PVD PROCESSES

Figure 24-5 illustrates a generalized PVD system with its three phases of vapor *emission* from a source, *transport* to, and *condensation* on the substrate. Also depicted are a number of system requirements to operate the process, as well as options that enable reactive deposition, a plasma-enhanced vapor, and ion bombardment of the growing film.

The most widely used commercial PVD processes are vacuum arc evaporation, electron beam evaporation, and high-rate magnetron sputtering. These are commonly referred to as ion-assisted processes and can be used to deposit coatings by nonreactive and reactive means. A comparison of the three processes for TiN deposition is presented in Table 24-6.

Reactive PVD

In reactive PVD, carbides, nitrides, oxides, carbonitrides, and many other compound types are deposited by introducing a reactive gas (simple hydrocarbons such as CH_4, C_2H_6, nitrogen,

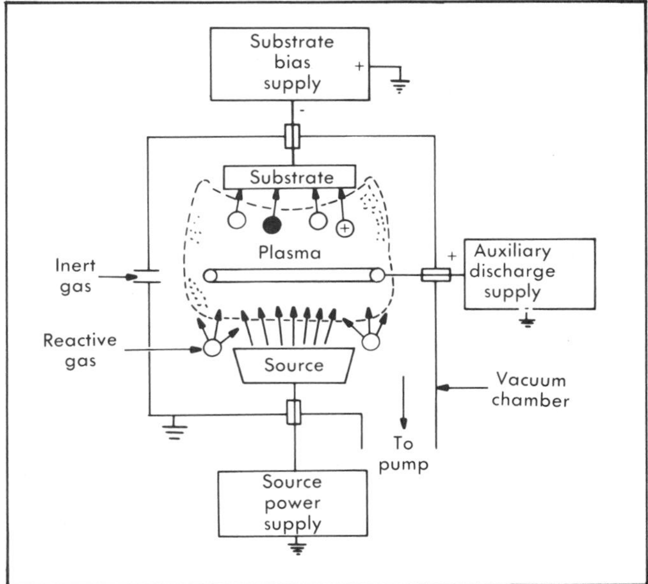

Fig. 24-5 Schematic of generalized PVD system. (*Multi-Arc Vacuum Systems, Inc.*)

oxygen, and other gases) into the physical vapor stream. Reactions between the gas and physical vapor can occur at the source surface, in transit, or at the substrate surface, as well as on the chamber walls and other surfaces.

In most processes, reactions at the vapor source are minimal owing to the nature of the process and/or the desires of the process designer. In vacuum arc evaporation, surface reactions often take place at the source; these reactions do not significantly affect the vaporization rates because most of the vapor release comes from beneath the source surface. With electron beam evaporation, the maximum operating pressure at the source is too low at typical evaporation rates for significant reactions to occur. For reactive planar magnetron sputtering, cathode vaporization rates are often reduced by a factor of 10 or more at a given power input when oxide, nitride, or carbide layers form on the sputtering surface.

To avoid such rate reductions, formation of reacted surface layers is often inhibited by enclosing the cathode in a housing having a vapor emission slot and by simultaneously directing reactive gases toward the substrates. Either the reactive gas flow or the cathode power level must then be controlled to maintain the proper balance of metal and gas reactants arriving at the substrates to achieve the desired film composition, while at the same time maintaining a sufficiently low concentration of reactive gas at the cathode to prevent the formation of rate-limiting reaction films. Automatic process control is required to achieve the continuous monitoring and short response times necessary to maintain the balance between metal and gas reactants at the substrates.

Reactions in transit are a minor effect because physical conservation laws prohibit most reactions in binary collisions leading to a single, final-state particle. There is a low probability of simultaneous multiparticle interactions at the pressures used.

The primary reaction site for most PVD processes is at the substrate. Generally, the flux of reactive gas constituents arriving at the substrates must be balanced with the physical

TABLE 24-6
Comparison of Ion-Assisted Vapor Deposition Processes

	Source Type		
	Arc	Electron Beam	Sputtering
Source location	Top/side/bottom	Bottom	Top/side/bottom
Number of sources	Multiple	1 (typical)	Multiple
Source to substrate distance, in. (mm)	4-20 (100-500)	8-28 (200-700)	2-8 (50-200)
Ionization (metal), %	≥50	1 to ≥10	≤1
Substrate heating and surface preparation	Metal ions	Radiant heat, electron bombardment, and argon ions	Radiant heat and argon ions
Substrate temperature, °F (°C)	360-1020 (180-550)	390-1020 (200-550)	570-1020 (300-550)
Substrate bias, -V	75-400	100-400	100-500
N_2 pressure @ stoichiometry	Wide range	Narrow range	Narrow range
Deposition rate, μin./min (μm/min)	Moderate, 1-11.8 (0.025-0.3)	High, ≤40 (1.0)	Low, ≤8 (0.2)
System throughput	High	Low	Very low
Alloy sources	No bulk depletion	Deplete high vapor pressure components	No bulk depletion
Operation complexity	Low	High	Moderate

(*Multi-Arc Vacuum Systems, Inc.*)

vapor flux to achieve the desired film composition. Process control is more easily achieved when excess reactive gas is rejected from the growing film. Films grown under the influence of substantial ion bombardment have been observed to reject excess reactive gas.[27]

Plasma Enhancement

Plasma enhancement refers to various methods of increasing the number of ions, electrons, fragmented molecules, and excited neutrons in a gas or vapor. To employ plasma enhancement, inert and/or reactive gases must be present at pressures suitable for maintaining a gas discharge. When used in conjunction with a negative substrate bias, the increased ion bombardment resulting from plasma enhancement is an important factor that generally results in improved film properties. In reactive PVD, plasma enhancement produces excited and ionized atoms, molecules, and molecular fragments that react more readily with the surface of the growing film.

The electrode attached to the auxiliary discharge supply (refer to Fig. 24-5) represents a plasma (or ionization) enhancement device. In its simplest form, a water-cooled anode accelerates secondary electrons from a planar magnetron or electron beam source to form an auxiliary gas discharge in the vapor transport region. Application of a high-frequency or RF voltage to such an electrode is another way to create a gas discharge. Another common method of plasma enhancement is to accelerate electrons from a thermionic filament through the transport region to an auxiliary anode.

The voltage required to set up an auxiliary gas discharge is generally 100-1000 V at pressures of 10 to 30 x 10^{-3} torr (1.34-4.0 Pa); voltages on the lower end of the range apply when there is a ready supply of initial electrons. Once started, the gas discharge clamps the voltage at a level determined by electrode geometry, surface composition, and gas pressure and composition. Attempts to increase voltage result in increased discharge current over several orders of magnitude at a relatively constant voltage.

Ion Bombardment

When energetic ion bombardment of a growing film is present, the process is referred to as ion-bombardment-assisted deposition, or simply ion-assisted deposition. Ion-assisted deposition processes are very successful in many applications because they produce films with extremely high adherence and good physical properties.

In most ion-assisted PVD processes, ions gain energy as a result of a negative bias voltage that is applied to the substrates. For nonconductive substrates or coatings, a high-frequency or RF bias voltage is applied; positive ions are accelerated to the substrates during the negative phase, and charging of the substrates is neutralized by electrons during the positive phase of each cycle.

In most PVD processes, the ion bombardment is supplied primarily by argon or reactive gas ions from a gas discharge. The level of ionization in these discharges is generally less than a few percent. With hollow cathode discharge (HCD) electron beam sources, in which the vaporized metal passes through the incident high-density electron beam, a very significant fraction of the metal vapor is ionized. Ionized fractions up to 15% have been reported for chromium.[28] With arc evaporation sources, vaporized metal passes through a region of extremely high-density electron flux that results in an ionization fraction of 35% for chromium.[29]

Ion Plating Processes

Ion plating is the name given by Mattox in 1964 to the PVD process illustrated in Fig. 24-6.[30] In this process, substrates are heated and sputtered clean by argon ions from a glow discharge that is supported by a negatively biased substrate; the negative bias is between -2000 to -5000 V. During deposition, molecules from a thermal evaporation source pass through the glow discharge on their way to the substrate. The vapor condensing on the substrate is accompanied by the bombardment of metal and argon ions as well as energetic molecules. This process produces highly adherent coatings with a relatively uniform thickness on irregularly shaped substrates due to gas scattering and sputtering-recondensation effects.

In addition to its use in the original ion plating process, ion bombardment is employed for surface cleaning and conditioning in many contemporary PVD processes. Although this procedure is not an economical substitute for thorough cleaning prior to loading in the coating chamber, it provides a very effective final touch that is an important factor in achieving good film adhesion. In the wake of a large number of PVD processes that utilize energetic particle bombardment of growing films, the term "ion plating" is gradually being replaced by the more general and descriptive term "ion-assisted deposition."

Another early ion-assisted deposition process is activated reactive evaporation (ARE), developed by Bunshah.[31] In this process, a glow discharge is supported between an auxiliary electrode and a conventional high-voltage, low-current electron beam evaporator, as illustrated in Fig. 24-7. The ARE process is further enhanced by applying a substrate bias; this arrangement, which provides separate means for creating ions and accelerating them to the substrate, is a form of ion plating. Various ion plating and ARE processes have been used to deposit a wide variety of compounds, including hard coatings, on an experimental and commercial basis.

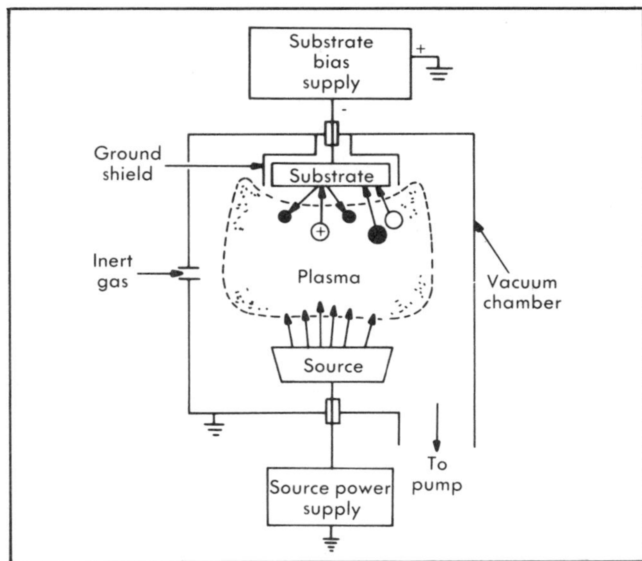

Fig. 24-6 Schematic of classical ion plating system. (*Multi-Arc Vacuum Systems, Inc.*)

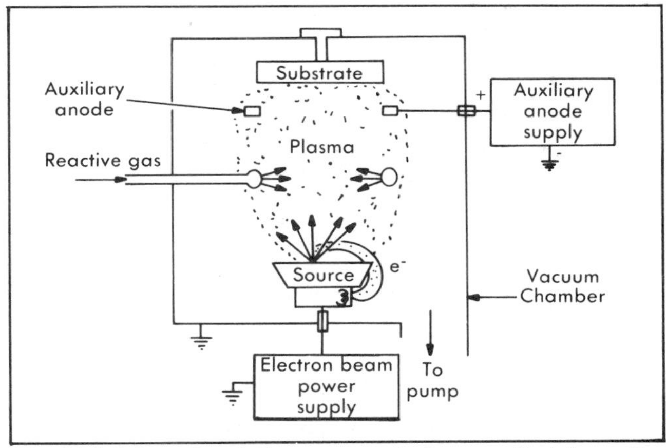

Fig. 24-7 Schematic of the components in an activated reactive evaporation (ARE) system. (*Multi-Arc Vacuum Systems, Inc.*)

Vacuum Arc Evaporation

In processes employing vacuum arc evaporation, illustrated in Fig. 24-8, metal vapor is evaporated from a cathode source as a result of intense localized heating by arc spots that move more or less randomly across the cathode surface. Distinguishing features of arc evaporation processes are that the vapor source remains solid while a major fraction of the metal vapor is ionized.

Because arc melting is highly localized, cathodes may be used in any orientation, allowing greater flexibility in designing systems with multiple sources for coating very large parts or large quantities of smaller parts. With judicious source placement, uniform coatings can be applied to a variety of substrates through the use of very simple fixturing and rotation devices.

The highly ionized metal vapor is used for substrate heating and surface preparation, as well as for coating purposes. After reaching a suitable vacuum level, approximately 1×10^{-5} torr (1.3×10^{-3} Pa), substrates are heated by metal-ion bombardment and by operating one or more arc sources while applying a voltage to the substrates. In addition to heating, the incident ions condition substrate surfaces to increase film adhesion by a combination of sputtering, surface mixing, and diffusion mechanisms.

Reactive deposition takes place while operating arc sources in the presence of a suitable reactive gas using substrate bias voltages from -75 to -400 V. As a result of the high percentage of metal ions, dense stoichiometric films with high adherence are deposited over a wide range of reactive gas pressures, substrate temperatures, and substrate voltages. Technologically successful TiN films have been deposited at N_2 pressures from 1 to 60×10^{-3} torr (0.13 to 8.0 Pa) and at temperatures as low as 455° F (180° C).

Electron Beam Reactive Deposition

The electron beam reactive deposition processes in current industrial use are thermionically-enhanced-triode ion plating, hollow cathode discharge, and coaxial. In each process, metal heated by an electron beam evaporates from a molten pool. The substrates are oriented to collect vapor from bottom-mounted sources. Substrate rotation is commonly used to increase the uniformity of the film thickness and part capacity. The reactive gas pressure must be closely controlled to achieve good film stoichiometry.

Thermionically enhanced. The thermionically-enhanced-triode ion plating process is shown schematically in Fig. 24-9.[32] After the chamber is pumped down to 1×10^{-5} torr (1.3 x

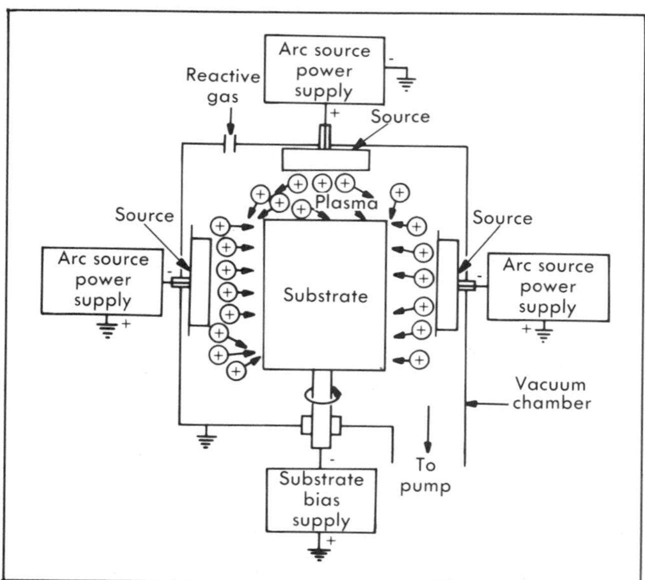

Fig. 24-8 Schematic of components in a vacuum arc reactive system. (*Multi-Arc Vacuum Systems, Inc.*)

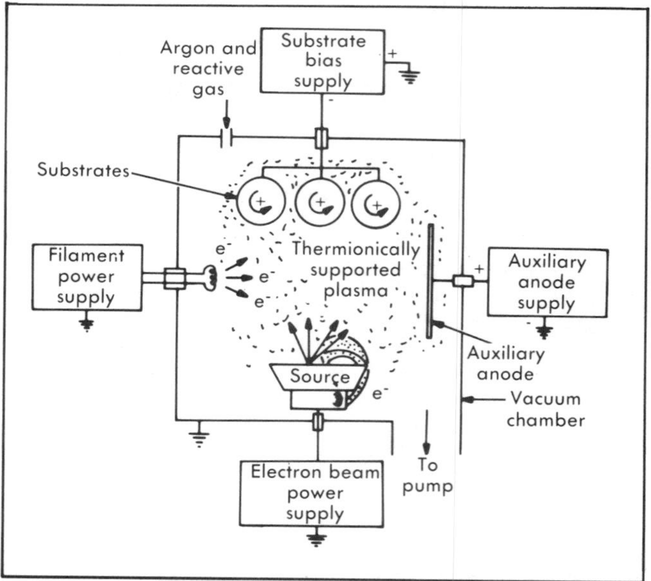

Fig. 24-9 Schematic of components in a thermionically-enhanced-triode ion plating system. (*Multi-Arc Vacuum Systems, Inc.*)

10^{-3} Pa), the substrates are heated to at least 570° F (300° C) by argon ion bombardment. An argon ion discharge, supported by a high-voltage substrate bias, is used to sputter clean the substrate surfaces; thermionic electrons accelerated from the hot filament to the auxiliary anode enhance the strength of this discharge. During coating, a bias of several hundred volts is applied to the substrates, and the desired reactive gases are introduced into the chamber. The electron beam source is a high-voltage, low-current unit that produces little ionization. The thermionic enhancement feature increases the ionization fraction (to as much as 3%), thereby improving adhesion and film quality.

Hollow cathode discharge. The hollow cathode discharge electron beam reactive deposition process, depicted in Fig. 24-10, is similar to the thermionically-enhanced-triode ion plating process. The hollow cathode discharge gun produces a very high-current electron beam that strikes the metal source at an energy on the order of 100 eV. Approximately 10-20% of the metal vapor is ionized as it passes through the incident electron flux.[33] The high ionization is beneficial to film growth.

Coaxial electron beam. The coaxial electron beam reactive deposition process (see Fig. 24-11)[34] differs from the thermionically enhanced and hollow cathode discharge processes previously described. First, the electron source contains a magnetically enhanced, thermionically supported discharge that emits a very high electron current along with ionized gas. Second, the substrates are heated by electron bombardment; the positive side of the discharge supply is connected to the substrates. Third, when the substrate temperature reaches about 570° F (300° C), electrons from the source are accelerated to the auxiliary anode, and the substrates are sputter cleaned with argon ions.

During cleaning, the substrates are biased negatively. A strong gas discharge is supported by the electron beam in conjunction with a magnetic field produced by the Helmholz coils. Finally, during coating, the electron beam is focused on the metal source. The substrates are biased to about -200 V, and reactive gas(es) is introduced into the electron source. The excited and ionized gases from the electron source along with a large percentage, greater than approximately 10%, of vaporized metal ions enhance the coating.

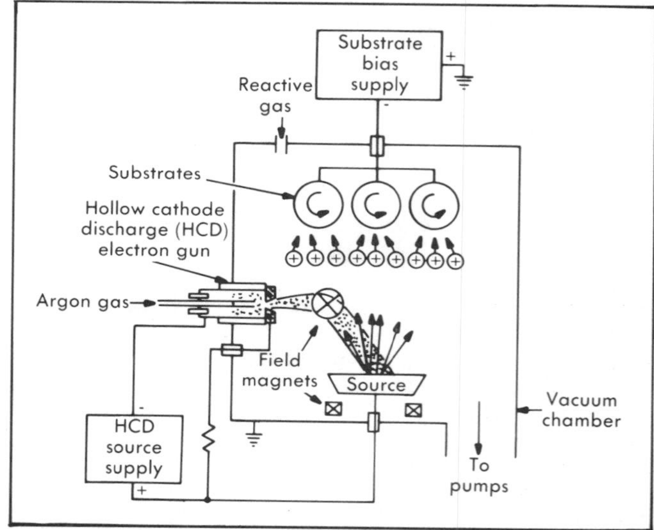

Fig. 24-10 Schematic of components in a hollow cathode discharge electron beam deposition system. (*Multi-Arc Vacuum Systems, Inc.*)

High-Rate Reactive Magnetron Sputtering

In high-rate reactive magnetron sputtering, metal vapor is sputtered from cathode sources by an intense inert gas discharge that is largely confined to a region within 4″ (100 mm) of the cathode source; magnetron sources may be used in any orientation. Compared to evaporation sources in general, the vaporization rates per unit area are relatively low; but large single sources can be used, as well as multiple sources, to compensate for the low rates. To maintain maximum sputtering

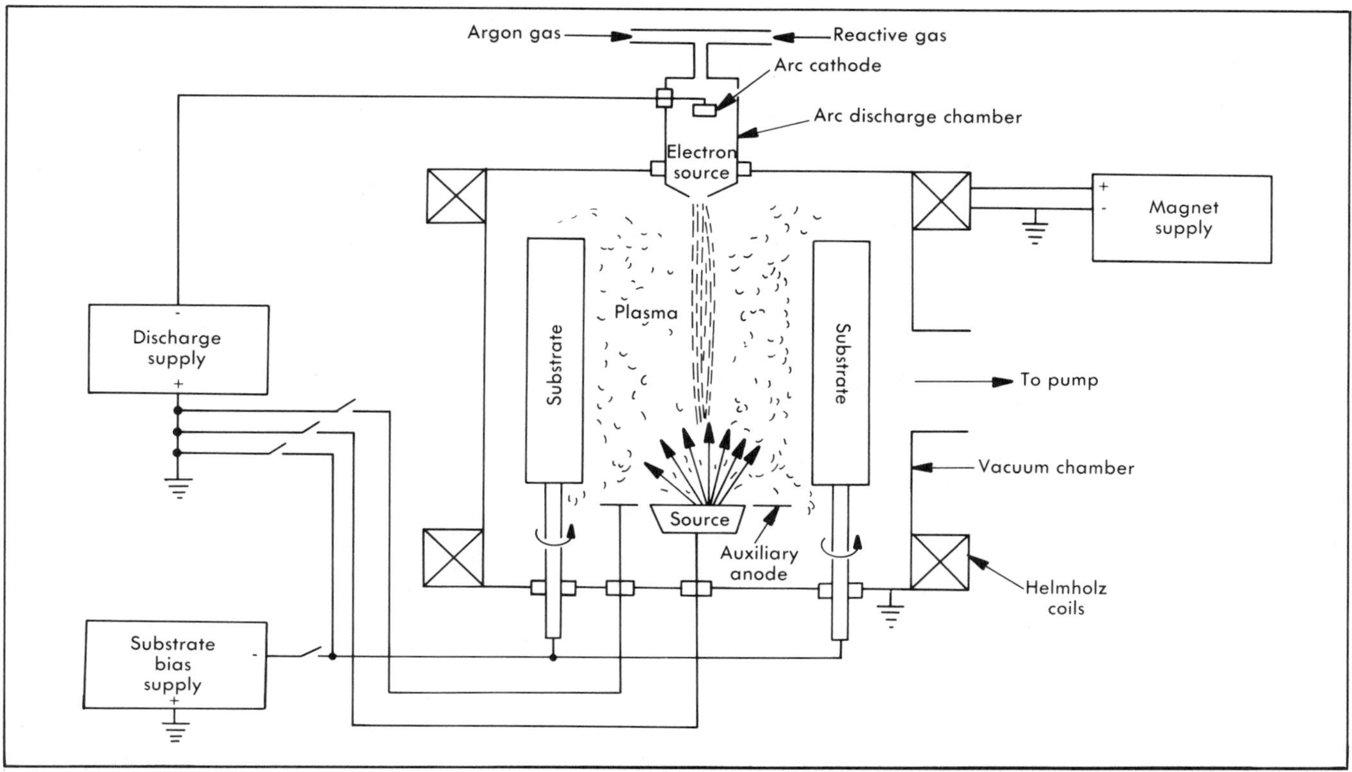

Fig. 24-11 Schematic of coaxial electron beam reactive deposition system. (*Multi-Arc Vacuum Systems, Inc.*)

rates during reactive deposition, the components and process controls must be arranged to prevent the formation of reaction products on the cathode surface. Magnetron sputtering is well suited for applying uniform coatings on large, planar surfaces and for use in continuous in-line systems.

The principles for magnetron sputtering are illustrated in Fig. 24-12.[35-37] The arrangement of the components and gas inlets maximizes the amount of reactive molecules that will react on the substrate or surrounding surfaces before reaching the cathode surface. It is also necessary to balance the flow rate of reactive gas with the metal vaporization rate to achieve film stoichiometry while maintaining a suitably clean cathode surface. Argon ion bombardment is frequently used to sputter clean the substrate surfaces, as well as to supply additional heat.

PROCESS CONSIDERATIONS

Although PVD methods typically require more expensive equipment, more parts preparation, more system maintenance, and greater processing times than many alternate processes, PVD is a viable coating process for hard coatings. In some instances, PVD is the only method capable of depositing a coating of the desired material. The overall cost of PVD can often be less than other coating methods when all factors are taken into account. The primary considerations that must be evaluated before selecting a PVD process include desired end result, substrate properties, cleaning and preparation, fixturing, temperature, and coating rates. When alternative means of achieving the end result are available, total costs must be compared among the alternatives; total cost would include equipment cost and operating environment.

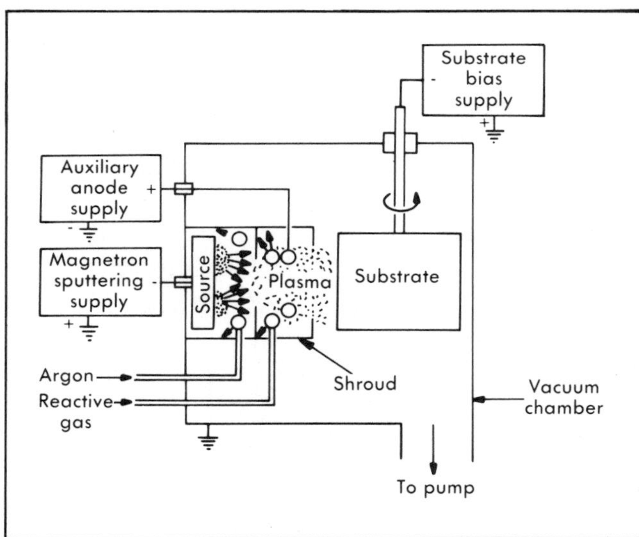

Fig. 24-12 Schematic of components in a high-rate reactive magnetron sputter system. (*Multi-Arc Vacuum Systems, Inc.*)

Substrates

The easiest substrates to coat are those that are electrically conductive and remain stable (minimum outgassing or decomposition of bulk material) at elevated temperatures. The level of difficulty and the number of process constraints increase when progressing from metal to glass to plastic substrates.

PHYSICAL VAPOR DEPOSITION OF HARD COATINGS

With metal substrates, certain materials, surface conditions, and assembling techniques must be avoided to achieve good adhesion and film properties. For example, cadmium and zinc alloys create problems during coating if the temperature exceeds 390° F (200° C) because of their high vapor pressures. Porous metals such as low-grade castings are generally difficult to coat because oils and contaminants remain entrapped in the pores. Burrs must be removed from the parts before coating to prevent exposure of uncoated metal when they are broken off later. Other surface conditions that reduce film adhesion are surface oxides, grinding burns, imbedded polishing compounds, and rust-preventive films. Parts that have been brazed or press fit together are often difficult to coat by PVD because of contaminants entrapped between the parts. Although some materials and assembly methods are undesirable for PVD coatings, many of the problems previously described can be overcome by thorough cleaning and operating parameter adjustments.

Cleaning and Preparation

Although substrate heating and ion bombardment is used for final surface preparation, the most important cleaning is done outside the vacuum coating chamber. The purpose of surface cleaning and preparation is to remove contaminants and conditions that would reduce adhesive strength. The choice of the cleaning method must take into consideration the type of soil or contaminant to be removed and how the cleaning method will affect part dimensions, surface finish, and surface appearance.

The basic cleaning procedures for most metals include degreasing, ultrasonic cleaning, water rinsing, and dewatering and drying. The parts should not be allowed to dry between the cleaning steps of the aqueous cleaning sequence. For additional information on the various cleaning methods, refer to Chapter 18, "Cleaning," of this volume.

Degreasing. Vapor degreasing in trichloroethylene or another strong solvent is used to remove soluble oils and grease. Some rust-inhibiting films leave insoluble residue on the surface of the part that must be removed by other methods. Ultrasonic cleaning in a heavy-duty aqueous detergent is an alternative degreasing method that has been used successfully.

Ultrasonic cleaning. Ultrasonic cleaning is performed in an aqueous detergent to remove soluble inorganic salts and solid residues.

Water rinsing. The detergent solution is rinsed off parts by dunking or spraying with clean water. Deionized water is preferred, but tap water is suitable for many ion plating processes.

Drying. Parts are often dewatered and dried using a two-tank Freon™ drying system. When used properly, this method does not leave any residue on the part surface.

Storing and handling. Cleaned parts may be stored up to 24 hours or longer if they are protected from airborne contaminants and exposure to humid air. Cleaned parts must never be handled with bare hands; gloves or other noncontaminating means must be used to avoid recontamination of cleaned surfaces.

Fixturing

Several types of fixturing devices are used in PVD; the three basic types are stationary, rotary, or rotary with planetary motion. The optimum fixture should:

- Hold parts so that they are exposed to the PVD vapor.
- Maximize the quantity of parts that can be simultaneously coated with acceptable uniformity, in a given length of time.
- Shield surfaces that must not be coated.
- Minimize detrimental effects due to the presence of fixture surfaces.

Temperature

The temperature requirements of a particular PVD process need to be matched with the temperature characteristics of the substrate. Excessive temperatures may change the desired properties of the substrate or may cause distortion or dimensional changes to the parts. For most PVD applications, a temperature of at least 355° F (180° C) is required for optimum film characteristics.

Coating Rates

Physical vapor deposition coating rates depend on vapor source characteristics, substrate properties, and the quantity of parts to be coated. Table 24-6 identifies the typical coating rates for the three main PVD methods. Physical vapor deposition coating rates are less than the rates of other coating processes such as electroplating or plasma spraying. This deficiency is partially compensated for because PVD films are usually much thinner than the films (coatings) of the two previously mentioned processes.

FILM EVALUATION

In a production environment, films are typically evaluated for visual defects, thickness, and adhesion. It should be pointed out that absolute film thickness is often not a very critical property. Visual defects such as bare spots, small voids, incorporated flakes, or debris can be observed with a stereo microscope having a magnification of 10 to 100 times.

Film thickness is generally measured by one of the following methods:

- Polished metallurgical microsections are used to microscopically observe the coating thickness on various part surfaces. This method is the most direct way to determine thickness uniformity.
- Beta (high-energy electron) backscatter instruments are used to measure the film thickness nondestructively. This is an indirect method that requires calibration with a known standard; substantial errors can be made in measuring the film thickness on curved surfaces if care is not exercised.
- A ball-crater instrument can be used to polish through the surface of a coating. The relationship between the diameter of the polishing ball, the maximum diameter that shows the effects of polishing, and the diameter of the substrate area that is exposed by polishing is used to calculate the thickness. Coatings that are up to 120 μin. (3 μm) thick can be measured with an accuracy of $\pm$4 μin. ($\pm$0.1 μm) without difficulty on relatively smooth, flat or cylindrical surfaces.
- X-ray fluorescence or energy-dispersive X-ray analysis in a scanning electron microscope can be used to measure film thickness.

The adhesion between coating and substrate is difficult to measure directly for highly adherent films; pull tests capable of measuring yield strengths that are typical of metals and PVD hard coatings on metals have not been developed. A commonly used indirect test is the manual stone abrasion test (SAT). In this test, a fine sharpening stone is rubbed back and forth across the coated surface, allowing the stone particles to make grooves in the surface by nonelastic deformation. The film is then inspected under a microscope to obtain adhesion information.

VACUUM METALLIZING

In vacuum metallizing, a metal or metal compound is evaporated at high temperature in a closed, evacuated chamber and then allowed to condense on a workpiece within the chamber. The workpiece is usually at room or a relatively low temperature. The coating can be deposited on flexible substrates and on discrete objects. When the coatings are deposited on rolled flexible substrates, the process is referred to as roll or semicontinuous metallizing; when deposited on discrete or solid objects, the process is called batch metallizing.

ROLL METALLIZING

Vacuum metallizing of flexible substrates, while involving the same vacuum deposition process as used in batch metallizing, requires additional process operations and controls. In general, no substrate cleaning or preparation is required, although certain film and paper grades are more suitable for roll metallizing. In the basic vacuum deposition process, the pressure in the vacuum chamber is reduced to a pressure of approximately 1×10^{-4} torr (0.013 Pa) or lower. The evaporation source heats the coating material to a temperature such that the vapor pressure exceeds the chamber pressure, permitting the coating material to be deposited on the substrate. The substrate is then transported above the evaporation source at a high rate of speed as it is rewound onto the rewind shaft. The substrate is normally held in contact with a chilled drum when it is exposed to the evaporation source to protect it from thermal shock due to infrared radiation from the source and the heat of condensation from the evaporant.

The deposited metal coating is approximately 1-2 μin. (0.03-0.05 μm) thick and closely duplicates the surface on which it is deposited. Aluminum is the most common metal deposited on flexible substrates. The characteristics of the metal coating are mainly determined by the rate of deposition and the operating pressure. The substrate temperature, the angle of deposition on the substrate, and several other factors also contribute to the final characteristics of the coating.

The adhesion of the deposited metal to the substrate is dependent upon the rate of deposition, the operating pressure, and the surface characteristics of the substrate. At pressures above 1×10^{-3} torr (0.13 Pa), the metal coating begins to degrade rapidly and becomes less adherent. At these high pressures, the coating also shows some reduction in reflectance and increased absorption due to changes in crystal structure. Metal adhesion and appearance are also dependent upon the speed at which the substrate is processed. Normal operating speeds for metallizing flexible substrates range from 500 to 1500 sfm (150 to 460 m/min), with some of the newer equipment capable of operating at speeds up to 2000 sfm (610 m/min).

Applications

The metallized products manufactured by roll metallizing are used in a wide range of applications. Products include hot stamping foils, laminated film products, capacitors, solar control products, reflective insulation, packaging materials, labels, decals, optical products, decorative materials, wrapping papers and films, protective fabrics, magnetic coatings, and many other types of functional products.

Many of the newer packaging materials are being produced by laminating metallized films to other, special functional films and papers to produce the desired appearance and properties. Metal thickness is a major factor in the performance of these packaging materials and must be accurately monitored and controlled to provide protection against oxygen, water, and light. Materials are being manufactured for packaging dried foods, snack foods, coffee, and many other products, and are currently being developed for packaging liquids and juices to further compete in the packaging market. Metallized films also reduce static in the packaging of powders and coffee, and prevent damage to electronic equipment due to static discharge.

Metallized paper products are beginning to compete with foil laminates and are used in a wide range of applications. Cigarette packaging, decals, labels, gum, and candy packaging are rapidly growing markets. These products can be produced by direct metallizing of precoated papers or by transfer metallizing. The latter process involves metallizing an untreated film that is laminated to the base paper and then delaminated for remetallization. Currently, transfer films can generally be reused in excess of 20 times, and in no case is it less than 15 times. This process produces a bright metal coating that is quite independent of many of the problems involving the direct metallizing of coated papers. Metallized films are also being used in solar applications to control light transmission, ultraviolet degradation, and selective wavelength reflection.

The primary growth of vacuum metallized products has been directly related to the increasing cost of aluminum foil and the concurrent development of more efficient vacuum evaporation process equipment. With the continuously improving economics of the vacuum metallizing process, the metallized products markets have been growing rapidly and should continue to increase in new nonaluminum foil markets.

Vacuum evaporation process applications are of interest in product areas other than the evaporation of metals onto flexible substrates. The use of dielectric materials for special optical properties should be expected to be developed for a wide range of special applications involving barrier and other characteristics. The vacuum deposition of dyes and organic compounds should also become important manufacturing processes as products could be developed for medical and other special applications.

Substrates

A wide variety of products are manufactured by metallizing flexible substrates, including plastic films, metallized papers, fabrics, and even sheet metals. The economics of the roll metallizing process is dependent on the length and width of the substrates. The length of the roll is inversely proportional to its thickness and proportional to its diameter. For film substrates, the roll diameter ranges from 14 to 36″ (356 to 915 mm); for paper substrates, from 30 to 48″ (760 to 1220 mm).

Polyester film is the most frequently used film for roll metallizing. Other films such as polypropylene, polyethylene, polyimide, polyamide, polyvinyl chloride, polystyrene, and polycarbonate are also frequently used, but do not have as many overall favorable characteristics as polyester film.

The films used in roll metallizing are generally available in widths of 48″ (1220 mm) or wider and have a number of the following characteristics:

- Low cost.
- Available in a wide range of thicknesses, 0.00006-0.007″ (1.5-175 μm).
- High tensile strength.
- High clarity.
- Low extensibility.
- Good temperature stability.
- Good surface smoothness and uniformity.
- Minimal residual volatile components.
- Low moisture absorption.

Special functional requirements dictate the choice of the film in the metallized product structure used in packaging, reflective insulation, solar control, and other specific functional products. In these products, the functional characteristics of the substrate, the metal coating, the adhesives, inhibitors, and organic dyes and coatings are used collectively to provide the overall functional characteristics of the finished product.

Papers are being used in great quantity as substrates for roll metallizing due to the increase in the cost of aluminum foil. Many metallized paper products are being developed and marketed by depositing thin aluminum coatings on precoated papers and glassines, or by transfer metallizing paper board, tissue, or other papers. Most of these products are manufactured and sold for their appearance, not for their functional barrier properties. Typical applications include beer, liquor, and food labels.

The type of paper selected is dependent upon basis weight of the paper (weight per ream or weight per square meter), surface smoothness, density, and cost. The amount of precoating required is a function of the quality of the paper. Dense, smooth papers require approximately 3 lb (1.4 kg) of coating per 3000 ft^2 (278 m^2) ream, while less dense, fibrous, highly absorbent papers require significantly more precoating to develop a suitably smooth surface for metallizing. As the weight of the precoating material increases, the coated paper becomes much less flexible, and the cost increases rapidly.

The metallizing of fabrics has been limited since the products in which they are used have been restricted to applications that primarily emphasized their decorative and not their functional properties. Nylon fabrics have been most frequently used for decorative applications. However, other fabrics are now being used more extensively in functional applications, such as in heat-reflecting draperies and shades, and in reflective clothing applications, such as for firefighters' suits and space suits. The fabrics that have been found to be most satisfactory for these functional applications have generally been woven from synthetic yarns.

Aluminum-coated steel products have been manufactured in production quantities, but the economics of the process has prevented them from being used competitively with other plated steel products. Special sheet metal products are being roll metallized for semiconductor applications, but these products are manufactured in limited volume.

Metallizing Operation

The substrates used in the metallizer must be loaded onto the unwind shaft and then carried through the system over the various rolls to the rewind shaft position. The substrate is next carefully aligned and attached to the rewind shaft core so that a minimal number of wrinkles or surface irregularities are present at the start of the rewinding. A small amount of the substrate is wound onto the rewind shaft to check that the roll is properly aligned and that the substrate is free of wrinkles, creases, or other defects that could adversely affect the quality of the rewound substrate.

The evaporation sources are then cleaned, loaded with metal, and checked for proper electrical operation prior to closing the chamber. The system is closed, the chamber evacuated, and the evaporation sources heated to vaporization temperature under a movable shutter. When the sources have reached proper temperature, the substrate is brought up to speed, and the movable shutter is opened, allowing the evaporation process to start. Evaporation continues until the roll is completely metallized, or until the operation is stopped for any operational problem.

At the end of the metallizing process, the shutter is closed, the substrate decelerated and stopped, and the sources left to cool briefly. After cooling, the chamber is vented and opened. The roll of substrate is then removed from the rewind shaft, and a new roll is installed on the unwind shaft. Before the new substrate roll is threaded through the system, the rolls in the metallizer must be cleaned. It is also necessary to do a certain amount of cleaning around the evaporation sources, the movable shutter, and other areas after each evaporation cycle to avoid operational problems in subsequent runs.

Each step in the process requires a certain amount of time. Table 24-7 lists the process steps sequentially, with an approximate time requirement for each step; the metallizing time includes both acceleration time and deceleration time. The wide variation in process time is due to techniques used, the amount of cleaning required for the run, the quantity of metal deposited, the type of evaporation sources used, and operator skill in overlapping source heatup and pump-down time.

Metallizing Equipment

Roll metallizing equipment is capable of handling a wide range of roll widths varying from 10 to 120″ (250 to 3050 mm). Two main equipment designs are generally used. In the first

TABLE 24-7
Roll Metallizing Cycle and Process Times

Process Step	Estimated Time,* minutes
1. Load substrate.	10-15
2. Prepare evaporation sources.	10-20
3. Evacuate chamber.	15-30
4. Heat sources.	5-10
5. Metallize substrate.	41-44
6. Cool sources.	2-5
7. Vent chamber to atmosphere.	2-3
8. Unload substrate.	3-5
9. Clean system.	5-15
	Total 93-147

(*Charles A. Baer Associates*)

* Based on a 40,000 ft (12 200 m) long roll of substrate that is metallized at 1000 fpm (305 m/min).

design, the unwind and rewind system is incorporated within the vacuum chamber. Several winding configurations are used with this first design, as illustrated in Fig. 24-13. The configuration in view *a* has a single large drum that permits better pre and postcooling than the configuration in view *b*, and it has a wider deposition angle. However, this configuration is capable of one-side metallizing only and may be limited to smaller roll sizes than the single small-drum configuration. View *b* shows a roll metallizer with a single small drum. This configuration is capable of handling larger rolls and permits a greater evaporation throw distance. It too is limited to one-side metallizing and has a narrower deposition angle than the single large-drum configuration. The double small-drum configuration is shown in view *c*. This configuration is capable of metallizing both sides of the substrate in one pass and of applying thick coatings. This configuration is limited to smaller roll diameters and has a narrow deposition angle. In all these configurations, the chambers are frequently subdivided to remove volatiles without contaminating the deposition zone.

In the second design, the substrate is carried into and then out of the vacuum chamber through a series of vacuum seals, as illustrated in Fig. 24-14. When metallizing substrates with this type of equipment, the process is generally referred to as "continuous" or "air-to-air" metallizing. This process is used for coating substrates when large production quantities are required with a constant film width and thickness. It is also useful when the substrates contain large quantities of volatile materials.

In addition to the vacuum chamber and the unwind and rewind system, the roll metallizer is equipped with evaporation sources and a pumping system. Since the speed and tension of flexible substrates are important operating parameters, a speed and tension control system is also incorporated.

Evaporation sources. The evaporation sources differ from sources used in batch metallizing in that the sources required for this process must be used for long intervals of time and provide high rates of evaporation. As most metallizing involves the evaporation of aluminum, the choice of evaporation source is a matter of choosing a material that is a sufficiently good refractory material to tolerate the required 2730° F (1500° C) temperature and will not have a high rate of reaction with the liquid aluminum at this temperature. The choice is generally limited to one basic type of source material, a boron nitride/titanium diboride composite.

While mixtures of carbon and graphite can be used as source materials, the major problem with their use is the rapid formation of aluminum carbide, a low vapor pressure reaction product that coats the surface of the liquid aluminum and rapidly reduces the rate of evaporation from the source. The rate of reaction between aluminum and graphite at this temperature is very high due to intergranular penetration. The use of additives such as titanium, zirconium, or other metals chosen from Group IV or Group V of the periodic table serves to minimize the formation of the aluminum carbide. For resistance-heated sources, the boron nitride composite materials are far superior to carbon/graphite sources.

In order to achieve good across-the-web uniformity with the evaporated metal coating, it is necessary to use a number of individual evaporation sources. The number of sources used is determined by the amount of metal that has to be evaporated, the distance from the source to the surface of the substrate, the deposition pressure, and the spacing between sources. As a good basic rule, the distance between sources should be approximately equal to the distance between the source and the substrate. For better uniformity, the distance between sources should be less than this guide distance.

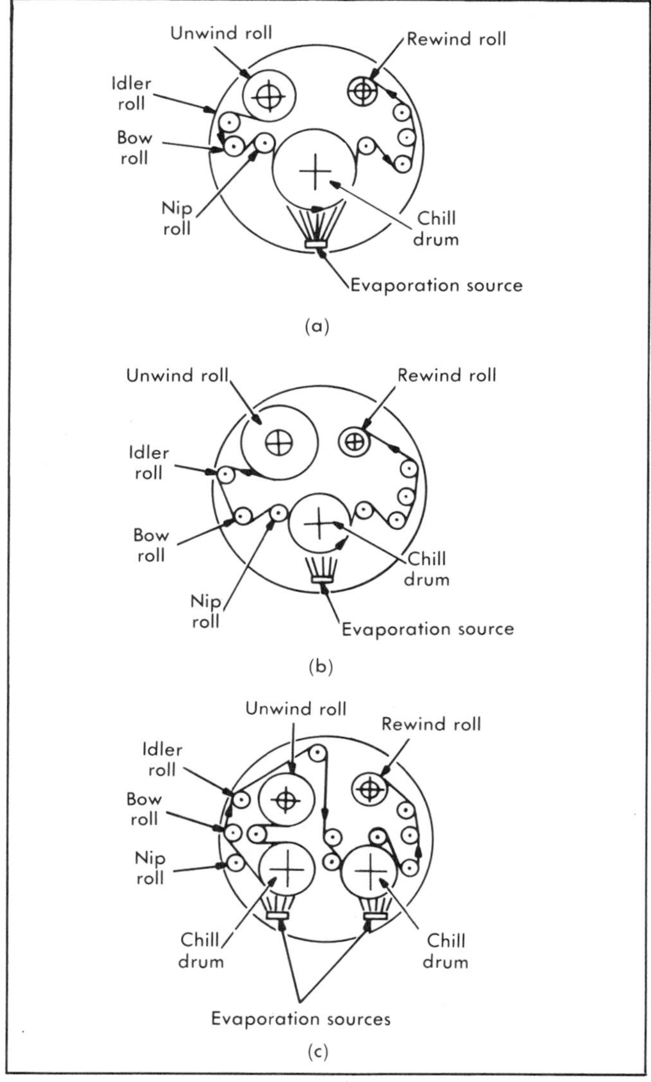

Fig. 24-13 Typical winding configurations for a semicontinuous metallizer: (a) single large drum, (b) single small drum, and (c) double small drum. (*Charles A. Baer Associates*)

Resistance-heated sources. Various types of evaporation sources have been used industrially over the past 30 years. The most frequently used evaporation source is a resistance-heated source of boron nitride/titanium diboride composite. This source is heated by means of a low-voltage transformer (5-15 V) or a saturable core reactor to provide the necessary power to evaporate the aluminum or metal that is being evaporated. The aluminum is fed to the source by means of a wire feed unit that continuously feeds a 0.06" (1.5 mm) diam wire into the cavity on the surface of the evaporation source. The wire feed rate is controlled and adjusted to match the operating temperature of the source to provide the needed evaporation rate.

The resistance sources are operated either in a parallel mode or by individual control. Evaporation sources were generally operated in parallel in the older equipment, but this method

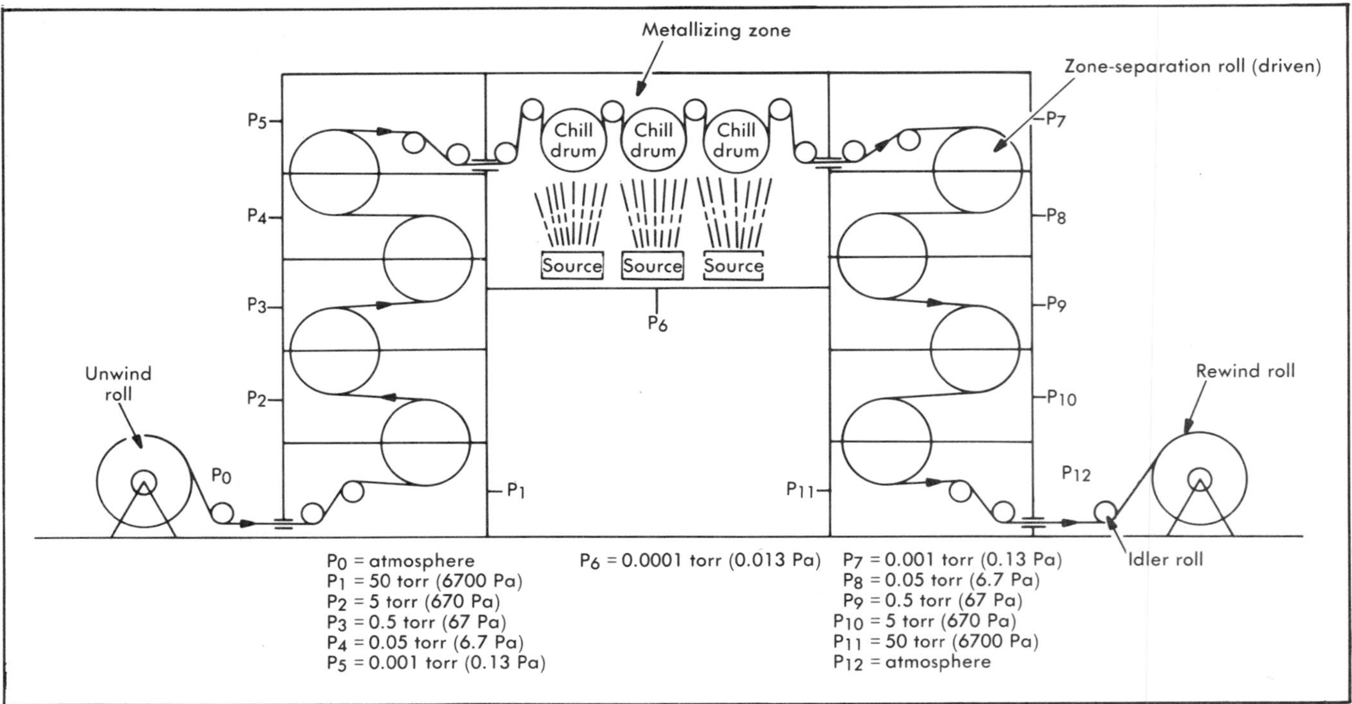

Fig. 24-14 Continuous or air-to-air metallizer for flexible substrates. (*Charles A. Baer Associates*)

introduced a number of problems that adversely affected the uniformity of the deposited metal. Any variation in the resistance of an individual source was a cause for a change in temperature. As evaporation rate is dependent upon temperature of the source and the source area, good uniformity from source to source could be achieved only through the use of sources operating at approximately the same temperature. To achieve good uniformity, it was necessary to use sources with essentially the same resistivity. To add to the problem, the absorption of aluminum into the sources and the amount of metal held within the cavity on the surface of the source were also factors that affected the evaporation rate from the source.

The use of individually controlled resistance-heated sources provides a far superior means of obtaining uniform metal thickness on the substrate when used in conjunction with in-chamber metal thickness detectors. Individual transformers, or saturable core reactors, with silicon-controlled rectifiers are now generally used for controlling the output power and the evaporation rate from the individual sources.

Induction-heated sources. Inductively heated sources are also used for production applications. These sources are generally carbon/graphite crucibles, 5-7″ (125-180 mm) in diameter and approximately 2″ (50 mm) in depth, that are heated by means of induction coils placed around the crucibles. The sources may be operated individually or in series when multiple sources are required. The individual control is by far the more satisfactory method to provide accurate control over the evaporation rate.

The induction power supply is a motor generator or solid-state system providing an output of 9600 Hz, with a power rating of 15 to 30 kW per source. The crucibles used are designed with a nominal 0.5″ (12 mm) thick wall, and the major portion of the heating of the source is from the direct heating of the crucible wall and the conduction of heat to the liquid metal within the crucible. At lower frequencies, or with thinner crucible walls, the power from the generator is coupled to the metal in the crucible, and excessive stirring occurs. The stirring can be sufficient to throw liquid metal from the crucible during operation, with possible damage to the substrate and to other components within the vacuum metallizer.

The induction-heated sources are initially loaded with a charge of aluminum and are not fed during the operating cycle. This procedure currently limits the evaporation time of the source. With the larger rolls of substrate now being processed, with 50,000 to 100,000 ft (15 240 to 30 500 m) of substrate per roll, the 5″ (125 mm) induction sources are limited to being used only when thin metal coatings are acceptable; otherwise, a double bank of sources must be used in order to obtain the quantity of metal necessary for the production run. New 7″ (180 mm) diam sources have been developed as they can supply sufficient metal to process the larger rolls of substrate.

Electron beam sources. Electron beam sources are also being used industrially, but only in limited number. The electron beam sources provide a unique means of evaporating metals without reaction between the source and the metal being vaporized as the container is generally a water-cooled copper hearth that prevents a reaction between the metal being evaporated and the hearth by freezing the metal at the interface. While electron beam sources eliminate one problem, they require the dissipation of a considerably higher amount of energy by conduction to the water-cooled hearth. A broader variety of materials can be deposited with an electron beam source than with the other evaporation sources; coating uniformity can also be controlled more quickly and easily.

For the electron beam sources to operate efficiently, they must be operated at a pressure of 1×10^{-5} torr (0.0013 Pa) or lower to eliminate and avoid arcing. The deposition process is performed in a process zone separated from the other sections of the metallizer by means of low-conductance seals through which the substrate is carried. For single-zone systems, the electron beam sources are quite unsatisfactory.

The electron beam sources can be operated with a fixed volume of melt or by means of a rod feed system. The latter involves the continuous feeding of a rod into the source to provide a continuing supply of metal for vaporization. Wire feed systems are also used when large volumes of metal must be continuously fed for long periods of time.

Comparative economics of the three sources show the resistance-heated sources to be the lowest in capital cost, induction-heated sources next, and the electron beam sources the most expensive.

Pumping system. Pumping systems for roll metallizing equipment must be specifically designed to provide the required pumping for the high gas loads involved. The design of the pumping system involves the selection of suitable mechanical pumps to handle the initial pumping from atmospheric pressure to approximately 20 torr (2700 Pa). The second stage of the pumping system is usually a mechanical booster-type system, or, in some applications, oil-vapor booster pumps. These pumps serve to handle the pumping during the drop in pressure from 20 to 0.1 torr (2700 to 13 Pa). When the pressure reaches 0.1 torr, the diffusion pumping system is activated.

Mechanical pumps. The selection of the mechanical pumps and mechanical boosters, or oil-vapor booster pumps, is determined by the pump-down cycle time desired. With the first stage of the pumping dependent upon the low pumping speeds of the mechanical pumps, nominally 150-900 cfm (4-25 m^3/min), it is necessary to install a sufficient number of mechanical pumps to evacuate the vacuum system to 20 torr (2700 Pa) in approximately 3 to 5 minutes.

The capacity of the booster pumps and the number installed is determined by the gas load during this stage of the roughing cycle and by the forepressure requirements of the diffusion pumps since the mechanical pumps and boosters are used as backing pumps for the diffusion pumping system under high-vacuum operation. The mechanical boosters range in capacity from 1300 to 14,000 cfm (37 to 396 m^3/min) per pump, and the evacuation of the vacuum chamber from 20 to 0.1 torr (2700 to 13 Pa) is accomplished in 1 to 2 minutes or less. The gas load handled by the boosters can be quite high when a paper or solvent-coated film is being processed. If inadequately dried substrates are metallized, the roughing cycle time increases.

Diffusion pumps. The diffusion pumping system must be quite large owing to the heavy evolution of gas from the surface of the substrate during the unwinding, the exposure of the substrate to the vapor stream, and the infrared radiation from the evaporation sources. A large volume of gas is given off from the surface of the substrate, from any residual solvent in or on the surface of the substrate, from trapped air between the layers of the substrate, and from absorbed moisture.

To provide additional pumping capability for condensable gases, cryogenic panels or coils are quite frequently placed within the vacuum chamber or directly over the diffusion pumps to condense these gases and hold them on the cryogenic surface. Since water vapor is the primary condensable gas present, it can be very conveniently and efficiently pumped by operating a cryogenic surface at approximately -180° F (-118° C) or lower. The low temperature can be achieved through the use of mechanical cascade refrigeration systems, or through the use of liquid nitrogen. At temperatures below -148° F (-100° C), the vapor pressure of water is 1 x 10^{-5} torr (0.0013 Pa) or lower and is for all practical purposes pumped from the system. At a liquid nitrogen temperature of -320° F (-195° C), the vapor pressure of water is many decades lower.

Cryogenic pumps are now being used along with the diffusion pumping systems in order to provide even greater pumping of a wider range of condensable gases. With the cryogenic pumps, operating in the range of -420 to -440° F (-250 to -260° C) and supplemented with activated charcoal, many of the residual gases, such as nitrogen, neon, argon, oxygen, and hydrogen, are pumped by condensing them on the cryogenic surfaces or by absorbing them onto the charcoal. The cryogenic pumps provide exceptionally high pumping capabilities and are very efficient additions to metallizing systems.

Through the use of high-efficiency pumping systems, the pump-down cycles for large roll metallizing systems are generally 15-30 minutes. With the shorter pumping cycles, more time is available for the metallizing cycle and for loading and unloading the systems, resulting in much higher productivity for the systems and a lower unit cost for manufacturing the end product.

Speed control. During the metallizing process, the speed should be controlled to obtain uniform coating thicknesses. Speed control is generally achieved through the use of adjustable-speed drives. Various types have been used, including variable-speed hydraulic drives, a-c drives, and d-c drives. The variable-speed d-c drives are most frequently used. With the more sophisticated equipment being currently manufactured, the drives are generally comprised of three or four motors, with one motor controlling the unwind speed and tension, one or two motors controlling the chilled drum (or drums), and one motor controlling the speed and tension on the rewind.

Tension control. Tension control is a major factor in obtaining an acceptable finished product. Variations in tension can introduce wrinkles, creases, fold-overs, and telescoping of the roll during rewinding. As the substrate is rewound under vacuum, with no layer of air between adjacent layers of substrate, any variation in thickness of the substrate is amplified in rewinding. These variations in plastic film are referred to as gauge bands; and unless special care is taken in adjusting tension, these gauge variations form hard bands in the rewound film and can result in stretching and distortion of the film. Tension is monitored and controlled through the use of load cells or dancer rolls mounted within the vacuum system, or through the use of control circuitry and tachometer generators on the various motors.

In order to isolate the tension for the unwind and the rewind sections, it is desirable to provide either a nip roll location or an S-wrap roll. The nip roll provides a positive isolation, while the S-wrap roll provides a degree of isolation dependent upon wrap angle and coefficient of friction between the substrate and the roll. The nip roll, while providing positive tension isolation, can introduce wrinkles or cause damage to the substrate by embedding contaminants in the substrate or by causing substrate perforation. The S-wrap roll minimizes physical damage to the substrate, but reduces tension isolation to a marked extent.

Evaporation Rate

A wide range of evaporation rates can be achieved by the various sources. Depending upon the type of metal that is to be evaporated, its vapor pressure, and the area of the evaporation source, quantities of metal ranging from fractions of a gram per minute to pounds per minute can be vaporized and deposited onto flexible substrates. Fig. 24-15 gives the evaporation rate for aluminum as a function of temperature. The maximum evaporation rate of other metals from a free liquid or solid surface can be determined by the following equation:

$$R = 5.83 \times 10^{-2} P \quad (1)$$

where:

R = evaporation rate, g/cm²/s
P = vapor pressure of metal at metal temperature, torr
M = molecular weight of the metal
T = temperature of metal, Kelvin

The vapor pressure of the metal deposited is dependent upon its temperature. As the temperature increases, the vapor pressure of the metal increases. Table 24-8 shows that there is little correlation between the melting point of a metal and its evaporation rate. In general, the metals with higher melting points are more difficult to vaporize.

BATCH METALLIZING

In batch metallizing, the parts are loaded on fixturing racks and then placed in the vacuum work chamber; as many as several thousand parts may be coated in one load. The smallest part size that can be processed is approximately 1/8″ (3 mm) in diameter by 1/4″ (6 mm) long; smaller parts are more difficult to grip individually. Maximum size is limited only by the size of the metallizer.[39] Coating rates are typically 50 μin./min (1.3 μm/min) and higher for aluminum, and from 5 to 20 μin./min (0.13 to 0.56 μm/min) for most other materials.

The substrates upon which metallic vapor can be deposited include glass, plastics, and metallic materials. The primary requirement of the material to be coated is that it be stable in vacuum. It must not evolve gas or vapor when exposed to the metal vapor.[40]

Coatings Deposited

Virtually all metals, many alloys and semimetallic elements, and innumerable compounds can be deposited by vacuum metallizing. A limiting factor is that as the vapor pressure of the coating material drops, vaporization becomes progressively more difficult because a better vacuum is required and the design and material problems of the vapor source become increasingly complex.

In practice, only a small group of coatings are commonly deposited by vacuum metallizing. Aluminum is the most widely used material because of its high vapor pressure, excellent reflectivity, and attractive appearance. After aluminum, some of the more commonly used materials, in order of decreasing vapor pressure, are selenium, cadmium, silicon monoxide, silver, and copper. Gold, chromium, nickel chromium, palladium, titanium, and magnesium fluoride are also deposited.

Coating Characteristics

Vacuum-metallized coatings are normally very thin and generally conform to the finish of the substrate. Decorative and barrier coatings are usually 0.4-4 μin. (0.01-0.1 μm) thick. Magnetic coatings are 4-5 μin. (0.1-0.13 μm) thick, and transparent conductive coatings are typically 0.2-20 μin. (0.005-0.5 μm) thick. To provide wear and corrosion protection, a protective coating or film may be applied over the metallic layer. Decorative products are generally coated with an organic resin, and optical applications are coated with silicon monoxide, magnesium fluoride, or similar transparent, dielectric coatings. Properly prepared deposits have well-ordered crystalline structures.

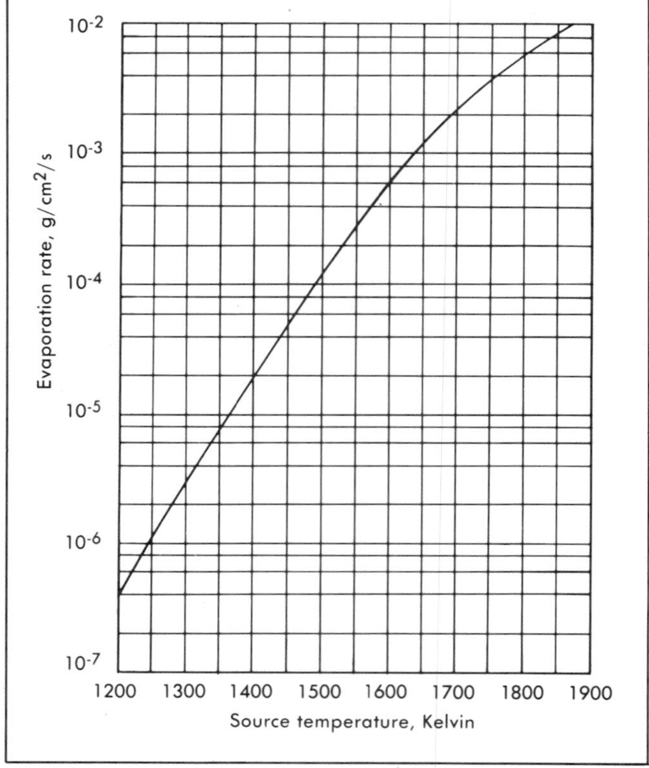

Fig. 24-15 Graph of the evaporation rate of aluminum as a function of temperature. (*Charles A. Baer Associates*)

TABLE 24-8
Vapor Pressure Data for a Select Group of Metals[38]

Metal	Melting Point, Kelvin	Vapor Pressure, torr: 0.001	0.01	0.1	1	10	100
		Temperature of Metal, Kelvin					
Aluminum	932	1370	1495	1650	1845	2090	2410
Copper	1357	1400	1530	1690	1890	2130	2460
Gold	1336	1545	1685	1850	2060	2330	2670
Chromium	2176	1545	1670	1830	2010	2260	2580
Silver	1234	1190	1300	1430	1600	1810	2090
Nickel	1725	1660	1800	1970	2180	2450	2790
Lead	601	900	990	1105	1250	1440	1700
Tin	505	1380	1510	1680	1885	2150	2500
Titanium	1945	1845	2000	2200	2440	2730	3120
Zinc	693	566	618	682	760	870	1010

VACUUM METALLIZING

Applications

Vacuum metallizing was first used as a laboratory method for depositing aluminum and silver on glass substrates. During World War II, the process was widely used for the large-scale production of coated optical instruments. Research also shows that vacuum-metallized coatings can be successfully deposited on plastics and metal castings and stampings to produce highly decorative results.

Vacuum metallizing is most commonly used to deposit decorative aluminum coatings onto plastics and metal substrates. The two types of decorative coatings are referred to as first-surface and second-surface coatings. First-surface coatings are deposited on the outer or front surface of the part and are generally used for applications that require wear or corrosion resistance. For typical application examples such as automotive instrument panels or television control bezels and knobs, a thin, transparent coating is applied over the aluminum to provide wear and corrosion resistance.

Second-surface coatings are deposited on the back side of a transparent material. These coatings can be used for both inside and outside applications because one side of the coating is protected by the transparent substrate and the other side is protected by a heavy coat of corrosion-resistant paint. Second-surface metallizing has been used by the automotive and appliance industries for many years. Typical applications include automotive instrument panels, control panels on washing machines and dryers, and panels for hi-fi radios and television sets. These coatings are also used on exterior automotive parts where the brilliant aluminum coating matches chromium-plated parts.

Vacuum metallizing is also used for depositing functional coatings in the optical, electrical/electronic, and corrosion-resistance fields. In the optical field, these coatings are selected primarily for their reflective or antireflective characteristics. They are used for mirrors varying from rearview mirrors and sealed-beam headlamps to optical components of scientific instruments, such as microscopes, monochromators, and telescopes. In other optical applications, the absorption property of the metal film is used, serving as light attenuators or neutral-density filters to reduce intensity levels. Typical product applications of these coatings are sunglasses and specially graded filters to provide uniform illumination of wide-angle lenses.

A variety of metals and alloys are deposited for use in the electrical/electronics industry as conductors, resistors, and capacitors. Cadmium and aluminum, in thicknesses of 0.25 to 1.0 mil (6 to 25 μm), are used for corrosion resistance.

Process Description

The vacuum metallizing process as used today for decorative purposes includes several steps in addition to the actual evaporation of the thin, brilliant metallic coating. In general, the process includes precleaning, precoating, loading, metallizing, and topcoating.

Precleaning. A clean surface is extremely important for the proper adhesion of coatings deposited by vacuum metallizing. Solvent cleaning and vapor degreasing are common methods used for substrates other than plastics. Plastics materials are generally given a precoat or base coat to ensure adherence and brilliance as well as a suitably smooth surface for optimum appearance. After cleaning, it is advisable to use cotton gloves when handling the parts.

When adhesion is critical, the parts can receive a final cleaning, referred to as glow-discharge cleaning, in the vacuum chamber. In glow-discharge cleaning, the chamber is pumped down to 0.1 to 0.01 torr (13.3 to 1.3 Pa), and 2000-10,000 V are applied across electrodes within the chamber. The current ionizes the residual gases and causes a glow to occur within the chamber. The gaseous ions bombard the surface and mechanically remove any residual or loosely adhering surface contaminants and volatile materials.

Loading. The parts to be metallized are normally clamped or clipped to a parts reel that will eventually be gear engaged on a planetary, rotating work carriage. From this point on, the parts usually remain affixed to the reel until they are removed for packaging. For some applications such as cosmetic closures, the parts are removed from the reel, placed on spindles and sprayed clear or gold, baked on line, and then packed.

Precoating. For plastics substrates used in decorative applications, an organic base coat is required to provide adhesion and a high-gloss surface, which serves as a base for the metallic film. Generally, lacquers are used to coat thermoplastics, and crosslinking systems to coat thermosetting plastics. The base coat also helps fill tiny cracks, scratches, and mold defects, while sealing pores and reducing outgassing. Outgassing increases as the temperature increases and pressure decreases. If this outgassing is not prevented, the pumping cycle time is greater, the deposition of the vaporized metal is impeded, and the resultant appearance and adhesion are adversely affected.

The base coat can be applied by several ways, depending on the size and shape of the parts to be coated. Some of the more common processes include flow coating, dipping, or spraying. The coatings are then cured in a convection or infrared oven.

Metallizing. The loaded reels are placed on a rotating jig or carriage; then the whole assembly is rolled into the vacuum chamber, and the door is closed. The multistage pumping system is activated, and the pressure within the chamber is brought down to 2×10^{-4} to 1×10^{-4} torr (0.067 to 0.013 Pa); under good conditions, pumping requires from 8 to 12 minutes. After the desired pressure has been attained, high-current, low-voltage power is applied to the heating filament. The power is gradually increased to first melt then evaporate the deposit metal. During evaporation, the carriage is rotated within the chamber to expose all the surfaces of the parts to the evaporated metal deposit. Complete metallizing takes approximately 10-60 seconds.

Topcoating. After metallizing, a topcoat is applied over the thin metal coating to provide protection from abrasion and corrosion. The coating may be the same synthetic material applied in the precoating cycle and may be left colorless or dyed. Dyeing can make the aluminum finish look like gold, bronze, copper, brass, or any other metallic hue. The topcoat is then baked in an oven to give the finished part an appearance that is comparable to electroplating. Following baking, the parts are removed from the reels and packaged.

Equipment

Batch metallizing equipment consists of the vacuum chamber, a part fixturing and rotating system, a pumping system, and a source heating system. A vacuum metallizer is essentially a horizontal steel tank, usually 5-7 ft (1.5-2.1 m) in diameter and 5-6 ft (1.5-1.8 m) in length, that has a full-opening hinged door at one end and the pumping equipment at the other end.

Fixturing devices may be either stationary or rotating racks on which the individual parts are attached by clips. The shape of the part determines whether stationary or rotating racks are required. Rotating racks are held by a carriage mechanism that rotates the entire workload about the central axis of the chamber. The carriage may also be stationary while the racks rotate.

The vacuum pumping system generally consists of a mechanical roughing pump, a mechanical blower, and a diffusion pump. The roughing pump reduces the chamber pressure from atmospheric pressure to 20 torr (2600 Pa); then the mechanical blower cuts in automatically and reduces the pressure to 5×10^{-2} torr (7 Pa). At this point, the diffusion pump is opened to the system to decrease the pressure to approximately 1×10^{-4} torr (1×10^{-2} Pa). Some systems include a cold trap that is cooled to -180° F (-118° C) or lower to pump water vapor.

For most materials deposited by vacuum metallizing, the simplest and most convenient heating source is a tungsten wire filament, which heats and holds the material to be evaporated. The source may also be a conical, coiled basket in which chunks of the metal are placed. Induction coils, coupled to metallic or carbon crucibles, are used to evaporate both conductive and nonconductive materials; and electron beam sources are used to evaporate high-temperature, low vapor pressure materials such as tungsten, or when extremely high deposition rates are required for aluminum.

SPUTTERING

Sputtering is the deposition of materials under vacuum onto prepared substrates to produce specific films for both decorative and functional applications. Sputtering differs from vacuum metallizing (evaporation) in that the material is removed from a solid cathode or target instead of being vaporized by a heating source. Rapidly moving gas ions in the vacuum chamber strike a negatively biased target causing metal atoms to be ejected through a transfer of momentum (Fig. 24-16). Subsequently, the ejected target atoms strike and adhere to the substrate surface forming a thin coating that has the same composition as the target. The process dates back to 1852 when W. R. Grove, while studying the electrical conductivity of gases, observed that metallic deposits formed on glass walls. However, it wasn't until 1950 that the ionization phenomenon was understood.

Because the coating material is passed into the vapor phase by a mechanical process rather than a chemical or thermal process, virtually any material is a candidate for coating.[41] Virtually all types of metals can be successfully sputtered, including chromium, stainless steel, titanium, aluminum, copper, brass, tungsten, molybdenum, gold, silver, and tantalum. Alloys and compounds can also be sputtered without altering their original compositions. New compounds can be created by sputtering with a gas background that reacts with a metal or metal alloy to form new materials such as oxides, carbides, and nitrides. Semiconducting and insulating materials may be sputtered by applying a radio frequency potential to the target.

Sputtered coatings can be deposited on both conductive and nonconductive substrates; some metals and nylons may require a primer or pretreatment before sputtering. Base coats are generally applied to the substrates when a bright finish is required. Most sputtered deposits are from 2 to 40 μin. (0.05 to 1 μm) thick.

ADVANTAGES AND LIMITATIONS

Sputtering is a nonpolluting, controllable process and a versatile tool for both decorative and functional finishing. As previously stated, a wide variety of metals and alloys can be successfully deposited. The coatings adhere better than conventional vacuum-metallized coatings. The primary limitation to sputtering is the slow deposition rate, but newer sputtering methods have resulted in faster rates.

APPLICATIONS

Sputtered coatings are used in decorative, decorative/functional, and functional applications. In decorative applications, the coatings are primarily for aesthetic purposes. Typical parts include toys, cosmetic caps, and picture frames. In decorative/functional applications, the coatings provide aesthetic value along with resistance to corrosion and impact, reflectivity, durability, and adhesion. These coatings are often used for appliance endcaps; automotive grills, wheelcovers, and hubcaps; fixtures for plumbing, marine, and electrical use; and knobs, buttons, and door hardware. Typical functional applications include electromagnetic interference shielding, semiconductor fabrication, transparent conductors, printed circuit boards, optical and magnetic storage media, electrical connections, architectural windows, solar control film, and barrier coats.

SPUTTERING METHODS

The sputtering process may be used as a method for either surface coating or etching. Some of the more common coating methods are planar diode, triode, magnetron, and ion gun sputtering. Direct current discharges are generally used for sputtering conductive substrates; a radio frequency (RF) potential must be applied to the target to sputter nonconducting substrates.[42] Reactive sputtering can be performed with the addition of small, controlled quantities of a reactive gas, such as oxygen, to the argon stream.

Planar Diode Sputtering[43]

A planar diode is a two-component system. The sputtering target is the cathode, and the substrate to be coated is the anode (see Fig. 24-17). When a negative potential of several hundred to a few thousand volts is applied to the cathode, a glow discharge (plasma) ignites within the vacuum chamber after the appropriate pressure level is reached. The plasma usually consists of argon gas ions.

With this simple arrangement, deposition rates are low, and it is difficult to avoid contamination in the film. In order to produce results that can be duplicated, it is necessary to control residual gas pressure. Because of the low deposition rates, the base pressure required before backfilling should be less than 1×10^{-7} torr (1.3×10^{-5} Pa). However, the low base pressure is not practical in a production environment. The advantages of diode sputtering are as follows:

- Refractory films can be deposited.
- Insulating films can be deposited.

The limitations of diode sputtering are:

- Good film adhesion is only possible with a few materials.

- Low base pressures are required.
- The source material must be available in sheet form.
- Special holders are required to maintain low substrate temperatures.
- The deposition rates are usually less than 4 +in./min (0.1 +m/min).

Triode Sputtering[44]

In triode sputtering, the apparatus consists of three electrodes, an anode and target, and an additional electron source (see Fig. 24-18). The chamber is held at the required pressure, and electrons are generated by a thermionically heated filament. The electrons are accelerated toward the anode where they ionize a larger portion of the gas molecules. The process that takes place in the triode system is known as electron-supported discharge

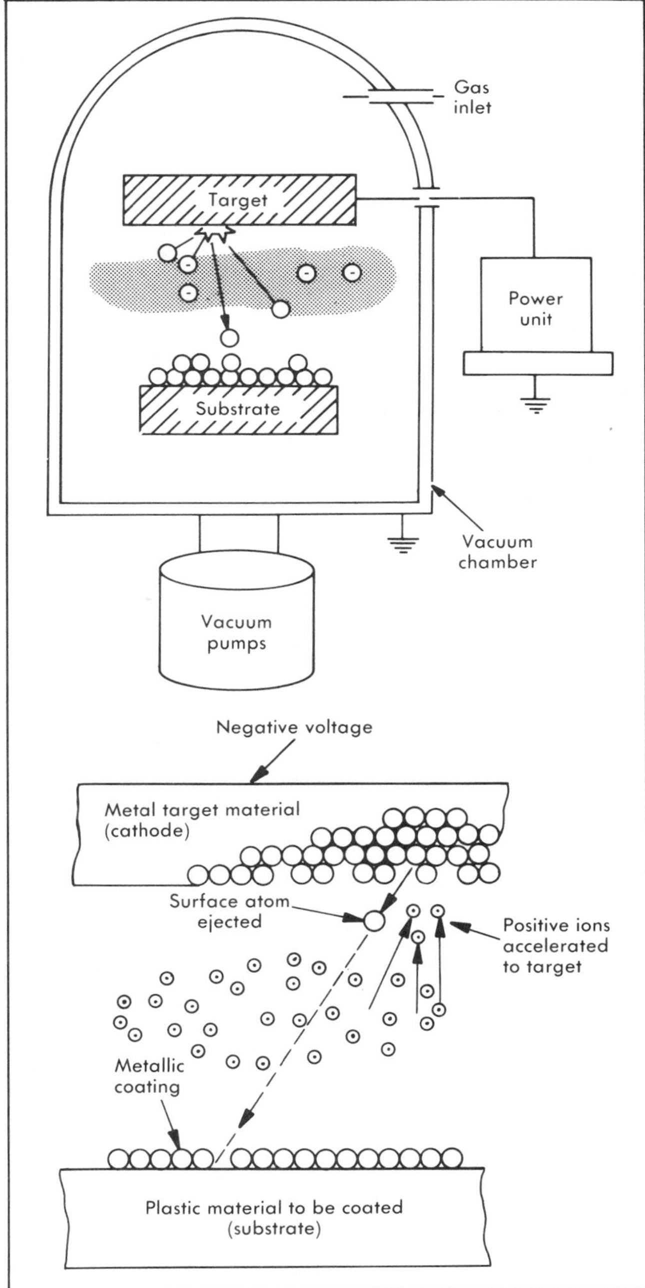

Fig. 24-16 Principles of sputtering operation. (*General Electric*)

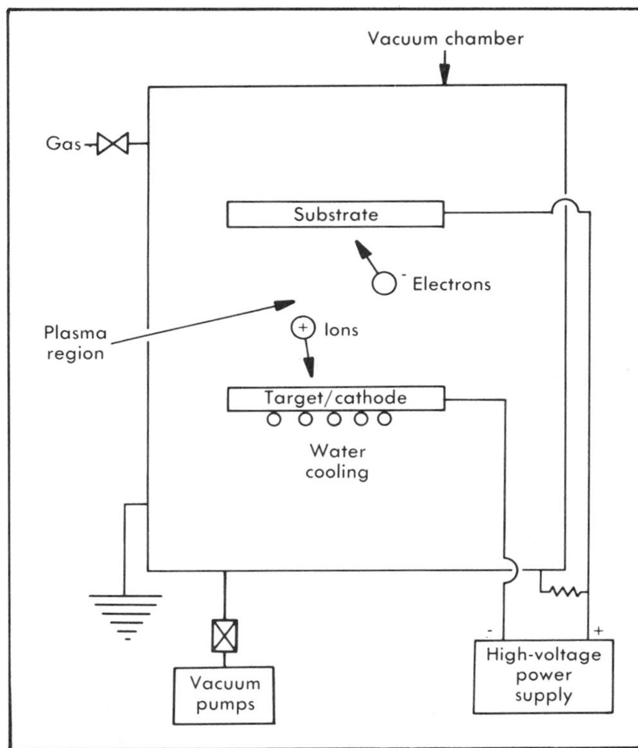

Fig. 24-17 Schematic of planar diode sputtering. (*Technical Enterprises, Inc.*)

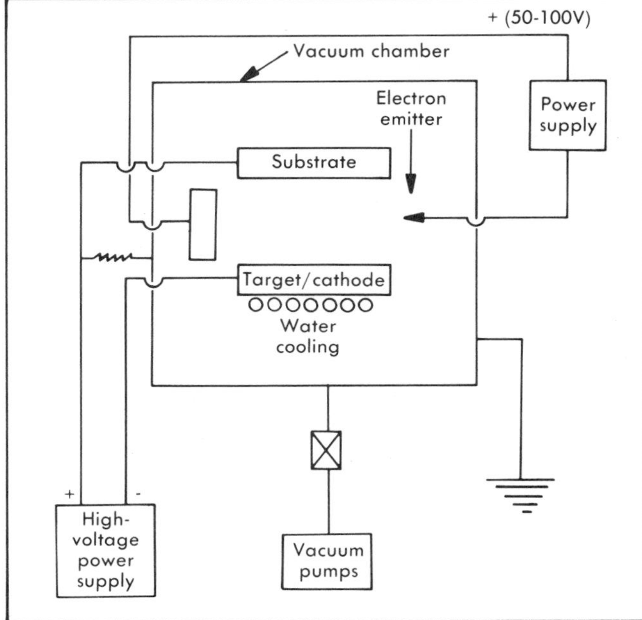

Fig. 24-18 Schematic of triode sputtering.

because it relies on the abundance of electrons generated by the heated filament to ensure sufficient ionizing collisions. The abundance of electrons allows the plasma to operate and sputter material at pressures considerably lower than diode sputtering.

The advantages of low-pressure triode sputtering over diode sputtering are:

- Higher deposition rates.
- Decreased ratio of sputtering gas molecules to sputtered atoms, which means that the film density and purity are enhanced.
- More consistent straight-line deposition from target to substrate, making it possible to use masks to define film patterns.

However, the use of the heated element as a source of free electrons creates the following limitations when compared with diode sputtering:

- The thermionic elements make reactive sputtering with chemically reactive gas impractical or even dangerous.
- The filament contributes to contamination and is subject to burnout.

Magnetron Sputtering

In magnetron sputtering, a magnetic field is applied over the sputtering target to confine the high-density plasma. The argon ions in the plasma are accelerated to the target, which is negatively biased. The collision of the ions with the cathode ejects particles of the target material with mean kinetic energies of 4 to 6 eV. The substrate to be coated is positioned in front of the target, and the particles that strike the substrate condense to form an adherent coating (see Fig. 24-19). A number of target configurations are commercially available, such as planar magnetron and cylindrical-post magnetron; planar targets are more suitable for in-line or semicontinuous process operations, and cylindrical targets are more suitable for batch processes.

High sputtering rates and efficiency are possible with magnetron sputtering because a greater percentage of the current flow is carried by the argon gas. Because of the higher sputtering rates, the sputtered material can be made to cover larger areas. Magnetron sputtering typically takes place at pressures midway between diode and triode sputtering methods.

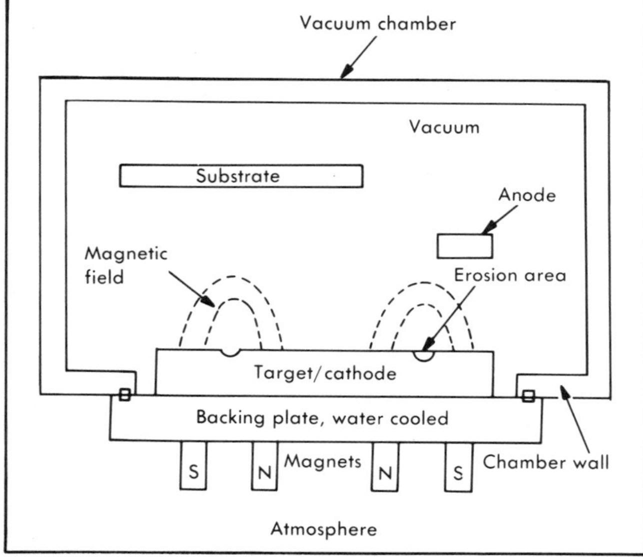

Fig. 24-19 Schematic of magnetron sputtering.

Ion Gun Sputtering

Ion gun sputtering is a method that extracts ions from a Kaufman source and impinges them on a target that becomes eroded; the eroded atoms are then deposited on the substrate. The Kaufman source uses a magnetically enhanced low-voltage discharge that is similar to a cold-cathode gage.

The ions are extracted through a series of screen grids and dumped into a low-gas-density chamber where they can be further accelerated at the target. Since the gas density or pressure is low, few if any collisions take place, resulting in higher energy ions striking the target. Because the background pressure is low, less argon is included in the coating. This method is used for sputtering, etching, and ion implantation; however, it has not been used for production applications.

PROCESS DESCRIPTION

Sputtering is performed on both rigid and flexible substrates. Rigid substrates are sputtered in batch or in-line multichamber equipment, whereas flexible substrates are sputtered in semicontinuous equipment.

Batch Process

A typical sputtering cycle consists of precleaning, racking, pumping down, backfilling with argon gas, sputtering, and unloading. For plastics and some metals, an organic base coat and topcoat may be applied as a separate operation or combined in a production cycle. The base coat helps level the surface of the part, thereby increasing brightness and improving adhesion of the deposit. The base coat also reduces the amount of outgassing that occurs under vacuum conditions. The topcoat, which is generally transparent and colorless, protects the metallic deposit from abrasion and improves corrosion resistance. The topcoat can also be tinted to provide a variety of color shades such as black chromium, gold, pewter, brass, or brushed aluminum.

Cleaning. Cleaning can be discussed only in general because different procedures are required for different circumstances. Some of the factors that influence the cleaning method used are substrate material, coating specifications, and the type of equipment that is readily available for cleaning. The main purpose of cleaning is to reduce surface contamination to an acceptable level. Since all chemical solutions leave some level of contamination on the surface, sputter cleaning is performed before sputtering.

Process. After the parts are loaded onto the racks and placed into the vacuum chamber, the vacuum chamber is closed, either by lowering the vessel onto the baseplate or by closing the door. The mechanical pump is then activated, and the system is pumped down to approximately 0.1 torr (13.3 Pa). At this point, the roughing pump is shut off and the high-vacuum valve opened to allow the high-vacuum pump to evacuate the system to less than 1×10^{-5} torr (1.3×10^{-3} Pa). After the required pressure has been attained, the chamber is backfilled with an inert gas that is easily ionized, such as argon, to a pressure range of 1×10^{-1} to 2×10^{-3} torr (13 to 0.26 Pa). Ionization occurs when a high negative voltage is applied to the target. The ionized gas molecules strike the target, causing atoms to be ejected. The ejected atoms eventually strike and adhere to the substrate surface. After the sputtering is completed, the chamber is vented to the atmosphere. The actual sputtering time depends on the required film thickness. Typical times range from 1 to 20 minutes.

Equipment. The sputtering system consists of vacuum chambers and pumping systems, as well as the sputtering sources, power supplies, and a feed and control system for the sputtering gas. The vacuum chamber in the batch process can be vertically or horizontally oriented, and is approximately 47-55″ (1200-1400 mm) in diameter and 59-87″ (1500-2200 mm) in height or length. Parts are loaded onto racks that can be rotated during sputtering to increase the uniformity of the deposited film. A mechanical pump and one or more high-vacuum pumps—such as diffusion, turbomolecular, cryogenic, or a combination of these—reduce the pressure to approximately 1×10^{-6} torr (1.3×10^{-4} Pa) prior to backfilling the chamber to a pressure of approximately 2×10^{-3} torr (0.26 Pa).

In-Line Sputtering

In-line sputtering, like roll sputtering, is a production-oriented process because large quantities of parts can be coated in relatively short times. Architectural glass, semiconductor wafers, flat-panel displays, and automotive grills are commonly sputtered in an in-line system.

A diagram of an in-line sputtering system is shown in Fig. 24-20. In this system, the parts are placed face up on coating platens and then fed into the entrance chamber; the entrance chamber cycles between atmospheric pressure and a vacuum environment. A transport system moves the platens from chamber to chamber. Generally, at least one conditioning chamber is required to allow outgassing of the substrates. The length of the conditioning chamber depends on the amount of outgassing required; outgassing is dependent on the type of material being coated, the exposed surface area, the length of time that the material is in the chamber, and the temperature of the substrate. Additional chambers can be incorporated in the system to ensure that residual gases and traces of the atmosphere are eliminated.

Before sputtering, the parts can be heated and/or sputter etched. Heating is used to remove gases that have been physically absorbed by the substrate surface, and sputter etching is used to remove thin oxide films or other surface contaminants. Planar magnetron sources are generally used because they have higher deposition rates than planar diode and triode sources.

The number of sputtering sources and the type of target material used vary with the individual coating process. The sputtering source can be oriented to sputter down, up, or sideways. In-line systems can be designed to accommodate multiple sources, either of the same material to increase the production rate or of different materials to deposit a multilayer coating. Following sputtering, the platens move from a vacuum environment to atmospheric pressure through a buffer-and-exit chamber. The main advantage of in-line systems is that the parts and the coating chamber are conditioned prior to sputtering. Another advantage is the ability to unload and reload a batch of parts while another batch is being sputtered.

Roll Sputtering

Roll sputtering, or semicontinuous sputtering, is a process used to deposit functional coatings on flexible substrates such as plastic films. The coating may be applied to either one or both sides of the substrate depending on the roll coater design. Typical applications of sputtered coatings on flexible substrates include transparent, electrically conductive films for liquid crystal displays, electroluminescent displays, touch panels, and membrane switches; electrostatic imaging film for document processors and copiers; electronic shielding; solar energy control films; and high-density magnetic recording media.

Process. In a typical roll coater, the roll of plastic film is mounted on the unwind mandrel, and then the entire roll is placed inside the vacuum chamber. The film is next threaded through idler, drive, and tension control rolls, passed through the sputtering regions, and attached to the rewind mandrel. The pumping system is activated, and the vacuum chamber is evacuated to a pressure of approximately 1×10^{-6} torr (1.3×10^{-4} Pa). After the required pressure has been attained, the chamber is backfilled with an inert gas (argon) to a pressure of approximately 2×10^{-3} torr (0.26 Pa).

During sputtering, the flexible substrate passes through a cloud of charged particles at a specific rate. Positively charged ions in the plasma (cloud of charged particles) strike the negatively biased target, causing the metal atoms to be ejected toward the substrate surface. The ejected metal atoms strike and adhere to the substrate surface, forming a thin coating. Electrons contained within the plasma help ionize the incoming argon gas, thereby increasing the deposition rate. After the entire roll is coated, the vacuum chamber is opened and the roll removed.

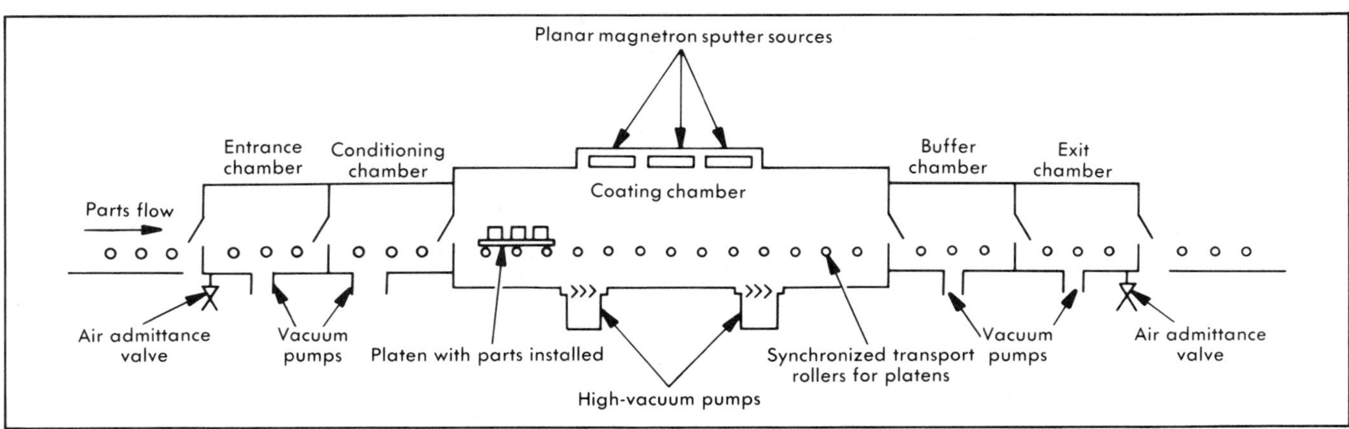

Fig. 24-20 Schematic of in-line magnetron sputtering system for coating architectural glass panels. (*Airco Temescal*)

REFERENCES

Equipment. A typical roll coater consists of a vacuum chamber, sputtering sources and targets, a pumping system, and a winding system. Most production roll coaters can handle rolls up to 80″ (2000 mm) wide and films up to 0.030″ (0.76 mm) thick. Both planar diode and magnetron sources can be used in roll coaters; however, magnetron sources are preferred because of the low heat load generated during sputtering. The sputtering source is generally oriented around a drum. Targets of pure elements, alloys, and compounds can be sputter deposited. New compounds can also be deposited by adding reactive gases into the vacuum chamber during sputtering. Multiple sources can be used to deposit multilayer coatings consisting of a variety of materials.

The pumping system of the roll coater must be large enough to handle the air that is released when the roll is unwound in the vacuum chamber during sputtering. To minimize outgassing (release of trapped air), materials with high vapor pressure may require a separate unrolling and then reloading onto another mandrel while in a vacuum. Outgassing can cause cloudy films and brittleness, and reduced adhesion, electrical conductivity, and reflectivity. The winding system feeds the film over the various rollers and in front of the sputtering source while maintaining a constant speed and tension until the entire roll is coated.

References

1. A. Kenneth Graham, "Plating Bath Compositions and Operating Conditions," *Electroplating Engineering Handbook*, 3rd ed. (New York: Van Nostrand Reinhold Co., 1971), Chap. 6.
2. John M. Blocher, Jr., "Chemical Vapor Deposition," *Metals Handbook*, 9th ed., vol. 5 (Metals Park, OH: American Society for Metals, 1982), pp. 381-386.
3. M. Bonetti-Lang, et. al., "Carbonitride Coatings at Moderate Temperatures Obtained from Organic C/N-Compounds," *Proceedings of the 8th International CVD Conference* (Pennington, NJ: The Electrochemical Society, 1981).
4. W. Hanni and H. E. Hintermann, "Chemical Vapor Deposition of Chromium," *Proceedings of the International Conference—Metallurgical Coatings*, 1976, Vol. II (Lausanne, Switzerland: Elsevier Sequoia, S.A.), pp. 107-114.
5. A. Brenner, "Electrolysis in Nonaqueous Systems in Delahay," *Advances in Electrochemistry and Electrochemical Engineering*, Vol. V (New York: John Wiley and Sons, Inc., 1967), p. 217.
6. Brenner, *loc. cit.*
7. M. Bonetti-Lang, *loc. cit.*
8. A. J. Caputo, "Fabrication and Characterization of Vapor-Deposited Niobium and Zirconium Carbides," *Proceedings of the International Conference—Metallurgical Coatings*, 1976, Vol. II (Lausanne, Switzerland: Elsevier Sequoia, S.A.), p. 50.
9. H. O. Pierson and A. W. Mullendore, "The Chemical Vapor Deposition of Boron from DiBorane," *Proceedings of the Seventh International Conference on Chemical Vapor Deposition* (Princeton, NJ: The Electrochemical Society, 1979), pp. 360-367.
10. G. Male and D. Salanoubat, "Preparation of Pyrolytic Boron Nitride (PBN) by C.V.D. and Reduced Pressure," *Proceedings of the Seventh International Conference on Chemical Vapor Deposition* (Princeton, NJ: The Electrochemical Society, 1979), pp. 391-397.
11. L. R. Newkirk, et. al., "Preparation of Fiber Reinforced Titanium DiBoride and Boron Carbide Composite Bodies," *Proceedings of the Seventh International Conference on Chemical Vapor Deposition* (Princeton, NJ: The Electrochemical Society, 1979), pp. 515-524.
12. Newkirk, *loc.cit.*
13. R. W. Kidd, M. F. Browning, and J. M. Rusin, "Chemically Vapor Deposited Coatings for Multibarrier Containment of Nuclear Waste," *Proceedings of the Seventh International Conference on Chemical Vapor Deposition* (Princeton, NJ: The Electrochemical Society, 1979), pp. 563-577.
14. B. Armas and M. Morales, "Chemical Vapor Deposition at Low Pressure in the Iron-Boron System," *Proceedings of the Seventh International Conference on Chemical Vapor Deposition* (Princeton, NJ: The Electrochemical Society, 1979), pp. 618-631.
15. Richard P. Vento, *Low Pressure-High Temperature Chemical Vapor Deposition—An Applications Guide*, SME Technical Paper No. AD83-870 (Dearborn, MI: Society of Manufacturing Engineers, 1983).
16. B. O. Seraphin, "Chemical Vapor Deposition of Thin Semiconductor Films for Solar Energy Conversion," *Proceedings of the International Conference—Metallurgical Coatings*, 1976, Vol. I (Lausanne, Switzerland: Elsevier Sequoia, S.A.), pp. 87-94.
17. E. Fitzer and D. Kehr, "Carbon, Carbide and Silicide Coatings," *Proceedings of the International Conference—Metallurgical Coatings*, 1976, Vol. I (Lausanne, Switzerland: Elsevier Sequoia, S.A.), pp. 55-67.
18. Fitzer, *loc. cit.*
19. D. M. Mattox, *Film Deposition Using Accelerated Ions*, Sandia Corporation Development Report No. SC-DR-281-63 (Oak Ridge, TN: Division of Technical Information Extension of Sandia Corp., November 1963).
20. M. J. Paleen, *Summary of Tests on Ivadizer® Aluminum Coated Fasteners*, MDC Report No. A5517 (St. Louis, MO: McDonnell Aircraft Co., April 4, 1979).
21. E. R. Fannin, "Aluminum Coated Fasteners by Ion Vapor Deposition," *Fastener Technology* (August 1978), pp. 25-29.
22. Military Specification MIL-C-83488, "Coating, Aluminum, Ion Vapor Deposition."
23. D. E. Muehlberger and J. J. Reilly, *Improved Equipment Productivity Increases Applications for Ion Vapor Deposition of Aluminum*, SAE Reprint No. 830691 (Warrendale, PA: Society of Automotive Engineers, 1983).
24. Military Specification MIL-C-5541, "Chemical Conversion Coating for Aluminum Alloys."
25. E. R. Fannin, *Ion Vapor Deposited Aluminum Coatings for Improved Corrosion Protection*, Presented at AGARD meeting in Florence, Italy, 26-28 September 1978, MCAir Report No. 78-007 (St. Louis, MO: McDonnell Aircraft Co., September 1978).
26. C. E. Toth, "Transition Metal Carbides and Nitrides," *Refractory Metals*, vol. 7 (New York: Academic Press, 1971).
27. J. M. E. Harper, H. T. G. Hentzell, and J. J. Cuomo, "A Quantitative Ion Beam Process Applied to the Deposition of Aluminum Nitride Thin Films," *Journal of Vacuum Science and Technology* (April-June 1984), p. 405.
28. S. Komiya and K. Tsuruoha, "Thermal Input to Substrate During Deposition by Hollow Cathode Discharge," *Journal of Vacuum Science and Technology* (January-February 1975), p. 589.
29. A. I. Vasin, A. M. Dorodonov, and V. A. Petrosov, "Vacuum Arc with a Distributed Discharge and an Expendable Cathode," *Soviet Technology Physics Letters* (December 1979), p. 634.
30. D. M. Mattox, "Film Deposition Using Accelerated Ions," *Electrochemical Technology* (September-October 1964), p. 1385.
31. R. F. Bunshah and A. C. Raghuram, "Activated Reactive Evaporation Process for High Rate Deposition of Compounds," *Journal of Vacuum Science and Technology* (November-December 1972), p. 1385.
32. A. Matthews and D. G. Teer, "Characteristics of a Thermionically-Assisted Ion Plating System," *Thin Solid Films* (June 1981), pp. 41-48.
33. S. Komiya, *loc. cit.*
34. E. Moll and H. Daxinger, "Method and Apparatus for Evaporating Materials in a Vacuum Coating Plant," U.S. Patent 4,197,175.
35. M. Scherer and P. Wirz, "Reactive High Rate D. C. Sputtering of Oxides," *Thin Solid Films* (December 1984).
36. S. Schiller, G. Beister, and W. Sieber, "D. C. Sputtering: Deposition Rate, Stoichiometry, and Features of TiO and TiN Films with Respect to the Target Mode," *Thin Solid Films*, Vol. III (1984), pp. 259-268.

37. S. Schiller, et. al., "Deposition of Hard, Wear Resistant Coatings by Reactive D. C. Plasmatron Sputtering," *Thin Solid Films* (December 1984).
38. R. E. Honig and D. A. Kramer, "Vapor Pressure Data for the Solid and Liquid Elements," *RCA Review*, vol. 30, no. 2 (June 1969), pp. 295-297.
39. David V. Rigney, "Vacuum Coating," *Metals Handbook*, 9th ed., vol. 5 (Metals Park, OH: American Society for Metals, 1982), p. 397.
40. *Ibid.*, p. 396.
41. John A. Thorton and Wolf-Dieter Munz, "Sputtering," *Metals Handbook*, 9th ed., vol. 5 (Metals Park, OH: American Society for Metals, 1982), p. 412.
42. Thorton, *loc. cit.*
43. Russel J. Hill, *Physical Vapor Deposition* (Berkeley, CA: Airco Temescal, 1976), pp. 104-105.
44. *Ibid.*, pp. 105-106.

Bibliography

Bunshah, R. F., et. al. *Deposition Technologies for Films and Coatings*. Park Ridge, NJ: Noyes Publications, 1982.

Chapman, B. *Glow Discharge Processes*. New York: John Wiley and Sons, 1980.

Stuart, R. V. *Vacuum Technology, Thin Films, and Sputtering*. New York: Academic Press, 1983.

Vossen, J. L., and Kern, W., eds. *Thin Film Processes*. New York: Academic Press, 1978.

SPECIAL PROCESSES

FLEXIBLE OVERLAYS

CHAPTER CONTENTS:

A new process has recently been developed for applying hard facing to workpiece surfaces that are susceptible to wear. The process consists of applying both a flexible clothlike material containing a hard, wear-resistant metal or ceramic powder and a cloth containing a brazing alloy to the surface of a workpiece. During the heating cycle, the metal powders are fused to the surface of the workpiece.

The flexible overlay process offers advantages over some of the more conventional methods of applying hard facing, which are discussed in Chapter 22, "Thermal Spraying and Hard Facing," of this volume. Table 25-1 compares the flexible overlay process (Conforma Clad®) to conventional hard-facing methods.

ADVANTAGES AND LIMITATIONS

The major advantage of the flexible overlay process is the ability to coat selected areas of the workpiece while retaining excellent dimensional control of the coating. Another important advantage is flexibility in the selection of coating materials since the process is not constrained by powder chemistry, particle shape, or size distribution. In addition, material utilization is high since there is no overspray and cloth remnants can be easily recycled with no loss in coating properties. Finally, the microstructure of the coating is homogeneous and does not contain the slag, entrapped oxide, or unmelted particles that are sometimes associated with thermal spray processes.

The major limitation of the flexible overlay process is its inability to lay down thin coatings. Coating thicknesses are usually 0.01 to 0.1" (0.25 to 2.5 mm). Another limitation of this new process involves the high temperatures required for the brazing cycle. Since the whole workpiece is heated, grain growth and loss of temper may occur. The brazing furnace capacity also limits the size of the workpiece that can be hard faced.

APPLICATIONS

Flexible overlays can be used for applications requiring high abrasion resistance, erosion resistance, and/or impact resistance. Examples of such applications are debarkers and chain saw teeth in the forest products industry. High-volume production of coated components for business machines is routinely handled through resistance fusion. In the oil drilling and mining industries, rock bits, attack points, driveshaft caps, and drill collars have been successfully coated. Other applications for flexible overlays include extrusion dies and screws, combustion fan liners, valves, turbine tips, rocker arms, agricultural fan blades, and wear tiles.

AVAILABLE COATINGS

Various composite coatings for hard-facing applications are available, with the hard particles generally being tungsten carbide and/or tungsten carbide with cobalt; the brazing alloy is generally nickel-based. By varying the proportion of brazing alloy and carbide, or by changing the carbide volume fraction using particle size control, coatings can be offered with hardnesses up to R_C 72. The performance characteristics of the coating can also be changed by varying the toughness of the carbide particles or by changing the brazing alloy composition. The intercarbide particle spacing is another variable that can be controlled to provide a coating with specific properties. Generally, for the low end of hardnesses (R_C 40-60), single-alloy coatings made from a cobalt-based or a nickel-based alloy are preferred. Within the composite coatings family, a chromium carbide hard particle coating is also offered, with a nickel-based alloy matrix. Experimental coatings have also been developed containing molybdenum tungsten carbide or tungsten titanium carbide.

PROCESS

The flexible overlay process, illustrated in Figs. 25-1 and 25-2, uses a two-cloth approach developed by Imperial Clevite, Inc. called Conforma Clad®. Figure 25-1 presents a breakdown of the process in intermediate steps. Figure 25-2 depicts how the metal-impregnated cloth is fused to the surface of the workpiece. A variation of the two-cloth technique uses a single cloth made from a mixture of brazing alloy and hard particle powders. Instead of the composite coating, which is comprised of hard particles dispersed within a brazing matrix, single-alloy coatings can also be used. The control of the coating geometry becomes more demanding during

Contributors of sections of this chapter are: ***Manek Dustoor****, Manager, P/M Research, Powder Metal Products Div., Imperial Clevite, Inc.;* ***John H. Eggleston****, Product Manager, Lindberg, A Unit of General Signal Corp.;* ***Robert D. Fisher****, Manager, Marketing Development, Lindberg, A Unit of General Signal Corp.;* ***Dr. Wesley H. Weisenberger****, President, Ion Implant Services.*

Reviewers of sections of this chapter are: ***Dr. Arnold H. Deutchman****, President, BeamAlloy Corp.;* ***Manek Dustoor****, Manager, P/M Research, Powder Metal Products Div., Imperial Clevite, Inc.;* ***Robert D. Fisher****, Manager, Marketing Development, Lindberg, A Unit of General Signal Corp.;* ***Robert F. Hochman****, Professor, Metallurgy Program, Georgia Institute of Technology;* ***Larry N. Moskowitz****, Amoco Research Center, Standard Oil Co.;* ***Robert J. Partyka****, Vice President, BeamAlloy Corp.;* ***Gilbert Saltzman****, Technical Director, Research, Metallurgical Industries, Inc.;* ***Dr. Wesley H. Weisenberger****, President, Ion Implant Services.*

CHAPTER 25

FLEXIBLE OVERLAYS

TABLE 25-1
Comparison of Hard-Facing Methods

Method	Dilution	Bond Strength, ksi (MPa)	Thickness, in. (mm)	Surface Finish	Porosity	Substrate Temperature, °F (°C)	Dimensional Control	Material Conservation	Coating Homo-geneity
Weld overlay	Medium-high	High	0.04 (1.0) and up	Poor	Low	High locally	Poor	Good	Good
Flame spray	Low	2-4 (15-30)*	0.02-0.1 (0.5-2.5)	Good	5%	1000 (540)	Fair	Poor	Fair
Spray and fuse	Low-medium	High	0.01-0.07 (0.8-1.8)	Very good	Low	1800-2000 (980-1095)	Fair	Fair	Fair
Plasma spray	Low	6-10 (40-70)*	0.005-0.02 (0.13-0.5)	Good	5-10%	300 (150) and up	Fair-good	Poor	Good
Plasma transferred arc	Medium	High	0.03 (0.75) and up	Fair	Low	High locally	Fair	Good	Fair
Detonation gun	Low	6 - >25 (41 - >172)†	0.002-0.01 (0.05-0.25)	Very good	Low	300 (150)	Good	Poor	Good
Cemented carbide (cobalt binder)	None	NA	0.1 (2.5) and up	Excellent	None	Varies with braze	Excellent	NA	Good
Conforma Clad	Low	High, >30 (>205)	0.01-0.1 (0.8-2.5)	Good	Low	1800-2150 (980-1175)**	Excellent	Excellent	Good

(*Imperial Clevite, Inc.*)

* Bond coats often needed for adequate bond strength.
** Except resistance fusion.
† Depends on coating material; most bond strengths are greater than 25 ksi (172 MPa).

fusion with the single-alloy materials than with the composite materials when dimensional control is dictated by the relatively inert refractory material, such as tungsten carbide. Flexible overlay coatings have been fused on plain-carbon steels, high-alloy steels, stainless steels, nickel-based superalloys, and cast irons.

Cloth Generation

The flexible overlay cloth is generated by mixing very low percentages, typically 3 to 6% by volume (0.3 to 0.6% by weight), of polytetrafluoroethylene (PTFE) with metal or ceramic powder particles. The PTFE is then formed into fibers by a proprietary process to develop a network that entraps the powder particles into a flexible sheet. The binder volume required in this process is lower than what is conventionally used in the manufacture of other, less flexible products, such as brazing tapes. The low binder volume allows for a higher particulate density, with fewer problems associated with binder removal and residue. The strength of the cloth can be varied from 5 to 200 psi (35 to 1379 kPa), depending on processing parameters, binder content, powder particle morphology, and particle size distribution. The density of the cloth can also be varied from about 45 to 70% of its theoretical value. The cloth thicknesses generally used range from 0.030 to 0.060″ (0.76 to 1.52 mm), although experimental samples have been made as thin as 0.002″ (0.05 mm).

Cleaning

The substrate preparation requirements are similar to those required for thermal-sprayed coatings. Grit blasting or chemical pickling is generally used to remove surface oxides. If necessary, a degreasing step is also included. On occasion, a part that is not heavily oxidized may require little or no surface preparation, depending on the fusion process variables that are used.

Application

The cloth is either cut and then placed over the part, or it is draped over the area to be coated and trimmed off. A proprietary adhesive is applied to the cloth to hold it in place. The application step can be labor intensive, depending on the complexity and accessibility of the area to be coated. However, a high degree of automation can be incorporated if justified by the production volume.

Fusion

Brazing is conducted in a conveyor belt furnace with a controlled atmosphere such as hydrogen (as indicated in Fig. 25-1), or in a vacuum furnace. The melting point of the brazing alloys typically used is in the range of 1900 to 2000° F (1040 to 1100° C). While these high temperatures are economical for certain component geometries and substrate materials, undesirable metallurgical changes or thermal distortion of the

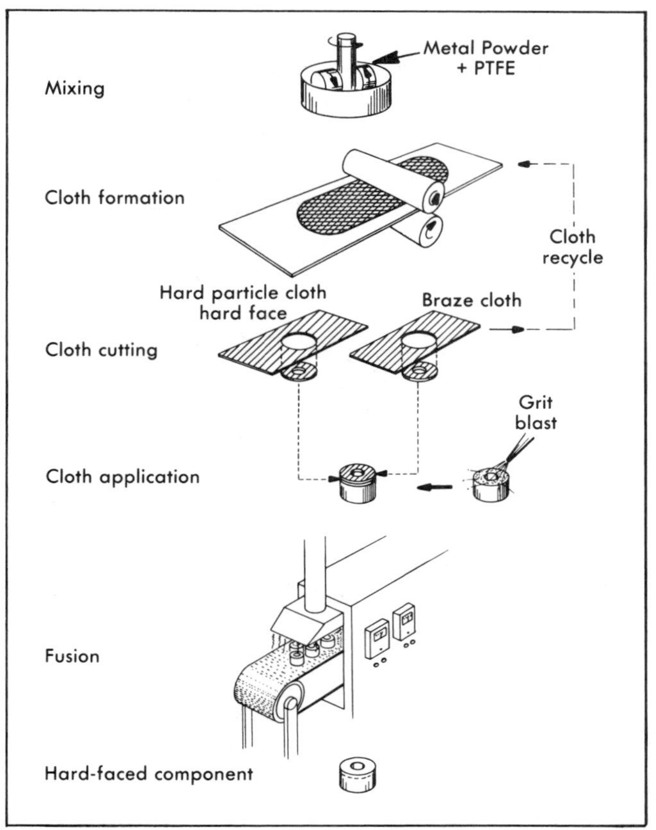

Fig. 25-1 Flexible overlay process (Conforma Clad®) for applying hard-facing coatings. (*Imperial Clevite, Inc.*)

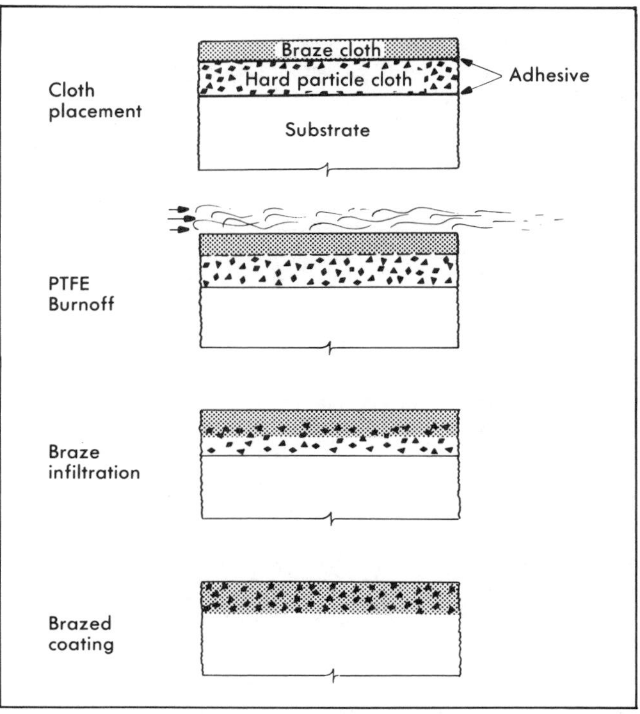

Fig. 25-2 Sequence involved in the infiltration of the brazing alloy in composite coatings. (*Imperial Clevite, Inc.*)

substrate material may occur as a result of the heating. Resistance fusion, generally applicable for small-area coatings, can be used to minimize metallurgical changes or thermal distortion. In resistance fusion, the thermal energy to melt the brazing alloy is generated by resistance heating and is similar to a resistance welding process. Lasers and induction heating coils have also been used to melt the brazing alloy.

The coated part can generally be used in the as-brazed condition since the surface finish is in the range of 100 to 1000 μin. (2.5 to 25 μm) rms. When necessary, standard grinding and lapping practices for carbide coatings can also be utilized to give a 3 μin. (0.08μm) rms finish.

STEAM TREATING

Treating parts in a steam atmosphere is an older process that is being increasingly used by present-day metalworkers owing to the escalating production costs of other processes. During steam treating, a tightly adherent oxide film (Fe_3O_4) is formed on the surface of ferrous parts. The film is usually a blue-black color and approximately 0.0002″ (5 μm) thick.

ADVANTAGES AND LIMITATIONS

Steam treating is a clean and safe process that requires only heat and steam to produce the coating on parts. Supervision and labor during the process are minimal. Further, steam treating can be combined with other heat processing operations.

The materials being treated must be capable of withstanding the processing temperatures, usually 400 to 1200° F (200 to 650° C). Since stainless steels do not react to the steam atmosphere, they cannot be treated by the process. Finally, steam treating is usually limited to parts that can be treated in a batch process.

APPLICATIONS

Steam treating can be used in both ferrous and nonferrous applications. In most nonferrous applications, the steam provides a protective atmosphere during annealing or aging operations. Generally, the natural color of the metal is retained.

Ferrous Materials

Almost all steam treating applications for ferrous parts stem from the benefits of the oxide film (Fe_3O_4) developed on the work surface. In general, steam treating can be used to provide improved lubricity, seal porous materials, provide a base for paint adhesion, develop black body conditions and interlaminar resistance, and provide corrosion resistance and the final product color. Steam treating is also used solely as a scale-free atmosphere for low-temperature annealing or stress relieving of in-process parts.

Lubricity. One of the best known steam treating applications is for high-speed cutting tools. The oxide film formed by the treatment holds the cutting lubricant and prevents chip buildup behind the cutting edge, extending tool life as much as four times between sharpening operations. Table 25-2 indicates the tool life improvement of various tools treated by steam. Most broaches and milling cutters require retreating after sharpening because the beneficial oxide coating has been ground away. Drills, however, do not normally need to be retreated after they have been sharpened. Steam treatment is also performed to impart better wear and wear-in properties to cast iron parts such as piston rings, valve tappets, and cam shafts.

Sealing of porosity. Steam treating is used widely by the powdered metal industry to build up iron oxide formations within the pores of low and medium-density sintered components. Items such as powdered metal air conditioning components and automotive shock absorber pistons benefit by reduced leakage of freon gas and hydraulic fluid respectively. In a few applications, steam treating has been employed to seal microscopic porosity in thin-walled iron castings.

Base for adhesion. Some spot-welded household appliance components are steam treated in preparation for painting. Steam treating prevents the entrapment of chemical solutions between sheet metal surfaces, solutions that could bleed out later and spoil the finish. The oxide coating on steel also provides an excellent surface for different bonding agents.

Black body conditions. Color television tube shadow masks are treated to create a black body condition on their surfaces. Steam treating improves radiation characteristics of the shadow masks, which aids heat dissipation. X-ray tube components are also processed by steam treating.

Interlaminar resistance. Steam treating has gained some acceptance as a means for developing interlaminar resistance and corrosion resistance for silicon-steel laminations. Loose laminations in both bundles and assemblies, such as die cast motor rotors, stators and relay cores, are successfully treated.

Corrosion resistance. Corrosion resistance varies from a slight improvement on mild steel and powdered iron parts (24 hours in a 5% salt-spray environment), to excellent improvement on cast iron products (160 hours in a salt-spray environment). The oxide coating is porous and therefore absorbs oil and proprietary rust inhibitors, which further improves corrosion resistance.

Final color. Products including hand tools, hand power tool components, industrial spark plug shells, carbide toolholders, saw blades, fasteners, and toy components are steam treated either for combination final temper and color or solely for color. Proprietary rust inhibitors enhance both the corrosion resistance and the color.

Nonferrous Materials

Steam is essentially nonreactive as an atmosphere when heat treating nonferrous materials, such as copper, brass, silver, and beryllium copper products. Accepted applications include electrical contacts, connectors, and instrument components made of copper, beryllium copper, and silver. Brass product applications include cosmetic cases, band instrument components, swivel faucets, gas appliance hoses, fittings for industrial gas accessories, and automotive tire stems.

TABLE 25-2
Steam Treating Tool Improvements

Tool	Application	Tool Life	
		Before Steam Treating	After Steam Treating
Drills	Drilling Bakelite plastic insulating blocks.	10 holes	25 holes
	Phenolic terminal plate.	1700 holes per grind	8500 holes per grind
	Drilling AISI 4340 steel 1″ (25 mm) thick.	17 holes	81 holes
Milling cutters	Milling 2 slots 21 15/16″ (557.2 mm) long x 0.130″ (3.3 mm) wide in 1020 steel.	150 cuts per grind	308 cuts per grind
	Slotting 1020 steel type bars.	2000 cuts per grind	7000 cuts per grind
Hobs	Cutting teeth on AISI 3140 steel forged gear.	62.5% increased life	
Broaches	Cutting AISI 1010 steel latch.	20 hours per grind	70 hours per grind
Punches	Cutting 1/2″ (13 mm) thick steel plate.	1000% increased life	
End mill tools	Cutting 8740 steel forgings.	30 pieces	200 pieces
Taps	Cutting SAE 52100 steel.	1800 pieces	3000 pieces
Saw blades	Cutting 3″ (75 mm) rods of 17a Cyclops austenitic steel.	100% life expectancy at 102 fpm (31 m/min)	120% at 112 fpm (34 m/min)

(*Lindberg, A Unit of General Signal Corp.*)

STEAM TREATING

PROCESS DESCRIPTION

Steam treating is performed in heat treating furnaces. The actual process cycle for ferrous parts is different than for nonferrous parts. Before loading the parts into the furnace, it is important that the parts are dry to produce a clean and scale-free coating.

Cleaning and Loading

For best surface appearance, the parts should be clean and bright prior to processing because the steam atmosphere does not clean the parts. Washing and rinsing or vapor degreasing is recommended for all items except powdered iron parts and laminations containing entrapped lubricant; powdered iron parts are usually cleaned using the burnoff method. Ferrous parts should also be free of rust because the ferrous oxide (Fe_3O_4) breaks down after a short period of time in locations that have rusted.

Parts may be dump, random, or stack loaded into the furnace. Line contact of round parts presents no problem, and bundles of electrical laminations are also successfully steam treated. If uniform color on ferrous materials is a requirement, pieces with flat surfaces must be separated. Delicate items such as color television tube shadow masks must be fixtured.

Ferrous Parts

After loading ferrous parts into the furnace, they are heated to 700° F (370° C) without steam. [If steam comes in contact with the parts before they reach 212° F (100° C), the moisture will form rust on the surface of the parts.] Steam is then injected into the work chamber for a period of time to purge all air from the chamber. (If air is present during steam treatment, the parts exhibit a nonuniform color, and the oxide that forms on the surface is nonadhering.) Following air purging, steam continues to flow while the work is heated to a specific temperature within the range of 800 to 1200° F (425 to 650° C). The parts are allowed to soak at this temperature from ½ to 3 hours. Soak time and temperature depend on the material and results desired. Most ferrous products can be unloaded at soak temperature without scaling. A diagram of a typical steam treatment cycle for ferrous metals is presented in Fig. 25-3.

The processing cycle for powdered iron parts varies considerably from wrought iron parts. Additional information on treating powdered iron parts can be obtained from the sources listed in the bibliography at the end of this chapter.

Nonferrous Parts

Processing nonferrous materials consists of loading the parts into the furnace and then heating to 300° F (150° C). Steam is injected for a period of time to remove all air from the furnace and to raise the temperature to between 400 and 1200° F (200 and 650° C). After the required heat treatment has been achieved, the parts are either removed from the furnace and water quenched, or they are cooled back to purge temperature while remaining in the furnace in a steam atmosphere. A diagram of this cycle is presented in Fig. 25-4.

EQUIPMENT

Parts are processed in heat treating furnaces specifically designed for use with steam as an atmosphere. The steam serves solely to react with or protect the work surfaces. The furnaces are usually heated by electrical resistance elements and can be designed to accommodate parts in batches or in continuous operation.

Batch Type

Most commercially available equipment is of the cylindrical top-loading pit-type design (see Fig. 25-5). Workspace sizes generally range from 12 to 50″ (305 to 1270 mm) in diameter and 15 to 96″ (380 to 2440 mm) deep. Steam for the atmosphere is supplied from plant steam or from a separate steam generator. The steam source pressure is usually 5 psi (34.5 kPa), and operating pressure within the work chamber is typically 4″ (102 mm) of water. Atmosphere circulating fans distribute the steam and heat uniformly.

Continuous Type

Continuous steam treating equipment is used in the electrical lamination industry. Some of the continuous steam treating units are in line following the annealing zones, with the work being moved through both the annealing and steam treating zones by a common mechanism. Other units are not used in conjunction with a heat treating line.

Currently, there are no standard preengineered continuous steam treating furnaces commercially available; consequently,

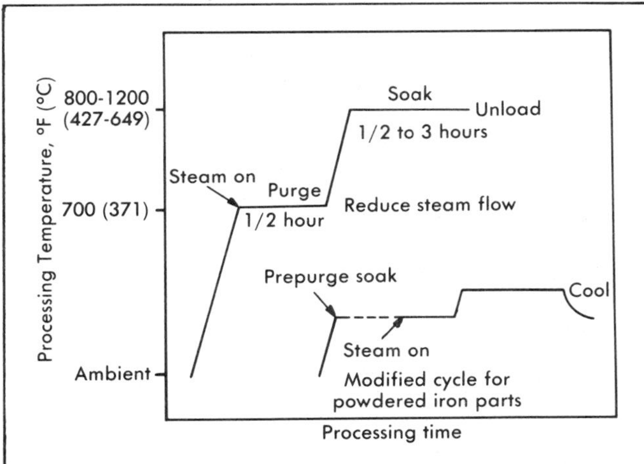

Fig. 25-3 Process cycle for steam treating ferrous parts. (*Lindberg, A Unit of General Signal Corp.*)

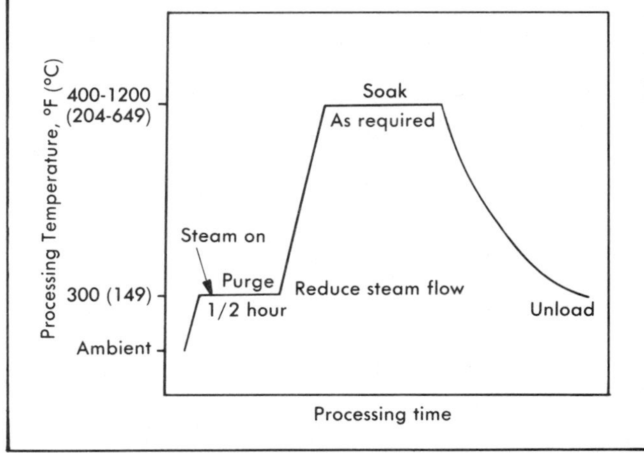

Fig. 25-4 Process cycle for steam treating nonferrous parts. (*Lindberg, A Unit of General Signal Corp.*)

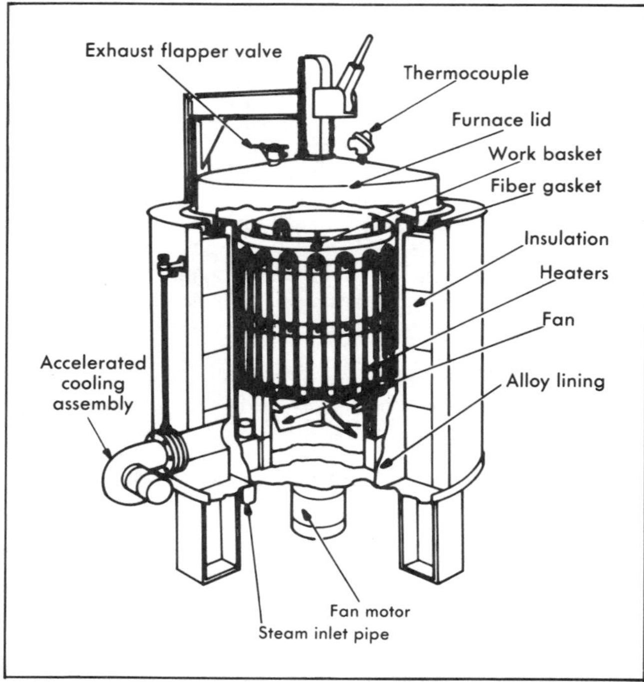

Fig. 25-5 Cutaway view of a batch-type steam treating furnace. *(Lindberg, A Unit of General Signal Corp.)*

these units must be designed for specific applications. Economic use of such equipment therefore depends upon high production rates, with little change in product configuration or steam treating time/temperature cycles.

SAFETY AND ENVIRONMENTAL CONSIDERATIONS

Steam treating presents very few safety hazards. One exception is possible explosions in batch-type furnaces loaded with oily work. Another exception concerns powdered iron products. During the steam treating process, both H_2 and O_2 are generated; normally their quantities are extremely small and do not present any hazards. With large, dense loads of powdered iron products, however, sufficient H_2 is generated to cause a minor explosion if the lid is opened when the furnace temperature is above 1000° F (538° C). This hazard can be eliminated by cooling the load in a steam atmosphere to 700° F (370° C) before removing it from the furnace. Components retaining 0.8 to 1.0% carbon after sintering should be handled in the same manner as powdered iron parts.

Environmental conditions may actually be improved by steam treating. Some blackening processes require toxic chemical solutions. Both the spent chemicals and the rinse water must be treated, or they must be removed by special and expensive removal methods. Similarly hazardous conditions exist with pickling solutions for nonferrous materials. The requirement for pickling, however, may be entirely eliminated or greatly reduced by using steam as a protective atmosphere while heat treating.

ION IMPLANTATION

Ion implantation is a process by which atoms of virtually any element can be injected into the near-surface region of any solid. The implantation process involves forming a beam of charged ions of the desired element and then accelerating them at high energies towards the surface of the solid, which is held under high vacuum. The atoms penetrate into the solid to a depth of 0.01 to 1.0 μin. (0.25 to 25 nm). This process differs from coating processes such as electroplating in that it does not produce a discrete coating; rather, it alters the chemical composition near the surface of the solid.[1] Some of the material properties influenced by ion implantation are conductivity, hardness, and wear and corrosion resistance.

PROCESS DESCRIPTION

In ion implantation, atoms of a desired element are ionized and accelerated using an electric field, then the ionized atoms are scanned across the solid to obtain a uniform deposition. The energy imparted to the ionized atoms determines the depth of their penetration into the solid; their concentration throughout the surface can be accurately predicted using rather well-documented theories and experimental data based on energy loss and stoppage of energetic particles in matter.

As ions enter the solid, they transfer energy through excitation of electrons and by collisions with the nuclei of the solid. Ions of elements with small nuclear charges lose most of their energy to the solid through electronic excitation, while heavier elements lose more energy to target atoms through collision cascades. As might be expected, all of the ions do not penetrate to the same depth. Some make collisions near the surface, while others go deeper, transferring energy along the way and eventually coming to rest. At some point between the surface and the deepest ion, there is a peak in the distribution of ions. The depth of this peak is referred to as the *projected range* (R_p). The cross section of ions coming to rest in a solid roughly follows a Gaussian distribution for specific species of ions implanted at specific energies; thus, range concentration curves can be described, and projected ranges calculated for given elements at various energies. A plot for boron ions implanted into silicon is presented in Fig. 25-6. The standard deviation of the projected range is ΔR_p.

Projected range data for ions penetrating crystalline solids such as silicon can be different than the calculated range concentration due to an effect called *channeling*. Channeling occurs when most of the ions are steered away from the closely packed rows or planes of atoms in the crystal, thereby avoiding collisions with the atoms until near the end of the path. Ions entering the solid in a direction that is parallel to a row or plane of atoms (channeling direction) will have a deeper distribution because only electronic collision will have occurred along their path. Since it is often important to control the depth of ion penetration, it is necessary to direct most of the ions in a channeling direction, or prevent channeling entirely, by tilting the crystal to a relatively nonchanneled direction before implanting.

ADVANTAGES AND LIMITATIONS

The successful use of ion implantation in semiconductor production has led to the development of equipment for use in other areas. For example, experiments have shown that ion

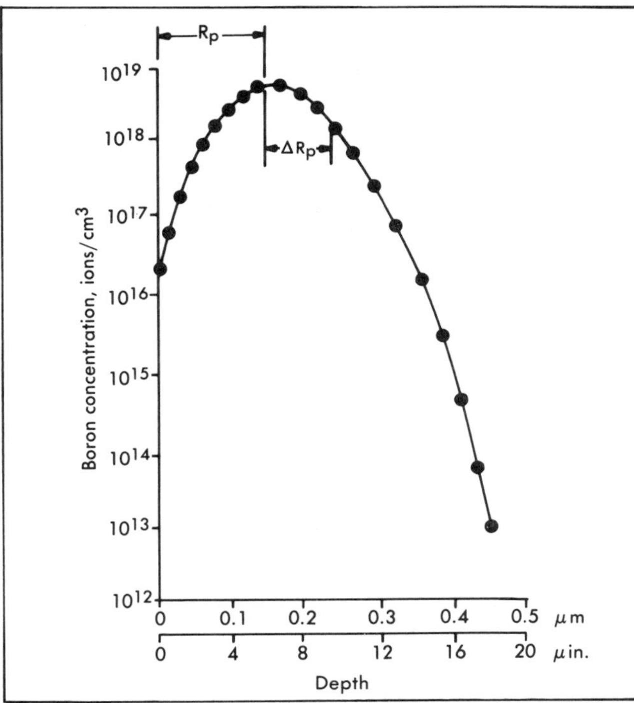

Fig. 25-6 Projected range plot of boron implanted in silicon; the dose of boron ions is $10^{14}/cm^2$, and the implant energy is 50 keV.

implantation can be used to improve the functional capability of materials in a variety of applications. A report by the National Materials Advisory Board cited the following advantages of ion implantation:[2]

- A variety of ion species can be implanted with the same basic apparatus. Almost all elements of the periodic table have been implanted.
- Ion implantation is a low-temperature process. It can often be added to the end of a production line without affecting existing operations.
- The surface of finished products can be treated without introducing significant dimensional changes and without changing bulk properties.
- The process is easily controlled through the electrical signals applied to the ion accelerator.
- Novel nonequilibrium structures and metallurgical phases with properties that cannot be duplicated in bulk material can be produced at the surface.
- Ion implantation creates no problems for disposal of waste products, as does electroplating.
- The absence of a discontinuous interface between the implanted surface layer and the bulk leads to excellent adhesion of the implanted layer.

In addition to the advantages previously described, ion implantation offers new opportunities for metallurgical research. Some of the unusual properties or characteristics that can be produced are as follows:

- New metallurgical phases with new properties can be formed. In certain cases, such as heavy implantations of tantalum in copper or phosphorus in iron, amorphous or glassy phases can be formed.
- If the implanted atoms are mobile, precipitates or inclusions can be formed. For example, implanted argon and helium atoms are insoluble in metals and may form bubbles; nitrogen in titanium may form titanium nitride.
- The composition of a surface layer can be changed by differential sputtering caused by the implanted ions.
- Damage and high concentrations of lattice defects, resulting from atomic displacements produced by the incident atoms, can change the chemical reactivity and mechanical characteristics of a treated surface.
- Implantation can enhance the diffusion of impurities already deposited in a substrate, presumably through the motion of the high concentrations of lattice defects produced by the incident ions.
- Cooperative effects of two implanted species can occur; for example, implantation of both molybdenum and sulfur into steels seem to have an effect similar to lubrication with molybdenum disulfide.
- Surface layers, either contaminants or deliberately deposited layers, can be driven into the substrate by impinging atoms.
- Surface layers with conventional (in the sense that they are the same as in bulk material of the same composition) chemical, optical, magnetic, and mechanical properties may be produced.

The primary limitation of ion implantation is the present cost of ion-implanting equipment capable of providing a few hundred kilovolts of acceleration potential. Ion implantation is also limited to line-of-sight applications; surfaces such as the interior of a gun barrel are difficult to implant. Further, mechanical manipulators are required to ensure a uniform deposition on all the exterior surfaces of a part. The depth of the implanted layer may also be a limiting factor in some applications because the implanted ions may diffuse away from the surface layer when the part is exposed to high temperatures.

APPLICATIONS

The major industrial applications of ion implantation techniques currently include fabrication of semiconductor devices (computer microcircuitry) and modification of the surface performance characteristics of tools, dies, and other metal components. At the same time, researchers are investigating other areas where ion implantation can be used.

Semiconductors

Ion implantation technology has been widely used in manufacturing integrated circuits. Minute quantities of elemental materials called *dopants* are implanted into silicon (a semiconductor) to create conductive layers; pure silicon is a material that has conductive characteristics somewhere between insulators like glass and conductors like metal. The introduction of "p-type" dopants (boron) creates a region with hole-type conductivity. Conversely, implanting "n-type" dopants (phosphorus or arsenic) creates electron-type conductivity. Combinations of "hole" and electron-rich regions can be used to produce devices such as transistors, metal oxide semiconductors (MOS), diodes, and capacitors. Sets of these devices are then connected to form integrated circuits (IC's). In all semiconductor devices, conduction is determined by the dominant dopant in each region. Since the product of p and n-type charges is a constant at any given temperature, regions of p or

ION IMPLANTATION

n-type dominance can be created if there is good control over the depth and concentration of dopant atoms introduced into the semiconductor material.

In the past, the more conventional methods of doping semiconductor wafers (i.e., adding elemental materials) involved thermal diffusions of dopant atoms into the silicon substrate. Ion implantation offers some major advantages over the "in-diffusion" doping techniques and hence has become the primary doping technique in the semiconductor industry. Implantation allows more precise control over the number of dopant atoms because the electrical charge deposited by the ion beam can be measured. Similarly, the uniformity of the doping across the semiconductor wafer and the depth profile of the dopant can be more accurately controlled with ion implantation than with other doping techniques that rely on temperature-dependent thermodynamic forces and kinetics.

The application of ion implantation in the semiconductor industry originated in the early 1970s as a technique for adjusting the threshold voltages of MOS devices.[3] The next major application was in the production of complimentary metal oxide semiconductor (CMOS) devices.

Among the more recent applications of ion implantation are the creation of "source" and "drain" regions in the latest high-speed memory and microprocessor devices. These heavily doped regions typically contain from 10^{15} to 10^{16} ions/cm^2. Ion implantation can produce n-doped and p-doped regions more shallow and more narrow than diffusion techniques can, thereby permitting smaller and faster devices to be produced. Ion implantation is used to manufacture virtually all state-of-the-art microprocessors and memory chips greater than 64k bits.

Ion implantation is also used for emitters and bases in bipolar semiconductor devices. The current gain of a bipolar device is a function of the ratio of the total charge of the emitter to the total charge of the base. Again, by precise ion implantation, the current gain can be accurately controlled. Numerous other applications for ion implantation have developed within the semiconductor industry. Resistors and resistor networks for both bipolar and MOS circuits are manufactured using ion implantation. The production of photoelectric diodes also involves ion implantation steps.

Metal Surface Modification

The science of metallurgy has developed a large body of knowledge demonstrating that corrosion, fatigue, and wear of metals and alloys are governed by characteristics and properties of the surface layers of the materials. This key surface layer is often only about 1 μin. (25 mm) thick. As a result, the technology of metallurgical surface treating addresses both the importance of the near-surface characteristics for wear, as well as the necessity to provide the surfaces of metallic components with special properties in order to fulfill specific applications. Ion implantation is a surface treatment technique that offers the following potential advantages:

1. A pure alloying element can be precisely introduced into the surface layers of a component in a reproducible way; no further grinding or straightening is required.
2. The difficulties of conventional surface treatment techniques are largely overcome because ion implantation can be performed at room temperature so the material need not be affected by heating or melting; there are no dimensional changes in the component; and the surface remains an integral part of the component, without the adhesion problems or the discontinuities associated with the interface of traditional coatings and metallic surfaces.
3. Ions implanted into the surface of metals can create regions of alloy concentration levels that cannot be achieved with conventional metal processing techniques. By controlling the temperature of the metal during implantation, ions can be implanted into structural positions that would have been impossible had the atoms been added at an earlier stage in the metal fabrication process.

As previously noted, ion implantation permits a metallurgist to create a new alloy on the surface of a finished component without the usual interface or adhesion problems. However, there are some limitations to metallurgical ion implantation. Because the ion-implanted layer is very shallow, current applications tend to be limited to components that are not subjected to high temperatures where implanted ions may diffuse away from the surface layer. Another limitation is the high cost of the equipment necessary for large-scale implantation into metallic components.

At present, the equipment designed to implant gaseous species such as nitrogen ions for antiwear applications seems to be the most practical and economical. In addition, the wear and fatigue performances of nitrogen-implanted steel and tungsten carbide seem to be comparable to those obtained through more conventional surface hardening processes. A compilation of successful applications of nitrogen implantation in a variety of metal components is presented in Table 25-3.[4–9]

Ion implantation has also been used to reduce corrosion and improve the hardness characteristics of metal components. Implantation for these purposes, however, is typically limited to small areas and is quite expensive. The solution to these limitations may lie in alternative methods to direct implantation, such as ion beam mixing or ion-beam-enhanced deposition (IBED). The latter is a process during which one metal surface is bombarded with ions simultaneous with the deposition from an evaporation or sputtering source. The ion beam interactions tend to alter the barrier layers (oxide layers) and allow intermixing of the substrate and deposited material. Also, the presence of lattice defects can aid in transporting atoms across metal interfaces. When large transport ratios are obtained, this process is more efficient than direct implantation.

Ion beam mixing of predeposited films is a process that allows the implantation of heavy metallic species using existing implantation equipment and gaseous sources. There is still a need to develop data on the beneficial effects brought about by various combinations of bombarding ions and diffused species in this process, but the process may be a more efficient, economic method than direct implantation. In many cases, it is possible to use the nonequilibrium nature of the ion beam mixing to override the normal solubility and alloy formation considerations. For example, argon bombardment has been found to cause enhanced diffusion of mutually insoluble iridium and gold.[10] Ion implantation and associated techniques will continue to have a significant impact on surface metallurgy and the performance improvement of metallic components.

Research and Development

Among the diverse research and development applications of ion implantation is the simulation of neutron damage in surfaces subject to the effect of nuclear decay. For example, nickel ions have been implanted into the surface of nuclear reactor materials, simulating the neutron collisions that would occur over several years or decades of neutron bombardment in the operation of the reactor. There is no other convenient way to accelerate the testing of the metals involved with currently available neutron sources.

TABLE 25-3
Performance of Ion-Implanted Components

Component	Material	Useful Life Increase
Paper slitters	Chrome steel	2x
Taps for phenolic resin	M2 tool steel	12x
Thread cutting dies	M2 tool steel	5x
Slitters for rubber	Tungsten carbide	12x
Wire dies for copper wire	Tungsten carbide	5x
Deep drawing dies	Tungsten carbide	2x
Wire dies for steel wire	Tungsten carbide	4x
Injection molding nozzle	Tool steel	2x
Swaging dies for steel	Tungsten carbide	2x
Forming tools	Carburized mild steel	3x
Bearings	AISI 52100 steel	2x
Punch and die sets	Tungsten carbide	6x
Punches for acetate	Chrome steel	2x
Mill rolls	H-13 steel	5x
Injection mold screws	Tool steel	10x
Drills for epoxy board	Tungsten carbide	2x

(*BeamAlloy Corp.*)

EQUIPMENT

As illustrated in Fig. 25-7, present-day semiconductor ion implanters have the following components:[11]

- An ion source.
- A magnetic mass-analyzing system to select the isotope of the atom to be implanted.
- An accelerating stage or stages that can bring the ions up to energies of hundreds of kilovolts.
- A beam scanner or "wobbler" to obtain beam uniformity.
- A target chamber that can be instrumented to handle large numbers of planar structures, rotate samples in and out of the beam, and make use of complicated masks.
- Electronics that provide readouts of mass analysis, beam current, beam profile, and vacuum conditions in the machine and at or near the implantation site.

Ion implantation systems designed for processing metal components are configured much like those designed for semiconductor fabrication. In the system most widely used for nitrogen implantation of metals, the magnetic mass-analyzing system is eliminated by using prepurified nitrogen gas in the ion source.

ENVIRONMENTAL AND SAFETY ISSUES

The use of high voltage in ion implantation presents the obvious electrical shock hazard. Typically the voltage is in kilovolts or hundreds of kilovolts at current levels that are extremely dangerous. All commercial machines contain extensive interlocks designed to reduce the risk of potential accidents. In addition, extensive safety procedures are necessary for the grounding of all implanter power supplies.

An advantage of ion implantation is that only small quantities of dangerous dopant materials are used. Small bottles of toxic gas are used instead of cylinders, but they still present a potential hazard and must be stored in specially vented cabinets. Extreme care must be used in following the procedures for installation and use of ion source materials.

Hazardous waste issues are relatively minor in the operation of ion implantation equipment, but expert waste handling firms are frequently used to dispose of such materials as contaminated

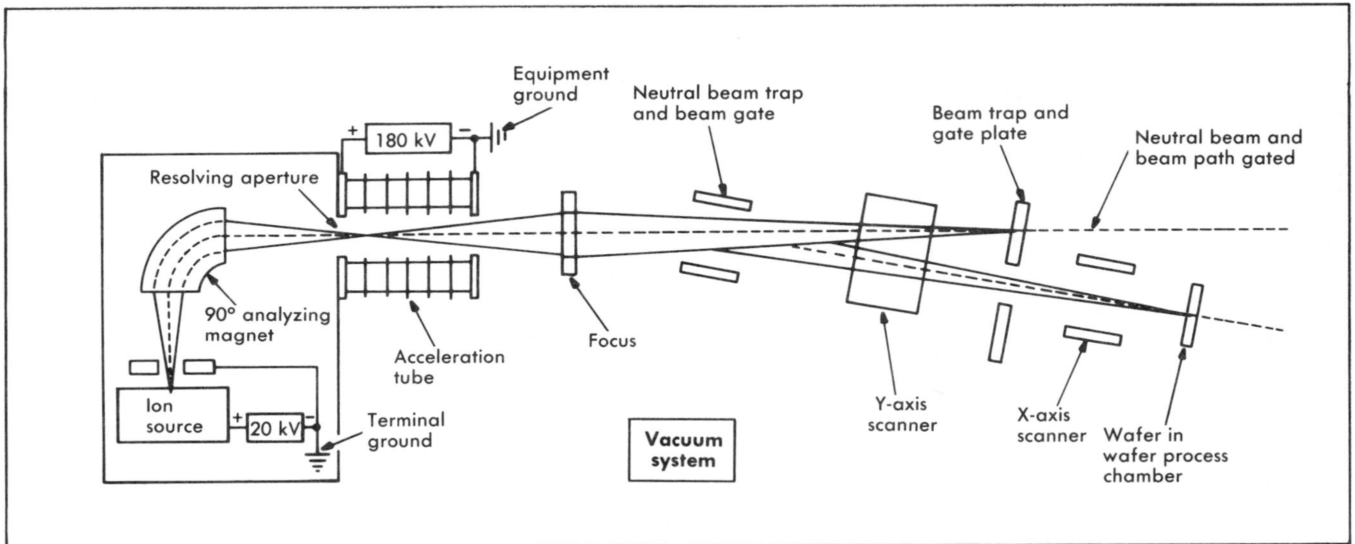

Fig. 25-7 Typical ion implantation equipment. (*Ion Implant Services*)

REFERENCES

oils from diffusion pumps and the residue from ion source cleaning. Proper venting and protective clothing are also mandatory when working with all hazardous waste materials associated with ion implantation.

Emissions into the atmosphere from the normal operation of ion implanters are typically in the one-part-per-billion range or below. Emissions should be monitored, but they are normally well below levels that pose a threat of atmospheric contamination.

References

1. National Materials Advisory Board, *Ion Implantation as a New Surface Treatment Technology*, NMAB-349 (Washington, DC: 1979), p. 3.
2. *Ibid.*, pp. 4-5.
3. M. R. MacPherson, "The Adjustment of MOS Transistor Threshhold Voltage by Ion Implantation," *Applied Physics Letters*, vol. 18, no. 11 (June 1971), pp. 502-504.
4. G. Dearnaley, "Ion Implantation for Improved Resistance to Wear and Corrosion," *Materials in Engineering Applications*, Vol. I (September 1978).
5. N. E. W. Hartley, et. al., *Proceeds from the International Conference on Applications of Ion Beams to Metals*, Albuquerque, NM, October 1973 (New York: Plenum Press, 1974), p. 123.
6. S. LoRusso, et. al., "Effect of Nitrogen Ion Implantation on the Unlubricated Sliding Wear of Steel," *Applied Physics Letters*, vol. 34, no. 10 (May 1979), pp. 627-629.
7. Robert N. Bolster and Irwin L. Singer, "Surface Hardness and Abrasive Wear Resistance of Ion-Implanted Steels," *American Society of Lubrication Engineers (ASLE) Transactions*, vol. 24, no. 4 (Park Ridge, IL: ASLE, 1980), pp. 526-532.
8. G. Dearnaley, ed., *New Uses of Ion Accelerators* (New York: Plenum Press, 1975), pp. 283-322.
9. A. H. Deutchman and R. J. Partyka, "Wear Resistant Performance of Ion Implanted Alloy Steels," *Proceedings of NATO Advanced Study Institute—Surface Engineering*, Les Arcs, France, July 1983 (The Netherlands: Nijhoff Publishers).
10. G. Dearnaley, "Techniques and Equipment for Ion Implantation in Metals," *Fourth International Conference on Ion Implantation*, Berchtesgaden, September 1982, vol. 11 (New York: Springer Verlag, 1983), p. 336.
11. National Materials Advisory Board, *op. cit.*, p. 33.

Bibliography

Eggleston, J. H., and Fisher, R. D. "Steam Treating of Iron Powder Metal Parts." *1982 National Powder Metallurgy Conference Proceedings*. Princeton, NJ: Metal Powder Industries Federation, 1982.

Eggleston, J. H., and Spangler, F. L. *Heat Treatment of Powdered Iron Products*. SME Technical Paper CM72-812. Dearborn, MI: Society of Manufacturing Engineers, 1972.

ORGANIC COATINGS

SECTION

5

COATING MATERIALS

Organic surface coatings are complex mixtures of materials that are designed to enhance the appearance of and/or to protect a substrate. The coating itself is normally composed of a number of ingredients including (1) the polymer (binder), which is designed to provide the major properties of the coating; (2) solvents, which are used to adjust the viscosity of the coating primarily for application; (3) pigments, which are designed to hide the substrate, provide decorative color, and enhance specific desired properties in the coating, such as corrosion resistance; and (4) additives, which include materials such as thickeners, flow agents, catalysts, inhibitors, and stabilizers.

Because of the complexity of a coating, the coating chemist should have thorough knowledge in a number of chemical disciplines such as physical chemistry, colloid chemistry, organic chemistry, inorganic chemistry, and polymer chemistry. As an increasing number of coating chemists possess this knowledge, the coatings industry is rapidly becoming highly scientific. Several universities offer specialized curricula for technical training.

The scientific approach to coatings development has resulted in major changes in coatings technology, including the advent of water-reducible coatings, powder coatings, and high-solids coatings. The coatings industry is an extremely important industry and, despite government restrictions on the use of solvents, manufactures excellent coating products for use in a variety of applications. This chapter is designed to introduce the reader to the coatings field and provide an overview of the materials used in developing coating products.

CHAPTER CONTENTS:

GLOSSARY OF TERMS

The following painting terms are abstracted from the glossary of *Understanding Paint and Painting Processes.*[1]

acetone A powerful ketone-type lacquer solvent.

acrylic A coating based on a polymer containing short-chain esters of acrylic and methacrylic acid. Acrylics are widely used as automotive topcoats. Their physical properties can be controlled in part by the choice of the alcohol used to make the ester.

active solvent A liquid that can dissolve a paint binder when used alone.

additive Any one of a number of special chemicals added to a paint to bring about special effects. Examples are plasticizers, light stabilizers, and fungicides.

adhesion The phenomenon by which one material is attached to another by means of surface attraction.

agglomerate Clumps of pigment crystals that have formed loose clusters containing entrapped air. Usually undesirable in paint, as they tend to settle out and have poor optical properties.

aliphatic solvent A type of solvent comprised mainly of straight-chain hydrocarbons. Examples are gasoline, kerosene, hexane, and naphtha.

alkyd A coating based on a polyester binder. The polyester binders are chemical combinations of molecules that contain more than one acid or alcohol group. Alkyds are widely used in water-based house paints and automotive primers.

anhydride A reactive form of dicarboxylic acid containing a monomer that has one mole of water removed. The major anhydride used in the synthesis of alkyds is phthalic anhydride.

antiskinning agents Chemicals added to a paint to help prevent the formation of a surface film on the paint.

aromatic A type of solvent based on benzene ring molecules. Aromatics are often used as diluents in acrylic lacquers. Typical examples are benzene, xylol, and toluol.

benzoic acid An aromatic monocarboxylic acid used in terminating chain growth in polyester or alkyd polymers. Also used in the manufacture of plasticizers.

beta rays Beams of electrons that can be used to cure certain kinds of paint.

binder The paint material that forms the film, so called because it binds the pigment and any additives present into a solid durable film. Also referred to as the resin.

branched polymer A polymer that has some branching along its backbone chain. An example is low-density polyethylene.

catalyst A chemical used to change the rate of a chemical reaction. Differs from a curing agent

Contributors of sections of this chapter are: ***Dr. John C. Graham****, Professor and Program Coordinator—Polymers and Coatings, Dept. of Interdisciplinary Technology, Eastern Michigan University;* ***Daniel C. Riter****, Product Manager, Powder Coatings Dept., Coatings Div., Ferro Corp.*

Reviewers of sections of this chapter are: ***John H. Daniel, Jr., Ph.D.****, Consultant;* ***John A. Gordon****, Adjunct Professor—Coatings Technology, Dept. of Interdisciplinary Technology, Eastern Michigan University;* ***Dr. John C. Graham****, Professor and Program Coordinator—Polymers and Coatings, Dept. of Interdisciplinary Technology, Eastern Michigan University;* ***Herman J. Lanson, Ph.D.****, Consultant;* ***Dr. George R. Patrick****, Industry Manager—Automotive Finishes, P & A Div., Pigments Dept., CIBA-GEIGY Corp.;* ***Joseph W. Prane****, Industrial Consultant.*

in that the catalyst is not itself chemically consumed in the reaction, while a curing agent is. Technically, catalysts that increase reaction rates are called accelerators; those that decrease reaction rates are called inhibitors or retarders.

cathodic protection The prevention of corrosion of a metal by electrically connecting it to a sacrificial anode. The anode is itself decomposed, and the object of interest is protected. The sacrificial anode must be replaced periodically.

coalescence The fusing or flowing together of liquid or solvent particles.

colloids Aggregates of molecules in solution (dispersion) resulting in particles having dimensions in the 0.001 millimicron to 1000 micron range.

condensation cure Any crosslinking process that liberates water and other simple molecules during the reaction.

conjugated double bond Two double bonds in alternate positions as indicated by the formula -CH=CH-CH=CH-.

copolymer A polymer comprised of two or more different monomer units.

critical pigment volume concentration (CPVC) The volume percent pigment in a coating in which the pigment particles are surrounded by resin so that no free surface pigment exists.

cure The process by which paint is converted from the liquid to the solid state.

Desmodur N® An aliphatic-type polyisocyanate commercially available from Mobay Chemicals.

diluent A liquid that extends a solution but definitely acts to weaken the solvent power of the active solvent.

double bond An unsaturated hydrocarbon of the type C_nH_n with the formula -C=C-, indicated by the suffix -ene.

drier A catalyst added to speed the cure of oil-based paints. Driers are often metal salts of carboxylic acids.

drying oil A water-insoluble liquid, usually obtained from a plant source, that reacts with oxygen (from the air) to form a crosslinked polymeric film.

electrocoating See "electrodeposition."

electrodeposition The process by which electrically charged paint is plated on conductive surfaces of the opposite charge.

electrolyte A substance that dissociates to some extent into two or more ions in water and other polar solvents. Solutions of electrolyte conduct electrical current and can be decomposed by it (electrolysis).

electron beam curing A system for curing paint films using the energy of an electron beam. The process lends itself to high-speed curing of paint on flat surfaces. Special paints must be used and personal shielding is required.

electron beam radiation Radiation generated from high-energy electrons that is used in crosslinking coating systems.

electrostatic spray The process by which paint particles are electrically charged and attracted to a substrate bearing an opposite charge.

emulsion polymerization The formation of a polymer in which the growing polymer molecules form droplets in the reaction medium. This situation arises when the solvent can dissolve the monomer, but not the polymer.

emulsion A class of colloidal dispersions containing two or more immiscible liquids such as oil in water. Emulsions are usually unstable and will separate into their components unless a stabilizing agent is present.

enamel A broad classification of free-flowing clear or pigmented varnishes, treated oils, or other forms of organic coatings that usually dry to a hard, glossy or semiglossy finish.

epoxy Synthetic resins formed by the condensation of epichlorohydrin and bisphenol-A.

exempt solvents Solvents that are not subject to air pollution legislation. Many alcohols, esters, some ketones, and mineral spirits are exempt. Aromatic and some ethylenic compounds are not exempt, and their use as solvents is therefore subject to regulation.

flash time The time between paint application and baking. Usually a considerable quantity of solvent is lost during this interval, and this solvent loss prevents popping problems in the oven.

functionality Ability of a compound to form covalent bonds.

gamma radiation High-energy radiation, similar to X-ray radiation, that is emitted by radioactive substances.

glass transition temperature The temperature at which polymer molecules are able to move fairly freely in the solid state.

hiding power The ability of a paint to mask the color or pattern of a surface. Usually expressed as square feet per gallon or square meters per liter.

high-solids paint Paint containing 35-80% solids. These products have become popular because of the reduction in solvent emissions associated with their use.

homopolymer A polymer containing only one kind of monomer.

inhibitor A chemical added to retard some particular reaction. Examples are antioxidants and antiskinning agents.

interfacial free energy The minimum amount of work required to create an interface between two immiscible materials.

latent solvent A liquid that cannot itself dissolve a binder but increases the tolerance of the paint for a diluent.

linear polymer A polymer containing little or no branching. Examples are high-density polyethylene and nitrocellulose or acrylic lacquers.

molecular weight The relative mass of a molecule in relation to that of a hydrogen atom. It is obtained by adding together the atomic weights indicated in the formula of the substance.

monomers Low-molecular-weight reactive materials that are used in the synthesis of polymers.

nonconjugated double bond Double bonds that are not in the relationship outlined under conjugated double bonds. They are indicated by the formula -C-C=C-C-C=C-C.

oil-based paints Paints with films that form solids by the air-induced crosslinking of certain unsaturated plant oils known as drying oils. Oxygen is consumed in the process.

paint A material that when applied as a liquid to a surface forms a solid film for the purpose of decoration and/or protection. Generally, a paint contains a binder(s), solvent(s), and a pigment(s). Often other materials are present to give special properties to the paint film. Examples of such additives are rust inhibitors, light stabilizers, and softening agents (plasticizers).

percent solids The percent mass of a paint due to its nonliquid components.

pigment Small particles added to the paint to influence properties such as color, corrosion resistance, and mechanical strength.

pigment volume concentration (PVC) The percent volume of a paint film occupied by the pigment.

plasticizer A low-molecular-weight material added to polymeric materials such as paints, plastics, or adhesives to improve their flexibility.

polyamides Polymeric compounds synthesized by the reaction of amine and carboxylic-containing compounds. They are sometimes amine terminated and used in the crosslinking of epoxide polymers.

BINDERS

polymers Large molecules built up by the combination of many small molecules.

primer A type of paint applied to a surface to increase its compatibility with the topcoat or to improve the corrosion resistance of the substrate.

refractive index The ratio of the velocities of light in a medium and in air under the same conditions. The result is that light passing from one medium to another is bent to some degree.

skinning The formation of a thin, tough film on the surface of a liquid paint film, usually due to reaction with the air or to rapid solvent loss.

styrene An unsaturated reactive monomer used extensively in the synthesis of polymers. It can also be used to thin out reactive polyesters with subsequent crosslinking in the ethylenic groups.

thermoplastic A type of polymer that softens and melts when heated and then resolidifies upon cooling. Thermoplastics generally have linear or branched structures.

thermosetting A type of polymer that does not soften appreciably when heated. Thermosets may char when heated in air. They are generally crosslinked polymers.

thixotropy The tendency for the viscosity of a liquid to be shear-rate dependent. When the liquid is rapidly shaken, brushed, or otherwise mechanically disturbed, the viscosity decreases rapidly. Thixotropic behavior is the result of molecules or particles in the liquid forming weakly associated structures that break apart when agitated.

throwing power The ability of an electrodeposition resin to coat recessed areas, usually measured by noting the coating distance up a cylindrical tube that is coated in an electrodeposition bath.

topcoat Usually the final paint film applied to a surface.

ultraviolet radiation High-energy short-wavelength radiation used in coatings to crosslink primarily acrylic and methacrylic systems by means of free-radical reactions.

UV stabilizers Chemicals added to paint to absorb the ultraviolet radiation present in sunlight. Ultraviolet radiation decomposes the polymer molecules in a paint film, and thus UV stabilizers are used to prolong paint life.

vehicle The combination of binder and solvents or diluents, which are used to put the binder in a liquid, usable form.

vinyl cure A curing process involving the crosslinking of vinyl groups.

vinyl toluene An unsaturated, aromatic monomeric compound reacted into oil-modified alkyds to modify its drying properties.

viscosity The property of liquid that enables it to resist flow, often measured by the time required for a given volume of liquid to flow through a small hole in the bottom of a cup under controlled conditions. A thick liquid-like molasses has a high viscosity.

volatile organic compounds (VOC) Volatile organic materials, such as solvents, that are present in many coating products.

BINDERS

Binders are defined as liquid polymeric or resinous materials that are used in coatings to hold the pigment and various additives together, to provide adhesion, and to supply the major properties of the coating. In the chemical industry, these materials are commonly known as polymers, the name *polymer* being derived from the Greek words *poly* meaning many and *meros* meaning repeating parts. Over the years, the word polymer has come to be used for any large molecules with or without simple repeating units. Another term often used synonymously with polymers in the coatings industry is *synthetic resin*. The term *vehicle* is also commonly used in the coatings industry. A vehicle is essentially a polymer dissolved in a suitable solvent. Because of the high viscosity of polymeric systems, most polymers used in the coatings industry are purchased from the raw material supplier in the form of vehicles.

CLASSIFICATIONS

As used in the coatings industry, polymers can be divided into various classes depending on the type of material and the type of crosslinking required to generate the final properties of the polymer. Polymers are most commonly used in such coating systems as lacquers, auto-oxidation film formers, nonauto-oxidation film formers, radical polymerization-curable film formers, and emulsion-type vehicles.

Lacquers

Lacquers are polymers dissolved in a solvent and require simple solvent evaporation to yield the final-film properties of the polymer. Because polymeric properties generally improve as the molecular weight of the polymer increases, lacquers are limited to polymers of reasonably high molecular weight. In addition, since high-molecular-weight polymers are very viscous in solvents, these materials are normally available at very low solids concentrations. Concentrations as low as 20% are not unusual. Since the volatile organic content of lacquers is high, these materials generate large amounts of organic volatiles, considered harmful to the environment, during film formation.

Auto-Oxidation Film Formers

Auto-oxidation film formers are low-molecular-weight polymers, primarily alkyds, natural oils, and/or epoxy esters, containing carbon-carbon double bonds that allow the materials to oxidize in the presence of atmospheric oxygen to form crosslinked systems. The concept of crosslinking is an especially useful concept in the coatings industry since it allows low-molecular-weight products with relatively poor properties to be crosslinked into high-molecular-weight materials with extremely good properties. In crosslinking, lower molecular-weight materials can be applied to the substrate at high-solids concentrations in solvents. Curing occurs on the substrate after the solvent has evaporated, producing high-molecular-weight materials with excellent properties.

Nonauto-Oxidation Film Formers

Nonauto-oxidation film formers are low-molecular-weight polymers containing functional groups that are useful in crosslinking reactions. Unlike the auto-oxidation film formers that react with oxygen, these materials react with a crosslinking agent to form high-molecular-weight products. Similar to the auto-oxidation film formers, these materials are usually of lower molecular weight and are available without solvent or at higher solids concentration in solvents. Baking alkyds using melamine or urea-formaldehyde resins as crosslinking agents are examples of polymeric materials of this type.

CHAPTER 26

BINDERS

Radical Polymerization-Curable Film Formers

Radical polymerization-curable film formers are low-molecular-weight materials supplied in reactive diluents that can be cured by exposure to radiation. This radiation can take the form of ultraviolet radiation (UV), electron beam radiation (EB), gamma radiation (γ), or other radiation sources. Since both the polymeric materials and diluents are reactive, these materials can be fully cured into highly crosslinked systems without the liberation of large amounts of volatile organic compounds. As such, radical polymerization-curable film formers are considered to be 100% nonvolatile. Materials normally found in this classification possess acrylic and methacrylic properties that permit chemical reactions to occur.

Emulsion-Type Film Formers

Emulsion-type film formers are extremely high-molecular-weight polymers that are dispersed as particles of polymers in a nonsolvent phase. Two types of emulsion systems exist: (1) the aqueous type, which is more commonly used in latex paints, and (2) the nonaqueous dispersion type, which is commonly abbreviated as NAD paint. In emulsion-type systems, the nonsolvent phase is considered to be the continuous phase of the media, and the polymeric particles are considered to be the dispersed phase of the media. Film formation in both instances occurs by the evaporation of the nonsolvent, either water or an organic nonsolvent, followed by coalescence of the polymeric particles into a continuous coating. Numerous polymers are useful in these systems, including acrylics and vinyls.

TYPES OF BINDERS

Various polymeric materials or binders of many chemical types are used in coatings, depending on the end use of the material or the curing cycle required for the system. The binders most commonly used are natural oils or vegetable oils, alkyds, polyesters, aminoplast resins, phenolic resins, polyurethane resins, epoxy resins, silicone resins, acrylic resins, vinyl resins, cellulosics, and fluorocarbons.

Natural Oils

Natural drying oils are polyunsaturated fatty acid esters of glycerol that can be crosslinked into protective coatings by reaction with oxygen in the air. In their natural form, oils are mixtures of different fatty materials containing different levels of unsaturation. The fatty acid compositions of various oils are presented in Table 26-1. Depending on the number of double bonds present in the mixture, oils will dry (cure) at various speeds. For example, oils with the greater number of double bonds will dry more rapidly than those with fewer double bonds. In addition to the number of bonds present, the relationship between double bonds and their locations in the molecule affects the rate of cure. Basically there are three types of double bonds that occur in natural oils: isolated double bonds, conjugated double bonds, and nonconjugated double bonds. On the basis of reactivity, conjugated double bonds react faster than nonconjugated double bonds, which react much faster than isolated double bonds.

Based on the composition of natural oils, the coatings industry has divided oils into drying, semidrying, and nondrying classes. These classifications are based on how rapidly or easily the oil reacts with oxygen present in the air.

Although the mechanism of drying or oxidative curing of oils is not well understood, it is known that certain divalent salts of carboxylic acids have a profound effect on the ability of unsaturated materials to react with oxygen and crosslink into cured products. These carboxylic acid salts, primarily salts of cobalt, zirconium, manganese, zinc, calcium and iron, are called driers and are basically catalysts, catalyzing the absorption of oxygen and/or the decomposition of peroxides. Each drier functions differently, and some are used in combination with others. The cobalt driers, which are essentially surface driers, are possibly the most potent driers and are normally used in combination with auxiliary driers or through driers.

Alkyd Resins

Although oils are quite flexible and cure by reacting with oxygen in the air, they lack the hardness and fast drying speed required in many applications where coatings are used. Alkyd resins, essentially polyesters modified by reaction with fatty acids, do not exhibit the hardness of polyesters, but have flexibility and drying properties superior to oils.

By virtue of the synthetic method used in their preparation, alkyds contain varying amounts of oils and/or fatty acids. The amount and type of oil or fatty acids present in alkyds determines the mechanism by which alkyds are generally crosslinked. The amount of oil is referred to as the oil length of the alkyd and relates to the percentage of fatty acid present in the alkyd itself. Short-oil alkyds, which contain low amounts of unsaturated groups and appreciable amounts of hydroxyl groups, are normally used in baking applications using crosslinking agents such as aminoplasts. Medium-oil alkyds can be used in both baking or air-dry applications; and flexible, long-oil alkyds, possessing appreciable amounts of unsaturation, are used primarily in air-drying systems such as house paints and enamels. The very long-oil alkyds, containing 75% or more of oil in the polymer composition, are used primarily by the printing ink industry where extremely rapid air drying is desired.

Although most alkyd systems have been used traditionally in solvent-soluble systems, the advent of water-soluble or water-reducible systems has changed the orientation of the alkyd market. Water-soluble or water-reducible alkyds are modified polyesters similar to the alkyds discussed previously. The only difference is that the water-reducible system contains carboxylic acid functionalities, which can be neutralized by bases such as ammonium hydroxide or amines to generate water-soluble polymers. As in the case of solvent-soluble alkyds, driers are used as catalysts in these water-soluble systems.

In recent years, very long-oil alkyds have been used as modifiers for latex-based paints in which the alkyd is post-emulsified into the latex. Modified latexes of this type show improved adhesion when applied over chalky substrates. Long-oil alkyds containing 60-75% oil are commonly used in architectural and maintenance coatings as brushing enamels and primers. Most of the long-oil alkyds are soluble in aliphatic solvents, permitting excellent brushing properties with good flow characteristics and easy cleaning.

The medium-oil alkyds, containing 45-60% oil, are primarily used in heavy-duty maintenance, automotive refinishing, and farm implement enamels. In some cases, they are modified with short-oil alkyds, depending on the individual application. The short-oil alkyds, which contain less than 45% oil, perform well in a force drying and baking system where they are cured by the use of aminoplast resins primarily of the urea-formaldehyde and melamine-formaldehyde types. Short-oil alkyds find extensive use as general-purpose industrial enamels and in metal decorating. Nonoxidizing alkyds are important for plasticizing cellulosic lacquers.

TABLE 26-1
Approximate Fatty Acid Composition of Selected Vegetable Oils, % by Weight

Fatty Acid	Double Bonds	Tung	Dehydrated Castor	Linseed	Safflower	Soya	Tall Oil Acids
Oleic	1	8	9	22	13	25	46
Linoleic	2	4	82	16	75	51	41
Linolenic	3	3	---	52	1	9	3
Eleostearic	3	80	---	---	---	---	---
Licanic	3	---	---	---	---	---	---
Ricionoleic	1	---	9	---	---	---	---
Palmitoleic	1	---	---	---	---	---	---
Arachidonic	4	---	---	---	---	---	---
Clupanodonic	5	---	---	---	---	---	---
Stearic	0	1	2	4	4	4	3
Palmitic	0	8	---	6	6	11	5
Myristic	0	---	---	---	---	---	---

In addition to the solvent-soluble alkyds and water-soluble alkyds, modified alkyds that exhibit rapid drying are also available. These materials are vinyl toluene (VT), styrene-modified alkyds, and benzoic-acid-modified alkyds. The benzoic acid modified, commonly known as chain-stopped alkyds, have better durability than other fast-drying alkyds, but exhibit poor recoatability.

Polyester Resins

Polyesters are the reaction products of polyfunctional carboxylic acids and polyfunctional alcohols. They are divided into saturated and unsaturated polyesters. Saturated polyesters are basically oil-free alkyds with crosslinking occurring after blending with co-reactants such as melamine-formaldehyde resins or isocyanate prepolymers. Unsaturated polyesters use a separate reaction, normally after the coating application, to convert the lower molecular-weight polyesters, as synthesized, into higher molecular-weight crosslinked products.

The saturated polyesters, basically the reaction products of anhydrides and alcohols, find only minimal use as adhesives; but when modified by the incorporation of natural oils, find extensive use in oil-modified alkyd resins as discussed in a previous section. Unsaturated polyesters are commonly used in manufacturing fiber-reinforced products such as sheet molding compounds (SMC). The unsaturated polyesters contain carbon-carbon double bonds in the polymer that react with unsaturated monomers to form crosslinked products. Styrene is by far the most common unsaturated monomer used in this application.

Unsaturated polyesters are especially prone to inhibition by oxygen during curing, and thus care must be taken to exclude oxygen. Oxygen exclusion can be accomplished either by adding wax, which migrates to the surface to exclude oxygen during the curing process, or by the use of modified polyesters.

Aminoplast Resins

Aminoplast resins are condensation products of urea, melamine, or benzoguanamine and formaldehyde. As initially synthesized, these materials are rather unstable and prone to self-condensation. They are stabilized quite readily, however, by reaction with alcohols to form alkylated aminoplast resins. In this form, aminoplast resins are used extensively as thermally initiated crosslinking agents for use with hydroxyl-containing polymers.

Since aminoplast resins are polyfunctional, they can react with hydroxyl functionalities in the polymer, forming extensively crosslinked systems. In the presence of an acid catalyst such as toluenesulfonic acid, crosslinking occurs rapidly at temperatures ranging from 212 to 300° F (100 to 150° C). Table 26-2 lists the various ureas and melamines that form aminoplast resins.

Polymers that use aminoplast resins as crosslinking agents include such materials as oil-modified alkyds, polyesters, acrylics, and epoxy resins. Since aminoplasts can be both solvent and water soluble, it is possible to incorporate them into water-reducible systems to provide crosslinking. Solubility in water and/or organic solvents depends to a great extent on the alcohol group used in stabilizing the aminoplast resin. For example, the fully methylated aminoplast resins are water soluble; however, as the alcohol group increases in size from ethyl to propyl to butyl, water solubility decreases and the materials become more soluble in common organic solvents such as alcohols. A great number of aminoplast crosslinkers possessing different solubilities, molecular weights, and degrees of functionality are commercially available.

Phenolic Resins

Phenolic resins, among the first synthetic resins ever produced for molding, were first studied in the late 1800s and prepared synthetically for commercial use in the early 1900s. Phenolic resins are synthesized by reacting phenols with formaldehyde under acidic or basic conditions. As with any reaction of this type, a large number of products are formed, ranging from dimers (two molecules hooked together) to highly crosslinked polymers that are of little use in surface coating applications. The phenols that are used in the reaction with formaldehyde include phenol, m- and p-cresol, bisphenol A, and p-tertiary butyl phenol, as well as other similar products.

When the reaction between the phenol and formaldehyde is carried out under acidic conditions, the product formed is called a novolac. This product is basically a low-molecular-weight polymer to which additional formaldehyde can be added to generate high-molecular-weight products, such as in the synthesis of molding resins. Under alkaline conditions, the reaction between phenols and formaldehyde generates resoles. After resoles are heated, they form hard, insoluble, infusible resins that are highly crosslinked.

TABLE 26-2
Partial List of Commercially Available Aminoplast Resins

Chemical Type	Trade Name	Manufacturer
Methylated melamine	Cymel 301	American Cyanamid Co.
Isobutylated melamine	Cymel 255	American Cyanamid Co.
Mixed isobutyl/methyl melamine	Cymel 1133	American Cyanamid Co.
Methylated melamine	Resimene 745	Monsanto Co.
Butylated melamine	Resimene 872	Monsanto Co.
Urea	Beckamine	Reichhold Chemicals, Inc.
Isobutylated urea	Beetle 1047	American Cyanamid Co.
Methylated urea	Beetle 55	American Cyanamid Co.
Methylated urea	2148	Cargill, Inc.

Although the early types of phenol-formaldehyde resins found only limited application in varnishes, recent modifications allow them to be used in more general applications. Specifically, resinous materials are used as modifying agents in phenol-formaldehyde resins to achieve improved solubility. Alkyl-substituted phenol-formaldehyde resins are dissolved in oil to generate varnishes and printing-ink bases. The most important application of phenolic resins is in the area of air drying and baking metal primers and spar varnishes.

Polyurethane Resins

By using polyfunctional reactants, it is possible to synthesize the polymeric materials known as polyurethanes. Polyfunctional reactants are formed by reacting compounds containing two or more isocyanate groups with compounds containing two or more hydroxyl groups. Polyurethane resins are especially noted for their abrasion resistance, toughness, and flexibility. They are resistant to chemicals, possess excellent electrical properties, and crosslink or cure at very low temperatures.

Polyisocyanates can also be used as crosslinking agents when the hydroxyl functionalities on the polymeric chains react with the isocyanate functionality to generate crosslinked species containing the polyurethane group. Polyfunctional isocyanates commonly used in this process include such materials as toluene diisocyanate (TDI), hexamethylene diisocyanate (HDI), isophorone diisocyanate (IPDI), dicyclohexyl methane diisocyanate (HMDI), diphenyl methane diisocyanate (MDI), and 2,2,4-trimethyl-1,6-hexane diisocyanate (TMDI). These materials are all used in the generation of polyurethanes or in the crosslinking of polymers containing hydroxyl functionality.

One major problem with using monomeric isocyanates is that of toxicity, a problem further compounded by the fact that these materials are also extremely volatile. Many attempts have been made to convert the commercially available monomeric isocyanates into higher molecular-weight compounds that would have lower volatility and be less dangerous in the workplace. Many of the commercially available materials such as Desmodur N® are higher molecular-weight isocyanates. These products are generally known as biurets and isocyanurates.

In almost all cases, isocyanate materials are too reactive to be used in the generation of single-package coatings. However, it is possible to block the isocyanate. The blocking material is liberated from the free isocyanate after heating to temperatures of approximately 200 to 250°F (93 to 120°C). With this technique, single-package urethanes are produced providing stability at room temperature but crosslinking at temperatures in excess of 212°F (100°C). The actual temperature at which unblocking occurs is greatly dependent on the blocking agent. Phenols, lactams, and oximes are commonly used in blocking isocyanates.

Urethane coatings can be divided into one-package prereacted, one-package moisture-cured, one-package heat-crosslinked, and two-package catalyzed urethanes, as well as two-package polyol polyurethanes.

One-package prereacted urethanes. One-package prereacted urethanes are characterized by the absence of any significant quantity of free-isocyanate groups. They are usually the reaction products of a polyisocyanate with polyfunctional alcohol esters of vegetable oils and are crosslinked via air oxidation. They are not true polyurethanes but are considered a special class of alkyds referred to as uralkyds.

One-package moisture-cured urethanes. One-package moisture-cured urethanes have free-isocyanate groups that are capable of conversion to useful coatings by the reaction of these groups with atmospheric moisture.

One-package heat-crosslinked urethanes. One-package heat-crosslinked urethanes cure by thermal unblocking of the urethane, regeneration of the isocyanate group, and subsequent reaction of the isocyanate group with polymers containing active hydrogen groups such as alcohols and amines. The blocking agent is normally an alcohol or phenol that is volatile and is released into the atmosphere as the system cures.

Two-package catalyzed urethanes. Two-package catalyzed urethane systems comprise two separate component packages: one package contains a prepolymer possessing free-isocyanate groups capable of reacting with materials in the second package, which consists of monomeric polyols or polyamines. These systems have limited pot life after the two components are mixed.

Two-package polyol polyurethanes. In the two-package polyol polyurethane system, one package contains a prepolymer polyisocyanate capable of forming useful coatings by combining with a substantial amount of the second package, which contains a polymer possessing active hydrogen groups with or without the benefit of a catalyst. The two-package polyol polyurethanes also possess a limited pot life, with crosslinking occurring normally at room temperature.

Epoxy Resins

Most of the epoxy resins used in this country are derived from a series of prepolymers that are based on bisphenol A and epichlorohydrin. A number of epoxy resins exist with different molecular weights and different epoxy contents. A list of some of the commercially available epoxy-resin prepolymers is provided in Table 26-3. Because of the inherent toxicity associated with epichlorohydrin, most manufacturers use the bisphenol A/epichlorohydrin (epoxy) prepolymers as listed in Table 26-3, converting them by a variety of processes into higher molecular-weight compounds or crosslinked systems. The important properties of cured epoxy resins include:

- Chemical resistance in corrosive environments.
- Excellent adhesion to a wide variety of materials including metal, wood, concrete, glass, ceramic, and many plastics.
- Low shrinkage during curing resulting in good dimensional stability and excellent adhesion.
- Ease of fabrication, which is inherent in systems that can be cured at room temperature.
- Good physical properties such as toughness, flexibility, and abrasion resistance.
- Excellent performance at elevated temperatures.

Epoxy resins, or at least the commercially available prepolymers, are readily converted into crosslinked products by the addition of curing agents such as polyfunctional amines and polyamides. Epoxy coating materials are available in either single-package or two-package systems. The single-package epoxy system includes epoxy esters, which are basically the reaction products of epoxides and unsaturated acids. The chemistry involved in the crosslinking of these materials is similar to that for alkyds and oils in which curing involves the reaction of oxygen in the air with the unsaturation of the epoxy ester system.

Two-package epoxy systems are essentially low-molecular-weight resins dissolved in a solvent and are crosslinked by the addition of curing agents into finished products. Liquid epoxy resins are used in the manufacture of solventless and high-solids epoxy coatings. For adequate application and cure, liquid epoxy resins require two-package spray units capable of accurately metering the components and controlling their temperature at suitable levels.

Epoxy-resin materials are used as primers in maintenance coating systems when comparatively mild corrosion resistance is required. They exhibit much better adhesion than alkyd

TABLE 26-3
Partial List of Commercially Available Epoxy Prepolymers

Numerical Value of X*	Epoxide Equivalent**	Product Designation†
0.15	185-192	EPON 828
0.4	230-280	EPON 834
1	290-335	EPON 836
2	450-550	EPON 1001
3	600-700	EPON 1002
4	850-975	EPON 1004
10	1600-2300	EPON 1007
13	2300-3800	EPON 1009

* An epoxy prepolymer is represented by:

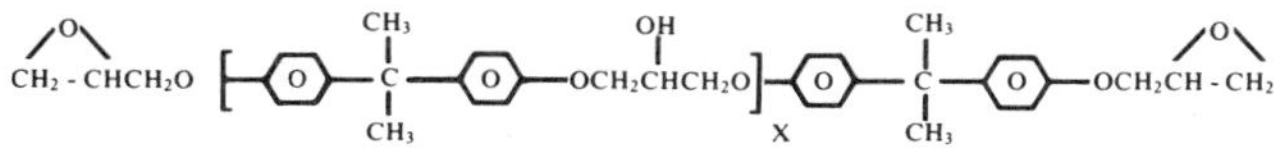

$$\overset{O}{CH_2-CHCH_2O}\left[-\bigcirc-\underset{CH_3}{\overset{CH_3}{C}}-\bigcirc-OCH_2\overset{OH}{C}HCH_2O\right]_x-\bigcirc-\underset{CH_3}{\overset{CH_3}{C}}-\bigcirc-OCH_2\overset{O}{CH-CH_2}$$

** The epoxide equivalent is defined as the weight of epoxy polymer that contains one mole of epoxy functionality.

† Shell Chemical Company

coatings and are widely used as primers for interior and exterior protection of structural steel and as exterior coatings for storage tanks in refineries and chemical plants. Epoxy-resin paints are not generally used as topcoats because they chalk severely, resulting in poor color and gloss retention.

High-molecular-weight solid epoxy materials with suitable co-reactants are also used in powder coating applications. Powder coating materials can be applied by dipping the preheated object into fluidized powder or by spraying the powder electrostatically. (Application methods are discussed in further detail in Chapter 27 of this volume.) Curing can be accomplished by heating the part to the required curing temperature. Epoxy powder materials are used in applications where severe conditions are encountered. Powder coating materials are discussed subsequently in this chapter.

Epoxy coatings are also available in water-reducible form. Aqueous systems available include:

- Water-reducible two-package room temperature curing systems.
- Single-package baking enamels for can coatings.
- Water-reducible electrodeposited systems.
- Epoxy powder aqueous suspensions.

Silicone Resins

Silicone resins differ considerably from the carbon-hydrogen polymers that have been discussed previously because they contain only silicon and oxygen in the basic building block of the polymer. They are used primarily for high-temperature applications, such as for ovens, stacks, exhaust systems, and space heaters. In recent years, however, silicone-modified resins have found application as decorative coatings when improved weather resistance is required. In addition, they possess other useful properties such as a low coefficient of friction (slip), water-repellent action, and antifoam properties.

Although a great number of silicone resins can be made commercially, the only useful forms are those that contain methyl and/or phenyl substituents, with the ratio of the silicon to the methyl and phenyl content determining the properties of the system. Heat resistance, for example, can be directly related to the percentage of silicon in the resin; properly formulated enamels can withstand temperatures of 480 to 660° F (250 to 350° C) for several thousand hours. Aluminum enamels, based on silicone resins, are serviceable up to 1020 or 1200° F (550 or 650° C).

Coatings based on silicone resins are virtually nonyellowing and nonchalking, and they are resistant to attack by most chemicals, dilute acids, and salts. Some silicone resins also have good compatibility with short and medium-length alkyds, phenolics, urea and melamine-formaldehyde resins, low-molecular-weight epoxy resins, cellulosics, acrylics, and polyesters. Many silicone resins are typically blended or co-reacted with these materials to form the final products. Air-drying alkyds reacted with special silicone resins are the basis for air-dry coatings having excellent durability.

The three types of silicone resins commercially produced are a two-component condensation-cured system, a one-component moisture-cured system, and a two-component vinyl-cured system. The most common examples of the two-component condensation silicone system are the RTV silicones, which are extremely useful in electrical encapsulation. The one-component moisture-cured systems, commonly used for bathroom caulking materials, generate either acetic acid or alcohols during the curing cycle and display excellent adhesion. The two-component vinyl-cure systems are normally used for encapsulation since no volatile products are released during curing; their one drawback is high cost.

Acrylic Resins

Acrylic resins were first produced commercially in the 1930s, but their use remained limited until the 1950s when the first acrylic emulsion polymer was produced. Since then, both emulsion and solution acrylics have established a solid reputation as coatings, especially for exterior applications.

Acrylics are divided into thermoplastic dispersions, thermoplastic solutions, and thermosetting acrylics. Although the word *acrylic* technically refers to polymers produced exclusively from acrylic monomers, over the years *acrylics* has come to include virtually any thermosetting or thermoplastic materials that are based on acrylic monomers or any other monomers that will easily copolymerize with them. The properties of acrylic polymers are greatly dependent on the monomers used in the synthesis of the polymer.

The major use of acrylics today is in the emulsion field. Basically, there are two different types of emulsions, the aqueous emulsions and the nonaqueous dispersions, which are commonly abbreviated NAD's. The aqueous emulsions are commonly known as latex paint vehicles. Both emulsions form continuous coatings by the evaporation of the nonsolvent. In the case of aqueous emulsions, the nonsolvent is water; in nonaqueous dispersions, the nonsolvent is organic. The process by which the continuous coating forms is called coalescence and is typical for both aqueous and nonaqueous systems.

Unlike solvent-based systems in which the polymer is soluble in the solvent and depends on crosslinking to generate high-molecular-weight materials with good properties, emulsion systems contain high-molecular-weight polymers dispersed as particles in the aqueous or nonaqueous phase. Consequently, no crosslinking or curing process is required to generate the final propertres. In fact, after most latexes coalesce, it is virtually impossible to redisperse the material in water. Unlike water-soluble polymers, which are always susceptible to attack by moisture, aqueous emulsion systems show little sensitivity to water once coalescence is complete.

Emulsion systems find widespread use in both architectural and industrial environments as base coats and topcoats. Nonaqueous dispersions have been used for over 10 years in the area of automotive finishes; however, such application is declining in favor of high-solids enamels.

Vinyl Resins

Vinyl resins are produced by the polymerization of certain monomers containing a double bond between adjacent carbon atoms; for example, the polymerization of vinyl chloride to form polyvinyl chloride. Although there are many resins and plastics that could fit into the vinyl classification of coating resins, a number of them, such as polyethylene, acrylic, and styrene-butadiene, are generally not included. Some of the common types of vinyl polymers and copolymers used in organic coatings include polyvinyl chloride, vinyl chloride-acetate copolymer, polyvinyl butyral, polyvinyl formal, and polyvinyl alcohol.

The resins produced from vinyl chloride have limited solubility in practical paint solvents, so they find limited use in solution-type coatings. However, there is a substantial, growing use of vinyl chloride resins in organosol and plastisol coatings. The resin is dispersed in a hydrocarbon such as kerosene (organosol) or a plasticizer such as dioctylphthalate (plastisol). After application, the coating is baked to fuse the dispersed particles into a continuous film. Polyvinyl organosols and plastisols are used as coatings for metal parts. One of the main advantages of these coatings is their excellent chemical resistance, with the chemical resistance of the plastisols varying according to the type and amount of plasticizer. Typical applications include dishwasher racks, steel window frames, steel container linings, automotive trim, and wire.

Vinyl resins are generally modified with ultraviolet light and/or heat stabilizers. When not stabilized properly, the resins deteriorate during baking and outdoor exposure, releasing small amounts of hydrochloric acid that react with the metal substrate and darken the coating. Baking-type vinyl coatings find extensive use in metal decoration and protection, including can linings, steel container linings, steel doors, and curtain walls. Vinyl resins are also used in the coating of steel sheet and coil, which are postformed into building components.

Cellulosics

The cellulosic polymers used in organic coatings are esters and ethers of cellulose. Perhaps the best known polymer to the average user of paints is nitrocellulose, which is used in many fast-drying lacquers. Other polymers include cellulose acetate, cellulose acetate butyrate, ethyl cellulose, methyl cellulose, and carboxymethyl cellulose. The last two polymers are water soluble and thus are not included among the lacquer resins. Cellulosic polymers are produced in different molecular weights and varying degrees of solubility.

The cellulosics are good film-forming materials and form the basis for fast-drying lacquers; however, they require modification for specific properties. The drying speed of the lacquers depends upon the rate of evaporation of the solvents and the ability of the coating to release the last few percent of solvent. In formulating cellulosic lacquers, various plasticizers and resins are incorporated to impart specific properties such as increased flexibility of the film, better adhesion, increased gloss, increased solids content (and thus better coating build), and improved durability.

Some of the best solvents for nitrocellulose lacquers are ketones, esters, and ether alcohols. These true solvents are blended with latent solvents and diluents primarily to reduce the cost of the lacquer thinner. Both aromatic and aliphatic hydrocarbons are used as diluents, and they are often good solvents for the plasticizer and resin portion of the lacquer. The lacquer solvent is thus a carefully balanced blend of true solvent, latent solvent, and diluent to obtain the best balance of solvency, evaporation rate, and cost.

Fluorocarbons

The fluorocarbon resins are unique materials that are noted for their lubricity and antistick properties as well as their weatherability, which is three to five times greater than other commonly used resins. The principal fluorocarbon polymers available today are prepared from the following monomers: tetrafluoroethylene, chlorotrifluoroethylene, vinyl fluoride, and vinylidene fluoride. These polymers are extremely inert materials and require special techniques in preparing their resins for application as coatings, generally as dispersions. The continuous coating is formed by fusing the dispersed particles of resin into a continuous film.

Tetrafluoroethylene and chlorotrifluoroethylene materials have found application as chemical-resistant coatings and linings when severe chemical environments justify the high cost. These coatings have exceptional release characteristics, and so tetrafluoroethylene (the highest melting polymer) has been used to line cooking and baking utensils. Vinyl fluoride and vinylidene fluoride have been finding some use in the long-term protection of metals outdoors, and they are unexcelled in their exterior durability. Fluorocarbon resins are also being used as mold-release agents and as high-temperature coatings.

PIGMENTS

Paint pigments are solid grains or particles of uniform and controlled size that are permanently insoluble in the vehicle of the coating.[2] They are generally dispersed in the vehicle by some type of grinding operation, usually performed by the paint manufacturer. Pigments are dispersed in a coating to hide or enhance the appearance or protection of a substrate. In addition, pigments have a positive effect on a number of properties including strength, adhesion, durability, gloss, flow, and corrosion resistance.

One very important factor in pigment dispersion is the pigment volume concentration (PVC).[3] The PVC is defined as the volume percent of the dry film that is actually pigment, as opposed to binder or additives. When the pigment particles are in contact with each other in the vehicle, the concentration is referred to as critical pigment volume concentration (CPVC). Exceeding the CPVC results in a deterioration in the resistance properties of the paint. The effects of PVC on various film properties are shown in Fig. 26-1. The pigment volume concentration ranges from zero in unpigmented films to over 90% in the case of zinc-rich primers.

PIGMENT CLASSES

The four commonly recognized classes of pigments are: (1) colored pigments; (2) white pigments, which include the primary and extender pigments; (3) metallic powders; and (4) functional pigments. The first class is available in both inorganic and organic compounds. The other classes are mostly inorganic materials. Functional pigments provide corrosion resistance, antifouling protection, slip resistance, or other desired properties. Table 26-4 summarizes the different types of pigments commonly being used by the paint industry.

PIGMENT PROPERTIES

The important pigment properties include tinting strength, light fastness, bleed characteristics, hiding power, transparency, particle size and shape, chemical and thermal stability, physical properties, rheological properties, and oil absorption.

Tinting Strength

The great majority of paints or pigmented coatings contain a white base, primarily titanium dioxide (TiO_2) to generate hiding, and are tinted with a colored pigment to produce the required color. The amount of pigment required to generate the color is a measure of the tinting strength of the pigment used. The tinting strength of a pigment is an important consideration in determining the cost involved in using a new pigment versus the pigment currently being used.

Light Fastness

Light fastness is a measure of the ability of a pigment to maintain its initial color and not fade or change in color when exposed to the weather, especially ultraviolet radiation from the sun.

Bleed Characteristics

Bleeding is the ability of a pigment in a coating to dissolve into an upper coat when a topcoat is applied. The dissolution is based on the solubility of the pigment in the solvent and the vehicle used for the topcoat. Some organic pigments are especially prone to bleeding.

Hiding Power

Hiding power is the ability of a pigment to hide the substrate and is usually expressed as square feet per gallon or square meters per liter. This ability depends to a great degree on the difference between the refractive indices of the resin and the pigment. The larger the difference, the better the hiding the pigment will produce. White, yellow, and orange pigments have the lowest indices of refraction, while black has the highest.

Chemical and Thermal Stability

Since certain pigments, especially those of lead and zinc, are prone to reaction with chemicals found in industrial environments, care must be exercised when using these pigments in certain environments. In addition, organic pigments show a pronounced tendency towards thermal decomposition and thus should be avoided in coatings that will be exposed to conditions of high heat.

Rheological Properties

The rheological (flow) properties of paints at the dispersion stage are critical to the development of pigment properties such as tinting strength, hiding power, and transparency. The pigment's particle shape, size, and surface characteristics play a major role in influencing the rheological properties of the coating material and therefore can effect the choice of a dispersion method. In addition, the pigments used in a paint can markedly influence the rheological properties of the paint at the application stage.

Oil Absorption

Oil absorption is the amount of polymer required to completely cover the surface of the pigment particle and is normally expressed as grams of oil required for 100 g (3.5 oz) of pigment. The coating chemist can convert the oil absorption value from a weight basis to a volume basis, changing this value to the critical pigment volume concentration (CPVC). The CPVC is a useful concept to the coatings chemist, especially since many properties change at or near the critical pigment volume concentration. Some of the properties that change depending upon the CPVC are blistering, rusting, staining, enamel holdout, tensile strength, and adhesion.

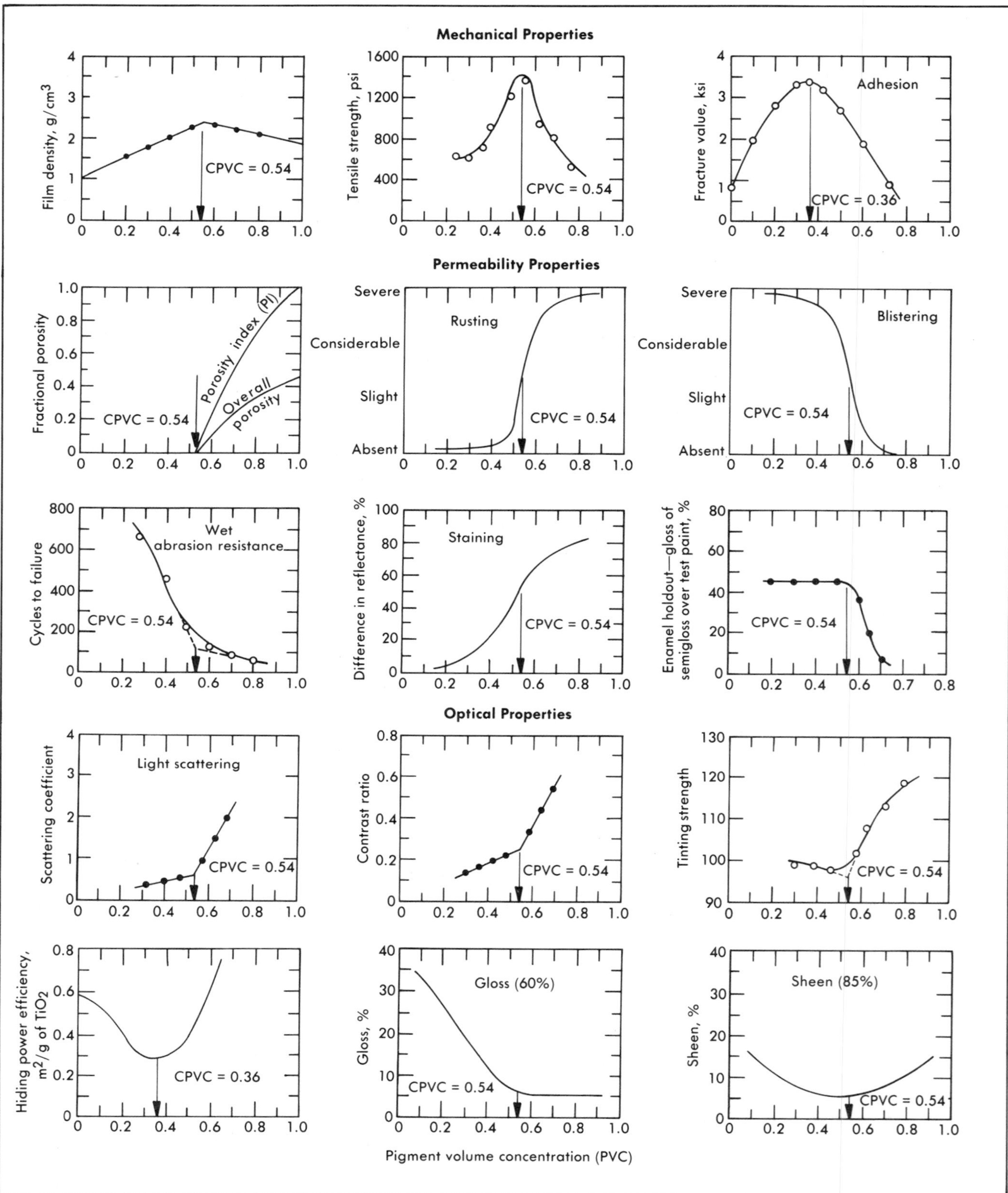

Fig. 26-1 The effects of pigment volume concentration on various coating properties.[4]

TABLE 26-4
Common Pigments and Extenders Used by the Coating Industry

White pigments	Red pigments	Metallics and miscellaneous types
Titanium dioxide (anatase and rutile)	Cadmium red	Aluminum powders
Zinc oxide	Benzimidazolone red	Bronze powders
White lead—basic carbonate	Toluidine red	Zinc dust
White lead—basic sulfate	Para red	Lead powder and flake
White lead—basic silicate		Nickel flake
Leaded zinc oxide	**Yellow and orange pigments**	Stainless steel
Antimony oxide	Lead chromate yellow	Cuprous oxide
Iron oxides	Chrome orange	
	Molybdate orange	**Extenders**
Red and brown iron oxides—natural	Zinc chromate yellow	Barytes
Red and brown iron oxides—synthetic	Cadmium yellow and orange	Blanc fixe
Yellow iron oxide—synthetic	Toluidine yellow	Calcium carbonate
Ochre		Silica—amorphous
Raw and burnt siennas	**Blue pigments**	Silica—diatomaceous
Raw and burnt umbers	Iron blue	Talc
Black iron oxide	Phthalocyanine blue	China clay
	Ultramarine blue	Mica
Green pigments	Indanthrone blue	Bentonite
Chrome green		
Chromium oxide	**Luminous and fluorescent pigments**	
Phthalocyanine green	Several colors	
Pigment green B		

SOLVENTS

Although solvents are defined as liquids that dissolve other substances, there are many materials used in the coating industry that are solvents for one type of resin but not solvents for another type of resin. In order to distinguish various solvents, the coatings chemist recognizes three types of solvent materials: (1) active solvents, (2) diluents, and (3) thinners. Active solvents are defined as liquids that dissolve binders and rapidly reduce the viscosity of the coating material. A diluent is a liquid that extends a solution but definitely acts to weaken the solvent power of the active solvent. The diluent must be completely miscible with the active solvent used. When using a diluent, it is quite common to have the viscosity actually increase as the solvency power of the active solvent is reduced. A thinner is a liquid that also extends a solution but does not materially impair the solvent power of the active solvent. It should be recognized that materials can function as active solvents, diluents, or thinners depending on the polymer system that is being dissolved.

In evaluating the solvent power of a particular material, the coatings chemist relies on information such as the solubility parameter of the solvent, the kauri-butanol value, and the aniline point of the solvent. The solubility parameter is a value assigned to a solvent, which is an attempt to place the solubility of a solvent on a sound thermodynamic basis. Generally speaking, this approach generates solubility parameters for all solvents that are known. Solubility parameters of various polymers are then predicted on the basis of which solvent or solvent blend will dissolve that particular polymer. For example, the solubility parameter of a solvent that will dissolve the polymer is assigned to that polymer. Extensive tables exist listing the solubility parameters of various solvents and polymers. In its simplest form, the coating chemist matches the solubility parameters of a solvent to a polymer in order to determine which solvent(s) will be effective in dissolving a particular polymer. The kauri-butanol value refers to the quantity of unknown solvent that can be added to a solution of kauri gum and butyl alcohol before the solution fails to be a solvent for the gum itself. The aniline point is the temperature at which a certain volume of a hydrocarbon will dissolve an equal volume of aniline oil.

Although there are many ways to classify solvents, solvents are generally divided into (1) hydrocarbons consisting of both aliphatic and aromatic hydrocarbons, (2) alcohols and amines, (3) active solvents such as esters and ketones, and (4) chlorinated solvents. The use of solvents in surface coatings is best illustrated by examining their function in various coating systems in which the selection of the solvent or solvent combination depends to a great extent on the type of resin, the method of application, and the drying time or the evaporation time required for the coating system. In addition, the solvent chosen must be suitable for the entire coating system, which might contain plasticizers and various additives such as pigments.

The application methods must also be considered in selecting the solvents or solvent combination. For example, paints applied by spraying require fast-evaporating solvents in order to ensure that a large proportion of the solvent is evaporated during the spraying process. The solvent evaporation prevents sagging of the film on vertical surfaces and allows quick setup of the coating. However, if the coating is to be applied by brushing, slow-evaporating solvents are required to ensure adequate flow and to avoid lap marks. Dipping or roll coating applications require slow-evaporating solvents so that solvent loss from the dipping bath or coater is minimized and coating flowout is enhanced.

Air-drying paints and fast-drying enamels are normally based on alkyd resins that require aromatic solvents and mineral spirits for the solvent combination. Aliphatic and aromatic solvents are commonly used. Baking enamels, primers, and automotive primers with alkyd resins are usually based on short-oil alkyd resins that require toluene and/or xylene as the solvent(s) for spray applications. Epoxy resins such as epoxy esters, which have been discussed previously, require both aromatic and aliphatic solvents for solubility. Nitrocellulose and vinyl resins normally require ketones to generate soluble systems. Latex paints, which are primarily waterborne, incorporate solvents as coalescing aids; glycol ethers are frequently used. A partial list of commercially available solvents with some of their properties is presented in Table 26-5.

TABLE 26-5
Partial List of Commercially Available Solvents

Chemical Name	Boiling Point, °F (°C)	Flash Point,* °F (°C)	Solubility Parameter
Pentane	97 (36)	0 (-18)	7
Hexane	152 (67)	0 (-18)	7.1
Lactol spirits	213 (101)**	20 (-7)	7.7
VM & P naphtha	252 (122)**	44 (7)	7.8
Mineral spirits	338 (170)**	95 (35)	7.8
Kerosene	410 (210)**	130 (54)	7.7
Odorless mineral spirits	362 (183)**	128 (53)	6.5
460 solvent	405 (207)**	150 (66)	7.5
Benzene	176 (80)	12 (-11)	9.2
Toluene	231 (111)	40 (4)	8.9
Xylene	280 (138)	80 (27)	8.9
Aromatic 100	330 (166)**	110 (43)	8.7
Aromatic 150	380 (193)**	150 (66)	8.7
Methanol	148 (64)	54 (12)	14.5
Ethanol	173 (78)	58 (14)	12.8
Isopropanol	180 (82)	53 (12)	11.5
Butanol	245 (118)	97 (36)	11.6
Ethyl acetate	170 (77)	24 (-4)	8.7
Isobutyl acetate	240 (116)	64 (18)	8.3
Methyl ethyl ketone	175 (79)	16 (-9)	9.3
Methyl isobutyl ketone	240 (116)	62 (17)	8.4
Diacetone alcohol	320 (160)	142 (61)	9.2

* Flash point tests are primarily performed in an open or closed cup.

** These materials are essentially mixtures of many different compounds and therefore do not have definite boiling points. The value given is approximate.

ADDITIVES

Additives are materials that, when added to a coating system in very small concentrations, exhibit a profound influence on the physical and chemical properties of the coating. Additives include such materials as surfactants, colloids and thickeners, biocides and fungicides, freeze/thaw stabilizers, coalescing agents, defoamers, plasticizers, flattening agents, flow modifiers, stabilizers, catalysts, and antiskinning agents. Each of these additives imparts special properties to the coating without which it would be impossible, in many cases, to develop a useable formulation.

SURFACTANTS

Surfactants are among the most versatile of the additives in the chemical industry, appearing in such diverse products as motor oils, pharmaceuticals, and detergents, as well as coatings. In spite of a wealth of experience in the field, the use of surfactants for a particular purpose remains more of an art than a science.

By definition, a surface-active agent (surfactant) is a substance that when present at a low concentration has the ability of adsorbing at the interface of a system and altering—usually reducing—the interfacial free energy. Surface-active agents have a characteristic molecular structure consisting of a group, known as the lyophilic group, that has very little attraction for the solvent, together with a group, called the lyophobic group, that has a very strong attraction for the solvent. When the solvent is water, these groups are referred to as hydrophilic or hydrophobic, respectively.

Literally thousands of surfactants have been synthesized. Of these, over 700 have been found to be of sufficient interest to be offered for sale in the United States by nearly 200 manufacturers. Surfactants are grouped into three classifications: (1) anionic surfactants, (2) cationic surfactants, and (3) nonionic surfactants. Each of these types of surfactants can be used alone or in combination with one of the others—except for the cationic and anionic surfactants, which are mutually antagonistic. In fact, when combinations of surfactants are used, such as anionic and nonionic or cationic and nonionic, the total effect is greater than the sum of the effects with separate surfactants.

Although it is possible for surfactants to affect many of the properties of a coating, the coatings industry is primarily concerned with (1) the ability of surfactants to affect the wetting of the substrate by the coating, (2) the effect of surfactants on pigment dispersions, and (3) the ability of the surfactant to stabilize an emulsion or dispersed system. Different materials that can function as surfactants are listed in Table 26-6.

TABLE 26-6
Partial List of Commercially Available Surfactants

Chemical Name	Typical Trade Names
Anionics	
Sodium stearate	Soap
Potassium lauryl sulfate	Conco Sulfate P
Sodium alkyl polyphosphate	Victawet 58B
Sodium dodecyl benzene sulfonate	LAS
Sodium dioctylsulfosuccinate	Aerosol OS
Sodium dioctylsulfosuccinate	Aerosol OT
Sodium ethoxylated and sulfated lauryl alcohol	Avirol 100-E
Sodium ethoxylated and sulfated alkyl phenol	Triton X-200
Cationics	
Ethoxylated fatty (soya) amine	Varonic L205
Stearylbenzyldimethylammonium chloride	Triton X-400
Lauryl pyridinium chloride	
Dodecyltrimethylammonium chloride	
Lauryl ether primary amine	Arosurf MG-70
Nonionics	
Sorbitan monolaurate	Span 20
Ethoxylated tridecyl alcohol	Renex 30
Ethoxylated nonylphenol	Triton N-101
Polyoxyethylene polyoxyropylene polyoxyethylene	Pluronic series
Alkylpolyoxyalkylene phosphate	Victawet 12
Diethanolamide of lauric acid 2/1	Emid 6540
Tertiary acetylenic glycol	Surfynol 104

PROTECTIVE COLLOIDS AND THICKENERS

Protective colloids and thickeners are essentially water-soluble resins that are used in emulsion-type paints at levels ranging from 0.1 to 1% by weight. They aid in pigment dispersion by increasing the viscosity of the paint system, preventing coagulation and settling, and controlling the flow and leveling of the final paint system. Materials commonly used as protective colloids and thickeners include starches, cellulosics, polyvinyl alcohols, sodium polyacrylates, and natural gums.

The disadvantages in using protective colloids and thickeners in coating formulations include the facts that they are water soluble, sometimes expensive, incompatible with the dry resin, and, in some cases, cause flocculation in coatings, especially if these colloids and thickeners possess a different charge than the surfactant used to disperse the resin. Colloids and thickeners are also prone to attack by bacteria and fungi.

BIOCIDES AND FUNGICIDES

Microbial growth can occur in paint systems when temperatures are between 72 and 100° F (22 and 38° C), when oxygen is readily available, and when sufficient water is available at neutral pH levels. The enzymes produced by microorganisms promote degradation of the stabilizers in the polymeric system.

Biocides and fungicides function in water-reducible coatings as inhibitors to prevent the formation of microbial growth. Specifically, biocides function to inhibit microbial growth in aqueous uncured systems, whereas fungicides inhibit microbial growth in dry paint systems. Materials commonly used as biocides and/or fungicides include mercuric compounds and sodium salts of pentachlorophenol. Table 26-7 lists some of the biocides and fungicides that are commercially available.

TABLE 26-7
Partial List of Commercially Available Biocides and Fungicides

Chemical Type	Trade Name	Manufacturer
Oxazolidine	Bioban CS-1135	Angus Chemical Co.
Barium metaborate	Busan 11	Buckman Laboratories, Inc.
Tin	Cotin 300	Cosan Chemical Corp.
Phenolic	Dowicide G	Dow Chemical Co.
Copper	Intercide	Interstab Chemicals
Mercurial	PMA-18	Nuodex, Inc.
Mercurial	Troysan PMO	Troy Chemical Corp.

FREEZE/THAW STABILIZERS

Freeze and/or thaw stabilizers, commonly of the glycol or glycol ether types, are useful in water-based systems in which freezing is particularly destructive to the emulsion. Freeze/thaw stabilizers essentially function to reduce the freezing point of the aqueous system much like antifreeze functions in the radiator of a car. A partial list of commercially available freeze/thaw stabilizers can be found in Table 26-8.

TABLE 26-8
Partial List of Commercially Available Freeze-Thaw Stabilizers

Ethylene glycol
Propylene glycol
Glycol ethers
Polyalkylene polyamines
Glyceryl monoricinoleate
Ureas
Thioureas

ADDITIVES

COALESCING AGENTS

Coalescing agents are essentially high-boiling solvents, normally of the glycol ether or polyethylene oxide type, that are added to water-soluble systems to aid in coalescing the polymer particles as the water evaporates from the system. Some of the common coalescing agents are listed in Table 26-9.

TABLE 26-9
Partial List of Commercially Available Coalescing Solvents

Chemical Type	Trade Name
Glycol ether ester	Butyl Carbitol Acetate
Glycol ether ester	Butyl Cellosolve Acetate
Glycol	Carbitol
Glycol ester	Carbitol Acetate
Hexylene glycol	Hexylene Glycol
Ether alcohol	Texanol

DEFOAMERS

Foaming is commonly caused by the presence of surfactants and various additives such as solvents and amines. Foaming causes cratering in the paint film and may occur during or after paint application. Defoamers function by displacing surfactants from the surface of the foam, thereby increasing the surface tension and breaking the foam. A list of some commercially available defoamers is presented in Table 26-10.

The addition of materials such as silicone oils or hydrocarbon solvents in small amounts (no more than 2% by weight) normally prevents foaming. Excessive amounts of a defoamer, especially in the case of silicone compounds, may result in cratering and poor intercoat adhesion.

PLASTICIZERS

A plasticizer is defined as a substance that is added to a polymer or a paint system to increase the flexibility of the finished product. In the coating industry, plasticizers are used to lower the glass transition temperature (T_g) of the polymer. Some materials commonly used as plasticizers in the coatings industry can be found in Table 26-11.

FLATTENING AGENTS

Flattening agents are materials that are basically added to a paint or coating to reduce the gloss or angular sheen of the dry coating. Materials commonly available that can be used as flattening agents include a great variety of synthetic silicas and zinc stearate. Table 26-12 is a list of some commercially available flattening agents.

FLOW MODIFIERS

Flow modifiers are used in coatings to impart thixotropy to the coating system. Thixotropy is the property possessed by certain gels or dispersions that enables the viscosity to decrease when the liquid is shaken, brushed, or mechanically disturbed, but increase when left in an undisturbed state. This property is necessary when the coating must be applied at low viscosity; yet after application, the coating must thicken rapidly to prevent sagging. Various materials can be added to paint systems to impart thixotropy. A partial list of commercially available flow modifiers is presented in Table 26-13. Some of these materials are similar to the protective colloids and thickeners previously discussed.

STABILIZERS

Polymers are prone to degradation under the influence of ultraviolet light and heat. To avoid degradation, stabilizers are added to coating systems containing polymers. Stabilizers come in many forms depending on the role they play in stabilizing the polymer. A list of commercially available stabilizers and their functions in coating systems is given in Table 26-14.

CATALYSTS

Catalysts are used in chemical applications to lower the energy at which chemical reactions, such as crosslinking, take place. Catalysts are normally metallic driers such as divalent salts of carboxylic acids, strong acids such as sulfonic acids, and weaker acids such as carboxylic acids, amines, and polyamides. The catalysts used in coating applications are fairly specific to the particular application involved. Driers are used exclusively in promoting the oxidative crosslinking of alkyds, epoxy esters, and other materials containing unsaturation in the form of natural oils. Sulfonic acids are commonly used in thermal crosslinking of polymers to promote the reaction between aminoplast crosslinkers and hydroxyl and carboxyl-containing polymers. Amines and polyamides are used with epoxy-resin systems. Table 26-15 is a partial list of commercially available catalysts.

ANTISKINNING AGENTS

Alkyd coatings and other coating materials containing unsaturated natural oils are especially prone to surface oxidation in open containers, leading to the formation of a polymeric skin on the surface of the paint. Antiskinning agents prevent the paint from forming a skin while in storage. Oxime and phenolic compounds, when used as additives, are especially helpful in preventing skinning in alkyd coating systems. Examples of some commercially available antiskinning agents are listed in Table 26-16.

TABLE 26-10
Partial List of Commercially Available Defoaming Agents

Trade Name	Manufacturer	Use
Byk 0	Byk-Mallinckrodt Co.	Solvent coatings
Byk 020	Byk-Mallinckrodt Co.	Water-reducible coatings
Byk W	Byk-Mallinckrodt Co.	Latex coatings
Busperse 54C	Buckman Lab	Latex coatings
Dapro DF-975	Daniel Products Co.	Water coatings
Foamaster B	Diamond Chemical Co.	All purpose
Nopco NXZ	Diamond Chemical Co.	All purpose
Defoamer 357	Hercules, Inc.	Latex coatings

TABLE 26-11
Partial List of Commercially Available Plasticizers

Chemical Type	Trade Name	Manufacturer
Abietate	Abalyn	Hercules, Inc.
Adipate	Santicizer 97	Monsanto Co.
Benzoate	Nuoplaz 1046	Nuodex, Inc.
Castor oil	Polyol 1066	Spencer-Kellogg, Div. of Textron, Inc.
Azaleic/alkyd	Paraplex RGA-2	Rohm & Haas Co.
Phthalates	Kodaflex DMP	Eastman Chemical Products
Phosphates	Santicizer 143	Monsanto Co.
Sebacates	Nuoplaz 6814	Nuodex, Inc.

TABLE 26-12
Partial List of Commercially Available Flattening Agents

Type	Trade Name	Manufacturer	Use
Silica	Flat-Ayd 3B	Daniel Products Co.	Alkyd coatings
Zinc stearate	59	Inmont Corp.	Lacquers
Silica	Syloid 74	Davison Specialty Chemicals	General purpose
Silicate	Zeolex 80	J.M. Huber Corp.	Latex coatings

TABLE 26-13
Partial List of Commercially Available Flow Modifiers

Chemical Type	Trade Name	Manufacturer
Hydroxyethyl cellulose	Natrosol 250	Hercules, Inc.
Nonionic polymer	Cyanamer P-350	American Cyanamid Co.
Hydroxypropyl methyl cellulose	Methocel	Dow Chemical Co.
Silicate	Attagel	Engelhard Industries
Carboxymethyl cellulose	CMC-12	Hercules, Inc.
Organosilicate	Bentone	NL Chemicals
	Raybo	Raybo Chemical Co.
Sodium polyacrylate	Acrysol GS	Rohm & Haas Co.
Castor oil derivative	Thixcin	NL Chemical

TABLE 26-14
Partial List of Commercially Available Stabilizers

Chemical Type	Trade Name	Manufacturer	Use
Organotin	837	Ferro Corp.	Heat stabilizer
Calcium/zinc	59VII	Ferro Corp.	Light and heat stabilizer
Organotin	Thermolite 12	M & T Chemicals, Inc.	Heat stabilizer
Phosphite	Thermolite 187	M & T Chemicals, Inc.	Auxiliary stabilizer
Lead salts	Halstab 30	Halstab, Div. Hammond Lead Products	Heat stabilizer
Sodium polyacrylate	Modicol VD	Diamond Shamrock Corp.	Chemical stabilizer
Epoxy resin	909	Ferro Corp.	Heat stabilizer

TABLE 26-15
Partial List of Commercially Available Catalysts

Chemical Type	Trade Name	Manufacturer
Cobalt acetate	none	Mooney Chemicals, Inc.
Cobalt carbonate	none	Mooney Chemicals, Inc.
Calcium octoate	Catalox	Ferro Corp.
Cobalt octoate	Hexogen	Interstab Chemicals, Inc.
Zinc octoate	Hexogen	Interstab Chemicals, Inc.
Zirconium octoate	Zirco	Interstab Chemicals, Inc.
Manganese octoate	Troychem	Troy Chemical Co.
Toluenesulfonic acid	PTSA	Various manufacturers

TABLE 26-16
Partial List of Commercially Available Antiskinning Agents

Type	Trade Name	Manufacturer	Use
Butyraldoxime	Coskin B	Cosan Chemical Corp.	General purpose
Methyl ethyl ketoxime	Aska	Interstab Chemicals, Inc.	General purpose
Methyl ethyl ketoxime	Exkin 2	Nuodex, Inc.	General purpose
Phenolic	Anti-Skin S	Troy Chemical Co.	Alkyd coatings

FORMULATIONS

Paint formulation is the art of bringing together a number of different materials to form a final product that will meet a particular need. Although there have been recent attempts to make this activity more scientific, the ability to formulate is still an art that is learned through years of experience rather than from any scientific principles that are taught in universities. When formulating a coating, the coatings chemist is guided by personal experience and any information on the various raw materials that may be available from the manufacturer.

When developing a formulation, the basic prerequisite is a broad knowledge of the properties of the raw materials that are used in coatings and how these materials interact with each other, as well as an understanding of all the components that make up a coating including the polymer, the pigments, the additives, and the solvents. The coatings chemist must also be aware of both the costs involved in putting these materials together and the processing costs. Other essential information includes the application method, the substrates on which the coating will be used, the kind of exposure that the paint will be required to endure, and the viscosity, evaporation rate, color, and hiding power required.

The coatings chemist must take into account the pigment volume concentration and the critical pigment volume concentration, discussed previously in this chapter, which define the amount of pigment plus extender that will be required for a particular application in order to optimize properties such as permeability, gloss, blistering, and rusting. In addition, the amount and the type of pigment has a profound effect on the rheological properties of the system. The amount of pigment used also has an effect on the gloss of the system. For example, if a high-gloss system is required, the pigment concentration must be kept fairly low so that the reflection of light from the surface will be that of the polymer or resin itself. Pigments must be selected to provide hiding of the substrate and the required color for the application. Further, pigment type and concentration affect durability, light fastness, chalking, and leveling properties of the system. The selection of resins and solvents is based on the type of coating properties required from the final coating and the curing conditions of the application method. Table 26-17 lists the common coating systems in use today along with their polymers.

TABLE 26-17
Types of Commonly Used Coating Systems

Coating System	Polymer
Air drying	Alkyds and epoxy esters are commonly used in conjunction with driers to promote oxidative crosslinking.
Baking coatings	Alkyds and epoxy esters are commonly used with aminoplast (melamines or ureas) crosslinkers and acid catalysts.
Two-part coatings	Urethane (isocyanate) type polymers with polyester crosslinkers. Tertiary amines can be used to catalyze the crosslinking reaction.

(continued)

TABLE 26-17—*Continued*

Coating System	Polymer
Latex coatings	Primarily vinyl, acrylic, or styrene polymer combinations. These coatings do not require crosslinking.
Powder coatings	Usually vinyl, polyester, or epoxy-type polymers requiring plasticizers but no crosslinking.
High-solids coatings	Low-solids-type polymers dissolved in strong solvents such as ketones to achieve higher solids level. Can also contain low-molecular-weight varieties of alkyds, epoxy esters, or polyesters.
Water-reducible coatings	Water-soluble versions of alkyds, epoxy esters, or polyesters. Curing can be accomplished by either oxidative crosslinking using driers or by thermal crosslinking using aminoplasts and acid catalysts.
Radiation-curable coatings	Primarily acrylic or methacrylic functional polymers diluted with low-molecular-weight reactive diluents (solvents) containing acrylic or methacrylic functionalities and photocatalysts to promote crosslinking. Epoxy functional polymers using photoionic catalysts have also been used.

Quality control for coating systems is a major responsibility of the coatings chemist and is essential in determining if the coating will meet the standards required in the application process and in the end use. Most of the test procedures have been standardized by the ASTM and can be found in parts 27, 28, and 29 of the *Annual Handbook of American Society for Testing and Materials (ASTM) Standards*. Specifically, these tests are designed to:

- Gather information regarding the coating during the development process.
- Determine if an experimental coating is ready for use as a product.
- Evaluate new raw materials to replace existing ones.
- Evaluate the suitability of the coating for the application.
- Verify the manufacturer's claims for the coating.

These tests are conducted by both the formulations chemist and the end user. Table 26-18 presents a summary of the tests. For additional information on these tests, refer to Chapter 30, "Testing, Troubleshooting and Safety."

TABLE 26-18
Common Organic Coating Tests

Test	Description
Color	Performed using an instrument or by comparison to a standard color.
Gloss	Comparison of coating to a standard gloss sample using a photoelectric glossmeter.
Hiding power	Determined by calculating the contrast ratio of the paint, which is the ratio of the light reflectance over a black surface divided by the light reflectance over a white surface.
Mar resistance	A test to determine the ability of the coating surface to resist defacing.
Chemical resistance	A series of tests designed to evaluate the resistance of the coating to a prescribed chemical. The coated sample may be immersed in the chemical or a drop may be placed on the sample.
Weathering resistance	Performed by exposing coated panels to outside exposure. Observations are made regarding appearance, gloss, chalking, cracking, peeling, mildew, rusting, fading, blistering, and yellowing. Accelerated weathering tests have also been developed.
Abrasion resistance	A test to determine the ability of the coating to resist wear when in contact with an abrasive. Abrasive wheels and sprayed abrasive materials are commonly used.
Humidity resistance	A test to determine the resistance of the coating to damage by water. Commonly performed in humidity cabinets.
Salt-spray resistance	A test used to evaluate the resistance of the coating to corrosion. Normally determined by observing the amount of rusting that develops from an X scratched through the coating to the metal surface.
Flexibility	A test to measure the ability of the coating to expand without rupturing. Normally performed

(continued)

TABLE 26-18—*Continued*

Test	Description
	by bending on a mandrel or by impact.
Density	A measure of the weight per unit volume; usually expressed in lb/gal.
Dispersion	A measure of the approximate particle size of the paint pigment. Normally determined as the fineness of grind using a Hegman gauge.
Viscosity	A measure of the resistance of the liquid coating material to shearing forces. Common procedures include ASTM Bubble test, Ford cups, Zahn cups, Stormer viscometer, and Brookfield viscometer.
Drying	A measure of the amount of time required for a coating to achieve a certain degree of dryness. By the finger test method, the extent of dryness is determined as wet, set-to-touch, tack-free, and dry.
Film thickness	A measurement on the wet or dry film using an appropriate measuring device. Test can be destructive or nondestructive depending on the test method selected.

CORROSION-PREVENTIVE MATERIALS

Corrosion is the process by which man-made materials assume their most stable state. In order for corrosion to occur, oxygen, an electrolyte dissolved in water, and a conductive material such as iron are required. The most common forms of corrosion involve the oxidation of aluminum and iron to their stable oxide forms. Corrosion itself actually involves an oxidation reaction occurring in concert with a reduction reaction.

Iron oxidation, which occurs primarily at the anode, involves the conversion of iron in its zero oxidation state to iron in its plus two (Fe^{+2}) or plus three (Fe^{+3}) oxidation state. The reduction, which occurs simultaneously at the cathode, involves the conversion of water to hydrogen and oxygenated products.

The corrosion-preventive function of a coating is to prevent the onset of corrosion or to control the spread of corrosion from the initial site of corrosion. Coatings have been developed that function as barrier coatings, conversion coatings, or sacrificial coatings. Sacrificial coatings are primarily inorganic zinc systems containing between 80 and 90% zinc. The zinc oxidizes preferentially to the iron, providing corrosion protection of the metal.

BARRIER COATINGS

The permeation of coatings by water requires the surface of the metal to be constantly exposed to moisture. If oxygen and electrolytes are available, rapid corrosion can form on the surface. However, organic coatings exhibit low permeability to ionic species (electrolytes) dissolved in water. Consequently, the concentration of the electrolyte at the surface is minimal, making it difficult to set up a conductive cell as long as the coating remains intact. Barrier coatings are normally optimized by applying thick films of materials to the substrates to be coated. No coating, however, is completely impermeable to moisture and any electrolytes that might be dissolved in the moisture, and corrosion will occur regardless of the thickness of the coating. The tendency for corrosion to occur is simply reduced as the thickness of the barrier coating is increased.

CONVERSION COATINGS AND CONVERSION-TYPE PIGMENTS

Inorganic phosphate coatings are commonly used to protect the substrate from corrosion. Phosphate coatings deposit an insoluble metal phosphate on the surface of the substrate that raises the oxidation potential of the surface and prevents corrosion. Materials commonly used are iron phosphates, zinc phosphates, and manganese phosphates; all three materials are only sparingly soluble in water and will raise the oxidation potential of the metal surface. On a cost/performance basis, the zinc phosphate coatings are much preferred since they provide long-term resistance to corrosion and an excellent surface for the adhesion of paints. In addition to phosphate conversion coatings, inorganic chromate conversion coatings are often used, primarily on zinc, aluminum, tin, and magnesium substrates. For additional information on conversion coatings, refer to Chapter 19, "Conversion Coatings and Anodizing."

Many paints use inorganic pigments that inhibit the corrosion of the metal by processes similar to those of conversion coatings. These inorganic pigments, primarily phosphates or chromates, passivate the surface of the metal. The solubility of inorganic pigments in water is important because a pigment must not be too soluble or it will be washed rapidly from the metal surface. However, if the pigment is too insoluble, there will not be enough pigment migrating to the surface in the form of an aqueous solution to passivate the metal.

Chromates are the most effective inorganic pigments, with zinc yellow or basic zinc chromate being the most commonly used. The various molybdates are fairly expensive but are also effective inhibitors. Compounds of lead and chromium are considered to possess a certain degree of toxicity, and hence efforts are under way to replace these traditional corrosion-inhibiting pigments with other materials. Table 26-19 lists commercially available corrosion-retardant pigments in use today.

TABLE 26-19
Partial List of Commercially Available Corrosion-Retardant Pigments

Lead silicochromate	Calcium borosilicate
Strontium chromate	Calcium-zinc molybdate
Zinc phosphate	Barium metaborate

RADIATION-CURABLE COATINGS

Radiation-curable coatings have been developed over the past fifteen years primarily as clear (nonpigmented) coatings for use on floor tile, beverage containers, building products, and packaging. These coatings are acrylic or methacrylic functional polymers diluted with reactive diluents containing the same type of functionality as the polymer. They are also epoxy functional polymers diluted with diluents containing epoxy functionality. Both types of coatings can be crosslinked under ultraviolet light, with the acrylic/methacrylic coatings using photoinitiators to promote radical crosslinking and the epoxy coatings using organic salts such as diphenyliodonium fluoroborate to initiate cationic crosslinking of the epoxy groups.

Curing is rapid, generally occurring within a few seconds for both types of coatings. Since the solvents contain reactive functionalities and are polymerized into the coating system during cure, these coatings are considered essentially nonpolluting. Care must be used in handling these materials, however, because they can be toxic and possibly cause allergic reactions. Table 26-20 lists some of the commonly used photoinitiators, while Table 26-21 lists some of the reactive diluents commonly used in these coatings.

TABLE 26-20
Commonly Used Photoinitiators

Chemical Name	Coating System
Benzophenone	Acrylic/methacrylic
Diethoxyacetophenone	Acrylic/methacrylic
Diphenyl iodonium fluoroborate	Epoxy
Diphenyl sulfonium fluoroborate	Epoxy

TABLE 26-21
Chemical Names of Commonly Used Reactive Diluents

Triethyleneglycol dimethacrylste
Vinyl pyrrolidone
Isoborneol acrylate
Trimethylolpropane trimethacrylste

VAPOR CURE COATINGS

Vapor cure coatings are the latest type of coatings produced by the coatings industry. These coatings utilize the rapid crosslinking reaction that occurs between isocyanates and polyester alcohols when catalyzed by tertiary amines. The earliest vapor cure coatings required premixing the isocyanate prepolymer and the polyester alcohol, applying this mixture to the part to be coated, and then exposing the coating to a low concentration of tertiary amine gas in an inert carrier gas. Recent advances in vapor cure coatings enable all the components to be mixed with a less volatile amine. Coatings prepared by this process show many, if not all, of the properties of standard urethane coatings.

POWDER COATINGS

Powder coatings are unique among all present-day compliance coatings in that they are dry, solid coating materials in contrast to the liquid materials of other coating technologies.[5] Prior to baking, powder coatings are finely pulverized, plastic compositions; consequently, the resins used in powder coatings are different from those in liquid paints. Liquid paints require resins that are soluble or miscible with solvents and/or water, whereas powder coatings require resins that are solid materials. These solid materials must be solid at ambient and elevated storage temperatures, but must be capable of melting sharply to low viscosities when heated. Powder coatings are available as thermoplastic and thermosetting resins.

In the early 1970s, the number of solid resin systems available to the powder coatings manufacturer was somewhat restricted; consequently, powder coating was limited in its ability to meet the diverse needs of the finishing industry. However, with the growth of powder coatings, the technology has rapidly expanded, resulting in many new resin systems and other compositional components. There are now available a wide range of powder coatings that will meet and often exceed the properties of most present-day solvent-borne and water-borne baking enamels.

Like all other industrial surface coatings, powder coatings are individually formulated to meet the industrial user's specific finishing needs. Individual formulation means matching color and film performance requirements, all within the particular restrictions of the finisher's operation. As with the selection of any industrial finish, therefore, a close relationship must be developed between the user of the coating and the coatings supplier so that exacting requirements are thoroughly understood and the correct type of finishing material is supplied.

THERMOPLASTIC POWDERS

A thermoplastic powder coating is one that melts and flows with the application of heat, but maintains the same chemical composition when it solidifies on cooling. Thermoplastic powder coatings are based on thermoplastic resins of high molecular weight. These tough and resistant resins tend to be difficult and also expensive to grind into the very fine particles necessary for the fusion of thin paintlike film thicknesses. Consequently, thermoplastic resin systems are used more as thick-film functional coatings and are applied mainly by the fluidized-bed application technique. Typical film thicknesses are 5-12 mils (0.13-0.30 mm) when applied by the fluidized-bed technique and 3-5 mils (0.08-0.13 mm) when applied using electrostatic spraying techniques. The various coating application techniques are discussed in greater detail in Chapter 27, "Application Methods."

The most commonly used thermoplastic powders are polyvinyl chloride, nylon, polyester, polyethylene, and polypropylene. Some of these powders require a primer to be applied to the substrate surface prior to coating. Table 26-22 summarizes the physical and coating properties of the common thermoplastic powders.

THERMOSETTING POWDERS

Thermosetting powder coatings are quite different from thermoplastic powder coatings in that they are based on lower molecular-weight solid resins, which, on melting and flowing, chemically crosslink within themselves or with other reactive components to form a higher molecular-weight reaction product. The coating film formed by this reaction is heat stable and will not soften back to a liquid on further exposure to heat. Powders based on these thermosetting resins can be ground into fine particle sizes, in the range of 0.001 to 0.0015" (0.03 to 0.038 mm) diameter. Due to the rheological characteristics of thermosetting resins, thin paintlike surface coatings, 1-2 mil (0.03-0.05 mm) thick, can be produced; their properties are usually better than the properties of liquid coatings.

The main types of resins used in thermosetting powder coatings are epoxy, polyester, and acrylic. Table 26-23 summarizes typical physical and coating properties of these powders.

TABLE 26-22
Physical and Coating Properties of Thermoplastic Powders

Property	PVC	Nylon-11	Polyester	Polyethylene	Polypropylene
Primer required	Yes	Yes	No	Yes	Yes
Melting point, °F (°C)	266-300 (130-150)	367 (186)	320-338 (160-170)	248-300 (120-130)	329-338 (165-170)
Typical preheat/ postheat, °F (°C)	446-554 (230-290)	482-590 (250-310)	482-572 (250-300)	392-446 (200-230)	428-482 (220-250)
Specific gravity	1.20-1.35	1.01-1.15	1.30-1.40	0.91-1.00	0.90-1.02
Adhesion* **	G-E	E	E	G	G-E
Gloss, Gardner 60° meter	40-90	20-95	60-95	60-80	60-80
Hardness, Shore D	30-55	70-80	75-85	30-50	40-60
Flexibility†	Pass	Pass	Pass	Pass	Pass
Resistance:*					
Impact	E	E	G-E	G-E	G
Salt spray	G	E	G	F-G	G
Weathering	G	G	E	P	P
Humidity	E	E	G	G	E
Acid‡	E	F	G	E	E
Alkali‡	E	E	G	E	E
Solvent‡	F	E	F	G	E

* E = excellent, G = good, F = fair, P = poor.
** With primer where indicated.
† No cracking, 1/8" (3 mm) diam mandrel bend.
‡ Inorganic, dilute.

BIBLIOGRAPHY

TABLE 26-23
Physical and Coating Properties of Thermosetting Powders

Property	Epoxy	Polyurethane[a]	Polyester[b]	Hybrid	Acrylic[a]
Film thickness, mils	1-4	1-3.5	1-10	1-10	1-2.5
(mm)	(0.02-0.10)	(0.02-0.09)	(0.02-0.25)	(0.02-0.25)	(0.02-0.06)
Cure type[c]	L, N, Q	N, Q	L, N	N	N
Adhesion[d]	E	G-E	G-E	G-E	G
Gloss, Gardner 60°	5-95	10-95	40-95	30-95	80-95
Hardness (pencil)	H-4H	H-4H	H-2H	H-2H	H-5H
Flexibility[d]	E	E	E	E	F
Resistance:[d]					
Impact	E	G-E	G-E	G-E	F
Overbake	F-G	G-E	E	E	G
Weathering	P	G-E	E	P-F	G-E
Acid[e]	E	G	G	G	F
Alkali[e]	G	F	F	G	G
Solvent	E	G	F-G	F	F
Corrosion	E	E	E	E	E

[a] Hydroxy functional—blocked isocyanate cure.
[b] TGIC (triglycidyl isocyanurate) cure.
[c] L = low, N = normal, Q = quick.
[d] E = excellent, G = good, F = fair, P = poor.
[e] Inorganic, dilute.

References

1. Gerald L. Schneberger, *Understanding Paint and Painting Processes*, 2nd ed. (Wheaton, IL: Hitchcock Publishing Co.), pp. 223-241.
2. Guy E. Weismantel, ed., *Paint Handbook* (New York: McGraw-Hill Book Co., 1981), pp. 1-18.
3. Schneberger, *op. cit.*, p. 7.
4. T. C. Patton, *Paint Flow and Pigment Dispersion*, 2nd ed. (New York: John Wiley and Sons—Interscience, 1979), p. 172.
5. Peter R. Gribble, *Powder Coatings Materials, You and Your Supplier*, SME Technical Paper FC81-447 (Dearborn, MI: Society of Manufacturing Engineers, 1981).

Bibliography

Banov, A., ed. *Paints and Coatings Handbook*. New York: McGraw Hill Book Co., 1982.

Gordon, J. L., and Prane, J. W., eds. *Non-Polluting Coatings and Coating Processes*. New York: Plenum, 1973.

Hess, M. *Paint Film Defects*. London: Chapman and Hall, 1979.

Martens, E. R. *Waterborne Coatings*. New York: Van Nostrand Reinhold, 1981.

Oil and Color Chemists Association. *Surface Coatings—Raw Materials*, vol. 1. New York: Chapman and Hall, 1983.

Roffey, C. G. *Photopolymerization of Surface Coatings*. New York: John Wiley and Sons—Interscience, 1982.

Rosen, M. J. *Surfactants and Interfacial Phenomenon*. New York: John Wiley and Sons—Interscience, 1978.

Snogren, R. D., ed. *Handbook of Surface Preparation*. New York: Palmerton, 1974.

Sward, G. G., ed. *Paint Testing Manual*. Philadelphia: American Society for Testing and Materials, 1972.

APPLICATION METHODS

The successful use of organic coatings depends to a great extent upon their correct application. The general requirement placed upon the various coating methods used is that they be capable of depositing a layer of the coating material approximately a thousandths of an inch thick. This layer must then be distributed with maximum uniformity and complete continuity.

The selection of an application method is determined by the type of coating material applied, the coating thickness and final properties required, the quantity of parts being coated, and the transfer efficiency of the application method. The transfer efficiency is the percentage of paint solids consumed that actually coats the merchandise. Transfer efficiencies vary from 30 to almost 100%; in other words, some application methods waste 70% of the consumed paint, while others are nearly waste-free.

The most widely used application methods are discussed subsequently; each has specific advantages and limitations. Being familiar with all these methods will assist the manufacturing engineer in selecting the proper method for a specific application. Although one process may be suitable for one installation, it may not be suitable for another. Table 27-1 summarizes the capabilities of the various application methods.

CHAPTER CONTENTS:

SUBSTRATE PREPARATION

The successful performance of a finish applied to any substrate is dependent upon the proper conditioning of the substrate surface. Failure to properly prepare the substrate can result in paint peeling, flaking, or blistering, underfilm rusting of ferrous metals, and underfilm corrosion of nonferrous metals.[1] Although in theory each surface should be chemically and physically clean before painting, in practice the choice of an adequate surface preparation or pretreatment is often dictated by the economics of the particular situation. To gain full advantage of the coating characteristics, surface preparation should be part of the finishing process. In general, substrate preparation serves to (1) remove soils or imperfections from the substrate, (2) create a surface susceptible to bonding, and (3) establish a chemical coating on the surface that will slow corrosion in case the coating film is damaged.

The soils to be removed may range from oil, grease, and wax-based soils to defects such as oxidation, rust, corrosion, heat scale, tarnish, and smut. Oils and greases are generally removed with alkaline or solvent-type cleaners, whereas natural conversion coatings are removed with acidic or chelated alkaline cleaners. For additional information on cleaning procedures and solvents, refer to Chapter 18, "Cleaning." Mechanical cleaning methods are discussed in Chapter 16, "Mechanical and Abrasive Deburring and Finishing."

With some cleaning techniques, a surface suitable for bonding may be produced while removing soils or imperfections; with other techniques, it may be necessary to roughen the surface. A corrosion-inhibiting surface can be produced by phosphating, chromating, or anodizing. These surface treatments are discussed in Chapter 19, "Conversion Coatings and Anodizing."

SPRAY COATING

Spray coating is the process of applying a liquid coating by causing the liquid coating material to be broken up or atomized into a fine mist or spray that is then deposited onto the part surface. The individual droplets of material flow together on the surface to form the coating film. The three basic

Contributors of sections of this chapter are: ***George E. F. Brewer****, Coating Consultants;* ***Glen L. Muir****, Sr. Project Design Engineer, Finishing Div., Graco, Inc.*

Reviewers of sections of this chapter are: ***George E. F. Brewer****, Coating Consultants;* ***Dr. Kenneth J. Coeling****, Manager, Divisional Engineering, Industrial-Commercial Div., DeVilbiss Co.;* ***Archie W. Garner****, Technical Manager, Glidden Coatings and Resins;* ***Lyle E. Gilbert****, Sales Engineer, MetoKote Corp.;* ***Rolf Gruener****, President, Koating Machinery Company, Inc.;* ***Jerry P. Hund****, Training Instructor, Training Div., Binks Mfg. Co.;* ***Carl P. Izzo****, Fellow Scientist, Research and Development Center, Westinghouse Electric Corp.;* ***Donald T. Jones****, Sr. Sales Engineer, Black Bros. Co.;* ***Harry M. Leister****, Group Lab Manager, Quaker Chemical Corp.;* ***Dean M. McCaskill****, Product/Marketing Manager—Powder, Finishing Equipment Div., Nordson Corp.;* ***Mark A. McQuaide****, Process Sales Engineer, Rapid Engineering, Inc.;* ***David Meynell****, Manager of Engineering, Durr Engineering and Management, Inc.;* ***Don Miller****, Sales Manager, Technical Products Div., George Koch Sons, Inc.;* ***Glen L. Muir****, Sr. Project Design Engineer, Finishing Div., Graco, Inc.;* ***Daniel C. Riter****, Product Manager, Powder Coatings Dept., Coatings Div., Ferro Corp.;* ***Ronald E. Schaefer****, Chief Chemist, MetoKote Corp.;* ***Ellsworth A. Stockbower****, Sales Manager, Autodeposition, AMCHEM Products, Inc.;* ***D. F. Stofleth****, Engineering Project Manager, Industrial Div., George Koch Sons, Inc.;* ***Gil Tredwell****, Sales Manager, Industrial Div., George Koch Sons, Inc.*

TABLE 27-1
Comparative Data on Application Methods for Organic Coatings

Method	Typical Coating Thickness,* mils (mm)	Transfer Efficiency, %	Equipment Cost	Operating Cost
Air spraying	1-3 (0.025-0.075)	30-40	Low	Low
Airless spraying	1-3 (0.025-0.075)	40-50	Low	Medium
Air-assisted airless spraying	0.5-4 (0.013-0.10)	40-50	Medium	Low
Electrostatic rotary atomization	0.5-4 (0.013-0.10)	80-95	High	Low
Electrostatic spraying	0.5-4 (0.013-0.10)	60-80	Medium	Low
Electrostatic powder spraying	1-3 (0.025-0.075)	95-100	Medium	Medium
Fluidized bed	1-10 (0.025-0.25)	100	Low	Low
Electrostatic fluidized bed	1-3 (0.025-0.075)	95-100	Medium	Medium
Electrocoating	0.5-1.4 (0.013-0.036)	100	High	High
Autodeposition	0.5-1 (0.013-0.025)	100	Medium	Medium
Dip coating	1-3 (0.025-0.075)	90-100	Medium	Low
Flow coating	1-3 (0.025-0.075)	95-100	Low	Low
Curtain coating	1-3 (0.025-0.075)	100	Medium	Medium
Roller coating	1-2 (0.025-0.050)	100	High	High

* Spray equipment is generally rated by the flow rate of the coating material rather than thickness. Coating thickness is dependent on material properties, so not all materials will be able to achieve the same thickness.

methods of breaking up or atomizing liquid coatings are: (1) conventional or air atomization, (2) airless or hydraulic atomization, and (3) rotary (centrifugal) atomization. There are also two adaptations of these basic methods that are gaining noted use today. In the newest adaptation, the primary atomization is with hydraulic pressure, while air pressure is used to control the spray pattern. This method is normally referred to as air-assisted airless. The application of an electrostatic charge to the basic methods of atomization can aid in the efficiency of the droplet transfer to the part being painted. This method is usually referred to as electrostatic spraying.

THE LIQUID COATING MATERIAL

Before a coating material is sprayed, it is necessary to make sure that the coating is well mixed and of the proper viscosity. Generally, small containers, less than 5 gal (18 L), can be mixed by hand using paddles. Large containers, holding more than 5 gal (18 L), should always be mixed using a mechanical mixer. Some materials require constant mixing to prevent settling.

Viscosity is resistance to the flow of the material and refers to the thickness or thinness of the material. The viscosity of the material is expressed in a term unit called poise (1 dyne sec/cm^2); however, a unit called centipoise is more commonly used. Materials of very low viscosities can cause runs and sags on the painted surfaces. The viscosity of the coating material is dependent on the amount of solvent present and the temperature of the material. The optimum application viscosity of the coating material is generally supplied by the material manufacturer.

In painting operations, there are two viscosities that affect the resulting finish of the material on the part: the viscosity of the material when it is atomized and the viscosity of the material when it reaches the part. The viscosity of the material when it is atomized is critical because it determines the fineness of the material particle. The viscosity of the material when it reaches the part can cause sags and runs, but it can be controlled by the quantity of material applied. The difference between these two viscosities is the result of solvent evaporation as the particle goes to the part.

The viscosity of a coating material is typically measured by a method that measures the time for a given volume of material to flow through a given orifice. In the paint industry, the Ford or Zahn cup method of viscosity measurement is widely used. The value obtained from these methods is then converted to a centipoise number for general reference. To make these measurements, a cup-like device with a hole in it is dipped into the coating material until it is completely full. It is then withdrawn, and the time from the moment the cup clears the surface of material until the time the solid stream of material breaks up or stops is recorded. Depending upon the type of cup used, this time can be converted into a different scale if required. Table 27-2 shows the relationship between the Ford and Zahn cup measurements.

If the viscosity must be adjusted, the operator can either add a proper solvent to lower the viscosity or add paint solids to raise the viscosity. Generally, a material is either thinned or used as is, as adding paint solids is not normally recommended. Table 27-3 lists some of the solvents that can be used in coating materials, but care should be taken to add only the solvent that is compatible with the material being used. Another method of reducing the viscosity is to heat the material. The addition of heat causes the viscosity to lower, but it is important to check and adjust the viscosity for the temperature at which the material will be used.

CONVENTIONAL AIR SPRAY

In conventional air spray, the material is usually supplied from a container in one of two ways. In the first way, the container may use pressure, up to 100 psi (690 kPa), to force the material to the spraying equipment via a pressure vessel or a pumping device. The other method of material supply uses a vacuum created by the spraying device to pull the material from

TABLE 27-2
Viscosity Conversion Table

	Viscosity Test				
Centipoise	Zahn #1, s	Zahn #2, s	Zahn #3, s	Ford #2, s	Ford #4, s
10	30	16			5
15	34	17			8
20	37	18		28	10
25	41	19		32	12
30	44	20		37	14
40	52	22		46	18
50	60	24		51	22
60	68	27		56	25
70	72	30		60	28
80	81	34		66	31
90	88	37	10	73	32
100		41	12	79	34
120		49	14		41
140		58	16		45
160		66	18		50
180		74	20		54
200		82	23		58
220			25		62
240			27		65
260			30		68
280			32		70
300			34		74
320			36		89
340			39		95
360			41		100
380			43		106
400			46		112
420			48		118
440			50		124
460			52		130

TABLE 27-3
Commonly Used Paint Solvents and Thinners

Aromatic	Alcohols
Benzene	Methanol
Toluene	Ethanol
Xylene	Isopropyl alcohol
Aromatic naphthas	n-Propyl alcohol
Aromatic petroleum solvents	n-Butyl alcohol
Others	Secondary butyl alcohol
	Amyl alcohol
Aliphatic	Cyclohexanol
Petroleum ether	
Lacquer diluent	**Acetates**
VM & P naphtha	Ethyl
Mineral spirits	Isopropyl
Odorless mineral spirits	n-Propyl
Kerosene	Secondary butyl
High-flash naphthas	n-Butyl
	Amyl
Glycol ethers	
Several commercial grades	

TABLE 27-3—*Continued*

Chlorinated solvents	Ketones
Methylene chloride	Acetone
1,1,1-Trichloroethane	Methyl ethyl ketone
Carbon tetrachloride	Methyl acetone
Ethylene dichloride	Methyl isobutyl ketone
Trichlorethylene	Diacetone
Perchlorethylene	Cyclohexanone
	Isophorone
Terpenes	Di-isobutyl ketone
Turpentine	
Dipentene	
Pine oil	

the container to the atomizing area. The method using pressure to force the material to the spray device is called pressure feed, and the method using the vacuum to draw the material is called suction feed. Generally, suction feed equipment is used for low-production jobs [up to 12 fl oz/min (355 mL/min)] when paint viscosity can be readily controlled on the small batches. Examples of such usage may be found in body shops or with most do-it-yourself-type spraying equipment. Pressure feed equipment is almost always used when high production and large volumes [12-30 fl oz/min (355-887 mL/min)] of material are involved. Pressure feed equipment can handle a much wider variety of coating viscosities and material volumes. Air pressures normally used vary from 10 to 100 psi (70 to 690 kPa).

In air-atomizing equipment, the material is supplied to the spraying device where the material is forced through the fluid nozzle in a stream. This stream of material is torn apart by an annular jet of air surrounding the paint nozzle, causing the material to become atomized into fine particles. Other air streams coming from extensions on the air cap or nozzle impinge on this stream of atomized particles, causing it to be spread into a desired pattern. The pattern generated is particular to the type of air cap being used. In most equipment, the air that controls the spray pattern is termed *shaping air* since it controls the shape of the material pattern. The material pattern can normally be adjusted from a small, round area to a flat, fan-type pattern; when the amount of air to the shaping area is increased, the pattern gets bigger.

A typical air-atomizing spray system consists of: (1) an air pressure source, probably a compressor; (2) an air regulator, to control the flow of air to the spray equipment; (3) an air line to the spray equipment; (4) a material supply, either suction or pressure feed, and a material supply line to the spray device; and (5) a spray device, usually a spray gun. The spray gun is generally categorized by the method used to control the air and fluid flow or by the design of the air nozzle.

A type of spray gun that controls only the fluid flow is known as a bleeder-type spray gun because the air constantly bleeds from the gun as it is being used. This type of equipment is common on small portable compressors since they need to maintain constant airflow to avoid possible damage. The other type of spray gun, known as a nonbleeder, controls the air and the fluid by the action of the trigger. These nonbleeder-type spray guns use a mechanical method to ensure that the air comes on before the fluid begins flowing and shuts off after the fluid flow has stopped. Known as lead-lag, this mechanical method helps the spray gun to keep clean and eliminates spitting.

SPRAY COATING

Spray Gun Nozzles

The air nozzles, referred to as air caps, are the most important part of the air spray gun. They direct the air to the material and cause atomization and pattern development. The two basic types of air spray systems used are external mixing and internal mixing. External mixing systems (see Fig. 27-1) mix the air and the fluid outside of the air cap. This type of cap is used on both bleeder and nonbleeder types of spray guns and can be either siphon or pressure fed. External mixing systems are the most common type used in production.

Internal mixing systems (see Fig. 27-2) mix the air and the fluid inside the air cap before being released. The air cap's exit-hole shape controls the pattern of the material spray, which cannot be varied with the gun controls. Internal mixing equipment must be pressure fed, and the air and fluid balance must be closely maintained. This type of air spray system is generally used for applying heavy-bodied coating materials since low pressures are normally used and the resulting finish is very coarse. Internal mixing equipment is hard to keep clean and requires additional maintenance.

Application Techniques

It is extremely important to follow proper application techniques in order to ensure a smooth, uniform film on the part. The application techniques used also affect the overall economy of the spray method.

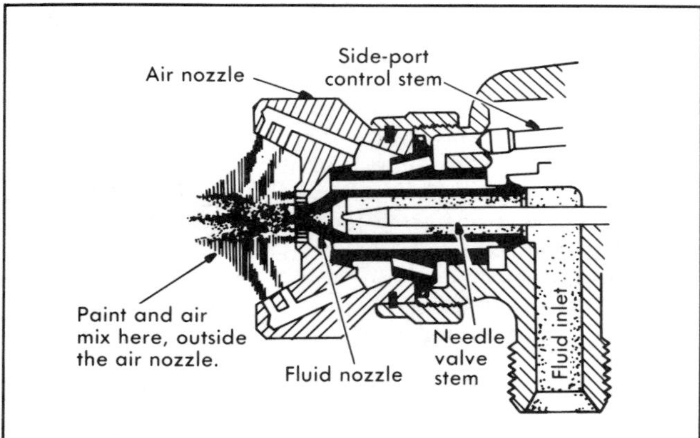

Fig. 27-1 Cross section of a spray gun with an external-mix nozzle. (*Binks Manufacturing Co.*)

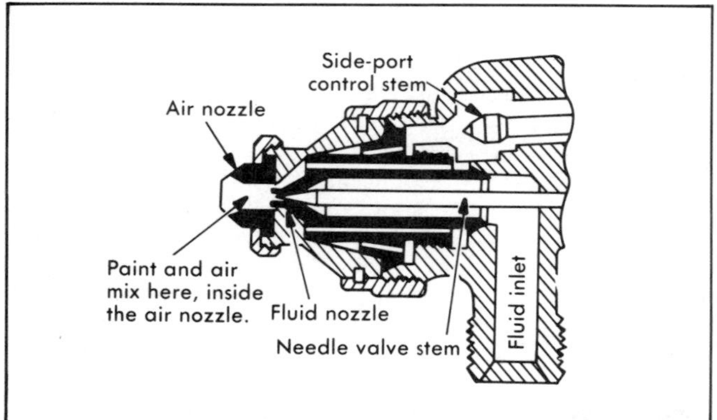

Fig. 27-2 Cross section of a spray gun with an internal-mix nozzle. (*Binks Manufacturing Co.*)

During spraying, the gun should always be held perpendicular and moved parallel to the part surface, as shown in Fig. 27-3. The best working distance is between 6 and 8" (150 and 200 mm). Uneven film thickness is often caused by moving the gun in an arc (see Fig. 27-4). Arcing deposits a heavy film when the gun is closest to the part and a light film when it is arced away to either side. Arcing is caused by using only the wrist or forearm to control the spray gun movement; the wrist, elbow, and shoulder must all be used.

The proper stroking speed allows a full wet-coat application with each stroke and helps to conserve the coating material. If the desired film thickness cannot be obtained in a single stroke or pass because of sagging, then two or more coats should be applied with a flashoff period between each coat. The spray movement should be at a comfortable rate. If the spray gun movement is excessive in order to keep from flooding the work, then the fluid nozzle is too large or the fluid pressure is too high. A good rule of thumb is to use the largest fluid nozzle size possible to achieve the fluid flow required and to use the lowest fluid pressure to achieve the best atomization.

Proper lapping (the distance between strokes) is essential in producing uniform film thickness. The overlap should be the minimum required to give the degree of uniformity required. A typical overlap is 50% for most applications. When painting large panels as depicted in Fig. 27-5, a single vertical stroke on the end of each panel ensures complete coverage and reduces waste. It is also essential to trigger the gun on as it approaches the part and then to release the trigger as the gun passes the end of the part (see Fig. 27-6).

Long panels should be sprayed in sections of 24 to 36" (600 to 900 mm) each. The edges of these sections should be overlapped 3-4" (75-100 mm). When spraying a level surface, the overspray

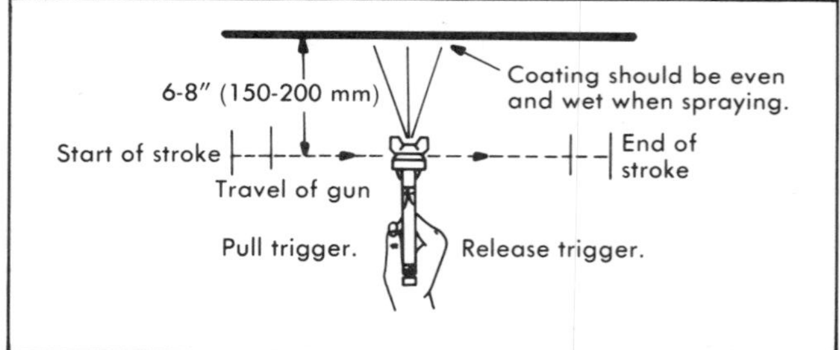

Fig. 27-3 During spraying, the spray gun should always be held perpendicular and moved parallel to the part surface. (*Binks Manufacturing Co.*)

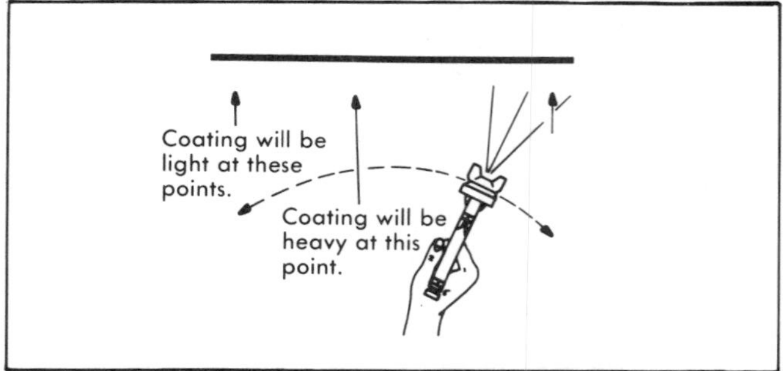

Fig. 27-4 Moving the gun in an arc results in a heavy film when the gun is closest to the surface and a light film when it is farthest from the surface. (*Binks Manufacturing Co.*)

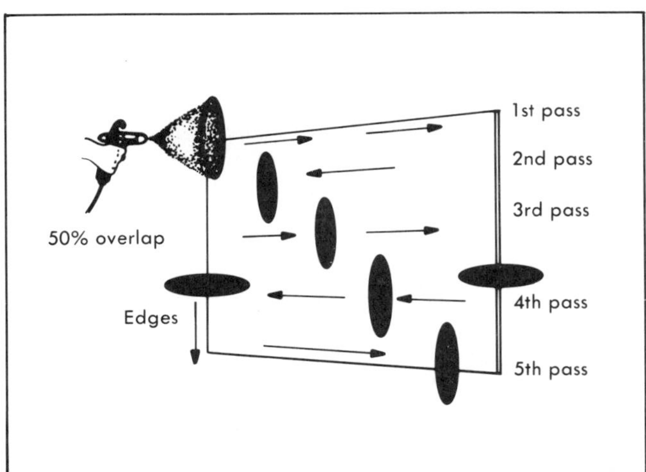

Fig. 27-5 When painting parts that require more than one pass, the typical overlap is 50%. (*Binks Manufacturing Co.*)

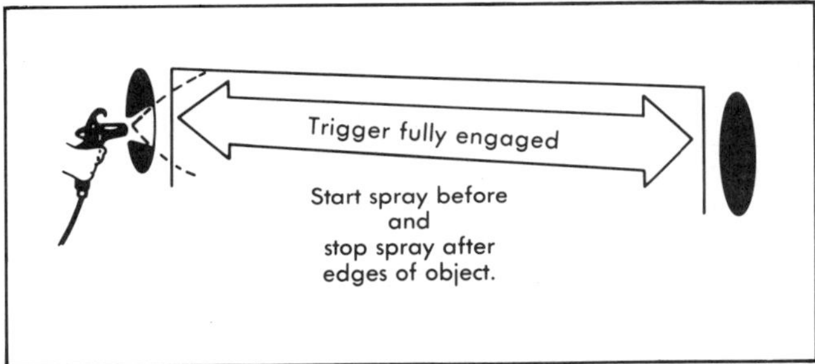

Fig. 27-6 The spray gun should be triggered just before it reaches the part and then released after it passes the part. (*Binks Manufacturing Co.*)

should be directed away from the surface previously painted to help reduce orange peel.

Troubleshooting

To ensure that the spray gun and nozzles are functioning correctly, a series of spray pattern checks should be made prior to painting the actual part. Figure 27-7 illustrates the common incorrect spray patterns due to a variety of causes. Table 27-4 lists the causes and corrections for these incorrect spray patterns. Paint film defects are discussed in Chapter 30, "Testing, Troubleshooting and Safety."

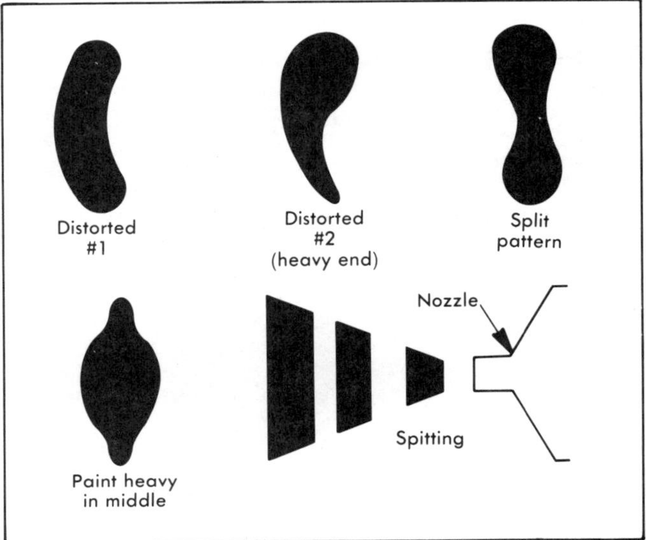

Fig. 27-7 Incorrect spray patterns of a conventional air spray gun. (*Binks Manufacturing Co.*)

TABLE 27-4
Common Spray Pattern Problems, Causes, and Corrections

Problems	Causes	Corrections
Distorted #1	Dried paint in one of the side portholes of air nozzle.	Dissolve paint in side porthole with thinner; do not probe in any of the holes with a tool harder than brass.
Distorted #2	Fluid buildup on side of fluid nozzle. Damaged fluid nozzle because spray gun was dropped.	Remove air nozzle and wipe off fluid nozzle. Replace damaged fluid nozzle.
Split pattern	Air pressure too high. Spray pattern too wide. Fluid pressure too low.	Reduce air pressure. Reduce fan width. Increase fluid supply.
Paint heavy in middle of pattern	Air pressure too low. Excessive fluid velocity or too much fluid.	Increase air pressure. Use smaller fluid nozzle orifice and/or lower fluid pressure.
Spitting	Air entering the fluid supply could be caused by: Fluid nozzle loose or not seating properly due to dirt. Loose packing nut, or missing or dried fluid packing. Fluid connection loose.	 Tighten fluid nozzle, or clean fluid nozzle seat area. Tighten packing nut, or replace missing or dried fluid packing. Tighten all fluid supply connections leading to spray gun.

AIRLESS (HYDRAULIC) SPRAYING

Airless spraying is a method of spray application that uses hydraulic pressure to atomize the fluid. The fluid is pumped to the spray device at high pressures, 500-4500 psi (3500-31 000 kPa), and then forced through a very small orifice at the spray nozzle. As the fluid is released from the nozzle at these high pressures, it is atomized into small droplets, resulting in a fine spray. The material is discharged at such a high velocity that after atomization the particles travel to the workpiece by their initial momentum. The pressures that are required to atomize various materials depend upon the material's viscosity and solids makeup, the distance of fluid travel, and the size of the orifice tip in the spray gun.

Advantages and Limitations

Airless spraying is generally used to apply a large volume of paint to a large area in a short period of time. Materials can be applied to the surface as fast as the operator can move and control the spray device. The degree of atomization is typically not as fine as in air spray, but fluid delivery is much higher. Because of this, airless spraying is generally not used for fine finish work or for coating metal substrates. The spray pattern is also very defined, which causes overlapping to be difficult.

Airless spraying typically is cleaner and faster than air spray due to its reduction in overspray and bounce-back. The main advantage of airless spraying is its ability to spray a variety of coating materials without reduction in viscosity; even high-viscosity materials can be sprayed successfully with little or no reduction in viscosity.

The main limitation with airless equipment is that, unlike air spray equipment, it can only vary the pressure and cannot be throttled. Greater operator skill is also necessary with airless equipment than with air equipment when spraying in difficult-to-reach areas because the full material flow comes from the spray device whenever the trigger is pulled. Table 27-5 compares conventional air spray with airless spray methods.

Spray Equipment

The airless spray gun is specifically designed for use with high fluid pressures. The pressurized fluid enters the gun either at the base of the handle or at the back of the nozzle. Some spray guns use a tungsten carbide ball and seat in the fluid shutoff as well as a tungsten carbide tip for maximum wear resistance and service life. A fluid filter may be inserted in the gun when fine filtration is required. Reversible tips are also available to help reduce downtime due to tip plugging.

Some spray guns contain a trigger-release knob for safety when removing or replacing the nozzle tip. Diffuser nuts may be installed behind the spray tip so that, when the tip is removed, any fluid accidentally released would not be harmful. Caution must be used when servicing airless spray equipment because the high internal pressures can cause serious injury.

Nozzle tips are selected by their orifice size and fan angle. The proper selection is determined by the fan width required for the specific application and by the desired amount of fluid delivery needed while providing adequate atomization. Typical orifice sizes range from 0.011 to 0.072″ (0.28 to 1.83 mm) diam, and fan angles range from 10 to 80°. Although orifices are given in diameter, it is generally understood that the given diameter means equivalent orifice size, which is used for flow rate considerations.

For light-viscosity materials, a small orifice is generally required; for heavy materials, the larger orifices are required. The equipment supplier can provide assistance in the proper choice of nozzle tips. Since some materials are hard to atomize adequately with a nozzle tip alone, a secondary orifice can be placed behind the nozzle tip (ahead of the tip in the fluid stream) to assist in the atomization. These additional orifices are referred to as preorifices. The preorifice size should be close to the nozzle tip size being used, but never smaller.

Application Techniques

The proper techniques for airless spraying are essentially the same as with air spray, as follows:

- Hold the gun perpendicular and parallel to the surface, the ideal distance being 12-14″ (300-355 mm).
- Move the gun at a steady rate to provide an even film.
- Lap the same distance when making successive passes; airless spray has quite a defined pattern and is therefore somewhat harder to lap than conventional air spray.
- Trigger each stroke in the same manner; airless guns cannot be feathered and require more precise trigger control.

Troubleshooting

The main operating problem associated with airless spraying is tip plugging, and its most common cause is foreign matter in the coating material. Strainers and filters help to remove this matter, but proper cleanup after use is the most important action that can be taken. Table 27-6 identifies most of the common operating problems along with their possible solutions.

Faulty airless spray patterns are readily recognizable as faulty conventional air spray patterns, but they are due to different causes. Figure 27-8 depicts a number of the faulty spray patterns, and Table 27-7 lists their possible causes and corrections. A table of frequently observed paint defects, with their causes, preventive measures, and corrections, is provided in Chapter 30, "Testing, Troubleshooting and Safety."

Safety

It is extremely important to exercise caution when operating airless equipment because of the high pressures that are developed. The high-pressure jet from the gun can cause serious injury. Medical attention is required for any injury because permanent damage can result. Some of the precautions to take when operating an airless system are:

- Make sure all fluid connections are tight before starting the fluid pump.
- Ensure that the gun and pump are grounded to prevent static buildup.
- Ensure that all fluid hoses are in good condition and free from cuts and sharp bends.
- Use only high-pressure hoses and fittings when modifying the paint distribution system.
- Handle the gun with care and watch where the gun is pointed.
- Relieve fluid pressure or disengage trigger before changing the nozzle.

ROTARY ATOMIZATION

Rotary atomizers differ greatly in their operation from that of either air spray or airless equipment. The atomization takes place by adding energy to the coating material through the high-speed rotation of a disc or cup-shaped part onto which the

TABLE 27-5
General Comparison of Conventional Spraying with Airless Spraying

Factor	Airless Spraying	Conventional Spraying
Means of atomization	High velocity of fluid using hydraulic pressure through small orifice.	Fluid stream torn apart by jets of compressed air.
Pattern control	Nozzle shape and size—must change nozzle to change pattern.	Control of air and fluid pressure provide complete control of pattern.
Air volume	None required.	4 to 20 cfm (0.1 to 0.5 m^3/min)
Air pressure requirements	Not applicable.	Medium to low air pressures best, 20-75 psi (138-520 kPa).
Fluid pressure requirements	600 to 4000 psi (4135 to 27 580 kPa).	Low pressures—generally to 18 psi (124 kPa) at nozzle.
Fluid delivery	Medium to high delivery. Provides fastest application speeds. Excellent for large areas.	Low to medium delivery. Usually not more than 32 fl oz/min. Less speed than airless, more control.
Air contamination	More overspray (material that misses the object) but less fog and rebound (material that bounces back from the surface).	Less overspray. More fog and rebound. Proportional to the atomizing pressure. Higher pressure, more fog.
Materials	Not all materials can be sprayed. Requires uniform fine grinds; particle size up to 0.008″ (0.20 mm) diam. Heavy-pigmented, fiber-filled, abrasive, or cohesive materials will not work. Viscosity of material can vary greatly.	Materials that flow can be sprayed. Rather narrow band of material viscosity can be sprayed (viscosity of some materials must be adjusted).
Material preparation	Requires considerable care in preparation to ensure proper patterns with no tip plugging.	Less care required. Follow material supplier's recommendations.
Maintenance	More required because higher pressure pumping equipment and smaller fluid tip orifices are required.	Less required because equipment is more basic.
Product contamination	No contamination from air line impurities.	Impurities in the air supply can spoil the finish.
Spraying advantage	Materials may be sprayed into cavities and corners with little rebound coming from the opening.	Difficult to spray into cavities and corners because of the large amounts of air required to atomize the materials; creates an air cushion that inhibits paint deposition.
Atomization	Generally courser atomization.	Fine atomization for all high-quality finishes.

(*Binks Manufacturing Co.*)

TABLE 27-6
Common Airless Gun Operating Problems, Causes, and Corrections

Problem	Possible Causes	Corrections
Spitting gun.	Air in system.	Inspect connections and siphon hose for leak.
	Dirty gun.	Disassemble and clean gun.
	Needle cartridge out of adjustment or damaged.	Inspect needle cartridge and adjust or replace.
	Broken or chipped needle seat.	Replace damaged needle seat.
	Packing nut too tight or not lubricated.	Loosen and lubricate packing.
	Twist tip or nozzle tip screen dirty.	Replace twist tip with standard airless tip. Clean nozzle tip screen.
Gun will not shut off.	Worn parts, broken or chipped needle seat.	Inspect spray gun. Replace defective parts.
	Needle cartridge out of adjustment.	Inspect needle cartridge and clean.
	Dirty gun.	Disassemble and clean spray gun.
	Packing gland or nut too tight.	Loosen gland or packing nut and lubricate needle.

(continued)

TABLE 27-6—*Continued*

Problem	Possible Causes	Corrections
Gun does not spray any fluid.	No paint. Plugged filters or tip. Broken needle in spray gun.	Check fluid supply. Clean filters or tip. Replace broken needle.

(*Binks Manufacturing Co.*)

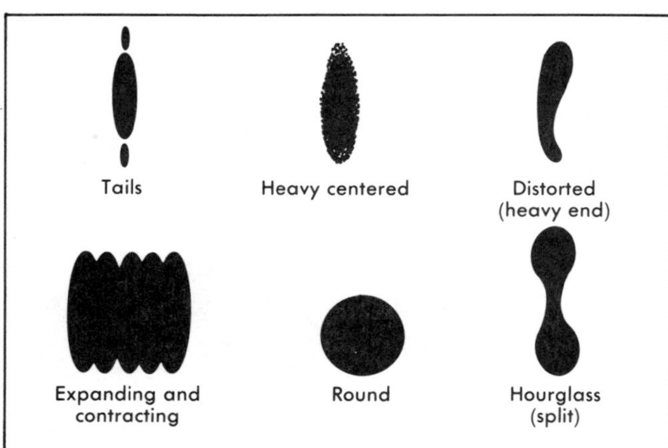

Fig. 27-8 Faulty spray patterns of an airless spray gun. (*Binks Manufacturing Co.*)

material is pumped. The material is pumped by a pressure source to a nozzle, either at the center of the rotating part or close to the center but off to the side. These methods of material feeding are referred to as center feed and side feed respectively.

As previously mentioned, there are two forms of rotating members. A flat platelike member is referred to as a disc and can vary from 4 to 14″ (100 to 350 mm) diam. It is normally rotated at 2000 to 20,000 rpm and can be driven by either an electric motor or an air turbine. A disc almost always has a side-feed type of material discharge. The material is pumped onto the surface of the rotating disc and is formed into a fine spray by the flinging action of the rotation. Figure 27-9 illustrates this type of an arrangement.

When the rotating member is in the shape of a cup or bowl, it is referred to as a bell atomizer. This type of atomizer also uses the rotational force to cause the atomization, but the bell arrangement directs the spray created to a smaller area. Bells use a ring of small, directed holes or air jets (referred to as

TABLE 27-7
Common Faulty Airless Spray Pattern Problems, Causes, and Corrections

Problems	Possible Causes*	Corrections*
Tails	Inadequate fluid pressure. Fluid not atomizing. Insufficient fluid velocity. Material too cohesive.	Increase fluid pressure. Change to smaller tip orifice size. Reduce fluid viscosity. Clean gun and filter(s). Reduce number of guns using pump. Install properly matched preorifice.
Heavy centered pattern	Worn tip. Fluid will not spray with airless.	Same as above. Reduce viscosity.
Distorted pattern	Plugged or worn nozzle tip.	Clean or replace nozzle tip.
Pattern expanding and contracting (surge)	Pulsating fluid delivery. Leak in suction tube. Pump capacity too low. Material too viscous.	Change to a smaller tip orifice size. Install pulsation chamber in system or drain existing one. Reduce number of guns. Remove restriction in system. Clean or remove screens or filters; use larger hose or pump if necessary. Inspect siphon tube and hose assembly for leak (sucking air). Reduce fluid viscosity.
Round pattern	Worn tip. Fluid too heavy for tip. Fluid will not spray with airless.	Replace tip. Increase fluid pressure. Thin material. Change nozzle tip. Install preorifice.
Hourglass pattern	Fluid too cohesive (adhesive-type materials).	Increase fluid pressure. Thin material. Install preorifice.

(*Binks Manufacturing Co.*)

* When multiple causes and corrections are listed for a particular airless spray pattern problem, more than one correction may be the solution to the cause.

shaping air) to control the pattern size and direction of the atomized particles. The bells are usually 2-4" (50-100 mm) diam and operate at speeds from 20,000 to 60,000 rpm. The bell can be either center feed or side feed. A handheld bell is shown in Fig. 27-10.

The rotary method of material atomization is the most efficient method of transforming the material into an atomized form. It requires little material reduction and can handle a variety of material viscosities, as well as a wide range of fluid flow rates. Typically, these rotary methods of atomization are coupled with an electrostatic charge to improve the transfer efficiency. Electrostatic atomization is discussed subsequently, in the section on electrostatic spraying.

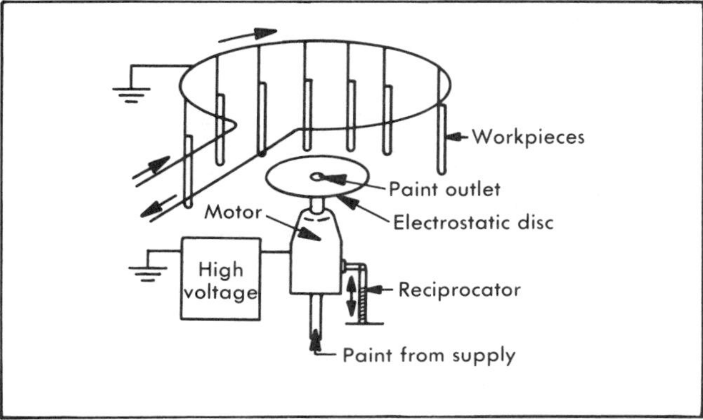

Fig. 27-9 Schematic of an atomization disc with an electrostatic charge.

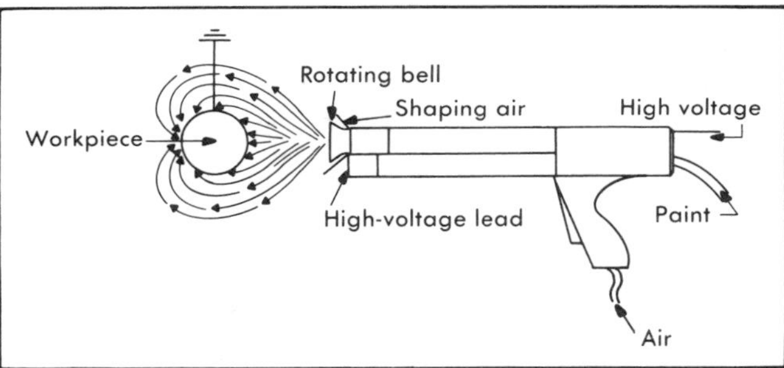

Fig. 27-10 Schematic of a handheld electrostatic bell atomizer.

AIR-ASSISTED AIRLESS SPRAY

An adaptation of airless spraying has recently become an accepted method of high transfer efficiency in spray coating. This method uses the principle of airless (hydraulic) atomization but adds a concentrated airflow to control the atomized particles. In general, a material can be broken into fine particles (atomized) at a relatively low pressure, 300-1000 psi (2000-7000 kPa). However, higher pressures are needed to complete the pattern forming. The air-assisted airless principle uses the low hydraulic pressure to provide the initial atomization, but then uses a low-pressure air to further atomize the tails and complete the pattern forming. This method also allows the pattern width to be adjusted somewhat by adding air to jets that act to compress the fan pattern.

Air-assisted airless spraying overcomes some of the problems of both air spray and airless spraying. The quality of the atomization approaches that of air spray but does not require the material reduction needed with air spray. The fluid delivery is higher than air spray, allowing faster coverage without bounce-back or overspray, which are common with conventional air spray. The use of low hydraulic pressure improves pump life, reduces maintenance cost, and is much safer to operate. Air-assisted airless spray produces a finish comparable to that of air spray with the speed of airless equipment.

HOT SPRAYING

Hot spraying is a technique in which the coating material is heated to a temperature of 120 to 160° F (50 to 70° C) before spraying. The increased temperature lowers the viscosity of the material, permitting materials to be atomized easier. The hot spraying technique can be used for all forms of spray application.

The coating material can be heated using electrically heated coils or a hot-water heat exchanger. Two types of heating systems are in common use, a recirculation system and a nonrecirculation system. In the recirculating system, the material is continually recirculated between the gun and the heater. In the nonrecirculating system, the material is only circulated through the heater one time, subjecting the material to cooling, which is dependent on hose length and application conditions.

When properly prepared materials are used, hot spraying has several advantages over unheated spraying. Since the viscosity is controlled by the temperature, little or no solvent is used for thinning, reducing labor costs, solvent costs, and solvent emissions. The coating material can also be stored in unheated areas or at a lower temperature than materials used for unheated methods. The disadvantages of the hot spraying method, however, are notable. A material heater and possibly a special gun are required, thereby increasing the initial cost. Application techniques are more critical with hot spraying than with unheated methods to minimize overspray losses as the cost of the material is higher than solvent-thinned material. In addition, many types of materials are affected by heat, so care must be used to limit the amount of heat used. The material supplier can provide additional information regarding the benefits and limitations of the hot spray method for certain coating materials.

ELECTROSTATIC SPRAYING

As the basic methods of material atomization have advanced, it has been found that the addition of an electrostatic charge to the atomized particles causes a dramatic increase in the material-to-part transfer efficiency. This application of electrostatic charges to the material particles causes them to act like small magnets when placed in the vicinity of a grounded object. During the spraying process, the part to be painted is grounded. As the material is sprayed at the part, the magnetic action of the charged particles causes the particles, which would normally be lost due to bounce-back or blow-by, to be attracted back to the part by actually wrapping the part in material particles. This phenomenon is known as *wrap* and is the prime force in the move to use electrostatics. By applying an electrostatic charge to the material particles, transfer efficiencies of 60 to 90% are possible.

Another phenomenon achieved with electrostatics is actual electrostatic atomization. When a high-voltage potential is created on a thin film of material and it is allowed to be free in atmosphere, the material, with all the same types of charges, tries to repel itself and forms a type of atomization. The particles formed by this atomization process maintain their charge, and the process may continue if the initial charge is high enough. An example of this type of setup is depicted in Fig. 27-11. This illustration shows the material being pumped over a knife-edge, creating the thin film required. A high-voltage charge is placed on the blade and thus to the material. As the material leaves the blade, it forms into small particles and falls to the part surface. This process uses no air or hydraulic forces to cause the atomization. The actual process of material atomization by pure electrostatics is not widely used, but the particular property of this like-charge repelling action is an advantage in using electrostatics. Electrostatic atomization is mainly a theoretical demonstration as it is extremely dependent upon material type, viscosity, fluid flow, and material distribution.

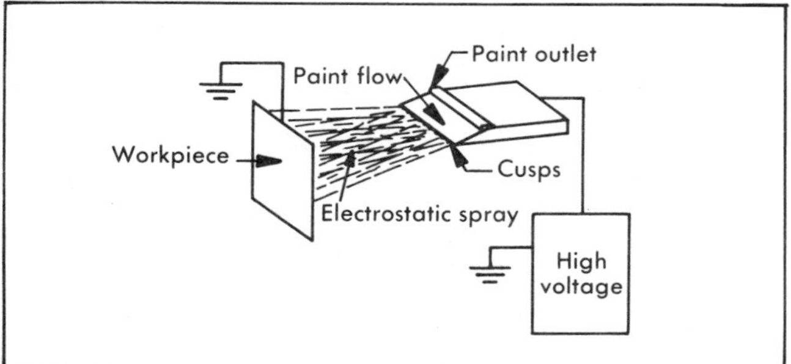

Fig. 27-11 Schematic of a blade-type electrostatic atomization spraying device.

There are a few limitations to the use of electrostatics. Electrostatic attraction is greater on outside edges and hole-edge areas due to what is known as edge phenomenon. It is known that magnetic forces are concentrated on the outside surfaces of an object and that any sharp edge is a collection point. This concentration of magnetic forces causes a heavy buildup of material on the edges of a part. The buildup can be controlled, however, by the application method and the charge used. Another problem associated with electrostatic application methods involves the Faraday cage effect. The Faraday cage effect is caused by the focused concentration of the charge resulting in low levels of material getting into recessed areas. Application methods can help in overcoming this problem, but normally a separate application using conventional air spray will ensure the proper coverage, while still maintaining the large savings in material for the major part of the application process.

Types of Electrostatic Systems

The application of electrostatics to any of the atomizing processes involves two types of systems. The first system applies the charge to the entire spraying device. The spraying device is then isolated from ground and, while operating, maintains the high-voltage charge. This system is normally referred to as a stiff or nonresistive type. The material is fed through the spraying device and picks up the charge; the electrostatic forces enhance the spray operation. This method requires a rather sophisticated control system due to the high-voltage potential present while the equipment is operating. It is normally limited to automatic-type systems because isolation of the spray area is possible.

The second method of electrostatic spraying in current use is referred to as a resistive system. In this system, the high voltage is very closely controlled and is fed through the spraying device, which is made of an insulative material and allowed to enter the material flow only at the point of atomization. This system uses a series of current-limiting resistors to control the current available to the spraying device, enhancing operator safety. The resistive systems are commonly used in both handheld and automatic equipment. The equipment used with normal solvent-based materials does not require isolation and is generally safe; additional equipment is required to maintain operator and system safety. Figure 27-12 is a sketch of a typical handgun installation.

The resistive system is a step in technological development to offer improved safety and reliability. The ability to control the current available to the spray device is critical. If the relative energy potential in a resistive system was compared to that for a nonresistive system, the difference could be easily seen. For example, if an energy potential having a value of one (1) was assigned to a resistive system, then a typical nonresistive system would have a value of 60,000. This difference demonstrates the need for system isolation in a nonresistive system.

The type of coating material used is critical as to the type and isolation of the system required. A complete review of the system requirements should be made with the material supplier and the equipment specialist. While electrostatic spray application is safe when properly used, care should be taken in the system installation and maintenance.

Application Techniques

Application techniques are somewhat simpler with electrostatic methods than with conventional or airless spraying. Lapping is less critical in applying an even material film, and, for many applications, careful attention to triggering is less necessary. Overspray is also reduced as a result of the improved efficiency. As previously mentioned, however, coverage in recessed areas and edge buildup must be controlled. Application methods for various part configurations can best be addressed by the equipment supplier.

Safety

Electrical shock and electrical discharge arcs are the major problems associated with electrostatic spraying. The operator can sustain a shock from the high voltage if contact is made with a portion of the power supply wiring not protected by a current-limiting device. Accidental shocks can be controlled by proper installation and maintenance of the equipment. Shocks from resistive equipment are not normally serious, but shocks from nonresistive systems contain so much energy that they are much more severe.

Electrical arc generation is the most prevalent problem in an electrostatic system because electrons generated by the electrostatic equipment will collect on any ungrounded, conductive object in the area of the spray equipment. When the electrons collect on the object, it can become charged. If this charge is not allowed to dissipate to ground through a grounding strap, an arc could be caused if the object or some other grounded object were brought near. This arc could be severe enough to cause a fire or serious injury to the operator. The best way to protect against arcing is to mechanically ground all objects in the spraying area and to employ good housekeeping practices.

Ventilation of the spray area is also important for both operator and operation safety, as well as for efficient spray operation. Removal of harmful vapors is critical, but too much ventilation will inhibit particle attraction formed by the electrostatic charge. Ventilation requirements must be reviewed and

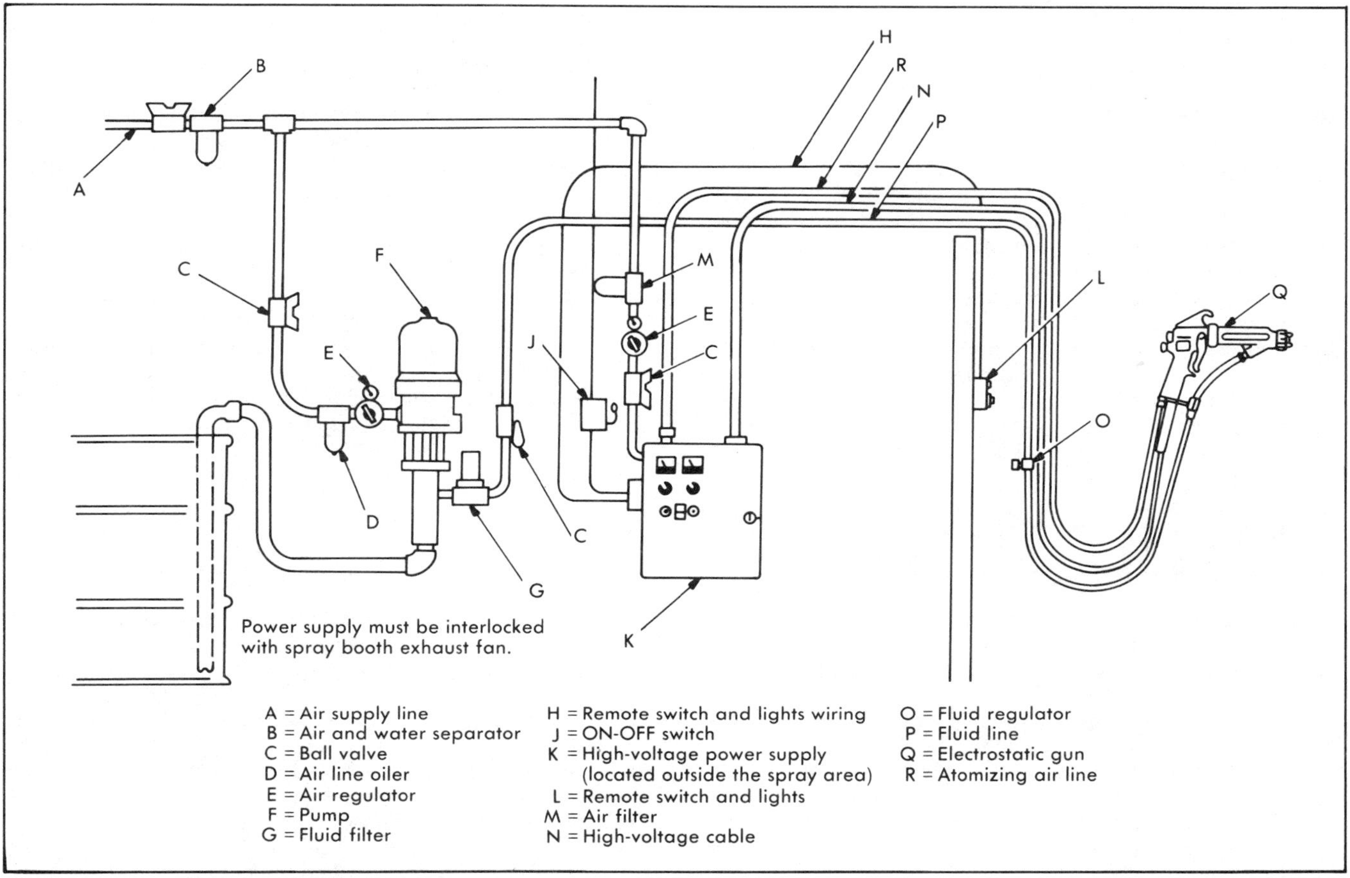

Fig. 27-12 Schematic of a resistive-type electrostatic spray system. (*Graco, Inc.*)

adjusted to maintain both a highly efficient spraying operation and a safe working environment. For additional information on safety procedures during electrostatic spraying, refer to the National Fire Protection Association's (NFPA) Standard No. 33, "Spray Application."

AUTOMATED SPRAYING

In manual spraying, the operator selects the system variables such as fluid flow rate, atomizing pressure, fan shape, paint temperature, and the sweep of the gun. In unmanned spraying processes, these parameters are preset to coat a preselected number or type of parts, with the operator checking that the parameters stay within the specified tolerances.

Two methods of unmanned spraying are in current use, automated booths and robotic systems. A part identification system is necessary for both. These systems size the part with mechanical fingers, limit switches, photocells, magnetic strips, bar codes, or by visual observation. A color-change system is also required for unmanned spraying processes; in it, a manual or automatic signal interacts with the color-change mechanism, which ejects the paint, cleans the paint line with solvent and air, and then refills the line with the newly selected paint. The time required for the color change varies with the paint viscosity and length of supply lines. In addition, unmanned systems require the parts to be hung uniformly since missing or improperly hung parts cannot be detected. Conveyors must also operate smoothly because these systems cannot compensate for swaying parts or other unforeseen motion.

Automated spraying methods ensure greater uniformity of finish from part to part as well as from day to day, while reducing the amount of manpower required. In addition, ventilation can be reduced; unmanned booths may require as little as 1/3 the amount of air as manned booths.

Automated Booths

Automated booths are used when a limited number of different parts have to be painted and when model changes are infrequent. They use spray painting guns that are mounted in either fixed positions or on reciprocating arms. In some instances, the arm tilts the gun to follow the contour of the part, using curved rails, cam-operated valves, or timers, as well as some programmable controls. Air, airless, or electrostatic spray guns can be used in these booths.

Theoretically, it is possible to set the equipment so that all the parts are coated acceptably. Practically, however, in order to reach certain areas of the part, other areas may receive an unnecessarily thick coat. It is therefore preferable to use a manual touch-up. In some applications, touch-up is performed before machine spraying.

The safety of automated booths is relatively high since all the operations are carried out within an enclosure. To guard against entering the booth during spraying, interlocks on doors and equipment are provided, as well as pressure-sensitive mats and photocells.

Robotic Spraying

Robotic spraying is performed using a six-axis servo-controlled robot that is capable of reproducing the wrist, arm, and waist movements of a human painter. Since the robot operates in the presence of flammable solvents, it is hydraulically activated. The volume of space that the robot can reach is called the working envelope.

Training a robot can be performed using continuous programming or point-to-point programming. In continuous programming, the robot's memory is activated, and then the robot arm is moved through the required motion pattern. On most robots, minor changes in the program can be made during each operating cycle until the optimum program has been achieved. Point-to-point programming is much slower than continuous programming and is generally only used when straight passes of the gun are required. In either case, the program can be speeded up or slowed down to meet the variations in the conveyor speed. However, care must be taken when increasing or decreasing the robot's speed so that the other parameters—such as proper spraying technique and, especially, the stroking speed (too fast = dry spray, too slow = runs and sags)—and the curing cycle are not adversely affected.

Robots are currently being used to apply topcoats to truck beds and steel office furniture, primers to truck cabs, hoods, and fenders, stains on wooden furniture, sound deadener on appliances, and porcelain enamel on bathtubs. Robots are most practical in coating applications that would be hazardous to a human painter, that are monotonous and repetitive, and that require the coating to be applied completely and uniformly. Air, airless, and electrostatic spray guns can be used with most robots.

POWDER COATING

Powder coating is a dry painting process in which powder particles are applied directly to the surface to be coated without the use of solvents or water. Each powder particle contains the resin, pigments, modifiers, and, if it is a reactive system, the curing agent. Most powders are formulated to provide the color and properties required by the manufacturer.

Two main types of powder coatings are used: thermosetting and thermoplastic. Thermosetting powder coatings chemically crosslink within themselves or with other reactive components to form a high-molecular-weight reaction product. The coating film formed by this reaction is heat stable and will not soften on further exposure to heat. Thermoplastic powder coatings do not chemically crosslink upon application of heat, but soften and level on the part in the oven. As the part cools outside the oven, the film hardens, but will soften upon application of sufficient heat.

The type of powder coating selected is based on application. Thermosetting powder coatings are generally used in decorative and protective applications or when comparatively thin coatings, 1-2 mils (0.03-0.05 mm) thick, are desired. Thermoplastic powder coatings are more suitable for items requiring a thicker coating, greater than 7 mils (0.18 mm) thick, and when extreme performance requirements must be met. Additional information on powder coating materials can be found in Chapter 26, "Coating Materials."

Some of the major advantages of powder coating are as follows:[2]

- Very little waste since overspray can be reused.
- No liquid solvent cost or handling.
- Lower fire hazard.
- Fewer toxicity problems.
- No air or water pollution (easy EPA compliance).
- No liquid mixing or pumping problems.
- No viscosities to maintain.
- Little or no makeup air is required for a powder booth.
- No flashoff space required.
- Lower oven exhaust rate.
- Less tendency to trap airborne dirt.
- Fewer shrink stresses developed upon curing.

The most common limitations of powder coating are problems with film appearance and powder handling. For example, powder coatings often have more "orange peel" defects than conventional films because of the high melt viscosity of the powder. Flow modifiers are used to achieve an acceptable smoothness for most applications. Changing colors may also present a problem because of the necessity to collect and reuse overspray if powder coating is to be economically feasible. To reuse overspray effectively, powder particles of each color must be kept separated if a booth is to be used for more than one color. To avoid handling problems, it is important to maintain a clean, dry air supply when transporting the powder particles.

APPLICATION METHODS

Several methods have been developed to apply powder coatings to the workpiece. The most commonly used method is electrostatic powder spraying. Some of the other methods include the fluidized bed, the electrostatic fluidized bed, and powder flocking. The particular method is selected based on the required coating properties and coating thickness, the size and shape of the parts, the rate of production, and the material handling techniques employed.

Electrostatic Powder Spraying

In electrostatic powder spraying, dry powder is pneumatically fed from a supply reservoir to a spray gun where a low-amperage, high-voltage charge is imparted to the powder particles (see Fig. 27-13).[3] The parts to be coated are electrically grounded so that the charged particles projected at them are firmly attracted to the part's surface and held there until melted and fused into a smooth finish in a baking oven.

Electrostatic powder spraying can be performed manually, or it can be a highly sophisticated automatic operation in which programmed robots can perform the spraying in booths. Because the powder is dry when sprayed, any overspray can be readily retrieved and recycled if only one color is used, regardless of whether the finishing system is manual or automated. Although several powder recycling methods are used, they all operate in a similar manner: the unused powder is separated from the airstream by various vacuuming and filtering methods, recycled, and returned to a feed hopper for reuse.

Spray gun. The design of the spray gun is an important factor in the spray system for achieving particle charging, particle deposition, and film build.[4] Figure 27-14 shows a cross section of a typical electrostatic powder spray gun.

Guns should possess some means whereby the shape of the spray pattern can be adjusted to that best suited to the material being applied and to the configuration of the part being coated. Spray patterns are generally cone or rod-shaped, and their dimensions can be controlled for diameter or width. Flat fan patterns can also be sprayed. The uniformity of the pattern is directly dependent on a constant-volume, puff-free delivery of powder by the feeder. In addition, the internal passages of the gun must be designed and manufactured free of irregularities and abrupt bends that can cause powder buildup and plugging of the gun passages; these flaws would create constrictions that adversely affect delivery of powder from the gun nozzle.

The gun must also house the charging electrode and provide the space and electrical insulation to permit the introduction of the cable carrying the high voltage to the electrode. The charging voltage must effect the highest particle charge but must be resistorized and insulated so as not to create potential shock hazards for the operator.

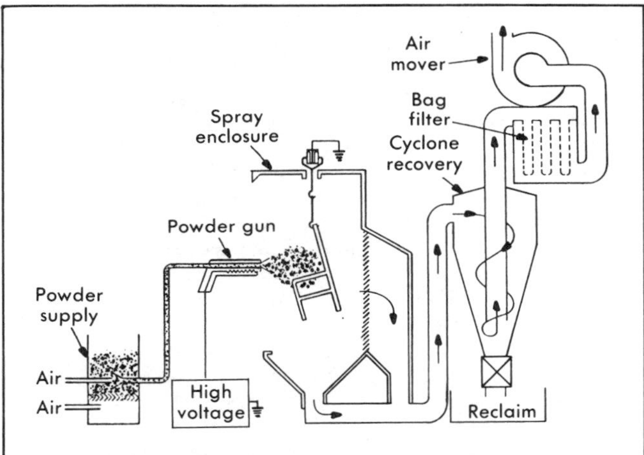

Fig. 27-13 Typical components in an electrostatic powder spray system.

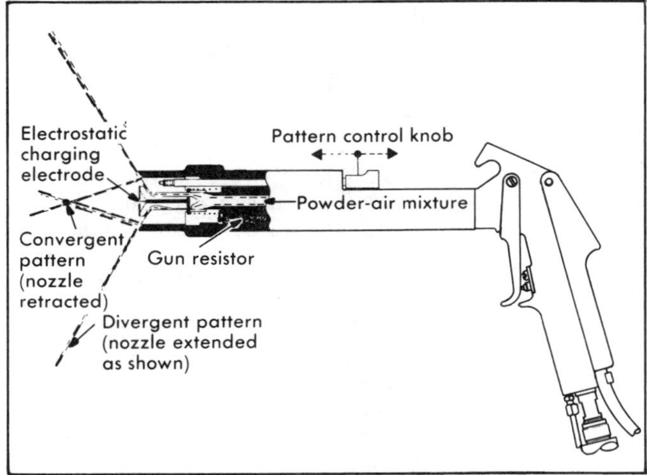

Fig. 27-14 Cross section of a typical electrostatic powder spray gun.

Most guns are designed to permit the highest possible delivery rate consistent with efficient deposition. If the powder flow rate is increased too much, the charge associated with the powder particles within the spray cloud generates a competitive electrostatic field that opposes the ion field being produced by the gun electrode. This opposing electrostatic space-charge field lowers the strength of the ion bombardment currents and, in turn, lessens the charge on the particles and their attraction to the part being painted. The low attraction also reduces the ratio of powder actually being deposited on the work to the total powder emerging from the gun, while increasing the amount of overspray powder being recirculated and recycled. Since excessive recycling of powder tends to degrade the particles in certain reclamation systems, it can degrade finishes and increase material costs.

These same guns are used in automated systems, but they are attached to a suitable support or manipulator. The guns are then appropriately positioned about the part or moved about over the surface of the part in accordance with a preestablished program. Since the part is exposed to the spray from a particular gun for a relatively short time, the guns are usually placed in series along the conveyor path. The guns can also be programmed to operate continuously.

Electrostatic discs. The electrostatic disc is a modification of the electrostatic powder spray gun and uses the same general methods for creating charged powder particles.[5] The disc is made of a nonconducting material, but has a coating on its surface that is highly resistive yet still capable of carrying the voltage applied at the center of the disc to the outer disc edge. The powder is delivered in an airstream to a cavity at the center of the disc. In this cavity, the powder encounters and is picked up by a series of spirally directed airstreams introduced tangentially into the cavity. The powder spins rapidly about the axis of the disc so that, as it is released from the cavity, it is spun or thrown outwardly past the disc edge toward the part.

As the powder passes through the high electrical field gradient at the disc edge, it becomes charged and is attracted to the parts. The parts are carried around the disc as they are moved about the loop in the conveyor, which permits the parts to be subjected to the powder distribution for a considerable period of time. Planar items can be moved about two successive discs with an intermediate index station if both sides are to be fully coated; multisurface parts can be rotated as they move about the disc. The disc can be moved up and down along the center axis of the loop to distribute the powder over parts of longer length.

Operating parameters. There are a few basic rules that can normally be applied when considering use of the electrostatic powder spraying method.[6] The parameters given are general and should be used as a guide in establishing exact operating parameters for any given application.

Spray gun distance. The distance from the spray gun to the part being coated is important. If the gun is too close, the air velocity associated with the powder transportation to the gun can overcome the force of the electrostatic field causing uneven deposition and increased overspray. If the gun is too far away, the charged particles can become influenced by the gravitational forces and drop before they reach the part or become entrained in the exhaust air and be drawn off before being deposited on the part.

A good working distance is generally in the range of 12 to 18″ (305 to 460 mm). This distance allows for the dissipation of the powder-transporting air and allows for efficient electrostatic field influence.

Applied voltage. The applied voltage is varied depending on the thickness of the film being applied and the ability of the powder to penetrate into recessed areas (Faraday areas). When thin films are desired along with high transfer efficiency, larger spray patterns and good wraparound, high-voltage potentials must be used. For penetration into hard-to-reach areas, low voltage potentials are used. Higher film builds are achieved through combinations of higher voltages and longer gun dwell times on the parts being sprayed, and sometimes by preheating the substrate. Typical voltages are between 60 and 100 kV d-c.

Volume resistivity. The volume resistivity of most dry powder materials can be changed by preheating the parts to be coated. With every 18° F (10° C) increase in part temperature, the volume resistivity changes by a factor of one or two. The conductivity of the powder particle is an inherent feature of the powder itself; the temperature of the substrate is unrelated to conductivity. The charge applied to the powder is a feature of the charging ability of the application. Preheating the substrate enhances the ability of the powder to achieve a higher film build than for unheated parts.

Temperature. In general, parts are electrostatically powder sprayed at room temperature for film thicknesses of 1 to 3 mils (0.03 to 0.08 mm). For thicker films up to 25 mils (0.6 mm) in one pass, the parts are heated above the melting point of the powder prior to spraying. During spraying, the powder particles adhere to the surface and melt, which allows the particles to continue to accumulate on the surface until spraying has stopped or the part has cooled.

Powder flow rate. The powder flow rate that has been found best for most applications is between 20 and 30 lb (9 and 13.6 kg) per hour per gun. At a powder flow rate of 30 lb/hr (13.6 kg/hr) and a deposition efficiency of 60% of the powder emerging from the nozzle, the gun operator can apply a 1.5 mil (0.04 mm) thick coating to 25 ft^2 (2.3 m^2) of surface per minute. If the flow rate is reduced to below 20 lb/hr (9 kg/hr), the coverage will be satisfactory, but the line speed must be reduced. If flow rates above 30 lb/hr (13.6 kg/hr) are used, the ratio of material usefully deposited to material emerging from the nozzle decreases. More oversprayed material must be recovered and recycled, increasing recovery system loading.

Fluidized Bed

Fluidized-bed coating is a method for applying thermosetting or thermoplastic materials in the form of fine powders to preheated metal parts. The powders are placed in the upper chamber of a dip tank. Pressurized air flows through a diffuser plate into a powder chamber, causing the powder to become suspended (fluidized) in the airstream. In this state, the air-powder mixture resembles a boiling liquid. The part to be coated is heated to a temperature above the powder's melting point and then immersed in the air-powder mixture (see Fig. 27-15). The powder particles that contact the hot surface begin to fuse and form a film on the surface. Uniform distribution of the particles over the surface is enhanced by vibrating the part while it is in the powder chamber. After the part is removed from the chamber, it is generally reheated to achieve good fusion and film properties; in the case of thermosetting powders, reheating is performed to cure the coating.

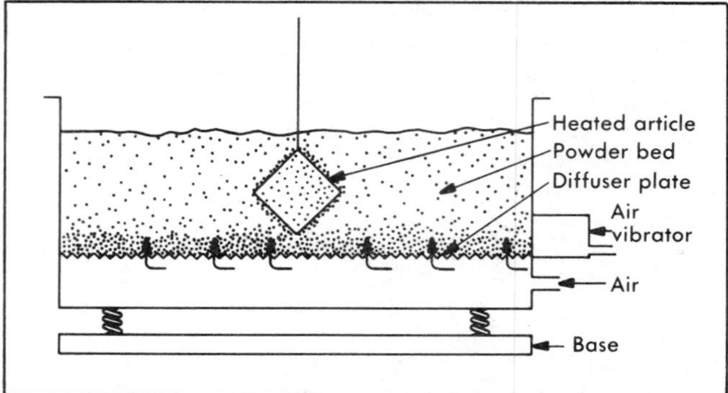

Fig. 27-15 Schematic diagram of a fluidized powder bed.

Advantages and limitations. Fluidized-bed coating is widely used to apply films in the range of 6 to 60 mils (0.15 to 1.50 mm) thick. Variables used to control film thickness are preheat oven temperature, dwell time in the preheat oven, dip time, powder density, and part motion. Flat areas, recessed areas, and edges are all coated during the process, thereby improving corrosion protection. Actual immersion times are relatively short, from 4 to 10 seconds; and the part does not have to be withdrawn at a specified rate as in dipping.

Although most parts can be coated with this method, parts with low mass are difficult to coat due to the rapid loss of heat between the preheat oven and the powder chamber. Films less than 6 mils (0.15 mm) thick are also difficult to obtain by this method. In addition, films are generally applied to both sides of the part rather than to just one side because masking costs would make fluidized-bed coating less competitive with the other powder coating methods. Fluidized-bed coating also requires a greater amount of powder than the other powder coating methods so that the parts can be completely immersed in the air-powder mixture.

Equipment. The minimum equipment required for fluidized-bed coating includes a preheat oven, a fluid-bed tank, and a pressurized air or inert gas source. In a production installation, however, additional equipment is required. Such equipment would include metal cleaning equipment, priming tanks (for thermoplastic powders), a tank exhaust fan and powder collector, a postheating oven, and an overhead parts conveyor.

Electrostatic Fluidized Bed

The electrostatic fluidized-bed coating method combines the principles of the fluidized-bed process with those of electrostatic deposition.[7] An illustration of a typical electrostatic fluidized bed is found in Fig. 27-16. The powder particles are placed in the fluidized bed, which has special charging electrodes built into the diffuser plate, to a depth of approximately 1 to 3″ (25 to 75 mm). When a high-voltage source of a given potential and polarity is applied, the powder particles become charged; the charged powder particles repel each other and form a cloud over the bed.

Parts to be coated are grounded and then transported over the cloud where the powder is electrostatically attracted to the parts. If the part is cold, the powder will collect on the surface until the accumulated surface charge is sufficient to repel further collection (in essence, the part is insulated). The part can then be transported to an oven where the collected powder is fused to the surface of the part. If the part is preheated and then transported over the particle cloud, the powder will fuse as it

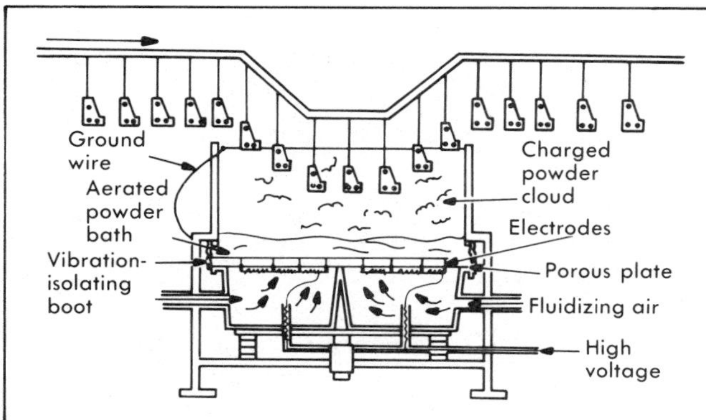

Fig. 27-16 Schematic of typical electrostatic fluidized bed.

collects. Since the fused particles lose their charge, the powder will continue to be attracted to the part as long as it is held in place over the cloud or until the powder ceases to fuse to the part due to cooling.

Not all powder materials that are normally associated with product finishes can be applied using the electrostatic fluidized-bed method. Some of the materials that have been successfully applied are epoxy, cellulose acetate butyrate, polyester, polypropylene, polyethylene, and acrylic. Polyvinyl chlorides and polyamides are not suited for this method. Powder particles should be as near to a uniform spherical shape as possible, and the optimum particle size should be in the range of 0.0008 to 0.004" (20 to 100 μm) diameter. For good electrostatic response, the powder material should have a bulk or volume resistivity of approximately 10^{10} Ω/cm^2.

Equipment. The electrostatic fluidized bed consists of a plenum chamber for the pressurized air as a base, an upper chamber to contain the coating powder, an air diffuser plate separating the two chambers, and an electrode grid system in the upper chamber to charge the powder particles. The plenum and upper chamber are usually made of electrical insulating materials such as rigid polyvinyl chloride or polyester fiberglass. Either high-density, porous polyethylene or porous ceramic materials are used for the diffuser plate. The supporting framework can be of steel construction, but must be electrically grounded in all cases. A compressor or a blower is used to supply air to the fluidized bed at a pressure of 2 to 20 psi (14 to 140 kPa) and a rate of 5 cfm (0.14 m^3/min). The air must be dry and free from contaminants such as oils or silicons. The power supply should be capable of providing a d-c potential of 30 to 100 kV and a current output of 200 to 500 μA.

Operating parameters.[8] As with the electrostatic spray process, there are a few basic rules that can normally be applied to the use of the electrostatic fluid bed. The parameters given are general and should only be used as a guide in establishing exact operating parameters for any given application:

- The thickness of the deposit is controlled by the high-voltage potential, substrate distance above the dry-powder fluidized-bed level, the transient time of the substrate through the charged cloud, and the substrate temperature.
- The charged cloud density is controlled by the high-voltage potential and the surface area of the grounded substrate passing over or through the cloud.
- A nominal optimum point for deposition of certain thermosetting materials on flat stock is 150 sfm (45 m/min) at 4" (100 mm) above the fluidized bed and at a potential of approximately 80 kV.

Powder Flocking

Powder flocking combines the fluidizing feature of the fluidized-bed process with the flexibility of the flocking process. The basic steps in this coating method are part preparation, preheating, spraying, and postheating. Generally, most parts are prepared using conventional methods such as cleaning and priming. The parts are then heated to a temperature slightly higher than the melting point of the coating material. Spraying of the powder onto the parts can be performed either manually or automatically. After spraying, the parts are heated to cause the coating to flow and then to cure the coating film. Overspray is collected and reprocessed for future use.

Equipment. The most basic piece of equipment is the manually operated flock gun (see Fig. 27-17). Like a liquid spray gun, the powder is placed in the cup of the spray gun. The incoming air is divided as it enters the gun; a portion of the air enters the cup to semifluidize and drive the powder out of the spray orifice while the other portion is used as diffusion or vehicle air.

Powder flocking can also be performed automatically. The equipment is designed to control the gun position, powder flow, and air pressure.

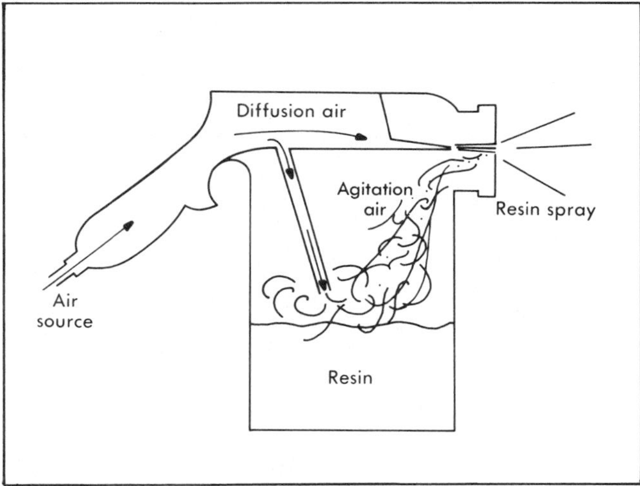

Fig. 27-17 Cross section of a manually operated flocking gun.

Advantages and limitations. Powder flocking has advantages over other powder coating methods that are often overlooked when choosing a powder coating method. One advantage is the low cost of coating equipment. Powder flocking is also capable of applying a full range of coating thicknesses, as well as providing good penetration into cavities; however, it is difficult to control the uniformity of film thickness.

Powder flocking has limitations due to its design. For example, the compressed air used to transport the powder cools the parts during coating, thereby requiring the part to be heated to a higher temperature than with the fluidized-bed method. Compared with the electrostatic powder spraying method, powder flocking does not provide a wrap effect when coating circular or irregularly shaped parts.

POWDER COATING

TROUBLESHOOTING

Defective powder coatings can often be attributed to low powder output, chargeability problems with the powder, or disturbances in the powder coating film. Table 27-8 lists some of the common problems that occur in powder coating along with possible causes and suggested solutions.

TABLE 27-8
Common Problems in Electrostatic Powder Spraying

Problem	Possible Causes	Suggested Solutions
	Powder Output	
Bad fluidizing properties in powder hopper.	Pressure of fluidizing air too low.	Adjust (increase) pressure of fluidizing air.
	Fluidized bed is blocked.	Clean or replace the fluidized bed; see instructions of equipment supplier.
	Humidity of compressed air too high.	Install an air drier with a corresponding oil microfilter or other suitable drying system.
	Humidity of the powder too high.	Check storage facilities. Powder shall be stocked at room temperature in closed packing (maximum humidity 75%).
	Free-flowing properties of the powder are bad.	Contact powder supplier.
Blocking in venturies and hoses.	Fusing of the powder in the venturi.	Clean or replace the venturi; see instructions of the equipment supplier. If necessary, reduce pressure of powder transport air.
	Fusing of the powder in the hoses.	Clean the hose by bending and breaking up the fused powder; if necessary, replace it.
	Humidity of compressed air too high.	Install an air drier with a corresponding oil microfilter or other suitable drying system.
	Bad free-flowing properties of the powder.	Contact powder supplier.
Blockage in the gun.	Fusing in the gun or gun outlet.	Clean the gun according to the instructions of equipment supplier. When blocking occurs, frequently check humidity of compressed air and the free-flowing properties of the powder.
	Blockage caused by contamination of the powder with dust of other coarse materials.	Clean the gun according to the instructions of equipment supplier and determine the reason for contamination.
	Powder Chargeability	
Insufficient wraparound.	Bad chargeability of the powder.	Adjust level of high-tension unit (increase voltage). If not possible, check equipment and guns according to instructions of the equipment supplier. Check if the powder takes an opposite static charge through friction in the hoses; if necessary, install hoses of other materials to avoid this charging.
	Insufficient ground contact.	Check and clean ground contacts; if necessary, use a suitable measuring instrument.
	Output of powder too low.	See Powder Output.
	Using an unsuitable powder.	Contact powder supplier.
Bad penetration.	Output of powder too low.	See Powder Output.
	Insufficient ground contact.	Check the ground contacts; if necessary, use a suitable measuring instrument.
	Powder cloud too wide.	Narrow powder cloud. If necessary, install a more suitable deflector or adjust air for cone adjustment.
	Voltage too high.	Adjust voltage; preheat part if necessary.

(continued)

TABLE 27-8—*Continued*

Problem	Possible Causes	Suggested Solutions
The powder falls down from the object.	Bad chargeability of the powder.	Adjust level of high-tension unit (increase voltage); if not possible, check equipment and guns according to instructions of equipment supplier. Check if the powder takes an opposite static charge in the hoses of other materials to avoid this charging.
	Powder output too high or the pressure for the transport air too high, which blows the powder from the object.	Reduce powder output and/or reduce pressure of the transport air.
	Unsuitable particle-size distribution of the powder or unsuitable powder type for the objects.	Contact powder supplier.
	Powder Coating Film	
Dust, precured, or other coarse material.	Dust or other coarse parts on the metal surface.	Check pretreatment.
	Dust or other coarse parts in powder.	Check powder and locate the cause of contamination; if necessary, clean up the installation and use fresh or sieved powder.
	Precured material from fusing in venturies, hoses, or the gun.	See Blocking in Venturies and Hoses *and* Blockage in the Gun.
	Precured powder particles from original powder that is stocked according to the instructions.	Install sieve. Contact powder supplier.
Matting of powder surface.	Contamination with other powder (based on other raw materials).	Clean up the installation; if necessary, contact powder supplier.
Orange peel.	Warming-up of the coated material is too slow or too fast.	Check curing cycle and curing oven; if necessary, contact powder supplier.
	Powder type too fast, or too coarse particle-size distribution.	Contact powder supplier.
Cratering.	Contamination with other powder (based on other raw materials).	Clean up the installation; if necessary, contact powder supplier.
	Bad pretreatment with remaining greases.	Check pretreatment; if necessary, contact pretreatment supplier.
	Contamination with incompatible materials such as silicones from the spraying area.	Check the presence of incompatible materials; if necessary, clean up the installation and contact powder supplier.
Pinholing.	Humidity of the powder too high.	Check storage facilities. Powder should be stocked at room temperature in closed packing (maximum humidity 75%).
	Air enclosure with hot-dip galvanized, cast iron, or cast aluminum material.	Preheat objects over 320° F (160° C) and cool down before application (only galvanized) or contact powder supplier who can recommend a specially developed powder.
	Gas enclosure and escaping due to chemical reaction.	Keep coating thickness below 100 microns; if necessary, contact powder supplier.

ELECTROCOATING

Electrocoating is a process for applying organic coatings through the use of film-forming organic macro-ions; the process resembles metal plating.[9] The part to be coated is electrically activated and then immersed in a bath of paint that has been given an electrical charge of the opposite polarity. The resin and pigment migrate to the part, and a uniform film is deposited. The part is then removed from the paint bath, rinsed to remove any excess material, and baked to cure the finish.

Electrocoating is not a panacea for the painting industry, but it does offer several advantages over other coating methods. Some of the advantages are as follows:[10]

- Approximately 90% of the paint adheres to the work.
- Water-based paints are used, resulting in fewer fire hazard and solvent air pollution problems.
- Sags, runs, and tears are eliminated.
- Uniform film thickness; film thickness can also be accurately controlled.
- Sharp edges, points, angles, and welding seams can be successfully painted.

Electrocoating is limited, however, to one-coat applications; and separate tanks are required for each color used. Large, flat surfaces can be electrocoated, but the equipment must be designed for the particular application.

THEORY

The deposition of a paint film by electrocoating is the result of four different processes that occur simultaneously: (1) electrophoresis, (2) electrolysis, (3) electro-osmosis, and (4) polarization. Electrophoresis is the movement of colloidal materials dispersed in a liquid medium under the influence of a potential gradient. Electrolysis is the formation of H_2, OH^-, and O_2 at the cathode and OH^+ at the anode as the result of an applied d-c potential. Electro-osmosis is the movement of the liquid phase under the influence of the potential gradient; in essence, electro-osmosis wrings the water from the deposited film. Polarization is the ability of the deposited film to exhibit electrical resistance, which permits a uniform coating over the entire surface.

When a part is immersed in the tank, the initial electrical resistance between the part and the counterelectrode is relatively low. As the film is deposited, the electrical resistance increases with the film thickness. Since the current seeks the path of least resistance, paint tends to deposit at the fastest rate in the areas that have the thinnest deposit. Deposition ceases when the thickness of the paint film has become uniform and the applied voltage is no longer sufficient to cause additional electrophoresis, electrolysis, and electro-osmosis.

The ability of the electrocoating process to form films of even thickness in intricate recessed areas is known as throwing power. Throwing power is proportional to the voltage; as the voltage increases, the throwing power increases. However, for each paint, there is a limit to the voltage that can be applied. Excessive voltage causes a film to rupture, resulting in a blemished appearance and a reduction in corrosion protection. Other variables that influence the throwing power are the bath conductivity and the electric equivalent weight of the paint.

Two methods are currently used in commercial electrocoating, anodic and cathodic. Although some manufacturers prefer anodic deposition, the cathodic method is becoming increasingly popular. In anodic electrocoating (see Fig. 27-18), the part to be coated is made the positively charged anode. When the part is immersed in the bath, negatively charged paint, pigment, and resin particles are deposited on the anode.

The dispersion reaction is represented by the equation:

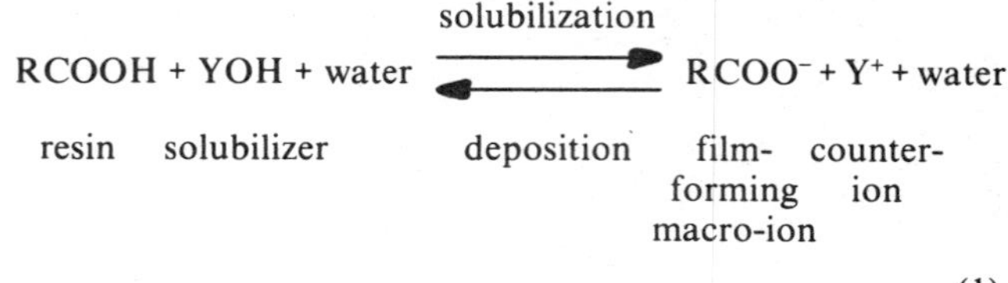

(1)

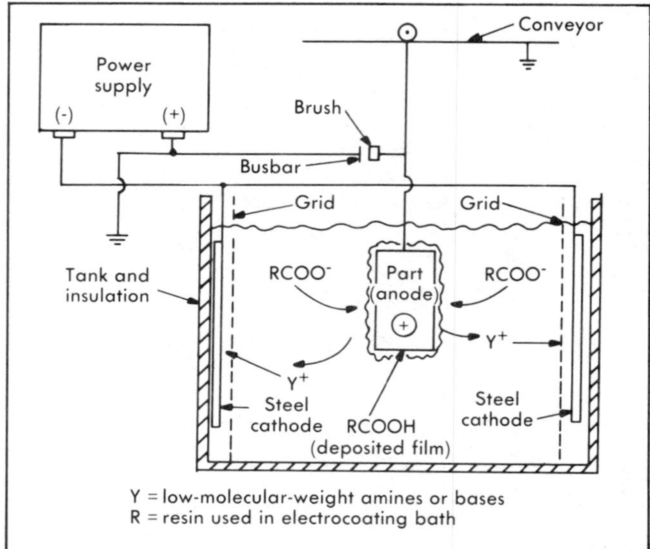

Fig. 27-18 Schematic of an anodic electrocoating reaction.

During deposition, however, metal ions start to dissolve from both the substrate and the pretreatment. These metallic ions mix with the paint particles, resulting in inclusions in the paint film. The initial dissolution at the substrate, which forms metallic salts, reduces the film's corrosion resistance and durability in external applications, as well as affects the color uniformity.

In cathodic electrocoating (see Fig. 27-19), the part is charged negatively while the paint pigment and resin particles are charged positively. The dispersion reaction is represented by the equation:

$$\underset{\text{resin}}{R_3N} + \underset{\text{solubilizer}}{HX} + \text{water} \underset{\text{deposition}}{\overset{\text{solubilization}}{\rightleftharpoons}} \underset{\text{film-forming macro-ion}}{R_3NH^+} + \underset{\text{counter-ion}}{X^+} + \text{water} \qquad (2)$$

Since the part is made the cathode, there is less of a tendency for metal to be dissolved and for metal ions to be included in the coating. Other advantages of cathodic electrocoating over anodic electrocoating are inherently better corrosion resistance and greater throwing power. Table 27-9 compares the properties and characteristics of anodically deposited coatings with cathodically deposited coatings.

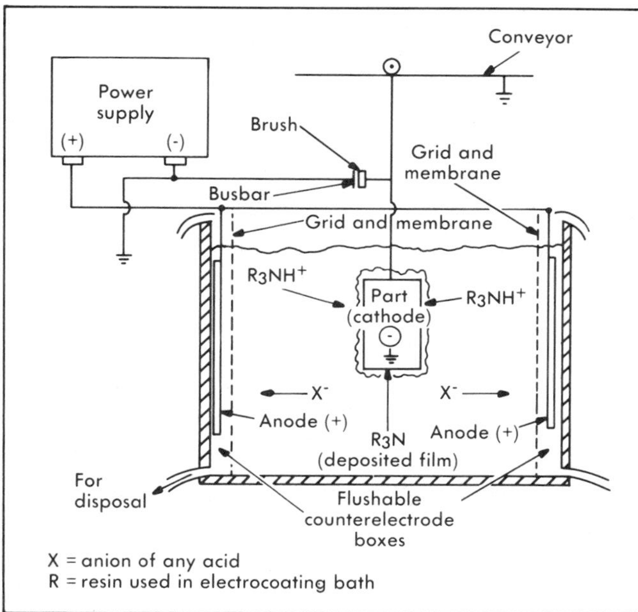

Fig. 27-19 Schematic of a cathodic electrocoating reaction.

TABLE 27-9
Comparison of Anodic and Cathodic Electrocoatings

Property	Anodic	Cathodic
Characteristics of deposited resin	Mild organic acids	Alkaline polymers
Substrate dissolution	Significant	Minimal
20° gloss (ASTM D 523), %	<50	40-85
60° gloss (ASTM D 523), %	15-85	30-90
Corrosion resistance (ASTM B 117), hr	96-336	500-1000
Detergent resistance (ASTM D 2248), hr	<100	500
Gloss retention (Florida exposure), %	20-60 at 9-12 months	60-80 at 9-12 months
Pencil hardness, H	2-3	4
pH	7-8.5	3-7.5
Voltage, V	150-250	200-450
Typical film thickness, mils (mm)	1-1.2 (0.025-0.03)	0.65-1.4 (0.016-0.036)
Color choice	Wide range	Limited

(*Durr Industries, Inc.*)

APPLICATIONS AND COATINGS APPLIED

Electrocoating was first used approximately 20 years ago for single coats to prime automotive bodies with alkyd and epoxy primers. Currently, electrocoating is used to apply primers and one-coat or topcoat finishes to a diverse range of automotive, furniture, appliance, industrial, and consumer products.

Primers are available as oleoresinous, epoxy, resinous polyols, and polybutadiene products.[11] Oleoresinous products are the least costly electrocoats. Because of their salt-spray protection, high intercoat adhesion to nearly any topcoat, and excellent impact resistance, they are popular choices for a wide variety of primer applications. They are not recommended for applications requiring alkali resistance or color fastness on overbake. Oleoresinous products are available in dark colors.

Epoxy primers possess all of the desirable properties of oleoresinous types plus offer improved salt-spray and alkali resistance, hydrolytic stability, and reduced discoloration on overbake. They are, however, higher in cost than oleoresinous primers. Epoxy primers are available in dark and earthtone colors.

Resinous polyols, also available in dark and earthtone colors, are applied at high voltages to achieve maximum throwing power in order to coat pieces with deep recessed areas. While their corrosion resistance and intercoat adhesion to topcoats are comparable to that of epoxy primers, their cost is slightly lower.

Polybutadiene primers offer perhaps the best corrosion resistance of any anionic primer for nonpretreated or poorly pretreated metal. They should be considered for applications in which corrosion protection is of utmost importance. They are available in dark colors.

Electrocoating products available for one-coat or topcoat applications include acrylics, which can be obtained in almost any color desired, and modified acrylics. The relatively wide color range and high gloss of acrylics make them the best choice for a decorative enamel. Other desirable enamel properties possessed by the acrylics include rapid film build, good stain and mar resistance, excellent color fastness on overbake, good salt-spray resistance, and humidity resistance. Cured acrylic films have excellent hardness, flexibility, and impact resistance. Their color and gloss retention in exterior applications is good when recommended metal substrates and pretreatments are used.

Modified acrylics combine many of the advantages of a corrosion-resistant primer and a decorative finish. They offer a film hardness of 2H (as measured by the pencil hardness scratch test described in Chapter 30), excellent salt-spray resistance when applied over pretreated steel, good alkali resistance, high impact resistance, good mar resistance, and high throwing power. Color and gloss ranges are similar to those of acrylic enamels.

EQUIPMENT

Electrocoating uses the concept of total system design by incorporating pretreatment equipment, dryoff oven, cooling equipment, electrocoating tank and allied equipment, rinsing equipment, and the bake oven in one line (see Fig. 27-20). The equipment required for cathodic electrocoating is similar to the equipment used in anodic electrocoating. Although a variety of conveyor techniques are suitable to transport the parts, the most common is the monorail. The most space and energy-saving equipment is incorporated into the vertical entry system.

ELECTROCOATING

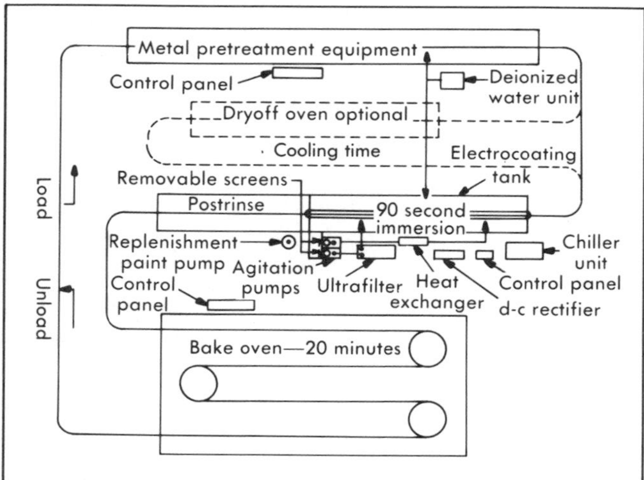

Fig. 27-20 Typical electrocoating system. (*George Koch Sons, Inc.*)

Pretreatment

The equipment used to pretreat the substrates prior to electrocoating is similar to that used for conventional painting methods. Depending on the quality of the finish required, three to nine pretreatment stages may be used. The first stage uses chemical techniques to remove surface soil, oils used in forming and drawing operations, and various other surface contaminants. After the parts have been thoroughly rinsed, they are given a conversion coating to provide additional corrosion resistance and a good base for paint adhesion. Both zinc and iron phosphate conversion coatings are commonly used. The parts are rinsed in deionized water to minimize the contaminants being carried into the paint tank.

Dryoff Oven

As indicated in Fig. 27-20, a dryoff oven is sometimes installed behind the pretreatment equipment to dry the parts before they enter the tank. The most commonly used ovens are direct gas-fired convection ovens. The temperature of the oven is generally maintained at 250 to 450° F (120 to 230° C), and the parts are dried for 5 to 10 minutes. Both the oven temperature and part drying time should be kept to a minimum since most parts have to be cooled after drying. If the design of the parts trap or pocket the rinse water, an air blowoff of the parts may be required prior to their entering the oven. The paint characteristics and cosmetic requirements determine whether or not a dryoff oven is required.

Cooling Equipment

Parts entering the electrocoating tank should have a temperature from 90 to 120° F (32 to 50° C). If there is insufficient time for the parts to cool between the dryoff oven and tank, a cooling tunnel or deionized quench should be installed.

Electrocoating Tank

The electrocoating tank is generally made from steel plate with an inner lining made of a chemically resistant dielectric material. The tank itself is usually electrically grounded, except in certain anodic electrocoating systems where the tank serves as the cathode. Depending on which electrocoating method is used, the part to be painted is either made the cathode or anode. In anodic electrocoating, the cathode plates are generally made from AISI Type 316 stainless steel; in cathodic electrocoating, the anode plates are made from carbon or stainless steel. Current flows from the power supply through the counterelectrode, through the paint to the part, and through the part back to the power supply. The size of the tank is directly related to the size and configuration of the parts being coated and to the length of time required for electrocoating (dwell time). Typical dwell times in the tank are from 90 to 180 s. Production tanks range in size from 200 to 120,000 gal (760 to 455 000 L).

The electrocoating section of the tank should be enclosed to prevent access to electrically dangerous areas, to confine vapors, and to minimize paint contamination from the plant environment. Additional equipment is included with the tank to provide electrical power, temperature control, filtration, agitation, and paint replenishment.

Electrical power. The power supply is generally a rectifier that converts alternating current into direct current. Unlike electroplating, electrocoating requires relatively high voltages and low currents; typical voltages are from 50 to 500 V. The current output is generally based on the maximum surface area of the parts being coated; a rule of thumb is 2 A/ft^2 (21.5 A/m^2).

Many systems are equipped with two small power supplies instead of one large power supply. The two power supplies are then interconnected to the work through a transfer device. Using two small power supplies permits coating to be started with one voltage and then increased to another. It also permits electrocoating to continue, on a limited scale, in case one of the power supplies is damaged.

Temperature control. Most electrocoating baths operate at temperatures between 60 and 90° F (15 and 32° C). During electrocoating, the temperature of the bath increases as a result of voltage and current flow, paint circulation, parts that have been through a dryoff oven, and high ambient temperatures. To remove the excess heat, a shell-and-tube or plate-and-frame heat exchanger is commonly used. Paint flows through the tubes, and chilled water is circulated on the shell side. The chilled water source may be a mechanical chiller, a cooling tower, or a well.

Filtration. Cartridge-type filters, capable of filtering from 10 to 100 microns, are commonly used to remove lint, solid impurities, and pigment agglomerates from the electrocoating bath. Normally, the entire bath should be filtered in 30 to 120 minutes. If the filter is located in line with the circulation pump, a manual or automatic bypass is required to maintain constant paint flow and to permit filter changing.

Agitation. Continuous agitation of any type of bath material is necessary to prevent paint solids from settling, provide cooling of the part being coated, and provide a constant supply of suitable paint to the areas being coated. In general, the higher the solids content of the bath, the higher the required tank agitation rate. As a rule of thumb, 6-8 tank volume changes (turnover) per hour are required; cathodic systems require more turnovers than anodic systems. Agitation can be provided by draft tubes, eductors, and perforated piping connected to circulating pumps.

Paint replenishment. As the parts are coated, paint solids are depleted from the bath. In order to control the level of paint solids, replenishment paint must be added to equal the depletion rate of paint solids. An ampere-hour device can be used to monitor the depletion rate and then activate the feed pumping system at required intervals based on the power consumed in applying the paint.

The replenishment material is normally supplied as either a one-package or two-package feed system. The one-package material contains all the components required to maintain the bath within specified parameters. As such, it controls the resin-crosslinker ratio, the pigment-binder ratio, and the solvent levels of the bath. It may also be fully solubilized or solubilizer deficient. The two-package system is supplied as a pigment concentrate and as a separate resin package. Both must be fed to the tank in specified proportions to properly maintain the bath. As with the one-package system, either of these components may be supplied either fully solubilized or solubilizer deficient.

Each method has advantages and disadvantages. The two-package system allows the pigment-binder ratio to be monitored more often. A pigment-binder drift can be easily corrected by addition of the appropriate package. Proper incorporation of feed materials into the electrocoat bath is facilitated by having a fully solubilized system; however, counterelectrode boxes are then required to properly control the bath solubilizer level.

If the pigment and resin are furnished by the paint manufacturer as separate components (two-package system), the paint replenishment system must be designed to mix them in the proper ratio. The mixing is usually done in a static mixer, and then the concentrated mixed paint is pumped to the tank. If the paint is furnished with the resin and pigment premixed (one-package system), then only a pump is required.

Rinsing Equipment

An excessive amount of the paint bath may be dragged out as the parts are removed from the electrocoating tank. To remove this excess paint, the parts are passed through a series of rinses using a permeate solution and deionized water; the deionized water rinse is used after two or three permeate rinses. The permeate solution, also called ultrafiltrate, contains water, solvents, low-molecular-weight resins, and salts. As the parts pass through the rinses, the permeate solution and undeposited paint flow back into the electrocoating tank where the solution is pumped past a semipermeable membrane (ultrafilter), which separates high-molecular-weight solutes and colloids from low-molecular-weight compounds (see Fig. 27-21). Approximately 1% of the volume of the bath that enters the ultrafiltration equipment passes through the membrane. The solutes and colloids are recirculated into the tank, while the solution that passes through the membrane is used as permeate.

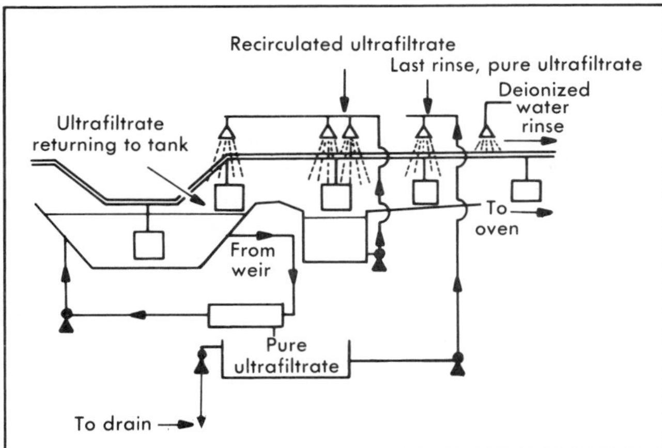

Fig. 27-21 Schematic layout of electrocoating tank with closed-loop ultrafiltration. (*Durr Industries, Inc.*)

Bake Ovens

Most electrocoating ovens are direct gas-fired convection ovens; some automotive companies use fresh-air-type heaters. Typical baking temperatures range from 350 to 450° F (175 to 230° C), with baking times from 15 to 30 minutes. Although there is less solvent in electrocoating paints than in other paints, a sufficient amount of fresh air must be supplied to prevent the volatiles from condensing and to keep them sufficiently above their lower explosion limit. In addition, cooling tunnels are sometimes required to reduce the part temperature as quickly as possible and to contain the smoke given off by some paints during the baking cycle. Steps must also be taken to control pollution from the exhaust.

OPERATING PARAMETERS

In order to obtain good films by electrocoating, it is necessary to monitor the bath parameters.[12] The primary parameters monitored are percent solids, solubilizer level, pH, conductivity, and bath temperature. These five parameters should be monitored on a daily basis by the tank operator, while the paint supplier should check them periodically. In addition, pigment-binder ratio and solvent analyses are performed by the paint supplier.

Percent Solids

Cathodic electrocoating systems typically operate in the 8 to 20% solids range. An increase in the bath solids results in:

- An increase in the concentration of solubilizer per gallon of bath. This increase can lead to postrinse washoff problems due to a high level of solubilizer in the permeate.
- An increase in ultrafilter area requirements because of an inverse relationship between bath solids and permeate flux rates.
- A need for higher capacity pumps to ensure proper agitation of the bath by increasing circulation.
- A reduction in voltage requirements.

Solubilizer Level

The level of solubilizer is normally defined in terms of milli-equivalents (MEQ's) per 100 g (3.5 oz) of bath solids. It may also be expressed as a function of bath volume or bath weight.

Excessively low MEQ levels can cause destabilization of the bath, resulting in settling problems, low film-rupture voltages (approximately 400 V), and poor product appearance. Excessively high levels increase power requirements, reduce coating efficiency and film build response, and cause gassing, which also creates poor product appearance. Low levels are easily corrected by solubilizer additions, but high levels require control by either counterelectrode boxes or ultrafiltration.

pH

The cited pH values for operational cathodic electrocoat systems cover the range from 3.0 to 7.5. Most of the apprehension concerning this parameter relates to potential damage to installed equipment due to the presence of an air-liquid interface. Of primary concern are areas that have turbulent flow, such as elbows and valves in the piping system, eductors, pump housings, heat exchangers, and the postrinse chamber. Corrosion failure of nonferrous parts can potentially occur for high pH systems. Anodic systems usually operate with a pH of 7.0 to 8.5.

Conductivity

Conductivity varies directly with percent solids, MEQ level, and bath temperature. The relationship between conductivity and percent of bath solids is shown in Fig. 27-22. If these parameters are established, conductivity should remain within a consistent range. If the bath becomes contaminated with soluble conductive materials or if any of the previously discussed parameters are on the high side, conductivity will exceed recommended limitations. Conductivity is normally controlled with counterelectrode boxes or ultrafiltration equipment. On some occasions, the inverse of conductivity, specific resistance, is monitored to control the electrocoating bath.

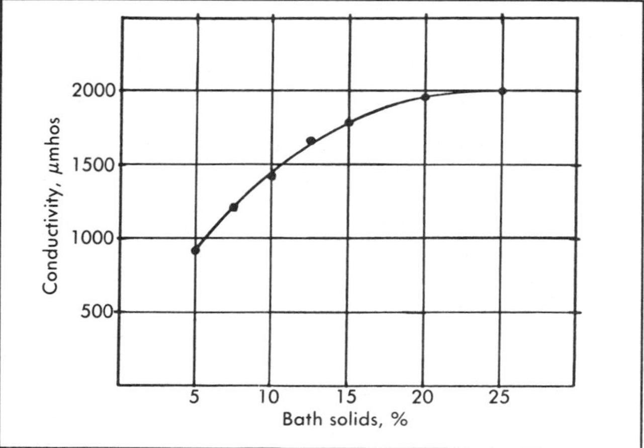

Fig. 27-22 Graph showing the relationship between conductivity and percent solids in an electrocoating bath.

Bath Temperature

Reported operating temperatures cover the 75° to 90°F (24 to 32°C) range. Once established, however, this parameter is controlled within a few degrees by the chiller and heat exchanger. The required cooling capacity is a function of the operating voltage and current, the heat input due to the circulation pumps, the temperature of incoming parts, and the ambient temperature. Higher operational bath temperatures offer approximately a 10% efficiency increase for the chiller and ultrafilter, and the potential for a corresponding reduction in heat exchanger requirements. Another benefit from higher bath temperatures is a reduced voltage requirement for a given film build. Higher operational bath temperatures increase bath conductivity and lower the film resistance of the deposited film. One important adverse effect of high bath temperatures is reduced throwing power.

Pigment-Binder Ratio

The pigment-binder ratio is normally a function of the paint formulation. It is determined by the performance properties required, such as color, gloss, detergent performance, and film hardness. Fluctuations are small, usually within a defined range, with little effect on other parameters. The pigment-binder ratio is controlled by proper changes in a one-package system or by the feed ratios of a two-package system.

Solvent Levels

Organic solvents are used in electrocoating systems for bath stability, film build characteristics, film flow and smoothness, and replenishment viscosity. The level of organic solvents is usually monitored by the paint supplier as part of a tank-return analysis program.

AUTODEPOSITION

Autodeposition is a coating process that has been in commercial use since 1973.[13] The film is deposited chemically, rather than electrolytically as in electrocoating. The film has a slightly textured, matte appearance. One of the most important characteristics of autodeposition is that it does not require a metal conversion coating because the adhesion of the deposited coating is extremely good due to the strong interaction between the metal and the organic film. The elimination of a phosphate stage leads to savings in floor space and operating costs.[14]

Some of the advantages of autodeposition are:[15]

- Minimum overall energy requirements for the complete coating process.
- Uniform, self-limiting film thickness.
- Uniform coverage on interior and exterior surfaces (excellent throwing power).
- Excellent adhesion and impact resistance, and good salt-spray resistance after baking.
- No solvents required, minimizing fire hazards.
- Minimum rack stripping required.
- Minimum air and water pollution.
- High system efficiency.
- High production rates.
- Low coating cost per unit area of steel coated.

However, autodeposition does have limitations. The color and polymer selection is limited because color capability is under development; currently, only black is available. In addition, only one company offers commercial autodeposition systems.

So far, autodeposited coatings have been commercially applied on automotive components such as engine mounts, lamp housing stampings, and axle housings.[16] The process is not limited, however, to the automotive industry. Autodeposition can be used whenever 200-500 hours of salt-spray resistance is required on steel substrates. The coating can serve as a primer or as a final finish on parts that are not highly visible after assembly. The quality level depends upon the type of steel substrate being coated and the process selection. Part size is not a critical factor.

AUTODEPOSITION BATH

The autodeposition bath is composed of a resin in the form of a latex (a water-dispersible resin), hydrofluoric acid, hydrogen peroxide, and deionized water. Theoretically, the percentage of solids in the bath may vary widely; but for commercial applications, the range is generally from 3 to 6% by volume.[17] The bath is acidic and operates at a pH of 2.6 to 3.5.

When a part is immersed in the bath, the proton attack on the substrate produces ionic species such as ferrous ions (Fe^{+2}). Hydrogen peroxide in the bath oxidizes the ferrous ions, producing ferric ions (Fe^{+3}). The ferric ions act as destabilizing agents for the colloidal latex particles in the bath, causing their deposition on the metal surface. As long as the environment at the liquid-metal interface contains the proper ions for producing destabilizing agents, the film continues to grow. Figure 27-23 shows the film buildup with respect to the immersion time for a typical latex coating. The film buildup reaches a dry-film thickness of about 0.2 mils (5 μm) in the first 30 seconds but slows down as the deposited particles begin to interfere with the diffusion of activators at the metal substrate. A typical film build is approximately 0.6 mils (15 μm) thick.

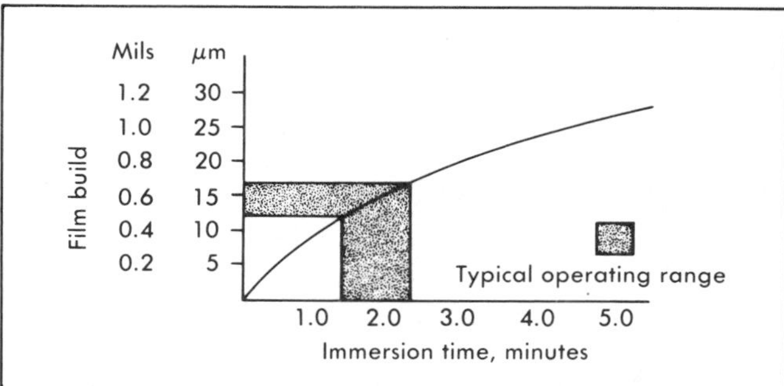

Fig. 27-23 Graph showing the film build versus the immersion time for cold rolled steel panels coated by autodeposition. (*AMCHEM Products, Inc.*)

PROCESS DESCRIPTION

At the most simplified level, autodeposition is a five-step process, plus includes curing in an oven. The five steps are cleaning, rinsing, coating, rinsing, and final rinsing.

Cleaning

The first step of the autodeposition process, cleaning, is common to all metal finishing processes. In the case of autodeposition, however, cleaning deserves particular attention because the chemical reactions that initiate resin deposition will not occur when an oily, greasy surface has formed a barrier. Heavy deposits block the surface from reacting with the constituents of the bath, and hence deposition may not occur. This requirement of not coating on extremely dirty surfaces could be viewed as a process advantage in the sense that a processor very quickly knows that a cleaning problem exists. Still, a superclean surface is not required for autodeposition because coating by this method will proceed on surfaces that are not water-break free or on surfaces that have a light iron oxide coating. Care must nevertheless be taken in line design to provide for adequate cleaning.

Cleaning is generally performed using a two-stage process. The first stage uses detergents and an alkaline cleaner to remove the bulk of the impurities by spraying; a typical spray cycle is one minute. The initial spray cleaning stage also protects the subsequent dip cleaning stage from a buildup of oils that would necessitate bath stabilization. A two-minute dip cleaning removes any remaining dirt, particularly in recessed areas. For some applications, the spray-dip sequence is changed to a dip-spray sequence to ensure adequate cleaning.

Rinsing

After the dirt is removed, the salts that remain on the metal from the cleaner baths have to be removed to protect the subsequent coating bath from building excess ionic contaminants. These salts are removed by water rinsing, which is usually performed by spray; unless the part being coated has interior metal surfaces that are inaccessible to spray rinsing, and then a dip rinse stage is required. Thus, the design of the rinse depends on the geometry of the parts to be processed.

The reason for rinsing prior to coating in an autodeposition process is to reduce the conductivity of the drippings from the metal parts to approximately 50 microhms. Since most plant water is substantially above this conductivity level, the process requires deionized water spraying; usually a short 5-10 second spray is sufficient.

Coating

Following rinsing is the coating cycle. The coating bath is normally maintained at 68°F (20°C) to help stabilize film buildup and ensure high quality.

Rinsing

The step after coating is a water rinse. Prior to rinsing, it is recommended that some air-dry time be provided to allow the coating reactions to continue, which helps to minimize resin loss from the parts. Currently, all rinsing in this stage is done by immersion because the wet deposited film is fairly soft and may be damaged by spray impingement. The rinse stage uses plant water and is generally overflowed to maintain the resin solids at approximately 1%.

Final Rinsing

The final rinse may contain small amounts of additives to improve corrosion resistance or to promote adhesion, depending on the type of resin used in the bath. A proprietary mixture of hexavalent chromium is generally used to improve corrosion resistance. (The hexavalent chromium mixture should never be mixed with the postcoating water rinse overflow before it has been reduced to trivalent chromium because the latex in the water rinse would coagulate, resulting in clogged transfer pipes.)

Recent developments have eliminated the need for the chromium mixture in the final rinse.[18] The rinses may now contain small amounts of nontoxic sealing agents that enhance the performance of the coating.

Curing

Curing is generally performed in a one or two-zone oven depending on the resin used. In a two-zone oven, the coated parts are heated in the first zone for 10 to 15 minutes at 230 to 248°F (110 to 120°C) to evaporate the water from the wet coating. More or less time may be required depending on the part geometry. Full curing occurs in the second zone, which is normally held at 284 to 356°F (140 to 180°C). Total curing time in this zone is usually 10-15 minutes, but the actual time depends on the thickness of the metal being processed. For lower curing resins, a single-zone oven is required. Oven temperatures are approximately 210-230°F (100-110°C), and cure time is 10-20 minutes.

EQUIPMENT

The nature of the chemical coating composition that is kept in the tank imposes several special requirements on the design of autodeposition equipment. For example, the tank should be lined, whether it is made from mild steel or stainless steel, to prevent a reaction between the tank and the bath. Free-standing plastic tanks could be used if the size of the line make them practical.

The tank should provide a means for bath agitation to hold the resin and pigment in suspension. To accomplish this agitation, variable-speed mixers with propellers are mounted along the sides of the tank. Typical rotational speeds of mixers for a production run are 150-200 rpm.

Although the autodeposition reaction is not exothermic, a heat exchanger or chiller is required to maintain the proper bath temperature. The bath temperature fluctuates mainly because of variations in plant temperatures. A means for soluble-iron removal must also be considered; the two most common methods are settling in a holding tank and bath stabilization.

The racks for parts should be made of mild steel and should be designed to minimize the area of the rack that comes in contact with the coating bath solution. After a period of time, the resinous material builds up on the racks, requiring cleaning. Although cleaning should normally be performed once per year, cleaning can be performed every two or three years for the low-temperature cure resins.

DIP, FLOW AND CURTAIN COATING

The common feature of dip, flow, and curtain coating methods is that the parts are coated with a large volume of coating material and the excess is allowed to drain off. In the case of curtain coating, however, the excess is the material that is not deposited on the part; whereas in the other two methods, the coating material contacts the surface and then flows off. Electrodeposition, autodeposition, and fluidized-bed powder coating (all of which have been discussed previously in this chapter) also use the drainoff principle.

DIP COATING

In its simplest form, dip coating is the complete immersion of a part in an open tank of a liquid coating material, the withdrawal of the part from the liquid, the supporting of the part over the tank or drainboard until it has completely drained, and finally the drying or curing of the coating. Although supporting and dipping the part manually is possible, mass production systems normally employ conveyors to transport and support the parts during the coating cycle. As shown in Fig. 27-24, a dip in the conveyor lowers the parts into the tank. During the period of time in which the part is immersed in the coating material, the conveyor moves evenly and smoothly to minimize withdrawal lines and improve film characteristics on the finished part. The controlled withdrawal also permits the meniscus of the reservoir to pull off the excess coating material, thereby reducing the drainage time. Since draining requires time, conveyorized systems incorporate a drainboard that extends under the conveyor to catch the excess material.

Large tanks and large quantities of coating material are required for mass production systems. To minimize solvent evaporation, tanks should have lids. An agitation system should also be incorporated in the tank to ensure uniform material distribution and to prevent settling or flotation.

Generally, dip-coated parts are hung from the conveyor so that the excess material flows to a point, as illustrated in Fig. 27-24. The heavy material accumulation at these drain points can be removed by wiping, pressurized air, or electrostatic detearing. Drain holes are normally added to the part during manufacturing because parts containing pockets or reentrant angles may trap the excess coating material as it drains off, thereby requiring special hanging arrangements. To prevent closed internal cavities from creating air pockets, which prevent the coating material from reaching the entire surface, the part should be oscillated during immersion. In some cases, pivoted hangers can be used that are cammed from one position to another, allowing a cavity to be flooded and subsequently drained. In addition to part fixturing recommendations, the temperature and viscosity of the coating material should be controlled to maintain proper flow characteristics.

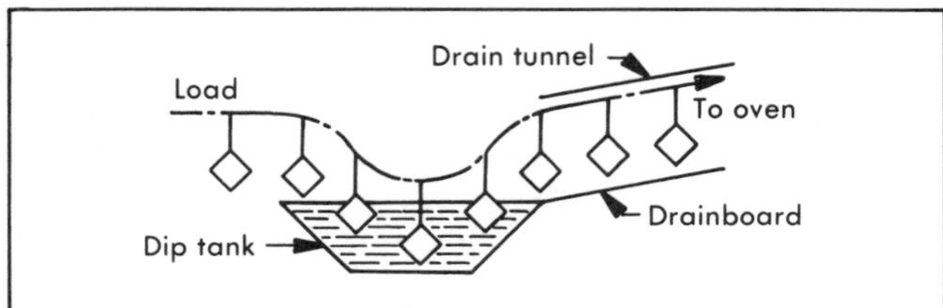

Fig. 27-24 Conveyor movement in dip coating.

Dip coating is widely used in industry for rough finishes, for applying a prime coat, and to coat the internal and external surfaces of a part. Dipping is generally efficient in its use of materials; up to 90% of the material is used, whereas only 50% of the material is used in spraying. Material loss is due to drag-out and drainage. One disadvantage of dip coating is that the coatings tend to be heavier on the bottom of the part than at the top. This difference in coating thickness is called wedging. To achieve more uniform film thicknesses, it is necessary to coordinate the solvent evaporation rate with the conveyor speed. Another disadvantage is that color changes can only be economically accomplished through the use of separate tanks.

FLOW COATING

In the flow coating process, the part is hung on a conveyor and carried through an enclosure, as illustrated in Fig. 27-25. Inside the enclosure, a series of nozzles connected to a pump that draws the coating material from a reservoir are suitably positioned with respect to the part. On some flow coaters, the

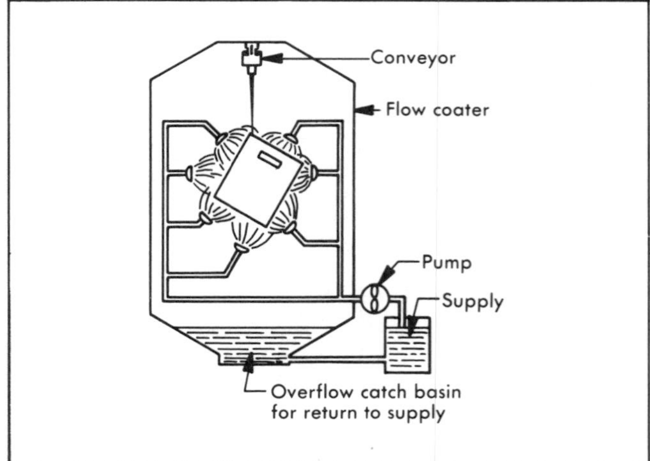

Fig. 27-25 Cross section of typical flow coating enclosure.

nozzles are mounted on a manifold tube that oscillates in the enclosure. The liquid material leaves the nozzles as a shower, coating the entire surface of the part. The excess material flows from the part surface to the bottom of the enclosure and then back to the supply reservoir. After passing through the shower, the part is carried by a conveyor over a drainboard until drainage stops and is then transported to the curing oven. In large industrial installations, extensive flow tunnels outside the application enclosure keep the area around the part saturated with solvent to improve flowout (drainage and uniform film formation).

Three types of flow coating are currently being used: high pressure, low pressure, and centrifugal. In the high-pressure method, the individual nozzles regulate the pressure of the coating material. In the low-pressure method, the pressure is regulated at the reservoir tank and rarely exceeds 10 psi (70 kPa). In the centrifugal method, the coating material flows out of nozzles mounted at the end of a rotating arm.

The viscosity and temperature of the coating material must be carefully controlled during flow coating. If the viscosity is too high, the coating material will not flow off properly, resulting in sags, beads, and blisters during baking. A viscosity that is too low results in excessive solvent loss and inadequate film thickness. The temperature of the coating material should be maintained at 70 to 90° F (21 to 32° C). If the operating temperature is too high, an excessive amount of solvent is lost before baking owing to evaporation. Low operating temperatures, however, require an increased use of solvents to control viscosity.

Flow coating can be used to economically coat both sides of a part; up to 95% of the coating material is used. Because of the placement of the nozzles, the coating material penetrates into difficult-to-reach areas. Compared with dipping, flow coating does not require large tanks or large quantities of coating material. However, flow coating does have certain limitations. For example, lacquers and other fast-drying coatings do not usually produce satisfactory films. Sharp corners on the parts may not be adequately coated because of flowaway or bubble shadows. The parts being coated must also be properly cleaned to ensure good adhesion and to prevent contaminants from entering the paint and being recirculated on other parts. Some of the parts being coated by flow coating are home appliance panels, shelves, and toys.

CURTAIN COATING

Curtain coating is a specialized type of flow coating in which the flat surface to be coated is passed through a continuous sheet or curtain of material. A typical curtain coater contains a coating system and a conveyor system. The coating system consists of a coating head, reservoir tank, variable-speed pump, filter, and a catch basin or return trough (see Fig. 27-26). The conveyor system consists of an infeed and an outfeed conveyor with a variable-speed drive; speeds range from 0 to 600 fpm. A heating system may also be incorporated in the curtain coater to maintain the coating material at a specified temperature.

The curtain of material can be produced by a pressure/gravity type of coating head or by a weir-type coating head. A pressure-type coating head is essentially a closed, airtight, V-shaped reservoir that has an adjustable slot across the bottom to provide a controlled, continuous curtain. Coating material is pumped from the reservoir tank and through a filter into the coating head until the material reaches the overflow level. At this point, the air bleeder valve is closed and the adjustable blades opened to the desired width, causing the coating material to flow out under pressure (see Fig. 27-27). If the bleeder valve is left open, the material flows out owing to gravity. In the weir-type coating head, the coating material is pumped into the coating head at a controlled rate through a distributor pipe. When the head is full, the material flows over the weir (knife plate), forming a continuous curtain. The catch basin collects and returns any excess coating material to the reservoir tank.

Curtain coating is most readily applied to flat surfaces and is widely used in the coating of sheets such as plywood and chipboard surfaces, floor tiles, cabinet doors, corrugated boxes, and mirror backings. Curtain coaters are capable of coating sheets in widths from 12 to 144″ (300 to 3650 mm). Some curtain coaters have been designed to coat rolls (webs) of materials; the coatings produced are smooth and uniform in thickness. Edges are not easily coated by this method if they are at right angles to the main surface.

Coating thicknesses from 0.5 to 100 mils (0.013 to 2.5 mm) can be applied by curtain coating, depending on the characteristics and flow rate of the coating material, conveyor speed, and

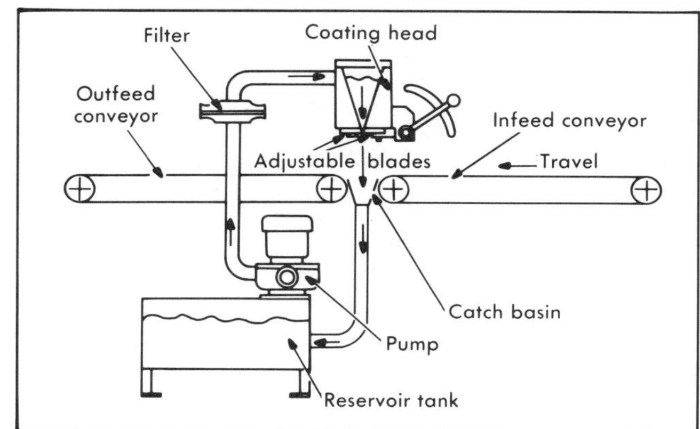

Fig. 27-26 Pressure head-type curtain coater designed for coating sheet products. (*Koating Machinery Company, Inc.*)

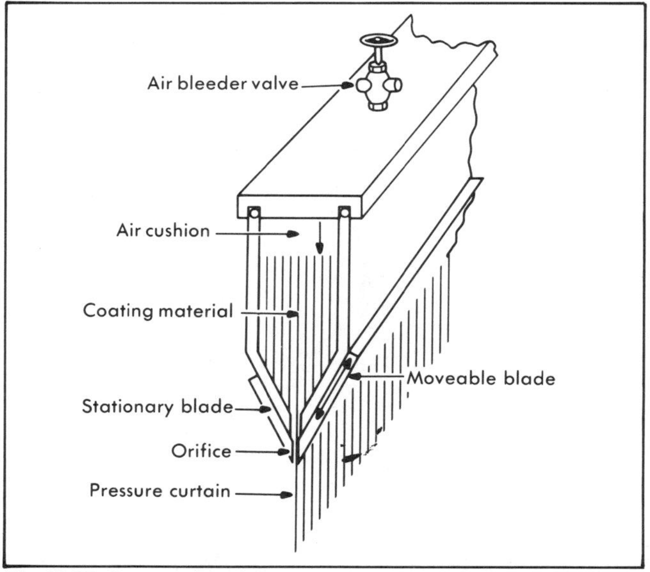

Fig. 27-27 Cross section of pressure-type coating head. (*Koch Technical Products Division*)

the type of parts being coated. The coating thickness is directly proportional to the flow rate and inversely proportional to the conveyor speed.

Coating materials used in curtain coating must be specifically formulated because not all materials form a continuous, break-free curtain. In addition, the viscosity and temperature of the coating material should be constantly monitored because curtain behavior is related to these factors. The coating material must also be carefully filtered because the head slot can be easily blocked, causing a split in the curtain and streaks on the surface.

ROLLER COATING

Roller coating is a high-speed machine painting process used for the continuous coating of flat panels or coils (webs) of metal, plastics, paper, film, and fabric. The process consists of transferring an organic coating from a revolving applicator roller to the surface of the material as it passes through the machine. Depending on the equipment design, the top, bottom, or both surfaces can be coated in one pass.[20] Most organic coating materials can be applied by roller coating. In addition, roll coaters are used to apply filler materials to porous or rough surfaces prior to applying the topcoat. Paints are usually thinned with slow-evaporating solvents and are applied at higher viscosities than in other application methods, permitting close control of paint flow and film thickness.[21]

When the material and the rollers travel in the same direction and at the same speed, the process is called direct roller coating. When the roller motion is in the opposite direction to the motion of the material, the process is called reverse roller coating.[22] When direct or reverse roll coaters are used to apply a coating to a continuous strip of metal, the process is commonly referred to as coil coating.

A direct roll coater consists of a coating roll, a doctor roll located next to the coating roll, and a feed roll beneath the coating roll (see Fig. 27-28).[23] The rolls are usually made from ethylene propylene rubber of various hardnesses to provide resiliency. The coating and feed rolls turn at the same surface speed, which is about six times the speed of the doctor roll, to obtain smooth coatings. The gap between the coating and feed rolls is usually adjusted to a distance a little less than the thickness of the material being coated to ensure a positive drive of the material through the machine. The gap between the doctor and coating rolls can be adjusted to vary the amount of coating material applied. Depending on the viscosity and other properties of the coating material, the speed of the coating and feed rolls can also be adjusted to vary the amount of coating material applied. The range of typical coatings applied by a direct roll coater is from 1 to 3 mils (0.025 to 0.075 mm) thick. Direct roll coaters can be connected in tandem to apply thicker coatings.

Another type of roll coater is the precision roll coater, which is a modified direct roll coater. Instead of the surface of the doctor roll being smooth, it has an engraved pattern on it that allows a definite deposit to remain on the coating roll. The doctor roll is also in direct contact with the coating roll and thus turns at the same speed. The pattern on the doctor roll determines, to a great extent, the thickness of the coating film on the substrate. The film thickness can also be varied slightly by adjusting the gap between the doctor and coating rolls. Film thickness is usually limited to 1 mil (0.025 mm); tandem precision roll coaters are commonly used for thicker films.

A reverse roll coater is essentially a direct roll coater with the bottom feed roll turning in the opposite direction to the coating roll (see Fig. 27-29). The opposite turning direction serves to wipe the coating material onto the substrate. The coating roll is also made from metal that has been chromium plated instead of being made from a resilient material; the doctor and feed rolls are generally made from a resilient material. Reverse roll coaters are commonly used for coating material in coil (web) form; the coil is held in contact with the coating roll by two smaller rolls, one on either side of the coating roll. The coating film applied by reverse roll coaters is thicker than the film applied by direct roll coaters.

A fourth type of roll coater, referred to as a combination roll coater, combines a direct roll coater with a reverse roll coater in the same machine frame.[24] At the first station (direct roll coater), an excess amount of coating material is applied to the surface. The second station wipes off the excess material from the surface and then applies a smooth, uniform film across the entire surface. The combination roll coater is commonly used to apply opaque base coats to hardboard surfaces.

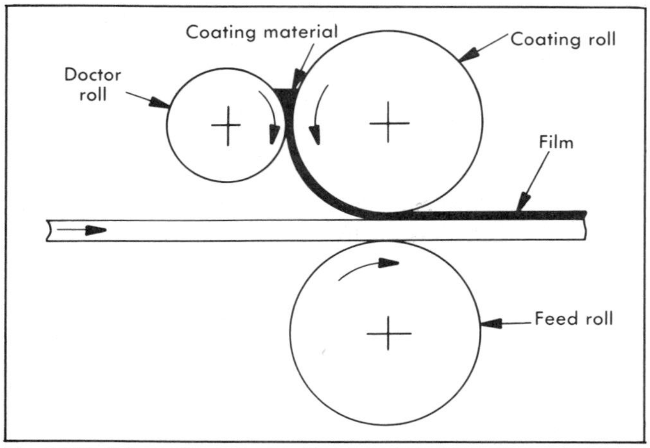

Fig. 27-28 Schematic of direct roll coater. (*The Black Bros. Co.*)

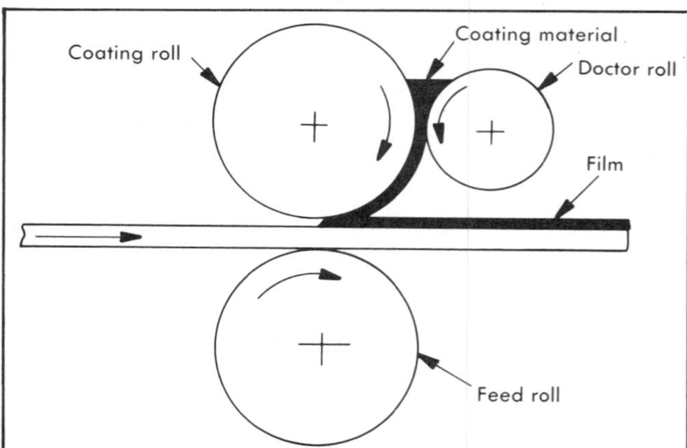

Fig. 27-29 Schematic of reverse roll coater. (*The Black Bros. Co.*)

BRUSHING AND ROLLING

The two most common manual coating methods are brushing and rolling. In brushing, a handle containing bristles on one end is dipped into the paint container and then applied to the surface using even strokes. Dabbing the brush on the side of the container helps to distribute the paint among the bristles. When applying the paint, the brush should be held at a 60° angle to the work. Several light strokes followed by moderate, even-pressure strokes are used to apply the paint to the surface. Long strokes in a crosswise direction and with minimal pressure are used to finish off the surface. The surfaces can be flat, round, curved, or irregular in shape.

The shape and size of the brush are mainly a matter of personal preference. Flat brushes are generally preferred for work on flat surfaces, whereas tapered brushes are used for trim work. Round or oval brushes are often preferred for rough surfaces, for painting rivet heads, and for constricted areas. The size of the brush must be adapted to the surface being painted.[19]

The bristles used in brushes are natural, synthetic, or a combination of the two. The best natural bristles are Chinese hog bristles because they are of the right length, have good resiliency, and are long lasting. The ends of hog bristles are forked, permitting more paint to be carried in the brush than by other bristles. In addition, the coating flows together smoothly with few brush marks. Horsehair is a less expensive natural bristle, but it does not perform as well as hog bristle. Another type of natural bristle is badger hair, which is generally used for varnishing brushes. For lettering and other fine line work, brushes are frequently made from sable or squirrel fur.

The most common synthetic bristles are made from nylon and are used for water-thinned coatings. Although nylon bristles are better than horsehair bristles, they are susceptible to damage from organic solvents. Brushes with mixed bristles combine the advantages and limitations of several types of bristles.

In rolling, a roller (rotatable cage) attached to a handle is used for applying paint to flat or continuous surfaces. The length of the roller is typically from 1.5 to 18″ (38 to 460 mm); selection of the proper roller length is determined by the area to be painted. The roller sleeve, mounted on the roller, is covered with fabrics that expand and contract as they are brought in contact with the surface. This expanding and contracting action transfers the paint from the sleeve to the surface. Paint is loaded onto the sleeve by rolling it back and forth in a holding tray; a secondary surface in the tray is used to even out the load. The paint may also be fed to the roller from a pressure tank through a hose to the handle. To obtain complete coverage, especially on rough surfaces, the roller must be worked in several directions back and forth. All final strokes should be made systematically in one direction.

Roller sleeves are plastic or fiberboard cylinders that are commonly covered with lamb's pelt (wool), mohair or angora, modified acrylic, or polyester fabrics. Nap lengths (length of fabric fibers) range from 0.2 to 1 1/4″ (5 to 30 mm); short naps are used when rolling smooth surfaces, while longer naps are used when rolling rough surfaces. The fabric from which the cover is made should be selected according to the surface profile of the work to be painted and the type of paint used.

Roller sleeves covered with lamb's wool are available in all nap lengths and are generally used when rolling semismooth or rough surfaces. Lamb's wool sleeves are widely used when rolling synthetic, solvent-borne paints, but are not recommended for use with waterborne paints. Mohair-covered sleeves typically have a short nap length and are used for rolling smooth surfaces; they are suitable for use with both solvent-borne and waterborne paints. Modified-acrylic-covered sleeves are available in all nap lengths and can also be used with both solvent-borne and waterborne paints; however, they are not recommended for use with lacquers. Polyester-covered sleeves are available in short and medium nap lengths and can be used for applying both solvent-borne and waterborne paints.

Rolling is recommended for applying paint to flat interior surfaces such as walls and ceilings, especially when the surfaces have uneven textures such as those of brick, plaster, and acoustical tile. Rolling is also used when painting open work such as wire-link fencing by using a long-nap roller sleeve.

AUXILIARY EQUIPMENT

In addition to the spray guns and coating tanks that are used for the various application methods, other equipment is required to produce and regulate the pressurized air or fluid, to control air pollution, and to transport the parts throughout the finishing line.

AIR COMPRESSORS

The air compressor provides the power to operate any compressed-air-operated system. The basis for the selection of an air compressor is the amount (volume) of airflow required, measured in cubic feet (meters) per minute, and the air pressure required, measured in psi or kPa.

The most common types of compressors used are diaphragm, rotary, and reciprocating or piston. In a diaphragm compressor, the air pressure is developed through the reciprocating or oscillating action of a flexible disc, normally made of metal or a rubber composition. The compressed air delivered with these compressors must be oil-free, but not moisture-free, and pressures are in the 30 to 40 psi (200 to 275 kPa) range. Diaphragm-type compressors are designed for occasional operation and are used by the homeowner, small painting contractor, or hobbyist.

Rotary compressors have become increasingly popular and are usually quieter, easier to install, and less expensive than reciprocating types. This type of compressor design is based on the operating principle of a rotor fitted with vanes eccentrically mounted in a housing. During operation, the vanes are held against the housing by centrifugal force. A lubricating fluid is used in these compressors for both sealing and cooling.

Piston-type compressors are the most common compressors used by the painting industry. The pressure chamber consists of a piston fitted with automotive-type piston rings, a cylinder, and intake and discharge valve assemblies. Compression can take place on either side of the piston or on both sides. When compression occurs on both sides of the piston, the compressor is referred to as a double-acting compressor. For most applications, single-acting compressors are widely used.

Single-acting compressors can be further divided into single stage, dual stage, or multistage; dual-stage and multistage compressors can be air or water cooled. When air is drawn from the atmosphere and compressed to a given pressure in a single stroke, the compressor is a single-stage unit. These compressors are capable of producing pressures up to 100 psi (690 kPa). Two-stage compressors compress the air to an intermediate pressure, approximately 40 psi (275 kPa), pass the air through an intercooler, and then compress the air to the final pressure, which is typically in the 100 to 200 psi (690 to 1380 kPa) range.

Two-stage compressors are recommended for the majority of production spray applications. The compressor output should be passed through a heat-exchange cooler to remove condensed water vapor. If the vapor is not removed, moisture in the line can damage the paint film. The compressor should also be equipped with a safety valve to release excessive pressures.

HOSES AND FITTINGS

Materials used for paint lines include black iron, stainless steel, plastics, fiber-reinforced plastics, and plastics-lined steel, depending on the nature of the paint. Special attention should be given to lines used for airless spray methods because they usually operate at pressures from 1000 to 4000 psi (6900 to 27 500 kPa). If the paint shoots out of a ruptured line, it could cause serious injury to personnel. Hoses should be made from materials that are resistant to paint solvents, especially at elevated temperatures, and should be capable of withstanding any unexpectedly high pressures that may occasionally occur.

Hose connections and fittings are available in various thread styles and sizes. The most common thread styles used in the painting industry are either national pipe straight (NPS) or national pipe tapered (NPT). The NPS style is preferred for most spray guns and air and fluid inlets, whereas the NPT style is the most common type used for piping. Sometimes special thread styles are used, such as joint industrial connections (JIC) for high-pressure hydraulic fittings.

To ensure adequate operating pressures, it is important to take into consideration the pressure loss that will occur as the paint flows through the lines and fittings. Pressure loss in the line depends upon the flow rate and viscosity of the paint and the inside diameter and length of the line; hoses or lines with a small ID and a long length have the greatest pressure loss. The pressure loss in the line can be estimated using the following equation:

$$P_L = \frac{0.0273 \times F \times V \times L}{D_I^4} \quad (3)$$

where:

P_L = pressure loss, psi
F = flow rate of the paint, gpm
V = viscosity of the paint, poises
L = length of the line, ft
D_I = inside diameter of the line, in.

To obtain the pressure loss in kPa, multiply the value in psi by 6.895. In addition, a pressure loss of 0.8 psi (5.5 kPa) occurs for every foot (0.3 m) of vertical rise. A pressure loss also occurs in line bends and fittings.

PUMPS

Pumps are used to move liquids from place to place in pretreatment and paint systems. For paint systems, the paint should be moved at a constant rate with minimal pressure surges. The four general pump classifications are: reciprocating piston, centrifugal, rotary, and pistonless.

The reciprocating piston pump is the type most commonly used in a material spraying system. It consists of the power section and the fluid section. The power section can be either an air-driven or hydraulic-motor-driven piston that is usually connected directly to the liquid-handling fluid piston (see Fig. 27-30). In some designs, the fluid piston can be driven by an electric motor through a piston rod and crankshaft linkage system.

As the fluid piston moves upward, it draws the coating material into the intake chamber through the foot valve. When the piston moves downward, the material is forced through the piston body and into the pressure-discharge chamber. Upward movement draws additional material into the intake chamber as well as pushes the material out of the discharge chamber and into the supply line. A constant supply of coating material is supplied to the lines on both the upstroke and downstroke.

The turbine centrifugal pump usually has its impeller and fixed vane section submerged in the liquid material. The rotating impellers receive the material centrally and discharge it peripherally (by centrifugal force) into the vanes. The vanes guide the material upward into successive impellers where additional pressure and velocity is imparted to the material until it is discharged. The magnitude of pressure depends on the number of impellers and fixed vane stages in the pump.

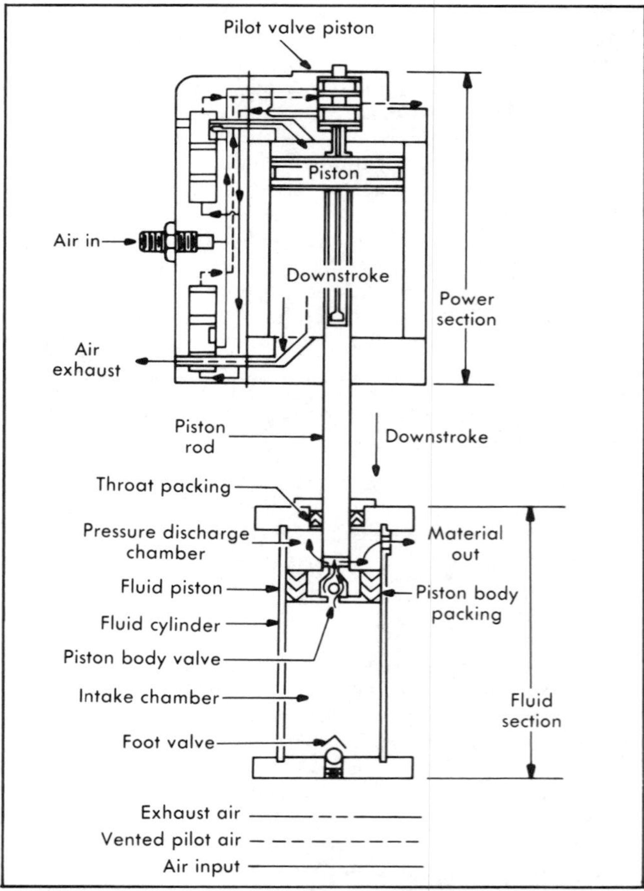

Fig. 27-30 Cross section of an air motor-driven reciprocating piston fluid pump. (*Binks Manufacturing Co.*)

Centrifugal pumps are similar to turbine centrifugal pumps, except for differences in the impeller-vane design.

A rotary pump is a positive-displacement pump consisting of a fixed casing containing gears, cams, screws, vanes, plungers, or similar elements, driven by a rotating driveshaft. Rotary pumps are not as accurate as piston pumps. Pistonless pumps utilize the direct pressure of gas or steam on the coating material.

REGULATORS AND FILTERS

In order to obtain optimum results from spray equipment, it is necessary to regulate the fluid pressure and air pressure, as well as remove contaminants from the coating material or air. A fluid pressure regulator is a device used to control the pressure or fluid flow to the spray guns. It is usually mounted on the discharge side of the fluid pump and as close to the spray gun(s) as possible. An air regulator is a mechanical device that reduces and maintains the main line air pressure at a specific operating pressure. It is usually mounted on the compressor or pressure tank.

Fluid filters are made of stainless steel wire mesh, as well as fibers that are resistant to organic solvents, resins, and additives. They are used to remove relatively large pigment particles or resin conglomerates that could clog the nozzle of the spray gun or adversely affect the surface finish of the coating. Most filters are capable of removing particles that are larger than 1-100 μm (0.00004-0.004") in size. The type of paint being applied determines the type of filter used.

Air filters are mechanical devices that remove solid particles from the air supply. Since they do not remove moisture, it is also necessary to include a drier in the system. The most common types of driers used are chemical, desiccant, and refrigeration types. Combination regulators and filters are designed to both control the air pressure and remove solid particles, entrained oil, and moisture from the airstream.

PAINTING BOOTHS

Spray booths are fire and health protection enclosures designed to provide a positive movement of air through the spray area. The air movement causes the solvent fumes to be carried to the atmosphere, and most solid particles are eliminated in the filter system. In addition, spray booths reduce air, water, and noise pollution. Further, the quality of the product's finish is improved. Spray booths can be categorized by their method of filtration and their construction.

The three main filtration methods are dry, water, and powder. Spray booths are of downdraft or sidedraft construction.[25] In downdraft booths, the air enters from the booth ceiling and exits through the floor. Downdraft booths are usually used for large, conveyor-moved objects. Since these booths are large, the operator can move around the part being painted. The excess paint is trapped in a water tank under the floor. Sidedraft booths have the air entering at one side of the booth and leaving at the other. Since the paint spray can only flow in one direction, the operator must stand upwind from the part being coated. Parts are usually rotated or indexed if all sides must be coated.

Dry Spray Booths

Dry booths have enclosed bottoms, tops, and sides and are generally used when paints are applied manually by air spray guns.[26] The backs of the booths, through which air is exhausted, are equipped with disposable filters, rolls of filter media, or staggered plates to catch most of the paint overspray. The exhaust rates must be high enough to draw the overspray away from the operator at a specified rate.

Water Spray Booths

The bottom, top, and two sides of water spray booths are generally enclosed. The basic method of operation involves a water curtain maintained by a pump that provides a continuous circulation of water. In some water booths, exhaust air is drawn through a water curtain that flows from an upper trough down over the surface facing the operator into a lower trough.[27] In downdraft water booths, the exhaust air (laden with paint overspray) is drawn through a metal floor grating into a pan or water trough below.

The water used in these booths is chemically treated to cause the water-entrapped overspray to either sink or float. After decanting, the sludge can be disposed of in accordance with existing waste handling regulations.

Powder Spray Booths

Powder overspray must be reclaimed and recycled to improve the efficiency of powder coating operations.[28] Powder coating booths have smooth sides with steep, hopperlike sloping bottoms that empty into collectors and an exhaust system designed to remove airborne suspended powder. The powder is drawn into a cylindrical chamber that has a centrifugal blower to force the powder to the outside walls where it collects and then falls through an opening in the cone-shaped bottom. The air flows through a filter at the top to remove any fine suspended powder particles. The reclaimed powder can then be blended with fresh material of a suitable color.

CONVEYORS

When production rates are low or if the parts are large and cumbersome to handle, parts can be painted and allowed to dry in place.[29] In most instances, however, the parts are moved from station to station on a finishing line by a conveyor. Conveyors are available in many sizes and shapes and can generally be classified as overhead, belt, or floor conveyors.

Overhead Conveyors

Most finishing lines use overhead monorail conveyors that are capable of going up and down inclines and around curves. The most commonly used monorail configuration is the I-beam. A continuous-link chain having pulleys designed to ride on the monorail has a drive mechanism and brackets for attaching the part hangers. The size of the chain varies depending on the amount of tension required to pull the parts through the finishing line.

The chain is moved along the I-beam by a drive mechanism consisting of a motor gearing unit, speed regulator, drive sprockets, and, frequently, a special chain with fingers to engage the conveyor chain while transmitting the drive force to a number of links simultaneously. A lubrication system applies oil or grease to the pulleys at prescribed intervals. A chain cleaner, which uses rotating brushes and air blasts, removes the paint residues that accumulate on the chain during painting.

For small, lightweight parts, the pulleys are mounted inside a continuous hollow tube that has a slot at or near its bottom for the part hangers. Some small-part conveyors use a steel cable in place of a chain.

Belt Conveyors

Belt conveyors are used when a large number of small parts are being finished. The belts are generally either a continuous metal screen or an open-mesh belt. Belt conveyors are generally designed to operate in one phase of the finishing operation, rather than operating in a continuous manner.

REFERENCES

Floor Conveyors

Floor conveyors are used for moving large objects such as automobile bodies, truck cabs, and appliances that require coating on the sides and the top. To carry the parts, a chain or cable moves a table, platform, or dolly having wheels or casters that are guided through the various stages of the operation by tracks. The drive chain or cable is mounted either on or below the floor level.

HANGERS AND RACKS

In production coating systems, a part hanger is an important tool and should be properly designed and used.[30] Most hangers are made from either stainless or carbon steel. Although stainless steel is more expensive than carbon steel, the additional cost can be justified because stainless steel possesses greater strength and is capable of withstanding repeated heating without damage.

Several hanger designs are in common use. A typical hanger used in the appliance-coating industry is illustrated in Fig. 27-31. For a high-production finishing process, parts can be prehung on a large hanger that is then hung on the conveyor. Small parts should be hung on a hanger in a step pattern from top to bottom to eliminate dripping onto finished parts. Add-on hangers should be used when the parts being coated have open areas. An example of a part with open areas is a picture frame. The add-on hanger would attach to the main hanger and allow smaller picture frames, for example, to be coated at the same time as the large picture frames. Magnetic hangers can be used to coat decorative steel parts that do not contain hanger holes or when hook-type hangers are not practical.

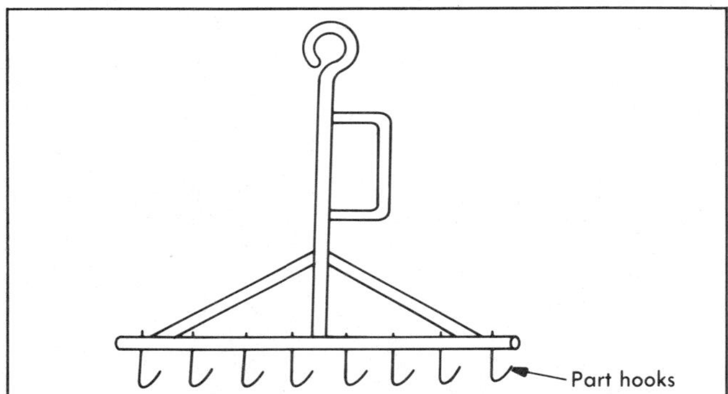

Fig. 27-31 Typical hanger used in coating appliance components.

References

1. Carl Izzo, "Paint Finishing Materials and Equipment Guide," *Products Finishing*, 1984 Directory (September 1983), p. 6.
2. Gerald L. Schneberger, ed., *Understanding Paint and Paint Processes*, 2nd ed. (Wheaton, IL: Hitchcock Publishing Co.), pp. 104-106.
3. The Powder Coating Institute, *Powder Coatings* (Greenwich, CT: 1983).
4. Dr. Emery P. Miller and Dr. David D. Taft, eds., *Fundamentals of Powder Coating* (Dearborn, MI: Society of Manufacturing Engineers, 1974), p. 24.
5. *Ibid.*, p. 47.
6. Daniel R. Savage, *The Application of Powder Coatings by Electrostatic Deposition Methods*, SME Technical Paper FC71-841 (Dearborn, MI: Association for Finishing Processes of SME, 1971).
7. Daniel R. Savage, *The Electrostatic Fluidized Bed—Theory, Design and Application*, SME Technical Paper FC72-935 (Dearborn, MI: Association for Finishing Processes of SME, 1972).
8. Savage, FC71-841, *loc. cit.*
9. George E. F. Brewer, ed., *Electrodeposition of Coatings* (Washington, DC: American Chemical Society, 1973).
10. Robert W. Whitehall, *Electrocoating Equipment*, SME Technical Paper FC75-564 (Dearborn, MI: Association for Finishing Processes of SME, 1975).
11. Dr. Edward L. Jozwiak, Jr., "Cathodic Instead of Anodic: Is There Justification?" *Manufacturing Engineering* (April 1981), pp. 131-133.
12. Dr. Rodney W. Stockstad, "Cathodic Electrocoats: Operational Considerations," *Finishing '83 Conference Proceedings*, October 10-13, 1983, Cincinnati, OH (Dearborn, MI: Association for Finishing Processes of SME, 1983), pp. 2-19 to 2-26.
13. Harry M. Leister, *Processing With Autodeposition*, SME Technical Paper FC81-248 (Dearborn, MI: Association for Finishing Processes of SME, 1981).
14. M. J. Johnson and N. R. Roobol, "An Evaluation of Autodeposition," *Plating and Surface Finishing* (July 1984), pp. 58-62.
15. Joseph W. Prane, "Organic Coatings Technology, A Review—Part II," *Metal Finishing* (October 1983), pp. 75-78.
16. Harry M. Leister, "Autodeposition of Organic Coatings," *Plating and Surface Finishing* (July 1982), pp. 46-48.
17. Johnson, *loc. cit.*
18. Ellsworth A. Stockbower, *Autodeposition of Organic Films—Current Developments*, SME Technical Paper FC84-884 (Dearborn, MI: Association for Finishing Processes of SME, 1984).
19. William F. Gross, *Applications Manual for Paint and Protective Coatings* (New York: McGraw-Hill Book Co., 1970), p. 182.
20. Melvin H. Sandler, et. al., "Painting," *Metals Handbook*, 9th ed., vol. 5 (Metals Park, OH: American Society for Metals, 1982), p. 482.
21. *Ibid.*, p. 483.
22. Izzo, *op. cit.*, p. 36.
23. Donald T. Jones, "Roll Coating Equipment—Applications and Limitations," *Radiation Curing V Proceedings*, September 23-25, 1980, Boston, MA (Dearborn, MI: Association for Finishing Processes of SME, 1980), pp. 401-412.
24. Donald T. Jones, "Latest Roll Coater Technology for Improved Panel Finishing," *Finishing '83 Conference Proceedings*, October 10-13, 1983, Cincinnati, OH (Dearborn, MI: Association for Finishing Processes of SME, 1983), pp. 3-15 to 3-23.
25. Schneberger, *op. cit.*, pp. 66-67.
26. Izzo, *op. cit.*, p. 20.
27. *Ibid.*, p. 24.
28. *Ibid.*, p. 25.
29. *Ibid.*, pp. 43-44.
30. Walter P. Sobas, *Component Hangers—Design and Application*, SME Technical Paper FC80-626 (Dearborn, MI: Association for Finishing Processes of SME, 1980).

Bibliography

Adams, John R. *Powder Coating Using the Flocking Process.* SME Technical Paper FC72-933. Dearborn, MI: Association for Finishing Processes of SME, 1972.

Anderson, R. K. *History: The Whys and Wheres of Powder Coating.* SME Technical Paper FC71-843. Dearborn, MI: Association for Finishing Processes of SME, 1971.

Case, Leo L. *Electrocoating Systems—Anodic, Cathodic or Both.* SME Technical Paper FC77-680. Dearborn, MI: Association for Finishing Processes of SME, 1977.

Keown, Robert H. "Powder Coating: The Total System Concept." *Finishing '83 Conference Proceedings.* October 10-13, 1983, Cincinnati, OH. Dearborn, MI: Association for Finishing Processes of SME, 1983.

Koch, Russell R. *Electrocoating Materials Today and Tomorrow.* SME Technical Paper FC75-563. Dearborn, MI: Association for Finishing Processes of SME, 1975.

Papp, Nick. *Ultrafiltration Applications in the Electrodeposition Process.* SME Technical Paper FC80-617. Dearborn, MI: Association for Finishing Processes of SME, 1980.

Phillips, Phillip G. *The Fluidized Bed Method of Powder Coatings.* SME Technical Paper FC71-844. Dearborn, MI: Association for Finishing Processes of SME, 1971.

Prane, Joseph W. "Organic Coatings Technology, A Review—Part III." *Metal Finishing* (October 1983), pp. 75-78.

CURING METHODS

While the terms *drying* or *baking* are commonly used in the painting industry when referring to curing, there is a distinction between the processes of drying (baking) and curing. A paint film can be dry, but not cured. Dry can refer to the film resulting from the loss of solvent from a paint; the solvent is gone, but the resin is unchanged. To be cured, however, the resin must be converted to a new resin. An incompletely cured film may be dry to the touch, but fail in use.

All organic coating films, as they occur on a surface as a finish, are polymers containing large numbers of molecular units of the same chemical form, bound together to create a solid, continuous film. To produce such film structures, it is necessary that the chemical resin components are in a form such as a liquid so that application is possible. Once the material is applied, it is allowed to level and convert to the desired solid film.

In some cases, the resins are applied as hot melts and simply allowed to flow, level, cool, and solidify as a film of material chemically like the original unmelted resin. In other cases, the resin materials are dissolved in a compatible solvent to produce a liquid material. This liquid material is applied as a film that returns to an uncured resin coating upon evaporation of the solvent component. The residual resin layer is then converted to the desired resin film by a further polymerization or crosslinking process called curing. Curing can be brought about by several methods, as discussed in this chapter.

CHAPTER CONTENTS:

AMBIENT TEMPERATURE DRYING

As previously mentioned, some organic coating films are applied as hot melts of the basic resin and are then solidified by cooling in air. Thermoplastic materials in this category include waxes, polyamides, polyethylenes, and bitumens. The article to be coated may be dipped in the hot melt, or the hot melt may be applied to the article in other ways. Thermoplastic coating materials may be prepared as powders and deposited by electrostatic spray or from a fluidized bed. Residual heat in the part or subsequently applied heat may be used to melt and fuse the particles into a continuous coating. Again, air cooling solidifies the film. In some cases, water quenching is used, especially for thick films.

Some organic coatings are solutions of resins in a solvent and are applied by brushing, spraying, or dipping. The coating films are formed rapidly, simply by evaporation of the solvent, leaving behind a solid film. The rate of drying is governed by the rapidity of the solvent evaporation. The last small percentage of solvent usually evaporates slowly; as a result, maximum hardness also develops slowly. Moderate heat can be applied to accelerate solvent evaporation to achieve maximum hardness quickly; essentially, however, the process remains one of air drying.

The process of air drying basically involves a physical change and is therefore reversible; formed films can be redissolved by the use of proper solvents. Nitrocellulose lacquers dry by solvent evaporation, as do other coatings—vinyls, vinyl copolymers, acrylic resins, shellac, styrene copolymers, and solutions of bitumens. Certain acrylic resins and styrene copolymers require baking. Other coating materials, such as drying oils and oleoresinous varnishes, are liquids or soft, sticky solids that form films by both solvent evaporation and combination with oxygen taken from the air. Oxygen take-up is a chemical process and irreversible.

Air drying is an energy-saving process and therefore quite desirable. Air-drying paints, however, require large quantities of solvents for reduction to a viscosity low enough to allow brushing, spraying, or other methods of application. Paints having high contents of volatile organic compounds have to be avoided because of environmental reasons. However, it is possible to replace some of the objectionable solvents with so-called "compliance solvents," like 1,1,1-trichloroethane and/or methylene chloride. The use of these chlorinated compounds is permissible in many areas where the ordinary organic solvents are restricted. Air-drying, waterborne emulsion paints are also being used.

CATALYTIC CURING

Properties of catalyzed conversion coatings are given in Table 28-1, and the useful lifetimes of coatings in various reagents are listed in Table 28-2. The polymerization or curing of many coating materials is brought about chemically. Two types of polymer linkage (linear and crosslinkage) are

***This chapter contributed by: George E. F. Brewer**, Coating Consultants.*
***Reviewers for sections of this chapter are: George E. F. Brewer**, Coating Consultants; **Horace H. Homer**, Consultant; **Dr. Emery P. Miller**, Consultant; **Dr. Norman R. Roobol**, GMI Engineering & Management Institute.*

CATALYTIC CURING

TABLE 28-1
Properties of Catalytically Cured and Other Heavy-Duty Organic Coatings

Coating Type	Primer	Inter-mediate Coat	No. of Coats*	Solvent Resistance**	Maximum Dry Heat Resistance, °F (°C)
Inorganic zinc	None	No	F1	E	750 (400)
Polyamide-cured epoxy	Inhibitive	No	P1,F2	Aliphatics, VG	250 (120)
	Zinc-rich epoxy	No	P1,F2	Aliphatics, E	250 (120)
	Inhibitive	Yes	P1,I1,F1	Aliphatics, E	250 (120)
	Zinc-rich epoxy	Yes	P1,I1,F1	Aliphatics, E	250 (120)
	Zinc-rich inorganic	Yes	P1,I1,F1	Aliphatics, VG	250 (120)
	Zinc-rich solvent epoxy	Yes	P2,I1,F1	Aliphatics, E	250 (120)
Amine-adduct-cured epoxy	Inhibitive	No	P1,F2	Aromatics, E	300 (150)
	Zinc-rich inorganic	No	P1,F2	Aromatics, E	300 (150)
	Zinc-rich epoxy	No	P1,F2	Aromatics, E	250 (120)
	Polyamide epoxy	No	P1,F2	Aliphatics, VG	250 (120)
Solventless epoxy	Zinc-rich epoxy or inorganic	No	P1,F2	Ketones, E	300 (150)
	Inhibitive	No	P1,F1	Ketones, E	300 (150)
Epoxy ester	Inhibitive	No	P1,F2	Aliphatics, G	250 (120)
	Epoxy ester	No	P1,F2	Aliphatics, VG	250 (120)
Vinyl	Inhibitive	Yes	P1,I1,F1	Alcohol, G	120 (50)
	Zinc-rich inorganic	Yes	P1,I1,F1	Alcohol, VG	120 (50)
	Zinc-rich epoxy	Yes	P1,I1,F1	Alcohol, VG	120 (50)
	Vinyl wash	Yes	P1,I1,F3	Alcohol, VG	120 (50)

(continued)

TABLE 28-1—*Continued*

Maximum Wet Heat Resistance,† °F (°C)	Abrasion Resistance	Flexibility and Impact Resistance	Fresh and Salt-Water Resistance†	Gloss and Color Retention†	Theoretical Coverage,†† ft^2 gal (m^2/L) at 1 mil (0.001″)
140 (60)	E	G	E	N.A.	960 (23.6)
150 (65)	E	VG	E	Chalks early	P660 (16.2) F660 (16.2)
150 (65)	E	VG	E	Chalks early	P855 (21.0) F660 (16.2)
150 (65)	E	VG	VG	Chalks early	P660 (16.2) I930 (22.8) F660 (16.2)
150 (65)	E	G	VG	Chalks early	P855 (21.0) I930 (22.8) F660 (16.2)
150 (65)	E	VG	E	Chalks early	P960 (23.6) I930 (22.8) F660 (16.2)
150 (65)	G	G	VG	Chalks, no erosion	P600 (14.7) F660 (16.2)
180 (80)	G	G	N.R.	Chalks early	P640 (15.7) F640 (15.7)
180 (80)	VG	Fr	E	Chalks, no erosion	P960 (23.6) F640 (15.7)
150 (65)	E	Fr	E	Chalks, no erosion	P855 (21.0) F640 (15.7)
150 (65)	E	VG	VG	Chalks, no erosion	P660 (16.2) F640 (15.7)
180 (80)	G	Pr	E	Chalks early	P855-960 (21.0-23.6) F1300 (31.9)
180 (80)	G	Pr	VG	Chalks early	P1300 (31.9) F1300 (31.9)
N.R.	Fr	G	Pr	Fr, chalks	P500 (12.3) F750 (18.4)
N.R.	Fr	VG	Fr	Fr, chalks	P660 (16.2) F750 (18.4)
120 (50)	Fr	E	E	---	P500 (12.3) I480 (11.8) F275 (6.8)
120 (50)	Fr	VG	E	E	P855 (21.0) I480 (11.8) F275 (6.8)
120 (50)	Fr	VG	E	E	P960 (23.6) I480 (11.8) F275 (6.8)
120 (50)	Fr	VG	E	E	P155 (3.8) I480 (11.8) F275 (6.8)

(continued)

TABLE 28-1—*Continued*

Coating Type	Primer	Inter-mediate Coat	No. of Coats*	Solvent Resistance**	Maximum Dry Heat Resistance, °F (°C)
	Vinyl	Yes	P1,I1,F3	Alcohol, VG	120 (50)
Chlorinated rubber	Inhibitive	No	P1,F2	Alcohol, G	190 (90)
	Inhibitive	Yes	P1,I1,F1	Alcohol, G	190 (90)
	Zinc-rich epoxy	Yes	P1,I1,F1	Alcohol, G	190 (90)
	Zinc-rich solvent epoxy	Yes	P2,I1,F1	Alcohol, G	190 (90)
Acrylic emulsion latex	Inhibitive	Yes	P1,I1,F1	Aliphatics, G	180 (80)
	Acrylic inhibitive	Yes	P1,I1,F1	Aliphatics, G	180 (80)
	Zinc-rich inorganic	Yes	P1,I1,F1	Aliphatics, G	180 (80)
	Zinc-rich epoxy	Yes	P1,I1,F1	Aliphatics, G	180 (80)
Silicone polyester	Inhibitive	No	P1,F2	Aliphatics, G	350 (175)
	Zinc-rich inorganic	Yes	P1,I1,F1	Aliphatics, F	250 (120)
Alkyd fortified	Inhibitive	No	P1,F1	Aliphatics, F	250 (120)
	Inhibitive	No	P1,F2	Aliphatics, F	250 (120)
	Inhibitive	No	P1,F2	N.R.	250 (120)
Coal-tar epoxy	Inhibitive	No	P1,F2	Aliphatics, VG	325 (165)
Polyamide cured	Zinc-rich epoxy	No	P1,F2	Aliphatics, VG	325 (165)

(continued)

TABLE 28-1—*Continued*

Maximum Wet Heat Resistance,† °F (°C)	Abrasion Resistance	Flexibility and Impact Resistance	Fresh and Salt-Water Resistance†	Gloss and Color Retention†	Theoretical Coverage,†† ft² gal (m²/L) at 1 mil (0.001″)
120 (50)	Fr	G	VG	E	P270 (6.6) I480 (11.8) F275 (6.8)
120 (50)	G	VG	VG	G	P480 (11.8) F560 (13.8)
120 (50)	G	VG	VG	G	P480 (11.8) I560 (13.8) F560 (13.8)
120 (50)	G	VG	E	G	P855 (21.0) I560 (13.8) F560 (13.8)
120 (50)	G	VG	E	G	P600 (14.7) I560 (13.8) F560 (13.8)
N.R.	Pr	E	G	E	P500 (12.3) I600 (14.7) F600 (14.7)
N.R.	Pr	E	G	E	P615 (15.1) I600 (14.7) F600 (14.7)
N.R.	Pr	E	G	E	P960 (23.6) I600 (14.7) F600 (14.7)
N.R.	Pr	E	G	E	P855 (21.0) I600 (14.7) F600 (14.7)
N.R.	Fr	E	Fr	E	P500 (12.3) F675 (16.6)
N.R.	G	VG	G	E	P960 (23.6) I480 (11.8) F675 (16.6)
N.R.	G	G	Fr	Fr	P780 (19.2) F740 (18.2)
N.R.	G	G	Fr	Fr	P850 (20.9) F680 (16.7)
N.R.	G	G	G	Fr	P500 (12.3) F570 (14.0)
180 (80)	G	Fr	VG	Chalks, no erosion	P660 (16.2) F1125 (27.6)
180 (80)	G	Fr	E	Chalks, no erosion	P855 (21.0) F1125 (27.6)

(continued)

TABLE 28-1—*Continued*

Coating Type	Primer	Inter-mediate Coat	No. of Coats*	Solvent Resistance**	Maximum Dry Heat Resistance, °F (°C)
	Inhibitive	No	P1,F2	Aromatics, VG	400 (205)
Amine-adduct cured	Zinc-rich epoxy	No	P1,F2	Aromatics, E	400 (205)

TABLE 28-2
Chemical Immersion Resistance of Catalytically Cured Coatings*

	Reagent and Concentration Percentage											
	Nitric Acid				Sulfuric Acid				Hydrochloric Acid			
Coating Type	Concen-tration	50	20	5	Concen-tration	50	20	5	Concen-tration	20	10	5
Polyamide-cured epoxy	3/4 h	6 1/3 h	23 3/4	14 d	2 h	21 d	122 d	122 d	70 1/2 h	21 d	49 d	71 d
Amine-adduct-cured epoxy	2 h	6 1/3 h	23 h	71 1/2 h	1/2 h	7 m	122 d	7 m	7d	21 d	7 m	28 d
Amine-cured epoxy (low mol. wt.)	3/4 h	6 1/3 h	23 3/4 h	28 d	1/2 h	7 m	122 d	7 m	7 d	36 d	7 m	7 m
Amine-cured coal-tar epoxy	3/4 h	30 h	7 m	7 m	1/2 h	7 m	7 m	7 m	7 m	7 m	7 m	7 m
Polyamide-cured coal-tar epoxy	3/4 h	6 1/3 h	7 m	154 d	2 h	7 m	7 m	7 m	21 d	7 m	7 m	7 m
Chlorosulforated polyethylene	1/3 h	71 1/3 h	7 m	7 m	7 d	7 m	7 m	7 m	5/6 h	7 m	7 m	7 m
Vinyl	1 1/2 h	71 2/3 h	7 m	7 m	7 d	7 m	7 m	7 m	21 d	7 m	7 m	7 m
Chlorinated rubber	5 3/4 h	22 h	23 h	23 1/4 h	7 d	21 d	21 1/2 h	72 h	29 3/4 h	23 h	22 1/2 h	22 1/2 h
Epoxy polyester	3/4 h	46 1/2 h	48 1/4 h	7 m	1/2 h	7 m	7 m	7 m	9 d	7 m	7 m	7 m

	Reagent and Concentration Percentage									
	Phosphoric Acid				Glacial Acetic Acid				Chromic Acid	Silage Acids
Coating Type	Concen-tration	50	20	5	Concen-tration	50	20	5	20	
Polyamide-cured epoxy	51 1/4 h	14 d	71 d	7 m	1 1/2 h	6 h	47 h	6 d	21 h	13 d
Amine-adduct-cured epoxy	99 1/2 h	7 d	49 d	7 m	6 h	21 h	47 h	6 d	21 h	34 d
Amine-cured epoxy (low mol. wt.)	7 d	122d	122d	7 m	21 1/4 h	21 h	47 h	6 d	21 h	13 d
Amine-cured coal-tar epoxy	58 d	122 d	122 d	7 m	21 1/4 h	6 d	48 d	7 m	90 d	7 m
Polyamide-cured coal-tar epoxy	14 d	14 d	28 d	193 d	21 1/4 h	21 h	6 d	70 d	6 d	70 d

(continued)

TABLE 28-1—*Continued*

Maximum Wet Heat Resistance,† °F (°C)	Abrasion Resistance	Flexibility and Impact Resistance	Fresh and Salt-Water Resistance†	Gloss and Color Retention†	Theoretical Coverage,†† ft² gal (m²/L) at 1 mil (0.001″)
180 (80)	G	Pr	VG	Chalks, no erosion	P660 (16.2) F1125 (27.6)
180 (80)	G	Pr	E	Chalks, no erosion	P855 (21.0) F1125 (27.6)

* P, primer; I, intermediate coat (may be high-build type); F, final coat.
** Pr, poor; Fr, fair; G, good; VG, very good; E, excellent.
† N.A., not applicable; N.R., not recommended.
†† Theoretical coverage does not allow for surface texture or irregularities, spray losses, or other types of waste.

Table 28-2—*Continued*

	Reagent and Concentration Percentage									
	Phosphoric Acid				Glacial Acetic Acid				Chromic Acid	Silage Acids
Coating Type	Concentration	50	20	5	Concentration	50	20	5	20	
Chlorosulforated polyethylene	7 m	7 m	122 d	71 d	74 1/4 h	19 d	7 m	100 d	27 h	7 m
Vinyl	7 m	7 m	7 m	7 m	1 1/2 h	6 d	6 d	20 d	6 d	48 d
Chlorinated rubber	21 d	7 d	14d	21 d	6 h	21 h	47 h	46 1/2 h	21 h	13 d
Epoxy polyester	154 d	7 m	7 m	7 m	1 3/4 h	13 d	20 d	70 d	48 d	90 d

	Reagent and Concentration Percentage							
	Sodium Hydroxide			Ammonium Hydroxide				Potassium Hydroxide
Coating Type	50	20	5	28	20	10	5	50
Polyamide-cured epoxy	7 m	7 m	7 m	21 d	21 d	21 d	21 d	7 m
Amine-adduct-cured epoxy	7 m	7 m	7 m	7 d	4 d	14 d	14 d	7 m
Amine-cured epoxy (low mol. wt.)	7 m	7 m	7 m	21 d	21 d	21 d	21 d	7 m
Amine-cured coal-tar epoxy	7 m	7 m	7 m	7 m	7 m	7 m	7 m	7 m
Polyamide-cured coal-tar epoxy	7 m	7 m	7 m	7 m	7 m	7 m	7 m	7 m
Chlorosulforated polyethylene	7 m	7 m	7 m	7 m	7 m	7 m	7 m	7 m
Vinyl	7 m	49 d	7 m	58 d	71 d	109 d	109 d	7 m
Chlorinated rubber	7 m	7 m	7 m	7 d	14 d	14 d	14 d	94 d
Epoxy polyester	7 m	67 1/2 h	67 1/2 h	66 h	44 h	70 h	70 h	44 h

* Tests discontinued after 7 months. h, hours; d, days; m, months.

possible, depending upon the type of chemical agent used to promote the polymerization reaction and the resin. Catalytic agents are used to trigger a reaction between molecules of resin materials to create larger molecules. Reactive agents, when used, themselves combine with one or more resin molecules to form larger molecules. Whether the polymerization is caused by catalysis or by reaction, heat is given off as the molecules crosslink, and the heat accelerates curing. For some types of metal protection, catalyzed conversion coatings are unsurpassed.

Epoxy materials are the most frequently used catalytic conversion coating materials for metals exposed to corrosive chemical environments. Other types of catalyzed conversion coatings include alkyd-urea, alkyd-triazines, two-package urethanes, and polyester finishes.

Several types of resins are marketed in the form of two-liquid components that are intended to be mixed and then applied and allowed to chemically react to form a film. These components are carefully metered and mixed because the rate at which they react is determined by the mix ratio. The allowable time before use is called pot life and varies from minutes to as long as hours. Widely used representatives of this group are the two-component epoxies and polyurethanes.

ELEVATED TEMPERATURE CURING

With some materials, the polymerization reaction can be accelerated by raising the temperature of the applied coating. The accelerated, molecular crosslinking produced by heating is often called baking. In addition to the actual polymerization of a film, a certain amount of drying by evaporation takes place as solvents are liberated. In the case of oil-based and oleoresinous paints, oxidation (another chemical polymerization process) also takes place. Oxidation likewise occurs to some degree with all alkyds. In fact, with most paints, the curing process is a combination of evaporation, oxidation, and polymerization. Heating increases the rate of all three processes.

TIME-TEMPERATURE RELATIONSHIP

While many chemical coatings are more rapidly polymerized by heating, there is a wide range of times and temperatures that will cause resin curing. For example, a curing cycle of 10 minutes at 500° F (250° C) may be equivalent to a cycle of 20 minutes at 350° F (175° C) for a specific coating. The general shape of a time and temperature curing curve for one material is shown in Fig. 28-1. It is the temperature of the paint film that is critical, rather than the temperature of the oven.

A given coating film usually has a certain optimum temperature-time curing cycle for which it exhibits the best final properties. The temperature should be maintained as constant as possible throughout the cycle. Minimum and maximum times are important considerations in selecting production conditions.

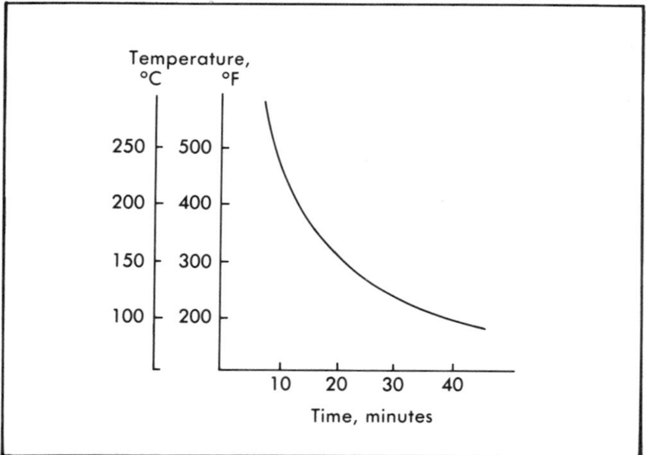

Fig. 28-1 Time-temperature curve for curing an organic film.

EFFECTS OF COATING MATERIAL

The nature of the coating material may also determine the nature of the heat application. Certain materials may be modified if subjected to the combustion products of organic fuels. If an organic fuel must be used, the heating unit should be an indirect type. With other materials, too much heat applied too rapidly may cause a dry surface skin to form over undried material beneath. The solvent in the undried material may then rupture the dried skin or cause blistering as it evaporates. Even if the finish is not damaged, the underlayer of material may never completely cure, thus making the finish highly susceptible to later damage.

SUBSTRATE EFFECTS

The nature of the substrate may also have a limiting effect on the range of temperatures or the type of heat that may be applied. Plastic substrates may soften and lose dimensional stability at higher temperatures (see Table 28-3). At still higher temperatures, sheet metal parts may warp and lose dimensional stability. Wood substrates may lose too much moisture or may even char if exposed to high temperatures. Coatings on cast materials should not be cured at temperatures higher than those to which the castings were originally exposed in a dryoff oven because pocketed gases trying to escape may cause the film to rupture.

CURING OVENS

Three means can be used to impart heat to the coatings applied to parts: conduction, convection, and radiation. Conduction is an important factor in the design of indirect-fired ovens and promotes the uniform temperature of metallic parts. Convection heating generally uses air or, infrequently, an inert gas as the medium for transferring heat from the heat source to the objects.

Convection (air movement) can be either natural, the result of temperature-caused density differences, or forced by fans. Most industrial convection ovens use a forced draft. Recirculating fans route the air or other gas through the heater, depositing convection currents on the work and recirculating all or part of the air back to the heater. Some or all of the used air may go through a heat exchanger where fresh air is warmed before entering the heater. Seals are provided at the entrance and exit openings of ovens to minimize heat losses.

Convection Ovens

Convection ovens are generally referred to as either low-velocity ovens, with air speeds to 4000 fpm (1220 m/min), or

TABLE 28-3
Temperature Sensitivity of Plastics

Plastics	Comments	Avoid the Following Solvents	Do Not Expose to Temperatures Above Approximately °F (°C)
Polyethylene	Treat surface before painting.		115 (45)
Vinyls		Ketones, esters, and aromatic hydrocarbons.	130 (55)
Cellulosics, acrylics		Ketones, esters, aromatics, and chlorocarbons.	150 (65)
Polystyrene		Aromatics and chlorocarbons.	
ABS (acrylonitrile butadiene styrene)		Ketones, esters, and some chlorocarbons.	165 (75)
Nylons	Good solvent resistance.		
Acetals	Roughen surface for painting; good solvent resistance.		185 (85)
Polypropylene	Good solvent resistance.		205 (95)
Polycarbonate	Sensitive to nearly all solvents.	Aromatics and chlorocarbons.	240 (115)

high-velocity ovens, with higher air speeds. The air stream is usually directed downward, entering the oven through overhead ducts.

In convection ovens, the heat transferred is proportional to the temperature drop between the supply ductwork and the recirculation ductwork. The magnitude of this temperature drop, or thermal head, determines the uniformity of the temperature in the oven. The higher the thermal head, the more difficult is the control problem. A low temperature drop requires heat to be transferred with a larger air volume, yielding better temperature uniformity in the oven and better heat transfer between the air and the parts.

With any type of oven, air must be exhausted from the oven enclosure and fresh air drawn in to remove the volatiles released from the coating film, to dilute the oven atmosphere, and to prevent the accumulation of an explosive concentration of vapors. In the case of an organic fuel oven, the fresh air is also required for the combustion of the fuel. A typical convection oven arrangement is illustrated in Fig. 28-2. Convection ovens are of two types, directly fired and indirectly fired.

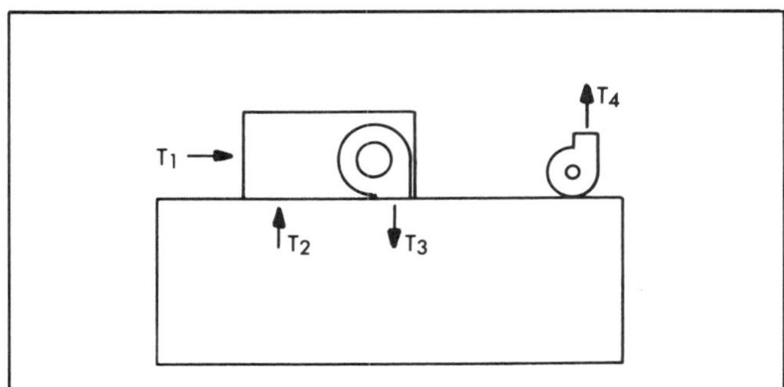

Fig. 28-2 A typical convection oven arrangement: T_1, intake air temperature; T_2, oven air temperature; T_3, recirculated air temperature; T_4, exhaust air temperature.

Direct-fired ovens. These ovens (see Fig. 28-3, view *a*) have the heating agent placed directly in the airstream. The heating agent can be a gas flame, an oil flame, or an electric resistance coil. These ovens are often used and are applicable whenever the products of combustion will not affect the coating or substrate.

Two modifications of direct-fired ovens are used to avoid contact of solvent fumes with the furnace flame. One modification is the baffle-type oven with recirculating system (see Fig. 28-3, view *b*). Another modification is a direct-fired oven with a heat exchanger system (view *c*).

Indirect-fired ovens. In these ovens, the heating device is not placed directly in the airstream (see Fig. 28-4). The flame or the electric coils are shielded from the airstream by an airtight housing to prevent any contamination or explosion hazard. Steam coils are essentially one type of indirect firing device.

In indirect firing, a heat exchanger introduces a degree of inefficiency into the circuit; however, a heat exchanger is required for some critical processes, even though it adds to the initial cost of the equipment. Steam coils are usually somewhat lower in initial cost than the other types of indirect heaters and thus should be considered when steam is available at the pressure needed for the temperatures required.

Zoned Ovens

Quality requirements for the coated products and/or economics often dictate the use of zoned ovens. In the first zone of such ovens, most of the paint solvent is evaporated. This arrangement facilitates vapor incineration if required. The second zone of the oven is heated to the required temperature for proper curing.

Explosion Limit Control

Drying ovens are usually operated at a lower explosion limit (LEL) of 25%; that is, a circulation of 10,000 ft^3 of air per gallon (75 m^3/L) of solvent evaporated. However, calculations show that operation at 50% LEL may bring heat savings as high as

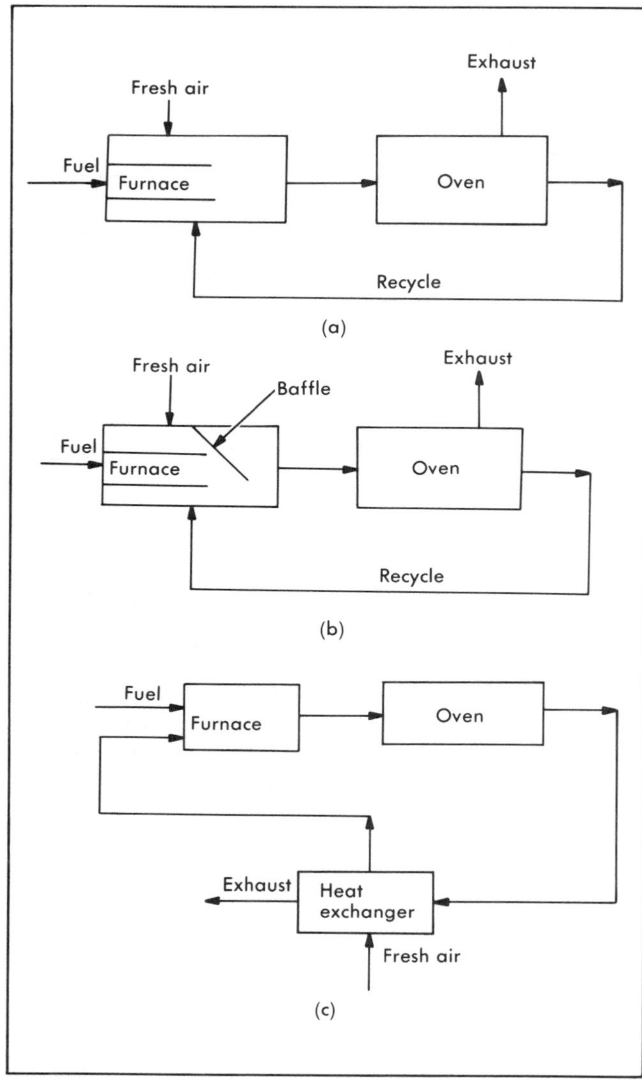

Fig. 28-3 Direct-fired ovens: (a) conventional, (b) baffle type, and (c) with heat exchanger.

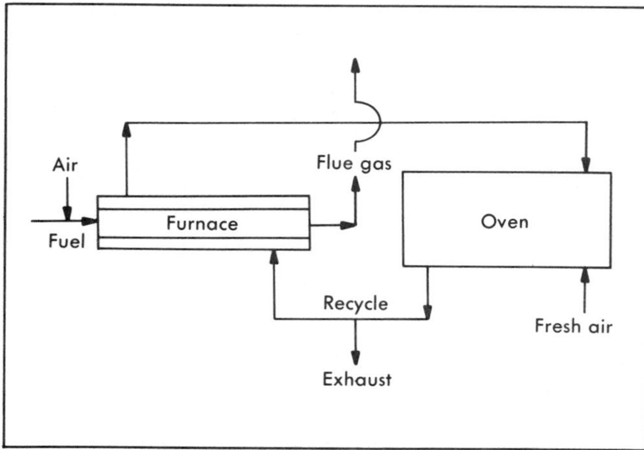

Fig. 28-4 Indirect-fired oven.

75%. The availability of continuous vapor-concentration indicators and controls makes it possible to operate ovens at 50% LEL.[1] The indicators are mounted at critical points and sound an alarm, control dampers, and turn off the flame when the preset danger level, usually 60% LEL, is reached.

HEATING REQUIREMENTS FOR CURING

In addition to bringing about the conversion of the resinous part of the paint, the curing process must also evaporate the solvent in the wet paint film. In the case of spray painting with solvent-borne paints, the wet paint film on products usually contains only 15 to 20% of the original solvent. In the case of waterborne paint, 50 to 60% of the solvent may still be in the paint film. Roller-applied coatings carry approximately 90% of the solvent into the curing area.

If heated air convection is used to accomplish the curing, enough heat has to be provided to heat the ventilating air, conveyor, hangers or other supports, the products, and the paint film. Sufficient heat is also needed to compensate for heat loss through the oven walls and through the entrance and exit openings of the oven. Infrared radiation curing primarily has to provide enough surface heat to dry or cure the paint coat sufficiently. However, some of the surface heat will be conducted into the ventilating air and the underlying products.

Heat Calculations

For conventional high-temperature heat polymerization systems, three basic components must be calculated to obtain the total heat required. The components are: the product load, the enclosure load, and the fresh-air load.

Product load. The first component of the total heat requirement is the heat required to bring the products plus hangers to temperature. For light-gage sheet metal parts, this amount of heat is usually the quantity required to bring the entire weight of the products to curing temperatures. However, for heavier metal thicknesses or for nonmetallic substrates, it may not be necessary to bring the entire mass to temperature because the product surfaces may be at the curing temperature range for a sufficiently long period to effect film conversion before the entire mass of the product is at uniform temperature.

The amount of heat required to bring the products to temperature can be calculated from the following equation:

$$Q = MCT \tag{1}$$

where:

Q = heat required, Btu/hr (J/hr)
M = mass of product introduced into oven per hour, lb (kg)
C = specific heat of the product material, Btu/lb/°F (J/kg/°C)
T = product temperature change required, °F (°C)

For metric usage, 1 Btu = 1055.056 J or 1.055 kJ, and 1 lb = 0.45 kg. The product mass, M, should include all associated hardware, such as hangers and conveyors.

Enclosure load. The second component in total heat requirement is the enclosure load, which is the heat loss through the oven enclosure. This heat loss can be calculated from the following equation:

$$Q = hAT \tag{2}$$

where:

Q = heat required, Btu/hr (J/hr)
h = coefficient of heat transfer through the enclosure walls and openings, Btu/hr/ft²/°F (J/hr/m²/°C)

A = the enclosure area, ft^2 (m^2)
T = temperature difference across the enclosure walls, °F (°C)

Fresh-air load. The third heat-load requirement is the heat required to bring incoming fresh air to oven temperature. This heat load can be calculated from the following equation:

$$Q = 1.08\ FT \qquad (3)$$

where:

Q = heat load, Btu/hr
F = fresh-air volume, cfm
T = thermal head (temperature difference), °F

For metric usage, equation (3) becomes:

$$Q = 20.12\ FT \qquad (4)$$

where:

Q = heat load, J/s
F = fresh-air volume, m^3/min
T = thermal head, °C

Size of Recirculating Fan

For a convection oven, the size of the recirculating fan required can be calculated from the following equation:

$$F = 1.08\ QT \qquad (5)$$

where:

F = fan volume requirement, cfm
Q = product load plus enclosure load, Btu/hr
T = thermal head, °F

For metric usage, equation (5) becomes:

$$F = 20.12\ QT \qquad (6)$$

where:

F = fan volume requirement, m^3/min
Q = product load plus enclosure load, J/s
T = thermal head, °C

Efficiency

The efficiency of any oven system may be calculated by dividing the product load by the total energy input. This efficiency is then a measure of the amount of total heat input that actually goes toward raising the temperature of the product. The efficiency of an oven is affected to a large degree by the amount of ventilation required for solvent vapor removal.

PROCESS SELECTION

Selection of the proper elevated temperature curing process for a specific application necessitates consideration of several requirements, including material, plant layout, fuel, and economic requirements. The first consideration in the selection of a curing process must be a consideration of the product itself. The substrate material as well as the coating material must be compatible with the characteristics of the elevated temperature curing technique. With respect to plant layout, oven enclosures can sometimes be made as weatherproof housings, thus allowing the ovens to be located outside, such as on the roof of the plant.

In the selection of a curing process, care must be taken to ensure that the proper type and quantities of fuel are available. Varying costs for fuels in different locations can determine which type of oven will be most economical. Curing of resins by baking is more expensive than simple air drying or forced drying at lower temperatures; however, it is sometimes capable of producing better final-film properties at an overall lower cost.

The initial cost of equipment for a given system is influenced by the type of insulated housing used, the type of heating equipment selected, the quantity of air circulated, the quantity of air exhausted, entry and exit sealing arrangements, and structural support schemes. Many of these factors will also affect the operating costs of the oven so that the lowest initial cost may not be the major consideration.

VAPOR INCINERATION

The solvent-laden exhaust from a curing oven may be burned in a suitably arranged incinerator. Such vapor incineration serves the dual purpose of complying with environmental regulations by lowering solvent emission and of generating heat by burning the normally wasted solvent vapors. Many plants, however, cannot use the waste heat; and without heat recovery, incineration costs are prohibitive. As a result, this method is not feasible for most paint lines, but it is being used with good results for high-volume coil coating lines.

Incinerators operate at approximately 1400° F (760° C), or at about 900° F (480° C) for catalytic systems. In all cases, the operating efficiency is about 90%; thus, solvent vapor emission is reduced to 10%.

Various designs of heat exchangers to be used with vapor incinerators are available using liquids or air as heat transfer media. Figure 28-5 is a schematic drawing of a vapor incinerator using a ceramic-wheel heat exchanger. Reference to the calculations provided with this illustration indicates that, whereas the operation of the bake oven requires 1.5 x 10^6 Btu/hr (1.58 x 10^6 kJ/hr), the incineration generates 0.7 x 10^6 Btu/hr (0.74 x 10^6 kJ/hr). Such recaptured energy can be used for producing hot water or plant heating, or for other purposes. Regenerative heat exchangers are available that use stoneware beds as the heat recovery medium. Temperature gradients across the beds are typically 1000° F (540° C).

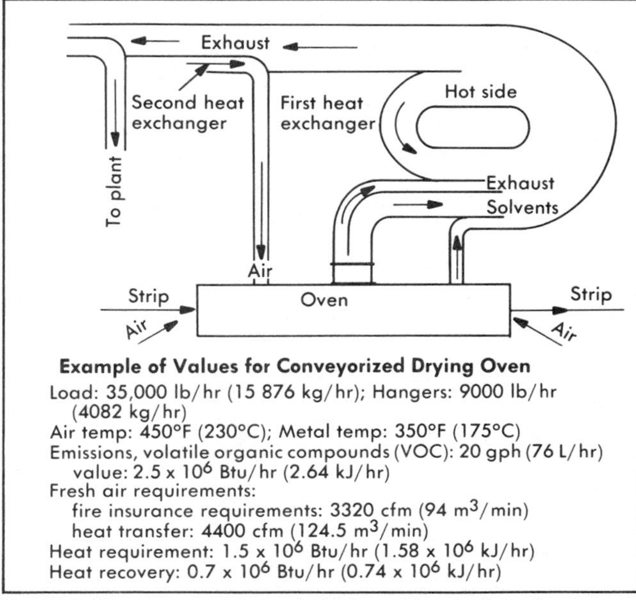

Exhaust
Second heat exchanger
First heat exchanger
Hot side
To plant
Exhaust
Solvents
Air
Strip
Oven
Strip
Air
Air
Example of Values for Conveyorized Drying Oven
Load: 35,000 lb/hr (15 876 kg/hr); Hangers: 9000 lb/hr (4082 kg/hr)
Air temp: 450°F (230°C); Metal temp: 350°F (175°C)
Emissions, volatile organic compounds (VOC): 20 gph (76 L/hr)
value: 2.5 x 10^6 Btu/hr (2.64 kJ/hr)
Fresh air requirements:
fire insurance requirements: 3320 cfm (94 m^3/min)
heat transfer: 4400 cfm (124.5 m^3/min)
Heat requirement: 1.5 x 10^6 Btu/hr (1.58 x 10^6 kJ/hr)
Heat recovery: 0.7 x 10^6 Btu/hr (0.74 x 10^6 kJ/hr)

Fig. 28-5 Heat generated by vapor incineration of paint solvents.

VAPOR CURING

A comparatively new technique for bringing about the polymerization of an applied coating is referred to as vapor curing. Amine-vapor-curable polyurethanes are the coating materials that are cured by this technique, but the method has had only limited application. Paints to be used are prepared by mixing a liquid polyol resin with a liquid polymeric, crosslinking isocyanate resin and all other ingredients required, like pigments. Two methods of curing these paints when they have been applied to product surfaces are: vapor permeation curing (VPC) and vapor injection curing (VIC).[2]

VAPOR PERMEATION CURING (VPC)

A curing line for the VPC coating system is designed specifically for a given application. It consists of four basic components: a tertiary amine gas generator, a curing chamber, a scrubber, and analytical instrumentation to monitor amine and oxygen levels. The process is amenable to almost all coating application methods, including gravure coating, curtain coating, conventional spray, electrostatic spray, and airless spray.

Inside the curing chamber, the coated substrate is subjected to a low concentration of a gaseous catalyst, such as triethylamine, in a carrier gas, such as nitrogen or carbon dioxide. Cure time and amine level depend largely upon the coating thickness. Thin films cure in 15 seconds at amine levels on the order of 2.5%. Such VPC coatings are now commercially applied and cured at line speeds as high as 125 fpm (38 m/min). Thicker films cure in 1 to 2 minutes at amine levels as low as 0.5%.

The catalyst environment is controlled using an amine analyzer. A circulation system and carefully designed containment system assure that a constant amine level is maintained in the chamber. Small amounts of amine that escape into the entry and exit containment chambers are vented to a scrubber. In many cases, depending upon the level of amine used and local regulations, a scrubber is not necessary. In some cases, particularly for thicker coatings such as automotive primers and topcoats, a 1 to 3 minute postflash at a low temperature is recommended to assist in the removal of residual solvents and to facilitate early handleability.

VAPOR INJECTION CURING (VIC)

A new vapor curing process has recently been developed that eliminates the need for a curing chamber. This process is called the vapor injection curing (VIC) coating process. It is particularly applicable to large automotive parts, such as reaction injection molding (RIM) fascias, sheet molding compound (SMC) hoods, grill panels, lift gates, and similar parts.

A schematic of the VIC process is shown in Fig. 28-6. The vaporous amine is generated in essentially the same manner as in vapor permeation curing. In the case of VIC coatings, however, the amine can be generated either in dry air or in an inert gas, such as nitrogen or carbon dioxide. This catalyst-containing carrier gas then serves essentially as a replacement for normal air, either as atomizing air in the case of conventional air spray or air-atomized electrostatic spray, or as the shaping air in such spray techniques as mechanically atomized electrostatic spray.

The VIC process is particularly applicable to heat-sensitive substrates, such as many plastics. It will easily cure any contour, such as three-dimensional plastic parts and coatings with high levels of pigmentation. The process is custom tailored to the specific needs of the applicator.

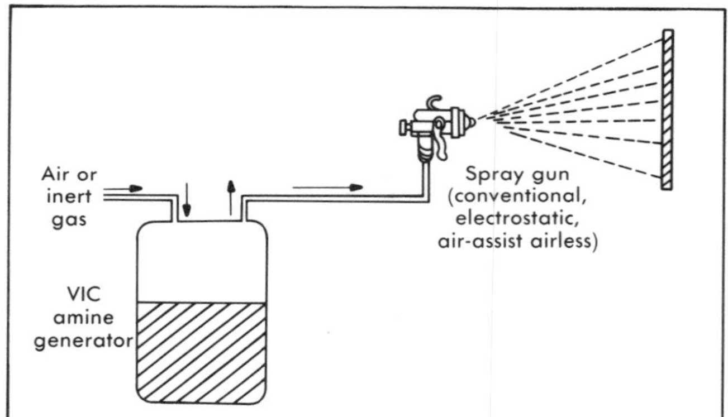

Fig. 28-6 Schematic drawing of the vapor injection curing (VIC) process. (*Ashland Chemical Co.*)

RADIATION CURING

Radiation curing is the process by which specifically formulated inks, coatings, and adhesives are dried and converted to a useable coating by means of high-energy electrons or short wavelengths of light. Several ranges of radiating energy are widely used for the drying and curing of paint. These include microwaves and infrared radiation having waves longer than visible light, ultraviolet radiation of shorter wavelength than visible light, and the still shorter waves of electron beams.

PROCESS ADVANTAGES AND LIMITATIONS

A major advantage of the radiation curing process is the short time required for curing. All the other curing processes previously discussed require curing times several orders of magnitude greater than that required by radiation curing. Short curing times are economically favorable in several ways. For example, short curing times allow reductions of in-process inventory and processing space, and they also allow simplified and improved line control.

The radiation curing process makes possible the maintenance of product dimensional stability, better quality control of finishes, lower utility costs, immediate startup and shutdown, and easier product handling. Some coatings contain little or no volatile organic components (VOC) and therefore do not contribute to air pollution. Solventless monomers can be used for zero VOC emission.

The radiation curing process is not always applicable to all substrate shapes. Radiation usually follows a straight line from the source to the part surface. As a result, reentrant angles and very deep recesses may cause insufficient curing. However, when high-speed electron curing is used, electron scattering can be depended on to provide a sufficiently uniform dosage to cure most injection molded or stamped metal contours encountered in production. The low impact strength of parts made from some plastics is another limitation to the use of radiation curing.

PROCESS PRINCIPLES

Radiation curing employs the physical effects of ionized particles and radiation (high-speed electrons, charged ions, gamma rays, X rays, and ultraviolet light) on polymers and monomers to generate free radicals that promote crosslinking in a properly formulated organic coating. From an industrial standpoint, high-speed electrons in the range of 300 kV and ultraviolet light are the only practical sources of radiation available for curing thin, liquid organic coatings when intensity, controllability, and economics are considered.

APPLICATIONS

For a coating to be radiation curable and also have commercially acceptable properties for the intended product, the unsaturation of the material should approximate one-half to three double bonds per 1000 molecular weight. Below this range, the coating cannot be suitably cured; above it, the film tends to be brittle. However, certain applications requiring high hardness can employ coatings with as much as five double bonds per 1000 molecular weight.

Radiation-curable coatings are formulated to be applied by one of the common industrial coating methods. All the standard pigments and fillers may be used to provide specific properties to the coating. Polymers that can be formulated to cure by crosslinking, such as polyesters, acrylics, silicones, and various combinations and modifications of these, are suitably cured by radiation. These resins are combined, when appropriate, with various monomers to make a complete coating binder.

Many coatings require the protection of an inert gas such as argon during the curing step to minimize oxygen inhibition and provide maximum exterior durability. The inert gas atmosphere also prevents ozone generation by the electron beams. Only polymers that cure by a free-radical mechanism are radiation curable.

With few exceptions, any substrate that is industrially coated and cured is amenable to having its coating cured by the radiation curing process. An exception is flint glass, which acquires a brown discoloration upon being irradiated; if necessary, however, the discoloration can be bleached out by moderate heating. Coating adhesion on organic materials is enhanced with radiation curing because of the promotion of chemical bonding between the substrate and the coating. Typical products for which the finish topcoat is radiation cured are exterior-exposure fiberboard, paper-coated plywood, interior particle board, aluminum residential siding, galvanized industrial siding, and coated coil products. Injection molded ABS, polypropylene automotive parts, and other plastic products are being radiation cured; however, many plastic parts become too embrittled to be radiation cured.

MICROWAVE CURING

Microwave curing is a relatively new heating process that uses wavelengths much longer than those in the visible light spectrum. Microwave curing offers the advantage of causing polymerization in a relatively fast time, with low power input. At present, microwave curing has its greatest application for finishes that mainly require evaporation for curing and for coating nonmetallic substrates.

INFRARED CURING

Infrared ovens use the radiant energy emitted from high-temperature sources in the form of waves above the wavelength of visible light; these waves are absorbed by and raise the temperature of the part and the coating film. Infrared waves may be generated by a variety of sources including quartz tubes, lamps, metal tubes, panels, wire-heated ceramic shapes, and gas-fired, porous ceramic plates through which gas travels and burns on the surface.

Not all infrared waves generated by the various sources are of the same wavelength; wavelength varies with the temperature of the source. At higher temperatures, more waves of shorter length are generated. The shorter wavelengths are reflected rather than absorbed by light-colored, bright surfaces. Thus, the degree of efficiency of infrared curing will vary with the combination of products heated, the paint colors, and the type of source used.

Exhaust temperatures from infrared ovens are comparatively low because air is relatively transparent to infrared energy. The exhaust temperature from an oven that is drying parts at 400° F (205° C) may be as low as 130° F (55° C). Under proper conditions, the infrared heating technique offers faster curing times than convection heating. The initial cost may also be somewhat lower than that of convection equipment. Infrared heating lends itself to rapid off-and-on operation, as well as adjustment, which is advantageous for some applications. Infrared equipment consists of two basic types: reflector and flat panel.

Reflector Heaters

Reflector-type heaters use a bulb, rod, tube, cone, or other means of heating. They depend upon a reflector to collect and concentrate the energy or to direct the energy towards the product. Reflector heaters have the disadvantage of nonuniformity, particularly when the heating source is close to the product. They also tend to be discolored by organic vapors, and the reflectors must be cleaned frequently. The reflector-type heaters have no thermal time lag; they are instantaneously on or off. Temperature control can only be accomplished by voltage regulation.

Flat-Panel Heaters

The flat-panel heater offers uniformity, but cannot concentrate or direct the infrared energy toward the product. The energy is given off in straight lines, but not necessarily perpendicular to the plane of the emitter. When the panel heater is close to the product, it is quite efficient. Some flat-panel heaters have a grid, like an egg crate, in front of the heat emitter that offers advantages in heat distribution.[3]

Infrared ovens, like other ovens, can be divided into successive zones to supply different amounts of energy. In addition, they can be zoned vertically to provide temperature control for various sections to ensure uniformity of product heating. Panel-type infrared units have a thermal time lag and can be cycled on and off without appreciable fluctuation of the source temperature.

ULTRAVIOLET (UV) CURING

Curing accomplished by using the ultraviolet portion of the electromagnetic spectrum is called UV curing. The use of

RADIATION CURING

ultraviolet radiation for curing is limited to clear or dye-containing coatings, or thin, 0.1 to 0.5 mil (0.0025 to 0.013 mm), pigmented coatings. Ultraviolet curing is particularly suited for clear-filler base coats on porous substrates, such as particle board and fiberboard. Curing times range from 10 seconds to several minutes, depending upon the ultraviolet source and the coating formulation. The coatings often have a catalyst added to accelerate curing.

Nature of Ultraviolet Light

Although the UV range of radiation extends from 100 to 400 nm (3.9 to 15.7 μin.), only the range from 200 to 400 nm (7.87 to 15.7 μin.) is effective in the UV curing of coatings. Ultraviolet wavelengths below 200 nm are absorbed by the gaseous molecules in air, resulting in ozone formation, and are therefore used only in a vacuum.

When passing through a nonabsorptive medium, such as a gas, liquid, or solid, a beam of UV light or photons will suffer a small loss of intensity. First, refractive index differences (variations in the speed of light) between the medium and its surroundings will result in light reflection at the phase boundaries. Second, inhomogeneities (mixtures) or thermal fluctuations in the bulk medium produce a small loss of power through light scattering. The effect of passing through an absorptive medium will be discussed later in this section.

Source of UV Light

The principal source of actinic radiation for processing UV-curable coatings is the medium-pressure mercury lamp. At pressures between one and two atmospheres, 14.7 and 29.4 psi (101 and 203 kPa), mercury vapor undergoes multiple electronic excitations. This excitation leads to the emission of discrete wavelengths of energy ranging from 185 to 1370 mm (7.3 to 54 μin.).

Excitation of mercury can be induced either through the use of microwave energy or by the more conventional method of applying a voltage across two electrodes. High-intensity lamps, such as pulsed xenon and other specialized light sources, are also available for excitation; they are designed to concentrate light output at preselected wavelengths.

Maximum curing efficiency is attained by concentrating the light at the substrate surface through focused lamp reflectors. This arrangement is essential when rapid line speed is desired because the rate of polymerization is proportional to the square root of the light intensity. With an unfocused light source or when curing beyond the reflector's focal point, the distance between substrate and light source is also important because the luminous flux or the rate of radiation is inversely proportional to the square of the distance from the light source; this relationship is derived for a point light source.

The special distribution of energy emitted by mercury lamp sources is available from the lamp manufacturer; each wavelength of energy is reported in watts. This data can then be used to optimize the photo response of a UV-curable coating by matching the absorption spectrum of a photoinitiator with the wavelength(s) of maximum photon output of the light source.

Some Aspects of UV Light Absorption

Absorption of UV light (photon) by an organic molecule results in raising the energy content of the molecule from its ground-state condition to an electronically excited state. The magnitude of this excitation energy for a single molecule is calculated by the following equation:

$$E_2 - E_1 = hF \tag{7}$$

where:

E_2 = final energy state, erg/photon
E_1 = initial energy state, erg/photon
h = Planck's constant (6.62×10^{-27} erg · s/photon)
F = frequency of absorbed radiation, Hz

Once in the excited state, the molecule has three general routes by which the excitation energy can be dissipated (see Fig. 28-7). It is the route that leads to chemical reaction products, route C, that provides a means for the UV curing of coatings.

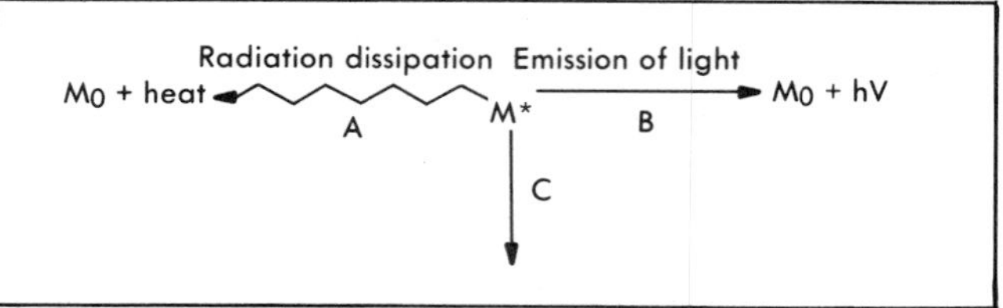

Fig. 28-7 Three routes for energy loss. Route C leads to the formation of chemical reaction products: M_0 = ground state of molecule; M* = excited state of molecule; hV = energy (erg/photon).

The process of converting light energy into chemical energy can be divided into the following three steps:

1. The absorption of light, producing an electronically excited state.
2. A photochemical process that leads to an electronically reactive excited state.
3. The initiation of a chemical reaction by the excited state produced in step 2.

Aromatic ketones containing the benzoyl group have been found to be most effective for initiating free-radical polymerization after the absorption of light energy. These compounds (chromophores) have two principal regions of absorption: 250 and 330 nm (9.84 and 13 μin.).

Polymerization Techniques

Three polymerization techniques are widely used for UV-curable coatings: free-radical addition polymerization, free-radical step-growth photopolymerization, and cationic chain-growth photopolymerization. The term *photoinitiator* as it relates to UV-curable compositions generally refers to the component(s) responsible for absorbing actinic radiation and generating a polymerization-initiating species through a photochemical process.

Free-radical addition polymerization. Of the vinyl-type monomers that undergo radical-addition-type polymerization, acrylates have been found to be best suited for UV-curable compositions. Not only do acrylates provide commercially acceptable cure speeds, but final-film properties can be enhanced by incorporating into these formulations urethane and other "toughening" components that have attached reactive acrylate groups. Methacrylates are slower curing than acrylates and therefore find little use when rapid curing is essential.

Free-radical step-growth photopolymerization. In the presence of a radical initiator, mercaptans add to olefins in a step-growth polymerization process. Vinyl ethers and the allyl groups are the preferred olefins for mercapto-addition polymerization. As with acrylates, urethane and other "toughening" structures can also be incorporated into these formulations.

Cationic chain-growth photopolymerization. Photocurable coatings based on the cationic polymerization process are not used in industry as much as the free-radical-based photocurable compositions. Some of the drawbacks to this technology are as follows:

1. A high concentration of ambient moisture can retard polymerization.
2. Inadvertant contamination of the coating with alkali also retards polymerization.
3. The urethane structure so widely incorporated in UV-curable, free-radical polymerization compositions (to toughen film properties) has the effect of retarding cationic polymerization.

The most successful cationic photopolymerizable systems are epoxy-based compositions with an aryldiazonium salt or a triarylsulfonium salt photocatalyst.

ELECTRON BEAM (EB) CURING

A radiation curing installation using electron beams (EB) generally consists of integrated line operations, including substrate preparation, coating, curing, further finishing if necessary, and preparation for distribution. The entire operation can be automated, with the required cure dose electronically coordinated with line speed.

Electron-accelerator positioning and part indexing may be used to take maximum advantage of process parameters. External shielding is required to protect against the X rays produced when high-speed electrons impinge on the substrate. Maximum shielding compactness is obtainable with lead and steel, but concrete vault construction with labyrinth accesses may also be used.

Costs are not sufficiently established by experience to determine which applications for radiation curing with electron beams will be economically competitive. It is clear, however, that high-speed, high-volume production installations to cure large panels, continuous sheet or strip, and relatively simple contoured parts are the first to take economic advantage of the process. Curing with EB is also useful for substrates that are temperature sensitive and for special applications in which the high cost of the equipment can be justified.

Principles of Operation

An accelerated electron can be described as a negatively charged energetic particle capable of directly breaking chemical bonds (ionizing molecules) contained in an irradiated medium. As one of the ways accelerated electrons differ from the highly energetic forms of electromagnetic (EM) radiation (X rays and gamma rays), EM radiation does not directly ionize matter upon impact. Instead, X rays and gamma rays are limited to producing ions by a secondary mechanism in which electrons are expelled from molecules upon impact. Also, ionizing electrons, unlike the electrons of EM radiation, have a mass in addition to a negative charge.

Because of the negative charge of accelerated electrons, they do not penetrate matter to the same depth as uncharged electrons of EM radiation. The kinetic energy of an electron is usually expressed in electron volts (eV). The 1.17 and 1.33 MeV gamma radiation emitted by Co^{60} penetrates 17″ (43.2 cm) of water or 1.6″ (4.1 cm) of lead while losing only 10% of its initial intensity. In contrast, radiating the same thickness of water or lead with charged electrons at 2 MeV results in complete absorption of the electrons.

The ionizing and chemical changes caused by absorption of an accelerated electron, as well as the penetrating properties, depend upon the kinetic energy of the individual electron. Modern accelerators are capable of producing electrons with values as high as a few hundred billion eV (BeV). For electron beam processing of a typical organic coating or pigmented film, electrons in the 150 keV to 300 keV range have been found to be most cost effective. Penetration through significantly thick substrates, as encountered in vulcanizing rubber, requires beam voltages of over 300 keV.

Sources of High-Energy Electrons

Negatively charged electrons can be obtained either by the natural emission of radioactive isotopes, such as Be^9, H^2, and C^{12}, or they may be artificially produced by electron beam accelerators. The use of natural sources of electrons (beta rays) is hampered by the low concentration of emitted electrons and the associated hazards in using a radioactive energy source. Artificial generation of electrons through the use of sophisticated electronics has proven highly desirable as very intense beams of electrons, with electron energies ranging up to several billion eV, are possible.

An EB accelerator may be thought of as a vacuum triode in which low-velocity electrons are produced at the surface of a cathode or a filament and then accelerated to the anode by application of a high negative voltage. A focusing arrangement (electric or magnetic field) is used to ensure maximum concentration of electrons at the intended surface. Individual electrons emitted from the cathode carry the full energy of the accelerating potential as they strike the anode. Electron density is controlled by the applied amperage.

The process of producing accelerated electrons is carried out in an evacuated chamber. To maintain the required vacuum while allowing for the emergence of electrons, a thin metallic "window" is used. This window is designed to minimize electron energy losses through absorption, while preventing the flow of ambient air into the chamber. The emerging electrons are directed into a curing compartment designed for maintaining an inert (oxygen-free) atmosphere. The radiation process is completely free of radioactivity; however, some X-radiation is formed as accelerated electrons collide with heavy atoms (metal). The use of lead shielding ensures the safety of operating personnel from exposure to harmful radiation.

In passing through the window of an EB accelerator, an electron experiences an approximate 10% loss in energy. Dense windows and low operating voltages tend to increase electron energy losses. The beam power of an electron source is calculated as the product of the accelerating voltage and the cathode current. Dosimetry measurements are used to determine the actual beam power striking the medium.

The linear filament type of electron beam accelerator appears to be emerging as the equipment of choice for generating the low-energy beams (150 to 300 keV) required to cure coatings and adhesives. This equipment is capable of producing a continuous stream of electrons across the length of the filament at intensities (beam current) high enough to achieve commercial line speeds that approach 1000 fpm (305 m/min). In equipment designed with one filament to cure a wide line, the filament is positioned across the direction of substrate travel. Additional filaments may be added in tandem, mounted across

BIBLIOGRAPHY

the width of the line. Multifilament devices are also available with the filaments mounted side by side, oriented parallel to the direction of substrate travel.

For electron voltages over 300 keV, the scanning electrode accelerator becomes an attractive alternative to the filament accelerator. In this more expensive unit, electrons are generated in a narrow beam and rapidly scanned (deflected) across the direction of substrate travel.

Electron Absorption

When an organic coating is exposed to a beam of accelerated electrons, electron kinetic energy is transferred to the absorbing medium. This encounter between energetic electrons and matter sets up a coulombic field that results in several pathways for producing radicals capable of initiating polymerization. The exact mechanism of energy transfer and the resulting energized molecular species produced are highly complex. Mass spectroscopy data has been used as a model to elucidate the condensed phase processes such as occur in a radiation-curable coating.

The preponderance of chemical evidence indicates that most of the chemical changes occurring under the influence of accelerated electrons are caused by free radicals. Ionic species are thought to play a minor role in effecting chemical reactions because they undergo rapid recombination reactions.

Chemistry of EB-Curable Coatings

Electron beam curing of coatings is effective for radical-type polymerization processes, but not with epoxy-based formulations that typically cure with UV light. Also, unlike UV processing in which curing can be carried out in air, EB curing requires an inert atmosphere (free of oxygen). This difference in curing requirements stems from the way energy is transferred in each of the two curing methods.

Under UV curing conditions, a disproportionately high amount of UV light is absorbed at the coating's surface. This effect ensures the formation of a high enough concentration of initiating radicals at the surface to overcome oxygen's retarding effect on polymerization. In EB curing, the concentration of initiating radicals tends to be more uniformly dispersed throughout the bulk of the coating. Once formed by EB radiation, the initiating radicals can be expected to combine with reactive resin monomers.

References

1. Frank Melette, "The 50% LEL Theory is Verified," *Industrial Finishing* (July 1977), pp. 10-12.
2. Michael G. Cobb, *The Development of Vapor Permeation Cure (VPC) Coatings for RIM and SMC Substrates*, presentation at AFP/SME clinic "Finishing Automotive Plastics," Dearborn, MI, October 23-25, 1984.
3. Harold D. Wells, *Infrared Versus Convective Ovens*, AFP/SME Technical Paper FC82-307 (Dearborn, MI: Society of Manufacturing Engineers, 1982).

Bibliography

Baer, George F. *Safety and Handling Considerations of UV Equipment*. AFP/SME Technical Paper FC84-986. Dearborn, MI: Society of Manufacturing Engineers, 1984.

___________. *UV Curing—An Overview*. AFP/SME Technical Paper FC83-248. Dearborn, MI: Society of Manufacturing Engineers, 1983.

Bayer, W. G. *Radiation Chemistry—An Overview*. AFP/SME Technical Paper FC79-203. Dearborn, MI: Society of Manufacturing Engineers, 1983.

Best, W. H. *Accelerated Drying Principles of Coatings*. AFP/SME Technical Paper FC79-478. Dearborn, MI: Society of Manufacturing Engineers, 1983.

Brann, Bill L., and Reed, J. Michael. *The Effects of Curing on the Physical Properties of UV Films*. AFP/SME Technical Paper FC84-998. Dearborn, MI: Society of Manufacturing Engineers, 1984.

Eeg, Marlene. *High-Intensity Infrared-Oven Selection and Application*. AFP/SME Technical Paper FC82-298. Dearborn, MI: Society of Manufacturing Engineers, 1983.

Karmann, Dr. Werner. *Radiation Curing Equipment*. AFP/SME Technical Paper FC83-269. Dearborn, MI: Society of Manufacturing Engineers, 1983.

Klein, Alan F. *Developments in Electron Beam Curing of Magnetic Media*. AFP/SME Technical Paper FC83-244. Dearborn, MI: Society of Manufacturing Engineers, 1983.

Knight, R. E. *Ultraviolet Equipment '83*. AFP/SME Technical Paper FC83-276. Dearborn, MI: Society of Manufacturing Engineers, 1983.

Koleske, J. V., and Mazzariello, R. G. *Cationic, Ultraviolet-Light Cure Technology*. AFP/SME Technical Paper FC84-976. Dearborn, MI: Society of Manufacturing Engineers, 1984.

Le, D. D. *A Method for Optimizing UV—Curing Conditions*. AFP/SME Technical Paper FC84-997. Dearborn, MI: Society of Manufacturing Engineers, 1984.

Nablo, Sam V. *Principles of Electron Beam Processing Equipment*. AFP/SME Technical Paper FC82-310. Dearborn, MI: Society of Manufacturing Engineers, 1983.

Radcure '84 Conference Proceedings, sponsored by AFP/SME, September 10-13, 1984, Atlanta, GA. Dearborn, MI: Society of Manufacturing Engineers, 1984.

Ross, Don, and Halvorson, Greg. *Drying and Curing Coatings with High Density Infrared*. AFP/SME Technical Paper FC80-539. Dearborn, MI: Society of Manufacturing Engineers, 1983.

Saraiya, S., and Hashimoto, Kelichi. *Overview of Radiation Curing Chemistry*. AFP/SME Technical Paper FC80-521. Dearborn, MI: Society of Manufacturing Engineers, 1980.

Tripp III, E. P. *Electron Beam Curing is Less Expensive Than You Think*. AFP/SME Technical Paper FC82-305. Dearborn, MI: Society of Manufacturing Engineers, 1983.

COATING SYSTEMS

Numerous organic coating systems are being used. They vary in size, complexity, and cost, from simple maintenance and touch-up facilities to large computer-controlled installations. Selection of a particular system for a specific application depends primarily upon the products to be coated, the coating materials, and production requirements. A comprehensive economic analysis is required before selecting a system. Adaptability and compatibility of the system with existing plant facilities must also be considered.

CHAPTER CONTENTS:

ECONOMIC ANALYSIS

A comprehensive economic analysis entails consideration of the cost of the coating materials, the comparative costs of various coating processes, energy costs, and environmental factors.

COST OF COATING MATERIALS

Coating materials are usually sold by weight, pounds (lb) or kilograms (kg); and liquid materials are often sold by volume, gallons (gal) or liters (L). However, neither the weight nor the volume of the paint are directly related to the cost of the applied coating.

Finished coatings are films of specified thicknesses applied to designated surface areas of products. Since *area* times *thickness* equals *volume*, the cost of the coating material depends only upon the price paid per volume of nonvolatile (NV) materials contained in the weight or volume of paint purchased.

To compare the costs of different coating materials, the following three factors must be considered:

1. The cost per volume of NV materials.
2. The efficiency of the paint application method, usually expressed as percent transfer efficiency (te).
3. The total applied cost, including all applicable costs such as cleanup, waste disposal, labor, energy, building, and amortization.

COST PER VOLUME OF NV MATERIAL

The data required to determine the cost per volume of NV material in a paint include the volume and weight of the nonvolatiles, of the volatile organic compounds (VOCs), and of the water and/or compliance solvents. The paint user can obtain these data from the vendor or by means of analysis.

It may be convenient to compile the data on a form, such as the one shown in Table 29-1. Crosschecks can be made by adding some of the vertically and horizontally arranged data. This same data can also be used for computing the compatibility of the paint with environmental regulations, which are discussed in Chapter 30.

As an example of paint costs, a certain liquid coating with 37.3% volume of NV materials costs \$12.55/gal (\$3.32/L). The paint is reduced to spray viscosity by the addition of 1 qt (0.95 L) of thinner per gallon (3.785 L) of paint, with the thinner costing \$1.56/gal (\$0.41/L). Thus, the cost of the paint with thinner is \$12.94 for five quarts or \$10.35/gal (\$2.73/L). With the addition of the thinner, the NV materials content within paint is reduced to 29.84% (see Table 29-2). Dividing \$10.35 by 0.2984 results in an NV cost of \$34.68/gal (\$9.16/L).

The cost per unit volume of paint solids (NV material) can be translated into cost per square foot or square meter for a given film thickness by use of the following equation:

$$\$/\text{ft}^2(\$/\text{m}^2) = \frac{\$/\text{gal (L) of NV x } t}{K} \qquad (1)$$

where:

t = dry film thickness, mil (μm)

$$K = \frac{1604 \text{ ft}^2 \text{ x mil}}{\text{gal of NV}} \left(\frac{1000 \text{ m}^2 \text{ x } \mu\text{m}}{\text{L of NV}}\right)$$

Example: The cost of paint per square foot with a film thickness of 1.5 mil, using a paint costing \$34.68/gal of NV, is:

$$\$/\text{ft}^2 = \frac{34.68 \text{ x } 1.5}{1604} = \$0.0324$$

TRANSFER EFFICIENCY VERSUS PAINT COST

A transfer efficiency (te) of 50% means that only one gallon of paint reaches the product for each two gallons used; the other gallon of paint is wasted (see Fig. 29-1). Transfer efficiencies for various application methods are presented in Table 29-3. In general terms, the waste factor is 100% divided by the percent te. Cost per square foot can be calculated from the following equation:

$$\$/\text{ft}^2 = \frac{100\%}{\%\text{ te}} \text{ x } \frac{\$/\text{gal of NV x } t}{K} \qquad (2)$$

where:

t and K are the same as for equation (1).

TOTAL APPLIED COST OF COATING

The total applied cost of a specific coating varies with different installations, depending primarily

This chapter contributed by: George E. F. Brewer, Coating Consultants.

CHAPTER 29

ECONOMIC ANALYSIS

TABLE 29-1
Data Needed for Cost and Environmental Calculations for Various Paints

Paint Component*	Volume/gal (L), %	Coating, lb/gal (kg/L)	Paint Density, lb/gal (kg/L)	Weight, %
NV				
VOC				
Water or compliance solvent				
	100			100

* NV = nonvolatile materials
VOC = volatile organic compounds

TABLE 29-2
Cost and Environmental Calculations for a Specific Paint

Paint Component*	Volume/gal (L), %	Coating, lb/gal (kg/L)	Paint Density, lb/gal (kg/L)	Weight, %
NV	29.84	3.43 (0.4)	11.33 (1.4)	40.2
VOC	70.16	4.73 (0.6)	6.96 (0.8)	59.8
Water or compliance solvent	None	None	None	None
	100	8.16 (1.0)	18.29 (2.2)	100

* NV = nonvolatile materials
VOC = volatile organic compounds

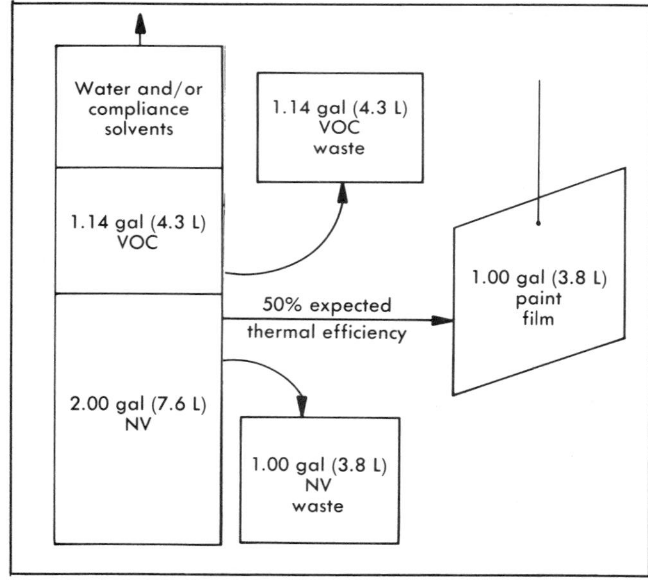

Fig. 29-1 Paint wasted as the result of a 50% transfer efficiency (te).

TABLE 29-3
Expected Transfer Efficiencies for Various Application Methods

Organic Coating Application Method	Expected Transfer Efficiency (te), %	Source of Data*
Air atomized, conventional	43	1
	50	2
	30-60	8
Air atomized, electrostatic	87	1
	68-87	9
Centrifugally atomized, electrostatic	93	1
	85-95	9
Pressure atomized, conventional	65-70	3
Pressure atomized, electrostatic	85-90	3
Dip, flow, and curtain coat	75-90	4
Coil and roll	90-98	4
	96-98	7
Electrocoat	90-96	4
	99	2
Powder coat	50-80	5
	98	2
	90-99	6

* 1. E P. Miller, SME Technical Paper FC73-553.
2. J. A. Antonelli, SME Technical Paper FC74-654.
3. W. H. Cobbs, Jr., Nordson Corp., direct communication.
4. F. Scofield, Wapora, Inc., EPA Contract 68-01-2656.
5. S. B. Levinson, *Journal of Paint Technology,* July 1972, pp. 35-36.
6. T. W. Seitz, Sherwin-Williams Co., direct communication—newer reuse methods.
7. M. Wismer, PPG Industries, direct communication.
8. J. A. Antonelli, Du Pont Co., direct communication—depending upon equipment and shape of products.
9. E. P. Miller, Ransburg Co., direct communication—depending upon product being coated.

upon the investment in coating equipment, the production rates, and the labor costs. A comparatively simple method of determining the total applied cost is to complete a form such as the one shown in Table 29-4. The form can be used to compare both the cost of paints from different suppliers and the cost of application by different methods.

RELATIVE COSTS OF VARIOUS COATING PROCESSES

The variety of production volumes and coating requirements in different plants makes it difficult, if not impossible, to realistically compare the relative costs of various coating processes. Manufacturers must therefore study their own specific applications.

Coating literature contains a number of good studies with respect to the costs of different coating methods, but inflation makes the dollar figures too time dependent to be of value.

However, some relative figures based on several studies are presented in Table 29-5. The comparative values in this tabulation are based on the assumption that the value of conventional solvent spray with air atomization is 100.

ENERGY CONSUMPTION

Variations in the size, weight, and shape of products, as well as differences in installations and other factors, do not allow absolute figures with respect to energy consumption for various coating processes. Valid comparisons for various paints and processes, however, are presented in Table 29-6, based on relative percentage numbers. In another study[2], assuming the energy consumption to be 100% for autodeposition, the value for electrocoats was found to be 154%; and for waterborne spray paints, 157%.

TABLE 29-4
Total Annual Coating Cost

Paint Variables	Supplier or Coating Method A	B	C	D
Coating cost, nonvolatiles, \$/gal (\$/L)				
Dry-film thickness, mil (μm)				
Transfer efficiency, %				
Applied cost, \$/ft² (\$/m²)				
Coated/year, ft² (m²)				
Annual cost of coating materials, \$				
Other Costs				
Material costs				
Labor cleanup costs, including disposal				
Energy costs				
Maintenance costs				
Depreciation costs				
Total annual costs, \$				
Total applied costs, \$/ft² (\$/m²)				

TABLE 29-6
Energy Consumptions for Various Paints and Coating Processes[1]

Requirement	Paint: Solvent	Paint: Urethane	Process: Powder Coat	Process: Electrocoat
	Energy Consumption, Btu/hr x 10^3 (kJ/hr x 10^3)			
Total electrical	920 (970)	920 (970)	850 (895)	990 (1045)
Power washer	5500 (5800)	5500 (5800)	5500 (5800)	5500 (5800)
Dryoff	540 (570)	540 (570)	540 (570)	
Air makeup	520 (550)	520 (550)	520 (550)	
Bake oven	1080 (1140)	390 (410)	780 (825)	785 (830)
Heat savings	250 (265)	20 (20)	20 (20)	20 (20)
Net energy	8310 (8765)	7850 (8280)	8170 (8620)	7255 (7655)
Relative percentage	100	94	98	87

TABLE 29-5
Relative Cost of Various Coating Methods, Including Materials, Energy, Labor, Maintenance, and Depreciation

	Coating Method							
Study*	Air Atomization	Electrostatic	Two-Component Spray	Waterborne	High Solids	Powder Coating	Electrocoat	Dip Coat
1	100	---	---	---	---	---	64	80
2	---	---	---	---	---	---	85	100
3	100	---	---	---	---	---	68	---
4	100	81	---	---	---	---	64	78
5**	100	---	117	---	---	108	52	---
6	100	---	---	113	99	87	---	---
7	100	---	---	113	102	92	---	---

* 1. Frangen, K.H.: *Flaeche & Farbe,* No. 3, pp. 9-12 (1964).
2. Mueller, W.: *Galvotechnik,* 57, 4 (1966).
3. Hestermann, G.: *Industrie-Lackier Betrieb,* 36, 2, pp. 55-62 (1968).
4. Frangen, K. H.: *Chemical Industrie,* 17, pp. 219-221 (April 1965).
5. Hutchinson, C.O.: National Paint and Coating Assn. (NPCA) Conference, Cincinnati, April 22, 1976.
** Without depreciation.
6. Lovano, S.: AFP/SME Technical Paper FC80-465.
7. Cole, G.E.: *Products Finishing,* pp. 80-86, 1984.

PAINT APPLICATION SYSTEMS

Equipment used for the application of organic coatings varies considerably in sophistication and cost from industry to industry, and even within the same industry. Details of several different systems are presented in this section.

TYPICAL AUTOMOTIVE PAINT SYSTEMS

Automobile bodies are generally made from cold rolled steel about 0.035″ (0.89 mm) thick. Some parts of the bodies that are particularly prone to corrode are made from steel galvanized on one side, steel coated on one side with Zincrometal, or steel coated on one side with zinc-rich paint, zinc-plated steel, or similar materials. The zinc-protected sides are usually the inner or invisible sides of the bodies.

Parts made from plastics, especially fiberglass, are being increasingly used for body components. Stainless steel and nickel-plated steel are used as trim and for windshield wipers. Aluminum parts have only limited use for automobile bodies. Except for parts painted by the autodeposition process, metal surfaces are generally prepared for organic coating by a five to nine-stage zinc phosphate treatment, as discussed in Chapter 19, "Conversion Coatings and Anodizing."

Prime Coat

Approximately 80% of all automotive painting starts with an electrocoat primer, usually applied by cathodic deposition. Further paint coats depend upon the location of the body component. The locations are generally divided into four groups: invisible, like the inside of the door; in doors, like the door jambs; outside, like the fenders, roof, and other components; and the so-called "highlights," like parts of the window frames and roof. The invisible parts receive no additional paint coat after the usually gray or black electrocoated primer.

Subsequent Coats

Visible indoor areas of automobile bodies receive a topcoat, usually of the same color as the overall body topcoat. The underside of the hood and inside of the engine compartment usually receive a sprayed-on topcoat of black alkyd or acrylic paint; thus, they carry a two-coat system.

Outside surfaces of the body receive a full or partial sprayed-on, sandable surface coat, applied on either the wet or incompletely baked electrocoat. The coat is then sanded, at least in the highlight areas. Next, the color topcoat, usually an acrylic resin, is sprayed on and baked. In many cases, a clear coat is sprayed over the color coat to provide "depth." In all, the outside areas of the body receive three or four paint coats.

Some automotive paint systems, not used extensively, apply paint coats on the outside first, like sprayed-on primer and/or powder coats, and then protect the inner surfaces by means of an electrocoat. On surfaces that are exposed to considerable damage from flying gravel, plastisol or plastisol mats are applied over the prime coat before the application of topcoats.

PAINTING AGRICULTURAL MACHINERY

Diverse methods are being used for painting agricultural machinery. Surface preparation ranges from sandblasting to five-stage iron phosphate treatment. Paints used range from air-drying alkyds to catalytically cured polyurethanes. The paints are applied by electrostatic, air-atomizing handguns or pressure guns in manned or unmanned booths. Some manufacturers use robots and computer control for paint application.

Tractors have a history of high-quality finishes. Even if the metal is 3 1/2″ (90 mm) thick in some places, it is generally prepared for painting with the same care as though it were sheet metal. Since the components of the tractor chassis are all primed, the system selected by one manufacturer for finish coating is as follows: three-stage iron phosphate treatment, wash and rinse, water blowoff, oven dry and cooldown, mask, spray paint, and oven bake.[3]

Robots were specified for this system to replace as many human painters as possible. Robotic water blowoff stations were required, with three robots specified for painting and two for water blowoff. Accomplishments with this system included:

1. About 95% of the manufacturer's two-wheel drive tractors and 90% of the four-wheel drive tractors can be accommodated.
2. A labor reduction of ten operators was attained.
3. Paint savings of about 13% were realized.
4. Energy usage was reduced by not having to heat the robot spray booth.
5. Overall quality was improved, and rework and repair reduced.

When a tractor is ready to be painted, a main computer (also known as the host) places the tractor into an area for subsequent painting. An operator looks at the tractor and punches in the paint code on the computer terminal. If the host and the human do not agree, everything stops until the problem is resolved. After agreement is reached, the tractor is entered into an area controlled by a programmable controller. This controller monitors the mechanical equipment in the paint department, but not the robots.

The mechanical equipment controller passes the paint code to another programmable controller that tells the robots what to paint and when. The system works on a first-in, first-out basis. The robots have their own memories; and based on the paint code, they select the appropriate programs for painting the tractors.

Because of the great number of different models to be painted and the fact that each robot has only 64 minutes of memory, it was not possible to paint all the manufacturer's tractors. This problem, however, was solved by dividing the tractors into three sections: front, middle, and rear. Most middle sections are common for all models. Some front sections are also common, but the rears are usually different. It became necessary to learn how to paint these sections individually and then to link their programs together to paint the total tractor.

AIRCRAFT PAINTING

Painting processes for aircraft vary widely, from the use of hand-sprayed alkyds to highly automated paint and process lines. Aircraft parts will corrode in service if cleaning and painting are not done correctly. Aircraft parts in general are designed with low safety factors (1.5), and corrosion must be prevented since it greatly reduces the fatigue life of highly stressed parts. If cleaning and painting are not done correctly before assembly, expensive rework of the aircraft will be required soon after it is put into service. It is almost impossible to determine how well paint is attached to the surface without destructive testing, and sample testing in a nonautomated system has little value. To minimize these problems, one aircraft manufacturer has developed the automated paint and process line (APPL) illustrated in Fig. 29-2.[4]

With this line, all fabricated metal parts are continually delivered without regard to part type or quantity. Sizes vary from very small parts moved in tote pans to large parts such as 48 ft (14.6 m) wing panels that are transported on special dollies. The production rate for 1979 was over a quarter of a million parts per month, and the finishing costs are estimated at $10-20 million per year for the 1980s.

Control System Design

Figure 29-3 is a block diagram of the computer control system for the automated paint and process line. This design centers around a hierarchy of minicomputers and microcomputers that allow the total control system to be split into various functional levels as follows:

- Level 1, the lowest level, consists of servomechanisms, mechanical actuators, and feedback devices that control and monitor the desired physical actions required.
- Level 2 consists of the amplifiers, drivers, and signal conditioners that convert the electronic signals of level 3 to the power signals of level 1. This level also converts the contact and encoder signals of level 1 to electronic signals for level 3.
- Level 3 consists of control microcomputers that contain the control programs necessary to convert level 4 commands to controlled physical action. This is the highest level that deals with all actuator and sensor actions required by the hardware.
- Level 3.5 consists of a communication microcomputer. It is necessary for coordination in the conveyor system because the level 3 microcomputers are distributed over a wide area for lower wiring costs and a single level 4 command may require action of more than one level 3 computer.
- Level 4 consists of minicomputers that contain the programs necessary to control subsystem (paint/process) operations according to manufacturing procedures. This level also interacts with production personnel via CRTs and monitors operations via various alarm methods.
- Level 5 consists of a supervisory minicomputer system that handles database storage and retrieval via communications from level 4, the system log, audit trails, management reports, software development, and program downloads to level 4 computers.
- Levels 6 and up are for future development that will coordinate plant and corporation-wide production and operation.

System Description

The automated paint and process line is composed of two major parts: a chemical process line consisting of 22 large tanks and a small-parts painting line housed in a two-story structure. Major components of the APPL are:

- A chemical processing line of twenty-two 50 ft (15.2 m) long, 15,000 gal (56 775 L) tanks, with computerized control of the tanks and feedback devices.
- A computer-controlled power supply (up to 37,000 A and to 48 V) for sulfuric, chromic, and phosphoric anodizing.
- A crane system with three 5 ton (4536 kg) microcomputer-controlled bridge cranes.
- A power-and-free conveyor system with microcomputer-controlled load-bar dispatching and identification.
- A paint enclosure system, 206 x 32 ft (62.8 x 9.8 m), with resources to maintain a safe painting environment.
- A paint curing oven, 24 x 19 ft (7.3 x 5.8 m), with two heated areas, 250 and 150° F (120 and 65° C).

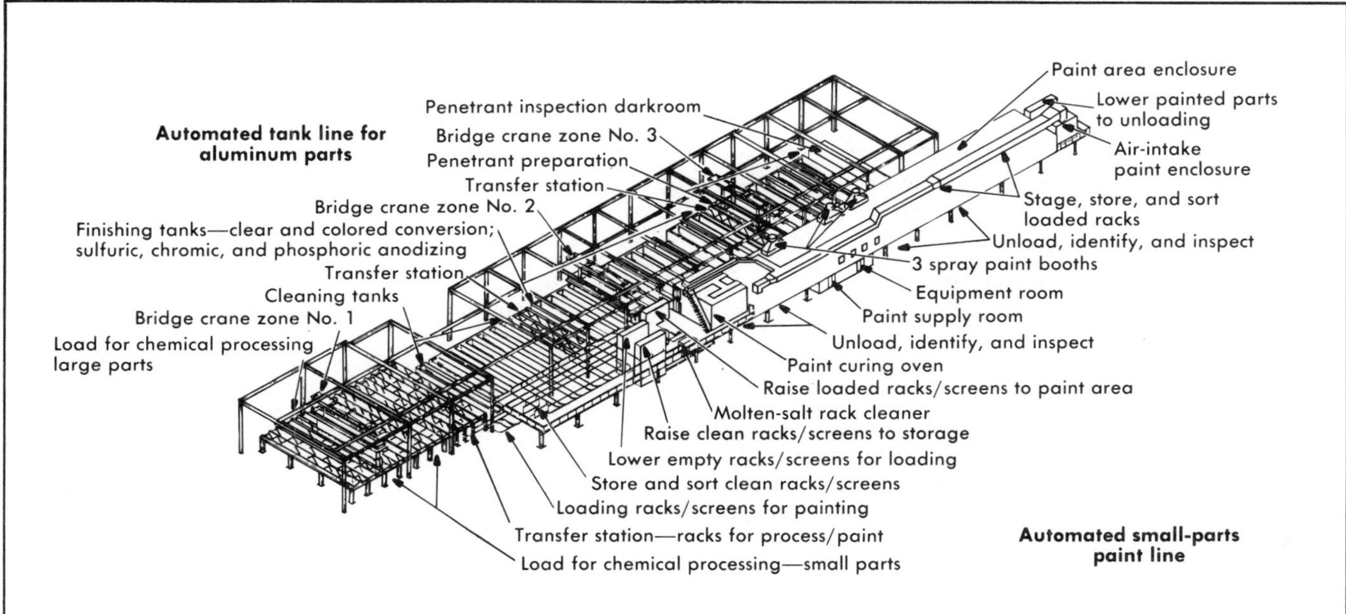

Fig. 29-2 Automated paint and process line. (*Lockheed Georgia Co.*)

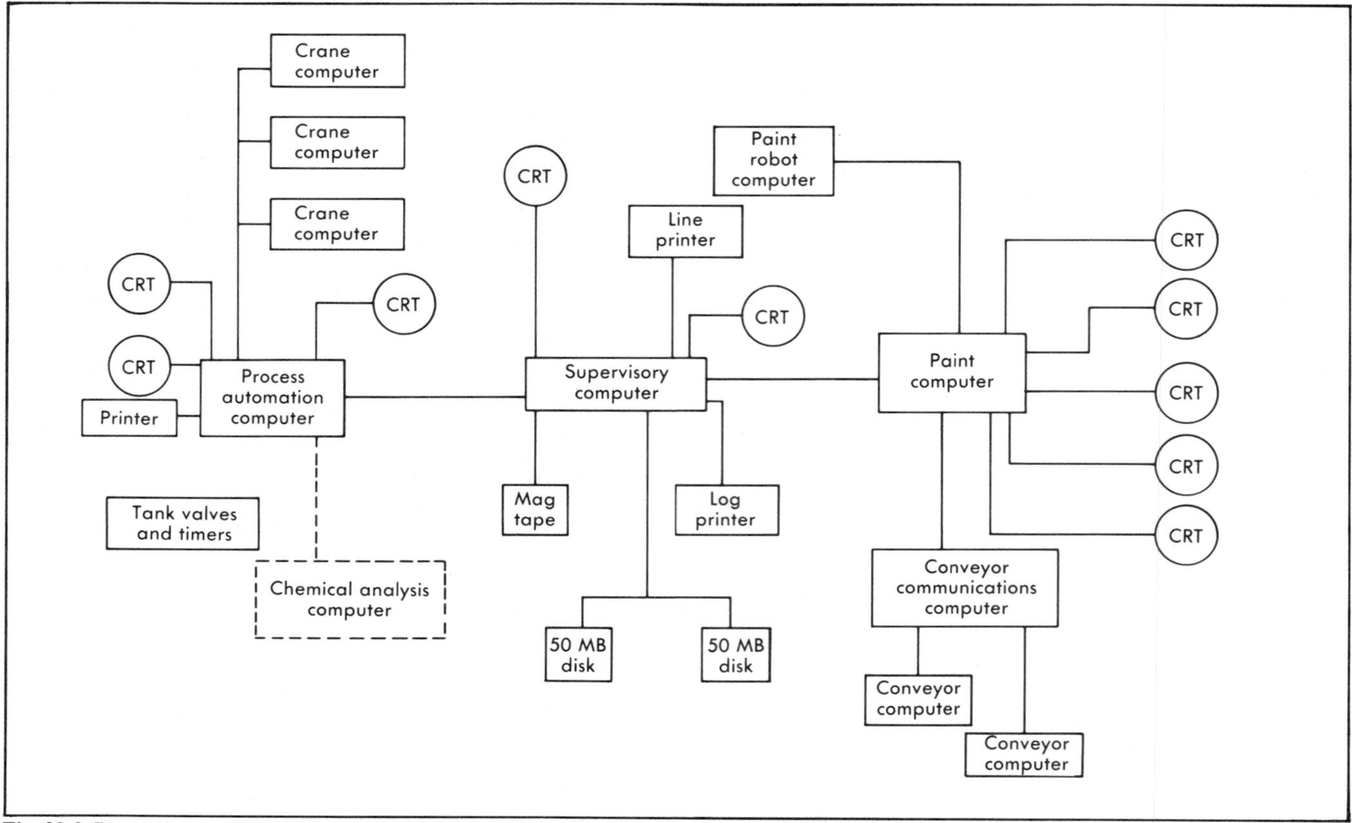

Fig. 29-3 Block diagram of computer system for automated paint and process line (see Fig. 29-2).

- Three water-washed paint booths housed within the paint enclosure.
- One microcomputer-controlled paint robot for automatic painting when possible.
- A paint mix, supply, and distribution system to supply the proper paint to each booth.
- A salt-bath cleaning system to remove old paint from racks before reuse.
- A master control system consisting of a supervisory computer, paint control computer, and process control computer.

Touch-Up and Repainting[5]

Hangars are sometimes used for the spray touch-up and repainting of the exteriors of large aircraft. These operations require control of the temperature and in the hangars, as well as control of coating thickness, drying time, and other variables. For painting over an existing coating, pretreatment operations, such as sanding and wiping with solvent, are necessary. When old paint is stripped, the metal is usually treated to form a chromate conversion coating for better adhesion of the primer.

Airflow is recommended throughout the entire hangar cross section, with inlet ports on an entire wall or ceiling and outlet ports on the opposing wall or floor. It is also desirable to have the air flow horizontally from the nose to the tail of the plane, thus minimizing turbulence and overspray deposition. Adequate filtration is required for both intake and exhaust air to ensure cleanliness.

Recirculation of some of the exhaust air, mixed with fresh air, is possible, but the air must be continuously monitored by means of a gas analyzer. The use of work-access stands allows painters to apply coatings without walking on surfaces of the airplane.

PAINTING APPLIANCES

Major household appliances, such as washers, driers, and refrigerators, have to be protected against alkaline detergents, high humidity, and household stains on their appearance surfaces. These surfaces are usually spray coated with high-solids paints that are hardened with melamine. Some of the functional steel areas may carry a sacrificial zinc coat, such as a galvanized coat, a zinc-rich primer, or an organo-zinc coat. These coats are applied on the coil stock before blanking and forming since the coats are weld-through primers. Zinc-rich prime coats can also be sprayed on after forming and even after welding. Areas exposed to humidity and water are usually electrocoated.

Some assembled appliance cabinets receive a seven-stage zinc phosphate metal preparation and are then prime coated inside and out by cathodic electrodeposition. The cabinets can also be spray primed with a thermosetting epoxy-resin-based paint, followed by a sprayed-on topcoat of acrylic melamine paint. Some appliances carry a powder coat, sprayed directly over the metal preparation, plus a decorative acrylic melamine coat.

The household appliance industry is a major user of thermosetting powder coatings, replacing porcelain enamel, liquid paints, and, in some cases, metal plating operations.[6] Reasons for the change to powder coatings include easier compliance with environmental and waste disposal laws, cost savings, and improved product quality.

Application methods for powder coatings are easily automated. Improved-design guns permit easier powder delivery and electrostatic charging of the powder coating. Redesigned reclamation systems with cartridge filters keep the reclaimed powder clean and maintain particle-size balance. Robots have replaced manual spray guns for touch-up on many application lines. Computers are being used to control both the powder output of electrostatic spray guns and their motion in respect to the parts being painted. Multicolor switching devices permit fast color changes at the guns.

FURNITURE PAINTING

Steel furniture for indoor use generally receives a three to five-stage iron phosphate metal preparation, plus a dip, spray, or electrodeposited prime coat. The topcoat is usually an alkyd or acrylic with the addition of crosslinkers. Steel outdoor furniture and steel doors usually receive a seven or nine-stage zinc phosphate treatment, plus a prime coat of epoxy-based spray paint or an electrocoat. The topcoats may be alkyds or polyesters, sometimes silicone modified. In some cases, powder coats are directly applied over the iron phosphate preparation.

To enhance the grain of wood furniture, a large number of different coats (up to 20) are brushed, sprayed, rolled, or wiped on. First, wipeable stains are used, usually followed by sanding; then sealers are applied. Additional coats are used to fill and color the grooves, and others to stain the top of the grain. Finally, clear coats are used to produce the depth of the finish. Frequent sanding between coats improves the surface appearance and product quality. Waterborne paints are frequently used, and ultraviolet radiation has been introduced as a curing method. Polyesters filled with wood flour are used to produce moldings and three-dimensional effects.

PAINTING PLASTICS

The introduction of plastics into industry in the mid-1800s marked the beginning of a new era in engineering and commodity substrates.[7] Over the years, plastics technology has grown into a major market. Plastics are currently being considered a practical and economical alternative for metal, metal alloys, glass, ceramic, and wood structures.

Plastics substrates are divided into three basic categories according to the properties of their polymers: (1) Thermosetting plastics consist of polymers that, when heated and formed into a part, are cured or crosslinked into a permanent shape. The crosslinking is irreversible; and although the part can be softened by heat, it cannot be returned to its original flowable state. (2) Thermoplastic substrates differ from the thermosetting by not crosslinking when heated. If the final part is reheated, the thermoplastic returns to its original flowable state. Repeated heating and cooling can cause degradation of the polymer. (3) Elastomers are plastics that exhibit flexibility and elastic deformation. Elastic deformation is the ability of a substrate to deform under stress and to recover when that stress is removed. In contrast, thermoplastics and thermoset plastics exhibit plastic deformation, the deforming under stress that will not recover when the load is removed. Natural rubber is an example of an elastomer. (Refer to Chapter 8, "Plastics and Composites," of this volume for a description of the different types of plastics being used and a summary of their properties.)

Plastics substrates can have in-mold color, texture, and a wide range of gloss; however, organic coatings produce all of these features with better consistency and quality. Organic coatings can be made in any color and held within the tolerances required by the various industries. Plastics substrates often have swirl patterns, porosity, knit lines, and other surface defects from processing; organic coatings fill and hide these defects. Organic coatings meet all of the appearance requirements of the manufacturers, as well as enhancing the quality of the product with the protective properties of the finish.

Painting Considerations

The solvent sensitivity, adhesion characteristics, and heat distortion temperature of plastics substrates make it more difficult to paint plastics than metal parts.[8] For example, quite often the best reducing solvents for paints are the most aggressive solvents in attacking sensitive plastics. Consequently, it is necessary to balance the strength of the solvent employed versus its evaporation rate to achieve an acceptable compromise for reducing the paint and yet not degrading the plastic. In some instances, it is possible to use an aggressive solvent in an air-atomized application as long as it is fast evaporating, used in moderation, and accompanied by a suitable amount of slower evaporating, less-aggressive solvent. Table 29-7 lists some of the solvent effects on several plastics substrates.

Adhesion characteristics change dramatically not only from plastic to plastic (for example, ABS to polyphenylene oxide), but also from one particular grade of plastics to another. In addition, plastics have a lower surface tension than metals that are normally painted and consequently are harder to wet and obtain adhesion on. Adhesion, however, can generally be achieved by using a polymer with a resin similar to the substrate. For example, polyester resins offer good adhesion to polyesters and polyester-based fiberglass-reinforced materials.

Baking temperatures for plastics are lower than baking temperatures for metals. Table 29-8 lists both the maximum baking temperatures of commonly used plastics substrates and the adhesion characteristics of both conventional coatings and low-VOC coatings.

Application Procedures

Applying organic coatings to plastics substrates requires knowledge of both the type of plastics and the coating being used. Procedures vary from substrate to substrate and from coating to coating. Before the coating is applied, proper substrate pretreatment is necessary to remove mold releases, oils, or other surface contaminants. A solvent wipe removes most contaminants, but caution should be exercised when selecting a solvent. Certain solvents degrade the plastic, causing loss of strength, blisters, or poor adhesion when coated. Sanding is another method occasionally used to ensure that the plastics surface is clean; it is also used to smooth rough substrates.

Spraying is the most popular method of applying coatings onto plastics substrates.[9] When electrostatic methods are used, it is necessary to render the part conductive using a hygroscopic ionized-salt solution that can be applied using dip or spray techniques. Roller coating is commonly used for coating flat surfaces and for producing woodgrain patterns on the surface. Flow coating has also been used to apply base coats and topcoats for vacuum-metallized plastics parts. Refer to Chapter 27, "Application Methods," for additional information on these coating methods.

TABLE 29-7
Solvent Effects on Some Sensitive Plastics

Solvents*	Evaporation Rate	ABS**	Acrylic**	Polycarbonate**	Styrene**
Ketones:					
Acetone	Very fast	1	2	1	1
Methyl ethyl ketone (MEK)	Fast	1	1	1	1
Methyl isobutyl ketone (MIBK)	Medium	1	2	1	1
Diisobutyl ketone (DIBK)	Medium	3	2	1	1
Cyclohexanone	Slow	1	1	1	1
Esters:					
Isopropyl acetate	Medium	1	2	1	1
n-Butyl acetate	Medium	1	2	1	1
Cellosolve acetate	Slow	1	1	1	1
Isobutyl isobutyrate	Slow	3	4	1	1
Aromatic hydrocarbons:					
Toluene/xylene	Fast to medium	1	4	1	1
Alcohols, glycol ethers, aliphatics (as a clear)		4	3-4	4	4

* As a group, ketones and esters are aggressive; aromatics are moderately aggressive; and aliphatics, alcohols, and glycol ethers are nonaggressive.
** 1 = aggressive; 4 = no effect.

TABLE 29-8
Adhesion and Paintability of Various Plastic Substrates

Substrate	Maximum Bake Temperature, °F (°C)	Adhesion	Solvent Sensitivity	Conventional Coating Type	Low-VOC Type
ABS	170 (77)	Easy	Poor	Lacquer Enamel Two component	Water UV High-solids two pack
Acrylic	180 (82)	Easy	Good	Lacquer Enamel Two component	
Cellulosics	220 (104)	Difficult	Good	Lacquer	
Polyolefins	250 (120)	Moderately difficult*	Good	Lacquer Two component Enamel	
Nylon	300 (150)	Easy	Good	Enamel	High solids/water
Polycarbonate	250 (120)	Easy	Poor	Lacquer Enamel	High solids Water/UV
Modified PPO	180 (82)	Fairly easy	Fair	Lacquer	Water/UV
Polystyrene	140 (60)	Easy	Very poor	Lacquer	Water

(continued)

TABLE 29-8—*Continued*

Substrate	Maximum Bake Temperature, °F (°C)	Adhesion	Solvent Sensitivity	Conventional Coating Type	Low-VOC Type
Vinyls	210 (100)	Fairly easy	Good	Enamel Lacquer	UV
Urethanes	250 (120)	Fairly easy	Good	Two-pack enamels	Water
				Enamels	High solids
Xenoy	170** (77)	Fairly easy	Fair	Enamels	High-solids two pack
				Two-pack enamels	
Polyesters	400 (200)	Fairly easy	Excellent	Enamels Two component	High solids High-solids two pack
Phenolic	500 (260)	Fairly difficult	Excellent	Enamels	High solids High-solids two pack

* Treatment required.
** Nonsupported.

OTHER PAINT APPLICATION SYSTEMS

Many other systems are being used for the application of organic coatings. The systems vary by industry, the products to be coated, the coating materials used, and the coating requirements. A few of the more common systems are discussed in this section.

Containers

Cans, drums, and other containers for nonfood merchandise receive an iron phosphate preparation, plus a sprayed-on coat of phenolic-resin paint or a clear coat. After baking, the coating shows high impact and chemical resistance. Other drums are sandblasted on the outside and sprayed with air-drying coats, usually of alkyd or polyester. Food-container paints, which include ultraviolet curing materials, are regulated by the U.S. Food and Drug Administration.

Electrical Equipment

Paints for motors, generators, and other electrical equipment have to be highly heat and corrosion resistant. A five-stage iron phosphate treatment is frequently used as metal preparation, followed by an anodic or cathodic, baked-on epoxy electrocoat. Polyester and acrylic coats are also used. Armatures often receive a baked-on phenolic spray coat. Frequently, a color coat is applied by spray over the prime coat.

Building Construction

For steel and aluminum doors, a phosphate metal preparation is often followed by an electrocoat, which in some cases is a metallic polyester. Insulated doors are spray coated with alkyd primer and then topcoated. Aluminum extrusions are often given a six-stage etching preparation, followed by an anodic electrocoat for complete coverage.

Steel or die-cast office enclosures receive an iron phosphate preparation, a sprayed-on alkyd primer, and a polyester topcoat. Higher durability is imparted by a zinc phosphate preparation, cathodic electrodeposition primer, and an acrylic-spray topcoat.

Catalytically cured urethane or polyester-epoxy resins are often sprayed or brushed on seamless flooring because of their resistance to wear and tear by traffic. Thick coats of fiberglass-filled paints are also sometimes used.

References

1. C. O. Hutchinson, "Electrocoat, Powder Coat, Radiate: Which and Why; Part 1: Electrocoat," National Paint and Coating Assn. (NPCA) Conference, April 22, 1976, Cincinnati, OH.
2. H. M. Leister, "Effect of Energy, Ecology, and Economics on Electrocoating," *Plating and Surface Finishing* (July 1982), pp. 46-48.
3. Robert E. Schuster, *Robotic Painting of Agricultural & Industrial Equipment—A User's Report*, AFP/SME Technical Paper FC81-421 (Dearborn, MI: Society of Manufacturing Engineers, 1981).
4. James A. Spruell, *Automatic Paint and Process Line*, AFP/SME Technical Paper FC80-624 (Dearborn, MI: Society of Manufacturing Engineers, 1980).
5. Charles B. Reymann, *Aircraft Painting Facility*, 28th National Society for the Advancement of Materials & Process Engineering (SAMPE) Symposium, April 12-14, 1983.
6. Peter R. Gribble, *Powder Coatings in the U.S. Major Appliance Industry*, AFP/SME Technical Paper FC83-606 (Dearborn, MI: Society of Manufacturing Engineers, 1983).
7. Arthur J. Kirby, "Coatings for Plastic: Advancing Technology and Versatility for the Plastic Industry," *Finishing '83 Conference Proceedings*, October 10-13, 1983, Cincinnati, OH (Dearborn, MI: Association for Finishing Processes of SME, 1983), pp. 13-73 to 13-81.
8. Ronald J. Lewarchik, *Low VOC Coatings for Automotive Plastics*, SME Technical Paper FC83-650 (Dearborn, MI: Association for Finishing Processes of SME, 1983).
9. Charles D. Storms, *Painting Plastics*, SME Technical Paper FC81-440 (Dearborn, MI: Association for Finishing Processes of SME, 1981).

CHAPTER 29

BIBLIOGRAPHY

Bibliography

Bergant, James M. *Appliance Finishing with an Industrial Robot*. AFP/SME Technical Paper FC81-441. Dearborn, MI: Society of Manufacturing Engineers, 1983.

Coutts, Garry D. *Developing a Paint System: An End User's Dilemma*. AFP/SME Technical Paper FC78-417. Dearborn, MI: Society of Manufacturing Engineers, 1983.

Goham, James E. *Decision Making Processes Employed in the Construction of a Contract Manufacturers Paint Finishing Facility*. AFP/SME Technical Paper FC80-616. Dearborn, MI: Society of Manufacturing Engineers, 1983.

Henke, Doyle W. *Implementation of Point-to-Point Spraying Robots for Painting Refrigerator Parts*. AFP/SME Technical Paper FC83-652. Dearborn, MI: Society of Manufacturing Engineers, 1983.

Mason, Mark R. *Robotic Spray Finishing of Heavy-Duty Truck Axles*. AFP/SME Technical Paper FC83-603. Dearborn, MI: Society of Manufacturing Engineers, 1983.

VanDyke, Harold R. *Robots in Paint Applications*. AFP/SME Technical Paper FC83-636. Dearborn, MI: Society of Manufacturing Engineers, 1983.

TESTING, TROUBLESHOOTING AND SAFETY

TESTING OF PAINT

CHAPTER CONTENTS:

Paint testing is divided into two distinct categories: testing of paint before application and testing of applied films. Seldom is a given paint tested for a single, well-defined property. Instead, testing is generally done to determine the performance of a paint under a specified set of circumstances. For example, a paint may be tested for resistance to outdoor exposure, immersion in water or chemical solutions, or other conditions.

In all cases, the supplier and the user of the paint must agree on the test procedures to be used, which should include specifications for the test equipment and all physical parameters. Most testing is done in accordance with specifications published by the American Society for Testing and Materials (ASTM). The tests for paints and paint films are discussed and illustrated in the ASTM *Paint Testing Manual*. A list of the more common ASTM tests is presented in Table 30-1.

TESTING PAINTS BEFORE APPLICATION

Prior to application, paints are often tested for content, physical properties, and optical properties.

Content and Physical Properties

Tests are used most extensively to determine the following content and properties of a paint:

- Amount of paint solids (nonvolatiles) in percentage by weight and/or percentage by volume.
- Weight of the paint and the paint solids, expressed in lb/gal or g/mL.
- Viscosity, a measure of the resistance of the liquid to shearing forces, which is generally determined by flow from a No. 4 Ford or No. 2 Zahn cup viscometers.
- Leveling characteristic (the ability to smooth out brush marks or furrows).
- Surface tension, with relation to the spreading of liquid paint.
- Particle size, measured by the use of sieves or microscopes.

Optical Properties

The optical properties of a paint are tested to determine its color. The color of a paint depends upon its spectral reflectance, the spectral composition of the light used, and the spectral sensitivity of the observer's eye. Fluorescence is the conversion of invisible light (such as ultraviolet) into visible light, thus increasing the apparent reflectance or luminance.

Color matching. Paint formulators are frequently called upon to match the color of a paint sample. It is, of course, possible to determine the exact pigment composition of a given paint and to duplicate it. Practically, however, it is often more economical and faster to match one paint/pigment composition with another, somewhat different composition. The two paints should match in both daylight and artificial light, and are tested for matching in a color booth.

Three characteristics usually considered in matching colors are color perception (hue), brightness, and saturation. Various instruments used to measure these characteristics include spectrophotometers and colorimeters.

Spectrophotometers. These instruments measure the reflected, absorbed, and transmitted wavelengths of light throughout the entire visual spectrum. Curves produced can be used for permanent reference.

Colorimeters. These instruments match a given color through the manipulation of filters. They measure hue, brightness, and saturation through the use of reference standards. Colorimeters are used extensively because of their ease and speed of operation, as well as their low cost.

Gloss. Measurement of gloss can be accomplished by visually comparing the distinctness of an image reflected by a sample surface with an image reflected by a known standard surface. More commonly, however, gloss is measured by a photoelectric glossmeter. With this instrument, gloss is determined as reflectance at a given angle of light incidence. Viewing angles of 85°, 60°, and 20° from the vertical are frequently used. The photoelectric cell in the instrument reports the intensity of reflected light as a percentage of the incident light.

Hiding power. This is the ability of a paint to obliterate color differences found on the substrate over which the paint is spread. Various devices and test methods have been designed and are in use for determining this capability. One method consists of calculating a contrast ratio: the ratio of light reflectance over a black surface divided by the light reflectance over a white surface. *Mass color* is a

***This chapter contributed by: George E. F. Brewer**, Coating Consultants.*

CHAPTER 30

TESTING OF APPLIED FILMS

TABLE 30-1
Partial List of ASTM Test Methods

Name of Test	ASTM Method
Color Matching by Eye	D 1729
Instrumental Color Measurement	D 2244
Colorfastness (Indoor)	D 2620
(Color) Bleed Resistance	D 868, D 969
Gloss Measurement	D 1471
Hiding Power	D 344, D 2805
Spot	D 1308
Natural Weathering (Test Fence)	D 1006
Artificial Weathering (Accelerated Weathering)	D 822
Abrasion Resistance	D 658, D 968
Water Immersion	D 870
Humidity	D 1735
Blister Resistance	D 714
Salt Spray (Fog)	B 117
Flexibility by Mandrel Test	D 522, D 1737
Impact	G 14
Resistance to Fungus Growth	D 2574, D 3273
Adhesion (by scratching and scraping the paint film)	D 2197
Density (Pounds-Per-Gallon Test)	D 1475
Fineness of Grind	D 1210
Viscosity by Rising Bubble or by Falling Ball of Plunger	D 1545
Viscosity by Efflux Viscometer	D 1200
Viscosity by Paddle Viscometer	D 562
Rotational Calibrated-Spring Viscometer	D 2196
Drying Rate by Finger-Test Method	D 1640
Percentage of Solids	D 2832
Magnetic or Electric Current	E 376
Destructive Film-Thickness-Testing Gauges	D 2691

term used to define the thickness of a coating needed to obscure the background colors. Tinting strength is a measurement of the power of a pigment to color a standard paint.

TESTING OF APPLIED FILMS

Most tests of applied paint films are carried out on samples of merchandise produced in the industrial plant or laboratory of the paint user. Paint manufacturers prepare sample pieces or test panels for their customers. These samples are made under strictly controlled conditions (for instance, by the use of automatic spraying equipment or other application methods); thus, the properties of the resulting coatings are reproducible. The tests used most extensively are for coating thickness, drying time, hardness, adhesion, flexibility, tensile strength and elongation, and chemical properties and resistance.

Coating Thickness

The thicknesses of both wet and dry films are often measured. The purpose of measuring wet films is to determine if they are sufficiently thick to develop the required dry-film thickness. Measurement of dry films is a check to ensure conformance with specifications.

Wet-film gages. Gages for determining the thickness of wet films cut through the films. Two types of gages used extensively are shown in Fig. 30-1. View *a* illustrates a tooth gage that is simply pressed into the wet film to measure its thickness. Wheel gages (view *b*) are rolled through the wet film to contact the base material in measuring film thickness.

Dry-film gages. A wide variety of gages are used for determining the thickness of dry films. Some gages, such as dial comparators or micrometers, require the removal of small areas of the film for measurement. Other gages are nondestructive and use magnetic or electrical functions, such as eddy currents, to measure the thickness. However, nondestructive gages are limited to the measurement of films on electrically conductive metals.

Drying Times

Laboratory test specifications exist for the many successive stages in the drying of paint films. Classifications for these stages include tack-free (surface dry), dry-to-touch, dry-through (dry to handle), and dry hard. Drying times are determined by noting the elapsed times until the paint film reaches the specified states of dryness.

Film Hardness

A widely accepted definition of hardness is resistance to indentation or scratching. In testing the hardness of paint films, related properties such as elasticity are also determined. A variety of devices are available to measure the hardness of paint films. Many of these devices are categorized as scratch or pendulum types.

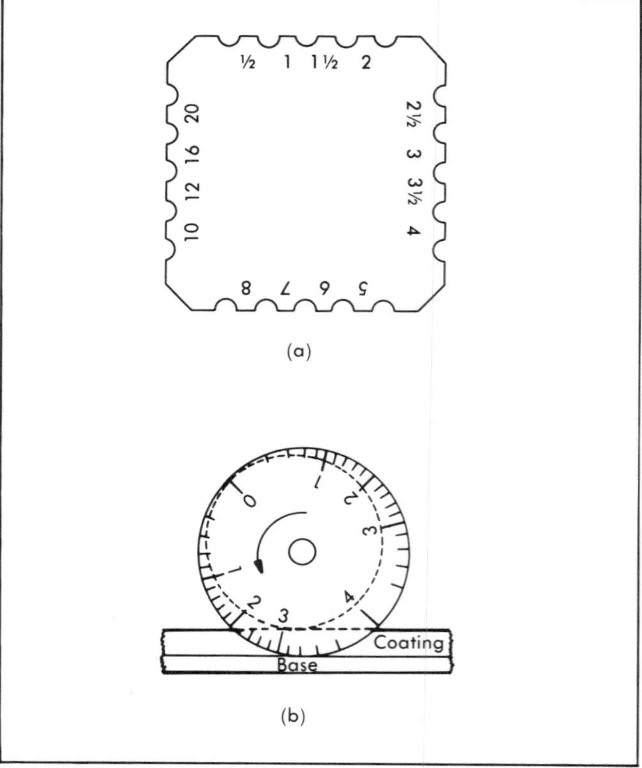

Fig. 30-1 Gages for measuring wet-film thickness: (a) tooth gage and (b) wheel gage.

Scratch tests. Devices for performing scratch tests on paint films include units having mechanically operated styli or knives. The pencil hardness method of scratch testing is used extensively. Pencils are available in 17 different grades of hardness ranging from 9H, the hardest, to 6B, the softest.

The following method of scratch testing with the pencil hardness method is common.[1] The wood of the pencil is stripped from the lead for a distance of about 1/4″ (6 mm) from the lead end, and the lead end is squared by applying a gentle rotary motion against 400-grit abrasive paper. Using successively harder pencils and holding them at an angle of 45°, the pencils are pushed forward against the paint film to be tested with a pressure just short of that required to break the lead. The hardness of the film is expressed as the hardness grade of the pencil just below the pencil that mars the film.

Pendulum-type tester. One type of pendulum tester is called the Sward Rocker. The mechanism of this device is calibrated to oscillate a cylindrical surface in contact with glass 50 times in 60 seconds before stopping. In testing paint films, the so-called "Rocker Hardness" is the number of oscillations before stopping with the cylindrical surface contacting the film.

Abrasion Resistance

Several properties of paint films are involved in the measurement of abrasion resistance. These include mar resistance, hardness, modulus of elasticity, and tensile strength. Various methods of testing for abrasion resistance include the use of falling abrasive and abrasive wheels or discs. The aircraft industry uses an erosion-resistance method of testing by exposure to rain, and traffic paints are tested by the use of crushed quartz or by rotating against rubber.

Falling abrasive. In this method of testing for abrasion resistance, sand or other abrasive from a hopper is allowed to impinge on a test panel, usually placed at an angle of 45° to the abrasive flow. Abrasion is generally expressed as the volume of abrasive required to wear through a unit thickness of coating. Another method is to measure the percent reduction in paint gloss caused by a given weight of impinging abrasive.

Abrasive wheels or discs. By rotating or moving abrasive wheels or discs rectilinearly, abrasion can be determined by measuring loss of material. One tester has a rotating table to which test panels can be clamped and that can be equipped with abrasive wheels of various grit sizes. Different loads can also be applied to the wheels.

Adhesion

Adhesion is defined in ASTM Designation D 907 as the state in which two surfaces are held together by interfacial forces that may consist of valence forces or interlocking action, or both. It would be difficult or even impossible to measure these forces; and if they could be measured, the data would have little practical meaning.

Difficulties in designing adhesion tests are many. First, every substrate has a profile that results in the creation of voids when a nonwetting liquid is applied (see Fig. 30-2). Second, careful studies have shown that the lift-off of coatings is due to cohesive failure.[2] In other words, a thin layer of coating remains on the substrate. Third, the substrate surface may carry oxides, scale, soil, or other impurities that can influence adhesion.

Many other factors, as well as substrate and paint properties, also influence adhesion. As a result, the types of tests used should be selected according to the modes of failure observed in service. Adhesion testing methods used most extensively include film removal, scraping, and scratching. Inertia tests, using vibrations to lift coatings, are rarely used.

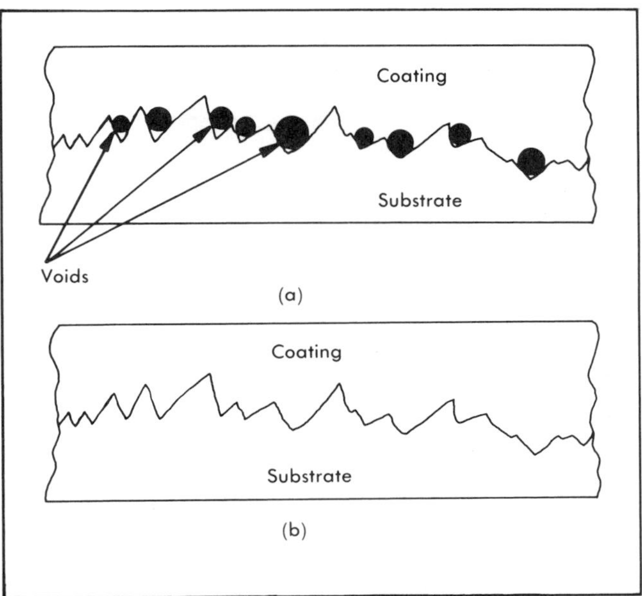

Fig. 30-2 Contact between coating and profiled surface on substrate with: (a) a nonwetting liquid and (b) an ideal wetting liquid.

Film removal. Methods of removing paint films for adhesion testing range from the application of pocket knives to mechanically operated cutting edges, blades, or points. Gages are used on some devices to measure the force needed for coating removal.

Scraping and scratching methods. In these methods of adhesion testing, a series of parallel cuts are made at 90° to each other to form a checkerboard-shaped grid (see Fig. 30-3). Then the percent of paint still remaining on the substrate is estimated.

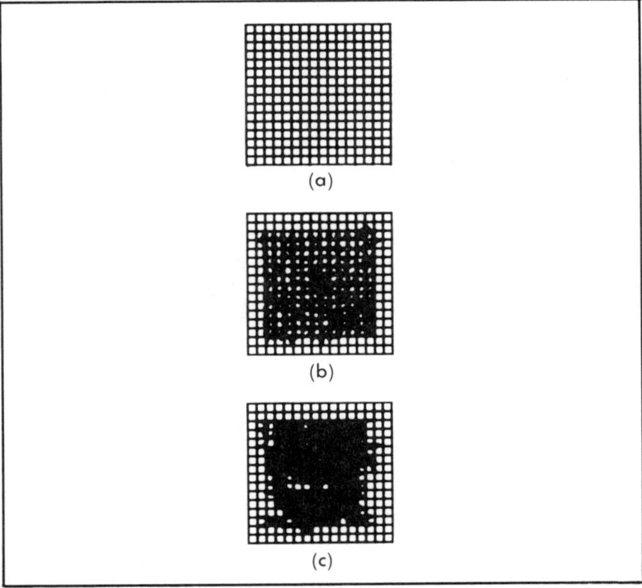

Fig. 30-3 Evaluation of crosshatch scratch tests showing various percentages of paint remaining: (a) 97%, (b) 70%, and (c) 25%.

CHAPTER 30

TESTING OF APPLIED FILMS

For some of these tests, additional diagonal (45°) cuts are made. In other tests, adhesive tape is pressed over the cut area and then lifted to remove loose coating.

Flexibility

Flexibility is sometimes defined as the capability of being bent, turned, or twisted without being broken and with or without returning to its former shape. The properties of adhesion and elongation of a paint film are definitely involved with flexibility. Humidity, temperature, and elongation time-rate also influence flexibility.

Mandrel tests for determining flexibility consist of bending and otherwise forming painted panels and then measuring the lengths of cracks produced. Conical mandrels are used for some tests. Impact tests apply falling weights, with the severity of the impact being measured in inch-pounds or newton-meters. In addition, deformation of the concave and convex sides of the substrate are checked. In cold-crack tests, successive cycles of heat, cold, and ambient temperature are applied. Most flexibility tests are performed on freshly prepared films and then repeated after exposure tests, discussed later in this section.

Tensile Strength and Elongation

While flexibility deals with films attached to substrates, tensile strength and elongation deal with properties of free (unsupported) films. Tensile strength is defined as the greatest longitudinal stress a substance can bear without tearing apart. Elongation is defined as the total deformation in the direction of load or per unit of length caused by a tensile force, or the maximum permanent stretch per unit of original length induced in a body by a force that causes it to break. The elongation of an organic coating is inversely proportional to the rate of strain, while the tensile strength is directly proportional.[3]

Tensile strengths and elongations of paint films are determined by using standard tension-testing machines, often under desired climatic conditions. Free films are prepared by casting or spraying coatings onto substrates that show little adhesion, such as polyethylene, Teflon, silvered glass, or fluorochemical-treated glass. Other substrates used, depending upon the dissolution of the substrate in water, include the gelatin side of photographic paper or an amalgamation of tinfoil or tinplate. The use of tin is least open to possible debate with respect to the paint film being identical with those formed and dried on steel.

Water Resistance

A chemical property of coatings that is of interest is resistance to water, including rain, condensation, and moisture in the substrate. Water permeability is tested by applying free films over the tops of jars or cups containing some drying agent. The weight increase is a measure of the water permeability of the film.

With one condensation tester, a room is maintained at a temperature of 77° F (25° C). A pan of water in the room, kept at 100° F (38° C), is covered with a test panel, coated side down. The temperature difference between the room and the water causes condensation on the coating.

Resistance of paints to moisture in the substrates is important when paints are applied to building walls. For such applications, moisture is measured and kept constant by the use of special hygrometers. In this way, the possibility of blistering or other damage to the paint is evaluated.

Chemical Resistance

To test paints for chemical resistance, it is necessary to expose the surfaces of the coatings to a hostile environment. Any changes in the coating may be considered as a basis for complaint. Possible changes include discoloration, loss of gloss, softening, swelling, and loss of adhesiveness. Test specifications have to be agreed upon between the seller and the buyer of the paint and then vigorously adhered to. Tests conducted to determine chemical resistance include spot, immersion, fire retardance, heat resistance, biological deterioration, and weathering tests.

Spot tests. For these tests, a selection is made from a large variety of chemicals to which the products may be exposed. For instance, appliances are tested for resistance to coffee, bacon grease, detergents, and other substances. Coatings for the transportation industry are tested by exposure to antifreeze, fluids, lubricating oils and greases, road oils, and similar products.

Immersion tests. This group of tests includes immersion in water, solutions of detergents in water, alkali and acid solutions, and solvents. The most severe test for outdoor coatings is the salt-fog exposure test, often called simply the salt-spray test.

The ASTM Method B 117 is used extensively for salt-fog exposure tests. Test pieces, usually 4 x 12″ (102 x 305 mm) test panels, are placed in a standard cabinet in a position 15° from the vertical and with their coated sides facing upward. The pieces or panels are exposed to the atomized fog from a solution containing 5% by weight of sodium chloride (NaCl) maintained at a temperature of 95° F (35° C). The concentration of the fog is described as 1-2 mL (0.034-0.068 fl oz) of salt solution to be collected per hour on every 80 cm^2 (12.4 $in.^2$) of horizontal surface. Failure of the specimen is described in different ways for different kinds of products; for instance, the number of hours of exposure that should cause rust on less than 5% of the surface. In variations of the salt-fog exposure test, acetic acid or copper-accelerated acetic acid is added to the salt solution.

Exposure of so-called "scribed panels" to salt fog is a method used extensively. A fairly large "X" is scratched through the coating, thus scoring the metal. The number of hours of exposure is specified, and failure is considered to be paint lift-off of more than 1/16″ (1.6 mm) from the sides of the scribed lines for a length of more than 1″ (25 mm).

The three-step practice at one automotive manufacturer consists of exposing scribed panels to:

1. Immersion in air that is saturated with a 5% NaCl solution for 15 minutes.
2. Exposure to air at a temperature of 72° F (22° C) for 1.25 hours, without rinsing or wiping.
3. Exposure to air at a temperature of 120° F (50° C), with 85-90% relative humidity, for 22.5 hours.

The procedure is repeated Monday through Friday, with the panels remaining in the humidity cabinet until the following Monday morning. Panels are exposed for 60 cycles of 24 hours each, and the paint lift-off is reported in millimeters, the total width perpendicular to the scribed lines.

Various products in different industries are required to pass salt-spray tests of 24, 96, 240, and 500 hours. For air conditioners and automobiles, 1000 hours may be specified. For neutral coatings, like epoxies, alkyds, and anodic electrocoats, a salt-spray exposure of 50 hours equals roughly one year in the field.

Fire retardance. Fire-retardant coatings are nonflammable and should be good heat insulators. Fire retardance tests imitate many natural situations, like dropping burning material on wooden shingles under precisely controlled conditions. In many of these tests, the duration of the flame and afterglow, and the weight loss are measures of the fire retardance.

Biological deterioration. Many liquid paints, particularly latex emulsions, are subject to attack by microorganisms. Tests for the effectiveness of antimicrobial agents and for the resistance of paint films against microbial attack have been developed. Field exposure tests are also used to determine the resistance of paint films to microbial damage.

Outdoor weathering tests. Test panels are exposed in various locations, notably Buffalo, NY and Florida, under precisely defined conditions. Specified, for example, are the construction of test racks and the angle of exposure. Most test racks are stationary, but some follow the sun. Other racks use mirrors to increase the intensity of the sun's rays, change the angle of the panel for continuous 90° incidence, and keep the test-panel temperature equal to a 45° exposure. Automotive paint systems are exposed in Florida with test panels facing south at a 45° angle for 18 months. No detectable change or deterioration in the coatings is allowable.

Weathering tests are evaluated by noting deteriorations in the appearance, gloss, darkening, fading, and yellowing of the paint films. Also observed is any cracking, peeling, flaking, chalking, blistering, or mildewing of the coatings. Deteriorations are generally reported using a scale from 10 (best) to 0 (complete failure).

Artificial weathering. A number of artificial weather machines of different designs are in use. They expose test panels to xenon or other lamps, water spray, varying humidity, different temperatures, a blend of gases like carbon dioxide (CO_2) and sulfur dioxide (SO_2), and other atmospheric phenomena. These exposures permit comparative scanning that, for a given set of conditions, may correlate results with those obtained from a specific natural environment.[4]

Specific Tests for Different Coatings

The ASTM *Paint Testing Manual* describes specific tests for different materials and coatings. These include oils, solvents, architectural paints, waxes and polishes, and marine paints. Information on paint testing by instrumental analysis, such as chromatography and atomic absorption, and the analysis of whole paint and paint components is presented. A list of organizations that issue paint testing specifications is also provided in this ASTM manual.

TROUBLESHOOTING

Unacceptable coatings can be traced to three different causes:

1. The substrate material.
2. The preparation of the surface to be coated.
3. The coating material.

In addition, the coating processes being used may be at fault. Some frequently observed paint defects, along with their causes, preventive methods, and means of repair, are presented in Table 30-2.

Production shutdowns because of failure to meet quality standards occur too frequently. While the causes for these shutdowns appear to happen suddenly, if samples are taken at regular intervals and inspected closely, it will usually be found that defects develop gradually.

TABLE 30-2
Frequently Observed Paint Defects, Their Causes, Preventive Methods, and Means of Repair

Paint Defect	Possible Causes	Preventive Measures	Methods of Repair
Orange peel	Improper oven flow.	Change flow.	Mild sanding and polishing.
	Fast surface drying.	Use paint thinner that evaporates more slowly.	Sand and repaint.
	Excess atomizing air.	Reduce air.	
	Gun too far away.	Move gun closer.	
	Not enough thinner.	Add thinner.	
Sagging (partial slipping of paint)	Too much thinner.	Reduce thinner.	Strip coat and repaint.
	Insufficient drying between coats.	Use thinner that evaporates faster.	
	Dip coats: exit from tank too fast.	Change cycle.	
	Gun too close.	Move gun away.	
	Coat too thick.	Reduce thickness.	

(continued)

TABLE 30-2—*Continued*

Paint Defect	Possible Causes	Preventive Measures	Methods of Repair
Flooding, mottling, floating, silking (pigment separation)	Poor paint agitation.	Improve agitation.	Repaint.
	Evaporation of thinner too slow.	Use thinner that evaporates faster.	
	Coat too thick.	Reduce thickness.	
	Low air pressure.	Increase pressure.	
	Gun too close.	Move gun away.	
Cratering, popping, pinholing, bubbling (indentations or eruptions in surface)	Oven temperature too high.	Reduce temperature.	Sand and repaint.
	Coat too thick.	Reduce thickness.	
	Gun too close.	Move gun away.	
	Impurities on substrate.	Clean substrate.	
	Water in solvent-borne paint.	Remove water.	
Cratering, fish-eyes	Oil, grease, or silicon on substrate or in paint.	Clean system.	Repaint.
		Use reducing agents to minimize defects.	
Wrinkling	Coat too thick.	Reduce thickness.	Sand and repaint.
	Drying too fast.	Reduce temperature.	
Dry spray (low gloss and roughness)	Insufficient thinner.	Add thinner.	Polish.
	Overspray on work.	Reduce overspray.	
	Booth draft too low.	Increase draft.	
	High gun pressure.	Reduce pressure.	
Blushing (milky-looking areas)	Booth humidity high.	Reduce humidity.	Repaint.
	Evaporation of thinner too fast.	Use thinner that evaporates more slowly.	
	Flashoff temperature too low.	Increase temperature.	
Light or dark areas	Usually result from repair or touch-up.	Improve procedures.	Repaint.
Cracking, crazing, checking	Overbaked film.	Change bake cycle.	Sand and repaint.
Chalking (powdery appearance)	Insufficient paint agitation.	Improve agitation.	Polish.
	Wrong solvent balance.	Check and correct.	
Water spotting	Surface tension of electrocoat rinse too high.	Treat rinse.	Polish.
Adhesion failure:			
Prime coat	Faulty surface preparation.	Improve process.	Strip and repaint.
Topcoat	Wrong thinner.	Change thinner.	Sand and repaint.
	Overbaked primer.	Reduce bake cycle.	
	Impurity on primer.	Clean system.	
Insufficient wraparound (electrostatic)	Solvent polarity high.	Reduce polarity.	Touch up.
	Poor grounding.	Improve grounding.	

(continued)

TABLE 30-2—*Continued*

Paint Defect	Possible Causes	Preventive Measures	Methods of Repair
Dirt in paint film	System dirty.	Clean lines. Check filters.	Sand and repaint.
Lint	Lint in contact with wet or semidry paint.	Keep booth and oven under positive pressure. Use lint-free work clothes.	Polish.

PINPOINTING TROUBLES

To pinpoint the cause of trouble, if there is a *single* cause of failure, the so-called "three test-panel method" may be helpful. This method, in its simplest form, consists of preparing three different test panels (see Table 30-3). If the paint is at fault, all three panels fail. If the steel is at fault, only panel A fails. If the user's surface preparation method is at fault, panels A and B fail.

TABLE 30-3
Three Test-Panel Method of Pinpointing Cause of Trouble

Test Panel	Steel for Panel	Surface Preparation of Panel	Paint for Panel
A	User	User	User
B	Standard*	User	User
C	Standard*	Standard*	User

* Available from surface preparation vendors.

PAINT STORAGE AND BAKING TEMPERATURE

A basic characteristic of organic coatings necessitates stability during storage at ambient temperature, 72° F (20° C). Paints are designed to undergo fundamental physical and/or chemical changes when signals, such as higher temperatures, are given; to prevent problems, paint should be stored at or near ambient temperature.

The evaporation of solvents can also cause problems. Vapors occupy about 1000 times the volume of liquids. As a result, if a coating entering the bake oven contains 99% nonvolatiles and 1% volatiles, by volume, then the evaporating vapor passing through the film will have a volume 10 times that of the film. To prevent problems, recommendations of the paint manufacturer with respect to the amount and types of thinners, baking temperatures, and length of bake should be adhered to strictly.

DEFECTIVE COATINGS

When defects have been found not to be connected with the substrate or the metal preparation, they are then due to either the coating composition or the processing of the coating. The decision to use a certain paint from a specific manufacturer is generally based on submission of samples, test panels, samples of coated products, and specifications by the paint manufacturer. Samples that pass inspection and tests should be kept for reference, together with all documents. At periodic intervals during production, new samples should be taken, sealed into tight containers, and stored for comparison with future deliveries.

When quality control personnel or a customer refuses products because of coating defects, test panels should be prepared from the currently used paint and from earlier, satisfactory deliveries. It is advisable to use test panels that have received the vendor's standard surface preparation. If the earlier delivered paint gives good test results while the currently delivered paint fails, then the cause of the defect is identified. If all test panels pass the tests, then the cause of complaint lies in the plant processing. A difficult situation is encountered if all test panels fail. This situation may be the result of deterioration of earlier paint samples, or the processing and testing of panels in the laboratory may be at fault.

COATING REMOVAL

Coating removal, also called stripping, is necessary to remove defective coatings from products and unwanted coatings from painting equipment, such as spray booths, drip sections, and hanging devices. One of the simplest methods for the removal of unwanted coatings is the covering of exposed surfaces with paper prior to painting and then removing the paper with waste paint on it after painting.

STRIPPABLE COATINGS

Spray booth walls, hangers, and other painting equipment that are indirectly painted can be initially coated with a paint that exhibits low adhesion but good cohesion and high flexibility. This coat, together with accumulated waste paint, is then peeled off.

BURNOFF STRIPPING

In this method of coating removal, parts to be stripped are placed in an oven and heated to approximately 1000° F (540° C). The organic coating exits as heavy fumes that require incineration. Paint pigments leave the flue as a fine ash or remain on the parts and in the oven to be brushed or vacuumed off.

MOLTEN SALT-BATH STRIPPING

A mixture of oxidizing inorganic salts is melted in a dip tank at approximately 900° F (480° C). After parts are immersed for a few minutes in the tank, the organic coating is rapidly oxidized or burned off. The parts are then lifted from the tank, cooled, and rinsed with water.

PHYSICAL METHODS OF STRIPPING

These methods of coating removal involve cutting through a paint film with wire brushes, scrapers, wet and dry sanding, or blasting the objects with abrasive grits, high-pressure water, or steam, as discussed in Chapter 16. These methods are used on parts that cannot withstand heat or chemicals, and on large or immovable structures. Rust, scale, and other contaminants are also removed by these physical methods.

BRUSHED-ON SOLVENT STRIPPERS

Viscous solvent-type strippers are brushed onto parts that do not lend themselves to other stripping methods. After one to two minutes, the loosened paint is rinsed off with pressurized water or steam.

FLUIDIZED-BED STRIPPING

Small, sandlike media is heated to a temperature from 850 to 1000°F (455 to 540°C) in a tank resembling the fluidized powder-bed equipment used for coating discussed in Chapter 27. The hot media is levigated by compressed air, and the parts to be stripped are lowered into the fluidized bed, using the media for heat transfer and for abrasion. During the first few minutes, a warm-up takes place, then heavy fumes are formed that may require incineration. During the last segment of the cycle, carbon dioxide (CO_2) is released. The stripping cycle typically lasts 30 minutes.

CRYOGENIC STRIPPING

A cryogenic liquid, such as nitrogen, having a boiling point of -320°F (-195°C), is sprayed onto the coated surface to create contraction of the metallic substrate and compression in the brittle coating. The cold part is then blasted with a nonabrasive plastic media of selected size and density to remove the coating and to scour any residue from the metal surface. The stripping cycle with this method usually requires less than 10 minutes. However, thin coatings less than 5 mils (0.13 mm) thick are difficult to remove. Cryogenic stripping is clean and fast, and the waste is uncontaminated, dry paint.

STEAM GUN STRIPPING

A hot alkaline stripper is sprayed over a painted surface. The solution valve of the steam gun is then closed, and high-pressure steam is used to lift off paint. A rinse with pressurized water removes the loosened paint.

HOT-FLOW STRIPPING

Large, essentially flat surfaces are flooded with hot stripping solution. A pan catches the solution flowing from the surfaces, and the solution is then pumped back into a heating tank for recirculation.

IMMERSION STRIPPING

Parts to be stripped are placed into expanded metal or other type baskets that are lowered into a tank containing hot stripping solution. The tanks are usually equipped with a circulation pump and an overflow dam to carry away loosened paint and soils. Immersion stripping, sometimes called chemical stripping, is based on the action of alkali or solvents. Alkali-based strippers are used hot or cold, while solvent-based strippers are used at ambient temperatures only.

SOLVENT STRIPPERS

Pure solvents are sometimes used as strippers, but solvent blends are used more extensively. Blends are based on ketones and saturated or aromatic hydrocarbons and are flammable. Chlorinated hydrocarbons are widely used, particularly methylene chloride and 1,1,1-trichloroethane, which are least objectionable and least flammable. All solvents are considered toxic to a degree, but these are also noncorrosive and easily handled. Thermosetting films and films on heat-sensitive or corrodible materials are frequently removed by use of solvent strippers.

WATERBORNE STRIPPERS

Hot waterborne strippers for ferrous metals are based on sodium hydroxide and usually contain trisodium phosphate and/or benzene sodium sulfonate. Hot strippers for nonferrous metals contain sodium metasilicate, detergents, and similar chemicals.

COLD ACID STRIPPERS

Cold acid strippers are essentially solvent strippers. They are usually based on methylene chloride, but also contain trichloracetic acid, formic acid, ethylene-diamine tetraacetic acid, some water-soluble solvents, and a small percentage of water. Phenolic compounds would aid in stripping but are ruled out by environmental regulations.

REGULATORY COMPLIANCE REQUIREMENTS

The finishing industry is required to comply with the Clean Air Act as administered by the Environmental Protection Agency (EPA) and the Occupational Safety and Health Act as administered by the Occupational Safety and Health Administration (OSHA). Regulations issued by EPA, OSHA, and state and local authorities establish limits for the following:

- Solvent emissions to the atmosphere.
- Levels of certain contaminants in liquid effluents.
- Restraints on the use of certain solvents and chemicals.
- Standards for minimizing hazards that might produce physical injuries to workers.
- Permissible levels of noise.

Due to regional variations that characterize state and local laws, only a rough outline of the general requirements for the finishing industry can be presented here. Environmental laws are in constant flux, and amendments are frequently issued. It is therefore incumbent on the reader to understand and comply with the rules and requirements of the latest laws and amendments.

SOLVENT EMISSION REGULATIONS

Solvents, called volatile organic compounds (VOC) by law, are photochemically reactive and give rise to objectionable ozone formation. As a result, these VOC emissions have to be held to a minimum. The accomplishable minimum VOC emission is called the reasonably available control technology (RACT). Large appliances, like washers, driers, and similar products, used to be spray painted with enamels containing 30% nonvolatiles (NV) and 70% volatile organic compounds (VOC) by volume, which is 0.3 gal (1.1 L) NV plus 0.7 gal (2.6 L) VOC. If it is assumed the solvent has a density of 7.25 lb/gal (0.87 kg/L), then the paint has 5.1 lb of VOC per gal of VOC + NV (0.61 kg/L). This unit, however, cannot be immediately related to the production of coatings; it first has to be converted to lb VOC/gal NV (kg/L), as follows:

$$\frac{5.1 \text{ lb VOC}}{0.3 \text{ gal NV}} = 17 \text{ lb VOC/gal NV } (2.04 \text{ kg/L})$$

Waterborne Paints

It seems reasonable that the paint discussed in the preceding paragraph can be replaced by a waterborne spray paint having a composition of 30% NV, 14% VOC, and 56% water, by volume. This content can be converted to the mass of VOC as follows:

0.14 gal (0.53 L) x 7.25 lb/gal (0.87 kg/L) = 1.0 lb (0.46 kg) of VOC, and:
1.0 lb (0.46 kg) of VOC divided by 0.44 gal (1.67 L) of NV gives VOC = 2.27 lb of VOC per gal (0.28 kg per L) of paint minus water; or 1.06 lb (0.46 kg) of VOC divided by 0.3 gal (1.14 L) of NV = 3.3 lb of VOC per gal (0.4 kg/L) of NV.

One gallon of nonvolatiles covers about 1604 ft^2 (149 m^2) of surface with a film thickness of 1 mil (0.03 mm). Thus, the ratio of VOC emissions for the waterborne paint versus the solvent-borne paint discussed is 3.3 lb (1.5 kg) of VOC (for the waterborne paint) divided by 17 lb (7.7 kg) of VOC (for the solvent-borne paint), which equals 0.19. In other words, the solvent emission with the waterborne paint is only 19% of that with the solvent-borne paint.

High Solids and Electrocoating

High-solids paint can be used for appliance painting with a reduction in VOC emissions. A typical high-solids paint may have a composition of 68% NV and 32% VOC, by volume, with 2.3 lb of VOC per gal (0.28 kg/L), NV + VOC. The 2.3 lb of VOC divided by 0.68 gal of NV equals 3.4 lb of VOC per gal of NV (0.4 kg/L). This amount represents an 80% reduction in VOC emissions compared with the conventional solvent-borne paint.

Electrocoating materials are another reasonable replacement for solvent-borne paints to reduce VOC emissions. Some electrocoating materials contain 1.8 lb of VOC per gal (0.22 kg/L), NV + VOC, or 2.4 lb of VOC per gal of NV (0.29 kg/L)—an 86% reduction in VOC emissions.

Emission Limitations

Regulators have decided that the use of paints containing 2.4 lb of VOC/gal (0.29 kg/L), minus water, is a reasonably available control technology (RACT). Through similar considerations in other finishing fields, the EPA proposed to state governments the solvent emission limitations listed in Table 30-4. If paints are used that contain more solvent (VOC) than mandated by law, the emitted vapors have to be reduced by 90% through vapor adsorption or incineration.

TABLE 30-4
Emission Limitations with Respect to Volatile Organic Compounds (VOC) for Surface Coatings

Affected Product or Facility	VOC, lb (kg) / VOC + NV, gal (L)
Automotive:	
Prime	1.9 (0.23)
Topcoat	2.8 (0.34)
Repair	4.8 (0.58)
Cans:	
Sheet and exterior	2.8 (0.34)
Interior	4.2 (0.50)
Side seam	5.5 (0.66)
Sealer	3.7 (0.44)
Coating:	
Coil	2.6 (0.31)
Fabric	2.9 (0.35)
Vinyl	3.8 (0.46)
Paper	2.9 (0.35)
Large Appliances:	
Prime and interior	2.4 (0.29)
Topcoat	3.7 (0.44)
Magnet Wire	Incineration

VOC = volatile organic compounds.
NV = nonvolatiles.

SELECTING PAINTS AND PROCESSES FOR COMPLIANCE

Any change to different coating materials or processes to meet emission limitations should be made by combining environmental desirability with lower cost and quality improvement whenever possible. This, however, may entail a sizeable capital investment.

Allowable Options

The range of organic volatiles per gallon of paint, minus water, was tabulated in 1979 and is presented as a bar graph in Fig. 30-4.[5] In addition, the coatings field was resurveyed in March 1981 and 10 important breakthroughs were noted and have been inserted on the graph. If a vertical line (parallel to the ordinate) is drawn intersecting the abscissa at the required VOC [for instance, 3.0 lb VOC/gal (VOC + NV) (0.36 kg/L)], then the right half of the graph is the area containing paints that generally have an unallowably high VOC content. It can be seen, however, that the VOC content of newly developed paints is in the allowable area, and thus the study of these materials is important to paint users.

Much progress has been made in the design of electrostatic spray equipment, which can increase transfer efficiency from 40% (for hand spray) to 95% (for centrifugal electrostatic atomization). In other words, a change of spray equipment can reduce the VOC emission by 50% and may make otherwise outlawed paints "equivalent" to allowable paints, as discussed later in this section.

Other ways of complying with solvent emission laws include the use of less than the allowable amount of VOC in one operation to compensate for excess VOC emission in another

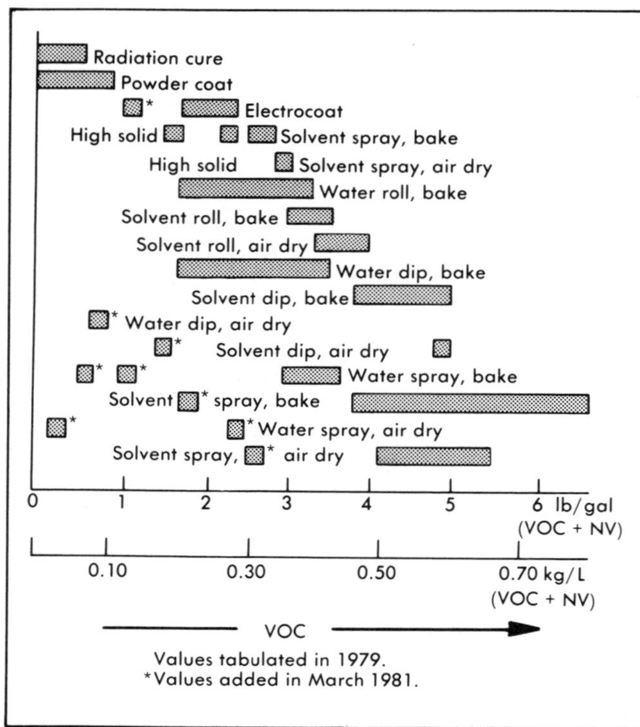

Fig. 30-4 Range of organic volatiles in paint, minus nonpolluting solvents.

operation. This approach is called the bubble concept or banking the saved amount of emission for future use, discussed later in this section. In addition, solvent emission can be reduced through incineration or carbon adsorption (add-on controls), autophoretic coating, and the use of radiation curing, as well as other methods. In all, numerous options are open to the finishing industry.

The selection process for coatings and processes that are in compliance with VOC emission regulations consists of three main steps and several substeps, as listed in Table 30-5. Computations required for these selection steps are facilitated by referring to the discussion in Chapter 29, "Coating Systems," and by the use of the following equation:

$$E = \frac{100}{te} \times \frac{A_p}{C} \times t \times B \qquad (1)$$

where:

E = weight of VOC emitted, kg or lb
te = transfer efficiency, %
A_p = area painted, m^2 or ft^2
C = 1000 m^2 x μm/L or 1604 ft^2 x mil/gal
t = film thickness, μm or mil
$B = \dfrac{\text{weight of VOC, kg or lb}}{\text{volume of NV, L or gal}}$

Equivalence Principle

Spray painting is the most widely used method of industrial painting. The transfer efficiency (te) of this method of painting is officially estimated at 40%, meaning that 60% of the expended paint is wasted as overspray. The transfer efficiency of spray painting can be markedly improved, however, through the application of an electric charge (see Table 30-6).

If, for instance, a certain paint applied at 40% te uses twice the allowable amount of VOC, then it will use not more than the allowable amount if applied at 80% te. In this case, the paint actually used has become "equivalent" to an allowable paint. In general, if conventional air-atomized spray is replaced by an electrostatic spray, the emission of VOC is reduced by approximately one half. The cost of buying paint is also reduced by one half.

TABLE 30-5
Steps in Selecting Coatings and Processes for Compliance with VOC Emission Regulations

Step	Action
1	Compute weight of VOC:
1a	Currently emitted.
1b	EPA mandated.
2	Compute weight of VOC reduced through:
2a	Transfer efficiency (equivalence).
2b	Pooling (bubble concept).
2c	Cooperation (banking).
2d	Different materials or processes.
3	Select most favorable option.

The Bubble Approach

Some proposed state laws define the bubble approach approximately as follows: "...compliance may be achieved on a source-wide basis...provided that...all applicable facilities are within the same source...The total VOC emissions do not exceed the sum of the allowable emissions from each facility..." Thus, the bubble approach offers the possibility of compensating for excess VOC emission from one operation by lowering the VOC emission from other operations to a level below the allowable limit.

Banking and Using VOC Emissions

The banking concept allows a company to store air quality credits earned through the emission of less than the allowable amount of VOC. The stored or banked credits can be used later by the company if needed to achieve compliance in new operations. The banked credits can also be given or even sold to other VOC-emitting parties. This offsetting policy seems to have been first used in Pennsylvania when the state agreed to

TABLE 30-6
Estimated Transfer Efficiencies for Painting Automobiles and Light-Duty Trucks*

Application Method	Transfer Efficiency (te), %
Air-atomized spray	40
Manual electrostatic spray	75
Automatic electrostatic spray	95
Electrodeposition	100

* Federal Register, October 5, 1970, p. 57792.

reduce hydrocarbon emissions from its highway paving operations to offset the expected pollution from a proposed Volkswagen plant.

Maryland's Economic and Community Development Department has proposed a statewide program for banking and marketing anti-air-pollution credits. This concept deserves attention as an effort to encourage industrial growth while still making progress toward clean air goals. Thus, emitting less than the allowable quantity of waste may create marketable assets and encourage the use of improved methods and materials.

Using Different Materials and Processes

In addition to the paints and processes identified in Fig. 30-4, consideration should be given to certain chlorinated hydrocarbons that can be used without VOC emission restrictions. Many states offer the possibility of replacing some of the VOC in a given formulation with chlorinated solvents. Thus, a paint that exceeds the allowable VOC content can be converted into a compliance paint, virtually without change in pigment or binder and even without change in application equipment. Other methods that should be seriously considered to accomplish compliance are the autodeposition materials and add-on controls, such as incineration and carbon adsorption.

REUSE AND REDUCTION OF WASTES

Reusing hazardous wastes means reduced disposal costs, reduced materials costs, avoidance of cradle-to-grave responsibilities, and a cleaner environment. Due to the high cost of paint, solvent, and oil, the used materials are a valuable resource. In addition, because of federal regulations, disposal costs for hazardous wastes will increase.[6]

EPA Regulations

The Environmental Protection Agency (EPA), under the 1976 Resource Conservation and Recovery Act (RCRA), has issued stringent regulations governing the management of hazardous wastes. In general, a hazardous waste has one or more of the following characteristics: toxicity, ignitability, corrosivity, or reactivity.

Effective November 19, 1980, the EPA established a cradle-to-grave management system for hazardous wastes. This system includes standards for generators and transporters of hazardous wastes, as well as for operators of facilities that store, treat, or dispose of hazardous wastes. A manifest system tracks the wastes from point of generation to point of disposal. Violations of these laws carry civil and criminal penalties of up to $25,000 per day and/or up to one-year imprisonment. Reusing hazardous wastes, however, can avoid disposal costs, as well as reduce material costs.

Waste Contents

The largest volume of finishing wastes are industrial sludges, solutions, and paints. Significant amounts of corrosives, waste oils, and solvents are also generated. Most wastes are disposed of in landfills, but some paint wastes are recyclable. Contacts with paint recycling companies should be made.

There are a number of companies seeking waste oils and solvents for purchase, recycling, and resale. Some acids and alkalis can be sold as feedstock for other industrial processes. Sludges and solutions, if they contain specialty metals or catalysts, can be reclaimed.

Reducing Wastes

For the immediate reduction of hazardous waste disposal costs at the plant level with no capital equipment expense, a program with the following procedures can be organized:

- *Separation and collection*. Waste paints, solvents, and oils are valuable resources. To realize their value, the wastes must be kept separate and uncontaminated. Once people are properly trained, proper separation should be no more expensive than present maintenance costs.
- *Reduce amount generated*. Common sense conservation and good housekeeping can reduce both the amount of virgin material used and the volume of waste generated. The life of oils and coolants can be extended through improved maintenance and analysis. Paint sludge can be reduced through improved paint transfer efficiency.
- *Reduce disposal quantity*. To lower disposal costs, minimize the number of drums. Make sure that all drums are full and that they do not contain garbage or water.
- *Qualify*. Determine if any waste can qualify as a *special waste*, which is significantly less expensive to dispose of than a *hazardous waste*.
- *Outright sale*. If wastes are separated and cleaned, it is possible to sell some of them. Local collectors purchase waste oils and halogenated solvents. For competitive bids, list wastes on the nearest industrial waste exchange.
- *Recycle*. Facilities that generate over 1000 gal (3785 L) per year of oils or halogenated solvents, 5000 gal (18 925 L) per year of nonhalogenated solvents, or 10,000 gal (37 850 L) of paint wastes should investigate recycling. Waste solvents can be recycled into solvent blends or back to pure solvents. If colors are kept separate, some paint sludge can be recycled to produce topcoats. Some mixed colors of paint sludge can be recycled to produce primer or chassis paint, where exact color is not critical.
- *Corrosive wastes*. These include paint stripper, chromic and mineral acids, spent pickle liquor, and strongly alkaline solutions. Corrosive wastes are either strongly acidic (pH 2) or strongly basic (pH 12.5). If a facility has both acidic and basic wastes, they can be neutralized on site. The salt resulting from the neutralization of an acid with a base is usually not considered hazardous.

  Under RCRA rules, neutralization was considered a treatment and required a permit. On November 17, 1980, however, the EPA ruled that owners and operators of wastewater treatment and elementary neutralization units are not required to obtain a RCRA permit. The waste corrosives and/or neutral salts can also be listed on a waste exchange as feedstock for other processes.
- *Waste oil*. Most waste oil should be burned on site for its heat content. Some waste oil can be sold to reclaimers or recycled on site. Because of the high cost of new oil, used oil is a valuable resource.
- *Solvents*. Solvents are divided into two categories: halogenated and nonhalogenated. The principal methods of their disposal are to burn them or to dump them in a landfill. However, because of the high cost of new solvent, used solvent has become a valuable resource too. To be recovered, waste solvents must be kept separate from each other and from other plant wastes. Bulk storage facilities for halogenated and nonhalogenated waste solvents may be needed.

Sludge Collection and Storage Procedures

Booth maintenance. To recycle paint sludge, neutral, clay-type booth chemicals should be used. These products have been used successfully for many years in production paint spray booths.

Drums. Paint sludge should be collected in clean, open-head drums or used paint drums. Covers should be kept on the drums to keep out trash and other contaminants.

Sludge collection. The sludge should be skimmed from the top of the water at the end of every eight-hour shift. Hardened paint from the sides of the booth and hook scrapings should be scrapped and not mixed with the skimmed sludge.

Sludge storage. Filled and covered drums of sludge can be stored outside. Freezing will not damage sludge, but excessive heat can prevent sludge from being reclaimed. In hot weather, therefore, the sludge should be stored inside or picked up and processed on a regular basis.

WASTE DISPOSAL REGULATIONS

The Resource Conservation and Recovery Act (RCRA) details a management program to control current and future waste disposal activities. For the purpose of this act, a solid waste includes solids, liquids, and contained gases. Hazardous waste is a subdivision of solid waste.

The producer of a waste that is designated or determined to be hazardous has cradle-to-grave responsibility for this waste and is required to:

- Obtain an Environmental Protection Agency (EPA) identification number (ID).
- Prepare a manifest for the off-site shipment of the hazardous waste, including a tracking system.
- Comply with pretransport requirements.
- Maintain records and prepare and submit reports (exception and annual).

The RCRA is administered by the United States Environmental Protection Agency (U.S. EPA). The EPA may delegate administration to states with programs that qualify for authorization. The federal program establishes the minimum requirements that a state must meet for approval, but states are free to impose more stringent regulations.

Identification of a RCRA Hazardous Waste

The EPA regulations detail identification of a RCRA hazardous waste. The substance meets a hazard characteristic, or it is listed on one of following four RCRA lists:

1. Ignitability; Listing Code = D001.
   - Liquid with a flashpoint of 60° C (140° F).
   - Nonliquid that is easily ignited and burns vigorously and persistently.
   - Compressed gas or oxidizer specified by the Department of Transportation (DOT).
2. Corrosivity; Listing Code = D002.
   - Aqueous, with a pH $\leq$ 2 or pH $\geq$ 12.5.
   - Liquid that corrodes steel 1/4″ (6.4 mm) per year.
3. Reactivity; Listing Code = D003.
   - Unstable; detonates or explodes; DOT forbidden explosive; reacts with water.
   - Forms toxic gases, vapors, or fumes.
4. EP (Extraction Procedure)—Toxicity; Listing Code = D004 to D017. Extract of solid waste containing any of eight elements or six pesticides at concentrations of 100 times the selected interim primary drinking-water standard.

Specifically Listed Hazardous Wastes

Specifically listed hazardous wastes are grouped in three categories, as follows: (Examples of hazardous wastes are presented in Table 30-7.)

1. Generic wastes—hazardous wastes from nonspecific sources: the F-List.
2. Process wastes—hazardous waste from specific sources: the K-List.
3. Discarded commercial chemical products, off-specification species, and spill residues thereof:
   - Acute hazardous wastes: includes mostly pesticides, organic chemicals, and certain heavy-metal compounds (the P-List).
   - Toxic wastes: mostly natural and synthetic organics like toluene and chlorinated solvents (the U-List).

TABLE 30-7
Solvents Listed as RCRA Hazardous Wastes

Solvent	Listing as RCRA Hazardous Wastes	
	F-List	U-List
Methylene chloride	F002	U080
1,1,1-trichloroethane	F002	U226
Trichloroethylene	F002	U228
Tetrachloroethylene	F002	U210
Toluene	F005	U220
Xylene	F003	U239
Acetone	F003	U002
n-Butyl alcohol	F003	U140
Isobutyl alcohol	F005	U031
Ethyl acetate	F003	U112
Methyl alcohol	F003	U154
Methyl ethyl ketone	F005	U159
Methyl isobutyl ketone	F003	U161

F002 The following spent halogenated solvents: tetrachloroethylene, methylene chloride, trichloroethylene, 1,1,1-trichloroethylene, chlorobenzene, 1,1,2-trichloro-1,2,2-trifluoroethane, o-dichlorobenzene, trichlorofluoromethane, and the still bottoms from the recovery of these solvents.

F003 The following spent nonhalogenated solvents: xylene, acetone, ethyl acetate, ethyl benzene, ethyl ether, methyl isobutyl ketone, n-butyl alcohol, cyclohexanone, methanol, and the still bottoms from the recovery of these solvents.

F005 The following spent nonhalogenated solvents: toluene, methyl ethyl ketone, carbon disulfide, isobutanol, pyridine, and the still bottoms from the recovery of these solvents.

U-List Discarded commercial chemical products, off-specification species, and spill residues thereof.

Halogenated and Nonhalogenated Solvents

Virtually all solvents are included in the RCRA, along with most of the priority pollutants from the Clean Water Act (65 chemical categories currently, covering 129 chemicals). Also, as discussed previously, regulated metals and pesticides from the interim primary drinking-water standard are included. Table 30-7 contains a sampling of typical halogenated and nonhalogenated solvents regulated under RCRA as hazardous wastes.

Solvent Handling and Waste Disposal

Industrial chemicals are strictly regulated to protect water and manage waste from industry by six major environmental statutes:

1. The Clean Air Act (CAA), which regulates discharges into ambient air.
2. The Clean Water Act (CWA), which regulates discharges and spills into surface water.
3. The Safe Drinking Water Act (SDWA).
4. The Toxic Substances Control Act (TSCA).
5. The Resource Conservation and Recovery Act (RCRA), which regulates discharges into ground water, some air emissions (such as from incinerators), and general waste management. Limited to solid and hazardous wastes.
6. The Comprehensive Environmental Response Compensation Act (CERCA) of 1980, also called the "Superfund."

The Superfund addresses both closed-site situations and releases of hazardous substances. Its principal focus is past management practices and present and future spills. A much broader definition of concerns includes releases of hazardous substances into the environment. Under the Superfund, the hazardous substance universe has been expanded to include:

- RCRA hazardous wastes (F-List, K-List, P-List, U-List, and characteristics of hazards).
- Sections 307 and 311 of the CWA.
- Section 112 of the CAA.
- Section 7 of the TSCA.

A continuing awareness of federal, state, and local regulations is essential to ensure present and future compliance when handling and disposing of all wastes. Illegal or improper disposal practices today can result in fines, litigation, and perhaps jail, along with cleanup costs at some future time. In addition, some chemicals not identified as hazardous today will be regulated in the near or distant future, with cleanup responsibilities for past practices a likely part of such regulations.

SAFETY OF PERSONNEL

Safety precautions with respect to various methods of applying paint are discussed in Chapter 29, "Coating Systems." As to the specific safety of personnel, the hazard and degree of exposure should first be determined. Suppliers of chemical materials provide safety data sheets listing flammability, reactivity, spill control, and suitable respiratory protection. These data sheets should be used by an industrial hygienist to develop safe procedures for the handling and the use of materials. Free or reasonably priced consultation is often available from state governments, funded by OSHA, and from workmen's compensation insurance carriers.

Industrial hygiene experts warn that an improper respirator gives employees a false sense of security, which may result in greater exposure to harmful materials. For instance, dust respirators do not protect against solvent vapors; usually, air-purifying masks or chemical cartridge respirators are required. The preferred means of minimizing toxic vapor exposure, however, include limitation of exposure duration and adequate exhaust ventilation.

As an example of adequate ventilation, the design of paint spray booths may be cited. For such equipment, the law demands solvent vapor concentrations not exceeding 50 to 200 parts per million by volume. One large spray booth emits 1245 lb (565 kg) of VOC per hour, using a solvent mixture having an average molecular weight of 82. Since 1 gram molecular weight of vapor equals 22.4 L, evaporation of 21 lb (9.5 kg) per minute creates 91 ft^3 (2.5 m^3) per minute. As a result, an airflow of 910,000 ft^3/min (25 480 m^3/min) is required.

References

1. W. T. Smith, "Hardness and Related Properties," *Paint Testing Manual*, 13th ed., G. G. Sward, ed. (Philadelphia: American Society for Testing and Materials, 1972), p. 283.
2. E. M. Corcoran, "Adhesion," *Paint Testing Manual*, 13th ed., G. G. Sward, ed. (Philadelphia: American Society for Testing and Materials, 1972), p. 314.
3. G. G. Schurr, "Tensile Strength and Elongation," *Paint Testing Manual*, 13th ed., G. G. Sward, ed. (Philadelphia: American Society for Testing and Materials, 1972), p. 338.
4. N. B. Garlock and G. G. Sward, "Artificial Weathering," *Paint Testing Manual*, 13th ed., G. G. Sward, ed. (Philadelphia: American Society for Testing and Materials, 1972), p. 405.
5. George E. F. Brewer, *The Technique of Complying with Paint Solvent Emission Regulations*, AFP/SME Technical Paper FC81-416 (Dearborn, MI: Society of Manufacturing Engineers, 1981).
6. Elaine M. Sloan, *Reusing Hazardous Wastes*, AFP/SME Technical Paper FC81-449 (Dearborn, MI: Society of Manufacturing Engineers, 1981).

Bibliography

Baer, George F. *Safety and Handling Considerations of UV Equipment*. AFP/SME Technical Paper FC84-986. Dearborn, MI: Society of Manufacturing Engineers, 1984.

Smith, Donald W. *Communicating Hazard with the Hazardous Materials Identification System*. AFP/SME Technical Paper FC84-985. Dearborn, MI: Society of Manufacturing Engineers, 1984.

INDEX

A

INDEX

B

D

F

G

I

L

M

N

INDEX

INDEX

T

U